PHYSICS for NEET

Volume 1

S A S Kirmani
Chairman
Erose Educational Infotech Limited
New Delhi

PHYSICS FOR NEET VOLUME 1

UNIVERSITIES PRESS (INDIA) PRIVATE LIMITED

Registered office
3-6-747/1/A & 3-6-754/1, Himayatnagar, Hyderabad 500 029, Telangana, India
info@universitiespress.com; www.universitiespress.com

Distributed by
Orient Blackswan Private Limited

Registered office
3-6-752 Himayatnagar, Hyderabad 500 029, Telangana, India

Other offices
Bengaluru, Chennai, Guwahati, Hyderabad, Kolkata,
Mumbai, New Delhi, Noida, Patna, Visakhapatnam

501752

ISBN: 978-93-89211-67-2

Cover and book design

Typeset in Times New Roman 10.5/12.5 by SRS Technologies, Puducherry

Printed in India by
Navya Printers, Hyderabad 500 013

Published by
Universities Press (India) Private Limited
3-6-747/1/A & 3-6-754/1, Himayatnagar, Hyderabad 500 029, Telangana, India

Endorsements

NEET Physics by SAS Kirmani is an all-inclusive book tailored for students preparing for the Medical Entrance Examination. Besides covering all the topics related to the NEET syllabus, the book has a unique 'Think and Grow' section that kindles logical thinking in the reader, which is essential for understanding the concepts of Physics.

Chandra Shekhara PS
Director, Marigold International School, Bengaluru
Former Principal, DPS, Damanjodi Orissa and DPS Ghana

The book will fulfill the much-felt need for a comprehensive book that addresses the specific needs of students taking up the NEET entrance examinations.
I wish the book all success.

Tarun Kumar Singh
Principal, Oil India H S School, Duliajan

This book will definitely help you to clear the NEET exam. With to-the-point explanations and an exhaustive range of problems, it will be of immense help in your mock sessions.

Sameer Gupta
Senior Faculty in Physics
Director, EDUNOVA
Ahmedabad | Vadodara | Dahod | Godhra

This book provides comprehensive material for a quick revision of all the important concepts of the subject. It is a useful study-aid for those preparing for competitive exams.

Dr Manish Tulsiani
Managing Director, ACDA Classes
Lucknow

With his lucid style and authentic approach, SAS Kirmani has put together in his book the essentials of higher secondary school Physics and made the subject appear less formidable for the NEET aspirants.

V K Sharma
HOD (Physics), Vasant Valley School
New Delhi

This book is one of the best collections of questions that cover all aspects of the subject the student has to be familiar with to appear for the NEET examination. The revision material and questions are well thought-out and would be of immense help to the students for practice sessions. I recommend this work to all NEET aspirants.

Dr Rajeev Tyagi
Principal, Mount Carmel School
Dwarka, New Delhi

The book is well-planned and executed. With its exhaustive solved questions with ample questions for practice, it will be very useful for students preparing for NEET.

Archna Deepak
Senior Physics Teacher
Delhi Public School
R K Puram, New Delhi

With incredible command over the subject, Shri Kirmani handles it in a manner that makes it appear so easy and comprehensible for the students. I am sure his book *Physics for NEET* will turn out to be a huge success, becoming a go-to book among the students preparing for JEE and NEET.

Gp Capt (R) M Anwar
Former Inspecting Officer, Sainik School Society
MOD, GOI
Former Principal and Director,
International Indian School,
Dammam, Saudi Arabia

The introduction to each chapter is unique; it is like giving inputs for a brain map diagram and helps to conceptualize and remember the whole chapter in a concise manner. Hence, the book is very useful for concept learning. I loved to read this book and recommend it to those who are keen to learn the key concepts of Physics and apply them to tackle NEET.

Dr Updesh Verma
(PhD, IIT Delhi)
Asst Professor of Physics
Manyavar Kanshiram Govt Degree College, Ghaziabad, UP

Table of Contents

Preface

It gives me immense pleasure in presenting this book to the students preparing for NEET and other similar examinations.

The book has two main objectives: first, to provide students a clear presentation of the basic concepts of physics; and second, to strengthen the understanding of the theoretical principles with a broad range of illustrations and examples of varying difficulty. To this end, I have used sound physical arguments and rationalised the problem-solving methods used to arrive at the correct answer. This logical approach to problem-solving will eventually help students to hone their technique and solve objective questions in optimal time. An assortment of problems from the previous years' NEET question papers has been included in the Exercises along with the Solutions. In addition, questions are provided under Chapter Test at the end of each chapter to help the student assess his/her application-level knowledge in the given topic.

The other useful feature of the book is a summary of necessary definitions, formulae and important notes under the heading "Points to Remember" at the end of each chapter. Physics is all about analytical reasoning and its concepts are readily applicable in our day-to-day life. To integrate the essentials of theory with practical everyday observations, another very special feature in the form of "Think and Grow" has been incorporated in each chapter. This will certainly arouse the interest of the reader.

I am highly indebted to the authors and publishers of the various books that I have freely consulted during the preparation of the manuscript.

I thank and acknowledge the following scholars for their suggestions, criticism and encouragement.

Chandra Shekhara	Director, Marigold International School Bengaluru Former Principal DPS Damanjodi Orissa and DPS Ghana
Tarun Kumar Singh	M.Sc. (Physics), M.Ed., Principal, Oil India HS School, Duliajan
Dr Vinita	Director, Erose Educational Infotech Ltd, New Delhi
Sameer Gupta	Senior Faculty, Physics, Director, EDUNOVA, Ahmedabad|Vadodra|Dahod|Godhra
Dr Manish Tulsiyan	Managing Director, ACDA classes, IIT-JEE, NEET Lucknow
V K Sharma	M.Sc., M.Phil., HOD (Physics), Vasant Valley School, New Delhi
Dr Rajeev Tyagi	Principal, Mount Carmel School, Dwarka, New Delhi
Arachna Deepak	Senior Physics Teacher, Delhi Public School, R K Puram, New Delhi
Capt M Anwar	Former Inspecting Officer, Sainik School Society, MOD, GOI Former Principal, Sainik School Balachadi Jamnagar, Former Principal and Director, International Indian School, Damman, Saudi Arabia
Dr Updesh Verma	Asst Professor of Physics – Manyavar Kanshiram Govt Degree College Nandgram, Ghaziabad, UP, Past Senate member, IIT Delhi Raman Postdoc Fellow and visiting Scientist at Princeton University, Princeton, New Jersey

I do not know how to adequately express my thanks to my wife Fatma Kirmani, for her constant motivation, support and long-term sacrifices. Also, my love and blessing to my little granddaughter Aaira Kirmani for her support and enthusiasm while I was writing the book. The book consumed lot of time which was exclusively dedicated to her.

I am also obliged to Anas, Neha and Aman for their transcription and the PATS team for proof reading and editing. I am grateful to my colleagues and friends who have helped me in bringing the book to fruition.

Finally, special thanks and recognition go to Thomas Mathew Rajesh and Ramesh M R of Universities Press (India) Private Limited, for their diligence and attention to detail.

I would be very pleased to receive suggestions and comments from the readers for further improvement of the book. These may please be sent to the publisher or directly to me at saskirmani@gmail.com

S A S KIRMANI

About the Author

SAS Kirmani is the Founder of Erose Educational Infotech Limited, a PAN India Educational and Research Company headquartered in New Delhi. He founded Erose in 1992 and is the chairman of the organisation. Under his leadership and vision, Erose has successfully guided and trained aspirants for competitive examinations, IIT-JEE, Medical and IAS. Kirmani articulated, designed and implemented the Global Education and Training model to prepare students for various entrance examinations, which has become a huge success. Under the model, he teaches students to focus on answering the maximum number of questions in optimal time by using the concepts for both straightforward application and lateral thinking.

An electrical engineer from HBTI Kanpur, Kirmani is on the advisory board of many schools and educational institutions to provide academic and administrative support. He was a member of the Draft Formulation Committee of Computer Literacy Awareness Program (CLASS) jointly sponsored by NCERT and Ministry of Education in the year 1984. In his initial years, he worked as senior physics faculty at Delhi Public School, R K Puram, New Delhi.

Units and Dimensions

Introduction

Physics describes the laws of nature. This description is quantitative and involves comparison and measurement of physical quantities. In physics, great importance is attached to measurements. This is also the reason that **physics is an exact science or the science of measurements.**

Therefore, to know and to understand a physical situation completely, one has to measure the quantities like mass, distance, time, acceleration, force, etc. All these quantities are known as physical quantities. So, the quantities which can be expressed in terms of laws of physics and can be measured directly or indirectly are called physical quantities.

The measurement of a physical quantity involves the following two steps:

(i) The selection of the appropriate unit
(ii) Finding the number of times that unit is contained in the physical quantity, i.e., numerical value.

For example, suppose the length of a room is 5 metres. Then 5 is the numerical value and metre is the unit of length. It is important to mention that the selected unit should have international acceptance. Otherwise, everyone will choose his own unit for that quantity. In that case, it would become difficult to communicate freely among the persons over the world. *The General Conference on Weights and Measures* in England has been given the authority to decide on units by international agreement.

In general,

Measure of a physical quantity

= *Numerical value* × *Size of its unit*

or $X = nu$

It should be remembered that $n\,u$ = constant.

Let n_1 is the numerical value of a physical quantity in unit u_1 and n_2 for unit u_2, then

$$n_1\ u_1 = n_2\ u_2$$

For example, l = 10m = 1000 cm.

Bigger the unit, smaller is the numerical value and vice-versa.

The magnitudes of physical quantities vary over a wide range. For example, the radius of an atomic nucleolus is 5.3×10^{-9} m whereas the radius of the earth is approximately 6.4×10^{6} m. To express the magnitudes of such widely varying quantities, we use certain standard prefixes for certain powers of ten. These are shown in Table 1.1.

Table 1.1 Prefix and symbols in terms of power of ten

Power of ten	Prefix	Symbol
18	Exa	e
15	Peta	p
12	tera	t
9	Giga	g
6	mega	m
3	Kilo	k
2	Hector	h
1	deka	da
–1	Deci	d
–2	Centi	c
–3	Milli	m
–6	Micro	μ
–9	Nano	n
–12	Pico	p
–15	Femto	f
–18	Atto	a

There are some practical units, frequently used for small and large quantities. These are:

(i) 1 fermi = 1 fm = 10^{-15} m
(ii) 1 X-ray unit = 1 xu = 10^{-13} m
(iii) 1 angstrom = 1 Å = 10^{-10} m
(iv) 1 micron = 1 µm = 10^{-6} m
(v) 1 mile = 1.6 km = 1.6×10^{3} m
(vi) 1 astronomical unit = 1 AU = 1.49×10^{11} m
(This is the average distance between the sun and the earth, i.e., equal to the radius of the earth's orbit.)

(vii) 1 Light year = 1 ly = 9.46×10^{15} m
This is the distance travelled by light in vacuum in one year.
(viii) 1 Parsec = 1 pc = 3.08×10^{16}m = 3.26 light year
(ix) 1 Barn = 1 b = 10^{-28} m^2
(x) 1 Are = 1 a = 100 m^2
(xi) 1 Tonne = 1 t = 1000 kg
(xii) 1 Chandrasekhar limit = 1.4 × Mass of the sun = 2.8×10^{30} kg

Fundamental and Derived Units

The physical quantities which do not depend upon other quantities are known as *fundamental quantities*. The internationally employed fundamental quantities are those of mass, length and time. Therefore, *the units of mass, length and time are called as fundamental units.* The units of the physical quantities which depend upon fundamental units or can be derived from fundamental units are known as *derived units* and quantities are known as *derived quantities*. Therefore, the units of all physical quantities which can be expressed in terms of fundamental units of mass, length and time are called as derived units.

Consider the following example:

$$\text{Speed} = \frac{\text{Distance}}{\text{Time}}$$

$$\therefore \quad \text{Unit of speed} = \frac{\text{Unit of distance}}{\text{Unit of time}}$$

$$= \frac{\text{metre}}{\text{second}} = \frac{\text{m}}{\text{s}} = \text{ms}^{-1}$$

The units of physical quantities like acceleration, momentum, force, work, power etc. can be expressed in terms of fundamental units of mass, length and time. Therefore, the units of all these physical quantities are known as derived units.

Principal System of Measurements

(i) CGS (Centimetre–Gram–Second) system.
(ii) FPS (Foot–Pound–Second) system.
(iii) MKS (Metre–Kilogram–Second) system.
(iv) SI (International system).

(i) CGS System: In this system, the unit of length is centimetre, that of mass, the gramme and that of time, the second.

(ii) FPS System: In this system, the units of length, mass and time are foot, pound and second respectively.

(iii) MKS System: In this system, the units of length, mass and time are metre, kilogram and second respectively.
1 metre is the distance that contains 1650763.73 wavelength of orange-red light of Kr-86.
1 kilogramme is the mass of 5.0188×10^{25} atoms of carbon-12.
1 second is the time in which a cesium atom vibrates 9192631770 times in an atomic clock.

(iv) SI System: In this system, there are seven fundamental quantities which are shown in Table 1.2.

Standard of Fundamental Units Mass, Length and Time

(i) Metre (m): 1 Metre is defined as the distance containing 1,650, 763.73 wavelengths of the orange light emitted by pure Krypton-86.

(ii) Kilogram (Kg): 1 Kilogram is defined as the mass of a platinum-iridium cylinder kept at International Bureau of Weights and Measures in Paris. In practice, the mass of 1 litre of water at 4°C is 1 Kilogram. On atomic scale, 1 Kilogram is equivalent to the mass of 5.0188×10^{25} atoms of $_6C^{12}$ (an isotope of carbon).

(iii) Second (s): Cesium-133 atom emits electromagnetic radiations of several wavelengths. A particular radiation is selected which corresponds to the transition

Table 1.2 Fundamental quantities used in different system of units with their units and symbols.

S. No	Fundamental quantities	Units and symbols in			
		CGS	FPS	MKS	SI
1	Mass	Gram (g)	Pound (lb)	Kilogram (kg)	Kilogram (kg)
2	Length	Centimetre (cm)	Foot (ft)	Metre (m)	Metre (m)
3	Time	Second (s)	Second (s)	Second (s)	Second (s)
4	Temperature	–	–	–	Kelvin (K)
5	Luminous intensity	–	–	–	Candela (cd)
6	Electric current	–	–	–	Ampere (A)
7	Amount of substance	–	–	–	Mole (mol)

between the two hyperfine levels of the ground state of Cs-133. Each radiation has a time period of repetition of certain characteristics. The duration of 9,192,631,770 time periods of the selected transition is defined as 1 s.

In SI, there are two supplementary units:

(i) Radian (rad): unit of plane angle

(ii) Steradian (sr): unit of solid angle.

Dimensions and Dimensional Formula

The fundamental quantities, also called base quantities, are represented by one-letter symbols. Generally, mass is denoted by M, length by L, time by T and electric current by I. The temperature is denoted by K, the luminous intensity by cd and the amount of substance is denoted by mol.

The **dimensions** of a derived unit are derived as the powers to which the fundamental units are raised in order to represent it. Thus, if a derived unit depends upon the *n*th power of a fundamental unit, it is said to be of *n* dimensions in that fundamental unit.

For example,

$$[\text{Velocity}] = \frac{[\text{Distance}]}{[\text{Time}]} = \frac{[\text{L}]}{[\text{T}]} = \left[\text{LT}^{-1}\right]$$

Hence, the dimensions of the unit of velocity are 1 in length and –1 in time. Since it is independent of mass, its dimension in mass is zero and we may, therefore, represent it by $[\text{M}^0\text{LT}^{-1}]$.

The physical quantity that is expressed in terms of the fundamental quantities is enclosed in square brackets simply as a reminder that the equation is among the dimensions and not among the magnitudes. Thus, the above equation may be written as [velocity] = LT^{-1}

A **dimensional formula** is an expression showing how and which of the fundamental units enter into the unit of a physical quantity.

Thus, the dimensional formula for velocity is M^0LT^{-1}.

But when we write it in the form $\text{V} = \text{M}^0\text{LT}^{-1}$, it is called dimensional equation for velocity.

There are two types of quantities, viz., variables and constants which may both be dimensional or non-dimensional (dimensionless). Thus, we have

i. *Dimensional variables*: These are quantities like acceleration, velocity, force, etc.

ii. *Non-dimensional variables*: These are quantities which are variables and yet have no dimension. Examples are specific gravity, strain, angle, etc. Such quantities are also called 'number'.

iii. *Dimensional constants*: These are quantities having a constant value and dimension. Examples are velocity of light in vacuum, gravitational constant, etc.

iv. *Non-dimensional constants:* These are merely numbers like π, 5, 9, etc.

To find out the dimensional formula for any physical quantity, we should know at least one equation in which it appears.

For examples the coefficient of viscosity of a medium can be obtained as follows:

From Stoke's formula, we have

$$F = 6\,\pi\eta r v$$

$$\Rightarrow \quad [\eta] = \frac{[F]}{6\pi r v} = \frac{\text{MLT}^{-2}}{\text{L.LT}^{-1}} = \text{ML}^{-1}\text{T}^{-1}$$

Illustration 1

Find the dimensional formulae of the following quantities:

(i) Coefficient of viscosity (η);

(ii) Young's modulus of elasticity (Y);

(iii) Momentum (P) and,

(iv) Angular momentum (***L***)

Solution

(i) The Stoke's law in viscosity is given by

$$F = 6\pi\eta r v$$

$$\Rightarrow \quad [\eta] = \frac{[F]}{[r][v]} \quad (6\,\pi \text{ is dimensionless})$$

$$\Rightarrow \quad [\eta] = \frac{\text{MLT}^{-2}}{\text{L.LT}^{-1}} = \text{ML}^{-1}\text{T}^{-1}$$

(ii) The Young's modulus of elasticity is defined as

$$Y = \frac{\text{Longitudinal stress}}{\text{Longitudinal strain}} = \frac{\text{F/L}}{\text{L/L}}$$

$$\Rightarrow \quad [Y] = \frac{[\text{F}][\text{L}]}{[\text{A}][\text{L}]} = \frac{\text{MLT}^{-2}.\text{L}}{\text{L}^2.\text{L}}$$

$$\Rightarrow \quad [Y] = \text{ML}^{-1}\text{T}^{-2}$$

(iii) The momentum P of a body is defined as the product of its mass (m) and velocity (v). That is

$$P = mv$$

$$\Rightarrow \quad [P] = [\text{MLT}^{-1}]$$

(iv) The angular momentum ***L*** is defined as the moment of momentum. That is,

$$\boldsymbol{L} = mvr$$

$$\Rightarrow \quad [\boldsymbol{L}] = [m]\,[r] = \text{M.LT}^{-1}.\text{L}$$

$$\Rightarrow \quad [\boldsymbol{L}] = \text{ML}^2\text{T}^{-1}$$

Illustration 2

The dimensions of the quantities in one (or more) of the following pairs are the same. Identify the pairs:

(a) Torque and work

(b) Angular momentum and work

(c) Energy and Young's modulus

(d) Light year and wave-length.

Solution

We have,

$$[\text{torque}] = ML^2T^{-2} \text{ and } [\text{work}] = ML^2T^{-2}$$

Since the dimensions of torque and work are the same, (a) is the correct answer.

Again [angular momentum] = ML^2T^{-1} and [work] = ML^2T^{-2}

Since the dimensions of these two quantities are not the same, (b) is wrong.

Again, [energy] = ML^2T^{-2} and [Young's modulus] = $ML^{-1}T^{-2}$

Since the dimensions of these two quantities are not the same, (c) is wrong.

Finally, light year is the distance travelled by light in one year and hence has the dimension. L. Wavelength is the minimum distance between two points of a medium which are in the same phase of vibration and hence has the dimension L. Therefore, both the light year and the wavelength have the same dimension and so, (d) is correct.

Hence, the pairs given by (a) and (d) have the same dimension.

Illustration 3

The equation of state of some gases can be expressed as

$$\left(P + \frac{a}{V^2}\right)(V - b) = RT$$

Here P is the pressure, V is the Volume, T the absolute temperature, and a, b and R constants. The dimensions of a are

(a) ML^5T^{-2} (b) $ML^{-1}T^{-2}$

(c) $M^0L^3T^0$ (d) $M^0L^6T^0$

Solution

Since in the given equation, $\frac{a}{V^2}$ is added to the pressure P, $\frac{a}{V^2}$ must have the dimension of pressure. This is so because only like quantities can be added or subtracted.

Hence, $\left[\frac{a}{V^2}\right] = [P]$

$$\Rightarrow \quad [a] = [P][V^2]$$

$$\Rightarrow \quad [a] = ML^{-1}T^{-2}\left(L^3\right)^2$$

$$\Rightarrow \quad [a] = ML^{-1}T^{-2}L^6$$

$$\Rightarrow \quad [a] = ML^5T^{-2}$$

Table 1.3 shows the dimensional formulae of some important physical quantities.

Table 1.3

Physical quantity	Dimensional formula
1. Area = Length × Breadth	$M^0L^2T^0$ or L^2
2. Volume = Length × Breadth × Height	$M^0L^3T^0$ or L^3
3. Velocity = Length/Time	M^0LT^{-1} or LT^{-1}
4. Acceleration = Velocity/Time	M^0LT^{-2} or LT^{-2}
5. Force = Mass × Acceleration	MLT^{-2}
6. Work = Force × Acceleration	ML^2T^{-2}
7. Kinetic energy = ½ mv^2	ML^2T^{-2}
8. Potential energy = mgh	ML^2T^{-2}
9. Couple = Force × Distance	ML^2T^{-2}
10. Torque = Force × Distance	ML^2T^{-2}
11. Power = Work/Time	ML^2T^{-3}
12. Radius of gyration (k)	M^0LT^0 or L
13. Moment of inertia = ½ mk^2	ML^2T^0 or ML^2
14. Density = Mass/Volume	$ML^{-3}T^0$ or ML^{-3}
15. Pressure = Force/Area	$ML^{-1}T^{-2}$
16. Stress = Force/Area	$ML^{-1}T^{-2}$
17. Strain = Length/Length	No dimension
18. Gravitational constant $G = F\,d^2/m_1m_2$	$M^{-1}L^3T^{-2}$
19. Surface tension = Force/Length	ML^0T^{-2} or MT^{-2}
20. Coefficient of viscosity	$ML^{-1}T^{-1}$
21. Frequency = 1/time	$M^0L^0T^{-1}$ or T^{-1}
22. Planck's constant $h = \frac{\text{Energy}\,(E)}{\text{Frequency}\,(\upsilon)}$	ML^2T^{-1}
23. Boltzmann's constant	$ML^2T^{-2}L^{-1}$
24. Angle = Arc/Radius	No dimensions

The analysis of the dimensions of the various physical quantities involved in a dimensional equation is known as dimensional analysis.

Its three main uses are the following:

(i) conversion of one system of units into another (Illustration 4)

(ii) checking the results arrived at (Illustration 5)

(iii) deriving a correct relationship between different physical quantities (Illustration 6).

Illustration 4

The SI and CGS units of energy are joule and erg, respectively. How many ergs are equal to one joule?

Solution

The dimensions of energy E is given by $[E] = ML^2 T^{-2}$

Now, 1 joule = (1 kg) (1 m)^2 (1 s)$^{-2}$

And 1 erg = (1g) × (1cm)^2 × (1 s)$^{-2}$

Hence,

$$\frac{1\text{ joule}}{1\text{ erg}} = \left(\frac{1\text{ kg}}{1\text{ g}}\right)\cdot\left(\frac{1\text{ m}}{1\text{ cm}}\right)^2\cdot\left(\frac{1\text{ s}}{1\text{ s}}\right)^{-2}$$

$$\Rightarrow \frac{1\text{ joule}}{1\text{ erg}} = \left(\frac{1000\text{ g}}{1\text{ g}}\right)\left(\frac{100\text{ cm}}{1\text{ cm}}\right)^2\left(\frac{1\text{ s}}{1\text{ s}}\right)^{-2}$$

$$\Rightarrow \frac{1\text{ joule}}{1\text{ erg}} = 10^7$$

$$\Rightarrow 1\text{ joule} = 10^7\text{ erg}$$

Illustration 5

Show that the expression $h = kT(r\,\rho\,g)$ for ascent h of a liquid of surface tension T and density ρ in a capillary tube of radius r is dimensionally correct.

Solution

First of all we write the dimensions of the terms used, i.e.,

h = [L], r = [L]

T = surface tension = [M T^{-2}]

ρ = [M L^{-3}] and g = [L T^{-2}]

Taking the dimensions of both sides, we have

$$[L] = \frac{[MT^{-2}]}{[L][ML^{-3}][LT^{-3}]} = [L]$$

As the dimensions of both sides of the above equation are same, the relation is correct.

Illustration 6

Using the method of dimensions, obtain an expression for

(i) the acceleration of the particle moving with a uniform speed v, in a circle of radius r;

(ii) the tension F in a uniform wire of radius r and mass m per unit length, rotating in its own plane with an angular velocity ω, about an axis passing through its centre and perpendicular to its plane and,

(iii) the mass M of a planet round which a satellite completes its orbit of radius r, in a time interval T.

Solution

(i) Let the acceleration f depends upon v^a and r^b so that

$$f = kv^a r^b \qquad (1)$$

where a and b are the dimensions of v and r, respectively, and k is a dimensionless constant.

$$\therefore \quad [f] = [v]^a[r]^b = [LT^{-1}]^a[L]^b$$

$$\Rightarrow \quad LT^{-2} = L^{a+b}T^{-a} \qquad (2)$$

Equating the dimensions of L and T on both sides of (2), we get

$$a + b = 1 \Rightarrow b = 1 - a$$
$$a = 2 \Rightarrow b = 1 - 2 = -1$$

Hence, $a = 2$ and $b = -1$

Putting these values of a and b in (1), we get

$$f = k\,v^2\,r^{-1}$$

$$\Rightarrow \quad f = k.\frac{v^2}{r}$$

(ii) Let $F \propto r^a\,m^b\,\omega^c$

$$\Rightarrow \quad F = k\,r^a\,m^b\,\omega^c \qquad (2)$$

where a, b and c are the dimensions of r, m and ω respectively and k is a dimensionless constant.

Taking dimensions on both sides of (2) and remembering that the tension F has the dimension of force, we get

$$[F] = [r]^a\,[m]^b\,[\omega]^c$$

$$MLT^{-2} = [L]^a[ML^{-1}]^b\left[\frac{LT^{-1}}{L}\right]^c$$

As m = mass per unit length and $\omega = \frac{v}{r}$

$$\Rightarrow \quad MLT^{-2} = M^bL^{a-b}T^{-c} \qquad (3)$$

Equating the dimensions of M, L and T on both sides of (3), we get

$$b = 1$$
$$a - b = 1 \Rightarrow a = 1 + b = 2$$
$$-c = -2 \Rightarrow c = 2$$

Putting these value of a, b and c in (2), we get

$$F = k\,r^2\,m\,\omega^2$$
$$\Rightarrow F = k\,m\,\omega^2\,r^2$$

(iii) Since the satellite revolves round the due to the gravitational attraction, the universal gravitational constant G will come in the expression.

Let $M \propto r^a\,T^b\,G^c$

$$M \propto r^a\,T^b\,G^c \qquad (4)$$

where a, b and c are the dimensions of r, T and G respectively and k is a dimensionless constant.

$$\therefore \quad [M] = [r]^a\,[T]^b\,[G]^c$$

$$[M] = [L]^a\,[T]^b\,[M^{-1}L^3T^{-2}]^c$$

$$\Rightarrow \quad ML^0T^0 = M^{-c}L^{a+3c}T^{b-2c} \qquad (5)$$

Equating the dimensions of M, L and T on both sides of (5), we get

$$-c = 1 \Rightarrow c = -1$$
$$a + 3c = 0 \Rightarrow a = -3c = 3$$
$$b - 2c = 0 \Rightarrow b = 2c = -2$$

Putting these values of *a*, *b* and *c* in (4), we get

$$M = k r^3 T^{-2} G^{-1}$$

$$\Rightarrow \quad M = k\frac{r^3}{T^2 G}$$

Errors in Measurement

Whenever a quantity is measured, there is always some error in its measurement because of the lack of perfection of the observer and the sensitivity limit (or the least count) of the measuring instrument. Hence, the true value of a physical quantity can never be ascertained.

If X be the true value of a physical quantity and $\bar{X}$ be the average of a equally reliable measurements $X_1, X_2, \ldots\ldots, X_i, \ldots, X_n$ of that physical quantity then $x_i = X_i - X$ is defined as the error in the *i*th observation and $\delta_i = X_i - \bar{X}$ is defined as the deviation (or residual) in the ith observation. As the true value of a physical quantity can never be ascertained, the error can never be ascertained and what we compute is the deviation.

Consider a physical quantity X that depends on three other quantities in the following manner:

$$X = K\frac{A^m B^n}{C^p}$$

Where *m*, *n*, *p* and *K* are constant.

Differentiating (1.4), we get

$$dX = K\left[m.\frac{A^{m-1}B^n dA}{C^p} + n.\frac{A^B B^{n-1} dB}{C^p} - \frac{rA^m B^n dC}{C^{r+1}}\right]$$

Dividing throughout by *X*, this gives

$$\frac{dX}{X} = m\frac{dA}{A} + n\frac{dB}{B} - r\frac{dC}{C}$$

Here, *dA*, *dB* and *dC* are the absolute errors in *A*, *B* and *C* respectively; $\frac{dA}{A}, \frac{dB}{B}$ and $\frac{dC}{C}$ are the relative errors in A, B and C respectively, and $\frac{dX}{X}$ gives the fractional error in *X*.

The maximum possible fractional error is obtained by taking all the signs as positive and hence, the maximum possible fractional error in X is,

$$\frac{dX}{X} = m\frac{dA}{A} + n\frac{dB}{B} + r\frac{dC}{C}$$

The maximum percentage error in obtained by multiplying the value by 100.

Illustration 7

A physical quantity *x* is calculated from the relation

$$x = \frac{a^2 b^3}{c\sqrt{d}}$$

If percentage error in *a*, *b*, *c* and *d* are 2%, 1%, 3% and 4% respectively, what is the percentage error in *x*.

Solution

$$\frac{\Delta x}{x} = \pm\left(2\frac{\Delta a}{a} + 3\frac{\Delta b}{b} + \frac{\Delta c}{c} + \frac{1}{2}\frac{\Delta d}{d}\right)$$

$$\therefore \quad \frac{\Delta x}{x} \times 100 = \pm\,(2 \times 2\% + 3 \times 1\% + 3\% + \frac{1}{2} \times 4\%)$$
$$= \pm(4\% + 3\% + 3\% + 2\%)$$
$$= \pm 12\%$$

Order of Magnitude

Very often, we come across a situation where a quick approximation is needed for a physical problem. Based on this information, we can decide whether a more precise calculation is necessary. Such an approximation is based on certain assumptions which can be modified depending upon the precision in the situation. To simplify this, we sometimes refer to an order of magnitude of certain quantity as the power of ten of a number which describes that physical quantity.

The order of magnitude gives a feel of largeness or smallness of a quantity.

Calculation of Order of Magnitude

Express each number in the form $a \times 10^b$ where $1 \le a < 10$ and *b* is a positive or negative integer. To get an approximate idea of the measure of the quantity, we may round off the number *a* to 1 if it is less than or equal to 5 and to 10 if it is greater than 5. Therefore the number can be expressed as 10^b, i.e., powers of 10.

Illustration 8

The distance from earth to nearest large galaxy Andromeda is 2×10^{22} m and the mean radius of earth is 6.4×10^6 m. Compare the order of the magnitude of the two.

Solution

Let us express the distance of the nearest galaxy as $a \times 10^b$ here $a = 2$ and $b = 22$, *a* is the less than 5, therefore we can round this off to 1.

Hence, order of magnitude is 10^{22}.

Similarly for radius of earth $a = 6.4$, $b = 6$ *a* could be rounded off to 10 as 6.4 is greater than 5. Therefore, order of magnitude becomes 10^7. Let us compare the order of magnitude in the given two cases $\frac{10^{22}}{10^7} = 10^{15}$.

It means the distance of the nearest galaxy from the earth is approximately 10^{15} times greater than the radius of earth.

Significant Figures

In measurement of certain quantities, the measured values are known only to the limits of experimental uncertainty. This uncertainty may be because of many factors such as the skill of the experimenter, quality of the apparatus etc. The number of significant figures is used in measurements to express this uncertainty.

For example, suppose we have to measure the area of a rectangular plate. If we claim that the area is 71.55 cm^2, our answer would be unjustifiable as it contains 4 significant figures, which is greater than the number of significant figures in either of the measured lengths. Then what are the significant figures in the calculation of area? The rule is that when multiplying or dividing several quantities, the number of significant figures in final calculation is the same number as the number of significant figures in the quantity having the lowest number of significant figure.

Now let us apply this rule to our area problem using a metre stick whose least count is 1mm (0.1 cm).

Let the measured value of length and breadth be respectively 13.5 cm and 5.3 cm. We can say that its length lies between 13.4 cm and 13.6 cm (13.5 ± 0.1). Similarly, breadth = (5.3 ± 0.1) cm. In the measurement of length we have 3 significant figures while in the measurement of breadth, we have 2.

Let us find the area of the plate by multiplying length and breadth.

$$\text{Area} = 13.5 \times 5.3$$
$$= 71.55 \text{ cm}^2$$

The calculated value is 71.55 cm^2 but the least value of significant figures is 2– that of breadth (5.3 cm).

Thus, we can claim that area is 71 cm^2. This value can vary between $13.4 \times 5.2 = 69.68$ and $13.6 \times 5.4 = 73.44$, i.e., 69 and 73.

Zeros Rule

Zero may or may not be a significant figure. If zeros are used to indicate the position of decimal point in a number, they are not counted in significant figures. For an example, number 0.007 and 0.00084 have respectively 1 and 2 significant figures. When the position of zeros comes after other digits there is possibility of ambiguity. Suppose the mass of an object is 2500 gram. This value is ambiguous because we cannot know if the last two zeros are being used to locate the decimal point or to represent significant figures. To sort out this ambiguity, we use scientific notation. The given mass 2500g could be expressed as 2.5×10^3 g if the measured value has 2 significant figures, 2.50×10^3 for 3 significant figures.

Precision and Accuracy of Measuring Instruments

We use various instruments to measure a physical quantity depending upon the precision we require. For example, we can measure the thickness of an object with the help of a metre scale as well as by Vernier Calipers. The difference lies in the least count of the two instruments. Therefore, the precision of a measuring instrument depends upon the least count of the instrument. The smaller the least count, the more precise the measured value when the numbers are added or subtracted. The number of decimal places in the result should be equal to the smallest number of decimal places of any term used in calculation.

SOLVED EXAMPLES

Example 1

The value of gravitational constant is 6.67×10^{-8} in CGS unit. What would be its numerical value in SI unit?

Solution

Dimension of

$$G = [\text{M}^{-1}\text{L}^3\,\text{T}^{-2}]$$

$$\therefore \quad n_2 = n_1 \left[\frac{\text{M}_1}{\text{M}_2}\right]^{-1} \left[\frac{\text{L}_1}{\text{L}_2}\right]^{3} \left[\frac{\text{T}_1}{\text{T}_2}\right]^{-2}$$

$$= 6.67 \times 10^{-8} \left[\frac{1\text{g}}{1\text{kg}}\right]^{-1} \left[\frac{1 \text{ cm}}{1 \text{ m}}\right]^{3} \left[\frac{1\text{sec}}{1\text{sec}}\right]^{-2}$$

$$= 6.67 \times 10^{-8} (10^{-3})^{-1} (10^{-2})^{3} (1)^{-2}$$

$$= 6.67 \times 10^{-11}$$

Example 2

Show dimensionally that the frequency n of the transverse waves in a string of length l and mass per unit length m under a tension T is given by

$$n = \frac{k}{l}\sqrt{\left(\frac{T}{m}\right)}$$

Solution

Here, $n = l^a\ T^b\ m^c$

or
$$n = kl^a\ T^b\ m^c$$

Taking the dimensions of both the sides, we have

$$[\text{T}^{-1}] = [\text{L}]^a [\text{M L T}^{-2}]^b\ [\text{M L}^{-1}]^c$$

or
$$\text{T}^{-1} = \text{L}^{a+b-c}\ \text{M}^{b+c}\ \text{T}^{-2b}$$

Comparing the powers of L,M and T on both sides, we get

$$a + b + c = 0,$$
$$b + c = 0$$
$$-2b = -1$$

On solving we get $a = -1$, $b = (1/2)$ and $c = (1/2)$

$$\therefore \quad n = k\,l^{-1}\ T^{1/2}\ m^{-1/2}$$

or
$$n = \frac{k}{l}\sqrt{\left(\frac{T}{m}\right)}$$

Example 3

A gas bubble, from an explosion under water, oscillates with a period proportional to $P^a d^b E^c$, where p is the static pressure, d is the density and E is the total energy of the explosion. Find the values of a, b and c.

Solution

Let the period of oscillation be t, then

$$t \propto p^a d^b E^c$$

$$\Rightarrow \quad t = K.p^a d^b E^c \qquad \text{(i)}$$

where K is a constant.

$$\Rightarrow \quad [t] = [p]^a [d]^b [E]^c$$

$$\Rightarrow \quad \mathrm{T} = [\mathrm{ML^{-1}T^{-2}}]^a [\mathrm{ML^{-3}}]^b [\mathrm{ML^2T^{-2}}]^c$$

$$\Rightarrow \quad \mathrm{M^0L^0T} = \mathrm{M}^{a+b+c}\mathrm{L}^{-a-3b+2c}.\mathrm{T}^{-2a-2c}$$

Now, equating the dimensions of M, L and T on both sides of (ii), we get

$$a + b + c = 0$$
$$-a - 3b + 2c = 0$$
$$-2a - 2c = 0$$

On Solving, we get

$$a = -\frac{5}{6}, b = \frac{1}{2}, c = \frac{1}{3}$$

Example 4

If the velocity of light c, gravitational constant G and Planck's constant h be chosen as fundamental units, find the dimensions of mass, length and time in the new system.

Solution

The dimensions of the velocity of light c, gravitational constant G and Planck's constant h in terms of mass M, length L and time T are given by

$$[c] = \mathrm{LT^{-1}} \qquad (1)$$
$$[G] = \mathrm{M^{-1}L^3T^{-2}} \qquad (2)$$
$$[h] = \mathrm{ML^2T^{-1}} \qquad (3)$$

Multiplying (1) and (3), we get

$$[c][h] = \mathrm{ML^3T^{-2}} \qquad (4)$$

Dividing (4) by (2), we get

$$\frac{[c][h]}{[G]} = \frac{\mathrm{ML^3T^{-2}}}{\mathrm{M^{-1}L^3T^{-2}}} = \mathrm{M^2}$$

$$\Rightarrow \quad [\mathrm{M}] = [c^{1/2} h^{1/2} c^{-1/2}] \qquad (5)$$

Multiplying (2) and (3), we get

$$[G][h] = \mathrm{M^1L^3T^{-2}.ML^2} = \mathrm{L^5T^{-3}} \qquad (6)$$

From (1) and (6)

$$\frac{[G][h]}{[c]^3} = \frac{\mathrm{L^5T^{-3}}}{\mathrm{L^3T^{-3}}} = \mathrm{L^2}$$

$$\Rightarrow \quad [\mathrm{L}] = \left[G^{1/2}h^{1/2}c^{3/2}\right] \qquad (7)$$

Putting the dimension of L in (1), we get

$$[\mathrm{L}][\mathrm{T^{-1}}] = [c]$$

$$\Rightarrow \quad [\mathrm{T}] = \frac{[\mathrm{L}]}{[c]} = \frac{[G^{1/2}h^{1/2}c^{-3/2}]}{[c]}$$

$$\Rightarrow \quad [\mathrm{T}] = \left[G^{1/2} \; h^{1/2} \; c^{-5/2}\right] \qquad (8)$$

Hence, form (5), (7) and (8), we have the dimensions of mass, length and time as

$$[\mathrm{M}] = [c^{1/2} \; h^{1/2} \; G^{-1/2}]$$
$$[\mathrm{L}] = [G^{1/2} \; h^{1/2} \; c^{-3/2}]$$
$$[\mathrm{T}] = [G^{1/2} \; h^{1/2} \; c^{-5/2}]$$

Example 5

For what value of n is the equation

$$\int \frac{dx}{\sqrt{2ax - x^2}} = a^n \sin^{-1}\left[1 - \frac{x}{a}\right] \text{ dimensionally correct?}$$

Solution

The given equation will be dimensionally correct only if the dimensions of both sides are the same.

An inspection of the left-hand side of the given equation shows that a and x have the same dimension as we are subtracting x^2 from $2ax$. Therefore, the L.H.S. of the given equation is dimensionless.

Now, since a and x have the same dimension, x/a is dimensionless and therefore, $\sin^{-1}[1 - x/a]$ is a non-dimensional constant.

Hence, the given equation will be dimensionally correct only if a^n is dimensionless and this will happen only if $n = 0$.

Example 6

Assuming that the period of vibration of a tuning fork depends upon the length l of its prongs, the density d and the Young's modulus Y of the material, find, by the method of dimensions, a formula for the period of vibration.

Solution

Let the period vibration be t. Then,

$$t \propto l^a \; d^b \; Y^c$$

$$\Rightarrow \quad t = k \; l^a \; d^b \; Y^c \qquad (1)$$

where, a, b and c are the dimensions of l, d and Y respectively and k is a dimensionless constant.

$$\therefore \quad [t] = [l]^a \; [d]^b \; [Y]^e$$

$$\Rightarrow \quad \mathrm{T} = [\mathrm{L}]^a [\mathrm{ML^{-3}}]^b [\mathrm{ML^{-1}T^{-2}}]^c$$

$$\Rightarrow \quad \mathrm{T} = \mathrm{M}^{b+c}\mathrm{L}^{a-3b-c}\mathrm{T}^{2c}$$

Equating the dimensions of M, L and T on both sides of the above equation, we get

$$b + c = 0$$
$$a - 3b - c = 0$$
$$-2c = 1$$

Solving these three equations, we get

$$a = 1,\ b = ½ \text{ and } c = -½$$

Putting these values of a, b and c in (1), we get

$$t = k\, l\, d^{½}\, Y^{½}$$

$$\Rightarrow \quad t = kl\sqrt{\frac{d}{Y}}$$

Example 7

The period of oscillation of a small drop of a liquid under surface tension depends upon the density ρ, radius r and surface tension s. Show, by the method of dimensions, that the period of oscillation

$$T \propto \sqrt{\frac{\rho r^3}{s}}$$

Solution

Let $T \propto \rho^a r^b s^c$

$$\Rightarrow \quad T = k\,\rho^a r^b s^c \tag{6}$$

where k is a dimensionless constant and a, b and c are the dimensions of ρ, r and s respectively.

Taking dimensions of both sides of (6), we get

$$[T] = [\rho]^a [r]^b [s]^c$$

$$\Rightarrow \quad T = [\mathrm{ML^{-3}}]^a [\mathrm{L}]^b [\mathrm{MT^{-2}}]^c$$

$$\Rightarrow \quad \mathrm{M^0L^0T = M^{a+c}L^{-3a+b}T^{-2c}} \tag{7}$$

Equating the dimensions of M, L and T on both sided of (7), we get

$$a + c = 0$$
$$-3a + b = 0$$
$$-2c = 1$$

Solving, we get, $a = ½$, $b = 3/2$, $c = -½$

Putting these values of a, b and c in (6), we get

$$T = k\rho^{½}\, r^{³⁄₂} s^{½}$$

$$\Rightarrow \quad T = k\sqrt{\frac{\rho r^3}{s}}$$

$$\Rightarrow \quad T \propto \sqrt{\frac{\rho r^3}{s}}$$

Example 8

Find the percentage error in specific resistance given by

$$\rho = \frac{\pi r^2 + R}{l}$$

where r is the radius having value (0.2 ± 0.02) cm, R, the resistance is (60 ± 2) ohm and l is the length (150 ± 0.1) cm.

Solution

$$\frac{\Delta\rho}{\rho} = \pm\left(2\frac{\Delta r}{r} + \frac{\Delta R}{R} + \frac{\Delta l}{l}\right)$$

$$= \pm\left(2\times\frac{0.02}{0.2} + \frac{2}{60} + \frac{0.1}{150}\right)$$

$$= \pm(0.2 + 0.0333 + 0.00066)$$

$$= +\,0.2340$$

Now, % error $= \pm\, 0.2340 \times 100$

$$= \pm\, 23.40\%$$

Example 9

The length and breadth of a rectangle are (5.7 ± 0.1) cm and (3.4 ± 0.2) cm. calculate area of the rectangle with error limits.

Solution

Here, $\quad l = (5.7 \pm 0.1)$ cm

$\quad b = (3.4 \pm 0.2)$ cm

Area $\quad A = l \times b$

$$= 5.7 \times 3.4 = 19.38 \text{ cm}^2$$

$$= 19.0 \text{ cm}^2$$

(Rounding off to two significant figures)

$$\frac{\Delta A}{A} = \pm\left(\frac{\Delta l}{l} + \frac{\Delta b}{b}\right)$$

$$= \pm\left(\frac{0.1}{5.7} + \frac{0.2}{3.4}\right)$$

$$= \pm\left(\frac{0.34 + 1.14}{5.7 \times 3.4}\right)$$

$$\frac{\Delta A}{A} = \pm\frac{1.48}{19.38}$$

$$\Delta A = \pm\frac{1.48}{19.38} \times A$$

$$= \pm\frac{1.48}{19.38} \times 19.38 = \pm 1.48$$

$$\Delta \mathrm{A} = \pm\, 1.5$$

(Rounding off to two significant figures)

$\therefore \quad$ Area = (19.0 ± 1.5) sq. cm.

Dimensional Analysis

To get an exact idea of the magnitude of any physical quantity, we need

(i) a standard or unit in which the quantity is measured and

(ii) a number that shows the number of times the physical quantity under question contains that unit. Larger the unit, the smaller the numerical value of the quantity.

(iii) The principle of homogeneity of dimensions states that the dimensions of the terms on the two sides of the expression must be the same.
Since pure numbers are dimensionless, the dimension of ½ gt^2 is the same as that of gt^2. Because of this, the equation $h = gt^2$ is also dimensionally correct. A dimensionally correct equation, therefore, need not be actually correct but, a dimensionally wrong equation is necessarily wrong.

Limitations of the dimensional analysis

(i) It gives no information about the pure numeric and non-dimensional constants involved in various physical relations and which, therefore, have to be determined by separate calculation or experiment.
For example, in the equation $S = ut + \frac{1}{2}at^2$, the constant $\frac{1}{2}$ cannot be obtained from dimensional analysis.

(ii) Its use is restricted solely to physical relationships which are power functions. Relationships involving exponential, logarithmic or trigonometric functions thus lie quite outside its purview.
For example, the formula for the refractive index of the material of the prism, *viz.*

$$\mu = \frac{\sin\frac{A+\delta m}{2}}{\sin\frac{A}{2}}$$ cannot be obtained by this method.

POINTS TO REMEMBER

1. If n_1 and n_2 are the numerical values of a physical quantity corresponding to the units u_1 and u_2 respectively then, $n_1 u_1 = n_2 u_2$
Hence, larger the unit the smaller the numerical value of the physical quantity.
2. The fundamental units cannot be derived from any other physical quantity or can be resolved into anything more fundamental.
All other units which can be derived from fundamental units are called derived units.
3. The following table shows the fundamental quantities used in different system of units with their units and symbols.

Fundamental quantities	Units and symbols in			
	CGS	FPS	MKS	SI
(i) Mass	gram (g)	Pound (lb)	Kilogram (kg)	Kilogram (kg)
(ii) Length	centimetre (cm)	Foot (ft)	Metre (m)	Metre (m)
(iii) Time	second (s)	second (s)	second (s)	second (s)
(iv) Temperature	–	–	–	kelvin (K)
(v) Luminous intensity	–	–	–	candela (cd)
(vi) Electric current	–	–	–	ampere (A)
(vii) Amount of substance	–	–	–	mole (mol)

4. In SI, there are two supplementary units
(i) Radian (rad) – unit of plane angle
(ii) Steradian (sr) – unit of solid angle
5. The fundamental quantities are also called the base quantities.
6. A dimensional formula is an expression showing how and which of the fundamental units enter into the unit of a physical quantity.
7. To find the dimensional formula for a physical quantity, we should know at least one equation in which it appears. For example the dimensional formula for the pressure can be found out from the following equation:

$$\text{Pressure} = \frac{\text{Force}}{\text{Area}} = \frac{MLT^{-2}}{L^2} = ML^{-1}T^{-2}$$

8. Physical quantities may be variables or constants and these may be both dimensional and non-dimensional. Therefore we can define following four types of physical quantities.
(i) Dimensional variables: These are quantities like acceleration, velocity, force etc.
(ii) Dimensional constants: These are quantities having a constant value and dimension. Examples are velocity of light in vacuum, gravitational constant, etc.
(iii) Non-dimensional variables: These are quantities which are variables and yet have no dimension. A few examples are specific gravity, strain, angle, etc. Such quantities are also called 'numerics'.
(iv) Non-dimensional constants: These are merely numbers like 3, 8, π, e, etc.
9. The principal of homogeneity of dimensions states that the dimensions of the terms of two sides of the expression must be the same.
10. The analysis of various physical quantities involved in a dimensional equation is known as dimensional analysis. Its three major uses are the following:

(i) Checking the dimensional homogeneity of equation
(ii) Conversion of one system of units into another
(iii) Establishing a relationship among different physical quantities.

11. If an equation is dimensionally correct, it does not mean that the equation is necessarily correct. But, if an equation is correct, it must be dimensionally correct.
For example, the dimensions of $\frac{1}{2}gt^2$ is same as that of gt^2.

12. Dimensional analysis gives no information about the pure numeric and non-dimensional constants involved in various physical relations. These constants, therefore have to be established by experiment.
The use of dimensional analysis is restricted to physical relationships which are power functions. Relationships involving trigonometric exponential and logarithmic functions cannot be used for analysis.

13. The magnitudes of physical quantities vary over a wide range. For example, the radius of an atomic nucleus is 5.3×10^{-9} m whereas the radius of the earth is approximately 6.4×10^6 m. To express the magnitudes of such widely varying quantities, we use certain standard prefixes for certain powers of ten.

14. In powers of ten method, each number is expressed as $a \times 10^b$, where $1 \le a < 10$, and b is an integer, positive or negative. To get an approximate idea of a number, the number a is rounded off to 1, if $a \le 5$ and to 10, if $a > 5$. The given number can then be approximated as 10^b.

15. Whenever a quantity is measured, there is always some error in its measurement because of the lack of perfection of the observer and the sensitivity limit (or the least count) of the measuring instrument. Hence, the true value of a physical quantity can never be ascertained.

16. If X be the true value of a physical quantity and $\overline{X}$ be the average of n equally reliable measurements $X_1, X_2, \ldots, X_i, \ldots, X_n$ of that physical quantity then $x_i = X_i - X$ is defined as the error in the ith observation and $\delta_i = X_i - \overline{X}$ is defined as the deviation (or residual) in the ith observation. As the true value of a physical quantity can never be ascertained, the error can never be ascertained and what we compute is the deviation.

17. Consider a physical quantity X that depends on three other quantities in the following way:

$$X = K\frac{A^m B^n}{C^p}$$

where m, n, p and K are constants.
Here dA, dB and dC are the absolute errors in A, B and C respectively.
$\frac{dA}{A}, \frac{dB}{B}$ and $\frac{dC}{C}$ are the relative errors in A, ,B and C respectively, and $\frac{dX}{X}$ gives the fractional error in X.

18. The maximum possible fractional error is obtained by taking all the signs as positive in (17). Hence, the maximum possible fractional error in X is,

$$\frac{dX}{X} = m\frac{dA}{A} + n\frac{dB}{B} + p\frac{dC}{C}.$$

EXERCISES

1. The expression $[ML^{-1}T^{-2}]$ does not represent
(a) pressure (b) stress
(c) Young's modulus (d) force.

2. Which is dimensionless ?
(a) Force/Acceleration (b) Velocity/Acceleration
(c) Volume/Area (d) Energy/Work

3. The velocity v of waves produced in water depends on their wavelength λ, the density of water ρ, and acceleration due to gravity g. The square of velocity is proportional to
(a) $\lambda^{-1}g^{-1}\rho^{-1}$ (b) λg
(c) $\lambda\rho g$ (d) $\lambda^2 g^{-2}\rho^{-1}$

4. The number of particles crossing a unit area perpendicular to x-axis in unit time is given by $n = -D\frac{n_2 - n_1}{x_2 - x_1}$ where n_1 and n_2 are number of particles per unit volume for the value of x meant for x_2 and x_1.
The dimensions of the diffusion constant D are
(a) $[M^0 L^1 T^{-1}]$ (b) $[M^0 L^2 T^{-1}]$
(c) $[M^1 L^1 T^{-1}]$ (d) $[M^1 L^2 T^{-2}]$

5. SI units of coefficient of viscosity are
(a) N m/s (b) N / m^2
(c) N m^2 s^{-1} (d) N s/m^2

6. Dimensions of thermal resistance are
(a) $M^{-1} L^{-2} T^3 K^{-1}$ (b) $M^{-1} L^{-2} T^3 K^{-2}$
(c) $M^{-1} L^{-2} T^3 K$ (d) $ML^{-2} T^3 K$

7. The time dependence of physical quantity P is given by $P = P_0 e^{-\alpha t^2 + \beta t + \gamma}$, where α, β, γ are constant and their dimensions are given by (where t is time)
(a) $M^0 L^0 T^{-2}$, $M^0 L^0 T^{-1}$, $M^0 L^0 T^0$
(b) $M^0 L^{-1} T^{-2}$, $M^0 L^0 T^{-1}$, $M^0 L^0 T$
(c) $M^0 L^0 T^{-1}$, $M L T^{-2}$, $M^0 L^0 T^{-1}$
(d) $M L T$, $M L T^0$, $M^0 L^0 T^0$

8. The de-Broglie wavelength associated with a particle of mass *m* and velocity *v* is *h/mv*. The dimensional formula of Planck's constant is
(a) $[M^2 L^2 T^{-1}]$ (b) $[M^2 L^2 T^{-3}]$
(c) $[M L^2 T^{-1}]$ (d) $[M^2 L^2 T^{-2}]$

9. The potential energy of a particle varies with distance *x* from a fixed origin as $V = \frac{A\sqrt{x}}{x+B}$ where A and B are constants. The dimensions of AB are
(a) $M L^{5/2} T^{-2}$ (b) $M^1 L^2 T^{-2}$
(c) $M^{3/2} L^{5/2} T^{-2}$ (d) $M^1 L^{7/2} T^{-2}$

10. If *e* is the charge, *V* is the potential difference, *T* is temperature, the units of $\frac{eV}{T}$ are the same as of
(a) Planck's constant (b) Stefan's constant
(c) Boltzmann constant (d) Gravitational constant.

Data for Questions 11 and 12

This section contains Statement-I and Statement-2. Of the four choices given here, choose the one that best describes the two statements
(a) Statement-1 is false, Statement-2 is true
(b) Statement-1 is true, Statement-2 is true; Statement-2 is a correct explanation for Statement-1
(c) Statement-1 is true, Statement-2 is true; Statement-2 is not a correct explanation for Statement-1
(d) Statement-1 is true, Statmetn-2 is false.

11. Statement–1: When we change the unit of measurement of a quantity, its numerical value changes.
and
Statement–2: Smaller the unit of measurement smaller is its numerical value.

12. Statement–1: Energy cannot be divided by volume
and
Statement–2: Dimensions of energy and volume are different.

13. The physical quantities not having same dimensions are
(a) torque and work
(b) momentum and Plank's constant
(c) stress and Young's modulus
d) speed and $(\mu_0 \varepsilon_0)^{-1/2}$

14. Identify the pair whose dimensions are equal:
(a) Torque and work (b) Stress and energy
(c) Force and stress (d) Force and work

15. Using mass (M), length (L), time (T) and electric current (A) as fundamental quantities the dimensions of permittivity will be
(a) $[M L T^{-1} A^{-1}]$ (b) $[M L T^{-2} A^{-2}]$
(c) $[M^{-1} L^{-3} T^4 A^2]$ (d) $[M^2 L^{-2} T^{-2} A^2]$

16. The terminal velocity v_r of a small steel ball of radius *r* falling under gravity through a column of a viscous liquid of coefficient of viscosity η depends on mass of the ball *m*, acceleration due to gravity *g*, coefficient of viscosity η and radius, *r*. Which of the following relations is dimensionally correct?
(a) $v_r \propto \frac{mgr}{\eta}$ (b) $v_r \propto mg\,\eta r$
(c) $v_r \propto \frac{mg}{r\eta}$ (d) $v_r \propto \frac{\eta mg}{r}$

17. The equation of state of some gases can be expressed as $\left(P + \frac{a}{V^2}\right) = \frac{R\theta}{V}$, where *P* is the pressure, *V* is the volume, θ the absolute temperature and a and *R* are constants. The dimensional formula of *a* is
(a) $M L^5 T^{-2}$ (b) $M^{-1} L^5 T^{-2}$
(c) $M L^{-1} T^{-2}$ (d) $M L^{-5} T^{-2}$

18. If the velocity of light (*c*), gravitational constant (G) and Planck's constant (*h*) are chosen as fundamental units, then the dimensions of mass in new system is:
(a) $c^{1/2} G^{1/2} h^{1/2}$ (b) $c^{1/2} G^{1/2} h^{-1/2}$
(c) $c^{1/2} G^{-1/2} h^{1/2}$ (d) $c^{-1/2} G^{1/2} h^{1/2}$

19. If *E* = energy, G = gravitational constant, *I* = impulse and *M* = mass, then dimensions of $\frac{GIM^2}{E^2}$ are same as that of
(a) time (b) mass
(c) length (d) force.

20. The refractive index of a material is given by the equation, $n = \frac{A+B}{\lambda^2}$, where A and B are constants. The dimensional formula for B is:
(a) $[M^0 L^2 T^{-1}]$ (b) $[M^0 L^{-2} T^0]$
(c) $[M^0 L^2 T^{-2}]$ (d) $[M^0 L^2 T^0]$

21. If F, T and V denote force, time and velocity respectively, then the dimension for mass 'M' is
(a) $F T V^{-1}$ (b) $F^{-1} T V$
(c) $F^{-1} T^{-1} V$ (d) $F T^{-1} V^{-1}$

22. The SI unit of gravitational potential is
(a) J (b) J kg^{-1}
(c) J kg (d) J kg^2

23. If $x = a - b$, the maximum percentage error in the measurement of *x* will be
(a) $\left(\frac{\Delta a}{a} + \frac{\Delta b}{b}\right) \times 100\%$
(b) $\left(\frac{\Delta a}{a} - \frac{\Delta b}{b}\right) \times 100\%$

(c) $\left(\frac{\Delta a}{a-b}+\frac{\Delta b}{a-b}\right)\times 100\%$

(d) $\left(\frac{\Delta a}{a-b}-\frac{\Delta b}{a-b}\right)\times 100\%$

24. Multiply 102.78 by 0.601 and express the result with correct number of significant figures.
(a) 61.77078 (b) 61.771
(c) 61.77 (d) 61.8

25. The percentage errors in measurement of mass and speed are 3% and 2% respectively. The error in kinetic energy will be,
(a) 6% (b) 7%
(c) 10% (d) 12%

26. The density of a cube is measured by measuring its mass and length of its sides. If the maximum error in the measurement of mass and length are 3% and 2% respectively, the maximum error in the measurement of density would be
(a) 12% (b) 14%
(c) 7% (d) 9%

27. Identify the pair which has different dimensions:
(a) Planck's constant and angular momentum
(b) Impulse and linear momentum
(c) Angular momentum and frequency
(d) Pressure and Young's modulus

28. Which one of the following is not related to time?
(a) Light year (b) Month
(c) Day (d) Time

29. A cube has a side of length 1.2×10^{-2} m. Calculate its volume
(a) 1.7×10^{-6} m^3 (b) 1.73×10^{-6} m^3
(c) 1.70×10^{-6} m^3 (d) 1.732×10^{-6} m^3

30. An equation is given here $\left(P+\frac{a}{V^2}\right)=b\frac{\theta}{V}$ where $P=$ Pressure, $V=$ Volume and $\theta=$ Absolute temperature. If a and b are constants, the dimensions of a will be
(a) $\left[ML^{-5}T^{-1}\right]$ (b) $\left[ML^{5}T^{1}\right]$
(c) $\left[ML^{5}T^{-2}\right]$ (d) $\left[M^{-1}L^{5}T^{2}\right]$

31. The equation of a stationary wave is $y=2A\sin\left(\frac{2\pi vt}{\lambda}\right)\cos\left(\frac{2\pi x}{\lambda}\right)$. Which of the statements is wrong?
(a) The unit of $v\,t$ is same as that of λ
(b) The unit of x is same as that of λ
(c) The unit of $2\pi v/\lambda$ is same as that of $2\pi x/\lambda t$
(d) The unit of v/λ is same as that of x/λ

32. The dimensional formula for magnetic flux is
(a) $ML^2T^{-1}A^3$ (b) $ML^0T^{-2}A^{-2}$
(c) $M^0L^{-2}T^{-2}A^{-2}$ (d) $ML^2T^{-2}A^{-1}$

33. The length and breadth of a metal sheet are 3.124 m and 3.002 m respectively. The area of this sheet up to four correct significant figures is
(a) 9.37 m^2 (b) 9.378 m^2
(c) 9.3782 m^2 (d) 9.37824 m^2

34. In a system of units, the units of mass, length and time are 1 quintal, 1 kilometre and 1 hour respectively. In this system 1 Newton force will be equal to
(a) 1 new unit (b) 129.6 new units
(c) 125.7 new units (d) 10^3 new units

35. Which of the following is a derived unit?
(a) Ampere (b) Candela
(c) Mole (d) Coulomb

36. Which pair do not have equal dimensions?
(a) Energy and torque
(b) Force and impulse
(c) Angular momentum and Planck's constant
(d) Elastic modulus and pressure.

37. The dimensions of pressure gradient are
(a) $[M\,L^{-2}\,T^{-2}]$ (b) $[M\,L^{-2}\,T^{-1}]$
(c) $[M\,L^{-1}\,T^{-2}]$ (d) $[M\,L^{-1}\,T^{-1}]$

38. The equation $\frac{dv}{dt}=At-Bv$ is describing the rate of change of velocity of a body falling from rest in a resisting medium. The dimensions of A and B are
(a) LT^{-3}, T (b) LT^{-3}, T^{-1}
(c) LT, T (d) LT, T^{-1}

39. If $x=ab$, the maximum percentage error in the measurement of x will be
(a) $\left(\frac{\Delta a}{a}\times 100\%\right)\times\left(\frac{\Delta b}{b}\times 100\%\right)$
(b) $\left(\frac{\Delta a}{a}\times 100\%\right)\div\left(\frac{\Delta b}{b}\times 100\%\right)$
(c) $\left(\frac{\Delta a}{a}-\frac{\Delta b}{b}\right)\times 100\%$
(d) $\left(\frac{\Delta a}{a}+\frac{\Delta b}{b}\right)\times 100\%$

40. A wire of length $l=(6\pm 0.06)$ cm and radius $r=(0.5\pm 0.005)$ cm, mass $m=(0.3\pm 0.003)$ gm. Maximum percentage error in density is
(a) 4 (b) 2
(c) 1 (d) 6.8

41. The velocity v of a particle at time t is given by, $v=\frac{a}{t}+\frac{bt}{t^2+c}$. the dimensions of a, b, c are respectively
(a) LT^{-2}, L, T (b) L, L, T^2
(c) L, LT, T^{-2} (d) L, L, LT^2

42. A cube has a side of length 1.2×10^{-2} m. Its volume with correct significant figures is
(a) 1.7×10^{-6} m^3 (b) 1.73×10^{-6} m^3
(c) 1.72×10^{-6} m^3 (d) 1.728×10^{-6} m^3

43. A quantity is represented by $X = M^a L^b T^c$. The % error in measurement of M, L and T are $\alpha\%$, $\beta\%$, $\gamma\%$ respectively. The % error in X would be
(a) $a\alpha - b\beta - c\gamma$ (b) $-a\alpha + b\beta + c\gamma$
(c) $a\alpha - b\beta + c\gamma$ (d) $a\alpha + b\beta + c\gamma$

44. The time period of a body under S.H.M. is represented by: $T = P^\alpha D^\beta S^\gamma$ where P is pressure, D is density and S is surface tension, then values of α, β and γ are
(a) $-\frac{3}{2}, \frac{1}{2}, 1$ (b) $1, 2, \frac{1}{3}$
(c) $-1, -2, 3$ (d) $\frac{1}{2}, \frac{-3}{2}, \frac{-1}{2}$

45. The dimensional formula of magnetic flux is
(a) $\left[M^0L^{-2}T^{-2}A^{-2}\right]$ (b) $ML^0T^{-2}A^{-2}$
(c) $\left[ML^2T^{-2}A^{-1}\right]$ (d) $\left[ML^2T^{-1}A^3\right]$

46. The time of oscillation T of a small drop of liquid depends on radius r, density ρ and surface tension S. The relation between them is given by
(a) $T \propto \sqrt{\frac{\rho r^3}{S}}$ (b) $T \propto \sqrt{\frac{S}{\rho r^3}}$
(c) $T \propto \sqrt{\frac{\rho r^3}{S^2}}$ (d) $T \propto \sqrt{\frac{S^2 r^3}{\pi}}$

Data for Questions 47 and 48

This section contains Statement-I and Statement-2. Of the four choices given here, choose the one that best describes the two Statements

(a) Statement-1 is false, Statement-2 is true
(b) Statements-1 is true, Statement-2 is true; Statements-2 is a correct explanation for Statements-1
(c) Statement-1 is true, Statement-2 is true; Statement-2 is not a correct explanation for statement-1
(d) Statement-1 is true, Statement-2 is false.

47. Statement-1: Dimension of coefficient of viscosity is $[M L^{-1} T^{-1}]$
and
Statement-2: Coefficient of viscosity is force acting per unit area per unit velocity gradient

48. Statement-1: The quantity $\left(1/\sqrt{\mu_0 \varepsilon_0}\right)$ is dimensionally equal to velocity and numerically equal to velocity of light.
and
Statement-2: μ_0 is permeability of free space and ε_0 is the permittivity of free space.

49. Out of the following pairs which one does not have identical dimensions is
(a) moment of inertia and moment of force
(b) work and torque
(c) angular momentum and Planck's constant
(d) impulse and momentum.

50. Dimensions of impulse are same as that of
(a) force (b) momentum
(c) energy (d) acceleration.

51. The dimensions of gravitational constant G are
(a) $[M L T^{-2}]$ (b) $[M L^3 T^{-2}]$
(c) $[M^{-1} L^3 T^{-2}]$ (d) $[M^{-1} L T^{2}]$

52. Which one of the following represents the correct dimensions of the coefficient of viscosity?
(a) $M L^{-1} T^{-1}$ (b) $M L T^{1}$
(c) $M L^{-1} T^{-2}$ (d) $M L^{-2} T^{-2}$

53. The dimensions of $\frac{e^2}{4\pi g \varepsilon_0 hc}$, where e, ε_0, h and c are electronic charge, electric permittivity, Planck's constant and velocity of light in vacuum respectively is
(a) $[M^0 L^0 T^0]$ (b) $[M^1 L^0 T^0]$
(c) $[M^0 L^1 T^0]$ (d) $[M^0 L^0 T^1]$

54. The dimensions of electric potential are
(a) $[M L^2 T^{-2} Q^{-1}]$ (b) $[M L T^{-2} Q^{-1}]$
(c) $[M L^2 T^{-1} Q]$ (d) $[M L^2 T^{-2} Q]$

55. A gas bubble, from an explosion under water, oscillates with a period T proportional to $p^a d^b E^c$, where p is the pressure, d is the density of water and E is the total energy of explosion. The value of a, b and c are
(a) $a = 1, b = 1, c = 2$ (b) $a = 1, b = 2, c = 1$
(c) $a=\frac{5}{6}, b=\frac{1}{2}, c=\frac{1}{3}$ (d) $a=-\frac{5}{6}, b=-\frac{1}{2}, c=\frac{1}{3}$

56. A force F is given by $F = at + bt^2$, where t is time. What are the dimensions of a and b?
(a) $[M L T^{-1}]$ and $[M L T^{-4}]$
(b) $[M L T^{-3}]$ and $[M L T^{-4}]$
(c) $[M L T^{-4}]$ and $[M L T^2]$
(d) $[M L^2 T^3]$ and $[M^{-1} L^2 T^1]$

57. The dimensions of kinetic energy is
(a) $[M^2 L T]$ (b) $[M L^2 T]$
(c) $[M L^2 T^{-2}]$ (d) $[M L^2 T^{-1}]$

58. A physical quantity A is related to four observables a, b, c and as following $A = \frac{a^2 b^3}{c\sqrt{d}}$ The percentage errors of measurement in a, b, c and d are 1%, 3%, 2% and 2% respectively. What is the percentage error in the quantity A?

(a) 12% (b) 7%
(c) 5% (d) 14%

59. A hypothetical experiment conducted to determine Young's formula gives $Y = \frac{\cos\theta\, T^x . \tau}{l^3}$. If Y = Young's modulus, T = time period, τ = torque and l = length, then the value of x is

(a) zero (b) 1
(c) 2 (d) 3

Questions Asked in NEET/CBSE Medicals

60. The unit of thermal conductivity is

(a) $\text{W m}^{-1}\text{K}^{-1}$ (b) J m K^{-1}
(c) $\text{J m}^{-1}\text{K}^{-1}$ (d) W m K^{-1}

61. In an experiment, the percentage of error occurred in the measurement of physical quantities A, B, C and D are 1%, 2%, 3% and 4% respectively. Then the maximum percentage of error in the measurement X, where $X = \frac{A^2 B^{1/2}}{C^{1/3} D^3}$, will be

(a) 10% (b) (3/13)%
(c) 16% (d) −10%

62. Which of the following dimensions will be the same as that of time?

(a) $\frac{L}{R}$ (b) $\frac{C}{L}$
(c) LC (d) $\frac{R}{L}$

63. The main scale of a vernier calipers has n divisions/cm. n divisions of the vernier scale coincide with (n - 1) division of main scale. The least count of the vernier calipers is,

(a) $\frac{1}{(n+1)(n-1)}$ cm (b) $\frac{1}{n}$ cm
(c) $\frac{1}{n^2}$ cm (d) $\frac{1}{n(n+1)}$ cm

64. A student measured the diameter of a small steel ball using a screw gauge of least count 0.001 cm. The main scale reading is 5 mm and zero of circular scale division coincides with 25 division above the reference level. If the screw gauge has a zero error of -0.004 cm, the correct diameter of the ball is

(a) 0.521 cm (b) 0.525 cm
(c) 0.053 cm (d) 0.529 cm

65. A physical quantity of the dimensions of length that can be formed out of c, G and $\frac{e^2}{4\pi\varepsilon_0}$ is [c is velocity of light, G is the universal constant of gravitational and e is charge]

(a) $c^2\left[G\frac{e^2}{4\pi\varepsilon_0}\right]^{1/2}$ (b) $\frac{1}{c^2}\left[\frac{e^2}{G4\pi\varepsilon_0}\right]^{1/2}$
(c) $\frac{1}{c}G\frac{e^2}{4\pi\varepsilon_0}$ (d) $\frac{1}{c}\left[G\frac{e^2}{4\pi\varepsilon_0}\right]^{1/2}$

66. Planck's constant (h), speed of light in vacuum (c) and Newton's gravitational constant (G) are three fundamental constant. Which of the following combinations of these has the dimension of length?

(a) $\frac{\sqrt{hG}}{c^{3/2}}$ (b) $\frac{\sqrt{hG}}{c^{5/2}}$
(c) $\sqrt{\frac{hc}{G}}$ (d) $\sqrt{\frac{Gc}{h^{3/2}}}$

67. If dimensions of critical velocity v_c of a liquid flowing through a tube are expressed as $[\eta^x \rho^y r^z]$ where η, ρ and r are the coefficient of viscosity of liquid, density of liquid and radius of the tube respectively, then the values of x, y and z are given by

(a) −1, −1, −1 (b) 1, 1, 1
(c) 1, −1, −1 (d) −1, −1, 1

68. If force (F), velocity (V) and time (T) are taken as fundamental units, then the dimensions of mass are

(a) $[FVT^{-1}]$ (b) $[FVT^{-2}]$
(c) $[FV^{-1}T^{-1}]$ (d) $[FV^{-1}T]$

69. In an experiment, four quantities a, b, c and d are measured with percentage error 1%, 2%, 3% and 4% respectively. Quantity P is calculated as follows $P = \frac{a^3 b^2}{cd}$. Percentage error in P is

(a) 7% (b) 4%
(c) 14% (d) 10%

70. The pair of quantities having same dimensions is

(a) Impulse and surface tension
(b) Angular momentum and work
(c) Work and torque
(d) Young's modulus and energy.

71. The damping force on an oscillator is directly proportional to the velocity. The units of the constant of proportionality are
(a) kg m s^{-1} (b) kg m s^{-2}
(c) kg s^{-1} (d) kg s

72. The dimensions of $(\mu_0 \varepsilon_0)^{-1/2}$ are
(a) $[L^{1/2}T^{-1/2}]$ (b) $[L^{-1}T]$
(c) $[LT^{-1}]$ (d) $[L^{1/2}T^{1/2}]$

73. The density of a material in CGS system of units is 4 g cm^{-3}. In a system of units in which unit of length is 10 cm and unit of mass is 100 g, the value of density of material will be
(a) 0.04 (b) 0.4
(c) 40 (d) 400

74. The dimensions of impulse are equal to that of
(a) pressure (b) linear momentum
(c) force (d) angular momentum.

75. The dimension of $\frac{1}{2}\varepsilon_0 E^2$, where ε_0 is permittivity of free space and E is electric field, is
(a) ML^2T^{-2} (b) $ML^{-1}T^{-2}$
(c) ML^2T^{-1} (d) MLT^{-1}.

76. A student measures the distance traversed in free fall of a body, initially at rest, in a given time. He uses this data to estimate *g*, the acceleration due to gravity. If the maximum percentage errors in measurement of the distance and the time are e^1 and e^2 respectively, the percentage error in the estimation of g is
(a) $e_2 - e_1$ (b) $e_1 + 2e_2$
(c) $e_1 + e_2$ (d) $e_1 - 2e_2$

77. If the dimensions of a physical quantity are given by $M^aL^bT^c$, then the physical quantity will be
(a) Velocity if $a = 1, b = 0, c = -1$
(b) Acceleration if $a = 1, b = 1, c = -2$
(c) Force if $a = 0, b = -1, c = -2$
(d) Pressure if $a = 1, b = -1, c = -2$

78. If the error in the measurement of radius of a sphere is 2%, then the error in the determination of volume of the sphere will be
(a) 8% (b) 2%
(c) 4% (d) 6%

79. Which two of the following five physical parameters have the same dimensions?
1. Energy density 2. Refractive index
3. Dielectric constant 4. Young's modulus
5. Magnetic field
(a) 1 and 4 (b) 1 and 5
(c) 2 and 4 (d) 3 and 5

80. Dimensions of resistance in an electrical circuit, in terms of dimension of mass M, of length L, of time T and of current I, would be
(a) $[ML^2T^{-2}]$ (b) $[ML^2T^{-1}I^{-1}]$
(c) $[ML^2T^{-3}I^{-2}]$ (d) $[ML^2T^{-3}I^{-1}]$

81. The velocity v of a particle at time t is given by $v = at + \frac{b}{t+c}$, where *a*, *b* and *c* are constants. The dimensions of *a*, *b* and *c* are
(a) [L], [LT] and $[LT^{-2}]$
(b) $[LT^{-2}]$, [L] and [T]
(c) $[L^2]$, [T] and $[LT^{-2}]$
(d) $[LT^{-2}]$, [LT] and [L]

82. The ratio of the dimensions of Planck's constant and that of moment of inertia is the dimensions of
(a) time (b) frequency
(c) angular momentum (d) velocity.

83. The dimensions of universal gravitational constant are
(a) $[M^{-1}L^3T^{-2}]$ (b) $[ML^2T^{-1}]$
(c) $[M^{-2}L^3T^{-2}]$ (d) $[M^{-2}L^2T^{-1}]$

84. The unit of permittivity of free space, ε_0 is
(a) coulomb/Newton-metre
(b) newton- $metre^2$/$coulomb^2$
(c) $coulomb^2$ / newton-$metre^2$
(d) $coulomb^2$/(newton-metre $)^2$

85. The dimensions of Planck's constant equals to that of
(a) energy (b) momentum
(c) angular momentum (d) power.

Answer Keys

EXERCISES

1. (d)	2. (d)	3. (b)	4. (b)	5. (d)	6. (c)	7. (a)	8. (c)	9. (d)	10. (c)
11. (d)	12. (a)	13. (b)	14. (a)	15. (c)	16. (c)	17. (a)	18. (c)	19. (a)	20. (d)
21. (a)	22. (b)	23. (c)	24. (d)	25. (b)	26. (d)	27. (c)	28. (a)	29. (b)	30. (c)
31. (d)	32. (d)	33. (b)	34. (b)	35. (d)	36. (b)	37. (a)	38. (b)	39. (d)	40. (a)

41. (b) 42. (a) 43. (d) 44. (a) 45. (c) 46. (a) 47. (b) 48. (c) 49. (a) 50. (b)
51. (c) 52. (a) 53. (a) 54. (a) 55. (d) 56. (b) 57. (c) 58. (d) 59. (a)

Questions Asked in NEET/CBSE Medicals

60. (a) 61. (c) 62. (a) 63. (c) 64. (d) 65. (d) 66. (a) 67. (c) 68. (d) 69. (c)
70. (c) 71. (c) 72. (c) 73. (c) 74. (b) 75. (b) 76. (b) 77. (d) 78. (d) 79. (a)
80. (c) 81. (b) 82. (b) 83. (a) 84. (c) 85. (c)

Solutions

1. (d)

$$F = ma$$

Dimension MLT^{-2}

Pressure, stress and Young's modules have the same dimensions.

$$P = \frac{F}{A}$$

$$\text{Dimension } \frac{MLT^{-2}}{L^2} = \left[ML^{-1}T^{-2}\right]$$

2. (d)
Energy and work has same units.
∴ Energy/work is dimensionless.

3. (b)
Given $v \propto \lambda^a \rho^b g^c$
Writing the dimensions of various quantity and equating the power of M, L and T,

$$a = 1, \quad b = 0, \quad g = 1$$

So, $\quad v \propto \lambda g$

4. (b)

$$D = \frac{n(x_2 - x_1)}{(n_2 - n_1)} \quad \text{(Numerically)}$$

$[n] = \left[L^{-2}T^{-1}\right]$ (∵ Number per unit area per sec)

$$\left[n_2 - n_1\right] = \left[L^{-3}\right]$$

$$\left[x_2 - x_1\right] = [L]$$

Substituting these, we get

$$[D] = \left[M^0 L^2 T^{-1}\right]$$

5. (d)

$$\frac{F}{A} = \eta \frac{dv}{dx}$$

$$\Rightarrow \quad \eta = \frac{Fdx}{Adv}$$

$$\Rightarrow \quad \eta = \frac{N}{m^2} \cdot \frac{m}{m/s} = Ns/m^2$$

6. (c)

$$\text{Thermal resistance} = \frac{\text{Temperature difference}}{\text{Rate of flow of heat}}$$

$$= \frac{\Delta T}{\left(\frac{\Delta Q}{\Delta T}\right)} = \frac{[K]}{\left(\frac{ML^2T^{-2}}{T}\right)}$$

$$= M^{-1}L^{-2}T^3K$$

7. (a)
$\alpha t^2, \beta t, \gamma$ all are dimensionless

$$\gamma = M^0L^0T^0$$

$$\beta t = M^0L^0T^0 \Rightarrow \beta = M^0L^0T^{0-1}$$

$$\alpha t^2 = M^0L^0T^0 \Rightarrow \alpha = M^0L^0T^{0-2}.$$

8. (c)

$$\lambda = \frac{h}{mv} \Rightarrow h = \lambda mv$$

$$h = \left[LMLT^{-1}\right] = \left[ML^2T^{-1}\right].$$

9. (d)

$$B = x = [L]; A\sqrt{x} = Vx \Rightarrow A = V\sqrt{x}$$

$$A = M^1L^2T^{-2}L^{1/2} = M^1L^{5/2}T^{-2}$$

$$AB = ML^{7/2}T^{-2}$$

10. (c)

$$\frac{eV}{T} = \frac{W}{T} = \frac{PV}{T} = R \text{ and } \frac{R}{N} = \text{Boltzmann constant.}$$

Alternatively

$$\frac{eV}{T} = \frac{[AT]\left[ML^2T^{-3}A^{-1}\right]}{[K]} = \left[ML^2T^{-2}K^{-1}\right]$$

and Boltzmann constant $\quad (k) = \frac{R}{N}$

(where N is Avogadro's number)

$$= \left[ML^2T^{-2}K^{-1}\right]$$

11. (d)
If u_1 and u_2 be the units to measure a quantity Q and n_1, n_2 be the numerical values respectively then

$$Q = n_1u_1 = n_2u_2$$

$$n \propto \frac{1}{u}$$

Smaller the unit of measurement, greater is the corresponding numerical value.

12. (a)
In division, there is no bar that quantities involved must have the same dimensions.

13. (b)

$$\text{Momentum} = \left[\text{MLT}^{-1}\right]$$

$$\text{Planck's constant} = \frac{E}{\nu} = \frac{\text{ML}^2\text{T}^{-2}}{\text{T}^{-1}} = \left[\text{ML}^2\text{T}^{-1}\right]$$

14. (a)
Torque and work have same dimensions.

15. (c)

$$F = \frac{1}{4\pi\varepsilon_0}\frac{q_1q_2}{r^2}$$

$$[F] = \left[\frac{1}{4\pi\varepsilon_0}\frac{q_1q_2}{r^2}\right]$$

$$[\varepsilon_0] = \left[\frac{q_1q_2}{r^2F}\right] = \left[\frac{\text{T}^2\text{A}^2}{\text{L}^2\text{MLT}^{-2}}\right]$$

$$= \left[\text{M}^{-1}\text{L}^{-3}\text{T}^4\text{A}^2\right]$$

16. (c)

$$\eta = \left[\text{ML}^{-1}\text{T}^{-1}\right], g = \left[\text{LT}^{-2}\right], v_r = \left[\text{LT}^{-1}\right]$$

$$\left[\frac{mgr}{\eta}\right] = \left[\frac{\text{MLT}^{-2}\text{L}}{\text{ML}^{-1}\text{T}^{-1}}\right] = \left[\text{L}^3\text{T}^{-1}\right]$$

$$[mg\eta r] = \left[\text{MLT}^{-2}\text{ML}^{-1}\text{T}^{-1}\text{L}\right] = \left[\text{M}^2\text{LT}^{-3}\right]$$

$$\left[\frac{mg}{r\eta}\right] = \left[\frac{\text{MLT}^{-2}}{\text{LML}^{-1}\text{T}^{-1}}\right] = \left[\text{LT}^{-1}\right] = v_r$$

17. (a)

$$\left[\frac{a}{V^2}\right] = [P]$$

$$\left[\text{ML}^{-1}\text{T}^{-2}\right] = \left[\frac{a}{\text{L}^6}\right]$$

$$\left[\text{ML}^5\text{T}^{-2}\right] = [a]$$

$$[a] = \left[\text{ML}^5\text{T}^{-2}\right]$$

18. (c)

$$c^{1/2}G^{1/2}h^{1/2} = \text{L}^{1/2}\text{T}^{-1/2}\text{M}^{-1/2}\text{L}^{3/2}\text{T}^{-1}\text{M}^{1/2}\text{LT}^{-1/2}$$
$$= \text{L}^3\text{T}^{-2}\text{M}^0$$

$$c^{1/2}G^{1/2}h^{-1/2} = \text{L}^{1/2}\text{T}^{-1/2}\text{M}^{-1/2}\text{L}^{-3/2}\text{T}^{-1}\text{M}^{-1/2}\text{L}^{-1}\text{T}^{-1/2}$$
$$= \text{LT}^{-1}\text{M}^{-1}$$

$$c^{1/2}G^{-1/2}h^{1/2} = \text{L}^{1/2}\text{T}^{-1/2}\text{M}^{1/2}\text{L}^{-3/2}\text{T}^{1}\text{M}^{1/2}\text{LT}^{-1/2}$$
$$= \text{L}^0\text{T}^0\text{M}$$

$\therefore$ Mass in new system is, $c^{1/2}G^{-1/2}h^{1/2}$.

19. (a)

$$GM^2 = \text{N.m}^2$$
$$I = \text{N.s}$$
$$E^2 = \text{N}^2.\text{m}^2$$
$$\Rightarrow \quad \frac{GM^2}{E^2} = \frac{\text{N.m}^2.\text{N.s}}{\text{N}^2\text{m}^2} = \text{s} = \text{time}$$

20. (d)
n is dimensionless. A and B have same dimensions and λ^2 has dimensions of L^2. Therefore B has dimensions of $L^2\left[\text{M}^0\text{L}^2\text{T}^0\right]$

21. (a)

$$F = \left[\text{MLT}^{-2}\right]$$
$$V = \left[\text{LT}^{-1}\right] \Rightarrow V^{-1} = \left[\text{L}^{-1}\text{T}\right]$$
$$FTV^{-1} = \left[\text{MLT}^{-2}\right][\text{T}]\left[\text{L}^{-1}\text{T}\right] = [\text{M}]$$

22. (b)

$$\text{Gravitational potential} = \frac{\text{Potential energy}}{\text{Mass}}$$
$$= \text{J/kg} = \text{J kg}^{-1}$$

23. (c)

$$x = a - b \quad \text{(i)}$$
$$\Delta x = \Delta a + \Delta b \quad \text{(ii)}$$

Dividing Eq. (ii) by Eq. (i),

$$\Rightarrow \quad \frac{\Delta x}{x}\times 100 = \left(\frac{\Delta a}{a-b} + \frac{\Delta b}{a-b}\right)\times 100\%$$

24. (d)

$$102.78 \times 0.601 = 61.77078$$

The answer should be in 3 significant figures hence corresponding to the minimum number, 61.77078 rounded to 3 significant figures is 61.8

25. (b)

$$\text{KE} = \frac{1}{2}mv^2$$

$$\frac{\Delta KE}{KE} = \frac{\Delta m}{m} + 2\frac{\Delta v}{v} \times 100$$

$$\frac{\Delta KE}{KE} \times 100 = \frac{\Delta m}{m} + 2\frac{\Delta v}{v} \times 100$$

$$= 3\% + 4\% = 7\%$$

26. (d)

Maximum error in mass $\left(\frac{\Delta m}{m}\right) = 3\% = \frac{3}{100}$ and maximum error in length $\left(\frac{\Delta l}{l}\right) = 2\% = \frac{2}{100}$

Maximum error in n the measurement of density

$$\frac{\Delta \rho}{\rho} = \frac{\Delta m}{m} + \left(3 \times \frac{\Delta l}{l}\right) = \frac{3}{100} + \left(3 \times \frac{2}{100}\right) = \frac{3}{100} + \frac{6}{100}$$

$$= \frac{9}{100} = 9\%$$

27. (c)

Angular momentum $= [ML^2T^{-1}]$, Frequency $= [T^{-1}]$

28. (a)

Light year is the unit of distance.

29. (b)

$$\text{Volume} = \left(1.2 \times 10^2\right)^3$$

$$= 1.728 \times 10^6 \, m^3$$

$$= 1.73 \times 10^6 \, m^3$$

30. (c)

Equation $\left(P + \frac{a}{V^2}\right) = b\frac{\theta}{V}$. Since $\frac{a}{V^2}$ is added to the pressure, there dimensions of $\frac{a}{V^2}$ and pressure (P) will be the same. And dimensions of $\frac{a}{V^2} = \frac{a}{[L^3]^2}[ML^{-1}T^{-2}]$

or $a = \left[ML^5T^{-2}\right]$

31. (d)

Option a, b, and c are dimensionally consistent. Option d is wrong.

$$\text{Dim } \frac{v}{\lambda} = \frac{LT^{-1}}{L} = T^{-1}$$

$$\text{Dim } \frac{x}{\lambda} = \frac{L}{L} = 1$$

32. (d)

$$\text{Magnetic flux } (\phi) = BA = \frac{F.A}{qv}$$

$$= \frac{\left[MLT^{-2}\right]\left[L^2\right]}{[AT]\left[LT^{-1}\right]} = \left[ML^2T^{-2}A^{-1}\right]$$

33. (b)

$$A = l \times b = 3.124 \times 3.002$$

$$= 9.378248 \text{ m}^2$$

Here length and breadth have four significant figures. Therefore, area of the sheet after rounding off to four significant figures is 9.378 m^2.

34. (b)

The dimensional formula of force

$$= [MLT^{-2}]$$

We know that $n_1u_1 = n_2u_2$

$$n_2 = n_1\left[\frac{M_1}{M_2} \times \frac{L_1}{L_2} \times \left(\frac{T_1}{T_2}\right)^{-2}\right]$$

$$= 1\left[\frac{1}{100} \times \frac{1}{100} \times \left(\frac{1 \text{ sec}}{1 \text{ hour}}\right)^{-2}\right]$$

$$= 1\left[\frac{1}{100} \times \frac{1}{100} \times (3600)^2\right]$$

$$= 129.6 \text{ new units.}$$

35. (d)

$$Q = It = \text{Ampere} \times \text{Second} = \text{Coulomb}$$

36. (b)

$$\text{Dimensions of force} = \left[MLT^{-2}\right]$$

$$\text{Dimensions of impulse} = \left[MLT^{-1}\right]$$

37. (a)

$$\text{Pressure gradient} = \frac{dP}{dx}$$

$$\text{Pressure} = \frac{F}{A} = \frac{\left[MLT^{-2}\right]}{\left[L^2\right]} = \left[ML^{-1}T^{-2}\right]$$

$$\text{Pressure gradient} = \frac{\left[ML^{-1}T^{-2}\right]}{[L]} = \left[ML^{-2}T^{-2}\right]$$

38. (b)

$$\left[\frac{dv}{dt}\right] = \frac{LT^{-1}}{T} = \left[LT^{-2}\right]$$

$$[At] = \left[LT^{-2}\right]$$

$$\Rightarrow \quad [A] = \left[LT^{-3}\right]$$

$$[Bv]=[LT^{-2}]$$

$$[B]=\frac{LT^{-2}}{LT^{-1}}=[T^{-1}]$$

39. (d)

$$x=ab \quad \ldots(i)$$

$$\Rightarrow \Delta x=b\Delta a+a\Delta b \quad \ldots(ii)$$

Divide Eq. (ii) by Eq. (i),

$$\Rightarrow \frac{\Delta x}{x}\times 100=\left(\frac{\Delta a}{a}+\frac{\Delta b}{b}\right)\times 100\%$$

40. (a)

$$\rho=\frac{m}{\pi r^2 l}$$

$$\frac{\Delta\rho}{\rho}=\frac{\Delta m}{m}+2\frac{\Delta r}{r}+\frac{\Delta l}{l}$$

$$\Delta m=0.003, \Delta r=0.005 \text{ cm}, \Delta l=0.06 \text{ cm}$$

$$\frac{\Delta\rho}{\rho}=\frac{4}{100} \therefore \frac{\Delta\rho}{\rho}\times 100=4\%$$

41. (b)

$$\frac{a}{t}=v\Rightarrow a=vt=[LT^{-1}.T]=[L]$$

c is added to t^2, $\therefore c=[T^2]$

$$bt=v(t^2+c)\Rightarrow b=\frac{v(t^2+c)}{t}=\frac{LT^{-1}.T^2}{T}=[L]$$

42. (a)

$$l=1.2\times 10^{-2}\,\text{m},$$

$$V=l^3=(1.2\times 10^{-2}\,\text{m})^3=1.728\times 10^{-6}\,\text{m}^3$$

l is given with 2 significant figure, so V should have only 2 significant figures.

$$V=1.7\times 10^{-6}\,\text{m}^3.$$

43. (d)

$$X=M^a L^b T^c$$

$$\frac{\Delta X}{X}\times 100=\left(a\frac{\Delta M}{M}+b\frac{\Delta L}{L}+c\frac{\Delta T}{T}\right)100$$

$$=(a\alpha+b\beta+c\gamma)\%$$

44. (a)

$$T=P^\alpha D^\beta S^\gamma$$

$$M^0L^0T^1=(ML^{-1}T^{-2})^\alpha(ML^{-3})^\beta(MT^{-2})^\gamma$$

$$=M^{\alpha+\beta+\gamma}L^{-\alpha-3\beta}T^{-2\alpha-2\gamma}$$

$$\therefore \quad \alpha+\beta+\gamma=0, -\alpha-3\beta=0, -2\alpha-2\gamma=1$$

On solving we get, $\alpha=\frac{-3}{2}, \beta=\frac{1}{2}, \gamma=1$

45. (c)

Magnetic flux, $\phi=BA=\left(\frac{F}{Il}\right)A$

$$=\frac{[MLT^{-2}][L^2]}{[A][L]}=[ML^2T^{-2}A^{-1}]$$

46. (a)

$$\sqrt{\frac{\rho r^3}{S}}=\sqrt{\frac{ML^{-3}L^3}{MT^{-2}}}=T.$$

47. (b)

Coefficient of viscosity is force acting per unit area per velocity gradient, i.e.,

$$[\eta]=\frac{[F]}{[A][(du/dr)]}=\frac{[MLT^{-2}]}{[L^2][LT^{-1}L^{-1}]}$$

$$=[ML^{-1}T^{-1}]$$

48. (c)

$$[\varepsilon_0]=[M^{-1}L^{-3}T^4A^2]$$

$$[\mu_0]=[MLT^{-2}A^{-2}]$$

So $$\left[\frac{1}{\sqrt{\mu_0\varepsilon_0}}\right]=\left[\frac{1}{\sqrt{[MLT^{-2}A^{-2}][M^{-1}L^{-3}T^4A^2]}}\right]$$

$$=\left[\frac{1}{\sqrt{L^{-2}T^2}}\right]=[LT^{-1}]=[\text{Velocity}]$$

Also, $$\frac{1}{\sqrt{\mu_0\varepsilon_0}}=\frac{1}{\sqrt{(\mu_0/4\pi)\times 4\pi\varepsilon_0}}$$

$$=\sqrt{\frac{9\times 10^9}{10^{-7}}}=\sqrt{9\times 10^{16}}$$

$$=3\times 10^8\,\text{m/s}$$

Therefore $\frac{1}{\sqrt{\mu_0\varepsilon_0}}$ has dimensions of velocity is numerically equal to the velocity of light.

49. (a)

Moment of inertia, $[I]=[mr^2]=[ML^2]$

Moment of force $(C)=rF$

$$[C]=[r][F]=[L][MLT^{-2}]=[ML^2T^{-2}]$$

$\therefore$ Henry has the dimensions of $\frac{ML^2}{Q^2}$.

50. (b)

Impulse = Change in momentum.

51. (c)

$$G = \frac{F.r^2}{m_1 m_2} = \frac{[MLT^{-2}][L^2]}{[M][M]} = [M^{-1}L^3T^{-2}]$$

52. (a)

$$n = \frac{F \times r}{A \times v} = [ML^{-1}T^{-1}]$$

53. (a)

$$\varepsilon_0 = \left[\frac{I^2T^4}{L^3.M}\right], I \text{ is current}$$

$$e^2 = [I^2T^2]$$

$$h = [ML^2T^{-1}]$$

$$c = [LT^{-1}]$$

$$\left[\frac{e^2}{4\pi\varepsilon_0 hc}\right] = \left[\frac{I^2T^2}{I^2T^4L^{-3}M^{-1}ML^2T^{-1}LT^{-1}}\right]$$

$$= [M^0L^0T^0]$$

54. (a)

Potential, $V = \frac{1}{4\pi\varepsilon_0}\frac{q}{r}$

$$\varepsilon_0 = Q^2T^2L^{-3}M^{-1}$$

$$[V] = \left[Q^{-2}T^{-2}L^3M\frac{Q}{L}\right] = [ML^2T^{-2}Q^{-1}]$$

55. (d)

$$T \propto p^a d^b E^c \Rightarrow T = kp^a d^b E^c$$

Using dimensional homogeneity,

$$[M^0L^0T^1] = [M^1L^{-1}T^{-2}]^a [M^1L^{-3}T^0]^b [M^1L^2T^{-2}]^c$$

$$\Rightarrow [M^0L^0T^1] = [M^{a+b+c}.L^{-a+3b+2c}.T^{-2a-2c}]$$

$$\Rightarrow a+b+c=0; -a-3b+2c=0; -2a-2c=1$$

Solving these three equations,

$$a = -\frac{5}{6}; b = \frac{1}{2}; c = \frac{1}{3}$$

56. (b)

$$[F] = [at] \text{ and also } [F] = [bt^2]$$

$$[MLT^{-2}] = [at]$$

$$\Rightarrow \quad [MLT^{-3}] = [a]$$

and $[MLT^{-2}] = [bt^2]$

$$[MLT^{-4}] = [b]$$

57. (c)

$$[MLT^2T^{-2}]$$

58. (d)

$$A = a^2b^3c^{-1}d^{\frac{1}{2}}$$

Percentage error in the quantity A
= 2 (Percentage error in a) + 3 (Percentage error in b)
+ 1 (Percentage error in c) + $\frac{1}{2}$ (Percentage error in d)

$$= (2 \times 1 + 3 \times 3 + 1 \times 2 + \frac{1}{2} \times 2)\%$$

$$= (2 + 9 + 2 + 1)\% = 14\%.$$

59. (a)

$$[Y] = [ML^{-1}T^{-2}]$$

$$[\tau] = [ML^2T^{-2}]$$

$$[l] = [L]$$

θ is dimensionless.

$$[ML^{-1}T^{-2}] = [T]^x[ML^2T^{-2}][L^{-3}]$$

$$= [M^1L^{2-3}T^{-2+x}]$$

Comparing the powers of T,

$$-2 = -2 + x$$

$$x = 0$$

Questions Asked in NEET/CBSE Medicals

60. (a)

$K = \frac{qx}{A(T_1 - T_2)t}$, where q is the amount of heat flow, x is the thickness of the slab, A is the area of cross-section, and t is the time taken.

$$K = \frac{Jm}{m^2K_s} = W\frac{1}{m}\frac{1}{K} = Wm^{-1}K^{-1}$$

61. (c)

$$X = \frac{A^2B^{1/2}}{C^{1/3}D^3}$$

Maximum percentage error in X

$$\left(\frac{dX}{X}\right) \times 100 = \left(2\frac{dA}{A} + \frac{1}{2}\frac{db}{B} + \frac{1}{3}\frac{dC}{C} + 3\frac{dD}{D}\right) \times 100$$

$$= 2 \times 1 + \frac{1}{2} \times 2 + \frac{1}{3} \times 3 + 3 \times 4 = 16\%$$

62. (a)

63. (c)
If *n*th division of vernier scale coincides with (*n* – 1) division of main scale,

$$n \text{ VSD} = (n-1) \text{ MSD}$$

$$\Rightarrow \quad 1 \text{ VSD} = \frac{(n-1)}{n} \text{MSD}$$

$$\therefore \quad \text{Least count} = 1 \text{ MSD} - 1 \text{ VSD}$$

$$= 1 \text{ MSD} - \frac{(n-1)}{n} \text{MSD}$$

$$= 1\text{MSD} - 1 \text{ MSD} + \frac{1}{n}\text{MSD}$$

$$= \frac{1}{n}\text{ MSD}$$

$$= \frac{1}{n} \times \frac{1}{n} = \frac{1}{n^2} \text{ cm} \quad \left[\therefore 1\text{MSD} = \frac{1}{n} \text{ cm}\right]$$

64. (d)
Diameter of the ball

$$= \text{MSR} + \text{CSR} \times (\text{Least count}) - \text{Zero error}$$

$$= 5 \text{ mm} + 25 \times 0.001 \text{ cm} - (-0.004 \text{ cm})$$

$$= 0.5 \text{ cm} + 25 \times 0.001 \text{ cm} - (-0.004) \text{ cm} = 0.529 \text{ cm}$$

65. (d)
Dimensions of

$$\frac{e^2}{4\pi\varepsilon_0} = \left[F \times d^2\right] = \left[\text{ML}^3\text{T}^{-2}\right]$$

Dimensions of $G = \left[\text{M}^{-1}\text{L}^3\text{T}^{-2}\right]$

Dimensions of $c = \left[\text{LT}^{-1}\right]$

$$l \propto \left(\frac{e^2}{4\pi\varepsilon_0}\right)^2 G^q c^r$$

$$\therefore \quad \left[\text{L}^1\right] = \left[\text{ML}^3\text{T}^{-2}\right]^p \left[\text{M}^{-1}\text{L}^3\text{T}^{-2}\right]^q \left[\text{LT}^{-1}\right]^r$$

On comparing both sides and solving, we get

$$p = \frac{1}{2}, q = \frac{1}{2} \text{ and } r = -2$$

$$\therefore \quad l = \frac{1}{c^2}\left[\frac{Ge^2}{4\pi\varepsilon_0}\right]^{1/2}$$

66. (a)
According to question

$$l \propto h^p c^q G^r$$

$$l = kh^p c^q G^r \qquad \text{(i)}$$

Writing dimensions of physical quantities on both sides,

$$\left[\text{M}^0\text{LT}^0\right] = \left[\text{ML}^2\text{T}^{-1}\right]^p \left[\text{LT}^{-1}\right]^q \left[\text{M}^{-1}\text{L}^3\text{T}^{-2}\right]^r$$

Applying the principal of homogeneity of dimensions, we get,

$$p - r = 0 \qquad \text{(ii)}$$

$$2q + q + 3r = 1 \qquad \text{(iii)}$$

$$-p - q - 2r = 0 \qquad \text{(iv)}$$

From Eqs. (ii), (iii) and (iv), we get

$$p = r = \frac{1}{2}, q = -\frac{3}{2}$$

Substituting in (i), we get $l = \frac{\sqrt{hG}}{c^{3/2}}$

67. (c)

$$\left[v_c\right] = \left[\eta^x \rho^y r^z\right] \text{(given)} \qquad \text{(i)}$$

Writing the dimensions of various quantities in Eq. (i), we get

$$\left[\text{M}^0\text{LT}^{-1}\right] = \left[\text{ML}^{-1}\text{T}^{-1}\right]^x \left[\text{ML}^{-3}\text{T}^0\right]^y \left[\text{M}^0\text{LT}^0\right]^z$$

$$= \left[\text{M}^{x+y}\text{L}^{-x-3y+z}\text{T}^{-x}\right]$$

Applying the principal of homogeneity of dimensions, we get

$$x + y = 0; -x - 3y + z = 1; -x = -1$$

On solving, we get $x = 1$, $y = -1$, $z = -1$

68. (d)
Let mass $m \propto F^a V^b T^c$ or $m = kF^a V^b T^c$ (i)
where k is a dimensionless constant and a, b and c are the exponents.
Writing dimensions on both sides, we get

$$\left[\text{ML}^0\text{T}^0\right] = \left[\text{MLT}^{-2}\right]^a \left[\text{LT}^{-1}\right]^b \left[\text{T}\right]^c$$

$$\left[\text{ML}^0\text{T}^0\right] = \left[\text{M}^a\text{L}^{a+b}\text{T}^{-2a-b+c}\right]$$

Applying the principle of homogeneity of dimensions, we get

$$a = 1 \qquad \text{(ii)}$$

$$a + b = 0 \qquad \text{(iii)}$$

$$-2a - b + c = 0 \qquad \text{(iv)}$$

Solving Eqs. (ii), (iii) and (iv), we get

$$a = 1, b = -1, c = 1$$

From Eq. (i), $[m] = \left[\text{FV}^{-1}\text{T}\right]$

69. (c)

As $P = \dfrac{a^3b^2}{cd}$

% error in P is

$$\frac{\Delta P}{P}\times 100 = \left[3\left(\frac{\Delta a}{a}\right)+2\left(\frac{\Delta b}{b}\right)+\frac{\Delta c}{c}+\frac{\Delta d}{d}\right]\times 100$$

$$= [3\times 1\% + 2\times 2\% + 3\% + 4\%] = 14\%$$

70. (c)

Impulse = Force × Time

$$= [\text{MLT}^{-2}][\text{T}] = [\text{MLT}^{-1}]$$

$$\text{Surface tension} = \frac{\text{Force}}{\text{Length}} = \frac{[\text{MLT}^{-2}]}{[\text{L}]} = [\text{ML}^0\text{T}^{-2}]$$

Angular momentum = Moment of inertia × Angular velocity

$$= [\text{ML}^2][\text{T}^{-1}] = [\text{ML}^2\text{T}^{-1}]$$

$$\text{Work} = \text{Force}\times\text{Distance} = [\text{MLT}^{-2}][\text{L}] = [\text{ML}^2\text{T}^{-2}]$$

$$\text{Energy} = [\text{ML}^2\text{T}^{-2}]$$

Torque = Force × Distance =

$$[\text{MLT}^{-2}][\text{L}] = [\text{ML}^2\text{T}^{-2}]$$

$$\text{Young's modulus} = \frac{\text{Force/Area}}{\text{Change in length/Original length}}$$

$$= \frac{[\text{MLT}^{-2}]/[\text{L}^2]}{[\text{L}]/[\text{L}]} = [\text{ML}^{-1}\text{T}^{-2}]$$

Hence, among the given pair of physical quantities work and torque have the same dimensions $[\text{ML}^2\text{T}^{-2}]$

71. (c)

Damping force, $F \propto v$ or $F = kv$

where k is the constant of proportionality

$$k = \frac{F}{v} = \frac{\text{N}}{\text{ms}^{-1}} = \frac{\text{kgms}^{-2}}{\text{ms}^{-1}} = \text{kgs}^{-1}$$

72. (c)

The speed of the light in vacuum is

$$c = \frac{1}{\sqrt{\mu_0\varepsilon_0}} = (\mu_0\varepsilon_0)^{-1/2}$$

$$[(\mu_0\varepsilon_0)^{-1/2}] = [c] = [\text{LT}^{-1}]$$

73. (c)

As $n_1u_1 = n_2u_2$

$$4\frac{\text{g}}{\text{cm}^3} = n_2\frac{100\text{ g}}{(10\text{ cm})^3} \Rightarrow n_2 = 40$$

74. (b)

Impulse = Force × Time

Therefore,

Dimensional formula of impulse = Dimensional formula of force × Dimensional formula of time $= [\text{MLT}^{-2}][\text{T}] = [\text{MLT}^{-1}]$ and dimensional formula of linear momentum $[p] = \text{MLT}^{-1}$

75. (b)

Energy density of an electric field E is

$$u_E = \frac{1}{2}\varepsilon_0 E^2$$

where ε_0 is permittivity of free space

$$u_E = \frac{\text{Energy}}{\text{Volume}} = \frac{\text{ML}^2\text{T}^{-2}}{\text{L}^3} = \text{ML}^{-1}\text{T}^{-2}$$

Hence, the dimension of $\frac{1}{2}\varepsilon_0 E^2$ is $\text{ML}^{-1}\text{T}^{-2}$

76. (b)

From the relation, $h = ut + \frac{1}{2}gt^2$

$$h = \frac{1}{2}gt^2 \Rightarrow g = \frac{2h}{t^2}\text{ (body initially at rest)}$$

Taking natural logarithm on both sides, we get $\ln g = \ln h - 2\ln t$

Differentiating, $\dfrac{\Delta g}{g} = \dfrac{\Delta h}{h} - 2\dfrac{\Delta t}{t}$

For maximum permissible error,

$$\text{or }\left(\frac{\Delta g}{g}\times 100\right)_{\max} = \left(\frac{\Delta h}{h}\times 100\right) + 2\times\left(\frac{\Delta t}{t}\times 100\right)$$

According to problem,

$$\frac{\Delta h}{h}\times 100 - e_1 \text{ and } \frac{\Delta t}{t}\times 100 = e_2$$

$$\text{Therefore, }\left(\frac{\Delta g}{g}\times 100\right)_{\max} = e_1 + 2e_2$$

77. (d)
Pressure

$$P = \frac{\text{Force}}{\text{Area}} = \frac{\text{Mass} \times \text{Acceleration}}{\text{Area}}$$

$$\therefore \quad [P] = \frac{\text{M}^1\text{LT}^{-2}}{\text{L}^2} = \left[\text{M}^1\text{L}^{-1}\text{T}^{-2}\right] = \text{M}^a\text{L}^b\text{T}^c$$

$$\therefore \quad a = 1, b = -1, c = -2$$

78. (d)

$$V = \frac{4}{3}\pi R^3; \ \ln V = \ln\left(\frac{4}{3}\pi\right) + \ln R^3$$

Differentiating, $\frac{dV}{V} = 3\frac{dR}{R}$

Error in the determination of the volume

$$= 3 \times 2\% = 6\%$$

79. (a)

$$[\text{Energy density}] = \left[\frac{\text{Work done}}{\text{Volume}}\right]$$

$$= \frac{\text{MLT}^{-2}\,\text{L}}{\text{L}^3} = \left[\text{ML}^{-1}\text{T}^{-2}\right]$$

$$[\text{Young's modulus}] = [\text{Y}] = \left[\frac{\text{Force}}{\text{Area}}\right] \times \frac{[l]}{[\Delta l]}$$

$$= \frac{\text{MLT}^{-2}}{\text{L}^2} \cdot \frac{\text{L}}{\text{L}} = \left[\text{ML}^{-1}\text{T}^{-2}\right]$$

The dimensions of 1 and 4 are the same.

80. (c)

Ohm's law, $V = RI$ or $R = \frac{V}{I}$

$$\text{Dimensions of } V = \frac{W}{q} = \frac{\left[\text{ML}^2\text{T}^{-2}\right]}{[\text{IT}]}$$

$$R = \frac{\left[\text{ML}^2\text{T}^{-2}/\text{IT}\right]}{[\text{I}]} = \left[\text{ML}^2\text{T}^{-3}\text{I}^{-2}\right]$$

81. (b)

$$v = at + \frac{b}{t+c}$$

As c is added to t, $[c] = [T]$

$$[at] = \left[\text{LT}^{-1}\right] \text{ or } [a] = \frac{\left[\text{LT}^{-1}\right]}{[\text{T}]} = \left[\text{LT}^{-2}\right]$$

$$\frac{[b]}{[T]} = \left[\text{LT}^{-1}\right] \therefore [b] = [\text{L}]$$

82. (b)

$$\frac{h}{I} = \frac{E\lambda}{c \times I} = \frac{\left[\text{ML}^2\text{T}^{-2}\right][\text{L}]}{\left[\text{LT}^{-1}\right] \times \left[\text{ML}^2\right]}$$

$$\frac{h}{I} = \left[\text{T}^{-1}\right] = \text{Frequency}$$

83. (a)
Gravitational constant G

$$= \frac{\text{Force} \times (\text{Distance})^2}{\text{Mass} \times \text{Mass}}$$

$$\therefore \text{ Dimensions of G} = \frac{\left[\text{ML}^2\text{T}^{-2}\right]\left[\text{L}^2\right]}{[\text{M}][\text{M}]}$$

$$= \left[\text{M}^{-1}\text{L}^3\text{T}^{-2}\right]$$

84. (c)
Force between two charges

$$F = \frac{1}{4\pi\varepsilon_0 r^2} \Rightarrow \varepsilon_0 = \frac{1}{4\pi}\frac{q^2}{Fr^2} = \text{C}^2/\text{N-m}^2$$

85. (c)
Dimensions of Planck's constant

$$h = \frac{\text{Energy}}{\text{Frequency}} = \frac{\left[\text{ML}^2\text{T}^{-2}\right]}{\left[\text{T}^{-1}\right]} = \left[\text{ML}^2\text{T}^{-1}\right]$$

Dimensions of angular momentum L

$$= \text{Moment of inertia (I)} \times \text{Angular velocity } (\omega)$$

$$= \left[\text{ML}^2\right]\left[\text{T}^{-1}\right] = \left[\text{ML}^2\text{T}^{-1}\right]$$

QUICK QUESTIONS

Quiz

1. The dimensions of a physical quantity is given, can it be unique?
2. Can there be a physical quantity which has no units and dimensions?
3. Can a physical quantity have a unit without having dimensions?
4. Dimensional analysis of the equations $(\text{velocity})^x = k(\text{pressure difference})^{3/2}\,(\text{density})^{-3/2}$ where k is a dimensionless constant, gives the value of x as?
5. What does the abbreviation 'SI' in SI units stand for?
6. Which of the following physical relations cannot be derived using the method of dimensional analysis – logarithmic, multiplication, division and exponential?
7. If pressure P, velocity V and time T are taken fundamental physical quantities, what is the dimensional formula for force?
8. Will the magnitude of a physical quantity change if there is a change in unit?
9. What is the ratio of 1 giga m to 1 nano m?
10. Arrange the following units of length in increasing order of magnitude: Angstrom, Fermi, X-ray unit, Light year, Astronomical unit.

Think and Grow

11. There are many misconceptions in writing the unit of a physical quantity, such as the unit of force, newton. Some people write Newton instead of newton; similarly some denote it by n while others use N. The other misconception is plurality like 100 m is denoted by 100 ms. Can you explain what is the correct position?
12. Is the time (t) vs. position (x) graph possible in nature as shown in the figure? What happens if the axes are interchanged?

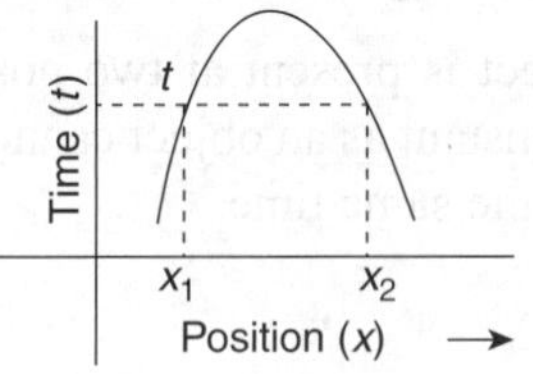

13. In defining the standard of length we have to specify the temperature at which the length has been taken as standard. Are we justified in defining the length as fundamental quantity as temperature has to be specific in choosing a standard?

Explanations

Quiz

1. No, it may not be unique. Surface tension, surface energy and force constant has the same dimension MT^{-2}.
2. Yes, the strain and refractive index.
3. Yes, angle with radian.
4. $x = 3$
5. It has been derived from French name ''Le System International d' units''.
6. Logarithmic and exponential
7. PV^2T^2

$$F = KP^aV^bT^c$$

$$MLT^{-2} = [M^{-1}LT^{-2}]^a[LT^{-}]^bT^c$$

Comparing the dimension

$$a = 1, -a + b = 1;$$
$$-2a - b + c = -2$$
$$b = 2, c = 2.$$

8. The magnitude will change; larger the unit, smaller will be the magnitude.
9. $\dfrac{10^9}{10^{-9}} = 10^8$
10. In increasing order (m),
(i) Fermi = 1 fm = 10^{-15}
(ii) X-ray unit = 1 xu = 10^{-13}
(iii) Angstrom = 1 Å = 10^{-10}
(iv) Astronomical unit = 1AU = 1.49×10^{11}
(v) Light year = 9.45×10^{15}

Think and grow

11. While writing the unit of a physical quantity the following conventions are adopted.
   (i) If a unit is named after a person, the unit is not written with first letter of the name as capital letter. For unit of force we should write newton not Newton.
   (ii) For a unit named after a person, the symbol is capital letter. For example N for newton, It is not n. Symbols of other units are not written in capital letter.
   (iii) The symbols of other units are not expressed in plural form.
   For example, the CGS unit of work is erg.
   100 unit is expressed as 100 erg, not 100 ergs.
   (iv) Not more than one oblique is used, for example, the unit of viscosity poise is written as 1 g/s cm or gs^{-1} cm^{-1} and not g/s/cm.

12. No, the object is present at two positions at the same time. At an instant as an object cannot be present at two positions at the same time.

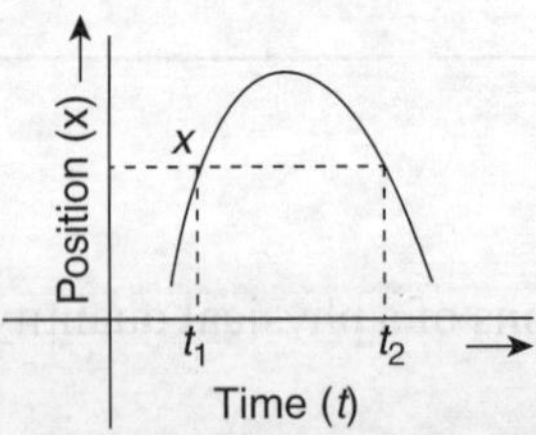

If we interchange the time and position axes the graph is possible as the object may be present at a position at two different times.

13. Yes, length is a fundamental unit. Standard metre length is defined as the distance travelled by light in vacuum during a time of 1/299792458 second. This arises from the established fact that speed of light in vacuum is exactly 299792458 m/s. Hence, standard metre is independent of temperature. But the length of an object varies with temperature and is given by the relation. $L_t = L_0(1+\alpha t)$.
Therefore we usually specify the temperature at which a particular measurement has been done.

CHAPTER TEST

1. Write dimensional formula for the intensity of radiation.
   (a) $M^1L^0T^3$ (b) $M^1L^0T^{-3}$
   (c) $M^1L^0T^{-2}$ (d) $M^1L^2T^{-3}$

2. The dimensional formula of physical quantity is $M^aL^bT^c$. Then that physical quantity is
   (a) Surface tension if $a=1$, $b=1$, $c=2$
   (b) Force if $a=1$, $b=1$, $c=2$
   (c) Angular frequency if $a=0$, $b=0$, $c=-1$
   (d) Spring constant if $a=1$, $b=-1$, $c=-2$

3. "Pascal – Second" has dimension of
   (a) Force (b) Energy
   (c) Pressure (d) Coefficient of viscosity

4. In the relation $y - a\cos(\omega t - kx)$, the dimensions formula for k is
   (a) $[M^0L^{-1}T^{-1}]$ (b) $[M^0LT^{-1}]$
   (c) $[M^0L^{-1}T^0]$ (d) $[M^0LT]$

5. Which one of the following does not have the same dimensions?
   (a) Work and energy
   (b) Angle and strain
   (c) Relative density and refractive index
   (d) Planck constant and energy.

6. The flux density of mass is defined as the amount of mass crossing unit area per unit time. The dimension of this quantity is
   (a) $ML^{-2}T^{-1}$ (b) ML^2T^{-1}
   (c) MLT^{-1} (d) $M^{-1}L^{-2}T$

7. Pressure gradient has the same dimension as that of
   (a) Velocity gradient (b) Potential gradient
   (c) Energy gradient (d) None of these.

8. The dimensions of electric potential are
   (a) $[ML^2T^{-2}Q^{-1}]$ (b) $[MLT^{-2}Q^{-1}]$
   (c) $[ML^2T^{-1}Q]$ (d) $[ML^2T^{-2}Q]$

9. Out of the following pairs, which one does not have identical dimensions?
   (a) Moment of inertia and moment of force
   (b) Work and torque
   (c) Angular momentum and Planck's constant
   (d) Impulse and momentum.

10. In a system of units, if force (F), acceleration (A) and time (T) are taken as fundamental units, then the dimensional formula of energy is
   (a) FA^2T (b) FAT^2
   (c) F^2AT (d) FAT

11. The dimensions of K in the equation $W = \frac{1}{2}Kx^2$ is

(a) $M^1L^0T^{-2}$ (b) $M^0L^1T^{-1}$

(c) $M^1L^1T^{-2}$ (d) $M^1L^0T^{-1}$

12. An object is moving through a liquid. The viscous damping force acting on it is proportional to the velocity. Then the dimension of constant of proportionality is

(a) $ML^{-1}T^{-1}$ (b) MLT^{-1}

(c) M^0LT^{-1} (d) ML^0T^{-1}

13. A force F is given by $F = at + bt^2$, where t is time. What are the dimensions of a and b?

(a) MLT^{-3} and ML^2T^{-4} (b) MLT^{-3} and MLT^{-4}

(c) MLT^{-1} and MLT^0 (d) MLT^{-4} and MLT^1

14. If the time period (T) of vibration of a liquid drop depends on surface tension (S), radius (r) of the drop, then the expression of T is

(a) $T = k\sqrt{\rho r^3 / S}$ (b) $T = k\sqrt{\rho^{1/2} r^3 / S}$

(c) $T = k\sqrt{\rho r^3 / S^{1/2}}$ (d) None of these.

15. The fundamental physical quantities that have same dimensions in the dimensional formulae of torque and angular momentum are

(a) Mass, time (b) Time, length

(c) Mass, length (d) Time, mole.

16. Which of the following pairs does not have similar dimensions?

(a) Stress and pressure

(b) Angle and strain

(c) Tension and surface tension

(d) Planck's constant and angular momentum.

17. The percentage errors in the measurement of mass and speed are 2% and 3% respectively. How much will be the maximum error in the estimation of the kinetic energy obtained by measuring mass and speed

(a) 11% (b) 8%

(c) 5% (d) 1%

18. If the time period of oscillation of a pendulum is measured as 2.5 second using a stop watch with the least count $\frac{1}{2}$ second, then the permissible error in the measurement is

(a) 10% (b) 30%

(c) 15% (d) 20%

19. The unit of percentage error is

(a) same as that of physical quantity

(b) different from that of physical quantity

(c) percentage error is unit-less

(d) Errors have got their own units, which are different from that of physical quantity measured.

20. The period of oscillation of a simple pendulum in the experiment is recorded as 2.63 s, 2.56 s, 2.42 s, 2.71 s and 2.80 s respectively. The average absolute error is

(a) 0.1 s (b) 0.11 s

(c) 0.01 s (d) 1.0 s

21. The number of significant figures in all given number 25.12, 2009, 4.156 and 1.217×10^{-4} is

(a) 1 (b) 2

(c) 3 (d) 4

22. If the length of rod A is 3.25 ± 0.01 cm and that of B is 419 ± 0.01 cm, then the rod B is longer than rod A by

(a) 0.94 ± 0.00 cm (b) 0.94 ± 0.01 cm

(c) 0.94 ± 0.02 cm (d) 0.94 ± 0.005 cm

23. Choose the incorrect statement out of the following:

(a) Every measurement by a measuring instrument has some error.

(b) Every calculated physical quantity that is based on measured values has some error.

(c) A measurement can have more accuracy but less precision and vice versa.

(d) The percentage error is different from relative error.

24. If the acceleration due to gravity is 10 ms^{-2} and the units of length and time are changed into kilometre and hour respectively, the numerical value of the acceleration is

(a) 3,60,000 (b) 72,000

(c) 36,000 (d) 1,29,600

25. With the usual notations, the following equation $S_t = u + \frac{1}{2}a(2t - 1)$ is

(a) only numerically correct

(b) only dimensionally correct

(c) Both numerically and dimensionally correct

(d) Neither numerically nor dimensionally correct.

26. A wire has a mass 0.3 ± 0.003 g, radius 0.5 ± 0.005 mm and length 6 ± 0.06 cm. The maximum percentage error in the measurement of its density is

(a) 1 (b) 2

(c) 3 (d) 4

27. The number of significant figures in 5418000.

(a) 4 (b) 5

(c) 6 (d) 7

28. In the relation $P = \frac{\alpha}{\beta} e^{\frac{\alpha Z}{k\theta}}$ P is pressure, Z is the distance, k is Boltzmann's constant and θ is the temperature. The dimensional formula of β will be

(a) $[M^0L^2T^0]$ (b) $[M^1L^2T^1]$

(c) $[M^1L^0T^{-1}]$ (d) $[M^0L^2T^{-1}]$

29. The density of a material in SI units is 128 kg m^{-3}. In certain units in which the unit of length is 25 cm and the unit of mass is 50 g, the numerical value of density of the material is
(a) 410 (b) 640
(c) 40 (d) 16

30. Newton / metre^2 is the unit of
(a) Energy (b) Momentum
(c) Force (d) Pressure

31. How many significant figures are there in 30.00?
(a) 1 (b) 2
(c) 3 (d) 4

32. Ampere-hour is a unit of
(a) Quantity of electricity
(b) Strength of electric current
(c) Power
(d) Energy.

33. Which of the following is not the unit of energy?
(a) Calorie (b) Joule
(c) Electron volt (d) Watt

34. Parsec is a unit of
(a) Distance (b) Velocity
(c) Time (d) Angle.

35. The dimensional formula for impulse is
(a) MLT^{-2} (b) MLT^{-1}
(c) ML^2T^{-1} (d) M^2LT^{-1}

36. The dimensions of specific heat are
(a) $M^1L^2T^{-2}\theta^{-1}$ (b) $L^2T^{-2}\theta^{-1}$
(c) $L^2T^{-2}\theta^1$ (d) $M^1L^2\theta^1$

37. If *E*, *M*, *L* and *G* denote energy, mass, angular momentum and gravitational constant respectively, then the quantity (EL^2 / M^2G^2) has the dimensions of
(a) Angle (b) Length
(c) Mass (d) Time

38. In CGS system, the magnitude of the force is 100 dyne. In another system where the fundamental physical quantities are kilogram, metre and minute, the magnitude of the force is
(a) 0.036 (b) 0.36
(c) 3.6 (d) 36

39. The dimensional formula of relative density is
(a) ML^{-3} (b) LT^{-1}
(c) MLT^{-2} (d) Dimensionless

40. Out of following four dimensional quantities, which one quantity is to be called a dimensional constant?
(a) Acceleration due to gravity
(b) Surface tension of water
(c) Weight of a standard kilogram mass
(d) The velocity of light in vacuum.

41. Which of the following quantities is dimensionless?
(a) Gravitational constant (b) Planck's constant
(c) Power of a convex lens (d) None.

42. $ML^3T^{-1}Q^{-2}$ is dimensions of
(a) Resistivity (b) Conductivity
(c) Resistance (d) None of these.

43. The resistance $R=\frac{V}{I}$ where $V=100\pm5$ volts and $I=10\pm0.2$ amperes. What is the total error in R?
(a) 5% (b) 7%
(c) 5.2% (d) $\frac{5}{2}$ %

44. If the error in the measurement of radius of a sphere is 2%, then the error in the determination of volume of the sphere will be
(a) 8% (b) 2%
(c) 4% (d) 6%

45. A thin copper wire of length *l* metre increases in length by 2% when heated through 10° C. What is the percentage increase in area when a square copper sheet of length *l* metre is heated through 10° C
(a) 4% (b) 8%
(c) 16% (d) None of the above.

Answer Keys

1. (b)	2. (c)	3. (d)	4. (c)	5. (d)	6. (a)	7. (d)	8. (a)	9. (a)	10. (b)
11. (a)	12. (d)	13. (b)	14. (a)	15. (c)	16. (c)	17. (b)	18. (d)	19. (c)	20. (b)
21. (d)	22. (c)	23. (d)	24. (d)	25. (c)	26. (d)	27. (a)	28. (a)	29. (c)	30. (d)
31. (d)	32. (a)	33. (d)	34. (a)	35. (b)	36. (b)	37. (a)	38. (c)	39. (d)	40. (d)
41. (d)	42. (a)	43. (b)	44. (d)	45. (a)					

Solutions

1. (b)

$$\text{Intensity of Radiation} = \frac{\text{Radiation Energy}}{\text{Area} \times \text{Time}}$$

$$\Rightarrow \quad I = \frac{[\text{ML}^2\text{T}^{-2}]}{[\text{L}^2 \times \text{T}]} = \text{ML}^0\text{T}^{-3}$$

2. (c)

$$\text{Angular frequency} = 2\pi \times \text{Frequency}$$
$$= [\text{T}^{-1}] = [\text{M}^0\text{L}^0\text{T}^{-1}]$$

$$a = 0, b = 0, c = -1$$

3. (d)

$\text{Nsm}^{-2} = \text{Nm}^{-2} \times \text{s}$ = Pascal-second

4. (c)

[Kx] = Dimension of ωt = (dimensionless)

Hence $\quad K = \frac{1}{x} = \frac{1}{\text{L}} = [\text{L}^{-1}]$

$$[K] = [\text{L}^{-1}]$$

5. (d)

[Planck constant] = $[\text{ML}^2\text{T}^{-1}]$ and

$[\text{Energy}] = [\text{ML}^2\text{T}^{-2}]$

6. (a)

$$\text{Flux density of mass} = \frac{\text{Amount of mass}}{\text{Area} \times \text{Time}}$$

$$\text{Flux density of mass} = \frac{[\text{M}]}{[\text{L}^2][\text{T}]} = [\text{ML}^{-2}\text{T}^{-1}]$$

7. (d)

$$\text{Velocity gradient} = \frac{v}{x} = \frac{[\text{LT}^{-1}]}{[\text{L}]} = [\text{T}^{-1}]$$

$$\text{Potential gradient} = \frac{V}{x} = \frac{[\text{ML}^2\text{T}^{-3}\text{A}^{-1}]}{[\text{L}]}$$
$$= [\text{MLT}^{-3}\text{A}^{-1}]$$

$$\text{Energy gradient} = \frac{E}{x} = \frac{[\text{ML}^2\text{T}^2]}{[\text{L}]} = [\text{MLT}^{-2}]$$

$$\text{And pressure gradient} = \frac{P}{x} = \frac{[\text{ML}^{-1}\text{T}^{-2}]}{[\text{L}]} = [\text{ML}^{-2}\text{T}^{-2}]$$

8. (a)

$$V = \frac{\text{W}}{\text{Q}} = [\text{ML}^2\text{T}^{-2}\text{Q}^{-1}]$$

9. (a) **10.** (b)

11. (a)

$$W = \frac{1}{2}kx^2 \Rightarrow [k] = \frac{[\text{W}]}{[x^2]} = \left[\frac{\text{ML}^2\text{T}^{-2}}{\text{L}^2}\right] = [\text{MT}^{-2}]$$

12. (d)

$$F \propto v \Rightarrow F = kv \Rightarrow [k] = \left[\frac{F}{v}\right] = \left[\frac{\text{MLT}^{-2}}{\text{LT}^{-1}}\right] = [\text{MT}^{-1}]$$

13. (b)

From the principal of dimensional homogeneity

$$[a] = \left[\frac{F}{t}\right] = [MLT^{-3}] \text{ and } [b] = \left[\frac{F}{t^2}\right] = [\text{MLT}^{-4}]$$

14. (a)

15. (c)

Torque = $[\text{ML}^2\text{T}^{-2}]$, Angular momentum = $[\text{ML}^2\text{T}^{-1}]$

So mass and length have the same dimensions.

16. (c)

Tension = $[\text{MLT}^{-2}]$, Surface Tension = $[\text{MT}^{-2}]$

17. (b)

$$E = \frac{1}{2}mv^2$$

$\therefore$ % error in KE

= 2% error in mass + 2 × 3 % error in velocity

= 2 + 2 × 3 = 8%

18. (d)

Here, $T = 2.5$ second and $\Delta T = \frac{1}{2}$ second

Therefore, the permissible error in the measurement of the time period is

$$\frac{\Delta T}{T} \times 100 = \frac{\left(\frac{1}{2}\right)}{(2.5)} \times 100 = 20\%$$

19. (c) **20.** (b)

21. (d)

The number of significant figures in all of the given numbers is 4.

22. (c) **23.** (d)

24. (d)

$$n_2 = n_1\left[\frac{L_1}{L_2}\right]^{-1}\left[\frac{T_1}{T_2}\right]^{-2} = 10\left[\frac{\text{m}}{\text{km}}\right]^{1}\left[\frac{\text{sec}}{\text{hr}}\right]^{-2}$$

$$n_2 = 10\left[\frac{\text{m}}{10^3\text{m}}\right]^{1}\left[\frac{\text{sec}}{3600\,\text{sec}}\right]^{-2} = 129600$$

25. (c)

We can derive this equation from equations of motion, so it is numerically correct.

S_t = distance travelled in tth second $= \dfrac{\text{Distance}}{\text{time}} = [\text{LT}^{-1}]$

u = velocity = $[\text{LT}^{-1}]$ and $\dfrac{1}{2}a(2t-1) = [\text{LT}^{-1}]$

As dimensions of each term in the given equation are same, the equation is also dimensionally correct.

26. (d)

$\therefore$ Density, $\rho = \dfrac{M}{V} = \dfrac{M}{\pi r^2 L}$

$$\Rightarrow \quad \frac{\Delta\rho}{\rho} = \frac{\Delta M}{M} + 2\frac{\Delta r}{r} + \frac{\Delta L}{L}$$

$$= \frac{0.003}{0.3} + 2\times\frac{0.005}{0.5} + \frac{0.06}{6}$$

$$= 0.01 + 0.02 + 0.01 = 0.04$$

$\therefore$ Percentage error $= \dfrac{\Delta\rho}{\rho}\times 100 = 0.04\times 100 = 4\%$

27. (a)

As all zeros to the right of the last non-zero digits are non-significant, hence there are four significant figures.

28. (a) **29.** (c)

30. (d)

31. (d)

We know that all zeros to the right of a decimal point are significant, if they are not followed by non-zero digits. Therefore 30.00 has four significant figures.

32. (a)

Charge = Current × Time

33. (d)

Watt is a unit of power.

34. (a)

Astronomical unit of distance

35. (b)

Impulse = Force × Time = $[\text{MLT}^{-2}][\text{T}] = [\text{MLT}^{-1}]$

36. (b) **37.** (a)

38. (c)

$$n_2 = n_1\left(\frac{\text{M}_1}{\text{M}_2}\right)^{1}\left(\frac{\text{L}_1}{\text{L}_2}\right)^{1}\left(\frac{\text{T}}{\text{T}_2}\right)^{-2}$$

$$= 100\left(\frac{\text{gm}}{10^3\text{ gm}}\right)^{1}\left(\frac{\text{cm}}{10^2\text{ cm}}\right)^{1}\left(\frac{\text{sec}}{60\,\text{sec}}\right)^{-2}$$

$$n_2 = \frac{3600}{10^3} = 3.6$$

39. (d)

$$\text{Relative density} = \frac{\text{Density of substance}}{\text{Density of water}} = [\text{M}^0\text{L}^0\text{T}^0]$$

40. (d)

41. (d)

$$[G] = [\text{M}^{-1}\text{L}^3\text{T}^{-2}];[h] = [\text{ML}^2\text{T}^{-1}]$$

$$\text{Power} = \frac{1}{\text{Focal length}} = [\text{L}^{-1}]$$

All quantities have dimensions.

42. (a)

$$\text{Resistivity } [\rho] = \frac{[R][A]}{[l]} \text{ where } [R] = [\text{ML}^2\text{T}^{-1}\text{Q}^{-2}]$$

$$[\rho] = [\text{ML}^3\text{T}^{-1}\text{Q}^{-2}]$$

43. (b)

$$\therefore \quad \left(\frac{\Delta R}{R}\times 100\right)_{\max} = \frac{\Delta V}{V}\times 100 + \frac{\Delta I}{I}\times 100\,]$$

$$= \frac{5}{100}\times 100 + \frac{0.2}{10}\times 100 = (5+2)\% = 7\%$$

44. (d)

As $V = \dfrac{4}{3}\pi r^2$

$$\frac{dV}{V} = 3\left(\frac{dr}{r}\right)$$

$\therefore$ Percentage error in determination of volume = 3 (Percentage error in measurement of radius)

= 3 (2%) = 6%

45. (a)

The percentage increase in length = 2%

Hence, percentage increases in area of square sheet

= 2 × 2% = 4%

CHAPTER 2

Kinematics

Introduction

Kinematics is that branch of mechanics in which we study ways to describe motion of objects independent of motion, and often even independent of the particular object that is moving.

Frame of Reference

The context of frame of reference comes to define the Rest or Motion state of body. Our general notion of rest and motion is – when a body does not change its position with respect to time it is said to be in rest. On the contrary, if it changes its position with respect to time we say that body is in motion.

Let us consider a situation: A group of three students set out on a picnic in a car. A teacher (observer) focuses on the motion of the car. The students, with respect to each other, find themselves in a state of rest. But, the students with respect to the observer are in motion, i.e., the position of the students are changing because of motion of the car. Therefore, rest or motion is a relative term and depends upon the position of the observer. In the language of physics, the observer is termed as frame of reference. Nothing is at an absolute state of rest or motion. The concept of rest or motion is meaningless without a frame of reference. Motion is a combined property of the object under observation and the observer.

There are many ways to locate the position of an object. A very convenient way is to choose Cartesian coordinate system, sometimes also called the rectangular coordinate system. A coordinate system used to specify the locations in space consists of

(1) A fixed reference point called origin
(2) A set of specified directions or axes with an appropriate scale. The other common coordinate system is polar coordinates system.

Position Vector

Motion involves change in position of a particle. The position of a particle with respect to a coordinate system is given by a vector quantity called position vector. It is given by the arrow from the origin of the coordinate system to the position of the particle. For example, when the particle is at A (in the given figure), its position vector is given by $\overrightarrow{r_A}$; when the particle is at B, its position vector is r_B. The position vector is also given in terms of the co-ordinates of the point (position). For example $\overrightarrow{r_A}$ can be written as,

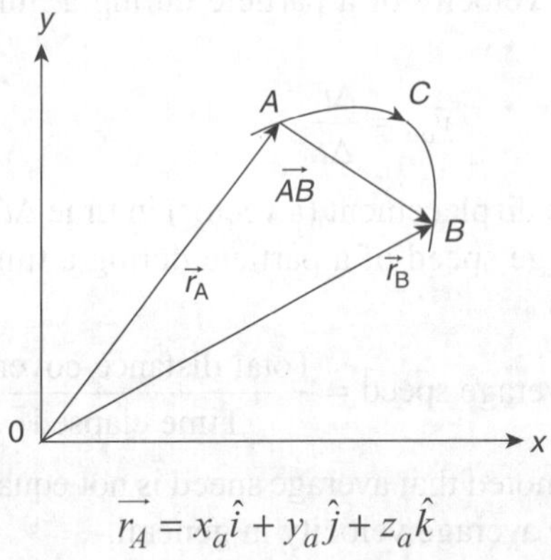

$$\overrightarrow{r_A} = x_a\hat{i} + y_a\hat{j} + z_a\hat{k}$$

where (x_a, y_a and z_a) are the co-ordinates of point A and $\hat{i}, \hat{j}, \hat{k}$ are unit vectors.

Distance and Displacement

The total length of the actual path covered by a particle is known as distance travelled by the particle whereas the directional distance between the initial and final position of the particle is called its displacement. In the above figure, let a particle move from A to B along the arc ACB.

Then,

$$\text{Distance} = \text{Length of path (arc) ACB}$$

$$\text{Displacement} = \overrightarrow{r_{AB}} = \overrightarrow{r_B} - \overrightarrow{r_A}$$

Hence, the displacement is a vector quantity which is given by the difference between the final and initial position vectors. In analytic representation, the displacement vector $\overrightarrow{r_{AB}}$ is given as,

$$\hat{r}_{AB} = (x_b - x_a)\hat{i} + (y_b - y_a)\hat{j} + (z_b - z_a)\hat{k}$$

The magnitude of $\overrightarrow{r_{AB}}$ equals the length of line AB and the direction is from A to B. The distance, on the other hand, is a scalar quantity having magnitude only.

Speed and Velocity

Rate of change of distance is called speed and the rate of change of displacement is known as velocity. The displacement being a vector quantity, the velocity of a particle is also a vector quantity. Speed, on the other hand, is a scalar quantity. The SI unit of both speed and velocity is ms^{-1} and their dimensional formula is $[LT^{-1}]$.

The speed of a particle changes if its magnitude is changed. The velocity of a particle changes if a change in either its magnitude or direction or both takes place. If the speed of a particle does not change during a time interval, it is known as uniform speed. Similarly, if the velocity remains constant during a time interval, it is called uniform velocity.

For varying speed and velocity, it is more meaningful to define average and instantaneous values of these quantities.

Average Velocity and Speed

The average velocity of a particle during a time interval Δt is defined by

$$\vec{v}_{av} = \frac{\Delta \vec{r}}{\Delta t}$$

where $\Delta \vec{r}$ is displacement (a vector) in time Δt.

The average speed of a particle during a time interval Δt is defined as

$$\text{Average speed} = \frac{\text{Total distance covered}}{\text{Time elapsed}}$$

It should be noted that average speed is not equal to the magnitude of the average velocity in general.

Illustration 1

A man walks on a straight road from his residence to his office 2.5 km away with a speed of 5 km/h. Finding the office closed, he immediately turns and walks back to his residence with a speed 0 7.5 km/h. What is the

(a) magnitude of average velocity and

(b) average speed of the man over the interval of time?

(i) 0 to 30 min (ii) 0 to 50 min (iii) 0 to 40 min

Solution

(a) Given that distance of office from the residence, $s = 2.5$ km.

Speed of the man for forward journey, $v_1 = 5$ km/h

$\therefore$ Time taken to reach the office

$$t_1 = \frac{s}{v_1} = \frac{2.5 \text{ km}}{5 \text{ km/h}} = \frac{1}{3} \text{ h} = 30 \text{ min}$$

Speed of the man for backward journey, $v_2 = 7.5$ km/h

$\therefore$ Time taken to return to residence

$$t_2 = \frac{s}{v_2} = \frac{2.5 \text{ km}}{7.5 \text{ km/h}} = \frac{1}{3} \text{ h} = 20 \text{ min}$$

(b) (i) Over the interval of time from 0 to 30 min

During this time interval, the man covers the distance from his residence up to his office, i.e., distance covered = 2.5 km.

Also displacement = 2.5 km.

Time taken = 30 min = $\frac{1}{2}$ h

$$\therefore \text{ Magnitude of average velocity } = \frac{\text{displacement}}{\text{time}} = \frac{2.5}{0.5} = 5 \text{ km/h}$$

$$\text{Average speed} = \frac{\text{distance}}{\text{time}} = \frac{2.5 \text{ km}}{0.5 \text{ h}} = 5 \text{ km/h}$$

(ii) Over the interval of time from 0 to 50 min

During this time interval, the man returns to his residence,

i.e., Displacement = 0 km, Distance covered = 5 km

$$\text{Time taken} = \frac{50}{60} = \frac{5}{6} \text{ h}$$

$$\therefore \text{ Magnitude of average velocity} = \frac{0 \text{ km}}{(5/6)\text{ h}} = 0 \text{ km/h}$$

and $$\text{Average speed} = \frac{5 \text{ km}}{(5/6)\text{ h}} = 6 \frac{\text{km}}{\text{h}}$$

(iii) Over the interval of time from 0 to 40 min

During the first 30 min of this interval, the man covers a distance of 2.5 km and in the next 10 min, he covers 2.5/2, i.e., 1.25 km in opposite direction.

$\therefore$ Displacement = 2.5 − 1.25 km

and distance covered = 2.5 + 1.25 − 3.75 km

time taken = 40 min $\frac{2}{3}$ h.

Hence, magnitude of average velocity = $\frac{1.25 \text{ km}}{(2/3)\text{ h}}$ = 1.875 km/h

$$\text{and average speed} = \frac{3.75 \text{ km}}{(2/3)\text{ h}} = 5.625 \text{ km/h}$$

Instantaneous Velocity and Speed

The velocity of a particle at any given instant of time is called the instantaneous velocity.

$$\text{Mathematically, instantaneous velocity} = \vec{v} \lim_{\Delta t \to 0} \frac{\Delta \vec{r}}{\Delta t} = \frac{\Delta \vec{r}}{dt}$$

i.e., differential coefficient of the position vector with respect to time gives the instantaneous velocity. The direction of $\vec{v}$ is the limiting direction that $\Delta \vec{r}$ takes as Δt approaches zero. This limiting direction is that of the tangent to the path of the particle at the given point.

The magnitude v of the instantaneous velocity is called the instantaneous speed or simply the speed of the particle. Hence,

$$\text{Speed} = v = |\vec{v}| = \left|\frac{dr}{dt}\right|$$

Speed, being the magnitude of a vector, is intrinsically positive.

Illustration 2

A particle moves alon a track such that at any instant its position coordinates are given by $x = a\cos\theta$ and $y = a\sin\theta$, a being a constant measured in metre and θ being the angle which the position vector makes with x-axis. Find

(a) the equation of trajectory of the particle. What is this trajectory?
(b) the distance travelled when $\theta = 60°$. Assume that at $t = 0$, particle lies on the x-axis.
(c) the average speed
(d) the magnitude of average velocity
(e) the magnitude of average acceleration.
(Assume that the time taken to complete one cycle of motion is T.)

Solution

(a) Since, $x = a\cos\theta$ ...(i)

$y = a\sin\theta$ ...(ii)

Squaring (i) and (ii) and adding, we get

$$x^2 + y^2 = a^2\left(\cos^2\theta + \sin^2\theta\right)$$

$$\Rightarrow \quad x^2 + y^2 = a^2 \quad \text{(Equation of trajectory)}$$

This equation is that of circle with center at (0, 0) and radius a.

(b) Distance travelled is the length of the actual path followed between the initial position A to the final position B.

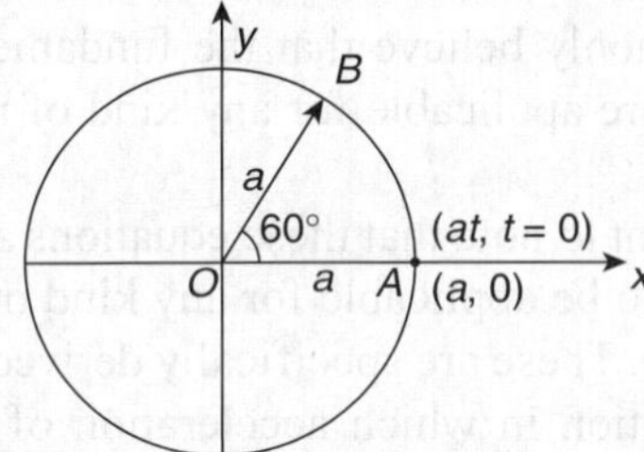

$$\text{Distance} = \text{Length of arc } AB = \frac{2\pi a}{6} = \frac{\pi a}{3}\text{ m}$$

(c) $\text{Average speed} = \dfrac{\text{Total Distance}}{\text{Total Time Taken}} = \dfrac{\overset{\frown}{AB}}{t_{AB}}$

Since T is the time taken to complete one cycle of motion, the time taken by particle to go from A to B is

$$t_{AB} = \frac{T}{6}$$

$$\Rightarrow \quad v_{av} = \frac{\pi a}{3} \times \frac{6}{T} = \frac{2\pi}{T}a$$

(d) Average velocity $\vec{v}_{av} = \dfrac{\Delta\vec{s}}{\Delta t}$

$$\Rightarrow \quad |\vec{v}_{av}| = \left|\frac{\Delta\vec{s}}{\Delta t}\right|$$

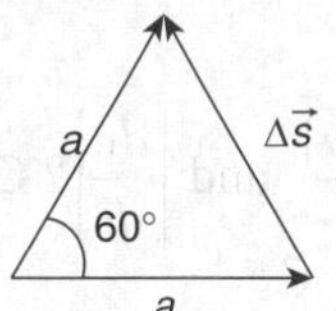

Here, $|\Delta\vec{s}| = \sqrt{a^2 + a^2 - 2a^2\cos(60°)} = |\Delta\vec{s}| = a$

$$\Rightarrow \quad |\vec{v}_{av}| = \frac{a}{\frac{T}{6}} = \frac{6a}{T}$$

(e) Average acceleration $\vec{a}_{av} = \dfrac{\Delta\vec{v}}{\Delta t}$

$$\Rightarrow \quad |\vec{a}_{av}| = \frac{|\Delta\vec{v}|}{\Delta t}$$

Further, $|\Delta\vec{v}| = \sqrt{v^2 + v^2 - 2v^2\cos(60°)} = v$

$$\Rightarrow \quad |\vec{a}_{av}| = \frac{v}{\frac{T}{6}} = \frac{6v}{T}$$

Acceleration

The rate of change of velocity of a particle is called its acceleration. It is a vector quantity and its units is m/s^2.

The acceleration could be uniform or variable. In the case of variable acceleration, we define average and instantaneous values of the acceleration.

$$\text{Average acceleration} = \frac{\Delta\vec{v}}{\Delta t}$$

$$\text{Instantaneous acceleration} = \vec{a} = \lim_{\Delta t \to 0}\frac{\Delta\vec{v}}{\Delta t} = \frac{d\vec{v}}{dt}$$

We will mainly be concerned with uniform acceleration only. A negative acceleration is called retardation.

Position, Velocity and Acceleration of a Particle in Three-dimensional Space

Consider a particle moving in space. Let at any instant of time its position vector be given by $\vec{r}$ such that,

$$\vec{r} = x\hat{i} + y\hat{j} + z\hat{k}$$

If $\vec{v}$ is the instantaneous velocity and $\vec{a}$ be the instantaneous acceleration of the particle, then

$$\vec{v} = \frac{d\vec{r}}{dt} = \dot{\vec{r}} = \hat{i}\left(\frac{dx}{dt}\right) + \hat{j}\left(\frac{dy}{dt}\right) + \hat{k}\left(\frac{dz}{dt}\right)$$

$$\Rightarrow \quad \vec{v} = v_x\hat{i} + v_y\hat{j} + v_z\hat{k}$$

Further, $\vec{a} = \dfrac{d\vec{v}}{dt} = \dot{\vec{v}} = \hat{i}\left(\dfrac{d^2x}{dt^2}\right) + \hat{j}\left(\dfrac{d^2y}{dt^2}\right) + \hat{k}\left(\dfrac{d^2z}{dt^2}\right)$

$$\Rightarrow \quad \vec{a} = a_x\hat{i} + a_y\hat{j} + a_z\hat{k}$$

Illustration 3

What is meant by $\frac{d|v|}{dt}$ and $\left|\frac{dv}{dt}\right|$? Can these be equal? Can

(i) $\frac{d|v|}{dt} = 0$ and $\left|\frac{dv}{dt}\right| \neq 0$

(ii) $\frac{d|v|}{dt} \neq 0$ and $\left|\frac{dv}{dt}\right| = 0$

Solution

$\frac{d|v|}{dt}$ means the time rate of change of speed, and

$\left|\frac{dv}{dt}\right|$ means the magnitude of acceleration.

These two can be equal in two situations:
When a particle moves with uniform velocity, and
When a particle moves with constant acceleration along a straight line.

(i) $\frac{d|v|}{dt} = 0 \Rightarrow$ Speed = Constant

$\left|\frac{dv}{dt}\right| \neq 0 \Rightarrow |\text{acceleration}| \neq 0$

It happens only when a particle moves with constant speed along a circular path of constant radius.

(ii) $\left|\frac{dv}{dt}\right| = 0 \Rightarrow |\text{acceleration}| = 0$

$\frac{d|v|}{dt} \neq \Rightarrow$ Speed $\neq$ Constant

For zero acceleration, speed cannot change.
Therefore, this case is not possible.

Illustration 4

A particle moves with a speed varying with distance as $v = \frac{1}{3}ks^3$ along a straight line. Calculate the average speed of the particle, when it reaches a distance S_0 (assume that it starts from $S = 0$).

Solution

Since, $$v_{av} = \frac{\int_0^{s_0} v\,ds}{\int_0^{s_0} ds}$$

$$\Rightarrow \quad v_{av} = \frac{k}{3s_0}\int_0^{s_0} s^3\,ds$$

$$\Rightarrow \quad v_{av} = \frac{k}{3s_0}\left[\left.\frac{s^4}{4}\right|_0^{s_0}\right]$$

$$\Rightarrow \quad v_{av} = \frac{ks_0^3}{12}$$

Motion in one dimension with uniform acceleration

The simplest motion to describe is motion of a particle along a straight line, called one-dimensional motion. The various quantities are related by the following equations, known as the equations of motion.

(i) $v = u + at$

(ii) $S = ut + \frac{1}{2}at^2$

(iii) $v^2 = u^2 + 2as$

where u = initial velocity of the particle
v = final velocity of the particle after time t
S = distance travelled in time t
a = uniform acceleration of the particle

(iv) The distance covered by the particle in the nth second

$$S_{\text{nth}} = S_n - S_{n-1}$$

$$= u + \frac{1}{2}a(2n - 1)$$

Note

Students commonly believe that the fundamental equations of kinematics are applicable for any kind of motion along a straight line.

It is important to note that these equations are not the general equations to be applicable for any kind of motion along the straight line. These are specifically derived for uniformly accelerated motion in which acceleration of the particle is constant, that is, acceleration is not a function of time, displacement or velocity.

Illustration 5

Can you analyse the motion of a particle travelling along x-axis such that its position at any instant of time t is given by $x_t = t^3 - 9t^2 + 6t$, where x_t is in metre and t is in second? Give reasonable mathematical arguments in support of your analysis.

Solution

When we are asked to analyse a particular motion whose mathematical function is given, we follow the following simple steps.

Step 1 Find the velocity as a function of time.

$$\frac{dx}{dt} = 3t^2 - 18t + 6$$

$$\Rightarrow \quad v = 3t^2 - 18t + 6 \quad \ldots\text{(i)}$$

Step 2 Always check the points (or corresponding time(s)) where this velocity is zero.

For this, equate (i) to zero.

$$\Rightarrow \quad 3t^2 - 18t + 6 = 0$$

$$\Rightarrow \quad t^2 - 6t + 2 = 0$$

$$\Rightarrow \quad t = \frac{6 \pm \sqrt{36-8}}{2}$$

$$\Rightarrow \quad t = \left(3 \pm \sqrt{7}\right)\text{s}$$

Step 3 We have got two values for the time when the velocity of the particle is zero.

$$t_1 = \left(3 - \sqrt{7}\right)\text{s} = 0.35\,\text{s} \quad \text{the earlier time}$$

$$t_2 = \left(3 + \sqrt{7}\right)\text{s} = 5.65\text{s} \quad \text{the later time}$$

[Suppose if any of these time values comes out to be negative, then just accept the positive value, as negative value of time has a lesser significance here: it corresponds to motion prior to the initial instant.]

Step 4 To find the significance of these two time values let's go further and calculate the acceleration also:

$$a = \frac{dv}{dt} = 6t - 18 \qquad \text{...(ii)}$$

Step 5 Then calculate the value of v and a, both at $t = 0$ i.e. initially. Let them be denoted by v_0 and a_0. So,

$$v_0 = 6\ \text{ms}^{-1} \quad \text{and} \quad a_0 = -18\ \text{ms}^{-2}$$

Here, we get a conclusion that, initially the particle has velocity and acceleration directed opposite to each other. This acceleration can then be regarded as the retardation offered to the motion of the body.

Step 6 Actually, what happens is that initially the particle moves with 6 ms^{-1} and experiences a retardation of 18 ms^{-2} (at $t = 0$) which decreases as time progresses and becomes zero at $t = 3$ s. But at $t_1 = 0.35$ s. The velocity of the particle becomes zero, but retardation remains finite such that the particle reverses its direction of motion and now it will be accelerated as both acceleration and velocity will be in the same direction. Further at a later time $t_2 = 5.65$ s its velocity again becomes zero.

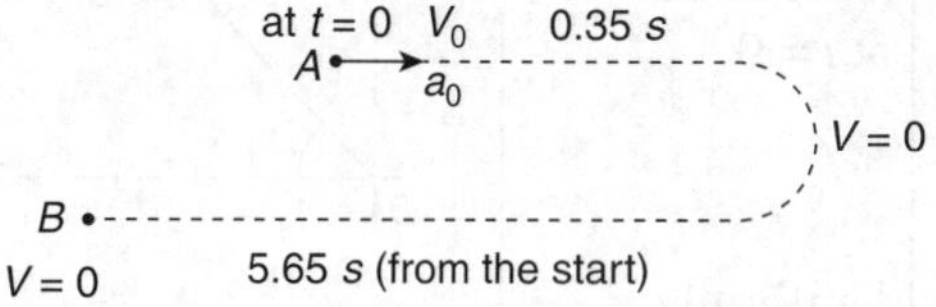

Step 7 The displacement of the particle in going from the first zero of velocity, i.e., at $t_1 = \left(3 - \sqrt{7}\right)\text{s}$ to the second zero of velocity i.e. at $t_2 = \left(3 + \sqrt{7}\right)\text{s}$ is

$$\Delta x = x_2 - x_1$$

$$\Rightarrow \quad \Delta x = \left[\left(3+\sqrt{7}\right)^3 - 9\left(3+\sqrt{3}\right)^2 + 6\left(3+\sqrt{7}\right)\right] - \left[\left(3-\sqrt{7}\right)^3 - 9\left(3-\sqrt{7}\right)^2 + 6\left(3-\sqrt{7}\right)\right]$$

$$\Rightarrow \quad \Delta x \approx -74\ \text{m}$$

Interpretation of Graphs

1. $\vec{x} - t$ graph

(a) The point on the graph is called an event.

(b) The average slope of the graph (or curve) gives average velocity.

(c) The instantaneous slope gives instantaneous velocity.

2. $\vec{v} - t$ graph

(a) The average slope of a curve in $\vec{v} - t$ graph gives average acceleration.

(b) The instantaneous slope of a curve in $\vec{v} - t$ graph gives instantaneous acceleration.

(c) The area under a curve in $\vec{v} - t$ graph gives the displacement.

The following are the graphs for one dimensional motion with uniform velocity or uniform acceleration.

S. No	Cases	v – t graph	s – t graph	Conclusion
1.	Uniform motion v = constant $a = 0$	V; V = constant; O; t	s; s = vt; θ; O; t	(1) Acceleration = slope of $v - t$ graph = 0 (2) Slope of $s - t$ graph = $v = \tan\theta$ (3) At $t = 0$, $s = 0$

(Continued)

S. No	Cases	v – t graph	s – t graph	Conclusion
2.	Uniformly accelerated motion with $u = 0$ and $s = 0$ at $t = 0$	$v = at$	$s = \frac{1}{2}at^2$	(1) $\tan\theta = a =$ constant (2) Slope of $s - t$ graph is zero at $t = 0$
3.	Uniformly accelerated motion with non-zero initial velocity $(u \neq 0)$	$v = u + at$	$s = ut + \frac{1}{2}at^2$	(1) $\tan\theta = a =$ constant (2) $u \neq 0$, so slope of $s - t$ graph at $t = 0$ is non zero and equals u. (3) Slope of $s - t$ graph is $\frac{ds}{dt} = u + at$, which is gradually increasing.
4.	Uniformly Accelerated motion with $u \neq 0$, and $s = s_0$ at $t = 0$	$v = u + at$	$s = s_0 + ut + \frac{1}{2}at^2$	(1) $\tan\theta = a =$ constant (2) Still slope is gradually increasing.
5.	Uniformly retarded motion with initial velocity u and final velocity zero.	$v = u - at$	t_0 $(< t)$	(1) $\tan\theta = -a =$ constant (2) Slope of $s - t$ graph at $t = 0$ is u. (3) Slope of $s - t$ graph at $t = t_0$ is zero.
6.	Uniformly retarded, then accelerated motion	t_0	$t = 0$	(1) $\tan\theta = -a =$ constant (2) In $s - t$ graph, slope or velocity first decreases, then increase with a reverse sign.

Relative Velocity

Consider a situation when two cars moving with different speeds along parallel paths cross each other. The car moving slowly forwards appears to be moving backward to an observer sitting in the second car moving faster.

In general, when an observer measures the velocity of a moving body, we get different results if one observer is moving relative to the other. The velocity of an object A as measured by an observer B is called the velocity of A relative to the observer B and is expressed as $\vec{v}_{A/B}$. We shall see that $\vec{v}_{A/B} = \vec{v}_A - \vec{v}_B$, when $\vec{v}_A$ and $\vec{v}_B$ are the velocities of A and B with respect to common reference frame (inertial).

Let $\vec{r}_A$ and $\vec{r}_B$ be the position vector of A and B with respect to the origin of an inertial frame of reference S. The relative separation $\vec{r}_{A/B}$ between the particles (position of A with respect to B) is obtained from the vector triangle OAB.

$$\vec{r}_B + \vec{r}_{A/B} = \vec{r}_A, \text{ or } \vec{r}_{A/B} = \vec{r}_A - \vec{r}_B$$

Here the three vectors can be resolved along the three coordinate axes.

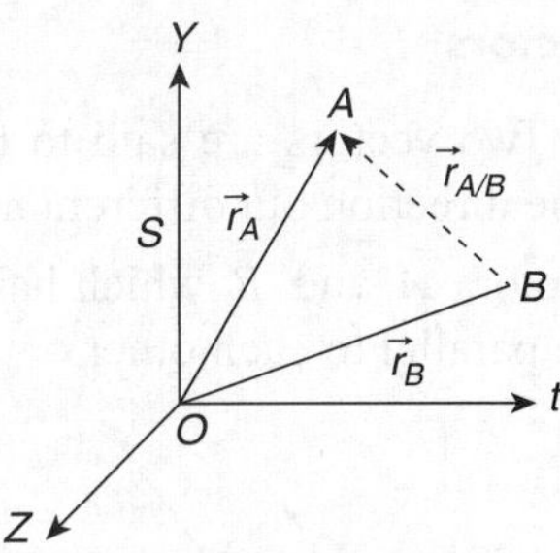

Now, differentiating both sides w.r.t time, we get

$$\frac{d}{dt}(\vec{r}_{A/B}) = \frac{d\vec{r}_A}{dt} - \frac{d\vec{r}_B}{dt}$$

or $\quad \vec{v}_{A/B} = \vec{v}_A - \vec{v}_B.$

Similarly $\vec{v}_{A/B}$, the velocity of B with respect to A is expressed as $\vec{v}_{B/A} = \vec{v}_B - \vec{v}_A$.

The following points regarding relative velocity must be noted.

1. If $\vec{v}_A = \vec{v}_B$, the two objects are said to be at relative rest, although both are in motion.
2. If the motion is considered in space, $\vec{r}_{A/B} = \vec{r}_A - \vec{r}_B$ is expressed in three dimensions, leading to

   $x_{A/B} = x_A - x_B \quad y_{A/B} = y_A - y_B \quad z_{A/B} = z_A - z_B$
3. For relative velocity,

   $$\vec{v}_{A/B} = \vec{v}_A - \vec{v}_B$$
   $$= (v_{Ax} - v_{Bx})\hat{i} + (v_{Ay} - v_{By})\hat{j} + (v_{Az} - v_{Bz})\hat{k}$$
   $$= (v_{A/B})_x \hat{i} + (v_{A/B})_y \hat{j} + (v_{A/B})_z \hat{k}.$$
4. $\vec{v}_{B/A} = -\vec{v}_{A/B}$ and $|\vec{v}_{A/B}| = |\vec{v}_{B/A}|$
5. If the direction of $\vec{v}_A$ and $\vec{v}_B$ make an angle θ

   $$|\vec{v}_{B/A}| = \sqrt{v_A^2 + v_B^2 - 2v_A v_B \cos\theta}$$

   and the direction of $\vec{v}_{B/A}$ is given by

   $$\phi = \tan^{-1}\left(\frac{v_B \sin\theta}{v_A - v_B\cos\theta}\right)$$

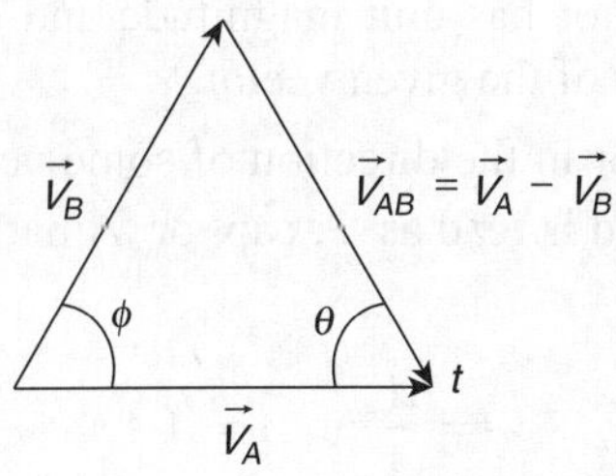

Illustration 6

Consider the situation shown in figure.

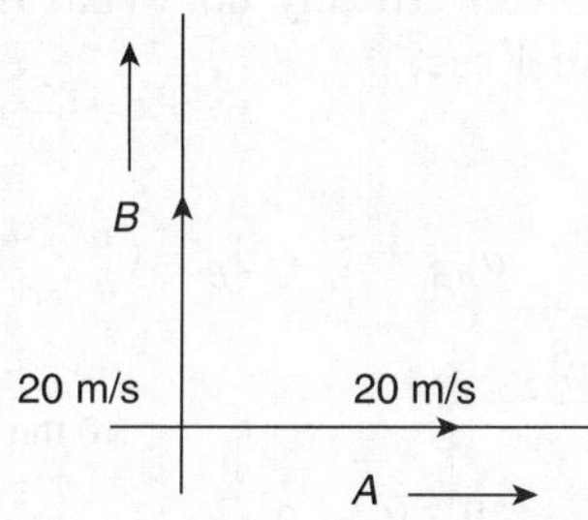

Solution

(i) The velocity of B with respect to A

$$\vec{V}_{AB} = \vec{V}_B - \vec{V}_A = 20\,\hat{j} - 20\hat{i}$$

(ii) The velocity of A with respect to B

$$\vec{V}_{AB} = \vec{V}_A - \vec{V}_B = 20\hat{i} - 20\,\hat{j}$$

Illustration 7

Objects A and B have velocities 10 m/s. A is moving along East while B is moving towards North from the same point as shown. Find the velocity of A relative to B ($\vec{V}_{AB}$).

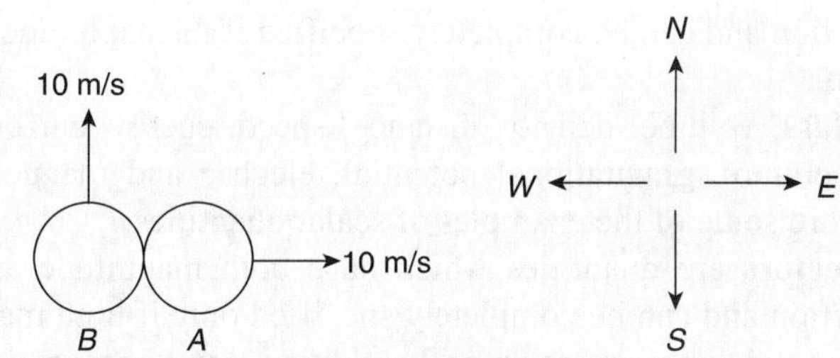

Solution

$$\vec{V}_{AB} = \vec{V}_A - \vec{V}_B$$

$$\therefore |\vec{V}_{AB}| = 10\sqrt{2}$$

$\vec{V}_A$ 10, $10\sqrt{2}$, 10 $-\vec{V}_B$

Note: $\quad |\vec{V}_A - \vec{V}_B| = \sqrt{V_A^2 + V_B^2 V_B \cos\theta}$

N, W, E, 45°, 45°, V_{AB}, S, A

Illustration 8

Two particles A and B are projected in air. A is thrown horizontally, B is thrown vertically up. What is the separation between them after 1 sec?

Solution

$$\vec{a}_{AB} = \vec{a}_A - \vec{a}_B = 0$$

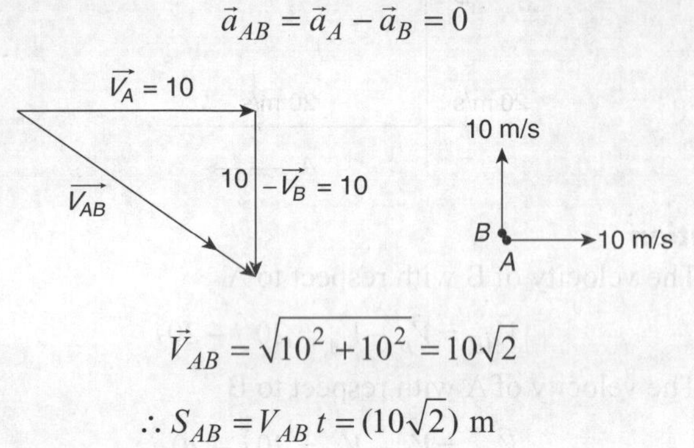

$$\vec{V}_{AB} = \sqrt{10^2 + 10^2} = 10\sqrt{2}$$

$$\therefore S_{AB} = V_{AB}\, t = (10\sqrt{2})\ \text{m}$$

Vectors

All physical quantities can be divided into two categories — **scalars** and **vectors**.

Scalars are quantities which have only magnitude but no direction and can be completely specified if its magnitude is given.

Mass, volume, density, distance, speed, energy, current, temperature, gravitational potential, electric and magnetic flux are some of the examples of scalar quantities.

Vectors are quantities which have both magnitude and direction and can be completely specified only if both magnitude and direction are given. In addition, a vector must also obey the rules of vector algebra.

Displacement, velocity, force, angular momentum, intensity of gravitational field, intensity of electric and magnetic fields, moment of a force are some of the examples of vector quantities.

A vector quantity is represented by putting an arrow over the symbol representing it. Hence, for example, $\vec{A}$ stands for a vector quantity.

Graphical representation of a vector

A vector is denoted by $\overrightarrow{OA}$, i.e., with an arrow over OA. The point O is known as origin while A as the terminus as shown in the figure below. The magnitude of the vector is given by the length OA.

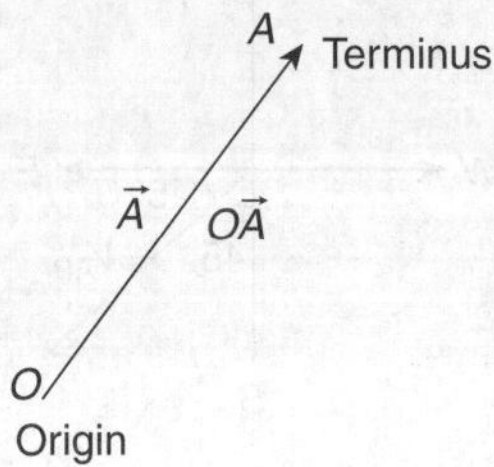

Some specific vectors

1. **Like vectors** Two vectors are said to be like vectors if they have same direction but different magnitude. Figure shows two vectors $\vec{A}$ and $\vec{B}$ which have different magnitude but are parallel to each other.

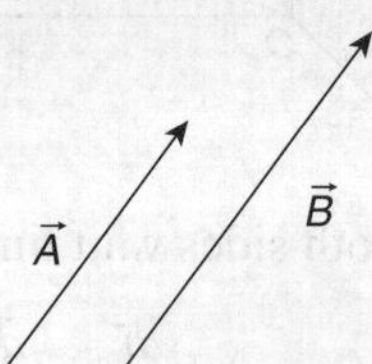

2. **Unlike vectors** The vectors having opposite direction and different magnitude are called unlike vectors. Figure shows two such vectors $\vec{A}$ and $\vec{B}$ which have different magnitudes and are antiparallel to each other.

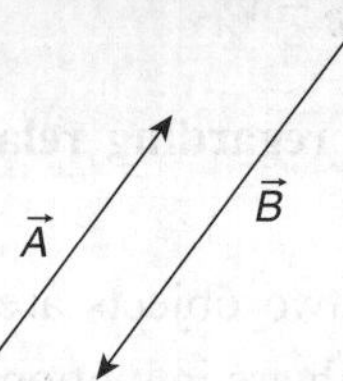

3. **Equal vectors** Two vectors are said to be equal, if they have the same magnitude and direction. Figure shows two vector $\vec{A}$ and $\vec{B}$ having the same magnitude and same direction and therefore

4. **Opposite vectors** The vectors having same magnitude but opposite direction, are known as opposite vectors. Figure shows two such vectors $\vec{A}$ and $\vec{B}$ having the same magnitude and opposite direction. Here, $\vec{A} = -\vec{B}$

5. **Null Vector** It is defined as a vector having zero magnitude. It has a direction which is indeterminate as its magnitude is zero.
6. **Unit Vector** A vector divided by its magnitude is called a unit vector along the direction of the vector. Obviously, the unit vector has unit magnitude and direction is the same as that of the given vector.

   A unit vector in the direction of some vector $\vec{A}$ is written as $\hat{A}$ and is read as 'A cap' or 'A hat'. Therefore, by definition

$$\hat{A} = \frac{\vec{A}}{A} \quad \text{or} \quad \vec{A} = \hat{A}\, A$$

Thus, any vector can be expressed as magnitude times the unit vector along its own direction.
Unit vectors along, x, y and z axes are represented by $\hat{i}$, $\hat{j}$ and $\hat{k}$ respectively.

7. Co-initial vectors Vectors are said to be co-initial, if they have a common initial point. In figure, $\vec{A}$ and $\vec{B}$ originating from the same point O as their origin are called co-initial vectors.

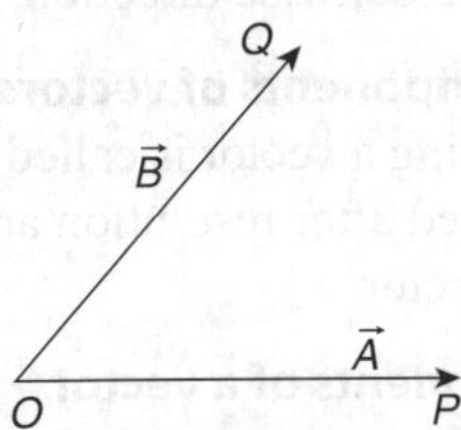

8. Co-linear Vectors Two vectors having equal or unequal magnitudes, which either act along the same line [Figure (i)] or along the parallel lines in the same direction [Figure (ii)] or along the parallel lines in opposite direction [Figure (iii)], are called collinear vectors. Like, unlike, equal and opposite vectors are collinear.

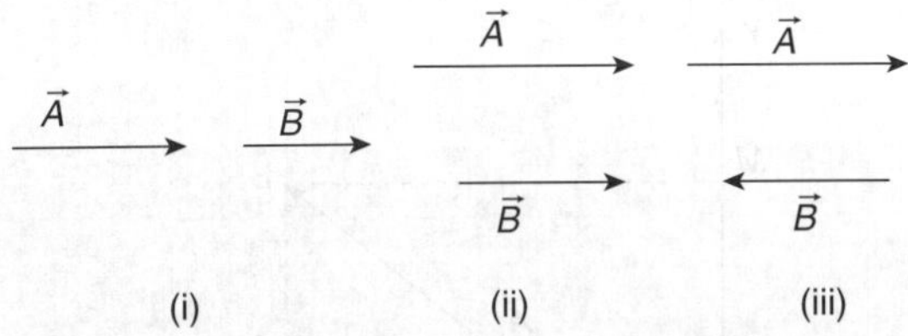

9. Coplanar vectors Vectors lying in the same plane are called coplanar vectors. Figure (i) shows three vectors $\vec{X}$, $\vec{Y}$ and $\vec{Z}$ along mutually ⊥ axes x, y and z respectively. These vectors are non-coplanar but the vectors $\vec{x}-\vec{y}$, $\vec{y}-\vec{z}$ and $\vec{z}-\vec{x}$ are coplanar in Figure (ii).

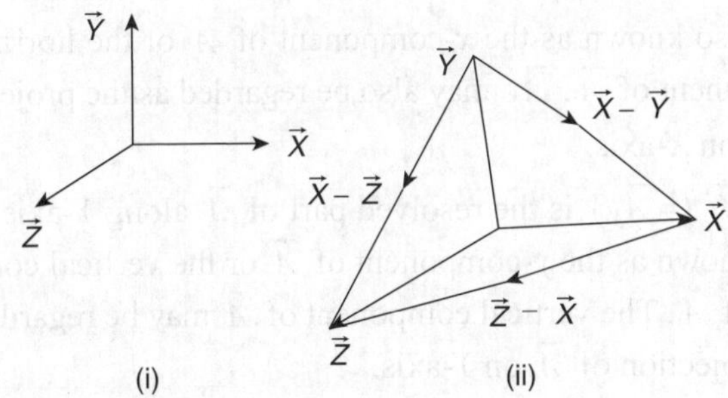

Addition of vectors

Because the vectors have directions also along with their magnitudes, they cannot be added according to simple rules of algebra as applied to scalars. In order to make it clear let us add two displacement vector $\vec{A}$ and $\vec{B}$ having different possible directions with respect to each other.

Case I: When two vectors $\vec{A}$ and $\vec{B}$ are in the same direction – If a body moves a distance of 1 cm first and then 2 cm, due east in both the cases and the two displacement vectors are represented by $\vec{A}$ and $\vec{B}$ (figure), then the sum of two displacement vectors, i.e., $\vec{A}+\vec{B}$, will be equal to a displacement of 3 cm along east.

$\vec{A}$ → $\vec{B}$ → = $\vec{C}=\vec{A}+\vec{B}$ East →

Case II: When two vectors $\vec{A}$ and $\vec{B}$ are in opposite directions – If the body moves a distance of 1 cm due west and then through 2 cm due east, (figure), then sum $(\vec{A}+\vec{B})$ of two displacement vectors will be equal to a displacement of 1 cm due east.

← $\vec{A}$ + $\vec{B}$ → = $\vec{D}=\vec{A}+\vec{B}$ East →

Case III: When two vectors are inclined to each other: We consider here the simple case of two mutually perpendicular vectors. Suppose the body moves a distance of 1 cm due east and then a distance of 2 cm due north. In figure given below the two displacement vectors are shown by $\overline{AB}$ and $\overline{BC}$, according to some scale. It is clear that the magnitude of the displacement vector $\overline{AC}$ is $\sqrt{5}$ cm and direction of displacement is due north of east measured in terms of θ.

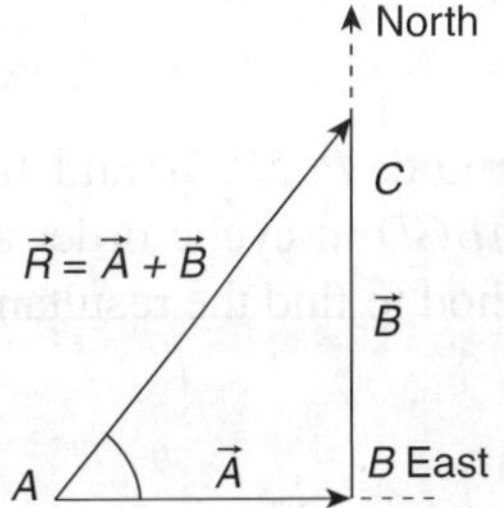

Therefore, if instead of moving a distance of 1 cm along AB and then a distance of 2 cm along BC, the body moves a distance of $\sqrt{5}$ cm along AC at an angle θ with AB, it reaches the same point. This means the displacement of $\sqrt{5}$ cm along AC produces the same effect as the displacements of 1 cm along AB and 2 cm along BC together. This implies that displacement vector $\overline{AC}$ is the sum or resultant of displacement vectors $\overline{AB}$ and $\overline{BC}$, i.e., $\overline{AB}+\overline{BC}=\overline{AC}$

If vectors $\overline{AB}$, $\overline{BC}$ and $\overline{AC}$ are represented by vectors $\vec{A}$, $\vec{B}$ and $\vec{R}$ respectively, then $\vec{A}+\vec{B}=\vec{R}$ where $\vec{R}$ is called as the resultant of vectors $\vec{A}$ and $\vec{B}$. Thus we conclude that

(i) the resultant of two or more vectors is a single vector which produces the same effect as the individual vectors do together.

(ii) vectors are not added algebraically but are added geometrically.

The triangle law of vector addition states that if three vectors $\vec{A}$, $\vec{B}$ and $\vec{C}$ can be represented in magnitude and direction by the sides of a triangle taken in order then,

$$\vec{A}+\vec{B}+\vec{C}=0$$

$$\Rightarrow \vec{A}+\vec{B}=-\vec{C}$$

$\vec{C}$ is the resultant of $\vec{A}$ and $\vec{B}$ and its magnitude is given by

$$C = |\vec{C}| = \sqrt{A^2 + B^2 + 2AB\cos\theta}$$

where θ is the angle between $\vec{A}$ and $\vec{B}$.

The resultant $\vec{C}$ is inclined with $\vec{A}$ by an angle α, where

$$\tan\alpha = \frac{B\sin\theta}{A + B\cos\theta}$$

for $A = B$,

$$\tan\alpha = \frac{\sin\theta}{1+\cos\theta} = \frac{2\sin\theta/2\cdot\cos\theta/2}{1+2\cos^2\frac{\theta}{2}-1} = \tan\theta/2$$

$$\Rightarrow \alpha = \theta/2$$

Hence, the resultant of two equal vectors bisects the angle between them. Further,

if $\theta = 0$, $C = (A + B)$ = Maximum value of C and
if $\theta = 180°$, $C = (A \sim B)$ = Minimum value of C.

For any other value of θ between 0° and 180°, C lies between $(A + B)$ and $(A - B)$.

Illustration 9

Four forces of magnitude P, $2P$, $3P$ and $4P$ act along the four sides of a square $ABCD$ in cyclic order as shown in figure. Use the vector method to find the resultant force.

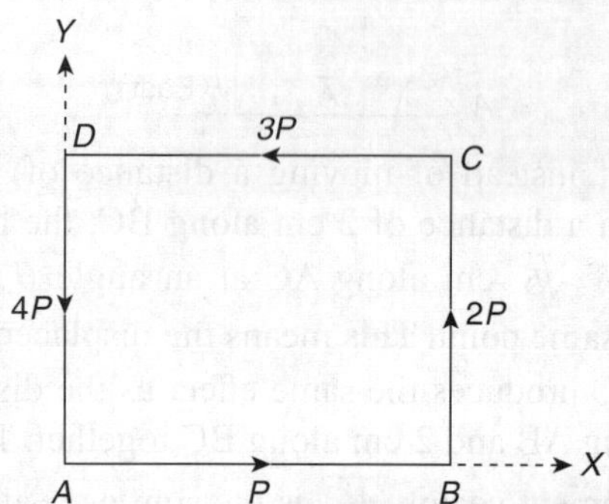

Solution

Let us assume X-axis along the side AB and Y-axis along the side AD of the square $ABCD$.

Since a parallel translation of a vector does not bring any change in it, we can assume that the force $3P$ is acting along the negative direction of X-axis and the force $4P$ is acting along the negative direction of Y-axis,

It follows from figure, that effectively a force $2P$ acts along the negative Y-direction,

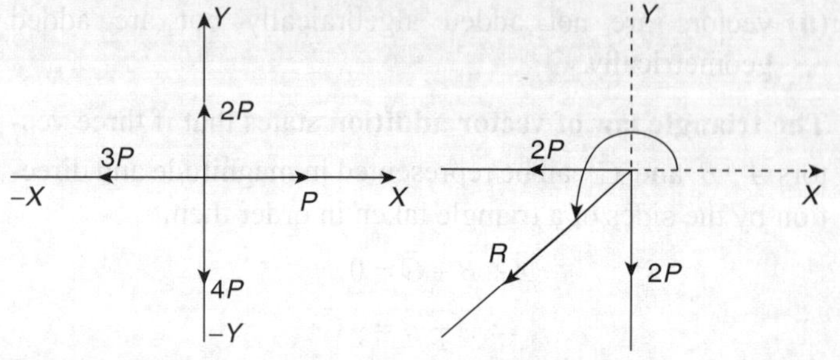

The resultant of these two equal forces $2P$ inclined at an angle of 90° is R, where $R = \sqrt{(2P)^2 + (2P)^2} = 2\sqrt{2}P$.

Since the forces are equal, their resultant R bisects the angle between them and hence, is inclined to either force at an angle of 45°.

Hence, the resultant of the four given forces is $2\sqrt{2}\,P$ and is inclined with the axis of X at an angle of 225° (180° + 45°) measured in the anticlockwise direction.

Resolution and components of vectors in a plane

The process of splitting a vector is called resolution of a vector. The parts obtained after resolution are known as *components* of the given vector.

Rectangular components of a vector

If the components of a given vector are perpendicular to each other, then they are called rectangular components.

Let us consider a vector $\vec{A}$ represented by $\overline{OP}$. Through the point O, draw two mutually perpendicular axes-X-axis and Y-axis. Let the vector $\vec{A}$ make an angle θ with the X-axis. From the point P, drop a perpendicular PN on X-axis.

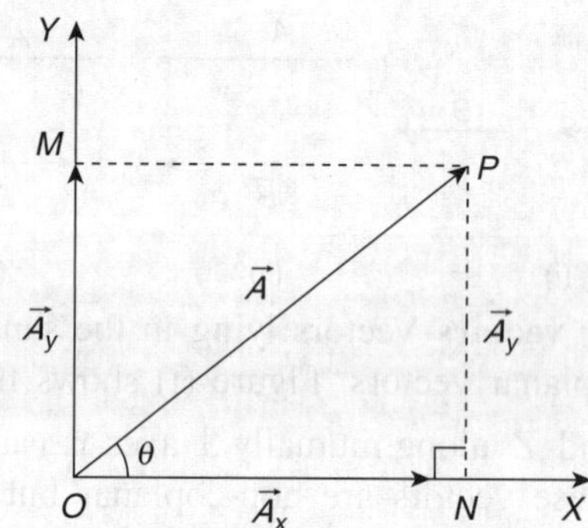

Now $\overline{ON}$ $(= \vec{A}_x)$ is the resolved part of $\vec{A}$ along X-axis. It is also known as the x-component of $\vec{A}$ or the horizontal component of $\vec{A}$. $\vec{A}_x$ may also be regarded as the projection of $\vec{A}$ on X-axis.

$\overline{OM}$ $(= \vec{A}_y)$ is the resolved part of $\vec{A}$ along Y-axis. It is also known as the y-component of $\vec{A}$ or the vertical component of $\vec{A}$. The vertical component of $\vec{A}$ may be regarded as the projection of $\vec{A}$ on Y-axis.

So, $\vec{A}_x$ and $\vec{A}_y$ are the rectangular components of $\vec{A}$.

Applying triangle law of vectors to the vector triangle ONP, we get

$$\vec{A}_x + \vec{A}_y = \vec{A}$$

In right-angled triangle ONP,

$$\cos\theta = \frac{A_x}{A} \Rightarrow A_x = A\cos\theta \quad \text{...(1)}$$

$$\sin\theta = \frac{A_y}{A} \Rightarrow A_y = A\sin\theta \quad \text{...(2)}$$

Squaring and adding (1) and (2), we get

$$A_x^2 + A_y^2 = A^2\cos^2\theta + A^2\sin^2\theta$$

$$A_x^2 + A_y^2 = A^2(\cos^2\theta + \sin^2\theta)$$

$$A_x^2 + A_y^2 = A^2$$

or $$A = \sqrt{A_x^2 + A_y^2}$$

$$\tan\theta = \frac{A_y}{A_x}$$

$$\theta = \tan^{-1}\frac{A_y}{A_x}$$

Illustration 10

Figure shows three vectors $\overrightarrow{OA}$, $\overrightarrow{OB}$ and $\overrightarrow{OC}$ having same magnitude. Find the direction of $\overrightarrow{OA} + \overrightarrow{OB} - \overrightarrow{OC}$

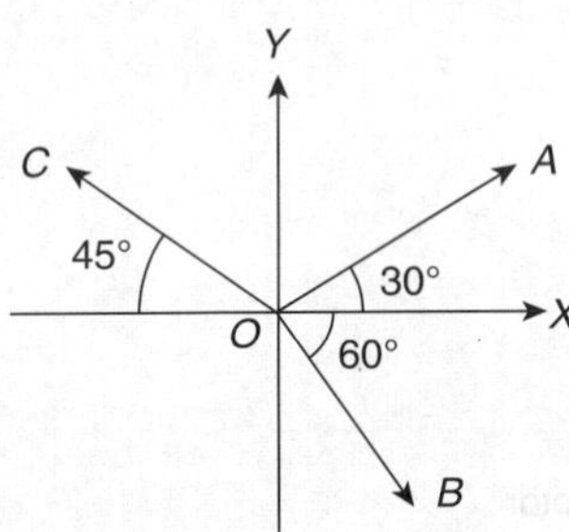

Solution

Let $OA = OB = OC = P$

x-component of $\overrightarrow{OA} = P\cos 30° = \dfrac{P\sqrt{3}}{2}$

x-component of $\overrightarrow{OB} = P\cos 60° = \dfrac{P}{2}$

x-component of $\overrightarrow{OC} = P\cos 135° = -\dfrac{P}{\sqrt{2}}$

$\therefore$ x-component of

$$(\overrightarrow{OA} + \overrightarrow{OB} - \overrightarrow{OC}) = \frac{P\sqrt{3}}{2} + \frac{P}{2} - \frac{P}{\sqrt{2}} = \frac{P}{2}(\sqrt{3} + 1 - \sqrt{2})$$

Similarly,

y-component of $\overrightarrow{OA} = P\cos 60° = \dfrac{P}{2}$

y-component of $\overrightarrow{OB} = P\cos 150° = -\dfrac{P\sqrt{3}}{2}$

y-component of $\overrightarrow{OC} = P\cos 45° = \dfrac{P}{\sqrt{2}}$

$\therefore$ y-component of

$$\left(\overrightarrow{OA} + \overrightarrow{OB} - \overrightarrow{OC}\right) = \frac{P}{2} + \left(-\frac{P\sqrt{3}}{2}\right) - \frac{P}{\sqrt{2}} = \frac{P}{2}(1 - \sqrt{3} - \sqrt{2})$$

Thus, angle of $\left(\overrightarrow{OA} + \overrightarrow{OB} - \overrightarrow{OC}\right)$ with the x-axis

$$= \tan^{-1}\frac{(P/2)(1 - \sqrt{3} - \sqrt{2})}{(P/2)(1 + \sqrt{3} + \sqrt{2})} = \tan^{-1}\frac{(1 - \sqrt{3} - \sqrt{2})}{(1 + \sqrt{3} - \sqrt{2})}$$

Illustration 11

Can a vector be zero if any of its components is not zero?

Solution

In terms of components in three dimensions, $\vec{A} = \vec{i}A_x + \vec{j}A_y + \vec{k}A_z$ with $A = \sqrt{A_x^2 + A_y^2 + A_z^2}$ so if any of A_x, A_y and A_z is not zero, the vector cannot be zero, i.e., for a vector to be zero (or null) all its component must be zero.

Illustration 12

The x and y components of vector $\vec{A}$ are 4 and 6 m respectively. The x and y components of vector $\vec{A} + \vec{B}$ are 10 and 9 m respectively. Calculate for the vector $\vec{B}$, the following: (a) its x and y components, (b) its length, and (c) the angle it makes with x-axis.

Solution

In terms of components,

$$\vec{A} + \vec{B} = (\vec{i}A_x + \vec{j}A_y) + (\vec{i}B_x + \vec{j}B_y)$$

i.e., $$\vec{A} + \vec{B} = \vec{i}(A_x + B_x) + \vec{j}(A_y + B_y)$$

According to given problem,

$$A_x + B_x = 10 \quad \text{and} \quad A_y + B_y = 9$$

and $$A_x = 4 \quad \text{and} \quad A_y = 6$$

So,

(a) $B_x = 10 - 4 = 6$ m and $B_y = 9 - 6 = 3$ m

(b) $B = \sqrt{B_x^2 + B_y^2} = \sqrt{6^2 + 3^2} = \sqrt{45}\text{ m} = 3\sqrt{5}\text{ m}$

(c) $\theta = \tan^{-1}\left[\dfrac{B_y}{B_x}\right] = \tan^{-1}\left[\dfrac{3}{6}\right] = \tan^{-1}\left[\dfrac{1}{2}\right] = 26.6°$

Component of a vector in three dimensions

A vector can be, in general, resolved into any three mutually perpendicular components. If a vector $\vec{A}$ makes angles α, β and γ with x, y and z axes respectively then,

$$\vec{A} = A\cos\alpha\,\hat{i} + A\cos\beta\,\hat{j} + A\cos\gamma\,\hat{k}$$

$$\Rightarrow \vec{A} = A_x\hat{i} + A_y\hat{j} + A_z\hat{k}$$

where,

$A_x = A\cos\alpha$, the component of $\vec{A}$ along x-axis,

$A_y = A\cos\beta$, the component of $\vec{A}$ along y-axis

$A_z = A\cos\gamma$, the component of $\vec{A}$ along z-axis.

Illustration 13

A particle is moving eastward with a velocity of 5 m/s. in 10 s, the velocity changes to 5 m/s northward. The average acceleration in this time is

(a) zero

(b) $\frac{1}{\sqrt{2}}$ ms^{-2} toward north-east

(c) $\frac{1}{\sqrt{2}}$ ms^{-2} towards south-west

(d) $\frac{1}{2}$ ms^{-2} towards north-west.

Solution

Let us consider the *x*-axis eastward and *y*-axis northward. Then, the initial velocity $\vec{v}_i$ is given by

$$\vec{v}_i = 5\hat{i}$$

and the final velocity $\vec{v}_f$, after time $t = 10$ s, is given by

$$\vec{v}_f = 5\hat{j}$$

The change in velocity is, then, $\vec{v}_f - \vec{v}_i$ and the acceleration, defined as the rate of change of velocity, is

$$\vec{a} = \frac{\vec{v}_f - \vec{v}_i}{t} = \frac{5\hat{j} - 5\hat{i}}{10} = \frac{\hat{j}}{2} - \frac{\hat{i}}{2}$$

$$\Rightarrow a = |\vec{a}| = \sqrt{\left(\frac{1}{2}\right)^2 + \left(\frac{1}{2}\right)^2} = \frac{1}{\sqrt{2}} \text{ms}^{-2}$$

If this acceleration makes an angle θ with the axis of *x* then

$$\tan\theta = \frac{1/2}{-1/2} = -1 = \tan 135°$$

$$\Rightarrow \theta = 135°$$

Thus, the average acceleration of the particle is $\frac{1}{\sqrt{2}}$ ms^{-2} along the north-west direction. Hence, the correct answer is (b).

Multiplication of Vectors

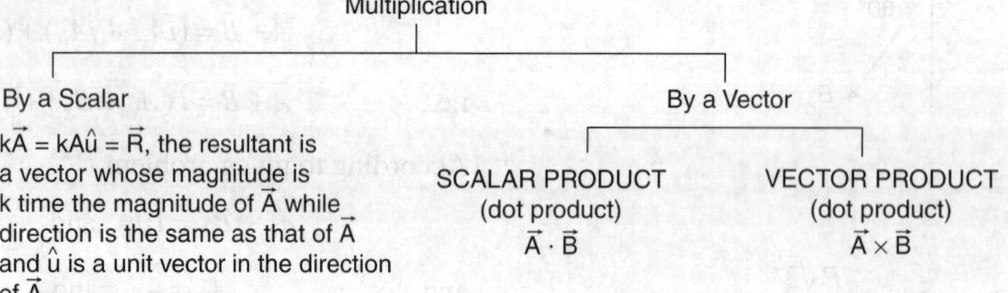

Scalar product of two vectors

The scalar product of two vectors in defined as scalar quantity having magnitude equal to the product of the magnitude of two vectors and the cosine of the smaller angle between them.

Mathematically, if θ is the angle between vectors $\vec{A}$ and $\vec{B}$, then $\vec{A}.\vec{B} = AB\ \cos\theta$

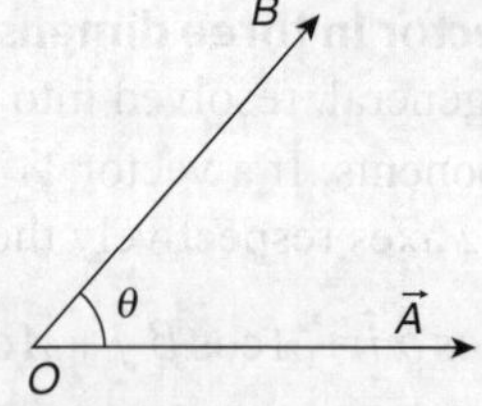

The above result can also be expressed as $\vec{A}.\vec{B} = A(B\cos\theta) = B(A\cos\theta)$

where $B\cos\theta$ is the magnitude of component of $\vec{B}$ along the direction of vector $\vec{A}$ and $A\cos\theta$ is the magnitude of component of $\vec{A}$ along the direction of vector $\vec{B}$. Therefore the dot product of two vectors can also be interpreted as the product of the magnitude of one vector and the magnitude of the component of other vector along the direction of first vector.

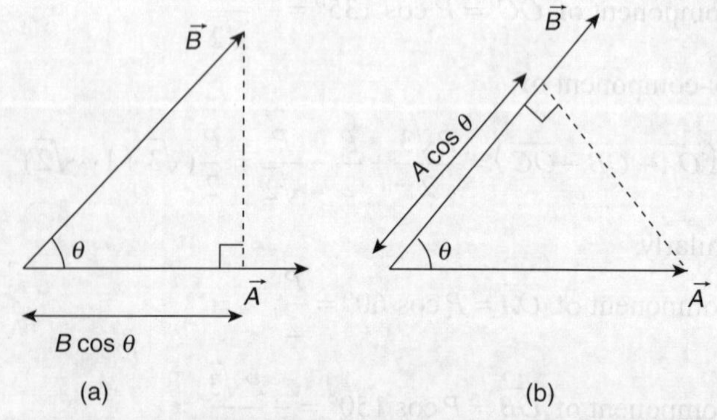

(iii) Dot product of two vectors can be positive or zero or negative depending upon whether θ is less than 90° or equal to 90° or $90° < \theta < 180°$.

(iv) Dot product of two vectors is always commutative i.e., $\vec{A}.\vec{B} = \vec{B}.\vec{A}$

$$\vec{A}.\vec{B} = AB\cos\theta$$

(θ is the angle between $\vec{A}$ and $\vec{B}$ measured in anti-clockwise direction)

and $\vec{B}.\vec{A} = BA\cos(-\theta) = AB \cos\theta$

Thus $\vec{A}.\vec{B} = \vec{B}.\vec{A}$

(v) The dot product of a vector with itself gives square of its magnitude, i.e.,

$$\vec{A}.\vec{A} = A\ A\cos 0° = A^2$$

(vi) The dot product of two mutually perpendicular vectors is zero, i,e., if two vectors $\vec{A}$ and $\vec{B}$ are perpendicular then

$$\vec{A}.\vec{B} = AB \cos 90° = 0$$

(vii) The dot product obeys the distributive law, i.e.,

$$\vec{A}.(\vec{B}+\vec{C}) = \vec{A}.\vec{B} + \vec{A}.\vec{C}$$

(viii) Two vectors are collinear, if their dot product is numerically equal to product their magnitudes, i.e., when $\theta = 0°$ or $180°$, $|\vec{A}.\vec{B}| = AB$

(ix) In case of unit vectors $\hat{i}, \hat{j}$ and $\hat{k}$, we have following two properties:

(a) $\hat{i}.\hat{i} = \hat{j}.\hat{j} = \hat{k}.\hat{k} = 1$ (as $\theta = 0°$)

(b) $\hat{i}.\hat{j} = \hat{j}.\hat{k} = \hat{k}.\hat{i} = 0$ (as $\theta = 90°$)

(x) Dot product of two vectors in terms of their rectangular components in three dimensions:

$$\vec{A}.\vec{B} = (A_x\hat{i} + A_y\hat{j} = A_z\hat{k}).(B_x\hat{i} + B_y\hat{j} + B_z\hat{k})$$

$$= A_x B_z + A_z B_z$$

(xi) Examples of some physical quantities which can be expressed as scalar product of two vectors:

(a) Work (W) is defined as the scalar product of force ($\vec{F}$) and the displacement ($\vec{s}$) i.e., W = $\vec{F}.\vec{s}$

(b) Power (P) is defined as the scalar product of force ($\vec{F}$) and the velocity ($\vec{v}$) i.e., $P = \vec{F}.\vec{v}$

(c) Magnetic flux (ϕ) linked with a surface is defined as the dot product of magnetic induction ($\vec{B}$) and area vector (A) i.e. $\phi = \vec{B}.\vec{A}$

(Magnetic flux ϕ is a part of XII class syllabus)

Illustration 14

Find the angle between the vectors

$$\vec{A} = \hat{i} + 2\hat{j} + 3\hat{k} \text{ and } \vec{B} = 2\hat{i} + 3\hat{j} + 4\hat{k}$$

Solution

We have, $\vec{A}.\vec{B} = AB \cos\theta$

$$\Rightarrow \cos\theta = \frac{\vec{A}.\vec{B}}{AB} \quad (1)$$

Now, $\vec{A}.\vec{B} = A_x B_x + A_y B_y + A_z B_z$

$$\Rightarrow \vec{A}.\vec{B} = 1\times 2 + 2\times 3 + 3\times 4$$

$$\Rightarrow \vec{A}.\vec{B} = 20 \quad (2)$$

Also, $A = \sqrt{A_x^2 + A_y^2 + A_z^2} = \sqrt{1^2+2^2+3^2} = \sqrt{14}$ (3)

and $B = \sqrt{B_x^2 + B_y^2 + B_z^2} = \sqrt{2^2+3^2+4^2} = \sqrt{29}$ (4)

Hence, $\cos\theta = \frac{\vec{A}.\vec{B}}{AB} = \frac{.20}{\sqrt{14}\sqrt{29}} = \frac{20}{\sqrt{406}}$

$$\Rightarrow \theta = \cos^{-1}\left(\frac{20}{\sqrt{406}}\right)$$

Illustration 15

The angle between two vectors = $-2\hat{i} + 3\hat{j} + \hat{k}$ and $\hat{i} + 2\hat{j} - 4\hat{k}$ is

(a) 0° (b) 60°
(c) 90° (d) 180°

Solution

Let $\vec{A} = -2\hat{i} + 3\hat{j} + \hat{k}$

and $\vec{B} = \hat{i} + 2\hat{j} - 4\hat{k}$

Then,

$$\vec{A}.\vec{B} = (-2\hat{i} + 3\hat{j} + \hat{k}).(\hat{i} + 2\hat{j} - 4\hat{k})$$

$$\Rightarrow \vec{A}.\vec{B} = -2\times 1 + 3\times 2 - 1\times 4 = 0$$

$$\Rightarrow \vec{A}.\vec{B} = 0$$

Since $\vec{A}.\vec{B}$ is zero, the given vectors are at right angle to each other. Hence, the correct answer is (c).

Illustration 16

Two billiard balls are rolling on a flat table. One has velocity components $\upsilon_x = 1$ m/s, $\upsilon_y = \sqrt{3}$ m/s and the other has components $\upsilon_x' = 2$ m/s and $\upsilon_y' = 2$ m/s. If both the balls start moving from the same point, the angle between their paths is

(a) 60° (b) 45° (c) 22.5° (d) 15°

Solution

$$v_1 = 1\,i + \sqrt{3}j \text{ and } v_2 = 2\,i + 2\,j$$

$$v_1.v_2 = (i+\sqrt{3}j).(2i+2j)$$

$$= 2 + 2\sqrt{3}$$

or $\upsilon_1\,\upsilon_2 \cos\theta = 2 + 2\sqrt{3}$

$$\cos\theta = \frac{1}{2}\times\frac{1}{\sqrt{2}} + \frac{\sqrt{3}}{2}\times\frac{1}{\sqrt{2}}$$

$$= \cos 60 \cos 45 + \sin 60 \sin 45$$

$$= \cos(60-45) = \cos 15$$

or $\theta = 15°$

Vector product of two vectors

(i) The vector product of two vectors is defined as a vector having magnitude equal to the product of the magnitudes of two vectors with the sine of angle between them and direction ⊥ to the plane containing the two vectors in accordance with right-handed screw rule or right-hand thumb rule.

(ii) Mathematically, if θ is the angle between vectors $\vec{A}$ and $\vec{B}$, then ...(i)

$$\vec{A}\times\vec{B} = AB \sin\theta\ \hat{n}$$

The direction of vector $\vec{A}\times\vec{B}$ is the same as that of unit vector $\hat{n}$. It is decided by any of the following two rules:

(a) Right-handed screw rule: Rotate a right-handed screw form vector $\vec{A}$ to $\vec{B}$ through the smaller angle between them; then the direction of motion of screw gives the direction of vector $\vec{A}\times\vec{B}$ [Figure (i)].

(b) Right-hand thumb rule: Bend the fingers of the right hand in such a way that they point in the direction of rotation and the thumb points in the direction of vector $\vec{A}\times\vec{B}$ [Figure (ii)]

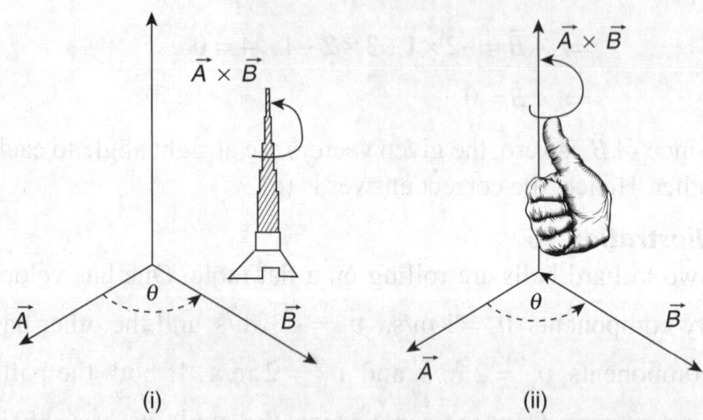

In case of vector $\vec{B}\times\vec{A}$, the direction of advancement of screw, on rotating it form vector $\vec{B}\times\vec{A}$, will be just opposite to that observed in the case of vector $\vec{A}\times\vec{B}$. Hence vector $\vec{B}\times\vec{A}$ is directed just in opposite direction to that of vector $\vec{A}\times\vec{B}$ i.e., $\vec{B}\times\vec{A} = -(\vec{A}\times\vec{B})$

(iii) The cross product of the two vectors does not obey commutative law. As discussed above $\vec{A}\times\vec{B} = -(\vec{B}\times\vec{A})$ i.e., $\vec{A}\times\vec{B} \neq (\vec{B}\times\vec{A})$

(vi) The cross product follows the distributive law, i.e.,

$$\vec{A}\times(\vec{B}+\vec{C}) = \vec{A}\times\vec{B} + \vec{A}\times\vec{C}$$

(v) The cross product of two vectors with itself is a NULL vector, i.e.,

$$\vec{A}\times\vec{A} = (A)(A)\sin 0°\ \hat{n} = 0.$$

(iv) The cross product of two vectors represents the area of the parallelogram formed by them.

The figure below shows a parallelogram PQRS whose adjacent sides $\overrightarrow{PQ}$ and $\overrightarrow{PS}$ are represented by vectors $\vec{A}$ and $\vec{B}$ respectively.

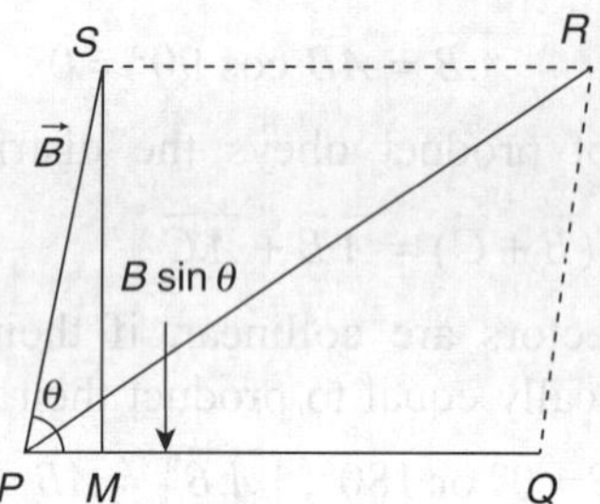

Now, area of parallelogram

$$= PQ \times SM = AB \sin\theta$$

Because the magnitude of vector $\vec{A}\times\vec{B}$ is $AB \sin\theta$, cross product of two vectors represents the area of parallelogram formed by it. It is worth noting that area vector $\vec{A}\times\vec{B}$ acts along the perpendicular to the plane of two vectors $\vec{A}$ and $\vec{B}$.

(vii) In case of unit vectors $\hat{i}, \hat{j}, \hat{k}$, we obtain following two important properties:

(a) $\hat{i}\times\hat{i} = \hat{j}\times\hat{j} = \hat{k}\times\hat{k} = (1)(1)\sin 0°\ (\hat{n}) = 0$

(b) $\hat{i}\times\hat{j} = (1)(1)\sin 90°\ (\hat{k}) = \hat{k}$

where, $\hat{k}$ is a unit vector ⊥ to the plane of $\hat{i}$ and $\hat{j}$ in a direction in which a right hand screw will advance, when rotated from $\hat{i}$ to $\hat{j}$.

Also, $-\hat{j}\times\hat{i} = -(1)(1)\sin 90°\ (-\hat{k}) = \hat{k}$

Similarly, $\hat{j}\times\hat{k} = -\hat{k}\times\hat{j} = \hat{i}$ and $\hat{k}\times\hat{i} = -\hat{i}\times\hat{k} = \hat{j}$

(viii) Cross product of two vectors in terms of their rectangular components:

$$\vec{A}\times\vec{B} = (A_x\hat{i} + A_y\hat{j} + A_z\hat{k})\times(B_x\hat{i} + B_y\hat{j} + B_z\hat{k})$$

$$= (A_yB_z - A_zB_y)\hat{i} + (A_zB_x - A_xB_z)\hat{j} + (A_xB_y - A_yB_x)\hat{k}$$

$$\begin{vmatrix} \hat{i} & \hat{j} & \hat{k} \\ A_x & A_y & A_z \\ B_x & B_y & B_z \end{vmatrix}$$

Illustration 17

Two vectors A and B lie in the same xy plane of a rectangular coordinate system. The vector A has components A_x and A_y along the x and y axes. The vector B has components B_x and B_y along the x and y axes. Find the magnitude and direction of $A \times B$.

Solution

Let $\boldsymbol{i}, \boldsymbol{j}$ and $\boldsymbol{k}$ be unit vectors along the x, y and z axes of a rectangular coordinate system. Then

$$\boldsymbol{A} = A_x\,\boldsymbol{i} + A_y\,\boldsymbol{j} \text{ and } \boldsymbol{B} = B_x\,\boldsymbol{i} + B_y\,\boldsymbol{j}$$

Now $\boldsymbol{A} \times \boldsymbol{B} = (A_x\,\boldsymbol{i} + A_y\boldsymbol{j}) \times (B_x\,\boldsymbol{i} + B_y\boldsymbol{j})$

$$= A_x B_x\,\boldsymbol{i}\times\boldsymbol{i} + A_x B_y\,\boldsymbol{i}\times\boldsymbol{j} + A_y B_x\boldsymbol{j}\times\boldsymbol{i} + A_y B_y\boldsymbol{j}\times\boldsymbol{j}$$

We know that

$$\boldsymbol{i}\times\boldsymbol{i} = \boldsymbol{j}\times\boldsymbol{j} = 0, \boldsymbol{i}\times\boldsymbol{j} = \boldsymbol{k} \text{ and } \boldsymbol{j}\times\boldsymbol{i} = -\boldsymbol{k}$$

$\therefore \quad \boldsymbol{A} \times \boldsymbol{B} = (A_x B_y - A_y B_x)\,\boldsymbol{k}$

Thus the vector $\boldsymbol{A} \times \boldsymbol{B}$ is in the direction of z-axis.

The magnitude of $\boldsymbol{A} \times \boldsymbol{B}$ is given by $|\boldsymbol{A} \times \boldsymbol{B}| = (A_x B_y - A_y B_x)$.

Illustration 18

If $A \times B = A \times C$, is it correct to conclude that $B = C$?

Solution

Given that $A \times B = A \times C$

$$\therefore \quad A\,B \sin\theta_1\,\hat{n}_1 = AC\sin\theta_2\hat{n}_2$$

where $\hat{n}_1$ and $\hat{n}_2$ are unit vectors in the direction of vectors $(\boldsymbol{A} \times \boldsymbol{B})$ and $(\boldsymbol{A} \times \boldsymbol{C})$ respectively.

Here, we consider the following four cases:

(i) If $\theta_1 = \theta_2$ and $\hat{n}_1 = \hat{n}_2$, then
$B = C$ and $\boldsymbol{B} = \boldsymbol{C}$

(ii) If $\theta_1 = \theta_2$ and $\hat{n}_1 \neq \hat{n}_2$, then
$B = C$ but $\boldsymbol{B} \neq \boldsymbol{C}$

(iii) If $\theta_1 \neq \theta_2$ and $\hat{n}_1 \neq \hat{n}_2$, then
$B \neq C$ and $\boldsymbol{B} \neq \boldsymbol{C}$

(iv) If $\theta_1 \neq \theta_2$ and $\hat{n}_1 = \hat{n}_2$, then
$B \neq C$ and $\boldsymbol{B} \neq \boldsymbol{C}$

Projectile Motion

When a body moves under an acceleration whose direction is along its initial line of motion, then the path described by the body is a straight line. This type of motion is called 'straight-line motion'. For example, when we throw a ball vertically upwards, or drop it downwards, then it moves in a straight line under the action of the earth's gravitational field.

But, when a body moves with an acceleration whose direction is different from the direction of the initial velocity, then both the magnitude and the direction of its velocity change continuously with time. Hence the body moves on a curved path in a plane. This type of motion is called 'planar motion'. Examples of planar motion are projectile motion and circular motion.

Motion in a Plane

Planar motion can be analyzed by resolving it along the two mutually perpendicular directions in x and y axes.

Consider a particle moving along a curve C in the $x-y$ plane. Let it be at a point P (x,y) which has a position vector $\vec{r}$ at any particular instant of time t. Let this position vector make an angle θ with the x – axis as shown in figure.

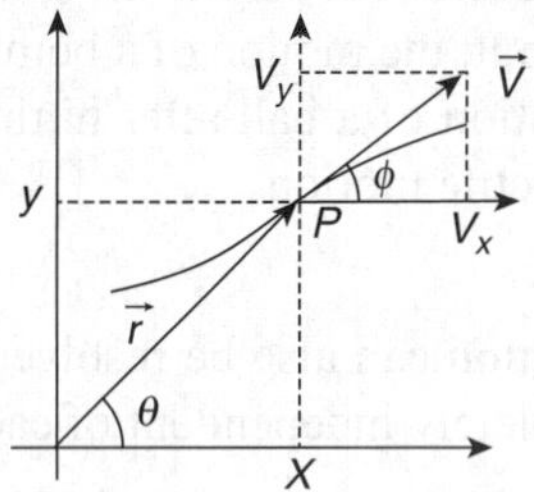

From figure, we get

$$x = r\cos\theta \quad \text{and} \quad y = r\sin\theta$$

Also, $\quad r = \sqrt{x^2 + y^2} \quad$ and $\quad \tan\theta = \dfrac{y}{x}$

Let $\vec{v}$ be the velocity of the particle at point P and let ϕ be the angle made by $\vec{v}$ with the x – axis. So, again we get

$$v_x = v\cos\phi \quad \text{and} \quad v_y = v\sin\phi$$

$$v = \sqrt{v_x^2 + v_y^2} \quad \text{and} \quad \tan\phi = \frac{v_y}{v_x}$$

A similar expression can also be arrived at for the components of acceleration a_x and a_y. Further, the equation of motion in vector form are

$$\vec{v} = \vec{u} + \vec{a}t$$

$$\vec{r} = \vec{u}t + \frac{1}{2}\vec{a}t^2$$

$$v^2 - u^2 = 2\vec{a}.\vec{r}$$

X-component of equation of motion	Y-component of equation of motion
$v_x = u_x + a_x t$	$v_y = u_y + a_y t$
$x = u_x t + \frac{1}{2}a_x t^2$	$y = u_y t + \frac{1}{2}a_y t^2$
$v_x^2 - u_x^2 = 2a_x x$	$v_y^2 - u_y^2 = 2a_y y$

Projectile Motion Under Gravity

When a body is thrown upward in a direction *different* from the vertical, due to gravity (which acts vertically downwards) it moves along a *curved* path in a vertical plane. Its path is referred to as the trajectory.

For studying projectile motion, the following assumptions must be kept in mind:

1. The particle remains close to the surface of the earth.
2. Acceleration due to gravity $(\vec{g})$ has a constant value throughout the entire motion.
3. No air drag is present.

The motion of a bullet shot from the gun, the motion of a rocket after burn-out, the motion of a bomb dropped from an aeroplane, the motion of a ball after hitting the bat, etc. are examples of projectile motion.

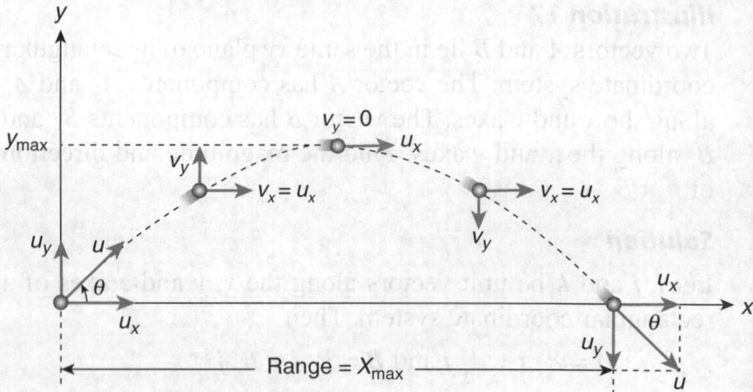

The velocity components are shown for various times : note that $v_y = 0$ at the top of the arc, or at y_{max}. The range (R) is the maximum horizontal distance, or x_{max}. Essentially, the ball travels up and then down as it simultaneously travels to the right.

The projectile motion can also be resolved into two simple motions, one horizontal and the other vertical motion. These two motions are completely independent of each other and the path of a projectile is totally determined by its initial velocity.

Motion along *X*-axis	Motion along *Y*-axis
The general equations for horizontal motion are given by $u_x = u\cos\theta,\ a_x = 0$ $v_x = u_x + 0(t) = u_x = u\cos\theta$ ...(i) i.e. the horizontal component of initial velocity is constant. $x = u_x t + \frac{1}{2}a_x t^2 = u\cos\theta t + \frac{1}{2}(0)t^2$ $= (u\cos\theta)t$ ...(iv)	The general equations for vertical motion are given by $u_y = u\sin\theta,\ a_y = -g$ $v_y = u_y + a_y t = u\sin\theta - gt$ ...(ii) $v_y^2 = u_y^2 + 2a_y y = u^2\sin^2\theta - 2gy$ ...(iii) $y = u_y t + \frac{1}{2}a_y t^2 = u\sin\theta t - \frac{1}{2}gt^2$ ...(v)

Trajectory

It is the path followed by the projectile, i.e., a curve joining position of the same particle in motion at different instants of time.

Substituting the value of t from Eq. (iv) in Eq. (v), we have,

$$y = u\sin\theta \times \frac{x}{u\cos\theta} - \frac{1}{2}\times g \times \left(\frac{x}{u\cos\theta}\right)^2$$

$$y = x\tan\theta - \frac{gx^2}{2u^2\cos^2\theta} \quad \text{(vi)}$$

This equation of trajectory represents a parabola; so the path of the projectile is a parabola.

The Eq. (vi) may be rewritten as

$$y = x\tan\theta - \frac{gx^2\tan\theta}{2u^2\cos\theta\sin\theta}$$

$$y = x\tan\theta\left[1 - \frac{x}{R}\right]$$

where $R = \dfrac{2u^2\sin\theta\cos\theta}{g}$

Maximum Height [*H*]

Maximum vertical displacement of projectile is called **Maximum Height**. At maximum height, $v_y = 0$.

Since, $v_y^2 - u_y^2 = 2a_y y$

$\Rightarrow \quad 0^2 - (u\sin\theta)^2 = 2(-g)H$

$\Rightarrow \quad H = \dfrac{u^2\sin^2\theta}{2g}$

Alternative method

At the highest point, the slope of the curve $\frac{dy}{dx} = 0$, hence from Eq. (vi)

$$0 = \tan\theta - \frac{g}{2u^2\cos^2\theta}.2x_m$$

$\therefore$ The corresponding x – coordinate, x_m is

$$x_m = \frac{u^2 \sin\theta\cos\theta}{g} = \frac{u^2\sin 2\theta}{2g} \quad \text{(vii)}$$

Putting x_m from Eq. (vii) in Eq. (iv), we get the time t_m to reach the maximum height, given by

$$t_m = \frac{x_m}{u\cos\theta} = \frac{u\sin\theta}{g} \quad \text{(viii)}$$

Now, putting t_m and x_m in Eq. (vi), we get the maximum height reached by the particle,

$$y_m = H = x_m(\tan\theta) - \frac{gx_m^2}{2u^2\cos^2\theta}$$

$$= \frac{u^2\sin^2\theta}{g} - \frac{u^2\sin^2\theta}{2g}$$

Hence, the maximum height,

$$H = \frac{u^2\sin^2\theta}{2g} \quad \text{(ix)}$$

Time of Flight [T]

The time of flight is the time taken by the projectile to strike the ground. At $t = T$, $y = 0$ as the projectile returns to the ground. Since

$$y = u_y t + \frac{1}{2}a_y t^2$$

$$\Rightarrow \quad 0 = (u\sin\theta)T + \frac{1}{2}(-g)T^2$$

$$\Rightarrow \quad T = \frac{2u\sin\theta}{g} \quad \{\because T \neq 0\}$$

Further, $T = t_{ascent} + t_{descent}$ and since no air drag exists so,

$$t_{ascent} = t_{descent}$$

$$\Rightarrow \quad t_{ascent} = t_{descent} = \frac{T}{2} = \frac{u\sin\theta}{g} \text{ or } T = \frac{2u\sin\theta}{g}$$

Range [R]

Maximum horizontal distance travelled is called **Range.**

At $\quad t = T,\ x = R$

Since $\quad x = u_x t + \frac{1}{2}a_x t^2$

and $\quad a_x = 0$ {$\because$ horizontal motion is non – accelerated}

$$\Rightarrow \quad R = (u\cos\theta)T$$

$$\Rightarrow \quad R = \frac{2}{g}(u\cos\theta)(u\sin\theta)$$

$$\Rightarrow \quad R = \frac{u^2\sin(2\theta)}{g}$$

Application of General Equations of a Projectile

(A) Projection from a height

Case I	Case II	Case III
$u_y = 0$, u_x, $t = 0$; u_x, v, v_y; u_x, v, v_y; u_x, v_y, v; y, x	u_y, u, $t = 0$, u_x; u_x, v, v_y; u_x, v, v_y; u_x, v_y, v; y, x	$t = 0$, θ, u_x, u_y; u_x, v, v_y; u_x, v, v_y; u_x, v_y, v; y, x

(Continued)

Horizontal projectile	**Projection at an angle θ above horizontal**	**Projection at an angle θ below horizontal**
$u_x = u,\ u_y = 0$ and $a_y = -g$ Horizontal motion: $x = ut$ Vertical motion: $-h = 0(t) - \frac{1}{2}gt^2$ From the above equations: $\Rightarrow \quad t = \sqrt{\frac{2h}{g}}$ $\Rightarrow \quad R = u \times t = u\sqrt{\frac{2h}{g}}$	$u_x = u\cos\theta,\ u_y = u\sin\theta$ and $a_y = -g$ Eqn. of horizontal motion: $x = u\cos\theta \times t$ Eqn. of vertical motion: $-h = u\sin\theta \times t - \frac{1}{2}gt^2$ From the above equations: $t = \frac{u\sin\theta}{g} \pm \sqrt{\frac{u^2\sin^2\theta}{g^2} + \frac{2h}{g}}$	$u_x = u\cos\theta,\ u_y = -u\sin\theta$ and $a_y = -g$ Eqn. of horizontal motion: $x = u\cos\theta t$ Eqn. of vertical motion: $-h = -u\sin\theta t - \frac{1}{2}gt^2$ From the above equations: $t = \frac{-2u\sin\theta \pm \sqrt{4u^2\sin^2\theta + 8gh}}{2g}$

(B) Projection from a moving body

Consider a boy throwing a ball from a moving trolley. Let the velocity of ball relative to boy be u.

Let v_{bt} be the velocity of the ball with respect to the trolley; then,

$$\vec{v}_{bt} = \vec{v}_b - \vec{v}_t$$

$$\vec{v}_b = \vec{v}_{bt} + \vec{v}_t$$

The above equation shows that absolute velocity of ball is vector sum of its velocity with respect to trolley and velocity of the trolley. Apply this equation to horizontal as well as vertical motion of the ball. Now consider following cases:

Case (i): Ball is projected in direction of motion of trolley Horizontal component of ball's velocity = $u\cos\theta + v$

Vertical component of ball's velocity = $u\sin\theta$

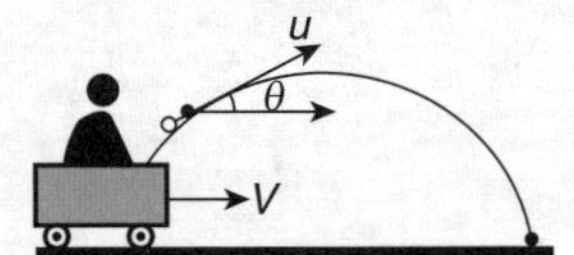

Horizontal component = $u\cos\theta + v$
Vertical component = $u\sin\theta$

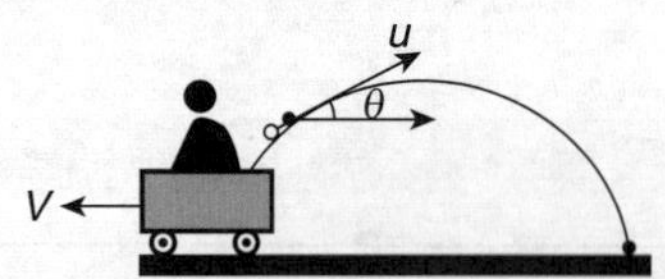

Horizontal component = $u\cos\theta - v$
Vertical component = $u\sin\theta$

Case (ii): Ball is projected opposite to direction of motion of trolley

Horizontal component of ball's velocity = $u\cos\theta - v$

Vertical component of ball's velocity = $u\sin\theta$

Case (iii): Similarly for a ball projected upwards from a upward moving platform

Horizontal component of ball's velocity = $u\cos\theta$

Vertical component of ball's velocity = $u\sin\theta + v$

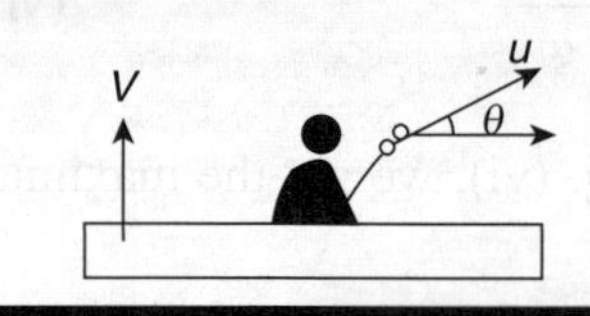

Horizontal component = $u\cos\theta$
Vertical component = $u\cos\theta + V$

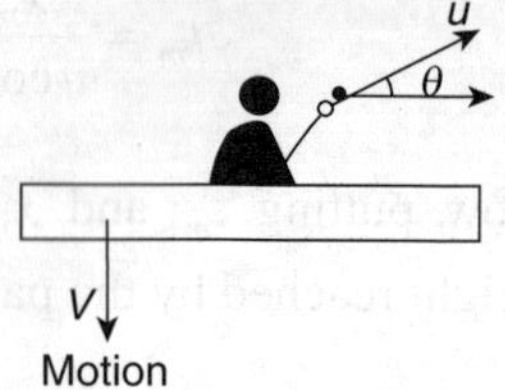

Horizontal component = $u\cos\theta$
Vertical component = $u\cos\theta - V$

Case (iv): For a downward moving platform,

Horizontal component of ball's velocity = $u\cos\theta$

Vertical component of ball's velocity = $u\sin\theta - v$

(c) Projection on an inclined plane

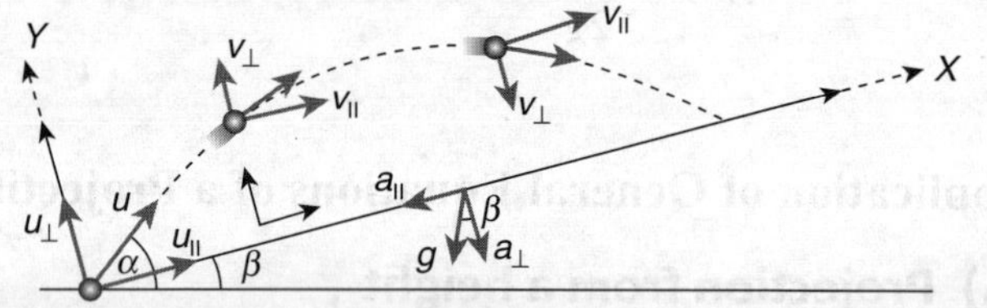

Consider a projectile launched with an initial velocity $\vec{u}$ at an angle α with horizontal on an inclined plane inclined at an angle $\beta(<\alpha)$ with the horizontal. Then $\vec{u}$ and $\vec{g}$ can both be resolved into two components, one along the incline and the other perpendicular to the incline.

- $u_{|} = u\cos(\alpha - \beta)$ along the incline along $+x$ axis.
- $u_\perp = u\sin(\alpha - \beta)$ along the incline along $+y$ axis.
- $a_{|} = g\sin\beta$ along $-x$ axis, acting as retardation to motion along the incline.
- $a_\perp = g\cos\beta$ along $-y$ axis, acting as retardation to motion perpendicular to the incline.

Time of flight

Time of flight, $T = \dfrac{2u_\perp}{a_\perp} = \dfrac{2u\sin(\alpha-\beta)}{g\cos\beta}$.

Range of projectile

$$R = u_\parallel T + \frac{1}{2}a_\parallel T^2$$

$$R = u\cos(\alpha-\beta)\times\frac{2u\sin(\alpha-\beta)}{g\cos\beta} - \frac{1}{2}g\sin\beta\times\left(\frac{2u\sin(\alpha-\beta)}{g\cos\beta}\right)^2$$

On solving we get,

$$R = \frac{2u^2\sin(\alpha-\beta)\cos\alpha}{g\cos^2\beta}$$

Condition for range to be maximum

Since $R = \dfrac{u^2}{g\cos^2\beta}\left[2\sin(\alpha-\beta)\cos\alpha\right]$

$\Rightarrow \quad R = \dfrac{u^2}{g\cos^2\beta}\left[\sin(\alpha-\beta+\alpha)+\sin(\alpha-\beta-\alpha)\right]$

$$[\because 2\sin A\cos B = \sin(A+B)+\sin(A-B)]$$

$\Rightarrow \quad R = \dfrac{u^2}{g\cos^2\beta}\left[\sin(2\alpha-\beta)-\sin\beta\right]$

For R to be maximum, $\sin(2\alpha-\beta)$ is maximum, i.e., 1

$\Rightarrow \quad \sin(2\alpha-\beta) = 1$

$\Rightarrow \quad 2\alpha-\beta = \dfrac{\pi}{2}$

$\Rightarrow \quad \alpha = \dfrac{\pi}{4}+\dfrac{\beta}{2}$

and maximum range is given by

$$R_{\max} = \frac{u^2}{g(1+\sin\beta)}$$

Maximum height

If h is maximum height above the incline, then at the maximum height, we have $u_\perp = 0$. So,

$$H = \frac{u_\perp^2}{2a_\perp} = \frac{u^2\sin^2(\alpha-\beta)}{2g\cos\beta}$$

Relative Motion Between the Two Projectiles

Let two particles A and B be projected from the ground with speeds u_A and u_B at angles θ_1 and θ_2 respectively as shown in figure.

The acceleration of both particles in y-coordinate is g (downwards). So relative acceleration between them in y – direction,

$$(a_{AB})_y = a_{Ay} - a_{By} = (-g)-(-g) = 0$$

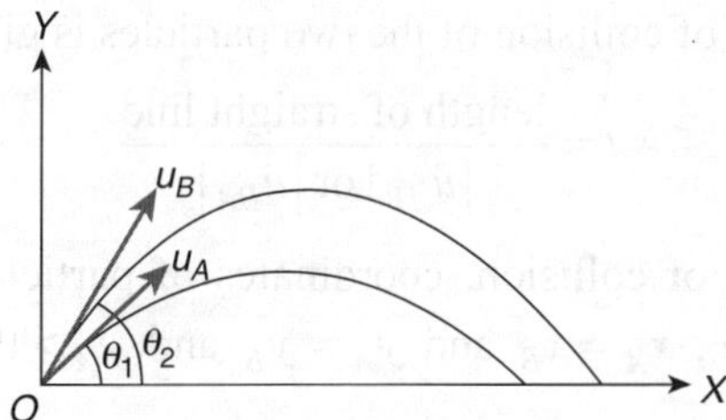

and acceleration of both particle in x – coordinate is also zero because

$a_{Ax} = a_{Bx} = 0$, i.e., the relative motion between the two particles is uniform. Similarly the components of relative velocity of two particles are

$$(u_{AB})_x = u_{Ax} - u_{Bx} = u_A\cos\theta_1 - u_B\cos\theta_2$$

and $\quad (u_{AB})_y = u_{Ay} - u_{By} = u_A\sin\theta_1 - u_B\sin\theta_2$

Hence, relative motion of particle A w.r.t. particle B is a straight line inclined at an angle.

$$\tan\theta = \frac{(u_{AB})_y}{(u_{AB})_x}$$

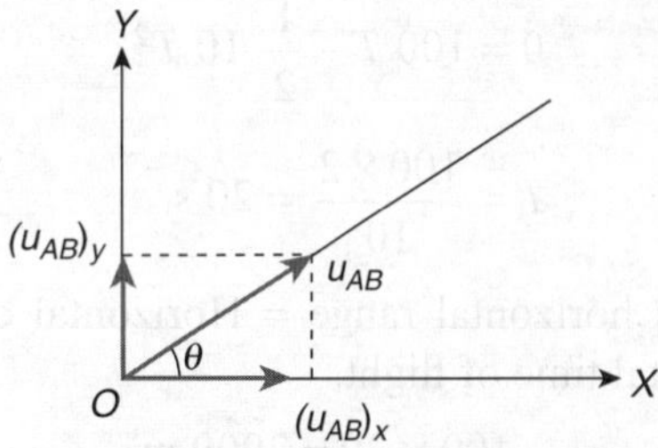

Condition of collision of two projectiles:

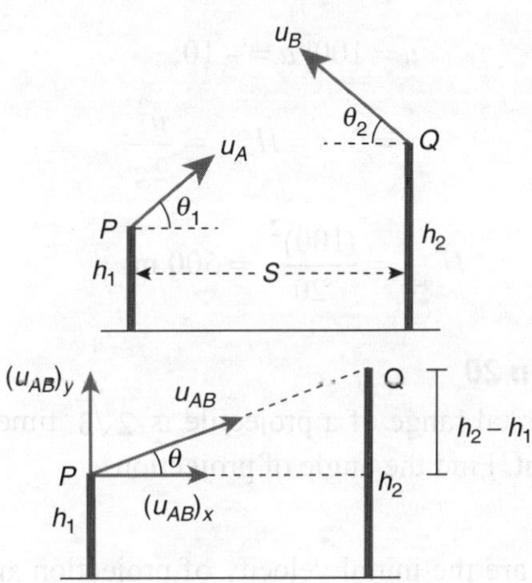

Let the two particles be projected simultaneously from two different heights h_1 and h_2 with speeds u_A and u_B at angles θ_1 and θ_2 respectively as shown in the given figure. Then, the two particles may collide in mid-air if the relative velocity of the two particles $\vec{u}_{AB}$ or $\vec{u}_{BA}$ is along the straight line PQ,

The time of collision of the two particles is given by,

$$t = \frac{\text{length of straight line}}{|\vec{u}_{AB}| \text{ or } |\vec{u}_{BA}|}$$

At the time of collision, coordinates of particle should be identical, i.e., $x_A = x_B$ and $y_A = y_B$ and $y_A > 0,\ y_B > 0$.

Illustration 19

The speed with which a bullet can be fired is $100\sqrt{2}$ m/s. Calculate the greatest distance to which it can be projected and also the maximum height to which it would rise.

Solution

The greatest horizontal range is achieved at an angle of projection of 45°.

Components of initial velocity $= 100\sqrt{2}\cos 45° = 100$ m/s.

Now, if T = time of flight, then considering the vertical motion of the bullet,

$$u = 100;\ a = -10;\quad s = 0;\quad t = T$$

Using $\quad s = ut + \frac{1}{2}at^2$

we get $\quad 0 = 100\,T - \frac{1}{2}\,10\,T^2$

$$\Rightarrow \quad T = \frac{100 \times 2}{10} = 20 \text{ s.}$$

$\therefore$ Maximum horizontal range = Horizontal component of velocity × Total time of flight.

$$100 \times 20 = 2000 \text{ m}$$

Again, if H_{max} be the maximum height to which the bullet rises, then

$$u = 100;\ a = -10;$$

$$v = 0;\ s = H_{max} = \frac{u^2}{2g}$$

$$\Rightarrow \quad H_{max} = \frac{(100)^2}{20} = 500 \text{ m.}$$

Illustration 20

The horizontal range of a projectile is $2\sqrt{3}$ times its maximum height. Find the angle of projection.

Solution

If u and α are the initial velocity of projection and angle of projection respectively, then

the maximum height attained $= \dfrac{u^2 \sin^2\alpha}{2g}$

and horizontal range $= \dfrac{2u^2 \sin\alpha\cos\alpha}{g}$

According to the problem, $\dfrac{2u^2\sin\alpha\cos\alpha}{g} = 2\sqrt{3}\left(\dfrac{u^2\sin^2\alpha}{2g}\right)$

$$\Rightarrow \quad \tan\alpha = \left(\frac{2}{\sqrt{3}}\right)$$

$$\Rightarrow \quad \alpha = \tan^{-1}\left(\frac{2}{\sqrt{3}}\right)$$

Illustration 21

A batsman hits a pitched cricket ball at a height of 1.2 m above the ground so that its angle of projection is 45° and its horizontal range is 110 m. The ball is lifted towards the left field line, where a fence of 7.5 m is located 100 m from the position of the batsman. Will the ball clear the fence?

Solution

Let us first calculate initial velocity of launch of projectile. Considering origin at O we have,

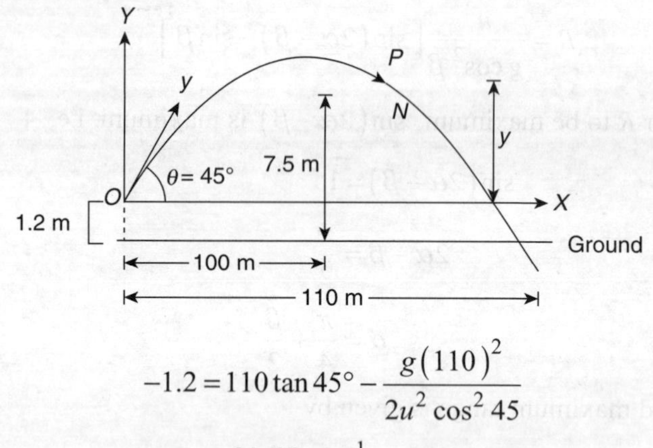

$$-1.2 = 110\tan 45° - \frac{g(110)^2}{2u^2\cos^2 45}$$

$$\Rightarrow \quad u = 32.66 \text{ ms}^{-1}$$

Let us calculate the vertical distance of ball from x-axis at the position of fence i.e. we are to find y for

$$x = 100 \text{ m},\ \theta = 45°,\ u = 32.66 \text{ ms}^{-1}$$

$$y = 100\tan 45° - \frac{9.8 \times (100)^2}{2 \times (32.66)^2 \cos^2 45°}$$

$$y = 100 - 91.87 = 8.13\text{m}$$

$\therefore$ Total height of ball from the ground $= 1.2 + 8.13 = 9.33$ m

Since 9.33 m > 7.5 m, therefore, the ball will clear the fence.

Illustration 22

Trajectories of two projectiles are shown in figure. Let T_1 and T_2 be the time periods and u_1 and u_2 their speeds of projection, then

(a) $T_2 > T_1$ (b) $T_1 = T_2$
(c) $u_1 > u_2$ (d) $u_1 < u_2$

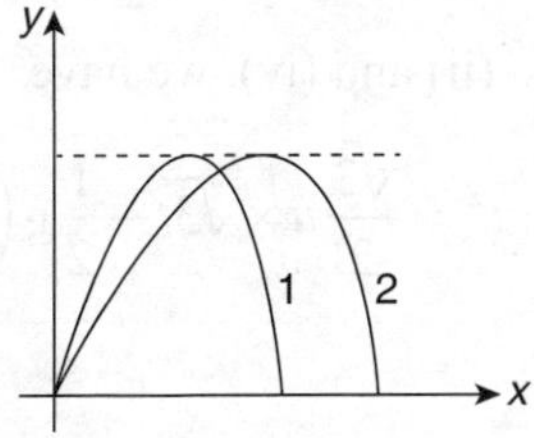

Solution

Maximum height and time of flight depends on the vertical component of initial velocity.

$$H_1 = H_2$$

$$\Rightarrow \quad u_{y_1} = u_{y_2}$$

Hence $\quad T_1 = T_2$

$$\text{Range, } R = \frac{u^2 \sin 2\theta}{g} = \frac{2(u\sin\theta)(u\cos\theta)}{g} = \frac{2u_x u_y}{g}$$

$$R_2 > R_1$$

$$u_{x_2} > u_{x_1}$$

or $\quad u_2 > u_1$

Illustration 23

A projectile has the same range R for two angles of projections. If T_1 and T_2 be the times of flight in the two cases, then

(a) $T_1T_2 \propto R$ (b) $T_1T_2 \propto R^2$
(c) $\frac{T_1}{T_2} = \tan\theta$ (d) $\frac{T_1}{T_2} = \cot\theta$

Solution

We know, $\quad R = \frac{u^2 \sin 2\theta}{g} = \frac{2u^2 \sin\theta\cos\theta}{g} \quad \ldots(1)$

Also $\quad R = R_{90^\circ-\theta}$.

Now, $T_1 = \frac{2u\sin\theta}{g}$ and $T_2 = \frac{2u\sin(90^\circ-\theta)}{g} = \frac{2u\cos\theta}{g}$

$$\therefore T_1T_2 = \frac{4u^2 \sin\theta\cos\theta}{g} = \left(\frac{2u^2 \sin\theta\cos\theta}{g}\right)\left(\frac{2}{g}\right) = \frac{2R}{g}$$

or $\quad T_1T_2 \propto R$

$$\frac{T_1}{T_2} = \frac{\frac{2u\sin\theta}{g}}{\frac{2u\cos\theta}{g}} = \tan\theta$$

Hence, options (a) and (c) are correct.

Illustration 24

A motorcycle stunt rider rides off the edge of a cliff. Just at the edge, his velocity is horizontal, with magnitude 9.0 m/s. Find the motorcycle's position, distance from the edge of the cliff, and velocity after 0.50 s.

Solution

The coordinate system is shown in the figure. We choose the origin at the edge of the cliff, where the motorcycle first becomes a projectile, so $x = 0$ and $y = 0$. The initial velocity is purely horizontal (that is, $\theta = 0$), so the initial velocity components are $u_x = 9.0\text{m/s}$ and $u_y = 0$.

The position of motorcycle at $t = 0.50$ s is given by

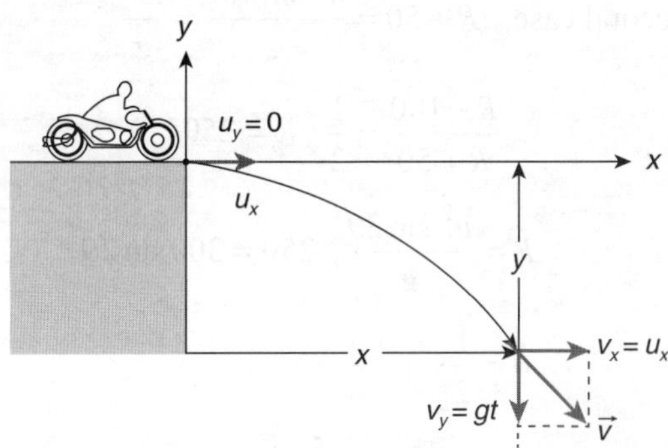

$$x = u_x t$$

$$= (9.0\text{ m/s})(0.50\text{ s})$$

$$= 4.5\text{ m}.$$

$$y = -\frac{1}{2}gt^2$$

$$= -\frac{1}{2}(9.8\text{ m/s}^2)(0.50\text{ s})^2$$

$$v_y = -gt = (-9.8\text{ m/s}^2)(0.50\text{ s}) = -4.9\text{ m/s}.$$

The negative value of y shows that at this time the motorcycle is below its starting position.

The distance of the motorcycle from the origin at this instant

$$r = \sqrt{x^2 + y^2}$$

$$= \sqrt{(4.5\text{ m})^2 + (-1.2\text{ m})^2}$$

$$= 4.7\text{ m}.$$

and the x and y components of velocity at time $t = 0.50$ s are

$$v_x = u_x = 9.0\text{ m/s},$$

$$v_y = -gt = (-9.8\text{ m/s}^2)(0.50\ s) = -4.9\text{ m/s}.$$

Note that the motorcycle has the same horizontal velocity u_x, as when it left the cliff at $t = 0$, but in addition there is a vertical velocity v_y directed downward. The net velocity at $t = 0.50\ s$ is 4.9 m/s.

Illustration 25

A projectile aimed at a mark which is in the horizontal plane falls 100 m short when the angle of projection is 15° and falls 50 m too far when the angle of projection is 45°. What should be the angle of projection so that the projectile hits the mark? Also find the corresponding velocity of projection.

Solution

Let R be the correct range of the projectile so that it hits the mark.

In the first case, $R-100=\dfrac{u^2\sin 30°}{g}=\dfrac{u^2}{2g}$

In the second case, $R+50=\dfrac{u^2\sin 90°}{g}=\dfrac{u^2}{g}$

$\therefore \quad \dfrac{R-100}{R+50}=\dfrac{1}{2},\ R=250$

Now, $\quad R=\dfrac{u^2\sin 2\theta}{g},\ 250=300\sin 2\theta$

$\therefore \quad \theta=\dfrac{1}{2}\sin^{-1}\dfrac{5}{6}$

and $\quad u^2=\dfrac{Rg}{\sin 2\theta}=\dfrac{2500}{5/6}$

or, $\quad u=10\sqrt{30}$ m/s

Illustration 26

A cannon fires two shells successively with a speed of $50\left(1+\sqrt{3}\right)$ m/s. The angles of projection are 60° and 30° respectively. What should be the time interval τ between the firings so that they collide?

Solution

Suppose that the angles of projection for the two shells are 60° and 30° respectively. Let the time of flight be $t+\tau$ and t, respectively, for the two shells before they collide. The time interval if the collision occurs at (x, y), we have for

1st shell: $\quad x=u\cos 60°\left(t+\tau\right)=\dfrac{u}{2}\left(t+\tau\right), \qquad$ (i)

$$y=u\sin 60°(t+\tau)-\frac{1}{2}g(t+\tau)^2=\frac{u\sqrt{3}}{2}(t+\tau)-\frac{1}{2}g(t+\tau)^2 \qquad \text{(ii)}$$

2nd shell: $\quad x=u\cos 30°t=\dfrac{\sqrt{3}\,ut}{2} \qquad$ (iii)

$$y=u\sin 30°t-\frac{1}{2}gt^2=\frac{ut}{2}-\frac{1}{2}gt^2 \qquad \text{(iv)}$$

From Eqs. (i) and (iii), we have,

$$\frac{t+\tau}{t}=\sqrt{3},$$

$\Rightarrow \quad t+\tau=\sqrt{3}t$

$\Rightarrow \quad \tau=\left(\sqrt{3}-1\right)t$

Similarly from Eqs. (ii) and (iv), we have

$$\frac{ut}{2}-\frac{1}{2}gt^2=\frac{\sqrt{3}}{2}u\times\sqrt{3}t-\frac{1}{2}g\left(\sqrt{3}t\right)^2$$

On solving, we get

$$t=\frac{u}{g}$$

$$\tau=\left(\sqrt{3}-1\right)t$$

$\Rightarrow \quad \tau=\left(\sqrt{3}-1\right)\times\dfrac{u}{g}=\left(\sqrt{3}-1\right)\dfrac{50\left(1+\sqrt{3}\right)}{10}=10$ s

Illustration 27

Two guns, situated on the top of a hill of height 10 m, fire one shot each with the same speed $5\sqrt{3}$ ms^{-1} at some interval of time. One gun fires horizontally and other fires upward at an angle of 60° with the horizontal. The shots collide in air at a point P. Find (a) the time-interval between the firings and (b) the coordinate of the point P. Take the origin of the coordinate system at the foot of the hill right below the muzzles and the trajectories in the X-Y plane. $\left(g=10\ \text{ms}^{-2}\right)$.

Solution

The shot-1 is fired with a speed $u_1=5\sqrt{3}$ m/s horizontally, while the shot-2 with (same) speed $u_2=5\sqrt{3}$ m/s upwards at an angle of 60° with the horizontal. The coordinates of the muzzles of the guns are (0,10 m).

Suppose the shots 1 and 2 take times t_1 and t_2 respectively before they collide at the point P.

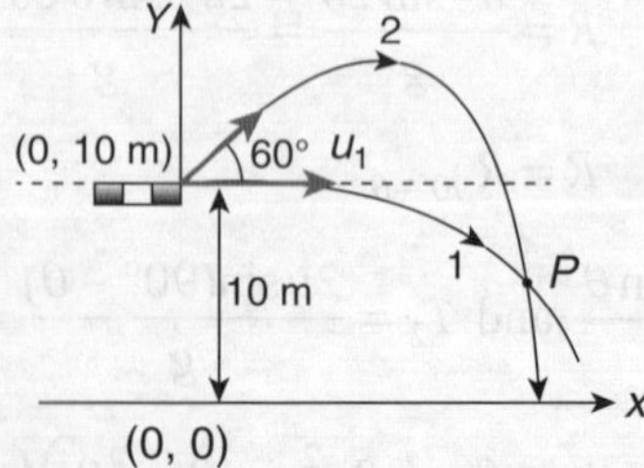

Now, we use the equation of motion $s=s_0+ut+\dfrac{1}{2}gt^2$.

For shot-1, we have

$$x_1=u_1t_1=5\sqrt{3}t_1 \qquad \ldots\text{(i)}$$

and $\quad y_1=10-\dfrac{1}{2}gt_1^2 \qquad \ldots$(ii)

For shot-2, we have

$$x_2=\left(u_2\cos 60°\right)t_2=\frac{5\sqrt{3}}{2}t_2 \qquad \ldots\text{(iii)}$$

and $\qquad y_2 = 10 + (u_2 \sin 60°)t_2 - \frac{1}{2}gt_2^2$

or $\qquad y_2 = 10 + \frac{15}{2}t_2 - \frac{1}{2}gt_2^2 \qquad \ldots\text{(iv)}$

At the point P, where shots collide, we have,

$$x_1 = x_2 \quad \text{and} \quad y_1 = y_2$$

Therefore, from Eqs. (i) and (iii) and from Eqs. (ii) and (iv), we have

$$5\sqrt{3}t_1 = \frac{5\sqrt{3}}{2}t_2$$

or, $\qquad t_2 = 2t_1 \qquad \ldots\text{(v)}$

and $10 - \frac{1}{2}gt_1^2 = 10 + \frac{15}{2}t_2 - \frac{1}{2}gt_2^2$ or $g\left(t_2^2 - t_1^2\right) = 15t_2 \ldots\text{(vi)}$

Solving Eqs. (v) and (vi), taking $g = 10$ ms^{-2}, we get

$$t_1 = 1 \text{ s and } t_2 = 2 \text{ s}$$

(a) The time-interval between the firings of the shots is

$$t_2 - t_1 = 2 - 1 = 1 \text{ s}$$

(b) Putting $t_1 = 1$ s and $g = 10$ ms^{-2} in Eqs. (i) and (ii), the coordinates of the point P are

$$(x_1, y_1) = \left(5\sqrt{3} \text{ m}, 5 \text{ m}\right)$$

Illustration 28

A particle is projected at an angle θ above the horizontal with an initial velocity of 105 m/s, from a platform moving horizontally with a constant speed of 49 m/s, as shown in the diagram.

(a) Find the total time of flight and the range of the projectile, as measured by an observer on the ground.

(b) For what value of θ, the angle of projection, will the projectile have the maximum range? What will be the value of this range? Ignore the height of the platform and use $g = 9.8$ m/s^2.

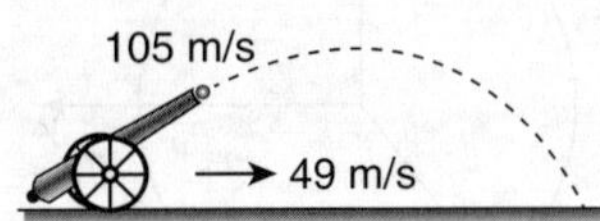

Solution

(a) Velocity of projection relative to the moving platform is $u = 105$ m/s at an angle θ with the horizontal.

The time of flight $T = \frac{2u\sin\theta}{g} = \frac{2 \times 105}{9.8}\sin 0$

$$= \frac{150}{7}\sin\theta \text{ s.} \qquad \ldots\text{(i)}$$

Horizontal component of the projectile w.r.t. ground is $(u\cos\theta + v)$, where v = speed of the platform.

Horizontal range as observed w.r.t the ground is

$$R = (u\cos\theta + v).T = (105\cos\theta + 49).\frac{150}{7}\sin\theta$$

$$= 150\sin\theta(15\cos\theta + 7)\text{m} \quad \ldots\text{(ii)}$$

(b) For the range R to be maximum,

$$\frac{dR}{d\theta} = 0$$

or $\qquad \frac{d}{d\theta}\left[150\sin\theta(15\cos\theta' + 7)\right] = 0$

or $\qquad 15\cos 2\theta + 7\cos\theta = 0$

Solving this we get $\cos\theta = \frac{3}{5}$; $\sin\theta = \frac{4}{5}$

$\therefore$ Required angle of projection for maximum rang is

$$\theta_{max} = \cos^{-1}\frac{3}{5} = 53°$$

Hence the maximum range is obtained by putting

$$\theta_{max} = 53° \text{ in Eq. (ii)}$$

$$R_{max} = 150\sin\theta_m(15\cos\theta_m + 7)$$

$$= 150 \times \frac{4}{5}\left(15 \times \frac{3}{5} + 7\right) = 1920 \text{ m}$$

Illustration 29

A projectile starts from the ground with velocity u making an angle of α with ground. The mean velocity between the starting point and the highest point of its path is

(a) $u\sqrt{1 + \cos^2\alpha}$ (b) $\frac{u}{2}\sqrt{1 - \cos^2\alpha}$

(c) $\frac{u}{2}\sqrt{1 - 3\cos^2\alpha}$ (d) $\frac{u}{2}\sqrt{1 + 3\cos^2\alpha}$

Solution

Here, displacement

$$s = \sqrt{(x^2 + y^2)} = \sqrt{H^2 + \left(\frac{R}{2}\right)^2}$$

where $H = \frac{u^2\sin^2\alpha}{2g}$, $R = \frac{u^2\sin 2\alpha}{g}$ and $t = \frac{u\sin\alpha}{g}$.

Mean velocity

$$= \frac{\sqrt{H^2 + \left(\frac{R}{2}\right)^2}}{t} = \frac{\sqrt{\left(\frac{u^2\sin^2\alpha}{2g}\right)^2 + \left(\frac{u^2\sin 2\alpha}{2g}\right)^2}}{\frac{u\sin\alpha}{g}}$$

$$= \frac{u}{2}\sqrt{1 + 3\cos^2\alpha}.$$

$\therefore$ (d) Ans.

Illustration 30

A particle is projected up an inclined plane of inclination α to the horizontal. If the particle strikes the plane horizontally then $\tan \alpha =$

(a) $\tan \beta$ (b) $2 \tan \beta$

(b) $3 \tan \beta$ (d) $\frac{1}{2} \tan \beta$

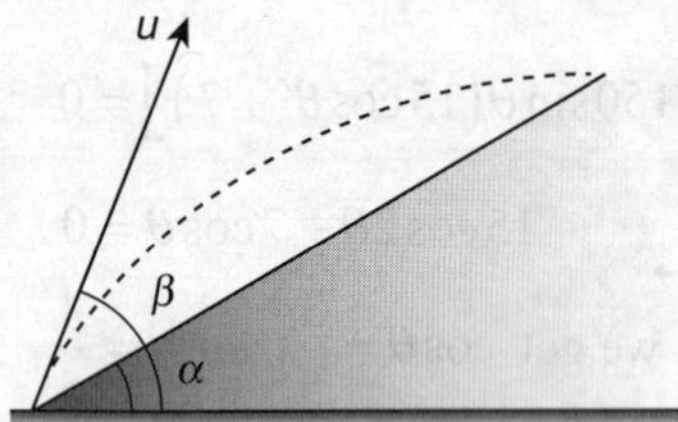

Solution

As the particle strikes the plane horizontally, it strikes the plane at the highest point of its trajectory.

$$\therefore \quad \tan \alpha = \frac{H}{R/2}$$

$$= \frac{u^2 \sin^2 \beta}{2g} \times \frac{g}{u^2 \sin \beta \cos \beta} = \frac{1}{2} \tan \beta$$

$\therefore$ Answer is (d)

Illustration 31

A particle is projected from a point O with velocity u in a direction making an angle α upward with the horizontal. At P, it is moving at right angles to its initial direction of projection. Its velocity at P is

(a) $u \tan \alpha$ (b) $u \cot \alpha$

(c) $u \operatorname{cosec} \alpha$ (d) $u \sec \alpha$

Solution

The horizontal component of velocity at P is given by

$$v_x = v \cos(90 - \alpha) = v \sin \alpha$$

Here $a_x = 0$, i.e., the horizontal component of velocity remains same.

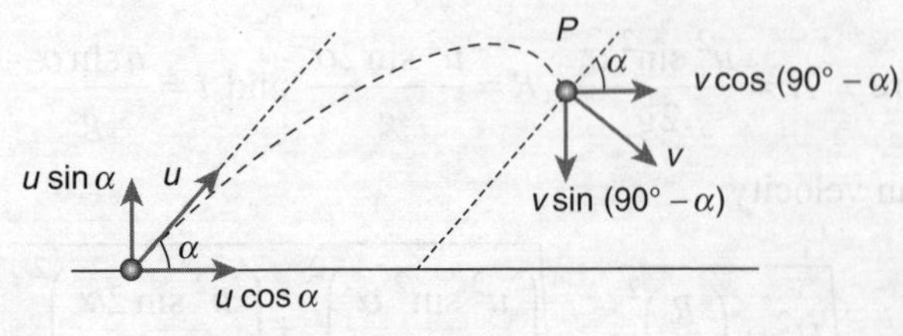

$$u_x = v_x$$

$$\Rightarrow \quad u \cos \alpha = v \sin \alpha$$

$$\Rightarrow \quad v = u \cot \alpha$$

$\therefore$ (b) Ans.

Uniform Circular Motion

When a particle moves in a circle with constant speed, the motion is called uniform circular motion.

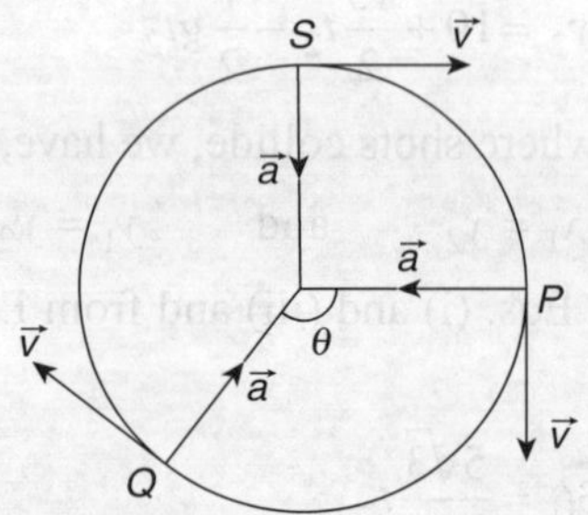

Consider a particle undergoing uniform circular motion. Let v be the speed of the particle and r be the radius of the circle. The velocity vector at any point on the circle is along the tangent at the point, and it changes continuously in direction but not in magnitude. Since the magnitude of $\vec{v}$ remains constant and only its direction changes, the acceleration must be perpendicular to the velocity (a component of acceleration along the velocity will change its magnitude). For uniform circular motion, the velocity is always tangent to the circle in the direction of motion, the acceleration $\vec{a}$ is always directed radially inward. Because of this, $\vec{a}$ is called centripetal (that is, centre-seeking) acceleration. The magnitude of this acceleration is given as v^2/r.

Angular velocity and angular acceleration

The rate of change of angle with respect to time is called angular velocity and is denoted by ω. It is a vector quantity and its direction is perpendicular to the plane of the circle.

Linear velocity = $r \times$ Angular velocity

The time for one complete revolution is called time period, T. We have,

$$2\pi = \omega T \Rightarrow T = 2\pi / \omega$$

The frequency f is defined as the number of revolutions per second.

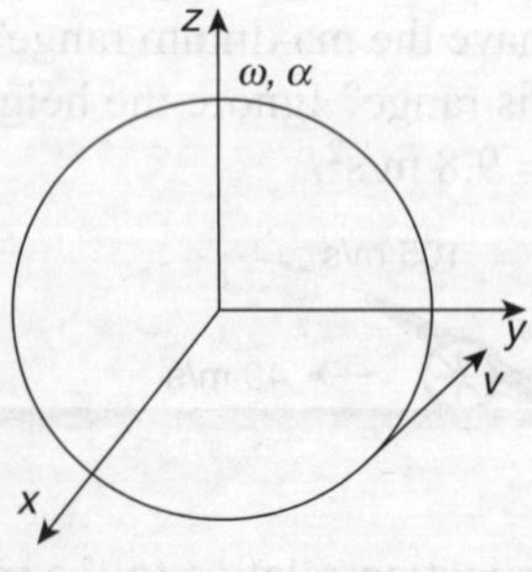

Therefore, $f = \frac{1}{T} = \frac{\omega}{2\pi}$

or $\omega = 2\pi f$

The rate of change of angular velocity is called angular acceleration. It is denoted by α and its unit is rad.s^{-2}. It is a vector quantity and its direction is perpendicular to the plane of the circle. Linear acceleration (a) = $r \times$ Angular acceleration (α).

Motion in three dimensions

The motion in three dimensions can be described in the same ways as motion in two dimensions. The vectors $\vec{r}, \vec{v}$ and $\vec{a}$ are interrelated and can be expressed in terms of their components as

$$\vec{r} = x\hat{i} + y\hat{j} + z\hat{k}$$

$$\vec{v} = v_x\hat{i} + v_y\hat{j} + v_z\hat{k}$$

$$\vec{a} = a_x\hat{i} + a_y\hat{j} + a_z\hat{k}$$

The motion along x-, y-, z- directions can be studied independently and later combined together. The values of these vectors at any time t can be obtained with the help of one-dimensional equations of motion.

Non-uniform circular motion

In a uniform circular motion, the particle moves along a circle with velocity of constant magnitude (i.e., constant speed). The direction of the velocity changes continuously. If the magnitude of the velocity also changes and the particle is made to move along the circle, then the motion is called non-uniform circular motion. The change in magnitude of the velocity occurs due to the tangential component of the acceleration.

In general, the acceleration of any particle can be written as,

$$\vec{a} = a_t\hat{t} + a_n\hat{n}$$

where t is a unit vector in the direction of velocity and $\hat{n}$ is a unit vector in the direction normal to the direction of $\vec{v}$, a_t and a_n are the components of $\vec{a}$ along $\hat{t}$ and $\hat{n}$. These are called tangential and normal components of the acceleration, respectively. The effect of a_t is to produce a change in the magnitude of the velocity while a_n produced a change in the direction only. Now, three cases arise.

(i) $a_t \neq 0$, $a_n = 0$. In this case, the direction remains constant but its magnitude changes. The resulting motion is along a straight line.

(ii) $a_t = 0$, $a_n \neq 0$. In this case the magnitude remains constant but the direction changes. If a_n is constant, then the rate of change of direction is constant and the resulting motion is a uniform circular motion.

(iii) $a_t \neq 0$, $a_n \neq 0$. In this case both the magnitude and the direction of the velocity change and the particle will move along a general curved path. If the particle is confined to move along a circle, the motion becomes non-uniform circular motion.

In terms of speed radius of curvature, a_t and a_n are given as,

$$a_t = \frac{dv}{dt}, a_n = \frac{v^2}{r}$$

Hence, $\quad a_t = 0$

$\Rightarrow \quad v = \text{constant}$

$\quad a_n = 0$

$\Rightarrow \quad r = \alpha \Rightarrow$ straight line motion.

Illustration 32

A particle is moving along a circle of radius r with constant speed v (figure). What is the change in velocity if the particle moves from the point P to point Q where $\angle POQ = \theta$?

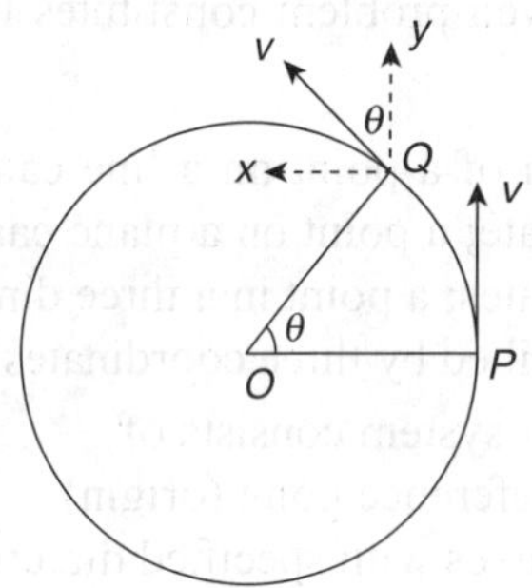

Solution

The velocity at any point of the path is always along the tangent to the path at that point. The velocity v at the point Q can be resolved into two mutually perpendicular components – one parallel to the velocity at P and is equal to $v \cos \theta$ and the other perpendicular to it and is $v \sin \theta$. Taking x and y axes along the directions shown, we can write

$$\vec{v}_P = \text{velocity at } P = v\,\hat{j}$$

and $\quad \vec{v}_Q = \text{velocity at } Q = v \sin \theta\, \hat{i} + v \cos\theta\, \hat{j}$

Therefore, the change in velocity is

$$d\vec{v} = \vec{v}_Q - \vec{v}_P = v \sin \theta \hat{i} + v(\cos\theta - 1)\,\hat{j}$$

$$\Rightarrow dv = |d\vec{v}| = \sqrt{v^2 \sin^2 \theta + v^2(\cos\theta - 1)^2}$$

$$\Rightarrow dv = \sqrt{v^2 \sin^2 \theta + v^2(\cos^2 \theta - 2\cos\theta + 1)}$$

$$\Rightarrow dv = \sqrt{2v^2(1 - \cos\theta)} = \sqrt{2v^2 . 2\sin^2 \theta / 2}$$

$$\Rightarrow dv = 2\, v \sin \theta / 2$$

POINTS TO REMEMBER

Concept of Motion

1. The concept of motion is a relative one and a body that may be in motion relative to one reference system, may be at rest relative to another. So, motion is not absolute, it is relative.
2. The motion of a particle is always described with reference system. A reference system is made by taking an arbitrary point as origin and imagining a coordinate system to be attached to it. This coordinate system chosen for a given problem constitutes the reference system for it.
3. The location of a point on a line can be described by one coordinate; a point on a plane can be described by two coordinates; a point in a three dimensional volume can be described by three coordinates.

   A coordinate system consists of
   (a) a fixed reference point (origin)
   (b) a set of axes with specified directions and scales
   (c) instructions that specify how to label a point in space relative to the origin and axes.

   For example,
   - Cartesian coordinate system (rectangular coordinate system): (x, y)

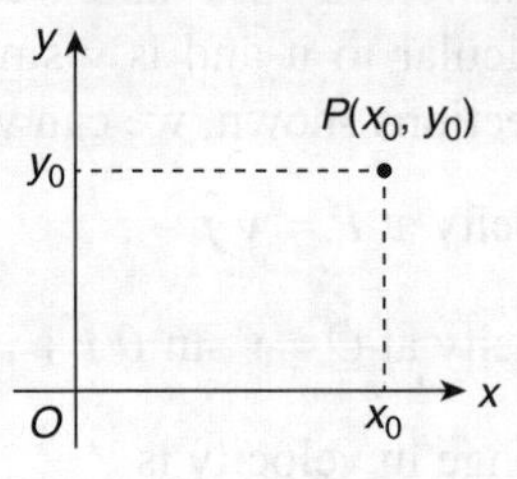

Cartesian coordinate system

   - Plane polar coordinates system : (r, θ)

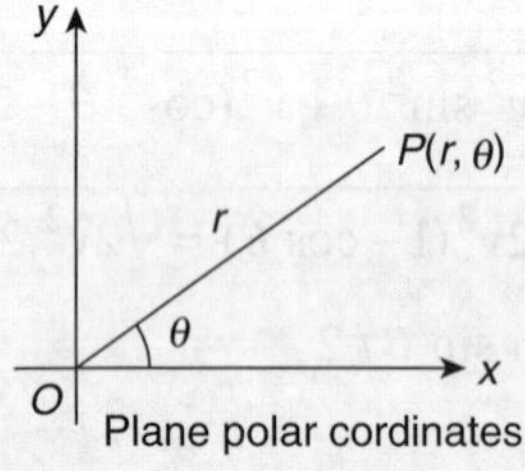

Plane polar cordinates

4. Distance is the actual path (length) covered by a moving particles (or object) in a given interval of time while displacement is the change in position, i.e., a vector joining on final position.
   (a) Distance ≥ |Displacement|.
   (b) For motion between two points, displacement is single-valued while distance depends on actual path and so can have many values.
   (c) For a moving particle distance can never decrease with time while displacement can.
   (d) For a moving particle distance can never be negative or zero while displacement can be.
   (e) In general, magnitude of displacement is not equal to distance. However, it can be so if the motion is along a straight line without change in direction.
5. For straight line motion, angle between velocity and acceleration should be either zero or π.
6. Average speed in different circumstances:
   (a) Bodies covering different distances with different speeds:

   Average speed,

$$v_{av} = \frac{\text{total distance covered}}{\text{total time taken}}$$

$$v_{av} = \frac{d}{t} = \frac{S_1 + S_2 + S_3 + \ldots\ldots\ldots}{t_1 + t_2 + t_3 + \ldots\ldots\ldots}$$

$$v_{av} = \frac{S_1 + S_2 + S_3 + \ldots\ldots\ldots}{\left(\frac{S_1}{v_1} + \frac{S_2}{v_2} + \ldots\ldots\right)}$$

   - Special case: If $S_1 = S_2 = S$; equal distances with different velocities

$$v_{av} = \frac{2S}{S\left(\frac{1}{v_1} + \frac{1}{v}\right)} = \frac{2v_1v_2}{v_1 + v_2}$$

   Or $\frac{2}{v_{av}} = \frac{1}{v_1} + \frac{1}{v_2}$ (Harmonic mean)

   - If a particle were to move along an N-sided regular polygon with speeds $v_1, v_2, \ldots v_n$, its average speed is

$$v_{av} = \frac{N}{\frac{1}{v_1} + \frac{1}{v_2} + \ldots\ldots + \frac{1}{v_n}}$$

Speed and Velocity

(a) If during motion, velocity remains constant throughout a given interval of time the motion is said to be uniform. For uniform motion,
Instantaneous velocity = Average velocity.
However, converse may or may not be true, i.e., if Average velocity = Instantaneous velocity, the motion may or may not be uniform.

(b) If velocity is constant, speed (= |velocity|) will also be constant. However, converse may or may not be true, i.e., if speed is constant, velocity may or may not be constant as velocity has direction in addition to magnitude which may or may not be change, e.g., in case of uniform circular motion.

(c) Velocity can be positive or negative as it is a vector but speed can never be negative as it is magnitude of velocity.

Acceleration

1. The direction of the acceleration vector depends on two things:
   (a) whether the object is speeding up or slowing down.
   (b) whether the object is moving in the positive or negative direction.
2. If the speed of a particle decreases with time, we say that it is decelerating, or it has deceleration or retardation. The acceleration of the body is opposite to that of velocity, and then the body decelerates.
3. If acceleration is zero, velocity will be constant and the motion will be uniform. However, if acceleration is constant, acceleration is uniform but motion is non-uniform and if acceleration is not constant both motion and acceleration are non-uniform.

Equations of Kinematics

1. The equations of kinematics for uniformly accelerating body are:
   (i) $v = u + at$
   (ii) $S = ut + \frac{1}{2}at^2$
   (iii) $v^2 - u^2 = 2aS$
   (iv) $S = \frac{u+v}{2}t$
2. Formula for finding the displacement of a body in the nth second: $S_n = u + a\left(n - \frac{1}{2}\right)$.

Some important facts about velocity and acceleration

1. If a particle is accelerated for a time t_1 with acceleration a_1 and for time t_2 with acceleration a_2 then average acceleration is

$$a_{av} = \frac{a_1t_1 + a_2t_2}{t_1 + t_2}$$

2. If a body starts from rest and moves with uniform acceleration then distance travelled by the body in t seconds is proportional to t^2 (i.e., $S \propto t^2$).
   So we can say that the ratio of distance covered in 1s, 2s, 3s is $1^2 : 2^2 : 3^2$ or 1 : 4 : 9.
3. A particle moving with uniform acceleration from A to B along a straight line has velocities v_1 and v_2 at A and B respectively. If C is the mid-point between A and B then velocity of the particle at C is equal to

$$v = \sqrt{\frac{v_1^2 + v_2^2}{2}}$$

4. A ball is dropped from a building of height h and it reaches after t seconds on ground. From the same building if two balls are thrown (one upwards and other downwards) with the same velocity u and they reach the ground after t_1 and t_2 seconds respectively then,

$$t = \sqrt{t_1t_2}$$

5. A particle is dropped vertically from rest from a height. The time taken by it to fall through successive distance of 1 m each will then be in the ratio of the difference in the square roots of the integers i.e.,

$$\sqrt{1} : (\sqrt{2} - \sqrt{1}) : (\sqrt{3} - \sqrt{2}) : ......$$

6. When a body is moving with a constant acceleration (constant retardation), the displacement-time graph is a (parabola) curve bending upwards (downwards).

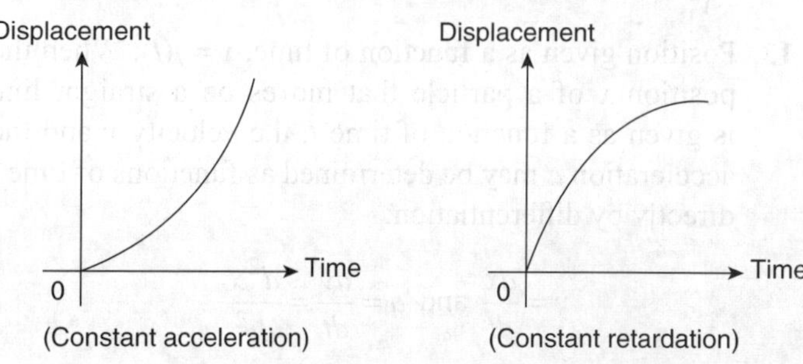

(Constant acceleration) (Constant retardation)

7. When a body returns towards the original point of reference while moving with uniform negative velocity, the displacement-time graph is oblique straight line AB making an angle $\theta > 90°$ with the time axis.

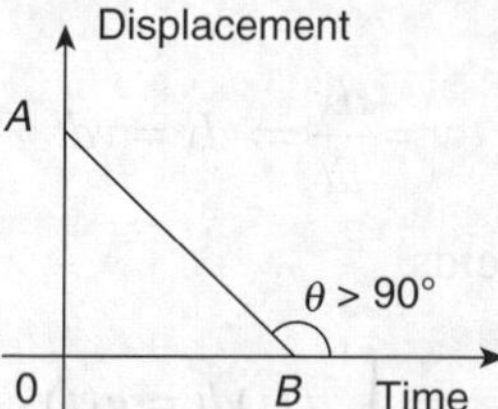

8. When a body is moving with increasing acceleration (decreasing acceleration), the velocity – time graph is a curve bending upwards (downwards).

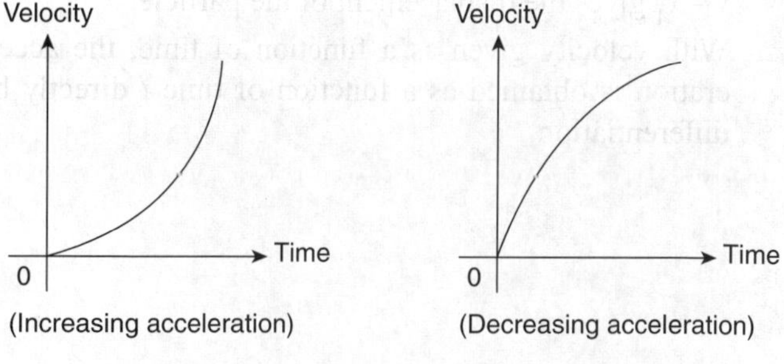

(Increasing acceleration) (Decreasing acceleration)

9. When acceleration is increasing (decreasing) and positive (negative), then acceleration-time graph:

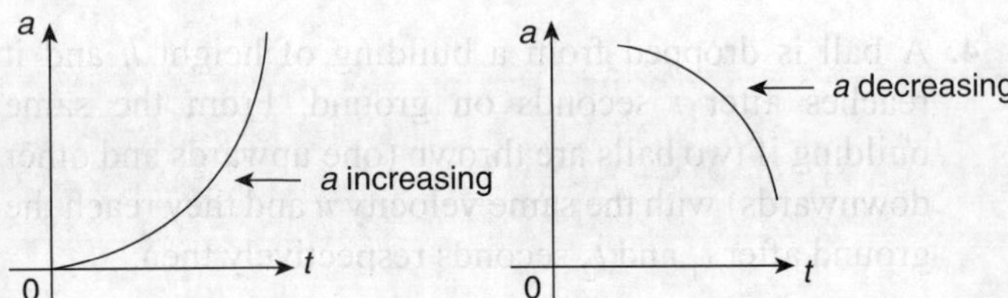

10. In a graph of position versus time, the slope of the curve is the velocity of the particle.

11. In a graph of velocity versus time,
 (a) the area under the curve between times t_1 and t_2 is the displacement $x_2 - x_1$.
 (b) the slope of the curve is the acceleration of the particle.

12. In a graph of acceleration versus time, the area under the curve between times t_1 and t is the change in velocity $v_2 - v_1$.

13. Position given as a function of time, $x = f(t)$. When the position x of a particle that moves on a straight line is given as a function of time t, the velocity v and the acceleration a may be determined as functions of time t directly by differentiation:

$$v = \frac{dx}{dt} \text{ and } a = \frac{dv}{dt} = \frac{d^2x}{dt^2}$$

14. Velocity given as a function of time, $v = f(t)$. When the velocity v of a particle that moves on a straight line is given as a function of time t, the position x and the acceleration a may be determined as functions of time t as follows:

$$v = \frac{dx}{dt} \Rightarrow dx = vdt$$

Integration yields,

$$x - x_1 = \int_{t_1}^{t} f(t)dt = g(t)$$

where x is the position of the particle at time t and x_1 is the position of the particle at time t_1. The difference $x - x_1$ gives the displacement of the particle.
With velocity given as a function of time, the acceleration is obtained as a function of time t directly by differentiation.

$$v = f(t)$$

$$a = \frac{dv}{dt}$$

15. Acceleration given as a function of time, $a = f(t)$. When the acceleration a of a particle that moves on a straight line is given as a function of time t, the position x and the velocity v may be determined as functions of time t as follows:

$$a = \frac{dv}{dt} \Rightarrow dv = adt$$

Then integration yields $v - v_1 = \int_{t_1}^{t} a\,dt$

where v is the velocity at time t and v_1 is the velocity at time t_1.

$$v = \frac{dx}{dt} \Rightarrow dx = v\,dt$$

Then integration yields $x - x_1 = \int_{t_1}^{t} v\,dt$

16. Acceleration given as a function of position, $a = f(x)$. When the acceleration a of a particle that moves on a straight line is given as a function of position x, the position x and the velocity v may be determined as functions of time t as follows:

$$\because \quad a = \frac{dv}{dt} = \frac{dv}{dx}\frac{dx}{dt} = v\frac{dv}{dx}$$

Separating variable, $vdv = adx$

Integration then yields $\frac{v^2}{2} - \frac{v_1^2}{2} = \int_{x_1}^{x} a\,dx$

Where v is a function of position x.

$$\because \quad v\frac{dx}{dt}$$

$$\Rightarrow \quad dt = \frac{dx}{v}$$

Then integration yields $t - t_1 = \int_{x_1}^{x} \frac{dx}{v}$.

EXERCISES

1. A locomotive begins to move with uniform acceleration at the moment when a boy runs past it with a uniform velocity of 2 ms^{-1}. The velocity of the locomotive when it overtakes the boy is
(a) 2 ms^{-1} (b) 4 ms^{-1}
(c) 6 ms^{-1} (d) 8 ms^{-1}

2. A body starts from rest and moves with constant acceleration. The ratio of distance covered by the body in *n*th second to that covered in *n* seconds is
(a) $1 : n$ (b) $\frac{2n-1}{n^2}$
(c) $\frac{n^2}{2n-1}$ (d) $\frac{2n-1}{2n^2}$

3. A train runs past a telegraph pole in 15 seconds, and through a tunnel, 450 metres long, in 45 seconds. When it meets another train 300 metres long both trains cross each other in 21 seconds, then the velocity of the second train is
(a) 10 ms^{-1} (b) 15 ms^{-1}
(c) 20 ms^{-1} (d) 30 ms^{-1}

4. Two spheres of same size, one of mass 2 kg and another of mass 4 kg, are dropped simultaneously from the top of Qutab minar (height = 72 m). When they are 1 m above the ground, the two spheres have the same
(a) momentum (b) kinetic energy
(c) potential energy (d) acceleration.

5. A body starts from the origin and moves along the *x*-axis such that the velocity at any instant is given by $4t^3 - 2t$, where *t* is in seconds and velocity in metre per second. What is the acceleration of the particle when it is 2 m from the origin?
(a) 28 ms^{-2} (b) 22 ms^{-2}
(c) 12 ms^{-2} (d) 10 ms^{-2}

6. The velocity of a body depends upon time according to the equation $v = 20 + 10t^2$. The body is undergoing
(a) zero acceleration (b) uniform acceleration
(c) uniform retardation (d) non-uniform acceleration.

7. For a particle moving along a straight line, the displacement *x* depends on time *t* as $x = \alpha t^3 + \beta t^2 + \lambda t + \delta$. The ratio of its initial acceleration to its initial velocity depends
(a) only on α and β (b) only on β and λ
(c) only on α and λ (d) only on α.

8. Three particles *A*, *B* and *C* are thrown from the top of a tower with the same speed. *A* is thrown straight up, *B* is thrown straight down and *C* is thrown horizontally. They hit the ground with speeds v_A, v_B and v_C respectively.
(a) $v_A = v_B = v_C$ (b) $v_B > v_C > v_A$
(c) $v_A = v_B > v_C$ (d) $v_A > v_B = v_C$

9. A particle starting from the origin (0,0) moves in a straight line in the (*x*, *y*). Its co-ordinates at a later time are $(3, \sqrt{3})$. The path of the particle makes with the *x*-axis an angle of
(a) 30° (b) 45°
(c) 60° (d) 0°

10. The velocity time graph for a straight line motion is shown in the figure. If distances travelled from *O* to *A*, *A* to *B* and *B* to *C* are in the ratio 1 : 2 : 3, then the ratio of maximum velocity to average velocity is

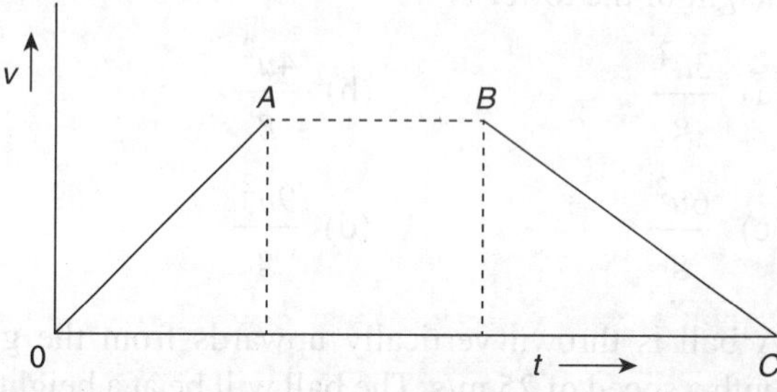

(a) 5/2 (b) 5/3
(c) 5/4 (d) 7/6

11. The *x* and *y* co-ordinates of a particle at any time *t* is given by $x = 7t + 4t^2$ and $y = 5t$ where *x* and *y* are in metres and *t* is in seconds. The acceleration of a particle at $t = 5$s is
(a) zero (b) 8 m/s²
(c) 20 m/s² (d) 40 m/s²

12. A particle goes along a quadrant circle of radius 15 cm with a constant speed 25 cm/s as shown. Find the average velocity over the interval *AB*.

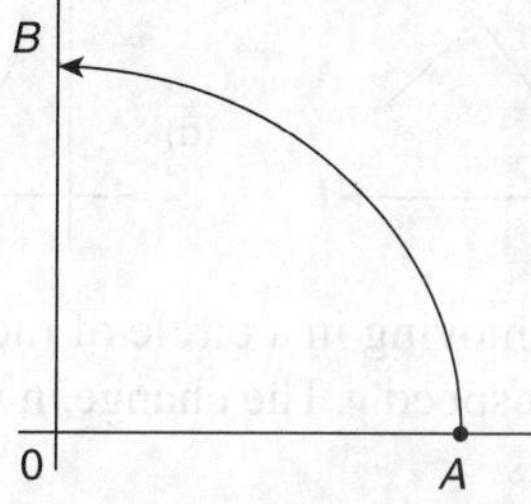

(a) 2.252 cm/s (b) 22.25 cm/s
(c) 4.49 cm/s (d) 3.14 cm/s

13. A point traversed half the distance with a velocity v_0. The remaining part of the distance was covered with velocity v_1 for half the time and with velocity v_2 for the other half of the time. The mean velocity of the point averaged over the whole time of motion is

(a) $\frac{v_0 + v_1 + v_2}{3}$ (b) $\frac{2v_0 + v_1 + v_2}{3}$

(c) $\frac{v_0 + 2v_1 + 2v_2}{3}$ (d) $\frac{2v_0(v_1 + v_2)}{2v_0 + v_1 + v_2}$

14. Two cars A and B, each having a speed of 30 km hr^{-1} are heading towards each other along a straight path. A bird that can fly at 60 km hr^{-1} flies off car A when the distance between the cars is 60 km, heads directly towards car B, on reaching B, the bird directly flies back to A and so forth, then the total distance the bird travels till the cars meet is

(a) infinite (b) 30 km
(c) 60 km (d) 120 km.

15. A stone thrown upwards with a speed u from the top of the tower reaches the ground with a velocity $3u$. The height of the tower is

(a) $\frac{3u^2}{g}$ (b) $\frac{4u^2}{g}$

(c) $\frac{6u^2}{g}$ (d) $\frac{9u^2}{g}$

16. A ball is thrown vertically upwards from the ground with a speed of 25 m/s. The ball will be at a height of 30 m above the ground after time

(a) 2.0 s only (b) 2.0 s and 3.0 s
(c) 3.0 s only (d) 1.0 s and 2.0 s

17. A particle is thrown above, then the correct $v-t$ graph will be

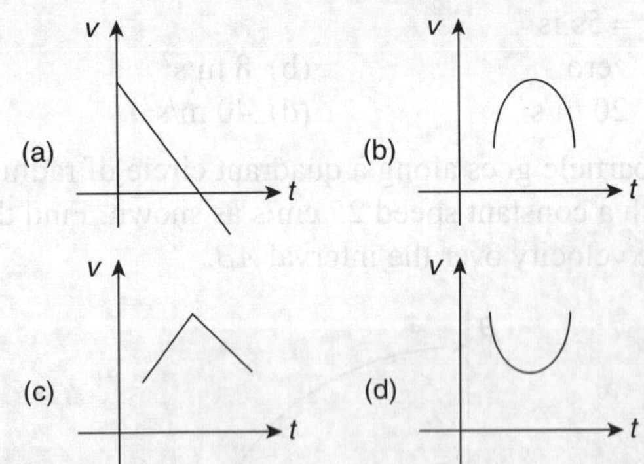

18. A particle is moving in a circle of radius r centred at O with constant speed v. The change in velocity in moving from A to B is

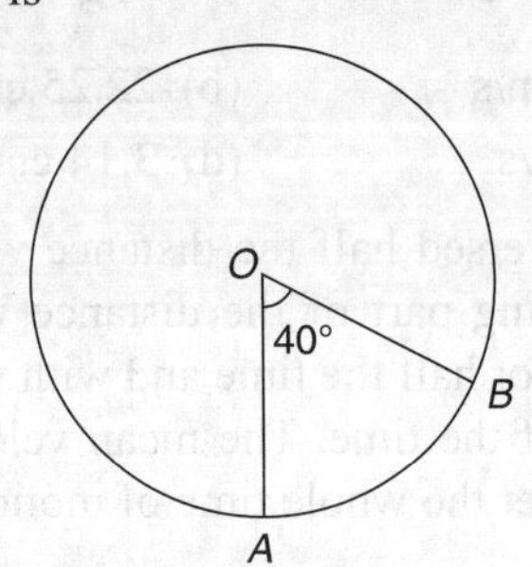

(a) $2v \cos 40°$ (b) $2v \sin 40°$
(c) $2v \cos 20°$ (d) $2v \sin 20°$

19. A cyclist moves in such a way that he takes 60° turn after 100 metres. What is the displacement when he takes seventh turn?

(a) 100 m (b) 200 m
(c) $100\sqrt{3}$ m (d) $100/\sqrt{3}$ m

20. A car runs at a constant speed on a circular track of radius 100 m, taking 62.8 s for every circular lap. The average velocity and average speed for each circular lap respectively is

(a) 0, 0 (b) 0, 10 m/s
(c) 10 m/s, 10 m/s (d) 10 m/s, 0

21. Which of the following options is correct for the object having a straight line motion represented by the following graph:

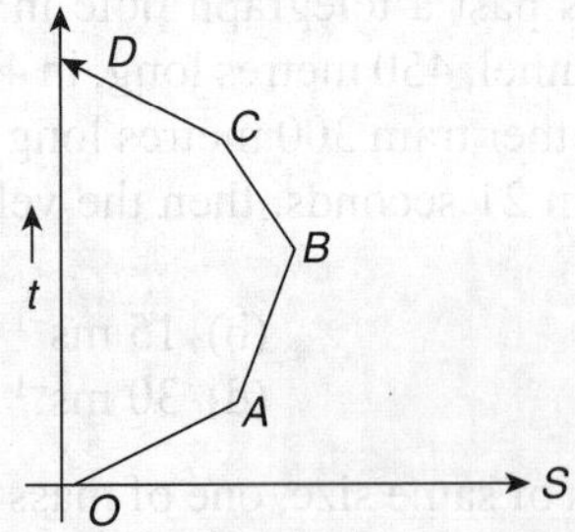

(a) The object moves with constantly increasing velocity from O to A and then it moves with constant velocity.
(b) Velocity of the object increases uniformly.
(c) Average velocity is zero.
(d) The graph shown is impossible.

22. A parachutist, after bailing out, falls 50 m without friction. When parachute opens, it decelerates at 2 m/s^2. He reaches the ground with a speed of 3 m/s. At what height, did he bail out?

(a) 91 m (b) 182 m
(c) 293 m (d) 111 m

23. A particle is moving eastward with a velocity of 5 m/s. In 10 seconds the velocity changes to 5 m/sec northward. The average acceleration in this time is

(a) $1/\sqrt{2}$ m/s^2 towards north-west
(b) $1/\sqrt{2}$ m/s^2 towards north-east
(c) 1/2 m/s^2 towards north-west
(d) 1/2 m/s^2 towards north.

24. A car is travelling at 20 m/s on a circular road of radius 100 m. It is increasing in speed at a rate of 3 m/s^2. Its acceleration is

(a) 3 m/s^2 (b) 5 m/s^2
(c) 30 m/s^2 (d) 7 m/s^2

25. An object is moving in a circle of radius 100 m with a constant speed of 31.4 m/s. What is its average speed for one complete revolution?

(a) Zero (b) 31.4 m/s

(c) 3.14 m/s (d) $\sqrt{2} \times 31.4$ m/s

26. From the top of a tower, a stone is thrown up and it reaches the ground in time t_1. A second stone is thrown down with the same speed and it reaches the ground in time t_2. A third stone is released from rest and it reaches the ground in time t_3.

(a) $t_3 = \frac{1}{2}(t_1 + t_2)$ (b) $t_3 = \sqrt{t_1 t_2}$

(c) $\frac{1}{t_3} = \frac{1}{t_2} - \frac{1}{t_1}$ (d) $t_3^2 = t_1^2 - t_2^2$

27. A balloon starts rising from the ground with an acceleration of 1.25 m/s^2. After 8 s, a stone is released from the balloon. The stone will

(a) cover a distance of 40 m

(b) have a displacement of 50 m

(c) reach the ground in 4 s

(d) begin to move down after being released.

28. Which graph pertains to uniform acceleration?

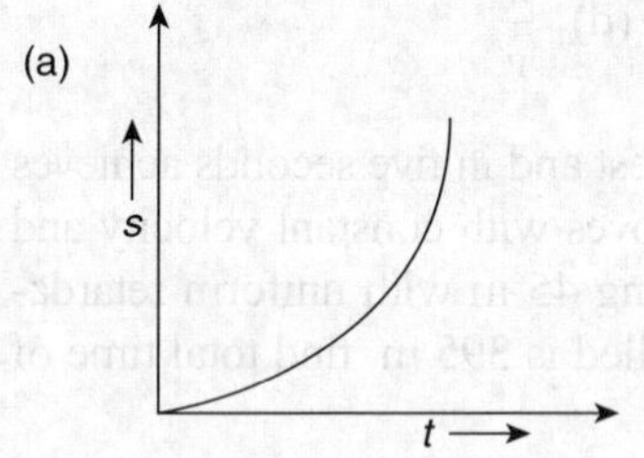

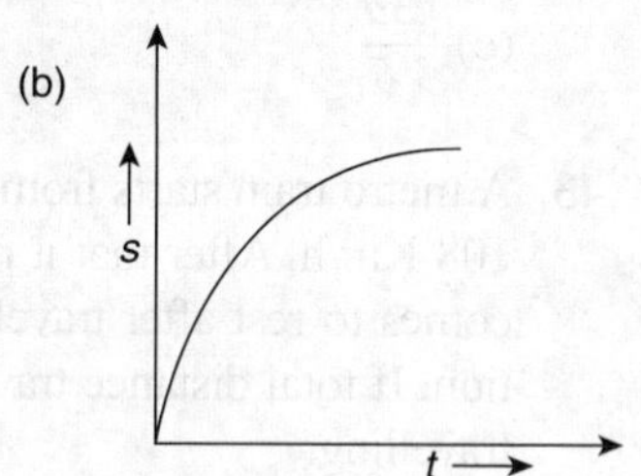

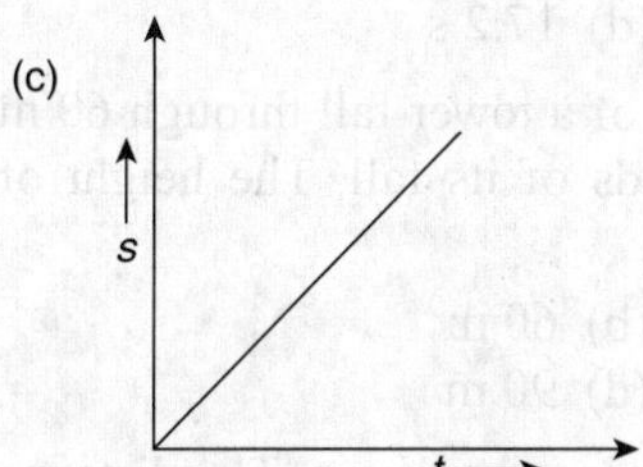

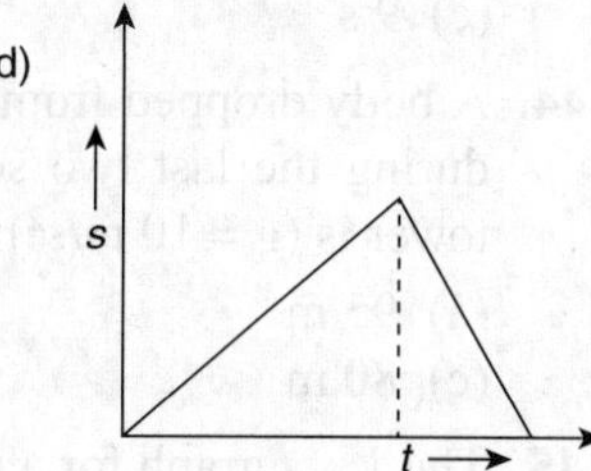

29. A frictionless wire AB is fixed on a sphere of radius R. A very small spherical ball slips on the wire. The time taken by this ball to slip from A to B is

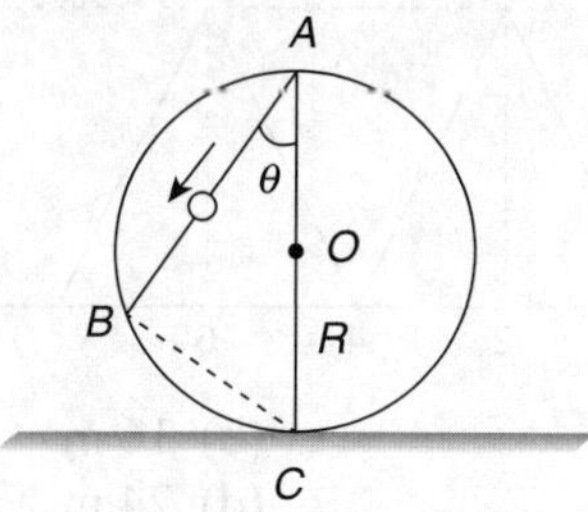

(a) $\frac{2\sqrt{gR}}{g\cos\theta}$ (b) $2\sqrt{gR}\,\frac{\cos\theta}{g}$

(c) $2\sqrt{\frac{R}{g}}$ (d) $\frac{gR}{\sqrt{g\cos\theta}}$

30. The numerical ratio of average velocity to average speed is

(a) always less than one (b) always equal to one

(c) always more than one (d) equal to or less than one.

31. A ball is dropped on a floor and bounces back to a height somewhat less than the original height, which of the curves depicts its motion correctly?

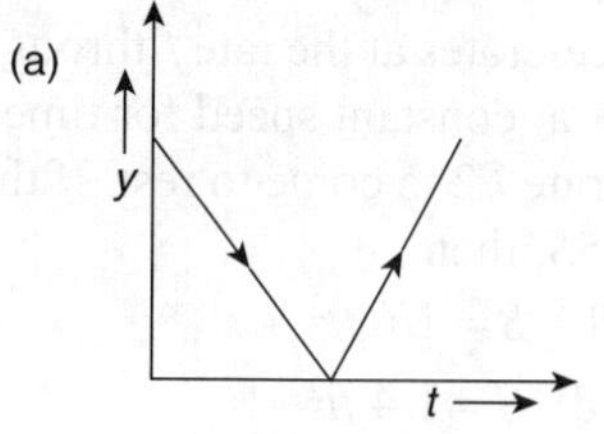

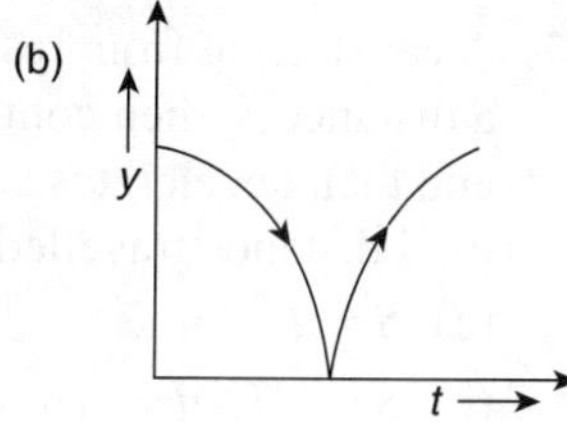

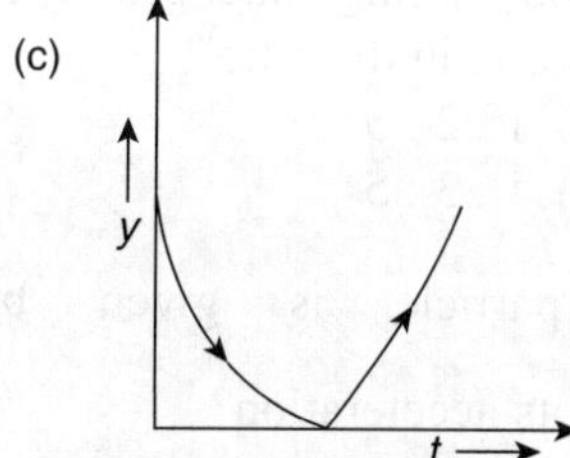

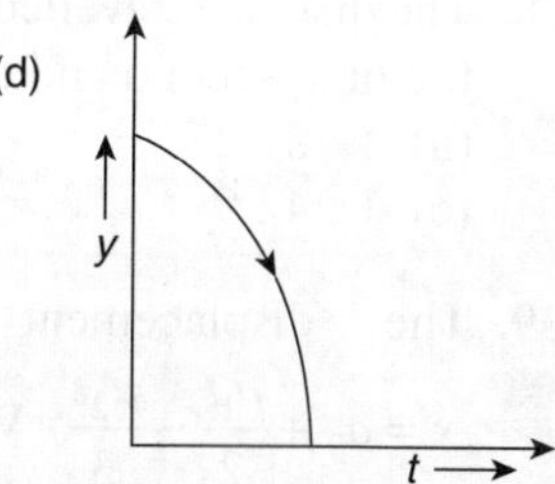

32. A body covers 200 cm in the first 2 seconds and 220 cm in the next 4 seconds. The velocity at the end of 7th second is

(a) 20 cm s^{-1} (b) 10 cm s^{-1}

(c) 115 cm s^{-1} (d) -10 cm s^{-1}

33. A stone falls freely from rest and the total distance covered by it in the last second of its motion equals $\frac{9}{25}$ of its motion, the stone remains in air for

(a) 4 s (b) 5 s

(c) 6 s (d) 7 s

34. The figure shows velocity time graph for a particle moving along a straight line. The distance covered by the particle in time interval $t = 1$ to $t = 8$ sec is

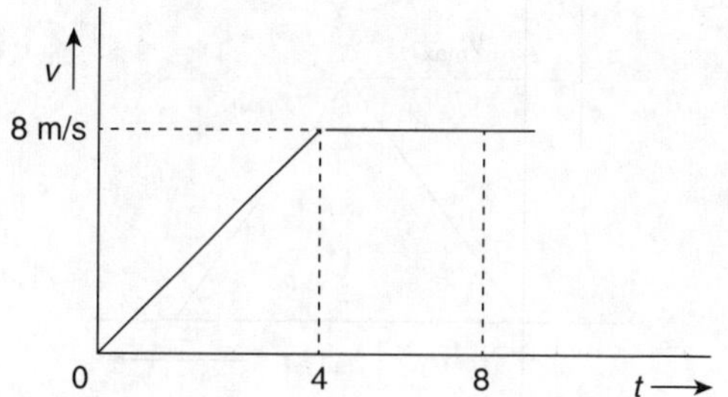

(a) 49 m (b) 32 m
(c) 47 m (d) 15 m

35. A boy moving with a uniform acceleration crosses a distance of 65 m in the 5th second and 105 m in the 9th second. How far will he go in 20 s?
(a) 2040 m (b) 240 m
(c) 2400 m (d) 2004 m

36. The relation between time t and distance x is $t = ax^2 + bx$, where a and b are constant. The acceleration is
(a) $-2abv^2$ (b) $2bv^3$
(c) $-2av^3$ (d) $2av^2$

37. A car starting from rest, accelerates at the rate f through a distance S, then continues at constant speed for time t and then decelerates at the rate $f/2$ to come to rest. If the total distance travelled is $15S$, then
(a) $S = ft$ (b) $S = 1/6\, ft^2$
(c) $S = 1/72\, ft^2$ (d) $S = 1/4\, ft^2$

38. The distance travelled by a body falling under gravity in the first, second and third second in the ratio
(a) 1 : 8 : 27 (b) 1 : 2 : 3
(c) 1 : 4 : 9 (d) 1 : 3 : 5

39. The displacement of particle is given by $x = a_0 + \frac{a_1 t}{2} - \frac{a_2 t}{3}$. What is its acceleration?
(a) $\frac{2a_2}{3}$ (b) $-\frac{2a_2}{3}$
(c) a_2 (d) zero

40. An object may have
(a) varying speed without having varying velocity
(b) varying velocity without having varying speed
(c) non-zero acceleration without having varying velocity
(d) None of these.

41. A car accelerates from rest at rate α for sometime after which it decelerates at rate β to come to rest. If the total time lapse is t seconds, then the total distance covered by the car is

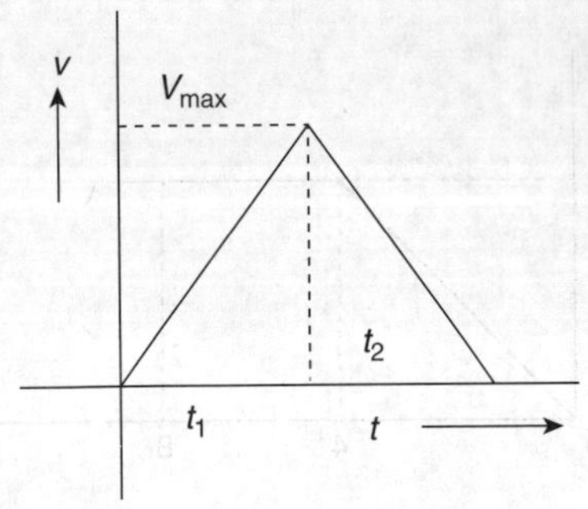

(a) $\frac{1}{2}\left(\frac{\alpha+\beta}{\alpha\beta}\right)t^2$ (b) $\frac{1}{2}\frac{\alpha\beta}{\alpha-\beta}t^2$
(c) $\frac{1}{2}\left(\frac{\alpha-\beta}{\alpha\beta}\right)t^2$ (d) $\frac{1}{2}\left(\frac{\alpha\beta}{\alpha+\beta}\right)t^2$

42. Three tortoises are located at the vertices of an equilateral triangle whose side equals a. They all start moving simultaneously with velocity v constant in modulus with the first tortoise heading continually for the second, the second for the third and the third for the first. They will meet after time

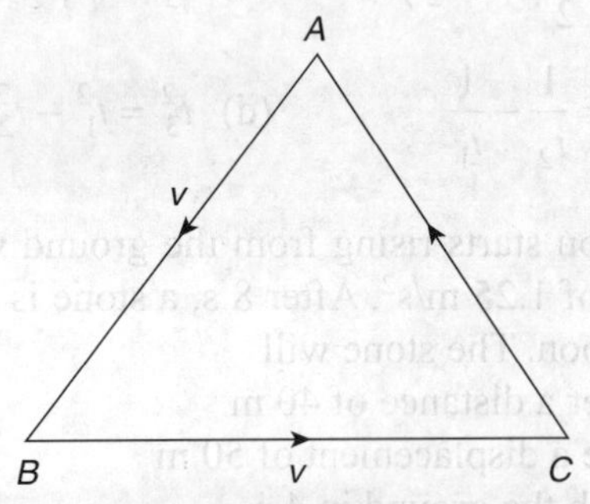

(a) $\frac{a}{3v}$ (b) $\frac{2a}{3v}$
(c) $\frac{3a}{v}$ (d) $\frac{a}{v}$

43. A metro train starts from rest and in five seconds achieves 108 Km/h. After that it moves with constant velocity and comes to rest after travelling 45 m with uniform retardation. If total distance travelled is 395 m, find total time of travelling.
(a) 12.2 s (b) 15.3 s
(c) 9 s (d) 17.2 s

44. A body dropped from top of a tower fall through 60 m during the last two seconds of its fall. The height of tower is ($g = 10$ m/s²)
(a) 95 m (b) 60 m
(c) 80 m (d) 90 m

45. The $V - t$ graph for a particle is as shown. The distance travelled in the first four seconds is

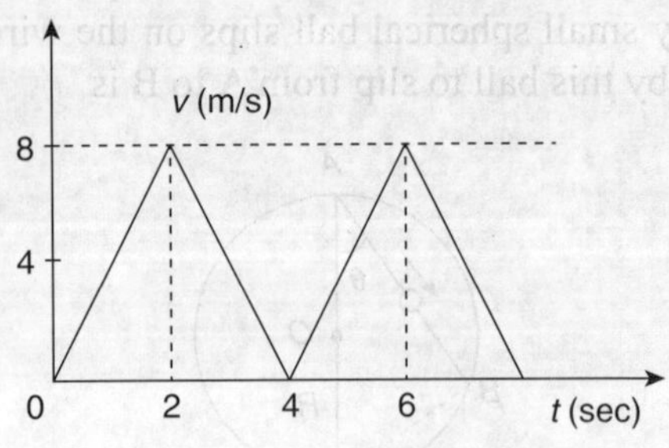

(a) 12 m (b) 16 m
(c) 20 m (d) 24 m

46. A body is thrown vertically upwards with a velocity v. The distance travelled by it in the fifth and the sixth seconds are equal. The velocity v is given by ($g = 10$ m/s^2)

(a) 24.5 m/s (b) 49.0 m/s
(c) 73.5 m/s (d) 98.0 m/s

47. The displacement (x) of a particle is related to time t as $x = at + bt^2 - ct^3$, where a, b and c are constants of motion. The velocity of the particle when its acceleration is zero is given by:

(a) $a+\frac{b^2}{c}$ (b) $a+\frac{b^2}{2c}$
(c) $a+\frac{b^2}{3c}$ (d) $a+\frac{b^2}{4c}$

48. The figure shows a particle moving along x-axis subjected to three periods of acceleration. Rank the periods according to the increases they produce

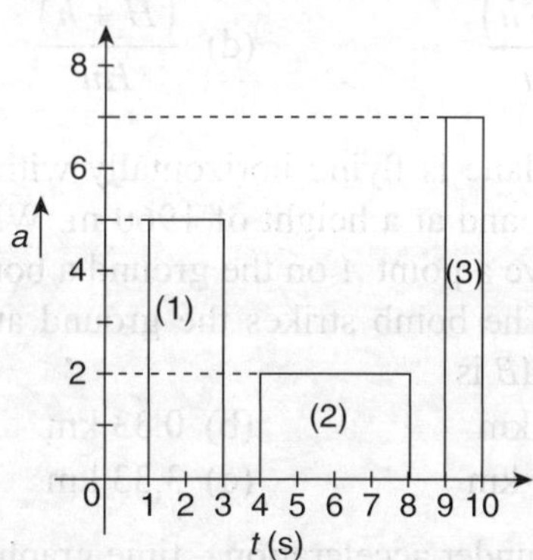

(a) (2) > (1) > (3) (b) (2) > (3) > (1)
(c) (1) > (3) > (2) (d) (1) > (2) > (3)

49. A ball is thrown vertically up with a velocity of 4.9 m/s. The ball is then collected by the person on ground after a time interval of

(a) 3.0 s (b) 2.0 s
(c) 1.0 s (d) 0.5 s

50. The position x (in metres) of a particle on the x-axis is given by

$$x = 5t^3 + 3t^2 - 9$$

where t is the time in seconds. What is the acceleration of the particle at $t = 2$ seconds (in metres/s^2)?

(a) 31 (b) 66
(c) 29 (d) 75

51. A car is travelling at 72 km/h and is 20 m from a barrier when the driver puts on the brakes. The car hits the barrier 2 seconds later. What is the magnitude of the constant declaration?

(a) 72 m/s^2 (b) 10 m/s^2
(c) 36 m/s^2 (d) 15 m/s^2

52. A man walks for 2 seconds at speed of 2.5 m/s and then runs for 4 seconds at a speed of 3.5 m/s. What is the average speed?

(a) 3.05 m/s (b) 2.75 m/s
(c) 3.16 m/s (d) 3.25 m/s

Direction Questions 53–56

53. Velocity of a projectile at height 15 m from ground is $\vec{v} = (20\hat{i} + 10\hat{j})$ m/s. Here $\hat{i}$ is in horizontal direction and $\hat{j}$ is vertically upwards. Then:
Speed with which particle is projected form ground is ________ m/s

(a) 30 (b) $20\sqrt{2}$
(c) $\sqrt{2}$ (d) $3\sqrt{40}$

54. Angle of projectile with ground is ________

(a) 45° (b) 30°
(c) 25° (d) 60°

55. Maximum height from the ground is ________ m

(a) 30 (b) 60
(c) 40 (d) 20

56. Horizontal range on the ground is ________ m

(a) 60 (b) 50
(c) 80 (d) 70

57. Three balls are projected upwards with the same initial speed at angles 25°, 45° and 65° with the horizontal. If R_1, R_2, R_3 are the horizontal in the three cases respectively, then

(a) $R_1 > R_2 > R_3$
(b) $R_3 > R_2 > R_1$
(c) $R_2 > R_3$ or R_1 and $R_3 = R_1$
(d) $R_2 < R_1 = R_3$ and $R_3 = R_1$

58. A car is moving horizontally along a straight line with constant speed 30 m/s. A projectile is to be fired from the moving car in such a way that it will return to the car after the car has moved 80 m. At what speed (relative to the car) must the projectile be fired (takes $g = 10$ m/s^2).

(a) 10 m/s (b) $10\sqrt{8}$ m/s
(c) $\frac{40}{3}$ m/s (d) None of these.

59. A body is initially at rest and it undergoes one-dimensional motion with constant acceleration. The power delivered to it at time t is proportional to

(a) $t^{1/2}$ (b) t
(c) $t^{3/2}$ (d) t^2

60. A boy begins to walk eastward along a street in front of his house, and the graph of his displacement from home is shown in the following figure. His average speed for the whole time interval is equal to

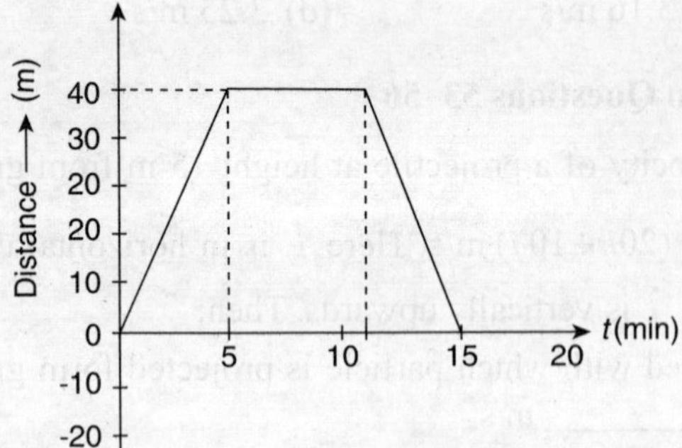

(a) 8 m/min (b) 6 m/min
(c) 8/3 m/min (d) 2 m/min

61. A ball is thrown vertically upwards with an initial velocity 1.4 m/s return in 2 seconds. The total displacement of the ball will be
(a) 22.4 m (b) zero
(c) 33.6 m (d) 44.8 m

62. A particle moves along a straight line such that its displacement at any time t is given by $s = t^3 - 6t^2 + 3t + 4$. The velocity when its acceleration is zero, is
(a) 2 m/s (b) 12 m/s
(c) – 9 m/s (d) 22 m/s

63. A ball is thrown vertically upwards. Which of the following graph represents velocity time graph of the ball during its flight ignoring its resistance of air?

(a)
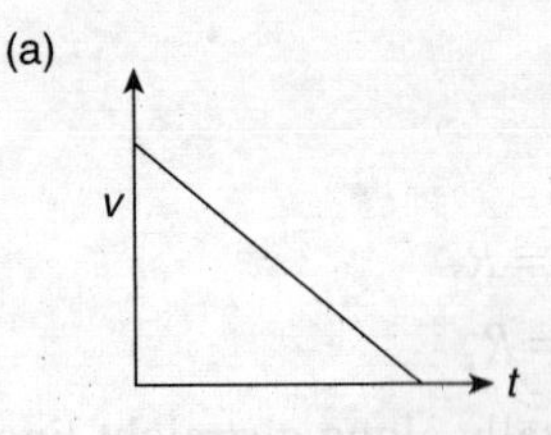

(b)
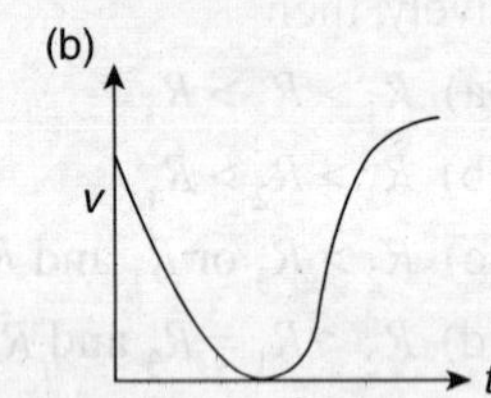

(c)
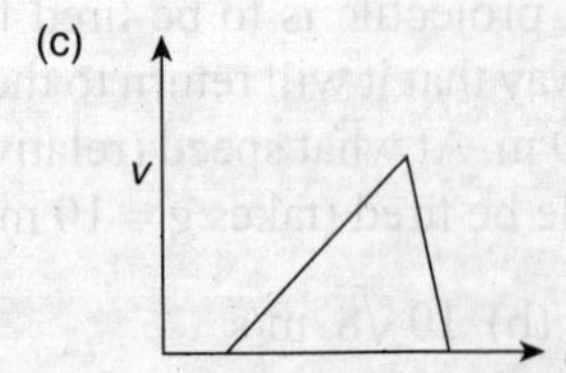

(d)
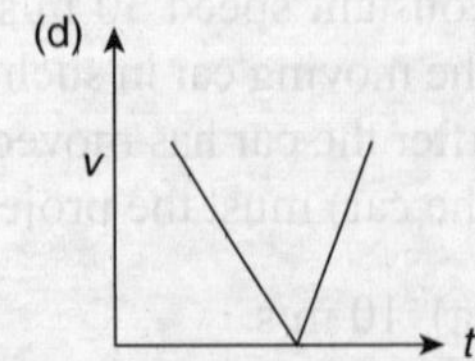

64. If the figure below represents a parabola, identify the physical quantities representing Y and X for constant acceleration

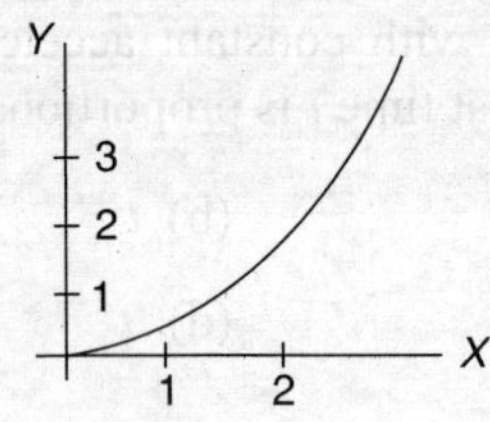

(a) X = time, Y = velocity
(b) X = velocity, Y = time
(c) X = time, Y = displacement
(d) X = time, Y = acceleration

65. Two bodies of different masses are dropped from heights of 16 m and 25 m respectively. The ratio of the time taken by them to reach the ground is
(a) $\frac{25}{16}$ (b) $\frac{5}{4}$
(c) $\frac{4}{5}$ (d) $\frac{16}{25}$

66. A man of height h walks in a straight path towards a lamp post of height H with uniform velocity u. Then the velocity of the edge of the shadow on the ground will be
(a) $\frac{hu}{(H-h)}$ (b) $\frac{Hu}{(H+h)}$
(c) $\frac{(H-h)}{Hu}$ (d) $\frac{(H+h)}{Hu}$

67. An aeroplane is flying horizontally with a velocity of 600 km/h and at a height of 1960 m. When it is vertically above a point A on the ground a bomb is released from it. The bomb strikes the ground at point B. The distance AB is
(a) 1200 km (b) 0.33 km
(c) 333.3 km (d) 3.33 km

68. The area under acceleration – time graph gives
(a) distance travelled (b) change in acceleration
(c) force acting (d) change in velocity

69. The magnitude of a vector, on the addition of two vectors $6\hat{i} + 7\hat{j}$ and $3\hat{i} + 4\hat{j}$, is
(a) $\sqrt{132}$ (b) $\sqrt{136}$
(c) $\sqrt{160}$ (d) $\sqrt{202}$

70. A car moves along a circular track of radius 10 km with a constant speed. If it takes 15 minutes to move from a point on the path to the diametrically opposite point, the ratio of distance covered to displacement is
(a) $\pi : 2$ (b) $2\pi : 3$
(c) $\pi : 4$ (d) $\pi : 5$

71. A bullet leaves a rifle with a muzzle velocity of 500 m/s. While accelerating through the barrel of the rifle, the bullet moves a distance of 1 m. The acceleration of the bullet (assuming uniform acceleration) is
(a) 2.5×10^5 m/s^2 (b) 1.25×10^5 m/s^2
(c) 3.75×10^5 m/s^2 (d) 5×10^5 m/s^2

72. A body falling from rest has a velocity v after it falls through a distance of h. The distance it has fallen when its velocity is $\frac{v}{2}$ is

(a) $\frac{h}{2}$ (b) $\frac{h}{4}$

(c) $\frac{h}{3}$ (d) $\frac{h}{\sqrt{2}}$

73. A stone is thrown with a velocity u making an angle θ with the horizontal. The horizontal distance covered by its fall to ground is maximum when the angle θ is equal to:

(a) 0° (b) 30°

(c) 45° (d) 90°

74. A ball is thrown upwards and it returns to the ground describing a parabolic path. Which of the following remains constant?

(a) Kinetic energy of the ball

(b) Speed of the ball

(c) Horizontal component of velocity

(d) Vertical component of velocity.

75. A ball is projected with $20\sqrt{2}$ m/s at angle 45° with horizontal. The angular velocity of the particle at highest point of its journey about point of projection is:

(a) 0.1 rad/s (b) 1 rad/s

(c) 2 rad/s (d) 0.4 rad/s

76. A body travels in a straight line from A to B. It travels half the time with velocity v_1 and the other half the time with velocity v_2. The average velocity is given by

(a) $\frac{2v_1v_2}{v_1+v_2}$ (b) $\frac{v_1+v_2}{v_1v_2}$

(c) $\frac{v_1+v_2}{2}$ (d) $\frac{v_1v_2}{v_1+v_2}$

77. A projectile is launched with an initial velocity v_0 = 2 m/s $\hat{i}$ + 3 m/s $\hat{j}$. What is the velocity at the top of the trajectory?

(a) 2 m/s (b) – 2 m/s

(c) 5 m/s (d) – 5 m/s

78. A particle travels half the distance with speed v_1 and the other half the distance with speed v_2. Average speed is

(a) $\frac{2v_1v_2}{v_1+v_2}$ (b) $\frac{v_1+v_2}{v_1v_2}$

(c) $\frac{v_1+v_2}{2}$ (d) $\frac{v_1v_2}{v_1+v_2}$

79. In projectile motion, if air resistance is ignored, the horizontal motion takes place with

(a) constant acceleration (b) constant velocity

(c) variable acceleration (d) constant retardation.

80. A projectile is thrown with an initial velocity of $(ai + bj)$ m/s. If the range of projectile is twice the maximum height reached by it, then

(a) $b = a/2$ (b) $b = a$

(c) $b = 2a$ (d) $b = 4a$

81. In a projectile motion the velocity

(a) is always perpendicular to the acceleration

(b) is never perpendicular to the acceleration

(c) is perpendicular to the acceleration for one instant only

(d) None of these.

82. A stone, projected so as to reach a height h passes points P and Q with velocities $\frac{v}{2}$ and $\frac{v}{3}$, where v is initial velocity with which the body is thrown. The distance between P and Q in terms of h is:

(a) $\frac{7}{36}h$ (b) $\frac{5}{36}h$

(c) $\frac{9}{36}h$ (d) $\frac{8}{36}h$

83. A train accelerates from rest at a constant rate 'α' for some time and then it retards to rest at the constant rate 'β'. If the total distance covered by the particle is x then the maximum velocity of the train is

(a) $\sqrt{\frac{3\alpha\beta x}{(\alpha+\beta)}}$ (b) $\sqrt{\frac{4\alpha\beta x}{\alpha+\beta}}$

(c) $\sqrt{\frac{2\alpha\beta x}{\alpha+\beta}}$ (d) $\sqrt{\frac{\alpha+\beta}{2\alpha\beta}x}$

84. A particle is constrained to move on a straight line path. It returns to the starting point after 10 s. The total distance covered by the particle during this time is 30 m. Which one of the following statements about the motion of the particle are true?

(a) Average speed of the particle is 3 m/s

(b) Displacement of the particle is 30 m

(c) Displacement of the particle is 0 m

(d) Both (a) and (c).

85. An airplane is in a level flight at 144 km/h at an altitude of 1000 m. How far from a given target should a bomb be released from it to hit the target?

(a) 570 m (b) 670 m

(c) 470 m (d) 450 m

86. Three particles, A, B and C are projected from the same point with same initial speeds making angles 30°, 45° and 60° respectively with the horizontal. Which of the following statements is correct?
(a) A, B and C have unequal ranges.
(b) Ranges of A and C are equal and less than that of B.
(c) Ranges of A and C are equal and greater than that of B.
(d) A, B and C have equal ranges.

87. A body of mass m_1 projected vertically upwards with an initial velocity v reaches a maximum height h. Another body of mass m_2 is projected along an inclined plane making an angle 30° with the horizontal and with speed u. The maximum distance travelled along the incline is
(a) $\frac{h}{2}$ (b) $4h$
(c) $2h$ (d) h

88. The correct statement from the following is:
(a) A body having zero velocity will not necessarily have zero acceleration.
(b) A body having zero velocity will necessarily have zero acceleration.
(c) A body having uniform speed can have only uniform acceleration.
(d) A body having non-uniform velocity will have zero acceleration.

89. Two stones are projected with the same speed but making different angles with the horizontal.
Their ranges are equal. If the angle of projection of one is $\pi/3$ and its maximum height is y_1, then the maximum height of the other will be:
(a) $3y_1$ (b) $2y_1$
(c) $y_1/2$ (d) $y_1/3$

90. A ball is thrown from a point with a speed v_o at an angle of projection θ. From the same point and at the same instant, a person starts running with a constant speed $\frac{v_0}{2}$ to catch the ball. Will the person be able to catch the ball? If yes, what should be the angle of projection?
(a) Yes, 60° (b) Yes, 30°
(c) No (d) Yes, 45°

91. A point moves with uniform acceleration and v_1, v_2 and v_3 denote the average velocities in there successive intervals of time t_1, t_2 and t_3, which one of the following relations is correct.
(a) $(v_1 - v_2):(v_2 - v_3) = (t_1 - t_2):(t_2 + t_3)$
(b) $(v_1 - v_2):(v_2 - v_3) = (t_1 + t_2):(t_2 + t_3)$
(c) $(v_1 - v_2):(v_2 - v_3) = (t_1 - t_2):(t_2 - t_3)$
(d) $(v_1 - v_2):(v_2 - v_3) = (t_1 - t_2):(t_2 - t_3)$

92. A particle experiences a constant acceleration for 20 s after starting from rest. If it travels a distance S_1 in the first 10 s and S_2 in next 10 s, then
(a) $S_1 = \frac{S_2}{4}$ (b) $S_1 = \frac{S_2}{3}$
(c) $S_1 = \frac{S_2}{2}$ (d) $S_1 = S_2$

93. A particle is moving with constant acceleration from A to B in a straight line AB. If u and v are the velocities at A and B respectively, then its velocity at the midpoint C will be
(a) $\left(\frac{u^2 + v^2}{2u}\right)$ (b) $\frac{u+v}{2}$
(c) $\frac{v-u}{2}$ (d) $\sqrt{\frac{u^2 + v^2}{2}}$

94. The displacement x of a particle along a straight line at time t is given by $x = a_0 + a_1 t + a_2 t^2$. The acceleration of the particle is
(a) a_2 (b) $2a_2$
(c) a_1 (d) $a_0 + a_1 + 2a_2$

95. The coordinates of a moving particle at any time are given by $x = at^2$, $y = bt^2$. The speed of the particle at any time is given by
(a) $2t\sqrt{a^2 + b^2}$ (b) $t\sqrt{a^2 + b^2}$
(c) $2t\sqrt{a^2 - b^2}$ (d) $2t(a+b)$

96. An electron starting from rest has a velocity that increase linearly with time that is $v = kt$, where $k = 2$ m/s². The distance travelled in the first 3 seconds will be
(a) 36 m (b) 27 m
(c) 16 m (d) 9 m

97. The initial velocity of a particle is u at $t = 0$ and acceleration is given by at. Which of the following relations is valid?
(a) $v = u$ (b) $v = u + at$
(c) $v = u + \frac{at^2}{2}$ (d) $v = u + at^2$

98. Two cars get closer by 8 m every second while traveling in the opposite directions. They get closer by 0.8 m while traveling in the same directions. What are the speeds of the two cars?

(a) 4 m/s and 4.4 m/s (b) 4.4 m/s and 3.6 m/s
(c) 4 m/s and 3.6 m/s (d) 4 m/s and 3 m/s

99. A river is flowing with a speed of 1 km/hr. A swimmer wants to go to point 'C' starting from 'A'. He swims with a speed of 5 km/hr, at an angle θ w.r.t. the river. If AB = BC = 400 m. Then the value of θ is:

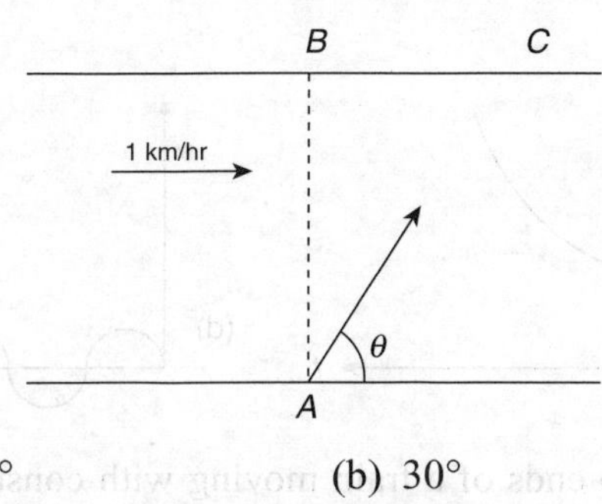

(a) 37° (b) 30°
(c) 53° (d) 45°

100. Which of the following would show the *v-t* graph for a body acceleration time graph is shown in figure?

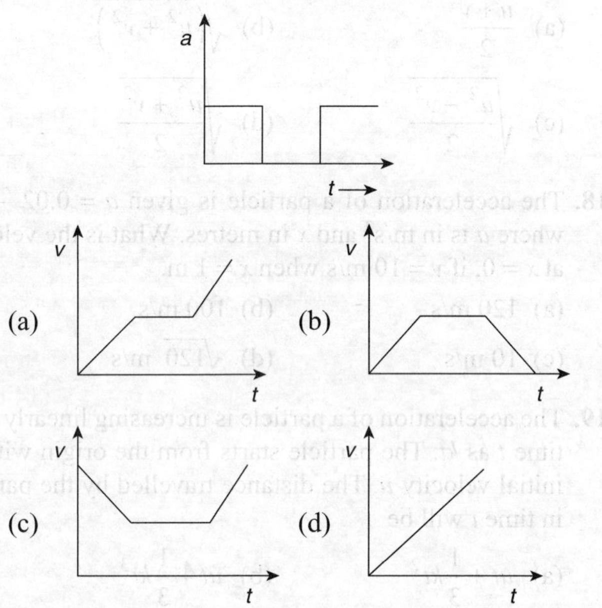

101. The speed versus time graphs for the two cars A and B shown in the figure. The figure shows that

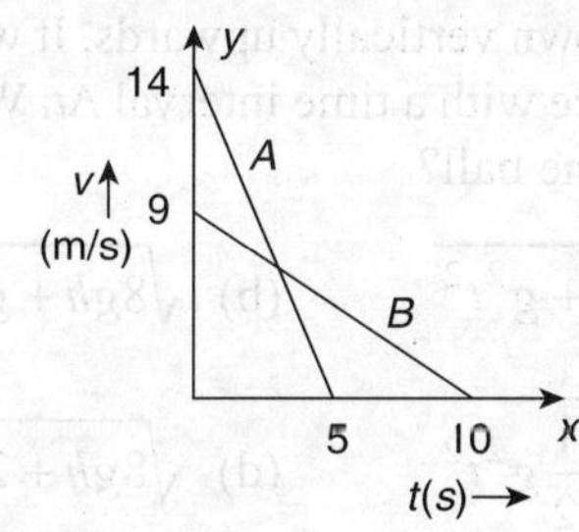

(a) Car B travelled farther
(b) Car A travelled farther
(c) Car B never meets with Car A on the road
(d) Both cars have the same retardation.

102. The displacement is given by, $x = 2t^2 + t + 5$, then acceleration at $t = 2$ s is
(a) 2 m/s^2 (b) 3 m/s^2
(c) 4 m/s^2 (d) 5 m/s^2

103. If $y = x - x^2$ is the equation for the path of a projectile, then which of the following is incorrect:
(a) Range = 1m
(b) Maximum height = 0.25 m
(c) Time of flight = 0.5 sec
(d) Angle of projection = 45°.

104. A particle is projected vertically upwards. It will be at 3/4th of its greatest height at time which are in the ratio is
(a) 4 : 1 (b) 1 : 3
(c) 2 : 1 (d) 1 : 2

105. A policeman in a jeep is chasing a dacoit on a straight road. The jeep is moving with a maximum uniform speed *v*. The dacoit rides on a motorcycle of his waiting friend when the jeep is at a distance *d* from him and the motorcycle starts with constant acceleration *a*. The dacoit will be caught if
(a) $v \le 2\sqrt{ad}$ (b) $v \le \sqrt{2ad}$
(c) $v \ge \sqrt{2ad}$ (d) $v \ge 2\sqrt{ad}$

106. The angle of projection, at which the horizontal range and maximum height of projectile are equal is:
(a) 45° (b) $\theta = \tan^{-1}(0.25)$
(c) $\theta = \tan^{-1} 4$ (d) 60°

107. A particle travels along a straight line such that the distance covered by it in any second, exceeds that in the previous second by 2 m. If it covers a distance of 1 m in the first second, find the distance travelled at the end of the *n* seconds.
(a) n (b) n^3
(c) $n^{3/2}$ (d) n^2

108. A rubber ball is dropped from a height of 2 m above the floor. Each time it strikes the ground, it rises to a height of 0.8 times the height from which it fell. Find the total distance covered by the ball before it stops.
(a) 18 m (b) 10 m
(c) 9 m (d) 180 m

109. Which of the following is correct?
1. When a body moves in a plane with constant acceleration in a direction different from initial velocity, the trajectory of the body is parabolic.
2. If θ is the angle of projection at which horizontal range and maximum height of a projectile are equal, then $\tan\theta = 4$.

3. Two bullets simultaneously fired with different speeds horizontally from the same place hit the ground at the same time

(a) 1 and 2 (b) 2 and 3
(c) 1,2 and 3 (d) 1 and 3

110. When particle moves with uniform acceleration, the distances described in consecutive equal intervals of time are in A.P. with common difference

(a) at^2 (b) $2at^2$
(c) $\frac{1}{2}at^2$ (d) $\frac{1}{4}at^2$

111. In car race, car A takes a time of t seconds less than car B at the finish and passes the finishing point with a velocity v m/s more than car B. Assuming that the cars starts from rest and travel with constant acceleration a_1 and a_2 respectively, what is the relation between v, t, a_1 and a_2?

(a) $v = t\left(\sqrt{a_1} + \sqrt{a_2}\right)$ (b) $v = t\left(\sqrt{a_1} - \sqrt{a_2}\right)$
(c) $v = t\left(\sqrt{a_1 + a_2}\right)$ (d) $v = t\left(\sqrt{a_1 a_2}\right)$

Direction Questions 112 – 115

Two inclined planes OA and OB having inclinations 30° and 60° with the horizontal respectively intersect each other at O, as shown in figure. A particle is projected from point P with velocity $u = 10\sqrt{3}$ m/s along a direction perpendicular to plane OA. If the particle strikes plane OB perpendicular at Q.

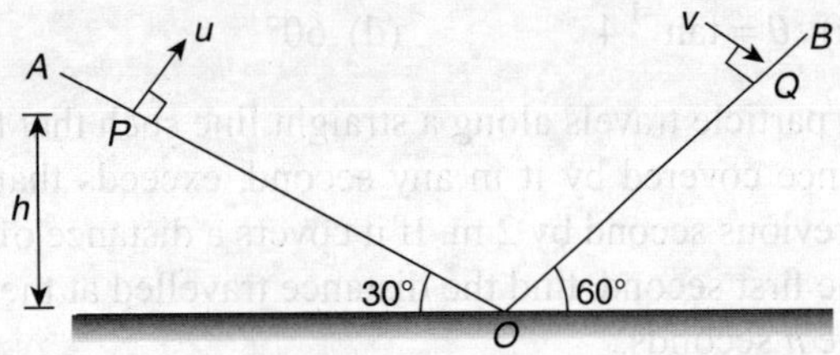

112. The time of flight is
(a) 3s (b) 2s
(c) 4s (d) 1s

113. Velocity with which the particle strikes the plane OB is
(a) 5 m/s (b) 6 m/s
(c) –10m/s (d) 8 m/s

114. Height h of point P from point O is
(a) 7 m (b) 5 m
(c) 3 m (d) 10 m

115. Distance PQ is
(a) 12.5 m (b) 15 m
(c) 10 m (d) 20 m

116. Which of the following curves do not represent motion in one dimension?

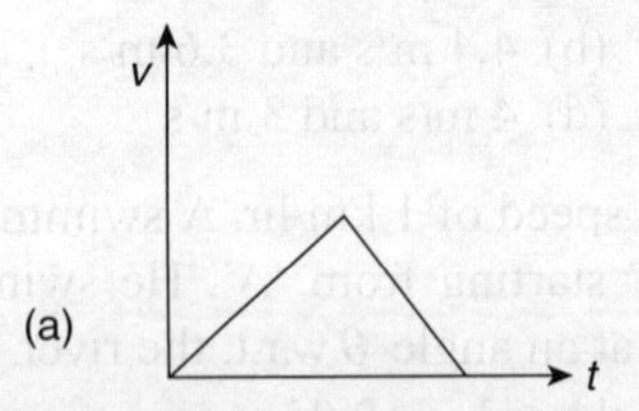

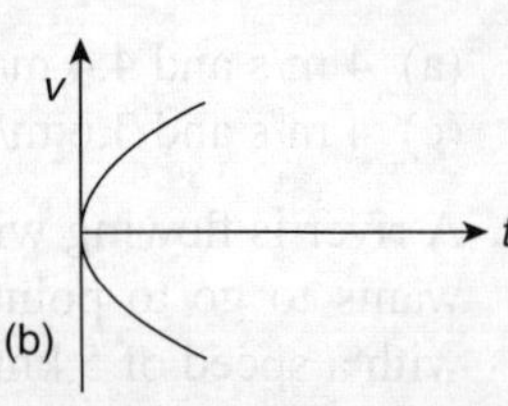

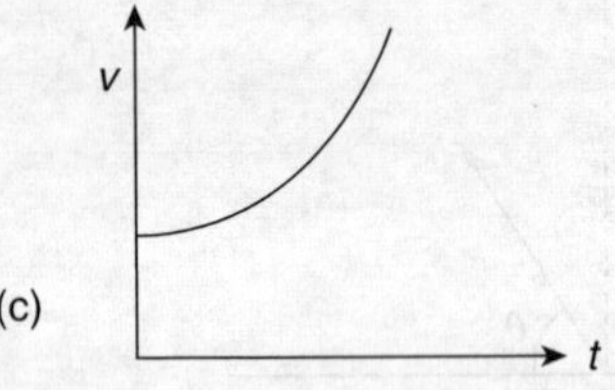

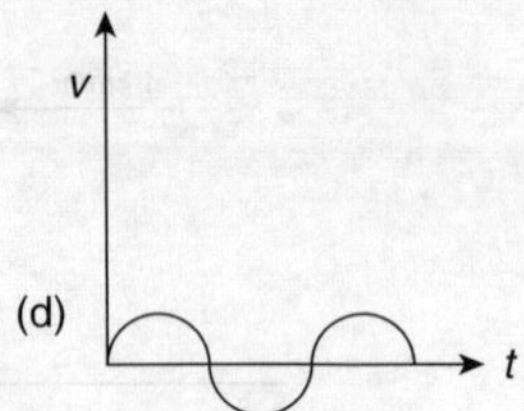

117. The two ends of a train moving with constant acceleration pass a certain point with velocities u and v. The velocity with which the middle point of the train passes the same point is

(a) $\frac{u+v}{2}$ (b) $\sqrt{\left(u^2+v^2\right)}$
(c) $\sqrt{\frac{u^2-v^2}{2}}$ (d) $\sqrt{\frac{u^2+v^2}{2}}$

118. The acceleration of a particle is given $a = 0.02 - 20x$ where a is in m/s^2 and x in metres. What is the velocity at $x = 0$, if $v = 10$ m/s when $x = 1$ m.

(a) 120 m/s (b) 100 m/s
(c) 10 m/s (d) $\sqrt{120}$ m/s

119. The acceleration of a particle is increasing linearly with time t as kt. The particle starts from the origin with an initial velocity u. The distance travelled by the particle in time t will be

(a) $ut + \frac{1}{3}kt^2$ (b) $ut + \frac{1}{3}kt^3$
(c) $ut + \frac{1}{6}kt^3$ (d) $ut + \frac{1}{2}kt^2$

120. A ball is thrown vertically upwards. It was observed at a height h twice with a time interval Δt. What is the initial velocity of the ball?

(a) $\frac{1}{2}\sqrt{8gh + g^2t^2}$ (b) $\sqrt{8gh + g^2t^2}$
(c) $\sqrt{8gh + \frac{1}{2}g^2t^2}$ (d) $\sqrt{8gh + 2g^2t^2}$

121. A particle moves in a straight line with a velocity $v_t = |(t-2)|$ m/s, where t is time in seconds. What is the distance covered by the particle in 4 s ?

(a) 2 m (b) 4 m
(c) 1 m (d) 8 m

122. Four persons are standing at the four corners of a square *PQRS* of side *a*. They simultaneously start running such that *P* runs towards *Q*, *Q* runs towards *R*, *R* runs towards *S* and *S* runs towards *P*, each with velocity *v*. They will meet at '*O*' after time *t* given by

(a) $\frac{2a}{v}$ (b) $\frac{a}{v}$

(c) $\frac{a}{\sqrt{2}v}$ (d) $\frac{3}{2}\frac{a}{v}$

123. A body starts from origin and moves along *x*-axis. At any instant velocity is $v_t = 4t^3 - 2t$ where *t* is in second and v_t is in m/s. What is the acceleration of the particle, when it is 2 m from the origin?

(a) 20 m/s^2 (b) 21 m/s^2
(c) 10 m/s^2 (d) 22 m/s^2

124. The displacement of a particle undergoing rectilinear motion along the *x*-axis is given by $x = (2t^3 - 21t^2 + 60t + 6)$ m. What is the acceleration of the particle when its velocity is zero?

(a) 36 m/s^2 (b) 8 m/s^2
(c) –8 m/s^2 (d) –18 m/s^2

125. A self-propelled vehicle of mass m whose engine delivers a constant power *P* has an acceleration $a = \frac{P}{mv}$. What is the distance travelled by it (assuming no friction) to increase the velocity of the vehicle from v_1 and v_2?

(a) $x = \frac{3m}{P}\left(v_2^3 - v_1^3\right)$ (b) $x = \frac{m}{3P}\left(v_2^3 - v_1^3\right)$

(c) $x = \frac{2P}{m}\left(v_2^3 - v_1^3\right)$ (d) $x = \frac{m}{2P}\left(v_2^3 - v_1^3\right)$

126. A particle moves such that its displacement as a function of time is given by $x^2 = t^2 + 1$. What is its acceleration?

(a) $\frac{1}{x^3}$ (b) $-\frac{t}{x^2}$

(c) $\frac{1}{x} - \frac{t}{x^3}$ (d) $\frac{1}{x} - \frac{t}{x^2}$

127. If a body falls in a vertical line from rest at *A*, then the times of descent through three equal distance *AB*, *BC*, *CD* are in the ratio

(a) $1:\sqrt{2}:\sqrt{3}$
(b) $1:\left(\sqrt{2}-\sqrt{1}\right):\left(\sqrt{3}-\sqrt{2}\right)$
(c) $1:\sqrt{3}:\sqrt{2}$
(d) $\sqrt{3}:\sqrt{2}:1$

128. A body is at rest at $x = 0$. At $t = 0$, it starts moving in the positive *x*-direction with a constant acceleration. At the same instant another body passes through $x = 0$ moving in the positive *x*-direction with a constant speed. The position of the first body is given x_1 (t) after time *t* and that of the second body by x_2 (t) after the same time interval. Which of the following graphs correctly describes $(x_1 - x_2)$ as a function of time *t*?

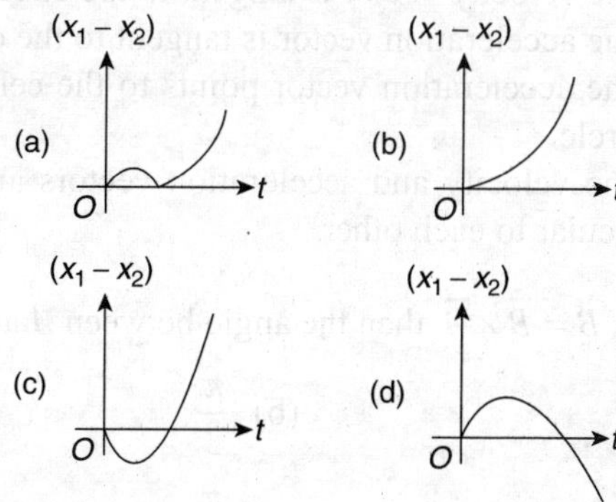

129. The velocity of a particle is $v = v_0 + gt + ft^2$. If its position is $x = 0$ at $t = 0$, then its displacement after unit time ($t = 1$) is

(a) $v_0 + \frac{g}{2} + f$ (b) $v_0 + 2g + 3f$

(c) $v_0 + \frac{g}{2} + \frac{f}{3}$ (d) $v_0 + g + f$

130. A particle located at $x = 0$ at time $t = 0$, starts moving along the positive *x*-direction with a velocity *v* that varies as $v = \alpha\sqrt{x}$. The displacement of the particle varies with time as

(a) t^3 (b) t^2
(c) t (d) $\sqrt{t}$

131. A ball is released from the top of a tower of height *h* metre. It takes *T* seconds to reach the ground. What is the position of the ball in $\frac{T}{3}$ seconds?

(a) $\frac{h}{9}$ m from the ground

(b) $\frac{7h}{9}$ m from the ground

(c) $\frac{8h}{9}$ m from the ground

(d) $\frac{17h}{18}$ m from the ground.

132. A projectile can have the same range *R* for two angles of projection. If t_1 and t_2 be the time of flights in the two cases, then the product of the two times of flight is proportional to

(a) $\frac{1}{R}$ (b) R

(c) R^2 (d) $\frac{1}{R^2}$

133. Which of the following statements is false for a particle moving in a circle with a constant angular speed?

(a) The velocity vector is tangent to the circle.

(b) The acceleration vector is tangent to the circle.

(c) The acceleration vector points to the centre of the circle.

(d) The velocity and acceleration vectors are perpendicular to each other.

134. If $\vec{A}\times\vec{B}=\vec{B}\times\vec{A}$ then the angle between A and B is

(a) π (b) $\frac{\pi}{3}$

(c) $\frac{\pi}{2}$ (d) $\frac{\pi}{4}$

135. The co-ordinates of a moving particle at any time t are given by $x=\alpha t^3$ and $y=\beta t^3$. The speed of the particle at time t is given by

(a) $3t\sqrt{\alpha^2+\beta^2}$ (b) $3t^2\sqrt{\alpha^2+\beta^2}$

(c) $t^2\sqrt{\alpha^2+\beta^2}$ (d) $\sqrt{\alpha^2+\beta^2}$

136. A boy playing on the roof of a 10 m high building throws a ball with a speed of 10 m/sec at an angle of 30° with the horizontal. How far from the throwing point will the ball be at the height of 10 m from the ground? [$g=10$ m/sec^2]

(a) 5.20 m (b) 4.33 m

(c) 2.60 m (d) 8.66 m

137. If the angle between the vectors $\vec{A}$ and $\vec{B}$ is θ, the value of the product $(\vec{B}\times\vec{A}).\vec{A}$ is equal to

(a) $BA^2\cos\theta$ (b) $BA^2\sin\theta$

(c) $BA^2\sin\theta\cos\theta$ (d) zero

138. A particle is projected with velocity $2\sqrt{gh}$ such that it just crosses two walls of height h and separated by h. Find the angle of projection.

(a) 15° (b) 75°

(c) 60° (d) 30°

139. For a projectile, (range)^2 is 48 times of (maximum height)^2 obtained. Find angle of projection.

(a) 60° (b) 30°

(c) 45° (d) 75°

140. In the arrangement shown in the figure, if v_1 and v_2 are instantaneous velocities of masses m_1 and m_2, respectively, and angle $ACB=2\theta$ at the instant then:

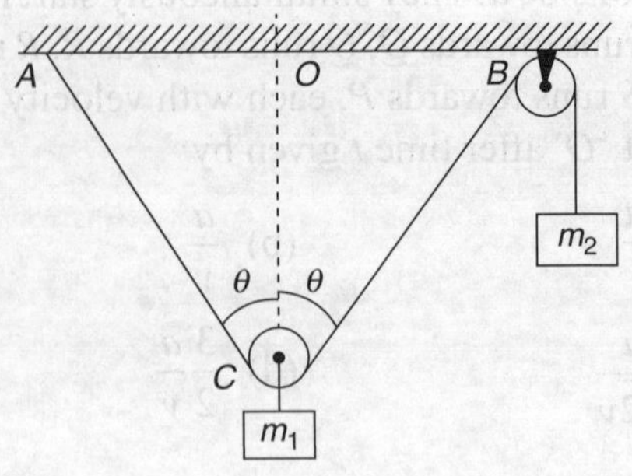

(a) $\theta=\cos^{-1}\frac{v_2}{2v_1}$ (b) $\theta=\cos^{-1}\frac{v_1}{2v_2}$

(c) $\theta=\tan^{-1}\frac{v_1}{2v_2}$ (d) $\theta=\sin^{-1}\frac{v_1}{v_2}$

141. A projectile has initially the same horizontal velocity as it would acquire if it had moved from rest with uniform acceleration of 3 m/s^2 for 0.5 minutes. If the maximum height reached by it is 80 m, then the angle of projection is ($g=10$ ms^{-1})

(a) $\tan^{-1}(3)$ (b) $\tan^{-1}\left(\frac{3}{2}\right)$

(c) $\tan^{-1}\left(\frac{4}{9}\right)$ (d) $\sin^{-1}\left(\frac{4}{9}\right)$

142. Let $\vec{r}_1(t)=3t\hat{i}+4t^2\hat{j}$ and $\vec{r}_2(t)=4t^2\hat{i}+3t\hat{j}$ represent the positions of particles 1 and 2, respectively as function of time t; $\vec{r}_1(t)$ and $\vec{r}_2(t)$ are in metre and t in second. The relative speed of the two particles at the instant $t=1$ s, will be

(a) 1 m/s (b) $3\sqrt{2}$ m/s

(c) $5\sqrt{2}$ m/s (d) $7\sqrt{2}$ m/s

143. The vectors $\vec{A}=(\hat{i}+\hat{j}-2\hat{k}),\vec{B}=(2\hat{i}+2\hat{j}-\hat{k})$ and $\vec{C}=(-\hat{i}+\alpha\hat{j}+\hat{k})$ are coplanar when the constant α is equal to

(a) $\frac{1}{3}$ (b) 1

(c) 3 (d) None of these.

144. An object falls from a bridge 45 m above the water level in a river. It falls directly into a boat moving with constant speed. The boat was 18 m away from the point of impact. What is the speed of the boat (in m/s) ?

(a) 6 (b) 9

(c) 12 (d) 15

145. A projectile is thrown from the surface of ground on Earth with velocity 16 m/s at an angle of 75° from the vertical. The projectile would be able to cover, approximately a horizontal distance of (in metres)

(a) 13 (b) 17
(c) 15 (d) 21

146. A bat and an insect are flying with velocities $\vec{V}_{BG}, \vec{V}_{IG}$ with respect to ground (in unit vector, $\hat{i}, \hat{j}$ notation) $\vec{V}_{BG} = 3.5\hat{i} + 9.2\hat{j}, \vec{V}_{IG} = -2.5\hat{i} + 1.8\hat{j}$. Calculate the velocity of the insect with respect to the bat.

(a) $1.5\hat{i} + 11.0\hat{j}$ (b) $-6\hat{i} - 7.4\hat{j}$
(c) $6\hat{i} - 11.0\hat{j}$ (d) $1.0\hat{i} + 7.4\hat{j}$

147. If $\vec{A} = \hat{i} + \hat{j}$ and $\vec{B} = \hat{i} + \hat{k}$, then $\vec{A} \times \vec{B}$ is

(a) $\hat{i} + \hat{j} + \hat{k}$ (b) $\hat{i} - \hat{j} + \hat{k}$
(c) $\hat{i} + \hat{j} - \hat{k}$ (d) $\hat{i} - \hat{j} - \hat{k}$

148. The (*x*, *y*, *z*) co-ordinates of two points *A* and *B* are given respectively as (0, 3, 1) and (– 2, 6, 4). The displacement vector from *A* to *B* may be given by

(a) $-2\hat{i} + 6\hat{j} + 4\hat{k}$ (b) $-2\hat{i} + 3\hat{j} + 3\hat{k}$
(c) $-2\hat{i} + 3\hat{j} + 5\hat{k}$ (d) $2\hat{i} - 3\hat{j} - 3\hat{k}$

149. An aircraft is flying at a height of 3400 m above the ground. If the angle subtended at a ground observation point by the aircraft positions 10 s apart is 30°, then the speed of the aircraft is

(a) 19.63 ms^{-1} (b) 1963 ms^{-1}
(c) 108 ms^{-1} (d) 196.3 ms^{-1}
(e) 10.8 ms^{-1}

150. A metal ring is held horizontally and bar magnet is dropped through the ring with its length along the axis of the ring. The acceleration of the falling magnet is

(a) equal to *g*
(b) less than *g*
(c) greater than *g*
(d) dependent on the conductivity of the ring.

151. Force acting on a particle is $\vec{F} = 6\hat{i} - 8\hat{j} + 10\hat{k}$ N accelerates it with $|a| = 1$ m/s². The mass of the body is

(a) $10\sqrt{2}$ kg (b) $2\sqrt{10}$ kg
(c) 10 kg (d) 20 kg

152. The component of vector $\vec{A} = a_x\hat{i} + a_y\hat{j} + a_z\hat{k}$ along the direction of $\hat{i} - \hat{j}$ is

(a) $a_x - a_y + a_z$ (b) $a_x - a_y$
(c) $\dfrac{a_x - a_y}{\sqrt{2}}$ (d) $a_x + a_y + a_z$

153. A body is projected from the earth at angle 30° with the horizontal with same initial velocity. If its range is 20 m, the maximum height reached by it is (in metres)

(a) $5\sqrt{3}$ (b) $\dfrac{5}{\sqrt{3}}$
(c) $\dfrac{10}{\sqrt{3}}$ (d) $10\sqrt{3}$

154. At a given instant of time, two particles are having the position vectors $4\hat{i} - 4\hat{j} + 7\hat{k}$ metres and $2\hat{i} + 2\hat{j} + 5\hat{k}$ metres respectively. If the velocity of the first particle be $0.4\hat{i}$ ms^{-1}, the velocity of second particle in metres per second, if they collide after 10 sec, is

(a) $6\left(\hat{i} - \hat{j} + \frac{1}{3}\hat{k}\right)$ (b) $0.6\left(\hat{i} - \hat{j} + \frac{1}{3}\hat{k}\right)$
(c) $6\left(\hat{i} + \hat{j} + \frac{1}{3}\hat{k}\right)$ (d) $0.6\left(\frac{1}{3}\hat{i} + \hat{j} + \frac{1}{3}\hat{k}\right)$

155. The horizontal and vertical displacement *x* and *y* of a projectile at a given time *t* are given by $x = 6t$ metre and $y = 8t - 5t^2$ metre. The range of the projectile in metre is

(a) 9.6 (b) 10.6
(c) 19.2 (d) 88.4

156. The equation of motion of a projectile are given by $x = 36t$ metre and $2y = 96t - 9.8t^2$ metre. The angle of projection is

(a) $\sin^{-1}(4/5)$ (b) $\sin^{-1}(3/5)$
(c) $\sin^{-1}(4/3)$ (d) $\sin^{-1}(3/4)$

157. The horizontal acceleration that should be given to a smooth inclined plant of angle $\sin^{-1}(1/l)$ to keep an object stationary on the plane, relative to the inclined plane is

(a) $\dfrac{g}{\sqrt{l^2 - 1}}$ (b) $g\sqrt{l^2 - 1}$
(c) $\dfrac{\sqrt{l^2 - 1}}{g}$ (d) $\dfrac{g}{\sqrt{l^2 + 1}}$

158. Two stones are projected with the same speed but making different angles with the horizontal. Their ranges are equal. If the angle of projection of one is $\dfrac{\pi}{3}$ and its maximum height is h_1. Then the maximum height of the other will be

(a) $3h_1$ (b) $2h_1$
(c) $\dfrac{h_1}{2}$ (d) $\dfrac{h_1}{3}$

159. At a height 0.4 m from the ground, the velocity of a projectile in vector form is $\vec{v} = (6\hat{i} + 2\hat{j})$ m/s. The angle of projection is ($g = 10$ m/s^2)

(a) 45° (b) 60°

(c) 30° (d) $\tan^{-1}\frac{3}{4}$

160. The horizontal and vertical displacement of a projectile at time t are $x = 36t$ and $y = 48t - 4.9t^2$ respectively. Initial velocity of the projectile in m/s is

(a) 15 (b) 30

(c) 45 (d) 60

161. Two stones are projected with the same speed but making different angles with the horizontal. Their horizontal range are equal. The angle of projection of one is $\frac{\pi}{3}$ and the maximum height reached by it is 102 m. Then the maximum height reached by the other in metre is

(a) 336 (b) 224

(c) 56 (d) 34

162. The sum of two vectors $\vec{A}$ and $\vec{B}$ is at right angles to their difference. Then

(a) $A = B$

(b) $A = 2B$

(c) $B = 2A$

(d) $\vec{A}$ and $\vec{B}$ have the same direction.

163. An aeroplane is flying horizontally at a height of 49 m with a velocity of 150 ms^{-1}. A bag containing food is to be dropped to the jawans on the ground. How far from them should be the bag dropped so that it directly reaches them?

(a) 750 m (b) 1500 m

(c) 2000 m (d) 1000 m

164. A stone is thrown at an angle θ to the horizontal reaches a maximum height H. Then the time of flight of stone will be

(a) $\frac{\sqrt{2H}}{g}$ (b) $2\sqrt{\frac{2H}{g}}$

(c) $\frac{2\sqrt{2H\sin\theta}}{g}$ (d) $\frac{\sqrt{2H\sin\theta}}{g}$

165. A train of 150 m length is going towards north direction at a speed of 10 ms^{-1}. A parrot files at a speed of 5 ms^{-1} towards south direction parallel to the railway track. The time taken by the parrot to cross the train is equal to

(a) 12 s (b) 8 s

(c) 15 s (d) 10 s

166. A cyclist is moving in a circular track of radius 80 m with a velocity $v = 36$ km/h. He has to lean from the vertical approximately through an angle (Takes g = 10 m/s^{-2})

(a) $\tan^{-1}(4)$ (b) $\tan^{-1}\left(\frac{1}{3}\right)$

(c) $\tan^{-1}\left(\frac{1}{4}\right)$ (d) $\tan^{-1}\left(\frac{1}{8}\right)$

167. The acceleration experienced by a moving boat after its engine is cut off, is given by $a = -kv^3$, where k is a constant. If v_0 is the magnitude of velocity at cut off, then the magnitude of the velocity at time t after the cut off is

(a) $\frac{v_0}{2ktv_0^2}$ (b) $\frac{v_0}{1+2ktv_0^2}$

(c) $\frac{v_0}{\sqrt{1-2kv_0^2}}$ (d) $\frac{v_0}{\sqrt{1+2ktv_0^2}}$

Questions Asked in NEET/CBSE Medicals

168. The speed of a swimmer in still water is 20 m/s. The speed of river water is 10 m/s and is flowing due east. If he is standing on the south bank and wishes to cross the river along the shortest path, the angle at which he should make his strokes w.r.t. north is, given by

(a) 45° west (b) 30° west

(c) 0° (d) 60° west

169. When an object is shot from the bottom of a long smooth inclined plane kept at an angle 60° with horizontal, it can travel a distance x_1 along the plane. But when the inclination is decreased to 30° and the same object is shot with the same velocity, it can travel x_2 distance. Then $x_1 : x_2$ will be

(a) $1:2\sqrt{3}$ (b) $1:\sqrt{2}$

(c) $\sqrt{2}:1$ (d) $1:\sqrt{3}$

170. Two particles A and B are moving in uniform circular motion in concentric circles of radii r_A and r_B with speed v_A and v_B respectively. Their time period of rotation is the same. The ratio of angular speed of A to that of B will be

(a) 1 : 1 (b) $r_A : r_B$

(c) $v_A : v_B$ (d) $r_B : r_A$

171. A particle starting from rest, moves in a circle of radius r. It attains a velocity of V_0 m/s in the nth round. Its angular acceleration will be

(a) $\frac{V_0}{n}$ rad/s^2 (b) $\frac{V_0}{2\pi nr^2}$ rad/s^2

(c) $\frac{V_0^2}{4\pi nr^2}$ rad/s^2 (d) $\frac{V_0^2}{4\pi nr}$ rad/s^2

172. The x and y coordinates of a particle at any time are $x = 5t - 2t^2$ and $y = 10t$ respectively, where x and y are in metres and t in seconds. The acceleration of the particle at $t = 2$ s is

(a) 5 ms^{-2} (b) –4 ms^{-2}

(c) –8 ms^{-2} (d) 0

173. In the given figure a = 15 ms^{-2} represents the total acceleration of a particle moving in the clockwise direction in a circle of radius R = 2.5 m at a given instant of time. The speed of the particle is

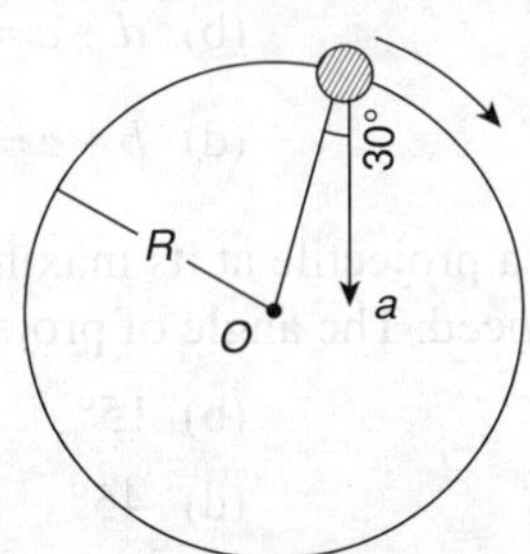

(a) 4.5 ms^{-1} (b) 5.0 ms^{-1}

(c) 5.7 ms^{-1} (d) 6.2 ms^{-1}

174. If the magnitude of sum of two vectors is equal to the magnitude of difference of the two vectors, the angle between these vectors is

(a) 45° (b) 180°

(c) 0° (d) 90°

175. A particle moves so that its position vector is given by $\vec{r} = \cos\omega t\hat{x} + \sin\omega t\hat{y}$, where ω is a constant. Which of the following is true?

(a) Velocity is perpendicular to $\vec{r}$ and acceleration is directed towards the origin.

(b) Velocity is perpendicular to $\vec{r}$ and acceleration is directed away from the origin.

(c) Velocity and acceleration both are perpendicular to $\vec{r}$.

(d) Velocity and acceleration, both are parallel to $\vec{r}$.

176. If vectors $\vec{A} = \cos\omega t\hat{i} + \sin\omega t\hat{j}$ and $\vec{B} = \cos\frac{\omega t}{2}\hat{i} + \sin\frac{\omega t}{2}\hat{j}$ are functions of time, then the value of t at which they are orthogonal to each other is

(a) $t = \frac{\pi}{\omega}$ (b) $t = 0$

(c) $t = \frac{\pi}{4\omega}$ (d) $t = \frac{\pi}{2\omega}$

177. The position vector of a particle $\vec{R}$ as a function of time is given by $\vec{R} = 4\sin(2\pi t)\hat{i} + 4\cos(2\pi t)\hat{j}$ where R is in metres, t is in seconds and $\hat{i}$ and $\hat{j}$ denote unit vectors along x-and y-directions, respectively. Which one of the following statements is wrong for the motion of particle?

(a) Magnitude of the velocity of particle is 8π metres/second.

(b) Path of the particle is a circle of radius 4 metres.

(c) Acceleration vector is along $-\vec{R}$.

(d) Magnitude of acceleration vector is $\frac{v^2}{R}$, where v is the velocity of particle.

178. A projectile is fired from the surface of the earth with a velocity of 5 ms^{-1} and angle θ with the horizontal. Another projectile fired from another planet with a velocity of 3 ms^{-1} at the same angle follows a trajectory which is identical with the trajectory of the projectile fired from the earth. The value of the acceleration due to gravity on the planet is (in ms^{-2}) is (Given g = 9.8 ms^{-2})

(a) 3.5 (b) 5.9

(c) 16.3 (d) 110.8

179. A particle is moving such that its position coordinates (x, y) are (2 m, 3 m) at time t = 0, (6 m, 7 m) at time t = 2 s and (13 m, 14 m) at time t = 5 s. Average velocity vector $(\vec{v}_{av})$ from t = 0 to t = 5 s is

(a) $\frac{1}{5}(13\hat{i} + 14\hat{j})$ (b) $\frac{7}{3}(\hat{i} + \hat{j})$

(c) $2(\hat{i} + \hat{j})$ (d) $\frac{11}{5}(\hat{i} + \hat{j})$

180. The velocity of a projectile at the initial point A is $(2\hat{i} + 3\hat{j})$ m/s. Its velocity (in m/s) at point B is

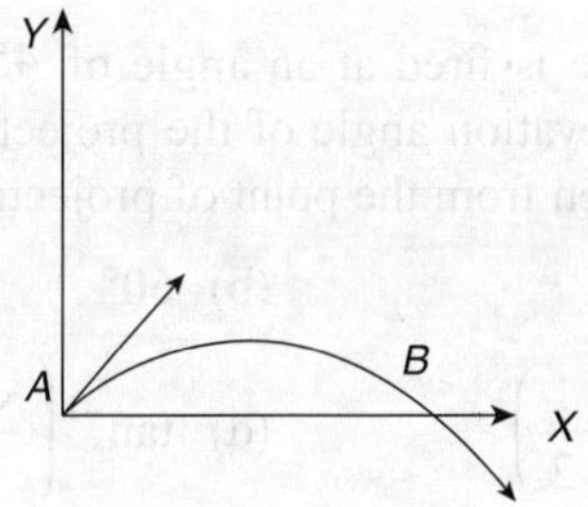

(a) $2\hat{i}-3\hat{j}$ (b) $2\hat{i}+3\hat{j}$
(c) $-2\hat{i}-3\hat{j}$ (d) $-2\hat{i}+3\hat{j}$

181. Vectors, $\vec{A}$, $\vec{B}$ and $\vec{C}$ are such that $\vec{A}.\vec{B}=0$ and $\vec{A}.\vec{C}=0$. Then the vector parallel to $\vec{A}$ is
(a) $\vec{A}\times\vec{B}$ (b) $\vec{B}+\vec{C}$
(c) $\vec{B}\times\vec{C}$ (d) $\vec{B}$ and $\vec{C}$

182. The horizontal range and the maximum height of a projectile are equal. The angle of projection of the projectile is
(a) $\theta=\tan^{-1}\left(\frac{1}{4}\right)$ (b) $\theta=\tan^{-1}(4)$
(c) $\theta=\tan^{-1}(2)$ (d) $\theta=45°$

183. A particle has initial velocity $\left(2\vec{i}+3\vec{j}\right)$ and acceleration $\left(0.3\vec{i}+0.2\vec{j}\right)$. The magnitude of velocity after 10 seconds will be
(a) $9\sqrt{2}$ units (b) $5\sqrt{2}$ units
(c) 5 units (d) 9 units

184. A particle moves in a circle of radius 5 cm with constant speed and time period $0.2\,\pi$ s. The acceleration of the particle is
(a) 15 m/s^2 (b) 25 m/s^2
(c) 36 m/s^2 (d) 5 m/s^2

185. A missile is fired for maximum range with an initial velocity of 20 m/s. If $g=10$ m/s^2, the range of the missile is
(a) 40 m (b) 50 m
(c) 60 m (d) 20 m

186. A body is moving with velocity 30 m/s towards east. After 10 seconds its velocity becomes 40 m/s towards north. The average acceleration of the body is
(a) 1 m/s^2 (b) 7 m/s^2
(c) $\sqrt{7}$ m/s^2 (d) 5 m/s^2

187. A projectile is fired at an angle of 45° with the horizontal. Elevation angle of the projectile at its highest point as seen from the point of projection, is
(a) 45° (b) 60°
(c) $\tan^{-1}\left(\frac{1}{2}\right)$ (d) $\tan^{-1}\left(\frac{\sqrt{3}}{2}\right)$

188. A particle has initial velocity $\left(3\hat{i}+4\hat{j}\right)$ and has acceleration $\left(0.4\hat{i}+0.3\hat{j}\right)$. Its speed after 10 is
(a) 7 units (b) $7\sqrt{2}$ units
(c) 8.5 units (d) 10 units

189. Six vectors, $\vec{a}$ through $\vec{f}$ have the magnitudes and directions indicated in the figure. Which of the following statements is true?

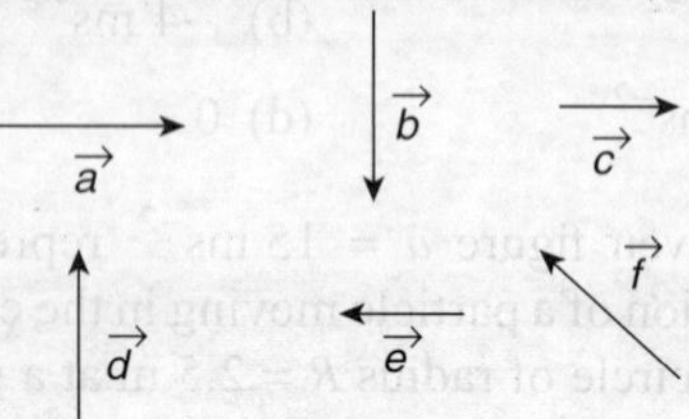

(a) $\vec{b}+\vec{c}=\vec{f}$ (b) $\vec{d}+\vec{c}=\vec{f}$
(c) $\vec{d}+\vec{e}=\vec{f}$ (d) $\vec{b}+\vec{e}=\vec{f}$

190. The speed of a projectile at its maximum height is half of its initial speed. The angle of projection is
(a) 60° (b) 15°
(c) 30° (d) 45°

191. A particle moves in *x-y* plane according to rule $x=a\sin\omega t$ and $y=a\cos\omega t$. The particle follows
(a) an elliptical path
(b) a circular path
(c) a parabolic path
(d) a straight line path inclined equally to a and *y*-axes

192. A particle shows distance-time curve as given in this figure. The maximum instantaneous velocity of the particle is around the point
(a) D (b) A
(c) B (d) C

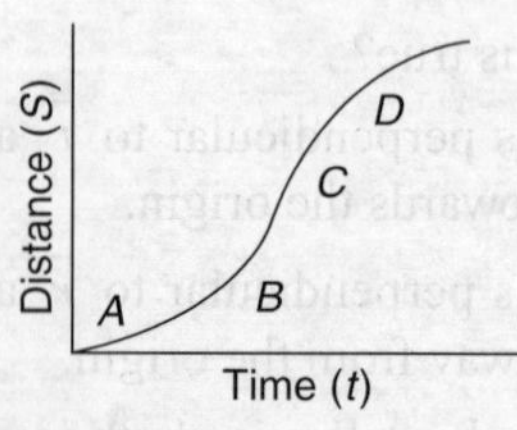

193. A particle of mass m is projected with velocity v making an angle of 45° with the horizontal. When the particle lands on the level ground the magnitude of the change in its momentum will be
(a) $mv\sqrt{2}$ (b) zero
(c) 2 mv (d) $mv/\sqrt{2}$

194. $\vec{A}$ and $\vec{B}$ are two vectors and θ is the angle between them, if $|\vec{A}\times\vec{B}| = \sqrt{3}(\vec{A}.\vec{B})$, the value of θ is

(a) 45° (b) 30°
(c) 90° (d) 60°

195. A particle starting from the origin (0, 0) moves in a straight line in the (x, y) plane. Its coordinates at a later time are $(\sqrt{3}, 3)$. The path the particle makes with the x-axis an angle of

(a) 45° (b) 60°
(c) 0° (d) 30°

196. For angles of projection of a projectile at angle $(45° - \theta)$ and $(45° + \theta)$, the horizontal range described by the projectile are in the ratio of

(a) 2 : 1 (b) 1 : 1
(c) 2 : 3 (d) 1 : 2

197. The vectors $\vec{A}$ and $\vec{B}$ are such that $|\vec{A}+\vec{B}| = |\vec{A}-\vec{B}|$. The angle between the two vectors is

(a) 45° (b) 90°
(c) 60° (d) 75°

198. Two boys are standing at the ends A and B of a ground where $AB = a$. The boy at B starts running in a direction perpendicular to AB with velocity v_1. The boy at A starts running simultaneously with velocity v and catches the other in a time t, where t is

(a) $\dfrac{a}{\sqrt{v^2 + v_1^2}}$ (b) $\dfrac{a}{v + v_1}$

(c) $\dfrac{a}{v - v_1}$ (d) $\sqrt{\dfrac{a^2}{v^2 - v_1^2}}$

199. A stone tied to the end of a string of 1 m long is whirled in a horizontal circle with a constant speed. If the stone makes 22 revolutions in 44 seconds, what is the magnitude and direction of acceleration of the stone?

(a) π^2 ms^{-2} and direction along the radius towards the centre

(b) π^2 ms^{-2} and direction along the radius away from the centre

(c) π^2 ms^{-2} and direction along the tangent to the circle

(d) $\pi^2/4$ ms^{-2} and direction along the radius towards the centre.

200. If the angle between the vectors $\vec{A}$ and $\vec{B}$ is θ, the value of the product $(\vec{B}\times\vec{A}).\vec{A}$ is equal to

(a) $BA^2 \sin\theta$ (b) $BA^2 \cos\theta$
(c) $BA^2 \sin\theta\cos\theta$ (d) zero

201. If a vector $2\hat{i} + 3\hat{j} + 8\hat{k}$ is perpendicular to the vector $4\hat{j} - 4\hat{i} + \alpha\hat{k}$, then the value of α is

(a) 1/2 (b) – 1/2
(c) 1 (d) – 1

202. If $|\vec{A}\times\vec{B}| = \sqrt{3}\vec{A}.\vec{B}$ then the value of $|\vec{A}+\vec{B}|$ is

(a) $(A^2 + B^2 + AB)^{1/2}$ (b) $\left(A^2 + B^2 + \dfrac{AB}{\sqrt{3}}\right)^{1/2}$

(c) $A + B$ (d) $(A^2 + B^2 + \sqrt{3}AB)^{1/2}$

203. The vector sum of two forces is perpendicular to their vector differences. In that case, the forces

(a) are equal to each other
(b) are equal to each other in magnitude
(c) are not equal to each other in magnitude
(d) cannot be predicted.

204. A particle moves along a circle of radius $\left(\dfrac{20}{\pi}\right)$ m with constant tangential acceleration. If the velocity of the particle is 80 m/s at the end of the second revolution after motion has begun, the tangential acceleration is

(a) 40 m/s^2 (b) 640 π m/s^2
(c) 160 π m/s^2 (d) 40 π m/s^2

205. A particle A is dropped from a height and another particle B is projected in horizontal direction with speed of 5 m/sec from the same height. Then correct statement is

(a) particle A will reach at ground first with respect to particle B
(b) particle B will reach at ground first with respect to particle A
(c) both particle will reach at ground simultaneously
(d) both particles will reach at ground with same speed.

206. An object of mass 3 kg is at rest. Now a force of $\vec{F} = 6t^2\hat{i} + 4t\hat{j}$ is applied on the object then velocity of object at $t = 3$ s is

(a) $18\hat{i} + 3\hat{j}$ (b) $18\hat{i} + 6\hat{j}$
(c) $3\hat{i} + 18\hat{j}$ (d) $18\hat{i} + 4\hat{j}$

207. If $|\vec{A}+\vec{B}| = |\vec{A}| + |\vec{B}|$ then angle between A and B will be

(a) 90° (b) 120°
(c) 0° (d) 60°

208. Two particle having mass M and m are moving in a circular path having radius R and r. If their time periods are same then the ratio of angular velocity will be

(a) $\frac{r}{R}$ (b) $\frac{R}{r}$
(c) 1 (d) $\sqrt{\frac{R}{r}}$

209. The width of river is 1 km. The velocity of a boat is 5 km/hr. The boat covered the width of river in 15 min. Then the velocity of river stream is

(a) 3 km/hr (b) 4 km/hr
(c) $\sqrt{29}$ km/hr (d) $\sqrt{41}$ km/hr

210. If two projectiles of same mass and with same velocity are thrown at an angle 60° and 30° with the horizontal, then which will remain same?

(a) Time of fight
(b) Range of projectile
(c) Maximum height acquired
(d) All of them.

211. A 500 kg car takes a round turn of radius 50 m with a velocity of 36 km/hr. The centripetal force is

(a) 1000 N (b) 750 N
(c) 250 N (d) 1200 N

212. A person aiming to reach exactly opposite point on the bank of a stream is swimming with a speed of 0.5 m/s at an angle of 120° with the direction of flow of water. The speed of water in the stream is

(a) 0.25 m/s (b) 0.5 m/s
(c) 1.0 m/s (d) 0.433 m/s

213. Two racing cars of masses m_1 and m_2 are moving in circles of radii r_1 and r_2 respectively. Their speeds are such that each makes a complete circle in the same time t. The ratio of the angular speeds of the first to the second car is

(a) $r_1 : r_2$ (b) $m_1 : m_2$
(c) 1:1 (d) $m_1 : m_2 : r_1 : r_2$

214. If a unit vector is represented by $0.5\hat{i} - 0.8\hat{j} + c\hat{k}$ then the value of c is

(a) $\sqrt{0.01}$ (b) $\sqrt{0.11}$
(c) 1 (d) $\sqrt{0.39}$

215. What is the value of linear velocity, if $\vec{r} = 3\hat{i} - 4\hat{j} + \hat{k}$ and $\vec{\omega} = 5\hat{i} - 6\hat{j} + 6\hat{k}$

(a) $4\hat{i} - 13\hat{j} + 6\hat{k}$ (b) $18\hat{i} + 13\hat{j} - 2\hat{k}$
(c) $6\hat{i} + 2\hat{j} - 3\hat{k}$ (d) $6\hat{i} - 2\hat{j} + 8\hat{k}$

216. A ball of mass 0.25 kg attached to the end of a string of length 1.96 m is moving in a horizontal circle. The string will break if the tension is more than 25 N. What is the maximum speed with which the ball can be moved?

(a) 5 m/s (b) 3 m/s
(c) 14 m/s (d) 3.92 m/s

217. Identify the vector quantity among the following.

(a) distance (b) angular momentum
(c) heat (d) energy

218. A body is whirled in a horizontal circle of radius 20 cm. It has an angular velocity of 10 rad/s. What is its linear velocity at any point on a circular path?

(a) 20 m/s (b) $\sqrt{2}$ m/s
(c) 10 m/s (d) 2 m/s

219. Preeti reached the metro station and found that the escalator was not working. She walked up the stationary escalator in time t_1. On other days, if she remains stationary on the moving escalator, then the escalator takes her up in time t_2. The time taken by her to walk up on the moving escalator will be

(a) $\frac{t_1 t_2}{t_2 - t_1}$ (b) $\frac{t_1 t_2}{t_2 + t_1}$
(c) $t_1 - t_2$ (d) $\frac{t_1 + t_2}{2}$

220. Two cars P and Q start from a point at the same time in a straight line and their positions are represented by $x_P(t) = (at + bt^2)$ and $x_Q(t) = (ft - t^2)$. At what time do the cars have the same velocity?

(a) $\frac{a-f}{1+b}$ (b) $\frac{a+f}{2(b-1)}$
(c) $\frac{a+f}{2(1+b)}$ (d) $\frac{f-a}{2(1+b)}$

221. If the velocity of a particle is $v = At + Bt^2$, where A and B are constant, then the distance travelled by it between 1s and 2s is

(a) $\frac{3}{2}A + \frac{7}{3}B$ (b) $\frac{A}{2} + \frac{B}{3}$

(c) $\frac{3}{2}A + 4B$ (d) $3A + 7B$

222. Aparticleofunitmassundergoesone-dimensionalmotion such that its velocity varies according to $v(x) = \beta x^{-2n}$, where β and n are constants and x is the position of the particle. The acceleration of the particle as a function of x, is given by

(a) $-2\beta^2 x^{-2n+1}$ (b) $-2n\beta^2 e^{-4n+1}$

(c) $-2n\beta^2 x^{-2n-1}$ (d) $-2n\beta^2 x^{-4n-1}$

223. A stone falls freely under gravity. It covers distance h_1, h_2 and h_3 in the first 5 seconds, the next 5 seconds and the next 5 seconds respectively. The relation between h_1, h_2 and h_3 is

(a) $h_2 = 3h_1$ and $h_3 = 3h_3$

(b) $h_1 = h_2 = h_3$

(c) $h_1 = 2h_2 = 3h_3$

(d) $h_1 = \frac{h_2}{3} = \frac{h_3}{5}$

224. The displacement x (in metres) of a particle of mass m (in kg) moving in one dimension under the action of a force, is related to time t (in sec) by $t = \sqrt{x} + 3$. The displacement of the particle when its velocity is zero, will be

(a) 4 m (b) 0 m (zero)

(c) 6 m (d) 2 m

225. The motion of a particle along a straight line is described by equation $x = 8 + 12t - t^3$ where x is in metre and t in second. The retardation of the particle when its velocity becomes zero is

(a) 24 ms^{-2} (b) zero

(c) 6 ms^{-2} (d) 12 ms^{-2}

226. A boy standing at the top of a tower of 20 m height drops a stone. Assuming $g = 10$ ms^{-2}, the velocity with which it hits the ground is

(a) 10.0 m/s (b) 20.0 m/s

(c) 40.0 m/s (d) 5.0 m/s

227. A particle covers half of its total distance with speed v_1 and the remaining half distance with speed v_2. Its average speed during the complete journey is

(a) $\frac{v_1 + v_2}{2}$ (b) $\frac{v_1 v_2}{v_1 + v_2}$

(c) $\frac{2v_1 v_2}{v_1 + v_2}$ (d) $\frac{v_1^2 v_2^2}{v_1^2 + v_2^2}$

228. A particle moves a distance x in time t according to equation $x = (t+5)^{-1}$. The acceleration of particle is proportional to

(a) $(\text{velocity})^{3/2}$ (b) $(\text{distance})^2$

(c) $(\text{distance})^{-2}$ (d) $(\text{velocity})^{2/3}$

229. A ball is dropped from a high rise platform at $t = 0$ starting from rest. After 6 seconds, another ball is thrown downwards from the same platform with a speed v. The two balls meet at $t = 18$ s. What is the value of v? (Take $g = 10$ m/s^2)

(a) 75 m/s (b) 55 m/s

(c) 40 m/s (d) 60 m/s

230. A particle starts its motion from rest under the action of a constant force. If the distance covered in first 10 seconds is S_1 and that covered in the first 20 seconds is S_2, then

(a) $S_2 = 3S_1$ (b) $S_2 = 4S_1$

(c) $S_2 = S_1$ (d) $S_2 = 2S_1$

231. A bus is moving with a speed of 10 ms^{-1} on a straight road. A scooter wishes to overtake the bus in 100 s. If the bus is at a distance of 1 km from the scooter, with what speed should the scooters chase the bus?

(a) 40 ms^{-1} (b) 25 ms^{-1}

(c) 10 ms^{-1} (d) 20 ms^{-1}

232. A particle moves in a straight line with a constant acceleration. It changes its velocity from 10 ms^{-1} while passing through a distance 135 m in t seconds. The value of t is

(a) 12 (b) 9

(c) 10 (d) 1.8

233. The distance travelled by a particle starting from rest and moving with an acceleration $\frac{4}{3}$ ms^{-2}, in the third second is

(a) $\frac{10}{3}$ m (b) $\frac{19}{3}$ m
(c) 6 m (d) 4 m

234. A particle moving along *x*-axis has acceleration *f*, at time *t*, given by $f = f_0\left(1-\frac{t}{T}\right)$, where f_0 and T are constants. The particle at $t = 0$ has zero velocity. In the time interval between $t = 0$ and the instant when $f = 0$, the particle's velocity $\left(v_x\right)$ is

(a) $\frac{1}{2}f_0T^2$ (b) f_0T^2
(c) $\frac{1}{2}f_0T$ (d) f_0T

235. A car moves from *X* to *Y* with a uniform speed v_u and returns to *X* with a uniform speed v_d. The average speed for this round trip is

(a) $\sqrt{v_u v_d}$ (b) $\frac{v_d v_u}{v_d + v_u}$
(c) $\frac{v_u + v_d}{2}$ (d) $\frac{2v_d v_u}{v_d + v_u}$

236. The position *x* of a particle with respect to time *t* along *x*-axis is given by $x = 9t^2 - t^3$ where *x* is in metres and *t* in seconds. What will be the position of this particle when it achieves maximum speed along the + *x* direction?

(a) 54 m (b) 81 m
(c) 24 m (d) 32 m

237. Two bodies *A* (of mass 1 kg) and *B* (of mass 3 kg) are dropped from heights of 16 m and 25 m, respectively. The ratio of the time taken by them to reach the ground is

(a) 4/5 (b) 5/4
(c) 12/5 (d) 5/12

238. A car runs at a constant speed on a circular track of radius 100 m, taking 62.8 seconds for every circular lap. The average velocity and average speed for each circular lap respectively is

(a) 10 m/s, 0 (b) 0, 0
(c) 0, 10 m/s (d) 10 m/s, 10 m/s

239. A particle moves along a straight line *OX*. At a time *t* (in seconds) the distance x (in metres) of the particle from *O* is given by $x = 40 + 12t - t^3$. How long would the particle travel before coming to rest?

(a) 16 m (b) 24 m
(c) 40 m (d) 56 m

240. A ball is thrown vertically upward. It has a speed of 10 m/sec when it reaches one half of its maximum height. How high does the ball rise? (Take $g = 10\text{ m/s}^2$)

(a) 10 m (b) 5 m
(c) 15 m (d) 20 m

241. The displacement *x* of a particle varies with time *t* as $x = ae^{-\alpha t} + be^{\beta t}$, where *a*, *b*, α and β are positive constant. The velocity of the particle will

(a) be independent of β
(b) drop to zero when $\alpha = \beta$
(c) go on decreasing with time
(d) go on increasing with time.

242. A man throws balls with the same speed vertically upwards one after the other at an interval of 2 seconds. What should be the speed of the throw so that more than two balls are in the sky at any time? (Given $g = 9.8\text{ m/s}^2$)

(a) more than 19.6 m/s
(b) at least 9.8 m/s
(c) any speed less than 19.6 m/s
(d) only with speed 19.6 m/s.

243. If a ball is thrown vertically upwards with speed *u*, the distance covered during the last *t* seconds of its ascent is

(a) ut (b) $\frac{1}{2}gt^2$
(c) $ut - \frac{1}{2}gt^2$ (d) $\left(u + gt\right)t$

244. A particle is thrown vertically upward. If its velocity at half of the height is 10 m/s, then the maximum height attained by it is ($g = 10\text{ m/s}^2$)

(a) 8 m (b) 20 m
(c) 10 m (d) 16 m

245. Motion of a particle is given by the equation $s = \left(3t^3 + 7t^2 + 14t + 8\right)$ m. The value of acceleration of the particle at $t = 1$ sec is

(a) 10 m/s^2 (b) 32 m/s^2
(c) 23 m/s^2 (d) 16 m/s^2

Answer Keys

EXERCISES

1. (b)	2. (b)	3. (b)	4. (d)	5. (b)	6. (d)	7. (b)	8. (a)	9. (c)	10. (b)
11. (b)	12. (b)	13. (d)	14. (c)	15. (b)	16. (b)	17. (a)	18. (d)	19. (a)	20. (b)
21. (c)	22. (c)	23. (a)	24. (b)	25. (b)	26. (b)	27. (c)	28. (a)	29. (c)	30. (d)
31. (b)	32. (b)	33. (b)	34. (c)	35. (c)	36. (c)	37. (c)	38. (d)	39. (b)	40. (b)
41. (d)	42. (b)	43. (d)	44. (c)	45. (b)	46. (b)	47. (c)	48. (d)	49. (c)	50. (b)
51. (b)	52. (c)	53. (b)	54. (a)	55. (d)	56. (c)	57. (c)	58. (c)	59. (b)	60. (b)
61. (b)	62. (c)	63. (d)	64. (c)	65. (c)	66. (a)	67. (d)	68. (d)	69. (d)	70. (a)
71. (b)	72. (b)	73. (c)	74. (c)	75. (b)	76. (c)	77. (a)	78. (a)	79. (b)	80. (c)
81. (c)	82. (b)	83. (c)	84. (d)	85. (a)	86. (b)	87. (c)	88. (a)	89. (d)	90. (a)
91. (b)	92. (b)	93. (d)	94. (b)	95. (a)	96. (d)	97. (c)	98. (b)	99. (c)	100. (a)
101. (a)	102. (c)	103. (c)	104. (b)	105. (c)	106. (c)	107. (d)	108. (a)	109. (c)	110. (a)
111. (d)	112. (b)	113. (c)	114. (b)	115. (d)	116. (b)	117. (b)	118. (d)	119. (c)	120. (a)
121. (b)	122. (b)	123. (d)	124. (d)	125. (b)	126. (a)	127. (b)	128. (c)	129. (c)	130. (b)
131. (c)	132. (b)	133. (b)	134. (a)	135. (b)	136. (d)	137. (d)	138. (d)	139. (b)	140. (a)
141. (c)	142. (c)	143. (d)	144. (a)	145. (a)	146. (b)	147. (d)	148. (c)	149. (d)	150. (b)
151. (a)	152. (c)	153. (b)	154. (d)	155. (a)	156. (a)	157. (a)	158. (d)	159. (c)	160. (d)
161. (d)	162. (a)	163. (b)	164. (b)	165. (d)	166. (d)	167. (d)			

Questions Asked in NEET/CBSE Medicals

168. (b)	169. (d)	170. (a)	171. (c)	172. (b)	173. (c)	174. (d)	175. (a)	176. (a)	177. (a)
178. (a)	179. (d)	180. (a)	181. (c)	182. (b)	183. (b)	184. (d)	185. (a)	186. (d)	187. (c)
188. (b)	189. (c)	190. (a)	191. (b)	192. (d)	193. (a)	194. (d)	195. (b)	196. (b)	197. (b)
198. (d)	199. (a, b)	200. (d)	201. (b)	202. (a)	203. (b)	204. (a)	205. (c)	206. (b)	207. (c)
208. (c)	209. (a)	210. (b)	211. (a)	212. (a)	213. (c)	214. (b)	215. (b)	216. (c)	217. (b)
218. (d)	219. (b)	220. (d)	221. (a)	222. (d)	223. (d)	224. (b)	225. (d)	226. (b)	227. (c)
228. (a)	229. (a)	230. (b)	231. (d)	232. (b)	233. (a)	234. (c)	235. (d)	236. (a)	237. (a)
238. (c)	239. (a)	240. (a)	241. (d)	242. (a)	243. (b)	244. (c)	245. (b)		

Solutions

1. (b)

Let the velocity of the locomotive be v when it overtakes after time t at a distance s then, the distance covered by the boy in time t,

$$s = ut = 2t$$

The distance covered by the locomotive in time t

$$s = \frac{1}{2}at^2$$

$$\frac{1}{2}at^2 = 2t \Rightarrow at = 4$$

Now, $v = u + at = 0 + 4 = 4$ m/s

2. (b)

The distance covered by body in n seconds $= \frac{1}{2}an^2\ (u = 0)$

The distance covered by the body in nth second $= \frac{1}{2}a(2n-1)$

$$\text{Ratio} = \frac{\frac{1}{2}a(2n-1)}{\frac{1}{2}an^2} = \frac{2n-1}{n^2}$$

3. (b)

Let the length of the first train be l_1 and velocity v_1. Then,

$$\frac{l_1}{v_1} = 15 \quad \text{(i)}$$

$$\frac{l_1 + 450}{v_1} = 45$$

$$\Rightarrow v_1 = \frac{l_1}{15} = \frac{l_1 + 450}{45} \quad \text{(ii)}$$

$l_1 = 225$ m and $v_1 = 15$ m/s

Let the velocity of the second train be v_2

Relative velocity $= v_1 + v_2 = 15 + v_2$

$$\frac{225+300}{15+v_2} = 21$$

$$\Rightarrow v_2 = 20 \text{ m/s}$$

4. (d)

When two spheres are dropped from rest, force of attraction due to earth acts on it. Under the force of attraction, they will acquire the same acceleration which is due to gravitational effect.
Also, acceleration due to gravity is independent of mass of the body. Hence, the two spheres have the same acceleration.

5. (b)

$$v = 4t^3 - 2t$$

$$a = \frac{dv}{dt} = 12t^2 - 2 \quad \text{(i)}$$

Now, $$\frac{ds}{dt} = 4t^2 - 2t$$

$$s = \int_0^t 4t^3 - 2t$$

$$2 = t^4 - t^2$$

$$t^4 - t^2 - 2 = 0$$

$$t^4 - 2t^2 + t^2 - 2 = 0$$

$$t^2(t^2-2) + 1(t^2-2) = 0$$

$$(t^2+1)(t^2-2) = 0$$

$$t^2 = -1 \text{ or } t^2 = 2$$

Neglecting imaginary value of t,
From Eq. (i)

$$a = 12t^2 - 2$$

$$= 12 \times 2 - 2 = 22 \text{ m/s}$$

6. (d)

$$v = 20 + 10t^2$$

Acceleration, $a = \frac{dv}{dt} = 20t$

Acceleration is a function of time therefore acceleration is non-uniform.

7. (b)

$$x = \alpha t^3 + \beta t^2 + \lambda t + \delta$$

$$v = \frac{dx}{dt} = 3\alpha t^2 + 2\beta t + \lambda$$

Velocity at $t = 0$ is λ

$$a = \frac{dv}{dt} = 6\alpha t + 2\beta$$

Acceleration at $t = 0$ is 2β

Ratio of a to v $= \frac{2\beta}{\lambda}$

Therefore, it depends upon β and λ

8. (a)

Let their velocity be u

$$v_A^2 = u^2 + 2gh \quad \text{(i)}$$

$$v_B^2 = u^2 + 2gh \quad \text{(ii)}$$

(u for B also remains same because when it comes back to tower, it has velocity u in downward direction)

$v_c^2 = 2gh$ (u for c is in horizontal direction)

Therefore, $v_A = v_B > v_C$

9. (c)

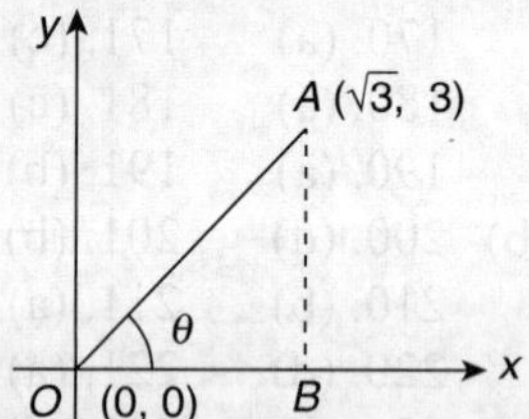

Draw the situation as shown OA represent the path of the particle starting from origin O (0,0). Draw a perpendicular from point A to x-axis. Let path of the particle makes an angle with the x-axis, then $\tan\theta$ = slope of line OA

$$= \frac{AB}{OB} = \frac{3}{\sqrt{3}} = \sqrt{3} \text{ or } \theta = 60°$$

10. (b)

Let S, $2S$ and $3S$ be the distance for motion OA, AB and BC respectively t_1, t_2 and t_3 be the time respectively

$$S = \frac{1}{2} v_{max} \times t_1 \quad \text{...(i)}$$

$$2S = v_{max} \times t_2 \quad \text{...(ii)}$$

$$3S = \frac{1}{2} v_{max} \times t_3 \quad \text{...(iii)}$$

$$v_{max} = \frac{S + 2S + 3S}{\frac{1}{2}t_1 + t_2 + \frac{1}{2}t_3} \quad \text{...(iv)}$$

$$v_{av} = \frac{S + 2S + 3S}{t_1 + t_2 + t_3} \quad \text{...(v)}$$

From (i) and (ii) $t_1 = t_2$

From (i) and (iii) $3t_1 = t_3$

From (iv) an (v)

$$\frac{v_{max}}{v_{av}} = \frac{t_1 + t_2 + t_3}{\frac{1}{2}t_1 + t_2 + \frac{1}{2}t_3} = \frac{t_1 + t_1 + 3t_1}{t_1 + \frac{1}{2}t_1 + \frac{3}{2}t_1} = \frac{5}{3}$$

11. (b)

$$x = 7t + 4t^2$$

$$v_x = \frac{dx}{dt} = 7 + 8t$$

$$a_x = \frac{dv_x}{dt} = 8$$

Similarly $y = 5t$

$$v_y = 5$$

$$a_y = 0$$

$$a = \sqrt{a_x^2 + a_y^2} = \sqrt{8^2} = 8$$

12. (b)

$$\vec{v}_{av} = \frac{\Delta \vec{s}}{\Delta \vec{t}}$$

Time $\Delta t = \frac{1}{4} \times \frac{2\pi r}{v} = \frac{1}{4} \times \frac{2\pi \times 15}{25} = \frac{3\pi}{10}$ s

Magnitude of displacement

$$= \Delta S = AB = \sqrt{(15)^2 + (15)^2} = 15\sqrt{2}$$

$$v_{av} = \frac{15\sqrt{2}}{3\pi / 10} = 22.25 \text{ cm/s}$$

13. (d)

Let total distance be = 4 s

For the second half the average velocity $v = \frac{v_1 + v_2}{2}$

$$v_{av} = \frac{\text{Total distance}}{\text{Total Time}}$$

Time t_1 for first half of the journey $= \frac{2s}{v_0}$

Time t_2 for second half of the journey $\frac{2s}{v}$

$$v_{av} = \frac{4s}{\frac{2s}{v_0} + \frac{2s}{v}}$$

$$= \frac{2v_0(v_1 + v_2)}{2 \times \left(v_0 + \frac{v_1 + v_2}{2}\right)}$$

$$= \frac{2v_0(v_1 + v_2)}{2v_0 + v_1 + v_2}$$

14. (c)

Their relative velocity = 30 + 30 = 60 km/hr^{-1}

The bird will keep on flying till the two cars will be

Time = 1 hr

The distance covered by bird also 60 km

15. (b)

$$v^2 = u^2 + 2gh$$

$$(3u)^2 = u^2 + 2gh$$

$$\Rightarrow h = \frac{8u^2}{2g} = \frac{4u^2}{g}$$

16. (b)

Let max height to which ball rises be *H*. Then,

$$v^2 = u^2 - 2gh$$

$$0 = (25)^2 - 2 \times 10H$$

$$\Rightarrow H = \frac{25 \times 25}{2 \times 10} = 31.25 \text{ m}$$

Let *t* be the time when the ball is at a height of 30 m

$$h = ut - \frac{1}{2}gt^2$$

$$30 = 25t - \frac{1}{2} \times 10t^2$$

$$\Rightarrow 6 = 5t - t^2$$

$$\Rightarrow t^2 - 5t + 6 = 0$$

$$(t-2)(t-3) = 0$$

$$t = 2, t = 3$$

The ball will be at a height of 30 m after 2*s* and next, after 3*s* while coming down after attaining maximum height.

17. (a)

Take initial position as origin and direction of motion as positive. As the particle is thrown with initial velocity; at highest point, its velocity is zero and then it returns to its reference position. This situation is best depicted in figure of option (a).

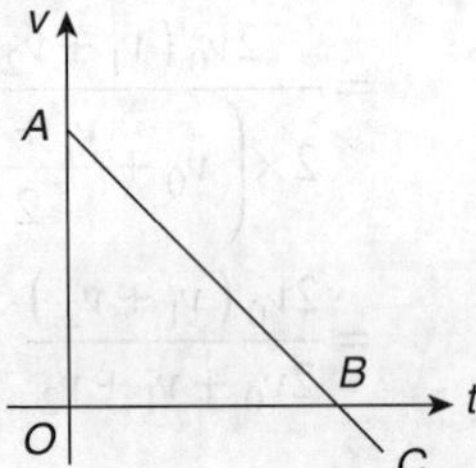

In the figure, AB part denotes upwards motion and BC part denotes downward motion.

18. (d)

$$\left|\vec{v_2}\right| = \left|\vec{v_1}\right|^2 = v$$

$$\Delta v = \vec{v_2} - \vec{v_1} = \vec{v_2} + \left(-\vec{v_1}\right)$$

Using law of parallelogram

$$\Delta v = \sqrt{\left(-v_1\right)^2 + v_2^2 - 2v_1v_2\cos\theta}$$

$$= \sqrt{v^2 + v^2 - 2v^2\cos\theta}$$

$$= \sqrt{2v^2\left(1 - \cos^2\theta/2 + \sin^2\theta/2\right)}$$

$$= \sqrt{2v^2 \times 2\sin^2\theta/2}$$

$$= 2v\sin\theta/2$$

$$= 2v\sin 20°$$

19. (a)

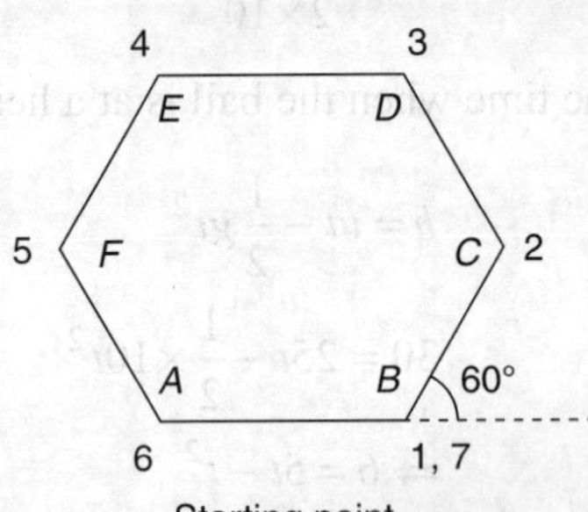

In 6 turns, each of 60, the cyclist traversed a regular hexagon path having each side 100 m. So, at 7th turn, he will be again at point B which is at a distance 100 m from starting point A. Hence, net displacement of cyclist is 100 m.

20. (b)

In completing a circular lap, car's displacement is zero. Average velocity is defined as the ratio of displacement to the time taken. Average speed of a particle in a given interval of time is defined as the ratio of distance travelled to the time taken. On a circular path, in completing on turn, the distance travelled is 2 while displacement is zero.

$$\text{Hence, average velocity} = \frac{\text{displacement}}{\text{time interval}} = \frac{0}{t} = 0$$

$$\text{Average speed} = \frac{\text{displacement}}{\text{time interval}} = \frac{2\pi r}{t}$$

$$= \frac{2 \times 3.14 \times 100}{62.8}$$

$$= 10 \text{ m/s}$$

21. (c)

Velocity is a vector quantity. Average velocity is defined as the displacement divided by time.
In the given graph, displacement is zero.

$$\text{Hence, average velocity} = \frac{\text{total displacement}}{\text{total time}} = \frac{0}{t} = 0$$

22. (c)

Velocity acquired by parachute after falling 50 m.

$$v^2 = \sqrt{2gh} = \sqrt{2 \times 10 \times 50} = 10\sqrt{10} \text{ m/s}$$

Let h be the height at which it bails out. Then,

$$v^2 = u^2 - 2ah$$

$$(3)^2 = \left(10\sqrt{10}\right)^2 - 2 \times 2 \times h$$

$$9 = 1000 - 4h$$

$$h = \frac{1000 - 9}{4} = 247.75 \text{ m}$$

Total height $= 50 + 247.5 = 297.5$

Closest answer is option (c).

23. (a)

$$\vec{a} = \frac{\Delta\vec{v}}{\Delta t} = \frac{\vec{v_2} - \vec{v_1}}{\Delta t}$$

$$|\vec{a}| = \frac{\sqrt{5^2 + 5^2}}{10} = \frac{5\sqrt{2}}{10} = \frac{1}{\sqrt{2}}$$

towards northwest

24. (b)

It is a case of non-uniform circular motion, therefore it has two components of acceleration – tangential and radial

$$a_r = \frac{u^2}{r} = \frac{20 \times 20}{100} = 4 \text{ m/s}^2$$

$$a_t = 3 \text{ m/s}^2$$

$$a = \sqrt{(4)^2 + (3)^2} = 5$$

25. (b)

Speed is a scalar quantity. The average speed of an object is defined as the distance (d) travelled divided by the time elapsed (t)

i.e., $$v = \frac{d}{t}$$

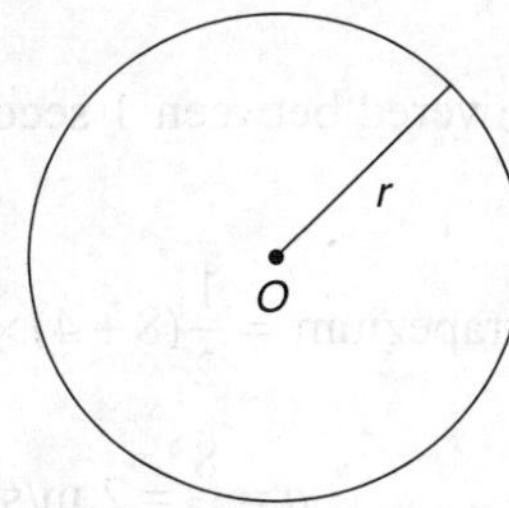

Circumference of a circle of radius r is $2\pi r$.

The time taken is $t = \frac{2\pi r}{\text{speed}} = \frac{2\times 3.14\times 100}{3.14} = 20$ s

Hence, average speed is

$$v_{av} = \frac{2\times 3.14\times 100}{20} = 31.4 \text{ m/s}$$

26. (b)

Assuming upward positive, downward negative:

$$-h = ut_1 - \frac{1}{2}gt_1^2 \quad \text{...(i)}$$

$$-h = ut_2 - \frac{1}{2}gt_2^2 \quad \text{...(ii)}$$

$$-h = -\frac{1}{2}gt_3^2 \quad \text{...(iii)}$$

From (i) and (ii),

$$ut_2 + ut_1 = \frac{1}{2}gt_2^2 - \frac{1}{2}gt_1^2$$

$$u(t_2 + t_1) = \frac{1}{2}g(t_1^2 - t_2^2)$$

$$u(t_2 + t_1) = \frac{1}{2}(t_1 - t_2)(t_2 + t_1)$$

$$u = \frac{1}{2}g(t_1 - t_2)$$

$$-\frac{1}{2}gt_3^2 = \frac{1}{2}g(t_1 - t_2)t_1 - \frac{1}{2}gt_1^2$$

$$-t_3^2 = t_1^2 - t_1t_2 - t_1^2$$

$$t_3^2 = t_1t_2$$

$$t_3 = \sqrt{t_1t_2}$$

27. (c)

Velocity of balloon after 8 seconds,

$$v = u + at$$

$$= 0 + 1.25\times 8 = 10 \text{ m/s}$$

The height reached by balloon,

$$h = \frac{1}{2}\times 1.25\times 8^2 = 40 \text{ m}$$

According to inertia, the initial velocity of stone will be – 10m/s.

$$h' = -10t + \frac{1}{2}\times 10t^2 = 40$$

$$\Rightarrow \quad t = 4, -2$$

28. (a)

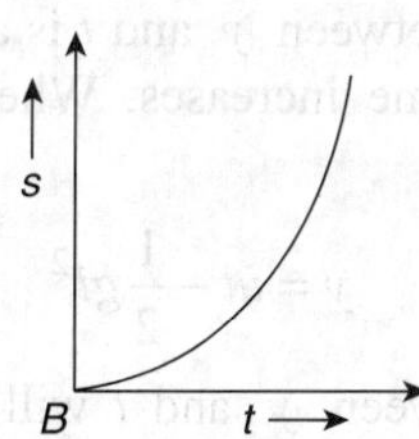

An object is said to be moving with uniform acceleration if its velocity changes by an equal amount in equal interval of time.

When a body is moving with a constant acceleration, the time-displacement graph is a curve with bend upwards.

29. (c)

Acceleration of body along AB is $g \cos \theta$

Distance travelled in time t sec = $AB = \frac{1}{2}(g\cos\theta)t^2$

From ΔABC, $AB = 2R\cos\theta; 2\cos\theta = \frac{1}{2}g\cos\theta t^2$

$$\Rightarrow t^2 = \frac{4R}{g} \text{ or } t = 2\sqrt{\frac{R}{g}}$$

30. (d)

We know that average speed

$$v_{av} = \frac{\text{length of path}}{\text{time interval } (t_2 - t_1)} \quad \text{...(i)}$$

and average velocity is defined as,

$$\vec{v}_{av} = \frac{\text{displacement}}{\text{time interval}} = \frac{\vec{r}_2 - \vec{r}_1}{t_2 - t_1} \quad \text{...(ii)}$$

But we know that distance is always greater than or equal to magnitude of displacement. So the average speed will always be greater than or equal to the magnitude of average velocity.

From Eqs. (i) and (ii)

$$\frac{\vec{v}_{av}}{v_{av}} = \frac{\text{displacement}}{\text{length of path (distance)}} \le 1$$

31. (b)

When a ball is dropped on a floor, then

$$y = \frac{1}{2}gt^2 \quad \text{...(i)}$$

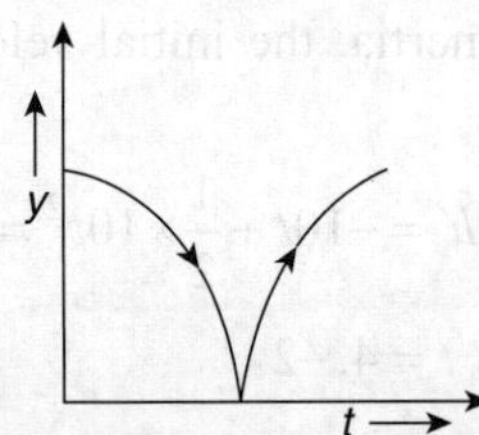

So, the graph between y and t is a parabola. Here, y decreases as time increases. When the ball bounces back then,

$$y = ut - \frac{1}{2}gt^2 \qquad \ldots\text{(ii)}$$

The graph between y and t will be parabola. Here, y increases when time increases. Hence, the required graph between y and t is shown in the figure.

32. (b)

$$200 = u \times 2 + \frac{1}{2}a \times (2)^2$$

$$200 = 2u + 2a$$

$$100 = u + a \qquad \ldots\text{(i)}$$

The distance covered in 6 seconds

$$420 = 6u + \frac{1}{2}a \times 6^2$$

$$70 = u + 3a \qquad \ldots\text{(ii)}$$

From (i) and (ii)

$$2a = -30$$

$$\Rightarrow \quad a = -15 \text{ cm/s}^2$$

From (i)

$$100 = u - 15$$

$$\Rightarrow \quad u = 115 \text{ cm/s}^2$$

Velocity at the end of 7 seconds

$$v = 115 + (-15 \times 7)$$

$$= 10 \text{ cm/s}$$

33. (b)

The distance travelled in tth second $= ut + \frac{1}{2}a(2t-1)$

$$\frac{9}{25}h = 0 + \frac{1}{2}g(2t-1) \qquad \ldots\text{(i)}$$

$$h = \frac{1}{2}gt^2 \qquad \ldots\text{(ii)}$$

From (i) and (ii)

$$\frac{25}{9} = \frac{t^2}{2t-1}$$

$$\Rightarrow \quad 9t^2 - 50t + 25 = 0$$

$$\Rightarrow \quad t = 5 \text{ or } \frac{5}{9}$$

34. (c)

The distance covered between 1 second and 8 seconds $= S_8 - S_1$

$$S_8 = \text{Area of trapezium} = \frac{1}{2}(8+4) \times 8 = 48 \text{ m}$$

Acceleration $\quad a = \frac{8}{4} = 2 \text{ m/s}^2$

Velocity after 1 second $= u + at$

$$= 0 + 2 \times 1$$

$$= 2 \text{ m/s}$$

Therefore, $S_1 = \text{Area of triangle} = \frac{1}{2} \times 2 \times 1 = 1 \text{ m}$

$$S_8 - S_1 = 48 - 1 = 47 \text{ m}$$

35. (c)

We have

$$s_t = u + \frac{a}{2}(2t-1)$$

At first $\quad s_t = 65 \text{ m}, t = 5 \text{ sec}$

$$65 = u + \frac{a}{2}(10-1)$$

$$\Rightarrow \quad 65 = u + \frac{9}{2a} \qquad \ldots\text{(i)}$$

Again, $\quad t = 9 \text{ sec and } s = 105 \text{ m}$

$$105 = u + \frac{a}{2}[17]$$

$$105 = u + \frac{17a}{2} \qquad \ldots\text{(ii)}$$

Solving Eqs. (i) and (ii)

$$105 - 65 = \frac{17}{2}a - \frac{9}{2}a$$

$$a = 10 \text{ m/s}^2$$

and $\quad u = 20 \text{ m/s}$

Now, from $\quad s = ut + \frac{1}{2}at^2$

Here, $u = 20$ m/s, $t = 20$ s and $a = 10 \text{ m/s}^2$

$$s = 20 \times 20 + \frac{1}{2} \times 10 \times (20)^2$$

$$= 400 + 5 \times 400$$

$$= 2400 \text{ m}$$

36. (c)

Given, $\quad t = ax^2 + bx$

Differentiating w.r.t to t,

$$1 = \frac{dt}{dt} = 2ax\frac{dx}{dt} + b\frac{dx}{dt}$$

$$v = \frac{dx}{dt} = \frac{1}{(2ax+b)}$$

Again, differentiating $w.r.t$ to t

$$\frac{d^2x}{dt} = \frac{-2a}{(2ax+b)^2} \cdot \frac{dx}{dt}$$

$$f = \frac{d^2}{dt^2} = \frac{-1}{(2ax+b)^2} \cdot \frac{2a}{(2ax+b)}$$

or $$f = \frac{-2a}{(2ax+b)^3} = -2av^3$$

37. (c)

The velocity – time graph for the given situation can be drawn as below

Magnitude of slope of $OA = f$

and slope of $BC = f/2$

$$v = \frac{f}{t_1} = f/2t_2$$

$$t_2 = 2t_1$$

In graph area of ΔOAD gives

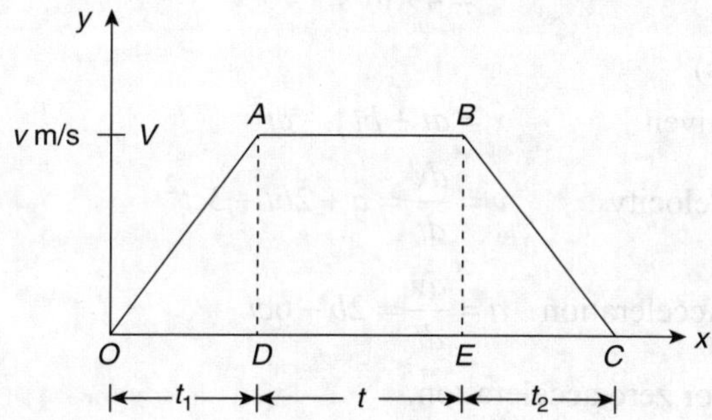

Distance $= S_1 = 1/2\, ft_1^2$ ...(i)

Area of rectangle $ABED$ gives distance travelled in time t

$$S_2 = (ft_1)t$$

Distance travelled in time $t_2 = S_3 = 1/2.f/2(2t_1)^2$

Thus, $S_1 + S_2 + S_3 = 15S$

$$S + (ft_1)t + ft_1^2 = 15S$$

$$S + (ft_1)t + 2S = 15S \quad \left(S = \frac{1}{2}ft_1^2\right)$$

$$(ft_1)t = 12S \quad \text{...(ii)}$$

$$S = \frac{1}{2}f(t_1)^2$$

$$= \frac{1}{2}f(t/6)^2 = \frac{1}{72}ft^2$$

38. (d)

$$s = ut + 1/2g(2n-1)$$

Where u is initial velocity, t is time, g is acceleration due to gravity and n is the particular second.

Given $u = 0$

For $n = 1, S = 0 + 1/2g(2-1) = g/2$ ...(i)

$n = 2, S_2 = 0 + 1/2g(2\times2-1) = 3g/2$ ...(ii)

$n = 3, S_3 = 0 + 1/2g(2\times3-1) = 5g/2$ ...(iii)

From Eqs. (i), (ii) and (iii),

$$S_1 : S_2 : S_3 = g/2 : 3g/2 : 5g/2$$

$$= 1:3:5$$

39. (b)

Here, $x = a_0 + \frac{a_1 t}{2} - \frac{a_2 t^2}{3}$

Velocity, $v = \frac{dx}{dt} = \frac{a_1}{2} - \frac{2a_2 t}{3}$

Acceleration, $a = \frac{dv}{dt} = \frac{d^2x}{dt^2} = -\frac{2a_2}{3}$

40. (b)

41. (d)

Let the car accelerate for time t_1 and decelerate for time t_2

$$\alpha = \frac{V_{max}}{t_1}$$

Similarly, $\beta = \frac{V_{max}}{t_2}$

$$t = t_1 + t_2$$

$$= \frac{V_{max}}{\alpha} + \frac{V_{max}}{\beta}$$

$$V_{max} = \frac{\alpha\beta t}{\alpha+\beta}$$

Total distance = Area under velocity time graph

$$= \frac{1}{2}V_{max}t$$

$$= \frac{1}{2}\frac{\alpha\beta}{\alpha+\beta}t^2$$

42. (b)

Velocity of A is v along AB. The velocity of B is along BC component of velocity v along BA

$$= v\cos 60° = \frac{v}{2}$$

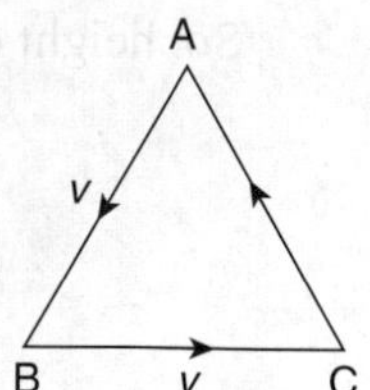

The relative velocity of tortoise A with respect B is

$$= v + \frac{v}{2} = \frac{3v}{2}$$

It means their distance is reduced by $\frac{3v}{2}$ unit/s

$$t = \frac{2a}{3v}$$

43. (d)

Given, $v = 108 \text{ km/h} = 30 \text{ m/s}$

$$v = u + at$$

$$30 = 0 + a \times 5$$

or $a = 6 \text{ m/s}^2$ $(\because u = 0)$

Thus, distance travelled by metro train in 5s is given by

$$s_1 = 1/2at^2 = 1/2 \times 6 \times (5)^2 = 75 \text{ m}$$

Distance travelled before coming to rest = 45 m

Thus, from third equation of motion,

$$0^2 = (30)^2 - 2a \times 45$$

$$a = \frac{30 \times 30}{2 \times 45} = 10 \text{ m/s}^2$$

Time taken in travelling 45 m is given by,

$$t_3 = \frac{30}{10} = 3 \text{ s}$$

Now, total distance = 395 m

or $75 + s_1 + 45 = 395 \text{ m}$

$$s_1 = 395 - 75 - 45 = 275 \text{ m}$$

$$t_2 = \frac{275}{30} = 9.2s$$

Thus, total time taken in whole journey

$$= t_1 + t_2 + t_3$$

$$= 5 + 9.2 + 3 = 17.2 \text{ s}$$

44. (c)

Let the body reach the ground in t seconds.

$\therefore$ Velocity of body after $(t - 2)$ seconds

$$v = u + gt = 0 + g(t - 2)$$

Distance covered in last seconds

$$h' = g(t-2) \times 2 + 1/2g(2)^2$$

$$60 = 20(t - 2) + 20$$

On solving, $t = 4$ s

So, height of tower is given by

$$h = 1/2gt^2 = 1/2 \times 10(4)^2 = 80 \text{ m}$$

45. (b)

Distance covered = Area enclosed by $v - t$ graph

= Area of triangle = $1/2 \times 4 \times 8 = 16$ m

46. (b)

Bu = 0
v = 0
AB in 5th sec
BA in 1 sec (6th sec)
A
1)
u

$$AB = u - \frac{1}{2}g(2t - 1)$$

$$= u - \frac{1}{2}g(2 \times 5 - 1)$$

$$= u - \frac{1}{2}g.9$$

$$BA = 0 + \frac{1}{2}g.1^2$$

These are equal.

$$\therefore \quad u - 9/2g = \frac{1}{2}g$$

$$u = 5 \times 9.84 \text{ m/s}$$

$$= 49 \text{ m/s}$$

47. (c)

Given $x = at + bt^2 - ct^3$

Velocity $v = \frac{dx}{dt} = a + 2bt - 3ct^2$

Acceleration $a = \frac{dv}{dt} = 2b - 6ct$

For zero acceleration,

$$2b - 6ct = 0 \text{ or } t = \frac{b}{3c}$$

The velocity at $t = \frac{b}{3c}$

$$v = a + 2b\left(\frac{b}{3c}\right) - 3c\left(\frac{b}{3c}\right)^2$$

$$= a + \frac{2b^2}{3c} - \frac{b^2}{3c} = a + \frac{b^2}{3c}$$

48. (d)

The area of $a - t$ represents change in velocity. Thus,

$$A_1 = 5 \times 2 = 10;$$

$$A_2 = 2 \times 4 = 8;$$

$$A_3 = 7 \times 1 = 7;$$

$$(1) > (2) > (3)$$

49. (c)
For displacement be zero, let $t = T$

$$0 = uT - \frac{1}{2}gT^2$$

$$T = \frac{2u}{g} = \frac{2\times 4.9}{9.8} = 1.0 \text{ s}$$

50. (b)

$$x = 5t^3 + 3t^2 - 9$$

Velocity, $v = \frac{dx}{dt} = 15t + 6t$

Acceleration, $a = \frac{dv}{dt} = 30t + 6$

At $t = 2$ $a = 30\times 2 + 6 = 66 \text{ m/s}^2$

51. (b)
Initial velocity

$$u = 72 \text{ km/h} \times \frac{5}{18} = 20 \text{ m/sec}$$

We have,

$$s = 20 \text{ m}$$

$$t = 2 \text{ sec}$$

$$s = ut + \frac{1}{2}at^2$$

$$20 = 20\times 2 + \frac{1}{2}\times a \times 4$$

$$20 - 40 = 2a \quad \Rightarrow a = -10 \text{ m/sec}^2$$

So, deceleration is 10 m/sec^2.

52. (c)
Distance travelled by walking.

$$= 2.5\times 2 = 5 \text{ m}$$

Distance travelled by running

$$= 3.5\times 4 = 14 \text{ m}$$

$$\text{Average speed} = \frac{\text{total distance}}{\text{total time}}$$

$$= \frac{14+5}{4+2} = \frac{19}{6} = 3.16 \text{ m/s}$$

53. (b)

$$v_x = 20 \text{ m/s}, v_y = 10 \text{ m/s}$$

$$v_y^2 = u_y^2 - 2gh$$

$$(10)^2 = u_y^2 - 2\times 10\times 15$$

$$u_y^2 = 400 \Rightarrow uy = 20 \text{ m/s}$$

$$u = \sqrt{u_x^2 + u_y^2} = \sqrt{(20)^2 + (20)^2}$$

$$= 20\sqrt{2} \text{ m/s}$$

54. (a)

$$u = 20i + 20j$$

$$\tan\theta = \frac{u_y}{u_x} = \frac{20}{20} = 1$$

$$\Rightarrow \quad \theta = 45°$$

55. (d)

$$H = \frac{u^2 \sin^2\theta}{2g} = \frac{(20\sqrt{2})^2 \times \left(\frac{1}{\sqrt{2}}\right)^2}{2\times 10} = 20 \text{ m}$$

56. (c)

$$R = \frac{u^2 \sin 2\theta}{g} = \frac{(20\sqrt{2})^2}{10} \sin 90° = 80 \text{ m}$$

57. (c)

58. (c)
Time taken by the car to cover 80 m

$$t = \frac{s}{v} = \frac{80}{30} = \frac{8}{3} \text{sec}$$

We have, $v = u + at$

$$u = ?\ v = 0, a = -g = -10 \text{ m/s}$$

and $t = \frac{8/3}{2} = \frac{4}{3}$ sec (since time of flight to reach the highest point is half of the total time of flight)

$$0 = u - 10\times\frac{4}{3}$$

$$u = \frac{40}{3} \text{ m/s}$$

59. (b)

$$a = \text{constant}$$

$$\frac{dv}{dt} = K$$

$$\int dv = \int K dt$$

$$v = Kt \quad \text{i.e., } v \propto t$$

But $P = F \times v$

So, $P \propto t$

60. (b)
Net distance = Distance from 0 to five seconds + Distance from 5 to 10 second + Distance from 10 to 15 seconds + Distance from 15 to 20 second

$$= 40 + 0 + 40 + 20 + 20 = 120 \text{ m}$$

Total time taken = 20 min

Hence, average speed $= \frac{120}{20} = 6$ m/min

61. (b)

$$u = 1.4 \text{ m/s}$$
$$T = 2$$

Displacement = Final position – Initial position
Since initial position and final position are same, displacement = 0

62. (c)

$$s = t^3 - 6t^2 + 3t + 4$$

$$v = \frac{ds}{dt} = 3t^2 - 12t + 3$$

$$a = \frac{dv}{dt} = 6t - 12$$

If $a = 0 \Rightarrow 6t - 12 = 0$

$$t = 2$$

$\therefore$ $v = 3\times(2)^2 - 12\times 2 + 3$

$$= 12 - 24 + 3 = -9 \text{ m/s}$$

63. (d)

As the ball goes upwards the velocity of ball decreases as time increases, at maximum height the velocity becomes zero.

64. (c)

Since body has uniform acceleration. velocity of the particle increases with time. Hence, this is displacement – time graph with X as time and Y as displacement axis.

65. (c)

$$h = 0 + \frac{1}{2}gt^2$$

$\Rightarrow$ $t^2 \propto h$

$\therefore$ $\frac{t_1}{t_2} = \sqrt{\frac{h_1}{h_2}} = \sqrt{\frac{16}{25}} = \frac{4}{5}$

66. (a)

Let in some time, t_1 distance covered by man and edge image be respectively x and y.

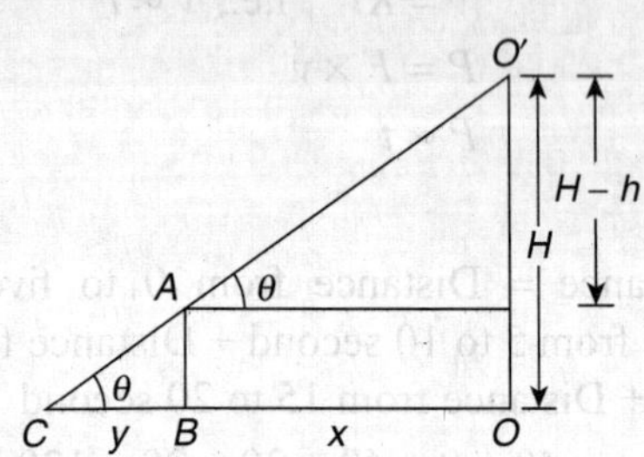

$$\tan\theta = \frac{H-h}{x} = \frac{H}{x+y}$$

$\Rightarrow$ $Hx = Hx + Hy - hx - hy$

$\Rightarrow$ $\frac{x}{y} = \frac{(H-h)}{h}$

Now $x = ut$ and $y = vt$

$\Rightarrow$ $\frac{x}{y} = \frac{u}{v}$

Hence, $\frac{u}{v} = \frac{(H-h)}{h}$

$\Rightarrow$ $\frac{u}{v} = \frac{hu}{(H-h)}$

67. (d)

$$h = \frac{1}{2}gt^2$$

$\Rightarrow$ $t = \sqrt{\frac{2\times 1960}{9.8}} = 20$ s

$\therefore$ $s = AB = ut = \frac{600\times 10^3 \times 20}{60\times 60} \times \frac{10}{3}\times 10^3 = 3.33$ km

68. (d)

The area under acceleration – time graph gives change in velocity.

69. (d)

70. (a)

Distance covered $= \pi r = \pi \times 10$ km

Displacement $= 2r = 2\times 10$ km

Ratio $= \frac{\pi\times 10}{2\times 10} = \pi : 2$

71. (b)

$v = 500$ m/s, $s = 1$ m, $u = 0 = ?$

$$v^2 - u^2 = 2as$$

$$(500)^2 - 0^2 = 2\times a\times 1$$

$\Rightarrow$ $a = 1.25\times 10^5 \text{ ms}^{-1}$

72. (b)

$$2gh = v^2 - 0^2 \Rightarrow v^2 = 2gh$$

$$\left(\frac{v}{2}\right)^2 = 2gh_1 \Rightarrow v^2 = 8gh_1$$

$\Rightarrow$ $8gh_1 = 2gh \Rightarrow h_1 = \frac{h}{4}$

73. (c)

Let angle of projection be θ. Then vertical component of velocity $= 40\sin\theta$

$$h = 4t - \frac{1}{2}gt^2$$

$$20.4 = 40\sin\theta \times 2 - \frac{1}{2}\times 9.8\times(2)^2$$

$$20.4 = 80\sin\theta - 19.6$$

$$20.4 + 19.6 = 80\sin\theta$$

$$40 = 80\sin\theta$$

$$\Rightarrow \sin\theta = \frac{1}{2} \Rightarrow \theta = 30°$$

74. (c)

75. (b)

Velocity at highest point $= u\cos 45°$

$$= 20\sqrt{2}\times\frac{1}{2}$$

$$= 20 \text{ m/s}$$

Maximum height $H = \dfrac{u^2\sin^2}{2g}$

$$= \frac{\left(20\sqrt{2}\right)^2\times\left(\frac{1}{\sqrt{2}}\right)^2}{2g} = 20 \text{ m}$$

$$\omega = \frac{v}{r} = \frac{20}{20} = 1 \text{ rad/s}$$

76. (c)

Average speed $= \dfrac{S_1 + S_2}{2t}$

$$S_1 = v_1 t, S_2 = v_2 t$$

$$\Rightarrow \quad v_{av} = \frac{v_1 t + v_2 t}{2t} = \frac{v_1 + v_2}{2}$$

77. (a)

$$v = \sqrt{2^2 + 3^2} = \sqrt{13}$$

$$\cos\theta = \frac{2}{\sqrt{13}}$$

Horizontal component of velocity

$$= v\cos\theta = \sqrt{13}\times\frac{2}{\sqrt{13}} = 2 \text{ m/s}$$

Otherwise also $v_x = 2$ m/s

78. (a)

Average speed $= \dfrac{2S}{t} = \dfrac{2S}{t_1 + t_2}$

$$t_1 = \frac{S}{v_1}, t_2 = \frac{S}{v_2}$$

$$t_1 + t_2 = \frac{S}{v_1} + \frac{S}{v_2} = \frac{S(v_1 + v_2)}{v_1 + v_2}$$

$$\frac{2S}{t_1 + t_2} = \frac{2v_1 v_2}{v_1 + v_2}$$

79. (b)

80. (c)

Range $R = \dfrac{2u^2\cos\theta.\sin\theta}{g}$

Maximum height, $H = \dfrac{u^2\sin^2\theta}{2g}$

$$R = 2H$$

$$\frac{2u^2\cos\theta.\sin\theta}{g} = \frac{2(u^2\sin^2\theta)}{2g}$$

$$\tan\theta = 2$$

$$\tan\theta = \frac{b}{a} = 2$$

$$\Rightarrow \quad b = 2a$$

81. (c) At highest point

82. (b)

$$v^2 = 2gh$$

$$\left(\frac{v}{2}\right)^2 - \left(\frac{v}{3}\right)^2 = 2gh'$$

$$v^2 = \frac{36}{5}\times 2gh'$$

$$2gh = \frac{36}{5}2gh' \Rightarrow h' = \frac{5}{36}h$$

83. (c)

$$v^2 - 0^2 = 2\alpha x_1$$

$$0 - v^2 = -2\beta x_2$$

$$v^2 = 2\beta x_2$$

$$x = x_1 + x_2$$

$$x = \frac{v^2}{2\alpha} + \frac{v^2}{2\beta}$$

$$x = \frac{v^2}{2}\left[\frac{\alpha+\beta}{\alpha\beta}\right]$$

$$\sqrt{\frac{2\alpha\beta}{\alpha+\beta}}x = v$$

84. (d)

$$\text{Average speed} = \frac{\text{Distance}}{\text{Time}} = \frac{30\text{ m}}{10\text{ s}} = 3\text{ m/s}$$

Displacement is zero, as the particle comes back to starting position.

85. (a)

$$v = 144\text{ km/h} = 40\text{ m/s}$$

$$h = \frac{1}{2}gt^2$$

$$1000 = \frac{1}{2}\times 10t^2$$

$$t = 10\sqrt{2}$$

$$s = ut = 40\times 10\sqrt{2} = 400\sqrt{2} \simeq 570\text{ m}$$

86. (b)

30° and 60° are supplementary, therefore their ranges are equal $R_A = R_C$

Maximum range occurs at 45° Therefore R_B is greatest and $R_A = R_C < R_B$

87. (c)

Maximum height attained by a body of mass m_1 is

$$h = \frac{u^2}{2g} \quad \ldots\text{(i)}$$

If S is the maximum distance travelled by the body of mass m_2 along the inclined plane is

$$S = \frac{v^2 - u^2}{2a} = \frac{0 - u^2}{2-(-2g\sin 30°)} = \frac{u^2}{g} \quad \ldots\text{(ii)}$$

From (i) and (ii), $S = 2h$

88. (a)

A body thrown up, on reaching maximum height, will have velocity zero but acceleration g.

89. (d)

$R_1 = R_2$ therefore if $\theta_1 = \pi/3$

$$\theta_2 = \left(\frac{\pi}{2} - \frac{\pi}{3}\right) = \pi/6$$

$$y_1 = \frac{u^2\sin^2\left(\frac{\pi}{3}\right)}{2g} = \frac{u^2}{2g}\left(\frac{\sqrt{3}}{2}\right)^2$$

$$y_2 = \frac{u^2\sin^2\left(\frac{\pi}{6}\right)}{2g} = \frac{u^2\left(\frac{1}{2}\right)^2}{2g}$$

$$\frac{y_2}{y_1} = \frac{1}{3} \Rightarrow y_2 = \frac{1}{3}y_1$$

90. (a)

Let θ be angle of projection

Range $= v_0\cos\theta\cdot T$

Where T is time of flight

During the time T the distance covered by the person $= \frac{v_0}{2}\cdot T$

$$V_0\cos\theta\, T = \frac{v_0}{2}T$$

$$\Rightarrow \cos\theta = \frac{1}{2} \Rightarrow \theta = 60°$$

91. (b)

Let u_1, u_2, u_3 and u_4 be the velocities at time $t = 0$, t_1, $(t_1 + t_2)$ and $(t_1 + t_2 + t_3)$ respectively and acceleration is a. Then,

$$v_1 = \frac{u_1 + u_2}{2}, v_2 = \frac{u_2 + u_3}{2} \text{ and } v_3 = \frac{u_3 + u_4}{2}$$

$$u_2 = u_1 + at_1$$

$$u_3 = u_1 + a(t_1 + t_2)$$

$$u_4 = u_1 + a(t_1 + t_2 + t_3)$$

$$v_1 - v_2 = \frac{u_1 - u_3}{2} = \frac{-a(t_1 + t_2)}{2}$$

$$v_2 - v_3 = \frac{-a(t_1 + t_2)}{2}$$

$$\frac{v_1 - v_2}{v_2 - v_3} = \frac{t_1 + t_2}{t_2 + t_3}$$

92. (b)

$$S_1 = \frac{1}{2}\times a\times 10^2 = 50a\text{ m}$$

$$S_1 + S_2 = \frac{1}{2}a\times 20^2 = 200a\text{ m}$$

$$\Rightarrow \quad S_2 = 150a\text{ m}$$

$$S_2 = 3S_1, S_1 = \frac{S_2}{3}$$

93. (d)

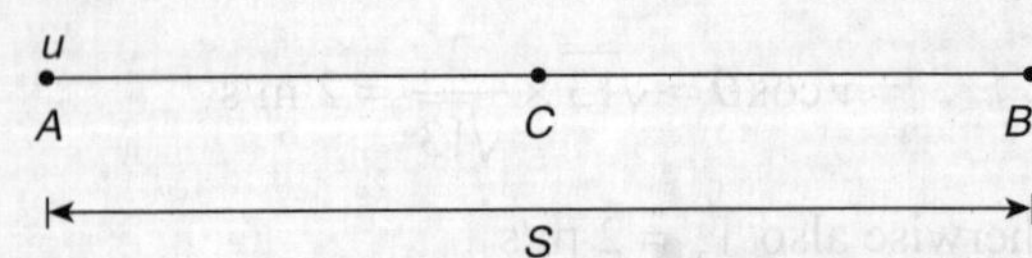

Let S be the distance between AB and a be the constant acceleration of a particle. Then,

$$v^2 - u^2 = 2aS$$

or $$aS = \frac{v^2 - u^2}{2} \qquad \ldots(i)$$

Let v_c be velocity of a particle at midpoint C

$$\therefore v_c^2 - u^2 = 2a(\frac{S}{2})$$

$$v_c^2 = u^2 + aS = u^2 + \frac{v^2 - u^2}{2}$$

$$v_c = \sqrt{\frac{u^2 + v^2}{2}}$$

94. (b)

$$x = a_0 + a_1 t + a_2 t^2$$

$$v = \frac{dx}{dt} = a_1 + 2a_2 t$$

$$a = \frac{dv}{dt} = \frac{d^2x}{dt^2} = 2a_2$$

95. (a)

$$\frac{dx}{dt} = 2at, \quad \frac{dy}{dt} = 2bt$$

Speed, $$v = \sqrt{\left(\frac{dx}{dt}\right)^2 + \left(\frac{dy}{dt}\right)^2} = 2t\sqrt{a^2 + b^2}$$

96. (d)

$$v = kt \qquad \frac{dS}{dt} = kt$$

$$\int_0^S dS = k\int_0^3 t dt = k \cdot \frac{t^2}{2}$$

$$S = k.\frac{9}{2}\text{m} = 9\text{ m as } k = 2\text{ m}$$

97. (c)

$$\frac{dv}{dt} = \text{acceleration} = at$$

$$\int_u^v dv = a\int_0^t t\,dt$$

$$v - u = \frac{at^2}{2}$$

$$v = u + \frac{at^2}{2}$$

98. (b)

Let the speed of cars be v_1 and v_2

$$v_1 + v_2 = 8 \qquad \ldots(i)$$

$$v_1 - v_2 = 0.8 \qquad \ldots(ii)$$

Solving (i) and (ii) we get,

$$2v_1 = 8.8$$

$$v_1 = 4.4\text{ m/s}$$

$$v_2 = 8 - 4.4 = 3.6\text{ m/s}$$

99. (c)

$$v_m \sin\theta = \frac{0.4}{t} \qquad (1)$$

$$v_m \cos\theta + v_r = \frac{0.4}{t} \qquad (2)$$

From (1) and (2),

$$v_m \sin\theta = v_m \cos\theta + v_r$$

$$5\sin\theta = 5\cos\theta + 1$$

$$5\sqrt{1 - \cos^2\theta} = 5\cos\theta + 1$$

$$25(1 - \cos^2\theta) = 25\cos^2\theta + 1 + 10\cos\theta$$

$$50\cos^2\theta + 10\cos\theta - 24$$

$$\cos\theta = \frac{3}{5}, -\frac{4}{5}$$

$$\theta = 53°$$

100. (a)

101. (a)

102. (c)

$$v = \frac{dx}{dt} = 4t + 1$$

$$a = \frac{d^2x}{dt^2} = 4\text{ m/s}^2 \text{ independent of time.}$$

103. (c)

104. (b)

If u is the initial velocity of a particle while going vertically upwards, then maximum height attained, $h = \frac{u^2}{2g}$

Let t be the time when particle reaches at a height $\left(\frac{3}{4}h\right)$

$$\therefore \quad \frac{3}{4}h = ut - \frac{1}{2}gt^2$$

$$\frac{3}{4}\left(\frac{u^2}{2g}\right) = ut - \frac{1}{2}gt^2 \qquad \left(\because h = \frac{u^2}{2g}\right)$$

$$\Rightarrow \quad t^2 - \frac{2u}{g}t + \frac{3}{4}\frac{u^2}{g^2} = 0$$

$$\Rightarrow \quad t = \frac{\frac{2u}{g} \pm \sqrt{\frac{4u^2}{g^2} - 4\times1\times\frac{3}{4}\frac{u^2}{g^2}}}{2}$$

$$\Rightarrow \quad t = \frac{u}{g} \pm \frac{u}{2g}$$

Taking negative sign, $t_1 = \frac{u}{g} - \frac{u}{2g} = \frac{u}{2g}$

Taking positive sign, $t_2 = \frac{u}{g} + \frac{u}{2g} = \frac{3u}{2g}$

$$\therefore \quad \frac{t_1}{t_2} = \frac{1}{3}$$

$$\Rightarrow \quad t_1 : t_2 = 1:3$$

105. (c)
Suppose the dacoit is caught at a time t seconds after the motorcycle starts.
The distance travelled by the motorcycle during this interval is

$$S = \frac{1}{2}at^2$$

During the interval, the jeep travels a distance

$$S + d = vt$$

$$\Rightarrow \quad \frac{1}{2}at^2 - vt + d = 0$$

or $$t = \frac{v \pm \sqrt{v^2 2ad}}{a}$$

The dacoit will be caught if t is real and positive

$$\Rightarrow \quad v^2 - 2ad \geq 0$$

$$\Rightarrow \quad v \geq \sqrt{2ad}$$

106. (c)

Max height, $H = \frac{u^2 \sin^2\theta}{2g}$ ...(i)

Range, $R = u\cos\theta \cdot \frac{2u\sin\theta}{g}$ ...(ii)

$$\frac{2u^2\cos\theta.\sin\theta}{g} = \frac{u^2}{2g}\sin^2\theta$$

$$\Rightarrow \quad \tan\theta = 4$$

$$\theta = \tan^{-1} 4$$

107. (d)
Distance covered in nth second

$$[1+(n-1)2]m = (2n-1)m$$

$$[\text{n}^{\text{th}}\text{term in A.P.}, T_n = a+(n-1)d]$$

$\therefore$ Total distance covered at the end of n seconds

$$= \Sigma(2n-1) = 2\Sigma n - \Sigma 1$$

$$= 2\left[\frac{n(n+1)}{2}\right] - n$$

$$= n(n+1) - n = n^2$$

108. (a)
Before the ball strikes the ground for the first time, it falls a distance of 2 m.
After the first strike, it rises and falls, each with a height of (0.8×2) m
After second strike, it rises and falls, through distances of 0.8 (0.8 m each and so on.
$\therefore$ Total distance covered

$$= 2 + 2\left[0.8\times2 + 0.8^2\times2 + ...\infty\right]$$

$$= 2 + 4\times0.8\left[1 + 0.8 + 0.8^2 + ...\infty\right]$$

$$= 2 + 3.2\left[\frac{1}{1-0.8}\right]\left[\because \text{In G.P.}, S_\infty = \frac{a}{a-r}\right]$$

$$= 2 + \frac{3.2}{0.2} = 2 + 16 = 18 \text{ m}$$

109. (c)

110. (a)
Let us consider equal intervals of time, each having value t.

u = initial velocity

a = uniform acceleration

Distance travelled during the first n intervals

$$S_1 = u(nt) + \frac{1}{2}(nt)^2 a$$

Distance travelled during the first $(n-1)$ intervals

$$S_2 = u(n-1)t + \frac{1}{2}a(n-t)^2 t^2$$

Distance travelled in the nth time interval

$$= S_2 - S_1 = ut + \frac{1}{2}a(2n-t)t^2$$

Putting $n = 1, 2, 3, ..,$ we get distance travelled in 1st, 2nd, 3rd,.... intervals to be $ut + \frac{1}{2}at^2, ut + \frac{3}{2}at^2, ut + \frac{5}{2}at^2, ..$
Thus, distance described in consecutive equal intervals of time are in A.P. with common difference at^2.

111. (d)

The distance covered by b the cars is same

$$S_1 = S_2 = S$$

If the cars take time t_1 and t_2 for the race and their velocities at the finish be v_1 and v_2, then

$$\frac{v_1}{2}t_1 = S \text{ and } \frac{v_2}{2}t_2 = S \qquad [\because S = v_{av}\cdot t]$$

$$\frac{v_1}{t_2} = \frac{v_2}{t_1} = \frac{v_1 - v_2}{t_2 - t_1} = \frac{v}{t}$$

$$\left(\frac{v}{t}\right)^2 = \frac{v_1}{t_2}.\frac{v_2}{t_1} = \frac{v_1}{t_1}.\frac{v_2}{t_2} = \frac{v}{t}$$

$$\Rightarrow \quad v = t\sqrt{a_1 a_2}$$

112. (b)

Let us choose the coordinate system x axis along OB and y axis along OA.

$$u_x = 10\sqrt{3}, u_y = 0$$

$$a_x = -g\sin 60° = -5\sqrt{3} \text{ m/s}^2$$

$$a_y = -g\cos 60° = -10\times\frac{1}{2} = -5 \text{ m/s}^2$$

At point Q, x component of velocity is zero. Therefore,

$$v_x = u_x + a_x t$$

$$0 = 10\sqrt{3} - 5\sqrt{3}t$$

$$\Rightarrow \quad t = 2 \text{ s}$$

113. (c)

At point Q,

$$v_y = u_y + a_y t$$

$$v_y = 0 - 5\times 2 = -10 \text{ m/s}$$

114. (b)

$$S_y = u_y t + \frac{1}{2}a_y t^2$$

$$= 0 - \frac{1}{2}\times 5\times 2^2 = -10 \text{ m}$$

$$\Rightarrow PO = 10 \text{ m}$$

$$h = 10\sin 30° = 10\times\frac{1}{2} = 5 \text{ m}$$

115. (d)

$$OQ = u_n t + \frac{1}{2}a_n t^2$$

$$= (10\sqrt{3})\times 2 + \frac{1}{2}(5\sqrt{3})\times 2^2 = 10\sqrt{3}$$

$$PQ = \sqrt{PO^2 + (OQ)^2} = \sqrt{(10)^2 + (10\sqrt{3})^2} = 20 \text{ m}$$

116. (b)

117. (b)

The length of train $= S$

$v =$ Final velocity after travelling a distance S

$u =$ Initial velocity

Let $a =$ acceleration of train

$$v^2 = u^2 + 2aS \Rightarrow a = \frac{v^2 - u^2}{2S}$$

Let v_1 = Velocity of the train after travelling a distance $\frac{S}{2}$

$$v_1^2 = u^2 + 2a\left(\frac{S}{2}\right)$$

$$\Rightarrow \quad v_1^2 = u^2 + \frac{v^2 - u^2}{2S}(S)$$

$$\Rightarrow \quad v_1^2 = u^2 + \frac{v^2 - u^2}{2}$$

$$\Rightarrow \quad v_1^2 = \frac{v^2 + u^2}{2}$$

$$\Rightarrow \quad v_1 = \sqrt{\frac{v^2 + u^2}{2}}$$

118. (d)

$$a = v\frac{dv}{dx} = 0.2 - 20x$$

$$\Rightarrow \quad vdv = (0.02 - 20x)dx$$

Integrating both sides,

$$\frac{v^2}{2}. = 0.02x - 10x^2 + C_1$$

given $\quad v = 10 \text{ m/s at } x = 1 \text{ m}$

$$50 = 0.02 - 10 + C_1$$

$$\Rightarrow \quad C_1 = 59.98$$

$$v^2 = 0.04x - 20x^2 + 119.9$$

Velocity when

$$x = 0$$

$$v = \sqrt{120} \text{ m/s}$$

119. (c)

$$a = kt$$

$$\frac{dv}{dt} = kt$$

$$dv = kt\,dt$$

Integrating,

$$v = \frac{kt^2}{2} + C_1$$

$$v = u \text{ at } t = 0$$

$$u = 0 + C_1$$

$$\Rightarrow \quad C_1 = u$$

$$v = \frac{dS}{dt} = \frac{kt^2}{2} + u$$

$$dS = \left(u + \frac{kt^2}{2}\right)dt$$

Integrating both sides,

$$S = ut + \frac{1}{6}kt^3 + C_2$$

at $t = 0, S = 0 \therefore C_2 = 0$

$$S = ut + \frac{1}{6}kt^3$$

120. (a)

Let the ball be at height h at time t_1 and time t_2.

$$h = ut_1 - \frac{1}{2}gt_1^2$$

and

$$h = ut_2 - \frac{1}{2}gt_2^2$$

$$\therefore ut_1 - \frac{1}{2}gt_1^2 = ut_2 - \frac{1}{2}gt_2^2$$

$$ut_2 - ut_1 = \frac{1}{2}gt_2^2 - \frac{1}{2}gt_1^2$$

$$\Rightarrow \quad u(t_2 - t_1) = \frac{1}{2}g(t_2 + t_1)(t_2 - t_1)$$

$$\Rightarrow \quad u = \frac{g}{2}(t_2 + t_1)$$

$$\Rightarrow \quad t_2 + t_1 = \frac{2u}{g} \qquad [\because t_2 - t_1 = t]$$

$$\Rightarrow \quad t + t_1 + t_1 = \frac{2u}{g}$$

$$\Rightarrow \quad 2t_1 = \frac{2u}{g} - t$$

$$\Rightarrow \quad t_1 = \frac{u}{g} - \frac{t}{2} = \frac{2u - gt}{2g}$$

Substituting in $h = ut_1 - \frac{1}{2}gt_1^2$

$$\Rightarrow \quad h = u\left(\frac{2u - gt}{2g}\right) - \frac{1}{2}g\frac{(2u - gt)^2}{4g^2}$$

$$\Rightarrow \quad h = \frac{8u^2 - 4ugt - 4u^2 - g^2t^2 + 4ugt}{8g}$$

$$\Rightarrow \quad h = \frac{4u^2 - g^2t^2}{8g}$$

$$\Rightarrow \quad u = \frac{1}{2}\sqrt{8gh + g^2t^2}$$

121. (b)

Graph of $v - t$.

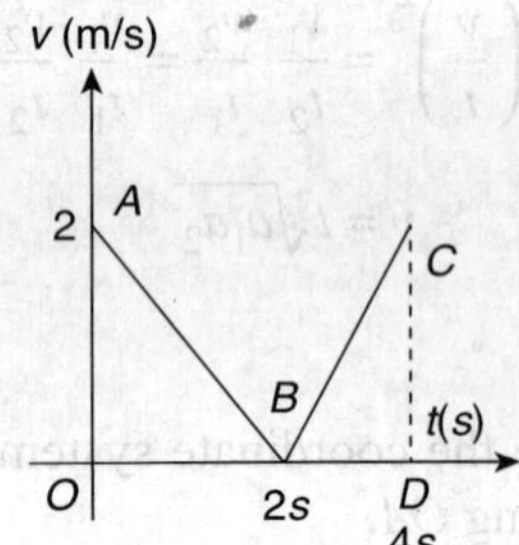

Area under the curve is the distance travelled.

$$S = ar(\Delta AOB) + ar(\Delta BCD)$$

$$= \frac{1}{2} \times 2 \times 2 + \frac{1}{2} \times 2 \times 2 = 4 \text{ m}$$

122. (b)

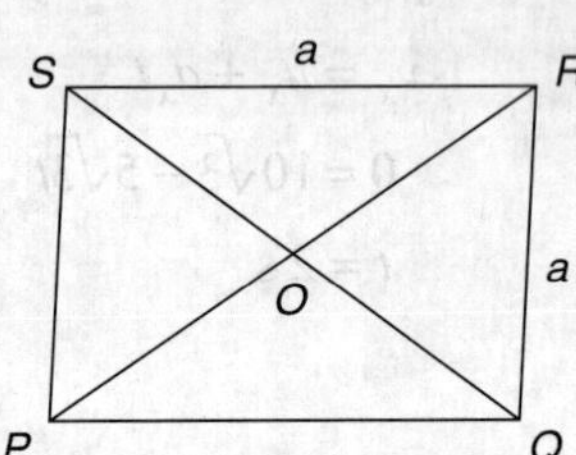

The four persons describe curvilinear motion to meet at O, the centre of the square.

Displacement $= PO = QO = RO = SO$

$$= \frac{1}{2}\sqrt{a^2 + a^2} = \frac{a}{\sqrt{2}}.$$

Velocity component with which each moves

$$= v\cos 45° = \frac{v}{\sqrt{2}}.$$

$$t = \frac{\frac{a}{\sqrt{2}}}{\frac{v}{\sqrt{2}}} = \frac{a}{v}$$

123. (d)

$$v_t = 4t^3 - 2t$$

$$\Rightarrow \quad \frac{dx_t}{dt} = 4t^3 - 2t$$

$$\Rightarrow \quad \int dx_t = \int 4t^3 dt - \int 2t\, dt$$

$$\Rightarrow \quad x_t = t^4 - t^2$$

Given $\quad x_t = 2$ m

$$t^4 - t^2 - 2 = 0$$

$$\Rightarrow \quad t = \sqrt{2} \text{ s}$$

$$a_t = \frac{dv_t}{dt} = 12t^2 - 2$$

$\therefore$ Acceleration at $t = \sqrt{2}$ s

$$a = 12(\sqrt{2})^2 - 2 = 24 - 2 = 22 \text{ m/s}^2$$

124. (d)

$$v = \frac{dx}{dt} = 6t^2 - 42t + 60$$

$$v = 0$$

$$\Rightarrow \quad t^2 - 7t + 10 = 0$$

$$\Rightarrow \quad t = 5 \text{ s or } t = 2 \text{ s}.$$

$$a = \frac{dv}{dt} = 12t - 42$$

at $\quad t = 5$ s; $a = 12 \times 5 - 42 = 18$ m/s^2

at $\quad t = 2$ s; $a = 24 - 42 = -18$ m/s^2

125. (b)

$$a = \frac{dv}{dt} \times \frac{dx}{dx}$$

$$\Rightarrow \quad \frac{vdv}{dx} = \frac{P}{mv}$$

$$\Rightarrow \quad \int_{v_1}^{v_2} v^2 dv = \int_0^x \frac{P}{m} dx$$

$$\Rightarrow \quad \left[\frac{v^3}{3}\right]_{v_1}^{v_2} = \frac{P}{m} x$$

$$\Rightarrow \quad x = \frac{m}{3P}(v_2^3 - v_1^3)$$

$$a = \frac{dv}{dt} \times \frac{dx}{dx}$$

$$= \frac{vdv}{dx}$$

126. (a)

$$x^2 = t^2 + 1$$

Differentiating,

$$2x\frac{dx}{dt} = 2t \Rightarrow v = \frac{dx}{dt} = \frac{t}{x}$$

$$a = \frac{dv}{dt} = \frac{x - t\dfrac{dx}{dt}}{x^2} = \frac{x - \dfrac{t^2}{x}}{x^2}$$

$$\Rightarrow \quad a = \frac{x^2 - t^2}{x^3} = \frac{1}{x^3} \qquad (\because x^2 - t^2 = 1)$$

127. (b)

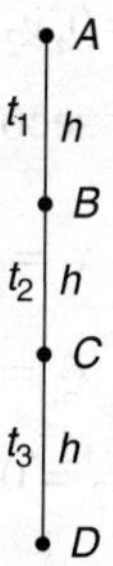

For $\quad AB = h = \frac{1}{2} g t_1^2 \quad$ ...(i)

For $\quad AC = 2h = \frac{1}{2} g(t_1 + t_2)^2 \quad$ ...(2)

For $\quad AD = 3h = \frac{1}{2} g(t_1 + t_2 + t_3)^2 \quad$ ...(3)

where t_1, t_2 and t_3 are times taken to cover the distances *AB*, *BC* and *CD* respectively.

From (1) and (2),

$$\frac{(t_1 + t_2)^2}{t_1^2} = 2$$

$$\Rightarrow \quad \left(1 + \frac{t_2}{t_1}\right)^2 = 2$$

$$\Rightarrow \quad 1 + \frac{t_2}{t_1} = \sqrt{2}$$

$$\Rightarrow \quad \frac{t_2}{t_1} = \sqrt{2} - 1$$

or $\quad \frac{t_1}{t_2} = 1 : (\sqrt{2} - 1)$

From (3) and (1),

$$\frac{(t_1 + t_2 + t_3)^2}{t^2} = 3$$

$$\Rightarrow \quad 1 + \frac{t_2}{t_1} + \frac{t_3}{t_1} = \sqrt{3}$$

$$\Rightarrow \quad 1 + \sqrt{2} - 1 + \frac{t_3}{t_1} = \sqrt{3}$$

$$\Rightarrow \quad \frac{t_1}{t_3} = \frac{1}{\sqrt{3}-\sqrt{2}}$$

$$\therefore \quad t_1 : t_2 : t_3 = 1:\left(\sqrt{2}-\sqrt{1}\right):\left(\sqrt{3}-\sqrt{2}\right)$$

128. (c)

$$u_1 = 0, v_1 = at$$
$$u_2 = 0, v_2 = v$$

(Let constant speed of second particle = v)

$$x_1 = \frac{1}{2}at^2$$
$$x_2 = vt$$

at $t = 0$; $x_1 = 0$ and $x_2 = 0 \Rightarrow x_1 - x_2 = 0$

at $t = t$;

$$x_1 - x_2 = \frac{1}{2}at^2 - vt = 0$$

$$\Rightarrow (vt - \frac{1}{2}at^2)^2 = 0$$

$$\Rightarrow \quad t = 0, \frac{2v}{a}$$

$x_1 - x_2$ is zero at $t = 0$,

After time at $t = \frac{2v}{a}, x_1 - x_2 = 0$,

In between $t = 0$ to $t = \frac{2v}{a}$, $x_1 - x_2$ is negative.

At time $t > \frac{2v}{a}$, $x_1 - x_2$ is positive.

Equation of $x_1 - x_2 = \frac{1}{2}at^2 - vt$ is a parabola.

The correct option is (c) .

129. (c)

$$v = v_0 + gt + ft^2$$

$$\frac{dx}{dt} = v_0 + gt + ft^2$$

$$\Rightarrow \quad \int_0^x dx = \int_0^t (v_0 + gt + ft^2)dt$$

$$\Rightarrow \quad x = v_0t + \frac{gt^2}{2} + \frac{ft^3}{3}$$

at $t = 1$, $\quad x = v_0 + \frac{g}{2} + \frac{f}{3}$.

130. (b)

$$v = \alpha\sqrt{x}$$

$$\frac{dx}{dt} = \alpha\sqrt{x}$$

$$\Rightarrow \quad \int_0^x \frac{dx}{\sqrt{x}} = \int_0^t \alpha dt$$

$$\Rightarrow \quad 2x^{1/2} = \alpha t$$

$$x = \left(\frac{\alpha}{2}\right)^2 t^2$$

$\therefore$ Displacement is proportional to t^2.

131. (c)

$$S = ut + \frac{1}{2}gt^2$$

$$\therefore \quad h = 0 + \frac{1}{2}gT^2$$

$$\Rightarrow \quad gT^2 = 2h$$

At $\quad t = \frac{T}{3}\text{s}, h_1 = \frac{1}{2}gt^2 = \frac{1}{2}g\left(\frac{T}{3}\right)^2$

$$\Rightarrow \quad h_1 = \frac{1}{18}gT^2$$

$$\Rightarrow \quad h_1 = \frac{1}{18}2h = \frac{h}{9} \text{ from the top.}$$

$\therefore$ Height from the ground $= h - \frac{h}{9} = \frac{8h}{9}$ m

132. (b)

From the given data, range is same for angles of projection θ and $(90° - \theta)$.

$\therefore$ Time of flight in first case $t_1 = \frac{2u\sin\theta}{g}$

Time of flight in second case $t_2 = \frac{2u\sin(90° - \theta)}{g}$

Product of time of flights

$$(t_1t_2) = \left(\frac{2u\sin\theta}{g}\right)\left(\frac{2u\sin(90° - \theta)}{g}\right)$$

$$= \frac{2}{g}\left(\frac{u^2\sin 2\theta}{g}\right)$$

$$t_1t_2 = \frac{2R}{g} \quad \left(\because R - \frac{u^2\sin 2\theta}{g}\right)$$

$\therefore$ t_1t_2 is proportional to R

133. (b)

The acceleration vectors acts along the radius of the circle. The given statement is false.

134. (a)

$$\vec{A}\times\vec{B}=\vec{B}\times\vec{A}$$

$$AB\ \sin\ \theta\ \hat{n}=AB\ \sin(-\theta)\ \hat{n}$$

$$\sin\theta=-\sin\theta$$

$$2\sin\theta=0$$

$$\Rightarrow \quad \theta=0,\pi,2\pi.....$$

$$\theta=\pi$$

135. (b)

$$x=\alpha t^3$$

$$\frac{dx}{dt}=3\alpha t^2$$

$$\Rightarrow \quad v_x=3\alpha t^2$$

Again $\quad y=\beta t^3$

$$\Rightarrow \quad \frac{dy}{dt}=v_y=3\beta t^2$$

$$v^2=\left\{v_x{}^2+v_y{}^2\right\}$$

$$v^2=(3\alpha t^2)^2+(3\beta t^2)^2$$

$$v^2=(3t^2)^2\left[\alpha^2+\beta^2\right]$$

$$v=3t^2\sqrt{\alpha^2+\beta^2}$$

136. (d)

Given height of building = 10 m

The ball projected from the roof of building will be back to roof height of 10 m after covering the maximum horizontal range.

Maximum horizontal range $R=\dfrac{u^2\sin 2\theta}{g}$

$$R=\frac{(10)^2\times\sin 60°}{10}=10\times 0.866$$

$$R=8.66\text{ m}$$

137. (d)

$$(\vec{B}\times\vec{A}).\vec{A}=BA\ \cos\theta\hat{n}.\ \vec{A}=0$$

Here $\hat{n}$ is perpendicular to both $\vec{A}$ and $\vec{B}$

Alternative: $(\vec{A}\times\vec{B}).\vec{A}$

Interchange the cross and dots,

$$(\vec{B}\times\vec{A}).\vec{A}=\vec{B}.(\vec{A}\times\vec{A})=0 \qquad (\because \vec{A}\times\vec{A}=0)$$

138. (d)

For vertically upward motion of a projectile,

$$y=(u\ \sin\ \alpha)t-\frac{1}{2}gt^2$$

or $\quad h=(u\ \sin\ \alpha)t-\dfrac{1}{2}gt^2$

$$gt^2-(2u\ \sin\ \alpha)t+2h=0$$

$$\therefore \quad t=\frac{2u\ \sin\ \alpha\pm\sqrt{(4u^2\ \sin^2\ \alpha)-8gh}}{2g} \qquad ...(i)$$

If two roots of the quadratic equation (i) are t_1,t_2, then

$$t_1=\frac{2u\ \sin\ \alpha+\sqrt{(4u^2\ \sin^2\ \alpha)-8gh}}{2g}$$

and $\quad t_2=\dfrac{2u\ \sin\ \alpha-\sqrt{(4u^2\ \sin^2\ \alpha)-8gh}}{2g}$

If particle crosses the wall at times t_1 and t_2 respectively, then the time of flight is

$$t=\sqrt{t_1t_2}$$

or $\quad t^2=t_1t_2$

$$\therefore \quad \left(\frac{2u\ \sin\ \alpha}{g}\right)^2=\frac{(2u\ \sin\ \alpha)^2-(4u^2\sin^2\alpha-8gh)}{4g^2}$$

or $\quad \dfrac{4u^2\sin^2\alpha}{4g^2}=\dfrac{8gh}{4g^2}$

or $\quad 2u^2\sin^2\alpha=gh \qquad (\because u=\sqrt{2gh})$

$$2(2gh)\sin^2\alpha=gh$$

$$\sin^2\alpha=\frac{1}{4}$$

or $\quad \sin\alpha=\dfrac{1}{2}$

$$\Rightarrow \quad \alpha=30°.$$

139. (b)

As shown in figure, distance OB is the range R. Thus applying angle of projection,

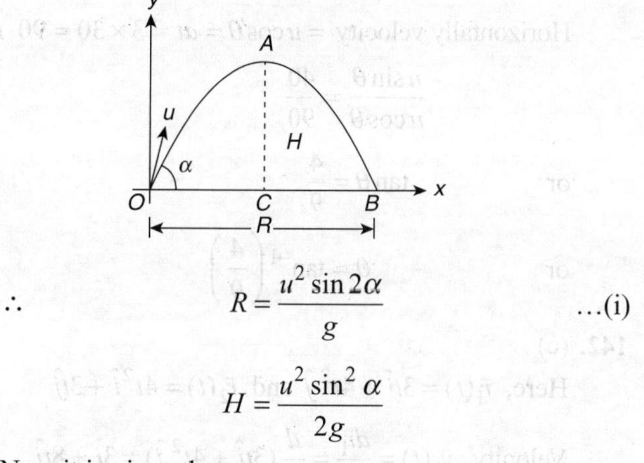

$$\therefore \quad R=\frac{u^2\sin 2\alpha}{g} \qquad ...(i)$$

$$H=\frac{u^2\sin^2\alpha}{2g}$$

Now it is given that,

$$(\text{Range})^2=48(\text{maximum height})^2$$

$$\therefore \quad \left(\frac{u^2 \sin 2\alpha}{g}\right)^2 = 48\left(\frac{u^2 \sin^2 \alpha}{2g}\right)^2$$

or $$\frac{u^2 \sin 2\alpha}{g} = 4\sqrt{3}\left(\frac{u^2 \sin^2 \alpha}{2g}\right)$$

or $$\frac{2\sin\alpha\cos\alpha}{4\sqrt{3}} = \frac{\sin^2 \alpha}{2}$$

or $$\tan\alpha = \frac{4}{4\sqrt{3}} = \frac{1}{\sqrt{3}}$$

$$\therefore \quad \alpha = 30°.$$

140. (a)

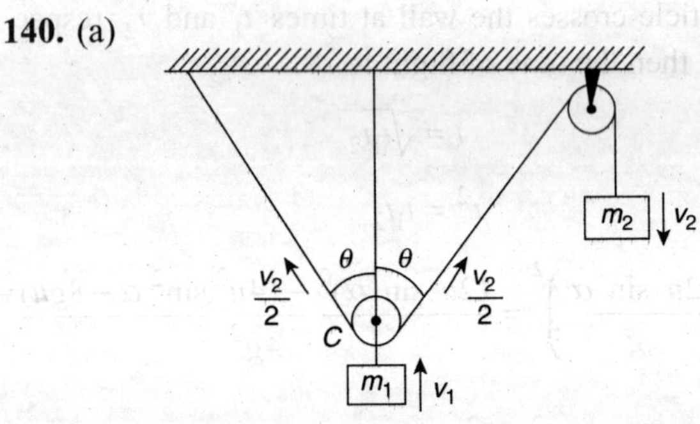

$$2v_1 \cos\theta = v_2$$

$$\cos\theta = \frac{v_2}{2v_1}$$

$$\theta = \cos^{-1}\frac{v_2}{2v_1}$$

141. (c)

$$H = \frac{u^2 \sin^2 \theta}{2g}$$

or $$80 = \frac{u^2 \sin^2 \theta}{2\times 10}$$

or $$u^2 \sin^2 \theta = 1600$$

or $$\sin\theta = 40 \text{ ms}^{-1}$$

Horizontally velocity $= u\cos\theta = at = 3\times 30 = 90 \text{ ms}^{-1}$

$$\frac{u\sin\theta}{u\cos\theta} = \frac{40}{90}$$

or $$\tan\theta = \frac{4}{9}$$

or $$\theta = \tan^{-1}\left(\frac{4}{9}\right)$$

142. (c)

Here, $\vec{r}_1(t) = 3t\hat{i} + 4t^2\hat{j}$ and $\vec{r}_2(t) = 4t^2\hat{i} + 3t\hat{j}$

Velocity, $\vec{v}_1(t) = \frac{d\vec{r}_1}{dt} = \frac{d}{dt}(3t\hat{i} + 4t^2\hat{j}) = 3\hat{i} + 8t\hat{j}$

$$\vec{v}_2(t) = \frac{d\vec{r}_2}{dt} = \frac{d}{dt}(4t^2\hat{i} + 4t\hat{j}) = 8t\hat{i} + 3j$$

The relative speed of particle 1 with respect to particle 2 is

$$\vec{v}_{12} = \vec{v}_1 - \vec{v}_2$$
$$= (3\hat{i} + 8t\hat{j}) - (8t\hat{i} + 3\hat{j}) = (3-8t)\hat{i} + (8t-3)\hat{j}$$

At $i = 1$s,

$$\vec{v}_{12} = (3-8)\hat{i} + (8-3)\hat{j} = -5\hat{i} + 5\hat{j}$$

$$|\vec{v}_{12}| = \sqrt{(-5)^2 + (5)^2} = \sqrt{25+25} = 5\sqrt{2} \text{ m/s}$$

143. (d)

Three vectors are co-planar when their triple product is zero.

$$\therefore A\times B\times C = \begin{vmatrix} 1 & 1 & -2 \\ 2 & 2 & -1 \\ -1 & \alpha & +1 \end{vmatrix} = 0$$

$$= 1(2+\alpha) - 1(2-1) - 2(2\alpha+2) = 0$$
$$= 2+\alpha-1-4\alpha-4 = 0$$
$$= -3\alpha - 3 = 0$$

or $$\alpha = -1$$

144. (a)

Time taken by the object to fall through 45 m is $t = \sqrt{\frac{2h}{g}}$

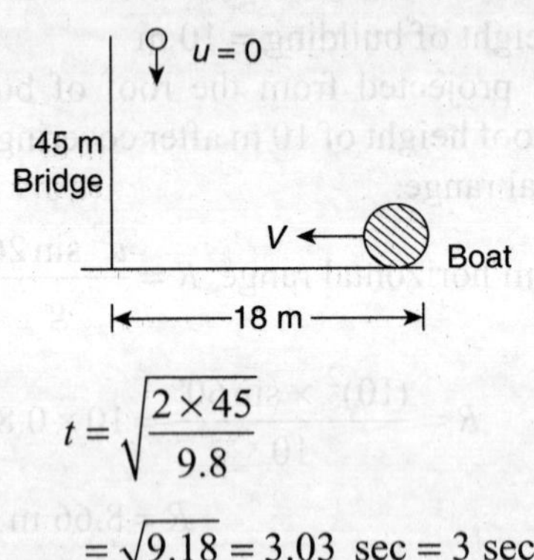

$$t = \sqrt{\frac{2\times 45}{9.8}}$$

$$= \sqrt{9.18} = 3.03 \text{ sec} = 3 \text{ sec}$$

If v is the speed of boat, then the distance travelled by boat in $t = 3$ sec is,

$$18 \text{ m} = v\times 3 \text{ sec}$$

$$v = 6 \text{ m/s}.$$

145. (a)

Range (horizontal distance) of projectile is

$$R = \frac{u^2 \sin 2\theta}{g}$$

$$= \frac{(16)^2 \times \sin 2(15^\circ)}{g}$$

$$= \frac{16 \times 16 \times \sin 30^\circ}{9.8}$$

$$= \frac{16 \times 16}{9.8} \times \frac{1}{2} = \frac{16 \times 8}{9.8} = \frac{128}{9.8} = 13.06 \text{ m}$$

$R = 13$ m.

146. (b)

Given: $\overrightarrow{V_{BG}} = 3.5\hat{i} + 9.2\hat{j}$

$$\overrightarrow{V_{IG}} = -2.5\hat{i} + 1.8\hat{j}$$

$$\overrightarrow{V_{IG}} = \overrightarrow{V_{IB}} + \overrightarrow{V_{BG}}$$

$$\overrightarrow{V_{IB}} = \overrightarrow{V_{IG}} - \overrightarrow{V_{BG}}$$

$$= (-2.5\hat{i} + 1.8\hat{j}) - (3.5\hat{i} + 9.2\hat{j})$$

$$= -6\hat{i} - 7.4\hat{j}$$

147. (d)

$$A \times B = \begin{vmatrix} \hat{i} & \hat{j} & \hat{k} \\ 1 & 1 & 0 \\ 1 & 0 & 1 \end{vmatrix}$$

$$= \hat{i} - \hat{j} - \hat{k}$$

148. (c)

Displacement vector from A to B

(i) Along x-axis from 0 to – 2 means $-2\hat{i}$

(ii) Along y-axis from 3 to 6 means $+3\hat{j}$

(iii) Along z-axis from – 1 to 4 means $+5\hat{k}$

Displacement = $-2\hat{i} + 3\hat{j} + 5\hat{k}$

149. (d)

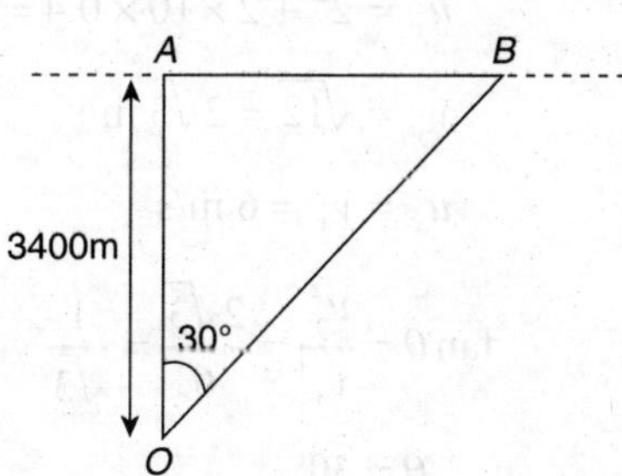

0 is the observation point on the ground. A and B are the positions of aircraft for which $\angle AOB = 30^\circ$. Time taken by aircraft from A to B is 10s.

ΔAOB

$$\tan 30^\circ = \frac{AB}{3400}$$

$$AB = 3400 \tan 30^\circ = \frac{3400}{\sqrt{3}} \text{ m}$$

$\therefore$ Speed of aircraft

$$v = \frac{AB}{10} = \frac{3400}{10\sqrt{3}} = 196.3 \text{ ms}^{-1}$$

150. (b)

Due to induction, the fall of the magnet is opposed by field induced in the metal ring acting in the upward direction.

Hence, the acceleration of the falling magnet is less than g.

151. (a)

$$|\vec{F}| = \sqrt{(6)^2 + (8)^2 + (10)^2} = 10\sqrt{2} \text{ N}$$

$$\therefore \text{ Mass} = \frac{|\vec{F}|}{|\vec{a}|} = \frac{10\sqrt{2}}{1} = 10\sqrt{2} \text{ kg}$$

152. (c)

Let $\vec{B} = \hat{i} - \hat{j}$

The component of $\vec{A}$ along $\vec{B}$ is

$$\vec{A}.\vec{B} = AB\cos\theta$$

$$\therefore \quad A\cos\theta = \frac{\vec{A}.\vec{B}}{B}$$

$$= \frac{(a_x\hat{i} - a_y\hat{j} - a_z\hat{k}) \cdot (\hat{i} - \hat{j})}{\sqrt{1+1}}$$

$$= \frac{a_x - a_y}{\sqrt{2}}$$

153. (b)

$$R = \frac{u^2 \sin 2\theta}{g}$$

$$u^2 = \frac{400}{\sqrt{3}}$$

$$H_{\max} = \frac{u^2 \sin^2\theta}{2g}$$

$$= \frac{400}{\sqrt{3}} \times \frac{1}{4} \times \frac{1}{20} = \frac{5}{\sqrt{3}} \text{ m}$$

154. (d)

$$\vec{r_1} = 4\hat{i} - 4\hat{j} + 7\hat{k}$$

$$\vec{r_2} = 2\hat{i} - 2\hat{j} + 5\hat{k}$$

$$\overrightarrow{r_{12}} = \vec{r_1} - \vec{r_2} = 2\hat{i} - 6\hat{j} + 2\hat{k} \quad \text{(i)}$$

$$\overrightarrow{v_1} = 0.4\hat{i}$$

Let $\overrightarrow{v_2} = a\hat{i} + b\hat{j} + c\hat{k}$

$$\overrightarrow{v_{12}} = \overrightarrow{v_1} - \overrightarrow{v_2} = (0.4 - a)\hat{i} - b\hat{j} - c\hat{k}$$

$$\overrightarrow{r_{12}} = \overrightarrow{v_{12}}t = [(0.4 - a)\hat{i} - b\hat{j} - c\hat{k}]10$$

$$(4 - 10a)\hat{i} - b\hat{j} - 10c\hat{k} \quad \text{(ii)}$$

Comparing the coefficients of $\hat{i}, \hat{j}$ and $\hat{k}$ with Eq. (i)

$$4 - 10a = 2 \Rightarrow a = \frac{1}{5} = 0.2$$

and $-10b = -6$

so, $b = \dfrac{3}{5} = 0.6$

and $c = -\dfrac{1}{5} = -0.2$

$$\overrightarrow{v_2} = 0.2\hat{i} + 0.6\hat{j} - 0.2\hat{k}$$

$$= 0.6\left(\frac{\hat{i}}{3} + \hat{j} - \frac{\hat{k}}{3}\right)$$

155. (a)

$$x = 6t, y = 8t - 5t^2$$

Eliminating t, we get

$$y = \frac{4x}{3} - \frac{5x^2}{36}$$

Put $y = 0$, $x\left(\dfrac{4}{3} - \dfrac{5x}{36}\right) = 0 \Rightarrow x = 9.6$

156. (a)

$$x = 36t$$

$$2y = 96t - 9.8t^2$$

or $y = 48t - \dfrac{9.8}{2}t^2$

$$y = u\sin\alpha t - \frac{1}{2}gt^2$$

$$x = u\cos\alpha$$

$$u\sin\alpha = 48, u\cos\alpha = 36$$

$$u = \sqrt{48^2 + 36^2} = 60$$

$$\sin\alpha = \frac{48}{u} = \frac{48}{60} = \frac{4}{5}$$

$$\Rightarrow \alpha = \sin^{-1}\left(\frac{4}{5}\right)$$

157. (a)

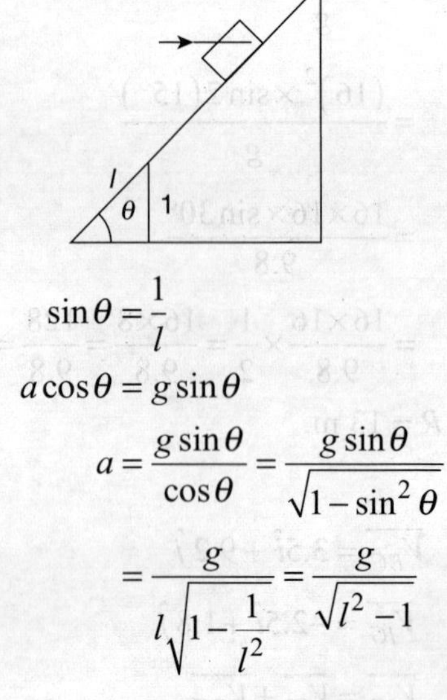

$$\sin\theta = \frac{1}{l}$$

$$a\cos\theta = g\sin\theta$$

$$a = \frac{g\sin\theta}{\cos\theta} = \frac{g\sin\theta}{\sqrt{1 - \sin^2\theta}}$$

$$= \frac{g}{l\sqrt{1 - \frac{1}{l^2}}} = \frac{g}{\sqrt{l^2 - 1}}$$

158. (d)

Since horizontal range is same for both the stones. It is only their angle of projection is θ or $90 - \theta$. So if

$$\theta_1 = \frac{\pi}{3}, \text{ then } \theta_2 = \frac{\pi}{6}$$

$$\frac{h_2}{h_1} = \frac{u^2\sin^2\theta_2}{2g} \times \frac{2g}{u^2\sin^2\theta_1}$$

$$= \frac{\sin^2\theta_2}{\sin^2\theta_1} = \frac{\sin^2\frac{\pi}{6}}{\sin^2\frac{\pi}{3}} = \frac{1}{3}$$

$$h_2 = \frac{h_1}{3}$$

159. (c)

$$u^2 = v^2 + 2gh$$

$$u_y^2 = v_y^2 + 2gh$$

or $u_y = 2^2 + 2 \times 10 \times 0.4 = 12$

$$v_y = \sqrt{12} = 2\sqrt{3} \text{ m/s}$$

and $u_x = v_x = 6 \text{ m/s}.$

$$\tan\theta = \frac{u_y^2}{v_y} = \frac{2\sqrt{3}}{6} = \frac{1}{\sqrt{3}}$$

$\Rightarrow$ $\theta = 30°$

160. (d)

$$u\cos\theta = 36, u\sin\theta = 48$$

$$\tan\theta = \frac{48}{36} = \frac{4}{3}$$

$$\Rightarrow \sin\theta = \frac{4}{5}$$

$$u\sin\theta = 48$$

$$u = 48 \times \frac{5}{4} = 60 \text{ ms}^{-1}$$

161. (d)
If two stones projected with same speed but at different angles has the same range, then sum of angles of projection is

$$\frac{\pi}{2} \Rightarrow \frac{\pi}{3} + \theta = \frac{\pi}{2} \Rightarrow \theta = \frac{\pi}{6}$$

$$H_{\text{max}} = \frac{u^2 \sin^2\theta}{2g}$$

$$102 = \frac{u^2}{2g}\left[\sin^2\frac{\pi}{3}\right]$$

$$102 = \frac{u^2}{2g}\left(\frac{3}{4}\right)$$

$$\Rightarrow \frac{u^2}{2g} = \frac{102\times 4}{3} = 136$$

$$H'_{\text{max}} = \frac{u^2}{2g}\sin^2\left(\frac{\pi}{6}\right)$$

$$= 136 \times \frac{1}{4} = 34 \text{ m}$$

162. (a)

$$(\vec{A}+\vec{B}).(\vec{A}-\vec{B}) = 0$$

$$\Rightarrow (\vec{A})^2 = (\vec{B})^2$$

$$\Rightarrow |A| = |B|$$

163. (b)

$$t = \sqrt{\frac{2h}{g}} = \sqrt{\frac{2\times 490}{9.8}} = 10 \text{ s}$$

$$R = 150 \times 10 = 1500 \text{ m}$$

164. (b)

$$H = \frac{u^2\sin^2\theta}{2g} \text{ and } T = \frac{2u\sin\theta}{g}$$

$$\therefore \quad \frac{T^2}{H} = \left(\frac{2u\sin\theta}{g}\right)^2 \times \frac{2g}{u^2\sin^2\theta} = \frac{8}{g}$$

or $$T^2 = \frac{8H}{g}$$

$$T = 2\sqrt{\frac{2H}{g}}$$

165. (d)

$$\text{Time} = \frac{\text{Length}}{\text{Velocity}} = \frac{150}{10+5} = 10 \text{ s}$$

166. (d)

$$v = 36 \text{ km/h} = 10 \text{ m/s}$$

$$\tan\theta = \frac{v^2}{rg} = \frac{10\times 10}{80\times 10} = \frac{1}{8}$$

$$\Rightarrow \quad \theta = \tan^{-1}\left(\frac{1}{8}\right)$$

167. (d)

$$a = -kv^3$$

$$\Rightarrow \quad \frac{dv}{dt} = -kv^3$$

On integrating,

$$\Rightarrow \quad \int_{v_0}^{v}\frac{dv}{v^3} = -\int_0^t kdt$$

$$\left[\frac{v^{-2}}{-2}\right]_{v_0}^{v} = -kt$$

$$\frac{v^{-2}}{-2} + \frac{v_0^{-2}}{2} = -kt$$

$$\frac{1}{v_0^2} + 2kt = \frac{1}{v^2}$$

$$\frac{1+2ktv_0^2}{v_0^2} = \frac{1}{v^2}$$

$$v = \frac{v_0}{\sqrt{1+2ktv_0^2}}$$

Questions Asked in NEET/CBSE Medicals

168. (b)

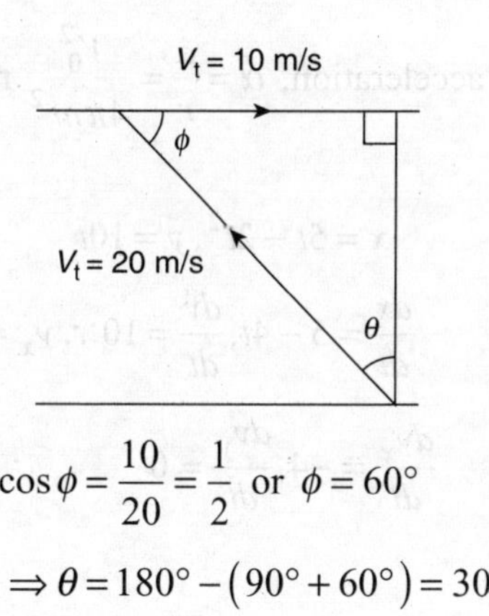

$$\cos\phi = \frac{10}{20} = \frac{1}{2} \text{ or } \phi = 60^\circ$$

$$\Rightarrow \theta = 180^\circ - (90^\circ + 60^\circ) = 30^\circ$$

169. (d)

$$v^2 - u^2 = 2ax$$

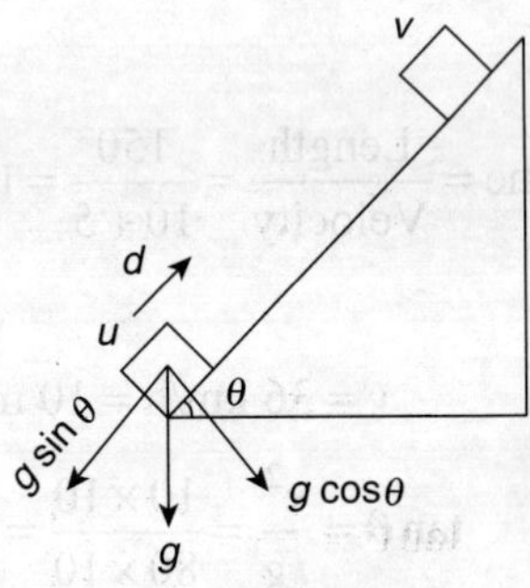

For Case I: $v^2 = u^2 - 2(g\sin\theta_1)x_1$

$[\because a = -g\sin\theta]$

$$x_1 = \frac{u^2}{2g\sin\theta_1} \qquad [\because v = 0]$$

For Case II:

$$v^2 = u^2 - (2g\sin\theta_2)x_2$$

$$x_2 = \frac{u^2}{2g\sin\theta_2}$$

$$\therefore \frac{x_1}{x_2} = \frac{\sin\theta_2}{\sin\theta_1} = \frac{\sin 30°}{\sin 60°} = \frac{1}{\sqrt{3}}$$

170. (a)

Time period, $T = \frac{2\pi}{\omega}$

As $T_A = T_B$

So, $\frac{2\pi}{\omega_A} = \frac{2\pi}{\omega_B}$ or $\omega_A : \omega_B = 1:1$

171. (c)

Distance travelled in nth round $= (2\pi r)n$

Using $v^2 = u^2 + 2as$ or, $V_0^2 = 0 + 2a(2\pi r)n$

$$a = \frac{V_0^2}{4\pi nr}$$

Angular acceleration, $\alpha = \frac{a}{r} = \frac{V_0^2}{4\pi nr^2}$ rad/s^2

172. (b)

$$x = 5t - 2t^2, y = 10t$$

$$\frac{dx}{dt} = 5 - 4t, \frac{dy}{dt} = 10 \therefore v_x = 5 - 4t, v_y = 10$$

$$\frac{dv_x}{dt} = -4, \frac{dv_y}{dt} = 0 \qquad \therefore a_x = -4, a_y = 0$$

Acceleration, $\vec{a} = a_x\hat{i} + a_y\hat{j} = -4\hat{i}$

$\therefore$ The acceleration of the particle at $t = 2$ s is -4 ms^{-2}.

173. (c)

Here, $a = 15$ ms^{-2},

$R = 2.5$ m

From figure,

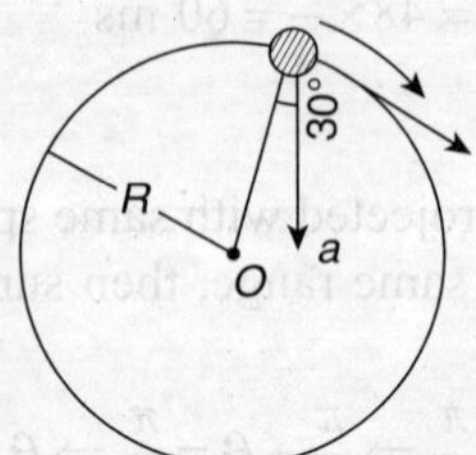

Radial acceleration, $a_R = a\cos 30° = 15 \times \frac{\sqrt{3}}{2}$ ms^{-2}

As we known, $a_R = \frac{v^2}{R} \Rightarrow v = \sqrt{a_R R}$

$$\therefore v = \sqrt{15 \times \frac{\sqrt{3}}{2} \times 2.5} = 5.69 \simeq 5.7 \text{ ms}^{-1}$$

174. (d)

Let the two vectors be $\vec{A}$ and $\vec{B}$

Then, magnitude of sum of $\vec{A}$ and $\vec{B}$,

$$|\vec{A} + \vec{B}| = \sqrt{A^2 + B^2 + 2AB\cos\theta}$$

and magnitude of difference of and

$$|\vec{A} - \vec{B}| = \sqrt{A^2 + B^2 - 2AB\cos\theta}$$

$$|\vec{A} + \vec{B}| = |\vec{A} - \vec{B}| \text{ (given)}$$

or $\sqrt{A^2 + B^2 + 2AB\cos\theta} = \sqrt{A^2 + B^2 - 2AB\cos\theta}$

$$4AB\cos\theta = 0$$

$$\because 4AB \neq 0,$$

$$\therefore \cos\theta = 0 \text{ or } \theta = 90°$$

175. (a)

Given, $\vec{r} = \cos\omega t\hat{x} + \sin\omega t\hat{y}$

$$\therefore \vec{v} = \frac{d\vec{r}}{dt} = -\omega\sin\omega t\hat{x} + \omega\cos\omega t\hat{y}$$

$$\vec{a} = \frac{d\vec{v}}{dt} = -\omega^2\cos\omega t\hat{x} - \omega^2\sin\omega t\hat{y} = -\omega^2\vec{r}$$

Since position vector $(\vec{r})$ is directed away from the origin, so, acceleration $(-\omega^2\vec{r})$ is directed towards the origin.

Also,

$$\vec{r}.\vec{v} = (\cos\omega t\hat{x} + \sin\omega t\hat{y}).(--\omega\sin\omega t\hat{x} + \omega\cos\omega t\hat{y})$$

$$= -\omega\sin\omega t\cos\omega t + \omega\sin\omega t\cos\omega t = 0$$

$$\Rightarrow \vec{r} \perp \vec{v}$$

176. (a)

Two vectors $\vec{A}$ and $\vec{B}$ are orthogonal to each other, if their scalar product is zero, i.e., $\vec{A}.\vec{B} = 0$

Here, $\vec{A} = \cos\omega t\hat{i} + \sin\omega t\hat{j}$

and $\vec{B} = \cos\frac{\omega t}{2}\hat{i} + \sin\frac{\omega t}{2}\hat{j}$

$$\therefore \vec{A}.\vec{B} = (\cos\omega t\hat{i} + \sin\omega t\hat{j}).\left(\cos\frac{\omega t}{2}\hat{i} + \sin\frac{\omega t}{2}\hat{j}\right)$$

$$= \cos\omega t\cos\frac{\omega t}{2} + si\omega t\sin\frac{\omega t}{2}$$

$$(\because \hat{i}.\hat{i} = \hat{j}.\hat{j} = 1 \text{ and } \hat{i}.\hat{j} = \hat{j}.\hat{i} = 0)$$

$$= \cos\left(\omega t - \frac{\omega t}{2}\right)$$

But $\vec{A}.\vec{B} = 0$ (as $\vec{A}$ and $\vec{B}$ are $\perp$ to each other)

$$\therefore \cos\left(\omega t - \frac{\omega t}{2}\right) = 0$$

$$\cos\left(\omega t - \frac{\omega t}{2}\right) = \cos\frac{\pi}{2}$$

$$= \omega t - \frac{\omega t}{2} = \frac{\pi}{2}$$

$$\frac{\omega t}{2} = \frac{\pi}{2} \Rightarrow t = \frac{\pi}{\omega}$$

177. (a)

Here, $\vec{R} = 4\sin(2\pi t)\hat{i} + 4\cos(2\pi t)\hat{j}$

The velocity of the particle is

$$\vec{v} = \frac{d\vec{R}}{dt} = \frac{d}{dt}[4\sin(2\pi t)\hat{i} + 4\cos(2\pi t)\hat{j}]$$

$$= 8\pi\cos(2\pi t)\hat{i} - 8\pi\sin(2\pi t)\hat{i}$$

Its magnitude is

$$|\vec{v}| = \sqrt{(8\pi\cos(2\pi t))^2 + (8\pi\sin(2\pi t))^2}$$

$$= \sqrt{64\pi^2} = 8\pi \text{ m/s}$$

178. (a)

The equation of trajectory is

$$y = x\tan\theta - \frac{gx^2}{2u^2\cos^2\theta}$$

where θ is the angle of projection and u is the velocity with which projectile is projected.

For equal trajectory and for same angles of projection,

$$\frac{g}{u^2} = \text{constant}$$

$$\frac{9.8}{5^2} = \frac{g'}{3^2}$$

where g' is acceleration due to gravity on the planet.

$$g' = \frac{9.8 \times 9}{25} = 3.5 \text{ ms}^{-2}$$

179. (d)

At time $t = 0$, the position vector of the particle is $\vec{r}_1 = 2\hat{i} + 3\hat{j}$

At time $t = 5$ s, the position vector of the particle is $\vec{r}_2 = 13\hat{i} + 14\hat{j}$

Displacement from $\vec{r}_1$ to $\vec{r}_2$ is

$$\Delta\vec{r} = \vec{r}_2 - \vec{r}_1 = (13\hat{i} + 14\hat{j}) - (2\hat{i} + 3\hat{j}) = 11\hat{i} + 11\hat{j}$$

$\therefore$ Average velocity

$$\vec{v}_{av} = \frac{\Delta\vec{r}}{\Delta t} = \frac{11\hat{i} + 11\hat{j}}{5 - 0} = \frac{11}{5}(\hat{i} + \hat{j})$$

180. (a)

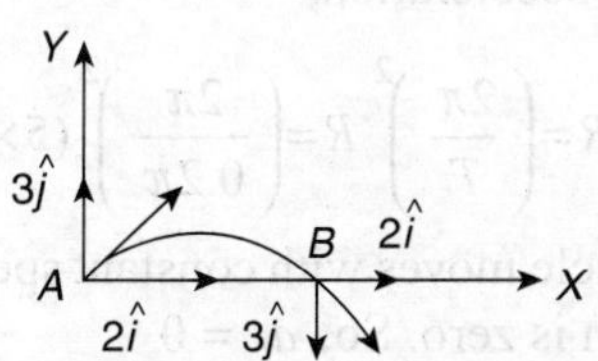

At point B, X-component of velocity remains unchanged while Y-component reverses its direction.

$\therefore$ The velocity of the projectile at point B is $2\hat{i} - 3\hat{j}$ m/s.

181. (c)

Vector triple product of three vectors $\vec{A}, \vec{B}$ and $\vec{C}$ is

$$\vec{A} \times (\vec{B} \times \vec{C}) = (\vec{A}.\vec{C})\vec{B} - (\vec{A}.\vec{B})\vec{C}$$

Given: $\vec{A}.\vec{B} = 0, \vec{A}.\vec{C} = 0$

$$\therefore \vec{A} \times (\vec{B} \times \vec{C}) = 0$$

Thus the vector $\vec{A}$ is parallel to vectors $\vec{B} \times \vec{C}$

182. (b)

Horizontal range, $R = \dfrac{u^2 \sin 2\theta}{g}$

where u is the velocity of projection and θ is the angle of projection

Maximum height, $H = \dfrac{u^2 \sin^2 \theta}{2g}$

Given, $R = H$

$$\therefore \frac{u^2 \sin 2\theta}{g} = \frac{u^2 \sin^2 \theta}{2g}$$

$$\frac{2u^2 \sin\theta \cos\theta}{g} = \frac{u^2 \sin^2 \theta}{2g}$$

$$\tan\theta = 4 \text{ or } \theta = \tan^{-1}(4)$$

183. (b)

Here, $u = 2\hat{i} + 3\hat{j}, a = 0.3\hat{i} + 0.2\hat{j}, t = 10$ s

As $\vec{v} = \vec{u} + \vec{a}t$

$$\therefore \vec{v} = (2\hat{i} + 3\hat{j}) + (0.3\hat{i} + 0.2\hat{j})(10)$$

$$= 2\hat{i} + 3\hat{j} + 3\hat{i} + 2\hat{j} = 5\hat{i} + 5\hat{j}$$

$$\therefore |\vec{v}| = \sqrt{(5)^2 + (5)^2} = 5\sqrt{2} \text{ units}$$

184. (d)

Here, radius, $R = 5$ cm $= 5 \times 10^{-2}$ m

Time period, $T = 0.2\pi$ s

Centripetal acceleration,

$$a_c = \omega^2 R = \left(\frac{2\pi}{T}\right)^2 R = \left(\frac{2\pi}{0.2\pi}\right)^2 (5 \times 10^{-2}) = 5 \text{ m/s}^2$$

As the particle moves with constant speed, its tangential acceleration is zero. So, $a_t = 0$

The acceleration of the particle is

$a = \sqrt{a_c^2 + a_t^2} = 5 \text{ m/s}^2$ directed towards centre.

185. (a)

Here, $u = 20$ m/s, $g = 10$ m/s^2

For maximum range, angle of projection is $\theta = 45°$

$$\therefore R_{\max} = \frac{u^2 \sin 90°}{g} = \frac{u^2}{g} \left(\because R = \frac{u^2 \sin 2\theta}{g}\right)$$

$$= \frac{(20\text{m/s})^2}{(10m/s^2)} = 40 \text{ m}$$

186. (d)

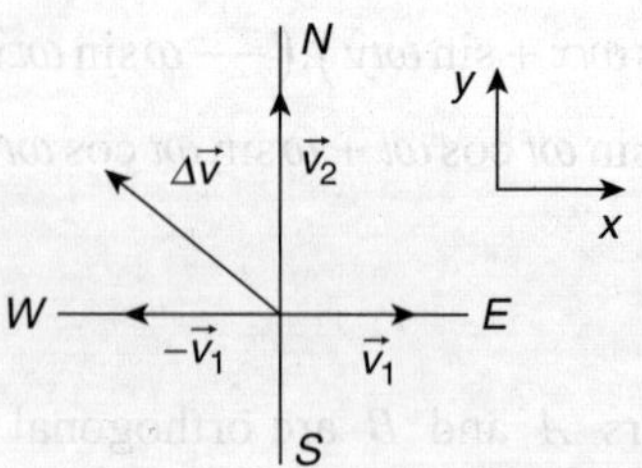

Velocity towards east direction, $v_1 = 30\hat{i}$ m/s

Velocity towards north direction, $v_2 = 40\hat{j}$ m/s

Change in velocity, $\Delta\vec{v} = \vec{v}_2 - \vec{v}_1 = (40\hat{j} - 30\hat{i})$

$$\therefore |\Delta\vec{v}| = |40\hat{j} - 30\hat{i}| = 50 \text{ m/s}$$

Average acceleration, $\vec{a}_{av} = \dfrac{\text{Change in velocity}}{\text{Time interval}}$

$$\vec{a}_{av} = \frac{\vec{v}_2 - \vec{v}_1}{\Delta t} = \frac{\Delta\vec{v}}{\Delta t}$$

$$|\vec{a}_{av}| = \frac{|\Delta\vec{v}|}{\Delta t} = \frac{50 \text{ m/s}}{10 \text{ s}} = 5 \text{ m/s}^2$$

187. (c)

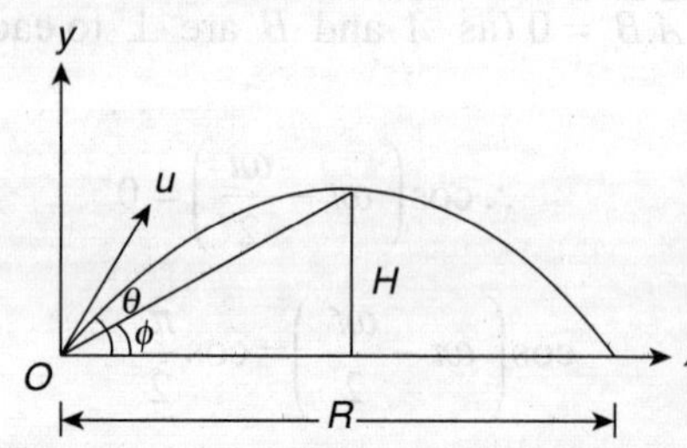

Let ϕ be elevation angle of the projectile at its highest point as seen from the point of projection O and θ be angle of projection with the horizontal.

From figure, $\tan\phi = \dfrac{H}{R/2}$ ...(i)

Maximum height, $H = \dfrac{u^2 \sin^2 \theta}{2g}$

Horizontal range, $R = \dfrac{u^2 \sin 2\theta}{g}$

Substituting these values of H and R in (i), we get

$$\tan\phi = \frac{\dfrac{u^2 \sin^2 \theta}{2g}}{\dfrac{u^2 \sin 2\theta}{2g}}$$

$$\tan\phi = \frac{\sin^2 \theta}{\sin 2\theta} = \frac{\sin^2 \theta}{2\sin\theta\cos\theta} = \frac{1}{2}\tan\theta$$

$$\therefore \tan\phi = \frac{1}{2}\tan 45° = \frac{1}{2}$$

$$\phi = \tan^{-1}\left(\frac{1}{2}\right)$$

188. (b)

Here, initial velocity, $\vec{u} = 3\hat{i} + 4\hat{j}$

Acceleration, $\vec{a} = 0.4\hat{i} + 0.3\hat{j}$ time, $t = 10$ s

Let $\vec{v}$ be velocity of a particle after 10 s.

Using, $\vec{v} = \vec{u} + \vec{a}t$

$$\therefore \vec{v} = \left(3\hat{i} + 4\hat{j}\right) + \left(0.4\hat{i} + 0.3\hat{j}\right)(10)$$

$$= 3\hat{i} + 4\hat{j} + 4\hat{i} + 3\hat{j} = 7\hat{i} + 7\hat{j}$$

Speed of the particle after 10 s $= |\vec{v}|$

$$= \sqrt{(7)^2 + (7)^2} = 7\sqrt{2} \text{ units}$$

189. (c)

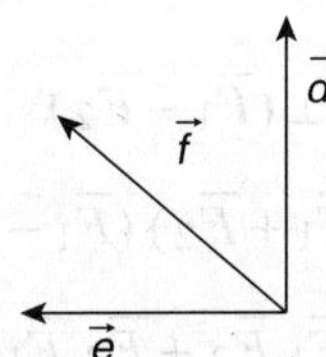

From figure, $\vec{d} + \vec{e} = \vec{f}$

190. (a)

Let v be velocity of a projectile at maximum height H.

$$v = u\cos\theta$$

Given, $v = \dfrac{u}{2}$

$$\therefore \frac{u}{2} = u\cos\theta \Rightarrow \theta = 60°$$

191. (b)

$$x = a\sin\omega t \Rightarrow \frac{x}{a} = \sin\omega t \quad \text{(i)}$$

$$y = a\cos\omega t \Rightarrow \frac{y}{a} = \cos\omega t \quad \text{(ii)}$$

Squaring and adding, Eqs. (i) and (ii)

$$\frac{x^2}{a^2} + \frac{y^2}{a^2} = 1 \quad \left(\because \cos^2\omega t + \sin^2\omega t = 1\right)$$

or $\quad x^2 + y^2 = a^2$

This is the equation of a circle. Hence, the particle follows a circular path.

192. (d)

Because the slope is highest at C,

$$v = \frac{ds}{dt} \text{ is maximum.}$$

193. (a)

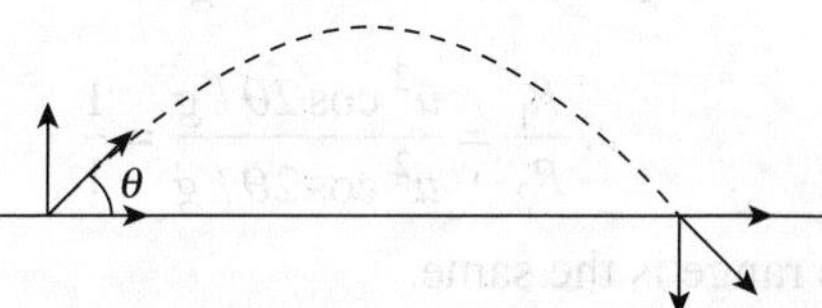

The horizontal momentum does not change. The change in vertical momentum is

$$mv\sin\theta - (-mv\sin\theta) = 2mv\frac{1}{\sqrt{2}} = \sqrt{2}mv$$

194. (d)

$$\left|\vec{A} \times \vec{B}\right| = \sqrt{3}(\vec{A}.\vec{B})$$

$$\therefore AB\sin\theta = \sqrt{3}AB\cos\theta$$

or, $\quad \tan\theta = \sqrt{3}$ or $\theta = \tan^{-1}\sqrt{3} = 60°$

195. (b)

Let θ be the angle which the particle makes with x-axis. From figure,

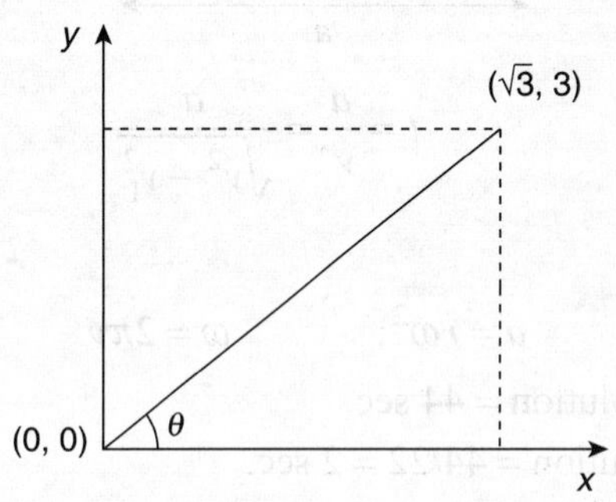

$$\tan\theta = \frac{3}{\sqrt{3}} = \sqrt{3}$$

or, $\quad \theta = \tan^{-1}\sqrt{3} = 60°$

196. (b)

Horizontal range $R = \dfrac{u^2\sin 2\theta}{g}$

For angle of projection $(45° - \theta)$, the horizontal range is

$$\therefore R_1 = \frac{u^2\sin[2(45° - \theta)]}{g} = \frac{u^2\sin(90° - 2\theta)}{g}$$

$$= \frac{u^2\cos 2\theta}{g}$$

For angle of projection $(45°+\theta)$, the horizontal range is

$$R_2 = \frac{u^2 \sin[2(45°+\theta)]}{g} = \frac{u^2 \sin(90°+2\theta)}{g} = \frac{u^2 \cos 2\theta}{g}$$

$$\therefore \frac{R_1}{R_2} = \frac{u^2 \cos 2\theta / g}{u^2 \cos 2\theta / g} = \frac{1}{1}$$

∴ The range is the same.

197. (b)

Let θ be angle between $\vec{A}$ and $\vec{B}$

$$|\vec{A}+\vec{B}| = |\vec{A}-\vec{B}|, \text{ then } |\vec{A}+\vec{B}|^2 = |\vec{A}-\vec{B}|^2$$

or $$(\vec{A}+\vec{B}).(\vec{A}+\vec{B}) = (\vec{A}--\vec{B}).(\vec{A}-\vec{B})$$

or $$\vec{A}.\vec{A}+\vec{A}.\vec{B}+\vec{B}.\vec{A}+\vec{B}.\vec{B} = \vec{A}.\vec{A}-\vec{A}.\vec{B}-\vec{B}.\vec{A}+\vec{B}.\vec{B}$$

or $4AB\ \cos\theta = 0$ or $\cos\theta = 0°$ or $\theta = 90°$

198. (d)

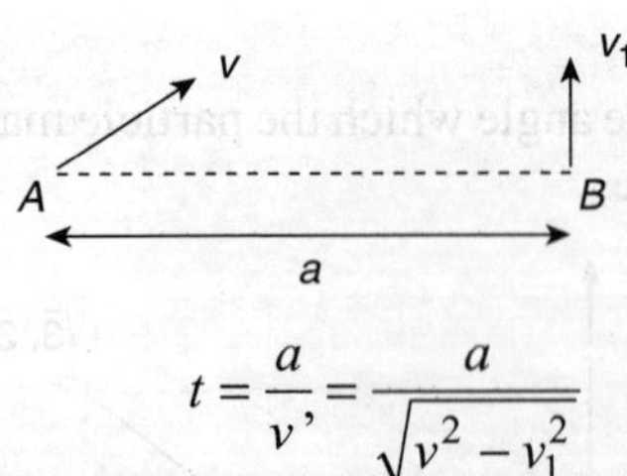

$$t = \frac{a}{v'} = \frac{a}{\sqrt{v^2 - v_1^2}}$$

199. (a, b)

$a = r\omega^2$; $\quad \omega = 2\pi\nu$

22 revolution = 44 sec.

1 revolution = 44/22 = 2 sec.

$\nu = 1/2$ Hz

$$a = r\omega^2 = 1 \times \frac{4\pi^2}{4} = \pi^2 \text{m/s}^2$$

Towards the centre, the centripetal acceleration is $-\omega^2 R$

and away from the centre, the centrifugal acceleration is $+\omega^2 R$

∴ (a) and (b) are correct as the directions are given.

200. (d)

Let $\vec{A}\times\vec{B} = \vec{C}$

The cross product of $\vec{B}$ and $\vec{A}$ is perpendicular to the plane containing $\vec{A}$ and $\vec{B}$, i.e., perpendicular to $\vec{A}$.

If a dot product of this cross product and $\vec{A}$ is taken, as the cross product is perpendicular to $\vec{A}$, $\vec{C}\times\vec{A} = 0$.

Therefore product of $(\vec{B}\times\vec{A}).\vec{A} = 0$.

201. (b)

$$\vec{a} = 2\hat{i}+3\hat{j}+8\hat{k}, \vec{b} = 4\hat{j}-4\hat{i}+\alpha\hat{k}$$

$$\vec{a}.\vec{b} = 0 \text{ if } \vec{a}\perp\vec{b}$$

$$(2\hat{i}+3\hat{j}+8\hat{k}).(-4\hat{i}+4\hat{j}+\alpha\hat{k}) = 0$$

or, $-8+12+8\alpha = 0 \Rightarrow 4+8\alpha = 0$

$$\Rightarrow \alpha = -1/2$$

202. (a)

$$|\vec{A}\times\vec{B}| = \sqrt{3}\vec{A}.\vec{B}$$

$$|\vec{A}||\vec{B}|\sin\theta = \sqrt{3}|\vec{A}||\vec{B}|\cos\theta$$

$$\tan\theta = \sqrt{3} \Rightarrow \theta = 60°$$

$$|\vec{A}+\vec{B}| = \sqrt{|\vec{A}|^2+|\vec{B}|^2+2|\vec{A}||\vec{B}|\cos\theta} = \left(A^2+B^2+AB\right)^{1/2}$$

203. (b)

Given: $(\vec{F}_1+\vec{F}_2)\perp(\vec{F}_1-\vec{F}_2)$

$$\therefore (\vec{F}_1+\vec{F}_2).(\vec{F}_1-\vec{F}_2) = 0$$

$$F_1^2 - F_2^2 - \vec{F}_1.\vec{F}_2 + \vec{F}_2.\vec{F}_1 = 0 \Rightarrow F_1^2 = F_2^2$$

i.e., F_1, F_2 are equal to each other in magnitude.

204. (a)

Given:

$$r = \frac{20}{\pi}\text{ m}, v = 80 \text{ m/s}, \theta = 2 \text{ rev} = 4\pi \text{ rad.}$$

From equation $\omega^2 = \omega_0^2 + 2\alpha\theta(\omega_0 = 0)$

$$\omega^2 = 2\alpha\times 4\pi$$

$$\omega^2 = 8\pi\alpha$$

$$\frac{v^2}{r^2} = 8\pi\alpha$$

$$v^2 = 8\pi\alpha r^2$$

$$(80)^2 = 8\pi ar$$

$$a = \frac{80\times 80\times\pi}{8\pi\times 20} = 40 \text{ m/s}^2$$

205. (c)

Both the particles have their initial velocity in vertical direction as zero.

206. (b)

Mass $m = 3$ kg. Force, $F = 6t^2\hat{i} + 4t\hat{j}$

$$\therefore \text{Acceleration } a = F/m = \frac{6t^2\hat{i} + 4t\hat{j}}{3} = 2t^2\hat{i} + \frac{4}{3}t\hat{j}$$

$$a = \frac{dv}{dt} = 2t^2\hat{i} + \frac{4}{3}t\hat{j}$$

$$\therefore dv = \left(2t^2\hat{i} + \frac{4}{3}t\hat{j}\right)dt$$

$$v = \int_0^3 \left(2t^2\hat{i} + \frac{4}{3}t\hat{j}\right)dt$$

$$= \frac{2}{3}t^3\hat{i} + \frac{4}{3}t^2\hat{j}_0^3 = 18\hat{i} + 6\hat{j}$$

207. (c)

$$\left|\vec{A}+\vec{B}\right| = \left|\vec{A}\right| + \left|\vec{B}\right| \text{ if } \vec{A} \parallel \vec{B}.\theta = 0°$$

208. (c)

$$\omega_1 = \frac{2\pi}{T_1}, \omega_2 = \frac{2\pi}{T_2}$$

Given, $T_1 = T_2$

Therefore $\frac{\omega_1}{\omega_2} = 1$

209. (a)

$$v_{\text{Resultant}} = \frac{1 \text{ km}}{1/4 \text{ hr}} = 4 \text{ km/hr}$$

$$\therefore v_{\text{River}} = \sqrt{5^2 - 4^2} = 3 \text{ km/hr}$$

210. (b)

$$\theta_2 = (90 - \theta_1) \quad (60° + 30° = 90°)$$

So range of projectile,

$$R_1 = \frac{v_0^2 \sin 2\theta}{g} = \frac{v_0^2 \sin\theta\cos\theta}{g}$$

$$R_2 = \frac{v_0^2 2\sin(90-\theta_1)\cos(90-\theta_1)}{g}$$

$$R_2 = \frac{v_0^2 2\cos\theta_1 \sin\theta_1}{g} = R_1$$

211. (a)

$$f_{\text{centripetal}} = \frac{mv^2}{R}; \; v = \left(36 \times \frac{5}{18}\right) \text{m/s}$$

$$f_{\text{centripetal}} = \frac{500 \times \left(36 \times \frac{5}{18}\right)^2}{50} = 1000 \text{ N}$$

212. (a)

Let v be the velocity of river water. As shown in figure,

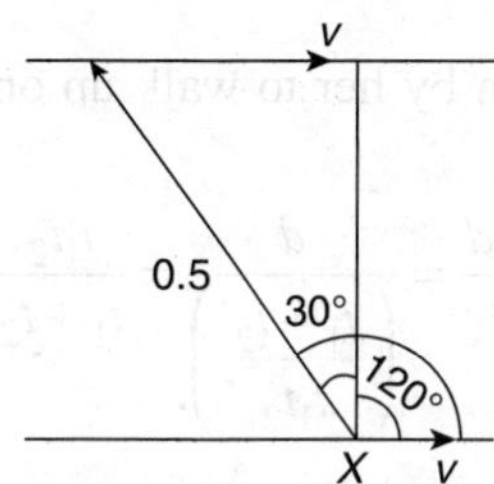

$$\sin 30° = \frac{v}{0.5}$$

or, $$v = 0.5 \sin 30°$$

$$= 0.5 \times (1/2) = 0.25 \text{ m/s}$$

213. (c)

$$t = \frac{2\pi}{\omega_1} = \frac{2\pi}{\omega_2} \Rightarrow \frac{\omega_1}{\omega_2} = \frac{1}{1}$$

214. (b)

For a unit vector $\hat{n}, |\hat{n}| = 1$

$$\left|0.5\hat{i} - 0.8\hat{j} + c\hat{k}\right| = 1^2 \Rightarrow 0.25 + 0.64 + c^2 = 1$$

or $$c = \sqrt{0.11}$$

215. (b)

$$\vec{v} = \vec{\omega} \times \vec{r} = \begin{vmatrix} \hat{i} & \hat{j} & \hat{k} \\ 5 & -6 & 6 \\ 3 & -4 & 1 \end{vmatrix} = 18\hat{i} + 13\hat{j} - 2\hat{k}$$

216. (c)

$$\frac{mv^2}{r} = 25; \; v = \sqrt{\frac{25 \times 1.96}{0.25}} = 14 \text{ m/s.}$$

217. (b)

218. (d)

Linear velocity $(v) = r\,\omega = 0.2 \times 10 = 2$ m/s.

219. (b)

Let v_1 be the velocity of Preeti on stationary escalator and d the distance travelled by her.

$$\therefore v_1 = \frac{d}{t_1}$$

Again, let v_2 be the velocity of escalator

$$\therefore v_2 = \frac{d}{t_2}$$

$\therefore$ Net velocity of Preeti on moving escalator with respect to the ground

$$v = v_1 + v_2 = \frac{d}{t_1} + \frac{d}{t_2} = d\left(\frac{t_1 + t_2}{t_1 t_2}\right)$$

The time taken by her to walk up on the moving escalator will be

$$t = \frac{d}{v} = \frac{d}{\left(\frac{t_1 + t_2}{t_1 t_2}\right)} = \frac{t_1 t_2}{t_1 + t_2}$$

220. (d)

Position of the car P at any time t, is $x_p(t) = at + bt^2$

$$v_p(t) = \frac{dx_p(t)}{dt} = a + 2bt$$

Similarly, for car Q

$$x_Q(t) = ft - t^2$$

$$v_Q(t) = \frac{dx_Q(t)}{dt} = f - 2t$$

$$\because v_p(t) = v_Q(t)$$

$$\therefore a + 2bt = f - 2t \text{ or, } 2t(b+1) = f - a$$

$$\therefore t = \frac{f - a}{2(1+b)}$$

221. (a)

Velocity of the particle is $v = At + Bt^2$

$$\frac{ds}{dt} At + Bt^2; \int ds = \int \left(At + Bt^2\right)dt$$

$$\therefore s = \frac{At^2}{2} + B\frac{t^3}{3} + C$$

$$S_1(t = 1s) = \frac{A}{2} + \frac{B}{3} + C; S_2(t = 2) = 2A + \frac{8}{3}B + C$$

$$S = S_2 - S_1$$

$$= \left(2A + \frac{8}{3}B + C\right) - \left(\frac{A}{2} + \frac{B}{3} + C\right) = \frac{3}{2}A + \frac{7}{3}B$$

222. (d)

According to question, velocity of unit mass varies as

$$v(x) = \beta x^{-2n} \quad \text{...(i)}$$

$$\frac{dv}{dx} = -2n\beta x^{-2n-1} \quad \text{...(ii)}$$

Acceleration of the particle,

$$a = \frac{dv}{dt} = \frac{dv}{dx} \times \frac{dx}{dt} = \frac{dv}{dx} \times v$$

Using Eqs. (i) and (ii), we get

$$a = \left(-2n\beta x^{-2n-1}\right) \times \left(\beta x^{-2n}\right) = -2n\beta^2 x^{-4n-1}$$

223. (d)

Distance covered by the stone in first 5 seconds

$$h_1 = \frac{1}{2}g(5)^2 = \frac{25}{2}g \quad \text{...(i)}$$

Let h_2 be the distance covered in next 5 seconds then $h_1 + h_2$ will be the distance covered in 10 seconds

$$h_1 + h_2 = \frac{1}{2}g(10)^2 = \frac{100}{2}g \quad \text{...(ii)}$$

Similarly, the distance covered in 15 seconds

$$h_1 + h_2 + h_3 = \frac{1}{2}g(15)^2 = \frac{225}{2}g \quad \text{...(iii)}$$

Subtract (i) from (ii), we get

$$\left(h_1 + h_2\right) - h_1 = \frac{100}{2}g - \frac{25}{2}g = \frac{75}{2}g$$

$$h_2 = \frac{75}{2}g = 3h_1$$

Subtract (ii) from (iii), we get

$$\left(h_1 + h_2 + h_3\right) - \left(h_2 + h_1\right) = \frac{225}{2}g - \frac{100}{2}g$$

$$h_3 = \frac{125}{2}g = 5h_1$$

From (i), (iv) and (v), we get

$$h_1 = \frac{h_2}{3} = \frac{h_3}{5}$$

224. (b)

Given $t = \sqrt{x} + 3 \Rightarrow \sqrt{x} = t - 3$

Squaring both sides, we get

$$x = (t-3)^2$$

Velocity, $v = \frac{dx}{dt} = \frac{d}{dt}(t-3)^2 = 2(t-3)$

Velocity of the particle becomes zero, when

$$2(t-3) = 0 \text{ or } t = 3 \text{ s}$$

At $t = 3$ s,

$$x = (3-3)^2 = 0 \text{ m}$$

225. (d)

Given: $x = 8 + 12t - t^3$

Velocity, $v = \frac{dx}{dt} = 12 - 3t^2$

When $v = 0$, $12 - 3t^2 = 0$ or $t = 2$ s

$$a = \frac{dv}{dt} = -6t$$

$$a\Big|_{t=2s} = -12 \text{ ms}^{-2}$$

Negative sign means retardation its value is 12 ms^2

226. (b)

$$\therefore v^2 = u^2 + 2gh$$

or $v = \sqrt{2gh} = \sqrt{2\times10\times20} = 20$ m/s $[\because u = 0]$

227. (c)

Let S be the total distance travelled by the particle.

Let t_1 be the time taken by the particle to cover first half of the distance. Then

$$t_1 = \frac{S/2}{v_1} = \frac{S}{2v_1}$$

Let t_2 be the time taken by the particle to cover remaining half of the distance. Then

$$t_2 = \frac{S/2}{v_2} = \frac{S}{2v_2}$$

Average speed,

$$v_{av} = \frac{\text{Total distance travelled}}{\text{Total time taken}}$$

$$= \frac{S}{t_1 + t_2} = \frac{S}{\frac{S}{2v_1} + \frac{S}{2v_2}} = \frac{2v_1v_2}{v_1 + v_2}$$

228. (a)

Distance, $x = (t+5)^{-1}$ ...(i)

Velocity, $v = \frac{dx}{dt} = \frac{d}{dt}(t+5)^{-1} = -(t+5)^{-2}$ ...(ii)

Acceleration, $a = \frac{dv}{dt} = \frac{d}{dt}[-(t+5)^{-2}] = 2(t+5)^{-3}$ ...(iii)

From Eq. (ii), we get

$$v^{3/2} = -(t+5)^{-3} \quad \text{...(iv)}$$

Substituting this in Eq. (iii), we get

Acceleration, $a = -2v^{3/2}$ or $a \propto (\text{velocity})^{3/2}$

From Eq. (i), we get

$$x^3 = (t+5)^{-3}$$

Substituting this in Eq. (iii), we get

Acceleration, $a = 2x^3$ or $a \propto (\text{distance})^3$

229. (a)

Let the two balls meet after t s at distance x from the platform.

For the first ball

Using $h = ut + \frac{1}{2}gt^2$

$$\therefore x = \frac{1}{2}\times10\times18^2 \quad \text{...(i)}$$

For the second ball

$$u = v,\ t = 12\text{ s},\ g = 10\text{ m/s}^2$$

Using $h = ut + \frac{1}{2}gt^2$

$$\therefore x = v\times12 + \frac{1}{2}\times10\times12^2 \quad \text{...(ii)}$$

From Eqs. (i) and (ii),

$$\frac{1}{2}\times10\times18^2 = 12v + \frac{1}{2}\times10\times(12)^2$$

or $12v = \frac{1}{2}\times10\times[(18)^2 - (12)^2]$

$$\frac{1}{2}\times10\times[(18+12)(18-12)]$$

$$12v = \frac{1}{2}\times10\times30\times6$$

or $v = \frac{1\times10\times30\times6}{2\times12} = 75$ m/s

230. (b)

Given $u = 0$

Distance travelled in 10 s, $S_1 = \frac{1}{2}a.10^2 = 50a$

Distance travelled in 20 s, $S_2 = \frac{1}{2}a.20^2 = 200a$

$$\therefore S_2 = 4S_1$$

231. (d)

Let v_s be the velocity of the scooter, $S = 1000$ m, $t = 100$

The velocity of the bus $= 10\text{ m s}^{-1}$

Time taken to overtake = 100 s

Relative velocity of the scooter with respect to the bus $= (v_s - 10)$

$$\therefore \frac{1000}{(v_s - 10)} = 100s \Rightarrow v_s = 20\text{ ms}^{-1}$$

232. (b)

$$v^2 - u^2 2as$$

Given $v = 20\text{ m s}^{-1}$, $u = 10\text{ m s}^{-1}$, $s = 135$ m

$$\therefore a = \frac{400-100}{2\times135} = \frac{300}{270} = \frac{10}{9}\text{ m/s}^2$$

$$v = u + at \Rightarrow t = \frac{v-u}{a} = \frac{10\text{ m/s}}{\frac{10}{9}\text{ m/s}^2} = 9\text{ s}$$

233. (a)

The distance travelled in nth second

As, $u = 0$

$$S_n = \frac{1}{2}a(2n-1)$$
$$= \frac{1}{2}\times\frac{4}{3}(2\times3-1)$$
$$= \frac{1}{2}\times\frac{4}{5}\times5$$
$$= \frac{10}{3}m$$

234. (c)

Given: At time $t = 0$, velocity, $v = 0$

Acceleration $f = f_0 = \left(1-\frac{t}{T}\right)$

At $f = 0$, $0 = f_0\left(1-\frac{t}{T}\right)$

Since f_0 is a constant,

$$\therefore 1-\frac{t}{T} = 0 \text{ or } t = T$$

Also, acceleration $f = \frac{dv}{dt}$

$$\therefore \int_0^{v_x} dv = \int_{t=0}^{t=T} f dt = \int_0^T f_0\left(1-\frac{t}{T}\right)dt$$

$$\therefore v_x = \left[f_0 t - \frac{f_0 t^2}{2T}\right]_0^T = f_0 T - \frac{f_0 T^2}{2T} = \frac{1}{2}f_0 T$$

235. (d)

$$\text{Average speed} = \frac{\text{Total distance travelled}}{\text{Total time taken}}$$

$$= \frac{s+s}{t_1+t_2} = \frac{2s}{\frac{s}{v_u}+\frac{s}{v_d}} = \frac{2v_u v_d}{v_d+v_u}$$

236. (a)

Given $x = 9t^2 - t^3$

Speed $v = \frac{dx}{dt} = \frac{d}{dt}\left(9t^2 - t^3\right) = 18t - 3t^2$

For maximum speed, $\frac{dv}{dt} = 0 \Rightarrow 18 - 6t = 0$

$$\therefore t = 3s$$

$$x_{max} = 9t^2 - t^3$$

$$\therefore x_{max} = 81 \text{ m} - 27 \text{ m} = 54 \text{ m}$$

237. (a)

Time taken by a body to fall from a height h is $t = \sqrt{\frac{2h}{g}}$

$$\therefore \frac{t_A}{t_B} = \frac{\sqrt{\frac{2h_A}{g}}}{\sqrt{\frac{2h_B}{g}}} = \sqrt{\frac{h_A}{h_B}} = \sqrt{\frac{16}{25}} = \frac{4}{5}$$

238. (c)

Distance travelled in one rotation (lap) $= 2\pi r$

$$\therefore \text{Average speed} = \frac{\text{distance}}{\text{time}} = \frac{2\pi r}{t}$$

$$= \frac{2\times3.14\times100}{62.8} = 10 \text{ ms}^{-1}$$

Net displacement in one lap = 0

$$\text{Average velocity} = \frac{\text{net displacement}}{\text{time}} = \frac{0}{t} = 0$$

239. (a)

$$x = 40 + 12t - t^3$$

$$\therefore \text{Velocity, } v = \frac{dx}{dt} = 12 - 3t^2$$

When particle come to rest, $\frac{dx}{dt} = v = 0$

$$12 - 3t^2 = 0 \Rightarrow t = 2$$

Distance travelled by the particle before coming to rest

$$\int_0^s ds = \int_0^2 v dt \text{ or } s = \int_0^2 \left(12 - 3t^2\right)dt = 12t - \frac{3t^3}{3}$$

$$s = 12\times2 - 8 = 24 - 8 = 16 \text{ m}$$

240. (a)

$$v^2 = u^2 2gh$$

After reaching maximum height velocity becomes zero.

$$0 = (10)^2 - 2\times10\times\frac{h}{2} \quad \therefore h = \frac{200}{20} = 10 \text{ m}$$

241. (d)

$$x = ae^{-\alpha t} + be^{\beta t}; \frac{dx}{dt} = -a\alpha e^{-\alpha t} + b\beta e^{\beta t}$$

$$v = -a\alpha e^{-\alpha t} + b\beta e^{\beta t}$$

For certain value of t, velocity will increases.

242. (a)

If we want that minimum three (more than two) balls remain in air, then time of flight of first ball must be greater than 4 sec.

$$T > 4 \text{ sec or } \frac{2u}{g} > 4 \text{ sec} \Rightarrow u > 19.6 \text{ m/s}$$

243. (b)

Let total height $= H$, Time of ascent $= T$

So, $H = uT - \frac{1}{2}gT^2$ (i)

Distance covered by ball in time $(T - t)$ sec.

$$h = u(T - t) - \frac{1}{2}g(T - t)^2 \quad \text{(ii)}$$

So distance covered by ball in last t sec.

$$h' = H - h$$

$$= \left[uT - \frac{1}{2}gT^2\right] - \left[u(T - t) - \frac{1}{2}g(T - t)^2\right] \quad \text{(iii)}$$

We have $0 = u - gt$ or $T = \frac{u}{g}$

Putting this value in Eq. (iii),

$$h' = \frac{1}{2}gt^2$$

244. (c)

$$10^2 = u^2 - 2g\frac{h}{2} \quad \ldots\text{(i)}$$

$$0 = u^2 - 2gh \quad \ldots\text{(ii)}$$

From (i) and (ii)

$$\Rightarrow 10^2 = \frac{2gh}{2} \Rightarrow h = 10 \text{ m}$$

245. (b)

$$\frac{ds}{dt} = 9t^2 + 14t + 14$$

$$\Rightarrow \frac{d^2s}{dt^2} = 18t + 14 = a$$

$$a_{t=1} = 18 \times 1 + 14 = 32 \text{ m/s}^2$$

QUICK QUESTIONS

Quiz

1. Can there be?

(i) A physical quantity which has direction but not a vector

(ii) A physical quantity which is either scalar or vector

(iii) A physical quantity which is neither a scalar nor vector

(iv) Two unequal vectors with their resultant zero

(v) Three non-coplanar vectors with resultant zero

(vi) Two vectors with their sum equal to their difference

(vii) Two vectors with the sum of their magnitude equal to their difference

(viii) A vector which can be divided by a vector

(ix) A motion in which average speed is lesser than the magnitude of average velocity

(x) A projectile motion under gravity following a parabolic path with velocity at the topmost point being zero.

2. If vector ***B*** is added to vector ***A***, under what circumstances does the resultant vector ***A*** + ***B*** have magnitude $A + B$? Under what circumstances is the resultant vector equal to zero?

3. If one component of a vector is not zero, can the magnitude of the vector be zero? Explain.

4. If ***A*** + ***B*** = 0, what can you say about the components of the two vectors?

5. Figure shows two vectors lying in $x - y$ plane. Determine the signs of (a) the x components of ***A*** and ***B***, (b) the y component of ***A*** and ***B***, and (c) the x and y components of ***A*** + ***B***.

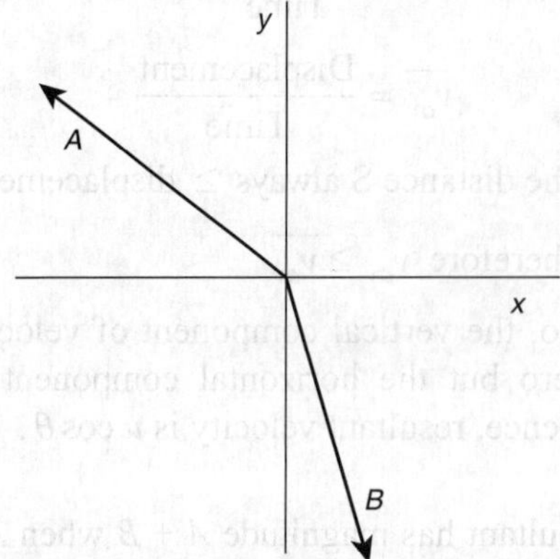

6. Under what circumstances can the component of a vector be equal to the magnitude of the vector?

7. (a) Can an object accelerate if its speed is constant?

(b) Can an object accelerate if its velocity is constant?

8. (a) Can a particle have a constant velocity and a varying speed?

(b) Can a particle have a constant speed and a varying velocity? Give examples of a positive answer and explain a negative answer.

9. Can a particle with constant acceleration ever stop and stay stopped?

Think and grow

10. Consider the following motions of an object in one dimension: (a) A ball is thrown directly upward; it rises to its highest point and falls back into the thrower's hand. (b) A race car starts from rest and speeds up to 120 m/s along a straight line. (c) A Voyager spacecraft drifts through empty space at constant velocity. Are there any instants of time in the motion of these objects at which the instantaneous velocity and the average velocity over the entire interval are the same? If so, identify the points (s).

11. A sky diver steps out of a stationary helicopter. A few seconds later, another sky diver steps out, so that they both fall along the same vertical line. Ignore air resistance, so that both sky divers fall with the same acceleration. Let us model the sky divers as particles under constant acceleration. Does the vertical separation distance between them stay the same? Does the difference in their speeds stay the same?

12. The gas pedal in an automobile is commonly called the accelerator. Are there any other controls that could also be considered accelerators?

Explanations

Quiz

1. Can there be?

(i) Yes, pressure

(ii) Yes, angular displacement

(iii) Yes, moment of inertia

(iv) No

(v) No, the third vector should be equal to the magnitude of the resultant of the other two vectors and in opposite direction in the same plane

(vi) Yes, if one of the vectors is null vector.

(vii) Yes, if the vectors are mutually perpendicular

(viii) No, vector can never be divided by a vector

(ix) No, $v_{av} = \dfrac{\text{Distance}}{\text{Time}}$

$\overline{v_{av}} = \dfrac{\text{Displacement}}{\text{Time}}$

The distance S always ≥ displacement

Therefore $v_{av} \geq \overline{v_{av}}$

(x) No, the vertical component of velocity becomes zero but the horizontal component is $u \cos\theta$. Hence, resultant velocity is $u \cos\theta$.

2. The resultant has magnitude $A + B$ when $\boldsymbol{A}$ is oriented in the same direction as $\boldsymbol{B}$. The resultant vector is zero only when the two vectors $\boldsymbol{A}$ and $\boldsymbol{B}$ have the same magnitude and are oriented in opposite directions.

3. No. The magnitude of a vector $\boldsymbol{A}$ is equal to $\sqrt{A_x^2 + A_y^2 + A_z^2}$. Therefore, if any component is nonzero, A cannot be zero.

4. The fact that $\boldsymbol{A} + \boldsymbol{B} = 0$ tells you that $\boldsymbol{A} = -\boldsymbol{B}$. Therefore, the components of the two vectors must have equal magnitudes and opposite signs: $A_x = -B_x, A_y = -B_y$, and $A_z = -B_z$.

5. (a) Because $\boldsymbol{A}$ is oriented toward the left, its x component is negative. Because $\boldsymbol{B}$ is oriented toward the right, its x component is positive. (b) Because $\boldsymbol{A}$ is oriented upward, its y component is positive. Because $\boldsymbol{B}$ is oriented downward, its y component is negative. (c) The resultant vector $\boldsymbol{R} = \boldsymbol{A} + \boldsymbol{B}$ lies in the third quadrant, so both x and y components are negative.

6. The only circumstances under which this can happen is if the vector is oriented along the positive direction for one of the three axes: x, y, or z. Then the magnitude of the vector is the same as the component associated with its direction and the other components are zero.

7. (a) Yes. An object moving along a curved path at constant speed is accelerating because the direction of the velocity vector is changing. (b) No. If the velocity is constant, there is no change in the velocity and the acceleration is zero.

8. (a) No. If the velocity is constant, the magnitude of the instantaneous velocity vector, which is speed, is constant. (b) Yes. An electron traveling around a curved path at a constant speed has a varying velocity because the direction of the velocity vector is changing.

9. No, for an accelerating particle to stop at all, the velocity and acceleration must have opposite signs, so that the speed is decreasing. In this case, the particle will eventually come to rest. If the acceleration remains constant, however, the particle must begin to move again, opposite to the direction of its original motion. If the particle comes to rest and then stays at rest, the acceleration has become zero at the moment the motion stops. This is the case for a braking cars – the acceleration is negative and goes to zero as the car comes to rest.

Think and grow

10. (a) The average velocity over the entire interval for the thrown ball is zero – the ball returns to the thrower's hand at the end of the time interval. There is one point at which the instantaneous velocity is zero – at the top of the motion. (b) The average velocity for the motion of the race car cannot be evaluated unambiguously with the information given, but its magnitude must be some value between 0 and 120 m/s. Because the magnitude of the instantaneous velocity of the car will have every value between 0 and 120 m/s at some time during the interval, there must be some instant at which the instantaneous velocity is equal to the average velocity over the entire interval. (c) Because the instantaneous velocity of the spacecraft is constant, its instantaneous velocity at any point of time and its average velocity over any time interval are the same.

11. At any given point of time, the speeds of the jumpers are different, because one had a head start over the other. In any time interval, however, each jumper increases his or her speed by the same amount, because they have the same acceleration. Thus, the difference in speeds remains the same. The first jumper will always be moving with a higher speed than the second. In a given time interval, then, the first jumper will have a larger displacement than the second. Thus, the distance of separation between them increases.

12. The gas pedal is called the accelerator because the common usage of the word acceleration refers to an increase in speed. The scientific definition, however, is that acceleration occurs whenever the velocity changes in any way. Thus, the brake pedal can also be considered an accelerator because it causes the car to slow down. The steering wheel is also an accelerator because it changes the direction of the velocity vector.

CHAPTER TEST

1. A car is travelling at 20 m/s on a circular road of radius 100 m. It is increasing in speed at a rate of 3 m/s^2. Its acceleration is
   (a) 3 m/s^2 (b) 5 m/s^2
   (c) 30 m/s^2 (d) 7 m/s^2

2. A car is moving horizontally along a straight line with constant speed 30 m/s. A projectile is to be fired from the moving car in such a way that it will return to the car after the car has moved 80 m. At what speed (relative to the cart) must the projectile be fired (takes g = 10 m/s^2).
   (a) $10\sqrt{8}$ m/s (b) $10\sqrt{8}$ m/s
   (c) $\frac{40}{3}$ m/s (d) None of these.

3. The area under acceleration-time graph gives
   (a) distance travelled (b) change in acceleration
   (c) force acting (d) change in velocity

4. A car moves along a circular track of radius 10 km with a constant speed. If it takes 15 minutes to move from a point on the path to the diametrically opposite point, the ratio of distance covered to displacement is
   (a) $\pi:2$ (b) $2\pi:3$
   (c) $\pi:4$ (d) $\pi:5$

5. A body falling from rest has a velocity v after it falls through a distance of h. The distance it has fallen when its velocity is $\frac{v}{2}$ is
   (a) $\frac{h}{2}$ (b) $\frac{h}{4}$
   (c) $\frac{h}{3}$ (d) $\frac{h}{\sqrt{2}}$

6. The correct statement from the following is:
   (a) A body having zero velocity will not necessarily have zero acceleration
   (b) A body having zero velocity will necessarily have zero acceleration
   (c) A body having uniform speed can have only uniform acceleration
   (d) A body having non-uniform velocity will have zero acceleration.

7. The acceleration of a particle is given $a = 0.02 - 20x$ where a is in m/s^2 and x in metres. What is the velocity at $x = 0$, if $v = 10$ m/s when $x = 1$ m.
   (a) 120 m/s (b) 100 m/s
   (c) 10 m/s (d) $\sqrt{120}$ m/s

8. A particle moves so that its position vector is given by $\vec{r} = \cos\omega t\hat{x} + \sin\omega t\hat{y}$, where ω is a constant. Which of the following is true?
   (a) Velocity is perpendicular to $\vec{r}$ and acceleration is directed towards the origin.
   (b) Velocity is perpendicular to $\vec{r}$ and acceleration is directed away from the origin.
   (c) Velocity and acceleration both are perpendicular to $\vec{r}$.
   (d) Velocity and acceleration both are parallel to $\vec{r}$

9. The velocity of a projectile at the initial point A is $(2\hat{i}+3\hat{j})$ m/s. Its velocity (in m/s) at point B is

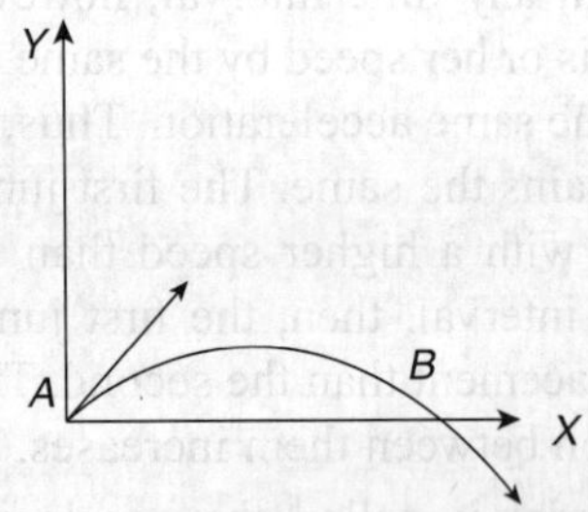

(a) $2\hat{i}-3\hat{j}$ (b) $2\hat{i}+3\hat{j}$
(c) $-2\hat{i}-3\hat{j}$ (d) $-2\hat{i}+3\hat{j}$

10. A body is moving with velocity 30 m/s towards east. After 10 seconds its velocity becomes 40 m/s towards north. The average acceleration of the body is
(a) 1 m/s^2 (b) 7 m/s^2
(c) $\sqrt{7}$ ms^2 (d) 5 m/s^2

11. The vector sum of two forces is perpendicular to their vector differences. In that case, the forces
(a) are equal to each other
(b) are equal to each other in magnitude
(c) are not equal to each other in magnitude
(d) cannot be predicted.

12. A particle A is dropped from a height and another particle B is projected in horizontal direction with speed of 5 m/sec from the same height. Then the correct statement is
(a) particle A will reach at ground first with respect to particle B
(b) particle B will reach at ground first with respect to particle A
(c) both particles will reach the ground simultaneously
(d) both particles will reach the ground with same speed.

13. Two particles having mass M and m are moving in a circular path having radius R and r respectively. If their time periods are same then the ratio of angular velocity will be
(a) $\dfrac{r}{R}$ (b) $\dfrac{R}{r}$
(c) 1 (d) $\sqrt{\dfrac{R}{r}}$

14. Identify the vector quantity among the following:
(a) Distance (b) Angular momentum
(c) Heat (d) Energy

15. Two cars P and Q start from a point at the same time in a straight line and their positions are represented by $x_P\ (t) = \left(at+bt^2\right)$ and $x_Q\ (t) = \left(ft-t^2\right)$. At what time do the cars have the same velocity?
(a) $\dfrac{a-f}{1+b}$ (b) $\dfrac{a+f}{2(b-1)}$
(c) $\dfrac{a+f}{2(1+b)}$ (d) $\dfrac{f-a}{2(1+b)}$

16. A particle starts its motion from rest under the action of a constant force. If the distance covered in first 10 seconds is S_1 and that covered in the first 20 seconds is S_2, then
(a) $S_2 = 3S_1$ (b) $S_2 = 4S_1$
(c) $S_2 = S_1$ (d) $S_2 = 2S_1$

17. A body starting from rest moves along a straight line with a constant acceleration. The variation of speed (v) with distance (S) is represented by the graph:

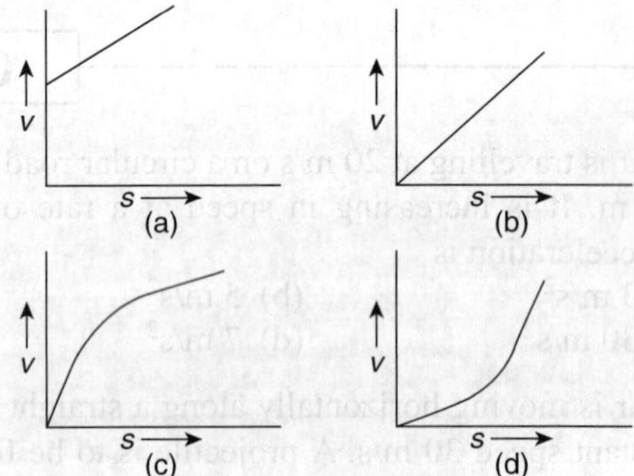

18. A ball is thrown vertically upwards. Which of the following plots represents the speed-time graph of the ball during its flight if the air resistance is not ignored ?

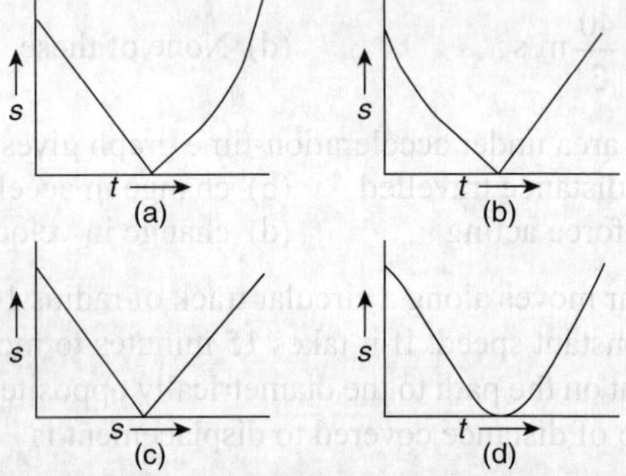

19. The resultant $\vec{R}$ of vector $\vec{P}$ and $\vec{Q}$ is perpendicular to $\vec{P}$ also $|\vec{P}| = |\vec{R}|$, the angle between $\vec{P}$ and $\vec{Q}$ is:
(a) 45° (b) 135°
(c) 225° (d) None of these

20. In the given figure, O is the centre of regular pentagon $ABCDE$. Five forces, each of magnitude F_0, are acted as shown in figure. The resultant force is:

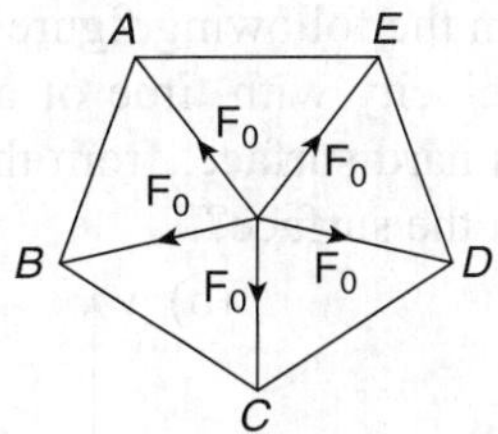

(a) $5\,F_0$ (b) $5\,F_0 \cos 72°$
(c) $5\,F_0 \sin 72°$ (d) Zero

21. Choose the correct option for $\vec{A} \times \vec{B} = \vec{C}$:
(i) $\vec{C}$ is perpendicular to $\vec{A}$
(ii) $\vec{C}$ is perpendicular to $\vec{B}$
(iii) $\vec{C}$ is perpendicular to $\left(\vec{A} + \vec{B}\right)$
(iv) $\vec{C}$ is perpendicular to $\left(\vec{A} \times \vec{B}\right)$
(a) Only (i) and (ii) are correct
(b) Only (ii) and (iv) are correct
(c) (i), (ii) and (iii) are correct
(d) All are correct.

22. A particle moves with constant speed v along a regular hexagon $ABCDEF$ in same order. The magnitude of average velocity for its motion from A to C is
(a) v (b) $\frac{v}{2}$
(c) $\frac{\sqrt{3}v}{2}$ (d) None of these.

23. A particle is moving on a straight line path with constant acceleration directed along the direction of instantaneous velocity. Which of following statements is/are false about the motion of particle?
(a) Particle may reverse the direction of motion
(b) Distance covered is not equal to magnitude of displacement
(c) The magnitude of average velocity is less than average speed.
(d) All the above.

24. A particle starts from rest with acceleration 2 m/s^2. The acceleration of the particle decreases down to zero uniformly during a time interval of 4 seconds. The velocity of the particle after 2 seconds is:
(a) 3 m/s (b) 4 m/s
(c) zero (d) 8 m/s

25. The acceleration of a train between two stations 2 km apart is shown in the figure. The maximum speed of the train is:

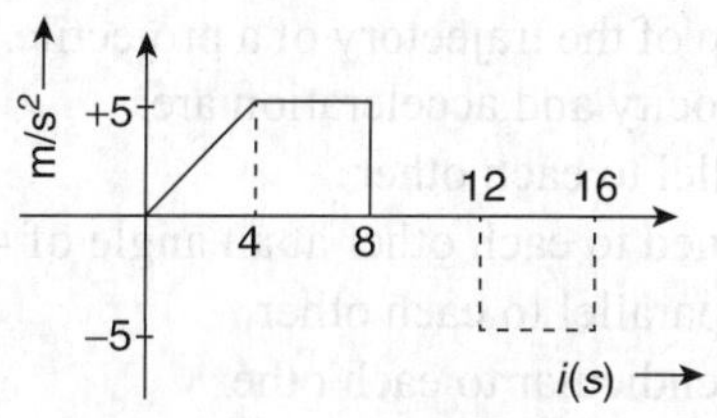

(a) 60 m/s (b) 30 m/s
(c) 120 m/s (d) 90 m/s

26. An object is thrown upwards with a velocity u, then the displacement – time graph is:

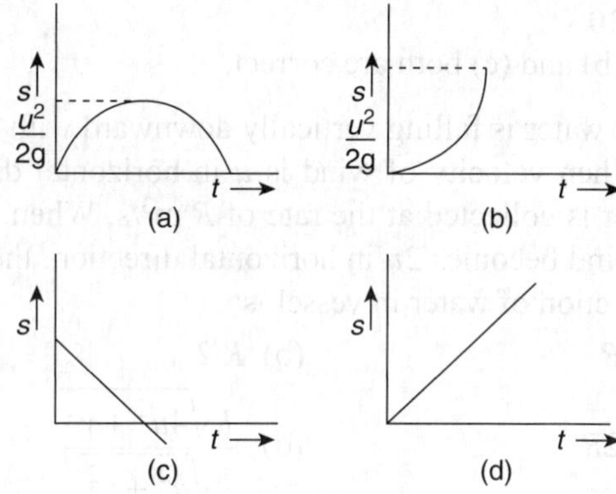

27. Two stones having different masses m_1 and m_2 respectively are projected at an angle α and with same velocity from same point. The ratio of their maximum heights is:
(a) 1 : 1 (b) 1 : tan α
(c) tan α : 1 (d) $\tan^2\alpha$: 1

28. Which of the following statements is incorrect?
(a) Component of a vector perpendicular to itself is zero
(b) Angle between two vectors can never be greater than 180°
(c) A vector cannot be divided by a vector
(d) Addition as well as subtraction of vectors is commutative.

29. A deer wants to save her life from a lion. The lion follows a path whose equation is $x^2 + y^2 = 16$. For saving life, the deer moves on a path whose equation is /are:
(a) $x^2 + y^2 = 4$ (b) $x^2 + y^2 = 16$
(c) $x^2 + y^2 - 64 = 0$ (d) Both (a) and (c) are correct

30. A ball slides down on a smooth inclined plane when released from the top, while another falls freely under gravity from the same point. Then
(a) falling body reaches the ground first
(b) sliding body reaches the ground first
(c) the velocity of falling body is more than that of sliding body
(d) the velocity of the sliding body is more than that of falling body

31. At the top of the trajectory of a projectile, the directions of its velocity and acceleration are:
(a) parallel to each other
(b) inclined to each other at an angle of 45
(c) anti parallel to each other
(d) perpendicular to each other.

32. A projectile is thrown at an angle of $\theta = 45°$ to the horizontal, reaches a maximum height of 16 m, then:
(a) its velocity at the highest point is zero
(b) its range is 64 m
(c) its range will decrease when it is thrown at an angle 30
(d) (b) and (c) both are correct.

33. Rain water is falling vertically downward with velocity v. When velocity of wind is u in horizontal direction, water is collected at the rate of R m^3/s. When velocity of wind becomes $2u$ in horizontal direction, the rate of collection of water in vessel is:
(a) R (b) $R/2$
(c) $2R$ (d) $\dfrac{R\sqrt{4u^2 + v^2}}{\sqrt{a^2 + v^2}}$

34. A particle moves along the positive branch of the curve $y = x^2/2$ where $x = t^2/2$, where x and y are measured in metres and t in seconds. At $t = 2$ sec, the velocity of the particle is:
(a) $(2\hat{i} - 4\hat{j})$ m/sec (b) $(2\hat{i} + 4\hat{j})$ m/sec
(c) $(2\hat{i} + 2\hat{j})$ m/sec (d) $(4\hat{i} - 2\hat{j})$ m/sec

35. A piece of marble is projected from earth's surface with velocity of 50 m/s, 2 second later, it just clears a wall 5 m high. What is the angle of projection?
(a) 45° (b) 30°
(c) 60° (d) None of these.

36. A river is flowing from west to east at a speed 15 m/s. A boy on the south bank of the river, capable of swimming at 30 m/s in still water, wants to swim and cross the river in the shortest time. He should swim in the direction
(a) due north (b) 30° east of north
(c) 30° west of north (d) 60° east of north.

37. If $x = a(\cos\theta + \theta\sin\theta)$ and $y = a(\sin\theta - \theta\cos\theta)$ and θ increases at uniform rate ω. The velocity of particle is:
(a) $a\omega$ (b) $\dfrac{a^2\theta}{\omega}$
(c) $\dfrac{a\theta}{\omega}$ (d) $a\theta\omega$

38. Which graph in the following figure best represents the variation of velocity with time of a ball that bounces vertically on a hard surface, from the moment when it rebounds from the surface?

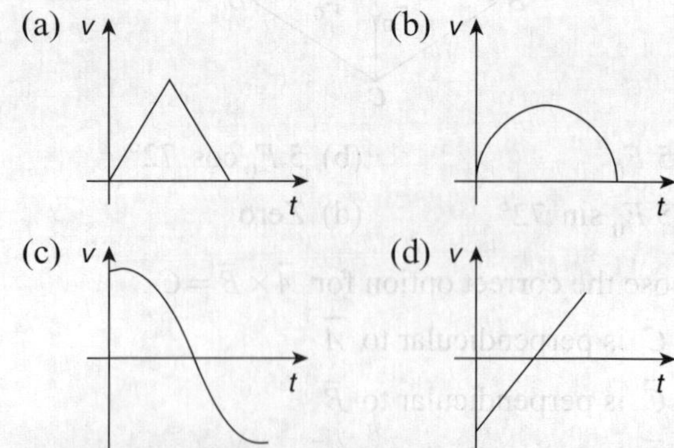

39. A time – velocity graph of two vehicles A and B starting from rest at the same time is given in the figure. The statement that can be deduced correctly from the graph is:

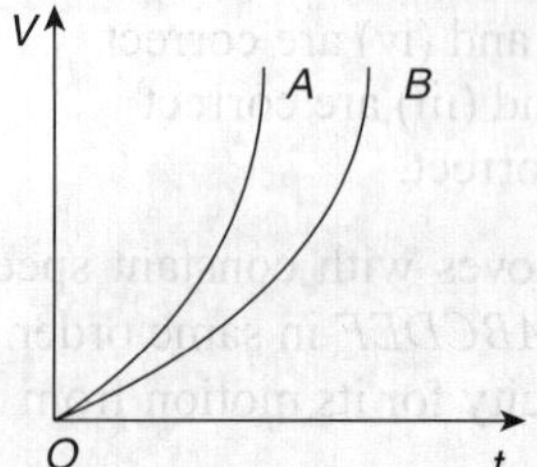

(a) Acceleration of A is greater than that of B
(b) Acceleration of B is greater than that of A
(c) Acceleration of A is increasing at a slower rate than that of B
(d) Velocity of B is greater than that of A.

40. A small object is dropped from the top of a building and it falls to the ground. As it falls, accelerating due to gravity, it passes a window. It has speed v_1 at the top of the window and speed v_2 at the bottom of the window. At what point does it have a speed $\dfrac{v_1 + v_2}{2}$? Neglect the air resistance.
(a) It depends on the height of the window or its distance from the top of the building.
(b) Above the centre point of the window
(c) Below the centre point of the window
(d) At the centre point of the window

41. A particle has an initial velocity of 9 m/s due east and a constant acceleration of 2m/s^2 due west. The distance covered by the particle in the fifth second of its motion is:
(a) 0 (b) 0.5 m
(c) 2 m (d) None of these.

42. Which of the following statements is correct?
(a) A projectile fired from the ground follows a parabolic path. The speed of the projectile at the highest point is zero.

(b) For angles of projection which exceed and fall short of 45° by equal amounts, the ranges are equal.
(c) The weight of a falling body is twice the weight of a body when it is at rest.
(d) The speed acquired by a falling body in vacuum is dependent on mass.

43. The greatest acceleration or deceleration that a train may have is 'a'. The minimum time in which the train can get from one station to the next at a distance s is:

(a) $\sqrt{\frac{s}{a}}$ (b) $\sqrt{\frac{2s}{a}}$

(c) $\frac{1}{2}\sqrt{\frac{s}{a}}$ (d) $2\sqrt{\frac{s}{a}}$

44. The x and y components of a vector are 3 and 7 units. The x and y components of vector $\vec{A}+\vec{B}$ are 9 and 10 units. Then vector $\vec{B}$ is

(a) $3\hat{i}+6\hat{j}$ (b) $6\hat{i}+3\hat{j}$

(c) $6\hat{i}+5\hat{j}$ (d) $6\hat{i}-5\hat{j}$

45. The sum of the magnitudes of two forces acting at a point is 18 and the magnitude of their resultant is 12. If the resultant is at 90° with the force of smaller magnitude, what are the magnitudes of forces?

(a) 7, 11 (b) 5, 13
(c) 8, 10 (d) 6, 12

Answer Keys

1. (b)	2. (c)	3. (d)	4. (a)	5. (b)	6. (a)	7. (d)	8. (a)	9. (a)	10. (d)
11. (b)	12. (c)	13. (c)	14. (b)	15. (d)	16. (b)	17. (c)	18. (d)	19. (b)	20. (d)
21. (d)	22. (c)	23. (d)	24. (a)	25. (b)	26. (a)	27. (d)	28. (c)	29. (d)	30. (a)
31. (d)	32. (d)	33. (a)	34. (b)	35. (b)	36. (a)	37. (d)	38. (d)	39. (a)	40. (b)
41. (b)	42. (b)	43. (d)	44. (b)	45. (b)					

Solutions

1. (b)
It is a case of non-uniform circular motion; therefore it has two components of acceleration tangential and radial.

$$a_r = \frac{u^2}{r} = \frac{20\times 20}{100} = 4\text{m/s}^2$$

$$a_t = 3\text{m/s}^2$$

$$a = \sqrt{(4)^2+(3)^2} = 5$$

2. (c)
Time taken by the car to cover 80 m

$$t = \frac{s}{v} = \frac{80}{30} = \frac{8}{3}\text{sec}$$

We have, $v = u + at$

$$u = ?\ v = 0, a = -g = -10 \text{ m/s}$$

and $t = \frac{8/3}{2} = \frac{4}{3}$sec (since time of flight to reach the highest point is half of the total time of flight)

$$\therefore \quad 0 = u - 10\times\frac{4}{3}$$

$$u = \frac{40}{3}\text{ m/s}$$

3. (d)
The area under acceleration-time graph gives change in velocity.

4. (a)

Distance covered $= \pi r = \pi \times 10$ km

Displacement $= 2r = 2\times 10$ km

$$\text{Ratio} = \frac{\pi\times 10}{2\times 10} = \pi : 2$$

5. (b)

$$2gh = v^2 - 0^2 \Rightarrow v^2 = 2gh$$

$$\left(\frac{v}{2}\right)^2 = 2gh_1 \Rightarrow v^2 = 8gh_1$$

$$\Rightarrow \quad 8gh_1 = 2gh \Rightarrow h_1 = \frac{h}{4}$$

6. (a)
A body thrown up, on reaching maximum height, has velocity zero but acceleration g.

7. (d)

$$a = v\frac{dv}{dx} = 0.2 - 20x$$

$$\Rightarrow \quad vdv = (0.02 - 20x)dx$$

Integrating both sides,

$$\frac{v^2}{2} = 0.02x - 10x^2 + C_1$$

given $\quad v = 10$ m/s at $x = 1$ m

$$50 = 0.02 - 10 + C_1$$

$$\Rightarrow \quad C_1 = 59.98$$

$$\therefore \quad v^2 = 0.04x - 20x^2 + 119.9$$

Velocity when

$$x = 0$$

$$v = \sqrt{120} \text{ m/s}$$

8. (a)

Given, $\vec{r} = \cos\omega t\hat{x} + \sin\omega t\hat{y}$

$$\therefore \vec{v} = \frac{d\vec{r}}{dt} = -\omega\sin\omega t\hat{x} + \omega\cos\omega t\hat{y}$$

$$\vec{a} = \frac{d\vec{v}}{dt} = -\omega^2\cos\omega t\hat{x} - \omega^2\sin\omega t\hat{y} = -\omega^2\vec{r}$$

Since position vector ($\vec{r}$) is directed away from the origin, so, acceleration ($-\omega^2\vec{r}$) is directed towards the origin.

Also,

$$\vec{r}.\vec{v} = (\cos\omega t\hat{x} + \sin\omega t\hat{y}).(--\omega\sin\omega t\hat{x} + \omega\cos\omega t\hat{y})$$

$$= -\omega\sin\omega t\cos\omega t + \omega\sin\omega t\cos\omega t = 0$$

$$\Rightarrow \vec{r} \perp \vec{v}$$

9. (a)

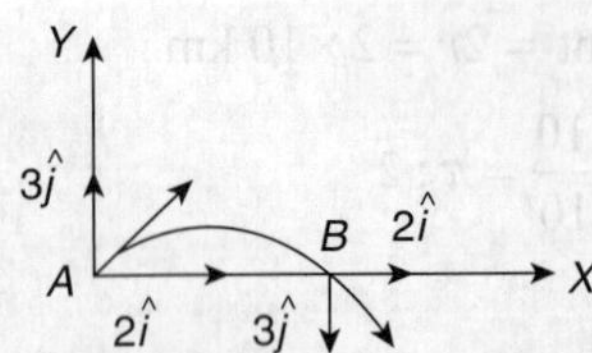

At point B, X Component of velocity remains unchanged while Y component reverses its direction.

$\therefore$ The velocity of the projectile at point B is $2\hat{i} - 3\hat{j}$ m/s.

10. (d)

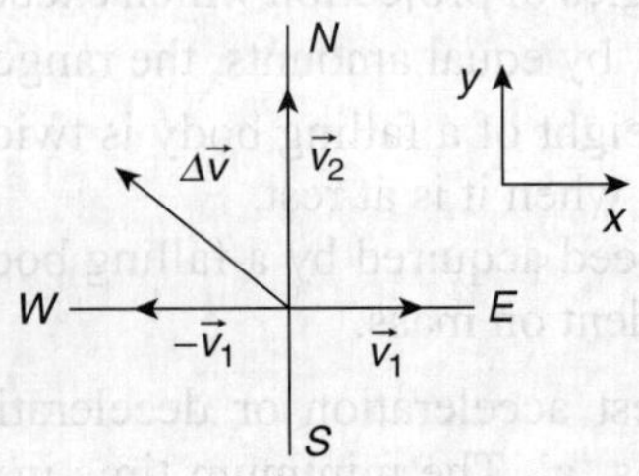

Velocity towards east direction, $\vec{v_1} = 30\hat{i}$ m/s

Velocity towards north direction, $\vec{v_2} = 40\hat{j}$ m/s

Change in velocity, $\Delta\vec{v} = \vec{v_2} - \vec{v_1} = (40\hat{j} - 30\hat{i})$

$$\therefore |\Delta\vec{v}| = |40\hat{j} - 30\hat{i}| = 50 \text{ m/s}$$

Average acceleration, $\vec{a}_{av} = \dfrac{\text{Change in velocity}}{\text{Time interval}}$

$$\vec{a}_{av} = \frac{\vec{v}_2 - \vec{v}_1}{\Delta t} = \frac{\Delta\vec{v}}{\Delta t}$$

$$|\vec{a}_{av}| = \frac{|\Delta\vec{v}|}{\Delta t} = \frac{50 \text{ m/s}}{10 \text{ s}} = 5 \text{ m/s}^2$$

11. (b)

Given: $(\vec{F}_1 + \vec{F}_2) \perp (\vec{F}_1 - \vec{F}_2)$

$$\therefore (\vec{F}_1 + \vec{F}_2).(\vec{F}_1 - \vec{F}_2) = 0$$

$$F_1^2 - F_2^2 - \vec{F}_1.\vec{F}_2 + \vec{F}_2.\vec{F}_1 = 0 \Rightarrow F_1^2 = F_2^2$$

i.e., F_1, F_2 are equal to each other in magnitude.

12. (c)

Both the particles have their initial velocity in vertical direction zero.

13. (c)

$$\omega_1 = \frac{2\pi}{T_1}, \omega_2 = \frac{2\pi}{T_2}$$

Given, $T_1 = T_2$

Therefore $\dfrac{\omega_1}{\omega_2} = 1$

14. (b)

15. (d)

Position of the car P at any time t, is $x_P(t) = at + bt^2$

$$v_P(t) = \frac{dx_P(t)}{dt} = a + 2bt$$

Similarly, for car Q

$$x_Q(t) = ft - t^2$$

$$v_Q(t) = \frac{dx_Q(t)}{dt} = f - 2t$$

$$\because v_p(t) = v_Q(t)$$

$$\therefore a + 2bt = f - 2t \text{ or, } 2t(b+1) = f - a$$

$$\therefore t = \frac{f-a}{2(1+b)}$$

16. (b)

Given $u = 0$

Distance travelled in 10 s, $S_1 = \frac{1}{2}a.10^2 = 50a$

Distance travelled in 20 s, $S_2 = \frac{1}{2}a.20^2 = 200a$

$$\therefore S_2 = 4S_1$$

17. (c)

From equation of motion, we have

$$v^2 = u^2 + 2as$$

Where, v is final velocity, u is initial velocity, a is acceleration and t is time.

Since, body starts from rest $u = 0$

$$v^2 = 2as$$

which is general equation of parabola $y^2 = 4ax$.

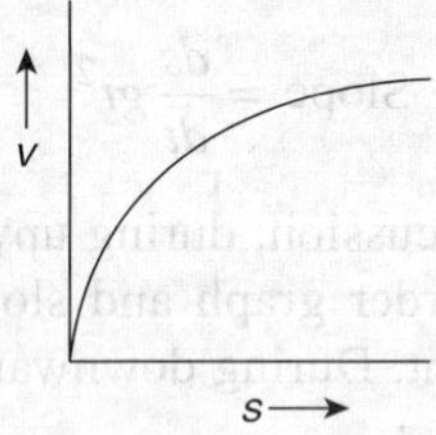

Hence, graph is parabolic and is symmetrical along s – axis.

18. (d)

From equation of motion, we have

$$v^2 = u^2 - 2gh$$

when $\quad u = 0, v^2 = -2gh$

As the ball is thrown upwards, its velocity decreases as height decreases.

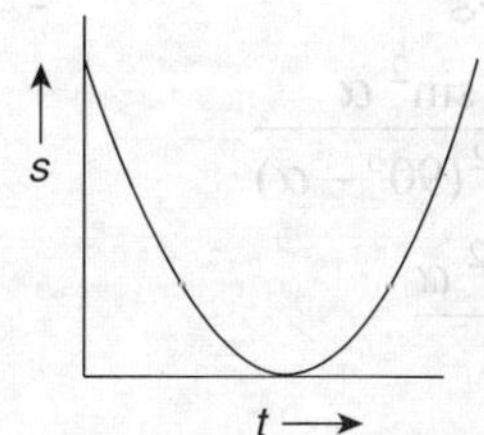

At the highest point, velocity is zero and after that, the body starts downwards motion; its speed increases.

19. (b)

$$\tan\theta = \frac{|\vec{R}|}{|\vec{P}|} = 1$$

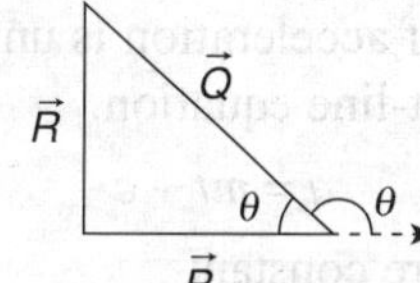

$$\theta' = 45°$$

$$\therefore \theta = 180° - 45° = 135°$$

Hence the answer is (b).

20. (d)

According to polygon law, resultant force will be zero.

Hence the answer is (d)

21. (d)

$$W = P\Delta V = 95 \times 10^3 \times (5 - 0) = 4.75 \times 10^5 \text{ J}$$

22. (c)

$$\text{Average velocity} = \frac{\text{Total displacement}}{\text{Total time}}$$

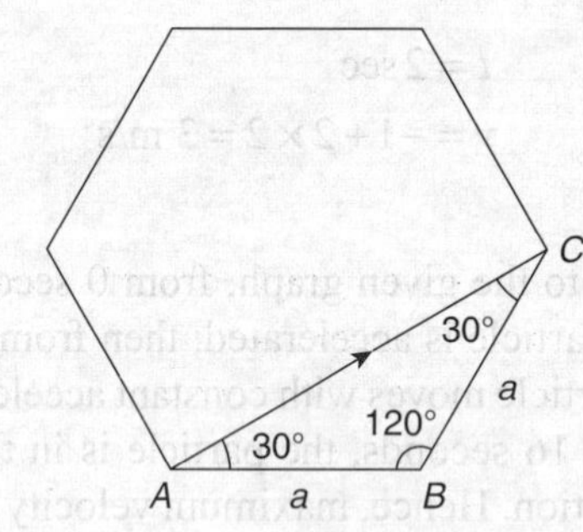

$$\vec{v}_{av} = \frac{AC}{t}$$

$$|\vec{v}_{av}| = \frac{AC}{t}$$

The time taken by particle to reach from A to C is

$$t = \frac{AB + BC}{v} = \frac{2a}{v}$$

In ΔABC, $\cos 120° = \frac{a^2 + a^2 - AC^2}{2a^2}$ or $-\frac{1}{2} = \frac{2a^2}{2a^2} - \frac{AC^2}{2a^2}$

$$\therefore \quad AC = \sqrt{3}a$$

$$\therefore \quad |\vec{v}_{av}| = \frac{\sqrt{3}a}{\frac{2a}{v}} = \frac{\sqrt{3}v}{2}$$

23. (d)
Since, acceleration is in the direction of instantaneous velocity, the particle always moves in forward direction.

24. (a)
Since variation of acceleration is uniform, acceleration obeys the straight-line equation.

$\therefore \qquad a = mt + c$

Where m and c are constant .

According to the problem, at $t = 0$, $a = 2$ m/s^2

$$\therefore 2 = m \times 0 + c \therefore c = 2 \qquad \text{...(i)}$$

$$\therefore a = mt + 2$$

When $t = 4$ sec, $a = 0$

$$\therefore 0 = m \times 4 + 2$$

$$\therefore m = -\frac{2}{4} = -\frac{1}{2} \qquad \text{...(ii)}$$

From eq. (i) and (ii)

$$a = -\frac{t}{2} + 2 \text{ or } \frac{dv}{dt} = -\frac{t}{2} + 2 \text{ or } \int_0^v dv = \int_0^t \left(-\frac{t}{2} + 2\right) dt$$

$$v = -\frac{t^2}{4} + 2t$$

At $\qquad t = 2$ sec

$$v = -1 + 2 \times 2 = 3 \text{ m/s}$$

25. (b)
According to the given graph, from 0 second to 8 seconds, the particle is accelerated; then from 8 to 12 second, the particle moves with constant acceleration. Then from 12 to 16 seconds, the particle is in the condition of deceleration. Hence, maximum velocity will be from 8 seconds to 12 seconds. During 0 to 4 seconds, then acceleration will be a function of time. The equation of straight line is

$$a = \frac{5}{4}t$$

$$\therefore \frac{dv}{dt} = \frac{5}{4}t$$

$$\therefore v = \int_0^t \frac{5}{4} t dt = \frac{5}{4}\frac{t^2}{2} = \frac{5}{8}t^2$$

The velocity $t = 4$ second is $u = 10$ m/s

The distance travelled during 4 to 8 sec is

$$s_2 = ut + \frac{1}{2}at^2 = 10 \times 4 + \frac{1}{2} \times 5 \times 4^2 = 40 + 40 = 80 \text{ m}$$

The velocity at $t = 8$ seconds is

$$v = 10 + 5 \times 4 = 30 \text{ m/s}$$

This is the maximum velocity.

Tricky approach: The area of $a - t$ graph gives change in velocity. The area of the graph from 0 to 8 second

$$= v - u = \frac{1}{2} \times 4 \times 5 + 4 \times 5 = 30$$

But $\qquad u = 0 \therefore v = 30$ m/s .

26. (a)
During upward motion

$$s = ut - \frac{1}{2}gt^2$$

$$\text{Slope } = \frac{ds}{dt} = u - gt^2$$

v = 0
at v = 0 A
Maximum height
g
$h = \frac{u^2}{2g}$
u
Ground
B B

During downward motion,

$$-s = \frac{1}{2}gt^2$$

(Displacement is in opposite direction)

$$\text{Slope } = \frac{ds}{dt} gt^2$$

From above discussion, during upward motion, graph is decreasing order graph and slope becomes zero a maximum height. During downward motion the slope becomes negative. This means from $t = 0$ to $t = \frac{u}{g}$ (upward motion), the graph is in decreasing order from maximum height to ground (downward motion) graph has negative slope, i.e., graph is increasing order graph.

27. (d)

$$H_1 = \frac{u^2 \sin^2 \alpha}{2g}, H_2 = \frac{u^2 \sin^2 (90° - \alpha)}{2g}$$

$$\frac{H_1}{H_2} = \frac{\sin^2 \alpha}{\sin^2 (90° - \alpha)}$$

$$\frac{H_1}{H_2} = \frac{\tan^2 \alpha}{1}$$

$$\therefore H_1 : H_2 = \tan^2 \alpha : 1$$

28. (c)

29. (d)
Here the lion moves on a circular path of radius 4 unit. If deer moves on the circular path of same radius, then deer may be killed by the lion.
So for saving her life, the deer must move on circular path of different radius.

30. (a)

31. (d)
The trajectory is shown in figure. From figure, at the top, velocity and gravitational acceleration are mutually perpendicular to each other.

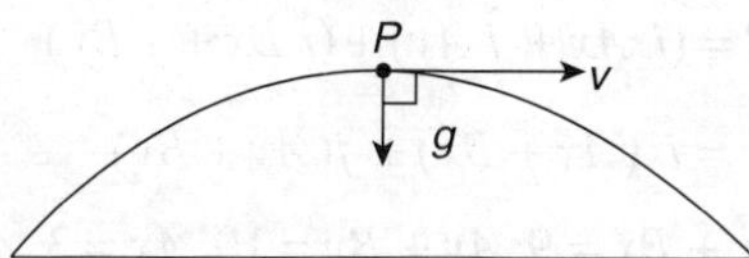

32. (d)
The maximum height is

$$H = \frac{u^2 \sin^2 \theta}{2g} \text{ or } 16 = \frac{u^2 \sin^2 45°}{19.6}$$

$$\therefore u = \sqrt{16 \times 2 \times 19.6}$$

But $R = \frac{u^2 \sin^2 \theta}{g} = \frac{16 \times 2 \times 19.6 \sin 90°}{9.8} = 64$ m

when $\theta = 45°, \sin 2\theta = 1$

But when $\theta = 30°, \sin 2\theta = \frac{\sqrt{3}}{2} = 0.866$

so, at $\theta = 30°$, range is lesser than that at $\theta = 45°$.

33. (a)
The rate of rain water in vessel depends upon vertical component of velocity. In this problem, the vertical component of velocity in both cases are same.

34. (b)

$$\because \text{Here, } x = \frac{t^2}{2}$$

$$\therefore v_x = \frac{dx}{dt} = t \;\therefore\; v_x = 2 \text{ m/s}$$

$$\because y = \frac{x^2}{2}$$

$$\therefore v_y = \frac{dy}{dt} = \frac{2x}{2}\frac{dx}{dt} \quad \therefore v_y = xv_x$$

$$= \frac{t^2}{2} v_x = \frac{(2)^2}{2}(2) \text{ m/s} = 4 \text{ m/s}$$

$$\therefore \vec{v} = v_x \hat{i} + v_x \hat{j} = (2\hat{i} + 4\hat{j}) \text{ m/s}$$

35. (b)
Horizontal component = u cos
Vertical component = $u \sin \theta$

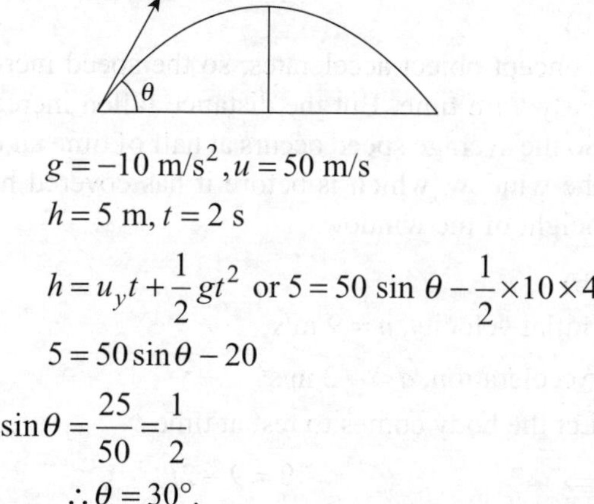

$$g = -10 \text{ m/s}^2, u = 50 \text{ m/s}$$

$$h = 5 \text{ m}, t = 2 \text{ s}$$

$$h = u_y t + \frac{1}{2} g t^2 \text{ or } 5 = 50 \sin \theta - \frac{1}{2} \times 10 \times 4$$

$$5 = 50 \sin \theta - 20$$

$$\sin \theta = \frac{25}{50} = \frac{1}{2}$$

$$\therefore \theta = 30°.$$

36. (a)

$$v \cos \theta = \frac{l}{t} \text{ or } t = \frac{l}{v \cos \theta}$$

When t is minimum $\cos \theta$ is maximum

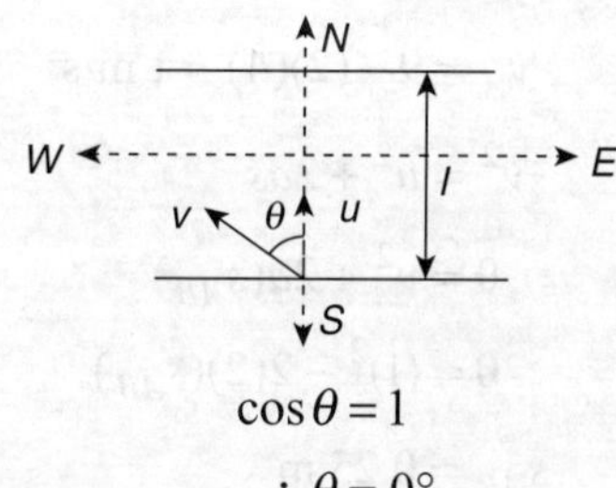

i.e., $\cos \theta = 1$

$$\therefore \theta = 0°$$

So, the boy should swim due north.

37. (d)

Here, $\omega = \frac{d\theta}{dt}$

$$\therefore v_x = \frac{dx}{dt} = a\left\{-\sin \theta \frac{d\theta}{dt} + \sin \theta \frac{d\theta}{dt} + \theta \frac{d(\sin \theta)}{dt}\right\}$$

$$= a\omega\{-\sin \theta + \sin \theta + \theta \cos \theta\} = a\omega\theta \cos \theta$$

$$v_y = \frac{dy}{dt} = a\left\{\cos \theta \frac{d\theta}{dt} - \cos \theta \frac{d\theta}{dt} + \theta\omega \sin \theta\right\}$$

$$= a\omega\{\theta \sin \theta\} = a\omega\theta \sin \theta$$

$$v = \sqrt{v_x^2 + v_y^2} = a\omega\theta$$

38. (d)
Throughout motion, the ball has acceleration g in downward direction; so slope of v – -t graph must be constant. Only possible option here is (d).

39. (a)

As slope of A is more than that of B hence option (a) is correct.

40. (b)

Concept object accelerates, so the speed increases linearly with time. But the distance fallen increases at t^2. So the average speed occurs at half of time taken to pass the window, which is before it has covered half of the height of the window.

41. (b)

Initial velocity, $u = 9$ m/s

Acceleration, $a = -2$ m/s^2

Let the body comes to rest at time t

$$\Rightarrow \quad 0 = 9 - 2t$$
$$t = 4.5 \text{ s}$$

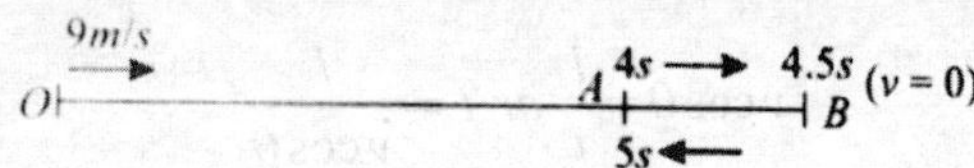

Velocity of body at A,

$$v_A = 9 - (2)(4) = 1 \text{ m/s}$$

Using $\quad v^2 = u^2 + 2as$

$$0 = v_A^2 + 2a(s_{AB})$$
$$0 = (1)^2 - 2(2)(s_{AB})$$
$$s_{AB} = 0.25 \text{ m}$$

Similarly, $\quad s_{BA} = 0.25$ m

$\therefore$ Distance covered by particle in fifth second of its motion,

$$s = s_{AB} + s_{BA} = 0.25 \text{ m} + 0.25 \text{ m} = 0.5 \text{ m}$$

42. (b)

43. (d)

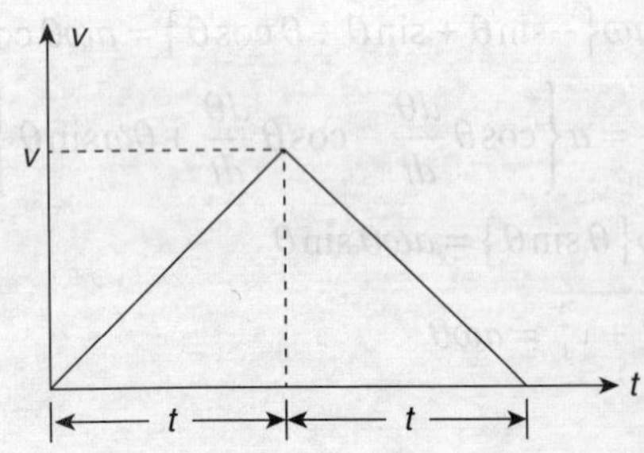

$$s = \frac{1}{2}(2t)v = vt$$
$$s = at^2$$
$$t = \sqrt{\frac{s}{a}}$$

Thus, time to go to another station,

$$2t = 2\sqrt{\frac{s}{a}}$$

44. (b)

$$\vec{A} + \vec{B} = (\hat{i}\ Ax + \hat{j}\ Ay) + (i\ Bx + \hat{j}\ By)$$
$$= \hat{i}\ (Ax + Bx) + \hat{j}(Ay + By)$$

Given $Ax + Bx = 9, Ay + By = 10, Ax = 3, Ay = 7$

Therefore, $\quad Bx = 9 - 3 = 6$

$By = 10 - 7 = 3$

vector $\vec{B} = 6\hat{i} + 3\hat{j}$

45. (b)

Let P be the smaller force, then it is given that

$$P + Q = 18 \qquad \text{...(1)}$$

$$R = \left[P^2 + Q^2 + 2PQ \cos\theta\right]^{1/2} = 12 \qquad \text{...(2)}$$

and

$$Q \sin\theta / [P + Q\cos\theta] = \tan\phi = \tan 90° = \infty$$

i.e., $$P + Q\cos\theta = 0 \qquad \text{...(3)}$$

Substituting the value of P from Eq. (1) in (3),

$$Q(1 - \cos\theta) = 18 \qquad \text{...(4)}$$

Now subtracting square of Eq. (2) from (1),

$$2PQ(1 - \cos\theta) = 18^2 - 12^2 = 180 \qquad \text{...(5)}$$

Dividing Eq. (5) by (4),

$$2P = 10, \text{ i.e., } P = 5$$

Substituting for P in Eq. (1),

$$Q = 13$$

So the magnitudes of forces are 5 and 13 units.

CHAPTER 3

Laws of Motion

Introduction

Force represents the interaction between two bodies. The term "force" appears both in "force laws" and the "laws of motion". The force laws give the forces acting on a body (system) and that exerted by another body (environment) while the laws of motion give the effect of the forces. One effect of the force is to alter the dimensions or shape of a body on which it acts; and the other effect is to alter of motion of the body.

An unbalanced force on an object causes the object to accelerate in the direction of the force.

The acceleration is proportional to the force and inversely proportional to the mass of the object.

Newton's Laws of Motion

First Law

"Every body continues in its state of rest or of uniform motion in a straight line, unless it is compelled to change its state by forces impressed on it".

It follows that once a body has been set in motion it is no longer necessary to exert a force on it to maintain it in motion.

Newton's first law gives a qualitative definition of force, or at least one aspect of the force concept as "that which changes the state of motion of body".

Second Law

"The rate of change of linear momentum of a body is directly proportional to the force acting on it and it takes place in the direction of the applied force".

Let $\vec{F}$ be the net force and $\vec{P}$ be the momentum, then

$$\vec{F} \propto \frac{d\vec{P}}{dt} = k\frac{d\vec{P}}{dt} \quad (k \text{ is a constant of proportionality})$$

$$\vec{F} = m\frac{d\vec{v}}{dt} + \vec{v}\frac{dm}{dt}, \text{ if } m \text{ is not constant}$$

$$\vec{F} = \frac{d\vec{P}}{dt} = m\frac{d\vec{v}}{dt} = m\vec{a}\text{ , if } m \text{ is a constant}$$

In the SI system, F is in Newton, m is in kg, and a is m/s^2. The vector equation

$\vec{F} = m\vec{a}$ may be written in terms of components, as

$$\Sigma F_x = ma_x, \Sigma F_y = ma_y, \Sigma F_z = ma_z$$

Third Law

To every action there is always an equal and opposite reaction, or the mutual action of two bodies upon each other are always equal and directed to contrary parts. Either of the forces may be considered the "action", the other as "reaction".

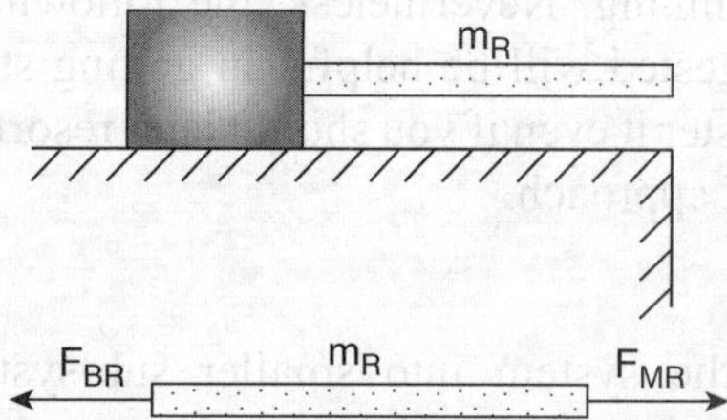

For example, consider a man pulling one end of a rope attached to a block, as shown in figure.

The Action–Reaction pairs of forces are indicated in the figure $\vec{F}_{MR}$ represents the force exerted by the man on the rope. A similar notation follows for other forces. The reaction force of $\vec{F}_{MR}$ is $\vec{F}_{RM}$ the force exerted by the rope on man. According to third law, $\vec{F}_{MR} = -\vec{F}_{RM}$, similarly $\vec{F}_{MR} = -\vec{F}_{RB}$. Force $\vec{F}_{MR}$ and $\vec{F}_{BR}$ are opposite in direction and have the same line of action. They do not constitute an action–reaction pair. This is because, both these forces act on the same body (the rope), while action–reaction pair necessarily act on different bodies.

Further, forces $\vec{F}_{MR}$ and $\vec{F}_{BR}$ are not necessarily equal in magnitude. If the block and rope are moving toward right (as shown in the figure) with increasing speed, the rope is not in equilibrium and F_{MR} is greater than F_{BR} . For a special case, when the rope is rest or in uniform motion, F_{MR} and F_{RB} are equal in magnitude ($F_{MR} = F_{BR}$, first law, and $F_{RB} = F_{BR}$, third law). Hence, F_{MR} is equal to F_{RB}. The rope can therefore be considered to transmit to the block, without change, the forces exerted on it by the man.

As another example, consider a body "A" suspended by means of a string from the ceiling (see figure). The body A, because of its weight W, exerts a force F on the lower of the string. The lower end of the string pulls the body up by a reaction force F. The upper end of the string pulls down the ceiling with a force F'. The string is therefore pulled in opposite direction, with forces F and F'. Since the string is at rest, we have $F = F'$. The force F acting on the string is called tension.

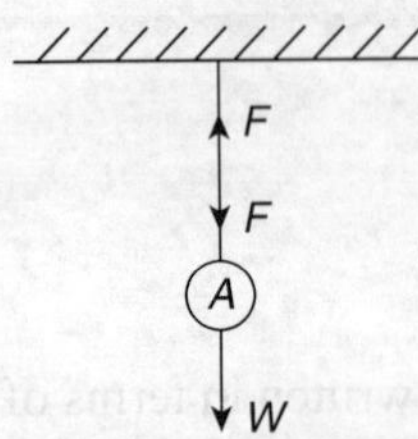

Application of Newton's Laws

It is neither possible nor desirable to follow rigidly a prescribed set of rules in solving problems involving Newton's laws. There are many ways of approaching a problem and the procedure we suggest is certainly not the only one. In fact, no cut-and-dry procedure can even substitute for an intelligent analytical thinking. Nevertheless, the following systematic method suggested will be helpful in getting started and we urge you master it even if you should later resort to short-cuts or a different approach.

Steps

1. Divide the system into smaller sub-system, each of which can be treated as point mass.
2. (i) Represent the body by a point or simply symbol.
   (ii) Draw the **Free-Body Diagram (FBD)** – The representation of all the forces acting on a body is known as its free-body diagram, i.e., draw all the forces acting on the chosen object, for example, the body's weight, the forces applied by the other objects with which the chosen object is in contact and non-contact forces (in addition to the weight), such as electric and magnetic forces, if any, acting on the body. Remember that while applying Newton's second law, we are concerned only with the external forces acting on the chosen object. The chosen body also exerts forces on its surrounding but such forces are irrelevant because they do not act on the chosen object. Indicate on the diagram the magnitude and direction of each force.

As an example, consider two blocks at rest on a table top. The free-body diagram for block A is shown in Figure (a). F_1 is the force exerted on block A by block B, and W_A is the force of gravity on A, called the weight. Similarly the free-body diagram for block B has been drawn in Figure (b). W_B is the force of gravity on block B, and F_2 is the force exerted by A on B. There are no other physical interactions on B.

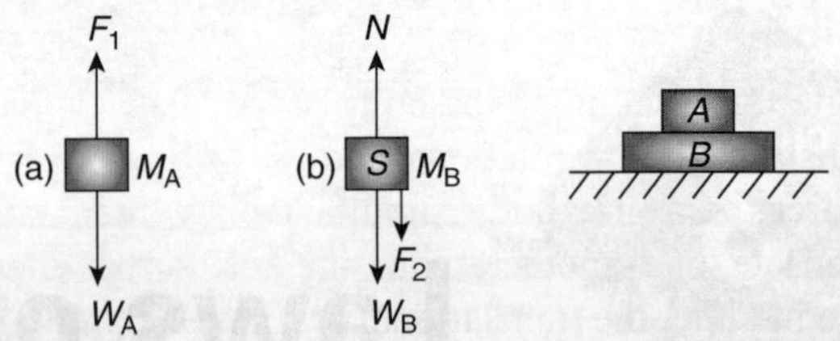

The following are some more examples of free-body diagram:

In this diagram, the object of interest is isolated from its surroundings and the interactions between the object and the surroundings are represented in terms of forces.

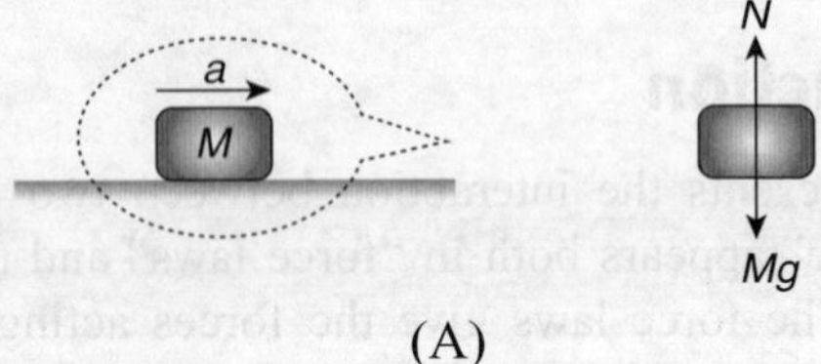

(A)

FBD of a block placed on a horizontal surface (Fig. A). Two vertical forces act on the block: The earth pulls the block downward by mg, and the surface pushes the block upward by N.

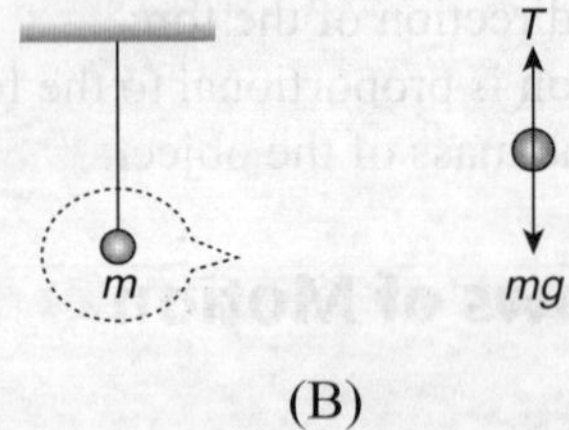

(B)

FBD of a ball suspended by a string (Fig. B). Two vertical forces act on the ball: The earth pulls the ball downward by mg and the string pulls the ball upwards by T.

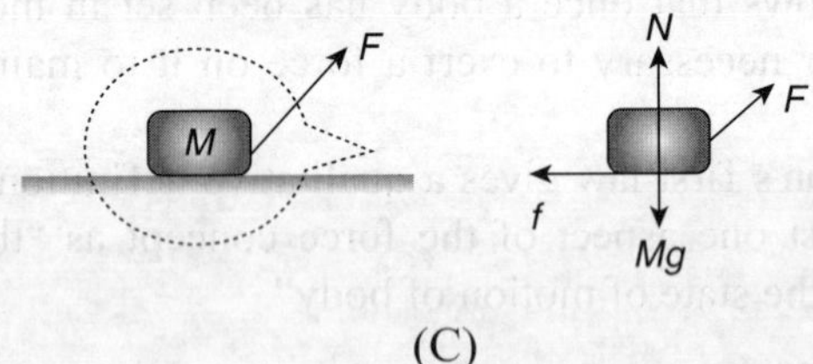

(C)

FBD of a block placed on a rough surface being pulled by an external force (Fig. C). There are four forces acting on the block: the gravitational pull mg; the normal reaction N; the external force F; and the tangential force of friction f.

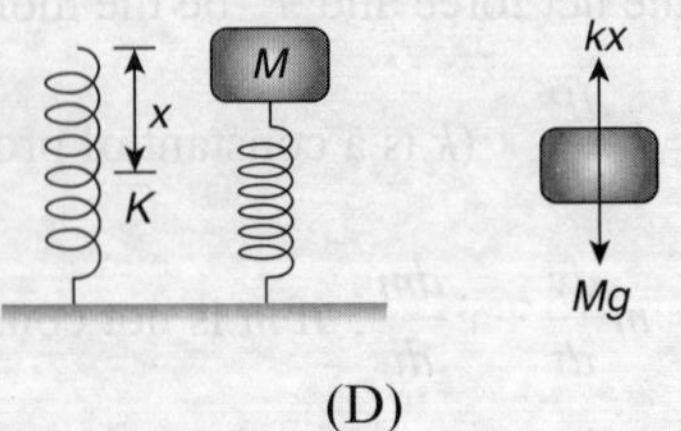

(D)

FBD of a block supported by a spring of stiffness k (Fig. D). Two vertical forces act on the block: the gravitational pull, and the spring force kx.

Equilibrium

A single force acting on the body produces a change in both its translational and rotational motion. However, when several forces act on a body simultaneously, their effects can compensate one another, with the result that there is no change in either the translational or rotational motion. In this situation, the body is said to be in equilibrium. It means, for the equilibrium of a body, (1) the body as a whole remains at rest or moves in a straight line with constant speed, (2) the body is either not rotating at all or rotating at a constant rate. Mathematically,

$$\Sigma\vec{F}=0,\ \text{1st condition of equilibrium}$$

$$\Sigma\vec{\tau}=0\ \text{2nd condition of equilibrium}$$

τ being the torque or moment of equilibrium

The first condition ensures that a body will be in translational equilibrium, the second, that it will be in rotational equilibrium. When both the conditions are satisfied, the body will be in complete equilibrium, which is Newton's first law of motion. In this chapter, we will discuss only translational equilibrium.

Illustration 1

An object is in equilibrium under four concurrent forces in the directions shown in figure. Find the magnitude of $\vec{F_1}$ and $\vec{F_2}$.

Solution

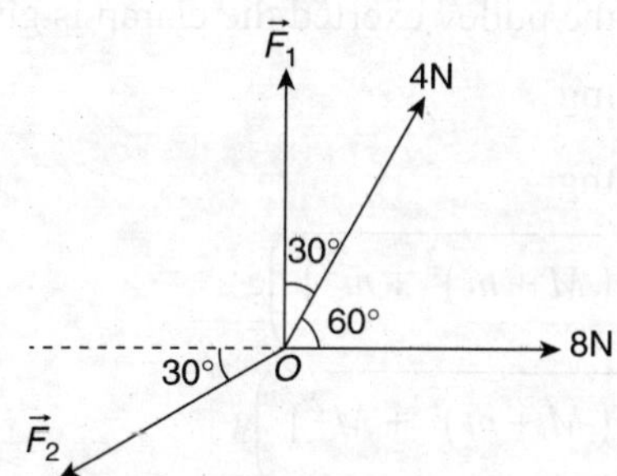

The object is in equilibrium. Hence,

$$\Sigma F_x=0\,,\ \Sigma F_y=0$$

$$\Sigma F_x=8+4\cos60°-F_2\cos30°=0$$

$$=8+2-F_2\frac{\sqrt{3}}{2}=0$$

$$F_2=\frac{20}{\sqrt{3}}\text{N}$$

$$\Sigma F_y=F_1+4\sin60°-F_2\sin30°=0$$

$$F_1+\frac{4\sqrt{3}}{2}-\frac{F_2}{2}=0$$

$$F_1=\frac{F_2}{2}-2\sqrt{3}=\frac{10}{\sqrt{3}}-2\sqrt{3}$$

$$F_1=\frac{4}{\sqrt{3}}\text{N}$$

Illustration 2

A worker applies a constant horizontal force with magnitude 10 N to a box with mass 20 kg resting on a level surface with negligible friction. What is the acceleration of the box?

Solution

$$\Sigma F_x=ma_x$$

$$ax=\frac{\Sigma F_x}{m}=\frac{10}{20}=0.50\text{ m/s}^2$$

Illustration 3

There forces act on a particle that moves with unchanging velocity $\vec{v}=2\text{m/s}\,i-7\text{ m/s}\,j$. Two of the forces are $\vec{F}_1\ 2Ni+3N\hat{j}-2N\hat{k}$ and $\vec{F}_2=-5N\hat{i}+8N\hat{j}-(-2N)\hat{k}$. What is the third force?

Solution

Since acceleration is zero $\Sigma F=0$

$$\vec{F}_1+\vec{F}_2+\vec{F}_3=0$$

$$F_3=-(\vec{F}_1+\vec{F}_2)$$

$$=3N\hat{i}-11\hat{j}+4N\hat{k}$$

Tension

Nature of the forces exerted by an ideal rope: An ideal rope is one which is perfectly flexible, inextensible and massless. No such rope or cord exists. But in many situations a real rope is very much close to the ideal. For a massless rope, the tension is the same at all points along the length of the rope irrespective of whether the rope is accelerated or in equilibrium.

Let us consider a short segment of a cord as shown in the figure. The tension on it from the neighbouring parts on the cord at the two ends are labeled T_1 and T_2; its weight is mg. Applying Newton's second law to the small segment, we have

$$\Sigma\vec{F}=m\vec{a}$$

$$\vec{T_1}+\vec{T_2}+m\vec{g}=m\vec{a}$$

But if mass m is negligibly small, we may discard both mg and ma terms, and the above equation reduces as $-\vec{T}_1=\vec{T}_2$ or in magnitude $T_1=T_2$.

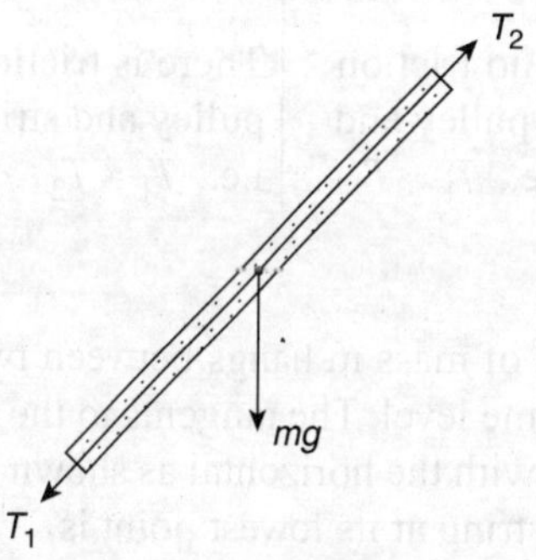

Then tension T has the same magnitude at both ends of the small segment. Moreover, the tension, defined as the force exerted by any cord segment on the neighbouring segment, is the same at all points along the cord, provided the cord's mass and weight are negligible.

The following are some of the examples of tension force when a body is connected by means of a string or a rope.

(i) If a string is ideal (inextensible and massless), then the magnitude of the accelerations of any number of masses connected through this string is always same.

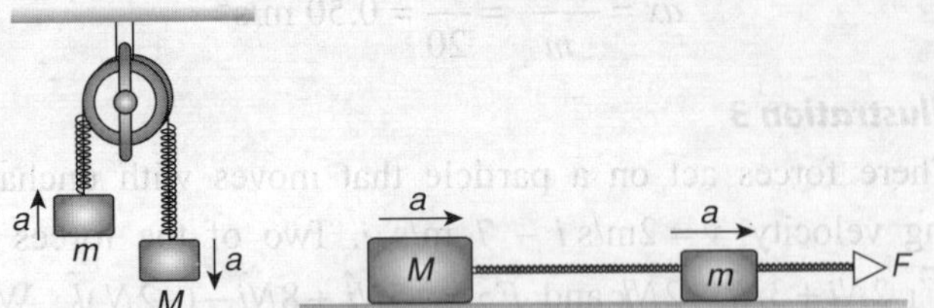

(ii) If a string is massless, the tension in it is same everywhere; on the other hand, if a string is not massless (i.e., its mass is not negligible), tension at different points may be different.

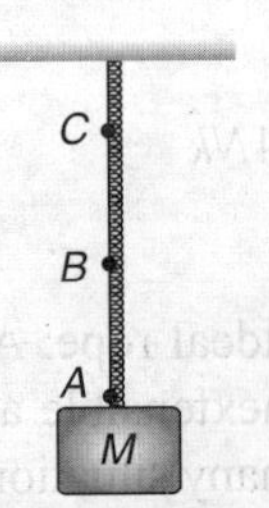

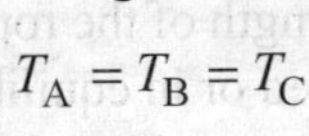

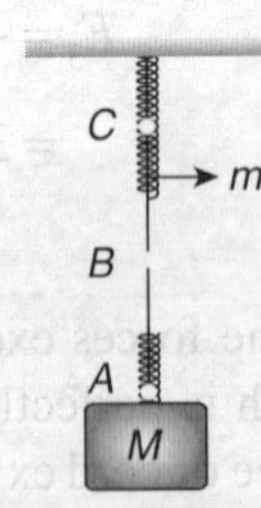

String is massless, i.e., $T_A = T_B = T_C$

String is not massless, i.e., $T_A \neq T_B \neq T_C$ and $T_A < T_B < T_C$

(iii) If friction is present between the pulley and the string, tension is different on the two sides of the pulley But if the friction is absent between the pulley and the string, tension will be same on both sides of the pulley, provided the pulley is massless.

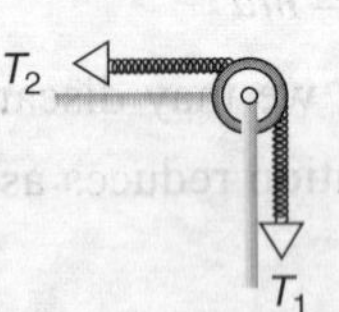

There is no friction between pulley and string, i.e., $T_1 = T_2$

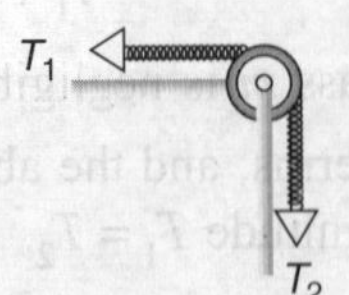

There is friction between pulley and string, i.e., $T_1 \neq T_2$

Illustration 4

A heavy string of mass m hangs between two fixed points A and B at the same level. The tangents to the string at A and B make angle θ with the horizontal as shown in the figure. The tension in the string at its lowest point is

(a) $\frac{mg}{2\sin\theta}$ (b) $\frac{mg}{2\cos\theta}$

(c) $\frac{mg}{2\tan\theta}$ (d) $\frac{mg}{2\cot\theta}$

Solution

Let T be the tension at each of point A and B. Then for equilibrium of the string

$$2T\sin\theta = mg$$

$$\Rightarrow \quad T = \frac{mg}{2\sin\theta}$$

Further if T' be the tension at the lowest point of the string for equilibrium of the half of string, balancing the horizontal force,

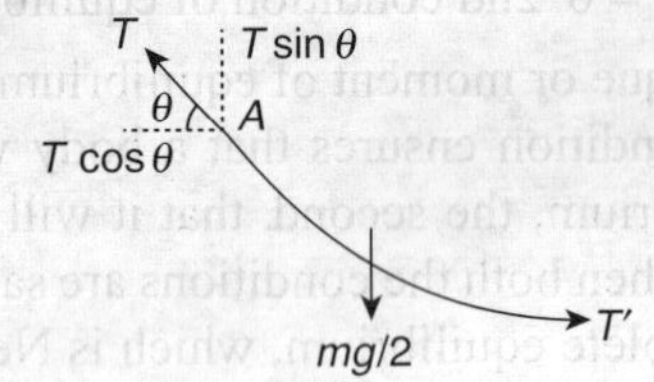

$$T' = T\cos\theta = \frac{mg}{2\sin\theta}\cos\theta = \frac{mg}{2\tan\theta}$$

Ans. (c)

Illustration 5

A string of negligible mass going over a clamped pulley of mass m supports a block of mass M as shown in the figure. The force on the pulley exerted the clamp is given by

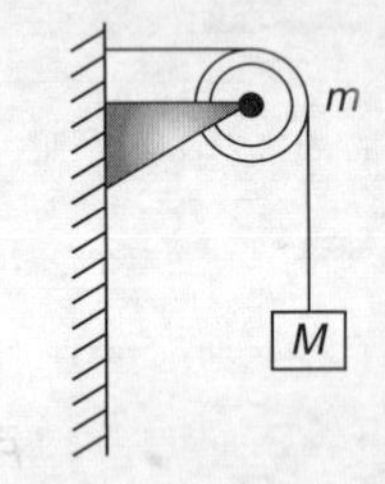

(a) $\sqrt{2}$ mg

(b) $\sqrt{2}$ mg

(c) $\left(\sqrt{[(M+m)^2 + m^2]}\right)$g

(d) $\left(\sqrt{[(M+m)^2 + M^2]}\right)$g

Solution

Drawing the FBD of the pulley and considering its equilibrium

F_x = horizontal component of the force by the clamp on the pulley = Mg and F_y = and vertical component of the force by the clamp on the pulley = $(M + m)\,g$

$$\therefore \quad F = \sqrt{F_x^2 + F_y^2} = \sqrt{(M+m)^2 g^2 + M^2 g^2}.$$

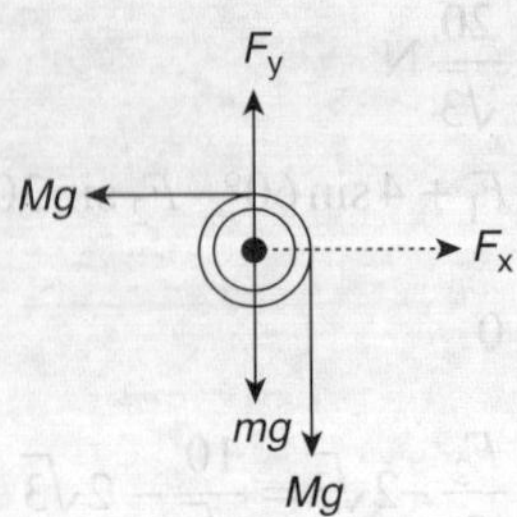

Ans. (d)

Illustration 6

Two masses m and M are attached with strings as shown. For the system to be in equilibrium we have

(a) $\tan\theta = 1 + \frac{2M}{m}$ (b) $\tan\theta = 1 + \frac{2m}{M}$

(c) $\tan\theta = 1 + \frac{M}{2m}$ (d) $\tan\theta = 1 + \frac{m}{2M}$

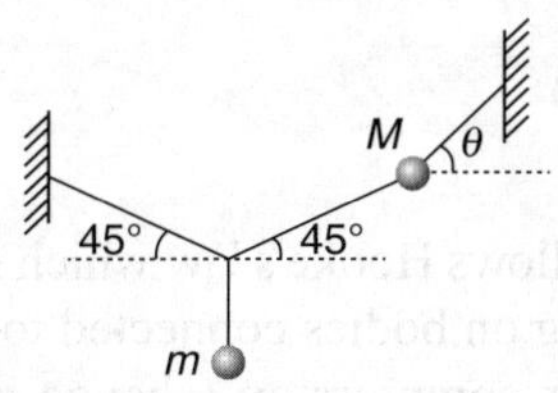

Solution

Draw the FBD of m and M and considering their equilibrium

$$mg = 2T\sin 45$$

$$\Rightarrow \quad mg = \sqrt{2}T$$

or $$T = \frac{mg}{\sqrt{2}} \quad \text{...(i)}$$

$$T_1\cos\theta = T\cos 45$$

$$\Rightarrow \quad T_1\cos\theta = \frac{T}{\sqrt{2}} = \frac{mg}{2} \quad \left\{\because T = \frac{mg}{\sqrt{2}}\right\}$$

Further, $Mg + T\sin 45° = T_1\sin\theta$

$$\Rightarrow \quad T_1\sin\theta = Mg + \frac{mg}{\sqrt{2}}\frac{1}{\sqrt{2}}$$

$$\Rightarrow \quad T_1\sin\theta = Mg + \frac{mg}{2} \quad \text{...(ii)}$$

$$\Rightarrow \quad \tan\theta = \frac{Mg + \frac{mg}{2}}{\frac{mg}{2}} = 1 + \frac{2M}{m} \quad \{\text{Divide (ii) by (i)}\}$$

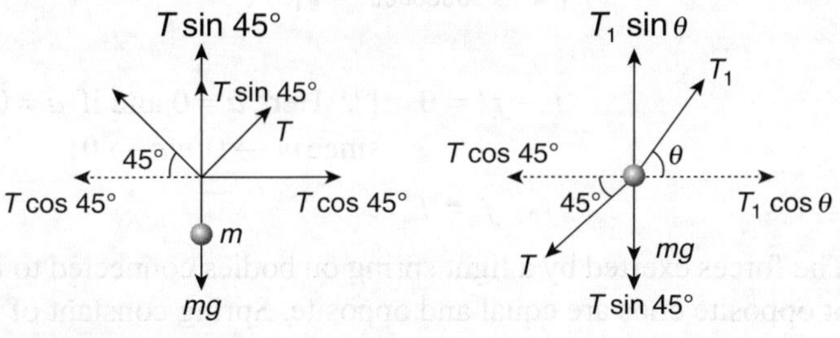

Ans. (a)

Spring and String Analogy

Consider a spring whose one end is fixed to the ceiling and the other end attached to a block at rest (Figure (a)). Since all the bodies are at rest, the vector sum of forces acting on a body is zero. The free-body diagram of forces acting on the block is shown in Figure (b)

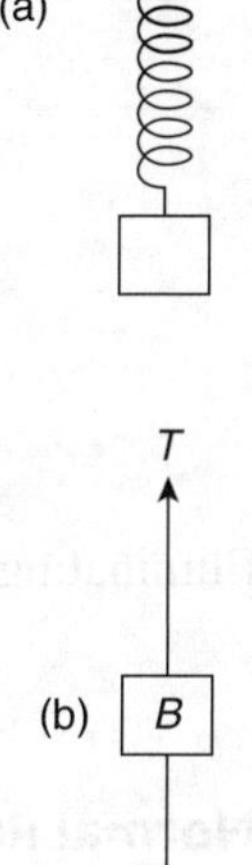

T is the tension in the stretched spring pulling the block up. W is pull of the earth acting vertically downward.

$\Sigma\vec{F}$, the resultant force on the block is given as $\Sigma\vec{F} = \vec{T} + \vec{W}$,

The block is at rest, $\Sigma\vec{F} = 0; \vec{T} = -\vec{W}$

T' = Force exerted by the block on the spring

P = Upward pull of the ceiling on the spring

W = Weight of the spring

Since the spring is at rest, $\Sigma\vec{F} = \vec{P} + \vec{T} + \vec{W} = 0$

$$\vec{T'} = -\vec{T} = \vec{W}$$

$$\Rightarrow \vec{P} = -(\vec{W} + \vec{w})(T' = T = W)$$

or $$P = W + w$$

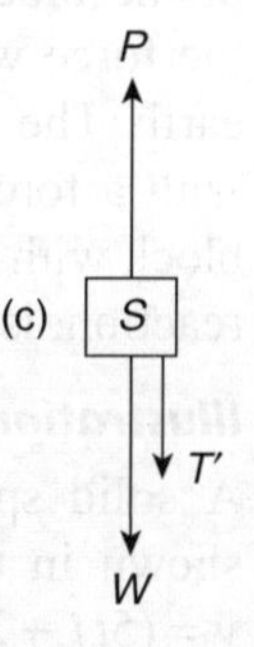

The ceiling therefore pulls up on the spring with a force whose magnitude is the sum of the weights of the block and spring. The force exerted by the spring on the ceiling is also equal to $(W + w)$.

In general, the spring exerts different forces on the bodies attached at its different ends, as P and T. In the special case in which the weight of the spring is negligible, $w = 0$ and $P = W = T$. Therefore a weightless spring or string may be considered to transmit a force from one end to the other without change.

Illustration 7

A heavy rope of length L, lying on a smooth horizontal surface, is pulled by a constant force F. What is the tension T in the rope at a distance l from the end where force F is applied?

Solution

Let AB be the rope of length L and C be the point such that CB equals l. The length of AC is $(L – l)$. Let m be the mass per unit length of the rope.

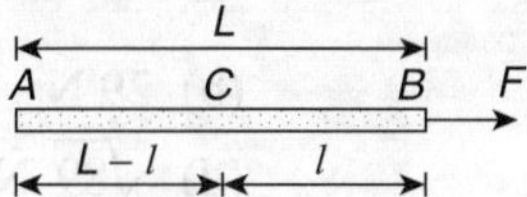

Part AC of the rope may be treated as a body of mass $(L – l)$ m to which the rope BC is tied. Let the tension between the rope AC and BC be T.

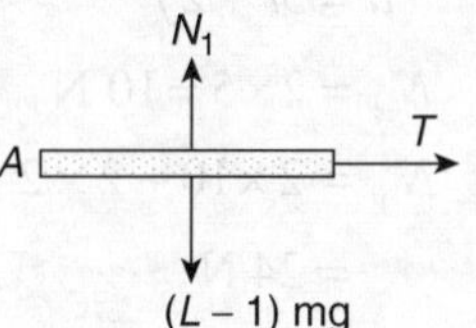

The free-body diagram of the two segments AC and CB are shown in the figure.

Applying Newton's law for the two segments in the horizontal plane, we get

$$F - T = mla \quad \ldots\text{(i)}$$

$$T = m\,(L - l)\,a \quad \ldots\text{(ii)}$$

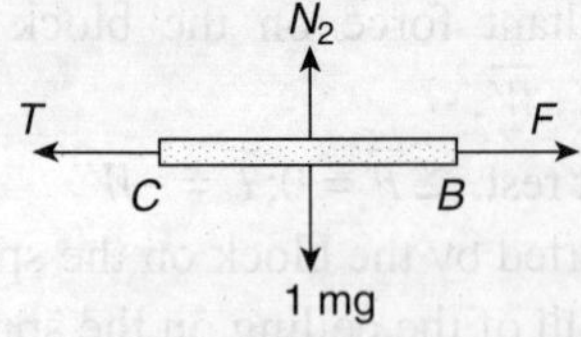

Eliminating ma from the two equations, we get

$$T = F\frac{(L - l)}{L}$$

Normal Reaction

Let a block B rest on a horizontal surface A. The block B is attracted by the earth with a force W. This force is called the weight of the block. The reaction to this force is the force with which the block attracts the earth. The block presses down on the table with a force $F = W$. The table at the same time pushes the block with a force $N = F = W$. The force N, called the normal reaction, is the reaction to the force F.

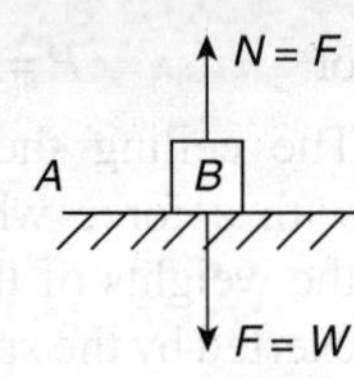

Illustration 8

A solid sphere of mass 2 kg is resting inside a cube as shown in the figure. The cube is moving with a velocity $\vec{v} = (5t\,\hat{i} + 2t\,\hat{j})$ m/s. Here t is the time in second. All surfaces are smooth. The sphere is at rest with respect to the cube. What is the total force exerted by the sphere on the cube (take $g = 10$ m/s^2)?

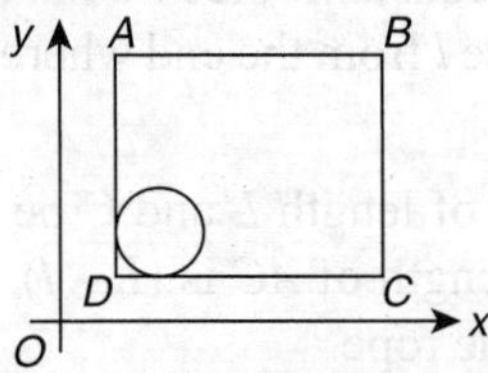

(a) $\sqrt{29}$ N (b) 29 N
(c) 26 N (d) $\sqrt{89}$ N.

Solution

$$\vec{v} = 5t\,\hat{i} + 2t\,\hat{j}$$

$$\therefore \quad \vec{a} = 5\hat{i} + 2\hat{j}$$

$$\therefore \quad N_x = 2 \times 5 = 10 \text{ N}$$

$$N_y = 2 \times 10 + 2 \times 2$$

$$= 24 \text{ N}$$

$$\therefore \quad \text{Total force} = \sqrt{N_x^2 + N_y^2}$$

$$= \sqrt{(10)^2 + (24)^2} = 26 \text{ N}.$$

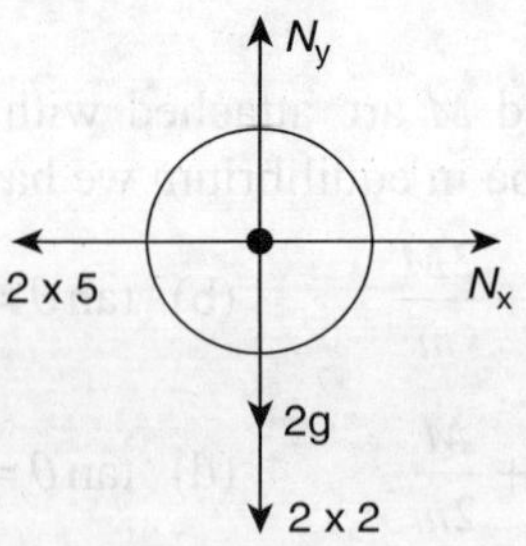

Ans. (c)

Elastic Force

An ideal spring follows Hooke's law which says that the force applied by a spring on bodies connected to it is proportional to its extension or compression (change in length from its natural length).

Consider a light spring which is connected to a vertical wall as shown in the figure. Suppose that it is pulled to the right by means of a force F, which causes the spring to get elongated by x over its natural length. Instantaneous rest to spring force f will be equal to applied force and in the opposite direction.

Then, $\quad f \propto -x$

$\Rightarrow \quad f = -kx$

where k is a constant that is characteristic of the spring, also known as the spring constant.

In this case, the force applied by the spring on the wall is also equal to F. Suppose that the force applied by the spring on the wall is f, then the wall will also exert a force f' on the spring. Consider the forces acting on the spring:

If the mass of spring is m and its acceleration a to the right,

$$f - f' = 0 \quad \{\because \text{ Here } a = 0 \text{ and if } a \neq 0 \text{ since } m \to 0, ma \to 0\}$$

$\Rightarrow \quad f = f'$

The forces exerted by a light spring on bodies connected to it at opposite ends are equal and opposite. Spring constant of a spring is inversely proportional to its relaxed length $k \propto \frac{1}{\ell}$.

Illustration 9

By applying a force F, the mass M is displaced vertically down by y from equilibrium position. Find the force F in terms of the force constant k of the spring and displacement y, for the cases (a) and (b) as shown in figure.

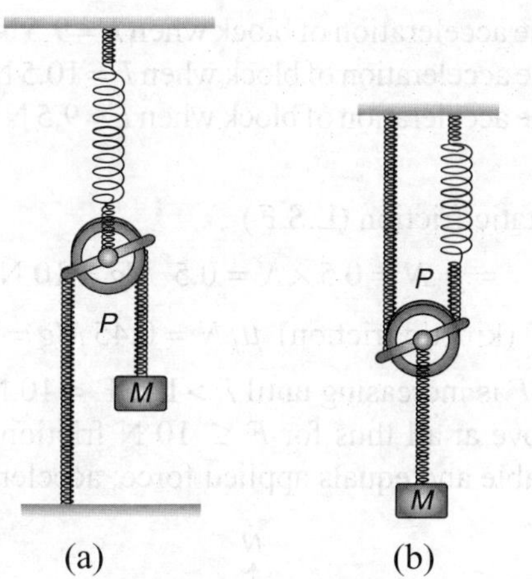

(a) (b)

Solution

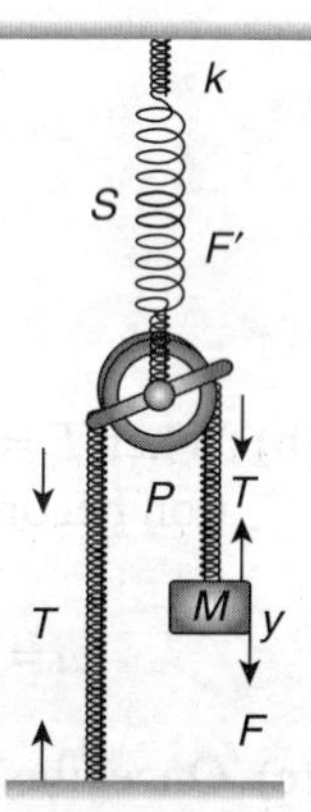

(a) Let T be additional tension created in string, due to application of F.
For the equilibrium of mass M, $F = T$ ...(1)
and for the equilibrium of pulley
$2T = F'$ ...(2)
Here F' is the additional force in the spring
But a due to shift of mass M by y, the spring stretches additionally by $(y/2)$
so $F' = k(y/2)$ ...(3)
So substituting F' from Eq. (3) in Eq. (2) and then T from Eq. (2) in Eq. (1), we get $F' = (k/4)y$.

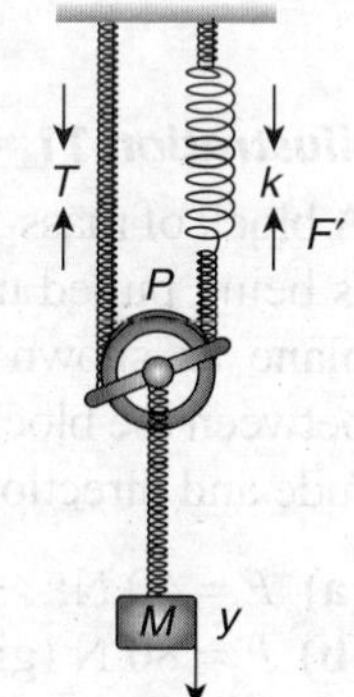

(b) As tension in massless string and spring will be same
$T = F'$ ...(1)
and for equilibrium of pulley,
$T + F' = F$ ...(2)

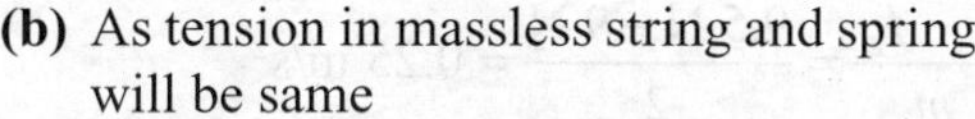
So from Eqs. (1) and (2), $F = 2F'$

Now if the mass M shifts by y the spring will stretch by $2y$ Therefore $F' = k(2y)$

So substituting F' from Eq. (4) in Eq. (3), we get, $F = (4k)y$

Frictional Forces

Friction is a force that opposes relative motion between bodies in contact. It can act between solids or, between a pair of fluids and between a solid and a fluid. Friction acting between two bodies in relative motion is known as kinetic friction while the force of friction acting between bodies, opposing the onset of relative motion is known as static friction.

The following points about force of friction are worth noting:

(i) If a body is at rest on a rigid horizontal surface and no pulling force is acting on it, force of friction on it is zero.

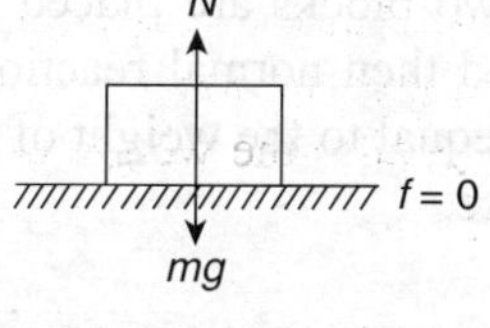

(ii) Now if a force is applied to pull the body and it does not move, the force of friction which acts on it is equal in magnitude and opposite in direction to the applied force, i.e., friction is a self-adjusting force. Since the body is at rest, the friction is called static friction.

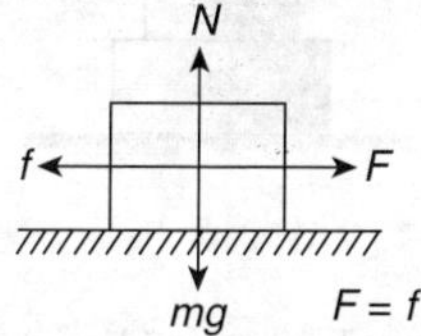

(iii) If the applied force is increased, the force of static friction also increases. If the applied force exceeds a certain (maximum) value, the body starts moving. This maximum force of static friction up to which body does not move is called limiting friction. Thus, the force of static friction is a self-adjusting force with an upper limit which is known as the limiting force of static friction.

(iv) This limiting force of friction (F_L) is found experimentally to be proportional to the force of normal reaction, i.e.,

$$f_L \propto N$$

or,
$$f_L = \mu_s N$$

where μ_s is a dimensionless constant called coefficient of static friction which depends upon the nature of surfaces in contact.

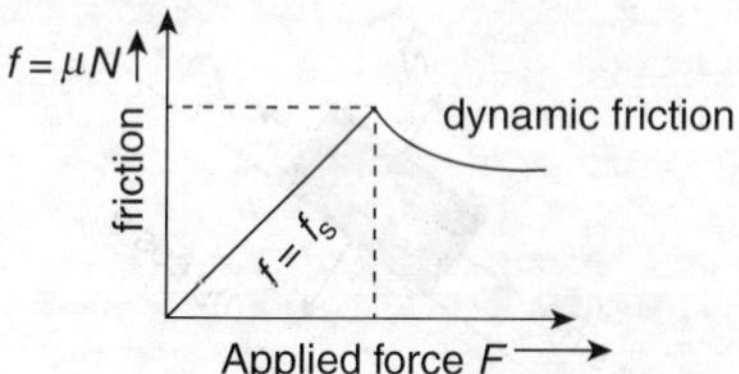

If the applied force is increased above its limiting value, it is observed that relative motion occurs. The force of friction opposing motion is known as kinetic friction (sliding friction).

$$f_K = \mu_K N$$

where μ_K is coefficient of kinetic friction and is less than μ_s.

(v) According to laws of friction, the force of kinetic friction is independent of the area of contact and independent of the relative velocities of the bodies.

The following points are worth remembering while solving problems on frictions:

(a) In problems if μ_s and μ_K are separately not given but only a single μ is given then use

$$f_K = \mu_K N = \mu N \text{ and } f_L = \mu_S N = \mu N$$

(b) If more than two blocks are placed one over the other on a the ground then normal reaction between the two blocks will be equal to the weight of the blocks over the common surface.

For example,

$N_1 = m_A g$ (Normal reaction between A and B)

$N_2 = (m_B + m_A)g$ (Normal reaction between B and C)

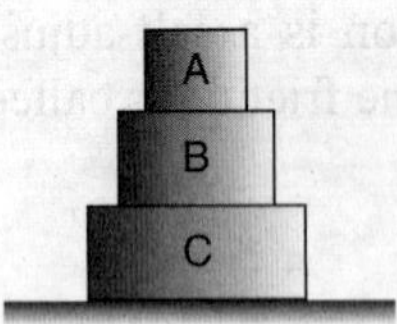

Angle of friction

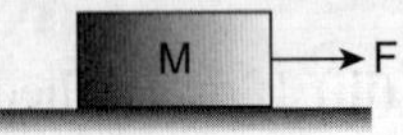

Suppose a body is placed on an inclined surface whose angle of inclination is $\theta\left(0° < \theta < \pi/2\right)$. The coefficient of friction between the body and the surface is μ_s.

Let the initial value of θ be 0°. If we slowly start increasing the value of θ, then at a particular value of θ (say ϕ), the block starts to move. This value of θ at which motion commences is called the angle of friction; at this point $mg \sin \theta = f$

When
$$\theta = \phi, mg\sin\theta = f_{\max},$$

or,
$$mg\sin\phi = \mu_s N$$
$$mg\sin\phi = \mu_s mg\cos\phi$$
$$\tan\phi = \mu_s$$
$$\phi = \tan^{-1}\mu_s$$

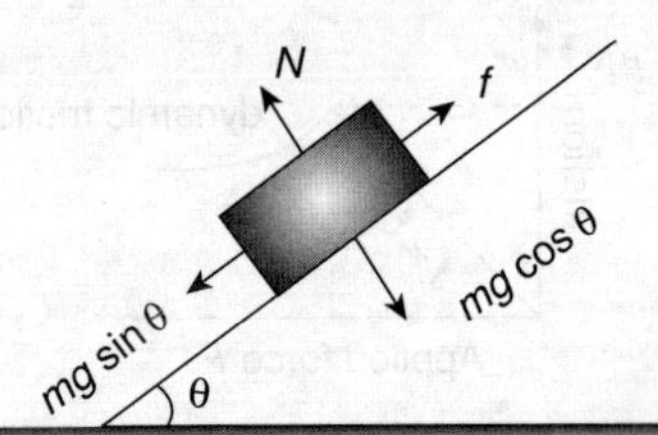

A block of mass m is placed on an incline whose inclination may be varied between 0 to $\pi/2$.

The angle of friction is that maximum angle of inclination of the plane at which a body placed at rest on the inclined plane is about to slide down.

When $\theta \le \phi$ the body is at rest. When the angle of inclination is more than the angle of friction $(\theta > \phi)$ the block accelerates down and if we wish to still keep it at rest an external force needs to be applied.

Illustration 10

A block of mass 2 kg is lying over a rough horizontal surface at rest with $\mu_s = 0.5$ and $\mu_k = 0.45$. A horizontal force F acts on block such that F increases from 5 N to 12 N continuously in certain time and then decreases from 12 N to 5 N continuously in certain time. [g = 10 m/s²]

(a) Find the acceleration of block when $F = 9.5$ N (increasing)
(b) Find the acceleration of block when $F = 10.5$ N (increasing)
(c) Find the acceleration of block when $F = 9.5$ N (decreasing).

Solution

Limiting static friction (L.S.F.)
$$= \mu_s N = 0.5 \times N = 0.5\ Mg = 10\text{ N}$$
$$f_K \text{ (kinetic friction) } \mu_k N = 0.45\ Mg = 9\text{ N}$$

(a) When F is increasing until $F >$ L.S.F. = 10 N, block does not move at all thus for $F \le 10$ N friction is static and adjustable and equals applied force; acceleration is zero.

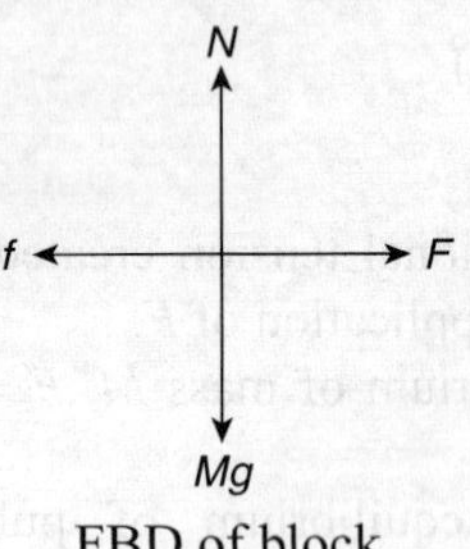

FBD of block

(b) When $F = 10.5$ N > L. S. F. (10 N) block moves and friction becomes kinetic $F = f_K$
$$a = \frac{F - f_K}{m} = \frac{10.5 - 9}{2} = \frac{1.5}{2} = 0.75\text{ m/s}^2$$

(c) Once the motion starts, friction becomes kinetic; the block will only be accelerating for $F > 9$ N, thus when force $F = 9.5$ N (decreasing) $F = f_K$
$$a = \frac{F - f_K}{m} = \frac{9.5\text{ N} - 9\text{ N}}{2} = 0.25\text{ m/s}^2.$$

Illustration 11

A block of mass 10 kg is placed at rest on an inclined plane and is being pulled upward with a force F parallel to the inclined plane as shown in figure. The coefficient of static friction between the block and the incline is $\mu_s = 0.3$. Find the magnitude and direction of the frictional force acting on the block if

(a) $F = 40$ N;
(b) $F = 80$ N (given sin 37° = 3/5, cos 37° = 4/5)

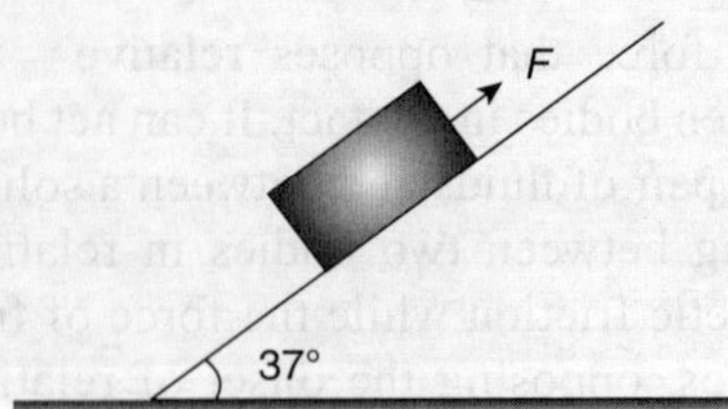

Solution

(a) Since $Mg \sin 37° > F$, the block has a tendency to slide downward in the absence of friction. Thus, the frictional force will act upward on the block to oppose the relative motion between the two.

Now to keep the block stationary, the force of friction should be

$$f = Mg \sin 37° - F$$
$$f = 60 - 40 = 20 \text{ N}$$

The maximum value of friction is

$$f_{\max} = \mu_s N = \mu_s mg \cos 37°$$
$$= (0.3)(10)(10)\left(\frac{4}{5}\right)$$
$$= 24\text{N}$$

Since $f < f_{\max}$, the frictional force acting on the block is $f = 20$ N (upward).

(b) Since, $Mg \sin 37° < F$, the block has a tendency to slide upward in the absence of friction. The friction force acts downward on the block. Now, to keep the block stationary the force of friction should be

$$f = F - Mg \sin 37°$$
$$f = 80 - 60 = 20 \text{ N.}$$

Since $f < f_{\max}$ the force of friction acting on the blocks is $f = 20$ N (downward).

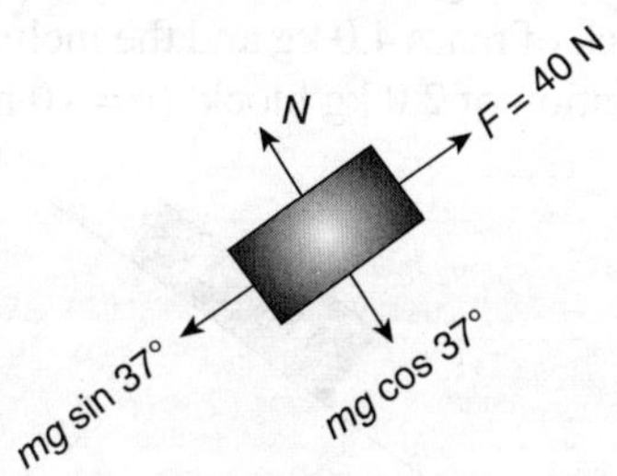

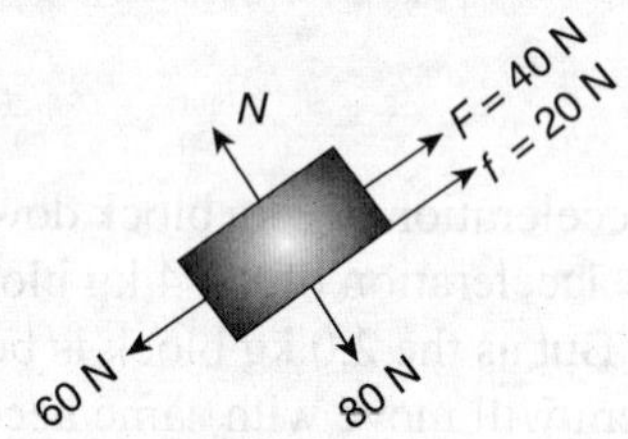

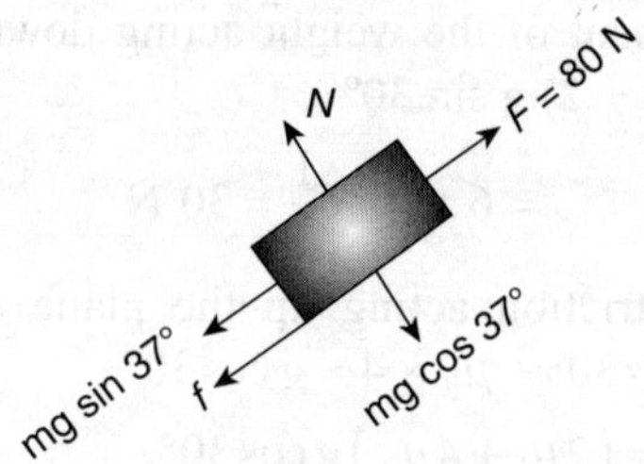

Illustration 12

A block of mass M is at rest on a rough floor. The coefficient of static friction between block and the surface is μ. A pulling force F is applied at angle θ with the horizontal as shown

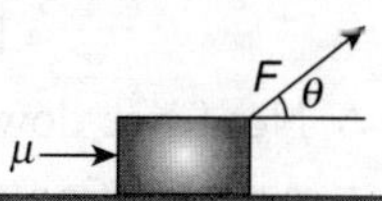

(a) For given θ, find the minimum value of F needed to move the block.

(b) If θ is also adjustable, find the minimum value of F needed to move the block.

Solution

(a) FBD of block

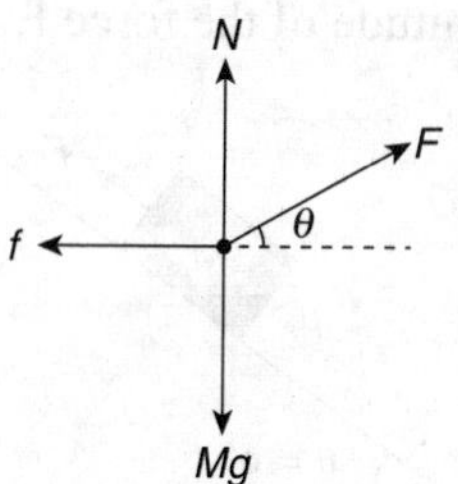

Breakup of forces along horizontal and vertical component

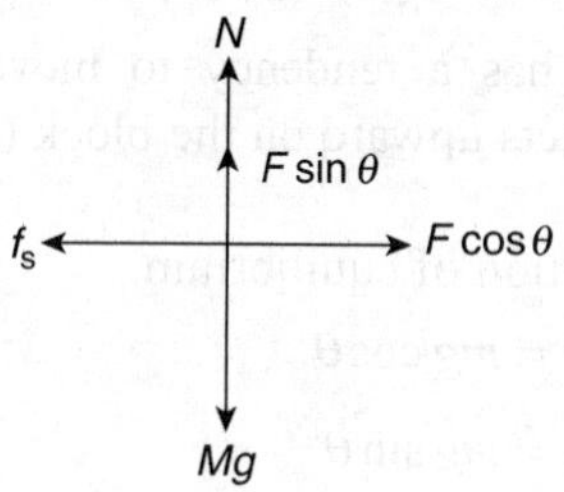

$$f_s \le \text{L.S.F.} = \mu N$$
$$= \mu(Mg - F\sin\theta)$$

For motion of block, net horizontal force > 0

$$F\cos\theta > \mu(Mg - F\sin\theta)$$
$$F(\cos\theta + \mu\sin\theta) > \mu Mg$$
$$F > \frac{\mu Mg}{\cos\theta + \mu\sin\theta}$$
$$F_{\min} > \frac{\mu Mg}{\cos\theta + \mu\sin\theta}$$

(b) $F = \dfrac{\mu Mg}{\cos\theta + \mu\sin\theta}$

$$\frac{\mu Mg}{\sqrt{1+\mu^2}\sin(\theta+\phi)}$$

where $\phi = \tan^{-1}\left(\frac{1}{\mu}\right)$

As $\sin(\theta = \phi) \le 1$

$$F \ge \frac{\mu Mg}{\sqrt{1+\mu^2}}$$

Thus minimum force required to move the block is $\dfrac{\mu Mg}{\sqrt{1+\mu^2}}$.

Illustration 13

A block of mass m is placed at rest on an inclined plane whose angle with the horizontal is more than the angle of friction $(\theta > \phi)$. An external force parallel to the inclined plane is applied in the upward direction to keep it in equilibrium. Find the magnitude of the force F.

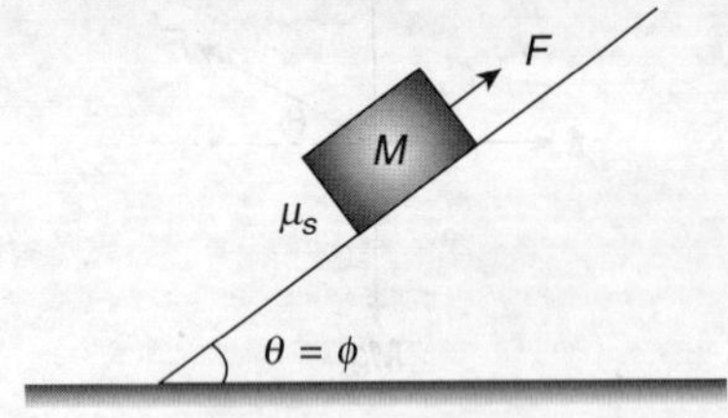

Solution

Since the block has a tendency to move downward, the force of friction acts upward on the block (see the free-body diagram).

Applying equation of equilibrium,

$$N = mg\cos\theta$$

$$F + f_{max} = mg\sin\theta$$

$$F = mg\sin\theta - f_{max}$$

$$= mg\sin\theta - \mu_s mg\cos\theta,$$

$$F = mg(\sin\theta - \mu_s\cos\theta).$$

Note that this is the minimum magnitude of force required to keep it in equilibrium. If we apply a force slightly more than this minimum, the block does not start moving up but the magnitude of the friction force gets reduced. It becomes equal to zero when the external force attains a value equal to $F = mg\sin\theta$, as shown in figure.

If the magnitude of force is increased further then the block has the tendency to move upward. The direction of frictional force gets reversed but the block does not start moving up unless the external force attains its maximum value. Balancing forces, we get,

$$F_{max} = mg\sin\theta + f_{max}$$

$$F_{max} = mg\sin\theta + \mu_s N$$

$$F_{max} = mg\sin\theta + \mu_s mg\cos\theta$$

$$F_{max} = mg(\sin\theta + \mu_s\cos\theta)$$

i.e., the block does not move (it remains stationary) if

$$F_{min} \le F \le F_{max}.$$

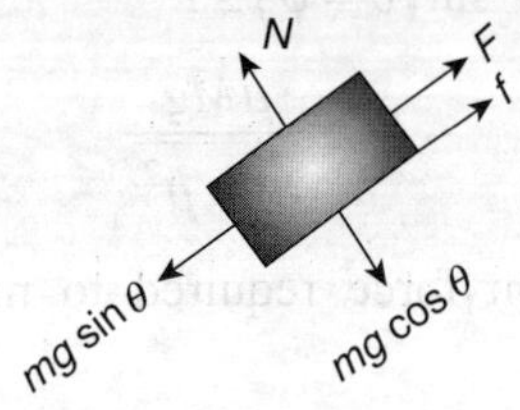

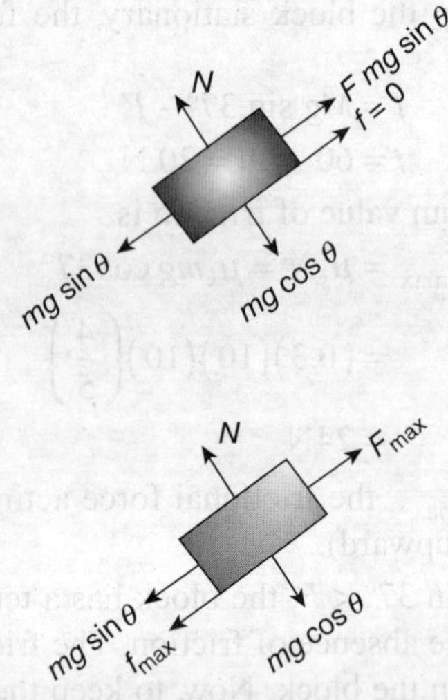

Illustration 14

The figure shows two blocks in contact sliding down an inclined surface of inclination 30º. The coefficient of friction between the block of mass 2.0 kg and the incline is $\mu_1 = 0.20$ and that between the block of mass 4.0 kg and the incline is $\mu_2 = 0.30$. Find the acceleration of 2.0 kg block. ($g = 10$ m/s^2).

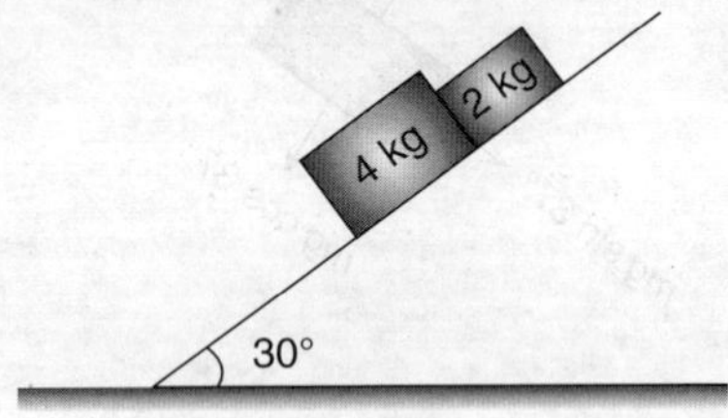

Solution

Since, $\mu_1 < \mu_2$ acceleration of 2 kg block down the plane will be more than the acceleration of the 4 kg block is allowed to move separately. But as the 2.0 kg block is behind the 4.0 kg block both of them will move with same acceleration, say a. Taking both the blocks as a system.

The component of the weight acting down the plane on the system = (4 + 2) g sin 30º

$$= 6 \times 10 \times \frac{1}{2} = 30 \text{ N}$$

The force of friction acting up the plane on the system $= \mu_1 \times 2 \times g\cos 30° + \mu_2 \times 4 \times g\cos 30°$

$$= (2\mu_1 + 4\mu_2)g\cos 30°$$

$$= (2 \times 0.2 + 4 \times 0.3) \times 10 \times 0.86$$

$$\approx 13.76 \text{ N}$$

∴ Net force down the plane $F = 30 - 13.76 = 16.24$ N

∴ Acceleration of both the blocks down the plane will be

$$a = \frac{F}{4+2} = \frac{16.24}{6} = 2.7 \text{ m/s}^2.$$

Frame of Reference

In order to study rest and motion, one needs to observe the position of bodies, which can be done only with respect to a frame of reference. The frame of reference consists of an observer, with a coordinate frame (Cartesian or, otherwise) to measure position and clocks to measure time. Frame of references are of two types.

Inertial frame

An inertial frame of reference is one in which Newton's first law of motion holds. A non-accelerating frame or all frames moving with uniform motion relative to an inertial frame are also inertial, i.e., in an inertial reference frame, a body which is not subject to any force will either stay at rest or move with a constant velocity.

Non-inertial frame

Frames of reference which are accelerating or rotating with respect to an inertial frame are non-inertial. In a non-inertial frame, Newton's second law of motion, $\vec{F} = m\vec{a}$ does not hold. For a non-inertial frame moving with acceleration relative to any inertial frame, the second law of motion takes the form:

$$\vec{F} - m\vec{\alpha} = m\vec{a}$$

where a is the acceleration of the body relative to the non-inertial frame. The term $-m\alpha$ is an example of a pseudo-force.

In general, the 'Second Law' takes the form: $\vec{F} + \vec{F}_P = m\vec{a}$, in a non-inertial frame.

Remember the following two important points regarding frame of reference:

(a) A frame can be approximately inertial in one context and non-inertial in another. The earth's frame of reference is approximately inertial for most laboratory scale phenomena but non-inertial for astronomical observations.

(b) Pseudo forces should be involved only in non-inertial frames such as accelerating or freely falling lift, etc.

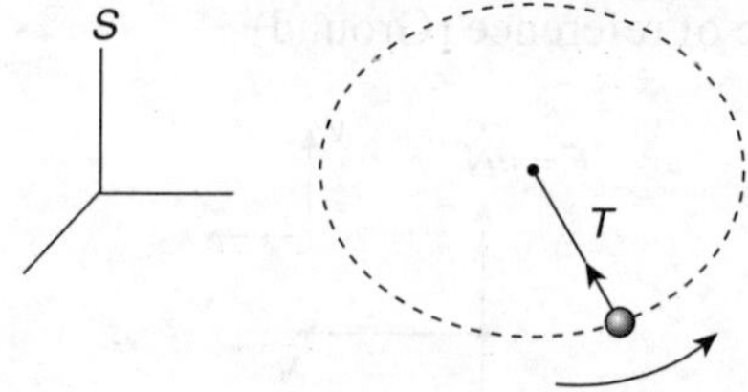

(i) relative to inertial frame of reference

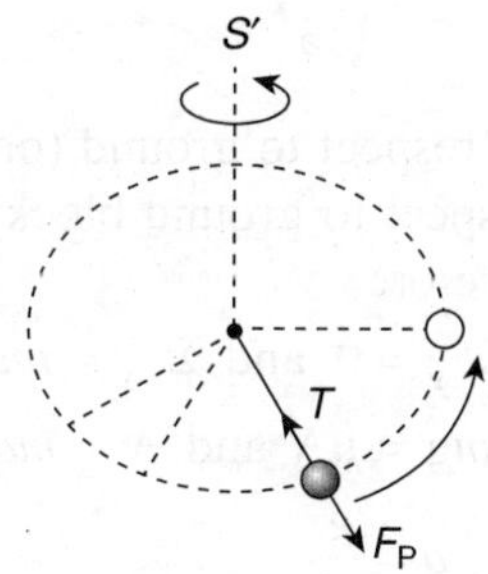

(ii) relative to non-frame of reference

Illustration 15

Apparent weight in an accelerating elevator: A man of mass M stands on a scale in an elevator, as shown in the figure. Find the scale reading when the elevator is accelerating with a_0 (i) upwards and (ii) downward ?

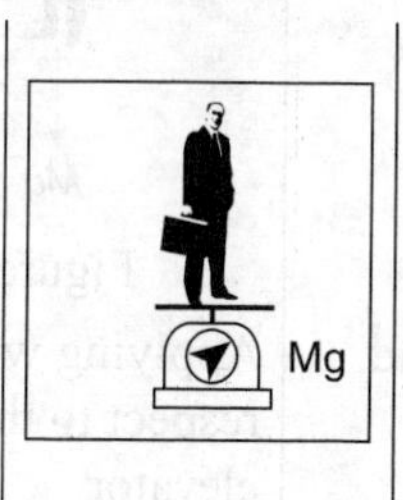

Solution

(a) Since the man is stationary relative to the elevator, it also accelerates up with a_0. The forces on the man are reaction N (exerted by the scale platform on which it rests) and Mg, the force of gravity. In the reference frame attached to the elevator an additional pseudo-force acts opposite to the acceleration of the frame. The free-body diagrams of the man w.r.t. ground and the elevator are shown in figure (b and c).

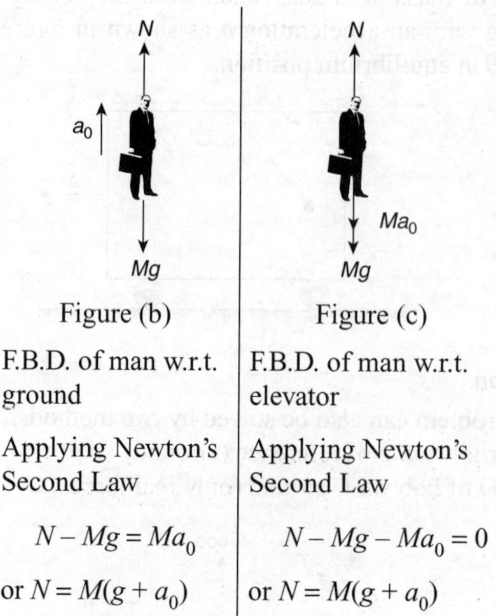

Figure (b)	Figure (c)
F.B.D. of man w.r.t. ground	F.B.D. of man w.r.t. elevator
Applying Newton's Second Law	Applying Newton's Second Law
$N - Mg = Ma_0$	$N - Mg - Ma_0 = 0$
or $N = M(g + a_0)$	or $N = M(g + a_0)$

The force exerted by the man on the scale is equal and opposite to N and determines the reading on the scale or the apparent weight. Thus, when the elevator accelerates up, the apparent weight of the block is greater than its true weight.

(b) When the elevator accelerates downward with a_0 the free-body diagrams can be drawn as shown in the figure.

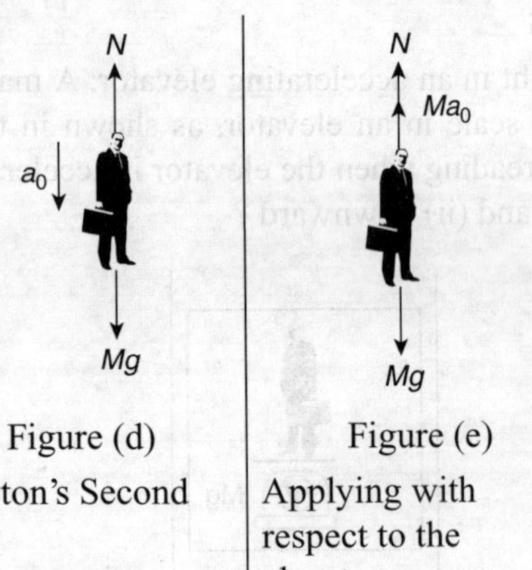

Figure (d)	Figure (e)
Newton's Second Law	Applying with respect to the elevator
$Mg - N = Ma_0$	$N + Ma_0 - Mg = 0$
$N - Mg = Ma_0$	$N - Mg - Ma_0 = 0$
or $N = M(g - a_0)$	or $N = M(g - a_0)$

Thus, when the elevator accelerates down, the apparent weight of the man is lesser than his true weight by Ma_0. If $a_0 = g$, as it would if the elevator were in free fall, the man would be apparently weightless.

Illustration 16

A bob of mass m is suspended from the ceiling of a train moving with an acceleration a as shown in figure. Find the angle θ in equilibrium position.

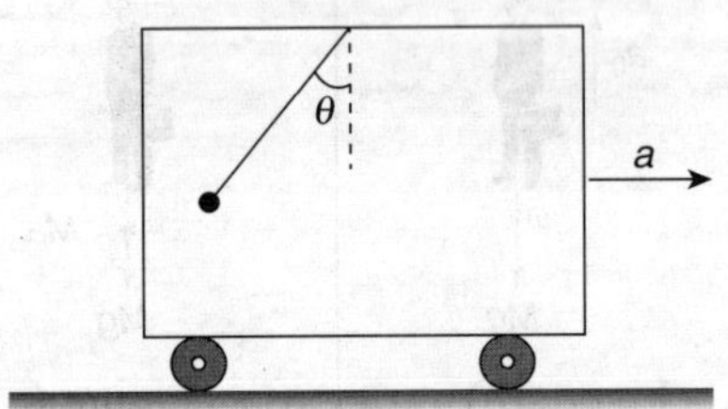

Solution

This problem can also be solved by two methods.

Inertial frame of reference (Ground)

FBD of bob w.r.t. ground (only real forces):

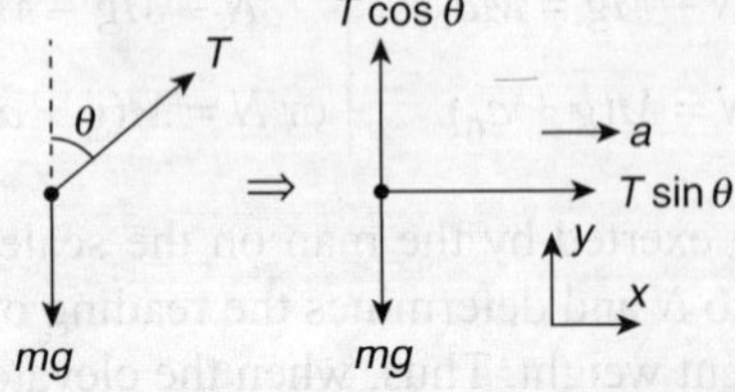

With respect to ground, bob is also moving with an acceleration a.

$\therefore \quad \Sigma F_x = 0 \Rightarrow T\sin\theta = ma \quad \ldots(i)$

and $\quad \Sigma F_y = 0 \Rightarrow T\cos\theta = mg \quad \ldots(ii)$

From Eqs. (i) and (ii), we get

$$\tan\theta = \frac{a}{g} \text{ or } \theta = \tan^{-1}\left(\frac{a}{g}\right)$$

Non-inertial frame of reference (Train)

FBD of bob w.r.t train. (Real forces + Pseudo force):

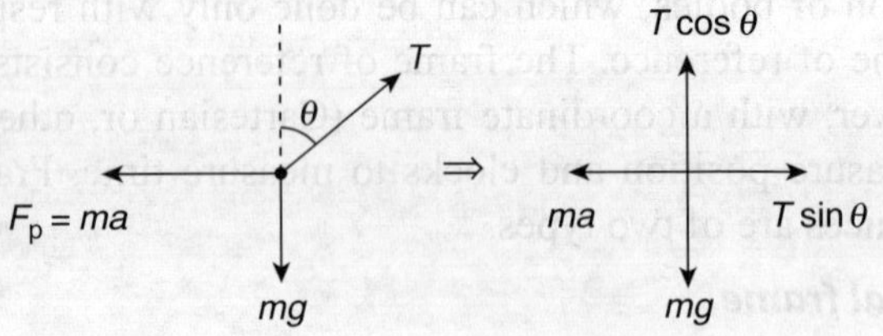

With respect to train, the bob is in equilibrium.

$\therefore \quad \Sigma F_x = 0 \Rightarrow T\sin\theta = ma \quad \ldots(iii)$

$\therefore \quad \Sigma F_y = 0 \Rightarrow T\cos\theta = mg \quad \ldots(iv)$

From Eqs. (iii) and (iv), we get the same result, i.e.,

$$\theta = \tan^{-1}\left(\frac{a}{g}\right)$$

Illustration 17

In the adjoining figure, the coefficient of friction between wedge (of mass M) and block (of mass m) is μ.

Find the minimum horizontal force F required to keep the block stationary with respect to the wedge.

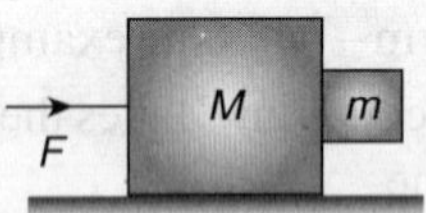

Solution

Such problems can be solved with or without using the concept of pseudo force. Let us solve the problem by both the methods.

a = acceleration of (wedge + block) in horizontal direction

$$= \frac{F}{M + m}$$

Inertial frame of reference (Ground)

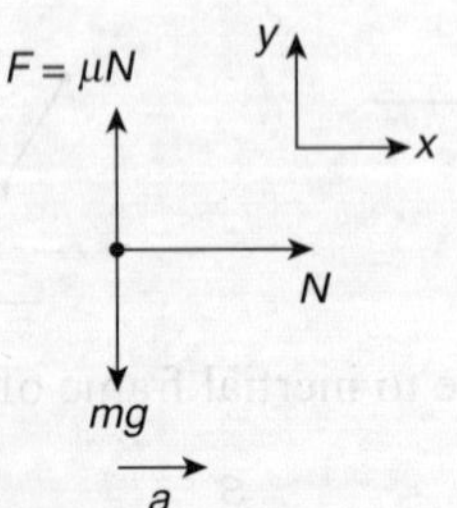

FBD of block with respect to ground (only real forces have to applied) with respect to ground block is moving with an acceleration a. Therefore,

$$\Sigma F_y = 0 \text{ and } \Sigma F_x = ma$$

$$mg = \mu N \text{ and } N = ma$$

$$\therefore \quad a = \frac{g}{\mu}$$

$$\therefore \quad F = (M + m)a = (M + m)\frac{g}{\mu}$$

Non-inertial frame of reference (Wedge)

$F = \mu N$

$F_P = ma$ N

mg

FBD of m with respect to wedge (Real + One pseudo force) with respect to wedge block is stationary.

$\therefore \quad \Sigma F_x = 0 = \Sigma F_y$

$\therefore \quad mg = \mu N \text{ and } N = ma$

$\therefore \quad a = \dfrac{g}{\mu}$

and $\quad F = (M + m)a$

$$= (M + m)\frac{g}{\mu}$$

From the above discussion, we can see that from both the methods, results are same.

Illustration 18

All surfaces are smooth in the adjoining figure. Find F such that block remains stationary with respect to the wedge.

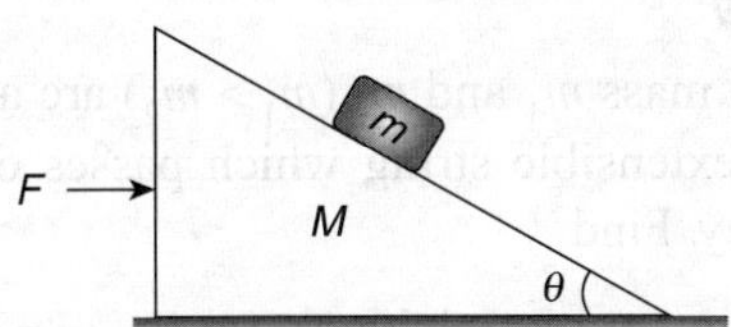

Solution

Acceleration of (block + wedge) $a = \dfrac{F}{(M + m)}$

Let us solve the problem by both the methods.

From inertial frame of ref. (Ground)	From non-inertial frame of ref. (Wedge)
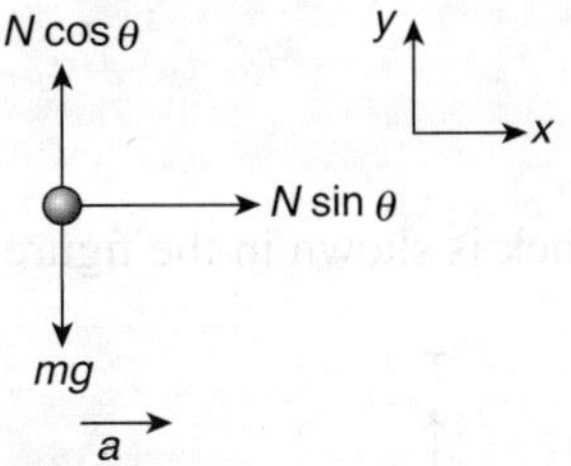	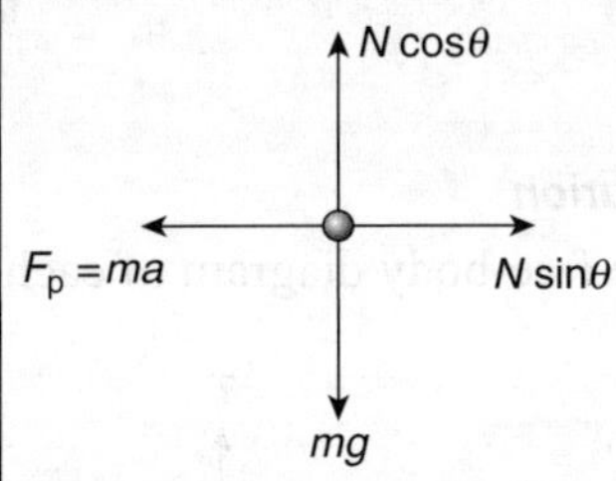
FBD of block w.r.t. ground (Apply real forces):	FBD of block w.r.t. wedge (real forces + pseudo force):
w.r.t. ground, the block is moving with an acceleration a	w.r.t. wedge, the block is stationary
$\therefore \sum F_y = 0$	$\therefore \sum F_y = 0$
$\Rightarrow N\cos\theta = mg$ ...(i)	$\Rightarrow N\cos\theta = mg$ ...(iii)
and $\sum F_x = ma$	$\sum F_x = 0$
$\Rightarrow N\sin\theta = ma$ ...(ii)	$\Rightarrow N\sin\theta = ma$ ...(iv)
From Eqs. (i) and (ii)	From Eqs. (iii) and (iv), we will get the same result
$a = g\tan\theta$ i.e., $\therefore F = (M + m)a$ $= (M + m)g\tan\theta$.	$F = (M + m)g\tan\theta$

Constraints motion

In a system composed of several bodies, the acceleration is often related by constraints. The equations of constraints can often be found by simple inspection. But the most general approach is to start with coordinate geometry. Constraint relation is a mathematical equation relating the kinematic parameters such as displacement x, velocity v, and acceleration a.

The procedure to determine constraint relation with this method is straightforward and given in the following steps:

Mathematical Procedure to Determine Constraint Relation:

Step I. Assume the direction of acceleration (or velocity) of each body (or particle),

Step II. Locate the position of each particle from a fixed point e.g. centre of pulley,

Step III. Identify the constraint and write down the equation of constraint in terms of distance assumed.

Many types of constraints are existent in various problems. But two general types of constraints can be identified.

(i) **Wedge type constraints:** When a body is placed on a wedge, the acceleration relationship between the wedge and the body can be found. Consider the case below. A and B are blocks, both of which can move (figure). From the diagram, it is clear that B is free to move along x-direction whereas along y-direction it is forced to stick to the surface of the wedge.

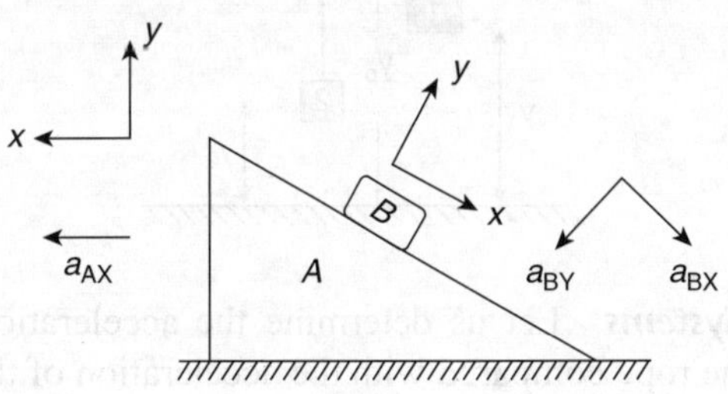

Therefore, there exists a constraint along y-direction of B. The relationship can be determined as below. Assume that "A" moves by a distance x in the backward direction. The new position of "A" and "B" are shown by the dotted lines in figure (b). They y-distance moved by $B =$ length DE. But $\frac{DE}{x} = \sin\theta$.

Therefore, $DE = x \sin\theta$

Differentiating twice, we have $a_{BY} = a_{AX} \sin\theta$.

This is a very important constraint and can be frequently used in many situations.

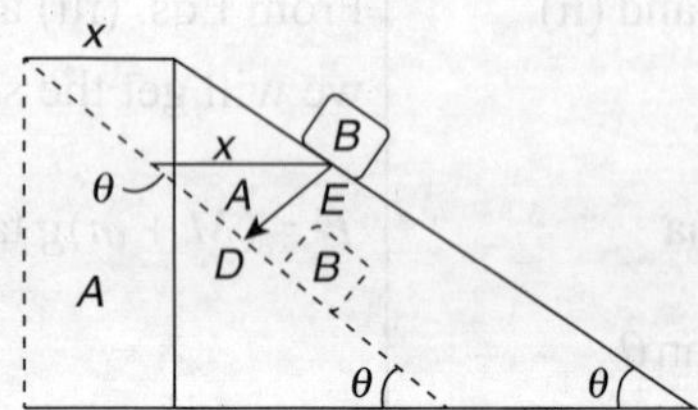

Therefore, we can summarize and say that if a wedge has the acceleration a along its x-direction as shown above, the block which is lying on the inclined surface of the wedge will have acceleration $a \sin\theta$ along its y-direction i.e., perpendicular to the inclined surface.

(ii) Masses and Pulley: Two masses are connected by a string which passes over a pulley accelerating upward at the rate a, as shown in figure. To find how the accelerations of the bodies are related, assume that there is no horizontal motion. The length of the string, 1, is constant. Hence, if y_P is measured to the centre of the pulley of radius R,

$$1 = \pi R + (y_p - y_1) + (y_p - y_2)$$

Differentiating twice with respect to time, we find the constraint conditions. Thus,

$$O = \frac{d^2 y_p}{dt^2} - \frac{d^2 y_1}{dt^2} + \frac{d^2 y_p}{dt^2} - \frac{d^2 y_2}{dt^2}$$

$$\frac{d^2 y_p}{dt^2} = \left(\frac{d^2 y_1}{dt^2} + \frac{d^2 y_2}{dt^2}\right) \text{ or } a = \frac{1}{2}(a_1 + a_2)$$

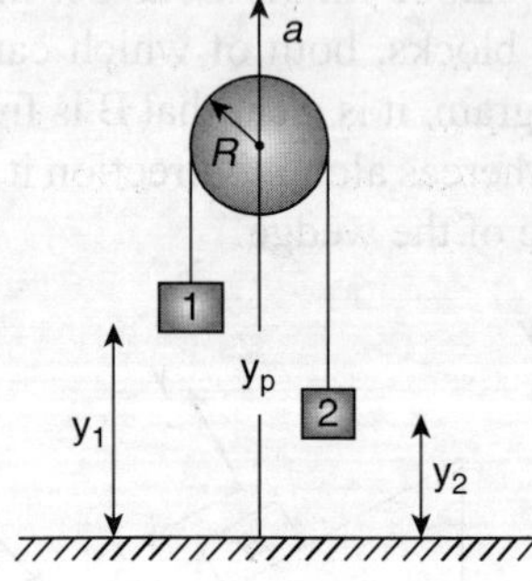

Pulley Systems Let us determine the acceleration of the end of the rope compared with the acceleration of the block.

Using the coordinates indicated, the length of the rope is given by

$$1 = x_2 + \pi R + (x_2 - h) + \pi R + (x_1 - h)$$

where R is the radius of the pulleys.

$$\text{Hence, } \frac{d^2 x_2}{dt^2} = -\frac{1}{2}\left(\frac{d^2 x_1}{dt^2}\right)$$

The block accelerates half as fast as the hand, and in the opposite direction.

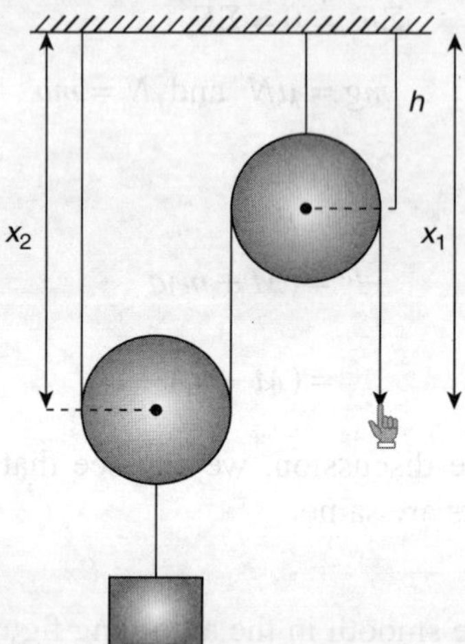

Illustration 19

Two blocks of mass m_1 and m_2 ($m_1 > m_2$) are attached at the ends of an inextensible string which passes over a smooth massless pulley. Find

(a) the acceleration of each block.

(b) the tension in the string.

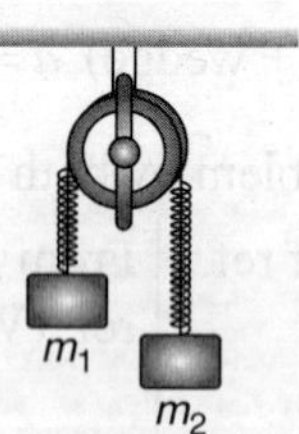

Solution

The free-body diagram of each block is shown in the figure.

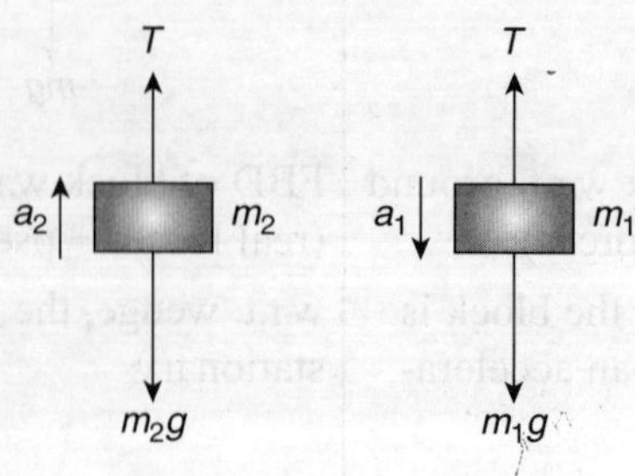

The block m_1 is assumed to be accelerating downward and the block m_2 is assumed to be accelerating upward.

This is merely an assumption and it does not imply the real direction of acceleration.

If the value of a_1 and a_2 come out to be positive then only the assumed directions are correct; otherwise the bodies accelerate in the opposite directions.

Since the pulley is smooth and massless, the tension in the string on both sides of the pulley is identical. Applying Newton's second law on

Block m_1: $m_1g - T = m_1a_1$ ...(i)

Block m_2: $-m_2g + T = m_2a_2$ ...(ii)

Here the number of unknowns is three: T, a_1 and a_2

The number of equations is two. We require one more equation to solve the problem. Note that the whenever one finds the number of equations is less than the number of unknowns, one may think about the constraint relations.

If y_0 be the distance of the ground from the centre of the pulley then the length of the string is:

$$(y_0 - y_1) + (y_0 - y_2) = \text{Constant.}$$

$$2y_0 - y_1 - y_2 = \text{Constant.}$$

Here y_1 and y_2 are the positions of m_1 and m_2 above the ground (in the upwards directions).

Differentiating twice w.r.t. time, we get ,

$$0 - \frac{d^2y_1}{dt^2} - \frac{d^2y_2}{dt^2} = 0$$

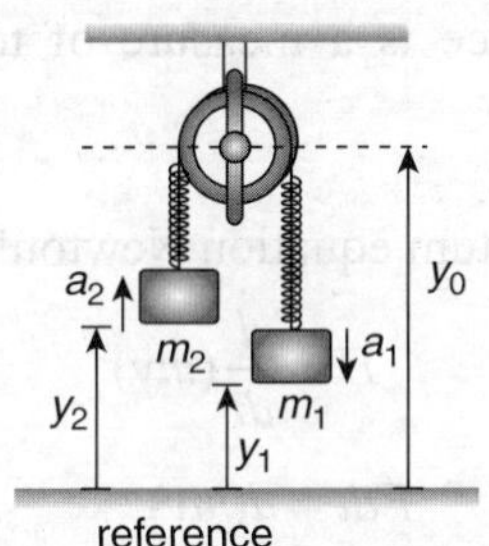

Now, $\frac{d^2y_1}{dt^2} = -a_1$ and $\frac{d^2y_2}{dt^2} = +a_2$

Thus, $a_1 - a_2 = 0$

$$a_1 = a_2 = a$$

Substituting in Eqs. (i) and (ii) and after solving them,

we get, $a = \left(\frac{m_1 - m_2}{m_1 + m_2}\right)g$

Also $$T = \left(\frac{2m_1m_2}{m_1 + m_2}\right)g.$$

Alternatively

One can also assume the distance of each block from the centre of pulley downwards, as shown in the figure. The length of string remains constant. Thus,

$$x_1 + x_2 = \text{constant}$$

Differentiating twice w.r.t. time, we get,

$$\frac{d^2x_1}{dt^2} + \frac{d^2x_2}{dt^2} = 0\,.$$

Since the block is assumed to be moving downward (x_1 is increases with time).

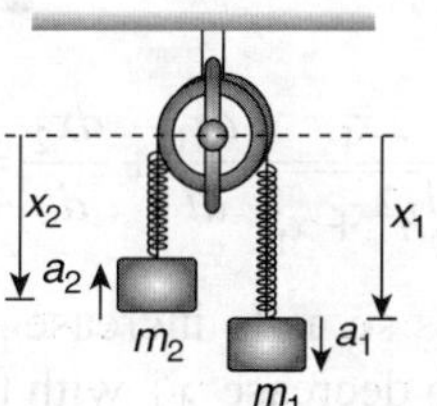

$\therefore$ $$\frac{d^2x_1}{dt^2} = +a_1 \quad (a_1 > 0)$$

and the block m_1 is assumed to be moving upward (x_2 is decreasing with time)

$\therefore$ $$\frac{d^2x_2}{dt^2} = -a_2\,,\ (a_2 > 0)$$

Thus $a_1 - a_2 = 0$

or, $a_1 = a_2 = a$ (say)

Substituting $a_1 = a_2 = a$ in Eqs. (i) and (ii) and solving them, we get

$$a = \left(\frac{m_1 - m_2}{m_1 + m_2}\right)g$$

and $$T = \left(\frac{2m_1m_2}{m_1 + m_2}\right)g.$$

As the bodies are connected, their acceleration will have the same magnitude here.

Illustration 20

In the figure, a ball and a block are joined together with an inextensible string. The ball can slide on a smooth horizontal surface. If v_1 and v_2 are the respective speeds of the ball and the block, then determine the constraint relation between the two.

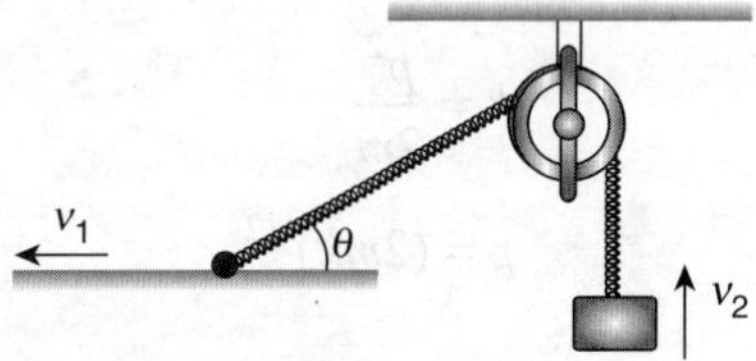

Solution

Distances are assumed from the centre of the pulley as shown in figure. The constraint relation implies that the length of string remains constant.

$$\sqrt{h^2 + x_1^2} + x_2 = \text{constant}$$

Differentiating both the sides w.r.t time, we get,

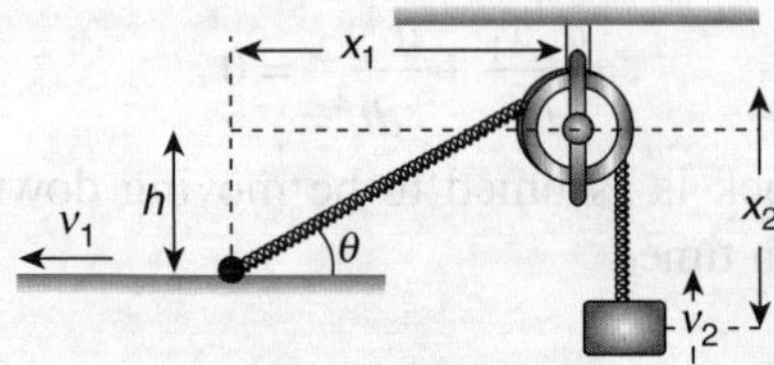

$$\frac{2x_1}{2\sqrt{h^2+x_1^2}}\frac{dx_1}{dt}+\frac{dx_2}{dt}=0$$

Since the ball moves so as to increase x_1 with time and the block moves so as to decrease x_2 with time, we can write

$$\frac{dx_1}{dt}=v_1(+ve)\ \frac{dx_2}{dt}=v_2(-ve)$$

By geometry, $\frac{x_1}{\sqrt{h^2+x_1^2}}=\cos\theta$

$$v_1\cos\theta - v_2 = 0$$

or, $$v_2 = v_1\cos\theta$$

Momentum, Impulse and Law of Conservation of Momentum

Momentum – Linear momentum of a particle is defined as product of mass m with its velocity $\vec{v}$. Further on we will use momentum. The Linear word is used to distinguish it with angular momentum which we learn in the chapter of Motion of System of Particles and Rigid Body.

Regarding momentum, the following points are worth remembering.

1. Momentum is a velocity quantity with dimensions $[MLT^{-1}]$ and SI unit Kg m/s.
2. Momentum depends upon frame of r reference.
3. Momentum $\vec{p}=m\vec{v}$, of a body changes either change in mass or velocity or both.
4. Two bodies having same mass and same speed moving in different direction will have different momentum.
5. $k=\frac{1}{2}mv^2=\frac{1}{2}m\frac{p^2}{m^2}$

$$k=\frac{p^2}{2m}$$

$$p=(2mk)^{1/2}$$

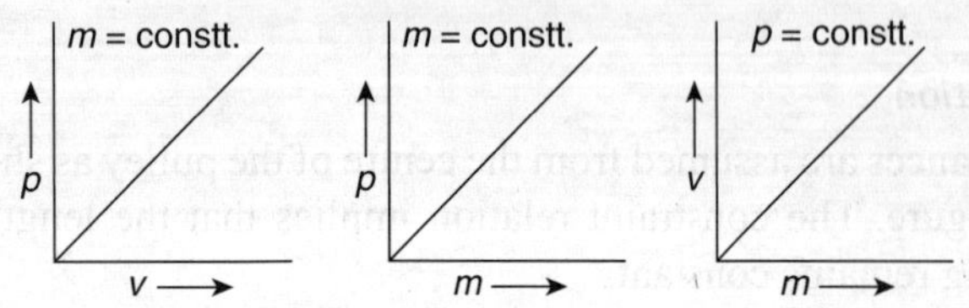

Impulse is a vector quantity having the direction, same as that of force. Its dimension is $[MLT^{-1}]$ and units ms^{-1}.

Impulse is equal to the area under F vs. t graph.

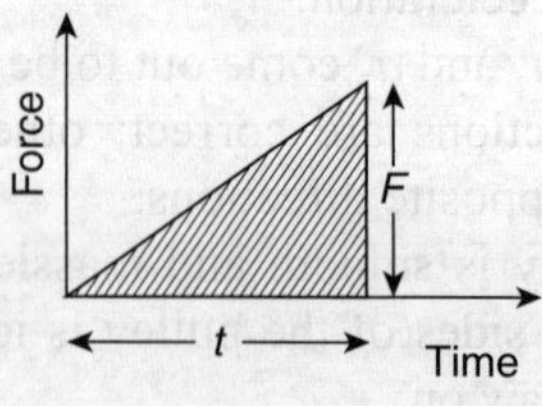

$$\text{Impulse}=\int_0^t Fdt = Fav.t$$

where Fav is the average force during the time interval.

6. A body cannot have momentum without possessing kinetic energy or vice-versa.
7. The graph between K vs. m, p vs. m and k vs. p are given below.

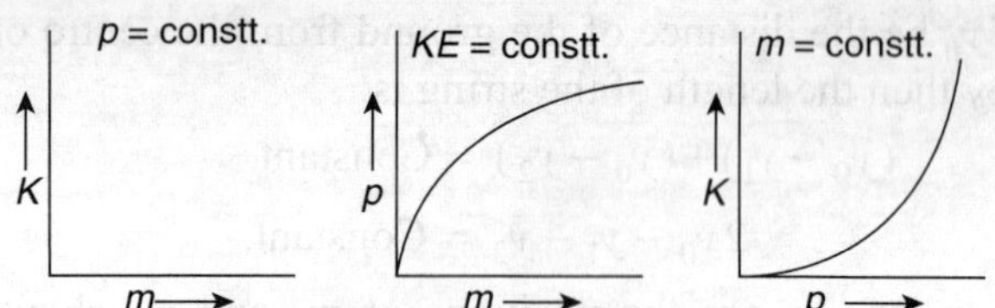

Impulse – When a large force acts on a body for a very small time, the product of force F and time Δt is known as impulse during time Δt

$$\text{Impulse}=\vec{F}.\Delta t$$

The impulsive force does not remain constant. Hence, Impulsive of a force is a measure of total effect of force

$$\text{Impulsive}=\int_0^t \vec{F}dt$$

Impulse momentum equation Newton's Law is

$$F=\frac{d}{dt}(mv)$$

or $$Fdt=d(mv)$$

$$\int_{t_1}^{t_2}Fdt=mv_2-mv_1$$

or $$\text{Impulse = Change in momentum}$$

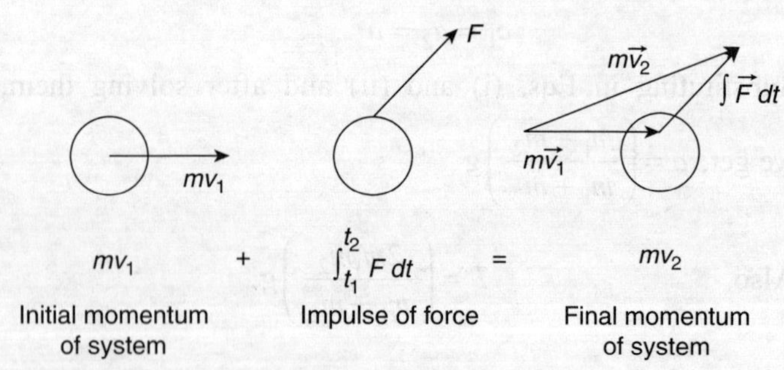

If the motion of bodies takes place in a plane, to simplify the calculation, we can resolve force and momentum and write the equation as

$$mv_1x+\int_{t_1}^{t_2}F_xdt=mv_2x$$

$$mv_1y + \int_{t_1}^{t_2} F_y dt = mv_2y$$

Law of conservation of momentum in an isolated system, the net (total) momentum of the system remains constant. In isolated system, no external force acts.

$$\vec{p}_1 + \vec{p}_2 + \vec{p}_3 .... = \text{Constant}$$

According to the equation for an isolated system, the momentum of individual particles may change but their sum total remains constant.

A very important point about law of conservation of momentum is that it is independent of frame of reference but linear momentum depends upon frame of reference. This law is universal. It equally applies to macroscopic as well microscopic system.

Applications of law of conservation of momentum

(a) Firing a bullet from a gun.

(i) If the bullet is considered as a system, the force exerted by the trigger will be external and momentum of the bullet will change from 0 to v. It is not a violation of conservation law of momentum.

(ii) If the bullet and gun is a system – In this case the force exerted by the trigger will be internal, therefore the net momentum $\vec{p} = \vec{p}_1 + \vec{p}_2$ where $\vec{p}_1$ = momentum of the bullet.
$\vec{p}_2$ = momentum of the gun.
Initially, bullet and the gun are at rest.
Therefore $\vec{p} = 0$

$$\vec{p}_1 + \vec{p}_2 = 0$$

$$\vec{p}_1 = -\vec{p}_2$$

It means if bullet acquires forward momentum, the gun will acquire backward momentum of same magnitude. Let m and $\vec{v}$ be the mass and velocity of the bullet while M and $\vec{V}$ be mass and velocity of the gun respectively.

$$m\vec{v} + M\vec{V} = 0$$

or

$$\vec{V} = -\left(\frac{m}{M}\right)\vec{v}$$

It means if the bullet moves forward, the gun recoils.

Motion of two masses connected by a spring

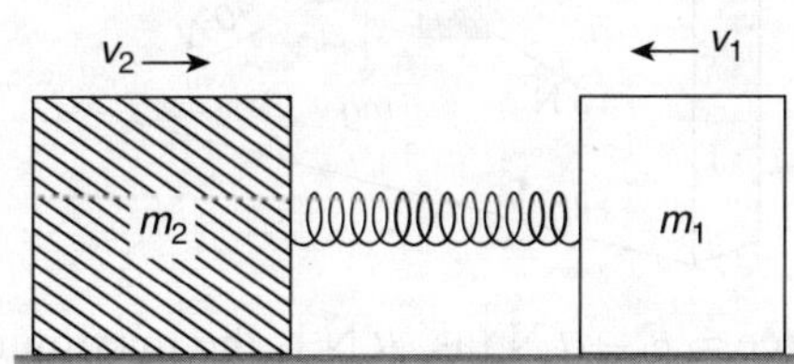

Consider the two blocks resting on a friction-less surface and connected by a mass-less spring. Let the spring be compressed and released from rest.

Then $\vec{p} = \vec{p}_1 + \vec{p}_2$ = Constant

Because external force = 0

$$\vec{p}_1 + \vec{p}_2 = \text{constant}$$

Initially both the blocks are at rest; therefore,

$$\vec{p}_1 + \vec{p}_2 = 0$$

$$\vec{p}_1 = -\vec{p}_2$$

We can conclude from this that at any point of time, for the system of two blocks, there is no external force.

In absence of external force the momentum of the system remains conserved.

$$m_1v_1 + m_2v_2 = 0$$

$$v_2 = -\frac{m_1v_1}{m_2}$$

This implies that if block A moves to right, block B will move to left.

Blocks have momentum equal in magnitude but opposite in direction.

Further, $m_1\vec{v}_1 + m_2\vec{v}_2 = 0$

$$\vec{v}_2 = -\left(\frac{m_1}{m_2}\right)\vec{v}_1$$

Higher block moves faster.

Illustration – Discuss the motion of the two boats.

Two boat of masses m_1 and m_2 are at rest. A person sailing in boat A pulls the other boat with a rope. In the absence of resistance of water, there is no external force acting in a two-boat system.

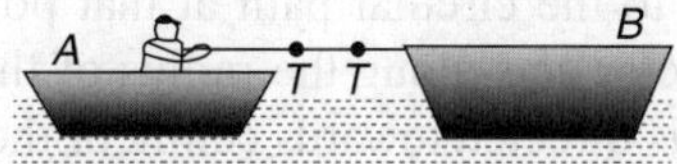

Uniform Circular Motion

When a particle moves in a circle with uniform speed, it is said that body is in uniform circular motion. According to Newton's first law of motion, some external force must be acting on the particle as the velocity is changing because of change in direction. Since, there is continuous change in direction the external force also must be acting continuously.

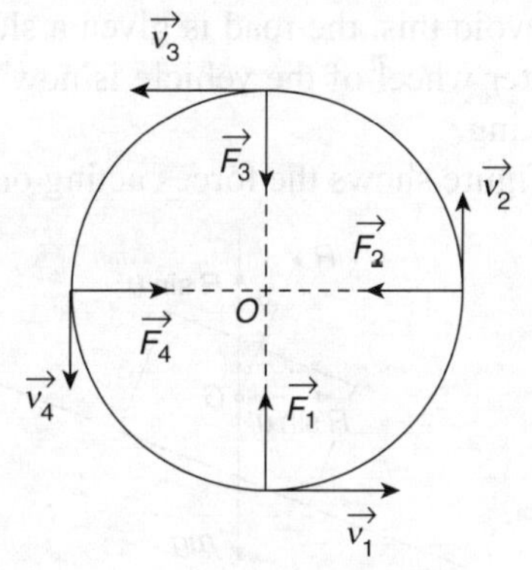

This external force is known as centripetal force.

Centripetal force

It is the force required to give circular motion to the particle or to a body. When a particle moves in a circle of radius r, its speed is constant but the direction of velocity is continuously changing. Thus the particle experiences an acceleration. The magnitude of the acceleration is v^2/r. This is always directed towards the centre of the circle. According to Newton's law, the particle must be under the action of an external force in order to have this acceleration. The external force which must be imposed on the body, so as to make it move with uniform speed around the circle is called centripetal force.

As $$F = ma = m\frac{v^2}{r}$$

the direction of F at any instant must be radially inward.

Examples of centripetal force

(1) For the motion of the earth around the sun, the centripetal force is provided by gravitational force

(2) For a stone whirled on the end of a string the centripetal force is provided by tension in the string

(3) For an amusement park patron pressed against the inner wall of a rapidly rotating circular ball, the centripetal force is provided by the normal reaction of the ball.

Important points

(i) An object moving on a circular path with uniform speed is always acted upon by a force directed towards the centre of the circle. This force is known as centripetal force.

(ii) In a circular motion, the velocity vector at any point is tangential to the circular path at that point. The acceleration vector acts along the radius of the circle at that point and is directed towards centre of the circle. This is called as centripetal acceleration.

(iii) The electrostatic force of attraction between the electron and the nucleus supplies the necessary centripetal force to keep the electron in its orbit.

(iv) When a stone is tied to one end of a rope and whirled in a horizontal circle, then the necessary centripetal force is supplied by the tension in the rope.

Banking of roads

When a vehicle goes around a curve on the road with excessive speed, the there is a tendency for the vehicle to overturn outwards. To avoid this, the road is given a slope rising outwards. The outer wheel of the vehicle is now raised. This is known as banking

The given figure shows the forces acting on a vehicle.

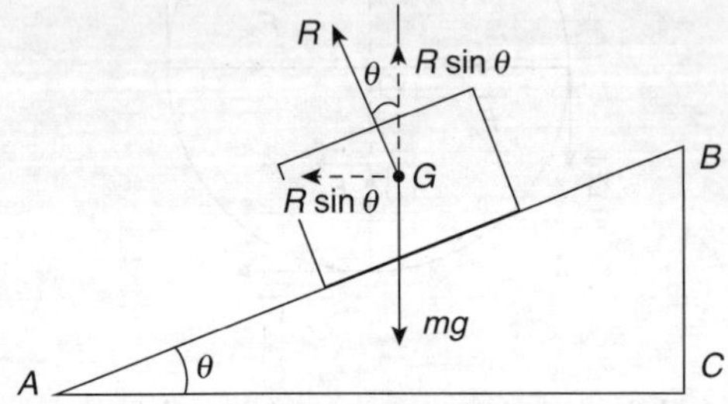

ABC is the section of a road having a slope θ. In the figure, R is the resultant normal reaction of the ground. This is resolved in two parts $R\cos\theta$ acting vertically and $R\sin\theta$ acting horizontally.

For vertical equilibrium,

$$R\cos\theta = mg \quad \text{(i)}$$

The horizontal component $R\sin\theta$ is the required centripetal force.

Hence, $$R\sin\theta = \frac{mv^2}{r} \quad \text{(ii)}$$

From Eqs. (i) and (ii), we get

$$\tan\theta = \frac{v^2}{rg} \quad \text{(iii)}$$

Expression (iii) gives the angle of banking required. This banking eliminates the lateral thrust in case of trains on rails or the friction in case of road vehicles when going around a curve.

SOLVED EXAMPLES

Example 1

A block of mass 2 kg is pushed against a rough vertical wall with a force of 40 N, the coefficient of static friction being 0.5. Another horizontal force of 15 N, is applied on the block in a direction parallel to the wall. Will the block move? If yes, in which direction and with what minimum acceleration? If no, find the frictional force exerted by the wall on the block.

Solution

The resultant of the external forces acting on the body, parallel to the surface of the wall is

$$F = \sqrt{20^2 + 15^2} = 25 \text{ N}$$

in a direction $\tan^{-1}\left(\frac{15}{20}\right) = 37°$ with the vertical.

The friction, by its very virtue of opposing the tendency, will act in a direction opposite to the resultant.

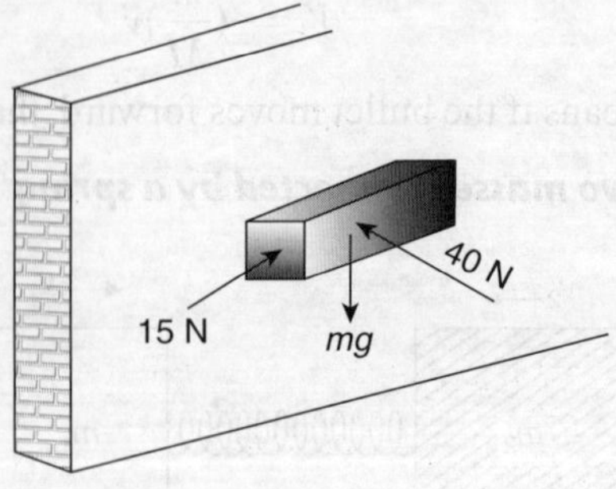

Now, net force = $F - \mu$ N (as, μ N is the maximum possible frictional force)

$$= 25 - 0.5 \times 40 = 5 \text{ N}$$

$\therefore$ Minimum acceleration is $\frac{5}{2} = 2.5 \text{ m/s}^2$

Example 2

Two masses m_1 and m_2 are connected by means of a light string that passes over a light pulley as shown in the figure. If $m_1 = 2$ kg and $m_2 = 5$ kg and a vertical force F is applied on the pulley, find the acceleration of the masses and that of the pulley when

(a) $F = 35$ N (b) $F = 70$ N

(c) $F = 140$ N.

Solution

Since string is massless and friction is absent, tension in the string is the same everywhere.

(a) Let acceleration of the pulley be a_p.

For a_p to be non-zero,

$$F = 2T \quad \text{...(i)}$$

also $T \geq m_1 g$

$\Rightarrow$ $T \geq 2g$ ...(ii)

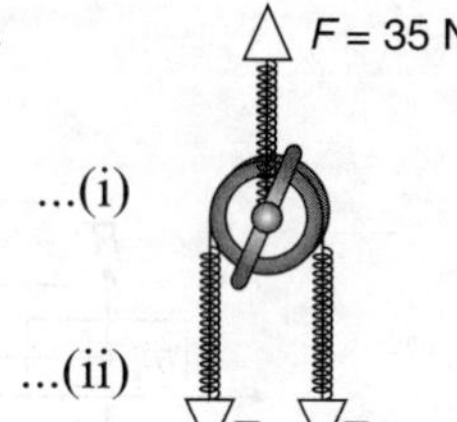

From (i) and (ii), we get,

$$F \geq 2 \times (2g)$$

$\Rightarrow$ $F \geq 40$ N

Therefore when $F = 35$ N

$$a_p = 0 \text{ and hence } a_1 = a_2 = 0$$

FBD of the pulley

(b) As mass of the pulley is negligible,

$$F - 2T = 0$$

or, $T = F/2$

or, $T = 35$ N

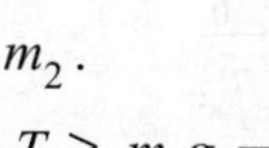

To lift m_2.

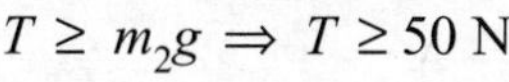

$$T \geq m_2 g \Rightarrow T \geq 50 \text{ N}$$

Therefore, block m_2 will not move.

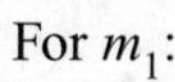

For m_1:

$$T - m_1 g = m_1 a_1$$

or, $35 = 2a_1$

or, $a_1 = \dfrac{35}{2}$ m/s^2

Constraint equation:

$$y_p + (y_p - y_1) = \text{Constant}$$

or, $2y_p - y_1 = c$

or, $2\dfrac{d^2 y_p}{dt^2} - \dfrac{d^2 y_1}{dt^2} = 0$

or, $a_p = \dfrac{a_1}{2} = \dfrac{35}{4}$ m/s^2.

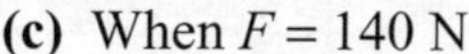

(c) When $F = 140$ N

$$T = 70 \text{ N}$$

or, $T - m_1 g = m_1 a_1$ ...(i)

or, $70 \text{ N} - 20 \text{ N} = 2 \times a_1$

or, $a_1 = 25$ m/s^2

$$T - m_2 g = m_2 a_2 \quad \text{...(ii)}$$

or, $70 \text{ N} - 50 \text{ N} = 5a_2$

or, $a_2 = 4$ m/s^2

Constraint equation

$$(y_p - y_2) + (y_p - y_1) = c$$

or, $2y_p - y_2 - y_1 = c$

$$\Rightarrow \quad 2\frac{d^2 y_p}{dt^2} - \frac{d^2 y_2}{dt^2} - \frac{d^2 y_1}{dt^2} = 0$$

$$\Rightarrow \quad a_p = \frac{a_1 + a_2}{2} = \frac{29}{2} \text{ m/s}^2.$$

Example 3

In the shown figure, the wedge A is fixed to the ground. The prism B of mass M and the block C of mass m are placed as shown. Find the acceleration of the block C w.r.t. block B when the system is set free. Neglect any friction.

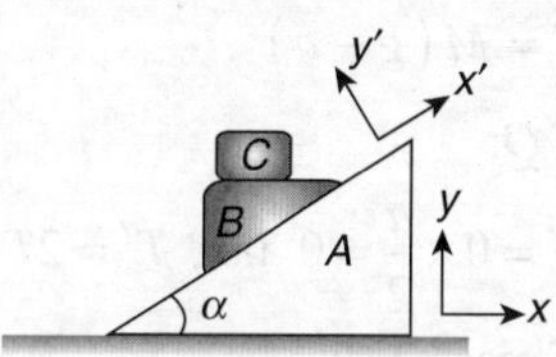

Solution

Let a be the acceleration of B. We consider the motion of C relative to B by applying pseudo force

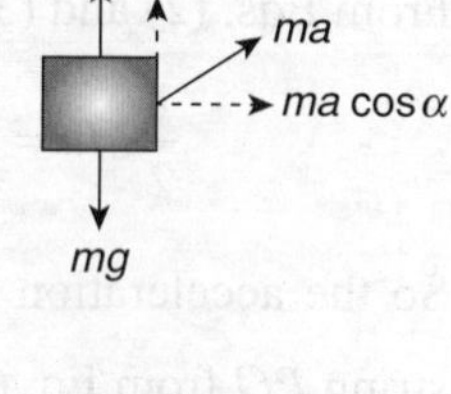

$mg - N - ma \sin\alpha = 0$, as C is always in contact with B.

or, $N = mg - ma \sin\alpha$ ...(1)

and $ma \cos\alpha = ma'$,

where a' = acceleration of C relative to B

$\Rightarrow$ $a' = a \cos\alpha$ ...(2)

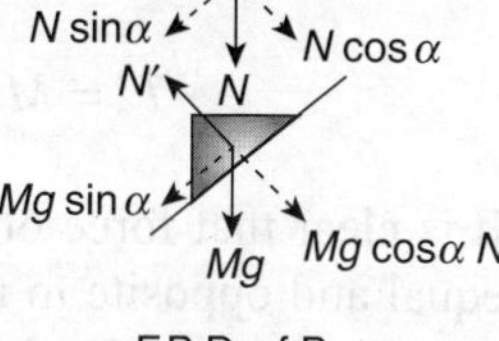

F.B.D. of B

From $Mg \sin\alpha + N \sin\alpha = Ma$

Putting the value of N from (1), we get,

$$\Rightarrow \quad Mg \sin\alpha + mg \sin\alpha - ma \sin^2\alpha = Ma \quad \text{...(3)}$$

$$\Rightarrow \quad a = \frac{(M+m)g\sin\alpha}{M + m\sin^2\alpha}$$

from Eq. (2),

$$a' = \frac{(M+m)g\sin\alpha\cos\alpha}{M + m\sin^2\alpha}.$$

Example 4

Consider the situation shown in Figure (a). Both the pulleys and the strings are light and all the surfaces are frictionless. Calculate (a) the acceleration of mass M, (b) tension in the string PQ and (c) force exerted by the clamp on the pulley.

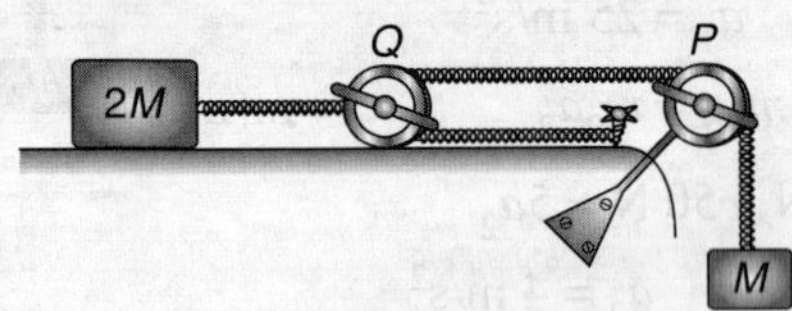

Solution

As pulley Q is not fixed, if it moves a distance d the length of string between P and Q will changes by $2d$, i.e., M will move $2d$. This in turn implies that if a is the acceleration of M the acceleration of Q and of $2M$ will be $(a/2)$.

Now, if we consider the motion of mass M, it is accelerated downward, so

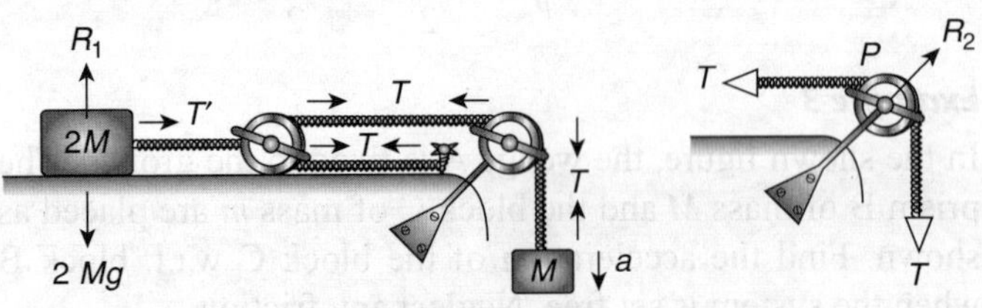

$$T = M(g-a) \qquad ...(1)$$

For the motion of Q:

$$2T - T' = 0 \times \frac{a}{2} = 0 \text{ i.e., } T' = 2T \qquad ...(2)$$

For the motion of mass $2M$:

$$T' = 2M\frac{a}{2}, \text{ i.e. } T' = Ma \qquad ...(3)$$

From Eqs. (2) and (3), as $T = \frac{1}{2}Ma$, Eq. (1) reduces to

$$\frac{1}{2}Ma = M(g-a) \text{ or } a = \frac{2}{3}g$$

So the acceleration of mass M is $\frac{2}{3}g$ while tension in the string PQ from Eq. (1) will be

$$T' = M\left(g - \frac{2}{3}g\right) = \frac{1}{3}Mg$$

It is clear that force on pulley exerted by the clamp will be equal and opposite to the resultant of the two forces of tension T, shown acting in figure (b) that are

$$(R_2) = \sqrt{T^2 + T^2} = \sqrt{2}T = \frac{\sqrt{2}}{3}Mg.$$

Example 5

A plank of mass m_1 with a bar of mass m_2 placed on it lies on a smooth horizontal plane. A horizontal force increasing with time t as $F = kt$ (k is a constant) is applied to the bar. Find how the acceleration of the plank a_1 and of the bar a_2 depend on t, if the coefficient of friction between the plank and the bar is equal to μ.

Solution

The situation is shown in Figure (a). Figure (b) shows the free-body diagrams of m_1 and m_2. It is obvious that if the frictional force is greater than the applied force, then both the blocks will move together. After some time, when the frictional force is less than applied force, then both the blocks will move with different accelerations.

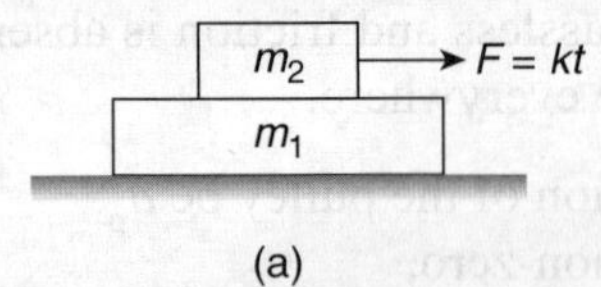

(a)

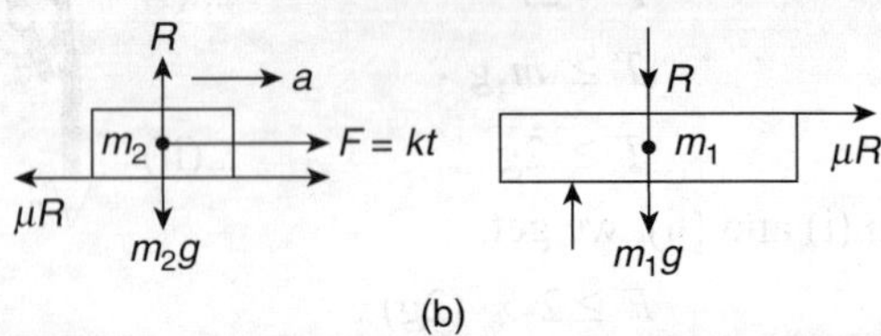

(b)

(1) Let both the blocks move together for a time t_0. In this case, acceleration $a = \frac{F}{m_1 + m_2}$

or, $$a = \frac{kt_0}{m_1 + m_2}.$$

From the free-body diagram of m_2, we have

$$F - \mu m_2 g = m_2 a$$

$$kt_0 - \mu m_2 g = \frac{m_2 k t_0}{m_1 + m_2}$$

$$t_0\left(k - \frac{m_2 k}{m_1 + m_2}\right) = \mu m_2 g$$

$$\therefore \quad t_0 = \frac{\mu m_2 g(m_1 + m_2)}{m_1 k}$$

Hence for $t \le t_0$, both the blocks will move together with acceleration.

(2) When applied force on m_2 is greater than frictional force: In this case, both the blocks will move with different accelerations.

Consider the free diagram of body m_2. We have,

$$(F - \mu m_2 g) = m_2 a_2 \text{ or, } (kt - \mu m_2 g) = m_2 a_2.$$

$$\therefore \quad a_2 = \frac{(kt - \mu m_2 g)}{m_2} \quad \text{(where } t \ge t_0\text{)}.$$

Similarly, for mass m_1, we have,

$$a_1 = (\mu m_2 g)/m_1 \text{ where } t \ge t_0.$$

Example 6

A body of mass 5×10^{-3} kg is launched up on a rough inclined plane making an angle of 30° with the horizontal. Obtain the co-efficient of friction between the body and the plane if the time of ascent is half of the time of descent.

Solution

The situation is shown in Figures (a) and (b).

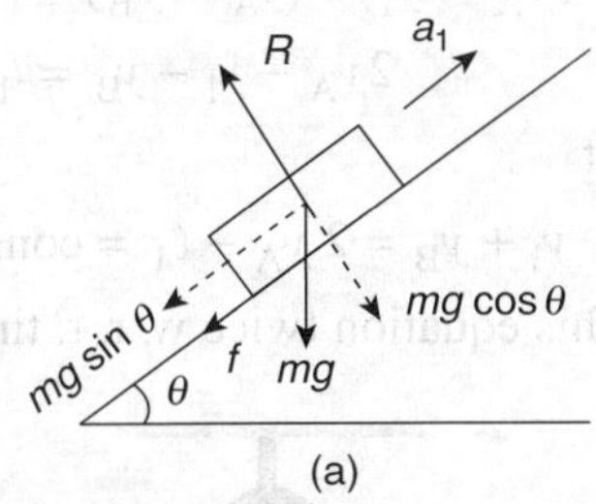

(a)

For upward motion,

$$mg\sin\theta + f = ma_1$$

or $$mg\sin\theta + \mu mg\cos\theta = ma_1 \quad ...(1)$$

$$\{\because f = \mu R = \mu\, mg\cos\theta\}$$

$$a_1 = g\sin\theta + \mu g\cos\theta$$

Let ℓ be the length of the plane. Suppose t_1 is the time taken by the block to move the length ℓ, then

$$\ell = 0 + \frac{1}{2}(g\sin\theta + \mu g\cos\theta)t_1^2 \quad ...(2)$$

$$(\because S = ut + \frac{1}{2}at^2)$$

Similarly, for downward motion

$$mg\sin\theta - f = ma_2$$

or $$mg\sin\theta - \mu mg\cos\theta = ma_2$$

$$a_2 = (g\sin\theta - \mu g\cos\theta) \quad ...(3)$$

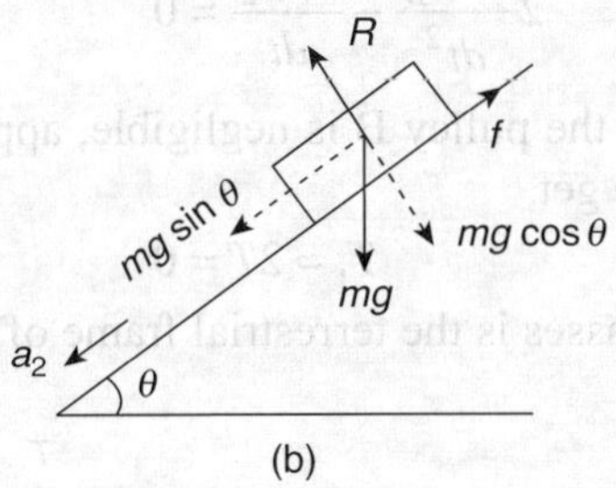

(b)

Let t_2 be the time taken for downward motion.

Hence $$\ell = \frac{1}{2}(g\sin\theta - \mu g\cos\theta)t_2^2 \quad ...(4)$$

From Eqs. (2) and (4), we get

$$\frac{1}{2}(g\sin\theta + \mu g\cos\theta)t_1^2 = \frac{1}{2}(g\sin\theta - \mu g\cos\theta)t_2^2$$

or $$(g\sin\theta + \mu g\cos\theta)\frac{t_2^2}{4} = (g\sin\theta - \mu g\cos\theta)t_2^2$$

$$g\sin\theta + \mu g\cos\theta = 4g\sin\theta - 4\mu g\cos\theta$$

$$5\mu g\cos\theta = 3g\sin\theta$$

$$\mu = \frac{3}{5}\tan\theta = 0.6\tan 30°$$

$$= 0.6\times\left(\frac{1}{\sqrt{3}}\right) = 0.346.$$

Example 7

In the figure, a painter of mass 100 kg stands in a crate C of mass 25 kg, and pulls himself and the crate up. If the painter exerts, an effective force of 450 N on the floor of the crate, find

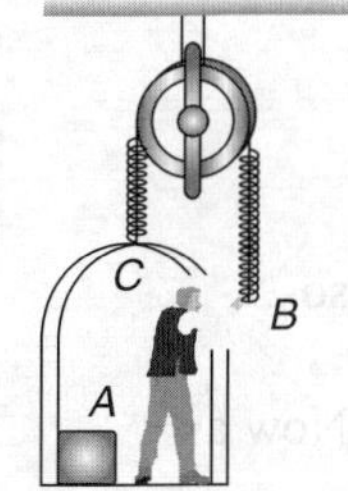

(a) The acceleration of the painter and
(b) The tension in the string, taking g = 10 m/s².

Solution

Suppose the man pulls the rope with a constant force $\vec{F}$ downwards, then the equation of motion for point B will be

$$F + m_B g - T = m_B a$$

i.e., $$F = T\ [\text{as } m_B = 0] \quad ...(1)$$

While for the painter,

$$R + T - mg = ma$$

i.e., $$R + T = m\,(g + a) \quad ...(2)$$

And for the system,

$$2T - (m + M)g = (m + M)a$$

i.e., $$2T = (m + M)\,(g + a). \quad ...(3)$$

Here, m = 100 kg, M = 25 kg, $R = A$ = 450 N

and g = 10 m/s².

So from Eqs. (2) and (3), we have,

$450 + T = 100(10 + a)$ and $2T = (100 + 25)(10 + a)$

i.e., $T = 550 + 100\,a$ and $2T = 125(10 + a)$

which on solving yields

a = 2 m/s² and $T = F$ = 750 N [as from Eq. (1) $F = T$]

Example 8

A small block starts sliding down an inclined plane subtending an angle θ with the horizontal. The coefficient of friction between the block and plane depends on the distance covered from rest along the plane as $\mu = \mu_0 x$ where μ_0 is a constant. Find:

(a) The distance covered by the block down the plane till it stops sliding and
(b) Its maximum velocity during this journey.

Solution

If a be the acceleration of the block sliding down the plane when it is at a distance x from the starting point, the equation of motion will be

$$ma = mg\sin\theta - \mu R$$

But $R = mg\cos\theta$ and $\mu = \mu_0 x$

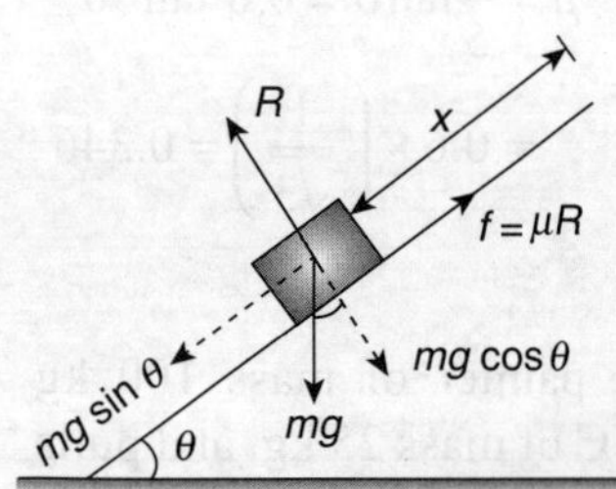

so $a = g\sin\theta - \mu_0 xg\cos\theta$ ...(1)

Now as $a = \frac{dv}{dt} = \frac{dv}{dx}.\frac{dx}{dt} = v\frac{dv}{dx}$

Equation (1) becomes

$$v\frac{dv}{dx} = g\sin\theta - \mu_0 xg\cos\theta$$

or $v\,dv = (g\sin\theta - \mu_0 xg\cos\theta)dx$

The above equation, on integration, yields

$$\frac{v^2}{2} = gx\sin\theta - \frac{1}{2}\mu_0 x^2 g\cos\theta + C$$

Now as at $x = 0$, $v = 0$ so, $C = 0$

so $v^2 = 2gx\sin\theta - \mu_0 x^2 g\cos\theta$ ...(2)

From Eq. (2), it is clear that

(a) v will be zero when

$$2gx\sin\theta - \mu_0 x^2 g\cos\theta = 0$$

i.e., $gx(2\sin\theta - \mu_0 x\cos\theta) = 0$

i.e., $x = 0$ or $x = 2\tan\theta/\mu_0$ ...(3)

(b) v will be maximum when

$$(dv/dx) = 0$$

But from Eq. (2),

$$v^2 = 2gx\sin\theta - \mu_0 x^2 g\cos\theta$$

i.e., $2v\left(\frac{dv}{dx}\right) = 2g\sin\theta - 2\mu_0 xg\cos\theta$

For $(dv/dx) = 0$, it yields $x = \tan\theta/\mu_0$ ...(4)

Substituting this value of x in Eq. (2), we get

$$v_{max}^2 = 2g\sin\theta\frac{1}{\mu_0}\tan\theta - \mu_0 g\cos\theta\frac{1}{\mu_0^2}\tan^2\theta$$

which on simplification gives

$$v_{max} = \sin\theta\sqrt{\frac{g}{\mu_0\cos\theta}} \quad ...(5)$$

Example 9

In the shown figure, masses of the pulleys and strings are negligible. Find the acceleration of the masses m_1 and m_2.

Solution

Let the lengths of the strings passing over A be ℓ_1 and of that passing over B be ℓ_2 from the ground level and y_A and y_B be the distance of the pulleys.

$\because$ $(y_A - y_1) + (y_A - y_B) = \ell_1$

$\Rightarrow$ $2y_A - y_1 - y_B = \ell_1$

$\because$ y_A is constant

$\therefore$ $y_1 + y_B = 2y_A - \ell_1 =$ constant

Differentiating this equation twice w. r. t. time, we get

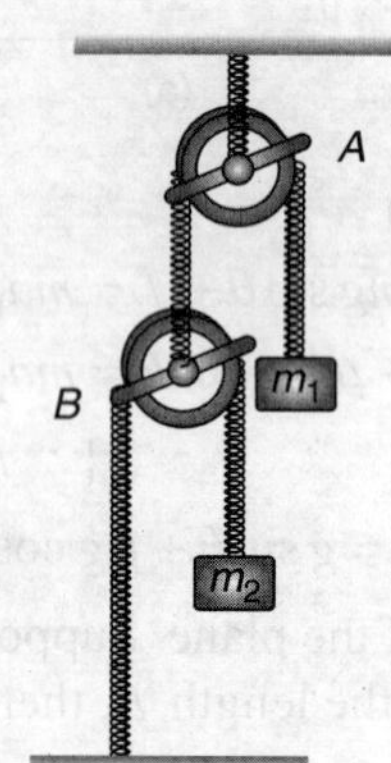

$$\frac{d^2y_1}{dt^2} + \frac{d^2y_B}{dt^2} = 0 \quad ...(1)$$

Similarly, $y_B + (y_B - y_2) = l_2$

$\Rightarrow$ $2y_B - y_2 =$ Constant

$\Rightarrow$ $2\frac{d^2y_B}{dt^2} - \frac{d^2y_2}{dt} = 0$ ...(2)

Since mass of the pulley B is negligible, applying Newton's second law we get,

$\Rightarrow$ $T_1 = 2T = 0$

FBD of the masses is the terrestrial frame of reference,

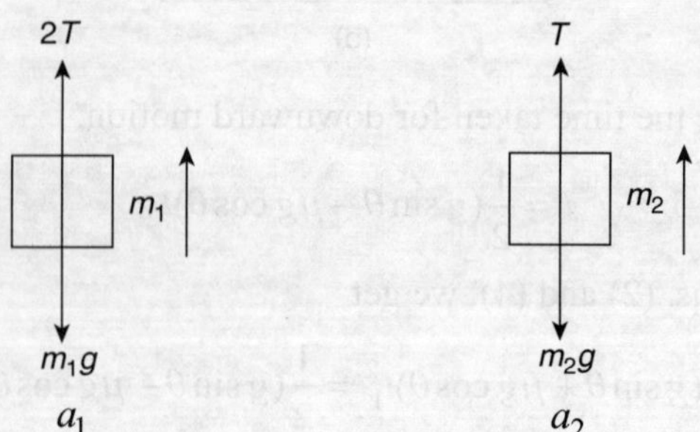

Equation of motion

$$-2T + m_1 g = m_1 a_1 \quad ...(i)$$

$$T - m_2 g = m_2 a_2 \qquad \text{...(ii)}$$

Solving Eqs. (i) and (ii), we get

$$\Rightarrow \qquad a_1 = \frac{m_1 - 2m_2}{m_1 + 2m_2} g$$

$$a_2 = \frac{2(m_1 - 2m_2)}{(m_1 + 2m_2)} g.$$

Example 10

For the arrangement shown in figure, the coefficient of friction between any two surfaces is 0.5. Which of the following graphs shows the correct variation of the frictional force f between the 2 kg block and the floor with the applied force F (Take leftward direction of f as positive)?

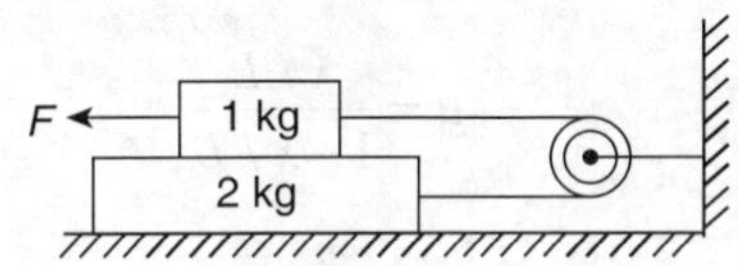

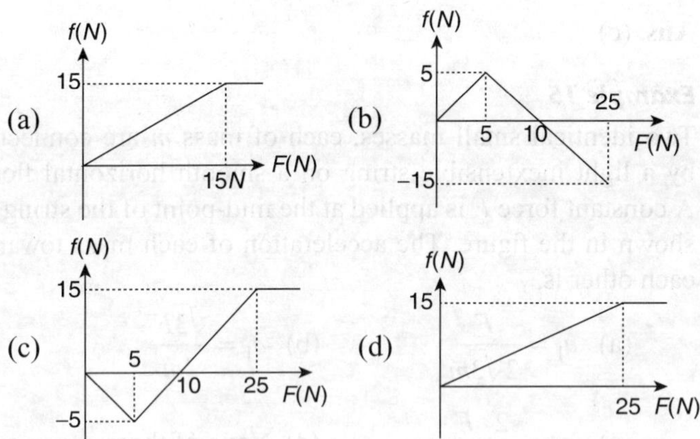

Solution

If $\qquad F \le 0.5 \times 1 \times 10$

$\Rightarrow \qquad F \le 5N, T = 0$

$\therefore \qquad f_1 = F$

$\therefore \qquad f = f_1 = F$

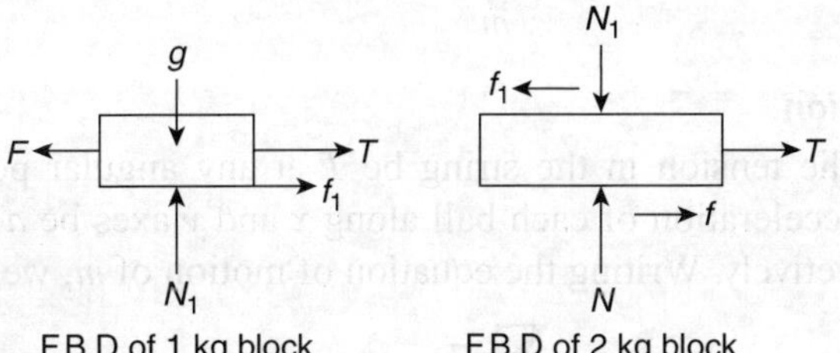

When $\qquad F > 5$ N,

Tension will arise in string and $f_1 = 5$ N, the limiting value. With the increase in F, tension will increase and f will decrease.

For $F > 10$ N, frictional force will change the direction to left and will continue to increase. For $F = 25$ N, f becomes maximum and then it will become constant.

Ans. (c)

Example 11

A block of mass $\sqrt{3}$ kg is placed on a rough horizontal surface whose coefficient of fiction is $\frac{1}{2\sqrt{3}}$. Then the maximum value of force F (shown in figure) such that block does not move ($g = 10$ m/s^2) is

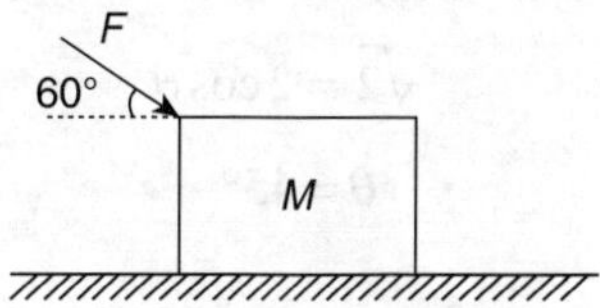

(a) 10 N (b) 20 N
(c) 12 N (d) None of these.

Solution

We have

$$N = 10\sqrt{3} + \frac{\sqrt{3}}{2}F \text{ and } F\cos 60° = \mu N$$

$$\Rightarrow \qquad \frac{F}{2} = \frac{1}{2\sqrt{3}}\left(10\sqrt{3} + \frac{\sqrt{3}}{2}F\right)$$

$$\Rightarrow \qquad F = 20 \text{ N.}$$

Ans. (b)

Example 12

Let R, N and f denote the magnitudes of the contact force, normal force and the friction force exerted by one surface on the other when kept in contact. If none of these is zero then which of the following is incorrect

(a) $R > N$ (b) $R > f$
(c) $N > f$ (d) $N - f < R < N + f$.

Solution

$$R = \sqrt{N^2 + f^2},\ f = \mu N$$

$\therefore \qquad R > N$ and $R > f$

R is the resultant of N and f. The magnitude of resultant is less than the sum and greater than the difference of the magnitudes of the constituent force,

$\therefore \qquad N - f < R < N + f$

Ans. (c)

Example 13

The pulleys and strings shown in the figure are smooth and of negligible mass. For the system to remain in equilibrium, the angle θ should be

(a) 0º (b) 30º
(c) 45º (d) 60º.

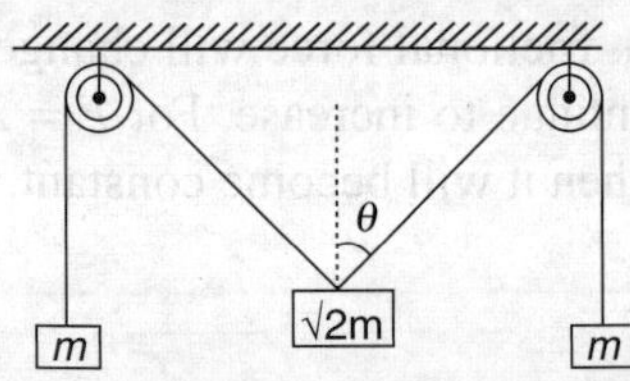

Solution

Drawing the F.B.D of the masses $\sqrt{2}m$ and m and considering their equilibrium $\sqrt{2}\,mg = 2T\cos\theta$

and $$mg = T$$

$\Rightarrow$ $$\sqrt{2} = 2\cos\theta$$

$$\theta = 45°$$

Ans. (c)

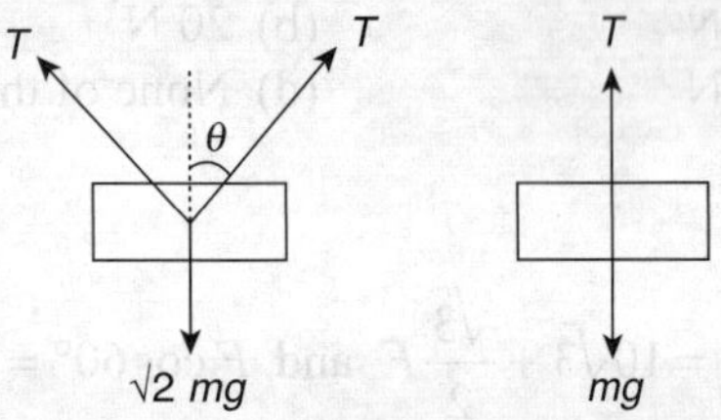

Example 14

A uniform chain is just at rest over a rough horizontal table with $\frac{1}{n}$th of its length hanging vertically over the edge of the table. The co-efficient of static friction between the chain and the table is,

(a) $\mu = \frac{1}{1-n}$ (b) $\mu = \frac{1}{1+n}$

(c) $\mu = \frac{n}{1-n}$ (d) $\mu = \frac{n}{1+n}$.

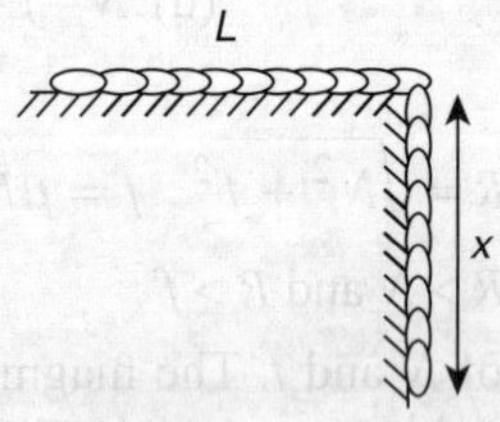

Solution

We see that a portion of the chain is lying on the table top. Let the mass of that portion be m_1. Let the mass of the remaining (hanging) portion of the chain be m_2. Since the chain is at the point of slipping, the weight of the hanging portion of the chain counterbalances the maximum static frictional force f_{max} between m_1 and the surface.

$\Rightarrow$ $$m_2 g = f_{max};\quad m_2 g = \mu N_1$$

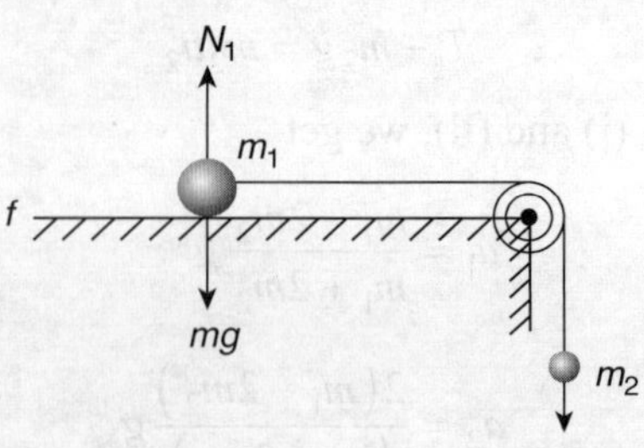

where $N_1 - m_1 g = 0$ for the equilibrium of the portion of chain lying as on the table.

$\Rightarrow$ $$m_2 g - \mu m_1 g = 0$$

$\Rightarrow$ $$\mu = \frac{m_2}{m_1} = \frac{\frac{M}{L}x}{\frac{M}{L}(L-x)}$$

$\Rightarrow$ $$\mu = \frac{x/L}{1 - x/L}$$

$\Rightarrow$ $$\mu = \frac{n}{1-n}$$

Ans. (c)

Example 15

Two identical small masses, each of mass m are connected by a light inextensible string on a smooth horizontal floor. A constant force F is applied at the mid-point of the string as shown in the figure. The acceleration of each mass towards each other is,

(a) $a_1 = \frac{F}{2\sqrt{3}m}$ (b) $a_1 = \frac{\sqrt{3}F}{2m}$

(c) $a_1 = \frac{2}{\sqrt{3}}\frac{F}{m}$ (d) None of these.

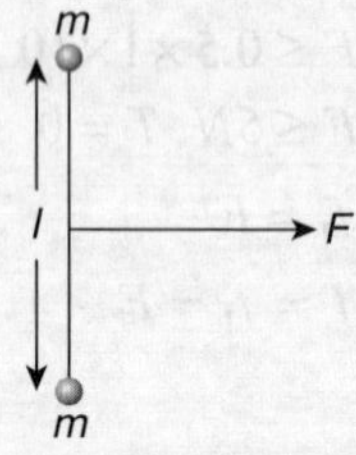

Solution

Let the tension in the string be T at any angular position, the acceleration of each ball along x and y axes be a and a_1 respectively. Writing the equation of motion of m, we obtain

$$\sum F_x = ma$$

$\Rightarrow$ $$T\cos\theta = ma \quad \text{...(i)}$$

$$\sum F_y = ma_1$$

$\Rightarrow \qquad T\sin\theta = ma_1 \qquad \ldots(ii)$

At point P, as it is accelerating with an acceleration a,

$$F - 2T\cos\theta = m_p a$$

where m_p = mass of the string at the point $P \cong 0$

$\Rightarrow \qquad F = 2T\cos\theta \qquad \ldots(iii)$

Divide (ii) and (i), then we get

$\Rightarrow \qquad \tan\theta = \dfrac{a_1}{a}$

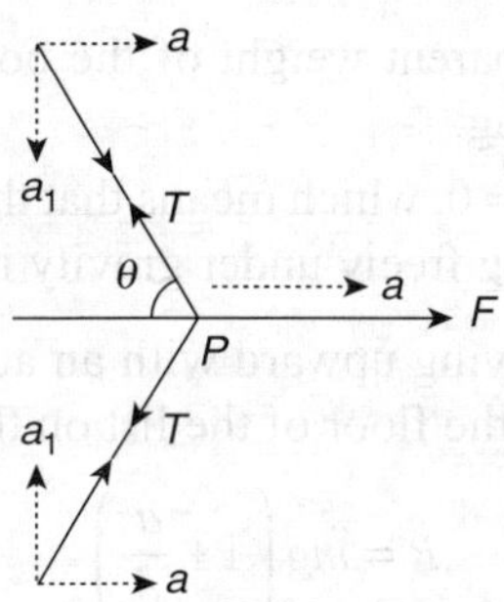

$\Rightarrow \qquad a_1 = a\tan\theta$

where $\qquad a = \dfrac{T\cos\theta}{m}$ from (i)

Putting $T = \dfrac{F}{2\cos\theta}$ from (iii), we obtain.

$$a_1 = \frac{F}{2m}\tan\theta, \text{ Putting } \theta = 30°$$

$$a_1 = \frac{F}{2\sqrt{3}m}.$$

Ans. (a)

Example 16

A man of 60 kg is standing on a horizontal conveyer belt. If the coefficient of static friction between his shoes and the belt is 0.2, to what value can the acceleration of the belt be increased so that the man continues to remain stationary relative to the belt?

(a) 1m/s² (b) 2 m/s²
(c) 3 m/s² (d) 4 m/s².

Solution

We know that $\mu_s = \dfrac{\text{Limiting friction}}{\text{Normal reaction}} = \dfrac{f}{mg}$

Since $\mu_s = 0.2$, the force of limiting friction between the man's shoes and belt is

$$f = \mu_s \times mg = 0.2 \times 60 \times 10 = 120 \text{ N}$$

The man will continue to remain stationary relative to the belt and will just start slipping on the belt if the acceleration of the belt (which equals his acceleration) is increased to a critical value a_c such that the force ma_c acting on him happens to equal the force of limiting friction f, i.e., if

$$ma_c = f \text{ or } a_c = \frac{f}{m} = \frac{120}{60} = 2 \text{ ms}^{-2}.$$

Ans. (b)

POINTS TO REMEMBER

1. Newton's three laws of motion are:

(i) Every body continues in its state of rest or of uniform motion along a straight line unless an external force acts on it.

(ii) The rate of change of momentum is directly proportional to the impressed force and takes place in the direction in which the force acts.

(ii) To every action, there is an equal and opposite reaction.

2. The property of a body to maintain its state of rest or the uniform motion along a straight line unless an external force acts on it, is called inertia. Newton's first law is, therefore, also called law of inertia.

3. The mass m of a body is a measure of its inertia. Larger the mass, more the time it takes to change its velocity by a given amount and vice-versa.

4. The momentum P of a body is given by the product of its mass m and velocity v and is, therefore, a vector quantity, i.e,

$$\vec{p} = m\vec{v}$$

The unit of momentum in SI, is kg-m/s.

5. The principle of conservation of linear momentum states that the total linear momentum of a system in any given direction remains conserved if no external force acts on it.

If u_1 and u_2 be the velocities of particles of masses m_1 and m_2 respectively along a given direction before collision, and v_1 and v_2 be their velocities along the same straight line after collision, then according to the principle of conservation of linear momentum,

$$m_1 u_2 + m_2 u_2 = m_1 v_1 + m_2 v_2$$

6. Newton's second law gives the definition of force and a way to measure it. The force is an external agent that changes or tends to change the state of body. If m be the mass of a body moving with an acceleration $\bar{a}$ under the action of a force $\vec{F}$ then,

$$\vec{F} = m\vec{a}$$

The unit of force in CGS, FPS, and SI system are dyne, pound and Newton respectively.

7. The impulse of a force $\vec{F}$ is defined as the product of the force and the time Δt during which it acts. Thus,

Impulse of a force = $\vec{F}.\Delta t$

The unit of impulse is Newton-second (N-s)

8. The second law also gives the principle of physical independence of forces. It states that if a number of forces acts on a body, then the effect produced by each force is independent of the effect produced by the other forces.

9. Action and reaction are nothing but forces. The first one to act is known as action and the other that comes into play as a result of the action is called reaction.

10. The forces of action–reaction appear simultaneously due to interaction between two bodies and reaction exists only so long as action is present.

11. The action and reaction always act on two different bodies and hence, in no case, can produce equilibrium.

12. Inertial frames are those which are either at rest or move with uniform velocity $\vec{v}$. Newton's laws are valid only in the inertial frames of reference.

13. Non-inertial frames are those which move with an acceleration or deceleration.

14. Newton's laws of motion are not valid if one is working from a non-inertial frame. If the frame is moving with respect to an inertial frame with an acceleration $\vec{a}_0$, one must assume the existence of a pseudo force $-m\vec{a}_0$ acting on the particle of mass m. Once this pseudo force is included, one can use Newton's laws in their usual form.

15. A body shows a weight only because the force of reaction acts on it. If a body falls freely under gravity, its apparent weight during its free fall is zero because, there is no force of reaction acting on it.

16. The apparent weight of a body in a moving lift is usually different from its actual weight. Hence, three cases may arise:

(i) If the lift is moving downward with an acceleration a, then the reaction of the floor of the lift on the body is

$$R = mg\left(1 - \frac{a}{g}\right)$$

$$\Rightarrow \qquad R < mg$$

That is, the apparent weight of the body is less than its actual weight mg.

If $a = g$, then $R = 0$, which means that the apparent weight of a body falling freely under gravity is zero.

(ii) If the lift is moving upward with an acceleration a, then the reaction of the floor of the lift on the body is

$$R = mg\left(1 + \frac{a}{g}\right)$$

$$\Rightarrow \qquad R > mg$$

That is, the apparent weight of the body is greater than its actual weight mg.

(iii) If the lift moves in either direction with uniform velocity, then $R = mg$, i.e., the apparent weight of the body is equal to its actual weight mg.

17. When several bodies are connected by unstretchable strings passing over several pulleys, the motion of the bodies takes place under certain constraints. The length of the string remains constant and the string is in state of tension. The tensions in different strings will be different.

The tension in different parts of a string will be the same if the string is light and the pulley is frictionless; otherwise it will be different.

18. When the motion of body is to be considered then all the forces acting on that body, and not the forces applied by that body, should be taken into account.

19. Whenever a spring is compressed or stretched by x, a force equal to $-Kx$ is developed in the spring that tends to ring the spring back to its normal length. The negative sign shows that the force developed is in a direction opposite to x. Here, K is the force constant of the spring and is defined as the force developed in the spring per unit compression or elongation.

EXERCISES

1. A block of mass m is supported by a string passing through a smooth peg as shown in the figure. Variation of tension in the string T as a function of θ best represented by

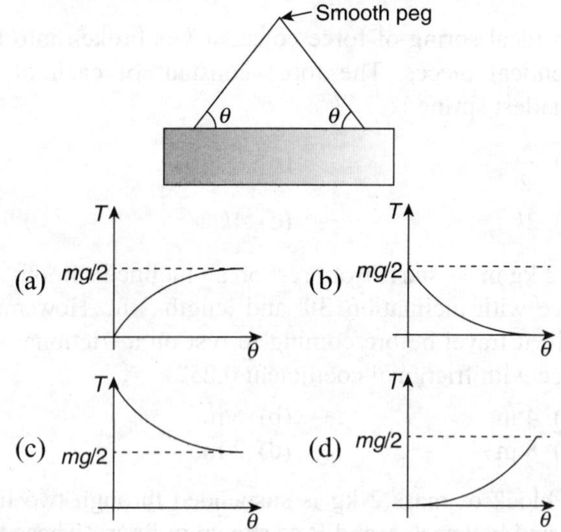

2. Physical independence of force is a consequence of
(a) Third law of motion (b) Second law of motion
(c) First law of motion (d) All of these laws.

3. Starting from rest, a body slides down a 45° inclined plane in twice the time it takes to slide down the same distance in the absence of friction. The coefficient of friction between the body and the inclined plane is
(a) 0.80 (b) 0.75
(c) 0.25 (d) 0.33

4. A person is sitting in a lift accelerating upwards. Measured weight of the person will be
(a) less than actual weight
(b) equal to actual weight
(c) more than actual weight
(d) None of these.

5. Which is a non-central force?
(a) Electrostatic force (b) Nuclear force
(c) Gravitational force (d) None of these.

6. In the arrangement shown in figure, the strings are light and inextensible. The surface over which blocks are placed is smooth. What is the acceleration of each block?

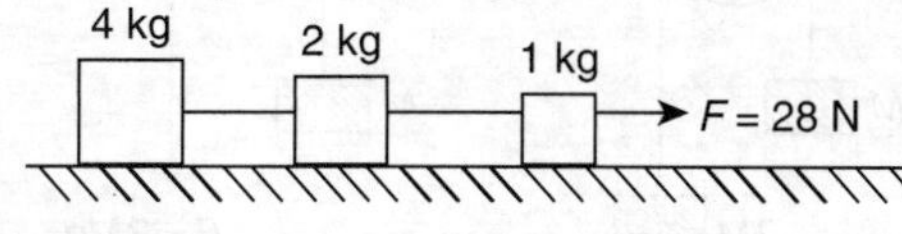

(a) 8 m/s^2 (b) 4 m/s^2
(c) 2 m/s^2 (d) 14 m/s^2

7. An ice cube is kept on an inclined plane of angle 30° coefficient of kinetic friction between block and inclined plane is $\frac{1}{\sqrt{3}}$.What is acceleration of block?
(a) zero (b) 2 m/s^2
(c) 1.5 m/s^2 (d) 5 m/s^2

8. A body of mass 5 kg starts from the origin with an initial velocity $\vec{u} = (30\hat{i} + 40\hat{j})$ m/s. If a constant force $(-6\hat{i} - 5\hat{j})$ N acts on the body, the time in which the y-component of the velocity becomes zero, is
(a) 5 s (b) 20 s
(c) 40 s (d) 80 s

9. If two forces are acting at a point such that the magnitude of each force is 2 N and the magnitude of their resultant is also 2 N, then the angle between the two force is
(a) 120° (b) 60°
(c) 90° (d) 0°

10. If two forces of 5 N each are acting along x and y axis, then the magnitude and direction of the resultant is
(a) $5\sqrt{2}, \frac{\pi}{3}$ (b) $5\sqrt{2}, \frac{\pi}{4}$
(c) $-5\sqrt{2}, \frac{\pi}{3}$ (d) $-5\sqrt{2}, \frac{\pi}{4}$

11. Natural length of spring is 60 cm and its spring constant is 4000 N/m. A mass of 2 kg is hung from it. The extension produced in the spring is (Take $g = 9.8$ m/s^2)
(a) 4.9 cm (b) 0.49 cm
(c) 9.4 cm (d) 0.94 cm

12. A body executing uniform circular motion. At any instant, its velocity and acceleration vector will be
(a) along the same direction
(b) in opposite direction
(c) normal to each other
(d) not related to each other.

13. The spring balance inside a lift suspends an object. As the lift begins to ascend, the reading indicated by the spring balance will
(a) increase
(b) decrease
(c) remain unchanged
(d) depend on the speed of ascend.

14. Block A of mass m and Block B of mass 2 m are placed on a fixed triangular wedge by means of a massless, inextensible spring and a frictionless pulley as shown in figure. The wedge is inclined at 45° to the horizontal on both the sides. If the coefficient of friction between

Block A and the wedge is $\frac{2}{3}$ and that between Block B and the wedges is $\frac{1}{3}$ and both the blocks A and B are released from rest, the acceleration of A will be

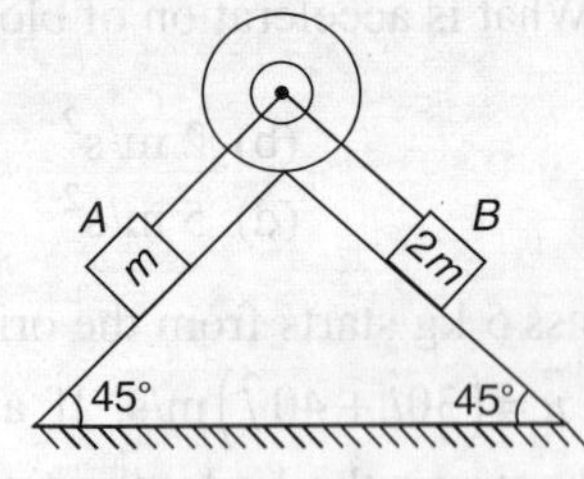

(a) −1 (b) 1.2
(c) 0.2 (d) zero.

15. The coefficient of limiting friction μ is defined as

(a) $\mu = \frac{R}{F}$ (b) $\mu = \sqrt{\frac{F}{R}}$

(c) $\mu = \frac{F}{R}$ (d) $\mu = \sqrt{\frac{R}{F}}$

16. A body of mass 12 kg is suspended by a coil spring of natural length 50 cm and spring constant 2×10^3 N/m. The length of the spring after extension will be

(a) 0.00588 m (b) 0.0588 m
(c) 0.5588 m (d) None of these.

17. Consider a car moving on a straight road with a speed of 100 m/s. The distance at which car can be stopped, is [$\mu_k = 0.5$]

(a) 800 m (b) 1000 m
(c) 100 m (d) 400 m

18. A block moving on a surface with velocity 20 m/s comes to rest because of surface friction over a distance of 40 cm. Taking $g = 10$ m/s^2, the coefficient of dynamic friction is

(a) 0.5 (b) 0.3
(c) 0.2 (d) 0.1

19. Two masses A and B of 15 kg and 10 kg are connected with a string passing over a frictionless pulley fixed at the corner of a table (as shown in figure). The coefficient of friction between the table and block is 0.4. The minimum mass of C that may be placed on A to prevent it from moving is

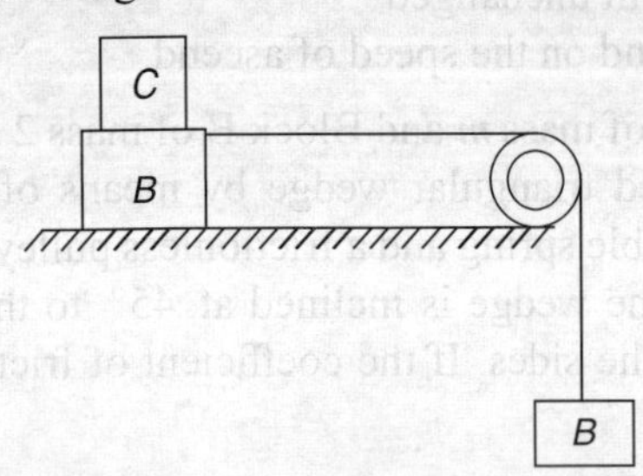

(a) 10 kg (b) 5 kg
(c) Zero (d) 15 kg

20. When a force of 0.1 N is applied, the spring is stretched by 1.5 cm. If the spring is cut into three parts and one part is stretched by 3 cm, find the force required for doing so.

(a) 0.2 N (b) 0.3 N
(c) 0.4 N (d) 0.6 N

21. An ideal spring of force constant k is broken into two identical pieces. The force constant of each of the smallest spring is

(a) $\frac{k}{2}$ (b) k
(c) $2k$ (d) $4k$

22. A 2 kg mass starts from rest on an inclined smooth surface with inclination 30° and length 2 m. How much will it travel before coming to rest on a frictional surface with frictional coefficient 0.25?

(a) 4 m (b) 6 m
(c) 8 m (d) 2 m

23. A block of mass 2 kg is suspended through two light spring balances A and B as shown in figure. These balances A and B will respectively read

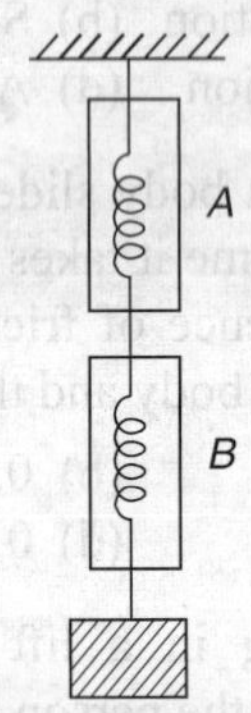

(a) 2 kg and zero (b) zero and 2 kg
(c) 2 kg and 2 kg (d) 1 kg and 1 kg.

24. A pulley has two different arrangements (i) and (ii) as shown. Neglecting the masses of the rope and the pulley, the ratio of the acceleration of the mass M in the arrangement (i) to that in the arrangement (ii) is

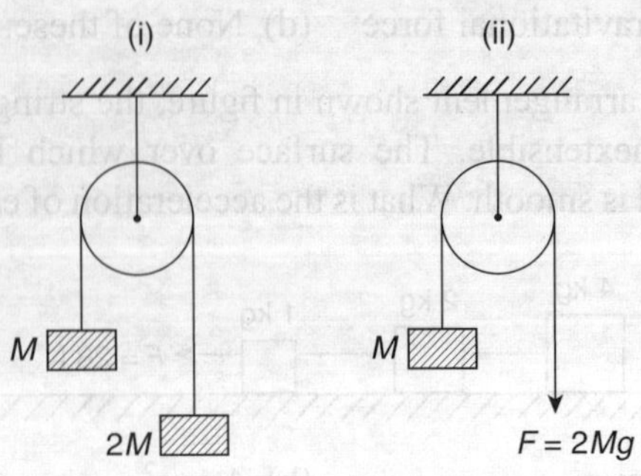

(a) 3 : 1 (b) 2 : 1
(c) 1 : 2 (d) 1 : 3

25. A 600 kg rocket is set for a vertical firing. If the exhaust speed is 1000 m s^{-1}, the mass of the gas ejected per second to supply the thrust needed to overcome the weight of rocket is

(a) 117.6 kg s^{-1} (b) 58.6 kg s^{-1}
(c) 6 kg s^{-1} (d) 76.4 kg s^{-1}

26. A man of mass $m = 55$ kg is climbing inside a wide vertical hollow pipe. After reaching some height h, he takes rest by pressing his shoulder and feet on the opposite sides of the pipe. If the two coefficients of static frictions are 1.1 and 0.7, what is the minimum horizontal push so that he does not fall?

(a) 700 N (b) 550 N
(c) 940 N (d) 860 N

27. A force of 6 N acts on a body at rest and of mass 1 kg. During this time, the body attains a velocity of 30 m/s. The time for which the force acts on the body is

(a) 7 seconds (b) 5 seconds
(c) 10 seconds (d) 8 seconds.

28. In the given figure, three blocks are connected by massless cords. Masses of the blocks are m, $3m$ and $5m$ and they are pulled by a force F and frictionless horizontal surface. If the tension $T_1 = 16$ N, then T_2 is equal to

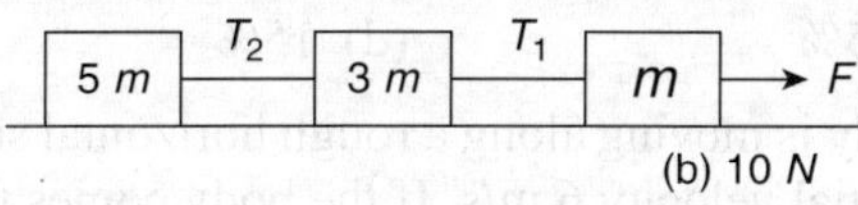

(b) 10 N

(a) 6 N (b) 10 N
(c) $\frac{80}{3}$ N (d) 18 N

29. A 1.5 kg ball drops vertically on a floor hitting with a speed of 25 m/s. It rebounds with an initial speed of 15 m/s. If the ball was in contact for only 0.03 second, the force exerted on the floor by the ball is

(a) 2000 N (b) 3000 N
(c) 3500 N (d) 4000 N

30. A particle of mass m is rotating in a horizontal circle of radius R and is attached to a hanging mass M as shown in the figure. The speed of rotation required by the mass m to keep M steady is

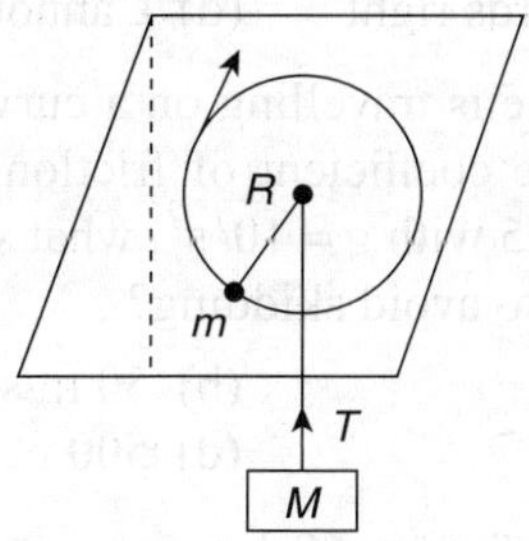

(a) $\sqrt{\frac{mgR}{M}}$ (b) $\sqrt{\frac{MgR}{m}}$
(c) $\sqrt{\frac{mg}{MR}}$ (d) $\sqrt{\frac{mR}{Mg}}$

31. An 80 kg person is parachuting and is experiencing a downward acceleration of 20.8 m/s^2. The mass of the parachute is 5 kg. The upward force on the open parachute is (Take $g = 9.8$ m/s^2)

(a) 595 N (b) 675 N
(c) 456 N (d) 925 N

32. Two block of masses M and m are connected by a string passing over a pulley as shown in the figure. The downward acceleration of the block with mass m is.

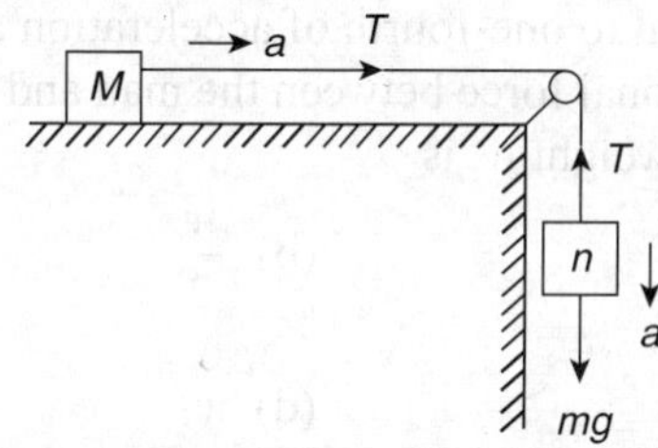

(a) $\frac{M}{(m+M)g}$ (b) $\frac{mg}{(m+M)}$
(c) $\frac{(m+M)}{mg}$ (d) $\frac{(m+M)}{mg}$

33. A force F_1 accelerates a particle from rest to a velocity v. Another force F_2 decelerates the same particle from v to rest. Then,

(a) F_2 is always greater than F_1
(b) F_2 is greater than F_1
(c) F_2 may be smaller than, greater than or equal to F_1
(d) F_2 cannot be equal to F_1

34. In the given figure the pulley is assumed massless and frictionless. If the friction force on the object of mass m is f, then its acceleration in terms of the force F will be equal to

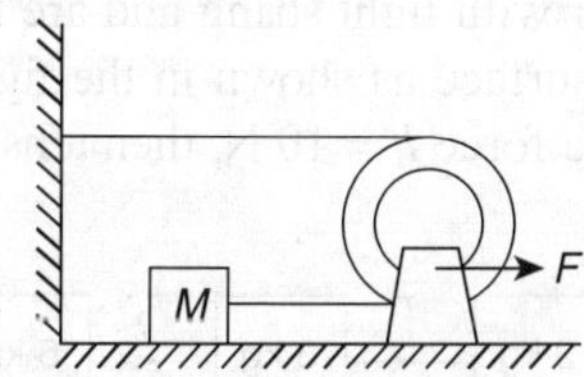

(a) $\frac{(F-f)}{M}$ (b) $\frac{(F/2-f)}{M}$
(c) $\frac{F}{2M}$ (d) None of these.

35. In the first second of its flight, a rocket ejects $\frac{1}{60}$ of its mass with a velocity of 2400 ms^{-1}. The acceleration of rocket is

(a) 19.6 $m s^{-1}$ (b) 30.2 $m s^{-1}$
(c) 40 $m s^{-1}$ (d) 49.9 $m s^{-1}$

36. A body of mass 5 kg explodes at rest into three fragments with masses in the ratio 1 : 1 : 3. The fragments with equal masses fly in mutually perpendicular directions with speed of 21 m/s. The velocity of heaviest fragment in m/s will be

(a) $7\sqrt{2}$ (b) $5\sqrt{2}$
(c) $3\sqrt{2}$ (d) $\sqrt{2}$

37. A man slides down on a telegraphic pole with acceleration equal to one-fourth of acceleration due to gravity. The frictional force between the man and pole, in terms of man's weight w, is

(a) $\frac{w}{4}$ (b) $\frac{w}{2}$
(c) $\frac{3w}{4}$ (d) w

38. The minimum force required to move a body up an inclined plane is three times the minimum force required to prevent it from sliding down the plane. If the coefficient of friction between the body and the inclined plane is $\frac{1}{2\sqrt{3}}$, the angle of the inclined plane is

(a) 60° (b) 45°
(c) 30° (d) 15°

39. A body is sliding down a rough inclined plane. The coefficient of friction between the body and the plane is 0.5. The ratio of the net force required for the body to slide down and the normal reaction of the body is 1 : 2. Then the angle of the inclined plane is

(a) 15° (b) 30°
(c) 45° (d) 60°

40. Three blocks of masses 2 kg, 3 kg and 5 kg are connected to each other with light string and are than placed on a frictionless surface as shown in the figure. The system is pulled by a force $F = 10$ N, then tension T_1

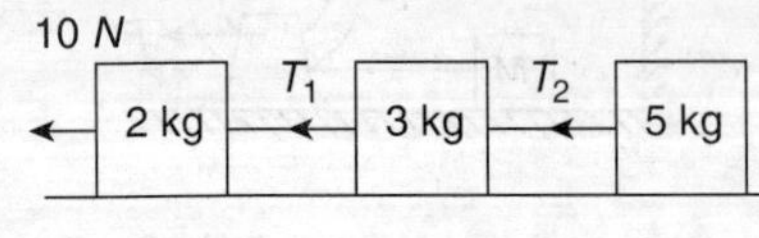

(a) 1 N (b) 5 N
(c) 8 N (d) 10 N

41. A ball is thrown up at an angle with the horizontal. Then the total change of momentum at the instant it returns to ground is

(a) Acceleration due to gravity × Total time of flight
(b) Weight of the ball × Half the time of flight
(c) Weight of the ball × Total time of flight
(d) Weight of the ball × Horizontal range.

42. The coefficient of friction between the tires and the road is 0.25. The maximum speed with which car can be driven around a curve of radius 40 m without skidding is (assume $g = 10$ m s^{-2})

(a) 40 $m s^{-2}$ (b) 20 $m s^{-2}$
(c) 15 $m s^{-2}$ (d) 10 $m s^{-2}$

43. If the coefficient of static friction between the tires and road is 0.5, what is the shortest distance in which an automobile can be stopped when travelling at 72 km/h?

(a) 50 m (b) 60 m
(c) 40.8 m (d) 80.16 m

44. A heavy uniform chain lies on horizontal table top. If the coefficient of friction between the chain and the table surface is 0.25, then the maximum fraction of the length of the chain that can hang over one edge of the table is

(a) 20% (b) 25%
(c) 35% (d) 15%

45. A body is moving along a rough horizontal surface with an initial velocity 6 m/s. If the body comes to rest after travelling a distance 9 m, then the coefficient of sliding friction will be

(a) 0.4 (b) 0.2
(c) 0.6 (d) 0.8

46. The given figure shows a wooden block on a horizontal plane at a rest being acted upon by three forces: $F_1 = 10$ N, $F_2 = 2$ N, and friction. If F_1 is removed, the resultant force acting on the block will be:

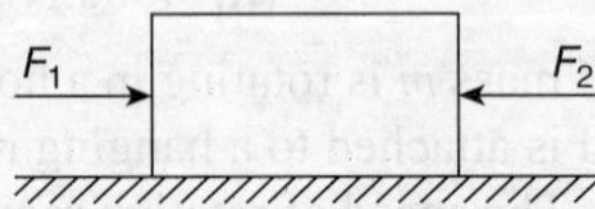

(a) 2 N towards left (b) 0 N
(c) 2 N towards right (d) Cannot be determined.

47. A motor cycle is travelling on a curved track of radius 500 m. If the coefficient of friction between the tires and road is 0.5 with $g = 10/s^2$, what should be the maximum speed to avoid skidding?

(a) 10 m/s (b) 50 m/s
(c) 250 m/s (d) 500 m/s

48. An ice cart of mass 60 kg rests on a horizontal snow patch with coefficient of static friction $\frac{1}{3}$. Assuming

that there is no vertical acceleration, find the magnitude of the maximum horizontal force required to move the ice cart. ($g = 9.8\ m\,s^{-2}$)

(a) 100 N (b) 110 N
(c) 209 N (d) 196 N

49. A uniform metal chain is placed on a rough table such that one end of it hangs down over the edge of the table. When one-third of its length hangs over the edge, the chain starts sliding. Then, the coefficient of static friction is

(a) $\frac{3}{4}$ (b) $\frac{1}{4}$
(c) $\frac{2}{3}$ (d) $\frac{1}{2}$

50. When milk is churned, cream gets separated due to

(a) centripetal force (b) centrifugal force
(c) frictional force (d) gravitational force.

51. A block rests on a rough plane whose inclination θ to the horizontal can be varied. Which of the following graphs indicates how the frictional force F between the block and the plane varies as θ is increased?

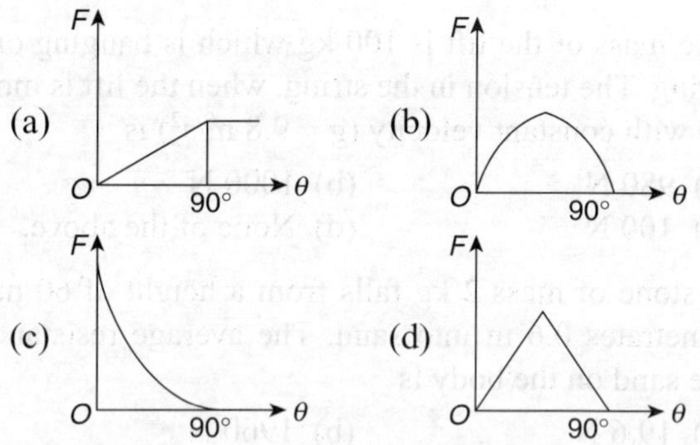

52. In the figure, the blocks A, B and C of mass m each have acceleration a_1, a_2 and a_3 respectively. F_1 and F_2 are external forces of magnitude 2 mg and mg respectively. Then,

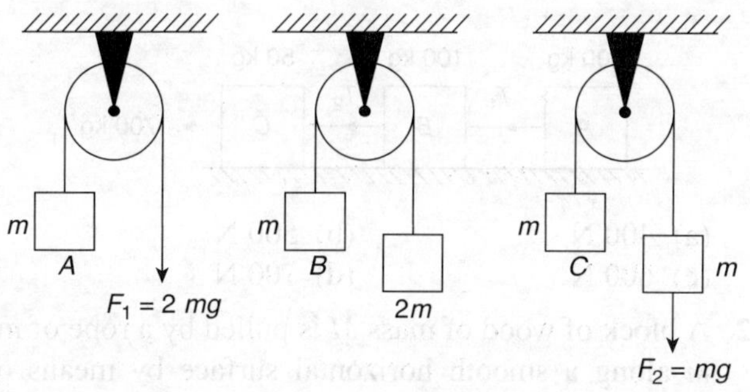

(a) $a_1 = a_2 = a_3$ (b) $a_1 > a_3 > a_2$
(c) $a_1 = a_2,\ a_2 > a_3$ (d) $a_1 > a_2,\ a_2 = a_3$

53. A force-time graph for a linear motion of a body is shown in the figure. The change in linear momentum between 0 and 7 s is

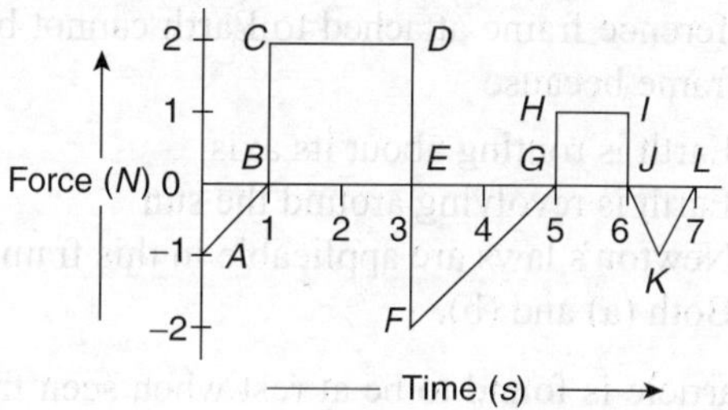

(a) 2 N-s (b) 3 N-s
(c) 4 N-s (d) 5 N-s

54. Figure shows two blocks A and B pushed against the smooth wall with the force F. The wall is smooth but the surface in contact of A and B is rough. Which of the following is true for the system of blocks?

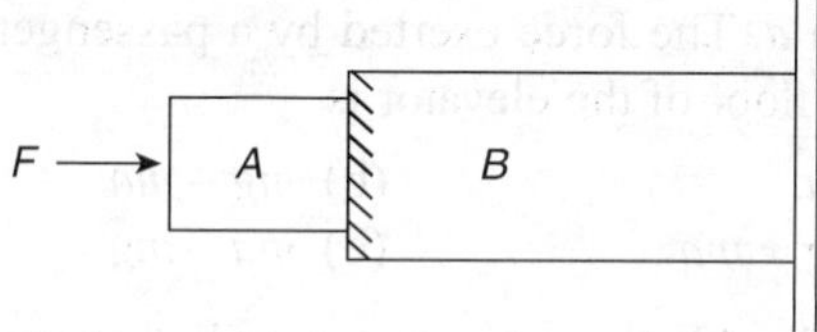

(a) The system cannot be in equilibrium.
(b) F should be equal to weight of A and B.
(c) F should be less than the weight of A and B.
(d) F should be more than the weight of A and B.

55. A weightless thread can bear tension up to 37 N. A stone of mass 500 g is tied to it and revolved in a circular path of radius 4 m in a vertical plane. If $g = 10\ m\,s^{-2}$, then the maximum angular velocity of the stone will be

(a) 2 rad s^{-1} (b) 4 rad s^{-1}
(c) 8 rad s^{-1} (d) 16 rad s^{-1}

56. You are on a frictionless horizontal plane. How can you get off it if no horizontal force is exerted by pushing against the surface

(a) by jumping
(b) by spitting and sneezing
(c) by rolling your body on the surface
(d) by running on the plane.

57. Pulling is easier than pushing because

(a) force of friction is in opposite direction
(b) when we pull a roller, the vertical component of the pulling force acts in the direction of weight
(c) it is possible in the case of roller only
(d) the vertical component of the pulling forces acts in the opposite direction of weight.

58. It is difficult to move a cycle with breaks on because

(a) rolling friction is more than sliding friction
(b) sliding friction opposes motion on road
(c) rolling friction opposes motion on road
(d) sliding friction is more than rolling friction.

59. A reference frame attached to Earth cannot be an inertial frame because

(a) Earth is routing about its axis
(b) Earth is revolving around the sun
(c) Newton's laws are applicable in this frame
(d) Both (a) and (b).

60. A particle is found to be at rest when seen from frame S_1 and moving with a constant velocity when seen from another frame S_2. Mark out the possible options.

(a) S_1 is an inertial and S_2 is non-inertial frame.
(b) Both frames are inertial
(c) Both the frames are non-inertial
(d) Either (b) or (c).

61. Consider an elevator moving downwards with an acceleration a. The force exerted by a passenger of mass m on the floor of the elevator is

(a) ma (b) $mg - ma$
(c) $mg + ma$ (d) $ma - mg$

62. A 30 kg block rests on a rough horizontal surface. A force of 200 N is applied on the body. The block acquires a speed of 4 m/sec starting from rest, in 2 seconds. What is the value of coefficient of friction?

(a) 0.185 (b) $\frac{\sqrt{3}}{10}$
(c) 0.476 (d) $\frac{10}{\sqrt{3}}$

63. A solid sphere of 2 kg is suspended from a horizontal beam by two supporting wires. The tension in each wire is approximately ($g = 10$ m/s^2).

(a) 10 N (b) 20 N
(c) 30 N (d) 5 N

64. A body of mass 4 kg moving on a horizontal surface with an initial velocity of 6 m/s comes to rest after 3 seconds. If one wants to keep the body moving on the same surface with the velocity of 6 m/s, the force required is

(a) zero (b) 4 N
(c) 8 N (d) 16 N

65. A uniform rope of length L resting on a frictionless horizontal surface is pulled at one end by a factor F. What is the tension in the rope at a distance l from the end where the force is applied?

(a) F (b) $\frac{F}{2}$
(c) $F\left(1+\frac{l}{L}\right)$ (d) $F\left(1-\frac{l}{L}\right)$

66. The elevator shown in figure is descending with an acceleration of 2 m/s^2. The mass of the block A = 0.5 kg. The force exerted by the block A on block B is

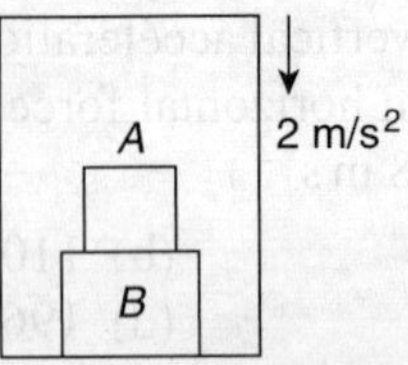

(a) 8 N (b) 6 N
(c) 4 N (d) 2 N

67. Two blocks of masses 2 kg and 1 kg are placed on a smooth horizontal table in contact with each other. A horizontal force of 3 Newton is applied on the first block so that the block moves with a constant acceleration. The force between the blocks would be

(a) 2 N (b) 1 N
(c) 3 N (d) zero

68. A monkey is descending from the branch of a tree with constant acceleration. If the breaking strength is 75% of the weight of the monkey, the minimum acceleration with which monkey can slide down without branch is

(a) g (b) $3g/4$
(c) $g/4$ (d) $g/2$

69. The mass of the lift is 100 kg which is hanging on the string. The tension in the string, when the lift is moving up with constant velocity ($g = 9.8$ m/s^2) is

(a) 980 N (b) 1000 N
(c) 100 N (d) None of the above.

70. A stone of mass 2 kg falls from a height of 60 m and penetrates 0.6 m into sand. The average resistance of the sand on the body is

(a) 19.6 N (b) 1960 N
(c) 787.6 N (d) 1979.6 N

71. Three blocks coupled together as shown in figure, are pulled by a force 700 N. The tension in the coupling between A and B is

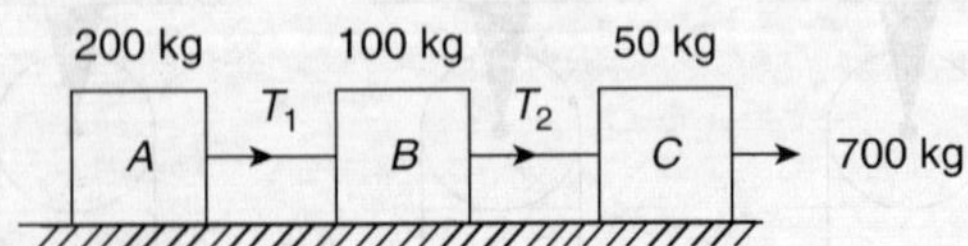

(a) 400 N (b) 500 N
(c) 600 N (d) 700 N

72. A block of wood of mass M is pulled by a rope of mass m along a smooth horizontal surface by means of a force p applied at the free end of the rope. The force exerted by the rope on the block is

(a) p (b) $\frac{pM}{M+m}$
(c) $\frac{pm}{M+m}$ (d) $\frac{PM}{m}$

73. A balloon with a total load of 2000 kg moves down with an acceleration 2.5 m/s^2. What load must be removed from it so that it goes up with the same acceleration? (g = 10 m/s^2)

(a) 500 kg (b) 200 kg
(c) 800 kg (d) 1200 kg

74. A force of 200 N acts on a body of mass 100 kg for 10 seconds. The distance travelled by the body in the next 10 sec is

(a) 100 m (b) 200 m
(c) 300 m (d) 400 m

75. A spring balance is adjusted to zero reading. Small particles, each having mass m, fall from a height h at the rate of n particles/sec and remain in the pan. The reading after one second is

(a) $mng + mn\sqrt{mgh}$ (b) $mn\sqrt{2gh}$
(c) mng (d) zero

76. Sand drops from a stationary hopper at 5 kg/s on a conveyor belt moving uniformly at 2 m/s. The force required to keep the belt moving is

(a) 10 N (b) 49 N
(c) 98 N (d) zero.

77. A 100 kg body is lowered from the top of a building by means of rope that can withstand a tension of 784 N. The maximum acceleration is

(a) g m/s^2 (b) $\frac{g}{5}$ m/s^2
(c) 9.8 m/s^2 (d) 7.84 m/s^2

78. The tension of the massless thread shown in the figure in newton is

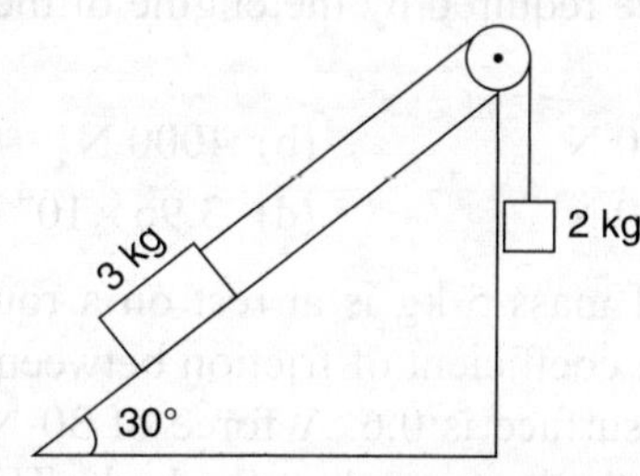

(a) g (b) 1.5 g
(c) 1.8 g (d) 2 g

79. A cart weighing 20 kg can roll without friction along a horizontal path. The cart carries a block of mass 2 kg. The coefficient of friction between block and cart is 0.25. A force of 19.6 N is applied as shown below. The acceleration of the cart is (g = 10 m/s^2)

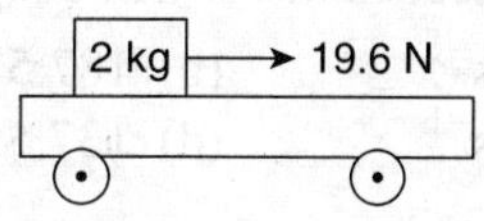

(a) 0.25 m/s^2 (b) 7.5 m/s^2
(c) 9.2 m/s^2 (d) zero

80. A certain force applied to a mass m_1 gives it acceleration of 15 m/s^2. The same force produces acceleration of 10 m/s^2 in another mass m_2. If two masses stick to each other, what is the acceleration produced by same force.

(a) 5 m/s (b) 1 m/s^2
(c) 6 m/s^2 (d) 25 m/s^2

81. The ratio between the acceleration of the two systems shown in figure is (g = 10 m/s^2)

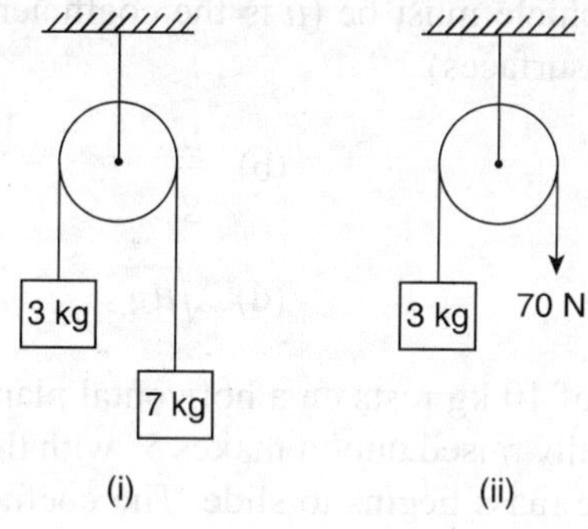

(a) 7 : 3 (b) 1 : 11
(c) 10 : 3 (d) 3 : 10

82. Two blocks A and B of masses 5 kg and 3 kg respectively are connected by light strings to a spring balance of mass 2 kg, free to slide on a smooth horizontal table. The tensions in the strings are as marked. Then

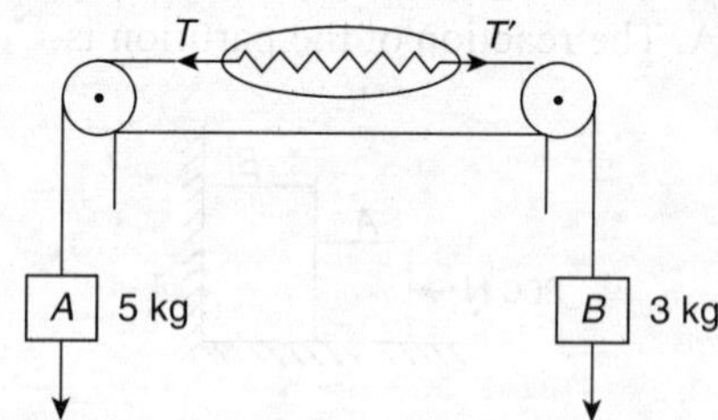

(a) T = 36.8 N (b) $\frac{T}{T'} = \frac{11}{9}$
(c) T' = 35.3 N (d) $T = T'$

83. The force of friction on a body of mass 50 kg on a rough surface ($\mu = 0.2$) when the body is pulled horizontally with a force of 20 N is

(a) 52 N (b) 20 N
(c) 124 N (d) 98 N

84. A body is pressed on a vertical wall by means of a force 60 N to just keep it in rest. The coefficient of friction between wall and body is 0.2. The frictional force on the body is

(a) $\frac{mg}{5}$ (b) 5 mg
(c) 12 N (d) 60 N

85. In the front plane surface of a vehicle, a body of mass m is placed as shown in the figure.

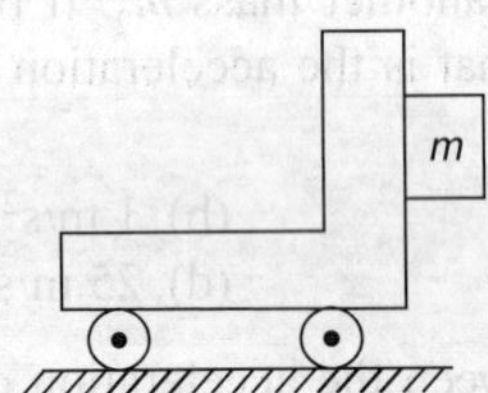

If it is not to fall or move in any way, the acceleration of the vehicle must be (μ is the coefficient of friction between surfaces)

(a) $\sqrt{\frac{g}{\mu}}$ (b) $\frac{\mu}{g}$

(c) $\frac{g}{\mu}$ (d) $\sqrt{\mu g}$

86. A mass of 10 kg rests on a horizontal plane. The plane is gradually raised until it makes 5° with the horizontal, when the mass begins to slide. The coefficient of friction between the surfaces is

(a) 0.17 (b) 0.34

(c) 0.087 (d) cos 5°

87. Two bodies A and B of masses 5 kg and 10 kg, in contact with each other, rest on a table against a rigid partition. The coefficient of friction between the bodies and the table is 0.15. A force of 200 N is applied horizontally at A. The reaction of the partition is

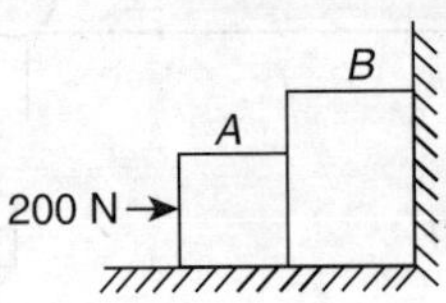

(a) 178 N (b) 185 N

(c) 192.5 N (d) 200 N

88. A man drives a motor cycle around a circular vertical wall of diameter 100 m. If the coefficient of friction between wall and cycle is 0.49, his speed to prevent himself from sliding down is

(a) 5 m/s (b) 20 m/s

(c) 40 m/s (d) $10\sqrt{10}$ m/s

89. Two trolleys of masses m and $3m$ are connected by a spring. They are compressed and released. They move off in opposite directions and come to rest after covering distances S_1 and S_2 respectively. Assuming coefficient of friction is uniform, the ratio of the distance $S_1 : S_2$ is

(a) 1 : 3 (b) 3 : 1

(c) 9 : 1 (d) 1 : 9

90. The coefficient of static and dynamic frictions are 0.5 and 0.45. After the body started moving by application of force F, the acceleration of the body is

(a) 0.5 m/s^2 (b) 0.47 m/s^2

(c) 0.49 m/s^2 (d) zero

91. When a force of 200 N is applied on a body on a rough surface, its acceleration is 3 m/s^2. When the force is increased to 300 N, the acceleration of the body is 5 m/s^2. The mass of the body is

(a) 60 kg (b) 63.3 kg

(c) 66.67 kg (d) 50 kg.

92. The figure below shows a man standing stationary with respect to a horizontal conveyor belt that is accelerating with 1 ms^{-2}. If the coefficient of static friction between the man's shoes and the belt is 0.2, up to what acceleration of belt can the man continue to be stationary relative to the belt? Mass of the man is = 65 kg.

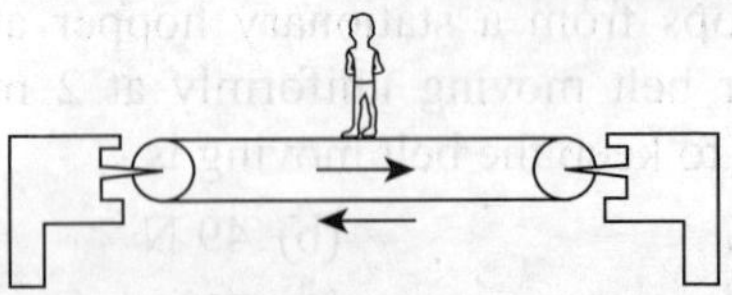

(a) 1.96 m/s^2 (b) 0.196 m/s^2

(c) 49 m/s^2 (d) $\frac{1}{49}$ m/s^2

93. An aeroplane required a takeoff speed of 72 km/hr, the run on the ground being 100 m. The mass of the plane is 10, 000 kg and coefficient of friction between the tyres of the plane and the ground is 0.2. Assume that plane accelerates uniformly during takeoff. What is the minimum force required by the engine of the plane for the take-off?

(a) 20, 000 N (b) 4000 N

(c) 19, 600 N (d) 3.96×10^4 N

94. A body of mass 5 kg is at rest on a rough horizontal plane. The coefficient of friction between the body and the plane surface is 0.6. A force of 30 N is applied at 30° above the horizontal on the body. The acceleration of the body is

(a) 1.12 m/s^2 (b) 2 m/s^2

(c) 5.88 m/s^2 (d) 0.12 m/s^2

95. A 5000 kg rocket is set for vertical firing. The exhaust speed is 800 m s^{-1}. To give an initial upward acceleration of 20 m s^{-2}, the amount of gas ejected per second to supply the needed thrust will be (g = 10 m s^{-2})

(a) 185.5 kg s^{-1} (b) 187.5 kg s^{-1}

(c) 127.5 kg s^{-1} (d) 137.5 kg s^{-1}

96. A block has been placed on an inclined plane. The slope angle θ of the plane is such that the block slides down the plane at constant speed. The coefficient of kinetic friction is

(a) $\sin\theta$ (b) $\cos\theta$
(c) g (d) $\tan\theta$

97. A sledge slides down a hill of height h and stops after covering a distance BC.

The coefficient of friction between sledge and hill is (AB = BC = s)

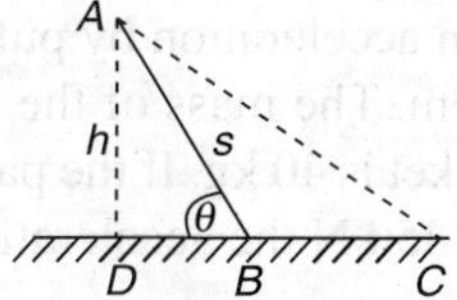

(a) $\frac{1}{hs}$ (b) $\frac{1}{s}$
(c) $\frac{1}{h}$ (d) $\frac{h}{s}$

98. The force parallel to the plane required to move a body of mass m up inclined plane of angle θ when the angle of repose is λ

(a) $\frac{mg\sin(\theta+\lambda)}{\cos\lambda}$ (b) $\frac{mg\sin(\theta-\lambda)}{\cos\lambda}$
(c) $\frac{mg\sin(\theta+\lambda)}{\sin\lambda}$ (d) $\frac{mg\sin(\theta-\lambda)}{\sin\lambda}$

99. A body just slides down an inclined plane, the top half of which is perfectly smooth and lower half rough. The angle of the plane is 30° with horizontal. The coefficient of friction μ of the lower half is

(a) $\frac{1}{\sqrt{3}}$ (b) $\frac{2}{\sqrt{3}}$
(c) $\frac{\sqrt{3}}{2}$ (d) 1

100. If a block moving up a plane at 30° with a velocity of 5 m/s stops after 0.5 sec then μ is

(a) 0.5 (b) 1.25
(c) 0.6 (d) None of these.

101. If a force equal to half the weight of the body can just move the body on a plane surface, the coefficient of frication is

(a) 0.2 (b) 1
(c) 0.25 (d) 0.5

102. How much force is required to push a 100 N wooden block up a frictionless plane inclined at 30° to the horizontal so that it can have an acceleration of 2.5 m/s^2 up the plane ($g = 10$ m/s^2)

(a) 25 N (b) 50 N
(c) 75 N (d) 100 N

103. A body is sliding down an inclined plane having coefficient of friction 0.5. If the normal reaction is twice that of the resultant downward force along the incline, the angle between the inclined plane and horizontal is

(a) 45° (b) 30°
(c) 15° (d) 60°

104. In a rocket, fuel burns at the rate of 1 kg/s. This fuel is ejected from the rocket with a velocity of 60 km/s. This exerts a force on the rocket equal to

(a) 6000 N (b) 60000 N
(c) 60 N (d) 600 N

105. A 45 kg load is being pulled vertically by a rope whose breaking strength is 495 N. With what maximum acceleration can the load be pulled up? ($g = 10$ m/s^2)

(a) 1 m/s^2 (b) 10 m/s^2
(c) 11 m/s^2 (d) 21 m/s^2

106. With what minimum acceleration can a fireman slide down a rope that has breaking strength as two-third of its weight?

(a) $\frac{2}{3}g$ (b) $\frac{1}{3}g$
(c) g (d) Zero

107. What force should be applied on a 5 kg body so that it has a downward acceleration of 4 m/s^2?

(a) 69 N upward (b) 69 N downward
(c) 29 N upward (d) 29 N downward

108. A 1000 kg lift is supported by a cable that can support 2000 kg. The shortest distance in which the lift can be stopped when it is descending with a speed of 2.5 m/s is ($g = 10$ m/s^2)

(a) $\frac{5}{4}$ m (b) $\frac{5}{8}$ m
(c) $\frac{5}{16}$ m (d) $\frac{5}{32}$ m

109. A body of mass 0.1 kg is suspended from a massless string. The point of suspension moves (i) up (ii) down with an acceleration of 5 m/s^2. If the tension in newtons in the string in the two cases are T_1 and T_2 respectively, then

(a) $T_1 = T_2$ (b) $T_2 - T_1 = 1$
(c) $T_1 - T_2 = 2$ (d) $T_1 - T_2 = 1$

110. A man slides down a rope of negligible mass, whose breaking strength is n times his weight ($n < 1$). What should be his maximum acceleration, so that the rope just breaks?

(a) $\frac{g}{1+n}$ (b) $g(1-n)$
(c) $\frac{g}{(1-n)}$ (d) $g(1+n)$

111. A balloon of gross weight W is falling vertically downward with constant acceleration a. The amount Q that must be thrown out in order to give the balloon an equal upward acceleration a is such that

(a) $Q=\frac{W}{2}$ (b) $Q=\frac{W(a-g)}{g}$
(c) $Q=2W\frac{g}{(a+g)}$ (d) $Q=\frac{W(a-g)}{a}$

112. An elevator starts from rest with a constant upward acceleration. It moves 2.0 m in the first 0.5 sec. A passenger in the elevator is holding a 3 kg package by a vertical spring. The tension in the spring is

(a) 30 N (b) 48 N
(c) 36 N (d) 78 N

113. A 2 kg mass pulls horizontally on a 3 kg mass by means of a lightly stretched spring. If at one instant, the 3 kg mass has acceleration of 1.8 m/s^2 towards 2 kg mass, the acceleration of 2 kg mass is

(a) 1.2 m/s^2 (b) 3.6 m/s^2
(c) 2.7 m/s^2 (d) zero

114. Two masses 2 kg and 1 kg are connected by an inextensible string passing over a smooth fixed pulley. The pressure in kg-wt on the axle of the pulley is

(a) 1.5 kg-wt (b) 3 kg-wt
(c) less than 3 kg w-t (d) more than 3 kg-wt

115. Three equal weight of mass 2 kg each are hanging on a string passing over a fixed pulley as shown in figure.

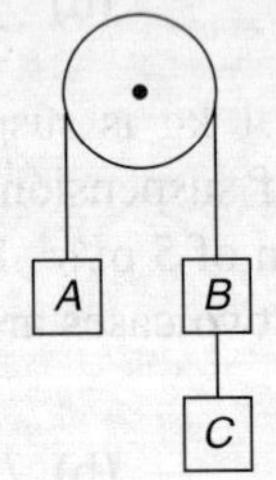

What is the tension in the string connecting the weights B and C?

(a) 3.3 N (b) 13.3 N
(c) 19.6 N (d) Zero

116. The system shown in figure is in equilibrium. The spring balance is calibrated in newtons. What weight does it record? ($g = 10$ m/s^2)

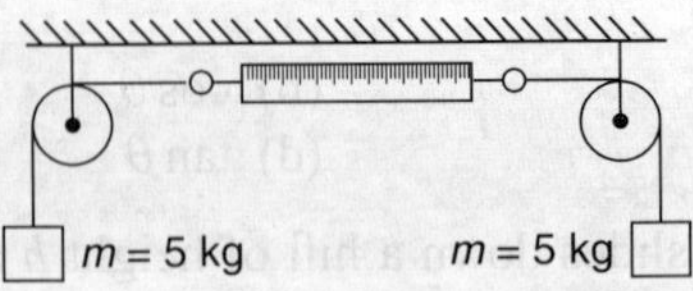

(a) 100 N (b) 25 N
(c) 50 N (d) Zero

117. A painter is raising himself and the basket on which he sits with certain acceleration by pulling a massless pulley-string system. The mass of the painter is 60 kg and mass of the basket is 40 kg. If the painter pulls the string with a force of 450 N the acceleration of the system is

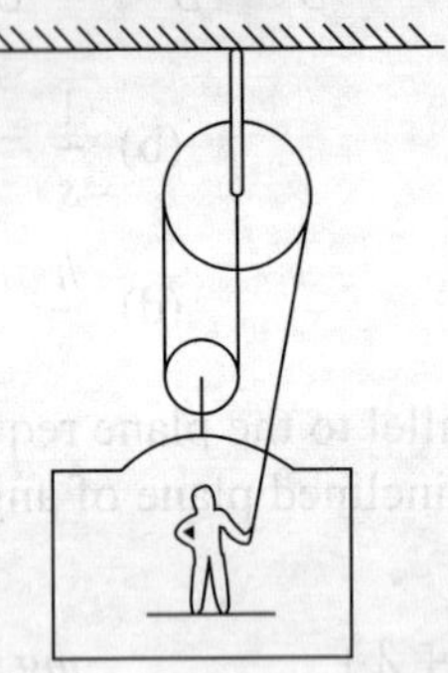

(a) 0.5 m/s^2 (b) 2 m/s^2
(c) 3.5 m/s^2 (d) 5 m/s^2

118. A 16 kg monkey wishes to raise a 20 kg mass by climbing up a rope that passes over a pulley attached to the mass. The minimum acceleration of the monkey up the rope must be

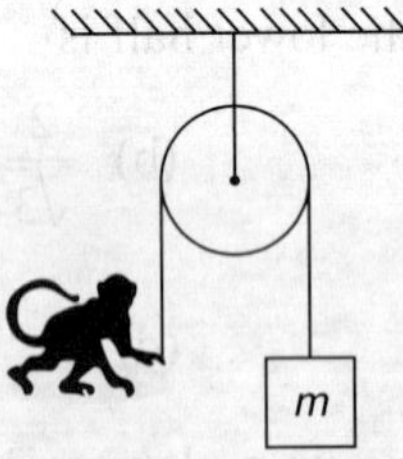

(a) g (b) $\frac{g}{2}$
(c) $\frac{g}{4}$ (d) $\frac{3g}{4}$

119. A block of W weighing 500 kgf is suspended by means of two strings P and Q as shown in figure. The tension in the string P is

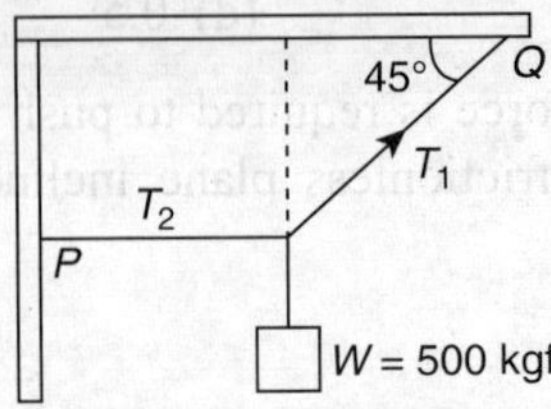

(a) 350 kgf (b) 250 kgf
(c) 500 kgf (d) 600 kgf

120. What is the Tension T in the string $ABCD$ if the weight 20 N is moving upward with a constant velocity?

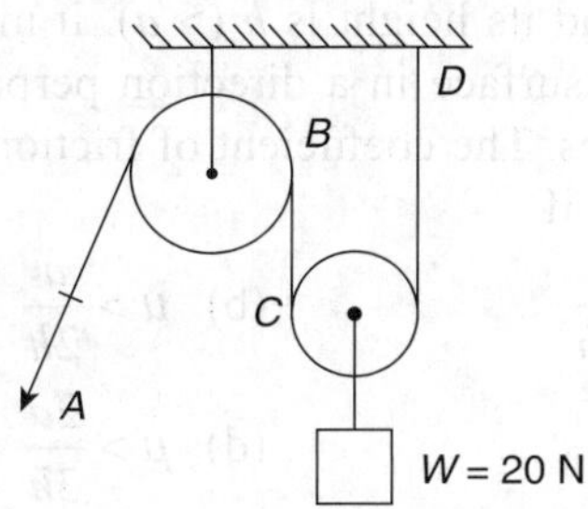

(a) $10\sqrt{2}$ N (b) 10 N
(c) 20 N (d) $\frac{10}{\sqrt{2}}$ N

121. A ball is suspended by a string from the mass of a boat which is moving with an acceleration of 4.9 m/s^2. By what approximate angle to the vertical the string will be inclined?

(a) 10° (b) 0°
(c) 30° (d) 45°

122. A pendulum is suspended from the ceiling of a carriage which is sliding freely down on a smooth inclined plane. The inclined plane makes an angle θ with the horizontal. The string makes an angle α with the vertical only when

(a) $\alpha > \theta$ (b) $\alpha < \theta$
(c) $\alpha = \theta$ (d) $\alpha = 0$

123. A mass M is placed on a very smooth wedge resting on a surface without friction. Once the mass is released, the acceleration to be given to the wedge so that M remains at rest is a where

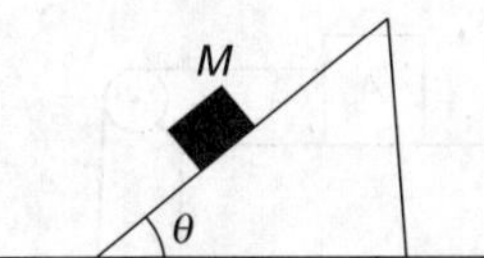

(a) a is applied to the left and $a = g \tan\theta$
(b) a is applied to the right and $a = g\tan\theta$
(c) a is applied to the left and $a = g\sin\theta$
(d) a is applied to the left and $a = g\cos\theta$.

124. A rope lies on a table such that a part of it hangs down the table, when the length of the hanging part is $\frac{1}{3}$ of the entire length, the rope begins to slide. The coefficient of friction between the rope and table is

(a) $\frac{1}{6}$ (b) $\frac{1}{3}$
(c) $\frac{2}{3}$ (d) $\frac{1}{2}$

125. A uniform chain of length l lies on a rough table with coefficient of friction μ. What maximum portion of its length can overhang from the edge of the table without sliding down?

(a) $\frac{l}{\mu}$ (b) $\frac{l}{\mu+1}$
(c) $\frac{l}{\mu+1}\mu$ (d) $\frac{l}{\mu-1}l$

126. A man fires a bullet of mass 200 g at a speed of 5 m/s. The gun is of one kg mass. By what velocity the gun rebounds backward?

(a) 1 m/s (b) 0.01 m/s
(c) 0.1 m/s (d) 10 m/s

127. A 40 kg slab rests on a frictionless floor. A 10 kg block rests on the slab. The coefficient of static friction between the block and the slab is 0.6 while the kinetic coefficient is 0.4 kg. The 10 kg block is acted upon by a horizontal force of 100 N. If $g = 9.8$ m/s^2, the resulting acceleration of the slab will be

(a) 1.52 m/s^2 (b) 1.96 m/s^2
(c) 1.47 m/s^2 (d) 0.98 m/s^2

128. Two blocks A and B, attached to each other by a mass-less spring, are kept on a rough horizontal surface ($\mu = 0.1$) and pulled by a force $F = 200$ N as shown in the figure. If at some instant, the 10 kg mass has acceleration of 12 m/s^2, what is the acceleration of the 20 kg mass?

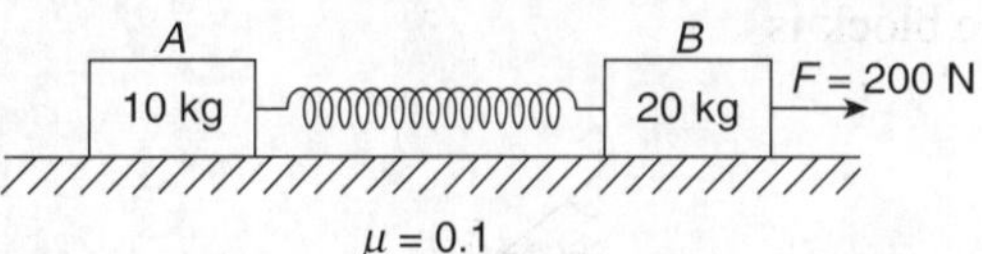

(a) 3.6 m/s^2 (b) 2.5 m/s^2
(c) 3.6 m/s^2 (d) 1.2 m/s^2

129. The coefficient of static friction between two blocks A and B is 0.65. What should be the minimum acceleration of the system shown so that block B does not fall?

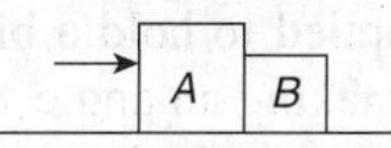

(a) 4.9 m/s^2 (b) 15.1 m/s^2
(c) 6.53 m/s^2 (d) 19.6 m/s^2

130. A piece of ice slides down a 45° incline in twice the time it takes to slide down a frictionless 45° incline. The coefficient of kinetic friction between the ice and the incline is

(a) 0.25 (b) 0.5
(c) 0.75 (d) 1.00

131. A body A of mass 1 kg rests on a smooth surface. Another body B of mass 0.2 kg is placed over A as shown in figure.

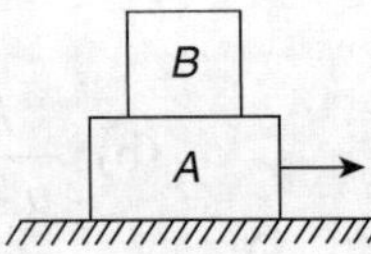

The coefficient of static friction between A and B is 0.15. The body B will begin to slide on body A if body A is pulled with a force greater than

(a) 1.764 N (b) 0.300 N
(c) 0.1764 N (d) It will not slide for any F.

132. A block A of mass 200 kg rests on a block B of mass 300 kg. A is tied with horizontal string to a wall. Coefficient of friction between A and B is 0.25 and that between B and floor is 0.2. The horizontal force F needed to move the block B is ($g = 10$ m/s^2)

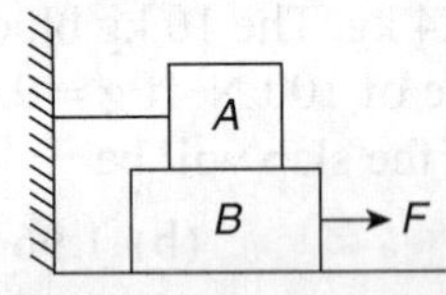

(a) 1500 N (b) 1100 N
(c) 550 N (d) 2200 N

133. A block of mass $\sqrt{3}$ kg is placed on a rough horizontal surface ($\mu = \dfrac{1}{2\sqrt{3}}$). A force F is applied as shown in figure. The minimum value of F required just to slide the block is

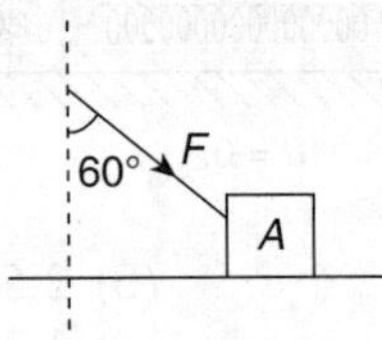

(a) $\dfrac{20}{3}$ N (b) 5 N
(c) 10 N (d) 20 N

134. A force F is applied to hold a block of mass m on an inclined plane making an angle θ with the horizontal. The force F is perpendicular to the plane. The coefficient of friction between the plane and the block is μ. The minimum force F necessary to keep the block at rest on the inclined plane is

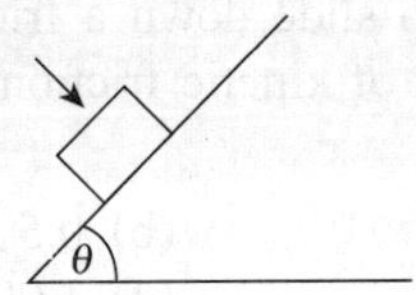

(a) $\dfrac{mg}{\mu}\sin\theta$ (b) $\dfrac{mg}{\mu}(\sin\theta - \mu\cos\theta)$
(c) $mg\sin\theta$ (d) $mg\cos\theta$

135. A rectangular block as a square base measuring sides as $a \times a$ and its height is h (> a). It moves on a rough horizontal surface in a direction perpendicular to one of the edges. The coefficient of friction is μ. The block will topple if

(a) $\mu > \dfrac{a}{4h}$ (b) $\mu > \dfrac{a}{2h}$
(c) $\mu > \dfrac{a}{h}$ (d) $\mu > \dfrac{2a}{3h}$

136. Two monkeys P and Q of masses M and m (> M) hold a light rope passing over a smooth fixed pulley. P and Q climb up the rope so that the acceleration of Q upwards is double than that of P downward. The tension in the rope is

(a) $\dfrac{2Mm}{M+m}g$ (b) $\dfrac{3Mm}{2M+m}g$
(c) $\dfrac{3Mm}{2M+2m}g$ (d) $\dfrac{3Mm}{M+2m}g$

137. A block can slide on a smooth inclined plane of inclination θ kept on the floor of the lift. When the lift is descending with a retardation a, the acceleration of the block relative to the inclined plane is

(a) $(g - a)$ (b) $g\sin\theta$
(c) $(g + a)\sin\theta$ (d) $(g - a)\sin\theta$

138. A block of mass m is placed over a block B of mass $2m$. The block B can move on a smooth horizontal surface. If the coefficient of friction between blocks A and B is μ, then the minimum force F required to initiate sliding motion in block A is

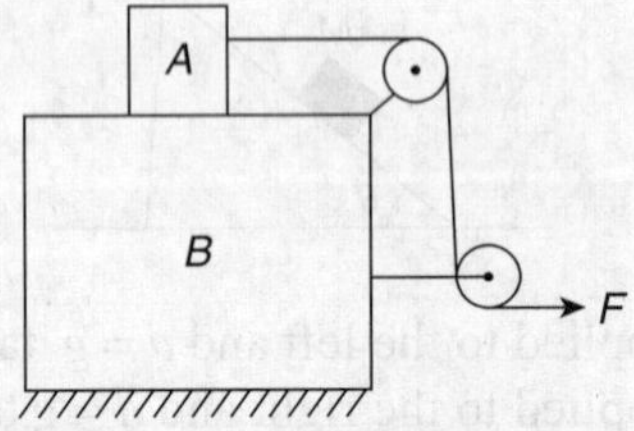

(a) $\dfrac{3}{2}\mu$ mg (b) $\dfrac{3\text{ mg}}{\mu}$
(c) μ mg (d) 3μ mg

139. A bullet is fired from a gun. The force on the bullet is given by $F = 600 - 2 \times 10^5\, t$ where, F is in newton and t in seconds. The force on the bullet becomes zero as soon as it leaves the barrel. What is the average impulse imparted to the bullet?

(a) 9 N-s (b) zero
(c) 1.8 N-s (d) 0.9 N-s

140. A body takes n times as much time to slide down a 45° rough incline as it takes to slide down a smooth 45° incline. The coefficient of fiction is

(a) 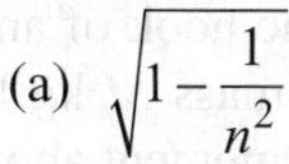 $\sqrt{1-\frac{1}{n^2}}$ (b) $1+\frac{1}{n^2}$

(c) $\mu=1-\frac{1}{n^2}$ (d) $\mu=\sqrt{1+\frac{1}{n^2}}$

141. Two masses A and B of 10 kg and 5 kg respectively are connected with a string passing over a frictionless pulley fixed at the corner of a table as shown. The coefficient of static friction between A and the table is 0.2. The minimum mass C that should be placed on 'A' to prevent it from moving is equal to

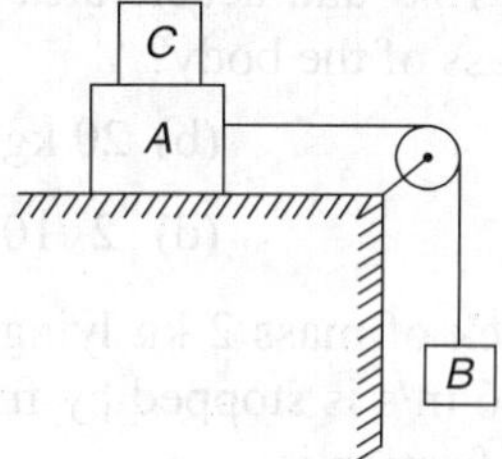

(a) 12 kg (b) 5 kg
(c) 10 kg (d) 15 kg

142. A body of mass m = 3.513 kg is moving along the x-axis with a speed of 5.00 m s^{-1}. The magnitude of its momentum is recorded as?

(a) 17.57 kg/m s^{-1} (b) 15.5 kg/m s^{-1}
(c) 13.52 kg/m s^{-1} (d) 10.57 kg/m s^{-1}

143. A block of mass m is connected to another block of mass M by a spring (massless) of spring constant k. The blocks are kept on a smooth horizontal plane. Initially the blocks are at rest and the spring is unstretched. Then a constant force F starts acting on the block of mass M to pull it. Find the force of the block of mass m.

(a) $\frac{MF}{(m+M)}$ (b) $\frac{mF}{M}$

(c) $\frac{(M+m)F}{m}$ (d) $\frac{mF}{(m+M)}$

144. A ball of mass 0.2 kg is thrown vertically upwards by applying a force by hand. If the hand moves 0.2 m while applying the force and the ball goes up to 2 m height further, find the magnitude of the force. Consider $g = 10\,\text{m/s}^2$.

(a) 22 N (b) 4 N
(c) 16 N (d) 20 N

145. A player caught a cricket ball of mass 150 g moving at a rate of 20 m/s. If the catching process is completed in 0.1 s, the force of the blow exerted by the ball on the hand of the player is equal to

(a) 300 N (b) 150 N
(c) 3 N (d) 30 N

146. Consider a car moving on a straight road with a speed of 100 m/s. The distance at which car can be stopped is [$\mu_k = 0.5$]

(a) 100 m (b) 400 m
(c) 800 m (d) 1000 m

147. A block is kept on a frictionless inclined surface with angle of inclination α. The incline is given an acceleration a to keep the block stationary. Then a is equal to

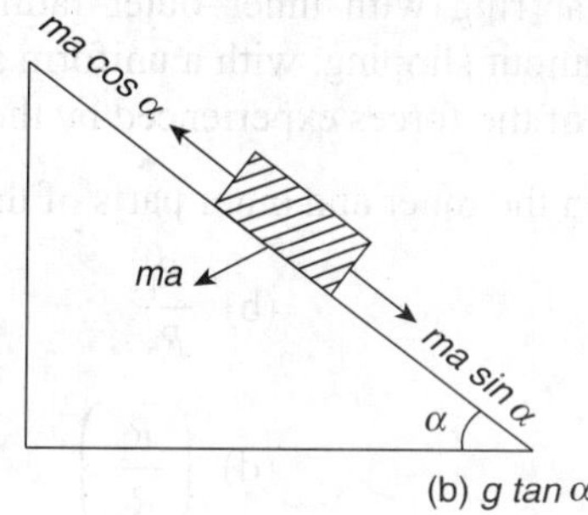

(b) $g \tan \alpha$

(a) g (b) $g \tan \alpha$

(c) $\frac{g}{\tan \alpha}$ (d) g cosec α

148. The particle of mass 0.3 kg is subjected to a force $F = -kx$ with k = 15 N/m. What will be its initial acceleration if it is related from a point 20 cm away from the origin?

(a) 5 m/s^2 (b) 10 m/s^2
(c) 3 m/s^2 (d) 15 m/s^2

149. A bullet fired into a fixed target loses half its velocity after penetrating 3 cm. How much further it will penetrate before coming to rest assuming that it faces constant resistance to motion?

(a) 1.5 cm (b) 1.0 cm
(c) 3.0 cm (d) 2.0 cm

150. The upper half of an inclined plane with inclination ϕ is perfectly smooth while the lower half is rough. A body starting from rest at the top will again come to rest at the bottom if the coefficient of friction for the lower half is given by

(a) 2 tan ϕ (b) tan ϕ
(c) 2 sin ϕ (d) 2 cos ϕ

151. A block of mass 2 kg is given a push horizontally and then the block starts sliding over a horizontal plane. The graph shows the velocity–time graph of the motion. The co-efficient of sliding friction between the plane and the block is

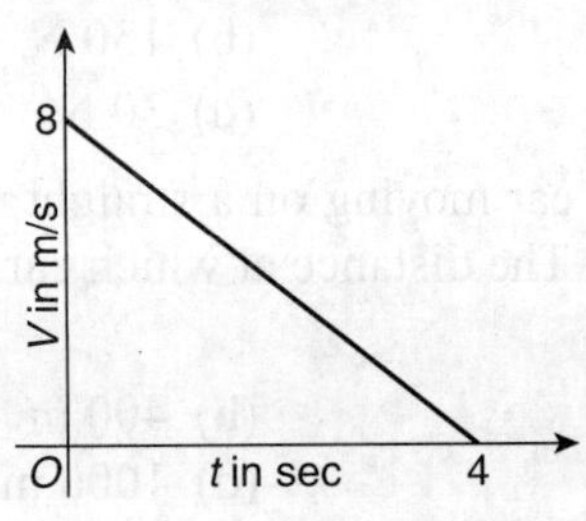

(a) 0.02 (b) 0.04
(c) 0.2 (d) 0.4

152. An annular ring with inner–outer radii R_1 and R_2 is rolling without slipping, with a uniform angular speed. The ratio of the forces experienced by the two particles situated on the inner and outer parts of the ring, $\frac{F_1}{F_2}$ is

(a) 1 (b) $\frac{R_1}{R_2}$

(c) $\frac{R_2}{R_1}$ (d) $\left(\frac{R_1}{R_2}\right)^2$

153. A block rests on a rough inclined plane making an angle of 30° with the horizontal. The coefficient of static friction between the block and the plane is 0.8. If the frictional force on the block is 10 N, the mass of the block (in kg) is (take $g = 10\ \text{ms}^{-2}$)

(a) 2.0 (b) 4.0
(c) 1.6 (d) 2.5

154. Two masses $m_1 = 5$ kg and $m_2 = 4.8$ kg tied to a string are hanging over a light frictionless pulley. What is the acceleration of the masses when left free to move? (Take $g = 9.8\ \text{m/s}^2$)

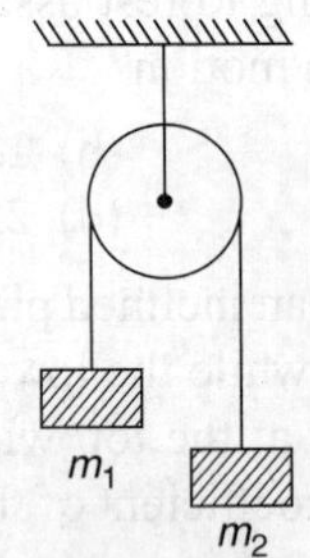

(a) $0.2\ \text{m/s}^2$ (b) $9.8\ \text{m/s}^2$
(c) $5\ \text{m/s}^2$ (d) $4.8\ \text{m/s}^2$

155. A mass of 1 kg is suspended by a thread. It is
(i) lifted up with an acceleration 4.9 m/s
(ii) lowered with an acceleration 4.9 m/s
The ratio of the tension is

(a) 1 : 3 (b) 1 : 2
(c) 3 : 1 (d) 2 : 1

156. A rocket with a lift-off mass 3.5×10^4 kg is blasted upwards with an initial acceleration of 10 m/s^2. Then the initial thrust of the blast is

(a) 3.5×10^5 N (b) 7.0×10^5 N
(c) 14.0×10^5 N (d) 1.75×10^5 N

157. A light spring balance hangs from the hook of another light spring balance and a block of mass M kg hangs from the former one. Then the true statement about the scale reading is

(a) both the scales read M kg each
(b) the scale of the lower one reads M kg and of the upper one zero
(c) the reading of the two scales can be M kg
(d) both the scales read $\frac{M}{2}$ kg.

158. A force vector applied on a mass is represented as $\vec{F} = 6\hat{i} - 8\hat{j} + 10\hat{k}$ and accelerates with 1 m/s^2. What will be the mass of the body?

(a) 10 kg (b) 20 kg
(c) $10\sqrt{2}$ kg (d) $2\sqrt{10}$ kg

159. A marble block of mass 2 kg lying on ice when given a velocity of 6 m/s is stopped by friction in 10 s. Then coefficient of friction is

(a) 0.02 (b) 0.03
(c) 0.06 (d) 0.01

160. A horizontal force of 10 N is necessary to just hold a block stationary against a wall. The coefficient of friction between the block and the wall is 0.2. The weight of the block is

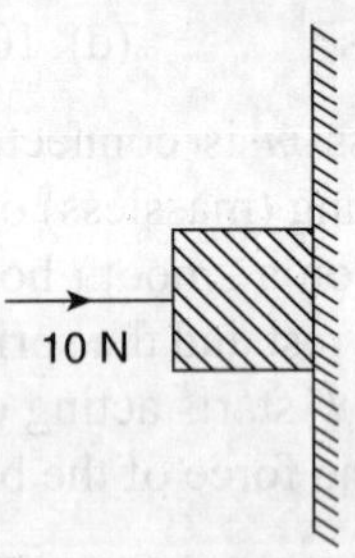

(a) 20 N (b) 50 N
(c) 100 N (d) 2 N

161. A spring balance is attached to the ceiling of a lift. A man hangs his bag on the spring and the spring reads 49 N, when the lift is stationary. If the lift moves downward with an acceleration of 5 m/s^2, the reading of the spring balance will be

(a) 24 N (b) 74 N
(c) 15 N (d) 49

162. Three forces start acting simultaneously on a particle moving with velocity $\vec{v}$. These forces are represented in magnitude and direction by the three sides of a triangle ABC (as shown). The particle will now move with velocity

(a) less than $\vec{v}$
(b) greater than $\vec{v}$
(c) $|\vec{v}|$ direction of the largest force BC
(d) $\vec{v}_1$ remaining unchanged.

163. Three identical blocks of masses $m = 2$ kg are drawn by a force $F = 10.2$ N with an acceleration of 0.6 m/ s^2 on a frictionless surface, then what is the tension (in N) in the string between the blocks B and C?

(a) 9.2 (b) 7.8
(c) 4 (d) 9.8

164. A light string passing over a smooth light pulley connects two blocks of masses m_1 and m_2 (vertically). If the acceleration of the system is $g/8$, then the ratio of the masses is

(a) 8 : 1 (b) 9 : 7
(c) 4 : 3 (d) 5 : 3

165. One end of a massless rope, which passes over a massless and frictionless pulley P is tied to a hook C while the other end is free. Maximum tension that the rope can bear is 960 N. With what value of maximum safe acceleration (in m s^{-2}) can a man of 60 kg climb on the rope?

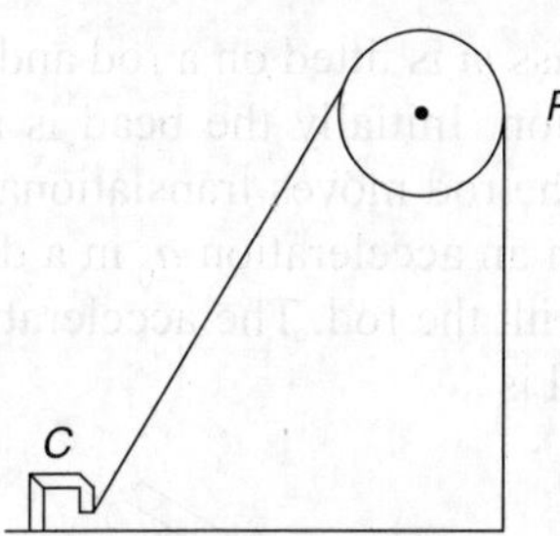

(a) 16 (b) 6
(c) 4 (d) 8

166. When forces F_1, F_2, F_3 are acting and particle of mass m such that F_2 and F_3 are mutually perpendicular, then the particle remains stationary. If the force F_1 is now removed then the acceleration of the particle is

(a) $\frac{F_1}{m}$ (b) $\frac{F_2F_3}{mF_1}$

(c) $\frac{(F_2 - F_3)}{m}$ (d) $\frac{F_2}{m}$

167. A lift is moving down with acceleration a. A man in the lift drops a ball inside the lift. The acceleration of the ball, as observed by the man in the lift and a man standing stationary on the ground are, respectively

(a) g, g (b) $g - a, g - a$
(c) $g - a, g$ (d) a, g

168. The minimum velocity (in m s^{-1}) with which a car driver must traverse a flat curve of radius 150 m and coefficient of friction 0.6 to avoid skidding is

(a) 60 (b) 30
(c) 15 (d) 25

169. A ship of mass 3×10^7 kg initially at rest, is pulled by a force of 3×10^4 N through a distance of 3 m. Assuming that the resistance due to water is negligible, the speed of the ship is

(a) 91.5 m/s (b) 60 m/s
(c) 0.1 m/s (d) 5 m/s

170. The string between blocks of mass m and $2m$ is massless and inextensible. The system is suspended by a massless spring as shown. If the string is cut, find the magnitudes of accelerations of mass $2m$ and m (immediately after cutting)

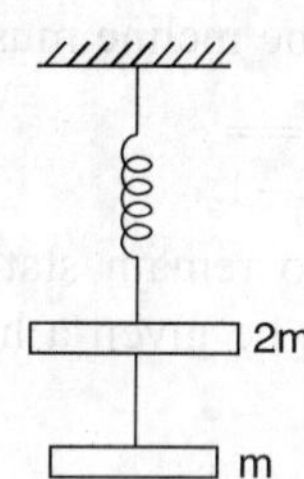

(a) g, g (b) $g, \frac{g}{2}$

(c) $\frac{g}{2}, g$ (d) $\frac{g}{2}, \frac{g}{2}$

171. Two particles of mass m each are tied at the ends of a light string of length $2a$. The whole system is kept on a frictionless horizontal surface with the string held tight so that each mass is at distance a form the centre P (as shown in the figure). Now, the mid-point of the string is pulled vertically upwards with a small but constant force F. As a result, the particle moves towards each other on the surface. The magnitude of acceleration, when the separation between them becomes $2x$, is

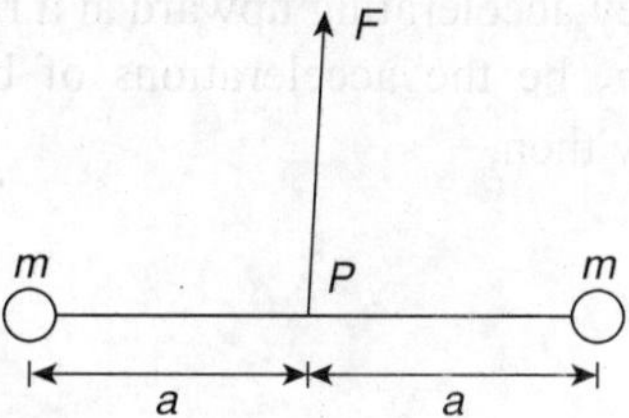

(a) $\frac{F}{2m}\frac{a}{\sqrt{a^2-x^2}}$ (b) $\frac{F}{2m}\frac{x}{\sqrt{a^2-x^2}}$

(c) $\frac{F}{2m}\frac{x}{a}$ (d) $\frac{F}{2m}\frac{\sqrt{a^2-x^2}}{x}$

172. A bowl of soup rests on a table in the dining car of train. If the acceleration of the train is a_0 in the forward direction then the angle which the surface of the soup makes with the horizontal is

(a) $\arctan\left(\frac{a_o}{g}\right)$ (b) $\arctan\left(\frac{g}{a_0}\right)$

(c) $\arctan\left(\frac{a_0}{\sqrt{a_0{}^2+g^2}}\right)$ (d) $\left(\frac{g}{\sqrt{a_0{}^2+g^2}}\right)$

173. A body is kept on a smooth inclined plane having an inclination of 1 in x. which of the following is incorrect?

(a) Slope of inclined plane is $\frac{1}{x}$

(b) Slope of inclined plane is $\frac{1}{\sqrt{x^2-1}}$

(c) For the body of mass m to remain stationary relative to the incline, the incline must offer a normal reaction of $mg\frac{x}{\sqrt{x^2-1}}$

(d) For the body to remain stationary relative to the incline, it must be given a horizontal acceleration of $\frac{g}{\sqrt{x^2-1}}$

174. A body of mass 2 Kg is moving towards east with a uniform speed of 2 ms^{-1}. A force of 3 N is applied to it towards north. The magnitude of the displacement of the body 2 s after the application of force is

(a) 4 m (b) 5 m
(c) 6 m (d) 7 m.

175. A particle of mass m is joined to very heavy mass M by a light spring passing over a light and frictionless pulley. Both the bodies are then set free. The total thrust on the pulley is

(a) mg (b) 2 mg
(c) 4 mg (d) much greater than mg

176. Two masses are connected by a string which passes over a pulley accelerating upward at a rate A as shown. If a_1 and a_2 be the accelerations of bodies 1 and 2 respectively then,

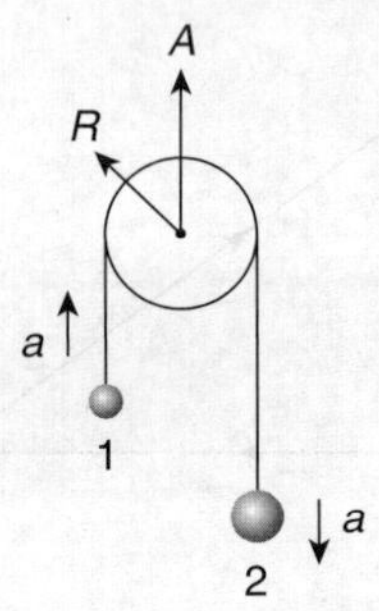

(a) $A = a_1 - a_2$ (b) $A = a_1 + a_2$

(c) $A = \frac{a_1 - a_2}{2}$ (d) $A = \frac{a_1 + a_2}{2}$

177. A mass m_1, placed on top of a trolley of mass m_3, is connected to another mass m_2 by means of a string passing over a smooth pulley as shown in the figure. The friction between the surfaces is negligible. For m_1 and m_2 not to move with respect to trolley, the horizontal force F to be applied on trolley is

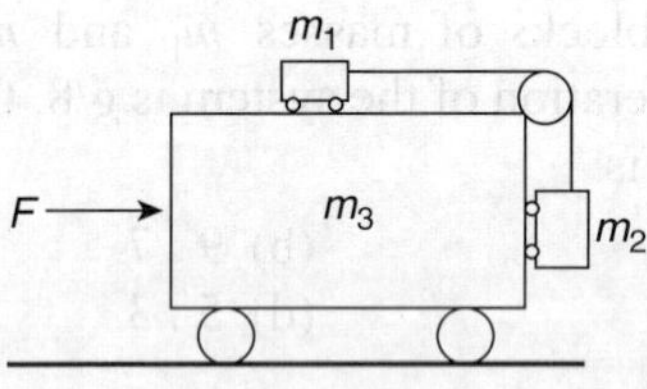

(a) $F = m_3\,g$
(b) $F = (m_1 + m_2)g$
(c) $F = (m_1 + m_2 + m_3)\frac{m_2 g}{m_1}$
(d) $F = m_1 g$

178. A bead of mass m is fitted on a rod and it can be moved without friction. Initially the bead is at the middle of the rod and the rod moves translationally in a horizontal plane with an acceleration a_0 in a direction forming an angle α with the rod. The acceleration of bead with respect to rod is

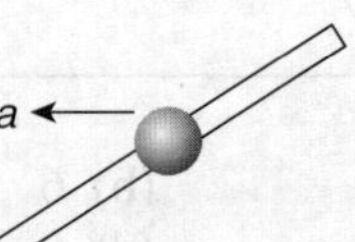

(a) $g \sin \alpha$ (b) $(g + a_0) \sin \alpha$
(c) $g \sin \alpha + a_0 \cos \alpha$ (d) $g \sin \alpha - a_0 \cos \alpha$

179. A pendulum of mass m hangs from a support to a trolley. The direction of the string when the trolley rolls up a plane of inclination α with acceleration a_0 is

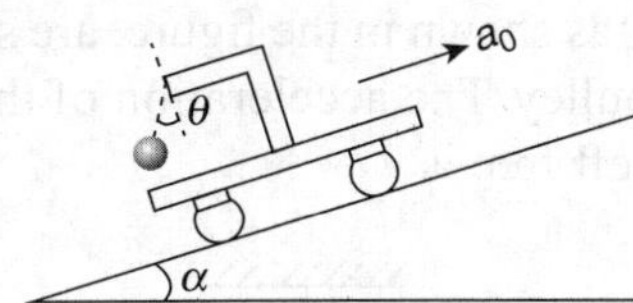

(a) $\theta = \tan^{-1}\alpha$ (b) $\theta = \tan^{-1}\left(\frac{a_0}{g}\right)$

(c) $\theta = \tan^{-1}\left(\frac{g}{a_0}\right)$ (d) $\theta = \tan^{-1}\left(\frac{a_0 + g\sin\alpha}{g\cos\alpha}\right)$

180. A bar of mass m resting on a smooth horizontal plane starts moving due to force $F = \frac{mg}{3}$ of constant magnitude. In the process of its rectilinear motion, the angle θ between the direction of this force and the horizontal varies as $\theta = Ks$, where K is constant. The velocity of bar as function of θ is

(a) $v = \sqrt{\frac{2g}{3}}\sin\theta$ (b) $v = \sqrt{\frac{2g}{3k}}\cos\theta$

(c) $v = \sqrt{\frac{2g}{3K}}\sin\theta$ (d) $v = \sqrt{\frac{2g}{3}}\cos\theta$

181. If the force on a rocket, moving with a velocity of 300 m/s is 210 N, then the rate of combustion of the fuel is

(a) 0.07 kg/s (b) 1.4 kg/s
(c) 0.7 kg/s (d) 10.7 kg/s

182. Beads A and B, each of mass m, are connected by a light inextensible cord. They are constrained to move on a frictionless ring in a vertical plane as shown. The beads are released from rest at the position shown. The tension in the cord just after the release is.

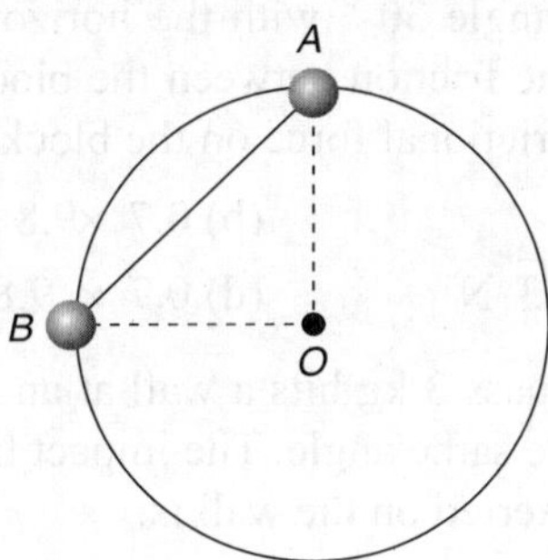

(a) $\sqrt{2}\,mg$ (b) mg

(c) $\frac{mg}{\sqrt{2}}$ (d) $2\,mg$

183. The system shown is just on the verge of slipping. The co-efficient of static friction between the block and the table top is:

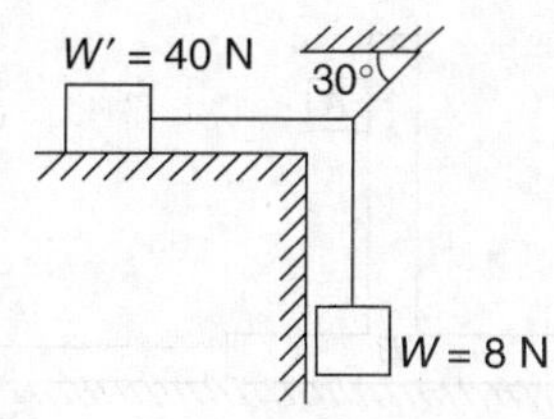

(a) 0.5 (b) 0.35
(c) 0.15 (d) 0.95

184. A prism of mass M is placed on a horizontal surface. A block of mass m slides on it, which in turn slides on a horizontal surface. Assume all surfaces to be frictionless. The acceleration of the block with respect to the prism is

(a) $g\sin\theta$
(b) $g\cos\theta$
(c) $\left(\frac{M+m}{M+m\sin^2\theta}\right)g\sin\theta$
(d) $\left(\frac{M+m}{M+m\sin^2\theta}\right)g\cos\theta$

185. A mass m is drawn up a smooth inclined plane of height h and length l by means of a string over a pulley fixed at the vertex of the plane as shown. A mass M hangs from the other end. For m to just reach the top of the plane, M must be detached when m has moved through a distance

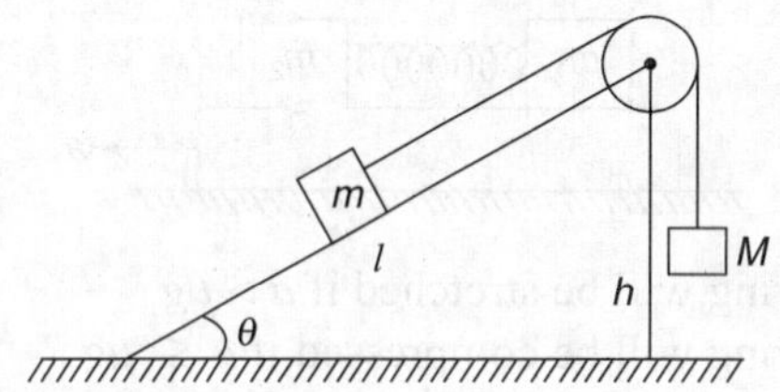

(a) $\frac{mh}{M}$ (b) $\frac{ml}{M}$

(c) $\frac{M}{M+m}\left(\frac{hl}{h+l}\right)$ (d) $\frac{M+m}{M}\left(\frac{hl}{h+l}\right)$

186. In the arrangement shown in the figure, the mass of the block B and A are $2m$ and $8m$ respectively. The surface between B and the floor is smooth. The block B is connected to block C by means of a pulley. If the whole system is released, then the minimum value of mass of the block C so that the block A remains stationary with respect to B is : (coefficient of friction between A and B is μ.)

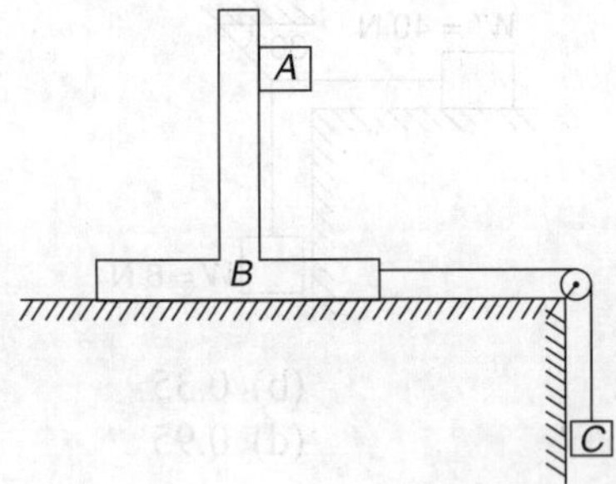

(a) $\frac{m}{\mu}$ (b) $\frac{2m}{\mu+1}$

(c) $\frac{10m}{1-\mu}$ (d) $\frac{10m}{\mu-1}$

187. A block of mass 1 kg lies on a horizontal surface in a truck. The coefficient of static friction between the block and the surface is 0.6. If the acceleration of the truck is $5 m/s^2$, the frictional force acting on the block is

(a) 3 N (b) 5 N

(c) 6 N (d) 7 N

188. Two blocks of masses m_1 and m_2 are connected with a massless unstretched spring and placed over a plank moving with acceleration a as shown in the figure. The coefficient of friction between the block and platform is μ. Then,

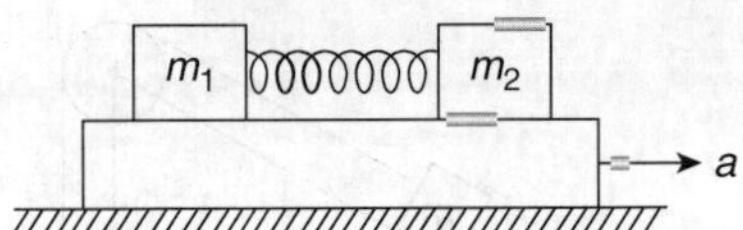

(a) spring will be stretched if $a > \mu g$

(b) spring will be compressed if $a \leq \mu g$

(c) spring will neither be compressed nor be stretched for $a \leq \mu g$

(d) spring will be in its natural length under all conditions.

189. A bead of mas m is located on a parabolic wire with its axis vertical and vertex directed downward as in figure and whose equation is $x^2 = 2\ ay$. If the coefficient of friction is μ, the highest distance above the x-axis at which the particle will be in equilibrium is

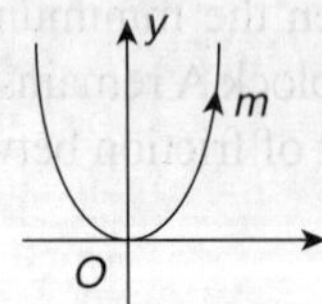

(a) μa (b) $\frac{\mu^2 a}{2}$

(c) $\frac{1}{4}\mu^2 a$ (d) $\frac{1}{2}\mu a$

190. Two masses, as shown in the figure, are suspended from a massless pulley. The acceleration of the system when masses are left free is

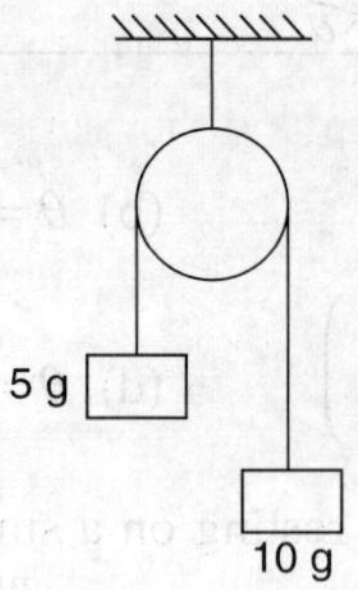

(a) $\frac{2g}{3}$ (b) $\frac{g}{3}$

(c) $\frac{g}{9}$ (d) $\frac{g}{7}$

191. The coefficient of friction between 4 kg and 5 kg is 0.2 and between 5 kg block and ground is 0.1 respectively. Choose the correct statements:

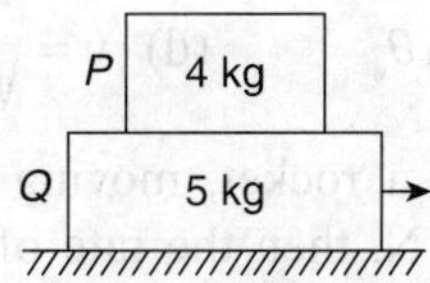

(a) Minimum force needed to cause system to move is 17 N.

(b) When force is 4 N static friction, at all surfaces is 4 N to keep system at rest.

(c) Maximum acceleration of 4 kg bock is $2 m/s^2$.

(d) Slipping between 4 kg and 5 kg blocks starts when F is 17 N.

192. A block of mass 2 kg rests on a rough inclined plane making an angle 30° with the horizontal. The coefficient of static friction between the block and the plane is 0.7. The frictional force on the block is

(a) 9.8 N (b) $0.7 \times 9.8 \times \sqrt{3}$ N

(c) $9.8 \times \sqrt{3}$ N (d) 0.7×9.8 N

193. A body of mass 3 kg hits a wall at an angle of 60° and returns at the same angle. The impact time was 0.2 sec. The force exerted on the wall is:

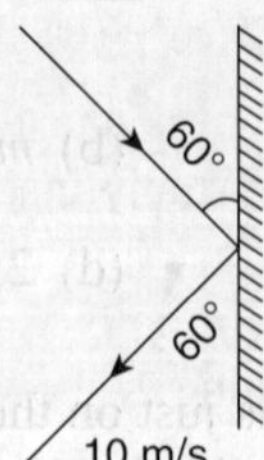

(a) $150\sqrt{3}$ N (b) $50\sqrt{3}$ N

(c) 100 N (d) $75\sqrt{3}$ N

194. A plank of mass $m_1 = 8$ kg with a bar of mass $m_2 = 2$ kg placed on its rough surface, lies on a smooth floor of elevator ascending with an acceleration $g/4$. The coefficient of friction is $\mu = 1/5$ between m_1 and m_2. A horizontal force $F = 30$ N is applied to the plank. Then the acceleration of bar and the plank in the reference frame of elevator are:

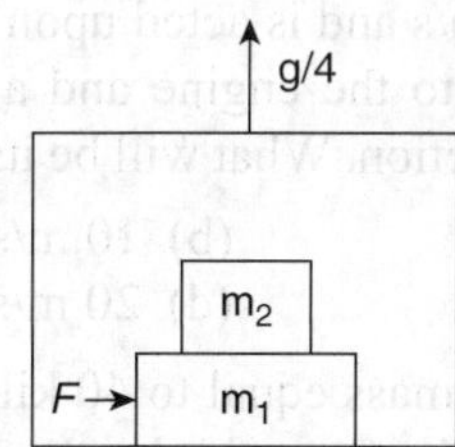

(a) 3.5 m/s^2 , 5 m/s^2 (b) 5 m/s^2, $\frac{50}{8}$ m/s^2

(c) 2.5 m/s^2, $\frac{25}{8}$ m/s^2 (d) 4.5 m/s^2 , 4.5m/s^2

195. A block of 0.1 is held against a wall applying a horizontal fore of 5 N on the block. If the coefficient of friction between block and the wall is 0.5, the magnitude of the frictional force acting on the block is:

(a) 2.5 N (b) 0.98 N
(c) 4.9 N (d) 0.49 N

196. A block of mass 1 kg lies on a horizontal surface in a truck. The coefficient of static friction between the block and the surface is 0.6. If the acceleration of the truck is 5 m/s^2, the frictional force acting on the body is:

(a) 5 N (b) 6 N
(c) 10 N (d) 15 N

197. In the shown mass–pulley system, pulleys and string are mass-less. If one end of the string is pulled by the force $F = mg$, the acceleration of the block will be:

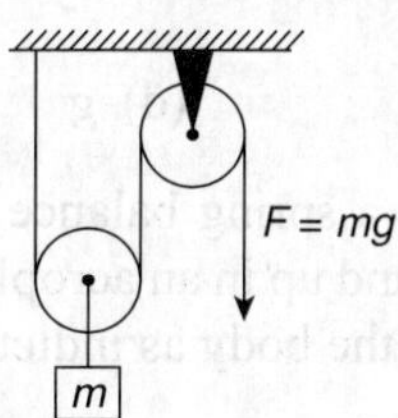

(a) g (b) 0
(c) g/2 (d) 3 g

198. A chain of length 5 m is placed on a horizontal surface as shown in figure. At any instant x is the length of the chain on rough surface and the remaining portion lies on smooth surface. Initially $x = 0$. A horizontal force P is applied to the chain (as shown in figure). In the duration when x changes from $x = 0$ to $x = 5$, for the chain to move with constant speed.

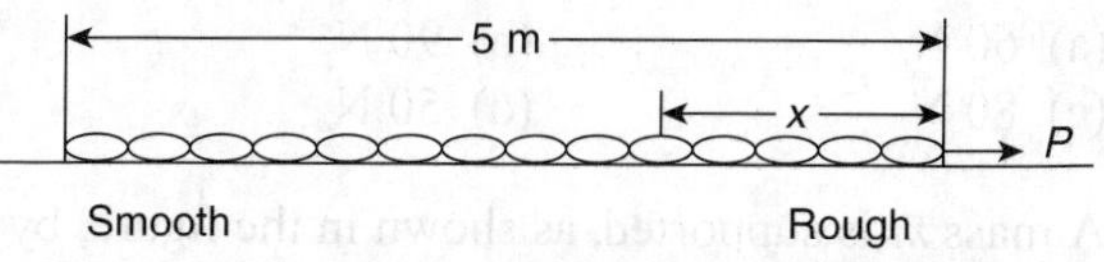

(a) the magnitude of P should increase with time
(b) the magnitude of P should decrease with time
(c) the magnitude of P should increase first and then decrease with time
(d) the magnitude of P should decrease first and then increase with time.

199. In the system shown in figure, the friction coefficient between ground and bigger block is μ. There is no friction between both the blocks. The string connecting both the block is light; all three pulleys are light and frictionless. Then the minimum limiting value of μ so that the system remains in equilibrium is

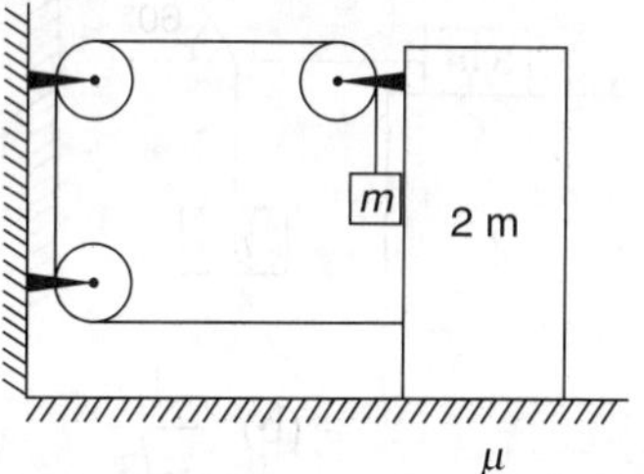

(a) $\frac{1}{2}$ (b) $\frac{1}{3}$

(c) $\frac{2}{3}$ (d) $\frac{3}{2}$

200. An insect crawls up a hemispherical surface very slowly as shown in figure. The coefficient of friction between the insect and the surface is 1/3. If the line joining the centre of the hemispherical surface to the insect makes an angle α with the vertical, the maximum possible value of α is given by

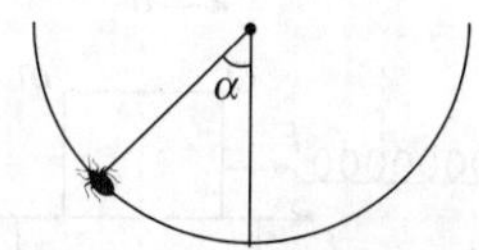

(a) $\cot \alpha = 3$ (b) $\tan \alpha = 3$
(c) $\sec \alpha = 3$ (d) $\operatorname{cosec} \alpha = 3$

201. Given $m_A = 20$ kg, $m_B = 10$ Kg, $m_c = 20$ kg. Between A and B, $\mu = 0.2$; between B and C, $\mu = 0.3$; and between C and ground, $\mu_3 = 0.1$. The least horizontal force F to start the motion of any part of the system of three blocks resting upon one another, as shown in the figure, is

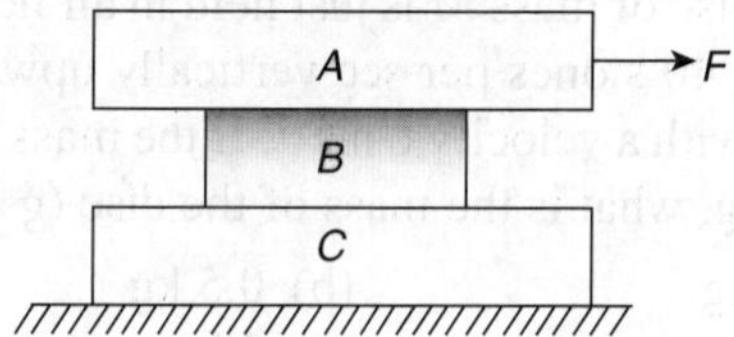

(a) 60 N (b) 90 N
(c) 80 N (d) 50 N

202. A mass m is supported, as shown in the figure, by ideal strings connected to a rigid wall and to a mass $3m$ at rest on a fixed horizontal surface. The string connected to larger mass is horizontal, that connected to smaller mass is vertical and the one connected to wall makes an angle 60° with horizontal. Then the minimum coefficient of static friction between the large mass and the horizontal surface that permits the system to remain in equilibrium in the situation shown is

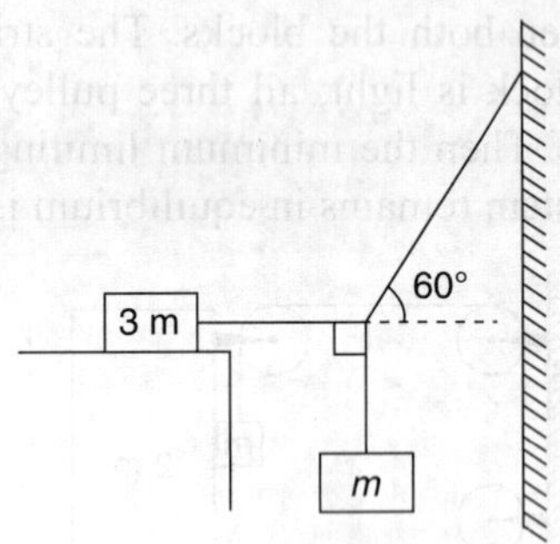

(a) $\frac{1}{\sqrt{3}}$ (b) $\frac{1}{3\sqrt{3}}$

(c) $\frac{\sqrt{3}}{2}$ (d) $\sqrt{\frac{3}{2}}$

203. A block P of mass m is placed on a horizontal frictionless plane. A second block of same mass m is placed on it and is connected to a spring of spring constant K. The two blocks are pulled by distance A. Block Q oscillates without slipping. What is the maximum value of frictional force between the two blocks?

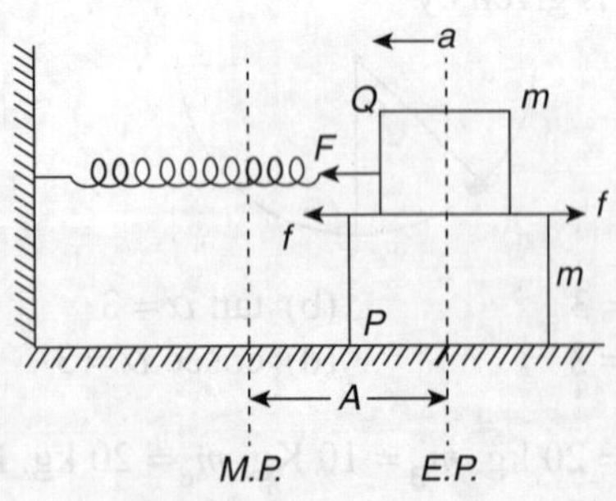

(a) $KA/2$ (b) KA
(c) $\mu_s mg$ (d) Zero

204. A solid disc of mass M is just held in air horizontally by throwing 40 stones per sec vertically upwards to strike the disc with a velocity 6 ms^{-1}. If the mass of each stone is 0.05 kg, what is the mass of the disc (g = 10 ms^{-2})

(a) 1.2 kg (b) 0.5 kg
(c) 20 kg (d) 3 kg

205. A sphere is accelerated upwards by a cord whose breaking strength is four times its weight. The maximum acceleration with which the sphere can move up without breaking the cord is

(a) g (b) $3g$
(c) $2g$ (d) $4g$

206. An army vehicle of mass 1000 kg is moving with a velocity of 10 m/s and is acted upon by a forward force of 1000 N due to the engine and a retarding force of 500 N due to friction. What will be its velocity after 10 s

(a) 5 m/s (b) 10 m/s
(c) 15 m/s (d) 20 m/s

207. A boy having a mass equal to 40 kilograms is standing in an elevator. The force felt by the feet of the boy will be greatest when the elevator (g = 9.8m/s^2)

(a) stands still
(b) moves downward at a constant velocity of 4 m/s
(c) accelerates downward with an acceleration equal to 4 m/s^2
(d) accelerates upward with an acceleration equal to 4m/s^2.

208. A block of mass 5 kg is moving horizontally at a speed of 1.5 m/s. A perpendicular force of 5 N acts on it for 4 sec. What will be the distance of the block from the point where the force started acting?

(a) 10 m (b) 8 m
(c) 6 m (d) 2 m

209. The ratio of the weight of a man in a stationary lift to when it is moving downward with uniform acceleration a is 3: 2. The value of a (g – Acceleration due to gravity of the earth) is

(a) $\frac{3}{2}g$ (b) $\frac{g}{3}$

(c) $\frac{2}{3}g$ (d) g

210. If a person with a spring balance and a body hanging form it goes up and up in an aeroplane, then the reading of the weight of the body as indicted by the spring balance will

(a) go on increasing
(b) go on decreasing
(c) first increase and then decrease
(d) remain the same.

211. A plumb line is suspended form a ceiling of a car moving with horizontal acceleration a. What will be the angle of its inclination with vertical?

(a) $\tan^{-1}(a/g)$ (b) $\tan^{-1}(g/a)$
(c) $\cos^{-1}(a/g)$ (d) $\cos^{-1}(g/a)$

212. A 1 kg stationary bomb is exploded in three parts having mass 1 : 1: 3 respectively. If the parts having same mass move in perpendicular direction with velocity 30 m/s, then the velocity of bigger part will be

(a) $10\sqrt{2}$ m/s (b) $\frac{10}{\sqrt{2}}$ m/s

(c) $15\sqrt{2}$ m/s (d) $\frac{15}{\sqrt{2}}$ m/s

213. A batsman hits back a ball of mass 0.15 kg straight in the direction of the bowler without changing its initial speed of 10 ms^{-1}. If the ball moves linearly, then the impulse imparted on it (in N-s) is

(a) 3.0 (b) 2.0
(c) 1.5 (d) 1.9
(e) 0

214. An aircraft is moving with a velocity of 300 ms^{-1}. If all the forces acting on it are balanced, then

(a) it still moves with the same velocity
(b) it will be just floating at the same point in space
(c) it will fall down instantaneously
(d) it will lose its velocity gradually
(e) it will explode.

215. A body of mass 1000 kg is moving horizontally with a velocity 50 m/s. A mass of 250 kg is added. Find the final velocity

(a) 40 m/s (b) 2.0 m/s
(c) $30\sqrt{2}$ m/s (d) 3.0 m/s

216. A stationary body of mass 3 kg explodes into three equal pieces. Two of the pieces fly off in two mutually perpendicular directions, one with a velocity of $3\hat{i}$ ms^{-1} and the other with a velocity of $4\hat{j}$ ms^{-1} if the explosion occurs in 10^{-4} s, the average force acting on the third piece in newton is

(a) $(3\hat{i} + 4\hat{j}) \times 10^{-4}$
(b) $(3\hat{i} - 4\hat{j}) \times 10^{-4}$
(c) $(3\hat{i} + 4\hat{j}) \times 10^{4}$
(d) $-(3\hat{i} + 4\hat{j}) \times 10^{4}$
(e) $(4\hat{i} + 3\hat{j}) \times 10^{4}$

217. In the first seconds of its fight, a rocket ejects 1/60 of its mass with a velocity of 2400 ms^{-1}. The acceleration of the rocket is

(a) 19.6 ms^{-2} (b) 30.2 ms^{-2}
(c) 40 ms^{-2} (d) 49.8 ms^{-2}

218. Which of the four arrangements in the figure correctly shows the vector addition of two forces $\vec{F_1}$ and $\vec{F_2}$ to yield the third force $\vec{F_3}$?

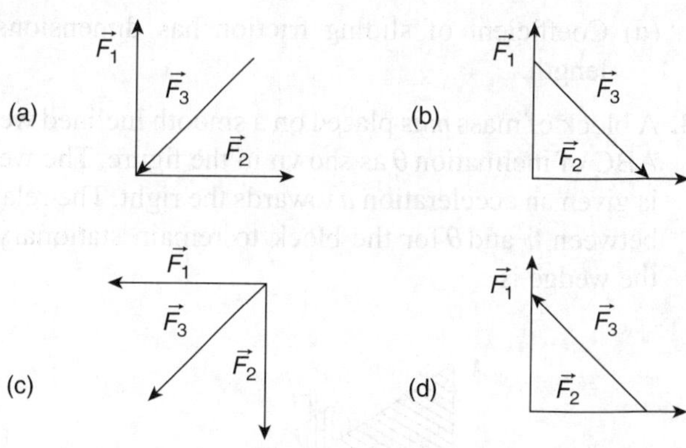

Questions Asked in NEET/CBSE Medicals

219. A block of mass 10 kg is in contact against the inner wall of a hollow cylindrical drum of radius 1 m. The coefficient of friction between the block and the inner wall of the cylinder is 0.1. The minimum angular velocity needed for the cylinder to keep the block stationary when the cylinder is vertical and rotating about its axis, will be $\left(g = 10 \text{ m/s}^2\right)$

(a) 10π rad/s (b) $\sqrt{10}$ rad/s

(c) $\frac{10}{2\pi}$ rad/s (d) 10 rad/s

220. A particle moving with velocity $\vec{v}$ is acted by three forces shown by the vectors triangle PQR. The velocity of the particle will

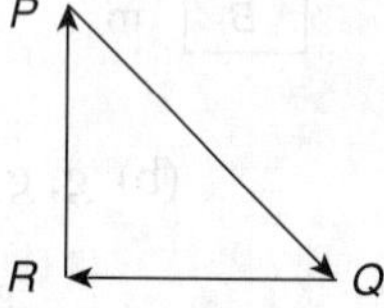

(a) change according to the smallest force $\overline{QR}$
(b) increase
(c) decrease
(d) remain constant.

221. An object flying in air with velocity $\left(20\hat{i} + 25\hat{j} - 12\hat{k}\right)$ suddenly breaks in two pieces whose masses are in the ratio 1 : 5. The smaller mass flies off with a velocity $\left(100\hat{i} + 35\hat{j} + 8\hat{k}\right)$. The velocity of the larger piece will be

(a) $4\hat{i} + 23\hat{j} - 16\hat{k}$ (b) $-100\hat{i} - 35\hat{j} - 8\hat{k}$
(c) $20\hat{i} + 15\hat{j} - 8\hat{k}$ (d) $-20\hat{i} - 15\hat{j} - 8\hat{k}$

222. Which one of the following statement is incorrect?

(a) Rolling friction is smaller than sliding friction.
(b) Limiting value of static friction is directly proportional to normal reaction.
(c) Frictional force opposes the relative motion.

(d) Coefficient of sliding friction has dimensions of length.

223. A block of mass m is placed on a smooth inclined wedge ABC of inclination θ as shown in the figure. The wedge is given an acceleration a towards the right. The relation between a and θ for the block to remain stationary on the wedge is

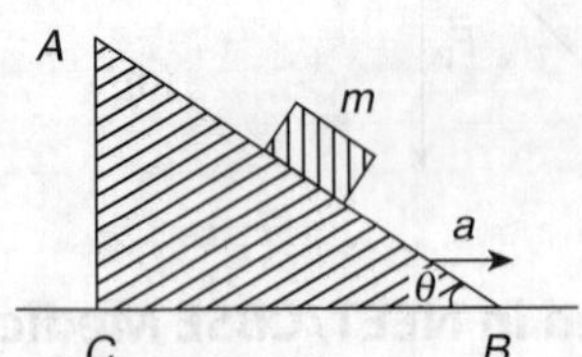

(a) $a = \dfrac{g}{\operatorname{cosec}\theta}$ (b) $a = \dfrac{g}{\sin\theta}$

(c) $a = g\cos\theta$ (d) $a = g\tan\theta$

224. Two blocks A and B of masses 3m and m respectively are connected by a massless and inextensible string. The whole system is suspended by a massless spring as shown in figure. The magnitudes of acceleration of A and B immediately after the string is cut, are respectively

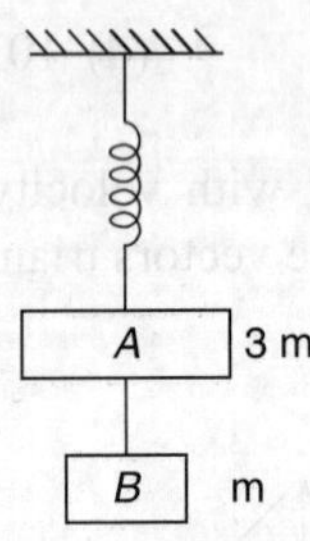

(a) $\dfrac{g}{3}, g$ (b) g, g

(c) $\dfrac{g}{3}, \dfrac{g}{3}$ (d) $g, \dfrac{g}{3}$

225. One end of string of length l is connected to a particle of mass m and the other end is connected to a small peg on a smooth horizontal table. If the particle moves in circle with speed v, the net force on the particle (directed towards centre) will be (T represents the tension in the string)

(a) $T + \dfrac{mv^2}{l}$ (b) $T - \dfrac{mv^2}{l}$

(c) zero (d) $T = \dfrac{mv^2}{l}$

226. A rigid ball of mass m strikes a rigid wall at 60° and gets reflected without loss of speed as shown in the figure. The value of impulse imparted by the wall on the ball will be

(a) mV (b) $2\,mV$

(c) $\dfrac{mV}{2}$ (d) $\dfrac{mV}{3}$

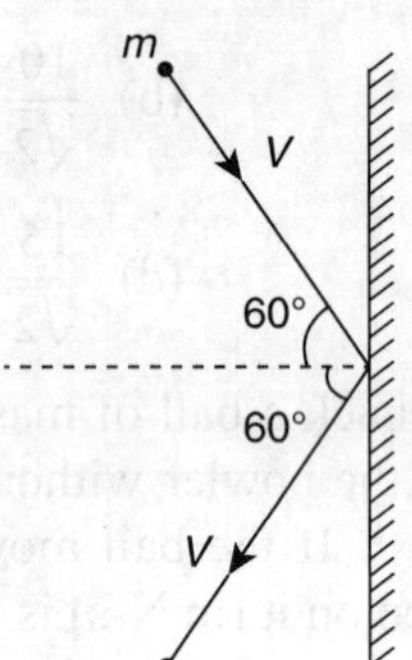

227. A car is negotiating a curved road of radius R. The road is banked at an angle θ. The coefficient of friction between the tires of the car and the road is μ_s. The maximum safe velocity on this road is

(a) $\sqrt{\dfrac{g}{R}\dfrac{\mu_s + \tan\theta}{1 - \mu_s\tan\theta}}$ (b) $\sqrt{\dfrac{g}{R^2}\dfrac{\mu_s + \tan\theta}{1 - \mu_s\tan\theta}}$

(c) $\sqrt{gR^2\dfrac{\mu_s + \tan\theta}{1 - \mu_s\tan\theta}}$ (d) $\sqrt{gR\dfrac{\mu_s + \tan\theta}{1 - \mu_s\tan\theta}}$

228. Two stones of masses m and 2m are whirled in horizontal circles, the heavier one in a radius r/2 and the lighter one in radius r. The tangential speed of lighter stone is n times that of the value of heavier stone when they experience same centripetal forces. The value of n is

(a) 4 (b) 1

(c) 2 (d) 3

229. A plank with a box on it at one end is gradually raised about the other end. As the angle of inclination with the horizontal reaches 30°, the box starts to slip and slides 4.0 m down the plank in 4.0 s. The coefficients of static and kinetic friction between the box and the plank will be, respectively,

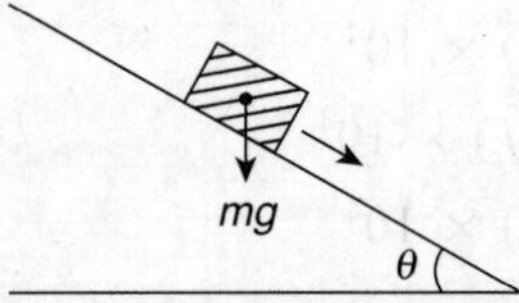

(a) 0.5 and 0.6 (b) 0.4 and 0.3

(c) 0.6 and 0.6 (d) 0.6 and 0.5

230. A cricketer catches a ball of mass 150 g in 0.1 sec moving with speed 20 m/s, then he experience force of

(a) 300 N (b) 30 N

(c) 3 N (d) 0.3 N

231. A system consists of three masses m_1, m_2 and m_3 connected by a string passing over a pulley P. The mass m_1 hangs freely and m_2 and m_3 are on a rough horizontal table (the coefficient of friction $= \mu$). The pulley is frictionless and of negligible mass. The downward acceleration of mass m_1 is
(Assume $m_1 = m_2 = m_3 = m$)

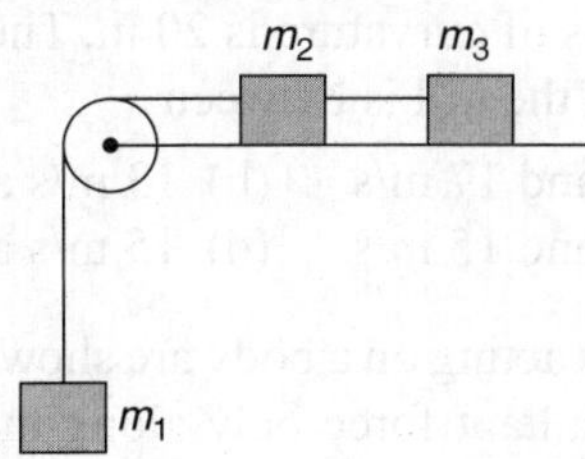

(a) $\dfrac{g(1-g\mu)}{9}$ (b) $\dfrac{2g\mu}{3}$
(c) $\dfrac{g(1-2\mu)}{3}$ (d) $\dfrac{g(1-2\mu)}{2}$

232. The force F acting on a particle of mass m is indicated by the force-time graph as shown. The change in momentum of the particle over the time interval from zero to 8 s is

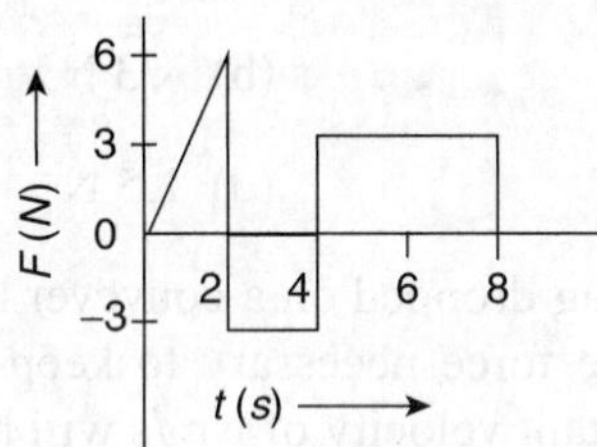

(a) 24 N-s (b) 20 N-s
(c) 12 N-s (d) 6 N-s

233. A balloon with mass m is descending down with an acceleration a (where $a < g$). How much mass should be removed from it so that it starts moving up with an acceleration a?

(a) $\dfrac{2ma}{g+a}$ (b) $\dfrac{2ma}{g-a}$
(c) $\dfrac{ma}{g+a}$ (d) $\dfrac{ma}{g-a}$

234. Three blocks with masses m, $2m$ and $3m$ are connected by strings, as shown in the figure. After an upward force F is applied on block m, the masses move upward at constant speed v.
What is the net force on the block of mass $2m$? (g is the acceleration due to gravity.)

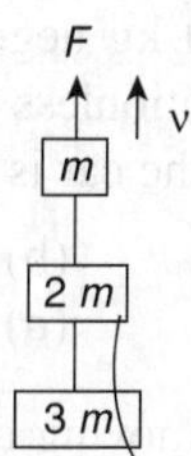

(a) 3 mg (b) 6 mg
(c) zero (d) 2 mg

235. An explosion breaks a rock into three parts in a horizontal plane. Two of them go off at right angles to each other. The first part of mass 1 kg moves with a speed of 12 m s^{-1} and the second part of mass 2 kg moves with 8 m s^{-1} speed. If the third part flies off with 4 m s^{-1} speed, then its mass is

(a) 7 kg (b) 17 kg
(c) 3 kg (d) 5 kg.

236. A trolley of mass 250 kg carrying a sand bag of 25 kg is moving uniformly with a speed of 35 kg km/h on a frictionless tracker. After some time, the sand starts falling out of a hole (in sand bag) on the trolley's floor at the rate of 0.05 kg/s. What is the speed of trolley after the entire sand bag is empty?

(a) 40 km/h (b) 35 km/h
(c) 48 km/h (d) None of these.

237. A car is moving in a circular horizontal track of radius 10 m with a constant speed of 1.0 m/s. A bob is suspended from the roof of the car by a light wire of length 1.0 m. The angle made by the wire with the vertical is

(a) $\pi/3$ (b) $\pi/6$
(c) $\pi/4$ (d) 0°

238. A person holding a rifle (mass of person and rifle together is 100 kg) stands on a smooth surface and fires 10 shots horizontally, in 5 s. Each bullet has a mass of 10 g with a muzzle velocity of 800 ms^{-1}. The final velocity acquired by the person and the average force exerted on the person are

(a) –0.08 ms^{-1},16 N (b) –0.08 ms^{-1},8 N
(c) –1.6 ms^{-1},16 N (d) –1.6 ms^{-1},8 N

239. A stone is dropped from a height h. It hits the ground with a certain momentum P. If the same stone is dropped from a height 100% more than the previous height, the momentum when it hits the ground will change by

(a) 68% (b) 41%
(c) 200% (d) 100%

240. A car of mass 1000 kg negotiates a banked curve of radius 90 m on a frictionless road. If the banking angle is 45°, the speed of the car is

(a) 20 m s^{-1} (b) 30 m s^{-1}
(c) 5 m s^{-1} (d) 10 m s^{-1}

241. A car of mass m is moving on a level circular track of radius R. If μ_s represents the static friction between the road and tires of the car, the maximum speed of the car in circular motion is given by

(a) $\sqrt{\mu_s mRg}$ (b) $\sqrt{\frac{Rg}{\mu_s}}$
(c) $\sqrt{\frac{mRg}{\mu_s}}$ (d) $\sqrt{\mu_s Rg}$

242. A person of mass 60 kg is inside a lift of mass 940 kg and presses the button on control panel. The lift starts moving upwards with an acceleration 1.0 m/s^2. If $g = 10 \text{ m s}^2$, the tension in the supporting cable is

(a) 8600 N (b) 9680 N
(c) 11000 N (d) 1200 N

243. A body of mass M hits normally a rigid wall with velocity V and bounces back with the same velocity. The impulse experienced by the body is

(a) MV (b) $1.5\ MV$
(c) $2\ MV$ (d) zero.

244. A conveyor belt is moving at a constant speed of 2 m s^{-1}. A box is gently dropped on it. The coefficient of friction between them is $\mu = 0.5$. The distance that the box will move relative to belt before coming to rest on it, taking $g = 10 \text{ m s}^{-2}$, is

(a) 0.4 m (b) 1.2 m
(c) 0.6 m (d) zero.

245. A block of mass m is in contact with the cart C as shown in the figure. The coefficient of static friction between the block and the cart is μ. The acceleration α of the cart that will prevent the block from falling satisfies

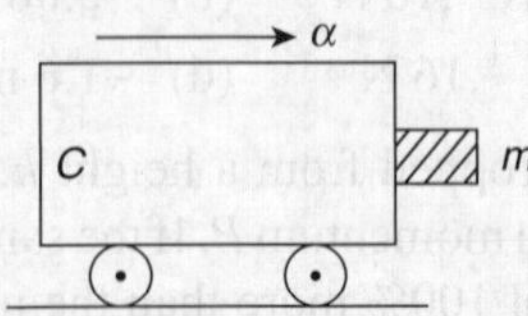

(a) $\alpha > \frac{mg}{\mu}$ (b) $\alpha > \frac{g}{\mu m}$
(c) $\alpha \geq \frac{g}{\mu}$ (d) $\alpha < \frac{g}{\mu}$

246. The mass of a lift is 2000 kg. When the tension in the supporting cable is 28000 N, then its acceleration is

(a) 4 m s^{-2} upwards (b) 4 m s^{-2} downwards
(c) 14 m s^{-2} upwards (d) 30 m s^{-2} downwards.

247. A body, under the action of a force $\vec{F} = 6\hat{i} - 8\hat{j} + 10\hat{k}$, acquires an acceleration of 1 m/ s^{-2}. The mass of this body must be

(a) 10 kg (b) 20 kg
(c) $10\sqrt{2}$ kg (d) $2\sqrt{10}$ kg

248. A roller coaster is designed such that riders experience "weightlessness" as they go round the top of a hill whose radius of curvature is 20 m. The speed of the car at the top of the hill is between

(a) 16 m/s and 17 m/s (b) 13 m/s and 14 m/s
(c) 14 m/s and 15 m/s (d) 15 m/s and 16 m/s

249. Three forces acting on a body are shown in the figure. To have the resultant force only along the y-direction, the magnitude of the minimum additional force needed is

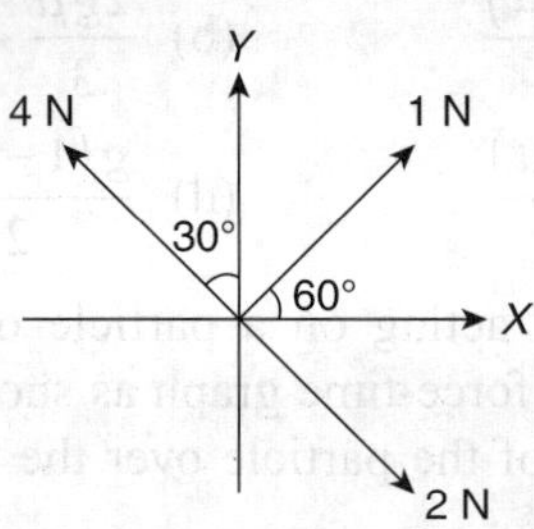

(a) $\frac{\sqrt{3}}{4}$ N (b) $\sqrt{3}$ N
(c) 0.5 N (d) 1.5 N

250. Sand is being dropped on a conveyer belt at the rate of M kg/s. The force necessary to keep the belt moving with a constant velocity of v m/s will be

(a) $\frac{Mv}{2}$ newton (b) zero
(c) Mv newton (d) $2\ Mv$ newton

251. A block B is pushed momentarily along a horizontal surface with an initial velocity V. If μ is the coefficient of sliding friction between B and the surface, block B will come to rest after a time

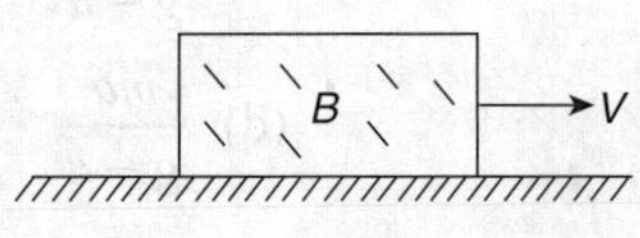

(a) $g\mu / V$ (b) g / V
(c) V / g (d) $V / (g\mu)$

252. A 0.5 kg ball moving with a speed of 12 m/s strikes a hard wall at an angle of 30° with the wall. It is reflected with the same speed at the same angle. If the ball is in contact with the wall for 0.25 seconds, the average force acting on the wall is

(a) 96 N (b) 48 N
(c) 24 N (d) 12 N

253. A block of mass m is placed on a smooth wedge of inclination θ. The whole system is accelerated horizontally so that the block does not slip on the wedge. The force exerted by the wedge on the block will be (g is acceleration due to gravity)

(a) $mg \cos \theta$ (b) $mg \sin \theta$
(c) mg (d) $mg/\cos \theta$

254. The coefficient of static friction, μ_s between block A of mass 2 kg and the table as shown in the figure is 0.2. What would be the maximum mass value of block B so that the two blocks do not move? The string and pulley are assumed to be smooth and mass less. ($g = 10$ m/ s^2)

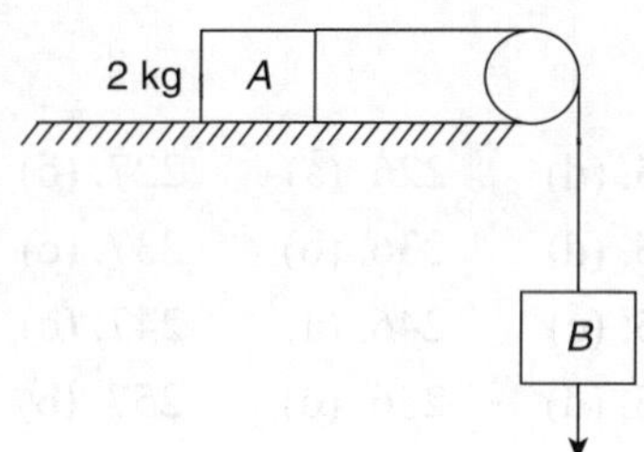

(a) 2.0 kg (b) 4.0 kg
(c) 0.2 kg (d) 0.4 kg

255. A man weight 80 kg. He stands on a weighing scale in a lift which is moving upwards acceleration of 5 m/ s^2. What would be the reading on the scale? ($g = 10$ m/ s^2)

(a) zero (b) 400 N
(c) 800 N (d) 1200 N

256. A monkey of mass 20 kg is holding a vertical rope. The rope will not break when a mass of 25 kg is suspended from it but will break if the mass exceeds 25 kg. What is the maximum acceleration with which the monkey can climb up along the rope? ($g = 10$ m/ s^2)

(a) 5 m/s^2 (b) 10 m/ s^2
(c) 25 m/ s^2 (d) 2.5 m/ s^2

257. Consider a lift of mass 1000 kg which is moving with acceleration of 1 m/ s^2 in upward direction. Then the tension developed in string which is connected to lift is

(a) 9800 N (b) 10, 800 N
(c) 11,000 N (d) 10,000 N

258. A block of mass 10 kg is placed on a rough horizontal surface having coefficient of friction $\mu = 0.5$. If a horizontal force of 100 N is acting on it then acceleration of the block will be

(a) 10 m/ s^2 (b) 5 m/ s^2
(c) 15 m/ s^2 (d) 0.5 m/ s^2

259. 250 N force is required to raise 75 kg mass from a pulley. If the rope is pulled 12 m then the load is lifted to 3 m; the efficiency of pulley system will be

(a) 25% (b) 33.3%
(c) 75% (d) 90%

260. Determine the magnitude and direction of the net force acting on a stone of mass 0.2 kg, just after it is dropped from the window of a train which is accelerating at 5 m/s^2? Take $g = 10$ m/s^2

(a) 2N downward
(b) $\sqrt{5}$ N making an angle of $\tan^{-1}(2)$ with train motion
(c) 2N upward
(d) $\sqrt{5}$ N making an angle of $\tan^{-1}(2)$ with vertical.

Answer Keys

EXERCISES

1. (c)	2. (c)	3. (b)	4. (c)	5. (b)	6. (b)	7. (a)	8. (c)	9. (a)	10. (b)
11. (a)	12. (c)	13. (a)	14. (d)	15. (c)	16. (c)	17. (b)	18. (a)	19. (a)	20. (d)
21. (c)	22. (a)	23. (c)	24. (d)	25. (c)	26. (a)	27. (b)	28. (b)	29. (a)	30. (b)
31. (a)	32. (b)	33. (c)	34. (c)	35. (c)	36. (a)	37. (c)	38. (c)	39. (c)	40. (b)
41. (c)	42. (d)	43. (c)	44. (a)	45. (b)	46. (b)	47. (b)	48. (d)	49. (d)	50. (b)
51. (b)	52. (b)	53. (a)	54. (a)	55. (b)	56. (b)	57. (d)	58. (d)	59. (c)	60. (d)
61. (b)	62. (c)	63. (b)	64. (c)	65. (d)	66. (c)	67. (b)	68. (c)	69. (a)	70. (d)
71. (a)	72. (b)	73. (c)	74. (b)	75. (a)	76. (a)	77. (b)	78. (c)	79. (a)	80. (c)
81. (d)	82. (c)	83. (b)	84. (c)	85. (c)	86. (c)	87. (a)	88. (d)	89. (c)	90. (c)
91. (d)	92. (a)	93. (d)	94. (a)	95. (b)	96. (d)	97. (d)	98. (a)	99. (a)	100. (c)
101. (d)	102. (c)	103. (a)	104. (b)	105. (a)	106. (b)	107. (c)	108. (c)	109. (d)	110. (b)

111. (c)	112. (d)	113. (c)	114. (c)	115. (b)	116. (c)	117. (c)	118. (c)	119. (c)	120. (b)
121. (c)	122. (c)	123. (a)	124. (d)	125. (c)	126. (a)	127. (d)	128. (b)	129. (b)	130. (c)
131. (a)	132. (a)	133. (d)	134. (b)	135. (c)	136. (d)	137. (c)	138. (a)	139. (d)	140. (c)
141. (d)	142. (a)	143. (d)	144. (d)	145. (d)	146. (d)	147(b)	148. (b)	149. (b)	150. (a)
151. (c)	152. (b)	153. (a)	154. (a)	155. (c)	156. (a)	157. (a)	158. (c)	159. (c)	160. (d)
161. (a)	162. (d)	163. (b)	164. (b)	165. (b)	166. (a)	167. (c)	168. (b)	169. (c)	170. (b)
171. (b)	172. (a)	173. (a)	174. (b)	175. (c)	176. (d)	177. (c)	178. (d)	179. (d)	180. (c)
181. (c)	182. (c)	183. (b)	184. (c)	185. (d)	186. (d)	187. (b)	188. (d)	189. (b)	190. (b)
191. (c)	192. (a)	193. (a)	194. (c)	195. (b)	196. (a)	197. (a)	198. (a)	199. (c)	200. (a)
201. (d)	202. (b)	203. (a)	204. (a)	205. (b)	206. (c)	207. (d)	208. (a)	209. (b)	210. (c)
211. (a)	212. (a)	213. (a)	214. (a)	215. (a)	216. (d)	217. (c)	218. (c)		

Questions Asked in NEET/CBSE Medicals

219. (d)	220. (d)	221. (a)	222. (d)	223. (d)	224. (a)	225. (d)	226. (a)	227. (d)	228. (c)
229. (d)	230. (b)	231. (c)	232. (c)	233. (a)	234. (c)	235. (d)	236. (b)	237. (c)	238. (b)
239. (b)	240. (b)	241. (d)	242. (c)	243. (c)	244. (a)	245. (c)	246. (a)	247. (c)	248. (c)
249. (c)	250. (c)	251. (d)	252. (c)	253. (d)	254. (d)	255. (d)	256. (d)	257. (b)	258. (b)
259. (c)	260. (a)								

Solutions

1. (c)

$$2T\sin\theta = mg$$

If $\theta \to 0 \quad T \to \infty$

and $\theta \to \dfrac{\pi}{2}, T \to \dfrac{mg}{2}$

2. (c)

First law states that the body will continue in its state of rest or uniform motion if no net force is acting on it.

3. (b)

The various forces acting on the body have been shown in the figure. The force on the body down the inclined plane in presence of friction μ is

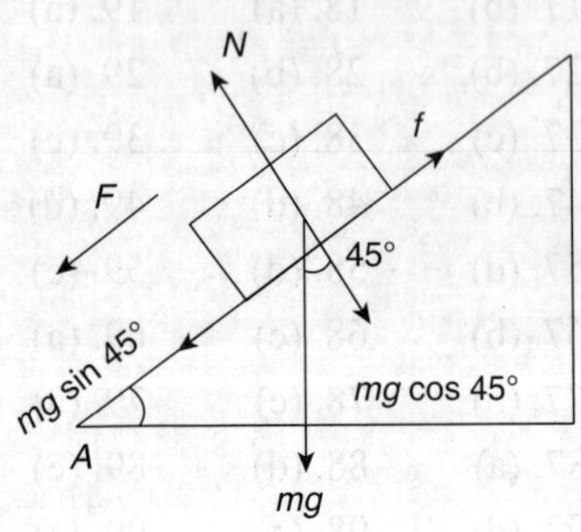

$$F = mg\sin\theta - f = mg\sin\theta - \mu N = ma$$

or $\quad a = g\sin\theta - \mu g\cos\theta$

Since block is at rest, initial velocity $u = 0$

$\therefore$ Time taken to slide down the plane

$$t_1 = \sqrt{\frac{2s}{a}} = \sqrt{\frac{2s}{g\sin\theta - \mu g\cos\theta}}$$

In absence of friction, time taken will be

$$t_2 = \sqrt{\frac{2s}{g\sin\theta}}$$

Given : $t_1 = 2t_2$

$\therefore t_1^2 = 4t_2^2 \quad$ or $\quad \dfrac{2s}{g(\sin\theta - \mu g\cos\theta)} = \dfrac{2s \times 4}{g(\sin\theta)}$

or $\sin\theta = 4\sin\theta - 4\mu\cos\theta$ or $\mu = \dfrac{3}{4}\tan\theta = 0.75$

4. (c)

Reaction by surface in lift is the apparent weight of person. When a person of mass m is placed in a lift, then actual weight of the person is mg. This acts on the weighing machine which offers a reaction R, given by the reading of the weighing machine. This reaction exerted by the surface of contact on the person and is apparent weight of the person.

When lift is accelerating upwards at the rate of a, then reaction

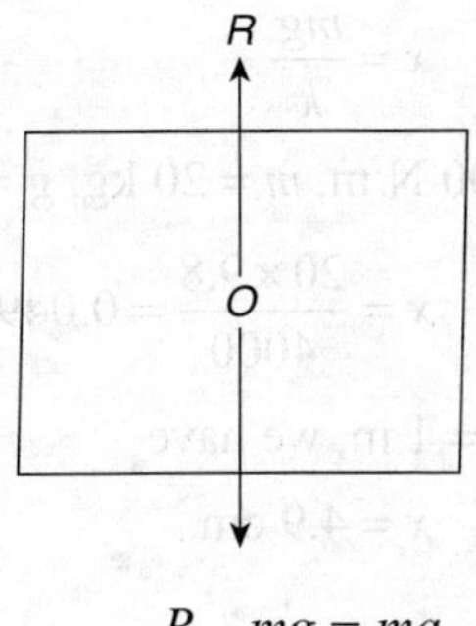

$$R - mg = ma$$
$$R = m(g + a)$$

Hence, apparent weight is greater than actual weight.

5. (b)

A force, which always acts toward or away from a fixed point, and whose magnitude depends only on the distance from that point is called a 'central force'. This point is called as 'centre of force'. Examples of central force are

(i) the gravitational force exerted on a particle by another particle which is stationary in an inertial frame of reference

(ii) the electrostatic force exerted on a charged particle

(iii) a particle attached to one end of a spring whose other end is stationary in an inertial frame of reference. On the other hand, nuclear forces are non-central forces.

6. (b)

Let a be the acceleration of each block and T_1 and T_2 be the tensions in the two strings as shown in figure. Taking the three blocks and the two strings as the system

Using, $\Sigma F_x = ma_x$

or $28 = (4 + 2 + 1)a$

or $a = \frac{28}{7} = 4 \text{ m/s}^2$.

7. (a)

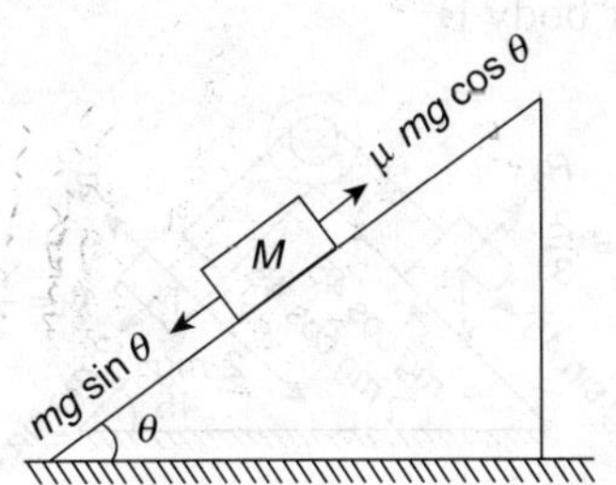

$$a = \frac{f_x - mg\sin\theta}{m}$$

$$= \frac{\mu_k mg\cos\theta - mg\sin\theta}{m}$$

$$= \frac{\frac{1}{\sqrt{3}} mg\cos 30° - mg\sin 30°}{m} \quad (\theta = 30°)$$

$$= \frac{\frac{1}{\sqrt{3}} g \times \frac{\sqrt{3}}{2} - g \times \frac{1}{2}}{1} = 0$$

8. (c)

From Newton's second law of motion,

Force = Mass × Acceleration

Acceleration along y-axis is

$$a_y = -\frac{5}{5} \text{ m/s}^2 = -1 \text{ m/s}^2$$

From equation of motion

$$v_y = u_y + a_y t$$

where u_y is the initial velocity along y-axis, t the time and v_y the final velocity along y-axis.

Given $v_y = 0, u_y = 40 \text{ m/s}, a_y = -1 \text{ m/s}$

$$0 = 40 - 1 \times t$$
$$t = 40 \text{ s.}$$

9. (a)

Let there be two vectors $\vec{A}$ and $\vec{B}$ where $|\vec{A}| = |\vec{B}|$ their sum is

$$\vec{R} = \vec{A} + \vec{B}$$

Taking self, product of both sides, we have

$$\vec{R}.\vec{R} = (\vec{A} + \vec{B}).(\vec{A} + \vec{B})$$
$$= A^2 + 2AB\cos\theta + B^2$$

Where θ is angle between $\vec{A}$ and $\vec{B}$.

Given, $|\vec{R}| = |\vec{A}| = |\vec{B}| = 2N$, therefore

$$(2)^2 = (2)^2 + 2 \times 2 \times 2\cos\theta + (2)^2$$

$$\Rightarrow \quad \cos\theta = -\frac{1}{2}$$

$$\theta = 120°$$

In this condition, the angle between the given vectors should be $120°$.

Alternative

If two vectors and their sum vector form a triangle: Given, two forces are of equal magnitude 2 N and the magnitude of their resultant is 2 N. Since these are equal in magnitude, the triangle is an equilateral one. In this case angle between the force is 120° (angle is measured anticlockwise)

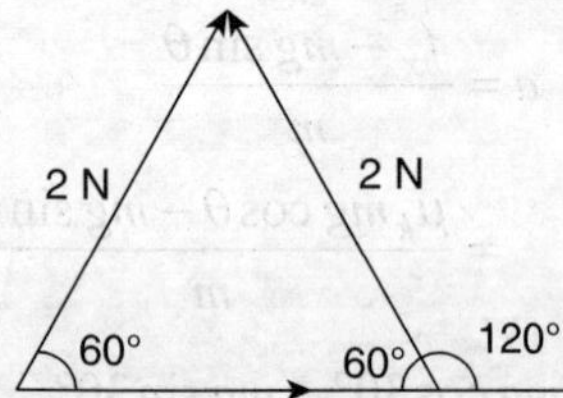

10. (b)

If a vector is resolved into two vectors whose combined effect is same as that of the given vector, then the resolved vectors are called components of the given vector. If θ is angle between vectors $\vec{A}$ and $\vec{B}$, we have

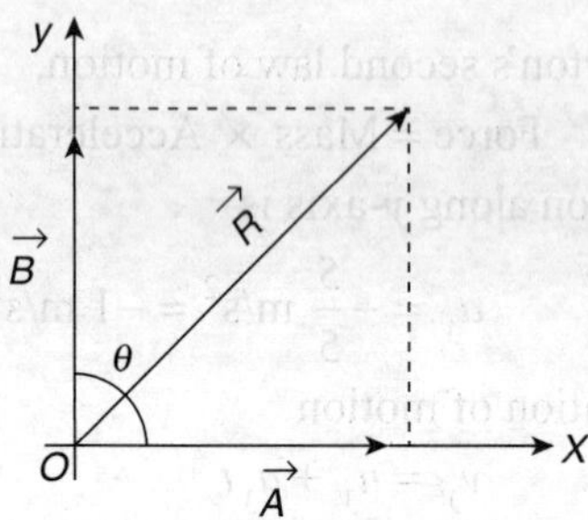

$$\left|\vec{R}\right| = \sqrt{A^2 + B^2 + 2AB\cos\theta}$$

Given, $A = B = 5, \theta = 90°$

$$\left|\vec{R}\right| = \sqrt{5^2 + 5^2 + 2 \times 25 \times \cos 90°}$$

$$\left|\vec{R}\right| = \sqrt{50}N = 5\sqrt{2}N$$

The direction of resultant is

$$\tan\theta = \frac{5\sin 90°}{5 + 5\cos 90°} = \frac{5}{5} = 1$$

$$\theta = 45° = \frac{\pi}{4}.$$

11. (a)

Restoring force on spring is equal to force due to gravity. Let extension produced in the spring when a mass of 20 kg is hung be x, then since the spring mass system executes SHM the restoring force acting on mass m is

$$F = -kx \quad \text{...(i)}$$

And force due to gravity is

$$F = mg \quad \text{...(ii)}$$

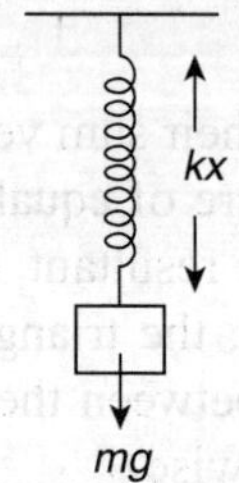

Equating Eqs. (i) and (ii), we get

$$kx = mg$$

$$x = \frac{mg}{k}$$

Given $k = 4000$ N/m, $m = 20$ kg, $g = 9.8$ m/s^2

$$x = \frac{20 \times 9.8}{4000} = 0.049 \text{ m}$$

Since 100 cm = 1 m, we have

$$x = 4.9 \text{ cm}.$$

12. (c)

An object moving in uniform circular motion is moving around the perimeter of the circle with a constant speed. While the speed of object is constant, its velocity is changing. Velocity, being a vector quantity, has a constant magnitude but a changing direction. The direction is always tangential to the circle and as the object turns the circle, the tangent line is always pointing in a new direction. Also, when it is moving in circular motion it is acted upon by a centripetal force directed towards the centre; hence acceleration is perpendicular to velocity.

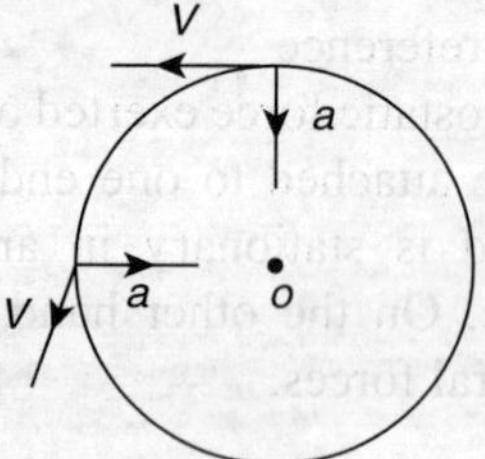

13. (a)

Let R be the reaction of base of lift and g the acceleration due to gravity, acting downwards. Their resultant provides the net acceleration to the lift. Therefore

$$R - mg = ma$$

$$R = m(g + a)$$

Also $\quad R = mg'$

$$g' = g + a$$

Therefore, net acceleration increases and hence, reading indicated by spring balance will increase.

14. (d)

The situation is as shown in the figure. The equation motion for body B

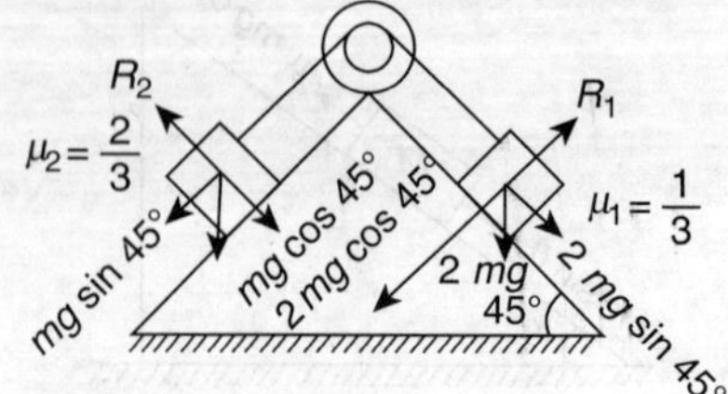

$$2mg\sin 45° - \mu_1 R_1 - T = 2ma$$

$$2mg\sin 45° - \frac{1}{3}2mg\cos 45° - T = 2ma$$

$$\Rightarrow \quad 2mg \times \frac{1}{\sqrt{2}} - \frac{1}{3} \times 2mg \times \frac{1}{\sqrt{2}} - T = 2ma \quad \text{...(i)}$$

In this problem, as $(m_B - m_A)g\sin\theta = \left(\frac{mg}{\sqrt{2}}\right)$ is lesser than $(\mu_B m_B - \mu_A m_A)g\cos\theta = \left(\frac{4mg}{3\sqrt{2}}\right)$ the masses will not move. Hence,

Acceleration of B = Acceleration A = 0.

15. (c)

We know the friction force

$$F = \mu R$$

So, the coefficient of limiting friction

$$\mu = \frac{F}{R}$$

Hence, R is normal reaction.

16. (c)

We know that

$$F = kx \quad (\text{Here } F = mg)$$

$$mg = kx, x = \text{extension in the spring}$$

or $$x = \frac{mg}{k}$$

$$\Rightarrow \quad x = \frac{12 \times 9.8}{2 \times 10^3}$$

or $$x = 58.8 \times 10^{-3}\text{ m}$$

$$\Rightarrow \quad x = 0.0588\text{ m}$$

Hence, the total length of the spring

$$= 0.50\text{ m} + 0.0588\text{ m} = 0.5588\text{ m}.$$

17. (b)

$$S = \frac{v^2}{2\mu_k g} = \frac{100 \times 100}{2 \times 0.5 \times 10}$$

$$= \frac{100 \times 100}{5 \times 2} = 1000$$

18. (a)

$$v^2 = u^2 - 2as$$

$$(0)^2 = (20)^2 - 2a \times 40$$

$$80a = 400$$

$$\Rightarrow \quad a = 5\text{ m/s}^2$$

$\therefore$ Coefficient of dynamic friction

$$\mu = \frac{a}{g} = \frac{5}{10} = \frac{1}{2} = 0.5$$

19. (a)

The following free-body diagram shows the various forces acting on the system. Let m be the minimum mass of block C and f_s be the maximum value of static friction.

For block 'A'

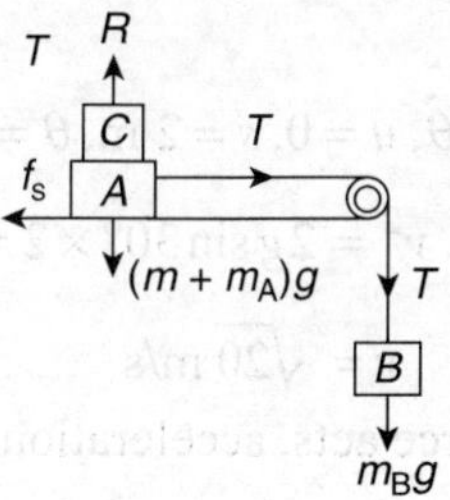

$$R = (m + m_A), f_s = T$$

$$\therefore \mu(m + m_A)g = T \quad \text{(i)}$$

For block 'B'

$$T = m_B g \quad \text{(ii)}$$

From Eqs. (i) and (ii), we get

$$m = \frac{m_B - \mu m_A}{\mu}$$

$$= \frac{10 - 0.4 \times 15}{0.4} = 10\text{ kg}$$

20. (d)

$$F = kx$$

$$0.1 = k(1.5 \times 10^{-2})$$

$$k = \frac{0.1}{(1.5 \times 10^{-2})}$$

When the spring is cut into three equal parts, then spring constant of each part will be $3k$. Force required

$$F' = 3k(x) = 3k(3 \times 10^{-2})$$

$$= \frac{0.1 \times 3}{1.5 \times 10^{-2}} \times 3 \times 10^{-2} = 0.6\text{ N}$$

21. (c)

When body of mass m is suspended from a spring of force constant k and stretched to distance x, then restoring force

$$F = -kx \quad \text{(i)}$$

When spring is broken to two identical springs $\left(\frac{x}{2}\right)$, then force constant of each spring is k', then

$$F' = k'\frac{x}{2} \quad \text{(ii)}$$

From Eqs. (i) and (ii), we get

$$kx = \frac{k'x}{2}$$

$$k' = 2k$$

22. (a)

When a body starts from rest, the initial velocity is zero. By equation of motion, we have

$$v^2 = u^2 + 2as$$

where u is the initial velocity, and v the final velocity. So we have

$$a = g\sin\theta,\ u = 0,\ s = 2\text{ m},\ \theta = 30°,\ g = 10\text{ m/s}^2$$

$$\therefore v^2 = 2g\sin 30° \times 2 = 20$$

or $\quad v = \sqrt{20}$ m/s

If frictional force acts, acceleration is given by $a' = \mu g$

$$\therefore v^2 = u^2 + 2a's'$$

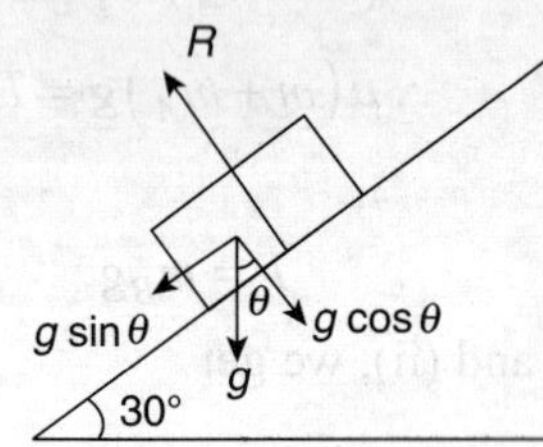

Here, $u = 0, a' = \mu g = 0.25 \times 10$ m/s²

$$v = \sqrt{20}\text{ m/s}$$

$$20 = 0 + 2 \times 0.25 \times 10 \times s'$$

or $\quad s' = \dfrac{20}{2 \times 0.25 \times 10} = 4$ m

23. (c)

The block will produce same extension in both the spring balanced A and B. Hence, both balances will read same value, 2 kg.

24. (d)

$$a_1 = \frac{(2M - M)g}{2M + M} = \frac{g}{3}$$

$$a_2 = \frac{2Mg - Mg}{M} = g$$

$$\frac{a_1}{a_2} = \frac{1}{3}$$

$\Rightarrow \quad a_1 : a_2 = 1 : 3$

25. (c)

Thrust is the force with which the rocket moves upward and is given by

$$F = u\frac{dm}{dt}$$

Thus mass of the gas ejected per second to supply the thrust needed to overcome the weight of the rocket is

$$\frac{dm}{dt} = \frac{F}{u} = \frac{m \times a}{u} \quad \text{or} \quad \frac{dm}{dt} = \frac{600 \times 10}{1000} = 6\text{ kg s}^{-1}$$

26. (a)

For equilibrium of the man

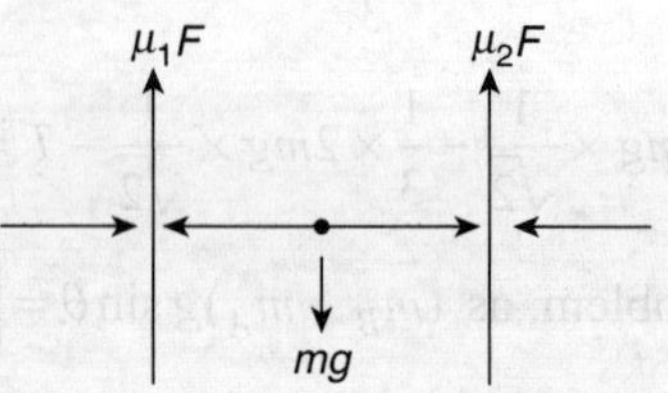

$$\mu_1 F + \mu_2 F = mg$$

$$\therefore F = \frac{mg}{\mu_1 + \mu_2} = \frac{55 \times 9.8}{1.1 \times 0.7} = 700\text{ N}$$

27. (b)

Force (F) = 6 N;
Initial velocity (u) = 0;
Mass (m) = 1 kg and final velocity (v) = 30 m/s

Therefore acceleration (a) $= \dfrac{F}{m} = \dfrac{6}{1} = 6$ m/s² and final velocity (v) $= 30 = u + at = 0 + 6 \times t$ or $t = 5$ seconds.

28. (b)

By Newton's second law,
$T_2 = (5m)a$ and $T_1 = (8m)a$

$$\frac{T_2}{T_1} = \frac{5}{8}$$

or $\quad T_2 = \dfrac{5}{8} \times T_1$

$$= \frac{5}{8} \times 16 = 10\text{ N}$$

29. (a)

$m = 1.5$ kg, $u = 25$ m/sec, $v = 15$ m/sec, $t = 0.03$ sec

$$F = \frac{m(v - u)}{t} = \frac{1.5[15 - (-25)]}{0.03}$$

$$F = \frac{1.5 \times 40}{0.03} = 2000\text{ N}$$

30. (b)

To keep the mass M steady, centripetal force balances by the weight of M.

$$\frac{mv^2}{R} = Mg$$

$$v^2 = \frac{MgR}{m}$$

$$v = \sqrt{\frac{MgR}{m}}$$

31. (a)

The net upward acceleration is 9.8 – 2.8 = 7 m/s^2
Total mass = 80 + 5 = 85 kg
So, net upward force is

$$F = 85 \times 7 = 595 \text{ N}$$

32. (b)

$$mg - T = ma \quad \ldots(1)$$

$$T = Ma \quad \ldots(2)$$

From Eqs. (i) and (ii)

$$mg = (m + M)a$$

$$a = \frac{mg}{m + M}$$

33. (c)

$F = ma$

where a is the acceleration.

$$a = \frac{\Delta v}{\Delta t}$$

Force depends upon acceleration and acceleration depends upon time. Even Δv is the same in both cases.

34. (c)

$$F - 2T = 0$$

$$T = \frac{F}{2}$$

∴ Acceleration $\quad a = \frac{T}{m} = \frac{F}{2m}$

T is the tension in the string.

35. (c)

Acceleration a = $\frac{1}{m}\left(\frac{dm}{dt}\right)u$

Rate of burning of fuel gases,

$$\frac{dm}{dt} = \frac{1/60 \text{ kg}}{1 \text{ s}} = \frac{1}{60} \text{kg/s}$$

$$m = 1 \text{ kg}, u = 2400 \text{ m/s}$$

$$\Rightarrow \quad a - \frac{1}{1} \times 2400 \times \frac{1}{60} = 40 \text{ m/s}^2$$

36. (a)

Since 5 kg body explodes into three fragments with masses in the ratio 1 : 1 : 3, the masses of fragments will be 1 kg, 1 kg and 3 kg respectively. The magnitude of resultant momentum of two fragments each of mass 1 kg, moving with velocity 21 m/s, in perpendicular directions is

$$\sqrt{(m_1v_1)^2 + (m_2v_2)^2}$$

$$m'v' = \sqrt{(21)^2 + (21)^2} = 21\sqrt{2} \text{ kg m/s}$$

According to law of conservation of linear momentum,

$$m_3v_3 = m'v' = 21\sqrt{2} \text{ or } 3v_3 = 21\sqrt{2}$$

or $\quad v_3 = 7\sqrt{2}$ m/s

37. (c)

Man is sliding down the telegraphic pole with acceleration $\frac{g}{4}$, so,

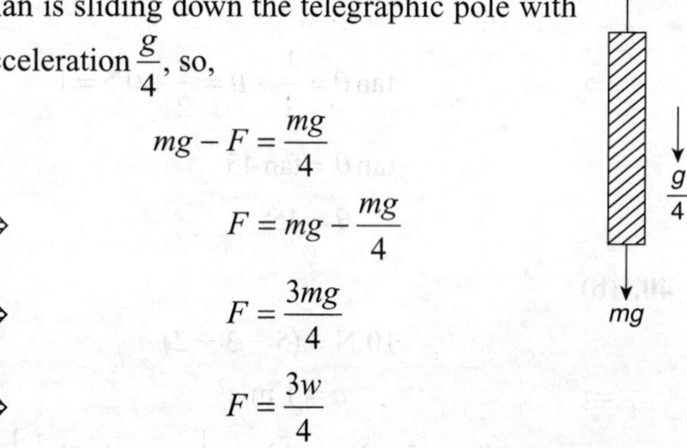

$$mg - F = \frac{mg}{4}$$

$$\Rightarrow \quad F = mg - \frac{mg}{4}$$

$$\Rightarrow \quad F = \frac{3mg}{4}$$

$$\Rightarrow \quad F = \frac{3w}{4}$$

38. (c)

Minimum force required to move a body up a rough inclined plane

$$F_1 = mg(\sin\theta + \mu\cos\theta)$$

Minimum force required to prevent the body from sliding down the rough inclined plane

$$F_2 = \mu mg\cos\theta$$

According to question,

$$F_1 = 3F_2$$

$$mg(\sin\theta + \mu\cos\theta) = 3(\mu mg\cos\theta)$$

$$(\sin\theta + \mu\cos\theta) = 3\mu\cos\theta$$

$$\sin\theta = 2\mu\cos\theta$$

$$\tan\theta = 2\mu$$

$$\tan\theta = 2 \times \frac{1}{2\sqrt{3}}$$

$$\tan\theta = \frac{1}{\sqrt{3}} = \tan 30°$$

$$\theta = 30°$$

39. (c)

$$\mu = 0.5, \frac{F}{R} = \frac{1}{2}$$

Net force acting downward = $mg\sin\theta - f$
where f = Friction force

$$= \mu R \ (R = \text{Normal reaction})$$

$$= \mu mg\cos\theta$$

$$\frac{F}{R} = \frac{1}{2}$$

$$\frac{mg\sin\theta - \mu mg\cos\theta}{mg\cos\theta} = \frac{1}{2}$$

$$\tan\theta - \mu = \frac{1}{2}$$

$$\Rightarrow \quad \tan\theta = \frac{1}{2} + \mu = \frac{1}{2} + 0.5 = 1$$

$$\tan\theta = \tan 45$$

$$\theta = 45°$$

40. (b)

$$10 \text{ N} = (5 + 3 + 2)\, a$$

$$\Rightarrow \quad a = 1 \text{ m/s}^2$$

$$T_1 = (5+3)a = 8 \text{ N} \quad \left[\because a = 1 \text{ m/s}^2\right]$$

$$T_2 = 5a = 5 \text{ N}$$

41. (c)

$$\Delta P = F\Delta t$$

$F = mg$ = weight of ball
t = time of flight

42. (d)

$$\mu mg = \frac{mv^2}{r}$$

$$\Rightarrow \quad v^2 = \mu mg = 0.25 \times 40 \times 10$$

$$\Rightarrow \quad v = 10 \text{ m/s}$$

43. (c)

$$F = ma$$

$$\mu mg = ma$$

$$\Rightarrow \quad a = \mu g$$

$$\Rightarrow \quad a = 4.9 \text{ m/s}^2$$

$$72 \text{ km/h} = 20 \text{ m/s}$$

$$2as = v^2 - u^2$$

$$\Rightarrow \quad 2 \times 4.9 \times s = (20)^2$$

$$s = 40.8 \text{ m}$$

44. (a)

Let M is the mass of the chain of length L. If y is the maximum length of chain which can hang outside the table without sliding, then for equilibrium of the chain, the weight of hanging part must be balanced by the force of friction on the portion of the table.

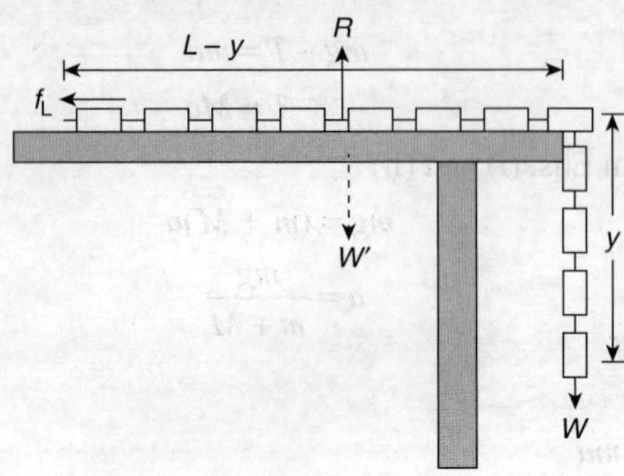

$$W = f_L \quad \ldots(1)$$

But from figure,

$$W = \frac{M}{L}yg \text{ and } R = W' = \frac{M}{L}(L - y)g$$

So that $f_L = \mu R = \mu \frac{M}{L}(L - y)g$

Substituting these values of W and f_L in Eqn. (i) we get

$$\mu \frac{M}{L}(L - y)g = \frac{M}{L}yg$$

or $\quad \mu(L - y) = y$ or $y = \dfrac{\mu L}{\mu + 1} = \dfrac{0.25L}{1.25} = \dfrac{L}{5}$

or $\quad \dfrac{y}{L} = \dfrac{1}{5} = \dfrac{1}{5} \times 100\% = 20\%$

45. (b)

Given: $u = 6$ m/s; $s = 9$ m; $v = 0$
Using $v^2 - u^2 = 2as$

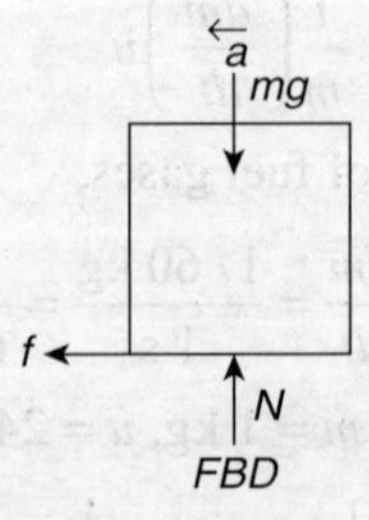

$$\Rightarrow \quad a = -\frac{u^2}{2s} = -\frac{6^2}{2 \times 9}$$

$$= -2 \text{ m/s}^2$$

$$\because \quad f = \mu mg = ma$$

$$\Rightarrow \quad \mu = \frac{a}{g} = \frac{2}{10} = 0.2 \quad (g = 10 \text{ m/s}^2)$$

46. (b)

$$F_1 = F_2 + F_r$$
$$10 = 2 + F_r$$
$$F_r = 8\text{ N (left)}$$

If F_1 is removed, the frictional force is acting opposite to applied force and this will be 2 N. Hence the forces are balanced and the block will remain at rest. The net force on it will be zero.

47. (b)

$$\frac{mv^2}{r} = \mu mg$$
$$v^2 = \mu rg$$
$$v^2 = 0.5\times 500\times 10$$
$$\Rightarrow \quad v = 50\text{ m/s}$$

48. (d)

$$(F_{\text{Static friction}})_{\text{max}} = \mu\,\text{mg}$$
$$= \frac{1}{3}\times 60\times 9.8 = 1.96\text{ N}$$

49. (d)

$$\mu\left(\frac{2}{3}m\right)g = \frac{m}{3}g$$
$$\mu = \frac{1}{2}$$

50. (b)

When milk is churned, cream gets separated due to centrifugal force.

51. (b)

As angle increases, the static friction which is balancing the weight component of the block along the incline also increases. So static friction will be $Mg\sin\theta$ and when the block starts sliding then it experiences kinetic friction given as $\mu Mg\cos\theta$.

52. (b)

In first case, the acceleration of block is g, in second case it is $g/3$ and in third case it is $g/2$. Hence option (b) is correct.

53. (a)

$$\Delta p = F.t$$

Now, $\quad F\,t = $ Area under the curve

Area under the curve

$$= \left(-\frac{1}{2}\times1\times1\right)+(2\times2)+\left(-\frac{1}{2}\times2\times2\right)+(1\times1)+\left(-\frac{1}{2}\times1\times1\right)$$

$$F.\,t = 2\text{ N-s}$$
$$\Delta p = 2\text{ N-s}$$

54. (a)

FBD of two body system

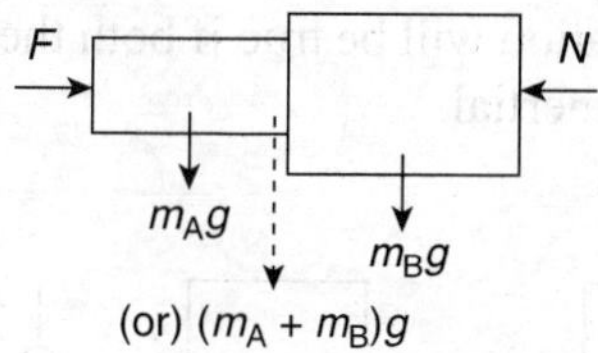

$\because$ $(m_A + m_B)g$ is acting downwards. There is no force acting upwards to balance this force.
$\therefore$ The system cannot be in equilibrium.

55. (b)

The tension in thread will be maximum when it is at the bottom of the circle.

$$\Rightarrow mr\omega^2 + mg = T_{\text{thread}}$$
$$0.5\times 4\times\omega^2 + 0.5\times 10 = 3$$
$$\omega^2 = 16\text{ rad/s} \Rightarrow \omega = 4\text{ rad/s}$$

56. (b)

By spitting or sneezing we get a momentum in opposite direction which will help us in getting off the plane. In all other cases we will slip on ice as there is no friction.

57. (d)

The vertical component of the pulling force acts in the opposite direction of weight.

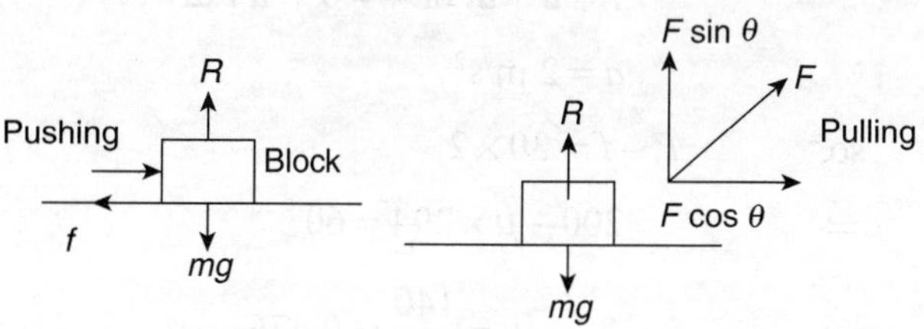

Here $f = \mu R = \mu mg$ $\qquad$ Here $f = \mu(mg - F\sin\theta)$

58. (d)

When brakes are on, the wheels of the cycle will slide on the road instead of rolling there. It means the sliding friction will come into play instead of rolling friction. The value of sliding friction is more than that of rolling friction.

59. (c)

An internal frame of reference is one in which law of inertia holds good, Newton's laws of motion are applicable equally.

If Earth is revolving around the sun or earth is rotating about its axis, then forces are acting on the earth and hence there will be acceleration of earth due to these forces.

That is why Earth cannot be an inertial frame of reference.

60. (d)

The observation will be true if both the frames are inertial or non-inertial.

61. (b)

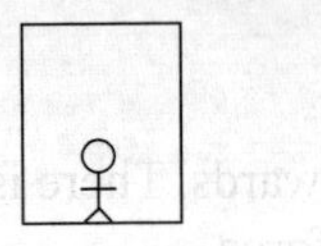

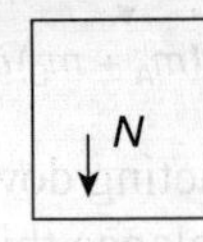

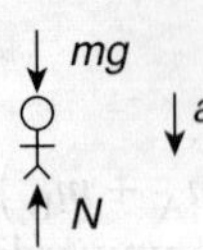

(a) System (b) FBD of lift and man

From FBD of man: $mg - N = ma$

$\Rightarrow \quad N = mg - ma$

The force exerted by passenger on the floor of the elevator $= mg - ma$.

62. (c)

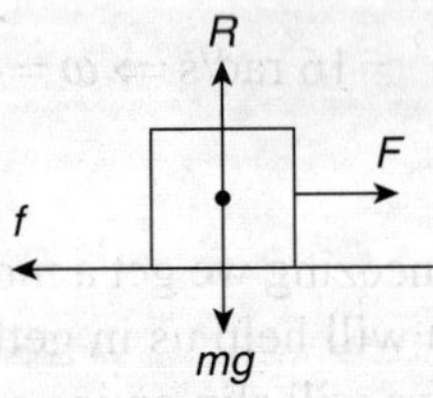

Equation of motion is $F - f = ma$

where $f = \mu R = \mu mg = \mu \times 30 \times 9.8 = 294\mu$

Acceleration of the body is

$$v = u + at \text{ or } 4 = 0 + a \times 2$$

$$a = 2 \text{ m/s}^2$$

so, $\quad F - f = 30 \times 2$

$\Rightarrow \quad 200 - \mu \times 294 = 60$

$\Rightarrow \quad \mu = \dfrac{140}{294} = 0.476$

63. (b)

$$2T\cos 60° = mg$$

or $\quad T = mg = 2 \times 10 = 20 \text{ N}$

64. (c)

Acceleration, $\quad a = \dfrac{v - u}{t} = \dfrac{0 - 6}{3} = -2 \text{ m/s}^2$

$$\text{Force} = m \times a = 4 \times 2 = 8 \text{ N}$$

65. (d)

Mass of the rope $= m \times L$

Where m = mass per unit length

Acceleration in the rope due to force F will be $a = \dfrac{F}{mL}$

Mass of rope of length $(L - l)$ will be $m(L - l)$.

Therefore, tension in the rope of length $(L - l)$ is equal to pulling force on it.

$$= m(L - l)a = m(L - l) \times \frac{F}{mL} = F\left(1 - \frac{l}{L}\right)$$

66. (c)

$$R = mg - ma = 0.5 \times 10 - 0.5 \times 2 = 5 - 1 = 4 \text{ N}$$

67. (b)

Let R be the reaction force between the block

Then acceleration

$$a = \frac{F}{m} = \frac{3}{3} = 1 \text{ m/s}^2$$

Now for 2 kg block

$$F - R = ma$$

$$3 - R = 2 \times 1$$

$\Rightarrow \quad R = 3 - 2 = 1$

68. (c)

In presence of friction $a = (g\sin\theta - \mu g\cos\theta)$

$\therefore$ Time taken to slide down the plane

$$t_1 = \sqrt{\frac{2s}{a}} \quad \sqrt{\frac{2s}{g(\sin\theta - \mu\cos\theta)}}$$

In absence of friction $t_2 = \sqrt{\dfrac{2s}{g\sin\theta}}$

$$t_1 = 2t_2$$

$$\therefore t_1^2 = 4t_2^2$$

or $\quad \dfrac{2s}{g(\sin\theta - \mu\cos\theta)} = \dfrac{2s \times 4}{g\sin\theta}$

$$\sin\theta = 4\sin\theta - 4\mu\cos\theta$$

$$\mu = \frac{3}{4}\tan\theta = \frac{3}{4} = 0.75 \quad (\because \theta = 45°)$$

69. (a)

$$T = m(g + a) = 100(9.8 + 0) = 980 \text{ N}$$

70. (d)

$$mg(h+s) = Fs$$
$$2\times 9.8(60.6) = F\times 0.6$$
$$F = 19.6\times 101 = 1979.6 \text{ N}$$

71. (a)

$$a = \frac{F}{m_1+m_2+m_3} = \frac{700}{350} = 2$$
$$T_1 = 200\times 2 = 400 \text{ N}$$

72. (b)

$$a = \frac{p}{M+m}$$

Force on the block $= Ma = \dfrac{Mp}{M+m}$

73. (c)

For downward motion: 2000 g – F (up thrust) = 2000 a

For upward motion

$$F - (2000 - M)g = (2000 - M)a$$

So, $\quad Mg = (4000 - M)\, 2.5$

$$M(10+2.5) = 4000\times 2.5$$
$$M = 800 \text{ kg}$$

74. (b)

$$a = \frac{200}{100} = 2 \text{ m/s}^2$$
$$s = \frac{1}{2}at^2 = \frac{1}{2}\times 2\times 100 = 100 \text{ m}$$

at the end of 10 sec

$$v = u + at$$
$$= 2\times 10 = 20 \text{ m/s}$$

$\therefore$ Distance travelled in next 10 sec $= 20\times 10 = 200$ m

75. (a)

Rate of change of momentum $= mn\sqrt{mgh}$

$\therefore$ Force $= mng + mn\sqrt{mgh}$

76. (a)

$$F = v\frac{dm}{dt} = 2\times 5 = 10 \text{ N}$$

77. (b)

$$T = 784 = 100(g-a) = 100(9.8 - a)$$
$$a = 9.8 - 7.84 = 1.96 \text{ m/s}^2 = g/5 \text{ m/s}^2$$

78. (c)

$$2g - T = 2a$$
$$T - 3g\sin\theta = 3a$$
$$2g - 3g\sin 30° = 5a = \frac{g}{2}$$
$$a = g/10 \quad \therefore T = 2g - 2a = 1.8\, g$$

79. (a)

Force of friction $= \mu mg = 0.25\times 2\times 10 = 5$ N

This is the force on the cart.

$\therefore$ Acceleration of cart $= \dfrac{5}{20} = 0.25$ m/s^2

80. (c)

$$F = m_1\times 15 = m_2\times 10 = (m_1+m_2)a$$
$$F = \left(\frac{F}{15}+\frac{F}{10}\right)a$$
$$\frac{1}{a} = \frac{1}{10}+\frac{1}{15} = \frac{1}{6}$$
$$a = 6 \text{ m/s}^2$$

81. (d)

Case I:

$$a_1 = \frac{(m_2-m_1)g}{m_1+m_2} = \frac{(7-3)g}{7+3} = 0.4\, g$$

Case II:

$$70 - 3g = 3a_2$$
$$a_2 = \frac{70-3\times 10}{3} = \frac{40}{3}$$
$$a_1 : a_2 = 4 : \frac{40}{3} = 3 : 10$$

82. (c)

$$5g - T = 5a$$
$$T - T' = 2a$$
$$T' - 3g = 3a$$

Adding $2g = 10a \quad \therefore a = \dfrac{g}{5}$

$$T' = 3(g+a) = 3\left(\frac{6g}{5}\right)$$
$$= 3.6\text{ g} = 3.6\times 9.8 = 35.3 \text{ N}$$
$$T = 5g - 5a = 5\left(g - \frac{g}{5}\right) = 4g = 39.2 \text{ N}$$

83. (b)

Frictional force is a self-adjusting force

$$\mu mg \text{ N} = 0.2\times 50\times 9.8 \text{ N}$$
$$20 \text{ N} < 98 \text{ N}$$

Frictional force = 20 N

84. (c)

$$F = \mu R = 0.2 \times 60 = 12 \text{ N}$$

85. (c)

Reaction of the body = ma

$\therefore$ Frictional force = μma

This acts upwards. For equilibrium, $\mu ma = mg$

$$a = \frac{g}{\mu}$$

86. (c)

$$\mu = \tan 5° = 0.087$$

87. (a)

$$\text{Reaction} = 200 - \mu \times 15 \times g = 200 - 0.15 \times 15g$$
$$= 200 - 22 = 178 \text{ N}$$

88. (d)

$$\mu \frac{mv^2}{r} = mg$$

$$\Rightarrow \quad 0.49 \times v^2 = 9.8 \times 50$$
$$v^2 = 100$$

or $\quad v = 10\sqrt{10}$ m/s

89. (c)

Since centre of mass remains at rest

$$mv_1 + 3mv_2 = 0$$
$$v_1 = -3v_2 \; \frac{v_1^2}{v_2^2} = \frac{S_1}{S_2} = \frac{9}{1}$$

90. (c)

Maximum force of friction = $\mu_s mg = 0.5$ mg

Frictional force when body moves = $\mu_k mg = 0.45$ mg

$\therefore$ Force accelerating the body = 0.5 mg – 0.45 mg

= 0.05 mg

$\therefore$ Acceleration = 0.05 g = 0.49 m/s^2

91. (d)

$$200 - F = m \times 3 \text{ and } 300 - F = m \times 5$$

$$\Rightarrow \quad \frac{200 - F}{300 - F} = \frac{3}{5}$$

$$1000 - 5F = 900 - 3F$$
$$2F = 100$$

or $\quad F = 50$ N = Frictional force.

$\Rightarrow \quad m = 50$ kg

92. (a)

$$\mu mg = ma = 0.2 \times 9.8 \Rightarrow a = 1.96 \text{ m/s}^2$$

93. (d)

The acceleration to provide a velocity of 72 km/h = 20 m/s

$$v^2 - u^2 = 2as$$

$u = 0$, $v = 20$ m/s and $s = 100$

$$v^2 = 2as$$
$$(20)^2 = 2a \times 100$$
$$2 \text{ m/s}^2 = a$$

Let F be the force, then

$$F - F_f = ma$$
$$F_f = \mu mg = 0.2 \times 10000 \times 9.8 = 19600 \text{ N}$$
$$F - 19600 = 10000 \times 2$$
$$F = 20000 + 196000$$
$$= 39600 \text{ N}$$
$$= 3.96 \text{ N}$$

94. (a)

$$\text{Frictional force} = \mu(mg - F\sin\theta)$$

$\therefore$ Force causing acceleration

$$= F\cos\theta - \mu(mg - F\sin\theta)$$
$$= 30 \times \frac{\sqrt{3}}{2} - 0.6\left(5 \times 9.8 - 30 \times \frac{1}{2}\right)$$
$$= 15\sqrt{3} - 20.4$$

$$\text{Acceleration} = \frac{15\sqrt{3} - 20.4}{5} = 3\sqrt{3} - 4.08 = 1.12 \text{ m/s}^2$$

95. (b)

$$\text{Thrust} = M(g + a) = u\frac{dm}{dt}$$

$$\frac{dm}{dt} = \frac{M(g+a)}{u} = \frac{5000(10+20)}{800} = 187.5 \text{ kg/s}$$

96. (d)

The acceleration is nullified by the force of kinetic friction/mass $mg\sin\theta$ is force downwards μ_k is the coefficient of kinetic friction $\mu_k mg\cos\theta$ is the force acting upwards

$$\therefore mg\sin\theta - \mu_k mg\cos\theta = \text{Mass} \times \text{Acceleration}$$

Acceleration = 0 as v is constant

$$\therefore mg\sin\theta = \mu_k mg\cos\theta$$

or, $\mu_k = \tan\theta$

97. (d)

$$mgh = \mu mgs$$
$$\mu = \frac{h}{s}$$

98. (a)

$$mg\sin\theta + \mu mg\cos\theta = mg(\sin\theta + \mu\cos\theta)$$
$$= mg(\sin\theta + \tan\lambda\cos\theta)$$

$$= mg\frac{(\sin\theta + \tan\lambda\cos\theta)}{\cos\lambda}$$

$$= mg\frac{\sin(\theta+\lambda)}{\cos\lambda}$$

99. (a)

Energy acquired in the upper half $= mg\frac{l}{2}\sin\alpha$

Energy wasted in the lower half $= \mu mg\cos\alpha\frac{l}{2}$

These are equal.

$$mg\frac{l}{2}\sin\alpha = \mu mg\cos\alpha\frac{l}{2}$$

$$\tan\alpha = \mu = \tan 30° = \frac{1}{\sqrt{3}}$$

100. (c)

$$v = u + at$$

$$0 = 5 - (g\sin\theta + \mu g\cos\theta)t$$

$$g\sin\theta + \mu g\cos\theta = \frac{5}{0.5} = 10$$

$$9.8[0.5 + \mu \times 0.866] = 10$$

$$\mu = 0.6 \text{ (approx.)}$$

101. (d)

$$F = \mu \text{ mg} = \frac{mg}{2}$$

So, $\mu = 0.5$

102. (c)

Let force be F

$$mg = 100 \Rightarrow m = \frac{100}{10}$$

$$m = 10 \text{ kg}$$

$$F - mg\sin 30° = ma$$

$$F - 100 \times \frac{1}{2} = 10 \times 2.5$$

$$F - 50 = 25$$

$$\Rightarrow F = 50 + 25 = 75 \text{ N}$$

103. (a)

$$\mu mg\cos\theta \times 2 = mg\sin\theta$$

$$0.5 \times 2\cos\theta = \sin\theta$$

$$\theta = 45°$$

104. (b)

Rate of burning of fuel $(\frac{dm}{dt})$ = 1 kg/s and velocity of ejected fuel $(v) = 60$ km/s $= 60 \times 10^3$ m/s.

Force = Rate of change of momentum

$$\frac{dp}{dt} = \frac{d(mv)}{dt} = v\frac{dm}{dt} = (60\times10^3)\times 1 = 60000 \text{ N}$$

105. (a)

$$T - mg = ma$$

$$495 - (45)(10) = 45\ a_{max}$$

$$\Rightarrow a_{max} = 1 \text{ m/s}^2$$

106. (b)

$$mg - T = ma$$

$$mg - \frac{2}{3}mg = ma$$

$$\Rightarrow a = \frac{g}{3}$$

107. (c)

$$mg - F = ma$$

$F = m(g - a) = 5(9.8 - 4) = 29$ N upwards.

108. (c)

Maximum retardation of the lift is

$$a = \frac{T - mg}{m}$$

$$= \frac{2000(10) - 1000(10)}{1000} = 10 \text{ m/s}^2$$

$u = 2.5$ m/s and an upward acceleration

$$a = 10 \text{ m/s}^2$$

$$0 = u^2 - 2as$$

$$\Rightarrow s = \frac{u^2}{2a} = \frac{(2.5)^2}{2\times10} = \frac{5}{16} \text{ m}$$

109. (d)

$$T_1 - mg = ma$$

$$mg - T_2 = ma$$

$$T_1 - T_2 = 2ma = 2(0.1)(5) = 1 \text{ N}$$

110. (b)

$$mg - T_{max} = ma_{max} \text{ and } T_{max} = nmg$$

$$a_{max} = g(1 - n)$$

111. (c)

If upthrust due to air is P, then

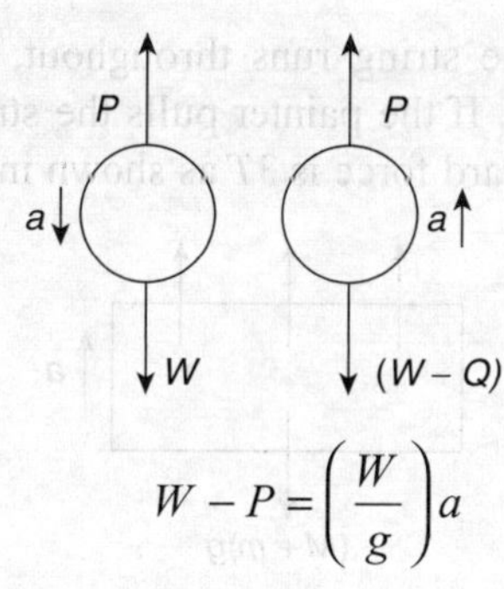

$$W - P = \left(\frac{W}{g}\right)a$$

$$P-(W-Q)=\left(\frac{W-Q}{g}\right)a$$

Adding $\quad \dfrac{Q}{a}=\dfrac{W}{g}+\dfrac{(W-Q)}{g}$

$$Q=\frac{2Wa}{(g+a)}$$

112. (d)

$s=\frac{1}{2}at^2$, the acceleration of elevator is

$$a=\frac{2s}{t^2}=\frac{2\times 2}{(0.5)^2}=16\text{ m/s}^2$$

As the lift is having an upward acceleration a,

$$T=m(g+a)=3(10+16)=78\text{ N}$$

113. (c)

$$T=m_1a_1=m_2a_2$$

$$2a_1=3(1.8)$$

$\Rightarrow \quad a_1=2.7\text{ m/s}^2$ towards m_2

114. (c)

$$T'=2T=2\left(\frac{2m_1m_2}{m_1+m_2}\right)g$$

$$T'=\frac{8}{3}g=2.67\text{ kg wt}$$

115. (b)

$$a=\frac{(2+2)-2}{2+2+2}g=\frac{g}{3}$$

For block C,

$$mg-T=ma$$

$$T=\left(mg-\frac{mg}{3}\right)=\left(\frac{2}{3}\right)mg=13.3\text{ N}$$

116. (c)

As $a=0$, so $T=\text{mg}$

The spring balanced reads the tension.

So, its reading is 5(10) = 50 N.

117. (c)

As the same string runs throughout, tension is same everywhere. If the painter pulls the string with a force T, then upward force is $3T$ as shown in figure.

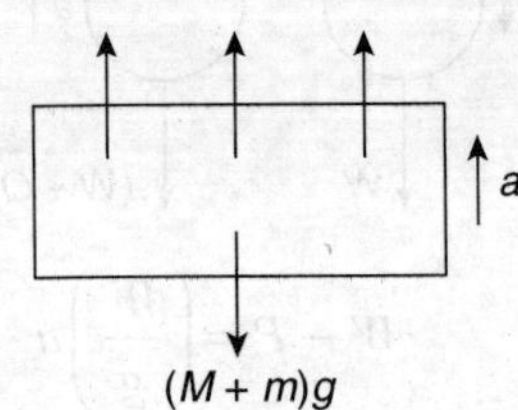

$$a=\frac{3T-(M+m)g}{M+m}$$

$$=\frac{3(450)-(60+40)(10)}{60+40}$$

$$=3.5\text{ m/s}^2$$

118. (c)

When the monkey of mass m climbs with upward acceleration a, the tension T in the string is

$$T-mg=ma$$

$\Rightarrow \quad T=ma+ma$

This tension T must be at least equal to the weight Mg of the block which the monkey wishes to raise.

Hence $T=mg+ma=Mg$

$$a=\frac{(M-m)}{m}g=\frac{(20-16)}{16}g=\frac{g}{4}$$

119. (c)

Let T_1 and T_2 be the tension in the strings P and Q respectively.

Then,

$$T_1\sin 45^\circ=500 \qquad \text{(i)}$$

$$T_1\cos 45^\circ=T_2 \qquad \text{(ii)}$$

From (i) and (ii),

$$\frac{T_2}{500}=\frac{T_1\cos 45^\circ}{T_1\sin 45^\circ}$$

$\Rightarrow \quad T_2=500$

120. (b)

If the tension in the lower string supporting the weight $w=20$ N is T' and tension in the upper string is T, then

$$T'=2T \text{ and } T'=W$$

Therefore, $\quad T=\dfrac{W}{2}=10\text{ N}$

121. (c)

$$\tan\theta=\frac{ma}{mg}=\frac{a}{g}=\frac{4.9}{9.8}=0.5$$

As $\tan 45^\circ=1; \tan 30=\dfrac{1}{\sqrt{3}}=0.58$

The angle θ is approximately 30°

122. (c)

When carriage is sliding down freely on inclined plane

$$a=g\sin\theta$$

The forces acting on the bob are

(i) mg vertically downward.

(ii) Tension T inclined at an angle with the vertical.

(iii) Pseudo force $mg\ \sin\theta$ along the plane in the upward direction. As the bob is in equilibrium in this frame, so

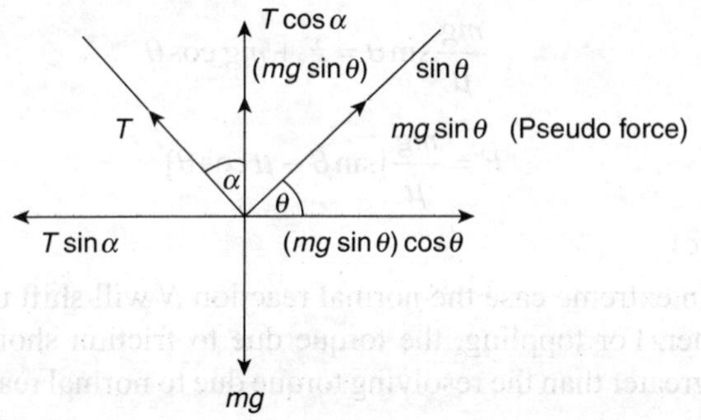

$$T\cos\alpha = mg - (mg\sin\theta)\sin\theta = mg\cos^2\theta \qquad \ldots\text{(i)}$$

$$T\sin\alpha = (mg\sin\theta).\cos\theta \qquad \ldots\text{(ii)}$$

Divide (ii) by (i), we get

$$\tan\alpha = \tan\theta$$

$$\alpha = \theta$$

123. (a)

The pseudo acceleration for the body $a' = a$

If the pseudo force $Ma\cos\theta = Mg\sin\theta$ then the body will be at rest, $a = g\tan\theta$

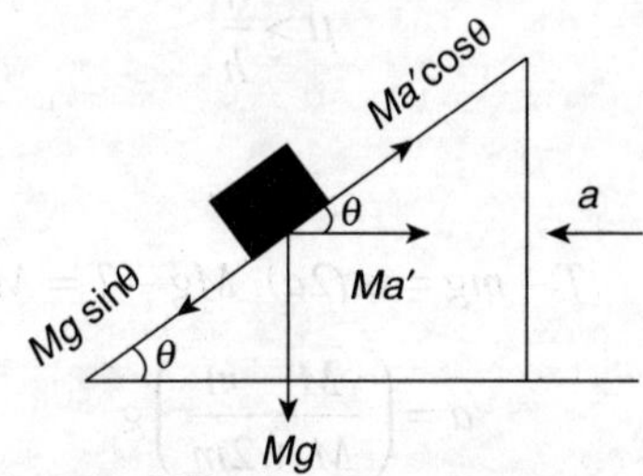

This horizontal acceleration should be applied to the wedge to the left.

124. (d)

$$\mu\left(\frac{2}{3}mg\right) = \frac{1}{3}mg$$

$$\Rightarrow \qquad \mu = \frac{1}{2}$$

125. (c)

$$\mu(l-x)mg = x\,mg$$

$$\Rightarrow \qquad x = \frac{\mu l}{\mu + l}$$

126. (a)

Mass of bullet $(m_1) = 200$ g $= 0.2$ kg; speed of bullet $(v_1) = 5$ m/sec and mass of gun $(m_2) = 1$ kg. Before firing, total momentum is zero.

$\therefore$ After firing total momentum is $m_1v_1 + m_2v_2$

From the law of conservation of momentum

$$m_1v_1 + m_2v_2 = 0$$

or
$$v_2 = \frac{-m_1v_1}{m_2} = \frac{-0.2\times 5}{1} = -1 \text{ m/s}$$

127. (d)

$$a_{\text{block}} = \frac{F - \mu_k m_{\text{block}}g}{m_{\text{block}}}$$

$$a_{\text{slab}} = \frac{\mu_k m_{\text{block}}g}{m_{\text{slab}}} = \frac{0.4\times 10\times 9.8}{40}$$

$$= 0.98 \text{ m/s}^2$$

128. (b)

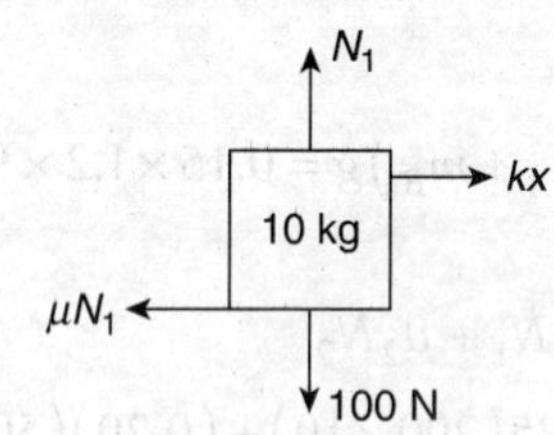

$$N_1 = 100$$

$$kx - \mu N_1 = 10\times 12$$

$$kx - 10 = 120$$

$$kx = 130$$

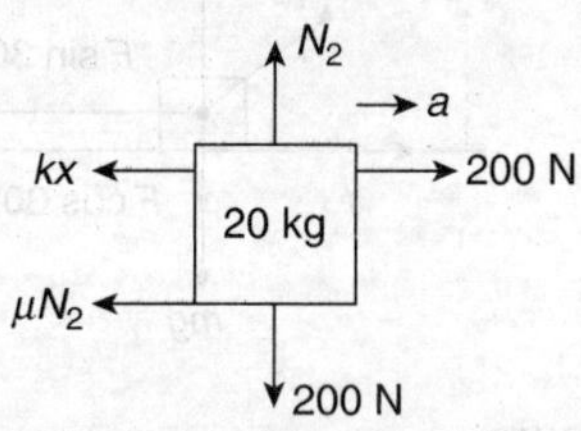

$$N_2 = 200$$

$$200 - kx - \mu N_2 = 20a$$

$$200 - 130 - (0.1\times 12) = 20a$$

$$70 - 20 = 20a$$

$$a = \frac{50}{20} = 2.5 \text{ m/s2}$$

129. (b)

The Block B is accelerated due to the normal contact force N exerted by the block A. If the acceleration of the system of two blocks is a, then

$$N = m_B a$$

And $f = \mu N = \mu m_B a$

As the force of friction must be able to balance the weight of the block B, so

$$\mu m_B a = m_B g$$

$$a = \frac{g}{\mu} = \frac{9.8}{0.65} = 15.1 \text{ m/s}^2$$

130. (c)

Distance covered $l = \frac{1}{2}at^2$

Hence the body takes times t_1 and t_2 under accelerations a_1 and a_2 then,

$$a_1 t_1^2 = a_2 t_2^2$$

$$g(\sin\theta)t_1^2 = g(\sin\theta - \mu\cos\theta)t_2^2$$

$$\theta = 45° \text{ and } t_2 = 2t_1$$

Therefore, $1 - \mu = \frac{1}{4}$

$\Rightarrow$ $\mu = 0.75$

131. (a)

$$F = \mu(m_A + m_B)g = 0.15 \times 1.2 \times 9.8 = 1.764 \text{ N}$$

132. (a)

$$F = \mu_1 N_1 + \mu_2 N_2$$
$$= 0.25(200 \times 10) + (0.20)(500 \times 10)$$
$$= 1500 \text{ N}$$

133. (d)

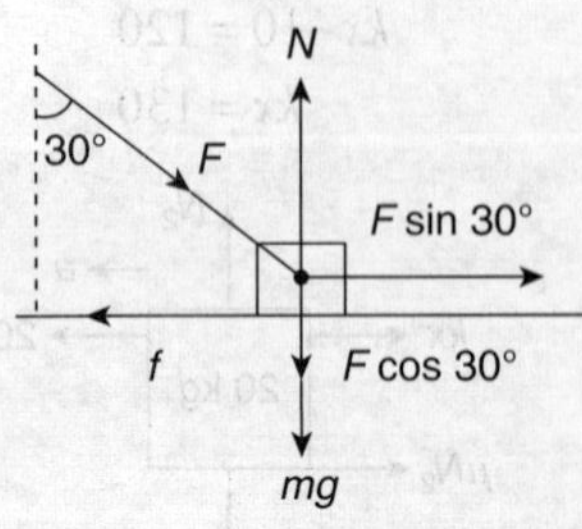

From figure

$$F\sin\theta = f = \mu N \quad \text{(i)}$$
$$F\cos\theta + mg = N \quad \text{(ii)}$$

From Eqs. (i) and (ii), we get

$$F\cos\theta + mg = \frac{F\sin\theta}{\mu}$$

Substituting the value, we get

$$F\cos 30° + \sqrt{3} \times 10 = \frac{F\sin 30°}{\frac{1}{2\sqrt{3}}}$$

$\Rightarrow$ $\frac{F}{2} + 10 = F$

$\Rightarrow$ $F = 20 \text{ N}$

134. (b)

Normal reaction N on the block is

$$N - F - mg\cos\theta = 0$$

$\Rightarrow$ $N = F + mg\cos\theta$ (i)

For equilibrium in the inclined plane

$$mg\sin\theta = \mu N \quad \text{(ii)}$$

$$\frac{mg}{\mu}\sin\theta = F + mg\cos\theta$$

$$F = \frac{mg}{\mu}[\sin\theta - \mu\cos\theta]$$

135. (c)

In extreme case the normal reaction N will shift to corner. For toppling, the torque due to friction should be greater than the resolving torque due to normal reaction.

$$f \times \frac{h}{2} > N \times \frac{a}{2}$$

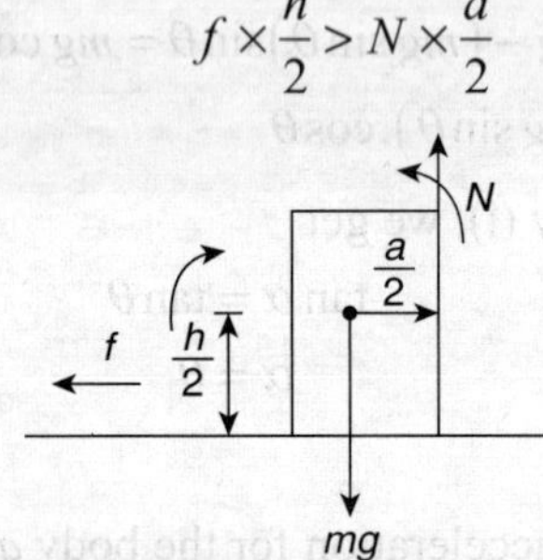

As $f = \mu N$

$$\mu > \frac{a}{h}$$

136. (d)

$$T - mg = m(2a), \; Mg - T = Ma$$

So, $a = \left(\frac{M - m}{M + 2m}\right)g$

$$T = M(g - a) = \frac{3Mm}{M + 2m}g$$

137. (c)

Lift is descending with a retardation a, so the lift is accelerating upward with acceleration a.

Effective acceleration due to gravity for the frame attached with lift is $(g + a)\sin\theta$.

138. (a)

When $F \geq F_c$ where F_c is critical (minimum) force to initiate sliding, then

$$a_A = \frac{F - \mu mg}{m}$$

$$a_B = \frac{\mu mg}{2m}$$

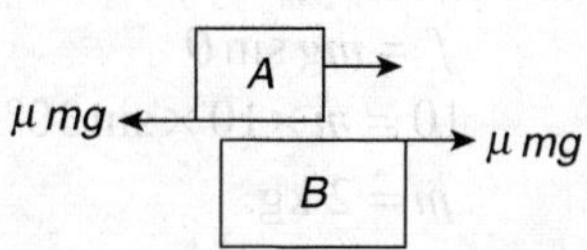

For A to slide over B, $a_A \geq a_B$

When sliding just begins $a_A = a_B$ gives

$$F = \frac{3}{2}\mu mg$$

139. (d)

When $F = 0$, $600 - 2\times10^5 t = 0$

$$\therefore t = \frac{600}{2\times10^5} = 3\times10^{-3}\,s$$

Now, impulse, $I = \int_0^t F dt = \int_0^t \left(600 - 2\times10^5 t\right) dt$

$$600t - 2\times10^5\frac{t^2}{2} = 600\times3\times10^{-3} - 10^5\times\left(3\times10^{-3}\right)^2$$

or, $I = 1.8 - 0.9 = 0.9$ N-s

140. (c)

$$\frac{1}{2}a_1t_1^2 = \frac{1}{2}a_2t_2^2$$

$$t_2 = nt_1 \quad a_1 = g\sin\theta$$

and $\quad a_2 = \left(g\sin\theta - \mu g\cos\theta\right)$

$$g\sin\theta = \left(g\sin\theta - \mu g\cos\theta\right)n^2$$

As $\quad \theta = 45°$

So, $\quad 1 - \mu = \frac{1}{n^2}$ or $\mu = 1 - \frac{1}{n^2}$

141. (d)

As there is no motion

$$T = m_B g = \mu\left(m_A + m_C\right)g$$

$$5 = 0.2(10 + m_C)$$

$$m_C = 15 \text{ kg}$$

142. (a)

Momentum is mv

$$m = 3.513 \text{ kg};\ v = 5.00 \text{ m/s}$$

$$mv = 17.57 \text{ ms}^{-1}$$

Because the values will be accurate up to second decimal place only, 17.565 = 17.57.

143. (d)

Acceleration of the system $a = \frac{F}{m + M}$

Force on block of mass $m = ma = \frac{mF}{(m + M)}$

144. (d)

Work done by hand = Potential energy of the ball

$$Fs = mgh$$

$$\Rightarrow \quad F = \frac{mgh}{s} = \frac{0.2\times10\times2}{0.2} = 20 \text{ N}.$$

145. (d)

Force × Time = Impulse = Change of momentum

$$\text{Force} = \frac{\text{Impulse}}{\text{Time}} = \frac{3}{0.1} = 30 \text{ N}.$$

146. (d)

Retardation due to friction = μg

$$v^2 = u^2 + 2as$$

$$0 = (100)^2 - 2(\mu g)s$$

or $\quad 2\mu gs = 100\times100$

$$s = \frac{100\times100}{2\times0.5\times10} = 1000 \text{ m}.$$

147. (b)

The incline is given an acceleration a. Acceleration of the block is to right. Pseudo acceleration a acts on the block to the left. Equate resolved parts of a and g along incline

$$ma\cos\alpha = mg\sin\alpha$$

or $\quad a = g\tan\alpha.$

148. (b)

$$F = -kx$$

or $\quad F = -15\times\left(\frac{20}{100}\right) = -3 \text{ N}$

Initial acceleration is overcome by retarding force.

or $\quad m\times(\text{acceleration } a) = 3$

$$a = \frac{3}{m} = \frac{0}{0.3}$$

$$= 10 \text{ m/ s}^2.$$

149. (b)

For first part of penetration, by equation of motion,

$$\left(\frac{u}{2}\right)^2 = (u)^2 - 2a(3)$$

or $$3u^2 = 24a \quad \ldots(i)$$

Further part of penetration

$$0 = \left(\frac{u}{2}\right)^2 - 2a$$

or $$u^2 = 8ax \quad \ldots(ii)$$

From (i) and (ii)

$$3 \times (8ax) = 24a$$

or $$x = 1 \text{ cm}.$$

150. (a)

For upper half smooth incline, component of g down the incline $= g \sin\phi$.

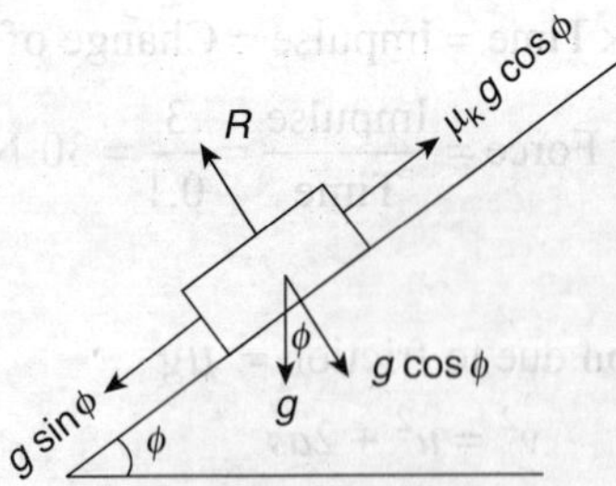

$$v^2 = 2(g \sin \phi)\ \frac{1}{2} \times l$$

For lower half rough incline, frictional retardation

$$= \mu_k g \cos\phi$$

$\therefore$ Resultant acceleration $= g\sin\phi - \mu_k g\cos\phi$

or $$0 = v^2 + 2(g\sin\phi - \mu_k g\cos\phi)\frac{1}{2} \times l$$

or $$0 = \sin\phi + \sin\phi - \mu_k \cos\phi$$

or $$\mu_k \cos\phi = 2\sin\phi$$

$$\mu_k = 2\tan\phi.$$

151. (c)

$$a = \frac{f}{m} - \frac{\mu(20)}{2} = 10\mu$$

From v–t graph,

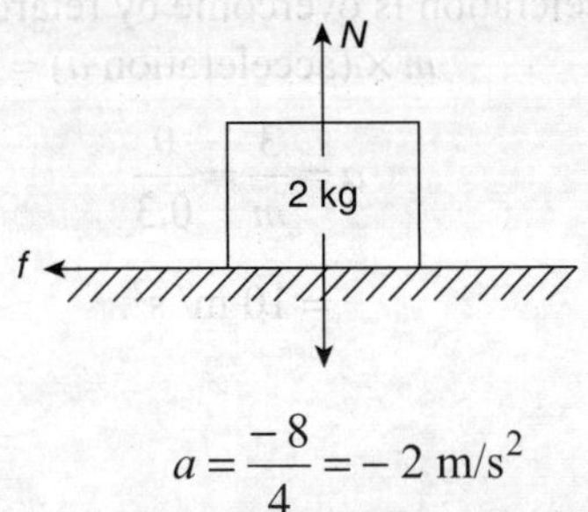

$$a = \frac{-8}{4} = -2 \text{ m/s}^2$$

$$10\mu = 2$$

$$\mu = \frac{2}{10} = 0.2$$

152. (b)

Centripetal force on particle $= mR\omega^2$

$$\frac{F_1}{F_2} = \frac{mR_1\omega^2}{mR_2\omega^2} = \frac{R_1}{R_2}$$

153. (a)

For equilibrium of block,

$$f = mg\sin\theta$$

$$10 = m \times 10 \times \sin 30°$$

or $$m = 2 \text{ kg}.$$

154. (a)

$$\frac{a}{g} = \frac{(m_1 - m_2)}{(m_1 + m_2)} = \frac{(5-4.8)}{(5+4.8)} = \frac{0.2}{9.8}$$

or $$a = g \times \frac{0.2}{9.8} = \frac{9.8 \times 0.2}{9.8} = 0.2 \text{ m/s}^2.$$

155. (c)

Upward acceleration, $ma = T_1 - mg$

$$T_1 = m(g+a)$$

Downward acceleration, $ma = mg - T_2$

or $$T_2 = m(g-a)$$

$$\frac{T_1}{T_2} = \frac{g+a}{g-a} = \frac{9.8+4.9}{9.8-4.9} = 3:1$$

156. (a)

Initial thrust = (Lift-off mass) × Acceleration

$$= (3.5 \times 10^4) \times 10 = 3.5 \times 10^5 \text{ N}.$$

157. (a)

Both the scales read M kg each.

158. (c)

Force ($\vec{F}$) $= 6\hat{i} - 8\hat{j} + 10\hat{k}$ and acceleration $(a) = 1$ m/s^2

$$\text{Mass } (m) = \frac{|\vec{F}|}{a} = \frac{|6\hat{i} - 8\hat{j} + 10\hat{k}|}{1}$$

$$= \sqrt{36 + 64 + 100} = \sqrt{200} = 10\sqrt{2} \text{ kg}$$

159. (c)

Frictional force provides the retarding force.

$$\mu\ mg = ma$$

or $$\mu = \frac{a}{g} = \frac{u/t}{g}$$

$$= \frac{6/10}{10} = 0.06.$$

160. (d)

Weight of the block is balanced by force of friction.

∴ Weight of the block = $\mu R = 0.2 \times 10 = 2\text{N}$.

161. (a)

When the lift is standing, $W_1 = mg$. When the lift descends with acceleration a, $W_2 = m(g-a)$

$$\frac{W_2}{W_1} = \frac{m(g-a)}{mg} = \frac{9.8-5}{9.8} = \frac{4.8}{9.8}$$

or $$W_2 = W_1 \times \frac{4.8}{9.8} = \frac{49 \times 4.8}{9.8} = 24 \text{ N}.$$

162. (d)

By triangle law of force, the particle will be in equilibrium under the three forces. Obviously, the resultant force on the particle will be zero. Consequently, the acceleration will be zero. Hence, the particle velocity remains unchanged at $\vec{v}$.

163. (b)

∵ Force = Mass × Acceleration

$$F - T_{AB} = ma$$

And $$T_{AB} - T_{BC} = ma$$

$$T_{BC} = F - 2ma$$

or $$T_{BC} = 10.2 - (2 \times 2 \times 0.6)$$

or $$T_{BC} = 7.8 \text{ N}.$$

164. (b)

$$\frac{a}{g} = \frac{(m_1 - m_2)}{(m_1 + m_2)}$$

$$\frac{1}{8} = \frac{(m_1 - m_2)}{(m_1 + m_2)}$$

or $$\frac{m_1}{m_2} = \frac{9}{7}.$$

165. (b)

$$T - 60g = 60a$$

or $$960 - (60 \times 10) = 60a$$

or $$60a = 360$$

or $$a = 6 \text{ ms}^{-2}.$$

166. (a)

F_2 and F_3 have resultant equivalent F_1

$$\text{Acceleration} = \frac{F_1}{m}.$$

167. (c)

For observer in the lift, acceleration = $(g - a)$,

For observer standing outside, acceleration = g.

168. (b)

For no skidding along curved track,

$$v = \sqrt{\mu R g}$$

$$v = \sqrt{0.6 \times 150 \times 10} = 30 \text{ m/s}$$

169. (c) $F = ma$

$$\Rightarrow \quad a = \frac{F}{m} = \frac{5 \times 10^4}{3 \times 10^7} = \frac{5}{3} \times 10^{-3} \text{ ms}^{-2}$$

Also $\quad v^2 - u^2 = 2as$

$$\Rightarrow \quad v^2 - 0^2 = 2 \times \frac{5}{3} \times 10^{-3} \text{ x } 3 = 10^{-2}$$

$$\Rightarrow \quad v = 0.1 \text{ ms}^{-1}$$

170. (b) By equilibrium of mass m, $T' + mg$ ...(i)

By equilibrium of mass $2m$, $T' + 2mg = T$ ...(ii)

From (i) and (ii), $T = 2mg + mg = 3mg$ ...(iii)

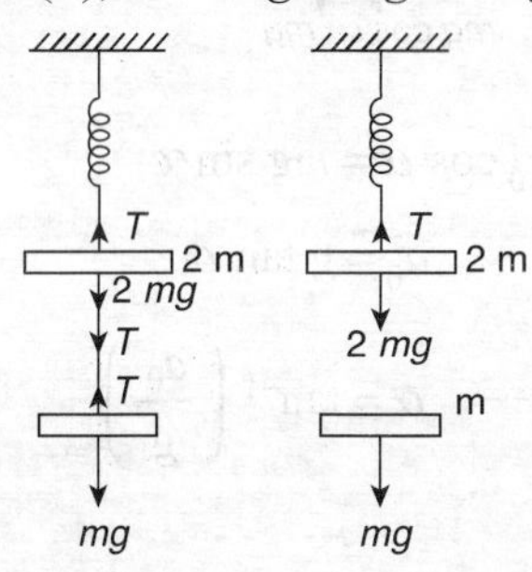

When the string is cut:

For mass m

$$F_{net} = ma_m \Rightarrow mg = ma_n \Rightarrow a_m = g$$

For mass $2m$:

$$T - 2\,mg = 2ma$$

$$3\,mg - 2\,mg = 2ma$$

$$\Rightarrow \quad a = \frac{g}{2} \text{ upward}$$

171. (b) The acceleration of mass m is due to the force $T\cos\theta$

$$T\cos\theta = ma$$

$$\Rightarrow \quad a = \frac{T\cos\theta}{m} \quad \text{...(i)}$$

Also, $\quad F = 2T\sin\theta$

$$\Rightarrow \quad T = \frac{F}{2\sin\theta} \quad \text{...(ii)}$$

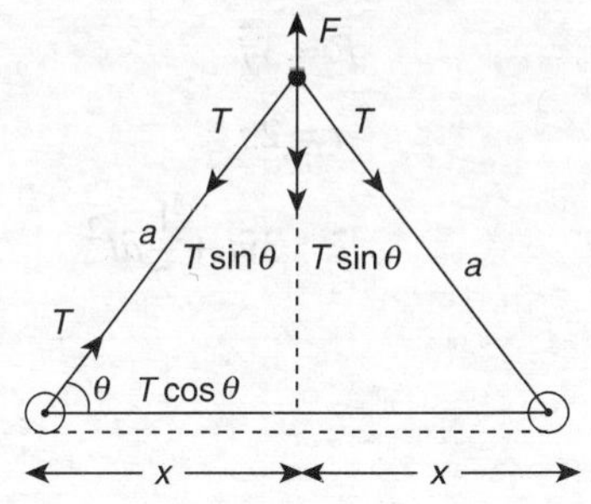

From (i) and (ii)

$$a=\left(\frac{F}{2\sin\theta}\right)\frac{\cos\theta}{m}=\frac{F}{2m\tan\theta}=\frac{F}{2m}\frac{x}{\sqrt{a^2-x^2}}$$

$$\left[\because \tan\theta=\frac{\sqrt{a^2-x^2}}{x}\right]$$

172. (a)

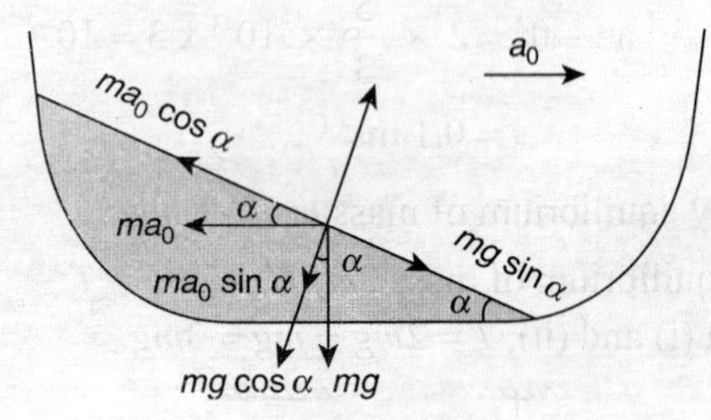

$$ma_0\cos\alpha = mg\sin\alpha$$

$$\Rightarrow \quad a_0 = g\tan\alpha$$

$$\Rightarrow \quad \alpha=\tan^{-1}\left(\frac{a_0}{g}\right)$$

173. (a)

$$A = g\tan\alpha$$

Since, $\quad \sin\alpha=\frac{1}{x}$

$$\Rightarrow \quad \tan\alpha=\frac{1}{\sqrt{x^2-1}}$$

$$a=\frac{g}{\sqrt{x^2-1}} \qquad [\text{using}(1)]$$

And $\quad A=\frac{mg}{cos\alpha}=\frac{mgx}{\sqrt{x^2-1}}$

174. (b)

$$m = 2\text{ kg}$$

since $\quad \vec{v}=2\hat{i}$

$$\vec{F}=3\hat{j}$$

$$t=2s$$

$$\vec{s}=3\vec{v}t+\frac{1}{2}\vec{a}t^2$$

$$\Rightarrow \vec{s}=\left(2\hat{i}\right)2+\frac{1}{2}\left(\frac{3}{2}\right)(4)\hat{j}$$

$$\Rightarrow \vec{s}=4\hat{i}+3\hat{j}$$

$$\Rightarrow |\vec{s}|=\sqrt{4^2+3^2}=5\text{ m}$$

175. (c)

$$\text{Thrust}=2T=\left(\frac{4mM}{M+m}\right)g$$

$$\Rightarrow \quad \frac{4mg}{1+\frac{m}{M}}$$

Since, $\quad \frac{m}{M}\rightarrow 0$

$\Rightarrow$ Thrust $\approx$ 4 mg

176. (d)

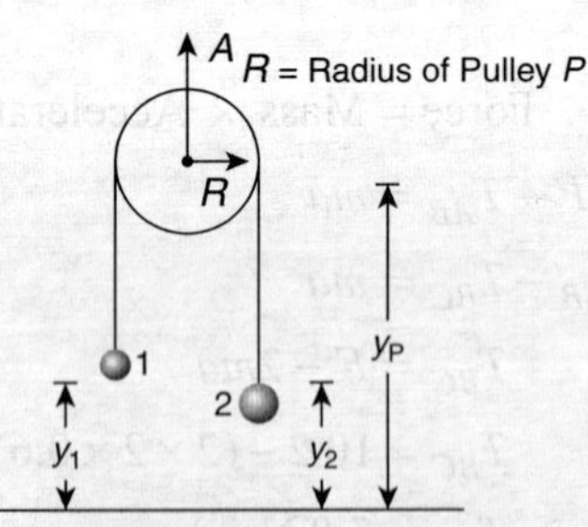

Since length of the string is constant,

$$Y_p-\ddot{y}_1+\pi R+(Y_p-y_2)=\text{Constant}$$

Take derivative of both sides twice w.r.t time t, we get

$$\Rightarrow \quad 2\,Y_p=\ddot{y}_1+\ddot{y}_2$$

$$\Rightarrow \quad A=\frac{a_1+a_2}{2}$$

177. (c)

$$a=\frac{F}{m_1+m_2+m_3}$$

For equilibrium $m_1a=m_2\,g$

$$\Rightarrow m_1\left(\frac{F}{m_1+m_2+m_3}\right)=m_2g$$

$$\Rightarrow F\frac{m_2g}{m_1}(m_1+m_2+m_3)$$

178. (d)

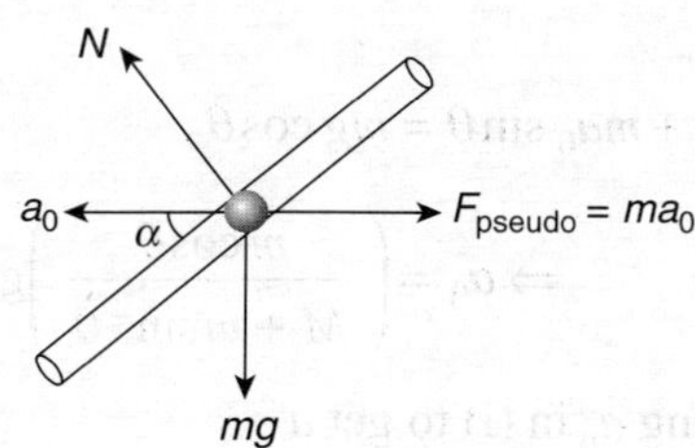

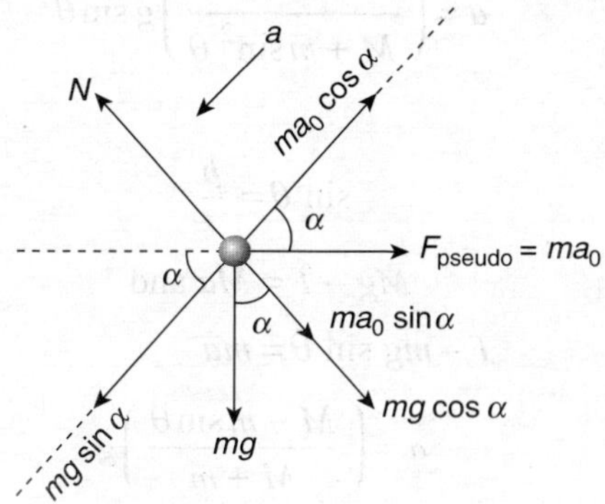

$mg \sin\alpha - ma_0 \cos\alpha = ma$

$$mg \sin\alpha - m_0 \cos\alpha = ma$$

$$\Rightarrow a = g \sin\alpha - a_0 \cos\alpha$$

179. (d)

$$T \cos\theta = mg \cos\alpha \text{ and} \quad \ldots\text{(i)}$$

$$T \sin\theta = ma_1 + mg \sin\alpha \quad \ldots\text{(ii)}$$

$$\Rightarrow \tan\theta = \frac{a_0 + g \sin\alpha}{g \cos\alpha}$$

180. (c)

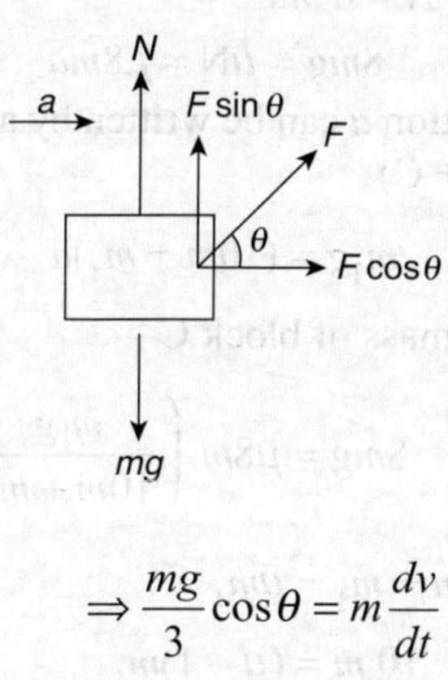

$$\Rightarrow \frac{mg}{3} \cos\theta = m\frac{dv}{dt}$$

$$\Rightarrow \frac{mg}{3} \cos(ks) = m\frac{dv}{ds}\frac{ds}{dt}$$

$$\Rightarrow \frac{mg}{3} \cos(ks) = mv\frac{dv}{ds}$$

$$\Rightarrow vdv = \frac{g}{3} \cos(ks)\, ds$$

$$\Rightarrow \frac{v^2}{2} = \frac{g}{3K} \sin(k\beta)$$

$$\Rightarrow v = \sqrt{\frac{2g}{3K}} \sin\theta$$

181. (c)

$$\text{Force} = \frac{d}{dt}(\text{momentum})$$

$$= \frac{d}{dt}(mv) = v\left(\frac{dm}{dt}\right) \Rightarrow 210 = 300\left(\frac{dm}{dt}\right)$$

$$\frac{dm}{dt} = \text{Rate of combustion} = \frac{210}{300} = 0.7 \text{ kg/s}$$

182. (c)

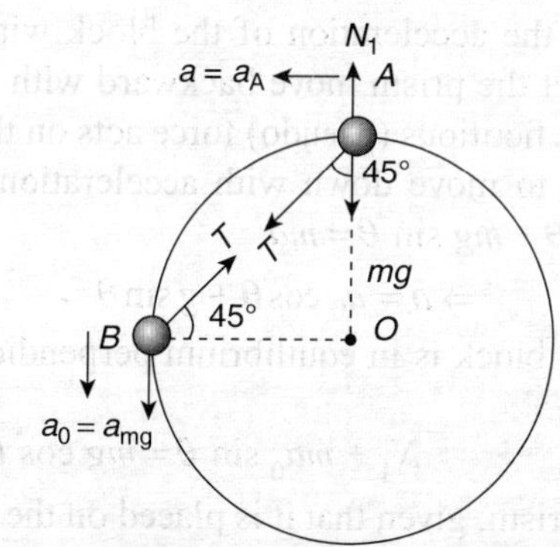

$$T \sin 45° = ma \text{ (For A)} \quad \ldots\text{(i)}$$

$$mg - T \sin 45° = ma \text{ (For B)} \quad \ldots\text{(ii)}$$

$$T = \frac{mg}{2 \sin 45} = \frac{mg}{\sqrt{2}}$$

183. (b)

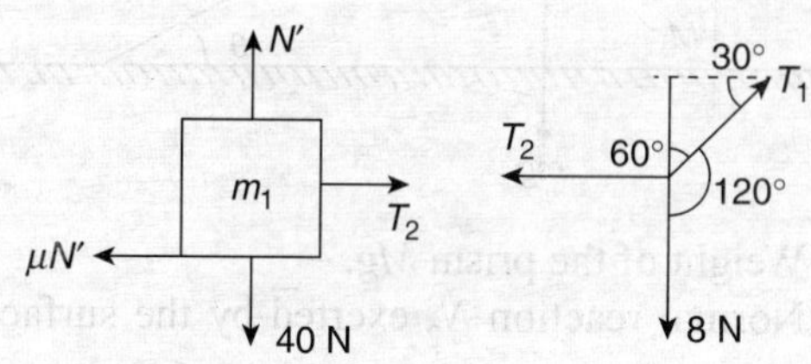

FBD of blocks are

$$N' = 40 \text{ N}$$

$$T_2 = 40\mu$$

$$\frac{8}{\sin(90° + 60°)} = \frac{T_1}{\sin 90°} = \frac{T_2}{\sin 120°}$$

$$16 = T_1 = \frac{2T_2}{\sqrt{3}}$$

$$\frac{2T_2}{\sqrt{3}} = 16$$

$$40\mu = \frac{\sqrt{3}\times 16}{2}$$

$$\mu = \frac{\sqrt{3}}{5} = 0.35$$

184. (c)

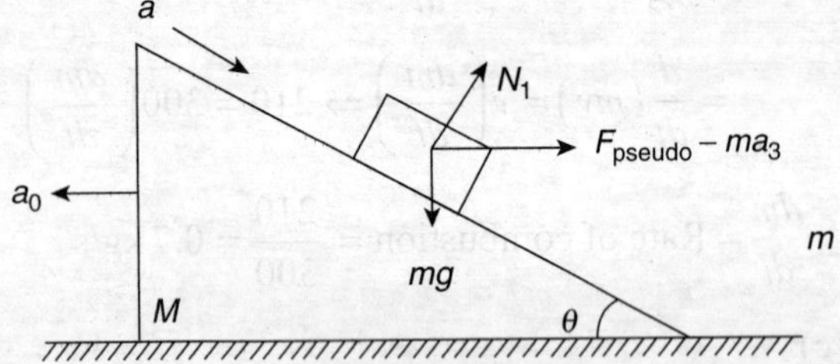

Let a be the acceleration of the block with respect to prism. Let the prism move backward with acceleration a_0; then a fictitious (pseudo) force acts on the mass. For the mass to move down with acceleration a, we have $ma_0 \cos\theta + mg \sin\theta = ma$

$$\Rightarrow a = a_0 \cos\theta + g\sin\theta \quad \text{...(i)}$$

Also, the block is in equilibrium perpendicular to then incline

i.e. $N_1 + ma_0 \sin\theta = mg\cos\theta$ ...(ii)

For the prism, given that it is placed on the ground with no fictitious force acting on it, the following forces are to be considered.

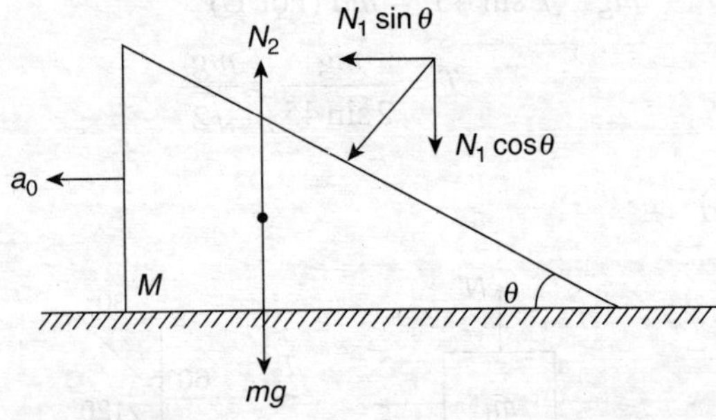

(a) Weight of the prism Mg.
(b) Normal reaction N_2 exerted by the surface on the prism.
(c) Normal reaction N_1 exerted by the block placed on the prism.

$N_1 \sin\theta$ is the component responsible for horizontal motion of the prism. So,

$$N_1 \sin\theta = Ma_0$$

$$\Rightarrow N_1 = \frac{Ma_0}{\sin\theta} \quad \text{...(iii)}$$

Substituting (iii) in (ii), we get

$$\frac{Ma_0}{\sin\theta} + ma_0\sin\theta = mg\cos\theta$$

$$\Rightarrow a_0 = \left(\frac{m\cos\theta}{M + m\sin^2\theta}\right) g\sin\theta \quad \text{...(iv)}$$

Substituting a_0 in (i) to get a

$$a = \left(\frac{M+m}{M + m\sin^2\theta}\right) g\sin\theta$$

185. (d)

$$\sin\theta = \frac{h}{\ell}$$

$$Mg - T = Ma \text{ and}$$

$$\Rightarrow \quad T - mg\sin\theta = ma$$

$$\Rightarrow \quad a = \left(\frac{M - m\sin\theta}{M+m}\right) g$$

Since $v^2 - 0^2 = 2as$, where s is the distance travelled by m before M is detached. After detaching M, m will not move the remaining distance of $(\ell - s)$ up the incline, under a retardation $g\sin\theta$.

$$\Rightarrow \quad 0^2 = v^2 - 2(g\sin\theta)(\ell - s)$$

$$\Rightarrow \quad 0^2 = 2as - 2g\sin\theta(\ell - s)$$

$$\Rightarrow \quad s = \frac{M+m}{M}\left(\frac{h\ell}{h+\ell}\right)$$

186. (d)

If the acceleration of 'C' is a

For block 'A' $N = 8\,ma$ ...(1)

$$8mg = \mu N = \mu 8ma \quad \text{...(2)}$$

and acceleration a can be written by an equation of system (A + B + C)

$$m_1 g = (10m + m_1)a$$

where m_1 is mass of block C

$$8mg = \mu 8m\left(\frac{m_1 g}{10m + m_1}\right)$$

$$10m + m_1 = \mu m_1$$

$$10\,m = (\mu - 1)m_1$$

$$\Rightarrow \quad m_1 = \frac{10m}{\mu - 1}$$

187. (b)

Maximum frictional force is $= \mu N = 0.6\times 1\times 10 = 6$ N

The truck is accelerating at 5 m/s^2.

Pseudo force = ma

$$= 5 \times 1 = 5\text{ N}$$

This force is less than max force of friction, so the friction force is 5 N.

188. (d)

Let the value of a be increased from zero. As long as $a \le \mu g$, there shall be no relative between m_1 and m_2 and platform, that is, m_1 and m_2 shall move with acceleration a.

As $a < \mu g$ the acceleration of m_1 and m_2 shall become μg each.

Hence at all instants, the velocity of m_1 and m_2 shall be same

$\therefore$ The spring shall always remain at its natural length

189. (b)

$$\tan\theta \le \mu$$

$$\tan\theta = \frac{dy}{dx} = \frac{x}{a} = \frac{\sqrt{2ay}}{a} = \sqrt{\frac{y}{a}}.2$$

$$\therefore \sqrt{\frac{2y}{a}} \le \mu \text{ or } y \le \frac{a\mu^2}{2}$$

190. (b)

The force equations are

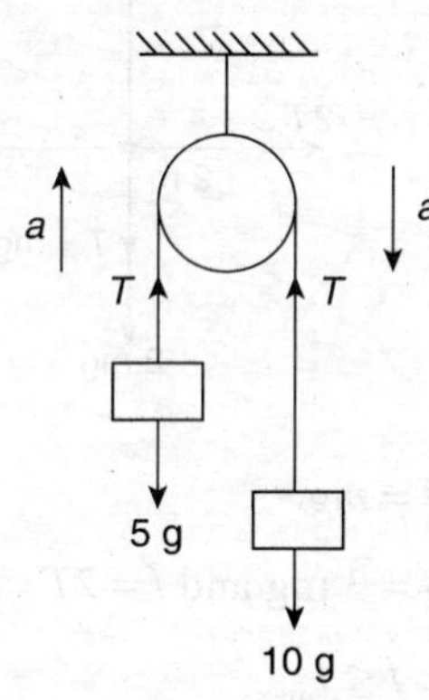

$$T - 5g = 5a,$$

$$10g - T = 10a$$

Adding, $\quad 10g - 5g = 15a$

or $\quad a = \frac{5g}{15} = \frac{g}{3}$

191. (c)

Block Q is moving due to force while block P, due to friction.

Friction direction on both the blocks are shown.

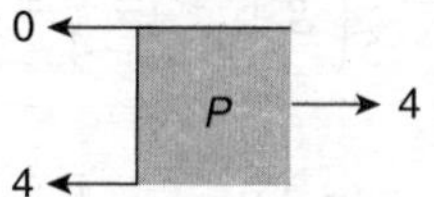

First block 'Q' will move and P will move with Q; so by FBD, taking P and Q as system

$$F - 9 = 0$$

$$\Rightarrow \quad F = 9\text{ N}$$

When applied force is 4 N as per the FBD

Q → 0

0 ← P → 4, 4 ←

4 kg block is moving due to friction and maximum friction force is 8 N.

So acceleration $= \frac{8}{4} = 2\text{ m/s}^2 = a_{max}$

Slipping will start when Q has +ve acceleration equal to maximum acceleration of P i.e. 2m/s^2.

$$F - 17 = 5 \times 2 \Rightarrow F = 27\text{ N}$$

192. (a)

(i) The force acting on the block along the incline to shift the block downwards

$$= mg\sin\theta = 2 \times 9.8 \times \sin 30° = 9.8\text{ N}$$

The maximum frictional force

$$= \mu_s mg\cos\theta = 0.7 \times 2 \times 9.8 \times \frac{\sqrt{3}}{2} = 11.8\text{ N}$$

Therefore, the frictional force is 9.8 N.

193. (a)

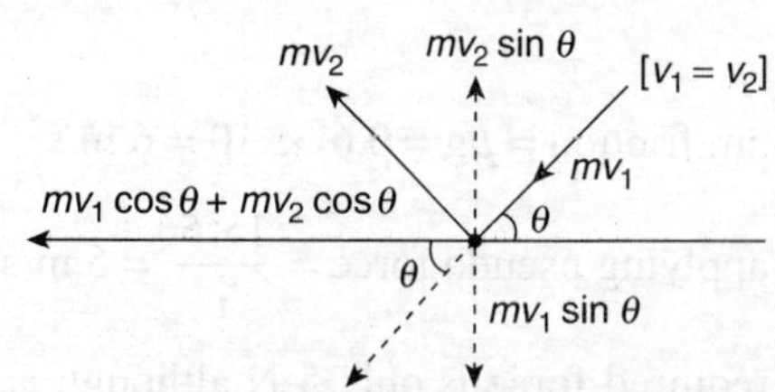

Change in momentum $= 2 \times 3 \times 10 \times \sin 60° = 60 \times \frac{\sqrt{3}}{2}$

Force = Change in momentum/Impact time

$$= \frac{30\sqrt{3}}{0.2} = 150\sqrt{3}\text{ N}$$

194. (c)

FBD in reference frame of the lift

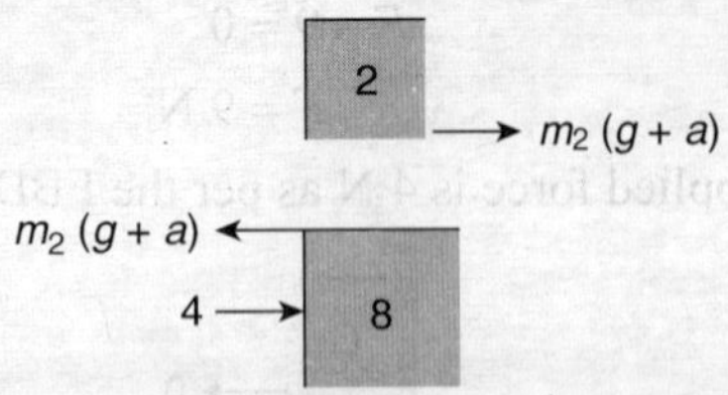

$$a_2 = \frac{1}{5}\left(g + \frac{g}{4}\right) = \frac{g}{4} = 2.5\text{ m/s}^2$$

$$a_8 = \frac{30 - \left[\mu.2\left(g + \frac{g}{4}\right)\right]}{8} = \frac{30 - \left[\frac{1}{5} \times 2 \times \frac{50}{4}\right]}{8} = \frac{25}{8}\text{ m/s}^2$$

195. (b)

The magnitude of the frictional force has to balance the weight 0.98 N acting downwards.

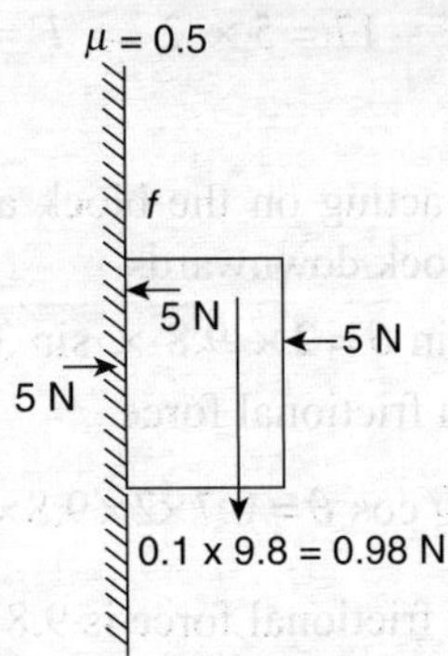

Therefore the frictional force = 0.98 N.

196. (a)

Maximum friction $= \mu g = 0.6 \times 10 = 6\text{ m/s}^2$

But by applying pseudo force $= \frac{1 \times 5}{1} = 5\text{ m/s}^2$

So the required force is only 5 N although *ma* friction available is 6 N.

197. (a)

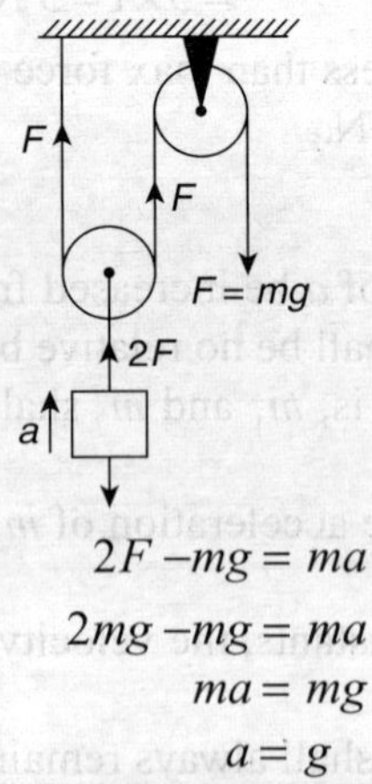

$$2F - mg = ma$$
$$2mg - mg = ma$$
$$ma = mg$$
$$a = g$$

198. (a)

For the chain to move with constant speed, *P* needs to be equal to frictional force on the chain as the length of chain on the rough surface increases. Hence the friction force,

$$f_k = \mu_k\text{N increases.}$$

199. (c)

In equilibrium,

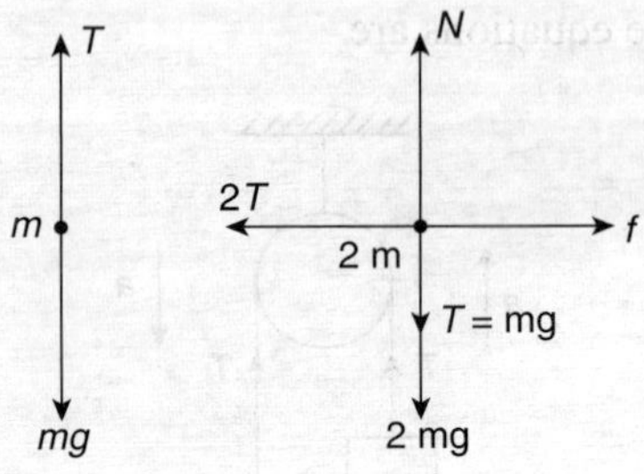

$$T = mg$$

$$N = 3\text{ mg and } f = 2T = 2\text{ mg}$$

In limiting case $f < f_{max}$

$$\Rightarrow \quad 2\text{ mg} \le \mu N \Rightarrow 2\text{ mg} \le 3\mu\text{mg}$$

$$\Rightarrow \quad \mu \ge \frac{2}{3}$$

200. (a)

The two forces acting at the insect are *mg* and *N*. Let us resolve *mg* into two components.

$mg\cos\alpha$ is balanced by *N*

$mg\sin\alpha$ is balanced by the frictional force.

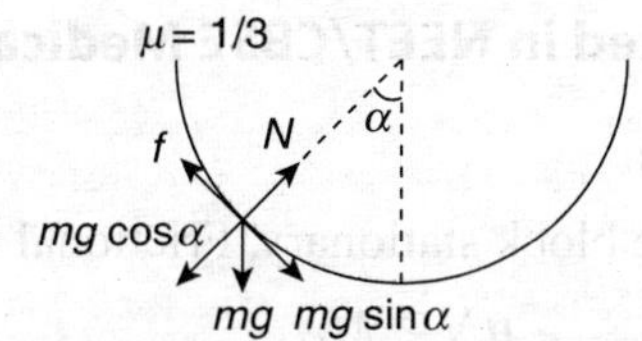

$$N = mg\cos\alpha \ s$$

$$f = mg\sin\alpha \text{ but } f = \mu N = \mu mg\cos\alpha$$

$$\mu mg\cos\alpha = mg\sin\alpha$$

$$\Rightarrow \quad \cot\alpha = \frac{1}{\mu} \Rightarrow \cot\alpha = 3$$

201. (d)

Limiting friction between A and B = 6 N
Limiting friction between B and C = 90 N
Limiting friction between C and ground = 50 N
Since limiting friction is least between C and ground, slipping will occur at first between C and ground. This will occur when $F = 50$ N.

202. (b)

At the instant 3 m is about to slip, the tension in all the strings are as shown

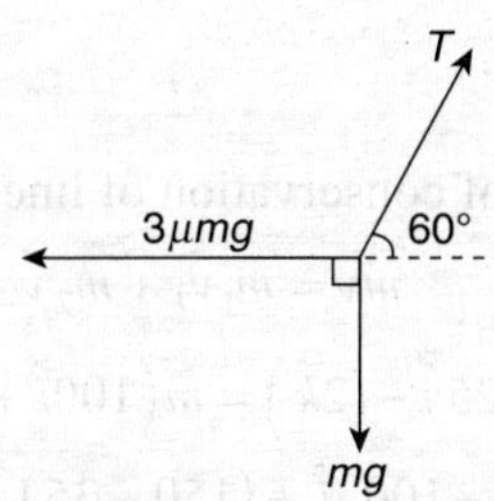

$$\therefore \mu mg = T\cos 60° \quad \ldots(i)$$

$$\text{and } mg = T\sin 60° \quad \ldots(ii)$$

$$\mu = \frac{1}{3\sqrt{3}}$$

203. (a)

The forces acting on the masses are shown.
Applying Newton's second law on mass Q, we get

$$F - f = ma \quad \ldots(i)$$

where a is acceleration at the extreme position.
Now, applying Newton's second law on mass P.

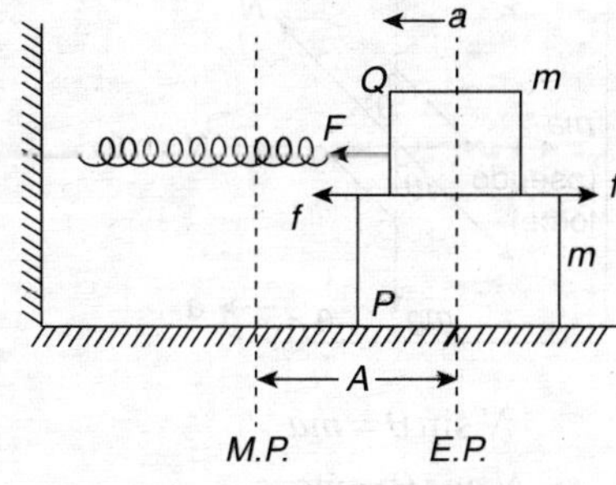

$$f = ma \quad \ldots(ii)$$

[acceleration is same as no slipping occurs between Q and P]

From (i) and (ii)

$$F = 2ma$$

$$\Rightarrow a = \frac{F}{2m} = \frac{KA}{2m} \quad [\because F = KA]$$

Substituting this value of a in Eq. (ii), we get

$$f = m \times \frac{kA}{2m} = \frac{kA}{2}$$

204. (a)

Weight of the disc will be balanced by the force applied by the bullet on the disc in vertically upward direction.

$$F = nmv = 40 \times 0.05 \times 6 = Mg$$

$$\Rightarrow M = \frac{40 \times 0.05 \times 6}{10} = 1.2 \text{ kg}$$

205. (b)

Here, the tension in the cord is given by

$$T = mg + ma$$

Here upwads acceleration = a,
Mass of sphere = m,

$$T = 4 \text{ mg}$$

$$\Rightarrow 4mg = mg + ma$$

$$\Rightarrow 3mg = mg \Rightarrow a = 3g$$

206. (c)

$$v = u + \frac{F}{m}t = 10 + \left(\frac{1000 - 500}{1000}\right) \times 10 = 15 \text{ m/s}$$

207. (d)

208. (a)

$$S_{\text{Horizontal}} = ut = 1.5 \times 4 = 6 \text{ m}$$

$$S_{\text{Vertical}} = \frac{1}{2}at^2 = \frac{1}{2}\frac{F}{m}t^2 = \frac{1}{2} 1 \times 16 = 8 \text{ m}$$

$$S_{\text{Net}} = \sqrt{6^2 + 8^2} = 10 \text{ m}$$

209. (b)

$$\frac{mg}{m(g-a)} = \frac{3}{2} \Rightarrow a = g/3$$

210. (c)

Initially due to upward acceleration, apparent weight of the body increases but then it decreases due to decrease in gravity.

211. (a)

$$T\cos\theta = mg$$
$$T\sin\theta = ma$$
$$\tan\theta = \frac{a}{g}$$
$$\therefore \theta = \tan^{-1}(a/g)$$

212. (a)

Apply conservation of linear momentum.
Total momentum before explosion = Total momentum after explosion

$$0 = \frac{m}{5}v_1\hat{i} + \frac{m}{5}v_2\hat{j} + \frac{3m}{5}\vec{v}_3$$
$$\frac{3m}{5}\vec{v}_3 = -\frac{m}{5}[v_1\hat{i} + v_2\hat{j}]$$
$$\vec{v}_3 = \frac{-v_1}{3}\hat{i} - \frac{v_2}{3}\hat{j}$$
$$\therefore v_1 = v_2 = 30 \text{ m/s}$$
$$\vec{v}_3 = -10\hat{i} - 10\hat{j}; v_3 = 10\sqrt{2} \text{ m/s}$$

213. (a)

Impulse = Change in momentum
$= 2\,mu = 2\times 0.15\times 10 = 3.0$ Ns

214. (a)

215. (a)

$$m_1 = 1000kg.v_1 = 50m/s, m_2 = 1250kg, v_2 = ?$$

From conservation law of linear momentum

$$m_1v_1 = m_2v_2$$
$$\Rightarrow 1000\times 50 = 1250\times v_2 \Rightarrow v_2 = 40 \text{ m/s}$$

216. (d)

According to law of conservation of momentum, the third piece has momentum $\frac{5}{4}$ m

Impulse = Average force × Time

$$\Rightarrow \text{Average force} = \frac{\text{Impluse}}{\text{time}}$$
$$\frac{\text{Change in momentum}}{\text{time}} = \frac{-(3\hat{i}+4\hat{j})\text{kg ms}^{-1}}{10^{-4}\text{s}}$$
$$= -(3\hat{i}+4\hat{j})\times 10^4 \text{N}$$

217. (c)

$$\text{Acceleration } a = \frac{1}{m}\left(\frac{-dm}{dt}\right)v_r = \frac{1}{1}\left(\frac{1}{60}\right)\times 2400 = 40 \text{ ms}^{-2}$$

218. (c)

Questions Asked in NEET/CBSE Medicals

219. (d)

To keep the block stationary, Frictional force ≥ Weight

$$\mu N \geq Mg$$

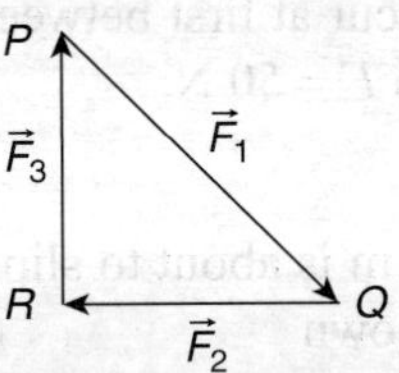

Here, $N = M\omega^2 r$

$r = 1$ m, $\mu = 0.1$

For minimum $\mu M\omega^2 r = Mg$

$$\omega = \sqrt{\frac{g}{\mu r}} = \sqrt{\frac{10}{0.1\times 1}} = 10 \text{ rad s}^{-1}$$

220. (d)

As per triangle law, $\vec{F}_1 + \vec{F}_2 + \vec{F}_3 = 0$ i.e., net force on the particle is zero. So, acceleration is also zero. Hence velocity of the particle will remain constant.

221. (a)

From the law of conservation of linear momentum

$$m\vec{v} = m_1\vec{v}_1 + m_2\vec{v}_2$$
$$\Rightarrow 6m(20\hat{i} + 25\hat{j} - 12\hat{k}) = m(100\hat{i} + 35\hat{j} + 8\hat{k}) + 5m\vec{v}_2$$
$$\Rightarrow 5\vec{v}_2 = (120-100)\hat{i} + (150-35)\hat{j} + (-72-8)\hat{k}$$
$$\Rightarrow 5\vec{v}_2 = 20\hat{i} + 115\hat{j} - 80\hat{k}$$
$$\Rightarrow \vec{v}_2 = 4\hat{i} + 23\hat{j} - 16\hat{k}$$

222. (d)

Coefficient of sliding friction has no dimension.

$$f = \mu_s N \Rightarrow \mu_s = \frac{f}{N}$$

223. (d)

In non-inertial frame,

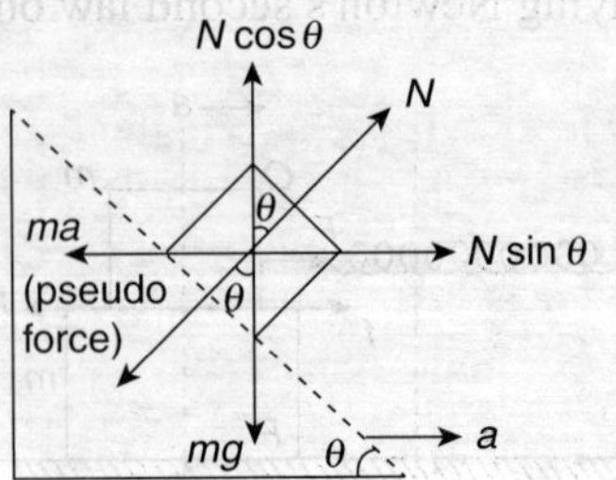

$$N\sin\theta = ma \quad \text{(i)}$$
$$N\cos\theta = mg \quad \text{(ii)}$$

From (i) and (ii)

$$\tan \theta = \frac{a}{g}$$

$$\Rightarrow a = g \tan\theta$$

224. (a)

Before the string is cut

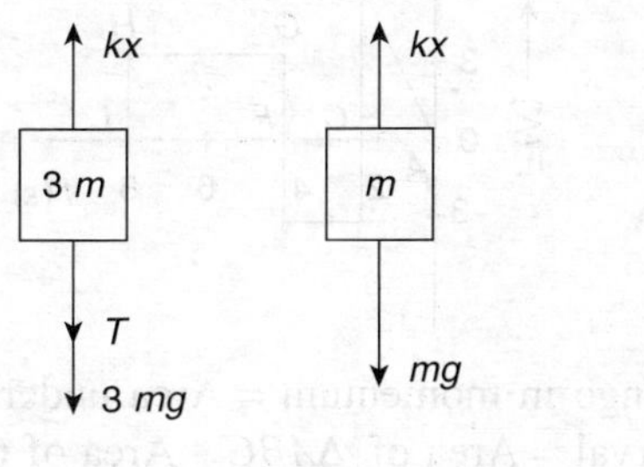

$$kx = T + 3 \text{ mg} \quad \text{(i)}$$

$$T = mg \quad \text{(ii)}$$

From Eqs. (i) and (ii)

$$Kx = 4 \text{ mg}$$

Just after the string is cut,

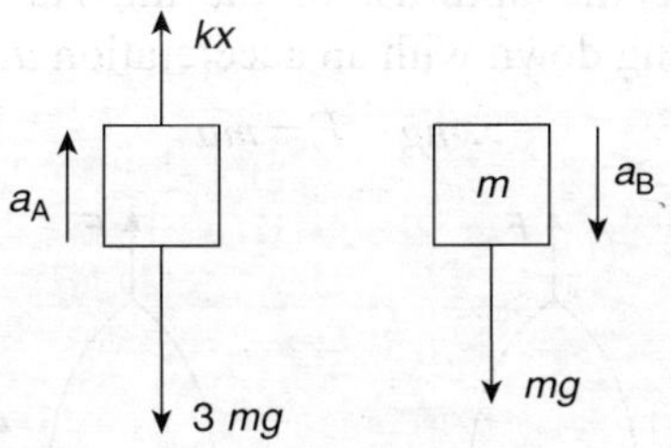

After the string is cut

$$T = 0$$

$$a_A = \frac{kx - 3mg}{3m}$$

$$a_A = \frac{4mg - 3mg}{3m}$$

$$= \frac{mg}{3m} = \frac{g}{3}$$

and also $a_B = g$

225. (d)

Centripetal force $\left(\frac{mv^2}{l}\right)$ is provided by tension. So, net force on the particle will be equal to tension T.

226. (a)

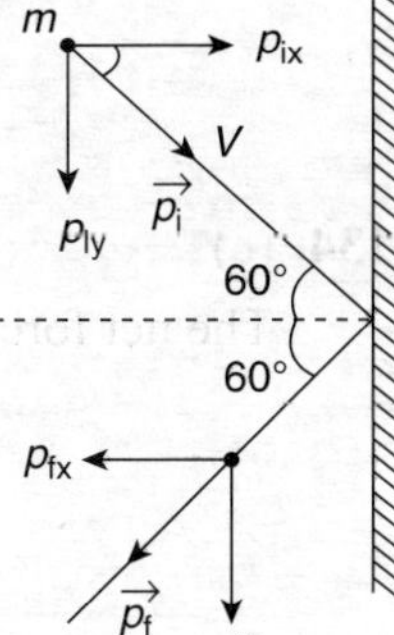

Given, $p_i = p_f = mV$

$$p_i x = mV \cos 60°$$

$$p_i y = mV \sin\theta$$

$$p_f x = -mV \cos 60°$$

$$p_f y = -mV \sin 60°$$

$$p_f y - p_i y = 0 \text{ (No change in momentum)}$$

$$mV \cos 60° - (-mV \cos 60°) = 2mV \cos 60°$$

$\because$ Impulse imparted by the well = Change in the momentum of the ball = mV.

227. (d)

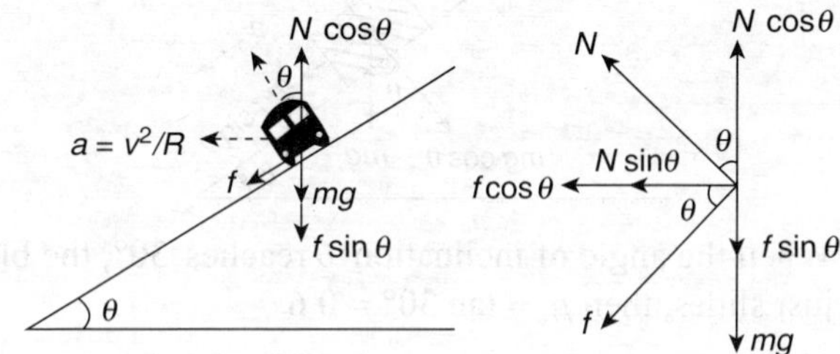

For vertical equilibrium on the road,

$$N \cos\theta = mg + f \sin\theta$$

$$mg = N \cos\theta - f \sin\theta \quad \text{(i)}$$

Centripetal force for safe turning,

$$N \sin\theta + f \cos\theta = \frac{mv^2}{R} \quad \text{(ii)}$$

From Eqs. (i) and (ii), we get

$$\frac{v^2}{Rg} = \frac{N \sin\theta + f \cos\theta}{N \cos\theta - f \sin\theta}$$

$$\Rightarrow \frac{v_{\max}{}^2}{Rg} = \frac{N \sin\theta + \mu_s N \cos\theta}{N \cos\theta - \mu_s N \sin\theta}$$

$$v_{\max} = \sqrt{gR \frac{\mu_s + \tan\theta}{1 - \mu_s \tan\theta}}$$

228. (c)

Let v be tangential speed of heavier stone. Then, centripetal force experienced by lighter stone is

$$(F_c)_{\text{lighter}} = \frac{m(nv)^2}{r}$$

and that of heavier stone is

$$(F_c)_{\text{heavier}} = \frac{2mv^2}{(r/2)}$$

But $(F_c)_{\text{lighter}} = (F_c)_{\text{heavier}}$ (given)

$$\therefore \frac{m(nv)^2}{r} = \frac{2mv^2}{(r/2)}$$

$$n^2\left(\frac{mv^2}{r}\right) = 4\left(\frac{mv^2}{r}\right)$$

$$n^2 = 4 \text{ or } n = 2$$

229. (d)

Let μ_s and μ_k be the coefficients of static and kinetic friction between the box and the plank respectively.

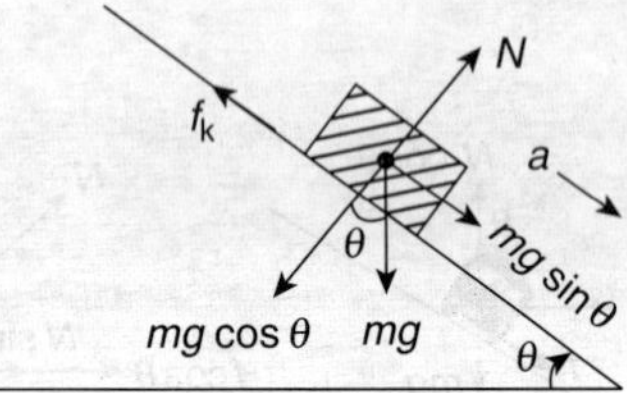

When the angle of inclination θ reaches 30°, the block just slides, then $\mu_s = \tan 30° = 0.6$

Acceleration a of the box $= \frac{2s}{t^2} = \frac{2\times4}{(4)^2} = \frac{1}{2}$ m/s^2

$mg \sin\theta = \mu_k\, mg \cos\theta = ma$

$g(\sin\theta - \mu_k \cos\theta) = a$

Solve to get $\mu_k = 0.5$

230. (b)

Impulse = Change in momentum

$$F.\Delta t = m.v; F = \frac{m.v}{\Delta t} = \frac{150\times10^{-3}\times20}{0.1} = 30 \text{ N}$$

231. (c)

Force of friction on mass $m_2 = \mu m_2 g$

Force of friction on mass $m_3 = \mu m_3 g$

Let a be common acceleration of the system.

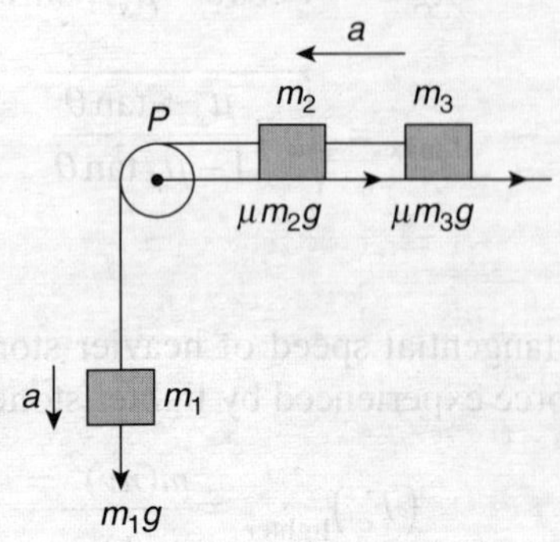

$$\therefore a = \frac{m_1 g - \mu m_2 g - \mu m_3 g}{m_1 + m_2 + m_3}$$

Here, $m_1 = m_2 = m_3 = m$

$$\therefore a = \frac{mg - \mu mg - \mu mg}{m + m + m} = \frac{mg - 2\mu mg}{3m} = \frac{g(1-2\mu)}{3}$$

232. (c)

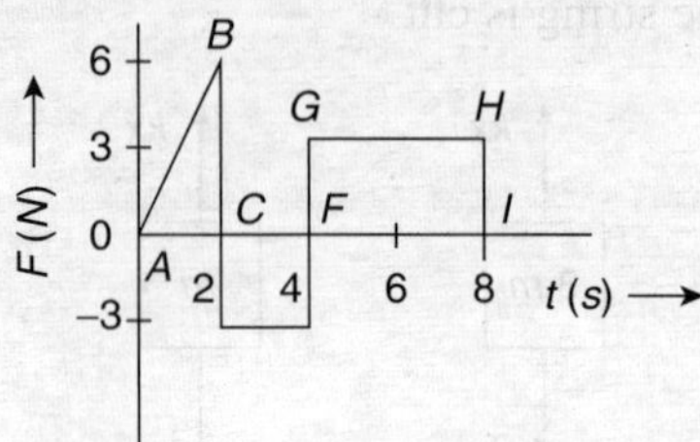

Change in momentum = Area under F-t graph in that interval = Area of ΔABC – Area of rectangle CDEF + Area of rectangle FGHI

$$= \frac{1}{2}\times2\times6 - 3\times2 + 4\times3 = 12 \text{ N s}$$

233. (a)

Let F be the upthrust of the air. As the balloon is descending down with an acceleration a,

$$\therefore mg - F = ma \quad \text{(i)}$$

Let mass m_0 be removed from the balloon so that it starts moving up with an acceleration a. Then,

$$F - (m - m_0)g = (m - m_0)a$$

$$F - mg + m_0 g = ma - m_0 a \quad \text{(ii)}$$

Adding Eqs. (i) and (ii), we get

$$m_0 g = 2ma - m_0 a; m_0 g + m_0 a = 2ma$$

$$m_0(g + a) = 2ma$$

$$m_0 = \frac{2ma}{a+g}$$

234. (c)

The net force is zero.

235. (d)

The situation is as shown in the figure.

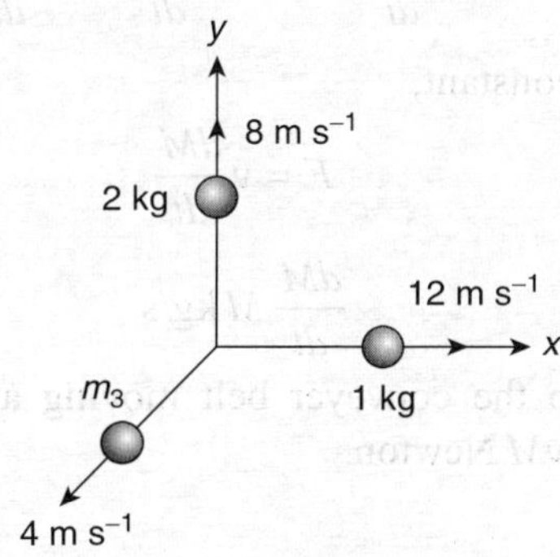

According to law of conservation of linear momentum,

$$\vec{p}_1+\vec{p}_2+\vec{p}_3=0 \quad \therefore \vec{p}_3=-\left(\vec{p}_1+\vec{p}_2\right)$$

Here, $\vec{p}_1=(1\text{ kg})(12\text{ ms}^{-1})\hat{i}=12\hat{i}\text{ kg ms}^{-1}$

$$\vec{p}_1=(2\text{ kg})(8\text{ ms}^{-1})\hat{j}=16\hat{j}\text{ kg ms}^{-1}$$

$$\therefore \vec{p}_3=-(12\hat{i}+16\hat{j})\text{ kg ms}^{-1}$$

The magnitude of p_3 is

$$p_3=\sqrt{(12)^2+(16)^2}=20\text{ kgms}^{-1}$$

$$\therefore m_3=\frac{p_3}{v_3}=\frac{20\text{ kgms}^{-1}}{4\text{ ms}^{-1}}=5\text{ kg}$$

236. (b)

Transportation of mass is taking place within the system (trolley + sand bag). No external force is acting on system, so momentum remains conserved. As mass of system is constant, there is no change in velocity of system.

237. (c)

Let θ is the angle made by the wire with the vertical.

$$\therefore \tan\theta=\frac{v^2}{rg}$$

Here, v = 10 m/s, r = 10 m, g = 10 m/ s^2

$$\therefore \tan\theta=\frac{\left(10\text{ m/s}^2\right)}{10\text{ m}\left(10\text{ m/s}^2\right)}=1$$

$$\theta=\tan^{-1}(1)=\frac{\pi}{4}$$

238. (b)

239. (b)

When a stone is dropped from a height h, it hits the ground with a momentum

$$P=m\sqrt{2gh} \qquad \text{(i)}$$

where m is the mass of the stone.

When the same stone is dropped from a height $2h$ (i.e. 100% of initial), then its momentum with which it hits the ground becomes

$$P'=m\sqrt{2g(2h)}=\sqrt{2}P \qquad \text{(ii)}$$

$$\%\text{ change in momentum}=\frac{P'-P}{P}\times 100\%$$

$$\frac{\sqrt{2}P-P}{P}\times 100\%=41\%$$

240. (b)

Here, m = 1000 kg, R = 90 m, $\theta = 45°$

For banking, $\tan\theta=\dfrac{v^2}{Rg}$

or $v=\sqrt{Rg\tan\theta}=\sqrt{90\times 10\times \tan 45°}=30\text{ ms}^{-1}$

241. (d)

Force of friction provides the necessary centripetal force.

$$f\le \mu_s N=\frac{mv^2}{R};\ v^2\le\frac{\mu_s RN}{m}$$

$$v^2\le\mu_s Rg \qquad [\because N=mg]$$

or $v\le\sqrt{\mu_s Rg}$

$\therefore$ The maximum speed of the car in circular motion is

$$v_{max}=\sqrt{\mu_s Rg}$$

242. (c)

Here, mass of a person, m = 60 kg; Mass of lift M = 940 kg,

$$a=1\text{ m/s}^2,\ g=10\text{ m/s}^2$$

Let T be the tension in the supporting cable.

$$\therefore T-(M+m)g=(M+m)a$$

$$T=(M+m)(a+g)=(940+60)(1+10)=11000N$$

243. (c)

Impulse = Change in linear momentum

= MV – (–MV) = 2 MV

244. (a)

Force of friction, $f = \mu$ mg

$$\therefore a=\frac{f}{m}=\frac{\mu mg}{m}=\mu g=0.5\times 10=5\text{ ms}^{-2}$$

Using $v^2-u^2=2aS$

$$0^2-2^2=2(-5)\times S\Rightarrow S=0.4\text{ m}$$

245. (c)

Pseudo force or fictitious force,

$$f_{fic} = m\alpha$$

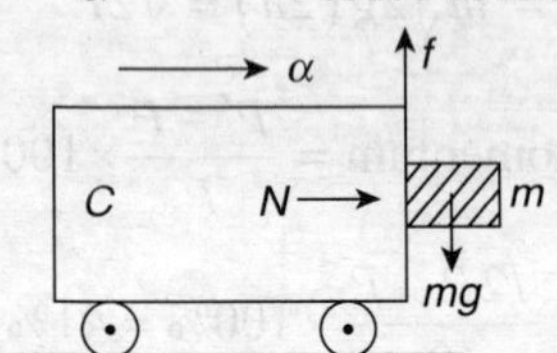

Force of friction, $f = \mu N = \mu m\alpha$

The block of mass m will not fall as long as

$$f \geq mg; \mu m\alpha \geq mg$$

$$\alpha \geq \frac{g}{\mu}$$

246. (a)

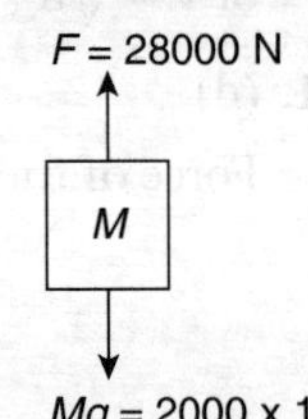

$$F - Mg = Ma$$

$$8000 = 2000a$$

$\therefore$ Acceleration is 4 ms^{-2} upwards

247. (c)

$$\vec{F} = 6\hat{i} - 8\hat{j} + 10\hat{k}$$

$$\left|\vec{F}\right| = \sqrt{36+64+100} = \sqrt{200}\ \text{N} = 10\sqrt{2}\ \text{N}$$

Acceleration, $a = 1$ ms^{-2}

$\therefore$ Mass $M = \frac{10\sqrt{2}}{1} = 10\sqrt{2}$ kg

248. (c)

$$mg = \frac{mv^2}{R} \Rightarrow v = \sqrt{Rg}$$

$$v = \sqrt{20\times 10} = \sqrt{200} = 14.1 \text{ m/s}$$

i.e., between 14 and 15 m/s.

249. (c)

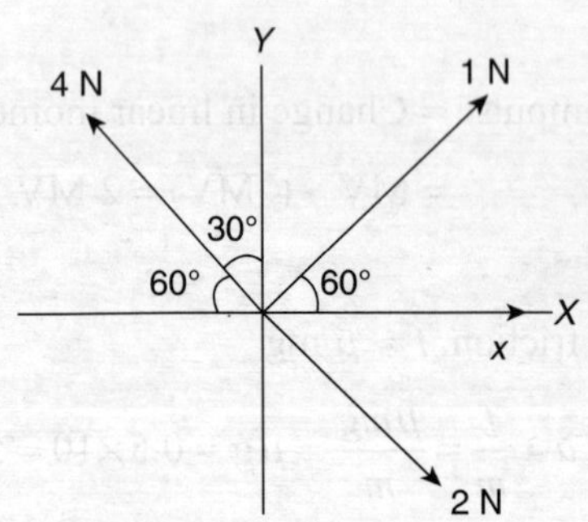

Taking x-components, the total should be zero.

$$1\times\cos 60° + 2\cos 60° + x - 4\cos 60° = 0$$

$$\therefore x = 0.5 \text{ N}$$

250. (c)

$$F = \frac{d}{dt}(Mv) = v\frac{dM}{dt} + M\frac{dv}{dt}$$

As v is a constant,

$$F = v\frac{dM}{dt}$$

But $\quad \frac{dM}{dt} M$ kg/s

$\therefore$ To keep the conveyer belt moving at v m/s, force needed = vM Newton.

251. (d)

Given $u = V$, final velocity = 0.

Using $v = u + at$

$\therefore 0 = V - at$ or, $-a = \frac{0-V}{t} = -\frac{V}{t}$

$f = \mu R = \mu mg$ (f is the force of friction)

$\therefore$ Retardation, $a = \mu g \therefore t = \frac{V}{a} = \frac{V}{\mu g}$

252. (c)

Components of momentum parallel to the wall add to each other and components of momentum in the perpendicular to the wall are opposite to each other. Therefore change of momentum is Final momentum – Initial momentum

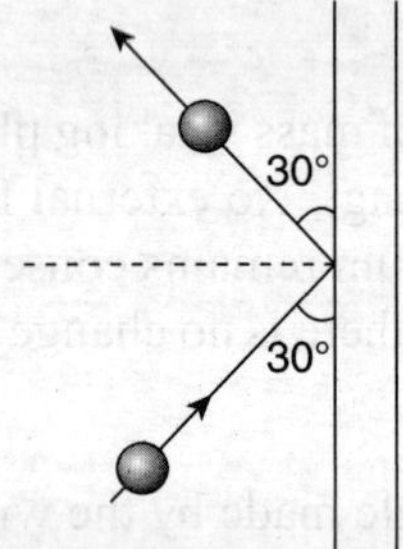

i.e., $mv\sin\theta$ after collision – ($-mv\sin\theta$ before collision)

$$F\times t = \text{Change in momentum} = 2mv\sin\theta$$

$$\therefore F = \frac{2mv\sin\theta}{t}$$

$$= \frac{2\times 0.5\times 12\times \sin 30°}{0.25} = 48\times\frac{1}{2} = 24 \text{ N}.$$

253. (d)

The wedge is given acceleration to the left.

$\therefore$ The block has a pseudo acceleration to the right, pressing against the wedge because of which the block is not moving.

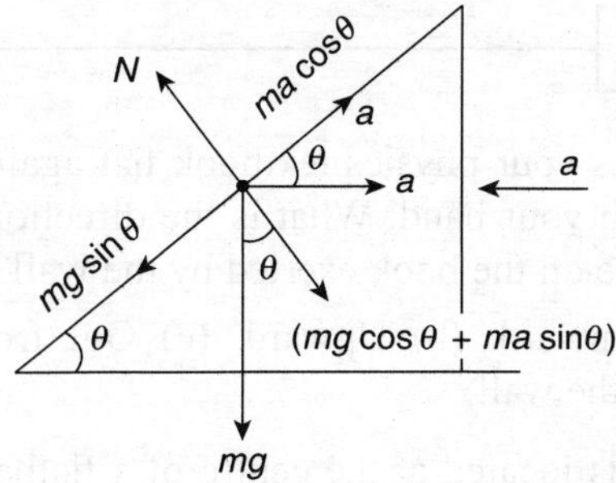

$$\therefore mg\sin\theta = ma\cos\theta \text{ or } a = \frac{g\sin\theta}{\cos\theta}$$

Total reaction of the wedge on the block is

$$N = mg\cos\theta + ma\sin\theta$$

$$\text{or } N = mg\cos\theta + \frac{mg\sin\theta.\sin\theta}{\cos\theta}$$

$$\text{or} \quad N = \frac{mg\left(\cos^2\theta + \sin^2\theta\right)}{\cos\theta} = \frac{mg}{\cos\theta}$$

254. (d)

Free-body diagram of two masses is

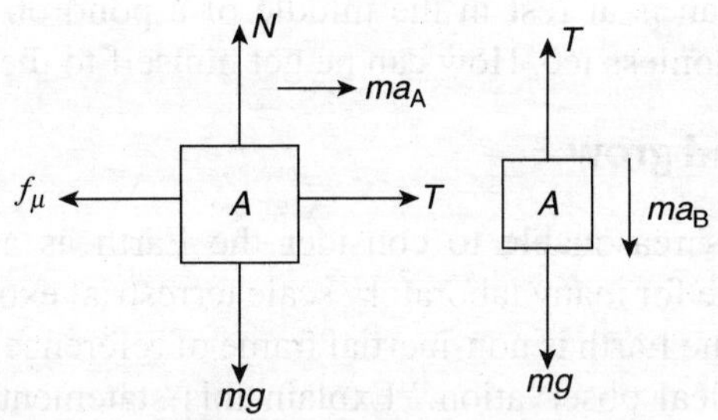

We get equations

$$T - f = ma \text{ or } T = \mu N_A \text{ (for } a = 0)$$

$$T = m_B g \text{ (for } a = 0)$$

$$\therefore \mu N_A = m_B g \Rightarrow m_B = \mu m_A = 0.2 \times 2 = 0.4 \text{ kg}$$

255. (d)

When the lift is accelerating upwards with acceleration a, then reading on the scale

$$R = m\,(g + a) = 80\,(10 + 5) \text{ N} = 1200 \text{ N}.$$

256. (d)

Let T be the tension in the rope when the monkey climbs up with an acceleration a. Then,

$$T - mg = ma$$

$$25g - 20g = 20a \Rightarrow a = \frac{5\times10}{20} = 2.5 \text{ m/s}^2$$

257. (b)

For a lift which is moving in upward direction with an acceleration a, the tension T developed in the string connected to the lift is given by $T = m\,(g + a)$.

Here, $m = 1000$ kg, $a = 1$ m/s^2, $g = 9.8$ m/s^2

$$\therefore T = 1000\,(9.8 + 1) = 10{,}800 \text{ N}.$$

258. (b)

$$m = 10 \text{ kg},$$
$$R = mg$$

$$\therefore \text{Frictional force} = f_k = \mu_k R = \mu_k mg = 0.5\times10\times10 = 50 \text{ N } [g = 10 \text{ m/s}^2]$$

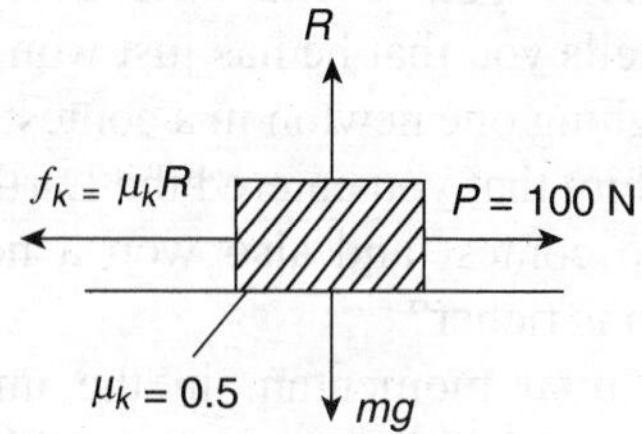

$\therefore$ Net force acting on the body

$$= F = P - f_k = 100 - 50 = 50 \text{ N}.$$

$\therefore$ Acceleration of the block $= a = F/m$

$$= 50/10 = 5 \text{ m/s}^2$$

259. (c)

$$\text{Load } W = Mg = 75 \times 10 = 750 \text{ N}$$

$$\text{Effort } (P) = 250 \text{ N}$$

$\therefore$ Mechanical advantage

$$= \frac{\text{load}}{\text{effort}} = \frac{W}{P} = \frac{750}{250} = 3$$

Velocity ratio

$$= \frac{\text{distance travelled by effort}}{\text{distance travelled by load}} = \frac{12}{3} = 4$$

$$\text{Efficiency, } \eta = \frac{\text{Mechanical advantage}}{\text{Velocity ratio}} = (3/4)\times 100 = 75\%$$

260. (a)

Just after the stone is dropped from the train, the only force acting on stone is gravity force. Hence, the net force acting on stone is mg (= 2N) just after it is dropped from the train.

QUICK QUESTIONS

Quiz

1. (i) Is it possible to have motion in the absence of a force? Is it possible to have force in the absence of motion?

   (ii) (a) If a single force acts on an object, does the object accelerate? (b) If an object experiences acceleration, is a force acting on it? (c) If an object experiences no acceleration, is no force acting on it?

   (iii) If an object experiences a net force in the x direction, does it move in the x direction?

   (iv) Suppose you are talking by interplanetary telephone to your friend, who lives on the Moon. He tells you that he has just won a piece of gold weighing one newton in a contest. Excitedly, you tell him that you entered the Earth version of the same contest and also won a newton of gold! Who is richer?

   (v) The total momentum in the universe remains constant. Is this statement true? If so, why?

   (vi) In a tug of war, three men pull on a rope to the left at A and three men pull to the right at B with forces of equal magnitude. Now a weight of 5 kg is hung vertically from the centre of the rope. (a) Can the men get the rope AB to be horizontal? If not, explain. If so, determine the magnitude of the forces required at A and B to do this.

   (vii) Is a 'single isolated force' possible in nature?

   (viii) Can a rocket operate in free space?

   (ix) Can a body in linear motion be in equilibrium?

   (x) A body is acted upon by a number of external forces. Can it remain at rest?

   (xi) Cite an example of (a) constant force, (b) time dependent force

   (xii) Can the tension in a string be different at different points?

   (xiii) Can the force of friction on a body be zero even if it is on a rough surface?

2. If a sports car collides head-on with a massive truck, which vehicle experiences the greater force? Which vehicle experiences the greater acceleration?

3. A large man and a small boy stand facing each other on frictionless ice. They put their hands together and push toward each other, so that they move away from each other. Who exerts the larger force? Who experiences the large acceleration? Who moves away with higher velocity? Who moves over a longer distance while their hands are in contact?

4. You press your physics textbook flat against a vertical wall with your hand. What is the direction of the friction force on the book exerted by the wall?

   (a) Downward. (b) Upward. (c) Out from the wall. (d) Into the wall.

5. A crate is located at the centre of a flatbed truck. The truck accelerates toward the east, and the crate moves with it, not sliding on the bed of the truck. What is the direction of the friction force exerted by the bed of the truck on the crate?

   (a) To the west. (b) To the east. (c) There is no friction force because the crate is not sliding.

6. Explain why

   (a) a horse cannot pull a cart and run in empty space,
   (b) passengers are thrown forward from their seats when a speeding bus stops suddenly,
   (c) it is easier to pull a lawn-mower than to push it,
   (d) a cricketer moves his hands backwards when holding a catch?

7. A man is at rest in the middle of a pond on perfectly frictionless ice. How can he get himself to the shore?

Think and grow

8. "It is reasonable to consider the Earth as an inertial frame for many laboratory scale terrestrial experiments, but the Earth is non-inertial frame of reference for astronomical observation." Explain this statement. Do you see how the same frame of reference is approximately inertial for one purpose, and non-inertial for the other?

9. In the absence of air friction, it is claimed that all objects fall with the same acceleration. A heavier object is pulled to the Earth with more force than a light object. Why does the heavier object not fall faster?

10. You are riding on a Ferris wheel that is rotating with constant speed. The car in which you ride always maintains its correct upward orientation – it does not invert. What is the direction of your centripetal acceleration when you are (a) at the top of the wheel? (b) at the bottom of the wheel? What is the direction of the normal force on you from the seat when you are (c) at the top of the wheel? (d) At the bottom of the wheel? (e) At which point, top or bottom, is the normal force the largest in magnitude?

11. The Copernican theory of the solar system is a model in which the planets are assumed to travel around the Sun in circular orbits. Historically, this was a break from the Ptolemaic theory, a model in which the Earth was at the

centre. When the Copernican theory was proposed, a natural question arose – what keeps the Earth and other planets moving in their paths around the sun? An interesting response to this question comes from Richard Feynman- "In those days, one of the theories proposed was that the planets went around because behind them there were invisible angels, beating their wings and driving the planets forward… It turns out that in order to keep the planets going around, the visible angels must fly in a different direction…" What did Feynman mean by this?

12. A person weighs a watch on a spring scale attached to the ceiling of an elevator. If the elevator accelerates, the spring scale reads a weight different from the true weight of the watch. Why?

Explanations

Quiz

1. (i) Motion requires no force. Newton's first law says an object in motion continues to be in motion by itself in the absence of external forces. It is possible for forces to act on an object with no resulting motion if the forces are balanced.

(ii) (a) If a single force acts on an object, it must accelerate. (b) If an object accelerates, at least one force must act on it. (c) If an object has no acceleration, we cannot conclude that no forces act on it. In this case, we can only say that the net force on the object is zero.

(iii) If the object begins at rest or is moving with a velocity with only an x component, the net force in the x direction will cause the object to move in the x direction. In any other case, the motion of the object will involve velocity components in directions other than x. Thus, the direction of the velocity vector will not generally be along the x axis. What we can say with certainly is that a net force in the x direction causes the object to accelerate in the x direction.

(iv) Because the value of g is smaller on the Moon than on the Earth, more mass of gold would be required to represent 1 newton of weight on the Moon. Thus, your friend on the Moon is richer roughly by a factor of 6.

(v) Since no external force can be applied on the universe, therefore total momentum is conserved.

(vi) Infinite tension is required to straighten the rope. So, it is impossible to straighten the rope.

(vii) A single isolated force is not possible. This follows from Newton's third law of motion, according to which to every action, there is an equal and opposite reaction. So, the forces must always exist in pairs. When we talk of a single force, we are considering only one aspect of mutual interaction.

(viii) Yes.

(ix) Yes, provided the vector sum of the forces acting upon the body is zero.

(x) If the vector sum of all the external forces is zero, then the body will remain at rest.

(xi) (a) Tension in a massless string (b) Force acting in a collision

(xii) Yes, when the string is not massless.

(xiii) Yes, when it is at rest and forces are acting on it.

2. The car and truck experience forces that are equal in magnitude but opposite in directions, according to Newton's third law. A calibrated spring scale placed between the colliding vehicles reads the same whichever way it faces. Because the car has the smaller mass, it experiences the greater acceleration.

3. According to Newton's third law, the force of the man on the boy and the force of the boy on the man are reaction forces, so they must be equal in magnitude. Both individuals experience the same magnitude of force, so the boy, with the smaller mass, experiences the larger acceleration. Both individuals experience acceleration over the same time interval. The larger acceleration of the boy results in his moving away from the interaction with a larger velocity. Again, because of the larger acceleration, the boy moves through a larger distance during the interval the hands are in contact.

4. (b). Friction forces are always parallel to the surfaces in contact, which, in this case, are the wall and the cover of the book. This tells us that the friction force is either upward or downward. Because the tendency of the book is to fall due to gravity, the friction force must be in the upward direction.

5. (b) The static friction force between the bottom surface of the crate and the surface of the truck bed is the net horizontal force on the crate that causes it to accelerate. It is in the same direction as the acceleration of the crate, to the east.

6. (a) No reaction is available in empty space.
   (b) Inertia of motion.
   (c) In the case of pull, the effective weight is reduced due to the vertical component of the pull. In the case of push, the vertical component increases the effective weight.
   (d) By increasing time, force is reduced.

7. If the man throws away his shirt in a direction opposite to the desired direction of motion, he can get himself to the shore.

Think and grow

8. On account of rotation of Earth about its axis, an object placed at the equator experiences an acceleration of $3.41\times10^{-2}\,ms^{-2}$. On account of revolution of Earth around the Sun, an object experiences an acceleration of $6\times10^{-3}\,ms^{-2}$. Since acceleration is present, strictly speaking, Earth is non-inertial frame. However, the accelerations on account of rotation and revolution are very small as compared to $9.8\,ms^{-2}$. Therefore, for all experiments performed in a laboratory, it is reasonable to consider the Earth as an inertial frame.

9. It is true that the heavier object is pulled with a larger force. The strength of the force is determined by the gravitational mass of the object. The resistance to the force and, therefore, to the change in motion of the object, is represented by the inertial mass. Thus, if an object has twice as much mass as another, it is pulled to the Earth with twice the force, but it also exhibits twice the resistance to having its motion changed. These factors cancel, so that the change in motion, the acceleration, is the same for all objects, regardless of mass.

10. The centripetal acceleration is always directed toward the center of the circle. Therefore, at the top (a), the centripetal acceleration is downward and at the bottom (b), it is upward. The normal force is always perpendicularly outward from the surface, so the normal force on you from the seat is upward in both situations: the top (c) and the bottom (d). (e) The normal force is largest at the bottom because it is in the same direction as the centripetal acceleration.

11. The question asked by those at the time of Copernicus indicates that they did not have proper understanding of inertia as Newton's first law. A that time in history, before Galileo and Newton, the interpretation was that motion was caused by force. This is different from our current understanding that changes in the state of motion are caused by force. Thus, it was natural for Copernicus's contemporaries to ask what force propelled a planet in its orbit. According to our current understanding, it is equally natural for us to realize that no force tangent to the orbit is necessary – the motion simply continues due to inertia.

    Thus, in Feynman's imagery, we do not need the angels pushing the planet from behind. We need the angels to push inward, to provide the centripetal acceleration associated with the orbital motion of the planet. Of course, the angels are not real, from a scientific point of view – they are a metaphor for the gravitational force.

12. An observer on the accelerating elevator is not in an inertial frame. We need to analyze this situation in an inertial frame, so let us imagine observing it from the stationary ground. We model the watch as a particle under a net force. The external forces acting on the watch are the downward gravitational force F_g and the upward force T exerted on it by the hook hanging from the bottom of the scale. Because the tension is the same everywhere in the hook supporting the watch, the hook pulls downward with a force of magnitude T on the spring scale. Thus, the tension T in the hook is also the reading of the spring scale.

    If the elevator is either at rest or moves at constant velocity, the watch is not accelerating and is a particle in equilibrium, which gives net force in vertical direction $\sum F = T - mg$

$$\Rightarrow T - mg = 0$$

$$T = mg$$

    If the elevator accelerates with acceleration $\vec{a}$ relative to an observer in an inertial frame outside the elevator we have,

$$\sum F = mg - T = ma$$

$$\Rightarrow T = mg - ma$$

    Thus, tension T is less than mg.

CHAPTER TEST

1. A ship of mass 3×10^7 kg, initially at rest, is pulled by a force of 5×10^4 N through a distance of 3 m. Assume that the resistance due to water is negligible. Determine the speed of the ship.
   (a) $0.5\ ms^{-1}$
   (b) $0.1\ ms^{-1}$
   (c) $5\ ms^{-1}$
   (d) $2\ ms^{-1}$

2. The force on a particle of mass 10 g is $(10\hat{i}+5\hat{j})$ N. If it starts from rest, what would be its position at time t = 5 seconds?

(a) $12500\hat{i}+6250\hat{j}$ (b) $2500\hat{i}+31250\hat{j}$
(c) $31250\hat{i}+25000\hat{j}$ (d) $2500\hat{i}+12500\hat{j}$

3. A bullet of mass 0.04 kg moving with speed of 90 m s^{-1} enters a heavy wooden block and is stopped after a distance 60 cm. What is the average resistive force exerted by the block on the bullet?

(a) 135 N (b) 540 N
(c) 270 N (d) 810 N

4. A stream of water flowing horizontally with a speed of 15 m s^{-1} gushes out of a tube of cross-sectional area 10^{-2} m^2 and hits at a vertical wall nearby. What is the force exerted on the wall by the impact of water, assuming it does not rebound?

(a) 2250 N (b) 5000 N
(c) 1250 N (d) 2500 N

5. A shell of mass 0.020 kg is fired by a gun of mass 100 kg. If the muzzle speed of the shell is 80 m s^{-1}, what is the recoil speed of the gun ?

(a) $^{-}$0.032 m s^{-1} (b) 0.064 m s^{-1}
(c) $^{-}$0.016 m s^{-1} (d) $-$0.016 m s^{-1}

6. A cart and a man move towards each other. The masses of a cart and man are 25 kg and 50 kg respectively. The velocity of the cart is 0.2 m s^{-1}. When the man approaches the cart, he jumps on to it. After this, the cart carrying the man acquires a velocity of 0.6 m s^{-1} in the direction in which the man was walking. What is the velocity with which the man was walking?

(a) 2.5 m s^{-1} (b) 1.5 m s^{-1}
(c) 2 m s^{-1} (d) 1 m s^{-1}

7. A body of mass 5 kg is hanging from a string of negligible weight. A horizontal force F is applied at a point of string due to which two parts of the string include an angle of 150°. If g = 10 ms^{-2}, find the force applied and also the tension in the upper part of the string.

(a) 68 N, 32.5 N (b) 57.5 N, 32.7 N
(c) 28.7 N, 57.7 N (d) 82.5 N, 41.25 N

8. A block of weight 10 N rests on a rough inclined plane of slope 3 in 5. Calculate the minimum force required to start moving the block up the inclined plane. Given: $\mu = 0.6$.

(a) 5.2 N (b) 24.4 N
(c) 10.8 N (d) 12.2 N

9. Determine the maximum acceleration of the train in which a box lying on its floor will remain stationary, given that the co-efficient of static friction between the box and the train's floor is 0.15. Take g = 10 m s^{-2}.

(a) 3.5 m s^{-2} (b) 2.5 m s^{-2}
(c) 1.25 m s^{-2} (d) 1.5 m s^{-2}

10. A ball is dropped from a spacecraft revolving around the earth at a height of 120 km. What will happen to the ball at that instant?

(a) It will fall down to the earth.
(b) It will go very far in space.
(c) It will move with the same speed, tangentially to the spacecraft.
(d) It will continue to move with the same speed along the original orbit of spacecraft.

11. A block slides down an inclined plane of slope angle θ with constant velocity. It is then projected up the same plane with an initial velocity v_0. How far up the incline will it move before coming to rest?

(a) $\dfrac{v_0^2}{4g\sin\theta}$ (b) $\dfrac{v_0^2}{2g\sin\theta}$
(c) $\dfrac{v_0^2}{\sin\theta}$ (d) $\dfrac{v_0^2}{\sin 2\theta}$

12. A block is about to slide down an inclined plane when its inclination to the horizontal is θ. If now a 5 kg weight is attached on the block:

(a) It is still be about to slide down the plane.
(b) It will not slide down the plane unless the inclination is increased.
(c) It will not slide down the plane unless the inclination is decreased.
(d) It will never slide down whatever be the inclination.

13. A boy of mass M is applying a horizontal force to slide a box of mass M'on a rough horizontal surface. The coefficient of friction between the shoes of the boy and the floor is μ and that between the box and the floor is μ'. In which of the following cases it is certainly not possible to slide the box?

(a) $\mu < \mu', M < M'$ (b) $\mu > \mu', M < M'$
(c) $\mu < \mu', M > M'$ (d) $\mu > \mu', M > M'$

14. A block A kept on an inclined surface just begins to slide if the inclination is 30°. The block is replaced by another block B and it is found that it is just begins to slide if the inclination is 40°.

(a) Mass of A > Mass of B
(b) Mass of A < Mass of B
(c) Mass of A = Mass of B
(d) All the three are possible.

15. A body of mass m is kept stationary on a rough inclined plane of angle of inclination θ. The magnitude of force acting on the body by the inclined plane is equal to:

(a) mg (b) $mg \sin \theta$
(c) $mg \cos \theta$ (d) None

16. A particle is moving on the inside of a smooth sphere of radius r describing a horizontal circle at a distance $r/2$ below the centre of the sphere. What is its speed?

(a) $\sqrt{5gr}$ (b) $\sqrt{4gr/3}$
(c) $\sqrt{3gr/2}$ (d) $\sqrt{\sqrt{3}gr}$

17. A ball weighing 10 g hits a hard surface vertically with a speed of 5 m/s and rebounds with the same speed. The ball remains in contact with the surface for (0.01) sec. The average force exerted by the surface on the ball is:

(a) 100 N (b) 10 N
(c) 1 N (d) 150 N

18. A body is moving down a long inclined plane of angle of inclination θ. The coefficient of friction between the body and the plane varies as $\mu = 0.1\,x$, where x is the distance moved down the plane. The body will have the maximum velocity when it has traveled a distance x given by:

(a) $x = 10 \tan \theta$ (b) $x = 5 \tan \theta$
(c) $x = \sqrt{2} \cot \theta$ (d) $x = \dfrac{\sqrt{10}}{\cot \theta}$

19. A man of mass m stands on a frame of mass M. He pulls on a light rope, which passes over a pulley. The other end of the rope is attached to the frame. For the system to be in equilibrium, what force must the man exert on the rope?

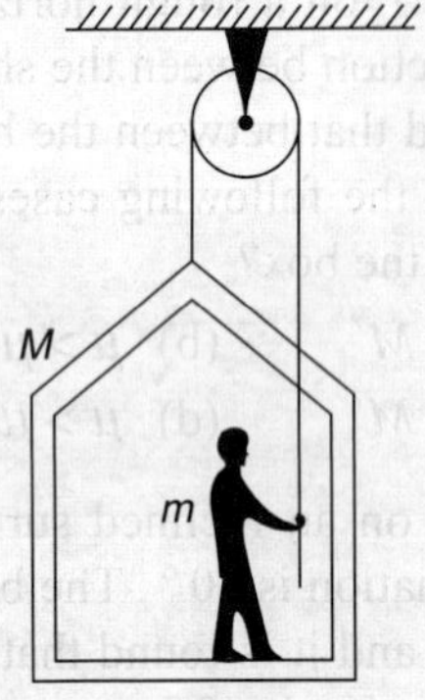

(a) $\dfrac{(M+m)g}{2}$ (b) $(M+m)g$
(c) $(M-m)g$ (d) $(M+2m)g$

20. A car starts from rest to cover a distance s. The coefficient of friction between the rod and the tyres is μ. The minimum time in which the car can cover the distance is proportional to:

(a) μ (b) $\sqrt{\mu}$
(c) $1/\mu$ (d) $1/\sqrt{\mu}$

21. A body of mass m starts sliding down an incline of 30° from rest. The body comes to rest just when it reaches the bottom. If the top half of the plane is perfectly smooth and the lower half is rough, find the force of friction:

(a) $\dfrac{mg}{4}$ (b) $\dfrac{mg}{\sqrt{3}}$
(c) mg (d) $\dfrac{mg}{\sqrt{2}}$

22. A mass of 0.5 kg is just able to slide down the slope of an inclined rough surface when the angle of inclination is 60°. The minimum force necessary to pull the mass up the incline along the line of greatest slope is:

(a) 20 N (b) 9 N
(c) 100 N (d) 1 N

23. A block a mass 10 kg is suspended through two light spring balances as shown in figure.

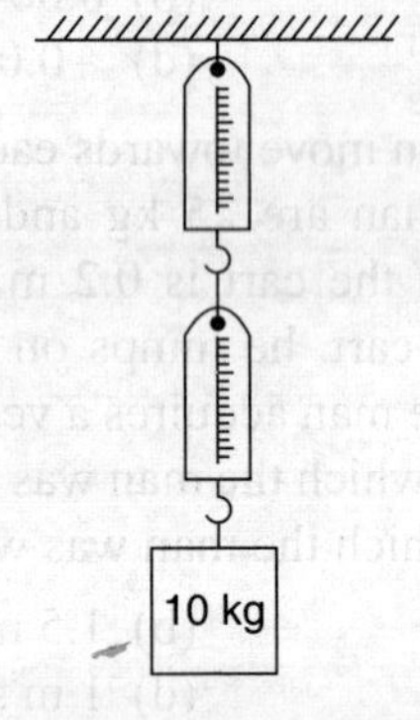

(a) Both the scales will read 10 kg
(b) Both the scales will read 5 kg
(c) The upper scale will read 10 kg and the lower zero.
(d) The reading may be anything but their sum will be 10 kg.

24. Two blocks of masses m_1 and m_2 are placed in contact with each other on a horizontal platform. The coefficient of friction between the platform and two blocks is the same. The platform moves with an acceleration. The force of interaction between the blocks is:

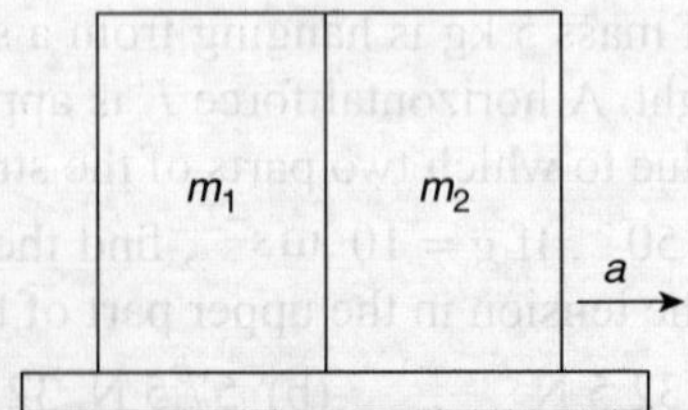

(a) Zero in all cases
(b) Zero only if $m_1 = m_2$
(c) Non-zero only if $m_1 > m_2$
(d) Non-zero only if $m_1 < m_2$

25. The human body can safely stand an acceleration 9 times that due to gravity which is 10 m/s². The minimum radius of curvature with which a pilot may safely turn a plane vertically upward at the end of a dive, when the plane's speed is 720 km/hr is:

(a) 500 m (b) 612 m
(c) 475 m (d) 323 m

26. A car moves on a circular road, describing equal angles about the centre in equal intervals of times. Which of the following statements about the velocity of car are not true?

(a) Velocity is constant
(b) Magnitude of velocity is constant but the direction of velocity changes
(c) Both magnitude and direction of velocity change
(d) Velocity is directed towards the centre of circle.

27. Two spheres of equal masses are attached to a string of length 2 m as shown in figure.

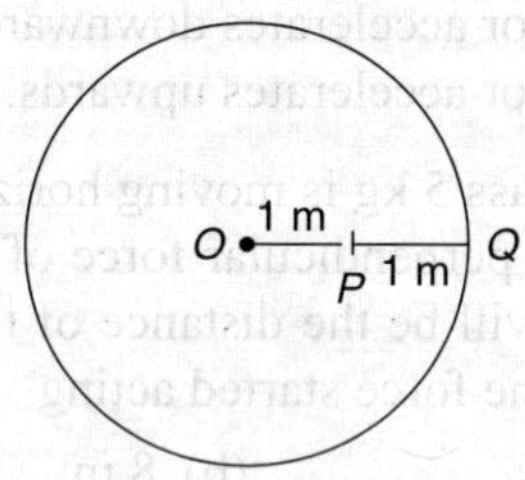

The string and spheres are then whirled in a horizontal circle about O at a constant rate. What is the value of the ratio?

(a) 1/2 (b) 2/3
(c) 3/2 (d) 2

28. A bird is sitting in a wire cage hanging from the spring balance. Let the reading of spring balance be W_1. If the bird flies about inside the cage, the reading of the spring balance is W_2. Which of the following is true?

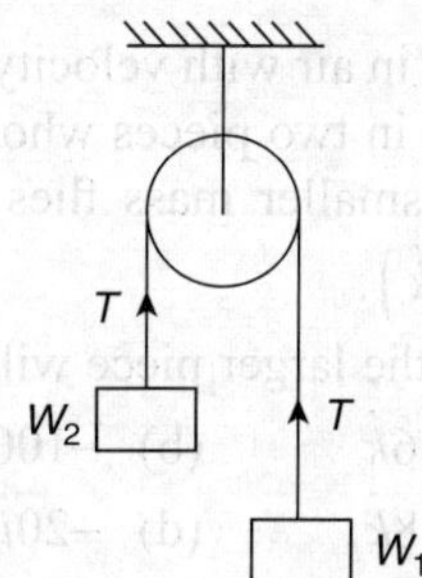

(a) $W_1 = W_2$
(b) $W_1 > W_2$
(c) $W_1 < W_2$
(d) Nothing definite can be predicted.

29. The linear momentum P of a particle varies with time as follows

$$P = a + bt^2$$

Where a and b are constants. The net force acting on the particle is

(a) proportional to t (b) proportional to t^2
(c) zero (d) constant

30. A ball of mass 150 g moving with an acceleration 20 m/s² is hit by a force, which acts on it for 0.1 sec. The impulsive force is

(a) 0.5 N (b) 0.1 N
(c) 0.3 N (d) 1.2 N

31. Which of the following sets of concurrent forces may be in equilibrium?

(a) $F_1 = 3$ N, $F_2 = 5$ N, $F_3 = 6$ N
(b) $F_1 = 3$ N, $F_2 = 5$ N, $F_3 = 15$ N
(c) $F_1 = 3$ N, $F_2 = 5$ N, $F_3 = 1$ N
(d) $F_1 = 3$ N, $F_2 = 5$ N, $F_3 = 9$ N

32. A block is given certain upward velocity along the incline of elevation α. The time of ascent to upper point was found to be half the time of descent to initial point. The co-efficient of friction between block and incline is:

(a) $0.5 \tan \alpha$ (b) $0.3 \tan \alpha$
(c) $0.6 \tan \alpha$ (d) $0.2 \tan \alpha$

33. The direction of three forces 1 N, 2 N and 3 N acting at a point are parallel to the sides of an equilateral triangle taken in order. The magnitude of their resultant is:

(a) $\sqrt{3}$ N (b) $\frac{\sqrt{3}}{2}$ N
(c) $\frac{3}{2}$ N (d) zero

34. Two blocks of masses 7 kg and 3 kg are connected by a spring of stiffness 1000 N/m and placed on a smooth horizontal surface. They are acted by horizontal forces $F_1 = 72$ N and $F_2 = 32$ N in opposite directions as shown in the figure. When accelerations of the blocks are equal and constant, the extension in cm of the spring is

$F_2 = 32$ N $k = 1000$ N/M $F_1 = 72$ N
7 Kg 3 Kg

(a) 4 cm (b) 5 cm
(c) 6 cm (d) 8 cm

35. Two identical small masses, each of mass m, are connected by light inextensible string on a smooth horizontal floor. A constant force F is applied at the mid-point of the string as shown in the figure. The acceleration of each mass towards each other at the instant when the string encloses an angle of 60° is

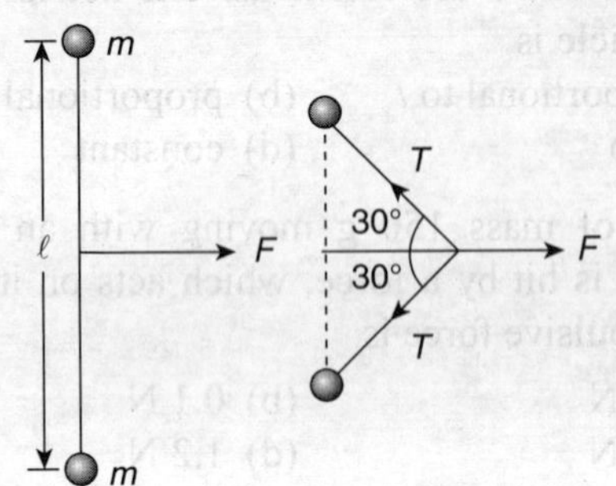

(a) $\frac{F}{2\sqrt{3}m}$ (b) $\frac{\sqrt{3}F}{2m}$
(c) $\frac{2F}{\sqrt{3}m}$ (d) None of these

36. Gravels are dropped on a conveyor belt at the rate of 0.5 kg s^{-1}. The extra force required in newton to keep the belt moving at 2 ms^{-1} is
(a) 1 (b) 2
(c) 4 (d) 0.5

37. A lift of mass 1000 kg starts moving from rest in the upward direction. The figure shows variation of speed of lift. If T_1, T_2, T_3 represent the tension in the cable supporting the lift, when moving with uniform acceleration, uniform velocity, and uniform retardation respectively, then

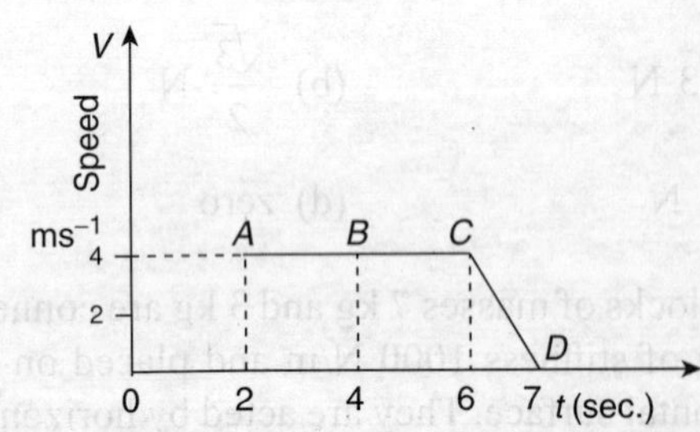

(a) $T_1 : T_2 : T_3 :: 1 : 1 : 1$
(b) $T_1 : T_2 : T_3 :: 6 : 5 : 3$
(c) $T_1 : T_2 : T_3 :: 3 : 5 : 6$
(d) $T_1 : T_2 : T_3 :: 6 : 5 : 6$

38. A chain is lying on a rough table with a fraction $\frac{1}{n}$ of its length hanging down from the edge of the table. If it is just on the point of sliding down from the table, then the coefficient of friction between the table and the chain is

(a) $\frac{1}{n}$ (b) $\frac{1}{n-1}$
(c) $\frac{1}{n+1}$ (d) $\frac{n-1}{n+1}$

39. Two weights W_1 and W_2 are attached to the ends of a string, which passes over a frictionless pulley. If the pulley is placed in a rocket accelerating upwards at 9.8 m s^{-2}, the tension in the string would be

(a) $2\frac{W_1W_2}{W_1+W_2}$ (b) $\frac{W_1W_2}{W_1+W_2}$
(c) $4\frac{W_1W_2}{W_1+W_2}$ (d) $\frac{W_1W_2}{2(W_1+W_2)}$

40. A monkey sits on the pan of spring scale kept in an elevator. The reading of the spring scale will be maximum when
(a) the elevator is stationary
(b) the elevator cable breaks and it falls freely towards earth.
(d) the elevator accelerates downwards.
(d) the elevator accelerates upwards.

41. A block of mass 5 kg is moving horizontally at a speed of 1.5 m/s. A perpendicular force of 5 N acts on it for 4 sec. What will be the distance of the block from the point where the force started acting
(a) 10 m (b) 8 m
(c) 6 m (d) 2 m

42. An object flying in air with velocity $(20\hat{i}+25\hat{j}-12\hat{k})$ suddenly breaks in two pieces whose masses are in the ratio 1: 5. The smaller mass flies off with a velocity $(100\hat{i}+35\hat{j}+8\hat{k})$.
The velocity of the larger piece will be
(b) $4\hat{i}+23\hat{j}-16\hat{k}$ (b) $-100\hat{i}-35\hat{j}-8\hat{k}$
(c) $20\hat{i}+15\hat{j}-8\hat{k}$ (d) $-20\hat{i}-15\hat{j}-8\hat{k}$

43. Which one of the following statement is incorrect?
(e) Rolling friction is smaller than sliding friction.
(f) Limiting value of static friction is directly proportional to normal reaction.
(g) Frictional force opposes the relative motion.
(h) Coefficient of sliding friction has dimensions of length.

44. Two stones of masses m and $2m$ are whirled in horizontal circles, the heavier one in a radius $r/2$ and the lighter one in radius r. The tangential speed of lighter stone is n times that of the value of heavier stone when they experience same centripetal forces. The value of n is

(a) 4 (b) 1
(c) 2 (d) 3

45. A system consists of three masses m_1, m_2 and m_3 connected by a string passing over a pulley P. The mass m_1 hangs freely and m_2 and m_3 are on a rough horizontal table (the coefficient of friction $= \mu$). The pulley is frictionless and of negligible mass. The downward acceleration of mass m_1 is

(Assume $m_1 = m_2 = m_3 = m$)

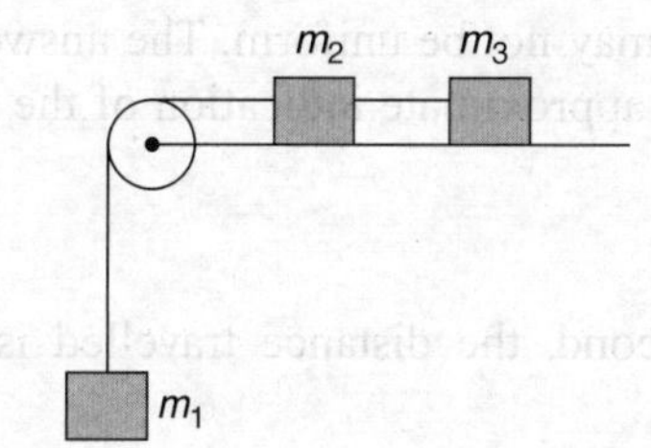

(a) $\frac{g(1-g\mu)}{9}$ (b) $\frac{2g\mu}{3}$
(c) $\frac{g(1-2\mu)}{3}$ (d) $\frac{g(1-2\mu)}{2}$

Answer Keys

1. (b)	2. (a)	3. (c)	4. (a)	5. (d)	6. (d)	7. (c)	8. (c)	9. (d)	10. (d)
11. (a)	12. (a)	13. (a)	14. (d)	15. (a)	16. (c)	17. (b)	18. (a)	19. (a)	20. (d)
21. (c)	22. (b)	23. (a)	24. (a)	25. (a)	26. (a,c, and d)	27. (b)	28. (b)	29. (a)	30. (c)
31. (a)	32. (c)	33. (a)	34. (c)	35. (b)	36. (a)	37. (b)	38. (b)	39. (c)	40. (d)
41. (a)	42. (a)	43. (d)	44. (c)	45. (c)					

Solutions

1. (b)

Acceleration, $a = \frac{\text{Force}}{\text{Mass}} = \frac{5\times10^4\ \text{N}}{3\times10^7\ \text{kg}}$

$= \frac{5}{3000}\ \text{m s}^{-2} = \frac{1}{600}\ \text{m s}^{-2}$

Now, $v^2 - 0^2 = 2\times\frac{1}{600}\times3 = \frac{1}{100}$

or $v = \sqrt{\frac{1}{100}}\ \text{m s}^{-1} = \frac{1}{10}\ \text{m s}^{-1} = 0.1\ \text{m s}^{-1}$

2. (a)

$F_x = 10\text{N},\ a_x = \frac{F_x}{m} = \frac{10\text{N}}{0.01\ \text{Kg}} = 1000\ \text{m s}^{-2}$

This is a case of constant acceleration in x - direction.

Using

$x = v_x(0)t + \frac{1}{2}a_x t^2 = 0 + \frac{1}{2}\times1000\times(5)^2\ \text{m} = 12500\ \text{m}$

Again, $a_y = \frac{F_y}{m} = \frac{5\text{N}}{0.01\ \text{kg}} = 500\ \text{m s}^{-2}$ and

$y = 0 + \frac{1}{2}\times500\times(5)^2\ \text{m} = 6250\ \text{m}$

So, the position of the particle at time 5 second is

$\vec{r} = (12500\hat{i} + 6250\hat{j})$ metre

3. (c)

$u = 90\ \text{m s}^{-1}, v = 0,\ s = 60\ \text{cm} = 0.6\ \text{m},\ a = ?$

We know that $v^2 - u^2 = 2as$

or $-90^2 = 2\times a\times0.6$

or $a = \frac{-90\times90}{2\times0.6}\ \text{m s}^{-2} = -6750\ \text{m s}^{-2}$

Retardation $= -a = 6750\ \text{m s}^{-2}$

$\therefore$ Retarding force $= 0.04\ \text{kg}\times6750\ \text{m s}^{-2} = 270\ \text{N}$

The actual resistive force and therefore, retardation of

the bullet may not be uniform. The answer is therefore merely an approximate indication of the average resistive force.

4. (a)

In one second, the distance travelled is equal to the velocity v.

$\therefore$ Volume of water hitting the wall per second, $V = av$ where a is the cross-sectional area of the tube and v is the speed of water coming out of the tube.

$$V = 10^{-2} \text{ m}^2 \times 15 \text{ m s}^{-1} = 15 \times 10^{-2} \text{ m}^3 \text{ s}^{-1}$$

Mass of water hitting the wall per second

$$= 150 \text{ kg s}^{-1} \times 15 \text{ m s}^{-1} = 2250 \text{ kg m s}^{-2} \text{ or } 2250 \text{ N}$$

Final momentum per second = 0

Force exerted by the wall $= 0 - 2250$ N $= -2250$ N

$\therefore$ Force exerted on the wall $= -(-2250)$N = 2250N

5. (d)

$$m = 0.02\text{kg}, M = 100\text{kg}, v = 80 \text{ m s}^{-1}, V = ?$$

$$V = -\frac{mv}{M} = -\frac{0.020 \text{ kg} \times 80 \text{ m s}^{-1}}{100 \text{ kg}} = -0.016 \text{m s}^{-1}$$

Negative sign indicates that the gun moves in a direction opposite to the direction of motion of the bullet.

6. (d)

Mass of cart, $m_1 = 25$kg ; Mass of man, $m_2 = 50$kg

Velocity of cart, $v_1 = -0.2$ m s^{-1}

Common velocity of cart and man, V = 0.6 m s^{-1}

Let v_2 be the velocity of man before jumping.

Applying law of conservation of momentum,

Total momentum before jumping = Total momentum after jumping

$$\therefore\ -25 \times 0.2 + 50 \times v_2 = (25 + 50)0.6 \text{ or } 50v_2 = 50 \text{ or}$$

$$v_2 = 1 \text{ m s}^{-1}$$

7. (c)

In the equilibrium position, the gravitational force mg acting on the body is balanced by tension T_2.

$$T_2 = mg = 5 \times 10\text{N} = 50 \text{ N}$$

Using Lami's theorem,

$$\frac{T_1}{\sin 90°} = \frac{F}{\sin 150°} = \frac{T_2}{\sin 120°}$$

or $$T_1 = \frac{T_2 \sin 90°}{\sin 120°} = \frac{50 \times 1 \times 2}{\sqrt{3}} \text{N}$$

or $$T_1 = \frac{100}{\sqrt{3}} \text{ N} = \frac{100\sqrt{3}}{3} \text{ N} = \frac{100 \times 1.732}{3} \text{ N} = 57.73 \text{ N}$$

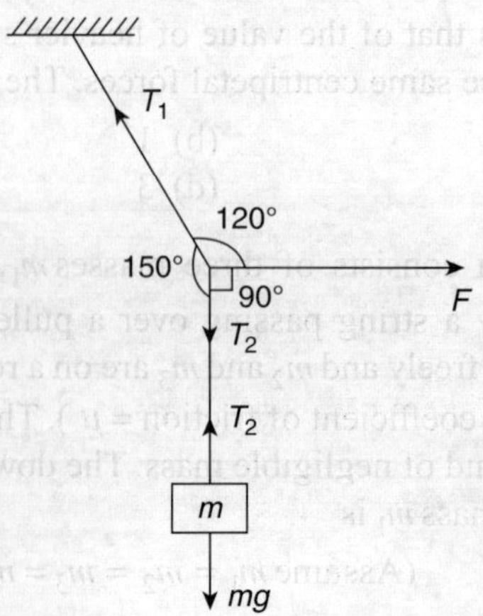

Again, $$F = \frac{T_2 \sin 150°}{\sin 120°} = \frac{T_2 \sin(180° - 30°)}{\sin(180° - 60°)}$$

or $$F = \frac{50 \sin 30°}{\sin 60°} \text{N} = \frac{50 \times \frac{1}{2}}{\frac{\sqrt{3}}{2}} \text{N} = \frac{50}{\sqrt{3}} \text{N}$$

$$= \frac{50\sqrt{3}}{3} \text{N} = \frac{50 \times 1.732}{3} \text{N} = 28.87 \text{ N}$$

8. (c)

$$mg = 10 \text{ N}, \sin\theta = \frac{3}{5}$$

$$\cos\theta = \frac{4}{5}$$

Let P be the required force.

Clearly, $P = mg\sin\theta + f = mg\sin\theta + \mu\, mg\cos\theta$

$$= mg(\sin\theta + \mu\cos\theta) = 10\left(\frac{3}{5} + 0.6 \times \frac{4}{5}\right)$$

$$= 10(0.6 + 0.48) \text{ N}$$

$$= 10.8 \text{ N}$$

9. (d)

$$f \le \mu_s R$$

or $$ma \le \mu_s mg$$

or $$a \le \mu_s g$$

$\therefore$ $$a_{\max} = \mu_s mg = 0.15 \times 10 \text{ m s}^{-2}$$

$$= 1.5 \text{ m s}^{-2}$$

10. (d)

As no external force is applied to the ball dropped from the spacecraft, it will continue to move with the same speed along the original orbit of the space craft.

11. (a)

As block slides with uniform velocity, its weight component $mg \sin\theta$ is balanced by friction between its surface and ground.

Thus we have,

$$mg\sin\theta = \mu mg \cos\theta$$

or

$$\mu = \tan\theta$$

When it is projected up with a velocity v_0, it will be retarded by both $g\sin\theta$ force and $\mu\ mg\ cos\theta$. It will now be retarded by $2g\sin\theta$, hence the distance covered by the particle along the plane is

$$s = \frac{v_0^2}{2a} = \frac{v_0^2}{4g\sin\theta}$$

12. (a) θ is the angle of repose and it is independent of the mass of the body placed on the inclined plane.

13. (a) The box can slide only when the force on it is more than the limiting friction between box and the floor. For this, the certain case will be when $\mu M < \mu' M'$

14. (d) As angle of repose is different for the two blocks. That means the friction coefficient is different in the two cases and angle of repose is independent of the mass of body placed on incline. So, all the cases are possible.

15. (a)

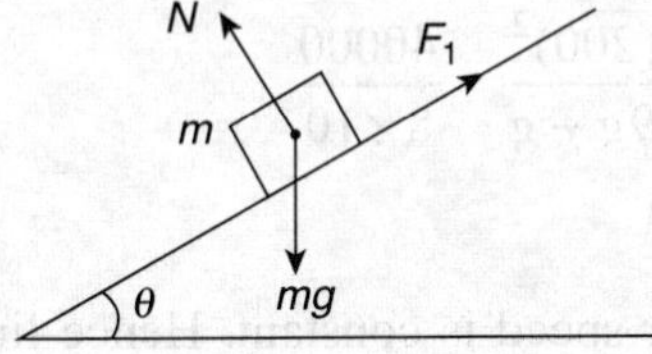

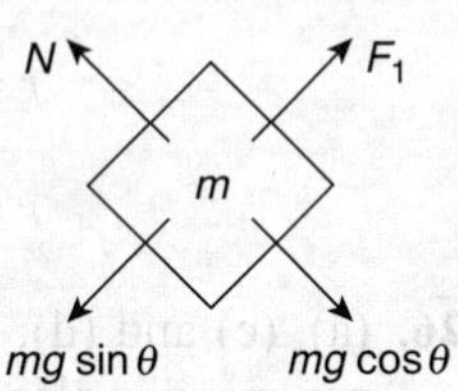

Since the body is at rest,

$$N = mg\cos\theta$$

$$F_r = mg\sin\theta$$

Magnitude of force acting on block by inclined surface,

$$F = \sqrt{N^2 + F_r^2}$$

$$= \sqrt{(mg\cos\theta)^2 + (mg\sin\theta)^2}$$

$$F = mg\sqrt{\cos^2\theta + \sin^2\theta}$$

$$F = mg$$

16. (c)

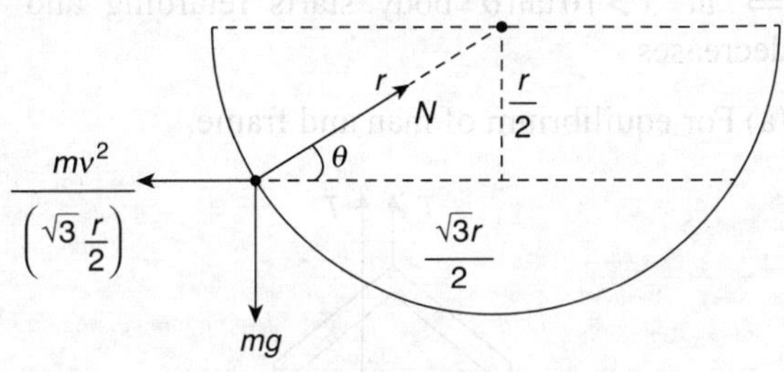

$$N\cos\theta = \frac{mv^2}{\frac{\sqrt{3}r}{2}} \quad ...(1)$$

$$N\sin\theta = mg \quad ...(2)$$

Dividing (1) by (2) $\tan\theta = \dfrac{mg}{\dfrac{2mv^2}{\sqrt{3}r}} = \dfrac{\sqrt{3}rg}{2v^2}$

$$v^2 = \frac{\sqrt{3}rg}{2\tan\theta}$$

From figure,

$$\tan\theta = \frac{\frac{r}{2}}{\sqrt{3}\frac{r}{2}} = \frac{1}{\sqrt{3}}$$

$$\Rightarrow \quad v^2 = \frac{\sqrt{3}rg}{2\left(\frac{1}{\sqrt{3}}\right)} = \frac{3rg}{2}$$

$$\Rightarrow \quad v = \sqrt{\frac{3rg}{2}}$$

17. (b)

$$F = \frac{m\left|\left(\vec{v} - \vec{u}\right)\right|}{t}$$

$$F = \frac{(10\times10^{-3}\,\text{kg})[5-(-5)]}{0.01}$$

$$F = \frac{10^{-2}\times10}{10^{-2}} = 10\text{ N}$$

18. (a) Downward acceleration of body is

$$a = g\sin g\theta - \mu g\cos\theta$$

$$a = g\left[\sin\theta - 0.1x\cos\theta\right]$$

Acceleration will remain $+ve$ up to $x = 10\tan\theta$

$\Rightarrow$ at $x > 10\tan\theta$ body starts retarding and speed decreases

19. (a) For equilibrium of man and frame,

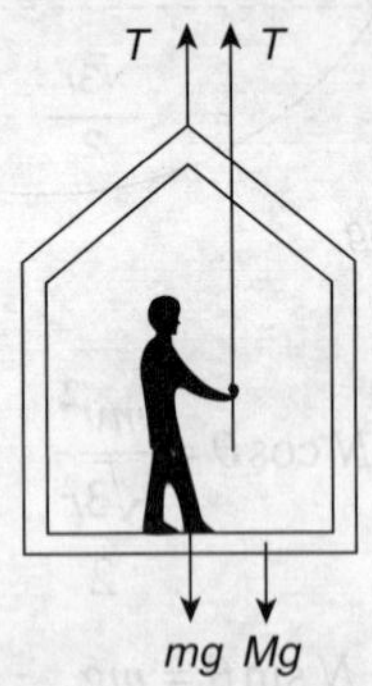

$$2T = (M+m)g$$
$$T = \frac{(M+m)g}{2}$$

20. (d)

$$s = \frac{1}{2}at^2$$

Maximum acceleration of car is

$$a = \mu g$$
$$s = \frac{1}{2}\mu g t^2$$
$$t^2 = \frac{2s}{\mu g} \Rightarrow t = \sqrt{\frac{2s}{\mu g}}$$
$$t \propto \frac{1}{\sqrt{\mu}}$$

21. (c) Let $2l$ be the length of inclined plane

Let f be the friction force in part BC

Work done against friction in part BC

$$w = fl$$

No work is done in covering upper part AB.

$$U = mgh$$
$$\sin 30° = \frac{h}{2l}$$
$$\Rightarrow \quad h = 2l\sin 30° = l$$
$$U = w$$
$$mgl = fl$$
$$f = mg$$

22. (b)

$$\mu\, mg\, \cos 60° = mg\sin 60°$$
$$\mu = \sqrt{3}$$

Now, if force F is applied to pull it along the incline,

$$F = mg\sin 60° + \mu mg\cos 60°$$
$$F = \frac{5\sqrt{3}}{2} + \frac{5\sqrt{3}}{2}$$
$$F = 5\sqrt{3}N = 8.66\text{ N} \approx 9\text{ N}$$

23. (a) Both the scales will read 10 kg as tension in ideal balances connected in series remain same.

24. (a) The force of friction between the blocks can be in vertical direction only. Since there is no net force or acceleration in this direction, force of interaction between the blocks is zero in all cases.

25. (a) Speed of plane = 720 km/h = 200 m/s

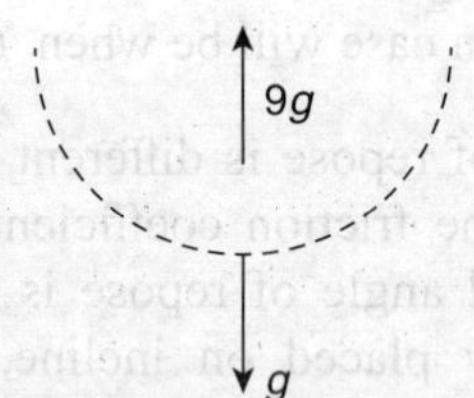

$$a = \frac{v^2}{r}$$
$$r = \frac{v^2}{a} = \frac{(200)^2}{9g - g} = \frac{40000}{8\times 10}$$
$$r = 500\text{ m}$$

26. (a), (c) and (d); Angular speed is constant. Hence linear speed is also constant, i.e., magnitude of velocity is constant, but direction is changing.

27. (b)

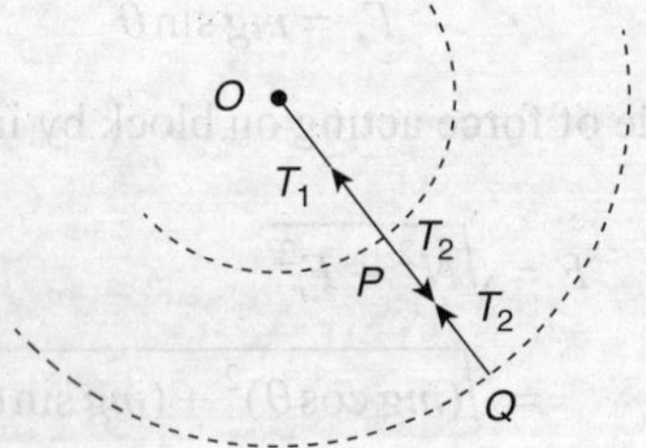

We know that centripetal force = $mr\omega^2$

For point $P, T_1 - T_2 = m\times 1\times\omega^2$ ...(i)

For point $Q, T_2 = m\times 2\times\omega^2$ ...(ii)

Adding Eq. (i) and Eq (iii), we get

$$T_1 = 3m\omega^2 \quad \ldots\text{(iii)}$$

Dividing Eq. (ii) by Eq. (iii), we have

$$\frac{T_2}{T_1} = \frac{2m\omega^2}{3m\omega^2} = \frac{2}{3}.$$

28. (b) In the second case, the weight of the bird will be excluded.

29. (a)

$$F = \frac{d}{dt}(p) = \frac{d}{dt}(a + b\,t^2) = 2\,b\,t \propto t.$$

30. (c) Force $F = ma = 0.15 \times 20 = 3$ N

Impulsive force $= F \times dt$

$= 3 \times 0.1 = 0.3$ N

31. (a) Here the set of forces $F_1 = 3$ N, $F_2 = 5$ N, $F_3 = 6$ N are in equilibrium because sum of F_1 and F_2 is greater than F_3 and their difference is less than F_3.

32. (c) Way up

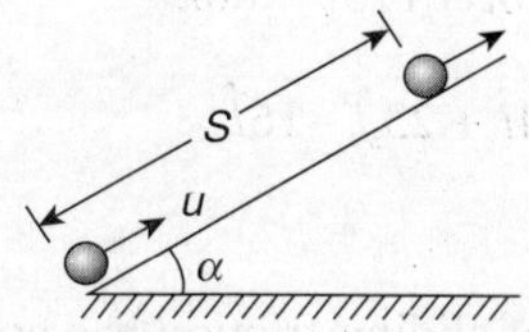

$$a_1 = -(g\sin\alpha + \mu g\cos\alpha)$$

$$S = ut_1 - \frac{1}{2}(g\sin\alpha + \mu g\cos\alpha)t_1^2 \quad \ldots(1)$$

$$S = \frac{1}{2}(g\sin\alpha + \mu g\cos\alpha)t_2^2 \quad \ldots(2)$$

Given $\quad t_1 = \frac{t_2}{2}$

From I equation while way up,

$$0 = u - (g\sin\alpha + \mu g\cos\alpha)t_1$$

$$u = (g\sin\alpha + \mu g\cos\alpha)t_1 \quad \ldots(3)$$

From (1) and (3), we get

$$S = \frac{1}{2}(g\sin\alpha + \mu g\cos\alpha)t_1^2$$

$$t_1 = \sqrt{\frac{2S}{g\sin\alpha + \mu g\cos\alpha}}$$

$$t_2 = \sqrt{\frac{2S}{g\sin\alpha - \mu g\cos\alpha}}$$

$$\Rightarrow \quad \frac{2S}{g\sin\alpha + \mu g\cos\alpha} = \frac{1}{4}\frac{2S}{g\sin\alpha - \mu g\cos\alpha}$$

$$4g\sin\alpha - 4\mu g\cos\alpha = g\sin\alpha + \mu g\cos\alpha$$

$$3g\sin\alpha = 5\mu g\cos\alpha$$

$$\mu = \frac{3}{5}\tan\alpha = 0.6\tan\alpha$$

33. (a)

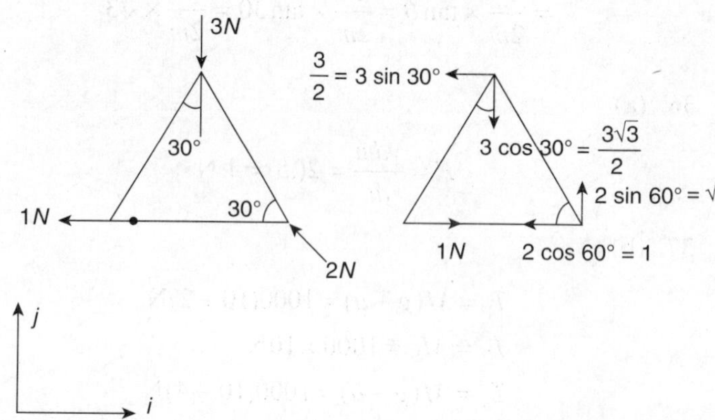

Resultant force on block is

$$\vec{F} = 1\hat{i} - 1\hat{j} - \frac{3}{2}\hat{i} + \sqrt{3}\hat{j}$$

$$\vec{F} = \frac{-3}{2}\hat{i} - \frac{\sqrt{3}}{2}\hat{j}$$

$$|\vec{F}| = \sqrt{\frac{9}{4} + \frac{3}{4}} = \sqrt{\frac{12}{4}} = \sqrt{3}\text{ N}$$

34. (c) Apply Newton's second law of motion on each mass

$$72 - T = 3a$$

$$T - 32 = 7a$$

On solving, $\quad a = 4\text{ ms}^{-2}$

$$T = 60N = kx$$

$$60 = 1000 \times x$$

$$x = \frac{60}{1000}\text{ m} = 6\text{ cm.}$$

35. (b)

$$F = 2T\cos\theta$$

$$T\sin\theta = m\,a;\; a = \frac{T\sin\theta}{m} = \frac{F}{2\cos\theta} \times \frac{\sin\theta}{m}$$

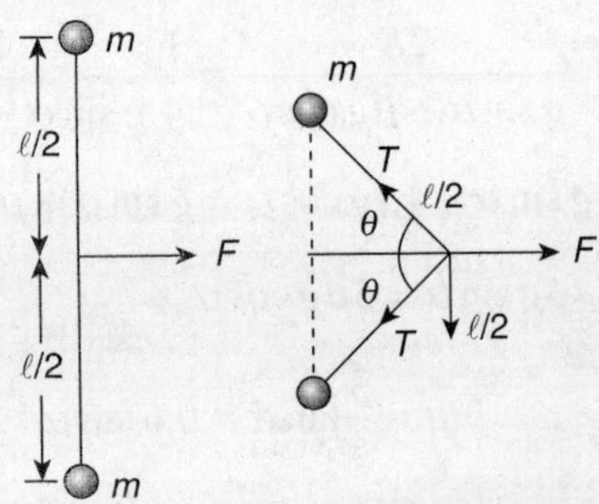

$$= \frac{F}{2m} \times \tan\theta = \frac{F}{2m} \times \tan 30 = \frac{F}{2m} \times \sqrt{3}$$

36. (a)

$$F = \frac{vdm}{dt} = 2(.5) = 1\text{ N}$$

37. (b)

$$T_1 = M(g+a) = 1000(10+2)\text{N}$$
$$T_2 = Mg = 1000 \times 10\text{N}$$
$$T_3 = M(g-a) = 1000(10-4)\text{N}$$

Ratio $T_1 : T_2 : T_3 :: 12 : 10 : 6$

38. (b)

Let l be the length of chain and m is its mass μ

$$\frac{m}{l}(l - \frac{l}{n})g = \frac{m}{l} \times \frac{l}{n} g, \mu = \frac{1}{n-1}$$

39. (c)

Tension in the string connecting masses, when the pulley is stationary

$$T = \frac{2m_1 m_2 g}{m_1 + m_2}$$
$$= \frac{2(m_1 g)(m_2 g)}{m_1 g + m_2 g}$$
$$T = \frac{2\ W_1 W_2}{W_1 + W_2}$$

When the pulley is accelerated upward with acceleration 'g' each weight is doubled as $W' = m(g+a)$

$$= m(g+g) = 2mg$$

$$\therefore \text{ Required tension } T' = \frac{2 \times 2W_1 \times 2W_2}{2W_1 + 2W_2}$$
$$= \frac{4\ W_1 W_2}{W_1 + W_2}$$

40. (d)

When elevator accelerates downward, apparent weight is $= M(g-a)$
When elevator accelerates upward apparent weight is $M\ (g + a)$. Obviously apparent weight is maximum if elevator accelerates upwards.

41. (a)

$$S_{\text{Horizontal}} = ut = 1.5 \times 4 = 6\text{ m}$$
$$S_{\text{Vertical}} = \frac{1}{2}at^2 = \frac{1}{2}\frac{F}{m}t^2 = \frac{1}{2}1 \times 16 = 8\text{ m}$$
$$S_{\text{Net}} = \sqrt{6^2 + 8^2} = 10\text{ m}$$

42. (a)

From the law of conservation of linear momentum

$$m\vec{v} = m_1\vec{v}_1 + m_2\vec{v}_2$$
$$\Rightarrow 6k\left(20\hat{i} + 25\hat{j} - 12\hat{k}\right) = k\left(100\hat{i} + 35\hat{j} + 8\hat{k}\right) + 5k\vec{v}_2$$
$$\Rightarrow 5\vec{v}_2 = (120 - 100)\hat{i} + (150 - 35)\hat{j} + (-72 - 8)\hat{k}$$
$$\Rightarrow 5\vec{v}_2 = 20\hat{i} + 115\hat{j} - 80\hat{k}$$
$$\Rightarrow \vec{v}_2 = 4\hat{i} + 23\hat{j} - 16\hat{k}$$

43. (d)

Coefficient of sliding friction has no dimension.

$$f = \mu_s N \Rightarrow \mu_s = \frac{f}{N}$$

44. (c)

Let v be tangential speed of heavier stone. Then, centripetal force experienced by lighter stone is

$$(F_c)_{\text{lighter}} = \frac{m(nv)^2}{r}$$

and that of heavier stone is

$$(F_c)_{\text{heavier}} = \frac{2mv^2}{(r/2)}$$

But $(F_c)_{lighter} = (F_c)_{heavier}$ (given)

$$\therefore \frac{m(nv)^2}{r} = \frac{2mv^2}{(r/2)}$$
$$n^2\left(\frac{mv^2}{r}\right) = 4\left(\frac{mv^2}{r}\right)$$
$$n^2 = 4 \text{ or } n = 2$$

45. (c)

Force of friction on mass $m_2 = \mu m_2 g$

Force of friction on mass $m_3 = \mu m_3 g$

Let a be the common acceleration of the system.

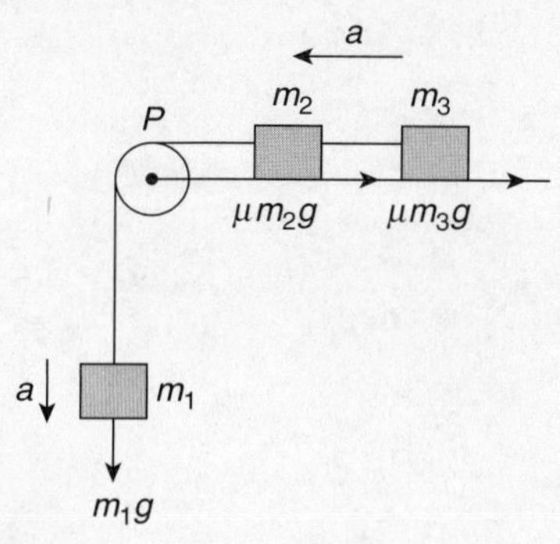

$$\therefore a = \frac{m_1 g - \mu m_2 g - \mu m_3 g}{m_1 + m_2 + m_3}$$

Here, $m_1 = m_2 = m_3 = m$

$$\therefore a = \frac{mg - \mu mg - \mu mg}{m + m + m} = \frac{mg - 2\mu mg}{3m} = \frac{g(1 - 2\mu)}{3}$$

Work, Power and Energy

Introduction

In daily life, the term work is applied to any activity requiring the exertion of muscular or mental effort. In Physics, the term is applied in a much more specific sense, and denotes an action which requires the application of a force to a body and displacement of that body.

Definition of Work

Let d be the distance moved by a body along a straight line, while acted upon by a constant force of magnitude F in the same direction as the motion. The work W done by the force is

$$W = Fd$$

However, the constant force acting on the body may not be in the direction in which the body moves. In the figure shown below, the constant force F makes an angle θ with the displacement.

The work done W by this force when its point of application undergoes a displacement d is defined as the product of the magnitude of displacement and the component of the force in the direction of the displacement.

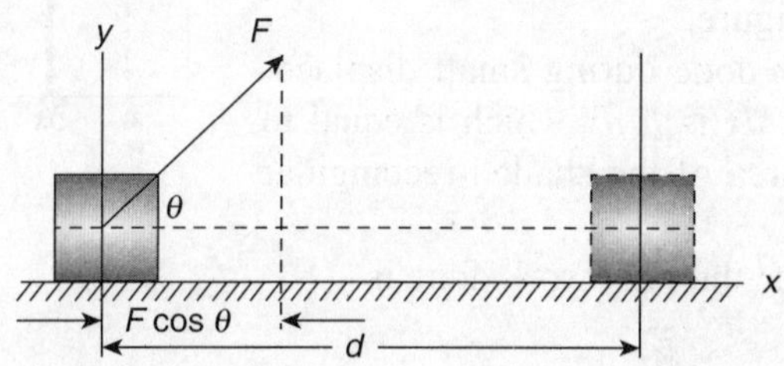

The component of F in the direction of d is $F \cos \theta$. Then

$$W = F \cos \theta\, d$$

An alternative interpretation of the above expression is that $d \cos \theta$ is the component of displacement in the direction of F.

Thus the work may also be defined as the product of magnitude of force and displacement in the direction of force. Comparing the expression, $W = Fd \cos \theta$ with the definition of the scalar product of two vectors, we observe that work can also be expressed as the scalar product of force and displacement vectors.

$$W = \vec{F}.\vec{d}$$

Work is a scalar quantity. However, it is an algebraic quantity; it can be positive or negative. When the component of force is in the same direction as the displacement, ($\theta = 0$) the work W is positive. If the force is at right angles ($\theta = 90°$) to the displacement it has no component in the direction of the displacement and work is zero. When it is opposite to the displacement, the work W is negative ($\theta = 180°$).

If θ is acute, work is positive.

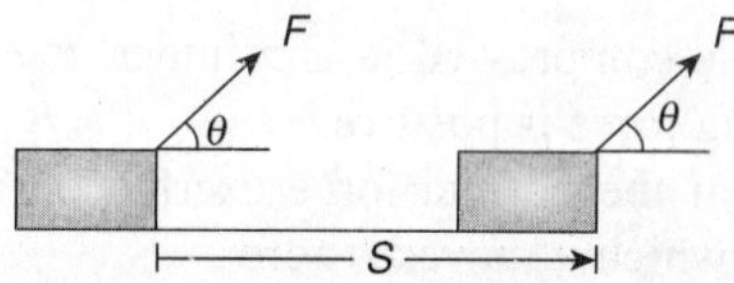

If $\theta = 90°$, force and displacement vectors are mutually perpendicular and $W = 0$.

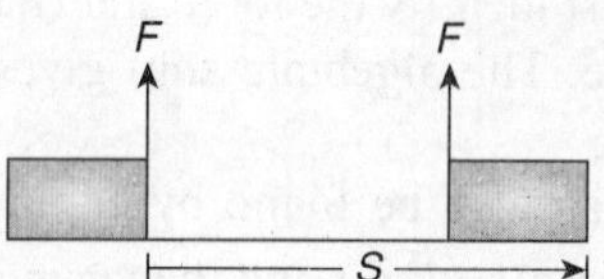

If θ is obtuse, work is negative.

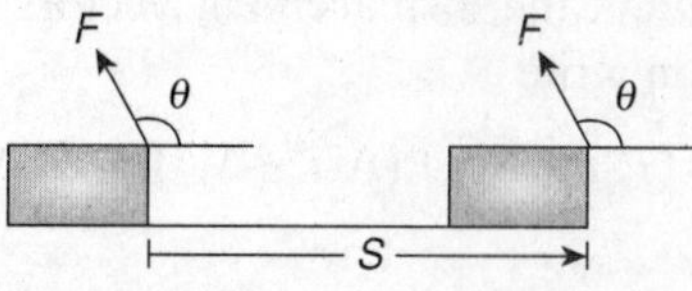

Illustration 1

Consider a block sliding down a rough inclined plane. The work done by gravity is positive (W_g) and tends to speed up the block, whereas the work (W_{f_r}) done by friction (direction opposite to the displacement) is negative.

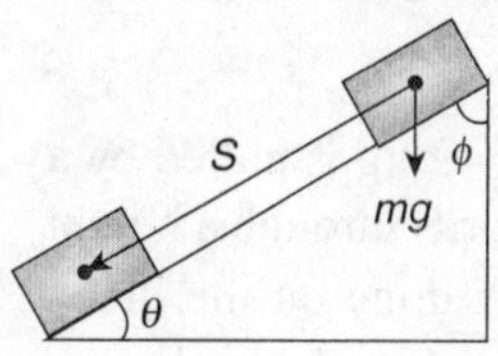

Solution

$$W_g = mg \cdot S\cos\phi = mg\ S\sin\theta \text{ is positive.}$$

$$W_{fr} = -\mu_K mg\cos\theta\ S \text{ is negative.}$$

The third force acting on the block is the normal reaction N which does zero work since its direction is perpendicular to the displacement.

The SI unit of work is joule (abbreviated as J) and is defined as the work done by a force of 1 N to produce a displacement of 1 m in the same direction.

Therefore, $1\text{ J} = 1\text{ N.m}$

(a) When a body is lifted, the work done by the lifting force is positive.

(b) The work of gravitational force on a body being lifted is negative.

(c) When a body slides on a stationary surface, the work of the frictional force exerted on the body is negative.

(d) When a spring is compressed or stretched by a force, the work done by the compressing or stretching force is positive while the work done by the elastic force is always negative.

(e) When gas is compressed in a cylinder, the work of the compressing force is positive.

(f) The work of the normal force exerted on a body by a surface on which it moves is zero.

(g) The work of the centripetal force on a body moving in a circle is also zero.

(h) In order to find the total work done on a body by several forces, we first identify the force and find the work done by each force. The algebraic sum gives the total work done.

(i) Total work can also be found by computing the vector sum of the force $(= \vec{F}_{net})$ and then using the relation:

$$W = \vec{F}_{net}.\vec{S}$$

(j) If the force and the displacement vectors lie in the x-y plane, we can write

$$W = \vec{F}.\vec{S} = (F_x\hat{i} + F_y\hat{j}).(\Delta x\hat{i} + \Delta y\hat{j}) = F_x\Delta x + F_y\Delta y$$

Illustration 2

You push your physics book 1.00 m along a horizontal table with a force of 2.00 N parallel to the plane of table. How much work does your 2.00 N force do on the book?

Solution

$$W = \vec{F}.\vec{s}$$

$$= Fs\cos\theta$$

$$= Fs = 2.00 \times 1.00 = 2.0\text{ Joule}$$

Illustration 3

A block of mass $m = 2$ kg is pulled by a force $F = 40$ N upwards through a height $h = 2$ m. Find work done on the block by the applied force F and weight mg. ($g = 10\text{ m/s}^2$)

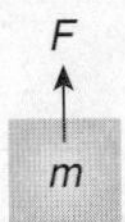

Solution

$$\text{Weight } mg = (2)(10) = 20\text{ N}$$

Work done by the applied force $W_F = F\,h\cos 0°$.

As the angle between force and displacement is 0°

or $$W_F = (40)(2)(1) = 80\text{ J}$$

Similarly, work done by its weight

$$W_{mg} = (mg)(h)\cos 180°$$

or $$W_{mg} = (20)(2)(-1) = -40\text{ J}$$

Illustration 4

A particle moves from a point $\vec{r}_1 = (5\text{ m})\hat{i} + (5\text{ m})\hat{j}$ to another point $\vec{r}_2 = (6\text{ m})\hat{i} + (4\text{ m})\hat{j}$ during which a certain force $\vec{F} = (5\text{ N})\hat{i} + (5\text{ N})\hat{j}$ acts on it. Find the work done by the force on the particle during the displacement.

Solution

Displacement vector

$$\vec{S} = \vec{r}_2 - \vec{r}_1 = (\hat{i} - \hat{j})\text{m}.$$

$$\text{Work done} = \vec{F}.\vec{S} = (5\hat{i} + 5\hat{j}).(\hat{i} - \hat{j})$$

$$= (5 - 5) = 0.$$

Work done by a variable force

The equation $W = \vec{F}.\vec{S} = FS\cos\theta$ is valid when F remains constant but when the force is variable, total work done is obtained by integration method using $W = \int dW = \int \vec{F}.d\vec{S}$.

(a) When the force is some function of position

Let a force $F = F(x)$ act along the x-axis as shown in the F–x graph in the figure.

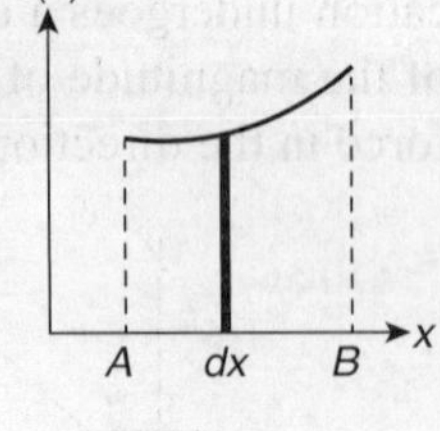

Work done during small displacement dx is $F\,dx$ which is equal to the area of the shaded rectangular strip.

Hence the total work done is

$$F = \int_A^B F\,dx = \text{Area under the } F-x \text{ graph}$$

An example of the variable force is the spring force, in which the force F is proportional to the extension x and is given by $F = -kx$, where k = force constant and the negative sign indicates opposite direction of spring force F and extension x.

$$\therefore \quad \text{Work done by spring} = \int \vec{F}.d\vec{x} = -\int_{x_1}^{x_f} kx\,dx$$

$$= -\frac{1}{2}k(x_f^2 - x_i^2).$$

If spring be initially relaxed, then work done by the spring (or, work done against the external agent) for an extension x is obtained by putting $x_i = 0$ and $x_f = x$ leading to

$$W = -\frac{1}{2}kx^2$$

Note that the work done by the spring is negative whereas the work done by the external agent producing the extension is positive. The two, however, are equal in quantity.

(b) When the force is time dependent

Let $F = F(t)$, then

$$W = \int \vec{F}.d\vec{x} = \int \vec{F}.\left(\frac{d\vec{x}}{dt}\right)dt = \int \vec{F}.\vec{v}dt, \text{ where } \vec{F} \text{ and } \vec{v}$$

are instantaneous force and velocity vectors expressed as functions of time.

Work Depends on the Reference Frame

Consider a situation in which a person pushes a block inside a moving train through a distance S. The work done on the block in the train's frame will be $\vec{F}.\vec{S}$. The work done on the block in the ground frame will be $\vec{F}.(\vec{S}+\vec{S}_0)$, where $\vec{S}_0$ is the displacement of the train relative to the ground.

Work done by friction

Work done by friction may be zero, positive or negative depending upon the situation.

(i) When the applied force is insufficient to overcome friction, work done by friction is zero.

(ii) When external force is greater than limiting friction, the work done by friction will be negative since friction and displacement are in opposite directions.

(iii) When force is applied on a block A placed over another block B, the work done by friction on the block B is positive since frictional force and displacements have the same direction.

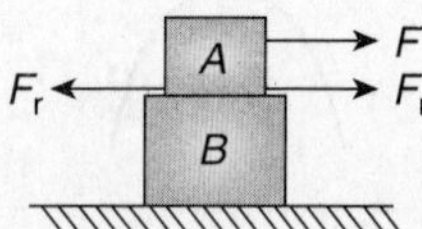

Let us consider the following situation to illustrate the work done by friction as observed in different frames: Suppose that an observer A and a block are stationary in a trolley accelerating in the horizontal direction. The horizontal surface of the trolley is sufficiently rough, so that the block does not slip.

(i) To an observer A, the work done by friction (between trolley and block) is zero.

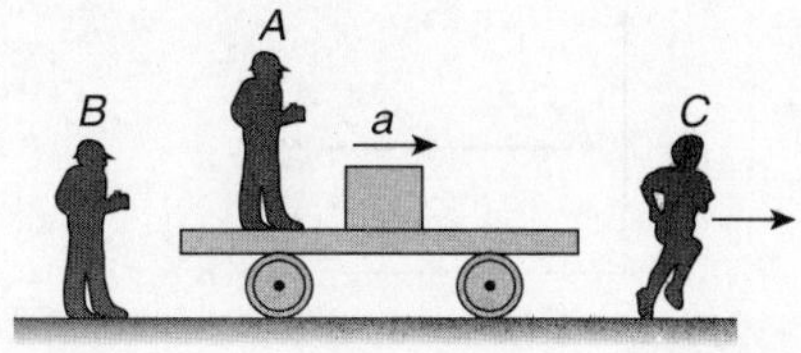

(ii) To an observer B, on the ground, work done by friction on the block is positive.

(iii) To an observer C, who is running faster than the trolley in the same direction, work done by friction is negative.

Conservative and Non-conservative Forces

In mechanics, we encounter two types of forces: conservative force and non-conservative force.

Conservative Force

A force is said to be conservative if the work by the force in moving a mass through any closed path is zero. Alternatively, if the net work done by the force in moving a mass between two points is path independent, i.e., the work done depends only on the location of two points and not on the path followed, the force may be termed conservative. Conservative forces are non-dissipative and can store work (as energy)

Examples of conservative forces are Gravitational force, Electrostatic force etc.

Non-conservative force

The forces which do not fulfill the above-mentioned conditions are non-conservative. For these, the energy is not stored but dissipated during displacement. Examples of non-conservative forces are friction, viscous forces etc.

Conservative force and potential energy

Potential energy of any object is the energy possessed by the object by virtue of its position or configuration. Each type of potential energy is associated with a corresponding conservative force and therefore it depends only on position. The potential energy is measured as the magnitude of the work done against the associated conservative force.

For every conservative force $F(x)$ (a function of position x) there exists an associated potential energy function $U(x)$. When conservative force does positive work on a system the potential energy of the system decreases and vice versa. Thus work done by conservative force:

$$F(x)\,\Delta x = -\Delta U$$

or,

$$F(x) = -\frac{\Delta U}{\Delta x}$$

In differential form,

$$F(x) = -\frac{dU}{dx}$$

Integrating both sides for a displacement from $x = x_1$ to $x = x_2$, we get,

$$U_2 - U_1 = -\int_{x_1}^{x_2} F(x)\,dx$$

Mechanical energy The following are the two kinds of mechanical energy.

(i) Kinetic energy: Energy possessed by a body because of its motion is known as kinetic energy and it is $= \frac{1}{2}mv^2$.

(ii) Potential energy: When work is done on a system, it gains energy. If the system stores this energy in such a way that it can be subsequently recovered, the system is said to possess potential energy. Examples of potential energy are gravitational, elastic (in stretched or expanded spring), and electrostatic (interaction between charges) potential energies.

Illustration 5

The potential energy of a diatomic molecule is expressed as $U = \frac{C}{x^{12}} - \frac{D}{x^6}$, where x is the separation between the atoms which constitute the molecule and C and D are positive constants. Find the separation between the atoms when the system is in equilibrium.

Solution

The interaction force between the atoms is given by

$$F_x = -\frac{dU}{dx} = -\frac{d}{dx}\left[\frac{C}{x^{12}} - \frac{D}{x^6}\right]$$

$$= \frac{12C}{x^{13}} - \frac{6D}{x^7}$$

At equilibrium, $\quad F_x = 0,\ x^6 = \frac{2C}{D}$

$$x = \left(\frac{2C}{D}\right)^{\frac{1}{6}}$$

Importance of potential energy

We know that for a conservative force there exists a potential energy function $U(x)$ which is a function of position x. Also for a conservative isolated system, the total mechanical energy is constant,

i.e., $\quad K + U = \text{Constant}$

or, $\quad \frac{1}{2}mv^2 + \frac{1}{2}kx^2 = \text{constant (for a spring-mass system)}$

$\therefore \quad \frac{d}{dx}\left(\frac{1}{2}mv^2 + \frac{1}{2}kx^2\right) = 0$

or, $\quad mv\frac{dv}{dx} + kx = 0$

or, $\quad ma + kx = 0 \qquad \left[\because a = \frac{dv}{dt} = v\frac{dv}{dx}\right]$

or, $\quad ma = -kx.$

But $\quad \frac{d}{dx}(\text{KE}) = ma,$

and $\quad \frac{d}{dx}(U) = kx.$

$\therefore \quad F = ma = -\frac{dU(x)}{dx}.$

Hence $F = -\frac{dU(x)}{dx}$, i.e., the force of interaction is the negative of the slope of potential energy function with respect to x.

Equilibrium and potential energy

In purely translational motion, a system is said to be in equilibrium if $\vec{F}_{\text{net}} = \vec{0}$.

For conservative forces, $F = -\frac{dU}{dr}$

Thus, for equilibrium under conservative forces, $\frac{dU}{dr} = 0$.

Hence the slope of U–r graph is zero at equilibrium position. This means that the potential energy is either a maximum or a minimum. Equilibrium states are of three types when both the conditions $\vec{F}_{\text{net}} = \vec{0}$ and $\frac{-dU}{dr} = 0$ are simultaneously satisfied. These are:

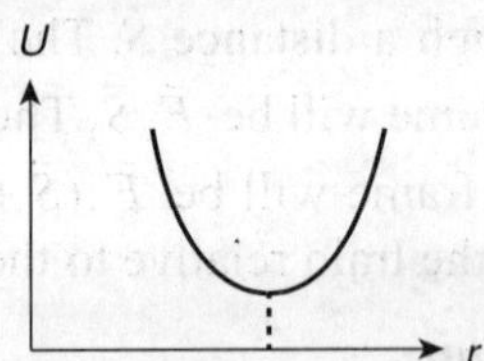

(i) Stable equilibrium if U is a minimum. Mathematically,

$$\frac{dU}{dr} = 0, \text{ and}$$

$$\frac{d^2U}{dr^2} > 0$$

$\Rightarrow$ Stable equilibrium.

(ii) Unstable equilibrium if U is a maximum.

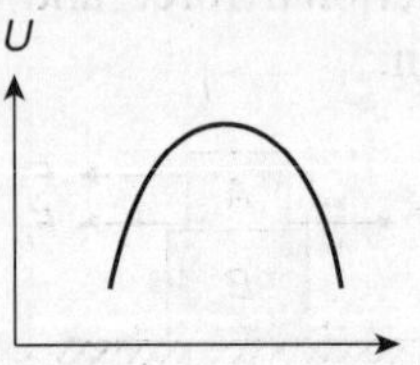

Mathematically, $\quad \frac{dU}{dr} = 0$, and

$$\frac{d^2U}{dr^2} < 0$$

$\Rightarrow$ Unstable equilibrium.

(iii) Neutral equilibrium if U is flat i.e. is independent of r.

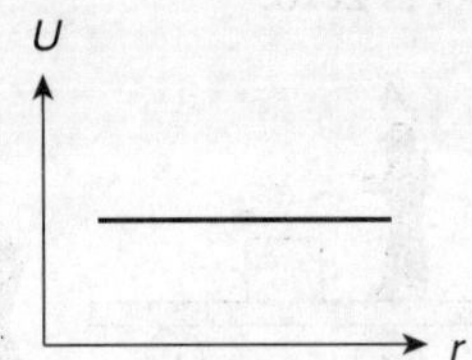

Work-Energy Theorem

This theorem gives a convenient approach to solve mechanics problems without going into the analytical details of Newton's laws of motion. The theorem states that: "When a body is subjected to a number of forces, both conservative and non-conservative, the increase in its kinetic energy is always equal to the total work done by all the forces which produce a given displacement." Thus

$$W_{\text{net}} = K_f - K_i = \Delta K = \text{Change in KE}$$

In the simplest case, let us consider a block of mass m moving along a straight line (x-axis), under the action of a constant force F.

$$F = m\frac{dv}{dt} = mv\frac{dv}{dx}$$

$$\therefore \quad \text{Work done, } W = \int F\,dx = \int mv\frac{dv}{dx}.dx$$

$$= \int_u^v mv\,\text{dv} = \frac{1}{2}m(v^2 - u^2) = \text{Change in KE}$$

This theorem can be generalized for a body moving in three dimensions when it is subjected to a number of forces (both conservative and non-conservative).

Consider a rigid body which is moving under the action of a number of component forces given by $\vec{F}_{\text{net}} = \vec{F}_1 + \vec{F}_2 + \vec{F}_3 + \ldots$

$$\text{But,} \quad \vec{F}_{\text{net}} = m\frac{d\vec{v}}{dt}$$

$$\therefore \quad \vec{F}_1 + \vec{F}_2 + \vec{F}_3 + \ldots = m\frac{d\vec{v}}{dt}$$

Let the body be displaced through $d\vec{r} = \vec{v}dt$, then the net work done is,

$$dW = \vec{F}_1.d\vec{r} + \vec{F}_2.d\vec{r} + \ldots = m\frac{d\vec{v}}{dt}.\vec{v}\,dt$$

$$= m\vec{v}.d\vec{v}$$

In rectangular coordinate system,

$$\vec{v} = v_x\hat{i} + v_y\hat{j} + v_z\hat{k}$$

$$\text{and} \quad d\vec{v} = dv_x\hat{i} + dv_y\hat{j} + dv_z\hat{k}.$$

$$\therefore \quad dW = \overrightarrow{F_{\text{net}}}.\overrightarrow{dr} = m\overrightarrow{v}.\overrightarrow{dv} = m(v_x dv_x + v_y dv_y + v_z dv_z).$$

If the body is displaced from point A to point B, the total work done will be

$$W = \int dW = m\left[\int_A^B v_x dv_x + \int_A^B v_y dv_y + \int_A^B v_z dv_z\right]$$

$$\text{or} \quad W = \frac{m}{2}\left[v_x^2 + v_y^2 + v_z^2\right]_A^B = \frac{1}{2}m\left(v_B^2 - v_A^2\right).$$

$$\text{Thus} \quad W = \int \overrightarrow{F}_{net}.\overrightarrow{dr} = K_f - K_i = \Delta\text{KE}.$$

How to apply Work-Energy theorem

(i) Draw that FBD and select all the forces acting on the particle (or the body).

(ii) Find the dot product $\vec{F}.d\vec{r}$ for all the forces where $d\vec{r}$ is the displacement vector.

(iii) Find the algebraic sum of the work done by all forces.

(iv) Equate the total work done with the final kinetic energy minus the initial kinetic energy $(= K_f - K_i = \Delta\text{KE})$.

(v) If W_{net} is positive, ΔKE must be positive.

If W_{net} is negative, ΔKE must be negative, i.e., kinetic energy has decreased.

Illustration 6

An object of mass m is tied to a string of length l and a variable force F is applied on it which brings the string gradually to an angle θ with the vertical. Find the work done by the force F.

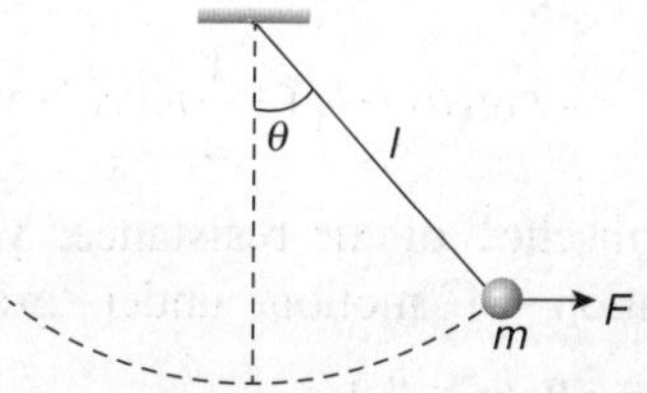

Solution

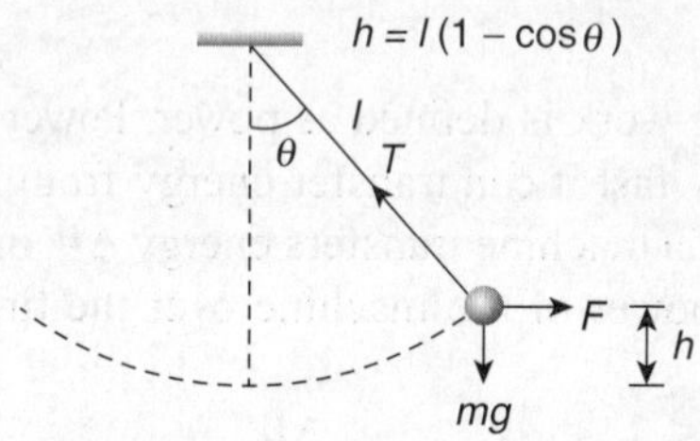

In this case three forces are acting on the object:

1. tension (T)
2. weight (mg) and
3. applied force (F)

Using Work–Energy theorem

$$W_{\text{net}} = \Delta\text{KE}$$

$$\text{or} \quad W_{\text{T}} + W_{\text{mg}} + W_{\text{F}} = 0 \quad \ldots(1)$$

$$\text{as} \quad \Delta\text{KE} = 0$$

$$\text{because} \quad K_{\text{i}} = K_{\text{f}} = 0$$

Further, $W_{\text{T}} = 0$, as tension is always perpendicular to displacement.

$$W_{\text{mg}} = -\,mgh \text{ or } W_{\text{mg}} = -\,mgl\,(1 - \cos\theta)$$

Substituting these values in Eq. (i), we get

$$W_{\text{F}} = mgl\,(1 - \cos\theta)$$

Illustration 7

A ball of mass m is projected from a height h_1 with speed v_1 in the vertically upward direction. If its speed reduces to v_2 at height h_2, find the work done on the ball by air resistance.

Solution

Work done by gravity is

$$W_g = -mg(h_2 - h_1) \quad [\because \text{Weight and displacement are opposite}]$$

Let the work done by air resistance be W_{air}.

From Work–Energy principle,

$$W_g + W_{air} = \frac{1}{2}mv_2^2 - \frac{1}{2}mv_1^2$$

or,
$$-mg(h_2 - h_1) + W_{air} = \frac{1}{2}m(v_2^2 - v_1^2)$$

$\therefore$
$$W_{air} = mg(h_2 - h_1) + \frac{1}{2}m(v_2^2 - v_1^2)$$

$$= mg(h_2 - h_1) - \frac{1}{2}m(v_1^2 - v_2^2)$$

Note: In the absence of air resistance, you will get the common equation of motion under gravity, which is $v_1^2 - v_2^2 = 2gh = 2g(h_2 - h_1)$.

Power

Rate of doing work is defined as power. Power of a machine indicates how fast it can transfer energy from itself to some other body. If a machine transfers energy ΔW in time Δt, then the average power of the machine over the time interval Δt is defined by,

$$P_{av} = \frac{\Delta W}{\Delta t}$$

The instantaneous power P is defined by,

$$P = \lim_{\Delta t \to 0} \frac{\Delta W}{\Delta t} = \frac{dW}{dt} = \frac{d}{dt}(\vec{F}.\vec{S})$$

$$= \vec{F}.\frac{d\vec{S}}{dt} = \vec{F}.\vec{V}$$

Power is a scalar quantity. The SI unit of power (Js^{-1}) is called watt. Instantaneous power of an automobile depends on both the force and the instantaneous velocity.

Illustration 8

A body of mass m is projected at an angle ϕ to the horizontal from the ground with a velocity u. Find (i) the mean power developed by gravity over the whole time of motion and (ii) the instantaneous power of gravity as a function of time.

Solution

(i) Mean power developed by gravity

$$= \frac{\text{Total work done}}{\text{Time of flight}} = \frac{0}{T} = \mathbf{0}.$$

(ii) Instantaneous power at time t is

$$\vec{F}.\vec{v} = mg\, v\cos(90° + \phi),$$

where ϕ is the angle between velocity vector at time t and the horizontal

$$\therefore\ P = -mgv\sin\phi = -mg(u\sin\theta - gt) = mg(gt - u\sin\theta).$$

Illustration 9

An automobile of mass m accelerates starting from rest, while the engine supplies constant power P. Show that

(a) the velocity is given as a function of time by $v = \left(\frac{2Pt}{m}\right)^{1/2}$.

(b) the position is given as a function of time by $s = \left(\frac{8P}{9m}\right)^{1/2} t^{3/2}$.

Solution

(a) Applying the Work–Energy theorem, we get

Total work done in time t = Total kinetic energy of automobile.

$$Pt = \frac{1}{2}mv^2$$

or
$$v = \sqrt{\frac{2Pt}{m}}.$$

(b) Using the expression for velocity $v = \frac{ds}{dt}$ in the above result, we get,

$$\frac{ds}{dt} = \sqrt{\frac{2P}{m}}\, t^{1/2}$$

Integrating from 0 to t, we get,

$$\int_0^s ds = \sqrt{\frac{2P}{m}} \int_0^t t^{1/2} dt \ \text{ or, } s = \left(\frac{8P}{9m}\right)^{1/2} t^{3/2}.$$

Motion in a Vertical Circle

This is a specific example of non-uniform circular motion. In this case, a particle of mass m is connected to a light inextensible string at one end and the other end is fixed at some point O. The particle is whirled in a vertical circle of radius r (= length of the string). The forces acting on the particle when the string is inclined at an angle θ with the vertical are

(i) Weight mg, vertically downward, and

(ii) Tension T, along the string.

Resolving mg along the radial and tangential direction, we see that the component $mg \sin\theta$ decelerates the motion and the net radial force $T - mg \cos \theta$ provides the required centripetal force.

If the speed of the particle at P be v,

we have, $$T - mg\cos\theta = \frac{mv^2}{r}$$

or $$T = \frac{mv^2}{r} + mg\cos\theta.$$

At the lowermost point A, the speed is maximum so that tension is maximum,

$$T_{\max} = \frac{mv_A^2}{r} + mg$$

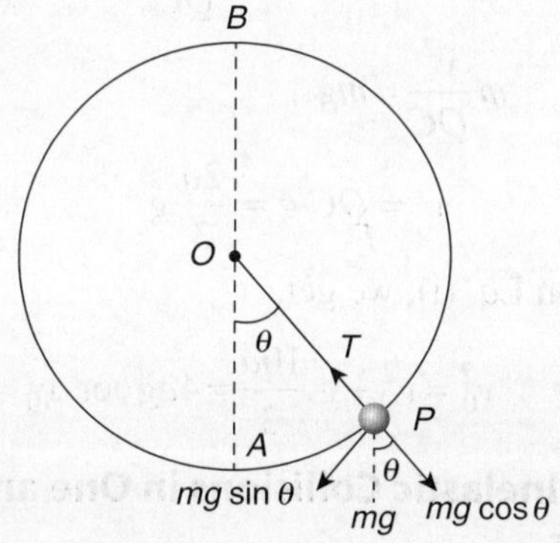

If the speed at the uppermost point be v_B (minimum speed), the tension in the string is minimum:

$$T_{\min} = \frac{mv_B^2}{r} - mg$$

Critical velocity

It is the minimum velocity imparted to the particle so that it completes the circle. As the particle goes up, the tension in the string decreases and the string has a tendency to slacken. This tendency is maximum at the topmost point B.

Now, $$T_B = \frac{mv_B^2}{r} - mg.$$

The minimum value of T_B is zero, which leads to the required minimum speed at B:

$$\frac{mv_B^2}{r} = mg, \text{ or } v_B = \sqrt{gr}.$$

This is the critical (or minimum) speed at the topmost point.

In order to find the corresponding critical velocity at the lowest point, we apply the Work–Energy theorem:

$$W_g = K_f - K_i$$

or, $$mg(2r) = \frac{1}{2}mv_A^2 - \frac{1}{2}mv_B^2.$$

Substituting $v_B = \sqrt{gr}$, we get $v_A = \sqrt{5gr}$.

This is the required critical velocity at the lowest point of the vertical circle.

Note: If the particle is attached with a light rod, the critical velocity at B and A will be zero and $2\sqrt{gr}$ respectively.

Illustration 10

A point mass m starts from rest and slides down the surface of a frictionless solid sphere of radius R as shown in the figure. At what angle will this body break off the surface of the sphere? Find the velocity with which it will break off.

Solution

Applying the principle of conservation of energy, we have at the points A and B,

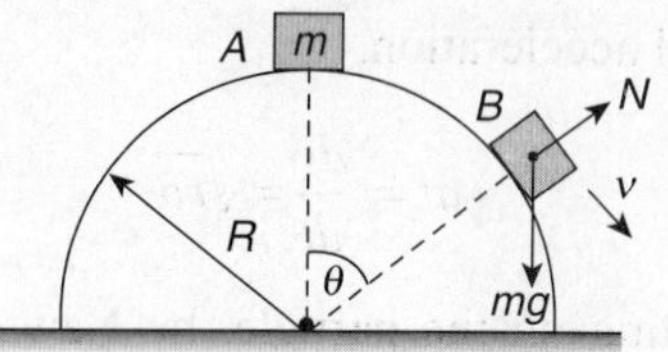

$$mgR\,(1 - \cos\theta) = \frac{1}{2}mv^2 \quad \text{...(i)}$$

Applying $F_{\text{net}} = ma$, we get,

$$mg\cos\theta - N = mv^2/R$$

$N = 0$ for break off.

$$\therefore \quad v = \sqrt{gR\cos\theta}. \quad \text{...(ii)}$$

Putting it in (i), we get,

$$\cos\theta = 2/3$$

Putting this in (ii), we get

$$v = \sqrt{\frac{2}{3}gR}$$

Illustration 11

A particle of mass m moves along a circle of radius r with a normal acceleration varying with time as $an = at^2$, where a is a constant. Find the time dependence of the power developed by all the forces acting on the particle, and the mean value of this power averaged over the first t seconds after the beginning of motion.

Solution

The diagram showing the particle's motion is shown here. The centre of the circle of rotation is chosen as the origin, with the particle's position being represented by P, at time t. We choose the x-axis along OP, and the y-axis perpendicular to it.

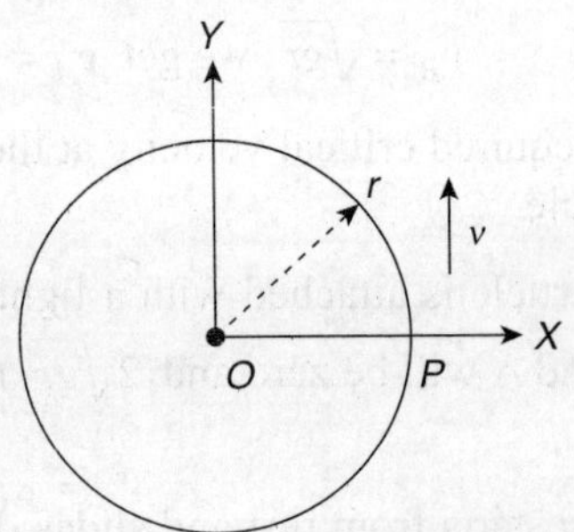

The velocity is $\vec{v} = v\hat{j}$

where, $\frac{v^2}{r} = a_n = at^2$

or, $v = \sqrt{ra}.t$

The tangential acceleration,

$$at = \frac{dv}{dt} = \sqrt{ra}$$

The force acting on the particle, by Newton's second law, equals

$$\vec{F} = m\vec{a} = m\left(-a_n\hat{i} + a_t\hat{j}\right)$$

The net power is

$$P = \vec{F}.\vec{v} = m(-a_n\hat{i} + a_t\hat{j}).v\hat{j}$$
$$= matv = m\sqrt{ra}.\sqrt{ra}.t$$

or, $P(t) = mrat.$

The average power,

$$(P)av = \frac{1}{t}\int_o^t P(t)\,dt = \frac{1}{t}\int_o^t mrat\,dt$$
$$= \frac{mrat}{2}$$

Illustration 12

A particle is hanging from a fixed point O by means of a string of length a. There is a small nail Q in the same horizontal line with O at a distance b ($b = a/3$) from O. Find the minimum velocity with which the particle should be projected so that it may make a complete revolution around the nail without the string being slackened.

Solution

Suppose that the velocity with which the ball is projected at A so as to just reach the highest point C, is v_0. Since the forces acting on the system are conservative and constraint forces, mechanical energy is conserved (between A and C):

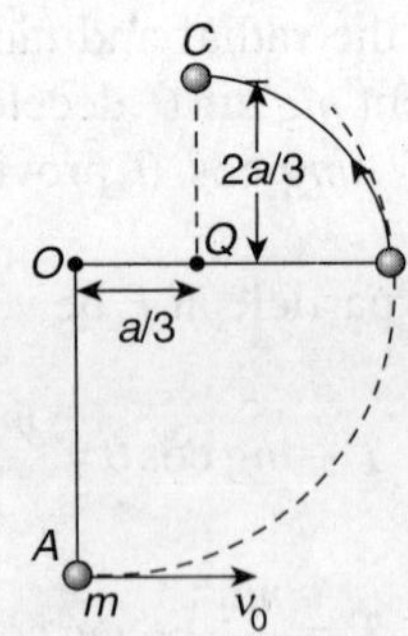

$$\frac{1}{2}mv_0^2 = \frac{1}{2}mv^2 + mg\left(a + \frac{2a}{3}\right) \quad \text{...(i)}$$

The string just gets slack at the highest point C, i.e. the tension is zero.

$\therefore$ The centripetal acceleration, $\frac{v^2}{QC}$ is caused by the weight

$$m\frac{v^2}{QC} = mg$$

or $$v^2 = QC \cdot g = \frac{2a}{3}g \quad \text{...(ii)}$$

Substituting in Eq. (i), we get,

$$v_0^2 = v^2 + g.\frac{10a}{3} = 4ag \text{ or } v_0 = 2\sqrt{ag}.$$

Elastic and Inelastic Collisions in One and Two Dimensions

Collisions

A collision is said to take place when either two bodies physically collide against each other or when the path of one body is changed by the influence of some force of the other body.

The result of collision:

(i) The momentum and kinetic energy of the interacting bodies many change
(ii) The interacting forces are internal (action–reaction); therefore, total momentum as well total energy is conserved.

We can broadly classify collision into two categories:

(1) Elastic
(2) Inelastic.

Elastic collision

A collision is said to be an elastic collision if both the kinetic energy and momentum are conserved in the collision.

During collision, the bodies are deformed but in elastic collision they regain their original shape completely. In elastic collision the interacting forces are conservative in nature.

Inelastic collision

A collision is said to be an inelastic collision if the kinetic energy is not conserved in the collision. However, the momentum is conserved.

The loss in kinetic energy appears in the form of heat, sound and light energy. Here, the forces are non-conservative. Most of the collisions between macroscopic bodies are inelastic.

Perfectly inelastic or plastic collision

If the two bodies stick together after the collision, the collision is said to be perfectly inelastic.

One-dimensional elastic collision (Head-on collision)

One-dimensional elastic collision is that elastic collision in which the colliding bodies move along the same straight line path before and after the collision.

Consider two bodies of masses m_1 and m_2 respectively with velocities $\vec{v}_{1i}$ and $\vec{v}_{2i}$ moving along the same straight line in the same direction and $|\vec{v}_{1i}|>|\vec{v}_{2i}|$

Since they are moving in the same direction along the same line and velocity of one body is greater than that of the other, they will collide after some time.

Let $\vec{v}_{1f}$ and $\vec{v}_{2f}$ be velocities of bodies respectively after the collision.

Applying the law of conservation of momentum,

Total momentum before collision = Total momentum after collision

$$\therefore \quad m_1v_{1i} + m_2v_{2i} = m_1v_{1f} + m_2v_{2f} \quad \ldots\text{(i)}$$

Kinetic energy before collision = Kinetic energy after collision

$$\frac{1}{2}m_1v_{1i}^2 + \frac{1}{2}m_2v_{2i}^2 = \frac{1}{2}m_1v_{1f}^2 + \frac{1}{2}m_2v_{2f}^2 \quad \ldots\text{(ii)}$$

From Eqs. (i) and (ii) we get

$$v_{1f} = \frac{m_1 - m_2}{m_1 + m_2}v_{1i} + \frac{2m_2}{m_1 + m_2}v_{2i} \quad \ldots\text{(iii)}$$

$$v_{2f} = \frac{m_2 - m_1}{m_1 + m_2}v_{2i} + \frac{2m_1}{m_1 + m_2}v_{1i} \quad \ldots\text{(iv)}$$

We can get Eq. (iv) by interchanging the value of m_1 and m_2 and vice versa in special cases.

Case I: Equal masses; $m_1 = m_2$

$$v_{1f} = \frac{m_1 - m_2}{m_1 + m_2}v_{1i} + \frac{2m_1}{m_1 + m_2}v_{2i} = v_{2i}$$

We find that velocity of body (1) after collision is equal to velocity of body (2) before collision.

$$v_{2f} = \frac{m_2 - m_1}{m_1 + m_2}v_{2i} + \frac{2m_1}{m_1}v_{1i}$$

Putting $m_2 = m_1$

$$v_{2f} = v_{1i}$$

The velocity of body (2) after collision is equal to the velocity of body (1) before collision.

To summarize, in a one-dimensional elastic collision of two bodies are of equal masses, they interchange their velocities.

Case II: When the body B is initially at rest, i.e., $v_{2i} = 0$

$$v_{1f} = \frac{m_1 - m_2}{m_1 + m_2}v_{1i}$$

$$v_{2f} = \frac{2m_1}{m_1 + m_2}v_{1i}$$

Further, two sub-cases arise when the body (2) is at rest.

Sub-case I: $m_2 << m_1$ i.e., the mass of the body (2) is negligible as compared to the mass of the body (1).

$$v_{1f} \approx v_{1i}$$

So, the velocity of the massive body after collision is nearly equal to after collision.

$$v_{2f} = 2v_{1i}$$

The velocity of the light body after collision becomes approximately twice the massive body before collision

Sub-case II: $m_1 << m_2$ i.e. when light body (1) collides with a massive body (2).

$$v_{1f} \approx -v_{1i}$$

$$v_{2f} \approx 0$$

The velocity of the massive body after collision is very small.

Elastic collision in two dimensions

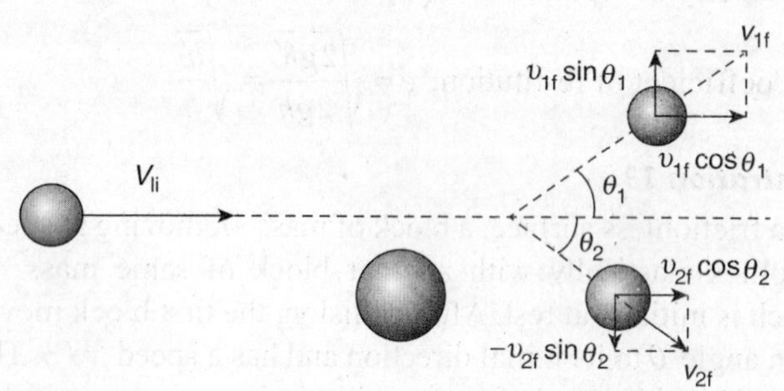

(a) Before the collision (b) After the collision

A body of mass m_1 is moving with velocity $\vec{v}_{1i}$ along x-axis. It suffers elastic collision with a stationary body of mass m_2. After collision, let body of mass m_1 be deflected at angle θ_1 with the initial direction. The body of mass m_2 is deflected at angle θ_2 with the initial direction. Let $\vec{v}_{1f}$ and $\vec{v}_{2f}$ be the velocities of bodies of masses m_1 and m_2 respectively after collision.

Applying the law of conservation of momentum to the x-component of motion, we get

$$m_1v_{1i} = m_1v_{1f}\cos\theta_1 + m_2v_{2f}\cos\theta_2 \quad \ldots\text{(i)}$$

For y-component of motion,

$$m_1v_{1f}\cos\theta_1 - m_2v_{2f}\cos\theta_2 = 0 \quad \ldots\text{(ii)}$$

The collision is elastic, so kinetic energy will be conserved.

$$\therefore \quad \frac{1}{2}m_1v_{1i}^2 = \frac{1}{2}m_1v_{1f}^2 + \frac{1}{2}m_2v_{2f}^2$$

or $$m_1 v_{1i}^2 = m_1 v_{1f}^2 + m_2 v_{2f}^2 \quad \text{...(iii)}$$

Equations (i), (ii) and (iii) can be solved for v_{1f} and v_{2f} if m_1, m_2, v_{1i}, θ_1 and θ_2 are known.

Coefficient of restitution

Newton, while experimenting on impact, found that the ratio of relative speed of separation after collision and the relative speed of approach before collision is a constant. This constant is called coefficient of restitution or coefficient of resilience. It is denoted by e. It is a measure of the degree of elasticity of a collision. Its value depends upon the nature of the colliding bodies.

$$e = \frac{|\vec{v}_{2f} - \vec{v}_{1f}|}{|\vec{v}_{1i} - \vec{v}_{2i}|}$$

For a perfectly elastic collision $e = 1$

For a perfectly inelastic collision $e = 0$

In general, $0 < e < 1$

Coefficient of restitution between a ball and the ground

Suppose that a ball falls from rest on to the ground from a height h and bounces to a height h'.

Velocity of approach $= \sqrt{2gh}$

Velocity of separation $= \sqrt{2gh'}$

Coefficient of restitution, $e = \sqrt{\dfrac{2gh'}{2gh}} = \sqrt{\dfrac{h'}{h}}$

Illustration 13

On a frictionless surface, a block of mass M, moving at speed v collies elastically with another block of same mass M, which is initially at rest. After collision, the first block moves at an angle θ to its initial direction and has a speed $v/3$. The second block's speed after the collision is

(a) $\dfrac{3}{\sqrt{2}}v$ (b) $\dfrac{\sqrt{3}}{2}v$

(c) $\dfrac{2\sqrt{2}}{3}v$ (d) $\dfrac{3}{4}v$

Solution

The situation is shown in the figure.

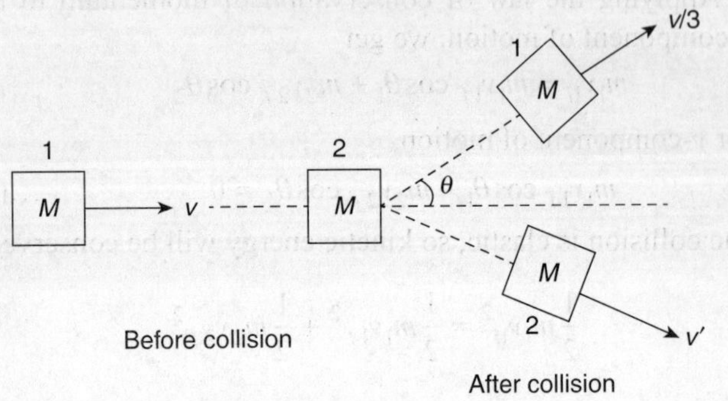

Let v' be speed of second block after the collision.

$$\frac{1}{2}Mv^2 + 0 = \frac{1}{2}M\left(\frac{v}{3}\right)^2 + \frac{1}{2}Mv'^2$$

$$\Rightarrow v^2 = \frac{v^2}{9} + v'^2 \Rightarrow v'^2 = v^2 - \frac{v^2}{9} = \frac{9v^2 - v^2}{9} = \frac{8}{9}v^2$$

$$v' = \sqrt{\frac{8}{9}v^2} = \frac{\sqrt{8}}{3}v = \frac{2\sqrt{2}}{3}v$$

Ans. (c)

Illustration 14

A body of mass 1 kg moving with a velocity of 12 m/s, collided with another body moving in the opposite direction with a velocity of 24 m/s. If the mass of the second body is 2 kg and if the coefficient of restitution is 2/3, then the velocities of bodies after collision is

(a) –28 m/s, –4 m/s (b) –4 m/s, –28 m/s

(c) 28 m/s, 4 m/s (d) 4 m/s, 28 m/s

Solution

$$m_1u_1 - m_2u_2 = m_1u_1 + m_2u_2$$

$$12 - 24 \times 2 = v_1 + 2v_2$$

$$12 - 48 = v_1 + 2v_2$$

$$-36 = v_1 + 2v_2$$

$$e = \frac{v_2 - v_1}{u_1 - u_2} = \frac{2}{3}$$

$$\frac{v_2 - v_1}{12 - (-24)} = \frac{2}{3}$$

$$\frac{v_2 - v_1}{36} = \frac{2}{3}$$

$$3v_2 = -12$$

$$v_2 = -4 \text{ m/s}$$

$$v_1 = v_2 - 24 = -4 - 24$$

$$v_1 = -28 \text{ m/s}$$

Ans. (a)

SOLVED EXAMPLES

Example 1

A force $F = a + bx$ acts on a particle in x direction where a and b are constants. Find the work done by this force during a displacement from $x = 0$ to $x = d$.

Solution

$$W = \int F.dx = \int_0^d (a + bx)\,dx$$

$$= \left[ax + \frac{1}{2}bx^2 \right]_0^d = ad + \frac{1}{2}bd^2$$

$$= \left(a + \frac{bd}{2} \right) d$$

Example 2

Sand drops vertically at the rate of 3 kg/s on to a conveyor belt moving horizontally with a velocity of 0.1 m/s. Calculate

(a) The power needed to keep the belt moving.
(b) The rate of kinetic energy supplied by the sand.

Solution

(a) $P = Fv = \frac{dp}{dt}.v = \frac{d}{dt}(m.v).v$

$$= v^2 \frac{dm}{dt} = (0.1)^2 \times 3 = \mathbf{0.03W}$$

(b) $\frac{dK}{dt} = \frac{d}{dt}\left(\frac{1}{2}mv^2\right) = \frac{1}{2}\left(\frac{dm}{dt}\right)v^2$

$$= \frac{1}{2} \times 3 \times (0.1)^2 = \mathbf{0.015\ W.}$$

Example 3

A smooth sphere of radius R is made to translate in a straight line with constant acceleration a. A particle kept on the top of sphere is released from there at zero velocity w.r.t. the sphere. Find the speed of the particle with respect to the sphere as a function of angle θ as it slides on the spherical surface.

Solution

Here the motion is described w.r.t the sphere frame which has uniform acceleration. Hence the work is done simultaneously by gravity and the pseudo force.

Work done by gravity

$$w_g = mg\,R\,(1 - \cos\theta)$$

Work done by pseudo force

$$w_p = ma\,R\sin\theta$$

Applying Work–Energy principle

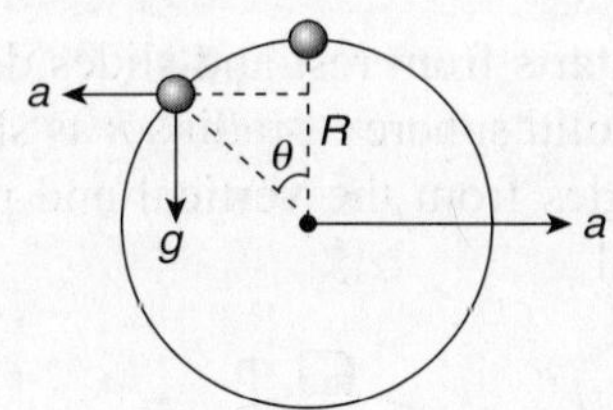

$$\frac{1}{2}mv^2 = maR\sin\theta + mgR\,(1 - \cos\theta)$$

or $$v = \sqrt{2R(a\sin\theta + g - g\cos\theta)}.$$

Example 4

A small block of mass m is kept on a rough inclined plane surface of inclination θ fixed in an elevator going up with uniform velocity v. Given that the block does not slide on the surface, find the work done by the force of friction on the block in time t.

Solution

Force due to friction = $mg\sin\theta$, up the plane.

Displacement = vt in the vertically upward direction. The angle between the force and the displacement vector is $(90° - \theta)$

$\therefore\quad W = \vec{F}.\vec{S}$ = mg $\sin\theta$ .vt . cos $(90° - \theta)$ = $\boldsymbol{mg\,vt\,\sin^2\theta}$.

Example 5

A person is pulling a mass m slowly on a rough hemispherical fixed surface from ground level up to the top of the hemisphere with the help of a light inextensible string as shown in the figure. The radius of the hemisphere is R. Find the work done by the tension in the string.

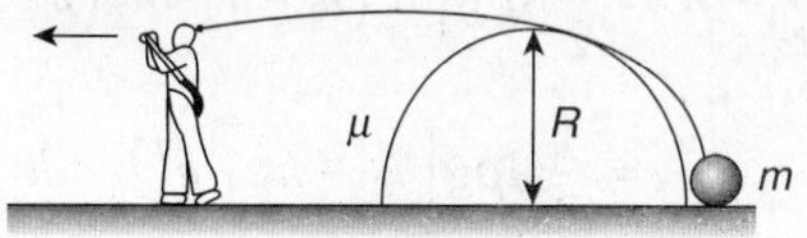

Solution

Since the mass m is being pulled slowly, the kinetic energy can be considered to be negligibly small. Hence change of KE is zero. From Work–Energy theorem, total work done by tension, gravity and friction must equal zero.

$$\therefore \qquad W_g + W_T + W_f = 0$$

T
N = mg sin θ
friction
mg
θ

Here $\quad W_g = -(mgR)$

$$W_f = -\int_0^{\pi/2} \mu mg\sin\theta.Rd\theta = \mu mgR[\cos\theta]_0^{\pi/2}$$

$$= -\mu mgR$$

$$\therefore W_T = -W_g - W_f = mgR + \mu\,mgR$$

$$= \mathbf{mgR\,(1 + \mu)}$$

Example 6

Two cylindrical vessels of equal cross-sectional area A contain water up to heights h_1 and h_2, with $h_1 > h_2$. The vessels are interconnected so that the levels in them become equal. Calculate the work done by the force of gravity during the process. The density of water is ρ.

Solution

Since volume remains constant,

$$A(h_1+h_2)=A(h+h)=2Ah$$

or, $$h=\frac{h_1+h_2}{2}.$$

Water level in one vessel falls from h_1 to h while it rises from h_2 to h in the other.

Work done by gravity in the first vessel is $W_1=+\frac{1}{2}Ah_1\rho g(h_1-h)$ and that in the second, it is $W_2=-\frac{1}{2}Ah_2\rho g(h-h_2)$, since the shift in the C.G. will be $\frac{1}{2}(h_1-h)$ and $\frac{1}{2}(h-h_2)$ respectively.

$\therefore$ Net work done will be,

$$W=W_1-W_2=\frac{1}{2}A\rho g\left[h_1(h_1-h)-h_2(h-h_2)\right]$$

$$=\frac{1}{2}A\rho g\left[h_1^2+h_2^2-h(h_1+h_2)\right]$$

$$=\frac{1}{2}A\rho g\left[h_1^2+h_2^2-\frac{(h_1+h_2)^2}{2}\right]=\frac{1}{4}A\rho g(h_1-h_2)^2.$$

Alternative method

Initial PE of arm of height h_1 = Weight $\times\frac{h_1}{2}$

$$=(Ah_1)(\rho)g\times\frac{h_1}{2}.$$

Final PE of Arm $h_1=Ah\rho g\times\frac{h}{2}$.

Work done by gravity for arm of height h_1,

$$w_1=\frac{A\rho g}{2}\left[h_1^2-h^2\right].\ (\text{+ve since } h_1>h)$$

Similarly for arm of height h_2,

$$w_2=\frac{A\rho g}{2}[h_2^2-h^2].\ (\text{–ve since } h_2<h_1)$$

Net work done $w=w_1+w_2$,

$$=\frac{A\rho g}{4}[h_1-h_2]^2$$

Example 7

Two bodies of masses m_1 and m_2 are placed on a horizontal table with coefficient of friction μ and joined by a spring. Initially, the spring has its natural length. Find the minimum force F which, when applied continuously, will make the other block m_2 just move. (k is the spring constant).

Solution

Motion of m_2 starts when $kx=\mu.m_2.g$, where x = extension in the spring

$$x=\mu m_2g/k$$

The minimum force will be such that m_1 has no kinetic energy. Applying Work–Energy principle on body of mass m_1,

$$\int_0^x(F_{\min}-\mu m_1g-kx)\,dx=0$$

$$\Rightarrow\quad F_{\min}x-\mu m_1gx-\frac{1}{2}kx^2=0$$

$$\Rightarrow\quad F_{\min}=\left[\mu m_1g+\frac{1}{2}kx\right]=\left[\mu m_1g+\frac{\mu m_2g}{2}\right]$$

$$\Rightarrow\quad F_{\min}=\mu m_1g+\frac{\mu m_2g}{2}.$$

Example 8

A block of mass m = 0.1 kg is released from a height of 4 m on a curved smooth surface. On the horizontal surface path AB is smooth and path BC is rough with coefficient of friction μ = 0.1. If the impact of block with the vertical wall at C be perfectly elastic, find the total distance covered by the block on the horizontal surface before coming to rest (take g = 10 ms^{-2}).

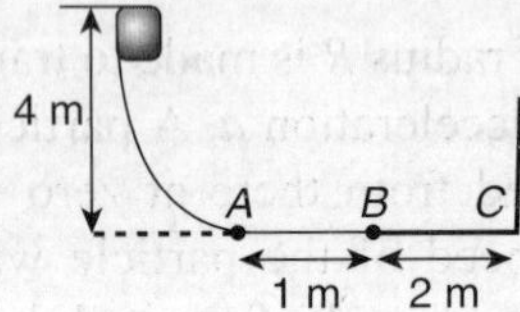

Solution

KE attained by block at $B=mgh$ = 4 J.

Work done by frictional force on path BC = (μ mg)(BC) = 0.2 J.

On the other horizontal surface, the block can make total forward and backward trips whose number is $\left(\frac{4}{0.2}\right)=20$ trips. If the last of these trips will end at point B,

Required distance = 20 × (AB + BC) – AB = **59 m.**

Example 9

A point mass m starts from rest and slides down the surface of a frictionless solid sphere of radius r as shown in the figure. Measure angles from the vertical and potential energy from the top. Find

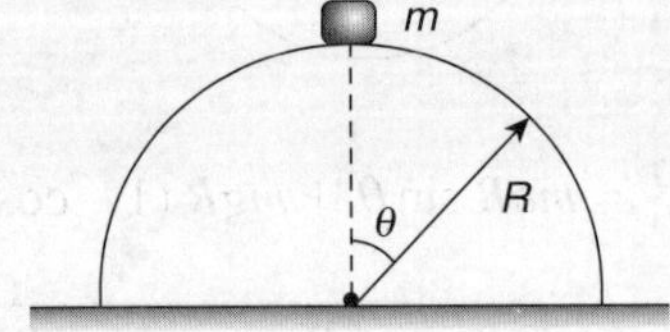

(a) the change in potential energy of the mass with angle.
(b) the kinetic energy as a function of angle.
(c) the radial and tangential acceleration as a function of angle.
(d) the angle at which the mass flies off the sphere.

Solution

(a) Consider the mass when it is at the point B.

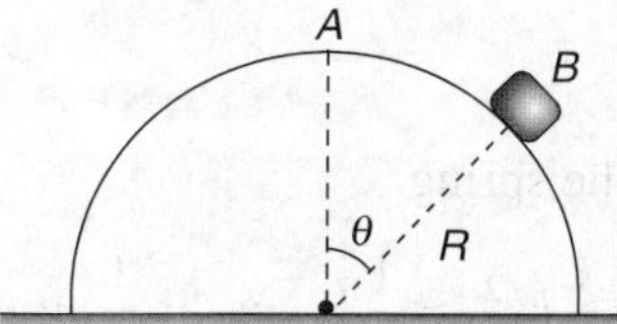

As the block shifts from A to B, its potential energy decreases by

$$mg \times \text{vertical displacement}$$
$$= mg(R - R\cos\theta)$$
$$= mgR(1 - \cos\theta)$$

$\therefore$ The change

$$\Delta U = -mgR(1 - \cos\theta).$$

Negative sign indicates that PE decreases as particle slides down.

(b) Apply the law of conservation of energy at points A and B.
Since the total mechanical energy is conserved,

$$(\text{KE} + \text{PE}) \text{ at } A = (\text{KE} + \text{PE}) \text{ at } B$$

$$\therefore \quad (\text{KE})_B = U_A - U_B = mgR(1 - \cos\theta)$$

(c) $\Delta \text{KE} = \frac{1}{2}mv^2$

$$\Rightarrow \quad \frac{1}{2}mv^2 = mgR(1 - \cos\theta) = 2mgR\sin^2\left(\frac{\theta}{2}\right)$$

$$\Rightarrow \quad v = 2\sqrt{gR}\,\sin\left(\frac{\theta}{2}\right).$$

The radial acceleration $a_r = v^2/R$

$$\Rightarrow \quad a_r = 4g\sin^2\left(\frac{\theta}{2}\right).$$

Also, in circular motion velocity is along the tangent, therefore

$$v_t = 2\sqrt{gR}\sin\left(\frac{\theta}{2}\right)$$

$$\because \quad a_t = \frac{d}{dt}(v_t)$$

$$\Rightarrow \quad a_t = \sqrt{gR}\cos\left(\frac{\theta}{2}\right)\frac{d\theta}{dt}$$

$$\Rightarrow \quad a_t = \sqrt{(g/R)}v\cos\left(\frac{\theta}{2}\right), \text{ as } \omega R = v$$

$$= \sqrt{gR}\cos\frac{\theta}{2}\frac{v}{R} = g\sin\theta \text{ putting } v = 2\sqrt{gR}\sin\frac{\theta}{2}$$

$$\Rightarrow \quad a_t = g\sin\theta.$$

(d) For circular motion

$$mg\cos\theta - N = \frac{mv^2}{R}$$

At the moment when the particle breaks off the sphere, the normal reaction $N = 0$

$$\Rightarrow \quad mg\cos\theta = \frac{mv^2}{R}$$

$$\Rightarrow \quad g\cos\theta = v^2/R$$

$$v = 2\sqrt{gR}\sin\frac{\theta}{2}$$

$$g\cos\theta = 4g\sin^2\left(\frac{\theta}{2}\right) = 2g(1 - \cos\theta),$$

$$\Rightarrow \quad \cos\theta = \frac{2}{3}$$

$$\Rightarrow \quad \theta = \cos^{-1}\left(\frac{2}{3}\right).$$

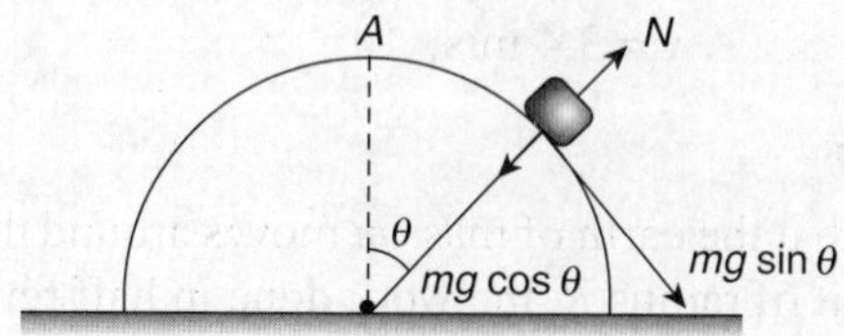

Example 10

A string with one end fixed on a rigid wall passing over a fixed frictionless pulley at a distance of 2 m from the wall has a point mass $M = 2$ kg attached to it at a distance of 1 m from the wall. A mass $m = 0.5$ kg attached at the free end is held at rest so that the string is horizontal between the wall and the pulley, and vertical beyond the pulley. What will be the speed with which the mass M will hit the wall when mass m released?

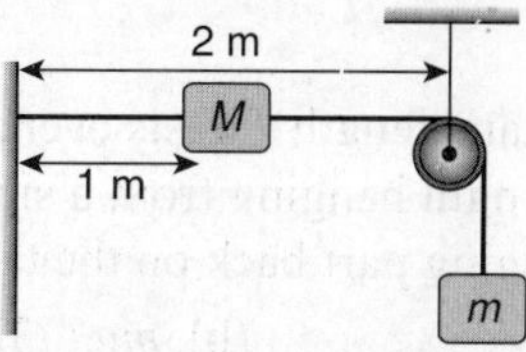

Solution

At the moment when the block M is just about to hit the wall, its velocity is along the horizontal, let the velocity of the block at that moment be v. Hence velocity of the block m is

$$v' = v\cos\theta.$$

If h be the displacement of the block m, then

$$(1+h)=\sqrt{2^2+1^2}=\sqrt{5}$$

$$\Rightarrow \quad h=\left(\sqrt{5}-1\right)\text{m}.$$

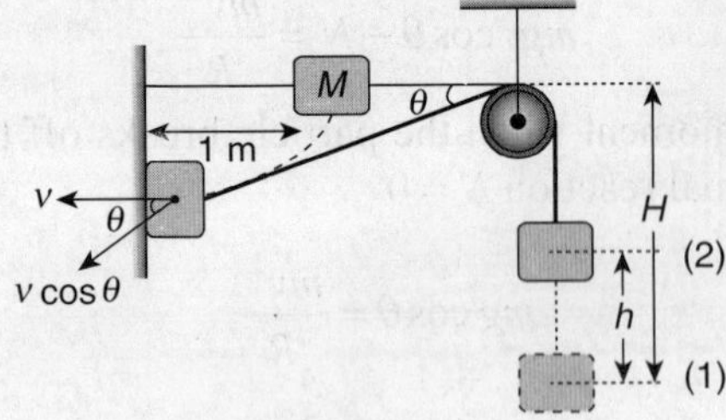

Conserving mechanical energy, we get

$$Mg(1)=\frac{1}{2}Mv^2+\frac{1}{2}m(v\cos\theta)^2+mgh$$

$$\Rightarrow \quad 0=\frac{1}{2}Mv^2+\frac{1}{2}mv^2\cos^2\theta-Mg+mgh$$

$$=\frac{1}{2}(2)v^2+\frac{1}{2}(0.5)v^2\left(\frac{2}{\sqrt{5}}\right)^2-2\times10+0.5\times10\times\left(\sqrt{5}-1\right)$$

$$=\frac{6}{5}v^2-20+6.18$$

$$\Rightarrow \quad v=3.4\text{ m/s}.$$

Example 11

Supposing that the earth of mass m moves around the sun in a circular orbit of radius R, the work done in half revolution is

(a) $\frac{mv^2}{R}\times\pi R$ (b) $\frac{mv^2}{R}\times 2R$

(c) Zero (d) None of these.

Solution

Force is radial and displacement is tangential. Hence $\vec{F}$ and $\vec{S}$ are always mutually perpendicular. Hence, work done = 0 for any amount of angular displacement during the orbital motion.

Ans. (c)

Example 12

A string of mass m and length ℓ rests over a frictionless table with 1/4 th of its length hanging from a side. The work done in bringing the hanging part back on the table is

(a) $mg\ell/4$ (b) $mg\ell/32$

(c) $mg\ell/16$ (d) None of these.

Solution

$$W=-\int_{\ell/4}^{0}\frac{mg}{\ell}x\,dx=\frac{mg}{2\ell}\left[x^2\right]_0^{\ell/4}=\frac{mg}{2\ell}\cdot\frac{\ell^2}{16}=\frac{mg\ell}{32}$$

Ans. (b)

Example 13

A weight mg is suspended from a spring. If the elongation in the spring is x_0, the elastic energy stored in it is

(a) $\frac{1}{2}mgx_0$ (b) $2\,mgx_0$

(c) mgx_0 (d) $\frac{1}{4}mgx_0$

Solution

Energy stored in the spring

$$U=\frac{1}{2}kx_0^2=\frac{1}{2}kx_0.x_0=\frac{1}{2}mgx_0.$$

Ans. (a)

Example 14

A constant retarding force is applied to stop a train. If its initial speed were doubled, the distance covered by the train by the time it stops will be (compared with the previous value)

(a) eight times (b) doubled

(c) half (d) four times.

Solution

From Work–Energy principle

$$F.S=-\frac{1}{2}mv^2,\ FS'=-\frac{1}{2}m(2v)^2$$

$$\therefore \quad \frac{S'}{S}=\frac{4}{1},\text{ or } S'=4S.$$

Ans. (d)

Example 15

Work done in time t on a body of mass m which is uniformly accelerated from rest to a speed v in time t_1 is given by

(a) $\frac{1}{2}m\frac{v}{t_1}t^2$ (b) $m\frac{v}{t_1}t^2$

(c) $\frac{1}{2}\left(\frac{mv}{t_1}t\right)^2 t^2$ (d) $\frac{1}{2}m\frac{v^2}{t_1^2}t^2$.

Solution

$$W=\int F.ds=\int ma\,vdt$$

$$=ma\int_0^t at\,dt=ma^2\frac{t^2}{2}=\frac{m}{2}\frac{v^2}{t_1^2}t^2.$$

Ans. (d)

Example 16

A ball is thrown up with a certain velocity at angle θ to the horizontal. The kinetic energy KE of the ball varies with horizontal displacement x as:

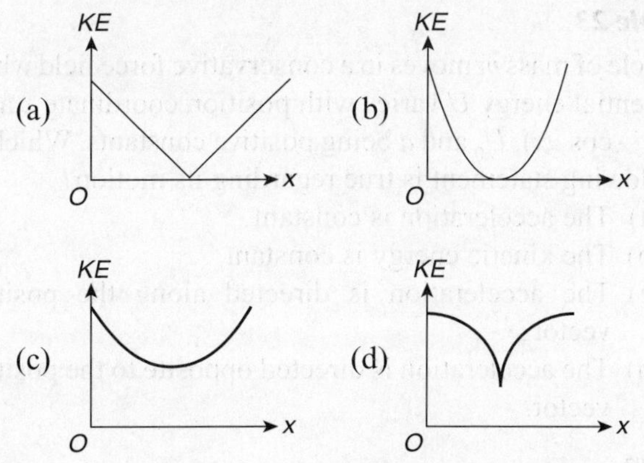

Solution

From Work–Energy principle

$$\frac{1}{2}mv^2 - \frac{1}{2}mu^2 = -mg\,y$$

where $$y = x\tan\theta - \frac{gx^2}{2u^2\cos^2\theta}.$$

Combining them,

$$\frac{1}{2}mv^2 = -\frac{1}{2}mu^2 - mg\left(x\tan\theta - \frac{gx^2}{2u^2\cos^2\theta}\right)$$

This equation has the form

KE $= a + bx + cx^2$, which has the nature of graph as in (**c**).

Ans. (c)

Example 17

A body of mass m_1 is projected vertically upwards with velocity v_1. Another body of same mass is projected at an angle of 45°. Both reach the same maximum height. What is the ratio of their kinetic energies at the point of projection?

(a) 1 (b) 1/2

(c) 1/3 (d) 1/4.

Solution

From Work–Energy principle,

$$-mgH = 0 - \frac{1}{2}mv_1^2 = -K_1$$

or $$K_1 = mgH$$

In the projectile motion,

$$-mgH = \frac{1}{2}m(v_2\cos 45^o)^2 - \frac{1}{2}mv_2^2 = \frac{K_2}{2} - K_2 = -\frac{K_2}{2}$$

$$\therefore \quad K_2 = 2mgH$$

$$\therefore \quad \frac{K_1}{K_2} = \frac{1}{2}.$$

Ans. (b)

Example 18

A simple pendulum has a string of length ℓ and bob of mass m. When the bob is at its lowest position, it is given the minimum horizontal speed necessary for it to move in a circular path about the point of suspension. The tension in the string at the lowest position of the bob is

(a) 3 mg (b) 4 mg

(c) 5 mg (d) 6 mg.

Solution

The critical velocity at the lowest point is $v = \sqrt{5g\ell}$. The centripetal force is given by $T - mg = \frac{mv^2}{r}$.

But $$v = \sqrt{5g\ell}$$

$$\therefore \quad T = mg + \frac{5mg\ell}{\ell} = 6\text{ mg}$$

Ans. (d)

Example 19

A particle of mass m is fixed to one end of a light rigid rod of length ℓ and rotated in a vertical circular path about its other end. The minimum speed of the particle at its highest point must be

(a) zero (b) $\sqrt{g\ell}$

(c) $\sqrt{1.5g\ell}$ (d) $\sqrt{2\,g\ell}$.

Solution

The rod will not slack at the highest point. Hence the minimum speed at the highest point is **zero.**

Ans. (a)

Example 20

A particle of mass m is fixed to one end of a light spring of force constant k and unstretched length ℓ. The system is rotated about the other end of the spring with an angular velocity ω, in gravity-free space (see figure). The increase in length of the spring will be

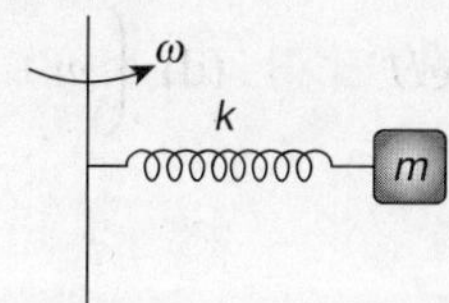

(a) $\frac{m\omega^2 l}{k}$ (b) $\frac{m\omega^2 l}{k - m\omega^2}$

(c) $\frac{m\omega^2 l}{k + m\omega^2}$ (d) none of these.

Solution

The required centripetal force is provided by the spring force. For extension x_0, $m\omega^2(1 + x_0) = kx_0$,

$$\therefore \quad x_0 = \frac{m\omega^2 l}{k - m\omega^2}.$$

Ans. (b)

Example 21

In the figure, a ball a is released from rest when the spring is at its natural (unstretched) length. For the block B of mass M to leave contact with the ground at some stage, the minimum mass of A must be

(a) $2M$
(b) M
(c) $M/2$
(d) a function of M and the force constant of the spring.

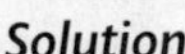

Solution

Applying Work–Energy principle,

$$mgx - \frac{1}{2}kx^2 \geq 0$$

$\therefore$ Minimum value of $x = 2\ mg/k$.

The block will leave contact with ground provided

$$kx = Mg$$

or

$$k.\frac{2mg}{k} = Mg$$

$$\therefore \quad m = \frac{M}{2}$$

Ans. (c)

Example 22

A man pulls a bucket of water from a well of depth H. If the mass of the rope and that of the bucket full of water are m and M respectively, then the work done by the man is

(a) $(m + M)gH$ (b) $\left(\frac{m}{2} + M\right) gH$

(c) $\left(\frac{m+M}{2}\right) gH$ (d) $\left(m + \frac{M}{2}\right) gH.$

Solution

$$w = w_1 + w_2 = mg\frac{H}{2} + MgH$$

$$= \left(\frac{m}{2} + M\right) gH$$

Ans. (b)

Example 23

A particle of mass m moves in a conservative force field where the potential energy U varies with position coordinate x as $U = U_0(1 - \cos ax)$, U_0 and a being positive constants. Which of the following statement is true regarding its motion?

(a) The acceleration is constant
(b) The kinetic energy is constant
(c) The acceleration is directed along the position vector
(d) The acceleration is directed opposite to the position vector.

Solution

$$F = -\frac{dU}{dx} = -aU_0 \sin ax$$

$$\therefore \text{Acceleration} = \frac{F}{m} = -\frac{aU_0}{m}\sin ax$$

$$= -\frac{a^2U_0}{m}x \text{ for small value of } x$$

$\therefore$ **Acceleration** $\propto -x$.

Ans. (d)

Example 24

A particle of mass m moves under the influence of the force $\vec{F} = a(\sin \omega t\ \hat{i} + \cos \omega t\ \hat{j})$ where a, ω are constants and t is time. The particle is initially at rest at the origin. The instantaneous power given to the particle is

(a) zero (b) $a^2 \sin \omega t / m\omega$

(c) $a^2 \cos \omega t / m\omega$ (d) $a^2(\sin \omega t + \cos \omega t)/m\omega$.

Solution

$$\vec{F} = m\frac{d\vec{v}}{dt} = a(\sin \omega t\,\hat{i} + \cos \omega t\,\hat{j})$$

$$\therefore \quad \vec{v} = \int_0^t \frac{a}{m}(\sin \omega t\,\hat{i} + \cos \omega t\,\hat{j})dt$$

$$= \frac{a}{m\omega}\left[(1 - \cos \omega t)\hat{i} + (\sin \omega t)\hat{j}\right]$$

$\therefore$ Instantaneous power $P = \vec{F}.\vec{v}$

$$= \frac{a^2}{m\omega}\left[\sin \omega t\,\hat{i} + \cos \omega t\,\hat{j}\right].\left[(1 - \cos \omega t)\hat{i} + \sin \omega t\,\hat{j}\right]$$

$$= \frac{a^2}{m\omega}\left[\sin \omega t \times (1 - \cos \omega t) + \sin \omega t \cos \omega t\right] = \frac{a^2}{m\omega}\sin \omega t$$

Ans. (b)

POINTS TO REMEMBER

Work = Force × Distance

Units of work

CGS system - erg.
MKS system - Joule
SI unit - Joule
1 Newton metre = 1 Joule
1 dyne centimetre = 1 erg.
1 J = 10^7 erg.

Power

It is the rate of doing work.

$$\text{Power} = \frac{\text{Work done}}{\text{Time}}$$

or
$$P = \frac{W}{T}$$

If F is constant, then $W = \vec{F}.\vec{S}$

$$\text{and } P = \frac{\vec{F}.\vec{S}}{t} = \vec{F}.\frac{\vec{S}}{t}$$
$$= \vec{F}.\vec{v}$$

Units of power

Watt and Kilowatt

1 Watt = 1 J/s
1 Kilowatt = 1000 watt

Horse Power is another unit of power

1 H.P = 746 watt

Energy

The capacity to do work is known as energy.

Various forms of energy

Internal energy, kinetic energy, potential energy, chemical energy, thermal energy, electrical energy, etc.

Kinetic energy

It is the energy possessed by a body by virtue of its motion.

$$\text{KE} = \frac{1}{2}mv^2$$

Work done by force acting on a body is equal to the change in KE of the body

$$W = \frac{1}{2}m\,v^2 - \frac{1}{2}mu^2$$

Potential energy

It is the energy possessed by a body by virtue of its position or distortion.

$$\text{PE} = mgh$$

A system is stable if the potential energy is minimum.

Law of conservation of energy

Energy can neither be created nor destroyed; it can be changed from one from to another.

In an isolated system, total energy is conserved.

If only conservative forces act, then the principal of energy conservation applies in the form.

$$\Delta K_e + \Delta U_g + \Delta U_e + \Delta U = 0$$

where
ΔK_e = change in kinetic energy
ΔU_g = change in gravitational potential energy
ΔU_e = change in elastic potential energy
ΔU = change in other forms of energy

If non-conservative forces act, then it assumes the form

$$\Delta K_e + \Delta U_g + \Delta U_e + \Delta U = W$$

where
W = Work done by non-conservative forces with which it is possible to identify a change of some form of energy, e.g., heat energy in case of friction.

Mass and energy

For a particle moving at the speed of light, the conventional laws of mechanics are valid if we define the mass of body as

$$\text{(i) } m = \frac{m_o}{\sqrt{1-\left(\frac{v}{c}\right)^2}}$$

where
m_o = rest mass
v = particle velocity
c = (velocity of light)

Relation between mass and energy

$$E = mc^2$$

It is called Einstein's equation
where E = Total energy of body
and m = relative mass of body

$$E_o = m_oC^2$$

where E_o = energy of particle at rest
and KE = $E - E_o$
$= (m - m_o)\,C^2$

EXERCISES

1. A ball is released from the top of a tower. The ratio of work done by force of gravity in first second and third second of the motion of ball is:

(a) 1 : 2 : 3 (b) 1 : 4 : 16
(c) 1 : 3 : 5 (d) 1 : 9 : 25

2. A particle is released from a height H. At a certain height h, its kinetic energy is two times its potential energy. The height and speed of particle at that instant are:

(a) $\frac{H}{3}, \sqrt{\frac{2gh}{3}}$ (b) $\frac{H}{3}, 2\sqrt{\frac{gh}{3}}$

(c) $\frac{2H}{3}, \sqrt{\frac{2gh}{3}}$ (d) None

3. The displacement of a body of mass 2 kg varies with time t as $s = t^2 + 2t$, where s is in metres and t is in seconds. The work done by all the forces acting on the body during the time interval $t = 2$ s to $t = 4$ s is:

(a) 36 J (b) 64 J
(c) 100 J (d) 120 J

4. Kinetic energy of a particle moving in a straight line varies with time t as $K = 4t^2$. The force acting on the particle:

(a) is constant
(b) is increasing
(c) is decreasing
(d) first increases and then decreases.

5. A ball of mass m is attached to one end of a light rod of length l, the other end of which is hinged. What minimum velocity v should be imparted to ball downwards, so that it can complete the circle?

(a) $\sqrt{gl}$ (b) $\sqrt{5gl}$
(c) $\sqrt{3gl}$ (d) $\sqrt{2gl}$

6. A particle of mass 2 kg starts moving in a straight line with an initial velocity of 2 m/s at a constant acceleration of 2 m/s^2. Then rate of change of kinetic energy:

(a) is four times the velocity at any moment
(b) is two times the displacement at any moment
(c) is four times the rate of change of velocity at any moment
(d) is constant throughout.

7. A particle moves on a rough horizontal ground with some initial velocity say v_0. If 3/4th of its kinetic energy is lost in friction in time t_0, then coefficient of friction between the particle and the ground is:

(a) $\frac{v_0}{2gt_0}$ (b) $\frac{v_0}{4gt_0}$

(c) $\frac{3v_0}{4gt_0}$ (d) $\frac{v_0}{gt_0}$

8. An object of mass m is allowed to fall from rest along a rough inclined plane. The speed of the object on reaching the bottom of the plane is proportional to:

(a) m° (b) m
(c) m^2 (d) m^{-1}

9. A vertical spring of force constant 100 N/m is attached with a hanging mass of 10 kg. Now an external force is applied on the mass so that the spring is stretched by an additional 2 m. The work done by the force F is: (g =10 m/s^2)

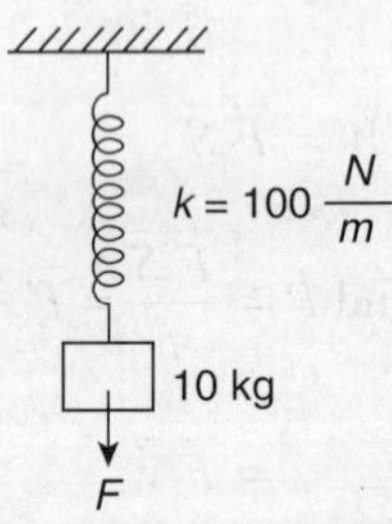

(a) 200 J (b) 400 J
(c) 450 J (d) 600 J

10. Two masses of 1 g and 4 g are moving with equal kinetic energies. The ratio of the magnitudes of their momenta is:

(a) 4 : 1 (b) $\sqrt{2}$: 1
(c) 1 : 2 (d) 1 : 16

11. A compressed spring has

(a) no energy stored in it
(b) negative mechanical energy stored in it
(c) positive mechanical energy stored in it
(d) kinetic energy stored in it.

12. If the force acting on a body is inversely proportional to speed, the kinetic energy of the body is

(a) constant
(b) directly proportional to time
(c) inversely proportional to time
(d) directly proportional to square of time.

13. A uniform force of 4 N acts on a body of mass 8 kg for a distance of 2.0 m. The KE acquired by the body is (g = 10 m/s^2)

(a) 8 J (b) 64 J
(c) 4 J (d) 160 J

14. A body falls freely under gravity. If its speed is v when it has lost an amount V of gravitational potential energy, then its mass is

(a) Vg/v^2 (b) V^2/g
(c) $2V/v^2$ (d) $2V/gv^2$

15. A uniform chain has a mass m and length l. It is placed on a frictionless table with a length l_0 hanging over the edge. The chain begins to slide down. Then the speed v with which the end slides away from the edge is given by

(a) $\sqrt{\frac{g}{l}(l^2 - l_0^2)}$ (b) $\sqrt{\frac{g}{l}(l - l_0)}$

(c) $\sqrt{\frac{g}{l}(l + l_0)}$ (d) $\sqrt{2g(l - l_0)}$

16. An elastic string of unstretched length L and force constant K is stretched by a small length x. It is further stretched by another small length y. The work done in second stretching is

(a) $\frac{1}{2}Ky^2$ (b) $\frac{1}{2}K(x^2 + y^2)$

(c) $\frac{1}{2}K(x + y)^2$ (d) $\frac{1}{2}ky(2x + y)$

17. A car starts from rest and attains a kinetic energy K by accelerating without slipping along a horizontal road in 20 seconds. If air resistance is neglected, the work done by external forces which accelerate the car will be

(a) zero (b) K
(c) $2K$ (d) $K/10$

18. A rod of mass m and length l is lying on a horizontal table. Work done in making it stand on one end will be

(a) mgl (b) $mgl/2$
(c) $mgl/4$ (d) $2mgl$

19. A mass of 20 kg moving with a speed of 10 m/s collides with another stationary mass of 5 kg. As a result of collision, the two masses stick together. The KE of composite mass will be

(a) 600 J (b) 800 J
(c) 1000 J (d) 1200 J

20. A car moving with a velocity of 50 km/hr can be stopped by brakes after at least 6 m. If the same car is moving at a speed of 100 km/hr, the minimum stopping distance is

(a) 12 m (b) 6 m
(c) 18 m (d) 24 m

21. A stationary particle explodes into two particles of masses m_1 and m_2, which move in opposite directions with velocities v_1 and v_2. The ratio of their kinetic energies (E_1/E_2) is

(a) m_2/m_1 (b) m_1/m_2
(c) 1 (d) m_1v_2/m_2v_1

22. When a force is applied on a moving body, its motion is retarded. Then the work done is

(a) positive (b) negative
(c) zero (d) positive and negative.

23. A gas expands from 5 litres to 105 litres at a constant pressure 100 N m^{-2}. The work done is

(a) 1 joule (b) 4 joule
(c) 8 joule (d) 10 joule.

24. The momentum of a body having kinetic energy E is doubled. The new Kinetic energy is

(a) E (b) $4E$
(c) $16E$ (d) $32E$

25. If distance is plotted against x-axis and kinetic energy against y-axis, then the slope of the graph so obtained is proportional to

(a) distance (b) kinetic energy
(c) velocity (d) acceleration.

26. A one kilowatt motor is used to pump water from a well 10 m deep. The quantity of water pumped out per second is nearly

(a) 1 kg (b) 10 kg
(c) 100 kg (d) 1000 kg

27. A load of mass M is moved up a smooth inclined plane of inclination θ, height h and length l. The work done is

(a) Mgl (b) Mgh

(c) $Mg \cos\theta$ (d) $\frac{M \tan\theta}{g}$

28. Which of the following statements s incorrect?

(a) The work done by sun in rotating planets around it is zero.
(b) Two vehicles having equal masses and equal speed moving in opposite directions possess equal kinetic energy.
(c) Potential energy arising from attractive forces is always positive.
(d) A particle is moving along the circumference of a circular track with variable speed. Some work is being done in this case.

29. A crane can lift a body of mass 100 kg vertically upwards with a constant speed of 5 ms^{-1}. If the value of acceleration due to gravity is 10 ms^{-2}, then the power of the crane is

(a) $100 \times 10 \times 5$ W (b) 100×5 W

(c) $\frac{100}{5}$ W (d) 100 W

30. Which of the following statements is incorrect?
(a) Kinetic energy may be zero, positive or negative.
(b) Potential energy may be zero, positive or negative.
(c) Power, Energy and Work are all scalars.
(d) Ballistic pendulum is a device for measuring the speed of bullets.

31. A body falling from height of 10 m rebounded from the floor and loses 20% of its energy after impact. It will rise
(a) 10 m (b) 8 m
(c) 5 m (d) 12 m

32. A sand bag of 10 kg mass is suspended by a 3 metre long weightless spring. A 0.2 kg mass bullet enters the bag with a velocity of 20 m/s and gets embedded into it. The loss in kinetic energy in the collision is
(a) 40.2 J (b) 38.2 J
(c) 49.2 J (d) 39.2 J

33. A 2 kg block collides with a horizontal weightless spring of force constant 2 N/m. The block compresses the spring 4 m from the rest position. Speed of the block at the instant of collision is
(a) 8 m/s (b) 16 m/s
(c) 12 m/s (d) 4 m/s

34. A 1 kg mass falls from a height of 10 m into a sand box. What is the speed of the mass just before hitting in sand box? If it travels a distance of 2 cm into the sand before coming to rest, average retarding force is
(a) 18 m/s and 8100 N (b) 16 m/s and 6400 N
(c) 14 m/s and 4900 N (d) 12 m/s and 3600 N

35. An object of mass m is tied to a string of length L and a variable horizontal force is applied on it, which starts at zero and gradually increases until the string makes an angle θ with the vertical. Work done by the force F is:

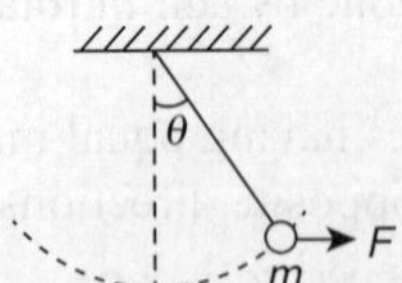

(a) $mgL(1 - \sin\theta)$ (b) mgL
(c) $mgL(1 - \cos\theta)$ (d) $mgL(1 + \cos\theta)$

36. In the given curved road, if particle is released from A then

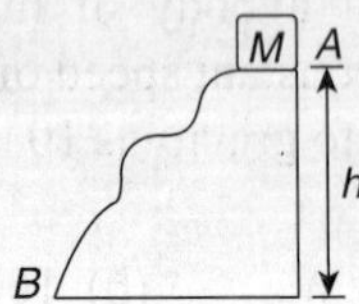

(a) kinetic energy at B must be mgh
(b) kinetic energy at B may be zero
(c) kinetic energy at B must be less than mgh
(d) kinetic energy at B must not be equal to zero.

37. Two unequal masses A and B moving along a straight line are brought to rest by equal retarding forces. If A takes twice the time as B but goes only 1/3 of the distance covered by B before coming to rest, the ratio of their velocity is
(a) 1 : 6 (b) 6 : 1
(c) 1 : 2 (d) 12 : 1

38. A body of mass 500 g is accelerated from a velocity $\left(3\hat{i} - 4\hat{j}\right)$ m s^{-1} to $\left(-6\hat{j} + 2\hat{k}\right)$ m s^{-1}. Find the work done
(a) 3.75 J (b) 0
(c) 4.75 J (d) 16 J

39. If bullet losses $\frac{1}{n}$ th of its velocity passing through each plank. How many such planks are required to stop the bullet?
(a) $\frac{n^2}{2n-1}$ (b) $\frac{2n^2}{(n-1)}$
(c) infinite (d) n

40. Two masses are connected by a light string which passes over a fixed pulley. The system is released from rest. Find the speed of block m_1 when it strikes the ground after failing a distance h. All surfaces are frictionless, as shown in figure

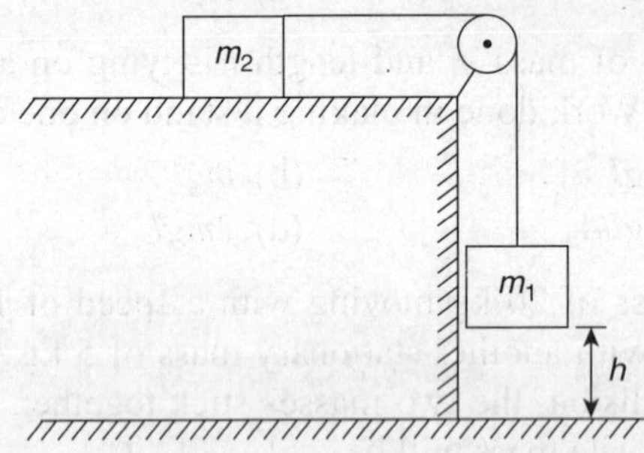

(a) $v = \sqrt{\frac{2m_2gh}{m_1 + m_2}}$ (b) $v = \sqrt{\frac{2m_1gh}{m_1 + m_2}}$
(c) $v = \sqrt{\frac{m_1 + m_2}{2m_1gh}}$ (d) $v = \sqrt{\frac{m_1 + m_2}{2m_2gh}}$

41. A particle of mass m is moving in a horizontal circle of radius r under centripetal force equal to $\frac{-k}{r^2}$, where k is a constant. The total energy of the particle is

(a) $\frac{-k}{2r^2}$ (b) $\frac{k}{2r}$

(c) $\frac{-k}{2r}$ (d) $\frac{-k}{r^2}$

Data for Questions 42 to 44

A body of mass 10 kg is moving on a horizontal surface by applying a force of 10 N in forward direction. If body moves with constant velocity,

42. The work done by applied force for a displacement of 2 m is

(a) 20 joule (b) 10 joule
(c) 30 joule (d) 40 joule

43. The work done by force of friction is

(a) – 20 joule (b) 10 joule
(c) 20 joule (d) –5 joule

44. The work done by normal reaction is

(a) 20 joule (b) 196 joule
(c) zero (d) None of these.

45. A body of mass m be dropped from a height h on a sand floor. If the body penetrates x cm into the sand, the average resistance offered by the sand to the body is

(a) $mg\frac{h}{x}$ (b) $mg\left(1+\frac{h}{x}\right)$

(c) $mgh+mgx$ (d) $mg\left(1-\frac{h}{x}\right)$

46. A particle of mass m is moving in a circular path of constant radius r such that centripetal acceleration a_c varying with time t is $a_c = k^2rt^2$ where k is a constant. What is the power delivered to the particle by the force attracting on it?

(a) $2mkr^2t$ (b) $mk^2r^2t^2$
(c) mk^2r^2t (d) mk^2rt^2

47. Which of the following is/are not conservative force?

(a) Gravitational force
(b) Electrostatic force in the coulomb field
(c) Frictional force
(d) All of the above.

48. Mark the correct option.

(a) The negative change in potential energy is equal to work done.
(b) Mechanical energy of a system remains constant.
(c) If internal forces are non-conservative, the net work done by internal forces must be zero.
(d) None of the above.

49. A shotput event was finished with a distance of 20 m. If the mass of shot is 5 kg and assuming that the shot is projected at an angle 45°, find the initial kinetic energy of the shot.

(a) 500 J (b) 490 J
(c) 450 J (d) 0 J

50. A bomb of mass 9 kg explodes into two pieces of mass of 3 kg and 6 kg. The velocity of 3 kg mass is 16 ms^{-1}. Kinetic energy of 6 kg mass particle is

(a) 288 J (b) 96 J
(c) 1025 J (d) 192 J

51. A stone of mass 1 kg falls on a spring from a height of 10 m. If the spring constant is 10 N/m, the compression of the spring is (g = 10 m/sec)

(a) $2\sqrt{5}$ m (b) $\sqrt{10}$ m
(c) $\left(1+\sqrt{21}\right)$ m (d) $5\sqrt{2}$ m

52. A block of mass m = 2 kg is pulled by a force F = 40 N upwards through a height h = 2 m. Find the work done on the block by the applied force F and by its weight mg. (g = 10 m/sec^2)

(a) 80 J, –40 J (b) 80 J, 40 J
(c) 40 J by each (d) –40 J by each

53. The block of mass m shown in figure, initially has a velocity v_0 to the right and its position is such that the spring exerts no force on it. The spring is not stretched or compressed. The block moves to the right through a distance before stopping in the dotted position shown. The spring constant is k and the coefficient of kinetic friction between the block and the table is μ. As the block moves the distance l, the total work done on the block is

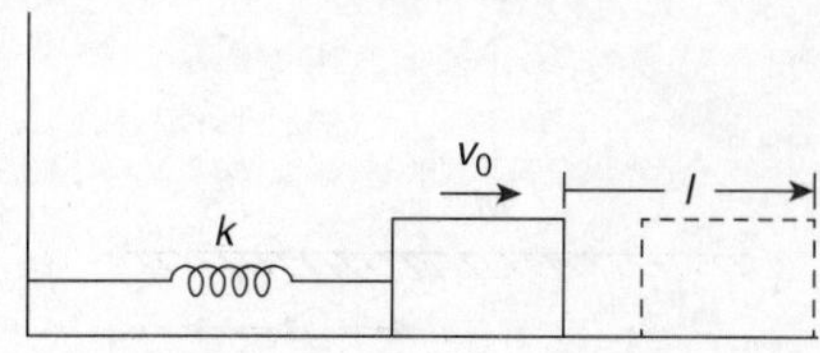

(a) 0 (b) $\frac{1}{2}kl^2$

(c) $-\mu mgl$ (d) $\left(\mu mgl+\frac{1}{2}kl^2\right)$

54. What work must be done to completely empty the water in a well? The area of cross-section of the well is 3.2 m^2, the water depth is 12 m and the water surface is at a depth of 18 m from the surface of the ground.

(a) 9.03×10^3 J (b) 9.31 J
(c) 9.03×10^5 J (d) 9.03×10^6 J

55. An aeroplane ascends and reaches a velocity of 360 km/h at an altitude of 5 km. By how many times is the work performed during ascent against the force of gravity more than that performed to increase the velocity of the aeroplane?

(a) 0 (b) 980
(c) 98 (d) 9.8

56. Two spheres of same size, one of mass 2 kg and another of mass 4 kg, are dropped simultaneously from the top of Qutab Minar (height = 72 m). When they are 1 m above the ground, the two spheres have the same

(a) momentum (b) kinetic energy
(c) potential energy (d) acceleration.

57. For inelastic collision between two spherical rigid bodies,

(a) the total kinetic energy is conserved
(b) the total mechanical energy is not conserved
(c) the linear momentum is not conserved
(d) the linear momentum is conserved.

58. A mass of 400 g and a mass of 100 g have same KE, then the ratio of their momentum will be

(a) 2 : 1 (b) 1 : 2
(c) 1 :3 (d) 3 : 1

59. A ball is dropped from height 20 m. If coefficient of restitution is 0.9, what will be the height attained after first bounce?

(a) 1.62 m (b) 16.2 m
(c) 18 m (d) 14 m

60. If reaction is R and coefficient of friction is μ, what is work done against friction in moving a body by distance d?

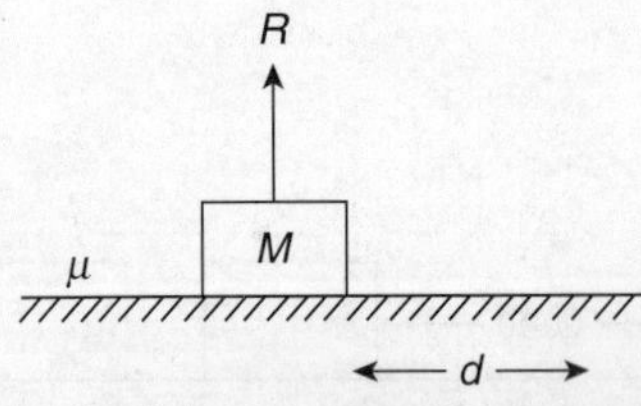

(a) $\frac{\mu Rd}{4}$ (b) $2\mu Rd$
(c) μRd (d) $\frac{\mu Rd}{2}$

61. A ball falls from 20 m height on floor and rebounds to 5 m. Time of contact is 0.02 sec. Find acceleration during impact.

(a) 1200 m/s^2 (b) 1000 m/s^2
(c) 2000 m/s^2 (d) 1500 m/s^2

62. Consider a collection of large number of particles, each with speed v. The direction of velocity is randomly distributed in the collection. The magnitude of the relative velocity between a pair of particles averaged over all the pairs in the collection is

(a) $\frac{4v}{\pi}$ (b) greater than $\frac{4v}{\pi}$
(c) less than $\frac{4v}{\pi}$ (d) zero.

63. A hockey player receives a corner shot at a speed of 15 m/s at an angle of 30° with the y-axis and then shoots the ball of mass 100 g along the negative x-axis with a speed of 30 m/s. If it remains in contact with the hockey stick for 0.01 s, the force imparted to the ball in the x-direction is

(a) 281.25 N (b) 187. N
(c) 562.5 N (d) 375 N

64. A bullet of mass 5 g is shot from a gun of mass 5 kg. The muzzle velocity of the bullet is 500 m/s. The recoil velocity of the gun is

(a) 0.5 m/s (b) 0.25 m/s
(c) 1 m/s (d) Data is insufficient.

65. A bomb is kept stationary at a point. It suddenly explodes into two fragments of masses 1 g and 3 g. The total KE of the fragments is 6.4×10^4 J. What is the KE of the smaller fragments?

(a) 2.5×10^4 J (b) 3.5×10^4 J
(c) 4.8×10^4 J (d) 5.2×10^4 J

66. A body moving with a velocity v, breaks up into two equal parts. One of the part retraces back with velocity v. Then velocity of the other part is

(a) v in forward direction
(b) $3v$ in forward direction
(c) v in backward direction
(d) $3v$ in backward direction.

67. In an inelastic collision, what is conserved?

(a) Kinetic energy (b) Momentum
(c) Both a and b (d) Neither a nor b.

68. If two particles collide at constant temperature then which of the following is conserved?

(a) Kinetic energy (b) Momentum
(c) Temperature (d) Velocity

69. Two masses of 1g and 4 g have same kinetic energy. What is the ratio of their momenta?

(a) $\frac{1}{2}$ (b) $\frac{1}{4}$
(c) 2 (d) 4

70. A spacecraft of mass M and moving with velocity v suddenly breaks in two pieces of same mass m. After the explosion, one of the mass m becomes stationary. What is the velocity of the other part of craft?

(a) $\frac{Mv}{M-m}$ (b) v

(c) $\frac{Mv}{m}$ (d) $\frac{M-m}{m}.v$

71. A particles moves with force $F = Cx$ from $x = 0$ to $x = x_1$. The work done is

(a) Cx_1^2 (b) $\frac{Cx_1^2}{2}$

(c) zero (d) Cx^3

72. If the linear momentum is increased by 50%, then kinetic energy will be increased by

(a) 50% (b) 20%

(c) 125% (d) None of these.

73. Which of the following statements are incorrect?

(i) If there were no friction, work need to be done to move a body up an inclined plane is zero.

(ii) If there were no friction, moving vehicles could not be stopped even by locking the brakes.

(iii) As the angle of inclination is increased, the normal reaction on the body placed on it increases.

(iv) A duster weighting 0.5 kg is pressed against a vertical board with a force of 11 N. If the coefficient of friction is 0.5, the work done in rubbing it upward through a distance of 10 cm is 0.55 J.

(a) (i) and (ii) (b) (i), (iii), (iv)

(c) (ii) and (iii) (d) All of these.

74. A long spring is stretched by 2 cm and its potential energy is U. If the spring is stretched by 10 cm, its potential energy will be

(a) $\frac{U}{5}$ (b) $\frac{U}{25}$

(c) $5U$ (d) $25U$

75. A 4 kg mass and a 1 kg mass are moving with the equal energies. The ratio of their momenta is

(a) 1: 2 (b) 1 : 1

(c) 2 : 1 (d) 4 : 1

76. A spring which is initially in its unstretched condition, is first stretched by a length x and then again by a further length x. The work done in the first case is W_1, and in the second case is W_2. Then

(a) $W_2 = W_1$ (b) $W_2 = 2W_1$

(c) $W_2 = 3W_1$ (d) $W_2 = 4W_1$

77. A mass m moves with a velocity v and collides inelastically with another identical mass. After collision, the first mass moves with velocity $\frac{v}{\sqrt{3}}$ in a direction perpendicular to the initial direction of motion. Find the speed of the second mass after collision.

(a) v (b) $\sqrt{3}v$

(c) $\frac{2}{\sqrt{3}}v$ (d) $\frac{v}{\sqrt{3}}$

78. A body of mass m is accelerated uniformly from rest to a speed v in a time T. The instantaneous power delivered to the body as a function of time, is given by

(a) $\frac{mv^2}{T^2}t$ (b) $\frac{mv^2}{T^2}t^2$

(c) $\frac{1}{2}\frac{mv^2}{T^2}t$ (d) $\frac{1}{2}\frac{mv^2}{T^2}t^2$

79. A particle of mass m moving with velocity v collides with a stationary particle of mass $2m$. Then speed of the system after collision is

(a) $2v$ (b) $\frac{v}{2}$

(c) $3v$ (d) $\frac{v}{3}$

80. A ball of mass 2 kg moving with velocity 3 m/s, collides with springs of nature/length 2 m and force constant 144 N/m. What will be length of compressed spring?

(a) 2 m (b) 1.5 m

(c) 1 m (d) 0.5 m

81. A bullet of mass 20 g and moving with 600 m/s collides with a block of mass 4 kg hanging with the string. What is velocity of bullet when it comes out of the block, if the block rises to height 0.2 m after collision

(a) 200 m/s (b) 150 m/s

(c) 400 m/s (d) 300 m/s

82. A steel ball of mass 5 g is thrown downward with velocity 10 m/s from height 19.5 m. It penetrates sand by 50 cm. The change in mechanical energy will be ($g = 10$ m/s^2)

(a) 1 J (b) 1.25 J

(c) 1.5 J (d) 1.75 J

83. A block C of mass m is moving with velocity v_0 and collides elastically with block A of mass m and connected to another block A of m and connected to another block B of mass 2 m through spring constant k. What is k if x_0 is compression of spring when velocity of A and B is same?

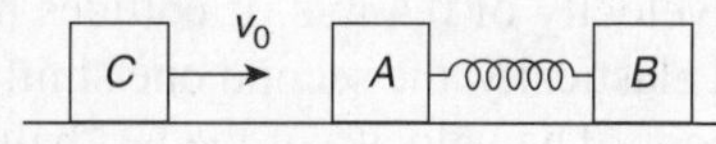

(a) $\dfrac{mv_0^2}{x_0^2}$ (b) $\dfrac{mv_0^2}{2x_0^2}$

(c) $\dfrac{3}{2}\dfrac{mv_0^2}{x_0^2}$ (d) $\dfrac{2}{3}\dfrac{mv_0^2}{x_0^2}$

84. As bomb at rest explodes into 3 parts of the same mass. The momentum of the 2 parts is $-2p\hat{i}$ and $p\hat{j}$. The momentum of the third part will have a magnitude of

(a) p (b) $\sqrt{3}p$
(c) $p\sqrt{5}$ (d) zero

85. A ball is hit at 45° to the horizontal with a kinetic energy E_k. The kinetic energy at the highest point is

(a) E_k (b) $\dfrac{E_k}{2}$
(c) $\dfrac{E_k}{\sqrt{2}}$ (d) zero

86. A 5 kg stationary bomb is exploded in three parts having mass 1 : 1 : 3 respectively. Parts having same mass move in perpendicular direction with velocity 39 m/s, then the velocity of the bigger part will be

(a) $\dfrac{15}{\sqrt{2}}$ m/s (b) $\dfrac{10}{\sqrt{2}}$ m/s
(c) $10\sqrt{2}$ m/s (d) $13\sqrt{2}$ m/s

87. When the speed of a moving body is doubled,

(a) its acceleration is doubled
(b) its momentum is doubled
(c) its kinetic energy is doubled
(d) its potential energy is doubled.

88. Consider elastic collision of a particle of mass m moving with a velocity u with another particle of the same mass at rest. After the collision, the projectile and the struck particle move in directions making angles θ_1 and θ_2 respectively with the initial direction of motion. The sum of the angles $\theta_1 + \theta_2$ is

(a) 45° (b) 90°
(c) 135° (d) 180°

89. If F is the force required to keep a train moving at a constant speed v. The power required is

(a) $\dfrac{1}{2}Fv^2$ (b) Fv^2
(c) $\dfrac{1}{2}Fv$ (d) Fv

90. Four smooth steel balls of equal mass at rest are free to move along a straight line without friction, the first ball is given a velocity of 0.4 m/s. It collides head-on with the second elastically, the second one similarly with the third and soon. The velocity of the last ball is

(a) 0.4 m/s (b) 0.2 m/s
(c) 0.1 m/s (d) 0.05 m/s

91. An object of mass $3M$ splits into three equal fragments. Two fragments have velocities $v\hat{j}$ and $v\hat{i}$. The velocity of the third fragment is

(a) $v(\hat{j}-\hat{i})$ (b) $v(\hat{i}-\hat{j})$
(c) $v(\hat{i}+\hat{j})$ (d) $\dfrac{v(\hat{i}+\hat{j})}{\sqrt{2}}$

92. A shell at rest at the origin explodes into three fragments of masses 1 kg, 2 kg and m kg. The 2 kg and 1 kg pieces fly off with speeds of 12 m/s along x-axis and 16 m/s along y-axis respectively. If the m kg piece flies off with a speed of 40 m/s, the total mass of the shell must be

(a) 3.7 kg (b) 4 kg
(c) 4.5 kg (d) 5 kg

93. A shell is fired from a cannon with a velocity v at an angle with the horizontal direction. At the highest point in its path, it explodes into two pieces of equal masses and one of the pieces retraces its path to the cannon. The speed of the other piece immediately after the explosion is

(a) $3v\cos\theta$ (b) $2v\cos\theta$
(c) $\dfrac{3}{2}v\cos\theta$ (d) $v\cos\theta$

94. A 16 kg block moving on a frictionless horizontal surface with a velocity of 4 m/s compresses an ideal spring and comes to rest. If the force constant of the spring be 100 N/m, then the spring is compressed by

(a) 1.6 m (b) 4 m
(c) 6.1 m (d) 3.2 m

95. A can filled with water is revolved in a vertical circle of radius 4 m and the water does not fall down. The time period for a revolution is about

(a) 2 s (b) 4 s
(c) 8 s (d) 10 s

96. The displacement x of a particle of mass 1.0 kg moving in one dimension, under the action of a constant force is related to time by the equation $t = \sqrt{x} + 3$ (in SI units). The work done by force in the first 10 seconds, in joules, is

(a) 169 (b) 98
(c) 49 (d) zero

97. A bullet of mass m moving with a speed u penetrates a block of wood of thickness x and emerges with a speed v. The force of resistances offered by the wood is given by

(a) $\frac{m}{x}\left(u^2 - v^2\right)$ (b) $\frac{m}{2x}\left(u^2 - v^2\right)$

(c) $\frac{m}{2x}\left(v^2 - u^2\right)$ (d) $\frac{m}{x}\left(v^2 - u^2\right)$

98. Solid line in the figure shows the potential energy $v(x)$ as a function of x of a particle confined to move along x-axis. Regions AB, BC, CD, EF, FG and GH are of equal distance. Rank the regions, AB, BC, CD and EF according to the magnitude of the force on the particle, greatest first.

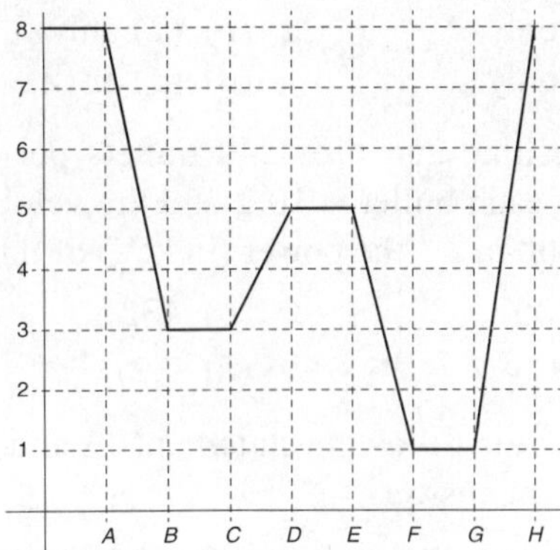

(a) AB > EF > CD > BC (b) CD > AB > EF > BC
(c) AB > EF > BC > CD (d) BC > EF > AB > CD

99. Two bodies with kinetic energies in the ratio of 4 : 1 are moving with equal linear momentum. The ratio of their masses is

(a) 4 : 1 (b) 1 : 1
(c) 1 : 2 (d) 1 : 4

100. The given diagram represents the potential energy curve of a particle in a field. The magnitude of the force on the particle will be maximum when its position coordinate 'x' is equal to

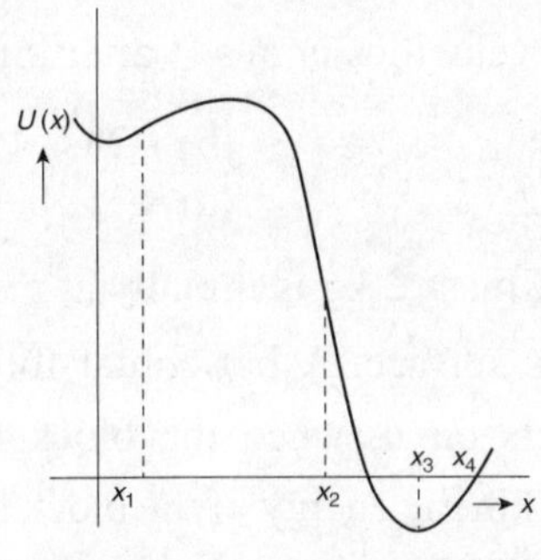

(a) x_1 (b) x_2
(c) x_3 (d) x_4

101. A bullet moving with a speed of 100 m s^{-1} can just penetrate two planks of equal thickness. Then, the number of such planks penetrated by the same bullet when the speed is doubled, will be

(a) 6 (b) 10
(c) 4 (d) 8

102. A mass of 1 kg is thrown up with a velocity of 100 m/s. After 5 seconds, it explodes into two parts. One part of mass 400 g comes down with a velocity 2 m/s. The velocity of other part is (take g = 10 ms^{-2})

(a) 40 m/s (b) 80 m/s
(c) 100 m/s (d) 60 m/s

103. A force of 1200 N acts on a 0.5 kg steel ball as a result of a collision lasting 25 ms. If the force is in a direction opposite to the initial velocity of 14 m/s, then the final speed of the steel ball would be

(a) 24 m/s (b) 35 m/s
(c) 12 m/s (d) 46 m/s

104. The kinetic energy of electron moving with the velocity 4×10^6 m/s, will be

(a) 600 eV (b) 50 eV
(c) 45 eV (d) 30 eV

105. A bucket full of water weighing 5 kg, is pulled from a well 20 m deep. There is a small hole in the bucket through which water leaks at a constant rate of 2.0 kg/m. The total work done in pulling the bucket up from the well is (g = 10 m/s^2)

(a) 600 J (b) 400 J
(c) 100 J (d) 500 J

106. For the same kinetic energy, the momentum is maximum for

(a) an electron (b) a proton
(c) a deuteron (d) an α particle.

107. There will be an increase in potential energy of the system if work is done upon the system by

(a) any conservative or non-conservative forces
(b) a non-conservative force
(c) a conservative force
(d) None of the above.

108. A river of salty water is flowing with a velocity 2 m/s. If the density of the water is 1.2 g/cc, then the kinetic energy of each cubic metre of water is

(a) 2.4 J (b) 24 J
(c) 2.4 kJ (d) 4.8 kJ

109. A ball is dropped from a height on a floor of coefficient of restitution e. The total distance covered by the ball just before second hit is

(a) $h\left(1-2e^2\right)$ (b) $h\left(1+2e^2\right)$

(c) $h\left(1+e^2\right)$ (d) $h2e^2$

110. A block of mass m = 25 kg sliding on a smooth horizontal surface with a velocity v = 3 m s^{-1} meets the spring of spring constant k = 100 N/m fixed at one end as shown in figure. The maximum compression of the spring and velocity of block as it returns to the original position respectively are

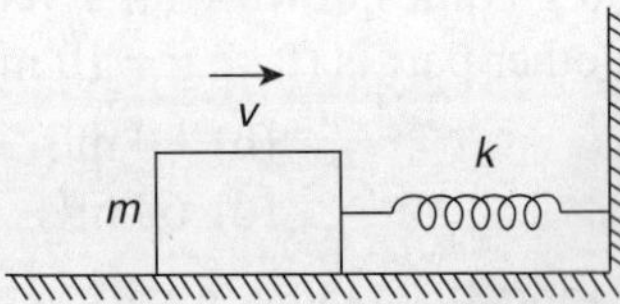

(a) 1.5 m, -3 m s^{-1} (b) 1.5 m, 0.01 m s^{-1}
(c) 1.0 m, 3 m s^{-1} (d) 0.5 m, 2 m s^{-1}

111. A rifle of 20 kg mass can fire 4 bullets per second. The mass of each bullet is 35×10^{-3} kg and its final velocity is 400 m s^{-1}. Then what force must be applied on the rifle so that it does not move backward while firing the bullets?

(a) 80 N (b) 28 N
(c) – 112 N (d) – 56 N

112. In two separate collisions, the coefficient of restitutions e_1 and e_2 are in the ratio 3 : 1. In the first collision, the relative velocity of separation to that of approach is 1 : 2. Then the ratio between relative velocity of approach and the relative velocity of separation in the second collision is

(a) 1 : 6 (b) 2 : 3
(c) 3 : 2 (d) 6 : 1

113. A bucket filled with water is tied to a rope of length 0.5 m and is rotated in a circular path in vertical plane. The least velocity it should have at the lowest point of circle so that water does not spill is (g = 10 ms^{-1})

(a) $\sqrt{5}$ m/s (b) $\sqrt{10}$ m/s
(c) 5 m/s (d) $2\sqrt{5}$ m/s

114. A motor is used to deliver water at a certain rate through a given horizontal pipe. To deliver n times the water through the same pipe in the same time, the power of the motor must be increased as follows:

(a) n times (b) n^2 times
(c) n^3 times (d) n^4 times

115. A bullet of mass 10 g is fired horizontally with a velocity 1000 m s^{-1} from a rifle situated at a height 50 m above the ground. If the bullet reaches the ground with a velocity 500 ms^{-1}, the work done against air resistance in the trajectory of the bullet is (g = 10 m/s^{-2})

(a) 50005 J (b) 3755 J
(c) 3750 J (d) 17.5 J

116. A man of 50 kg is standing at one end on a boat of length 25 m and mass 200 kg. If he starts running and when he reaches the other end, he has a velocity 2 m s^{-1} with respect to the boat. The final velocity of the boat is (in m s^{-1})

(a) $\frac{2}{5}$ (b) $\frac{2}{3}$
(c) $\frac{8}{5}$ (d) $\frac{8}{3}$

117. For a system to follow the law of conservation of linear momentum during a collision, the condition is

1. total external force acting on the system is zero
2. total external force acting on the system is finite and time of collision is negligible
3. total internal force acting on the system is zero.

(a) (1) only (b) (2) only
(c) (3) only (d) (1) or (2)

118. The machine gun fires 240 bullets per minute. If the mass of each bullet is 10 g and the velocity of the bullets is 600 m s^{-1} the power (in kW) of the gun is

(a) 43200 (b) 432
(c) 72 (d) 7.2

119. Consider the following statements A and B and identify the correct answer.

(A) In an elastic collision, if a body suffers a head-on collision with another of same mass at rest, the first body comes to rest while the other starts moving with the velocity of the first one.
(B) Two bodies of equal mass suffering a head-on elastic collision merely exchange their velocities.

(a) Both A and B are true
(b) Both A and B are false
(c) A is true but B is false
(d) A is false but B is true

120. A 2 kg ball moving at 24 m s^{-1} undergoes inelastic head-on collision with a 4 kg ball moving in the opposite direction at 48 m s^{-1}. If the coefficient of restitution is $\frac{2}{3}$, their velocities in m s^{-1} after impact are

(a) –56, –8 (b) –28, –4
(c) –14, –2 (d) –7, –1

121. A block of mass 2 kg is initially at rest on a horizontal frictionless surface. A horizontal force $\vec{F}=\left(9-x^2\right)\hat{i}$ newton acts on it, when the block is at x = 0. The maximum kinetic energy of the block between x = 0 and x = 3 m in joule is

(a) 24 (b) 20
(c) 18 (d) 15

122. Consider the following statements A and B and identify the correct answer given below.

(A) A body initially at rest is acted upon by a constant force. The rate of change of its kinetic energy varies linearly with time.

(B) When a body is at rest, it must be in equilibrium.

(a) A and B are correct
(b) A and B are wrong
(c) A is correct and B is wrong
(d) A is wrong and B is correct.

123. A body of mass M kg is on the top point of a smooth hemisphere of radius 5 m. It is released to slide down the surface of the hemisphere. It leaves the surface when velocity is 5 m/s. At this instant, the angle made by the radius vector of the body with the vertical is (acceleration due to gravity = 10 m s^{-1})

(a) 30° (b) 45°
(c) 60° (d) 90°

124. A body of mass 4 kg is moving with momentum of 8 kg m s^{-1}. A force of 0.2 N acts on it in the direction of motion of the body for 10 seconds. The increases in kinetic energy in joule is

(a) 10 (b) 8.5
(c) 4.5 (d) 4

125. A body is moving up an inclined plane of angle θ with an initial kinetic energy E. The coefficient of friction between the plane and the body is μ. The work done against friction before the body comes to rest is

(a) $\dfrac{\mu \cos\theta}{E\cos\theta + \sin\theta}$ (b) E

(c) $\dfrac{\mu E\cos\theta}{\mu\cos\theta - \sin\theta}$ (d) $\dfrac{\mu E\cos\theta}{\mu\cos\theta + \sin\theta}$

126. A body of mass 2 kg starts from and moves with uniform acceleration. It acquires a velocity 20 m s^{-1} in 4 s. The power exerted on the body in 2 s in watt is

(a) 50 (b) 100
(c) 150 (d) 200

127. An engine pumps up 100 kg of water through a height of 10 m in 5 s. Given that the efficiency of engine is 60%. If g = 10 ms^{-2}, the power of the engine is

(a) 3.3 kW (b) 0.33 kW
(c) 0.033 kW (d) 33 kW

128. When a spring is stretched by a distance x, it exerts a force, given by

$$F = (-5x - 16x^3) \text{ N}$$

The work done, when the spring is stretched from 0.1 m to 0.2 m is

(a) 8.1×10^{-2} J (b) 12.2×10^{-2} J
(c) 8.7×10^{-1} J (d) 12.2×10^{-1} J

129. 4 m^3 of water is to be pumped to a height of 20 m and forced into a reservoir at a pressure of 2×10^5 N/m^2. The work done by the motor is (external pressure = 10^5 N/m^2)

(a) 8×10^5 J (b) 16×10^5 J
(c) 12×10^5 J (d) 32×10^5 J

130. A body moves a distance of 10 m along a straight line under the action of a force of 5 N. If work done is 25 J, then angle between the force and direction of motion of the body will be

(a) 75° (b) 60°
(c) 45° (d) 30°

131. A body is initially at rest. It undergoes one-dimensional motion with constant acceleration. The power delivered to it at time t is proportional to

(a) $t^{1/2}$ (b) t
(c) $t^{3/2}$ (d) t^2

132. A block of mass m initially at rest is dropped from a height h on to a spring of force constant k. If the maximum compression in the spring is x, then:

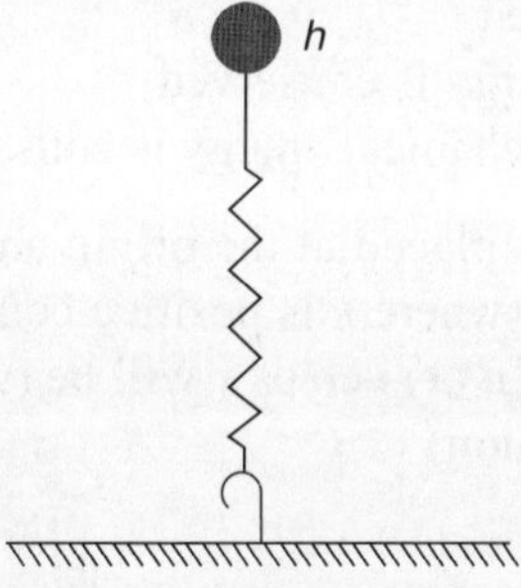

(a) $mgh = \dfrac{1}{2kx^2}$

(b) $mg(h+x) = \dfrac{1}{2}kx^2$

(c) $mgh = \dfrac{1}{2}k(x+h)^2$

(d) $mg(h+x) = \dfrac{1}{2}k(x+h)^2$

133. A ball of mass m moves with speed v and strikes a wall having infinite mass and it returns with the same speed. Then the work done by the ball on the wall is

(a) zero (b) mv
(c) $\dfrac{m}{v}$ (d) $\dfrac{v}{m}$

134. A body of mass M moves with velocity v and collides elastically with another body of mass m ($M >> m$) at rest then the velocity of body of mass m is

(a) v (b) $2v$
(c) $\dfrac{v}{2}$ (d) zero

135. A 30 g bullet initially travelling 500 m/s penetrates 12 cm into a wood block. What average force does it exert?

(a) 3125 N (b) 2825 N
(c) 4125 N (d) None of these.

136. A 10 kg object collides with stationary 5 kg object and after collision they stick together and move forward with velocity 4 ms^{-1}. What is the velocity with which the 10 kg object hits the second one?

(a) 4 ms^{-1} (b) 6 ms^{-1}
(c) 10 ms^{-1} (d) 12 ms^{-1}

137. When a force is applied on a moving body, its motion is retarded. Then the work done is

(a) positive (b) negative
(c) zero (d) positive and negative.

138. In elastic collision,

(a) both momentum and kinetic energies are conserved
(b) both momentum and kinetic energies are not conserved
(c) only energy is conserved
(d) only mechanical energy is conserved.

139. A particle is placed at the origin and a force $F = kx$ is acting on it (where k is positive constant). If $u(0) = 0$, the graph of $u(x)$ versus x will be (where u is potential energy function)

(a)
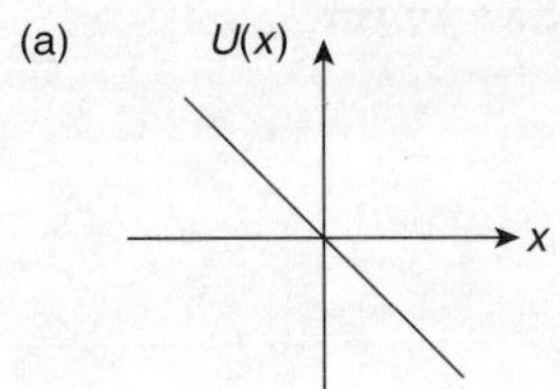

(b)
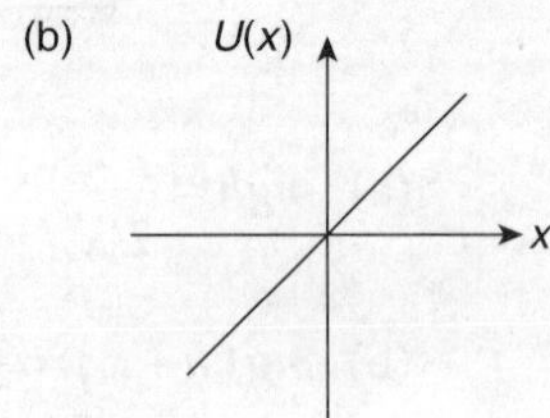

(c)
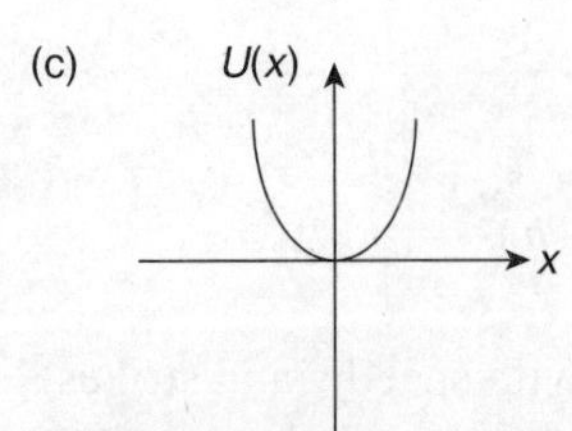

(d)
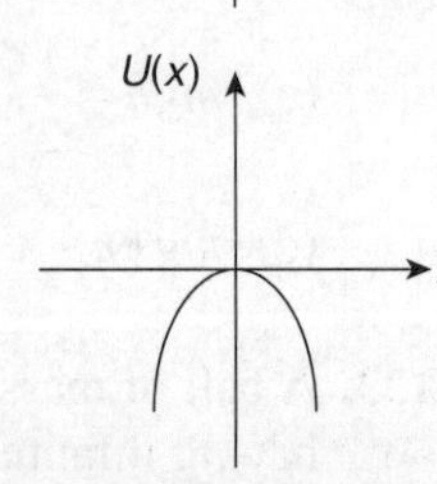

140. If $\vec{F} = (60\hat{i} + 15\hat{j} - 3\hat{k})$ N and $\vec{v} = (2\hat{i} - 4\hat{j} + 5\hat{k})$ m/s, then instantaneous power is

(a) 195 watt (b) 45 watt
(c) 75 watt (d) 100 watt

141. A block of mass 0.50 kg is moving with a speed of 2.00 m s^{-1} on a smooth surface. It strikes another mass of 1.00 kg and then they move together as a single body. The energy loss during the collision is

(a) 1.00 J (b) 0.67 J
(c) 0.34 J (d) 0.16 J

142. An athlete in the Olympic Games covers a distance of 100 m in 10 s. His kinetic energy can be estimated to be in the range

(a) 2×10^5 J – 3×10^5 J (b) 20,000 J – 50, 000 J
(c) 2,000 J – 5,000 J (d) 200 J – 500 J

143. A 2 kg block slides on a horizontal floor with a speed of 4 m/s. It strikes an uncompressed spring, and compresses it till the block is motionless. The kinetic friction force is 15 N and spring constant is 10,000 N/m. The spring compresses by

(a) 5.5 cm (b) 2.5 cm
(c) 11.0 cm (d) 8.5 cm

144. Which of the following holds good when two balls of masses m_1 and m_2 undergo elastic collision?

(a) When $m_1 < m_2$ and m_2 is at rest, there will be maximum transfer of momentum.
(b) When $m_1 > m_2$ and m_2 is at rest, after collision the ball of mass m_2 moves with four times the velocity of m_1.
(c) When $m_1 = m_2$ and m_2 is at rest, there will be maximum transfer of KE
(d) When collision is oblique and m_2 is at rest with $m_1 = m_2$, after collision the ball moves in opposite direction.

145. The potential energy of a 1 kg particle free to move along the x-axis is given by

$$V(x) = \left(\frac{x^4}{4} - \frac{x^2}{2}\right) \text{J}$$

The total mechanical energy of the particle is 2 J. Then, the maximum speed in (m/s) is

(a) 2 (b) $\frac{3}{\sqrt{2}}$
(c) $\sqrt{2}$ (d) $\frac{1}{\sqrt{2}}$

146. A particle is projected making an angle of 45° with horizontal having kinetic energy K. The kinetic energy at highest point will be

(a) $\frac{K}{\sqrt{2}}$ (b) $\frac{K}{2}$
(c) $2K$ (d) K

147. A particle of mass 100 g is thrown vertically upwards with a speed of 5 m/s. The work done by the force of gravity during the time the particle goes up is

(a) –0.5 J (b) 1.25 J
(c) –1.25 J (d) 0.5 J

148. A bomb of mass 16 kg at rest explodes into two pieces of masses 4 kg and 12 kg. The velocity of the 12 kg mass is 4 m s^{-1}. The kinetic energy of the other mass is

(a) 288 J (b) 192 J
(c) 96 J (d) 144 J

149. A mass of M kg is suspended by a weightless string. The horizontal force that is required to displace it until the string makes an angle of 45° with the initial vertical directions is

(a) $Mg(\sqrt{2}-1)$ (b) $\frac{Mg}{\sqrt{2}}$
(c) $Mg\sqrt{2}$ (d) $Mg(\sqrt{2}+1)$

150. A mass m moves with a velocity v and collides inelastically with another identical mass. After collision, the first mass moves with velocity $\frac{v}{\sqrt{3}}$ in a direction perpendicular to the initial direction of motion. Find the speed of the second mass after collision.

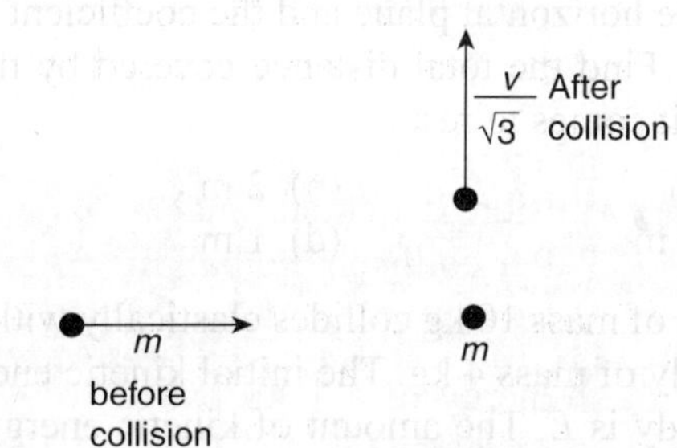

(a) v (b) $\frac{v}{\sqrt{3}}$
(c) $\frac{2}{\sqrt{3}}v$ (d) $\sqrt{3}v$

151. The block of mass M moving on the frictionless horizontal surface collides with the spring of spring constant K and compresses it by length L. The maximum momentum of the block after collision is

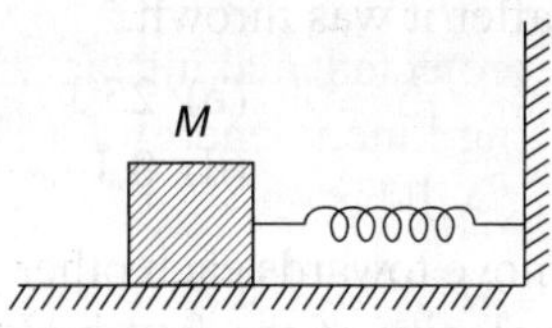

(a) $\frac{KL^2}{2M}$ (b) $\frac{ML^2}{K}$
(c) $\sqrt{MKL}$ (d) zero

152. A spherical ball of mass 20 kg is stationary at the top of a hill of height 100 m. It rolls down a smooth surface to the ground, then climbs up another hill of height 30 m and finally rolls down to a horizontal base at height of 20 m above the ground. The velocity attained by the ball is

(a) 10 m s^{-1} (b) 10 $\sqrt{30}$ m s^{-1}
(c) 40 m s^{-1} (d) 20 m s^{-1}

153. The law of conservation of linear momentum is a logical consequence of Newton's

(a) first law (b) second law
(c) third law (d) All the three laws.

154. A particle moves in a straight line with retardation proportional to its displacement. Its loss of kinetic energy for any displacement x is proportional to

(a) x^2 (b) e^2
(c) $\log_e x$ (d) x

155. A particle is acted upon by a force of constant magnitude, which is always perpendicular to the velocity of the particle. The motion of the particle takes place in a plane. It follows that

(a) its kinetic energy is constant
(b) it moves in a straight line
(c) its velocity is constant
(d) its acceleration is constant.

156. A force $\vec{F} = (r\hat{i} + 3\hat{j} + 2\hat{k})$ N is applied over a particle which displaces it from its origin to the point $\vec{r} = (2\hat{i} - \hat{j})$ m. The work done (in J) on the particle is

(a) + 13 (b) +10
(c) +7 (d) −7

157. A uniform chain of length 2 m is kept on a table such that a length of 60 cm hangs freely from the edge of the table. The total mass of the chain is 4 kg. What is the work done in pulling the entire chain on the table?

(a) 3.6 J (b) 7.2 J
(c) 1200 J (d) 120 J

158. An automobile travelling with a speed of 60 km/h, can brake to stop within a distance of 20 m. If the car is going twice as fast, 120 km/h, the stopping distance will be

(a) 60 m (b) 20 m
(c) 40 m (d) 80 m

159. Two springs A and B having spring constant K_A and K_B ($K_A = 2K_B$) are stretched by applying force of equal magnitude. If energy stored in spring A is E_A then energy stored in B will be

(a) $2E$ (b) $E_A / 4$
(c) $E_A / 2$ (d) $4E_A$

160. Consider the following two statements:

A. Linear momentum of a system of particles is zero.
B. Kinetic energy of a system of particles is zero.

Then,
(a) A does not imply B and B does not imply A
(b) A implies B but B does not imply A
(c) A does not imply B but B implies A
(d) A implies B and B implies A.

161. A body is moved along a straight line by a machine delivering a constant power. The distance moved by the body in time t is proportional to

(a) $t^{1/2}$ (b) $t^{1/4}$
(c) $t^{3/4}$ (d) $t^{3/2}$

162. Speeds of two identical cars are u and $4u$ at a specific instant. The ratio of the respective distances in which the two cars are stopped from that instant is

(a) 1 : 1 (b) 1 : 8
(c) 1 : 4 (d) 1 : 16

163. If body loses half of its velocity on penetrating 3 cm in a wooden block, then how much will it penetrate more before coming to rest?

(a) 3 cm (b) 4 cm
(c) 1 cm (d) 2 cm

164. In a one-dimensional elastic collision, the relative velocity of approach before collision is equal to

(a) sum of the velocities of the bodies
(b) e times the relative velocity of separation after collision
(c) $1/e$ times the relative velocity of separation after collision
(d) relative velocity of separation after collision.

165. A particle of mass m is moving in a horizontal circle of radius r with a centripetal force $\left(\frac{-1}{r^2}\right)$. The total energy is

(a) $\frac{-4}{r}$ (b) $\frac{-2}{r}$
(c) $\frac{-1}{r}$ (d) $\frac{-1}{2r}$

166. A bomb of mass 12 kg exploded into two pieces of masses 4 kg and 8 kg. If the velocity of mass 8 kg is 6 m/s, the kinetic energy of mass 4 kg will be

(a) 48 J (b) 32 J
(c) 24 J (d) 288 J

167. A running man has half the kinetic energy as that of a boy of half his mass. The man speeds up by 1 m/sec so as to have the same kinetic energy as the boy. What were the original speeds of man and boy respectively?

(a) 2.4, 4.8 m/s (b) 2.4, 3.4 m/s
(c) 3.4, 4.8 m/s (d) 3.4, 6.8 m/s

168. A bullet having a speed of 100 m/sec crashes through a plank of wood. After passing through the plank, its speed is 80 m/s. Another bullet of the same mass and size, but travelling at 80 m/s is fired at the plank. The speed of the second bullet after travelling through the plank is (assume that resistance of the plank is independent of the speed of the bullet)

(a) $10\sqrt{7}$ m/sec (b) $20\sqrt{7}$ m/sec
(c) $30\sqrt{7}$ m/sec (d) $20\sqrt{5}$ m/sec

169. A bullet of mass 50 g is fired from a gun of mass 2 kg. If the total kinetic energy produces is 2050 J, the energy of the bullet and the gun separately is

(a) 200 J, 5 J (b) 2000 J, 50 J
(c) 5 J, 200 J (d) 50 J, 2000 J

170. A particle of mass 1 g executes an oscillatory motion on the concave surface of a spherical dish of radius 2 m placed on horizontal plane. The motion of the particle begins from a point on the dish at a height of 1 cm from the horizontal plane and the coefficient of friction is 0.01. Find the total distance covered by the particle before it comes to rest.

(a) 4 m (b) 2 m
(c) 1.5 m (d) 1 m

171. A body of mass 10 kg collides elastically with a stationary body of mass 4 kg. The initial kinetic energy of the first body is E. The amount of kinetic energy left over with the first body is

(a) $\frac{3}{7}E$ (b) $\frac{4}{7}E$
(c) $\frac{9}{49}E$ (d) $\frac{40}{49}E$

172. A man, standing on a cart at rest throws a stone weighing 2 kg in a horizontal direction. The cart begins to roll back with a velocity of 0.1 m/s. If the cart, with the man, weighs 100 kg, find the kinetic energy of the stone 0.5 seconds after it was thrown.

(a) 15 J (b) 25 J
(c) 49 J (d) 0 J

173. Two bodies move towards each other and collide inelastically. The velocity of the first body is u_1= 2 m/s and that of the other is u_2= 4 m/s before impact. The total velocity of the bodies after collision is in the direction of the first body and is equal to 1 m/s. By how many times does the kinetic energy of the first body exceed that of the second one?

(a) 1.25 (b) 0.25
(c) 0.75 (d) None of these.

174. A massless platform is kept on a light elastic spring as shown in fig. When sand particle of 0.1 kg mass is dropped on the pan from a height of 0.24 m, the particle strikes the pan, and the spring compresses by 0.01 m. From what height should the particle be dropped to cause a compression of 0.04 m?

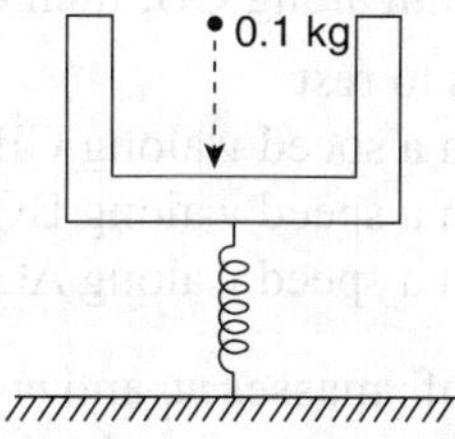

(a) 0.096 m (b) 0.296 m
(c) 0.02 m (d) 0.048 m

175. A body is dropped from a height h. The coefficient of restitution is e. Find the height to which the body rebounds after nth rebounds

(a) 0 (b) $h_n = he^{2n}$
(c) $h_n = he^{2n+1}$ (d) $h\left(e^2\right)^2$

176. A body is dropped from height h. Find the total distance travelled by the body before coming to rest.

(a) $h\left[\frac{1+e^2}{1-e^2}\right]$ (b) $\frac{he^2}{1+e^2}$
(c) he^2 (d) $h\left[\frac{1-e^2}{1+e^2}\right]$

177. As shown in the figure, six steel balls of identical size are lined up along a straight frictionless groove. Two similar balls moving with speed v along the groove collide with this row on the extreme left end. Then,

(a) one ball from the right end will move on with speed $2v$ and all the remaining ball will be rest.
(b) two balls from the extreme right end will move on with speed v and the remaining balls will be at rest.
(c) all the balls will start moving to the right with speed $v/8$.
(d) all the six balls originally at rest will move on with speed $v/6$ and the incident balls will comes to rest.

178. Two spheres of masses m_1 and m_2 moving with velocities u_1 and u_2 collide. The velocities after impact are v_1 and v_2. If the loss of kinetic energy of each sphere is same, then $u_1 + u_2$ is equal to

(a) $v_1 + v_2$ (b) $-\left(v_1 + v_2\right)$
(c) $v_1 - v_2$ (d) $v_2 - v_1$

179. An object initially at rest explodes into 3 fragments A, B and C. The momentum of A is P and that of B is $\sqrt{3}P$ where P is a positive number. The momentum of C will be

(a) $\left(1+\sqrt{3}\right)P$ in a direction making with that of A
(b) $\left(1+\sqrt{3}\right)P$ in a direction making with that of B
(c) $2P$ in a direction making with that of A
(d) $2P$ in direction making with that of B.

180. A ball 'A' moving with a speed of 90 m/s collides directly with another identifiable ball 'B' moving with a speed v in the opposite direction. 'A' comes to rest after collision. If the coefficient of restitution is 0.8, the speed of 'B' before collision is

(a) 10 m/s (b) 81 m/s
(c) 22.5 m/s (d) 90 m/s

181. A 3 kg ball strikes a heavy rigid wall a speed of 10 m/s at an angle $60°$. It gets reflected with the same speed and angles as shown here. If the ball is in contact with the wall for 0.20 sec, what is the average force exerted on the ball by the wall?

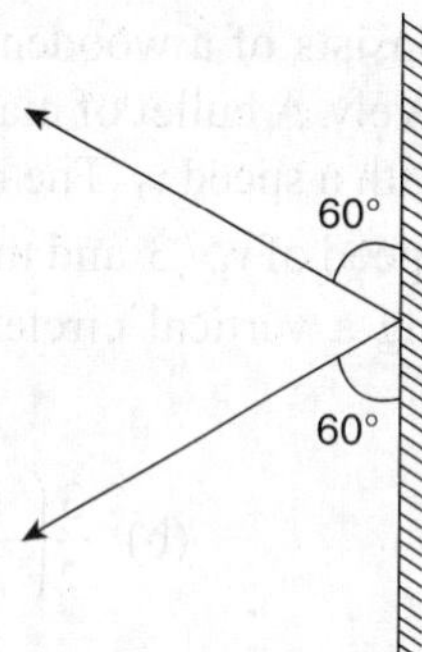

(a) 300 N (b) 150 N
(c) 150 N (d) 0

182. A body of mass M and moving with velocity v makes a head-on elastic collision with another stationary body of mass m. If $A = \frac{m}{M}$, then the ratio (f) of the loss of energy of M to its initial energy will be

(a) $f = A\left(A+1\right)^2$ (b) $f = \frac{A}{\left(A+1\right)^2}$
(c) $f = \frac{vA}{\left(A+1\right)^2}$ (d) $f = \frac{4A}{\left(A+1\right)^2}$

183. A plastic ball is dropped from a height of 1 m and rebounds several times from the floor. If 1.3 sec elapses from the moment it is dropped to the second impact with the floor, what is the coefficient of restitution?

(a) 0.93 (b) 0.94
(c) 0.94 (d) 0.95

184. A ball of mass m moving with a velocity collides against a stationary ball of mass m. The two balls stick together during collision. If E be the initial kinetic energy, then the loss of kinetic energy in the collision is

(a) E (b) $\frac{E}{2}$
(c) $\frac{E}{3}$ (d) $\frac{E}{4}$

185. Consider a one-dimensional elastic collision between a given two incoming bodies A and B, initially at rest. How would you choose the mass of B in comparison to the mass of A in order than B should recoil with greatest momentum?

(a) $m_B >> m_A$ (b) $m_B = m_A$
(c) $m_A >> m_B$ (d) None of these.

186. A child is sitting on a swing. Its minimum and maximum heights from the ground are 0.75 m and 2 m respectively. Its maximum speed will be

(a) 10 m/s (b) 5 m/s
(c) 8 m/s (d) 15 m/s

187. A pendulum consists of a wooden bob of mass m and length l respectively. A bullet of mass m_1 is fired towards the pendulum with a speed v_1. The bullet emerges out of the bob with a speed of $v_1/3$ and the bob just completes the motion along a vertical circle. Then v_1 is (Assume $g = 10\ \text{m/sec}^2$)

(a) $\left(\frac{m}{m_1}\right)\sqrt{5gl}$ (b) $\frac{3}{2}\left(\frac{m}{m_1}\right)\sqrt{5gl}$

(c) $\frac{2}{3}\left(\frac{m}{m_1}\right)\sqrt{5gl}$ (d) $\left(\frac{m}{m_1}\right)\sqrt{gl}$

188. A chain is held on a frictionless horizontal table with $\left(\frac{1}{n}\right)$ th of its length hanging over edge. If the chain has a length l and mass m then the work required to pull the hanging part back on to the table is

(a) $\frac{MgL}{n^2}$ (b) $\frac{MgL}{n}$

(c) $\frac{MgL}{2n^2}$ (d) $mgLn^2$

189. A train of mass 100 metric tons is drawn up an incline of 1 in 49 at the rate of 360 km/hour by an engine. If the resistance due to friction is 10 N per metric ton, then the power of the engine is

(a) 2×10^4 watt (b) 5×10^4 watt
(c) 2.5×10^6 watt (d) 2.1×10^6 watt

190. Three particles, each of mass m, are located at the vertices of an equilateral triangle ABC. They start moving with equal speeds v, each along the medians of the triangle and collide at its centroid G. A comes to rest and B retraces its path along GB; then C

(a) also comes to rest
(b) moves with a speed v along CB
(c) moves with a speed v along BG
(d) moves with a speed v along AG.

191. Two blocks of masses m_1 and m_2 are connected by spring of spring constant k. Initially, the spring is at its natural length. The coefficient of friction between the block and the surface is μ. What minimum constant force has to be applied in the horizontal direction to the block of mass l_1, in order to shift the other block?

(a) $F = \mu\left(m_1 + \frac{m_2}{2}\right)g$ (b) $F = \frac{\mu}{2}(m_1 + m_2)g$

(c) $F = \mu\left(\frac{m_1}{2} + m_2\right)g$ (d) $F = \mu\left(\frac{m_1 m_2 g}{k}\right)^2$

192. A body projected up has a potential energy of 1080 J and a kinetic energy 880 J at a point P in its path vertically upwards. If its mass is 10 kg, what is the maximum height reached by it? What is the height of P expressed as a fraction of the maximum height reached?

(a) 10 m, $\frac{49}{27}H$ (b) 0 m, 10 H

(c) 20 m, $\frac{27}{49}H$ (d) 49 m, 27 H

193. A wooden block of mass 10 g is dropped from the top of a tower 100 m high. Simultaneously a bullet of mass 10 g is fired from the foot of the tower vertically upwards with a velocity of 100 m/s. If the bullet gets embedded in the block, how high will be rise above the tower before it starts falling?

(a) 80 m (b) 85 m
(c) 75 m (d) 10 m

194. A car weighing 2×10^3 kg and moving 20 m/sec along a main road collides with a lorry of mass 8×10^3 kg which emerges at 5 m/sec from a cross road at right angle to the main road. If the two vehicles lock, what will be their velocity after the collision?

(a) 4 m/s, 45° with cross road
(b) $4/\sqrt{2}$ m/s, 60° with cross road
(c) $4\sqrt{2}$ m/s, 45° with main road
(d) $4\sqrt{2}$ m/sec, 45° with cross road

195. A ball moving with a velocity of 6 m/sec strikes an identical stationary ball. After collision each ball moves at an angle of 30° with the original line of motion. What are the speeds of the balls after the collision?

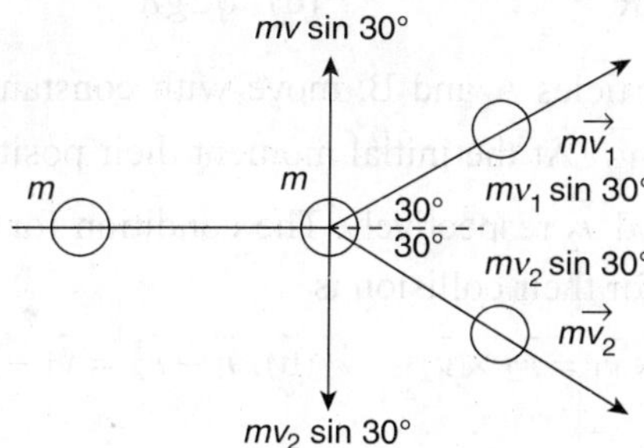

(a) $\frac{\sqrt{3}}{2}$ m/s (b) 3 m/s
(c) $2\sqrt{3}$ m/s (d) $\sqrt{3}$ m/s

196. Two pendulums each of length l are initially situated in a horizontal and other in vertical position respectively. The first pendulum is released and strikes the second. Assume that the collision is completely inelastic and neglect the mass of the string and any frictional effects. How high does the centre of mass rise after the collision?

(a) $l\left(\frac{m_1}{m_1+m_2}\right)^2$ (b) $l\left(\frac{m_1}{m_1+m_2}\right)$
(c) $l\left(\frac{m_1+m_2}{m_2}\right)^2$ (d) $l\left(\frac{m_2}{m_1+m_2}\right)^2$

197. A small ball A slides down the quadrant of a circle as shown in figure and hits the ball B of equal mass which is initially at rest. Find the velocity of the balls after collision. Neglect the effect of friction and assume the collision to be elastic

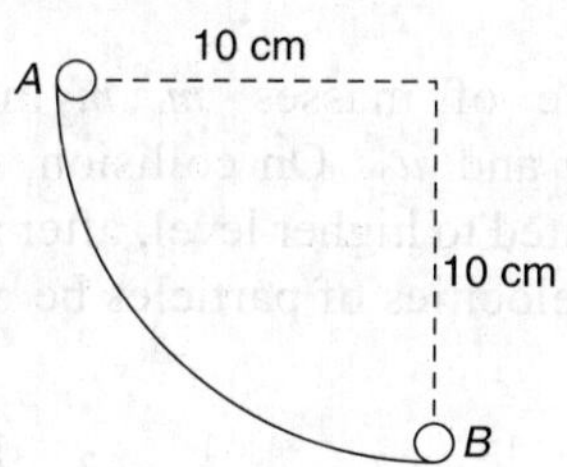

(a) 70 cm/s (b) 280 cm/s
(c) 0 (d) 140 cm/s

198. A bullet travelling horizontally with a velocity of 600 m/s and weighing 0.25 N strikes a wooden block of weight 50 N resting on a rough horizontal floor. The coefficient of friction between the floor and the block $\mu = 0.5$. Find the distance through which the block is displaced from its initial position. Assume that bullet, after striking the block, remains buried in the block.

(a) 5 m (b) 0.90 m
(c) 0.5 m (d) 0 m

Questions Asked in NEET/CBSE Medicals

199. Body A of mass $4m$ moving with speed u collides with another body B of mass $2m$, at rest. The collision is head-on and elastic in nature. After collision, the fraction of energy lost by the colliding body A is

(a) 5/9 (b) 1/9
(c) 8/9 (d) 4/9

200. A force $F = 20 + 10y$ acts on a particle in y-direction where F is in newton and in metre. Work done by this force to move the particle from $y = 0$ to $y = 1$ m is

(a) 20 J (b) 30 J
(c) 5 J (d) 25 J

201. A mass m is attached to a thin wire and whirled in a vertical circle. The wire is most likely to break when

(a) inclined at an angle of 60° from vertical
(b) the mass is at the highest point
(c) the wire is horizontal
(d) the mass is at the lowest point

202. A particle of mass $5m$ at rest suddenly breaks on its own into three fragments. Two fragments of mass m each move along mutually perpendicular direction with speed v each. The energy released during the process is

(a) $\frac{3}{5}mv^2$ (b) $\frac{5}{3}mv^2$
(c) $\frac{3}{2}mv^2$ (d) $\frac{4}{3}mv^2$

203. A moving block having mass m, collides with another stationary block having mass $4m$. The lighter block comes to rest after collision. When the initial velocity of the lighter block is v, then the value of coefficient of restitution (e) will be

(a) 0.5 (b) 0.25
(c) 0.8 (d) 0.4

204. A body initially at rest and sliding along a frictionless track from a height h (as shown in the figure) just completes a vertical circle of diameter AB = D. The height h is equal to

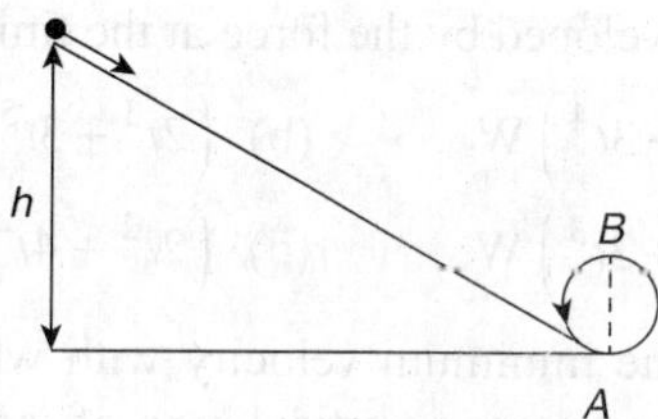

(a) $\frac{3}{2}D$ (b) D
(c) $\frac{7}{5}D$ (d) $\frac{5}{4}D$

205. Consider a drop of rain water having mass 1 g falling from a height of 1 km. It hits the ground with a speed of 50 m s^{-1}. Take g constant with a value 10 m s^{-2}. The work done by the (i) gravitational force and the (ii) resistive force of air is

(a) (i) 1.25 J (ii) –8.25 J (b) (i) 100 J (ii) 8.75 J
(c) (i) 10 J (ii) –8.75 J (d) (i) –10 J (ii) –8.25 J

206. A bullet of mass 10 g moving horizontally with a velocity of 400 m s^{-1} strikes a wooden block of mass 2 kg, which is suspended by light inextensible string of length 5 m. As a result, the centre of gravity of the block is found to rise a vertical distance of 10 cm. The speed of the bullet after it emerges out horizontally from the block will be

(a) 100 m s^{-1} (b) 80 m s^{-1}
(c) 120 m s^{-1} (d) 160 m s^{-1}

207. Two identical balls A and B having velocities of 0.5 m s^{-1} and –0.3 s^{-1} m respectively collide elastically in one dimension. The velocities of B and A after the collision respectively will be

(a) –0.5 m s and 0.3 m s^{-1}
(b) 0.5 m s and –0.3 m s^{-1}
(c) –0.3 m s and 0.5 m s^{-1}
(d) 0.3 m s and 0.5 m s^{-1}

208. A particle moves from a point $(-2\hat{i}+5\hat{j})$ to $(4\hat{j}+3\hat{k})$ when a force of $(4\hat{i}+3\hat{j})$ N is applied. How much work has been done by the force?

(a) 8 J (b) 11 J
(c) 5 J (d) 2 J

209. A particle of mass 10 g moves along a circle of radius 6.4 cm with a constant tangential acceleration. What is the magnitude of this acceleration if the kinetic energy of the particle becomes equal to 8×10^{-4} J by the end of the second revolution after the beginning of the motion?

(a) 0.18 m/s^2 (b) 0.2 m/s^2
(c) 0.1 m/s^2 (d) 0.15 m/s^2

210. A body of mass 1 kg begins to move under the action of a time dependent force $\vec{F}=\left(2t\hat{i}+3t^2\hat{j}\right)$N, where $\hat{i}$ and $\hat{j}$ are unit vectors along x and y axis. What power will be developed by the force at the time t?

(a) $\left(2t^3+3t^4\right)$ W (b) $\left(2t^3+3t^5\right)$ W
(c) $\left(2t^2+3t^3\right)$ W (d) $\left(2t^2+4t^4\right)$ W

211. What is the minimum velocity with which a body of mass m must enter a vertical loop of radius R so that it can complete the loop?

(a) $\sqrt{3gR}$ (b) $\sqrt{5gR}$
(c) $\sqrt{gR}$ (d) $\sqrt{2gR}$

212. Two particles A and B, move with constant velocities $\vec{v_1}$ and $\vec{v_2}$. At the initial moment their position vectors are $\vec{r_1}$ and $\vec{r_2}$ respectively. The condition for particles A and B for their collision is

(a) $\vec{r_1}\times\vec{v_1}=\vec{r_2}\times\vec{v_2}$ (b) $\vec{r_1}-\vec{r_2}=\vec{v_1}-\vec{v_2}$
(c) $\dfrac{\vec{r_1}-\vec{r_2}}{\left|\vec{r_1}-\vec{r_2}\right|}=\dfrac{\vec{v_2}-\vec{v_1}}{\left|\vec{v_2}-\vec{v_1}\right|}$ (d) $\vec{r_1}.\vec{v_1}=\vec{r_2}.\vec{v_2}$

213. The heart of a man pumps 5 litres of blood through the arteries per minute at a pressure of 150 mm of mercury. If the density of mercury be 13.6×10^3 kg/m^3 and $g=10$m/s^2 then the power (in watt) is

(a) 3.0 (b) 1.50
(c) 1.70 (d) 2.35

214. A ball is thrown vertically downwards from a height of 20 m with an initial velocity v_0. It collides with the ground, loses 50 percent of its energy in collision and rebounds to the same height. The initial velocity v_0 is (Take g = 10 m s^{-2})

(a) 28 m s^{-1} (b) 10 m s^{-1}
(c) 14 m s^{-1} (d) 20 m s^{-1}

215. A particle of mass m is driven by a machine that delivers constant power k watts. If the particle starts from rest, the force on the particle at time t is

(a) $\sqrt{2mk}t^{-1/2}$ (b) $\frac{1}{2}\sqrt{mk}t^{-1/2}$
(c) $\sqrt{\frac{mk}{2}}t^{-1/2}$ (d) $\sqrt{mk}t^{-1/2}$

216. Two particle of masses m_1, m_2 move with initial velocities u_1 and u_2. On collision, one of the particles get excited to higher level, after absorbing energy ε. If final velocities of particles be v_1 and v_2 then we must have

(a) $\frac{1}{2}m_1u_1^2+\frac{1}{2}m_2u_2^2-\varepsilon=\frac{1}{2}m_1v_1^2+\frac{1}{2}m_2v_2^2$

(b) $\frac{1}{2}m_1^2u_1^2+\frac{1}{2}m_2^2u_2^2+\varepsilon=\frac{1}{2}m_1^2v_1^2+\frac{1}{2}m_2^2v_2^2$

(c) $m_1^2u_1+m_2^2u_2-\varepsilon=m_1^2v_1+m_2^2v_2$

(d) $\frac{1}{2}m_1u_1^2+\frac{1}{2}m_2u_2^2=\frac{1}{2}m_1v_1^2+\frac{1}{2}m_2v_2^2-\varepsilon$

217. Two similar springs P and Q have spring constants K_P and K_Q, such that $K_P > K_Q$. They are stretched first by the same amount (case a), then by the same force (case b). The work done by the springs W_P and W_Q are related as, in case (a) and case (b) respectively,

(a) $W_P > W_Q; W_Q > W_P$ (b) $W_P < W_Q; W_Q < W_P$

(c) $W_P = W_Q; W_P > W_Q$ (d) $W_P = W_Q; W_P = W_Q$

218. A body of mass ($4m$) is lying in x-y plane at rest. It suddenly explodes into three pieces. Two pieces, each of mass (m) move perpendicular to each other with equal speeds (v). The total kinetic energy generated due to explosion is

(a) mv^2 (b) $\frac{3}{2}mv^2$

(c) $2mv^2$ (d) $4mv^2$

219. A uniform force of ($3\hat{i} + \hat{j}$) newton acts on a particle of mass 2 kg. Hence the particle is displaced from position ($2\hat{i} + \hat{k}$) metre to position ($4\hat{i} + 3\hat{j} - \hat{k}$) metre. The work done by the force on the particle is

(a) 13 J (b) 15 J
(c) 9 J (d) 6 J

220. A particle with total energy E is moving in a potential energy region $U(x)$. Motion of the particle is restricted to the region when

(a) $U(x) < E$ (b) $U(x) = 0$

(c) $U(x) \leq E$ (d) $U(x) > E$

221. One coolie takes 1 minute to raise a suitcase through a height of 2 m but the second coolie takes 30 s to raise the same suitcase to the same height. The powers of two coolies are in the ratio

(a) 1 : 3 (b) 2 : 1
(c) 3 : 1 (d) 1 : 2

222. The potential energy of a particle in a force field is $U = \frac{A}{r^2} - \frac{B}{r}$ where A and B are positive constants and r is the distance of the particle from the centre of the field. For stable equilibrium, the distance of the particle is

(a) $\frac{B}{2A}$ (b) $\frac{2A}{B}$

(c) $\frac{A}{B}$ (d) $\frac{B}{A}$

223. Two spheres A and B of masses m_1 and m_2 respectively collide. A is at rest initially and B is moving with velocity v along x-axis. After collision B has velocity $\frac{v}{2}$ in a direction perpendicular to the original direction. The mass A moves after collision in the direction

(a) same as that of B
(b) opposite to that of B

(c) $\theta = \tan^{-1}\left(\frac{1}{2}\right)$ to the x-axis

(d) $\theta = \tan^{-1}\left(-\frac{1}{2}\right)$ to the x-axis

224. A car of mass m starts from rest and accelerates so that the instantaneous power delivered to the car has a constant magnitude P_0. The instantaneous velocity of this car is proportional to

(a) $t^2 P_0$ (b) $t^{1/2}$

(c) $t^{-1/2}$ (d) $\frac{t}{\sqrt{m}}$

225. The potential energy of a system increases if work is done

(a) upon the system by a non-conservative force
(b) by the system against a conservative force
(c) by the system against a non-conservative force
(d) upon the system by a conservative force.

226. A body projected vertically from the earth reaches a height equal to earth's radius before returning to the earth. The power exerted by the gravitational force is greatest

(a) at the highest position of the body.
(b) at the instant just before the body hits the earth.
(c) It remains constant all through.
(d) at the instant just after the body is projected.

227. A particle of mass m is released from rest and follows a parabolic path as shown. Assuming that the displacement of the mass from the origin is small, which graph correctly depicts the position of the particle as a function of time?

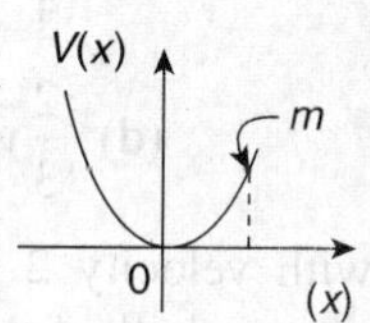

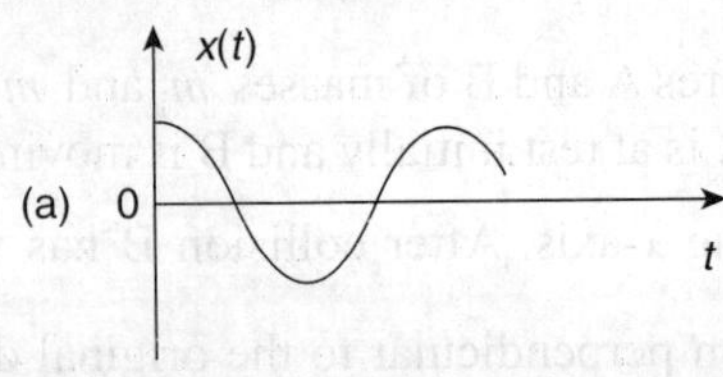

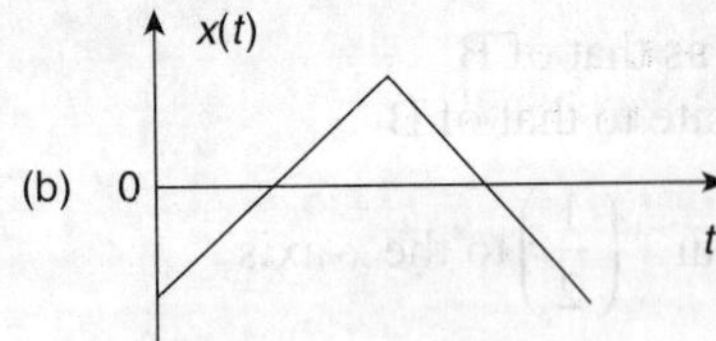

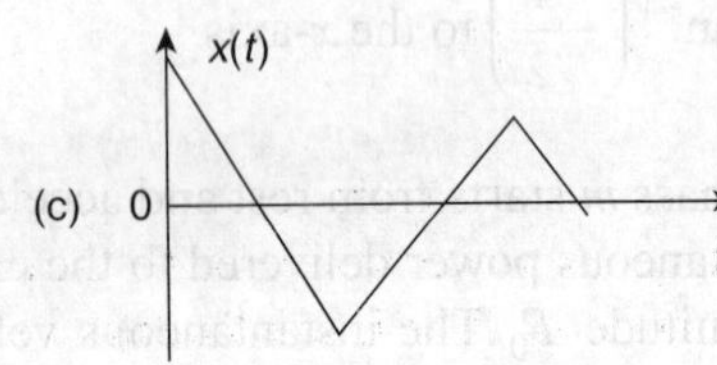

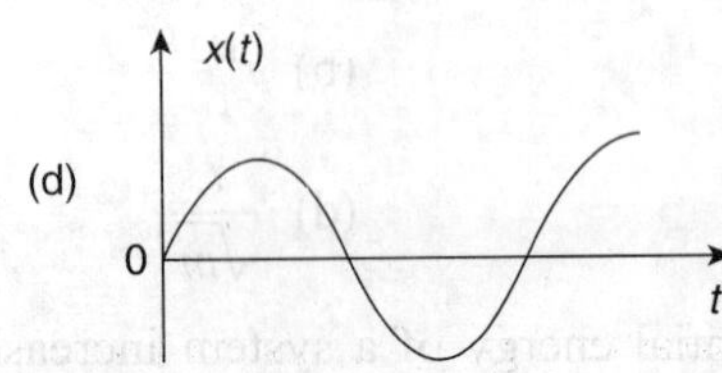

228. Force F on a particle moving in a straight line varies with distance d as shown in figure. The work done on the particle during its displacement of 12 m is

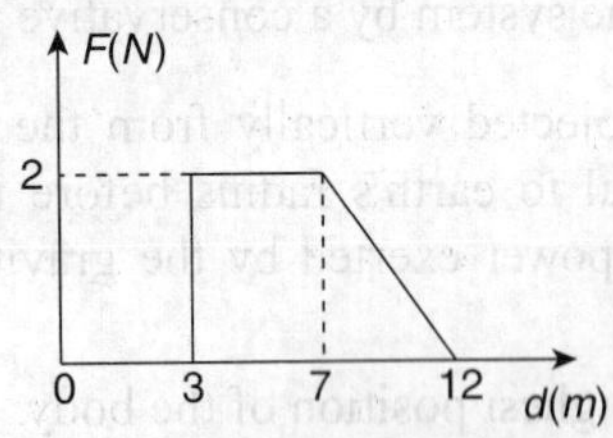

(a) 18 J (b) 21 J
(c) 26 J (d) 13 J

229. A mass m moving horizontally (along the x-axis) with velocity v collides and sticks to a mass of $3m$ moving vertically upward (along the y-axis) with velocity $2v$. The final velocity of the combination is

(a) $\frac{3}{2}v\hat{i}+\frac{1}{4}v\hat{j}$ (b) $\frac{1}{4}v\hat{i}+\frac{3}{2}v\hat{j}$

(c) $\frac{1}{3}v\hat{i}+\frac{2}{3}v\hat{j}$ (d) $\frac{2}{3}v\hat{i}+\frac{1}{3}v\hat{j}$

230. A ball moving with velocity 2 m/s collides head on with another stationary ball of double the mass. If the coefficient of restitution is 0.5, then their velocities (in m/s) after collision will be

(a) 0, 1 (b) 1, 1
(c) 1, 0.5 (d) 0, 2

231. An engine pumps water through a hose pipe. Water passes through the pipe and leaves it with a velocity of 2 m/s. The mass per unit length of water in the pipe is 100 kg/m. What is the power of the engine?

(a) 400 W (b) 200 W
(c) 100 W (d) 800 W

232. A particle of mass M, starting from rest, undergoes uniform acceleration. If the speed acquired in time T is V, the power delivered to the particle is

(a) $\frac{MV^2}{T}$ (b) $\frac{1}{2}\frac{MV^2}{T^2}$

(c) $\frac{MV^2}{T^2}$ (d) $\frac{1}{2}\frac{MV^2}{T}$

233. A block of mass M is attached to the lower end of a vertical spring. The spring is hung from a ceiling and has force constant value k. The mass is released from rest with the spring initially unscratched. The maximum extension produced in the length of the spring will be

(a) $2Mg/k$ (b) $4Mg/k$
(c) $Mg/2k$ (d) Mg/k

234. A body of mass 1 kg is thrown upwards with a velocity 20 m/s. It momentarily comes to rest after attaining a height of 18 m. How much energy is lost due to air friction? ($g = 10\text{ m/s}^2$)

(a) 30 J (b) 40 J
(c) 10 J (d) 20 J

235. An explosion blows a rock into three parts. Two parts go off at right angles to each other. These two are, 1 kg first part moving with a velocity of 12 m s^{-1} and 2 kg second part moving with a velocity 8 m s^{-1}. If the third part flies off with a velocity of 4 m s^{-1}, its mass would be

(a) 7 kg (b) 17 kg
(c) 3 kg (d) 5 kg

236. An engine pumps water continuously through a hose. Water leaves the hose with a velocity v and m is the mass per unit length of the water jet. What is the rate at which kinetic energy is imparted to water?

(a) mv^3 (b) $\frac{1}{2}mv^3$

(c) $\frac{1}{2}m^2v^2$ (d) $\frac{1}{2}mv^3$

237. A shell of mass 200 g is ejected from a gun of mass 4 kg by an explosion that generates 1.05 kJ of energy. The initial velocity of the shell is

(a) 40 m s^{-1} (b) 120 m s^{-1}

(c) 100 m s^{-1} (d) 80 m s^{-1}

238. Water falls from a height of 60 m at the rate of 15 kg/s to operate a turbine. The losses due to frictional forces are 10% of energy. How much power is generated by the turbine? ($g = 10$ m/s^2)

(a) 12.3 kW (b) 7.0 kW

(c) 8.1 kW (d) 10.2 kW

239. A vertical spring with force constant k is fixed on a table. A ball of mass m at a height h above the free upper end of the spring falls vertically on the spring so that the spring is compressed by a distance d. The net work done in the process is

(a) $mg(h+d)-\frac{1}{2}kd^2$ (b) $mg(h-d)-\frac{1}{2}kd^2$

(c) $mg(h-d)+\frac{1}{2}kd^2$ (d) $mg(h+d)+\frac{1}{2}kd^2$

240. 300 J of work is done in sliding a 2 kg block up an inclined plane of height 10 m. Work done against friction is (Take $g = 10$ m/s^2)

(a) 100 J (b) 200 J

(c) 100 J (d) zero

241. The potential energy of a long spring when stretched by 2 cm is U. If the spring is stretched by 8 cm, the potential energy stored in it is

(a) $U/4$ (b) $4U$

(c) $8U$ (d) $16U$

242. A body of mass 3 kg is under a constant force which causes a displacement s in metres in it, given by the relation $s = \frac{1}{3}t^2$, where t is in seconds. Work done by the force in 2 second is

(a) $\frac{19}{5}$ J (b) $\frac{5}{19}$ J

(c) $\frac{3}{8}$ J (d) $\frac{8}{3}$ J

243. A bomb of mass 30 kg at rest explodes into two pieces of masses 18 kg and 12 kg. The velocity of 18 kg mass is 6 m s^{-1}. The kinetic energy of the other mass is

(a) 324 J (b) 486 J

(c) 256 J (d) 524 J

244. A force F acting on an object varies with distance x as shown here. The force is in N and x in m. The work done by the force in moving the object from $x = 0$ to $x = 6$ m is

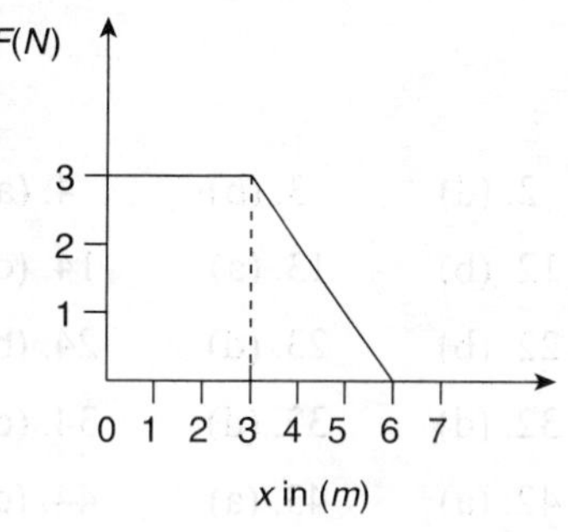

(a) 18.0 J (b) 13.5 J

(c) 9.0 J (d) 4.5 J

245. A particle of mass m_1 is moving with a velocity v_1 and another particle of mass m_2 is moving with a velocity v_2. Both of them have the same momentum but their different kinetic energies are E_1 and E_2 respectively. If $m_1 > m_2$ then

(a) $E_1 < E_2$ (b) $\frac{E_1}{E_2} = \frac{m_1}{m_2}$

(c) $E_1 > E_2$ (d) $E_1 = E_2$

246. A ball of mass 2 kg and another of mass 4 kg are dropped together from a 60 feet tall building. After a fall of 30 feet each towards earth, their respective kinetic energies will be in the ratio of

(a) $\sqrt{2}:1$ (b) $1:4$

(c) $1:2$ (d) $1:\sqrt{2}$

247. A mass of 0.5 kg moving with a speed of 1.5 m/s on a horizontal smooth surface, collides with a nearly weightless spring of force constant $k = 50$ N/m. The maximum compression of the spring would be

(a) 0.15 m (b) 0.12 m

(c) 1.5 m (d) 0.5 m

248. When a long spring is stretched by 2 cm, its potential energy is U. If the spring is stretched by 10 cm, the potential energy stored in it will be

(a) $U/5$ (b) $5U$
(c) $10U$ (d) $25U$

249. A stationary particle explodes into two particles of masses m_1 and m_2 which move in opposite directions with velocities v_1 and v_2. The ratio of their kinetic energies E_1 / E_2 is

(a) m_2 / m_1 (b) m_1 / m_2
(c) 1 (d) $m_1 v_2 / m_2 v_1$

250. If kinetic energy of a body is increased by 300%, then percentage change in momentum will be

(a) 100% (b) 150%
(c) 265% (d) 73.2%

Answer Keys

EXERCISES

1. (c)	2. (d)	3. (b)	4. (a)	5. (b)	6. (a)	7. (a)	8. (a)	9. (b)	10. (c)
11. (c)	12. (b)	13. (a)	14. (c)	15. (a)	16. (d)	17. (b)	18. (b)	19. (b)	20. (d)
21. (a)	22. (b)	23. (d)	24. (b)	25. (d)	26. (b)	27. (b)	28. (c)	29. (a)	30. (a)
31. (b)	32. (d)	33. (d)	34. (c)	35. (c)	36. (a)	37. (a)	38. (a)	39. (a)	40. (b)
41. (c)	42. (a)	43. (a)	44. (c)	45. (b)	46. (c)	47. (c)	48. (a)	49. (b)	50. (d)
51. (c)	52. (a)	53. (d)	54. (d)	55. (d)	56. (d)	57. (d)	58. (a)	59. (b)	60. (c)
61. (d)	62. (a)	63. (d)	64. (a)	65. (c)	66. (b)	67. (b)	68. (b)	69. (a)	70. (c)
71. (b)	72. (c)	73. (b)	74. (d)	75. (c)	76. (c)	77. (c)	78. (a)	79. (d)	80. (b)
81. (a)	82. (b)	83. (d)	84. (c)	85. (b)	86. (d)	87. (b)	88. (b)	89. (d)	90. (a)
91. (c)	92. (a)	93. (a)	94. (a)	95. (b)	96. (b)	97. (c)	98. (a)	99. (d)	100. (b)
101. (d)	102. (c)	103. (d)	104. (c)	105. (a)	106. (d)	107. (c)	108. (c)	109. (b)	110. (a)
111. (d)	112. (d)	113. (c)	114. (c)	115. (c)	116. (a)	117. (a)	118. (c)	119. (a)	120. (a)
121. (c)	122. (c)	123. (c)	124. (c)	125. (b)	126. (b)	127. (a)	128. (a)	129. (c)	130. (b)
131. (b)	132. (b)	133. (a)	134. (b)	135. (a)	136. (b)	137. (b)	138. (a)	139. (d)	140. (b)
141. (b)	142. (c)	143. (a)	144. (c)	145. (b)	146. (b)	147. (c)	148. (a)	149. (a)	150. (c)
151. (c)	152. (c)	153. (b)	154. (a)	155. (a)	156. (c)	157. (a)	158. (d)	159. (a)	160. (c)
161. (d)	162. (d)	163. (c)	164. (d)	165. (c)	166. (d)	167. (a)	168. (b)	169. (b)	170. (d)
171. (c)	172. (c)	173. (b)	174. (c)	175. (b)	176. (a)	177. (b)	178. (b)	179. (d)	180. (a)
181. (b)	182. (d)	183. (a)	184. (b)	185. (a)	186. (b)	187. (b)	188. (c)	189. (d)	190. (c)
191. (b)	192. (c)	193. (c)	194. (c)	195. (c)	196. (a)	197. (d)	198. (b)		

Questions Asked in NEET/CBSE Medicals

199. (c)	200. (d)	201. (d)	202. (d)	203. (b)	204. (d)	205. (c)	206. (c)	207. (b)	208. (c)
209. (c)	210. (b)	211. (b)	212. (c)	213. (c)	214. (d)	215. (c)	216. (a)	217. (a)	218. (b)
219. (c)	220. (c)	221. (d)	222. (b)	223. (d)	224. (b)	225. (b)	226. (b)	227. (a)	228. (d)
229. (b)	230. (a)	231. (a)	232. (d)	233. (a)	234. (d)	235. (d)	236. (d)	237. (c)	238. (c)
239. (a)	240. (c)	241. (d)	242. (d)	243. (b)	244. (b)	245. (a)	246. (c)	247. (a)	248. (d)
249. (a)	250. (a)								

Solutions

1. (c)

The distance travelled in nth second

$$S_n = u + \frac{1}{2}a(2n-1) \quad (u=0)$$

$$S_1 = \frac{1}{2}\times g(2\times 1-1) = \frac{1}{2}g$$

$$S_2 = \frac{1}{2}\times g(2\times 2-1) = \frac{3}{2}g$$

$$S_3 = \frac{1}{2}\times g(3\times 2-1) = \frac{5}{2}g$$

Work done $= mgs$

$$W_1;W_2;W_3 = mg\frac{1}{2}g : mg\frac{3}{2}g; mg\frac{5}{2}g$$

$$W_1;W_2;W_3 = 1:3:5$$

2. (d)

KE of the particle at height h is

$$= \frac{1}{2}mv^2$$

$$= \frac{1}{2}m\times 2g(H-h)$$

$$= mg(H-h)$$

PE at height $h = mgh$

According to the problem,

$$mg(H-h) = 2mgh$$

$$H-h = 2h$$

$$\Rightarrow h = \frac{H}{3}$$

Now,
$$v^2 = 2g(H-h)$$

$$= 2g(3h-h)$$

$$= 4gh$$

$$v = 2\sqrt{gh}.$$

3. (b)

$$s = t^2 + 2t$$

$$v = \frac{ds}{dt} = 2t+2$$

$$a = \frac{dV}{dt} = 2$$

Force $= ma = 2\times 2 = 4\ \text{m/s}^2$

Now, displacement in 2s is $= (2)^2 + 2\times 2$

$$= 8\ \text{m}$$

Displacement in 4s is $= (4)^4 + 4\times 2$

$$= 24\ \text{m}$$

$$W = Fds = 4\times(24-8)$$

$$= 64\ \text{J}$$

4. (a)

$$k = 4t^2$$

$$\frac{1}{2}mv^2 = 4t^2$$

$$v^2 = \frac{8}{m}t^2$$

$$v = \sqrt{\frac{8}{m}}\,t$$

$$v = k't$$

$$\frac{dv}{dt} = k'$$

Force $= ma = mk'$

Mass is constant k' is constant, therefore force is constant.

5. (b)

6. (a)

Work–Energy theorem

$$W = \Delta\text{KE}$$

$$\int Fds = \Delta\text{KE}$$

$$F = ma = 2\times 2 = 4$$

Therefore,
$$4s = \Delta\text{KE}$$

$$4\frac{ds}{dt} = \frac{d\text{ KE}}{dt}$$

$$4v = \frac{d\text{ KE}}{dt}$$

Rate of change of KE is 4 times velocity.

7. (a)

Initial KE $= \frac{1}{2}m{v_0}^2$

Let velocity after time be v then

$$\frac{1}{2}m{v_0}^2 - \frac{1}{2}mv^2 = \frac{3}{4}\times\frac{1}{2}m{v_0}^2 \quad \text{(as per question)}$$

$$\Rightarrow \quad v^2 - {v_0}^2 - \frac{3}{4}{v_0}^2 = \frac{1}{4}{v_0}^2$$

$$v = \frac{1}{2}v_0$$

Now,
$$v = v_0 - at_0$$

$$v = v_0 - \mu g t_0 \quad (a = \mu g)$$

$$\mu = \frac{v_0 - v}{gt_0} = \frac{v_0 - \frac{v_0}{2}}{gt_0}$$

$$= \frac{v_0}{2gt_0}$$

8. (a) Speed is independent of mass

9. (b)

Initial stretch of the spring is x

$$F = kx$$

$$mg = kx \Rightarrow x = \frac{10 \times 10}{100} = 1\text{m}$$

Work done $= \frac{1}{2}kx_2^2 - \frac{1}{2}kx_1^2$

$$= \frac{1}{2} \times 100(3^2 - 1^2) \qquad (x_2 = 1 + 2)$$

$$= \frac{1}{2} \times 100 \times 8$$

$$= 400 \text{ J}$$

10. (c)

$$\frac{1}{2}m_1v_1^2 = \frac{1}{2}m_2v_2^2$$

$$\Rightarrow \quad \frac{v_1}{v_2} = \sqrt{\frac{m_2}{m_1}}$$

$$\frac{m_1v_1}{m_2v_2} = \frac{m_1}{m_2} \times \sqrt{\frac{m_2}{m_1}}$$

$$= \frac{\sqrt{m_1}}{m_2} = \sqrt{\frac{1}{4}} = \frac{1}{2}$$

11. (c)

12. (b)

$$F \propto \frac{1}{v} \Rightarrow F = \frac{k}{v}$$

$$\Rightarrow m\frac{dv}{dt} = \frac{k}{v}$$

$$mv\,dv = k\,dt$$

$$m\int_0^v v\,dv = \int_0^t k\,dt$$

$$\frac{1}{2}mv^2 = kt$$

$$\text{KE} \propto t$$

13. (a)

ΔKE = Work done by the force

$= Fs$

$= 4 \times 2$

$= 8$J

14. (c)

Gain in KE = Loss in PE

$$\frac{1}{2}mv^2 = V$$

$$\Rightarrow \quad m = \frac{2V}{v^2}$$

15. (a)

Let the mass per unit length of the chain be $m = \frac{M}{l}$

Potential energy of the hanging part $= ml_0g\left(H - \frac{l_0}{2}\right)$

(H is the height of the table from ground)

Potential energy of the part $(l - l_0)$ which is lying on table $= m(l - l_0)gH$

Potential energy of the chain when its end slides away from the edge $= ml\,g\left(H - \frac{l}{2}\right)$

Applying conservation of Mechanical energy

$$= \frac{1}{2}mlv^2 + ml\,g\left(H - \frac{l}{2}\right)$$

$$= ml_0g\left(H - \frac{l_0}{2}\right) + m(l - l_0)gH$$

$$= \frac{1}{2}lv^2 = l_0gH - \frac{l_0^2}{2}g + lgH - l_0gH - lgH + g\frac{l^2}{2}$$

$$\frac{1}{2}lv^2 = g\frac{l^2}{2} - g\frac{l_0^2}{2}$$

$$lv^2 = g(l^2 - l_0^2)$$

$$v^2 = \frac{g}{l}(l^2 - l_0^2)$$

$$v = \sqrt{\frac{g(l^2 - l_0^2)}{l}}$$

16. (d)

Work done $= \frac{1}{2}k(x + y)^2 - \frac{1}{2}kx^2$

$$= \frac{1}{2}k(x + y + x)(x + y - x)$$

$$= \frac{1}{2}ky(2x + y)$$

17. (b)

The work done = change in KE

18. (b)

Work done = ΔPE + ΔKE

The centre of gravity is at the midpoint of the rod, therefore $\Delta PE = mg\frac{l}{2}$

19. (b)

Let v_1 and v_2 be the velocity of two masses respectively and v be the velocity of combined mass, then

$$m_1v_1 + m_2v_2 = (m_1 + m_2)v$$

$$20 \times 10 + 5 \times 0 = (20 + 5)v$$

$$200 = 25v$$

$$v = \frac{200}{25} = 8 \text{ m/s}$$

$$KE = \frac{1}{2}(m_1 + m_2)v^2$$

$$= \frac{1}{2} \times 25 \times 8 \times 8$$

$$= 800 \text{ J}$$

20. (d)

$$S_1 = 6; S_2 = ?$$

Let stopping force be F, then

$$FS_1 = \frac{1}{2}mv_1^2 \quad \ldots\text{(i)}$$

$$FS_2 = \frac{1}{2}mv_2^2 \quad \ldots\text{(ii)}$$

From (i) and (ii)

$$\frac{S_2}{S_1} = \frac{v_2^2}{v_1^2} = \left(\frac{100 \times 1000}{60 \times 60}\right)^2 / \left(\frac{50 \times 1000}{60 \times 60}\right)^2$$

$$\frac{S_2}{S_1} = 4$$

$$S_2 = 4 \times 6$$

$$= 24 \text{ m}$$

21. (a)

Using conservation of momentum

$$m_1v_1 - m_2v_2 = 0$$

$$m_1v_1 = m_2v_2$$

$$\frac{m_1}{m_2} = \frac{v_2}{v_1}$$

$$\frac{E_1}{E_2} = \frac{\frac{1}{2}m_1v_1^2}{\frac{1}{2}m_2v_2^2}$$

$$= \frac{m_1}{m_2} \times \left(\frac{v_1}{v_2}\right)^2$$

$$= \frac{m_1}{m_2} \times \left(\frac{m_2}{m_1}\right)^2$$

$$= \frac{m_2}{m_1}$$

22. (b)

$$W = \vec{F}.\vec{S}$$

$$= Fs \cos\theta$$

$$= Fs \cos 180° = -Fs$$

23. (d)

$$W = P\Delta V$$

$$= 100 \times (105 - 5) \times 10^{-3}$$

$$= 100 \times 100 \times 10^{-3}$$

$$= 10 \text{ J}$$

24. (b)

$$E_1 = E; \; E_2 = ?$$

$$p_1 = mv_1; p_2 = mv_2$$

$$mv_2 = 2mv_1$$

$$v_2 = 2v_1$$

$$\frac{E_2}{E_1} = \frac{\frac{1}{2}mv_2^2}{\frac{1}{2}mv_1^2} = \left(\frac{v_2}{v_1}\right)^2 = 2^2 = 4$$

$$E_2 = 4E$$

25. (d)

KE = Work done

KE = Fs = mas

KE = mas

KE $\propto a$

where a is acceleration

26. (b)

The energy required to pump m kg of water = mgh

Energy/second =1000 J/s

$$m \times 10 \times 10 = 1000$$

$$\Rightarrow m = 10 \text{ kg}$$

27. (b)

$$W = \Delta PE$$

$$= Mgh$$

28. (c)

29. (a)

$KE = \frac{1}{2}mv^2$. It is square of velocity, so it can never be negative.

30. (a)

31. (b)

$$\frac{\frac{1}{2}mv_2^2}{\frac{1}{2}mv_1^2} = \frac{v_2^2}{v_1^2} = \frac{2gh_2}{2gh_1} = \frac{h_2}{h_1} = 0.8$$

$$h_2 = h_1 \times 0.8 = 10 \times 0.8 = 8 \text{ m}$$

32. (d)

$$0.2 \times 20 = 10 \times v$$

$$v = 0.4 \text{ m/s}$$

$$\text{Initial KE} = \frac{1}{2} \times 0.2 \times 400 = 40 \text{ J}$$

$$\text{Final KE} = \frac{1}{2} \times 10 \times 0.16 = 0.8 \text{ J}$$

So loss in KE in collision $= (40 - 0.8) = 39.2$ J

33. (d)

$$\frac{1}{2}mv^2 = \frac{1}{2}kx^2$$

$$\frac{1}{2} \times 2 \times v^2 = \frac{1}{2} \times 2 \times 16$$

$$v^2 = 16$$

$$v = 4 \text{ m/s}$$

34. (c)

$$v = \sqrt{2gh} = \sqrt{2 \times 9.8 \times 10} = 14 \text{ m/s.}$$

$$\text{Average retarding force} = \frac{mgh}{s}$$

$$= \frac{1 \times 9.8 \times 10}{2 \times 10^{-2}} = 4900 \text{ N}$$

35. (c)

Work done = PE

$$= mg(L - L\cos\theta)$$

$$= mgL(1 - \cos\theta)$$

36. (a)

Gain in KE = Loss in PE

$$= mgh$$

37. (a)

$$F.s. = \frac{1}{2}mv^2 \quad \text{or} \quad \frac{mv}{t}.s = \frac{1}{2}mv^2$$

$$\text{or} \quad \frac{s}{t} \propto v \quad \Rightarrow \quad \frac{v_A}{v_B} = \frac{s_A}{t_A} \times \frac{t_B}{s_B}$$

$$= \frac{s_B t_B}{2t_B \times 3s_B} = \frac{1}{6}$$

38. (a)

$$\text{Work done} = \frac{1}{2}mv^2 - \frac{1}{2}mu^2$$

$$u = 3\hat{i} - 4\hat{j}$$

$$u^2 = (3\hat{i} - 4\hat{j}) \cdot (3\hat{i} - 4\hat{j})$$

$$= 9 + 16 = 25$$

$$v^2 = \bar{v} \cdot \bar{v}$$

$$= (-6\hat{j} + 2\hat{k}) \cdot (-6\hat{j} + 2\hat{k})$$

$$= 36 + 4 = 40$$

$$W = \frac{1}{2} \times \frac{500}{1000}(40 - 25) = \frac{15}{4} = 3.75 \text{ J}$$

39. (a)

Consider a bullet of mass m is moving with initial velocity u and after passing through is plane of width x its velocity decreases to v.

$$\text{Given } u - v = \frac{1}{n}.u \Rightarrow v = u - \frac{u}{n}$$

$$v = u\left(\frac{n-1}{n}\right).$$

Let F be the retarding force applied by each plank by using Work–Energy theorem

$$F.x = \frac{1}{2}mu^2 - \frac{1}{2}mv^2$$

$$\text{or} \quad F.x = \frac{1}{2}mu^2\left(1 - \frac{(n-1)^2}{n^2}\right)$$

$$\text{or} \quad = \frac{1}{2}mu^2\left(\frac{2n-1}{n^2}\right) \quad \text{...(i)}$$

Let P be the total number of planks required to stop the bullet. Then total distance traveled by the bullet before coming to rest is Px.

Again applying Work–Energy theorem,

$$F(Px) = \frac{1}{2}mu^2 - 0$$

Using Eq. (i), we get

$$P = \frac{n^2}{2n-1}$$

40. (b)

Considering the system of m_1, m_2 and earth, there is no friction (non-conservative force)

The k mechanical energy of the system is conserved. Hence, $\Delta K + \Delta U = 0$. If v is the velocity of both masses when m strikes the ground is

$$\Delta K = \frac{1}{2}m_1v^2 + \frac{1}{2}m_2v^2 = 0$$

$$\text{and} \quad \Delta U = -m_1gh$$

$$\Rightarrow \quad \frac{1}{2}m_1v^2 + \frac{1}{2}m_2v^2 - m_1gh = 0$$

$$\therefore \quad v = \sqrt{\frac{2m_1gh}{m_1+m_2}}$$

41. (c)

Centripetal force $= \frac{mv^2}{r} = \frac{k}{r^2}$ (in magnitude)

$$\text{KE} = \frac{1}{2}mv^2 = \frac{k}{2r}$$

$$\text{PE} = U = -\int_{\infty}^{r} \frac{-k}{r^2}.dr = \frac{-k}{r}$$

$$\text{TE} = \text{PE} + \text{KE} = \frac{k}{2r} - \frac{k}{r} = \frac{-k}{2r}$$

42. (a)

The work done by the applied force $= \vec{F}.\vec{S}$

$$= Fs\cos\theta \quad (\theta = 0)$$

$$= 10 \times 2 \times 1$$

$$= 20 \text{ J}$$

43. (a)

Net work done = Work done by applied force + Work done by friction + Work done by normal reaction + Work done by gravity

$$0 = 20 + W_f + 0 + 0$$

$$\Rightarrow W_f = -20 \text{ J}$$

44. (c)

Work done by normal reaction is zero because the force is perpendicular to the surface.

45. (b)

If body strikes the floor with a velocity v then

$$mgh = \frac{1}{2}mv^2 \quad \ldots\text{(i)}$$

With this velocity v, when body passes through the sand floor it comes to rest after travelling a distance x

Body
h
v
F
O
x
sand floor

Let F be the resisting for acting on the body.

Then the resultant force $= F - mg$.

By using Work–Energy theorem,

$$\Rightarrow \quad (F - mg)x = \frac{1}{2}mv^2 - 0 \quad \text{using (i)}$$

or $\quad F.x = mgh + mgx$

or $\quad F = mg\left(1 + \frac{h}{x}\right)$

46. (c)

$$a_c = \frac{v^2}{r} = k^2rt^2$$

$$\text{KE} = \frac{1}{2}mv^2 = \frac{1}{2}mk^2r^2t^2$$

According to Work–Energy theorem change in kinetic energy is equal to work done.

$$W = \frac{1}{2}mk^2r^2t^2;$$

$$P = \frac{dW}{dt} = \frac{1}{2}mk^2r^2(2t)$$

$$P = mk^2r^2t$$

47. (c)

48. (a)

49. (b)

The range is maximum for the angle of projection 45°.

$$R_{\max} = \frac{u^2}{g}$$

$$\therefore \quad u^2 = g.R_{\max}$$

$$\text{KE} = \frac{1}{2}mu^2 = \frac{1}{2}mg\,R_{\max}$$

$$= \frac{1}{2} \times 5 \times 9.8 \times 20 = 490 \text{ J}$$

50. (d)

$$m_1v_1 + m_2v_2 = 0$$

$$3 \times 16 + 6\,(v_2) = 0$$

$$v_2 = -8 \text{ m/s}$$

$$\text{KE} = \frac{1}{2} \times 6 \times 64 = 192 \text{ J}$$

51. (c)

$$mg(h + x) = \frac{1}{2}kx^2$$

$$1 \times 10[10 + x] = \frac{1}{2} \times 10 \times x^2$$

$$5x^2 - 10x - 100 = 0$$

$$x^2 - 2x - 20 = 0$$

$$x = \frac{2 + \sqrt{4 + 4 \times 1 \times 20}}{2 \times 1}$$

$$= \frac{2 + \sqrt{84}}{2} = (1 + \sqrt{21}) \text{ m}$$

52. (a)

Weight $mg = 2(10) = 20$ N
Work done by the applied force $W_F = Fh \cos 0°$.
As the angle between force and displacement is 0°.

or $\quad W_F = 40\ (2)\ (1) = 80$ J.

Similarly, work done by its weight

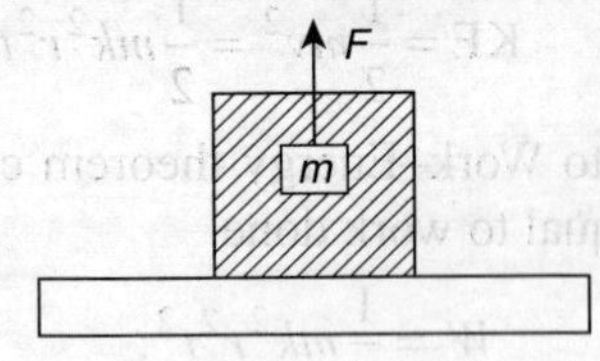

$$W_{mg} = (mg)h \cos 180°$$

or $\quad W_{mg} = 20\ (2)\ (-1) = -40$ J.

53. (d)

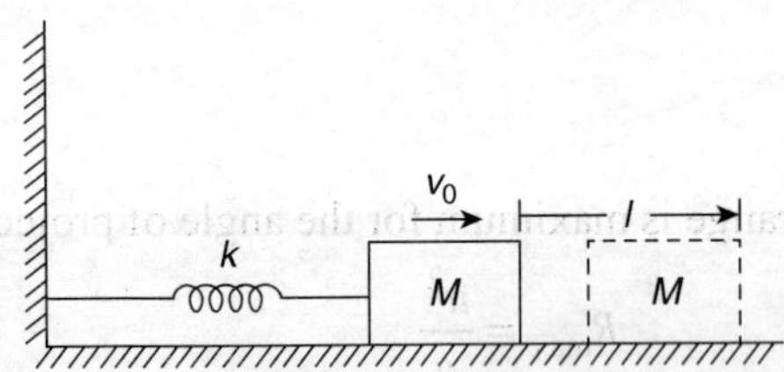

Total work done on block $= (\mu mgl + \frac{1}{2}kl^2)$

54. (d)

Mass of water = Volume × Density = $3.2 \times 12 \times 10^3$ kg
Work done = mg (average height through which water must be lifted)

$$= 3.2 \times 12 \times 10^3 \times 9.8(18 + 6)$$
$$= 3.2 \times 12 \times 10^3 \times 9.8(24)$$
$$= 384 \times 98 \times 240 = 9031680 \text{ J}$$
$$= 9.03 \times 10^6 \text{ J}$$

55. (d)

$$\frac{\text{Work done during ascent}}{\text{Work done for increasing velocity}} = \frac{mgh}{\frac{1}{2}mv^2}$$
$$= \frac{2gh}{v^2}$$
$$= \frac{2 \times 9.8 \times 5000}{\left(360 \times \frac{5}{18}\right)^2}$$
$$= 9.8$$

56. (d)

Both spheres are falling under the same gravitational acceleration.

57. (d)

In an inelastic collision, the particles do not regain their shape and size completely after collision. Some fraction of mechanical energy is retained by the colliding particles in the form of deformed potential energy. Thus, the kinetic energy of particles no longer remains conserved. However, in the absence of external forces, law of conservation of linear momentum still holds good.

58. (a)

Let m_1 and m_2 be the masses of the lighter and the heavier bodies and v_1 and v_2 be their respective velocities. Since, they have equal kinetic energies,

$$\frac{1}{2}m_1v_1{}^2 = \frac{1}{2}m_2v_2{}^2 \quad \text{...(i)}$$

If p_1 and p_2 are the respective momentums of the lighter and heavier bodies, then

$$\frac{P_1}{P_2} = \frac{m_1v_1}{m_2v_2} = \sqrt{\frac{(m_1v_1)^2}{(m_2v_2)^2}}$$
$$= \sqrt{\frac{\frac{1}{2}m_1v_1^2}{\frac{1}{2}m_2v_2^2}} \times \sqrt{\frac{2m_1}{2m_2}} \quad \text{...(ii)}$$

Using Eq. (i), we have

$$\frac{P_1}{P_2} = \sqrt{\frac{2m_1}{2m_2}} = \sqrt{\frac{m_1}{m_2}}$$
$$\frac{P_2}{P_1} = \sqrt{\frac{m_2}{m_1}}$$

Here, $m_2 = 400$g, $m_1 = 100$g

$$\frac{P_2}{P_1} = \sqrt{\frac{400}{100}} = \frac{\sqrt{4}}{1} = \frac{2}{1}$$
$$P_2 : P_1 = 2:1$$

59. (b)

If a body falls from height h, then from equation of motion we know that it will hit the ground with a velocity say $u = \sqrt{2gh}$, which is also the velocity of approach here now. If after collision it gains a height h_1 then again by equation of motion $v = \sqrt{2gh_1}$ which is also the velocity of separation. So, by definition of e,

$$e = \sqrt{\frac{2gh_1}{2gh}}$$

or $\quad h_1 = e^2h.$

Here, $h = 20$ m, $e = 0.9$

∴ Height attained after first bounce

$$h_1 = (0.9)^2 \times 20$$

$$= 0.9 \times 0.9 \times 20 = 16.2 \text{ m}$$

60. (c)

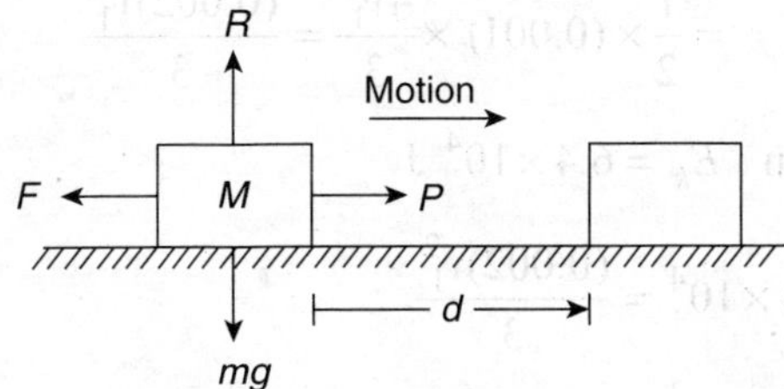

As shown a block of mass M is lying over a rough horizontal surface. Let μ be the coefficient of kinetic friction between the two surfaces in contact. The forces of friction between the block and horizontal surface is given by

$$F = \mu R = \mu Mg$$

To move the block without acceleration the force (P) required will be just equal to the force of friction, i.e.,

$$P = F = \mu R$$

If d is the distance moved, then work done is given by

$$W = P \times d = \mu R d$$

61. (d)

According to Newton's second law of motion, force acting on a body is equal to the rate of change of momentum during impact.

$$F = \frac{\Delta P}{\Delta t}$$

Also $\quad F = ma$

$$\therefore \quad ma = \frac{P_2 - P_1}{\Delta t}$$

or $\quad a = \dfrac{mv_2 - (-mv_1)}{m\Delta t} = \dfrac{v_2 + v_1}{\Delta t}$

$$\therefore \quad a = \frac{\sqrt{2 \times 10 \times 20} + \sqrt{2 \times 10 \times 5}}{0.02}$$

$$= \frac{20 + 10}{0.02} = 1500 \text{ m/s}^2$$

62. (a)

The below figure shows that each particle is moving with speed v, but in different directions.

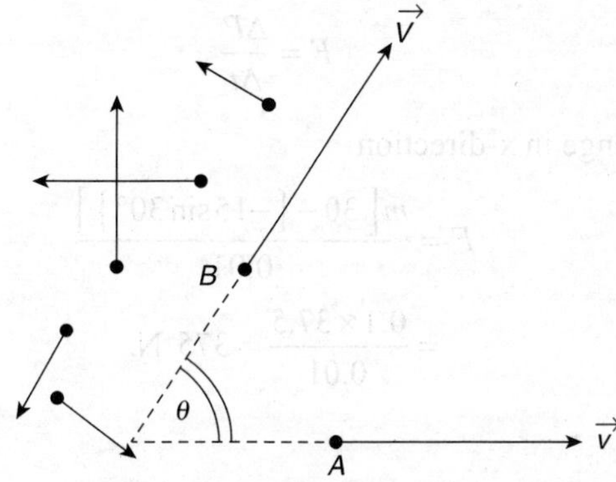

Consider any two particles having angle θ between directions of their velocities is

Then, $\quad \vec{v}_{rel} = \vec{v} - \vec{v}_A$

i.e., $\quad v_{rel} = \sqrt{v^2 + v^2 - 2v.v\cos\theta}$

or $\quad v_{rel} = \sqrt{2v^2(1 - \cos\theta)} = 2v\sin\left(\dfrac{\theta}{2}\right)$

So averaging v_{rel} over all pairs

$$\vec{v}_{rel} = \frac{\int_0^{2\pi} v_{rel} d\theta}{\int_0^{2\pi} d\theta} = \frac{\int_0^{2\pi} 2v\sin\left(\frac{\theta}{2}\right) d\theta}{\int_0^{2\pi} d\theta}$$

$$= \frac{2v \times 2\left[-\cos\frac{\theta}{2}\right]_0^{2\pi}}{2\pi}$$

or $\quad \vec{v} = \dfrac{4v}{\pi}$

63. (d)

If a constant force $\vec{F}$ is applied on a body for a short interval of time Δt; then the impulse of this force is $F \times \Delta t$.

Since, impulse = change in momentum (Δp)

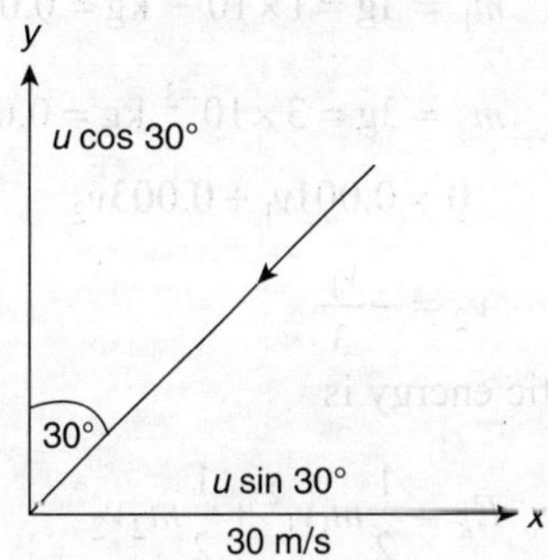

$$F \times \Delta t = \Delta P$$

$$F = \frac{\Delta P}{\Delta t}$$

Change in x-direction

$$F = \frac{m[30 - \{-15\sin 30°\}]}{0.01}$$

$$= \frac{0.1 \times 37.5}{0.01} = 375 \text{ N.}$$

64. (a)

Momentum remains conserved.

Let u_1, u_2 be their initial velocities and v_1, v_2 be their final velocities. From law of conservation of momentum,

$$m_1u_1 + m_2u_2 = m_1v_1 + m_2v_2$$

Given, $m_1 = 5g = 5 \times 10^{-3}$ kg, $m_2 = 5$ kg

$u_1 = 0, u_2 = 0$ (at rest), $v_1 = 500$ m/s

$$m_1 \times 0 + m_2 \times 0 = 5 \times 10^{-3} \times 500 + 5 \times v_2$$

$$v_2 = -\frac{2.5}{5} = -0.5 \text{ m/s}$$

Negative sign of v_2 indicates that v_2 moves in opposite direction.

65. (c)

Let m_1, m_2 be mass of first and second fragment respectively and v_1, v_2 be their velocities after explosion. From conservation of momentum,

$$Mv = m_1v_1 + m_2v_2$$

Where, M is mass of bomb before explosion and v its velocity.

Since, bomb is stationary, hence $v = 0$

$$0 = m_1v_1 + m_2v_2$$

Given, $m_1 = 1\text{g} = 1 \times 10^{-3} \text{ kg} = 0.001 \text{ kg}$

and $m_2 = 3\text{g} = 3 \times 10^{-3} \text{ kg} = 0.003 \text{ kg}$

$\Rightarrow$ $0 = 0.001v_1 + 0.003v_2$

$$v_2 = -\frac{v_1}{3} \quad \text{...(i)}$$

Total kinetic energy is

$$E_k = \frac{1}{2}m_1v_1^2 + \frac{1}{2}m_2v_2^2$$

$$= \frac{1}{2} \times (0.001)v_1^2 + \frac{1}{2} \times (0.003)v_2^2 \quad \text{...(ii)}$$

From Eq. (i), $v_2 = -\frac{v_1}{3}$

$$E_k = \frac{1}{2} \times (0.001)v_1^2 + \frac{1}{2} \times (0.003) \times \left(-\frac{v_1}{3}\right)^2$$

$$= \frac{1}{2} \times (0.001)\left[v_1^2 + 3 \times \frac{v_1^2}{9}\right]$$

$$= \frac{1}{2} \times (0.001) \times \frac{4v_1^2}{3} = \frac{(0.002)v_1^2}{3} \quad \text{...(iii)}$$

Given $E_k = 6.4 \times 10^4$ J

$$6.4 \times 10^4 = \frac{(0.002)v_1^2}{3}$$

$$v_1^2 = 96 \times 10^6 = 9.6 \times 10^7 \text{ m/s}$$

Hence, kinetic energy of smaller fragment is

$$E_k' = \frac{1}{2}m_1v_1^2$$

$$= \frac{1}{2} \times (0.001) \times 9.6 \times 10^7 = 4.8 \times 10^4 \text{ J.}$$

66. (b)

Momentum is conserved.

Let M be the mass of body moving with velocity v and m be the mass of each broken part, the velocity of one part which retraces back is v and that of second part is v'.

Momentum before breaking = Momentum after breaking

$$Mv = m(-v) + mv'$$

$$v' = \frac{Mv + mv}{m}$$

Since, $M = 2m$,

Therefore $v' = \frac{(2m + m)}{m}v = 3v$

Thus, the second part moves with a velocity three times more than the initial as velocity in forward direction.

67. (b)

68. (b)

69. (a)

The relation between kinetic energy (K) and momentum P is given by

$$K = \frac{P^2}{2m}$$

Given, $m_1 = 1\text{g} = 1 \times 10^{-3} \text{ kg} = 0.001 \text{ kg}$

$$m_2 = 4g = 4 \times 10^{-3} \text{ kg} = 0.004 \text{ kg}$$

$$K_1 = K_2$$

$$\frac{P_1^2}{2m_1}=\frac{P_2^2}{2m_2}$$

$$\Rightarrow \quad \frac{P_1}{P_2}=\sqrt{\frac{m_1}{m_2}}=\sqrt{\frac{0.001}{0.004}}=\frac{1}{2}$$

70. (c)

Momentum remains conserved in explosion.
Let velocity of masses after explosion be v_1, and v_2. Then from law of conservation of momentum we have, Momentum before explosion = Momentum after explosion.

$$Mv=m_1v_1+m_2v_2$$

$$m_1=m_2=m, v_2=0$$

$$Mv=mv_1+m\times 0$$

$$\Rightarrow \quad v_1=\frac{Mv}{m}.$$

71. (b)

The work done is measured by the product of force (F) and displacement (x) in the direction of force

$$dW=F.dx$$

Given, $F=Cx$

$$W=\int_0^{x_1} Cxdx$$

Using $\int x^n dx=\frac{x^{n+1}}{n+1}.x^n$

$$W=c\left[\frac{x^2}{2}\right]_0^x=\frac{1}{2}Cx_1^2$$

72. (c)

We know $P^2=2mK$ ...(i)

When linear momentum is increased by 50%, then

$$p'=\frac{150}{100}P=\frac{3}{2}P$$

$$(p')^2 2mK'$$

$$\left(\frac{3}{2}P\right)^2=2mK'$$

$$\frac{9}{4}P^2=2mK'$$

From Eqs. (i) and (ii), we get

$$\frac{9}{4}\times 2mK=2mK'$$

$$K'=\frac{9}{4}K$$

Increase in kinetic energy

$$\Delta K=\frac{9}{4}K-K=\frac{5}{4}K$$

Hence, % increase in kinetic energy.

$$=\frac{\left(\frac{5}{4}\right)K}{K}\times 100\%=\frac{5}{4}\times 100\%=125\%$$

73. (b)

(i) If a body is moved up an inclined plane, then the work done against friction force is zero as there is no friction. But a work has to be done against the gravity. So Statement (i) is incorrect.

(ii) If there were no friction, moving vehicles could not be stopped by locking the brakes. This statement is correct.

(iii) In this situation the normal reaction is by

$$R=mg\cos\alpha$$

If α increases then the value of cos α decreases and hence, normal reaction R also decreases. So this statement is also incorrect.

(iv) When the duster is rubbing upward then an external force is applied and its value is

$$F'=0.5g+\mu R=0.5\text{ g}+0.5\times 11$$
$$=(0.5\times 10+5.5)\text{N} \quad (R=11\text{ N})$$

or $F'=10.5\text{ N}$

Hence, work done in rubbing the duster through a distance of 10 cm is

$$W=F'\times s=10.5\times\frac{10}{100}=1.05\text{ J}$$

74. (d)

We know the potential energy of a stretched spring is

$$U=\frac{1}{2}kx^2$$

Here, k is spring constant and x is elongation in spring. But here $x=2$ cm

$$U=\frac{1}{2}k(2)^2=\frac{1}{2}k\times 4 \quad \text{...(i)}$$

If $x=10\text{cm}$

then $U'=\frac{1}{2}k(10)^2$

$$U'=\frac{1}{2}k\times 100 \quad \text{...(ii)}$$

From Eq. (i) and Eq. (ii), we get

$$\frac{U'}{U} = \frac{\frac{1}{2}k \times 100}{\frac{1}{2}k \times 4} = 25$$

$$U' = 25U.$$

75. (c)

$$p = \sqrt{2mK}$$

Here, $m_1 = 4\text{kg}, m_2 = 1\text{kg}$

and $K_1 = K_2 = K$

If K is constant, then

$$p \propto \sqrt{m} \Rightarrow \frac{P_1}{P_2} = \sqrt{\frac{m_1}{m_2}}$$

or $\frac{P_1}{P_2} = \sqrt{\frac{4}{1}} \Rightarrow \sqrt{\frac{P_1}{P_2}} = \frac{2}{1}$

or $P_1 : P_2 = 2:1$

76. (c)

$$W_1 = \frac{1}{2}kx^2 \quad \ldots\text{(i)}$$

The work done in stretching the spring by a further length x.

$$W_2 = \frac{1}{2}k(2x)^2 - \frac{1}{2}(kx^2)$$

$$= \frac{1}{2}k \times 4x^2 - \frac{1}{2}kx^2$$

$$W_2 = 3 \times \frac{1}{2}kx^2 \quad \ldots\text{(ii)}$$

From Eqs. (i) and (ii), we have

$$W_2 = 3W_1$$

77. (c)

In x-direction

$$mu_1 + 0 = 0 + mv_x$$

$$mv = mv_x$$

$$v_x = v$$

In y-direction

$$0 + 0 = m\left(\frac{v}{\sqrt{3}}\right) - mv_y$$

$$v_y = \frac{v}{\sqrt{3}}$$

$\therefore$ Velocity of second mass after collision

$$v' = \sqrt{\left(\frac{v}{\sqrt{3}}\right)^2 + v^2} = \sqrt{\frac{4}{3}v^2}$$

$$v' = \frac{2}{\sqrt{3}}v.$$

78. (a)

$$F = ma = \frac{mv}{T} \qquad \left(\because \frac{v-0}{T} = a\right)$$

Instantaneous power $= Fv$

$$= mav$$

$$= \frac{mv}{T}.at = \frac{mv}{T}\frac{v}{T}.t$$

$$= \frac{mv^2}{T^2}.t$$

79. (d)

Since linear momentum is conserved,

$$mv = (2m + m)v'$$

$$\Rightarrow \quad v' = \frac{v}{3}$$

80. (b)

Suppose spring is compressed by a length x.

i.e., $\frac{1}{2}mv^2 = \frac{1}{2}kx^2$

Given, $m = 2\text{kg}, v = 3\text{m/3}, k = 144 \text{ N/m}$

Thus, $\frac{1}{2} \times 2 \times (3)^2 = \frac{1}{2} \times 144 \times x^2$

or $9 = 72x^2$

Thus $x = \sqrt{\frac{9}{72}} = \frac{1}{2\sqrt{2}}$ m

Thus length of compressed spring is given by

$$= 2 - \frac{1}{2\sqrt{2}} \approx 1.5 \text{ m}.$$

81. (a)

According to law of conservation of linear momentum,

$$m_1v_1 = m_1v + m_2v_2$$

Where v is the velocity of bullet after collision and v_2 is the velocity of block

$$0.02 \times 600 = 0.02v + 4v_2$$

$$0.02v = 12 - 8 \qquad (v_2 = \sqrt{2gh})$$

$$v = \frac{4}{0.02} = 200 \text{ m/s}$$

82. (b)

The change in mechanical energy

$$\Delta U = mg(h + x) + \frac{1}{2}mv^2$$

Given, $m = 5 \text{ g} = 5\times10^{-3} \text{ kg} = 0.05 \text{ kg}, h = 19.5 \text{ m}$

$$x = 50 \text{ cm} = 0.5 \text{ m}, v = 10 \text{ m/s}, g = 10 \text{ m/s}^2$$

Thus, $\Delta U = 0.005\times10(19.5+0.5)+\frac{1}{2}\times0.005\times(10)^2$

$$= 1+0.25 = 1.25 \text{ J}$$

$$\Delta U = 1.25 \text{ J.}$$

83. (d)

Using conservation of linear momentum, we have

$$mv_0 = mv+2mv \Rightarrow v = \frac{v_0}{3}$$

$$\Delta PE = -\Delta KE$$

$$\frac{1}{2}kx_0^2 = \frac{1}{2}mv_0^2 - \frac{1}{2}(3m)\left(\frac{v_0}{3}\right)^2$$

$$\Rightarrow \quad \frac{1}{2}kx_0^2 = \frac{1}{2}mv_0^2 - \frac{1}{2}\frac{mv_0^2}{3}$$

and $$kx_0^2 = mv_0^2 - \frac{mv_0^2}{3}$$

$$k = \frac{2mv_0^2}{3x_0^2}.$$

84. (c)

The momentum of third part, will be equal to the resultant of moment of two parts.

$$P_3 = \sqrt{P_1^2+P_2^2} = \sqrt{(-2P)^2+(P)^2}$$

$$= \sqrt{4P^2+P^2} = P\sqrt{5}.$$

85. (b)

At highest point, there is only horizontal component velocity.

$\therefore$ KE at highest point

$$= \frac{1}{2}m(u\cos\theta)^2 = \frac{1}{2}m(u\cos45°)^2$$

$$= \frac{1}{2}m\left(\frac{u}{\sqrt{2}}\right)^2 = \frac{1}{2}\left(\frac{1}{2}mu^2\right) = \frac{E_k}{2}.$$

86. (d)

Since, 5 kg mass explodes into three pieces of masses 1:1:3,

$$m_1 = 1 \text{ kg}, m_2 = 1 \text{ kg}, m_3 = 3 \text{ kg}$$

Resultant momentum

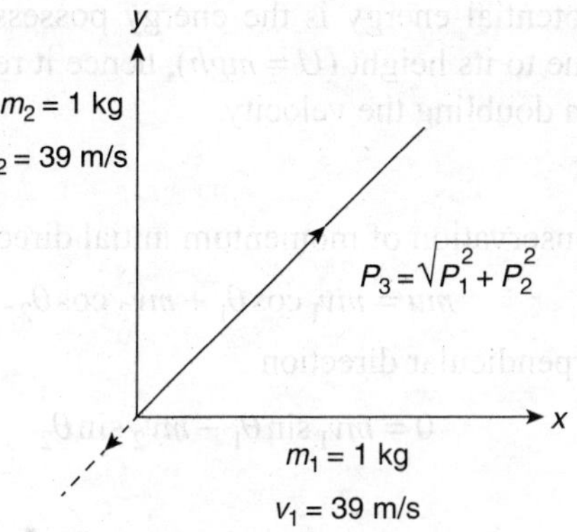

$$P_3 = \sqrt{P_1^2+P_2^2}$$

$$mv_3 = \sqrt{(m_1v_1)^2+(m_2v_2)^2}$$

$$3v_3 = \sqrt{(1\times39)^2+(1\times39)^2} = 39\sqrt{2}$$

$$v_3 = \frac{39\sqrt{2}}{3} = 13\sqrt{2} \text{ m/s.}$$

87. (b)

(a) Acceleration is proportional to rate of change of velocity.

$$a = \frac{dv}{dt} = \frac{v_2 - v_1}{t}$$

Given, $v_2 = 2v, v_1 = v$

$$a = \frac{2v-v}{t} = \frac{v}{t}$$

(b) Momentum (P) is defined as product of mass (m) and velocity (v)

$$P = mv$$

$$P = mv_1, P_2 = m(2v_1)$$

$$\frac{P^2}{P^2} = \frac{mv_1}{2mv_1} = \frac{1}{2}$$

$$P_2 = 2P_1$$

(c) Kinetic energy is the energy possessed by a body due to its velocity.

$$K = \frac{1}{2}mv^2$$

$$K_1 = \frac{1}{2}mv_1^2$$

$$K_2 = \frac{1}{2}m(2v_1)^2$$

$$\frac{K_1}{K_2} = \frac{v_1^2}{(2v_1)^2} = \frac{1}{4}$$

$$K_2 = 4K_1$$

Hence, kinetic energy becomes four times.

(d) Potential energy is the energy possessed by body due to its height ($U = mgh$), hence it remains same on doubling the velocity.

88. (b)

By conservation of momentum initial direction

$$mu = mv_1 \cos\theta_1 + mv_2 \cos\theta_2 \quad \text{...(i)}$$

In perpendicular direction

$$0 = mv_1 \sin\theta_1 - mv_2 \sin\theta_2 \quad \text{...(ii)}$$

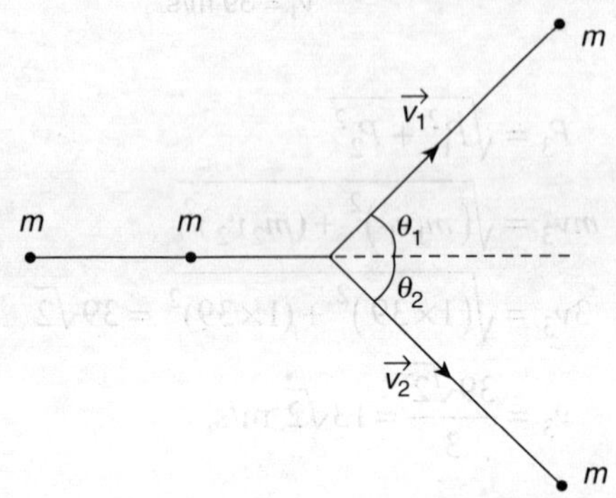

Since the collision is elastic, the KE is also conserved

$$\frac{1}{2}mu^2 = \frac{1}{2}mv_1^2 + \frac{1}{2}mv_2^2 \quad \text{...(iii)}$$

$$u^2 = v_1^2 + v_2^2$$

Squaring Eqs. (i) and (ii) adding

$$mu^2 = m^2(v_1 \cos\theta_1 + v_2 \cos\theta_2)^2 + m^2(v_1 \sin\theta_1 - v_2 \sin\theta_2)^2$$

$$u^2 = v_1^2 + v_2^2 + 2v_1v_2 \cos\theta_1 \cos\theta_2 - 2v_1v_2 \sin\theta_1 \sin\theta_2$$

$$= v_1^2 + v_2^2 + 2v_1v_2 \cos(\theta_1 + \theta_2)$$

Using Eq. (iii),

$$2v_1v_2 \cos(\theta_1 + \theta_2) = 0$$

Since $v_1v_2 \neq 0$

Hence, $\cos(\theta_1 + \theta_2) = 0$

$$\theta_1 + \theta_2 = 90°$$

When two identical particles collide elastically and obliquely, with one being at rest, then they fly off in mutually perpendicular directions.

89. (d)

$$p = \frac{W}{t} = \frac{F \times x}{t} = F \times v$$

90. (a)

In elastic collision of two particles of same mass, the masses exchange velocities after collision. Hence, the velocity of the last ball will be 0.4 m/s.

91. (c)

Using conservation of momentum;

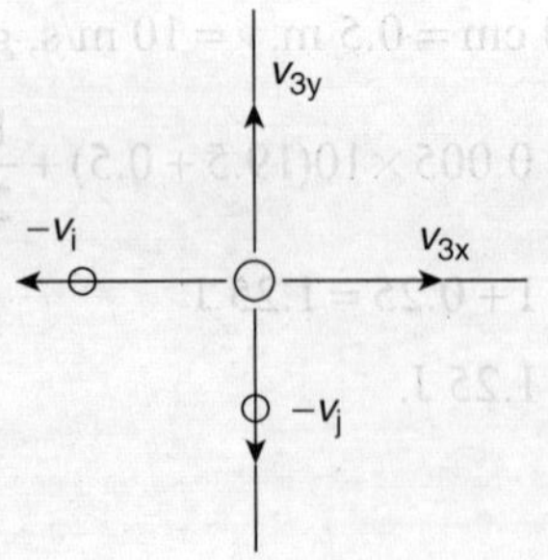

$$mv_{3x} - mvi = 0$$

$$mv_{3x} - mvj = 0$$

$$\Rightarrow \quad v_{3x} = vi$$

and $\quad v_{3y} = vj$

$$v_3 = vi + vj$$
$$= v(i + j)$$

92. (a)

From principle of conservation of momentum,

$$P_x - P_m \cos\theta = 0$$
$$P_y - P_m \sin\theta = 0$$

$$P_m = \sqrt{P_x^2 + P_y^2}$$

$$m \times 40 = \sqrt{(2 \times 12)^2 + (1 \times 16)^2} = 29$$

$$m = 0.7$$

Hence, total mass of the shell

$$M = 1 + 2 + 0.7 = 3.7 \text{ kg}$$

93. (a)

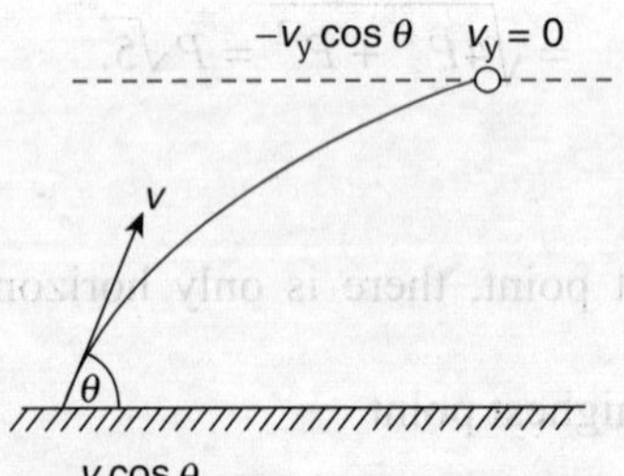

At the highest point, vertical component of velocity is zero. Let the shell be fired from the cannon with a velocity v at an angle θ with the horizontal. At highest point, the horizontal component of velocity is $v \cos\theta$. Since, momentum remains conserved

Momentum before explosion = Momentum after explosion

$$mv\cos\theta = -\frac{mv\cos\theta}{2} + \frac{mv'}{2}$$

where v' is velocity of second part. So $v' = 3v\cos\theta$.

94. (a)

Kinetic energy of the block is

$$K = \frac{1}{2}mv^2$$

This kinetic energy is equal to the work done by the block before coming to rest. The work done in compressing the spring through a distance x from its normal length is

$$W = \frac{1}{2}kx^2$$

$$\frac{1}{2}mv^2 = \frac{1}{2}kx^2$$

$$x = v\sqrt{\frac{m}{k}}$$

given, $v = 4$ m/s, $m = 16$ kg, $k = 100$ N/m

$$x = 4 \times \sqrt{\frac{16}{100}} = 1.6 \text{ m}$$

95. (b)

$$mg = \frac{mv^2}{r}$$

$$\Rightarrow \quad v^2 = rg \qquad \ldots\text{(i)}$$

Circumference of circle is $2\pi r$.

Time for a revolution $= \dfrac{2\pi r}{v}$

Putting the value of v from Eq. (i), we get

$$T = \frac{2\pi r}{\sqrt{gr}} = 2\pi\sqrt{\frac{r}{g}}$$

Here, $r = 4$ m, $g = 9.8$ m/s^2

$$T = 2\pi\sqrt{\frac{4}{9.8}}$$

$$T = \frac{4\pi r}{\sqrt{9.8}} = 4 \text{ s.}$$

96. (b)

$$t = \sqrt{x} + 3$$

$$x = (t-3)^2$$

Velocity, $v = \dfrac{dx}{dt} = 2(t-3)$

Acceleration, $a = \dfrac{dv}{dt} = 2$

$\therefore$ Force, $F = ma = 1 \times 2 = 2$ N

Displacement in 10 seconds

$$s = (10-3)^2 = 49 \text{ m}$$

Work done $= F.s = 2 \times 49 = 98$ J

97. (c)

$$v^2 - u^2 = 2ax$$

$$a = \frac{1}{2x}(v^2 - u^2)$$

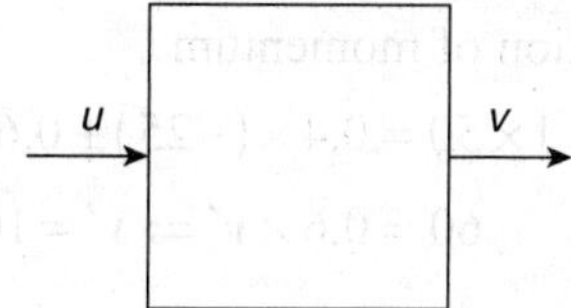

$$F = \frac{m}{2x}(v^2 - u^2)$$

98. (a)

The magnitude of force is given by

$$F = \frac{dU}{dx}$$

The greatest slope is of AB and then of EF

Thus AB > EF > CD > BC

99. (d)

$$\text{KE} = \frac{p^2}{2m} \Rightarrow \frac{\text{KE}_1}{\text{KE}_2} = \frac{m_2}{m_1} = \frac{4}{1}$$

or $$\frac{m_1}{m_2} = \frac{1}{4}$$

100. (b)

The magnitude of force is given by

$$|F| = \frac{dU}{dx}$$

Thus maximum force will occur at x_2.

101. (d)

$$\because \quad v^2 - u^2 = 2as \qquad (\text{as } v = 0)$$

$$a = -\frac{u^2}{2s}$$

Resistance offered by the plank and is constant.

$$u^2 \propto s$$

$$\left.\begin{aligned} u_1^2 &\propto 2t \\ u_2^2 &\propto nt \end{aligned}\right\}$$ where n is no. of planks and t is thickness of plank

$$\frac{u_2^2}{u_1^2} = \frac{n}{2}$$

$$\Rightarrow \quad \frac{(2u_1)^2}{u_1^2} = \frac{n}{2} \qquad (\because u_2 = 2u_1)$$

$$n = 8.$$

102. (c)

Velocity after 5 s, $v = u - gt$

$$= 100 - 10 \times 5 = 50 \text{ m/s}$$

By conservation of momentum

$$1 \times 50 = 0.4 \times (-25) + 0.6 \times v'$$

$$60 = 0.6 \times v' \Rightarrow v' = 100 \text{ m/s upwards}$$

103. (d)

Given: F = 1200 N, $m = 0.5$kg

$$t = 25 \times 10^{-3} \text{ sec}, u = -14 \text{m/ sec}$$

($\because$ direction is opposite to force)

So, Impulse = $F \times t = m(v - u)$

$$1200 \times 25 \times 10^{-3} = 0.5(v + 14)$$

$$v + 14 = \frac{1200 \times 25 \times 10^{-3} \times 10}{5} = 60$$

$$v = 60 - 14 = 46 \text{ m/ sec}$$

104. (c)

Kinetic energy, $\text{KE} = \frac{1}{2}mv^2$

$$= \frac{1}{2} \times 9.1 \times 10^{-31} \times (4 \times 10^6)^2$$

$$= \frac{9.1 \times 10^{-31} \times 16 \times 10^{12}}{2}$$

$$= 72.8 \times 10^{12} \times 10^{-31} \text{J}$$

$$= \frac{72.8 \times 10^{12} \times 10^{-31}}{1.6 \times 10^{-19}} \text{eV}$$

$$= 45.5 \text{ eV} \approx 45 \text{ eV}$$

105. (a)

$$M = 5 \text{ kg}, h = 20 \text{ m}$$

Since the water leaks at a constant of 0.2 kg/m, suppose x be the distance pulled up, then the reduced mass, $m = 0.2x$.

$\therefore$ Effective mass= $(M - m) = (5 - 0.2x)$

$\therefore$ Total work done, $W = \int_{x=0}^{x=20} F.dx$

$$= \int_{x=0}^{x=20} (M - m)gdx$$

$$= \int_{x=0}^{x=20} (5 - 0.2) \times 10dx$$

$$= \left[50x - 2.\frac{x^2}{2}\right]_0^{20}$$

$$= 1000 - 400 = 600 \text{ J}$$

106. (d)

Momentum, $P = \sqrt{2mE}$ where E is the kinetic energy. Because for same KE, the mass of α-particle is maximum and hence momentum is maximum.

107. (c)

108. (c)

Given, velocity of river, $(v) = 2$ m/s

Density of water (ρ)= 1.2 g/cc

Mass of each cubic metre

$$m = \frac{1.2 \times 10^{-3}}{(10^{-2})^3} = 1.2 \times 10^3 \text{ kg}$$

$\therefore$ Kinetic energy $= \frac{1}{2}mv^2$

$$= \frac{1}{2} \times 1.2 \times 10^3 \times (2)^2$$

$$= 2.4 \times 10^3 \text{ J} = 2.4 \text{ kJ}$$

109. (b)

Total distance travelled by the ball before its second hit is

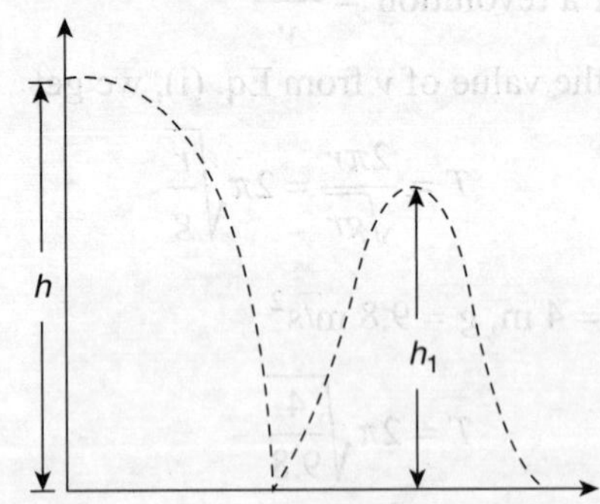

$$H = h + 2h_1$$

$$= h\left[1 + 2e^2\right] \qquad \left[\because h_1 = he^2\right]$$

110. (a)

When the block strikes the spring, the kinetic energy of the block converts into potential energy of spring, i.e.,

$$\frac{1}{2}mv^2 = \frac{1}{2}kx^2$$

where x is compression in spring.

$$x = \sqrt{\frac{mv^2}{k}}$$

$$= \sqrt{\frac{25 \times (3)^2}{100}} = \sqrt{\frac{9}{4}} = \frac{3}{2} = 1.5 \text{ m}$$

When the block returns to its original position, again potential energy converts into kinetic energy of the block, so velocity of the block is the same as before but its sign changes as it goes to mean position.

Hence, $v = -3 \text{ ms}^{-1}$

111. (d)

From conservation of momentum

$$Mv + (4mv) = 0$$

$$\Rightarrow \quad v = -\frac{4mv}{M} = \frac{-4 \times 35 \times 10^{-3} \times 400}{20}$$

$$= -2.8 \text{ms}^{-1}$$

Force applied on the rifle

$$F = \frac{Mv}{t} = -\frac{20 \times 2.8}{1}$$

$$= -56 \text{ N}.$$

112. (d)

Coefficient of restitution of momentum

$$e_1 = \left(\frac{\text{Relative velocity of separation}}{\text{Relative velocity of approach}}\right)_1$$

$$= \frac{1}{2}$$

and $$e_2 = \left(\frac{\text{Relative velocity of separation}}{\text{Relative velocity of approach}}\right)_2$$

given $$\frac{e_1}{e_2} = \frac{3}{1} \Rightarrow \frac{e_2}{e_1} = \frac{1}{3}$$

$$\Rightarrow \left(\frac{\text{Relative velocity of separation}}{\text{Relative velocity of approach}}\right)_2 = \frac{1}{3} \times \frac{1}{2} = \frac{1}{6}$$

$$\left(\frac{\text{Relative velocity of approach}}{\text{Relative velocity of separation}}\right) = \frac{6}{1}$$

113. (c)

For water not to spill out of the bucket,

$$v_{\min} = \sqrt{5gR}$$

$$= \sqrt{5 \times 10 \times 0.5} = 5 \text{ m/s}.$$

114. (c)

If the motor pumps water (density $= \rho$) continuously through a pipe of area of cross-section A with velocity v, the mass flowing out per second

$$m = Av\rho \quad \text{...(i)}$$

Rate of increase of kinetic energy

$$= \frac{1}{2}mv^2 = \frac{1}{2}(Av\rho)v^2 \quad \text{...(ii)}$$

Mass m, flowing out per second, can be increased to m' by increasing v to v', then power increases from P to P'

$$\frac{P'}{P} = \frac{\frac{1}{2}A\rho(v')^3}{\frac{1}{2}A\rho v^3} \text{ or } \frac{P'}{P} = \left(\frac{v'}{v}\right)^3$$

Now, $$\frac{m'}{m} = \frac{A\rho v'}{A\rho v} = \frac{v'}{v}$$

As $$m' = nm, v' = nv$$

$$\frac{P'}{P} = n^3$$

$$\Rightarrow \quad P' = n^3 P.$$

115. (c)

From equation of motion,

$$v^2 = u^2 - 2as$$

$$(500)^2 = (1000)^2 - 2 \times a \times s$$

$$s = \frac{(1000)^2 - (500)^2}{2a} = \frac{375000}{a}$$

$\therefore$ work done against air resistance

$$= Fs = ma \times a$$

$$= \frac{10}{1000} a \times \frac{375000}{a} = 3750 \text{ J}.$$

116. (a)

Let v be the velocity of the boat with respect to the water, then from conservation of linear momentum

$$(200 + 50)v + 50 \times 2 = 50 \times 0 + 200 \times 0$$

$$250v = -100$$

$$v = \frac{-100}{250} = \frac{-2}{5} \text{ m/s}.$$

117. (a)

From Newton's second law,

$$F = \frac{dp}{dt}$$

If $F = 0$, then $\frac{dp}{dt} = 0$

$\Rightarrow p$ = constant.

118. (c)

Work done by the gun per minute = Total kinetic energy of the bullets

$$= n\frac{1}{2}mv^2$$

$$= 240\times\frac{1}{2}\times 10\times 10^{-3}(600)^2$$

$$= 120\times 10^{-2}\times 600\times 600$$

$$\text{Power of gun} = \frac{\text{Work done}}{\text{Time taken}}$$

$$= \frac{120\times 10^{-2}\times 600\times 600}{60} = 72000 \text{ W}$$

$$= 72 \text{ kW}$$

119. (a)

Both statements A and B are true.

120. (a)

$$v_1' = \left[\frac{m_1 - em_2}{m_1+m_2}\right]v_1 - \left[\frac{(1+e)m_2}{m_1+m_2}\right]v_2$$

$$= \left[\frac{2-\left(\frac{2}{3}\right)4}{2+4}\right]24 - \left[\frac{\left(1+\frac{2}{3}\right)4}{2+4}\right]48$$

$$= -56 \text{ m/s}$$

$$v_2' = \left[\frac{(1+e)m_1}{m_1+m_2}\right]v_1 - \left[\frac{m_2 - em_1}{m_1+m_2}\right]v_2$$

$$= \left[\frac{\left(1+\frac{2}{3}\right)2}{2+4}\right]24 - \left[\frac{4-\left(\frac{2}{3}\right)2}{2+4}\right]48$$

$$= -8 \text{ m/s.}$$

121. (c)

$$\text{KE} = \int F dx$$

$$= \int_{x=0}^{3}(9-x^2)dx$$

$$= 9[x]_{x=0}^{3} - \left[\frac{x^3}{3}\right]_{x=0}^{3}$$

$$= 27 - 9 = 18 \text{ joule}$$

122. (c)

$$\text{KE} = \frac{1}{2}mv^2 = \frac{1}{2}m(at)^2\frac{1}{2}ma^2t^2$$

Rate of change of KE,

$$\frac{dK}{dt} = \frac{d}{dt}\left[\frac{1}{2}ma^2t^2\right] = ma^2t$$

$$\frac{dK}{dt} \propto t$$

So, statement A is correct.

When the body is at rest then it may be or may not be in equilibrium, so statement B is wrong.

123. (c)

Suppose the body leaves the surface of the hemisphere at point P.

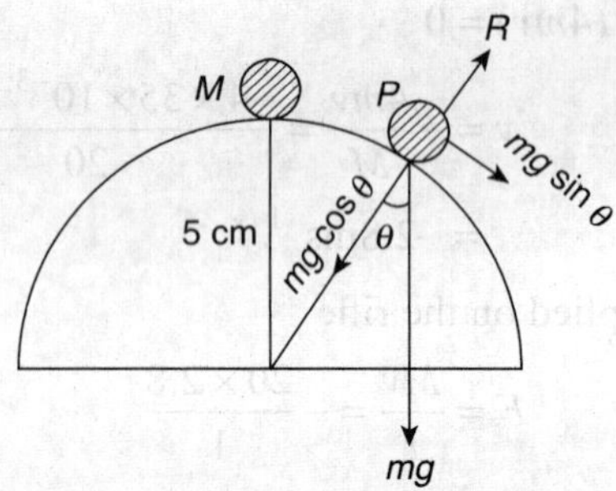

Let at point P radius vector of the body makes an angle θ.

$$mg\cos\theta - R = \frac{mv^2}{r}$$

When the body leaves the surface then $R = 0$.

$$mg\cos\theta = \frac{mv^2}{r}$$

$$\cos\theta = \frac{v^2}{rg} = \frac{(5)^2}{5\times 10} = \frac{25}{50} = \frac{1}{2}$$

$$\cos 60° = \frac{1}{2}$$

$$\theta = 60°.$$

124. (c)

$M = 4$ kg, $P = 8$ kg ms

$F = 0.2$ N, $t = 10$ sec.

From Newton's second law,

Force = Rate of change of momentum

$$F = \frac{P_2 - P_1}{t}$$

$$0.2 = \frac{mv-8}{10}$$

$$2 = 4\times v - 8$$

$$4v = 10 \Rightarrow v = \frac{5}{2} \text{ m/s}$$

Final momentum $P_2 = mv = 4\times\frac{5}{2} = 10$ kg-m/s

Increase in KE $= \dfrac{P_2^2 - P_1^2}{2m} = \dfrac{(10)^2 - (8)^2}{2\times 4}$

$= \dfrac{100-64}{8} = \dfrac{36}{8} = 4.5$ J.

125. (b)

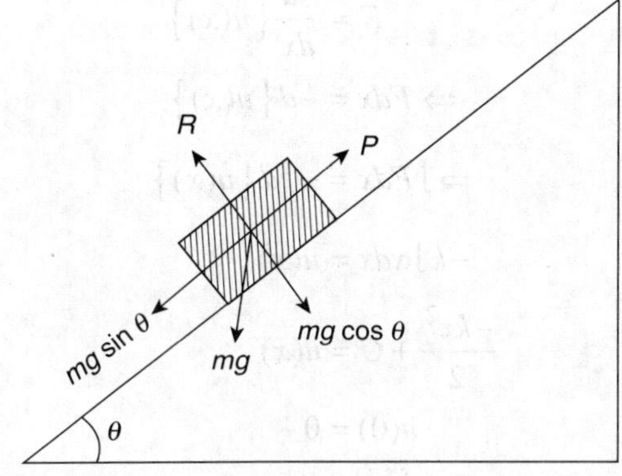

Acceleration of body up a rough inclined plane

$$a = g(\mu\cos\theta + \sin\theta)$$

From equation of motion

$$v^2 = u^2 - 2as$$

$$0 = u^2 - 2as$$

$$u^2 = 2as$$

Initial KE, $E = \dfrac{1}{2}mu^2$

$$E = \frac{1}{2}(2mas)$$

$$s = \frac{E}{ma}$$

Work done $W = F.s$

$$= (mg\sin\theta + \mu R)\frac{E}{ma}$$

$$= (mg\sin\theta + \mu mg\cos\theta).\frac{E}{m(\mu\cos\theta + \sin\theta)g}$$

$$= g(\sin\theta + \mu\cos\theta).\frac{E}{g(\mu\cos\theta + \sin\theta)} = E.$$

126. (b)

$m = 2$ kg, u = 0, $v = 20$ m/s, $t = 4$ sec

Acceleration $a = \dfrac{v-u}{t} = \dfrac{20-0}{4} = 5$ m/s^2

$\therefore$ Force applied $F = ma = 2\times 5 = 10$ N

Velocity at the end of 2 sec

$$v = u + at = 0 + 5\times 2 = 10 \text{ m/s}$$

$\therefore$ Power $P = Fv = 10\times 10 = 100$ watt

127. (a)

$$P_d = F.V.$$

$$V = \frac{10\text{ m}}{5\text{ s}} = 2\text{m/s}$$

$$F = 100\times 10 = 1000\text{ N}$$

$\Rightarrow$ $P_d = 2000$ N

$$\frac{60}{100}P_E = P_d$$

$\Rightarrow$ $P_E = \dfrac{10,000}{3} = 3.3$ kW

128. (a)

$$dW = F.dx$$

Applied force $= -F$

$\Rightarrow$ $dW = -(-5x - 16x^3)dx$

$$\int_0^w dW = \left[16\int_{0.1}^{0.2} x^3 dx + 5\int_{0.1}^{0.2} x dx\right]$$

$$= \left[4x^4 + \frac{5}{2}x^2\right]_{0.1}^{0.2}$$

$$= 8.1\times 10^{-2}\text{ J}$$

129. (c)

Total work done = Work done in lifting water 20 m + Work done in changing pressure.

$\Rightarrow$ $W_T = W_l + W_r$

$$W_r = \Delta p.V.$$

$$\Delta p = 2\times 10^5 - 10^5 = 10^5\text{ N/m}^2$$

$$W_1 = F.x$$

$$F = mg = \rho Vg$$

$$= 10^3\text{ kg/m}^3 \times 4\text{m}^3 \times 10\text{ m/s}^2$$

$$= 4\times 10^4\text{ N}$$

where ρ is density of water V is volume of water

$$x = 20\text{ m}$$

$\Rightarrow$ $W_r = 4\times 10^5$

$$W_1 = 8\times 10^5$$

$\Rightarrow$ $W_T = 12\times 10^5$ J

130. (b)

(5 cos θ) . 10 = 50 cos θ =25

$\Rightarrow \quad \cos\theta = \frac{1}{2}$

$\Rightarrow \quad \theta = 60°$

131. (b)

$p = F.v$

$v = u + at$

$\because \quad u = 0$

$\Rightarrow \quad v = at \Rightarrow P \propto t$

132. (b)

Loss in gravitational PE = Gain in elastic PE

$$mg(h+x) = \frac{1}{2}kx^2$$

133. (a)

The work done by the ball on the wall is equal to zero.

134. (b)

Let after the collision, velocity of m be v_1 and M be v_2, and let v_1 and v_2 be in the same direction.

Now, $\quad Mv = Mv_1 + mv_2 \quad$ ...(i)

(Conservation of momentum)

$v_1 - v_2 = v \quad (e = 1)$

$\Rightarrow \quad Mv_1 - Mv_2 = Mv \quad$ ...(ii)

Adding these two equations, we get

$$2Mv = (M+m)v_1$$

$$v_1 = \frac{2M}{M+m}v \quad [\because M \gg m]$$

$$\Rightarrow \quad v_1 \simeq 2v \quad \left(\because \frac{M}{M+m} \simeq 1\right)$$

135. (a)

Work done change in KE

$\Rightarrow \quad F.s = \frac{1}{2}mu^2$

$\Rightarrow \quad F \times 12 \times 10^{-2} = \frac{1}{2} \times 30 \times 10^{-3} \times (500)^2$

$\Rightarrow \quad F = \frac{500 \times 500}{8} = 3125 \text{ N}$

136. (b)

$10 \cdot v = 15 \times 4$

$\Rightarrow \quad v = 6 \text{ m/s}$

137. (b)

Direction of force and displacement are opposite.

∴ Work is negative.

138. (a)

139. (d)

$$F = \frac{-d}{dx}\{u(x)\}$$

$$\Rightarrow Fdx = -d\{u(x)\}$$

$$\Rightarrow \int Fdx = -\int d\{u(x)\}$$

$$-k\int xdx = u(x)$$

$$\frac{-kx^2}{2} + C = u(x)$$

$$u(0) = 0$$

$$\Rightarrow C = 0$$

$$\therefore u(x) = \frac{-k}{2}x^2$$

140. (b)

$$P = \vec{F}.\vec{v} = (60\hat{i} + 15\hat{j} - 3\hat{k}).\left(2\hat{i} - 4\hat{j} + 5\hat{k}\right)$$

$$= 120 - 60 - 15 = 45 \text{ watts.}$$

141. (b)

By the law of conservation of momentum

$$Mu = (M+m)\,v$$

$$0.50 \times 2.00 = (1 + 0.50)v, \quad \frac{1.00}{1.50} = v$$

Initial $\quad \text{KE} = \left(\frac{1}{2}\right) \times 0.50 \times (2.00)^2 = 1.00 \text{ J}$

Final $\quad \text{KE} = \frac{1}{2} \times 1.50 \times = \frac{1.00}{3.00} = 0.33 \text{ J}$

∴ Loss of energy = 1.00 – 0.33 = 0.67 J

142. (c)

$\bar{v} = \frac{v}{2}$ is average velocity

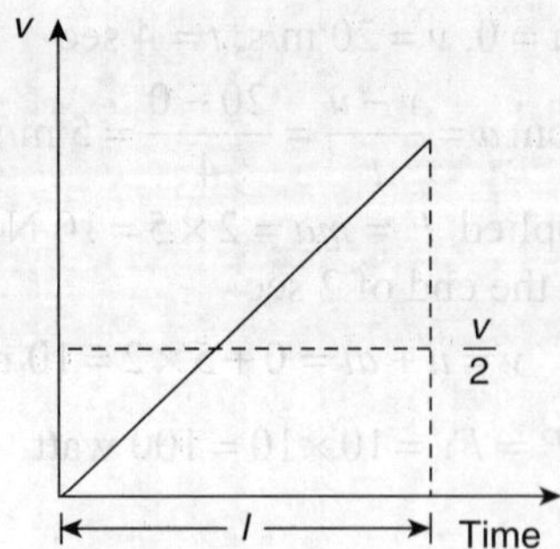

$S = 100\text{m}, t = 10$ s

$\therefore\ v_{average} = \left(\frac{v}{2}\right) = 10$ m/s

Assuming an athlete has about 50 to 100 kg, his kinetic energy would have been $\frac{1}{2}mv_{av}^2$.

$$\left(\frac{1}{2}\right)mv_{av}^2 = \left(\frac{1}{2}\right)\times 50\times 100 = 2500 \text{ J}$$

For 100 kg,

$$\left(\frac{1}{2}\right)\times 100\times 100 = 5000 \text{ J}$$

It could be in the range 2000 to 5000 J

143. (a)

Let the spring be compressed by x.
Initial kinetic energy of mass = potential energy of the spring + work done due to friction.

$$\frac{1}{2}\times 2\times 4^2 = \frac{1}{2}\times 10000\times x^2 + 15x$$

or $\quad 5000x^2 + 15x - 16 = 0$

or $\quad x = 0.055 \text{ m} = 5.5 \text{ cm}$

144. (c)

145. (b)

Total energy $E_T = 2$ J it is fixed.
For maximum speed, kinetic energy is maximum, the potential energy should therefore be maximum.

$\because \quad V(x) = \frac{x^4}{4} - \frac{x^2}{2}$

or $\quad \frac{dV}{dx} = \frac{4x^3}{4} - \frac{2x}{2}$

$$= x^3 - x = x(x^2 - 1)$$

For V to be minimum, $\frac{dV}{dx} = 0$

$$x(x^2 - 1) = 0$$

$$x = 0, \pm 1$$

At $x = 0, V(x) = 0$

At $x = \pm 1, V(x) = -\frac{1}{4}$ J

$\therefore$ (Kinetic energy)$_{max} = E_T - V_{min}$

or (Kinetic energy)$_{max} = 2 - \left(-\frac{1}{4}\right) = \frac{9}{4}$ J

or $\quad \frac{1}{2}mv_m^2 = \frac{9}{4}$

or $\quad v_m^2 = \frac{9\times 2}{m\times 4} = \frac{9\times 2}{1\times 4}$

$$v_m^2 = \frac{9}{2}$$

$$v_m^2 = \frac{3}{\sqrt{2}} \text{ m/s}$$

146. (b)

Kinetic energy of the ball = K
Angle of projection $(\theta) = 45°$
Velocity of the ball at the highest point = $v\cos\theta$

$$= v\ \cos 45° = \frac{v}{\sqrt{2}}$$

Therefore kinetic energy of the ball

$$\frac{1}{2}m\times\left(\frac{2}{\sqrt{2}}\right)^2 = \frac{1}{4}mv^2 = \frac{K}{2}$$

147. (c)

$$F = mg$$

$$mg = \Delta\text{KE} = 0 - \frac{1}{2}mv^2$$

$$(\text{Work done}) = \left(\frac{1}{2}\times\frac{100}{1000}\times 5^2\right)$$

$$= \frac{1}{2}\times\frac{100}{1000}5\times 5$$

$$= \frac{25}{20} = \frac{5}{4} = -1.25 \text{ J}$$

148. (a)

Linear momentum is conserved

$$0 = m_1v_1 + m_2v_2 = (12\times 4) + (4\times v_2)$$

or $\quad 4v_2 = -48$

$\Rightarrow \quad v_2 = -12$ m/s

$\therefore$ Kinetic energy of mass $m_2 = \frac{1}{2}m_2v_2^2$

$$= \frac{1}{2}\times 4\times(-12)^2 = 288 \text{ J.}$$

149. (a)

Work done in displacement is equal to gain in potential energy of mass.

Work done = $F \times 1 \sin 45°$

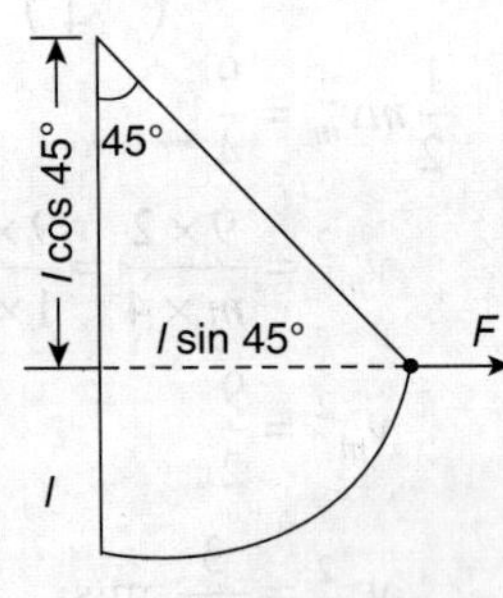

$$= \frac{Fl}{\sqrt{2}}$$

Gain in potential energy

$$= Mg\,(1 - 1 \cos 45°)$$

$$= mgl\left(1 - \frac{1}{\sqrt{2}}\right)$$

$$\frac{Fl}{\sqrt{2}} = \frac{Mgl(\sqrt{2}-1)}{\sqrt{2}}$$

or $\quad F = Mg(\sqrt{2}-1)$

150. (c)

Let v_1 = Speed of second mass, after collision momentum is conservation

Along x-axis, $\quad mv_1 \cos\theta = mv \quad \ldots$(i)

Along y- axis, $\quad mv_1 \sin\theta = \frac{mv}{\sqrt{3}} \quad \ldots$(ii)

From (i) and (ii)

$$\therefore\ (mv_1 \cos\theta)^2 + (mv_1 \sin\theta)^2 = (mv)^2 + \left(\frac{mv}{\sqrt{3}}\right)^2$$

$$m^2 v_1^2 = \frac{4m^2v^2}{3}$$

or $\quad v_1 = \frac{2}{\sqrt{3}}v$

151. (c)

Elastic energy stored in spring = $\frac{1}{2}KL^2$

$\therefore$ Kinetic energy of block E = $\frac{1}{2}KL^2$

Since $\quad P^2 = 2ME$

$$P = \sqrt{2ME}$$

$$= \sqrt{\frac{2M \times KL^2}{2}} = \sqrt{MKL^2}$$

$$= \sqrt{MK}L.$$

152. (c)

Here, $H = 100$ m, $h_2 = 20$ m

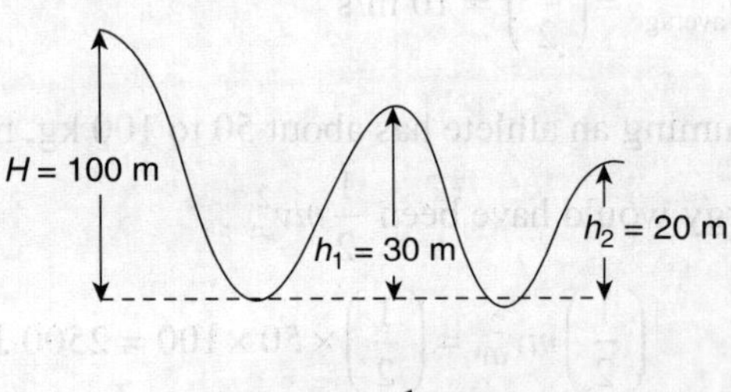

$$mgH = \frac{1}{2}mv^2 + mgh_2$$

or $\quad mg(H - h_2) = \frac{1}{2}mv^2$

or $\quad v^2 = 2g(H - h_2)$

$$= 2 \times 10 \times (100 - 20)$$

$$= 2 \times 10 \times 80$$

$$v = \sqrt{1600} = 40 \text{ m/s}$$

153. (b)

154. (a)

Given: Retardation $\propto$ Displacement

or $\quad \frac{dv}{dt} = kx$

$$\left(\frac{dv}{dx}\right)\left(\frac{dv}{dt}\right) = kx$$

$$dv(v) = kxdx$$

$$\int_{v_1}^{v_2} vdv = k\int_0^x xdx$$

$$\frac{mv_2^2}{2} - \frac{mv_1^2}{2} = \frac{mkx^2}{2}$$

$$K_2 - K_1 = \frac{mk}{2}.x^2$$

Therefore, loss of KE is proportional to x^2.

155. (a)

No work is done when a force of constant magnitude always acts at right angles to the velocity of a particle when the motion of the particle takes place in a plane. Hence KE of the particle remains constant.

156. (c)

Work done = $\vec{F}\ .\ \vec{r}$

or work done = $((5\hat{i} + 3\hat{j} + 2\hat{k})(2\hat{i} - \hat{j})$

$$= 10 - 3 = 7 \text{ J}.$$

157. (a)

The centre of mass of the hanging part is at 0.3 m from table.

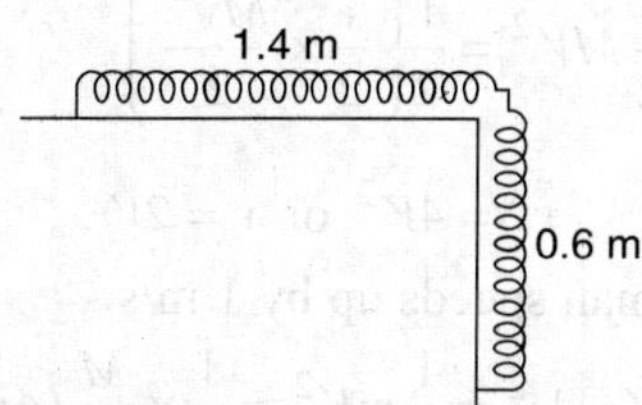

Mass of hanging part $= \frac{4 \times 0.6}{2} = 1.2$ kg

$$W = mgh$$

$$= 1.2 \times 10 \times 0.3 = 3.6 \text{ J.}$$

(C.G lies at 0.3 m below hanging part)

158. (d)

From Work–Energy theorem

$$(-F)\ (s) = (-F)(s) = 0 - \frac{1}{2}mu^2$$

$$\Rightarrow Fs = \frac{1}{2}mu^2$$

where

F = retarding force

m = mass of automobile

s = stopping distance

u = initial velocity of automobile

Since retarding force and mass of automobile are constant

$$s \propto u^2$$

$$s_1 = ku_1^2$$

$$s_2 = ku_2^2$$

$$\frac{s_2}{s_1} = \frac{u_2^2}{u_1^2}$$

$$\Rightarrow \quad s_2 = \left(\frac{u_2}{u_1}\right)^2 s_1$$

$$= \left(\frac{120}{60}\right)^2 (20 \text{ m})$$

$$\Rightarrow \quad s_2 = 4 \times 20 = 80 \text{ m.}$$

159. (a)

$$\text{Energy} = \frac{1}{2}Kx^2 = \frac{1}{2}\frac{F^2}{K}$$

$$\frac{K_A}{K_B} = 2$$

$$\therefore \frac{E_A}{E_B} = \frac{1}{2} \text{ or } E_B = 2E_A$$

160. (c)

A system of particles implies that one is discussing total momentum and total energy.

1(a) explodes

m m
1 (a) u ← • → u 1 (b)

Total momentum = 0

But total kinetic energy $= 2\left(\frac{1}{2}\right)mu^2$

But if total kinetic energy = 0,

Velocities are zero.

Here, A is true but B is not true; A does not imply B, but B implies A.

161. (d)

$$\text{Power} = \frac{\text{Work}}{\text{Time}} = \frac{\text{Force} \times \text{Distance}}{\text{Time}}$$

$$= \text{Force} \times \text{Velocity}$$

$\therefore$ Force $\times$ Velocity = Constant (k)

or $\quad (ma)\ (at) = k$

or $\quad a = \left(\frac{k}{mt}\right)^{1/2}$

$$s = \frac{1}{2}at^2$$

$$s = \frac{1}{2}\left(\frac{k}{mt}\right)^{1/2} .t^2 = \frac{1}{2}\left(\frac{k}{m}\right)^{1/2} .\ t^{3/2}$$

or s is proportional to $t^{3/2}$

162. (d)

$$\frac{s_1}{s_2} = \left(\frac{u_1}{u_2}\right)^2$$

$$= \left(\frac{4}{4u}\right)^2 = 1:16.$$

163. (c)

$$(-F)(s) = \frac{1}{2}m\left(\frac{v}{2}\right)^2 - \frac{1}{2}mv^2$$

$$= \frac{1}{2}\left(\frac{3}{4}v^2\right)m$$

$$\Rightarrow \frac{v^2}{4} = \left(\frac{2F}{3m}\right)s$$

$$\Rightarrow \frac{v}{2} = \sqrt{\left(\frac{2F}{3m}\right)s}$$

$\because$ The retarding force is same, once again using Work–Energy theorem, we get

$$(-F)(s_2) = 0 - \frac{1}{2}m\left(\frac{v}{2}\right)^2$$

$$\Rightarrow \quad (F)(s_2) = \frac{1}{2}m\left(\frac{2F}{3m}\right)s$$

$$\Rightarrow \quad s_2 = \frac{1}{3}s = \frac{1}{3}(3\text{cm}) = 1\text{ cm}$$

164. (d)

165. (c)

Centripetal force $= \dfrac{mv^2}{r} = \dfrac{-1}{r^2}$

$$mv^2 = \frac{1}{r}$$

$$\text{KE} = \frac{1}{2}mv^2$$

$$= \frac{1}{2r}$$

$$\text{PE} = -2\text{KE}$$

$$= \frac{-1}{r}$$

$$\text{TE} = \text{PE} + \text{KE}$$

$$= \frac{-1}{r} + \frac{1}{2r} = \frac{-1}{r}$$

166. (d)

$$m_1v_1 = m_2v_2$$

$$4 \times v_1 = 8 \times 6$$

$$v_1 = \frac{8 \times 6}{4} = 12\text{ms}^{-1}$$

KE of mass 4 kg $= \frac{1}{2} \times mv^2 = \frac{1}{2} \times 4 \times 12^2 = 288$ J

167. (a)

Let M and m be the mass of the man and boy respectively.

If V and v be their speeds then

$$\frac{1}{2}MV^2 = \frac{1}{2}\left(\frac{1}{2}mv^2\right) \text{ given } m = \frac{M}{2}$$

$$\frac{1}{2}MV^2 = \frac{1}{2}\left(\frac{1}{2} \times \frac{Mv^2}{2}\right)$$

$$v^2 = 4V^2 \text{ or } v = 2V.$$

When the man speeds up by 1 m/s

$$\frac{1}{2}M(V+1)^2 = \frac{1}{2}mV^2 = \frac{1}{2} \times \frac{M}{2}(4v^2)$$

or $\quad (V+1)^2 = 2V^2 \Rightarrow V^2 - 2V - 1 = 0$

On solving $V = 2.4$ m/sec (velocity of man)

$v = 4.8$ m/sec (velocity of boy)

168. (b)

Let F be the resistance force offered by the plank of width x.

In case of the first bullet

$$F.x = \frac{1}{2}m(100^2 - 80^2)$$

$$= \frac{1}{2}m(180 \times 20) \quad \ldots\text{(i)}$$

In case of second bullet, let v be the final velocity of bullet. Then

$$F.x = \frac{1}{2}m(80^2 - v^2) \quad \ldots\text{(ii)}$$

$$W_{\text{first bullet}} = W_{\text{second bullet}}$$

So, $\quad \frac{1}{2}m(180 \times 20) = \frac{1}{2}m(80^2 - v^2)$

$$v^2 = 6400 - 3600 = 2800$$

$$v = \sqrt{7 \times 400} = 20\sqrt{7} \text{ m/sec.}$$

169. (b)

$$\text{KE} = \frac{1}{2}mv^2 \quad \text{(where } v \text{ is constant)}$$

When bullet is fired from gun, the bullet and the gun moves with the same momentum but in opposite direction.

Hence $\quad \dfrac{(\text{KE})_B}{(\text{KE})_G} = \dfrac{M_G}{M_B}$

$$\Rightarrow \frac{(\text{KE})_B + (\text{KE})_G}{(\text{KE})_G} = \frac{M_G + M_B}{M_B}$$

$$\Rightarrow (\text{KE})_G = \frac{\text{Total KE} \times M_B}{M_G + M_B}$$

$$= \frac{2050\text{ J} \times 50\text{ gm}}{2050\text{ gm}} = 50\text{ J}$$

$$(\text{KE})_B = 2050 - 50 = 2000\text{ J}$$

170. (d)

Consider the disc as inclined plane. Using law of conservation of energy,

Loss in PE = Work done against friction

$$mgh = \mu\, mg\, \cos\,\theta .\, s$$

$$s = \frac{h}{\mu\cos\theta} = \frac{10^{-2} \times 200}{0.01 \times 99}$$

$$= 1.005\text{ m} = 1\text{ m}$$

171. (c)

$$\frac{1}{2}m_1 u_1{}^2 = E$$

$$v_1 = \frac{m_1 - m_2}{m_1 + m_2}.u_1 + \frac{2m_2 u_2}{m_1 + m_2}$$

$$v_1 = \left(\frac{10-4}{10+4}\right)u_1 + 0 \qquad (\because u_2 = 0)$$

$$= \frac{6}{14}u_1 = \frac{3}{7}u_1$$

$$\frac{1}{2}m_1 v_1{}^2 = \frac{1}{2}m_1\left(\frac{9}{49}u_1{}^2\right) = \frac{9}{49}E.$$

172. (c)

Momentum of stone = Momentum of cart

$$2v = 100 \times 0.1$$

So, $v = 5$ m/sec

Horizontal velocity of stone $= 5\text{ m/s} = v_x$

Vertical velocity of the stone after 0.5 sec $= gt$

$$= 9.8 \times \frac{1}{2} = 4.9\text{m/s} = v_y$$

$\therefore$ Kinetic energy of stone $= \frac{1}{2}mv^2$

$$= \frac{1}{2}m(v_x^2 + v_y^2)$$

$$= \frac{1}{2} \times 2\left[(5)^2 + (4.9)^2\right] = 49\text{ J}$$

173. (b)

$$m_1 u_1 + m_2 u_2 = (m_1 + m_2)v$$

$$2m_1 - 4m_2 = (m_1 + m_2) \times$$

So, $m_1 = 5m_2$

$$\frac{(\text{KE})_I}{(\text{KE})_{II}} = \frac{\frac{1}{2}m_1 u_1^2 - \frac{1}{2}m_2 u_2^2}{\frac{1}{2}m_2 u_2^2} = \frac{\frac{m_1}{m_2}u_1^2 - u_2^2}{u_2^2}$$

$$= \frac{5 \times 2^2 - 4^2}{4^2} = 0.25$$

So, the kinetic energy of the first body is 0.25 times that of the second.

174. (c)

If the particle is dropped from a height h and the spring is compressed by y, loss in PE of the particle will be $mg(h+y)$ (and not mgh) while gain in classic potential energy of the spring will be $1/2\,ky^2$, so by conservation of mechanical energy

$$mg(h+y) = 1/2ky^2$$

As here particle and spring remains same,

$$\frac{h_1 + y_1}{h_2 + y_2} = \left(\frac{y_1}{y_2}\right)^2$$

$$[\text{Here } h_1 = 0.24, y_1 = 0.01, h_2 = ?\, y_2 = 0.04]$$

$$\Rightarrow \left[\frac{0.24 + 0.01}{h_2 + 0.04}\right] = \left(\frac{0.01}{0.04}\right)^2 = 0.016\text{ m}$$

$$\Rightarrow h_2 = 0.02$$

175. (b)

Let the body fall freely from a height h.

Let it rise to a height h after 1st rebound

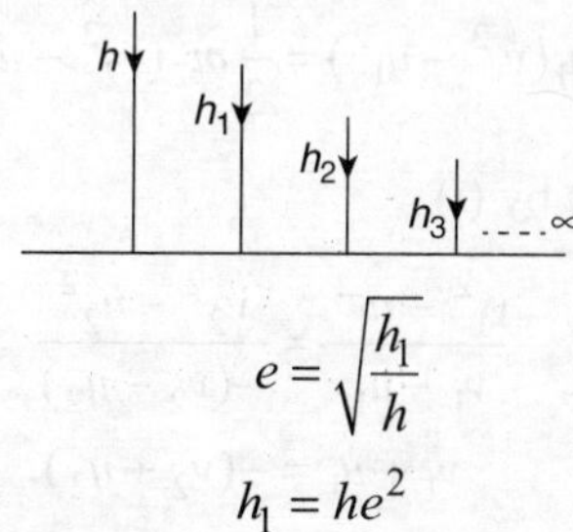

$$e = \sqrt{\frac{h_1}{h}}$$

$$h_1 = he^2$$

After 2nd rebound,

$$h_2 = h,e^2 = he^2.e^2$$

$$= he^4 = (he^2)^2$$

After 3rd rebound,

$$h_3 = h_2 e^2 = h\left(e^2\right)^2.e^2$$

$$= h(e^3)^2$$

$\therefore$ After nth rebounce,

$$h_n = he^{2n}$$

176. (a)

$$d = h + 2h_1 + 2h_2 + 2h_3 + \ldots\infty$$
$$= h + 2he^2 + 2he^4 + 2he^6 + \ldots\infty$$
$$= h + 2he^2(1 + e^2 + e^4 + \ldots\infty)$$
$$= h + 2he^2\left(\frac{1}{1-e^2}\right) = h\left[1 + \frac{2e^2}{1-e^2}\right]$$
$$= h\left[\frac{1 - e^2 + 2e^2}{1-e^2}\right]$$
$$d = h\left[\frac{1+e^2}{1-e^2}\right]$$

177. (b)

Since the collisions are elastic, both momentum and KE must be conserved.

178. (b)

$$m_1v_1 + m_2v_2 = m_1u_1 + m_2u_2$$

(Applying law of conservation of momentum)

$$m_1(v_1 - u_1) = -m_2(v_2 - u_2) \quad \ldots\text{(i)}$$

Loss in KE of first body = Loss in KE of second body

$$\frac{1}{2}m_1v_1^2 - \frac{1}{2}m_1u_1^2 = \frac{1}{2}m_2v_2^2 - \frac{1}{2}m_2u_2^2$$
$$\frac{1}{2}m_1(v_1^2 - u_1^2) = \frac{1}{2}m_2(v_2^2 - u_2^2)$$

Dividing (ii) by (i)

$$\frac{v_1^2 - u_1^2}{v_1 - u_1} = \frac{v_2^2 - u_2^2}{-(v_2 - u_2)}$$
$$v_1 + u_1 = -(v_2 + u_2)$$
$$u_1 + u_2 = -v_1 - v_2$$

Hence, $u_1 + u_2 = -(v_1 + v_2)$.

179. (d)

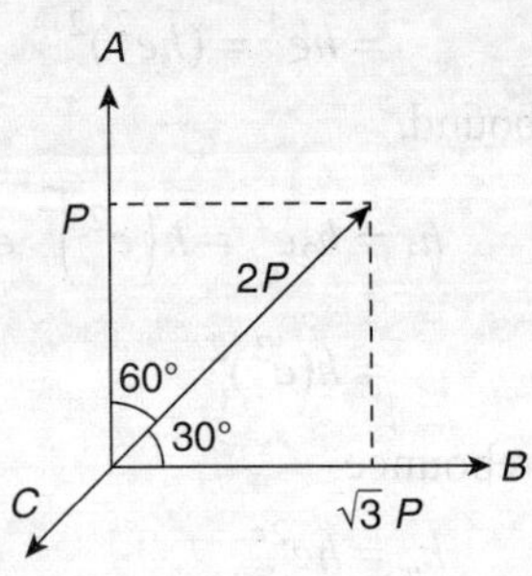

Momentum of C is $2P$ in a direction making 150° with that of B.

180. (a)

$$m.90 - mv = m.v_2$$
$$90 - v = v_2$$
$$e = \frac{v_2 - v_1}{u_1 - u_2}$$
$$0.8 = \frac{90 - v - 0}{90 + v}$$
$$72 + 0.8v = 90 - v$$
$$1.8v = 18$$
$$v = 10\text{m/sec.}$$

181. (b)

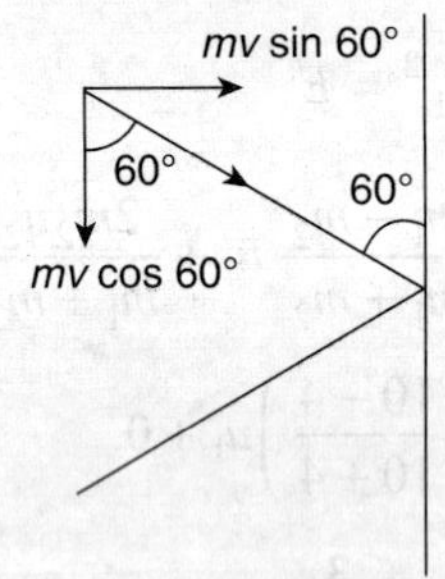

$$\Delta P = 2mv\sin 60$$
$$F = \frac{\Delta p}{\Delta t} = \frac{2mv\sin 60}{\Delta t} = \frac{2\times3\times10\times\frac{\sqrt{3}}{2}}{0.2}$$
$$= 150\sqrt{3}\text{ N}$$

182. (d)

$$v_1 = \left(\frac{M_1 - m_2}{M_1 + m_2}\right)u_1 + \frac{2M_2u_2}{M_1 + m_2}$$
$$= \left(\frac{M - m}{M + m}\right)u + 0$$
$$\text{KE}_i = \frac{1}{2}Mu^2$$
$$\text{KE}_f = \frac{1}{2}M\left(\frac{M-m}{M+m}\right)^2$$
$$\frac{\text{KE}_f}{\text{KE}_i} = \left(\frac{M-m}{M+m}\right)^2$$
$$\frac{\text{KE}_i - \text{KE}_f}{\text{KE}_i} = \frac{4Mm}{(M+m)^2} = \frac{4A.M^2}{M^2(1+A)^2}$$
$$= \frac{4A}{(1+A)^2}$$

183. (a)

The time elapsed from the moment it is dropped to the second impact with the floor.

$$t=\sqrt{\frac{2h}{g}}(1+2e)$$

where h is the initial height of the body from the ground.

$$1.3=\sqrt{\frac{2}{9.8}}(1+2e)$$

$$(1+2e)=1.3\sqrt{4.9}=2.877$$

On solving, we get

$$e=0.93$$

184. (b)

Applying the law of conservation of momentum

$$mv_1+0=2mv_2$$

$$v_2=\frac{v_1}{2}$$

$$\text{Loss in KE}=\frac{1}{2}mv_1^2-\frac{1}{2}(2m)v_2^2$$

$$=\frac{1}{2}mv_1^2-\frac{1}{2}(2m)\left(\frac{v_1^2}{4}\right)$$

$$=\frac{1}{2}mv_1^2\left(1-\frac{1}{2}\right)=\frac{E}{2}$$

185. (a)

In one-dimensional elastic collision

$$v_B=\frac{2m_Au_A}{m_A+m_B}+\frac{m_B-m_A}{m_A+m_B}u_B$$

Since, $u_B=0$

$$v_B=\frac{2m_Au_A}{m_A+m_B};\ p_B=m_Bv_B$$

$$p_B=\frac{2m_Au_Am_B}{m_B\left(1+\frac{1}{k}\right)}=\frac{2p_A}{1+\frac{1}{k}}$$

Where, $k=\frac{m_B}{m_A}$

To have greater momentum k should be more, i.e.,

$$m_B>>m_A.$$

186. (b)

Maximum drop in PE = Maximum gain in KE

$$\text{mg}(2-0.75)=\frac{1}{2}mv^2\Rightarrow v=\sqrt{2g(1.25)}=5\text{ m/s.}$$

187. (b)

Applying law of conservation of momentum

$$m_1v_1=mv+\frac{m_1v_1}{3}$$

$$v=\frac{m_1}{m}\times\frac{2v_1}{3}$$

To describe a vertical circle v should equal to $\sqrt{5gl}$.

Here $\sqrt{5gl}=\frac{m_1}{m}\times\frac{2}{3}v_1$

$$v_1=\frac{3}{2}\frac{m}{m_1}\sqrt{5gl}$$

188. (c)

The length of the hanging part $=\frac{L}{n}$

Mass of the hanging part $=\frac{M}{n}$

The weight of hanging part $\frac{mg}{n}$ acts at the centre of gravity G and is at a distance $h=\frac{L}{2n}$ below the surface of the table.

The work done in pulling the hanging part on to the table

$$=\frac{M}{n}\times g\times h$$

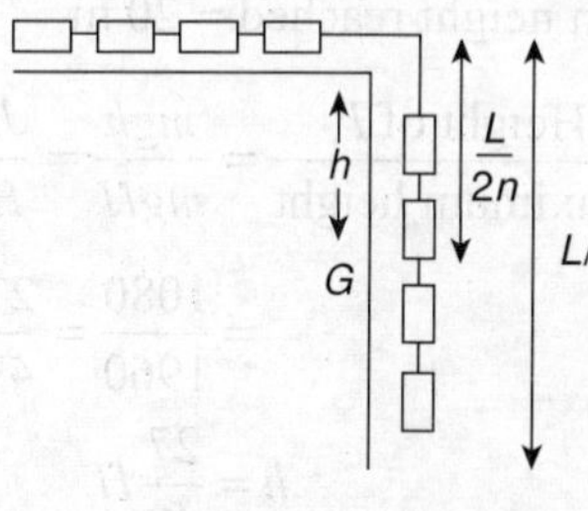

$$=\frac{M}{n}\times g\times\frac{L}{2n}=\frac{MgL}{2n^2}$$

189. (d)

Mass of the train = 100 metric tons

$$=100\times10^3\text{ kg}=10^5\text{ kg}$$

Velocity = 360km/h = 100m/sec

Frictional force $(f)=100\times10=1000\text{N}$

$$\sin\theta=\frac{1}{49}$$

Power required to move the body along the inclined plane in upward direction is

$$P = (mg\sin\theta + f)v$$

$$= (10^5 \times 9.8 \times \frac{1}{49} + 10^3) \times 100$$

$$= (10^4 \times 2 + 10^3) \times 100$$

$$= 21 \times 10^3 \times 10^2 = 2.1 \times 10^6 \text{ watt}$$

190. (c)

191. (b)

Block m_2 will move when spring pulls it to right by force equal to force of friction.

192. (c)

Total energy at any point = 1080 + 880

= 1960 J

= Initial kinetic energy

$\therefore$ Initial kinetic energy $= \frac{1}{2}mv^2 = \frac{1}{2} \times 10 \times v^2$

$\therefore$ Velocity of projection, $v = \sqrt{\frac{1960}{5}} = \sqrt{392}$

$= 14\sqrt{2}$ m/sec

Total energy is equal to the potential energy at the highest point.

$$1960 \text{ J} = mgh = 10 \times 9.8 \times h$$

$$h = 20 \text{ m}$$

$\therefore$ Maximum height reached = 20 m

$$\frac{\text{Height of } P}{\text{Maximum height}} = \frac{mgh}{mgH} = \frac{h}{H}$$

$$= \frac{1080}{1960} = \frac{27}{49}$$

$$\therefore \quad h = \frac{27}{49}H$$

193. (c)

The bullet and block will meet after time t, then

$$\frac{1}{2}gt^2 + ut - \frac{1}{2}gt^2 = 100$$

$$t = \frac{h}{u} = \frac{100}{100} = 1 \text{ s}$$

During this time distance travelled by the block

$$s_1 = \frac{1}{2}gt^2 = \frac{1}{2} \times 10 \times 1 = 5\text{m}$$

Distance travelled by bullet

$$s_2 = 100 - s_1 = 95 \text{ m}$$

Velocity of bullet before collision

$$u_2 = u - gt$$

or $\quad u_2 = 100 - 10 = 90$ m/s

Velocity of block before collision $u_1 = gt = 10$ m/sec

Using law of conservation of momentum,

$$(m_1u_1 + m_2u_2) = (m_1 + m_2)$$

$$\therefore 0.01(-10 + 90 = 0.02v)$$

(Consider velocity in upward direction as positive)

$$\Rightarrow \quad v = \frac{80}{2} = 40 \text{ m/sec}$$

After collision, the system moves with a velocity of 40 m/s in the upward direction

Maximum height reached by the block

$$= \frac{v^2}{2g} = \frac{40 \times 40}{2 \times 10} = 80 \text{ m}$$

Height reached above the tower

$$= 80 - s_1 = 80 - 5 = 75$$

194. (c)

The situation is shown in Figures (a), (b) and (c).

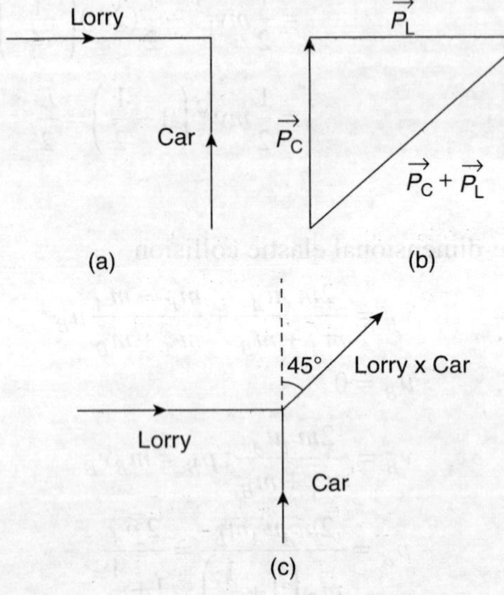

Total momentum before impact = $P_C + P_L$

$$(P_C + P_L) = \sqrt{(2 \times 10^3 \times 20)^2 + (8 \times 10^3 \times 5)^2}$$

$$= 40 \times 10^3 \times \sqrt{2} \text{ kg m/sec}$$

Direction of momentum with main road

$$\tan\theta = \frac{P_L}{P_C} = \frac{8 \times 10^3 \times 5}{2 \times 10^3 \times 20} = 1 \text{ or } \theta = 45°$$

After collision, two vehicles moved together at an angle of 45° with the main road. Let v be their common velocity. Applying the law of conservation of momentum, we have

$$10^3 \times 40 \times \sqrt{2} = \left(2\times 10^3 + 8\times 10^3\right)v'$$

$$\Rightarrow \quad 40\times 10^3 \times \sqrt{2} = 10\times 10^3 \times v'$$

or $v' = 4\sqrt{2}$ m/s at 45° with main road.

195. (c)

Applying the law of conservation of momentum in perpendicular to the initial line of motion,

$$0 = mv_1 \sin 30° - mv_2 \sin 30°$$

$$= v_1 = v_2 \quad \text{...(i)}$$

Now, in line of motion,

$$(m\times v) + m\times 0 = mv_1 \cos 30° + mv_2 \cos 30° \quad \text{....(ii)}$$

Put in (i) and (ii)

or $\quad mv = 2mv_1 \cos 30°$

or $\quad m\times 6 = 2\times m\times v_1 \times \dfrac{\sqrt{3}}{2}$

or $\quad v_1 = \dfrac{6}{\sqrt{3}} = 2\sqrt{3}\text{m/s.}$

196. (a)

Applying the law of conservation of momentum

$$m_1 v_1 = (m_1 + m_2)v \quad \text{...(i)} \ (v_2 = 0)$$

Here the PE $m_1 gl$ has been converted into KE is

$$m_1 gl = \frac{1}{2} m_1 v_1$$

or $\quad v_1 = \sqrt{2gl}$

Substituting the value v_1 from Eq. (ii) in Eq. (i),

$$m_1\sqrt{2gl} = v(m_1 + m_2), \text{ or } v = \left(\frac{m_1}{m_1 + m_2}\right)\sqrt{2gl}$$

Now let centre of mass rises through a height h after collision. In this case, the kinetic energy $(m_1 + m_2)$ system is converted into the PE at maximum height h.

$$\frac{1}{2}(m_1 + m_2)v^2 = (m_1 + m_2)gh$$

$$\frac{1}{2}(m_1 + m_2)\left\{\frac{m_1}{m_1 + m_2}\right\}^2 .2gl = (m_1 + m_2)gh$$

$$h = l\left(\frac{m_1}{m_1 + m_2}\right)^2$$

197. (d)

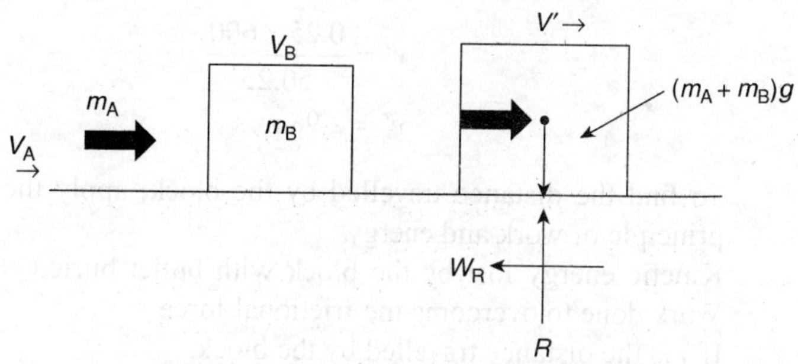

Let m be the mass of each ball. Further, let, u_1 be the velocities of ball A just before the collision. When ball slides a quadrant, its potential energy is converted into kinetic energy. Hence,

$$mgh = \frac{1}{2} m u_1^2$$

or $\quad u_1 = \sqrt{2gh}$

or $\quad u_1 = \sqrt{2\times 980\times 10} = 140$ cm/s

Let after collision, v_1 and v_2 be the velocities of ball A and B respectively. Applying the law of conservation of linear momentum,

$$mu_1 = mv_1 + mv_2$$

or $\quad u_1 = v_1 + v_2 \quad$ ...(i)

Again it is elastic collision $e = 1$

or $\quad v_1 - v_2 = u_2 - u_1 = -u_1 \quad (\because u_2 = 0)$

or $\quad v_1 - v_2 = -u_1 \quad$ ...(ii)

Solving Eqs. (i) and (ii),

$$v_1 = 0 \text{ and } v_2 = 140 \text{ cm/s}$$

This shows that ball A comes to rest while ball B moves with a velocity 140 cm/sec.

198. (b)

Velocity of the bullet before impact $v_a = 600$ m/s

Velocity of the block before impact $v_b = 0$

Mass of the bullet $m_a = \dfrac{0.2}{g}$ kg

Mass of the block $m_b = \dfrac{50}{g}$ kg

Let us assume that the bullet, after striking the block, remains buried in the block and both move with a common velocity v. Applying the principle of conservation of momentum,

$$m_a v_a + m_b v_b = (m_a + m_b)v'$$

$$\frac{0.25}{g}(600) + 0 = \frac{(50 + 0.25)}{g} v'$$

$$v' = \frac{0.25 \times 600}{50.25}$$

$$v' = 2.98 \text{ m/s}$$

To find the distance travelled by the block, apply the principle of work and energy.
Kinetic energy lost by the block with bullet buried = Work done to overcome the frictional force
If s is the distance travelled by the block,

$$(m_a + m_b)\frac{(v')^2}{2} = (\mu R)s$$

where $R = g(m_a + m_b)$

$$(m_a + m_b)\frac{(v')^2}{2} = \mu g(m_a + m_b)s$$

$$s = \frac{(v')^2}{2g\mu}$$

$$s = \frac{(2.98)^2}{2 \times 9.8 \times 0.5}$$

$$s = 0.90 \text{ m}$$

Questions Asked in NEET/CBSE Medicals

199. (c)

According to conservation of momentum,

$$4mu_1 = 4mv_1 + 2mv_2 \Rightarrow 2(u_1 - v_1) = v_2 \quad \ldots\text{(i)}$$

From conservation of energy.

$$\frac{1}{2}(4m)u_1^2 = \frac{1}{2}(4m)v_1^2 + \frac{1}{2}(2m)v_2^2$$

$$\Rightarrow 2(u_1^2 - v_1^2) = v_2^2 \quad \ldots\text{(ii)}$$

$$3v_1 = u_1 \quad \ldots\text{(iii)}$$

$$\frac{\Delta K}{K_i} = \frac{K_i - K_f}{K_i} = \frac{\frac{1}{2}(4m)u_1^2 - \frac{1}{2}(4m)v_1^2}{\frac{1}{2}(4m)u_1^2} \quad \ldots\text{(iv)}$$

Substituting (iii) in (iv), we get $\frac{\Delta K}{K_i} = \frac{8}{9}$

200. (d)

Given : $F = 20 + 10\,y$

Work done, $W = \int F.dy$

$$= \int_0^1 (20 + 10y)\,dy = \left[20y + \frac{10}{2}y^2\right]_0^1 = 20 + \frac{10}{2} = 25\text{J}$$

201. (d)

In vertical circular motion, tension in the wire is maximum at the lowermost point, so the wire is most likely to break when the mass is at the lowest point.

202. (d)

Let the speed of the third fragment of mass $3m$ be v'.

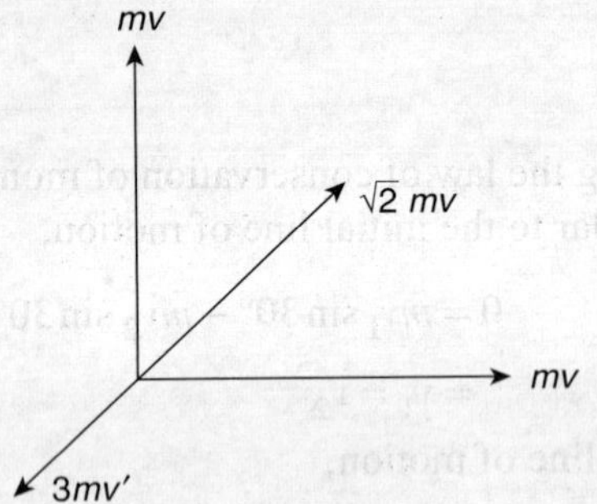

From law of conservation of linear momentum,

$$mv' = \sqrt{2}mv$$

$$\Rightarrow v' = \frac{\sqrt{2}v}{3} \quad \ldots\text{(i)}$$

$\therefore$ Energy released during the process is,

$$\text{KE} = 2\left(\frac{1}{2}mv^2\right) + \frac{1}{2}(3m)v'^2$$

$$= mv^2 + \frac{1}{2}(3m)\frac{2v^2}{9}$$

(Using Eq. (i))

$$= mv^2 + \frac{mv^2}{3} = \frac{4}{3}mv^2$$

203. (b)

Let final velocity of the block of mass, $4\text{m} = v'$
According to law of conservation of linear momentum,

$$mv + 4m \times 0 = 4mv' + 0 \Rightarrow v' = v/4$$

Coefficient of restitution,

$$e = \frac{\text{Relative velocity of separation}}{\text{Relative velocity of approach}} = \frac{v/4}{v} = 0.25$$

204. (d)

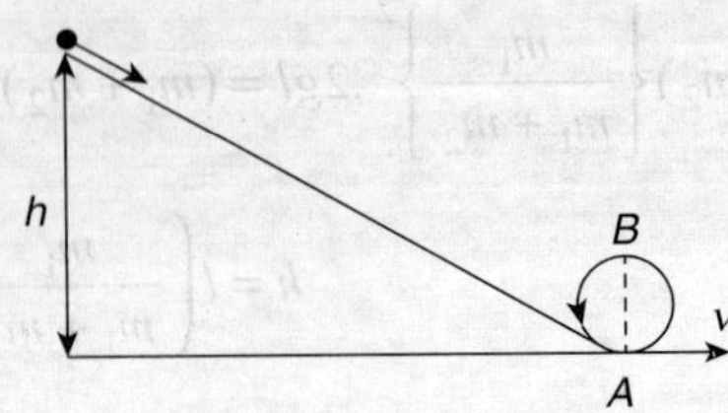

Initial speed = 0
At point A, speed = v.
As track is frictionless, the total mechanical energy remains constant.

$$\sum(\text{KE}+\text{PE})_i = \sum(\text{KE}+\text{PE})_f$$

$$0 + mgh = \frac{1}{2}mv^2 + 0 \text{ or } h = \frac{v^2}{2g}$$

For completing the vertical circle, $v \geq \sqrt{5gR}$

$$\therefore h = \frac{5gR}{2g} = \frac{5}{2}R = \frac{5}{4}D$$

205. (c)

Here, $m = 1\text{ g} = 10^{-3}\text{kg}$, $h = 1\text{ km} = 1000\text{ m}$, $v = 50\text{ ms}^{-1}$, $g = 10\text{ ms}^{-2}$.

(i) The work done by the gravitational force

$$= mgh = 10^{-3} \times 10 \times 1000 = 10\text{J}$$

$$\therefore W_g + W_r = \frac{1}{2}mv^2 - 0 \quad \text{(Work–Energy theorem)}$$

$$10 + W_r = \frac{1}{2} \times 10^{-3} \times 50 \times 50$$

or $\quad W_r = -8.75\text{J}$

206. (c)

Mass of bullet, $m = 10\text{ g} = 0.01\text{ kg}$
Initial speed of bullet, $u = 400\text{ ms}^{-1}$
Mass of block, $M = 2\text{ kg}$
Length of string, $l = 5\text{ m}$
Speed of the block after collision = v_1
Speed of the bullet on emerging from block, $v = ?$

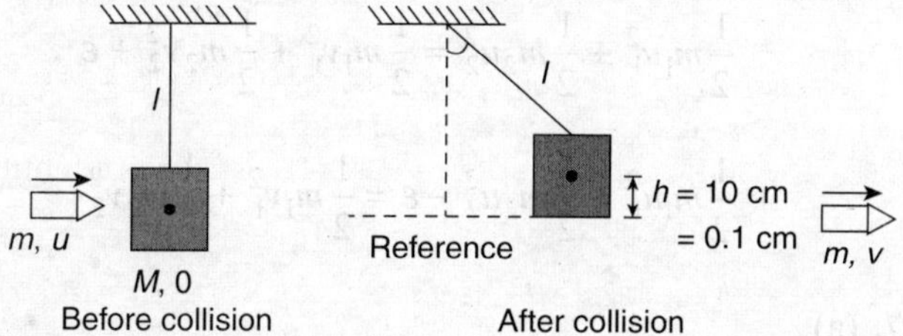

Using energy conservation principal for the block,

$$(\text{KE}+\text{PE})_{Reference} = (\text{KE}+\text{PE})_h$$

$$\Rightarrow \frac{1}{2}Mv_1^2 = Mgh \text{ or } v_1 = \sqrt{2gh}$$

$$v_1 = \sqrt{2 \times 10 \times 0.1} = \sqrt{2}\text{ms}^{-1}$$

Using momentum conservation principle for block and bullet system,

$$(M \times 0 + mu)_{\text{Before collision}} = (M \times v_1 + mv)_{\text{After collision}}$$

$$\Rightarrow 0.01 \times 400 = 2\sqrt{2} + 0.01 \times v$$

$$\Rightarrow v = \frac{4 - 2\sqrt{2}}{0.01} = 117.150\text{ ms}^{-1} \approx 120\text{ ms}^{-1}$$

207. (b)

Masses of the balls are same and collision is elastic, so their velocities will be interchanged after collision.

208. (c)

Here,

$$\vec{r}_1 = \left(-2\hat{i} + 5\hat{j}\right)m, \vec{r}_2 = \left(4\hat{j} + 3\hat{k}\right)m$$

$$\vec{F} = (4\hat{i} + 3\hat{j})N, W = ?$$

Work done by force F in moving from $\vec{r}_1$ to $\vec{r}_2$,

$$W = \vec{F}.\left(\vec{r}_2 - \vec{r}_1\right) \Rightarrow W = (4\hat{i} + 3\hat{j}).(4\hat{j} + 3\hat{k} + 2\hat{i} - 5\hat{j})$$

$$= (4\hat{i} + 3\hat{j}).(2\hat{i} - \hat{j} + 3\hat{k}) = 8 + (-3) = 5\text{J}$$

209. (c)

Here, $m = 10\text{ g} = 10^{-2}\text{ kg}$,
$R = 6.4\text{ cm} = 6.4 \times 10^{-2}\text{ m}$, $K_f = 8 \times 10^{-4}\text{ J}$

$$K_i = 0, a_t = ?$$

Using Work–Energy theorem,
Work done by all the forces = Change in KE

$$W_{tangential\ force} + W_{centripetal\ force} = K_f - K_i$$

$$= a_t = \frac{K_f}{4\pi Rm} = \frac{8 \times 10^{-4}}{4 \times \frac{22}{7} \times 6.4 \times 10^{-2} \times 10^{-2}}$$

$$= 0.099 \approx 0.1\text{ ms}^{-2}$$

210. (b)

Here, $\vec{F} = (2t\hat{i} + 3t^2\hat{j})N, m = 1\text{ kg}$

Acceleration of the body, $\vec{a} = \frac{\vec{F}}{m} = \frac{(2t\hat{i} + 3t^2\hat{j})N}{1\text{kg}}$

Velocity of the body at time t,

$$\vec{v} = \int \vec{a}dt = \int (2t\hat{i} + 3t^2\hat{j})dt = t^2\hat{i} + t^3\hat{j}\text{ms}^{-1}$$

$\therefore$ Power developed by the force at time t,

$$\text{P} = \vec{F}.\vec{v} = (2t\hat{i} + 3t^2\hat{j}).\left(t^2\hat{i} + t^3\hat{j}\right)\text{W}$$

$$= (2t^3 + 3t^5)\text{W}$$

211. (b)

212. (c)

Let the particles A and B collide at time t. At the time of collision, their position vector should be same.

$$\vec{r}_1 + \vec{v}_1 t = \vec{r}_2 + \vec{v}_2 t \quad \ldots\text{(i)}$$

$$\vec{r}_1 - \vec{r}_2 = \vec{v}_2 t - \vec{v}_1 t = \left(\vec{v}_2 - \vec{v}_1\right) t$$

$$\Rightarrow \quad \left|\vec{r}_1 - \vec{r}_2\right| = \left|\vec{v}_2 - \vec{v}_1\right| t \text{ or } t = \frac{\left|\vec{r}_1 - \vec{r}_2\right|}{\left|\vec{v}_2 - \vec{v}_1\right|}$$

Substituting this value of t in Eq. (i), we get

$$\vec{r}_1 - \vec{r}_2 = \left(\vec{v}_2 - \vec{v}_1\right) \frac{\left|\vec{r}_1 - \vec{r}_2\right|}{\left|\vec{v}_2 - \vec{v}_1\right|}$$

or
$$\frac{\vec{r}_1 - \vec{r}_2}{\left|\vec{r}_1 - \vec{r}_2\right|} = \frac{\left(\vec{v}_2 - \vec{v}_1\right)}{\left|\vec{v}_2 - \vec{v}_1\right|}$$

213. (c)

Here, volume of blood pumped by man's heart,

$V = 5 \text{ litres} = 5\times10^{-3}\text{ m}^{-3}$

$t = 1 \text{ min} = 60 \text{ s}$

$P = 150 \text{ mm of Hg} = 0.15 \text{ m of Hg}$

$P = h\rho g$

$= (0.15\text{ m})(13.6\times10^3 \text{kg/m}^3)(10\text{m/s}^2)$

$= 20.4\times10^3 \text{ N/m}^2$

$\therefore$ Power of the heart $= \dfrac{PV}{t}$

$$= \frac{\left(20.4\times10^3 \text{N/m}^2\right)\left(5\times10^{-3}\text{m}^3\right)}{60s} = 1.70\text{W}$$

214. (d)

$$\frac{1}{2}mv^2 - \frac{1}{2}mv_0^2 = mgh$$

or
$$v^2 - v_0^2 = 2gh$$

Now, when the ball collides with the ground, 50% of its energy is lost and it rebounds to the same height h.

$$\therefore \frac{50}{100}\left(\frac{1}{2}mv^2\right) = mgh$$

$$\frac{1}{4}v^2 = gh \text{ or } v^2 = 4gh$$

Substituting this value of v^2 in Eq. (i), we get

$$4gh - v_0^2 = 2gh$$

or
$$v_0^2 = 4gh - 2gh = 2gh \text{ or } v_0 = \sqrt{2gh}$$

Here, $g = 10 \text{ ms}^{-2}$ and $h = 20$ m

$$\therefore v_0 = \sqrt{2\left(10\text{ ms}^{-2}\right)(20\text{ m})} = 20\text{ ms}^{-1}$$

215. (c)

$$P = k \quad \text{(Power is constant)}$$

$$\frac{dW}{dt} = k; \quad dW = kdt$$

Integrating both sides, $\displaystyle\int_0^W dW = \int_0^t kdt$

$$\Rightarrow W = kt$$

Using Work–Energy theorem,

$$W = \frac{1}{2}mv^2 - \frac{1}{2}m(0)^2$$

$$kt = \frac{1}{2}mv^2$$

$$v = \sqrt{\frac{2kt}{m}}$$

Acceleration of the particle, $a = \dfrac{dv}{dt}$

$$a = \sqrt{\frac{k}{2mt}}$$

Force on the particle, $F = ma = \sqrt{\dfrac{mk}{2t}} = \sqrt{\dfrac{mk}{2}}t^{-1/2}$

216. (a)

Using conservation of energy,

$$\frac{1}{2}m_1u_1^2 + \frac{1}{2}m_2u_2^2 = \frac{1}{2}m_1v_1^2 + \frac{1}{2}m_2v_2^2 + \varepsilon$$

$$\therefore \quad \frac{1}{2}m_1u_1^2 + \frac{1}{2}m_2u_2^2 - \varepsilon = \frac{1}{2}m_1v_1^2 + \frac{1}{2}m_2v_2^2$$

217. (a)

Here, $K_P > K_Q$

Case (a): Elongation (x) in each spring is same.

$$W_P = \frac{1}{2}K_Px^2, W_Q = \frac{1}{2}K_Qx^2 \;\therefore W_P > W_Q$$

Case (b): Force is same.

So, $x_1 = \dfrac{F}{K_P}$ and $x_2 = \dfrac{F}{K_Q}$

$$W_P = \frac{1}{2}K_Px_1^2 = \frac{1}{2}\frac{F^2}{K_P}$$

$$\therefore W_Q = \frac{1}{2}K_Q x_2^2 = \frac{1}{2}\frac{F^2}{K_Q} \Rightarrow W_P < W_Q$$

218. (b)

Let $\vec{v}'$ be velocity of third piece of mass $2m$. Initial momentum, $\vec{p}_i = 0$

Final momentum, $\vec{p}_f = mv\hat{i} + mv\hat{j} + 2m\vec{v}'$

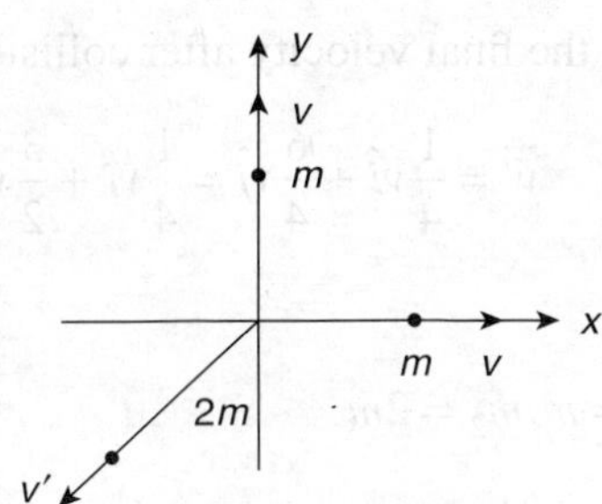

According to law of conservation of momentum

$$\vec{p}_f = \vec{p}_i$$

$$mv\hat{i} + mv\hat{j} + 2m\vec{v}' = 0$$

$$\vec{v}' = -\frac{v}{2}\hat{i} - \frac{v}{2}\hat{j}$$

The magnitude of v' is

$$v' = \sqrt{\left(-\frac{v}{2}\right)^2 + \left(-\frac{v}{2}\right)^2} = \frac{v}{\sqrt{2}}$$

Total kinetic energy

$$= \frac{1}{2}mv^2 + \frac{1}{2}mv^2 + \frac{1}{2}(2m)v'^2$$

$$= \frac{1}{2}mv^2 + \frac{1}{2}mv^2 + \frac{1}{2}(2m)\left(\frac{v}{\sqrt{2}}\right)^2$$

$$= mv^2 + \frac{mv^2}{2} = \frac{3}{2}mv^2$$

219. (c)

Here, $\vec{F} = (3\hat{i} + \hat{j})N$

Initial position, $\vec{r}_1 = (2\hat{i} + \hat{k})m$

Final position, $\vec{r}_2 = (4\hat{i} + 3\hat{j} - \hat{k})m$

Displacement, $\vec{r} = \vec{r}_2 - \vec{r}_1$

$$\vec{r} = (4\hat{i} + 3\hat{j} - \hat{k})m - (2\hat{i} + \hat{k})m = 2\hat{i} + 3\hat{j} - 2\hat{k}m$$

Work done, $W = \vec{F}.\vec{r} = (3\hat{i} + \hat{j}).(2\hat{i} + 3\hat{j} - 2\hat{k})$

$$= 6 + 3 = 9\text{J}$$

220. (c)

221. (d)

Power, $P = \dfrac{\text{Work done}}{\text{Time taken}}$

Here work done (= mgh) is same is for both cases.

$$\therefore \frac{P_1}{P_2} = \frac{t_2}{t_1} = \frac{30s}{1\,\text{min}} = \frac{30s}{60s} = \frac{1}{2}$$

222. (b)

Here, $U = \dfrac{A}{r^2} - \dfrac{B}{r}$

For equilibrium, $\dfrac{dU}{dr} = 0$

$$\therefore -\frac{2A}{r^3} + \frac{B}{r^2} = 0 \text{ or } \frac{2A}{r^3} = \frac{B}{r^2} \text{ or } r = \frac{2A}{B}$$

For stable equilibrium, $\dfrac{dU^2}{dr^2} > 0$

$$\frac{d^2U}{dr^2} = \frac{6A}{r^4} - \frac{2B}{r^3}$$

$$\left.\frac{d^2U}{dr^2}\right|_{r=(2A/B)} = \frac{6AB^2}{16A^4} - \frac{2B^4}{8A^3}\frac{B^4}{8A^3} > 0$$

So for stable equilibrium, the distance of the particle is $\dfrac{2A}{B}$

223. (d)

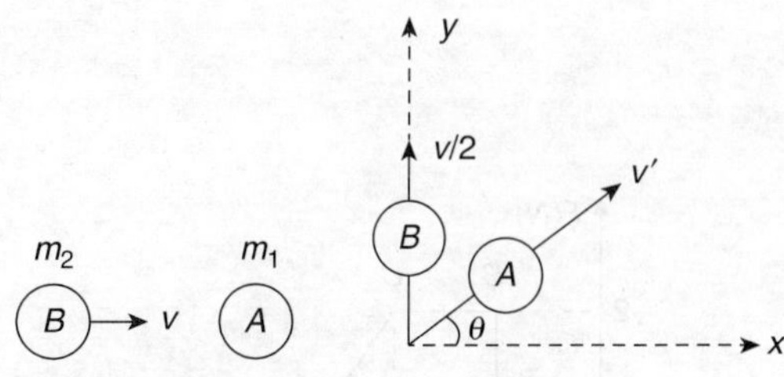

Applying to law of conservation of liner momentum along x-axis,

$$m_1 \times 0 + m_2 \times v = m_1 v' \cos\theta$$

$$m_2 v = m_1 v' \cos\theta$$

$$\Rightarrow \quad \cos\theta = \frac{m_2 v}{m_1 v'} \quad \text{....(i)}$$

Applying to law of conservation of liner momentum along y-axis,

$$m_1 \times 0 + m_2 \times 0 = m_1 v' \sin\theta + m_2 \frac{v}{2} - m_2 \frac{v}{2} = m_1 v' \sin\theta$$

$$\Rightarrow \quad \sin\theta = -\frac{m_2 v}{2m_1 v'} \quad \ldots\text{(ii)}$$

Dividing (ii) by (i), we get,

$$\tan\theta = -\frac{1}{2} \text{ or } \theta \tan^{-1}\left(-\frac{1}{2}\right) \text{ to the } x\text{-axis}$$

224. (b)

$$P_0 = Fv$$

$$\therefore F = ma = m\frac{dv}{dt}$$

$$\therefore P_0 = mv\frac{dv}{dt} \text{ or } \therefore P_0 dt = mvdv$$

Integrating both sides, we get $\int_0^t P_0 dt = m\int_0^t vdv$

$$P_0 t = \frac{mv^2}{2}$$

$$v = \left(\frac{2P_0 t}{m}\right)^{1/2} \text{ or } v \propto \sqrt{t}$$

225. (b)

226. (b)

Power, $\vec{F}.\vec{v} = Fv\cos\theta$

Just before hitting the earth $\theta = 0°$. Hence, the power exerted by the gravitational force is greatest at the instant just before the body hits the earth.

227. (a)

228. (d)

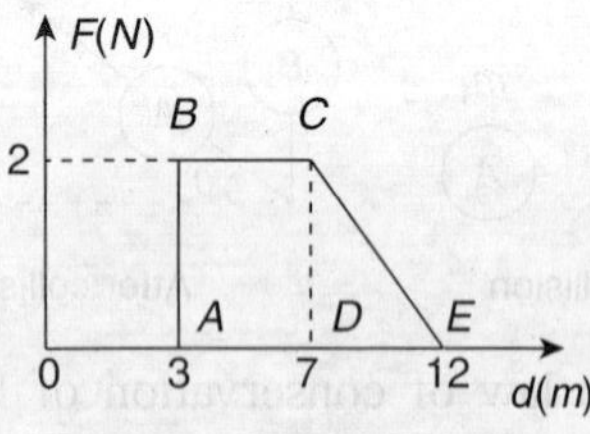

Work done = Area under (F–d) graph = Area of rectangle ABCD + Area of triangle DCE

$$= 2\times(7-3) + \frac{1}{2}\times 2\times(12-7) = 8 + 5 = 13\text{J}$$

229. (b)

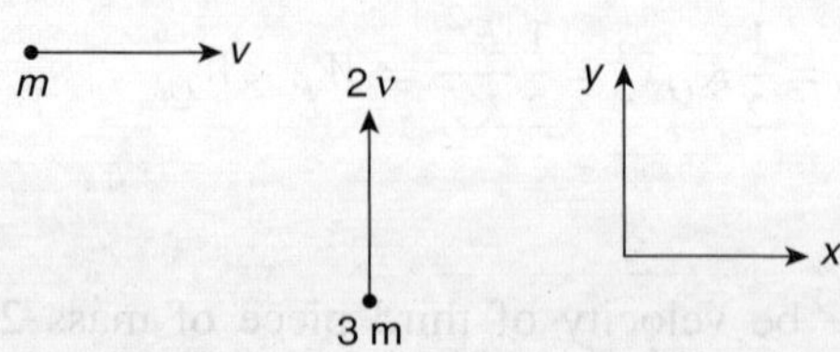

Applying to conservation of momentum,

$$mv\hat{i} + (3m)2v\hat{j} = (m + 3m)\vec{v'}$$

where $\vec{v'}$ is the final velocity after collision

$$\vec{v'} = \frac{1}{4}v\hat{i} + \frac{6}{4}v\hat{j} = \frac{1}{4}v\hat{i} + \frac{3}{2}v\hat{j}$$

230. (a)

Here, $m_1 = m, m_2 = 2m$

$$u_1 = 2\text{ m/s}, u_2 = 0$$

Coefficient of restitutions, $e = 0.5$

Let v_1 and v_2 be their velocities after collision.

Applying the law of conservation of linear momentum,

$$m_1u_1 + m_2u_2 = m_1v_1 + m_2v_2$$

$$\therefore m\times 2 + 2m\times 0 = m\times v_1 + 2m\times v_2$$

or $2m = mv_1 + 2mv_2$ or $2 = (v_1 + 2v_2)$ ... (i)

$$e = \frac{v_2 - v_1}{u_1 - u_2}$$

or $\quad e(u_1 - u_2) = v_2 - v_1$

$$\Rightarrow 0.5(2-0) = v_2 - v_1 \quad \ldots\text{(ii)}$$

Solving Eqs. (i) and (ii), we get

$$v_1 = 0\text{ m/s}, v_2 = 1\text{ m/s}$$

231. (a)

Mass per unit length of water = m

Mass flowing per second = mv

Power, P = kinetic energy/s = $\frac{1}{2}mvv^2$

$$= \frac{1}{2}mv^3$$

$$= \frac{1}{2}\times 100\times(2)^3$$

$$= 400\text{ W}$$

Velocity of water, $v = 2$ m/s

Power of the engine, $P = \mu v^3 = (100\text{ kg/m})(2\text{ m/s})^3$

$$= 800\text{ W}$$

232. (d)

Power delivered in time T is

$$P = F.V = Ma\ V$$

or $$P = MV\frac{dV}{dT} \Rightarrow PdT = MVdV$$

$$\Rightarrow PT = \frac{MV^2}{2}$$

$$\Rightarrow P = \frac{1}{2}\frac{MV^2}{T}$$

233. (a)

$$Mgx = \frac{1}{2}kx^2 \Rightarrow x = \frac{2Mg}{k}$$

234. (d)

Initial velocity $u = 20$ m/s; $m = 1$ kg

Kinetic energy = Maximum potential energy

Initial kinetic energy $= \frac{1}{2} \times 1 \times 20^2 = 200$ J

Mgh (max) $= 200 \Rightarrow 1 \times 10 \times h = 200$

$h = 20$ m (without air friction)

Actually the height is $h' = 18$ m

$\therefore$ Loss of energy due to air friction

$= mgh - mgh'$

$= 200 - 1 \times 10 \times 18 = 20$ J

235. (d)

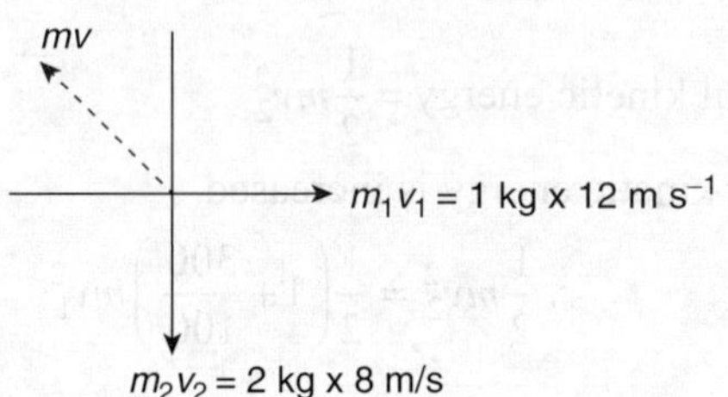

Total momentum of the two pieces 1 kg and 2 kg

$$= \sqrt{12^2 + 16^2} = 20\text{kgms}^{-1}$$

The magnitude of third piece is equal to 20 kg m/s and in the direction opposite to the resultant of these two momenta.

Velocity = 4 ms^{-1}

$\therefore$ Mass of the 3rd piece $= \frac{mv}{v} = \frac{20}{4} = 5$kg

236. (d)

Velocity of water is v, mass flowing per unit length is m.

$\therefore$ Mass flowing per second $= mv$

$\therefore$ KE per second

$$= \frac{1}{2}(mv)v^2 = \frac{1}{2}mv^3$$

237. (c)

$$mv = Mv' \Rightarrow v' = \left(\frac{m}{M}\right)v$$

Total KE of the bullet and gun = $= \frac{1}{2}mv^2 + \frac{1}{2}Mv'^2$

$$= \frac{1}{2}mv^2 + \frac{1}{2}M.\frac{m^2}{M^2}v^2$$

$$= \frac{1}{2}mv^2\left\{1 + \frac{m}{M}\right\}$$

$$= \left(\frac{1}{2} \times 0.2\right)\left(1 + \frac{0.2}{4}\right)v^2 = 1.05 \times 1000\text{J}$$

$$\Rightarrow v^2 = \frac{4 \times 1.05 \times 1000}{0.1 \times 4.2} = (100)^2$$

$$\therefore v = 100\text{ms}^{-1}$$

238. (c)

Mass of water falling/second = 15 kg/s

$h = 60$ m, $g = 10$ m/s^2, Loss = 10%

Power generated $= 15 \times 10 \times 60 \times 0.9 = 8100$ W $= 8.1$ kW.

239. (a)

Net work done $= W_{mg} + W_{\text{spring}}$

$$= mg\ (h + d) - \frac{1}{2}kd^2$$

240. (c)

Loss in potential energy $= mgh = 2 \times 10 \times 10 = 200$J

Gain in kinetic energy = Work done = 300 J

$\therefore$ Work done against friction = 300 – 200 = 100 J

241. (d)

Potential energy of spring $= \frac{1}{2}kx^2$

or, $\frac{U_1}{U_2} = \left(\frac{x_1}{x_2}\right)^2$ or, $\frac{U_1}{U_2} = \left(\frac{2}{8}\right)^2$

or, $\frac{U_1}{U_2} = \frac{1}{16}$ or, $U_2 = 16U_1 = 16U$. $(\because U_1 = U)$

242. (d)

$$s = \frac{t^2}{3}$$

$$\frac{ds}{dt} = \frac{2t}{3}$$

$$\frac{d^2s}{dt^2} = \frac{2}{3}$$

Work done, $W = \int Fds = \int m\frac{d^2s}{dt^2}ds$

$$= \int m\frac{d^2s}{dt^2}\frac{ds}{dt}dt = \int_0^2 3\times\frac{2}{3}\times\frac{2t}{3}dt = \frac{4}{3}\int_0^2 tdt$$

$$= \frac{4}{3}\left|\frac{t^2}{2}\right|^2 = \frac{4}{3}\times 2 = \frac{8}{3}j$$

243. (b)

$$30\times 0 = 18\times 6 + 12\times v$$

(Momentum conservation $v = -9$ m/s)
Negative sign indicates that both fragments move in opposite direction.

KE of 12 kg $= \frac{1}{2}mv^2 = \frac{1}{2}\times 12\times(-9)^2$

$$= \frac{1}{2}\times 12\times 81 = 486 \text{ J}$$

244. (b)

Work done = Area under F – x graph = Area of trapezium

$$\frac{1}{2}\times(6+3)\times 3 = \frac{9\times 3}{2} = 13.5 \text{ J}$$

245. (a)

Kinetic energy $= \frac{p^2}{2m}$

$$\therefore \frac{E_1}{E_2} = \frac{p_1^2/2m_1}{p_2^2/2m_2} \Rightarrow \frac{E_1}{E_2} = \frac{m_2}{m_1}$$

$\therefore E_1 < E_2$ [as $m_1 < m_2$]

246. (c)

Ratio of their kinetic energy

$$\frac{KE_1}{KE_2} = \frac{(1/2)m_1v_1^2}{(1/2)m_2v_2^2}$$

They acquire the same velocity as the height fall is same.

$$\frac{KE_1}{KE_2} = \frac{m_1}{m_2} = \frac{2}{4} = \frac{1}{2}$$

247. (a)

$$\frac{1}{2}mv^2 = \frac{1}{2}kx^2$$

$$\Rightarrow x = \sqrt{\frac{mv^2}{k}} = \sqrt{\frac{0.5\times(1.5)^2}{50}} = 0.15 \text{ m}$$

248. (d)

$U = -kx^2$, k = Spring constant

$$\frac{U_1}{U_2} = \frac{x_1^2}{x_2^2} = \frac{4}{100} \Rightarrow U_2 = 25U_1$$

249. (a)

$$m_1v_1 = m_2v_2$$

$$\frac{E_1}{E_2} = \frac{(1/2)m_1v_1^2}{(1/2)m_2v_2^2} = \frac{m_1^2v_1^2}{m_2^2v_2^2}\cdot\frac{m_2}{m_1} = \frac{m_2}{m_1}$$

250. (a)

$\therefore$ Initial kinetic energy $= \frac{1}{2}mv_1^2$

$\therefore$ Final kinetic energy $= \frac{1}{2}mv_2^2$

Initial kinetic energy is increased

$$\therefore \frac{1}{2}mv_2^2 = \frac{1}{2}\left(1+\frac{300}{100}\right)mv_1^2$$

$\Rightarrow v_2 = 2v_1$ or $v_2/v_1 = 2$ ...(i)

$$\therefore \frac{p_2}{p_1} = \frac{mv_2}{mv_1} = \frac{v_2}{v_1} = 2$$

$$\therefore p_2 = 2p_1$$

QUICK QUESTIONS

Quiz

1. Does kinetic energy depend upon the direction of motion?

2. Can kinetic energy be negative?

3. Does a single external force acting on a particle necessarily change its kinetic energy?

4. Can the kinetic energy of a system be increased or decreased without applying an external force on the system?

5. It is possible to exert a force on a body, thereby doing work on the body but without changing its kinetic energy?

6. Can a body have energy without having momentum?

7. Can a body have momentum without having energy?

8. A meteorite burns in the atmosphere before it reaches earth's surface. What happens to its momentum?

9. A particle of mass m has momentum p. What is its kinetic energy?

10. What is the momentum of a body of mass whose kinetic energy is E?

11. What is the type of collision when a negatively charged body collides with a positively charged particle?

12. A heavy stone is lowered to the ground. Is the work done by the applied force positive or negative?

13. Fast moving neutrons can be slowed down by passing them through heavy water. Comment on the statement.

14. Consider a tug-of-war in which two teams pulling on the rope are equally matched so that no motion takes place. Is there any work done on the rope? On the pullers? On the ground?

15. Give two examples of conservation force.

16. When is the exchange of energy maximum during an elastic collision of two bodies?

17. The casing of a rocket burns up due to friction. At whose expenses is the heat energy acquired for the burning obtained? The rocket or the atmosphere?

18. When a ball is thrown up in the air, its momentum first decreases and then increases. Does it not violate the law of conservation of momentum?

Think and Grow

19. In many situations, friction forces tend to reduce the kinetic energy of a body. But sometimes, there are situations where the kinetic energy of the body because of friction increases. Cite a situation where the kinetic energy increases.

20. Soft steel can be made red hot by repeated hammering on it. This does not go well with hard steel. Think of the reason for this difference.

21. A car strikes a tree and comes to rest. The total energy of the car system is now less than before. How did the energy get transferred from the system and in what form does it remain in the system?

22. A toaster is turn on. Discuss the different forms of energy transfer occurring in the coils of the toaster.

23. A bird is held in a light container which is completely closed. Can we judge whether the bird is resting or flying by weighing the containers?

Explanations

Quiz

1. No, it is a scalar quantity.

2. No, $KE = \frac{1}{2}mv^2$ velocity v appears in square, so it will always be positive.

3. No, example when the force is perpendicular to displacement as in circular motion.

4. Yes, by doing work from internal forces, basically other forms of energy are converted into work. In an adiabatic system the kinetic energy of the system increases during expansion at the expense of its internal energy. In an inelastic collision the kinetic energy decreases.

5. Yes, when a body is dragged on a rough horizontal surface. The applied force is neutralized by the force of friction and body moves.

6. Yes, The relation between kinetic energy and momentum is $KE = \frac{p^2}{2m}$ if $p = 0$, KE = 0, but Energy E = KE +PE and when KE = 0, E = PE.

Potential energy PE may or may not be zero, so a body may or may not have energy even when momentum is zero.

7. Yes, E = KE + PE

E = 0, means either KE or PE both are zero, or PE is negative of kinetic energy. Therefore kinetic energy may or may not be zero. That means it may have momentum.

8. The momentum of meteorite is transferred to air molecules by air resistance. Therefore the momentum of meteorite plus air molecules remains constant.

9. $\frac{p^2}{2m}$

10. $\sqrt{2mE}$

11. Perfectly inelastic

12. Negative

13. The masses of neutrons and heavy water are almost equal. Therefore the maximum exchange of energy takes place.

14. There is no motion; therefore the point of application of the forces from the puller has no displacement. Hence, no work is done on the rope. Because of the same reason, there is no work done on the pullers as well on the ground.

15. (i) Gravitational force

(ii) Electrostatic force

16. When collision bodies have same mass

17. The rocket provides the necessary energy.

18. No, the momentum of the system remains conserved. Here, the system consists of ball, earth and air.

Think and Grow

19. If a box is lying on the bed of a truck and the truck accelerates to west, the friction force on the box by the truck also acts in west direction assuming that the box does not slip.

20. In the case of soft steel, the energy is transferred by doing work on the soft steel. The continuous hammering results in the deformation of steel. Therefore, point of application of the force is displaced and positive work is done on the soft steel.

In the case of hard steel, because of less deformation, less work is done on it. Therefore soft steel is better than hard steel in absorbing energy due to hammering. A good amount of energy is transferred in the case of soft steel; its temperature rises considerably and it becomes red hot.

21. Energy is transferred out of system in a number of ways. The car did some work on the tree during collision deforming the tree. During the crash, the energy may transfer in the form of sound wave.

After the car has come to rest, the kinetic energy of the system becomes zero. The internal energy of the car may rise because of increased temperature after collision.

22. Let us assume the coils as a system. The energy change in the system is internal energy because the temperature of the coil rises. The energy transfer into the system is in the form of electrical energy. Energy transfers out of system in the form of electromagnetic radiation as the coil is hot and glowing. Energy is also transferred in the form of heat from the coil to the surrounding.

23. No, according to law of conservation of momentum, in absence of any external force, the net momentum of a system remains conserved. Now considering the bird + container as a system of constant mass the force which the bird exerts in flying is an internal force. The internal force does not change the momentum of an isolated system, so the weight of the system cannot change and we cannot judge whether bird is flying or resting.

CHAPTER TEST

1. A cube of mass M starts at rest from point 1 at a height 4 R, where R is the radius of the circular track. The cube slides down the frictionless track and around the loop. The force which the track exerts on the cube at point 2 is:

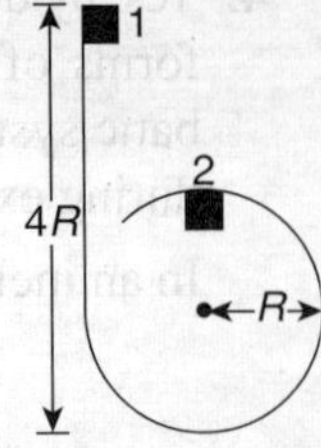

(a) 3 Mg
(b) Mg
(c) 2 Mg
(d) Cube will not reach the point 2

2. The work done in moving a particle from a point (1, 1) to (2, 3) in a plane and in a force field with potential $U = \lambda\ (x + y)$ is

(a) 0 (b) λ
(c) $3\ \lambda$ (d) $-3\ \lambda$

3. A particle of mass m is fixed to one end of a light spring of force constant k and unstretched length l. The system is rotated about the other end of the spring with an angular velocity ω, in gravity-free space. The increase in length of the spring will be

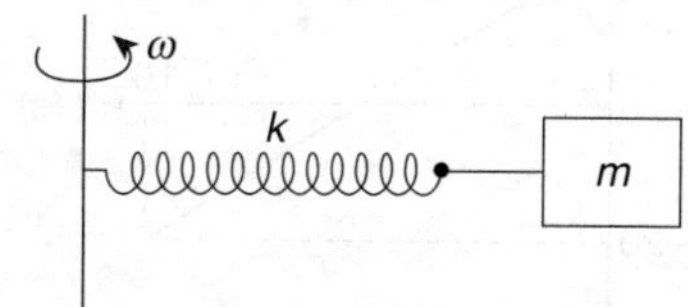

(a) $\dfrac{m\omega^2 l}{k}$ (b) $\dfrac{m\omega^2 l}{k - m\omega^2}$
(c) $\dfrac{m\omega^2 l}{k + m\omega^2}$ (d) None of these.

4. A spring which is initially in its unstretched condition, is first stretched by a length x and then again by a further length x. The work done in the first case is W_1 and in the second case is W_2 then:

(a) $W_2 = W_1$ (b) $W_2 = 2W_1$
(c) $W_2 = 3W_1$ (d) $W_2 = 4W_1$

5. A particle of mass 0.1 kg is subjected to a force which varies with distance as shown in figure. If it starts its journey from rest at x=0, its velocity at x = 12 m is:

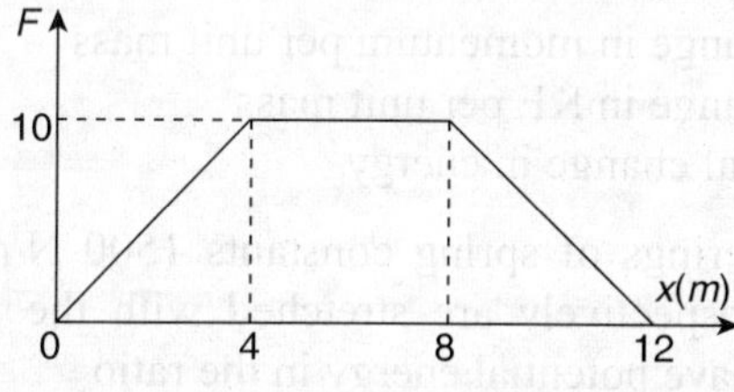

(a) 0 m/s (b) $20\sqrt{2}$ m/s
(c) $20\sqrt{3}$ m/s (d) 40 m/s

6. A motor is to be used to lift water from a depth of 10 m and discharging it on ground level at a speed 10 m/s. If 2 kg of water is lifted per second and the efficiency of motor is 40%, the minimum power of the motor should be nearly:

(a) 1.5 HP (b) 1 HP
(c) 0.67 HP (d) 2 HP

7. A block of mass 4 kg is resting on a horizontal table and a force of 10 N is applied on it in the horizontal direction. The coefficient of kinetic friction between the block and the table is 0.1. The work done in $\sqrt{20}$ s

(a) by the applied force is 150 J
(b) by the friction force is 60 J
(c) by the applied force and net force are 150 J and 90 J respectively
(d) All are correct.

8. A car of mass 500 kg is driven with acceleration I m/s^2 along a straight level road against constant external resistance of 1000 N. When the velocity is 5 m/s, the rate at which the engine is working is

(a) 5 kW (b) 7.5 kW
(c) 2.5 kW (d) 10 kW

9. In the figure shown, the net work done by the tension when the bigger block of mass M touches the ground is:

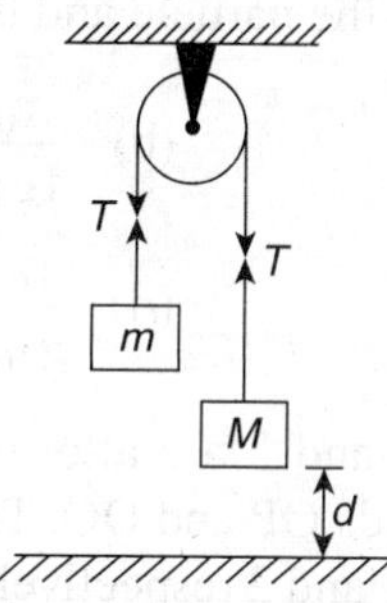

(a) $+Mgd$ (b) $-(M + m)\ gd$
(c) $-Mgd$ (d) Zero

10. Work done in time t on a body of mass m which is accelerated from rest to a speed v in time t_1 as a function of time t is given by:

(a) $\dfrac{1}{2} m \dfrac{v}{t_1} t^2$ (b) $m \dfrac{v}{t_1} t^2$
(c) $\dfrac{1}{2} (\dfrac{mv}{t_1})^2 t^2$ (d) $\dfrac{1}{2} m \dfrac{v^2}{t_1^2} t^2$

11. In a simple pendulum, the breaking strength of the string is double the weight of the bob. The bob is released from rest when string is horizontal. The string breaks when it makes an angle θ with the vertical:

(a) $\theta = \cos^{-1}(1/3)$ (b) $\theta = 60^0$
(c) $\theta = \cos^{-1}(2/3)$ (d) $\theta = 0$

12. A particle of mass 0.1 kg travels along a space curve with velocity given by $4\hat{i} + 16\hat{k}$ m/s. After some time, its velocity becomes $8\hat{i} + 20\hat{j}$ m/s due to the action of a conservative force. The work done on the particle during this interval of time is:

(a) 0.32 J (b) 6.9 J
(c) 9.6 J (d) 0.96 J

13. A 15 gm ball is shot from a spring gun whose spring has a force constant of 600 N/m. The spring is compressed

by 5 cm. The greatest possible horizontal range of the ball for this compression is: (Take $g = 10$ m/s^2)

(a) 6.0 m (b) 12.0 m
(c) 10.0 m (d) 8.0 m

14. Water is pumped from a depth of 10 m and delivered through a pipe of cross-section 10^{-2} m^2 up to a height of 10 m. If it is needed to deliver a volume of 0.2 m^3 per second, the power required will be:

(a) 19.6 kW (b) 9.8 kW
(c) 39.2 kW (d) 4.9 kW

15. A particle moves on a rough horizontal ground with some initial velocity, say v_0. If 3/4 th of its kinetic energy is lost in friction in time t_0, then coefficient of friction between the particle and the ground is:

(a) $\frac{v_0}{2g t_0}$ (b) $\frac{v_0}{4g t_0}$

(c) $\frac{3v_0}{4g t_0}$ (d) $\frac{v_0}{g t_0}$

16. Two particles 1 and 2 are allowed to descend on two frictionless chords OP and OQ. The ratio of the speeds of the particles 1 and 2 respectively, when they reach on the circumference is:

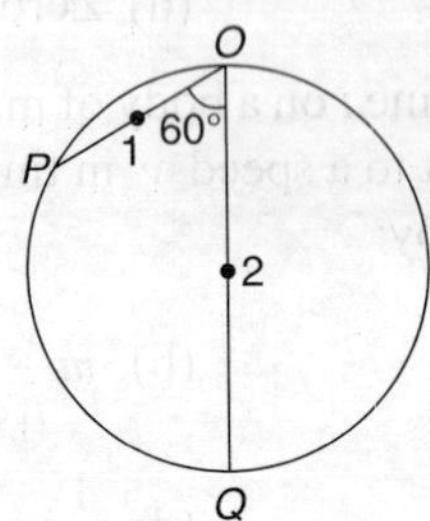

(a) $\frac{1}{4}$ (b) $\frac{1}{2}$

(c) 1 (d) $\frac{1}{2\sqrt{2}}$

17. A particle of mass m describes a circle of radius r. The centripetal acceleration of the particle is $4/r^2$. What will be the momentum of the particles?

(a) $2m/r$ (b) $2m/\sqrt{r}$
(c) $4m/\sqrt{r}$ (d) $4m/r$

18. A block of mass m slides down a rough inclined plane of inclination θ with horizontal with zero initial velocity. The coefficient of friction between the block and the plane is μ with $\theta > \tan^{-1}(\mu)$. Magnitude of rate of work done by the force of friction at time t is

(a) $\mu m g^2 t \sin\theta$

(b) $m g^2 t(\sin\theta - \mu\cos\theta)$

(c) $\mu m g^2 t \cos\theta(\sin\theta - \mu\cos\theta)$

(d) $\mu m g^2 t \cos\theta$

19. Velocity-time graph of a particle moving in a straight line is as shown in figure. Mass of the particle is 2 kg. Work done by all the forces acting on the particle in time interval between $t = 0$ to $t = 10$ s is:

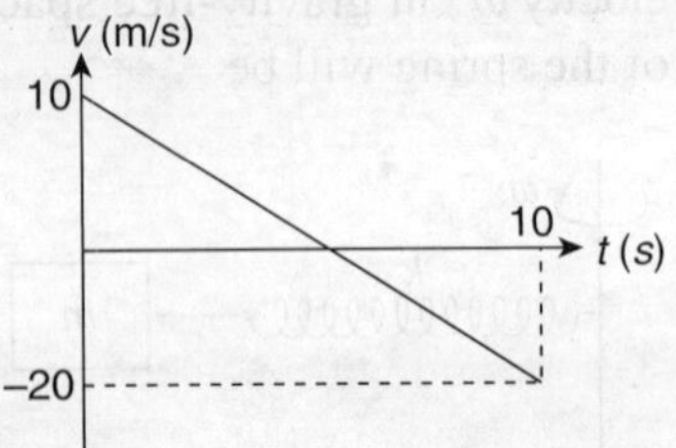

(a) 300 J (b) -300 J
(c) 400 J (d) -400 J

20. A stone tied to a string of length l is whirled in a vertical circle with the other end of the string at the centre. At a certain instant of time, the stone is at its lowest position and has a speed u. The magnitude of the change in its velocity at it reaches a position where the string is horizontal is

(a) $\sqrt{(u^2 - 2gl)}$ (b) $\sqrt{2gl}$

(c) $\sqrt{(u^2 - gl)}$ (d) $\sqrt{(2u^2 - gl)}$

21. The area of the acceleration-displacement curve of a body gives

(a) impulse
(b) change in momentum per unit mass
(c) change in KE per unit mass
(d) total change in energy.

22. Two springs of spring constants 1500 N/m and 3000 N/m respectively are stretched with the same force. They have potential energy in the ratio

(a) 4 : 1 (b) 1 : 4
(c) 2 : 1 (d) 1 : 2

23. A positive dependent force $F = 7 - 2x + 3x^2$ newton acts on a small body of mass 2 kg and displaces it from $x = 0$ to $x = 5$. The work done in joules is

(a) 70 (b) 270
(c) 35 (d) 135

24. When a spring is stretched by 2 cm, it stores 100 J of energy. If it is stretched further by 2 cm, the stored energy will be increased by

(a) 100 J (b) 200 J
(c) 300 J (d) 400 J

25. Which one of the following is not a conservative force?

(a) Gravitational force
(b) Electrostatic force between two charges
(c) Magnetic force between two magnetic dipoles
(d) Frictional force.

26. A ball is projected vertically upwards with a certain initial speed. Another ball of the same mass is projected at an angle of 60^0 with the vertical with the same initial speed. At highest points of their journey, the ratio of their potential energies will be

(a) 1 : 1 (b) 2 : 1
(c) 3 : 2 (d) 4 : 1

27. A body of mass (4*m*) is lying in x-y plane at rest. It suddenly explodes into three pieces. Two pieces each of mass (*m*) move perpendicular to each other with equal speeds (*v*). The total kinetic energy generated due to explosion is

(a) $2\,mv^2$ (b) $4\,mv^2$
(c) mv^2 (d) $\frac{3}{2}\,mv^2$

28. A sphere of mass *m*, moving with velocity *V*, enters a hanging bag of sand and stops. If the mass of the bag is *M* and it is raised by height *h*, then the velocity of the sphere is

(a) $\frac{M+m}{m}\sqrt{2gh}$ (b) $\frac{M}{m}\sqrt{2gh}$
(c) $\frac{m}{M+m}\sqrt{2gh}$ (d) $\frac{m}{M}\sqrt{2gh}$

29. If a lighter body (mass M_1 and velocity V_1) and a heavier body (mass M_2 and velocity V_2) have the same kinetic energy, then

(a) $M_2V_2 < M_1V_1$ (b) $M_2V_2 = M_1V_1$
(c) $M_2V_1 = M_1V_2$ (d) $M_2V_2 > M_1V_1$

30. A particle is released from a height *S*. At a certain height, its kinetic energy is three times its potential energy. The height and speed of the particle at that instant are, respectively,

(a) $\frac{S}{4}, \frac{3gS}{2}$ (b) $\frac{S}{4}, \frac{\sqrt{3gS}}{2}$
(c) $\frac{S}{2}, \frac{\sqrt{3gS}}{2}$ (d) $\frac{S}{2}, \sqrt{\frac{3gS}{2}}$

31. A particle is moving in a circular path of radius a under the action of an attractive potential $U = -\frac{k}{2r^2}$. Its total energy is

(a) $\frac{k}{3a^2}$ (b) Zero
(c) $-\frac{3}{2}\frac{k}{a^2}$ (d) $-\frac{k}{4a^2}$

32. A simple pendulum is released from A as shown. If *m* and *l* represent the mass of the bob and length of the pendulum, respectively, the gain in kinetic energy at B is

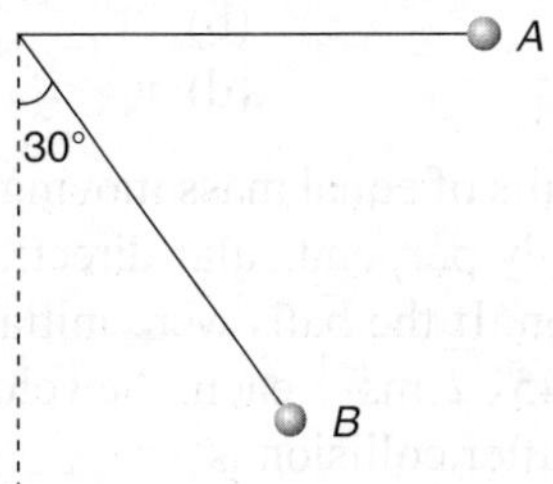

(a) $\frac{mgl}{2}$ (b) $\frac{mgl}{\sqrt{2}}$
(c) $\frac{\sqrt{3}}{2}mgl$ (d) $\frac{2}{\sqrt{3}}mgl$

33. A machine which is 75 percent efficient uses 12 joules of energy in lifting up a 1 kg mass through a certain distance. The mass is then allowed to fall through that distance. The velocity at the end of its fall is (in ms^{-1})

(a) $4\sqrt{3}$ (b) $4\sqrt{2}$
(c) $3\sqrt{2}$ (d) 3

34. A force $(4\hat{i} + \hat{j} - 2\hat{k})$ N acting on a body maintains its velocity at $(2\hat{i} + 2\hat{j} + 3\hat{k})ms^{-1}$. The power exerted is

(a) 4 W (b) 5 W
(c) 2 W (d) 8 W

35. Three objects A, B and C are kept in a straight line on a frictionless horizontal surface. These have masses m, 2*m* and *m*, respectively. The object A moves towards B with a speed 9 m/s and makes an elastic collision with it. Thereafter, B makes completely inelastic collision with C. All motions occur on the same straight line. Find the final speed (in m/s) of the object C.

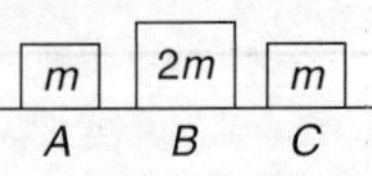

(a) 3m/s (b) 4 m/s
(c) 5m/s (d) 1 m/s

36. A particle of mass *m* moving in the x direction with speed 2*v* is hit by another particle of mass 2*m* moving in the y direction with speed *v*. If the collision is perfectly inelastic, the percentage loss in the energy during the collision is close to

(a) 44% (b) 50%
(c) 56% (d) 62%

37. A particle of mass *m* moving eastward with a speed *v* collides with another particle of the same mass moving northward with the same speed *v*. The two particles coalesce on collision. The new particle of mass 2*m* will move in the north-easterly direction with a velocity

(a) $v/2$ (b) $2v$
(c) $v/\sqrt{2}$ (d) v

38. Two putty balls of equal mass moving with equal velocity in mutually perpendicular directions, stick together after collision. If the balls were initially moving with a velocity of $45\sqrt{2}$ ms^{-1} each, the velocity of their combined mass after collision is

(a) $45\sqrt{2}$ ms^{-1} (b) 45 ms^{-1}
(c) 90 ms^{-1} (d) $22.5\sqrt{2}$ ms^{-1}

39. An open knife-edge of mass *m* is dropped from a height *h* on a wooden floor. If the blade penetrates up to a depth *d* into the wood, the average resistance offered by the wood to the knife edge is

(a) mg (b) $mg(1-\frac{h}{d})$
(c) $mg(1+\frac{h}{d})$ (d) $mg(1+\frac{h}{d})^2$

40. A time dependent force $F= 6t$ acts on a particle of mass 1 kg. If the particle starts from rest, the work done by the force during the first 1 sec will be

(a) 18 J (b) 4.5 J
(c) 22 J (d) 9 J

41. If speed is plotted against x-axis and kinetic energy against y-axis, then the graph obtained has a shape similar to that of

(a) circle (b) ellipse
(c) hyperbola (d) parabola

42. If two electrons are found to come closer to each other, then the potential energy of the electrons

(a) becomes zero (b) increases
(c) decreases (d) becomes infinite.

43. A 20 newton stone falls accidentally from a height of 4 m on to a spring of stiffness constant 1960 nm^{-1}. The maximum distance to which the spring shall suffer compression is:

(a) $\frac{2}{7}$ m (b) $\frac{3}{7}$ m
(c) $\frac{4}{6}$ m (d) 0.3295 m

44. The temperature at a bottom of a high water fall is higher than that at the top because

(a) by itself heat flow from higher to lower temperature
(b) the difference in height causes a difference in pressure
(c) thermal energy is transformed into mechanical energy
(d) mechanical energy is transformed into thermal energy.

45. In figure, a carriage P is pulled up from A to B. The relevant coefficient of friction is 0.40. The work done will be:

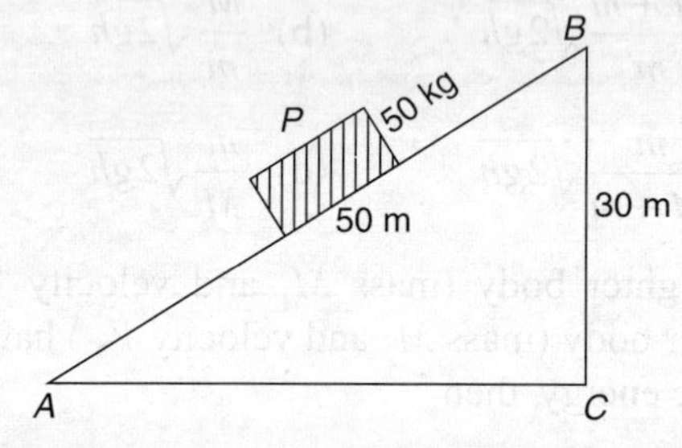

(a) 10 kJ (b) 23 kJ
(c) 25 kJ (d) 28 kJ

Answer Keys

1. (a)	2. (d)	3. (b)	4. (c)	5. (d)	6. (c)	7. (d)	8. (b)	9. (d)	10. (d)
11. (c)	12. (d)	13. (c)	14. (c)	15. (a)	16. (b)	17. (b)	18. (c)	19. (a)	20. (d)
21. (c)	22. (c)	23. (d)	24. (c)	25. (d)	26. (d)	27. (d)	28. (a)	29. (d)	30. (d)
31. (b)	32. (c)	33. (c)	34. (a)	35. (b)	36. (c)	37. (c)	38. (b)	39. (c)	40. (b)
41. (d)	42. (b)	43. (a)	44. (d)	45. (b)					

Solutions

1. (a)

Potential energy of cube at position 1, (considering the lower most portion of circular track as reference point)

$$U_1 = Mg(4R) = 4MgR$$

At position 2,

$$U_1 = U_2 + K$$

$$4MgR = 2MgR + \frac{1}{2}Mv^2$$

$$4MgR = Mv^2$$

$$v^2 = 4gR$$

$$N = \frac{Mv^2}{R} - Mg$$

$$N = \frac{M}{R}(4Rg) - Mg = 3Mg$$

2. (d)

$$U = \lambda(x+y)$$

$$\overline{F} = -\left[\frac{\partial U}{\partial x}\hat{i} + \frac{\partial U}{\partial y}\hat{j}\right]$$

$$\overline{F} = -\lambda\hat{i} - \lambda\hat{j}$$

$$\overline{F} = -\lambda\hat{i} - \lambda\hat{j}$$

$$w = \overline{F}.\overline{x}$$

$$W = (-\lambda\hat{i} - \lambda\hat{j}).[(2-1)\hat{i} + (3-1)\hat{j}]$$

$$W = (-\lambda\hat{i} - \lambda\hat{j}).(\hat{i} + 2\hat{j})$$

$$W = -\lambda - 2\lambda$$

$$W = -3\lambda$$

3. (b)

Let increase in length of spring be x

At equilibrium,

$$kx = m\omega^2 x + m\omega^2 l$$

$$kx - m\omega^2 x = m\omega^2 l$$

$$x = \frac{m\omega^2 l}{k - m\omega^2}$$

4. (c)

Work done to stretch a length x of spring,

$$W_1 = \frac{1}{2}kx^2$$

Work done to stretch the spring to length $2x$,

$$W = \frac{1}{2}k(2x)^2 = 4W_1$$

Work done to stretch by a further length x,

$$W_2 = W - W_1 = 3W_1$$

$$\Rightarrow \quad W_2 = 3W_1$$

5. (d)

Area under F-x graph,

$$A = \frac{1}{2}(4)(10) + (4\times10) + \frac{1}{2}\times4\times10$$

$$W = 20 + 40 + 20$$

From Work–Energy theorem,

$$W = \frac{1}{2}mv^2 \text{ (as body starts from rest)}$$

$$80 = \frac{1}{2}\times0.1\times v^2$$

$$v^2 = 1600$$

$$v = 40 \text{ m/s}$$

6. (c)

$$P_{out} = \frac{mgh}{t} = \frac{2\times10\times10}{1} = 200\text{ W}$$

$$\eta = \frac{P_{out}}{P_{in}}$$

$$\frac{40}{100} = \frac{200}{P_{in}}$$

$$P_{in} = 500 \text{ W}$$

As $\quad 1\text{ HP} = 746\text{ W}$

$$\Rightarrow \quad 500\text{ W} = \frac{500}{746}\text{ HP} = 0.67\text{ HP}$$

7. (d)

8. (b)

For motion of car we use

$$F - 1000\text{ N} = 500\times1$$

$$F = 1500\text{ N}$$

$$P = F.v$$

$$= 1500\times5$$

$= 7500$ W

$= 7.5$ kW

9. (d)

Work done by tension = zero as string is mass-less and friction-less and hence, it can never do work.

10. (d)

$$W = FS = ma.s$$

$$W = ma\frac{1}{2}at^2$$

As $a = \frac{v}{t_1}$

$\Rightarrow$ $W = \frac{1}{2}m\frac{v^2}{t_1^2}t^2$

11. (c)

12. (d)

13. (c)

$$\frac{1}{2}kx^2 = \frac{1}{2}mv^2$$

$$600 \times (0.05)^2 = (15 \times 10^{-3})v^2$$

$$v^2 = 100$$

$$v = 10 \text{ m/s}$$

Maximum range, $R = \frac{v^2}{g} = \frac{(10)^2}{10} = 10$ m

14. (c)

$$P = \frac{W}{t} = \frac{mgh}{t}$$

$t = 1$s

$V_\rho = 0.2 \text{ m}^3 \times 1000 \text{ kg/m}^3$

$m = 200$ kg

$$P = \frac{200 \times 9.8 \times (10+10)}{1}$$

$P = 200 \times 9.8 \times 20$

$P = 39200$ W $= 39.2$ kW

15. (a)

16. (b)

17. (b)

Centripetal acceleration

$$a_c = \frac{v^2}{r} = \frac{4}{r^2}$$

$$v = \frac{2}{\sqrt{r}}$$

Momentum of particle,

$P = mv$

$$P = \frac{2m}{\sqrt{r}}$$

18. (c)

Rate of doing work = Power

$P = F.\, v$

$$P = \mu mg\cos\theta \times [0 + (g\sin\theta - \mu g\cos\theta)t]$$

$$P = \mu mg^2 t\cos\theta(\sin\theta - \mu\cos\theta)$$

19. (a)

$$W = \Delta KE$$

$$= \frac{1}{2}m(v_2^2 - v_1^2)$$

$$= \frac{1}{2} \times 2 \times ((-20)^2 - (10)^2) = 300 \text{ J}$$

20. (d)

21. (c)

Area of acceleration – Displacement curve gives change in KE per unit mass

$$\frac{1}{2}m(v^2 - u^2) = F.S = \frac{mdv}{dt} \times s$$

$$\therefore \quad \frac{\text{Change in KE}}{\text{Mass}} = \frac{dv}{dt} \times s$$

22. (c)

$$U = \frac{F^2}{2k} \Rightarrow \frac{U_1}{U_2} = \frac{k_2}{k_1} \text{ (Force is constant)}$$

$$\therefore \quad \frac{U_1}{U_2} = \frac{3000}{1500} = \frac{2}{1}$$

23. (d)

$$W = \int_0^5 F dx = \int_0^2 (7 - 2x + 3x^2)dx = [7x - x^2 + x^3]_0^5$$

$35 - 25 + 125 = 135$ J

24. (c)

$$100 = \frac{1}{2}kx^2 \text{ (Given)}$$

$$W = \frac{1}{2}k(x_2^2 - x_1^2) = \frac{1}{2}k[(2x)^2 - x^2]$$

$$= 3 \times (\frac{1}{2}kx^2) = 3 \times 100 = 300 \text{ J}$$

25. (d)

26. (d)

27. (d)

Initial momentum = $P_i = 0$

Final momentum $P_f = 0 = mv\hat{i} + mv\hat{j} + \vec{P_3}$

$$\Rightarrow \quad P_3 = mv\sqrt{2}$$

$$\text{Total KE} = \frac{P_3^2}{2\times 2m} + \frac{1}{2}mv^2 + \frac{1}{2}mv^2$$

$$= \frac{2m^2v^2}{4m} + mv^2 = \frac{3}{2}mv^2$$

28. (a)

29. (d)

$p = \sqrt{2mE}$. If kinetic energy are equal then $p \alpha \sqrt{m}$

i.e., heavier body possesses large momentum

As $M_2 > M_1$. Therefore, $M_2V_2 > M_1V_1$.

30. (d)

31. (b)

$$F = \frac{\partial u}{\partial r} = \frac{K}{r^3}$$

Since it is performing circular motion

$$F = \frac{mv^2}{r} = \frac{K}{r^3}$$

$$mv^2 = \frac{K}{r^2}$$

$$\Rightarrow \quad KE = \frac{1}{2}mv^2 = \frac{K}{2r^2}$$

Total energy = PE + KE

$$= -\frac{K}{2r^2} + \frac{K}{2r^2} = \text{Zero}$$

32. (c)

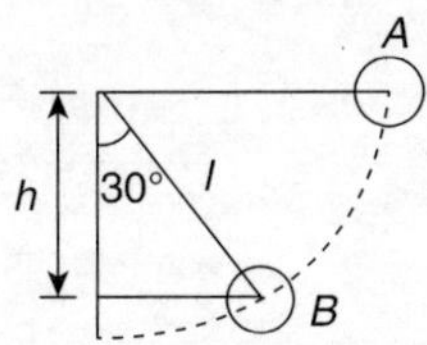

Vertical height = $h = l\cos 30^0$

Loss of potential energy = mgh

$$= mgl\cos 30^0 = \frac{\sqrt{3}}{2}mgl$$

$$\therefore \text{ Kinetic energy} = \frac{\sqrt{3}}{2}mgl$$

33. (c)

Potential energy of a body = 75% of 12 J

$$mgh = 9J \Rightarrow h = \frac{9}{1\times 10} = 0.9 \text{ m}$$

Now when this mass is allowed to fall, then it acquires velocity

$$v = \sqrt{2gh} = \sqrt{2\times 10\times 0.9} = \sqrt{18} \text{ m/s}$$

34. (a)

Here, Force $\vec{F} = (4\hat{i} + \hat{j} - 2\hat{k})$N

Velocity, $\vec{v} = (2\hat{i} + 2\hat{j} + 2\hat{k})$ ms^{-1}

Power,

$P = \vec{F}\vec{v} = (4\hat{i} + \hat{j} - 2\hat{k}).(2\hat{i} + 2\hat{j} + 3\hat{k}) = (8 + 2 - 6)$ W

= 4 W

35. (b)

$$v_2 = \frac{2m_1u_1}{m_1 + m_2} = \frac{2\times m\times 9}{m + 2m} = 6 \text{ m/s}$$

After elastic collision B collides with C with velocity of 6 m/s. Now, collision between B and C is perfectly inelastic.

By the law of conservation of momentum

$$2m\times 6 + 0 = 3m\times v_{sys}$$

$$\Rightarrow \quad v_{sys} = 4 \text{ m/s.}$$

36. (c)

37. (c)

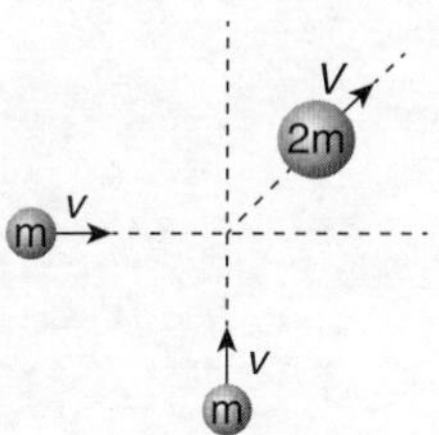

Initial momentum of the system

$$\vec{P_1} = mv\hat{i} + mv\hat{j} \Rightarrow |\vec{P_1}| = \sqrt{2}mv$$

Final momentum of the system = 2 mV

By the law of conservation of momentum

$$\sqrt{2}mv = 2mV \Rightarrow V = \frac{v}{\sqrt{2}}$$

38. (b)

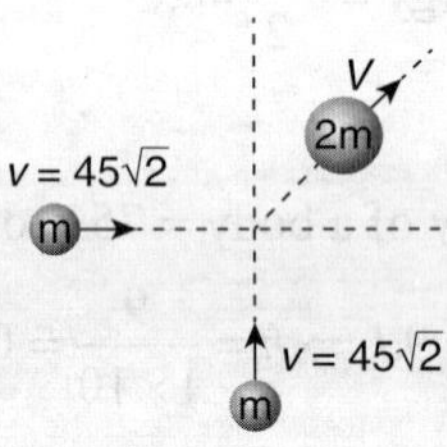

Initial momentum

$$\overline{P} = m45\sqrt{2}\hat{i} + m45\sqrt{2}\hat{j}$$

$$\Rightarrow |\overline{P}| = m \times 90$$

Final momentum $2m \times V$

$2m \times V = m \times 90$

$V = 45$ m/s.

39. (c)

Let the blade stop at depth d into the wood

$$v^2 = u^2 + 2aS$$

$$\Rightarrow 0 = (\sqrt{2ghg})^2 + 2(g-a)d$$

on solving $a = \left(1 + \frac{h}{d}\right)g$

So the resistance offered by the wood $= mg\left(1 + \frac{h}{d}\right)$

40. (b)

$$a = a = \frac{F}{m} = \frac{6t}{1} = 6t$$

$$\frac{dv}{dt} = 6t$$

$$dv = 6tdt$$

$$\int_0^v dv = 6\int_0^t tdt$$

$$v = 6\left[\frac{t^2}{2}\right]_0^1 = 3$$

$$W = \Delta KE = K_F - K_i$$

$$= \frac{1}{2}(1)(3)^2 = 4.5 \text{ J}$$

41. (d)

It is a case of $y \propto x^2$.

42. (b)

The work done against the force of repulsion is stored as potential energy.

43. (a)

$Mg = 20$ N, $h = 4$ m, PE = 80 J

$$80 \text{ J} = \frac{1}{2}kx^2$$

$$x^2 = \frac{80 \times 2}{1960} = \frac{16}{196}$$

$$x = \frac{4}{14} \text{ m or } x = \frac{2}{7} \text{ m}$$

44. (d)

45. (b)

Work done against gravity = 50 × 10 × 30 J

= 15 kJ

Work done against friction

$$= 0.4 \times 50 \times 10 \times \frac{4}{5} \times 50 \text{ J} = 8 \text{ kJ}$$

Total work done = 15 kJ + 8 kJ = 23 kJ

Motion of System of Particles and Rigid Body

Introduction

In a system of extended bodies, there is one special point whose motion represents the motion of the system or the body in the case of translatory motion. This point is called centre of mass, i.e., centre of mass of a body or system is a point where the whole mass of body or system may be supposed to be concentrated for the purpose of translational motion.

Centre of Mass

The concept of centre of mass is very useful in solving many problems, in particular, those concerned with collision of particles. In the following discussion, we are going to mathematically define the centre of mass first and later on, justify this definition by considering the motion of the centre of mass of the system of particles under the action of forces.

Position of the centre of mass (C.M.) of a large number of particles

For a system of n particles whose position vectors are $\vec{r}_1, \vec{r}_2, ..., \vec{r}_n$ as shown in figure, the position vector of the centre of mass $\vec{r}_{cm}$ is defined as

$$\vec{r}_{\text{cm}} = \frac{m_1\vec{r}_1 + m_2\vec{r}_2 \ldots + m_n\vec{r}_n}{m_1 + m_2 + \ldots + m_n}$$

or,

$$\vec{r}_{\text{cm}} = \frac{\sum m_i\vec{r}_i}{M}$$

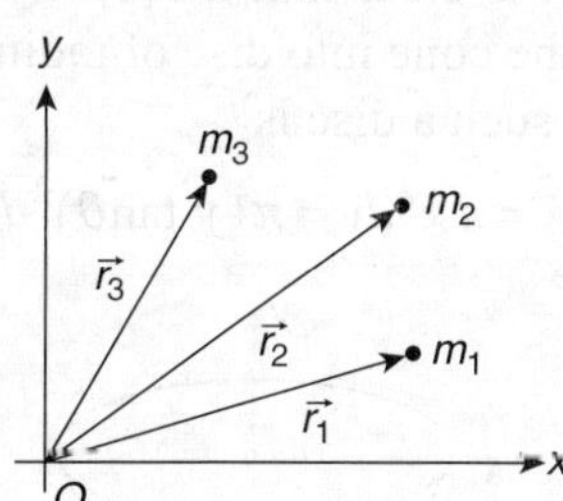

Here, $M = \sum m_i$ is the total mass of the system. The components of $\vec{r}_{\text{cm}}$ are defined as:

$$X_{\text{cm}} = \frac{\sum m_i x_i}{M};\ Y_{\text{cm}} = \frac{\sum m_i y_i}{M};\ Z_{\text{cm}} = \frac{\sum m_i z_i}{M}.$$

Position of centre of mass of continuous mass distributions

The centre of mass of a continuous mass distribution is defined as:

$$\vec{r}_{\text{cm}} = \frac{1}{M}\int \vec{r}\,dm$$

where $\vec{r}$ is the position vector on an element dm of the mass distribution.

The co-ordinates of the centre of mass are:

$$x_{\text{cm}} = \frac{1}{M}\int x\,dm$$

$$y_{\text{cm}} = \frac{1}{M}\int y\,dm$$

$$z_{\text{cm}} = \frac{1}{M}\int z\,dm$$

Position of centre of mass of two particles

Centre of mass of two particles of masses m_1 and m_2 separated by a distance d lies between the two particles. Taking moments about A,

$$r_{\text{cm}} = \frac{m_1 \times 0 + m_2 \times d}{m_1 + m_2} = \frac{m_2}{(m_1 + m_2)} \times d \text{ from } A.$$

Taking moment about the C.M. O, we have,

$$m_1 r_{\text{cm}} + m_2 \times (d - r_{\text{cm}}) = 0$$

$$\Rightarrow \quad m_2 d - (m_1 + m_2) \times r_{\text{cm}} = 0$$

$$\Rightarrow \quad r_{\text{cm}} = \frac{m_2 d}{m_1 + m_2}$$

r_1 C.M. r_2

A • ——— • ——— • B

m_1 ← r_{cm} → O m_2

← d →

Note

1. Moment can be taken about any convenient point.
2. If $m_1 = m_2$, $r_1 = r_2 = \frac{d}{2}$, C.M. lies midway between the two particles of equal masses.

If $m_1 < m_2$ then $r_1 > r_2$; and if $m_1 > m_2$ then $r_1 < r_2$ i.e., C.M. is nearer to the particle having a larger mass.

Illustration 1

Two particles of masses 2 kg and 4 kg are located at $x = 0$ and $x = 3$ m. Find the position of their centre of mass.

Solution

Since both the particles lie on x-axis, the C.M. will also lie on x-axis. Let the C.M. be located at a distance x from the origin. Taking moments about A,

$m_1 = 2$ kg   C.M. $m_2 = 4$ kg
A   B
$r_1 = x$ $r_2 = 3 - x$

$$r_{cm} = \frac{2\times 0 + 4\times 3}{6}$$

$$= 2 \text{ m from } A.$$

Thus the C.M. of the two particles located at $x = 2$ m.

Illustration 2

Find the centre of mass of the four point-masses as shown in the figure

Solution

The coordinates of C.M. can be calculated as follows:

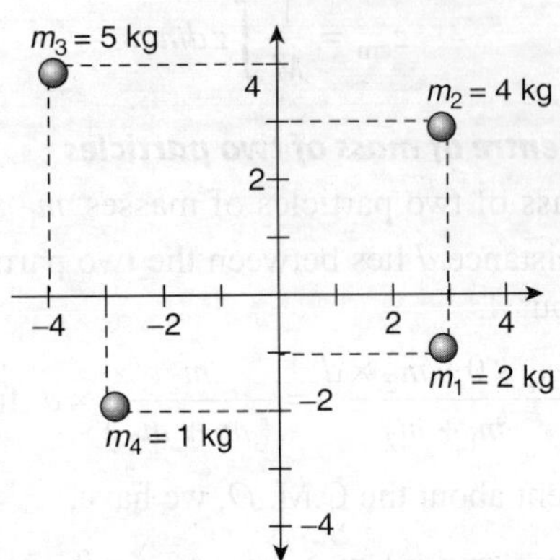

$$X_{cm} = \frac{x_1m_1 + x_2m_2 + x_3m_3 + x_4m_4}{m_1 + m_2 + m_3 + m_4}$$

$$= \frac{2\times 3 + 4\times 3 + 5\times(-4) + 1(-3)}{12}$$

$$= -\frac{5}{12}\text{ m} = -0.42 \text{ m}$$

$$Y_{cm} = \frac{m_1y_1 + m_2y_2 + m_3y_3 + m_4y_4}{m_1 + m_2 + m_3 + m_4}$$

$$= \frac{2(-1) + 4\times 3 + 5\times 4 + 1\times(-2)}{12}$$

$$= \frac{28}{12} = 2.3 \text{ cm}$$

The position vector of C.M. $\vec{r} = x_{cm}\hat{i} + y_{cm}\hat{j}$.

$$\vec{r} = \left(-0.42\hat{i} + 2.3\hat{j}\right) \text{ m}$$

Illustration 3

A thin rod of length $3L$ is bent at right angles at a distance L from one end. Locate the C.M. with respect to the corner. Take $L = 1.2$ m.

Solution

The centre of mass of each arm is at its mid-point. The C.M. of the two arms can be found by treating each arm as a point particle at its own C.M. From the figure, we see that

$$x_1 = \frac{L}{2},\ y_1 = 0 \text{ and } x_2 = 0,\ y_2 = L$$

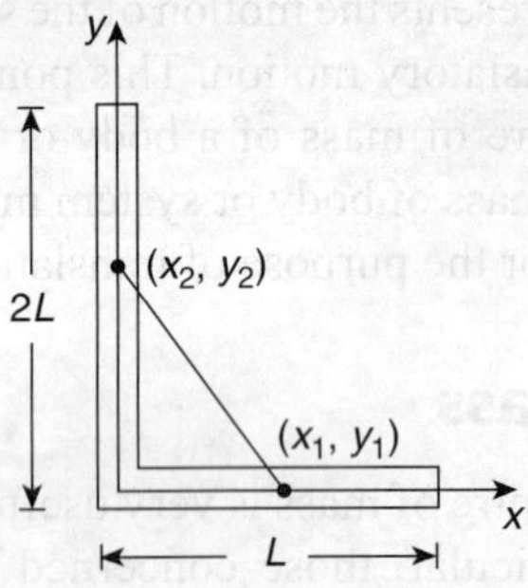

If we take $m_1 = m$ then $m_2 = 2$ m, we get

$$X_{cm} = \frac{m_1x_1 + m_2x_2}{3m} = \frac{m\left(\frac{L}{2}\right) + 2m(0)}{3m} = \frac{L}{6} = \frac{1.2}{6} = 0.2 \text{ m}$$

$$Y_{cm} = \frac{m_1y_1 + m_2y_2}{3m} = \frac{m(0) + 2m(L)}{3m} = \frac{2L}{3} = \frac{2\times 1.2}{3} = 0.8 \text{ m}$$

The position vector of C.M. is $\vec{r}_{cm} = 0.2\hat{i} + 0.8\hat{j}$ m

Illustration 4

Find the centre of mass of a uniform solid cone of height h and semi-angle θ as shown in figure.

Solution

Let us consider the apex of the cone as the origin and its axis along the y-axis. It is clear that the C.M. will lie along the y-axis. We divide the cone into disc of radius x and thickness dy. The volume of such a disc is

$$dV = \pi x^2 dy = \pi(y\tan\theta)^2 dy$$

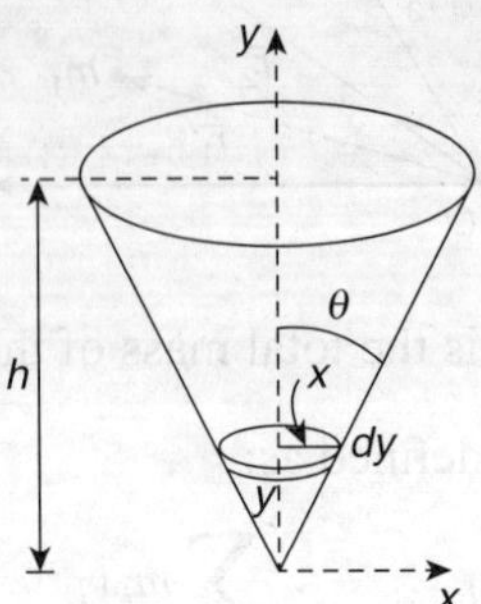

The mass of the disc is, $dm = \rho dy$ and the total mass of the cone

$$M = \int_0^h dm = \int_0^h \rho\pi(y\tan\theta)^2 dy$$

$$= \rho\pi\tan^2\theta\int_0^h y^2 dy$$

$$= \pi\rho\tan^2\theta\left[\frac{h^3}{3}\right] \quad \ldots(1)$$

The position of C.M. is given by

$$y_{cm} = \frac{\int y dm}{M}$$

$$\therefore y_{cm} = \frac{1}{M}\pi\rho\tan^2\theta\int_0^h y^3 dy$$

$$= \frac{1}{M}\pi\rho\tan^2\theta\left[\frac{h^4}{4}\right] \quad \ldots(2)$$

Using Eqs. (1) and (2), we find,

$$y_{cm} = \frac{3h}{4}$$

Motion of the Centre of Mass

Let us consider the motion of a system consisting of n particles of masses $m_1, m_2, \ldots, m_n$. The position of C.M. is defined by

$$r_{cm} = \frac{m_1\vec{r}_1 + m_2\vec{r}_2 + \ldots + m_n\vec{r}_n}{m_1 + m_2 + m_3 + \ldots + m_n} \quad \ldots(1)$$

or, $$M\vec{r}_{cm} = m_1\vec{r}_1 + m_2\vec{r}_2 + \ldots + m_n\vec{r}_n \quad \ldots(2)$$

where $$M = m_1 + m_2 + \ldots + m_n$$

Differentiating Eq. (2) w.r.t. time, we have

$$M\frac{d}{dt}\vec{r}_{cm} = m_1\frac{d}{dt}\vec{r}_1 + m_2\frac{d\vec{r}_2}{dt} + \ldots + m_n\frac{d\vec{r}_n}{dt}$$

$$M\vec{v}_{cm} = m_1\vec{v}_1 + m_1\vec{v}_2 + \ldots + m_n\vec{v}_n \quad \ldots(3)$$

or $$\vec{v}_{cm} = \frac{m_1\vec{v}_1 + m_2\vec{v}_2 + \ldots + m_n\vec{v}_n}{M} = \frac{\sum_{i=1}^{n} m_i\vec{v}_i}{M}$$

Further, $m\vec{v}$ represents the momentum of a particle P of mass m and velocity $\vec{v}$. Defining the momentum associated with the C.M. of the system $\vec{P}_{cm} = M\vec{v}_{cm}$, we get,

$$\vec{P}_{cm} = \vec{p}_1 + \vec{p}_2 + \ldots + \vec{p}_n = \sum_{i=1}^{n}\vec{P}_i$$

Differentiating Eq. (3) with respect to time t, we have,

$$m\frac{d\vec{v}_{cm}}{dt} = m_1\frac{d\vec{v}_1}{dt} + m_2\frac{d\vec{v}_2}{dt} + \ldots + m_n\frac{d\vec{v}_n}{dt}$$

$$M\vec{a}_{cm} = m_1\vec{a}_1 + m_2\vec{a}_2 + \ldots + m_n\vec{a}_n \quad \ldots(4)$$

$$a_{cm} = \frac{m_1\vec{a}_1 + m_2\vec{a}_2 + \ldots + m_n\vec{a}_n}{M} = \frac{\sum_{i=1}^{n} m_i\vec{a}_i}{M}$$

Newton's second law of motion states that $\vec{F} = m\vec{a}$, and so Eq. (4) can be rewritten as

$$\vec{F}_{cm} = \vec{F}_1 + \vec{F}_2 + \ldots + \vec{F}_n = \sum_{i=1}^{n}\vec{F}_i$$

Thus, we can conclude that for translation,

i. A body (system of particles) can be treated as a single particle of mass equal to itself and located at the centre of mass.

ii. Its motion is governed by the net external force acting on the system.

iii. Dynamical quantities like linear momentum may be associated with the centre of mass.

Important points regarding centre of mass

1. The centre of mass of a body may lie within or outside the body.

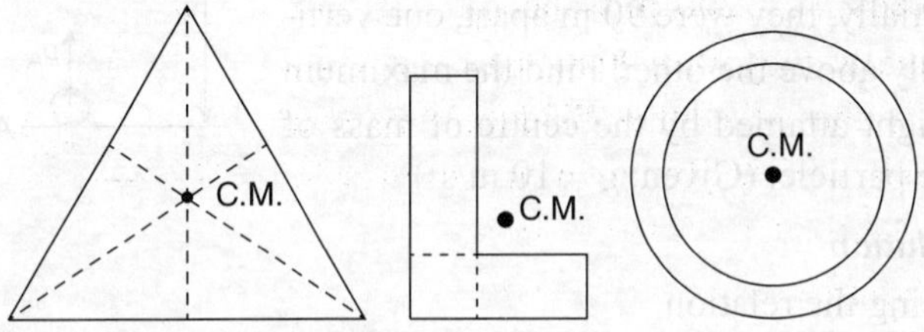

2. For symmetric bodies distribution of mass coincides with centre of symmetry or geometrical centre.
3. The center of mass of a system of extended bodies is defined by Eq. (1) where $\vec{r}_1, \vec{r}_2, \ldots, \vec{r}_n$ represent the C.M. of the respective bodies.
4. A frame of reference in which the C.M. of an isolated (i.e. a system not subjected to any external force) system of particles is at rest is known as the center of mass frame of reference. In this frame of reference, the position vector of C.M. is zero velocity and hence momentum associated with C.M. is zero.

Illustration 5

Two blocks of masses 5 kg and 2 kg are placed on a frictionless surface and connected by a spring. An external impulse imparts a velocity of 14 m/s to the heavier block in the direction of lighter one. Deduce

(1) the velocity gained by the centre of mass and

(2) the separate velocities of the two blocks with respect to the center of mass (C) frame just after the impulse is delivered.

Solution

Let us calculate by taking the direction of impulse imparted as positive.

(1) By definition

$$\vec{v}_{cm} = \frac{m_1 v_1 + m_2 v_2}{m_1 + m_2}$$

Here, $v_1 = 14$ m/s; $v_2 = 0$, $m_1 = 5$ kg and $m_2 = 2$ kg.

So, $$v_{cm} = \frac{5 \times 14 + 2(0)}{7} = 10 \text{ m/s}.$$

(2) The C-frame is one in which C.M. is at rest. So the velocity of the heavier block in this frame just after the impulse is delivered as

$$v_1' = v_1 - v_{cm} = 14 - 10 = 4 \text{ m/s}$$

and the lighter block is

$$v_2'' = v_2 - v_{cm} = 0 - 10 = -10 \text{ m/s}$$

i.e., in the C-frame, the block of 5 kg and 2 kg will approach each other with speed 4 m/s and –10 m/s respectively.

Illustration 6

Two particles A and B of mass 1 kg and 2 kg respectively, are projected towards each other as shown in figure with speeds $u_A = 200$ m/s and $u_B = 50$ m/s. Initially, they were 90 m apart, one vertically above the other. Find the maximum height attained by the centre of mass of the particle. (Given: $g = 10$ m/s²)

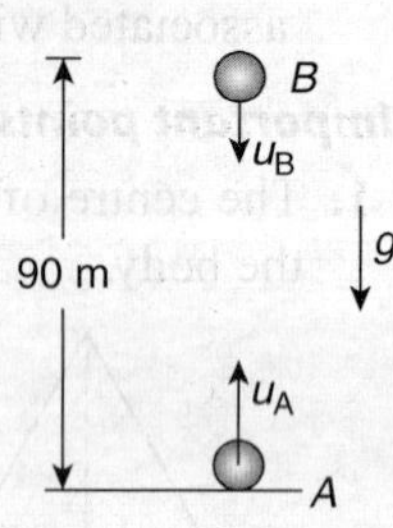

Solution

Using the relation,

$$m_A r_A = m_B r_B$$

$$r_A = 2r_B \quad \ldots(1)$$

and $$r_A + r_B = 90 \text{ m} \quad \ldots(2)$$

On solving these two equations,

$$r_A = 60 \text{ m and } r_B = 30 \text{ m}$$

i.e., C.M. is at height 60 m from the ground at time $t = 0$.

Further, $$a_{cm} = \frac{m_A a_A + m_B a_B}{m_A + m_B}$$

$$= \frac{1(g) + 2(g)}{3} = g \text{ (downwards)}$$

As $$a_A = a_B = g \text{ (downwards)},$$

$$u_{cm} = \frac{m_A u_A + m_B u_B}{m_A + m_B}$$

$$= \frac{1(200) - 2(50)}{1+2} = \frac{100}{3} \text{ m/s (upwards)}.$$

Let h be the height attained by C.M. beyond 60 m.

$$v_{cm}^2 = u_{cm}^2 + 2a_{cm} h$$

$$0 = \left(\frac{100}{3}\right)^2 - 2(10)h$$

$$h = 55.55 \text{ m}$$

Therefore, maximum height attained by the centre of mass is

$$H = 60 + 55.55 = 115.55 \text{ m}$$

Torque

The torque due to a force (or forces) describes the tendency of the force to cause rotation or alter the state of rotational motion of the body on which it acts. Torque due to a force $\vec{F}$ acting at A about an axis is given by,

$$\vec{\tau} = \vec{r}_A \times \vec{F}$$

where $\vec{r}_A$ is the position vector of the point A where the force F acts on the body. The origin of coordinates is chosen to lie on the axis. Newton's second law leads to the expression,

$$\tau = I\alpha$$

where I may be treated like a scalar for symmetric bodies and the axis of rotation is parallel to an axis of symmetry.

Angular Momentum and Conservation

Angular momentum of a rigid body undergoing pure rotation about a fixed axis is given by

$$\vec{L} = I\vec{\omega}$$

where I is the moment of inertia of the body about the axis of rotation and ω, its angular velocity. It is related to torque much like linear momentum is related to force:

$$\vec{F} = \frac{d\vec{p}}{dt}$$

$$\vec{\tau} = \frac{d\vec{L}}{dt}$$

Angular momentum is a vector quantity and it is directed along $\vec{\omega}$, provided I can be treated as a scalar (i.e., the rotation is along one of the axes of symmetry).

Angular momentum is conserved if the net torque acting on a body about the given axis is zero.

$$\vec{\tau} = \frac{d\vec{L}}{dt} = 0$$

$$\vec{L} = I\vec{\omega}$$

Angular Impulse

The torque due to an impulsive force when integrated over the time of action of the force gives a quantity known as angular impulse. Mathematically, it is useful in the solution of problems involving impulsive forces acting on extended bodies, tending to cause rotation and is given as $\vec{J}_\theta = \int \vec{\tau}\,dt$.

Illustration 7

A symmetrical body is rotating about its axis of symmetry, its moment of inertia about the axis of rotation being 1 kg-m^2 and its rate of rotation 2 rev./sec. (a) What is its angular momentum? (b) What additional work will have to be done to double its rate of rotation?

Solution

(a) As the body is rotating about its axis of symmetry, the angular momentum vector coincides with the axis of rotation.

$\therefore$ Angular momentum $L = I\omega$ ...(i)

Kinetic energy of rotation $E = \frac{1}{2}I\omega^2$

or $2E = I\omega^2$ from which, we get,

$\therefore$ $I\omega = \left(\sqrt{2IE}\right)$ ...(ii)

From Eqs. (i) and (ii), $L = \sqrt{2IE}$ ...(iii)

$\omega = 2$ rev/sec $= 2\times 2\pi$ or 4π radian s^{-1}

$\therefore$ $E = \frac{1}{2}\times 1\times (4\pi)^2 = 8\pi^2$ joules.

Now, $L = \sqrt{(2IE)} = \sqrt{(2\times 1\times 8\pi^2)}$

$= \sqrt{(16\pi^2)} = 4\pi$ joule $= 12.57$ kg.m^2/s

(b) When the rate of rotation is doubled, i.e,. 4 rev/sec or 8π radians/sec the kinetic energy of rotation is given by

$$E = \frac{1}{2}\times 1\times (8\pi)^2 = 32\pi^2 \text{ joules}$$

Additional work required = Final KE of rotation – Initial KE of rotation

$$= 32\pi^2 - 8\pi^2 = 24\pi^2 = 236.8 \text{ joules}$$

Illustration 8

A uniform flat disc of mass M and radius R rotates about a horizontal axis through its centre with angular speed ω_0. (a) What is its kinetic energy? Its angular momentum? (b) A chip of mass m breaks off the edge of the disc at an instant such that the chip rises vertically about the point at which it broke off. How high above the point does it rise before starting to fall? (c) What is the final angular momentum and energy of the disc?

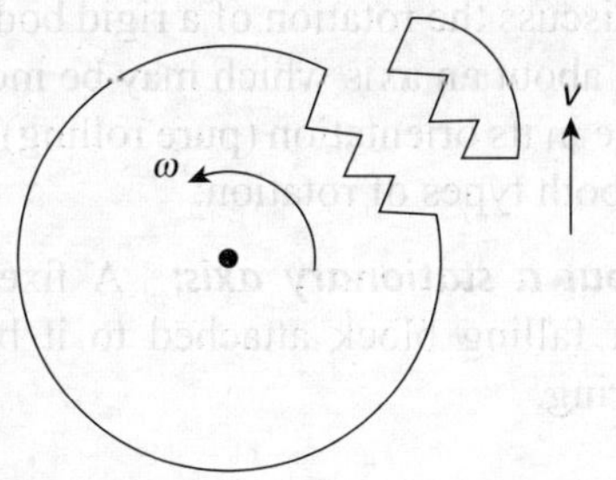

Solution

(a) The kinetic energy of rotation T_R is given by

$$T_R = \frac{1}{2}I\omega^2$$

Here, $I = \frac{1}{2}MR^2$ and $\omega = \omega_0$

$$T_R = \frac{1}{2}MR^2\omega_0^2$$

The angular momentum, $L = I\omega$

or $L = \frac{1}{2}MR^2\omega_0$

(b) We know that $v = r\omega$.

or $v = R\omega_0$

If the particle rises to a height h, then

$$v^2 = 2gh \text{ or } h = \frac{v^2}{2g} = \frac{R^2\omega_0^2}{2g}$$

(c) If the angular momentum of remainder be L_R, then applying the law of conservation of angular momentum we have

$$L_R + mR^2\omega_0 = \frac{1}{2}MR^2\omega_0$$

$$L_R = \left(\frac{1}{2}M - m\right)R^2\omega_0^2$$

Similarly, applying the law of conservation of energy, we get

$$(T_R)_R + \frac{1}{2}mR^2\omega_0^2 = \frac{1}{4}MR^2\omega_0^2$$

$$(T_R)_R = \frac{1}{2}\left[\left(\frac{1}{2}M - m\right)\right]R^2\omega^2$$

Illustration 9

A uniform rod of mass M and length a lies on a smooth horizontal plane. A particle of mass m moving at a speed v perpendicular to the length of the rod strikes it at a distance $\frac{a}{4}$ from the centre and stops after the collision. Find (a) the velocity of the centre of the rod and (b) the angular velocity of the rod after the collision.

Solution

We take the rod and the particle together as a single system. As there is no resultant force on the system as a whole, the linear momentum of the system will remain constant. Further, there is no resultant external torque on the system and so the angular momentum of the system about any point remains constant.

Suppose the velocity of the centre of the rod is V and the angular velocity about the centre is ω (see figure) after the collision.

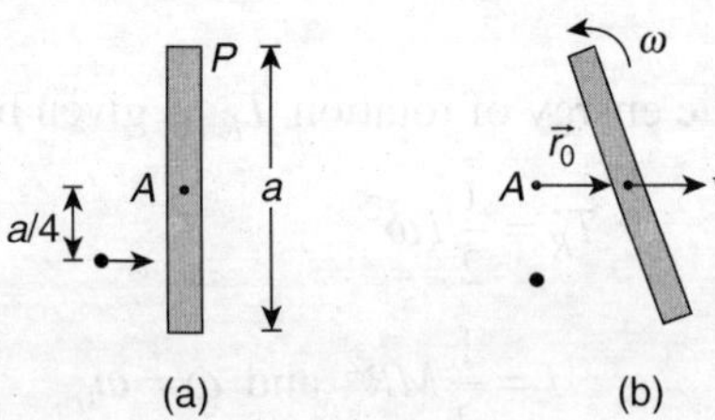

(a) The linear momentum before the collision is mv and that after the collision is MV. Thus,

$$mv = MV, \text{ or } V = \frac{m}{M}v$$

(b) Consider the angular momentum of 'the rod plus the particle' system about the mid-point of the rod, A. Initially the rod is at rest. The angular momentum of the particle about A is

$$L = mv\left(\frac{a}{4}\right)$$

After the collision, the particle comes to rest.

Thus, $\quad \vec{L} = \vec{L}_{cm}$

Hence the angular momentum of the rod about A is

$$L = I\omega = \frac{Ma^2}{12}\omega$$

Thus, $\quad \dfrac{Mva}{4} = \dfrac{Ma}{12}\omega$ or, $\omega = \dfrac{3mv}{Ma}$

Illustration 10

A disc of mass M and radius R is rolling with angular speed ω on a horizontal plane as shown in the figure.

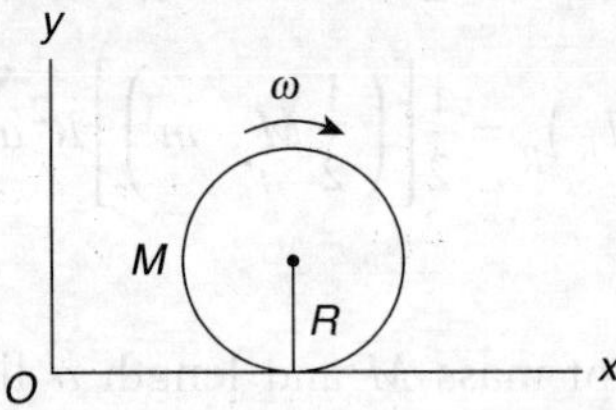

The magnitude of angular momentum of the disc about the origin O is

(a) $\left(\frac{1}{2}\right)MR^2\omega$ (b) $MR^2\omega$

(c) $\left(\frac{3}{2}\right)MR^2\omega$ (d) $2MR^2\omega$

Solution

The angular momentum of a body $\vec{L}$ may be expressed as the sum of two parts,

(a) one arising from the motion of the centre of mass of the body $\left(\vec{L}_{\text{orbital}}\right)$

(b) the other from the motion of the body with respect to its centre of mass $\left(\vec{L}_{\text{spin}}\right)$

ie., $\quad \vec{L}_{\text{total}} = \vec{L}_{C.M} + \vec{r}_{C.M} \times \vec{p}$

$\Rightarrow \quad \vec{L}_{\text{total}} = \vec{L}_{C.M.} + M\left(\vec{r}_{C.M} \times \vec{v}_{C.M}\right)$

For this problem,

$$L_{C.M} = I\omega = \frac{1}{2}MR^2\omega \text{ and}$$

$$M\left(\vec{r}_{C.M} \times \vec{v}_{C.M}\right) = MRv_{C.M.} = MR(R\omega)$$

$\Rightarrow \quad M\left(\vec{r}_{C.M} \times \vec{V}_{C.M}\right) = MR^2\omega$

$\Rightarrow \quad L_{\text{total}} = \dfrac{1}{2}MR^2\omega + MR^2\omega = \dfrac{3}{2}MR^2\omega$

Ans. (c)

Rigid Body

A body with a definite and unchanging shape and size is called a rigid body.

In practice, such a body does not exist. A real body undergoes changes in shape and size under the action of external forces. If however, these forces are small enough and we are dealing with solid bodies, changes in shape or size are insignificant. Under these circumstances, we may regard the body as a rigid body.

The motion of a rigid body in three dimensions in general is very complicated. The analysis of this motion can be done in terms of two simple types of motion: translation and rotation.

We define these two types of motion as below:

(a) Translation

A rigid body is said to undergo pure translation if it moves such that all its particles undergo the same displacement during any interval of time.

(b) Rotation

A rigid body is said to undergo pure rotation if it moves such that its constituent particles move along circular arcs, the centres of which lie on a straight line perpendicular to the plane of the arcs. This straight line is known as the axis of rotation, and the rigid body is said to undergo pure rotation about this axis.

Axis of rotation

The axis of rotation may be fixed in space (stationary axis) or moving (changing in position or direction). In this unit, we are going to discuss the rotation of a rigid body about a fixed axis, and also about an axis which may be moving but without any change in its orientation (pure rolling). The following figures show both types of rotation:

Rotation about a stationary axis: A fixed pulley with a falling block attached to it by means of a string.

Rotation about a translating axis: A cylinder rolling on an inclined plane.

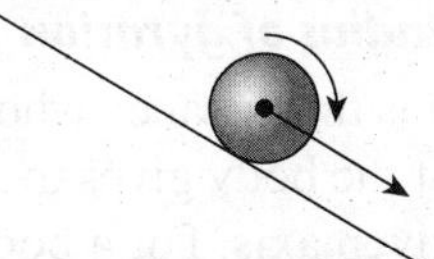

Equilibrium of Rigid Bodies

For equilibrium of rigid bodies, the linear acceleration of its centre of mass is zero and also the angular acceleration of the body about any fixed axis in the frame of reference is zero.

For the equilibrium of a rigid body, it is not necessary that the rigid body is at rest. However, if the rigid body is at rest, then the equilibrium of the rigid body is called static equilibrium.

For equilibrium of rigid bodies the following two conditions should be satisfied.

1. The translational motion of the centre of mass of a rigid body is governed by the following equation.

$$\sum F_{\text{ext.}} = \frac{d}{dt}(p)$$

Here, $\sum F_{\text{ext.}}$ is the sum of all the external forces acting on the rigid body.

For equilibrium, p must have a constant value.

$$\therefore \quad \frac{d}{dt}(\vec{p}) = 0$$

$$\therefore \quad \sum \overrightarrow{F_{\text{ext.}}} = 0$$

This vector equation is equivalent to the three scalar equations

$$\sum F_x = 0, \sum F_y = 0 \text{ and } \sum F_z = 0$$

"The vector sum of all the external forces acting on the rigid body must be zero".

2. The rotational motion of a rigid body is governed by the following equation.

$$\sum \vec{\tau}_{\text{ext.}} = \frac{d\vec{L}}{dt}$$

Here $\sum \vec{\tau}_{\text{ext.}}$ represents the sum of all the external torque acting on the body

For equilibrium, $\vec{L}$ must have a constant value.

$$\therefore \quad \frac{d}{dt}\vec{L} = 0$$

$$\therefore \quad \sum \vec{\tau}_{\text{ext.}} = 0$$

This vector equation can be written as three scalar equations.

$$\sum \tau_x = 0, \sum \tau_y = 0, \sum \tau_z = 0$$

"The vector sum of all the external torques acting on the rigid body must be zero"

Analogy between Translational and Rotational Kinematics

Linear motion of particle	Rotational motion (of particle about fixed axis)
$\vec{v} = \frac{d\vec{r}}{dt}$	$\vec{\omega} = \frac{d\theta}{dt}\hat{n}$
$\vec{a} = \frac{d\vec{v}}{dt}$	$\vec{\alpha} = \frac{d\vec{\omega}}{dt}\hat{n}$
$v = u + at$	$\omega = \omega_o + \alpha t$
$x = ut + \frac{1}{2}at^2$	$\theta = \omega_0 t + \frac{1}{2}\alpha t^2$
$v^2 = u^2 + 2ax$	$\omega^2 = \omega_0^2 + 2\alpha\theta$

Illustration 11

A wheel rotates for 5 second with constant angular acceleration and describes 100 radians during the time. It then rotates with constant angular velocity and during the next 5 seconds describes 80 radians. Find the angular acceleration.

Solution

For the last five seconds, the wheel describes 80 radians. Let the constant angular velocity be ω. Then

$$\omega = \frac{80}{5} = 16 \text{ radians /sec}$$

This is angular velocity at the end of first interval. Let ω_0 be the angular velocity at the start. Further, let α be the angular acceleration. So, for the first five seconds, we have

Initial velocity ω_0, Acceleration α, Time 5 sec. and Final velocity 16 radians/sec.

The wheel describes 100 radians during this time

$$16 = \omega_0 + 5\alpha$$

$$100 = 5\omega_0 + \frac{1}{2}\alpha \times 25 \quad \text{...(1)}$$

$$80 = 5\,\omega_0 + 25\alpha \quad \text{...(2)}$$

From Eqs. (1) and (2), we get

$$20 = -\frac{25}{2}\alpha \text{ or } \alpha = -\frac{40}{25} = -\frac{8}{5} = -1.6 \text{ rad/sec}^2$$

Illustration 12

Starting from rest, a fan takes 10 seconds to attain the maximum speed of 600 rpm (revolutions per minute). Assuming constant acceleration, find the time taken by the fan in attaining half the maximum speed.

Solution

Let the angular acceleration be α.

$$\omega = \omega_o + \alpha t$$

$$600 \text{ rev/min} = 0 + \alpha 10 \quad \ldots\text{(i)}$$

Let t be the time taken in attaining the speed of 300 rev/min which is half the maximum.

Then,

$$300 \text{ rev/min} = 0 + \alpha t \quad \ldots\text{(ii)}$$

Dividing (i) by (ii), we get,

$$2 = \frac{10}{t}$$

or, $$t = 5 \text{ s}$$

Moment of Inertia

It is the property of a body due to which it opposes any change in its state of rest or of uniform rotation. Analytically for a particle of mass m rotating in circle of radius r, moment of inertia about the axis of rotation is given as

$$I = mr^2 \quad \ldots(1)$$

(B) Discrete system (C) Continuous body

So, for a body made up of number of particles of masses m_1, m_2..... etc., at positions r_1, r_2..... etc., respectively from the axis of rotation:

$$I = m_1r_1^2 + m_2r_2^2 + ... = \Sigma mr^2 \quad \ldots(2)$$

While for a continuous distribution of mass, treating the element of mass dm at position r as particle

$$dI = dm\, r^2, \text{ i.e., } I = \int r^2 dm \quad \ldots(3)$$

Regarding moment of inertia, it is worth noting that

(1) It has dimensions $\left[ML^2\right]$ and SI unit kg m^2.

(2) Moment of inertia for a given body depends on the axis of rotation. So if axis of rotation changes, usually moment of inertia will change

(3) Moment of inertia for a given axis depends on mass, shape and size of the body. If same mass is casted in different shapes with same axis, moment of inertia will be different.

Radius of gyration

It is the distance whose square when multiplied with the mass of the body gives the moment of inertia of the body about the given axis. For a body mass M, if k is radius of gyration

$$I = Mk^2 \text{ or } k = \sqrt{(I/M)} \quad \ldots(4)$$

Through this concept, a real body is replaced by a point mass for dealing its rotational motion, e.g., in case of a disc rotating about an axis through its centre of mass and perpendicular to its plane.

$$k = \sqrt{\frac{I}{M}} = \sqrt{\frac{(1/2)MR^2}{M}} = \frac{R}{\sqrt{2}}$$

So, instead of disc we can assume a point mass M at a distance $(R/\sqrt{2})$ from the axis of rotation for dealing the rotational motion of the disc.

Kinetic Energy and Moment of Inertia

Suppose that a rigid body is undergoing pure rotation about a fixed axis in an inertial frame of reference, with an angular velocity ω. The particles of the rigid body are numbered from 1, 2, 3, ..., n. Let the mass of the ith particle be given by m_i and its distance from the axis of rotation be given by ri, then the speed of the particle is: $v_i = \omega r_i$ and its kinetic energy is:

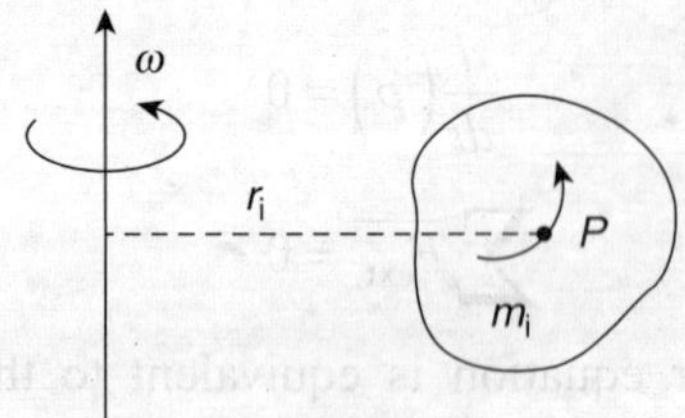

$$\text{KE} = \frac{1}{2}m_i v_i^2 = \frac{1}{2}m_i\left(\omega r_i\right)^2$$

The kinetic energy of the entire body is:

$$\text{KE} = \sum_i \frac{1}{2}m_i\,(\omega r_i)$$

$$= \frac{1}{2}\left(\sum_i m_i r_i^2\right)\omega^2 = \frac{1}{2}I\omega^2$$

Illustration 13

Three point of masses 2 kg, 3 kg and 7 kg respectively are located at the vertices of an equilateral triangle of side 6 m. What is the moment of inertia of the system about an axis along an altitude of the triangle passing through the 2 kg mass?

Solution

Moment of inertia about altitude AD is

$$I = m_1 \times \left(\text{distance of } m_1 \text{ from } AD\right)^2 + m_2 \times \left(\text{distance of } m_2 \text{ from } AD\right)^2 + m_3 \times \left(\text{distance of } m_3 \text{ from } AD\right)^2$$

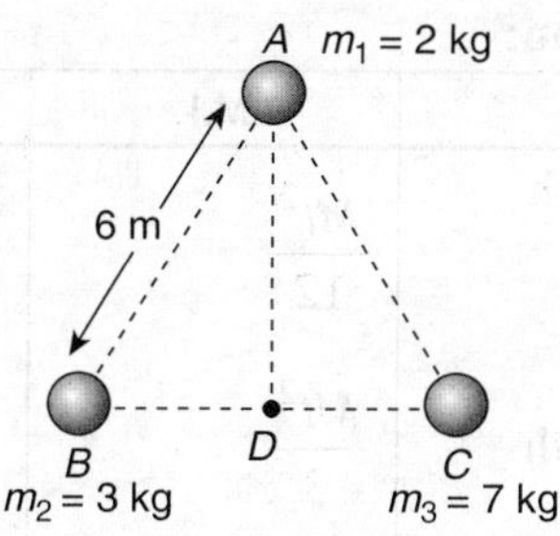

$$= m_1 \times 0 + m_2 \times (BD)^2 + m_3 \times (CD)^2$$

$$= 0 + m_2 \times \left(\frac{a}{2}\right)^2 + m_3\left(\frac{a}{2}\right)^2$$

$$= (m_2 + m_3)\frac{a^2}{4} = (3+7)\frac{6^2}{4} = 90 \text{ kg m}^2$$

Illustration 14

Find the radius of gyration of a uniform rod of length ℓ and mass m about an axis passing through the centre of mass of the rod and making an angle θ with the rod.

Solution

The moment of inertia of the rod about the given axis can be easily found by integration:

$$I = \int_{-\frac{\ell}{2}}^{\frac{\ell}{2}} \left(\frac{m}{\ell}dx\right)(x\sin\theta)^2 = \frac{1}{12}m\ell^2\sin^2\theta = mk^2$$

Parallel Axis Theorem

The moment of inertia of a rigid body about any given axis equals the moment of inertia about a parallel axis passing through its centre of mass plus the moment of inertia of a particle of equal mass about the given axis:

$$I_{\text{axis}} = I_{C.M.} + Md^2$$

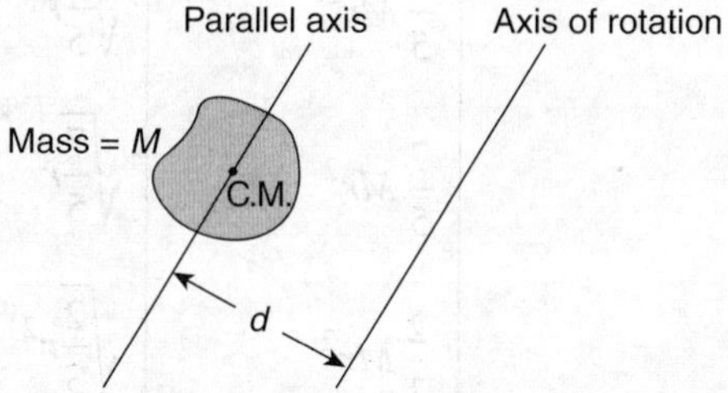

Here, M is the mass of the body and d is the distance between the two axes.

Illustration 15

Four solid spheres, each of mass m and radius r are located with their centres on four corners of a square of side a. What is the moment of inertia of the system about an axis (a) along any diagonal and (b) along any side of the square?

Solution

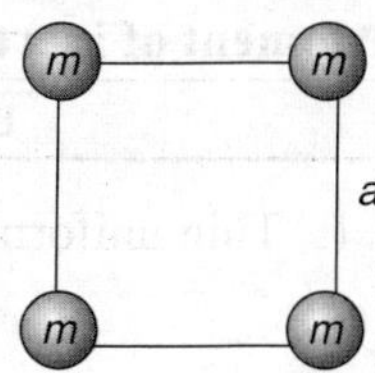

(a) Refer to the adjacent figure. The moment of inertia of spheres A and B about their common diameter $AB = \frac{2}{5}mr^2$ each. Also, the moment of inertia of spheres C and D about an axis passing through their centre and parallel to $AB = \frac{a}{\sqrt{2}}$. From the parallel axis theorem, the moment of inertia of spheres C and D about diagonal AB is

$$\frac{2}{5}mr^2 + m(CO)^2 = \frac{2}{5}mr^2 + m\left(\frac{a}{\sqrt{2}}\right)^2 = \frac{2}{5}mr^2 + \frac{1}{2}ma^2$$

Hence the moment of inertia (MI) of the system of four spheres about diagonal AB is

I_{AB} = MI of C about AB + MI of B about AB + MI of C about AB + MI of D about AB.

$$= \frac{2}{5}mr^2 + \frac{2}{5}mr^2 + \frac{2}{5}mr^2 + \frac{1}{2}ma^2 + \frac{2}{5}mr^2 + \frac{1}{2}ma^2$$

$$= \frac{8}{5}mr^2 + ma^2 = m\left(\frac{8r^2}{5} + a^2\right)$$

(b) Moment of inertia of sphere A about side AD = Moment of inertia of sphere D about side $AD = \frac{2}{5}mr^2$. Using the parallel axis theorem, moment of inertia of sphere C about AD = moment of inertia of sphere B about $AD = \frac{2}{5}mr^2 + ma^2$. Hence the moment of inertia of the system of four spheres about side AD is I_{AD} = MI of A about AD + MI of D about AD + MI of B about AD + MI of C about AD

$$= \frac{8}{5}mr^2 + 2ma^2 = m\left(\frac{8r^2}{5} + 2a^2\right)$$

Perpendicular Axis Theorem

The moment of inertia of a plane lamina about an axis perpendicular to the plane of the lamina is equal to the sum of its moments of inertia about two mutually perpendicular concurrent axes passing through the plane of the lamina:

$$I_z = I_x + I_y$$

Moment of inertia of bodies of regular shapes about particular axes of rotation:

Body	Axis passing through	M.I.	K
1. Thin uniform rod of length l	(i) Its C.M. and perpendicular to its length	$\frac{Ml^2}{12}$	$\frac{l}{2\sqrt{3}}$
	(ii) One end and perpendicular to its length	$\frac{Ml^2}{3}$	$\frac{l}{\sqrt{3}}$
2. Thin rectangular lamina of length l and breadth b	(i) Its C.M. perpendicular to its length	$\frac{Ml^2}{12}$	$\frac{l}{2\sqrt{3}}$
	(ii) Its C.M. and perpendicular to its breadth	$\frac{Mb^2}{12}$	$\frac{b}{2\sqrt{3}}$
	(iii) Its C.M. and perpendicular to its plane	$\frac{M\left(l^2+b^2\right)}{12}$	$\frac{\sqrt{l^2+b^2}}{2\sqrt{3}}$
3. Circular ring of radius r	(i) Its C.M. and perpendicular to its plane	Mr^2	r
	(ii) Its diameter	$\frac{Mr^2}{2}$	$\frac{r}{\sqrt{2}}$
	(iii) About a tangent its plane	$\frac{3}{2}Mr^2$	$\sqrt{\frac{3}{2}}r$
	(iv) About a tangent perpendicular to its plane.	$2Mr^2$	$\sqrt{2}r$
4. Circular disc of radius r	(i) Its C.M. and perpendicular to its plane	$\frac{Mr^2}{2}$	$\frac{r}{\sqrt{2}}$
	(ii) About a diameter	$\frac{Mr^2}{4}$	$\frac{r}{\sqrt{2}}$
	(iii) About a tangent in its plane	$\frac{5}{4}Mr^2$	$\frac{\sqrt{5}}{2}r$
	(iv) About a tangent perpendicular to its plane	$\frac{3}{2}Mr^2$	$\sqrt{\frac{5}{2}}r$
5. Solid sphere of radius r	(i) About a diameter	$\frac{2}{5}Mr^2$	$\sqrt{\frac{2}{5}}r$
	(ii) About a tangent	$\frac{7}{5}Mr^2$	$\sqrt{\frac{7}{5}}r$
6. Thin spherical shell of radius r	(i) About a diameter	$\frac{2}{3}Mr^2$	$\sqrt{\frac{2}{3}}r^2$
	(ii) About a tangent	$\frac{5}{3}Mr^2$	$\sqrt{\frac{5}{3}}r$

Body	Axis passing through	M.I.	K
7. Hollow cylinder of length l and radius r	(i) Its C.M. and perpendicular to its end faces	Mr^2	r
	(ii) Its C.M. perpendicular to its length	$M\left(\frac{r^2}{2}+\frac{l^2}{12}\right)$	$\sqrt{\left(\frac{r^2}{2}+\frac{l^2}{12}\right)}$
8. Solid cylinder of length l and radius r	(i) Its C.M. and perpendicular to its end faces	$\frac{Mr^2}{2}$	$\frac{r}{\sqrt{2}}$
	(ii) Its C.M. and perpendicular to its length	$M\left(\frac{l^2}{12}+\frac{r^2}{4}\right)$	$\sqrt{\frac{l^2}{12}+\frac{r^2}{4}}$

Illustration 16

The moment of inertia of a uniform circular disc of mass M and radius R about its axis is $\frac{1}{2}MR^2$. What is the moment of inertia of the disc about (i) a diameter (ii) an axis tangential to its circumference?

Solution

The moment of inertia of the disc about its axis is $\frac{1}{2}MR^2$.

Let us consider two perpendicular diameters, one along the x-axis and the other along the y-axis.

Then, $I_x = I_y$

(i) According to the theorem of perpendicular axes, the moment of inertia of the disc about an axis passing through its centre and normal to its plane is given by

$$I_C = I_x + I_y \text{ or } \frac{1}{2}MR^2 = 2I_x \text{ or } I_x = \frac{1}{4}MR^2$$

(ii) Since the disc is uniform, its centre of mass coincides with its centre. Therefore, the moment of inertia of the disc about an axis tangential to its circumference is

$$I_t = I_{C.M.} + Md^2$$
$$= I_{C.M.} + MR^2$$
$$= \frac{1}{2}MR^2 + MR^2$$
$$= \frac{3}{2}MR^2$$

Illustration 17

Let I be the moment of inertia of a uniform square plate about an axis AB that passes through its centre and is parallel to two of its sides. CD is a line in the plane of the plate that passes through the centre of the plate and makes an angle θ with AB. The moment of inertia of the plate about the axis CD is then equal to

(a) I (b) $I\sin^2\theta$

(c) $I\cos^2\theta$ (d) $I\cos^2\left(\frac{\theta}{2}\right)$

Solution

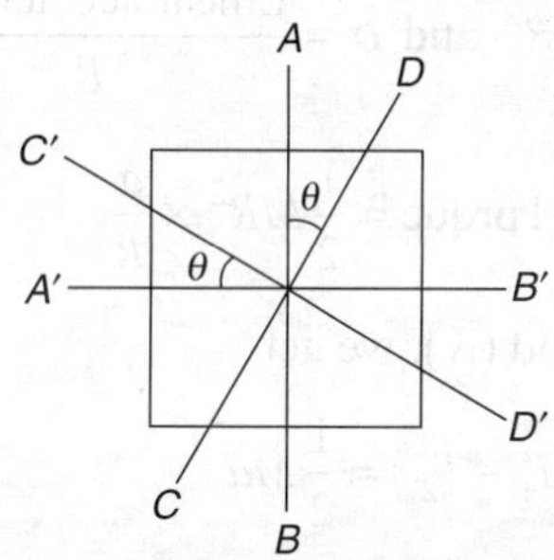

$$I_{AB} = I_{A'B'} = I \text{ and } I_{CD} = I_{C'D'}$$

If I_0 be the moment of inertia of the square plate about an axis passing through O and perpendicular to the plate, then by the perpendicular axis theorem

$$I_0 = I_{AB} + I_{A'B'} = 2I_{AB} \quad \dots (1)$$

or $$I_0 = I_{CD} + I_{C'D'} = 2I_{CD} \quad \dots (2)$$

From ((1) and (2))

$$I_{CD} = I_{AB} = I$$

Ans. (a)

Illustration 18

Two masses m_1 = 15 kg and m_2 = 10 kg are attached to the ends of a cord which passes over the pulley of an Atwood's machine. The mass of the pulley is M = 10 kg and its radius is R = 0.1 m. Calculate the tension in the cord, the acceleration a of the system and the number of revolutions made by the pulley at the end of 2 seconds from the start. There is sufficient friction to prevent slipping between the cord and the pulley. (Take g = 9.8 m/s²)

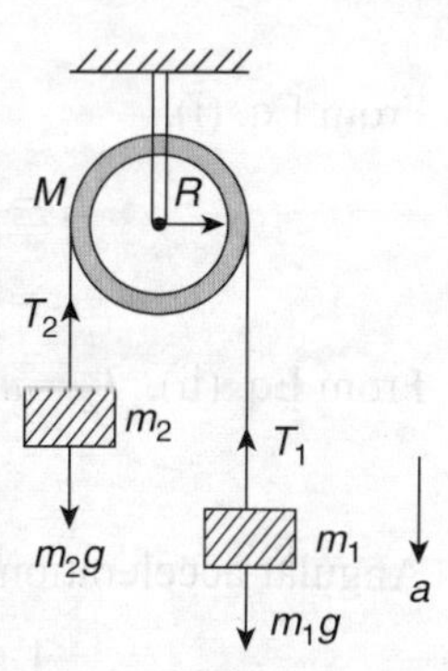

Solution

The figure shows an Atwood's machine. As the pulley has a finite mass, the two tensions T_1 and T_2 are not equal. If a be the linear acceleration of the system, then according to Newton's laws, we get,

$$m_1 g - T_1 = m_1 a \quad \text{...(i)}$$

$$T_2 - m_2 g = m_2 a \quad \text{...(ii)}$$

Net torque on the pulley in the clockwise direction,

$$= (T_1 - T_2)R \quad \text{...(iii)}$$

We know that,

Torque = Moment of inertia × Angular acceleration

$$= I \times \alpha$$

where, $I = \frac{1}{2}MR^2$ and $\alpha = \frac{\text{Linear acceleration}}{R} = \frac{a}{R}$

$$\text{Torque} = \frac{1}{2}MR^2 \times \frac{a}{R} \quad \text{...(iv)}$$

From Eqs. (iii) and (iv), we get

$$(T_1 - T_2) = \frac{1}{2}Ma \quad \text{...(v)}$$

Adding Eqs. (i) and (ii), we have

$$(m_1 - m_2)g = (m_1 + m_2)a + (T_1 - T_2)$$

Substituting the value of $(T_1 - T_2)$ from Eq. (v), we get,

$$(m_1 - m_2)g = (m_1 + m_2)a + \frac{1}{2}Ma$$

$$\therefore \quad a = \frac{(m_1 - m_2)g}{m_1 + m_2 + \frac{1}{2}M} \quad \text{...(vi)}$$

Given that,

$$m_1 = 15 \text{ kg}, m_2 = 10 \text{ kg and } M = 10 \text{ kg}$$

$$\therefore \quad \alpha = \frac{(25-10)\times 9.8}{15+10+5}.$$

From Eq. (i),

$$T_1 = m_1 g - m_1 a = m_1(g - a)$$
$$= 15(9.8 - 1.63) = 122.55 \text{ N}$$

From Eq. (ii), $T_2 = m_2(g + a)$

$$= 10 \times (9.8 + 1.63) = 114.3 \text{ N}$$

Angular acceleration $\alpha = 16.3$ radian/s^2

$$\theta = \frac{1}{2}(16.3)(2)^2 = 32.6 \text{ radians}$$

$$\therefore \text{ Number of revolutions } = \frac{32.6}{2\pi} = \mathbf{5.2}.$$

Illustration 19

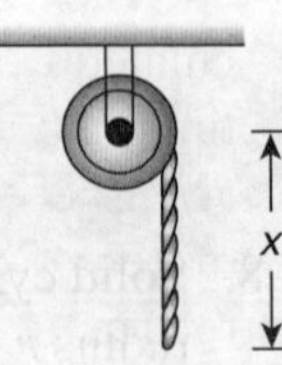

A uniform cylinder of radius R and mass M can rotate freely about a stationary horizontal axis O (see figure). A thin cord of length ℓ and mass m is wound on the cylinder in a single layer. Find the angular acceleration of the cylinder as a function of the length x of the hanging part of the cord. The wound part of the cord may be supposed to have its centre of gravity on the cylinder-axis.

Solution

Let m' be the mass of hanging part of the cord.

$$m' = \frac{m}{\ell}x$$

Let the cord be descending with an acceleration a. Then

$$\left(\frac{m}{\ell}xg\right) - T = \left(\frac{m}{\ell}x\right)a \quad \text{...(i)}$$

$$\tau = TR = I\alpha$$

where $\quad I = \frac{MR^2}{2} + \frac{m}{\ell}(\ell - x)R^2$ and $\alpha = \frac{a}{R}$

$$\therefore \quad TR = \left[\frac{MR^2}{2} + \frac{m}{\ell}(\ell - x)R^2\right]\frac{a}{R}$$

or $$T = \left[\frac{M}{2} + \frac{m}{\ell}(\ell - x)\right]a \quad \text{...(ii)}$$

From Eqs. (i) and (ii) we get

$$\left(\frac{m}{\ell}xg\right) - \left[\frac{M}{2} + \frac{m}{\ell}(\ell - x)\right]a = \left(\frac{m}{\ell}x\right)a$$

or $$\left(\frac{mxg}{\ell}\right) = \left[\frac{m}{\ell}x + \frac{M}{2} + \frac{m}{\ell}(\ell - x)\right]a$$

Solving we get,

$$a = \frac{mxg}{\ell\left(m + \frac{1}{2}M\right)}$$

Rolling Motion

When a disc rolls without sliding, there is no relative motion between contact point of the disc and ground. A rolling disc may be assumed to rotate about contact point, which we call instantaneous centre of rotation. A rolling disc may be compared with a block at rest on a surface. The magnitude of friction force on static block can vary from 0 to maximum value $F_{max} = \mu_s N$, where μ_s is coefficient of static friction. Hence in case of rolling motion the magnitude of friction force is impending on N, normal reaction.

When sliding is impending, the friction force can be calculated from N. When the disc slides and rotates at the same time, the force of friction has magnitude $F_k = \mu_k N$, where μ_k is coefficient of kinetic friction. In this, acceleration of center of mass of disc and angular acceleration are not related, i.e., a is not equal to $r\alpha$.

In brief, following three cases are possible:

Rolling, no sliding

$$F \le \mu_s N \quad a_{C.M.} = r\alpha$$

Rolling, sliding impending

$$F = \mu_s N \quad a_{C.M.} = \alpha r$$

Rotating and sliding

$$F = \mu_k N \quad a_{C.M.} \text{ and } \alpha \text{ are independent}$$

Note: When it is not clear whether a disc slides it is assumed that the disc rolls without sliding. If friction force is found to be less than or equal to $\mu_s N$, our assumption is correct. If friction force is found to be larger than $\mu_s N$, the assumption is incorrect and the problem should be started again, assuming the motion to be rotating and sliding.

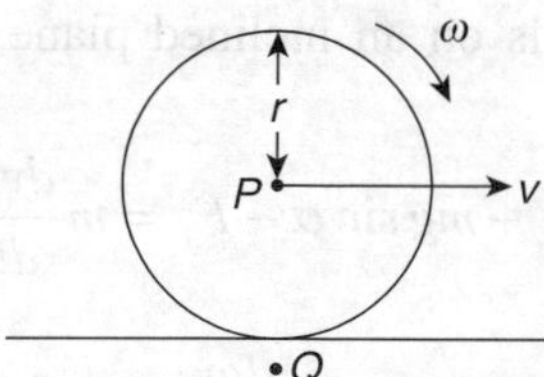

The given figure shows a cylinder rolling on a stationary horizontal surface: $v_P = v - \omega r$

$$v_Q = 0$$

$$v_{PQ} = v_P - v_Q = 0 \text{ or } v = \omega r = 0$$

In addition, the tangential relative acceleration.

$$\frac{dv_{PQ}}{dt} = a_{PQ} = 0 = \frac{dv}{dt} = \frac{d\omega}{dt}.r$$

for pure rolling to be sustained.

Kinetic energy in rolling motion

The kinetic energy of a body undergoing pure rolling is given by the sum of two terms –

(a) The kinetic energy of translation of the centre of mass.

(b) The kinetic energy of rotation about an axis through the centre of mass.

Thus, $$\text{KE} = \frac{1}{2}mv_{cm}^2 + \frac{1}{2}I_{cm}\omega^2$$

along with the constraint appropriate for rolling with or without slipping.

Illustration 20

A cylinder is released from rest at the top of an incline of inclination θ and length ℓ. If the cylinder rolls without slipping, what will be its speed when it reaches the bottom?

Solution

Let the mass of the cylinder be m and its radius r. Suppose the linear speed of the cylinder when it reaches the bottom is v. As the cylinder rolls without slipping, its angular speed about its axis is $\omega = v / r$. The kinetic energy at the bottom will be

$$K = \frac{1}{2}I\omega^2 + \frac{1}{2}mv^2$$

$$= \frac{1}{2}\left(\frac{1}{2}mr^2\right)\omega^2 + \frac{1}{2}mv^2 = \frac{1}{4}mv^2 + \frac{1}{2}mv^2 = \frac{3}{4}mv^2$$

This should be equal to the loss of potential energy $mg\ell \sin\theta$

Thus, $$\frac{3}{4}mv^2 = \text{mg}\,\ell \sin\theta \text{ or, } v = \sqrt{\frac{4}{3}g\ell \sin\theta}$$

Analogy between translational and rotational dynamics

Translational motion	Rotational motion
Displacement = s Velocity = υ Acceleration = α Inertia = m Force = f Momentum = $m\upsilon$ Power = $f\upsilon$	Angular displacement = θ Angular velocity = ω Angular acceleration = α Moment of inertia = I Torque = τ Angular momentum = $I\omega$ Rotational power = $\tau\omega$
Kinetic energy = $\frac{1}{2}mv^2$	Rotational KE = $\frac{1}{2}I\omega^2$

Illustration 21

State whether the statement given below is true or false giving reason in brief.

"A ring of mass 0.3 kg and radius 0.1 m and a solid cylinder of mass 0.4 kg and of the same radius are given the same kinetic energy and released simultaneously on a flat horizontal surface such that they begin to roll as soon as released towards a wall which is at the same distance from the ring and cylinder. The rolling friction in both the cases is negligible. The cylinder will reach the wall first."

Solution

In case of rolling as

$$K = \frac{1}{2}M\upsilon^2 + \frac{1}{2}I\omega^2 \text{ with } \upsilon = r\omega$$

So $$K = \frac{1}{2}M\upsilon^2[1 + I/Mr^2]$$

Now as for ring $I = Mr^2$

$$K_R = M\upsilon_R^2, \text{ i.e., } \upsilon_R = \sqrt{K_R/0.3}$$

And for cylinder, $I = (1/2)Mr^2$

$$K_C = (3/4)Mv_C^2,$$

$$\text{i.e., } v_C = \sqrt{(4K_C/3\times 0.4} = \sqrt{K_C/0.3}$$

Now according to given problem $K_R = K_C$, so $v_R = v_C$ and as the motion is uniform, both will reach the wall simultaneously; i.e., the given statement is false.

Frictional Forces Acting During Pure Rolling

Since the relative velocity at the point of contact between two rolling surfaces is zero, the only type of friction that may act during pure rolling is static friction. This force acts as a 'constraint force' maintaining rolling.

Rolling on a horizontal surface

If a body is undergoes pure rolling on a flat horizontal surface and no additional forces (or torques) are acting on it except contact forces and gravity (weight), no frictional force acts upon the body.

$$v_c = r\omega = \text{constant}$$

$$F_f = 0 \text{ (no friction)}$$

On the other hand, if a force F acts on the body the conditions to be satisfied are

(i) For Linear Motion

$$F - F_f = m\frac{dv_c}{dt}$$

(ii) For Rotation

$$F.b + F_f r = I\frac{d\omega}{dt}$$

or, $$F\frac{b}{r} + F_f = m\frac{k^2}{r^2}\frac{d\omega}{dt}r$$

and the equation of the constraint $v_c = r\omega$.

After solving we get, $F_f = -\left(\dfrac{b/r - k^2/r^2}{1 + k^2/r^2}\right)$

where k = the radius of gyration and it is assumed that the centre of mass coincides with the geometrical centre of the body.

Illustration 22

A force F acts tangentially at the highest point of a sphere of mass m kept on a rough horizontal plane. If the sphere rolls without slipping, find the acceleration of the centre of the sphere.

Solution

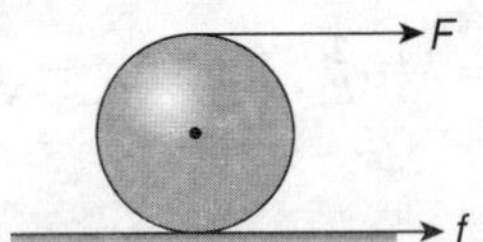

The situation is as shown in the figure. As the force F rotates the sphere, the point of contact has a tendency to slip towards the left, so the static friction on the sphere will act towards right.

Let r be the radius of the sphere and a be the linear acceleration of the centre of the sphere. The angular acceleration about the centre of the sphere is $\alpha = \frac{a}{r}$, as there is no slipping. For the linear motion of the centre

$$F + f = ma \quad \ldots\text{(i)}$$

and for the rotational motion about the centre,

$$Fr - fr = I\alpha = \left(\frac{2}{5}mr^2\right)\left(\frac{a}{r}\right)$$

or, $$F - f = \frac{2}{5}ma \quad \ldots\text{(ii)}$$

From (i) and (ii),

$$2F = \frac{7}{5}ma \text{ or, } a = \frac{10F}{7m}$$

Rolling on an inclined plane

It is clear from the above discussion that the direction and magnitude of friction are determined by forces and torques acting on the body, other than contact forces.

If the body rolls on an inclined plane of inclination α, then:

$$F + mg\sin\alpha - F_f = m\frac{dv_c}{dt}$$

$$F.b + F_f r = I\frac{d\omega}{dt} = mk^2\frac{d\omega}{dt}$$

or, $$F\frac{b}{r} + F_f = \frac{mk^2}{r^2}\left(\frac{d\omega}{dt}r\right)$$

Using the constraint, $v_c = \omega r$, we get

$$F_f = \frac{F\left(\dfrac{k^2}{r^2} - \dfrac{b}{r}\right)}{1 + \dfrac{k^2}{r^2}} + \frac{\dfrac{k^2}{r^2}mg\sin\alpha}{1 + \dfrac{k^2}{r^2}}$$

If $F = 0, F_f = \dfrac{\dfrac{k^2}{r^2}}{1 + \dfrac{k^2}{r^2}}mg\sin\alpha$, acting backward.

Note that this expression does not depend on whether the body rolls up or down the inclined plane. To remind you once again, if it is not clear in a certain situation as to whether it is the case of pure rolling or not, it is advisable to assume that pure rolling occurs. One should apply the rolling constraint and calculate the force of friction, and the normal reaction. The magnitude of the force of static friction always satisfies

$$F_f = \mu_s N$$

where μ_s is the coefficient of static friction between the two surfaces.

Illustration 23

A sphere of mass m rolls without slipping on an inclined plane of inclination θ. Find the linear acceleration of the sphere and the force of friction acting on it. What should be the minimum coefficient of static friction to support pure rolling?

Solution

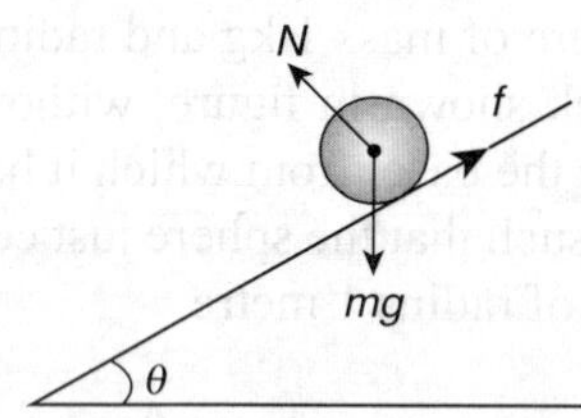

Suppose the radius of the sphere is r. The forces acting on the sphere are shown in figure. They are (a) weight mg, (b) normal force N and (c) friction f.

Let the linear acceleration of the sphere down the plane be a. The equation for the linear motion of the centre of mass is

$$mg\sin\theta - f = ma \qquad \text{...(i)}$$

As the sphere rolls without slipping, its angular acceleration about the centre is a/r. The equation of rotational motion about the centre of mass is,

$$fr = \left(\frac{2}{5}mr^2\right)\left(\frac{a}{r}\right)$$

or,
$$f = \frac{2}{5}ma \qquad \text{...(ii)}$$

From (i) and (ii),

$$a = \frac{5}{7}g\sin\theta \text{ and } f = \frac{2}{7}mg\sin\theta$$

The normal force is equal to $mg\cos\theta$ as there is no acceleration perpendicular to the incline. The maximum friction that can act is, therefore, $\mu mg\cos\theta$, where μ is the coefficient of static friction. Thus, for pure rolling

$$\mu mg\cos\theta \geq \frac{2}{7}mg\sin\theta$$

or,
$$\mu \geq \frac{2}{7}\tan\theta$$

The minimum coefficient of static friction to support pure rolling $= \frac{2}{7}\tan\theta$.

Friction and Rolling

(a) Pure rolling and static friction go hand-in-hand. If pure rolling occurs, there is no relative velocity between the surfaces at the point of contact and therefore, only static friction may act between the bodies.
It does not necessarily reach its limiting value. The correct expression is:

$$\left|F_{fr}\right| \leq \mu_s N \qquad \text{... (i)}$$

where F_{fr} represents the force of static friction, N, the normal reaction, and μ_s, the coefficient of static friction between the surfaces.
If it is known that pure rolling occurs, F_{fr} and N should be treated as unknown quantities which must be calculated from Newton's laws. Equation (i) then provides a consistency check for the problem.

(b) Another question that is frequently asked, quite often at the point where the FBD is being drawn, is the direction of the force of static friction.
The answer is: assume it to be an unknown vector, i.e., let its component along the surface, along the chosen direction, be F_{fr}. If $F_{fr} > 0$ as a result of the solution, the choice made is correct, otherwise it needs to be reversed.

(c) If it is known that the surfaces are sliding, friction is kinetic and its direction is such as to oppose relative velocity between the surfaces. Its magnitude is $\mu_k N$.

(d) In problems where it is not clear as to whether friction is static or kinetic, a safe method is to assume that it is static – solve accordingly and check for consistency:
If the check fails for a set of surfaces/bodies, assume that F_{fr} is kinetic, i.e., $F_{fr} = \mu_k N$ but along the proper direction that emerged as a result of the previous solution (static assumption). The process is repeated till the result is found to be consistent.

(e) Are there any shortcuts?
No, there aren't any universal "shortcuts". A "shortcut" may work in a few simple cases, but it would fail in other cases. It is best to avoid these.

(f) There do exist problems which cannot be solved entirely on the basis of Newton's laws alone, with or without constraints. Most of these problems are outside the purview of the syllabus.

(g) As for rolling friction, we have ignored it because it is not so large as to be of much consequence here.

Instantaneous Axis of Rotation

A rigid body that is in motion in general, undergoes a combination of translation and rotation. Rolling motion, for example, is a combination of translation and rotation. For such motion, the axis of rotation changes in position and /or orientation.

The instantaneous velocity of any particle, P, of the rigid body, can be written as:

$$\vec{v}_P = \vec{v}_0 + \vec{\omega} \times \vec{r}_P$$

where $\vec{v}_0$ is the velocity of the origin, located on the axis of rotation, and $\vec{r}_P$ is the position vector of the particle P about O. Here, $\vec{\omega}$ is the angular velocity of rotation of the rigid body about the given axis.

It can be proved that, for any moving rigid body, there exists a straight line passing through the body, for which, the velocity of the particles lying on it is zero. This line is known as the instantaneous axis of rotation, because the instantaneous velocity of any particle of the body may be written in the form:

$$\vec{v}_P = \vec{\omega} \times \vec{r}_P$$

where $\vec{r}_P$ is the position vector of this particle with respect to a point on this line.

This line, therefore, can be regarded as an axis of rotation; but only for an instant.

SOLVED EXAMPLES

Example 1

A particle of mass m is projected at time $t = 0$ from a point P with a speed v_0 at an angle of 45° to the horizontal. Find the magnitude and the direction of the angular momentum of the particle about the point P at time t.

Solution

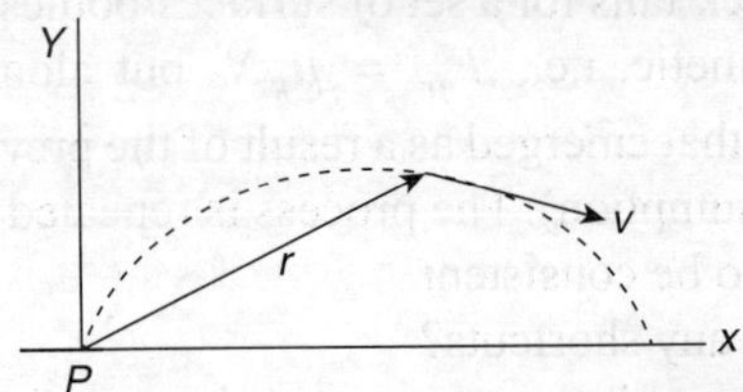

Let P be the origin, x-axis along the horizontal and y-axis along the vertically upward direction

For horizontal motion,

$$v_x = v_0 \cos 45° = \frac{v_0}{\sqrt{2}}$$

and $$x = v_x t = \frac{v_0 t}{\sqrt{2}}$$

For vertical motion,

$$v_y = v_0 \sin 45° - gt = \frac{v_0}{\sqrt{2}} - gt$$

and $$y = \left(v_0 \sin 45^0\right)t - \frac{1}{2}gt^2$$

$$= \frac{v_0}{\sqrt{2}}t - \frac{1}{2}gt^2$$

The angular momentum of the particle at time t about the origin is

$$\vec{L} = \vec{r} \times \vec{p} = \left(x\hat{i} + y\hat{j}\right) \times m\left(v_x\hat{i} + v_y\hat{j}\right)$$

$$= -\frac{mv_0}{2\sqrt{2}} gt^2 \hat{k}$$

where. $\hat{i} \times \hat{j} = \hat{k}$, which is out of the plane of the paper.

Thus, the angular momentum of the particle is $\frac{mv_0}{2\sqrt{2}} gt^2$ **in the negative z-direction** i.e., perpendicular to the plane of motion, going into the plane.

Example 2

A small solid sphere of mass 1 kg and radius 0.2 metre rolls down along a track shown in figure, without slipping. Find the height h above the base, from which it has to start rolling down the incline, such that the sphere just completes the vertical circular loop of radius 1 metre.

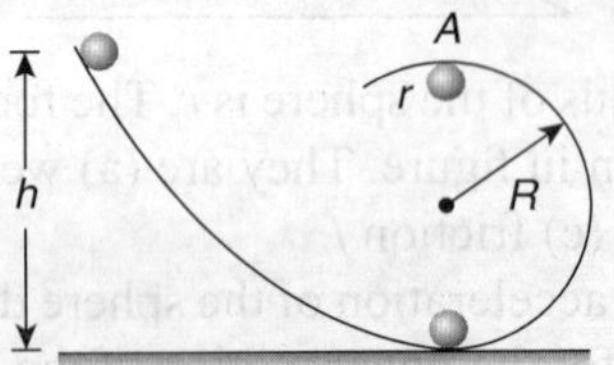

Solution

The situation is as shown in figure. Let m and r be the mass and radius of solid sphere. When the sphere reaches the point A, it descends through a vertical distance $(h - 2R + r)$, where R is the radius of circular loop. At A, the loss of potential energy

$$= mg(h - 2R + r) \quad \ldots\text{(i)}$$

Total kinetic energy of the sphere at A

$$= \text{KE of the rotation} + \text{KE of translation of sphere}$$

$$= \frac{1}{2}I\omega^2 + \frac{1}{2}mv^2$$

$$= \frac{1}{2}\left(\frac{2}{5}mr^2\right)\left(\frac{v^2}{r^2}\right) + \frac{1}{2}mv^2$$

$$= \frac{7}{10}mv^2 \quad \ldots\text{(ii)}$$

For just completing the vertical circular loop

$$\frac{mv^2}{(R - r)} = mg$$

or $$v^2 = (R - r)g$$

Now total kinetic energy of sphere at $A = \frac{7}{10}mv^2$

$$= \frac{7}{10} m(R-r)g$$

Loss of potential energy = Gain of KE

$$mg(h-2R+r) = \frac{7}{10} m(R-r)g$$

Solving we get,

$$h = \left(\frac{27}{10} \times R - \frac{17}{10} r\right)$$

Substituting the given values, we get

$$h = \left(\frac{27}{10} \times 1 - \frac{17}{10} \times 0.2\right) = 2.36 \text{ m}$$

Example 3

A carpet of mass M made of inextensible material is rolled along its length in the form of a cylinder of radius R and is kept on a rough floor. The carpet starts rolling without slipping on the floor when a negligibly small push is given to it. Calculate the horizontal velocity of the axis of cylindrical part of the carpet when its radius reduces to $R/2$.

Solution

The centre of mass of the rolled carpet lies on the axis of the cylinder. As the carpet rolls and its radius reduces from R to $R/2$, its centre of mass also comes down from a height R to $R/2$. Thus, there is a loss in the potential energy of the carpet.

If L be the length of the rolled carpet and ρ be its density, then

Initial mass of the rolled carpet = $M = (\pi R^2)L\rho$

And final mass of the rolled carpet = $M' = \left(\pi \frac{R^2}{4}\right).L\rho$

$$= M' = \frac{M}{4}$$

Loss in potential energy of the carpet is

$$E_P = MgR - \frac{M}{4} g \frac{R}{2} = \frac{7}{8} MgR$$

As the carpet unrolls, its centre of mass moves in a straight line. Thus, the carpet has both translational and rotational kinetic energy energies. The total kinetic energy, when its radius reduces to $R/2$ is

$$E_k = \frac{1}{2} M'v^2 + \frac{1}{2} I\omega^2$$

where v is the linear velocity of the centre mass and I is its moment of inertia about the axis of rotation.

As there is no sliding,

From the principle of conservation of energy, we have

$E_k = E_P$

$$\frac{1}{2} M'v + \frac{1}{2} I\omega^2 = \frac{7}{8} MgR$$

$$\frac{1}{2} \cdot \frac{M}{4} v^2 + \frac{1}{2} \cdot \left[\frac{1}{2} \frac{M}{4} \cdot \left(\frac{R}{2}\right)^2\right] \frac{v^2}{\left(\frac{R}{2}\right)^2} = \frac{7}{8} MgR$$

$$\frac{3}{2} v^2 = 7gR$$

$$v = \sqrt{\frac{14gR}{3}}$$

Example 4

A cylinder of mass m is suspended through two strings wrapped around it as shown in figure. Find (a) the tension T in the string and (b) the speed of the cylinder as it falls through a distance h.

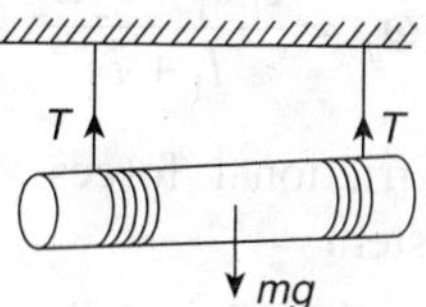

Solution

The portion of the strings between the ceiling and the cylinder is at rest. Hence the points of the cylinder where the strings leave it are at rest. The cylinder is thus rolling without slipping on the strings. Suppose the centre of the cylinder falls with an acceleration a. The angular acceleration of the cylinder about its axis is $\alpha = \frac{a}{r}$, as the cylinder does not slip over the strings.

The equation of motion for the centre of mass of the cylinder is

$$mg - 2T = ma \quad \text{...(i)}$$

and for motion about the centre of mass, it is

$$2Tr = \left(\frac{1}{2} mr^2 \alpha\right) = \frac{1}{2} mra$$

or,
$$2T = \frac{1}{2} ma \quad \text{...(ii)}$$

From (i) and (ii),

$$a = \frac{2}{3} g \text{ and } T = \frac{mg}{6}$$

As the centre of the cylinder starts moving from rest, the velocity after it has fallen through a distance h is given by

$$v^2 = 2\left(\frac{2}{3} g\right) h$$

or,
$$v = \sqrt{\frac{4gh}{3}}$$

Example 5

Two horizontal discs rotate freely about a common vertical axis passing through their centres. The moments of inertia of the discs relative to their axis are equal to I_1 and I_2, and their angular velocities are ω_1 and ω_2. The upper disc falls on the lower disc and after some time, both discs begin to rotate as a single body due to friction. Find

(a) the steady-state angular velocity of the discs, and

(b) the work performed by the frictional forces in the process.

Solution

(a) Since there is no external torque on the system consisting of both the discs, angular momentum of the system is conserved.

$$I_1\omega_1 + I_2\omega_2 = (I_1 + I_2)\omega_f$$

$$\Rightarrow \qquad \omega_f = \frac{I_1\omega_1 + I_2\omega_2}{I_1 + I_2}$$

(b) Work done by frictional forces = Change in kinetic energy of the system

$$W_{\text{friction}} = \text{Change in kinetic energy} = KE_f - KE_i$$

$$\text{or,} \quad W_{fr} = \frac{1}{2}(I_1 + I_2)\omega_f^2 - \left(\frac{1}{2}I_1\omega_1^2 + \frac{1}{2}I_2\omega_2^2\right)$$

$$= \frac{1}{2}(I_1 + I_2)\frac{(I_1\omega_1 + I_2\omega_2)^2}{(I_1 + I_2)^2} - \frac{1}{2}\left(I_1\omega_1^2 + I_2\omega_2^2\right).$$

$$W_{fr} = \frac{1}{2}\frac{(I_1\omega_1 + I_2\omega_2)^2}{I_1 + I_2} - \frac{1}{2}\left(I_1\omega_1^2 + I_2\omega_2^2\right)$$

$$= -\frac{I_1 I_2}{2(I_1 + I_2)}(\omega_1 - \omega_2)^2$$

which is very similar to the expression for a perfectly inelastic collision.

Example 6

A uniform sphere of mass m and radius R starts rolling without slipping down an inclined plane. Find the time dependence of the angular momentum of the sphere relative to the initial point of contact. How will the result be affected in the case of a perfectly smooth inclined plane? Here, θ = Angle of inclination of the plane.

Solution

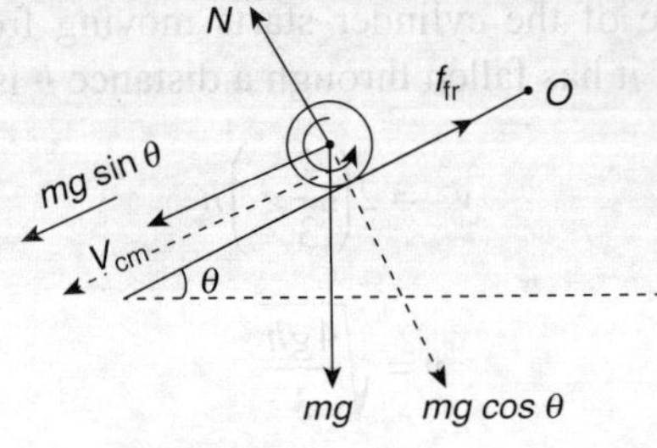

Let O be the initial point of contact.

For linear motion, we write,

$$mg\sin\theta - f_{fr} = ma_{\text{cm}}$$

For rotation,

$$\tau = f_{fr}R = I\alpha, \text{ where } \alpha = \frac{a_{\text{cm}}}{R}$$

$$\therefore \qquad mg\sin\theta - \frac{I\alpha}{R} = ma_{\text{cm}}$$

$$\text{or,} \qquad mg\sin\theta - \frac{Ia_{\text{cm}}}{R^2} = ma_{\text{cm}}$$

$$\text{or,} \qquad mg\sin\theta = \left(m + \frac{I}{R^2}\right)a_{cm} = \left(m + \frac{I}{R^2}\right)\frac{dv_{\text{cm}}}{dt}$$

$$mg\sin\theta \cdot dt = \left(m + \frac{I}{R^2}\right)dv_{\text{cm}}$$

Integrating,

$$\text{or,} \qquad v_{\text{cm}} = \frac{mg\sin\theta \cdot t}{m + \frac{I}{R^2}}$$

Using the expression for angular momentum $(L_{\text{axis}} = L_{\text{spin}} + L_{\text{orbital}})$

$$L_0 = L_{cm} + mv_{cm}R = I\omega + mv_{cm}R = \frac{Iv_{cm}}{R} + mv_{cm}R$$

$$L_0 = v_{cm}R\left(m + \frac{I}{R^2}\right) = \frac{mg\sin\theta \cdot t}{m + \frac{I}{R^2}}R\left(m + \frac{I}{R^2}\right)$$

$$= (mg\sin\theta)Rt$$

If the plane is perfectly smooth, there is no rotation.

$$\therefore \qquad L_0 = mv_{cm}R = mg\sin\theta Rt$$

Example 7

A sphere, a disc and a ring of the same mass and radius are allowed to roll down an inclined plane simultaneously from the same height without slipping. Prove that the sphere reaches down first, the disc next and the ring, the last.

Solution

We know that acceleration a of the body rolling down an inclined plane is given by

$$a = \frac{g\sin\theta}{\left(1 + \frac{K^2}{R^2}\right)} \qquad \ldots\text{(i)}$$

We also know that,

Moment of inertia of solid sphere

$$I_1 = MK_1^2 = \frac{2}{5}MR^2$$

$$\therefore \quad K_1^2 = \frac{2}{5}R^2 \qquad \text{...(ii)}$$

Moment of inertia of disc

$$I_2 = MK_2^2 = \frac{MR^2}{2}$$

$$\therefore \quad K_2^2 = \frac{R^2}{2} \qquad \text{...(iii)}$$

Moment of inertia of ring

$$I_3 = MK_3^2 = MR^2$$

$$\therefore \quad K_3^2 = R^2 \qquad \text{... (iv)}$$

Substituting these values in Eq. (i), we get

Acceleration a_1 for solid sphere,

$$\frac{g\sin\theta}{\left(1+\frac{2}{5}\right)} = \frac{5}{7}g\sin\theta$$

Acceleration a_2 for disc

$$\frac{g\sin\theta}{\left(1+\frac{1}{2}\right)} = \frac{2}{3}g\sin\theta$$

Acceleration a_3 for ring

$$\frac{g\sin\theta}{(1+1)} = \frac{g}{2}\sin\theta$$

$$\therefore \quad a_1 > a_2 > a_3$$

Thus solid sphere reaches down first, the disc next and the ring, last.

Example 8

A uniform sphere of mass m and radius r rolls without slipping down from the top of a larger fixed sphere of radius R. Find the angular momentum acquired by the smaller sphere about the centre of the larger sphere as well as its KE at the moment when it breaks off from the larger sphere. The initial velocity of the sphere is negligible.

Solution

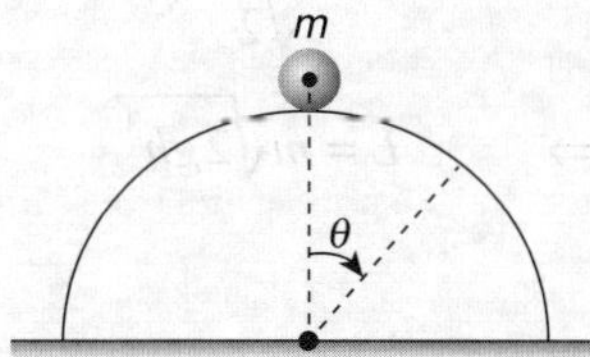

The centre of mass of the sphere moves in a vertical circle of radius $(R+r)$.

Considering the dynamics of circular motion of the centre of mass of the sphere

$$mg\cos\theta - N = \frac{mv^2}{R+r}$$

When it breaks off, $N=0$

At the point where the smaller sphere breaks off,

$$mg\cos\theta = \frac{m\omega^2 r^2}{R+r} \quad (\because v = \omega r) \qquad \text{(i)}$$

Since friction is static and gravity is conservative (other forces being constraint forces), mechanical energy is conserved.

$$mg(R+r) = mg(R+r)\cos\theta + \frac{1}{2}mv^2 + \frac{1}{2}I\omega^2$$

or $$mg(R+r) = mg(R+r)\cos\theta + \frac{7}{10}m\omega^2 r^2 \qquad \text{(ii)}$$

or $$mg(R+r) = m\omega^2 r^2 + \frac{7}{10}m\omega^2 r^2$$

or $$\omega = \sqrt{\frac{10}{17}g\frac{(R+r)}{r^2}}$$

The angular momentum of the sphere is

$$L = L_{\text{orbital}} + I\omega$$

$$= mv(R+r) + \frac{2}{5}mr^2\omega$$

$$= m\omega r(R+r) + \frac{2}{5}m\omega r^2$$

$$= m\omega r\left(R + \frac{7}{5}r\right)$$

$$= m\sqrt{\frac{10}{17}g(R+r)}\left(R+\frac{7}{5}r\right)$$

Kinetic energy of the sphere is

$$\text{KE} = \frac{1}{2}mv^2 + \frac{1}{2}I\omega^2$$

$$= \frac{1}{2}mr^2\omega^2 + \frac{1}{2}\frac{2}{5}mr^2w^2$$

$$= \frac{7}{10}mr^2\omega^2 = \frac{7}{10}mr^2 . \frac{10}{17}g\frac{(R+r)}{r^2}$$

$$= \frac{7}{17}\text{mg}(R+r)$$

Example 9

Two uniform thin rods A and B of length 0.6 m each and of masses 0.01 kg and 0.02 kg respectively are rigidly jointed, end to end. The combination is pivoted at the lighter end P as shown in the figure such that it can freely rotate about the point P in a vertical plane. A small particle of mass 0.05 kg, moving horizontally strikes the lower end of the combination and sticks to it. What should be the velocity of the particle so that the system just rises to a horizontal position?

Solution

The situation is as shown in figure. Let ℓ be the length of each rod and m_A and m_B their respective masses.

The torque about pivot is zero. Therefore angular momentum of the system is conserved, i.e.,

$$mv(2\ell) = I\omega \quad \text{... (i)}$$

where I is the moment of inertia of the system about point P and ω is the angular velocity of combination just after the collision. Now we shall calculate the value of I.

Moment of inertia of rod A about P, i.e., I_A is given by

$$I_A = \frac{1}{3} m_A \ell^2$$

$$= \frac{1}{3} \times 0.01 \times (0.6)^2 = 1.2 \times 10^{-3} \text{ kg-m}^2$$

The moment of inertia of rod B about P is

$$I_B = \frac{1}{12} m_B \ell^2 + m_B \left(\frac{3\ell}{2}\right)^2$$

(by the theorem of parallel axes)

$$\frac{28}{12} m_B \ell^2 = \frac{28}{12} \times 0.02 \times (0.6)^2$$

$$= 1.68 \times 10^{-2} \text{ kg-m}^2$$

The moment of inertia of particle about P is

$$I_C = m(\ell + \ell)^2 = 4m\ell^2$$

$\therefore$ Moment of inertia I of the system about P is given by

$$I = I_A + I_B + I_C = 9 \times 10^{-2} \text{ kg-m}^2 \quad \text{...(ii)}$$

Substituting the value of I in Eq. (i) we get

$$0.05v \times (2 \times 0.6) = (9 \times 10^{-2})\omega$$

or $$\omega = \frac{0.05 \times (2 \times 0.6)}{(9 \times 10^{-2})} \times v = \frac{2}{3} v \quad \text{...(iii)}$$

When the rod moves, the rotational kinetic energy of the system is converted into gravitational potential energy, i.e.,

$$\frac{1}{2} I\omega^2 = (m_A + m_B + m) g Y_{C.M.} \quad \text{...(iv)}$$

where $Y_{C.M.}$ is the distance of the centre of mass of system from P. $Y_{C.M.}$ is given by

$$Y_{C.M.} = \frac{m_A y_1 + m_B y_2 + m y_3}{(m_A + m_B + m)}$$

or $$Y_{C.M.} = \frac{0.01 \times 0.3 + 0.02 \times 0.9 + 0.05 \times 1.2}{0.01 + 0.02 + 0.05} = \frac{81}{80}$$

Substituting the values in Eq. (iv), we get

$$\frac{1}{2} \times (9 \times 10^{-2}) \left(\frac{2}{3} v\right)^2 = \frac{(0.01 + 0.02 + 0.05) \times 9.8 \times 81}{80}$$

Solving this equation for v, we get,

$$v = 6.3 \text{ m/s}$$

Example 10

A particle of mass m is projected with velocity v making an angle of 45° with the horizontal. The magnitude of the angular momentum of the projectile about the point of projection when the particle is at its maximum height h is

(a) Zero (b) $\frac{mv^3}{4\sqrt{2g}}$

(c) $\frac{mv^3}{\sqrt{2g}}$ (d) $m\sqrt{2gh^3}$

Solution

$$L = mv_x h$$

$$\Rightarrow \quad L = m(v\cos 45°)\frac{v^2 \sin^2 45°}{2g}$$

$$\Rightarrow \quad L = \frac{mv^3}{4\sqrt{2g}}$$

Further $L = mv_x h$

$$\Rightarrow \quad L = m(v\cos 45°)h$$

But $$h = \frac{v^2 \sin^2 45°}{2g}$$

$$\Rightarrow \quad v = 2\sqrt{gh}$$

$$\Rightarrow \quad L = \frac{mv}{\sqrt{2}} 2\sqrt{gh}\, h$$

$$\Rightarrow \quad L = m\sqrt{2gh^3}$$

Ans. (b) and (d)

Example 11

A uniform solid cylinder of mass M and radius R is resting on a horizontal platform (which is parallel to X-Y plane) with its axis along the Y-axis and free to roll on the platform. The platform is given a motion in X-direction given by $x = A\cos\omega t$. There is no slipping between the cylinder and the platform. The maximum torque acting on the cylinder as measured about its centre of mass

(a) $\frac{1}{2}MRA\omega^2$ (b) $MRA\omega^2$

(c) $2mRA\omega^2$ (d) $mR\omega A^2\cos^2\omega t$

Solution

$$\vec{\tau} = I\alpha = \frac{1}{2}MR^2 \times \left(\frac{\text{linear acceleration}}{R}\right)$$

Acceleration $= -\omega^2 A\sin\omega t$ or

$$(\text{linear acceleration})_{\max} = \omega^2 A$$

$$\therefore \quad (\tau)_{\max} = \frac{1}{2}\frac{MR^2\omega^2}{R} = \frac{1}{2}MRA\omega^2$$

Ans. (a)

Example 12

A metal ball of mass m is put at the point A of a loop track and the vertical distance of A from the lower most point of track is 8 times the radius R of the circular part. The linear velocity of ball when it rolls of the point B to a height R in the circular track will be

(a) $[10gR]^{1/2}$ (b) $7\left[\frac{gR}{10}\right]^{1/2}$

(c) $\left[\frac{7gR}{5}\right]^{1/2}$ (d) $[5gR]^{1/2}$

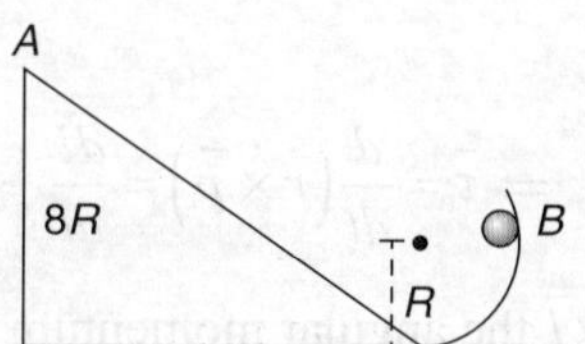

Solution

Applying the conservation of energy at points A and B, we have

$$mg(8R) - \frac{mv^2}{2} + \frac{1}{2}I\omega^2 + mgR$$

or $$mg(8R) = \frac{mv^2}{2} + \frac{1}{2}\left(\frac{2}{5}mR^2\right)\left(\frac{v}{R}\right)^2 + mgR$$

$$= \frac{7}{10}mv^2 + mgR$$

$$mg(8R - R) = \frac{7}{10}mv^2 \; [10gR]^{1/2}$$

$$v = [10gR]^{1/2}$$

Ans. (a)

Example 13

A uniform solid cylinder of mass m can rotate freely about its axis which is kept horizontal. A particle of mass m_0 hangs from the end of a light string wound round the cylinder. When the system is allowed to move, the acceleration with which the particle descends is

(a) $\frac{2m_0 g}{m_0 + 2m}$ (b) $\frac{2m_0 g}{m + 2m_0}$

(c) $\frac{m_0 g}{m + m_0}$ (d) $\frac{2mg}{m + 2m_0}$

Solution

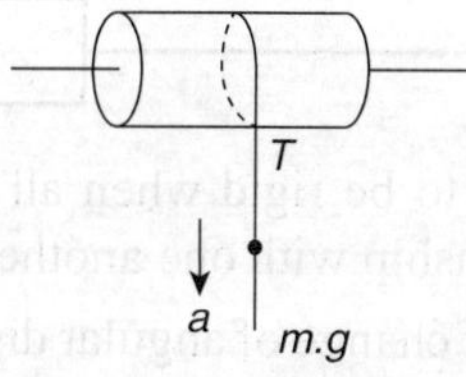

Suppose that the tension in the string is T.

Then, $\quad = m_0 g - T = m_0 a$

$$\therefore \quad T = m_0(g - a)$$

where a = acceleration

Further, $T.r = I\alpha$

α = angular acceleration and $T.r$ = moment of force acting on cylinder

or $$T = \frac{I\alpha}{r}$$

$$= \left(\frac{mr^2}{2}\right)\left(\frac{\alpha}{r}\right) = \frac{mr\alpha}{2} = \frac{ma}{2}$$

$$\therefore \quad a = \frac{2T}{m} = \frac{2m_0(g-a)}{m}$$

Solving for a, we get

$$a = \left(\frac{2m_0 g}{m + 2m_0}\right)$$

Ans. (b)

Example 14

A smooth uniform rod of length ℓ and mass M has two identical beads of negligible size, each of mass m, which can slide freely along the rod. Initially the two beads are at the centre of the rod and the system is rotating with angular velocity ω_0 about a fixed vertical axis perpendicular to rod and passing through the midpoint of the rod. When the beads reach the ends of the rod, the angular velocity of the system is

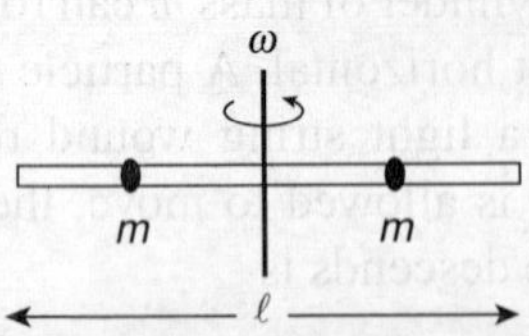

(a) ω_0 (b) $\left(\frac{M}{M+6m}\right)\omega_0$

(c) $\left(\frac{M}{M+3m}\right)\omega_0$ (d) $\left(\frac{M}{M-6m}\right)\omega_0$

Solution

Applying the conservation of angular momentum, we have

$$I_0\omega_0 = I\omega$$

$$I_0 = \frac{ML^2}{12} \text{ and } I = \frac{ML^2}{12} + 2m\left(\frac{L}{2}\right)^2$$

$$\therefore \quad \frac{ML^2}{12}\omega_0 = \left[\frac{ML^2}{12} + 2m\left(\frac{L}{2}\right)^2\right]\omega$$

$$\omega = \left(\frac{M}{M+6m}\right)\omega_0$$

Ans. (b)

POINTS TO REMEMBER

1. A body is said to be rigid when all its particles have a definite relationship with one another.
2. The time rate of change of angular displacement is called angular velocity, i.e.,
$$\text{Angular velocity} = \vec{\omega} = \frac{d\theta}{dt}$$
3. The time rate of change of angular velocity is called angular acceleration, i.e.,
$$\text{Angular acceleration} = \vec{\alpha} = \frac{d\vec{\omega}}{dt} = \frac{d^2\theta}{dt^2}$$
4. The following relationships exist between the initial angular velocity ω_0, final angular velocity ω, angular acceleration α, time t and the angular displacement θ:
$$\omega = \omega_0 + \alpha t$$
$$\omega^2 = \omega_0^2 + 2\alpha\theta$$
$$\theta = \omega_0 t + \frac{1}{2}\alpha t^2$$
If the body has angular deceleration instead of angular acceleration, then in the above equations, α has to be replaced by $-\alpha$.
5. Just as a force is necessary to bring about a change in the translational motion of a particle along a straight-line, a torque is necessary to cause a rotational motion of a particle about a given axis and is given by
$$\vec{\tau} = \vec{r} \times \vec{F}$$
where $\vec{r}$ the position is vector of the particle with respect to the origin of the given inertial frame and $\vec{F}$ is the force acting on it.
6. $\vec{\tau} = \vec{r} \times \vec{F}$ can also be written as
$$\vec{\tau} = \vec{r} \times \frac{d\vec{p}}{dt}, \text{ where } \vec{p} = \text{liner momentum}$$
$$\Rightarrow \vec{\tau} = \frac{d}{dt}\left(\vec{r} \times \vec{p}\right) = \frac{d\vec{l}}{dt}$$
where $\vec{l} = \vec{r} \times \vec{l}$ the angular momentum of the particle. Thus, the rate of change of angular momentum of a particle about a given origin is equal to the torque acting on that particle about that origin.
7. If a system consists of a number of particles, then torque takes the form
$$\vec{\tau} = \frac{d\vec{L}}{dt}$$

where $$\vec{L} = \sum \vec{l}_i$$

8. The moment of inertia of a particle of mass m about a given axis at a distance r from it is given by $I = mr^2$
If a body consists of a number of particles, the moment of inertia of the whole body about the given axis is

$$I = \sum m_i r_i^2$$

where r_i is the distance of ith particle of mass m_i from the axis of rotation. The moment of inertia is written, in short, as M.I. Its unit is kg-m^2.

9. The moment of inertia plays the same role in rotational motion as is played by mass in linear motion.

10. The angular momentum of a rigid body and the torque acting on it are related to its moment of inertia by the following equations:

$$L = I\omega$$

$$\tau = I\alpha$$

11. The kinetic energy of a body rotating about a given axis is

$$\text{KE} = \frac{1}{2} I\omega^2$$

12. If an external force $\vec{F}$ acting on a body is such that its line of action passes through the C.M. of the body, the motion of the body is one of pure translation and is governed by the following equation:

$$\vec{F} = M\overline{a_{\text{cm}}}$$

where M is the mass of the body and $\overline{a_{\text{cm}}}$ is the acceleration of its C.M.
If external forces acting on a body do not pass through its C.M., they produce both translation and rotation. The net external force produces translation of the C.M. of the body according to $\vec{F} = M\overline{a_{\text{cm}}}$.
Besides translation of the C.M., the resultant force also forms a couple about an axis passing through the C.M. of the body. The effect of this couple is to produce a rotation of the body about this axis according to the following equation:

$$\vec{L} = I_{cm}\vec{\omega}$$

$$\vec{\tau} = \frac{d\vec{L}}{dt}$$

Where I_{cm} is the M.I. of the body about an axis of rotation passing through its C.M.

The kinetic energy of a body for such a motion is equal to the sum of its translational kinetic energy and rotational kinetic energy,

i.e., $$\text{KE} = \frac{1}{2} mv_{\text{cm}}^2 + \frac{1}{2} + I_{\text{cm}}\omega^2$$

13. In a rigid body, there are a number of particles of different masses which are at different distances from the axis of rotation. The moment of inertia in such a case is given by

$$I = \sum m_i r_i^2$$

Let us assume that the total mass $M = \sum m_i$ of the body is concentrated at a point whose distance K from the axis of rotation is such that

$$I = \sum m_i r_i^2 = MK^2$$

Then, K is known as the radius of gyration of the body about that given axis.

14. Note that the moment of inertia of a body has a meaning only when the axis of rotation is given.

15. If I_0 be M.I. of a body about an axis passing through its centre of gravity (C.G.), then its moment of inertia about another axis parallel to the former axis and at a distance a from it given by $I = I_0 + Ma^2$
This is known as the theorem of parallel axes.

16. Let x and y be two mutually perpendicular axes in the plane of the body and the z-axis be perpendicular to the plane of the body passing through the point of intersection of the x- and y-axis. Let I_x, I_y, and I_z be the moments of inertia of the body about x, y and z axes respectively. Then, the theorem of perpendicular axes states that $I_z = I_x + I_y$

17. If a body of radius r rolls down an inclined plane, inclined at an angle θ to the horizontal, then its acceleration down the plane is given by

$$a = \frac{g\sin\theta}{1 + \frac{K^2}{r^2}}$$

where K is the radius of gyration of the body.

18. When it is said that a body rolls without slipping, it means that its C.M. moves forward and covers a linear distance while the body itself continues to rotate about a horizontal axis passing through its C.M. and the instantaneous linear velocity of its point in contact with the surface on which it rolls is zero.

EXERCISES

1. Consider a two particle system with particle having masses, m_1 and m_2. If the first particle is pushed towards the centre of mass through a distance d, by what distance should the second particle be moved, so as to keep the centre of mass at the same position?

(a) d (b) $\frac{m_2}{m_1}d$

(c) $\frac{m_1}{m_1+m_2}.d$ (d) $\frac{m_1}{m_2}.d$

2. A T-shaped object with dimensions shown in the figure, is lying on a smooth floor. A force $\vec{F}$ is applied at the point P parallel to AB, such that the object has only translational motion without rotation. Find the location of P with respect to C.

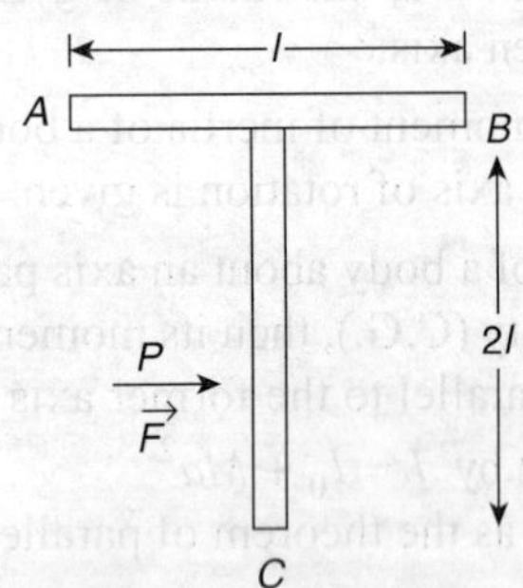

(a) $\frac{4}{3}l$ (b) l

(c) $\frac{3}{4}l$ (d) $\frac{3}{2}l$

3. A disc is rotating with an angular velocity ω_0. A constant retarding torque is applied on it to stop the disc. The angular velocity becomes $\frac{\omega_0}{2}$ after n rotations. How many more rotations will it make before coming to rest?

(a) n (b) $2n$

(c) $\frac{n}{2}$ (d) $\frac{n}{3}$

4. The magnitude of torque on a particle of mass 1 kg is 2.5 N-m about the origin. If the force acting on it is 1 N and the distance of the particle from the origin is 5m, then the angle between the force and the position vector (in radian) is

(a) $\frac{\pi}{8}$ (b) $\frac{\pi}{4}$

(c) $\frac{\pi}{3}$ (d) $\frac{\pi}{6}$

5. Seven identical circular planar discs, each of mass M and radius R are welded symmetrically as shown in the figure. The moment of inertia of the arrangement about an axis normal to the plane and passing through the point P is

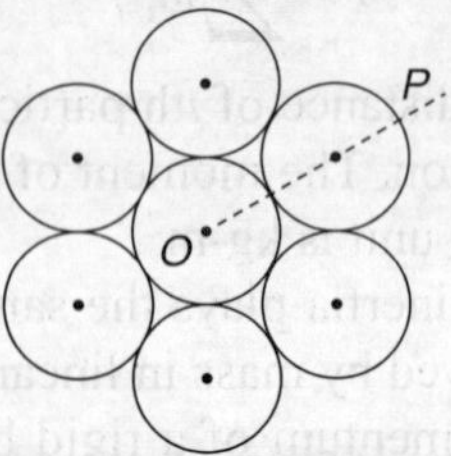

(a) $\frac{181}{2}MR^2$ (b) $\frac{19}{2}MR^2$

(c) $\frac{55}{2}MR^2$ (d) $\frac{73}{2}MR^2$

6. A particle undergoes uniform circular motion. About which point on the plane of the circle, will the angular momentum of the particle remain conserved?

(a) Centre of circle
(b) On the circumference of the circle
(c) Inside the circle
(d) Outside the circle.

7. A disc of mass M and radius R is rolling with angular speed ω on a horizontal plane as shown. The magnitude of angular momentum of the disc about the origin O is

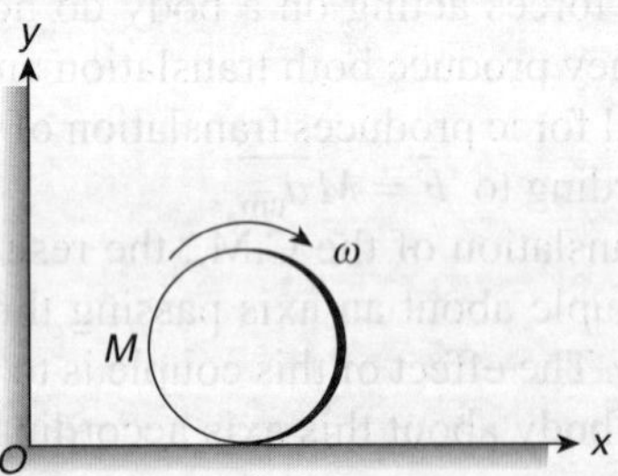

(a) $\left(\frac{1}{2}\right)MR^2\omega$ (b) $MR^2\omega$

(c) $\left(\frac{3}{2}\right)MR^2\omega$ (d) $2MR^2\omega$

8. A mass m is moving with a constant velocity along a line parallel to the X-axis, away from the origin. Its angular momentum with respect to the origin

(a) is zero
(b) remains constant
(c) goes on increasing
(d) goes on decreasing.

9. The moment of inertia of a body does not depend on:
(a) the mass of the body
(b) the angular velocity of the body
(c) the axis of rotation of the body
(d) the distribution of the mass in the body.

10. From a circular disc of radius R and mass $9M$, a small disc of radius $R/3$ is removed from the disc. The moment of inertia of the remaining disc about an axis perpendicular to the plane of the disc and passing through O is

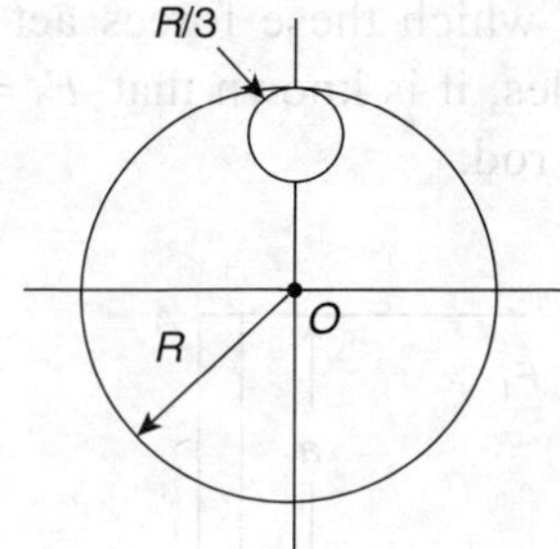

(a) $4MR^2$ (b) $\frac{40}{9}MR^2$
(c) $10MR^2$ (d) $\frac{37}{9}MR^2$

11. Two equal and opposite forces act on a rigid body at a certain distance. Then
(a) the body is in equilibrium
(b) the body will rotate about its centre of mass
(c) the body may rotate about any point other than its centre of mass
(d) the body cannot rotate about its centre of mass.

12. If the earth shrinks to half of its radius without change in mass, the duration of the day will be
(a) 48 hours (b) 24 hours
(c) 12 hours (d) 6 hours.

13. A circular platform is free to rotate in a horizontal plane about a vertical axis passing through its centre. A tortoise is sitting at the edge of the platform. Now, the platform is given an angular velocity ω_0. When the tortoise move along a chord of the platform with a constant velocity, the angular velocity of the platform $\omega(t)$ will vary with time t as

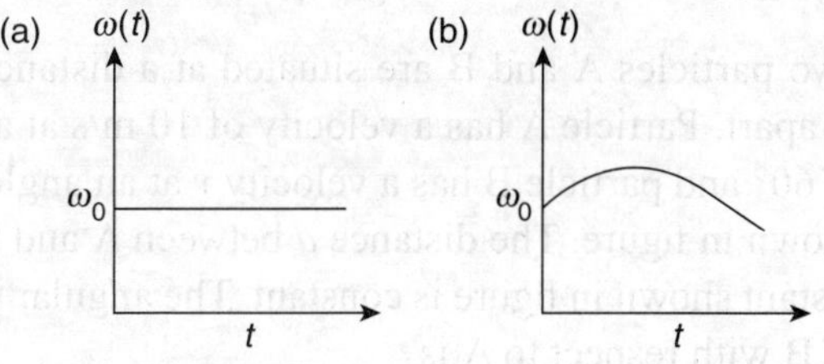

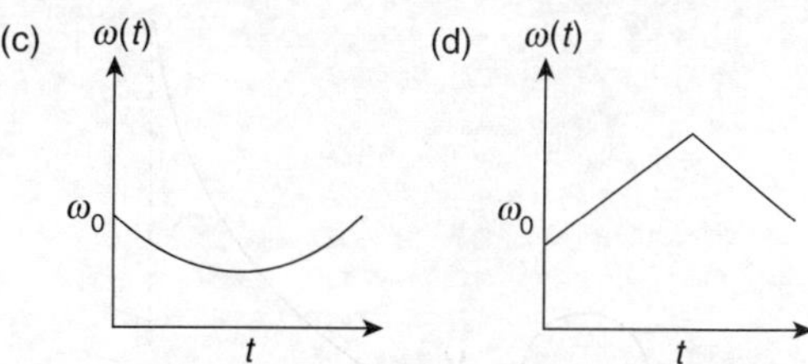

14. A thin wire of length L and uniform linear mass density is bent into a circular loop with centre at O as shown. The moment of inertia of the loop about the axis XX′ is

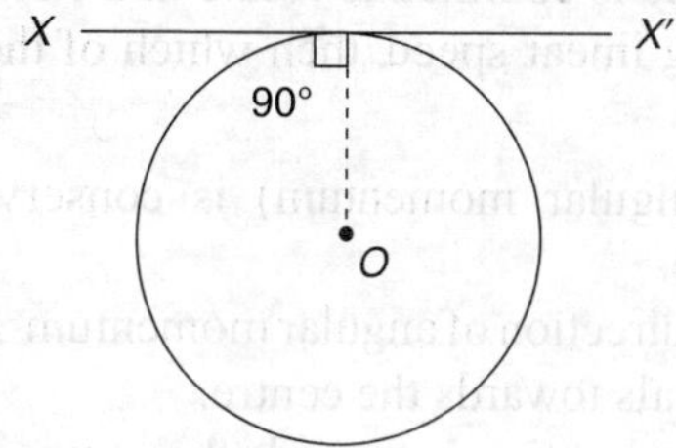

(a) $\frac{\rho L^3}{8\pi^2}$ (b) $\frac{\rho L^3}{16\pi^2}$
(c) $\frac{5\rho L^3}{16\pi^2}$ (d) $\frac{3\rho L^3}{8\pi^2}$

15. A cubical block of side L rests on a rough horizontal surface with coefficient of friction μ. A horizontal force F is applied on the block as shown. If the coefficient of friction is sufficiently high so that the block does not slide before toppling, the minimum force required to topple the block is

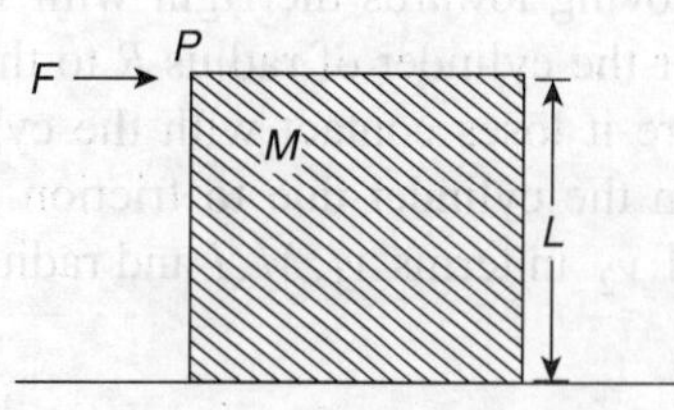

(a) infinitesimal (b) $\frac{mg}{4}$
(c) $\frac{mg}{2}$ (d) $mg(1-\mu)$

16. A small object of uniform density rolls up a curved surface with an initial velocity v. It reaches up to maximum height of $\frac{3v^2}{4g}$ with respect to the initial position. The object is

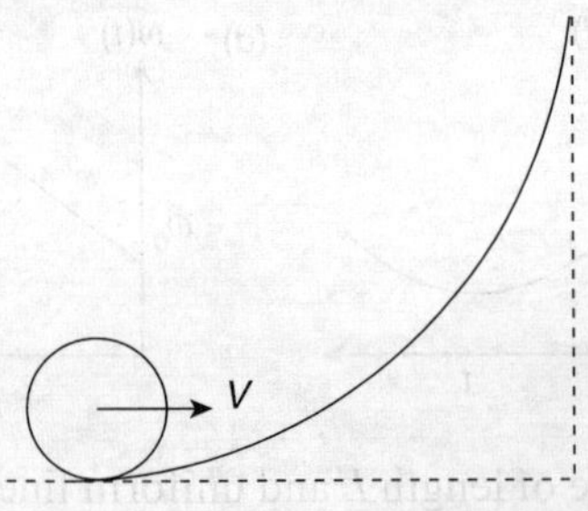

(a) ring (b) solid sphere
(c) hollow sphere (d) disc.

17. If a particle is confined to rotate in a circular path with decreasing linear speed, then which of the following is correct?

(a) $\vec{L}$ (angular momentum) is conserved about the centre.
(b) Only direction of angular momentum $\vec{L}$ is conserved
(c) It spirals towards the centre.
(d) Its acceleration is towards the centre.

18. A solid sphere of mass M and radius R having moment of inertia I about its diameter is recast into a solid disc of radius r and thickness t. The moment of inertia of the disc about an axis passing the edge and perpendicular to the plane remains I. Then R and r are related as

(a) $r=\sqrt{\frac{2}{15}}R$ (b) $r=\frac{2}{\sqrt{15}}R$

(c) $r=\frac{2}{15}R$ (d) $r=\frac{\sqrt{2}}{15}R$

19. The axis of the cylinder shown in figure is fixed and its initial angular velocity is zero. A block of mass m is initially moving towards the right with velocity v_1. It passes over the cylinder of radius R to the dotted position. Before it loses contact with the cylinder it stops slipping on the cylinder due to friction. Calculate the final speed v_2 in terms v_1, M, I and radius R.

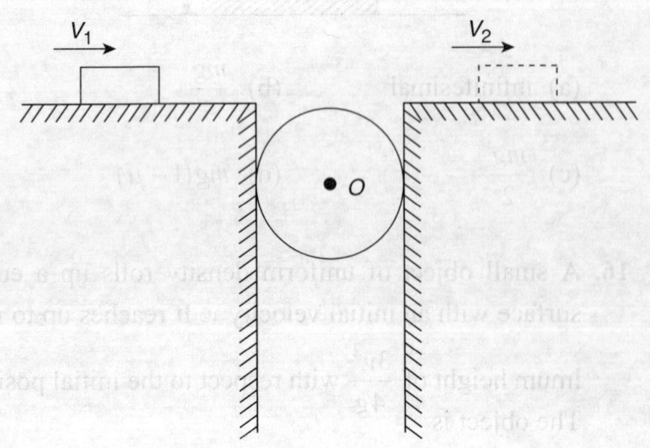

(a) $v_2=\frac{v_1}{(1+I/mR^2)}$ (b) $v_2=\frac{2v_1}{(1+I/mR^2)}$

(c) $v_2=\frac{2v_1}{(2+I/mR^2)}$ (d) $v_2=\frac{v_1}{(2+I/mR^2)}$

20. A uniform rod AB of mass m = 1.0 kg moves translation ally with acceleration 2.0 m/s^2 due to two anti-parallel forces F_1 and F_2. The distance between the points at which these forces act is equal to a = 20 cm. Besides, it is known that F_2 = 5.0N. Find the length of the rod.

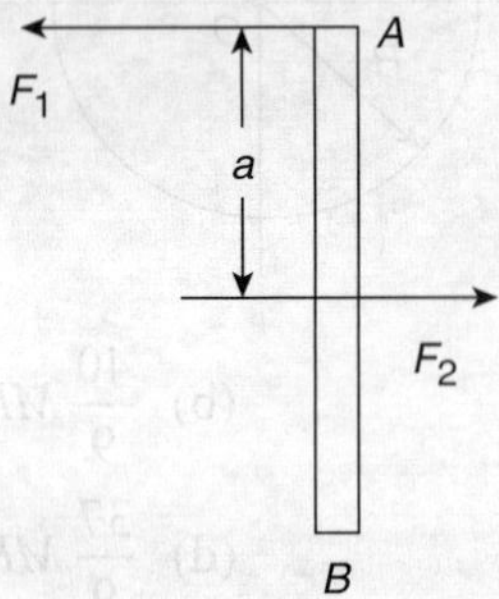

(a) 4 m (b) 3 m
(c) 2 m (d) 1 m

21. A bowling ball is thrown down in such a way that it slides with speed v_0 initially without rolling. Find the velocity of the ball when it starts rolling on the surface.

(a) $v_f=\frac{1}{7}v_0$ (b) $v_f=\frac{2}{7}v_0$

(c) $v_f=\frac{5}{7}v_0$ (d) $v_f=\frac{3}{7}v_0$

22. A particle moves in a circle with constant angular velocity ω about a point P on its circumference. The angular velocity of the particle about the centre C of the circle is

(a) 2 ω (b) $\frac{\omega}{2}$
(c) ω (d) not constant.

23. Two particles A and B are situated at a distance d = 2 m apart. Particle A has a velocity of 10 m/s at an angle of 60° and particle B has a velocity v at an angle 30° as shown in figure. The distance d between A and B at the instant shown in figure is constant. The angular velocity of B with respect to A is

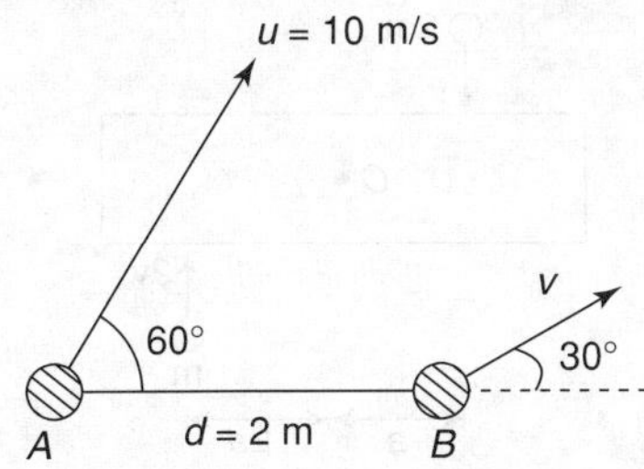

(a) $5\sqrt{3}$ rad/s (b) $\frac{5}{\sqrt{3}}$ rad/s

(c) $10\sqrt{3}$ rad/s (d) $\frac{10}{\sqrt{3}}$ rad/s

24. A uniform rod of length L and mass m is held vertically on smooth horizontal surface. At $t = 0$, it is slightly disturbed so that it starts slipping on the smooth surface. Find the velocity v_0 of centre of mass when rod makes an angle θ with the horizontal.

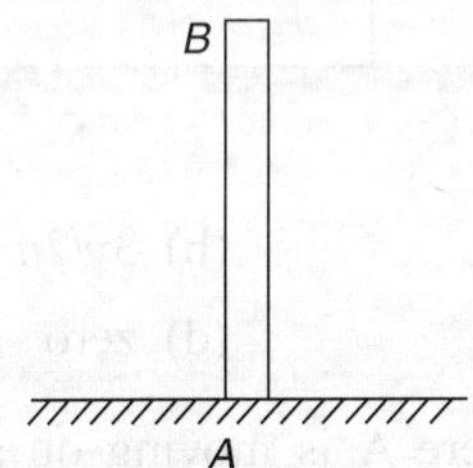

(a) $v_0 = \sqrt{\frac{12(1-\sin\theta)gL}{1+3\cos^2\theta}\left(\frac{7\cos\theta}{2}\right)}$

(b) $v_0 = \sqrt{\frac{12(1-\sin\theta)gL}{1+3\cos^2\theta}\left(\frac{5\cos\theta}{2}\right)}$

(c) $v_0 = \sqrt{\frac{12(1-\sin\theta)gL}{1+3\cos^2\theta}\left(\frac{1\cos\theta}{2}\right)}$

(d) $v_0 = \sqrt{\frac{12(1-\sin\theta)gL}{1+3\cos^2\theta}\left(\frac{3\cos\theta}{2}\right)}$

25. A billiards ball, initially at rest, is given a sharp impulse by a cue. The cue is held a distance h above centre line. The ball leaves the cue with a speed v_0 and acquired a final speed of $9/7v_0$. Find the value of h in terms of R, where R is the radius of ball.

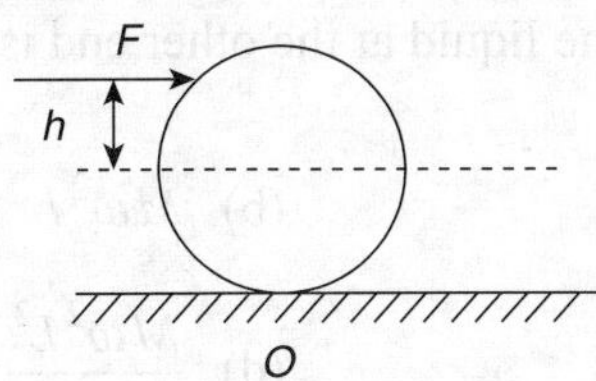

(a) $h = \frac{3}{5}R$ (b) $h = \frac{4}{5}R$

(c) $h = \frac{1}{5}R$ (d) $h = \frac{2}{5}R$

26. A solid sphere, ring and disc, all having same mass and radius, are sliding from top of an inclined plane having inclination θ. The least time will be taken for reaching the bottom by

(a) the solid sphere
(b) the ring
(c) the disc
(d) all will take the same time.

27. A wheel of radius R rolls on the ground with a uniform velocity v. The relative acceleration of top most point of the wheel with respect to the bottom most point is

(a) $\frac{v^2}{R}$ (b) $\frac{2v^2}{R}$

(c) $\frac{v^2}{2R}$ (d) $\frac{4v^2}{R}$

28. A thin smooth rod of length L and mass M is rotating freely with angular speed ω_0 about an axis perpendicular to the rod and passing through its centre. Two beads of mass m and negligible size are at the centre of the rod initially. The beads are free to slide along the rod. The angular speed of the system, when the beads reach the opposite ends of the rod, will be

(a) $\frac{M\omega_0}{M+3m}$ (b) $\frac{M\omega_0}{M+m}$

(c) $\frac{M\omega_0}{M+2m}$ (d) $\frac{M\omega_0}{M+6m}$

29. Let I be the moment of inertia of a uniform square plate about an axis AB that passes through its centre and is parallel to two of its sides. CD is a line in the plane of the plate that passes through the centre of the plate and makes an angle with AB. The moment of inertia of the plate about the axis CD is then equal to

(a) I (b) $I\sin^2\theta$

(c) $I\cos^2\theta$ (d) $I\cos^2(\theta/2)$

30. A thin circular ring of mass M and radius r is rotating about its axis with a constant angular velocity ω. Two objects, each of mass m, are attached gently to the opposite ends of a diameter of the ring. The wheel now rotates with an angular velocity

(a) $\omega M/(M+m)$ (b) $\omega(M-2m)/(M+2m)$
(c) $\omega M/(M+2m)$ (d) $\omega(M+2m)/M$

31. A homogeneous solid cylindrical roller of radius R and mass m is pulled on a cricket pitch by a horizontal force. Assuming rolling without slipping, angular acceleration of the cylinder is

(a) $\frac{F}{2mR}$ (b) $\frac{2F}{3mR}$

(c) $\frac{3F}{2mR}$ (d) $\frac{F}{3mR}$

32. Two solid cylinders P and Q of same mass and same radius start rolling down a fixed inclined plane from the same height at the same time. Cylinder P has most of its mass concentrated near its surface, while Q has most of its mass concentrated near the axis. Which statement(s) is (are) correct?

(a) Both cylinders P and Q reach the ground at the same time

(b) Cylinder P has larger linear acceleration than cylinder Q

(c) Both cylinders reach the ground with same translational kinetic energy

(d) Cylinder Q reaches the ground with larger angular speed.

33. A rod of length 50 cm is pivoted at one end. It is raised such that if makes an angle of 30° from the horizontal as shown and released from rest. Its angular speed when it passes through the horizontal (in rad s^{-1}) will be (take $g = 10$ ms^{-2})

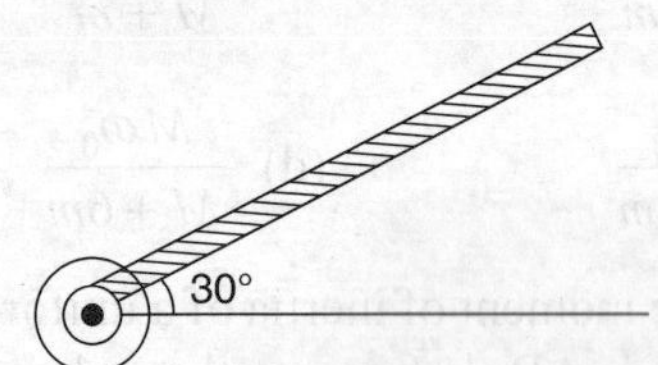

(a) $\frac{\sqrt{30}}{2}$ (b) $\sqrt{30}$

(c) $\frac{\sqrt{20}}{3}$ (d) $\sqrt{\frac{30}{2}}$

34. A uniform bar of length $6a$ and mass 8 m lies on a smooth horizontal table. Two point masses m and $2m$ moving in the same horizontal plane with speed $2v$ and v respectively, strike the bar [as shown in the figure] and are stuck to the bar after collision. Denoting angular velocity (about the centre of mass), total energy and centre of mass velocity by ω, E and v_c respectively, which of the following is incorrect after collision?

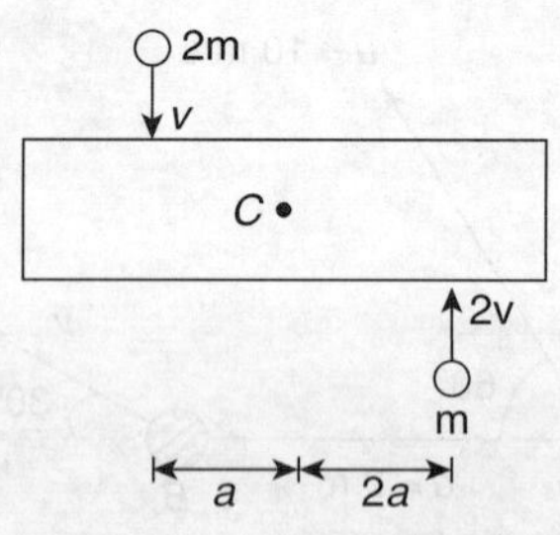

(a) $v_c = 0$ (b) $\omega = \frac{3v}{5a}$

(c) $\omega = \frac{v}{5a}$ (d) $E = \frac{3}{5}mv^2$

35. A cubical block of side a is moving with velocity v on a horizontal smooth plane as shown. It hits a ridge at point O. The angular speed of the block after it hits O is

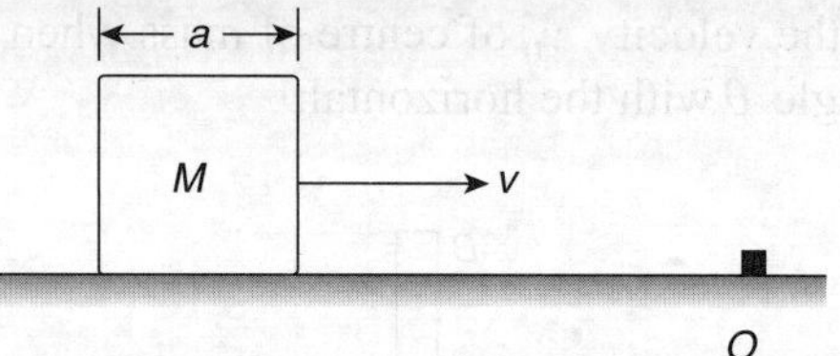

(a) $3v/4a$ (b) $3v/2a$

(c) $\sqrt{3}/\sqrt{2}a$ (d) zero

36. A smooth sphere A is moving on a frictionless horizontal plane with angular velocity ω and centre of mass velocity v. It collides elastically and head on with an identical sphere B at rest. Neglect frication everywhere. After the collision their angular speeds are ω_A and ω_B respectively. Then,

(a) $\omega_A < \omega_B$ (b) $\omega_A = \omega_B$

(c) $\omega_A = \omega$ (d) $\omega_B = \omega$

37. A particle moves in a circular path with decreasing speed. Choose the correct statement.

(a) Angular momentum remains constant.

(b) Acceleration (a) is towards the centre.

(c) Particle moves in a spiral path with decreasing radius.

(d) The direction of angular momentum remains constant.

38. A tube of length L is filled completely with an incompressible liquid of mass M and closed at both the ends. The tube is then rotated in a horizontal plane about one of its ends with a uniform angular velocity ω. The force exerted by the liquid at the other end is

(a) $\frac{M\omega^2 L}{2}$ (b) $M\omega^2 L$

(c) $\frac{M\omega^2 L}{4}$ (d) $\frac{M\omega^2 L^2}{2}$

39. Which of the following statements is incorrect?
(a) During rolling, the force of frication acts in the same direction as the direction of motion of the C.M. of the body
(b) The instantaneous speed of the point of contact during rolling is zero.
(c) The instantaneous acceleration of the point of contact during rolling is zero.
(d) A wheel moving down a perfect frictionless inclined plane will undergo slipping (not rolling) motion.

40. If there is a change of angular momentum from 1 J to 4 J in 4 sec, then the torque is
(a) 3/4 J (b) 1 J
(c) 5/4 J (d) 4/3 J

41. A thin rod of length L is lying along the x-axis with its ends at $x = 0$ and $x = L$. Its linear density (mass/length) varies with x as $k\left(x/L\right)^n$ where n can be zero or any positive number. If the position $x_{C.M.}$ of the centre of mass of the rod is plotted against n, which of the following graphs best approximates the dependence of $x_{C.M.}$ on n?

(a)
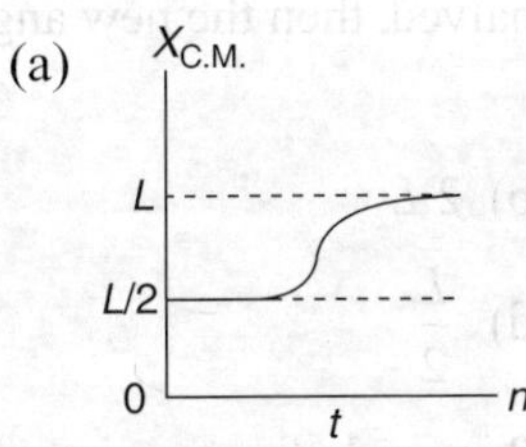

(b)
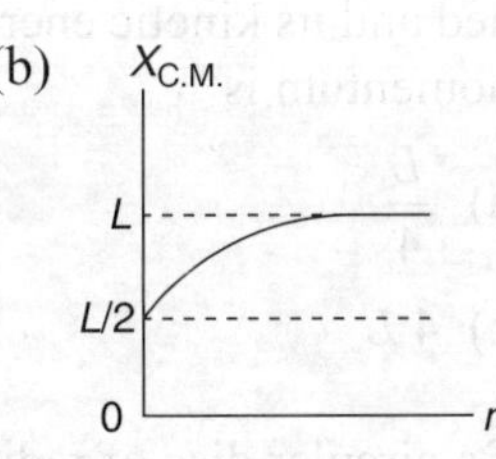

(c)
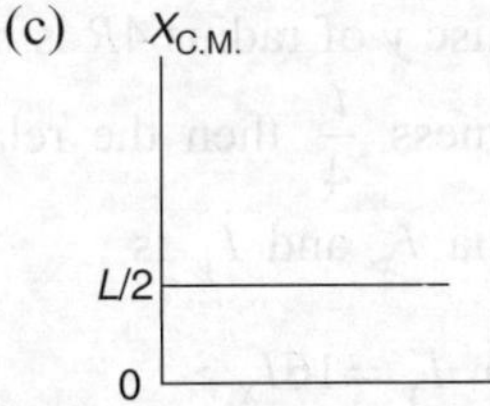

(d)
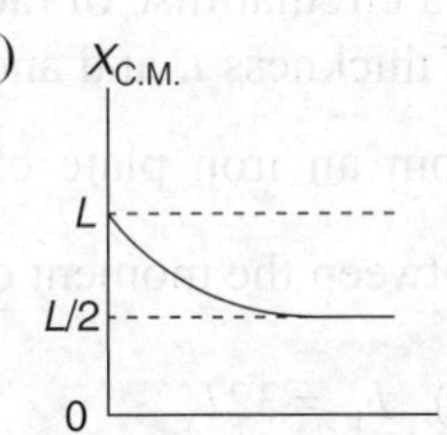

42. Consider a uniform square plate of side a and mass m. The moment of inertia of this plate about an axis perpendicular to its plane and passing through one of its corners is
(a) $\frac{2}{3}ma^2$ (b) $\frac{5}{6}ma^2$
(c) $\frac{1}{12}ma^2$ (d) $\frac{7}{12}ma^2$

43. Two bodies A and B are attracted towards each other due to gravitation. Given that A is much heavier than B. Which of the following correctly describes the relative motion of the centre of mass of the bodies?
(a) It moves towards A.
(b) It moves towards B.
(c) It moves perpendicular to the line joining the particles.
(d) It remains at rest w.r.t. A as well as B.

44. Angular momentum of the particle rotating with a central force is constant due to
(a) constant torque
(b) constant force
(c) constant linear momentum
(d) zero torque.

45. A round uniform body of radius R, mass M and moment of inertia I rolls down (without slipping) an inclined plane making an angle θ with the horizontal. Then its acceleration is
(a) $\dfrac{g\sin\theta}{1-\left(\dfrac{MR^2}{I}\right)}$ (b) $\dfrac{g\sin\theta}{1+\left(\dfrac{I}{MR^2}\right)}$
(c) $\dfrac{g\sin\theta}{1+\left(\dfrac{MR^2}{I}\right)}$ (d) $\dfrac{g\sin\theta}{1-\left(\dfrac{I}{MR^2}\right)}$

46. A circular disc of radius R is removed from a bigger circular disc of radius $2R$ such that the circumferences of the discs coincide. The centre of mass of the new disc is α/R from the centre of the bigger disc. The value of α is
(a) $\frac{1}{4}$ (b) $\frac{1}{3}$
(c) $\frac{1}{2}$ (d) $\frac{1}{6}$

47. Four point masses, each of value m, are placed at the corner of a square ABCD of side l. The moment of inertia of this system about an axis through A and parallel to BD is
(a) ml^2 (b) $2\,ml^2$
(c) $\sqrt{3}\,ml^2$ (d) $3\,ml^2$

48. The density of a rod gradually decreases from one end to the other. It is pivoted at an end so that it can move about a vertical axis through the pivot. A horizontal force F is applied on the free end in the direction perpendicular to the rod. The quantities that do not depend on which end of the rod is pivoted are:
(a) Angular acceleration
(b) Angular velocity when the rod completes one rotation
(c) Angular momentum when the rod completes one rotation
(d) Torque of the applied force.

49. A force of $-F\hat{K}$ acts on O, the origin of the co-ordinate system. The torque about the point (1, −1) is

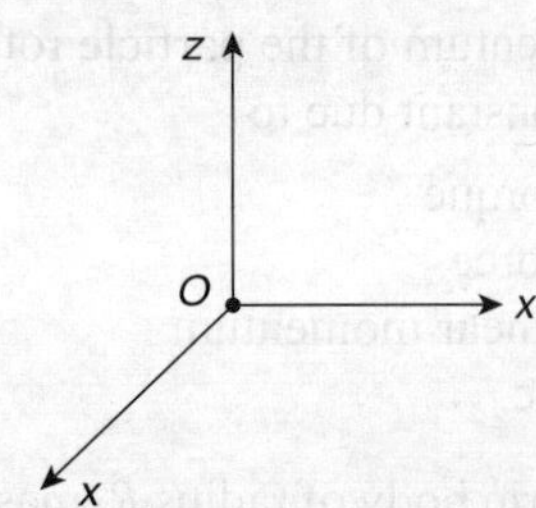

(a) $-F(\hat{i}-\hat{j})$ (b) $F(\hat{i}-\hat{j})$
(c) $-F(\hat{i}+\hat{j})$ (d) $F(\hat{i}+\hat{j})$

50. A rope of negligible mass is wound around a hollow cylinder of mass 3 kg and radius 40 cm. If the rope is pulled with a force of 30 N, what is the linear acceleration of the rope? Assume that there is no slipping.

(a) 10 m/s^2 (b) 5 m/s^2
(c) 25 m/s^2 (d) 15 m/s^2

51. Three identical particles, each of mass 1 kg, are placed touching each other with their centers on a straight line. Their centers are marked A, B and C respectively. The distance of centre of mass of the system from A is

(a) $\frac{AB+AC+BC}{3}$ (b) $\frac{AB+AC}{3}$
(c) $\frac{AB+BC}{3}$ (d) $\frac{AC+BC}{3}$

52. A body A of mass M while falling vertically downwards under gravity breaks into two parts; a body B of mass $\frac{1}{3}M$ and body C of mass $\frac{2}{3}M$. The centre of mass of bodies B and C taken together shifts compared to that of body A towards

(a) body C
(b) body B
(c) It depends on height of breaking
(d) It does not shift.

53. The moment of inertia of uniform semicircular disc of mass M and radius r about a line perpendicular to the plane of the disc through the centre is

(a) Mr^2 (b) $\frac{1}{2}Mr^2$
(c) $\frac{1}{4}Mr^2$ (d) $\frac{2}{5}Mr^2$

54. One solid sphere A and another hollow sphere B are of same mass and same outer radii. Their moment of inertia about their diameters are respectively I_A and I_B such that

(a) $I_A = I_B$ (b) $I_A > I_B$
(c) $I_A < I_B$ (d) $\frac{I_A}{I_B} = \frac{d_A}{d_B}$

where d_A and d_B are their densities.

55. A solid sphere is rotating in free space. If the radius of the sphere is increased keeping mass same, which one of the following will not be affected?

(a) Moment of inertia
(b) Angular momentum
(c) Angular velocity
(d) Rotational kinetic energy.

56. Three particles of masses 2 kg each are placed such that one lies on negative x-axis, 2nd one lies on negative y-axis and the third one lies on positive z-axis at distance 2 m, 3 m, and 1 m respectively from the origin. Then the square of the distance of centre of mass of the system from the origin is

(a) 1.55 m^2 (b) $\sqrt{1.55}$ m^2
(c) 1.55 m (d) 1.25 m^2

57. A particle performing uniform circular motion has angular momentum L. If its angular frequency is doubled and its kinetic energy halved, then the new angular momentum is

(a) $\frac{L}{4}$ (b) $2L$
(c) $4L$ (d) $\frac{L}{2}$

58. If a circular disc of radius R is made from an iron plate of thickness t, and another disc y of radius $4R$ is made from an iron plate of thickness $\frac{t}{4}$ then the relation between the moment of inertia I_x and I_y is

(a) $I_y = 32I_x$ (b) $I_y = 16I_x$
(c) $I_y = I_x$ (d) $I_y = 64I_x$

59. A particle of mass m moves along line PC with velocity v as shown. What is the angular momentum of the particle about P?

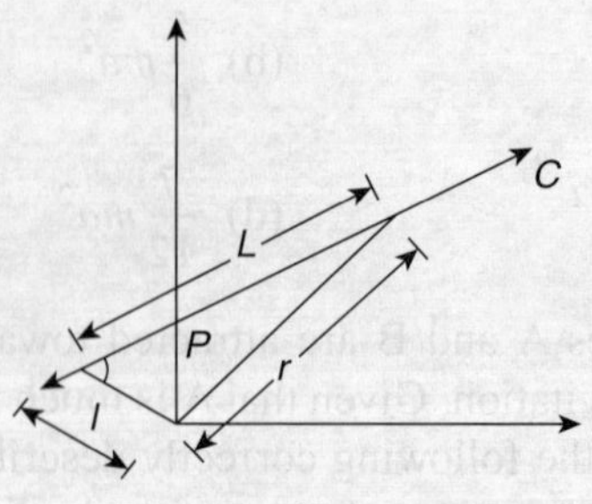

(a) mvl (b) mvl
(c) mvr (d) zero

60. Moment of inertia of a circular wire of mass M and radius R about its diameter is

(a) $\frac{MR^2}{2}$ (b) MR^2

(c) $2MR^2$ (d) $\frac{MR^2}{4}$

61. A solid sphere, a hollow sphere and a ring are released from top of an inclined plane (frictionless) so that they slide down the plane. The maximum acceleration down the plane is for (no rolling)

(a) solid sphere (b) hollow sphere
(c) ring (d) All same.

62. Initial angular velocity of a circular disc of mass M is ω_1. Then two small spheres of mass m are attached gently to two diametrically opposite points on the edges of the disc. What is the final angular velocity of the disc?

(a) $\left(\frac{M+m}{M}\right)\omega_1$ (b) $\left(\frac{M+m}{m}\right)\omega_1$

(c) $\left(\frac{M}{M+4m}\right)\omega_1$ (d) $\left(\frac{M}{M+2m}\right)\omega_1$

63. Two identical particles move towards each other with velocity $2v$ and v respectively. The velocity of centre of mass is

(a) v (b) $\frac{v}{3}$

(c) $\frac{v}{2}$ (d) zero.

64. The moment of inertia of a rod about an axis through its centre and perpendicular to it is $\frac{1}{2}ML^2$ (where M is the mass and L, the length of the rod). The rod is bent in the middle so that the two halves make an angle of 60°. The moment of inertia of the bent rod about the same axis would be

(a) $\frac{1}{48}ML^2$ (b) $\frac{1}{12}ML^2$

(c) $\frac{1}{24}ML^2$ (d) $\frac{ML^2}{8\sqrt{3}}$

65. A particle of mass m moves in the xy plane with a velocity v along the straight line AB. If the angular momentum of the particle with respect to origin O is L_A when it is at A and L_B when it is at B, then

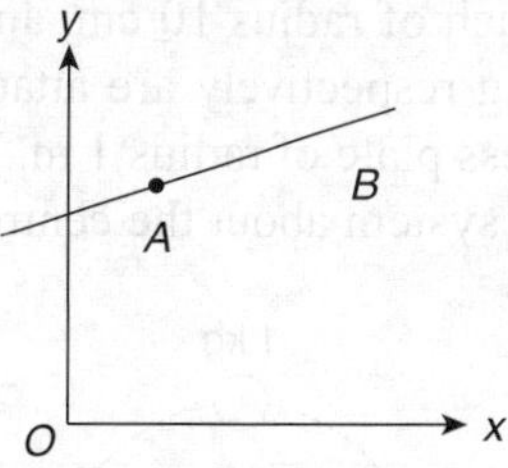

(a) $L_A > L_B$

(b) $L_A = L_B$

(c) the relationship between L_A and L_B depends upon the slope of the line AB.

(d) $L_A < L_B$

66. Two particles of masses 4 kg, 8 kg, are separated by a distance of 12 m. If they are moving towards each other under the influence of a mutual force of attraction, then the two particles will meet each other at a distance of – from 8 kg mass.

(a) 6 m (b) 2 m
(c) 4 m (d) 8 m

67. A disc is rolling on the inclined plane, what is the ratio of its rotational KE to the total KE?

(a) 1 : 3 (b) 3 : 1
(c) 1 : 2 (d) 2 : 1

68. A round disc of moment of inertia I_2 about its axis perpendicular to its plane and passing through its centre is placed over another disc of moment of inertia I_1 rotating with an angular velocity ω about the same axis. The final angular velocity about the combination of discs is

(a) $\frac{I_2\omega}{I_1+I_2}$ (b) ω

(c) $\frac{I_1\omega}{I_1+I_2}$ (d) $\frac{(I_1+I_2)\omega}{I_1}$

69. A sphere of radius R is rolling without sliding. What is the ratio of rotational kinetic energy to the total kinetic energy associated with body (radius of gyration of body is K)?

(a) $\frac{2}{7}$ (b) $\frac{2}{5}$

(c) 1 (d) $\frac{1}{2}$

70. A coin of mass 4.8 kg and radius 1 m is rolling on a horizontal surface without sliding with angular velocity 600 rotations/min. What is total KE of the coin?

(a) 360 J (b) 1440 π^2 J

(c) 4000 π^2 J (d) 600 π^2 J

71. Four balls, each of radius 10 cm, and mass 1 kg, 2 kg, 3 kg and 4 kg respectively are attached to the periphery of massless plate of radius 1 m. What is moment of inertia of the system about the centre of plate?

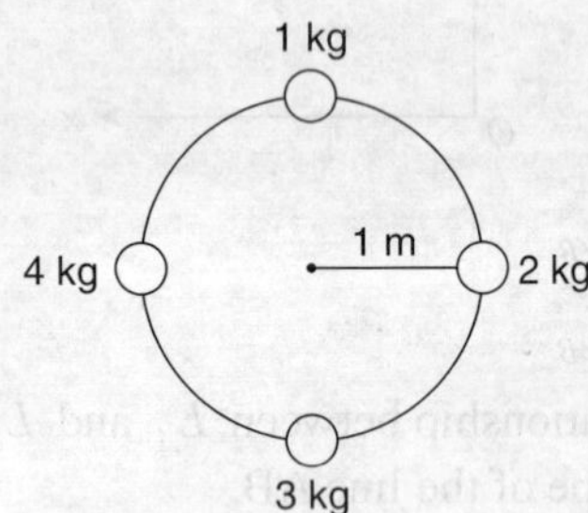

(a) 12.04 kg m^2 (b) 10.04 kg m^2
(c) 11.50 kg m^2 (d) 5.04 kg m^2

72. Two rings of radius R and nR made up of same material have the ratio of moment of inertia about an axis passing through their centre is 1 : 8. The value of n is

(a) 2 (b) $2\sqrt{2}$
(c) 4 (d) $\frac{1}{2}$

73. A cylinder rolls down an inclined plane of inclination 30°. The acceleration of cylinder is

(a) $\frac{g}{3}$ (b) g
(c) $\frac{g}{2}$ (d) $\frac{2g}{3}$

74. A solid sphere of radius R is rolling with velocity v on a smooth plane. The total kinetic energy of sphere is

(a) $\frac{7}{10}mv^2$ (b) $\frac{3}{4}mv^2$
(c) $\frac{1}{2}mv^2$ (d) $\frac{1}{4}mv^2$

75. The distance of the centre of mass of the T-shaped plate from O is

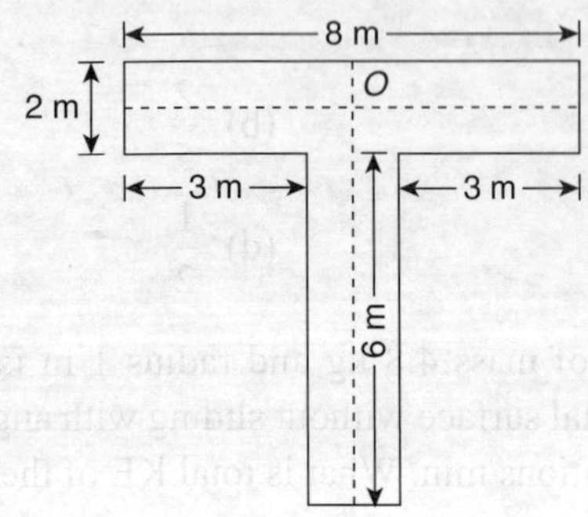

(a) 7 m (b) 2.7 m
(c) 4 m (d) 1 m

76. A couple is acting on a two particle system. The resultant motion will be
(a) purely rotational motion
(b) purely linear motion
(c) both (a) and (b)
(d) neither (a) nor (b).

77. The ratio of moments of inertia of circular ring and a circular disc having the same mass and radii about an axis passing through the centre and perpendicular to its plane is
(a) 1 : 1 (b) 2 : 1
(c) 1 : 2 (d) 4 : 1

78. Two spheres of equal masses, one of which is a thin spherical shell and the other a solid, have the same moment of inertia about their respective diameters. The ratio of their radii will be

(a) 5 : 7 (b) 3 : 5
(c) $\sqrt{3}:\sqrt{5}$ (d) $\sqrt{3}:\sqrt{7}$

79. The acceleration of the centre of mass of a uniform solid disc rolling down an inclined plane of angle α is

(a) $g \sin \alpha$ (b) $\frac{2}{3} g \sin \alpha$
(c) $\frac{1}{2} g \sin \alpha$ (d) $\frac{1}{3} g \sin \alpha$

80. A thin wire of mass M and length L is bent to form a circular ring. The moment of inertia of this ring about its axis is:

(a) $\frac{1}{4\pi^2}ML^2$ (b) $\frac{1}{12}ML^2$
(c) $\frac{1}{3\pi^2}ML^2$ (d) $\frac{1}{\pi^2}ML^2$

81. Three identical rods, each of length x, are joined to form a rigid equilateral triangle. Its radius of gyration about an axis passing through a corner and perpendicular to the triangle is

(a) $\frac{x}{\sqrt{3}}$ (b) $\frac{x}{2}$
(c) $\sqrt{\frac{3}{2}}x$ (d) $\frac{x}{\sqrt{2}}$

82. A solid sphere is rolling without slipping on a horizontal surface. The ratio of its rotational kinetic energy to its translational kinetic energy is

(a) $\frac{2}{9}$ (b) $\frac{2}{7}$
(c) $\frac{2}{5}$ (d) $\frac{7}{2}$

83. A solid sphere, disc and solid cylinder all of the same mass and made of the same material are allowed to roll down (from rest) on an inclined plane. Then

(a) the solid sphere reaches the bottom first
(b) the solid sphere reaches the bottom last
(c) the disc will reach the bottom first
(d) all reach the bottom at the same time.

84. Which one of the following is true?

(a) Momentum is conserved in all collisions and kinetic energy is conserved in elastic collisions.
(b) Momentum is conserved in all collisions but not kinetic energy.
(c) Both momentum and kinetic energy are conserved in elastic collisions.
(d) Neither momentum nor kinetic energy is conserved in elastic collisions.

85. What is moment of inertia in terms of angular momentum (L) and kinetic energy (K)?

(a) $\dfrac{L^2}{K}$ (b) $\dfrac{L^2}{2K}$

(c) $\dfrac{L}{2K^2}$ (d) $\dfrac{L}{2K}$

86. A small disc of radius 2 cm is cut form a disc of radius 6 cm. If the distance between their centers is 3.2 cm, what is the shift in the centre of mass of the disc?

(a) 0.4 cm (b) 2.4 cm
(c) 1.8 cm (d) 1.2 cm

87. In a system of particles 8 kg mass subjected to a force of 16 N along positive y-axis and another 8 kg mass is subjected to a force of 8 N along positive x-axis. What is the magnitude of acceleration of centre of mass and the angle made by it with x-axis?

(a) $\dfrac{\sqrt{5}}{2}$ m/s^2, $\theta = 45°$

(b) $3\sqrt{5}$ m/s^2 $\theta = \tan^{-1}\left(\dfrac{2}{3}\right)$

(c) $\dfrac{\sqrt{5}}{2}$ m/s^2, $\theta = \tan^{-1}(2)$

(d) 1 m/s^2, $\theta = \tan^{-1}\sqrt{3}$

88. A particle performs uniform circular motion with an angular momentum L. If the frequency of particle's motion is doubled and its K.E is halved, the angular momentum becomes

(a) $\dfrac{L}{2}$ (b) $2L$

(c) $4L$ (d) $\dfrac{L}{4}$

89. If the external forces acting on a system have zero resultant, the centre of mass

(a) may move but not accelerate
(b) may not move
(c) must not move
(d) None of the above.

90. A circular table rotates about a vertical axis with a constant angular speed ω. A circular pan rests on the rotating table and rotates along with the table. The bottom of the pan is covered with a uniform thick layer of ice which also rotates with the pan. The ice starts melting. The angular speed of the rotating table:

(a) remains the same
(b) decreases
(c) increases
(d) may increase or decrease depending upon the thickness of ice layer.

91. A couple produces

(a) a pure linear motion
(b) a pure rotational motion
(c) no motion
(d) both linear and rotational motion.

92. A body is rolling down an inclined plane. Its translational and rotational kinetic energies are equal. The body is a

(a) solid sphere (b) hollow sphere
(c) solid cylinder (d) hollow cylinder.

93. Two rods each of mass m and length l are joined at the centre to form a cross. The moment of inertia of this cross about an axis passing through the common centre of the rods and perpendicular to the plane formed by them, is

(a) $\dfrac{ml^2}{12}$ (b) $\dfrac{ml^2}{6}$

(c) $\dfrac{ml^2}{3}$ (d) $\dfrac{ml^2}{2}$

94. If the torque is zero, what will be the value of angular momentum?

(a) Constant in magnitude but changing in direction
(b) Changing in magnitude but constant in direction
(c) Constant in both magnitude and direction
(d) Zero.

95. A dancer is standing on a stool rotating about the vertical axis passing through its centre. She pulls her arms towards the body reducing her moment of inertia by factor of n. The new angular speed of turn table is proportional to

(a) n^1 (b) n^{-1}

(c) n^0 (d) n^2

96. A disc like reel with massless thread unrolls itself while falling vertically downwards. The acceleration of its fall is

(a) g (b) $\frac{g}{2}$

(c) zero (d) $\left(\frac{2}{3}\right)g$

97. A rod AB of mass M, length L is lying on a horizontal frictional surface. A particle of mass m travelling along the surface hits the end A of the rod with a velocity v_0 in a direction perpendicular to AB. The collision is completely elastic. After the collision, the particle comes to rest. The ratio of $\frac{m}{M}$ is

(a) $\frac{\omega^2 L^2}{9v_0^2}$ (b) $\frac{9v_0^2}{\omega^2 L^2}$

(c) $\frac{9v_0}{\omega L}$ (d) $\frac{\omega L}{9v_0}$

98. An equilateral triangle ABC is cut from a thin solid sheet of wood (see figure). D, E and F are the mid-points of its sides as shown and G is the centre of the triangle. The moment of inertia of the triangle about an axis passing through G and perpendicular to the plane of the triangle is I0. If the smaller triangle DEF is removed from ABC, the moment of inertia of the remaining figure about the same axis is I. Then

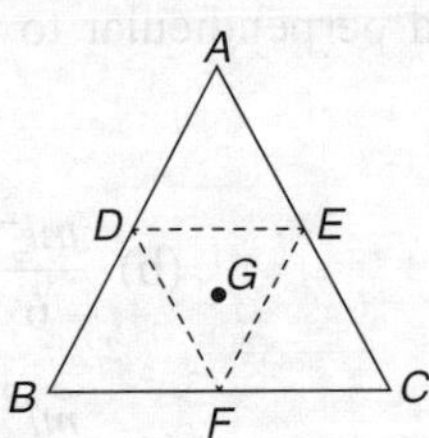

(a) $I = \frac{9}{16}I_0$ (b) $I = \frac{3}{4}I_0$

(c) $I = \frac{I_0}{4}$ (d) $I = \frac{15}{16}I_0$

99. The centre of mass of three particles of masses 1 kg, 2 kg and 3 kg is at (3, 3, 3) with reference to a fixed co-ordinate system. Where a fourth particle of mass 4 kg should be placed so that the centre of mass of the system of all particle shifts to a point (1, 1, 1)?

(a) (-1, -1, -1) (b) (-2, -2 , -2)
(c) (2, 2, 2) (d) (1, 1, 1)

100. If momenta of two particles of a system are given by $\vec{p}_1 = 2\hat{i} - \hat{j} + 3\hat{k}$ and $\vec{p}_2 = -\hat{i} + 2\hat{j} + 3\hat{k}$, then what is the angle made by the direction of motion of the system with $x-$axis?

(a) $\cos^{-1}\left(\frac{1}{6}\right)$ (b) $\cos^{-1}\sqrt{\frac{36}{38}}$

(c) 45° (d) $\cos^{-1}\sqrt{\frac{1}{38}}$

101. A uniform rod of length l and mass m is free to rotate in a vertical plane about A. The rod initially in horizontal position is released. The initial angular acceleration of the rod is (Moment of inertia of rod about A is $\frac{Ml^2}{3}$)

(a) $\frac{3g}{2l}$ (b) $\frac{2l}{3g}$

(c) $\frac{3g}{2l^2}$ (d) $mg\frac{l}{2}$

102. Choose the correct statement / statement from the following:

S_1: The angular-momentum of a body is always constant.

S_2: The direction of the angular momentum vector and the angular velocity vector are always same.

S_3: The direction of torque vector is always along the direction of change in angular momentum vector.

(a) S_1 and S_2 (b) S_2 and S_3

(c) S_1 only (d) S_3 only

103. Two particle of masses m_1 and m_2 ($m_1 > m_2$) initially at rest, move towards each other under an inverse square law force of attraction. Pick out the correct statement about the centre of mass (COM) of the system

(a) The COM moves towards m_1

(b) The COM moves towards m_2

(c) The COM remains at rest.

(d) The motion of COM is acceleration.

104. The angular acceleration a of a spinning top as a function of t is: $\alpha = 3t^2 + 5t$. At $t = 0$, the angular velocity $\omega_0 = 10$ rad/s and angular position $\theta_0 = 8$ rad. The angular position as a function of time t is given by which of the following expressions?

(a) $\frac{1}{4}t^4+\frac{5}{6}t^3+10t+8$ (b) $\frac{5}{6}t^4+\frac{1}{4}t^3+\frac{2}{5}t+8$

(c) $2t^4+3t^3+5t+8$ (d) $\frac{1}{5}t^4+\frac{3}{4}t^3+6t^2+8$

105. An 80 kg man is riding on a small 40 kg cart at a speed of 4 m/s. He jumps off the card with zero horizontal speed. What is the resulting change in the speed of the cart (in metres/s)?

(a) 4 (b) 8
(c) 12 (d) 0

106. Figure shows two particles, each having momentum $\overline{P}$. Which of the points O_1 and O_2 shown in the figure, can be chosen as origin so that the two particles have the same angular momentum vector, $\overline{L}$ about the origin?

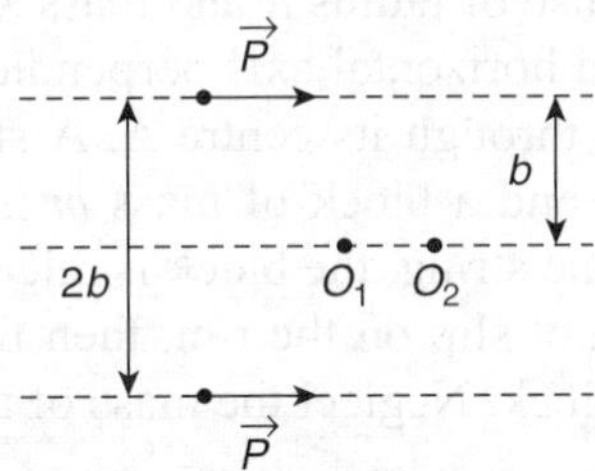

(a) O_1 (b) O_2
(c) Either of the two. (d) Neither of the two.

107. With O as the origin of the coordinate axis, the x and y-coordinates of the centre of mass of the system of particles shown in the figure may be given as

(a) $\left(-\frac{b}{2},O\right)$ (b) $\left(-\frac{b}{2},b\right)$

(c) $\left(-\frac{b}{3},b\right)$ (d) $\left(-\frac{2b}{5},b\right)$

Here m and $2\,m$ represent the masses of the particle.

108. A thin uniform circular disc of mass m and radius R is rotating in horizontal plane about an axis passing through its centre and perpendicular to the plane with an angular velocity ω. Another disc of same dimensions but of mass $\frac{1}{4}m$ is placed gently on the first disc co-axially. The angular velocity of the system is

(a) $\sqrt{2}\omega$ (b) $\frac{4}{5}\omega$

(c) $\frac{3}{4}\omega$ (d) $\frac{1}{3}\omega$

109. What is the magnitude of torque acting on a particle moving in the xy-plane about the origin, if its angular momentum is 4.0 $\sqrt{t}$ kg m²/s?

(a) $8\ t^{3/2}$ (b) $4.0/\sqrt{t}$

(c) $2.0/\sqrt{t}$ (d) $3/2\ \sqrt{t}$

110. A straight rod of length L has one of its ends at the origin and the other at $x=L$. If the mass per unit length of the rod is given by Ax where A is constant, where is its mass centre?

(a) $L/3$ (b) $L/2$
(c) $2L/3$ (d) $3L/4$

111. Four spheres each of diameter $2a$ and mass M are placed with their centres on the four corners of a square of side b. The moment of inertia of the system about any side of the square will be

(a) $\frac{8}{5}Ma^2+2Mb^2$ (b) $\frac{8}{5}Ma^2+Mb^2$

(c) $\frac{8}{5}Ma^2+4Mb^2$ (d) $\frac{8}{5}Ma^2+16Mb^2$

112. A ring of radius r and mass m rotates about an axis passing through its centre and perpendicular to its plane, its kinetic energy is

(a) $mr\omega$ (b) $\frac{1}{2}mr\omega^2$

(c) $mr\omega^2$ (d) $\frac{1}{2}mr^2\omega^2$

113. The moment of inertia of a thin circular disc about an axis passing through its centre and perpendicular to its plane is I. Then, the moment of inertia of the disc about an axis parallel to its diameter and touching the edge of the rim is

(a) I (b) $2I$

(c) $\frac{3}{2}I$ (d) $\frac{5}{2}I$

114. Two bodies of 6 kg and 4 kg masses have their velocities $5\hat{i}-2\hat{j}+10\hat{k}$, $10\hat{i}-2\hat{j}+5\hat{k}$ and $10\hat{i}-2\hat{j}+5\hat{k}$ respectively. Then the velocity of their centre of mass is

(a) $5\hat{i}-2\hat{j}-8\hat{k}$ (b) $7\hat{i}+2\hat{j}-8\hat{k}$
(c) $7\hat{i}-2\hat{j}+8\hat{k}$ (d) $5\hat{i}-2\hat{j}+8\hat{k}$

115. A uniform rod of length $8a$ and mass $6m$ lies on a smooth horizontal surface. Two points masses m and $2m$ moving in the same plane with speed $2v$ and v respectively strike the rod perpendicularly at distance a and $2a$ from the midpoint of the rod in opposite directions and stick to the rod. The angular velocity of the system immediately after collision is

(a) $\frac{6v}{32a}$ (b) $\frac{6v}{33a}$

(c) $\frac{6v}{40a}$ (d) $\frac{6v}{41a}$

116. The centre of mass of three particles of masses 1 kg, 2 kg and 3 kg is at (2, 2, 2). The position of the fourth mass of 4 kg to be placed in the system so that the new centre of mass is at (0, 0, 0) is

(a) (−3, −3, −3) (b) (−3, 3, −3)
(c) (2, 3, 3) (d) (2, −2, 3)

117. The instantaneous velocity of a point B of the given rod of length 0.5 m is 3 m/s in the represented direction. The angular velocity of the rod for minimum velocity of end A is

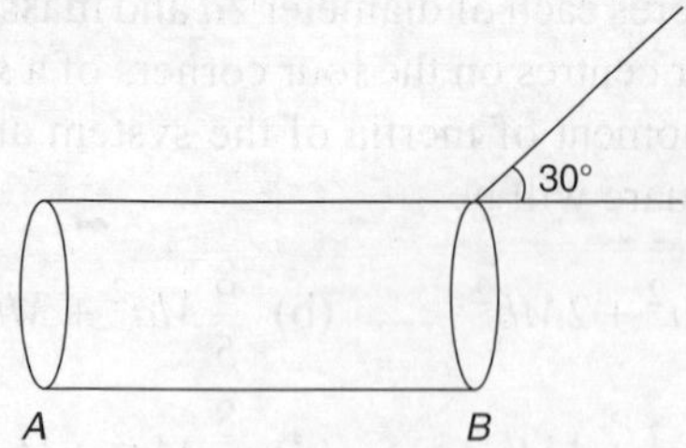

(a) 1.5 rad/s (b) 5.2 rad/s
(c) 2.5 rad/s (d) None of these.

118. Identify the increasing order of the angular velocities of the following:

1. Earth rotating about its own axis
2. Hour's hand of a clock
3. Second's hand of a clock
4. Flywheel of radius 2 m making 300 rpm.

(a) 1, 2, 3, 4 (b) 2, 3, 4, 1
(c) 3, 4, 1, 2 (d) 4, 1, 2, 3

119. Two particles of equal mass have velocities $\vec{v_1} = 4\hat{i}$ and $\vec{v_2} = 4\hat{j}\ ms^{-1}$. The first particle has acceleration $a_1 = (5\hat{i} + 5\hat{j})ms^{-1}$ while the acceleration of the other particle is zero. The centre of mass of the two particles moves in a path of:

(a) straight line (b) parabola
(c) circle (d) ellipse

120. A disc of mass 2 kg and radius 0.2 m is rotating with angular velocity 30 rad/s. What is angular velocity, if a mass of 0.25 kg is put on periphery of the disc?

(a) 24 rad/s (b) 36 rad/s
(c) 15 rad/s (d) 26 rad/s

121. A disc is rolling, the velocity of its centre of mass is v_{cm}. Which one will be correct?

(a) The velocity of highest point is 2 v_{cm} and at point of contact is zero.
(b) The velocity of highest point is v_{cm} and at point of contact is v_{cm}.
(c) The velocity of highest point is 2 v_{cm} and point of contact is v_{cm}.
(d) The velocity of highest point is 2 v_{cm} and point of contact is 2 v_{cm}.

122. For the adjoining diagram, the correct relation between I_1, I_2 and I_3 is, (*I*-moment of inertia)

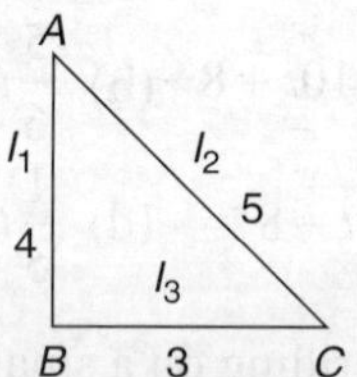

(a) $I_1 > I_2$ (b) $I_2 > I_1$
(c) $I_3 > I_1$ (d) $I_3 > I_2$

123. For a hollow cylinder and a solid cylinder rolling without slipping on an inclined plane, which of these reaches earlier?

(a) solid cylinder
(b) hollow cylinder
(c) both simultaneously
(d) Can't say anything.

124. A uniform disc of radius *R* and mass *A* is free to rotate about a fixed horizontal axis perpendicular to its plane and passing through its centre A. A string is wrapped over its rim and a block of mass *m* is attached to the free end of the string, the block is rcleased from rest. If string does not slip on the rim, then find the acceleration of the block. Neglect the mass of the string.

(a) $\frac{mg}{m+M}$ (b) $\frac{2mg}{m+M}$
(c) $\frac{2mg}{2m+M}$ (d) $\frac{3mg}{3m+M}$

125. Three identical metal balls, each of radius *r* are placed touching each other on a horizontal surface such that an equilateral triangle is formed when centres of three balls are joined. The centre of the mass of the system is located at

(a) line joining centres of any two balls
(b) centre of one of the balls
(c) horizontal surface
(d) point of intersection of the medians.

126. The moment of inertia of a disc of mass *M* and radius *R* about an axis, which is tangential to the circumference of the disc and parallel to its diameter is

(a) $\frac{5}{4}MR^2$ (b) $\frac{2}{3}MR^2$
(c) $\frac{3}{2}MR^2$ (d) $\frac{4}{5}MR^2$

127. Find the torque of a force $\overline{F} = 3\hat{i} + \hat{j} + 5\hat{k}$ acting at the point $\vec{r} = 7\hat{i} + 3\hat{j} + 5\hat{k}$

(a) $-21\hat{i} + 4\hat{j} + 4\hat{k}$ (b) $-14\hat{i} + 34\hat{j} - 16\hat{k}$
(c) $14\hat{i} - 38\hat{j} + 16\hat{k}$ (d) $4\hat{i} + 4\hat{j} + 6\hat{k}$

128. The centre of mass of system of particles does not depend on

(a) position of the particles
(b) relative distances between the particles
(c) masses of the particles
(d) forces acting on the particle.

129. A uniform rod is kept vertically on a horizontal smooth surface at a point O. If it is rotated slightly and released, it falls down on the horizontal surface. The lower end will remain:

(a) at O
(b) at a distance less than l/2 from O
(c) at a distance l/2 from O
(d) at a distance larger than l/2 from O.

130. A plank P is placed on a solid cylinder *S*, which rolls on a horizontal surface. The two are of equal mass. There is no slipping at any of the surfaces in contact. The ratio of the kinetic energy of *P* to the kinetic energy of *S* is

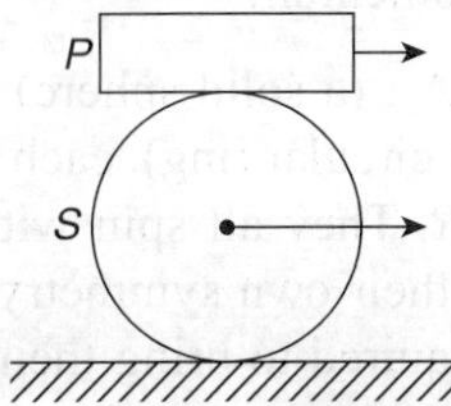

(a) 1:1 (b) 2:1
(c) 8:3 (d) 11:8

131. What is the torque of the force $\vec{F} = 2\hat{i} - 3\hat{j} + 4\hat{k}$ N acting at the point $\vec{r} = 3\hat{i} + 2\hat{j} + 3\hat{k}$ m about the origin?

(a) $-6\hat{i} + 6\hat{j} - 12\hat{k}$ (b) $-17\hat{i} + 6\hat{j} + 13\hat{k}$
(c) $6\hat{i} - 6\hat{j} + 12\hat{k}$ (d) $17\hat{i} - 6\hat{j} - 13\hat{k}$

132. A solid spherical ball rolls on a table. The ratio of its rotational kinetic energy to total kinetic energy is

(a) 1/2 (b) 1/6
(c) 7/10 (d) 2/7

133. In a rectangle ABCD (BC = 2AB). The moment of inertia is minimum along axis through

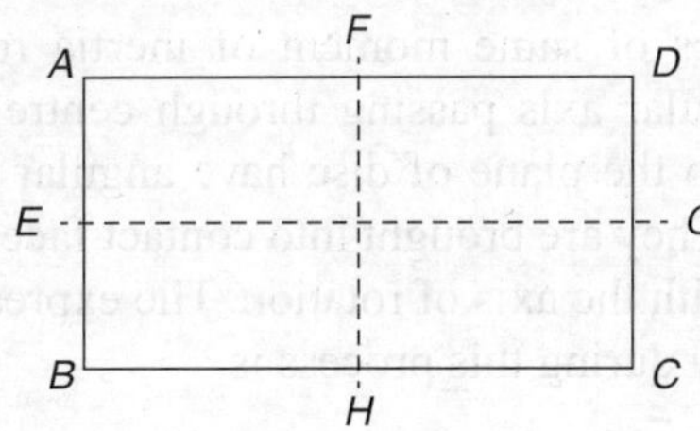

(a) BC (b) BD
(c) HF (d) EG

134. The angular velocity of the engine (and hence of the wheel) of a scooter is proportional to the petrol input per second. The scooter is moving on a frictionless road with uniform velocity. If the petrol input in increased by 1 0%, the linear velocity of the scooter is increased by

(a) 50% (b) 10%
(c) 20% (d) 0%

135. The speed of a homogeneous solid sphere after rolling down an inclined plane of vertical height *h* from rest without sliding is

(a) $\sqrt{\frac{10}{7}gh}$ (b) $\sqrt{gh}$
(c) $\sqrt{\frac{6}{5}gh}$ (d) $\sqrt{\frac{4}{3}gh}$

136. If a sphere is rolling, the ratio of the translational energy to total kinetic energy is given by

(a) 7: 10 (b) 2: 5
(c) 10: 7 (d) 5 : 7

137. A particle of mass $m = 5$ is moving with a uniform speed $v = 3\sqrt{2}$ in the XOY plane along the line Y = X + 4. The magnitude of the angular momentum of the particle about the origin is

(a) 60 units (b) $40\sqrt{2}$ units
(c) zero (d) 7.5 units.

138. A fly wheel rotating about fixed axis has a kinetic energy of 360 joule when its angular speed is 30 radian/sec. The moment of inertia of the wheel about the axis of rotation is

(a) 0.6 kg m^2 (b) 0.15 kg m^2
(c) 0.8 kg m^2 (d) 0.75 kg m^2

139. The moment of inertia of a body about a given axis is 1.2 kg m^2. Initially, the body is at rest. In order to produce a rotational kinetic energy of 1500 joule, an angular acceleration of 25 radian/sec^2 must be applied about that axis for a duration of

(a) 4 s (b) 2 s
(c) 8 s (d) 10 s

140. Moment of inertia of a uniform circular disc about a diameter is *I*. Its moment of inertia about an axis perpendicular to its plane and passing through a point on its rim will be

(a) 5*I* (b) 3*I*
(c) 6*I* (d) 4*I*

141. A solid homogenous sphere of mass *M* and radius *r* is moving on a rough horizontal surface, partly rolling and partly sliding. During this kind of motion of the sphere,

(a) the total kinetic energy is conserved
(b) the angular momentum of the sphere about the point of contact with the plane is conserved
(c) only the rotational kinetic energy about the centre of mass is conserved
(d) angular momentum about the centre of mass is conserved.

142. A ring of mass m and radius r rotates about an axis passing through its centre and perpendicular to its plane with angular velocity ω. Its kinetic energy is

(a) $\frac{1}{2}mr^2\omega^2$ (b) $mr\omega^2$
(c) $mr^2\omega^2$ (d) $\frac{1}{2}mr\omega^2$

143. Two coaxial discs, having moments of inertia I_1 and $\frac{I_1}{2}$ are rotating with respective angular velocities ω_1 and $\frac{\omega_1}{2}$ about their common axis. They are brought in contact with each other and thereafter thcy rotate with a common angular velocity. If E_f and E_t are the final and initial total energies, then $\left(E_f - E_t\right)$ is

(a) $-\frac{I_1\omega_1^2}{24}$ (b) $-\frac{I_1\omega_1^2}{12}$
(c) $\frac{3}{8}I_1\omega_1^2$ (d) $\frac{I_1\omega_1^2}{6}$

144. Only under gravity and some initial impulse, a sphere cannot roll without sliding on

(a) a smooth horizontal surface
(b) a smooth inclined surface
(c) a rough horizontal surface
(d) a rough inclined surface.

Questions Asked in NEET/CBSE Medicals

145. A solid cylinder of mass 2 kg and radius 4 cm is rotating about its axis at the rate of 3 rpm. The torque required to stop it after 2π revolutions is

(a) 2×10^6 N m (b) 2×10^{-6} N m
(c) 2×10^{-3} N m (d) 2×10^{-4} N m

146. A disc of radius 2 m and mass 100 kg rolls on a horizontal floor. Its centre of mass has speed of 20 cm/s. How much work is needed to stop it?

(a) 1 J (b) 3 J
(c) 30 k J (d) 2 J

147. A solid cylinder of mass 2 kg and radius 50 cm rolls up an inclined plane of angle inclination 30°. The centre of mass of cylinder has speed of 4 m/s. The distance travelled by the cylinder on the incline surface will be (Take $g = 10$ m/s^2)

(a) 2.2 m (b) 1.6 m
(c) 1.2 m (d) 2.4 m

148. A solid sphere is in rolling motion. In rolling motion a body possesses translational kinetic energy (K_t) as well as rotational kinetic energy (K_r) simultaneously. The ratio $K_t : (K_t + K_r)$ for the sphere is

(a) 7: 10 (b) 5 : 7
(c) 10: 7 (d) 2: 5

149. A solid sphere is rotating freely about its symmetry axis in free space. The radius of the sphere is increased keeping its mass same. Which of the following physical quantities would remain constant for the sphere?

(a) Angular velocity
(b) Moment of inertia
(c) Rotational kinetic energy
(d) Angular momentum

150. Three objects, A : (a solid sphere), B : (a thin circular disk) and C : (a circular ring), each have the same mass M and radius R. They all spin with the same angular speed w about their own symmetry axes. The amounts of work (W) required to bring them to rest, would satisfy the relation

(a) $W_C > W_B > W_A$ (b) $W_A > W_B > W_C$
(c) $W_B > W_A > W_C$ (d) $W_A > W_C > W_B$

151. The moment of the force, $\vec{F} = 4\hat{i} + 5\hat{j} - 6\hat{k}$ at (2, 0, -3), about the point (2, –2, –2), is given by

(a) $-8\hat{i} - 4\hat{j} - 7\hat{k}$ (b) $-4\hat{i} - \hat{j} - 8\hat{k}$
(c) $-7\hat{i} - 8\hat{j} - 4\hat{k}$ (d) $-7\hat{i} - 4\hat{j} - 8\hat{k}$

152. A rope is wound around a hollow cylinder of mass 3 kg and radius 40 cm. What is the angular acceleration of the cylinder if the rope is pulled with a force of 30 N?

(a) 0.25 rad s^{-2} (b) 25 rad s^{-2}
(c) 5 m s^{-2} (d) 25 m s^{-2}

153. Two discs of same moment of inertia rotating about their regular axis passing through centre and perpendicular to the plane of disc have angular velocities w_1 and w_2. They are brought into contact face to face coinciding with the axis of rotation. The expression for loss of energy during this process is

(a) $\frac{1}{4}I(w_1 - w_2)^2$ (b) $I(w_1 - w_2)^2$
(c) $\frac{1}{8}I(w_1 - w_2)^2$ (d) $\frac{1}{2}I(w_1 + w_2)^2$

154. Two rotating bodies A and B of masses m and $2m$ with moments of inertia I_A and I_B ($I_B > I_A$) have equal kinetic energy of rotation. If L_A and L_B be their angular momenta respectively, then

(a) $L_A = \frac{L_B}{2}$ (b) $L_A = 2L_B$
(c) $L_B > L_A$ (d) $L_A > L_B$

155. A solid sphere of mass m and radius R is rotating about its diameter. A solid cylinder of the same mass and same radius is also rotating about its geometrical axis with an angular speed twice that of the sphere. The ratio of their kinetic energies of rotation ($E_{sphere}/E_{cylinder}$) will be

(a) 2: 3 (b) 1 : 5
(c) 1 :4 (d) 3: 1

156. A light rod of length l has two masses m_1 and m_2 attached to its two ends. The moment of inertia of the system about an axis perpendicular to the rod and passing through the centre of mass is

(a) $\frac{m_1 m_2}{m_1 + m_2} l^2$ (b) $\frac{m_1 + m_2}{m_1 m_2} l^2$
(c) $(m_1 + m_2) l^2$ (d) $\sqrt{m_1 m_2} l^2$

157. A disc and a sphere of same radius but different masses roll off on two inclined planes of the same altitude and length. Which one of the two objects gets to the bottom of the plane first?

(a) Both reach at the same time
(b) Depends on their masses
(c) Disc
(d) Sphere

158. From a disc of radius R and mass M, a circular hole of diameter R, whose rim passes through the centre, is cut. What is the moment of inertia of the remaining part of the disc about a perpendicular axis passing through the centre?

(a) $11\ MR^2/32$ (b) $9\ MR^2/32$
(c) $15\ MR^2/32$ (d) $13\ MR^2/32$

159. A uniform circular disc of radius 50 cm at rest is free to turn about an axis which is perpendicular to its plane and passes through its centre. It is subjected to a torque which produces a constant angular acceleration of 2.0 rad s^{-2}. Its net acceleration in m s^{-2} at the end of 2.0 s is approximately

(a) 6.0 (b) 3.0
(c) 8.0 (d) 7.0

160. Point masses m_1 and m_2 are placed at the opposite ends of a rigid rod of length L, and negligible mass. The rod is to be set rotating about an axis perpendicular to it. The position of point P on this rod through which the axis should pass so that the work required to set the rod rotating with angular velocity ω_0 is minimum, is given by

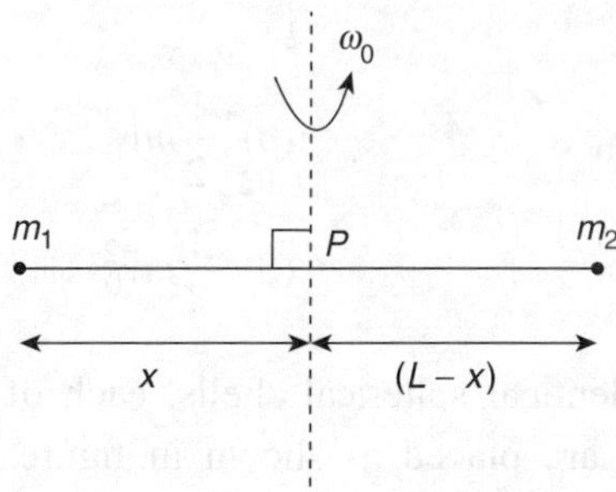

(a) $x = \frac{m_2}{m_1} L$ (b) $x = \frac{m_2 L}{m_1 + m_2}$
(c) $x = \frac{m_1 L}{m_1 + m_2}$ (d) $x = \frac{m_1}{m_2} L$

161. An automobile moves on a road with a speed of 54 km h^{-1}. The radius of its wheels is 0.45 m and the moment of inertia of the wheel about its axis of rotation is 3 kg m^2. If the vehicle is brought to rest in 15 s, the magnitude of average torque transmitted by its brakes to the wheel is

(a) 10.86 kg m^2 s^{-2} (b) 2.86 kg m^2 s^{-2}
(c) 6.66 kg m^2 s^{-2} (d) 8.58 kg m^2 s^{-2}

162. A force $\vec{F} = \alpha\hat{i} + 3\hat{j} + 6\hat{k}$ is acting at a point $\vec{r} = 2\hat{i} - 6\hat{j} - 12\hat{k}$. The value of α for which angular momentum about origin is conserved is

(a) zero (b) 1
(c) –1 (d) 2

163. A rod of weight W is supported by two parallel knife edges A and B and is in equilibrium in a horizontal position. The knives are at a distance d from each other. The centre of mass of the rod is at distance x from A. The normal reaction on A is

(a) $\frac{W(d-x)}{x}$ (d) $\frac{W(d-x)}{d}$
(c) $\frac{Wx}{d}$ (d) $\frac{Wd}{x}$

164. A mass m moves in a circle on a smooth horizontal plane with velocity v_0 at a radius R_0. The mass is attached to a string which passes through a smooth hole in the plane as shown. The tension in the string is increased gradually and finally m moves in a circle of radius $\frac{R_0}{2}$. The final value of the kinetic energy is

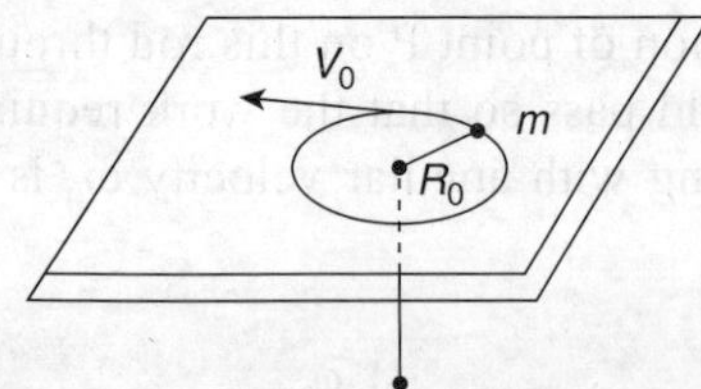

(a) $2mv_0^2$ (b) $\frac{1}{2}mv_0^2$

(c) mv_0^2 (d) $\frac{1}{4}mv_0^2$

165. Three identical spherical shells, each of mass m and radius r are placed as shown in figure. Consider an axis XX′ which is touching the two shells and passing through the diameter of the third shell. Moment of inertia of the system consisting of these three spherical shells about XX′ axis is

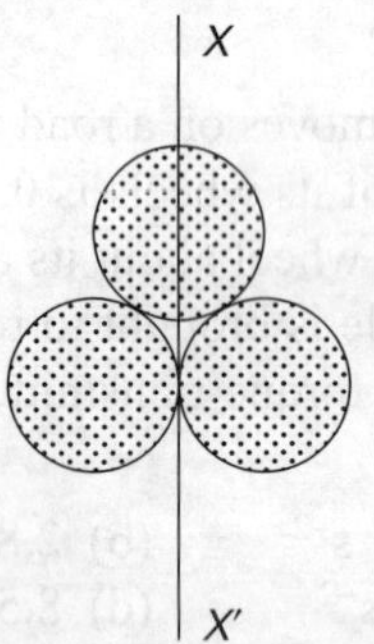

(a) $\frac{16}{5}mr^2$ (b) $4mr^2$

(c) $\frac{11}{5}mr^2$ (d) $3mr^2$

166. A solid cylinder of mass 50 kg and radius 0.5 m is free to rotate about the horizontal axis. A massless string is wound round the cylinder with one end attached to it and other hanging freely. Tension in the string required to produce an angular acceleration of 2 revolutions s^{-2} is

(a) 25 N (b) 50 N

(c) 78.5 N (d) 157 N

167. The ratio of the accelerations for a solid sphere (mass m and radius R) rolling down an incline of angle θ without slipping and slipping down the incline without rolling is

(a) 5 : 7 (b) 2 : 3

(c) 2 : 5 (d) 7:5

168. A rod PQ of mass M and length L is hinged at end P. The rod is kept horizontal by a massless string tied to point Q as shown in figure. When the string is cut, the initial angular acceleration of the rod is

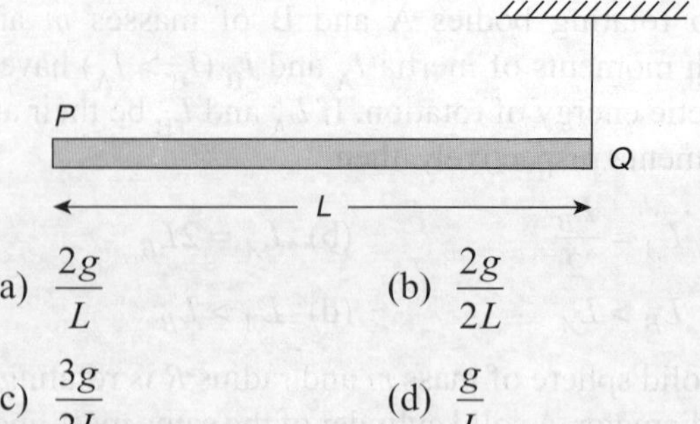

(a) $\frac{2g}{L}$ (b) $\frac{2g}{2L}$

(c) $\frac{3g}{2L}$ (d) $\frac{g}{L}$

169. A small object of uniform density rolls up a curved surface with an initial velocity v. It reaches up to a maximum height of $\frac{3v^2}{4g}$ with respect to the initial position. The object is a

(a) hollow sphere (b) disc

(c) ring (d) solid sphere.

170. Two discs are rotating about their axes, normal to the discs and passing through the centres of the discs. Disc D_1 has 2 kg mass and 0.2 m radius and initial angular velocity of 50 rad s^{-1}. Disc D_2 has 4 kg mass, 0.1 m radius and initial angular velocity of 200 rad s^{-1}. The two discs are brought in contact face to face, with their axes of rotation coincident. The final angular velocity (in rads^{-1}) of the system is

(a) 60 (b) 100

(c) 120 (d) 40

171. When a mass is rotating in a plane about a fixed point, its angular momentum is directed along

(a) a line perpendicular to the plane of rotation

(b) the line making an angle of 45° to the plane of rotation

(c) the radius

(d) the tangent to the orbit.

172. Two persons of masses 55 kg and 65 kg respectively, are at the opposite ends of a boat. The length of the boat is 3.0 m and weighs 100 kg. The 55 kg man walks up to the 65 kg man and sits with him. If the boat is in still water the centre of mass of the system shifts by

(a) 3.0 m (b) 2.3 m

(c) zero (d) 0.75 m

173. ABC is an equilateral triangle with O as its centre. $\overrightarrow{F_1}, \overrightarrow{F_2}$ and $\overrightarrow{F_3}$ represent three forces acting along the sides AB, BC and AC respectively. If the total torque about O is zero then the magnitude of $\overrightarrow{F_3}$ is

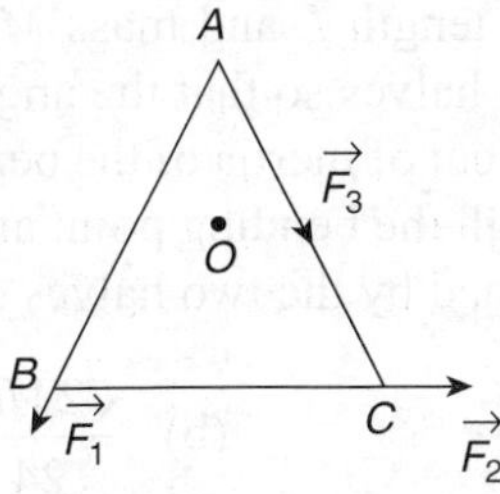

(a) $F_1 + F_2$ (b) $F_1 - F_2$

(c) $\frac{F_1 + F_2}{2}$ (d) $2(F_1 + F_2)$

174. A circular platform is mounted on a frictionless vertical axle. Its radius $R = 2$ m and its moment of inertia about the axle is 200 kg m^2. It is initially at rest. A 50 kg man stands on the edge of the platform and begins to walk along the edge at the speed of 1 m s^{-1} relative to the ground. Time taken by the man to complete one revolution is

(a) πs (b) $3\pi/2$s

(c) 2πs (d) $\pi/2$s

175. The moment of inertia of a uniform circular disc is maximum about an axis perpendicular to the disc and passing through

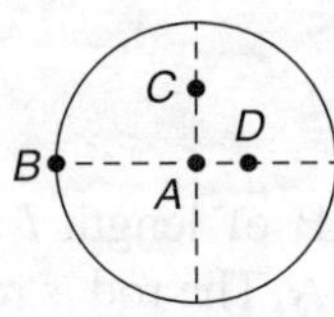

(a) B (b) C

(c) D (d) A

176. Three masses are placed on the x-axis: 300 g at origin, 500 g at $x = 40$ cm and 400 g at $x = 70$ cm. The distance of the centre of mass from the origin is

(a) 40 cm (b) 45 cm

(c) 50 cm (d) 30 cm

177. The instantaneous angular position of a point on a rotating wheel is given by the equation $\theta(t) = 2t^3 - 6t^2$. The torque on the wheel becomes zero at

(a) $t = 1$ s (b) $t = 0.5$ s

(c) $t = 0.25$ s (d) $t = 2$ s

178. The moment of inertia of a thin uniform rod of mass M and length L about an axis passing through its midpoint and perpendicular to its length is I_0. Its moment of inertia about an axis passing through one of its ends and perpendicular to its length is

(a) $I_0 + ML^2/2$ (b) $I_0 + ML^2/4$

(c) $I_0 + 2ML^2$ (d) $I_0 + ML^2$

179. A small mass attached to a string rotates on a frictionless table top as shown. If the tension in the string is increased by pulling the string causing the radius of the circular motion to decrease by a factor of 2, the kinetic energy of the mass will

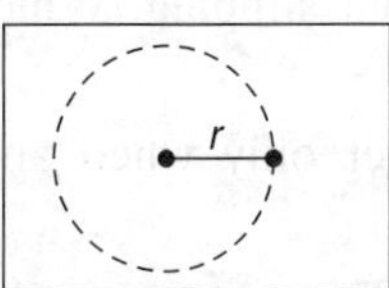

(a) decrease by a factor of 2

(b) remain constant

(c) increase by a factor of 2

(d) increase by a factor of 4.

180. A circular disk of moment of inertia I_t is rotating in a horizontal plane about its symmetry axis, with a constant angular speed ω_i. Another disk of moment of inertia I_b is dropped coaxially onto the rotating disk. Initially, the second disk has zero angular speed. Eventually both the disks rotate with a constant angular speed ω_f. The energy lost by the initially rotating disc to friction is

(a) $\frac{1}{2}\frac{I_b^2}{(I_t + I_b)}\omega_i^2$ (b) $\frac{1}{2}\frac{I_t^2}{(I_t + I_b)}\omega_i^2$

(c) $\frac{I_b - I_t}{(I_t + I_b)}\omega_i^2$ (d) $\frac{1}{2}\frac{I_b I_t}{(I_t + I_b)}\omega_i^2$

181. Two particles which are initially at rest move towards each other under the action of their internal attraction. If their speeds are v and $2v$ at any instant, then the speed of centre of mass of the system will be

(a) $2v$ (b) zero

(c) $1.5v$ (d) v

182. A gramophone record is revolving with an angular velocity ω. A coin is placed at a distance r from the centre of the record. The static coefficient of friction is μ. The coin will revolve with the record if

(a) $r = mg\omega^2$ (b) $r < \frac{\omega^2}{\mu g}$

(c) $r \le \frac{\mu g}{\omega^2}$ (d) $r \ge \frac{\mu g}{\omega^2}$

183. From a circular disc of radius R and mass $9M$, a small disc of mass M and radius $R/3$ is removed concentrically. The moment of inertia of the remaining disc about an axis perpendicular to the plane of the disc and passing through its centre is

(a) $\frac{40}{9}MR^2$ (b) MR^2

(c) $4MR^2$ (d) $\frac{4}{9}MR^2$

184. A solid cylinder and a hollow cylinder, both of the same mass and same external diameter, are released from the same height at the same time on an inclined plane. Both roll down without slipping. Which one will reach the bottom first?

(a) Both together only when angle of inclination of plane is 45°
(b) Both together
(c) Hollow cylinder
(d) Solid cylinder.

185. A thin circular ring of mass M and radius r is rotating about its axis with constant angular velocity ω. Two objects each of mass m are attached gently to the opposite ends of a diameter of the ring. The ring now rotates with angular velocity given by

(a) $\frac{(M+2m)\omega}{2m}$ (b) $\frac{2M\omega}{M+2m}$
(c) $\frac{(M+2m)\omega}{M}$ (d) $\frac{M\omega}{M+2m}$

186. A spherical solid ball of 1 kg mass and radius 3 cm is rotating about an axis passing through its centre with an angular velocity of 50 radian/s. The kinetic energy of rotation is

(a) 4500 J (b) 90 J
(c) 910 J (d) 9/20 J

187. If $\vec{F}$ is the force acting on a particle having position vector $\vec{r}$ and $\vec{\pi}$ be the torque of this force about the origin, then

(a) $\vec{r}.\vec{\pi} > 0$ and $\vec{F}.\vec{\pi} > 0$
(b) $\vec{r}.\vec{\pi} = 0$ and $\vec{F}.\vec{\pi} = 0$
(c) $\vec{r}.\vec{\pi} = 0$ and $\vec{F}.\vec{\pi} \neq 0$
(d) $\vec{r}.\vec{\pi} \neq 0$ and $\vec{F}.\vec{\pi} = 0$

188. Four identical thin rods, each of mass M and length l, form a square frame. Moment of inertia of this frame about an axis through the centre of the square and perpendicular to its plane is

(a) $\frac{2}{3}Ml^2$ (b) $\frac{13}{3}Ml^2$
(c) $\frac{1}{3}Ml^2$ (d) $\frac{4}{3}Ml^2$

189. Two bodies of mass 1 kg and 3 kg have position vectors $-2\hat{i} + 2\hat{j} + \hat{k}$ and $-3\hat{i} - 2\hat{j} + \hat{k}$, respectively. The centre of mass of this system has a position vector

(a) $-2\hat{i} - \hat{j} + \hat{k}$ (b) $2\hat{i} - \hat{j} - 2\hat{k}$
(c) $-\hat{i} + \hat{j} + \hat{k}$ (d) $-2\hat{i} + 2\hat{k}$

190. A thin rod of length L and mass M is bent at its mid-point into two halves so that the angle between them is 90°. The moment of inertia of the bent rod about an axis passing through the bending point and perpendicular to the plane defined by the two halves of the rod is

(a) $\frac{ML^2}{6}$ (b) $\frac{\sqrt{2}ML^2}{24}$
(c) $\frac{ML^2}{24}$ (d) $\frac{ML^2}{12}$

191. The ratio of the radii of gyration of a circular disc to that of a circular ring, each of same mass and radius, around their respective axes is

(a) $\sqrt{2}:1$ (b) $\sqrt{2}:\sqrt{3}$
(c) $\sqrt{3}:\sqrt{2}$ (d) $1:\sqrt{2}$

192. A particle of mass m moves in the XY plane with a velocity v along the straight line AB. If the angular momentum of the particle with respect to origin 0 is L_A when it is at A and L_B when it is at B, then

(a) $L_A = L_B$
(b) the relationship between L_A and L_B depends upon the slope of the line AB
(c) $L_A < L_B$
(d) $L_A > L_B$

193. A uniform rod AB of length l and mass m is free to rotate about point A. The rod is released from rest in the horizontal position. Given that the moment of inertia of the rod about A is $ml^2/3$, the initial angular acceleration of the rod will be

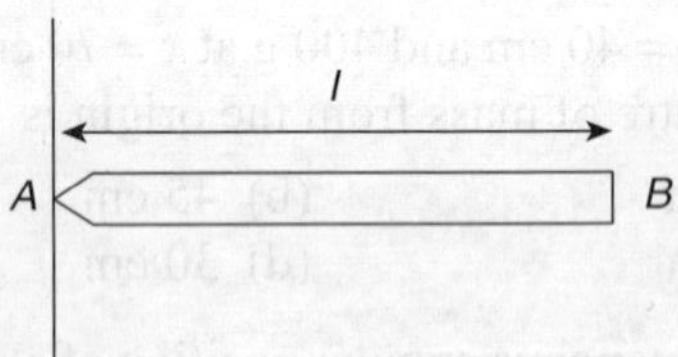

(a) $\frac{mgl}{2}$ (b) $\frac{3}{2}gl$
(c) $\frac{3g}{2l}$ (d) $\frac{2g}{3l}$

194. A wheel has angular acceleration of 3.0 rad/sec^2 and an initial angular speed of 2.00 rad/sec. In a time of 2 sec, it has rotated through an angle (in radian) of

(a) 10 (b) 12
(c) 4 (d) 6

195. The moment of inertia of a uniform circular disc of radius R and mass M about an axis touching the disc at its diameter and normal to the disc is

(a) $\frac{1}{2}MR^2$ (b) MR^2

(c) $\frac{2}{5}MR^2$ (d) $\frac{3}{2}MR^2$

196. A tube of length L is filled completely with an incompressible liquid of mass M and closed at both the ends. The tube is then rotated in a horizontal plane about one of its ends with a uniform angular velocity w. The force exerted by the liquid at the other end is

(a) $\frac{ML^2\omega^2}{2}$ (b) $\frac{ML\omega^2}{2}$

(c) $\frac{ML^2\omega}{2}$ (d) $ML\omega^2$

197. The moment of inertia of a uniform circular disc of radius R and mass M about an axis passing from the edge of the disc and normal to the disc is

(a) MR^2 (b) $\frac{1}{2}MR^2$

(c) $\frac{3}{2}MR^2$ (d) $\frac{7}{2}MR^2$

198. A drum of radius R and mass M, rolls down without slipping along an inclined plane of angle θ. The frictional force

(a) dissipates energy as heat
(b) decreases the rotational motion
(c) decreases the rotational and translational motion
(d) converts translational energy to rotational energy.

199. Two bodies have their moments of inertia I and $2I$ respectively about their axis of rotation. If their kinetic energies of rotation are equal, their angular velocity will be in the ratio

(a) 2 : 1 (b) 1 : 2

(c) $\sqrt{2}$: 1 (d) 1 : $\sqrt{2}$

200. Three particles, each of mass m gram, are situated at the vertices of an equilateral triangle ABC of side l cm (as shown in the figure). The moment of inertia of the system about a line AX perpendicular to AB and in the plane of ABC, in gram-cm^2 units will be

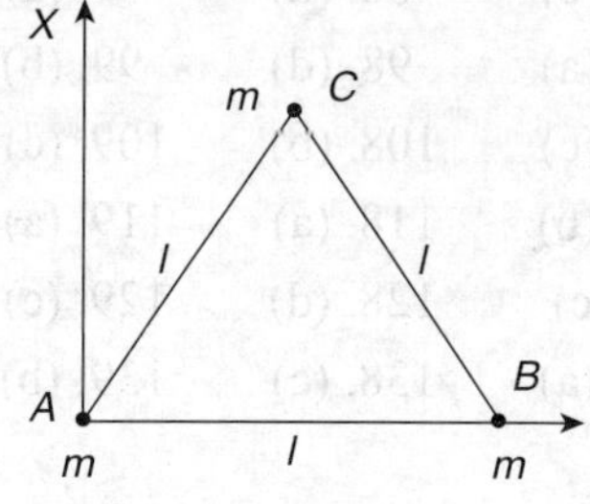

(a) $\frac{3}{4}ml^2$ (b) $2ml^2$

(c) $\frac{5}{4}ml^2$ (d) $\frac{3}{2}ml^2$

201. Consider a system of two particles having masses m_1 and m_2. If the particle of mass m_1 is pushed towards the centre of mass of the particles through a distance d, by what distance would be the particle of mass m_2 move so as to keep the centre of mass of the particles at the original position?

(a) $\frac{m_1}{m_1 + m_2}d$ (b) $\frac{m_1}{m_2}d$

(c) d (d) $\frac{m_2}{m_1}d$

202. A wheel having moment of inertia 2 kg m^2 about its vertical axis rotates at the rate of 60 rpm about this axis. The torque which can stop the wheel's rotation in one minute would be

(a) $\frac{2\pi}{15}$ Nm (b) $\frac{\pi}{12}$ Nm

(c) $\frac{\pi}{15}$ Nm (d) $\frac{\pi}{18}$ Nm

203. A round disc of moment of inertia I_2 about its axis perpendicular to its plane and passing through its centre is placed over another disc of moment of inertia I_1 rotating with an angular velocity ω about the same axis. The final angular velocity of the combination of discs is

(a) $\frac{I_2\omega}{I_1 + I_2}$ (b) ω

(c) $\frac{I_1\omega}{I_1 + I_2}$ (d) $\frac{(I_1 + I_2)\omega}{I_1}$

204. The ratio of the radii of gyration of a circular disc about a tangential axis in the plane of the disc to a circular ring of the same radius about a tangential axis in the plane of the ring is

(a) 2 : 3 (b) 2 : 1

(c) $\sqrt{5} : \sqrt{6}$ (d) 1: $\sqrt{2}$

205. A ball rolls without slipping. The radius of gyration of the ball about an axis passing through its centre of mass is K. If radius of the ball be R, then the fraction of total energy associated with its rotational energy will be

(a) $\frac{k^2 + R^2}{R^2}$ (b) $\frac{k^2}{R^2}$

(c) $\frac{k^2}{k^2 + R^2}$ (d) $\frac{R^2}{k^2 + R^2}$

206. A solid cylinder of mass M and radius R rolls without slipping down an inclined plane of length L and height h. What is the speed of its centre of mass when the cylinder reaches its bottom?

(a) $\sqrt{2gh}$ (b) $\sqrt{\frac{3}{4}gh}$
(c) $\sqrt{\frac{4}{3}gh}$ (d) $\sqrt{4gh}$

207. A thin circular ring of mass M and radius r is rotating about its axis with a constant angular velocity ω. Four objects, each of mass m, are kept gently at the opposite ends of two perpendicular diameters of the ring. The angular velocity of the ring will be

(a) $\frac{M\omega}{4m}$ (b) $\frac{M\omega}{M+4m}$
(c) $\frac{(M+4m)\omega}{M}$ (d) $\frac{(M-4m)\omega}{M+4m}$

208. A rod of length 3 m and its mass per unit length is directly proportional to distance x from one of its end. Then its centre of gravity from that end will be at

(a) 1.5 m (b) 2 m
(c) 2.5 m (d) 3.0 m.

209. Consider a contact point P of a wheel which rolls on the ground without slipping. Then value of displacement of point P when the wheel completes a half-rotation (if radius of wheel is 1 m) is

(a) 2 m (b) $\sqrt{\pi^2+4}m$
(c) π m (d) $\sqrt{\pi^2+2}m$

210. A solid sphere of radius R is placed on a smooth horizontal surface. A horizontal force F is applied at height h from the lowest point. For the maximum acceleration of centre of mass, which is correct?

(a) $h = R$
(b) $h = 2R$
(c) $h = 0$
(d) No relation between h and R.

211. A disc is rotating with angular speed ω. If a child sits on it, what is conserved?

(a) linear momentum (b) angular momentum
(c) kinetic energy (d) potential energy.

212. A circular disc is to be made by using iron and aluminium so that it acquires maximum moment of inertia about geometrical axis. It is possible by using

(a) aluminium at interior and iron surrounding it
(b) iron at interior and aluminium surrounding it
(c) iron and aluminium layers in alternate order
(d) sheet of iron at both external surfaces and aluminium sheet as internal layers.

Answer Keys

EXERCISES

1.(d)	2. (a)	3. (d)	4. (d)	5. (a)	6. (a)	7. (c)	8. (b)	9. (b)	10. (a)
11. (b)	12. (d)	13. (b)	14. (d)	15. (c)	16. (d)	17. (b)	18. (b)	19. (a)	20. (d)
21. (c)	22. (a)	23. (b)	24. (c)	25. (b)	26. (d)	27. (b)	28. (d)	29. (a)	30. (c)
31. (b)	32. (d)	33. (b)	34. (b)	35. (a)	36. (c)	37. (d)	38. (a)	39. (c)	40. (a)
41. (b)	42. (a)	43. (d)	44. (d)	45. (b)	46. (b)	47. (d)	48. (d)	49. (d)	50. (a)
51. (b)	52. (d)	53. (b)	54. (c)	55. (b)	56. (a)	57. (a)	58. (d)	59. (d)	60. (a)
61. (d)	62. (c)	63. (c)	64. (b)	65. (b)	66. (c)	67. (a)	68. (c)	69. (a)	70. (b)
71. (b)	72. (a)	73. (a)	74. (a)	75. (b)	76. (b)	77. (b)	78. (c)	79. (b)	80. (a)
81. (a)	82. (c)	83. (a)	84. (a)	85. (b)	86. (a)	87. (c)	88. (d)	89. (a)	90. (b)
91. (b)	92. (d)	93. (b)	94. (c)	95. (a)	96. (d)	97. (a)	98. (d)	99. (b)	100. (d)
101. (a)	102.(a)	103. (c)	104. (a)	105. (d)	106. (c)	107. (c)	108. (b)	109. (c)	110. (b)
111. (a)	112. (d)	113. (d)	114. (c)	115. (d)	116. (a)	117. (b)	118. (a)	119. (a)	120. (a)
121. (a)	122. (b)	123. (a)	124. (c)	125. (d)	126. (a)	127.(c)	128. (d)	129. (c)	130. (c)
131. (d)	132. (d)	133. (d)	134. (d)	135. (a)	136. (d)	137. (a)	138. (c)	139. (b)	140. (c)
141. (b)	142. (a)	143. (a)	144. (b)						

Questions Asked in NEET/CBSE Medicals

145. (b)	146. (b)	147. (d)	148. (b)	149. (d)	150. (a)	151. (d)	152. (b)	153. (a)	154. (c)
155. (b)	156. (a)	157. (d)	158. (d)	159. (c)	160. (b)	161. (c)	162. (c)	163. (b)	164. (a)
165. (b)	166. (d)	167. (a)	168. (c)	169. (b)	170. (b)	171. (a)	172. (c)	173. (a)	174. (c)
175. (a)	176. (a)	177. (a)	178. (b)	179. (d)	180. (d)	181. (b)	182. (c)	183. (a)	184. (d)
185. (d)	186. (d)	187. (b)	188. (d)	189. (a)	190. (d)	191. (d)	192. (a)	193. (c)	194. (a)
195. (d)	196. (b)	197. (c)	198. (d)	199. (c)	200. (c)	201. (b)	202. (c)	203. (c)	204. (c)
205. (c)	206. (c)	207. (b)	208. (b)	209. (b)	210. (d)	211. (b)	212. (a)		

Solutions

1. (d)

Let m_2 be moved by x so as to keep the centre of mass at the same position.

$\therefore \quad m_1 d + -m_2(-x) = 0$

or $\quad m_1 d = m_2 x$

or $\quad x = \dfrac{m_1}{m_2}.d$

2. (a)

It is a case of translational motion without rotation. The force should act at the centre of mass

$$l_{C.M.} = \frac{(m\times 2l)+(2m\times l)}{m+2m} = \frac{4}{3}l$$

Alternative

For pure translatory notion, net torque about centre of mass should be zero.

Thus, $\vec{F}$ is applied at centre of mass of system.

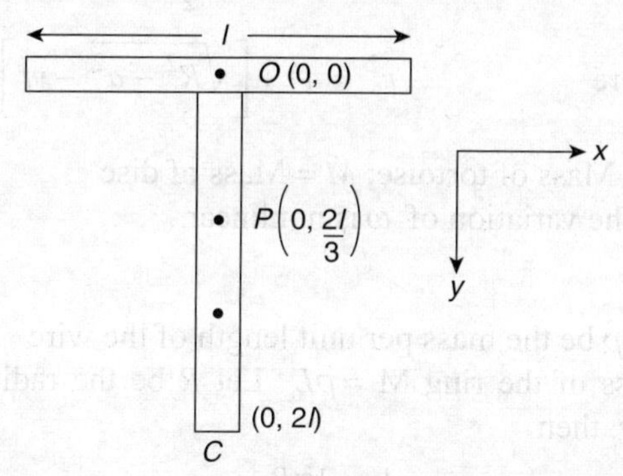

$$OP = \frac{m_1\times 0 + m_2\times l}{m_1+m_2}$$

where m_1 and m_2 are masses of horizontal and vertical section of the object. Assuming object is uniform,

$$m_2 = 2m_1 \Rightarrow OP = \frac{2l}{3}$$

$$\therefore \quad PC = \left(2l - \frac{2l}{3}\right) = \frac{4l}{3}$$

3. (d)

Retarding torque is constant. Therefore, angular retardation say α will also be constant. Applying

$$\omega^2 = \omega_0^2 - 2\alpha\theta$$

We get, $\quad \left(\dfrac{\omega_0}{2}\right)^2 = \omega_0^2 - 2\alpha\theta_1 \quad \ldots(i)$

and $\quad 0 - \left(\dfrac{\omega_0}{2}\right)^2 = 2\alpha\theta_2 \quad \ldots(ii)$

Solving Eqs. (i) and (ii),

$$\theta_2 = \frac{\theta_1}{3}$$

Therefore, the disc will make $\dfrac{n}{3}$ more rotations before coming to rest.

4. (d)

$m = 1$ kg, $|\tau| = 2.5$ N-m, $F = 1$ N and $r = 5$ m

We know that torque $\quad |\tau| = rF\sin\theta$

$$\Rightarrow 2.5 = 5\times 1\times \sin\theta$$

$$\Rightarrow \sin\theta = \frac{1}{2}$$

or $\quad \theta = \dfrac{\pi}{6}$ rad

5. (a)

From theorem of parallel axis,

$$I = I_{cm} + 7M(3R)^2$$

$$= \left[\frac{MR^2}{2} + 6\times\left\{\frac{MR^2}{2} + M(2R)^2\right\}\right] + 7M(3R)^2$$

$$= \frac{181MR^2}{2}$$

6. (a)

In uniform circular motion, the only force acting on the particle is centripetal (towards centre). Torque of this force about the centre is zero. Hence, angular momentum about centre remains conserved.

7. (c)

$$L_0 = L_{cm} + m(r\times v)$$

Angular momentum about O = Angular momentum about *C.M.* + Angular momentum of *C.M.* about origin.

$$\therefore L_0 = I\omega + MRv$$

$$= \frac{1}{2}MR^2\omega + MR(R\omega) = \frac{3}{2}MR^2\omega$$

8. (b)

$$|v| = v = \text{constant and } |r| = r$$

Angular momentum of the particle about origin O will be given by

$$L = \vec{r}\times\vec{p} = m(\vec{r}\times\vec{v})$$

or $$|L| = L = mrv\sin\theta = mv(r\sin\theta) = mvh$$

Now, *m*, *v* and *h* are constants.

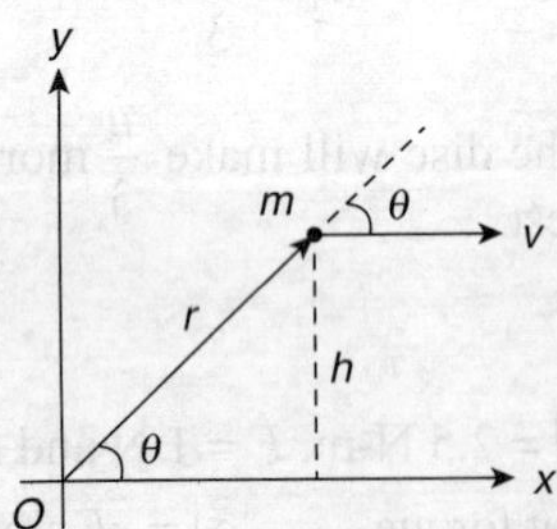

Therefore, angular momentum of particle about origin will remain constant. The direction of $\vec{r}\times\vec{v}$ also remains the same (negative *z*).

9. (b)

10. (a)

Let I be the moment of Inertia of the reaming portion, I_1 for whole portion and I_2 for removed portion then

$$I = I_1 - I_2$$

or $$I = \frac{1}{2}(9M)(R^2) - \left[\frac{1}{2}m\left(\frac{R}{3}\right)^2 + \frac{1}{2}m\left(\frac{2R}{3}\right)^2\right] \quad ...(i)$$

Here, $$m = \frac{9M}{\pi R^2}\times\pi\left(\frac{R}{3}\right)^2 = M$$

The value of m in Eq. (i)

We get, $I = 4MR^2$

11. (b)

Net force on centre of mass is zero, i.e., the centre of mass cannot move at all. Hence, the body will rotate about centre of mass only.

12. (d)

$$I_2\omega_2 = I_1\omega_1$$

$$\frac{2}{5}M\left(\frac{R}{2}\right)^2 = \frac{2}{5}MR^2\omega_1$$

$$\omega_2 = 4\omega_1$$

$$\omega_2 = 4\times\frac{2\pi}{24}$$

$$\frac{2\pi}{T_2} = \frac{2\pi}{6}$$

$$\Rightarrow \quad T_2 = 6 \text{ hours}$$

13. (b)

Since there is no external torque, angular momentum remains conserved. As moment of inertia initially decreases and then increases, ω will increase initially and then decrease. Moment of inertia when tortoise is at B,

$$I = mr^2 + \frac{Mr^2}{2}$$

where $$r^2 = a^2 + \left[\sqrt{R^2 - a^2} - vt\right]^2;$$

m = Mass of tortoise; M = Mass of disc

$\therefore$ The variation of ω is nonlinear.

14. (d)

Let ρ be the mass per unit length of the wire

Mass of the ring M $= \rho L$. Let R be the radius of the ring, then

$$L = 2\pi R$$

or $$R = \frac{L}{2\pi}$$

Moment of inertia about an axis passing through O and parallel to XX′ will be

$$= \frac{1}{2}MR^2$$

Therefore, moment of inertia about XX′ (from parallel axis theorem) will be

$$I_{XX'} = \frac{1}{2}MR^2 + MR^2 = \frac{3}{2}MR^2$$

Substituting values of M and R,

$$I_{XX'} = \frac{3}{2}(\rho L)\left(\frac{L^2}{4\pi^2}\right) = \frac{3\rho L^3}{8\pi^2}$$

15. (c)

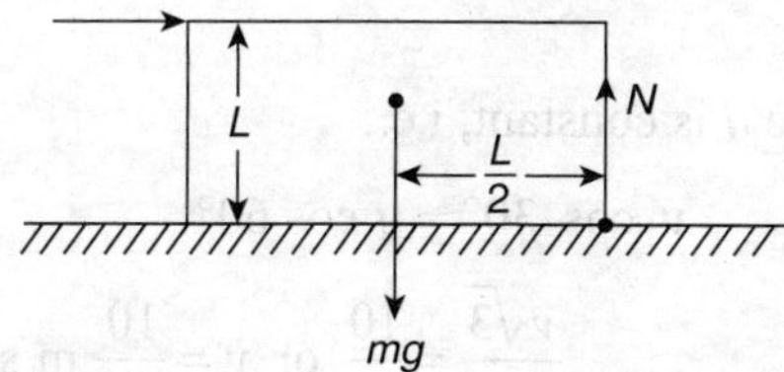

The applied force shifts the normal reaction to one corner as shown. At this situation, the cubical block starts toppling about O. Taking torque about O

$$F \times L = mg \times \frac{L}{2},$$

$$\Rightarrow \qquad F = \frac{mg}{2}$$

16. (d)

By the concept of energy conservation

$$\frac{1}{2}mv^2 + \frac{1}{2}I\omega^2 = mg\left(\frac{3v^2}{4g}\right)$$

For rolling motion $v = R\omega$

$$\therefore \quad \frac{1}{2}mv^2 + \frac{1}{2}I\frac{v^2}{R^2} = \frac{3}{4}mv^2$$

$$\therefore \quad \frac{1}{2}I\frac{v^2}{R^2} = \frac{3}{4}mv^2 - \frac{1}{2}mv^2 = \frac{1}{4}mv^2$$

$$I = \frac{1}{2}mR^2.$$

17. (b)

$$\left|\vec{L}\right| = mvr$$

Since v is changing (decreasing), L is not conserved in magnitude. Since it is given that the particle is confined to rotate in a circular path, it cannot have spiral path. Since the particle has two accelerations a_c and a_t the net acceleration is not towards the centre.
The direction of $\vec{L}$ remains same even when the speed decreases.

18. (b)

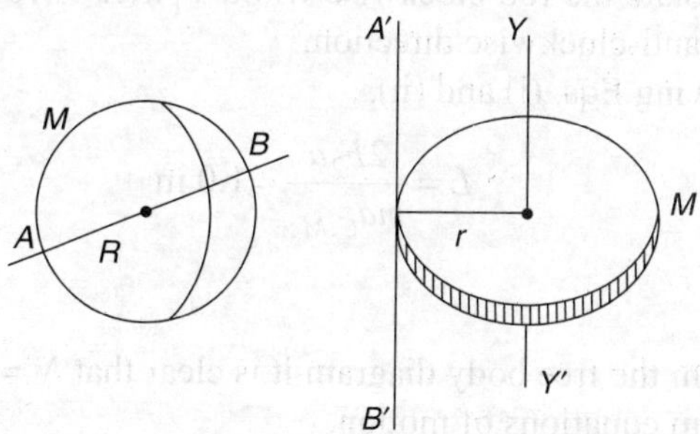

$$I_{AB} = \frac{2}{5}MR^2 = I \qquad \text{...(i)}$$

$$I_{A'B'} = I_{YY'} + Mr^2 = \frac{1}{2}Mr^2 + Mr^2 = \frac{3}{2}Mr^2$$

But $I_{A'B'} = \frac{3}{2}Mr^2 = I,$ ...(ii)

From (i) and (ii),

$$\frac{2}{5}MR^2 = \frac{3}{2}Mr^2 \Rightarrow r = \frac{2}{\sqrt{15}}R$$

19. (a)

Let friction force f act for time Δt.

$$f\Delta t = mv_1 - mv_2 \qquad \text{...(i)}$$

$$Rf\Delta t = I\omega \qquad \text{...(ii)}$$

But since block stops slipping before it looses contact with the cylinder,

$$v_2 = \omega R \qquad \text{...(iii)}$$

After solving Eqs. (i), (ii) and (iii)

$$v_2 = \frac{v_1}{(1 + I/mR^2)}$$

20. (d)

Since rod moves translationally, the angular acceleration α is zero.

$a_{C.M.}$ = Acceleration of centre of mass = 2 m/s^2

a = Distance between the forces = 20 cm = 0.2 m

From equation of motion,

$$F_2 - F_1 = m\,a_{C.M.} \qquad \text{...(i)}$$

The torque about the C.M. is

$$F_2(L/2 - a) - F_1\frac{L}{2} = 0 \qquad \text{...(ii)}$$

The torques act in the opposite direction since F_2 tries to rotate the rod clockwise while F_1 tries to rotate it in the anti-clockwise direction.
Solving Eqs. (i) and (ii),

$$L = \frac{2F_2 a}{m a_{C.M.}} = 1.0 \text{ m}$$

21. (c)

From the free body diagram it is clear that $N = mg$.
From equations of motion,

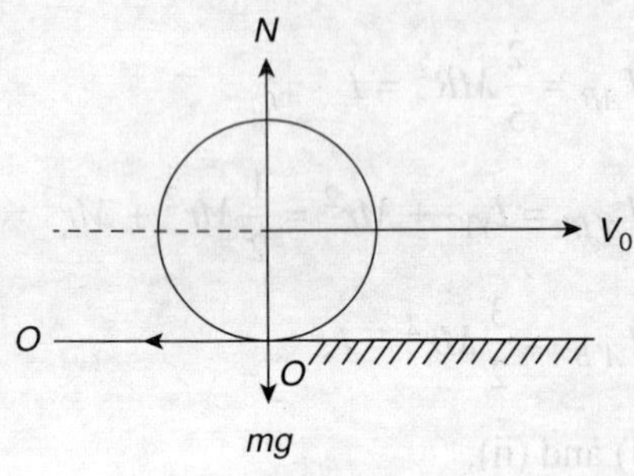

$$m\, a_{C.M.} = -\mu\, mg \quad \ldots\text{(i)}$$

$$I_{C.M.}\alpha = \mu\, mg\, R \quad \ldots\text{(ii)}$$

R = radius of the ball

$$\Rightarrow \quad a_{C.M.} = -\mu g$$

$$\Rightarrow \quad \alpha = \frac{\mu\, mgR}{I_{C.M.}} = \frac{\mu\, mgR}{\frac{2}{5}mR^2} = \frac{5\mu g}{2R} \quad \ldots\text{(iii)}$$

After time t, the velocity v_f of centre of ball and angular velocity ω will be

$$v_f = v_0 - \mu g t$$

$$\omega = 0 + \frac{5\mu g}{2R}t \quad \ldots\text{(iv)}$$

If it starts rolling after time t, then

$$v_f - \omega R = 0 \quad \ldots\text{(v)}$$

After solving Eqs. (iii), (iv) and (v)

$$v_f = \frac{5}{7}v_0$$

22. (a)

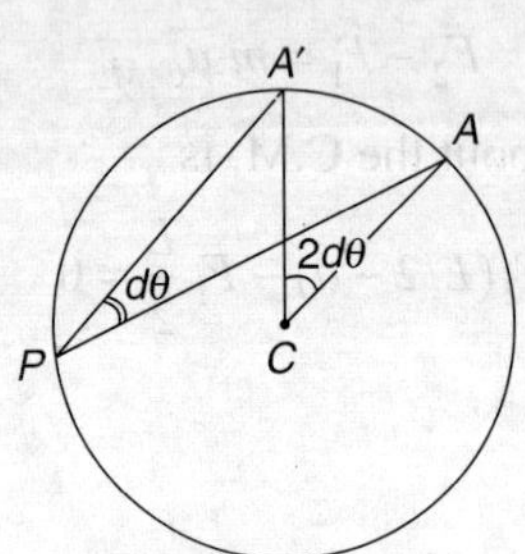

Let the particle moves from A to A′ in time dt.

If $\angle A'PA = d\theta$

Then

$$\angle A'PA = 2(d\theta)$$

$$\omega_P = \frac{d\theta}{dt} = 2\omega$$

$$\omega_C = 2\frac{d\theta}{dt} = 2\omega$$

$\therefore$ Angular velocity about $C = 2\omega$

23. (b)

Distance d is constant, i.e.,

$$v \cos 30° = u \cos 60°$$

or
$$\frac{v\sqrt{3}}{2} = \frac{10}{2} \text{ or } v = \frac{10}{\sqrt{3}} \text{ m/s}$$

Magnitude of angular velocity of 'B' with respect to 'A' is

$$\omega = \frac{u \sin 60° - v \sin 30°}{d} = \frac{(10)\frac{\sqrt{3}}{2} - \frac{10}{\sqrt{3}}.\frac{1}{2}}{2}$$

$$= \frac{5\sqrt{3} - \frac{5}{\sqrt{3}}}{2} = \frac{5}{\sqrt{3}} \text{ rad/s.}$$

24. (c)

Since external force, namely, mg and N act in vertical direction only, the centre of mass will move in the vertical direction. End A of the rod can only slip in the horizontal direction. The instantaneous axis rotation will be at point O when rod makes an angle θ.

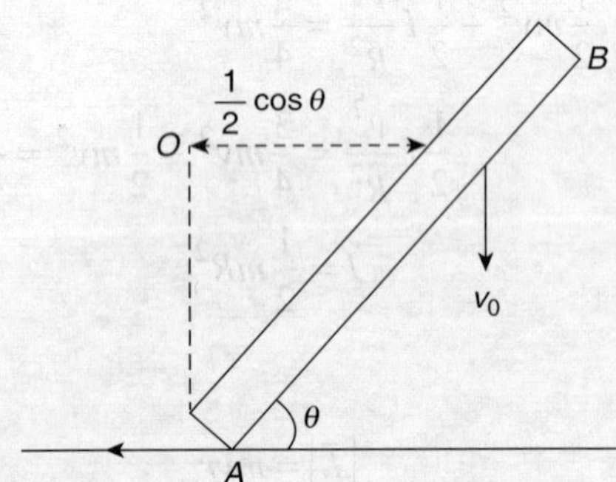

Conserving mechanical energy,

$$\frac{mgL}{2}(1 - \sin\theta) = \frac{1}{2}I_0\omega^2$$

in which $I_0 = I_{C.M.} + md^2$

$$= \frac{mL^2}{2} + \frac{mL^2}{4}\cos^2\theta$$

$$= \frac{mL^2}{12}(1 + 3\cos^2\theta)$$

$$\Rightarrow \quad \omega = \sqrt{\left(\frac{1-\sin\theta}{1+3\cos^2\theta}\right)\frac{12g}{L}}$$

The v_0 of the centre of mass is given by

$$v_0 = \left(\frac{\omega L}{2}\cos\theta\right)$$

$$= \sqrt{\frac{12(1-\sin\theta)gL}{1+3\cos^2\theta}}\left(\frac{\cos\theta}{2}\right)$$

25. (b)

Let F act for time Δt. Then

$$v_{C.M.} = v_0 = \frac{F\Delta t}{M} \text{ and } \omega = \frac{(F\Delta t)h}{2/5MR^2}$$

$$\Rightarrow \quad \omega = \frac{5v_0}{2R^2}h$$

After the application of impulse, angular momentum about O will be conserved.

$$\Rightarrow \quad L_i = L_f$$

$$\Rightarrow \quad Mv_0R + I_{C.M.}\omega = Mv_fR + I_{C.M.}\frac{v_f}{R}$$

$$\Rightarrow \quad Mv_0R + \frac{2}{5}MR^2 \times \frac{5v_0h}{2R^2} = \left(MR + \frac{2}{5}\frac{MR^2}{R}\right)\frac{9}{7}v_0$$

$$\Rightarrow \quad MR + Mh = \frac{9}{5}MR$$

$$\Rightarrow \quad h = \frac{4}{5}R$$

26. (d)

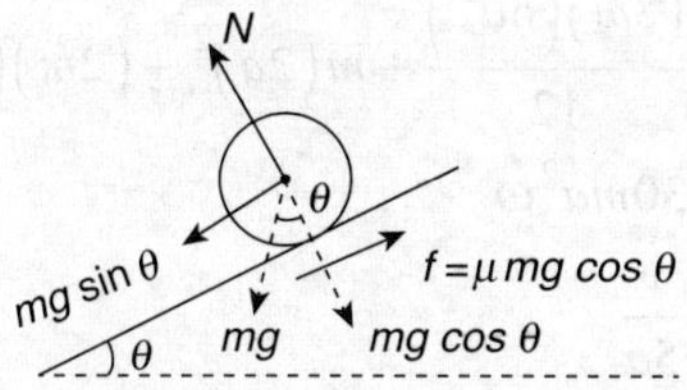

Let frictional coefficient be μ.
For all body, $mg \sin\theta - \mu mg \cos\theta = ma$

$$g(\sin\theta - \mu\cos\theta) = a$$

(Since they are sliding).
Hence $a_{sph} = a_{disc} = a_{ring}$.

27. (b)

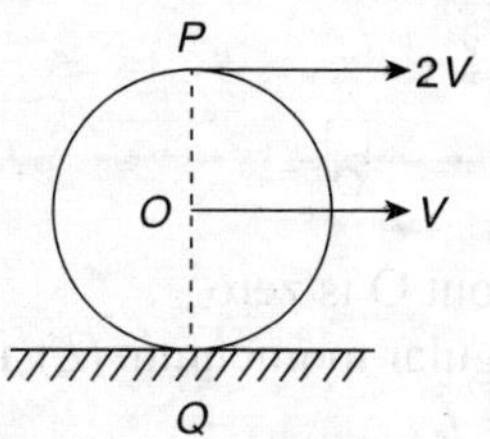

$$v_r = v_{PQ} = 2v$$

$$a_r = \frac{v_r^2}{2R} = \frac{4v^2}{2R} = \frac{2v^2}{R}$$

28. (d)

As there is no external torque on system, angular momentum of the system is conserved.

$$\Rightarrow I_i\omega_i = I_f\omega_f$$

Initially,
The M.I of beads when they reach opposite ends of the rod

$$= m\left(\frac{L}{2}\right)^2 + m\left(\frac{L}{2}\right)^2$$

$$= \frac{mL^2}{2}$$

The M.I of the rod + beads system

$$= \frac{ML^2}{12} + \frac{mL^2}{2} = \frac{ML^2 + 6mL^2}{12}$$

So, final angular speed of system is

$$\Rightarrow \omega = \frac{\frac{ML^2}{12}\cdot\omega_0}{\left(\frac{ML^2 + 6mL^2}{12}\right)} = \frac{M\omega_0}{M+6m}$$

29. (a)

$$A'B' \perp AB \text{ and } C'D' \perp CD$$

From symmetry $I_{AB} = I_{A'B'}$

and $\qquad I_{CD} = I_{C'D'}$

From theorem of perpendicular axes,

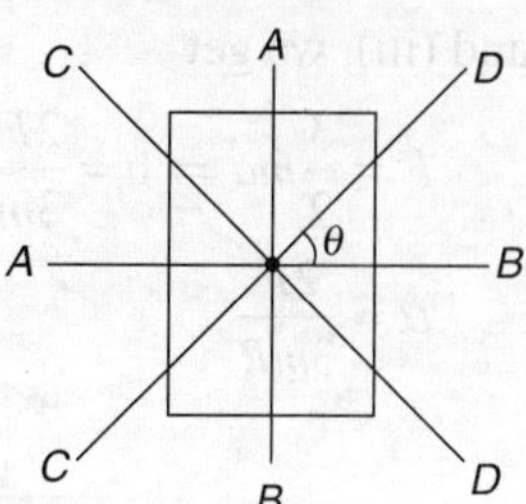

Alternatively,
the relation between I_{AB} and I_{CD} should be true for all values of θ

At $\qquad \theta = 0, I_{CD} = I_{AB}$

Similarly, at $\theta = \pi/2, I_{CD} = I_{AB}$ (by symmetry)

Keeping these things in mind, only option (a) is correct.

30. (c)

$$I_1\omega_1 = I_2\omega_2$$

$$\therefore \quad \omega_2 = \frac{I_1}{I_2}\omega = \left(\frac{Mr^2}{Mr^2 + 2mr^2}\right)\omega$$

$$= \left(\frac{M}{m+2m}\right)\omega$$

31. (b)

When force F is applied at the centre of roller of mass m as shown in the figure.

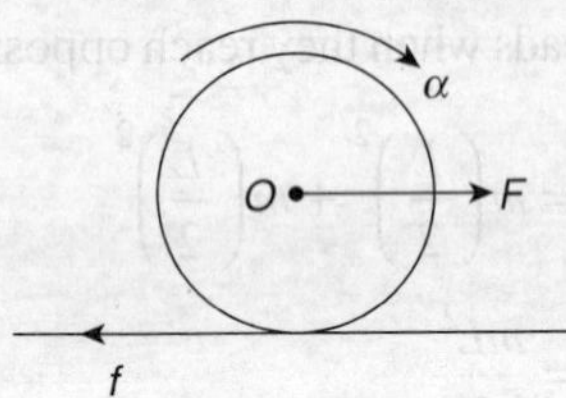

Its acceleration is

$$\frac{(F-f)}{m} = a \qquad \ldots\text{(i)}$$

where, f = Force of friction and
m = Mass of roller.
Torque on roller is provided by friction f and it is

$$\tau = fR = I\alpha \qquad \ldots\text{(ii)}$$

where, I = Moment of inertia of solid cylindrical roller $= mR^2/2$
and α = angular acceleration of cylinder = a/R

Therefore, $\tau = \frac{mR^2}{2}\cdot\frac{a}{R} = \frac{maR}{2}$

From Eq. (ii), $(\tau = fR)$

$$f = \frac{ma}{2} \qquad \ldots\text{(iii)}$$

From Eq. (i) and (iii), we get

$$F = \frac{3}{2}ma \Rightarrow a = \frac{2F}{3m}$$

So, $\alpha = \frac{2F}{3mR}$

32. (d)

$$I_P > I_Q \qquad (I = \varepsilon mR^2)$$

In case of pure rolling, $a = \frac{g\sin\theta}{1 + I/mR^2}$

$a_Q > a_P$ as its moment of inertia is less. Therefore, Q reaches first with more linear speed and more translational kinetic energy.

Further, $\omega = \frac{v}{R}$ or $\omega \propto v$

$\therefore \omega_Q > \omega_P$ as $v_P > v_Q$

33. (b)

Loss of potential energy of rod = Gain in kinetic energy

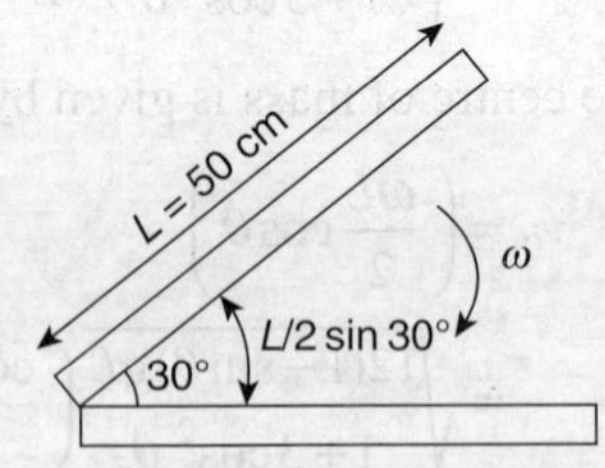

$$\therefore \Delta PE = \frac{1}{2}I\omega^2$$

$$\Rightarrow Mg \times \frac{L}{2}\sin 30° = \frac{1}{2} \times I \times \omega^2$$

$$\Rightarrow \quad Mg\frac{L}{2} \times \frac{1}{2} = \frac{1}{2} \times I \times \omega^2 \Rightarrow \frac{MgL \times 2}{4 \times I} = \omega^2$$

$$\sqrt{\frac{MgL}{4 \times \frac{ML^2}{3} \times \frac{2}{1}}} = \omega \qquad \left(I = \frac{ML^2}{3}\right)$$

$$\omega = \sqrt{30} \text{ rad/s}$$

34. (b)

$$P_i = 0$$

$$\therefore P_f = 0 \text{ or } v_c = 0$$

$$L_i = L_f \text{ or } (2mv)a + (2mv)(2a) = I\omega$$

(Conservation of momentum)

$$\text{Here, } I = \frac{(8m)(6a^2)}{12} + m(2a)^2 + (2m)(a)^2 = 30ma^2$$

$$6mv_a = 30ma^2\omega$$

$$\omega = \frac{v}{5a}$$

$$\text{Further, } E = \frac{1}{2}I\omega^2 = \frac{1}{2} \times (30ma^2)\left(\frac{v}{5a}\right)^2 = \frac{3mv^2}{5}$$

35. (a)

$$r = \sqrt{2}\frac{a}{2} \text{ or } r^2 = \frac{a^2}{2}$$

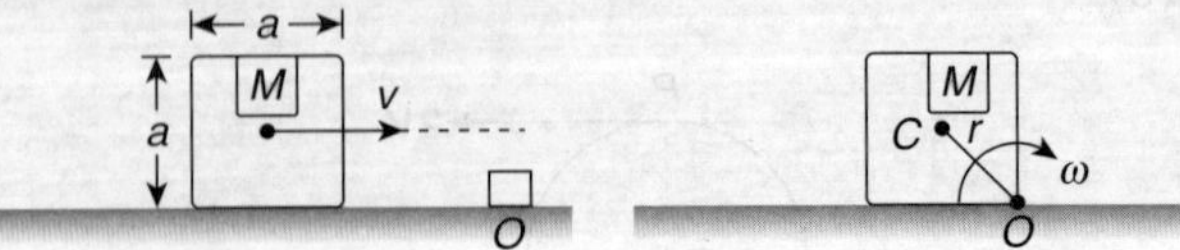

Net torque about O is zero
Therefore, angular momentum (L) about O will be conserved, or $L_i = L_f$

$$Mv\left(\frac{a}{2}\right)=I_0\omega=\left(I_{C.M.}+Mr^2\right)\omega$$

$$=\left\{\left(\frac{Ma^2}{6}\right)+M\left(\frac{a^2}{2}\right)\right\}\omega$$

$$=\frac{2}{3}Ma^2\omega\Rightarrow\omega=\frac{3v}{4a}$$

36. (c)

Since, it is head-on elastic collision between two identical spheres, they will exchange their linear velocities. A comes to rest and B starts moving with linear velocity v. As there is no friction anywhere, torque on both the spheres about their centre of mass remains zero and their angular velocities also remain unchanged. Therefore,

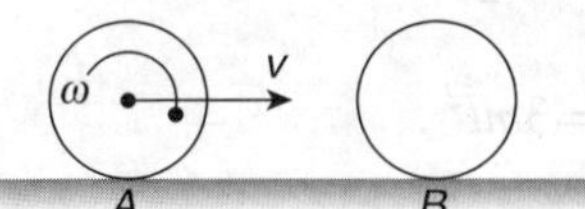

Before collision

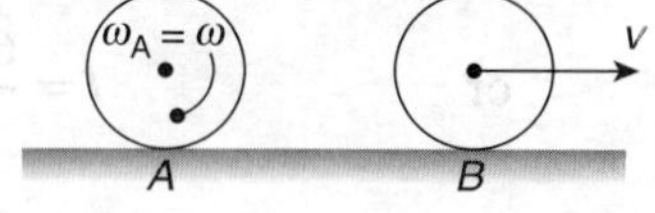

After collision

$$\omega_A=\omega \text{ and } \omega_B=0$$

37. (d)

$$\vec{L}=m\left(\vec{r}\times\vec{v}\right)$$

Direction of $\vec{r}\times\vec{v}$, hence the direction of angular momentum remains the same.

38. (a)

Mass of the element dx is $m=\frac{M}{L}dx$.

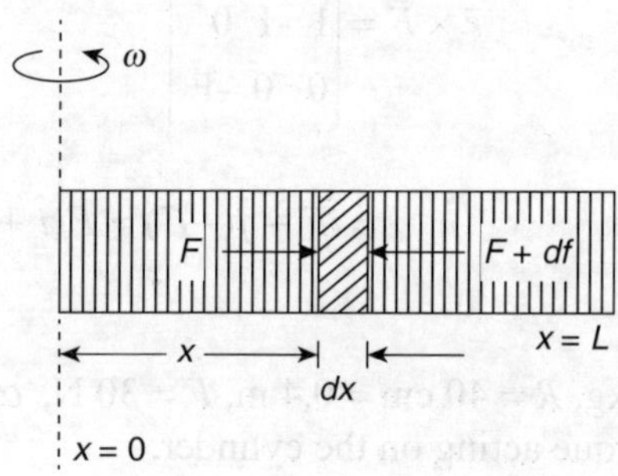

This element needs centripetal force for rotation.

$$\therefore\quad dF=mx\omega^2=\left(\frac{M}{L}x\omega^2dx\right)$$

$$\therefore\quad F=\int_0^L dF=\frac{M}{L}\cdot\omega^2\int_0^L xdx=\frac{M\omega^2L}{2}$$

This is the force exerted by the liquid at the other end.

39. (c)

40. (a)

The torque is the rate of change of angular momentum

$\therefore$ Torque = (4 J – 1 J) /4

$$=\frac{3}{4}\text{ J}$$

41. (b)

$$x_{C.M.}=\frac{\int_0^L\left(\frac{K}{L^n}.x^n.dx\right)x}{\int_0^L\frac{K}{L^n}.x^n.dx}$$

$$\Rightarrow\quad x_{C.M.}=\frac{\int_0^L x^{n+1}dx}{\int_0^L x^n dx}=\frac{L^{n+2}}{n+2}.\frac{(n+1)}{L^{n+1}}$$

$$\Rightarrow\quad x_{C.M.}=\frac{L(n+1)}{n+2}$$

42. (a)

For a rectangular sheet, moment of inertia passing through O perpendicular to the plate is

$$I_o=m\left(\frac{a^2+b^2}{12}\right)$$

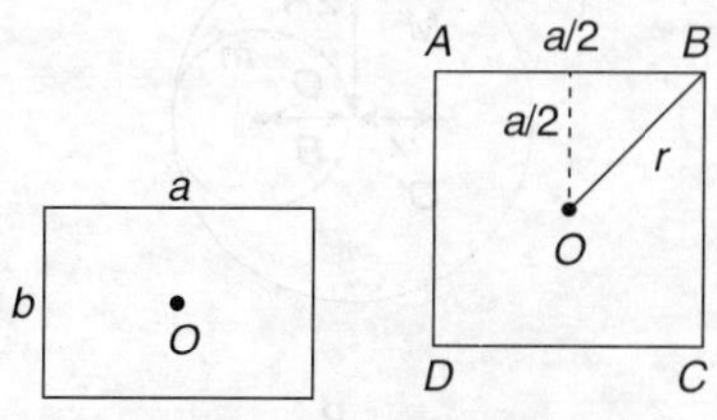

For square plate, it is $\frac{ma^2}{6}$.

$$r=\sqrt{\frac{a^2}{4}+\frac{a^2}{4}}=\frac{a}{\sqrt{2}}\therefore r^2=\frac{a^2}{2}$$

I about B parallel to the axis through O is

$$I_o+md^2=\frac{ma^2}{6}+\frac{ma^2}{2}=\frac{4ma^2}{6}$$

$$I=\frac{2}{3}ma^2$$

43. (d)

The two bodies will move towards the common centre of mass. But the location of the centre of mass will remain unchanged.

44. (d)

Central forces pass through axis of rotation, so torque is zero. If no external torque is acting on a particle, the angular momentum of a particle is constant.

45. (b)

Acceleration of a uniform body of radius R and mass M and moment of inertia I rolling down (without sliding) an inclined plane making an angle θ with the horizontal is given by

$$a = \frac{g\sin\theta}{1+\dfrac{I}{MR^2}}$$

46. (b)

$$(M'+m) = M = \pi(2R)^2.\sigma$$

where σ mass per unit area

$$\therefore \quad m = \pi R^2.\sigma, M' = 3\pi R^2\sigma.$$

$$\frac{3\pi R^2\sigma.x + \pi R^2\sigma.R}{M} = 0$$

because for the full disc, the centre of mass is at centre O.

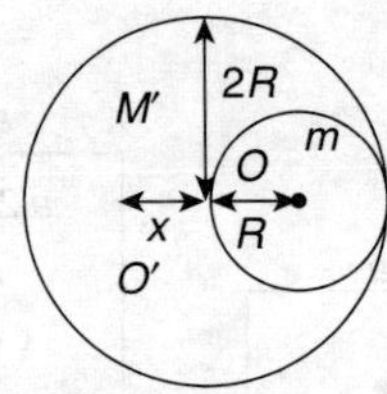

$$x = -\frac{R}{3} = \alpha R$$

$$\therefore \quad |\alpha| = \left|\frac{-1}{3}\right| \Rightarrow |\alpha| = \frac{1}{3}$$

The centre of mass is at $\frac{R}{3}$ to the left on the diameter of the original disc.

47. (d)

$$AO\cos 45° = \frac{l}{2} \therefore AO \times \frac{1}{\sqrt{2}} = \frac{l}{2}$$

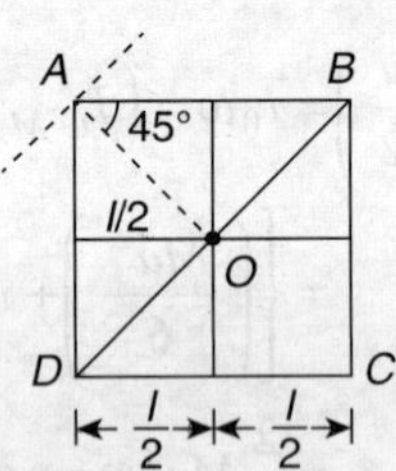

or $\quad AO = \frac{l}{\sqrt{2}}$

$$I = I_D + I_B + I_C$$

or $\quad I = \frac{2ml^2}{2} + m\left(\frac{2l}{\sqrt{2}}\right)^2$

$$I = \frac{2ml^2}{2} + \frac{4ml^2}{2}$$

or $\quad I = \frac{6ml^2}{2} = 3ml^2.$

48. (d)

For this rod, as linear density varies from one end to another, its moment of inertia also changes. If its pivoted end is changed, all physical quantities that depend upon moment of inertia will also change. Hence only option (d) is correct.

49. (d)

Torque $\quad \vec{\tau} = \vec{r} \times \vec{F}$

$$\vec{F} = -F\hat{k}, \vec{r} = \hat{i} - \hat{j}$$

$$\therefore \quad \vec{r} \times \vec{F} = \begin{vmatrix} \hat{i} & \hat{j} & \hat{k} \\ 1 & -1 & 0 \\ 0 & 0 & -F \end{vmatrix}$$

$$= \hat{i}F - \hat{j}(-F) = F(\hat{i} + \hat{j})$$

50. (a)

$M = 3$ kg, $R = 40$ cm $= 0.4$ m, $F = 30$ N, $\alpha = ?$, $a = ?$

The torque acting on the cylinder,

τ = force × radius of the cylinder = 30 N × 0.4 m = 12 N m

The moment of inertia of the hollow cylinder about its axis,

$$I = MR^2 = 3\text{ kg} \times (0.4)^2\text{ m}^2 = 0.48\text{ kg m}^2$$

Now, $\tau = I\alpha$, where α is angular acceleration.

$\therefore$ The angular acceleration of the cylinder,

$$\alpha = \frac{\tau}{I} = \frac{12}{0.48}\text{ rad s}^{-2} = 25\text{ rad s}^{-2}$$

The linear acceleration of the rope,

$$a = R\alpha = 0.4 \times 25 \text{ m s}^{-2} = 10 \text{ m s}^{-2}$$

51. (b)

$$R_{C.M.} = \frac{1\times 0 + 1\times AB + 1\times AC}{1+1+1} = \frac{AB+AC}{3}$$

52. (d)

The centre of mass of bodies B and C taken together does not shift as no external force is applied horizontally.

53. (b)

$$I = \frac{(\text{Mass of semicircular disc})\times r^2}{2} = \frac{Mr^2}{2}$$

Alternative

The mass of complete (circular) disc is

$$M + M = 2M$$

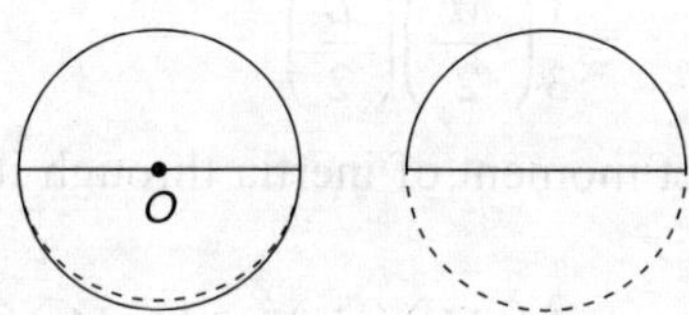

The moment of inertia of disc about the given axis is

$$I = \frac{2Mr^2}{2} = Mr^2$$

Let the moment of inertia of semicircular disc is I_1. The disc may be assumed as a combination of two semicircular parts.

Thus, $I_1 = I - I_1$

$\therefore$ $I_1 = \frac{I}{2} = \frac{Mr^2}{2}$.

54. (c)

For solid sphere,

$$I_A = \frac{2}{5}MR^2$$

For hollow sphere,

$$I_B = \frac{2}{3}MR^2$$

$$\therefore \quad \frac{I_A}{I_B} = \frac{2MR^2}{5} \times \frac{3}{2MR^2} = \frac{3}{5}$$

or $I_A < I_B$

55. (b)

Free space implies that no external torque is operating on the sphere; internal changes are responsible for increase in radius of sphere. Here the law of conservation of angular momentum applies to the system.

Alternative

Given a solid sphere is rotating in free space, so there is no external torque, i.e.,

$$\tau = 0$$

and the rate of change of angular momentum is equal to the torque.

Then, $\frac{dL}{dt} = 0.$

$\Rightarrow$ $dL = 0.$

$\Rightarrow$ $L = \text{constant}.$

But $L = I\omega,$

If we increase the radius of the sphere then its moment of inertia $\left(I = \frac{2}{5}MR^2\right)$ increases.

From $L = I\omega$

its angular velocity decreases.

Also, we know that the rotational kinetic energy,

$$K_r = \frac{L^2}{2I}$$

Hence, K_r decreases as I increases.

56. (a)

According to given data,

$$m_1 = 2kg; r_1 = -2i$$

$$m_2 = 2kg; r_2 = -3j$$

$$m_3 = 2kg; r_3 = k$$

$$r_{C.M.} = \frac{m_1r_1 + m_2r_2 + m_3r_3}{m_1+m_2+m_3}$$

$$r_{C.M.} = \frac{2(-2\hat{i}) + 2(-3\hat{j}) + 2(\hat{k})}{6}$$

$$r_{C.M.} = \frac{2\hat{i} - 3\hat{j} + \hat{k}}{3}$$

$$|r_{C.M.}| = \sqrt{x^2+y^2+z^2} = \sqrt{\frac{4}{9}+\frac{9}{9}+\frac{1}{9}} = \sqrt{\frac{14}{9}}$$

$$|r_{C.M.}|^2 = \frac{14}{9} = 1.55 \text{ m}^2$$

57. (a)

$$L = \frac{2K}{\omega}$$

or $$\frac{L_1}{L_2} = \frac{K_1}{K_2} \times \frac{\omega_2}{\omega_1} = 2 \times 2 = 4$$

$$\therefore \quad L_2 = \frac{L_1}{4} = \frac{L}{4}$$

58. (d)

Mass of disc $x = (\pi R^2 t)\sigma$ where $\sigma =$ density

$$\therefore \quad I_x = \frac{MR^2}{2} = \frac{(\pi R^2 t\sigma)R^2}{2}$$

$$= \frac{\pi R^4 t\sigma}{2}$$

Similarly, $I_y = \frac{(\text{Mass})\,(4\text{R})^2}{2} = \frac{\pi(4R)^2}{2} \cdot \frac{t}{4} \cdot \sigma \times 16R^2$

or $$I_y = 32\pi R^4 t\sigma$$

$$\frac{I_x}{I_y} = \frac{\pi R^4 \sigma t}{2} \times \frac{1}{32\pi R^4 t\sigma} = \frac{1}{64}$$

$$I_y = 64 I_x$$

59. (d)

The particle moves with linear velocity v along line PC. The line of motion is through P.
Hence, angular momentum is zero.

60. (a)

A circular wire behaves likes a ring. Therefore, moment of inertia about its diameter is

$$= \frac{MR^2}{2}$$

61. (d)

The bodies slide along inclined plane. They do not roll. Acceleration for each body down the plane = $g \sin \theta$. It is the same for each body.

62. (c)

Angular momentum of the system is conserved.

$$\therefore \quad \frac{1}{2}MR^2\omega_1 = 2mR^2\omega + \frac{1}{2}MR^2\omega$$

$$\Rightarrow \quad M\omega_1 = (4m + M)\omega$$

$$\omega = \frac{M\omega_1}{M + 4m}$$

63. (c)

$$v_C = \frac{m_1 v_1 + m_2 v_2}{m_1 + m_2}$$

$$v_C = \frac{m(2v) + m(-v)}{m + m} = \frac{v}{2}$$

64. (b)

Since rod is bent at the middle, each part of it will have same length $\left(\frac{L}{2}\right)$ and mass $\left(\frac{M}{2}\right)$ as shown.

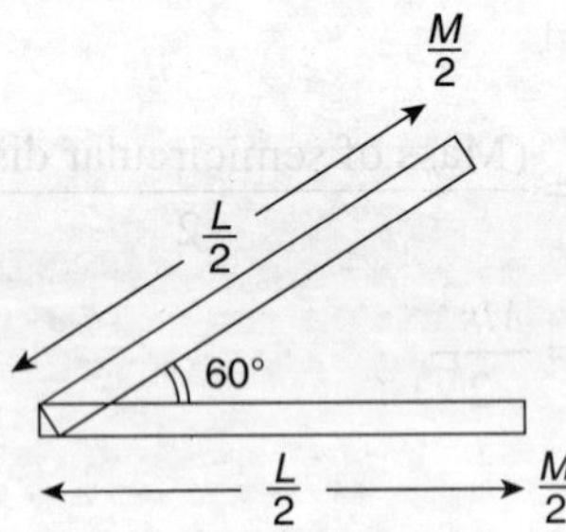

Moment of inertia of each part through its one end

$$= \frac{1}{3}\left(\frac{M}{2}\right)\left(\frac{L}{2}\right)^2$$

Hence, net moment of inertia through its middle point O is

$$= \frac{1}{3}\left(\frac{M}{2}\right)\left(\frac{L}{2}\right)^2 + \frac{1}{3}\left(\frac{M}{2}\right)\left(\frac{L}{2}\right)^2$$

$$= \frac{1}{3}\left(\frac{ML^2}{8} + \frac{ML^2}{8}\right) = \frac{ML^2}{12}$$

65. (b)

From the definition of angular momentum,

$$\vec{L} = \vec{r} \times \vec{P} = rmv \sin\phi(\hat{k})$$

Therefore, the magnitude of L is

$$L = rmv \sin\phi = mvd$$

where $d = r\sin\phi$ is the distance of closest approach of the particle to the origin. As d is same for both the particles, $L_A = L_B$.

66. (c)

We know that the two particles will meet at their centre of mass.

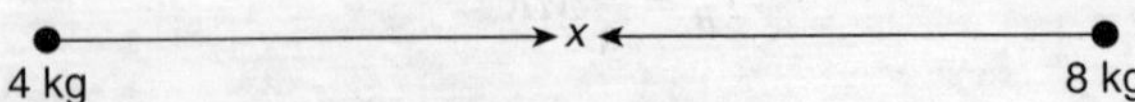

Distance of centre of mass from 8 kg

$$\frac{8 \times 0 + 4 \times 12}{12} = 4 \text{ m}$$

67. (a)

The rotational KE of the disc is

$$K_{\text{rot}} = \frac{1}{2}I\omega^2 = \frac{1}{2}\left(\frac{1}{2}MR^2\right)\omega^2$$

$$= \frac{1}{4}MR^2\omega^2$$

The translational kinetic energy is

$$K_{\text{trans}} = \frac{1}{2}Mv_{cm}^2$$

where v_{cm} is the linear velocity of its centre of mass.

Now, $v_{\text{cm}} = R\omega$,

$$K_{trans} = \frac{1}{2}MR^2\omega^2$$

Thus, $K_{total} = \frac{1}{2}MR^2\omega^2 + \frac{1}{2}MR^2\omega^2$

$$= \frac{3}{4}MR^2\omega^2$$

$$\frac{K_{rot}}{K_{total}} = \frac{\frac{1}{4}MR^2\omega^2}{\frac{3}{4}MR^2\omega^2} = \frac{1}{3}$$

68. (c)

When no external torque acts on a system of particles, then the total angular momentum of the system remains always constant.

The angular momentum of a disc of moment of inertia I_1 and rotating about its axis with angular velocity ω is

$$L_1 = I_1\omega.$$

When a round disc of moment of inertia I_2 is placed on the first disc, then angular momentum of the combination is

$$L_2 = (I_1 + I_2)\omega'.$$

In absence of any external torque, angular momentum remains conserved, i.e.,

$$L_1 = L_2$$

$$I_1\omega = (I_1 + I_2)\omega'$$

$$\Rightarrow \quad \omega' = \frac{I_1\omega}{I_1 + I_2}$$

69. (a)

When a body rolls down on an inclined plane, it makes rotational as well as translational motions. Thus, it is associated with rotational and translational kinetic energies. Hence, total kinetic energy of sphere

$$K = K_{rot} + K_{trans}$$

$$= \frac{1}{2}I\omega^2 + \frac{1}{2}mv^2$$

$$= \frac{1}{2}.\frac{2}{5}mr^2\left(\frac{v^2}{r^2}\right) + \frac{1}{2}mv^2$$

$$= \frac{1}{5}mv^2 + \frac{1}{2}mv^2$$

$$= \frac{7}{10}mv^2$$

Also, $K_{rot} = \frac{1}{2}I\omega^2 = \frac{1}{5}mv^2$

Hence, $\frac{K_{rot}}{K} = \frac{\frac{1}{5}mv^2}{\frac{7}{10}mv^2} = \frac{2}{7}$

70. (b)

Angular velocity is given by

$$\omega = 600rp$$

$$= \frac{600 \times 2\pi}{60} \text{ rad/s}$$

$$= 20\pi \text{ rad/s.}$$

Kinetic energy of coin which is due to rotation and translation is

$$K = \frac{1}{2}I\omega^2 + \frac{1}{2}mv^2$$

$$= \frac{1}{2} \times \frac{1}{2}mr^2\omega^2 + \frac{1}{2}m(\omega r)^2$$

$$= \frac{1}{4} \times 4.8 \times (1)^2(20\pi)^2 + \frac{1}{2}(4.8)(20\pi \times 1)^2$$

$$= 480\pi^2 + 960\pi^2$$

$$K = 1440\pi^2 \text{ J}$$

71. (b)

Moment of inertia of the system about the centre of plate is given by

$$I = \left[\frac{2}{5} \times 1 \times (0.1)^2 + 1 \times (1)^2\right] + \left[\frac{2}{5} \times 2 \times (0.1)^2 + 2 \times (1)^2\right]$$

$$+ \left[\frac{2}{5} \times 3 \times (0.1)^2 + 3 \times (1)^2\right] + \left[\frac{2}{5} \times 4 \times (0.1)^2 + 4 \times (1)^2\right]$$

$$= 1.004 + 2.008 + 3.012 + 4.016$$

$$= 10.04 \text{ kgm}^2.$$

72. (a)

If M is mass of ring and R its radius, then moment of inertia of ring is given by

$$I = MR^2.$$

Since both the rings are of same material, their densities will be equal. Hence,

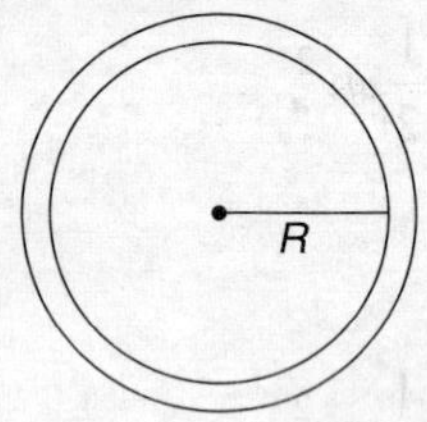

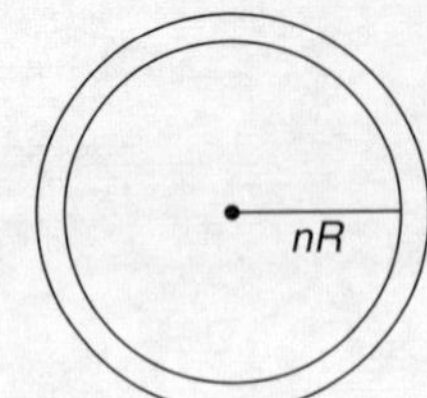

$$d = \frac{M}{V} = \frac{M}{2\pi RA} = \frac{M'}{2\pi(nR)A}$$

$\Rightarrow$ $M' = nM.$

Ratio of moment of inertia is

$$\frac{I_1}{I_2} = \frac{MR^2}{M'(nR)^2}$$

$$\Rightarrow \quad \frac{1}{8} = \frac{MR^2}{nMn^2R^2}$$

$$\Rightarrow \quad \frac{1}{8} = \frac{1}{n^3}$$

$$\Rightarrow \quad \frac{1}{n} = \frac{1}{2}$$

$$\therefore \quad n = 2.$$

73. (a)

Let the plane be inclined at an angle θ and the cylinder rolls down the plane. Then the acceleration of the cylinder of mass m, radius R, and I as moment of inertia is given by:

$$a = \frac{g\sin\theta}{\left(1 + \frac{I}{mR^2}\right)}$$

Moment of inertia (I) of a cylinder $= \frac{mR^2}{2}$

$$\therefore \quad a = \frac{g\sin\theta}{\left(1 + \frac{\frac{mR^2}{2}}{mR^2}\right)}$$

$$= \frac{2}{3}g\sin 30°$$

$$\Rightarrow \quad a = \frac{g}{3}.$$

74. (a)

A body of spherical section, whose centre of mass is C, as shown, rolls down a plane, which is inclined at an angle θ with the horizontal. The total kinetic energy is

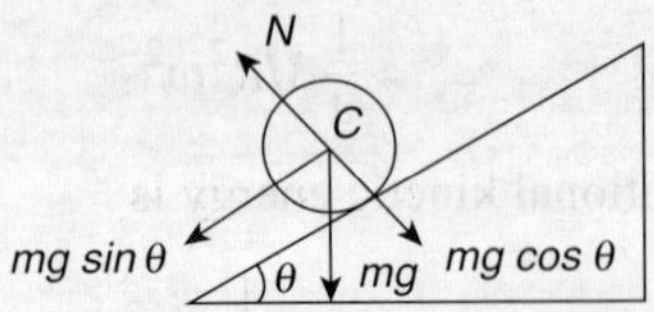

$$(\text{KE})_T = \text{Linear KE} + \text{Rotational KE}$$

$$= \frac{1}{2}mv^2 + \frac{1}{2}I\omega^2$$

where I is moment of inertia and ω the angular velocity.

$$I = \frac{2}{5}mR^2$$

where m is mass, R the radius and ω the angular velocity.

$$\therefore \quad \text{KE} = \frac{1}{2}mv^2 + \frac{1}{2}\left(\frac{2}{5}mR^2\right)\left(\frac{v}{R}\right)^2$$

$$\text{KE} = \frac{1}{2}mv^2 + \frac{1}{5}mv^2$$

$$\text{KE} = \frac{7}{10}mv^2$$

75. (b)

Co-ordinates of centre of mass is given by

$$X_{C.M.} = \frac{m_1x_1 + m_2x_2}{m_1 + m^2}$$

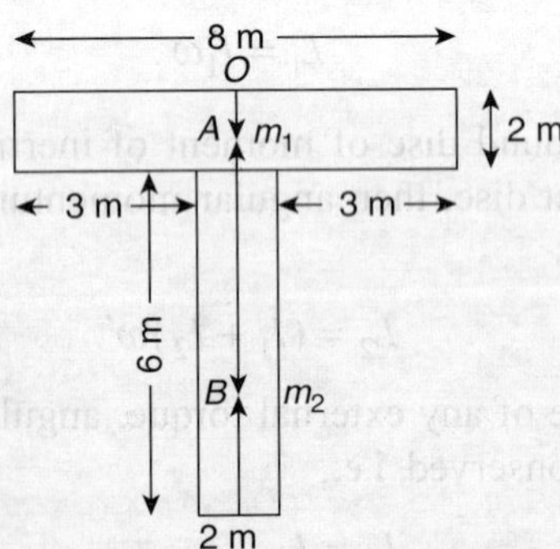

Taking parts A and B as two bodies of the same system

$$m_1 = l \times b \times \sigma = 8 \times 2 \times \sigma = 16\sigma$$

$$m_2 = l \times b \times \sigma = 6 \times 2 \times \sigma = 12\sigma$$

Choosing O as origin,

$$x_1 = 1\text{ m}, x_2 = 2 + 3 = 5\text{ m}$$

$$X_{C.M.} = \frac{16\sigma \times 1 + 12\sigma \times 5}{16\sigma + 12\sigma}$$

$$= \frac{19}{7} = 2.7\text{ m from O.}$$

76. (b)

77. (b)

Moment of inertia quantifies the rotational inertia of an object.

Moment of inertia of circular ring about an axis passing through its centre and perpendicular to its plane is

$$I_{ring} = MR^2.$$

Similarly, for a circular disc about the same axis of rotation,

$$I_{disc} = \frac{1}{2}MR^2$$

Hence, $\dfrac{I_{ring}}{I_{disc}} = \dfrac{MR^2}{\frac{1}{2}MR^2} = \dfrac{2}{1}.$

78. (c)

Let r_1 and r_2 be the radii of thin and solid spherical shell respectively.

∴ Moment of inertia of thin spherical shell about its diameter

$$I = \frac{2}{3}Mr_1^2 \quad \text{...(i)}$$

And moment of inertia of the solid sphere is

$$I = \frac{2}{5}Mr_2^2 \quad \text{...(ii)}$$

According to question, the masses and moment of inertia for both the bodies are equal. Then from Eqs. (i) and (ii),

$$\frac{2}{3}Mr_1^2 = \frac{2}{5}Mr_2^2 \Rightarrow \frac{r_1^2}{r_2^2} = \frac{3}{5}$$

$$\Rightarrow \quad \frac{r_1}{r_2} = \sqrt{\frac{3}{5}} \Rightarrow r_1 : r_2 = \sqrt{3} : \sqrt{5}.$$

79. (b)

Here acceleration of the body which is rolling down an inclined plane of angle α is

$$a = \frac{g\sin\alpha}{1+\dfrac{K^2}{r^2}}$$

Now, given the body is a uniform solid disc.

$$\frac{K^2}{r^2} = \frac{1}{2} \quad \text{(i)}$$

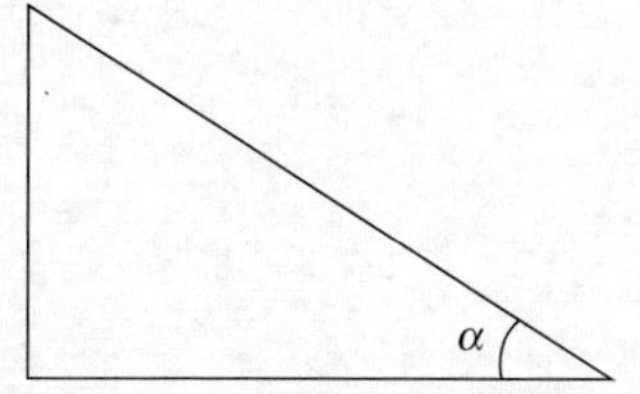

Hence, from Eq. (i),

$$a = \frac{g\sin\alpha}{1+\frac{1}{2}} = \frac{2}{3}g\sin\alpha$$

80. (a)

A thin wire of length L is better to form a circular ring.

∴ $2\pi r = L$ (where r is the radius of the ring)

or $r = \dfrac{L}{2\pi}$

Hence, the moment of inertia of the ring about its axis

$$= M\left(\frac{L}{2\pi}\right)^2 = \frac{ML^2}{4\pi^2}.$$

81. (a)

Clearly, the whole mass of the equilateral triangle acts at O. So the distance OP is the radius of gyration of the system.

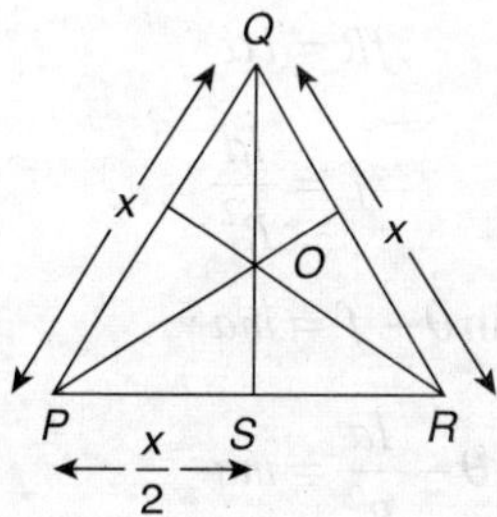

Now, in ΔPSQ,

$$x^2 = QS^2 + \left(\frac{x}{2}\right)^2$$

$$\Rightarrow \quad QS^2 = x^2 - \frac{x^2}{4} = \frac{3x^2}{4} QS$$

or $$= \sqrt{3}\frac{x}{2}.$$

Hence, the distance,

$$OQ = \frac{\sqrt{3}x}{2} \times \frac{2}{3}.$$

$$\Rightarrow \quad OQ = \frac{x}{\sqrt{3}}$$

But the distance OP, OQ and OR are the same.

So, $OP = \dfrac{x}{\sqrt{3}}.$

82. (c)

The kinetic energy of the solid sphere,

$$K = \frac{1}{2}Mv^2 \quad \text{...(i)}$$

The rotational kinetic energy,

$$K_r = \frac{1}{2} I\omega^2.$$

Here, $$K_r = \frac{1}{2} \times \frac{2}{5} MR^2 \times \frac{v^2}{R^2}$$

Since, $I = \frac{2}{5} MR^2$ for solid sphere and $\omega = \frac{v}{R}$.

$$\therefore \quad K_r = \frac{1}{5} Mv^2 \quad \text{...(ii)}$$

From Eqs. (i) and (ii),

$$\frac{K_r}{K} = \frac{\frac{1}{5}Mv^2}{\frac{1}{2}Mv^2}$$

$$\Rightarrow \quad \frac{K_r}{K} = \frac{2}{5}.$$

83. (a)

$$fR = I\alpha$$

$$\Rightarrow \quad f = \frac{Ia}{R^2}$$

$$mg\sin\theta - f = ma$$

$$\Rightarrow \quad mg\sin\theta - \frac{Ia}{R^2} = ma$$

$$\Rightarrow \quad a = \frac{mg\sin\theta}{m + \frac{1}{R^2}} \quad \text{.... (i)}$$

For cylinder/ disc $I = \frac{1}{2} mR^2 = 0.5mR^2$

For solid sphere $I = \frac{2}{5} mR^2 = 0.4mR^2$

a_{sphere} is greater than $a_{cylinder/disc}$ from Eq. (i) because I is less for sphere.

$\therefore$ t_{sphere} is less than $t_{\text{cylinder/disc}}$. $(\because t = \sqrt{2s/a})$

84. (a)

In collisions (elastic, plastic and inelastic), momentum is conserved. ($\because$ If no external force is acting, then momentum is constant.) In elastic collisions, kinetic energy is same before and after collision.

$\therefore$ Kinetic energy is conserved in elastic collisions.

85. (b)

Angular momentum of a rigid body about a fixed axis is

$$L = I\omega$$

Kinetic energy of body is given by

$$K = \frac{1}{2} I\omega^2 = \frac{1}{2} I \times \left(\frac{L}{I}\right)^2 = \frac{L^2}{2I}$$

$$\Rightarrow \quad I = \frac{L^2}{2K}.$$

86. (a)

The situation can be shown as:

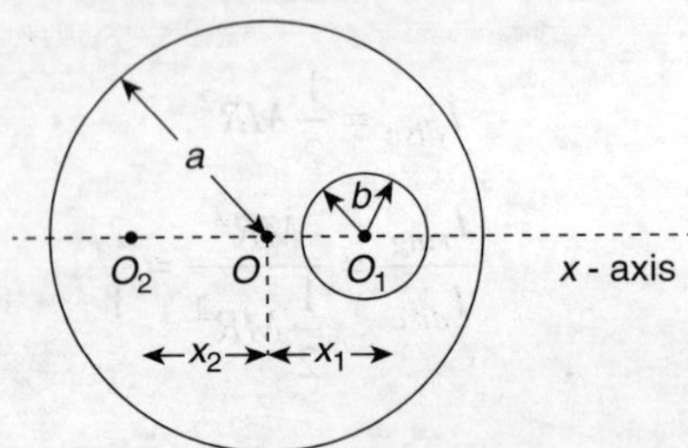

Let the radius of complete disc be a and that of small disc be b. Also, let centre of mass now shift to O_2 at a distance x_2 from original centre.

The position of new centre of mass is given by

$$x_{cm} = \frac{-\sigma\pi b^2 x_1}{\sigma\pi a^2 - \sigma\pi b^2}$$

Hence, $$x_{cm} = \frac{-\sigma \times \pi(2)^2 \times 3.2}{\sigma \times \pi \times (6)^2 - \sigma \times \pi \times (2)^2}$$

$$= -\frac{12.8\pi}{32\pi} = -0.4 \text{ cm.}$$

87. (c)

$$F_{ext} = \sqrt{F_x^2 + F_y^2} = \sqrt{16^2 + 8^2} = 8\sqrt{5} \text{ N};$$

$$M = m_1 + m_2 = 8 + 8 = 16;$$

$$F_{C.M.} = Ma_{C.M.}$$

$$|a_{C.M.}| = \frac{|F_{C.M.}|}{M} = \frac{8\sqrt{5}}{16} = \frac{\sqrt{5}}{2} \text{ m/s}^2$$

$a_{C.M.}$ lies in the direction of F_{ext}

Angle made by F_{ext} with x-axis $= \tan^{-1}\left|\frac{F_y}{F_x}\right|$

$$= \tan^{-1}\frac{16}{8} = \tan^{-1}(2)$$

88. (d)

Angular momentum $L = I\omega$.

$$\Rightarrow \quad L = \frac{2K}{\omega^2}.\omega = \frac{2K}{\omega}$$

New angular momentum

$$L' = \frac{2\left(\frac{K}{2}\right)}{2\omega} = \frac{1}{4}\left(\frac{2K}{\omega}\right) = \frac{L}{4}.$$

89. (a)

According to equation of motion of the centre of mass

$$M\vec{a}_{C.M.} = \vec{F}_{ext}$$

If $\quad \vec{F}_{ext} = 0,\ \vec{a}_{C.M.} = 0$

i.e., $\quad v_{C.M.} = \text{constant}.$

i.e., If no external force acts on a system, the velocity of its centre of mass remains constant. Thus, centre of mass may move but not accelerate.

90. (b)

As ice melts, due to centrifugal force in the frame of pan, water moves outward and total moment of inertia increases; hence to conserve angular momentum angular speed decreases.

91. (b)

Two equal and opposite, parallel forces not acting along the same line from a couple. A couple produces only rotation.

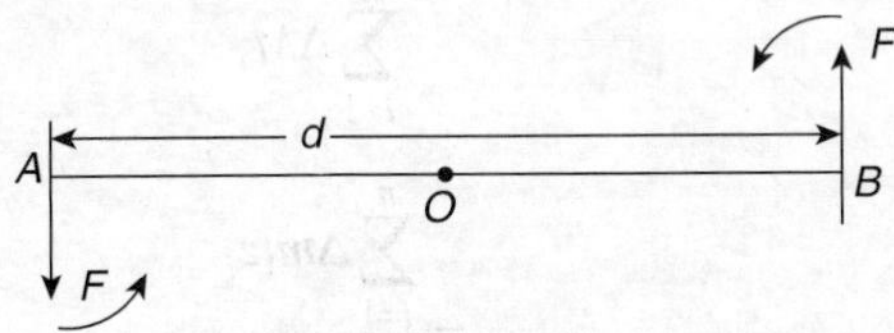

Couple can only be balanced by an equal and opposite couple. In the given figure, AB is a bar fixed at O. At ends A and B two equal and opposite forces are applied. The forces (F) cannot produce translational motion as their sum is zero. But they have turning effects in direction and hence form a couple, which rotates the bar.

92. (d)

$$\frac{1}{2}mv^2 = \frac{1}{2}I\omega^2$$

$$m\omega^2 r^2 = I\omega^2;\ I = mr^2$$

$\Rightarrow$ Moment of inertia of hollow cylinder about its axis.

93. (b)

Moment of inertia of the rod $= \dfrac{ml^2}{12}$

$\therefore$ Moment of inertia of the mass $= \dfrac{ml^2}{12} + \dfrac{ml^2}{12} = \dfrac{ml^2}{6}$

94. (c)

From law of conservation of angular momentum, if no external torque is acting on a body ($\tau = 0$) then the angular momentum ($L = I\omega$) or, in other words, the moment of momentum of body remains constant.

95. (a)

Angular momentum of a rotating system is conserved. As no external torque is acting the angular momentum (L) is conserved.

$$L = I\omega = \text{constant}$$

where $I = mr^2$ is moment of inertia and ω the angular velocity.

Here, $\quad I_1 = I, I_2 = \dfrac{I}{n}, \omega_1 = \omega$

$$L = I_1\omega_1 = I_2\omega_2$$

$$I\omega = \frac{I}{n}\omega_2$$

$$\Rightarrow \quad \omega_2 = n\omega$$

Hence, angular velocity increases by a factor of n.

96. (d)

We know that rotational force acts on the thread when unrolling. Suppose a is acceleration of fall of the thread. Then net force acting downwards, balances the force due to tension (T) in the thread.

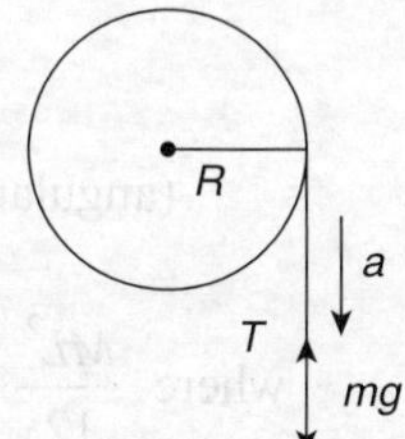

$$mg - T = ma$$

or $\quad mg - ma = T \quad \ldots(i)$

Also, torque (also known as momentum or couple acts on the system)

τ = Force $\times$ perpendicular distance from axis of rotation

$$\tau = \tau \times R$$

By Eq. (i), $\quad \tau = m(g-a)\times R \quad \ldots(ii)$

Suppose I is moment of inertia of reel and α the angular acceleration, then torque is

$$\tau = I\alpha \quad \ldots(iii)$$

where $I = \frac{1}{2}mR^2, \alpha = \frac{a}{R}$

$\therefore \quad \tau = \frac{1}{2}mR^2 \times \frac{a}{R} = \frac{mRa}{2}$ ...(iv)

Comparing Eqs. (ii) and (iv),

$$\tau = m(g-a)R = \frac{mRa}{2}$$

or $g - a = \frac{a}{2}$

or $a = \frac{2}{3}g.$

97. (a)

Both angular and linear momentums remain conserved

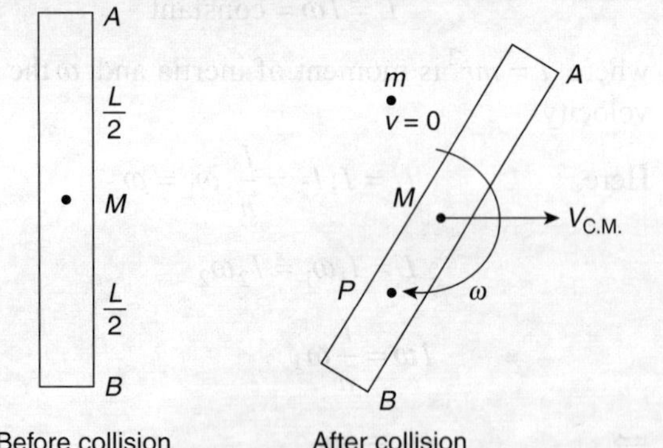

$mv_0 = Mv_{C.M.}$ (linear momentum is conserved)...(i)

$$mv_0\frac{L}{2} = \frac{ML^2}{12}\omega$$

(angular momentum is also conserved) ...(ii)

where $\frac{ML^2}{12}$ is the moment of inertia of the rod about the axis of rotation.

Since, collision is completely elastic, kinetic energy is also conserved. Thus,

$$\frac{1}{2}mv_0{}^2 = \frac{1}{2}Mv_{C.M.}^2 + \frac{1}{2}\left(\frac{ML^2}{12}\right)\omega^2 \quad ...(iii)$$

By Eqs. (i) and (ii)

$$v_{C.M.} = \frac{1}{6}\omega L$$

Putting this value in Eq. (iii),

$$\frac{1}{2}mv_0{}^2 = \frac{1}{2}M\left(\frac{1}{36}\omega^2L^2\right) + \frac{1}{2}M\left(\frac{1}{12}\omega^2L^2\right)$$

$$= \frac{1}{18}M\omega^2L^2$$

or $\frac{m}{M} = \frac{\omega^2L^2}{9v_0}.$

98. (d)

Suppose M is the mass and a is side of larger triangle, then $\frac{M}{4}$ and $\frac{a}{2}$ will be mass and side length of smaller triangle.

$$\frac{I_{removed}}{I_{original}} = \frac{K\frac{M}{4}(\frac{a}{2})^2}{KM(a)^2}$$

$$I_{removed} = \frac{I_0}{16}$$

So, $I = I_0 - \frac{I_0}{16} = \frac{15}{16}I_0$

99. (b)

Centre of mass of a solid body is given by

$$x_{C.M.} = \frac{\sum_{i=1}^{n}\Delta m_1 x_i}{\sum_{i=1}^{n}\Delta M_i},$$

$$y_{C.M.} = \frac{\sum_{i=1}^{n}\Delta m_1 y_i}{\sum_{i=1}^{n}\Delta M_i},$$

$$z_{C.M.} = \frac{\sum_{i=1}^{n}\Delta m_1 z_i}{\sum_{i=1}^{n}\Delta M_i},$$

$$1\times x_1 + 2\times x_2 + 3\times x_3 = (1+2+3)3 \quad ....(i)$$

and $x_1 = x_2 = x_3 = 3$

$$x_{C.M.} = y_{C.M.} = z_{C.M.} = 1$$

$$1(1+2+3+4) = 1x_1 + 2x_2 + 3x_3 + 4x_4 \quad ....(ii)$$

Solving Eqs. (i) and (ii)

$$4x_4 = 10 - 18$$

$\Rightarrow \quad x_4 = -2$

Similarly, $y_4 = -2, z_4 = -2.$

The fourth particle must be placed at the point (−2,−2,−2).

100. (d)

We know that, $p_{C.M.} = p_1 + p_2$

$$\Rightarrow \quad 2\hat{i} - \hat{j} + 3\hat{k} - \hat{i} + 2\hat{j} + 3\hat{k} = \hat{i} + \hat{j} + 6\hat{k}$$

Angle made by $p_{C.M.}$ with $x-axis$

$$= \cos^{-1}\frac{A_x}{|A|} = \cos^{-1}\frac{1}{\sqrt{38}}$$

101. (a)

The moment of inertia of the uniform rod about an axis through one end and perpendicular to length

$$I = \frac{ml^2}{3}$$

where m is mass of rod and l its length.

Torque $(\tau = I\alpha)$ acting on centre of gravity of rod is given by

$$\tau = mg\frac{l}{2}$$

or $$I\alpha = mg\frac{l}{2}$$

or $$\frac{ml^2}{3}\alpha = mg\frac{l}{2}$$

$$\alpha = \frac{3g}{2l}.$$

102. (a)

Angular momentum $\vec{L} = I\vec{\omega}$, and so $\vec{L}$ and $\vec{\omega}$ are along the same direction.

Also $$\vec{\tau} = \frac{\Delta\vec{L}}{\Delta t},$$

$$\vec{\tau}\,\Delta t = \Delta\vec{L}$$

103. (c)

When no external force acts on the system, initially velocity of centre of mass is zero and so the centre of mass remains at rest.

104. (a)

Given: $$\frac{d\omega}{dt} = 3t^2 + 5t$$

$$\therefore \quad \int_{10}^{\omega} d\omega = \int_{10}^{\omega}(3t^2 + 5t)dt$$

or $$\omega - 10 = t^3 + \frac{5t^2}{2}$$

$$\omega = t^3 + \frac{5t^2}{2} + 10$$

We can write:

$$\frac{d\theta}{dt} = t^3 + \frac{5}{2}t^2 + 10$$

or $$\int_8^{\theta} d\theta = \int_0^{t}(t^3 + \frac{5}{2}.t^2 + 10)dt$$

$$\theta - 8 = \frac{t^4}{4} + \frac{5t^3}{6} + 10t$$

$$\theta = \frac{t^4}{4} + \frac{5t^3}{6} + 10t + 8$$

105. (d)

By conservation of linear momentum,

$$(80 + 40) \times 4 = 40 \times 4 + 80v$$

$$\Rightarrow \quad v = 4 \text{ m/s}$$

The change in speed of the cart

$$= 4 - 4 = 0.$$

106. (c)

Taking the angular momentum about O_1,

$$L_1 = Pb - Pb = 0$$

Taking the angular momentum about O_2,

$$L_2 = Pb - Pb = 0$$

The angular momentum is same, i.e., zero for the two particles at O_1 and O_2.

107. (c)

$$\bar{x} = \frac{m(-2b) + 2m(-b) + m \times 0 + 2m(b)}{m + 2m + m + 2m}$$

$$= -\frac{b}{3}$$

and $\bar{y} = +b$

$$\therefore \quad (x, y) = \left(-\frac{b}{3}, b\right)$$

108. (b)

According to angular momentum conservation law

$$I_1\omega_1 = I_2\omega_2$$

M.I. of disc

$$=\frac{1}{2}mR^2$$

$$\frac{1}{2}mR^2\omega_1=\frac{1}{2}\left(m+\frac{1}{4}m\right)R^2\omega_2$$

$$m\omega_1=\frac{5}{4}m\cdot\omega_2$$

$$\omega_2=\frac{4}{5}\omega_1$$

109. (c)

Given: $L=4\sqrt{t}$

$$\tau=\frac{dL}{dt}=4.\frac{1}{2}t^{\frac{1}{2}-1}=\frac{2}{\sqrt{t}}$$

110. (b)

Let M be the mass of the rod, then mass per unit length, i.e.,

$$\frac{M}{L}=A$$

Consider an elemental disc at a distance x from the origin of thickness dx

∴ Mass of the elemental disc

$$dm=\frac{M}{L}dx=Adx$$

∴ Position of centre of mass

$$x_{C.M.}=\frac{\int_0^L xdm}{\int_0^L dm}=\frac{\int_0^L xAdx}{M}$$

$$=\frac{\int_0^L x\times\frac{M}{L}.dx}{M}=\frac{1}{L}\left[\frac{x^2}{2}\right]_0^L$$

$$=\frac{1}{L}\left[\frac{L^2}{2}\right]=\frac{L}{2}$$

111. (a)

I = M.I. of A about CD
+ M.I. of B about CD
+ M.I. of C about CD
+ M.I. of D about CD

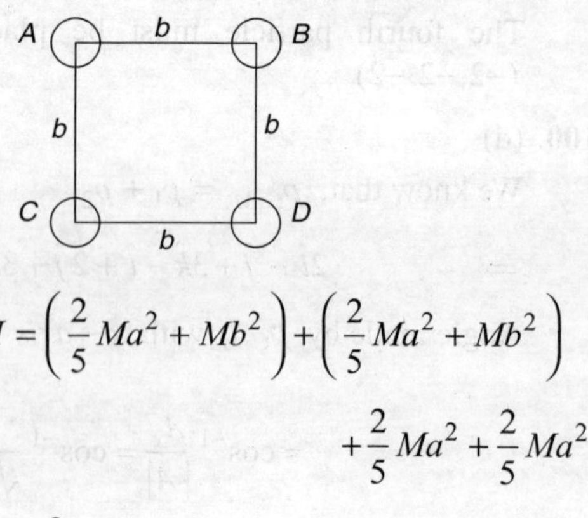

$$I=\left(\frac{2}{5}Ma^2+Mb^2\right)+\left(\frac{2}{5}Ma^2+Mb^2\right)$$

$$+\frac{2}{5}Ma^2+\frac{2}{5}Ma^2$$

$$I=\frac{8}{5}Ma^2+2Mb^2$$

112. (d)

Rotational

$$\text{KE}=\frac{1}{2}I\omega^2$$

Moment of inertia of circular ring of mass m and radius r about an axis passing through its centre and perpendicular to its plane is

$$I_{ring}=mr^2$$

$$\text{KE}=\frac{1}{2}mr^2\omega^2$$

113. (d)

Moment of inertia of a circular disc about an axis passing through centre of gravity and perpendicular to its plane

$$I=\frac{1}{2}MR^2 \quad \text{...(i)}$$

From Eq. (i),

$$MR^2=2I$$

Then, moment of inertia of disc about tangent in a plane

$$=\frac{5}{4}MR^2$$

$$=\frac{5}{4}(2I)$$

$$=\frac{5}{2}I.$$

114. (c)

$$m_1=6\text{ kg}, m_2=4\text{ kg}$$

$$\vec{v}=\frac{m_1\vec{v}_1+m_2\vec{v}_2}{m_1+m_2}$$

$$= \frac{6(5\hat{i} - 2\hat{j} + 10\hat{k}) + 4(10\hat{i} - 2\hat{j} + 5\hat{k})}{6+4}$$

$$= \frac{70\hat{i} - 20\hat{j} + 80\hat{k}}{10}$$

$$\therefore \quad \vec{v} = 7\hat{i} - 2\hat{j} + 8\hat{k}.$$

115. (d)

Applying conservation of angular momentum about point O,

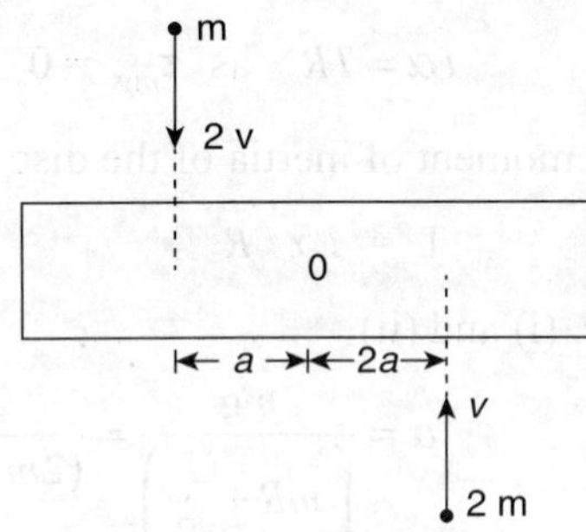

$$m(a)(2v) + 2m(2a)(v) = I\omega$$

or $\quad \omega = \frac{6mav}{I} \quad$ ...(i)

Now, $\quad I = \frac{6m(8a)^2}{12} + m(a)^2 + 2m(2a)^2$

$$= 32ma^2 + ma^2 + 8ma^2 = 41ma^2$$

$\therefore$ From Eq. (i),

$$\omega = \frac{6mav}{41ma^2} = \frac{6v}{41a}.$$

116. (a)

$$m_1 = 1\text{ kg}, m_2 = 2\text{ kg}, m_3 = 3\text{ kg}$$

Position of centre of mass

$$= (2, 2, 2)$$

$$m_4 = 4\text{ kg}$$

New position of centre of mass (0, 0, 0)
For initial position

$$X_{C.M.} = \frac{m_1x_1 + m_2x_2 + m_3x_3}{m_1 + m_2 + m_3}$$

$$2 - \frac{m_1x_1 + m_2x_2 + m_3x_3}{1+2+3}$$

$$\therefore \quad m_1x_1 + m_2x_2 + m_3x_3 = 12.$$

Similarly,

$$m_1y_1 + m_2y_2 + m_3y_3 = 12$$

and $\quad m_1z_1 + m_2z_2 + m_3z_3 = 12.$

For new position,

$$X'_{cm} = \frac{m_1x_1 + m_2x_2 + m_3x_3 + m_4x_4}{m_1 + m_2 + m_3 + m_4}$$

$$0 = \frac{12 + 4x_4}{1+2+3+4}$$

$$4x_4 = -12$$

$$x_4 = -3$$

Similarly $\quad y_4 = -3, z_4 = -3.$

$\therefore$ Position of fourth mass $= (-3. -3, -3)$.

117. (b)

If rod is rotated about end A, then vertical component of velocity v_1 of end A will be zero.

$$\therefore \quad \omega = \frac{v\cos 60^\circ}{l} = \frac{\sqrt{3}v}{2l}$$

$$= \frac{\sqrt{3} \times 3}{2 \times 0.5} = 5.2\text{ rad/s.}$$

118. (a)

1. Angular velocity of earth

$$\omega_1 = \frac{2\pi}{T} = \frac{2\pi}{24 \times 60 \times 60} = \frac{2\pi}{86400}\text{ rad/s.}$$

2. Angular velocity of hours hand of clock

$$\omega_2 = \frac{2\pi}{T} = \frac{2\pi}{12 \times 60 \times 60} = \frac{2\pi}{43200}\text{ rad/s}$$

3. Angular velocity of seconds hand of a clock

$$\omega_3 = \frac{2\pi}{T} = \frac{2\pi}{1 \times 60} = \frac{2\pi}{60}\text{ rad/s}$$

4. Angular velocity of flywheel

$$\omega_4 = 2\pi n = 2\pi \times \frac{300}{60} = 2\pi \times 5\text{ rad/s.}$$

119. (a)

$$\vec{v}_{C.M.} = \frac{m_1\vec{v}_1 + m_2\vec{v}_2}{m_1 + m_2} = \frac{\vec{v}_1 + \vec{v}_2}{2}$$

and $\quad \vec{a}_{C.M.} = \frac{m_1\vec{a}_1 + m_2\vec{a}_2}{m_1 + m_2} = \frac{\vec{a}_1 + 0}{2} = \frac{\vec{a}_1}{2}$

The centre of mass of two particles will move with the mean velocity of two particles having common acceleration $\frac{\vec{a}}{2}$. Hence, path of $C.M.$ will be a straight line.

120. (a)

In the absence of external torque on a system of particles, angular momentum of the system remains constant, i.e.,

$$\tau = 0$$

$$\Rightarrow \quad \frac{dL}{dt} = 0$$

$$\Rightarrow \quad L = I\omega = \text{constant}$$

$$I_1\omega_1 = I_2\omega_2$$

$$\frac{1}{2}Mr^2\omega_1 = \frac{1}{2}(M+2m)r^2\omega_2 \quad \text{...(i)}$$

Given, $\quad M = 2\text{ kg}, m = 0.25\text{ Kg}, r = 0.2\text{ m}$

$$\omega_1 = 30\text{ rad/s.}$$

Thus, after putting the given values in Eq. (i)

$$\frac{1}{2}\times 2\times(0.2)^2\times 30 = \frac{1}{2}(2+2\times 0.25)(0.2)^2\times\omega_2$$

$$\omega_2 = 24\text{ rad/s.}$$

121. (a)

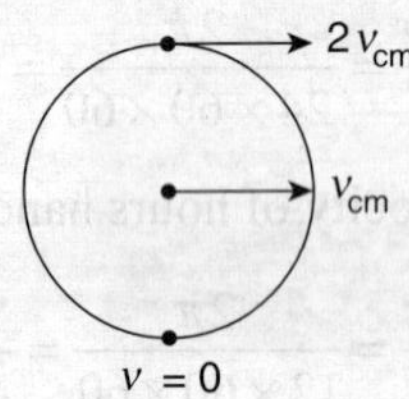

122. (b)

As effective distance of mass from BC is greater than the effective distance of mass from AB, $I_2 > I_1$

123. (a)

Solid sphere reaches the bottom first because for solid cylinder $\frac{K^2}{R^2} = \frac{1}{2}$

and for hollow cylinder $\frac{K^2}{R^2} = 1$

Acceleration down the inclined plane $\propto \frac{1}{K^2/R^2}$

Solid cylinder has greater acceleration, so it reaches the bottom first.

124. (c)

Since string does not slip on the disc, tangential acceleration of the point on the rim which is in contact with the string is equal to the acceleration of the block. Let angular acceleration of the disc about axis be α; hence acceleration of the block $a = \alpha R$

F.B.D of the block F.B.D of the Disc.

$$\Rightarrow \quad mg - T = m\,\alpha R,\ as = \alpha R \quad \text{....(i)}$$

Torque on the disc is

$$\tau = \tau_{Tension} + \tau_{mg}$$

$$\Rightarrow \quad I\alpha = TR \quad \text{as } \tau_{mg} = 0$$

Where I = moment of inertia of the disc about its axis.

$$\Rightarrow \quad T = I\alpha/R \quad \text{...(ii)}$$

Using Eqs. (i) and (ii)

$$\alpha = \frac{mg}{\left(mR + \frac{I}{R}\right)} = \frac{2mg}{(2m+M)R}$$

as $\quad I = \frac{MR^2}{2}$

Hence $\quad a = \alpha R = \frac{2mg}{2m+M}$

125. (d)

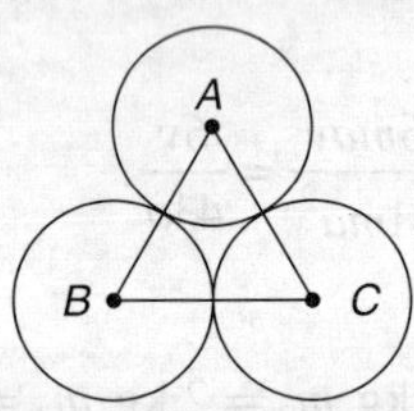

Centre of mass of each ball lies on the centre.

$\Rightarrow$ Centre of mass of combined body will be at the centroid of equilateral triangle.

126. (a)

Moment of inertia of a disc about its diameter $= \frac{1}{4}MR^2$

Using theorem of parallel axes,

$$I = \frac{1}{4}MR^2 + MR^2 = \frac{5}{4}MR^2$$

127. (c)

Force $\left(\vec{F}\right) = -3\hat{i} + \hat{j} + 5\hat{k}$ and distance of the point $\left(\vec{r}\right) = 7\hat{i} + 3\hat{j} + \hat{k}$

$$\text{Torque } \vec{\tau} = \vec{r} \times \vec{F} = \begin{vmatrix} \hat{i} & \hat{j} & \hat{k} \\ 7 & 3 & 1 \\ -3 & 1 & 5 \end{vmatrix} = 14\hat{i} - 38\hat{j} + 16\hat{k}$$

128. (d)

The resultant of all forces, on any system of particles, is zero. Therefore their centre of mass does not depend upon the forces acting on the particles.

129. (c)

As no external horizontal force is there on the rod, its centre of mass falls vertically. So the lower end will be at a distance 1/2 from O after it falls on ground.

130. (c)

If the cylinder moves at speed v, then the plank will move at speed $2v$ as there is no slipping anywhere. The kinetic energy of the plank is $(1/2)\,M(2v)^2$ and that of cylinder in pure rolling is $(3/4)Mv^2$. Thus option (c) is correct.

131. (d)

Force $\left(\vec{F}\right) = 2\hat{i} - 3\hat{j} + 4\hat{k}$ N and distance of the point from origin $(r) = 3\hat{i} + 2\hat{j} + 3\hat{k}$

Torque $\quad \vec{\tau} = \vec{r} \times \vec{F}$

$$= \left(3\hat{i} + 2\hat{j} + 3\hat{k}\right) \times \left(2\hat{i} - 3\hat{j} + 4\hat{k}\right)$$

$$= \begin{vmatrix} \hat{i} & \hat{j} & \hat{k} \\ 3 & 2 & 3 \\ 2 & -3 & 4 \end{vmatrix}$$

$$= 17\hat{i} - 6\hat{j} - 13\hat{k}$$

132. (d)

Linear KE of ball $= \dfrac{1}{2}mv^2$ and

Rotational KE of ball $= \dfrac{1}{2}I\omega^2 = \dfrac{1}{2}\left(\dfrac{2}{5}mr^2\right)\omega^2$

$$= \frac{1}{5}mv^2$$

Therefore total KE $= = \dfrac{1}{2}mv^2 + \dfrac{1}{5}mv^2 = \dfrac{7}{10}mv^2$

And ratio of rotational KE and total KE

$$= \frac{(1/5)mv^2}{(7/10)mv^2} = \frac{2}{7}$$

133. (d)

The moment of inertia is minimum about EG because mass distribution is at minimum distance from EG.

134. (d)

As the scooter is moving on a smooth road, no external force is acting on it. So internal forces cannot change the speed of centre of mass of scooter.

135. (a)

PE = total KE

$$mgh = \frac{7}{10}mv^2, v = \sqrt{\frac{10gh}{7}}$$

136. (d)

Total kinetic energy $= E_{trans} + E_{rot}$

$$= \frac{1}{2}mv^2 + \frac{1}{2}I\omega^2 = \frac{1}{2}mv^2 + \frac{1}{2} \times \left(\frac{2}{5}mr^2\right)\omega^2$$

$$= \frac{1}{2}mv^2 + \frac{1}{5}mv^2 = \frac{7}{10}mv^2$$

$$\therefore \frac{E_{trans}}{E_{total}} = \frac{\frac{1}{2}mv^2}{\frac{7}{10}mv^2} = \frac{5}{7}$$

137. (a)

$$\vec{L} = \vec{r} \times \vec{p}$$

$Y = X + 4$ line has been shown in the figure.

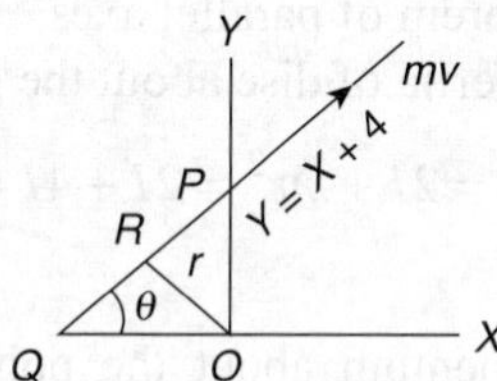

When $X = 0$,
$Y = 4$, So OP = 4.
The slope of the line can be obtained by comparing with the equation of line

$$y = mx + c$$

$$m = \tan\theta = 1 \Rightarrow \theta = 45°$$

$$\angle OQP = \angle OPQ = 45°$$

If we draw a line perpendicular to this line,
Length of the perpendicular = OR

$$\rightarrow OR = OP\sin 45° - 4\frac{1}{\sqrt{2}} = \frac{4}{\sqrt{2}} = 2\sqrt{2}$$

Angular momentum of particle going along this line

$$= r \times m = 2\sqrt{2} \times 5 \times 3\sqrt{2} = 60 \text{ units}$$

138. (c)

$$\text{KE} = \frac{1}{2}I\omega^2$$

$$I = \frac{2K.E.}{\omega^2} = \frac{2\times 360}{30\times 30} = 0.8 \text{ kg m}^2$$

139. (b)

$$I = 1.2 \text{ kg m}^2,\ E_r = 1500 \text{ J}$$

$$\alpha = 25 \text{ rad/s}^2,\ \omega_1 = 0,\ t = ?$$

As $\quad E_r = \frac{1}{2}I\omega^2, \omega = \sqrt{\frac{2E_r}{I}}$

$$\omega = \sqrt{\frac{2\times 1500}{1.2}} = 50 \text{ rad/s}$$

From $\quad \omega_2 = \omega_1 + \alpha t$

$$50 = 0 + 25t, \text{ or } t = 2 \text{ s}$$

140. (c)

Moment of inertia of uniform circular disc about diameter = I

According to theorem of perpendicular axes,

Moment of inertia of disc about axis

$$= 2I = \frac{1}{2}mr^2$$

Applying theorem of parallel axes,

Moment of inertia of disc about the given axis

$$= 2I + mr^2 = 2I + 4I = 6I$$

141. (b)

Angular momentum about the point of contact with the surface includes the angular momentum about the centre. Because of friction, linear momentum will not be conserved.

142. (a)

Kinetic energy $= \frac{1}{2}I\omega^2$, and for ring $I = mr^2$

Hence $\text{KE} = \frac{1}{2}mr^2\omega^2$

143. (a)

Initial kinetic energy of the given system,

$$\text{KE}_i = \frac{1}{2}I_1\omega_1^2 + \frac{1}{2}\left(\frac{I_1}{2}\right)\left(\frac{\omega_1}{2}\right)^2$$

$$= \left(\frac{1}{2}+\frac{1}{16}\right)I_1\omega_1^2 = \frac{9}{16}I_1\omega_1^2 \qquad \text{(i)}$$

Now, using angular momentum conservation law

Initial angular momentum = Final angular momentum

$$I_1\omega_1 + \left(\frac{I_1}{2}\right)\left(\frac{\omega_1}{2}\right) = I_1\omega' + \frac{I_1}{2}\omega'$$

$$\Rightarrow \frac{5}{4}\omega_1 = \frac{3}{2}\omega' \text{ or } \omega' = \frac{5}{6}\omega_1$$

Now, final kinetic energy (after contact)

$$\text{KE}_f = \frac{1}{2}I_1\omega'^2 + \frac{1}{2}\left(\frac{I_1}{2}\right)\omega'^2$$

$$= \frac{1}{2}I_1\left(\frac{5}{6}\omega_1\right)^2 + \frac{1}{4}I_1\left(\frac{5}{6}\omega_1\right)^2$$

$$= \left(\frac{25}{72}+\frac{25}{144}\right)I_1\omega_1^2$$

$$= \frac{25}{48}I_1\omega_1^2$$

Hence, change in KE,

$$\Delta\text{KE} = \text{KE}_f - \text{KE}_i$$

$$= \frac{25}{48}I_1\omega_1^2 - \frac{9}{16}I_1\omega_1^2 \qquad \text{(From Eq. (i))}$$

$$\Delta\text{KE} = -\frac{1}{24}I_1\omega_1^2$$

144. (b)

A sphere cannot roll on a surface where no torque is present to provide the rotational motion in accordance with the linear motion of sphere due to gravity. Hence only option (b) is the case among other given cases where sphere cannot roll without sliding.

Questions Asked in NEET/CBSE Medicals

145. (b)

Given: Mass $M = 2$ kg, Radius $R = 4$ cm

Initial angular speed

$$\omega_0 = 3rpm = 3\times\frac{2\pi}{60} \text{ rad/s} = \frac{\pi}{10} \text{ rad/s}$$

We know that, $\omega^2 = \omega_0^2 + 2\alpha\theta$

$$\Rightarrow 0 = \left(\frac{\pi}{10}\right)^2 + 2\times\alpha\times 2\pi\times 2\pi$$

$$\Rightarrow \alpha = \frac{-1}{800} \text{ rad/s}^2$$

Moment of inertia of a solid cylinder,

$$I = \frac{MR^2}{2} = \frac{2\times\left(\frac{4}{100}\right)^2}{2} = \frac{16}{10^4}$$

Torque $\tau = I\alpha = \left(\frac{16}{10^4}\right)\times\left(\frac{1}{800}\right) = -2\times10^{-6}$ Nm

146. (b)

Required work done $= -(K_f - K_i) = 0 + K_i = K_i$

$$= \frac{1}{2}I\omega^2 + \frac{1}{2}mv^2 = \frac{1}{2}\left(\frac{1}{2}mR^2\right)\omega^2 + \frac{1}{2}mv^2$$

$$= \frac{3}{4}mv^2 = \frac{3}{4}\times100\times\left(20\times10^{-2}\right)^2 = 3\text{ J}$$

147. (d)

Using law of conservation of energy,

$$mgh = \frac{1}{2}mv_{cm}^2 + \frac{1}{4}mr^2\omega^2$$

$$\Rightarrow mgs\sin30° = \frac{1}{2}mv_{cm}^2 + \frac{1}{4}mv_{cm}^{\ 2}$$

$$\Rightarrow s = \frac{3}{4}\frac{v^2}{g\sin30°} \Rightarrow s = \frac{3}{4}\times\frac{4^2}{10\times\frac{1}{2}}$$

$$\Rightarrow s = \frac{3\times4\times2}{10} = \frac{12}{5} = 2.4\text{ m}$$

148. (b)

Translational kinetic energy,

$$K_t = \frac{1}{2}mv^2$$

Rotational kinetic energy, $K_r = \frac{1}{2}I\omega^2$

$$\therefore K_t + K_r = \frac{1}{2}mv^2 + \frac{1}{2}I\omega^2$$

$$= \frac{1}{2}mv^2 + \frac{1}{2}\left(\frac{2}{5}mr^2\right)\left(\frac{v}{2}\right)^2$$

$$\left[\because I = \frac{2}{5}mr^2\text{ (for sphere)}\right]$$

$$\therefore K_t + K_r = \frac{7}{10}mv^2$$

So, $\frac{K_t}{K_t + K_r} = \frac{5}{7}$

149. (d)

As there is no external torque acting on a sphere, $\tau_{ex} = 0$

So, $\frac{dL}{dt} = \tau_{ex} = 0$, L = constant

So angular momentum remains constant.

150. (a)

Work done required to bring an object to rest $\Delta W = \Delta KE$

$$\Delta W = \frac{1}{2}I\omega^2;$$

where I = moment of inertia

For same $\omega, \Delta W \propto I$

For a solid sphere, $I_A = \frac{2}{5}MR^2$

For a thin circular disk, $I_B = \frac{1}{2}MR^2$

For a circular ring, $I_C = MR^2$

$\therefore I_C > I_B > I_A \therefore W_C > W_B > W_A$

151. (d)

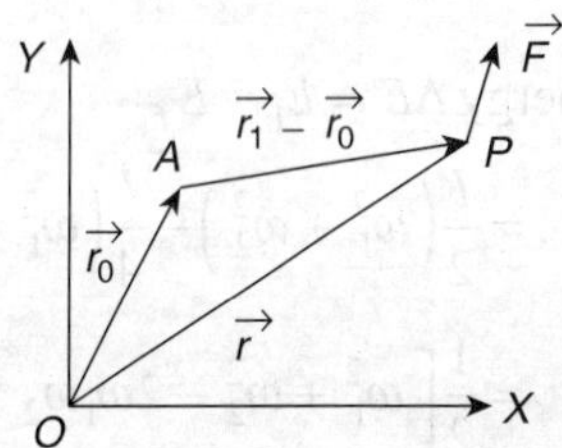

Moment of the force is,

$$\vec{\tau} = \left(\vec{r} - \vec{r_0}\right)\times\vec{F}$$

Here, $\vec{r_0} = 2\hat{i} - 2\hat{j} - 2\hat{k}$

and $\vec{r} = 2\hat{i} + 0\hat{j} - 3\hat{k}$

$$\therefore \vec{r} - \vec{r_0} = \left(2\hat{i} + 0\hat{j} - 3\hat{k}\right) - \left(2\hat{i} - 2\hat{j} - 2\hat{k}\right)$$

$$= 0\hat{i} + 2\hat{j} - \hat{k}$$

$$\therefore \vec{\tau} = \begin{vmatrix}\hat{i} & \hat{j} & \hat{k}\\ 0 & 2 & -1\\ 4 & 5 & -6\end{vmatrix} = -7\hat{i} - 4\hat{j} - 8\hat{k}$$

152. (b)

$m = 3$ kg, $r = 40$ cm $= 40\times10^{-2}$ m, $F = 30$ N

Moment of inertia of hollow cylinder about its axis

$$= mr^2 = 3\text{ kg}\times\left(0.4\right)^2\text{ m}^2 = 0.48\text{ kg m}^2$$

The torque, $\tau = I\alpha$

where I = moment of inertia,

α = angular acceleration

In the given case, $\tau = rF$, as the force is acting perpendicularly to the radial vector.

$$\therefore \alpha = \frac{\tau}{I} = \frac{Fr}{mr^2} = \frac{F}{mr} = \frac{30}{3 \times 40 \times 10^{-2}} = \frac{30 \times 100}{3 \times 40}$$

$\alpha = 25 \text{ rad s}^{-2}$

153. (a)

Initial angular momentum $= I\omega_1 + I\omega_2$

Let ω be angular speed of the combined system. Final angular momentum $= 2I\omega$

$\therefore$ According to conservation of angular momentum

$$I\omega_1 + I\omega_2 = 2I\omega \text{ or } \omega = \frac{\omega_1 + \omega_2}{2}$$

Initial rotational kinetic energy

$$E = \frac{1}{2} I\left(\omega_1^2 + \omega_2^2\right)$$

Final rotational kinetic energy

$$E_f = \frac{1}{2}(2I)\omega^2 = \frac{1}{2}(2I)\left(\frac{\omega_1 + \omega_2}{2}\right)^2 = \frac{1}{4} I\left(\omega_1 + \omega_2\right)^2$$

$\therefore$ Loss of energy $\Delta E = E_i - E_f$

$$= \frac{I}{2}\left(\omega_1^2 + \omega_2^2\right) - \frac{I}{4}\left(\omega_1^2 + \omega_2^2 + 2\omega_1\omega_2\right)$$

$$= \frac{1}{2}\left[\omega_1^2 + \omega_2^2 - 2\omega_1\omega_2\right] = \frac{1}{4}\left(\omega_1 - \omega_2\right)^2$$

154. (c)

Given, $m_A = m, m_B = 2m$

Both bodies A and B have equal kinetic energy of rotation

$$K_A = K_B \Rightarrow \frac{1}{2} I_A \omega_A^2 = \frac{1}{2} I_B \omega_B^2$$

$$\Rightarrow \frac{\omega_A^2}{\omega_B^2} = \frac{I_B}{I_A} \quad \text{(i)}$$

Ratio of angular moment a,

$$\frac{L_A}{L_B} = \frac{I_A \omega_A}{I_B \omega_B} = \frac{I_A}{I_B} \times \sqrt{\frac{I_B}{I_A}} \quad \text{[Using Eq. (i)]}$$

$$= \sqrt{\frac{I_A}{I_B}} < 1 \quad (\because I_B > I_A)$$

$\because \quad L_B > L_A$

155. (b)

$$\frac{E_{\text{Sphere}}}{E_{\text{Cylinder}}} = \frac{\frac{1}{2} I_s \omega_s^2}{\frac{1}{2} I_c \omega_c^2} = \frac{I_s \omega_s^2}{I_c \omega_c^2}$$

Here, $\quad I_s = \frac{2}{5} MR^2, I_c = \frac{1}{2} MR^2, \omega_c = 2\omega_s$

$$\frac{E_{\text{Sphere}}}{E_{\text{Cylinder}}} = \frac{\frac{2}{5} mR^2 \times \omega_s^2}{\frac{1}{2} mR^2 \times \left(2\omega_s\right)^2} = \frac{4}{5} \times \frac{1}{4} = \frac{1}{5}$$

156. (a)

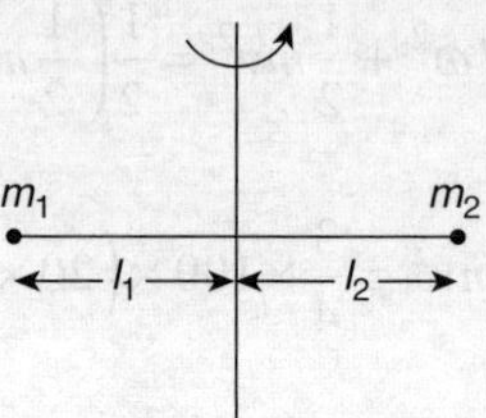

Here, $l_1 + l_2 = l$

Centre of mass of the system,

$$l_1 = \frac{m_1 \times 0 + m_2 \times l}{m_1 + m_2} = \frac{m_2 l}{m_1 + m_2}$$

$$l_2 = l_1 = \frac{m_1 l}{m_1 + m_2}$$

Required moment of inertia of the system,

$$I = m_1 l_1^2 + m_2 l_2^2 = \left(m_1 m_2^2 + m_2 m_1^2\right) \frac{l^2}{\left(m_1 + m_2\right)^2}$$

$$= \frac{m_1 m_2 \left(m_1 + m_2\right) l^2}{\left(m_1 + m_2\right)^2} = \frac{m_1 m_2}{m_1 + m_2} l^2$$

157. (d)

Time taken by the body to reach the bottom when it rolls down on an inclined plane without slipping is given by

$$t = \sqrt{\frac{2l\left(1 + \frac{k^2}{R^2}\right)}{g \sin\theta}}$$

Since g is constant and l, R and $\sin\theta$ are same for both

$$\therefore \frac{t_d}{t_s} = \frac{\sqrt{1 + \frac{k_d^2}{R^2}}}{\sqrt{1 + \frac{k_s^2}{R^2}}} = \sqrt{\frac{1 + \frac{R^2}{2R^2}}{1 + \frac{2R^2}{5R^2}}} \quad \left(\because k_d = \frac{R}{\sqrt{2}}, k_s = \sqrt{\frac{2}{5}} R\right)$$

$$= \sqrt{\frac{3}{2} \times \frac{5}{7}} = \sqrt{\frac{15}{14}} \Rightarrow t_d > t_s$$

Hence, the sphere gets to the bottom first.

158. (d)

Mass per unit area of disc

$$= \frac{M}{\pi R^2}$$

Mass of removed portion of disc,

$$M' = \frac{M}{\pi R^2} \times \pi \left(\frac{R}{2}\right)^2 = \frac{M}{4}$$

Moment of inertia of removed portion about an axis passing through centre of disc O and perpendicular to the plane of disc,

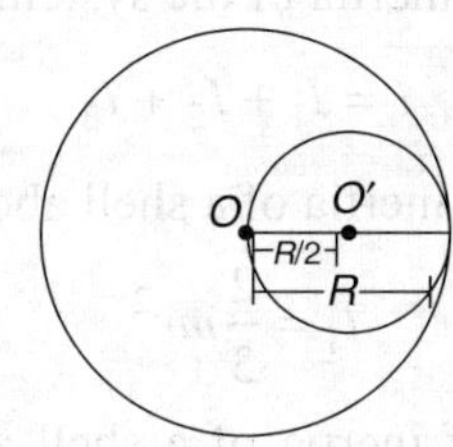

$$I'_O = I_{O'} + M'd^2$$

$$= \frac{1}{2} \times \frac{M}{4} \times \left(\frac{R}{2}\right)^2 + \frac{M}{4} \times \left(\frac{R}{2}\right)^2$$

$$= \frac{MR^2}{32} + \frac{MR^2}{16} = \frac{3MR^2}{32}$$

The moment of inertia of complete disc about centre O before removing the portion of the disc $I_O = \frac{1}{2}MR^2$

So, moment of inertia of the disc with removed portion is

$$I = I_O - I'_O = \frac{1}{2}MR^2 - \frac{3MR^2}{32} = \frac{13MR^2}{32}$$

159. (c)

Given, $r = 50$ cm $= 0.5$ m, $\alpha = 2.0$ rad s^{-2}, $\omega_0 = 0$

At the end of 2 s,

Tangential acceleration, $a_t = r\alpha = 0.5 \times 2 = 1$ m s^{-2}

Radial acceleration, $a_r = \omega^2 r = (\omega_0 + \alpha t)^2 r$

$$= (0 + 2 \times 2)^2 \times 0.5 = 8 \text{ m s}^{-2}$$

∴ Net acceleration,

$$a = \sqrt{a_t^2 + a_r^2} = \sqrt{1^2 + 8^2} = \sqrt{65} \approx 8 \text{ m s}^{-2}$$

160. (b)

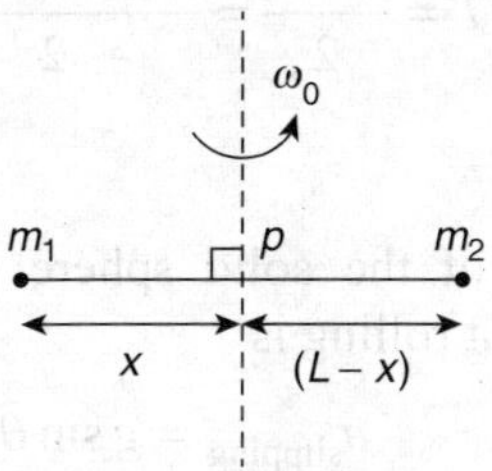

Moment of inertia of the system about the axis of rotation (through point P) is

$$I = m_1 x^2 + m_2 (L - x)^2$$

By work–energy theorem,

Work done to set the rod rotating with angular velocity ω_0 = Increase in rotational kinetic energy

$$W = \frac{1}{2} I \omega_0^2 = \frac{1}{2}\left[m_1 x^2 + m_2 (L - x)^2\right]\omega_0^2$$

For W to be minimum, $\frac{dW}{dx} = 0$

$$\frac{1}{2}\left[2m_1 x + 2m_2 (L - x)(-1)\right]\omega_0^2 = 0$$

or $\quad m_1 x - m_2 (L - x) = 0 \quad (\because \omega_0 \neq 0)$

or $\quad (m_1 - m_2)x = m_2 L$ or $x = \frac{m_2 L}{m_1 + m_2}$

161. (c)

Here, speed of the automobile,

$$v = 54 \text{ km h}^{-1} = 54 \times \frac{5}{18} \text{ m s}^{-1} = 15 \text{ m s}^{-1}$$

Radius of the wheel of the automobile, $R = 0.45$ m

Moment of inertia of the wheel about its axis of rotation, $I = 3$ kg m^2

Time in which the vehicle brought to rest, $t = 15$s

The initial angular speed of the wheel is

$$\omega_i = \frac{v}{R} = \frac{15\text{ms}^{-1}}{0.45\text{m}} = \frac{1500}{45} \text{ rad s}^{-1} = \frac{100}{3} \text{ rad s}^{-1}$$

and its final angular speed is

$$\omega_f = 0 \quad \text{(as the vehicle comes to rest)}$$

∴ The angular retardation of the wheel is

$$\alpha = \frac{\omega_f - \omega_i}{t} = \frac{0 - \frac{100}{3}}{15s} = -\frac{100}{45} \text{ rad s}^{-2}$$

The magnitude of required torque is

$$\tau = I|\alpha| = (3 \text{ Kg m}^2)\left(\frac{100}{45} \text{ rad s}^{-2}\right)$$

$$= \frac{20}{3} \text{ Kg m}^2\text{s}^{-2} = 6.66 \text{ Kg m}^2\text{s}^{-2}$$

162. (c)

For the conservation of angular momentum about origin, the torque $\vec{\tau}$ acting on the particle will be zero.

By definition, $\vec{\tau} = \vec{r} \times \vec{F}$

Here, $\vec{r} = 2\hat{i} - 6\hat{j} - 12\hat{k}$ and $\vec{F} = \alpha\hat{i} + 3\hat{j} + 6\hat{k}$

$$\therefore \vec{\tau} = \begin{vmatrix} \hat{i} & \hat{j} & \hat{k} \\ 2 & -6 & -12 \\ \alpha & 3 & 6 \end{vmatrix}$$

$$= \hat{i} = (-36+36) - \hat{j}(12+12\alpha) + \hat{k}(6+6\alpha)$$

$$= -\hat{j}(12+12\alpha) + \hat{k}(6+6\alpha)$$

But $\vec{\tau} = 0$

$\therefore 12 + 12\alpha = 0$ or $\alpha = -1$

163. (b)

Situation is shown in figure.

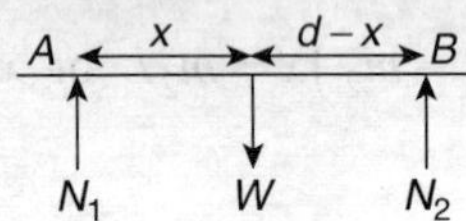

N_1 = Normal reaction on A

N_2 = Normal reaction on B

W = Weight of the rod

In vertical equilibrium

$$N_1 + N_2 = W \quad \text{(i)}$$

Torque balance about centre of mass of the rod,

$$N_1 x = N_2 (d - x)$$

Putting value of N_2 from Eq. (i)

$$N_1 x = (W - N_1)(d - x)$$

$$\Rightarrow N_1 x = Wd - Wx - N_1 d + N_1 x$$

$$\Rightarrow N_1 d = W(d - x)$$

$$\therefore N_1 = \frac{W(d-x)}{d}$$

164. (a)

According to law of conservation of angular momentum

$$mvr = mv'r'$$

$$v_0 R_0 = v\left(\frac{R_0}{2}\right); v = 2v_0 \quad \text{...(i)}$$

$$\therefore \frac{K_0}{K} = \frac{\frac{1}{2}mv_0^2}{\frac{1}{2}mv^2} = \left(\frac{v_0}{v}\right)^2$$

or $$\frac{K}{K_0} = \left(\frac{v_0}{v}\right)^2 = (2)^2 \quad \text{(Using (i))}$$

$$K = 4K_0 = 2mv_0^2$$

165. (b)

Net moment of inertia of the system,

$$I = I_1 + I_2 + I_3$$

The moment of inertia of a shell about its diameter,

$$I_1 = \frac{2}{3}mr^2$$

The moment of inertia of a shell about its tangent is given by

$$I_2 = I_3 = \frac{2}{3}mr^2 + mr^2 = \frac{5}{3}mr^2$$

$$\therefore I = 2 \times \frac{5}{3}mr^2 + \frac{2}{3}mr^2 = \frac{12mr^2}{3} = 4mr^2$$

166. (d)

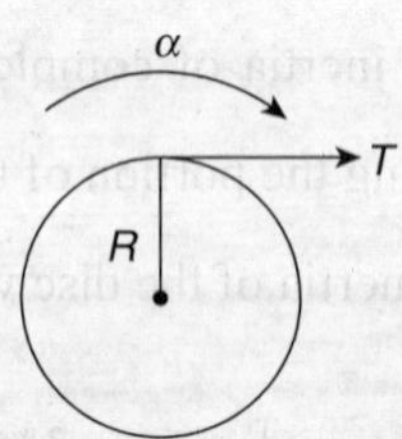

Here, mass of the cylinder, $M = 50$ kg

Radius of the cylinder, $R = 0.5$ m

Angular acceleration, $\alpha = 2 \text{ rev s}^{-2}$

$$= 2 \times 2\pi \text{ rad s}^{-2} = 4\pi \text{ rad s}^{-2}$$

Torque, $\tau = TR$

Moment of inertia of the solid cylinder about its axis,

$$I = \frac{1}{2}MR^2$$

$\therefore$ Angular acceleration of the cylinder,

$$\alpha = \frac{\tau}{I} = \frac{TR}{\frac{1}{2}MR^2}$$

$$T = \frac{MR\alpha}{2} = \frac{50 \times 0.5 \times 4\pi}{2} = 157 \text{ N}$$

167. (a)

Acceleration of the solid sphere slipping down the incline without rolling is

$$a_{\text{slipping}} = g\sin\theta \quad \text{....(i)}$$

Acceleration of the solid sphere rolling down the incline without slipping is

$$a_{\text{rolling}} = \frac{g\sin\theta}{1+\frac{k^2}{R^2}} = \frac{g\sin\theta}{1+\frac{2}{5}}$$

$$\left(\because \text{For solid sphere, } \frac{k^2}{R^2} = \frac{2}{5}\right) = \frac{5}{7} g\sin\theta \quad \text{(ii)}$$

From Eqs. (i) and (ii)

$$\frac{a_{\text{rolling}}}{a_{\text{slipping}}} = \frac{5}{7}$$

168. (c)

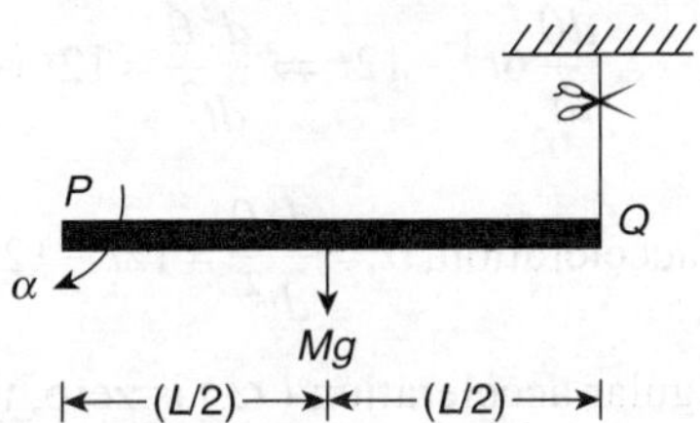

When the string is cut, the rod will rotate about P. Let α be initial angular acceleration of the rod. Then,

$$\text{Torque, } \tau = I\alpha = \frac{ML^2}{3}\alpha \quad \text{(i)}$$

(Moment of inertia of the rod about one end $= \frac{ML^2}{3}$)

$$\text{Also, } \tau = Mg\frac{L}{2} \quad \text{(ii)}$$

Equating (i) and (ii), we get

$$Mg\frac{L}{2} = \frac{ML^2}{3}\alpha \text{ or } \alpha = \frac{3g}{2L}$$

169. (b)

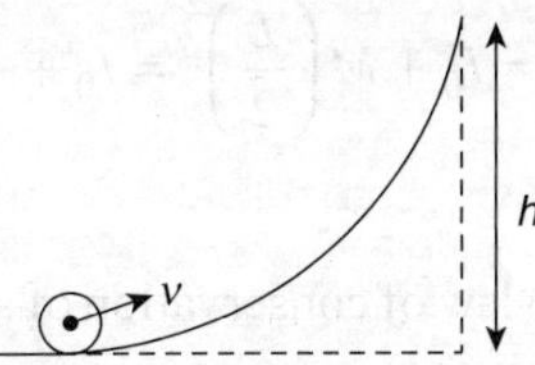

The kinetic energy of the rolling object is converted into potential energy at height

$$h = \frac{3v^2}{4g}$$

So by the law of conservation of mechanical energy, we have

$$\frac{1}{2}Mv^2 + \frac{1}{2}I\omega^2 = Mgh$$

$$\frac{1}{2}Mv^2 + \frac{1}{2}I\left(\frac{v}{R}\right)^2 = Mg\left(\frac{3v^2}{4g}\right)$$

$$\frac{1}{2}I\frac{v^2}{R^2} = \frac{3}{4}Mv^2 - \frac{1}{2}Mv^2$$

$$\frac{1}{2}I\frac{v^2}{R^2} = \frac{1}{4}Mv^2 \text{ or } I = \frac{1}{2}MR^2$$

Hence, the object is disc.

170. (b)

Moment of inertia of disc D_1 about an axis passing through its centre and normal to its plane is.

$$I_1 = \frac{MR^2}{2} = \frac{(2\text{ kg})(0.2\text{ m})^2}{2} = 0.04\text{ kgm}^2$$

Initial angular velocity of disc $D_1, \omega_1 = 50\text{ rads}^{-1}$.

Moment of inertia of disc D_2 about an axis passing through its centre and normal to its plane is

$$I_2 = \frac{(4\text{ kg})(0.1\text{ m})^2}{2} = 0.02\text{ kgm}^2$$

Initial angular velocity of disc $D_2, \omega_2 = 200\text{ rads}^{-1}$

Total initial angular momentum of the two discs is

$$L_i = I_1\omega_1 + I_2\omega_2$$

When two discs are brought in contact face to face (one on the top of the other) and their axes of rotation coincide, the moment of inertia I of the system is equal to the sum of their individual moment of inertia.

$$I = I_1 + I_2$$

Let ω be the final angular speed of the system.

The final angular momentum of the system is

$$L_f = I\omega = (I_1 + I_2)\omega$$

By law of conservation of angular momentum, we have

$$L_i = L_f$$

$$I\omega_1 + I_2\omega_2 = (I_1 + I_2)\omega$$

$$\omega = \frac{I_1\omega_1 + I_2\omega_2}{I_1 + I_2}$$

$$= \frac{(0.04\text{Kg m}^2)(50\text{ rad s}^{-1}) + (0.02\text{ Kg m}^2)(200\text{ rad s}^{-1})}{(0.04 + 0.02)\text{ Kg m}^2}$$

$$= \frac{(2+4)}{0.06}\text{ rad s}^{-1} = 100\text{ rad s}^{-1}$$

171. (a)

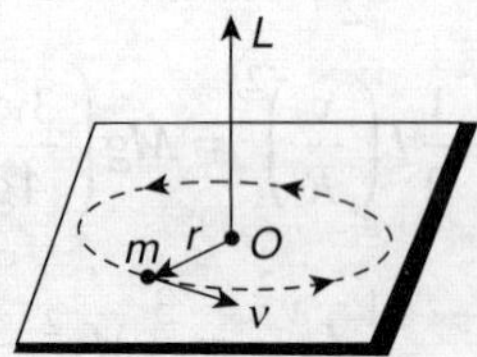

When a mass is rotating in a plane about a fixed point, its angular momentum is directed along a line perpendicular to the plane of rotation.

172. (c)

As no external force acts on the system, centre of mass will not shift.

173. (a)

Let x be the distance of centre O of equilateral triangle from each side.
Total torque about O = 0

$$\Rightarrow F_1x+F_2x-F_3x=0 \text{ or } F_3=F_1+F_2$$

174. (c)

As the system is initially at rest, initial angular momentum $F_i=0$.
According to the principle of conservation of angular momentum,
Final angular momentum, $F_f=0$
∴ Angular momentum of platform = Angular momentum of man in opposite direction of platform.

$$mvR=I\omega$$

or $\omega=\frac{mvR}{I}=\frac{50\times 1\times 2}{200}=\frac{1}{2}\text{ rad s}^{-1}$

Angular velocity of man relative to platform is

$$\omega_r=\omega+\frac{v}{R}=\frac{1}{2}+\frac{1}{2}=1\text{ rad s}^{-1}$$

Time taken by the man to complete one revolution is

$$T=\frac{2\pi}{\omega_r}=\frac{2\pi}{1}=2\pi s$$

175. (a)

According to the theorem of parallel axes, $I=I_{C.M.}+Ma^2$
As a is maximum for point B, I is maximum about B.

176. (a)

300 g 500 g 400 g
O x
40 cm
70 cm

The distance of the centre of mass of the system of three masses from the origin O is

$$X_{C.M.}=\frac{m_1x_1+m_2x_2+m_3x_3}{m_1+m_2+m_3}$$

$$=\frac{300\times 0+500\times 40+400\times 70}{300+500+400}$$

$$=\frac{500\times 40+400\times 70}{1200}=\frac{400[50+70]}{1200}$$

$$=\frac{50+70}{3}=\frac{120}{3}=40\text{ cm}$$

177. (a)

Given: $\theta(t)=2t^3-6t^2$

$$\therefore \frac{d\theta}{dt}6t^2-12t\Rightarrow\frac{d^2\theta}{dt^2}=12t-12$$

Angular acceleration, $\alpha=\frac{d^2\theta}{dt^2}=12t-12$

When angular acceleration (α) is zero, then the torque on the wheel becomes zero $(\because \tau=I\alpha)$

$$\Rightarrow 12t-12=0 \text{ or } t=1\text{ s}$$

178. (b)

According to the theorem of parallel axes, the moment of inertia of the thin rod of mass M and length L about an axis passing through one of the ends is

$$I=I_{C.M.}+Md^2$$

where $I_{C.M.}$ is the moment of inertia of the given rod about an axis passing through its centre of mass and perpendicular to its length and d is the distance between two parallel axes.

Here, $I_{C.M.}=I_0, d=\frac{L}{2}$

$$\therefore I=I_0+M\left(\frac{L}{2}\right)^2=I_0+\frac{ML^2}{4}$$

179. (d)

According to law of conservation of angular momentum

$$mvr=mv'r'$$

$$vr=v'\left(\frac{r}{2}\right)$$

$$v'=2v \quad \text{(i)}$$

$$\therefore \frac{K}{K'}=\frac{\frac{1}{2}mv^2}{\frac{1}{2}mv'^2}=\left(\frac{v}{v'}\right)^2$$

or $\frac{K'}{K} = \left(\frac{v}{v'}\right)^2 = (2)^2 \Rightarrow K' = 4K$ (Using (i))

180. (d)

As no external torque is applied to the system, the angular momentum of the system remains conserved.

$$\therefore L_i = L_f$$

According to given problem,

$$I_t\omega_i = (I_t + I_b)\omega_f$$

or $$\omega_f = \frac{I_t\omega_i}{(I_t + I_b)} \quad \text{(i)}$$

Initial energy,

$$E_i = \frac{1}{2}I_t\omega_i^2 \quad \text{(ii)}$$

Final energy,

$$E_f = \frac{1}{2}(I_t + I_b)\omega_f^2 \quad \text{(iii)}$$

Substituting the value of from Eq. (i) in Eq. (iii) we get, Final energy,

$$E_f = \frac{1}{2}(I_t + I_b)\left(\frac{I_t\omega_i}{I_t + I_b}\right)^2 = \frac{1}{2}\frac{I_t^2\omega_i^2}{2(I_t + I_b)} \quad \text{(iv)}$$

Loss of energy, $\Delta E = E_i - E_f$

$$= \frac{1}{2}I_t\omega_i^2 - \frac{1}{2}\frac{I_t^2\omega_i^2}{2(I_t + I_b)} \quad \text{(Using (ii) and (iv))}$$

$$= \frac{\omega_i^2}{2}\left(I_t - \frac{I_t^2}{(I_t + I_b)}\right) = \frac{\omega_i^2}{2}\left(\frac{I_t^2 + I_bI_t - I_t^2}{(I_t + I_b)}\right)$$

$$= \frac{1}{2}\frac{I_bI_t}{(I_t + I_b)}\omega_i^2$$

181. (b)

As no external force is acting on the system, the centre of mass must be at rest $v_{C.M.} = 0$

182. (c)

The coin will revolve with the record, if force of friction ≥ centripetal force

$$\mu mg \geq mr\omega^2$$

or $$r \leq \frac{\mu g}{\omega^2}$$

183. (a)

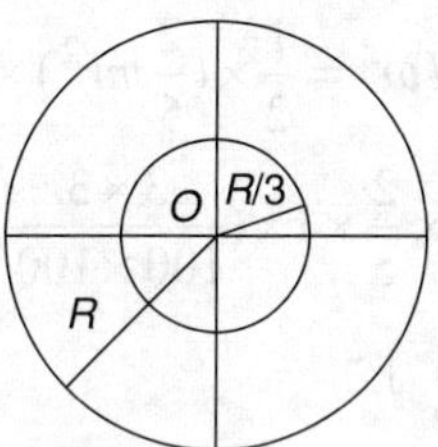

Mass of the disc = $9M$
Mass of removed portion of disc = M
The moment of inertia of the complete disc about an axis passing through its centre O and perpendicular to its plane is

$$I_1 = \frac{9}{2}MR^2$$

Now, the moment of inertia of the removed portion of the disc

$$I_2 = \frac{1}{2}M\left(\frac{R}{3}\right)^2 = \frac{1}{18}MR^2$$

Therefore, moment of inertia of the remaining portion of disc about O is

$$I = I_1 - I_2 = 9\frac{MR^2}{2} - \frac{MR^2}{18} = \frac{40MR^2}{9}$$

184. (d)

Time taken to reach the bottom of inclined plane,

$$t = \sqrt{\frac{2l\left(1 + \frac{K^2}{R^2}\right)}{g\sin\theta}}$$

Here, l is length of incline plane

For solid cylinder $K^2 = \frac{R^2}{2}$

For hollow cylinder $K^2 = R^2$
Hence, solid cylinder will reach the bottom first.

185. (d)

As no external torque is acting about the axis, angular momentum of system remains conserved.

$$I_1\omega_1 = I_2\omega_2$$

$$\Rightarrow \omega_2 = \frac{I_1\omega_1}{I_2} = \frac{Mr^2\omega}{(M + 2m)r^2} = \frac{M\omega}{(M + 2m)}$$

186. (d)

$$=\frac{1}{2}I\omega^2=\frac{1}{2}\times(\frac{2}{5}mr^2)\times\omega^2$$

$$=\frac{1}{2}\times\frac{2}{5}\times1\times(\frac{3\times3}{100\times100})\times50\times50$$

$$=\frac{9}{20}\text{ J}$$

187. (b)

Torque is always perpendicular to $\vec{F}$ as well as $\vec{r}$.

$\vec{r}\cdot\vec{\tau}=0$ as well as $\vec{F}\cdot\vec{\tau}=0$

188. (d)

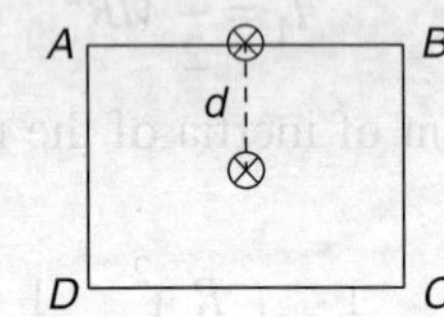

Moment of inertia for the rod AB rotating about an axis through the mid-point of AB perpendicular to the plane of the paper is $\frac{Ml^2}{12}$

$\therefore$ Moment of inertia about the axis through the centre of the square and parallel to this axis,

$$I=I_0+Md^2=M\left(\frac{l^2}{12}+\frac{l^2}{4}\right)=\frac{Ml^2}{3}$$

For all the four rods, $I=\frac{4}{3}Ml^2$

189. (a)

$\vec{r}_1=\hat{i}+2\hat{j}+\hat{k}$ for $M_1=1$ kg

$\vec{r}_2=-3\hat{i}-2\hat{j}+\hat{k}$ for $M_2=3$ k

$$r_{C.M.}=\frac{\Sigma m_i r_i}{\Sigma m_i}$$

$$\Rightarrow r_{C.M.}=\frac{\left(1\hat{i}+2\hat{j}+1\hat{k}\right)\times1+\left(-3\hat{i}-2\hat{j}+\hat{k}\right)\times3}{4}$$

$$\Rightarrow r_{C.M.}=\frac{-8\hat{i}-4\hat{j}+4\hat{k}}{4}=-2\hat{i}-\hat{j}+\hat{k}$$

190. (d)

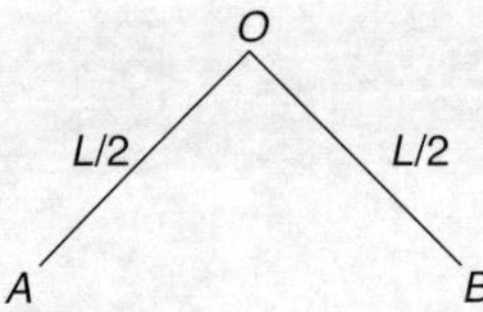

Total mass = M, total length = L

Moment of inertia of OA about O = Moment of inertia of OB about O.

$$\Rightarrow M.I._{\text{total}}=2\times\left(\frac{M}{2}\right)\left(\frac{L}{2}\right)^2\cdot\frac{1}{3}=\frac{ML^2}{12}$$

191. (d)

M.I. of a circular disc, $Mk^2=\frac{M\cdot R^2}{2}$

M.I. of a circular ring = MR^2

$\therefore$ Ratio of their radius of gyration $=\frac{1}{\sqrt{2}}:1$ or $1:\sqrt{2}$

192. (a)

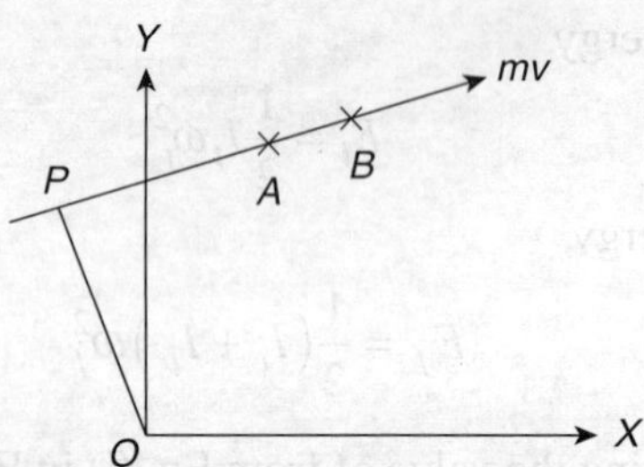

Moment of momentum is angular momentum. OP is the same whether the mass is at A or B.

$\therefore\ L_A=L_B$

193. (c)

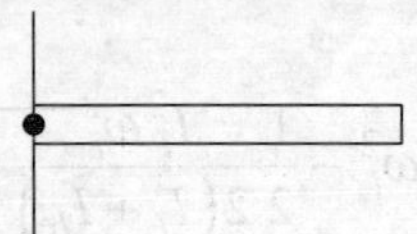

Torque about A,

$$\tau=mg\times\frac{l}{2}=\frac{mgl}{2}$$

Also $\qquad\tau=I\alpha$

$\therefore$ Angular acceleration,

$$\alpha=\frac{\tau}{I}=\frac{mgl/2}{ml^2/3}=\frac{3}{2}\frac{g}{l}$$

194. (a)

Given: Angular acceleration, $\alpha=3$ rad/s^2

Initial angular velocity $\omega_i=2$ rad/s

Time $t=2$ s

Using, $\theta=\omega_i t+\frac{1}{2}\alpha t^2$

$\therefore\theta=2\times2+\frac{1}{2}\times3\times4=4+6=10$ radians.

195. (d)

Moment of inertia of a uniform circular disc about an axis through its centre and perpendicular to its plane is

$$I_C = \frac{1}{2}MR^2$$

$\therefore$ Moment of inertia of a uniform circular disc about an axis touching the disc at its diameter and normal to the disc is I.

By the theorem of parallel axes,

$$I = I_C + Mh^2 = \frac{1}{2}MR^2 + MR^2 = \frac{3}{2}MR^2$$

196. (b)

The centre of the tube will be at length $L/2$. So radius $r = L/2$.

The force exerted by the liquid at the other end = Centrifugal force

$$\text{Centrifugal force} = Mr\omega^2 = M\left(\frac{L}{2}\right)\omega^2 = \frac{ML\omega^2}{2}$$

197. (c)

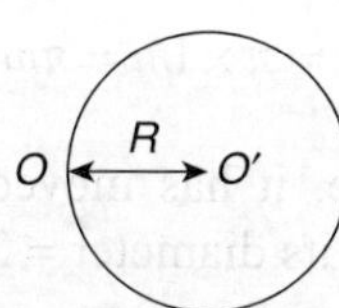

$$\text{M.I. of disc about its normal} = \frac{1}{2}MR^2$$

$$\text{M.I. about its one edge} = MR^2 + \frac{MR^2}{2}$$

(Perpendicular to the plane)

$$\text{Moment of inertia} = \frac{3}{2}MR^2$$

198. (d)

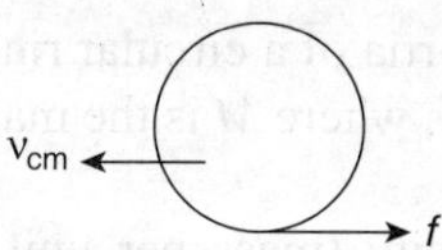

Required frictional force converts some part of translational energy into rotational energy.

199. (c)

$$\text{KE} = \frac{1}{2}I\omega^2$$

$$\therefore \frac{1}{2}I_1\omega_1^2 = \frac{1}{2}\cdot 2I_1\omega_2^2$$

$$\frac{\omega_1^2}{\omega_2^2} = \frac{2}{1} \Rightarrow \frac{\omega_1}{\omega_2} = \frac{\sqrt{2}}{1}$$

200. (c)

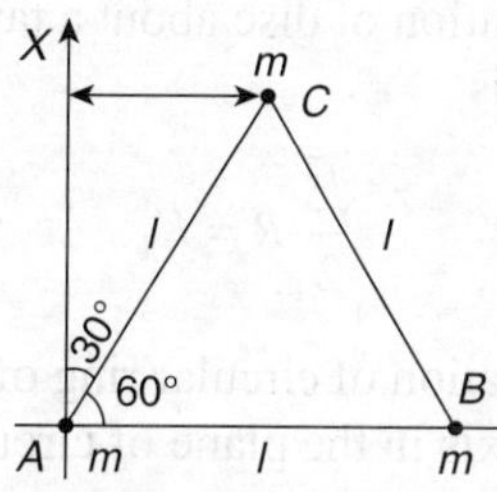

The moment of inertia of the system

$$= m_A r_A^2 + m_B r_B^2 + m_C r_C^2$$

$$= m_A(0)^2 + m(l)^2 + m(l\sin 30°)^2$$

$$= m_1^2 + ml^2 \times (1/4) = (5/4)ml^2$$

201. (b)

$$\text{C.M.} = \frac{m_1x_1 + m_2x_2}{m_1 + m_2} \quad \text{(i)}$$

After changing position of m_1, for keeping the position of C.M. same

$$\text{C.M.} = \frac{m_1(x_1 - d) + m_2(x_2 + d_2)}{m_1 + m_2}$$

$$0 = \frac{m_1d + m_2d_2}{m_1 + m_2}$$

[Substituting value of C.M. from (i)]

$$\Rightarrow d_2 = \frac{m_1}{m_2}d$$

202. (c)

$$\omega_f = \omega_i - \alpha t \Rightarrow 0 = \omega_i - \alpha_t$$

$\therefore \alpha = \omega_i/t$, where α is retardation.

The torque on the wheel is given by

$$\tau = I\alpha = \frac{I\omega}{t} = \frac{I\cdot 2\pi\upsilon}{t} = \frac{2\times 2\times\pi\times 60}{60\times 60}$$

$$\tau = \frac{\pi}{15}\text{ Nm}$$

This is the torque required to stop the wheel in 1 min. (or 60 s).

203. (c)

Applying conservation of angular momentum,

$$I_1\omega = (I_1 + I_2)\omega_1 \text{ or } \omega_1 = \frac{I_1}{(I_1 + I_2)}\omega$$

204. (c)

Radius of gyration of disc about a tangential axis in the plane of disc is

$$\frac{\sqrt{5}}{2}R = K_1$$

Radius of gyration of circular ring of same radius about a tangential axis in the plane of circular ring is

$$K_2 = \sqrt{\frac{3}{2}}R \therefore \frac{K_1}{K_2} = \frac{\sqrt{5}}{\sqrt{6}}$$

205. (c)

$$\text{Total energy} = \frac{1}{2}I\omega^2 + \frac{1}{2}mv^2 = \frac{1}{2}mv^2\left(1 + K^2/R^2\right)$$

$$\text{Required fraction} = \frac{K^2/R^2}{1 + K^2/R^2} = \frac{K^2}{R^2 + K^2}$$

206. (c)

Potential energy of the solid cylinder at height $h = Mgh$
KE of centre of mass when it reaches the bottom

$$= \frac{1}{2}Mv^2 + \frac{1}{2}I\omega^2 = \frac{1}{2}Mv^2 + \frac{1}{2}Mk^2v^2/R^2$$

$$= \frac{1}{2}Mv^2\left(1 + \frac{k^2}{R^2}\right)$$

For a solid cylinder $\frac{k^2}{R^2} = \frac{1}{2} \therefore K.E. = \frac{3}{4}Mv^2$

$$\therefore Mgh = \frac{3}{4}Mv^2, v = \sqrt{\frac{4}{3}gh}$$

207. (b)

According to conservation of angular momentum, L = $I\omega$ = constant.
Therefore, $I_2\omega_2 = I_1\omega_1$

$$\text{or } \omega_2 = \frac{I_1\omega_1}{I_1} = \frac{Mk^2\omega}{(M + 4m)k^2} = \frac{M\omega}{M + 4m}$$

208. (b)

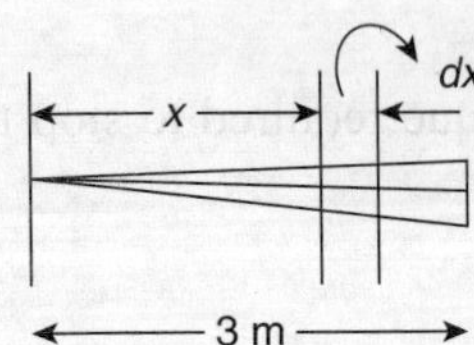

Let us consider an elementary length dx at a distance x from one end.
Its mass $= k.x.dx$

[k = proportionality constant]
Then centre of gravity of the rod x_c is given by

$$x_c = \frac{\int_0^3 kxdx \cdot x}{\int_0^3 kxdx} = \frac{\int_0^3 x^2dx}{\int_0^3 xdx} = \frac{\left.\frac{x^3}{3}\right|_0^3}{\left.\frac{x^2}{2}\right|_0^3}$$

$$\text{or} \quad x_c = \frac{27/3}{9/2} = 2$$

$\therefore$ Centre of gravity of the rod will be at distance of 2 m from one end.

209. (b)

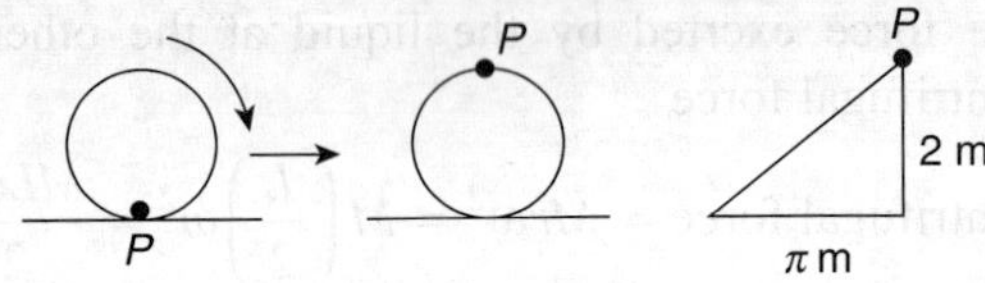

In half-rotation point P has moved horizontally.

$$\frac{\pi d}{2} = \pi r = \pi \times 1m = \pi m. \qquad [\because \text{radius} = 1\text{ m}]$$

In the same time, it has moved vertically a distance which is equal to its diameter = 2 m.

$\therefore$ Displacement of P $= \sqrt{\pi^2 + 2^2} = \sqrt{\pi^2 + 4}m$

210. (d)

Since there is no friction at the contact surface (smooth horizontal surface) there will be no rolling. Hence, the acceleration of the centre of mass of the sphere will be independent of the position of the applied force F. Therefore, there is no relation between h and R.

211. (b)

Law of conservation of angular momentum.

212. (a)

Moment of inertia of a circular ring about its geometrical axis is MR^2, where M is the mass and R is the radius of the ring.
Since the density (mass per unit volume) for iron is more than that of aluminum, the proposed rings made of iron should be placed at a greater radius to get more value of MR^2. Hence to get maximum moment of inertia for the circular disc, aluminium should be placed at interior and iron at the exterior position.

QUICK QUESTIONS

Quiz

1. Imagine an object rotating in space with no motion of its centre of mass. Each particle in the object is moving in a circular path; therefore there is momentum associated with the motion of each particle. As its centre of mass is at rest, the object has no linear momentum. What is the quantity of motion associated with its rotation?

2. Which of the following is conserved in an isolated system?

   (i) Total energy
   (ii) Linear momentum
   (iii) Angular momentum

3. Both torque and work are the products of force and distance. How are they different?

4. Does moment of inertia have a fixed value? What about torque?

5. Does the internal forces contribute to change the position of centre of mass?

6. What is the total kinetic energy of a rigid body such as cylinder which is rolling on a rough surface without slipping?

7. When an object is rolling, what is the direction of velocity of the points on the objects?

8. Is the moment of inertia is a scalar or vector quantity?

9. Why does rotational velocity of a star greatly increase when it collapses under gravitational pull and becomes a neutron star?

10. In case of rolling bodies, static friction is required for the body to rotate; otherwise it will slip. You know friction is a dissipative force, how is the law of conservation of mechanical energy justified?

Think and grow

11. A wheel of radius R is rotating about a fixed axis. Answer the following.

    (a) Do all points on the wheel have same angular speed?
    (b) Do they all have the same tangential speed?
    (c) If the angular speed is constant and equal to ω, describe the tangential speeds and total translational acceleration of the points located at $r = 0$ and $r = R/2$ from the centre of the wheel.

12. A skater moves along a straight line on a frictionless ice surface. Afterwards, he holds of a pole in the middle of the ice and rotates around the pole. Thus there is a switch over from translational motion to rotational motion. Is the angular momentum of the skater conserved?

13. Consider the wheel the tire of a truck moving to the right. Pieces of mud are stuck between the treads. As the mud is dislodged from the treads and gets free of tire, it is thrown in the direction of tangential velocity vector. All of these vectors have a velocity component in the forward direction, the direction in which the truck is moving. So why do trucks have mud flaps behind the tires.

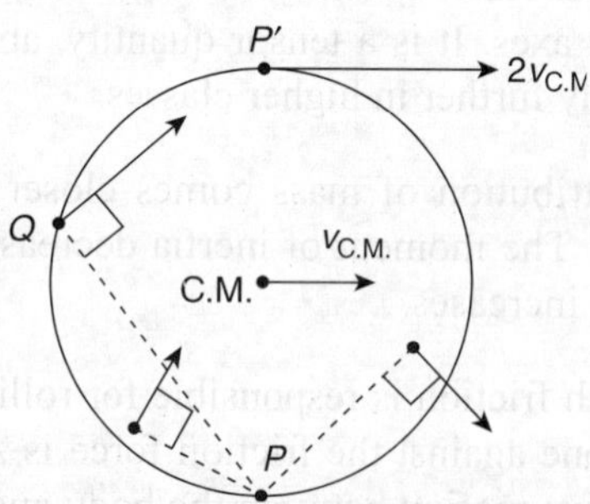

14. A thin wheel can stay upright in vertical position for a considerable time when rolled with a considerable velocity. On the contrary, it falls from the upright position at slight disturbance when stationary. Why?

Explanations

Quiz

1. Angular Momentum

2. All the above are conserved in an isolated system.

3. There are two major differences between work and torque. The displacement in the expression for work is parallel to force whereas position vector in the torque expression is perpendicular to the force. The second difference is that in the case of work, the work is done only when there is the displacement of the point of application of force. On the other hand, a force applied at a perpendicular distance from a rotating axis results in a torque, whether there is motion or not.

4. No, the moment of inertia depends upon axis of rotation, similarly for torque also.

5. No.

6. It is the sum of the rotating kinetic energy $\frac{1}{2}I_{cm}w^2$ about the centre of mass of the body and translational kinetic energy $\frac{1}{2}Mv_{em}^2$ of the centre of the mass.

Total kinetic energy $= \frac{1}{2}I_{cm} + \frac{1}{2}Mv_{em}^2$

7. It is perpendicular to an axis through the instantaneous point of contact.

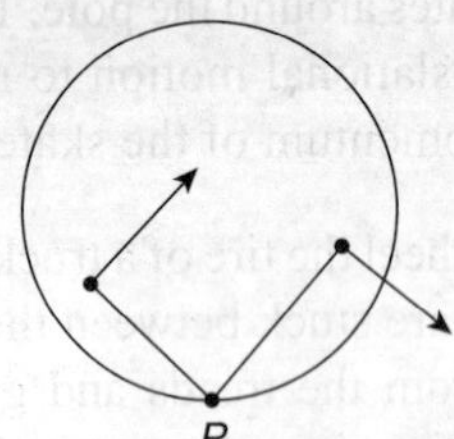

8. It is neither vector nor scalar, it is a tensor quantity because clockwise and anticlockwise is not specified. It is not a scalar quantity as it has different values for different axes. It is a tensor quantity, about which you will study further in higher classes.

9. The distribution of mass comes closer to the axis of rotation. The moment of inertia decreases and angular velocity increases.

10. Although friction is responsible for rolling motion, the work done against the friction force is zero as there is no relative motion between the body and surface at the point of contact.

Think and grow

11. (a) Yes, they all have same angular speed. All points on the wheel make same angular distance at the same time.
(b) No, points farther from the rotation axis have greater tangential speeds.
(c) The point at $r = 0$ has zero tangential speed and zero tangential acceleration. Point at $r = R/2$ has a tangential speed $\omega R/2$ and total translation acceleration ω^2 R/2. Tangential acceleration is zero at all points because ω is constant. A point at $r = R$ has a tangential speed $v = R\omega$ and total translational acceleration $\omega^2 R$.

12. Yes, even when the skater is not moving in circular path before grabbing the pole, he has angular momentum (A particle moving a straight line has angular momentum about any axis away from the path of the particle). After he grabs the pole, the force on the skater is along his arm and it is parallel to the position vector from the pole to his body. This force exerts no torque on the skater so he can be realized as an isolated system and angular momentum is conserved.

13. Consider a piece of mud released from the tyre immediate left of the point P in the figure. Although this mud piece has a component of velocity in forward direction, it is smaller than v_{cm}. Thus as tire moves horizontally, the mud, relative to the truck, moves backward.

14. A rolling wheel has a certain angular momentum with torque perpendicular to the axis of rotation. When there is a slight change in angular momentum, it makes the wheel unstable and causes it to fall.

The larger the angular momentum of the wheel, the lesser will be the effect of a given torque in changing the direction of the axis of the rolling wheel and more stable will be the orientation. This is the reason a rolling wheel is more stable than a stationary wheel.

CHAPTER TEST

1. When a torque acting upon a system is zero, which of the following will be constant?
(a) Force
(b) Linear momentum
(c) Angular momentum
(d) Linear impulse.

2. The moment of inertia of a solid sphere about an axis parallel to its diameter and at a distance of x from it, is I(x)'. Which one of the graphs represents the variation of I(x) with x correctly?

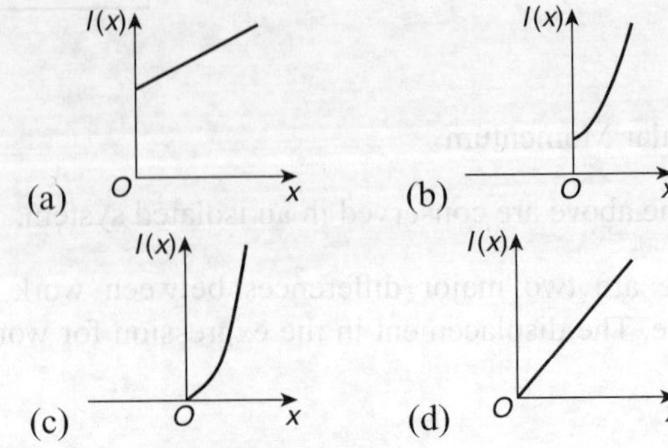

3. The moment of inertia of a body does not depend upon
   (a) the angular velocity of the body
   (b) the mass of the body
   (c) the distribution of mass in the body
   (d) the axis of rotation of the body.

4. Which of the following has largest moment of inertia?
   (a) A ring about its axis perpendicular to its plane
   (b) A disc about its axis perpendicular to its plane
   (c) A solid sphere
   (d) A bar magnet.

5. A body of m mass slides down an incline and reaches the bottom with a velocity v. If the same mass were in the form of a ring which rolls down this incline, the velocity of the ring at bottom would have been
   (a) v (b) $\sqrt{2}v$
   (c) $\frac{1}{\sqrt{2}}v$ (d) $\sqrt{2/5}.v$

6. A particle of mass m is moving in a plane along a circular path of radius r. Its angular momentum about the axis of rotation is L. The centripetal force acting on the particle is
   (a) L^2/mr (b) L^2m/r
   (c) L^2/mr^3 (d) L^2/mr^2

7. A thin circular ring of mass M and radius r is rotating about an axis passing through its centre and perpendicular to its plane with a constant angular velocity ω. Two objects, each of mass m, are attached gently to the opposite ends of a diameter of the ring. The ring now rotates with angular velocity
   (a) $\frac{\omega(M-2m)}{M+2m}$ (b) $\omega M(M-m)$
   (c) $\frac{\omega(M+2m)}{M}$ (d) $\frac{\omega M}{M+2m}$

8. A solid iron sphere A rolls down an inclined plane, while an identical sphere B slides down the plane in a frictionless manner. At the bottom of the inclined plane, the total kinetic energy of sphere A is
   (a) less than that of B
   (b) equal to that of B
   (c) more than that of B
   (d) sometimes more and sometimes less.

9. Two particles of mass m_1 and m_2 ($m_1 > m_2$) attract each other with a force inversely proportional to the square of the distance between them. The particles are initially held at rest and then released. Which one is correct?
   (a) The C.M. moves towards m_1
   (b) The C.M. moves towards m_2
   (c) The C.M. remains at rest
   (d) The C.M. moves at right angles to the line joining m_1 and m_2.

10. A raw egg and a hard-boiled egg are made to spin on a table with the same angular speed about the same axis. The ratio of the time taken by the two to stop is
   (a) = 1 (b) < 1
   (c) > 1 (d) None of the above.

11. The angular velocity of a body changes from ω_1 to ω_2 without applying torque but by changing moment of inertia. The ratio of initial radius of gyration to the final radius of gyration is
   (a) $\omega_2 : \omega_1$ (b) $\omega_2^2 : \omega_1^2$
   (c) $\sqrt{\omega_2} : \sqrt{\omega_1}$ (d) $1/\omega_2 : 1/\omega_1$

12. Two circular discs A and B are of equal masses and thickness but made of metals with density d_A and d_B ($d_A > d_B$). If their moments of inertia about an axis passing through the centre and normal to the circular faces be I_A and I_B then
   (a) $I_A = I_B$ (b) $I_A > I_B$
   (c) $I_A < I_B$ (d) $I_A \geq I_B$

13. A circular disc rolls down an inclined plane without slipping. What fraction of its total energy is translational?
   (a) $\frac{1}{2}$ (b) $\frac{1}{3}$
   (c) $\frac{2}{3}$ (d) $\frac{1}{4}$

14. A block Q of mass M is placed on a horizontal frictionless surface AB and a body P of mass m is released on its frictionless slope. As P slides by a length L on this slope of inclination θ, the block Q would slide by a distance

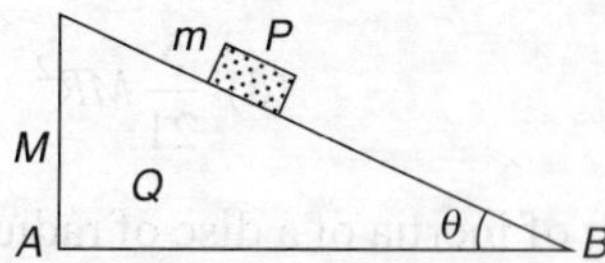

   (a) $(m/M)\ L \cos\theta$ (b) $m\ L/(M+m)$
   (c) $(M+m)/(m\ L\cos\theta)$ (d) $(m\ L\cos\theta)/(m+M)$

15. Without weighing, how will you distinguish between the two identical balls of same material, but one being solid and the other one being hollow?
   (a) by rolling them down an inclined plane in air
   (b) by determining their moment of inertia about the centre
   (c) by spinning them by equal torque
   (d) All the above.

16. A mass M is moving with a constant velocity parallel to x-axis. Its angular momentum with respect to origin
   (a) is zero (b) remains constant
   (c) goes on increasing (d) goes on decreasing.

17. If the moment of inertia of a disc about an axis tangentially and parallel to its surface be I, then what will be the moment of inertia about the axis tangential but perpendicular to the surface?

(a) $\frac{6}{5}I$ (b) $\frac{3}{4}I$

(c) $\frac{3}{2}I$ (d) $\frac{5}{4}I$

18. Three point-masses, each of mass m. are placed at the corners of an equilateral triangle of side b. The moment of inertia of the system about an axis coinciding with one side of the triangle is

(a) $3mb^2$ (b) mb^2

(c) $(3/4)mb^2$ (d) $(2/3)mb^2$

19. Two spheres each of mass M and radius $R/2$ are connected with a mass-less rod of length $2R$ as shown in figure.

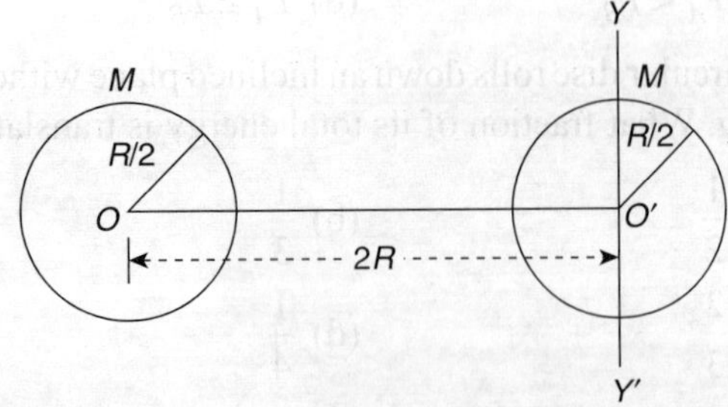

What will be the moment of inertia of the system about an axis passing through the centre of one of the spheres and perpendicular to the rod?

(a) $\frac{21}{5}MR^2$ (b) $\frac{2}{5}MR^2$

(c) $\frac{5}{2}MR^2$ (d) $\frac{5}{21}MR^2$

20. The moment of inertia of a disc of radius R and mass M about a diameter as axis is

(a) MR^2 (b) $\frac{MR^2}{2}$

(c) $\frac{MR^2}{4}$ (d) $\frac{5}{4}MR^2$

21. A constant torque acting on a uniform circular wheel changes its angular momentum from A_0 to $4A_0$ in 4 seconds. The magnitude of this torque is

(a) $3A_0/4$ (b) A_0

(c) $4A_0$ (d) $12A_0$

22. A particle of mass $m = 5$ is moving with a uniform speed $v = 3\sqrt{2}$ in the XOY plane along the line $Y = X + 4$. The magnitude of the angular momentum about origin is

(a) zero (b) 60 unit

(c) 7.5 unit (d) $40\sqrt{2}$

23. A uniform solid sphere rolls on a horizontal surface at 20 m/s. It then rolls up an incline having an angle of inclination of 30^0 with the horizontal. If the friction losses are negligible, the value of height h above the ground where the ball stops is

(a) 14.3 m (b) 28.6 m

(c) 57.2 m (d) 9.8 m

24. A wheel is rotating at 900 rpm about its axis. When the power is cut off, it comes to rest in one minute. The angular retardation (assuming to be uniform) in rad/sec^2 is

(a) $\pi/2$ (b) $\pi/4$

(c) $\pi/6$ (d) $\pi/8$

25. The moment of inertia of a thin uniform circular disc about one of its diameter is I. The moment of inertia about an axis perpendicular to the circular surface and passing through its centre is

(a) $\sqrt{(2I)}$ (b) $2I$

(c) $I/2$ (d) $I/\sqrt{2}$

26. Three bricks, each of length L and mass M are arranged as shown from the wall. The distance of the centre of mass of the system from the wall is

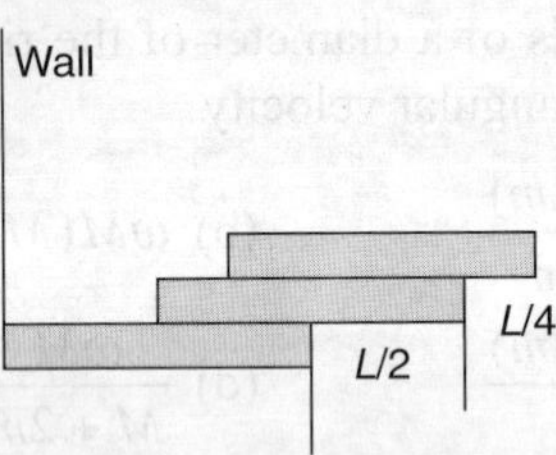

(a) $L/4$ (b) $L/2$

(c) $(3/2)L$ (d) $(5/6)L$

27. Three identical spheres, each of mass 1 kg are kept as shown in figure, touching each other, with their centres on a straight line. If their centres are marked P, Q, R respectively, the distance of centre of mass of the system from P is

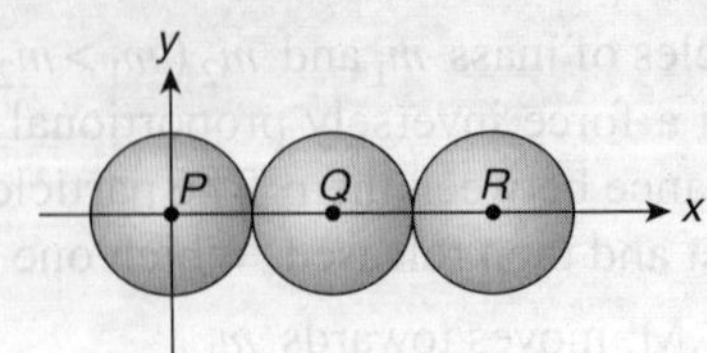

(a) $\frac{PQ+PR+QR}{3}$ (b) $\frac{PQ+PR}{3}$

(c) $\frac{PQ+QR}{3}$ (d) $\frac{PR+QR}{3}$

28. Centre of mass of 3 particles 10 kg, 20 kg and 30 kg is at (0,0,0). Where should a particle of mass 40 kg be placed so that the combined centre of mass will be at (3,3,3)?

(a) (0, 0, 0) (b) (7.5, 7.5, 7.5)
(c) (1, 2, 3) (d) (4, 4, 4)

29. When a ceiling fan is switched on, it makes 10 revolutions in the first 3 seconds. Assuming a uniform angular acceleration, how many rotation it will make in the next 3 seconds?

(a) 10 (b) 20
(c) 30 (d) 40

30. A circular platform is free to rotate in a horizontal plane about a vertical axis passing through its centre. A tortoise is sitting at the edge of the platform. Now the platform is given an angular velocity ω_0. When the tortoise moves along a chord of the platform with a constant velocity (with respect to the platform), the angular velocity of the platform $\omega(t)$ will vary with time t as

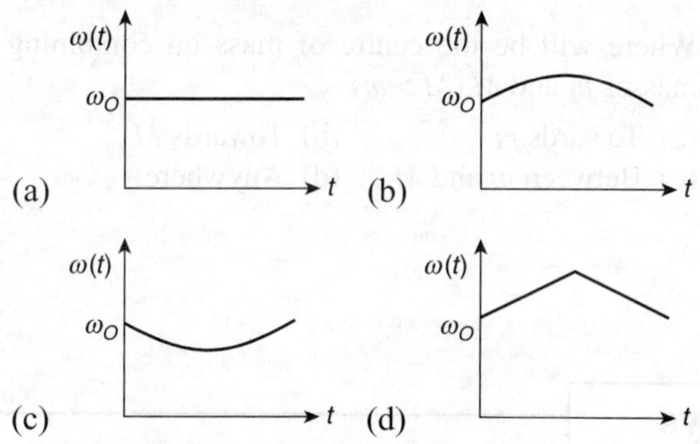

31. Turning effect is produced by

(a) tangential component of force
(b) radial component of force
(c) transverse component of force
(d) None of these.

32. A constant torque of 31.4 N-m is exerted on a pivoted wheel. If angular acceleration of wheel is 4πrad/sec^2, then the moment of inertia of the wheel is

(a) 2.5 kg-m^2 (b) 3.5 kg-m^2
(c) 4.5 kg-m^2 (d) 5.5 kg-m^2

33. A fly wheel of moment of inertia 3×10^2 kg-m^2 is rotating with uniform angular speed of 4.6 rad s^{-1}. If a torque of 6.9×10^2 N m retards the wheel, then the time in which the wheel comes to rest is

(a) 1.5 s (b) 2 s
(c) 0.5 s (d) 1 s

34. The total kinetic energy of a body of mass 10 kg and radius 0.5 m moving with a velocity of 2 m/s without slipping is 32.8 joule. The radius of gyration of the body is

(a) 0.25 m (b) 0.2 m
(c) 0.5 m (d) 0.4 m

35. Two rigid bodies A and B rotate with rotational kinetic energies E_A and E_B respectively. The moments of inertia of A and B about the axis of rotation are I_A and I_B respectively. If $I_A = I_B/4$ and $E_A = 100\ E_B$ the ratio of angular momentum (L_A) of A to the angular momentum (L_B) of B is

(a) 25 (b) 5/4
(c) 5 (d) ¼

36. The moment of inertia of a flywheel having kinetic energy 360 J and angular speed of 20 rad/s is

(a) 18 kg m^2 (b) 1.8 kg m^2
(c) 2.5 kg m^2 (d) 9 kg m^2

37. Two solid cylinders P and Q of same mass and same radius start rolling down a fixed inclined plane from the same height at the same time. Cylinder P has most of its mass concentrated near its surface, while Q has most or its mass concentrated near the axis. Which statement (s) is (are) correct?

(a) Both cylinders P and Q reach the ground at the same time.
(b) Cylinder P has larger linear acceleration than cylinder C.
(c) Both cylinder P and Q reach the ground with same translational kinetic energy.
(d) Cylinder Q reaches the ground with larger angular speed.

38. A cord is wound round the circumference of wheel of radius r. The axis of the wheel is horizontal and moment of inertia about it is I. A weight mg is attached to the end of the cord and falls from the rest. After falling through a distance h, the angular velocity of the wheel will be

(a) $\sqrt{\frac{2gh}{I+mr}}$ (b) $[\frac{2mgh}{I+mr^2}]^{-1/2}$

(c) $[\frac{2mgh}{I+mr^2}]^{1/2}$ (d) $\sqrt{2gh}$

39. One-quarter sector is cut from a uniform circular disc of radius R. This sector has mass M. It is made to rotate about a line perpendicular to its plane and passing through the centre of the original disc. Its moment of inertia about the axis of rotation is

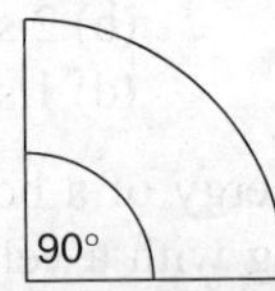

(a) $\frac{1}{2}MR^2$ (b) $\frac{1}{4}MR^2$

(c) $\frac{1}{8}MR^2$ (d) $\sqrt{2}MR^2$

40. Let I be the moment of inertia of a uniform square plate about an axis AB that passes through its centre and is parallel to two of its sides. CD is a line in the plane of the plate that passes through the centre of the plate and makes an angle θ with AB. The moment of inertia of the plate about the axis CD is then equal to

(a) I (b) $I\sin^2\theta$

(c) $I\cos^2\theta$ (d) $I\cos^2\frac{\theta}{2}$

41. A homogeneous solid cylindrical roller of radius R and mass M is pulled on a cricket pitch by a horizontal force. Assuming rolling without slipping, angular acceleration of the cylinder is

(a) $\frac{3F}{2mR}$ (b) $\frac{F}{3mR}$

(c) $\frac{2F}{3mR}$ (d) $\frac{F}{2mR}$

42. A dancer on ice spins faster when she folds her arms. This is due to

(a) increase in energy and increase in angular momentum

(b) decrease in friction at the skates

(c) constant angular momentum and increase in kinetic energy

(d) increase in energy and decrease in angular momentum.

43. Identify the correct statement for the rotational motion of a rigid body.

(a) Individual particles of the body do not undergo accelerated motion.

(b) The centre of mass of the body remains unchanged.

(c) The centre of mass of the body moves uniformly in a circular path.

(d) Individual particles and centre of mass of the body undergo an accelerated motion.

44. Moment of inertia of a sphere of mass M and radius R is I. Keeping M constant, if a graph is plotted between I and R, then its form would be

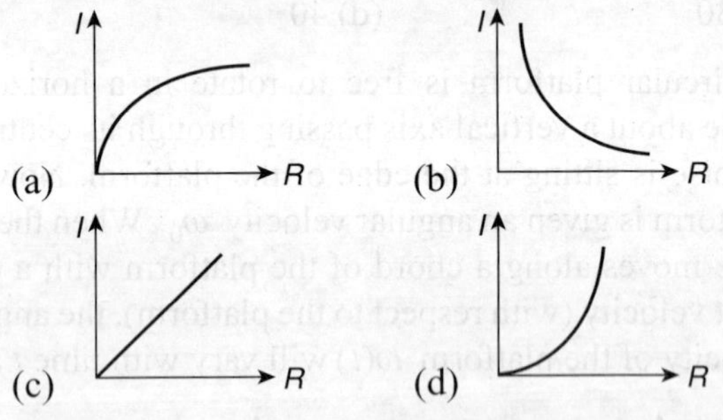

45. Where will be the centre of mass on combining two masses m and M ($M > m$)

(a) Towards m (b) Towards M

(c) Between m and M (d) Anywhere.

Answer Keys

1. (c)	2. (b)	3. (a)	4. (a)	5. (c)	6. (c)	7. (d)	8. (b)	9. (c)	10. (b)
11. (c)	12. (c)	13. (c)	14. (d)	15. (d)	16. (b)	17. (a)	18. (c)	19. (a)	20. (c)
21. (a)	22. (b)	23. (b)	24. (a)	25. (b)	26. (d)	27. (b)	28. (b)	29. (c)	30. (b)
31. (c)	32. (a)	33. (b)	34. (d)	35. (c)	36. (b)	37. (d)	38. (c)	39. (a)	40. (a)
41. (c)	42. (c)	43. (b)	44. (d)	45. (b)					

Solutions

1. (c)

2. (b)

$$I_x = \frac{2}{5}mR^2 + mx^2 \Rightarrow \text{parabola opening upward.}$$

3. (a)

4. (a)

5. (c)

For sliding, $a = g\sin\theta$. Hence, the velocity v is the given by

$$v^2 = 0 + 2(g\sin\theta)\times l \qquad \ldots(1)$$

For rolling, acceleration down the inclined plane is given by

$$a = \frac{g\sin\theta}{(1+\frac{K^2}{R^2})} = \frac{1}{2}g\sin\theta \quad (\because \; K^2 = R^2)$$

In case of a ring

$$V^2 = 2\times(\frac{1}{2}g\sin\theta)\times l = \frac{v^2}{2}$$

$\therefore$ $$V = v/\sqrt{2}$$

6. (c)

$L = m\,v\,r$ or $v = L/mr$

$$\text{Centripetal force} = \frac{mv^2}{r} = \frac{m(L/mr)^2}{r}$$

$$= \frac{L^2}{mr^3}$$

7. (d)

We know that the angular momentum is conserved

$$M.R^2\omega = (M+2m)R^2\omega'$$

$$\omega' = \frac{M\omega}{(M+2m)}$$

8. (b)

9. (c)

10. (b)

11. (c)

$$I_1\omega_1 = I_2\omega_2$$

or $$MK_1^2\omega_1 = MK_2^2\omega_2$$

$$\frac{K_1}{K_2} = \sqrt{\frac{\omega_2}{\omega_1}}$$

12. (c)

$$I_A = \frac{m_A r_A^2}{2} \text{ and } I_B = \frac{m_B r_B^2}{2}$$

$\therefore$ $$\frac{I_A}{I_B} = \frac{r_A}{r_B} \quad (\because m_A = m_B) \qquad \ldots(1)$$

Now, $$m_A = \pi r_A^2 t d_A$$

$$m_B = \pi r_B^2 t d_B$$

So, $$\pi r_A^2 t d_A = \pi r_B^2 t d_B$$

or $$\frac{r_A^2}{r_B^2} = \frac{d_B}{d_A} \qquad \ldots(2)$$

From Eqs. (1) and (2)

$$\frac{I_A}{I_B} = \frac{d_B}{d_A}$$

As $d_A > d_B$, hence $I_A < I_B$

13. (c)

The kinetic energy of translational motion is given by

$$(\text{KE})_t = \frac{1}{2}mv^2 = \frac{1}{2}mr^2\omega^2$$

The rotational kinetic energy is given by

$$(\text{KE})_t = \frac{1}{2}I\omega^2 = \frac{1}{2}(\frac{1}{2}mr^2)\omega^2$$

$$= \frac{1}{4}mr^2\omega^2$$

$\therefore$ Total $$(\text{KE})_T = \frac{1}{2}mr^2\omega^2 + \frac{1}{4}mr^2\omega^2 = \frac{3}{4}mr^2\omega^2$$

The ratio of translational kinetic energy to total kinetic energy is given by

$$\frac{(\text{KE})_t}{(\text{KE})_T} = \frac{(1/2)mr^2\omega^2}{(3/4)mr^2\omega^2} = \frac{2}{3}$$

14. (d)

Here, the centre of mass of the system remains unchanged. When the mass moves forward by a distance $L\cos\theta$, let the mass $(m+M)$ move by a distance x in the backward direction. Hence,

$$(M+m)x - mL\cos\theta = 0$$

$\therefore$ $$x = (mL\cos\theta)/(M+m)$$

15. (d)

16. (b)

17. (a)

According to the theorem of parallel axes, the moment of inertia of the disc about an axis tangentially and parallel to the surface is given by

$$I = I_{11} = \frac{MR^2}{4} + MR^2 = \frac{5}{4}MR^2$$

The moment of inertia of the disc about an axis tangential but perpendicular to the surface is given by

$$I' = I_\perp = \frac{MR^2}{2} + MR^2 = \frac{3}{2}MR^2$$

$$= \frac{6}{5}[\frac{5}{4}MR^2] = \frac{6}{5}I$$

18. (c)

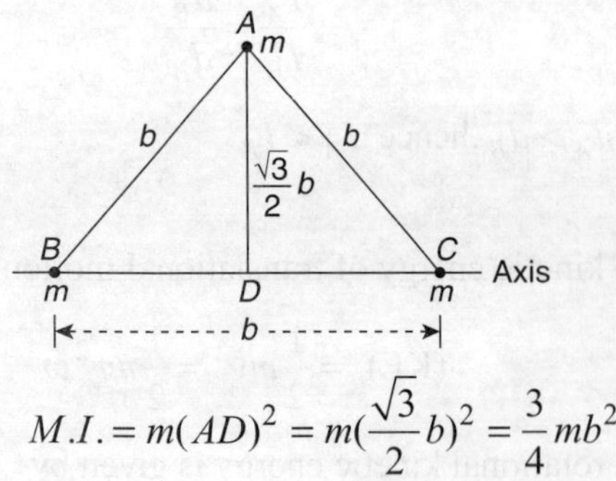

$$M.I. = m(AD)^2 = m(\frac{\sqrt{3}}{2}b)^2 = \frac{3}{4}mb^2$$

19. (a)

$$I = \frac{2}{5}M\left(\frac{R}{2}\right)^2 + M(2R)^2 + \frac{2}{5}M\left(\frac{R}{2}\right)^2$$

$$= \frac{MR^2}{5} + 4MR^2 = \frac{21}{5}MR^2$$

20. (c)

By the theorem of perpendicular axes,

$$I_z = I_x + I_y$$

or $$\frac{MR^2}{2} = I_x + I_x = 2I_x \quad (\because I_y = I_x)$$

$$\therefore \quad I_x = \frac{MR^2}{4}$$

21. (a)

$$\Delta L = 4A_0 - A_0 = 3A_0 \text{ and } \Delta T = 0$$

$$\therefore \quad \tau = \frac{\Delta L}{\Delta T} = \frac{3A_0}{4}$$

22. (b)

$$L = m\,v\,r = 5 \times 3\sqrt{2} \times \frac{4}{\sqrt{2}} \text{ unit}$$

$$= 60 \text{ unit.}$$

23. (b)

Applying the law of conservation of energy, we have

$$\frac{1}{2}mv^2 + \frac{1}{2}I\omega^2 = m\,g\,h$$

$$\frac{1}{2}mv^2 + \frac{1}{2}(\frac{2}{5}mr^2)\omega^2 = m\,g\,h$$

$$\frac{1}{2}mv^2 + \frac{1}{5}mv^2 = m\,g\,h$$

$$h = \frac{7}{10} \times \frac{v^2}{r} = \frac{7 \times (20)^2}{10 \times 9.8} \, 28.6 \text{ m}$$

24. (a)

$n = 900/60 = 15$ rps

$\omega_0 = 2\pi n = 2\pi \times 15 = 30\pi$

$\omega = \omega_0 + \alpha t$ or $0 = 30\pi + \alpha \times 60$

$\alpha = \frac{-30\pi}{60} = -\frac{\pi}{2}$ or retardation $\frac{\pi}{2}$

25. (b)

According to the theorem of perpendicular axes,

$I_z = I_x + I_y = I + I = 2I$

26. (d)

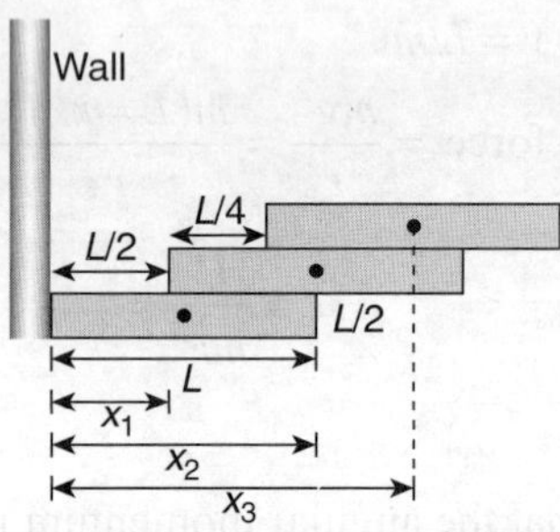

From figure $\quad x_1 = \frac{L}{2}, x_2 = \frac{L}{2} + \frac{L}{2} = L$

$$x_3 = \frac{L}{2} + \frac{L}{4} + \frac{L}{2} = \frac{5L}{4} \text{ v}$$

$$x_{cm} = \frac{m_1x_1 + m_2x_2 + m_3x_3}{m_1 + m_2 + m_3}$$

$$= \frac{M \times \frac{L}{2} + M \times L + M \times \frac{5L}{4}}{M + M + M} = \frac{\frac{11}{4}ML}{3M} = \frac{11L}{12}$$

27. (b)

$$x_{cm} = \frac{1 \times 0 + 1 \times PQ + 1 \times PR}{1+1+1} = \frac{PQ + PR}{3} \text{ and } y_{cm} = 0$$

28. (b)

$$X = \frac{m_{1x_1} + m_2x_2 + m_3x_3 + m_4x_4}{m_1 + m_2 + m_3 + m_4}$$

$$X = \frac{0 + 40x_4}{100} \Rightarrow 3 = \frac{40x_4}{100}$$

$$x_4 = \frac{300}{40} = 7.5$$

Similarly $y_4 = 7.5$ and $z_4 = 7.5$

29. (c)

Angle turned in three second $\theta_{3s} = 2\pi \times 10 = 20\pi$ rad

From $\theta = \omega_0 t + \frac{1}{2}\alpha t^2 \Rightarrow 20\pi = 0 + \frac{1}{2}\alpha \times (3)^2$

$\Rightarrow \alpha = \frac{40\pi}{9}$ rad / s^2

Now, angle turned in 6 sec from the start

$\theta_{6s} = \omega_0 t + \frac{1}{2}\alpha t^2 \Rightarrow 0 + \frac{1}{2} \times \left(\frac{40\pi}{9}\right) \times 6 = 80\pi$ rad

$\therefore$ angle turned between $t = 3$s to $t = 6$s

$\theta_{last\ 3s} = \theta_{6s} - \theta_{3s} = 80\pi - 20\pi = 60\pi$

Number of revolutions $= \frac{60\pi}{2\pi} = 30$ rev.

30. (b)

31. (c)

$$\bar{\tau} = \bar{r} \times \bar{F} = rF\sin\phi \;\therefore\; \tau = rF\sin\phi$$

$F\sin\phi$ = transverse component of force
$F\cos\phi$ = radial component of force

32. (a)

$I = \frac{\tau}{\alpha} = \frac{31.4}{4\pi} = 2.5$ kg m^2.

33. (b)

34. (d)

Total kinetic energy $= \frac{1}{2}mv^2(1 + \frac{K^2}{R^2}) = 32.8$ J

$\Rightarrow \frac{1}{2} \times 10 \times (2)^2 (1 + \frac{K^2}{(0.5)^2}) = 32.8 \Rightarrow K = 0.4$ m

35. (c)

Rotational kinetic energy $E = \frac{L^2}{2I} \therefore L = \sqrt{2EI}$

$$\Rightarrow \frac{L_A}{L_B} = \sqrt{\frac{E_A}{E_B} \times \frac{I_A}{I_B}} = \sqrt{100 \times \frac{1}{4}} = 5$$

36. (b)

Given: kinetic energy $K = 360$ J
Angular speed $\omega = 20$ rad/s

$$\therefore K = \frac{1}{2}I\omega^2$$

where I = month of inertia

$$\Rightarrow I = \frac{2K}{\omega^2} = \frac{2 \times 360}{20 \times 20} = 1.8 \text{ kg m}^{-2}.$$

37. (d)

38. (c)

We know $\quad v = \sqrt{\frac{2gh}{1 + \frac{k^2}{r^2}}}$

$$\omega = \frac{v}{r} = \sqrt{\frac{2gh}{r^2 + k^2}}$$

$$\omega = \sqrt{\frac{2mgh}{mr^2 + mk^2}} = \sqrt{\frac{2mgh}{mr^2 + I}}$$

39. (a)

Mass of the entire disc is 4 M and its moment of inertia about given axis would be $\frac{1}{2}(4M)R^2$.

For the given section, the moment of inertia about the same axis would be one-quarter of this, i.e., $\frac{1}{2}MR^2$.

40. (a)

41. (c)

42. (c)

43. (b)

44. (d)

$$I = \frac{2}{5}MR^2 \quad \therefore I \,\alpha\, R^2$$

This relation shows that graph between I and R will be parabola symmetric to I-axis.

45. (b)

Centre of mass always lies towards heavier mass.

Gravitation

Introduction

Gravitation is a fundamental force of nature. It is a non-contact force which determines the motion of planets in the solar system, the motion of the Sun in our Milky Way, the motion of our galaxy and so on in the universe.

Newton discovered in 1687 that the interaction which makes an apple fall from a tree also keeps the planets in the solar system in their orbits around the sun. This was the beginning of the celestial mechanics which has developed so extensively that it allows us to determine how to put a satellite into any desired orbit around the earth or to choose just the right trajectory to send a spacecraft to another planet.

Newton's Law of Gravitation

"Every particle of matter in the universe attracts every other particle with a force that is directly proportional to the product of the masses of the particles and inversely proportional to the square of the distance between them."

m_1 F F m_2
A B
r

Thus the magnitude of the gravitational attraction on either particle having masses m_1 and m_2 and distance apart being r is given by

$$F = G\frac{m_1 m_2}{r^2}$$

Here G is a fundamental physical constant, called the universal gravitational constant.

In SI units $G = 6.67 \times 10^{-11}$ Nm²/kg².

Dimensions of G are $M^{-1}L^3T^{-2}$.

Gravitational Force in Vector form

The mutual forces of gravitational attraction between A and B are equal in magnitude but in opposite directions (along the line joining them).

Thus force on B due to A is

$$\vec{F} = -\frac{Gm_1m_2}{r^2}\hat{r},$$

where $\hat{r}$ is a unit vector along the line from A to B.

A attracts B and B attracts A.

Important points regarding gravitational force

(i) This force is always attractive in nature.

(ii) This force is independent of the nature of the surrounding medium.

(iii) Gravitation is a conservative force, i.e., the work done by gravitational force is independent of path and the work done in a closed path is zero.

(iv) The law governing the gravitational interaction holds over a wide range of distances from inter-atomic to inter-planetary distances.

(v) The relative strengths of gravitational, electrostatic and nuclear forces can be expressed as: gravitational force << electrostatic force << nuclear force. The range of the gravitational force >> range of nuclear force.

Mass and Density of Earth

Mass of earth

The relation between acceleration due to gravity g and gravitation G is given by

$$g = \frac{GM}{R^2} \text{ or } M = \frac{gR^2}{G} \quad \ldots(1)$$

Where R is the radius of earth and M is its mass.

We know that,

$$G = 6.67\times10^{-11}\,\text{N-m}^2/\text{kg}^2 \text{ and } g = 9.8\,\text{m/s}^2$$

$$\therefore\ M = \frac{9.8\times(6.4\times10^6)^2}{6.67\times10^{-11}} = 6.018\times10^{24}\,\text{kg}$$

Density of earth

The density is defined as

$$\rho = \frac{\text{Mass}}{\text{Volume}} = \frac{M}{(4/3)\pi R^3}$$

or

$$\rho = \frac{3M}{4\pi R^3} \quad \ldots(2)$$

Substituting the value of M from Eq. (1) in Eq. (2), we get

$$\rho = \frac{3\times(gR^2/G)}{4\pi R^3} = \frac{3g}{4\pi RG} \quad \ldots(3)$$

Substituting the values, we get

$$\rho = \frac{3\times 9.8}{4\times(3.14)\times(6.4\times10^6)\times(6.67\times10^{-11})^2}$$
$$\approx 5.5\times10^3\ \text{kg/m}^3$$

Illustration 1

Calculate the mass and density of earth, given that gravitational constant $= 6.67\times10^{-11}$ N-m²/kg², the radius of earth $= 6.37\times10^6$ metre and $g = 9.8$ m/s².

Solution

The gravitational attraction of earth on a body of mass m on the surface of earth is given by

$$mg = \frac{GMm}{R^2}$$

Where g is acceleration due gravity.

$$\therefore\quad M = \frac{g\,R^2}{G} = \frac{9.8\times(6.37\times10^6)^2}{6.67\times10^{-11}}$$
$$= 6\times10^{24}\ \text{kg.}$$

If ρ be the density of earth, then

$$M = \frac{4}{3}\pi R^3\times\rho$$

or
$$\rho = \frac{3M}{4\pi R^3} = \frac{3\times(6\times10^{24})}{4\times3.14\times(6.67\times10^6)^3}$$
$$= 5.5\times10^3\ \text{kg/m}^3$$

Gravitational Field

A useful way to describe non-contact forces acting on a material particle is in terms of a physical quantity called "field". For example when a body is placed near earth's surface, it experiences a force which is known as its weight. This interacting force is greater when the body is close to the earth and decreases as the body moves away. We say that the earth has a gravitational field and it becomes weaker as one moves farther away.

The **intensity of the gravitational field** at a point is defined as "force per unit test mass" at that point. It is referred to as simply the gravitational field :

$$\vec{g} = \frac{\vec{F}}{m_0},$$

where m_0 = test mass, which is so small that it does not distort the field that is being measured.

In SI units $\vec{g}$ is expressed as N/kg or m/s² and has its dimensions LT⁻².

The field at P due to a mass m placed at a point O is $\frac{\vec{F}}{m_0}$, where

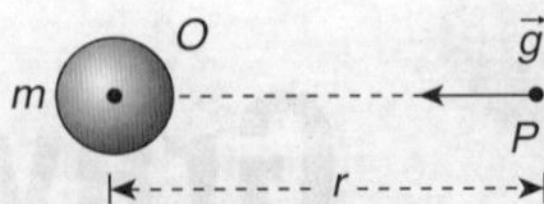

$\vec{F} = -\frac{Gmm_0}{r^2}\hat{r}$, and $\hat{r}$ is unit vector along $\overrightarrow{OP}$.

$$\therefore\ \vec{g} = -\frac{Gm}{r^2}\hat{r}.$$

Variation of Acceleration due to Gravity

The acceleration of a freely falling body, which can be measured on the surface of the earth, and which is conventionally referred to as the "acceleration due to gravity", varies from place to place on the earth's surface. In the discussion that follows, we calculate the apparent variation due to the effect of the earth's rotation (i.e., due to pseudo-force acting on a freely falling body), and the actual variation in the earth's gravitational field which depends on the location of the body (i.e., height as well as depth of the body relative to the earth's surface) and the shape of the earth.

Let us consider the earth as a uniform solid sphere of mass M and radius R. The acceleration due to gravity, $|\vec{g}|$, on the surface is :

$$|\vec{g}| = \frac{F}{m_0} = \frac{GMm_0}{R^2 m_0} = \frac{GM}{R^2}$$

$$g = \frac{GM}{R^2} = g_0 \text{ (on the surface of earth).}$$

(i) At a height h, above the surface of earth

$$g_h = \frac{F}{m_0} = \frac{GMm_0}{m_0 r^2} = \frac{GM}{(R+h)^2},\ (r = R+h)$$

In terms of g_0,

$$g_h = \frac{GM}{R^2\left(1+\frac{h}{R}\right)^2} = \frac{g_0}{\left(1+\frac{h}{R}\right)^2} \quad \ldots(1)$$

$\therefore\ g_h$ decreases with height h.

However, if $h << R$, this expression can be expanded by binomial theorem:

$$g_h = g_0\left(1+\frac{h}{R}\right)^{-2} \approx g_0\left(1-\frac{2h}{R}\right). \quad \ldots(2)$$

Can we use Eq. (2) to find the value of g_h at height $h=\frac{R}{2}$? Yes, before using binomial theorem, but not after that.

Illustration 2

A man weighs 720 N at the surface of earth. How much will he weigh at a height equal to half the radius of earth, above the surface of earth?

Solution

Acceleration due to gravity at a height h is given by

$$g_h = \frac{gR^2}{(R+h)^2}.$$

$$\frac{g_h}{g} = \left(\frac{R}{R+\frac{R}{2}}\right)^2 = \frac{4}{9}\quad\left(h=\frac{R}{2}\right)$$

$$W_h = mg_h = m\left(\frac{4}{9}g\right) = mg\times\frac{4}{9}.$$

$$720\times\frac{4}{9} = 320 \text{ N}.$$

(ii) Variation of g with depth

Consider a point inside the earth at a depth x, whose distance from the centre of earth is r. Let the value of g be g_x, so that

$$g_x = \frac{GM'}{r^2},$$

where $M' = \frac{Mr^3}{R^3}$ is the mass of the inner core of earth of radius r

$$g_x = \frac{GM}{r^2}\frac{r^3}{R^3} = \frac{GM}{R^3}r = \frac{GM}{R^2}\frac{r}{R} = g_0\left(1-\frac{x}{R}\right)$$

Thus the value of g_x decreases linearly with x and reduces to zero at the centre of the earth.

The variation of g with the distance r from the centre of the earth is shown in the graph.

Although $g_0 = 9.8$ m/s^2 is taken as the acceleration due to gravity at the surface of the earth, the actual value varies from place to place.

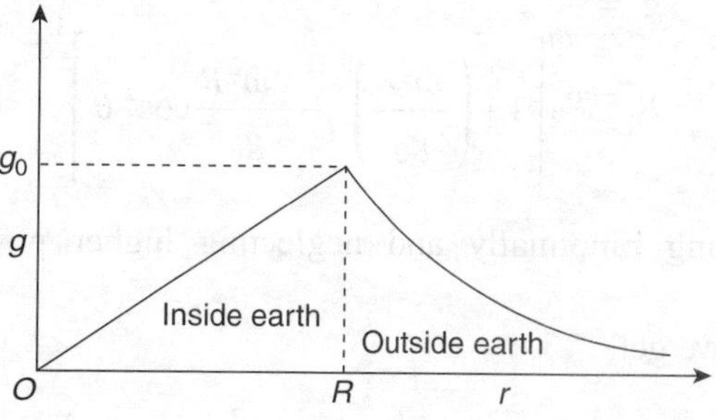

(iii) Variation of g due to rotation of earth

The measured acceleration of a freely falling body is, however, different from the actual acceleration due to gravity (i.e., the intensity of the earth's gravitational field). This is due to pseudo-force acting on a freely falling body in the non-inertial frame of the earth.

The earth spins about its polar axis with an angular frequency $\omega = 2\pi/T$, where T = 24 hours = 86400 seconds. Any object placed on its surface experiences a centrifugal force (pseudo-force) which will produce an apparent change in the measured value of the free-fall acceleration.

We consider an object of mass m at a point P whose latitude is θ. With the rotation of earth, this object also rotates with the same angular frequency ω about the polar axis, in the circle of radius $r = R\cos\theta$, where r is the distance of this object from the polar axis. In the earth's frame, the centrifugal force F_{CP} acts on the object in the direction $\overrightarrow{CP}$. The effective gravitational force is given by

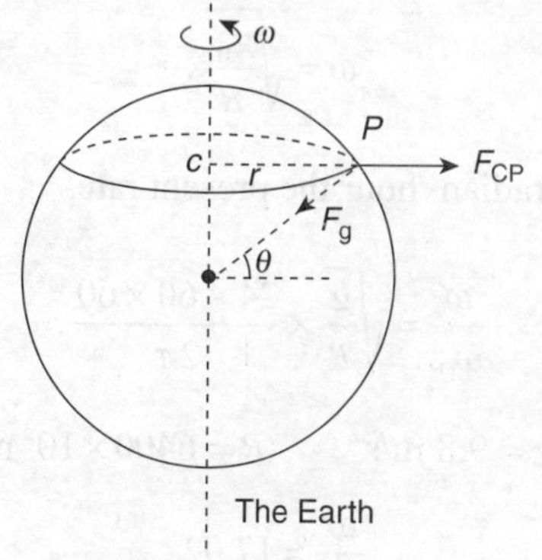

$$\vec{F}'_g = \vec{F}_g + \vec{F}_{CP}.$$

$$\therefore\ F'_g = \sqrt{F_g^2 + F_{CP}^2 - 2F_gF_{CP}\cos\theta}.$$

Here $F_g = mg_0$, $F_{CP} = m\omega^2 r$.

$$F'_g = \sqrt{(mg_0)^2 + (m\omega^2 r)^2 - 2m^2 g_0\omega^2 R\cos^2\theta}.$$

If g' be the measured value of the acceleration of a freely falling body at P, then

$mg' = F_g'$.

$g' = \frac{F_g'}{m}$.

$\therefore g' = g_0\left[1 + \left(\frac{\omega^2 r}{g_0}\right)^2 - \frac{2\omega^2 R}{g_0}\cos^2\theta\right]^{1/2}$.

Expanding binomially and neglecting higher powers of $\frac{\omega^2 R}{g_0}$, we get,

$$g' = g_0\left(1 - \frac{1}{2}\frac{2\omega^2}{g_0}R\cos^2\theta\right)$$

$$\Rightarrow g' = g_0\left(1 - \frac{\omega^2}{g_0}R\cos^2\theta\right)$$

or $g' = g_0 - \omega^2 R\cos^2\theta$ (approximately).

Illustration 3

How long would a day be if the earth were rotating so fast that objects at the equator were weightless ?

Solution

Due to rotation of the earth, effective value of g in given by

$$g_{eff} = g_0 - \omega^2 R\cos^2\theta.$$

At the equator latitude $\theta = 0$, hence

$$g_{eff} = g_0 - \omega^2 R$$

For weightlessness, $g_{eff} = 0$, hence

$$\omega = \sqrt{\frac{g_0}{R}}.$$

But $\omega_0 = \frac{2\pi}{24}$ radian /hour the present rate.

Hence, $\frac{\omega}{\omega_0} = \sqrt{\frac{g}{R}} \times \frac{24 \times 60 \times 60}{2\pi}$.

Substituting $g = 9.8\ \text{m/s}^2$, $R = 6400 \times 10^3\ \text{m}$,

We get $\frac{\omega}{\omega_0} = 17.02$.

Hence the earth should rotate approximately 17 times faster and the length of the day would be $\frac{24}{17}\ \text{h} = 1.41\ \text{h}$.

Work and Potential Energy

Let us consider the situation given in the figure. Suppose that there is a gravitation field due to a system of mass. If some external agent were to take a test mass **slowly** along path (1) and do some work in going from A to B against the gravitational field, then exactly the same amount of work will be done on the external agent when the particle returns to its previous position A.

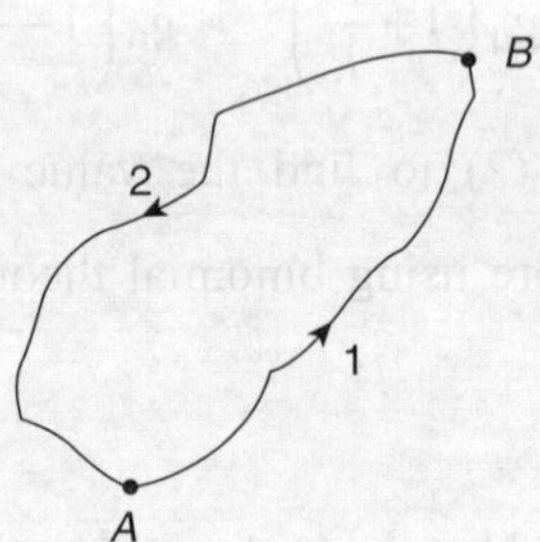

In other words we can say that the work that was done (by an external agent) in taking the particle from A to B was stored by the field as energy and the field returned the same amount of work (the stored energy) when the particle is returned slowly to its initial position A. Work done by the external agent in moving the test mass slowly against the gravitational attraction produced by particle of mass M is

$$\int dW_{ext} = -\int dW_{gr}$$

$$= \int \vec{F}.d\vec{r} = GMm\int_A^B \frac{dr}{r^2} \quad [\because W_{ext} + W_{gr} = \Delta KE = 0]$$

Thus the work done $= GMm\left(\frac{1}{r_A} - \frac{1}{r_B}\right)$.

This capacity of the field to do external work in configuration B (when the particle is at B relative to A) is called the energy. Since this is due to configuration, (position dependent) it is known as **potential energy**.

Thus the energy stored in the field

$U_{BA} = -$ Work done by an external agent against the gravitational field

$$= -W_{gr} = -\int \vec{F}_{gr}.d\vec{r}$$

Potential Energy

In order to define potential energy in absolute terms, we need a reference point (a fixed point) with respect to which the work done has to be defined. Conventionally, this reference point is chosen to be at infinity since the field at infinity is zero.

This choice works well for mass distribution whose field at infinity becomes zero.

However, it does not work for infinite line or a plane of infinite extension.

Thus, the potential energy of a mass m at a point P in a gravitational field will be

$$U_P = -\int_{\infty}^{P} \vec{F}_{gr}.d\vec{r}$$

Potential energy for a two-particle system

Consider a point mass m_1 located at the origin of coordinates. The potential energy of another point mass m_2 placed at point $P(\vec{r})$ is, according to the above equation,

$$U_{12} = -\int_{\infty}^{P} \vec{F}_{gr} \cdot d\vec{r}$$

$$= -\int_{\infty}^{P}\left(-\frac{Gm_1m_2}{r^3}\vec{r}\right).d\vec{r}$$

$$= +Gm_1m_2 \int_{\infty}^{P(r)} \frac{r\,dr}{r^3} \qquad [\because \vec{r}\cdot d\vec{r} = r\,dr]$$

$$= +Gm_1m_2\left[-\frac{1}{r}\right]_{\infty}^{r} = -\frac{Gm_1m_2}{r}.$$

Thus, the interaction (potential) energy

$$U_{12} = -\frac{Gm_1m_2}{r_{12}},$$

where r_{12} = separation between the two particles.

In general, the potential energy of a test mass m in the gravitational field produced by a system of point mass m_1, m_2 ... can be obtained by the principle of superposition leading to

$$U_{total} = U_1 + U_2 + U_3 + \ldots$$

$$= -\frac{Gm_1m}{r_1} - \frac{Gm_2m}{r_2} - \frac{Gm_3m}{r_3} - \ldots$$

$$= -Gm\left(\frac{m_1}{r_1} + \frac{m_2}{r_2} + \ldots\right)$$

$$= -Gm\,\Sigma\frac{m_i}{r_i}.$$

In the same way, the potential energy of a system of point masses m_1, m_2, m_3 ... will be

$$U_{total} = U_{12} + U_{13} + U_{14} + \ldots$$
$$+ U_{23} + U_{24} + \ldots$$
$$+ U_{34} + \ldots$$

$$= \sum_{i \neq j} U_{ij} = -G\sum_{i}\sum_{\substack{j \\ i \neq j}} \frac{m_im_j}{r_{ij}}.$$

Gravitational potential

Gravitational potential V at a point P in the gravitational field is the potential energy per unit test mass (m_0) located at that point, i.e.,

$$V(\vec{r}) = V_P = \frac{U_P}{m_0}.$$

The potential at point $P(x, y, z)$ due to a point mass m at the origin O will be

$$V_P = \frac{U_P}{m_0}, \quad \text{(where } m_0 \text{ is the test mass)}$$

$$= -\frac{1}{m_0}\int \vec{F}_{gr}.d\vec{r} = -\frac{1}{m_0}\int_{\infty}^{r} -\frac{Gmm_0}{r^2}dr$$

$$= -\frac{Gm}{r}$$

Gravitational potential is a scalar quantity and is always negative (like the potential energy) and has the maximum value (= zero) at infinity.

The total potential at a point due to a system of particles is simply the algebraic sum of the potential produced by them independently.

$$\therefore \quad V = V_1 + V_2 + V_3 + \ldots$$

$$= \sum V_i = -G\sum\frac{m_i}{r_i}.$$

For continuous mass distribution,

$$V = -G\int\frac{dm}{r}.$$

We obtained the gravitational potential by integrating the field from infinity up to r given by the expression

$$V(r) = -\int_{\infty}^{r}\vec{g}.d\vec{r}.$$

We can get the field by differentiating the potential, if the potential is known.

$$\int_{\infty}^{r}\vec{g}.d\vec{r} = -V_r \text{ and } \frac{d}{dr}(V_r) = -g_r.$$

Illustration 4

Find the amount of work done in displacing a body of mass m very slowly from the earth's surface to a point beyond the earth's gravitational field.

Solution

Since the kinetic energy has been assumed to be negligible, the work done is equal to the change in its potential energy. Thus, work done by an external agent is the change in the potential energy.

$$\therefore \quad W = \Delta U = U_f - U_i = U_{r=\infty} - U_{r=R}$$

$$= 0 - \left(-\frac{GMm}{R}\right) = \frac{GMm}{R} = \frac{gR^2 m}{R} = mgR$$

Illustration 5

For a certain mass distribution, the gravitational field g as a function of r is expressed as $g = -\frac{K}{r}$. Taking the reference point at $r = r_0$ with potential $V = V_0$, find the potential V in terms of r.

Solution

The relation connecting the field g and potential V is

$$g = -\frac{dV}{dr}$$

$$\therefore \quad -\frac{dV}{dr}$$

or $$\int_{v=v_0}^{v} dV = K \int_{r=r_0}^{r} \frac{dr}{r}$$

$$\therefore \quad V - V_0 = \text{[illegible]}$$

$$\Rightarrow \quad V = V_0 + K.$$

Illustration 6

A particle of mass M is placed at the centre of a uniform spherical shell of mass $2M$ and radius R. The gravitational potential on the surface of the shell is

(a) $-\frac{GM}{R}$ (b) $-\frac{3GM}{R}$

(c) $-\frac{2GM}{R}$ (d) zero

Solution

Potential is a scalar quantity. Hence,

$$V_{total} = V_{shell} + V_{particle}$$

$$= -\frac{G2M}{R} - \frac{GM}{R} = -\frac{3GM}{R}.$$

Ans. (b)

Gravitational Potential (V) and Field ($\vec{g}$)

Body		Potential	Field
Ring Mass = M Radius = R	R, O, x	$V = \frac{GMx}{\sqrt{R^2 + x^2}}$	$\vec{g} = -\frac{GMx}{\left(R^2 + x^2\right)^{3/2}}\hat{i}$
Disc Mass = M Radius = R	R, O, x	$V = G2\pi\sigma x - \sqrt{R^2 + x^2}$	$\vec{g} = -\frac{2GM}{R^2}\left(1 - \frac{x}{\sqrt{R^2 + x^2}}\right)\hat{i}$
Spherical shell Mass = M Radius = R	O, x	$V = -\frac{GM}{x}$, for $x > R$ $V = -\frac{GM}{R}$, on the shell $V = -\frac{GM}{R}$, for $x < R$.	$\vec{g} = -\frac{GM}{x^2}\hat{x}$ $\vec{g} = -\frac{GM}{R^2}\hat{x}$ $\vec{g} = \vec{0}$
Solid sphere Mass = M Radius = R		$V = -\frac{GM}{x}$, for $x > R$ $V = -\frac{GM}{R}$, for $x = R$ $V = -\frac{GM}{2R^3}\left(3R^2 - x^2\right)$, for $x < R$	$\vec{g} = -\frac{GM}{x^2}\hat{x}$ $\vec{g} = -\frac{GM}{R^2}\hat{x}$ $\vec{g} = -\frac{GMx}{R^3}\hat{x}$

Illustration 7

Find the strength of the gravitational field at a point along the axis of a thin rod of length L and mass M at a distance d from one end.

Solution

Linear density of mass of the rod is $\lambda = \dfrac{M}{L}$.

Mass of a small element dx is $dm = \lambda dx$.

Field at P due to this mass is

$$|dg| = \frac{G\lambda dx}{x^2}$$

$$\therefore \quad g = G\lambda \int_d^{L+d} \frac{dx}{x^2} = G\lambda \left[-\frac{1}{x} \right]_d^{L+d}$$

$$= G\lambda \left[\frac{1}{d} - \frac{1}{L+d} \right]$$

$$= \frac{GM}{L} \frac{L}{d(L+d)} = \frac{GM}{d(L+d)}.$$

The field is directed towards the rod along the axis.

Illustration 8

Two bodies of masses m_1 and m_2 are placed at a distance r apart. Show that the position where the gravitational field due to them is zero, the gravitational potential is given by $V = -\dfrac{G}{r}\left[m_1 + m_2 + 2\sqrt{m_1 m_2} \right]$.

Solution

Let the field be zero at a distance x from m_1, hence

$$\frac{Gm_1}{x^2} = \frac{Gm_2}{(r-x)^2}$$

or
$$(r-x)\sqrt{m_1} = x\sqrt{m_2}$$

or
$$x = \frac{\sqrt{m_1}\, r}{\sqrt{m_1} + \sqrt{m_2}}.$$

Similarly,

$$(r-x) = \frac{\sqrt{m_2}\, r}{\sqrt{m_1} + \sqrt{m_2}},$$

Hence, the required gravitational potential

$$V = -\frac{Gm_1}{x} - \frac{Gm_2}{(r-x)}$$

$$= -\frac{G\left(\sqrt{m_1} + \sqrt{m_2}\right)}{r}\left[\sqrt{m_1} + \sqrt{m_2} \right]$$

$$= -\frac{G}{r}\left(m_1 + m_2 + 2\sqrt{m_1 m_2} \right).$$

Illustration 9

From a large uniform solid sphere of radius R and mass M, a small spherical part of radius r is removed. Find the gravitational potential at a point P on the line OCP, where OP = x.

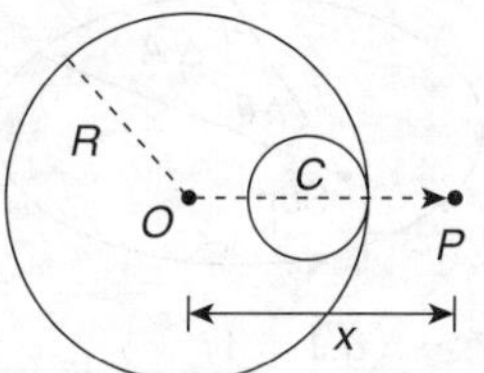

Solution

The potential at P due to the spherical part which is removed is

$$V_r = -\frac{Gm}{PC} = -\frac{G\dfrac{Mr^3}{R^3}}{x - (R-r)}.$$

The potential at P due to the complete solid sphere is $V_R = -\dfrac{GM}{x}$.

Let the required potential at P due to the remaining part be V, hence using the superposition principle

$$V_R = V + V_r$$

or
$$V = V_R - V_r = -\frac{GM}{x} - \left(-\frac{GMr^3}{R^3(x-R+r)} \right)$$

$$= -GM\left[\frac{1}{x} - \frac{1}{(x-R+r)} \frac{r^3}{R^3} \right].$$

Kepler's Laws of Planetary Motion

In the solar system, all planets move around the sun in different orbits with different times of revolution. Similarly the planets have their satellites, orbiting around them. These satellites are natural (like the moon) as well as artificial. All these orbital motions, of planets around the sun, or of satellites around their planets, obey certain laws known as Kepler's laws of planetary motion.

Kepler's Laws are:

1. **Law of Orbits**: All planets move in elliptical orbits with the sun at one focus.
2. **Law of Areas**: A line joining any planet to the sun sweeps out equal areas in equal times.

$\therefore$ Areal velocity $= \dfrac{dA}{dt} = \text{constant}$.

This law is basically the principle of *conservation of angular momentum* as shown below:

The area $\Delta A = \dfrac{1}{2}r.r\Delta\theta$, and the areal velocity

$$\frac{dA}{dt} = \lim_{\Delta t \to 0}\frac{\Delta A}{\Delta t} = \frac{1}{2}r^2\frac{d\theta}{dt} = \frac{1}{2}r^2\omega$$

$$\therefore \quad m\frac{dA}{dt} = \frac{1}{2}mr^2.\omega$$

$$= \frac{1}{2}(I\omega)$$

where $mr^2 = I$; the moment of inertia of the planet about the axis through the sun at that instant. Since $\dfrac{dA}{dt}$ is constant, the angular momentum $I\omega$ remains conserved. This law is true since gravitational force $(\vec{f})$ is a *central force* and the torque $\vec{r}\times\vec{f}$ is zero, hence $\vec{L}$, the angular momentum is conserved.

3. **Law of Periods**: The square of the period of any planet (time of revolution of the planet around the sun) is proportional to the cube of the semi-major axis of the elliptical orbit.

Thus, $T^2 \propto a^3$

$$\Rightarrow \quad \frac{T_1^2}{a_1^3} = \frac{T_2^2}{a_2^3},$$

where T_1 and T_2 are the periods of the two planets and a_1 and a_2 are the corresponding semi-major axes of the two elliptical paths. Using circular approximation for planetary motion,

$$\frac{GMm}{r^2} = \frac{mv^2}{r} = m\omega^2 r = m\frac{4\pi^2}{T^2}.r$$

$$\therefore \quad T^2 = \frac{4\pi^2}{GM}.r^3$$

or $\quad T^2 \propto r^3$.

Illustration 10

The period of revolution of an earth satellite close to the surface of earth is 90 minutes. The time period of another satellite in an orbit at a distance of three times the radius of earth from its surface will be

(a) $90\sqrt{8}$ min (b) 360 min
(c) 720 min (d) 270 min

Solution

According to Kepler's third law,

$$T^2 \propto r^3$$

$$\Rightarrow \quad \frac{T^2}{T_0^2} = \left(\frac{3R+R}{R}\right)^3 = 64$$

$$\therefore \quad T = \sqrt{64\times 90^2} =$$

Ans. (c)

Illustration 11

A planet is suddenly stopped in its orbit which is supposed to be circular. Show that it would fall into the sun in a time $\dfrac{\sqrt{2}}{8}$ times the period of the planet's revolution.

Solution

Orbital velocity of the planet is $v_0 = \dfrac{2\pi r}{T} = \sqrt{\dfrac{GM}{r}}$, where M = mass of the sun, and r = radius of orbit

$$\therefore \quad T^2 = \frac{4\pi^2 r^3}{GM}. \qquad \ldots(1)$$

When the planet stops, it would fall into the sun due to the sun's gravitational pull.

Let v be the velocity at a distance x from the sun. Conserving energy we get,

$$\frac{1}{2}mv^2 - \frac{GMm}{x} = 0 - \frac{GMm}{r}$$

or $$\left(-\frac{dx}{dt}\right)^2 = \frac{2GM}{r}\left(\frac{r-x}{x}\right) \quad \left[\because v = -\frac{dx}{dt}\right]$$

or $$\frac{dx}{dt} = \sqrt{\frac{2GM}{r}}\sqrt{\frac{r-x}{x}}$$

$$\therefore \quad dt = -\sqrt{\frac{r}{2GM}}\int_r^0\sqrt{\frac{x}{r-x}}\,dx.$$

Substituting $x = r\sin^2\theta$ and integrating, we get

$$t = \sqrt{\frac{r}{2GM}}\frac{\pi r}{2} = \sqrt{\frac{\pi^2 r^3}{8GM}}$$

$$\therefore \quad \frac{t}{T} = \sqrt{\frac{\pi^2 r^3}{8GM}\times\frac{GM}{4\pi^2 r^3}} = \sqrt{\frac{1}{32}} = \frac{\sqrt{2}}{8}$$

$$\therefore \quad t = \left(\frac{\sqrt{2}}{8}\right)T.$$

Alternative method

This problem can be also be solved using Kepler's laws.

During the fall of the planet directly on to the sun along a straight line, we may consider the straight line path as an extreme case of ellipse with minor axis = 0 and major axis

$2a = r$, or $a = \frac{r}{2}$.

Now, $T^2 = kr^3$ and

$$T_1^2 = ka^3$$

$$\therefore \quad \left(\frac{T_1}{T}\right)^2 = \left(\frac{a}{r}\right)^3 = \left(\frac{1}{2}\right)^3 = \frac{1}{8}$$

$\therefore T_1 = \frac{T}{\sqrt{8}}$ = time of revolution in this special case.

The required time of fall is half of T_1, hence

$$T' = \frac{T_1}{2} = \frac{T}{2\sqrt{8}} = \frac{\sqrt{2}}{8}T.$$

Motion of Satellites

Let us consider the motion of a (artificial) satellite of mass m which moves around the planet (earth) in approximately circular orbit. The gravitational attraction $\frac{GMm}{r^2}$ provides the necessary centripetal force for circular motion.

Orbital Speed (v_0)

$$\frac{mv_0^2}{r} = \frac{GMm}{r^2}$$

$$\therefore \quad v_0 = \sqrt{\frac{GM}{r}} = \sqrt{\frac{g_0R^2}{r}}$$

where g_0 = gravity at the surface of earth, and R = radius of earth.

If g = acceleration due to gravity at the circular orbit of radius r, then

$$v_0 = \sqrt{gr}. \qquad \left[\because \; g = \frac{GM}{r^2}\right]$$

If h = height of satellite above earth surface,

$$r = R + h$$

$$\therefore \quad v_0 = \sqrt{g_0 R^2/(R+h)} = \sqrt{g_0R}\left(1+\frac{h}{R}\right)^{-\frac{1}{2}}$$

or $$v_0 = \sqrt{g_0R}\left(1-\frac{h}{2R}\right), \text{ for } h << R.$$

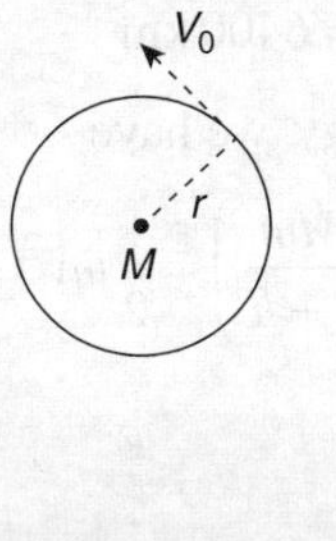

Neglecting h w.r.t. $2R$, the orbital speed close to the earth's surface,

$$v_0 \approx \sqrt{g_0R}.$$

Angular speed ω

For circular orbit, angular speed

$$\omega = \frac{v_0}{r}.$$

Putting $$v_0 = \sqrt{\frac{g_0R^2}{r}},$$

$$\omega = \sqrt{\frac{g_0R^2}{r^3}} = \sqrt{\frac{GM}{(R+h)^3}}.$$

Time of revolution

$$T = \frac{2\pi}{\omega}.$$

Putting $v_0 = \sqrt{\frac{g_0R^2}{r^3}}$, $T = 2\pi\sqrt{\frac{r^3}{g_0R^2}}$.

Higher the value of the radius of the orbit (r), the time of revolution is also higher. For artificial satellites close to earth, time of revolution is nearly 86 minutes whereas for the moon, it is around 27.3 days.

Potential Energy (U) of a satellite

The gravitational potential energy of a system of two masses m_1 and m_2 with separation r is $-\frac{Gm_1m_2}{r}$.

Hence the gravitational potential energy of planet-satellite system is $U_{gr} = -\frac{GMm}{r}$ where M and m are the masses of the planet and the satellites respectively. Here the negative sign for U_{gr} is due to attractive nature of the gravitational force. Although U_{gr} is the gravitational potential energy of the planet satellite system, we take, for simplicity, U_{gr} as the potential energy of the satellite (in the field of the planet).

Putting $GM = g_0R^2$, the potential energy of the satellite,

$$U_{gr} = \frac{-mg_0R^2}{r} = \frac{-mg_0R^2}{(R+h)} = -\frac{mg_0R}{\left(1+\frac{h}{R}\right)}.$$

For $h << R$,

$$U_{gr} = -mg_0R\left(1+\frac{h}{R}\right)^{-1}$$

$$= -mg_0R\left(1-\frac{h}{R}\right).$$

Neglecting h w.r.t. R,

$$U_{gr} = -mg_0R.$$

It is important to note that potential energy of the satellite is always negative. It increases with height h and attains the maximum value (=0) at infinity, which is the point beyond the gravitational field of the planet.

Kinetic Energy (E_K) of the Satellite

Since the orbital speed of the satellite in its orbit of radius r around the planet of mass M is given by

$$v_0 = \sqrt{\frac{GM}{r}}.$$

Hence its kinetic energy

$$E_k = \frac{1}{2}mv_0^2 = \frac{1}{2}\frac{GMm}{r} = \frac{1}{2}|U(r)|.$$

This result that the kinetic energy is half the magnitude of the potential energy is true for any circular orbit in an inverse-square force field.

Total energy of the satellite

$$E_{total} = E_P + E_K$$

$$= -\frac{GMm}{r} + \frac{GMm}{2r} = -\frac{GMm}{2r}.$$

Note that since the total energy of the planet–satellite system is negative $\left(= -\frac{GMm}{2r}\right)$, the system is called a bound system. The minimum amount of energy required to separate the given system for infinite separation will be $+\frac{GMm}{2r}$.

This energy $\frac{GMm}{2r}$ is also known as the **Binding Energy** of the system.

The following figure is a plot of potential energy function $U(r)$ versus r with the choice $U(r) = 0$ at $r = \infty$. This function begins at $U = -\frac{GMm}{R}$ at the earth's surface $(r = R)$ and increases as r increases; approaching zero at $r = \infty$.

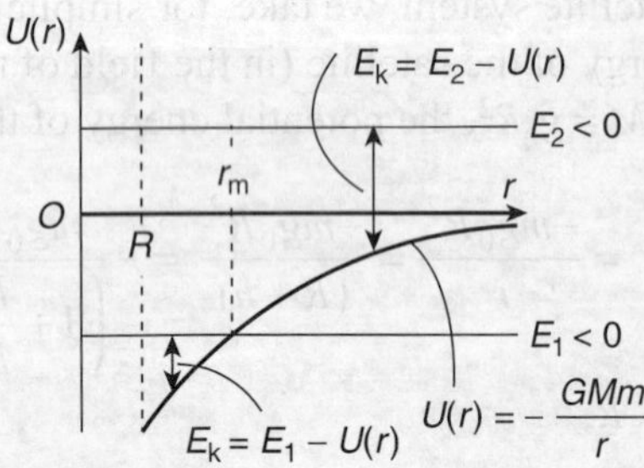

Two possible values for the total energy E are indicated on this graph. A negative total energy $(E_1 < 0)$ merely means that the kinetic energy at the earth's surface is less than $\frac{GMm}{R}$, so that it is never greater in magnitude than the negative potential energy. Thus for $E_{total} < 0$, the total energy line intersects the potential energy curve at some maximum separation r_m and the system is bound. On the other hand, for $E_{total} > 0$, there will be no such intersection and the system is unbound.

Geostationary Satellites

These are artificial satellites of earth which appears stationary to an observer on the earth's surface. Such a satellite is known as geostationary or geosynchronous satellite. If the satellite has to appears fixed at a position above a certain place on the earth, then it should co-rotate with the earth in such a manner that its orbital period around the earth is precisely equal to the rotational period of the earth itself. Clearly, $T = 24$ hours for a geostationary satellite. The height of geostationary satellite above earth surface is given by

$$h = \left[\frac{gR^2T^2}{4\pi^2}\right]^{1/3} - R$$

where, $R = 6.4 \times 10^6$ m, $g = 9.8$ m/s^2

$T = 24$ hours$= 24 \times 60 \times 60$ sec

$\therefore \quad h = 3.6 \times 10^7$ m $= 36000$ km

So, the height of the satellite above the earth surface is 36,000 km. When a satellite is placed in an orbit at this height above the equator, it will appear stationary.

Illustration 12

An artificial satellite is moving in a circular orbit around earth with a speed equal to half the magnitude of escape velocity from earth.

(a) Determine the height of the satellite above the earth's surface,

(b) If the satellite is stopped suddenly in its orbit and allowed to fall freely on the earth

Find the speed with which it hits the surface of earth. $\left[g = 9.8 \text{ m/s}^2 \text{ and } R_e = 6400 \text{ km}\right]$

Solution

(a) Let $h =$ height of satellite from the surface of earth.

$\therefore$ Orbital velocity $v_0 = \sqrt{\frac{GM}{r}} = \sqrt{\frac{GM}{R_e + h}}$.

Given, $\quad v_0 = \frac{1}{2}v_e = \frac{1}{2}\sqrt{\frac{2GM}{R_e}}$

$$\therefore \quad \frac{GM}{2R_e} = \frac{GM}{R_e + h}$$

$$\therefore \quad h = R_e = 6400 \text{ km}.$$

(b) Conserving energy, we have

$$0 + \left(-\frac{GMm}{R_e + h}\right) = \frac{1}{2}mv^2 + \left(-\frac{GMm}{R_e}\right)$$

$$\therefore \quad v^2 = 2GM\left(\frac{1}{R_e} - \frac{1}{R_e + h}\right)$$

$$= 2GM\left(\frac{1}{R_e} - \frac{1}{2R_e}\right) = \frac{GM}{R_e}$$

$$= gR_e$$

$$\therefore \quad v = \sqrt{gR_e} = \sqrt{9.8 \times 6.4 \times 10^6} = 7.9\,\text{km/s}.$$

Illustration 13

Two satellites of same mass are launched in the same orbit round the earth so as to orbit in opposite directions. They collide inelastically and stick together as wreckage. Find the total energy of the system

(a) just before the collision, and
(b) just after the collision.

Also describe the subsequent motion of the wreckage.

Solution

The total energy $(\text{KE} + \text{PE})$ of a satellite moving in an orbit of radius r is given by

$$E = -\frac{GMm}{2r}.$$

(a) Total energy just before collision is

$$E_i = E_1 + E_2 = 2E = -\frac{GMm}{r}.$$

(b) Conserving linear momentum

$$mv_0 - mv_0 = 2mv$$

or $\quad v = 0$.

Thus the wreckage comes to rest just after collision and the energy of the wreckage will be totally potential, given by

$$E_f = -\frac{GM(2m)}{r} = -\frac{2GMm}{r}.$$

In the subsequent motion, the stationary system in its orbit will radially move towards the earth under gravity.

Illustration 14

The distance (R) of a satellite from the centre of a spinning planet is such that it appears to be stationary with respect to an observer on its surface. The rate of spin of the planet is now suddenly halved. By what factor should the distance of the satellite from the centre of the planet be changed for the satellite to appear sationary relative to the surface of the planet?

(a) $(2)^{1/3}$ times (b) $(2)^{3/2}$ times

(c) $(2)^{2/3}$ times (d) 4 times

Solution

$$T^2 \propto r^3$$

$$\therefore \quad T \propto r^{3/2}$$

or $\quad \omega \propto r^{-3/2}$.

or $\quad r \propto \omega^{-2/3}$

Here $\quad \frac{r_i}{r_f} = \left(\frac{\omega_i}{\omega_f}\right)^{-2/3} = \left(\frac{\omega_f}{\omega_i}\right)^{2/3} = \left(\frac{1}{2}\right)^{2/3}$

$\Rightarrow \quad \Rightarrow \quad r = 2^{2/3}$ times initial radius.

Ans. (c)

Data for Illustrations 15 and 16

Two satellites S_1 and S_2 revolve round a planet in coplanar circular orbits in the same sense. Their periods of revolutions are 1 hour and 8 hour respectively. The radius of the orbit of $S_1 = 10^4$ km. When S_2 is closed to S_1

Illustration 15

The speed of S_2 relative to S_1 is

(a) $\pi \times 10^4$ km/h (b) $2\pi \times 10^4$ km/h

(c) $\frac{\pi}{3} \times 10^4$ km/h (d) $\frac{\pi}{2} \times 10^4$ km/h

Solution

Let the radius of the orbit of $S_2 = r_2$ km.

$$T^2 = \frac{4\pi^2}{GM} r^3$$

Hence $T_1^2 = \frac{4\pi^2}{GM} r_1^3$ and $T_2^2 = \frac{4\pi^2}{GM} r_2^3$

$$\therefore \quad \frac{T_1^2}{T_2^2} = \frac{r_1^3}{r_2^3} \text{ or } \frac{1}{64} = \left(\frac{10^4\text{ km}}{r_2\text{ km}}\right)^3$$

or $\quad \frac{1}{4} = \frac{10^4}{r_2}$ or $r_2 = 4 \times 10^4$ km

Now $\quad v_1 = \frac{2\pi r_1}{T_1} = \frac{2\pi \times 10^4}{1} = 2\pi \times 10^4$ km per hour

and $\quad v_2 = \frac{2\pi r_2}{T_2} = \frac{2\pi \times 4 \times 10^4}{8} = \pi \times 10^4$ km per hour

$$|v_2 - v_1| = \pi \times 10^4 \text{ km per hour}$$

Ans. (a)

Illustration 16

The angular speed of S_2 as actually observed by an astronaut in S_1.

(a) $\frac{\pi}{2}$ rad/h (b) $-\frac{\pi}{3}$ rad/h

(c) $\frac{\pi}{4}$ rad/h (d) $-\frac{\pi}{2}$ rad/h

Solution

$$\omega_2 = \frac{|v_2 - v_1|}{(r_2 - r_1)} = -\frac{\pi \times 10^4}{(4\times 10^4 - 10^4)} = -\frac{\pi}{3} \text{ radian/hour}$$

Ans. (b)

Escape Velocity

It is the minimum velocity v_e given to an object so that it escapes the gravitational pull of the planet. As the object moves away the kinetic energy is reduced and the potential energy increases from negative value to zero at infinity. Thus using energy conservation

$$(E_k + E_p)_{\text{at surface}} = (E_k + E_p)_{\text{at infinity}}$$

$$\therefore \quad \frac{1}{2}mv_e^2 + \left(-\frac{GMm}{R}\right) = 0 + 0 = 0$$

$$\Rightarrow \quad v_e = \sqrt{\frac{2GM}{R}}.$$

Illustration 17

A rocket is fired with a speed $v = 2\sqrt{\frac{GM}{R}}$ near the earth's surface in the vertically upward direction.

(a) Will it escape earth's gravity?

(b) Find the speed of the rocket in interstellar space if it escapes earth's gravity.

Solution

(a) The minimum speed to escape earth's gravity, v_e, is obtained by energy conservation principle.

We have $\frac{1}{2}mv_e^2 - \frac{GMm}{R} = 0$

$$\Rightarrow \quad v_e = \sqrt{\frac{2GM}{R}}.$$

Here the speed of projection

$$v = 2\sqrt{\frac{GM}{R}} > v_e.$$

$\therefore$ The rocket will escape earth's gravity.

(b) The potential energy beyond earth's field in the interstellar space is zero. Hence conserving energy,

We have $\frac{1}{2}mv^2 - \frac{GMm}{R} = \frac{1}{2}mv'^2$ where v' is the speed in the interstellar space beyond earth's gravity.

$$\text{or} \quad v'^2 = v^2 - \frac{2GM}{R} = 4\frac{GM}{R} - 2\frac{GM}{R}$$

$$\Rightarrow \quad v' = \sqrt{\frac{2GM}{R}}.$$

It is not dependent on the direction in which the projectile is fired.

Illustration 18

A body is projected vertically upward from the surface of earth with a speed kV_e, where $k < 1$ and V_e is the escape speed. Neglecting air resistance, find the maximum height measured from the centre of the earth to which the rocket will rise. Express your answer in terms of the radius of the earth $= R$. Also find the height h above earth's surface if $k = \frac{1}{9}$.

Solution

Applying the energy conservation principle,

$$\frac{1}{2}mv^2 + \left(-\frac{GMm}{R}\right) = 0 + \left(-\frac{GMm}{r}\right)$$

Given that $v = kv_e = k\sqrt{\frac{2GM}{R}}$, we have

$$\frac{1}{2}mk^2\frac{2GM}{R} - \frac{GMm}{R} = -\frac{GMm}{r}$$

$$\text{or} \quad \frac{1}{r} = \frac{1}{R} - \frac{k^2}{R} = \frac{1-k^2}{R}$$

$$\therefore \quad r = \frac{R}{1-k^2}$$

$$\text{If} \quad k = \frac{1}{9},\ r = \frac{81}{80}R$$

$$\therefore \quad h = r - R = \frac{R}{80}$$

SOLVED EXAMPLES

Example 1

A mass M is to be divided into two parts m_1 and m_2 such that the gravitational attraction between the two parts is maximum for a given separation. Find the ratio $\frac{m_1}{m_2}$. Assume that both the parts are spheres.

Solution

Let $m_1 = m$ and $m_2 = M - m$.

At a given separation r the force of attraction between them is

$$F = \frac{Gm(M-m)}{r^2} = \frac{G}{r^2}(Mm - m^2).$$

For F to be maximum,

$$\frac{dF}{dm} = 0$$

or $$\frac{G}{r^2}(M - 2m) = 0$$

or $$M - 2m = 0$$

$$\therefore \quad m = \frac{M}{2}.$$

Hence $m_1 = m_2 = \frac{M}{2}$, or $\frac{m_1}{m_2} = 1:1$, the mass has to be divided into two equal parts for maximum force of attraction.

Example 2

Three particles, each of mass m, are placed at the vertices of an equilateral triangle of side a.

(i) At which point is the gravitational field due to the given system zero?

(ii) Find the potential energy of the system when a fourth particle of equal mass m is placed at the point of zero field determined in (i).

Solution

(i) The field at the centre of symmetry due to a given mass distribution is always zero. For a two-particle system the field is zero at the mid-point. For a three particle system the field **is zero at the centroid** of the equilateral triangle. A uniform circular wire will have the field equal to zero at its centre.

(ii) Potential energy of the system with three particles at the vertices and fourth at the centroid is

$$U = U_{12} + U_{13} + U_{14} + U_{23} + U_{24} + U_{34}$$

$$= (U_{12} + U_{13} + U_{23}) + (U_{14} + U_{24} + U_{34})$$

$$= \left(-\frac{3Gm^2}{a}\right) + \left(-\frac{3Gm^2}{r}\right)$$

$$= -\frac{3Gm^2}{a} - \frac{3Gm^2}{a/\sqrt{3}} = -\frac{3Gm^2}{a}(1 + \sqrt{3}).$$

Example 3

If the radius of the moon be $\frac{1}{4}$th the radius of earth and its mass $\left(\frac{1}{80}\right)$th the mass of earth, find the value of $|\vec{g}|$ on the surface of moon. Also find the escape velocity on the surface of moon if it is 11.2 km/s on the surface of the earth. $\left[g \text{ on earth} = 9.8 \text{ m/s}^2\right]$

Solution

$$g = \frac{GM}{R^2}$$

$$\therefore \quad \frac{g_m}{g_e} = \frac{M_m}{M_e} \times \left(\frac{R_e}{R_m}\right)^2 = \frac{1}{80} \times (4)^2 = \frac{1}{5}$$

$$\therefore \; g \text{ on the surface of moon} = \frac{g_e}{5} = \frac{9.8}{5} = 1.96 \text{ m/s}^2$$

Again, escape velocity $= \sqrt{\frac{2GM}{R}}$

$$\therefore \quad \frac{v_m}{v_e} = \sqrt{\frac{M_m}{M_e} \times \frac{R_e}{R_m}} = \sqrt{\frac{1}{80} \times 4} = \frac{1}{\sqrt{20}}.$$

$$\therefore \text{ Escape velocity from moon} = \frac{v_e}{\sqrt{20}} = \frac{11.2}{\sqrt{20}} = 2.5 \text{ km/s}.$$

Example 4

The period of revolution of moon around the earth is 27.3 days and the radius of its circular orbit is 3.9×10^5 km. Estimate the mass of the earth. $\left[G = 6.67 \times 10^{-11} \text{ Nm}^2 \text{ kg}^{-2}\right]$.

Solution

The centripetal force required by the moon is provided by gravitational attraction between the moon and the earth.

Hence $$\frac{mv^2}{r} = \frac{GMm}{r^2}.$$

or, $$\frac{1}{r}\left(\frac{2\pi r}{T}\right)^2 = \frac{GM}{r^2}$$

$$\therefore \quad M = \frac{4\pi^2 r^3}{GT^2} \quad (1)$$

Here $r = 3.9 \times 10^5 \text{ km} = 3.9 \times 10^8$ m

$$T = 27.3 \text{ days} = 27.3 \times 24 \times 3600 \text{ secs.}$$

Substituting the values in Eq. (1),

$$M = \frac{4 \times (3.14)^2 \times (3.9 \times 10^8)^3}{(6.67 \times 10^{-11}) \times (27.3 \times 24 \times 3600)^2}$$

$$= 6.31 \times 10^{24} \text{ kg}$$

Example 5

A planet of mass m revolves around the sun in elliptical orbit with the sun at one of the foci. The maximum and minimum distance from the sun are r_1 and r_2 respectively. Find the angular momentum of the planet relative to the centre of the sun.

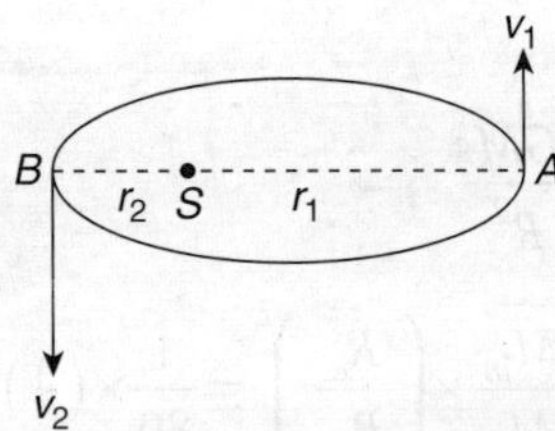

Solution

Since the gravitational force field is conservative, we conserve the total mechanical energy.

Hence, $\frac{1}{2}mv_1^2 - \frac{GMm}{r_1} = \frac{1}{2}mv_2^2 - \frac{GMm}{r_2}$.

or $$v_2^2 - v_1^2 = 2GM\left(\frac{r_1 - r_2}{r_1 r_2}\right). \quad ...(1)$$

As the force acting on the planet is a central force (directed along the line joining the sun and the planet), we conserve angular momentum.

Hence, $mv_1r_1 = mv_2r_2$.

or $$v_2 = \frac{v_1 r_1}{r_2}.$$

Substituting v_2 in Eq. (1) we get,

$$v_1 = \sqrt{\frac{2GMr_2}{r_1(r_1+r_2)}}.$$

$\therefore$ The angular momentum $L = mv_1r_1 = m\sqrt{\frac{2GMr_1r_2}{(r_1+r_2)}}$.

Example 6

A spaceship is launched into a circular orbit close to the earth's surface. What additional speed should now be imparted to the spaceship in that orbit to overcome the gravitational pull $\left(R = 6400\,\text{km}, g = 9.8\,\text{m/s}^2\right)$?

Solution

Orbital speed close to the earth's surface is

$$v_0 = \sqrt{\frac{GM}{R}} = \sqrt{gR}. \quad \left[\because GM = gR^2\right]$$

The escape speed v_e is given by

$$v_e = \sqrt{\frac{2GM}{R}} = \sqrt{2gR}.$$

Hence the additional speed required is

$$v = v_e - v_0 = \sqrt{gR}\left(\sqrt{2}-1\right)$$

$$= \sqrt{9.8\times 6400\times 10^3}\left(\sqrt{2}-1\right)$$

$$= 3.28\,\text{km/s}.$$

Example 7

Neptune is at an average distance of 4.5×10^9 km from the sun. Estimate the length of the Neptunian year given that the earth is 1.50×10^8 km from the sun on the average.

Solution

Using Kepler's third law,

$$\frac{T_E^2}{r_E^3} = \frac{T_N^2}{r_N^3}$$

Here $T_E = 1$ year, $r_E = 1.50\times 10^8$ km, $r_N = 4.5\times 10^9$ km

$$\therefore \quad \frac{1}{\left(1.50\times 10^8\right)^3} = \frac{T_N^2}{\left(4.5\times 10^9\right)^3}$$

or $$T_N^2 = (30)^3$$

$$\Rightarrow \quad T_N = 30\sqrt{30} \text{ years} = 164 \text{ years}.$$

The observed value of T_N (1 Neptunian year) = 164.8 years (Earth-year).

Example 8

If a tunnel is dug across the earth along its diameter and a block is released from the surface, find the speed of the block when it reaches the centre of the earth.

Solution

Here we conserve the total mechanical energy in earth's gravity.

$$\Rightarrow \quad (\text{KE}+\text{PE})_{\text{at } R} = (\text{KE}+\text{PE})_{\text{at } r<R}$$

$$\therefore \quad 0 + \left(-\frac{GMm}{R}\right) = \frac{1}{2}mv^2 + \left[\frac{-GMm\left(3R^2 - r^2\right)}{2R^3}\right].$$

At the centre of the earth, $r = 0$

$$\therefore \quad -\frac{GMm}{R} = \frac{1}{2}mv^2 - \frac{3}{2}\frac{GMm}{R}$$

or $$\frac{1}{2}v^2 = \frac{GM}{2R}$$

$$\Rightarrow \quad v = \sqrt{\frac{GM}{R}} = \sqrt{gR}.$$

Example 9

The masses and radii of a double star are M, R and $2M$ and $R/2$ respectively. Their centers are at distance d apart. What is the minimum speed with which an object should be projected from the midpoint of the line joining their centres so as to escape to infinity. ($d >> 3R/2$)

Solution

Since the gravitational field is conservative, we have

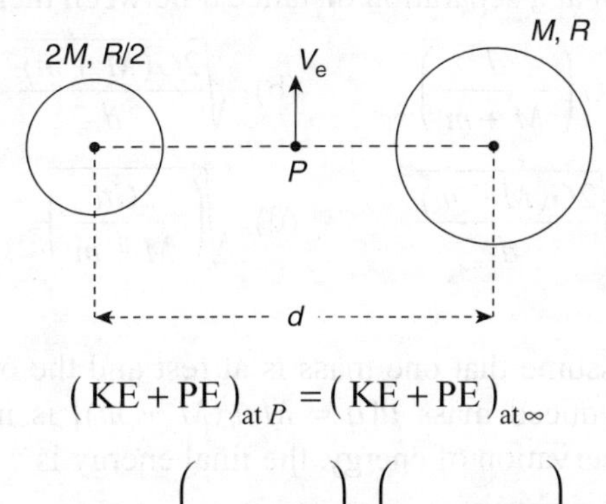

$$(\text{KE}+\text{PE})_{\text{at } P} = (\text{KE}+\text{PE})_{\text{at } \infty}$$

$$\Rightarrow \frac{1}{2}mv_e^2 + \left(-\frac{GMm}{\frac{d}{2}}\right) + \left(-\frac{2GMm}{\frac{d}{2}}\right) = 0$$

$$\therefore \quad v_e = 2\sqrt{\frac{3GM}{d}}.$$

Example 10

A particle is projected upward from the surface of earth (radius = R) with a speed equal to the orbital speed of a satellite near the earth's surface. The height to which it would rise is

(a) $\sqrt{2}R$ (b) $\frac{R}{\sqrt{2}}$

(c) R (d) $2R$

Solution

Orbital speed near the earth's surface is $v_0 = \sqrt{\frac{GM}{R}}$

Conserving energy,

$$\frac{1}{2}mv_0^2 - \frac{GMm}{R} = -\frac{GMm}{R+h}$$

or $$\frac{1}{2}\frac{GM}{R} - \frac{GM}{R} = -\frac{GM}{R+h} = -\frac{GM}{2R}$$

$$\Rightarrow \quad h = R.$$

Ans. (c)

Example 11

The minimum energy required to launch a satellite of mass m from the surface of earth of radius R in a circular orbit at an altitude $2R$, ignoring the earth's rotation, is (mass of earth is M)

(a) $\frac{5GmM}{6R}$ (b) $\frac{2GmM}{3R}$

(c) $\frac{GmM}{2R}$ (d) $\frac{GmM}{3R}$

Solution

Total energy of the satellite at earth is the potential energy $= -\frac{GMm}{R} = E_i$

The total energy of the satellite at a height $2R$ in the orbit of radius $3R$ is $-\frac{GMm}{2(3R)} = E_f$

Hence the additional energy required is

$$\Delta E = E_f - E_i = \frac{GMm}{R} - \frac{GMm}{6R} = \frac{5GMm}{6R}.$$

Ans. (a)

Example 12

If the period of revolution of an artificial satellite just above the earth's surface is T and the density of earth is ρ, then ρT^2

(a) is a universal constant whose value is $\frac{3\pi}{G}$

(b) is a universal constant whose value is $\frac{3\pi}{2G}$

(c) is proportional to radius of earth R

(d) is proportional to square of the radius of earth R^2

Here G = universal gravitational constant

Solution

Near the earth's surface, $v_0 = \sqrt{\frac{GM}{R}} = \frac{2\pi R}{T}$

$$\therefore \quad T^2 = \frac{4\pi^2 R^3}{GM} = \frac{4\pi^2 R^3}{G.\frac{4}{3}\pi R^3 \rho} = \frac{3\pi}{G\rho}$$

or $$\rho T^2 = \frac{3\pi}{G} = \text{Universal constant.}$$

Ans. (a)

Example 13

Imagine a light planet revolving around a very massive star in a circular orbit of radius R with a period of revolution T. If the gravitational force of attraction between the planet and the stat is proportional to $R^{-5/2}$, then T^2 is proportional to

(a) R^3 (b) $R^{7/2}$

(c) $R^{3/2}$ (d) $R^{3.75}$

Solution

Gravitational force between star and planet = Centripetal force on the planet

$$\therefore \quad \frac{GMm}{R^{5/2}} = \frac{mv^2}{R} \quad \text{or} \quad v^2 = \frac{GM}{R^{3/2}}$$

Let T be the time period of the planet around the star, then $T = 2\pi R/v$

Now $T^2 = \frac{4\pi^2 R^2}{v^2} = \frac{4\pi^2 R^2 \times R^{3/2}}{GM} = \frac{4\pi^2 R^{7/2}}{GM}$

$\therefore \quad T^2 \propto R^{7/2}$

Ans. (b)

Example 14

The magnitude of the gravitational field at distances r_1 and r_2 from the centre of a uniform sphere of radius R and mass M are F_1 and F_2 respectively. Then

(a) $\frac{F_1}{F_2} = \frac{r_1}{r_2}$ if $r_1 < R$ and $r_2 < R$

(b) $\frac{F_1}{F_2} = \frac{r_1^3}{r_2^3}$ if $r_1 > R$ and $r_2 > R$

(c) $\frac{F_1}{F_2} = \frac{r_1}{r_2}$ if $r_1 > R$ and $r_2 > R$

(d) $\frac{F_1}{F_2} = \frac{r_1^2}{r_2^2}$ if $r_1 < R$ and $r_2 < R$

Solution

Inside the uniform solid sphere,

$$F = (GM/R) \times r$$

$$\therefore \quad F_1 = \frac{GM}{R} r_1 \text{ and } F_2 = \frac{GM}{R} r_2$$

or

$$\frac{F_1}{F_2} = \frac{r_1}{r_2}$$

Ans. (a)

Example 15

Two particles of mass m and M respectively are initially at rest and infinitely separated from each other. Due to mutual interaction, they approach each other. Their relative velocity of approach at a separation distance d between them is

(a) $2G\left(\frac{d}{M+m}\right)$ (b) $\sqrt{\frac{2G(M+m)}{d}}$

(c) $\sqrt{\frac{2G(M-m)}{d}}$ (d) $\sqrt{\left(\frac{2Gd}{M+m}\right)}$

Solution

Here we assume that one mass is at rest and the other mass equal to reduced mass $\mu[\mu = Mm/(M + m)]$ is in motion. Using conservation of energy, the final energy is

$$\frac{1}{2}\mu v^2 - \frac{GMm}{d} = 0$$

or $$v^2 = \frac{2GMm}{\mu d} = \frac{2GMm(M+m)}{mMd} = \frac{2G(M+m)}{d}$$

$$v = \sqrt{\left[\frac{2G(M+m)}{d}\right]}$$

Ans. (b)

POINTS TO REMEMBER

1. Gravitation is a universal property of all matter.
2. Newton's law of gravitation states that any two particles attract each other with a force given by
$$F = G.\frac{m_1 m_2}{r^2} \qquad \text{(A)}$$
where m_1 and m_2 are the gravitational masses of the two particles and r is the distance between them. G is a universal constant, called gravitational constant, and its value is experimentally found to be 6.67×10^{-11} N-m^2/kg^2 (or m^3/kg-s^2).
3. The mass defined by Eq. (A) is known as gravitation mass whereas mass defined by Newton's second law of motion is known as inertial mass and therefore we use a common word "mass" for both inertial and gravitational masses.
4. Equation (A) holds only for point masses. The extended bodies behave for an external point, as if their entire mass were concentrated at their centres of gravity.
5. In our daily life, we do not observe this gravitational force between two bodies given by Eq. (A), because of the negligibly small value of G.
6. The gravitational field of a particle of mass m is a region of space where any other particle will experience a force. The gravitational field strength (or intensity of gravitational field) at a point is the force acting on a particle of unit mass placed at that point. It is a vector quantity. The gravitational field strength at a distance r from a point mass m is given by
$$g = -\frac{Gm}{r^2}$$
or
$$\vec{g} = -\frac{Gm}{r^3}\vec{r}$$

The negative sign shows that the gravitational field is always attractive.

7. The gravitational potential (V) at a point is equal to the amount of work that has to be done to bring a particle of unit mass from infinity to that point. It is a scalar quantity.
The gravitational potential at a distance r from a point mass m is given by

$$V = \frac{Gm}{r}$$

In bringing the unit mass from infinity, work is done by the system and hence its potential energy decreases. Therefore, negative sign is used.

8. The potential at a point does not depend on the actual path followed in bringing the particle of unit mass from infinity up to that point. Thus gravitational force is a conservative force.

9. Gravitional potential in the case of a solid sphere at the centre $V_{\text{centre}} = -\frac{3}{2}\frac{GM}{R}$

10. The gravitational energy of a particle of mass m at a point where the gravitational potential is
$V = -\frac{GM}{R}$ is $U = -\frac{GMm}{R}$

11. For points lying outside or on the surface of the spherical shell or solid sphere, its entire mass may be assumed to be concentrated at its centre.

12. The weight of a body is the force with which the earth attracts it towards its centre.
The force of attraction between the earth of mass M and radius R and a particle of mass m at a height h above the surface of the earth is given by

$$F = \frac{GMm}{(R+h)^2} \approx \frac{GMm}{R^2} \text{ as } h << R$$

If g be the acceleration of the particle then

$$g = \frac{F}{m} = \frac{GM}{R^2}$$

This acceleration is known as acceleration due to gravity.

13. Since the earth is not a perfect sphere, its radius at the pole and the equator are different. If R_p and R_e be the radii of the earth at the pole and at the equator and g_p and g_e be the corresponding values of acceleration due to gravity at those places, then

$$g_p = \frac{GM}{R_p^2} \text{ and } g_e = \frac{GM}{R_e^2}$$

As, $R_p < R_e, g_p > g_e$

14. Because of the rotation of the earth around its axis, there is an apparent change in the value of acceleration due to gravity.
The acceleration due to gravity at a place having latitude λ is given by

$$g' = g - \omega^2 R \cos^2 \lambda$$

At the equator, $\lambda = 0 \Rightarrow g' = g - \omega^2 R$
At the poles, $\lambda = 90° \Rightarrow g' = g$
Thus, the acceleration due to gravity is minimum at the equator and maximum at the poles.

15. If h is not small compared to R, then

$$g = \frac{GM}{(R+h)^2}$$

Hence, the value of acceleration due to gravity at a point varies with h.

16. Whether we go above the surface of earth or below the surface of earth the acceleration due to gravity decreases.

Above the surface of earth $g' = g\left(1 - \frac{2h}{R}\right)$. Below the surface of earth $g' = g\left(1 - \frac{d}{R}\right)$

17. Following are Kepler's laws of planetary motion:
   a. All planets move in elliptical orbits with the sun at one of its foci.
   b. The radius vector from the sun to the planet sweeps out equal area in equal interval of time. In other words, the areal velocity of the radius vector joining the sun to the planet is a constant.
   c. The square of the time period of the planet round the sun is directly proportional to the cube of the semi-major axis of the ellipse. If a be the semi-major axis of the ellipse, then, $T^2 \propto a^3$

18. The minimum velocity with which a body projected from the surface of a planet goes out of its gravitational field and never comes back to it, is known as escape velocity of that planet.
If M be the mass and R be the radius of the planet, then escape velocity of that planet is given by

$$v_e = \sqrt{\frac{2GM}{R}} = \sqrt{2gR}$$

where $g = \frac{GM}{R^2}$ is the acceleration due to gravity of that planet.
For the earth, $v_e = 11.2$ km/s
For the moon, $v_e = 2.38$ km/s
The low value of escape velocity on the moon explains the absence of atmosphere on its surface.

19. A satellite is a small body revolving around a large body under the gravitational attraction of the latter. The necessary centripetal force is provided by the gravitational force of attraction between the large body and the satellite.
The speed at which rotational equilibrium is attained is called orbital speed of satellite and is given by

$$v_0 = \sqrt{\frac{GM}{R}} = \sqrt{gR}$$

where M and R are the mass of the planet and radius of the elliptical orbit.

20. For a satellite to escape from its orbit, its velocity should be increased by a factor of $\sqrt{2}$.

21. A geostationary satellite is one whose time period is 24 hour and it moves in the same sense as the earth so as to be always overhead at a particular place on the equator. As seen from the earth, the satellite will appear to the stationary.
The height of a geostationary satellite above the surface of the earth is 3.6×10^4 km.

22. Movement of planets and satellite

$$G.M = \frac{4\pi^2 r^3}{T^2}$$

M = Mass of the sun (for planet) or of earth (for an earth satellite)
r = Radius of the orbit (or greater semi-axis if the orbit is elliptical)
T = Period of revolution.

23. Artificial satellite in a geostationary orbit

$$r^3 = G.\frac{M.T^2}{4\pi^2}$$

or

$$r = \left[\frac{T^2 gR^2}{4\pi^2}\right]^{\frac{1}{3}}$$

M = mass of earth $(5.98\times10^{24}\ kg)$
T = Period of rotation of earth and revolution of satellite (86400 sec)
r = Radius of the fixed orbit around the earth
If we substitute the values we find r = 42,250 km.

24. Height of geostationary satellite

$$h = R_0 - R$$

$$= (42{,}250 - 6380)\text{ km}$$

$$= 35{,}870\text{ km.}$$

25. Energy of the planet and satellite

$$E = \text{K.E.} + \text{P.}$$

$$= -G\frac{M.m}{2r}$$

M = Mass of sun (for planets) or of earth (for earth satellite)
m = Mass of earth or satellite
r = Radius of orbit

EXERCISES

1. If the gravitational force of attraction between any two bodies is proportional to R^3 instead of R^2, and directly proportional to the product of masses of the two bodies, then
(a) the weight of the bodies on earth's surface will be more than their present value
(b) the weight will be same as their present value
(c) the period of the planet round the sun will be such that $T \propto R^2$, where R is the mean distance
(d) $T^2 \propto R^{4.5}$

2. The moon would escape if its speed were increased by
(a) 39% (b) 42%
(c) 47% (d) 49%

3. What would be duration of a year if the distance between the earth and the sun gets doubled?
(a) 1.92 years (b) 2.83 years
(c) 7.23 years (d) 4.23 years

4. The magnitude of gravitational field at distance r_1 and r_2 from the centre of a uniform sphere of radius R and mass M are I_1 and I_2 respectively. Find the ratio of (I_1 / I_2) if (i) $r_1 > R$ and $r_2 > R$ and (ii) $r_1 < R$ and $r_2 < R$.
(a) $\left(\frac{r_2}{r_1}\right)^2, \left(\frac{r_1}{r_2}\right)^3$ (b) $\left(\frac{r_2}{r_1}\right)^3, \left(\frac{r_2}{r_1}\right)$
(c) $\left(\frac{r_2}{r_1}\right)^2, \left(\frac{r_1}{r_2}\right)$ (d) $\left(\frac{r_2}{r_1}\right), \left(\frac{r_1}{r_2}\right)^3$

5. Two planets have same density but different radii. The acceleration due to gravity would be

(a) same on both the planets
(b) greater on the smaller planet
(c) greater on the larger planet
(d) dependent of the distance of planet from the sun.

6. If a planet of given density were made larger, its force of attraction for an object on its surface would increase because of the planet's greater mass but would decrease because of the greater distance from the object to the centre of the planet. Which effect predominates?

(a) Increase in mass
(b) Increase in radius
(c) Both affect attraction equally
(d) None of these.

7. When astronauts return from the moon they have to live for some days in a caravan with the air pressure inside lower than that outside. The reason for this is that

(a) it is the easier for them to move about
(b) the men have to get used to earth
(c) they need to get used to normal air pressure gradually
(d) in the event of a leak, any bacteria would be forced back inside.

8. The time period of a satellite in a circular orbit of radius R is T. The radius of the orbit in which time period is $8T$ is

(a) $2R$ (b) $3R$
(c) $4R$ (d) $5R$

9. The acceleration due to gravity g and mean density of the earth ρ are related by which of the following relations? (where G is the gravitation constant and R is radius of the earth)

(a) $\rho = \frac{4\pi g R^2}{3G}$ (b) $\rho = \frac{4\pi g R^3}{3G}$

(c) $\rho = \frac{3g}{4\pi G R}$ (d) $\rho = \frac{3g}{4\pi G R^3}$

10. If both the mass and radius of the earth decrease by 1% the value of

(i) acceleration due to gravity would decrease by nearly 1%
(ii) acceleration due to gravity would increase by 1%
(iii) escape velocity from the earth's surface would decrease by 1%
(iv) the gravitational potential energy of a body on earth's surface will remain unchanged.

(a) (i) and (ii) are correct
(b) (ii) is correct
(c) (ii) and (iv) are correct
(d) None of these is correct.

11. How much faster than its present rate ω should the earth rotate about its axis in order that the weight of the body on the equator may be zero? Radius of earth = 6400 km.

(a) 2ω (b) 15ω
(c) 17ω (d) 19ω

12. A geostationary satellite orbits around the earth in a circular orbit of radius 36,000 km. Then, the time period of a spy satellite orbiting a few hundred kilometres above the earth's surface $(R_{\text{earth}} = 6400 \text{ km})$ will approximately be

(a) $\frac{1}{2}$ hr (b) 1 hr
(c) 2 hr (d) 4 hr

13. With what angular velocity should the earth rotate so that a body at the latitude of 45° would be weightless?

(a) $\omega = \sqrt{\frac{2g}{3R}}$ (b) $\omega = \sqrt{\frac{g}{3R}}$

(c) $\omega = \sqrt{\frac{2g}{R}}$ (d) $\omega = \sqrt{\frac{3g}{R}}$

14. If the radius of the earth were to shrink by one percent, its mass remaining the same, the acceleration due to gravity on the earth's surface would

(a) decrease by 2% (b) remain unchanged
(c) increase by 2% (d) decrease by 9.8%.

15. The weight of a body decreases by 51% of the weight on the earth's surface at a height of

(a) $\frac{10}{7}R$ (b) $\frac{7}{10}R$

(c) $\frac{3}{7}R$ (d) $\frac{2}{7}R$

16. At what height above the earth's surface is the acceleration due to gravity 4% less than its value at the surface of the earth?

(a) $\frac{R}{25}$ (b) $\frac{R}{40}$

(c) $\frac{R}{50}$ (d) $\frac{R}{96}$

17. At what depth below the earth's surface does the acceleration due to gravity fall to 2% of its value at the earth's surface?

(a) $\frac{R}{50}$ (b) $\frac{R}{25}$

(c) $\frac{49R}{50}$ (d) $\frac{4R}{25}$

18. If the radius of the earth decreases by 2% and the mean density of the earth increases by 2%, the percentage change in the orbital velocity of a satellite revolving close to the earth's surface will be

(a) 4% increase (b) 2% decrease
(c) 1% decrease (d) 1% increase.

19. The period of the moon's rotation round the earth is nearly 29 days. If the moon's mass were 2-fold its present value, and all other things remain unchanged, the period of moon's rotation would be nearly

(a) $29\sqrt{2}$ days (b) $\frac{29}{\sqrt{2}}$ days
(c) 29×2 days (d) 29 days.

20. A satellite of mass m is orbiting in a circular path of radius r around the earth. If its kinetic energy is E, then its angular momentum is

(a) $\left(2mr^2E\right)^{1/2}$ (b) $\left(\frac{1}{2}mr^2E\right)^{1/2}$
(c) $\left(\frac{2}{3}mr^2E\right)^{1/2}$ (d) $\left(mr^2E\right)^{1/2}$

21. If the potential energy is at infinity, the total energy of an orbiting satellite is

(a) negative of its kinetic energy
(b) positive of its kinetic energy
(c) positive of its potential energy
(d) negative of its potential energy.

22. P is a point at a distance r from the centre of a solid sphere of radius R_0. The gravitational potential at P is V. If V is plotted as a function of r, then the curve representing the plot correctly is:

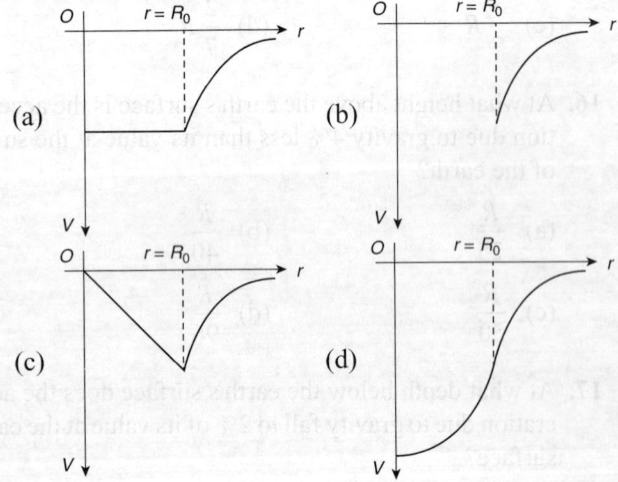

23. A satellite revolving around earth loses some energy due to collision. What would be the effect on its velocity and distance from the centre of earth?

(a) Velocity increases and distance decreases
(b) Both velocity and distance increase
(c) Both velocity and distance decrease
(d) Velocity decreases and distance increases.

24. Two identical trains P and Q move with equal speeds on parallel tracks along the equator. P moves from east to west and Q from west to east. Then,

(a) train P exerts greater force on track
(b) train Q exerts greater force on track
(c) both exert equal force on track
(d) data is insufficient to arrive at a conclusion.

25. A particle of mass m is projected vertically. A uniform gravitational field $\vec{g}$ is acted vertically downwards. The most appropriate graph between magnitude of potential energy U and height h (<< radius of earth) is (assume U to be zero on surface of earth):

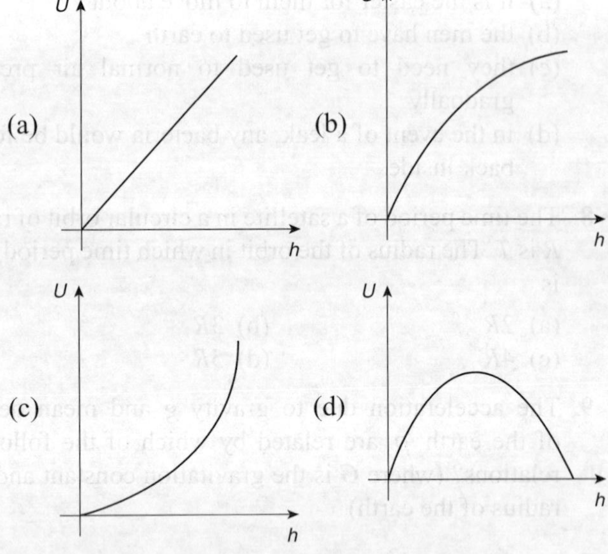

26. A satellite is to be stationed in an orbit such that it can be used for relay purpose (such a satellite is called a geostationary satellite). The conditions such a satellite should fulfill is/are:

(a) Its orbit must lie in equatorial plane.
(b) Its sense of rotation must be from east to west.
(c) Its orbital radius must be 44900 km.
(d) Its orbit must be elliptical.

27. Consider a planet in some solar system which has a mass double the mass of the earth and density equal to the average density of the earth. An object weighing W on earth will weigh:

(a) W (b) $2W$
(c) $W/2$ (d) $2^{1/3}\,W$

28. Figure shows the variation of energy with the orbit radius r of a satellite in a circular motion. Mark the correct statement:

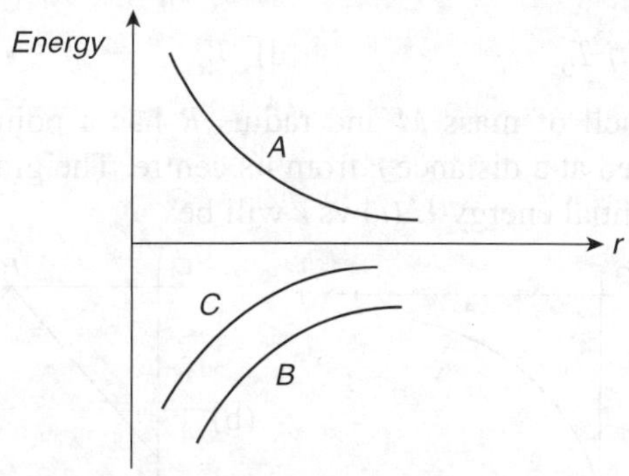

(a) C shows the total energy, B the kinetic energy and A the potential energy of the satellite
(b) A shows the kinetic energy, B the total energy and C the potential energy of the satellite
(c) A and B are the kinetic and potential energies and C the total energy of the satellite
(d) C and A are the kinetic and potential energies respectively and B the total energy of the satellite.

29. The largest and the shortest distance of the earth from the sun are r_1 and r_2. Its distance from the sun when it is at perpendicular to the major-axis of the orbit drawn around the sun is

(a) $\dfrac{r_1+r_2}{4}$ (b) $\dfrac{r_1+r_2}{r_1-r_2}$
(c) $\dfrac{2r_1r_2}{r_1+r_2}$ (d) $\dfrac{r_1+r_2}{3}$

30. A sphere of mass M and radius R_2 has a concentric cavity of radius R_1 as shown in figure. The force F exerted by the sphere on a particle of mass m located at a distance r from the centre of sphere varies as $(0 \le r \le \infty)$

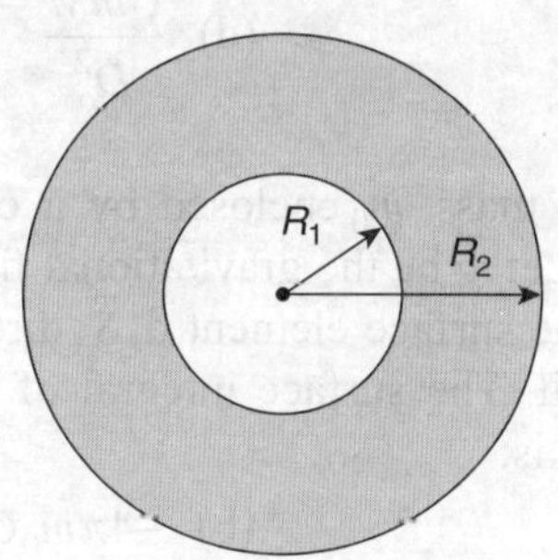

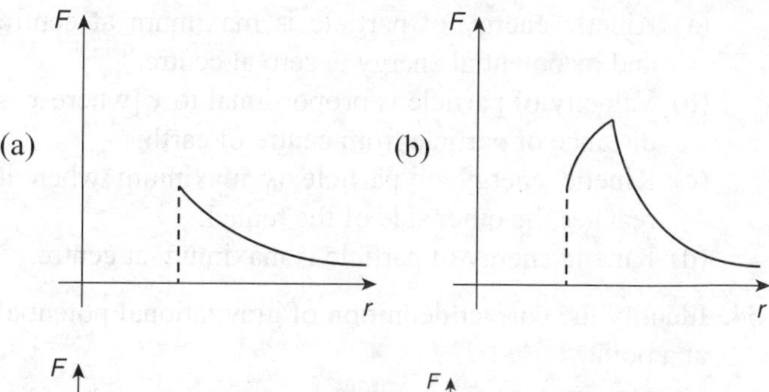

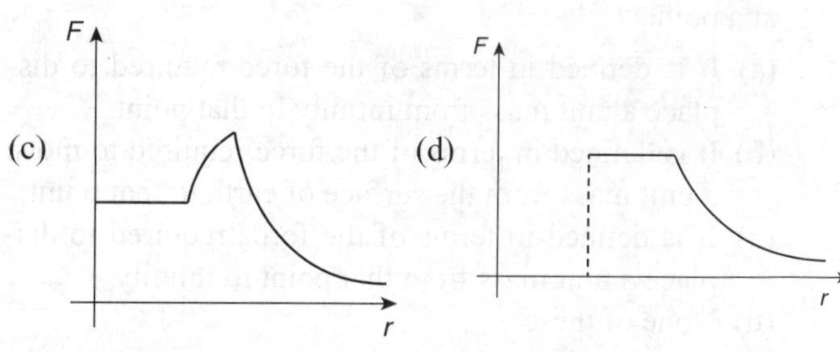

31. If we consider a solid sphere of mass M and radius R, then the potential energy of gravitational interaction of matter forming this solid sphere is:

(a) $\dfrac{GM^2}{R}$ (b) $-\dfrac{1}{2}\dfrac{GM^2}{R}$
(c) $-\dfrac{3}{5}\dfrac{GM^2}{R}$ (d) $-\dfrac{2}{3}\dfrac{GM^2}{R}$

32. A satellite is orbiting round the earth. While in orbit a small part separates from the satellite. The separated part
(a) falls directly to the earth
(b) moves in a spiral path and reaches the earth after few revolutions about the earth.
(c) continues to move in the same orbit
(d) moves gradually farther from the earth.

33. A tunnel is made inside earth passing through centre of earth. A particle is dropped from the surface of earth. Select the correct statement:

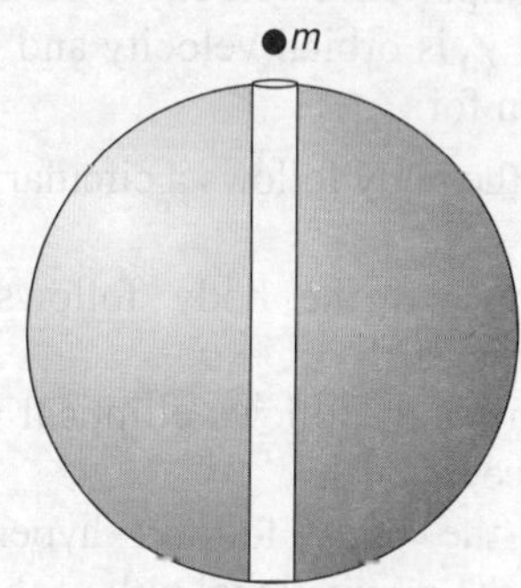

(a) Kinetic energy of particle is maximum at centre and its potential energy is zero at centre.
(b) Velocity of particle is proportional to x [where x is distance of particle from centre of earth.
(c) Kinetic energy of particle is maximum when it reaches the other side of the tunnel.
(d) Kinetic energy of particle is maximum at centre.

34. Identity the correct definition of gravitational potential at a point.
(a) It is defined in terms of the force required to displace a unit mass from infinity to that point.
(b) It is defined in terms of the force required to move a unit mass from the surface of earth to that point.
(c) It is defined in terms of the force required to displace a unit mass from that point to infinity.
(d) None of these.

35. If the sun were suddenly replaced by a black hole of one solar mass, what would happen to the earth's orbit immediately after the replacement?
(a) The earth would spiral into the black hole.
(b) The radius of the earth's orbit would be unchanged, but the period of the earth's motion would increase.
(c) The radius of the earth's orbit would be unchanged, but the period of the earth's motion would decrease.
(d) Neither the radius of the orbit nor the period would change.

36. A satellite is revolving round the earth in an orbit of radius r with time period T. If the satellite is revolving round the earth in an orbit of radius $r+\Delta r\,(\Delta r << r)$ with period $T+\Delta T\,(\Delta T << T)$ then,

(a) $\frac{\Delta T}{T}=\frac{3}{2}\frac{\Delta r}{r}$ (b) $\frac{\Delta T}{T}=\frac{2}{3}\frac{\Delta r}{r}$
(c) $\frac{\Delta T}{T}=\frac{\Delta r}{r}$ (d) $\frac{\Delta T}{T}=-\frac{\Delta r}{r}$

37. A body is imparted a velocity v from the surface of the earth. If v_0 is orbital velocity and v_e be the escape velocity, then for:
(a) $v=v_0$, the body follows a circular track around the earth.
(b) $v>v_0$ but $<v_e$ the body follows elliptical path around the earth
(c) $v<v_0$ the body follows elliptical path and returns to surface of earth
(d) $v<v_e$, the body follows hyperbolic path and escapes the gravitational pull of the earth.
(a) A, B (b) B, C
(c) A, B, C (d) A, B, C, D

38. The orbital period of revolution of a planet round the Sun is T_0. Suppose we make a model of solar system scaled down in the ratio η but of materials of the same mean density as the actual material of planet and the Sun has. The new orbital period is:
(a) ηT_0 (b) $\eta^2 T_0$
(c) $\eta^3 T_0$ (d) T_0

39. A shell of mass M and radius R has a point mass m placed at a distance r from its centre. The gravitational potential energy $U(r)$ vs r will be:

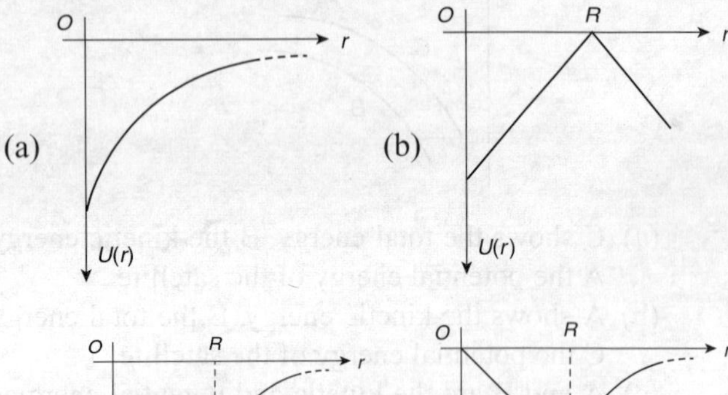

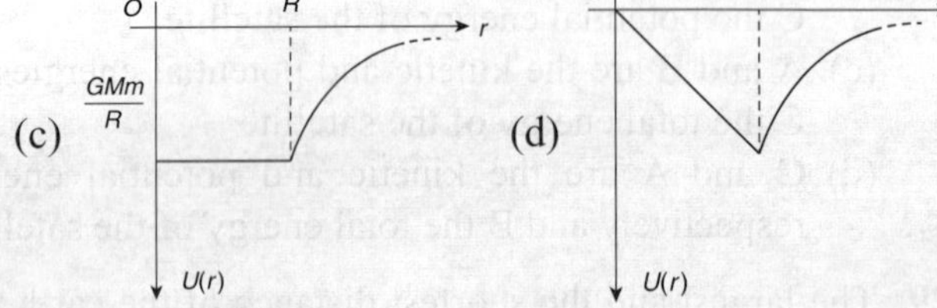

40. If the time of revolution of a satellite is T, then kinetic energy is proportional to:
(a) $1/T$ (b) $1/T^2$
(c) $1/T^3$ (d) $T^{-2/3}$

41. A spherical planet in space has a mass M_0 and diameter D_0. A particle of mass m falling freely near the surface of this planet will experience an acceleration due to gravity which is equal to:
(a) $\frac{GM_o}{D_o^2}$ (b) $\frac{4GmM_o}{D_o^2}$
(c) $\frac{4GM_o}{D_o^2}$ (d) $\frac{GmM}{D_o^2}$

42. Consider a mass m_0 enclosed by a closed imaginary surface S. Let $\vec{g}$ be the gravitational field intensity due to m_0 at the surface element $d\vec{S}$ directed as outward normal to it. The surface integral of the gravitational field over S is:
(a) $-m_0G$ (b) $-4\pi m_0G$
(c) $-\frac{m_0G}{4\pi}$ (d) None of these.

43. A solid sphere of uniform density and radius R exerts a gravitational force of attraction F_1 on a particle placed at P. The distance of P from the centre of the sphere is $2R$. A spherical cavity of radius $\frac{R}{2}$ is now made in the sphere. The sphere (with cavity) exerts gravitational force F_2 on the same particle at P. The radio $\frac{F_1}{F_2}$ is:

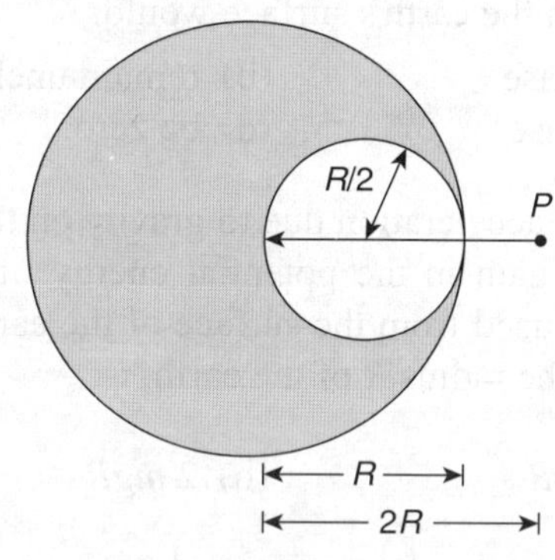

(a) $\frac{9}{7}$ (b) $\frac{7}{9}$
(c) $\frac{1}{2}$ (d) $\frac{4}{3}$

44. Two satellites are orbiting around the earth in circular orbits of the same radius. The mass of satellite A is five times greater than the mass of satellite B. Their periods of revolution are in the ratio:

(a) 1 : 1 (b) 1 : 10
(c) 5 : 1 (d) 1 : 5

45. When a planet moves around the sun,

(a) the angular momentum remains conserved
(b) the angular speed remains constant
(c) the linear momentum remains constant
(d) the linear velocity remains constant.

46. The value of universal gravitational constant G depends upon

(a) nature of material of two bodies
(b) heat content of two bodies
(c) acceleration of two bodies
(d) None of these.

47. In an elliptical orbit under gravitational force, in general,

(a) tangential velocity is constant
(b) angular velocity is constant
(c) radial velocity is constant
(d) areal velocity is constant.

48. If R be the radius of earth and g the value of acceleration due to gravity on the surface of earth then, the minimum speed with which the body must be thrown so that it reaches $R/4$ above the surface of earth is:

(a) $\sqrt{g}$ (b) $\sqrt{R}$
(c) $\sqrt{gR}$ (d) $\sqrt{\frac{2}{5}gR}$

49. The ratio of the kinetic energy required to be given to a satellite so that it escapes the gravitational field of earth to the kinetic energy required to put the satellite in a circular orbit just above the free surface of earth is:

(a) 1 (b) 2
(c) 3 (d) 5

50. If the value of acceleration due to gravity on the surface of earth is 10 ms^{-2}, then a girl can jump safely from a height of 2 m. If she has to jump on a planet where acceleration due to gravity is 0.2 m s^{-2}, then the safe height will be :

(a) 0.2 m (b) 10 m
(c) 2 m (d) 100 m.

51. The period of Moon's rotation around earth is nearly 29 days. If Moon's mass were 2-fold its present value, and all other things remains unchanged, the period of Moon's rotation would be nearly:

(a) $29\sqrt{2}$ days (b) $\frac{29}{\sqrt{2}}$ days
(c) 29×2 days (d) 29 days.

52. Two satellites A and B go around the earth in circular orbits at heights R_A and R_B respectively from the surface of earth. Assume the earth to be a uniform sphere of radius R_E. The ratio of the magnitudes of the velocity, $\frac{\upsilon_A}{\upsilon_B}$ of the satellite is:

(a) $\sqrt{\frac{R_B}{R_A}}$ (b) $\sqrt{\frac{R_B + R_E}{R_A + R_E}}$
(c) $\frac{R_B + R_E}{R_A + R_E}$ (d) $\left(\frac{R_A}{R_{AB}}\right)^2$.

53. Mars has about $\frac{1}{10}$ th as much mass as the earth and half as great a diameter. The acceleration of falling body on Mars is nearly:

(a) 9.8m s^{-2} (b) 1.96 m s^{-2}
(c) 3.92 m s^{-2} (d) 4.9 m s^{-2}

54. R is the radius of the earth and ω is its angular velocity and g_p is the value of g at poles. The effective value of g at the latitude $\lambda = 60°$ will be equal to

(a) $g_p - \frac{1}{4}R\omega^2$ (b) $g_p - \frac{3}{4}R\omega^2$

(c) $g_p - R\omega^2$ (d) $g_p + \frac{1}{4}R\omega^2$

55. A particle on earth's surface is given a velocity equal to its escape velocity. Its total mechanical energy will be

(a) negative (b) positive
(c) zero (d) infinite.

56. Three uniform spheres of mass M and radius R each are kept in such a way that each of them touches the other two. The magnitude of the gravitational force on any of the spheres due to the other two is:

(a) $\frac{\sqrt{3}}{4}\frac{GM^2}{R^2}$ (b) $\frac{3}{2}\frac{GM^2}{R^2}$

(c) $\frac{\sqrt{3}GM^2}{R^2}$ (d) $\frac{\sqrt{3}GM^2}{2R^2}$

57. The acceleration due to gravity on the moon is only one sixth that of earth. If the earth and moon are assumed to have the same density, the ratio of the radii of moon and earth will be:

(a) $\frac{1}{6}$ (b) $\frac{1}{(6)^{1/3}}$

(c) $\frac{1}{36}$ (d) $\frac{1}{(6)^{2/3}}$

58. For a given density of planet, the orbital period of a satellite near the surface of planet of radius R is proportional to:

(a) $R^{1/2}$ (b) $R^{3/2}$
(c) $R^{-1/2}$ (d) R^0

59. Which of the following is incorrect?

(a) The universe on the large scale is shaped by electromagnetic forces.
(b) The universe on the atomic scale is shaped by electromagnetic forces.
(c) The universe on the nuclear scale is shaped by electromagnetic force.
(d) The universe on the large scale is shaped by gravitational forces.

60. If G is the universal gravitational constant and ρ is the uniform density of a spherical planet, then shortest possible period of rotation of the planet can be:

(a) $\sqrt{\frac{\pi G}{2\rho}}$ (b) $\sqrt{\frac{3\pi G}{\rho}}$

(c) $\sqrt{\frac{\pi}{6G\rho}}$ (d) $\sqrt{\frac{3\pi}{G\rho}}$

61. If the radius of the earth were to shrink by one percent, its mass remaining the same, the acceleration due to gravity on the earth's surface would:

(a) decrease (b) remain unchanged
(c) increase (d) be zero.

62. If g is the acceleration due to gravity on the earth's surface, the gain in the potential energy of an object of mass m raised from the surface of the earth to a height equal to the radius R of the earth, is:

(a) $\frac{1}{2}mgR$ (b) $2\ mgR$

(c) mgR (d) $\frac{1}{4}mgR$

63. A simple pendulum has a time period T_1 when on the earth's surface and T_2 when taken to a height R above the earth's surface, where R is the radius of the earth. The value of T_2/T_1 is

(a) 1 (b) $\sqrt{2}$
(c) 4 (d) 2

64. The radius of the earth is 6400 km and g is 10 m s^{-2}. What should be the angular speed of earth so that 5 kg body placed at the equator of earth weighs zero ?

(a) $\frac{1}{800}$ rad s^{-1} (b) $\frac{1}{80}$ rad s^{-1}

(c) 5 rad s^{-1} (d) 10 rad s^{-1}

65. What keeps an earth's satellite moving in its orbit?

(a) Gravitational attraction between satellite and Earth
(b) Ejection of gases from the exhaust of the satellite
(c) Burning of fuel
(d) Gravitational attraction of Sun.

66. Two identical copper spheres, each of radius R, are in contact with each other. The gravitational force of attraction between two copper spheres is proportional to:

(a) R (b) $2R$
(c) R^2 (d) R^4.

67. If the change in the value of g at a height x ($<< R$) above the surface of the earth is same as at a depth y ($<< R$) below it, then
(a) $y = 2x$ (b) $x = 2y$
(c) $x = 3y$ (d) $y = 9.8x$.

68. Which of the following statements regarding an Earth's satellite is true?
(a) It has a powerful engine.
(b) It is a freely falling body.
(c) It is responsible for cloud formation.
(d) It sends cosmic rays into Earth's atmosphere.

69. A satellite is moving round the Earth in a circular orbit. The force acting on satellite is:
(a) centripetal only, provided by Earth's gravitational pull.
(b) centrifugal only, due to orbital motion.
(c) zero, because centripetal and centrifugal forces balance out.
(d) tangential to the orbit, provided by a machine fitted to the satellite.

70. A hollow sphere is held suspended. Sand is now poured into it in stages. The centre of gravity of the sphere with the sand:
(a) rises continuously
(b) remains unchanged in the process
(c) First rises and then falls to the original position
(d) First falls and then rises to the original position.

71. X and Y are two points at respective distances R and $2R$ from the centre of Earth, where R is greater than the radius of Earth. The gravitational potential at X is – 800 kJ kg^{-1}. When a 1 kg mass is taken from X to Y, the work done on the mass is
(a) 400 kJ (b) 200 kJ
(c) – 200 kJ (d) 80 kJ

72. The weight of an object in a coal mine, sea level and at the top of the mountain are respectively W_1, W_2 and W_3. Then
(a) $W_1 < W_2 > W_3$ (b) $W_1 = W_2 = W_3$
(c) $W_1 < W_2 < W_3$ (d) $W_1 > W_2 > W_3$

73. The escape velocity from earth for a rocket is 11.2 km s^{-1} Ignoring the air resistance, the escape velocity of 10 mg of sand from earth will be:
(a) 0.112 km s^{-1} (b) 11.2 km s^{-1}
(c) 1.12 km s^{-1} (d) None.

74. When a body is taken from equator to the poles, its weight
(a) remains the same
(b) increases
(c) decreases
(d) increases at North pole and decreases at South.

75. Earth is flattened at the poles and bulges at the equator. This is due to the fact that
(a) the earth revolves around the Sun in an elliptical orbit
(b) the angular velocity of spinning about its axis is more at the equator.
(c) the centrifugal force is more at the equator than at poles.
(d) None of these.

76. The mass of a planet is 6 times that of earth. Its radius is twice that of earth. It the escape velocity from earth is v, then the escape velocity from the planet is
(a) $\sqrt{3}\ v$ (b) $\sqrt{2}\ v$
(c) v (d) $\sqrt{5}\ v$

77. Two earth-satellites are revolving in the same circular orbit round the centre of the earth. They must have the same:
(a) mass (b) angular momentum
(c) kinetic energy (d) velocity

78. Three particles each having a mass of 100 g are placed on the vertices of an equilateral triangle of side 20 cm. The work done in increasing the side of this triangle to 40 cm is: $\left(G = 6.67 \times 10^{-11} \dfrac{\text{N-m}^2}{\text{kg}^2}\right)$
(a) 6.0×10^{-12} J (b) 2.25×10^{-10} J
(c) 4.0×10^{-11} J (d) 6.0×10^{-15} J

79. The minimum energy required to launch a satellite of mass m from the surface of earth of radius R in a circular orbit at an altitude $2R$ is: (mass of earth is M)
(a) $\dfrac{5GmM}{6R}$ (b) $\dfrac{2GmM}{3R}$
(c) $\dfrac{GmM}{2R}$ (d) $\dfrac{GmM}{3R}$

80. A planet is moving in an elliptical path around the sun as shown in figure. If speed of the planet in positions P and Q are v_1 and v_2 respectively with SP = r_1 and SQ = r_2, then v_1/v_2 is equal to:

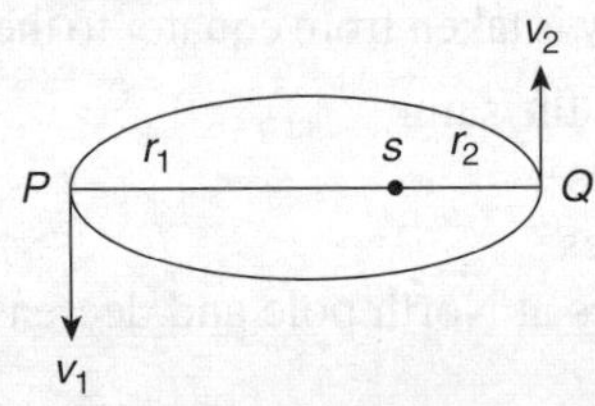

(a) $\frac{r_1}{r_2}$ (b) $\frac{r_2}{r_1}$

(c) constant (d) $\left(\frac{r_1}{r_2}\right)^2$

81. The potential at the surface of a planet of mass M and radius R is assumed to be zero. Choose the most appropriate option:

(a) The potential at infinity is $\frac{GM}{R}$

(b) The potential at the center of plane is $-\frac{GM}{2R}$

(c) Both (a) and (b) are correct.

(d) Both (a) and (b) are wrong.

82. If the distance between the earth and the sun were half its present value, the number of days in a year would have been:

(a) 64.5 (b) 129

(c) 182.5 (d) 730

83. Imagine a light planet revolving around a very massive star in a circular orbit of radius R with a period of revolution T. If the gravitational force of attraction between the planet and the star is proportional to $R^{-5/2}$, then:

(a) T^2 is proportional to R^2

(b) T^2 is proportional to $R^{7/2}$

(c) T^2 is proportional to $R^{3/2}$

(d) T^2 is proportional to $R^{3.75}$

84. If the gravitational force between two objects were proportional to $1/R$ (and not as $1/R^2$), where R is the distance between them, then a particle in a circular path (under such a force) would have its orbital speed v, proportional to

(a) R (b) R_0 (independent of R)

(c) $1/R^2$ (d) $1/R$

85. A satellite in force-free space sweeps stationary interplanetary dust at a rate of $dM/dt = \alpha v$, where M is mass and v is the speed of satellite and α is a constant. The acceleration of satellite is

(a) $\frac{-\alpha v^2}{2M}$ (b) $-\alpha v^2$

(c) $\frac{-2\alpha v^2}{M}$ (d) $\frac{-\alpha v^2}{M}$

86. The escape velocity from earth is 11.2 km/s. If a body is to be projected in a direction making an angle 45° to the vertical, then the escape velocity is

(a) 11.2×2 km/s (b) 11.2 km/s

(c) $11.2/\sqrt{2}$ km/s (d) $11.2\sqrt{2}$ km/s

87. A satellite A of mass m is at a distance of r from the surface of the earth. Another satellite B of mass $2m$ is at a distance of $2r$ from the earth's centre. Their time periods are in the ratio of

(a) 1: 2 (b) 1: 16

(c) 1: 32 (d) $1: 2\sqrt{2}$

88. The mean radius of earth is R, its angular speed on its own axis is ω and the acceleration due to gravity earth's surface is g. What will be the radius of the orbit of a geostationary satellite?

(a) $\left(R^2 g/\omega^2\right)^{1/3}$ (b) $\left(Rg/\omega^2\right)^{1/3}$

(c) $\left(R^2\omega^2/g\right)^{1/3}$ (d) $\left(R^2 g/\omega\right)^{1/3}$

89. The satellite of mass m is orbiting around the earth in a circular orbit with a velocity v. What will be its total energy?

(a) $(3/4)mv^2$ (b) $(1/2)mv^2$

(c) mv^2 (d) $-(1/2)mv^2$

90. A planet is moving in an elliptical orbit around the sun. If T, V, E and L stand respectively for its kinetic energy, gravitational potential energy, total energy and magnitude of angular momentum about the centre of force, which of the following is correct?

(a) T is conserved

(b) V is always positive

(c) E is always negative

(d) L is conserved but direction of vector L changes continuously.

91. For a satellite, its escape velocity is 11 km/s. If the satellite is launched at an angle of 60° with the vertical, then escape velocity will be

(a) 11 km/s (b) $11\sqrt{3}$ km/s

(c) $\frac{11}{\sqrt{3}}$ km/s (d) 33 km/s

92. A particle of mass m is located at a distance r from the centre of a shell of mass M and radius R. The force between the shell and mass is $F(r)$. The plot of $F(r)$ vs r is:

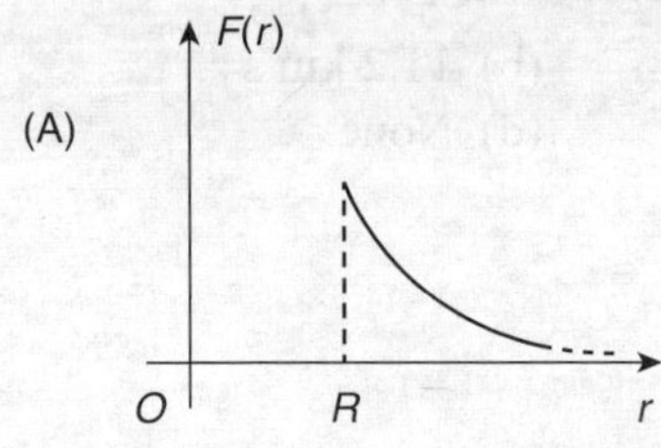

(B)

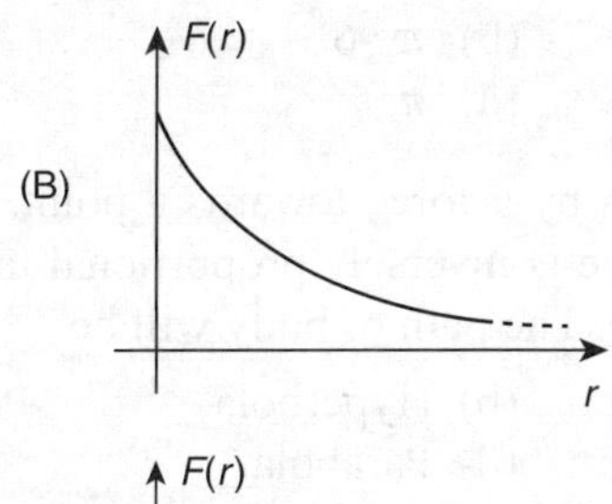

(C)

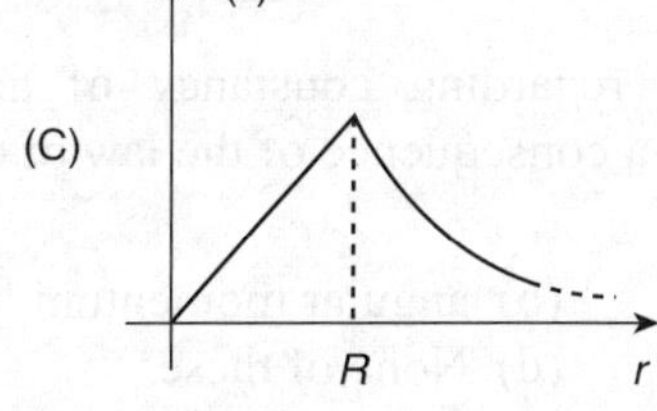

(D)

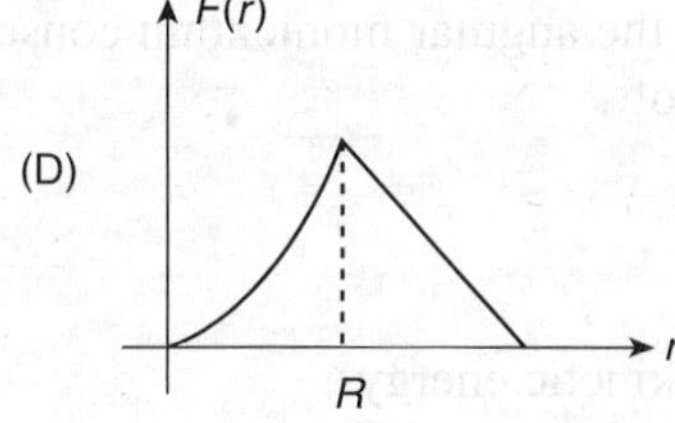

93. At what depth below the surface of the earth, acceleration due to gravity g will be half its value 1600 km above the surface of the earth?

(a) $4.2 \times {}^{6}$ m (b) $3.19 \times {}^{6}$ m

(c) 1.59×0^{6} m (d) None of these.

94. The acceleration of a body due to attraction of the earth (radius R) at a distance $2R$ from the surface of the earth is (g = acceleration due to gravity at the surface of the earth)

(a) $g/9$ (b) $g/3$

(c) $g/4$ (d) g

95. If the earth loses its gravity, then for a body

(a) Weight becomes zero, but not the mass

(b) Mass becomes zero, but not the weight

(c) Both mass and weight become zero

(d) Neither mass nor weight become zero.

96. A person will get more quantity of matter in kg-wt. at

(a) Poles (b) At latitude of 60°

(c) Equator (d) Satellite

97. Assuming earth to be a sphere of a uniform density, what is the value of gravitational acceleration in a mine 100 km below the earth's surface (R = 6400 km)

(a) 9.66 m/s^2 (b) 7.64 m/s^2

(c) 5.06 m/s^2 (d) 3.10 m/s^2

98. Two planets of radii in the ratio 2: 3 are made from the material of density in the ratio 3: 2. Then the ratio of acceleration due to gravity g_1 / g_2 at the surface of the two planets will be

(a) 1 (b) 2.25

(c) 4/9 (d) 0.12

99. Assume that the acceleration due to gravity on the surface of the moon is 0.2 times the acceleration due to gravity on the surface of the earth. If R_e is the maximum range of a projectile on the earth's surface, what is the maximum range on the surface of the moon for the same velocity of projection?

(a) 0.2 R_e. (b) 2 R_e

(c) 0.5 R_e (d) 5 R_e

100. Acceleration due to gravity on moon is 1/6 of the acceleration due to gravity on earth. If the ratio of densities of earth (ρ_e) and moon (ρ_m) is $\left(\frac{\rho_e}{\rho_m}\right) = \frac{5}{3}$ then radius of moon R_m in terms of R_e will be

(a) $\frac{5}{18} R_e$ (b) $\frac{1}{6} R_e$

(c) $\frac{3}{18} R_e$ (d) $\frac{1}{2\sqrt{3}} R_e$

101. Weight of a body is maximum at

(a) Moon (b) Poles of earth

(c) Equator of earth (d) Centre of earth.

102. Two balls of same size but with the density of one being greater than that of the other are dropped from the same height. Then which ball will reach the earth first (air resistance is negligible)?

(a) Heavy ball

(b) Light ball

(c) Both simultaneously

(d) Will depend upon the density of the balls.

103. From a solid sphere of mass M and radius R, a spherical portion of radius $\frac{R}{2}$ is removed. Taking gravitational potential the potential $V = 0$ at $r = \infty$, the centre of the cavity thus formed is (G = Gravitational constant)

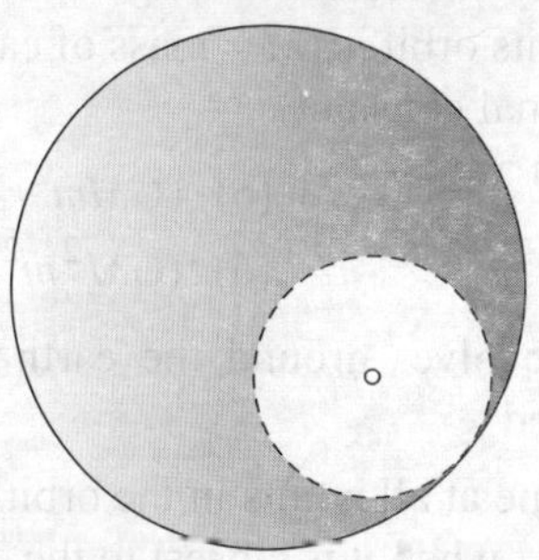

(a) $\frac{-GM}{2R}$ (b) $\frac{-GM}{R}$

(c) $\frac{-2GM}{3R}$ (d) $\frac{-2GM}{R}$

104. Escape velocity of a body of 1kg mass on a planet is 100 m/sec. Gravitational potential energy of the body at the planet is

(a) –5000 J (b) –1000 J
(c) –2400 J (d) 5000 J

105. A body is orbiting very close to the earth surface with kinetic energy KE. The energy required to completely escape from it is

(a) $\sqrt{2}KE$ (b) KE
(c) $KE/\sqrt{2}$ (d) None of these.

106. A point mass is placed inside a thin spherical shell of radius R and mass M at a distance $R/2$ from the centre of the shell. The gravitational force exerted by the shell on the point mass is

(a) $\frac{GM}{2R^2}$ (b) $-\frac{GM}{2R^2}$
(c) Zero (d) $\frac{GM}{4R^2}$

107. Four particles each of mass M, are located at the vertices of a square with side L. The gravitational potential due to this at the centre of the square is

(a) $-\sqrt{32}\frac{GM}{L}$ (b) $-\sqrt{64}\frac{GM}{L^2}$
(c) Zero (d) $\sqrt{32}\frac{GM}{L}$

108. Two satellite of masses m_1 and $m_2 (m_1 > m_2)$ are revolving round the earth in circular orbits of radius r_1 and $r_2 (r_1 > r_2)$ respectively. Which of the following statements is true regarding their speeds v_1 and v_2?

(a) $v_1 = v_2$ (b) $v_1 < v_2$
(c) $v_1 > v_2$ (d) $\frac{v_1}{r_1} = \frac{v_2}{r_2}$

109. A satellite of mass m is revolving in circular orbit of radius r round the earth. Its angular momentum w.r.t. the centre of its orbit is (M = mass of earth, G = universal gravitational constant)

(a) $(GMmr)^{1/2}$ (b) $(GMm^2r)^{1/2}$
(c) $(GMm^2r^2)^{1/2}$ (d) $(GM^2m^2r)^{1/2}$

110. A satellite revolves around the earth in an elliptical orbit. Its speed

(a) is the same at all points in the orbit
(b) is greatest when it is closest to the earth
(c) is greatest when it is farthest from the earth
(d) goes on increasing or decreasing continuously depending upon the mass of the satellite.

111. For a geostationary satellite, the angular speed in units of radian per hour is

(a) $\pi/12$ (b) $\pi/6$
(c) $\pi/2$ (d) π

112. A body is acted upon by a force towards a point. The magnitude of the force is inversely proportional to the square of the distance. The path of body will be

(a) Ellipse (b) Hyperbola
(c) Circle (d) Parabola.

113. Kepler's second law regarding constancy of aerial velocity of a planet is a consequence of the law of conservation of

(a) energy (b) angular momentum
(c) linear momentum (d) None of these.

114. In planetary motion, the angular momentum conservation leads to the law of

(a) Orbits
(b) Areas
(c) Periods
(d) Conservation of kinetic energy
(e) Conservation of linear energy.

115. The condition for a uniform spherical mass m of radius r to be a black hole is [G = gravitional constant and g = accleration due to gravity]

(a) $(2Gm/r)^{1/2} \le c$ (b) $(2Gm/r)^{1/2} = c$
(c) $(2Gm/r)^{1/2} \ge c$ (d) $(gm/r)^{1/2} \ge c$

116. If r denotes the distance between the sun and the earth, then the angular momentum of the earth around the sun is proportional to

(a) $r^{3/2}$ (b) r
(c) $\sqrt{r}$ (d) r^2
(e) r^3

117. If mass of a satellite is doubled and time period remain constant the ratio of orbit in the two cases will be

(a) 1:2 (b) 1:1
(c) 1:3 (d) None of these.

118. A satellite is launched into a circular orbit of radius R around the earth. A second satellite is launched into an orbit of radius $(1.01)R$. The period of the second satellite is larger than that of the first one by approximately.

(a) 0.5% (b) 1.0%
(c) 1.5% (d) 3.0%

119. Two planets at mean distance d_1 and d_2 from the sun and their frequencies n_1 and n_2 respectively, then

(a) $n_1^2 d_1^2 = n_2 d_2^2$ (b) $n_2^2 d_2^3 = n_1^2 d_1^3$
(c) $n_1 d_1^2 = n_2 d_2^2$ (d) $n_1^2 d_1 = n_2^2 d_2$

120. An artificial satellite moving in a circular orbit around the earth has a total (kinetic + potential) energy E_0. Its potential energy is

(a) $-E_0$ (b) $1.5E_0$
(c) $2E_0$ (d) E_0

121. Two balls, each of radius R, equal mass and density are placed in contact, then the force of gravitation between them is proportional to

(a) $F \propto \frac{1}{R^2}$ (b) $F \propto R$
(c) $F \propto R^4$ (d) $F \propto \frac{1}{R}$

122. If the earth stops rotating, the value of g at the equator

(a) increases (b) decreases
(c) no effect (d) None of these.

123. A satellite moves in elliptical orbit about a planet. Its maximum and minimum velocities of satellites are $3 \times {}^4$ m/s and $1 \times {}^3$ m/s respectively. What is the minimum distance of satellite from planet if maximum distance is 4×0^4 km?

(a) 4×10^3 km (b) 3×10^3 km
(c) $\frac{4}{3}\times10^3$ km (d) 1×10^3 km

124. The value of escape velocity on a certain planet is 2 km/s. Then, the value of orbital speed for a satellite orbiting close to its surface is

(a) 12 km/s (b) 1 km/s
(c) $\sqrt{2}$ km/s (d) $2\sqrt{2}$ km/s

125. Acceleration due to gravity g for a body of mass m on earth's surface is proportional to (Radius of earth = R, mass of earth = M)

(a) $\frac{M}{R^2}$ (b) m^0
(c) mM (d) $\frac{1}{R^{3/2}}$

126. A satellite is orbiting around the earth with orbital radius R and time period T. The quantity which remains constant is

(a) $\frac{T}{R}$ (b) $\frac{T^2}{R}$
(c) $\frac{T^2}{R^2}$ (d) $\frac{T^2}{R^3}$

127. The escape velocity for a body of mass 1 kg from the earth's surface is 11.2 Km s^{-1}. The escape velocity for a body of mass 100 kg would be

(a) 11.2×10^2 Km s^{-1} (b) 112 Km s^{-1}
(c) 11.2 Km s^{-1} (d) 11.2×10^{-2} Km s^{-1}

128. The escape velocity from earth is v_{es}. A body is projected with velocity $2v_{es}$. With what constant velocity will it move in the linear planetary space?

(a) v_{es} (b) $3v_{es}$
(c) $\sqrt{3}v_{es}$ (d) $\sqrt{5}v_{es}$

129. Two satellite A and B have ratio of masses 3: 1 in circular orbits of radii r and $4r$. The ratio of total mechanical energy of A and B is

(a) 1:3 (b) 3:1
(c) 3:4 (d) 12:1

130. A satellite A of mass m is at a distance of r from the surface of the earth. Another satellite B of mass $2m$ is at a distance of $2r$ from the earth's centre. Their time periods are in the ratio of

(a) 1:2 (b) 1:16
(c) 1:32 (d) $1: 2\sqrt{2}$

131. A ball is dropped from a satellite revolving around the earth at a height of 120 km. The ball will

(a) continue to move with same speed along a straight line tangentially to the satellite at that time
(b) continue to move with the same speed along the original orbit of satellite
(c) fall down to earth gradually
(d) go far away in space.

132. If a body is raised from the surface of the earth up to height R, what is the change in potential energy?

(a) mgR (b) $\frac{3}{2}\ mgR$
(c) $\frac{mgR}{2}$ (d) $\frac{mgR}{4}$

133. The ratio $\frac{g}{g_h}$, where g and g_h are the acceleration due to gravity at the surface of the earth and at a height h and above the earth's surface respectively, is

(a) $\left(1+\frac{h}{R}\right)^2$ (b) $\left(1+\frac{R}{h}\right)^2$
(c) $\left(\frac{R}{h}\right)^2$ (d) $\left(\frac{h}{R}\right)^2$

134. If the density of a small planet is the same as that of earth, while the radius of the planet is 0.2 times that of the earth, the gravitational acceleration of the surface of that planet is

(a) 0.2 g (b) 0.4 g
(c) 2 g (d) 4 g.

135. A satellite moves around the earth in a circular orbit of radius R making one revolution per day. A second satellite moving in a circular orbit, moves around the earth once in 8 days. The radius of the orbit of the second satellite is

(a) 8 R (b) 4 R
(c) 2 R (d) R.

136. Kepler's second law (law of areas) is nothing but a statement of
(a) work energy theorem
(b) conservation of linear momentum
(c) conservation of angular momentum
(d) conservation of energy.

137. In an elliptical orbit under gravitational force, in general
(a) tangential velocity is constant
(b) angular velocity is constant
(c) radial velocity is constant
(d) areal velocity is constant.

138. A person sitting in a chair in a satellite feels weightless because
(a) the earth does not attract the objects in a satellite
(b) the normal force by the chair on the person balances the earth's attraction
(c) the normal force is zero
(d) the person in satellite is not accelerated.

139. A geostationary satellite is orbiting the earth at a height of 6 R above the surface of earth R being the radius of earth. The time period of another satellite at a height of 2.5 R from the surface of earth, is

(a) 10 hour (b) $(6/\sqrt{2})$ hour
(c) 6 hour (d) $6\sqrt{2}$ hour.

140. A body weighs w newton at the surface of the earth. Its weight at a height equal to half the radius of the earth, will be

(a) $\frac{w}{2}$ (b) $\frac{2w}{3}$
(c) $\frac{4w}{9}$ (d) $\frac{8w}{27}$

141. An asteroid of mass m is approaching earth, initially at a distance of 10 R_E with speed v_i. It hits the earth with a speed v_f, then

(a) $v_f^2 = v_i^2 + \frac{2Gm}{M_E R}\left(1 - \frac{1}{10}\right)$
(b) $v_f^2 = v_i^2 + \frac{2GM_E}{R_E}\left(1 + \frac{1}{10}\right)$
(c) $v_f^2 = v_i^2 + \frac{2GM_E}{R_E}\left(1 - \frac{1}{10}\right)$
(d) $v_f^2 = v_i^2 + \frac{2Gm}{R_E}\left(1 - \frac{1}{10}\right)$

142. A rocket is sent vertically up with a velocity v less than the escape velocity from the earth. Taking M and R as the mass and radius of earth, the maximum height h attained by the rocket is given by the following expression

(a) $v^2R^2/(2GR - Mv)$ (b) $v^2R^2/(2GM + v^2R)$
(c) $v^2R^2/(2GM - v^2R)$ (d) $v^2R^2/(2Gv - RM)$

143. Two satellites A and B go round a planet in circular orbits having $4R$ and R respectively. If the speed of the satellite A is 3 v, the speed of satellite B will be

(a) $4v/3$ (b) $3v/2$
(c) $6v$ (d) $2v$.

144. If the speed of the rotation of the earth increases, then the weight of the body
(a) increases
(b) decreases
(c) remains unchanged
(d) may increases or decrease.

145. The radius to which the earth has be compressed so that the escape velocity is increased 10 times is

(a) 6.4 km (b) 64 km
(c) 640km (d) 4800 km

146. A satellite is launched into a circular orbit of radius R around earth while a second satellite is launched into an orbit of radius 1.02 R. The percentage difference in time periods is

(a) 0.7 % (b) 1.0 %
(c) 1.5 % (d) 3.0 %

147. A body is projected up with a velocity equal to $\left(\frac{3}{4}\right)$th the escape velocity from the surface of earth. The height it reaches is:

(a) $\frac{10R}{9}$ (b) $\frac{9}{7}R$
(c) $\frac{9}{8}R$ (d) $\frac{10R}{3}$

148. The height of the point vertically above the earth's surface at which the acceleration due to gravity become 1% of its value at the surface is

(a) 8 R (b) 9 R
(c) 10 R (d) 20 R

149. Find the ratio of acceleration due to gravity g at depth d and at height h, where $d = 2h$.

(a) 1:1 (b) 1:2
(c) 2:1 (d) 1:4

150. If the radius of earth of R then the height h at which the value of g becomes one-fourth, will be

(a) $\frac{R}{8}$ (b) $\frac{3R}{8}$

(c) $\frac{3R}{4}$ (d) $\frac{R}{2}$

151. A body is orbiting around earth at a mean radius which is two times as great as the parking orbit of a satellite. The period of body is

(a) 4 days (b) 16 days

(c) $2\sqrt{2}$ days (d) 64 days.

152. The escape velocity of a projectile on the earth's surface is 11.2 kms^{-1}. A body is projected out with thrice this speed. The speed of the body far away from the earth will be

(a) 22.4 km s^{-1} (b) 31.7 km s^{-1}

(c) 33.6 km s^{-1} (d) None of these.

153. The energy required to take a satellite to a height h above earth's surface (radius of earth = 6.4×10^3 km) is E_1 and kinetic energy required for the satellite to be in a circular orbit at this height is E_2. The value of h for which E_1 and E_2 are equal, is

(a) 1.28×10^4 km (b) 6.4×10^3 km

(c) 3.2×10^3 km (d) 1.6×10^3 km

154. When earth moves around the sun, the quantity which remains constant is

(a) angular velocity (b) kinetic energy

(c) potential energy (d) aerial velocity.

155. A satellite is orbiting around the earth. By what percentage should we increase its velocity so as to enable it escape away from the earth

(a) 41.4% (b) 50%

(c) 82.8% (d) 100%

156. If the earth is at one-fourth of its present distance from the sun, the duration of the year will be

(a) half the present year

(b) one-eight the present year

(c) one-fourth the present year

(d) one-sixth the present year.

157. The radius of a planet is $\frac{1}{4}$ of earth's radius and its acceleration due to gravity is double that of the earth's acceleration due to gravity. How many times will the escape velocity at the planet's surface be as compared to its value on earth's surface?

(a) $\frac{1}{\sqrt{2}}$ (b) $\sqrt{2}$

(c) $2\sqrt{2}$ (d) 2

158. The escape velocity from earth is 11.2 km/s. Another planet is having mass 1000 times, and radius 10 times that of earth, then escape velocity at that planet is

(a) 1120 km/s (b) 1.12 km/s

(c) 112 km/s (d) None of these.

159. A body of mass m is taken from the surface of the earth to a height equal to the radius R of the earth. The change in gravitational potential energy is

(a) 3 mgR (b) mgR

(c) $\frac{mgR}{2}$ (d) $\frac{mgR}{4}$

160. A planet is revolving around the sun in elliptical orbit. Its closest distance from the sun is r and the farthest distance is R. If the orbital velocity of the planet closet to the sun be v_0 then what is the velocity at the farthest point?

(a) $\frac{v_0 r}{R}$ (b) $\frac{v_0 R}{r}$

(c) $v_0\left(\frac{r}{R}\right)^{1/2}$ (d) $v_0\left(\frac{R}{r}\right)^{1/2}$

161. Acceleration due to gravity is g on the surface of the earth. Then the value of the acceleration due to gravity at a height of 32 km above earth's surface is

(a) 0.99 g (b) 0.8 g

(c) 1.01 g (d) 0.9 g

162. If the mass of Moon is $\frac{1}{90}$ of earth's mass, its radius is $\frac{1}{3}$ of earth's radius and if g is acceleration due to gravity on earth, then the acceleration due to gravity on moon is

(a) $\frac{g}{3}$ (b) $\frac{g}{90}$

(c) $\frac{g}{10}$ (d) $\frac{g}{9}$

163. A satellite is moving with a constant speed v in circular orbit around the earth. An object of mass m is ejected from the satellite such that it just escapes from the gravitational pull of the earth. At the time of ejection, the kinetic energy of the object is

(a) R^2 (b) mv^2

(c) $2mv^2$ (d) R^{-4}

164. Two spherical bodies of mass M and $5M$ and radii R and $2R$ respectively are released in free space with initial separation between their centres equal to 12 R. If they attract each other due to gravitational force only, then the distance covered by the smaller body just before collision, is

(a) 1.5 R (b) 2.5 R

(c) 4.5 R (d) 7.5 R

165. A satellite orbiting the earth in a circular orbit of radius R completes on revolution is $3h$. If orbital radius of geostationary satellite is 36,000 km, orbital radius of earth is

(a) 6000 km (b) 9000 kg
(c) 12000 km (d) 15000 km.

166. Two planets are at mean distance d_1, and d_2 from sun and their periods are T_1 and T_2 respectively. Then

(a) $n_1^2 d_1^2 = n_2^2 d_2^2$ (b) $n_2^2 d_2^3 = n_1^2 d_1^3$
(c) $n_1 d_1^2 = n_2 d_2^2$ (d) $n_1^2 d_1 = n_2^2 d_2$
(e) $n_1 d_1 = n_2 d_2$

167. A simple pendulum has a time period T_1 when on earth's surface, and T_2 when taken at a height $2R$ above the earth's surface where R is the radius of earth. The value of T_1 / T_2 is:

(a) $\frac{1}{9}$ (b) $\frac{1}{3}$
(c) $\sqrt{3}$ (d) 3

168. A research satellite of mass 200 kg circles the earth in an orbit of average radius of $\frac{3R}{2}$ where R is the radius of the earth. Assuming the gravitational pull on a mass of 1 Kg on the earth's surface to be 10 N, the pull on the satellite will be

(a) 880 N (b) 889 N
(c) 890 N (d) 882 N

169. If the earth were to suddenly contract to half the present radius (without any external torque acting on it), by how much would the day be decreased?

(a) 8 hrs (b) 6 hrs
(c) 4 hrs (d) 2 hrs

170. Observe on the 10th, 5th and ground floor of a tall building measure the velocity of a certain rain drop by some accurate method. Surprisingly, the velocity of the raindrop measured by the three observers was found to be the same. This is because,

(a) there is no gravitational force acting on the drop
(b) gravitational force on the raindrop is balanced by the force produced by the surrounding air
(c) gravitational force on the raindrop is balanced by the upward force of attraction produced by the cloud
(d) data is insufficient to predict.

171. The acceleration of a body due to attraction of the earth at the distance $2R$ from the surface of the earth is

(a) $\frac{g}{9}$ (b) $\frac{g}{3}$
(c) $\frac{g}{4}$ (d) g

172. The acceleration due to gravity near the surface of a planet of radius R and density d is proportional to:

(a) $\frac{d}{R^2}$ (b) dR^2
(c) dR (d) $\frac{d}{R}$

173. A satellite is moving around the earth with a speed v in a circular orbit of radius r. If the orbit radius is decreased by 1%, the speed of the satellite will

(a) increase by 1% (b) increase by 0.5%
(c) decrease by 1% (d) decrease by 0.5%

174. This question contains Statement-1 and Statement-2. Of the four choices given after the statements, choose the one that best describes the two statements.

Statement 1

For a mass M kept at the centre of a cube of side a,, the flux of gravitational field passing through its sides in 4πGM.

Statement 2

If the direction of a field due to point source is radial and its dependence on the distance r from the source is given as $\frac{1}{r^2}$, its flux through a closed surface depends only on the strength of the source enclosed by the surface and not on the size or shape of the surface.

(a) Statement-1 is true, Statement -2 is true; Statement-2 is a correct explanation for Statement-1.
(b) Statement-1 is true, Statement-2 is true; Statement-2 is not a correct explanation for Statement-1.
(c) Statement-1 is true, Statement-2 is false.
(d) Statement-1 is false, Statement-2 is true.

175. A planet in a distant solar system is 10 times more massive than the earth and its radius is 10 times smaller. Given that the escape velocity from the earth is 11 km/sec, the escape velocity from the surface of the planet would be

(a) 0.11 km/sec (b) 1.1 km/sec
(c) 11 km/sec (d) 110 km/sec

176. A particle of mass 10 g is kept on the surface of a uniform sphere of mass 100 kg and radius 10 cm. Find the work to be done against the gravitational force between them to take the particle far away from the sphere. ($G = 6.67 \times {}^{-11}$ Nm^2/kg^2)

(a) 6.67×0^{-9} J (b) 6.67×0^{-10} J
(c) 13.34×0^{-10} J (d) 3.33×0^{-10} J

177. Average density of the earth

(a) is directly proportional to g
(b) is inversely proportional to g
(c) does not depend on g
(d) is a complex function of g.

178. The change in the value of g at a height h above the surface of earth is the same as at a depth d below the surface of earth. When both d and h are much smaller than the radius of the earth, then which of the following is correct?

(a) $d = 2h$ (b) $d = h$

(c) $d = \frac{h}{2}$ (d) $d = \frac{3h}{2}$

179. Suppose the gravitational force varies inversely as the nth power of distance, then the time period of a planet in circular orbit of radius R around the sun will be proportional to

(a) $R^{\left(\frac{n+1}{2}\right)}$ (b) $R^{\left(\frac{n-1}{2}\right)}$

(c) R^n (d) $R^{\left(\frac{n-2}{2}\right)}$

180. If g is the acceleration due to gravity on the earth's surface, the gain in the potential energy of an object of mass m raised from the surface of the earth to a height equal to the radius R of the earth is

(a) 2 mgR (b) $\frac{1}{2}mgR$

(c) $\frac{1}{4}mgR$ (d) mgR

181. The time period of an earth-satellite in circular orbit is independent of

(a) the mass of the satellite
(b) radius of its orbit
(c) both the mass and radius of the orbit
(d) neither the mass of the satellite nor the radius of its orbit.

182. A satellite of mass m revolves around the earth of radius R at a height x from its surface. If g is the acceleration due to gravity on the surface of the earth, the orbital speed of the satellite is

(a) gx (b) $\frac{gR}{R-x}$

(c) $\frac{gR^2}{R+x}$ (d) $\left(\frac{gR^2}{R+x}\right)^{1/2}$

183. The escape velocity for a body projected vertically upwards from the surface of the earth 11 km/sec. If the body is projected at an angle of 45° ith the vertical, the escape velocity will be

(a) $11\sqrt{2}$ (b) 22 km/sec

(c) 11 km/sec (d) $\frac{11}{\sqrt{2}}$ m/sec

184. Two spherical bodies of mass M and $5M$ and radii R and $2R$ respectively are released in free space with initial separation between their centres equal to 9 R. If they attract each other due to gravitational force only, then the distance covered by the smaller body just before collision is

(a) 2.5 R (b) 4.5 R
(c) 7.5 R (d) 1.5 R

185. The time period of a satellite of earth is 5 hour. If the separation between the earth and the satellite is increased to 4 times the previous value, the new period will become

(a) 10 hour (b) 80 hour
(c) 40 hour (d) 20 hour

186. The escape velocity of a body depends upon mass as

(a) m^0 (b) m^1
(c) m^2 (d) m^3

187. The kinetic energy needed to project a body of mass m from the earth's surface (radius R) to infinity is

(a) $\frac{mgR}{2}$ (b) 2 mgR

(c) mgR (d) $\frac{mgR}{4}$

188. Energy required to move a body of mass m from an orbit of radius $2R$ to $3R$ is

(a) $\frac{GMm}{12R^2}$ (b) $\frac{GMm}{3R^2}$

(c) $\frac{GMm}{8R}$ (d) $\frac{GMm}{6R}$

189. If suddenly the gravitational force of attraction between the earth and a satellite revolving around it becomes zero, then the satellite will

(a) continue to move in its orbit with same velocity
(b) move tangentially to the original orbit with the same velocity
(c) become stationary in its orbit
(d) move towards the earth.

Questions Asked in NEET/CBSE Medicals

190. The work done to raise a mass m from the surface of the earth to a height h, which is equal to the radius of the earth, is

(a) $\frac{3}{2}mgR$ (b) mgR

(c) $2mgR$ (d) $\frac{1}{2}mgR$

191. A body weights 200 N on the surface of the earth. How much will it weigh half-way down to the centre of the earth?

(a) 100 N (b) 150 N
(c) 200 N (d) 250 N

192. The time period of a geostationary satellite is 24 h, at a height $6R_E$ (R_E is radius of earth) from surface of earth. The time period of another satellite whose height is 2.5 R_E from surface will be

(a) $6\sqrt{2}$ h (b) $12\sqrt{2}$ h
(c) $\frac{24}{2.5}$ h (d) $\frac{12}{2.5}$ h

193. The kinetic energies of a planet in an elliptical orbit about the Sun, at positions A, B and C are K_A, K_B and K_C, respectively. AC is the major axis and SB is perpendicular to AC at the position of the Sun S as shown in the figure. Then,

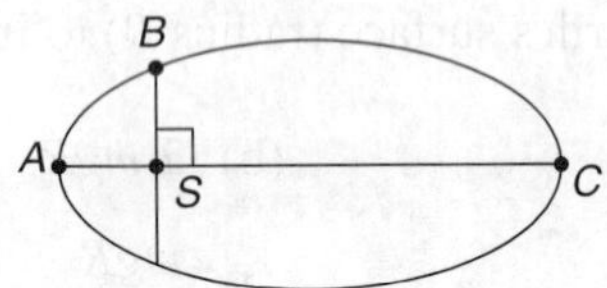

(a) $K_A < K_B < K_C$ (b) $K_A > K_B > K_C$
(c) $K_B < K_A < K_C$ (d) $K_B > K_A > K_C$

194. If the mass of the sun were ten times smaller and the universal gravitational constant were ten times larger in magnitude, which of the following is not correct?

(a) Raindrops will fall faster.
(b) Walking on the ground would become more difficult.
(c) Time period of a simple pendulum on the earth would decrease.
(d) g on the earth will not change.

195. The acceleration due to gravity at a height 1 km above the earth is the same as at a depth d below the surface of earth. Then,

(a) $d = 1$ km (b) $d = \frac{3}{2}$ km
(c) $d = 2$ km (d) $d = \frac{1}{2}$ km

196. Two astronauts are floating in gravitational free space after having lost contact with their spaceship. The two will

(a) move towards each other.
(b) move away from each other.
(c) will become stationary
(d) keep floating at the same distance between them.

197. Starting from the centre of the earth having radius R, the variation of g (acceleration due to gravity) is shown by

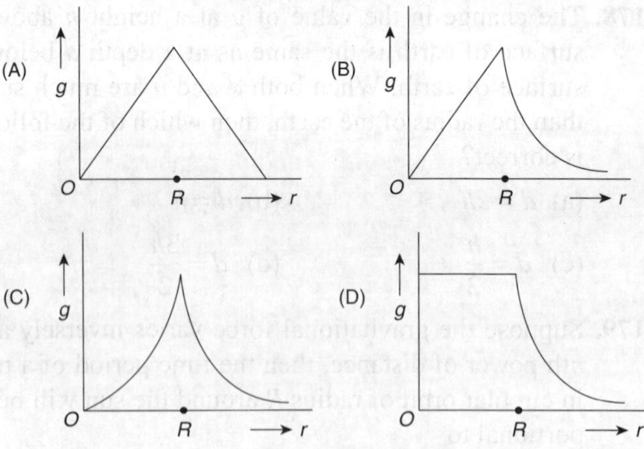

198. A satellite of mass m is orbiting the earth (of radius R) at a height h from its surface. The total energy of the satellite in terms of g_0, the value of acceleration due to gravity at the earth's surface is

(a) $\frac{mg_0R^2}{2(R+h)}$ (b) $-\frac{mg_0R^2}{2(R+h)}$
(c) $\frac{2mg_0R^2}{R+h}$ (d) $-\frac{2mg_0R^2}{R+h}$

199. At what height from the surface of earth the gravitation potential and the value of g are 5.4×10^7 J kg^{-1} and 6.0 m s^{-2} respectively? Take the radius of earth as 6400 km.

(a) 1400 km (b) 2000 km
(c) 2600 km (d) 1600 km

200. The ratio of escape velocity at earth (v_e) to the escape velocity at a planet (v_p) whose radius and mean density are twice as that of earth is

(a) 1: 4 (b) 1: $\sqrt{2}$
(c) 1: 2 (d) 1: $2\sqrt{2}$

201. A remote-sensing satellite of earth revolves in a circular orbit at a height of 0.25×10^6 m above the surface of earth. If earth's radius is 6.38×10^6 m and $g = 9.8$ m s^{-2}, then the orbital speed of the satellite is

(a) 9.13 km s^{-1} (b) 6.67 km s^{-1}
(c) 7.76 km s^{-1} (d) 8.56 km s^{-1}

202. A satellite S is moving in an elliptical orbit around the earth. The mass of the satellite is very small compared to the mass of the earth. Then,

(a) the liner momentum of S remains constant in magnitude
(b) the acceleration of S is always directed towards the centre of the earth
(c) the angular momentum of S about the centre of the earth changes in directions but its magnitude remains constant
(d) the total mechanical energy of S varies periodically with time.

203. Kepler's third law states that square of period of revolution (T) of a planet around the sun, is proportional to third power of average distance r between sun and planet, i.e., $T^2 = Kr^3$. Here K is constant. If the masses of sun and planet are M and m respectively then as per Newton's law of gravitation force of attraction between them is $F = \dfrac{GMm}{r^2}$, here G is gravitational constant. The relation between G and K is described as

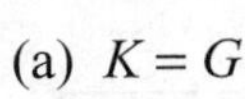
(a) $K = G$ (b) $K = \dfrac{1}{G}$

(c) $GK = 4\pi^2$ (d) $GMK = 4\pi^2$

204. A black hole is an object whose gravitational field is so strong that even light cannot escape from it. To what approximate radius would earth (mass = 5.98×10^{24} kg) have to be compressed to be a black hole?

(a) 10^{-9} m (b) 10^{-6} m

(c) 10^{-2} m (d) 100 m

205. Dependence of intensity of gravitational field (E) of earth with distance (r) from centre of earth is correctly represented by

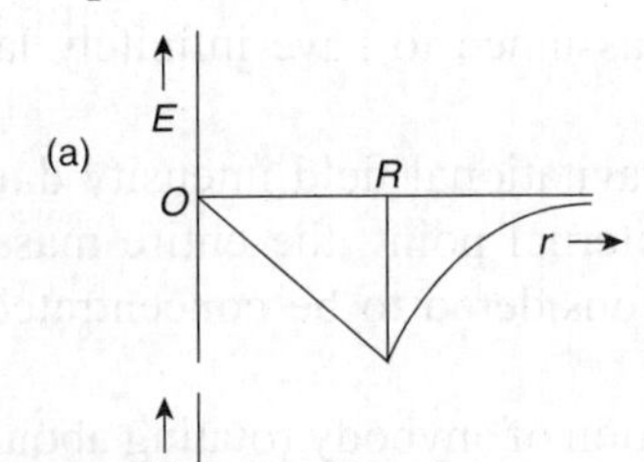

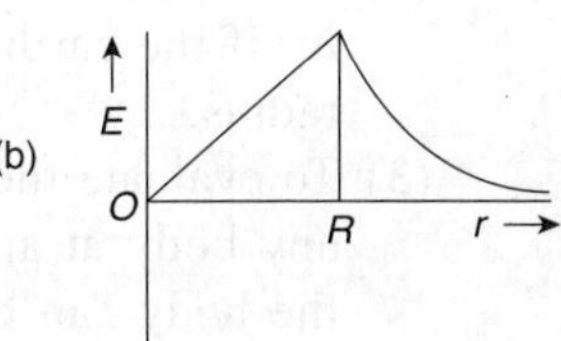

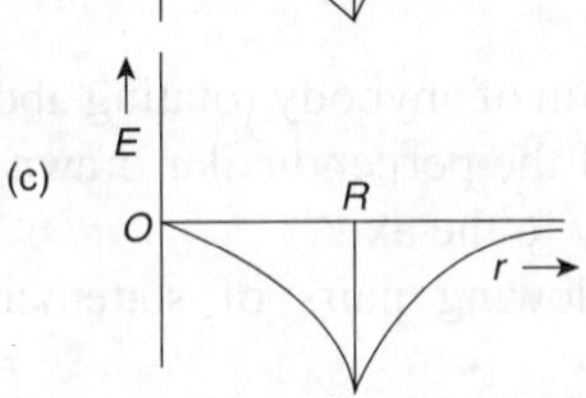

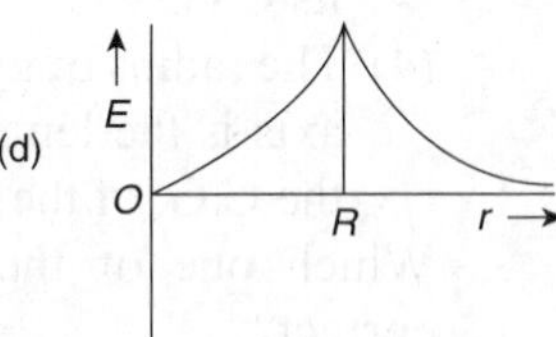

206. Infinite number of bodies, each of mass 2 kg are situated on x-axis at distance 1 m, 2 m, 4 m, 8 m, ..., respectively, from the origin. The resulting gravitational potential due to this system at the origin will be

(a) $-\dfrac{4}{3}$G (b) –4G

(c) –G (d) $-\dfrac{8}{3}$

207. A body of mass m is taken from the earth's surface to the height equal to twice the radius (R) of the earth. The change in potential energy of body will be

(a) $3mgR$ (b) $\dfrac{1}{3}mgR$

(c) $mg\,2R$ (d) $\dfrac{2}{3}mgR$

208. The radius of a planet is twice the radius of earth. Both have almost equal average mass-densities. V_P and V_E are escape velocities of the planet and the earth, respectively. Then

(a) $V_P = 1.5V_E$ (b) $V_P = 2V_E$

(c) $V_E = 3V_P$ (d) $V_E = 1.5V_P$

209. A particle of mass m is kept at rest at a height $3R$ from the surface of earth, where R is radius of earth and M is mass of earth. The minimum speed with which it should be projected, so that it does not return, is (g is acceleration due to gravity on the surface of earth)

(a) $\left(\dfrac{GM}{2R}\right)^{1/2}$ (b) $\left(\dfrac{gR}{4}\right)^{1/2}$

(c) $\left(\dfrac{2g}{R}\right)^{1/2}$ (d) $\left(\dfrac{GM}{R}\right)^{1/2}$

210. The height at which the weight of a body becomes $\left(\dfrac{1}{16}\right)$th, its weight on the surface of earth (radius R), is

(a) $5R$ (b) $15R$

(c) $3R$ (d) $4R$

211. A spherical planet has a mass M_P and diameter D_P. A particle of mass m falling freely near the surface of this planet will experience an acceleration due to gravity, equal to

(a) $\dfrac{4GM_p}{D_p^2}$ (b) $\dfrac{GM_pm}{D_p^2}$

(c) $\dfrac{GM_p}{D_p^2}$ (d) $\dfrac{4GM_pm}{D_p^2}$

212. A geostationary satellite is orbiting the earth at a height of $5R$ above the surface of the earth, R being the radius of the earth. The time period of another satellite in hours at a height of $2R$ from the surface of the earth is

(a) 5 (b) 10

(c) $6\sqrt{2}$ (d) $\dfrac{6}{\sqrt{2}}$

213. If v_e is escape velocity and v_0 is orbital velocity of a satellite for an orbit close to the earth's surface, then these are related by

(a) $v_0 = \sqrt{2}v_e$ (b) $v_0 = v_e$

(c) $v_e = \sqrt{2v_0}$ (d) $v_e = \sqrt{2}v_0$

214. The amount of work done in bringing a mass from the surface of earth on one side to a point diametrically opposite on the other side?

(a) $\sqrt{\dfrac{GM}{R}}$ (b) Zero

(c) $\sqrt{\dfrac{2GM}{R}}$ (d) $2mgR$

215. A planet moving along an elliptical orbit is closest to the sun at a distance r_1 and farthest away at a distance of r_2. If v_1 and v_2 are the linear velocities at these points respectively, then the ratio $\frac{v_1}{v_2}$ is

(a) $(r_1/r_2)^2$ (b) r_1/r_2
(c) $(r_2/r_1)^2$ (d) r_1/r_2

216. A particle of mass m is thrown upwards from the surface of the earth, with a velocity u. The mass and the radius of the earth are, respectively, M and R. G is gravitational constant and g is acceleration due to gravity on the surface of the earth. The minimum value of u so that the particle does not return to earth, is

(a) $\sqrt{\frac{2GM}{R^2}}$ (b) $\sqrt{\frac{2GM}{R}}$
(c) $\sqrt{\frac{2gM}{R^2}}$ (d) $\sqrt{2gR^2}$

217. A particle of mass M is situated at the centre of a spherical shell of same mass and radius a. The magnitude of the gravitational potential at a point situated at $a/2$ distance from the centre, will be

(a) $\frac{GM}{a}$ (b) $\frac{2GM}{a}$
(c) $\frac{3GM}{a}$ (d) $\frac{4GM}{a}$

218. The radii of circular orbits of two satellites A and B of the earth, are $4R$ and R, respectively. If the speed of satellite A is $3V$, then the speed of satellite B will be

(a) $\frac{3V}{4}$ (b) $6V$
(c) $12V$ (d) $\frac{3V}{2}$

219. A man of 50 kg is standing in a gravity-free space at a height of 10 m above the floor. He throws a stone of 0.5 kg mass downwards with a speed 2 m/s. When the stone reaches the floor, the distance of the man above the floor will be

(a) 9.9 m (b) 10.1 m
(c) 10 m (d) 20 m

220. The additional kinetic energy to be provided to a satellite of mass m revolving around a planet of mass M, to transfer it from a circular orbit of radius R_1 to another of radius $R_2 (R_2 > R_1)$ is

(a) $GmM\left(\frac{1}{R_1^2}-\frac{1}{R_2^2}\right)$ (b) $GmM\left(\frac{1}{R_1}-\frac{1}{R_2}\right)$
(c) $2GmM\left(\frac{1}{R_1}-\frac{1}{R_2}\right)$ (d) $\frac{1}{2}GmM\left(\frac{1}{R_1}-\frac{1}{R_2}\right)$

221. The dependence of acceleration due to gravity g at a distance r from the centre of the earth, assumed to be a sphere of radius R and of uniform density is as shown in figures.

The correct figure is

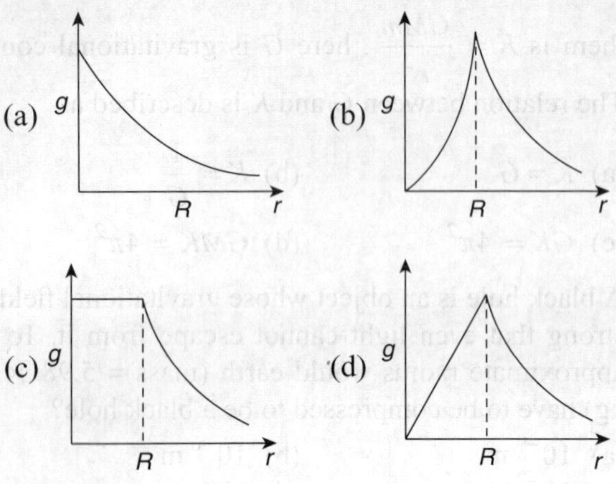

222. (1) Centre of gravity (C.G.) of a body is the point at which the weight of the body acts
(2) Centre of mass coincides with the centre of gravity if the earth is assumed to have infinitely large radius.
(3) To evaluate the gravitational field intensity due to any body at an external point, the entire mass of the body can be considered to be concentrated at its C.G.
(4) The radius of gyration of anybody rotating about an axis is the length of the perpendicular drawn from the C.G. of the body to the axis.

Which one of the following pairs of statements is correct?

(a) (4) and (1) (b) (1) and (2)
(c) (2) and (3) (4) (3) and (4)

223. The figure shows elliptical orbit of a planet m about the sun S. The shaded area SCD is twice the shaded area SAB. If t_1 is the time for the planet to move from C to D and t_2 is the time to move from A to B then

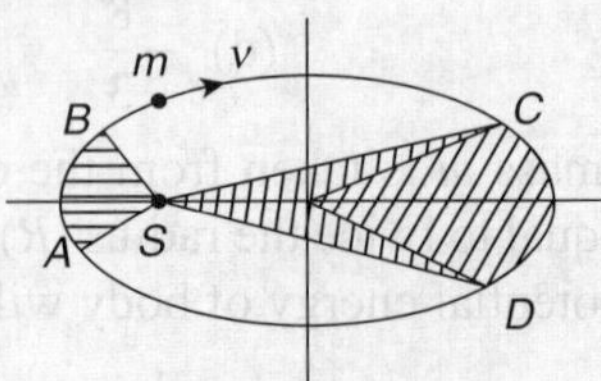

(a) $t_1 = 4t_2$ (b) $t_1 = 2t_2$
(c) $t_1 = t_2$ (d) $t_1 > t_2$

224. Two satellite of earth, S_1 and S_2 are moving in the same orbit. The mass of S_1 is four times the mass of S_2. Which one of the following statements is true?

(a) The potential energies of earth and satellite in the two cases are equal.
(b) S_1 and S_2 are moving with the same speed.
(c) The kinetic energies of the two satellite are equal.
(d) The time period of S_1 is four times that of S_2

225. The earth is assumed to be a sphere of radius R. A platform is arranged at a height R form the surface of the earth. The escape velocity of a body from this platform is fv, where v is its escape velocity from the surface of the earth. The value of f is

(a) 1/2 (b) $\sqrt{2}$
(c) $1/\sqrt{2}$ (d) 1/3

226. Imagine a new planet having the same density as that of earth but 3 times bigger than the earth in size. If the acceleration due to gravity on the surface of earth is g and that on the surface of the new planet is g', then

(a) $g' = g/9$ (b) $g' = 27g$
(c) $g' = 9g$ (d) $g' = 3g$

227. For a satellite moving in an orbit around the earth, the ratio of kinetic energy to potential energy is

(a) 1/2 (b) $1/\sqrt{2}$
(c) 2 (d) $\sqrt{2}$

228. The density of a newly discovered planet is twice that of earth. The acceleration due to gravity at the surface of the planet is equal to that at the surface of the earth. If the radius of the earth is R, the radius of the planet would be

(a) $2R$ (b) $4R$
(c) $\frac{1}{4}R$ (d) $\frac{1}{2}R$

229. Two spheres of mass m and M are situated in air and the gravitational force between them is F. The space around the masses is now filled with a liquid of specific gravity 3. The gravitational force will now be

(a) 3 F (b) F
(c) $F/3$ (d) $F/9$

230. The acceleration due to gravity on the planet A is 9 times the acceleration due to gravity on planet B. A man jumps to a height of 2 m on the surface of A. What is the height of jump by the same person o the planet B?

(a) (2/9) m (b) 18 m
(c) 6 m (d) (2/3) m

231. A body of mass m is placed on earth's surface which is taken from earth surface to a height of $h = 3R$, then change in gravitational potential energy is

(a) $\frac{mgR}{4}$ (b) $\frac{2}{3}mgR$
(c) $\frac{3}{4}mgR$ (d) $\frac{mgR}{2}$

232. With what velocity should a particle be projected so that its height becomes equal to radius of earth?

(a) $\left(\frac{GM}{R}\right)^{1/2}$ (b) $\left(\frac{8GM}{R}\right)^{1/2}$
(c) $\left(\frac{2GM}{R}\right)^{1/2}$ (d) $\left(\frac{4GM}{R}\right)^{1/2}$

233. For a planet having mass equal to mass of the earth but radius one-fourth of the radius of earth, the escape velocity will be

(a) 11.2 km/s (b) 22.4 km/s
(c) 5.6 km/s (d) 44.8 km/s

234. Gravitational force is required for

(a) stirring of liquid (b) convection
(c) conduction (d) radiation.

235. For a body of weight 72 N moves from the surface of earth at a height half of the radius of earth, the gravitational force exerted on it will be

(a) 36 N (b) 32 N
(c) 144 N (d) 50 N

236. The escape velocity of a sphere of mass m is given by (G = Universal gravitational constant; M_e = Mass of the earth and R_e = Radius of the earth)

(a) $\sqrt{\frac{2GM_e m}{R_e}}$ (b) $\sqrt{\frac{2GM_e}{R_e}}$
(c) $\sqrt{\frac{GM_e}{R_e}}$ (d) $\sqrt{\frac{2GM_e + R_e}{R_e}}$

237. The escape velocity of a body on the surface of the earth is 11.2 km/s. If the earth's mass increases to twice its present value and radius of the earth become half, the escape velocity becomes

(a) 22.4 km/s (b) 44.8 km/s
(c) 5.6 km/s (d) 11.2 km/s

238. The period of revolution of planet A around the sun is 8 times that of B. The distance of A from the sun is how many times greater than that of B from the sun?

(a) 4 (b) 5
(c) 2 (d) 3

239. What will be the formula of mass of the earth in terms of g, R and G?

(a) $G\frac{R}{g}$ (b) $g\frac{R^2}{G}$
(c) $g^2\frac{R}{G}$ (d) $G\frac{g}{R}$

240. A ball is dropped from a spacecraft revolving around the earth at a height of 120 km. What will happen to the ball?

(a) It will fall down to the earth gradually.
(b) It will go very far in the space.
(c) It will continue to move with the same speed along the original orbit of spacecraft.
(d) It will move with the same speed, tangentially to the spacecraft.

241. The acceleration due to gravity g and mean density of the earth ρ are related by which of the following relations? (where G is the gravitational constant and R is the radius of the earth)

(a) $\rho=\dfrac{3g}{4\pi GR}$ (b) $\rho=\dfrac{3g}{4\pi GR^3}$

(c) $\rho=\dfrac{4\pi gR^2}{3G}$ (d) $\rho=\dfrac{4\pi gR^3}{3G}$

242. Two particles of equal mass go around a circle of radius R under the action of their mutual gravitational attraction. The speed v of each particle is

(a) $\dfrac{1}{2}\sqrt{\dfrac{Gm}{R}}$ (b) $\sqrt{\dfrac{4Gm}{R}}$

(c) $\dfrac{1}{2R}\sqrt{\dfrac{1}{Gm}}$ (d) $\sqrt{\dfrac{Gm}{R}}$

243. The earth (mass = 6×10^{24} kg) revolves around the sun with an angular velocity of 2×10^{-7} rad/s in a circular orbit of radius 1.5×10^{8} km. The force exerted by the sun on the earth, in newton, is

(a) 36×10^{21} (b) 27×10^{39}
(c) zero (d) 18×10^{25}

244. The radius of Earth is about 6400 km and that of Mars is 3200 km. The mass of the earth is about 10 times mass of Mars. An object weights 200 N o the surface of Earth. Its weight on the surface of Mars will be

(a) 20 N (b) 8 N
(c) 80 N (d) 40 N

245. The distance of two planets from the sun are 10^{13} m and 10^{12} m respectively. The ratio of time periods of the planets is

(a) $\sqrt{10}$ (b) $10\sqrt{10}$
(c) 10 (d) $1/\sqrt{10}$

Answer Keys

EXERCISES

1. (c)	2. (b)	3. (b)	4. (c)	5. (c)	6. (a)	7. (c)	8. (c)	9. (c)	10. (c)
11. (c)	12. (c)	13. (c)	14. (c)	15. (c)	16. (c)	17. (c)	18. (c)	19. (d)	20. (a)
21. (a)	22. (d)	23. (a)	24. (a)	25. (a)	26. (a)	27. (d)	28. (c)	29. (c)	30. (b)
31. (c)	32. (c)	33. (d)	34. (d)	35. (d)	36. (a)	37. (d)	38. (d)	39. (c)	40. (b)
41. (c)	42. (b)	43. (a)	44. (a)	45. (a)	46. (d)	47. (d)	48. (d)	49. (b)	50. (d)
51. (d)	52. (b)	53. (b)	54. (a)	55. (c)	56. (a)	57. (a)	58. (d)	59. (a)	60. (d)
61. (c)	62. (a)	63. (d)	64. (a)	65. (a)	66. (d)	67. (a)	68. (b)	69. (a)	70. (d)
71. (a)	72. (a)	73. (b)	74. (b)	75. (c)	76. (a)	77. (d)	78. (a)	79. (a)	80. (b)
81. (c)	82. (b)	83. (b)	84. (b)	85. (d)	86. (b)	87. (d)	88. (a)	89. (d)	90. (c)
91. (a)	92. (a)	93. (a)	94. (a)	95. (a)	96. (d)	97. (a)	98. (a)	99. (d)	100. (a)
101. (b)	102. (c)	103. (b)	104. (a)	105. (b)	106. (c)	107. (a)	108. (b)	109. (b)	110. (b)
111. (a)	112. (a)	113. (b)	114. (b)	115. (c)	116. (c)	117. (b)	118. (c)	119. (b)	120. (c)
121. (c)	122. (a)	123. (c)	124. (c)	125. (a)	126. (d)	127. (c)	128. (c)	129. (d)	130. (d)
131. (c)	132. (c)	133. (a)	134. (a)	135. (b)	136. (c)	137. (d)	138. (c)	139. (d)	140. (c)
141. (c)	142. (c)	143. (c)	144. (b)	145. (b)	146. (d)	147. (b)	148. (b)	149. (a)	150. (c)
151. (c)	152. (b)	153. (c)	154. (d)	155. (a)	156. (b)	157. (a)	158. (c)	159. (b)	160. (a)
161. (a)	162. (c)	163. (b)	164. (d)	165. (b)	166. (b)	167. (b)	168. (b)	169. (b)	170. (a)

171. (a) 172. (c) 173. (b) 174. (a) 175. (d) 176. (a) 177. (a) 178. (a) 179. (a) 180. (b)
181. (a) 182. (d) 183. (c) 184. (c) 185. (c) 186. (a) 187. (c) 188. (d) 189. (b)

Questions Asked in NEET/CBSE Medicals

190. (d) 191. (a) 192. (a) 193. (b) 194. (d) 195. (c) 196. (a) 197. (b) 198. (b) 199. (c)
200. (d) 201. (c) 202. (b) 203. (d) 204. (c) 205. (a) 206. (b) 207. (d) 208. (b) 209. (a)
210. (c) 211. (a) 212. (c) 213. (d) 214. (b) 215. (b) 216. (b) 217. (c) 218. (b) 219. (b)
220. (d) 221. (d) 222. (a) 223. (b) 224. (b) 225. (c) 226. (d) 227. (a) 228. (d) 229. (b)
230. (b) 231. (c) 232. (a) 233. (b) 234. (b) 235. (b) 236. (b) 237. (a) 238. (a) 239. (b)
240. (c) 241. (a) 242. (a) 243. (a) 244. (c) 245. (b)

Solutions

1. (c)

Normal weight $w = mg = \dfrac{GMm}{R^2}$

When gravitational force of attraction is proportional to R^3 instead of R^2

The new weight $w' = \dfrac{GMm}{R^3} \Rightarrow w' < w,$

As $w' = \dfrac{GMm}{R^3} = \dfrac{mv^2}{R} \Rightarrow v = \sqrt{\dfrac{GM}{R^2}}$

and $T = \dfrac{2\pi R}{v} = \dfrac{2\pi R^2}{\sqrt{GM}}$

or $T \propto R^2$

2. (b)

Let M = Mass of the earth

r = Radius of the circular orbit

v_0 = Orbital velocity of the moon

$$\frac{mv_0^2}{r} = \frac{GMm}{r^2}$$

$$v_0 = \sqrt{\frac{GM}{r}}$$

Escape velocity of the moon

$$v_e = \sqrt{\frac{2GM}{r}}$$

% ncrease in the velocity of the moon

$$= \frac{v_e - v_0}{v_0} \times 10^2$$

$$= \frac{\sqrt{\frac{2GM}{r}} - \sqrt{\frac{GM}{r}}}{\sqrt{\frac{GM}{r}}} \times 100$$

$$= (1.414 - 1) \times 100$$

$$= 0.414 \times 100$$

3. (b)

$$T_1^2 = \frac{4\pi^2 R^3}{Gm}$$

$$T_2^2 = \frac{4\pi^2 (2R)^3}{Gm}$$

$$T_2^2 = 8T_1^2$$

$$T_2^2 = 8 \qquad [\because T_1 = 1 \text{ year}]$$

$$T_2 = 2.83 \text{ years.}$$

4. (c)

In case spherical volume distribution, as

$$I = \frac{GM}{r^2} \text{ for } r > R$$

and $$I = \frac{GM}{r^3} \text{ for } r < R$$

(i) As both r_1 and r_2 are greater than R.

$$\frac{I_1}{I_2} = \frac{(GM / r_1^2)}{(GM / r_2^2)} = \left(\frac{r_2}{r_1}\right)^2$$

(ii) As both r_1 and r_2 are lesser than R.

$$\frac{I_1}{I_2} = \frac{(GM/r_1^3)\,r_1}{(GM/r_2^3)\,r_2} = \frac{r_1}{r_2}$$

5. (c)

$$g = \frac{GM}{R^2} = G\frac{4}{3}\frac{\pi R^3 \rho}{R^2} = \frac{4}{3}\pi G\rho R$$

$$g \propto R$$

6. (a)

Let R be the original radius of planet. Attraction on a body of mass m placed on its surface will be

$$F = \frac{GMm}{R^2}$$

$R' = 2R$, then mass of the planet becomes

$$M' = \frac{4}{3}\pi(2R)^3\rho = 8\times\frac{4}{3}R^3\rho = 8M$$

7. (c)

On the surface of earth, the atmospheric pressure is quite high. The astronauts will feel great discomfort if they move on earth immediately after coming back from Moon. To avoid it, they need to get used to normal air pressure gradually. That is why they have to live for some days in a caravan with the air pressure lower than outside.

8. (c)

$$R_2 = R_1\left(\frac{T_2}{T_1}\right)^{2/3} = R\left(\frac{8T}{T}\right)^{2/3} = 4R$$

9. (c)

As $\quad g = \dfrac{GM}{R^2} = \dfrac{G(4/3)\pi R^3\rho}{R^2}$

or $\quad \rho = \dfrac{3g}{4\pi GR}$

10. (c)

As $\quad g = \dfrac{GM}{R^2}$

$$g' = G\frac{(0.99M)}{(0.99R)^2} = 1.01\,\text{g}$$

Thus, acceleration due to gravity would increase by 1 %

Escape velocity $v_e = \sqrt{\dfrac{2GM}{R}}$ will remain unchanged.

Potential energy $= \dfrac{-GMm}{R}$ will remain the same.

11. (c)

The value of g at latitude θ is

$$g_\theta = g - R\omega^2\cos^2\theta$$

At equator $\theta = 0° \quad \therefore \cos\theta = 1$

$$g_0 = g - R\omega^2$$

Weight of the body at the equator will be zero if ω change to ω_1.

$$mg_0 = 0 \text{ or } g_0 = 0$$

$$0 = g - R\omega_1^2$$

$$\omega_1 = \sqrt{\frac{g}{R}}$$

$$\frac{\omega_1}{\omega} = \sqrt{\frac{g}{R}}\times\frac{1}{\omega}$$

Using given values

$$\omega = \frac{2\pi}{T} = \frac{2\times 3.142}{24\times 3600}$$

$$= 7.27\times10^{-5}\text{ rad/sec}$$

$$\frac{\omega_1}{\omega} = \sqrt{\frac{9.8}{6.4\times10^6}}\times\frac{1}{7.27\times10^{-5}} = 17$$

$$\omega_1 = 17\omega$$

12. (c)

$$T^2 \propto R^3 \text{ with } R_e = 6400\text{ km}$$

$$\frac{T^2}{(24)^2} = \left(\frac{6400}{36000}\right)^3$$

$\Rightarrow \quad T \simeq 1.7$ hr

For spy satellite, R is slightly greater than R_e.

$\Rightarrow \quad T_S\,T > \therefore T_S = 2$ hr

13. (c)

$$g_\phi = g - R\omega^2\cos^2\phi$$

$$0 = g - R\omega^2\cos^2 45°$$

$$\omega = \sqrt{\frac{2g}{R}}$$

14. (c)

$$g=\frac{GM}{R^2}$$

$$\frac{\Delta g}{g}=-2\left(\frac{\Delta R}{R}\right)=-2\left(-\frac{1}{100}\right)$$

$$=\frac{2}{100}=2\%$$

We know, $g=\dfrac{GM}{R^2}$

When the radius shrinks 1% the new radius becomes $\left(\dfrac{99}{100}R\right)$

Hence, $g'=\dfrac{GM}{\left(\dfrac{99}{100}R\right)^2}$

$$=\left(\frac{100}{99}\right)^2\frac{GM}{R^2}$$

15. (c)

$$\frac{g_2}{g_1}=\left(\frac{R}{R+h}\right)^2$$

$$\frac{49}{100}=\left(\frac{R}{R+h}\right)^2$$

$$\frac{R}{R+h}=\frac{7}{10}\Rightarrow h=\frac{3}{7}R$$

16. (c)

$$\frac{\Delta g}{g}=\frac{-2h}{R}$$

$$\frac{-4}{100}=\frac{-2h}{R}$$

17. (c)

$$\frac{g_d}{g}=1-\frac{d}{R}$$

$$\frac{2}{100}=1-\frac{d}{R}$$

$$d=\frac{49R}{50}$$

18. (c)

$$v=\sqrt{\frac{GM}{R}}=\sqrt{\frac{4}{3}\frac{G\pi R^3}{R}.\rho}$$

$$v\propto R\sqrt{\rho}$$

$$\frac{\Delta v}{v}=\frac{\Delta R}{R}\frac{1}{2}\frac{\Delta\rho}{\rho}$$

$$=\frac{-2}{100}+\frac{1}{2}\times\frac{2}{100}=\frac{-1}{100}$$

19. (d)

The period of rotation does not depend upon mass.

20. (a)

The kinetic energy of satellite is

$$E=\frac{1}{2}mv^2=\frac{1}{2}mr^2\omega^2 \quad \text{...(i)}$$

Where v and ω are the linear and angular velocities of the orbiting satellite.

Since the gravitational force supplies the necessary centripetal force,

$$F=\frac{GMm}{r^2}=mr\omega^2$$

where M is the mass of the planet.

$$\frac{GM}{r}=r^2\omega^2 \quad \text{...(ii)}$$

Substituting the value of $r^2\omega^2$ in Eq. (i),

$$E=\frac{GMm}{2r}$$

Further, angular momentum $L=mvr=mr^2\omega$

$$L^2=m^2r^4\omega^2=m^2r^2(r^2\omega^2)$$

$$L^2=m^2r^2\left(\frac{GM}{r}\right) \quad \text{[From Eq. (ii)]}$$

$$L^2=2mr^2\left(\frac{2GMm}{2r}\right)$$

$$L^2=2mr^2E$$

$$L=(2mr^2E)^{1/2}$$

21. (a)

22. (d)

Gravitational potential at a distance r outside a solid sphere varies inversely with r and inside it is given by the expression $\left(KQ/2R^3\right)\left(3R^2 - r^2\right)$. Hence option (d) is correct.

23. (a)

As it loses some energy, it will be attracted toward the earth and distance decreases. Due to work done by gravity, its interaction energy decreases and kinetic energy increases. Hence option (a) is correct.

24. (a)

As train P is moving opposite to the direction of rotation of earth, its speed relative to earth's rotation axis is less and it experiences less centripetal force for its revolution along equator. Hence the normal reaction on this train by the track will be more compared to train Q, which is moving in the direction of rotation of earth.

25. (a)

Above the surface of earth, for small altitudes, gravitational potential varies linearly with height. Hence option (a) is correct.

26. (a)

A geostationary satellite must have an angular speed in circular orbit which is equal to that of earth's rotation in same direction. For the satellite to appear at rest from earth's surface it must be in equatorial plane. Hence option (a) is correct.

27. (d)

If the density of planet is equal to the average density of earth and mass is double then its radius must be $2^{1/3} R$ so that its volume is double that of earth. The value of acceleration due to gravity at the surface of the planet can be compared with that at earth as $g_p = G(2M)/\left(2^{1/3}R\right)^2 = 2^{1/3} g_E$. Hence option (d) is correct.

28. (c)

The potential, kinetic and total energy of a satellite varies inversely with the orbit radius of the satellite and the relation we know is TE = (1/2) PE = –KE. Hence, option (c) is correct.

29. (c)

30. (b)

Inside the cavity, the acceleration due to gravity is zero and outside the shell, it varies inversely proportional to r^2 and in the annular region it increases in some non-linear manner. Hence option (b) is correct.

31. (c)

Considering an elemental shell of radius r and width dr as shown:

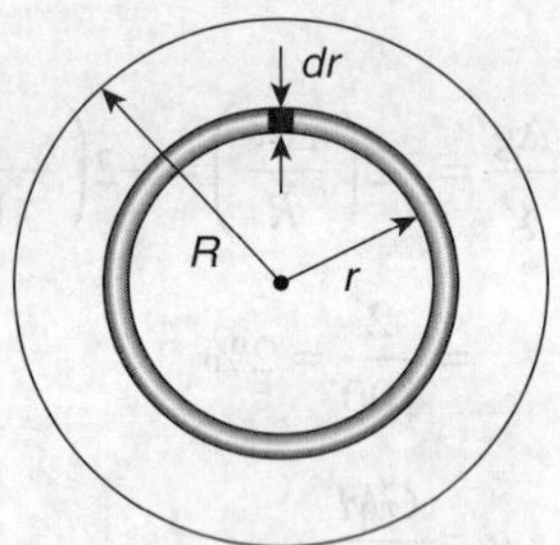

If inside mass is m then interaction energy of shell with mass is

$$dU = -\frac{Gmdm}{r}$$

$$\Rightarrow \quad dU = -\frac{G\left(\frac{4}{3}\pi r^3 \rho\right)\left(4\pi r^2 dr \rho\right)}{r}$$

$$\Rightarrow \quad dU = -\frac{16\pi^2 G\rho^2}{3} r^4 dr$$

Net interaction energy is

$$\Rightarrow \quad U = -\frac{16}{3}\pi^2 G\left(\frac{M}{\frac{4}{3}\pi R^3}\right)^2 \int_0^R r^4 dr$$

$$\Rightarrow \quad U = -\left(\frac{16}{3}\pi^2 G\right)\left(\frac{M^2}{\frac{16}{9}\pi^2 R^6}\right)\left(\frac{R^5}{5}\right)$$

$$\Rightarrow \quad U = -\frac{3}{5}\frac{GM^2}{R}$$

32. (c)

As the separated part has same speed as that of satellite, it will continue to move in the same orbit.

33. (d)

Kinetic energy is maximum at centre and potential energy is minimum.

34. (d)

In all the options given, the term force is used. But gravitational potential at a point is defined in terms of the work done in displacing a unit mass from infinity to that point without change in kinetic energy.

35. (d)

As the mass of earth remain same, nothing would happen because all gravitational forces remain same.

36. (a)

From Kepler's third law

$$T^2 = kr^3$$

$$\Rightarrow \quad 2\frac{\Delta T}{T} = 3\frac{\Delta r}{r}$$

$$\Rightarrow \quad \frac{\Delta T}{T} = \frac{3}{2}\frac{\Delta r}{r}$$

37. (d)

38. (d)

As from Kepler's law we have

$$T_0^2 = 4\pi^2\left(\frac{r^3}{GM}\right)$$

and

$$T^2 = 4\pi^2\left(\frac{(\eta r)^3}{G\left(\rho 4/3\pi(\eta R)^3\right)}\right)$$

$$= 4\pi^2\frac{r^3}{GM} = T_0^2$$

Therefore, $T = T_0$

39. (c)

For a shell at all interior points, the gravitational potential is constant and outside it varies inversely with the distance from centre. Hence, option (c) is correct

40. (b)

Orbital velocity

$$v = \frac{2\pi r}{T}$$

Hence, $$\text{KE} \propto \frac{1}{T^2}$$

41. (c)

Let g_0 to the acceleration due to gravity on surface, We use

$$mg_0 = \frac{GM_0 m}{\left(\frac{D_0}{2}\right)}$$

$$g_0 = \frac{4GM_0}{D_0^2}$$

42. (b)

Surface integral is

$$\phi = \oint g \cdot ds$$

If we consider S is a symmetric sphere of radius R, we use

$$\phi = \oint \frac{Gm_0}{R^2}\cdot ds = -\frac{Gm_0}{R^2}\cdot 4\pi R^2 = -4\pi G m_0$$

As the result is independent of R, for all surface of any shape this result is valid and this is also called Gauss's law for gravitational field.

43. (a)

Initially force on particle P due to sphere is

$$F_1 = m\cdot\frac{GM}{(2R)^2} = \frac{GMm}{4R^2}$$

After capacity is created by removing mass

$$m_{\text{cavity}} = \frac{M}{4/3\pi R^3}\times\frac{4}{3}\pi\left(\frac{R}{2}\right)^3 = \frac{M}{8}$$

The force on particle P is

$$F_2 = F_1 - F_{\text{cavity}}$$

$$= \frac{GMm}{4R^2} - m\frac{G(M/8)}{(3R/2)^2}$$

$$= \frac{GMm}{R^2}\left[\frac{1}{4} - \frac{1}{18}\right]$$

$$= \frac{7}{36}\frac{GMm}{R^2}$$

$$\Rightarrow \quad \frac{F_1}{F_2} = \frac{9}{7}$$

44. (a)

$$T^2 \propto R^3$$

Time period is independent of mass.

45. (a)

Kepler's second law

46. (d)

47. (d)

48. (d)

Using conservation of mechanical energy on the surface of earth and at a height $R/4$ from the surface of earth.

$$\text{KE}_1 + \text{PE}_1 = \text{KE}_2 + \text{PE}_2$$

Let the minimum velocity be v

$$\frac{1}{2}mv^2 + \left(-\frac{GMm}{R}\right) = 0 + \left(-\frac{GMm}{R+\frac{R}{4}}\right)$$

$$\frac{1}{2}mv^2 = \frac{GMm}{R} - \frac{4GMm}{5}$$

$$v^2 = \frac{2GM}{5R}$$

$$= \frac{2gR^2}{5R} = \frac{2}{5}gR \ (GM = gR^2)$$

$$v = \sqrt{\frac{2}{5}gR}$$

49. (b)

$$\frac{\text{KE of satellite to escape}}{\text{KE of satellite to put I circular orbit}} = \frac{\frac{1}{2}mv_e^2}{\frac{1}{2}mv_0^2}$$

$$= \frac{mv_e^2}{mv_0^2}$$

$$v_e = \sqrt{\frac{2GM}{R}}$$

$$v_0 = \sqrt{\frac{GM}{R}}$$

$$\frac{v_e^2}{v_0^2} = 2$$

50. (d)

$$mg_e h_e = mg_p h_p$$

$$h_p = \frac{g_e h_e}{g_p}$$

$$= \frac{10 \times 2}{0.2} = 100 \text{ m}$$

51. (d)

$$T^2 = \frac{4\lambda^2}{GM} r^3$$

where T is time period, M is mass of earth.
Time period is independent of mass of Moon.

52. (b)

$$v_A = \sqrt{\frac{GM}{R_E + R_A}}$$

$$v_B = \sqrt{\frac{GM}{R_E + R_B}}$$

$$\frac{v_A}{v_B} = \sqrt{\frac{R_E + R_B}{R_E + R_A}}$$

53. (b)

$$g_e = \frac{GMe}{R_e}$$

$$g_m = \frac{GMm}{R_m}$$

$$\frac{g_m}{g_e} = \frac{M_m}{M_e} \times \frac{R_e}{R_m} = \frac{1}{10} \times 2 = \frac{1}{5}$$

$$g_m = g_e \times \frac{1}{5} = \frac{9.8}{5} = 1.96$$

54. (a)

The variation of g with latitude ϕ is given by

$$g' = g - R\omega^2 \cos^2 \phi$$

At pole, $\phi = 90°$

$$g' = g - R\omega^2 \cos 90°$$

$$g' = g - 0 = g$$

$$g_p = g$$

$$g_\lambda = g_p - R\omega^2 \cos^2 60°$$

$$= g_p - R\omega^2 \left(\frac{1}{2}\right)^2$$

$$= g_p - \frac{1}{4}R\omega^2$$

55. (c)

Total mechanical energy = KE + PE

$$= \frac{1}{2}mv_e^2 + \frac{-GMm}{R}$$

$$= \frac{1}{2}m\left(\frac{\sqrt{2GM}}{R}\right)^2 - \frac{GMm}{R}$$

$$= \frac{GMm}{R} - \frac{GMm}{R}$$

$$= 0$$

56. (a)

Force due to C_1 on C_2

$$F_1 = F = \frac{Gmm}{4R^2} = \frac{Gm^2}{4R^2}$$

Force due to C_3 on C_2

$$F_2 = F = \frac{Gmm}{4R^2} = \frac{Gm^2}{4R^2}$$

Two equal forces acing on C_2 at angle $60°$

$\therefore$ The resultant force $= 2F \cos 30°$

$$= 2 \times \frac{Gm^2}{4R^2} \cos 30°$$

$$= \sqrt{3}\frac{GM^2}{4R^2}$$

57. (a)

$$g_e = \frac{GM_e}{R_e^2}$$

$$g_m = \frac{GMm}{R_m^2}$$

$$\frac{g_m}{g_e} = \frac{M_m}{M_e^2} \times \frac{R_e^2}{R_m^2}$$

$$\frac{1}{6} = \frac{\frac{4}{3}\lambda R_m^2}{\frac{4}{3}\lambda R_e^2} \times \frac{R_e^2}{R_{m2}}$$

$$= \frac{1}{6} = \frac{R_m}{R_e}$$

58. (d)

If R is fixed the time period will remain constant.

59. (a)

60. (d)

$$T^2 = \frac{4\pi^2}{GM} r^3$$

$$T^2 = \frac{4\pi^2 r^3}{G \times \frac{4}{3}\pi r^3 \rho}$$

$$T^2 = \frac{3\pi}{G\rho}$$

$$T = \sqrt{\frac{3\pi}{G\rho}}$$

61. (c)

Acceleration due to gravity g is inversely proportional to square of radius. Therefore, as radius is decreasing g will increase.

62. (a)

The potential energy on the surface of earth $= -\frac{GMm}{R}$

The PE at a height from the surface of earth $= -\frac{GMm}{2R}$

$$\text{Gain in PE} = -\frac{GMm}{2R} - \left(-\frac{GMm}{R}\right)$$

$$= \frac{GMm}{2R}$$

We know $g = \frac{GMm}{R^2}$

Therefore, $\frac{GMm}{2R} = \frac{1}{2} mgR$

63. (d)

$$T_1 = 2\lambda\sqrt{\frac{l}{g}}$$

$$T_2 = 2\lambda\sqrt{\frac{l}{g'}}$$

g' at a height R from the surface of earth $= g' = \frac{GM}{4R^2}$

$$\frac{g'}{g} = \frac{GM}{4R^2} \times \frac{R^2}{GM} = \frac{1}{4}$$

$$\frac{T_2}{T_1} = \sqrt{\frac{g}{g'}} = \sqrt{4} = 2$$

64. (a)

$$g_e = g - R\omega^2 \cos^2 \phi$$

at equator $\phi = 0 \Rightarrow \cos\phi = 1$

$$g_e = g - R\omega^2$$

$$0 = 10 - 6400 \times 10^3 \omega^2$$

$$\omega^2 = \frac{10}{6400 \times 10^3}$$

$$\omega = \frac{1}{800} \text{ rad/s}^{-1}$$

65. (a)

66. (d)

67. (a)

The value of acceleration due to gravity at a depth d

$$g_d = g\frac{(R-d)}{R}$$

The volume of acceleration due to gravity at a height h,

$$g_h = g\left(\frac{R}{R+h}\right)^2$$

As per problem $g_d = g_h$

$$g_d = g\left(\frac{R-y}{R}\right); g_h = g\left(\frac{R}{R+x}\right)$$

$$g\left(\frac{R-y}{R}\right) = g\left(\frac{R}{R+x}\right)^2$$

$$1-\frac{y}{R} = \frac{1}{\left(1+\frac{x}{R}\right)^2}$$

$$1-y/R = \left(1+\frac{x}{R}\right)^{-2}$$

Since $x \leq R$,

$$\left(1+\frac{x}{R}\right)^{-2} = 1-\frac{2x}{R}$$

$$1-\frac{y}{R} = 1-\frac{2x}{R}$$

$$\Rightarrow y = 2x$$

68. (b)

69. (a)

70. (d)

71. (a)

The gravitational potential is the gravitational potential energy for 1 kg mass.

The gravitational potential energy at x is

$$= -\frac{GMm}{R} = -800 \text{ KJ}$$

The gravitational potential energy at y is

$$= -\frac{GMm}{2R} = -\frac{800}{2} = -400$$

Therefore work done = – 400 – (–800) = 400 KJ

72. (a)

The weight at sea level is w_2. As we go above the surface of earth or below the surface of earth, the value of acceleration due to gravity decreases, Hence weight

$$w_1 < w_2 > w_3$$

73. (b)

74. (b)

75. (c)

76. (a)

Escape velocity $v_e = \sqrt{\frac{2GM}{R}}$

Let v_1 and v_2 be the escape velocity for earth and planet respectively.

Then $\frac{v_2}{v_1} = \sqrt{\frac{M_2}{M_1} \times \frac{R_1}{R_2}}$

$$\frac{v_2}{v_1} = \sqrt{6 \times \frac{1}{2}} = \sqrt{3}$$

$$v_2 = v_1\sqrt{3} = v\sqrt{3}$$

77. (d)

They may have different masses, therefore, Angular momentum and kinetic energy may be different but velocity depends on the orbit.

78. (a)

$$\text{PE} = -\frac{Gm_1m_2}{r}$$

$$\text{PE of the system} = -\frac{3Gm^2}{r}$$

$$\text{PE}_1 = -\frac{3Gm^2}{r}$$

$$\text{PE}_2 = -\frac{3Gm^2}{(r+0.4)}$$

Work done = Change in PE (assuming $\Delta kE = 0$)

$$PE_2 - PE_1 = -\frac{3Gm^2}{r+0.4} - \left(-\frac{3Gm^2}{r}\right)$$

$$= -3Gm^2\left(-\frac{1}{r+0.4} + \frac{1}{r}\right)$$

$$= \frac{3GM^2 \times 0.4}{r(r+0.4)}$$

$$\text{Substituting the values} = \frac{3 \times 6.67 \times 10^{-11} \times (0.1)^2 \times 0.4}{0.2 \times 0.6}$$

$$= 6.67 \times 10^{-12} \text{ J}$$

The nearest option is (a)

79. (a)

The required minimum energy is ΔKE + ΔPE

$\Delta\text{KE} = \frac{1}{2}mv_0^2$ where v_0 is orbital velocity

$$= \frac{1}{2}m\left(\sqrt{\frac{GM}{3R}}\right)^2 = \frac{1}{6}\frac{GMm}{R}$$

$$\Delta\text{PE} = -\frac{GMm}{3R} - \left(-\frac{GMm}{R}\right)$$

$$= \frac{2}{3}\frac{GMm}{R}$$

$$\Delta KE + \Delta PE = \frac{1GMm}{6R} + \frac{2GMm}{3R}$$

$$= \frac{5GMm}{6R}$$

80. (b)

Angular momentum is constant, therefore,

$$m_1 v_1 r_1 = m_2 v_2 r_2$$

$$\frac{v_2}{v_1} = \frac{r_1}{r_2}$$

81. (c)

82. (b)

$$T^2 \propto R^3 \text{ (Kepler's Law)}$$

$$\frac{T_2^2}{T_1^2} = \frac{R_2^3}{R_1^3}$$

$$\left(\frac{T_2}{T_1}\right)^2 = \left(\frac{R_2}{R_1}\right)^3 = \left(\frac{1}{2}\right)^3 = \frac{1}{8}$$

$$\frac{T_2}{T_1} = \frac{1}{2\sqrt{2}}$$

$$T_2 = \frac{365}{2\sqrt{2}} = 129$$

83. (b)

For stable circular orbit of planet, we use

$$mR\omega^2 \propto R^{-5/2}$$

$$\Rightarrow \qquad T^2 \propto R^{7/2}$$

84. (b)

Centripetal force (F) $= \dfrac{mv^2}{R}$ and the gravitational force

(F) $= \dfrac{GMm}{R^2} = \dfrac{GMm}{R}$

(where $R^2 \rightarrow R$). Since $\dfrac{mv^2}{R} = \dfrac{GMm}{R}$ therefore

$v = \sqrt{GM}$. Thus velocity v is independent of R.

85. (d)

Rate of change of mass $= \dfrac{dm}{dt} = \alpha v$

Retarding force = Rate of change of momentum

= Velocity $\times$ Rate of change in mass $= -v \times \dfrac{dM}{dt}$

$= -v \times \alpha v = -\alpha v^2$ (Minus sign of v due to deceleration)

Therefore, acceleration $= -\dfrac{\alpha v^2}{M}$

86. (b)

Escape velocity does not depend on the angle of projection.

87. (d)

Time period does not depend on the mass.

As $\qquad T^2 \propto r^3$

$$\frac{T_A}{T_B} = \frac{r^{3/2}}{2^{3/2} r^{3/2}} = 1 : 2\sqrt{2}$$

88. (a)

$$\frac{GMm}{r^2} = m\omega^2 r \Rightarrow r^3 = \frac{GM}{\omega^2} = \frac{gR^2}{\omega^2}$$

$$r = \left(gR^2 / \omega^2\right)^{1/3}$$

89. (d)

$$\text{Total energy} = -\text{KE} = -\frac{1}{2}mv^2$$

90. (c)

In a circular or elliptical orbital motion, torque is always acting parallel to velocity. So, angular momentum is conserved. In attractive field, potential energy and the total energy is negative. Kinetic energy increases with increase in velocity. If the motion is in a plane, the direction of L does not charge.

91. (a)

Since escape velocity $\left(v_e = \sqrt{2gR_e}\right)$ is independent of angle of projection, so it will not change.

92. (a)

Inside the shell, there is zero gravity. Hence, no force will act and outside the shell acceleration due to gravity is inversely proportional to r^2. Hence, option (A) is correct.

93. (a)

Radius of earth $R = 6400$ km $\therefore \; h = \dfrac{R}{4}$

Acceleration due to gravity at a height h

$$g_h = g\left(\frac{R}{R+h}\right)^2 = g\left(\frac{R}{R+\frac{R}{4}}\right)^2 = \frac{16}{25}g$$

At depth d value of acceleration due to gravity

$$g_d = \frac{1}{2}g_h$$

$$\Rightarrow g_d = \frac{1}{2}\left(\frac{16}{25}\right)g \Rightarrow g\left(1 - \frac{d}{R}\right) = \frac{1}{2}\left(\frac{16}{25}\right)g$$

By solving we get $d = 4.3 \times 10^6$ m.

94. (a)

$$g' = g\left(\frac{R}{R+h}\right)^2 = g\left(\frac{R}{R+2R}\right)^2 = \frac{g}{9}$$

95. (a)

96. (d)

Because the body weighs zero in satellite.

97. (a)

$$g' = g\left(1-\frac{d}{R}\right) = 9.8\left(1-\frac{100}{6400}\right) = 9.66 \text{ m/s}^2$$

98. (a)

$$\frac{g_1}{g_2} = \frac{\rho_1}{\rho_2}\times\frac{R_1}{R_2} = \frac{3}{2}\times\frac{2}{3} = 1$$

99. (d)

Range of projectile $R = \dfrac{u^2 \sin 2\theta}{g}$

If u and θ re constant then $R \propto \dfrac{1}{g}$

$$\frac{R_m}{R_e} = \frac{g_e}{g_m} \Rightarrow \frac{R_m}{R_e} = \frac{1}{0.2} \Rightarrow R_m = \frac{R_e}{0.2} \Rightarrow R_m = 5R_e.$$

100. (a)

$$g = \frac{4}{3}\pi G\rho R \Rightarrow g \propto \rho R \Rightarrow \frac{g_e}{g_m} = \frac{\rho_e}{\rho_m}\times\frac{R_e}{R_m}$$

$$\Rightarrow \frac{6}{1} = \frac{5}{3}\times\frac{R_e}{R_m} \Rightarrow R_m = \frac{5}{18}R_e$$

101. (b)

102. (c)

Since acceleration due to gravity is independent of mass, time is also independent of mass (or density) of an object.

103. (b)

Potential at internal point of solid sphere at a distance r

$$V = \frac{-GM}{R}\left[\frac{3}{2} - \frac{r^2}{2R^2}\right]$$

at $$r = \frac{R}{2}$$

$$V_1 = -\frac{GM}{R}\left[\frac{3}{2} - \frac{R^2}{8R^2}\right] = -\frac{11}{8}\frac{GM}{R}$$

Because of sphere removed,

$$V_2 = \frac{3}{2}\cdot\frac{G\dfrac{M}{8}}{\dfrac{R}{2}} = \frac{3}{8}\frac{GM}{R}$$

Net potential, $V = V_1 + V_2 = -\dfrac{11}{8}\dfrac{GM}{R} + \dfrac{3}{8}\dfrac{GM}{R}$

$$= -\frac{GM}{R}.$$

104. (a)

$$v_e = \sqrt{\frac{2GM}{R}} = 100 \Rightarrow \frac{GM}{R} = 5000$$

Potential energy $U = -\dfrac{GMm}{R} = -5000$ J.

105. (b)

Kinetic energy = – (Total energy)

Negative energy means the satellite is bound to the earth by an attractive force. Hence to remove satellite, the energy required is (+ K.E.).

106. (c)

Gravitational field inside hollow sphere will be zero.

107. (a)

Potential at the centre due to single mass $= \dfrac{-GM}{L/\sqrt{2}}$

Potential at the centre due to all four masses

$$= -4\frac{GM}{L/\sqrt{2}} = -4\sqrt{2}\frac{GM}{L}$$

$$= -\sqrt{32}\times\frac{GM}{L}.$$

108. (b)

$v = \sqrt{\dfrac{GM}{r}}$ if $r_1 > r_2$ then $v_1 < v_2$

Orbital speed of satellite does not depend upon the mass of the satellite.

109. (b)

110. (b)

111. (a)

$$\omega = \frac{2\pi}{T} = \frac{2\pi}{24} = \frac{\pi}{12}$$

112. (a)

When a body is acted on by the force towards a point and the magnitude of force is inversely proportional to the square of distance, it means it obeys inverse square law and represents ellipse. For example path of the planet around the sun and the force that acts between sun and planet are proportional to $\frac{1}{r^2}$.

113. (b)

$$\frac{dA}{dt}=\frac{L}{2m}=\text{constant}$$

114. (b)

115. (c)

Escape velocity for that body $v_e=\sqrt{\frac{2Gm}{r}}$

v_e should be more than or equal to speed of light

$$\sqrt{\frac{2Gm}{r}}\geq c$$

116. (c)

Angular momentum of the earth around the sun is $L=M_E v_0 r$

$$\Rightarrow \qquad L=M_E\sqrt{\frac{GM_s}{r}}r$$

$$\Rightarrow \qquad L=\left[M_E^2 GM_s r\right]^{1/2}$$

where, M_E = Mass of the earth

M_s = Mass of the sun

r = Distance between the sun and the earth

$$L\propto\sqrt{r}$$

117. (b)

Mass of satellite does not affect its orbital radius.

118. (c)

In the problem, orbital radius is increased by 1%.

Time period of satellite $T\propto r^{3/2}$

Percentage change in time period

$=\frac{3}{2}$ (% change in orbital radius) $=\frac{3}{2}$ (1%) = 1.5 %

119. (b)

$$\frac{T^2}{R^3}=\frac{T^2}{d^3}=\frac{1}{n^2d^3}=\text{Constant}$$

$$n_1^2 d_{12}^3=n_2^2 d_2^3$$

120. (c)

Potential energy $=2\times$ (Total energy) $=2E_0$

Because we know $U=\frac{-GMm}{r}$ and $E_0=\frac{-GMm}{2r}$

121. (c)

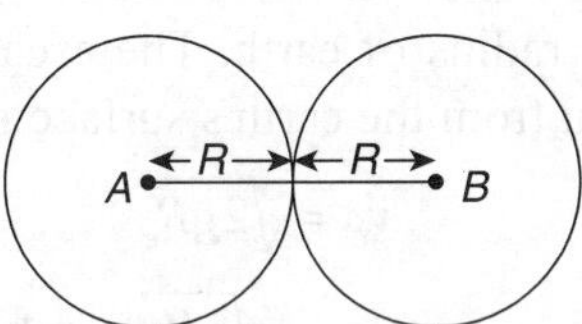

Let masses of two balls be $m_1=m_2=m$ and the density be ρ. Distance between their centres = AB = $2R$. Thus, the magnitude of the gravitational force F that two balls separated by a distance $2R$ exert on each other is

$$F=G\frac{(m)(m)}{(2R)^2}$$

$$=G\frac{m^2}{4R^2}=G\frac{\left(\frac{4}{3}\pi R^3\rho\right)^2}{4R^2}$$

$$F\propto R^4$$

122. (a)

The earth possesses rotational motion about an axis through its poles. The value of acceleration due to gravity at a place (at given latitude) is affected due to its rotational motion. If earth ceases to rotate, the weight of a body at equator will increase. However, there will be no effect on the weight at poles. The effect of rotation of the earth on acceleration due to gravity is to decrease its value. Therefore if the earth stops rotating, the value of g will increase.

123. (c)

If no external torque acts on a system, then angular momentum of the system does not change.

If $\qquad \tau=0$

$$\Rightarrow \qquad \frac{dL}{dt}=0$$

$$L=\text{constant}$$

Hence,

$$mv_{max}r_{min}=mv_{min}r_{max}$$

$$\Rightarrow \qquad r_{min}=\frac{v_{min}\times r_{max}}{v_{max}}$$

$$=\frac{1\times10^3\times4\times10^4}{3\times10}$$

$$=\frac{4}{3}\times10^3\text{ km}$$

124. (c)

The orbital velocity of a satellite close to the earth is

$$v_0 = \sqrt{gR_e} \quad \text{...(i)}$$

Where R_e is radius of earth. The escape velocity for a body thrown from the earth's surface is

$$v_e = \sqrt{2gR_e} \quad \text{...(ii)}$$

Thus, $$\frac{v_0}{v_e} = \frac{\sqrt{gR_e}}{\sqrt{2gR_e}} = \frac{1}{\sqrt{2}}$$

$$v_e = \sqrt{2}v_0$$

Given $v_e = 2$ km/s

$$v_e = \frac{v_e}{\sqrt{2}} = \frac{2}{\sqrt{2}} = \sqrt{2} \text{ km/s}$$

125. (a)

According to law of gravitation, the force of attraction acting on the body due to earth is given by

$$F = G\frac{Mm}{R^2} \quad \text{...(i)}$$

The acceleration due to gravity g in the body arises due to the force F from Newton's second law of motion, we have

$$F = mg \quad \text{...(ii)}$$

From Eqs. (i) and (ii) we get

$$mg = G\frac{Mm}{R^2}$$

$$\Rightarrow \quad g = \frac{GM}{R^2} \Rightarrow g \propto \frac{M}{R^2}$$

126. (d)

From Kepler's third law of planetary motion, the square of the period of revolution (T) of any planet around the sun is directly proportional to the cube of the semi major-axis of its elliptical orbit (R), that is

$$T^2 \propto R^3$$

$$\Rightarrow \quad \frac{T^2}{R^3} = \text{Constant}$$

127. (c)

At a certain velocity of projection, the body will go out of the gravitational field of earth and never return to earth. This initial velocity is called escape velocity.

$$V_e = \sqrt{2gR}$$

where g is acceleration due to gravity and R is the radius. As is clear from above formula, that escape velocity does not depend upon mass of body. Hence, it will be same for a body of 100 kg as for 1 kg body.

128. (c)

The kinetic energy required for body to escape the gravitational field is $k_1 = \frac{1}{2}mv_{es}^2$

The energy with which the body is projected is $k_2 = \frac{1}{2}m(2v_{es})^2$

$$\frac{1}{2}mv^2 = \frac{1}{2}m(2v_{es})^2 - \frac{1}{2}mv_{es}^2$$

$$\frac{1}{2}mv^2 = \frac{1}{2}m3v_{es}^2$$

$$\Rightarrow \quad v^2 = 3v_{es}^2$$

$$\Rightarrow \quad v = \sqrt{3}v_{es}$$

129. (d)

A satellite revolving around the earth has potential as well as kinetic energy. It has potential energy because it remains in the gravitational field and kinetic because it is in motion.

Hence, total energy is

$$E = U + k = \frac{-GMm}{R} + \frac{1}{2}\frac{GMm}{R}$$

$$E = -\frac{1}{2}\frac{GMm}{R}$$

For satellites A and B,

$$\frac{m_A}{m_B} = \frac{3}{1} \text{ and } \frac{r_A}{r_B} = \frac{r}{4r}$$

$$\frac{E_A}{E_B} = \frac{\frac{m_A}{r_A}}{\frac{m_B}{r_B}} = 3 \times 4 = \frac{12}{1} = 12$$

130. (d)

$$T^2 \propto R^3$$

$$\Rightarrow \quad \left(\frac{T_A}{T_B}\right)^2 = \left(\frac{R_A}{R_B}\right)^3$$

$$\Rightarrow \quad \left(\frac{T_A}{T_B}\right)^2 = \left(\frac{r}{2r}\right)^3 = \frac{1}{8}$$

$$\Rightarrow \quad \frac{T_A}{T_B} = \frac{1}{2\sqrt{2}}$$

131. (c)

Satellite is revolving around the earth in an orbit with a velocity v. When a ball is dropped from the satellite, the ball will attain the same velocity v. This velocity is sufficient to move in the same orbit of the satellite.

132. (c)

Let M be mass of earth, and m = mass of the body. The work required to raise the body at a height h is

$$W = GMm\left(\frac{1}{R} - \frac{1}{R+h}\right) = \frac{GMmh}{R(R+h)}$$

$\because \quad GM = gR^2$

$$W = \frac{gR^2mh}{R(R+h)} = \frac{mgh}{\left(1+\frac{h}{R}\right)}$$

Given $h = R$ and by substituting, we get

$$W = \frac{mgR}{\left(1+\frac{R}{R}\right)} = \frac{1}{2}mgR = \Delta U$$

$\therefore$ The change in potential energy $= \frac{1}{2}mgR$

133. (a)

$$g = \frac{GM}{R^2}$$

and $\quad g_h = \frac{GM}{R^2\left(1+\frac{h}{R}\right)^2}$

$$\frac{g}{g_h} = \left(1+\frac{h}{R}\right)^2$$

134. (a)

$$g = \frac{GM}{R^2} = \frac{G}{R^2}.\frac{4}{3}\pi R^3\rho = \frac{4G\pi}{3}R\rho$$

$\Rightarrow \quad g \propto R$

$$\frac{g}{g_1} = \frac{R}{R_1}$$

$$\frac{g}{g_1} = \frac{R}{0.2R} (\because R_1 = 0.2R)$$

$$g_1 = 0.2g$$

135. (b)

$\because \quad T^2 \propto R^3$

$\Rightarrow \quad \left(\frac{T_1}{T_2}\right)^2 = \left(\frac{R_1}{R_2}\right)^3 \quad$ ...(i)

Given $\quad R_1 = R;$

$T_1 = 1$ day and $T_2 = 8$ days

Substituting in Eq. (i) we will get

$R_2 = 4$ R

136. (c)

From Kepler's second law, the area swept out by planet per unit time is constant. Areal velocity, $\frac{dA}{dt} = \frac{L}{2m}$ where L is angular momentum.

$\therefore$ Kepler's second law, that areal velocity is constant means the angular momentum is conserved.

137. (d)

From Kepler's second law, areal velocity is constant.

138. (c)

The normal force exerted by the person on the floor of a lift moving downwards with an acceleration a is given by

$$N = m(g - a)$$

If $\quad a = g$, then $N = 0$

The same reasoning can be applied for the satellite orbiting around the earth.

139. (d)

From Kepler's third law:

$\because \quad T^2 \propto R^3$

Given $\quad R_1 = 6R + R = 7R$

$R_2 = 2.5R + R = 3.5R$

and $\quad T_1 = 24$ h and $T_2 = ?$

$$\left(\frac{T_2}{T_1}\right)^2 = \left(\frac{R_2}{R_1}\right)^3$$

$$\Rightarrow \left(\frac{T_2}{24}\right)^2 = \left(\frac{3.5R}{7R}\right)^3 = \frac{1}{8}$$

$$T_2^2 = \frac{(24)^2}{8}$$

$$\Rightarrow T_2 = \frac{24}{\sqrt{8}} = \frac{12}{\sqrt{2}} = 6\sqrt{2} \text{ hour}$$

140. (c)

The value of acceleration due to gravity at a height h above the earth's surface,

$$g_h = \frac{g}{\left(1+\frac{h}{R}\right)^2}$$

Given, $\quad h = \frac{R}{2} \Rightarrow g_h = \frac{g}{\left(1+\frac{1}{2}\right)^2} = \frac{4}{9}g$

$$\Rightarrow mg_h = \frac{4}{9}mg$$

$$\Rightarrow \qquad W' = \frac{4}{9}\text{ W}$$

141. (c)

Using law of conservation of energy,

$$K_i + U_i = K_f + U_f$$

$$\frac{1}{2}mv_i^2 - \frac{GM_E m}{10R_E} = \frac{1}{2}mv_f^2 - \frac{GM_E m}{R_E}$$

$$\Rightarrow \qquad v_f^2 = v_i^2 + \frac{2GM_E}{R_E}\left(1 - \frac{1}{10}\right)$$

142. (c)

Using law of conservation of energy,

$$K_i + U_i = K_f + U_f$$

$$\frac{1}{2}mv^2 - \frac{GMm}{R} = 0 - \frac{GMm}{(R+h)}$$

$$\Rightarrow \qquad \frac{1}{2}mv^2 = \frac{GMm}{R} - \frac{GMm}{R+h}$$

$$\Rightarrow \qquad \frac{1}{2}mv^2 = \frac{GMmh}{R(R+h)}$$

$$\Rightarrow \qquad Rv^2(R+h) = 2GMh$$

$$\Rightarrow \qquad v^2R^2 + v^2Rh = 2GMh$$

$$\Rightarrow \qquad h\left(-v^2R + 2GM\right) = +v^2R^2$$

$$h = \frac{v^2R^2}{2GM - v^2R}$$

143. (c)

From Kepler's third law, we have

$$T^2 \propto R^3$$

$$\left(\frac{T_A}{T_B}\right)^2 = \left(\frac{R_A}{R_B}\right)^3$$

$$\Rightarrow \qquad \frac{T_A}{T_B} = \left(\frac{4R}{R}\right)^{3/2} = 4^{3/2}$$

$$\because \qquad v = \frac{2\pi R}{T} \Rightarrow v \propto \frac{R}{T}$$

$$\frac{v_A}{v_B} = \frac{R_A}{R_B} \times \frac{T_B}{T_A}$$

$$\frac{3v}{v_B} = \frac{4R}{R} \times \frac{1}{4^{3/2}} = \frac{1}{\sqrt{4}} = \frac{1}{2}$$

$$v_B = 6v$$

144. (b)

Variation of acceleration due to gravity with the rotation of earth is given by $g' = \left(g - R\omega^2\right)$

It ω increases, then g' decreases.

$\because$ Weight $= mg'$

$\therefore$ Weight also decreases.

145. (b)

$$v_e = \sqrt{\frac{2GM}{R}}$$

$$v_e \propto \frac{1}{\sqrt{R}}$$

$$\frac{v_e}{v_e'} = \sqrt{\frac{R'}{R}} = \frac{1}{10}$$

$$R' = \frac{R}{100} = \frac{6400}{100} = 64\text{ km}$$

146. (d)

$$T^2 \propto R^3$$

$$\left(\frac{T_2}{T_1}\right)^2 = \left(\frac{R_2}{R_1}\right)^3 = (1.02)^3 = 1.06$$

$$\frac{T_2}{T_1} = 1.03$$

$$\frac{T_2 - T_1}{T_1} = 0.03$$

$$\frac{T_2 - T_1}{T_1} \times 100 = 3\%$$

147. (b)

Escape velocity for the earth is $v_e = \sqrt{\frac{2GM}{R}}$

Given the velocity projection of the body

$$\Rightarrow \qquad v = \frac{3}{4}v_e = \frac{3}{4}\sqrt{\frac{2GM}{R}}$$

Total energy on earth = Total energy at maximum height h

$$\frac{1}{2}mv^2 + \left(\frac{-GMm}{R}\right) = 0 + \left(\frac{-GMm}{R+h}\right)$$

$$\frac{1}{2}m\frac{9}{16}.\frac{2GM}{R} - \frac{GMm}{R} = \frac{-GMm}{R+h}$$

$$\frac{9}{16} - 1 = \frac{-R}{R+h}$$

$$h = \frac{9}{7}R$$

148. (b)

$$g_h = \left(\frac{R}{R+h}\right)^2 g$$

$$\frac{g_h}{g} = \left[\frac{1}{1+\frac{h}{R}}\right]^2 = \frac{g/100}{g}$$

$$\frac{1}{1+\frac{h}{R}} = 0.1$$

$$h = 9R.$$

149. (a)

$$g_d = g\left[1-\frac{d}{R_e}\right]$$

$$g_h = g\left[1-\frac{2h}{R_e}\right]^2$$

Now, $d = 2h$

$$g_d = g_h$$

150. (c)

Let ρ be the average density of earth. Now, $g = \frac{GM}{R^2}$.

$$M = \rho\left(\frac{4}{3}\pi R^3\right)$$

$$\Rightarrow \quad g = \frac{4}{3}G\rho\pi R$$

Now at a depth h, the gravitational field is only due to mass in $(R-h)$.

$$\Rightarrow \quad \frac{g}{4} = \frac{4}{3}G\rho\pi(R-h)$$

$$\Rightarrow \quad R = 4(R-h)$$

$$\Rightarrow \quad h = \frac{3R}{4}$$

151. (c)

$$\frac{T^2}{R^3} = \text{constant}$$

$$\Rightarrow \quad \frac{T_1^2}{R_1^3} = \frac{T_2^2}{R_2^3}$$

$$\Rightarrow \quad T_1^2 = \left(\frac{R_1}{R_2}\right)^3 (T_2)^2$$

Here, $T_2 = 1$ day, $R_1 = 2R_2$

$$\Rightarrow T_1 = 2\sqrt{2} \text{ days}$$

152. (b)

$$\frac{1}{2}mv_e^2 = \frac{GMm}{R}$$

$$\frac{GMm}{R} + \text{KE}_{\text{final}} = \frac{1}{2}m(3v_e)^2$$

$$\Rightarrow \quad \text{KE}_{\text{final}} = \frac{1}{2}m(2\sqrt{2}v_e)^2$$

$$\Rightarrow \quad \text{Final velocity} = 2\sqrt{2}v_e$$

$$v_e = 11.2 \text{ km/s}$$

$$\Rightarrow \quad \text{Final velocity} = 31.7 \text{ km/s}$$

153. (c)

$$U_{\text{surface}} + E_1 = U_h$$

KE of satellite is zero at earth surface and at height h

$$-\frac{GM_e m}{R_e} + E_1 = -\frac{GM_e m}{(R_e+h)}$$

$$E_1 = GM_e m\left(\frac{1}{R_e} - \frac{1}{R_e+h}\right)$$

$$E_1 = \frac{GM_e m}{(R_e+h)} \times \frac{h}{R_e}$$

Gravitational attraction $F_G = ma_c = \frac{mv^2}{(R_e+h)}$

$$\Rightarrow \frac{mv^2}{(R_e+h)} = \frac{GM_e m}{(R_e+h)^2}$$

$$mv^2 = \frac{GM_e m}{(R_e+h)}$$

$$E_1 = E_2$$

$$\frac{h}{R_e} = \frac{1}{2} \Rightarrow h = \frac{R_e}{2} = 3200 \text{ km.}$$

154. (d)

Kepler's second law: planets sweep out equal area in equal intervals of time, i.e., areal velocity is constant, when earth moves around the sun.

155. (a)

Velocity of geosynchronous satellite, $v_s = \sqrt{gR_e}$

Velocity of escape $= \sqrt{2gR_e}$

$$\text{Percentage of change} = \frac{\sqrt{2gR_e} - \sqrt{gR_e}}{\sqrt{gR_e}} \times 100$$

$$= 41.4\%$$

156. (b)

$$T_1^2 = \left(\frac{R_1}{R_2}\right)^3 T_2^2$$

$$R_1 = \frac{R_2}{4}$$

$$T_2 = 1 \text{ year}$$

$$\Rightarrow \quad T_1^2 = \frac{1}{64}$$

$$\Rightarrow \quad T_1 = \frac{1}{8}$$

157. (a)

$$v_e = \sqrt{2gR}$$

$$\frac{(v_e)_{\text{planet}}}{(v_e)_{\text{earth}}} = \frac{\sqrt{g_p R_p}}{\sqrt{g_e R_e}}$$

$$= \sqrt{\frac{2\times 1}{4}} = \frac{1}{\sqrt{2}}$$

158. (c)

Gravity on planet,

$$g_p = \frac{Gm_p}{R_p^2} = \frac{1000\, GM_e}{100\, R_e^2}$$

$$\Rightarrow \quad g_p = 10 g_e$$

$$(v_e)_{\text{earth}} = \sqrt{2gR_e}$$

$$(v_e)_{\text{planet}} = \sqrt{2g_p R_p}$$

$$= \sqrt{2\times 10 g_e \times 10 R_e}$$

$$= 10\sqrt{2g_e R_e}$$

$$= 10 \times 11.2 = 112 \text{ km/s}$$

159. (b)

$$\text{PE} = mgh = mgR$$

160. (a)

Angular momentum is conserved

$$\Rightarrow \quad mv_0 r = mvR$$

$$\Rightarrow \quad v = \frac{v_0 r}{R}$$

161. (a)

$$g = \frac{GM}{R^2}$$

$$g' = \frac{GM}{(R+32)^2}$$

$$\Rightarrow \quad g' = \frac{R^2}{(R+32)^2} g = 0.99 \text{ g}$$

$$g' = 0.99 \text{ g}$$

162. (c)

$$g = \frac{GM_e}{R_e^2}$$

$$g_m = \frac{GM_m}{R_m^2} = \frac{GM_e}{90.\dfrac{R_e^{\,2}}{9}} = \frac{g}{10}$$

163. (b)

At height r from centre of earth, orbital velocity

$$= \sqrt{\frac{GM}{r}}$$

By energy conservation,

$$\text{KE of } m + \left(-\frac{GMm}{r}\right) = 0 + 0$$

(At infinity PE = KE = 0)

$$\text{KE of } m = \frac{GMm}{r} = \left(\sqrt{\frac{GM}{r}}\right)^2 m = mv^2$$

164. (d)

The two bodies are attracted to each other due to gravitation. If they move towards each other under mutual interaction, they will meet at the centre of mass of the system of the two particles.

$$L = 12R - R - 2R = 9R$$

$\therefore$ The distance covered by the smaller body just before collision,

$$x_{\text{cm}} = \frac{m_2 L}{m_1 + m_2} = \frac{5M \times 9R}{M + 5M} = 7.5R$$

165. (b)

$$T^2 \propto R^3$$

$$\left(\frac{T_2}{T_1}\right)^2 = \left(\frac{R_2}{R_1}\right)^3$$

$$\Rightarrow \quad R_2^{\,3} = \left(\frac{T_2}{T_1}\right)^2 R_1^3$$

$$\Rightarrow \quad R_2 = \left(\frac{T_2}{T_1}\right)^{2/3} R_1$$

Given: $R_1 = 36000$ km; $T_1 = 24h$; $T_2 = 3h$.

$$\Rightarrow \quad R_2 = \left(\frac{3}{24}\right)^{2/3} (36000) \text{ km} = 9000 \text{ km}$$

166. (b)

$$T^2 \propto d^3$$

or $\frac{1}{n^2} \propto d^3 \Rightarrow n^2 d^3 \Rightarrow$ Constant.

$$n_2^2 d_2^3 = n_1^2 d_1^3$$

167. (b)

$$T \propto \frac{1}{\sqrt{g}}$$

$$\frac{T_1}{T_2} = \sqrt{\frac{g_2}{g_1}} = \frac{R}{R+h}$$

$$= \frac{R}{R+2R} = \frac{1}{3}$$

168. (b)

Gravitational acceleration at a height h is given by

$$g_h = g\left(\frac{R}{r}\right)^2 = g\left(\frac{R}{3R/2}\right)^2 = \frac{4}{9}g$$

$$....(R+h=r=\frac{3R}{2})$$

$\therefore$ Gravitational pull on the satellite of mass 200 Kg is

$$= mg_h = m\left(\frac{4}{9}g\right) = \frac{4}{9}\times 200\times 10$$

$$= 888.9\text{N} = 889 \text{ N}$$

169. (b)

Since there is no external torque, angular momentum remains conserved.

$$I_1\omega_1 = I_2\omega_2$$

$$I_1\left(\frac{2\pi}{T_1}\right) = I_2\left[\frac{2\pi}{T_2}\right]$$

$$T_2 = \left[\frac{I_2}{I_1}\right]T_1 = \left[\frac{\frac{2}{5}MR_2^2}{\frac{2}{5}M\text{ R}_1^2}\right]T_1$$

$$T_2 = \frac{T_1}{4} = \frac{24}{4}\text{hrs} = 6 \text{ hrs.}$$

170. (a)

Gravitational force is neutralized by the upward viscous drag $6\pi\eta rv$ and upthrust.

171. (a)

$$g_h = \left(\frac{R}{R+h}\right)^2 g = \left(\frac{R}{R+2R}\right)^2 g = \frac{g}{9}$$

172. (c)

$$g = \frac{GM}{R^2} = \frac{G.\frac{4}{3}\pi R^3 d}{R^2} = \frac{4}{3}\pi G\, Rd$$

$$g \propto R.d.$$

173. (b)

Orbital velocity $v_0 = v = R\sqrt{\frac{R}{R+h}} = R\sqrt{\frac{G}{r}}$

$$\frac{\Delta v}{v} = -\frac{1}{2}\frac{\Delta r}{r}$$

Thus, if r is decreased by 1%, v will increase by 0.5%.

174. (a)

Let A be the Gaussian surface enclosing charge Q.

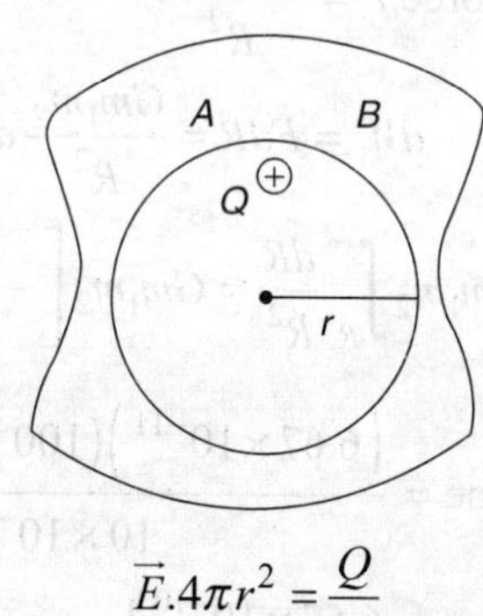

$$\vec{E}.4\pi r^2 = \frac{Q}{\varepsilon_0}$$

$$\vec{E} = \frac{Q}{4\pi\varepsilon_0 r^2}$$

Flux, $\phi = \vec{E}.4\pi r^2 = \frac{Q}{\varepsilon_0}$

Every line passing through A has to pass through B, whether B is a cube of any surface.

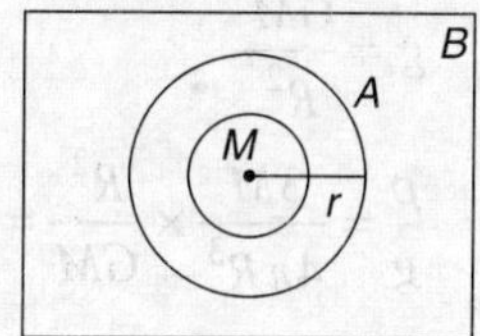

It is only for Gaussian surface, the lines of field should be normal.

Assuming the mass is a point mass,

$$\vec{I}, \text{ gravitational field } = -\frac{GM}{r^2}$$

Flux, $\phi_g = \left|\vec{I}.4\pi r^2\right| = \frac{4\pi r^2.GM}{r^2} = 4\pi GM$

Here B is a cube. As explained earlier, whatever be the shape, all the lines passing through A are passing through B, although all the lines are not normal.

Statement-2 is correct because when the shape of the earth is spherical, area of the Gaussian surface is $4\pi r^2$. This ensures the inverse square law.

175. (d)

$$v_e = \sqrt{\frac{2GM}{R}}$$

For the earth $v_e = 11$ km/sec

Mass of the planet = $10\ M_e$

Radius of the planet = $\frac{R}{10}$

$$v_e = \sqrt{\frac{2GM\times 10}{\frac{R}{10}}} = 10\times 11 = 110 \text{ km/sec}$$

176. (a)

Gravitational force $F = \frac{Gm_1m_2}{R^2}$

$$dW = FdR = \frac{Gm_1m_2}{R^2}dR$$

$$\therefore \int_0^w dW = Gm_1m_2\int_R^\infty \frac{dR}{R^2} = Gm_1m_2\left[-\frac{1}{R}\right]_R^\infty = \frac{Gm_1m_2}{R}$$

$$\text{Work done} = \frac{(6.67\times 10^{-11})(100)(100\times 10^{-3})}{10\times 10^{-2}}$$

Work done = 6.67×10^{-10} J

177. (a)

$$\text{Density} = \frac{\text{Mass of the earth}}{\text{Volume of the earth}}$$

$$\rho = \frac{M}{\left(\frac{4}{3}\right)\pi R^3} = \frac{3M}{4\pi R^3}$$

$$g = \frac{GM}{R^2}$$

$$\frac{\rho}{g} = \frac{3M}{4\pi R^3}\times\frac{R^2}{GM} = \frac{3}{4\pi RG}$$

$$\Rightarrow \rho = \frac{3}{4\pi RG}g$$

$\therefore$ Average density is directly proportional to g.

178. (a)

At height, $g_h = g\left(1-\frac{2h}{R}\right)$ where $h << R$

$$g - g_h = \frac{2hg}{R}$$

$$\Delta g_h = \frac{2hg}{R} \quad \ldots\text{(i)}$$

At depth, $g_d = g\left(1-\frac{d}{R}\right)$ where $d << R$

or $$g - g_d = \frac{dg}{R}$$

$$\Delta g_d = \frac{dg}{R}$$

Given, $\Delta g_h = \Delta g_d$

Therefore, from Eqs. (i) and (ii),

$$\frac{2hg}{R} = \frac{dg}{R}$$

$$\Rightarrow \quad d = 2h$$

179. (a)

For a motion of a planet in a circular orbit

Centripetal force = Gravitational force

$$mR\omega^2 = \frac{GMm}{R^n}$$

$$\omega = \sqrt{\frac{GM}{R^{n+1}}}$$

So, $$T = \frac{2\pi}{\omega} = 2\pi\sqrt{\frac{R^{n+1}}{GM}} = \frac{2\pi}{\sqrt{GM}}R^{\frac{(n+1)}{2}}$$

$\therefore$ T is proportional to $R^{\frac{(n+1)}{2}}$

180. (b)

Force on object $= \frac{GMm}{x^2}$ at x from the centre of the earth.

$$\text{Work done} = \frac{GMm}{x^2}dx$$

$$\int \text{Work done} = GMm\int_R^{2R}\frac{dx}{x^2}$$

$$\text{PE gained} = GMm\left[-\frac{1}{x}\right]_R^{2R}$$

$$= \frac{GMm\times 1}{2R}$$

$$\text{Gain in PE} = \frac{1}{2}mR\left(\frac{GM}{R^2}\right) = \frac{1}{2}mgR$$

181. (a)

For a satellite

Centripetal force = Gravitational force

$$mR\omega^2 = \frac{GmM_e}{r^2}$$

$$\omega = \sqrt{\frac{GM_e}{r^3}} = \sqrt{\frac{GM_e}{(R_e+h)^3}}$$

$$T = 2\pi\sqrt{\frac{(R_e+h)^3}{GM_e}}$$

$\therefore$ T is independent of mass (m) of satellite.

182. (d)

For a satellite

Centripetal force = Gravitational force

$$\frac{mv_0^2}{(R+x)} = \frac{GMm}{(R+x)^2}$$

$$v_0^2 = \frac{GM}{R+x} = \frac{gR^2}{R+x}$$

$$v_0 = \sqrt{\frac{gR^2}{R+x}} \text{ i.e., } \left(\frac{gR^2}{R+x}\right)^{1/2}$$

183. (c)

The escape velocity of a body does not depend on the angle of projection from the earth. It is 11 km/sec.

184. (c)

Let the spheres collide after time t. During this time, let the smaller sphere cover distance x_1 and bigger sphere cover x_2 towards the point of collision.

The gravitational force

$$F(x) = \frac{GM \times 5M}{(12R - x_1)^2}$$

Acceleration of smaller body

$$a_1(x) = \frac{GM \times 5M}{(12R - x_1)^2}$$

Acceleration of bigger body

$$a_2(x) = \frac{GM}{(12R - x_1)^2}$$

From equation of motion

$$x_1 = \frac{1}{2}a_1(x)t^2$$

$$x_2 = \frac{1}{2}a_2(x)t^2$$

$$\Rightarrow \quad \frac{x_1}{x_2} = \frac{a_1(x)}{a_2(x)} = 5$$

$$\Rightarrow \quad x_1 = 5x_2$$

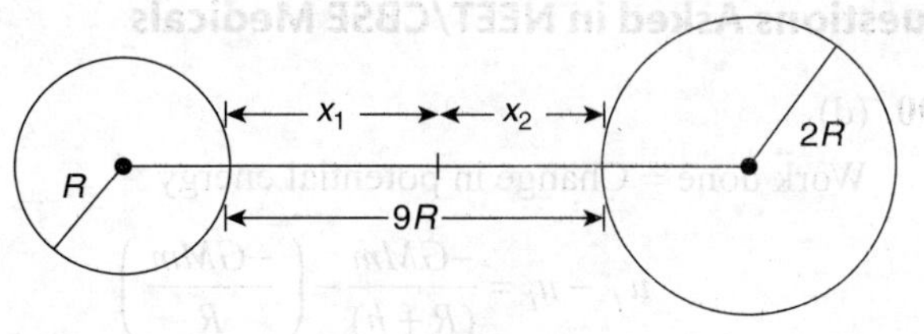

$$x_1 + x_2 = 9R$$

$$x_1 + \frac{x_1}{5} = 9R$$

$$x_1 = \frac{45R}{6} = 7.5R$$

Therefore, the two spheres collide when the smaller sphere covers the distance of 7.5 R.

185. (c)

According to Kepler's law,

$$T^2 \propto r^3$$

$$\left(\frac{T_1}{T_2}\right)^2 = \left(\frac{r_1}{r_2}\right)^3 = \left(\frac{1}{4}\right)^3 = \frac{1}{64}$$

$$\frac{T_1}{T_2} = \frac{1}{8}$$

$$T_2 = 8T_1 = 8 \times 5 = 40 \text{ hours}$$

186. (a)

Escape velocity $= \sqrt{2gR} = \sqrt{\dfrac{2GM_e}{R}}$

Escape velocity does not depend on mass of body which escapes. It depends on M_e, the mass of earth.

187. (c)

Escape velocity $v_e = \sqrt{2gR}$,

$$\text{KE} = \frac{1}{2}mv_e^2 = \frac{1}{2}m(2gR) = mgR$$

188. (d)

$$\text{Energy} = (\text{P.E})_{3R} - (\text{P.E})_{2R}$$

$$= -\frac{GMm}{3R} - \left(-\frac{GMm}{2R}\right)$$

$$\text{Energy} = \frac{GMm}{6R}$$

189. (b)

The centripetal and centrifugal forces disappear, the satellite has the tangential velocity and it will move in a straight line.

Questions Asked in NEET/CBSE Medicals

190. (d)

Work done = Change in potential energy

$$u_f - u_i = \frac{-GMm}{(R+h)} - \left(\frac{-GMm}{R}\right)$$

where M is the mass of earth and R is the radius of earth.

$$\therefore W = GMm\left[\frac{1}{R} - \frac{1}{(R+h)}\right]$$

Now, $h = R$

$$\therefore W = GMm\left[\frac{1}{R} - \frac{1}{2R}\right] = \frac{GMm}{2R}$$

$$\Rightarrow W = \frac{mgR}{2} \qquad \left[\because g = \frac{GM}{R^2}\right]$$

191. (a)

Acceleration due to gravity at a depth d,

$$g_d = g\left(1 - \frac{d}{R}\right)$$

For $d = R/2 \Rightarrow g_d = g\left(1 - \frac{R/2}{R}\right) = \frac{g}{2}$

Required weight $W' = mg_d = \frac{mg}{2} = \frac{W}{2} = \frac{200}{2} = 100$ N

192. (a)

Time period of geostationary satellite is,

$$T = 2\pi\sqrt{\frac{a^3}{GM}} \Rightarrow T^2 \propto a^3$$

$$\therefore \frac{T_1^2}{T_2^2} = \frac{a_1^3}{a_2^3} \Rightarrow \frac{(24)^2}{T_2^2} = \frac{(7R_E)^3}{(3.5R_E)^3}$$

$$\Rightarrow T_2^2 = \frac{(24)^2 \times (3.5)^3}{(7)^3}$$

$$\Rightarrow T_2 = \frac{\sqrt{(24)^2}}{\sqrt{8}} = 6\sqrt{2}\text{ h.}$$

193. (b)

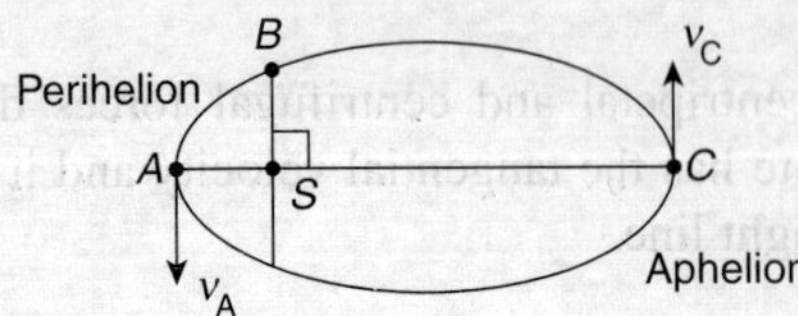

Point A is perihelion and C is aphelion.

So, $\qquad v_A > v_B > v_C$

As kinetic energy $K = 1/2\ mv^2$ or $K \propto v^2$

So, $\qquad K_A > K_B > K_C$

194. (d)

If universal gravitational constant becomes ten times, then $G' = 10\ G$.

So, acceleration due to gravity increases. That is, (d) is the wrong option.

195. (c)

The acceleration due to gravity at a height h is given as

$$g_d = g\left(1 - \frac{2h}{R_e}\right)$$

where R_e is radius of earth.

The acceleration due to gravity at a depth d is given as

$$g_d = g\left(1 - \frac{d}{R_e}\right)$$

Given, $\qquad g_h = g_d$

$$\therefore g\left(1 - \frac{2h}{R_e}\right) = g\left(1 - \frac{d}{R_e}\right)$$

$$\therefore d = 2h = 2 \times 1 = 2\text{ km} \qquad (\therefore h = 1\text{ km})$$

196. (a)

Since two astronauts are floating in gravitational free space. The only force acting on the two astronauts is the gravitational pull of their masses, $F = \frac{Gm_1m_2}{r^2}$ which is attractive in nature.

Hence, they move towards each other.

197. (b)

Acceleration due to gravity is given by

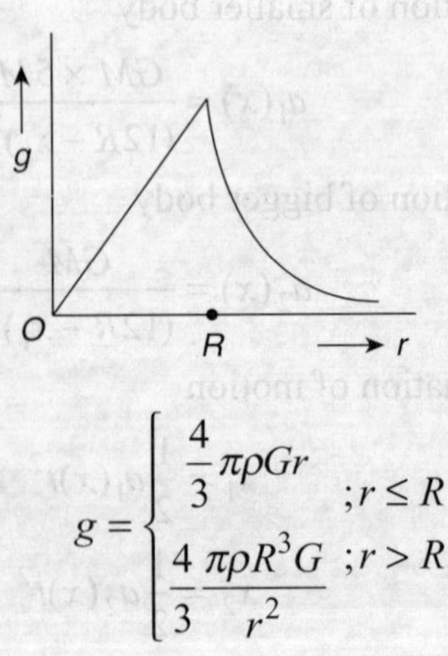

$$g = \begin{cases} \frac{4}{3}\pi\rho G r & ; r \le R \\ \frac{4}{3}\frac{\pi\rho R^3 G}{r^2} & ; r > R \end{cases}$$

198. (b)

Total energy of satellite at height h from the earth surface,

$$E = \text{PE} + \text{KE} = -\frac{GMm}{(R+h)} + \frac{1}{2}mv^2 \qquad \ldots\text{(i)}$$

Also, $\frac{mv^2}{(R+h)}+\frac{GMm}{(R+h)^2}$ or, $v^2=\frac{GM}{R+h}$ ...(ii)

From Eqs. (i) and (ii),

$$E=-\frac{GMm}{(R+h)}+\frac{1}{2}\frac{GMm}{(R+h)}=-\frac{1}{2}\frac{GMm}{(R+h)}$$

$$=-\frac{1}{2}\frac{GM}{R^2}\times\frac{mR^2}{(R+h)}$$

$$=-\frac{mg_0R^2}{2(R+h)}\qquad\left(\because g_0=\frac{GM}{R^2}\right)$$

199. (c)

Gravitation potential at a height h from the surface of earth, $V_h=-5.4\times10^7\,\text{Jkg}^{-1}$ At the same point acceleration due to gravity,

$$g_h=6\text{ ms}^{-2}$$

$$R=6400\text{ km}=6.4\times10^6\text{ m}$$

We know, $V_h=-\frac{GM}{(R+h)}$

$$g_h=\frac{GM}{(R+h)^2}=-\frac{V_h}{R+h}\Rightarrow R+h=-\frac{V_h}{g_h}$$

$$\therefore h=-\frac{V_h}{g_h}-R=-\frac{(-5.4\times10^7)}{6}-6.4\times10^6$$

$$=9\times10^6-6.4\times10^6=2600\text{ km}$$

200. (d)

As escape velocity, $v=\sqrt{\frac{2GM}{R}}$

$$=\sqrt{\frac{2G}{R}\cdot\frac{4\pi R^3}{3}\rho}=R\sqrt{\frac{8\pi G}{3}\rho}$$

$$\therefore\frac{v_e}{v_p}=\frac{R_e}{R_p}\times\sqrt{\frac{\rho_e}{\rho_p}}=\frac{1}{2}\times\sqrt{\frac{1}{2}}=\frac{1}{2\sqrt{2}}$$

$$(\because R_p=2R_e\text{ and }\rho_p=2\rho_e)$$

201. (c)

The orbital speed of the satellite is

$$v_o=R\sqrt{\frac{g}{(R+h)}}$$

where R is the earth's radius, g is the acceleration due to gravity on earth's surface and h is the height above the surface of earth.

Here, $R=6.38\times10^6$ m, $g=9.8\text{ ms}^{-2}$ and $h=0.25\times10^6$

$$\therefore v_o=(6.38\times10^6\text{ m})\sqrt{\frac{(9.8\text{ms}^{-2})}{(6.38\times10^6\text{ m}+0.25\times10^6\text{ m})}}$$

$$=7.76\times10^3\text{ ms}^{-1}=7.76\text{ kms}^{-1}$$

202. (b)

The gravitational force on the satellite S acts towards the centre of the earth, so the acceleration of the satellite S is always directed towards the centre of the earth.

203. (d)

Gravitational force of attraction between sun and planet provides centripetal force for the orbit of planet.

$$\therefore\frac{GMm}{r^2}=\frac{mv^2}{r};v^2=\frac{GM}{r}\quad\text{...(i)}$$

Time period of the planet is given by

$$T=\frac{2\pi r}{v},T^2=\frac{4\pi^2r^2}{v^2}=\frac{4\pi^2r^2}{\left(\frac{GM}{r}\right)}\quad\text{(Using (i))}$$

$$T^2=\frac{4\pi^2r^2}{GM}\quad\text{... (ii)}$$

According to question,

$$T^2=Kr^3$$

Comparing Eqs. (ii) and (iii), we get

$$K=\frac{4\pi^2}{GM},\therefore GMK=4\pi^2$$

204. (c)

Light cannot escape from a black hole,

$$v_{sec}=c$$

$$\sqrt{\frac{2GM}{R}}=c\text{ or }R=\frac{2GM}{c^2}$$

$$R=\frac{2\times6.67\times10^{-11}\text{Nm}^2\text{kg}^{-2}\times5.98\times10^{24}\text{ kg}}{(3\times10^8\text{ ms}^{-1})^2}$$

$$=8.86\times10^{-3}\text{ m}\propto10^{-2}\text{ m}$$

205. (a)

For a point inside the earth, i.e., $r<R$

$$E=-\frac{GM}{R^3}r$$

where M and R be mass and radius of the earth respectively.

At the centre, $r=0$

$$\therefore E = 0$$

For a point outside the earth, i.e., $r > R$

$$E = -\frac{GM}{r^2}$$

On the surface of the earth i.e. $r = R$

$$E = -\frac{GM}{R^2}$$

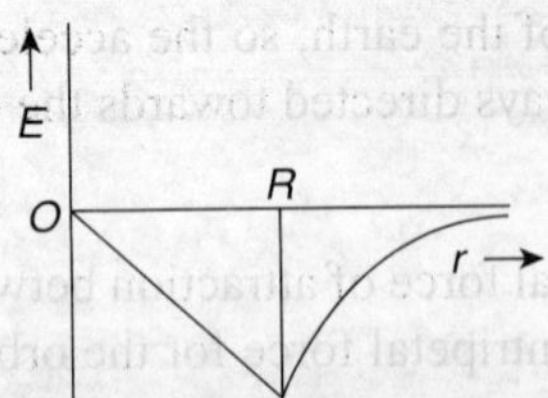

The variation of E with distance r from the centre is as shown in the figure.

206. (b)

The resulting gravitational potential at the origin O due to each of mass 2 kg located at positions as shown in figure is

O — 2 kg at $x = 1$, 2 kg at $x = 2$, 2 kg at $x = 4$, 2 kg at $x = 8$ ($x = 0$ at O)

$$V = -\frac{G\times 2}{1} - \frac{G\times 2}{2} - \frac{G\times 2}{4} - \frac{G\times 2}{8} - .....$$

$$= -2G\left[1+\frac{1}{2}+\frac{1}{4}+\frac{1}{8}+.....\right] = -2G\left[\frac{1}{1-\frac{1}{2}}\right]$$

$$= -2G\left[\frac{2}{1}\right] = -4G$$

207. (d)

Gravitational potential energy at any point at a distance r from the centre of the earth is,

$$U = -\frac{GMm}{r}$$

where M and m be masses of the earth and the body respectively.

At the surface of the earth, $r = R$

$$\therefore U_i = -\frac{GMm}{R}$$

At a height h from the surface,

$$r = R + h = R + 2R \qquad (h = 2R(\text{Given}))$$

$$= 3R$$

$$\therefore U_f = -\frac{GMm}{3R}$$

Change in potential energy, $\Delta U = U_f - U_i$

$$= -\frac{GMm}{3R} - \left(-\frac{GMm}{R}\right) = \frac{GMm}{R}\left(1-\frac{1}{3}\right)$$

$$= \frac{2}{3}\frac{GMm}{R} = \frac{2}{3}mgR \qquad \left(\because g = \frac{GM}{R^2}\right)$$

208. (b)

Here, $R_P = 2R_E, \rho_E = \rho_P$

Escape velocity of the earth,

$$V_E = \sqrt{\frac{2GM_E}{R_E}} = \sqrt{\frac{2G}{R_E}\left(\frac{4}{3}\pi R_E^3\rho_E\right)} = R_E\sqrt{\frac{8}{3}\pi G\rho_E} \quad ...(i)$$

Escape velocity of the planet,

$$V_P = \sqrt{\frac{2GM_P}{R_P}} = \sqrt{\frac{2G}{R_P}\left(\frac{4}{3}\pi R_P^3\rho_P\right)} = R_P\sqrt{\frac{8}{3}\pi G\rho_P} \quad ...(ii)$$

Divide (i) by (ii), we get

$$\frac{V_E}{V_P} = \frac{R_E}{R_P}\sqrt{\frac{\rho_E}{\rho_P}} = \frac{R_E}{2R_E}\sqrt{\frac{\rho_E}{\rho_E}} = \frac{1}{2} \text{ or } V_P = 2V_E$$

209. (a)

The minimum speed with which the particle should be projected from the surface of the earth so that it does not return is known as escape speed and it is given by,

$$v_e = \sqrt{\frac{2GM}{(R+h)}}$$

Here, $h = 3R$

$$\therefore v_e = \sqrt{\frac{2GM}{(R+3R)}} = \sqrt{\frac{2GM}{4R}} = \sqrt{\frac{GM}{2R}}$$

210. (c)

Acceleration due to gravity at a height h from the surface of earth is

$$g' = \frac{g}{\left(1+\frac{h}{R}\right)^2} \quad (i)$$

where g is the acceleration due to gravity at the surface of each and R is the radius of earth.

Multiplying m (mass of the body) on both sides, in (i) we get

$$mg' = \frac{mg}{\left(1+\frac{h}{R}\right)^2}$$

$\therefore$ Weight of body at height h, $W' = mg'$. Weight of body at surface of earth, $W = mg$

$$\therefore W' = \frac{1}{16}W \qquad \therefore \frac{1}{16} = \frac{1}{\left(1+\frac{h}{R}\right)^2}$$

$$\left(1+\frac{h}{R}\right)^2 = 16 \text{ or } 1+\frac{h}{R} = 4$$

or $$\frac{h}{R} = 3 \text{ or } h = 3R$$

211. (a)

Gravitational force acting on particle of mass m is

$$F = \frac{GM_p m}{(D_p/2)^2}$$

Acceleration due to gravity experienced by the particle is

$$g = \frac{F}{m} = \frac{GM_p}{(D_p/2)^2} = \frac{4GM_p}{D_p^2}$$

212. (c)

According to Kepler's third law $T \propto r^{3/2}$

$$\therefore \frac{T_2}{T_1} = \left(\frac{r_2}{r_1}\right)^{3/2} = \left(\frac{R+2R}{R+5R}\right)^{3/2} = \frac{1}{2^{3/2}}$$

Since $T_1 = 24$ hours,

$$\frac{T_2}{24} = \frac{1}{2^{3/2}} \text{ or } T_2 = \frac{24}{2^{3/2}} = \frac{24}{2\sqrt{2}} = 6\sqrt{2} \text{ hours}$$

213. (d)

Escape velocity, $$v_e = \sqrt{\frac{2GM}{R}} \quad \ldots\text{(i)}$$

where M and R are the mass and radius of the earth respectively.

The orbital velocity of a satellite close to the earth's surface is

$$v_o = \sqrt{\frac{GM}{R}} \quad \ldots\text{(ii)}$$

From (i) and (ii), we get $v_e = \sqrt{2}v_o$

214. (b)

Gravitational field due to the thin spherical shell

Inside the shell, i.e., (for $r < R$)

$$F = 0$$

On the surface of the shell, i.e., (for $r = R$)

$$F = \frac{GM}{R^2}$$

Outside the shell, i.e., (For $r > R$)

$$F = \frac{GM}{r^2}$$

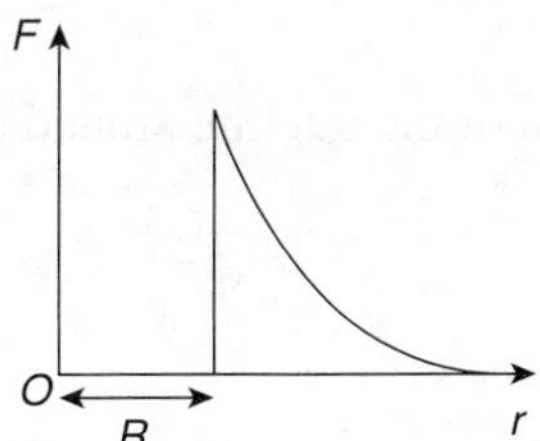

The variation of F with distance r from the centre is as shown in the figure.

215. (b)

According to the law of conservation of angular momentum $L_1 = L_2$

$$mv_1r_1 = mv_2r_2 \Rightarrow v_1r_1 = v_2r_2 \text{ or } \frac{v_1}{v_2} = \frac{r_2}{r_1}$$

216. (b)

According to law of conservation of mechanical energy,

$$\frac{1}{2}mu^2 - \frac{GMm}{R} = 0 \text{ or } u^2 = \frac{2GM}{R}$$

$$u = \sqrt{\frac{2GM}{R}} = \sqrt{2gR} \qquad \left(\because g = \frac{GM}{R^2}\right)$$

217. (c)

Here, Mass of a particle = M

Mass of a spherical shell = M

Radius of a spherical shell = a

Let O be centre of the spherical shell.

Gravitational potential at point P due to particle O is

$$V_1 = -\frac{GM}{a/2}$$

Gravitational potential at point P due to spherical shell is

$$V_2 = -\frac{GM}{a}$$

Hence, total gravitational potential at point P is

$$V = V_1 + V_2$$

$$\frac{-GM}{a/2} + \left(\frac{-GM}{a}\right) = \frac{-2GM}{a} - \frac{GM}{a} = \frac{-3GM}{a}$$

$$|V| = \frac{3GM}{a}$$

218. (b)

Orbital speed of the satellite around the earth is

$$v=\sqrt{\frac{GM}{r}}$$

For satellite A

$$r_A=4R, v_A=3V$$

$$v_A=\sqrt{\frac{GM}{r_A}} \quad \ldots\text{(i)}$$

For satellite B

$$r_B=R, v_B=?$$

$$v_B=\sqrt{\frac{GM}{r_B}} \quad \ldots\text{(ii)}$$

Dividing Eq. (ii) by Eq. (i), we get

$$\frac{v_B}{v_A}=\sqrt{\frac{r_A}{r_B}} \text{ or } v_B=v_A\sqrt{\frac{r_A}{r_B}}$$

Substituting the given values, we get

$$v_B=3V\sqrt{\frac{4R}{R}} \text{ or } v_B=6V$$

219. (b)

Since the man is in gravity-free space, force on man + stone system is zero. Therefore centre of mass of the system remains at rest. Let the man go x m above when the stone reaches the floor, then

$$M_{\text{man}}\times x=M_{\text{stone}}\times 10$$

$$x=\frac{0.5}{50}\times 10 \text{ or } x=0.1 \text{ m}$$

Therefore, final height of man above floor = 10 + x = 10 + 0.1 = 10.1 m

220. (d)

221. (d)

The acceleration due to gravity at a depth d below surface of earth is

$$g'=\frac{GM}{R^2}\left(1-\frac{d}{R}\right)=g\left(1-\frac{d}{R}\right)$$

$$g'=0 \text{ at } d=R.$$

i.e., acceleration due to gravity is zero at the centre of earth.

Thus, the variation in value of g with r is

For, $r>R$

$$g'=\frac{g}{\left(1+\frac{h}{R}\right)^2}=\frac{gR^2}{r^2}\Rightarrow g'\propto\frac{1}{r^2}$$

Here, $R+h=r$

For $r<R,\ g'=g\left(1-\frac{d}{R}\right)=\frac{gr}{R}$

Here, $R-d=r$

$$\Rightarrow g'\propto r$$

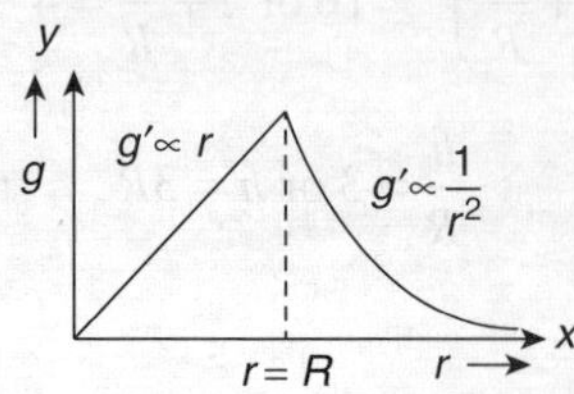

Therefore, the variation of g with distance from centre of the earth will be as shown in the figure.

222. (a)

223. (b)

Equal areas are swept in equal time. As it is given that area SCD = 2SAB.

The time taken to go from C to D, $t_1=2t_2$ where t_2 is the taken to go from A to B.

224. (b)

The satellite of mass m is moving in a circular orbit of radius r.

$\therefore$ Kinetic energy of the satellite, $K=\dfrac{GMm}{2r}$ ...(i)

Potential energy of the satellite, $U=\dfrac{-GMm}{r}$ ...(ii)

Orbital speed of satellite, $v=\sqrt{\dfrac{GM}{r}}$ ...(iii)

Time-period of satellite, $T=\left[\left(\dfrac{4\pi^2}{GM}\right)r^3\right]^{1/2}$ ...(iv)

Given $m_{S_1}=4m_{S_2}$

Since M, r is same for both the satellite S_1 and S_2

$\therefore$ From Eq. (ii), we get $U\propto m$

$$\therefore \frac{U_{S_1}}{U_{S_2}}=\frac{m_{S_1}}{m_{S_2}}=4 \text{ or } U_{S_1}=4U_{S_2}$$

Option (a) is wrong.

From (iii), since v is independent of the mass of a satellite, the orbital speed is same for both satellites S_1 and S_2

Hence option (b) is correct.

From (i), we get $K\propto m$

$$\therefore \frac{K_{S_1}}{K_{S_2}}=\frac{m_{S_1}}{m_{S_2}}=4 \text{ or } K_{S_1}=4K_{S_2}$$

Hence option (c) is wrong.
From (iv), since T is independent of the mass of a satellite, time period is same for both the satellite S_1 and S_2
Hence option (d) is wrong.

225. (c)

Escape velocity of the body from the surface of earth is $v = \sqrt{2gR}$

For escape velocity of the body from the platform, potential energy + kinetic energy = 0

$$-\frac{GMm}{2R} + \frac{1}{2}mv^2 = 0$$

$$\Rightarrow fv_{\text{escape}} = \sqrt{\frac{GM}{R^2}.R} = \sqrt{gR} = f$$

From the surface of the earth, $v_{\text{escape}} = \sqrt{2gR}$

$$\therefore fv_{\text{escape}} = \frac{v_{\text{escape}}}{\sqrt{2}} \therefore f = \frac{1}{\sqrt{2}}$$

226. (d)

$$g = \frac{GM}{r^2} = \frac{G}{r^2}\left(\frac{4}{3}\times\pi r^3\rho\right) = \frac{4}{3}\times\pi\rho Gr$$

$$\frac{g'}{g} = \frac{3R}{R} \Rightarrow g' = 3g$$

227. (a)

$$-\frac{GMm}{R^2} + m\omega^2 R = 0 \Rightarrow \frac{GMm}{R^2} = m\omega^2 R$$

$$\text{K.E.} = \frac{1}{2}I\omega^2 = \frac{1}{2}mR^2\omega^2 = \frac{GMm}{2R}$$

$$\text{P.E.} = -\frac{GMm}{R}$$

$$\therefore KE = \frac{|PE|}{2} \text{ or } \frac{KE}{|PE|} = \frac{1}{2}$$

228. (d)

From equation of acceleration due to gravity.

$$g_e = \frac{GM_e}{R_e^2} = \frac{G(4/3)\pi R_e^3}{R_e^2}\rho_e$$

$$g_e \propto R_e\rho_e$$

Acceleration due to gravity of planet $g_p \propto R_p\rho_p$

$$R_e\rho_e = R_p\rho_p \Rightarrow R_e\rho_e = R_p 2\rho_e$$

$$\Rightarrow R_p = \frac{1}{2}R \qquad (\therefore R_e = R)$$

229. (b)The gravitational force does not depend upon the medium in which the objects are placed.

230. (b)

The velocity of the mass while reaching the surface of both the planets will be same.

i.e., $\sqrt{2g'h'} = \sqrt{2gh}$

$$\sqrt{2\times g'\times h'} = \sqrt{2\times 9g\times 2} \Rightarrow 2h' = 36$$

$$\Rightarrow h' = 18 \text{ m}$$

231. (c)

Gravitational potential energy on earth's surface $= -\frac{GMm}{R}$, where M and R are the mass and radius of the earth respectively, m is the mass of the body and G is the universal gravitational constant.
Gravitational potential energy at a height $h = 3R$

$$= -\frac{GMm}{R+h} = -\frac{GMm}{R+3R} = -\frac{GMm}{4R}$$

$\therefore$ Change in potential energy

$$= -\frac{GMm}{4R} - \left(-\frac{GMm}{R}\right)$$

$$= \frac{GMm}{R+3R} = \frac{GMm}{4R} + \frac{GMm}{R} = \frac{3}{4}\frac{GMm}{R} = \frac{3}{4}mgR$$

232. (a)

Use $v^2 = \dfrac{2gh}{1+\dfrac{h}{R}}$ given $h = R$

$$\therefore v = \sqrt{gR} = \sqrt{\frac{GM}{R}}$$

233. (b)

$$v_e = \sqrt{2gR} = \sqrt{\frac{2GM}{R}} \because R_P = \frac{1}{4}R_e$$

$$v_P = 2v_e = 2\times 11.2 = 22.4 \text{ km/s}$$

234. (b)

235. (b)

$$F_{\text{surface}} = G\frac{Mm}{R_e^2}$$

$$F_{R_{e/2}} = G\frac{Mm}{(R_e + R_e/2)^2} = \frac{4}{9}\times F_{\text{surface}}$$

$$= \frac{4}{9}\times 72 = 32 \text{ N}$$

236. (b)

237. (a)

Escape velocity of a body (v_e) = 11.2 km/s; New mass of the earth M_e'=2 M_e and new radius of the earth R_e' = 0.5 R_e.

$$\text{Escape velocity } (v_e) = \sqrt{\frac{2GM_e}{R_e}} \propto \sqrt{\frac{M_e}{R_e}}$$

$$\text{Therefore} \quad \frac{v_e}{v_e'} = \sqrt{\frac{M_e}{R_e} \times \frac{0.5R_e}{2M_e}} = \sqrt{\frac{1}{4}} = \frac{1}{2}$$

$$\text{or} \quad v_e' = 2v_e = 22.4 \text{ km/s}$$

238. (a)

Period of revolution of planet. A, (T_A)=8 T_B. According to Kepler's third law of planetary motion $T^2 \propto R^3$

$$\text{Therefore} = \left(\frac{r_A}{r_B}\right)^3 = \left(\frac{T_A}{T_B}\right)^2 = \left(\frac{8T_B}{T_B}\right)^2 = 64$$

$$\text{or} \quad \frac{r_A}{r_B} = 4 \text{ or } r_A = 4r_B$$

239. (b)

240. (c)

Since no external torque is applied, according to law of conservation of angular momentum, the ball will continue to move with the same angular velocity along the original orbit of the spacecraft.

241. (a)

Acceleration due to gravity (g)

$$= G \times \frac{M}{R^2} = G\frac{(4/3)\pi R^3 \times \rho}{R^2} = G \times \frac{4}{3}\pi R \times \rho$$

$$\text{or } \rho = \frac{3g}{4\pi GR}$$

242. (a)

The force between the two masses, $F = -G\dfrac{mm}{4R^2}$

This force will provide the necessary centripetal force for the masses to go around a circle. Then

$$\frac{Gmm}{4R^2} = \frac{mv^2}{R} \Rightarrow v^2 = \frac{Gm}{4R} \Rightarrow v = \frac{1}{2}\sqrt{\frac{Gm}{R}}$$

243. (a)

$$\text{Mass } (m) = 6 \times 10^{24} \text{ kg}$$

Angular velocity (ω) = 2×10^{-7} rad/s and radius (r) = 1.5×10^8 km = 1.5×10^{11} m.

Force exerted on the earth = $mR\omega^2$

$$= \left(6 \times 10^{24}\right) \times \left(1.5 \times 10^{11}\right) \times \left(2 \times 10^{-7}\right)^2$$

$$= 36 \times 10^{21} \text{ N}$$

244. (c)

Radius of earth (R_e) = 6400 km; Radius of Mars (R_m) = 3200 km; Mass of Earth (M_e) = 10 Mm and weight of the object on Earth (W_e) = 200 N

$$\frac{W_m}{W_e} = \frac{mg_m}{mg_e} = \frac{M_m}{M_e} \times \left(\frac{R_e}{R_m}\right)^2 = \frac{1}{10} \times (2)^2 = \frac{2}{5}$$

$$\text{or} \quad W_m = W_e \times \frac{2}{5} = 200 \times 0.4 = 80 \text{ N}$$

245. (b)

Distance of two planets from sun, $r_1 = 10^{13}$ m and $r_2 = 10^{12}$ m

Relation between time period (T) and distance of the planet from the sun is $T^2 \propto r^3$ or $T \propto r^{3/2}$

$$\text{Therefore} = \frac{T_1}{T_2} = \left(\frac{r_1}{r_2}\right)^{3/2} = \left(\frac{10^{13}}{10^{12}}\right)^{3/2} = 10^{3/2}$$

$$= 10\sqrt{10}$$

QUICK QUESTIONS

Quiz

1. Why is most of the mass of a flywheel concentrated at its rim?

2. When a labourer cuts down a tree, he makes a cut on the side facing the direction in which he wants it to fall. Explain.

3. Why spokes are provided in a bicycle wheel?

4. Two satellites of equal masses are orbiting at different heights. Will their moments of Inertia be the same or different?

5. Can the mass of a body be assumed to be concentrated at the centre of mass for purpose of calculating its moment of Inertia?

6. An astronaut in a satellite releases a screwdriver into space. Will the screwdriver fall on earth?

7. How is the gravitational force different from electric and magnetic force while all are governed by inverse square law?

8. You know escape velocity V_e is given by $V_e = \sqrt{\frac{2GM}{R}}$ and orbital velocity $V_0 = \sqrt{\frac{GM}{R}}$ where R is the radius of earth and V_0 is orbital velocity for a satellite which is very close to the surface of earth. What will be the path of a satellite in the following situations?
   (i) If the velocity V of the satellite is greater than orbital velocity and less than escape velocity ($V_0 < V < V_e$)
   (ii) The velocity V is less than orbital velocity ($V < V_0$)
   (iii) The velocity V is greater than escape velocity ($V > V_e$)

9. A comet is in a highly elliptical orbit around the sun with a time period 90 days. Which of the following statements is true about the possibility of a collision between the comet and earth?
   (i) Collision is possible
   (ii) Collision is not possible
   (iii) More information is needed to determine whether collision is possible or not.

10. What are the algebraic signs of the following energies for a satellite moving around the earth?
    (i) Kinetic energy (ii) Potential energy
    (iii) Total energy

11. Why does the negative total energy of a satellite in orbit isignify?

12. Tycho Brahe (1546–1601) the Danish astronomer made accurate astronomical measurement and provided the basis for the laws of planetary motion. He observed around 750 stars. What were the devices and instruments he used for this?

Think and grow

13. An observer on earth measures the time period of two satellites, one moving from east to west and the other moving from west to east in a circular orbit near the equatorial plane of earth. Would the two time periods be the same or different ?

14. A block slides down a frictionless ramp and a sphere also rolls without sliding down a ramp of same, inclination and same height. The block and sphere have the same mass and descent through the same height.
    (a) In descent, is the work done by gravitational force on the block less than, greater than or equal to the work done by gravitational force on the sphere?
    (b) Which object has more translational kinetic energy?
    (c) Which object has more speed down the ramp?

15. A satellite of low-earth orbit experiences a small drag force from the earth's atmosphere. What happens to the speed of the satellite because of this drag force?

16. Can a satellite stay in stable orbit in a plane not passing through centre of earth?

17. According to Kepler's first law, all planets move in elliptical orbits. This is very intimately related to the fact that the gravitational force is inversely proportional to the square of distance between the planet and the sun. Suppose, instead of being inversely proportional to square of distance, the gravitational force is inversely proportional to the cube of the distance between planet and sun. This allows us to define some new types of orbits. What about the second and third law. Will they change if force depends on the inverse of cube of the distance between planet and sun?

18. The earth is closer to sun when it is winter in the northern hemisphere than in summer. In which month, July or January, does the earth move longer distance in its orbit?

19. In case of satellite motion, angular momentum is constant. Angular momentum orbital velocity $L = mvr =$

Constant, $v \propto \frac{1}{r}$, $v_0 = \sqrt{\frac{GM}{r}}$ and according to this relation, $v_0 \propto \frac{1}{\sqrt{r}}$. Does not the two results contradict to each other?

20. Imagine a group of scientists trapped on a floating ice berg near the north pole. They have a transmitter with which they can fix their position with "the aid of a geo-synchronous satellite". Can a satellite in polar orbit be geosynchronous?

Explanations

Quiz

1. To increase the moment of inertia. As a result, gain or loss of rotational energy brings about a comparatively small change in angular velocity. This keeps the motion nearly uniform.

2. The weight of the tree provides a torque about the point where cut is made. This causes rotation of the tree about the cut.

3. This is to increase the moment of inertia.

4. Different. The distances are different.

5. No, because moment of inertia depends upon distribution of mass also along with axis of rotating.

6. No, because of inertia of motion. The screwdriver acquires the same velocity as that of satellite. As the radius of orbit is independent of mass of the satellite, the screwdriver will continue move along with the satellite in the same orbit. It will not fall. It will become a satellite of earth.

7. Gravitational force is always attractive. The other major difference is that it is independent of medium between the particles.

8. (i) It will follow an elliptical path.
   (ii) It will follow a parabolic path.
   (iii) It will follow a hyperbolic path.

9. Collision is not possible. If we calculate the semi-major axis of the elliptical orbit of the comet, it comes out to be 1.2×10^{11}m. This is smaller than the earth–sun distance, so the comet cannot collide with the earth.

10. (i) Positive (ii) Negative (iii) Negative

11. It means that the satellite is bound to the earth by an attractive force and energy is needed to knock it from the orbit and to remove it to infinity. This energy is called binding energy.

12. Sextant and compass. The telescope had not been invented at that time.

Think and grow

13. The earth is rotating on its axis from west to east. Therefore, the velocity of a satellite V_0 moving towards east with respect to earth (velocity V_e) would be $(V_0 - V_e)$ and for the satellite moving towards west would be $(V_0 + V_e)$. So the period of the two satellite as observed by an observed by an observer on earth would be $T_1 = (\frac{2\pi R}{V_0 - V_e})$, $T_2 = (\frac{2\pi R}{V_0 + V_e})$
    Hence, the time period of the two satellites will be different. The time period of the satellite moving from west to east will be greater than the satellite moving from east to west.

14. (a) In both the cases loss of potential energy is same. Hence, the work done by gravitational forces in the two cases are same.
    (b) In the case of the block the entire gravitational potential energy has been transformed into translational kinetic energy while in case of sphere, the potential energy is transformed to partly translational and partly rotation energy. Therefore the translational kinetic energy for the block is more than that of sphere.
    (c) The speed of the block down the ramp is more than that of the sphere as greater the translational kinetic energy, greater will be speed.

15. Because of the air drag, some of the mechanical energy will be dissipated and there will be loss of total energy of the satellite.

    Total Energy $E = -\frac{GMm}{2r}$, with the result r will decrease and the satellite will follow a spiral path towards the earth.

    Now, Potential energy $= -\frac{GMm}{r}$

    and Kinetic energy $= \frac{GMm}{2r}$

    So, with decrease in r potential and kinetic energy will increase but as a whole the kinetic energy increases at a greater rate. Hence speed of the satellite increases.

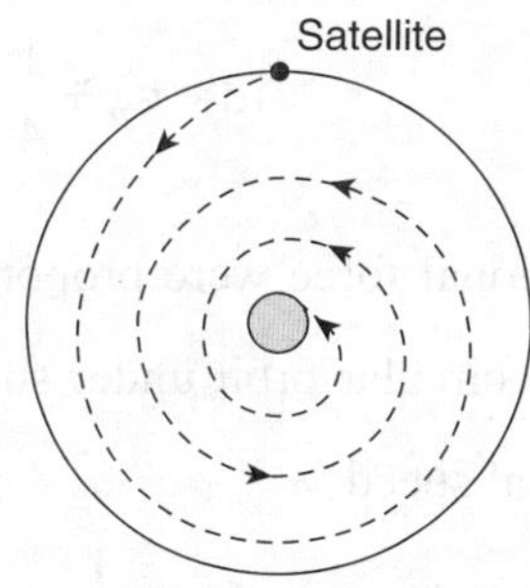

16. No, for orbital motion of a satellite the required centripetal force is provided by the gravitational force between the earth and satellite. If the plane of orbit is not passing through the centre of earth, the gravitational force F will not be directed towards the centre of the orbit O, and the orbit will not be stable. Then, there will be a component of the gravitational force perpendicular to the plane of the orbit. This force will pull the satellite till the plane of the satellite passes through the centre of the earth.

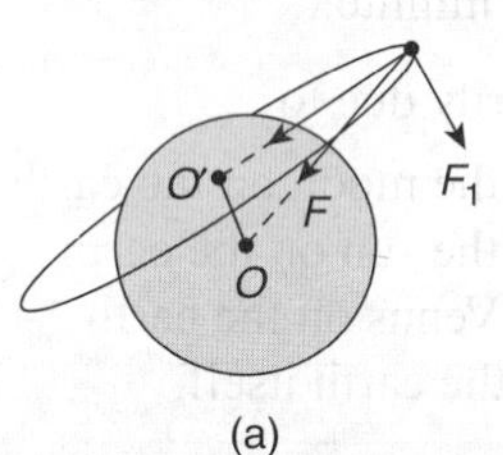

(a)

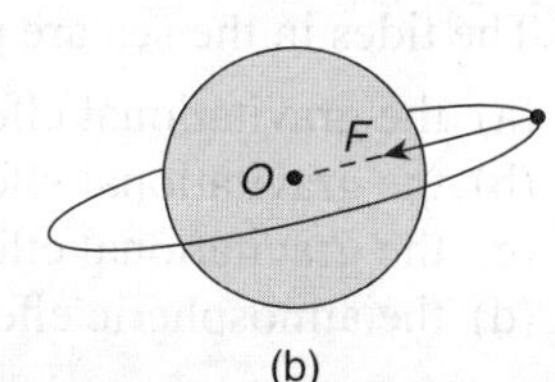

(b)

17. Kepler's second law will not change as it is a consequence of law of conservation of angular momentum. There is no external torque. Kepler's third law depends upon inverse square nature of the force.

Suppose force is inversely proportional to the cube of earth sun distance.

$$F = \frac{GmM}{R^3}$$

$$F = \frac{mv^2}{R}$$

$$\frac{mv^2}{R} = \frac{GmM}{R^3}$$

$$v^2 = \frac{GmM}{R^2}$$

$$\left(\frac{2\pi R}{T}\right)^2 = \frac{GM}{R^2}$$

$$\frac{4\pi^2 R^2}{T^2} = \frac{GM}{R^2}$$

$$T^2 = \frac{4\pi^2}{GM} R^4$$

$$T^2 \propto R^4$$

Time period will be proportional to square of distance between sun and planet.

18. The orbit of the earth around the sun is elliptical. According to Kepler's second law, in accordance with law of conservation of momentum, the earth moves faster. Therefore when it is closer to the sun in January, it moves faster in January than it is in July.

19. No, $v \alpha \frac{1}{r}$ holds for different points in the orbit while

$v_0 \alpha \frac{1}{\sqrt{r}}$ is applicable for different orbits.

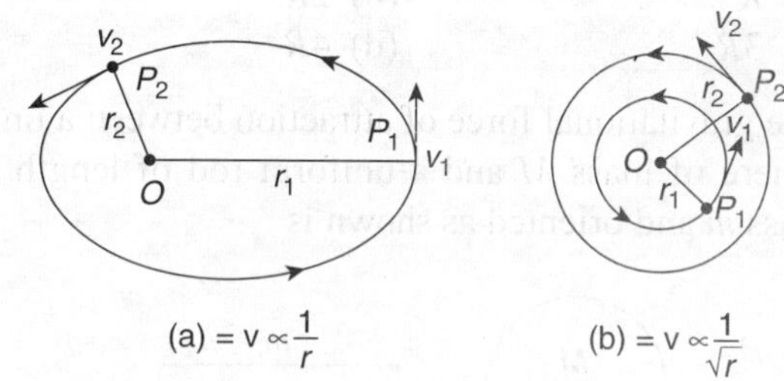

(a) = $v \propto \frac{1}{r}$ (b) = $v \propto \frac{1}{\sqrt{r}}$

20. A geosynchronous satellite is one whose period of revolution around the earth as rotational period of earth. It stays at one location of earth's surface for all the time. This results in no east–west motion relative to the observer to continue at a particular location.

CHAPTER TEST

1. An earth satellite can revolve around the earth only in such an orbit whose plane coincides with
   (a) the plane of some latitude circle of earth
   (b) a vertical plane not containing the centre of earth
   (c) the plane of tropic of cancer
   (d) the plane of any great circle around the earth.

2. Two identical copper spheres, each of radius R, are in contact with each other. The gravitational force of attraction between two copper spheres is proportional to
   (a) R (b) $2R$
   (c) R^2 (d) R^4

3. A body is detached gently from the outer wall of a satellite circling round earth. This body will
   (a) continue to move along with the satellite
   (b) fall to the earth
   (c) fall to the moon
   (d) follow an irregular path.

4. At sea level, a body will have minimum weight at
(a) pole (b) equator
(c) 42° south latitude (d) 37° north latitude.

5. If the earth shrinks, what will be the effect on the value of acceleration due to gravity g?
(a) It will decrease. (b) It will increase.
(c) It will become zero. (d) It will become infinite.

6. If the value of acceleration due to gravity on the surface of Earth is 10 in ms^{-2}, then a girl can jump safely from a height of 2 m. If she has to jump on a planet where acceleration due to gravity is it 0.2 ms^{-2}, then the safe height will be
(a) 0.2 m (b) 10 m
(c) 2 m (d) 100 m

7. If there were a small gravitational effect, then which of the following forces will undergo a change?
(a) Viscous force (b) Electrostatic force
(c) Magnetic force (d) Archimedes' uplift.

8. The value of g at a certain height h above the surface of Earth is $\frac{x}{4}$ where x is the value of g at the surface of Earth. The height h is
(a) R (b) $2R$
(c) $3R$ (d) $4R$

9. The gravitational force of attraction between a uniform sphere of mass M and a uniform rod of length l and mass m and oriented as shown is

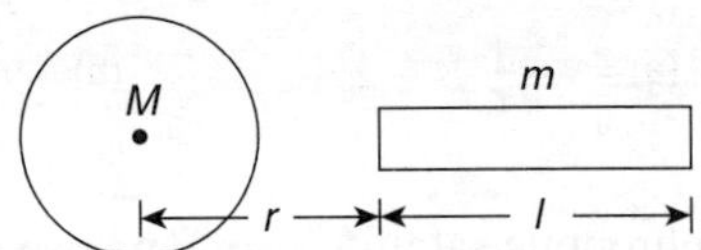

(a) $\frac{GMm}{r(r+l)}$ (b) $\frac{GM}{r^2}$
(c) $Mmr^2 + l$ (d) $(r^2 + l)nM$

10. Suppose that the acceleration of free fall at the surface of a distant planet were found to be equal to that at the surface of the earth. If the diameter of the planet were twice the diameter of earth, then the ratio of mean density of the planet to that of the earth would be
(a) 4 : 1 (b) 2 : 1
(c) 2 : 1 (d) 1 : 2

11. R is the radius of Earth and ω is its angular velocity and g_p is the value of g at poles. The effective value of g at the latitude $\lambda = 60°$ will be equal to
(a) $g_p - \frac{1}{4}R\omega^2$ (b) $g_p - \frac{3}{4}R\omega^2$
(c) $g_p - R\omega^2$ (d) $g_p + \frac{1}{4}R\omega^2$

12. If the gravitational force were proportional to $\frac{1}{r}$, then a particle in a circular orbit under such a force would have its original speed
(a) independent of r (b) $\propto \frac{1}{r}$
(c) $\propto \frac{1}{r^2}$ (d) $\propto r^2$

13. A particle on earth's surface is given a velocity equal to its escape velocity. Its total mechanical energy with zero potential energy reference at infinite separation will be
(a) negative (b) positive
(c) zero (d) infinite.

14. The tides in the sea are primarily due to
(a) the gravitational effect of the moon on the earth
(b) the gravitational effect of the sun on the earth
(c) the gravitational effect of Venus on the earth
(d) the atmospheric effect of the earth itself.

15. Two identical thin rings, each of radius R, are coaxially placed at a distance R. If the rings have a uniform mass distribution and each has mass m_1 and m_2 respectively, then the work done in moving a mass m from the centre of one ring to that of the other is
(a) Zero (b) $\frac{Gm(m_1 - m_2)(\sqrt{2}-1)}{\sqrt{2}R}$
(c) $\frac{Gm\sqrt{2}(m_1 - m_2)}{R}$ (d) $\frac{Gmm_1(\sqrt{2}+1)}{m_2 R}$

16. At point P ($R\sqrt{3}$, 0, 0) lies on the axis of a ring of mass M and radius R. The ring is located in y-z plane with its centre at origin O. A small particle of mass m starts from P and reaches O under gravitational attraction only. Its speed at O will be
(a) $\sqrt{\frac{GM}{R}}$ (b) $\sqrt{\frac{Gm}{R}}$
(c) $\sqrt{\frac{GM}{\sqrt{2}R}}$ (d) $\sqrt{\frac{Gm}{\sqrt{2}R}}$

17. A particle is projected vertically upwards from the surface of earth (radius R_e) with a kinetic energy equal to half of the minimum value needed for it to escape. The height to which it rises above the surface of earth is
(a) R_e (b) $2R_e$
(c) $3R_e$ (d) $4R_e$

18. A research satellite of mass 200 kg circles the earth in an orbit of average radius $2R/2$, where R is the radius of the earth. Assuming the gravitational pull on a mass of 1 kg on earth's surface to be 10 N, the pull on the satellite will be

(a) 880 N (b) 889 N
(c) 890 N (d) 982 N

19. The masses and radii of the earth and the moon are M_1, R_1 and M_2, R_2 respectively. Their centres are at distance d apart. The minimum speed with which a particle of mass m should be projected from a point midway between the two centres so as to escape to infinity is

(a) $v = \sqrt{\dfrac{4g(M_1 + M_2)}{d}}$ (b) $v = \sqrt{\dfrac{4G(M_1 + M_2)}{d}}$

(c) $v = \sqrt{4G(M_1 M_2)}$ (d) $v = \sqrt{4G(M_1 + M_2)}$

20. A planet moving along an elliptical orbit is closest to the sun at a distance r_1 and farthest away at a distance of r_2. If v_1 and v_2 are the linear velocities at these points respectively, then the ratio v_1 / v_2 is

(a) r_1 / r_2 (b) r_2 / r_1

(c) $(r_1 / r_2)^2$ (d) $(r_2 / r_1)^2$

21. Force between two objects of equal masses is F. If 25 % mass of one object is transferred to the other object, then the new force will be

(a) $\dfrac{F}{4}$ (b) $3\dfrac{F}{4}$

(c) $\dfrac{15}{16}F$ (d) F

22. Which force in nature exits everywhere ?

(a) Nuclear force (b) Electromagnetic force
(c) Weak force (d) Gravitation

23. One can easily "weigh the earth" by calculating the mass of earth using the formula (in usual notation)

(a) $\dfrac{G}{g} R_E^2$ (b) $\dfrac{g}{G} R_E^2$

(c) $\dfrac{g}{G} R_E$ (d) $\dfrac{g}{G} R_E^3$

24. The centripetal force acting on a satellite orbiting round the earth and the gravitational force of earth acting on the satellite both equal F. The net force on the satellite is

(a) Zero (b) F
(c) $F\sqrt{2}$ (d) $2F$

25. Mass M is divided into two parts, xM and $(1 - x)M$. For a given separation, the value of x for which the gravitational attraction between the two pieces becomes maximum is

(a) 1/2 (b) 3/5
(c) 1 (d) 2

26. The force of gravitation is

(a) Repulsive (b) Electrostatic
(c) Conservative (d) Non-conservative.

27. Read the following statements:

S_1: An object shall weigh more at the poles than at the equator when weighed by using a physical balance.
S_2: It shall weigh the same at the poles and the equator when weighed by using a physical balance.
S_3: It shall weigh the same at the poles and the equator when weighed by using a spring balance.
S_4: It shall weigh more at the poles than at the equator when weighed using a spring balance. Which of the above statements is/are correct?

(a) S_1 and S_2 (b) S_1 and S_4
(c) S_2 and S_3 (d) S_2 and S_4

28. If M be the mass of the earth and R its radius, the ratio of the gravitational acceleration and the gravitational constant is

(a) $\dfrac{R^2}{M}$ (b) $\dfrac{M}{R^2}$

(c) MR^2 (d) $\dfrac{M}{R}$

29. If the acceleration due to gravity, g, is 10 m/s^2 at the surface of the earth (radius 6400 km), then at a height of 1600 km the value of g will be (in m/s^2)

(a) 5 (b) 6.4
(c) 7.5 (d) 2.5

30. The radius of the earth is 6400 km and g = 10 m/sec^2. In order that a body of 5 kg weighs zero at the equator, the angular speed of the earth is

(a) 1/80 radian/sec (b) 1/400 radian/sec
(c) 1/800 radian/sec (d) 1/1600 radian/sec.

31. If the earth rotates faster than its present speed, the weight of an object will

(a) increase at the equator but remain unchanged at the poles
(b) decrease at the equator but remain unchanged at the poles
(c) remain unchanged at the equator but decrease at the poles
(d) remain unchanged at the equator but increase at the poles.

32. If the density of earth increased 4 times and its radius become half of what it is, our weight will

(a) be four times its present value
(b) be doubled
(c) remain same
(d) be halved.

33. The change in the gravitational potential energy when a body mass m is raised to a height nR above the surface of the earth (here R is the radius of the earth)

(a) $(\frac{n}{n+1})mgR$ (b) $(\frac{n}{n-1})mgR$
(c) $nmgR$ (d) $\frac{mgR}{n}$

34. The ratio of the radii of planets A and B is k_1 and the ratio of acceleration due to gravity on them is k_2. The ratio of escape velocities from them will be

(a) k_1k_2 (b) $\sqrt{k_1k_2}$
(c) $\sqrt{\frac{k_1}{k_2}}$ (d) $\sqrt{\frac{k_2}{k_1}}$

35. The satellite of mass m revolving in a circuit orbit of radius r around the earth has kinetic energy E. Then its angular momentum will be

(a) $\sqrt{\frac{E}{mr^2}}$ (b) $\frac{E}{2mr^2}$
(c) $\sqrt{2Emr^2}$ (d) $\sqrt{2Emr}$

36. Rockets are launched in eastward direction to take advantage of
(a) the clear sky on eastern side
(b) the thinner atmosphere on this side
(c) earth's rotation
(d) earth's tilt.

37. A satellite moves in a circle around the earth. The radius of this circle is equal to one-half of the radius of the moon's orbit. The satellite completes one revolution in

(a) $\frac{1}{2}$ lunar month
(b) $\frac{2}{3}$ lunar month
(c) $2^{-3/2}$ lunar month
(d) $2^{3/2}$ lunar month.

38. A geostationary satellite is revolving around the earth. To make it escape from gravitational field of earth, its velocity must be increased
(a) 100% (b) 41.4%
(c) 50% (d) 59.6%

39. The time period T of the moon of planet Mars (mass M_m) is related to its orbital radius R as (G=Gravitational constant)

(a) $T^2 = \frac{4\pi^2 R^3}{GM_m}$ (b) $T^2 = \frac{4\pi^2 GR^3}{M_m}$
(c) $T^2 = \frac{2\pi^2 GR^3}{M_m}$ (d) $T^2 = 4\pi M_m GR^3$

40. The distance of Neptune and Saturn from the sun is nearly 10^{13} and 10^{12} m respectively. Assuming that they move in circular orbits, their periodic times will be in the ratio
(a) $\sqrt{10}$ (b) 100
(c) $10\sqrt{10}$ (d) $1/\sqrt{10}$

41. A planet moving around the sun sweeps area A_1 in 2 days, A_2 in 3 days and A_3 in 6 days. Then the relation between A_1, A_2 and A_3 is

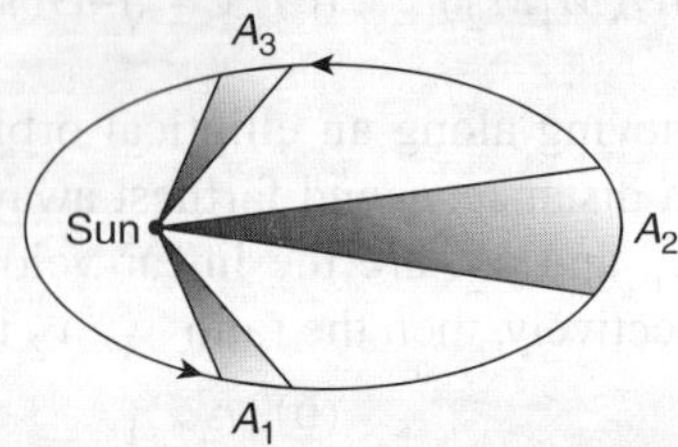

(a) $3A_1 = 2A_2 = A_3$ (b) $2A_1 = 3A_2 = 6A_3$
(c) $3A_1 = 2A_2 = 6A_3$ (d) $6A_1 = 3A_2 = 2A_3$

42. Which of the following astronomers first proposed that sun is static and earth goes around the sun?
(a) Copernicus (b) Kepler
(c) Galileo (d) None

43. If the distance between the earth and the sun becomes half its present value, the number of days in a year would have been
(a) 64.5 (b) 129
(c) 182.5 (d) 730

44. A particle is moving with a uniform speed in a circular orbit of radius R with a central force inversely proportional to the nth power of R. If the period of rotation of the particle is T then,

(a) $T \propto R^{\frac{n}{2}+1}$ (b) $T \propto R^{(n+1)/2}$
(c) $T \propto R^{n/2}$ (d) $T \propto R^{3/2}$ for any n

45. The mass and the diameter of a planet are three times the respective values for earth. The period of oscillation of a simple pendulum on the earth is 2s. The period of oscillation of the same pendulum on the planet would be

(a) $\frac{2}{\sqrt{3}}$ s (b) $2\sqrt{3}$s
(c) $\frac{\sqrt{3}}{2}$ s (d) $\frac{3}{2}$ s

Answer Keys

1. (d)	2. (d)	3. (a)	4. (b)	5. (b)	6. (d)	7. (d)	8. (a)	9. (a)	10. (d)
11. (a)	12. (a)	13. (c)	14. (a)	15. (b)	16. (a)	17. (a)	18. (b)	19. (b)	20. (b)
21. (c)	22. (d)	23. (b)	24. (b)	25. (a)	26. (c)	27. (d)	28. (b)	29. (b)	30. (c)
31. (b)	32. (b)	33. (a)	34. (b)	35. (c)	36. (c)	37. (c)	38. (b)	39. (a)	40. (c)
41. (a)	42. (a)	43. (b)	44. (b)	45. (b)					

Solutions

1. (d)

This is a case of a geostationary satellite

2. (d)

$$F = G\frac{M \times M}{(2R)^2} \text{ or } F = G\frac{(\frac{4}{3}\pi R^3 \times p)^2}{(2R)^2}$$

or $F \propto R^4$

3. (a)

The gravitational force between the body and Earth will provide the necessary centripetal force.

4. (b)

$$g' = g - R\omega^2 - mR\omega^2 \cos^2 \phi$$

$$mg' = mg - mR\omega^2 \cos^2 \phi$$

Out of the four given places, the value of ϕ is least at the equator. So, the term $mR\omega^2 \cos^2 \phi$ will have maximum value at the equator. Thus, mg' will be least.

5. (b)

$$g = \frac{GM}{R^2}$$

If the earth shrinks, its mass remains unchanged and its radius decreases. So the value of acceleration due to gravity increases.

6. (d)

7. (d)

Archimedes' uplift = Weight of liquid displaced. If the gravitational effect were smaller, then the weight of the liquid displaced will be smaller.

8. (a)

$$x = \frac{GM}{R^2}$$

Again, $\frac{x}{4} = \frac{GM}{(R+h)^2}$ or $x = GM(\frac{2}{R+h})^2$

$$\therefore \frac{1}{R^2} = (\frac{2}{R+h})^2$$

or $\frac{1}{R} = \frac{2}{R+h}$ or $R + h = 2R$

or $h = R$.

9. (a)

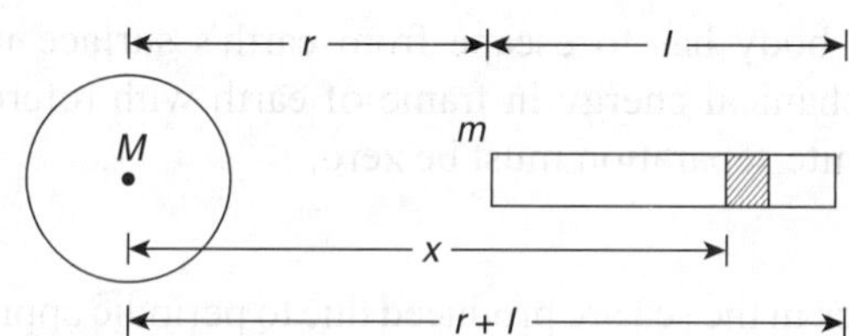

Mass per unit length $= \frac{m}{l}$

Mass of section of length $dx = \frac{m}{l}dx$

Let dF be the gravitational force of attraction between this section and sphere.

Then, $$dF = GM\frac{\left(\frac{m}{l}dx\right)}{x^2}$$

$$\int_r^{r+l} \frac{GMm}{lx^2}dx = \frac{GMm}{l}\int_r^{r+l}\frac{1}{x^2}dx$$

$$= \frac{GMm}{l}\int_r^{r+l} x^{-2}dx = \frac{GMm}{l}\left[\frac{x^{-2+1}}{-2+1}\right]_r^{r+l}$$

$$= -\frac{GMm}{l}[x^{-1}]_r^{r+l} = -\frac{GMm}{l}\left[\frac{1}{x}\right]_r^{r+l}$$

$$= \frac{GMm}{r(r+1)}$$

If $r >> l$, then $F = \frac{GMm}{r^2}$

10. (d)

$$g_p = g_c;\ \frac{GM_p}{R_p^2} = \frac{GM_e}{R_e^2};\ \frac{4}{3}\pi R_p d_p = \frac{4}{3}\pi R_e d_e$$

or $$2R_e d_e = R_e d_e, \frac{d_p}{d_e} = \frac{1}{2}$$

11. (a)

$$g' = g - R\omega^2 \cos^2 \lambda$$

$$= g_p - R\omega^2 \cos^2 60^0 = g_p - \frac{R\omega^2}{4}$$

12. (a)

If the gravitational force is inversely proportional to r, this force will provide the necessary centripetal force for circular motion of the particle. Then the radius r in denominator of both terms will cancel and speed will be independent of r.

13. (c)

If a body has to escape from earth's surface its total mechanical energy in frame of earth with reference at infinite separation must be zero.

14. (a)

Tides in the sea are produced due to periodic approaches of moon at different locations above the sea on earth due to the pull of gravitational attraction of moon.

15. (b)

16. (a)

By energy conservation, we use

$$\frac{1}{2}m(0)^2 - \frac{GMm}{\sqrt{(\sqrt{3}R)^2 + R^2}} = \frac{1}{2}mv^2 - \frac{GMm}{R}$$

$$-\frac{GMm}{R} = \frac{1}{2}mv^2 - \frac{GMm}{R}$$

$$v = \sqrt{\frac{GM}{R}}$$

17. (a)

By law of conservation of energy
Total surface = Potential energy at a height h
Energy at the surface of earth

$$\frac{1}{2}\frac{GMm}{R_e} - \frac{GMm}{R_e} = 0 - \frac{GMm}{R_e + h}$$

As velocity at maximum height is zero

$$h = R_e$$

18. (b)

Pull on satellite at height $h = mg\left(\frac{R}{R+h}\right)^2$

But $h = R/2$

$\therefore$ Pull of satellite $= 200 \times 10 \times \left(\frac{2R}{3R}\right)^2$

19. (b)

$$\frac{1}{2}mv^2 = \frac{GM_1 m}{d/2} + \frac{GM_2 m}{d/2}$$

20. (b)

The angular momentum of a planet about the sun is constant and hence vr = constant (mass of the planet is fixed).

21. (c)

$$F = \frac{GM_1M_2}{R} = \frac{G}{R} \times 100 \times 100$$

$$\frac{G}{R} = \frac{F}{10000}$$

$$F' = \frac{G}{R} 125 \times 75 = \frac{G}{R} \times 9375$$

$$F' = \frac{F}{10000} \times 9375$$

$$F' = \frac{15}{16}F$$

22. (d)

23. (b)

$mg = \frac{GM_E m}{R_E^2}$; where M_E and R_E is the mass and radius of the earth respectively. $M_E = \frac{g}{G} R_E^2$

24. (b)

Actually, gravitational force provides the centripetal force.

25. (a)

$$F \propto xm \times (1-x)m = xm^2(1-x)$$

For maximum force $\frac{dF}{dx} = 0$

$$\Rightarrow \frac{dF}{dx} = m^2 - 2xm^2 = 0 \Rightarrow x = 1/2$$

26. (c)

27. (d)

S_2 is correct because whatever be the g the same force is acting on both the pans. The value of g is greater at the poles. Therefore mg at the poles is greater. Hence, an object will weigh more at the poles if a spring balance is used. S_4 is correct. S_2 and S_4 are correct.

28. (b)

Acceleration due to gravity $g = \frac{GM}{R^2} \therefore \frac{g}{G} = \frac{M}{R^2}$

29. (b)

The value of g at height h from the surface of earth

$$g' = g\left(\frac{R}{R+h}\right)^2 = 10\left(\frac{6400}{6400+1600}\right)^2 = 6.4 \text{ ms}^{-2}$$

30. (c)

For the condition of weightlessness at equator,

$$\omega = \sqrt{\frac{g}{R}} \therefore \omega = \sqrt{\frac{1}{640\times 10^3}} = \frac{1}{800} \text{ rad/s}$$

31. (b)

$$g' = g - \omega^2 R\cos^2\lambda$$

Rotation of the earth results in the decreased weight apparently. This decrease in weight is not felt at the poles as the angle of latitude is 90^0.

32. (b)

$$G \propto \rho R$$

33. (a)

34. (b)

$$v = \sqrt{2gR} \Rightarrow \frac{v_A}{v_B} = \sqrt{\frac{g_A}{g_B}\times\frac{R_A}{R_B}} = \sqrt{k_1 \times k_2}$$

35. (c)

Angular momentum $L = mvr = \sqrt{2mE}.r = \sqrt{2mEr^2}$

36. (c)

Because earth rotates from west to east direction.

37. (c)

Time period of revolution of moon around the earth = 1 lunar month

$$\frac{T_s}{T_m} = \left(\frac{r_s}{r_m}\right)^{3/2} = \left(\frac{1}{2}\right)^{3/2} \Rightarrow T_s = 2^{-3/2} \text{ lunar month}$$

38. (b)

$$v_e = \sqrt{2}v_0 = 1.414v_0$$

Fractional increases in orbital velocity $\left(\frac{\Delta v}{v}\right)$

$$= \frac{v_e - v_0}{v_0} = 0.414$$

Percentage increase = 41.4 %

39. (a)

Time period $T = \frac{2\pi R}{\sqrt{\frac{GM_m}{R}}} = \frac{2\pi R^{3/2}}{\sqrt{GM_m}}$

where the symbols have their meaning as given in the question.

Squaring both sides, we get

$$T^2 = \frac{4\pi^2 R^3}{GM_m}$$

40. (c)

$$\frac{T_1}{T_2} = \left(\frac{R_1}{R_2}\right)^{3/2} = \left(\frac{10^{13}}{10^{12}}\right)^{3/2} = (1000)^{1/2} = 10\sqrt{10}$$

41. (a)

When a planet revolves around the sun, its real velocity is constant

$$\therefore \frac{A_1}{t_1} = \frac{A_2}{t_2} = \frac{A_3}{t_3}$$

$$\frac{A_1}{2} = \frac{A_2}{3} = \frac{A_3}{6} \Rightarrow 3A_1 = 2A_2 = A_3$$

42. (a)

43. (b)

According to Kepler's third law, the ratio of the squares of the periods of any two planets revolving about the sun is equal to the ratio of the cubes of their average distance from the sun, i.e.,

$$\left(\frac{T_1}{T_2}\right)^2 = \left(\frac{r_1}{r_2}\right)^3 = \left[\frac{r_1}{\frac{1}{2}r_1}\right]^3 = 8 \Rightarrow \frac{T_1}{T_2} = 2\sqrt{2}$$

$$T_2 = \frac{T_1}{2\sqrt{2}} = \frac{365 \text{ days}}{2\sqrt{2}} = 129 \text{ days}$$

44. (b)

45. (b)

CHAPTER 7

Properties of Bulk Matter

Introduction

All materials become distorted in some way under the action of external forces (extensional or compressional), and many of them regain their original size and shape when the extensional or compressional forces are removed. Such materials are said to be **elastic**. From an atomic viewpoint, elastic behaviour has its origin in the forces that atoms or molecules exert on each other.

Elasticity and Fluids

The interatomic forces that hold the atoms of a solid together are very strong and considerable external force has to be applied to stretch a solid object. The intermolecular forces arise due to two main causes:

(i) The potential energy of the molecules due to electrical interactions with the surrounding molecules.

(ii) The thermal energy of the molecules which depends on the temperature of the material.

The properties of a substance in its three different phases (solid, liquid or gas) are determined by the relative magnitudes of these two energies.

In this chapter we shall confine ourselves to the directly measurable quantities such as stress, strain, moduli of elasticity, strain energy etc., and shall not attempt detailed molecular explanation of the observed behaviour under external forces.

Stress

Consider a bar of uniform cross-sectional area A subjected to equal and opposite forces F at its ends (Fig. a). The bar is said to be under tension and each portion of the bar is in equilibrium as well. Consider a section shown by the dotted line which is perpendicular to the length of the bar. The portion at the right of the section must be pulling on the portion at the left with a force of equal magnitude F, and vice versa since every portion of the bar is in equilibrium. These pulls are uniformly distributed over the entire cross-section as indicated by short arrows (Fig. b) The Stress at the section is defined as the ratio of the force F to the area A:

$$\text{Stress} = \frac{F}{A}$$

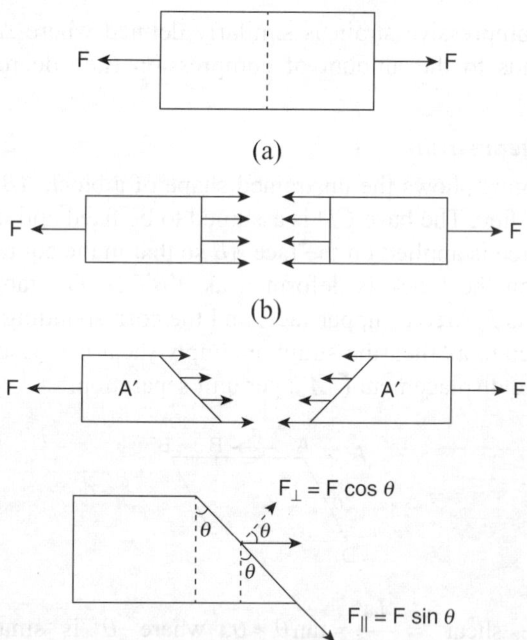

This stress is called a tensile stress or a normal stress. The stress at the inclined section can be resolved into:

$$\text{Normal stress} = \frac{F_\perp}{A'} = \frac{F\cos\theta}{A'} \text{ and}$$

$$\text{Tangential stress or Shearing stress} = \frac{F_\parallel}{A'} = \frac{F\sin\theta}{A'}$$

Units of Stress

(i) The SI unit of stress is N.m^{-2} which is also called pascal.

$1 \text{ pascal} = 1\,\text{Pa} = 1\,\text{N.m}^{-2}$

(ii) The CGS unit of stress is dyne. cm^{-2}.

(iii) The FPS unit of stress is pound per square inch $\left(\text{lb.in}^{-2}\right)$ which is commonly known as psi.

Strain

A body that is subjected to stress suffers some change in its dimension or shape. The relative change in its dimension or shape is called strain. Strains are basically of three types.

(i) Tensile or linear strain or longitudinal strain

It is defined as the ratio of the increase in length (ΔL) to the original length (L_0).

$$\therefore \quad \text{Tensile strain} = \frac{\text{Change in length}}{\text{Original length}}$$

$$= \frac{L - L_0}{L_0} = \frac{\Delta L}{L_0}$$

The compressive strain is similarly defined where ΔL corresponds to the amount of compression (i.e., decrease in length)

(ii) Shear strain

The figure shows the unstrained shape of a block *ABCD* by dotted line. The base *CD* is assumed to be fixed and tangential force is applied on the face *AB* so that in the equilibrium position the block is deformed as *A′B′CD*. The tangential stress is *F*/Area of upper face, and the corresponding strain, referred to as shearing strain or simply shear is measured by relative displacement (*AA′*) per unit separation.

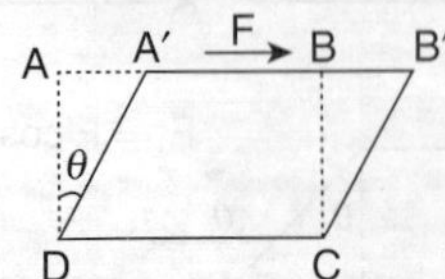

Hence shear $= \frac{AA'}{AD} = \tan\theta \approx \theta$, where θ is small and expressed in radian.

(iii) Volume Strain

When a body is subjected to compressive forces from all directions so that the size of every dimension (length, width and depth) decreases leading to a decrease in volume, the relative change in volume is defined as *volume strain* $= \frac{\Delta V}{V}$

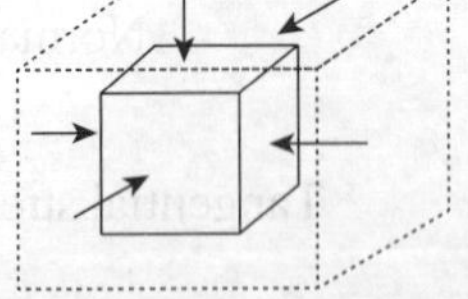

The strain produced by hydrostatic pressure is an example of volume strain.

Note that unlike stress, all the types of strains are pure numbers (having no unit or dimensions) because they are determined by the ratio of two lengths, or two volumes.

Hooke's Law

It is a common observation that when the stretching face F is increased (up to a certain limit) the extension ΔL also increases proportionately. The variation of $F \sim \Delta L$ is represented by a straight line. Based on this observation Robert Hooke, founder of the Royal Society, deduced the following relation in 1676 A.D. which is known as Hooke's Law:

"The force or a tension in a wire is proportional to the extension if the proportional limit is not exceeded".

Thus $\quad F \propto \Delta x$

or $\quad F = k.\Delta x,$

where k is commonly called the force constant or spring constant.

Hooke's Law is generalised as:

"For small deformation, the stress in a body is proportional to the corresponding strain", i.e., Stress $\propto$ Strain

or $\quad \text{Stress} = \text{E} \times \text{Strain}$

where the proportionality constant,

$\text{E} = \frac{\text{Stress}}{\text{Strain}}$ is known as modulus of elasticity.

Moduli of Elasticity

Depending upon the nature of stress exerted on the body, the modulus of elasticity is defined corresponding to the three types of strains:

(i) Young's Modulus (Y) for Longitudinal Strain

For small change in length of a wire, the ratio of tensile (or longitudinal stress) to the corresponding longitudinal strain is called the Young's modulus (Y) of the material of the wire.

Thus

$$Y = \frac{F/A}{\Delta L / L_0} = \frac{F}{A} \cdot \frac{L_0}{\Delta L}$$

If a wire of length L_0 and uniform radius r is clamped at one end and load $W = Mg$ is attached at the free end, then

$$F = Mg \quad \text{and} \quad A = \pi r^2$$

These substitutions lead to

$$Y = \frac{Mg}{\pi r^2} \cdot \frac{L_0}{\Delta L}$$

(ii) Bulk Modulus (B) for Volume Strain

For small change in volume of a body under uniform normal pressure, the ratio of pressure (or stress) to the corresponding volume strain is called the Bulk Modulus (B) of the material of the body.

Thus $\quad B = \frac{\text{Stress}}{\text{Volume strain}} = \frac{P}{-\frac{\Delta V}{V}}$

Note that the negative sign corresponds to decrease in volume associated with increase in pressure and vice versa.

The expression for B can also be expressed in the differential form:

$$B = \underset{\Delta V \to 0}{Lt} \frac{-\Delta P}{\Delta V} V = -V \frac{dP}{dV}$$

Here B is positive and $\frac{dP}{dV}$ is normally negative.

(iii) Compressibility

The reciprocal of the bulk modulus is referred to as *compressibility*, thus

$$\text{Compressibility}(K) = \frac{1}{B}$$

(iv) Modulus of Rigidity (η) for Shear Strain

For small shear in a block, the ratio of the shearing stress (or tangential stress) to the corresponding shear strain is called the Modulus of Rigidity (η) of the material of the body.

Thus $\quad \eta = \frac{\text{Tangential Stress}}{\text{Shear}} = \frac{F/A}{\theta}$

Poisson's Ratio (σ)

It is observed that extension in the length of a wire is accompanied with contraction in its radius. In other words longitudinal strain and lateral (or perpendicular) strain occur simultaneously. For small extension, the ratio of lateral strain to the longitudinal strain is a constant for the material of the wire and is called *"Poisson's Ratio"* (σ). If L_0 and D_0 indicate the initial values of the length and diameter respectively and ΔL and ΔD be the corresponding changes, then Poisson's ratio,

$$\sigma = \frac{\frac{-\Delta D}{D_0}}{\frac{\Delta L}{L_0}} = \frac{L_0}{D_0} \cdot \frac{\Delta D}{\Delta L}$$

The negative sign indicates that for extension (ΔL) there is contraction (ΔD) so that σ is positive.

For many common materials σ has a value between 0.1 and 0.3.

The constants Y, B, η and σ are inter-related by the following equation

$$Y = 3B(1-2\sigma) = 2\eta(1+\sigma).$$

Illustration 1

A stone of mass 0.5 kg is attached to one end of a 0.8 m long aluminum wire of diameter 0.7 mm and suspended vertically. The stone is now rotated in a horizontal plane at a rate such that the wire makes an angle of 85° with the vertical. Find the increase in the length of the wire. (Y of aluminum $= 7\times10^{10}\ \text{N/m}^2$, $\sin 85° = 0.9962$, $\cos 85° = 0.0872$).

Solution

The wire generates a cone and the system is referred to as a *conical pendulum*. The forces acting on the stone are

(i) Tension T is in the wire

(ii) Weight mg acting vertically downward.

Resolving T along the vertical and horizontal, the vertical component $T\cos\theta$ balances the weight mg and the horizontal component $T\sin\theta$ provides the required centripetal force $m\omega^2 r$, where r is the radius of the horizontal circle and ω is the angular frequency $\left(\frac{2\pi}{t} = 2\pi f\right)$.

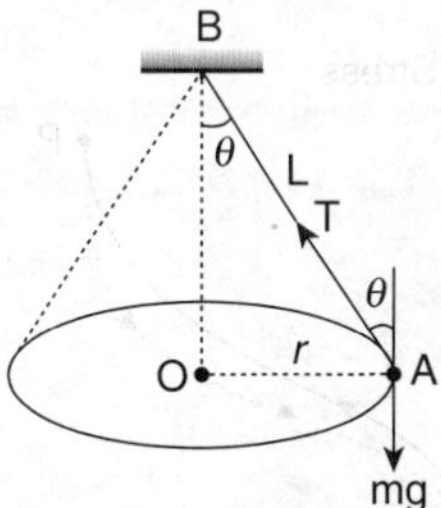

Thus $\quad T\sin\theta = m\omega^2 r$, and $T\cos\theta = mg$

$\therefore \quad \text{Tension } T = \frac{mg}{\cos\theta} = \frac{0.5\times9.8}{0.0872} = 56.19\text{N}$

Now, $\quad Y = \frac{F}{A}\frac{L}{\Delta L}$, or, $\Delta L = \frac{FL}{AY}$.

$$= \frac{56.19\times0.8}{3.14\times(0.35\times10^{-3})^2\times7\times10^{10}}$$

$$= 1.67\times10^{-3}\,m$$

$$= 1.67\ \text{mm}.$$

Proportionality Limit and Elastic Limit

According to Hooke's Law, stress is proportional to the strain. This law, in reality, is observed to be valid only upto a certain limit, known as the *proportionality limit.*

The stress-strain graph in the figure given here is a straight line up to a point P and afterwards the graph deviates from straight-line (shown by dotted line) and assumes a curved nature shown by the segment PQ.

The point P on the graph where the sample wire begins to deviate from the straight-line behaviour is called the "proportionality Limit". Beyond P, there exists a point Q on the graph such that the stress-strain linear relation is no longer valid in the region PQ but the object will return to its original size and shape when the stress is gradually reduced to zero from Q. The point Q is referred to as the elastic limit or the yield point. If the stress exceeds the elastic limit (point Q) the object no longer regains its original size and shape when the stress is removed; *the object remains permanently deformed.*

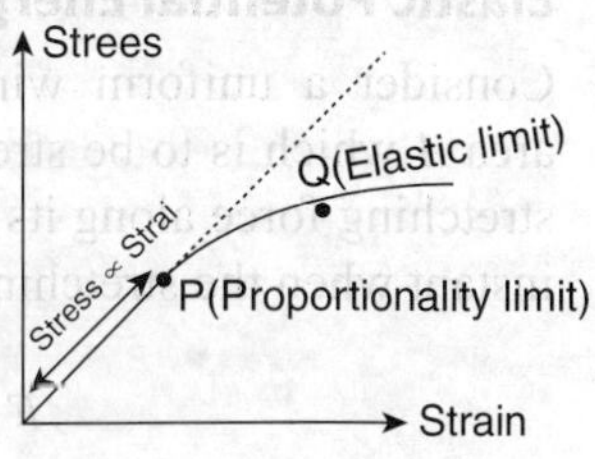

Elastic Hysteresis

The adjoining figure shows a typical stress-strain curve for a sample of hardened rubber which has been stretched to over seven times its original length. The material is elastic in the sense that on removal of the load, the rubber regains its original length but the stress-strain curve is not retraced. On decreasing the load the stress-strain curve follows the dotted curve shown in the figure. This lack of coincidence of the curves for increasing and decreasing stress is known as Elastic Hysteresis.

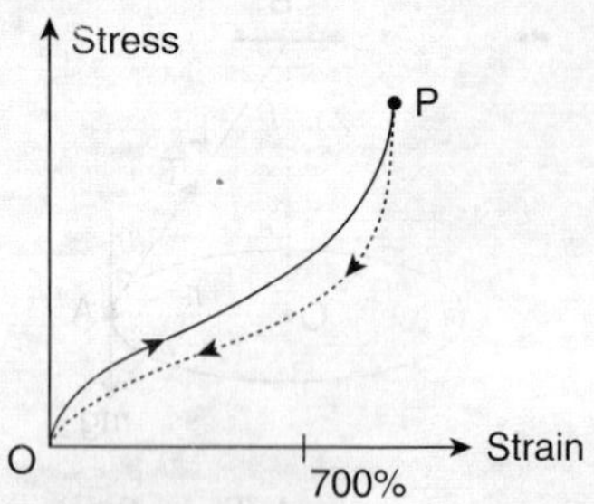

Force Constant (k) for a spring and a wire

For a spring, the stretching force (F) is directly proportional to the extension (or stretch) Δx, so that

$$F = k\Delta x \quad \ldots(1)$$

The constant k is known as spring constant or force constant for the spring.

Similar relation holds for a wire under the proportionality limit for which the Young's modulus

$$Y = \frac{F/A}{\Delta L/L_0}$$

$$\text{Stretching force } F = YA\frac{\Delta L}{L_0} = \left(\frac{YA}{L_0}\right)\Delta L \quad \ldots(2)$$

Comparing Eqs. (1) and (2), we conclude that the force constant (k) for a given wire can be expressed as

$$k = \frac{YA}{L_0}$$

Thus $\quad kL_0 = \text{constant}$

or $\quad k \propto \frac{1}{L_0}.$

Elastic Potential Energy

Consider a uniform wire of length L_0 and cross-section area A which is to be stretched *slowly* by an amount x_0 by a stretching force along its length. Let x be the extension at an instant when the stretching force is

$$F = \frac{YA}{L_0}x = kx$$

Work done for very small extension dx is

$$dW = Fdx = kxdx$$

$\therefore$ Total work done for the extension from $0 \to x_0$ will be

$$W = \int dW = \int_0^{x_0} kxdx = \frac{1}{2}kx_0^2 = \frac{1}{2}(kx_0)(x_0)$$

or $\quad W = \frac{1}{2}\text{stretching force} \times \text{stretch}$

This amount of work done by an external force is stored in the wire in the form of ***elastic potential energy*** *U*.

The total work done $W = U$ can also be expressed as

$$\frac{1}{2}\times\text{stress}\times\text{strain}\times\text{volume}.$$

$$U = \frac{1}{2}\cdot F\cdot x_0 = \frac{1}{2}\frac{F}{A}\cdot\frac{Ax_0}{L_0}\cdot L_0$$

$$= \frac{1}{2}\left(\frac{F}{A}\right)\left(\frac{x_0}{L_0}\right)(AL_0) = \frac{1}{2}\cdot\text{stress}\cdot\text{strain}\cdot\text{volume}$$

Thus the *elastic energy density*

$$u = \frac{U}{V} = \frac{1}{2}\text{stress}\cdot\text{strain}$$

The result derived above can be represented graphically as shown:

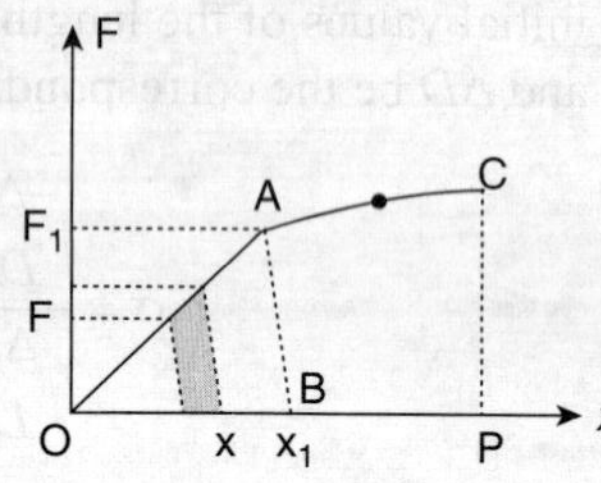

The extension (x) and the stretching force (F) are plotted along the x- and y-axes in the figure. The stretching force F_1 produces an extension x_1. At some stage before the extension x_1 is reached, let the restoring force in the wire be F. The work done (or energy stored in the wire) for small extension Δx is $F.\Delta x$ as shown by the shaded area

Hence the total work done for extension x_1 is

$$W_1 = \Sigma F\cdot\Delta x = \int_0^{x_1} Fdx = \text{area of the triangle } OAB$$

$$= \frac{1}{2}\cdot OB\cdot AB = \frac{1}{2}F_1x_1$$

$$= \frac{1}{2}\times\text{stretching force}\times\text{stretch}$$

This result agrees with the result obtained earlier. In terms of force constant (k), $F = kx$ and $W_1 = \int_0^{x_1} kx\,dx = \frac{1}{2}kx_1^2$

Note that work = area under $F - x$ graph can be used for both linear as well as non-linear (non-elastic) part of the curve. The energy stored in the wire for extension from O to P is still equal to the area $OACP$.

Thermal Stress and Strain

We know about thermal expansion in solids due to rise in its temperature. If a body is allowed to expand or contract freely with the rise or fall of its temperature, no restoring force or stress is developed. But some stress (known as thermal stress) is developed if the body is prevented from contracting or expanding. The corresponding strain is called thermal strain or temperature induced strain.

Let L_0 = initial length of a rod
A = cross-sectional area
Y = Young's modulus of the material of the rod
α = coefficient of linear expansion of the material of the rod

If the temperature is increased by $\Delta\theta$, the corresponding increase in its length (during free expansion) would be $\Delta L = L_0\alpha\Delta\theta$.

But when the rod is prevented from expanding, the thermal strain developed is $\frac{\Delta L}{L_0} = \alpha\Delta\theta$ and the corresponding thermal stress is $Y \times \text{strain} = Y\alpha\Delta\theta$.

The force exerted by the rod on the two rigid support at its ends is given by

$$F = \text{Stress.Area} = YA\alpha\Delta\theta.$$

Illustration 2

Two rods of different metals having the same area of cross-section A, are placed between the two massive walls as shown in the figure. The first rod has a length ℓ_1, coefficient of linear expansion α_1 and Young's modulus Y_1. The corresponding quantities for second rod are ℓ_2, α_2 and Y_2. The temperature of both the rods is now raised by T.

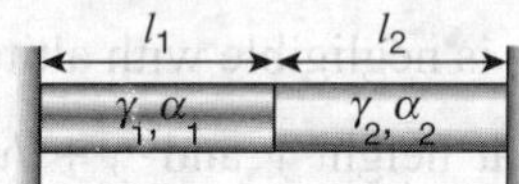

(i) Find the force with which the rods act on each other (at higher temperature) in terms of the given quantities.

(ii) Also find the lengths of the rods at higher temperature

Solution

(i) The coefficient of linear expansion

$$\alpha = \frac{1}{L}\frac{\Delta L}{\Delta T},$$

or, $$\Delta L = \alpha L \Delta T$$

$\therefore$ $$(\Delta L)_1 = \alpha_1\ell_1 T$$

$(\Delta L)_2 = \alpha_2\ell_2 T$, since it is given that $\Delta T = T$

$\therefore$ Total increase in length

$$\Delta L = (\Delta L)_1 + (\Delta L)_2 = (\alpha_1\ell_1 + \alpha_2\ell_2)T \quad \ldots\text{(i)}$$

As the walls are rigid, the length does not change but get compensated by the stress developed and the corresponding decrease in lengths will be

$$(\Delta L)_1 = \frac{F\ell_1}{Y_1 A}, \quad (\Delta L)_2 = \frac{F\ell_2}{Y_2 A}$$

$\therefore$ Total decrease in length

$$\Delta L = (\Delta L)_1 + (\Delta L)_2 = \frac{F}{A}\left(\frac{\ell_1}{Y_1} + \frac{\ell_2}{Y_2}\right) \quad \ldots\text{(ii)}$$

Equating (i) and (ii),

$$\frac{F}{A}\left(\frac{\ell_1}{Y_1} + \frac{\ell_2}{Y_2}\right) = (\ell_1\alpha_1 + \ell_2\alpha_2)T$$

Hence the required force,

$$F = \frac{(\ell_1\alpha_1 + \ell_2\alpha_2)AT}{\frac{\ell_1}{Y_1} + \frac{\ell_2}{Y_2}}.$$

(ii) Length of the rod at the higher temperature = Initial length + Thermal expansion – Contraction due to force F

$\therefore$ Length of the first rod

$$= \left(\ell_1 + \ell_1\alpha_1 T - \frac{F\ell_1}{AY_1}\right)$$

Similarly length of the second rod

$$= \left(\ell_2 + \ell_2\alpha_2 T - \frac{F\ell_2}{AY_2}\right)$$

Note that the total length remains unchanged.

Fluid Statics

Matter is commonly divided into three phases: solid, liquid and gas. Solids and liquids (also called condensed matter) have a certain set of properties in common: they are relatively incompressible and their densities remain relatively constant with temperature variation (provided the pressure is kept constant). Gases are easily compressible and their density changes substantially with temperature provided the pressure is held constant.

We assign the word Fluid, collectively for liquids and gases from a different view: both can flow and assume the shape of any container that holds them. In this chapter we first study the properties of fluids at rest and the laws governing them and later we will discuss the different laws governing the fluids in motion.

Pressure and Density

Fluids are unable to sustain shear when a force parallel to its surface is applied. Under static condition the only force component that needs to be considered is one that acts in a direction perpendicular to the surface of the fluid.

The forces between the interior and exterior of the fluid are everywhere normal to the fluid boundary and are independent of the shape of the fluid. The magnitude of the *normal force per unit surface area is called pressure.* A small element of surface area can be represented by a vector $\overrightarrow{\Delta A}$ along the outward normal. The force $\overrightarrow{\Delta F}$ exerted by the fluid lying within the bounding surface depends on the fluid pressure p according to

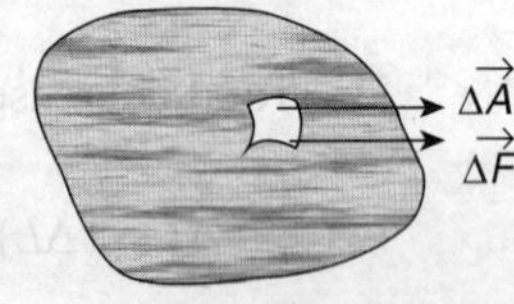

$$\overrightarrow{\Delta F} = p\,\overrightarrow{\Delta A}$$

Since $\overrightarrow{\Delta F}$ and $\overrightarrow{\Delta A}$ are parallel, we can write the pressure in terms of scalar relationship

$$p = \frac{\Delta F}{\Delta A}.$$

Density

The density ρ is defined as *mass per unit volume:*

$$\rho = \lim_{\Delta v \to 0} \frac{\Delta m}{\Delta v} = \frac{dm}{dv}.$$

Like pressure, density also has no directional property and is a *scalar* for a homogeneous medium. Density is simply mass divided by volume it occupies $(\rho = m / v)$.

Variation of Pressure with Depth

Consider a liquid at rest in a container (see figure). Let us select a cylindrical portion of the liquid of cross-sectional area A and height dy. The downward force on the top of the cylinder due to the surrounding fluid is PA and upward force on the bottom is $(P + \Delta P)A$. The gravitational force on the element is Δmg. Since the cylinder is in equilibrium, the net force on it is zero. Thus

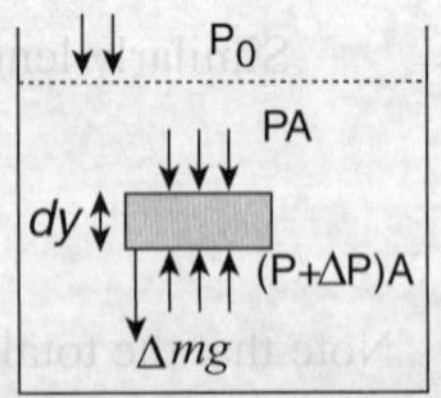

$$(P + \Delta P)A - PA = \Delta mg$$

$$= \rho A g \Delta y$$

$$\Delta \rho A = \rho A g \Delta y$$

$$\Rightarrow \quad \frac{dp}{dy} = \rho g$$

The above expression is true even if the density is non-uniform. We now consider the density to be uniform. The vessel is open to atmosphere at the top and therefore the pressure at the surface is P_0. To find the pressure at depth h, we integrate the above expression.

$$\int_{P_0}^{P} dP = \int_{0}^{h} \rho g dy$$

$$p = P_0 + \rho g h$$

Standard Atmospheric Pressure

If a standard atmosphere supports a column of height of exactly 76 cm of mercury, the standard atmospheric pressure can be expressed in pascals.

$$p_{\text{atm}} = h\rho g$$

Here, $h = 76\,\text{cm} = 0.76$ m

$$\rho_{\text{Hg}} = 13.6 \times 10^3\ \text{kg/m}^3, \text{ and } g = 9.8\ \text{m/s}^2$$

$$\therefore \quad p_{\text{atm}} = 0.76 \times 13.6 \times 10^3 \times 9.8\ \text{N/m}^2$$

$$= 1.013 \times 10^5\ \text{Pa} \quad (\text{or, } 101.3\ \text{kPa})$$

Changes in atmospheric pressure are observed as changes in the height of mercury column. One standard atmosphere is a unit of air pressure expressed in height units as

1 atm = 76 cm of Hg = 760 mm of Hg

In honour of Torricelli, a pressure supporting 1 mm of mercury is given the name **torr**:

1 mm Hg = 1 torr and
1 atm = 760 torr.

Variation of Pressure in the Atmosphere

The pressure of a gas in a small vessel is taken as uniform since its density is very small and the difference in pressure at two nearby points is negligibly small. However, this is not true if the height $(h_2 - h_1)$ is very large. The expression for the variation of pressure in the atmosphere is a bit difficult to derive. The reason is that gas (or air) is compressible and its density ρ varies with altitude. In order to determine the density $\rho(y)$ as a function of altitude y, we assume that

(i) the temperature of air remains constant with altitude, and

(ii) variation of g is negligible with altitude.

Let the pressure at height y and $y + dy$ be p and $p + dp$ respectively.

$$\therefore \quad dp = -\rho g\, dy$$

or, $$\frac{dp}{dy} = -\rho g.$$

(here the negative sign indicates that the pressure p decreases as height y increases)

Since density (ρ) is proportional to the pressure (p), we have

$$\frac{p}{p_0} = \frac{\rho}{\rho_0}, \quad \text{or} \quad \rho = \rho_0 \frac{p}{p_0}$$

$$\therefore \quad \frac{dp}{dy} = -g\rho_0 \frac{p}{p_0}$$

or,
$$\frac{dp}{p} = -\frac{g\rho_0}{p_0} dy \quad \ldots\text{(a)}$$

Integrating (a), we have

$$\int_{p_0}^{p} \frac{dp}{p} = -\frac{g\rho_0}{p_0} \int_0^h dy$$

or,
$$\ln \frac{p}{p_0} = -\frac{g\rho_0}{p_0} h$$

or,
$$p = p_0 e^{-\left(\frac{g\rho_0}{p_0}\right)h} \quad \ldots\text{(b)}$$

which can be expressed as

$$p = p_0 \cdot e^{-h/a} \quad \ldots\text{(c)}$$

where
$$a = \frac{p_0}{g\rho_0}$$

Assuming $p_0 = 1.01 \times 10^5$ pa, $\rho_0 = 1.21$ kg/m^3 and $g = 9.80$ m/s^2, we obtain

$$a = 8.55 \text{ km}.$$

Thus atmospheric pressure decreases exponentially with height [equation (c)] as shown in the adjoining graph.

Note that the constant a gives the difference in altitude over which the pressure drops by a factor of e. In other words, the pressure drops by a factor of 10 when the height increases by $a \ln 10 = a \times 2.30 = 20$ km. Thus at the heights $h = 20$ km, 40 km, above the sea level, the pressure would be 0.1 atm, 0.01 atm and so on.

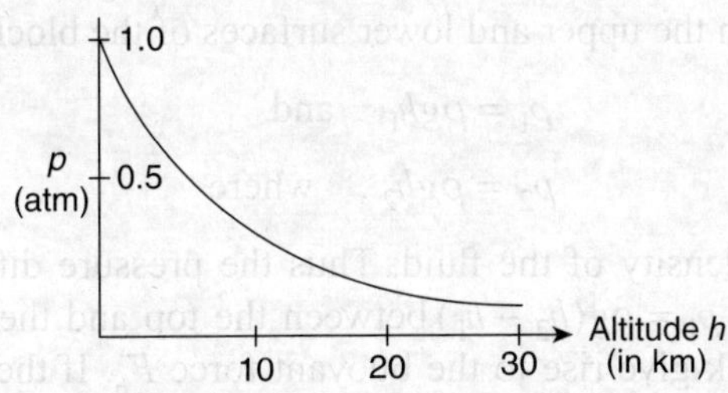

Illustration 3

A U-tube open to the atmosphere at both ends contains some mercury. A quantity of water is carefully poured into the left arm of the U-tube until the vertical height of water column is 15.0 cm.

(a) What is the gauge-pressure at the water-mercury interface?

(b) Calculate the vertical height h from the free surface of mercury to the free surface of water column.

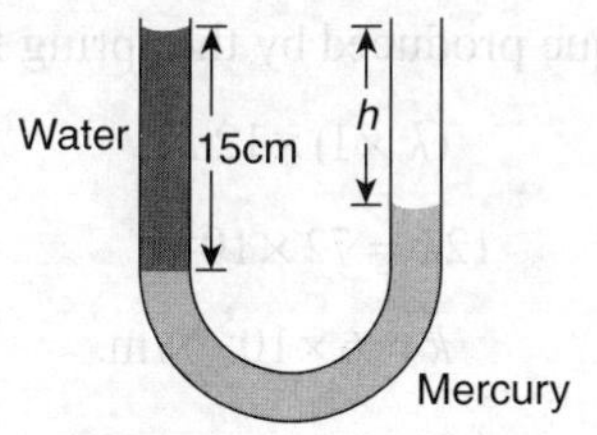

Solution

(a) Gauge pressure (which is the pressure above the atmospheric pressure) at water-mercury interface is

$$P = h_\omega \rho_\omega g = (15 \times 10^{-2})(1000)(10)$$
$$= 1500 \text{ Pa}.$$

(b) Equating the pressure at the same horizontal level of mercury in the two arms of the U-tube,

$$P_{atm} + h_\omega \rho_\omega g = p_{atm} + h_m \rho_m g$$

or,
$$15 \times 1000 = (15 - h) \times 13.6 \times 10^3$$

or,
$$h = 13.9 \text{ cm}.$$

Illustration 4

A barrier AB of length 12m is hinged at A. At the lower end a horizontal spring keeps the barrier closed. The height of water is 6m and the width of the barrier is 5m. Water level is 4m below the hinge A. If elongation of the spring to keep the barrier closed is 1m, find its spring constant (k). Neglect atmospheric pressure. ($g = 10$ m/s^2)

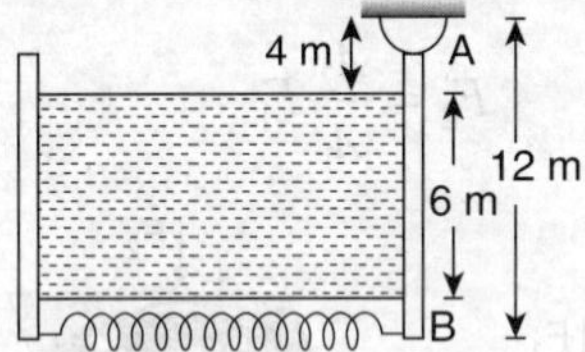

Solution

The fluid pressure at a distance x below the hinge is

$$p = (x - 4)\rho g$$

If b be the width of the barrier, the force acting on a thin area bdx is

$$F = p \cdot b\,dx = (x - 4)\rho g\, b\, dx$$

The torque of F about A is

$$d\tau = F.x = \rho g\, b(x - 4)x \cdot dx$$

Total deflecting torque due to fluid pressure is

$$\tau = \rho g b \int_4^{10} (x - 4)x \cdot dx$$

$$= 1000 \times 10 \times 5 \left[\frac{x^3}{3} - 2x^2\right]_4^{10}$$

$$= 72 \times 10^5 \text{ Nm}.$$

The restoring torque produced by the spring force is

$$(k \times 1) \times 12.$$

Hence, $\quad 12k = 72 \times 10^5,$

$\therefore \quad k = 6 \times 10^5$ N/m.

Pascal's Law

We know that fluid pressure depends on the depth h and is given by $p = p_0 + h\rho g$, where p_0 is the atmospheric pressure above the free surface of the liquid. A consequence of this fact used in hydraulic jack or lift is referred to as ***Pascal's law*** which states that:

"Pressure applied to a confined fluid (enclosed in a chamber) at rest is transmitted undiminished to every portion of the fluid and the walls of the containing vessel".

A well known application of Pascal's law is the hydraulic *lift of jack* to support or lift heavy objects as shown in the adjoining figure.

A piston with small cross-sectional area A_1 exerts a force F_1 on the surface of a liquid. The applied pressure $P = \dfrac{F_1}{A_1}$ is transmitted through the liquid in the connecting pipe to a larger piston of area A_2. The applied pressure is the same in both the cylinders, hence

$$P = \frac{F_1}{A_1} = \frac{F_2}{A_2},$$

or, $\quad F_2 = \dfrac{A_2}{A_1} F_1$

$A_2 >> A_1$, so $F_2 >> F_1$ and a small force F_1 acting on a small area A_1 results in a larger force F_2 acting on a larger area A_2. Thus, hydraulic lift is a force multiplying device with a multiplication factor equal to the ratio of the cross-sectional areas of the two pistons.

Illustration 5

A hydraulic press shown in the given figure consists of a larger piston of diameter 35 cm at a height of 1.5 m relative to a smaller piston of diameter 10 cm. The mass of the smaller piston is 20 kg. What is the force exerted on the load by the larger piston. The density of oil in the press is 0.75×10^3 kg/m^3 and $g = 9.8$ m/s^2.

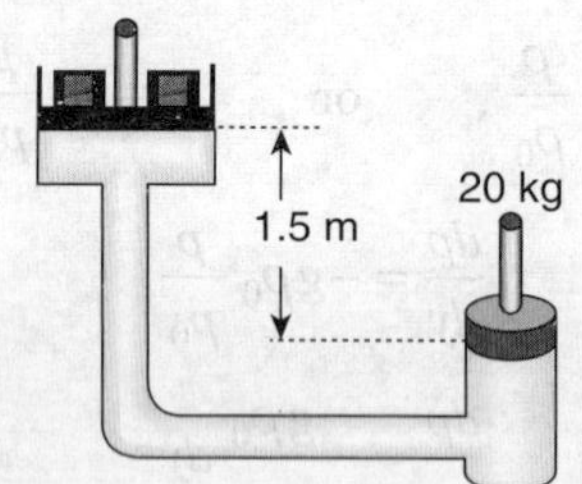

Solution

Fluid pressure on the smaller piston

$$P_1 = \frac{mg}{A} = \frac{20 \times 9.8}{\pi \times (5 \times 10^{-2})^2} \text{ N/m}^2$$

Similarly fluid pressure on the larger piston,

$$P_2 = \frac{F}{\pi (17.5 \times 10^{-2})^2} \text{ N/m}^2.$$

The difference between the two pressure is balanced by $h\rho g$, where $h = 1.5$ m and density of fluid $\rho = 0.75 \times 10^3$ kg/m^3

$$\therefore \quad \frac{20 \times 9.8}{\pi \times (5 \times 10^{-2})^2} - \frac{F}{\pi \times (17.5 \times 10^{-2})^2} = 1.5 \times 750 \times 9.8.$$

Simplifying we get,

$$F = 1.3 \times 10^3 \text{ N}$$

Buoyancy: Archimedes Principle

When an object is placed in a fluid, it will either sink or float. A similar effect is observed in gases as well. A falling object is sinking in the atmosphere, and objects such as hot-air balloons float.

Objects float because they are buoyant, or are buoyed up and thus require an upward net force. The upward force resulting from an object being wholly or partially immersed in a fluid is called the **buoyant force**.

Consider a buoyant object (piece of cork or wood) being held under the fluid surface as shown in the figure. The fluid pressure on the upper and lower surfaces of the block are

$$p_1 = \rho g h_1 \quad \text{and}$$

$$p_2 = \rho g h_2, \quad \text{where}$$

ρ is the density of the fluid. Thus the pressure difference, $\Delta p = p_2 - p_1 = \rho g (h_2 - h_1)$ between the top and the bottom of the block give rise to the buoyant force F_b. If the area of the top and bottom of the block is A, we have

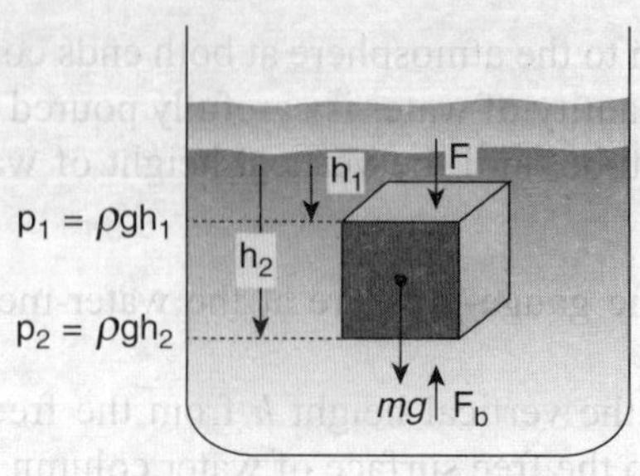

$$F_b = (\Delta p)A = \rho g(h_2 - h_1)A$$
$$= \rho gV,$$

where V is the volume of the block, which is also the volume of the fluid displaced by the block. But ρV = mass of the fluid displaced by the block. Thus the buoyant force

$$F_b = m_{fluid} \times g = \text{weight of fluid diplaced.}$$

This general result is known as **Archimedes Principle** as stated below:

A body immersed wholly or partially in a fluid is buoyed up by a force equal in magnitude to the weight of the volume it displaces.

Illustration 6

What fraction of the total volume of an iceberg floating is sea-water is exposed. Given: density of ice = 917 kg m^{-3} and density of sea-water = 1024 kgm^{-3}.

Solution

The weight of the iceberg acting vertically downward is

$$W_{ice} = \rho_{ice} V_{ice} g,$$

where V_{ice} is the volume of the iceberg. If V_w be the volume of the submerged part of the iceberg, the buoyant force F_b is the weight of the displaced seawater

$$\therefore \quad F_b = \rho_w V_w g$$

Since the iceberg is in equilibrium,

$$W_{ice} = F_b$$

or, $\quad \rho_{ice} \cdot V_{ice} g = \rho_w V_w \cdot g$

$$\therefore \quad \frac{V_w}{V_{ice}} = \frac{\rho_{ice}}{\rho_w} = \frac{917}{1024} = 0.896 = 89.6\%$$

$\therefore$ % of the iceberg exposed is 10.4%.

Stability of a Floating Body

The stability of a floating body depends on the positions of the effective point of application of the buoyant force and the centre of mass of the body. The buoyant force acts at the centre of mass of the displaced fluid which is known as the centre of buoyancy. When the floating body rests in equilibrium, the centre of mass of the body G and the centre of buoyancy B lie along a vertical axis of the body as shown in figure (a).

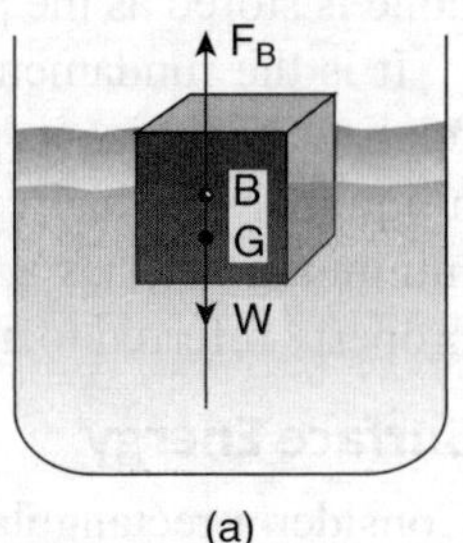

(a)

However, if the body tilts on one side, the line of action of the buoyant force F_B intersects the axis of the body at a point M, called the metacentre as shown in figure (b). If G is below M, the torque will restore the body to its stable equilibrium position.

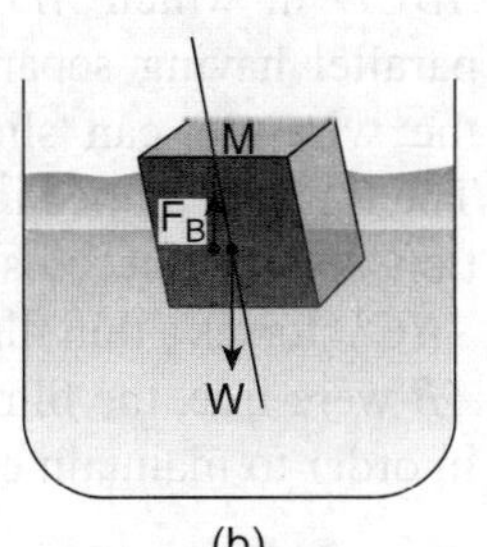

(b)

If G is above M, the torque will tend to rotate the body away from its equilibrium position and the body will be unstable.

Illustration 7

A uniform rod of length 6m is made of a material of specific gravity $= 25/36$. One end of the rod is tied to a string 5m long. The other end of the string is tied to the floor of a water-pool 10m deep. Find the length of the rod lying above the water surface.

Solution

Let a length x of the rod remains exposed. The free body diagram is shown in the figure. The forces acting are

(i) Weight of the rod $W = mg = AL\rho g$

(ii) Buoyant force $F_B = A(L-x)\rho_\omega g$ and

(iii) Tension in the string T.

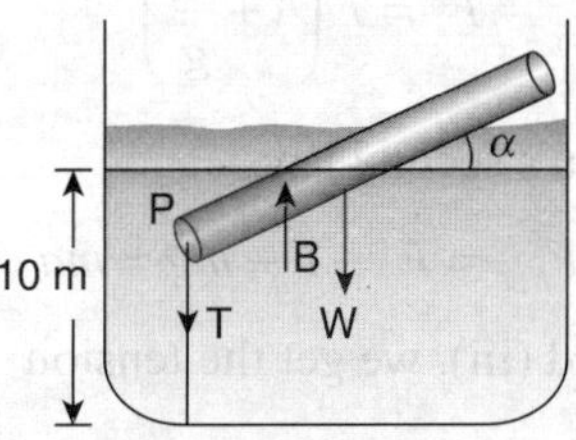

Since the system is in equilibrium, the moment of the forces about the point P must be zero. Hence

$$W \cdot \frac{L}{2}\cos\alpha - F_B \frac{(L-x)}{2}\cos\alpha = 0$$

or, $\quad AL\rho g \frac{L}{2} = A(L-x)\rho_\omega g \frac{(L-x)}{2}$

or, $\quad L^2\rho = (L-x)^2 \rho_\omega$

Substituting $\quad L = 6m,\ \rho = \frac{25}{36},\ \rho_\omega = 1$

$$6^2 . \frac{25}{36} = (6-x)^2 \times 1$$

$$\therefore \quad x = 1.0 \text{ m}.$$

Illustration 8

The tension in a light and inextensible string holding a solid block below the free surface of a liquid (of density grater than that of solid) is T_0 when the system is at rest relative to the ground. If the system has an upward acceleration a, what will be the tension in the string?

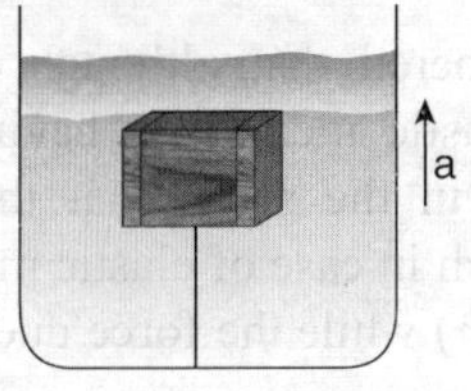

Solution

Figure (a) shown the free body diagram of the block when the system is at rest.

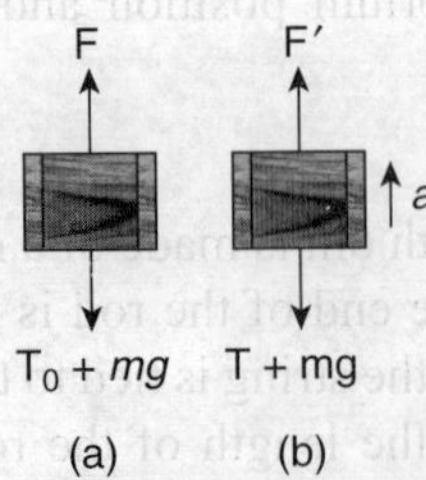

The buoyant force (or the upthrust) F is balanced by the tension T_0 plus the weight mg of the block. Thus

$$F = T_0 + mg \quad \text{...(i)}$$

When the lift accelerates upward, g_{eff} become $(g + a)$ instead of g. Hence the buoyant force becomes

$$F' = F\left(1 + \frac{a}{g}\right) \quad \text{... (ii)}$$

Now, from Newton's second law,

$$F_{net} = F' - T - mg = ma \quad \text{... (iii)}$$

Solving (i), (ii) and (iii), we get the tension

$$T = T_0\left(1 + \frac{a}{g}\right)$$

Surface Tension

It is a common observation that a small drop of a liquid e.g., mercury or water droplet on lotus leaves assumes a *spherical shape*. The spherical shape has a special feature. "It has a minimum surface area for a given volume".

We thus conclude that

"A liquid surface always tends to contract so as to have a minimum surface area". This contracting property of liquid-surface is known as SURFACE TENSION.

Quantitatively, the surface Tension (T) of a liquid surface is defined as the *force per unit length* on a line drawn on the surface directed perpendicular to the line and tangential to the liquid surface. Thus surface tension $T = \frac{F}{\ell}$ or $F = T.\ell$

The SI unit of Surface Tension is newton per metre ($1\ \text{N} \cdot \text{m}^{-1}$). This unit is not commonly used; the usual unit is the *cgs* unit, the dyne per centimetre (1 dyne. cm^{-1}). The conversion factor is

$$1\,\text{N.m}^{-1} = 1000\,\text{dyne.cm}^{-1}$$

It must be remembered that although the liquid surface behaves like an elastic membrane having the contracting property, it differs in the sense that the stretching force increases with stretch in case of elastic membrane (like rubber in which $F = kx$) while the force due to surface tension ($F = T.\ell$) remains constant and does not depend on the amount of stretch. What actually happens during increase in area of the liquid surface is that more interior molecules are brought to the surface for which work must be done against the inter-molecular forces. Surface tension is thus basically a molecular phenomenon.

EXPLANATION OF SURFACE TENSION AS A MOLECULAR PHENOMENON

Molecules of a liquid attract each other with a force called the *cohesive* force (the force between the molecules of the *same* substance). The cohesive force between the two molecules is appreciable within a certain small distance (called the *molecular range* or the *radius of molecular activity*) which is of the order of 10^{-9} metre, consisting of about 10 to 15 molecular diameter. These forces are electrical in nature because the gravitational forces are far too small to produce the observed surface phenomena.

Consider a molecule A well within the liquid. The resultant cohesive force on it is zero since the molecule is being equally attracted from all directions by the molecules lying in the *sphere of influence* (a sphere whose radius is equal to the molecular range). When the molecule is brought from A to B (very close to the free surface), its sphere of influence falls partly outside the liquid and it will experience net inward force. A molecule on the surface will experience a maximum cohesive force since it interacts with air molecules from above and liquid molecules from below. This *asymmetric force distribution* is responsible for surface tension.

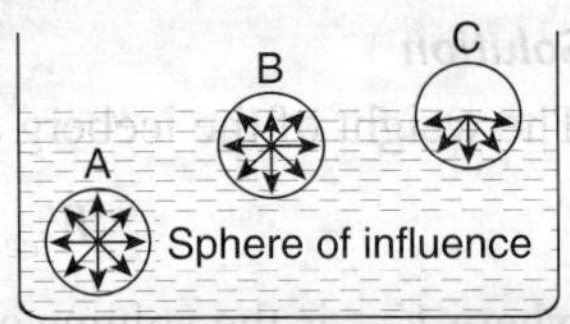

Thus a molecule on the surface experiences a downward drag which tends to bring it inside the liquid. In other words when the surface area of a liquid increases, more molecules are brought from within to the surface and work has to be done against the inward cohesive force. The amount of work done is stored as the potential energy of the surface.

It is the fundamental law of nature that a system always tends to minimize its potential energy and thus the surface of a liquid always has a tendency to reduce its surface area to a minimum, i.e., has a tendency to contract, giving rise to the property referred to as surface tension.

Surface Energy

Consider a rectangular wire frame $ABCD$ in which AD and BC are parallel having separation ℓ and the wire AB can slide smoothly. The frame is dipped in a soap solution so that the closed boundary $ABCD$ forms a thin film. If the wire AB were free, the film would contract by a force $= 2T\ell$ and in order to maintain equilibrium, an external force $F = 2T\ell$

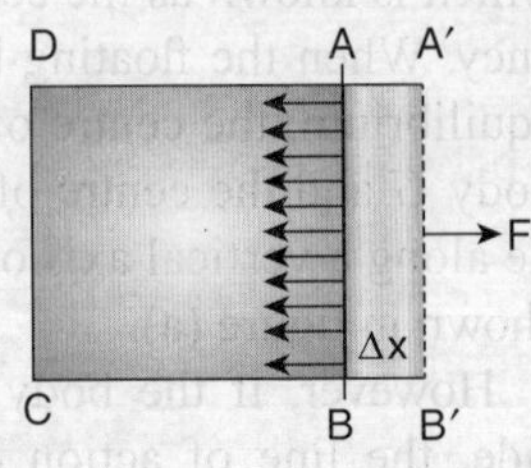

must be applied. Note that the factor 2 appears because the soap film has two surfaces.

Let a small displacement Δx be given to the movable wire so that AB is shifted to $A'B'$ and the film expands isothermally (at constant temperature). Work done by the external force F is

$$W = F.\Delta x = 2T\ell.\Delta x = T(2\ell.\Delta x)$$

or $$W = T.\Delta A \quad \text{... (i)}$$

where $\Delta A = 2\ell.\Delta x$ = increase in surface area of the film.

According to Eq. (i), the amount of work done in increasing the surface area of liquid surface is the product of surface tension and increase in surface area.

From Eq. (i),

$$T = \frac{W}{\Delta A} = \text{Work done per unit increase in surface area.}$$

Thus the surface tension has the SI unit, joule per square metre ($1\ \text{Jm}^{-2}$).

The work done per unit area in the surface is called the *surface energy*. In the isothermal expansion of the film, some heat H is also absorbed and constitutes a part of surface energy. Thus surface energy

$$E = T + H,$$

where H is the heat absorbed per unit increase in surface area.

Thus surface tension is numerically equal to the mechanical part of surface energy which is also called the free surface energy.

Adhesion, Cohesion, and Angle of Contact

Adhesive forces (or adhesion) are attractive forces between unlike molecules (i.e., between molecules of different substance like glass and water molecules).

Cohesive forces (or Cohesion) are attractive forces between like molecules (i.e., between molecules of water, between molecules of glass)

Cohesive forces hold a substance together, and adhesive forces hold different substances together (adhesives such as glue hold things together).

Wetting of a surface. If the adhesive forces between the molecules of the liquid and those of the surface (water on glass) are greater than the cohesive forces among the molecules of liquid, the liquid wets the surface.

On the other hand, if the cohesive forces are greater (between molecule of mercury) than the adhesive forces (between mercury and glass), the liquid does not wet the surface. Formation of water beads on a newly waxed car is a good example of non-wetting of surface.

Angle of Contact

When a liquid is in contact with a plane glass plate, the liquid assumes a curved shape. The curves *PAB* are called capillary curves. The tangent to the capillary curve at the point of contact P makes an angle θ with the solid (glass) *surface inside the liquid* which is called the angle of contact. This angle is acute ($\theta < 90^o$) for soild-liquid pair which wet the surface and it is obtuse ($\theta > 90^o$) for the pair (e.g. mercury and glass) which do not wet. Thus the angle of contact is a relative measure of the effects produced by the cohesive and adhesive forces. For water-glass pair $\underset{(water-glass)}{f_{adhesive}} > \underset{(water-water)}{f_{cohesive}}$ and θ is acute.

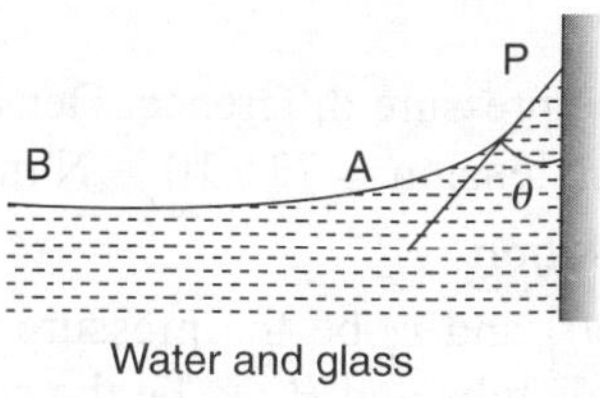

Water and glass

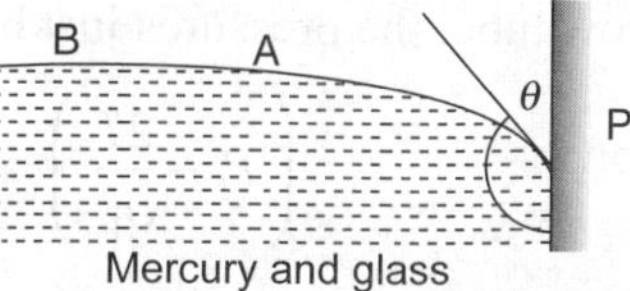

Mercury and glass

Similarly for mercury glass pair $\underset{(Hg-galss)}{f_{adhesive}} < \underset{(Hg-Hg)}{f_{cohesive}}$ and θ is obtuse.

Pressure Difference (or Excess Pressure) Across a Curved Membrane (or Film)

$$p = 2T\left(\frac{1}{R_1} + \frac{1}{R_2}\right)$$

Illustration 9

Two soap bubbles coalesce to form a single bubble without any leakage of air. If V be the change in volume of the enclosed air and S the total change in surface area of film, show that

$$3PV + 2ST = 0$$

where P is the atmospheric pressure and T is the surface tension of the soap solution.

Solution

Let a and b be the radii of the two bubble and c be the radius of the resulting bubble, we have

$$P_1V_1 + P_2V_2 = P_3V_3$$

or, $$\left(P + \frac{4T}{a}\right)\frac{4}{3}\pi a^3 + \left(P + \frac{4T}{b}\right)\frac{4}{3}\pi b^3 = \left(P + \frac{4T}{c}\right)\frac{4}{3}\pi c^3$$

Now $$S = 8\pi(a^2 + b^2 - c^2)$$

and $$V = \frac{4}{3}\pi(a^3 + b^3 - c^3)$$

Rearranging the terms,

$$4T\left(a^2 + b^2 - c^2\right) = P(c^3 - a^3 - b^3)$$

$$\Rightarrow \quad 3PV + 2ST = 0$$

Illustration 10

A barometer contains two uniform capillaries of radii 1.44×10^{-3} m and 7.2×10^{-4} m. If the height of liquid in narrow tube is 0.2m more than that in wider tube, calculate the true pressure difference. Density of liquid $=10^3$ kg/m^3, surface tension $=72\times10^{-3}$ N/m and $g=9.8$ m/s^2.

Solution

If P_1 and r_1 be the pressure and radius of meniscus in the wide tube and P_2, r_2 be the corresponding values in the narrow tube, the pressures just below their meniscus will be

$$\left(P_1-\frac{2T}{r_1}\right) \text{ and } \left(P_2-\frac{2T}{r_2}\right).$$

The difference between these two pressure will be balanced by the pressure $h\rho g$ where h is the difference in the heights of liquid columns.

$$\therefore \quad P_1-P_2=h\rho g-2T\left(\frac{1}{r_2}-\frac{1}{r_1}\right)$$

Assuming the angle of contact $\theta=0$, r_1 and r_2 will also be the radii of the two tubes

$\therefore$ Pressure difference

$$=0.2\times10^3\times9.8-2\times(72\times10^{-3})\left(\frac{10^4}{7.2}-\frac{10^3}{1.44}\right)$$

$$=1960-100=1860 \text{ Pa}.$$

Difference of Level in a Vertical U-Tube with Unequal Radii

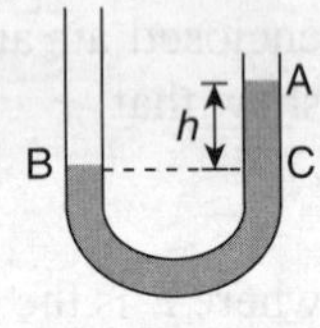

Let R and r be the radii of the two arms of a U-tube which is held vertically and contains a liquid whose surface tension is T. The liquid rises in the narrow tube due to capillary action so that the difference of levels in these be h. If p_0 = atmospheric pressure, then the pressure just below the meniscus at B is $p_0-\frac{2T}{R}$ and that just below that meniscus A is $p_0-\frac{2T}{r}$.

Since the points B and C are in the same horizontal level, we have pressure at B = pressure at C

$$\text{or} \quad p_0-\frac{2T}{R}=p_0-\frac{2T}{r}+h\rho g$$

$$\therefore \quad 2T\left(\frac{1}{r}-\frac{1}{R}\right)=h\rho g.$$

This equation can be used to find the surface tension T of the liquid.

Work Done in Blowing a Soap Bubble

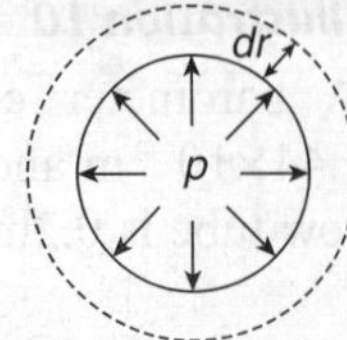

Let r be the radius of the bubble at an instant during blowing. The excess pressure inside will be

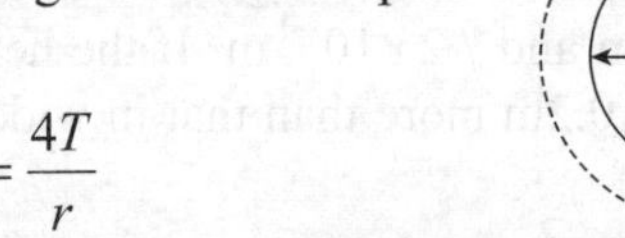

$$p=\frac{4T}{r}$$

which acts radially outward. Work done for small increment dr

$$dW=F\,dr=\left(\frac{4T}{r}\right)4\pi r^2\cdot dr$$

Total work done,

$$W=\int dW=4T\cdot4\pi\int_0^R r\,dr=8\pi TR^2$$

Alternatively, the same result can also be obtained from energy conservation:

$$\begin{aligned}\text{Work done} &= \text{Increase in the surface energy}\\ &= \text{Surface tension} \times \text{Increase in surface area}\\ &= T\times(A_{final}-A_{initial})\\ &= T\times(2\times4\pi R^2-0)\\ &= 8\pi TR^2.\end{aligned}$$

Raise of Liquid in a Capillary Tube

When an open tube of a very fine (narrow) capillary is dipped in a liquid (water), a rise of liquid is observed. This is a striking effect of surface tension, known as capillarity or capillary rise.

(i) The word capillary is derived from a Latin word meaning "hairlike".

(ii) Our smallest blood vessels are called capillaries and are so narrow that blood cells must pass through them in single file.

(iii) Capillary action is important in liquid transport. Common examples are the absorption of water by paper napkins, the wicking of oil in oil lamps, the distribution of water and nutrients in plants, and the holding of water in the soil.

(iv) Had there been no capillary action, the ground would be pretty dry, down to the water table.

(v) Less desirable results of capillary action are the wetting of concrete blocks in building walls and the rise of water in cracks and joints.

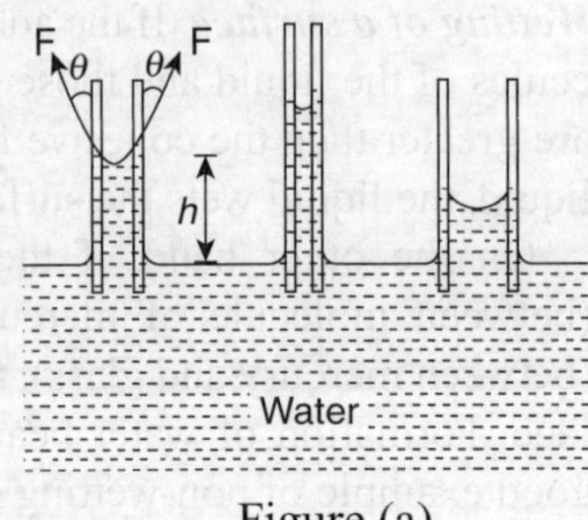

Figure (a)

Consider a tube of small radius r held vertically with one end submerged in a liquid that wets the walls of the tube. Let h be the height to which the liquid rises above the surface of the surrounding liquid due to the combined effect of surface tension and adhesion both.

Essentially, adhesion draws the water up the sides of the tube, and cohesion (surface tension) then pulls the column upward.

At equilibrium, the vertically upward component of surface tension force and the weight of the liquid column along vertically downward must be equal in magnitude. The force due to surface tension is

$$F = T \cdot L = T(2\pi r)$$

where $L = 2\pi r$ since the liquid is in contact with the tube at all points around its circumference. The vertical component of this force is

$$F\cos\theta = T(2\pi r)(\cos\theta) \quad \ldots\text{(i)}$$

The weight of the liquid column is

$$W = mg = \rho Vg = \rho(\pi r^2 h)g \quad \ldots\text{(ii)}$$

where V is the volume of the liquid column and ρ is the density of the liquid. Equating (i) and (ii), we get

$$T.2\pi r.\cos\theta = \rho.\pi r^2 h.g$$

or
$$T = \frac{rh\rho g}{2\cos\theta} \quad \ldots\text{(iii)}$$

If the liquid in the meniscus is also taken into account, assuming it to be hemispherical, the weight of the liquid column is $\pi r^2\left(h + \frac{r}{3}\right)\rho g$ and Eq. (iii) is modified as

$$T = \frac{r\left(h + \frac{r}{3}\right)\rho g}{2\cos\theta} \quad \ldots\text{(iii a)}$$

Note that $r/3 << h$ and may be considered as negligible.

Force Between two Plates Separated by a Thin Layer of a Liquid

If two glass plates are well wetted and placed together with their wet surfaces in contact, it is observed that, although they slide over each other comparatively easily, it requires considerable force to draw them apart normally with a risk of breakage as well. The force required for this normal detatchment is a measure of attractive force produced by surface tension and can be evaluated as given below:

Let A = wetted area of each plate
B = circumference of wetted area
d = separation between the parallel plates.

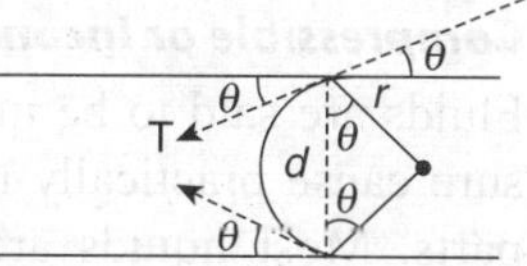

Normal component of surface tensional force around the periphery is $B.T.\sin\theta$.

The pressure within the film (being on the convex side of the liquid film) will be less than that outside by an amount $T\left(\frac{1}{r} - \frac{1}{R}\right)$ where r and R are the principal radii of curvature of the meniscus. From the given figure, $d = 2r\cos\theta$, and $R >> r$, the resultant excess pressure force acting downward on the upper plate in $AT/r = \frac{2AT}{d}\cos\theta$

$\therefore$ Resultant attraction

$$F = BT\sin\theta + \frac{2AT}{d}\cos\theta.$$

For most liquids which wet the plates, $\theta \approx 0$, and with sufficient accuracy the force F is given by

$$F = \frac{2AT}{d}.$$

Force to Pull a Plate From a Liquid Surface

Consider a plate placed on the surface of a liquid which is raised vertically upward. It is observed that a layer of the liquid will adhere to the plate and will thus be elevated. Considerable normal pull is required for its detatchment which can be evaluated as follows:

Figure (a) shows the situation when the liquid having been raised to a maximum amount h is about to break away.

Let A = area of the plate
p_0 = atmospheric pressure.

The pressure just below the plate (within the liquid will be $p_0 - h\rho g$, where ρ is the density of the liquid.

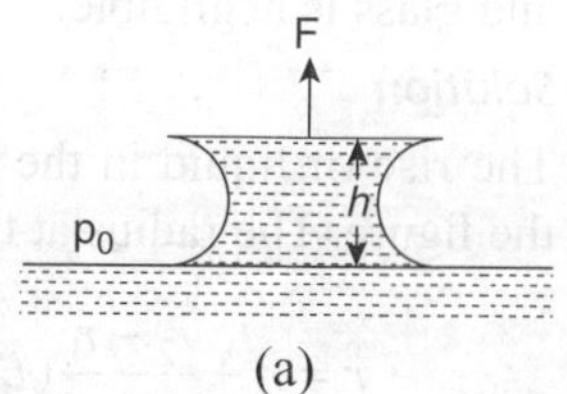

(a)

The downward force on the plate is

$$F_1 = p_0 A,$$

and the upward force is

$$F_2 = (p_0 - h\rho g)A$$

Hence the force required to pull the plate away from the liquid, is

$$F = p_0 A - (p_0 - h\rho g)A = Ag\rho h \quad \ldots\text{(i)}$$

To find h, consider the horizontal equilibrium of the slice of liquid of length L (figure (b).

The forces acting to the right are $2TL$, acting tangentially to the meniscus and $PLh = Lh\left(p_0 - \frac{1}{2}h\rho g\right)$ is due to hydrostatic pressure on the vertical rectangular section. The force due to atmospheric pressure is $p_0.Lh$, acting towards the left. Thus, for horizontal equilibrium

(b)

$$p_0 Lh = Lh\left[p_0 - \frac{1}{2}\rho gh\right] + 2TL$$

or, $$\frac{1}{2}\rho g h^2 = 2T;\quad h = 2\sqrt{\frac{T}{\rho g}} \quad \text{...(ii)}$$

Substituting the value of h in (i), we have

$$F = 2Ag\rho\sqrt{\frac{T}{\rho g}} = 2A\sqrt{T\rho g} \quad \text{...(iii)}$$

Illustration 11

A conical glass capillary tube of length 0.1m has diameters 10^{-3}m and 5×10^{-4}m respectively at its ends. When it is just immersed in a liquid at 0° C with larger radius in contact with it, the liquid rises to 8×10^{-2}m in the tube. In another cylindrical glass capillary tube, when immersed in the same liquid at 0°C, the liquid rises to 6×10^{-2}m high. The rise of liquid in the second capillary is only 5.5×10^{-2}m when the liquid is at 50° C. Find the rate at which the surface tension changes with temperature assuming the change to be linear. The density of liquid is $\left(\frac{1}{14}\right)\times 10^4\, kg/m^3$ and the angle of contact is zero. Effect of temperature on the density of liquid and glass is negligible.

Solution

The rise of liquid in the conical glass capillary is shown in the figure. The radius at the meniscus will be

$$r = r_1 + \frac{r_2 - r_1}{L}(L - h)$$

$$= (2.5\times 10^{-4}) + \frac{(2.5\times 10^{-4})}{0.1}(0.1 - 0.08)$$

$$= 3.0\times 10^{-4}\, m.$$

The surface tension at 0°C is

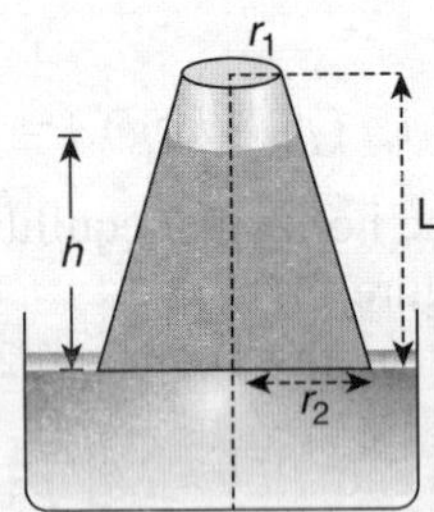

$$T_0 = rh\rho g/2$$

$$= \frac{(3.0\times 10^{-4})(8\times 10^{-2})(1/14\times 10^4)(9.8)}{2}$$

$$= 0.084\ \text{N/m}.$$

For the second tube let h_0 and h_{50} be the capillary rise at 0°C and 50°C respectively, then

$$\frac{T_{50}}{T_0} = \frac{h_{50}}{h_0} = \frac{5.5\times 10^{-2}}{6.0\times 10^{-2}} = \frac{11}{12}$$

$$\therefore \quad T_{50} = \frac{11}{12}\times T_0 = \frac{11}{12}\times 0.084 = 0.077\ \text{N/m}$$

Given that the change of surface tension with temperature is linear, we can write

$$T_\theta = T_0(1 + \alpha\theta)$$

where α is the temperature coefficient of surface tension. Hence the rate of change of T with θ is

$$\frac{dT}{d\theta} = \alpha T_0 = \frac{T_\theta - T_0}{\theta - 0}$$

$$= \frac{T_{50} - T_0}{50 - 0} = \frac{0.077 - 0.084}{50}$$

$$= -1.4\times 10^{-4}\ \text{N/m°C}.$$

Fluid Dynamics

Liquids and gases are collectively referred to as "Fluids" because they can flow. For example, water may flow smoothly and slowly in a quiet stream or violently over a water fall or in a *tsunami*. The air may form a gentle breeze or a raging tornado. To deal with such diversity, we first discuss some of the basic types of fluid flow and their characteristics.

Steady flow

The flow of fluid is said to be steady if the velocity of the fluid particles at any point is constant as time passes. In the given figure fluid particles flow with a velocity of $v_1 = 2\,\text{ms}^{-1}$ past point A and at another point B, the fluid velocity is $v_2 = 0.5\text{ms}^{-1}$. In steady flow, every particle passing through these points has these same velocities (in magnitude and direction) for all time during the flow.

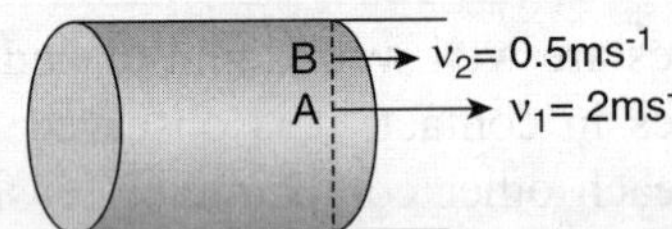

Unsteady or Turbulent Flow

Fluid flow is said to be unsteady or turbulent when the velocities of the fluid particles at any point changes erratically both in magnitude and direction as time passes. Turbulent flow occurs when there exist sharp obstacles or bends in the path of a fast moving fluids, as in the rapids.

Compressible or Incompressible Fluid Flow

Fluids are said to be incompressible if the changes in pressure cause practically no change in fluid density at various parts. Most liquids are nearly incompressible. To a good approximation, flow of liquids takes place in incompressible manner. In contrast, gases are highly compressible.

Viscous or Non-viscous Flow

It is our common observation that honey does not flow readily whereas water flows more readily. This difference exists due to a special property of fluid-flow, known as ***viscosity***

or internal friction in fluid flow. Honey is said to have large viscosities and water has comparatively smaller viscosity. The flow of a viscous fluid is an *energ-dissipating process*. Although no real fluid has zero viscosity at normal temperature, some fluids have negligibly small viscosities. *A fluid is said to be ideal if it is incompressible and non-viscous.*

Streamlines

During steady fluid-flow, the trajectories of the fluid particles are in general curved paths, known as streamlines. A streamline is a line drawn in the fluid such that the tangent to the streamline at any point is parallel to the fluid velocity at that point. The fluid velocity can vary (in both magnitude and direction) from point to point along a ***streamline***, but at any given point, the velocity remains constant in time during steady flow. Hence steady flow is also called ***streamline flow.***

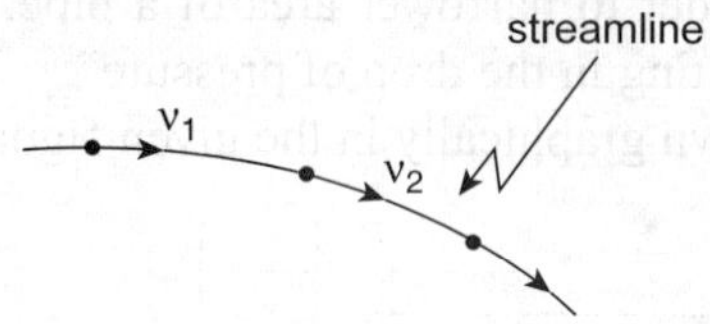

Equation of Continuity

The analysis of fluid-flow is not done in terms of Newton's laws but in terms of two basic conservation principles:

(i) The conservation of mass, and
(ii) The conservation of energy.

Assuming that there are no losses of fluid within the tube, the mass of the fluid flowing into the tube in a given time must be equal to the mass of the fluid flowing out of the tube in the same time (Principle of mass conservation).

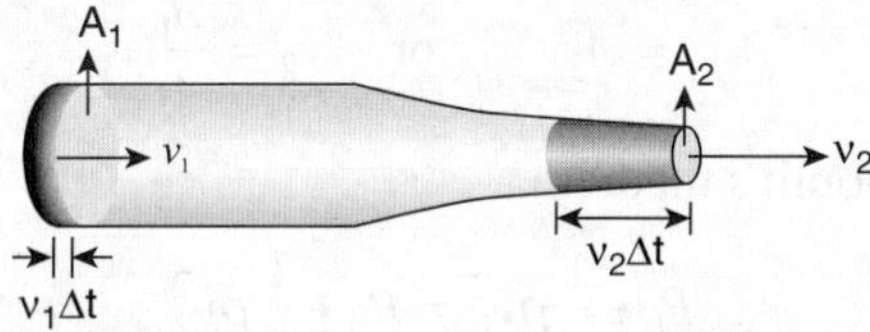

In the given figure, the mass (Δm_1) entering the tube in a short time (Δt) is

$$\Delta m_1 = \rho_1 \Delta V_1 = \rho_1 (A_1 v_1 \Delta t)$$

where A_1, is the cross-sectional area of the tube at the entrance, and in time Δt, a fluid particle moves a distance equal to $v_1 \Delta t$. Similarly, the mass leaving the tube in the same time interval is

$$\Delta m_2 = \rho_2 \Delta V_2 = \rho_2 (A_2 v_2 \Delta t)$$

Assuming that there are no places between the entry and exit points to add or remove fluid, the mass is conserved, i.e., $\Delta m_1 = \Delta m_2$, which leads to

$$\rho_1 A_1 v_1 = \rho_2 A_2 v_2$$

or,
$$\rho A v = \text{constant}$$

This general result is called the **equation of continuity.**

For an incompressible fluid, the density ρ is constant, so

$$A_1 v_1 = A_2 v_2$$

or,
$$Av = \text{constant.}$$

Bernoulli's Equation

For steady flow of an incompressible and non-viscous fluid, the speed (v), the pressure (p), and the elevation (h) are related by an equation discovered by Daniel Bernoulli, which is referred to as Bernoulli's Equation (given by Eq. (iii) on the next page.)

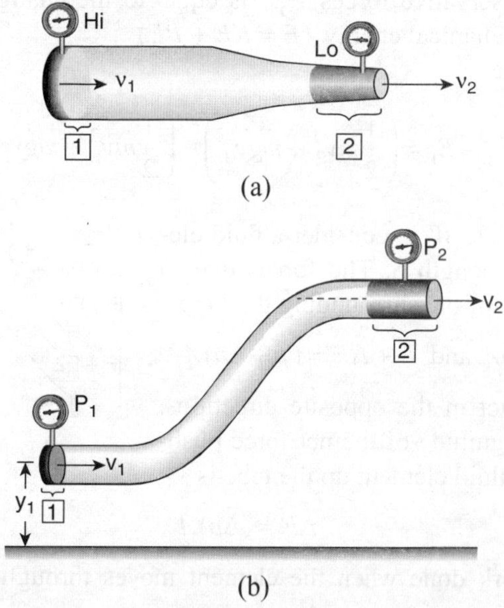

Before we derive Bernoulli's equation, let us consider two observations about a moving fluid. In figure (a), the fluid flowing in a horizontal tube encounters constriction (a region of reduced cross-sectional area), the pressure of the fluid drops. In this case the fluid speeds up (or accelerates), consistent with the equation of continuity $(A_1 v_1 = A_2 v_2)$. According to Newton's second law, the accelerating fluid must be subjected to an ***unbalanced force*** produced due to ***pressure difference*** between regions 1 and 2 $(p_1 > p_2)$. In figure(b) the fluid moves to a higher elevation in uniform tube, the pressure of the lower level is observed to be greater than the pressure at the higher level. The basis for this observation is like that of static fluid $(\Delta P = y\rho g)$. Here the velocities in regions 1 and 2 are same but elevations are different $(y_1 < y_2)$; consequently, the pressures are unequal $(p_1 > p_2)$.

Derivation of Bernoulli's Equation

Consider a tube of non-uniform cross-section through which an ideal fluid (non-viscous and incompressible) flows in streamline manner. Let the speed, pressure and elevation of an element of mass m in region 1 be v_1, p_1, and y_1 respectively and let these values in region 2 be v_2, p_2 and y_2.

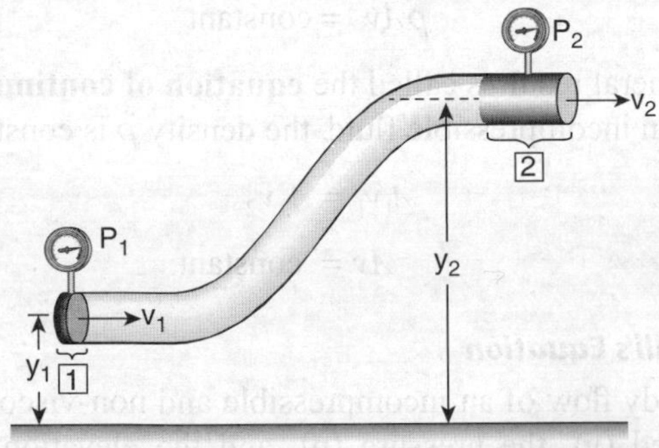

We know that the pressure within a fluid arises due to collisional forces, which are in fact, *non-conservative*. According to *work-energy theorem*, the net work done on an object by non-conservative forces W_{nc} is equal to the change in the total mechanical energy $(E = KE + PE)$.

Hence,

$$W_{nc} = E_2 - E_1 = \left(\frac{1}{2}mv_2^2 + mgy_2\right) - \left(\frac{1}{2}mv_1^2 + mgy_1\right) \quad \ldots\text{(i)}$$

To evaluate W_{nc}, consider a fluid element of length S. The forces due to pressure of the surrounding fluid are

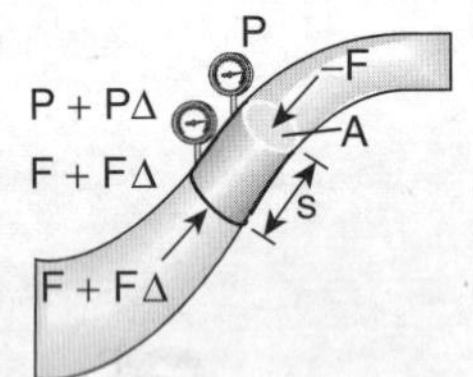

$$F = pA \text{ and } F + \Delta F = (p + \Delta p)A$$

which act in the opposite directions. The magnitude of the net force pushing the fluid element up the tube is

$$\Delta F = (\Delta p)A$$

The work done when the element moves through its own length S is

$$dW = (\Delta F)S = (\Delta p)As = (\Delta p)V$$

where V is the volume of the fluid element.

Hence the total work done on the fluid element in moving if from region 1 to region 2 is

$$W_{nc} = \int dW = \int_{p_2}^{p_1} V.dp = (p_1 - p_2)V \quad \ldots\text{(ii)}$$

Equating (i) and (ii), we get

$$(p_1 - p_2)V = \left(\frac{1}{2}mv_2^2 + mgy_2\right) - \left(\frac{1}{2}mv_1^2 + mgy_1\right)$$

Putting $m/V = \rho$, the density of the fluid and rearranging the terms, we get

$$p_1 + \frac{1}{2}\rho v_1^2 + \rho g y_1 = p_2 + \frac{1}{2}\rho v_2^2 + \rho g y_2$$

Since the points 1 and 2 were selected arbitrarily, we can write

$$p + \frac{1}{2}\rho v^2 + \rho g y = \text{constant} \quad \ldots\text{(iii)}$$

at all points in the flow. The expression given in (iii) is known as **Bernoulli's equation**.

Drop of Pressure when Fluid Moves from Broader to Narrower Region of a Horizontal Pipe

Equation of continuity of an incompressible fluid is

$$Av = \text{constant}$$

So, velocity increases as area decreases.

From Bernoulli's equation for a horizontal pipe (h = constant),

$$p + \frac{1}{2}\rho v^2 = \text{constant}$$

So, pressure will be small where the velocity is large. Combining these two results we conclude that "when a liquid flows from wider to narrower area of a pipe, fluid velocity increases resulting in the drop of pressure".

This is shown graphically in the given figure.

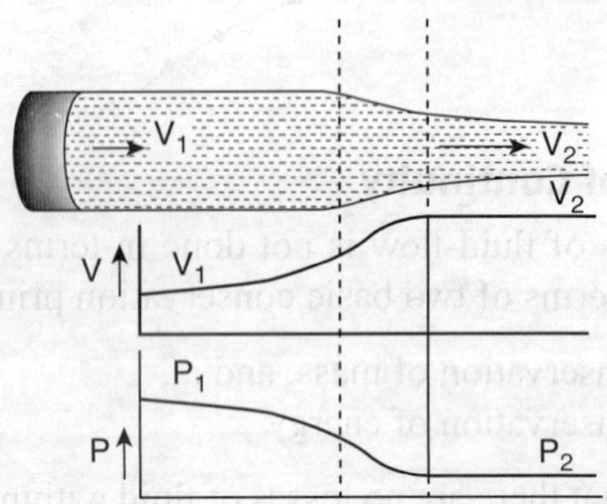

Let A_1, A_2 be the cross-sectional areas at points 1 and 2 where the fluid velocities are v_1 and v_2 and pressures are P_1 and P_2 respectively .

$$\therefore \quad A_1v_1 = A_2v_2, \quad \text{or} \quad v_2 = \frac{A_1}{A_2}v_1$$

From Bernoulli's theorem,

$$P_1 + \frac{1}{2}\rho v_1^2 = P_2 + \frac{1}{2}\rho v_2^2$$

or,
$$P_1 - P_2 = \frac{1}{2}\rho(v_2^2 - v_1^2)$$

$$= \frac{1}{2}\rho v_1^2\left(\frac{A_1^2}{A_2^2} - 1\right) \quad \ldots\text{(i)}$$

If the pressure difference corresponds to a height h of the same liquid,

$$P_1 - P_2 = h\rho g \quad \ldots\text{(ii)}$$

From (i) and (ii),

$$\frac{1}{2}\rho v_1^2\left[\left(\frac{A_1}{A_2}\right)^2 - 1\right] = h\rho g$$

or, $$v_1 = \left[\frac{2gh}{\left(\frac{A_1}{A_2}\right)^2 - 1}\right]^{1/2}$$

This principle is used to measure flow speed in a pipe by a **Venturimeter**.

The flow rate (or volume flowing per second) can be obtained as

$$Q = \frac{dV}{dt} = A_1 v_1 = A_1 \sqrt{\frac{2gh}{\left(\frac{A_1}{A_2}\right)^2 - 1}}.$$

Efflux Speed

Consider a vessel containing a liquid up to some height H. A small hole is punched in the wall at a depth h below the free surface. The speed of the liquid coming out of the hole is called the efflux speed.

Taking the level of the orifice (hole) O as reference level (or zero level of potential energy) and applying Bernoulli's equation to the points within and outside O, we have,

$$p_0 + h\rho g = p_0 + \frac{1}{2}\rho v^2,$$

where p_0 is the atmospheric pressure

or, $$v = \sqrt{2gh}.$$

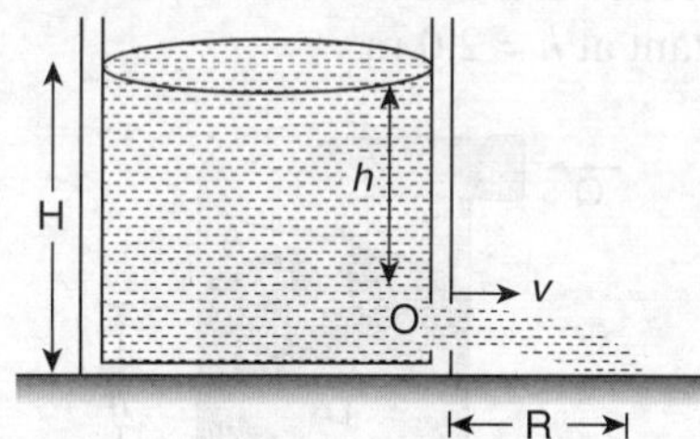

The same result can also be derived by considering a tube of flow, starting at the liquid surface and ending at the orifice. Assuming the speed at the free surface to be negligible, we have

$$0 + \rho gh + p_0 = \frac{1}{2}\rho v^2 + 0 + p_0$$

or, $$v = \sqrt{2gh}.$$

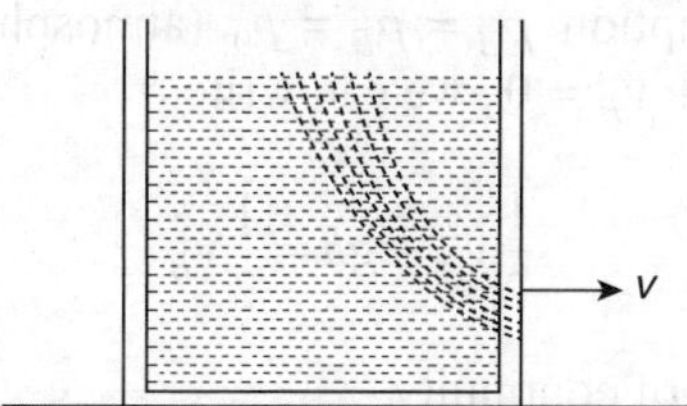

Note that this expression is the same as the liquid will acquire in falling vertically through a height h.

Horizontal Range (R)

Let the liquid jet strike the ground at a horizontal distance R. If t be the time of fall, then for motion along the vertical

$$H - h = \frac{1}{2}gt^2, \quad \text{or} \quad t = \sqrt{\frac{2(H-h)}{g}}$$

and for motion along the horizontal,

$$R = vt = \sqrt{2gh}.\sqrt{\frac{2(H-h)}{g}}.$$

or, $$R = 2\sqrt{h(H-h)}.$$

From this expression it is clear that for the same range R, the hole has to be punched at a depth h and $(H - h)$ from the free surface. These two points are at the same distance below the top and above the bottom.

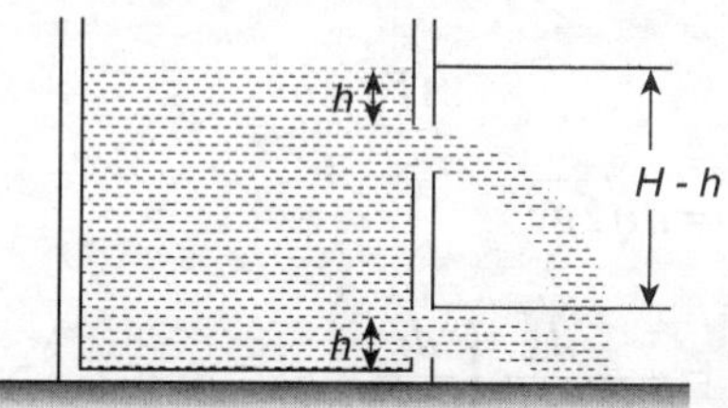

Condition for range R to be maximum

Since, $$R = 2\sqrt{h(H-h)}$$

or, $$R^2 = 4(hH - h^2)$$

$\therefore$ R is maximum, if

$$\frac{dR^2}{dh} = 0$$

or, $$H - 2h = 0$$

or, $$h = \frac{H}{2}.$$

The maximum value of range is

$$R_{max} = 2\sqrt{h(H-h)} = 2\sqrt{\frac{H}{2} \times \frac{H}{2}} = H.$$

Illustration 12

A tank is filled with a liquid upto a height H. A hole is punched at the bottom of the tank. Determine the time for the tank to be emptied. The cross-sectional area of the tank and the hole are A and a respectively.

Solution

Let the height of the liquid level at any time be x, so that the efflux speed $v = \sqrt{2gx}$.

Equation of continuity, $AV = av$ can be expressed as

$$-A\frac{dx}{dt} = a\sqrt{2gx}$$

or, $$-\frac{dx}{\sqrt{x}} = \frac{a}{A}\sqrt{2g}dt$$

Integrating, we get

$$\int_0^t dt = -\frac{A/a}{\sqrt{2g}}\int_H^0 \frac{dx}{\sqrt{x}}$$

or, $$t = \frac{A}{a}\sqrt{\frac{2H}{g}}$$

Illustration 13

A tank is filled up to a height H and a narrow hole is punched at its bottom. If t_1 and t_2 be the time taken for the tank to be emptied for its first and second half respectively, determine the ratio t_1/t_2.

Solution

If A and a be the cross-section of the tank and the hole, we have

$$AV_1 = av_2$$

or, $$-A\frac{dy}{dt} = a\sqrt{2gy}$$

∴ $$t = -\frac{A}{a\sqrt{2g}}\int\frac{dy}{\sqrt{y}}$$

$$t_1 = -\frac{A}{a\sqrt{2g}}\int_H^{H/2}\frac{dy}{\sqrt{y}} = \frac{A\cdot 2}{a\sqrt{2g}}\left[\sqrt{H} - \sqrt{\frac{H}{2}}\right]$$

$$= \frac{A}{a}\sqrt{\frac{2H}{g}}\left(1 - \frac{1}{\sqrt{2}}\right).$$

Similarly $$t_2 = -\frac{A}{a\sqrt{2g}}\int_{H/2}^{0}\frac{dy}{\sqrt{y}} = \frac{A}{a}\sqrt{\frac{H}{g}}$$

∴ $$\frac{t_1}{t_2} = \sqrt{2}\left(1 - \frac{1}{\sqrt{2}}\right) = (\sqrt{2} - 1).$$

Venturimeter

Venturimeter is used for measuring the velocity of a fluid. Two pressure gauges G_1 and G_2 measure the pressure in the pipe. One is placed in the main pipe and the other at a constriction inserted in it. The area of cross-section are A_1 and A_2 where the corresponding velocity of fluid flow are v_1 and v_2 repectively, as shown in the figure.

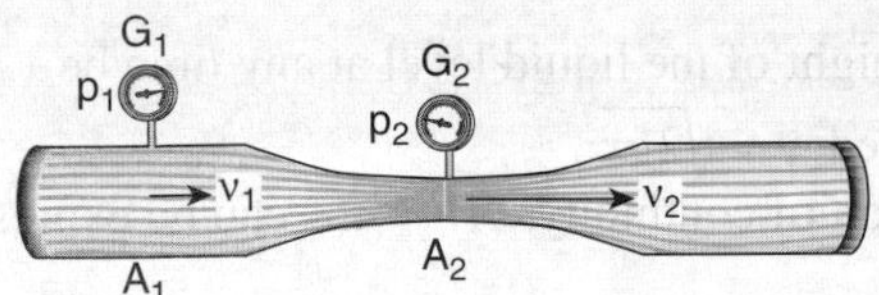

From the equation of continuity,

$$A_1v_1 = A_2v_2 \quad \text{or} \quad v_2 = \frac{A_1}{A_2}v_1$$

If the pipe is horizontal, Bernoulli's theorem gives

$$\frac{1}{2}\rho v_1^2 + p_1 = \frac{1}{2}\rho v_2^2 + p_2$$

or, $$\frac{1}{2}\rho v_1^2 + p_1 = \frac{1}{2}\rho\left(\frac{A_1}{A_2}v_1\right)^2 + p_2$$

Solving for v_1, we get

$$v_1 = A_2\left[\frac{2(p_1 - p_2)}{\rho(A_1^2 - A_2^2)}\right]^{1/2}.$$

This gives the fluid velocity in the main pipe. The flow rate or the volume flowing per second is given by

$$Q = A_1v_1 = A_1A_2\left[\frac{2(p_1 - p_2)}{\rho(A_1^2 - A_2^2)}\right]^{1/2} = K\left[(p_1 - p)\right]^{1/2}$$

where K is a constant depending on the size of the pipe and the density of the fluid. Thus by measuring p_1 and p_2 and calculating K, both the fluid velocity and the flow rate are determined

Illustration 14

A stream of water of diameter $d = 0.1$ m flows steadily from a tank of diameter $D = 1.0$ m as shown in figure. Find the flow rate Q needed from the inflow pipe if the water level h remains constant at $h = 2.0$ m.

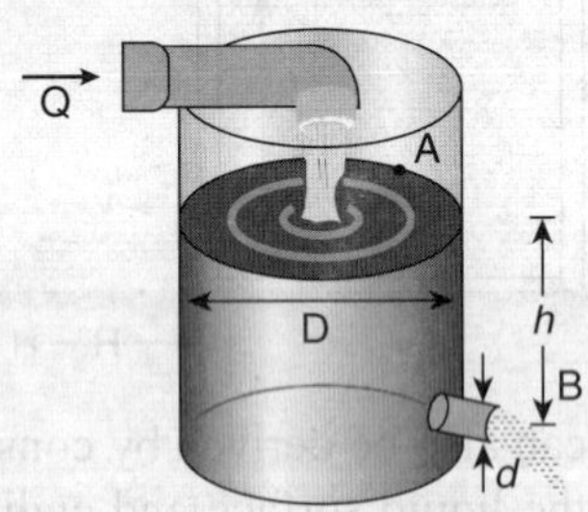

Solution

Assuming steady, inviscid, incompressible flow, we have from Bernoulli's principle,

$$p_A + \frac{1}{2}\rho v_A^2 + \rho g y_A = p_B + \frac{1}{2}\rho v_B^2 + \rho g y_B$$

With the assumption $p_A = p_B = p_0$ (atmospheric) $y_A = h$, and $y_B = 0$, we can write

$$\frac{1}{2}v_A^2 + gh = \frac{1}{2}v_B^2 \qquad \text{...(i)}$$

From equation of continuity,

$$A_1v_A = A_2v_B$$

or, $$\frac{\pi}{4}D^2 v_A = \frac{\pi}{4}d^2 v_B$$

Hence, $$v_A = \left(\frac{d}{D}\right)^2 v_B \qquad \ldots\text{(ii)}$$

From (i) and (ii),

$$v_B^2 = v_A^2 + 2gh = \left(\frac{d}{D}\right)^4 v_B^2 + 2gh$$

or, $$v_B = \sqrt{\frac{2gh}{1-\left(\frac{d}{D}\right)^4}} = \sqrt{\frac{2\times 9.81\times 2.0}{1-(0.1/1)^4}} = 6.26 \text{ m/s}$$

Thus flow rate $Q = A_1 v_A = A_2 v_B = \frac{\pi}{4}(0.1)^2 \times 6.26$

$$= 0.0492 \text{ m}^3/\text{s}.$$

Illustration 15

A cylindrical vessel of base area A has a small hole of cross-section a punched near its base. At time $t = 0$, water is supplied into the vessel at a constant rate α m^3/s. Find

(a) the maximum water level h_{max} in the vessel

(b) the time t when water level becomes $h(< h_{max})$.

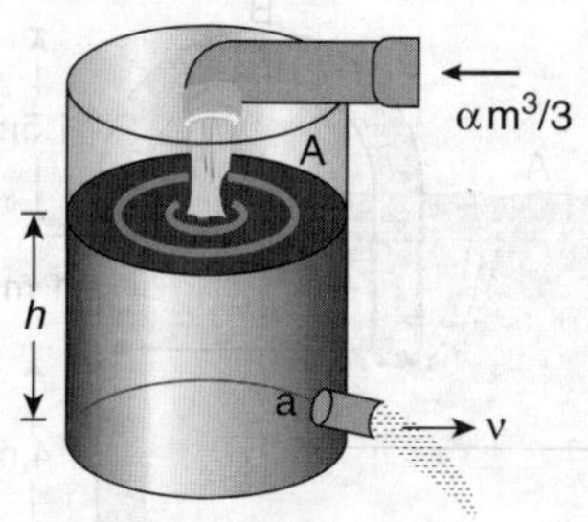

Solution

(a) Water level will have maximum height when *inflow rate = outflow rate* and there will be no further change in level.

$\therefore \quad \alpha = av$

or, $\alpha = a\sqrt{2gh_{max}} \quad [\because v = \sqrt{2gh}]$

or, $h_{max} = \alpha^2 / 2ga^2$

(b) Let the water level be y at time t.

$$\therefore \quad A\left(\frac{dy}{dt}\right) = \alpha - av = \alpha - a\sqrt{2gy}$$

Here $\frac{dy}{dt}$ is positive as y increases with time and instantaneous efflux velocity $v = \sqrt{2gy}$. Rearranging the above equation

$$\int_0^h \frac{dy}{\alpha - a\sqrt{2gy}} = \frac{1}{A}\int_0^t dt.$$

Integrating under the given limits, we get the required time,

$$t = \frac{A}{ag}\left[\frac{\alpha}{a}\ln\left(\frac{\alpha - a\sqrt{2gh}}{\alpha}\right) - \sqrt{2gh}\right]$$

This gives the time t as a function of h. For any value $h(\le h_{max})$, the corresponding time t can be evaluated.

Illustration 16

The shape of an ancient water clock jug is such that water level descends at a constant rate at all times. If the water level falls by 4 cm every hour, determine the shape of the jar, i.e., specify x as a function of y. The radius of the drain hole is 2mm and can be assumed to be very small compared to x.

Solution

Let the height of water level above the origin be y, so the efflux speed through the drain hole is $\sqrt{2gy}$. The volume flowing per second is

$$V = av = a\sqrt{2gy}$$

or, $$\pi x^2\left(-\frac{dy}{dt}\right) = a\sqrt{2gy} \qquad \ldots\text{(i)}$$

Given: $$\left(-\frac{dy}{dt}\right) = 4 \text{ cm/hour} = \frac{4\times 10^{-2}}{3600}\text{ m/s}$$

$$= 1.11\times 10^{-5} \text{ m/s}$$

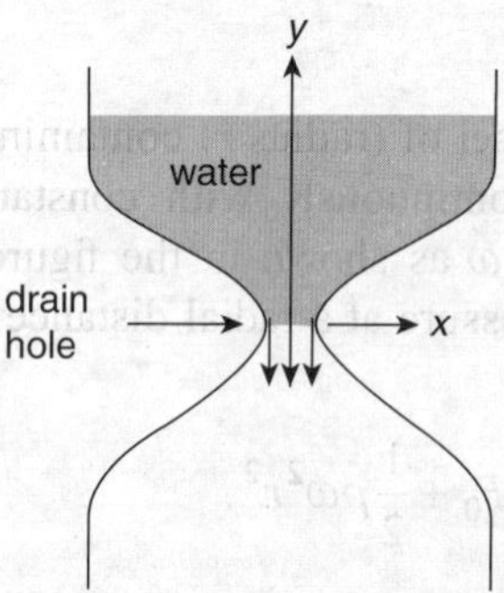

and the cross-section area of the drain hole $a = \pi r^2 = \pi\times(2\times 10^{-3})^2$

These substitutions in (i) leads to

$$\pi(x^2)(1.11\times 10^{-5}) = \pi(4\times 10^{-6})\sqrt{20y}$$

or, $$y = 0.4x^4$$

This expresses y as an explicit function of x.

Illustration 17

What work should be done in order to expel total amount of water contained in a horizontally fixed cylinder in time t by means of a constant force F acting on the piston? The total volume of water contained in the cylinder is V, cross-sectional area of the orifice is 'a'. Density of water is ρ and friction and viscosity are negligibly small.

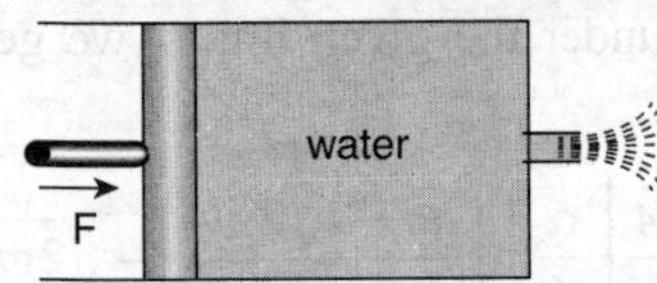

Solution

Applying Bernoulli's equation inside water near the piston and the orifice, we have

$$\frac{F}{A}=\frac{1}{2}\rho v^2,$$

or, $$F=\frac{1}{2}\rho A v^2$$

If L be the total initial length of the water column,

volume $V = Ax$,

total time $t=\frac{V}{av}$

and work done $W = Fx$

$$=\left(\frac{1}{2}\rho A v^2\right)\frac{V}{A}$$

$$=\frac{1}{2}\cdot\rho\left(\frac{V}{at}\right)^2\cdot V$$

$$=\frac{1}{2}\rho V^3/a^2t^2.$$

Illustration 18

A cylindrical vessel of (radius r) containing a liquid spins continuously with constant angular velocity ω as shown in the figure. Show that the pressure at a radial distance r from the axis is

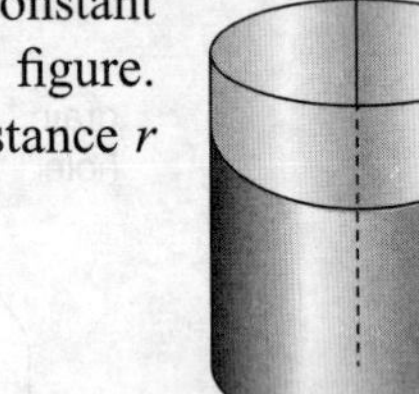

$$P=P_0+\frac{1}{2}\rho\omega^2r^2,$$

where P_0 = atmospheric pressure.

Solution

Consider *a* particle of the fluid at a point $P(x, y)$ w.r.t. the coordinate axes as shown in the figure. The forces acting on this particle are $m\omega^2x$ (the centrifugal force) and the weight mg.

The net force F acting at P should be perpendicular to the free surface, so that

$$\tan\theta=\frac{m\omega^2x}{mg}=\frac{\omega^2x}{g}$$

or, $$\frac{dy}{dx}=\frac{\omega^2x}{g}\quad\left[\because \text{slope}=\tan\theta=\frac{dy}{dx}\right]$$

or, $$y=\frac{\omega^2}{2g}x^2.$$

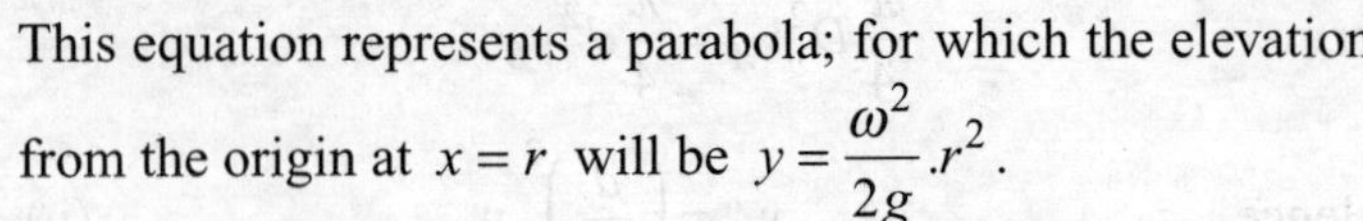

This equation represents a parabola; for which the elevation from the origin at $x = r$ will be $y=\frac{\omega^2}{2g}.r^2$.

$\therefore$ Pressure $P(r)=P_0+\rho gy=P_0+\frac{\rho\omega^2r^2}{2}$.

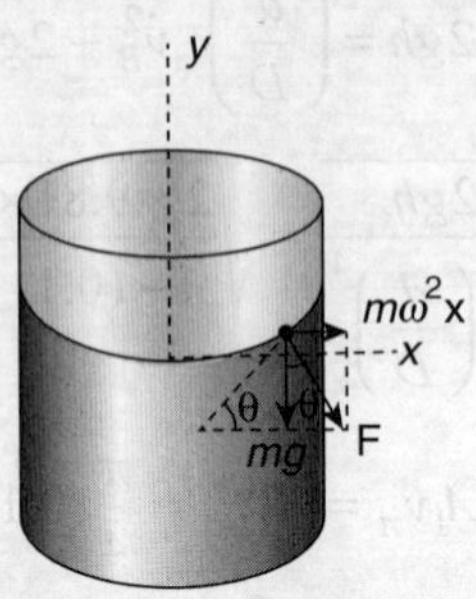

Illustration 19

A siphon tube is discharging a liquid of specific gravity 0.9 from a reservoir as shown in the figure. Find

(a) the speed of liquid through the siphon,

(b) the pressure at the highest point B.

(c) the pressure at point C.

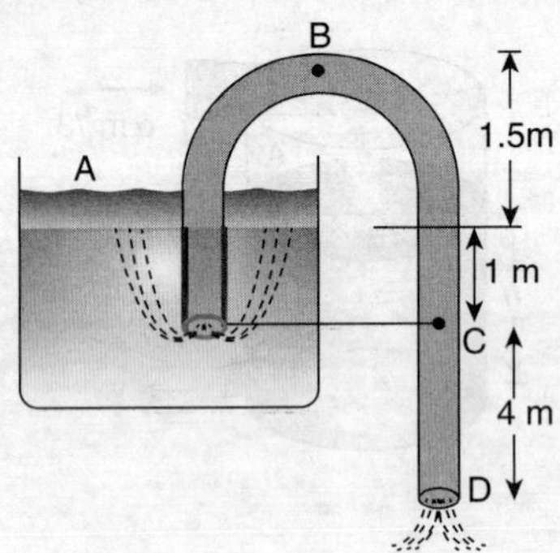

Solution

(a) Applying Bernoulli's principle,

$$\frac{1}{2}\rho v^2=\rho gh$$

$\therefore$ Speed of liquid through the siphon

$$v=\sqrt{2gh}=\sqrt{2\times10\times(4+1)}=10\text{ m/s}$$

(b) From Bernoulli's principle at A and B,

$$p_A+0+0=p_B+\frac{1}{2}\rho v^2+\rho g(1.5)$$

or, $$p_B=p_A-\frac{1}{2}\rho v^2-\rho g\times1.5$$

$$=1.01\times10^5-\frac{1}{2}\times900\times10^2-1.5\times900\times10$$

$$=4.25\times10^4\text{ Pa}.$$

(c) Similarly for points at A and C,

$$p_A = p_C + \frac{1}{2}\rho v^2 - \rho g(1.0)$$

or, $$p_C = p_A - \frac{1}{2}\rho v^2 + \rho g$$

$$= 1.01\times 10^5 - \frac{1}{2}\times 900\times (10)^2 + 900\times 10$$

$$= 6.5\times 10^4 \text{ Pa}.$$

Critical Velocity: Reynold's Number

For streamline (or steady) flow of a liquid through a tube the average velocity is normally small. But with gradual increase in flow velocity (by increasing the pressure difference) the flow is observed to be streamline upto a maximum velocity v_c, known as the ***critical velocity***, and afterwards (for $v > v_c$) the flow no longer remains streamline and becomes turbulent. Osborne Reynolds showed, by experiment that critical velocity (v_c) is given by the equation

$$v_c = \frac{k\eta}{\rho r}$$

where k = Reynolds number
ρ = density of the liquid
η = viscosity of the liquid, and
r = radius of the tube.

For narrow tubes, Reynold's number is approximately 1000.

Viscosity

A fluid is said to be ideal if it is incompressible and has no viscosity (non-viscous). The term viscosity refers to "internal friction or resistance in fluid flow due to relative motion between its different layers". This resistance (or viscosity) is caused by short-range cohesive forces in liquids and by collisions between the molecules in gases.

In an ideal fluid, there is no viscosity to hinder or resist the fluid layers as they slide past one another. Flow of liquid through a pipe of uniform cross-section may be considered as flow of different co-axial layers. In an ideal fluid, every layer moves with the same velocity, even the layer next to the wall [figure (a)]. In a real fluid (where viscosity is present), the fluid layers have different velocities (figure b). The fluid on the axis has the greatest velocities, which successively decreases as one moves towards the wall and the layer next to the wall does not move at all because it is held tightly by intermolecular forces. So strong are these forces that if the tube (solid surface) moves, the adjacent fluid layer moves along with it and remains at rest relative to the moving surface.

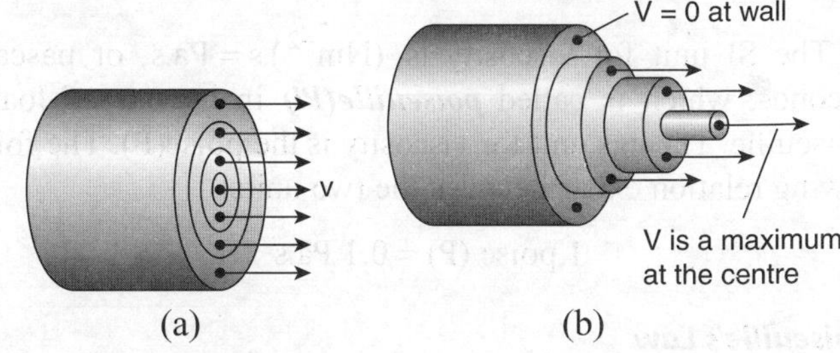

(a) (b)

Viscosity Coefficient

In order to introduce a quantitative measure of viscosity, we consider the layered motion (or the laminar flow) of a viscous fluid between two parallel plates P and Q. The upper plate Q is free to move while the lower plate P is stationary. A tangential force F makes the plate Q to move with velocity v relative to P. All the intermediate layers in between P and Q slide over each other. As each layer moves, it is subjected to viscous forces from its neighbours. The aim of the force F is to compensate for the effect of these viscous drags, so that any layer can move uniformly (with a constant velocity). It is observed that the tangential force F required to move the fluid layer at a constant speed depends upon the following factors:

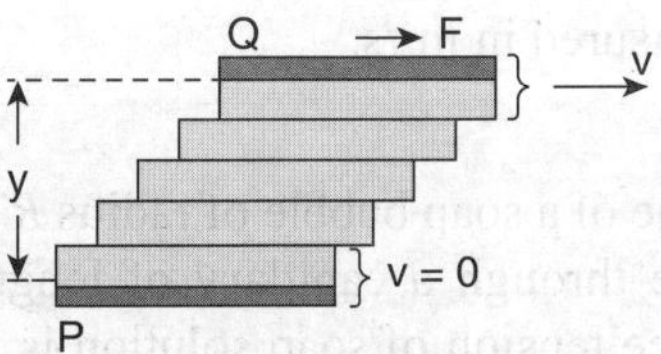

$F \propto A$ (area of contact)
$\propto v$ (velocity of layer)
$\propto 1/y$ (where y = separation between the top and bottom plates).

$$\therefore \quad F \propto \frac{Av}{y}$$

or, $$F = \eta\frac{Av}{y} \quad \ldots\text{(i)}$$

where the proportionality constant η is called the ***coefficient of viscosity*** or simply the ***viscosity***.

Equation (i) can be expressed in differential form as

$$F = \eta A\left(\frac{dv}{dy}\right) \quad \ldots\text{(ii)}$$

where $\frac{dv}{dy}$ is referred to as ***velocity gradient***. From (ii), we define the coefficient of viscosity as "numerically equal to the tangential stress so as the maintain unit velocity gradient in a direction perpendicular to the direction of motion of the fluid."

The SI unit for viscosity is $(\text{Nm}^{-2}).\text{s} = \text{Pa.s}$, or pascal seconds, which is called ***poiseuille(Pl)*** in honour of Joan Poiseuille. The cgs unit for viscosity is the poise(P). The following relation exists between the two units:

$$1 \text{ poise (P)} = 0.1 \text{ Pa.s}$$

Poiseuille's Law

The flow rate (or volume flowing per second) of a viscous fluid through a horizontal tube depends on the properties of the fluid, dimensions of the pipe, and on the pressure difference between the ends of the pipe. For streamline flow through a uniform tube, the flow rate is given by

$$Q = \frac{\pi(P_1 - P_2)r^4}{8\eta L} \quad \text{...(iii)}$$

where $(P_1 - P_2)$ = pressure difference between the ends of the tube

r = radius of the tube

L = length of the tube and

η = viscosity of the fluid.

The relation expressed in (iii) is known as ***Poiseuille's law*** in which Q is measured in m^3/s.

Illustration 20

Find the lifetime of a soap bubble of radius R connected with the atmosphere through a capillary of length ℓ and inner radius r. Surface tension of soap solution is T and viscosity of air is η.

Solution

Pressure difference across the two ends of the capillary tube at an instant when the radius of the bubble is x is given by

$$p = \frac{4T}{x}$$

According to Poiseuille's equation, flow rate,

$$-\frac{d}{dt}\left(\frac{4}{3}\pi x^3\right) = \frac{\pi p r^4}{8\eta\ell} = \frac{\pi \cdot \frac{4T}{x} r^4}{8\eta\ell}$$

or, $$-4\pi x^2 \frac{dx}{dt} = \frac{\pi T r^4}{2\eta\ell x}$$

or, $$dt = -(4\pi x^2\, dx)\frac{2\eta\ell x}{\pi T r^4}$$

$\therefore$ Total life time of the bubble

$$t = \int dt = -\frac{8\eta\ell}{Tr^4}\int_R^0 x^3\, dx$$

$$= \frac{2\eta\ell R^4}{Tr^4}.$$

Motion in a Viscous Medium: Stokes Law

When an object moves through a fluid, the layers of the fluid adjacent to the body tends to move with a velocity equal to that of the body while the fluid far away from it remains practically at rest. This sets up a relative motion between various layers of the fluid which is opposed by the forces due to viscosity. The viscous force exerted by the fluid on the object tends to slow it down.

For a small sphere of radius r, moving slowly with a speed v, the magnitude of the viscous force is given by

$$F = 6\pi\eta r v \quad \text{...(iv)}$$

Equation (iv) is known as Stokes Law.

Terminal Velocity

For a sphere moving under gravity through a viscous medium the viscous force $F = 6\pi\eta r v$ acts on the body in addition to the weight W and buoyant force B. Initially when $v = 0$, the net force $(W - B)$ acts on the body and it accelerates, consequently the velocity increases and the viscous force proportional to the velocity of the sphere develops. This viscous force increases with the increase in velocity and at a certain instant, the viscous force becomes equal to $(W - B)$. The net force then becomes zero and the sphere continues to move with a constant velocity which is known as the ***terminal velocity***.

Expression For Terminal Velocity

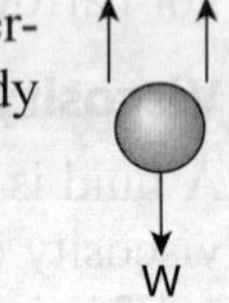

When the spherical body of radius r moves with terminal velocity v_0 in a viscous medium, free body diagram gives us

$$B + F = W$$

or, $$\frac{4}{3}\pi r^3 \sigma g + 6\pi\eta r v_0 = \frac{4}{3}\pi r^3 \rho g \quad \text{...(v)}$$

where σ = density of the viscous medium

ρ = density of spherical body

and η = viscosity of the viscous medium

Simplifying (v), we get

$$v_0 = \frac{2}{9}\frac{r^2(\rho - \sigma)g}{\eta} \quad \text{...(vi)}$$

SOLVED EXAMPLES

Example 1

Two rods of equal cross-section, one of copper and the other of steel, are joined to form a composite rod of 2.0m. At 20°C the length of the copper rod is 0.5m. When the temperature is raised to 120°C the length of the composite rod increases to 2.002 m. If the composite rod is fixed between two rigid walls and is thus not allowed to expand, it is found that the lengths of the two component rods also do not change with the increase in temperature. Calculate the Young's modulus and the coefficient of linear expansion of steel. Given: $Y_{copper} = 1.3\times10^{13}$ N/m^2, coefficient of linear expansion of copper $= 1.6\times10^{-5}$/°C.

Solution

Increase in length of Cu-rod is

$$(\Delta L)_{Cu} = \alpha L\Delta T = \alpha_{Cu}\times0.5\times(120-20) = 50\cdot\alpha_{Cu}$$

Similarly for steel rod,

$$(\Delta L)_{Steel} = \alpha L\Delta T = \alpha_s\times1.5\times100 = 150\alpha_s$$

Total change in length of the composite rod,

$$50\alpha_{Cu} + 150\alpha_s = 0.002$$

or, $$50\times1.6\times10^{-5} + 150\alpha_s = 0.002$$

or, $$\alpha_s = 0.8\times10^{-5}/°C$$

Since there is no change in the length of the component rods, the changes in length due to stress developed is balanced by that due to thermal expansion.

Hence, $$(Y\times strain)_{Cu} = (Y\times strain)_s$$

or, $$Y_{Cu}\times\alpha_{Cu}\times\Delta T = Y_s\times\alpha_s\times\Delta T$$

or, $$Y_s = Y_{Cu}\times\left(\frac{\alpha_{Cu}}{\alpha_s}\right) = 1.3\times10^3\times\frac{1.6\times10^{-5}}{0.8\times10^{-5}}$$

∴ $$Y_s = 2.6\times10^3\ \text{N/m}^2.$$

Example 2

A rod AD consisting of three segments AB, BC and CD joined together is hanging vertically from a fixed support at A. The lengths of the segments are respectively 0.1m, 0.2m and 0.15m. The cross-sectional area of the rod is uniform ($=10^{-1}$m^2). A weight of 10 kg is hung from D. Calculate the displacements of point B, C and D. Neglect the weight of rod and assume that $Y_{AB} = 2.5\times10^{10}$ N/m^2, $Y_{BC} = 4.0\times10^{10}$ N/m^2 and $Y_{CD} = 1.0\times10^{10}$ N/m^2.

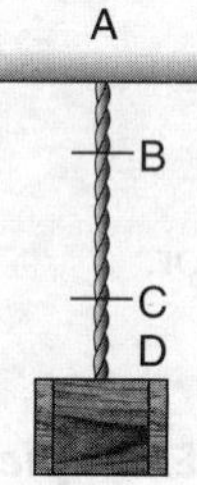

Solution

The extensions produced in the system is shown in the figure. We know that extension

$$\ell = \frac{Mg}{YA}\cdot L$$

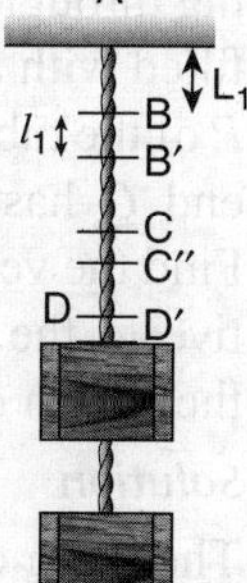

∴ for AB, $$\ell_1 = \frac{10\times9.8\times0.1}{2.5\times10^{10}\times10^{-4}}$$

$$= 3.92\times10^{-6}\ \text{m}.$$

Similarly for BC,

$$\ell_2 = \frac{10\times9.8\times0.2}{4.0\times10^{10}\times10^{-4}} = 4.9\times10^{-6}\ \text{m}$$

and for CD, $$\ell_3 = \frac{10\times9.8\times0.15}{1.0\times10^{10}\times10^{-4}} = 14.7\times10^{-6}\ \text{m}$$

Displacement of point B is

$$BB' = \ell_1 = 3.92\times10^{-6}\ \text{m}$$

Displacement of point C is

$$\ell_1 + \ell_2 = 3.92\times10^{-6} + 4.90\times10^{-6}$$

$$= 8.82\times10^{-6}\ \text{m}.$$

Similarly displacement of point D is

$$\ell_1 + \ell_2 + \ell_3 = 23.52\times10^{-6}\ \text{m}.$$

Example 3

A rod of length L and uniform cross-section A is tied to one end of a light string of length ℓ. The other end of the string is tied to the floor of a water pool whose depth is d. ρ_r and ρ_w are the densities of the material of the rod and water respectively. In the stable position as shown in the figure, find

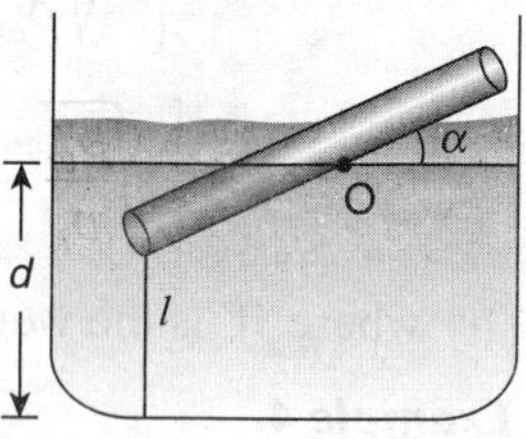

(a) the length of the rod above the water level
(b) the inclination of the rod with the horizontal
(c) the tension in the string.

Solution

(a) Let a length x of the rod remain outside water and the rod make an angle α with the horizontal in its equilibrium position. The forces acting on the rod are shown in the free-body-diagram

(i) Weight of the rod
$W = mg = AL\rho_r g$ acting at the $C.G.$ of the rod where $AG = \frac{L}{2}$.

(ii) Buoyant force $F_B = m'g = A(L-x)\rho_\omega g$ acting at the centre of buoyancy B, where $AB = (L-x)/2$.

(iii) Tension T in the string
For rotational equilibrium, the torque of the force must be zero. Taking the torque about A,

$$W\frac{L}{2}\cos\alpha - F_B\cdot\frac{(L-x)}{2}\cos\alpha = 0$$

or, $$\frac{1}{2}AL^2\rho_r g\cdot = \frac{1}{2}A(L-x)^2\rho_\omega g$$

Solving , we get $x=\left[1-\sqrt{\rho_r/\rho_\omega}\right]L$

(b) From figure, $\sin\alpha=\dfrac{AC}{AO}=\dfrac{d-\ell}{L-x}$

$$=\frac{(d-\ell)}{L}\sqrt{\frac{\rho_\omega}{\rho_r}}.$$

(c) For transnational equilibrium,

$$T+W-F_B=0$$

or, $T=F_B-W$

$$=A(L-x)\rho_W g-AL\rho_r g$$

$$=Ag\left[L\sqrt{\frac{\rho_r}{\rho_\omega}}\rho_\omega-L\rho_r\right]=AL\rho_r g\left[\sqrt{\rho_\omega/\rho_r}-1\right]$$

$$=W\left[\sqrt{\frac{\rho_\omega}{\rho_r}}-1\right]$$

where W is the weight of the rod.

Example 4

A cubical block of wood of each edge 3 cm floats in water. The lower surface of the cube just touches the free end of a vertical spring fixed at the bottom of the vessel. Find the maximum weight that can be placed on the block without wetting it. Given: density of wood = 800 kg/m^3, spring constant $k=50$ N/m and $g=10\,\text{m/s}^2$.

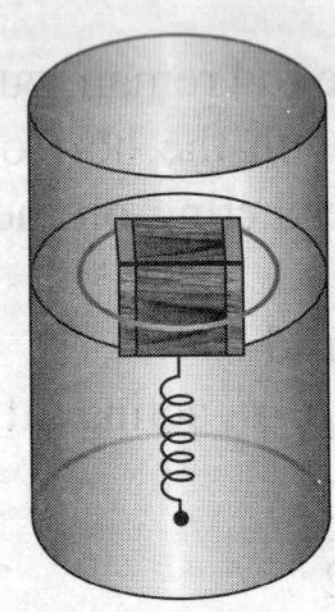

Solution

Let x be the height of block inside water.

$\therefore$ Buoyant force F_B = weight of block

or, $A\cdot x\rho_{water}=v\rho_{wood}$

W + W₀

W₀

kx

F$_B$

$\therefore$ $x=\dfrac{L^3}{L^2}\dfrac{\rho_{wood}}{\rho_{water}}=(3\,\text{cm})\times0.8=2.4\,\text{cm}$

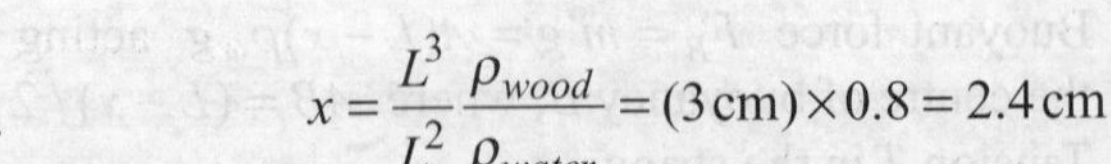

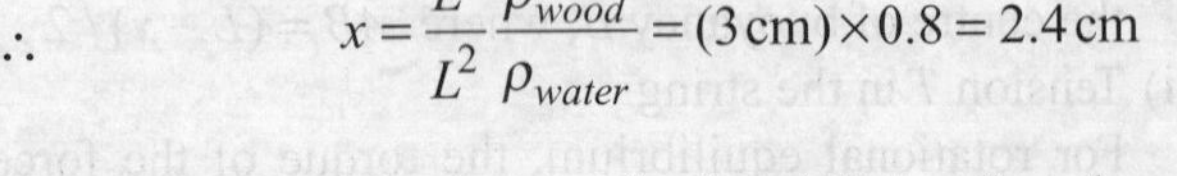

If W be the maximum required weight so that the cubical block just gets completely immersed and the spring gets compressed by $(L-x)=(3-2.4)=0.6\,\text{cm}$.

From the free body diagram,

$$W+W_0=F_B+kx\text{, where}$$

W_0 = Weight of the cubical block

$\therefore$ $W+L^3\rho_{wood}\cdot g=L^3\rho_{water}\cdot g+kx$

or, $W+27\times10^{-6}\times800\times10$

$$=27\times10^{-6}\times10^3\times10+50\times0.6\times10^{-2}.$$

Simplifying we get,

$$W=0.354\,\text{N}.$$

Example 5

A vessel of cross-sectional area A has a narrow horizontal tube of length ℓ and internal radius r. The free surface of the liquid contained in the vessel and the free end of the tube are open to the atmosphere. Find the time for the liquid level to fall from $3h$ to $2.5h$.

Solution

Let the length of the liquid level at any time t be x so that the pressure difference across the tube is $p=x\rho g$, where ρ is the density of the liquid. According to Poiseuille's equation, flow rate

$$\frac{dV}{dt}=\frac{\pi pr^4}{8\eta\ell}$$

or, $$-A\frac{dx}{dt}=\frac{\pi x\rho gr^4}{8\eta\ell}$$

or, $$dt=-\frac{8\eta\ell A}{\pi r^4\rho g}\frac{dx}{x}$$

Integrating, $$t=\int_0^t dt=-\frac{8\eta\ell A}{\pi r^4\rho g}\int_{3h}^{2.5h}\frac{dx}{x}$$

or, $$t=\frac{8\eta\ell A}{\pi r^4\rho g}ln1.2.$$

Example 6

A horizontal oriented tube PQ of length ℓ rotates with a constant angular velocity ω about a stationary vertical axis AB passing through the end P. The tube is filled with an ideal fluid. The end P of the tube is open and the other end Q has a very small orifice. Find the velocity of the fluid relative to the tube as a function of the length of liquid column h.

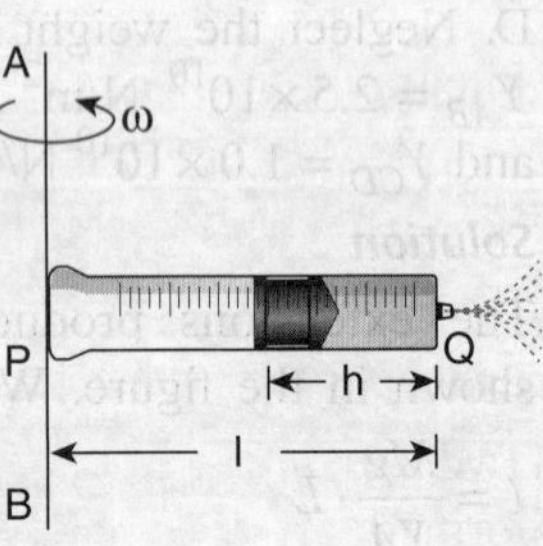

Solution

The water contained is the tube is thrown away from the axis due to the centrifugal force. For a small element dx at a distance x from the axis of rotation, the centrifugal force is

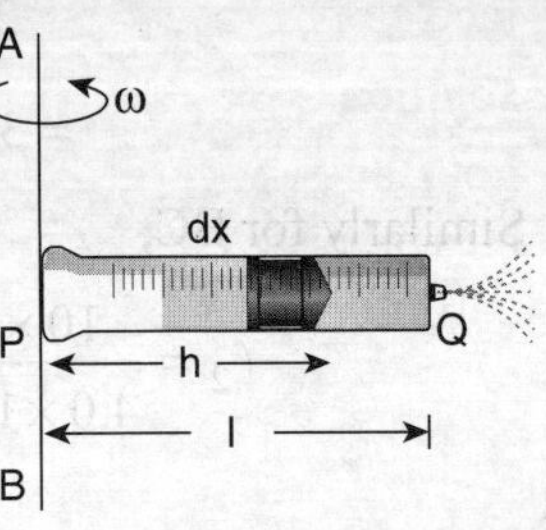

$$dm\cdot\omega^2\cdot x = dm\cdot a = dm\cdot v\frac{dv}{dx}$$

or,
$$\omega^2 x dx = v\,dv$$

Integrating, $\omega^2\int_{(\ell-h)}^{\ell} x\,dx = \int_0^v v\,dv$

or,
$$\frac{\omega^2}{2}\left[\ell^2-(\ell-h)^2\right]=\frac{v^2}{2}$$

Simplifying, we get

$$v=\omega\sqrt{2\ell h-h^2}$$
$$=\omega h\sqrt{\frac{2\ell}{h}-1}.$$

Example 7

A soap bubble is blown at the end of a uniform tube of length ℓ and internal diameter *D*. If the viscosity of air is η and surface tension of soap solution be *T*, find the time taken by the bubble to reduce its radius from r_1 and r_2.

Solution

Let the radius of the bubble at time t be r. The volume it contained $=V=\frac{4}{3}\pi r^3$

∴ Volume flowing, per second (according to Poiseuille's equation)

$$-\frac{dV}{dt}=\frac{\pi PR^4}{8\eta\ell}$$

or,
$$-4\pi r^2\frac{dr}{dt}=\frac{\pi D^4 T}{32\eta\ell r}\quad\left[\because P=\frac{4T}{r}\right]$$

or,
$$dt=-\frac{128\eta\ell r^3}{TD^4}dr$$

Integrating,

$$t=\frac{32\eta\ell}{TD^4}\left(r_1^4-r_2^4\right).$$

Example 8

Two separate air bubbles (radii 2×10^{-3}m and 4×10^{-3}m) formed of the same soap solution (surface tension $=0.07$ N/m) touch each other to form a double bubble. Find the radius and the sense of curvature of the internal film surface common to both the bubbles.

Solution

Let r and R be the radii of the two bubbles where $r<R$. If p_0 be the external pressure, the pressure inside are

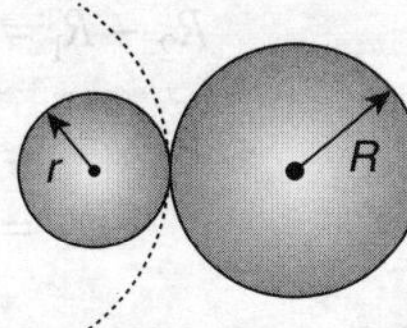

$$P_1=p_0+\frac{4T}{r}\text{ and }P_2=p_0+\frac{4T}{R}$$

Here, $P_1>P_2$ since $r<R$ and thus $P_1-P_2=\frac{4T}{R_0}$, where R_0 is the radius of the interface having its concavity towards the smaller bubble. Substituting the values of P_1 and P_2, we get

$$\left(p_0+\frac{4T}{r}\right)-\left(p_0+\frac{4T}{R}\right)=\frac{4T}{R_0}$$

Simplifying, we get

$$R_0=\frac{Rr}{R-r}=\frac{4\times10^{-3}\times2\times10^{-3}}{(4-2)\times10^{-3}}$$
$$=4\times10^{-3}\text{ m}.$$

Example 9

A vessel filled with air under pressure p_0 contains a soap bubble of diameter *d*. The air pressure having been reduced n-fold isothermally, the diameter of the bubble is observed to increase η fold. Find the surface tension of the soap solution.

Solution

Since the process is isothermal, Boyle's law will be applicable.

∴
$$P_1V_1=P_2V_2$$

or,
$$\left(p_0+\frac{4T}{R_i}\right)\times\frac{4}{3}\pi R_i^3=\left(\frac{p_0}{n}+\frac{4T}{R_f}\right)\times\frac{4}{3}\pi R_f^3$$

where, $R_i=R=\frac{d}{2}$ and $R_f=\eta R=\frac{\eta d}{2}$

These substitutions give us

$$\left(p_0+\frac{8T}{d}\right)\left(\frac{d}{2}\right)^3=\left(\frac{p_0}{n}+\frac{8T}{\eta d}\right)\left(\frac{\eta d}{2}\right)^3$$

or,
$$p_0\left(1-\frac{\eta^3}{n}\right)=\frac{8T}{d}\left(\eta^2-1\right)$$

or,
$$T=\frac{p_0 d}{8}\frac{\left(1-\frac{\eta^3}{n}\right)}{(\eta^2-1)}.$$

Example 10

The diameter of a gas bubble formed at the bottom of a deep tank is $4.0\,\mu m$. When the bubble slowly rises to the surface, its diameter increases *n* times where $n=1.1$. Find the depth of the tank. Assume atmospheric pressure is $1.01\times10^5\,Pa$; Surface tension of water $=72\times10^{-3}$ N/m.

Solution

Assuming isothermal expansion, we have

$$p_1V_1=p_2V_2$$

or,
$$\left(p_0+h\rho g+\frac{4T}{d}\right)\frac{4}{3}\pi\left(\frac{d}{2}\right)^3=\left(p_0+\frac{4T}{nd}\right)\frac{4}{3}\pi\left(\frac{nd}{2}\right)^3$$

or, $$\left(p_0 + h\rho g + \frac{4T}{d}\right) = \left(p_0 + \frac{4T}{nd}\right)n^3$$

or, $$h = \frac{1}{\rho g}\left[p_0(n^3-1) + \frac{4T}{d}(n^2-1)\right]$$

$$= \frac{1}{10^3 \times 10}\left[1.01\times10^5((1.1)^3-1) + \frac{4\times72\times10^{-3}}{4\times10^{-6}}((1.1)^2-1)\right]$$

$$= 4.86 \text{ m}.$$

Example 11

A mercury drop shaped as a round tablet of radius R and thickness h is located between two horizontal glass plates. Assuming that $h << R$, find the mass m of a block that has to be placed on the upper plate so as to reduce the separation between the plates by a factor of n. Assume that angle of contact $= \theta$.

Solution

The volume of mercury between the glass plates initially is approximately

$$V = \pi R^2 h.$$

On placing the weight mg, let R' be new radius and the thickness h reduces to $h' = \frac{h}{n}$.

Hence, $$V = \pi R'^2 h' = \pi R'^2 \cdot \frac{h}{n}$$

$$R' = \sqrt{n}R$$

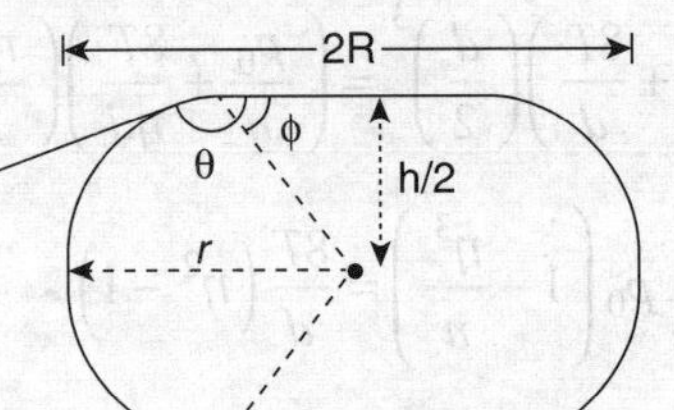

Example 12

If r be the radius of curvature of the vertical section of the drop, then

$$\frac{h/2}{r} = \sin\phi = |\cos(90"+\phi)| = |\cos\theta|$$

or, $$r = \frac{h}{2|\cos\theta|}.$$

∴ Excess pressure inside the drop $= \frac{T}{r}$

$$= \frac{2T|\cos\theta|}{h}$$

∴ Total force exerted on the upper glass plate by the mercury drop is

$$F = \pi R^2 \times \text{excess pressure}$$

or, $$m'g = \pi R^2 \times \frac{2T|\cos\theta|}{h}$$

In equilibrium, the weight of the plate $(= m'g)$ is balanced by F. When an additional weight mg is placed on the glass plate so that R changes to $R' = \sqrt{n}R$, we have,

$$(m+m')g = \frac{2\pi R'^2 T|\cos\theta|}{h/n}$$

or, $$mg = 2\pi T|\cos\theta|\left[\frac{nR'^2}{h} - \frac{R^2}{h}\right]$$

or, $$m = \frac{2\pi R^2 T|\cos\theta|}{gh} \times (n^2-1).$$

Example 13

A drop of water falls in air with uniform velocity. Find the difference between radii of curvature of its uppermost and lowermost points of the drop. The thickness of the drop along the vertical joining the two points is 2.3 mm. and surface tension of water $= 72\times10^{-3}$ N/m.

Solution

Let R_1 and R_2 be the radii of curvature at the uppermost and lowermost points with their centre at O_1 and O_2 as shown in the figure.

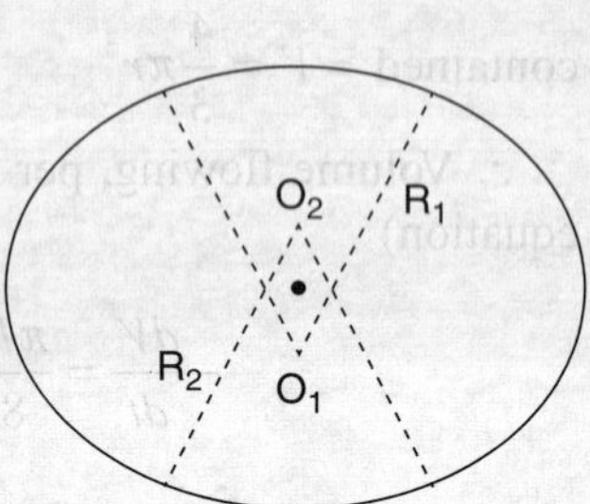

The pressure just inside the drop at the upper surface is

$P_1 = p_0 + \frac{2T}{R_1}$ and that inside the lower surface is $P_2 = p_0 + \frac{2T}{R_2}$.

Again, $$p_1 = p_2 + h\rho g$$

or, $$2T\left(\frac{1}{R_1} - \frac{1}{R_2}\right) = h\rho g$$

or, $$R_2 - R_1 = \frac{R_1 R_2 h\rho g}{2T}.$$

Since $h = 2.3$ mm, the drop is very small so that we can assume $R_1R_2 = \left(\frac{h}{2}\right)^2$. These substitutions give us

$$R_2 - R_1 = \left(\frac{h}{2}\right)^2 h\rho g/2T$$

$$= \frac{h^3\rho g}{8T}$$

$$= \frac{(2.3\times10^{-3})^3 \times 10^3 \times 10}{8\times 72\times10^{-3}}$$
$$= 0.21\,\text{mm}.$$

Example 14

Two soap bubbles of radii *a* and *b* coalesce isothermally to form a single bubble of radius *c* without any leakage of air. If the external pressure be *P*, show the surface tension *T* of the soap solution is given by

$$T = \frac{P(c^3 - a^3 - b^3)}{4(a^2 + b^2 - c^2)}$$

Solution

The pressure inside the two bubbles before they combine are

$$P_1 = \left(P + \frac{4T}{a}\right),\quad P_2 = \left(P + \frac{4T}{b}\right)$$

As the temperature of the enclosed air remains constant (isothermal) we can apply Boyle's law which leads to

$$P_1V_1 + P_2V_2 = P_3V_3$$

or,

$$\left(P + \frac{4T}{a}\right)\cdot\frac{4}{3}\pi a^3 + \left(P + \frac{4T}{b}\right)\cdot\frac{4}{3}\pi b^3 = \left(P + \frac{4T}{c}\right)\cdot\frac{4}{3}\pi c^3$$

or, $$P(a^3 + b^3 - c^3) + 4T(a^2 + b^2 - c^2) = 0$$

Note that of the two expressions contained within the brackets, one is positive and the other must be negative. The second term is positive since surface area has to decrease.

$$\therefore\quad 4T(a^2 + b^2 - c^2) = P(c^3 - a^3 - b^3)$$

or, $$T = \frac{P(c^3 - a^3 - b^3)}{4(a^2 + b^2 - c^2)}.$$

SOLVED OBJECTIVE PROBLEMS

Problem 1

A uniform metal wire fixed at one end is gradually stretched a little beyond its elastic limit and then the stretching force is removed. It will

(a) remain stretched and will not contract
(b) lose its elastic property completely
(c) contract, but its final length will be greater than its initial length
(d) contract upto the proportional limit.

Solution

When the stretching force is increased upto the proportional limit. Hooke's law is valid. Beyond the proportional limit and upto the elastic limit, Hooke's law is not valid but the elastic property is maintained. Beyond the elastic limit the wire does not regain the original length and remains slightly elongated.

Ans. (c)

Problem 2

A uniform rod of length ℓ and density ρ is being pulled on a smooth horizontal floor with an acceleration a. The stress at the transverse cross-section at the mid point of the rod has the magnitude

(a) $\frac{1}{2}\ell\rho a$ (b) $\ell\rho a$
(c) $\frac{1}{4}\ell\rho a$ (d) $\frac{3}{4}\ell\rho a$.

Solution

The force developed at the mid-point of the rod perpendicular to the cross-section (area = A)

$$F = \frac{1}{2}Ma = \frac{1}{2}\ell A\rho a$$

$$\therefore\quad \text{Stress} = \frac{F}{A} = \frac{1}{2}\ell\rho a$$

Ans. (a)

Problem 3

A metal wire of uniform cross-section *a*, length ℓ and Young's modulus *Y* behaves as a spring of spring constant *k*, where

(a) $k = YA/L$ (b) $k = YL/A$
(c) $k = YA/2L$ (d) $k = 2YA/L$.

Solution

If the stretching force be *F*, the Young's modulus of the material of the wire is

$$Y = \frac{F/A}{\Delta\ell/L} = \frac{FL}{\Delta\ell A}$$

or, $$F = \frac{YA}{L}\Delta\ell = k\Delta\ell$$

Ans. (a)

Problem 4

A uniform rod of mass *m*, length *L*, area of cross-section *A* and Young's modulus *Y* hangs from a rigid support. Its elongation under its own weight will be

(a) *zero* (b) $mgL/2YA$
(c) $mgL/\gamma A$ (d) $2mgL/\gamma A$.

Solution

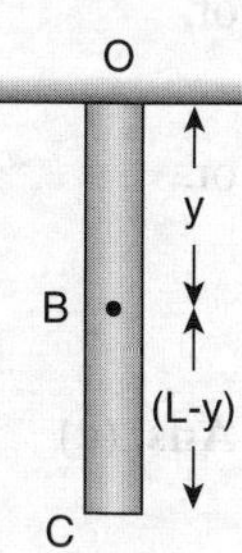

Linear mass density $\lambda = \frac{m}{L}$

Tension at section *B* is

$$F = \text{weight of part } BC$$
$$= \lambda(L - y)g$$

$\therefore$ Elongation of an element dy at *B*

$$= \frac{F}{AY}dy = \frac{\lambda(L-y)g}{AY}dy$$

$$\text{Total elongation} = \frac{\lambda g}{AY}\int_0^L (L-y)\,dy$$

$$= \frac{\lambda g}{AY}\frac{L^2}{2} = \frac{mgL}{2AY}.$$

Ans. (b)

Problem 5

The length of a uniform metal wire is observed to be ℓ_1 and ℓ_2 under the stretching forces F_1 and F_2. The natural length of the wire is

(a) $\sqrt{\ell_1\ell_2}$ (b) $\sqrt{\dfrac{\ell_1+\ell_2}{2}}$

(c) $\dfrac{\ell_1F_2-\ell_2F_1}{F_2-F_1}$ (d) $\dfrac{\ell_1F_2+\ell_2F_1}{F_1+F_2}$.

Solution

If L be the natural length,

$$F_1 = k\Delta L = k(\ell_1 - L)$$

and $$F_2 = k(\ell_2 - L)$$

$\therefore$ $$\frac{F_1}{F_2} = \frac{\ell_1 - L}{\ell_2 - L}$$

Simplifying, we get

$$L = \frac{\ell_1F_2 - \ell_2F_1}{F_2 - F_1}.$$

Ans. (c)

Problem 6

A vessel contains oil of density $800\,\text{kg/m}^3$ over mercury of density $13600\,kg/m^3$. A solid homogenous sphere floats such that the interface of the two liquids divides the sphere in two equal halves. The density of the material of the sphere in g/cm^3 is

(a) 7200 (b) 6.4

(c) 7.2 (d) 6400

According to the law of floatation, weight of sphere = buoyant force

or, $$v\cdot\rho_{solid}\,g = \frac{v}{2}\rho_1 g + \frac{v}{2}\rho_2 g$$

or, $$\rho_{solid} = \frac{\rho_1}{2} + \frac{\rho_2}{2} = \frac{800+13600}{2}$$

$$= 7200\ \text{kg/m}^3 = 7.2\ \text{g/cm}^3.$$

Ans. (c)

Problem 7

A hemispherical portion of radius R is removed from the bottom of a cylinder of radius R. The volume of the remaining cylinder is v and its mass is M. It is suspended by a string in a liquid of density ρ where it is at rest in the vertical position. The free surface of the liquid is at a height h above the upper surface of the cylinder. The force exerted by the liquid on the bottom of the cylinder is

(a) $Mg + \pi R^2 h\rho g$ (b) $Mg - v\rho g$

(c) $(v + \pi R^2 h)\rho g$ (d) Mg.

Solution

The buoyant force on the cylinder is equal to the force on its bottom (F) minus the force on the top of the cylinder.

Thus $$v\rho g = F - \pi R^2 h\rho g$$

or, $$F = (v + \pi R^2 h)\rho g.$$

Ans. (c)

Problem 8

Consider a liquid placed in a beaker which is accelerating horizontally with an acceleration a_0. The free surface of the liquid will

(a) remain horizontal

(b) be inclined to the horizontal where $\theta = \tan^{-1}\dfrac{a_0}{g}$, as shown in figure(i)

(c) be inclined to the horizontal where $\theta = \tan^{-1}\dfrac{a_0}{g}$, as shown in figure(ii).

(d) oscillate between the shapes shown in (i) and (ii).

Solution

Consider a cylindrical element of fluid with its axis horizontal and cross-sectional area ΔA. Forces acting on sections A and B are $P_1\Delta A$ towards right and $P_2\Delta A$ towards left. Hence, according to Newton's 2nd law,

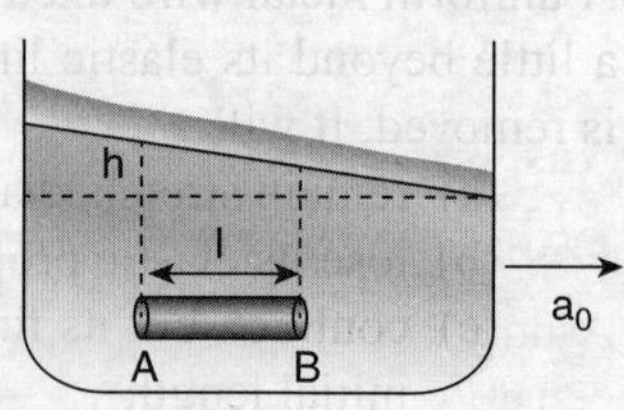

$$F_{net} = P_1\Delta A - P_2\Delta A = ma_0.$$

where the mass $m = v\rho = \Delta A \cdot \ell\rho$.

$\therefore \quad (P_1 - P_2)\Delta A = \Delta A\ell\rho a_0$

or, $\quad h\rho g = \ell\rho a_0$

or, $\quad \frac{h}{\ell} = \frac{a_0}{g}$

or, $\quad \tan\theta = \frac{a_0}{g}$, as in option (b)

Ans. (b)

Problem 9

A tank containing a liquid has two identical holes on opposite sides as shown in the figure. The free surface of the liquid is exposed to the atmosphere. The vertical separation between the two holes is h. As the liquid is expelled from the two holes, the tank will experience a net horizontal force which is proportional to

(a) h^2 (b) $h^{3/2}$

(c) h (d) $\sqrt{h}$.

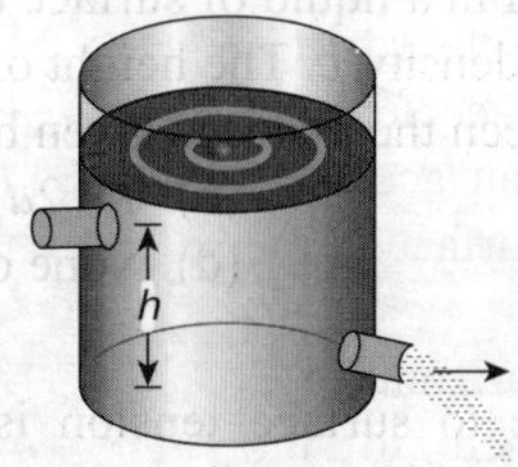

Solution

The efflux speed through the upper and lower holes are $v_1 = \sqrt{2gx}$ and $v_2 = \sqrt{2g(h+x)}$. If ρ be the density of the liquid, mass expelled per second through them will be $av_1\rho$ and $av_2\rho$ and the corresponding momentum discharged per second are $a\rho v_1^2$ and $a\rho v_2^2$ respectively. According to Newton's 2nd Law, these are the forces exerted on upper and lower holes respectively, along the horizontal

$\therefore$ Net force exerted on the tank

$$= a\rho(v_2^2 - v_1^2) = a\rho[2g(h+x) - 2gx]$$

$$= 2a\rho gh.$$

Ans. (c)

Problem 10

A uniform rod of length L and density ρ is placed in a wide tank containing a liquid of density ρ_0. The depth of the liquid in the tank is $\frac{L}{2}$. The rod rests in equilibrium with its lower end in contact with the bottom of the tank and its inclination with the horizontal base is $\tan^{-1}\frac{3}{4}$. The ratio ρ_0/ρ is

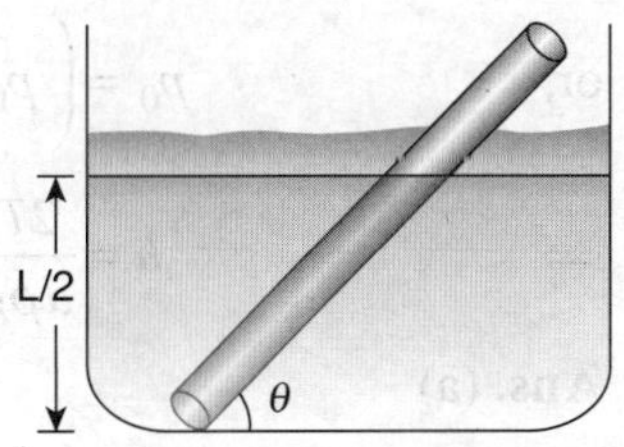

(a) 2 : 1 (b) 6 : 5

(c) 36 : 25 (d) None of these.

Solution

In the given figure,

$AB = L$, $AC = L/2$.

Let $AD = \ell$

The weight W acts at C and buoyant force F_B at mid-point of AD.

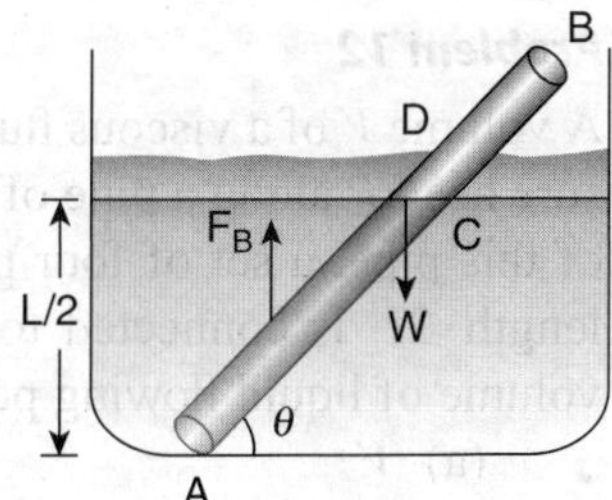

Taking the moment of W and F_B about A,

$$W.AC.\cos\theta = F_B.\frac{AD}{2}\cos\theta$$

or, $\quad aL\rho g \cdot \frac{L}{2} = a \cdot AD \cdot \rho_0 g \cdot \frac{AD}{2}$

or, $\quad \frac{\rho_0}{\rho} = \left(\frac{L}{AD}\right)^2 = (2\sin\theta)^2 \quad \left[\because \sin\theta = \frac{L/2}{AD}\right]$

$$= \left(2 \times \frac{3}{5}\right)^2 = \frac{36}{25}.$$

Ans. (c)

Problem 11

Three uniform tubes of length L, $2L$ and $3L$ are connected in series. Their radii are r, $\frac{r}{2}$ and $\frac{r}{3}$ respectively. When the flow through the system is streamline, the ratio of the pressure difference across the three tubes will be

(a) 1 : 1 : 1 (b) 1 : 4 : 9

(c) 1 : 16 : 81 (d) 1 : 32 : 243.

Solution

The flow rate according to Poiseuille's equation is

$$Q = \frac{\pi Pr^4}{8\eta\ell} = \frac{P}{(8\eta\ell/\pi r^4)} = \frac{P}{R}$$

where $\quad R = 8\eta\ell/\pi r^4$

Since the flow rate is same,

$$\frac{P_1}{R_1} = \frac{P_2}{R_2} = \frac{P_3}{R_3}$$

$\therefore \quad P_1 : P_2 : P_3 = R_1 : R_2 : R_3$

$$= \frac{\ell_1}{r_1^4} : \frac{\ell_2}{r_2^4} : \frac{\ell_3}{r_3^4}$$

$$= \frac{L}{r^4} : \frac{2L}{r^4/16} : \frac{3L}{r^4/81}$$

$$= 1:32:243.$$

Ans. (d)

Problem 12

A volume V of a viscous fluid flows per second due to a pressure head P along a pipe of diameter d and length ℓ. Instead of this pipe, a set of four pipes each of diameter $d/2$ and length 2ℓ is connected to the same pressure head P. The volume of liquid flowing per second now is

(a) V (b) $V/4$
(c) $V/8$ (d) $V/16$.

Solution

According to Poiseuille's equation, flow rate

$$V = \frac{\pi P r^4}{8\eta\ell} = \frac{P}{(8\eta\ell/\pi r^4)} = \frac{P}{R}$$

where R is the initial resistance. The resistance of each of the 4 pipes is

$$r = \frac{8\eta(2\ell)}{\pi(r/2)^4} = \frac{8\eta\ell}{\pi r^4} \times 32 = 32R$$

$\therefore$ Equivalent resistance of four such pipes connected in parallel is

$$R_{eq} = \frac{r}{4} = \frac{32R}{4} = 8R$$

$\therefore$ New flow rate $= \frac{P}{R_{eq}} = \frac{P}{8R} = \frac{1}{8}(V).$

Ans. (c)

Problem 13

A large drop of oil (density $0.8\,g/cm^3$ and viscosity η_0) floats up through a column of another liquid (density $1.2\,g/cm^3$ and viscosity η_L). Assuming that the two liquids do not mix, the velocity with which the oil drop rises will depend on

(a) η_0 only (b) η_L only
(c) both on η_0 and η_L (d) neither η_0 nor η_L.

Solution

Viscous force develops when different layers of the fluid move with different velocities. No such motion exists within the oil drop. Hence the motion in the viscous medium depends only on η_L.

Ans. (b)

Problem 14

Two rain drops of radii r_1 and r_2 reaching the ground with terminal velocities have their linear momenta p and $32p$. The ratio r_2/r_1 will be

(a) 2 : 1 (b) 1 : 2
(c) 2 : 3 (d) 3 : 2.

Solution

The terminal velocity $v \propto r^2$ and the mass of the drop $m \propto r^3$. Hence the momentum $p = mv \propto r^5$.

Given that $p_2/p_1 = \frac{32}{1}$

$$\therefore \quad \left(\frac{r_2}{r_1}\right)^5 = (2)^5$$

or, $r_2/r_1 = 2:1$.

Ans. (a)

Problem 15

Two parallel glass plates are held vertically at a small separation d and dipped in a liquid of surface tension T, angle of contact $\theta = 0$ and density ρ. The height of water that climbs up in the gap between the plates is given by

(a) $2T/d\rho g$ (b) $T/2d\rho g$
(c) $T/d\rho g$ (d) None of these.

Solution

Upward force due to surface tension is balanced by the weight of the liquid which rises in the gap, so

$$2T.b = b\,dh\rho g,$$

where, b = width of the plates

$$\therefore \quad h = \frac{2T}{d\rho g}.$$

Alternative Method

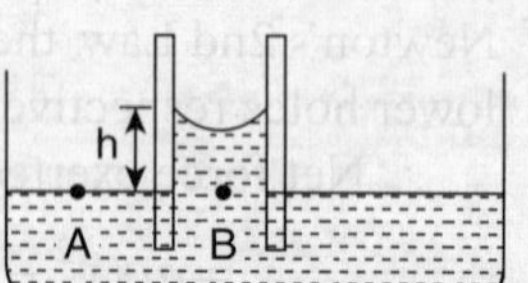

The meniscus between the plates has cylindrical shape with radius $r = \frac{d}{2}$.

The pressure just inside the meniscus is $p_0 - T\left(\frac{1}{r} + \frac{1}{\infty}\right) = p_0 - \frac{2T}{d}$,

Now, $p_A = p_B$

or, $$p_0 = \left(p_0 - \frac{2T}{d}\right) + h\rho g$$

$$\Rightarrow \quad h = \frac{2T}{d\rho g}.$$

Ans. (a)

Problem 16

An air bubble rises uniformly through a liquid column of density ρ. At some instant it is at a depth h below the free surface of water and its radius r. If T be the surface tension of the liquid, the pressure of enclosed air in the bubble exceeds the atmospheric pressure by an amount

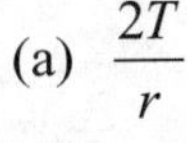

(a) $\frac{2T}{r}$

(b) $\frac{2T}{r}+h\rho g$

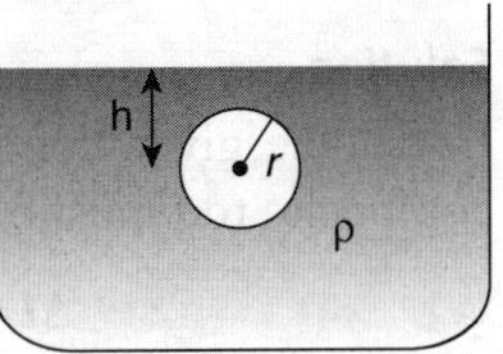

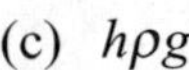

(c) $h\rho g$

(d) None of these.

Solution

Pressure outside the bubble, $p_0 = p_{atm} + h\rho g$

Pressure inside the bubble

$$p_i = p_0 + \frac{2T}{r} = p_{atm} + h\rho g + \frac{2T}{r}$$

$$\therefore \quad p_i - p_{atm} = h\rho g + \frac{2T}{r}.$$

Ans. (b)

Problem 17

A uniform capillary tube of inner radius 1.0mm is held vertically such that one end is just dipped in water (surface tension = 0.07N/m) and a length of 1cm is outside water. Which of the following options is true?

(a) The liquid will keep on overflowing from the open end as the tube length is insufficient

(b) The liquid will rise up to the top and will assume flat surface

(c) The liquid will rise the full exposed length and the radius of its meniscus will be equal to the radius of the capillary (= 1.0mm)

(d) The liquid will rise the full length of the tube and will assume a meniscus of radius = 1.4mm.

Solution

If θ be the angle of contact between the liquid and the glass, the capillary radius and meniscus radius are expressed

as $\quad r = R\cos\theta$

and $\quad 2\pi rT\cos\theta = \pi r^2 h'\rho g$

$$\therefore \quad R = \frac{2T}{\rho g h'} = \frac{2\times 0.07}{10^3\times 9.8\times(1\times 10^{-2})}$$

$$= 1.4\times 10^{-3}\text{ m} = 1.4\text{ mm}$$

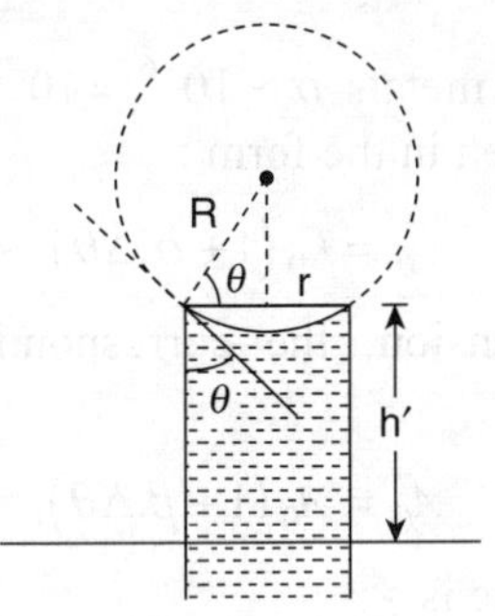

Ans. (d)

Problem 18

A vertical glass capillary tube of radius r open at both ends contains some water (surface tension T and density ρ). If L be the length of the water column, then

(a) $L = \frac{4T}{r\rho g}$ (b) $L = \frac{2T}{r\rho g}$

(c) $L = \frac{T}{4r\rho g}$ (d) $L = \frac{T}{2r\rho g}$.

Solution

The pressure inside the water column at the uppermost and lowermost points are $\left(p_0 - \frac{2T}{r}\right)$ and $\left(p_0 + \frac{2T}{r}\right)$, where p_0 is the atmospheric pressure.

$$\therefore \quad \left(p_0 + \frac{2T}{r}\right) - \left(p_0 - \frac{2T}{r}\right) = L\rho g$$

or, $$\frac{4T}{r} = L\rho g$$

Ans. (a)

THERMAL EXPANSION: SOLIDS AND LIQUIDS

It is observed that both solids and liquids undergo thermal expansion, i.e., their size (length, area or volume) increases as temperature increases. This is due to a corresponding decrease in the density. Metals, for instance, expand with temperature and this expansion is observed to be approximately linear with temperature:

$$\Delta\ell = \ell_0(1+\alpha.\Delta\theta) \qquad \ldots(i)$$

where $\Delta\ell = \ell_\theta - \ell_0$, is the increase in the length of the metal rod over its length at $0°\text{C}(\ell_0)$, $\Delta\theta =$ the increase in temperature, $\alpha =$ the coefficient of linear expansion of the material. For most metals $\alpha \sim 10^{-6} = 10^{-5}\,K^{-1}$. Equation (i) is frequently written in the form :

$$\ell_\theta = \ell_0\left(1+\alpha.\Delta\theta\right) \qquad \text{...(ii)}$$

For surface expansion, the corresponding coefficient is defined to be β,

$$A_\theta = A_0\left(1+\beta.\Delta\theta\right) \qquad \text{...(iii)}$$

and that for volume is γ:

$$V_\theta = V_0\left(1+\gamma.\Delta\theta\right) \qquad \text{...(iv)}$$

By considering the area of a small square plate of dimensions $\ell_0 \times \ell_0$ or, the volume of a small cube of material of side ℓ_0 it is possible to arrive at the useful result :

$$\alpha = \frac{\beta}{2} = \frac{\gamma}{3}, \qquad \text{...(v)}$$

Cavities and holes in solids also expand in the same proportion as if they were filled with the material.

For liquids, one defines an apparent (γ_a) and a real (γ_r) coefficient of thermal expansion, since linear and surface expansions do not carry any meaning. If the coefficient of thermal expansion of the material of the container is γ_g, then:

$$\gamma_a = \gamma_r - \gamma_g \qquad \text{...(iv)}$$

This is because expansion of the material of the container causes expansion of the cavity and this leads to lowering of the liquid level and thus, an apparent contraction. For most liquids, $\gamma_r \sim 10^{-4} - 10^{-3}\,K^{-1}$ or so.

Density varies in the same manner, except that it decreases with a rise in temperature. Since γ is very small for both solids and liquids, one ignores higher order k terms (γ^2 and smaller) to get:

$$\rho_\theta = \rho_0\left(1-\gamma.\Delta\theta\right) \qquad \text{...(vii)}$$

where γ is the coefficient of volume expansion (real for liquid).

Thermal expansion has lot of uses as well as unwanted side effects. Thermal expansion of mercury is used to construct the mercury-in-glass thermometer; temperature controllers made of bimetallic strips work on the same principle as well. Thermal expansion has to be properly accounted for in the construction of buildings, bridges, railroad tracks, aircrafts, computer chips and spacecraft.

Illustration 21

Two bars 1 and 2 of the same cross-sectional area having length ℓ_1 and ℓ_2, and made of different materials are clamped between two rigid walls. The materials of the bars have their Young's moduli Y_1 and Y_2 and linear coefficients of thermal expansion α_1 and α_2 respectively. What must be the relationship between these constants so that heating the bars does not change the position of the boundary between them ?

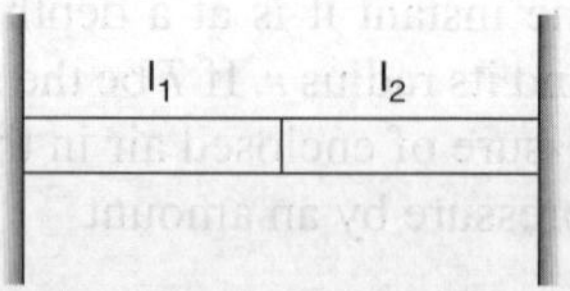

Solution

Under the given condition, the change in the length of the composite rod must be zero.

$\therefore \qquad \Delta L = 0$

or $\qquad \Delta(\ell_1+\ell_2) = 0$

or $\qquad \Delta\ell_1 + \Delta\ell_2 = 0 \qquad$ ...(i)

and, $\qquad \Delta\ell_1 = \Delta\ell_2 = 0 \qquad$ ...(ii)

Now, $\qquad \Delta\ell_1 = 0 = \alpha_1\ell_1\theta - \dfrac{F\ell_1}{AY_1}$

Similarly $\qquad \Delta\ell_2 = 0 = \alpha_2\ell_2\theta - \dfrac{F\ell_2}{AY_2}$,

where θ is the rise in temperature.

$$\therefore \qquad \frac{\alpha_1}{\alpha_2} = \frac{Y_2}{Y_1}$$

The Three Scales of Temperature

There are many scales of temperature. Among the useful scales of temperature are the Celsius, the Fahrenheit and the Kelvin (or the Absolute) scales.

The thermodynamic scale of temperature uses one 'fixed point' to define the temperature: the *triple point* of water. This is the temperature at which water vapour, pure water and melting ice, are all in equilibrium with each other. This temperature is defined as 273.16K. Using a constant-volume gas thermometer, the gas pressure p_0 is measured at the triple point of water (273.16 K). If the pressure is p at an unknown temperature T, then by definition,

$$T = \frac{p}{p_0} \times 273.16\text{ K}$$

The Celsius (C) and Fahrenheit (F) scales are defined by using two fixed points:

The *ice point* refers to the melting point of pure ice at a pressure of 760 mm Hg and it is defined to be 0° C or 32° F. The *steam point* refers to the boiling point of pure water or the temperature of steam under a pressure of 760 mm Hg and is defined to be 100° C or 212° F. It should be noted that if different types of thermometers are used to measure temperature, they only agree at the fixed points 273.16 K on the thermodynamic scale, and 0° C and 100° C on the Celsius scale. The absolute zero of temperature is 0 K (–273° C approximately). The conversion between all the three scales may be effected by the formula:

$$\frac{C}{5}=\frac{K-273}{5}=\frac{F-32}{9}$$

Calorimetry

Specific Heat

It is defined as the quantity of heat required to raise the temperature of 1 gram of a substance by 1°C.

Thermal capacity

It is defined as the quantity of heat required to raise the temperature of a substance by 1°C.

Calorie

It is defined as the quantity of heat required to raise the temperature of 1 gram of water from 14.5°C to 15.5°C, at a pressure of one standard atmosphere (1.01325×10^5 Pa). The mean calorie is defined as (1/100)th of the heat required to raise the temperature of 1 gram of water from 0°C to 100°C under a under a pressure of one atmosphere.

The basic equation of calorimetry is

$$\Delta Q = ms\Delta\theta = C\cdot\Delta\theta$$

where, $\Delta Q =$ heat supplied

$m =$ mass

$s =$ specific heat

$\Delta\theta =$ rise in the temperature

$C = ms =$ heat capacity of the mass

Since heat is a form of energy, one uses the same unit for heat and mechanical energy. Therefore, one calorie, the quantity of heat required to increase the temperature of 1 gram of water from 14.5° C to 15.5° C, can be expressed in Joule (J):

$$1 \text{ calorie} = 4.18 \text{ J}$$

Latent Heat

Many times it is observed that the supply of heat to a substance does not lead to any change in its temperature but rather to a change of phase: water at 100°C starts to boil under normal atmospheric pressure as heat is supplied to it. The steam that issues is also at 100° C, even though heat has been supplied to boil the water from 100° C to steam at 100° C. This heat, which is required to change the phase of a substance (solid-liquid or liquid-gas) is known as the latent heat. The latent heat is proportional to the mass of the substance, and the specific latent heat is the property of the substance itself:

$$\Delta Q = mL$$

where $m =$ mass of the substance at the phase transition,

$L =$ latent of the substance at the phase transition.

The latent heat of evaporation for water is 540 cal g^{-1} or 2.26×106 Jkg^{-1}. The latent heat of fusion for water (ice – water) is 80 cal g^{-1} or, 3.34×10^2 Jkg^{-1}.

Specific heats of solids and liquids are measured using calorimeter by the method of mixtures. The fundamental principle of calorimetry is sometimes stated in the following form:

Heat lost = Heat gained.

For an isolated system, the total amount of heat lost by the hot bodies is always equal to the total amount of heat gained by the colder bodies. This principle is frequently used to determine the final equilibrium temperature. It must be understood that the principle of calorimetry is a restatement of the law of conservation of energy and is true for all times.

Illustration 22

An insulated copper calorimeter of mass 102.5 g (specific heat capacity 0.5 cal/g°C) contains 200 g of water at 20⁰C. Steam at a temperature of 120°C is passed through the calorimeter at a rate of 10 g/min.

Find

(i) after what time the temperature of the calorimeter and its contents reaches 100 °C.

(ii) the total mass of the calorimeter and its contents when the water begins to boil.

Assume that the ambient pressure is equal to the atmospheric pressure and that the latent heat of vaporization of water is 540 cal/g.

Solution

At first the temperature of the calorimeter and its contents increases from 20 °C. Suppose that after x min. (when $10\,x$ g of steam has been passed), the temperature of the calorimeter and its contents reaches 100 ° C.

(i) Heat gained by calorimeter = (102.5) (0.1) (100 – 20) = 820 cal

Heat gained by water in calorimeter = 200 × 1 × (100 – 20) = 16000 cal cal.

Heat lost by steam condensed in calorimeter $=(10x)\left[(540)+0.5(120-100)\right]=5500x$

Since other mechanisms of heat transfer lead to negligible heat loss (or gain),

we have,

$$820+1600=5500\,x$$

$$\text{or,}\quad x=\frac{16820}{5500}=3.06 \text{ min.}$$

(ii) Total mass of the calorimeter and its contents $=(102.5+200+3.06\times10)\ \text{g}=333.1\ \text{g}.$

Illustration 23

A block of ice at –10°C is slowly heated and converted to steam at 100°C. Which of the following curves represents the phenomenon qualitatively?

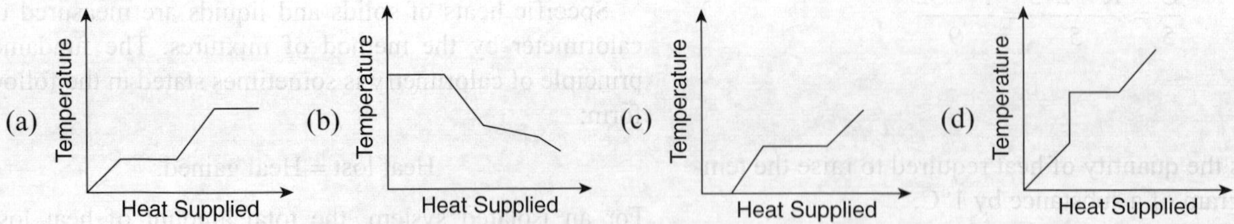

Solution

The scheme of the process is given below:

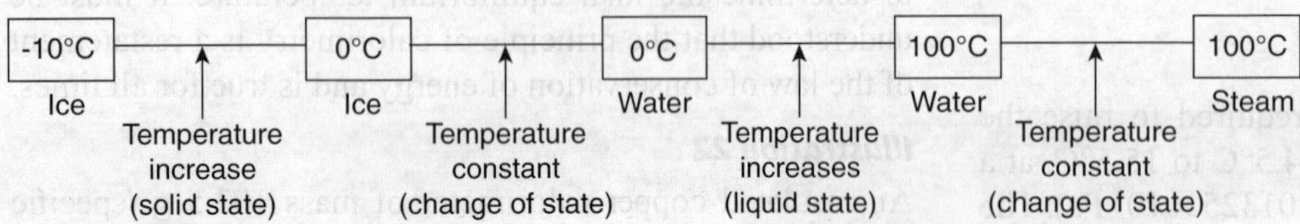

Ans. (a)

Heat Transfer

Heat can be transferred from one place to another by conduction, convection and radiation. During *conduction*, the molecules (or atoms) of the conducting medium maintain their mean positions and exchange energy through collisions (or through some coupling). This is the process by which a copper vessel conducts heat.

Convection is different in the sense that the flow of heat is accompanied by the flow of mass (or hot material). This is the process that occurs when a vessel containing water is heated from the bottom: the hotter water at the bottom is lighter and rises upward enabling cooler water to flow down and get heated, in turn. This happens only in the presence of gravity, and so, weaker gravitational fields lead to weaker convection currents. It is this very process that decides the weather and, as we all know, it is very complicated.

The third process is *radiation*. Radiation does not involve any transmitting medium; it can take place in vacuum. A hot body radiates its energy into space at a fixed rate which depends on the temperature and other properties of the body. We are going to study conduction and radiation and not convection, as the latter is more complicated and requires advanced mathematics.

Conduction

Suppose that a homogeneous uniform cylindrical rod is connected to a source of heat at temperature θ_1 on the left side and a sink (a large body at lower temperature) maintained at θ_2 on the other, as shown in the diagram.

Source of heat → Thermal current → Sink (ℓ; θ_1; $\theta_1 > \theta_2$; θ_2)

Suppose that the sides of the rod are insulated so that thermal energy does not flow out of the sides. Then, the rate of flow of heat through the rod, which occurs from the end at higher temperature to that at lower temperature, is equal to:

$$\frac{dQ}{dt} = \frac{KA}{\ell}(\theta_1 - \theta_2) \quad \text{...(i)}$$

where A is the cross-sectional area of the rod, and ℓ is the length of the rod. This expression is experimentally observed to be true, provided that the temperature differences are not extremely large. The quantity K is known as the thermal conductivity.

If, however, the temperature θ varies continuously, the expression (i) may be rewritten as:

$$\frac{dQ}{dt} = \text{thermal current}$$

$$= -KA\frac{d\theta}{dx} \quad \text{...(ii)}$$

θ(x) → Sink; x; x = 0; x = L

The minus sign indicates that $\frac{dQ}{dt}$ is positive if $\frac{d\theta}{dx}$ is negative, i.e., the temperature is decreasing along x. The quantity $\left(\frac{d\theta}{dx}\right)$ is also known as the 'temperature gradient'. The unit of thermal conductivity is watt (metre-Kelvin)$^{-1}$.

Another way of writing Eq. (ii) is:

J_Q = thermal current density

$$= \frac{1}{A}\frac{dQ}{dt} = -K\frac{d\theta}{dt} \quad \text{...(iii)}$$

Frequently, the thermal current is constant throughout the rod (or whatever body is used). This fact can be utilized to find the dependence of $\theta(x)$. It should be emphasized that thermal currents are similar to electrical currents and many of the expressions can be written by exploiting this. Thermal resistances add in series just like electrical resistances, thermal currents add up at junctions and Kirchoff's rules commonly applied to circuits are also applicable to thermal 'circuits' as well. Temperature differences play the role of emf. The thermal resistivity is:

$$R_H = \frac{\ell}{KA} \quad \text{...(iv)}$$

The study of thermal conduction in spherical and cylindrical geometries can be accomplished through the use of calculus.

Illustration 24

Two spheres of thermal capacity C and 2C, at temperature of 200° C and 300° C respectively are joined by a conducting rod of thermal resistance R_{Th}. Find the temperatures of the spheres as a function of time. Assume that the spheres can exchange heat by conduction only and there are no other mechanisms for heat loss. [Given: $R_{Th}.C = 2.5$ hour and the thermal capacity of the rod is negligible]

Solution

Suppose that at time t, the temperatures of A and B are θ_1 and θ_2.

$$\therefore \quad C\frac{d\theta_1}{dt} = -\frac{\theta_1 - \theta_2}{R_{Th}}$$

and

$$2C\frac{d\theta_2}{dt} = +\frac{\theta_1 - \theta_2}{R_{Th}}$$

$$\therefore \quad \frac{d\theta_1}{dt} = -\frac{(\theta_1 - \theta_2)}{R_{Th}C} \quad \text{...(i)}$$

$$\frac{d\theta_2}{dt} = +\frac{(\theta_1 - \theta_2)}{2R_{Th}C} \quad \text{...(ii)}$$

Subtracting, we get

$$\frac{d(\theta_1 - \theta_2)}{dt} = -\frac{3}{2}\frac{(\theta_1 - \theta_2)}{R_{Th}C}$$

or

$$\frac{d\theta}{dt} = -\frac{\theta}{\tau} \quad \text{...(iii)}$$

where θ is the difference in temperature, and

$$\tau = \frac{2}{3}R_{Th}C = \frac{2}{3} \times \frac{5}{2}\text{ hour} = 100\text{ min}$$

Solving (iii), we get

$$(\theta_1 - \theta_2) = \theta = -100 \cdot e^{-t/100\text{ min}} \quad \text{...(iv)}$$

Since, heat gained = heat lost, we get

$$C - 200 + 2C - 300 = C\theta_1 + 2C\theta_2$$

or

$$\theta_1 + 2\theta_2 = 800 \quad \text{...(v)}$$

Solving (iv) and (v), we get

$$\theta_1 = \frac{800}{3} - \frac{200}{3}e^{-t/100\text{ min}}$$

$$\theta_2 = \frac{800}{3} + \frac{100}{3}e^{-t/100\text{ min}}$$

Illustration 25

Two moles of a diatomic ideal gas at a temperature of 700 K is placed in a thermally insulated cylinder with a massless piston tightly fitting inside it. The bottom of the cylinder is fitted with a thin, weakly conducting slab of thickness w. It is observed that the piston moves with a speed :

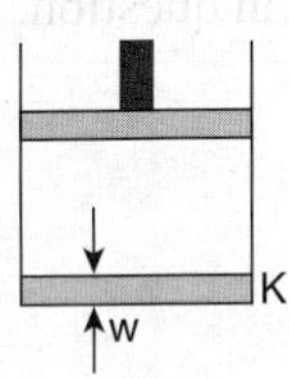

$$V = \text{constant}\, e^{-\alpha t}$$

Find the thermal conductivity of the slab. Assume that the cross-section of the cylinder is A and the temperature and pressure of the external surroundings are 300 K and p_0 respectively.

Solution

Thermal current through the slab is,

$$\frac{dQ}{dt} = -KA\frac{T - T_0}{w} \quad \text{...(i)}$$

Pressure of the enclosed gas,

$$p = p_0 \quad \text{...(ii)}$$

Since the process is isobaric,

$$\frac{dQ}{dt} = nC_p\frac{dT}{dt} = 2 \cdot \frac{7}{2}R\frac{dT}{dt} = 7R\frac{dT}{dt}$$

$$\therefore \quad 7R\frac{dT}{dt} = -KA\frac{T - 300}{w}$$

Radiation

Most of the thermal energy comes to us, directly or indirectly, from the sun. Solar energy traverses 150 × 10^6 km of space, travelling along straight lines just like light. The optical shadow of a building also coincides with its thermal shade. All these point to the fact that thermal energy is radiated; experiments bear out the fact that thermal radiation is also electromagnetic in nature. Black bodies are observed to be better absorbers (as well as emitters) of thermal radiation. Experiments performed on radiating bodies soon led one to the idea of a perfectly black body, i.e., one which absorbs all the radiation incident on it, and reflects and transmits none.

Experimentally, a good approximation to a black body is a spherical cavity within another material (e.g. porcelain). The inner surface of the cavity is blackened with soot and a small hole is made, connecting it with the exterior. The entire body is maintained at constant temperature (T). The radiation emerging from the hole is blackbody radiation; its character does not vary with the nature of the surface of the cavity. This is so because the radiation emerging from any area is made up partly of the radiation emitted by that area, and partly of the radiation from other area elements reflected at the area in question.

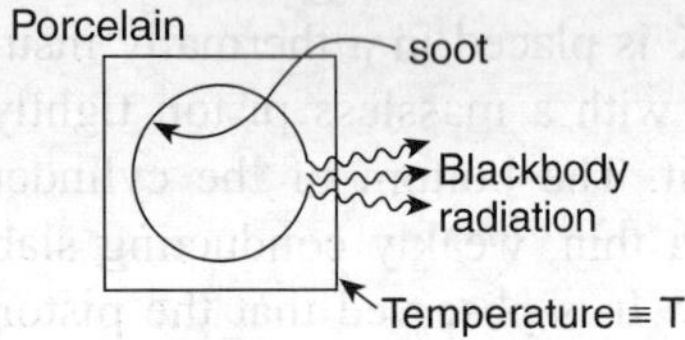

When the relative intensities of the radiation are plotted as a function of the wavelength a curve similar to the one shown in the adjacent diagram is obtained.

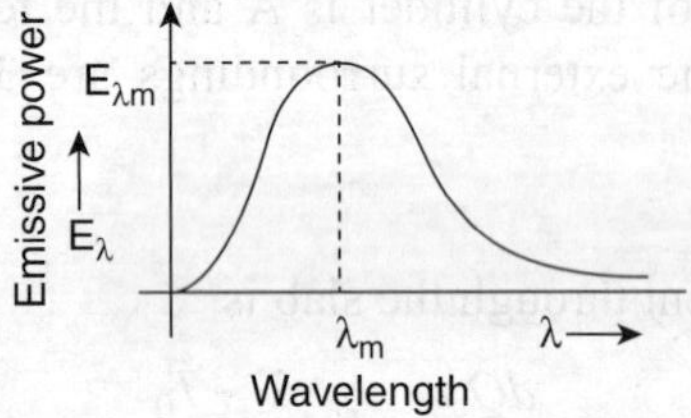

More specifically, one cannot speak of the intensity of radiation at a particular wavelength, as that would be meaningless, due to the fact that the spectrum is continuous. However, we can always speak of the intensity of radiation within a given band of wavelengths, λ to $\lambda + \Delta\lambda$. For a narrow band ($\Delta\lambda \to 0$), one can define the emissive power (E_λ):

$E_\lambda\, \Delta\lambda$ = Power (i.e. energy/sec) radiated by a blackbody per unit area within the wavelength band λ to $\lambda + \Delta\lambda$.

$\therefore$ E_λ = emissive power of a blackbody

$$= \frac{\text{Power radiated in the wavelength band } \lambda \text{ to } \lambda + \Delta\lambda \text{ per unit surface area of the blackbody}}{\Delta\lambda}$$

The practical unit of E_λ is watt per metre^2 per nanometer.

The total area under the curve represents the total power radiated by the blackbody over all wavelengths. Note that the curve has a maximum at wavelength λ_m; the emissive power reaches reaching its maximum value E_{λ_m}.

Laws of Blackbody Radiation

The curve shown in the figure (for emissive power) can only be explained in full by invoking Planck's quantum theory, this explanation is beyond the scope of this discussion. However, certain general laws were discovered before the curve was fully explained:

i) Stefan's Law

The total area under the E_λ versus λ curve, represents the total power radiated per unit area by a blackbody at temperature T. This is given by:

$$E = \sigma T^4,$$

where σ is a constant known as Stefan's constant.

$$\sigma = 5.67 \times 10^{-8}\ \text{Wm}^{-2}\ \text{K}^{-4}.$$

It gives the total radiant power per unit area of a blackbody at temperature T (in kelvin).

ii) Wien's Displacement Law

If λ_m is the wavelength corresponding to the peak of the curve for a blackbody at temperature T,

$$\lambda_m T = b\ (a \text{ constant}) = 2.9 \times 10^{-3}\ \text{m K}.$$

The dotted curve shown in the adjacent diagram shows the join of the maxima for the blackbodies radiating at different temperatures. A plot of λ_m vs T shows that λ_m is inversely proportional to T.

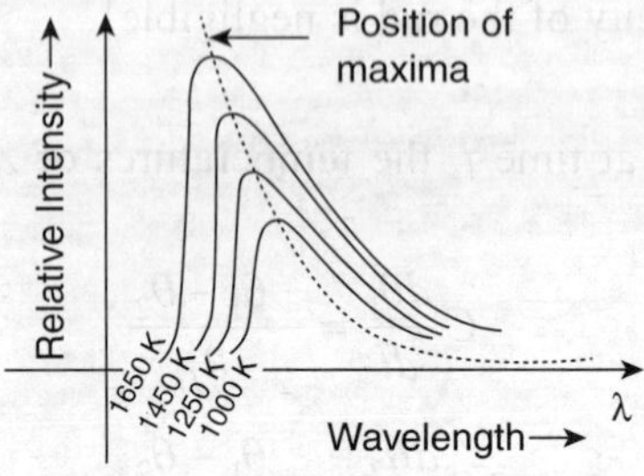

iii) If E_{λ_m} is the height of the curve corresponding to λ_m, then

$$E_{\lambda_m} \propto T^5$$

iv) The curve showing the variation of E_λ with λ at a constant temperature T obeys the Planck formula:

$$E_\lambda = \frac{c_1}{\lambda^5 (e^{c_2/\lambda T} - 1)}$$

where c_1, c_2 are constants.

Prevost's Theory of exchanges

Prevost, in 1792, applied the idea of dynamic equilibrium to radiation. He stated that a body radiates heat at a rate which depends on the nature of its surface and its temperature; and that it absorbs heat at a rate depending on the nature of its surface and the temperature of its surroundings. When a body is at constant temperature in vacuum, it is losing heat to its surroundings by radiation as well as gaining heat from its surroundings by radiation, at equal rates.

Kirchhoff's Law

Most ordinary bodies are not blackbodies, they are coloured. They absorb certain wavelengths much better than others. If a body absorbs a fraction a_λ of the radiation falling on it in the narrow band between λ and $\lambda + \Delta\lambda$, then a_λ is referred to as the spectral absorption factor of the body, for the wavelength λ.

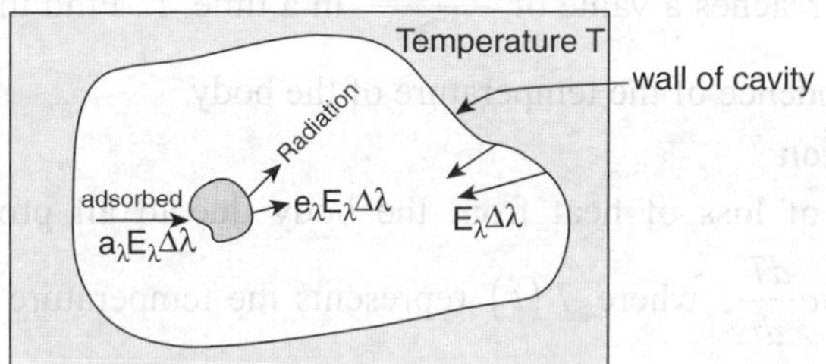

Assume that such a body is placed within a cavity at a temperature T. The radiation falling on the body in the band λ to $\lambda + \Delta\lambda$ has the power: $E_\lambda\ \Delta\lambda$, where E_λ is the emissive power of a blackbody at T in the neighborhood of λ. The body inserted within the enclosure absorbs a fraction a_λ of this. This is also equal to the power radiated, in equilibrium (i.e. when the body also reaches the temperature T of the enclosure).

$\therefore$ Power radiated in the band λ to $\lambda + \Delta\lambda$ per unit area by this body

$$= a_\lambda\ E_\lambda\ \Delta\lambda.$$

The spectral emissivity of body,

$$e_\lambda = \frac{\text{Power radiated by the body in } \lambda \text{ to } \lambda + \Delta\lambda}{\text{Power radiated by a blackbody of identical shape and size in } \lambda \text{ to } \lambda + \Delta\lambda}$$

(where both the bodies are at the same temperature)

$$= \frac{a_\lambda E_\lambda \Delta\lambda}{E_\lambda \Delta\lambda} = a_\lambda$$

This is Kirchhoff's law. The spectral emissivity (e_λ) of a body at a certain wavelength λ equals its spectral absorption factor (a_λ) for the same wavelength. It is not easy to demonstrate because the spectral emissivity of different coloured pigments vary substantially with temperature.

Illustration 26

A homogeneous metallic sphere of radius 10 cm is heated to a temperature of 727° C, and spun about its diameter at 1000 rpm in gravity free space. The temperature of its surrounding space is maintained at 27 °C .It is observed that the sphere speeds up because of thermal contraction due to radiative cooling. Find the change in its spin-rate after a time interval of 12.6 seconds, assuming that the sphere radiates like a black body. Given that for the material of the sphere, coefficient of linear expansion $\alpha - 1.2\times10^{-5}$ / °C, specific gravity $\rho = 7.2$, specific heat $= 0.09919$ cal/g. °C, and Stefan's constant $\sigma = \frac{17}{3}\times10^{-8}\ \text{W/m}^2\text{K}^4$.

Solution

Rate of loss of heat due to radiation is approximately

$$= \left|\frac{dQ}{dt}\right| = m.s.\frac{dQ}{dt} = \sigma.4\pi R_0^2\left(\theta^4 - \theta_0^4\right),$$

where θ = absolute temperature of the sphere = 1000 K and θ_0 = absolute temperature of the environment $= 273 + 27 = 300$K

$$\therefore \left|\frac{d\theta}{dt}\right| = \frac{\sigma.4\pi R_0^2\left(\theta^4 - \theta_0^4\right)}{\text{ms}}$$

$$= \frac{\sigma.4\pi R_0^4\left(\theta^4 - \theta_0^4\right)}{\frac{4}{3}\pi R_0^3 \rho s}$$

$$= \frac{3\sigma}{\rho s R_0}\left(\theta^4 - \theta_0^4\right)$$

$$= \frac{3\times\frac{17}{3}\times10^{-8}\left[(1000)^4 - (300)^4\right]}{7.2\times10^3\times0.09919\times4200\times0.1}$$

$$= \frac{17}{4.2\times7.2}\ \text{K/s}$$

Given that total number of seconds = 12.6

$$\therefore \qquad \delta\theta = \frac{17\times12.6}{4.2\times7.2} = \frac{17}{2.4}\text{K}$$

Since angular momentum is conserved,

$I\omega$ = constant, where

I = moment of inertia $= \frac{2}{5}\text{MR}^2$

$$\therefore \qquad \left|\frac{d\omega}{\omega}\right| = \left|\frac{dI}{I}\right| = 2\left|\frac{\delta R}{R}\right| = 2\alpha\delta\theta$$

$$= 2\times1.2\times10^{-5}\times\frac{17}{2.4} = 17\times10^{-5}$$

Initial spin rate = 1000 rpm

$$\therefore \qquad \text{Change in spin rate} = \left|\frac{d\omega}{\omega}\right|\times1000\,\text{rpm}$$

$$= 0.17\ \text{rpm}.$$

Newton's law of cooling

Newton investigated the heat lost by a body in air, and found that 'the rate of loss of heat is proportional to the excess temperature over the surrounding'. This is known as Newton's law of Cooling, and is true for temperature differences of 20–30°C in still air; but over a much wider range of temperatures in cases of forced convection, i.e., when there is a draught. With natural convection, a different empirical law proposed by Dulong and Petit holds over a wider range of temperatures, which is beyond the scope of this discussion. Newton's law is an empirical law, i.e., it is demonstrated by experiment: there is no simple theoretical derivation.

If the rate of loss of heat from a body be represented by $-\frac{dQ}{dt}$ then:

$$\frac{dQ}{dt} = -k(\theta - \theta_0) \quad \text{...(i)}$$

where k is constant and θ, θ_0 are the temperatures of the body and its surroundings.

If the thermal capacity of the body is C, then one can write,

$$C\frac{d\theta}{dt} = -k(\theta - \theta_0)$$

or

$$C\frac{dx}{dt} = -kx,$$

where $x = \theta - \theta_0$ is the excess temperature of the body over that of its surroundings.

$$\therefore \quad \frac{dx}{dt} = -\frac{k}{C}x \quad \text{...(ii)}$$

and this may be solved to give,

$$x(t) = x(0)e^{-kt/C}$$
$$= x(0)e^{-t/\tau}$$

where $\tau = C/k$, has dimensions of time.

This variation is analogous to an exponential decay (just like that in radioactivity) and x may be plotted as a function of time:

It shows all the characteristics of exponential decays like half-life, etc.

Illustration 27

A body of mass m and specific heat C is at an initial temperature T_0 which is greater than the temperature T_e of the surroundings. Suppose also that the temperature of the body reaches a value of $\frac{T_e + T_0}{2}$ in a time τ. Find the time dependence of the temperature of the body.

Solution

Rate of loss of heat from the body due to all processes $= -mc\frac{dT}{dt}$, where $T(t)$ represents the temperature of the body at time t.

$$\therefore \quad T(0) = T_0 \text{ (given)}$$

By Newton's law of cooling,

$$-mc\frac{dT}{dt} = \lambda(T - T_e)$$

$$\therefore \quad \frac{dT}{dt} = -\frac{\lambda}{mc}(T - T_e) \quad \text{...(i)}$$

Solving this, we get

$$T - T_e = (T_0 - T_e)e^{-\frac{\lambda}{mc}t} \quad \text{...(ii)}$$

It is given that the temperature becomes:

$$T(t = \tau) = \frac{T_e + T_0}{2}.$$

Putting this is (ii)

$$T(t) = T_e + (T_0 - T_e)e^{-\frac{t}{\tau}(\ln 2)} \quad \text{...(iii)}$$

This gives the time dependence of the temperature of the body.

POINTS TO REMEMBER

1. A substance that can flow is called a fluid. It includes both liquids and gases.
2. The free surface of a liquid at rest is horizontal.
3. The density of a substance is defined as the mass per unit volume of that substance. Its unit is kg/m^3.
4. The relative density of a substance is defined as the ratio of the density of that substance to that of water at 4°C. As this is the ratio of two densities, it has no unit.
5. The pressure at any point is defined as the normal force per unit area surrounding that point. It is a scalar quantity. The SI unit of pressure is N/m, called Pascal and abbreviated as Pa.
6. The pressure due to a liquid column of height h and density ρ is given by

$$P = h\rho g$$

7. The pressure difference between two points in a liquid which is separated by a vertical distance of h is given by

$$P_1 - P_2 = h\rho(g \pm a)$$

where a is the acceleration of the liquid in the vertically upward direction.
Thus, the pressure difference between two points in a liquid separated by a vertical distance increases if the liquid is accelerated in the upward direction and decreases if the liquid is accelerated in the downward direction (or equivalently, decelerated in the upward direction).
If $a = 0$

$$P_1 - P_2 = h\rho g$$

8. When a liquid is accelerated in the horizontal direction, the free surface of the liquid is no longer horizontal, but is inclined to the horizontal by an angle θ where

$$\tan\theta \frac{a_0}{g}$$

where a_0 is the horizontal acceleration of the liquid. This is shown in Fig 7.1.

9. The pressure difference between two points in a liquid separated by a distance l in the same horizontal level (Fig. 7.1) is

$$P_1 - P_2 = l\rho a_0$$

where a_0 is the horizontal acceleration of the liquid.
If a_0 =0, then $P_1 = P_2$, i.e., the pressure in a liquid at all points in the same level are equal.

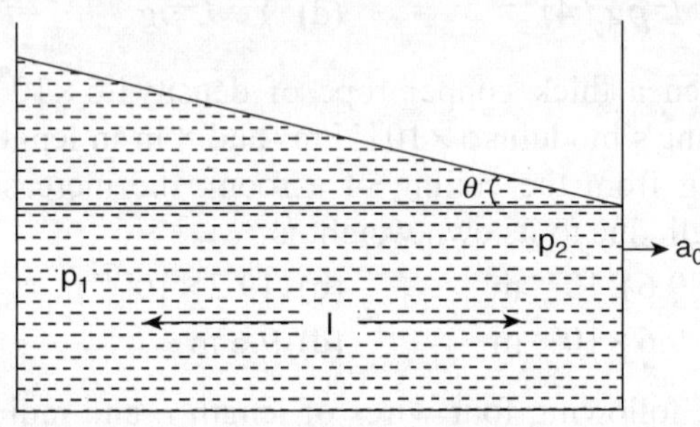

Fig. 7.1

10. At any point in a liquid, the pressure exerted by the liquid is the same in all directions.

11. If the pressure in a liquid is changed at any point, the change is transmitted to every point of the liquid without being diminished in magnitude. This is known as Pascal's law.

12. Archimedes' principle states that when a body is partially or wholly immersed in a liquid at rest, the liquid exerts an upward force of buoyancy on the body, equal to the weight of the displaced liquid.
Because of this upward force of buoyancy on the body, the apparent weight of a body in the liquid is reduced and the decrease in weight is equal to the weight of the liquid displaced.

13. A body floating in a fluid is in stable equilibrium if the centre of gravity of the displaced fluid, called the centre of buoyancy, is vertically above the centre of gravity of the body.

14. When a solid body is dipped into a liquid, it experiences an upward force of buoyancy due to the liquid. If this force of buoyancy is equal to the weight of the solid, the solid will remain in equilibrium. This is known as floatation. When the density of the solid is less than that of the liquid, the solid floats with a part of it in the liquid. The fraction dipped is such that the weight of the displaced liquid equals the weight of the solid.

15. All liquids we deal with are assumed to be incompressible.

16. In the streamline flow of a fluid, velocity of fluid particles reaching a particular point is the same at all time. Thus, each particle follows the same path as taken by its predecessor, passing through that point. The streamline flow is also known as steady flow.

17. If the velocities of different particles of the fluid passing through the same point is different and change erratically with time, the flow of the fluid is said to be turbulent.

18. In the case of flow of a liquid through a tube, the total volume of the liquid going into the tube through any cross-section is equal to the total volume coming out of the same tube from any other cross-section in the same time because, the liquids are incompressible.
If A_1 and A_2 be the areas of cross-section of the tube at two different places where the speeds of the liquids are v_1 and v_2 respectively, then

$$A_1 v_1 = A_2 v_2$$

This is known as equation of continuity and states that the product of the area of cross-section and the speed remains the same at all points of a tube.

19. When an incompressible and non-viscous fluid flows in streamline motion from one place to another, then at every point of its path, the total energy per unit volume (i.e., pressure energy P + kinetic energy $\frac{1}{2}\rho v^2$ + potential energy ρgh) is a constant. Thus,

$$P + \frac{1}{2}\rho v^2 + \rho gh = \text{a constant}$$

This equation is known as Bernoulli's theorem and can also be written as

$$\frac{P}{\rho g} + \frac{1}{2}\frac{v^2}{d} + h = a \text{ constant}$$

The quantity $\frac{P}{\rho g}$ is known as pressure head, $\frac{v^2}{2g}$ as velocity head and h as the gravitational head.

20. Torricelli's theorem states that the velocity of efflux of a liquid flowing through an orifice is equal to that which a body would acquire in falling freely from the free surface of the liquid to the orifice. This velocity is given by

$$v = \sqrt{2gh}$$

where h is the depth of the orifice from the free surface of the liquid.

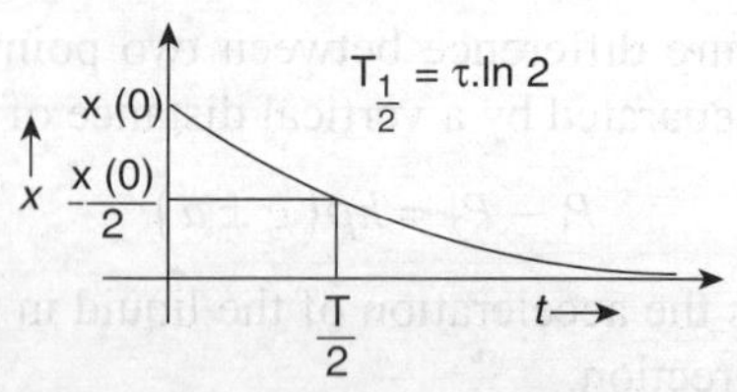

EXERCISES

1. Young's modulus is defined as
   (a) the ratio of linear strain to the normal stress
   (b) the ratio of normal stress to strain
   (c) product of linear strain and normal stress
   (d) square of the ratio of normal stress to linear strain.
2. The longitudinal extension of any elastic material is very small. In order to have an appreciable change, the material must be in the form of a
   (a) thin block of any cross-section
   (b) thick block of any cross-section
   (c) long thin wire
   (d) short thin wire.
3. Which of the following substance has the highest elasticity?
   (a) Steel (b) Copper
   (c) Rubber (d) Sponge
4. Shearing strain is expressed by
   (a) angle of shear
   (b) angle of twist
   (c) decrease in volume
   (d) increase in volume.
5. A cable that can support a load W is cut into two equal parts. The maximum load that can be supported by either part is
   (a) $\frac{W}{4}$ (b) $\frac{W}{2}$
   (c) W (d) $2\,W$
6. The effect of temperature on the value of Young's modulus of elasticity for various substances in general is,
   (a) it increases with increase in temperature
   (b) it remains constant
   (c) it decreases with rise in temperature
   (d) it sometimes increases and sometimes decreases with temperature.
7. The following four wires are made of the same material. Which of these will have the largest extension when the same tension is applied
   (a) length 50 cm and diameter 0.5 mm
   (b) length 100 cm and diameter 1 mm
   (c) length 200 cm and diameter 2 mm
   (d) length 300 cm and diameter 3 mm.
8. Two rods of different materials having coefficient of linear expansion α_1, α_2 and Young's modulii Y_1 and Y_2 respectively are fixed between two rigid massive walls. The rods are heated such that they undergo the same increase in temperature. There is no bending of rods. If $\alpha_1 : \alpha_2 = 2:3$,the thermal stresses developed in the two rods are equal provided $Y_1 : Y_2$ is equal to
   (a) 2 : 3 (b) 1 : 1
   (c) 3 : 2 (d) 4 : 9
9. An Indian rubber cord L metre long and area of cross-section A m^2 is suspended vertically. Density of rubber is ρkg/m^3 and Young's modulus of rubber is Y newton/m^2. If the rod extends by l metre under its own weight, then extension l is
   (a) $L^2\rho g / Y$ (b) $L^2\rho g / 2Y$
   (c) $L^2\rho g / 4Y$ (d) $Y / L^2\rho g$
10. When a thick copper rope of density 1.5×10^3 kg/m^3, Young's modulus 5×10^6 N/m^2 and 8 m in length when hung from the ceiling of a room, the increase in its length due to its own weight is
    (a) 9.6×10^{-5} m (b) 19.2×10^{-7} m
    (c) 9.6×10^{-2} m (d) 9.6 m
11. The following four wires of length L and radius r are made of the same material. Which of these will have the largest extension when the same tension is applied
    (a) $L = 50$ cm $r = 0.25$ mm
    (b) $L = 100$ cm $r = 0.5$ mm
    (c) $L = 200$ cm $r = 1$ mm
    (d) $L = 300$ cm $r = 1.5$ mm
12. A wire of length l and cross-sectional area A is made of a material of Young's modulus Y. If the wire is stretched by an amount x, the work done is
    (a) $\frac{YAx^2}{2L}$ (b) $\frac{YAx}{2L^2}$
    (c) $\frac{YAx}{2L}$ (d) $\frac{YAx^2}{L}$

13. Energy stored in stretching a string per unit volume is

(a) $\frac{1}{2}\times$ strees $\times$ strain (b) strees $\times$ strain

(c) $y\times(\text{strain})^2$ (d) $\frac{1}{2}\times y\times(\text{stress})^2$

14. The work done in deforming a body is given by

(a) strees $\times$ strain (b) $\frac{1}{2}\times(\text{strees}\times\text{strain})$

(c) $\frac{\text{stress}}{\text{strain}}$ (d) $\frac{\text{strain}}{\text{stress}}$

15. Two rods of different materials having coefficient of thermal expansion α_1 and α_2 and Young's moduli Y_1 and Y_2 respectively are fixed between two rigid walls. The rods are heated such that they undergo the same increase in temperature. There is no bending of rods. If $\frac{\alpha_1}{\alpha_2}=\frac{2}{3}$ and stresses developed in the two rods are equal, then $\frac{Y_1}{Y_2}$ is

(a) $\frac{3}{2}$ (b) 1

(c) $\frac{2}{3}$ (d) $\frac{1}{2}$

16. The mean distance between the atoms of iron is 3×10^{-10} m and interatomic force constant for iron is 7 N/m. The Young's modulus of elasticity for iron is

(a) 2.33×10^{5} N/m² (b) 23.3×10^{10} N/m²
(c) 233×10^{10} N/m² (d) 2.33×10^{10} N/m²

17. A wire elongates by l mm when a load W is hanged from it. If the wire goes over a pulley and two weights, W each, are hung at the two ends, the elongation of the wire will be (in mm).

(a) $\frac{l}{2}$ (b) l

(c) $2l$ (d) zero

18. A wire of length L and radius r fixed at one end and a force F applied to the other end produces an extension l. The extension produced in another wire of the same material of length 2 L and radius $2r$ by a factor 2 F, is

(a) l (b) $2l$
(c) $4l$ (d) $l/2$

19. Two wires of same radius and length are subjected to the same load. One wire is of steel and the other is of copper. If Young's modulus of steel is twice than that of copper, then the ratio of elastic energy stored per unit volume of steel to that of copper wire is:

(a) 2 : 1 (b) 1 : 2
(c) 1 : 4 (d) 4 : 1

20. Young's modulus of the material of a wire is Y. On pulling the wire by a force F, the increase in its length is x. The potential energy of the stretched wire is

(a) $\frac{1}{2}Fx$ (b) $\frac{1}{2}Yx$

(c) $\frac{1}{2}Fx^2$ (d) None of the above.

21. Adiabatic modulus of elasticity of a gas is $2.1\times10^{5}\,\text{N/m}^2$. What will be its isothermal modulus of elasticity? $\left(\frac{C_p}{C_v}=1.4\right)$

(a) $1.2\times10^{5}\,\text{N/m}^2$ (b) $4\times10^{5}\,\text{N/m}^2$
(c) $1.5\times10^{5}\,\text{N/m}^2$ (d) $1.8\times10^{5}\,\text{N/m}^2$

22. The bulk modulus of water is $2\times10^{9}\,\text{N/m}^2$. The change in pressure required to increase the density of water by 0.1% is

(a) $2\times10^{4}\,\text{N/m}^2$ (b) $2\times10^{6}\,\text{N/m}^2$
(c) $2\times10^{8}\,\text{N/m}^2$ (d) $2\times10^{9}\,\text{N/m}^2$

23. The increase in length of a wire on stretching is 0.025%. If its Poisson's ratio is 0.4, Then the percentage increase in diameter is

(a) 0.01 (b) 0.02
(c) 0.03 (d) 0.04

24. If S is stress and Y is Young's modulus of material of a wire, the energy stored in the wire per unit volume is

(a) $\frac{2Y}{S}$ (b) $\frac{S}{2Y}$

(c) $2S^2Y$ (d) $\frac{S^2}{2Y}$

25. A wire fixed at the upper end stretches by length l by applying a force F. The work done in stretching is

(a) $\frac{F}{2l}$ (b) Fl

(c) $2Fl$ (d) $\frac{Fl}{2}$

26. The Poisson 's ratio of material is 0.4. If a force is applied to a wire of this material, there is a decrease of cross-sectional area by 2%. The percentage increase in its length is

(a) 3% (b) 2.5%
(c) 1 % (d) 0.5 %

27. The length of an elastic string is a metre when the longitudinal tension is 4 N and b metre when the tension is 5 N. The length of the string (in m) when the longitudinal tension is 9 N is
(a) $a - b$ (b) $5b - 4a$
(c) $2b - \frac{a}{2}$ (d) $4a - 3b$

28. The elastic energy stored in a wire of Young's modulus Y is
(a) $\frac{1}{2} Y \times \text{stress} \times \text{strain} \times \text{volume}$
(b) $\frac{(\text{stress})^2 \times \text{volume}}{2Y}$
(c) $\text{stress} \times \text{strain} \times \text{volume}$
(d) $Y \times \frac{(\text{stress})^2}{\text{volume}}$

29. A wire suspended vertically from one of its ends is stretched by attaching a weight of 200 N to the lower end. The weight stretches the wire by 1 mm. Then the elastic energy stored in the wire is
(a) 0.2 J (b) 10 J
(c) 20 J (d) 0.1 J

30. Shearing stress is related to
(a) length (b) volume
(c) shape (d) area

31. A copper wire of length 4.0 m and area of cross section 1.2 cm^2 is stretched with a force of 4.8×10^3N. If Young's modulus for copper is 1.2×10^{11}N/m^2. The increase in length of the wire will be
(a) 2.66 cm (b) 2.66 mm
(c) 1.33 cm (d) 1.33 mm

32. A wire of length L and area of cross-section A is stretched through a distance x metre by applying a force F along length, then the work done in this process is (Y is Young's modulus of the material)
(a) $\frac{1}{2}(A.L)\left(\frac{Yx}{L}\right)\left(\frac{x}{L}\right)$ (b) $(A.L)(Y.L)\left(\frac{x}{L}\right)$
(c) $2(A.L)(Y.L)\left(\frac{x}{L}\right)$ (d) $3(A.L)(Y.L)\left(\frac{x}{L}\right)$

33. A wire can be broken by applying a load of 200 N. The force required to break another wire of the same length and same material, but double in diameter, is
(a) 200 N (b) 400 N
(c) 600 N (d) 800 N

34. A cube of side 40 mm has its upper face displaced by 0.1 mm by a tangential force of 8 kN, the shearing modulus of cube is
(a) 2×10^9 Nm^{-2} (b) 4×10^9 Nm^{-2}
(c) 8×10^9 Nm^{-2} (d) 16×10^9 Nm^{-2}

35. A work of 2×10^{-2} J is done on a wire of length 50 cm and area of cross-section 0.5 mm^2. If the Young's modulus of the material of the wire is 2×10^{10} N/m^2. Then the wire must be
(a) elongated to 50.1414 cm
(b) contracted by 2.0 mm
(c) stretched by 0.707 mm
(d) None of these.

36. When a weight of 5 kg is suspended from a copper wire of length 30 m and radius 0.5 m, the length of the wire increases by 2.4 cm. If the radius is doubled, the extension produced is
(a) 1.2 cm (b) 0.6 cm
(c) 0.3 cm (d) 0.15 m

37. Two bodies of masses 1 kg and 2 kg are connected by a steel wire of cross-section 2 cm^2 going over a smooth pulley (as shown in figure). The longitudinal strain in the wire is (Take $g = 10$ m/s^2, $Y = 2 \times 10^{11}$ N/m^2)

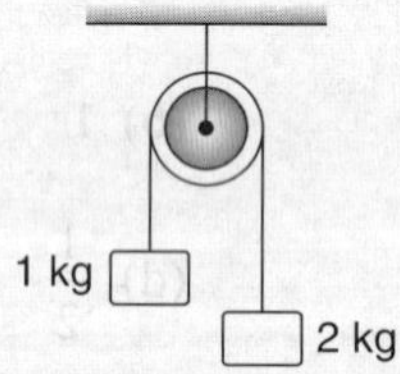

(a) 3.3×10^{-7} (b) 3.3×10^{-6}
(c) 2×10^{-6} (d) 4×10^{-6}

38. A three-block system is shown in the figure. Each has mass 3 kg. A string connected to P and Q are of equal cross-section and Young's modulus of 0.005 cm^2 and 2×10^{11} N/m^2 respectively. If friction is neglected, the longitudinal strain in A and B are:

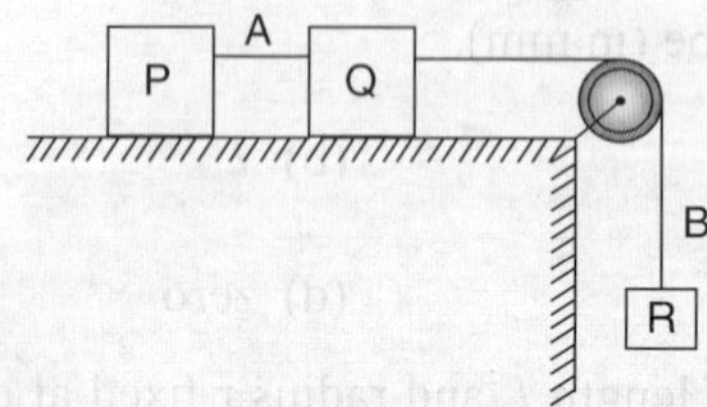

(a) $2.5 \times 10^{-4}, 1 \times 10^{-4}$ (b) $1 \times 10^{-4}, 2 \times 10^{-4}$
(c) $0.2 \times 10^{-4}, 2 \times 10^{-4}$ (d) None of these.

39. When tension in a metal wire is T_1, its length was l_1 and when tension is T_2, the length is l_2. Its unstretched length is:
(a) $\sqrt{l_1 l_2}$ (b) $\frac{l_1 + l_2}{2}$
(c) $\frac{(l_1 T_2 - l_2 T_1)}{T_2 - T_1}$ (d) $\frac{l_1 T_2 + T_1 l_2}{T_1 + T_2}$

40. What will happen if a metal wire is stretched a little beyond its elastic limit (or yield point) and released?
(a) It loses its elastic property completely.
(b) It does not contract.
(c) It contracts, but its final length will be greater than its initial length.
(d) It contracts only up to its length at the elastic limit.

41. Young's modulus is defined for:
(a) solid (b) liquid
(c) gas (d) All of these.

42. The bulk modulus for an incompressible liquid is:
(a) ∞ (b) 0
(c) 1 (d) 2

43. The value of modulus of rigidity for liquids is:
(a) ∞ (b) 0
(c) 1 (d) 2

44. A steel ring of radius r and cross-sectional area A is filled on to a wooden disc of radius R ($R > r$). If Young's modulus be Y, then the force with which the steel ring is expanded, is
(a) $AY\frac{R}{r}$ (b) $AY\left(\frac{R-r}{r}\right)$
(c) $\frac{Y}{A}\frac{(R-r)}{r}$ (d) $\frac{Yr}{AR}$

45. Elongation-load graph within elastic limit is shown in figure, of three wires A, B and C made of same material and of same length. The thickest wire is:

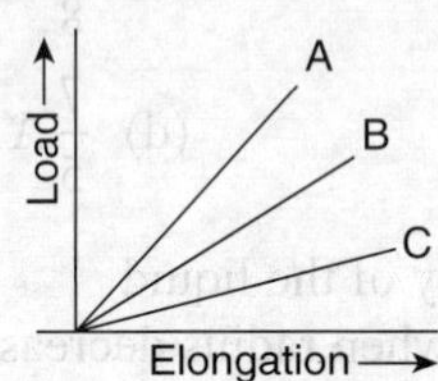

(a) A (b) B
(c) C (d) None of these.

46. A steel rod [Young's modulus= 2×10^{11} Nm^{-2}] has an area of cross-section 3×10^{-4} m^2 and length 1m. A force of 6×10^4 N stretches it axially. The elongation of the rod is
(a) 10^{-4} m (b) 5×10^{-3}
(c) 10^{-3} m (d) 5×10^{-2} m

47. A copper wire of length 1 m and a steel wire of length 0.5 m are fastened end to end and stretched by a load M. They have the same area of cross-section. Their Young's modulii are 1×10^{11} Nm^{-2} and 2×10^{11} Nm^{-2} respectively. If the copper wire stretches by 1 mm, the total extension of the combined wire is
(a) 1.25 cm (b) 12.5 cm
(c) 0.125 cm (d) zero

48. The potential energy U between the two molecules as a function of the distance x between them is shown in figure.

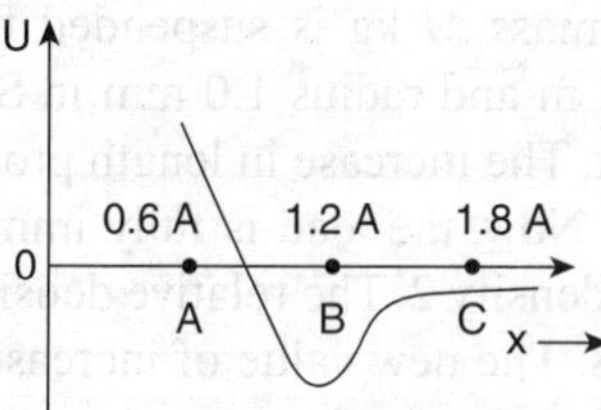

A, B and C are points for which x = 0.6 Å, 1.2 Å and 1.8 Å. At A, B and C force between the two molecules respective is
(a) zero, attractive, repulsive
(b) attractive, zero, repulsive
(c) zero, repulsive, attractive
(d) repulsive, zero, attractive.

49. One end of a uniform wire of length L and of weight w is attached rigidly to a point in the roof and a weight w_1 is suspended from its lower end. If s is the area of cross-section of the wire, the stress in the wire at height $\left(\frac{3L}{4}\right)$ from its lower end is
(a) $w_1 s$ (b) $\left(w_1+\frac{w}{4}\right)s$
(c) $\frac{\left(w_1+\frac{w}{4}\right)}{s}$ (d) $\frac{(w_1+w)}{s}$

50. A material has a Poisson's ratio 0.20. If a uniform rod of it suffers a longitudinal strain of 2×10^{-3}, then the percentage changes in volume is
(a) +0.28 (b) +0.12
(c) –0.12 (d) – 0.28

51. Find the amount of work done in twisting a steel wire of radius 1mm and length 25 cm through an angle of 45°, the modulus of rigidity of steel being 8×10^{10} N/m^2.
(a) 17×10^{-2} J (b) 20×10^{-2} J
(c) 15.5×10^{-2} J (d) 18×10^{-2} J

52. Young's moduli of two wires A and B are in the ratio 7 : 4. Wire A is 2 m long and has radius R. Wire B is 1.5 m long and has radius 2 mm. If the two wires stretch by the same length for a given load, then the value of R is close to
(a) 1.3 mm (b) 1.5 mm
(c) 1.9 mm (d) 1.7 mm

53. A steel wire having a radius of 2.0 mm, carrying a load of 4 kg, is hanging from a ceiling. Given that g = 3.1 πms^{-2}, what will be the tensile stress that would be developed in the wire?

(a) 6.2×10^6 Nm^{-2} (b) 5.2×10^6 Nm^{-2}
(c) 3.1×10^6 Nm^{-2} (d) 4.8×10^6 Nm^{-2}

54. A load of mass M kg is suspended from a steel wire of length 2 m and radius 1.0 mm in Searle's apparatus experiment. The increase in length produced in the wire is 4.0 mm. Now, the load is fully immersed in a liquid of relative density 2. The relative density of the material of load is 8. The new value of increase in length of the steel wire is
(a) zero (b) 5.0 mm
(c) 4.0 mm (d) 3.0 mm

55. The pressure of a medium is changed from 1.01×10^5 Pa to 1.165×10^5 Pa and change in volume is 10% keeping temperature constant. The bulk modulus of the medium is
(a) 204.8×10^5 Pa (b) 102.4×10^5 Pa
(c) 51.2×10^5 Pa (d) 1.55×10^5 Pa

56. The adjacent graph shows the extension (Δl) of a wire of length 1 m suspended from the top of a roof at one end and with a load w connected to the other end. If the cross-sectional area of the wires is 10^{-6} m^2, calculate from the graph the Young's modulus of the material of the wire.

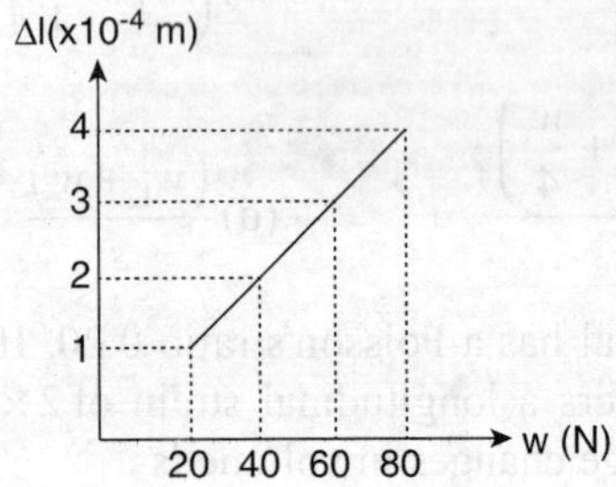

(a) 2×10^{11} N/m^2 (b) 2×10^{-11} N/m^2
(c) 3×10^{12} N/m^2 (d) 2×10^{13} N/m^2

Fluids

57. By sucking through a straw, a student can reduce the pressure in his lungs to 750 mm of Hg (density = 13.6 g cm^3). Using the straw, he can drink water from a glass up to a maximum depth of
(a) 10 cm (b) 75 cm
(c) 136 cm (d) 1.36 cm

58. A steel ball is dropped in oil, then
(a) the ball attains constant velocity after some time
(b) the ball stops
(c) the speed of ball will keep on increasing
(d) None of these.

59. There is a small bubble at one end and bigger bubble at other end of rod. What will happen?

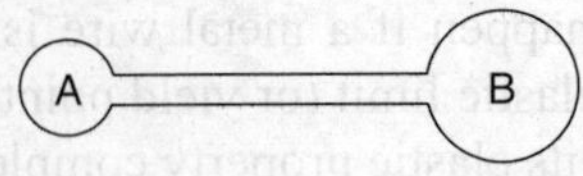

(a) Smaller bubble will grow until it collapses
(b) Bigger bubble will grow until it collapses
(c) Remain in equilibrium
(d) None of these.

60. A sphere of mass M and radius R is dropped in a liquid, then terminal velocity of sphere is proportional to
(a) R (b) $\frac{1}{R}$
(c) R^2 (d) $\frac{1}{R^2}$

61. Two drops of equal radius coalesce to form a bigger drop. What is the ratio of surface energy of bigger drop of smaller one?
(a) $2^{1/2}:1$ (b) $1:1$
(c) $2^{1/3}:1$ (d) None of these.

62. One drop of soap bubble of diameter D breaks into 27 drops having surface tension T. The change in surface energy is
(a) $2\pi TD^2$ (b) $4\pi TD^2$
(c) πTD^2 (d) $8\pi TD^2$

63. Two capillaries of lengths L and $2L$ and of radii R and $2R$ respectively are connected in series. The net rate of flow of fluid through them will be
(a) $\frac{8}{9}X$ (b) $\frac{9}{8}X$
(c) $\frac{5}{7}X$ (d) $\frac{7}{5}X$

64. Critical velocity of the liquid
(a) decreases when radius decreases
(b) increases when radius increases
(c) decreases when density decreases
(d) increases when density increases.

65. The rate of flow of liquid through an orifice of a tank does not depend upon
(a) the size of orifice
(b) density of liquid
(c) the height of fluid column
(d) acceleration due to gravity.

66. The rate of flow of liquids in a tube of radius r, length l, whose ends are maintained at a pressure difference P is $V = \frac{\pi QPr^4}{\eta l}$, where η is coefficient of viscosity and Q is

(a) 8 (b) $\frac{1}{8}$

(c) 16 (d) $\frac{1}{16}$

67. The terminal velocity v_r of a small steel ball of radius r falling under gravity through a column of a viscous liquid of coefficient of viscosity η depends on mass of the ball m, acceleration due to gravity g, coefficient of viscosity η and radius r. Which of the following relations is dimensionally correct?

(a) $v_r \propto \frac{mgr}{\eta}$ (b) $v_r \propto mg\eta r$

(c) $v_r \propto \frac{mg}{r\eta}$ (d) $v_r \propto \frac{\eta mg}{r}$

68. A tank is filled with water of density 1g per cm^3 and oil of density 0.9 g per cm^3. The height of water layer is 100 m and of the oil of layer is 400 cm. If g = 980 cm/s^2, then the velocity of efflux from an opening in the bottom of the tank is:

(a) $\sqrt{900 \times 980}$ cm/s (b) $\sqrt{1000 \times 980}$ cm/s

(c) $\sqrt{920 \times 980}$ cm/s (d) $\sqrt{950 \times 980}$ cm/s

69. Water is flowing through a pipe of uniform cross-section. At some point the pipe becomes narrow and the cross-section is haived. The speed of water is

(a) reduced to zero
(b) decreased by factor of 2
(c) increased by factor of 2
(d) unchanged.

70. In streamline flow of liquid, the total energy of liquid is constant at:

(a) all points (b) inner points
(c) outer points (d) None of these.

71. An iceberg is floating in water. The density of ice in the iceberg is 917 kg/m^3 and the density of water is 1024 kg/m^3. What percentage fraction of the iceberg would be visible?

(a) 5% (b) 10%
(c) 12% (d) 8%

72. A water tank, open to the atmosphere, has a leak in it in the form of a circular hole, located at a height h below the open surface of water. The velocity of the water coming out of the hole is

(a) $\sqrt{\frac{gh}{2}}$ (b) $\sqrt{gh}$

(c) $\sqrt{2gh}$ (d) $2\sqrt{gh}$

73. A horizontal pipe of cross sectional diameter 5 cm carries water at velocity of 4 m/s. The pipe is connected to a smaller pipe with a cross-sectional diameter 4 cm. The velocity of water through the smaller pipe is

(a) 6.25 m/s (b) 5.0 m/s
(c) 3.2 m/s (d) 2.56 m/s

74. Energy needed in the breaking of a drop of radius R into n drops of radius r is given by:

(a) $(4\pi r^2 n - 4\pi R^2)S$ (b) $\left(\frac{4}{3}\pi r^3 n - \frac{4}{3}\pi R^3\right)S$

(c) $(4\pi R^2 - 4\pi r^2)nS$ (d) $(4\pi r^2 n - 4\pi r^2)P$

75. A thread is tied slightly loose to a wipe frame as in figure and the frame is dipped into a soap solution and taken out. The frame is completely covered with the film. When the portion A is punctured with a pin, the thread

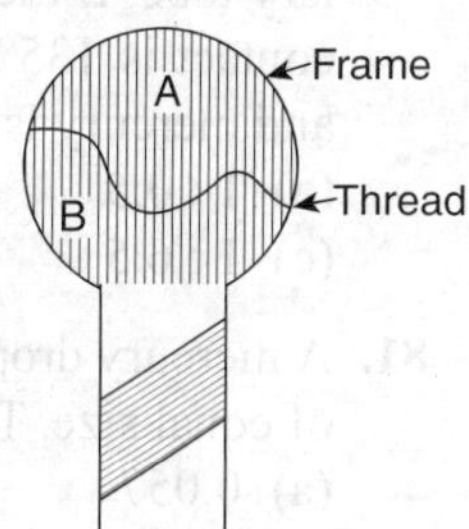

(a) becomes concave towards A
(b) becomes convex towards A
(c) either (a) or (b) depending on the size of A with respect to B
(d) remains in the initial position.

76. Water is flowing smoothly through a pipe as shown in the figure. Rank the four numbered sections of the pipe according to the water pressure P, greater first

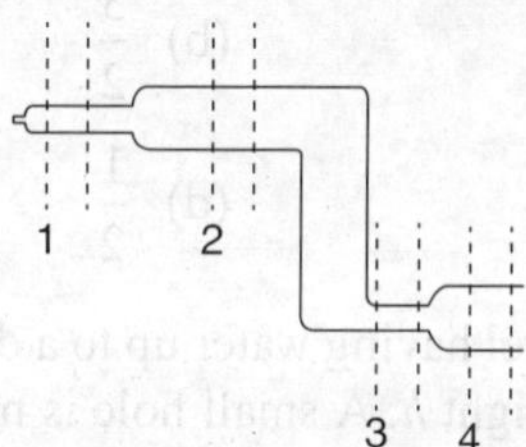

(a) $P_1 < P_2 < P_3 < P_4$ (b) $P_1 > P_2 > P_3 > P_4$
(c) $P_4 < P_3 < P_2 < P_1$ (d) $P_4 < P_1 < P_2 < P_3$

77. A water film is formed between two parallel wires of 10 cm length. The distance of 0.5 cm between the wires is increased by 1 mm. What will be the work done? (Surface tension of water = 72 N/m)

(a) 288 erg (b) 144 erg
(c) 72 erg (d) 36 erg

78. Viscosity is most closely related to

(a) friction
(b) adhesive molecular force
(c) cohesive molecular force
(d) Bernoulli's theorem.

79. The radii of the two columns in U-tube are r_1 and r_2. When a liquid of density ρ is filled in it the level difference of liquid in two arms is h. The surface tension of liquid is:

(a) $\frac{\rho g h r_1 r_2}{2(r_2 - r_1)}$ (b) $\frac{\rho g h(r_2 - r_1)}{2 r_1 r_2}$

(c) $\frac{2(r_2 - r_1)}{\rho g h r_1 r_2}$ (d) $\frac{\rho g h}{2(r_2 - r_1)}$

80. Water rises to a height of 10 cm in a capillary tube and mercury falls to a depth of 3.42 cm in the same capillary tube. If the density of mercury is 13.6 and angle of contact is 135°. The ratio of surface tension for water and mercury is :

(a) 1 : 0.5 (b) 1 : 3
(c) 1 : 6.5 (d) 1.5 : 1

81. A mercury drop of radius 1 cm is sprayed into 10^6 drops of equal size. The energy expanded in joule is:

(a) 0.057 (b) 5.7
(c) 5.7×10^{-4} (d) 5.7×10^{-6}

82. Tanks A and B open at the top contain two different liquids up to certain height in them. A hole is made to the wall of each tank at a depth h from the surface of the liquid. The area of the hole is A, is twice that in B. If the liquid mass flux through each hole in equal. Then the ratio the densities of the liquids respectively is :

(a) $\frac{2}{1}$ (b) $\frac{3}{2}$

(c) $\frac{2}{3}$ (d) $\frac{1}{2}$

83. A water barrel having water up to a depth d is placed on a table of height h. A small hole is made on the wall of the barrel at s bottom. If the stream of water coming out of the hole falls on the ground at a horizontal distance R from the barrel, then the value of d is:

(a) $\frac{4h}{R^2}$ (b) $4hR^2$

(c) $\frac{R^2}{4h}$ (d) $\frac{h}{4R^2}$

84. At what speed, the velocity head of water is equal to pressure head 40 cm of Hg?

(a) 10.3 m/s (b) 2.8 m/s
(c) 5.6 m/s (d) 8.4 m/s

85. A soap bubble A of radius 0.03 and another bubble B radius 0.04 m are brought together so that the combined bubble has a common interface of radius r, then the value of r is

(a) 0.24 m (b) 0.48 m
(c) 0.12 m (d) None of these.

86. An air bubble of radius 10^{-2} m is rising up at a steady rate of 2×10^{-3} m/s through a liquid of density 1.5×10^3 kg/m^3, the coefficient of viscosity neglecting the density of air, will be ($g = 10$ m/s^2)

(a) 23.2 units (b) 83.5 units
(c) 334 units (d) 167 units

87. Water is flowing in a pipe of diameter 4 cm with a velocity 3 m/s . The water then enters into a pipe of diameter 2 cm. The velocity of water in the other pipe is

(a) 3 m/s (b) 6 m/s
(c) 12 m/s (d) 8 m/s

88. The surface tension of a liquid is 5 N/m. If a film is held on a ring of area 0.02 m^2, its total surface energy is about

(a) 2×10^{-2} J (b) 2.5×10^{-2} J
(c) 2×10^{-1} J (d) 3×10^{-1} J

89. A viscous fluid is flowing through a cylindrical tube. The velocity distribution of the fluid is best represented by the diagram

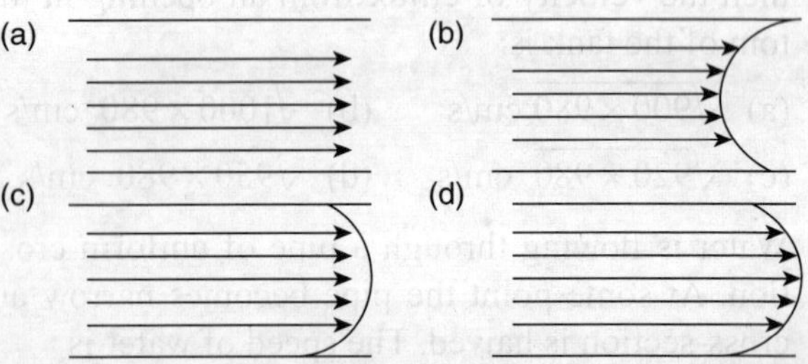

90. If the work done in blowing a bubble of volume V is W, then the work done in blowing a soap bubble of volume 2 V will be

(a) W (b) 2 W
(c) $\sqrt{2}W$ (d) $4^{1/3}$ W

91. A rain drop of radius 0.3 mm has terminal velocity 1 m/sec and the viscosity of air is 18×10^{-5} poise. The viscous force on the drop is

(a) 16.95×10^{-9} N (b) 1.695×10^{-9} N
(c) 10.17×10^{-9} N (d) 101.74×10^{-9} N

92. Pressure inside two soap bubbles are 1.01 and 1.02 atmosphere. Ratio between their volumes is

(a) 102 : 101 (b) $(102)^3 : (103)^3$
(c) 8 : 1 (d) 2 : 1

93. When the free surface area of a liquid is increased, the surface tension of the liquid

(a) increase (b) decrease
(c) remains unchanged (d) None of these.

94. Pressure inside the two soap bubbles is 1.01 and 1.02 atmosphere. Ratio between their volumes is

(a) 102 : 101 (b) $(102)^3 : (101)^3$
(c) 8 : 1 (d) 2 : 1

95. The surface tension of soap solution is 25×10^{-3} N/m. The excess pressure inside a soap bubble of diameter 1 cm is
(a) 40 pa (b) 20pa
(c) 10 pa (d) 5 pa

96. The shape of a liquid drop becomes spherical due to
(a) surface tension of the liquid
(b) density of the liquid
(c) viscosity of the liquid
(d) temperature of the air.

97. A manometer connected to a closed tap reads 4.5×10^5 Pa. When the tap is opened the reading of the manometer falls to 4.5×10^5 Pa. Then the velocity of flow of water is
(a) 7 ms^{-1} (b) 8 ms^{-1}
(c) 9 ms^{-1} (d) 10 ms^{-1}

98. The neck and the bottom of a bottle are 3 cm and 15 cm in radius respectively. If the cork is pressed with a force 12 N in the neck of the bottle, then force exerted on the bottom of the bottle is
(a) 30 N (b) 150 N
(c) 300 N (d) 600 N

99. A metal plate of area $10^3 cm^2$ rest on a layer of oil 6 mm thick. A tangential force of 10^{-2} N is applied on it to move it with a constant velocity of 6 cm^{-1}. The coefficient of viscosity of the liquid is
(a) 0.1 poise (b) 0.5 poise
(c) 0.7 poise (d) 0.9 poise

100. Sixty-four spherical rain drops of equal size are falling vertically through air with a terminal velocity 1.5 ms^{-1}. If these drops coalesce to form a big spherical drop, then terminal velocity of big drop is
(a) 8 ms^{-1} (b) 16 ms^{-1}
(c) 24 ms^{-1} (d) 32 ms^{-1}

101. A cylinder of height 20 m is completely filled with water. The velocity of efflux of water (in ms^{-1}) through a hole on the side wall of the cylinder near its bottom, is
(a) 10 (b) 20
(c) 25.5 (d) 5

102. A liquid flows through a pipe of non-uniform cross-section. If A_1 and A_2 are the cross-sectional areas of the pipe at two points, the ratio of velocities of the liquid at these points will be
(a) A_1A_2 (b) $\frac{A_1}{A_2}$
(c) $\frac{A_2}{A_1}$ (d) $\frac{1}{A_1A_2}$

103. Water rises in a capacity tube to a height h. Choose the false statement regarding a capillary rise from the following:
(a) On the surface of Jupiter, height will be less than h
(b) In a lift moving up with constant acceleration, height is less than h
(c) On the surface of moon, the height is more than h
(d) In a lift moving down with constant acceleration, height is less than h.

104. A jar is filled with two non-mixing liquids 1 and 2 having densities ρ_1 and ρ_2, respectively. A solid ball, made of a material density ρ_3, is dropped in the jar. It comes to equilibrium in the position shown in the figure.

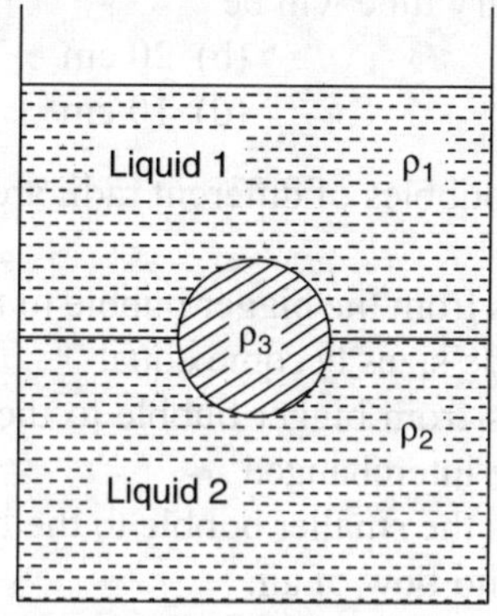

Which of the following is true for ρ_1, ρ_2 and ρ_3?
(a) $\rho_1 > \rho_2 > \rho_3$ (b) $\rho_1 < \rho_2 < \rho_3$
(c) $\rho_1 < \rho_3 < \rho_2$ (d) $\rho_3 < \rho_1 < \rho_2$

105. A capillary (A) is dipped in water. Another identical tube (B) is dipped in a soap-water solution. Which of the following shows the relative nature of the liquid columns in the two tubes?

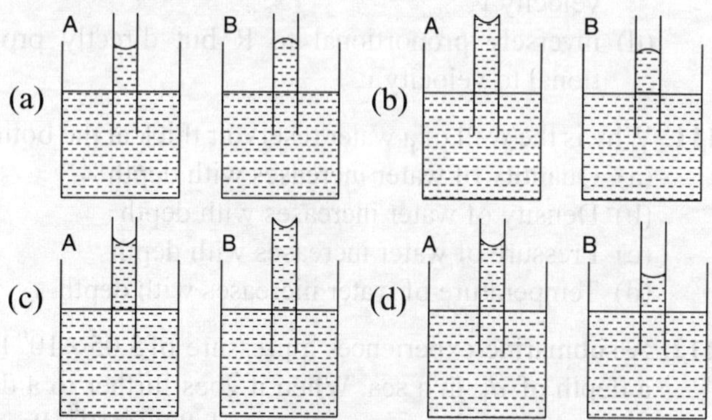

106. A spherical solid ball of volume V is made of material of density ρ_1. It is falling through a liquid of density ($\rho_2 < \rho_1$). Assume that the liquid applies a viscous force on the ball that is proportional to the square of its speed v, i.e., $F_{viscous} = -kv^2 (k > 0)$. The terminal speed of the ball is
(a) $\frac{Vg(\rho_1 - \rho_2)}{k}$ (b) $\sqrt{\frac{Vg(\rho_1 - \rho_2)}{k}}$
(c) $\frac{Vg\rho_1}{k}$ (d) $\sqrt{\frac{Vg\rho_1}{k}}$

107. If the terminal speed of a sphere of gold (density = 19.5 kg/m^3) is 0.2 m/s in a viscous liquid (density = 1.5 kg/m^3) then the terminal velocity of a silver sphere (density = 10.5 kg/m^3) of the same size in the same liquid is
(a) 0.2 m/s (b) 0.4 m/s
(c) 0.133 m/s (d) 0.1 m/s

108. A 20 cm long capillary tube is dipped in water. The water rises up to 8 cm. If the entire arrangement is put in a freely falling elevator, the length of water column in the capillary tube will be
(a) 4 cm (b) 20 cm
(c) 8 cm (d) 10 cm

109. If two soap bubbles of different radii are connected to a tube,
(a) air flows from the bigger bubble to the smaller bubble till the sizes become equal
(b) air flows from bigger bubble to the smaller till the sizes are interchanged
(c) air flows the smaller bubble to the bigger
(d) there is no flow of air.

110. Spherical balls of radius R are falling in a viscous fluid of viscosity η with a velocity v. The retarding viscous force acting on the spherical ball is
(a) directly proportional to R but inversely proportional to v
(b) directly proportional to both radius R and velocity v.
(c) inversely proportional to both radius R and velocity v.
(d) inversely proportional to R but directly proportional to velocity v.

111. Why is the wall of a water reservoir thick at the bottom?
(a) Quantity of water increases with depth
(b) Density of water increases with depth
(c) Pressure of water increases with depth
(d) Temperature of water increases with depth.

112. A submarine experiences a pressure of 5.05×10^6 Pa at a depth of d_1 in a sea. When it goes further to a depth of d_2, it experiences a pressure of 8.08×10^6 Pa, then $d_2 - d_1$ is approximately (density of water = 10^3 kg/m^3 and acceleration due to gravity = 10 ms^{-2})
(a) 500 m (b) 400 m
(c) 600 m (d) 300 m

113. A wooden block, with a coin placed on its top, floats in water as shown in figure. The distance l and h are shown there. After sometime, the coin falls into the water. Then

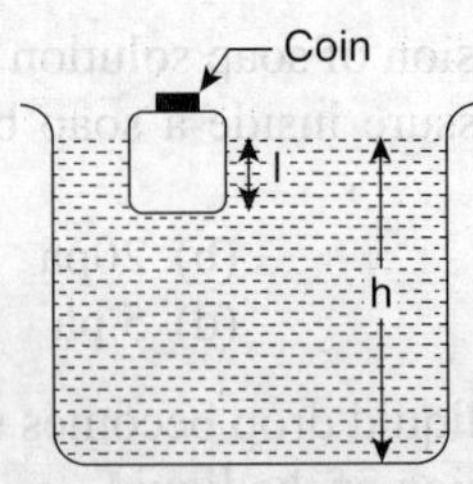

(a) l decreases and h increases
(b) l increases and h decreases
(c) both l and h increase
(d) both l and h decrease.

114. A vessel contains oil (density =0.8 g/cm^3) over mercury (density = 13.6 g/cm^3). A homogeneous sphere floats with half its volume immersed in mercury and the other half in oil. The density of the material of the sphere in g/cm^3 is
(a) 3.3 (b) 6.4
(c) 7.2 (d) 12.8.

115. A tank is filled with water up to a height H. Water is allowed to come out of a small hole P in one of the wall at a depth h below the surface of water. The water stream strikes the ground at a distance x from the tank x is given by

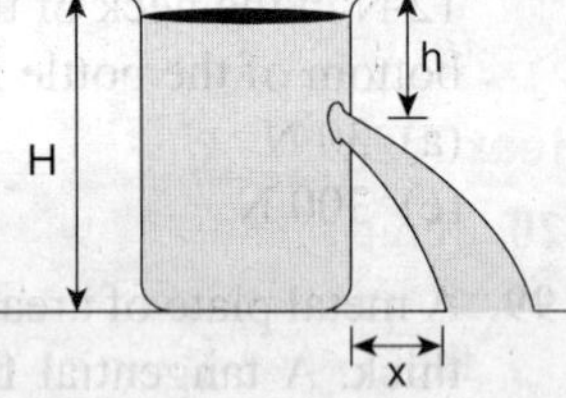

(a) $x = \sqrt{h(H-h)}$ (b) $x = \sqrt{h(H-h)/2}$
(c) $x = 2\sqrt{h(H-h)}$ (d) $x = 4\sqrt{h(H-h)}$.

116. The expression for the height of capillary rise between two parallel plates dipping in a liquid of density ρ separated by a distance d is (The surface tension of the liquid is T)
(a) $h = \dfrac{T}{\rho g d}$ (b) $h = \dfrac{2T}{\rho g d}$
(c) $h = \dfrac{T}{2\rho g d}$ (d) None of these.

117. A solid is completely immersed in a liquid. The force exerted by the liquid on the solid will
(a) increase if it is pushed deeper inside the liquid
(b) change if its orientation is changed keeping it inside
(c) decrease if it is taken partially out of the liquid
(d) be in the horizontal direction only.

118. From amongst the following curves, which one shows the variation of the velocity v with time t for a small sized spherical body falling vertically in a long column of viscous liquid?

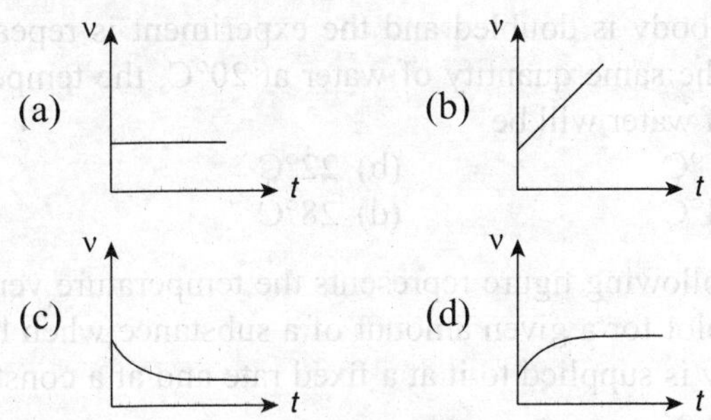

119. The journey of a small spherical solid ball dropped in a viscous liquid is best described by

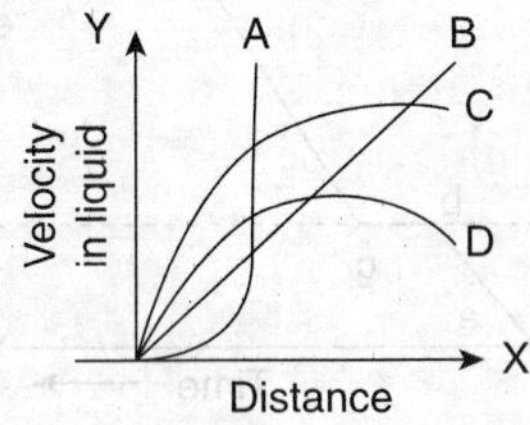

(a) curve A (b) curve B
(c) curve C (d) curve D

Heat Transfer

120. Temperature of two stars are in ratio 3 : 2. If wavelength of maximum intensity of first body is 4000Å, what is the corresponding wavelength of second body?
(a) 9000 Å (b) 6000 Å
(c) 2000 Å (d) 8000 Å

121. If boiling point of water is 95°F what will be reduction at Celsius scale?
(a) 7°C (b) 65°C
(c) 63°C (d) 35°C

122. A composite rod made of copper ($\alpha = 1.8 \times 10^{-5}\,K^{-1}$) and steel ($\alpha = 1.2 \times 10^{-5}\,K^{-1}$) is heated, then
(a) it bends with steel in concave side
(b) it bends with copper on concave side
(c) it does not expand
(d) data is insufficient.

123. A black body at a temperature of 227°C radiates heat at the rate of 20 cal $m^{-2}\,s^{-1}$ when its temperature rises to 727°C, the rate of heat radiated will be
(a) 40 cal $m^{-2}\,s^{-1}$ (b) 160 cal $m^{-2}\,s^{-1}$
(c) 320 cal $m^{-2}\,s^{-1}$ (d) 640 $m^{-2}\,s^{-1}$

124. 50 g of ice at 0°C is mixed with 50 g of water at 80°C final temperature of the mixture will be
(a) 0°C (b) 40°C
(c) 60°C (d) 4°C

125. Which of the following graphs correctly represents the relation between E and in T, where E is the amount of radiation emitted per unit time from unit area of body and T is the absolute temperature?

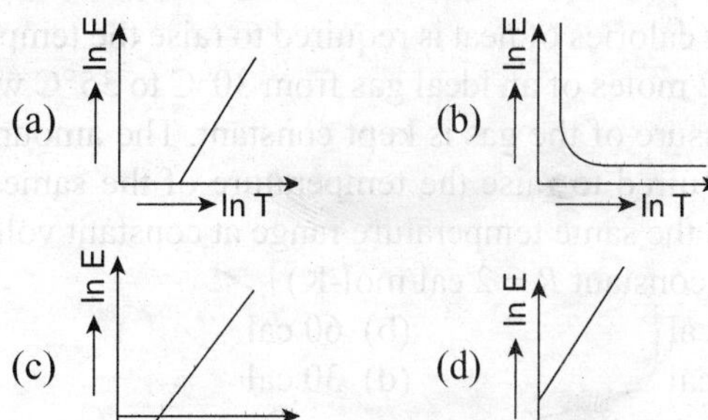

126. In a refrigerator the defrosting is done when frosting of ice takes place in the freezer compartment. What is correct to say?
(a) It is done to save mechanical damage to freezer
(b) It is done to reduce the temperature by melting ice
(c) It is done to pump out heat more efficiently from refrigerator
(d) It is done to save electricity.

127. The radiation from the sun, incident normally at the surface of the earth is 20 kcal/m^2 min. What would be the radiant energy, incident normally on the earth if the sun had a temperature, twice of the present one?
(a) 80 kcal/m^2 min (b) 320 kcal/m^2 min
(c) 160 kcal/m^2 min (d) 40 kcal/m^2 min

128. Three objects colored black, grey and white can withstand hostile conditions at 2800°C. These objects are thrown into furnace where each of them attains a temperature of 2000°C. Which object will glow brightest?
(a) The white object
(b) Black object
(c) All glow with equal brightness
(d) Grey object

129. A wire 3 m in length and 1 mm diameter at 30°C is kept in a low temperature at –170°C and is stretched by hanging a weight of 10 kg at one end. The change in length of the wire is [$Y = 2 \times 10^{11}\,N/m^2, g = 10 m/s^2$ and $\alpha = 1.2 \times 10^{-5}$ /°C]
(a) 5.2 mm (b) 2.5 mm
(c) 52 mm (d) 25 mm

130. An iron bar of length 10 m is heated from 0°C to 100°C. If the coefficient of linear thermal expansion of ions is 10×10^{-6} /°C the increase in the length of bar is
(a) 0.5 cm (b) 1.0 cm
(c) 1.5 cm (d) 2.0 cm

131. The radiation energy density per unit wavelength at a temperature T has a maximum at a wavelength λ_0. At temperature 2 T, it will have a maximum at a wavelength
(a) $4\lambda_0$ (b) $2\lambda_0$
(c) $\frac{\lambda_0}{2}$ (d) $\frac{\lambda_0}{4}$

132. Seventy calories of heat is required to raise the temperature of 2 moles of an ideal gas from 30°C to 35°C while the pressure of the gas is kept constant. The amount of heat required to raise the temperature of the same gas through the same temperature range at constant volume is (Gas constant $R = 2$ cal/mol-K)
(a) 70 cal (b) 60 cal
(c) 50 cal (d) 30 cal

133. Solar radiation emitted by the sun resembles that emitted by a black body at a temperature of 6000 K. Maximum intensity is emitted at a wavelength of about 4800 Å. If the sun were to cool down from 6000 K to 3000 K, then the peak intensity would occur at a wavelength
(a) 4800 Å (b) 9600 Å
(c) 7200 Å (d) 6400 Å

134. Air is pumped into an automobile tube up to a pressure of 200 k Pa in the morning when the air temperature is 22°C. During the day, temperature rises to 42°C and the tube expanded by 2%. The pressure of the air in the tube at this temperature, will be approximately
(a) 212 k Pa (b) 209 k Pa
(c) 206 k Pa (d) 200 k Pa

135. Two rods (one semi-circular and other straight) of same material and of same cross-sectional area are joined as shown in the figure. The points A and B are maintained at different temperatures. The ratio of the heat transferred through a cross-section of a semi-circular rod to the heat transferred through a cross-section of the straight rod in a given time, is

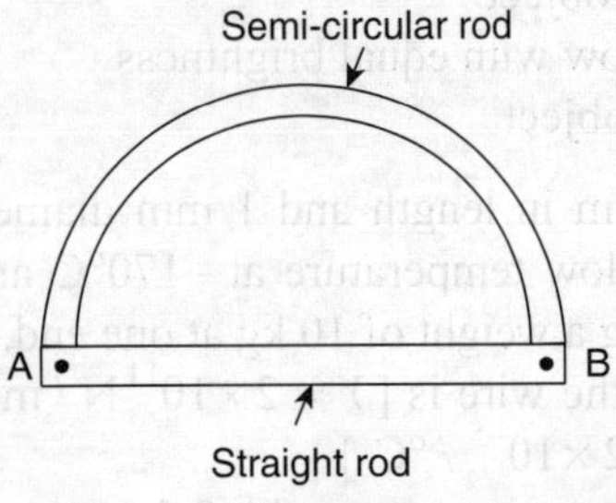

(a) $2:\pi$ (b) $1:2$
(c) $\pi:2$ (d) $3:2$

136. Two rods of the same length and material transfer a given amount of heat in 12 s when they are joined lengthwise. They will transfer the same amount of heat, in the same conditions, in
(a) 1.5 s (b) 3 s
(c) 24 s (d) 48 s

137. The total energy radiated from a black body source is collected for one minute and is used to heat a quantity of water. The temperature of water, is found to increase from 20°C to 20.5°C. If the absolute temperature of the black body is doubled and the experiment is repeated with the same quantity of water at 20°C, the temperature of water will be
(a) 21°C (b) 22°C
(c) 24°C (d) 28°C

138. The following figure represents the temperature versus time plot for a given amount of a substance when heat energy is supplied to it at a fixed rate and at a constant pressure.

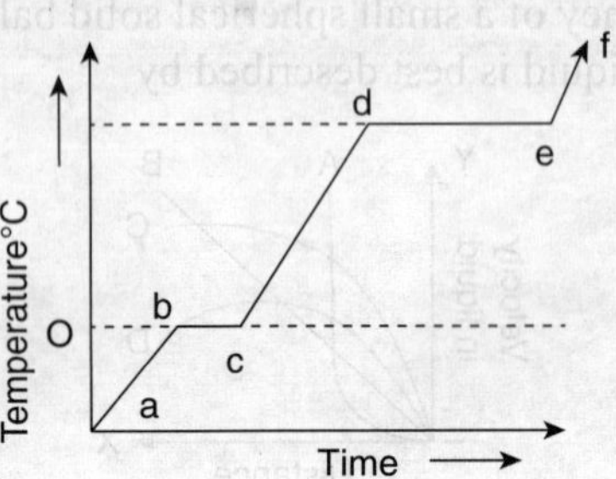

Which parts of the above plot represent a phase change?
(a) a to b and e to f (b) b to c and c to d
(c) d to e and e to f (d) b to c and d to e

139. A brass sphere immersed completely in water at 0°C weights 1.5 kg. If water is heated to 50°C, with sphere immersed in it all through, the weight of sphere now is
(a) 1.5 kg (b) > 1.5 kg
(c) < 1.5 kg (d) None of these.

140. Total energy emitted by a perfectly black body is directly proportional to T^n, where n is
(a) 1 (b) 2
(c) 3 (d) 4

141. Heat travels through vacuum by
(a) radiation (b) conduction
(c) convection (d) None of these.

142. An iron piece falls from a height of 1 km on the ground. If all the energy is converted into heat, the temperature of the iron piece will be (specific heat of iron = 0.1 k cal/kg °C)
(a) 0.233°C (b) 2.33°C
(c) 23.8°C (d) 233°C

143. A black body radiates energy at the rate of E watt/m² at a high temperature T K. When the temperature is reduced to $\left(\frac{T}{2}\right)$ K, the radiant energy is
(a) $\frac{E}{2}$ (b) 2 E
(c) $\frac{E}{4}$ (d) $\frac{E}{16}$

144. The relation between the coefficient of real expansion (γ_r) and coefficient of apparent expansion (γ_a) of a liquid and the coefficient of linear expansion (α_g) of the material of the container is
(a) $\gamma_r = \alpha_g + \gamma_a$ (b) $\gamma_r = \alpha_g + 3\gamma_a$
(c) $\gamma_r = 3\alpha_g + \gamma_a$ (d) $\gamma_r = 3(\alpha_g + \gamma_a)$

145. Two identical bodies have temperature 277°C and 67°C. If the surrounding temperature is 27°C, the ratio of loss of heats of the two bodies during the same interval of time is (approximately)
(a) 4 : 1 (b) 7 : 1
(c) 12 : 1 (d) 19 : 1

146. A black body of mass 34.88 g and surface area 19.2 cm^2 is at an initial temperature of 400 K. It is allowed to cool inside an evacuated enclosure kept at constant temperature 300 K. The rate of cooling is 0.04°C per second. The specific heat of the body in Jkg^{-1} K^{-1} is (Stefan's constant $\sigma = 5.73 \times 10^{-8}$ Wm^{-2}K^{-4})
(a) 2800 (b) 2100
(c) 1400 (d) 1200

147. The temperature of 5 moles of a gas at a constant volume is changed from 100°C to 120°C. The change in internal energy is 80 J. The total heat capacity of the gas at constant volume will be J/K
(a) 8 (b) 4
(c) 0.8 (d) 0.4

148. When temperature of a black body increases, it is observed that the wavelength corresponding to maximum energy changes from 0.26 m to 0.13 μ m. The ratio of the emissive power of the body at the temperature is
(a) $\frac{16}{1}$ (b) $\frac{4}{1}$
(c) $\frac{1}{4}$ (d) $\frac{1}{16}$

149. The coefficient of apparent expansion of a liquid, when determined using two different vessels A and B, are γ_1 and γ_2 respectively. If the coefficient of linear expansion of the vessel A is α the coefficient of linear expansion of vessel β is
(a) $\frac{\alpha\gamma_1 + \gamma_2}{\gamma_1 + \gamma_2}$ (b) $\frac{\gamma_1 - \gamma_2}{2\alpha}$
(c) $\frac{\gamma_1 - \gamma_2 + \alpha}{3}$ (d) $\frac{\gamma_1 - \gamma_2}{3} + \alpha$

150. A body cools from 60°C to 50°C in 10 minutes. If the room temperature is 25°C and assuming Newton's law of cooling to hold good, the temperature of the body at the end of the next 10 minutes will be
(a) 45°C (b) 42.85°C
(c) 40°C (d) 38.5°C

151. The ratio of the coefficient of thermal conductivity of two different materials is 5 : 3. If the thermal resistance of the rods of the same thickness of these materials is same, then the ratio of the length of these rods will be
(a) 3 : 5 (b) 5 : 3
(c) 3 : 4 (d) 3 : 2

152. Two circular discs A and B with equal radii are blackened. They are heated to same temperature and are cooled under identical conditions. What inference do you draw from their cooling curves?

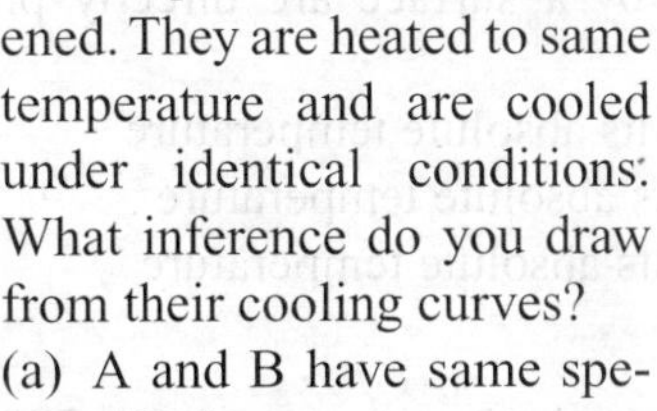

(a) A and B have same specific heats
(b) Specific heat of A is less
(c) Specific heat of B is less
(d) Nothing can be said.

153. The snow on the mountains does not melt all at once when it is heated by the sun because it
(a) becomes very hard
(b) reflects most of the heat from the sun
(c) has low specific heat capacity
(d) has a high latent heat of fusion.

154. The colour of a star is an indication of its
(a) weight (b) distance
(c) surface temperature (d) size.

155. A liquid cools from 50°C to 45°C in 5 minutes and from 45°C to 41.5°C in the next 5 minutes. The temperature of its surrounding is
(a) 27°C (b) 40.3°C
(c) 23.3°C (d) 33.3°C

156. The ratio of radii of two spheres of same material is 1 : 4. The ratio of their heat capacities will be
(a) $\frac{1}{64}$ (b) $\frac{1}{32}$
(c) $\frac{1}{2}$ (d) $\frac{1}{4}$

157. At 127°C a body radiates energy 10^3 J/s. At what temperature would it radiate energy 10^7 J/s.
(a) 400 K (b) 4000 K
(c) 80000 K (d) 40000 K

158. The wavelength of maximum energy, released during an atomic explosion was 2.93×10^{-10} m. Given that the Wien's constant is 2.93×10^{-3} m-K. The maximum temperature attained must be of the order

(a) 10^{-7} K (b) 10^{7} K
(c) 10^{-3} K (d) 5.86×10^{7} K

159. A gas in an air tight container is heated from 25°C to 90°C. The density of gas will
(a) increase slightly
(b) remain the same
(c) increase considerably
(d) decrease slightly.

160. Radiations emitted by a surface are directly proportional to
(a) fourth power of its absolute temperature
(b) third power of its absolute temperature
(c) twice power of its absolute temperature
(d) its temperature.

161. A perfect body emits radiation at temperature 300 K. If it radiates 16 times this amount, then the temperature will be
(a) 300 K (b) 600 K
(c) 1200 K (d) 4800 K

162. An object is cooled from 75°C to 65°C in two minutes in a room temperature at 30°C. The time taken to cool the same object from 55°C to 45°C in the same room (in minute) is
(a) 7 (b) 6
(c) 5 (d) 4

163. An ice box made of styrofoam (thermal conductivity = 0.01 J^{-1} s^{-1} K^{-1}) is used to keep liquids cool. It has a total wall area including lid of 0.8 m^2 and wall thickness of 2.0 cm. A bottle of water is placed in the box and filled with ice. If the outside temperature is 30°C, the rate of flow of heat into the box is (in J s^{-1})
(a) 16 (b) 14
(c) 12 (d) 10

164. The temperature at which centigrade and Fahrenheit scales give same reading is
(a) –40° (b) 40°
(c) –30° (d) 30°

165. A brass rod of length 500 mm and diameter 300 mm is joined to a steel rod of same length and diameter at 50°C. If the coefficients of linear expansion of brass and steel are $2.5\times10^{-5}\,°C^{-1}$ and $1.25\times10^{-5}\,°C^{-1}$ then change in the length of the combined rod at 200°C is
(a) 2.4 mm (b) 2.8 mm
(c) 3.2 mm (d) 3.6 mm

166. A copper block of mass 4 kg is heated in a furnace to a temperature 425°C and then placed on a large ice block. The mass of ice that will melt in this process will be (specific heat of copper = 500 $Jkg^{-1}\,°C^{-1}$ and heat of fusion of ice = 336 kJ kg^{-1})
(a) 0.5 kg (b) 1 kg
(c) 1.5 kg (d) 2.5 kg

167. A block of ice at –10°C is slowly heated and converted to steam at 100°C. What of the following curves represents this phenomenon qualitatively?

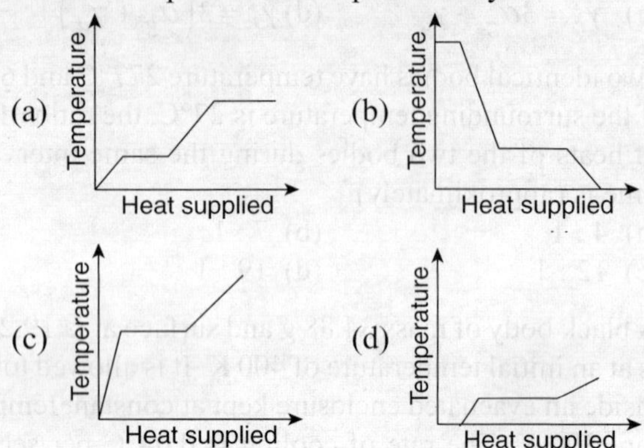

168. One mole of an ideal gas at an initial temperature of T K does 6 R joules of work adiabatically. If the ratio of the specific heats of this gas at constant pressure and at constant volume is 5/3, then final temperature of the gas will be
(a) $(T-4)$ K (b) $(T+4)$ K
(c) $(T-2.4)$ K (d) $(T+2.4)$ K

169. In which mode of transmission, the heat waves travel along straight line with the speed of light?
(a) Thermal radiation (b) Forced convection
(c) Natural convection (d) Thermal conduction

170. Consider a compound slab consisting of two materials having equal length, thickness and thermal conductivities K and 2 K respectively. The equivalent thermal conductivity of the slab is
(a) $\sqrt{2}$ K (b) 3 K
(c) $\frac{4}{3}$ K (d) $\frac{2}{3}$ K

171. An ideal black body at room temperature is thrown into a furnace. It is observed that
(a) it is the darkest body at all times
(b) it cannot be distinguished at all times
(c) initially it is the darkest body and later it becomes brightest
(d) initially it is darkest body and later it cannot be distinguished.

172. If temperature of the sun were to increases from T to 2 T and its radius from R to 2 R, then how many times the radiant energy will be received on Earth?
(a) 4 (b) 16
(c) 32 (d) 64

173. A black body has a wavelength of λ and temperature 2000 K. Its corresponding wavelength at 3000 K will be

(a) $\frac{2\lambda}{3}$ (b) $\frac{3\lambda}{2}$

(c) $\frac{4\lambda}{9}$ (d) $\frac{9\lambda}{4}$

174. One kg of ice at 0°C is mixed with 1 kg of water at 10°C. The resulting temperature will be

(a) between 0°C and 10°C
(b) equal to 0°C
(c) less than 0°C
(d) greater than 10°C.

175. The ratio of densities of two substance is 2 : 3 and their specific heats are in the ratio 3 : 4. The ratio of their thermal capacities for unit volume is

(a) 1 : 1 (b) 1 : 2
(c) 2 : 1 (d) 8 : 9

176. Which one of the following would raise the temperatures of 20 gm of water at 30°C most when added to water? (specific heat of copper = 0.1 cal/gm°C)

(a) 20 gm of water at 40°C
(b) 80 gm of water at 27°C
(c) 1000 gm of copper at 25°C
(d) 4 gm of water at 80°C

177. 10 gm of ice at –20°C is dropped into a calorimeter containing 10 gm of water at 10°C.The specific heat of water is twice that of ice. When equilibrium is reached, the calorimeter will contain

(a) 10 gm of ice and 10 gm of water
(b) 20 gm of water
(c) 5 gm of ice and 15 gm of water
(d) 20 gm of ice

178. Two liquids, A and B are at 32°C and 24°C respectively. When mixed in equal masses the temperature of the mixture is found to be 28°C. Their specific heats are in the ratio of

(a) 3 : 2 (b) 2 : 3
(c) 1 : 1 (d) 4 : 3

179. The temperature of equal masses of three different liquids A, B and C are 12°C, 19°C and 28°C respectively. The temperature when A and B are mixed is 16°C and when B and C are mixed, it is 23°C. Approximately what will be the temperature when A and C are mixed

(a) 22°C (b) 17°C
(c) 24°C (d) 14°C

180. Boiling water is changing into steam. Under this condition the specific heat of water is

(a) zero (b) one
(c) infinite (d) less than one.

181. Two copper cylinders of mass 100 gm and 200 gm respectively and of same height are heated at 80°C. If they are placed vertically on two large blocks of ice, then the ratio of their depths of penetration in the ice assuming no heat loss is

(a) 1 : 1 (b) 1 : 2
(c) 2 : 1 (d) 2 : 3

182. If there are no heat losses, the heat released by the condensation of x gm of steam at 100°C into water at 100°C can be used to convert y gm of ice at 0°C into water at 100°C. Then the ratio $y : x$ is nearly

(a) 1 : 1 (b) 2 : 1
(c) 2.5 : 1 (d) 3 : 1

183. A cold coke bottle is left open on the pan of a balance and its weight observed from time to time, the weight

(a) increases
(b) decreases
(c) increases, reaches a maximum and then starts decreasing
(d) remains stationary.

184. A spherical iron ball is placed on a large block of dry ice at 0°C. The ball sinks into the ice until it is half submerged. What was the temperature of iron ball?
Density if iron = 7675 kg m^{-3}
Density of ice = 921 kg m^{-3}
Specific heat of iron = 504 J kg^{-1} K^{-1}
Specific latent heat of fusion ice = 336×10^3 J kg^{-1}

(a) 30°C (b) 35°C
(c) 40°C (d) 45°C

185. 200 gm of water and an equal volume of another liquid of mass 250 gm are placed in turn in the same calorimeter of mass 100 gm and specific heat capacity 420 J kg^{-1} K^{-1}. The liquids are continuously stirred and are allowed to cool from 60°C to 40°C in 180 second and 140 seconds respectively. The temperature of surrounding is 20°C. The specific heat of liquid is

(a) 2576 J kg^{-1} (b) 1875 J kg^{-1}
(c) 1570 J kg^{-1} (d) 1976 J kg^{-1}

186. Newton's law of cooling is used in laboratory for the determination of

(a) specific heat of gases
(b) latent heat of gases
(c) specific heat of liquids
(d) latent heat of liquids.

187. A cylinder of radius R made of material of thermal conductivity k_1 is surrounded by a cylindrical shell of inner radius R and outer radius $2R$ made of material of thermal conductivity k_2. The two ends of the combined system are maintained at two different temperatures. There is no loss of heat across the cylindrical surface and the system is in steady state. The effective thermal conductivity of the system is

(a) $k_1 + k_2$ (b) $\frac{(k_1k_2)}{(k_1+k_2)}$

(c) $\frac{k_1+3k_2}{4}$ (d) $\frac{(3k_1+3k_2)}{4}$

188. A slab consists of two parallel layers of two different materials of same thickness and having thermal conductivities k_1 and k_2. The equivalent thermal conductivity of the slab is

(a) $k_1 + k_2$ (b) k_1k_2

(c) $\frac{k_1+k_2}{k_1k_2}$ (d) $\frac{2k_1k_2}{k_1+k_2}$

189. Heat is flowing through two cylindrical rods of same material. The diameter of the rod is in the ratio 1 : 2 and the lengths in the ratio of 2 : 1. If the temperature difference between the ends be the same, then the ratio of rate of flow of heat through them will be

(a) 1 : 8 (b) 8 : 1
(c) 1 : 1 (d) 2 : 1

190. Two identical objects A and B are at temperature T_A and T_B respectively. Both objects are placed in a room with perfectly absorbing walls maintained at a temperature $T(T_A > T_B)$. The objects A and B attain the temperature T eventually. Select the correct statements from the following:

(a) A only emits radiation, while B only absorbs it until both attain the temperature T.
(b) A loses more heat by radiation than it absorbs, while B absorbs more radiation than it emits until they attain temperature T.
(c) Both A and B only absorbs radiation but do not emit it until they attain the temperature T.
(d) Each object continuous to emit and absorb radiation even after attaining the temperature T.

191. A slab of ice has one half covered with black cloth and the other half with white cloth. This is then placed in sunlight. After some time the pieces of cloth are removed. Which of the following statements is correct?

(a) Ice has melted equally under both pieces
(b) More ice has melted under white cloth
(c) More ice has melted under black cloth
(d) It will depend upon the medium in which ice is placed.

192. Two walls of thickness d_1 and d_2 and thermal conductivities k_1 and k_2 are in contact. In the steady-state if the temperatures at the outer surface are T_1 and T_2, the temperature at the common wall is

(a) $\frac{k_1T_1d_2 + k_2T_2d_1}{(k_1d_2 + k_2d_1)}$ (b) $\frac{k_1T_1 + k_2T_2}{(d_2 + d_1)}$

(c) $\left(\frac{k_1d_1 + k_2d_2}{T_1 + T_2}\right)T_1T_2$ (d) $\frac{(k_1T_1d_1 + k_2T_2d_2)}{(k_1d_1 + k_2d_2)}$

193. Two rods of same length and area of cross-section A_1 and A_2 have their ends at the same temperature. If k_1 and k_2 are their thermal conductivities, C_1 and C_2 are their specific heats and d_1 and d_2 are their densities, then the rate of flow of heat is the same in both ends, if

(a) $\frac{A_1}{A_2} = \frac{k_1}{k_2}$ (b) $\frac{A_1}{A_2} = \frac{k_1C_1d_1}{k_2C_2d_2}$

(c) $\frac{A_1}{A_2} = \frac{k_1 2C_1d_1}{k_2C_2d_2}$ (d) $\frac{A_1}{A_2} = \frac{k_2}{k_1}$

194. The temperature of a body is increased from 27°C to 127°C. The radiation emitted by it increases by a factor of

(a) $\frac{256}{81}$ (b) $\frac{15}{9}$

(c) $\frac{4}{5}$ (d) $\frac{12}{27}$

195. On investigation of light from three different stars A, B and C, it was found that the intensity of red colour is maximum in the spectrum of A, the intensity of blue colour is maximum in B, and in C, the intensity of yellow colour is maximum. From these observations it can be concluded that

(a) the temperature of A is maximum, B is minimum and C is intermediate
(b) the temperature of A is maximum, C is minimum and B is intermediate
(c) the temperature of B is maximum, B is minimum and C is intermediate
(d) the temperature of C is maximum, B is minimum and A is intermediate.

196. According to Wien's law

(a) $\lambda_m T = \text{Constant}$ (b) $\frac{\lambda_m}{T} = \text{Constant}$

(c) $\frac{T}{\lambda_m} = \text{Constant}$ (d) $T + \lambda_m = \text{Constant}$

197. Certain substances emit only the wavelength λ_1, λ_2, λ_3 and λ_4 when it is at high temperature. When this substance is at a colder temperature, it will absorb only the following wavelengths

(a) λ_1 (b) λ_2
(c) λ_1 and λ_2 (d) λ_1, λ_2, λ_3 and λ_4

198. A bucket full of hot water is kept in a room and it cools from 75°C to 70°C in T_1 minutes from 70°C to 65°C in T_2 minutes and from 65°C to 60°C in T_3 minutes. Then
(a) $T_1 = T_2 = T_3$ (b) $T_1 < T_2 < T_3$
(c) $T_1 > T_2 > T_3$ (d) $T_1 < T_2 > T_3$

199. The temperature of a furnace is 2327°C and the intensity is maximum in its radiation spectrum nearly at 1200 Å. If the intensity in the spectrum of a star is maximum nearly at 4800 Å, then the surface temperature of the star is
(a) 8400°C (b) 7200°C
(c) 6500 K (d) 5900°C

200. The most appropriate material for a cooking pot is the one having
(a) high specific heat and low conductivity
(b) high specific heat and high conductivity
(c) low specific heat and low conductivity
(d) low specific heat and high conductivity.

201. Solar radiation emitted by the sun resembles that emitted by black body at a temperature of 6000 K. Maximum intensity is emitted at a wavelength of about 4800 Å. If the sun were to cool down from 6000 K to 3000 K, then the peak intensity would occur at a wavelength of
(a) 4800 Å (b) 9600 Å
(c) 2400 Å (d) 19200 Å

202. Ice starts forming in a lake with water at 0°C when the atmospheric temperature is –10°C. If the time taken for the first 1 cm of ice to be formed is 7 hrs, then the time taken for the thickness of ice to change from 1 cm to 2 cm is
(a) 7 hours (b) 14 hours
(c) 21 hours (d) 3.5 hours

203. Which of the following statements is true/correct?
(a) During clear nights, the temperature rises steadily upwards near the ground level.
(b) Newton's law of cooling, an approximate form of Stefan's law, is valid for only natural convection
(c) The total energy emitted by a black body per unit time per unit area is proportional to the square of its temperature in the Kelvin scale.
(d) Two spheres of the same material have radii 1 m and 4 m and temperature 4000 K and 2000 K respectively. The energy radiated per second by the first sphere is greater than that generated per second by the second sphere.

204. In a 10 m deep lake, the bottom is at a constant temperature of 4°C. The air temperature is constant at –4°C. The thermal conductivity of ice is three times that of water. Neglecting the expansion of water in freezing, the maximum thickness of ice will be
(a) 2.5 m (b) 5 m
(c) 6 m (d) 7.5 m

205. The power radiated by a black body is P and it radiates maximum energy around the wavelength λ_0. If the temperature of the black body is now changed, so that it radiates maximum energy around the wavelength $\frac{3\lambda_0}{4}$, the power radiated by it will increase by a factor of
(a) $\frac{4}{3}$ (b) $\frac{16}{9}$
(c) $\frac{64}{27}$ (d) $\frac{256}{81}$

206. When water is heated from 0°C to 10°C its volume
(a) increases
(b) decreases
(c) does not change
(d) first decreases, then increases.

207. Two spheres of the same radii are made from the same material. One is hollow and the other is solid. If they are heated from 20°C to 100°C,then
(a) both will expand equally
(b) hollow sphere will expand more
(c) solid sphere will expand more
(d) None of these.

208. On heating liquid of cubical expansion γ in a container having coefficient of linear expansion $\frac{\gamma}{3}$, the level of liquid in the container will
(a) rise (b) fall
(c) remain stationary (d) be unpredictable.

209. Two rods of lengths l_1 and l_2 are made of materials whose coefficients of linear expansion are α_1 and α_2. If the difference between the two lengths is independent of temperature
(a) $\frac{l_1}{l_2} = \left(\frac{\alpha_1}{\alpha_2}\right)$ (b) $\frac{l_1}{l_2} = \left(\frac{\alpha_2}{\alpha_1}\right)$
(c) $l_1^2\alpha_1 = l_2^2\alpha_2$ (d) $\alpha_1^2 l_1 = \alpha_2^2 l_2$

210. Two metal strips that constitute a thermostat must necessarily differ in their
(a) mass
(b) length
(c) resistivity
(d) coefficient of linear expansion.

211. Assuming the sun to be a spherical body of radius R at a temperature of T K, evaluate the total radiant power incident on the earth, at a distance r from the sun.
(a) $4\pi r_0^2 R^2 \sigma T^4 / r^2$ (b) $\pi r_0^2 R^2 \sigma T^4 / r^2$
(c) $r_0^2 R^2 \sigma T^4 / 4\pi r^2$ (d) $R^2 \sigma T^4 / r^2$

212. The system shows a system of two concentric spheres of radii r_1 and r_2 and kept at temperature T_1 and T_2 respectively. The radial state of flow of heat in a substance between the two concentric spheres is proportional to

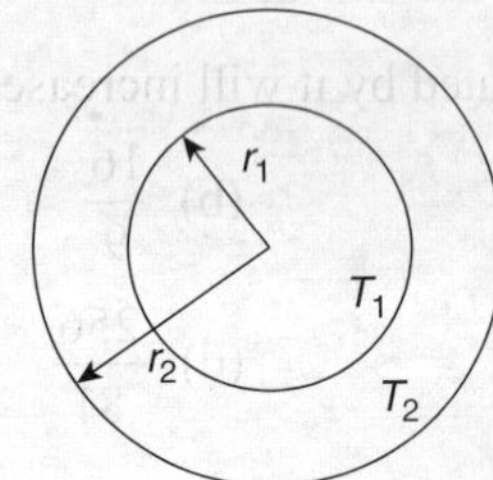

(a) $\frac{(r_2 - r_1)}{r_1 r_2}$ (b) $\frac{r_2}{r_1}$
(c) $\frac{r_1 r_2}{(r_2 - r_1)}$ (d) $(r_2 - r_1)$

213. A metallic ball and highly stretched spring are made of the same material and have the same mass. They are heated so that they melt, the latent heat required
(a) are the same for both
(b) is greater for the ball
(c) is greater for the spring
(d) for the two may or may not be the same depending upon the metal.

214. A radiation of energy E falls normally on a perfectly reflecting surface. The momentum transferred to the surface is
(a) E/c (b) $2E/c$
(c) Ec (d) E/c^2

215. The temperature of the two outer surfaces of a composite slab, consisting of two materials having coefficient of thermal conductivity K and $2K$ and thickness X and $4X$ respectively are T_1 and T_2 $(T_2 > T_1)$. The rate of heat transfer through the slab in a steady state is $\left(\frac{A(T_2 - T_1)K}{x}\right)f$ with f equal to

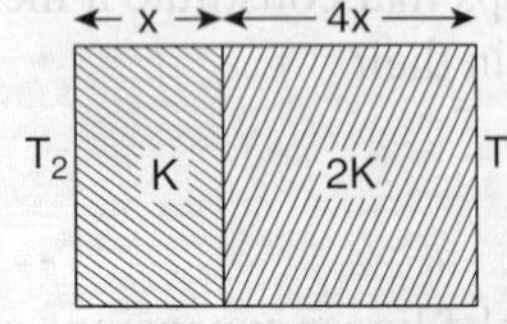

(a) 1 (b) 1/2
(c) 2/3 (d) 1/3

216. According to Newton's law of cooling, the rate of cooling of a body is proportional to $(\Delta\theta)^n$, where $\Delta\theta$ is the difference of the temperature of the body and the surroundings and n is equal to
(a) two (b) three
(c) four (d) one.

217. Which of the following is more close to a black body?
(a) Black board paint (b) Green leaves
(c) Black holes (d) Red roses

218. If 1 g of steam is mixed with 1 g of ice, then resultant temperature of the mixture is
(a) 100°C (b) 230°C
(c) 270°C (d) 50°C

219. The radiant energy from the sun, incident normally at the surface of earth is 20 kcal/m² min. What would have been the radiant energy, incident normally on the earth if the sun had a temperature twice that of the present one?
(a) 320 kcal/m² min (b) 40 kcal/m² min
(c) 160 kcal/m² min (d) 80 kcal/m² min

220. A black body is at a temperature of 500 K. It emits energy at a rate which is proportional to
(a) $(500)^3$ (b) $(500)^4$
(c) 500 (d) $(500)^2$.

221. A beaker full of hot water is kept in a room. If it cools from 80°C to 75°C in t_1 minutes, from 75°C to 70°C in t_2 minutes and from 70° C to 65° C in t_3 minutes, then
(a) $t_1 < t_2 < t_3$ (b) $t_1 > t_2 > t_3$
(c) $t_1 = t_2 = t_3$ (d) $t_1 < t_2 = t_3$.

222. Heat is flowing through two cylindrical rods of the same material. The diameters of the rods are in the ratio 1 : 2 and the lengths in the ratio 2 : 1. If the temperature difference between the ends is same, then ratio of the rate of flow of heat through them will be
(a) 2 : 1 (b) 8 : 1
(c) 1 : 1 (d) 1 : 8.

223. If the temperature of the sun is doubled, the rate of energy received on earth will be increased by a factor of
(a) 2 (b) 4
(c) 8 (d) 16

224. A Centigrade and a Fahrenheit thermometer are dipped in boiling water. The water temperature is lowered until the Fahrenheit thermometer registers 140° F. What is the fall in temperature as registered by the centigrade thermometer?
(a) 80° C (b) 60° C
(c) 40° C (d) 30° C

225. Thermal capacity of 40 g of aluminium ($s = 0.2$cal/g K) is
(a) 168 J/K (b) 672 J/K
(c) 840 J/K (d) 33.6 J/K

226. 10 gm of ice cubes at 0°C are released in tumbler (water equivalent 55 g) at 40°C. Assuming that negligible heat is taken from the surroundings, the temperature of water in the tumbler becomes nearly ($L = 80$ cal/g)
(a) 31°C (b) 22°C
(c) 19°C (d) 15°C

227. For the measurement of temperature of the order of 400°C, the preferred temperature will be
(a) Hg thermometer
(b) alcohol thermometer
(c) radiation thermometer
(d) thermocouple.

228. For measuring temperature near absolute zero the thermometer used is
(a) thermoelectric thermometer
(b) radiation thermometer
(c) magnetic thermometer
(d) resistance thermometer.

229. The mode of transfer heat which requires no medium, is called as
(a) radiation (b) conduction
(c) combustion (d) convection.

230. If K and σ respectively are the thermal and electrical conductivities of a metal absolute temperature T then
(a) $\frac{K}{\sigma T}$ = constant (b) $\frac{K}{\sigma}$ = constant
(c) $\frac{K}{T}$ = constant (d) $\frac{\sigma}{KT}$ = constant

231. Though air is bad conductor, a body kept in air losses heat quickly. This is due to
(a) conduction (b) convection
(c) radiation (d) None of these.

232. In a closed room, heat transfer takes place by:
(a) conduction (b) convection
(c) radiation (d) All of these.

233. A body of length 1m having cross-sectional area 0.75 m^2 has thermal conductivity 6000 L/s then the temperature difference if $K = 200$ Jm^{-1}K^{-1}, will be
(a) 20°C (b) 40°C
(c) 80°C (d) 100°C

234. A black body is maintained at 27°C and 927°C. What will be the ratio of radiation emitted?
(a) 1 : 4 (b) 1 : 16
(c) 1 : 16 (d) 1 : 256

235. One end of a metal rod is kept in steam. In steam state, the temperature gradient $\left(\frac{d\theta}{dx}\right)$
(a) may be variable (b) must be constant
(c) must be variable (d) None of these.

236. If in two identical containers, equal quantity of ice melts completely in 30 and 20 minutes respectively, then the ratio of the thermal conductivities of the material of two containers is
(a) 1: 1 (b) 1 : 2
(c) 3 : 2 (d) 2 : 3

237. Three slabs are placed in contact in order 1, 2 and 3 as shown in figure. The thickness of the slabs are x_1, x_2 and x_3 and made of materials of thermal conductivities K_1, K_2 and K_3 respectively. Assume that there is no less of heat due to radiation. At steady state, the combination will behave as a single material, if thermal conductivity K will be

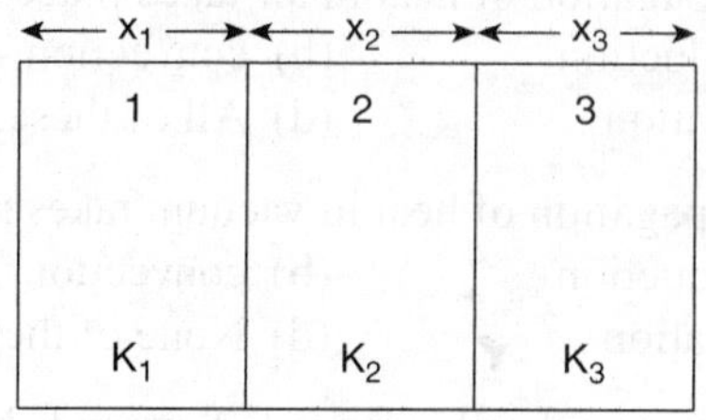

(a) $K = \dfrac{(x_1 + x_2 + x_3)}{\left(\dfrac{x_1}{K_1} + \dfrac{x_2}{K_2} + \dfrac{x_3}{K_3}\right)}$

(b) $K = \dfrac{2(x_1 + x_2 + x_3)}{\dfrac{x_1}{K_1} + \dfrac{x_2}{K_2} + \dfrac{x_3}{K_3}}$

(c) $K = \dfrac{(x_1 + x_2 + x_3)}{\left(\dfrac{x_1}{K_1} - \dfrac{x_2}{K_2} + \dfrac{x_3}{K_3}\right)}$

(d) None of these.

238. Choose the correct option: heat is flowing through a conductor length l from $x = 0$ to $x = l$. Assume thermal resistance per unit length is uniform.

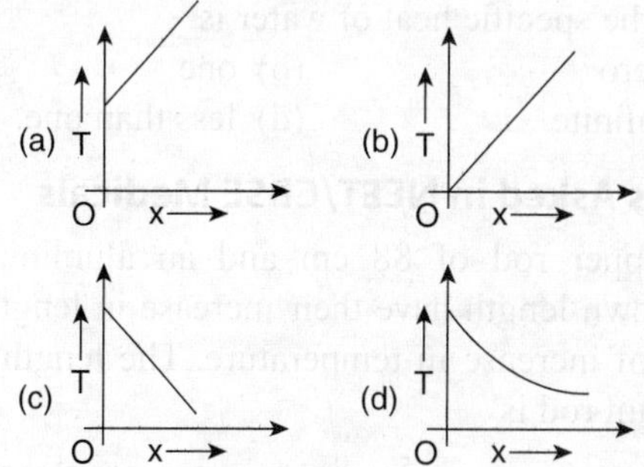

239. If the thermal conductivity of the material of the rod of length l is K, then the rate of heat flow through a tapering rod, tapering from radius r_1 and r_2, if the temperature of the ends are maintained at T_1 and T_2 is
(a) $\dfrac{\pi K r_1 r_2 (T_1 + T_2)}{l}$ (b) $\dfrac{\pi K r_1 r_2 (T_1 - T_2)}{l}$
(c) $\dfrac{\pi K r_1 r_2 (T_1 - T_2)}{2l}$ (d) $\dfrac{\pi K r_1^2 (T_1 - T_2)}{2l}$

240. Which process has maximum rate of heat transfer?
(a) Conduction
(b) Convection
(c) Radiation
(d) In all the above heat is transferred with the same velocity.

241. The propagation of heat in air takes place through
(a) conduction (b) convection
(c) radiation (d) All of these.

242. The propagation of heat in vacuum takes place by
(a) conduction (b) convection
(c) radiation (d) None of these.

243. When the metal rod is heated, it expands because
(a) the size of its atoms increases
(b) the distance among its atoms increases
(c) atmospheric air rushes into it
(d) The actual cause is still unknown.

244. If same amount of heat is supplied to two identical spheres (one is hollow and other is solid), then
(a) the expansion in hollow is greater than the solid
(b) the expansion in hollow is same as that in solid
(c) the expansion in hollow is lesser than the solid
(d) the temperature of both must be same.

245. A metal ball immersed in water weight W_1 at 0°C and W_2 at 50°C. The coefficient of cubical expansion of metal is less than that of water. Then
(a) $W_1 < W_2$ (b) $W_1 > W_2$
(c) $W_1 = W_2$ (d) data is not sufficient.

246. Boiling water is changing into steam. Under this condition, the specific heat of water is
(a) zero (b) one
(c) infinite (d) less than one.

Questions Asked in NEET/CBSE Medicals

247. A copper rod of 88 cm and an aluminium rod of unknown length have their increase in length independent of increase in temperature. The length of an aluminium rod is

$$(\alpha_{Cu} = 1.7\times10^{-5}\,\text{K}^{-1}, \alpha_{Al} = 2.2\times10^{-5}\,\text{K}^{-1})$$

(a) 68 cm (b) 6.8 cm
(c) 113.9 cm (d) 88 cm

248. A small hole of area of cross-section 2 mm² is present near the bottom of a fully filled open tank of height 2 m. Taking $g = 10$ m/s², the rate of flow of water through the open hole would be nearly
(a) 6.4×10^{-6} m³/s (b) 12.6×10^{-6} m³/s
(c) 8.9×10^{-6} m³/s (d) 2.23×10^{-6} m³/s

249. When a block of mass M is suspended by a long wire of length L, the length of the wire becomes $(L + l)$. The elastic potential energy stored in the extended wire is
(a) $\frac{1}{2}Mgl$ (b) Mgl
(c) Mgl (d) $\frac{1}{2}Mgl$

250. A soap bubble, having radius of 1 mm, is blown from a detergent solution having a surface tension of 2.5×10^{-2} N/m. The pressure inside the bubble equals at a point Z_0 below the free surface of water in a container. Taking $g = 10$ m/s², density of water = 10^3 kg/m³, the value of Z_0 is
(a) 0.5 cm (b) 100 cm
(c) 10 cm (d) cm

251. Two small spherical metal balls, having equal masses, are made from materials of densities ρ_1 and ρ_2 $(\rho_1 = 8\rho_2)$ and have radii of 1 mm and 2 mm respectively. They are made to fall vertically (from rest) in a viscous medium whose coefficient of viscosity equals η and whose density is $0.1\ \rho_2$. The ratio of their terminal velocities would be
(a) $\frac{79}{72}$ (b) $\frac{19}{36}$
(c) $\frac{39}{72}$ (d) $\frac{79}{36}$

252. The stress-strain curves are drawn for two different materials X and Y. It is observed that the ultimate strength point and the fracture point are close to each other for material X but are far apart for material Y. We can say that materials X and Y are likely to be (respectively)
(a) ductile and brittle (b) brittle and ductile
(c) brittle and plastic (d) plastic and ductile.

253. A small sphere of radius r falls from rest in a viscous liquid. As a result, heat is produced due to viscous force. The rate of production of heat when the sphere attains its terminal velocity, is proportional to
(a) r^3 (b) r^2
(c) r^5 (d) r^4

254. Two wires are made of the same material and have the same volume. The first wire has cross-sectional area A and the second wire has cross-sectional area $3A$. If the length of the first wire is increased by Δl on applying a force F, how much force is needed to stretch the second wire by the same amount?
(a) $9F$ (b) $6F$
(c) $4F$ (d) F

255. The power radiated by a black body is P and it radiates maximum energy at wavelength, λ_0. If the temperature of the black body is now changed so that it radiates maximum energy at wavelength) $\frac{3}{4}\lambda_0$, the power 4 radiated by it becomes n_p. The value of n is

(a) $\frac{3}{4}$ (b) $\frac{4}{3}$
(c) $\frac{256}{81}$ (d) $\frac{81}{256}$

256. Two rods A and B of different materials are welded together as shown in figure. Their thermal conductivities are K_1 and K_2. The thermal conductivity of the composite rod will be

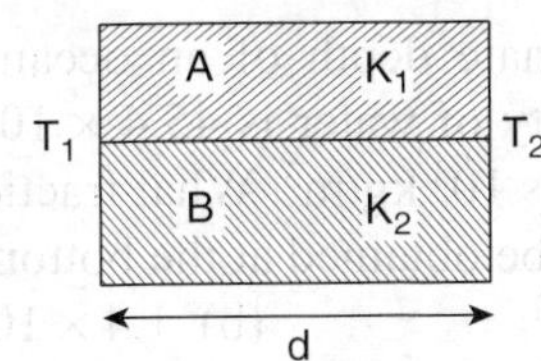

(a) $\frac{3(K_1+K_2)}{2}$ (b) K_1+K_2
(c) $2(K_1+K_2)$ (d) $\frac{K_1+K_2)}{2}$

257. A spherical black body with a radius of 12 cm radiates 450 watt power at 500 K. If the radius were halved and the temperature doubled, the power radiated in watt would be
(a) 450 (b) 1000
(c) 1800 (d) 225

258. A U tube with both ends open to the atmosphere, is partially filled with water. Oil, which is immiscible with water, is poured into one side until it stands at a distance of 10 mm above the water level on the other side. Meanwhile the water rises by 65 mm from its original level (see diagram). The density of the oil is

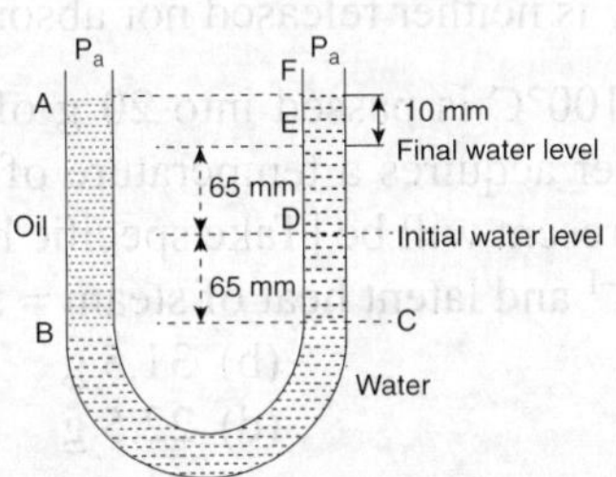

(a) 425 kg m^{-3} (b) 800 kg m^{-3}
(c) 928 kg m^{-3} (d) 650 kg m^{-3}

259. The bulk modulus of a spherical object is B. If it is subjected to uniform pressure p, the fractional decrease in radius is
(a) $\frac{B}{3p}$ (b) $\frac{3p}{B}$
(c) $\frac{P}{3B}$ (d) $\frac{P}{B}$

260. A rectangular film of liquid is extended from (4 cm × 2 cm) to (5 cm × 4 cm). If the work done is 3×10^{-4} J, the value of the surface tension of the liquid is
(a) 0.250 N m^{-1} (b) 0.125 N m^{-1}
(c) 0.2 N m^{-1} (d) 8.0 N m^{-1}

261. Three liquids of densities p_1, p_2 and p_3 (with $p_1 > p_2 > p_3$), having the same value of surface tension T, rise to the same height in three identical capillaries. The angles of contact θ_1, θ_2 and θ_3 obey
(a) $\pi/2 > \theta_1 > \theta_2 > \theta_3 \geq 0$
(b) $0 \leq \theta_1 < \theta_2 < \theta_3 < \pi/2$
(c) $\pi/2 < \theta_1 < \theta_2 < \theta_3 < \pi$
(d) $\pi > \theta_1 > \theta_2 > \theta_3 > \pi/2$

262. Two identical bodies are made of a material for which the heat capacity increases with temperature. One of these is at 100°C, while the other one is at 0°C. If the two bodies are brought into contact, then, assuming no heat loss, the final common temperature is
(a) 50°C
(b) more than 50°C
(c) less than 50°C but greater than 0°C
(d) 0°C

263. A body cools from a temperature 3 T to 2 T in 10 minutes. The room temperature is T. Assume that Newton's law of cooling is applicable. The temperature of the body at the end of next 10 minutes will be
(a) $\frac{7}{4}T$ (b) $\frac{3}{2}T$
(c) $\frac{4}{3}T$ (d) T

264. Coefficient of linear expansion of brass and steel rods are α_1 and α_2. Lengths of brass and steel rods are l_1 and l_2 respectively. If $(l_1 - l_2)$ is maintained same at all temperatures, which one of the following relations holds good?
(a) $\alpha_1^2 l_2 = \alpha_2^2 l_1$ (b) $\alpha_1 l_1 = \alpha_2 l_2$
(c) $\alpha_1 l_2 = \alpha_2 l_1$ (d) $\alpha_1 l_2^2 = \alpha_2 l_1^2$

265. A piece of ice falls from height h so that it melts completely. Only one-quarter of the heat produced is absorbed by the ice and all energy of ice gets converted into heat during its fall. The value of h is [latent heat of ice 3.4×10^5 J/kg and g = 10 N/kg]
(a) 136 km (b) 68 km
(c) 34 km (d) 544 km

266. A black body is at a temperature of 5760 K. The energy of radiation emitted by the body at wavelength 250 nm is U_1, at wavelength 500 nm is U_2 and that at 1000 nm

is U_3. Wien's constant, $b = 2.88 \times 10^6$ nm K. Which of the following is correct?
(a) $U_1 > U_2$ (b) $U_2 > U_1$
(c) $U_1 = 0$ (d) $U_3 = 0$

267. Two non-mixing liquids of densities ρ and $n\rho$ ($n > 1$) are put in a container. The height of each liquid is h. A solid cylinder of length L and density d is put in this container. The cylinder floats with its axis vertical and length pL ($p < 1$) in the denser liquid. The density d is equal to
(a) $\{2 + (n - 1)p\}\rho$ (b) $\{1 + (n - 1)p\}\rho$
(c) $\{1 + (n + 1)p\}\rho$ (d) $\{2 + (n + 1)p\}\rho$

268. The cylindrical tube of a spray pump has radius R, one end of which has n fine holes, each of radius r. If the speed of the liquid in the tube is V, the speed of the ejection of the liquid through the holes is
(a) $\frac{VR^2}{n^3r^2}$ (b) $\frac{V^2R}{nr}$
(c) $\frac{VR^2}{n^2r^2}$ (d) $\frac{VR^2}{nr^2}$

269. Water rises to a height h in capillary tube. If the length of capillary tube above the surface of water is made less than h, then
(a) water rises up to a point a little below the top and stays there.
(b) water does not rise at all.
(c) water rises up to the tip of capillary tube and then starts overflowing like a fountain.
(d) water rises up to the top of capillary tube and stays there without overflowing.

270. The value of coefficient of volume expansion of glycerin is 5×10^{-4} K^{-1}. The fractional change in the density of glycerin for a rise of 40° C in its temperature, is
(a) 0.025 (b) 0.010
(c) 0.015 (d) 0.020

271. The Young's modulus of steel is twice that of brass. Two wires of same length and of same area of cross section, one of steel and another of brass are suspended from the same roof. If we want the lower ends of the wires to be at the same level, then the weights added to the steel and brass wires must be in the ratio of
(a) 4 : 1 (b) 1 : 1
(c) 1 : 2 (d) 2 : 1

272. The two ends of a metal rod are maintained at temperatures 100°C and 110°C. The rate of heat flow in the rod is found to be 4.0 J/s. If the ends are maintained at temperatures 200°C and 210°C, the rate of heat flow will be
(a) 8.0 J/s (b) 4.0 J/s
(c) 44.0 J/s (d) 16.8 J/s

273. A wind with speed 40 m/s blows parallel to the roof of a house. The area of the roof is 250 m^2. Assuming that the pressure inside the house is atmospheric pressure, the force exerted by the wind on the roof and the direction of the force will be
($P_{air} = 1.2$ kg/m^3)
(a) 2.4×10^5 N, upwards
(b) 2.4×10^5 N, downwards
(c) 4.8×10^5 N, downwards
(d) 4.8×10^5 N, upwards

274. The approximate depth of an ocean is 2700 m. The compressibility of water is 45.4×10^{-11} Pa^{-1} and density of water is 10^3 kg/m^3. What fractional compression of water will be obtained at the bottom of the ocean?
(a) 1.2×10^{-2} (b) 1.4×10^{-2}
(c) 0.8×10^{-2} (d) 1.0×10^{-2}

275. Copper of fixed volume V is drawn into wire of length l. When this wire is subjected to a constant force F, the extension produced in the wire is Δl. Which of the following graphs is a straight line?
(a) Δl versus $1/l$ (b) Δl versus $l2$
(c) Δl versus $1/l2$ (d) Δl versus l

276. A certain number of spherical drops of a liquid of radius r coalesce to form a single drop of radius R and volume V. If T is the surface tension of the liquid, then
(a) energy $= 4VT\left(\frac{1}{r} - \frac{1}{R}\right)$ is released
(b) energy $= 3VT\left(\frac{1}{r} + \frac{1}{R}\right)$ is absorbed
(c) energy $= 3VT\left(\frac{1}{r} + \frac{1}{R}\right)$ is released
(d) energy is neither released nor absorbed.

277. Steam at 100°C is passed into 20 g of water at 10°C. When water acquires a temperature of 80°C, the mass of water present will be [Take specific heat of water = 1 cal g^{-1} °C^{-1} and latent heat of steam = 540 cal g^{-1}]
(a) 24 g (b) 31.5 g
(c) 42.5 g (d) 22.5 g

278. Certain quantity of water cools from 70°C to 60°C in the first 5 minutes and to 54°C in the next 5 minutes. The temperature of the surroundings is
(a) 45°C (b) 20°C
(c) 42°C (d) 10°C

279. A piece of iron is heated in a flame. It first becomes dull red then becomes reddish yellow and finally turns to white hot. The correct explanation for the above observation is possible by using
(a) Kirchhoff's law (b) Newton's law of cooling
(c) Stefan's law (d) Wien's displacement law.

280. The wettability of a surface by a liquid depends primarily on
(a) density
(b) angle of contact between the surface and the liquid
(c) viscosity
(d) surface tension.

281. The following four wires are made of the same material. Which of these will have the largest extension when the same tension is applied?
(a) length = 200 cm, diameter = 2 mm
(b) length = 300 cm, diameter = 3 mm
(c) length = 50 cm, diameter = 0.5 mm
(d) length = 100 cm, diameter = 1 mm

282. If the ratio of diameters, lengths and Young's modulus of steel and copper wires shown in the figure are p, q and s respectively, then the corresponding ratio of increase in their lengths would be

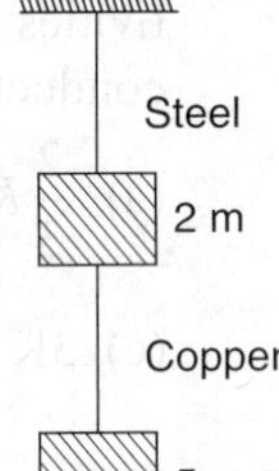

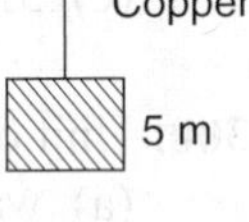

(a) $\dfrac{5q}{(7sp^2)}$ (b) $\dfrac{7q}{(5sp^2)}$
(c) $\dfrac{2q}{(5sp)}$ (d) $\dfrac{7q}{(5sp)}$

283. Two metal rods 1 and 2 of same lengths have same temperature difference between their ends. Their thermal conductivities are K_1 and K_2 and cross-sectional areas A_1 and A_2, respectively. If the rate of heat conduction in 1 is four times that in 2, then
(a) $K_1A_1 = 4K_2A_2$ (b) $K_1A_1 = {}_2K_2A_2$
(c) $4K_1A_1 = K_2A_2$ (d) $KA_1 = K_2A_2$

284. A fluid is in streamline flow across a horizontal pipe of variable area of cross-section. For this which of the following statements is correct?
(a) The velocity is maximum at the narrowest part of the pipe and pressure is maximum at the widest part of the pipe.
(b) Velocity and pressure both are maximum at the narrowest part of the pipe.
(c) Velocity and pressure both are maximum at the widest part of the pipe.
(d) The velocity is minimum at the narrowest part of the pipe and the pressure is minimum at the widest part of the pipe.

285. The density of water at 20° C is 998 kg/m^3 and at 40° C is 992 kg/m^3. The coefficient of volume expansion of water is
(a) 3×10^{-4}/°C (b) 2×10^{-4}/°C
(c) 6×10^{-4}/°C (d) 10^{-4}/°C

286. If the radius of a star is R and it acts as a black body, what would be the temperature of the star, in which the rate of energy production is Q?
(a) $\dfrac{Q}{4\pi R^2\sigma}$ (b) $\left(\dfrac{Q}{4\pi R^2\sigma}\right)^{-1/2}$
(c) $\left(\dfrac{4\pi R^2 Q}{\sigma}\right)^{1/4}$ (d) $\left(\dfrac{Q}{4\pi R^2\sigma}\right)^{1/4}$
(σ stands for Stefan's constant)

287. Liquid oxygen at 50 K is heated to 300 K at constant pressure of 1 atm. The rate of heating is constant. Which one of the following graphs represents the variation of temperature with time?

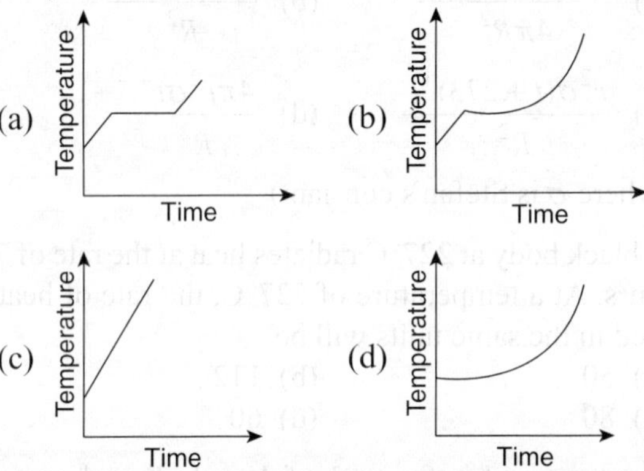

288. A slab of stone of area 0.36 m^2 and thickness 0.1 m is exposed on the lower surface to steam at 100° C. A block of ice at 0° C rests on the upper surface of the slab. In one hour 4.8 kg of ice is melted. The thermal conductivity of slab is (Given latent heat of fusion of ice = 3.36 × 105 J kg^{-1})
(a) 1.24 J/m/s/°C (b) 1.29 J/m/s/°C
(c) 2.05 J/m/s/°C (d) 1.02 J/m/s/°C

289. A cylindrical metallic rod in thermal contact with two reservoirs of heat at its two ends conducts an amount of heat Q in time t. The metallic rod is melted and the material is formed into a rod of half the radius of the original rod. What is the amount of heat conducted by the new rod, when placed in thermal contact with the two reservoirs in time t?
(a) $\dfrac{Q}{4}$ (b) $\dfrac{Q}{16}$
(c) $2Q$ (d) $\dfrac{Q}{2}$

290. The total radiant energy per unit area, normal to the direction of incidence, received at a distance R from the centre of a star of radius r, whose outer surface radiates as a black body at a temperature T K is given by

(a) $\frac{\sigma r^2 T^4}{R^2}$ (b) $\frac{\sigma r^2 T^4}{4\pi R^2}$

(c) $\frac{\sigma r^4 T^4}{R^4}$ (d) $\frac{4\pi\sigma r^2 T^4}{R^2}$

(where σ is Stefan's constant)

291. Assuming the sun to have a spherical outer surface of radius r, radiating like a black body at temperature t°C, the power received by a unit surface, (normal to the incident rays) at a distance R from the centre of the sun is

(a) $\frac{r^2\sigma(t+273)^4}{4\pi R^2}$ (b) $\frac{16\pi^2 r^2 \sigma t^4}{R^2}$

(c) $\frac{r^2\sigma(t+273)^4}{R^2}$ (d) $\frac{4\pi r^2 \sigma t^4}{R^2}$

(where σ is Stefan's constant)

292. A black body at 227°C radiates heat at the rate of 7 cals/cm^{2}s. At a temperature of 727°C, the rate of heat radiated in the same units will be

(a) 50 (b) 112
(c) 80 (d) 60

293. The two ends of a rod of length L and a uniform cross-sectional area A are kept at two temperatures T_1 and T_2 ($T_1 > T_2$). The rate of heat transfer, $\frac{dQ}{dt}$ through the rod in a steady state is given by

(a) $\frac{dQ}{dt} = \frac{k(T_1 - T_2)}{LA}$ (b) $\frac{dQ}{dt} = kLA(T_1 - T_2)$

(c) $\frac{dQ}{dt} = \frac{kA(T_1 - T_2)}{L}$ (d) $\frac{dQ}{dt} = \frac{kL(T_1 - T_2)}{A}$

294. On a new scale of temperature (which is linear) and called the W scale, the freezing and boiling points of water are 39°W and 239° respectively. What will be the temperature on the new scale, corresponding to a temperature of 39°C on the Celsius scale?

(a) 200°W (b) 139°W
(c) 78°W (d) 117°W

295. A black body is at 727° C. It emits energy at a rate which is proportional to

(a) $(1000)^4$ (b) $(1000)^2$
(c) $(727)^4$ (d) $(727)^2$

296. A black body at 1227° C emits radiation: with maximum intensity at a wavelength of 5000 Å. If the temperature of the body is increased by 1000° C, the maximum intensity will be observed at

(a) 3000 Å (b) 4000 Å
(c) 5000 Å (d) 6000 Å.

297. Which of the following rods, (given radius r and length l) each made of the same material and whose ends are maintained at the same temperature will conduct mos heat?

(a) $r = r_0, l = l_0$ (b) $r = 2r_0, l = /l_0$
(c) $r = r_0, l = 2/l_0$ (d) $r = 2r_0, l = 2/l_0$.

298. If λ_m denotes the wavelength at which till radiative emission from a black body at temperature T K is maximum, then

(a) λ_m, α T^4 (b) λ_m is independent of T
(c) λ_m α T (d) λ_m αT^{-1}

299. Consider a compound slab consisting of two different materials having equal thicknesses and thermal conductivities K and $2K$, respectively. The equivalent thermal conductivity of the slab is

(a) $\frac{2}{3}K$ (b) $\sqrt{2}K$

(c) 3K (d) $\frac{4}{3}K$

300. Unit of Stefan's constant is

(a) watt m^2 K^4 (b) watt m^2/K^4
(c) watt/m^{2}K (d) watt/m^2/K^4

301. Consider two rods of same length and different specific heats (S_1, S_2), conductivities (K_1, K_2) and area of cross-sections (A_1, A_2) and both having temperatures T_1 and T_2 at their ends. If rate of loss of heat due to conduction is equal, then

(a) $K_1A_1 = K_2A_2$ (b) $\frac{K_1A_1}{S_1} = \frac{K_2A_2}{S_2}$

(c) $K_2A_1 = K_1A_2$ (d) $\frac{K_2A_1}{S_2} = \frac{K_1A_2}{S_1}$

302. For a black body at temperature 727°C, its radiating power is 60 watt and temperature of surrounding is 227°C. If temperature of black body is changed to 1227°C then its radiating power will be

(a) 304 W (b) 320 W
(c) 240 W (d) 120 W

303. Which of the following is best close to an ideal black body?

(a) black lamp
(b) cavity maintained at constant temperature
(c) platinum black
(d) a lump of charcoal heated to high temperature.

304. The Wien's displacement law expresses the relation between

(a) wavelength corresponding to maximum energy and temperature
(b) radiation energy and wavelength
(c) temperature and wavelength
(d) colour of light and temperature.

305. A cylindrical rod having temperature T_1, and T_2 at its end. The rate of flow of heat is Q_1 cal/sec. If the linear dimension is doubled keeping temperature constant, then rate of flow of heat Q_2 will be

(a) $4Q_1$ (b) $2Q_1$
(c) $Q_1/4$ (d) $Q_1/2$

306. A black body has maximum wavelength λ at 2000 K. Its corresponding wavelength at 3000 K will be

(a) $\frac{3}{2}\lambda_m$ (b) $\frac{2}{3}\lambda_m$
(c) $\frac{16}{81}\lambda_m$ (d) $\frac{81}{16}\lambda_m$

Answer Keys

EXERCISES

1. (b)	2. (c)	3. (a)	4. (a)	5. (c)	6. (c)	7. (a)	8. (c)	9. (b)	10. (c)
11. (a)	12. (a)	13. (a)	14. (b)	15. (a)	16. (d)	17. (b)	18. (a)	19. (b)	20. (a)
21. (c)	22. (b)	23. (a)	24. (d)	25. (d)	26. (b)	27. (b)	28. (b)	29. (d)	30. (d)
31. (d)	32. (a)	33. (d)	34. (a)	35. (a)	36. (b)	37. (a)	38. (b)	39. (c)	40. (a)
41. (a)	42. (a)	43. (b)	44. (b)	45. (a)	46. (c)	47. (c)	48. (d)	49. (c)	50. (b)
51. (c)	52. (d)	53. (c)	54. (d)	55. (d)	56. (a)	57. (c)	58. (a)	59. (b)	60. (c)
61. (c)	62. (a)	63. (a)	64. (c)	65. (d)	66. (b)	67. (c)	68. (c)	69. (c)	70. (a)
71. (b)	72. (c)	73. (a)	74. (a)	75. (a)	76. (b)	77. (b)	78. (a)	79. (a)	80. (c)
81. (a)	82. (a)	83. (c)	84. (a)	85. (c)	86. (d)	87. (c)	88. (c)	89. (c)	90. (d)
91. (d)	92. (c)	93. (c)	94. (c)	95. (b)	96. (a)	97. (d)	98. (c)	99. (a)	100. (c)
101. (b)	102. (c)	103. (d)	104. (c)	105. (d)	106. (b)	107. (d)	108. (c)	109. (c)	110. (b)
111. (c)	112. (d)	113. (d)	114. (c)	115. (c)	116. (b)	117. (c)	118. (b)	119. (c)	120. (b)
121. (d)	122. (b)	123. (c)	124. (a)	125. (c)	126. (c)	127. (b)	128. (b)	129. (a)	130. (b)
131. (c)	132. (c)	133. (b)	134. (b)	135. (a)	136. (d)	137. (d)	138. (d)	139. (b)	140. (d)
141. (a)	142. (c)	143. (d)	144. (c)	145. (b)	146. (c)	147. (b)	148. (d)	149. (d)	150. (b)
151. (b)	152. (b)	153. (d)	154. (c)	155. (d)	156. (a)	157. (a)	158. (b)	159. (d)	160. (a)
161. (b)	162. (d)	163. (c)	164. (a)	165. (b)	166. (d)	167. (a)	168. (a)	169. (a)	170. (c)
171. (c)	172. (d)	173. (a)	174. (b)	175. (b)	176. (d)	177. (a)	178. (c)	179. (a)	180. (c)
181. (a)	182. (d)	183. (c)	184. (c)	185. (a)	186. (c)	187. (c)	188. (d)	189. (a)	190. (a,c)
191. (c)	192. (a)	193. (d)	194. (a)	195. (c)	196. (a)	197. (d)	198. (b)	199. (c)	200. (d)
201. (b)	202. (c)	203. (b)	204. (d)	205. (d)	206. (d)	207. (a)	208. (c)	209. (b)	210. (d)
211. (b)	212. (c)	213. (a)	214. (b)	215. (d)	216. (d)	217. (a)	218. (a)	219. (a)	220. (b)
221. (a)	222. (d)	223. (d)	224. (c)	225. (d)	226. (b)	227. (d)	228. (c)	229. (a)	230. (a)
231. (b)	232. (b)	233. (b)	234. (d)	235. (b)	236. (d)	237. (a)	238. (c)	239. (b)	240. (c)
241. (d)	242. (c)	243. (b)	244. (a)	245. (a)	246. (c)				

Questions Asked in NEET/CBSE Medicals

247. (a)	248. (b)	249. (d)	250. (d)	251. (d)	252. (b)	253. (c)	254. (a)	255. (c)	256. (d)
257. (c)	258. (c)	259. (c)	260. (b)	261. (b)	262. (b)	263. (b)	264. (b)	265. (a)	266. (b)
267. (b)	268. (d)	269. (d)	270. (d)	271. (d)	272. (b)	273. (a)	274. (a)	275. (b)	276. (c)
277. (d)	278. (a)	279. (d)	280. (b)	281. (c)	282. (b)	283. (a)	284. (a)	285. (a)	286. (d)
287. (a)	288. (a)	289. (b)	290. (a)	291. (c)	292. (b)	293. (c)	294. (d)	295. (a)	296. (a)
297. (b)	298. (d)	299. (a)	300. (d)	301. (a)	302. (b)	303. (b)	304. (a)	305. (b)	306. (b)

Solutions

1. (b)

2. (c)

$$Y = F\,L/Al \text{ or } l = F\,L/AY$$

3. (a)

4. (a)

5. (c)
The breaking stress does not depend on the length of the cable. Moreover, it also does not depend on the area of cross-section.

6. (c)
We know that inter-atomic binding energy decreases with temperature. So, more strain is produced at higher temperature. As a result, Y decreases.

7. (a)

$$l\alpha(L/A)$$

8. (c)

$$\text{Thermal stress} = Y\alpha(\Delta T)$$
$$(\text{Thermal stress})_1 = Y_1\alpha_1\Delta T$$
$$(\text{Thermal stress})_2 = Y_2\alpha_2\Delta T$$
$$Y_1\alpha_1\Delta T = Y_2\alpha_2\Delta T$$
$$\frac{Y_1}{Y_2} = \frac{\alpha_2}{\alpha_1} = \frac{3}{2}$$

9. (b)
Weight of the cord

$$W = V\rho g = AL\rho g$$
$$\text{Stress} = \frac{W}{A} = L\rho g$$

For the purpose of extension, the weight acts at the centre of gravity of the bar. Hence the upper half length (L/2) will be stretched.

$$\text{Strain} = \frac{l}{(L/2)} = \frac{2l}{L}$$
$$\text{Now, } Y = \frac{\text{Stress}}{\text{Strain}} = \frac{L\rho g \times L}{2l} = \frac{L^2\rho g}{2l}$$

or $\quad l = L^2\rho g/2Y$

10. (c)

$$\Delta L = \left(\frac{1}{2}\rho g L^2\right)/Y$$
$$= \frac{1}{2}\times\frac{(1.5\times10^3)\times64\times10}{5\times10^6}$$
$$= 9.6\times10^{-2}$$

11. (a)
We know that

$$Y = \frac{F}{\pi r^2}\times\frac{L}{l}$$

Here Y, F and π are same for all the four wires. So l will be maximum for a wire for which L/r^2 is maximum.

12. (a)

$$Y = \frac{FL}{Al} \text{ or } F = \frac{YAl}{L}$$
$$dW = Fdl = \frac{YA}{L}ldl$$
$$W = \int_0^x \frac{YA}{L}ldl = \frac{YA}{L}\left[\frac{l^2}{2}\right]_0^x = \frac{YAx^2}{2L}$$

13. (a)
Total work done in stretching a string

$$= \frac{1}{2}\times\text{Stress}\times\text{Strain}$$

Hence, the work done per unit volume is

$$\frac{1}{2}(\text{Stress}\times\text{Strain}).$$

This work is stored as the potential energy in the string.

14. (b)
When a body is deformed, work is done against the interatomic deforming forces which is stored as elastic potential energy.
Let L be the length of the body, A the area of cross-section and l the increase in length.

$$\text{Stress} = \frac{F}{A},\ \text{Strain} = \frac{l}{L}$$

Force necessary to deform the body is

$$F = \frac{YA}{L}.l$$

If body deformed by a distance dl, then

$$\text{Work done} = F\times dl = \frac{YA}{L}.ldl$$
$$W = \int_0^l \frac{YA}{L}ldl = \frac{YA}{L}\left[\frac{l^2}{2}\right]_0^l = \frac{1}{2}YA\frac{l^2}{L}$$
$$= \frac{1}{2}\left(Y\frac{l}{L}\right)\left(\frac{l}{L}\right)(AL)$$
$$= \frac{1}{2}(\text{stress}\times\text{strain})\times\text{volume}$$

Hence, work done for unit volume is

$$W = \frac{1}{2}\times\text{stress}\times\text{strain}$$

15. (a)
Stresses developed in the two rods are equal.

But $\quad \text{Stress} = Y\times\text{Strain}$

$\Rightarrow \quad \text{Stress} = Y\times\frac{x}{L}$ and elongation $= \alpha LT$

$$\frac{x_1}{L_1}\times Y_1 = \frac{x_2}{L_2}\times Y_2$$

$$\frac{\alpha_1 L_1 T}{L_1}\times Y_1 = \frac{\alpha_2 L_2 T}{L_2}\times Y_2$$

$$\Rightarrow \quad \frac{\alpha_1}{\alpha_2} = \frac{Y_2}{Y_1}$$

$$\Rightarrow \quad \frac{Y_1}{Y_2} = \frac{3}{2}$$

16. (d)

$$Y = \frac{k}{r_0} = \frac{7}{3\times 10^{-10}} = 2.33\times 10^{-10}\ \text{N/m}^2$$

17. (b)

$$Y = \frac{Force\times L}{A\times l} = \frac{WL}{Al}$$

$$l = \frac{WL}{AY}$$

Due to pulley arrangement, the length of wire is $\frac{L}{2}$ on each side and so the elongation will be $\frac{l}{2}$. For both sides, elongation = l.

18. (a)

Young's modulus = Y

$$(\text{Elongation in Wire})_1 = l = \frac{FL}{\pi r^2 Y}$$

$$(\text{Elongation in Wire})_2 = x = \frac{2F\times 2L}{\pi(2r)^2\times Y}$$

$$= \frac{FL}{\pi r^2 Y} = l$$

$$x = l$$

19. (b)

Elastic energy stored per unit volume

$$u = \frac{1}{2}\times \text{Stress}\times\text{Strain} = \frac{1}{2}\times\frac{\text{Stress}^2}{Y}$$

Load and radius are same for both the wires.

i.e., $\quad \text{Stress} = \frac{F}{A} = \frac{F}{\pi r^2}$ is same (constant)

$$u \propto \frac{1}{Y} \Rightarrow \frac{u_1}{u_2} = \frac{Y_2}{Y_1} = \frac{Y_2}{2Y_2} = \frac{1}{2}$$

$$\frac{u_1}{u_2} = \frac{1}{2}$$

20. (a)

The potential energy of the stretched wire

$$= \frac{1}{2}\times\text{Stress}\times\text{Strain}\times\text{Volume}$$

$$\Rightarrow \quad U = \frac{1}{2}\times\frac{F}{A}\times\frac{x}{L}\times AL$$

$$\Rightarrow \quad U = \frac{1}{2}Fx$$

Alternative Solution

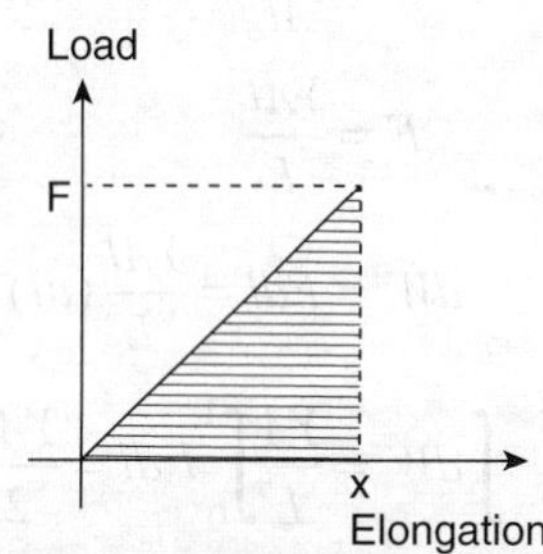

The potential energy of the stretched wire

= Area under load versus elongation curve

$$= \frac{1}{2}\times F\times x = \frac{1}{2}Fx$$

21. (c)

$$\gamma = \frac{C_p}{C_v} = \frac{Y_{\text{adiabatic}}}{Y_{\text{isothermal}}}$$

$$\Rightarrow \quad 1.4 = \frac{2.1\times 10^5\ \text{N/m}^2}{Y_{\text{isothermal}}}$$

$$\Rightarrow \quad Y_{\text{isothermal}} = \frac{2.1}{1.4}\times 10^5\ \text{N/m}^2$$

⇒ Isothermal modulus of elasticity $= 1.5\times 10^5\ N/m^2$

22. (b)

$$\text{Bulk modulus} = k = \frac{p}{\frac{\Delta v}{v}} = \frac{pv}{\Delta v}$$

$$\text{Given} \quad \frac{\Delta p}{p}\times 100 = 0.1$$

$$\frac{\Delta p}{p} = 0.001$$

$$\Rightarrow \quad p = k.\frac{\Delta v}{v} = k.\frac{\Delta p}{p}$$

$$\Rightarrow \quad p = 2\times 10^9\times 0.001 = 2\times 10^6\ \text{N/m}^2.$$

23. (a)

$$\sigma = \frac{\text{Lateral strain}}{\text{Longitudinal strain}} = \frac{\frac{\Delta D}{D}}{\frac{\Delta L}{L}}$$

$$\frac{\Delta D}{D} = \sigma\left(\frac{\Delta L}{L}\right) = 0.4 \times 0.025\% = 0.01\%$$

24. (d)

Energy stored per unit volume $= \frac{1}{2} \times \text{Stress} \times \text{Strain}$

$$= \frac{\text{Stress} \times \text{Stress}}{2Y} = \frac{S^2}{2Y}$$

25. (d)

Young's modulus $Y = \frac{FL}{Al}$

$$F = \frac{YAl}{L} \quad \ldots\text{(i)}$$

or $$dW = Fdl = \frac{YAl}{L}(dl)$$

or $$\int dW = \frac{YA}{L}\int_0^1 l\,dl = \frac{YAl^2}{2L}$$

or Work done $= \frac{YAl^2}{2L}$ ...(ii)

From Eqs. (i) and (ii), we get

Work done $= \frac{Fl}{2}$.

26. (b)

Poisson ratio $(\sigma) = \frac{\text{Lateral strain}}{\text{Longitudinal strain}} = \frac{\frac{\Delta D}{D}}{\frac{\Delta L}{L}}$

where L stands for length and D for diameter of the wire.

Here cross-sectional area of wire $A = \pi\left(\frac{D}{2}\right)^2$

On differentiating, $\Delta A = \frac{2\pi D.\Delta D}{4} = \frac{\pi D.\Delta D}{2}$

$$\frac{\Delta A}{A} = \frac{2\Delta D}{D}$$

Therefore, $$\sigma = \frac{\frac{\Delta A}{A}}{\frac{2\Delta L}{L}}$$

or $$\frac{\Delta L}{L} = \frac{1}{2\sigma}\frac{\Delta A}{A} = \frac{1}{2 \times 0.4} \times 2\% = 2.5\%$$

27. (b)

Let the original unstretched length of the elastic spring is L.

By definition, $$Y = \frac{\frac{F}{A}}{\frac{\Delta L}{L}} = \frac{F.L}{A.\Delta L}$$

$$\Delta L = \frac{F.L}{AY}$$

and $$L + \Delta L = L\left(1 + \frac{F}{AY}\right)$$

In the first case, $L + \Delta L = a$, when $F = 4N$

$$L\left(1 + \frac{4}{AY}\right) = a \quad \ldots\text{(i)}$$

In the second case, $L + \Delta L = b$, when $F = 5N$

$$L\left(1 + \frac{5}{AY}\right) = b \quad \ldots\text{(ii)}$$

In the third case, $L + \Delta L = x(say)$, when $F = 9N$

$$L\left(1 + \frac{9}{AY}\right) = x$$

From Eqs. (i) and (ii),

$$\frac{1 + \frac{5}{AY}}{1 + \frac{4}{AY}} = \frac{b}{a}$$

or $$1 + \frac{5}{AY} = \frac{b}{a} + \frac{4}{AY}.\frac{b}{a}$$

$$\frac{1}{AY}\left(5 - \frac{4b}{a}\right) = \frac{b}{a} - 1 = \frac{b-a}{a}$$

or $$AY = \frac{5a - 4b}{b - a}$$

Substituting the value of AY in Eq. (i),

$$L = \frac{a}{1 + \frac{4(b-a)}{5a - 4b}}$$

Hence, from Eq. (iii)

$$x = L\left(1 + \frac{9}{AY}\right)$$

$$x = (5a - 4b)\left[1 + \frac{9(b-a)}{5a - 4b}\right] = 5b - 4a$$

28. (b)

$$E = \frac{1}{2} \times \text{Stress} \times \text{Strain} \times \text{Volume}$$

$$= \frac{1}{2} \times \frac{(\text{Stress})^2 \times \text{Volume}}{\text{Young's modulus}}$$

$$= \frac{(\text{Stress})^2 \times \text{Volume}}{2 \times Y}$$

29. (d)

Elastic energy per unit volume $=\frac{1}{2}\times\text{Stress}\times\text{Strain}$

$$\text{Elastic energy}=\frac{1}{2}\times\text{Stress}\times\text{Strain}\times\text{Volume}$$

$$=\frac{1}{2}\times\frac{F}{A}\times\frac{\Delta L}{L}\times(AL)=\frac{1}{2}\times F.\Delta L$$

$$=\frac{1}{2}\times 200\times 10^{-3}=0.1\text{ J.}$$

30. (d)

$$\text{Stress}=\frac{\text{Force}}{\text{Area}}$$

31. (d)

$$\Delta L=\frac{FL}{AY}=\frac{4.8\times 10^{3}\times 4}{1.2\times 10^{-4}\times 1.2\times 10^{11}}$$

$$\Rightarrow\quad \Delta L=\frac{4.8\times\times 4\times 10^{-4}}{1.2\times 1.2}$$

$$=13.3\times 10^{-4}\text{ m}=1.33\text{mm}$$

32. (a)

Work done, W = Change in potential energy.

$$W=\frac{1}{2}\times\text{Stress}\times\text{Strain}\times\text{Volume}$$

$$=\frac{1}{2}\times\frac{Yx}{L}\times\frac{x}{L}\times AL$$

33. (d)

$$\text{Breaking stress}=\frac{FA}{A}=\frac{200}{\pi\frac{d^2}{4}}$$

$$\text{For second wire}=\frac{F}{\frac{4\pi d^2}{4}}$$

Since they both are made of same material, breaking stress must be equal.

$$\Rightarrow\quad \frac{200}{\frac{\pi d^2}{4}}=\frac{F}{\frac{4\pi d^2}{4}}$$

$$F=800\text{ N}$$

34. (a)

$$\text{Shearing modulus}=\frac{\text{Shear stress}}{\text{Shear strain}}$$

$$=\frac{\frac{8\times 10^{3}}{(4\times 10^{-2})^2}}{\left(\frac{0.1}{40}\right)}$$

$$=2\times 10^{9}\text{ N/m}^2$$

$\therefore$ Shearing modulus $=2\times 10^{9}\text{ N/m}^2$

35. (a)

$$W=\frac{1}{2}\times\frac{F}{A}\times\frac{\Delta l}{l}\times A\times l$$

$$=\frac{1}{2}\times\frac{F/A}{\Delta l/l}\times\left(\frac{\Delta l}{l}\right)^2\quad Al=\frac{YA(\Delta l)^2}{2l}$$

$$\text{or}\quad(\Delta l)^2=\frac{2lW}{YA}=\frac{2\times 50\times 10^{-2}\times 2\times 10^{-2}}{2\times 10^{10}\times 0.5\times 10^{-6}}$$

$$=2\times 10^{-6}\text{ m}^2$$

$$\Delta l=\sqrt{2}\times 10^{-3}\text{ m}=0.1414\text{cm.}$$

Total length of the wire after elongation.

$$=(50+0.1414)\text{cm}=50.1414\text{cm.}$$

36. (b)

$$Y=\frac{FL}{\pi r^2 l}$$

$$\Rightarrow\quad l=\frac{FL}{\pi r^2 Y}$$

$$l\propto\frac{1}{r^2}$$

If the radius is doubled, the extension will become $\frac{1}{4}$th

i.e., $\frac{2.4}{4}=0.6$ cm

37. (a)

Let's find tension T in the string

$$2g-T=2a \quad\text{(i)}$$

$$T-g=a \quad\text{(ii)}$$

From (i) and (ii)

$$T=\frac{40}{3}\text{N}$$

$$\text{Strain}=\frac{\text{Stress}}{Y}=\frac{\text{Force}}{Y\text{ Area}}$$

$$=\frac{40}{3\times 2\times 10^{11}\times 2\times 10^{-4}}=3.3\times 10^{-7}$$

38. (b)

Hint Find out the tension in portions A and B proceed as above.

39. (c)

If R is spring constant and L is the original length then

For 1st case - $T_1=R(l_1-L)$

For 2nd case - $T_2=R(l_2-L)$

$$T_1l_2-T_1L=T_2l_1-T_2L\ ;\ T_1l_2-T_2l_1=T_1L-T_2L,$$

$$L=\left(\frac{T_1l_2-T_2l_1}{T_1-T_2}\right)=\frac{T_2l_1-T_1l_2}{T_2-T_1}$$

40. (a)

41. (a)

42. (a)

43. (b)

44. (b)

$$Y=\frac{FL}{Al}$$

$$F=\frac{YAl}{L}=\frac{YA\times 2\pi(R-r)}{2\pi r}=\frac{YA(R-r)}{r}$$

45. (a)

46. (c)

$$\Delta l=\frac{FL}{YA}=\frac{6\times 10^4\times 1}{2\times 10^{11}\times 3\times 10^{-4}}=10^{-3}\,\text{m}$$

47. (c)

$$\Delta l=\frac{Fl}{AY}$$

Since $\frac{F}{A}$ is identical for both the wires.

$$\frac{\Delta l_{1(Cu)}}{\Delta l_{2(steel)}}=\frac{l_1}{Y_1}\cdot\frac{Y_2}{l_2}$$

$$\Delta l_2=\Delta l_1\cdot\frac{l_2}{l_1}\cdot\frac{Y_1}{Y_2}$$

$$=(1\text{mm})\left(\frac{0.5}{1}\right)\left(\frac{1\times 10^{11}}{2\times 10^{11}}\right)$$

$$=0.25\text{mm}$$

Total extension = 1 + 0.25 = 1.25 mm = 0.125 cm

48. (d)

At A, potential energy is positive. So, force is repulsive at C, potential energy is negative. So, force is attractive at B. Slope of graph is zero. $\left[F=-\frac{dV}{dr}\right]$

49. (c)

Weight attached at the end + Weight of the $\frac{3}{4}$ of the length. Hence stress $=\frac{F}{A}=\frac{\left(w_1+\frac{3w}{4}\right)}{s}$

50. (b)

$$\sigma=\frac{\text{Lateral strain}}{\text{Longitudinal strain}}$$

or Lateral strain, $\frac{dr}{r}=-0.2\times 2\times 10^{-3}=-0.40\times 10^{-3}$

Now $\quad V=\pi r^2 l$

$$dV=\pi r^2 dl+2\pi rl.dr$$

Now $\quad \frac{dV}{V}=\frac{\pi r^2 dl}{\pi r^2 l}+\frac{2\pi rldr}{\pi r^2 l}=\frac{dl}{l}+\frac{2dr}{r}$

$$=2\times 10^{-3}-2\times 0.40\times 10^{-3}=1.2\times 10^{-3}$$

Again, $\quad \frac{dV}{V}\times 100=1.2\times 10^{-3}\times 100=+0.12.$

51. (c)

The work done by torque τ in twisting through θ is $\tau\theta$. Hence, the work done for producing a total twist θ_0 is given by,

$$W=\int_0^{\theta}\tau\theta=\frac{1}{2}\tau\theta_0^2$$

$$\tau=\frac{\pi\eta r^4}{2l}$$

$$\therefore\quad W=\frac{1}{2}\frac{\pi\eta r^4}{2l}\theta_0^2 \qquad \ldots\text{(i)}$$

$\eta=8\times 10^{10}\,Pa, r=10^{-3}\,m, l=0.25m$ and $\theta_0=\frac{\pi}{4}$

Substituting these values in Eq. (i),

$$W=\frac{1}{2}\times\frac{\pi\times 8\times 10^{10}\times(10^{-3})^4}{2\times 0.25}\times\left(\frac{\pi}{4}\right)^2$$

$$=15.5\times 10^{-2}\,\text{J}$$

52. (d)

When a wire is stretched, then change in length of wire is

$$\Delta l=\frac{Fl}{\pi r^2 Y}\text{, where Y is its Young's modulus.}$$

Here, for wires A and B,

$$l_A=2\text{m}, l_B=1.5\text{m}$$

$$\frac{Y_A}{Y_B}=\frac{7}{4}, r_B=2mm=2\times 10^{-3}\,m \text{ and } \frac{F_A}{F_B}=1$$

As, it is given that $\Delta l_A=\Delta l_B$

$$\Rightarrow \frac{F_A l_A}{\pi r_A^2 Y_A} = \frac{F_B l_B}{\pi r_B^2 Y_B}$$

$$\Rightarrow r_A^2 = \frac{F_A}{F_B} \cdot \frac{l_A}{l_B} \cdot \frac{Y_B}{Y_A} \cdot r_B^2$$

$$= 1 \times \frac{2}{1.5} \times \frac{4}{7} \times 4 \times 10^{-6}\ \text{m} = 3.04 \times 10^{-6}\ \text{m}$$

or $\quad r_A = 1.7 \times 10^{-3}\ \text{m}$

or $\quad r_A = 1.7\ \text{mm}$

53. (c)

Radius of wire, $r = 2\ \text{mm} = 2 \times 10^{-3}\ \text{m}$

Weight of load, $m = 4\ \text{kg}$, $g = 3.1\pi \text{ms}^{-2}$

$$\therefore \quad \text{Tensile stress} = \frac{\text{Force}(F)}{\text{Area}(A)} = \frac{mg}{\pi r^2}$$

$$= \frac{4 \times 3.1 \times \pi}{\pi \times \left(2 \times 10^{-3}\right)} = 3.1 \times 10^6\ \text{Nm}^{-2}$$

54. (d)

When load M is attached to wire, extension in length of wire is

$$\Delta l_1 = \frac{Mgl}{A.Y} \qquad \text{(i)}$$

where, Y is the Young's modulus of the wire.

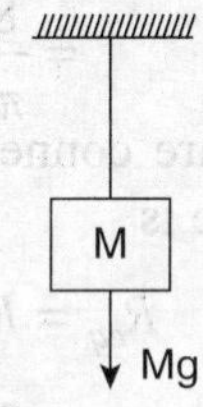

When load is immersed in liquid of relative density 2, increase in length of wires as shown in the figure is

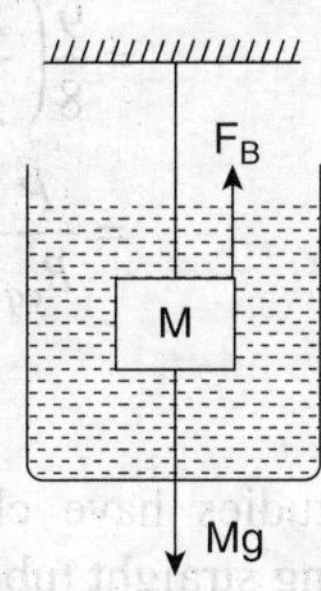

$$\Delta l_2 = \frac{(Mg - F_B)l}{A.Y}$$

where, F_B = Buoyant force

$$\therefore \Delta l_2 = \frac{\left(Mg - Mg \cdot \frac{\rho_l}{\rho_b}\right) l}{A.Y} \left[\because F_B = V\rho_l g = \frac{Mg}{\rho_b} \rho_l g\right]$$

Here given that, $\frac{\rho_l}{\rho_b} = \frac{2}{8} = \frac{1}{4}$

So, $$\Delta l_2 = \frac{\left(\frac{3}{4} Mg\right) l}{A.Y} \qquad \text{(ii)}$$

Dividing Eqs. (ii) by (i), we get

$$\frac{\Delta l_2}{\Delta l_1} = \frac{3}{4}$$

$$\Rightarrow \quad \Delta l_2 = \frac{3}{4} \times \Delta l_1$$

$$= \frac{3}{4} \times 4\text{mm} = 3\text{mm}$$

55. (d)

From the definition of bulk modulus, $B = \frac{-dp}{(dV/V)}$

Substituting the values, we have

$$B = \frac{(1.165 - 1.01) \times 10^5}{(10/100)} = 1.55 \times 10^5\ \text{Pa}$$

56. (a)

$$\Delta l = \left(\frac{l}{YA}\right) \cdot w$$

Graph is a straight line passing through origin (as shown in question also), the slope of which is $\frac{l}{YA}$

$$\text{Slope} = \left(\frac{l}{YA}\right)$$

$$Y = \left(\frac{l}{A}\right)\left(\frac{1}{\text{slope}}\right) = \left(\frac{1.0}{10^{-6}}\right)\frac{(80-20)}{(4-1)\times 10^{-4}}$$

$$= 2.0 \times 10^{11}\ \text{N/m}^2$$

FLUIDS

57. (c)

$$P_A - P_L = \rho_w g H_w$$

$$g\rho_m (760\ \text{mm} - 750\ \text{mm}) = \rho_w g H_w$$

$$13.6 \times 1\text{cm} = 1 \times H_w$$

$$H_w = 13.6\text{cm}$$

58. (a)

According to Stokes law, when a spherical body falls through a viscous fluid, it experiences a viscous force. The magnitude of the viscous force increases with the increase in velocity of the body falling under the action of its weight. As a result, the viscous force soon balances the driving force and the body starts moving with a constant velocity, known as its terminal velocity.

59. (b)

Excess pressure inside the bubble is given by $P + \frac{4T}{r}$. Smaller the radius r, the larger is the excess pressure P. It means, the pressure of air is more in bubble A than in bubble B. So, the air will go from bubble A to bubble B and B will grow until both bubbles collapse.

60. (c)

If a sphere of mass M and radius R is dropped in a liquid, its weight $Mg = \frac{4}{3}\pi R^3 \rho g$ acts vertically downwards. While upthrust $\frac{4}{3}\pi r^3 \sigma g$ and viscous force $6\pi\eta Rv$ acts vertically upwards. Initially, the body will be accelerated down. At a certain instant when viscous force F will balance the net downward force, acceleration will become zero and the body will fall with constant velocity. This constant velocity is called terminal velocity

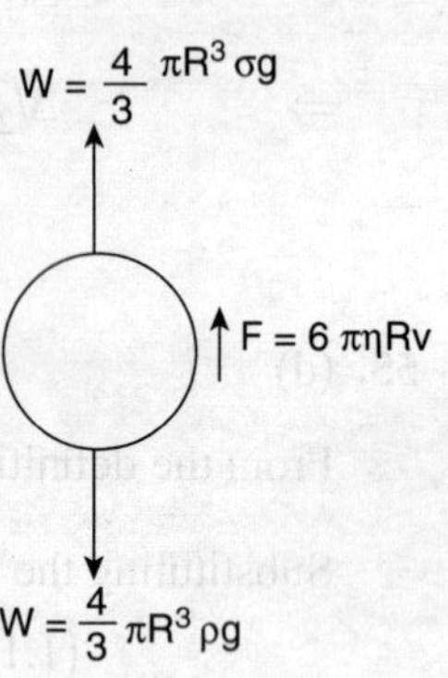

$$6\pi\eta R v_T = \frac{4}{3}\pi R^3(\rho - \sigma)g$$

i.e., $$v_T = \frac{2}{9}R^2\left(\frac{\rho - \sigma}{\eta}\right)g,$$

or $$v_T \propto R^2$$

61. (c)

Let r be the radius of each smaller drop and R be the radius of bigger drop.

$$\frac{4}{3}\pi r^3 \times 2 = \frac{4}{3}\pi R^3$$

or $$R^3 = 2r^3$$

$\Rightarrow$ $$R = (2)^{1/3} r$$

$\therefore$ Surface energy $\propto$ Surface area.

62. (a)

Surface tension (T) of a liquid is equal to the work (W) required to increase the surface area (ΔA) of the liquid film by unity at constant temperature.

$$W = T\Delta A = \text{surface energy.}$$

Also, volume of bigger drop = 27 × volume of smaller drop.

i.e., $$V = 27v$$

where V is the volume of bigger drop of diameter D and v is the volume of smaller drop of diameter d.

$$\frac{4}{3}\pi\left(\frac{D}{2}\right)^3 = 27 \times \frac{4}{3} \times \pi\left(\frac{d}{2}\right)^3$$

$\Rightarrow$ $$\frac{D}{2} = 3 \times \frac{d}{2}$$

$\Rightarrow$ $$d = \frac{D}{3}$$

$\therefore$ Radius of smaller drop, $r = \frac{d}{2} = \frac{D}{6}$

$\therefore$ Change in surface energy

$$= T(A_2 - A_1)$$
$$= T(27 \times 4\pi r^2 - 4\pi R^2)$$
$$= 4\pi T\left[27\left(\frac{D}{6}\right)^2 - \left(\frac{D}{2}\right)^2\right]$$
$$= 4\pi T\left[\frac{3D^2}{4} - \frac{D^2}{4}\right] = 2\pi D^2 T$$

63. (a)

If l and r be the length and radius of the tube and P the pressure difference, then from Poiseuille's formula, volume of liquid flowing per second is given by

$$X = \frac{\pi \text{P} r^4}{8\eta l}$$

Fluid resistance $$= \frac{8\eta l}{\pi r^4}$$

When capillaries are connected in series then equivalent fluid resistance is

$$R_{eq} = R_1 + R_2$$
$$= \frac{8\eta L}{\pi R^4} + \frac{8\eta(2L)}{\pi(2R)^4}$$
$$= \frac{9}{8}\left(\frac{8\eta L}{\pi R^4}\right)$$

Net rate of flow $$= \frac{P}{R_{eq}} = \frac{P}{\frac{9}{8}\left(\frac{8\eta L}{\pi R^4}\right)} = \frac{8}{9}X$$

64. (c)

Experimental studies have characterized the critical velocity for a long straight tube in the form

$$V_{\text{critical}} = \frac{N_R \eta}{2\rho r}$$

which depends upon viscosity, density and radius. Here R is Reynold's number.

Since $V_{\text{critical}} \propto \frac{1}{\text{density}}$, hence

Critical velocity decreases when density increases.

$$\underset{\text{Pressure energy}}{P_1} + \underset{\substack{\text{Kinetic energy}\\\text{Per unit volume}}}{\frac{1}{2}\rho V_1^2} + \underset{\substack{\text{Potential energy}\\\text{Per unit volume}}}{\rho g h_1}$$

$$= P_2 + \frac{1}{2}\rho V_2^2 + \rho g h_2$$

Thus, Bernoulli's equation is a form of conservation of energy.

65. (d)
The rate of flow of liquid through an orifice depends upon three factors: orifice size, atomizing surface area, liquid characteristics.

66. (b)
According to the Poiseuille's law

$$V = \frac{\pi r^4 p}{8\eta l} \quad \ldots\text{(i)}$$

where V is a volume of the liquid, per time t.
Comparing Eq. (i) with given equation, we get.

$$Q = \frac{1}{8}$$

67. (c)

$$v_r = \mathrm{LT}^{-1}$$

$$\text{Load} = mg$$

$$F = \mathrm{MLT}^{-2}$$

$$\eta = \mathrm{ML}^{-1}\mathrm{T}^{-1}$$

$$r = \mathrm{M}^0\mathrm{L}^1\mathrm{T}^0$$

$$\therefore \left[\mathrm{M}^0\mathrm{LT}^{-1}\right] = \left[\mathrm{MLT}^{-2}\right]^a\left[\mathrm{ML}^{-1}\mathrm{T}^{-1}\right]^b\left[\mathrm{M}^0\mathrm{L}^1\mathrm{T}^0\right]^c$$

$$\Rightarrow a + b = 0; a - b + c = 1; -2a - b = 1$$

Solving , $a = 1; b = -1; c = -1$

$$v_r \propto \frac{mg}{\eta r}$$

68. (c)
Let d_w = density of water and d_0 = density of oil.
Pressure at the bottom of the tank $= d_w g\, h_w + d_0 g h_0$
Let this pressure be equivalent to pressure due to the water of height h.

$$d_w g\, h = d_w g\, h_w + d_0 g\, h_0$$

$$\Rightarrow \quad h = h_w + \frac{d_0 h_0}{d_w} = 100 + \frac{0.9 \times 400}{1} = 460$$

$\therefore$ From Toricelli's theorem, we have $v = \sqrt{2gh}$

$$v = \sqrt{2 \times 980 \times 460} = \sqrt{920 \times 920} \text{ cm/s}$$

69. (c)
From the equation of continuity,

$$Av = \text{Constant}$$

$$A_1 v_1 = A_2 v_2$$

$$A_1 v_1 = \left(\frac{A_1}{2}\right) v_2$$

$$\Rightarrow \quad v_2 = 2v_1.$$

70. (a)
The total energy of liquid is constant at all points in a stream line flow of liquid.

71. (b)
The fraction of ice floating in water be n, then

$$nV\rho g = Vdg$$

Where ρ is the density of water and d is the density of ice,

$$n = \frac{d}{\rho} = \frac{917}{1024} = \frac{9}{10}$$

The fraction of ice visible

$$= 1 - \frac{9}{10} = \frac{1}{10} = 10\%$$

72. (c)
The hole will act like an orifice.
For orifice, $v = \sqrt{2gh}$.

73. (a)
From continuity equation,

$$\therefore \quad A_1 v_1 = A_2 V_2$$

$$\pi(2.5)^2 \times 4 = \pi(2.0)^2 \times v_2$$

$$\Rightarrow \quad v_2 = 6.25\text{m/s}$$

74. (a)
Energy needed = Work done
In splitting, the volume remains unchanged

$$\frac{4}{3}\pi R^3 = \frac{4}{3}\pi r^3 \times n$$

or

$$r = \frac{R}{(n)^{1/3}}$$

Energy needed = W = Surface tension × Change in area.

$$= S\left[4\pi r^2 n - 4\pi R^2\right]$$

75. (a)
Liquid surfaces tend to remain in its smaller possible surface area. When the surface A is punctured, the

surface area of B will be reduced. Hence, the thread becomes concave towards A.

76. (b)

$\because$ Pressure $P = \dfrac{F}{A} \Rightarrow P \propto \dfrac{1}{A}$

$\because$ $A_1 < A_3 < A_2 < A_4$

$\Rightarrow$ $P_1 > P_3 > P_2 > P_4.$

77. (b)

Work done, $W = 2T \times \Delta A$

$$T = 72\text{N/m}$$
$$= \frac{72\times10^5\text{ dyne}}{100\text{ cm}}$$
$$= 72\times10^3\text{ dyne/cm}$$
$$\Delta A = (1\text{ mm})(10\text{ cm})$$
$$= \left(\frac{1}{10}\text{cm}\right)(10\text{ cm}) = 1\text{cm}^2$$
$$W = 2\times72\times10^3\times1 = 144\times10^3\text{ erg}$$

78. (a)

Viscosity is most closely related to friction between the layers of liquid.

79. (a)

$$H = \frac{2T}{\rho r_2 g}$$

and $h + H = \dfrac{2T}{\rho r_1 g}$

$$h = \frac{2T}{\rho g}\left(\frac{1}{r_1} - \frac{1}{r_2}\right)$$

or $T = \dfrac{\rho g h_1 r_2}{2(r_2 - r_1)}$

80. (c)

$$\frac{T_{water}}{T_{mercury}} = \frac{\rho_w h_w(\cos\theta)_m}{\rho_m h_m(\cos\theta)_w}$$
$$= \frac{1\times10\times\cos135°}{13.6\times(-3.42)\times\cos0°}$$
$$= \frac{10}{13.6\times3.42\times1.414} = \frac{1}{6.5}$$

81. (a)

Volume of mercury drop = Volume of 10^6 droplets of radius r.

$$\frac{4}{3}\pi R^3 = 10^6\times\frac{4}{3}\pi r^3$$
$$R^3 = 10^6 r^3$$
$$R = 10^2.r$$
$$r = \frac{R}{100}$$

Energy expanded $= T\Delta A$

$$= T\left[10^6(4\pi r^2) - 4\pi R^2\right]$$
$$= 4\times3.14\times460\times10^{-3}\left[10^6\times\frac{R^2}{10^4} - R^2\right]$$
$$= 4\times3.14\times460\times10^{-3}\times99\times10^{-4}$$
$$= 5.7\times10^{-2}\text{ J} = 0.057\text{ J}$$

82. (a)

From the principle of continuity

$$A_1v_1 = A_2v_2$$
$$2Av_1 = A.v_2$$
$$\frac{v_2}{v_1} = 2$$

Liquid mass flux = Mass flowing per unit area per second.

$$= \frac{\text{Mass}}{\text{Area}\times\text{Time}}$$
$$= \frac{\text{Mass}}{\text{Area}\times\text{Distance}}\times\frac{\text{Distance}}{\text{Time}}$$
$$= \frac{\text{Mass}}{\text{Volume}}\times\text{Velocity}$$
$$= \text{Density}\times\text{Velocity} = \rho v$$

given $\rho_1v_1 = \rho_2v_2$ or $\dfrac{\rho_1}{\rho_2} = \dfrac{v_1}{v_2} = 2$

83. (c)

Velocity of efflux $v = \sqrt{2gd}$

Time taken by the water to reach the ground from a vertical height h is

$$t = \sqrt{\frac{2h}{g}} \quad \left(\because h = 0 + \frac{1}{2}gt^2\right)$$

Horizontal range $R = vt$

$$R = \sqrt{2gd}\sqrt{\frac{2h}{g}}$$
$$R = 2\sqrt{hd}$$

or $d = \dfrac{R^2}{4h}$

84. (a)

d_w = Density of water.

d_m = Density of mercury.

$$\frac{1}{2}d_w v^2 = d_m g \times 0.4$$

Specific gravity of Hg, $\frac{d_m}{d_w} = 13.6$

$\Rightarrow$ $v^2 = 2\times 13.6\times 9.8\times 0.4$

$\Rightarrow$ $v \cong 10.3$ m/s

85. (c)

Excess pressure in each of the two bubbles will be $\frac{4T}{0.04}$ and $\frac{4T}{0.03}$, where T is surface tension. Let the radius of the interface formed be r. Then excess pressure on the concave side compared to the convex side will be $\frac{4T}{r}$.

The difference in pressures on the two sides of the common surface is $\frac{4T}{0.03} - \frac{4T}{0.04}$.

$$\frac{4T}{r} = \frac{4T}{0.03} - \frac{4T}{0.04}$$

$\Rightarrow$ $\frac{1}{r} = \frac{0.01}{12\times 10^{-4}}$

$\Rightarrow$ $r = 0.12$ m

86. (d)

The bubble is moving with a constant velocity, So,

Buoyant force = Viscous force

$$\rho V g = 6\pi\eta vr$$

$$\rho\times\frac{4}{3}\pi r^3\times g = 6\pi\eta vr$$

$\Rightarrow$ $\eta = \frac{2\rho r^2 g}{9v}$

$\Rightarrow$ $\eta = \frac{2\times 1.5\times 10^3\times 10^{-4}\times 10}{9\times 2\times 10^{-3}}$

$$= \frac{1000}{6} \cong 167$$

87. (c)

In a steady flow,

$$A_1 v_1 = A_2 v_2$$

$$\pi\frac{d_1^2}{4}v_1 = \pi\frac{d_2^2}{4}v_2$$

$$v_1\times d_1^2 = v_2\times d_2^2$$

$$3\times(0.04)^2 = v_2\times(0.02)^2$$

$\Rightarrow$ $v_2 = 12$ m/s

88. (c)

$$T = \frac{E}{2A}$$

$\Rightarrow$ $E = 2A\times T$

$$= 2\times 0.02\times 5 = 2\times 10^{-1}\text{ J}$$

89. (c)

If a viscous fluid through a cylindrical tube, then the velocity distribution of the fluid is given by the Hazen Poiseuille law:

$$u = U_{\max}\times\left[1-\left(\frac{r}{R}\right)^2\right]$$

The velocity profile varies parabolically.

90. (d)

$$W = T(\Delta A)$$

$$\Delta A = 4\pi r^2$$

$\Rightarrow$ $V = \frac{4}{3}\pi r^3$

$\Rightarrow$ $r = \left(\frac{3V}{4\pi}\right)^{1/3}$

Now, radius of second bubble of volume 2 V be r_1.

Now, $2V = \frac{4}{3}\pi r_1^3$

$\Rightarrow$ $r_1 = 2^{1/3}r.$

$$W = T\times 4\pi r^2$$

$$W' = T\times 4\pi\times 4^{1/3}r^2$$

$\Rightarrow$ $W' = 4^{1/3}W.$

91. (d)

$$F = 6\pi\eta vr$$

$\Rightarrow$ $F = 6\pi\times 1.8\times 10^{-5}\times 1\times 3\times 10^{-4}$

$$= 101.74\times 10^{-9}\text{ N}$$

92. (c)

Atmospheric pressure = 1 atm

Pressure in two soap bubbles are 1.01 and 1.02 atm.

So, excess pressure are 0.01 and 0.02 atm

$$P_{excess} = \frac{4T}{R} \Rightarrow 0.01:0.02 = \frac{1}{R_1}:\frac{1}{R_2}$$

$\Rightarrow$ $1:2 = R_2:R_1$

Volumes $= R_1^3 : R_2^3 = 8:1$

93. (c)

Surface tension is defined as energy expanded per unit area of the surface. Surface tension depends upon (i) nature of the liquid (ii) nature of the surrounding matter

(iii) the kinetic energy and hence the temperature of the liquid molecules.

Even though the definition contains area, it does not depend on area of free surface, so the surface tension of the liquid remains unchanged.

94. (c)

Pressure inside the soap bubble

$$P = P_0 + \frac{4T}{R}$$

Where, P_0 is the atmospheric pressure.

Thus, $$P - P_0 = \frac{4T}{R}$$

For the two bubbles of volume V_A and V_B

$$P_A - P_0 = \frac{4T}{R} = 1.01 - 1.00 = 0.01$$

$$P_B - P_0 = \frac{4T}{R_B} = 1.02 - 1.00 = 0.02$$

($\because$ Atmospheric pressure $P_0 = 1$ atm)

$$\frac{R_A}{R_B} = 2 \text{ and } \frac{V_A}{V_B} = \frac{\frac{4}{3}\pi R_A^3}{\frac{4}{3}\pi R_B^3} = \left(\frac{R_A}{R_B}\right)^3 = 8$$

95. (b)

$$P = \frac{4T}{R} = \frac{8T}{d}$$

$$\Rightarrow \quad P = \frac{8 \times 25 \times 10^{-3}}{10^{-2}} = 20\text{Pa}.$$

96. (a)

Due to surface tension, an isolated drop of liquid takes nearly a spherical shape.

97. (d)

$$\frac{1}{2}\rho v^2 = (p_1 - p_2)$$

$$\Rightarrow \quad \frac{1}{2} \times 10^3 \times v^2 = 4.5 \times 10^5 - 4 \times 10^5$$

$$\Rightarrow \quad v = 10 \text{ m/s}$$

98. (c)

Pressure is constant at neck and bottom.

P = Pressure on neck

$$= \frac{12}{\pi \times (3 \times 10^{-2})^2}$$

The same pressure will be transferred to bottom = P

Force on bottom = $P \times$ Area of Bottom

$$= \frac{12}{\pi \times (3 \times 10^{-2})^2} \times \pi \times (15 \times 10^{-2})^2$$

$$= 300\text{N}$$

99. (a)

$$F = -\eta A \frac{dv}{dy}$$

$$\Rightarrow \quad \frac{F}{A} = -\eta \frac{dv}{dy}$$

The body is travelling in a straight line with constant velocity. So,

$$\frac{dv}{dy} = \frac{v}{y}$$

$$\Rightarrow \quad \frac{F}{A} = \eta \frac{v}{y}$$

$$F = 10^{-2}\,N = 10^3$$

$$A = 10^3 \text{ cm}^2$$

$$v = 6 \text{ cm/s}$$

$$y = \frac{6}{10} \text{ cm}$$

$$\Rightarrow \quad \frac{10^3}{10^3} = \eta \frac{6}{\frac{6}{10}}$$

$$\Rightarrow \quad \eta = 0.1 \text{ poise}$$

100. (c)

Volume is conserved in the process. Let r be the radius of smaller sphere and R be the radius of bigger sphere.

Now, $$64 \times \left(\frac{4}{3}\pi r^3\right) = \frac{4}{3}\pi R^3$$

$$\Rightarrow \quad R = 4r$$

$$v_T \propto r^2$$

$$\frac{v_1}{r_1^2} = \frac{v_2}{r_2^2}$$

$$\Rightarrow \quad \frac{1.5}{r^2} = \frac{v_2}{16r^2}$$

$$\Rightarrow \quad v_2 = 24\text{m/s}$$

101. (b)

$$v_e = \sqrt{2gh} = \sqrt{2 \times 10 \times 20} = 20\text{m/s}$$

102. (c)

$$A_1v_1 = A_2v_2$$

$$\Rightarrow \quad \frac{v_1}{v_2} = \frac{A_2}{A_1}$$

103. (d)

$$h = \frac{2T}{\rho rg}$$

$$h \propto \frac{1}{g}$$

$$\because \quad g_{pole} > g_{equator}$$

$$h_{pole} < h_{equator}$$

104. (c)

The liquid 1 is over liquid 2. Therefore $\rho_1 < \rho_2$.
If ρ_3 had been greater than ρ_2, it would not be partially inside but anywhere inside liquid 2, if

$$\rho_3 = \rho_2$$

or it would have sunk totally, if ρ_3 had been greater than ρ_2.

$$\rho_1 < \rho_3 < \rho_2.$$

105. (d)

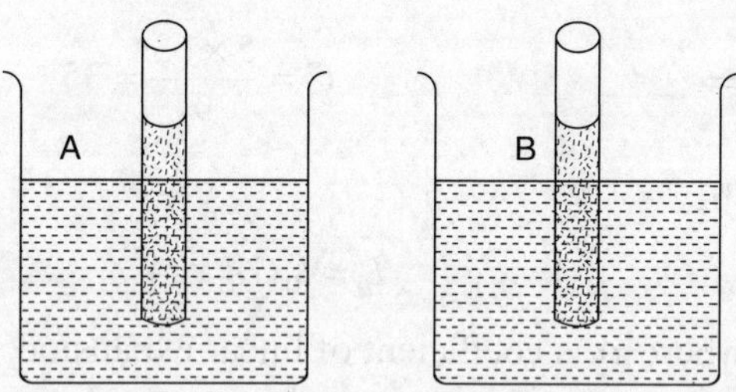

The force acting upwards $2\pi rT \simeq h\pi r^2\rho g$; the force acting down is $T\rho h$ without making finer corrections. Soap reduces the surface tension of water. The height of liquid supported decreases. But it is also a wetting agent. Therefore, the meniscus will not be convex as in mercury.

106. (b)

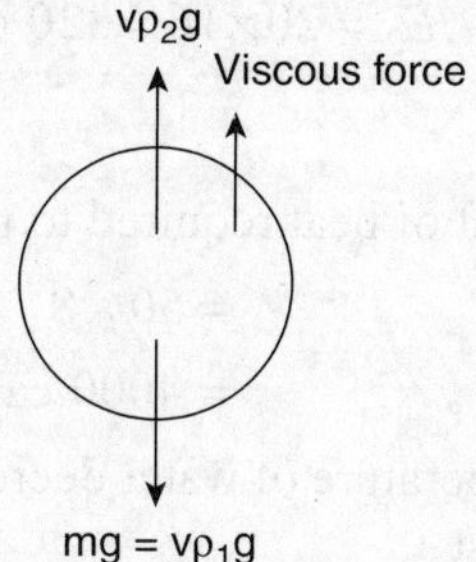

The forces acting on the solid ball when it is falling through a liquid are mg downwards, thrust by Archimedes principle upwards and the force due to the force of friction also acting upwards. The viscous force rapidly increases with velocity, attaining a maximum value when the ball attains the terminal velocity. Then the acceleration is zero.

$$mg - V\rho_2 g - kv^2 = ma$$

Where V is volume, v is the terminal velocity.
When the ball is moving with terminal velocity

$$a = 0.$$

Therefore,

$$V\rho_1 g - V\rho_2 g - kv^2 = 0$$

$$\Rightarrow \quad v = \sqrt{\frac{Vg(\rho_1 - \rho_2)}{k}}$$

107. (d)

Terminal velocity = v.
Viscous force upwards = Weight of sphere downwards

$$\text{or} \quad 6\pi\eta rv = \left(\frac{4}{3}\pi r^3\right)(\rho - \sigma)g$$

For gold and silver spheres falling in viscous liquid,

$$\frac{v_g}{v_s} = \frac{\rho_g - \sigma}{\rho_s - \sigma} = \frac{19.5 - 1.5}{10.5 - 1.5} = \frac{18}{9} = \frac{2}{1}$$

$$\text{or} \quad v_s = \frac{v_g}{2} = \frac{0.2}{2} = 0.1 \text{m/s}$$

108. (c)

In a freely falling elevator $g = 0$
Water will rise to the fall length, i.e., 20 cm to tube.

109. (c)

Pressure inside the bubble = $P_0 + \frac{4T}{r}$

Smaller the radius, greater will be the pressure.
Air flows from higher pressure to lower pressure. Hence air flows from the smaller bubble to the bigger.

110. (b)

Retarding viscous force = $6\pi\eta Rv$.
Obviously option (b) holds good.

111. (c)

A torque acts on the wall of the dam trying to make it topple. The bottom is made very board so that the dam will be stable.

112. (d)

Pressure inside a fluid volume open to atmosphere is

$$p = p_0 + h\rho g$$

where, p = pressure at depth h, h = depth

$$\rho = \text{density of fluid}$$

and g = acceleration due to gravity.

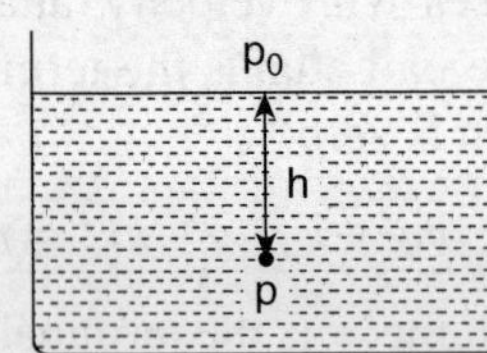

In problem given,
when $h = d_1$, pressure $p_1 = 5.05\times10^6$ Pa and when $h = d_2$, pressure $p_2 = 8.08\times10^6$ Pa
So, we have

$$p_1 = p_0 + d_1\rho g = 5.05\times10^6$$

and
$$p_2 = p_0 + d_2\rho g = 8.08\times10^6$$

$$\Rightarrow \quad p_2 - p_1 = (d_2 - d_1)\rho g = 3.03\times10^6$$

$$\Rightarrow \quad d_2 - d_1 = \frac{3.03\times10^6}{\rho g}$$

Given $\rho = 10^3\ \text{kgm}^{-3}$

and $g = 10\ \text{ms}^{-2}$

$$d_2 - d_1 = \frac{3.03\times10^6}{10^3\times10}$$
$$= 303\ \text{m} \approx 300\ \text{m}$$

113. (d)
l will decrease because the block moves up and h will decrease because the coin will displace the volume of water (V_1) equal to its own volume when it is in the water. However, when the coin is on the block it will displace the volume of water (V_2) whose weight is equal to the weight of coin and since density of coin is greater than the density of water, $V_2 < V_2$.

114. (c)
Let V be the volume of sphere, of which half is immersed in oil and the other half in mercury. Then for condition of floating

$$V\times\rho\times g = \frac{V}{2}\times13.6g + \frac{V}{2}\times0.8g$$

Where ρ is density of material of sphere.

$$\rho = \frac{13.6}{2} + \frac{0.8}{2} = 6.8 + 0.4 = 7.2\ \text{gm/cc}$$

115. (c)
$$v_{efflux} = \sqrt{2gh}$$

$$(H-h) = \frac{1}{2}gt^2, t = \frac{\sqrt{2(H-h)}}{g}$$

$$x = v_{eff}\times t = \sqrt{2gh}\times\frac{\sqrt{2(H-h)}}{g}$$
$$= 2\sqrt{(H-h)h}$$

116. (b)

117. (c)

118. (b)

119. (c)
As the body is dropped in a liquid, the velocity increases, viscous force which is velocity dependent also grows and a stage is reached when viscous force F equals the net weight, i.e., $w - u$. The body attains a steady velocity called terminal velocity, therefore, curve C is correct choice.

HEAT TRANSFER

120. (b)

$$\lambda_{m_1}T_1 = \lambda_{m_2}T_2$$

$$\lambda_{m_2} = \lambda_{m_1}\frac{T_1}{T_2}$$

$$= 4000\times\frac{3}{2} = 6000\ \text{Å}$$

121. (d)

$$\frac{C-0}{100} = \frac{F-32}{180}$$

$$\Rightarrow \quad \frac{C}{5} = \frac{95-32}{9}$$

$$\Rightarrow \quad C = \frac{5\times63}{9} = 35$$

122. (b)

$$l_t = l_0(1+\alpha t)$$

where α is coefficient of linear expansion.
l_t is proportional to α and $\alpha_c > \alpha_s$
Therefore, the expansion of copper rod is more than the expansion of steel rod. Hence, the rod bends with copper on concave side.

123. (c)

$$\frac{E_2}{E_1} = \frac{T_2^4}{T_1^4} = \left(\frac{1000}{500}\right)^4$$

$$E_2 = 20\times16 = 320\ \text{cal m}^{-2}s^{-1}$$

124. (a)
The amount of heat required to melt 50 g of ice

$$= 50\times8$$
$$= 4000\ \text{cal}$$

When temperature of water decreases from 80°C to 0°C the heat lost

$$= 50\times80$$
$$= 4000\ \text{cal}$$

Therefore, the temperature of mixture will be 0°C.

125. (c)

$$E = \sigma T^4$$

Taking log on both sides

$$L_n\ E = \log\ \sigma + 4\ L_n\ T$$

When $\quad T = 0$

$$L_n\ E = L_n\ \sigma$$

The body radiates (the net radiation) when the temperature of the body greater than surrounding.

126. (c)
It is done to pump up heat more efficiently.

127. (b)

$$E = 20 \times (2)^4$$

$$= 320\ \text{k cal/m}^2\ \text{min}$$

128. (b)

$$E = e\sigma T^4$$

As the temperature of all the bodies are same, the one with greatest emissivity will radiate more.
The emissivity of black body is greatest.

129. (a)
The length will change because of two factors:
(1) There will be contraction as the temperature has reduced
(2) There will be elongation as the wire is stretched a weight

The expansion

$$\Delta l_1 = l_1 \alpha\ \Delta t$$

$$= 3 \times 1.2 \times 10^{-5} \times [(30 - (-170))]$$

$$= 3 \times 1.2 \times 10^{-5} \times 200 = 7.2 \times 10^{-3}\ \text{m}$$

The compression

$$\Delta l_2 = \frac{l \times \text{stress}}{Y} = \frac{l \times \text{Force}}{Y \times \text{Area}}$$

$$= \frac{3 \times 10 \times 10}{2 \times 10^{11} \times \pi (0.5)^2 \times 10^{-6}} = \frac{6}{\pi} \times 10^{-3}$$

The net change in length

$$\Delta l_1 - \Delta l_2 = \left(7.2 - \frac{6}{\pi}\right) \times 10^{-3}\ \text{m}$$

$$= 5.2\text{mm}$$

130. (b)

$$\Delta l = l\ \alpha\ \Delta T$$

$$\Delta l = 10 \times 10 \times 10^{-6} \times 100$$

$$= 10^{-2}\ \text{m}$$

$$= 1\ \text{cm}$$

131. (c)

$$\lambda_0 T = \lambda' 2T$$

$$\Rightarrow \quad \lambda' = \frac{\lambda_0}{2}$$

132. (c)

$$Q = mC_p \Delta T$$

$$70 = 2C_p \times 5$$

$$\Rightarrow \quad C_p = 7$$

Now, $\quad C_v = C_p - R = 7 - 2 = 5$

$$Q' = 2 \times 5 \times 5$$

$$= 50 cal$$

133. (b)

$$\lambda_{m_1} T_1 = \lambda_{m_2} T_2$$

$$\lambda_{m_2} = \lambda_{m_1} \frac{T_1}{T_2}$$

$$= 4800 \times \frac{6000}{3000} = 9600\ \text{Å}$$

134. (b)

$$\frac{P_2 V_2}{T_2} = \frac{P_1 V_1}{T_1}$$

$$P_2 = \frac{P_1 V_1 \times T_2}{T_1 \times V_2}$$

$$= \frac{200 \times 100V \times 315}{295 \times 102V} = 209\ \text{kPa}$$

135. (a)
Rate of heat transfer,

$$\mu = \frac{KA(\theta_1 - \theta_2)}{l}$$

If the length of rod is l, the length of semicircular rod will be $\frac{\pi l}{2}$.

$$\mu \propto \frac{1}{l}$$

Therefore, the ratio is $l : \frac{\pi l}{2}$

$$= 2 : \pi$$

136. (d)
Let l be the length of the rod and A the area cross-section, Then, in the first case, the rods joined in series

$$Q_1 = \frac{KA(\theta_1 - \theta_2)t_1}{l}$$

When the rods are joined in parallel

$$Q_2 = \frac{K2A(\theta_1 - \theta_2)t_2}{l}$$

According to question, $Q_2 = Q_1$

$$\frac{t_1}{t_2} = 4 \Rightarrow t_1 = 12\times 4 = 48 \text{ s}$$

137. (d)

$$H = m\times 1\times(20.5 - 20) = \frac{m}{2}$$

$$\frac{H'}{H} = (2)^4 \Rightarrow H' = 2^4 H = 16H = 16\times 0.5$$

∴ The rise in temperature = 8°C.
∴ Final temperature = 20 + 8 = 28°C.

138. (d)
During phase change the temperature remains constant.

139. (b)
When water is heated the volume of the sphere increases due to expansion, but at the same time, the density of water decreases. The weight of the water displaced by sphere $= V\rho g$
Where ρ is the density of water. The density factor is more dominating than volume hence, overall $V\rho g$ (upthrust force) decreases and weight of the sphere in water increases.

140. (d)

$$E \propto T^4$$

Therefore, $n = 4$

141. (a)
Heat travels through vacuum by radiation.

142. (c)

$$ms\Delta T = mgh$$

$$\Delta T = \frac{gh}{s}$$

$$= \frac{10\times 1000}{0.1\times 1000\times 4.2}$$

$$\approx 23.8°C$$

143. (d)

$$\frac{E_2}{E_1} = \left(\frac{T_2}{T_1}\right)^4 = \left(\frac{T}{2T}\right)^4$$

$$= \frac{1}{16}$$

$$E_2 = \frac{E_1}{16}$$

144. (c)

$$\gamma_a = \gamma_r - \gamma_g$$

$$\gamma_a = \gamma_r - 3\alpha_g$$

$$\Rightarrow \quad \gamma_r = \gamma_a + 3\alpha_g$$

145. (b)
Let T_1 and T_2 be the temperature of the bodies in Kelvin

Then, $T_1 = 277 + 273 = 550\text{K}$

$$T_2 = 67 + 273 = 340\text{K}$$

Surrounding temperature

$$T_0 = 27 + 273 = 300\text{K}$$

Rate of heat transfer

$$H \propto (T^4 - T_0{}^4)$$

The ratio of heat transfer

$$\frac{H_1}{H_2} = \frac{(550)^4 - (300)^4}{(340)^4 - (300)^4}$$

$$= \frac{\left(\frac{550}{300}\right)^4 - 1}{\left(\frac{340}{300}\right)^4 - 1} \approx 7$$

146. (c)
Energy radiated per second

$$\sigma A(T^4 - T_0{}^4) = ms\frac{d\theta}{dt}$$

$$\Rightarrow \quad s = \frac{\sigma A(T^4 - T_0{}^4)}{ms\frac{d\theta}{dt}}$$

$$= \frac{(5.73\times 10^{-8})(19.2\times 10^{-4})\times(400^4 - 300^4)}{(34.38\times 10^{-3})(0.04)}$$

$$= 1400 \text{ JKg}^{-1}\text{K}^{-1}$$

147. (b)

$$\Delta U = ms\Delta T$$

or $$ms = \frac{\Delta U}{\Delta T} = \frac{80}{20}$$

$$= 4 \text{ J/K}$$

148. (d)

$$\frac{\lambda_{m_2}}{\lambda_{m_1}} = \frac{T_1}{T_2} = \frac{0.13}{0.26} = \frac{1}{2}$$

Now, $\frac{E_1}{E_2}=\left(\frac{T_1}{T_2}\right)^4=\left(\frac{1}{2}\right)^4=\frac{1}{16}$

149. (d)

$$\gamma_r=\gamma_a+\gamma_{vessel}$$

For vessel A,

$$\gamma_r=\gamma_1+3\alpha$$

For vessel B,

$$\gamma_r=\gamma_2+3\alpha'$$

Equating the above,

$$\gamma_2+3\alpha'=\gamma_1+3\alpha$$

$$\alpha'=\frac{\gamma_1-\gamma_2}{3}+\alpha$$

150. (b)

Newton's Law of cooling is

$$\frac{d\theta}{dt}\propto(\theta-\theta_0)$$

where θ_0 is iron temperature

$$\theta=\frac{\theta_1+\theta_2}{2}=\frac{60+50}{2}=55$$

$$\frac{60-50}{10}=K(55-25)$$

$$\Rightarrow K=\frac{1}{30}$$

Let the temperature in next 10 minutes be θ_3

$$\frac{50-\theta_3}{10}=K\left(\frac{50+\theta_3}{2}-25\right)$$

$$50-\theta_3=\frac{1}{3}\times\frac{\theta_3}{2}$$

Simplifying the above, we get $\theta_3=42.85°\text{C}$

151. (b)

Thermal resistance $=\frac{d}{KA}$

$$R_1=R_2$$

$$\frac{d_1}{K_1A}=\frac{d_2}{K_2A}$$

$$\Rightarrow\frac{K_1}{K_2}=\frac{d_1}{d_2}=\frac{5}{3}$$

152. (b)

Heat lost $=ms\frac{d\theta}{dt}$ per second

$$\frac{d\theta}{dt}\propto\frac{1}{s}$$

Rate of cooling is faster for A, therefore, the specific of A is less than B.

153. (d)

The latent heat of fusion of ice is large. It is 80 cal/g.

154. (c)

The temperature of white stars is the largest. Red stars are colder and the temperature of blue star is in between that of white star and red star.

155. (d)

$$\frac{50-45}{5}=K\left(\frac{50+45}{2}-\theta_0\right)$$

$$\frac{45-41.5}{5}K=\left(\frac{45+41.5}{2}-\theta_0\right)$$

On simplification the above

$$\theta_0=33.3°\text{C}$$

156. (a)

Thermal capacity $=ms$

The ratio $=\frac{m_1s_1}{m_2s_2}$ $(s_1=s_2=s)$

$$=\frac{\frac{4}{3}\pi R^3ds}{\frac{4}{3}\pi(4R)^3ds}=\left(\frac{1}{4}\right)^3=\frac{1}{64}$$

157. (a)

$$\frac{E_2}{E_1}=\left(\frac{T_2}{T_1}\right)^4$$

$$\frac{10^7}{10^3}=\left(\frac{T_2}{400}\right)^4$$

$$(10)^4=\left(\frac{T_2}{400}\right)^4$$

$$T_2=400\times10=400\text{ K}$$

158. (b)

$$\lambda_mT=b=\text{Wien's constant}$$

$$T=\frac{b}{\lambda_m}=\frac{2.93\times10^{-3}}{2.93\times10^{-10}}=10^7\text{ K}$$

159. (d)

In spite of the gas being in air-tight container, the gas will expand because of expansion of glass. Therefore, the density of gas will decrease.

160. (a)

$$E \propto T^4 \quad \text{(Stefan's Law)}$$

161. (b)

$$\left(\frac{T_2}{T_1}\right)^4 = \frac{E_2}{E_1} = 16$$

$$\frac{T_2}{T_1} = 2 \Rightarrow T_2 = 2 \times 300 = 600\text{K}$$

162. (d)

$$\frac{75-65}{2} = K\left(\frac{75+65}{2} - 30\right)$$

$$\left(\frac{55-45}{t}\right) = K\left(\frac{55+45}{2} - 30\right)$$

Solve the above two equations to find t

$t = 4$ minute

163. (c)

$$H = \frac{KA(\theta_1 - \theta_2)}{l}$$

$$= \frac{0.01 \times 0.8(30-0)}{2 \times 10^{-2}} = 12 \text{ J s}^{-1}$$

164. (a)

$$\frac{F-32}{9} = \frac{\theta}{5}$$

Let the temperature be θ,

Then $$\frac{\theta - 32}{9} = \frac{\theta}{5}$$

$$\Rightarrow \quad \theta = -40°\text{C}$$

165. (b)

Total change in length = Change in length of brass rod + Change in length of steel rod

$$\Delta l = \Delta l_1 + \Delta l_2$$

$$\Delta l_1 = l_1 \alpha_1 (T_2 - T_1)$$

$$= 500 \times 2.5 \times 10^{-5} \times (200 - 50)$$

$$= 1.875 \text{ mm}$$

$$\Delta l_2 = 500 \times 1.5 \times 10^{-5} \times (200 - 50)$$

$$= 0.9375 \text{ mm}$$

$$\Delta l = 1.875 + 0.9375$$

$$= 2.8 \text{ mm}$$

A better way of solving the problem is

$$\Delta l = \Delta l_1 (T_2 - T_1)(\alpha_1 + \alpha_2)$$

$$= 500 \times (250 - 50)(2.5 \times 10^{-5} + 1.25 \times 10^{-5})$$

$$= 2.8 \text{ mm}$$

166. (d)

The heat required to melt the ice is provided by the copper block. Let the mass of ice melt be $= m_2$

Then, $$m_2 L = m_1 s (T_2 - T_1)$$

$$m_2 \times 336 \times 10^3 = 4 \times 500 \times (425 - 0)$$

$$m_2 = \frac{4 \times 500 \times 425}{336 \times 10^3} = 2.5 \text{ kg}$$

167. (a)

The processes are as follows:

(i) The temperature of ice rises from –10°C to 0°C
(ii) Change of state at 0°C from ice to water
(iii) The temperature of water rises from 0°C to 100°C
(iv) Change of state at 100°C from 100°C water to 100°C steam.

The graph which depicts the above process is (a)

168. (a)

$$W_{ad} = \frac{nR(T_1 - T_2)}{\gamma - 1}$$

$$6R = \frac{R(T - T_2)}{\frac{5}{3} - 1}$$

$$6 = \frac{(T - T_2)}{2} \times 3$$

$$\Rightarrow \quad T - T_2 = 4$$

$$T_2 = (T - 4)\text{K}$$

169. (a)

Thermal radiation travels in a straight line with the speed of light.

170. (c)

$$K_{eq} = \frac{l_1 + l_2}{\frac{l_1}{k_1} + \frac{l_2}{k_2}}$$

$$= \frac{2l}{l\left(\frac{1}{K} + \frac{1}{2K}\right)} = \frac{4}{3}\text{K}$$

171. (c)

Kirchhoff's Law - A good absorber is also a good emitter. Initially it absorbs all the energy incident on it. Hence, it is darkest. Later on when the temperature of the black body becomes the temperature of the furnace, the black body radiates maximum energy and it will become brightest.

172. (d)

$$E = \sigma T^4$$

where E is emissive power and defined as energy per unit area per second. Therefore, E' energy per second is AE.

$$E' = \sigma A T^4$$

$$= \sigma (4\pi R^2) T^4$$

$$\frac{E_2'}{E_1'} = \left(\frac{T_2}{T_1}\right)^4 \left(\frac{R_2}{R_1}\right)^2$$

$$= (2)^4 \times (2)^2 = 64$$

173. (a)

$$\frac{\lambda_2}{\lambda_1} = \frac{T_1}{T_2} = \frac{200}{300}$$

$$\lambda_2 = \frac{2}{3}\lambda$$

174. (b)
The amount of heat required to melt 1 kg ice = mL

$$= 1 \times 33 \times 10^3 \text{ J/kg}$$

The heat lost by water in changing temperature from

$$10 \text{ to } 0°C = ms\ \Delta T$$

$$= 1 \times 4.2 \times 10^3 \times 10 = 42 \times 10^3 \text{ J}$$

The heat released by water is not sufficient to melt the entire ice. Therefore, the resulting temperature will be 0°C.

175. (b)
Thermal capacity = ms

$$\frac{c_1}{c_2} = \frac{m_1 s_1}{m_2 s_2}$$

$$= \frac{2}{3} \times \frac{3}{4} = \frac{1}{2}$$

176. (d)
Heat lost = Heat gained
Let t be the temperature of the mixture

then $\quad m_2 s_2 (t_2 - T) = m_1 s_1 (T - 30)$

$$m_2 s_2 (t_2 - T) = 20 \times 1(T - 30)$$

Substituting the values of $m_2 s_2$ and t_2 as given in the option and comparing, we find T is greatest for option (d).

177. (a)
Let specific heat of ice and water be respectively s_1 and s_2 heat lost by 10 g of water in reaching the temperature from 10°C to 0°C

$$= 10 s_2 \times 10$$

$$= 100 s_2 = 200 s_1 \quad (\because\ s_2 = 2 s_1)$$

The heat gained by ice in acquiring a temperature of 0°C is

$$10 s_1 \times 20 = 200 s_1$$

The heat lost by water is just sufficient to raise the temperature ice from –20°C to 10°C.

Therefore, calorimeter will contain 10 g of ice and 10 g of water.

178. (c)

$$m_A s_A (32 - 28) = m_B s_B (28 - 24)$$

$$m_A = m_B$$

$$\Rightarrow \quad \frac{s_A}{s_B} = \frac{4}{4} = 1:1$$

179. (a)
When A and B are mixed

$$m s_A (16 - 12) = m s_B (19 - 16)$$

$$\Rightarrow \quad \frac{s_A}{s_B} = \frac{3}{4}$$

Similarly, when B and C are mixed

$$\frac{s_B}{s_C} = \frac{(28 - 23)}{(23 - 19)} = \frac{5}{4}$$

$$\frac{s_A}{s_C} = \frac{s_A}{s_B} \times \frac{s_B}{s_C} = \frac{3}{4} \times \frac{5}{4} = \frac{15}{16}$$

When A and C are mixed.
Let temperature be T, then

$$s_A (T - 12) = s_C (28 - T)$$

$$\frac{s_A}{s_C} = \frac{28 - T}{T - 12} = \frac{15}{16}$$

$$T = 20.25$$

The nearest option is (a).

180. (c)

$$Q = ms\Delta T$$

$$s = \frac{Q}{m\Delta T}$$

$$\Delta T = 0$$

Therefore, specific heat of water s is infinite.

181. (a)
The ratio of amount of heat released by copper cylinders of masses 100 g and 200 g

$$= \frac{100 \times s \times 80}{200 \times s \times 80} = \frac{1}{2}$$

Now, $\quad \frac{m_1}{m_2} = \frac{A_1 h\rho}{A_2 h\rho} = \frac{100}{200}$

$$\Rightarrow \quad \frac{A_1}{A_2} = \frac{1}{2}$$

Let h_1 and h_2 be the depth of penetration, then

$$\frac{A_1 h_1 \rho}{A_2 h_2 \rho} = \frac{1}{2}$$

$$\frac{h_1}{h_2} = 2 \times \frac{1}{2} = 1$$

The ratio is 1 : 1

182. (d)

$$x \times 540 = y \times 80 + y \times 100$$

$$\Rightarrow \quad y/x = 3$$

$$y : x = 3 : 1$$

183. (c)

184. (c)

$$V \times 7675 \times 504 \times T = 336 \times 10^3 \times \frac{V}{2} \times 921$$

$$\Rightarrow \quad T = \frac{336 \times 10^3 \times 921}{7675 \times 504 \times 2} = 40°\text{C}$$

185. (a)

Newton's law of cooling

Rate of cooling $\propto$ Difference in temperature of the body and surroundings

First case with water

$$\frac{(0.2 \times 4200 + 0.1 \times 420) \times 40}{180} = k\left(\frac{60+40}{2} - 20\right) \quad \ldots\text{(i)}$$

Second case with liquid

$$\frac{(0.2 \times s + 0.1 \times 420) \times 40}{140} = k\left(\frac{60+40}{2} - 20\right) \quad \ldots\text{(ii)}$$

$\Rightarrow$ from Eqs. (i) and (ii),

$$s = 2390 \text{ J kg}^{-1}$$

The closest option is (a)

186. (c)

187. (c)

Let us consider the thermal resistance of inner and outer cylindrical shell as R_1 and R_2.

The temperature at the two ends are same. Therefore, the thermal resistances are in parallel

$$R_{eq} = \frac{R_1 R_2}{R_1 + R_2} = \frac{\dfrac{l}{k_1 A_1} \times \dfrac{l}{k_2 A_2}}{\dfrac{l}{k_1 A_1} + \dfrac{l}{k_2 A_2}}$$

Equating this to $\dfrac{l}{k\pi(2R)^2}$

Where k is equivalent thermal capacity of the system.

$$\frac{l}{k\pi(2R)^2} = \frac{\dfrac{l \times l}{k_1 A_1 \times k_2 A_2}}{\dfrac{l}{k_1 A_1} + \dfrac{l}{k_2 A_2}}$$

Put $\quad A_1 = \pi R^2$

$$A_2 = \pi(2R)^2 - \pi R^2 = 3\pi R^2$$

$$\Rightarrow \quad k = \frac{3k_2 + k_1}{4}$$

188. (d)

Let R and k be the equivalent thermal resistance and conductivity

$$\frac{1}{R} = \frac{1}{d/kA} \quad \ldots\text{(i)}$$

Then $\quad R_1 = \dfrac{1}{d/k_1 A/2}$

$$R_2 = \frac{1}{d/k_2 A/2}$$

$$\frac{1}{R} = \frac{1}{R_1} + \frac{1}{R_2}$$

$$\frac{1}{\dfrac{d}{A} \cdot \dfrac{1}{k}} = \frac{2}{\dfrac{d}{A}\left(\dfrac{1}{k_1} + \dfrac{1}{k_2}\right)} \quad \ldots\text{(ii)}$$

Comparing Eq. (ii) with Eq. (i)

$$k = \frac{2k_1 k_2}{k_1 + k_2}$$

189. (a)

$$\because \quad H_1 = \frac{kA_1}{l}(T_1 - T_2)$$

$$H_2 = \frac{k 4A_1 (T_1 - T_2)}{l/2} \quad (A_2 = 4A_1)$$

$$\Rightarrow \quad H_1 : H_2 :: 1 : 8$$

190. (a and c)

According to radiation theory every object emits radiation continuously. If the two bodies are at temperature T_A and T_A $(T_A > T_B)$

$$E_A = \sigma T_A{}^4; E_B = \sigma T_B{}^4$$

Radiation emitted by A is absorbed by B. Similarly radiation from B is absorbed by A

and $\quad T_A > T_B$

Therefore $\quad E_A > E_B$

Hence, the body A loses more heat than it absorbs.

191. (c)

Black bodies are better absorber of heat than white bodies.

192. (a)

$$H_1 = \frac{k_1 A(T_1 - T_0)}{d_1}$$

$$H_2 = \frac{k_2 A(T_0 - T_2)}{d_2}$$

where T_0 is the temperature at the common wall.

$$\frac{k_1 A(T_1 - T_0)}{d_1} = \frac{k_2(T_0 - T_2)}{d_2}$$

$$\frac{k_1(T_1 - T_0)}{d_1} = \frac{k_2}{d_2}(T_0 - T_2)$$

$$\Rightarrow \quad \frac{k_1 T_1}{d_1} - \frac{k_1 T_0}{d_1} = \frac{k_2 T_0}{d_2} - \frac{k_2 T_2}{d_2}$$

$$\left(\frac{k_1}{d_1} + \frac{k_2}{d_2}\right) T_0 = \frac{k_1 T_1}{d_1} + \frac{k_2 T_2}{d_2}$$

$$\Rightarrow \quad T_0 = \frac{k_1 T_1 d_2 + k_2 T_2 d_1}{k_1 d_2 + k_2 d_1}$$

193. (d)

$$H_1 = \frac{k_1 A_1 (T_1 - T_2)}{l}$$

$$H_2 = \frac{k_2 A_2 (T_1 - T_2)}{l}$$

$$H_1 = H_2 \Rightarrow \frac{A_1}{A_2} = \frac{k_2}{k_1}$$

194. (a)

$$\frac{E_2}{E_1} = \frac{\sigma T_2^4}{\sigma T_1^4} = \left(\frac{127 + 273}{27 + 273}\right)^4$$

$$\left(\frac{400}{300}\right)^4 = \frac{256}{81}$$

195. (c)

According to Wien's law

$$\lambda_m T = \text{Constant}$$

$$\lambda_{mr} T_A = \lambda_{mb} T_B = \lambda_{my} T_C$$

Where λ_m is the wavelength corresponding to maximum intensity.

Since $\lambda_{mb} < \lambda_{my} < \lambda_{mr}$ therefore $T_B > T_C > T_A$

196. (a)

$$\lambda_m T = \text{Constant}$$

where λ_m is the wavelength corresponding to maximum emissive power.

197. (d)

Kirchoff's law states that a substance in the vapour state at a given temperature emitting light of a particular wavelength would absorb light of the same wavelength at that temperature when exposed to white light from a source at a higher temperature, i.e., an emitter of a certain wavelength is also an absorber of the same wavelength.

198. (b)

Let the room temperature be T_0, then

$$\frac{dT}{dt} = -k(T_{\text{body}} - T_{\text{surrounding}})$$

where dT = Change in temperature
dt = Time for cooling

$$\left(\frac{75 - 70}{T_1}\right) = -k(72.5 - T_0) \quad \text{...(i)}$$

$$\left(\frac{70 - 65}{T_2}\right) = -k(67.5 - T_0) \quad \text{...(ii)}$$

$$\left(\frac{65 - 60}{T_3}\right) = -k(62.5 - T_0) \quad \text{...(iii)}$$

From Eqs. (i),(ii) and (iii),

$$\frac{5(72.5 - T_0)}{T_1} = \frac{5(67.5 - T_0)}{T_2} = \frac{5(62.5 - T_0)}{T_3}$$

$$\Rightarrow \quad T_1 < T_2 < T_3$$

199. (c)

Wien's law $\lambda_m T$ = Constant

$$\lambda_m T_1 = \lambda_{m_2} T_2$$

$$T_2 = \frac{\lambda_m T}{\lambda_{m_2}}$$

$$= \frac{12000 \times (2327 + 273)}{4800}$$

$$= \frac{12000 \times 2600}{4800}$$

$$= 6500 \text{ K}$$

200. (d)

Low specific heat helps in raising the cooking pot to a particular temperature with less amount of heat and high thermal capacity helps in transferring a particular amount of heat in lesser time.

201. (b)

λ_m corresponding to 3000 K is

$$= \frac{6000 \times 4800}{3000} = 9600 \text{ Å}$$

202. (c)

Let t be the time taken to increase the thickness ice from y_1 to y_2.

Then $$t = \frac{\rho L}{2k\theta}(y_2^2 - y_1^2)$$

where ρ is the density of ice, L is the latent heat of ice, K is the thermal conductivity of ice; let θ be the atmospheric temperature below 0°C.

$$t_1 = \frac{1}{2}\frac{\rho L}{k\theta}(1^2 - 0)$$

$$t_2 = \frac{1}{2}\frac{\rho L}{k\theta}(2^2 - 1^2)$$

$$\frac{t_2}{t_1} = \frac{3}{1} \Rightarrow t_2 = 7 \times 3 = 21 \text{ hours}$$

203. (b)

204. (d)

205. (d)

$$\lambda_{m_1} T_1 = \lambda_{m_2} T_2$$

$$\frac{T_2}{T_1} = \frac{\lambda_{m_1}}{\lambda_{m_2}} = \frac{\lambda_0 \times 4}{3\lambda_0}$$

$$\frac{T_2}{T_1} = \frac{4}{3}$$

$$\frac{P_2}{P_1} = \left(\frac{T_2}{T_1}\right)^4 = \left(\frac{4}{3}\right)^4 = \frac{256}{81}$$

206. (d)

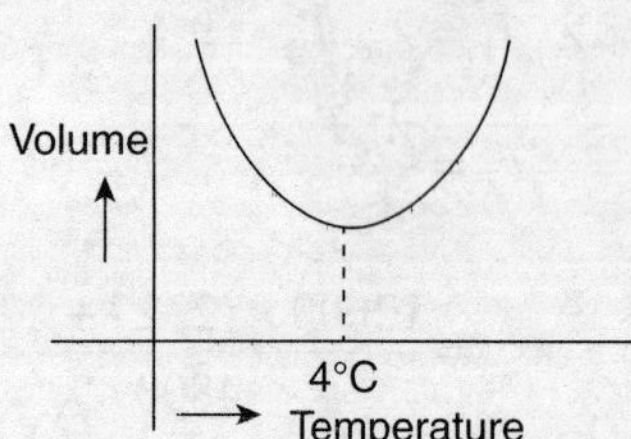

The above graph shows anomalous behavior of water at 4°C. The density of water is maximum and volume is minimum at 4°C.

207. (a)

208. (c)

209. (b)

$$l_1' = l_1(1 + \alpha_1 \Delta T)$$

$$l_2' = l_2(1 + \alpha_2 \Delta T)$$

$$l_2' - l_1' = (l_2 - l_1) + (l_2\alpha_2\Delta T - l_1\alpha_1\Delta T)$$

If $l_2' - l_1'$ is constant,

$$l_1\alpha_1\Delta T = l_2\alpha_2\Delta T$$

$$\Rightarrow \quad \frac{l_1}{l_2} = \frac{\alpha_2}{\alpha_1}$$

210. (d)

211. (b)

Energy radiated by the sun's surface per second is

$$P = \sigma \times 4\pi R^2 \times T^4$$

The emissive power at earth's surface $= \dfrac{\sigma 4\pi R^2 T^4}{4\pi r^2}$

The earth's surface which receives this energy is only half of the total surface area of the earth and its projection is πr_0^2 where r_0 is the radius of the earth.

The radiant power incident on the earth $= \dfrac{\sigma R^2 T^4}{r^2} \times \pi r_0^2$

$$= \frac{\pi r_0^2 R^2 \sigma T^4}{r^2}$$

The option (d) is in any case ruled out as it is dimensionally not satisfied.

212. (c)

Rate of flow of heat,

$$H = \frac{T_1 - T_2}{R}$$

(Analogy with current voltage and resistance relation $I = \dfrac{V}{R}$)

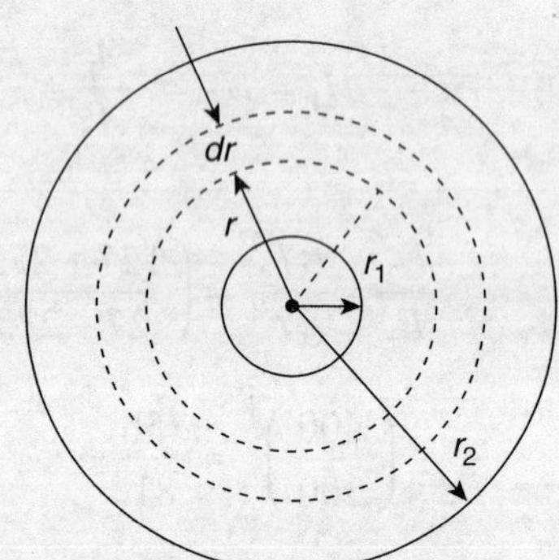

Resistance, $R = \displaystyle\int_{r_1}^{r_2} dR = \int_{r_1}^{r_2} \frac{dr}{k4\pi r^2}$

$$= \frac{1}{4\pi k}\left[\frac{1}{r_1} - \frac{1}{r_2}\right] = \frac{r_2 - r_1}{4\pi k r_1 r_2}$$

$$H = \frac{T_1 - T_2}{(r_2 - r_1)} 4\pi k r_1 r_2$$

Therefore, rate of heat flow is proportional to $\dfrac{r_1 r_2}{r_2 - r_1}$

213. (a)

Latent heat is independent of configuration. Ordered energy spent in stretching the spring will not contribute to heat, which is disordered kinetic energy of molecules of substance.

214. (b)

The initial momentum of the surface, $p_1 = \dfrac{E}{c}$

According to the question, the surface is perfectly reflecting.

So final momentum, $p_2 = -\frac{E}{c}$

∴ Change in momentum

$$\Delta p = p_2 - p_1$$

$$= -\frac{E}{c} - \frac{E}{c}$$

$$= -\frac{2E}{c}$$

So momentum transferred to the surface,

$$|\Delta p| = \frac{2E}{c}$$

215. (d)
Let T be the temperature of interface, then

$$H_2 = \frac{KA(T_2 - T)}{x}$$

$$H_1 = \frac{2KA(T - T_1)}{4x}$$

In steady state, the rate heat flow is constant. Therefore,

$$\frac{KA(T_2 - T)}{x} = \frac{2KA(T - T_1)}{4x}$$

$$\Rightarrow \quad T = \frac{2T_2 + T_1}{3}$$

$$H = \frac{KA(T_2 - T)}{x}$$

$$= \frac{KA}{x}\left(T_2 - \frac{2T_2 + T_1}{3}\right)$$

$$= \frac{KA(T_2 - T_1)}{3x}$$

Comparing this expression with the expression given in question,

$$\frac{A(T_2 - T_1)Kf}{x}$$

$$\therefore \quad f = 1/3$$

216. (d)
The rate of cooling is directly proportional to the difference in temperature of the body and surrounding

$$\frac{dQ}{dt} \propto \Delta\theta$$

The given relation is $\frac{dQ}{dt} \propto (\Delta\theta)^n$

$$\Rightarrow \quad n = 1$$

217. (a)

218. (a)

219. (a)

$$E = \sigma T^4 = 20. \quad T' = 2T$$

$$E' = \sigma(2T)^4 = 16\sigma T^4$$

$$= 16 \times 20 = 320 \text{ kcal/m}^2\text{min}$$

220. (b)
Temperature of black body (T') = 500 K. Therefore total energy emitted by the black body

$$(E) \propto T^4 \propto (500)^4$$

221. (a)
The rate of cooling is directly proportional to the temperature difference of the body and the surrounding.

222. (d)
Ratio of diameters of rod = 1 : 2 and ratio of their length 2 : 1.

The rate of flow of heat, $(R) = \frac{KA\Delta T}{l} \propto \frac{A}{l}$

Therefore, $\frac{R_1}{R_2} = \frac{A_1}{A_2} \times \frac{l_2}{l_1} = \left(\frac{1}{2}\right)^2 \times \frac{1}{2} = \frac{1}{8}$

or $R_1 : R_2 = 1 : 8$

223. (d)
Amount of energy radiated $\propto T^4$

224. (c)
Here, $F = 140°$

Using $\frac{F - 32}{180} = \frac{C}{100}$

$$\frac{140 - 32}{180} = \frac{C}{100} \Rightarrow C = 60°C$$

we get, fall in temperature = 40°C

225. (d)
Thermal capacity = ms = 40×0.2

$$= 8 \text{cal/K} = 33.6 \text{J/K}$$

226. (b)
Let the final temperature be T
Heat required by ice = mL + $m \times s \times (T - 0)$

$$= 10 \times 80 + 10 \times 1 \times T$$

Heat lost by water = $55 \times (40 - T)$
By using law of calorimetry,
Heat gained = Heat lost

$$800 + 10T = 55 \times (40 - T)$$

$$\Rightarrow T = 21.54°C = 22°C$$

227. (d)

228. (c)

229. (a)

230. (a)

231. (b)

232. (b)

233. (b)

234. (d)

235. (b)

In steady state,

$$\frac{\Delta H}{\Delta t} = -KA\frac{d\theta}{dx}$$

Here, $\frac{\Delta H}{\Delta t}$ = constant

$$\therefore \frac{d\theta}{dx} = \text{constant}$$

236. (d)

$$\frac{K_1 A(\theta - 0)t_1}{d} = mL = \frac{K_2 A(\theta - 0)t_2}{d}$$

$$K_1 t_1 = K_2 t_2$$

$$\frac{K_1}{K_2} = \frac{t_2}{t_1} = \frac{20}{30} = \frac{2}{3}$$

237. (a)

Let temperature of faces A, B, C and D be $\theta_1, \theta_2, \theta_3$ and θ_4. In steady state, same quantity of heat will flow from A to D per unit area per sec.

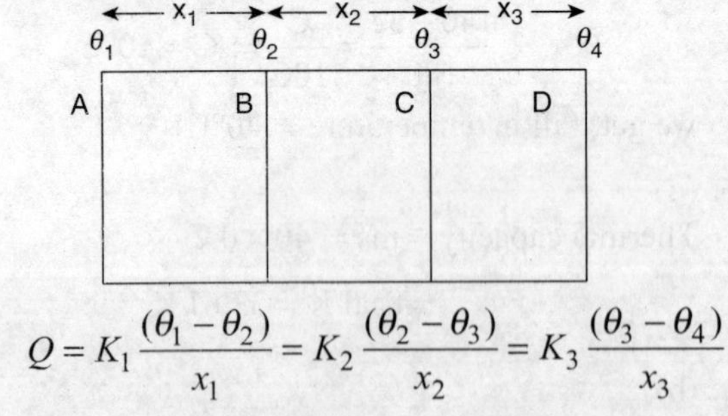

$$Q = K_1\frac{(\theta_1 - \theta_2)}{x_1} = K_2\frac{(\theta_2 - \theta_3)}{x_2} = K_3\frac{(\theta_3 - \theta_4)}{x_3}$$

Let the equivalent conductivity = K and thickness = $x_1 + x_2 + x_3$

$$Q = K_1\frac{(\theta_1 - \theta_4)}{(x_1 + x_2 + x_3)}$$

$$\therefore \frac{(x_1 + x_2 + x_3)}{K} = \frac{\theta_1 - \theta_4}{Q}$$

$$= \frac{\theta_1 - \theta_2}{Q} + \frac{\theta_2 - \theta_3}{Q} + \frac{(\theta_3 - \theta_4)}{Q}$$

$$Q = K = \frac{(x_1 + x_2 + x_3)}{\left(\frac{x_1}{K_1} + \frac{x_2}{K_2} + \frac{x_3}{K_3}\right)}$$

238. (c)

Heat in a conductor flows from higher to lower temperature. Hence, temperature at $x = 0$ is greater than $x = l$. So, (a) and (b) are wrong. Thermal resistance per unit length is constant so, temperature gradient $\left(\frac{dT}{dx}\right) = -H\lambda$.

Thus, temperature will decrease linearly with x. So, (c) is correct and (d) is wrong.

239. (b)

Let 'O' be the apex

$$\tan\theta = \frac{r_1}{x_1} = \frac{t_2}{x_2}$$

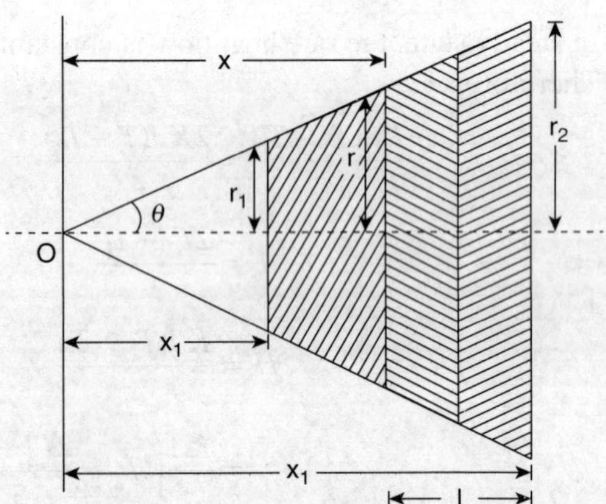

Let us consider a strip at a distance x of radius 'r'

$$x_2 - x_1 = l$$

Let rate of heat flow $= \frac{Q}{t} = -K\pi r^2\frac{dT}{dx}$

$$= -\pi K\frac{r_2^2}{x_2^2}x^2\frac{dT}{dx}$$

$$\int_{x_1}^{x_2}\frac{dx}{x^2} = -\int_{T_1}^{T_2}\frac{\pi K r_2^2}{x_2^2\left(\frac{Q}{t}\right)}dT$$

After solving $\frac{Q}{t} = \frac{\pi K r_1 r_2 (T_1 - T_2)}{l}$

240. (c)

241. (d)

242. (c)

243. (b)

244. (a)

245. (a)

246. (c)

Questions Asked in NEET/CBSE Medicals

247. (a)

$$l'_{cu} + l'_{Al} = l_{cu} + l_{Al} \Rightarrow \Delta l_{cu} = -\Delta l_{Al} \text{ or}$$

$$\Delta l_{Cu}\alpha_{Cu}\Delta T = -l_{Al}\alpha_{Al}\Delta T$$

$$|l_{Al}| = \frac{l_{Cu}\alpha_{Cu}}{\alpha_{Al}} = \frac{88\times 1.7\times 10^{-5}}{2.2\times 10^{-5}} = 68 \text{ cm}$$

248. (b)

According to Torricelli's theorem, Velocity, $v = \sqrt{2gh} = \sqrt{2\times 10\times 2} = 6.32$ m/s.

From equation of continuity,

Volume of liquid flowing per second, $Q = Av = 2\times 10^{-6}\times 6.32 = 12.6\times 10^{-6}\text{ m}^3/\text{s}$

249. (d)

$$\text{Stress} = \frac{F}{A} = \frac{Mg}{A}$$

$$\text{Strain} = \frac{\Delta L}{L} = \frac{L+l-L}{L} = \frac{l}{L}$$

Energy stored in the wire is,

$$U = \frac{1}{2}\times \text{Stress}\times\text{Strain}\times\text{Volume}$$

$$= \frac{1}{2}\times\frac{Mg}{A}\times\frac{l}{L}\times A\times L = \frac{1}{2}Mgl$$

250. (d)

The pressure at point Z_0 below the surface of water, $P_{Z_0} = P_0 + \rho g Z_0$

Pressure inside a soap bubble, $P = P_0 + \frac{4T}{R}$

$$P_{Z_0} = P$$

$$P_0 + \frac{4T}{R} = P_0 + \rho g Z_0$$

$$Z_0 = \frac{4T}{R\rho g} = \frac{4\times 2.5\times 10^{-2}}{1\times 10^{-3}\times 10^3\times 10}$$

$$1\times 10^{-2}\text{ m} = 1\text{ cm}$$

251. (d)

$$\text{Terminal velocity, } v = \frac{2r^2(\rho-\sigma)g}{9\eta}$$

Ratio of terminal velocity of spherical metal balls,

$$\frac{v_1}{v_2} = \frac{\frac{2}{9}(1)^2(8\rho_2 - 0.1\rho_2)}{\frac{2}{9}(2)^2(\rho_2 - 0.1\rho_2)}$$

$$\Rightarrow \quad \frac{v_1}{v_2} = \frac{7.9\rho_2}{4(0.9\rho_2)} = \frac{79}{36}$$

252. (b)

253. (c)

The viscous drag force, $F = 6\pi\eta r v$;

where v = terminal velocity

$\therefore$ The rate of heat generated = Power = Force × Terminal velocity

$$\Rightarrow \text{Power} = 6\pi\eta r v.v = 6\pi\eta r v^2 \quad \text{...(i)}$$

$$\because \text{Terminal velocity } v = \frac{2r^2(\rho-\sigma)g}{9\eta};$$

$$v \propto r^2$$

$$\text{Power} = 6\pi\eta r\left[\frac{4r^4(\rho-\sigma)^2}{81\eta^2}g^2\right]$$

or Power $\propto r^5$

254. (a)

$$\text{Young's modulus, Y} = \frac{Fl}{A\Delta l}$$

For wire 1,

$$\Delta l = \left(\frac{F}{AY}\right)3l \quad \text{(i)}$$

For wire 2, let F' force is applied

$$\frac{F'}{3A} = Y\frac{\Delta l}{l}$$

$$\Rightarrow \quad \Delta l = \left(\frac{F'}{3AY}\right)l \quad \text{(ii)}$$

From Eqs. (i) and (ii),

$$\left(\frac{F}{AY}\right)3l = \left(\frac{F'}{3AY}\right)l \Rightarrow F' = 9F$$

255. (c)

From Wien's law's, $\lambda_{max}T$ = constant. So, $\lambda_{max_1}T_1 = \lambda_{max_2}T_2$

$$\Rightarrow \lambda_0 T = \frac{\lambda_0}{4}T' \Rightarrow \frac{T'}{T} = \frac{4}{3}$$

According to Stefan–Boltzmann law, energy emitted in unit time by a black body is AeσT^4. This is the power radiated.

$$P \propto T^4$$

$$\text{So,} \quad \frac{P_2}{P_1} = \left(\frac{T'}{T}\right)^4 \Rightarrow n = \left(\frac{4}{3}\right)^4 = \frac{256}{81}$$

256. (d)

Equivalent thermal conductivity of the composite rod in parallel combination will be,

$$K = \frac{K_1A_1 + K_2A_2}{A_1 + A_2} = \frac{K_1 + K_2}{2}$$

257. (c)
According to Stefan-Boltzmann law, rate of energy radiated by a black body is given as

$$E = \sigma A T^4 = \sigma 4 p R^2 T^4$$

Given $E_1 = 450$ W, $T_1 = 500$ K, $R_1 = 12$ cm

$$R_2 = \frac{R_1}{2}, T_2 = 2T_1, E_2 = ?$$

$$\frac{E_2}{E_1} = \frac{\sigma 4\pi R_2^2 T_2^4}{\sigma 4\pi R_1^2 T_1^4} = \left(\frac{R_2}{R_1}\right)^2 \left(\frac{T_2}{T_1}\right)^4 = 4$$

$$E_2 = E_1 \times 4 = 450 \times 4 = 1800 \text{ W}$$

258. (c)
Pressure at point C,

$$P_C = P_a + \rho_{water} g h_{water}$$

where $h_{water} = CE = (65 + 65)$ mm $= 130$ mm
Pressure at point B, $P_B = P_a + \rho_{oil} g h_{oil}$
where $h_{oil} = AB = (65 + 65 + 10)$ mm $= 140$ mm
In liquid pressure is same at same liquid level,

$$P_B = P_C \Rightarrow \rho_{oil} g h_{oil} = \rho_{power} g h_{water}$$

$$\rho_{oil} = \frac{130 \times 10^3}{140} = \frac{13}{14} \times 10^3 = 928.57 \text{ kgm}^{-3}$$

259. (c)
Bulk modulus B is given as

$$B = \frac{-pV}{\Delta V} \quad \text{(i)}$$

The volume of a spherical object of radius r is given as

$$V = \frac{4}{3}\pi r^3, \Delta V = \frac{4}{3}\pi\left(3r^2\right)\Delta r$$

$$-\frac{V}{\Delta V} = \frac{\frac{4}{3}\pi r^3}{\frac{4}{3}\pi 3r^2 \Delta r} \text{ or } -\frac{V}{\Delta V} = \frac{r}{3\Delta r}$$

Put this value in Eq. (i), we get

$$B = -\frac{pr}{3\Delta r}$$

Fractional decreases in radius is

$$-\frac{\Delta r}{r} = \frac{p}{3B}$$

260. (b)
Work done = Surface tension of film $\times$ Change in area of the film
or, $W = T \times \Delta A$
Here, $A_1 = 4\text{cm} \times 2\text{cm} = 8\text{cm}^2$

$$A_2 = 5\text{cm} \times 4\text{cm} = 20\text{cm}^2$$

$$\Delta A = 2(A_2 - A_1) = 24\text{cm}^2 = 24 \times 10^{-4} \text{m}^2$$

$$W = 3 \times 10^{-4} \text{J}, T = ?$$

$$T = \frac{W}{\Delta A} = \frac{3 \times 10^{-4}}{24 \times 10^{-4}} = \frac{1}{8} = 0.125 \text{Nm}^{-1}$$

261. (b)
Capillary rise, $h = \dfrac{2T\cos\theta}{r\rho g}$

For given value of T and r, $h \propto \dfrac{\cos\theta}{\rho}$

Also, $h_1 = h_2 = h_3$ or $\dfrac{\cos\theta_1}{\rho_1} = \dfrac{\cos\theta_2}{\rho_2} = \dfrac{\cos\theta_3}{\rho_3}$

Since, $\rho_1 > \rho_2 > \rho_3$, so $\cos\theta_1 > \cos\theta_2 > \cos\theta_3$
For $0 \le \theta < \pi/2, \theta_1 < \theta_2 < \theta_3$
Hence, $0 \le \theta_1 < \theta_2 < \theta_3 < \pi/2$

262. (b)
Since heat capacity of material increases with increase in temperature, body at 100° C has more heat than body at 0° C. Therefore, final common temperature of the system will be closer to 100° C.

$$\therefore T_C > 50° \text{ C}$$

263. (b)
According to Newton's Law of cooling,

$$\frac{dT}{dt} = K(T - T_s)$$

Therefore,

$$\frac{dT_1}{dt} = K(T_1 - T_s) \text{ and } \frac{dT_2}{dt} = K(T_2 - T_s)$$

Here, $T_s = T, T_1 = \dfrac{3T + 2T}{2} = 2.5T$

and $\dfrac{dT_1}{dt} = \dfrac{3T - 2T}{10} = \dfrac{T}{10}$

$$T_2 = \frac{2T + T'}{2} \text{ and } \frac{dT_2}{dt} = \frac{2T - T'}{10}$$

So, $\dfrac{T}{10} = K(2.5T - T)$ (i)

$$\frac{2T - T'}{10} = K\left(\frac{2T + T'}{2} - T\right) \quad \text{(ii)}$$

Solving Eq. (i) and Eq. (ii)

$$T' = \frac{3}{2}T$$

264. (b)
Linear expansion of brass $= \alpha_1$
Linear expansion of steel $= \alpha_2$

Length of brass rod = l_1
Length of steel rod = l_2
On increasing the temperature of the rods by ΔT , new length would be

$$l_1' = l_1(1+\alpha_1\Delta T) \quad \text{(i)}$$

$$l_2' = l_2(1+\alpha_2\Delta T) \quad \text{(ii)}$$

Subtracting Eq. (i) from Eq. (ii), we get

$$l_2' - l_1' = (l_2 - l_1) + (l_2\alpha_2 - l_1\alpha_1)\Delta T \quad \text{(iii)}$$

According to Eq. (iii),

$$l_2' - l_1' = l_2 - l_1 \quad \text{(for all temperatures)}$$

$$l_2\alpha_2 - l_1\alpha_1 = 0 \text{ or } l_1\alpha_1 = l_2\alpha_2$$

265. (a)
Gravitational potential energy of a piece of ice at a height $(h) = mgh$
Heat absorbed by the ice to melt completely

$$\Delta Q = \frac{1}{4}mgh \quad \text{(i)}$$

Also, $$\Delta Q = mL \quad \text{(ii)}$$

From Eqs. (i) and (ii), $mL\frac{1}{4}mgh$ or $h = \dfrac{4L}{g}$

Here $L = 3.4\times10^5\,\text{Jkg}^{-1}, g = 10\text{Nkg}^{-1}$

$$h = \frac{4\times3.4\times10^5}{10} = 4\times34\times10^3 = 136\,\text{km}$$

266. (b)
According to Wien's displacement law

$$\lambda_m = \frac{b}{T} = \frac{2.88\times10^6\ \text{nmK}}{5760\text{K}} = 500\,\text{nm}$$

Clearly from graph, $U_1 < U_2 > U_3$

267. (b)
d = density of cylinder
A = area of cross-section of cylinder
Using law of floatation
Weight of cylinder = Upthrust by two liquids

$$L\times A\times d\times g = n\rho\times(pL\times A)g + \rho(L - pL)Ag$$

$$d = np\rho + \rho(1-p) = (np+1-p)\rho$$

$$d = \{1+(n-1)p\}\rho$$

268. (d)
Let the speed of the ejection of the liquid through the holes be v. Then according to the equation of continuity,

$$\pi R^2 V = n\pi r^2 v \text{ or } v = \frac{\pi R^2 V}{n\pi r^2} = \frac{VR^2}{nr^2}$$

269. (d)
Water will not overflow but will change its radius of curvature.

270. (d)
Let ρ_0 and ρ_T be densities of glycerin at 0°C and T°C respectively. Then,

$$\rho_T = \rho_0(1-\gamma\Delta T)$$

where γ is the coefficient of volume expansion of glycerin and ΔT is rise in temperature.

$$\frac{\rho_T}{\rho_0} = 1-\gamma\Delta T \text{ or } \gamma\Delta T = 1 - \frac{\rho_T}{\rho_0}$$

Thus, $$\frac{\rho_0-\rho_T}{\rho_0} = \gamma\Delta T$$

Here, $\gamma = 5\times10^{-4}K^{-1}$ and $\Delta T = 40° = 40$ K

$\therefore$ The fractional change in the density of glycerin

$$= \frac{\rho_0-\rho_T}{\rho_0} = \gamma\Delta T = (5\times10^{-4}K^{-1}) = (40K) = 0.020$$

271. (d)
Let L and A be length and area of cross-section of each wire. In order to have the lower ends of the wires to be at the same level (i.e., same elongation is produced in both wires), let weights W_s and W_b be added to steel and brass wires respectively. Then, by definition of Young's modulus, the elongation produced in the steel wire is

$$\Delta L_s = \frac{W_s L}{Y_s A}$$

and that in the brass wire is $\Delta L_b = \dfrac{W_b L}{Y_b A}$

But $$\Delta L_s = \Delta L_b \quad \text{(given)}$$

$$\therefore \frac{W_s L}{Y_s A} = \frac{W_b L}{Y_b A} \text{ or } \frac{W_s}{W_b} = \frac{Y_s}{Y_b}$$

$$\frac{Y_s}{Y_b} = 2 \quad \text{(given)}$$

Therefore $$\frac{W_s}{W_b} = \frac{2}{1}$$

272. (b)

273. (a)

274. (a)
Depth of ocean, $d = 2700$ m
Density of water, $\rho = 10^3\,\text{kgm}^{-3}$
Compressibility of water, $K = 45.4\times10^{-11}\,\text{Pa}^{-1}$

$$\frac{\Delta V}{V} = ?$$

Excess pressure at the bottom, $\Delta P = \rho g d$

$$= 10^3 \times 10 \times 2700 = 27 \times 10^6\ Pa$$

We know, $B = \dfrac{\Delta P}{(\Delta V / V)}$

$$\left(\frac{\Delta V}{V}\right) = \frac{\Delta P}{B} = K.\Delta P \quad \left(\therefore K = \frac{1}{B}\right)$$

$$= 45.4 \times 10^{-11} \times 27 \times 10^6 = 1.2 \times 10^{-2}$$

275. (b)

As $V = Al$

where A is the area of cross-section of the wire.

Young's modulus, $Y = \dfrac{(F/A)}{(\Delta l / l)} = \dfrac{Fl}{A\Delta l}$

$$\Delta l = \frac{Fl}{YA} = \frac{Fl^2}{YV} \quad \text{(Using (i))}$$

$$\Delta l \propto l^2$$

Hence, the graph between Δl and l^2 is a straight line.

276. (c) $\therefore$ Volume of n droplets = Volume of big drop

$$n \times \frac{4}{3}\pi r^3 = \frac{4}{3}\pi R^3 \Rightarrow n = \frac{R^3}{r^3} \qquad \text{(i)}$$

Volume of big drop, $V = \dfrac{4}{3}\pi R^3$ (ii)

Initial surface area of n droplets,

$$A_i = n \times 4\pi r^2 = \frac{R^3}{r^3} \times 4\pi r^2 \qquad \text{from (i)}$$

$$= 4\pi \frac{R^3}{r} = \left(\frac{4}{3}\pi R^3\right)\frac{3}{r} = \frac{3V}{r} \qquad \text{from (ii)}$$

Final surface area of big drop

$$A_f = 4\pi R^2 = \left(\frac{4}{3}\pi R^3\right)\frac{3}{R} = \frac{3V}{R} \qquad \text{from (ii)}$$

Decrease in surface area

$$\Delta A = A_i - A_f = \frac{3V}{r} - \frac{3V}{R} = 3V\left(\frac{1}{r} - \frac{1}{R}\right)$$

$\therefore$ Energy released = Surface tension $\times$ Decrease in surface area

$$= T \times \Delta A = 3VT\left(\frac{1}{r} - \frac{1}{R}\right)$$

277. (d)

Specific heat of water, $s_w = 1\ \text{cal}\ \text{g}^{-1}\ {}^\circ\text{C}^{-1}$

Latent het of steam, $L_s = 540\ \text{cal}\ \text{g}^{-1}$

Heat lost by m g of steam at 100° C to change into water at 80° C is

$$Q_1 = mL_s + ms_w \Delta T_w$$
$$= m \times 540 + m \times 1 \times (100 - 80)$$
$$= 540\ \text{m} + 20\ \text{m} = 560\ \text{m}$$

Heat gained by 20 g of water to change its temperature from 10° C to 80° C is

$$Q_2 = m_w s_w \Delta T_w = 20 \times 1 \times (80 - 10) = 1400$$

$\therefore$ 560m = 1400 or m = 2.5 g

Total mass of water present

$$= (20 + \text{m})\ \text{g} = (20 + 2.5)\ \text{g} = 22.5\ \text{g}$$

278. (a)

Let T_s be the temperature of the surroundings.

According to Newton's Law of cooling

$$\frac{T_1 - T_2}{t} = K\left(\frac{T_1 + T_2}{2} - T_s\right)$$

For first 5 minutes,

$$T_1 = 70°C, T_2 = 60°C, t = 5 \text{ minutes}$$

$$\frac{70 - 60}{5} = K\left(\frac{70 + 60}{2} - T_s\right)$$

$$2 = K\ (65 - T_s) \qquad \text{(i)}$$

For next 5 minutes,

$$T_1 = 60°C, T_2 = 54°C, t = 5 \text{ minutes}$$

$$\therefore \quad \frac{60 - 54}{5} = K\left(\frac{60 + 54}{2} - T_s\right)$$

$$\frac{6}{5} = K(57 - T_s) \qquad \text{(ii)}$$

Divide Eq. (i) by Eq. (ii), we get

$$\frac{5}{3} = \frac{65 - T_s}{57 - T_s v} \Rightarrow T_s = 45^\circ\ \text{C}$$

279. (d)

According to Wien's displacement law

$$\lambda_m T = \text{constant}$$

$$\lambda_m = \frac{\text{constant}}{T}$$

So when a piece of iron is heated, λ_m decreases, i.e., λ_m of the heated object will change from longer wavelength (red) to that of shorter (reddish yellow) wavelength and when the temperature is sufficiently high and all wavelength are emitted, the color will become white.

280. (b)

The wetting ability of a surface by a liquid depends primarily on angle of contact between the surface and the liquid.

281. (c)

Young's modulus,

$$Y = \frac{FL}{A\Delta L} = \frac{4FL}{\pi D^2 \Delta L} \text{ or } \Delta L = \frac{4FL}{\pi D^2 Y}$$

where F is the force applied, L is the length, D is the diameter and ΔL is the extension of the wire respectively.

For all the four wires, Y, F (= tension) are the same.

$$\Delta L \propto \frac{L}{D^2}$$

For (a) $\quad \frac{L}{D^2} = \frac{200\text{cm}}{(0.2\text{cm})^2} = 5\times10^3\ \text{cm}^{-1}$

For (b) $\quad \frac{L}{D^2} = \frac{300\text{cm}}{(0.3\text{cm})^2} = 3.3\times10^3\ \text{cm}^{-1}$

Similarly,

For (c) $\quad \frac{L}{D^2} = 20\times10^3\ \text{cm}^{-1}$.

For (d) $\quad \frac{L}{D^2} = 10\times10^3\ \text{cm}^{-1}$

Hence, ΔL is maximum for (c).

282. (b)

As $\quad Y = \frac{FL}{A\Delta L} = \frac{4FL}{\pi D^2 \Delta L}$

$$\Delta L = \frac{4FL}{\pi D^2 Y}$$

$$\frac{\Delta L_S}{\Delta L_C} = \frac{F_S}{F_C}\frac{L_S}{L_C}\frac{D_C^2}{D_S^2}\frac{Y_C}{Y_S}$$

Here, $\quad F_S = (5m+2m)g = 7mg$

$F_C = 5mg$

$\frac{L_S}{L_C} = q, \frac{D_S}{D_C} = p, \frac{Y_S}{Y_C} = s$

$$\frac{\Delta L_S}{\Delta L_C} = \left(\frac{7\text{mg}}{5\text{mg}}\right)(q)\left(\frac{1}{p}\right)^2\left(\frac{1}{s}\right) = \frac{7q}{5p^2 s}$$

283. (a)

Let L be length of each rod.

Rate of heat flow in rod 1 for the temperature difference ΔT is

$$H_1 = \frac{K_1 A_1 \Delta T}{L}$$

$$H_2 = \frac{K_2 A_2 \Delta T}{L}$$

Given, $H_1 = 4H_2$

$$\frac{K_1 A_1 \Delta T}{L} = 4\frac{K_2 A_2 \Delta T}{L}; K_1 A_1 = 4K_2 A_2$$

284. (a)

According to equation of continuity, Av = constant

Therefore, velocity is maximum at the narrowest part and minimum at the widest part of the pipe. According to Bernoulli's theorem, for a horizontal pipe.

$$P + \frac{1}{2}\rho v^2 = \text{constant}$$

Hence, when a fluid flows across a horizontal pipe of variable area of cross-section, its velocity is maximum and pressure is minimum at the narrowest part and vice versa.

285. (a)

As $\quad \rho_{T_2} = \frac{\rho_{T_1}}{(1+\gamma\Delta T)} = \frac{\rho_{T_1}}{1+\gamma(T_2-T_1)}$

Here, $\quad T_1 = 20°\text{C}, T_2 = 40°\text{C}$

$$\rho_{20} = 998\text{kg/m}^3, \rho_{40} = 992\text{kg/m}^3$$

$$992 = \frac{998}{1+\gamma(40-20)}$$

$$992 = \frac{998}{1+20\gamma} \text{ or } 992(1+20\gamma) = 998$$

$$= 3\times10^{-4}\ /°\text{C}$$

286. (d)

According to Stefan's law, $Q = \sigma A T^4$

or $\quad T = \left(\frac{Q}{\sigma A}\right)^{1/4} = \left(\frac{Q}{\sigma 4\pi R^2}\right)^{1/4}$

287. (a)

Temperature of liquid oxygen will first increase in the same phase. Then the liquid oxygen will change to gaseous phase during which temperature will remain constant. After that the temperature of oxygen in gaseous state will increase.

288. (a)

Heat flowing through the slab in t s is

$$Q = \frac{KA(T_1-T_2)t}{L} = \frac{K\times0.36\times(100-0)\times3600}{0.1}$$

$$= \frac{K\times0.36\times100\times3600}{0.1} \quad \text{(i)}$$

or $Q = m_{ice}L_f = 4.8\times3.36\times10^5 \quad$ (ii)

From (i) and (ii), we get

$$\frac{K\times0.36\times(100-0)\times3600}{0.1} = 4.8\times3.36\times10^5$$

$$K = \frac{4.8\times3.36\times10^5\times0.1}{0.36\times100\times3600} = 1.24\text{J/m/s/°C}$$

289. (b)

$$Q = \frac{KA(T_1 - T_2)t}{L}$$

Area of cross-section of new rod

$$A' = \pi\left(\frac{R^2}{2}\right)^2 = \frac{\pi R^2}{4} = \frac{A}{4}$$

$AL = A'L'$ (Volume remains constant)

where L' is the length of the new rod

$$\text{or } L' = L\frac{A}{A'}$$

$$= 4L$$

$$Q' = \frac{KA'(T_1 - T_2)t}{L'}$$

$$Q' = \frac{K(A/4)(T_1 - T_2)t}{4L} = \frac{1}{16}\frac{KA(T_1 - T_2)t}{L} = \frac{1}{16}Q$$

290. (a)

According to the Stefan–Boltzmann law,

$$P = \sigma 4\pi r^2 T^4$$

where, r = radius of the star, σ = Stefan's constant

The radiant power per unit area received at a distance R from the centre of a star is

$$S = \frac{P}{4\pi R^2} = \frac{\sigma 4\pi r^2 T^4}{4\pi R^2} = \frac{\sigma r^2 T^4}{R^2}$$

291. (c)

Power P radiated by the sun with its surface temperature $(t + 273)$ K is given by Stefan–Boltzmann law, where r is the radius of the sun and the sun is treated as a black body where $e = 1$.

The radiant power per unit area received by the surface at a distance R from the centre of the sun is given by

$$S = \frac{P}{4\pi R^2} = \frac{\sigma 4\pi r^2 (t+273)^4}{4\pi R^2} = \frac{r^2\sigma(t+273)^4}{R^2}$$

292. (b)

Rate of heat radiated at (227 + 273) K = 7 cals/(cm^{2}s)

Let rate of heat radiated at (727 + 273) K = x cals/(cm^{2}s)

By Stefan's law, $7 \propto (500)^4$ and $x \propto (1000)^4$

$$\frac{x}{7} = 2^4 \Rightarrow x = 7\times 2^4 = 112\text{cals/(cm}^2\text{s)}$$

293. (c)

$$\frac{dQ}{dt} = \frac{kA}{L}(T_1 - T_2)$$

294. (d)

$$39°\text{C} = 39\times 2 + 39 = (78 + 39)°\text{W}$$

$$= 117°\text{W}$$

295. (a)

According to Stefan's law, rate of energy radiated $E \propto T^4$

where T is the absolute temperature of a black body.

$$E \propto (727 + 273)^4 \text{ or } E \propto [1000]^4$$

296. (a)

According to Wien's displacement law,

$$\lambda_{\max} T = constnt$$

$$\frac{\lambda_{\max_1}}{\lambda_{\max_2}} = \frac{T_2}{T_1}$$

$$\text{or} \quad \lambda_{\max_2} = \frac{\lambda_{\max_1}\times T_1}{T_2} = \frac{5000\times 1000}{2500} = 3000\text{Å}$$

297. (b)

Heat conducted

$$= \frac{KA(T_1 - T_2)t}{l} = \frac{K\pi r^2(T_1 - T_2)t}{l}$$

The rod with the maximum ratio of A/l will conduct most. Here the rod with $r = 2r_0$ and $l = l_0$ will conduct most.

298. (d)

Wien's displacement law

$$\lambda_m T = \text{constant}, \ \lambda_m \propto T^{-1}$$

299. (a)

The slabs are in series.

$$\Rightarrow \quad \frac{l}{AK_{\text{effective}}} = \frac{l}{A.K} + \frac{l}{A2K}$$

$$\Rightarrow \quad \frac{1}{K_{\text{effective}}} = \frac{1}{K} + \frac{1}{2K} = \frac{3}{2K}$$

$$K_{\text{effective}} = \frac{2K}{3}$$

300. (d)

Unit of Stefan's constant is watt/m^{2}K^4

301. (a)

$$\text{Rate of heat loss in rod 1} = Q_1 = \frac{K_1A_1(T_1 - T_2)}{l_1}$$

$$\text{Rate of heat loss in rod 2} = Q_2 = \frac{K_2A_2(T_1 - T_2)}{l_2}$$

$$Q_1 = Q_2 \quad \text{(given)}$$

$$\frac{K_1A_1(T_1 - T_2)}{l_1} = \frac{K_2A_2(T_1 - T_2)}{l_2}$$

$$K_1A_1 = K_2A_2 \quad (\therefore l_1 = l_2)$$

302. (b)

Radiating power of a black body

$$= E_0 = \sigma(T^4 - T_0^4)A$$

where σ is known as the Stefan– Boltzmann constant. A is the surface area of the black body, T is temperature of the black body and T_0 is the temperature of the surrounding.

$$60 = \sigma\left(1000^4 - 500^4\right)$$

$$[T = 727°C = 727 + 273 = 1000K, T_0 = 227°C = 500K].$$

In the second case, $T = 1227°C = 1500$ K and let E' be the radiating power.

$$E' = \sigma\left(1500^4 - 500^4\right) \quad \text{(ii)}$$

From (i) and (ii) we have

$$\frac{E'}{60} = \frac{1500^4 - 500^4}{1000^4 - 500^4} = \frac{15^4 - 5^4}{10^4 - 5^4} = \frac{50000}{9375}$$

$$E' = \frac{50000}{9375} \times 60 = 320\text{W}$$

303. (b)

304. (a)

$$\lambda_{max} T = \text{Constant}$$

305. (b)

$$\text{Heat flow rate } \frac{dQ}{dt} = \frac{KA(T_1 - T_2)}{L} = Q$$

When linear dimensions are double

$$A_1 \propto r_1^2, L_1 = L$$

$$A_2 \propto 4r_1^2, L_2 = 2L \text{ so } Q_2 = 2Q_1$$

306. (b)

According to Wien's law,

$$\lambda_m T = \text{Constant}$$

$$\lambda' = (2/3)\lambda_m$$

QUICK QUESTIONS

Quiz

1. What is the origin of the saying, that "it is only the tip of an iceberg"?

2. Why are sleepers used below the rails?

3. A soft plastic bag weighs the same when empty or when filled with air at atmospheric pressure. Why?

4. When the same bag is weighed in vacuum, would the weight be the same?

5. How it is possible that a sadhu can stretch out for a nap on a bed of nails or thorns?

6. While drinking a liquid through a straw basically you reduce the pressure in your mouth and because atmospheric pressure is greater than mouth pressure, the liquid is pushed up. Could you use a straw to sip the liquid on the moon?

7. What kind of elasticity can be defined in the following activity?

(i) Blowing a balloon

(ii) Bending of beam

(iii) Cutting a piece of paper

(iv) Propagation of mechanical waves in a fluid.

8. How can mosquitoes and other insects live on water surface? And why do they sink when some kerosene oil is sprinkled on water?

9. The density of water is more than the density of air. Then explain why the clouds containing water droplets do not fall and instead continue to float.

10. The surface layer of a liquid is frequently likened to a stretched rubber film. In what respect does this analogy disagree with reality?

11. Why it is not advisable to sail two boats closely parallel to each other?

12. As water exits a faucet at a low setting so that a streamline flow of water results in the stream becoming narrower as it descends. Explain the reason.

13. On a winter night, you feel warmer when clouds cover the sky than when sky is clear. Explain.

14. What is wrong in the statement, ''A body at high temperature contains more heat"?

15. Distinguish between calorie and Calorie?

16. Can the specific heat of a substance be

(i) Zero (ii) Infinite (iii) Negative

Think and grow

17. A blackened platinum wire, when gradually heated, first appears dull red, then blue and finally white. Explain

18. A man is sitting in a boat which is floating in a pond. The man drinks some water from the pond. What will happen to level of water?

19. A bar of aluminium floats in mercury and the level is marked on the aluminium bar. Now the aluminum–mercury system is raised. Think of the level of mercury on the floating aluminium bar. Will it be higher, lower or at the same level as before?
Can the temperature of the system be raised to a point at which the aluminum bar sinks?

20. A home owner is painting a ceiling and a drop of paint falls from the brush onto an operating light bulb. The bulb breaks. Why?

21. Think! why in summer a piece of metal feels hotter than a piece of wood while in winter it is opposite although the temperature of both is same either in summer or winter. At what temperature the two would feel equally hot or cold?

22. Think and comment on the statement "Black body radiation is white"?

Explanations

Quiz

1. The density of ice is about 90% that of sea water and 90% of a floating iceberg is under water. Only 10% of it visible above water. Because of this reason we call it "tip of the iceberg".

2. This increases the cross sectional area. This, in turn, reduces the pressure on the rails.

3. Let us assume that the weight of empty bag is W and volume of the bag is V. When the bag is filled with air at N T P of density ρ its weight becomes $W + V\rho g$. Now when the bag filled with air is weighed in air, the thrust of the air is $V\rho g$ and the weight $W = W + V\rho g - V\rho g = W$.
The weight of the bag remains same whether it is filled with air or weight in air.

4. If the bas is weighed in vacuum, its weight is W when empty and when filled with air its weight will be $W + V\rho g$. In vacuum, there will be not be any upthrust force. So in vacuum, an air filled bag will weigh more than an empty bag.

5. The area of contact in lying posture is maximum but the pressure due to weight will be minimum and so they are not hurt.

6. No, moon has no atmosphere. There will be no atmospheric pressure to push the liquid up.

7. (i) Bulk (ii) Young's modulus
(iii) Shear (iv) Bulk

8. Their weight is balanced by the surface tension of water. When kerosene oil is sprinkled, due to contamination of surface, the surface tension decreases sharply and now water surface be no longer be able to support the weight of mosquitoes and insects. So they sink.

9. We know terminal velocity $v_t = \frac{2}{9}r^2(\frac{\rho - \tau}{g})g$

i.e., $v_t \propto r^2$. Due to small size of droplets in the clouds, they fall very slowly towards earth and appear to float in the sky.

10. When an elastic rubber film is stretched, the force of tension depends on the amount of deformation of the film. The force of surface tension is determined by the properties of the liquid only and does not change with increase of its surface.

11. When two boats move side by side in the same direction, the water in the region between them moves faster than on rear side. Therefore, in accordance with Bernoulli's principle, the pressure between the two is reduces. Because of this pressure difference, the boats will be pulled towards each other.

12. As the water exits, its speed increases. According to equation of continuity, Area x velocity (Av) remains constant at all points. As velocity increases, the stream must become narrower.

13. During daytime heat radiations coming from the sun are absorbed by the earth whereas at night, the earth radiates this heat. The temperature of sun is 6000 K. Therefore according to Wien's law $\lambda_m T$ = constant and the wavelength emitted by earth will be much longer than that that incident on it. Clouds have the property of passing

radiations of shorter wavelength and reflecting of larger wavelength. The heat emitted from the earth will not pass through the clouds. It is trapped between the earth and the cloud. Hence warmer nights.

In absence of clouds the energy emitted by the earth will be radiated in the sky. Hence, cool nights.

14. The concept of heat is the energy in transit. It depends upon the transfer of energy.

15. 'calorie' is the unit of heat energy whereas Calorie is the unit of energy value of food used in nutrition.

1 Cal = 1000 cal

16. (i) Yes,

$$\text{Specific heat } c = \frac{\Delta Q}{m\Delta T}$$

$c = 0$ is possible if either $\Delta Q = 0$

or $\Delta T = \infty$.

But ΔT cannot be ∞, so $\Delta Q = 0$

This means that specific heat is zero if the change is adiabatic such as during shaking of a liquid or compression of a gas.

(ii) Yes, for c to be infinity either $Q = \infty$ or $\Delta T = 0$

But Q cannot be infinity, when $\Delta T = 0$ the change is isothermal.

In state conversion $\Delta Q \neq 0$ but $\Delta T = 0$

(iii) Negative specific heat means that with increases in temperature ΔT, positive heat will be released and ΔQ is negative.

This happens in case of saturated vapors. Specific heat of saturated vapour is negative.

Think and grow

17. According to Wien's law $\lambda_m T$ = constant. When a black body is heated, λ_m decreases. It means, with increase in temperature the wavelength corresponding to maximum radiation shifts towards shorter wavelength. Therefore the colour of the heated objects change from longer to shorter wavelength, i.e., red to blue.

At a sufficiently high temperature all the wave lengths are emitted and the colour becomes white.

18. If man drinks m mass of water from the pond, the weight of the system forming man and boat will increase by mg and the system will displace additional mg weight for floating. Therefore, decrease in mg is compensated by the same weight mg because of water displaced from the system.

19. The density of aluminium 2700 kg/m^3 is less than density of mercury 13600 kg/m^3. The aluminium bar floats in mercury with 2700/13600$=\frac{1}{5}$ = 20% fraction submerged in mercury. The volume expansion coefficient of aluminium = 3 [14 × 10^{-6}(0°C)] $= 42\times10^{-6}\,C^{-1}$ is less than that of mercury 1182 × 10^{-6}(°C)$^{-1}$.

As the temperature increases, the density of both metals increases. But the density of mercury increases more. Therefore the fraction ρ_{Al}/ρ_{Hg} increases. As a result, a greater portion of the bar is immersed. The level of the mercury on the bar is higher than before and covers up the mark. In theoretical terms, the temperature of the system could be raised up to the point at which the density of the expanded mercury becomes equal to the density of aluminium and aluminium would then sink. Calculation shows that the temperature at which this occurs is far above the boiling point of mercury.

20. The glass envelope of the bulb receives energy on the inside surface by radiation from the hot filament. In addition because the bulb contains gas, the glass envelope receives energy by the movement of hot gas near the filament to the colder gas. Thus the glass can become very hot. If a drop of paint relatively colder falls onto the glass, the portion of glass envelope suddenly becomes colder than the other portions. This region contracts and can cause thermal stress and it might break the glass.

21. The feeling of hot or cold depends upon conductivity along with temperature. The thermal conductivity of metal is much more than the thermal conductivity of wood.

In summer the room temperature is more than that of body temperature 30°C. So on touching a metal, the heat will flow from the piece to our body. Since the conductivity of metal is much more than wood, the metal will conduct much more than wood, Therefore, the metal will appear hotter than wood.

In winter, the room temperature is lower than of our body so on touching a metal heat will flow from our body to the piece. But due to large thermal conductivity of the metal heat will flow from our body to metal therefore, it will appear colder than wood.

If the room temperature and body temperature are equal, there will not be any transfer of heat between our body and metal as well as wood. Therefore both will appear equally hot or cold.

22. The white radiation is composed of all possible wavelengths. A black body absorbs radiations of all wavelengths, so on heating to a suitable temperature it will emit radiations of all wavelengths. So we conclude that black body radiation is white.

CHAPTER TEST

1. A small hollow sphere, which has a small hole in it, is immersed in water to a depth of 0.5 m before any drop penetrates into it. If surface tension for water is 0.073 N/m, the radius of the hole is
(a) 0.03 mm (b) 0.06 mm
(c) 0.09 mm (d) 0.15 mm

2. Two spheres of the same material but of radii 0.01 m and 0.02 m are dropped one by one in the same viscous fluid. Their terminal velocities respectively will be in the ratio
(a) 1 : 1 (b) 1 : 2
(c) 1 : 4 (d) 4 : 1

3. The net rate at which heat is lost by a body due to radiation does not depend upon
(a) temperature of the body
(b) temperature of the surrounding
(c) material of the body
(d) nature of its surface.

4. The density of water a 40°C is 1000.0 kg/m^3 and at 1000°C it is 958.4 kg/m^3. The cubic expansivity of water between these temperatures is
(a) 4.5×10^{-3} / K (b) 5.4×10^{-5} / K
(c) 4.5×10^{-4} / K (d) 5.4×10^{-6} / K

5. Three metal rods of the same material and identical in all respects are joined as shown in the figure. The temperatures at the ends are maintained as indicated. Assuming no loss of heat from the curved surfaces of the rods, the temperature at the junction X would be

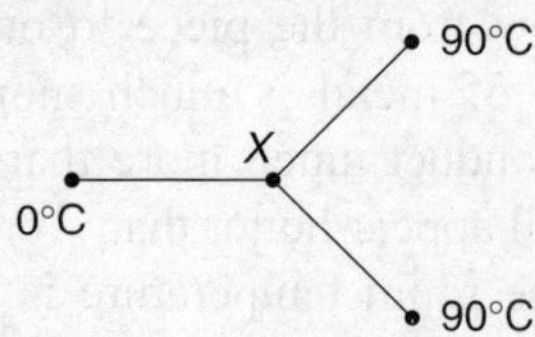

(a) 45° C (b) 60° C
(c) 30° C (d) 20° C

6. The planets with radii R_1, R_2 have densities ρ_1, ρ_2 respectively. Their atmospheric pressures are p_1, p_2 respectively. Therefore, the ratio of masses of their atmospheres, neglecting variation of g within the limits of atmosphere is
(a) $p_1 R_2 \rho_1 / p_2 R_1 \rho_2$ (b) $p_1 R_2 \rho_2 / p_2 R_1 \rho_1$
(c) $p_1 R_1 \rho_1 / p_2 R_2 \rho_2$ (d) $p_1 R_1 \rho_2 / p_2 R_2 \rho_1$

7. A large tank has two holes in the wall. Hole 1 has a square cross-section with side l and is cut at a depth h from the top. Hole 2 has a circular cross section of radius r at a depth of $4h$ from the top. When the tank is completely filled, the quantities of water flowing out through the holes per second are the same. Then, l is equal to
(a) $\frac{r}{\sqrt{2\pi}}$ (b) $r\sqrt{2\pi}$
(c) $\frac{r}{\sqrt{\pi}}$ (d) $r\sqrt{\pi}$

8. A body of density ρ enters a tank of water of density $\rho_1 (\rho_1 > \rho)$ after falling through a height h. The maximum depth to which it sinks in water is
(a) $h\rho_1 / (\rho_1 - \rho)$ (b) $h\rho / (\rho_1 - \rho)$
(c) $h\rho / \rho_1$ (d) $h\rho_1 / \rho$

9. Five identical balls made of different materials (Lead, Brass, Iron, Zinc and Copper) are heated up to 100°C in boiling water. The heat contained in all the balls will be
(a) equal
(b) different
(c) equal in zinc and lead ball
(d) equal in copper and iron ball.

10. Three cubes A, B, and C are connected to a horizontal pipe in which ideal liquid is flowing. The radii of the tubes A, B and C at the junction are respectively 2 cm, 1 cm, and 2 cm. It can be said:

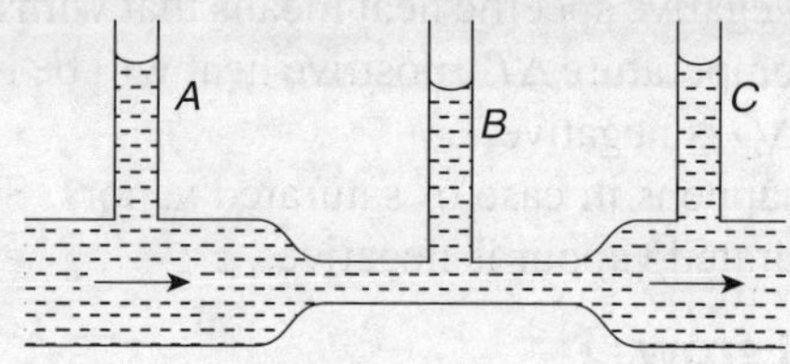

(a) The height of the liquid in the tube A is maximum.
(b) The height of the liquid in the tube A and B is same.
(c) The height of the liquid in the tube A, B and C is same.
(d) The height of the liquid in the tube A and C is same.

11. The raindrops falling from the sky neither hit us hard nor make holes on the ground because they move with
(a) constant acceleration
(b) variable acceleration
(c) variable speed
(d) constant terminal velocity.

12. Residents of those multi-storeyed buildings which are far removed from a municipal water tank generally find it difficult to get water on the top floor even if it is below the level of water in the main water tank. This is due to
(a) the fact that g decreases with altitude
(b) lesser value of atmospheric pressure at higher altitudes
(c) loss of pressure when water is flowing
(d) leakage of water.

13. A liquid flows through a horizontal hose pipe of radius 0.5 cm at a speed of 1 ms^{-1}. If the liquid is to come out of the pipe with a speed of 5 ms^{-1}, then the diameter of the nozzle should be
(a) $\frac{1}{\sqrt{5}}$ cm (b) $\sqrt{5}$ cm
(c) $(\sqrt{5}+1)$ cm (d) $(\sqrt{5}-1)$ cm

14. The equation of continuity
(a) is applicable to incompressible, non-viscous liquids only
(b) expresses the law of conservation of mass in fluid dynamics
(c) is applicable only when there are sinks but no sources
(d) is applicable only when there are sources but no sinks.

15. When there are no external forces, the shape of a liquid drop is determined by
(a) surface tension of the liquid
(b) density of liquid
(c) viscosity of liquid
(d) temperature of air only.

16. An iron needle slowly placed on the surface of water floats on it because
(a) when inside in water, it will displace water more than its weight
(b) the density of the material of needle is less than that of water
(c) of its surface tension
(d) of its shape.

17. A drop of oil is placed on the surface of water. Which of the following statements is correct?
(a) It will remain on it as a sphere
(b) It will spread as a thin layer
(c) It will partly be as spherical droplets and partly as thin film
(d) It will float as a distorted drop on the water surface.

18. Droplets of a liquid are usually more spherical in shape than larger drops of the same liquid because
(a) force of surface tension is equal and opposite to the force of gravity
(b) force of surface tension predominates the force of gravity
(c) force of gravity predominates the force of surface tension
(d) force of gravity and force of surface tension act in the same direction and are equal.

19. Two liquids of densities d_1 and d_2 are flowing in identical capillaries under same pressure difference. If t_1 and t_2 are the time taken for the flow of equal quantities of liquids, then the ratio of coefficients of viscosities of liquids must be
(a) $\frac{d_1 d_2}{t_1 t_2}$ (b) $\frac{d_1 t_1}{d_2 t_2}$
(c) $\frac{d_1 t_2}{d_2 t_1}$ (d) $\sqrt{\frac{d_1 t_1}{d_2 t_2}}$

20. What change in surface energy will be noticed when a drop of liquid of radius R splits up into 1000 droplets of radius r? Surface tension is T.
(a) $4\pi R^2 T$ (b) $7\pi R^2 T$
(c) $16\pi R^2$ (d) $36\pi R^2 T$

21. Work done to blow a bubble of volume V is W. The work done in blowing a bubble of volume $2V$ will be
(a) $(2)^{1/3} W$ (b) $2W$
(c) $(4)^{1/3} W$ (d) $4W$

22. Under a constant pressure head, the rate of flow of orderly volume flow of liquid through a capillary tube is V. If the length of the capillary is doubled and the diameter of the bore is halved, the rate of flow would become
(a) $V/4$ (b) 16 V
(c) $V/8$ (d) $V/32$

23. 16 cm^3 of water flows per second through a capillary tube of radius a cm and of length l cm when connected to a pressure head of h cm of water. If a tube of the same length and radius, $a/2$ cm is connected to the same pressure head, the quantity of water flowing through the tube per second will be
(a) 16 cm^3 (b) 4 cm^3
(c) 1 cm^3 (d) 8 cm^3

24. An incompressible fluid flows steadily through a cylindrical pipe which has radius $2R$ at point A and radius R at point B further along the flow direction. If the velocity at point A is v, its velocity at a point B will be
(a) $2v$ (b) v
(c) $v/2$ (d) $4v$

25. From amongst the following curves, which figure shows the variation of the velocity v with time t for a small-sized spherical body falling vertically in a long column of viscous liquid?

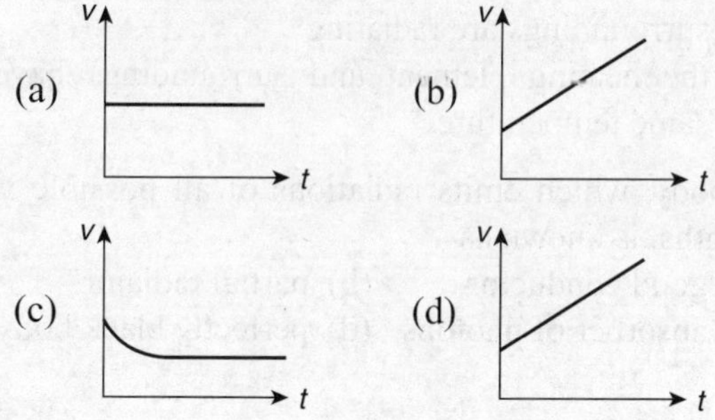

26. There is a dip in the capillary tube when the angle of the contact, θ is
(a) $\theta = 90°$ (b) $\theta = 0°$
(c) $\theta > 90°$ (d) $\theta < 90°$ but not zero.

27. A black body is heated from 27°C to 927°C. What will be the ratio of radiations emitted?
(a) 1 : 4 (b) 1 : 16
(c) 1 : 64 (d) 1 : 256

28. A hot liquid takes 5 minutes to cool from 70°C to 60°C. How long will it take to cool from 60°C to 50°C ?
(a) 5 minutes
(b) more than 5 mintues
(c) less than 5 minutes
(d) less or more than 5 minutes depending on the nature of liquid.

29. Figure shows the spectral energy density

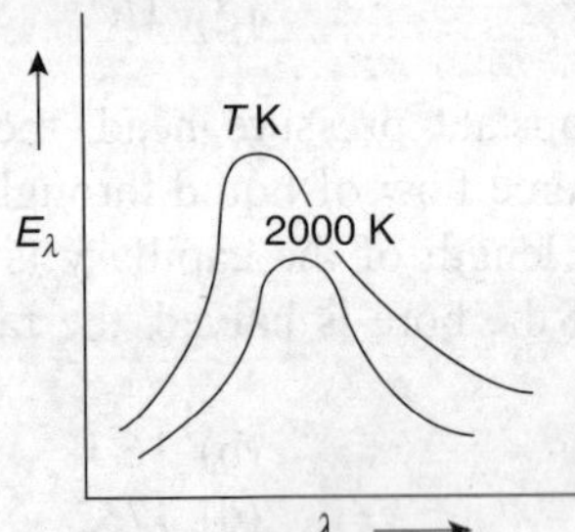

distribution E_λ of a black body at two different temperatures. If the areas under the curves are is the ratio 16 : 1, the value of temperature T is
(a) 32,000 K (b) 16,000 K
(c) 8,000 K (d) 4,000 K

30. An ideal black body at room temperature is thrown into a furnace. It is observed that
(a) initially it is the darkest body and at later times the brightest
(b) it is the darkest body at all times
(c) it cannot be distinguished at all times
(d) initially it is darkest body and at later times it cannot be distinguished.

31. When the energy supply being cut-off, an electric heater element cools down to temperature of its surroundings, but it will not cool further because
(a) supply is cut off
(b) it is made of metal
(c) surroundings are radiating
(d) the heating element and surroundings have the same temperature.

32. A body, which emits radiations of all possible wavelengths, is known as
(a) good conductor (b) partial radiator
(c) absorber of photons (d) perfectly black body.

33. Heat is flowing through a rod of length 50 cm and area of cross-section 5 cm^2. Its ends are respectively at 25°C and 125°C. The coefficient of thermal conductivity of the material of the rod is 0.092 kcal /m × s ×°C. The temperature gradient of the rod is
(a) 2°C/cm (b) 2°C/m
(c) 20°C/cm (d) 20°C/m

34. The energy spectrum of a black body exhibits maximum energy around a wavelength λ_0. The temperature of the black body is now changed such that the energy is maximum around a wavelength $3\lambda_0/4$. The power radiated by the black body will now increase by a factor of
(a) 256/81 (b) 64/27
(c) 16/9 (d) 4/3

35. Mud houses are cooler in summer and warmer in winter because
(a) mud is bad conductor of heat
(b) mud is superconductor of heat
(c) mud is good conductor of heat
(d) None of these.

36. The ratio of thermal conductivity of two rods of different materials is 5 : 4. The two rods of same area of cross-section and same thermal resistance will have the lengths in the ratio
(a) 5 : 4 (b) 1 : 9
(c) 9 : 1 (d) 4 : 5

37. Hooke's law essentially defines
(a) stress (b) strain
(c) yield point (d) elastic limit.

38. The longitudinal extension of any elastic material is very small. In order to have an appreciable change, the material must be in the form of a
(a) thin block of any cross-section
(b) thick block of any cross-section
(c) long thin wire
(d) short thin wire.

39. Which of the following substance has the highest elasticity?
(a) Steel (b) Copper
(c) Rubber (d) Sponge.

40. Shearing strain is expressed by
(a) angle of shear (b) angle of twist
(c) decrease in volume (d) increase in volume.

41. When a body undergoes a linear tensile strain, it experiences a lateral contraction also. The ratio of lateral contraction to longitudinal strain is known as
(a) Young's modulus (b) Bulk modulus
(c) Poisson's ratio (d) Hooke's law.

42. If a gas is heated at constant pressure, its isothermal compressibility
(a) remains constant
(b) increases linearly with temperature
(c) decreases linearly with temperature
(d) decreases inversely with temperature.

43. A long string is stretched by 2 cm and the potential energy is V. If the spring is stretched by 10 cm, its potential energy will be
(a) $V/25$ (b) $V/5$
(c) $5V$ (d) $25V$

44. A wire of length L and radius r is fixed at one end and a force F applied to the other end produces an extension l. The extension produced in another wire of the same material of length $2L$ and radius $2r$ by a force 2F is
(a) l (b) $2l$
(c) $l/2$ (d) $4l$

45. Two bars A and B of circular cross-section and same volume and made of the same material are subjected to tension. If the diameter of A is half that of B and if the force applied to both the rods is the same, the ratio of extension of A to that of B will be
(a) 16 (b) 15
(c) 8 (d) 24

Answer Keys

1. (a)	2. (c)	3. (c)	4. (c)	5. (b)	6. (d)	7. (b)	8. (b)	9. (b)	10. (d)
11. (d)	12. (c)	13. (a)	14. (a)	15. (a)	16. (c)	17. (b)	18. (b)	19. (b)	20. (d)
21. (c)	22. (d)	23. (c)	24. (d)	25. (d)	26. (d)	27. (d)	28. (b)	29. (d)	30. (d)
31. (d)	32. (d)	33. (a)	34. (a)	35. (a)	36. (a)	37. (d)	38. (c)	39. (a)	40. (a)
41. (c)	42. (a)	43. (d)	44. (a)	45. (a)					

Solutions

1. (a)
Excess of pressure inside a water drop = $2T/R$.
At depth h, pressure is $h\rho g$

2. (c)

$$\text{Terminal velocity} = v_T = \frac{2g(\rho-\sigma)r^2}{9\eta}$$

$$v_T \propto r^2$$

3. (c)

4. (c)

5. (b)

$$(90-\theta)+(90-\theta)=(\theta-0)$$

$$\Rightarrow \quad 3\theta = 180^0$$

$$\theta = 60^0$$

6. (d)

Use expression for $g = G\dfrac{(\frac{4}{3}\pi R^3\rho)}{R^2}$ and pressure

$$p = \frac{mg}{4\pi R^2}$$

7. (b)
The rates of flow of water (volume per unit time) through the holes 1 and 2 are respectively A_1v_1 and A_2v_2 whereas the speed v_1 and v_2 are $(2g \times h)^{1/2}$ and $(2g \times 4h)^{1/2}$. Also note that the areas A_1 and A_2 are l^2 and πr^2 respectively. Equate the rates to get the results.

8. (b)
When the body falls through a height h, it loses potential energy equal to mgh which is also the gain of kinetic energy, m being the mass of the body. This energy is utilized in taking the body to a depth of x, say against the upward force. The net upward force = Upthrust – Weight = $V\rho_1 g - V\rho g$ where V is the volume of the body. Equating the work done against this force to the gain of kinetic energy, that is, the loss of potential energy, one gets $(V\rho_1 g - V\rho g)x = V\rho gh$. Here, the mass in is written as $V\rho$.

9. (b) **10.** (d)

11. (d) **12.** (c)

13. (a)
Equation of continuity

$$A_1v_1 = A_2v_2$$

14. (a) **15.** (a)

16. (c) **17.** (b)

18. (b) **19.** (b)

20. (d)
We know that

$$W = 4\pi R^2[n^{1/3} - 1]T$$
$$= 4\pi R^2[(1000)^{1/3} - 1]\,T = 4\pi R^2 \times 9T$$
$$= 36\pi R^2 T$$

21. (c)

$$W_1 = 4\pi r_1^2 T \text{ and } W_2 = 4\pi r_2^2 T$$

$$\frac{W_2}{W_1} = \frac{r_2^2}{r_1^2} = \left[\left(\frac{r_2^3}{r_1^3}\right)^{1/3}\right]^2$$

$$= \left(\frac{V_2}{V_1}\right)^{2/3} = (2)^{2/3} = (4)^{1/3}$$

22. (d)

We know that $V = \dfrac{\pi \text{Pr}^4}{8\eta l}$

In second case $V_1 = \dfrac{\pi(\text{P})(\text{r}/2)^4}{8\eta(2l)} = \dfrac{V}{32}$

23. (c)

$$16 = \frac{\pi h a^4}{8\eta l}$$

$$V = \frac{\pi h(a/2)^4}{8\eta l} = \frac{\pi h a^4}{8\eta l} \times \frac{1}{16}$$

$$= 16 \times \frac{1}{16} = 1 \text{ cm}^3$$

24. (d)
We know that $A_1 v_1 = A_2 v_2$

$$4\pi(2R)^2 v = 4\pi R^2 v_1$$
$$v_1 = 4v$$

25. (d) **26.** (d)

27. (d)

28. (b)
We know that rate of loss of heat ∝ temperature difference

$$ms\frac{(70-60)}{5} \propto \left(\frac{70+60}{2} - T_0\right) \propto (65 - T_0)$$

$$\text{and } ms\frac{(60-50)}{t} \propto \left(\frac{60+50}{2} - T_0\right) \propto (55 - T_0)$$

This show that t should be greater than 5 minutes.

29. (d)

$$E_\lambda \propto T^4$$

$$\frac{(E_\lambda)_T}{(E_\lambda)_{2000\,K}} = \frac{T^4}{(2000)^4} = \frac{16}{1}$$

$$\left(\frac{T}{2000}\right) = \left(\frac{16}{1}\right)^{1/4} = 2$$

$$T = 4000 \text{ K}$$

30. (d)
Initially the body absorbs heat till it comes to the temperature of furnace. Then the rate of emission becomes equal to the rate of absorption. So, initially the body is darkest (being black body) and at later times, it cannot be distinguished.

31. (d)
When the energy is supplied to the electric heater element, it gets heated up and starts radiating heat to the surroundings due to temperature difference. When energy supply is cut-off, the heater element does not get heated. After sometime, its temperature becomes equal to that of surroundings. So, the heater does not cool further.

32. (d)
When a body emits radiations of all possible wavelengths, it is known as perfectly black body.

33. (a)
Temperature gradient

$$\frac{dT}{dx} = \frac{100°\text{C}}{50\,\text{cm}} = 2°\text{C/cm}$$

34. (a)

35. (a)
Mud is a bad conductor of heat. So it prevents flow of heat between the surrounding and inside.

36. (a)
Given that,

$$\frac{l_1}{K_1 A_1} = \frac{l_2}{K_2 A_2}$$

$$\frac{l_1}{l_2} = \frac{K_1 A_1}{K_2 A_2} = \frac{K_1}{K_2} \quad (\because A_1 = A_2)$$

$$\frac{l_1}{l_2} = \frac{5}{4}$$

37. (d)

38. (c)

$$Y = F\,L/Al \text{ or } l = FL/AY$$

39. (a)

40. (a)

41. (c)

42. (a)

43. (d)

$$V = \frac{1}{2}k(x)^2 = \frac{1}{2}k(2)^2$$

$$k = \frac{2V}{4} = \frac{V}{2}$$

$$V' = \frac{1}{2}k(10)^2 = \frac{1}{2}\times\left(\frac{V}{2}\right)(10)^2 = 25V$$

44. (a)

$$Y = \frac{F/\pi r^2}{l/L} = \frac{FL}{\pi r^2 l} \quad or \quad l = \frac{FL}{\pi r^2 Y}$$

Let l be the extension produced when a force 2 F is applied. Then,

$$l' = \frac{(2F)(2L)}{\pi(2r)^2 Y} = \frac{FL}{\pi r^2 Y} = l$$

45. (a)

$$Y = \frac{mgL}{\pi r^2 l} = \frac{mgL \times \pi r^2}{\pi^2 r^4 l} = \frac{MgV}{\pi r^4 l}$$

Here Y, volume V and mg are constants.

Hence $r^4 l$ = constant

$$r_1^4 l_1 = r_2^4 l_2 \text{ or } (l_1 / l_2) = (r_2 / r_1)^4$$

$$l_1 / l_2 = (2r / r)^4 = 16$$

Thermodynamics

Introduction

Thermodynamics deals with the relationship between heat, work and other forms of energy. It does not go into the detailed microscopic structure of a system but deals only with its macroscopic properties.

Thermometry

Thermal Equilibrium and Temperature

An important physical principle that is the basis of thermometry is the 'Zeroth law of thermodynamics'. It states that **two bodies, which are separately in thermal equilibrium with a third, will also be in thermal equilibrium with each other, if they are brought into thermal contact with each other.** Thermal equilibrium refers to the fact that the bodies don't exchange heat if they are placed in contact with each other, i.e., one does not become hotter and the other colder with time. Another important observation is that the flow of thermal energy between bodies is governed by the hotness or coldness of the bodies, which is quantitatively measured by temperature. These statements are sufficient to establish the existence of a physical quantity called 'temperature' which is associated with thermal equilibrium.

Thermodynamics

The study of energy relations and the manner in which energy changes take place are based on two general laws, namely, the first law and the second law of thermodynamics.

System

A system is defined as a region in space or a quantity of matter bounded by some closed surface on which the attention is focused.

Surroundings

Anything outside the system which can exchange energy and matter with the system and has a direct bearing on the behaviour of the system is termed the surroundings

Open system

A system which can exchange matter and energy with the surroundings is known as an open system, for example an air compressor.

Closed system

A system which can exchange only energy (not matter) with the surroundings is called a closed system. A gas cylinder fitted with a movable piston and heated by a burner is an example of a closed system.

Isolated system

A system which is not influenced in any way by its surroundings is an isolated system.

Thermodynamic State

When a system is in equilibrium with its surroundings, then thermodynamic coordinates Pressure (P), Volume (V) and temperature (T) of the system describe the thermodynamic state of the system

Process

Change of state of a system by altering the thermodynamic coordinates is known as process.

Reversible process

A reversible process is one that, by a differential change in the environment, can be made to retrace its path.

Quasi-static process

These are processes that are carried out slowly enough so that the system passes through a continuous sequence of equilibrium states. A quasi-static process may or may not be reversible.

Indicator diagram

The graphical representation of change of state of a gas by a thermodynamic process is called an indicator diagram. Indicator diagram is generally plotted as pressure vs. volume of a gas. The figure shows an indicator diagram in pressure and volume.

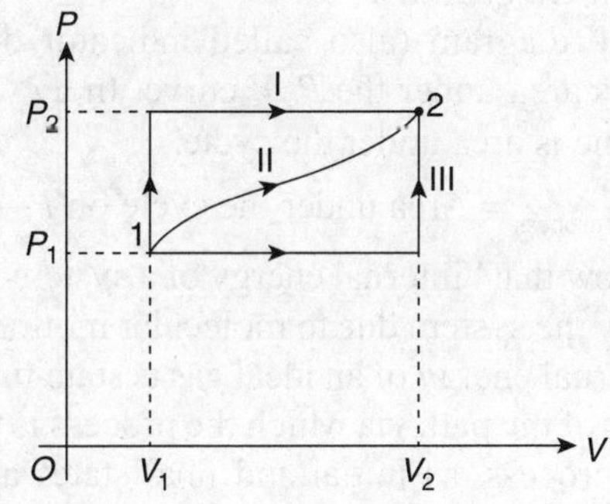

The indicator diagram shows the change of state 1 to 2, i.e., (P_1, V_1, T_1) to (P_2, V_2, T_2) through three different processes I, II and III. In path I, the initial volume V_1 is kept constant and pressure is raised to P_2; then pressure is kept constant and volume is changed to V_2.

In path II, pressure and volume are changed simultaneously along the curve shown, to attain state 2.

In path III, at first pressure is kept constant and volume is changed to V_2; then volume is kept constant and pressure is raised to P_2.

There are infinite ways in which the state of a system can be changed. When a system passes from one state to another state through a process, there are intermediate states. An indicator diagram represents the focus of all intermediate states between the initial and final state.

Quantities Involved in First Law of Thermodynamics

First law of thermodynamics is nothing but a restatement of the principle of conservation of energy. This law is the relation between Heat, Work and Internal Energy of a system. We know that heat is energy in transit across a boundary by virtue of temperature difference. If a gas is taken from state 1 to state 2 due to expansion via two paths, 1–A–2 and 1–B–2, then heat transfer in the two cases will be different, i.e., $(\Delta Q)_A \neq (\Delta Q_B)$, i.e., heat is path dependent between the two states.

We know that the heat required to change the temperature of a given gas at constant pressure is different from that required to change the temperature of the gas to the same extent at constant volume.

$(\Delta Q)_v = nC_V\Delta T$ if volume is kept constant

$(\Delta Q)_P = nC_P\Delta T$ if pressure is kept constant

$(\Delta Q)_{\text{cyclic process}}$ = Area under the cycle in the P–V diagram.

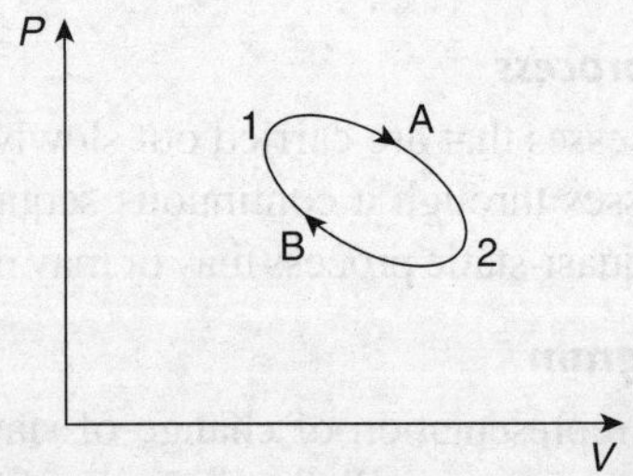

Similarly, work done in the two process between states 1 and 2 will be different, i.e., $(\Delta W)_A \neq (\Delta W)_B$ and thus work is also a path dependent quantity.

On a P–V diagram (also called indicator diagram), the work done is area under the P–V curve. In a cyclic process, the work done is area under the cycle.

$(\Delta W)_{\text{cyclic process}}$ = Area under the cycle on P–V diagram

Also, we know that "internal energy of a system is the energy possessed by the system due to molecular motion and configuration. Internal energy of an ideal gas is state function and is independent of the path via which the process is taking place. In a cyclic process, as initial and final states are the same, there is no net change in internal energy and $(\Delta U)_{\text{cyclic}} = 0$.

Also in any process,

$$\Delta U = nC_v\Delta T$$

Since $C_v = \dfrac{R}{\gamma - 1}$; between initial state 1 and final state 2;

$$\Delta U = \frac{nR\Delta T}{\gamma - 1} = \frac{P_2V_2 - P_1 - V_1}{\gamma - 1} a$$

Sign convention

Heat: When heat is supplied to the gas, it is taken as positive and when gas rejects heat it is taken as negative.

Work: When work is done by the gas, it is taken as positive; and when work is done on the gas, it is taken as negative.

Internal Energy: Increase in internal energy is taken as positive while decrease is taken as negative.

First Law of Thermodynamics

According to this law, heat given to a system (ΔQ) is equal to the sum of work done (ΔW) by the system and increase in its internal energy (ΔU), i.e.,

$$\Delta Q = \Delta U + \Delta W$$

This law is the consequence of many past experiments in which it was observed that if a process is taking place between two states i and f via many paths, in each case the quantity ($\Delta Q - \Delta W$) was the same though ΔQ and ΔW for different paths were different. It means that quantity ($\Delta Q - \Delta W$) depends only on state of gas and is path independent. Later, this quantity was termed as change in internal energy (ΔU) and U is a function of state of gas.

Path dependence of work and heat

The thermodynamic state of an ideal gas requires the specification of two variables: say temperature (T) and pressure (p). If these two variables are given, for a given mass of gas, the third variable (V: volume) may be calculated using the equation of state:

$$pV = nRT \quad \text{...(i)}$$

In order to represent all the equilibrium states of a given mass of an ideal gas, a p-V diagram is used. The work done by the gas in going from the state A to the state B by a reversible process is:

$$W_{AB} = \int_A^B dW = \int_A^B pdV \quad \text{...(ii)}$$

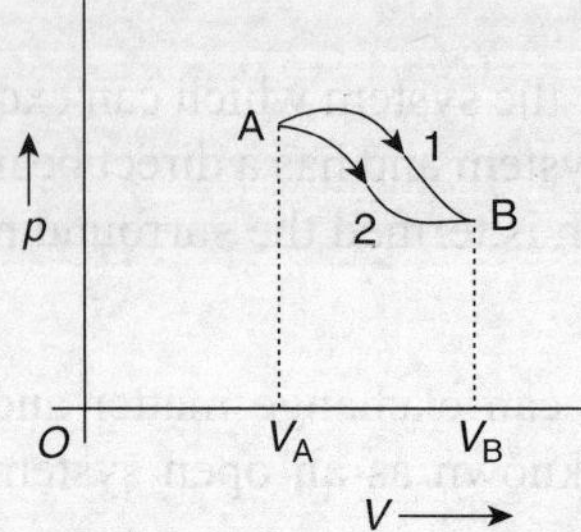

A reversible process is one in which the gas undergoes a slow transition such that it is almost always in equilibrium.

The quantity W_{AB} depends on the path taken by the process as it moves from A to B, which is given by the function $p(V)$. It is simple to compute this from the p-V diagram as this is the area under the curve. The work done in taking the system from A to B via path 1 is greater than that in path 2:

$$\Delta W_{AB}^{(1)} = \int_{A \text{ (path1)}}^{B} pdV > \int_{A \text{ (path2)}}^{B} pdV = \Delta W_{AB}^{(2)} \quad \text{...(iii)}$$

The change in internal energy (U) can be easily calculated from knowledge of the end points A and B:

$$\Delta U_{AB} = \int_{A}^{B} dU = U_B - U_A$$

since U is a state function, i.e., it depends only on the state of the gas. For an ideal gas, U is a function of temperature (T) alone,

$$\therefore \quad \Delta U_{AB} = U(T_B) - U(T_A) \quad \text{...(iv)}$$

The first law of thermodynamics, when applied to the process AB (path 1) gives:

$$\Delta Q_{AB}^{(1)} = \Delta U_{AB} + \Delta W_{AB}^{(1)}$$

where the superscript (1) indicates that the heat and the work done absorbed are both path dependent, and are computed, in this case on path 1. It is useful to integrate equation (ii) and remember the results for different types of processes; so we define the processes first and then state the results for each of these. We will restrict our study of processes to those that involve ideal gases.

Molar heat capacity of gases

The calorimetry of gases is different from that of liquids and solids, in that the gases are compressible. The heat capacity of a gas depends not only on the ambient pressure but also on the process undergone by the gas.

Considering 1 mole of an ideal gas,

$$dQ = dU + pdV \quad \text{...(i)}$$

We know that dU does not depend on the process, but dQ and $dW (= pdV)$ do.

$$\therefore \quad C = \left(\frac{dQ}{dT}\right)_{\text{proc}} = \frac{dU}{dT} + p\left(\frac{dV}{dT}\right)_{\text{proc}} \quad \text{...(ii)}$$

The subscript 'proc' in the term $\left(\frac{dV}{dT}\right)_{\text{proc}}$ indicates that it is dependent on the process. Two very useful and important definitions are the molar heat capacity at constant volume (C_v) and that at constant pressure (C_p) for ideal gases.

For a constant volume process, the gas is not allowed to expand; so the mechanical work done, $dW = pdV = 0$.

$$\therefore \quad C_v = \left(\frac{dQ}{dT}\right)_V = \frac{dU}{dT} = \frac{f}{2}R \quad \text{...(iii)}$$

assuming that the molecules have f degrees of freedom contributing to the internal energy.

For a constant pressure process, p = constant.

i.e., $$\frac{RT}{V} = \text{constant}$$

or, $$V \propto T$$

Taking the natural logarithm of both sides and differentiating,

$$\frac{dV}{V} = \frac{dT}{T}; \quad \left(\frac{dV}{dT}\right)_p = \frac{V}{T}$$

$$\therefore \quad C_p = \frac{dU}{dT} + p\left(\frac{dV}{dT}\right)_p$$

$$= C_v + p.\frac{V}{T} \quad \left(\because C_V = \frac{dU}{dT}\right)$$

$= C_v + R$ from the equation of state for 1 mole : $pV = RT$.

$$\therefore \quad C_p - C_v = R \quad \text{...(iv)}$$

This is also known as Mayer's relation.

$$\therefore \quad C_p = \left(\frac{f}{2} + 1\right)R \quad \text{...(v)}$$

The ratio of the heat capacity at constant pressure to that at constant volume is important and is defined as γ.

$$\gamma = \frac{C_p}{C_v} = \frac{f/2+1}{f/2} = 1 + \frac{2}{f} \quad \text{...(vi)}$$

The above relation also holds for mixtures of gases; the only caveat is that the number of degrees of freedom is replaced by the average number of degrees of freedom per molecule. For a mixture of ideal monoatomic and diatomic gases in the molar ratio 2 : 3 (say),

$$f_{av} = \frac{2f_1 + 3f_2}{5} = 4.2$$

The calculations for C_p, C_v and γ can be easily done using Eqs. (iii), (v) and (vi).

Illustration 1

Find the ratio of specific heats (C_p / C_v) of a mixture of He and N_2 in the molar ratio 1 : 1.

Solution

To find the ratio of specific heats of a mixture, we must first find the specific heats, C_p and C_v for the mixture.

The energy required to increase the temperature of 1 mole of the mixture by 1 K is equal to the sum of the energy required by He and H_2 present within the mixture. This is a result of the fact that energy is a scalar and the first law of thermodynamics holds.

$$C_p^{mix} = f_{He}C_p^{He} + f_{N_2}C_p^{N_2}$$

where f stands for molar fraction.

$$= \frac{1}{2}C_p^{He} + \frac{1}{2}C_p^{N_2}$$

$$= \frac{1}{2}\times\frac{5}{2}R + \frac{1}{2}\times\frac{7}{2} = 3R$$

Similarly, $$C_v^{mix} = f_{He}C_v^{He} + f_{N_2}C_v^{N_2}$$

$$= \frac{1}{2}\times\frac{3}{2}R + \frac{1}{2}\times\frac{5}{2}R = 2R$$

$$\therefore \quad \gamma^{mix} = \frac{C_p^{mix}}{C_v^{mix}} = \frac{\frac{5}{2}+\frac{7}{2}}{\frac{3}{2}+\frac{5}{2}} = \frac{3}{2} = 1.5$$

Illustration 2

An ideal monoatomic gas undergoes the following processes.

(a) $p = p_0 + \alpha V$

(b) $pV^3 = \text{constant}$

Find the molar heat capacity of the gas in each of these processes.

Solution

(a) To find the molar heat capacity of a gas undergoing a process, we use the first law to write

$$dQ = dU + pdV$$

$$C_{process} = \left(\frac{dQ}{dT}\right)_{process} = \left(\frac{dU}{dT}\right) + p\left(\frac{dV}{dT}\right)_{process}$$

$$= C_V + p\left(\frac{dV}{dT}\right)_{process} \quad \ldots\text{(i)}$$

For the process (a), we express V as a function of T:

$$pV = RT \quad \text{(for 1 mole)}$$

or $$\left(p_0 + \alpha V\right)V = RT \quad \ldots\text{(ii)}$$

which is an implicit relation for V. It is possible to solve it, but we won't. We will differentiate (ii) w.r.t. T, which leads to

$$\left(p_0 + 2\alpha V\right)\left(\frac{dV}{dT}\right)_{process} = R$$

or $$\left(\frac{dV}{dT}\right)_{process} = \frac{R}{p_0 + 2\alpha V}$$

$$\therefore \quad C_{process} = C_V + \left(\frac{p_0+\alpha V}{p_0+2\alpha V}\right)R \quad \text{[from (i)]}$$

which depends on the molar volume V.

$$\therefore \quad C_{process} = \frac{3}{2}R + \left(\frac{p_0+\alpha V}{p_0+2\alpha V}\right)R$$

(b) In this process, $pV^3 = \text{constant} = A$ (say)

and, $pV = RT$ (for an ideal gas)

Combining these two,

$$\frac{A}{V^2} = RT$$

Differentiating w.r.t. T,

$$-2\frac{A}{V^3}\left(\frac{dV}{dT}\right)_{process} = R$$

or $$\left(\frac{dV}{dT}\right)_{process} = -\frac{RV^3}{2A} = -\frac{RV^3}{2pV^3} = -\frac{R}{2p}$$

$$\therefore \quad C_{process} = C_V + p\left(-\frac{R}{2p}\right) \quad \text{(from (i))}$$

$$= \frac{3}{2}R - \frac{1}{2}R = R$$

Note: In (b) the molar heat capacity does not depend on the molar volume (or pressure or temperature). It is a constant. This is true for all polytropic processes.

Some Well-defined Processes

Many of the common reversible processes in thermodynamics have certain characteristics which are useful to remember, since these processes are repeated over and over again. The important characteristics are the heat absorbed (ΔQ), the work done by the gas (ΔW), the change in internal energy (ΔU) and the molar heat capacity (C). We will assume throughout that the process involves 1 mole of an ideal gas.

Isochoric process

An isochoric process is a process that takes place at constant volume.

Equation of the process: $V = \text{constant}; dV = 0$

Work done in the process: $\Delta W = 0$

Internal energy change: $\Delta U = C_v(T_2 - T_1)$

Heat absorbed: $\Delta Q = \Delta U + \Delta W = \Delta U$

The p-V diagram of the process is: V = constant line shown on the graph is an example. On the p-T diagram this translates into: $\frac{p}{T}$ = constant, which is a straight line extrapolating through the origin.

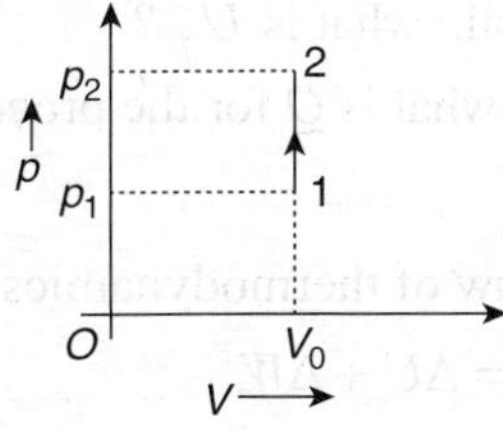

Isobaric process

An isobaric process, as the name implies, is a process that takes place at constant pressure.

Equation of the process P = constant; $dP = 0$

Work done in the process $\Delta W = P(V_2 - T_1)$

Internal energy change $\Delta U = C_v(T_2 - T_1)$

Heat absorbed $\Delta Q = \Delta U + \Delta W$

$$= C_v(T_2 - T_1) + P(V_2 - V_1)$$

$$= C_v(T_2 - T_1) + R(T_2 - T_1)$$

$$= (C_v + R)(T_2 - T_1)$$

$$= C_p(T_2 - T_1), \left(\text{Meyer's relation: } C_p = C_V + R\right)$$

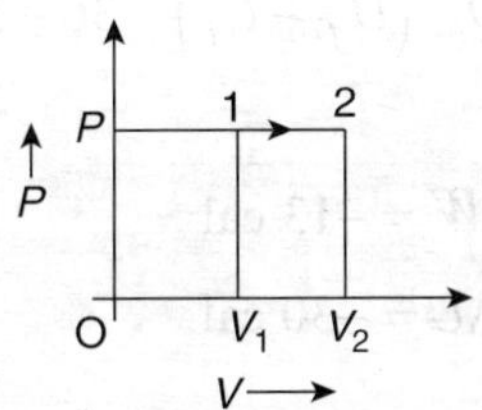

On the V-T diagram, the process takes the form of a straight line that extrapolates to the origin.

Work done in isothermal process

In an isothermal process, the temperature of the system remains constant $(\Delta T = 0)$.

Therefore the entire heat supplied to system is spent in doing work against surrounding. If the volume of the system changes from V_1 to V_2

Work done, $W = \int_{V_1}^{V_2} PdV$

$$W = \int_{V_1}^{V_2} \frac{nRT}{V} dV$$

(From ideal gas equation $PV = nRT$)

Substitute the value of P)

$$W = nRT\left[\log_e V\right]_{V_1}^{V_2}.$$

$$W = nRT(\log_e V_2 - \log_e V_1)$$

$$= nRT \log_e \frac{V_2}{V_1}$$

Boyle's Law $P_1V_1 = P_2V_2$ (constant temperature)

$$W = nRT \log_e \frac{P_1}{P_2}$$

As $P_1V_1 = P_2V_2 = nRT$

Work done W could also be expressed as

$$W = P_1V_1 \log_e \frac{V_2}{V_1}$$

$$= P_2V_2 \log_e \frac{V_2}{V_1}$$

Heat supplied in isothermal process

In isothermal process, change in internal energy $\Delta U = 0$

Using first law of thermodynamics

$$dQ = dU + dW$$

$$dQ = dW$$

or Work done = Heat supplied. Therefore, the work done in isothermal process equal to heat supplied to system

Adiabatic process

An adiabatic process is one in which no thermal energy is exchanged with the surroundings throughout the process: the process is either too fast or the system is sufficiently insulated.

Equation of the process is $dQ = 0$

This can be rewritten in terms of p, V or, V, T (or T, p):

$$dQ = dU + p\,dV$$

or, $$0 = C_v dT + \frac{RT}{V} dV$$

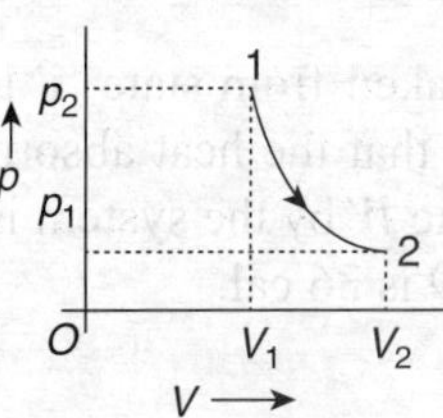

or $$\frac{dT}{T} + \frac{R}{C_v} \cdot \frac{dV}{V} = 0$$

or $$d(\log_e TV^{R/C_v}) = 0$$

or $TV^{\gamma-1}$ = constant, where $\gamma = C_p/C_v$

This can be rewritten to give: pV^{γ} = constant.

Work done in adiabatic process

In adiabatic process $\Delta Q = 0$. Therefore from first law of thermodynamics.

$$dW = -dU$$

$$PdV = -nC_V dT$$

$$= -\frac{nR}{\gamma - 1} dT$$

$$dW = \int_{V_1}^{V_2} pdV = -\int_{T_1}^{T_2} \frac{nR}{\gamma - 1} dT$$

$$dW = -\frac{nR}{\gamma - 1}(T_2 - T_1)$$

$$= \frac{nR}{\gamma - 1}(T_1 - T_2)$$

$$= \frac{P_1V_1 - P_2V_2}{\gamma - 1}$$

$$(P_1V_1 = nRT_1; P_2V_2 = nRT_2)$$

The internal energy in adiabatic process

In adiabatic system, the heat supplied to the system is zero ($dQ = 0$). That means work done is at the expense of internal energy. Therefore, the internal energy is negative of workdone.

An interesting property of an adiabatic curve is that it is always steeper than the isotherm passing through a common point.

There are many other processes as well. In fact, one can have infinite kinds of processes, one for each function $p(V)$. An interesting type of process is a polytropic process: pV^n = constant. (n is not necessarily equal to γ, and is known as the polytropic exponent).

Illustration 3

When a system is taken from state 'i' to state 'f' along the path iaf, it is found that the heat absorbed by the system is 50 cal and work done W by the system is equal to 20 cal and along the path ibf, Q is 36 cal.

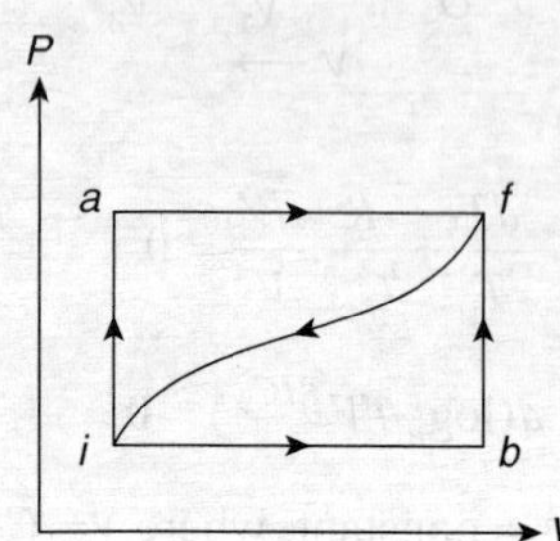

(a) What is W along the path ibf?

(b) If $W = -13$ cal., for the curved return path fi, what is Q for this path?

(c) When $U_1 = 10$ cal., what is U_f ?

(d) If $U_b = 22$ cal., what is Q for the processes bf and ib?

Solution

According to first law of thermodynamics,

$$\Delta Q = \Delta U + \Delta W$$

or $$\Delta Q = U_f - U_i + W$$

U_f = internal energy in final state and U_i is internal energy in initial state.

For Path iaf,

$$\Delta Q = 50 \text{ cal and } \Delta W = 20 \text{ cal}$$

Because

$$U_f - U_i = \Delta Q - \Delta W = 50 - 20 = 30 \text{ cal}$$

Here, it should be remembered that the change in internal energy between i and f state remains the same 20 cal whatever path is fallowed.

(a) For path ibf,

$$Q = 36 \text{ cal and } \Delta U = U_f - U_i = \text{cal.}$$

Therefore,

$$\Delta W = \Delta Q - (U_f - U_i) = 36 - 30 = 6 \text{ cal}$$

(b) For path fi,

$$\Delta W = -13 \text{ cal}$$

$$\Delta U = -30 \text{ cal}$$

Therefore,

$$\Delta Q = \Delta W + (U_i - U_f)$$

$$= -13 - 30 = -43 \text{ cal}$$

(c) $$U_f = 10 \text{ cal}$$

$$\Delta U = U_f - U_i = 30$$

$$U_f = 30 + U_i = 30 + 10 = 40 \text{ cal}$$

(d) For process bf, volume is constant that means workdone is zero.

Therefore,

$$\Delta Q = \Delta U = U_f - U_b = 40 - 22 = 18 \text{ cal}$$

For path ib,

$$\Delta Q = Q_{ibf} - Q_{bf} = 36 - 18 = 18 \text{ cal}$$

Illustration 4

One mole of a gas is isothermally expanded at 27 °C till the volume is doubled. Then it is adiabatically compressed to its original volume. Find the total work done.

($\gamma = 1.4$ and $R = 8.4$ joule/mole/°K.)

Solution

In case of isothermal expansion, the work done is given by

$$dW_1 = RT.\ln\left(V_f / V_i\right)$$

Here $R = 8.4$, $T = 27°C = 273 + 27 = 300°K$

$$\left(V_f / V_i\right) = 2$$

or $$dW_1 = 8.4 \times 300 \times \ln(2)$$

$$= 8.4 \times 300 \times 0.693$$

$$\simeq 1746 \text{ J}$$

Now the gas is adiabatically compressed to its original volume. Initially the temperature of the gas is 300 K and at the end of adiabatic compression, the temperature becomes T_2, $2V_1$. For an adiabatic process.

$$T_1V_1^{\gamma-1} = T_2V_2^{\gamma-1}$$

or $$300 \times (2V_i)^{\gamma-1} = T_2 \times (V_i)^{\gamma-1}$$

or $$300 \times (2)^{1.4-1} = T_2 (1)^{1.4-1}$$

or $$T_2 = 300 \times (2)^{0.4} = 395 \text{ K}$$

Work done during adiabatic process is

$$W_2 = \frac{nR}{(\gamma-1)}(T_2 - T_1)$$

$$= \frac{8.4}{(1.4-1)}(395 - 300)$$

$$= -1995 \text{ J}$$

Total workdone $= 1746 - 1995$

$$= -249 \text{ J}$$

Illustration 5

The *P-V* diagram for a cyclic process is a triangle ABC drawn in order. The coordinates of A, B and C are (4, 1), (2, 4) and (2, 1) respectively. The coordinates are in order *P*, *V*. Pressure is in N m^{-2} and volume is in litre. Calculate the work done during the process from A to B, B to C and C to A. Also calculate the work done in the complete cycle.

Solution

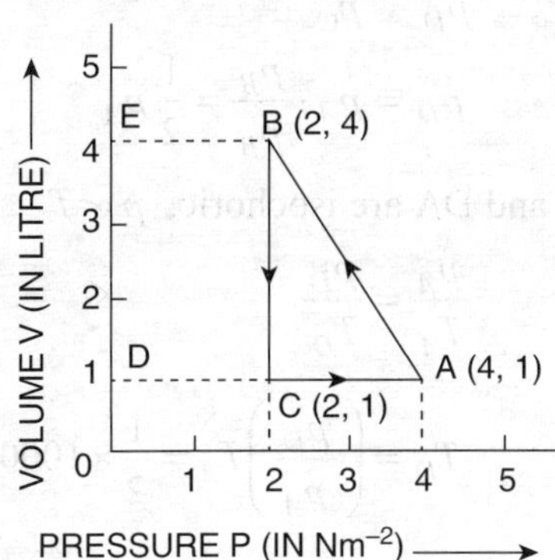

(i) Let W_{AB} be the work done during the process from A to B.

Then, W_{AB} = area of trapezium ABEDA = $\frac{1}{2}$ (DA + EB) CB

$$= \frac{1}{2}(4+2)\,\text{Nm}^{-2} \times 3 \times 10^{-3}\,\text{m}^3 = 9 \times 10^{-3}\,\text{J}$$

It is a case of an expanding gas. So, work done is positive.

(ii) Let W_{BC} be the work done during the process from B to C.

Then W_{BC} = area of rectangle BCDEB

$$DC \times DE$$

$$= 2\ \text{Nm}^{-2} \times 3 \times 10^{-3}\,\text{m}^{-3}$$

$$= 6 \times 10^{-3}\,\text{J}$$

In this case, the gas is being compressed. So, the work done is negative.

(iii) Let W_{CA} be the work done during the process from C to A.

Then, W_{CA} = area CAD = zero

(iv) Work done during the cycle ABCA

$$= W_{AB} + W_{BC} + W_{CA}$$

$$= 9 \times 10^{-3}\,\text{J} + \left(-6 \times 10^{-3}\,\text{J}\right) + 0 = 3 \times 10^{-3}\,\text{J}$$

Illustration 6

The pressure in a monatomic gas increases linearly from p_0 to $2p_0$ when its volume increases from V_0 to $2.5V_0$. Calculate the following:

(a) work done by the gas,
(b) increase in the internal energy,
(c) amount of heat supplied,
(d) molar heat capacity for the process.

Take $p_0 = 4 \times 10^5\, p_a, V_0 = 0.2\,\text{m}^3.$

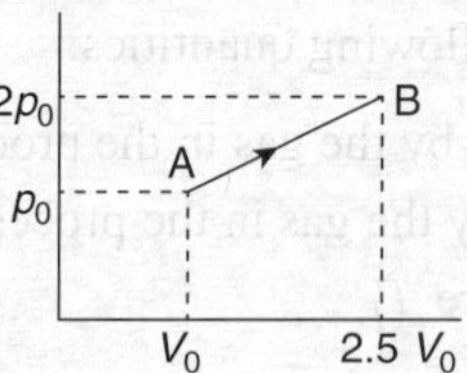

Solution

The straight line plot AB represents a linear increase in pressure on the *p-V* diagram.

(a) Work done by the gas

W = Area under the p-*V* diagram

$$= \frac{1}{2}(p_0 + 2p_0)(2.5V_0 - V_0)$$

$$= \frac{1}{2}(3p_0)(1.5V_0) = \frac{9}{4}p_0V_0$$

$$= \frac{9}{4} \times 4 \times 10^5 \times 0.2 = 1.8 \times 10^5 \text{ J}$$

(b) $$\Delta U = nC_V \Delta T = n \times \frac{3}{2} R \times (T_2 - T_1)$$

$$= n \times \frac{3}{2} R \times \left[\frac{p_2 V_2}{nR} - \frac{p_1 V_1}{nR} \right]$$

$$= \frac{3}{2} \left[2p_0 \cdot \frac{5}{2} V_0 - p_0 V_0 \right] = 6p_0 \times V_0$$

$$= 6 \times 4 \times 10^5 \times 0.2 = 4.8 \times 10^5 \text{ J}$$

(c) According to 1st law of thermodynamics

$$\Delta Q = \Delta U + \Delta W$$

$$= 4.8 \times 10^5 + 1.8 \times 10^5 = 6.6 \times 10^5 \text{ J}$$

(d) For the given process

$$\Delta Q = nC\Delta T$$

Illustration 7

One mole of a diatomic gas is taken through the cycle shown in the figure:

$A \to B$: adiabatic expansion

$B \to C$: cooling at constant volume

$C \to D$: adiabatic compression

$D \to A$: heating at constant volume.

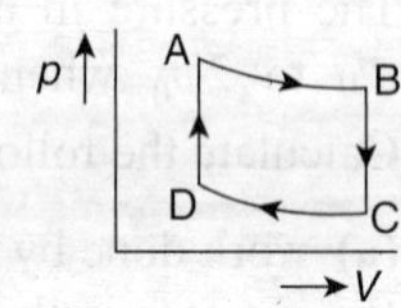

The pressure and temperature at A, B etc. are denoted by p_A, T_A, p_B, T_B etc. respectively. Given that $T_A = 1000$ K, $p_B = \frac{2}{3} p_A$ and $p_C = \frac{1}{3} p_A$.

Calculate the following quantities:

(i) the work done by the gas in the process $A \to B$,

(ii) the heat lost by the gas in the process $B \to C$,

(iii) the temperature T_D.

Given that $\left(2/3\right)^{2/7} = 0.90$.

Solution

(i) We use the same technique as for analysing a closed cyclic process. The end points of process $A \to B$ are specified in terms of pressure and temperature. For an adiabatic process,

$$pV^\gamma = \text{Constant, where } \gamma = \frac{C_p}{C_V} = \frac{7}{5}$$

and, $pV = RT$.

$$\therefore \quad p \times \left(\frac{RT}{p} \right)^\gamma = \text{Constant}$$

or, $$\frac{T^\gamma}{p^{\gamma-1}} = \text{constant, in terms of } p \text{ and } T.$$

Applying this equation,

$$\frac{T_A^\gamma}{p_A^{\gamma-1}} = \frac{T_B^\gamma}{p_B^{\gamma-1}}$$

or, $$(1000)^{7/5} \times \left(\frac{p_B}{p_A} \right)^{2/5} = T_B^{7/5}$$

$$\Rightarrow \quad T_B = 1000 \times \left(\frac{2}{3} \right)^{2/7} = 900 \text{ K}.$$

$\therefore$ Work done by the gas, $W_{AB} = -\Delta U = -C_V (T_B - T_A)$

$$= \frac{5}{2} R \times 100 = 250R$$

(substitute the value of R in the end)

(ii) The heat lost by the gas in the process $B \to C$:

$$\Delta Q_{BC} = C_V (T_B - T_C)$$

Now, $$p_C = \frac{1}{3} p_A, p_B = \frac{2}{3} p_A, \& \ pV = RT$$

$$\Rightarrow \quad \frac{p_B}{T_B} = \frac{p_C}{T_C}$$

$$\therefore \quad T_C = \frac{p_C}{p_B} \times T_B = \frac{\frac{1}{3} p_A}{\frac{2}{3} p_A} \times T_B = \frac{1}{2} \times 900 \text{ K} = 450 \text{ K}$$

$$\therefore \quad \Delta Q_{BC} = \frac{5}{2} R (900 - 450) = 1125R$$

(iii) Since AB and CD are adiabatic processes,

$$p_A V_A^\gamma = p_B V_B^\gamma$$

and $$p_D V_D^\gamma = p_C V_C^\gamma$$

Dividing, we get

$$\therefore \quad \frac{p_A}{p_D} = \frac{p_C}{p_C} \qquad [\because V_A = V_D, V_B = V_C]$$

$$\therefore \quad p_D = p_A \cdot \frac{p_B}{p_B} = \frac{1}{2} p_A$$

Since BC and DA are isochoric, $p \propto T$

$$\therefore \quad \frac{p_A}{T_A} = \frac{p_D}{T_D}$$

or, $$T_D = \left(\frac{p_D}{p_A} \right) T_A = \frac{1}{2} \times 1000 = 500 \text{ K}$$

Cyclic Processes

Thermodynamic cycles or cyclic processes are used by heat engines which either convert heat into work or, when working in reverse, pump heat from lower temperature to higher temperature (e.g. air conditioners, refrigerators). A cyclic process is either a closed process or, as is commonly the case, consists of several processes such that together they constitute a closed cycle (see the diagram below). Heat is absorbed in some processes, and rejected (i.e., evolved) in others. The net thermal energy absorbed is:

$$Q_{net} = Q_{abs} - Q_{rej} = W$$

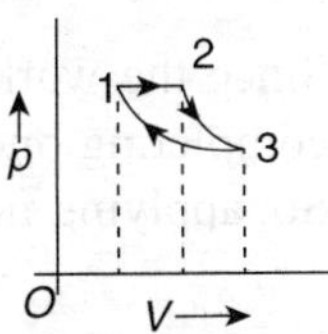

The work done in the cycle is zero, since the change in internal energy is zero in a closed cycle.

Second Law of Thermodynamics

The first law of thermodynamics is basically a statement of law of conservation of energy. This law does not restrict the types of energy conversions. Contrary to first law, only certain types of energy conversions are observed to take place. There are so many examples in which mechanical energy is completely converted into heat. There are also examples where heat energy is partially converted into heat. But till now there is no device which can convert heat into mechanical energy. The answer to this question lies in the second Law of thermodynamics.

The Kelvin-Plank statement of the second Law of thermodynamics

"It is impossible to construct a device that, operating in a cycle, produces no effect other than the absorption of energy from a reservoir and the performance of an equal amount of work".

The spirit of this law is that it is impossible to construct a device which converts whole of the heat energy into work i.e., it is impossible to get 100% efficiency.

Classius statement of the second law of thermodynamics

"It is impossible to make heat flow from a body at a lower temperature to a body at a higher temperature."

Equivalence of the two statements

At first instance, it appears as if the two statements of the second law of thermodynamics are non-coherent. But, in fact the two statements are coherent. Any violation of the first will violate the second.

The second law of thermodynamics is applicable to only a cyclic process in which the system returns to its initial state after completing the cycle.

Reversible and Irreversible Process

Reversible process

It is a process which can be retraced in opposite direction so that the system and surroundings pass through exactly the same state at every stage as in direct process. It means some heat is absorbed in the direct process from the surroundings, and the same amount of heat is rejected to surroundings in the reverse process. Similarly if some work is done on the system in the direct process, the same amount of work is done in the indirect process. At the conclusion of reversible process, the system and surrounding should restore its initial stage.

The following conditions should be satisfied in order that a process to be reversible.

(1) The process should be quasistatic. It should remain in mechanical, thermal as well as chemical equilibrium with surroundings at each stage.

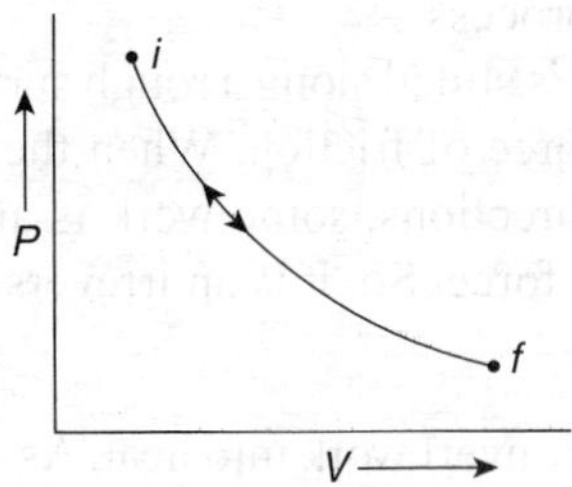

(2) There should be no dissipative forces (friction, viscosity, electrical resistance etc).

Examples

(1) When current is passed through an electrolyte, ions are formed. The positive ions move towards the cathode while the negative ions move towards anode. If the direction of electric current is reversed, the direction of motion of ions will also be reversed. So, electrolysis is a reversible process. Here, it is assumed that the electrical resistance offered by an electrolyte is zero.

(2) When an elastic spring is compressed gradually, some work is done on the spring. When the applied force is reduced, the spring expands and does nearly the same amount of work as was initially done upon it. So, slow compression or extension of a spring is a reversible process.

Irreversible process

It is a process which is not exactly reversed, i.e., the system does not pass through the same intermediate state as in the direct process.

Every process in nature is an irreversible process. Some examples are as follows:

(1) All chemical reactions are irreversible because they involve changes in the internal structure of the constituents. The constituents can be restored only when energy is supplied from outside. Rusting of iron is one example of a chemical reaction. When iron rusts, iron is converted to iron oxide. Iron and oxygen can be separated only by supplying energy from outside. So, rusting of iron is an irreversible process.

(2) The decay of organic matter is an irreversible process.

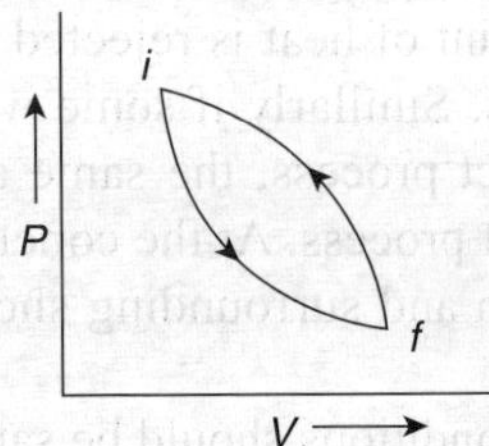

(3) When an electric current flows through a conductor, some heat is produced. If the electric current is made to flow in the reverse direction, some heat will be again produced. So, flow of current through a conductor is an irreversible process.

(4) When a body is slided along a rough surface, work is done against the force of friction. When the body is slided in the reverse directions, some work is again done against the frictional force. So, it is an irreversible process.

Heat engine

It is very easy to convert work into heat. As an example, when two bodies are rubbed against each other, heat is produced. However, for the conversion of heat into work, we require a device. This device is called a heat engine.

Heat engine is a device used for converting heat energy into mechanical energy.

There are two types of heat engines – external combustion engine and internal combustion engine.

Steam engine is an example of external combustion engine. In such an engine, the fuel is burnt outside the working portion of the engine.

Internal combustion engines are used in motor cars and aeroplanes. Petrol engine and Diesel engine are examples of internal combustion engines. In such engines, the fuel is burnt within the working portion of the engine.

A heat engine consists essentially of the following parts:

(i) **Source** or **Heat reservoir:** It is the supplier of heat energy

(ii) **Sink** or **Cold reservoir:** That heat which has not been converted into work is rejected to the sink.

(iii) **Working substance:** It absorbs a certain quantity of heat from the source, converts a part of it into work and rejects the remaining heat to the sink. It is taken through a cycle of operations. At the end of each cycle, the working substance is restored to its original condition and there is no change in its internal energy. Since the cycle of operations can be repeated indefinitely, the heat engine gives a continuous supply of work.

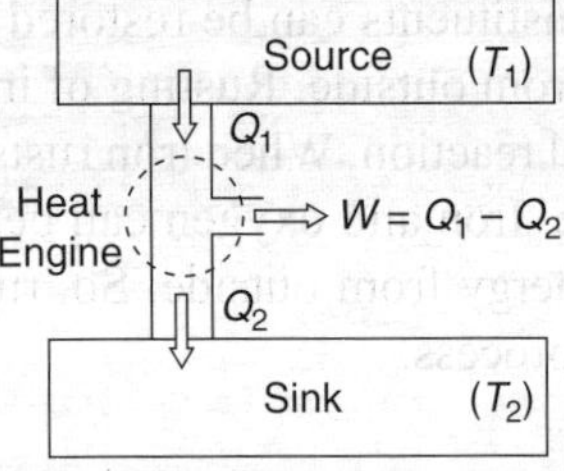

Block diagram of a heat engine

Thermal efficiency of heat engine

It is defined as the ratio of the net external work done by the engine during one cycle to the heat absorbed from the source during that cycle. It is denoted by η.

The work W obtained in each cycle is called the output while the heat Q_1 extracted from the source in one cycle is called the input.

$$\therefore \quad \text{Efficiency, } \eta = \frac{\text{output}}{\text{input}} = \frac{W}{Q_1}$$

Since the working substance returns to its initial state after completing one cycle, there is no change in internal energy. So, applying first law of thermodynamics, we get

$$Q_1 - Q_2 = W$$

Here, Q_1, Q_2 and W are all measured in the same units

$$\eta = \frac{Q_1 - Q_2}{Q_1} = 1 - \frac{Q_2}{Q_1}$$

Discussion

(1) For a given value of Q_1, the smaller the value of Q_2, the higher is the efficiency of the heat engine.

(2) The value of η will be one, i.e., efficiency of the heat engine will be 100% if the value of Q_2 is zero. In such case, the whole of the heat extracted from the source will be converted into mechanical work. No heat will be rejected to the sink. But this is impossible. Thus, we conclude that efficiency cannot be 100%.

(3) For the sake of discussion, let us discuss the purely hypothetical case of an engine whose efficiency is greater than 1. Such a heat engine will extract heat both from the source and the sink.

(4) In actual practice, the efficiency of heat engine does not exceed 55%, i.e., at the most only 55% of the total heat energy extracted from the source is converted into mechanical work. Total efficiency of a steam engine varies from 12% to 17%. The maximum efficiency of a diesel engine is 40% while that of a petrol engine is only up to 26%.

(5) The low efficiency of a heat engine is not due to the bad designing of a heat engine but it is the consequence of a certain law of nature. It was shown by a French engineer Sadi Carnot that there is something in nature which forbids the complete conversion of heat into work.

Refrigerator (heat pump)

An ideal refrigerator may be regarded as a Carnot's ideal heat engine working in the reverse order.

In an actual refrigerator, the vapours of some low boiling point liquid act as the working substance. The working substance absorbs a certain quantity of heat Q_2 from the cold body or sink at lower temperature T_2. In a household refrigerator, the ice cubes in the freezer compartment and food constitute the cold body. A certain amount of work W is performed by the compressor of the refrigerator on the working substance. The compressor is operated by an electric motor.

The quantity of heat T_2 is rejected to the hot body (atmosphere) at temperature T_1K by the radiator.

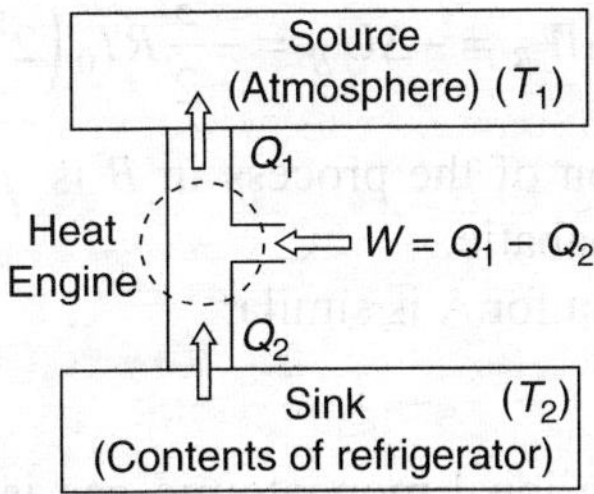

Coefficient of performance

It measures the efficiency of a refrigerator.

It is defined as the ratio of the quantity of heat extracted per cycle from the contents the refrigerator to the mechanical work W done by the external agency to do so.

It is denoted by β or K or ω.

$$\therefore \quad \beta = \frac{Q_2}{W}$$

Smaller the amount of mechanical work done in removing heat Q_2, greater will be the coefficient of performance.

Applying first law of thermodynamics,

$$W = Q_1 - Q_2$$

$$\therefore \quad \beta = \frac{Q_2}{Q_1 - Q_2} \quad \ldots(1)$$

This expression may be put in another form also.

$$\beta = \frac{1}{\frac{Q_1}{Q_2} - 1}$$

But $\quad \frac{Q_1}{Q_2} = \frac{T_1}{T_2}$

$$\therefore \quad \beta = \frac{1}{\frac{T_1}{T_2} - 1} \quad \ldots(2)$$

or $\quad \beta = \frac{T_2}{T_1 - T_2}$

Discussion

(i) In actual practice, β varies from 2 to 6: for an actual refrigerator, the value of β is less than that calculated from Eq. (1) or (2).

(ii) Lesser the difference in temperature between the cooling chamber and the atmosphere, higher is the coefficient of performance of the refrigerator.

(iii) In a heat engine, the efficiency can never exceed 100%. But in the case of a refrigerator, the coefficient of performance may be much higher than 100%.

(iv) As the refrigerator works, T_2 continues to decrease due to formation of too much ice. There is practically no change in T_1. This decreases the value of β. However, if the refrigerator is defrosted, T_2 shall increase and consequently the value of β also increases. So, it is necessary to defrost the refrigerator.

Illustration 8

An automobile engine absorbs 1600 J from a hot reservoir and expels 1000 J to a cold reservoir in each cycle. (a) What maximum work is done in each cycle? (b) What is the efficiency of the cycle?

Solution

(a) $W = Q_1 - Q_2 = 1600 - 1000 = 600$ J

(b) $\eta = \frac{W}{Q_1} = \frac{600}{1600} = 0.375 = 37.5\%$

Illustration 9

The temperatures T_1 and T_2 of two heat reservoirs in the ideal Carnot engine are 1500 °C and 500 °C respectively. Which of these, increasing T_1 by 100 °C or decreasing T_2 by 100 °C, would result in a greater improvement in the efficiency of the engine?

Solution

The efficiency of a heat engine is given by

$$\eta = 1 - \frac{Q_2}{Q_1} = \frac{Q_1 - Q_2}{Q_1}$$

When Q_1 is increased by 100 °C,

$$\eta_1 = \frac{(1600 - 500)}{(1600 + 273)}$$

$$= \frac{1100}{1873} = 59\% \quad \ldots(i)$$

When Q_2 is decreased by 100 °C,

$$\eta_2 = \frac{(1500 - 400)}{(1500 + 273)}$$

$$= \frac{1100}{1773} = 62\% \quad \ldots(ii)$$

From Eqs. (i) and (ii), it is clear that decreasing Q_2 results in greater improvement in efficiency.

Illustration 10

A refrigerator, whose coefficient of performance K is 5, extracts heat from the cooling compartment at the rate of 250 J per cycle.

(a) How much work per cycle is required to operate the refrigerator cycle?

(b) How much heat per cycle is discharged to the room which acts as the high temperature reservoir?

Solution

(a) As coefficient of performance of a refrigerators is defined as $K = Q_2/W$,

$$W = \frac{Q_2}{K} = \frac{250}{5} = 50 \text{ J}$$

(b) As $Q_1 = Q_2 + W$

So, $Q_1 = 250 + 50 = 300$ J

SOLVED EXAMPLES

Example 1

Two chambers A and B, each of volume V_0, each contain 1 mole of a monoatomic ideal gas, and are separated from each other by a sliding adiabatic partition (of negligible mass). The partition is now slowly displaced in the direction of B (by an external force) until the volume of B equals $V_0/2$. If the entire system is insulated, find

(i) the temperature in each of the vessels A and B,

(ii) the work done on the gas in B,

(iii) the equation of the process in A and B.

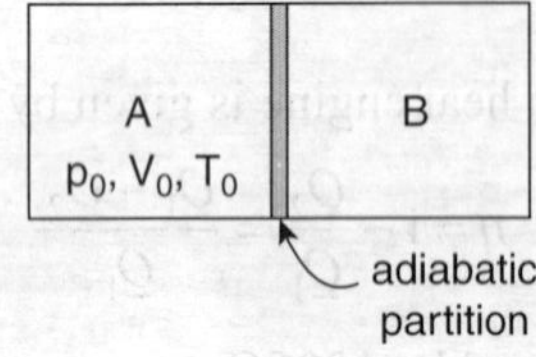

Solution

(i) Since the partition is slowly displaced and both the chambers are insulated, the process in *A* and *B* are adiabatic and quasi-static.
Since the gas in monatomic, $\gamma = 5/3$.
For the gas in B,

$$TV^{\gamma-1} = \text{constant}$$

$$\Rightarrow \quad T_{Bf}\left(\frac{V_0}{2}\right)^{2/3} = T_0 \times V_0^{2/3} \quad \left(\gamma - 1 = \frac{5}{3} - 1 = \frac{2}{3}\right)$$

$$\therefore \quad T_{Bf} = 2^{2/3}.T_0$$

For the gas in A,

$$T_{Af}\left(\frac{3}{2}V_0\right)^{2/3} = T_0\left(V_0\right)^{2/3}$$

$$\therefore \quad T_{Af} = \left(\frac{2}{3}\right)^{2/3} T_0 .$$

(ii) Since the process is adiabatic,

$$\Delta Q_B = 0 = \Delta U_B + \Delta W_B$$

But $\quad \Delta U_B = C_V \Delta T = \frac{3}{2}R\left(2^{2/3} - 1\right)T_0$, and

$$\Delta W_B = -\Delta U_B = -\frac{3}{2}RT_0\left(2^{2/3} - 1\right).$$

(iii) The equation of the process in B is $p_B V_B^{5/3} = p_0 V_0^{5/3}$, which is adiabatic.
The equation for A is similar.

Example 2

Two moles of an ideal monoatomic gas is taken through a cycle ABCA as shown in the *p-T* diagram. During the process AB, pressure and temperature of the gas vary such that pT^2 = constant. If $T_1 = 400$ K, calculate:

(a) the work done on or by the gas in each process,

(b) the heat absorbed or released by the gas in each process,

(c) the efficiency of the cycle.

[Assume: $ln2 = 0.7$]

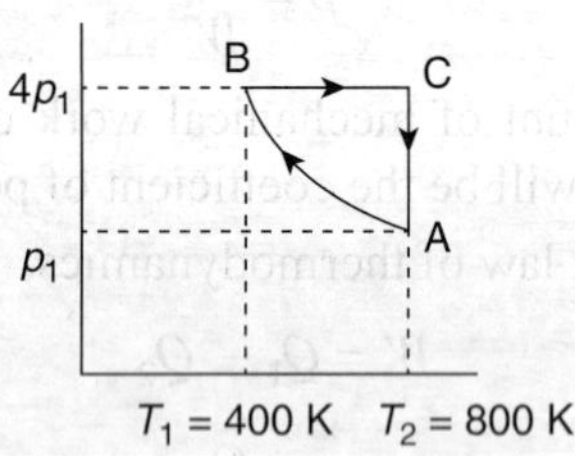

Solution

(a) During the process AB, pT^2 = constant $= A$

$$\therefore \quad T = \sqrt{A/p} = \frac{a}{\sqrt{p}}, \text{ where } \sqrt{A} = a$$

According to the gas equation,

$$pV = nRT$$

or

$$V = \frac{nRT}{p} = \frac{naR}{p^{3/2}}$$

$$\therefore \quad dV = -\frac{3}{2}\frac{naR}{p^{5/2}}dp$$

$\therefore$ Work done on the gas in the process AB

$$W_{AB} = \int pdV = \int_{p_1}^{4p_1}\left(-\frac{3}{2}\right)\frac{naR}{p^{3/2}}dp$$

$$= \left(-\frac{3}{2}naR\right)\times(-2)\left[\frac{1}{\sqrt{p}}\right]_{p_1}^{4p_1}$$

$$= -3naR\left[\frac{1}{\sqrt{p_1}} - \frac{1}{2\sqrt{p_1}}\right]$$

$$= 3nR\left[\frac{a}{\sqrt{p_1}} - \frac{a}{2\sqrt{p_1}}\right] = -3nR\left[T_2 - T_1\right]$$

$$= -3 \times 2 \times R\left[800 - 400\right] = -2400\text{R}$$

Work done by the gas during the process BC:

$$W_{BC} = \int pdV = 4p_1\left[V_C - V_B\right]$$

$$= 2R\left(T_2 - T_1\right)$$

$$= 2R\left(800 - 400\right) = 800 \text{ R}$$

Process CA is isothermal expansion, for which the work done

$$W_{CA} = nRT\ln\frac{V_A}{V_C} = nRT\ln\frac{p_C}{p_A}$$

$$= 2R \times 800 \times \ln 4$$

$$= 3200R\ln 2$$

$$= 2240\,R$$

(b) For the process AB,

$$\Delta U = nC_V\Delta T = 2\times\frac{3}{2}R\times\left(T_1 - T_2\right)$$

$$= 3\text{ R}\times\left(400 - 800\right) = -1200\text{ R}$$

$\therefore$ From the first law,

$$\left(\Delta Q\right)_{AB} = \Delta U + \Delta W = -1200R - 2400R$$

$$= \left(-3600R\right), \text{ heat is released}$$

For the process BC: It is an isobaric process

$$\therefore \quad \left(\Delta Q\right)_{BC} = nC_p\Delta T$$

$$= 2\times\frac{5}{2}R\times\left(T_C - T_B\right)$$

$$= 5R\times\left(800 - 400\right)$$

$= 2000$ R, so heat is absorbed

For the process CA: It is an isothermal process

$$\therefore \quad \left(\Delta Q\right)_{CA} = \Delta U + \Delta W = 0 + 2240\text{R}$$

$= 2240$ R, so heat is absorbed

These results are given in the following table.

Process	ΔW	ΔU	ΔQ
AB	$-2400\,R$	$-1200\,R$	$-3600\,R$
BC	$800\,R$	$1200\,R$	$2000\,R$
CA	$2240R$	0	$3240R$
Total	$640R$	0	$640R$

(c) Efficiency of the cycle

$$\eta = \frac{\text{Work done}}{\text{Total heat absorbed}}$$

$$= \frac{3200\,R\ln 2 - 1600R}{3200R\ln 2 + 2000R}$$

$$= \frac{16\ln 2 - 8}{16\ln 2 + 10} = \frac{3.2}{21.2} = \frac{8}{53} = 15\%\text{, approximately.}$$

Example 3

A horizontal insulated cylindrical vessel of length 2ℓ is separated by a thin insulating piston into two equal parts, each of which contains n moles of an ideal diatomic gas at a temperature T. The left part is in contact with a thermostat (a device which maintains a constant temperature). When an amount of heat Q is supplied to the gas in the right part, the piston is displaced to the left by a distance $x = \ell/2$. Determine the heat Q' given away to the thermostat at temperature T by the gas on the left side. Assume that $K\ell^2 = 1.25RT$ where K is spring constant

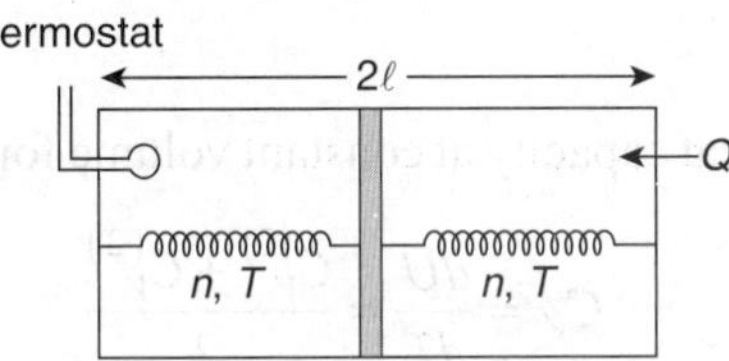

Solution

Total heat given to the system is $\left(Q - Q'\right)$. From first law,

$$Q - Q' = \Delta W + \Delta U, \quad \ldots\text{(i)}$$

where

$\Delta W =$ total work done by the gas in both the chambers, and

$\Delta U =$ change in internal energy of both the gases.

Now, $\quad \Delta W =$ sum of potential energy stored in the springs

$$= 2\times\frac{1}{2}K\left(\frac{\ell}{2}\right)^2 = \frac{1}{4}K\ell^2 \quad \ldots\text{(ii)}$$

The temperature of the left part is maintained constant, so $\left(\Delta U\right)_{\text{left}} = 0$. Hence

$$\Delta U = \left(\Delta U\right)_{\text{right}} = nC_V\Delta T$$

$$= \frac{5}{2}nR\Delta T \quad \ldots\text{(iii)}$$

Pressure on the right side = Pressure on the left

or $$\frac{nR\left(T + \Delta T\right)}{A\left(\ell + \frac{\ell}{2}\right)} = \frac{nRT}{A\left(\ell - \frac{\ell}{2}\right)} + \frac{2K\left(\frac{\ell}{2}\right)}{A}$$

Simplifying, we get

$$\Delta T = \frac{3K\ell^2}{2nR} + 2T$$

$$\therefore \quad Q' = Q - \Delta W - \Delta U$$

$$= Q - \frac{1}{4}K\ell^2 - \frac{5}{2}nR\left(\frac{3K\ell^2}{2nR} + 2T\right)$$

$$= Q - 4K\ell^2 - 5nRT$$

$$= Q - 4 \times 1.25RT - 5nRT$$

$$= Q - (n+1)5RT$$

Example 4

An equimolar mixture of a monoatomic and diatomic gas is subjected to continuous expansion such that

$$dQ = \frac{1}{3}dU + \frac{1}{2}dW.$$

(i) Find the equation of the process in terms of the variables P and V.

(ii) Calculate the molar heat capacity of the gas during the process.

Solution

The molar heat capacity at constant volume for the mixture is:

$$C_V = \frac{dU}{dT} = \frac{C_V^{(1)} + C_V^{(2)}}{2}$$

$$= \frac{\frac{3}{2}R + \frac{5}{2}R}{2} = 2R$$

Equation of process: $dQ = \frac{1}{3}dU + \frac{1}{2}dW$ ...(i)

First law: $dQ = dU + dW$ ...(ii)

$\therefore$ $$dU + dW = \frac{1}{3}dU + \frac{1}{2}dW$$

or $$\frac{2}{3}dU = -\frac{1}{2}dW \quad \text{...(iii)}$$

or $$\frac{2}{3}C_V dT = -\frac{1}{2}pdV$$

or $$\frac{4}{3}RdT = -\frac{1}{2}pdV \quad \text{...(iv)}$$

Gas equation gives, $pV = RT$ (for 1 mole)

or $$pdV + Vdp = RdT \quad \text{...(v)}$$

From (iv) and (v) we get,

$$\frac{4}{3}(pdV + Vdp) = -\frac{1}{2}pdV$$

or $$\frac{11}{8}pdV + Vdp = 0$$

Dividing by pV,

$$\frac{dp}{p} + \frac{11}{8}\frac{dV}{V} = 0$$

or $$d\left(\ln pV^{11/8}\right) = 0$$

$\therefore$ $$pV^{11/8} = \text{Constant} \quad \text{...(vi)}$$

which is the required equation of state.

Molar heat capacity for the process can be calculated from

$$CdT = dQ = dU + dW$$

or $$C = \frac{dU}{dT} + \frac{dW}{dT}$$

$$= \frac{dU}{dt} - \frac{4}{3}\frac{dU}{dT} \quad \text{[from Eq. (iii)]}$$

$$= -\frac{1}{3}\frac{dU}{dT} = -\frac{1}{3} \cdot C_V$$

$$= -\frac{2}{3}R$$

Example 5

One mole of a monoatomic gas is taken through the cycle shown in the pV diagram. For this cycle, calculate

(i) the heat absorbed / rejected by the gas.
(ii) the temperatures at A, B, C, D in the cycle.
(iii) the work done in each process.
(iv) the efficiency of the cycle.

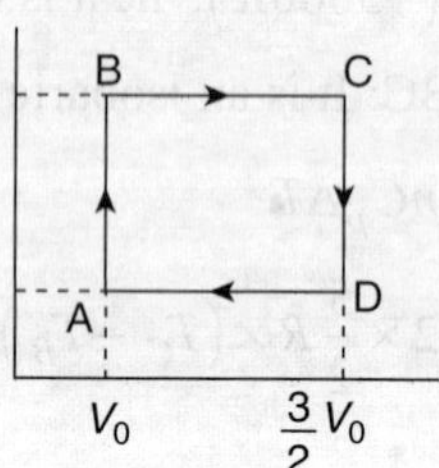

Solution

In discussing this problem, we are going to show you a way of organising your calculations so that you can minimise your errors. We will make a table, and fill in the entries which are most easily calculated.

To calculate ΔU, we will need the temperatures at A, B, C, D since $\Delta U = C_V \Delta T = C_V\left(T_{\text{final}} - T_{\text{initial}}\right)$

In some cases ΔQ is known, e.g., for adiabatic process, isobaric process, etc. In such cases, we don't calculate ΔW separately but use the first law.

(i) $T_A = \frac{p_0 V_0}{R} = T_0$ (say)

$$T_B = \frac{2p_0V_0}{R} = 2T_0, \quad [\because p \propto T, \text{ as } V = \text{constant}]$$

$$T_C = \frac{3}{2}T_B = 3T_0 \quad [\because V \propto T, \text{as } p = \text{constant}]$$

$$T_D = \frac{1}{2}T_C = \frac{3}{2}T_0$$

For AB, $\Delta U = C_V\left(T_B - T_A\right) = \frac{3}{2}RT_0$

$\Delta W = 0$ ($\because$ it is a constant volume process)

For CD and DA we calculate similarly. The table is shown below:

Process	ΔU	ΔW	ΔQ
AB	$\frac{3}{2}RT_0$	0	$\frac{3}{2}RT_0$
BC	$\frac{3}{2}RT_0$	RT_0	$\frac{5}{2}RT_0$
CD	$-\frac{9}{4}RT_0$	$-\frac{3}{4}RT_0$	$-\frac{9}{4}RT_0$
DA	$-\frac{3}{4}RT_0$	$-\frac{1}{2}RT_0$	$-\frac{5}{4}RT_0$
Net	0	$+\frac{1}{2}RT_0$	$+\frac{1}{2}RT_0$

The table is now "checked". If all the ΔU are added up (for a closed cycle), you should get 0: Since U is a 'state' function and in our case, the material undergoes a 'closed cycle', the net change in U is zero. We add up the ΔW column. It gives the area enclosed by the cycle on the p-V diagram. Obviously, it should be:

$$p\times\frac{1}{2}V_0=\frac{1}{2}RT_0$$

Now, we add up the ΔQ column. The result should be equal to that of the ΔW column. For a closed cycle, the first law $(\Delta Q=\Delta U+\Delta W)$ is always valid.

Now, we find the efficiency of the cycle:

$$\eta=\frac{\Delta W^{\text{net}}}{\Delta Q^{\text{net}}_{+\text{ve}}}$$

where

$$\Delta Q^{\text{net}}_{+\text{ve}}=\frac{3}{2}RT_0+\frac{5}{2}RT_0=4RT_0$$

$$\therefore\quad \eta=\frac{\frac{1}{2}RT_0}{4RT_0}=\frac{1}{8}=12.5\%$$

Practice this method for solving cyclic processes.

The answers to (i) are in the ΔQ column, those to (ii) have been calculated, and (iii) are in the ΔW column. The efficiency is calculated to be 12.5%.

Example 6

Two cylinders A and B fitted with pistons contain equal amounts of an ideal diatomic gas at 300 K. The piston in A is free to move, while that in B is held fixed. The same amount of heat is given to the gas in each cylinder. If the rise in temperature of the gas in A is 30 K, then the rise in temperature of the gas in B is

(a) 30 K (b) 18 K
(c) 50 K (d) 42 K

Solution

For A, pressure is constant and for B, volume is constant.

Hence, $$(dQ)_A=nC_p\Delta T=n\times\frac{7}{2}R\times(30K)$$

$$(dQ)_B=nC_v\Delta T=n\times\frac{5}{2}R\times\Delta T$$

$$\because\quad (dQ)_A=(dQ)_B$$

$$\Rightarrow\quad \Delta T=42\text{ K}$$

Ans. (d)

Example 7

A monatomic ideal gas, initially at temperature T_1, is enclosed in a cylinder fitted with a frictionless piston. The gas is allowed to expand adiabatically to a temperature T_2 by releasing the piston suddenly. If L_1 and L_2 are the lengths of the gas column before and after expansion respectively, then T_1/T_2 is given by

(a) $\left(\frac{L_1}{L_2}\right)^{2/3}$ (b) $\frac{L_1}{L_2}$

(c) $\frac{L_2}{L_1}$ (d) $\left(\frac{L_2}{L_1}\right)^{2/3}$

Solution

For adiabatic expansion

$$T_1V_1^{\gamma-1}=T_2V_2^{\gamma-1}$$

$$T_1(AL_1)^{5/3-1}=T_2(AL_2)^{\frac{5}{3}-1}$$

$$\Rightarrow\quad \frac{T_1}{T_2}=\left(\frac{L_2}{L_1}\right)^{2/3}$$

Ans. (d)

Example 8

Starting with the same initial conditions, an ideal gas expands from volume V_1 and V_2 in three different ways. The work done by the gas is W_1 if the process is purely isothermal, W_2 if it is purely isobaric and W_3 if it is purely adiabatic. Then,

(a) $W_2>W_1>W_3$ (b) $W_2>W_3>W_1$
(c) $W_1>W_2>W_3$ (d) $W_1>W_3>W_2$

Solution

Three processes are shown in the adjoining p–V diagram. The area under the curves represents the work done,

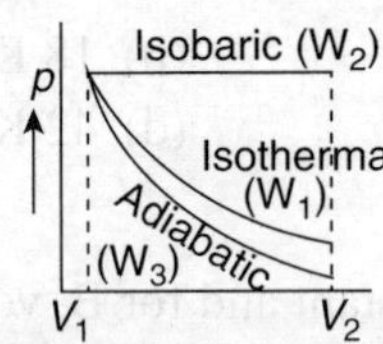

Hence, $W_2 > W_1 > W_3$

Ans. (a)

Example 9

An ideal gas is taken through the cycle $A \to B \to C \to A$, as shown in the figure. If the net heat supplied to the gas in the cycle is 5 J, the work done by the gas in the process $C \to A$ is

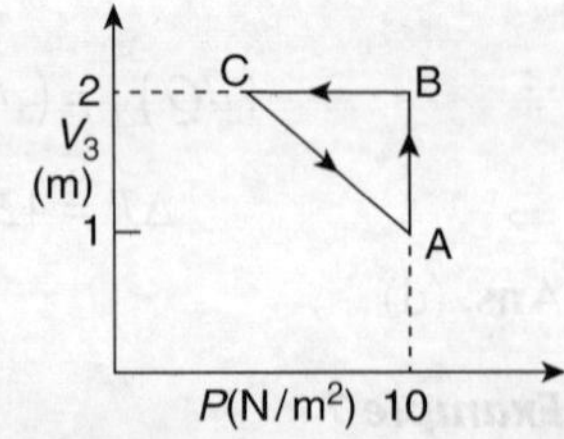

(a) – 5 J (b) – 10 J

(c) – 15 J (d) – 20 J

Solution

In a closed cycle,

$$\Delta Q = \Delta W = W_{AB} + W_{BC} + W_{CA}$$

or, $$5\text{ J} = 10(2-1)\text{ J} + 0 + W_{CA}$$

$\therefore$ $$W_{CA} = -5\text{ J}$$

Ans. (a)

Example 10

Figure shows five processes occurring with an ideal gas. Regarding the behaviour of the changes in internal energy of the gas in these processes, mark the correct statement.

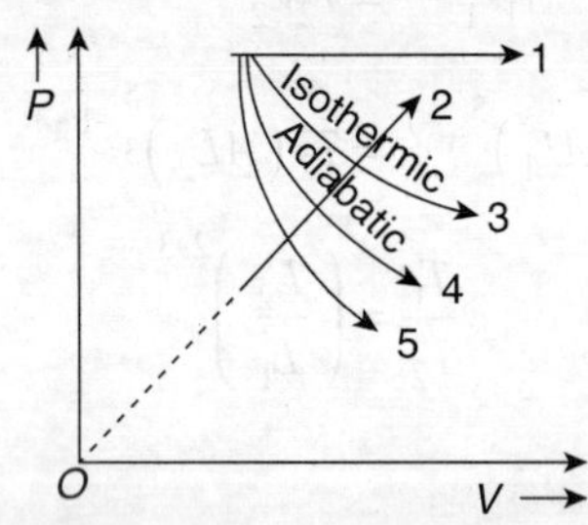

(a) In 1, 2 and 3 it grows and in 4, 5 it diminishes.

(b) In 1, it remains constant, in all others it decreases.

(c) In 1 and 2 it grows, in 3 it stays constant and in 4 and 5 it diminishes.

(d) In 1 and 2 it diminishes, in 3 it is constant and in 4 and 5 it grows.

Solution

For an ideal gas $PV = \mu RT$

For process represented by graph 1,($V \propto T$ at constant pressure)

$\therefore$ T increases, dT is +ve

Since $dU = C_V\, dT$,

dU is positive

So U increases.

In process 2, P increases, V also increases,

$$\left(\frac{PV}{T} = \text{constant}\right)$$

$\therefore$ T must also increase; dT = positive

$\therefore$ $$dU = C_V\, dT,$$

$$= +\text{ve}, \therefore U \text{ increases}$$

In process represented by 3 (isothermal process) T is constant

$$dU = C_V dT = 0$$

In an adiabatic process represented by 4 $\quad dW$ is +ve

$$dQ = 0, \therefore\ dU + dW = 0 \qquad dU \text{ is –ve}$$

Similarly, in 5 also U decreases.

Ans. (c)

Example 11

Six moles of an ideal gas perform a cycle shown in figure.

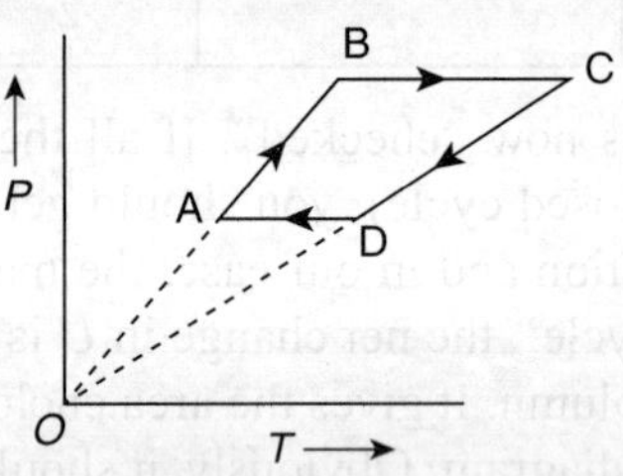

If the temperature: $T_A = 600$ K, $T_B = 800$ K, $T_C = 2200$ K and $T_D = 1200$ K, the work done per cycle is

(A) 20 kJ (B) 30 kJ

(C) 40 kJ (D) 60 kJ

Solution

Processes A to B and C to D are parts of straight line graphs of the form

$$y = mx$$

$$P = \frac{nR}{V}T\ (n = 6)$$

$P \propto T$, therefore volume remains constant.

So no work is done during processes from A to B and C to D, $W_{AB} = W_{CD} = 0$

$$W_{BC} = P_2(V_C - V_B) = 6R(T_C - T_B)$$

$$= 6R\,(2200 - 800) = 6R \times 1400 = 6 \times 8.31 \times 1400$$

$$= 50 \times 1400$$

$$W_{DA} = P_1(V_A - V_D) = 6R(600 - 1200) = -50(600)\text{ J}$$

Work done during the cycle from ABCD is

$W_{AB} + W_{BC} + W_{CD} + W_{DA}$
$= 0 + 50 \times 1400 + 0 - 50 \times 600 \text{ J} = 50(800) \text{ J} = 40 \text{ kJ}$

Ans. (c)

Example 12

n_1 and n_2 mols of two ideal gases of thermodynamic constants γ_1 and γ_2 respectively are mixed, $\frac{C_P}{C_V}$ for the mixture is

(A) $\frac{\gamma_1 + \gamma_2}{2}$

(B) $\frac{n_1\gamma_1 + n_2\gamma_2}{n_1 + n_2}$

(C) $\frac{n_1\gamma_2 + n_2\gamma_1}{n_1 + n_2}$

(D) $\frac{n_1\gamma_1(\gamma_2 - 1) + n_2\gamma_2(\gamma_1 - 1)}{n_1(\gamma_2 - 1) + n_2(\gamma_1 - 1)}$

Solution

$$n = n_1 + n_2 \; ; \; U = U_1 + U_2$$

$$nC_V = n_1C_{V_1} + n_2C_{V_2} = n_1\frac{R}{\gamma_1 - 1} + n_2\frac{R}{\gamma_2 - 1}$$

$$\frac{C_P}{C_V} = \gamma$$

$$nC_P = n_1\gamma_1\frac{R}{\gamma_1 - 1} + n_2\gamma_2\frac{R}{\gamma_2 - 1}$$

$$\gamma = \frac{nC_P}{nC_V} = \frac{C_P}{C_V} = \frac{\frac{n_1\gamma_1 R}{\gamma_1 - 1} + n_2\gamma_2\frac{R}{\gamma_2 - 1}}{\frac{n_1 R}{(\gamma_2 - 1)} + \frac{n_2 R}{(\gamma_1 - 1)}}$$

$$= \frac{n_1\gamma_1(\gamma_2 - 1) + n_2\gamma_2(\gamma_1 - 1)}{n_1(\gamma_2 - 1) + n_2(\gamma_1 - 1)}$$

Ans. (d)

POINTS TO REMEMBER

Thermodynamics: It is the branch of physics which deals with the conversion of heat into mechanical energy and vice-versa.

Thermodynamic State: It refers to a state of a body (or the system) that is completely defined by pressure, volume and temperature of the body (or the system).

Isotherms are the curves (graphs) between pressure and volume at constant temperature.

Zeroth law of thermodynamics

According to the zeroth law, when two bodies A and B are in thermal equilibrium with another body C, then bodies A and B will also be in thermal equilibrium with each other.

Work done in different processes

S. No.	Process	Const.	Work done ΔW
1.	Free expansion	U	0 as $P = 0$
2.	Isobaric	P	$P(V_f - V_i)$ or $nR\,(T_f - T_i)$
3.	Isochoric	V	0 as $\Delta V = 0$
4.	Isothermal	T	$nRT\log_e\left(\frac{V_f}{V_i}\right)$ or $nRT\log_e\left(\frac{P_i}{P_f}\right)$

(Continued)

S. No.	Process	Const.	Work done ΔW
5.	Adiabatic	Q	$\frac{nR}{(\gamma - 1)}(T_i - T_f)$ or $\frac{1}{(\gamma - 1)}\left[P_iV_i - P_fV_f\right]$

First law of thermodynamics

First law of thermodynamics is a statement of conservation of energy in thermodynamic process. According to it, heat given to a system (ΔQ) is equal to the sum of increase in its internal energy (ΔU) and the work done (ΔW) by the system against the surroundings, i.e.,

$$\Delta Q = \Delta U + \Delta W$$

(1) It defines internal energy U as a thermodynamic function of state, which is path independent, although it depends on initial and final positions of the state. According to the first law, change in internal energy $\Delta U\left[= U_f - U_i = \Delta Q + (-\Delta W)\right]$ is same for all possible paths between given points although ΔQ and ΔW are different for different paths.

(2) It makes no distinction between work and heat.

The first law of thermodynamics applied to different processes

S. No.	Process	ΔQ	ΔU	ΔW
1.	Cyclic	ΔW	0	= Area of cycle
2.	Isochoric	ΔU	$nC_V dT$	0
3.	Isothermal	ΔW	0	$nRT \log_e\left(\frac{V_f}{V_i}\right)$
4.	Adiabatic	0	$-\Delta W$	$\frac{nR(T_i - T_f)}{(\gamma - 1)}$
5.	Isobaric	$nC_P\Delta T$	$nC_V\Delta T$	$P(V_f - V_i) = nR(T_f - T_i)$

Heat Capacity (*C*)

$$C = \frac{\Delta Q}{\Delta t}$$

where

ΔQ = amount of heat absorbed

Δt = rise in temperature

SI unit of heat capacity: JK^{-1}, (J/°K)

Specific heat is the amount of heat required to raise the temperature of 1 kg of substance by 1°C or by 1°K .

SI unit: Joules per kilogram per Kelvin ($Jkg^{-1}K^{-1}$).

The specific heat of water is maximum.

Molar specific heat is the amount of heat required to raise the temperature of 1 mole of substance by 1°K .

SI unit – J $mole^{-1}K^{-1}$

Molar specific heat at constant volume (C_V)

It is the amount of heat required to raise the temperature of one mole of a gas through 1 K, keeping its volume constant.

SI units – J $mole^{-1}K^{-1}$

Specific heat at constant pressure (C_P)

It is the amount of heat required to raise the temperature of one mole of gas through 1°C or 1 K, keeping its pressure constant.

SI units – J $mole^{-1}K^{-1}$

C_p is always greater than C_v

Relation between C_P and C_V

$$C_P - C_V = R$$

Second law of thermodynamics

According to the Kelvin–Plank statement of the second law of thermodynamics, "It is impossible to construct a device that, operating in a cycle, produces no effect other than the absorption of energy from a reservoir and the performance of an equal amount of work".

Classius statement of the second Law of thermodynamics "It is impossible to make heat flow from a body at a lower temperature to a body at a higher temperature."

Reversible process: It is a process which can be retraced in opposite direction so that the system and surroundings pass through exactly the same state at every stage as in direct process. It means that as some heat is absorbed in the direct process from the surroundings, so the same amount of heat is rejected to surroundings in the reverse process.

Irreversible process: It is a process which is not exactly reversed, i.e., the system does not pass through the same intermediate state as in the direct process.

Heat engine

Heat engine is a device which converts heat into work continuously working through a cycle.

The efficiency of ideal engine (Carnot engine)

$$\eta = \frac{W}{Q_1} = \frac{Q_1 - Q_2}{Q_1}$$

$$= 1 - \frac{T_2}{T_1}$$

where Q_1 = Heat absorbed by the engine

Q_2 = Heat rejected to the surrounding

T_1 = Source temperature

T_2 = Sink temperature

Refrigerator (heat pump) - coefficient of performance (CoP)

It is defined as the ratio of the quantity of heat extracted per cycle from the contents of the refrigerator to the mechanical work *W* done by the external agency in doing so.

It is denoted by β or K

CoP of Refrigerator β

$$\beta = \frac{Q_2}{W} = \frac{Q_2}{Q_1 - Q_2}$$

$$\beta = \frac{T_2}{T_1 - T_2}$$

EXERCISES

1. An ideal engine exhausting heat at 27° C is to have 25% efficiency. It must take heat at
(a) 127° C (b) 227° C
(c) 327° C (d) 673° C

2. An ideal gas A and a real gas B have their volumes increased from V to $2V$ under isothermal conditions. The increase in internal energy
(a) will be zero in both A and B
(b) will be zero in both the gases
(c) of B will be more than that of A
(d) of A will be more than that of B.

3. P–V plots for two gases during adiabatic processes are shown in the figure. Plots 1 and 2 should correspond respectively to

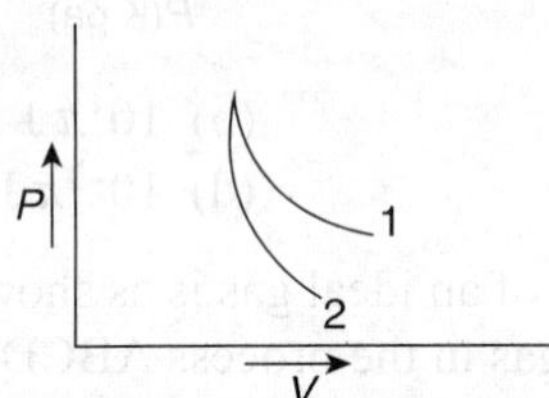

(a) He and O_2 (b) O_2 and He
(c) He and Ar (d) O_2 and N_2

4. In the indicator diagram shown, the net amount of work will be

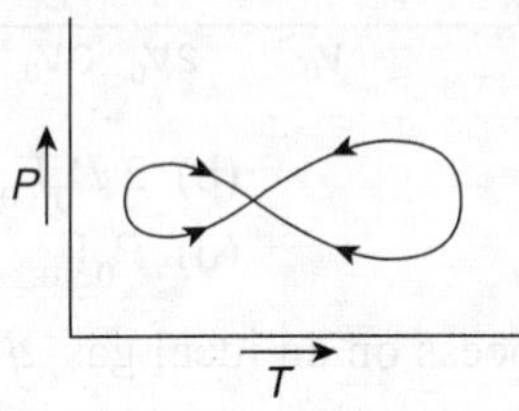

(a) zero (b) positive
(c) negative (d) Data insufficient.

5. Which of the following is incorrect regarding the first law of thermodynamics?
(a) It is not applicable to any cyclic process.
(b) It is a restatement of the principle of conservation of energy.
(c) It introduces the concept of the internal energy.
(d) It introduces the concept of the entropy.

6. A Carnot engine takes 3×10^6 Cal of heat from a reservoir at 627° C and gives it to a sink at 27° C. The work done by the engine is:
(a) 4.2×10^6 J (b) 8.4×10^6 J
(c) 16.8×10^6 J (d) zero

7. An ideal monatomic gas is taken around the cycle ABCDA as shown in the P-V diagram. The work done during cycle is given by:

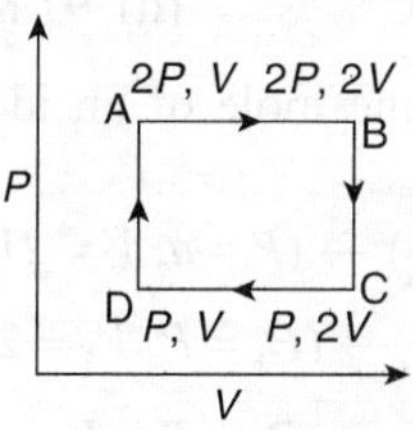

(a) (1/2) PV (b) PV
(c) 2PV (d) 4PV

8. The efficiency of a Carnot engine operating with reservoir temperature of 100° C and – 23° C will be
(a) $\frac{373+250}{373}$ (b) $\frac{373-250}{373}$
(c) $\frac{100+23}{100}$ (d) $\frac{100-23}{100}$

9. A balloon that is initially flat, is inflated by filling it from a tank of compressed air. The final volume of the balloon is 5 m^3. The barometer reads 95 kPa. The work done in this process is:
(a) 3.75×10^5 J (b) 4.75×10^7 J
(c) 4.75×10^3 J (d) 4.75×10^5 J

10. One mole of an ideal gas is enclosed in a conducting vertical cylinder under a light piston. If the volume of the gas is increased isothermally n times, then the work done in increasing the volume is: [Assume atmospheric pressure is P_0 and temperature is T_0]
(a) $-RT_0 \log_e n$
(b) $RT_0 \log_e n$
(c) $RT_0 \log_e n - (n-1)RT_0$
(d) $RT_0 \log_e n - (n-1)RT_0$

11. We consider a thermodynamic system. If ΔU represents the increase in its internal energy and W the work done by the system, which of the following statements is true?
(a) $\Delta U = -W$ in an isothermal process
(b) $\Delta U = W$ in an isothermal process
(c) $\Delta U = -W$ in an adiabatic process
(d) $\Delta U = W$ in an adiabatic process

12. In a polytrophic process, PV^n = constant:
(a) If $n = 1$, process is isothermal
(b) If $n = \infty$, process is isochoric
(c) If $n = 0$, process is isobaric
(d) All the above.

13. The (W/Q) of a Carnot engine is 1/6, now the temperature of sink is reduced by 62° C, then this ratio becomes twice, therefore the initial temperature of the sink and source are respectively
(a) 13° C, 67° C (b) 37° C, 99° C
(c) 67° C, 33° C (d) 97 K, 37 K

14. A process on one mole of an ideal gas is defined as follows:

$$(P_A, V_A, T_A) \rightarrow (P = a, V = 2V_A, T = T_A)$$
$$\rightarrow (P_1 = P, V_1 = 2V_A, T_1 = b)$$
$$\rightarrow (P_2 = P_A, V_2 = V_A, T_2 = c)$$

The values of a, b and c are:
(a) $\frac{P_A}{2}, T_A, T_A$ (b) P_A, T_A, V_A
(c) $2P_A, T_A, T_A$ (d) $P_A, 2T_A, T_A$

15. In an adiabatic expansion, a gas does 25 J of work while in an adiabatic compression 100 J of work is done on a gas. The change of internal energy in the two process respectively are:
(a) 25 J and –100 J (b) –25 J and 100 J
(c) –25 J and –100 J (d) 25 J and 100 J

16. If a gas is compressed adiabatically:
(a) the internal energy of gas increases
(b) the internal energy of gas decreases
(c) the internal energy of gas does not change
(d) the work done is positive.

17. In an isothermal process, the volume of the ideal gas is halved. One can say that
(a) internal energy of the system decreases
(b) work done by the gas is positive
(c) work done by the gas is negative
(d) pressure does not change.

18. Which of the following best represents the process of above problem?

(a) 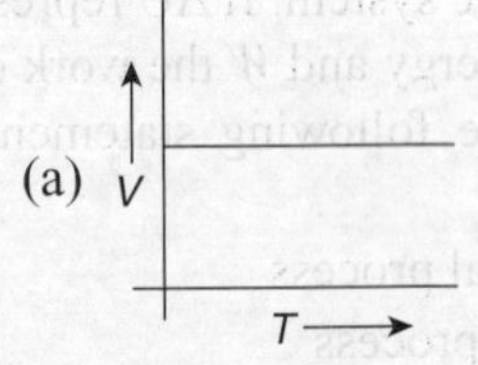

(b) 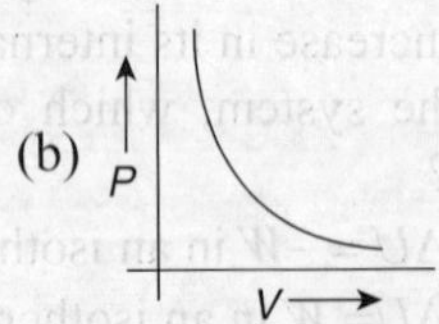

(c) 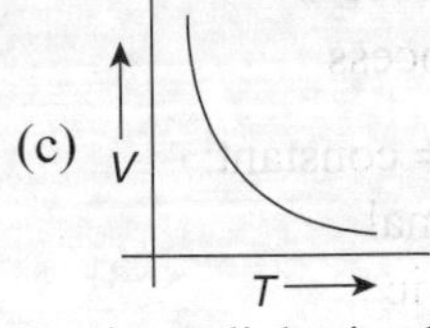

(d) All of these.

19. During adiabatic change, specific heat is
(a) zero (b) greater than zero
(c) less than zero (d) infinity.

20. A Carnot engine has an efficiency of 40% when its sink is at temperature of 300 K. By what temperature should the source be raised so that the efficiency may become 70%?
(a) 300 K (b) 400 K
(c) 500 K (d) 600 K

21. Heat energy absorbed by the system in going through a cyclic process is

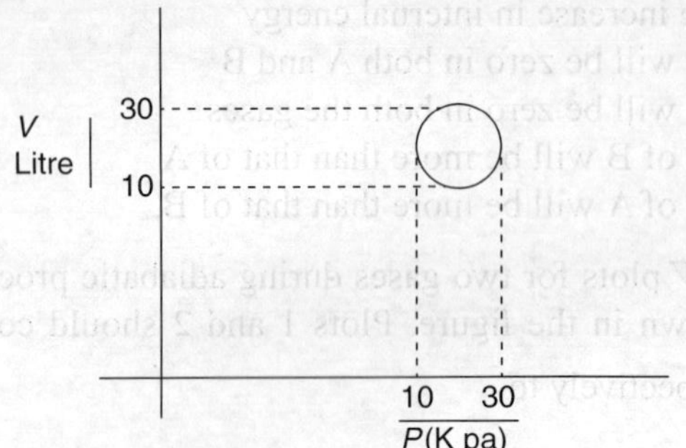

(a) $10^7 \pi$ J (b) $10^4 \pi$ J
(c) $10^2 \pi$ J (d) $10^{-3} \pi$ J

22. P-V diagram of an ideal gas is as shown in figure. Work done by the gas in the process ABCD is

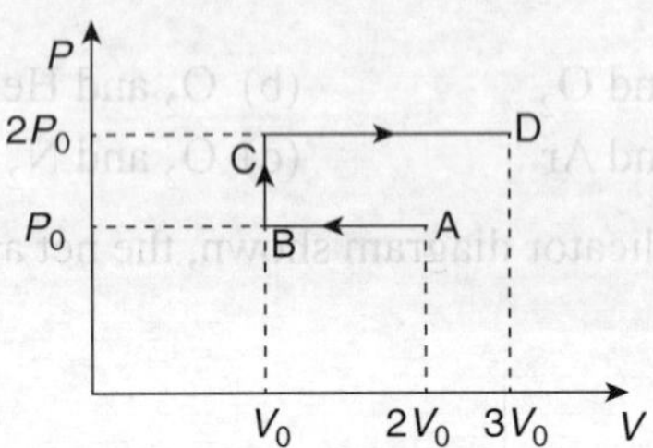

(a) $4\ P_0V_0$ (b) $2\ P_0V_0$
(c) $3\ P_0V_0$ (d) P_0V_0

23. In a given process on an ideal gas, $dW = 0$ and $dQ < 0$; then for the gas
(a) the temperature will decrease
(b) the volume will increase
(c) the pressure will remain constant
(d) the temperature will increase.

24. Heat is not being exchanged in a body. If its internal energy is increased, then
(a) its temperature will increase
(b) its temperature will decrease
(c) its temperature will remain constant
(d) None of these.

25. The internal energy of a gas during isothermal expansion
(a) increases (b) decreases
(c) becomes zero (d) remains constant.

26. When two bodies A and B are in thermal equilibrium,
(a) the kinetic energies of all the molecules of A and B will be equal

(b) the potential energies of all the molecules of A and B will be equal
(c) the internal energies of two bodies will be equal
(d) the average kinetic energy of the molecules of the two bodies will be equal.

27. If C_P and C_V denote the specific heats of nitrogen per unit mass at constant pressure and constant volume respectively, then:
(a) $C_P - C_V = R/28$ (b) $C_P - C_V = R/14$
(c) $C_P - C_V = R$ (d) $C_P - C_V = 28\,R$

28. When a system is taken from state i to state f along the path iaf, it is found that $Q = 50$ cal and $W = 20$ cal. Along the path ibf $Q = 36$ cal. W along the path ibf is:

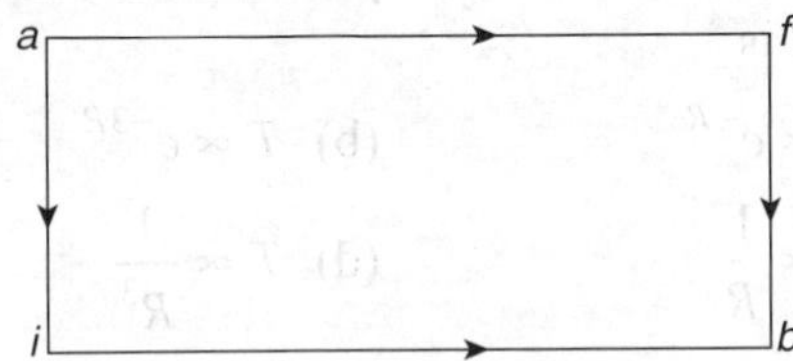

(a) 6 cal (b) 16 cal
(c) 66 cal (d) 14 cal

29. Which of the following statements is correct for any thermodynamic system?
(a) The internal energy changes in all processes
(b) Internal energy and entropy are state functions
(c) The change in entropy can never be zero
(d) The work done in an adiabatic process is always zero.

30. During an adiabatic process, the pressure of a gas is found to be proportional to the cube of its absolute temperature. The ratio C_P/C_V for the gas is:
(a) 4/3 (b) 2
(c) 5/3 (d) 3/2

31. Which of the following parameters does not characterize the thermodynamic state of matter?
(a) Temperature (b) Pressure
(c) Work (d) Volume

32. What work will be done, when 3 moles of an ideal gas are compressed to half the initial volume at a constant temperature of 300 K?
(a) –5188 J (b) 5000 J
(c) 5188 J (d) –5000 J

33. During an isothermal expansion of an ideal gas,
(a) its internal energy decreases
(b) its internal energy does not change
(c) the work done by the gas is equal to the quantity of heat absorbed by it
(d) Both (b) and (c) are correct.

34. The P-V diagram shows that two adiabatic parts for the same gas intersect two isothermals at T_1 and T_2. How are the ratios (V_a/V_d) and (V_b/V_c) related to each other?

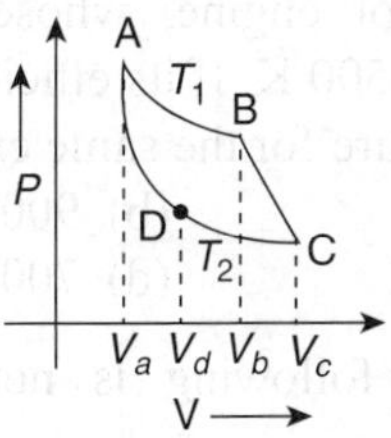

(a) $\frac{V_a}{V_d} = \frac{V_c}{V_d}$ (b) $\frac{V_a}{V_d} = \frac{V_b}{V_c}$
(c) $\frac{V_a}{V_d} = 2\frac{V_c}{V_b}$ (d) $\frac{V_a}{V_d} = \frac{1}{2}\frac{V_b}{V_c}$

35. A given system undergoes a change in which the work done by the system equals the decrease in its internal energy. The system must have undergone an
(a) isothermal change (b) adiabatic change
(c) isobaric change (d) isochoric change.

36. The molar specific heat of mixture at constant volume, if one mole of a monoatomic gas is mixed with three moles of diatomic gas is:
(a) 3.33 R (b) 2.25 R
(c) 1.15 R (d) 6.72 R

37. In a given process for ideal gas, $dW = 0$ and $dQ > 0$. Then for the gas:
(a) the volume remains constant
(b) the volume will increase
(c) temperature will increase
(d) Both (a) and (c) are correct.

38. When a gas is compressed to half its initial volume,
(a) final pressure is more if the change is isothermal
(b) final pressure is more if the change is adiabatic
(c) work done is less if the change is adiabatic
(d) work done is more if the change is isothermal.

39. A quantity of heat Q is supplied to a mono-atomic ideal gas which expands at constant pressure. The fraction of heat that goes in to work done by the gas is
(a) 2/5 (b) 3/5
(c) 2/3 (d) 1

40. An engine operates between temperatures of 900 K and T_2K and another engine between T_2K and 400 K. For both to do equal work value of T_2 will be
(a) 600 K (b) 625 K
(c) 650 K (d) 750 K

41. A sample of gas expands from volume V_1 to V_2. The amount of work done by the gas is greatest, when the expansion is
(a) adiabatic (b) equal in all cases
(c) isothermal (d) isobaric.

42. An ideal Carnot engine, whose efficiency is 40%, receives heat at 500 K. If its efficiency is 50%, then the intake temperature for the same exhaust temperature is
(a) 800 K (b) 900 K
(c) 600 K (d) 700 K

43. Which of the following is not thermo dynamical function?
(a) Enthalpy (b) Work done
(c) Gibb's energy (d) Internal energy

44. 110 joule of heat is added to a gaseous system whose internal energy is 40 J, then the amount of external work done is
(a) 150 J (b) 70 J
(c) 110 J (d) 40 J

45. A perfect gas contained in a cylinder is kept in vacuum. If the cylinder suddenly bursts, then the temperature of the gas
(a) remains constant (b) becomes zero
(c) increases (d) decreases.

46. A thermodynamic process is shown in the figure. The pressure and volumes corresponding to some points in the figure are

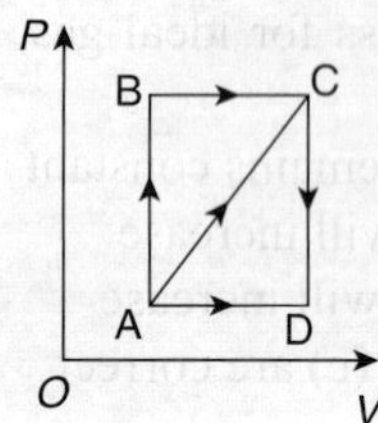

$P_A = 3 \times 10^4$ Pa; $V_A = 2 \times 10^{-3}$ m³;
$P_B = 8 \times 10^4$ Pa; $V_D = 5 \times 10^{-3}$ m³.

In the process AB, 600 J of heat is added to the system and in process BC, 200 J of heat is added to the system. The change in internal energy of the system in the process AC would be
(a) 560 J (b) 800 J
(c) 600 J (d) 640 J

47. First law of thermodynamics is consequence of conversation of
(a) work (b) energy
(c) heat (d) All of these.

48. Ideal gas undergoes an adiabatic change in its state from (P_1, V_1, T_1) to (P_2, V_2, T_2). The work done (W) in the process is (μ = number of moles; C_p and C_v are molar specific heats of gas)
(a) $W = \mu(T_1 - T_2) C_p$ (b) $W = \mu(T_1 - T_2) C_\upsilon$
(c) $W = \mu(T_1 + T_2) C_p$ (d) $W = \mu(T_1 + T_2) C_\upsilon$

49. Air in a cylinder is suddenly compressed by a piston, which is then maintained at the same position. With the passage of time,
(a) the pressure decreases
(b) the pressure increases
(c) the pressure remains the same
(d) the pressure may increase or decrease depending upon the nature of the gas.

50. An adiabatic process occurs at constant
(a) temperature (b) pressure
(c) heat (d) temperature and pressure.

51. Consider a spherical shell of radius R at temperature T. The black body radiation inside it can be considered as an ideal gas of photons with internal energy per volume $u = \frac{U}{V} \propto T^4$ and pressure $p = \frac{1}{3}\left(\frac{U}{V}\right)$. If the shell now undergoes an adiabatic expansion, the relation between T and R is
(a) $T \propto e^{-R}$ (b) $T \propto e^{-3R}$
(c) $T \propto \frac{1}{R}$ (d) $T \propto \frac{1}{R^3}$

52. A scientist says that the efficiency of his heat engine which works at source temperature 127° C and sink temperature 27° C is 26%; then
(a) it is impossible
(b) it is possible but less probable
(c) it is quite probable
(d) data are incomplete.

53. The work of 146 kJ is performed in order to compress one kilo mole of a gas adiabatically and in the process, the temperature of the gas increase by 7° C. The gas is ($R = 8.3$ J mol⁻¹ K⁻¹)
(a) triatomic
(b) a mixture of monatomic and diatomic
(c) monatomic
(d) diatomic.

54. 5.6 Litres of helium gas at STP is adiabatically compressed to 0.7 litre. Taking the initial temperature to be T_1, the work done in the process is
(a) $\frac{9}{8}RT_1$ (b) $\frac{3}{2}RT_1$
(c) $\frac{15}{8}RT_1$ (d) $\frac{9}{2}RT_1$

55. Which is incorrect?
(a) In an isobaric process, $\Delta P = 0$
(b) In an isochoric process, $\Delta W = 0$
(c) In an isothermal process, $\Delta T = 0$
(d) In an isothermal process, $\Delta Q = 0$

56. A Carnot engine absorbs an amount Q of heat from a reservoir at an absolute temperature T and rejects heat to a sink at a temperature of $T/3$. The amount of heat rejected is

(a) $Q/4$ (b) $Q/3$
(c) $Q/2$ (d) $2Q/3$

57. Choose the incorrect statement from the following
S1: The efficiency of a heat engine can be 1, but the coefficient of performance of a refrigerator can never be infinity.
S2: The first law of thermodynamics is basically the principle of conservation of energy.
S3: The second law of thermodynamics does not allow several phenomena consistent with the first law.
S4: A process, whose sole result is the transfer of heat from a colder object to a hotter object is impossible.
(a) S1 (b) S3
(c) S2 (d) S4

58. Two moles of ideal helium gas are in a rubber balloon at 30 °C. The balloon is fully expandable and can be assumed to require no energy in its expansion. The temperature of the gas in the balloon is slowly changed to 35 °C. The amount of heat required in raising the temperature is nearly (take $R = 8.31$ J / mol. K)
(a) 62 J (b) 104 J
(c) 124 J (d) 208 J

59. The efficiency of Carnot engine is 50% and temperature of sink is 500 K. If temperature of source is kept constant and its efficiency raised to 60%, then the required temperature of sink will be
(a) 100 K (b) 600 K
(c) 400 K (d) 500 K

60. Two cylinders A and B, fitted with pistons, contain equal amounts of an ideal diatomic gas at 300 K. The piston of A is free to move while that of B is held fixed. The same amount of heat is given to the gas in each cylinder. If the rise in temperature of the gas in A is 30 K, then the rise in temperature of the gas in B is
(a) 30 K (b) 18 K
(c) 50 K (d) 42 K

61. A Carnot engine whose sink is at a temperature of 300 K has an efficiency of 40%. By how much should the temperature of the source be increased so as to increase the efficiency to 60%?
(a) 380 K (b) 325 K
(c) 250 K (d) 275 K

62. An electrical refrigerator abstracts 2000 calories from ice trays. The coefficient of performance is 4. Then the work done by motor in calories is
(a) 8000 (b) 4
(c) 2000 (d) 500

63. In a thermodynamics process, pressure of a fixed mass of a gas is changed in such a manner that the gas molecules gives out 30 joules of heat and 10 joules of work is done on the gas. If the initial internal energy of the gas was 40 joules, then the final internal energy will be
(a) zero (b) 80 joules
(c) 20 joules (d) –20 joules.

64. An ideal gas heat engine operates in a Carnot cycle between 227° C and 127° C. It absorbs 6 kcal at the higher temperature.
The amount of heat (in kcal) converted into work is equal to
(a) 4.8 (b) 3.5
(c) 1.6 (d) 1.2

65. What is the nature of change in internal energy in the following thermodynamic processes shown in figure?

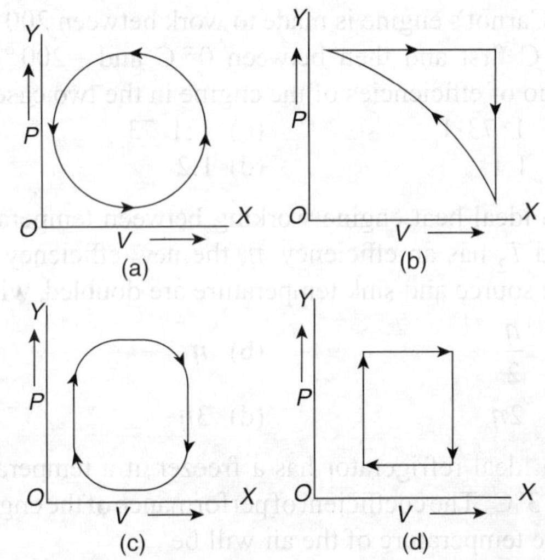

(a) ΔU is + in all cases
(b) ΔU is – in all cases
(c) ΔU zero in all cases
(d) ΔU is + in (a) and (c) and – in (b) and (d).

66. In figure, V_c is

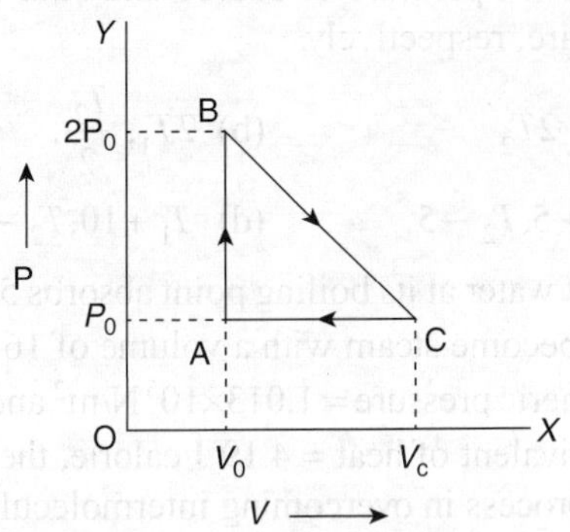

(a) $3\ V_0$ (b) $2\ V_0$
(c) V_0 (d) None of the above.

67. When an ideal gas placed in vacuum inside a container is allowed to expand then
(a) internal energy decreases, temperature increases
(b) internal energy increases, temperature decreases
(c) both internal energy and temperature decreases
(d) neither internal energy nor temperature decreases.

68. The temperature of 5 moles of gas which was held at constant volume was changed from 100 °C to 120 °C. The change in the internal energy of the gas was found to be 80 joule, the total heat capacity of the gas at constant volume will be equal to
(a) 8 joule/K (b) 0.8 joule/K
(c) 4.0 joule/K (d) 0.4 joule/K

69. In a given process on *n* ideal gas, $dW = 0$ and $dQ < 0$. Then for the gas,
(a) the temperature will decrease
(b) the volume will increase
(c) the pressure will remain constant
(d) the temperature will increase.

70. A Carnot's engine is made to work between 200 ° C and 0 ° C first and then between 0 ° C and – 200 ° C. The ratio of efficiencies of the engine in the two cases is
(a) 1:73·1 (b) 1:1·73
(c) 1:1 (d) 1:2

71. An ideal heat engine working between temperature T_1 and T_2 has an efficiency η, the new efficiency if both the source and sink temperatures are doubled, will be
(a) $\dfrac{\eta}{2}$ (b) η
(c) 2η (d) 3η

72. An ideal refrigerator has a freezer at a temperature of –13°C. The coefficient of performance of the engine is 5. The temperature of the air will be
(a) 325 ° C (b) 325 K
(c) 39 ° C (d) 320 ° C

73. The efficiency of Carnot's heat engine is 0.5 when the temperature of the source T_1 and that of sink is T_2. The efficiency of another Carnot's heat engine is also 0.5. The temperature of source and sink of the second engine are, respectively,
(a) $2T_1, 2T_2$ (b) $2T_1, \dfrac{T_2}{2}$.
(c) $T_1 + 5, T_2 - 5$ (d) $T_1 + 10, T_2 - 10$

74. 1 cm³ of water at its boiling point absorbs 540 calories of heat to become steam with a volume of 1671 cm³. If the atmospheric pressure = 1.013×10^5 N/m² and the mechanical equivalent of heat = 4.19 J/calorie, the energy spent in this process in overcoming intermolecular forces is
(a) 540 cal (b) 40 cal
(c) 500 cal (d) zero.

75. A monoatomic gas is supplied the heat Q very slowly keeping the pressure constant. The work done by the gas will be
(a) $\dfrac{2}{3}Q$ (b) $\dfrac{3}{5}Q$
(c) $\dfrac{2}{5}Q$ (d) $\dfrac{1}{5}Q$

76. An engineer claims to have made an engine delivering 10 kW power with fuel consumption of 1 g/s. The calorific value of the fuel is 2 k cal g. The claim of the engineer
(a) is valid
(b) is invalid
(c) depends on engine design
(d) depends on the load.

77. Half mole of an ideal monatomic gas is heated at constant pressure of 1 atm from 20 °C to 90 °C. Work done by gas is close to
(a) 73 J (b) 291 J
(c) 581 J (d) 146 J

78. A thermodynamic system is taken through the cycle PQRSP process. The net work done by the system is

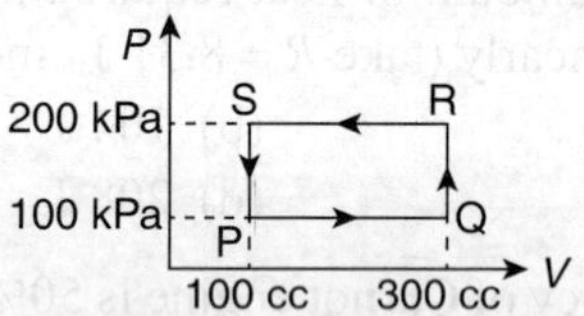

(a) 20 J (b) –20 J
(c) 400 J (d) –374 J

79. An ideal monatomic gas is taken through the thermodynamics states A → B → C → D via the paths shown in the figure If U_A, U_B, U_C and U_D represent the internal energy of the gas in states A, B, C and D respectively, then which of the following is not true?

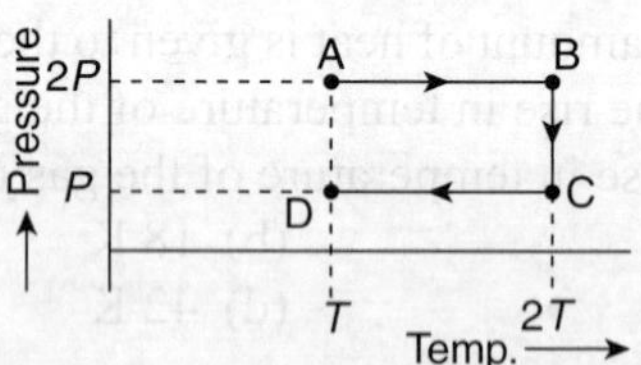

(a) $U_A = U_D$ (b) $U_B < U_A$
(c) $U_B = U_C$ (d) $U_C > U_D$

80. Six moles of O_2 gas is heated from 20 °C to 35 °C at constant volume. Its specific heat capacity at constant pressure is 8 cal /mol-K and R = 8.31 J/mol-K. What is the change in internal energy of gas?
(a) 180 cal (b) 300 cal
(c) 360 cal (d) 540 cal

81. An ideal gas is taken from point A to the point B as shown in P-V diagram keeping temperature constant. The work done in the process is

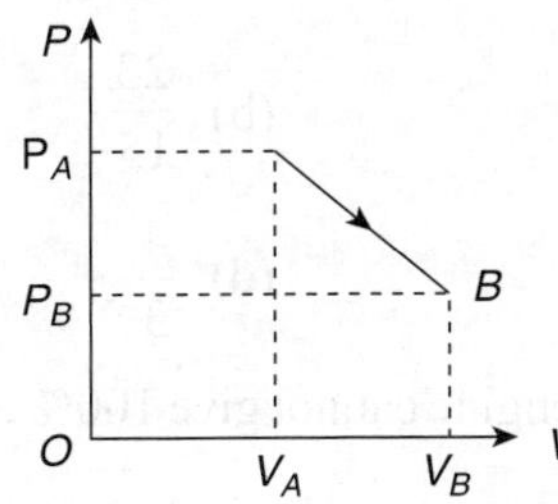

(a) $(P_A - P_B)(V_B - V_A)$

(b) $\frac{1}{2}(P_B - P_A)(V_B + V_A)$

(c) $\frac{1}{2}(P_B - P_A)(V_B - V_A)$

(d) $\frac{1}{2}(P_B + P_A)(V_B - V_A)$

82. 1 mole of a monatomic gas is first heated at constant volume at 100 °C. The change in internal energy of the gas is Δu_1. Next it is heated at constant pressure at 100 °C, the change in internal energy of the gas is Δu_2. Then the ratio is $\frac{\Delta u_1}{\Delta u_2}$

(a) 1 (b) $\frac{3}{5}$

(c) $\frac{5}{3}$ (d) uncertain.

83. If the degrees of the freedom of a gas are f, then the ratio of its specific heats $\frac{C_P}{C_V}$ is given by

(a) $1+\frac{2}{f}$ (b) $1-\frac{2}{f}$

(c) $1+\frac{1}{f}$ (d) $1-\frac{1}{f}$

84. The temperature of the system decreases in the process of

(a) free expansion (b) adiabatic expansion

(c) isothermal expansion (d) isothermal compression.

85. Two cylinders A and B fitted with pistons contain equal number of moles of an ideal monatomic gas at 400 K. The piston of A is free to move while that of B is held fixed. Same amount of heat energy is given to the gas in each cylinder. If the rise in temperature of the gas in A is 42 K, then the rise in temperature of B is $(\gamma = 5/3)$

(a) 25.2 K (b) 35 K

(c) 42 K (d) 70 K

86. The ratio of specific heats of a gas is γ. The change in internal energy of one mole of the gas, when the volume changes from V to $2V$ at constant pressure P is

(a) $\frac{\gamma-1}{PV}$ (b) PV

(c) $\frac{PV}{\gamma-1}$ (d) $\frac{PV}{\gamma}$

87. A given mass of a gas is compressed isothermally until its pressure is doubled. It is then allowed to expand adiabatically until its original volume is restored and its pressure is then found to be 0.75 of its initial pressure. The ratio of the specific heats of the gas is approximately.

(a) 1.20 (b) 1.41

(c) 1.67 (d) 1.83

88. The pressure and density of a given mass of a diatomic gas $\left(\gamma = \frac{7}{5}\right)$ change adiabatically from (P, d) to (P', d'). If $\frac{d'}{d} = 32$, then $\frac{P'}{P}$ is (γ ratio of specific heats)

(a) $\frac{1}{128}$ (b) $\frac{1}{64}$

(c) 64 (d) 128

89. An ideal refrigerator has a freezer at a temperature of -13 °C. The coefficient of performance of the engine is 5. The temperature of the air (at which heat is rejected) will be

(a) 325 °C (b) 325 K

(c) 39 °C (d) 320 °C

90. A P-V diagram of a system undergoing thermodynamics transformation is shown in figure. The work done by the system is going from A→B→ C is 30 J and 40 J respectively. If heat is given in the system, the change in internal energy between A and C is

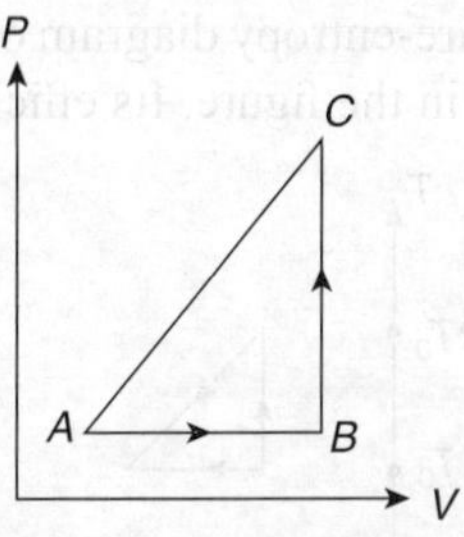

(a) 10 J (b) 70 J

(c) 84 J (d) 134 J

91. One mole of an ideal gas at an initial temperature of T K does 6 R joules of work adiabatically. If the ratio of specific heats of this gas at constant pressure and at constant volume is 5/3, then final temperature of the gas will be

(a) $(T-4)$ K (b) $(T+4)$ K

(c) $(T-2.4)$K (d) $(T+2.4)$ K

92. Ten moles of an ideal gas at constant temperature 500 K is compressed from 50 L to 5 L. Work done in the process is

(a) -1.2×10^4 J (b) -2.4×10^4 J
(c) -4.8×10^4 J (d) -9.6×10^4 J

93. A Carnot engine having an efficiency of $\eta = \frac{1}{10}$ as heat engine, is used as a refrigerator. If the work done on the system is 10 J, the amount of energy absorbed from the reservoir at lower temperature is

(a) 99 J (b) 90 J
(c) 1 J (d) 100 J

94. A system performs work ΔW when an amount of heat ΔQ is added to the system. The corresponding change in the internal energy is ΔU. A unique function of the initial and final states (irrespective of the mode of change) is

(a) ΔQ (b) ΔW
(c) ΔU and ΔQ (d) ΔU

95. A system goes from A to B via two processes I and II as shown in the figure. If ΔU_1 and ΔU_2 are the changes in the internal energies in the process I and II respectively, then

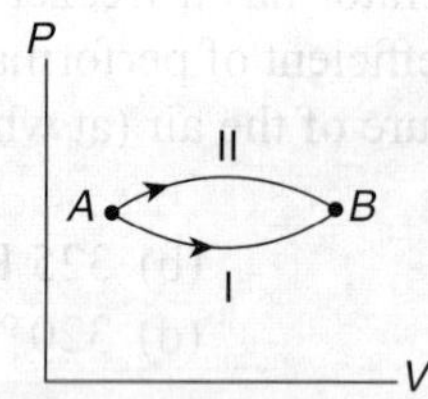

(a) $\Delta U_1 = \Delta U_2$
(b) relation between ΔU_1 and ΔU_2 cannot be determined
(c) $\Delta U_2 > \Delta U_1$
(d) $\Delta U_2 < \Delta U_1$

96. The temperature-entropy diagram of a reversible engine cycle is given in the figure. Its efficiency is

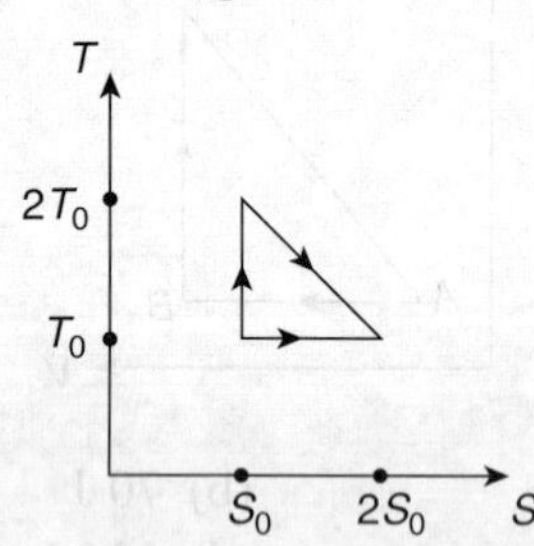

(a) $\frac{1}{2}$ (b) $\frac{1}{4}$
(c) $\frac{1}{3}$ (d) $\frac{2}{3}$

97. One mole of ideal monatomic gas $\left(\gamma = \frac{5}{3}\right)$ is mixed with one mole of diatomic gas $\left(\gamma = \frac{7}{5}\right)$. What is γ for the mixture? γ denotes the ratio of specific heat at constant pressure to specific heat at constant volume.

(a) $\frac{3}{2}$ (b) $\frac{23}{15}$
(c) $\frac{35}{23}$ (d) $\frac{4}{3}$

98. Even Carnot engine cannot give 100% efficiency because we cannot

(a) prevent radiation
(b) find ideal sources
(c) reach absolute zero temperatures
(d) eliminate friction.

99. One mole of an ideal gas, at an initial temperature of T K, does 6 R joule of work adiabatically. If the ratio of specific heats of this gas at constant pressure and at constant volume is 5/3, the final temperature of gas will be

(a) $(T + 2.4)$ K (b) $(T - 2.4)$ K
(c) $(T + 4)$ K (d) $(T - 4)$ K

100. A thermodynamic system is taken from state A to B along ACB and is brought back to A along BDA as shown in the P–V diagram. The net work done during the complete cycle is given by the area

(a) P_1A C B P_2 P_1 (b) A C B B′ A′ A
(c) A C B D A (d) A D B B′ A′ A.

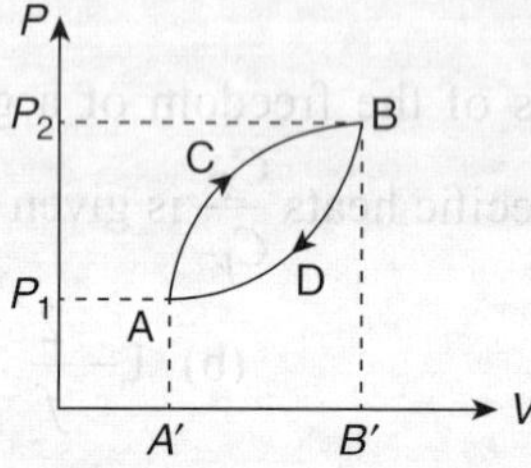

101. In a thermodynamic process, the pressure of a mixed mass of the gas is changed in such a manner that the gas releases 20 J of heat and 8 J of work is done on the gas. If the initial internal enrgy of the gas was 30 J, then the final internal energy will be

(a) 2 J (b) 42 J
(c) 18 J (d) 58 J

102. A gas is compressed at a constant pressure of 50 N/m^2 from volume of 10 m^3 to a volume of 4 m^3. Energy of 100 J as then added to the gas by heating. Its internal energy is

(a) increased by 400 J (b) increased by 200 J
(c) increased by 100 J (d) decreased by 200 J.

103. A frictionless heat engine can be 100% efficient only if its exhaust temperature is

(a) 0 °C
(b) 0 K
(c) equal to its input temperature
(d) half of its input temperature.

104. One of the most efficient engines ever developed operates between 2100 K and 700 K. Its actual efficiency is 40%. What percentage of its maximum possible efficiency is this?

(a) 40% (b) 60%
(c) 66.67% (d) 33.37%

105. If the ratio of specific heat of a gas at constant pressure to that at constant volume is γ, the change in internal energy of the mass of gas, when the volume changes from V to $2V$ at constant pressure P is

(a) $R/(\gamma - 1)$ (b) PV
(c) $PV/(\gamma - 1)$ (d) $\gamma PV/(\gamma - 1)$

106. In a thermodynamic system, if the working substance is an ideal gas, its internal energy is in the form of

(a) kinetic energy only
(b) kinetic and potential energy
(c) potential energy
(d) None of these.

107. Figure shows a cyclic process a b c a for one mole of an ideal gas. If ab is an

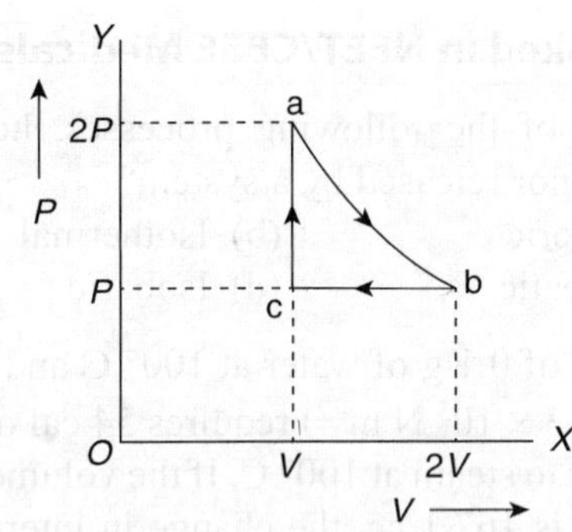

isothermal process, then which one is the P-T diagram for the cyclic process?

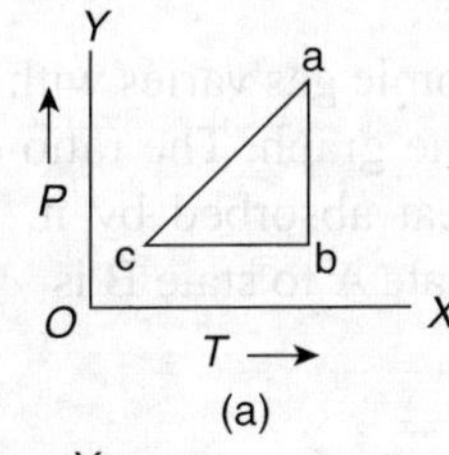

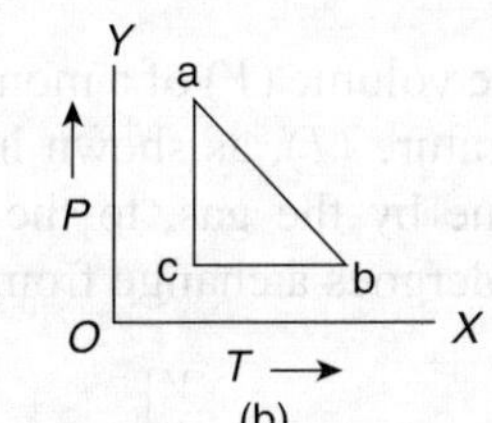

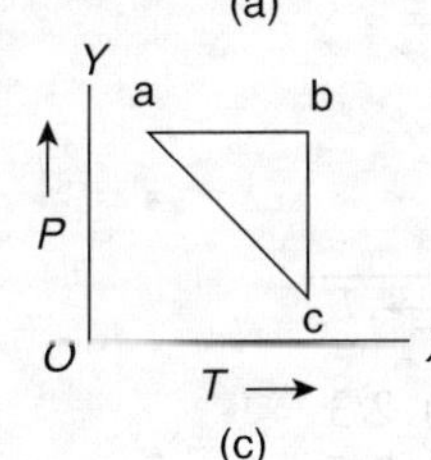

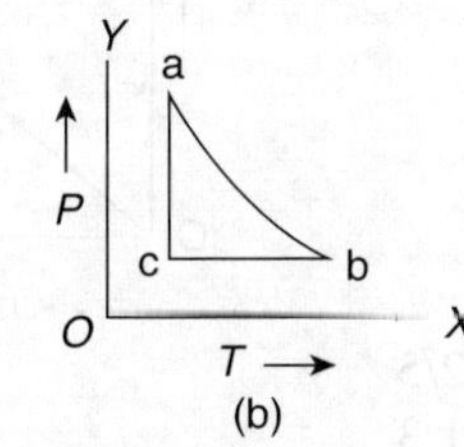

108. In the indicator diagram, the net amount of work done is

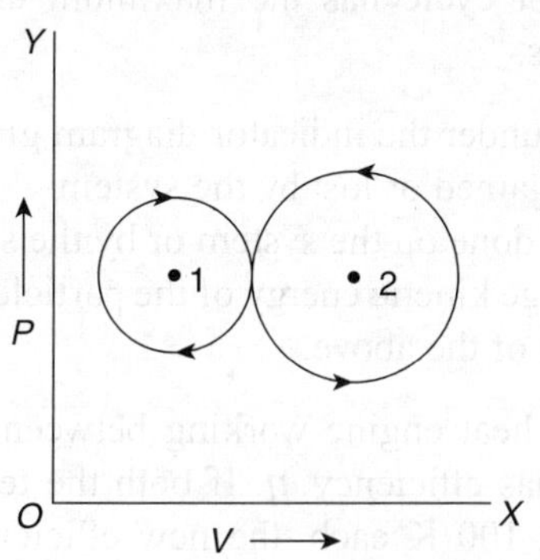

(a) negative (b) positive
(c) zero (d) infinity.

109. Which of the following processes is reversible?

(a) Transfer of heat by conduction
(b) Transfer of heat by radiation
(c) Isothermal compression
(d) Electrical heating of a nichrome wire.

110. A cyclic process is shown in the below P–T diagram. Which of the curves show the same process on V–T diagram?

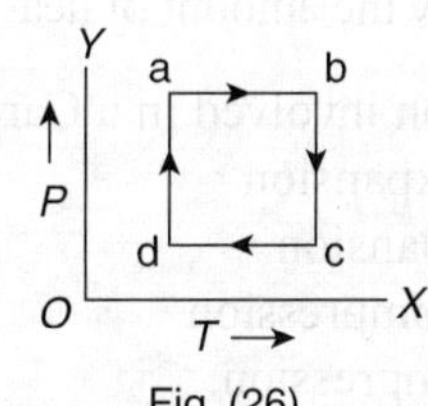

Fig. (26)

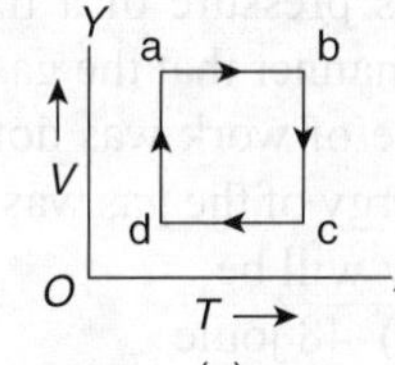

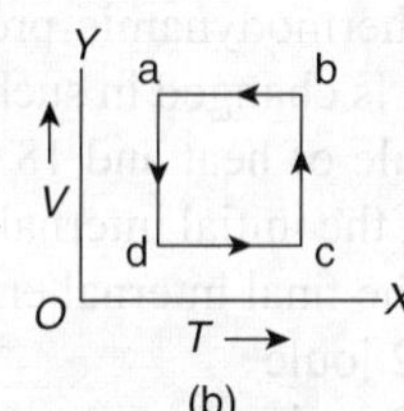

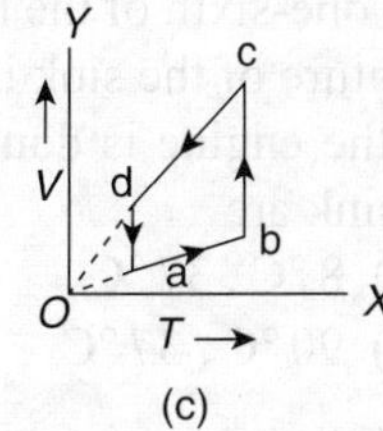

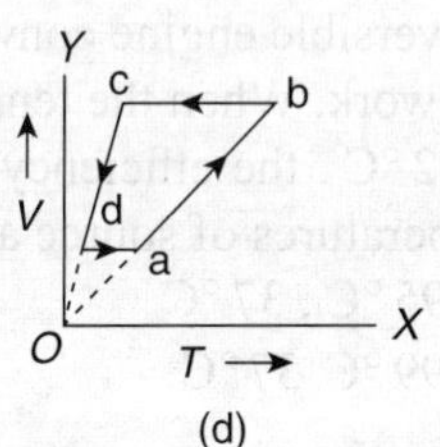

111. Which statement is incorrect?

(a) All reversible cycles have same efficiency.
(b) Reversible cycle has more efficiency than an irreversible one.

(c) Carnot cycle is a reversible one
(d) Carnot cycle has the maximum efficiency in all cycles.

112. The area under the indicator diagram gives
(a) heat gained or lost by the system
(b) work done on the system or by the system
(c) average kinetic energy of the particles of the system
(d) None of the above.

113. An ideal heat engine working between temperature T and T' has efficiency η. If both the temperatures are raised by 100 K each, the new efficiency of the heat engine will be
(a) equal to η
(b) greater than η
(c) less than η
(d) greater or less than η depending upon the nature of the working substances.

114. In a Carnot heat engine, the temperature of the working substance at the end of the cycle is
(a) equal to that at the beginning
(b) more than that at the beginning
(c) less than that at the beginning
(d) determined by the amount of heat rejected at the sink.

115. The first operation involved in a Carnot cycle is
(a) isothermal expansion
(b) adiabatic expansion
(c) isothermal compression
(d) adiabatic compression.

116. In a thermodynamic process pressure of a fixed mass of gas is changed in such a manner that the gas releases 30 joule of heat and 18 joule of work was done on the gas. If the initial internal energy of the gas was 60 joule, then the final internal energy will be
(a) 32 joule (b) 48 joule
(c) 72 joule (d) 96 joules.

117. A reversible engine converts one-sixth of the heat input into work. When the temperature of the sink is reduced by 62 °C, the efficiency of the engine is doubled. The temperatures of source and sink are
(a) 95 °C, 37 °C (b) 8 °C, 37 °C
(c) 99 °C, 37 °C (d) 90 °C, 37 °C

118. The following are the P–V diagrams for cyclic processes for a gas. In which of these processes is heat absorbed by the gas?

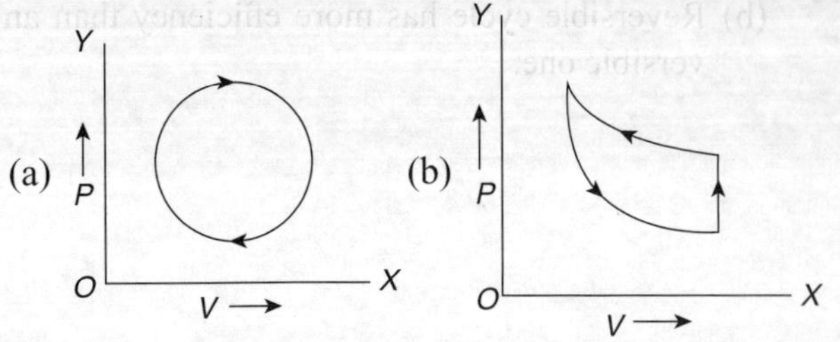

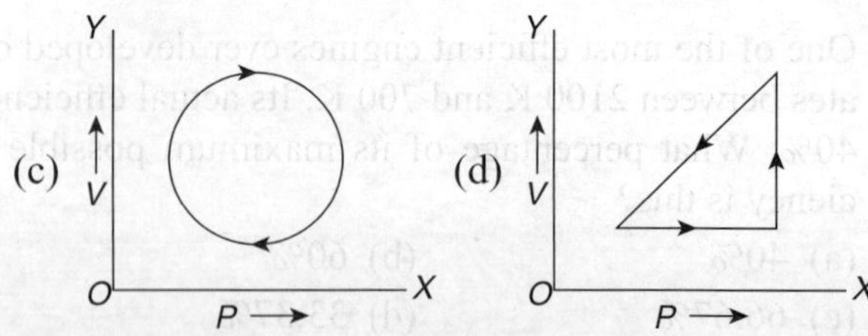

119. The indicator diagrams representing minimum and maximum amounts of work done are respectively
(a) c and a (b) a and c
(c) b and a (d) d and b.

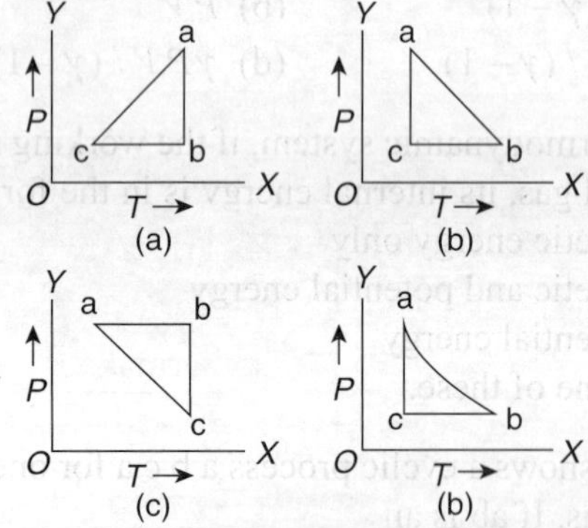

Questions Asked in NEET/CBSE Medicals

120. In which of the following processes, heat is neither absorbed nor released by a system?
(a) Isochoric (b) Isothermal
(c) Adiabatic (d) Isobaric

121. A sample of 0.1 g of water at 100° C and normal pressure (1.013×10^5 N m^{-2}) requires 54 cal of heat energy to convert to steam at 100° C. If the volume of the steam produced is 167.1 cc, the change in internal energy of the sample, is
(a) 104.3 J (b) 208.7 J
(c) 42.2 J (d) 84.5 J

122. The volume (V) of a monatomic gas varies with its temperature (T), as shown in the graph. The ratio of work done by the gas, to the heat absorbed by it, when it undergoes a change from state A to state B is

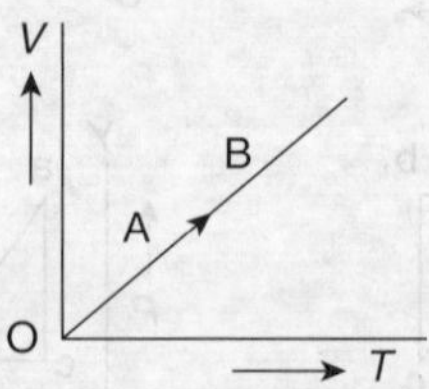

(a) 2/5 (b) 2/3
(c) 1/3 (d) 2/7

123. The efficiency of an ideal heat engine working between the freezing point and boiling point of water, is
(a) 26.8% (b) 20%
(c) 625% (d) 12.5%

124. Thermodynamic processes are indicated in the following diagram.

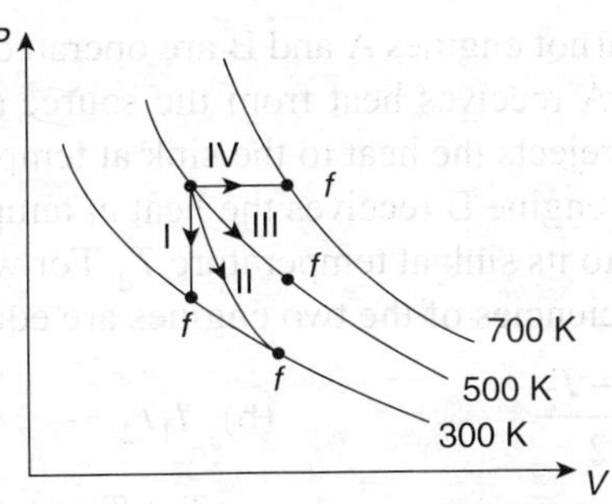

Match the following.

Column-1	Column-2
P. Process I	A. Adiabatic
Q. Process II	B. Isobaric
R. Process III	C. Isochoric
S. Process IV	D. Isothermal

(a) P → C, Q → A, R → D, S → B
(b) P → C, Q → D, R → B, S → A
(c) P → D, Q → B, R → A, S → C
(d) P → A, Q → C, R → D, S → B

125. A Carnot engine having an efficiency of 1/10 as heat engine is used as a refrigerator. If the work done on the system is 10 J, the amount of energy absorbed from the reservoir at lower temperature is
(a) 90 J (b) 99 J
(c) 100 J (d) 1 J

126. The temperature inside a refrigerator is t_2°C and the room temperature is t_1°C. The amount of heat delivered to the room for each joule of electrical energy consumed ideally will be

(a) $\frac{t_1}{t_1 - t_2}$ (b) $\frac{t_1 + 273}{t_1 - t_2}$

(c) $\frac{t_2 + 273}{t_1 - t_2}$ (d) $\frac{t_1 + t_2}{t_1 + 273}$

127. A gas is compressed isothermally to half its initial volume. The same gas is compressed separately through an adiabatic process until its volume is again reduced to half. Then,
(a) compressing the gas isothermally or adiabatically will require the same amount of work
(b) which of the case (whether compression through isothermal or through adiabatic process) requires more work will depend upon the atomicity of the gas
(c) compressing the gas isothermally will require more work to be done
(d) compressing the gas through adiabatic process will require more work to be done.

128. A refrigerator works between 4 °C and 30 °C. It is required to remove 600 calories of heat every second in order to keep the temperature of the refrigerated space constant. The power required is (Take 1 cal = 4.2 Joules)
(a) 236.5 W (b) 2365 W
(c) 2.365 W (d) 23.65 W

129. An ideal gas is compressed to half its initial volume by means of several processes. Which of the process results in the maximum work done on the gas?
(a) Isochoric (b) Isothermal
(c) Adiabatic (d) Isobaric

130. The coefficient of performance of a refrigerator is 5. If the temperature inside freezer is –20 °C, the temperature of the surroundings to which it rejects heat is
(a) 11 °C (b) 21 °C
(c) 31 °C (d) 41 °C

131. A Carnot engine, having an efficiency of 1/10 as heat engine, is used as a refrigerator. If the work done on the system is 10 J, the amount of energy absorbed from the reservoir at lower temperature is
(a) 90 J (b) 1 J
(c) 100 J (d) 99 J

132. One mole of an ideal diatomic gas undergoes a transition from A to B along a path AB as shown in the figure. The change in internal energy of the gas during the transition is

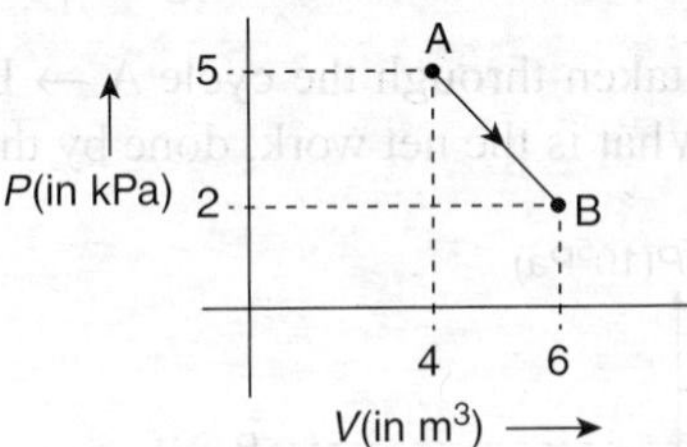

(a) 20 J (b) –12 kJ
(c) 20 kJ (d) –20 kJ

133. Figure below shows two paths that may be taken by a gas to go from a state A to a state C. In process AB, 400 J of heat is added to the system and in process BC, 100 J of heat is added to the system. The heat absorbed by the system in the process AC will be

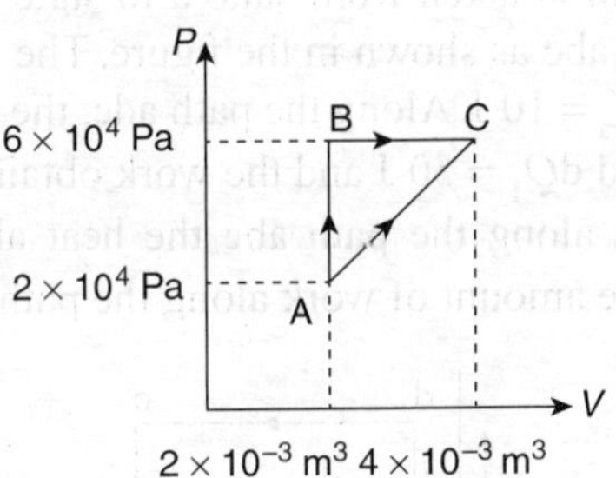

(a) 460 J (b) 300 J
(c) 380 J (d) 500 J

134. A thermodynamic system undergoes cyclic process ABCDA as shown in figure. The work done by the system in the cycle is

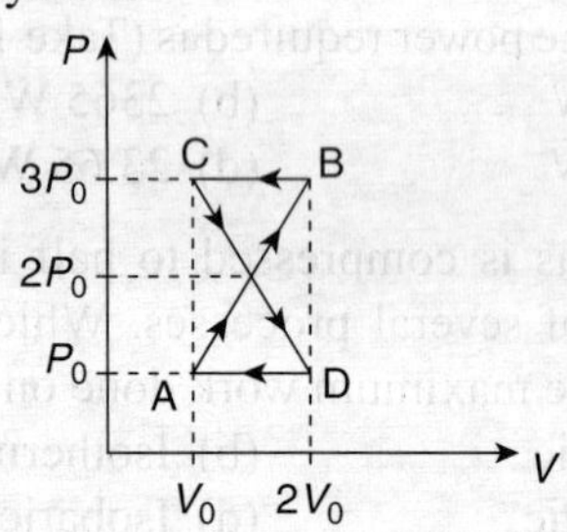

(a) P_oV_o (b) $2P_oV_o$
(c) $\frac{P_0V_0}{2}$ (d) zero

135. In the given (V–T) diagram, what is the relation between pressures P_1 and P_2?

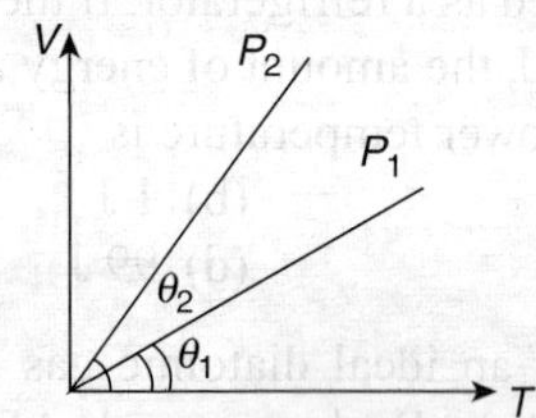

(a) $P_2 < P_1$ (b) Cannot be predicted
(c) $P_2 = P_1$ (d) $P_2 > P_1$

136. A gas is taken through the cycle A → B → C → A as shown. What is the net work, done by the gas?

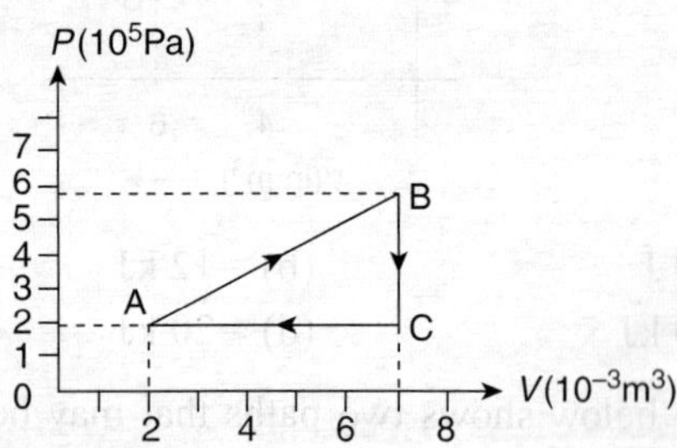

(a) Zero (b) –2000 J
(c) 2000 J (d) 1000 J

137. A system is taken from state a to state c by two paths adc and abc as shown in the figure. The internal energy at a is $U_a = 10$ J. Along the path adc, the amount of heat absorbed $dQ_1 = 50$ J and the work obtained $dW_1 = 20$ J whereas along the path abc the heat absorbed $dQ_2 = 36$ J. The amount of work along the path abc is

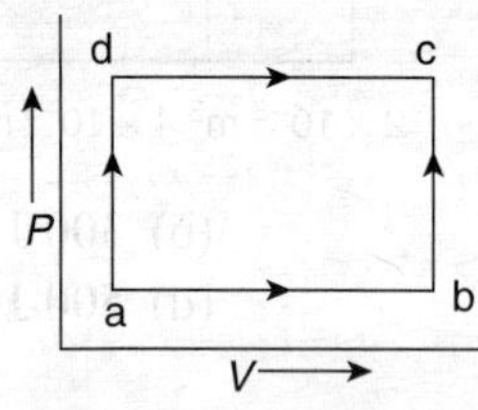

(a) 10 J (b) 12 J
(c) 36 J (d) 6 J

138. Two Carnot engines A and B are operated in series. The engine A receives heat from the source at temperature T_1 and rejects the heat to the sink at temperature T. The second engine B receives the heat at temperature T and rejects to its sink at temperature T_2. For what value of T the efficiencies of the two engines are equal?

(a) $\frac{T_1 - T_2}{2}$ (b) T_1T_2
(c) $\sqrt{T_1T_2}$ (d) $\frac{T_1 + T_2}{2}$

139. A thermodynamic system is taken through the cycle ABCD as shown in figure. Heat rejected by the gas during the cycle is

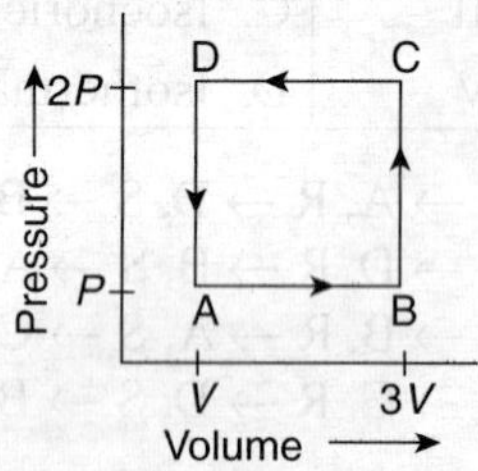

(a) $2PV$ (b) $4PV$
(c) $1/2\,PV$ (d) PV

140. One mole of an ideal gas goes from an initial state A to final state B via two processes: It first undergoes isothermal expansion from volume V to $3V$ and then its volume is reduced from $3V$ to V at constant pressure. The correct P-V diagram representing the two processes is

(a) 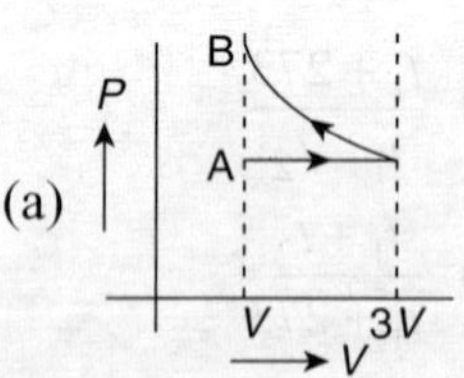

(b) 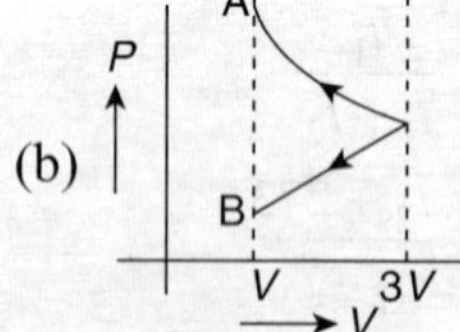

(c) 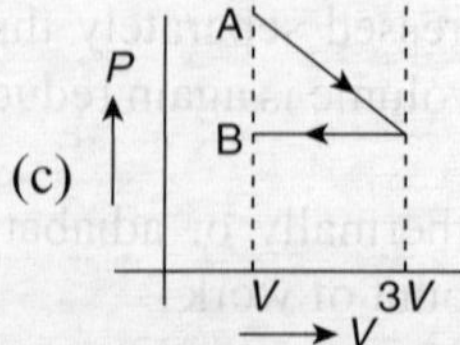

(d) 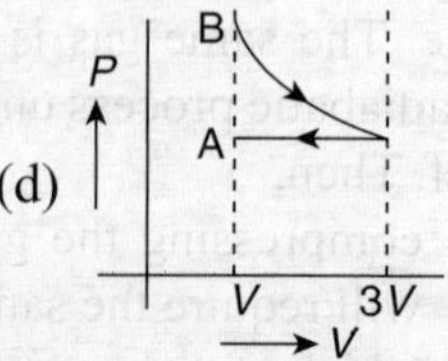

141. An ideal gas goes from state A to state B via three different processes as indicated in the P-V diagram. If Q_1, Q_2, Q_3 indicate the heat absorbed by the gas along the three processes and ΔU_1 ΔU_2, ΔU_3 indicate the change in internal energy along the three processes respectively, then

(a) $Q_1 > Q_2 > Q_3$ and $\Delta U_1 = \Delta U_2 = \Delta U_3$
(b) $Q_3 > Q_2 > Q_1$ and $\Delta U_1 = \Delta U_2 = \Delta U_3$

(c) $Q_1 = Q_2 = Q_3$ and $\Delta U_1 > \Delta U_2 > \Delta U_3$
(d) $Q_3 > Q_2 > Q_1$ and $\Delta U_1 > \Delta U_2 > \Delta U_3$

142. During an isothermal expansion, a confined ideal gas does –150 J of work against its surroundings. This implies that
(a) 150 J of heat has been removed from the gas
(b) 300 J of heat has been added to the gas
(c) no heat is transferred because the process is isothermal
(d) 150 J of heat has been added to the gas.

143. When 1 kg of ice at 0° C melts to water at 0° C, the resulting change in its entropy, taking latent heat of ice to be 80 cal/°C, is
(a) 273 cal/K (b) 8×10^4 cal/K
(c) 80 cal/K (d) 293 cal/K

144. If ΔU and ΔW represent the increase in internal energy and work done by the system respectively, in a thermodynamic process, which of the following is true?
(a) $\Delta U = -\Delta W$, in an adiabatic process
(b) $\Delta U = \Delta W$, in an isothermal process
(c) $\Delta U = \Delta W$, in an isochoric process
(d) $\Delta U = -\Delta W$, in an isobaric process

145. If C_P and C_V denote the specific heats per unit mass of an ideal gas molecular weight M, then
(a) $C_P - C_V = R/M^2$ (b) $C_P - C_V = R$
(c) $C_P - C_V = R/M$ (d) $C_P - C_V = MR$
where R is the molar gas constant.

146. The internal energy change in a system that has absorbed 2 kcal of heat and done 500 J of work is
(a) 6400 J (b) 5400 J
(c) 7900 J (d) 8900 J

147. In thermodynamic processes which of the following statements is not true?
(a) In an isochoric process pressure remains constant.
(b) In an isothermal process temperature remains constant.
(c) In an adiabatic process PV^γ = constant.
(d) In an adiabatic process the system is insulated from the surroundings.

148. If Q, E and W denote respectively the heat added, change in internal energy and the work done in a closed cyclic process, then
(a) $E = 0$ (b) $Q = 0$
(c) $W = 0$ (d) $Q = W = 0$

149. An engine has an efficiency of 1/6. When the temperature of sink is reduced by 62 °C, its efficiency is doubled. Temperature of the source is
(a) 37° C (b) 62° C
(c) 99° C (d) 124° C.

150. A Carnot engine whose sink is at 300 K has an efficiency of 40%. By how much should the temperature of source be increased so as to increase its efficiency by 50% of original efficiency?
(a) 380 K (b) 275 K
(c) 325 K (d) 250 K

151. An ideal gas heat engine operates in a Carnot cycle between 227° C and 127° C. It absorbs 6×10^4 cal of heat at higher temperature. Amount of heat converted to work is
(a) 4.8×10^4 cal (b) 6×10^4 cal
(c) 2.4×10^4 cal (d) 1.2×10^4 cal

Answer Keys

EXERCISES

1. (a)	2. (b)	3. (b)	4. (c)	5. (a, d)	6. (b)	7. (b)	8. (b)	9. (d)	10. (b)
11. (c)	12. (d)	13. (b)	14. (a)	15. (b)	16. (a)	17. (c)	18. (a)	19. (a)	20. (c)
21. (c)	22. (c)	23. (a)	24. (a)	25. (d)	26. (c)	27. (a)	28. (a)	29. (b)	30. (d)
31. (c)	32. (a)	33. (d)	34. (b)	35. (b)	36. (b)	37. (d)	38. (b)	39. (a)	40. (c)
41. (d)	42. (c)	43. (b)	44. (b)	45. (a)	46. (a)	47. (b)	48. (b)	49. (a)	50. (c)
51. (c)	52. (a)	53. (d)	54. (a)	55. (d)	56. (b)	57. (a)	58. (d)	59. (c)	60. (d)
61. (c)	62. (d)	63. (c)	64. (d)	65. (c)	66. (b)	67. (d)	68. (c)	69. (a)	70. (b)
71. (b)	72. (c)	73. (a)	74. (c)	75. (c)	76. (b)	77. (b)	78. (b)	79. (b)	80. (d)
81. (d)	82. (b)	83. (a)	84. (b)	85. (d)	86. (c)	87. (b)	88. (d)	89. (c)	90. (a)
91. (a)	92. (d)	93. (b)	94. (d)	95. (a)	96. (c)	97. (b)	98. (c)	99. (d)	100. (c)
101. (c)	102. (a)	103. (b)	104. (b)	105. (c)	106. (a)	107. (a)	108. (a)	109. (c)	110. (c)
111. (a)	112. (b)	113. (c)	114. (a)	115. (a)	116. (b)	117. (c)	118. (a, d)	119. (b)	

Questions Asked in NEET/CBSE Medicals

120. (c)	121. (b)	122. (a)	123. (a)	124. (a)	125. (a)	126. (b)	127. (d)	128. (a)	129. (c)
130. (c)	131. (a)	132. (d)	133. (a)	134. (d)	135. (a)	136. (d)	137. (d)	138. (c)	139. (a)
140. (d)	141. (a)	142. (d)	143. (d)	144. (a)	145. (c)	146. (c)	147. (a)	148. (a)	149. (c)
150. (d)	151. (d)								

Solutions

1. (a)

$$\eta = 1 - \frac{Q_2}{Q_1} = 1 - \frac{T_2}{T_1}$$

$$\frac{25}{100} = 1 - \frac{(273+27)}{T_1}$$

$$\frac{1}{4} = 1 - \frac{300}{T_1}$$

$$\Rightarrow \frac{300}{T_1} = 1 - \frac{1}{4}$$

$$\frac{300}{T_1} = \frac{3}{4}$$

$$T_1 = 400 \text{ K}$$

$$T_1 = 400 - 273 = 127°\text{C}$$

2. (b)
Under isothermal conditions, there is no change in internal energy.

3. (b)
In adiabatic process, slope of PV-graph

$$\frac{dP}{dV} = -\gamma \frac{P}{V} \Rightarrow |\text{Slope}| \propto \gamma$$

From the given graph $(\text{Slope})_2 > (\text{Slope})_1 \Rightarrow \gamma_2 > \gamma_1$
Therefore 1 should correspond to $O_2\,(\gamma = 1.4)$ and 2 should correspond to He $(\gamma = 1.66)$

4. (c)

5. (a, d)

6. (b)

$$\frac{Q_2}{Q_1} = \frac{T_2}{T_1}$$

where Q_1 = Heat absorbed from reservoir
Q_2 = Heat rejected to sink
T_1 = Reservoir temperature
T_2 = Sink temperature

$$\frac{Q_2}{3\times10^6} = \frac{(273+27)}{(273+627)}$$

$$\frac{Q_2}{3\times10^6} = \frac{300}{900}$$

$$Q_2 = 3\times10^6 \times \frac{1}{3} = 10^6 \text{ cal}$$

$$W = Q_1 - Q_2 = 3\times10^6 - 10^6$$

$$= 2\times10^6 \text{ cal}$$

$$W = 2\times10^6 \times 4.2$$

$$= 8.4\times10^6 \text{ J}$$

7. (b)
Area within indicator diagram = Area of rectangle on the PV diagram = $(2P - P)(2V - V) = PV$

8. (b)
Reservoirs temperature $(T_1) = 100°\text{C} = 373$ K and

$$T_2 = -23°\text{C} = 250\text{K}$$

The efficiency of a Carnot engine

$$\eta = \frac{T_1 - T_2}{T_1} = \frac{373-250}{373}$$

9. (d)

$$W = PdV$$

$$= 95\times10^3\,(5-0)$$

$$= 95\times5\times10^3$$

$$= 475\times10^3$$

$$= 4.75\times10^5 \text{ J}$$

10. (b)
The work done by the gas in isothermal process

$$W = nRT \log_e \frac{V_2}{V_1}$$

$$n = 1, T = T_0, \frac{V_2}{V_1} = n$$

$$W = RT_0 \log_e n$$

11. (c)

According to first law of thermodynamics,

$$\Delta Q = \Delta U + W$$

For an adiabatic process, $\Delta Q = 0$

$$\therefore \Delta U = -W$$

12. (d)

13. (b)

$$\frac{1}{6} = 1 - \frac{T_2}{T_1} \text{ or } \frac{5}{6} = \frac{T_2}{T_1}$$

Now, $$\frac{1}{3} = 1 - \frac{T_2 - 62}{T_1} = 1 - \frac{5}{6} + \frac{62}{T_1}$$

$$T_1 = 372 \text{ K} ; \ T_2 = 310 \text{ K}$$

$$\Rightarrow \quad T_1 = (372 - 273); \ T_2 = (310 - 273)$$

$$T_1 = 99°\text{C and } T_2 = 37°\text{C}$$

14. (a)

Ideal gas equation for one mole of a gas $= \frac{PV}{T} =$ constant

$$\frac{P_A V_A}{T_A} = \frac{a \times 2V_A}{T_A} = \text{K} \quad \text{(i)}$$

$$\Rightarrow a = \frac{P_A}{2}$$

$$\frac{P \times 2V_A}{b} = \frac{P_A V_A}{c} = \text{K} \quad \text{(ii)}$$

$$\frac{P_A V_A}{T_A} = \frac{P_A V_A}{c} = \text{K}$$

$$\Rightarrow c = T_A$$

But $$P = a$$

Therefore $$\frac{a \times 2V_A}{b} = \frac{P_A V_A}{T_A} = k$$

$$\frac{P_A \times 2V_A}{2b} = \frac{P_A V_A}{T_A}$$

$$\frac{1}{b} = \frac{1}{T_A}$$

$$\Rightarrow b = T_A$$

15. (b)

Applying first law of thermodynamics,

$$dQ = dU + dW$$

For adiabatic process $dQ = 0$.

$$dU = -dW$$

$$= -25 \text{ J (expansion)}$$

$$dU = -dW = -(-100)$$

(dW is –ve for compression)

$$= 100 \text{ J}$$

16. (a)

$$dQ = dU + dW$$

$$0 = dU + dW$$

$$dU = -dW \ (dW \text{ is –ve})$$

$$dU = -(-dW)$$

dU is positive. Therefore internal energy increases.

17. (c)

18. (a)

19. (a)

20. (c)

$$\eta_1 = 1 - \frac{300}{T_1}$$

$$\frac{40}{100} = 1 - \frac{300}{T_1}$$

$$\frac{300}{T_1} = 1 - \frac{40}{100}$$

$$\frac{300}{T_1} = \frac{60}{100}$$

$$\Rightarrow T_1 = 500 \text{ K}$$

$$\eta_2 = 1 - \frac{300}{T_1'}$$

$$\frac{70}{100} = 1 - \frac{300}{T_1'}$$

$$\frac{300}{T_1'} = 1 - \frac{70}{100}$$

$$\frac{300}{T_1'} = \frac{30}{100}$$

$$T_1' = 1000 \text{ K}$$

$$T_1' - T_1 = 1000 - 500 = 500 \text{ K}$$

21. (c)

Area of the circle

$$\pi r^2 = \pi \frac{d^2}{4}$$

$$= \frac{\pi \times 20 \times 20}{4}$$

$$= \pi 10^2$$

22. (c)

$$W_{AB} = P_0(v_0 - 2v_0)$$
$$= -P_0V_0$$
$$W_B = 0 \text{ (Isochoric process)}$$
$$W_{CD} = 2P_0(3v_0 - v_0)$$
$$= 4P_0V_0$$
$$W_{\text{net}} = -P_0V_0 + 0 + 4P_0V_0$$
$$= 3P_0V_0$$

23. (a)

We know that, $dQ = dU + dW$

When $dW = 0, dQ = dU$

As $dQ < 0$, therefore $dU < 0$

Internal energy decreases. So, the temperature will decreases.

24. (a)

With rise in temperature, internal energy also increases.

25. (d)

During isothermal expansion, temperature remains constant. Therefore, internal energy remains constant

26. (c)

27. (a)

$$C_P - C_V = R$$

where C_P and C_V are molar specific heats. The given specific heats are per gram. Therefore per mole = 28 C_P and 28 C_V

$$28C_P - 28C_V = R$$
$$C_P - C_V = \frac{R}{28}$$

28. (a)

$$dQ = dU + dW$$
$$50 = dU + 20$$
$$\Rightarrow \quad dU = 50 - 20 = 30 \text{ J}$$

Along ibf

$$dQ = dU + dW$$
$$36 = 30 + dW$$
$$dW = 6 \text{ cal}$$

29. (b)

30. (d)

$$P \propto T^3 \Rightarrow PT^{-3} = \text{Constant} \quad \text{(i)}$$

Adiabatic relations $P^{1-\gamma}T^{\gamma} = \text{Constant}$

$$PT^{\frac{\gamma}{1-\gamma}} = \text{Constant} \quad \text{(ii)}$$

Comparing Eq. (i) with Eq. (ii)

$$\frac{\gamma}{1-\gamma} = -3$$
$$-3 + 3\gamma = \gamma$$
$$2\gamma = 3 \Rightarrow \gamma = 3/2$$

31. (c)

32. (a)

The work done in isothermal process

$$W = nRT \log_e \frac{V_2}{V_1}$$
$$= 3 \times 8.31 \times 300 \log_e \frac{1}{2}$$
$$= -5188 \text{ J}$$

33. (d)

34. (b)

Point B and C lie on the same adiabatic curve, therefore

$$T_1V_a^{\gamma-1} = T_2V_c^{\gamma-1}$$
$$\frac{T_1}{T_2} = \frac{V_c^{\gamma-1}}{(V_b)^{\gamma-1}} \quad \text{(i)}$$

Similarly A and D lie on the same adiabatic curve

$$\frac{T_1}{T_2} = \frac{(V_d)^{\gamma-1}}{(V_a)^{\gamma-1}} \quad \text{(ii)}$$

From (i) and (ii)

$$\frac{V_c}{V_b} = \frac{V_d}{V_a} \Rightarrow \frac{V_a}{V_d} = \frac{V_b}{V_c}$$

35. (b)

In adiabatic change $\Delta Q = 0$

$$\Delta U = -\Delta W$$

36. (b)

$$C_V = \frac{n_1C_{V_1} + n_2C_{V_2}}{n_1 + n_2}$$
$$C_V = \frac{1 \times \frac{3}{2}R + 3 \times \frac{5}{2}R}{1+3} = \frac{9}{4}R$$
$$= 2.25R$$

37. (d)

$$\text{Work done} = \int PdV = 0$$
$$\Rightarrow dV = 0$$

Volume is constant

$$dQ = dU + dW$$
$$dQ = dU + 0$$
$$dQ = dU$$

Since $dQ > 0 \Rightarrow dU > 0$

$dV > 0$ means temperature increases

38. (b)

39. (a)

$$dQ_P = dU + dW$$
$$dQ_V = dU$$
$$dQ_P = nC_P dT$$
$$dQ_V = nC_V dT$$
$$dW = dQ - dV$$
$$= nC_P dT - nC_V dT$$
$$\frac{dW}{dQ_P} = \frac{nC_P dT - nC_V dT}{nC_P dT}$$
$$\frac{C_P - C_V}{C_P} = 1 - \frac{C_V}{C_P}$$
$$= 1 - \frac{\frac{3}{2}R}{\frac{5}{2}R} = 1 - \frac{3}{5}$$
$$= \frac{2}{5}$$

40. (c)

$$900 - T_2 = T_2 - 400$$
$$2T_2 = 1300$$
$$T_2 = 650$$

41. (d)

During expansion, work is performed by the gas. The isobaric expansion is represented by the horizontal straight line AB_1. Since the adiabatic curve is steeper than the isothermal curve, the adiabatic expansion curve (AB_3) must lie below the isothermal curve (AB_2) as shown in the given figure.

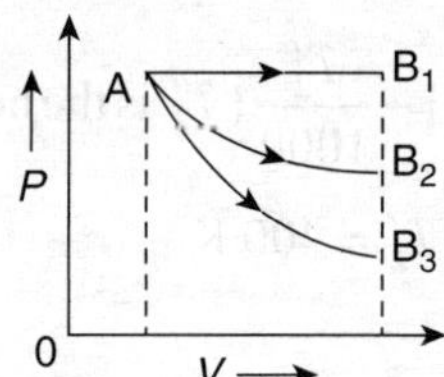

Since area under AB_1 is maximum the work done is maximum in case of isobaric expansions.

42. (c)

Efficiency of Carnot engine $(\eta_1) = 40\% = 0.4$;

Heat intake = 500 K and

New efficiency $(\eta_2) = 50\% = 0.5$

The efficiency $(\eta) = 1 - \frac{T_2}{T_1}$ or $\frac{T_2}{T_1} = 1 - \eta$

For first case, $\frac{T_2}{500} = 1 - 0.4$ or $T_2 = 300$ K

For second case, $\frac{300}{T_1} = 1 - 0.5$ or $T_1 = 600$ K

43. (b)

44. (b)

$$\Delta Q = \Delta U + \Delta W$$
$$\Rightarrow \Delta W = \Delta Q - \Delta U = 110 - 40 = 70 \text{ J}$$

45. (a)

During free expansion of a perfect gas, no work is done and also no heat is supplied from outside. Therefore, there is no change in internal energy. Hence, temperature remains constant.

46. (a)

Since AB is an isochoric process, no work is done. BC is isobaric process

$$W = P_B \times (V_D - V_A) = 240 \text{ J}$$

Therefore $\Delta Q = 600 + 200 = 800$ J

Using $\Delta Q = \Delta U + \Delta W$

$$\Rightarrow \Delta U = \Delta Q - \Delta W = 800 - 240 = 560 \text{ J}$$

47. (b)

Conservation of energy

48. (b)

Work done during an adiabatic change is

$$W = \frac{\mu R(T_1 - T_2)}{\gamma - 1}$$
$$\gamma = \frac{C_p}{C_v}$$
$$\therefore \gamma - 1 = \frac{C_p - C_v}{C_v} = \frac{R}{C_v}$$
$$\therefore W = \frac{\mu R(T_1 - T_2)C_v}{R} = \mu(T_1 - T_2)C_v$$

49. (a)

Due to compression, the temperature of the system increases to a very high value. This causes the flow of heat from system to the surrounding, thus decreasing the temperature. This decrease in temperature results in decrease in pressure.

50. (c)

In adiabatic process, there is no heat transfer between system and surrounding.

51. (c)

$$P=\frac{1}{3}\left(\frac{U}{V}\right)\propto T^4$$

For an ideal gas $PV=nR'T$

$$\therefore P=\frac{nRT}{V}\quad (R\text{ –molar gas constant})$$

$$\therefore \frac{nRT}{V}\propto T^4\Rightarrow\frac{nR}{V}\propto T^3$$

$$\therefore \frac{1}{V}\propto T^3\Rightarrow\frac{1}{\frac{4\pi}{3}R^3}\propto T^3$$

$$T^3\propto\frac{1}{R^3}\Rightarrow T\propto\frac{1}{R}$$

52. (a)

A Carnot engine has maximum efficiency

$$\eta=\frac{400-300}{400}\times 100\%=25\%$$

The maximum achievable efficiency is 25%. Hence it is impossible to achieve 26% efficiency.

53. (d)

In adiabatic process $\Delta Q=0$ and work is done on the system so internal energy of the system increases

$$\therefore \Delta U=\Delta W$$

$$\Rightarrow n\times C_v\times\Delta T=\Delta W$$

$$\Rightarrow n\times\left(\frac{R}{\gamma-1}\right)\times 7=146\times 10^3$$

$$\Rightarrow 10^3\times\frac{8.3}{(\gamma-1)}\times 7=146\times 10^3$$

$\Rightarrow$ on solving we get $\gamma=1.4$ the gas is diatomic.

54. (a)

$$\text{Number of moles of He}=\frac{5.6}{22.4}=\frac{1}{4}$$

$$\text{Now } T_1(5.6)^{\gamma-1}=T_2(0.7)^{\gamma-1}$$

$$T_1=T_2\left(\frac{1}{8}\right)^{2/3}$$

$$4T_1=T_2$$

$$\text{Work done}=-\frac{nR[T_2-T_1]}{\gamma-1}=\frac{\frac{1}{4}R[3T_1]}{\frac{2}{3}}=\frac{9}{8}RT_1$$

55. (d)

In isothermal process $\Delta Q\neq 0$

56. (b)

$$\because \eta=1-\frac{T_2}{T_1}=\frac{W}{Q_1}=\frac{Q_1-Q_2}{Q_1}$$

where $Q_1=$ heat absorbed, $Q_2=$ heat rejected

$$\Rightarrow 1-\frac{T/3}{T}=\frac{W}{Q_1}\Rightarrow\frac{2}{3}=\frac{W}{Q_1}=\frac{Q_1-Q_2}{Q_1}$$

$$\Rightarrow\frac{2}{3}=1-\frac{Q_2}{Q_1}\Rightarrow\frac{Q_2}{Q_1}=\frac{1}{3}\Rightarrow Q_2=\frac{Q_1}{3}=\frac{Q}{3}$$

57. (a)

Efficiency of heat engine, $\eta=1-\dfrac{T_2}{T_1}$

For $\eta=1\,(100\%)$ either, $T_1=\infty$ or $T_2=0$ K

As source at infinite temperature or sink at 0 K are not attainable, a heat engine cannot have efficiency 1.

58. (d)

$$\Delta Q=nC_P\Delta T$$

$$=2\left(\frac{3}{2}R+R\right)\Delta T$$

$$=2\left[\frac{3}{2}R+R\right]\times 5$$

$$=2\times\frac{5}{2}\times 8.31\times 5$$

$$=208\text{ J}$$

59. (c)

Efficiency (η) of a Carnot engine is given by

$\eta=1-\dfrac{T_2}{T_1}$ where T_1 is the temperature of the source and T_2 is the temperature of the sink.

Here, $T_2=500$ K

$$\therefore 0.5=1-\frac{500}{T_1}\Rightarrow T_1=1000\text{ K}$$

Now, $\eta=0.6=1-\dfrac{T_2'}{1000}$ (T_2' is the new sink temperature)

$$\Rightarrow T_2'=400\text{ K}$$

60. (d)

In both cylinders A and B, the gases are diatomic $(\gamma=1.4)$. Piston A is free to move as it is an isobaric process. Piston B is fixed as it is an isochoric process. If same amount of heat ΔQ is given to both then

$$(\Delta Q)_{\text{isobaric}} = (\Delta Q)_{\text{isochoric}}$$

$$\Rightarrow nC_p(\Delta T)_A = nC_v(\Delta T)_B$$

$$\Rightarrow (\Delta T)_B = \frac{C_p}{C_v}(\Delta T)_A = \gamma(\Delta T)_A$$

$$= 1.4 \times 30 = 42$$

61. (c)

$$\frac{40}{100} = 1 - \frac{300}{T_1} \text{ or } T_1 = 500 \text{ K}$$

For 60% efficiency

$$\frac{60}{100} = 1 - \frac{300}{T_1} \text{ or } T_1 = 750 \text{ K}$$

So the temperature should be increased 750 – 500 = 250 K

62. (d)

$$\beta = \frac{Q_1}{Q_1 - Q_2} = \frac{Q_1}{W}$$

$$W = \frac{Q_1}{\beta} = \frac{2000}{4} = 500 \text{ calories}$$

63. (c)

Given that $dQ = -30$ J and $dW = -10$ J

$U_i = 40$ J and let final internal energy $= U_f$.

So, $$dQ = U_f - U_i + dW$$

$$-30 = U_f - 40 - 10$$

$$U_f = 20 \text{ J}$$

64. (d)

Efficiency of Carnot engine

$$= \frac{W}{Q_1} = 1 - \frac{T_2}{T_1}$$

$$\frac{W}{6} = 1 - \frac{400}{500} = \frac{1}{5}$$

$$\Rightarrow W = \frac{6}{5} = 1.2 \text{ kcal}$$

65. (c)

In all cases, the curves are closed and represent cyclic changes. Therefore U is constant and $\Delta U = 0$

66. (b)

Process AB is isochoric. So,

$$PV = nRT$$

$$\frac{P_0}{2P_0} = \frac{T_A}{T_B} \text{ or } T_B = 2T_A$$

BC is isothermal process, i.e.,

$$T_B = T_C = 2T_A$$

CA represents an isobaric process. Hence

$$\frac{V_0}{T_A} = \frac{V_C}{T_C} \text{ or } V_C = \left(\frac{T_C}{T_A}\right)V_0 = \left(\frac{2T_A}{T_A}\right)V_0$$

or $$V_C = 2V_0$$

67. (d)

68. (c)

$$dU = nC_v\Delta T$$

$$80 = 5 \times C_v \times 20$$

$$C_v = \frac{80}{100} = 0.8 \text{ J/mole K}$$

or Heat capacity

$$= n\,C_v = 5 \times 0.8 = 4.0 \text{ J/K}$$

69. (a)

We know that, $dQ = dU + dW$

When $dW = 0$, $dQ = dU$

As $dQ < 0$, therefore $dU < 0$

i.e., internal energy decreases.

So, the temperature will decrease.

70. (b)

In first case,

$$\eta_1 = \left(1 - \frac{T_2}{T_1}\right) \times 100$$

$$= \left(1 - \frac{0 + 273}{200 + 273}\right) \times 100$$

$$= \frac{200}{473} \times 100$$

In second case,

$$\eta_2 = \left(1 - \frac{-200 + 273}{0 + 273}\right) \times 100$$

$$= \left(1 - \frac{73}{273}\right) \times 100$$

$$= \left(\frac{200}{473}\right) \times 100$$

$$\frac{\eta_1}{\eta_2} = \frac{273}{473} = \frac{1}{1 \cdot 73} = 1 : 1 \cdot 73.$$

71. (b)

In first case $\eta_1 = \dfrac{T_1 - T_2}{T_1}$

In second case $\eta_2 = \frac{2T_1 - 2T_2}{2T_1} = \frac{T_1 - T_2}{T_1} = \eta.$

72. (c)

Coefficient of performance

$$K = \frac{T_2}{T_1 - T_2}$$

$$\Rightarrow 5 = \frac{(273-13)}{T_1 - (273-13)} = \frac{260}{T_1 - 260}$$

$$\Rightarrow 5T_1 - 1300 = 260$$

$$\Rightarrow 5T_1 = 1560$$

$$\Rightarrow T_1 = 312 \text{ K} \rightarrow 39° \text{ C}.$$

73. (a)

Efficiency of Carnot's heat engine,

$$\eta = 1 - \frac{T_2}{T_1}$$

For first Carnot's heat engine, $0.5 = 1 - \frac{T_2}{T_1}$

For second heat engine, $0.5 = 1 - \frac{T_2'}{T_1'}$

Efficiency remains the same when both T_1 and T_2 are increased by same factor.

Therefore, the temperature of source and sink of second engine are $T_1' = 2T_1, T_2' = 2T_2$.

74. (c)

$$\Delta Q = \Delta U + \Delta W$$

$$\therefore \Delta U = \Delta Q - \Delta W = 540 - \frac{P(V_2 - V_1)}{J}$$

$$= 540 - \frac{1.013 \times 10^5 \times \left[(1671-1) \times 10^{-6}\right]}{4.2}$$

$$= 540 - 40 = 500 \text{ calories}$$

75. (c)

$$\Delta Q = \Delta U + \Delta W$$

$$\Rightarrow \Delta W = (\Delta Q)_P - \Delta U$$

$$= (\Delta Q)_P \left[1 - \frac{(\Delta Q)_V}{(\Delta Q)_P}\right] \quad (\Delta U = (\Delta Q)_V$$

$$= (\Delta Q)_P \left[1 - \frac{C_V}{C_P}\right] = Q\left[1 - \frac{3}{5}\right] = \frac{2}{5}Q$$

$\because (\Delta Q)_P = Q$ and $\gamma = \frac{5}{3}$ for monatomic gas.

76. (b)

Input energy $= \frac{1 \text{ g}}{\text{s}} \times \frac{2 \text{ kcal}}{\text{g}} = 2$ kcal/s.

Output energy $= 10$ kW $= 10$ kJ/S $= \frac{10}{4.2}$ kcal/s.

$$\Rightarrow \eta = \frac{\text{output energy}}{\text{input energy}} = \frac{10}{4.2 \times 2} > 1, \text{ it is impossible.}$$

77. (b)

$$WD = P\Delta V = nR\Delta T = \frac{1}{2} \times 8.31 \times 70 = 291 \text{ J}$$

78. (b)

Work done by the system = Area under P-V diagram

$$= (300\text{–}100)10^{-6} \times (100\text{–}200) \times 10^3 = -20 \text{ J}$$

79. (b)

Internal energy of a gas is

$$U = \frac{3}{2}nRT$$

For a given number of moles of the same gas, U depends only on T

Therefore U_B at $2T < U_A$ at temperature T is a wrong statement.

80. (d)

$$dU = nC_V dT$$

$$C_V = C_p - R = 8 - \frac{8.31}{4.2} = 6.02$$

$$dU = 6 \times 6.02 \times (35 - 20)$$

$$\approx 540 \text{ cal}$$

81. (d)

The work done is area of the trapezium

$$W = \frac{1}{2}(P_B + P_A)(V_B - V_A)$$

82. (b)

$$\frac{\Delta U_1}{\Delta U_2} = \frac{nC_V\Delta T}{nC_P\Delta T} = \frac{\frac{3}{2}R}{\frac{5}{2}R} = \frac{3}{5}$$

83. (a)

$$C_V = \frac{1}{2}fR$$

$$C_P = \frac{1}{2}fR + R$$

$$\frac{C_P}{C_V} = \frac{\frac{1}{2}fR + R}{\frac{1}{2}fR} = \frac{f+2}{f} = 1 + \frac{2}{f}$$

84. (b)

The internal energy decreases as the work is done by the system and no heat is supplied to the system.

85. (d)

$$dQ = nC_P\Delta T_1 = nC_V\Delta T_2$$
$$= C_P\Delta T_1 = C_V\Delta T_2$$
$$\frac{C_P}{C_V}\times\Delta T_1 = \Delta T_2$$
$$\frac{5}{3}\times 42 = \Delta T_2$$
$$70\text{ K} = \Delta T_2 \therefore T_2 = 70\text{ K}$$

86. (c)

87. (b)

Isothermal process

$$P_1V_1 = P_2V_2 \text{ or } P_1V_1 = 2P_1V_2$$
$$V_2 = \frac{P_1V_1}{2P_1} = \frac{V_1}{2}$$

Adiabatic process

$$PV^\gamma = \text{constant}$$
$$P_2V_2^\gamma = P_1V_1^\gamma$$
$$\frac{P_2}{P_1}\left(\frac{V_1}{V_2}\right)^\gamma$$
$$\frac{0.75P}{2P} = \left(\frac{V_1}{2\times V_1}\right)^\gamma$$
$$0.375 = \left(\frac{1}{2}\right)^\gamma$$

Solving, we get $\gamma = 1.41$

88. (d)

$$PV^\gamma = K$$
$$P\left(\frac{m}{d}\right)^\gamma = K$$
$$\Rightarrow \left(\frac{P}{d}\right)^\gamma = K$$
$$\frac{P'}{P} = \frac{d'}{d} = (32)^{7/5} = 128$$

89. (c)

Coefficient of Performance (CoP)

$$= \frac{T_2}{T_1 - T_2}$$
$$5 = \frac{(273-13)}{T_1 - 260}$$
$$5 = \frac{260}{T_1 - 260}$$
$$T_1 = 312\text{ K}$$
$$= 312 - 273 = 39\,°\text{C}$$

90. (a)

$$\Delta U = \Delta Q - \Delta W = 40 - 30 = 10\text{ J}$$

91. (a)

$$W = \frac{nR(T_1 - T_2)}{(\gamma - 1)}$$
$$6R = \frac{R(T - T_2)}{\left(\frac{5}{3} - 1\right)} = \frac{R(T - T_2)}{2/3}$$
$$\frac{2}{3}\times 6 = T - T_2$$
$$\Rightarrow T_2 = (T - 4)\text{K}$$

92. (d)

$$W_{\text{iso}} = nRT\log_e\frac{V_2}{V_1}$$
$$= 10\times 8.31\times 500\log_e\frac{5}{50}$$
$$= 8.31\times 5\times 10^4\log_e\frac{1}{10} = -9.6\times 10^4\text{ J}$$

93. (b)

$$\eta = 1 + \frac{Q_2}{Q_1}$$
$$\frac{1}{10} = 1 - \frac{Q_2}{Q_1}$$
$$\frac{Q_2}{Q_1} = \frac{9}{10}$$
$$W = Q_1 - Q_2 = 10$$
$$\frac{10}{9}Q_2 - Q_2 = 10$$
$$Q_2\times\frac{1}{9} = 10$$
$$Q_2 = 90\text{ J}$$

94. (d)

Change in internal energy (ΔU) depends upon initial and final state of the function while ΔQ and ΔW are path different.

95. (a)

$$\Delta U_1 = \Delta U_2$$

as change in internal energy is independent of path.

96. (c)

$$Q_1 = T_0S_0 + \frac{1}{2}T_0S_0 = \frac{3}{2}T_0S_0$$

$$Q_2 = T_0(2S_0 - S_0) = T_0 S_0$$
$$Q_3 = 0$$

Efficiency $\eta = 1 - \dfrac{Q_2}{Q}$

$$= 1 - \frac{T_0 S_0}{\frac{3}{2} T_0 S_0} = 1 - \frac{2}{3} = \frac{1}{3}$$

97. (b)

$$\frac{n_1 y_1 + n_2 y_2}{n_1 + n_2} = \frac{\frac{5}{3} + \frac{7}{5}}{2} = \frac{23}{15}$$

98. (c)

Efficiency $\eta = 1 - \dfrac{T_2}{T_1}$

For efficiency to be 100%, temperature of the sink, T_2 should be zero which cannot be achieved.

99. (d)

Work done in adiabatic process is given as

$$W = \frac{-1}{\gamma - 1}(P_f V_f - P_i V_i)$$

$$6R = \frac{-1}{5/3 - 1} R(T_f - T_i)$$

$$\Rightarrow T_f - T_i = -4 \therefore T_f = (T - 4)K$$

100. (c)

The work done in the cyclic process is equal to the area enclosed by the *P-V* diagram. Hence, the net work done during the complete cycle is given by the area A C B D A.

101. (c)

$$dU = dQ - dW$$
$$= -20\text{ J} + 8\text{ J} = -12\text{ J}$$

Hence final value of $U = 30\text{ J} - 12\text{ J} = 18\text{ J}$

102. (a)

$$dQ = dU + dW = dU + p\,dV$$
$$dU = dQ - p\,dV$$

Here, $dQ = 100\text{ J}, p = 50\text{ N/m}^2$ and $dV = -6\text{ m}^3$

$$dU = 100 + 50 \times 6 = 400\text{ J}$$

103. (b)

104. (b)

105. (c)

Change in internal energy is given by

$$\Delta U = n\,C_v\,\Delta T$$

$$= n\left(\frac{R}{\gamma - 1}\right)(2T - T) - n\left(\frac{RT}{\gamma - 1}\right)$$

$$= \frac{PV}{(\gamma - 1)} \qquad (\because nRT = PV)$$

106. (a)

Ideal gas possesses only kinetic energy.

107. (a)

In the figure, a b is isothermal and b c is isobaric process. The system returns to its original state from c to a.

108. (a)

It is obvious that process 1 is clockwise, therefore W_1 is positive. On the other hand, process 2 is anticlockwise and work done W_2 is negative. As $W_2 > W_1$ the net work done is negative.

109. (c)

Isothermal compression is reversible

110. (c)

111. (a)

112. (b)

The area under the indicator diagram gives the work done on the system or by the system.

113. (c)

$$\eta = \frac{T_1 - T_2}{T_1}$$

$$\eta' = \frac{(T_1 + 100) - (T_2 + 100)}{(T_1 + 100)} = \frac{T_1 - T_2}{(T_1 + 100)}$$

$$\therefore \eta' < \eta$$

114. (a)

In Carnot cycle, the temperature of working substance at the end of the cycle is equal to that at the beginning.

115. (a)

The first operation involved in a Carnot cycle is isothermal expansion.

116. (b)

$$dU = dQ - pdV = 30 - 18 = 12\text{ J}$$

As heat is released, so internal energy decreases from 60 J to $60 - 12 = 48$ J.

117. (c)

The efficiency of Carnot's cycle is given by

$$\eta = \frac{T_1 - T_2}{T_1}$$

where T_1 = Temperature of source, and

T_2 = Temperature of sink

Given that, $\eta = \frac{1}{6}$ for T_2 and $\eta = \frac{2}{6}$ for $T_2 - 62$

$$\frac{1}{6} = \frac{T_1 - T_2}{T_1} \quad ...(1)$$

and $$\frac{2}{6} = \frac{T_1 - (T_2 - 62)}{T_1} \quad ...(2)$$

From Eq. (1) $6T_1 - 6T_2 = T_1$

or $5T_1 = 6T_2$ or $T_2 = (6/5)T_2$

From Eq. (2) $2T_1 = 6T_1 - 6T_2 + 372$

or $6T_2 = 4T_1 + 372$

or $6T_2 = (25/4)T_2 + 372$

or $\frac{6}{5}T_2 = 372$ or $T_2 = 310\text{ K} = 37°\text{C}$

Now $T_1 = \frac{6}{5} \times 310 = 372\text{ K} = 99°\text{C}$

118. (a) and (d)

In a cyclic process, heat is absorbed by the gas when the work done by it is positive, i.e., work is done by the gas. Work is done on the gas if the closed curve in *P-V* diagram is clockwise with *P* on Y-axis and *V* on X-axis. On the other hand, the work-done is positive for anti-clockwise curve *P* on X-axis and *V* on Y-axis.

119. (b)

The work done can be obtained by the area enclosed between *P-V* curve

Questions Asked in NEET/CBSE Medicals

120. (c)

Adiabatic process is the process in which no exchange of heat energy takes place between the gas and the surroundings, i.e., $\Delta Q = 0$

121. (b)

Using first law of thermodynamics, $\Delta Q = \Delta U + \Delta W$

$\Rightarrow 54 \times 4.18 = \Delta U + 1.013 \times 10^5 (167.1 \times 10^{-6} - 0.1)$

$\Rightarrow \Delta U = 208.7\text{ J}$

122. (a)

Given process is isobaric.

$\therefore dQ = nC_p dT$ where C_p is specific heat at constant pressure.

or $$dQ = n\left(\frac{5}{2}R\right)dT$$

Also, $$dW = PdV = nRdT \quad (\because PV = nRT)$$

$$\text{Required ratio} = \frac{dW}{dQ} = \frac{nRdT}{n\left(\frac{5}{2}R\right)dT} = \frac{2}{5}$$

123. (a)

Efficiency of an ideal heat engine,

$$\eta = \left(1 - \frac{T_2}{T_1}\right)$$

Freezing point of water = 0°C = 273 K

Boiling point of water = 100°C = (100 + 273) K = 373 K

T_2 : Sink temperature = 273 K

T_1 : Source temperature = 373 K

$$\%\eta = \left(1 - \frac{T_2}{T_1}\right) \times 100 = \left(1 - \frac{273}{373}\right) \times 100$$

$$= \left(\frac{100}{373}\right) \times 100 = 26.8\%$$

124. (a)

In process I, volume is constant

$\therefore$ Process I $\to$ Isochoric; $P \to C$

As slope of curve II is more than the slope of curve III.

Process II $\to$ Adiabatic and

Process III $\to$ Isothermal

$\therefore Q \to A, R \to D$

In process IV, pressure is constant

Process IV $\to$ Isobaric; S $\to$ B

125. (a)

The relation between coefficient of performance and efficiency of Carnot engine is given as

$$\beta = \frac{1 - \eta}{\eta}$$

Given $\eta = \frac{1}{10}, W = 10\text{ J}$

$$\beta = \frac{1 - \frac{1}{10}}{\frac{1}{10}} = \frac{9}{10}.10 = 9$$

Since, $\beta = \frac{Q_2}{W}$, where Q_2 is the amount of energy absorbed from the reservoir

$\therefore Q_2 = \beta W = 9 \times 10 = 90\text{ J}$

126. (b)

Temperature inside refrigerator = t_2°C

Room temperature = t_1°C

For refrigerator,

$$\frac{\text{Heat given to high temperature}(Q_1)}{\text{Heat taken from lower temperature}(Q_2)} = \frac{T_1}{T_2}$$

$$\frac{Q_1}{Q_2} = \frac{t_1 + 273}{t_2 + 273}$$

$$\Rightarrow \frac{Q_1}{Q_1 - W} = \frac{t_1 + 273}{t_2 + 273} \text{ or } 1 - \frac{W}{Q_1} = \frac{t_2 + 273}{t_1 + 273}$$

or
$$\frac{W}{Q_1} = \frac{t_1 - t_2}{t_1 + 273}$$

The amount of heat delivered to the room for each joule of electrical energy ($W = 1$ J)

$$Q_1 = \frac{t_1 + 273}{t_1 - t_2}$$

127. (d)

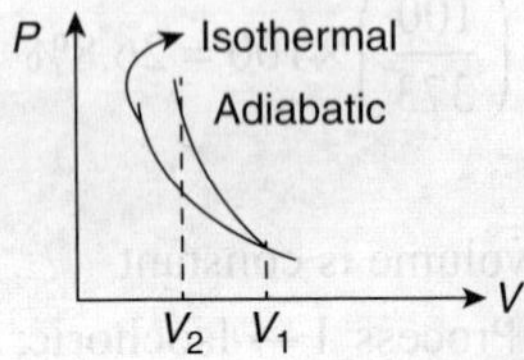

$$V_1 = V, V_2 = V/2$$

On *P*-*V* diagram, Area under adiabatic curve > Area under isothermal curve. So, compressing the gas through adiabatic process, will require more work to be done.

128. (a)

Given $\quad T_2 = 4°\text{C} = 277 \text{ K}$

$T_1 = 30°\text{C} = 303 \text{ K}$

$Q_2 = 600$ cal per second

Coefficient of performance,

$$\alpha = \frac{T_2}{T_1 - T_2} = \frac{277}{303 - 277} = \frac{277}{26}$$

Also, $\quad \alpha = \frac{Q_2}{W}$

$\therefore$ Work to be done per second = Power required

$$= W = \frac{Q_2}{\alpha} = \frac{26}{277} \times 600 \text{ cal per second}$$

$$= \frac{26}{277} \times 600 \times 4.2 \text{ J per second} = 236.5 \text{ W}$$

129. (c)

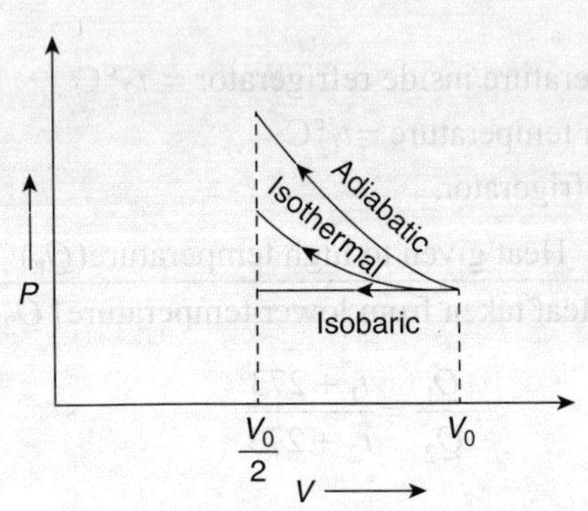

Work done on the gas = Area under *P*-*V* curve. As area under the *P*-*V* curve is maximum for adiabatic process, so work done on the gas is maximum for adiabatic process.

130. (c)

The coefficient of performance of a refrigerator is

$$\alpha = \frac{T_2}{T_1 - T_2}$$

where T_1 and T_2 are the temperature of hot and cold reservoirs (in Kelvin) respectively.

Here, $\quad \alpha = 5, T_2 = -20° \text{ C} = -20 + 273 \text{ K} = 253 \text{ K}$

$$T_1 = ?$$

$$\alpha = \therefore 5 = \frac{253}{T_1 - 253}$$

$$5T_1 - 5(253) = 253$$

$$5T_1 = 253 + 5(253) = 6(253)$$

$$T_1 = \frac{6}{5} \times 253 = 303.6 \text{ K}$$

$$= 303.6 - 273$$

$$= 30.6° \text{ C} \approx 31° \text{ C}$$

131. (a)

132. (d)

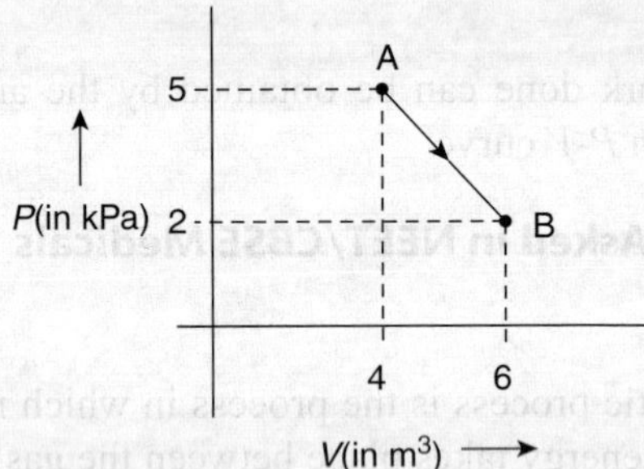

We know $\Delta U = nC_v \Delta T$

$$= n\left(\frac{5R}{2}\right)(T_B - T_A) \quad \left[\text{for diatomic gas, } C_v = \frac{5R}{2}\right]$$

$$= \frac{5nR}{2}\left(\frac{P_B V_B}{nR} - \frac{P_A V_A}{nR}\right) \quad [\because PV = nRT]$$

$$= \frac{5}{2}(P_B V_B - P_A V_A)$$

$$= \frac{5}{2}\left(2 \times 10^3 \times 6 - 5 \times 10^3 \times 4\right)$$

$$= \frac{5}{2}(-8 \times 10^3) = -20 \text{ kJ}$$

133. (a)

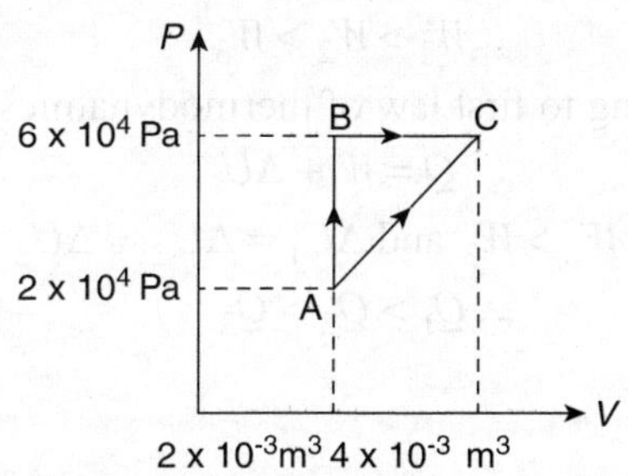

As initial and final points are same so $\Delta U_{ABC} = \Delta U_{AC}$

AB is isochoric process.

$$\Delta W_{AB} = 0$$

$$\Delta Q_{AB} = \Delta U_{AB} = 400 \text{ J}$$

BC is isobaric process.

$$\Delta Q_{BC} = \Delta U_{BC} + \Delta W_{BC}$$

$$100 = \Delta U_{BC} + 6\times10^4\left(4\times10^{-3} - 2\times10^{-3}\right)$$

$$100 = \Delta U_{BC} + 12\times10$$

$$\Delta U_{BC} = 100 - 120 = -20 \text{ J}$$

As, $\Delta U_{ABC} = \Delta U_{AC}$

$$\Delta U_{AB} + \Delta U_{BC} = \Delta Q_{AC} - \Delta W_{AC}$$

$$400 - 20 = \Delta Q_{AC} - \frac{1}{2}\left(2\times10^4 + 6\times10^4\right)\times2\times10^{-3}$$

$$380 = \Delta Q_{AC} - 80$$

$$\Delta Q_{AC} = 460 \text{ J}$$

134. (d)

In a cyclic process, work done is equal to the area under the cycle and is positive if the cycle is clockwise and negative if cycle is anticlockwise.

As is clear from the figure,

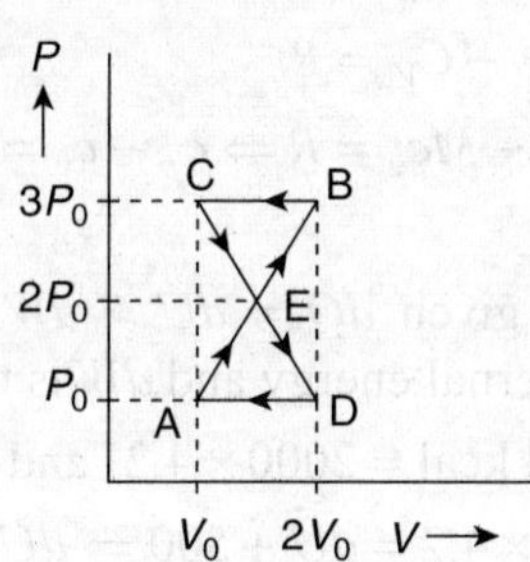

$$W_{AEDA} = +\text{ Area of } \Delta AED = +\frac{1}{2}P_0V_0$$

$$W_{BCEB} = -\text{Area of } \Delta BCE = -\frac{1}{2}P_0V_0$$

The net work done by the system is

$$W_{net} = W_{AEDA} + W_{BCEB}$$

$$= +\frac{1}{2}P_0V_0 - \frac{1}{2}P_0V_0 = \text{zero}$$

135. (a)

According to ideal gas equation,

$$PV = nRT$$

or $$V = \frac{nRT}{P}$$

For an isobaric process, P = Constant and $V \propto T$

Therefore, V-T graph is a straight line passing through origin. Slope of this line is inversely proportional to P.

In the given figure,

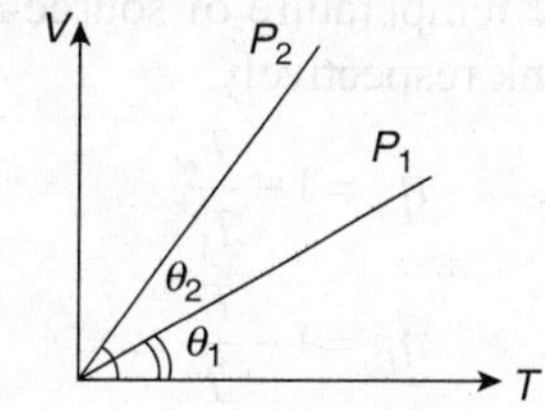

$$(\text{Slope})_2 > (\text{Slope})_1 \therefore P_2 < P_1$$

136. (d)

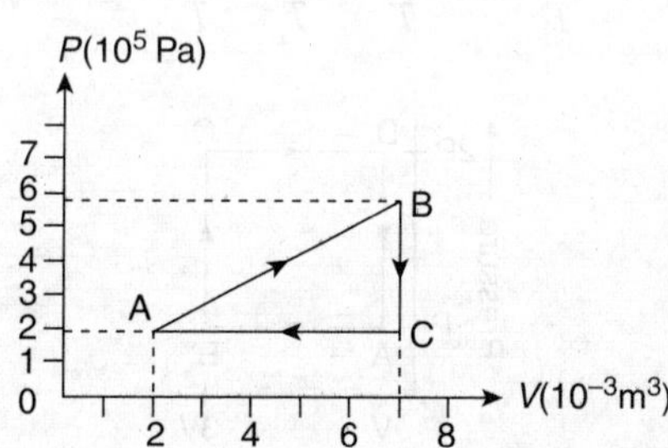

In a cyclic process, work done is equal to the area under the cycle and is positive if the cycle is clockwise and negative if the cycle is anticlockwise.

$\therefore$ The net work done by the gas is

$$W = \text{Area of the cycle ABCA}$$

$$= \frac{1}{2}\times(7-2)\times10^{-3}\times(6-2)\times10^5$$

$$= \frac{1}{2}\times5\times10^{-3}\times4\times10^5 = 10\times10^2 \text{ J} = 1000 \text{ J}$$

137. (d)

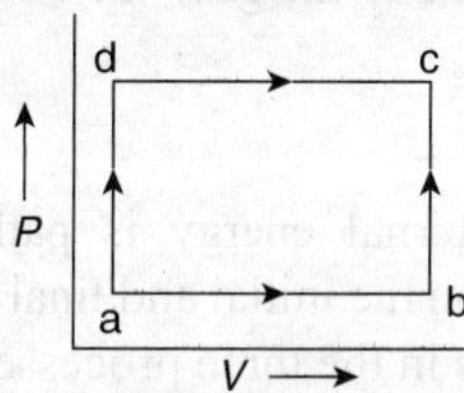

According to the first law of thermodynamics,

$$dQ = dU + dW$$

Along the path adc, change in internal energy,

$$dU_1 = dQ_1 - dW_1 = 50 \text{ J} - 20 \text{ J} = 30 \text{ J}$$

Along the path abc, change in internal energy,

$$dU_2 = dQ_2 - dW_2;$$

$$dU_2 = 36\text{ J} - dW_2$$

As change in internal energy is path independent,

$$\therefore dU_1 = dU_2 \Rightarrow 30\text{ J} = 36\text{ J} - dW_2$$

$$dW_2 = 36\text{ J} - 30\text{ J} = 6\text{ J}$$

138. (c)

Efficiency of a Carnot engine

$$\eta = 1 - \frac{T_2}{T_1}$$

where T_1 is the temperature of source and T_2 is the temperature of sink respectively.

For engine A, $\eta_A = 1 - \frac{T}{T_1}$

For engine B, $\eta_B = 1 - \frac{T_2}{T}$

Given, $\eta_A = \eta_B$

$$\therefore 1 - \frac{T}{T_1} = 1 - \frac{T_2}{T} \Rightarrow \frac{T}{T_1} = \frac{T_2}{T} \text{ or } T = \sqrt{T_1 T_2}$$

139. (a)

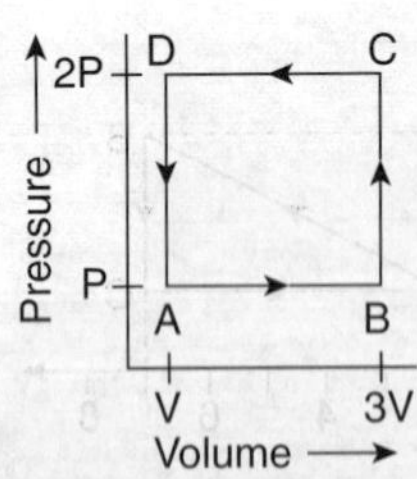

In a cyclic process, $\Delta U = 0$

In a cyclic process, work done is equal to the area under the cycle and is positive if the cycle is clockwise and negative if anticlockwise.

$\therefore \Delta W = -$ Area of rectangle ABCD $= -P(2V) = -2PV$

According to first law of thermodynamics

$$\Delta Q = \Delta U + \Delta W \text{ or } \Delta Q = \Delta W \qquad (\text{As } \Delta U = 0)$$

i.e, heat supplied to the system is equal to the work done

$$\Delta Q = \Delta W = -2PV$$

$\therefore$ Heat rejected by the gas $= 2PV$

140. (d)

141. (a)

Change in internal energy is path independent and depends only on the initial and final states. As the initial and final states in the three processes are same,

$$\Delta U_1 = \Delta U_2 = \Delta U_3$$

Work done, $W =$ Area under P-V graph

The area under curve 1 > area under curve 2 > area under curve 3.

$$\therefore W_1 > W_2 > W_3$$

According to first law of thermodynamics,

$$Q = W + \Delta U$$

As $W_1 > W_2 > W_3$ and $\Delta U_1 = \Delta U_2 = \Delta U_3$

$$\therefore Q_1 > Q_2 > Q_3$$

142. (d)

143. (d)

Heat required to melt 1 kg ice at 0°C to water at 0°C is

$$Q = m_{ice} L_{ice} = (1\text{ kg})(80\text{ cal/g})$$

$$= (1000\text{ g})(80\text{ cal/g}) = 8 \times 10^4\text{ cal}$$

Change in entropy, $\Delta S = \frac{Q}{T} = \frac{8 \times 10^4\text{ cal}}{(273\text{ K})} = 293\text{ cal K}^{-1}$

144. (a)

According to first law of thermodynamics

$$\Delta Q = \Delta U + \Delta W$$

where,

$\Delta Q =$ Heat supplied to the system

$\Delta U =$ Increase in internal energy of the system

$\Delta W =$ Work done by the system

For an adiabatic process

$$\Delta Q = 0 \qquad \therefore \Delta U = -\Delta W$$

For an isothermal process

$$\Delta U = 0$$

$$\therefore \Delta Q = \Delta W$$

145. (c)

Let C_V and C_P be molar specific heats of the ideal gas at constant volume and constant pressure, respectively, then

$$C_P = Mc_p \text{ and } C_V = Mc_v$$

$$C_P - C_V = R$$

$$\therefore Mc_p - Mc_v = R \Rightarrow c_p - c_v = R/M$$

146. (c)

Heat energy given $dQ = dU + dW$ where dU is the change in internal energy and dW is the work done.

Given $dQ = 2$ kcal $= 2000 \times 4.2$J and $dW = 500$ J

$$\therefore 2000 \times 4.2 = dU + 500 \Rightarrow dU = 7900\text{ J}$$

147. (a)

In isochoric process, volume is constant. If pressure is constant, it is an isobaric process.

148. (a)

Internal energy depends only on the initial and final states of temperature and not on the path. In a cyclic process, as internal and final states are the same, change in internal energy is zero.

149. (c)

Efficiency of an engine, $\eta = 1 - \frac{T_2}{T_1}$

where T_1 is the temperature of the source and T_2 is the temperature of the sink.

$$\therefore \frac{1}{6} = 1 - \frac{T_2}{T_1} \text{ or } \frac{T_2}{T_1} = \frac{5}{6} \qquad \text{(i)}$$

When the temperature of the sink is decreased by 62° C (or 62 K), efficiency becomes double. Since the temperature of the source remains unchanged

$$\therefore 2 \times \frac{1}{6} = 1 - \frac{(T_2 - 62)}{T_1} \text{ or } \frac{1}{3} = 1 - \frac{(T_2 - 62)}{T_1}$$

$$\text{or, } \frac{2}{3} = 1 - \frac{T_2 - 62}{T_1} \text{ or } 2T_1 = 3T_2 - 186 \qquad \text{(ii)}$$

$$\text{or } 2T_1 = 3\left[\frac{5}{6}\right]T_1 - 186 \qquad \text{[using (i)]}$$

$$2T_1 = \frac{5}{2}T_1 - 186 \text{ or, } \frac{T_1}{2} = 186$$

$$\text{or } T_1 = 372 \text{ K or } (372 - 273) = 99°\text{C}$$

150. (d)

Efficiency of a Carnot engine, $\eta = 1 - \frac{T_2}{T_1}$

$$\text{or, } \frac{40}{100} = 1 - \frac{T_2}{T_1}$$

$$\Rightarrow \frac{T_2}{T_1} = 1 - \frac{2}{5} = \frac{3}{5}$$

$$\therefore T_1 = \frac{5}{3} \times T_2 = \frac{5}{3} \times 300 = 500 \text{ K.}$$

Increase in efficiency = 50% of 40% = 20%

New efficiency, $\eta' = 40\% + 20\% = 60\%$

$$\therefore \frac{T_2}{T_1'} = 1 - \frac{60}{100} = \frac{2}{5}$$

$$T_1' = \frac{5}{2} \times T_2 = \frac{5}{2} \times 300 = 750 \text{ K.}$$

Increase in temperature of source = $T_1' - T_1$

= 750 – 500 = 250 K.

151. (d)

$$1 - \frac{T_2}{T_1} = 1 - \frac{Q_2}{Q_1}$$

$$\Rightarrow 1 - \frac{400}{500} = 1 - \frac{Q_2}{6 \times 10^4}$$

$$\Rightarrow \frac{4}{5} = \frac{Q_2}{6 \times 10^4} \Rightarrow Q_2 = 4.8 \times 10^4 \text{ cal.}$$

Net heat converted into work

$$= 6.0 \times 10^4 - 4.8 \times 10^4 = 1.2 \times 10^4 \text{ cal.}$$

QUICK QUESTIONS

Quiz

1. What is an energy reservoir?
2. What is a quasi-static process?
3. The sun is continually radiating heat energy. This means the temperature of earth should continuously rise. Then how come we say that temperature in a region has a particular value?
4. Imagine you have 100 gram each of iron, glass and water and all three at 20° C.
   (a) Rank them from lowest to highest temperature after 80 J of energy is added to each
   (b) Rank them from lowest to greatest amount of energy transfer by heat if the temperature of each is raised to 25° C.
5. Differentiate between state variables and transfer variables in thermodynamics.
6. If friction could be totally removed from a Carnot of engine, would it be 100% efficient?
7. On what factors does latent heat depend?

Think and grow

8. The equation $\Delta Q = ms\Delta T$ relates energy transfer to an object of mass to specific heat and resulting change in temperature ΔT . In fact the energy transfer on ΔQ could be made by any method other than heat. Think and give two more examples in which this equation could be used to find the change in temperature ΔT of an object due to energy transfer process other than heat.
9. Imagine you are in an empty space holding a thermometer. Does the thermometer measure the temperature of empty space? If not so, what does it measure?
10. If you are in front of a fire with closed eyes you can feel warmth in your eye lids. Now, you put on eye glasses and repeat the activity your eyelids will not feel so warm. What is the reason?
11. In a sweltering summer your air conditioning system is not operating. In your kitchen you have a working refrigerator and ice box full of ice. Which one should you open and leave it to cool the room more efficiency?

Explanations

Quiz

1. An energy reservoir is a source of internal energy that is considered to be so large that a finite transfer of energy does not change its temperature.
2. A process which proceeds slowly to allow the system to be always in state of thermal equilibrium.
3. The energy is also radiated by earth. When the rate of energy transfer due to these exchanges is equal, the temperature of the earth becomes constant.
4. (a) The order is water, glass, iron from highest specific heat to lowest specific heat.
   (b) For iron, glass and water for a given temperature, the rise in energy transfer by heat is proportional to specific heat.
5. State variables are characteristics of a system in a thermal equilibrium, while transfer variables are characteristics of a process in which energy is transferred between the system and its surrounding.
6. No, friction is one factor which reduces efficiency because even for Carnot engine, efficiency is always less than 100%.
7. It depends upon two factors:
   (i) Nature of the phase change
   (ii) Substance.

Think and grow

8. Example 1 – Toaster
   The temperature of the electric coil rises. The transfer mechanism here is electric transmission of energy through the power cable.
   Example 2 – Microwave
   The temperature of milk in a microwave oven increases due to the absorption of microwaves. In this case, the energy transfer mechanism is through electromagnetic radiation.
9. The thermometer does not measure the temperature of empty space because temperature is not defined for vacuum. The temperature is defined in terms of internal energy, which is related to the energy of motion of molecules. There are no molecules in vacuum.
   The thermometer in empty space measures the transfer of energy of the radiating objects because stationary thermometers require contact with molecules.
   If the thermometer is near a star, its temperature reading may go up very high as it absorbs energy by electromagnetic radiation. If it is away from the star its reading will go down as releases more radiation into the environment than it absorbs from the environment.
10. The warmth which we feel is due to electromagnetic radiation from the fire. A large portion of this radiation is in the form of infrared radiation. Our eyelids are very sensitive to infrared radiation. On the other hand, glass is largely opaque to infrared radiation. Thus, when you put on the eye glasses you block major portion of the radiation reaching to your eyelids and you feel relaxed.
11. The air in the kitchen is the high temperature reservoir. If the refrigerator door is left open, the energy would be drawn in from the kitchen air and passed through the refrigerator and transferred back to air. As a result the kitchen would be warmer due to addition of energy which is supplied through electric power.
    However, when the ice box is opened, the energy would be supplied to the ice and raise its temperature and cause it to melt. The transfer of energy from air to ice will decrease the temperature of air. Hence it would be more effective to open the ice box.

CHAPTER TEST

1. A vessel containing 5 litres of a gas at 0.8 m pressure is connected to an evacuated vessel of volume 3 litres. The resultant pressure inside will be (assuming whole system to be isolated)
   (a) 4/3 m (b) 0.5 m
   (c) 2.0 m (d) 3/4 m
2. An ideal gas, initially at 10 atm and 300 K, is permitted to expand adiabatically until its volume doubles. Then which of the following statements will be true?
   (a) If the gas be monatomic, the final temperature will be 189 K.
   (b) If the gas be diatomic, the final pressure will be 7.6 atm.
   (c) If the gas be now compressed isothermally back to its original volume, the net work done on the gas will be zero.
   (d) The work done in the adiabatic expansion will be the same for monatomic and diatomic gases.
3. A system performs work ΔW when an amount of heat ΔQ is added to the system. The corresponding change in the internal energy is ΔU. A unique function of initial and final states, irrespective of the mode of change, is:

(a) ΔW (b) ΔQ
(c) ΔW and ΔQ (d) ΔU

4. In a thermodynamic process pressure of fixed mass of gas is changed in such a way that the gas molecules give out 30 J of heat and 10 J of work is done on the gas. If the internal energy of the gas was 40 J, the final internal energy will be:
(a) Zero (b) 80 J
(c) 20 J (d) –20 J

5. For which combination of working temperature the efficiency of Carnot's engine is highest
(a) 80 K, 60 K (b) 100 K, 80 K
(c) 60 K, 40 K (d) 40 K, 20 K

6. A gas is compressed at a constant pressure of 50 N/m^2 from a volume of 10 m^3 to a volume of 4 m^3. Energy of 100 J is then added to the gas by heating. Its internal energy is
(a) increased by 400 J (b) increased by 200 J
(c) increased by 100 J (d) decreased by 200 J.

7. In an adiabatic change, between the system and surroundings,
(a) there is no transfer of heat; hence the temperature remains constant
(b) there is transfer of heat in such a way that the temperature remains constant
(c) there is no transfer of heat but the temperature may vary
(d) there is free transfer of heat as well as variation in the temperature.

8. The property of the system that does not change during an adiabatic change is:
(a) Temperature (b) Volume
(c) Pressure (d) Quantity of heat

9. An ideal gas A and a real gas B have their volumes increased from V to 2 V under isothermal conditions. The increase in internal energy
(a) will be same in both A and B
(b) will be zero in both the gases
(c) of B will be more than that of A
(d) of A will be more than that of B.

10. 1 g of an ideal gas expands isothermally; heat flow will be
(a) from the gas to outside atmosphere
(b) from outside atmosphere to the gas
(c) zero
(d) Both (a) and (b).

11. The pressure – volume graph of an ideal gas is shown in Figure. The adiabatic processes are:

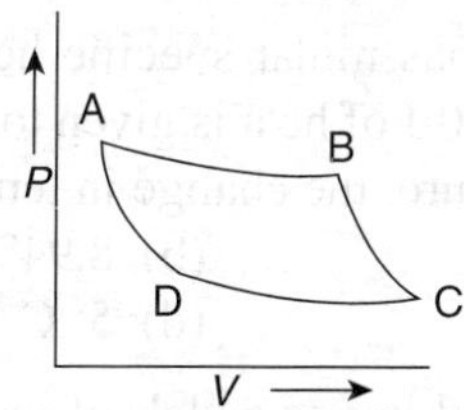

(a) AB and BC (b) AB and CD
(c) AD and BC (d) BC and CD

12. Two samples of air A and B having same composition and initially at the same temperature and pressure are compressed from a volume V to $V/2$, the sample A isothermally and the sample B adiabatically. The final pressure of
(a) A is greater than B
(b) A is lesser than B
(c) A is equal to that of B
(d) A is twice that of B.

13. One mole of O_2 gas having a volume equal to 22.4 litres at 0° C and 1 atmospheric pressure in compressed isothermally so that its volume reduces to 11.2 litres. The work done in this process is
(a) 1672.5 J (b) 1728 J
(c) –1728 J (d) –1572.5 J

14. The most efficient heat engine would operate by taking heat at a particular temperature and
(a) converting it all into work
(b) converting some of it into work and rejecting the rest at a lower temperature
(c) converting some of it into work and rejecting the rest at the same temperature
(d) converting some of it into work and rejecting the rest at a higher temperature.

15. The most efficient heat engine is one which is
(a) reversible
(b) irreversible
(c) driven by high-speed diesel
(d) driven by oil.

16. The ratio of the velocity of sound in a gas to the velocity of adiabatic exhaust of this gas jet (ratio of specific heats γ) from a vessel through a small hole into vacuum is given by:
(a) $\frac{1}{\sqrt{\gamma-1}}$ (b) $\frac{1}{\sqrt{2(\gamma-1)}}$
(c) $\frac{1}{\gamma}$ (d) $\sqrt{\frac{\gamma-1}{\gamma}}$

17. An ideal gas has molar specific heat $5R/2$ at constant pressure. If 300 J of heat is given to two moles of gas at constant pressure, the change in temperature is
(a) 7.22° C (b) 8.94° C
(c) zero (d) 5° C

18. During the melting of a slab of ice at 273 K at atmospheric pressure
(a) positive work is done by the ice-water system on the atmosphere
(b) positive work is done on the ice-water system by the atmosphere
(c) internal energy of ice-water system decreases
(d) None of the above.

19. An ideal gas is taken through the cycle A → B → C → A, as shown in the figure. If the net heat supplied to the gas in the cycle is 5 J, the work done by the gas in the process C → A is:

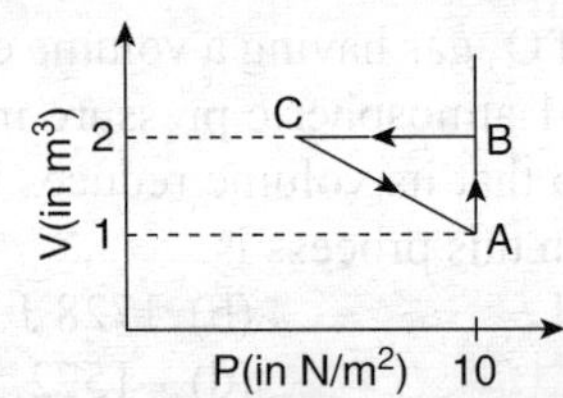

(a) –5 J (b) –10 J
(c) –15 J (d) –20 J

20. A Carnot engine operates with source at 127° C and sink at 27°C. If the source supplies 40 kJ of heat energy, the work done by the engine is:
(a) 30 kJ (b) 10 kJ
(c) 4kJ (d) 1 kJ

21. When heat Q is supplied to a diatomic gas of rigid molecules, at constant volume, its temperature increases by ΔT. The heat required to produce the same change in temperature, at a constant pressure, is
(a) $\frac{2}{3}Q$ (b) $\frac{5}{3}Q$
(c) $\frac{3}{2}Q$ (d) $\frac{7}{5}Q$

22. n molecules of an ideal gas with constant volume heat capacity C_V undergo an isobaric expansion by certain volume. The ratio of the work done in the process, to the heat supplied is,
(a) $\frac{4nR}{C_V + nR}$ (b) $\frac{4nR}{C_V - nR}$
(c) $\frac{nR}{C_V - nR}$ (d) $\frac{nR}{C_V + nR}$

23. The given diagram shows four processes, i.e. isochoric, isobaric, isothermal and adiabatic. The correct order of the processes, in the same order is given by

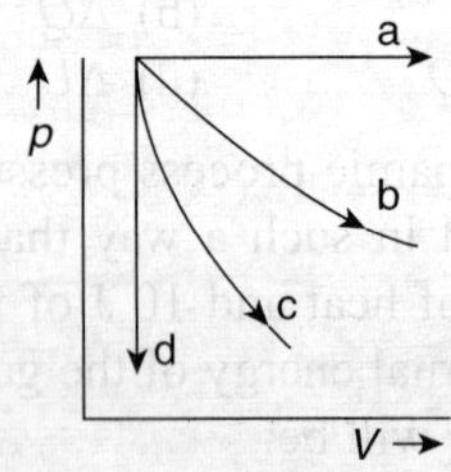

(a) d a b c (b) a d b c
(c) d a c b (d) a d c b

24. In a process, temperature and volume of one mole of an ideal monatomic gas are varied according to the relation $VT = k$, where k is a constant. In the process, the temperature of the gas is increased by ΔT. The amount of heat absorbed by gas is (where, R is gas constant)
(a) $\frac{1}{2}kR\Delta T$ (b) $\frac{2k}{3}\Delta T$
(c) $\frac{1}{2}R\Delta T$ (d) $\frac{3}{2}R\Delta T$

25. A rigid diatomic ideal gas undergoes an adiabatic process at room temperature. The relation between temperature and volume for this process is TV^x = constant, then x is
(a) $\frac{2}{5}$ (b) $\frac{2}{3}$
(c) $\frac{5}{3}$ (d) $\frac{3}{5}$

26. An ideal gas goes from the state i to the state f as shown in figure. The work done by the gas during the process

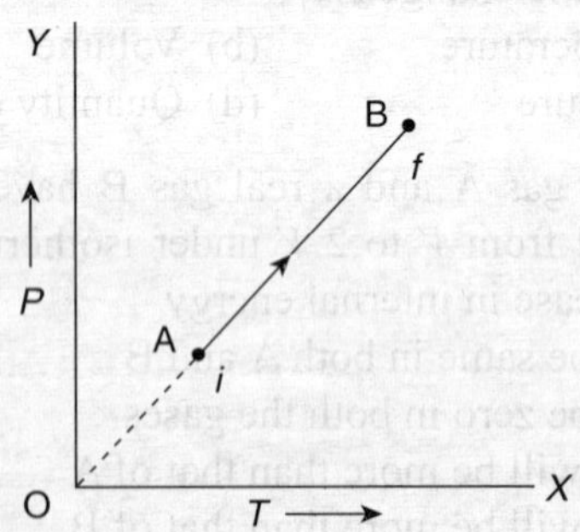

(a) is positive
(b) is negative
(c) is zero
(d) cannot be obtained from this information.

27. Half-mole of an ideal monatomic gas is heated at constant pressure of 1 atm from 20° C to 90° C . Work done by the gas is close to (take gas constant, $R = 8.31$ J/mol–K)
(a) 291 J (b) 581 J
(c) 146 J (d) 73 J

28. A monatomic gas of n moles is heated from temperature T_1 to T_2 under two different conditions (i) at constant volume and (ii) at constant pressure. The change in internal energy of the gas is
(a) more for (i)
(b) more for (ii)
(c) same in both cases
(d) independent of number of moles.

29. Two Carnot engines, A and B, are operated in series. The first one, A receives heat $T_1 = 600$ K and rejects to a reservoir at temperature T_2. The second engine B receives heat rejected by the first engine and in turn rejects to a heat reservoir at $T_3 = 400$ K. Calculate the temperature T_2 if the work outputs of the two engines are equal.
(a) 600 K (b) 500 K
(c) 400 K (d) 300 K

30. A gas can be taken from A to B via two different processes ACB and ADB.

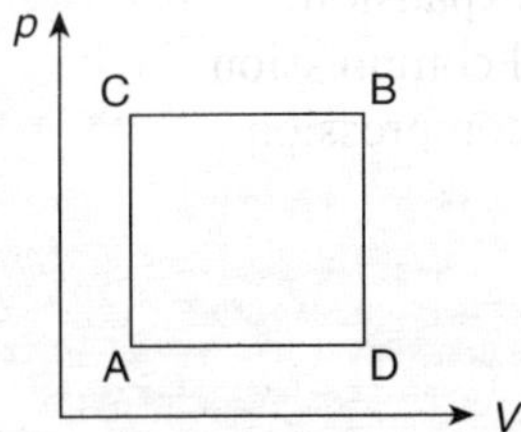

When path ACB is used, 60 J of heat flows into the system and 30 J of work is done by the system. If path ADB is used, work done by the system is 10 J. The heat flow into the system in path ADB is
(a) 80 J (b) 40 J
(c) 100 J (d) 20 J

31. Two moles of an ideal monatomic gas occupies a volume V at 27° C. The gas expands adiabatically to a volume 2 V. Calculate (i) the final temperature of the gas and (ii) change in its internal energy.
(a) (i) 195 K (ii) 2.7 kJ
(b) (i) 189 K (ii) 2.7 kJ
(c) (i) 195 K (ii) –2.7 kJ
(d) (i) 189 K (ii) –2.7 kJ

32. One mole of diatomic ideal gas undergoes a cyclic process ABC as shown in figure. The process BC is adiabatic. The temperatures at A, B and C are 400 K, 800 K and 600 K, respectively. Choose the correct statement.

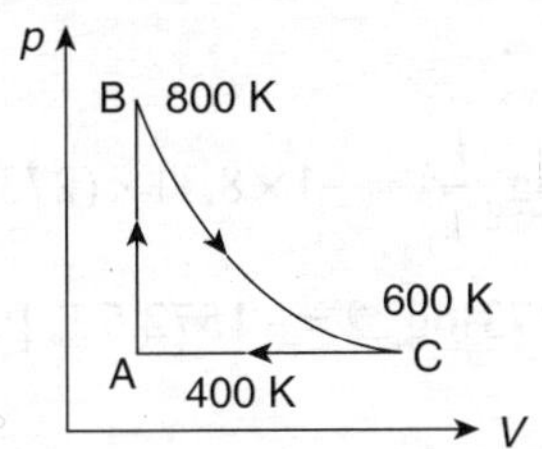

(a) The change in internal energy in whole cyclic process is 250 R.
(b) The change in internal energy in the process CA is 700 R.
(c) The change in internal energy in the process AB is –350 R.
(d) The change in internal energy in the process BC is –500 R.

33. Two cylinders A and B, fitted with pistons, contain equal amounts of an ideal diatomic gas at 300 K. The piston of A is free to move, while that of B is held fixed. The same amount of heat is given to the gas in each cylinder. If the rise in temperature of the gas in A is 30 K, then the rise in temperature of the gas in B is
(a) 30 K (b) 18 K
(c) 50 K (d) 42 K

34. A Carnot engine has an efficiency of 1/6. When the temperature of the sink is reduced by 62° C, its efficiency is doubled. The temperatures of the source and the sink are respectively,
(a) 62° C, 124° C (b) 99° C, 37° C
(c) 124° C, 62° C (d) 37° C, 99° C

35. A sample of an ideal gas is taken through the cyclic process abca as shown in the figure. The change in the internal energy of the gas along the path ca is –180 J. The gas absorbs 250 J of heat along the path ab and 60 J along the path bc. The work done by the gas along the path abc is

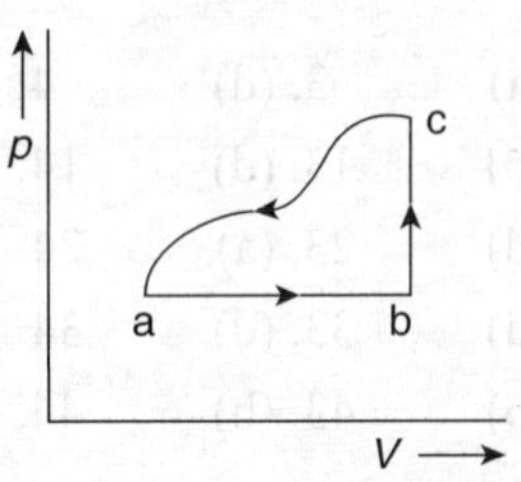

(a) 120 J (b) 130 J
(c) 100 J (d) 140 J

36. Which of the following cannot determine the state of a thermodynamic system?
(a) Pressure and volume
(b) Volume and temperature
(c) Temperature and pressure
(d) Any one of pressure, volume or temperature.

37. In a given process for an ideal gas, $dW = 0$ and $dQ < 0$. Then for the gas
(a) The temperature will decrease
(b) The volume will increase
(c) The pressure will remain constant
(d) The temperature will increase.

38. Temperature is a measurement of coldness or hotness of an object. This definition is based on
(a) Zeroth law of thermodynamics
(b) First law of thermodynamics
(c) Second law of thermodynamics
(d) Newton's law of cooling.

39. The gas law $\frac{PV}{T}$ = Constant is true for
(a) isothermal changes only
(b) adiabatic changes only
(c) both isothermal and adiabatic changes
(d) neither isothermal nor adiabatic changes.

40. A container that suits the occurrence of an isothermal process should be made of
(a) Copper (b) Glass
(c) Wood (d) Cloth.

41. A thermos flask made of stainless steel contains several tiny lead shots. If the flask is quickly shaken, up and down several times, the temperature of lead shots
(a) increases by adiabatic process
(b) increases by isothermal process
(c) decreases by adiabatic process
(d) remains same.

42. In changing the state of a gas adiabatically from an equilibrium state A to another equilibrium state B, an amount of work equal to 22.3 J is done on the system. If the gas is taken from state A to B via a process in which the net heat absorbed by the system is 9.35 calories, the net work done by the system in latter case will be
(a) 5.9 J (b) 16.9 J
(c) 9.3 J (d) 4.6 J

43. In which process the PV indicator diagram is a straight line parallel to volume axis?
(a) Isothermal (b) Isobaric
(c) Irreversible (d) Adiabatic

44. In an isochoric process if $T_1 = 27°C$ and $T_2 = 127°C$, then P_1/P_2 will be equal to
(a) 9/59 (b) 2/3
(c) 3/4 (d) None of these.

45. The first operation involved in a Carnot cycle is
(a) isothermal expansion
(b) adiabatic expansion
(c) isothermal compression
(d) adiabatic compression.

Answer Keys

1. (b)	2. (a)	3. (d)	4. (a)	5. (d)	6. (a)	7. (c)	8. (d)	9. (c)	10. (b)
11. (c)	12. (b)	13. (d)	14. (b)	15. (a)	16. (b)	17. (a)	18. (b)	19. (a)	20. (b)
21. (d)	22. (d)	23. (a)	24. (c)	25. (a)	26. (c)	27. (a)	28. (c)	29. (b)	30. (b)
31. (d)	32. (d)	33. (d)	34. (b)	35. (b)	36. (d)	37. (a)	38. (a)	39. (c)	40. (a)
41. (a)	42. (b)	43. (b)	44. (c)	45. (a)					

Solutions

1. (b)

$0.8 \times 5 = P \times (3+5) \Rightarrow P = 0.5$ m

2. (a) **3.** (d)

4. (a)

5. (d)

$\eta = 1 - \frac{T_2}{T_1}$; for η to be max ratio $\frac{T_2}{T_1}$ should be minimum.

6. (a) **7.** (c)

8. (d)

9. (c)
No change in the thermal energy of ideal gas, but for real gas internal energy increases because work is done against intermolecular forces.

10. (b) **11.** (c)

12. (b)

13. (d)

$$W = -\mu RT \log_e \frac{V_2}{V_1} = -1 \times 8.31 \times (273+0) \log_e \left(\frac{22.4}{11.2}\right)$$

$$= -8.31 \times 273 \log_e 2 = -1572.5 \text{ J} \quad [\because \log_e 2 = 0.693]$$

14. (b)

15. (a)

16. (b)

17. (a)

18. (b)

19. (a)

20. (b)

21. (d)

22. (d)

23. (a)

24. (c)

25. (a)

26. (c)

We know that, $PV = nRT$. It is obvious that if V is constant, then a graph between P and T will be a straight line passing through origin. Here, volume is constant. Therefore, work done is zero.

27. (a)

28. (c)

Change in internal energy $dQ = \mu C_V \Delta T$

It does not depend upon type of process. Actually, it is a state function.

29. (b)

30. (b)

31. (d)

32. (d)

33. (d)

34. (b)

35. (b)

36. (d)

The state of a thermodynamic system cannot be determined by a single variable (P or V or T).

37. (a)

38. (a)

39. (c)

40. (a)

An isothermal process that takes place at constant temperature, must be carried out in a vessel with conducting wall so that heat generated should go out at once.

41. (a)

42. (b)

According to I law of thermodynamics,

$\Delta Q = \Delta U + \Delta W$, in adiabatic process $\Delta Q = 0$

$0 = \Delta U - \Delta W$ (Work done on the system is – ve)

$$\Delta U = +\Delta W = +22.3$$

[Because work is done on the system the internal energy increases]

In 2 process $\quad \Delta Q = \Delta U + \Delta W$

$$9.35 \times 4.18 = 22.3 + \Delta W$$

Work done by system $\Delta W = 16.95$

43. (b)

44. (c)

At constant volume $P \propto T$

$$\Rightarrow \frac{P_1}{P_2} = \frac{T_1}{T_2}$$

$$\Rightarrow \frac{P_1}{P_2} = \frac{300}{400} = \frac{3}{4}$$

45. (a)

Behaviour of Perfect Gas and Kinetic Theory

Introduction

The behaviour of gases, in kinetic theory, is explained by considering the motion of their molecules. This molecular motion is governed by the pressure, volume and temperature of a gas and describes the state, i.e., physical condition of the system. A relation among these macroscopic properties is called **equation of state**. If any two of the values are known, the third has a unique value and can be determined.

Ideal Gas Equation

For a system containing N particles of a gas, when macroscopic properties Pressure (P), Volume (V) and Temperature (T) were studied through experiments it was found that:

$V \propto N$ if P and T are constant.

$V \propto (1/P)$ if N and T are constant.

and $V \propto T$ if P and N are constant.

On combining all these,

$$V \propto \frac{NT}{P}, \text{ i.e., } \frac{PV}{NT} = \text{const.} = k \quad \ldots(1)$$

The constant k is called Boltzmann's constant and has value 1.38×10^{-23} J/K.

Number of moles, $n = \dfrac{m}{M} = \dfrac{N}{N_A}$, $\quad N = nN_A$

So Eq. (1) can be written as,

$$\frac{PV}{nT} = N_A k = R \text{ with } R = N_A k$$

i.e., $$PV = nRT \text{ with } R = 8.31 \text{ J/mol K} \quad \ldots(2)$$

Equation (2) is called equation of state for an ideal gas where R is universal gas constant.

Important Points

While using gas equation, it is worthy to note that:

1. Zero Kelvin is a unique temperature at which pressure, volume and translational kinetic energy of an ideal gas would reach zero.
2. NTP (STP) refers to normal temperature and pressure and implies 0°C and 1 atm, i.e., 273 K and 1.01×10^5 N/m^2.
3. The temperature in gas equation is always in K (never in °C).
4. One mole of any substance at any pressure, temperature or volume always contains 6.02×10^{23} molecules, called **Avogadro's number** (N_A).
5. The volume of one mol of a gas at NTP (or STP) is 22400 cc or 22.4 litre.
6. Whatever be the process, in equilibrium state, ideal gas always satisfies the equation $PV = nRT$.
7. According to the gas equation,

$$PV = \left[\frac{m}{M}\right] RT,$$

i.e., $$P = \left[\frac{R}{M}\right] \frac{m}{V} T$$

or $$\frac{P}{\rho T} = \frac{R}{M} = \text{Const.} \quad \ldots(3)$$

This is gas equation for an i deal gas in terms of density ρ.

8. According to the gas equation,

$$P = \frac{nRT}{V} = \frac{nN_A kT}{V} \quad (R = N_A k)$$

$$P = \frac{N}{V} kT \quad (nN_A = N)$$

$$\Rightarrow \quad P = n' kT \quad \ldots(4)$$

This is gas equation for an ideal gas in terms of number of molecules per unit volume, $n' (= N/V)$.

9. If n_1 mol of a non-reactive gas in thermodynamic state (P_1, V_1, T_1) are mixed with n_2 mol of another non-reactive gas at (P_2, V_2, T_2) and the resultant mixture is at (P, V, T), then by **conservation of mass,**

$$n_1 + n_2 = n$$

i.e., $$\frac{P_1V_1}{T_1} + \frac{P_2V_2}{T_2} = \frac{PV}{T} \quad \left[\text{as } n = \frac{PV}{RT}\right] \quad \ldots(5)$$

Avogadro's number and Avogadro hypothesis

An Italian scientist Ameodeo Avogadro stated that equal volumes of gas at the same pressure and temperature contain equal number molecules. This statement is called Avogadro's hypothesis.

The number of molecule in a mole is known as Avogadro's number, N_A.

$$N_A = 6.023 \times 10^{23} \text{ molecules/ mole}$$

Thus in n moles of gas total number of molecules of the gas are

$$N = nN_A$$

or number of moles of a gas can be given as

$$n = \frac{N}{N_A} = \frac{m'N}{m'N_A} = \frac{m}{M}$$

Here m' is the mass of each molecule, m is the total mass of gas and M is the mass of 1 mole of molecules of a gas, i.e., its molecular mass,

From gas law,

$$PV = n\,R\,T$$

or $$PV = \frac{n}{N_A}RT = \frac{m}{M}RT$$

or $$PV = n\,k\,T$$

where $k = \frac{R}{N_A}$ is the Botzmann's constant and has the value

$k = 1.38 \times 10^{-23}$ J/K

Kinetic Theory of Gases

The pressure of a gas is due to the molecules bombarding the walls of its container. Whenever a molecule bounces off a wall, its momentum at right angles to the wall is reversed; the force which it exerts on the wall is equal to the rate of change of momentum. The average force per unit area on the walls, also known as the pressure, is due to the collision of all the molecules with the walls. The following assumptions are made so that calculations can be simplified:

Assumptions

(i) The molecules are perfectly elastic and undergo random motion, and collisions with the walls of the container.

(ii) The force of attraction between molecules is negligible, and collision between molecules may be ignored.

(iii) The molecules are very small in size, i.e., the total volume of the molecules is much smaller that the total volume of the gas.

(iv) The collisions of the molecules with the walls are instantaneous, i.e., the time of collision is much smaller than the time interval between two collisions.

Consider a cube of side ℓ, containing N molecules of an ideal gas, each of mass m. A typical molecule of this gas will have velocity:

$$\vec{c}_1 = c_{1x}\hat{i} + c_{2y}\hat{j} + c_{1z}\hat{k}$$

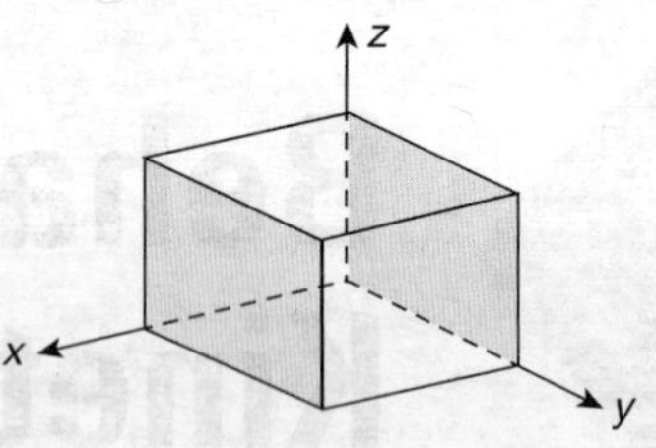

where, c_{1x}, c_{1y}, c_{1z} represent the components of the velocity $\vec{c}_1$ along O_x, O_y, O_z. Consider the walls YZ, which is normal to the x-axis. The molecule collides with YZ wall, and the x-component of its momentum gets reversed. The change in momentum Δp on impact is:

$$\Delta p = 2mc_{1x}$$

The time, Δt taken by the molecule to collide with the wall X' and return is:

$$\Delta t = \frac{2\ell}{c_{1x}}$$

$\therefore$ The average force exerted on this wall X due to this molecule is:

$$F_{av,x} = \frac{mc_{1x}^2}{\ell}$$

The net force on this wall due to all molecules is:

$$F_{av,x} = \frac{m\left(c_{1x}^2 + c_{2x}^2 + c_{3x}^2 + ... + c_{nx}^2\right)}{\ell}$$

$\therefore$ The pressure,

$$p_x = \frac{F_{av_1x}}{\ell^2} = \frac{m\left(c_{1x}^2 + c_{2x}^2 + c_{3x}^2 + ... + c_{nx}^2\right)}{V} \quad ...(i)$$

where $V = \ell^3$, represents the volume.

Similarly,

$$p_y = \frac{m\left(c_{1y}^2 + ... + c_{ny}^2\right)}{V} \quad ...(ii)$$

and $$p_z = \frac{m\left(c_{1z}^2 + ... + c_{nz}^2\right)}{V} \quad ...(iii)$$

$$c_1^2 = c_{1x}^2 + c_{1y}^2 + c_{1z}^2$$

Since the gas is homogeneous and isotropic, there is no reason why $p_x \neq p_y$ and $p_x \neq p_z$ and so we will assume that $p_x = p_y = p_z = p$.

$$\therefore \quad 3p = p_x + p_y + p_z = \frac{m\left(c_1^2 + c_2^2 + ... + c_n^2\right)}{V} \quad ...(iv)$$

$$\left(c_1^2 = c_{1x}^2 + c_{1y}^2 + c_{1z}^2\right)$$

Since $\frac{1}{2}mc^2$ is the kinetic energy of the molecules, we get,

$$3p = \frac{2}{V} \times \left(\frac{1}{2}mc_1^2 + \frac{1}{2}mc_2^2 + ... + \frac{1}{2}mc_n^2 \right)$$

$$= \frac{2}{V}.U_t \text{ where } U_t = \text{ the internal translational kinetic energy of the gas.}$$

$$\therefore p = \frac{2}{3}\frac{U_t}{V}$$

Further, $pV = \frac{2}{3}U_t = (k_BT).N$ (from gas laws) ...(v)

$$\therefore \bar{u}_t = \frac{U_t}{N} = \frac{3}{2}k_BT$$

which gives the average translational KE per molecule.

The root-mean-square (r.m.s) speed is defined by the expression:

$$\bar{c}^2 = \frac{c_1^2 + c_2^2 + ... + c_n^2}{N} \quad ...(vi)$$

and $\sqrt{\bar{c}^2}$ (r.m.s speed) $= \sqrt{\frac{c_1^2 + c_2^2 + ... + c_n^2}{N}}$ ...(vii)

Using the definition, we can write, the average translational KE,

$$\bar{u}_t = \frac{1}{2}m\bar{c}^2 \quad ...(viii)$$

The expression for pressure is:

$$p = \frac{N}{V}k_BT = n'k_BT,$$

where $n' = \frac{N}{V}$ = no of molecules per unit volume

= number density of the molecules.

$$= n'.\left(\frac{2}{3}\bar{u}_t\right)$$

$$= n'.\left(\frac{2}{3}.\frac{1}{2}m\bar{c}^2\right)$$

or, $p = \frac{1}{3}n'm\bar{c}^2$

The r.m.s. speed, $\sqrt{\bar{c}^2} = \sqrt{\frac{3p}{\nu m}}$ ...(ix)

$$= \sqrt{\frac{3p}{\rho}} \quad ...(ix)$$

where $\rho = n'm$ = Mass of gas per unit volume = Density of the gas

$$= \sqrt{\frac{3RT}{M}} \quad ...(x)$$

Various Speeds of Gas Molecules

1. **Root mean square speed:** It is defined as the square root mean of squares of the speed of different molecules,

$$v_{rms} = \sqrt{\left(v_1^2 + v_2^2 + ...\right)/N}$$

and according to the kinetic theory of gases,

$$v_{rms} = \sqrt{\frac{3PV}{M}} = \sqrt{\frac{3RT}{M}} = \sqrt{\frac{3kT}{m}}$$

where M is the molecular weight while m the mass of a single molecule.

2. **Most probable speed:** It is the speed which maximum number of molecules in a gas have at constant temperature and is given by

$$v_{mp} = \sqrt{\frac{2RT}{M}} = \left(\sqrt{\frac{2}{3}}\right)v_{rms} = 0.815v_{rms}$$

3. **Average speed:** It is the arithmetic mean of the speeds of molecules in a gas at a given temperature,

$$v_{av} = \left(v_1 + v_2 + v_3 + ...\right)/N$$

and according to kinetic energy of gases,

$$v_{av} = \sqrt{\frac{8RT}{\pi M}} = \left(\sqrt{\frac{8}{3\pi}}\right)v_{rms} = 0.92v_{rms}$$

From expressions (i), (ii) and (iii) it is evident that

$$v_{rms} > v_{av} > v_{mp}$$

Equipartition of Energy

An interesting conclusion from the kinetic theory of gases is derived from Eqs. (i) and (v) :

$$m\left(c_{1x}^2 + c_{2x}^2 + ... + c_{nx}^2\right) = p_xV$$

$$= pV$$

$$= Nk_BT$$

or,

$$\frac{\frac{1}{2}m\left(c_{1x}^2 + ... + c_{nx}^2\right)}{N} = \frac{1}{2}k_BT$$

or,

$$\in_{[x]} = \frac{1}{2}k_BT \quad ...(xi)$$

where $\in_{[x]}$ represents the average contribution of the translational kinetic energy emerging from motion along the x-direction only. This value is independent of direction i.e.,

$$\in_{[x]} = \in_{[y]} = \in_{[z]}, \quad \text{...(xii)}$$

A conclusion is derived from the identity of the three pressures. It can be generalized and it leads to what is known as the principle of **equipartition of energy.** This principle states that the average energy due to thermal motion for each degree of freedom is equal to:

$\in = \frac{1}{2}k_BT$, at the temperature T, Here, k_B is Boltzmann's constant.

Internal Energy of an Ideal Gas

The internal energy of an ideal gas is the total kinetic energy of random motion of the molecules of the gas, as the interaction between its molecules is zero, i.e., there is no potential energy of interaction.

The kinetic energy is, in general, due to three different types of motion: translational, rotational and vibrational motion.

The terms refer, respectively, to the energy associated with most motion of the centre of mass, rotation about the centre of mass and vibration of atoms within a molecule about their mean positions. At ordinary temperature, the vibrational states are not excited and most gas molecules behave like rigid bodies. Therefore, the internal energy (U) is due to translation and rotation alone, for most gases.

The principle of equipartition of energy applies equally well to rotational motion. Each term involving an independent contribution to the total energy from an independent type of motion is known as a 'degree of freedom'. Each degree of freedom contributes $\frac{1}{2}k_BT$ to the average energy of the molecule. The total internal energy of the gas is equal to the average energy per molecule multiplied by the total number of molecules. Hence, we need to only count the number of degrees of freedom to determine the total contribution of energy.

There are three translational degrees of freedom associated with motion along x, y, z axes, which are independent. For *monatomic* gases, there are no rotational degrees of freedom. For *diatomic* molecules, there are two rotational degrees of freedom, corresponding to rotation about the two axes perpendicular to the axis of the molecule. For *linear triatomic* molecules, like carbon dioxide, the number of degrees of freedom associated with rotation is two. For *non-linear triatomic* and *polyatomic* molecules, this becomes three, as there are three separate axes about which rotation can take place. The situation is summarised in the table below:

Type of molecule	No. of degrees of freedom			Average energy per molecule	Total internal energy (U) per mole
	Translational	Rotational	Total		
Monoatomic	3	0	3	$\frac{3}{2}k_BT$	$\frac{3}{2}RT$
Diatomic	3	2	5	$\frac{5}{2}k_BT$	$\frac{5}{2}RT$
Triatomic linear	3	2	5	$\frac{5}{2}k_BT$	$\frac{5}{2}RT$
Triatomic non-linear	3	3	6	$3k_BT$	$3RT$

If the total number of degree of freedom is f, then,

$$U = \frac{f}{2}RT$$

An important quantity is: $C_v = \frac{dU}{dT} = \frac{f}{2}R$, and it will be defined in the next subsection.

Specific heat in terms of degree of freedom

C_V : For a gas at temperature T, the internal energy

$$U = \frac{1}{2}nRT, \ \Delta U = \frac{f}{2}nR\Delta T \quad \text{(i)}$$

and $$(\Delta Q)_V = nC_V\Delta T = \Delta U \quad \text{(ii)}$$

From Eqs. (i) and (ii) $C_V = \frac{1}{2}fR$

C_P : From the Mayer's relation $C_p - C_v = R$

$$\Rightarrow C_P = C_V + R = \frac{f}{2}R + R = \left(\frac{f}{2}+1\right)R$$

Ratio of C_P and C_V (γ): $\gamma = \frac{C_P}{C_V} = \frac{\left(\frac{f}{2}+1\right)}{\frac{f}{2}R} = 1 + \frac{2}{f}$

Illustration 1

In a certain region of outer space there are only 100 molecules per cm^3 on an average. The temperature there is about 3 K. What is the average pressure of this gas?

Solution

$$P = \frac{1}{3}\rho v_{rms}^2$$

The density of the gas is $\rho = mN$ where m is the mass of the gas molecules and N is the number of molecules in 1 m^3 of the gas.

Here,

$$N = 100 \times 10^6$$

or

$$P = \frac{1}{3} \times \left(100 \times 10^6\right) m v_{rms}^2$$

As we known, rms velocity of the gas molecules is given as

$$v_{rms} = \sqrt{\frac{3RT}{M}} = \sqrt{\frac{3kT}{m}} \quad \left[\text{As } M = mN_A \text{ and } \frac{R}{N_A} = k\right]$$

So, we have $P = \frac{1}{3} \times \left(100 \times 10^6\right) \times 3kT$

$$\left(100 \times 10^6\right) kT$$

$$= \left(100 \times 10^6\right)\left(1.38 \times 10^{-23}\right)(3)$$

$$= 41.5 \times 10^{-16}$$

$$= 4.15 \times 10^{-15} \text{ Pa}$$

Illustration 2

A vessel of volume 0.2 m^3 contains hydrogen gas at temperature 300 K and pressure 1 bar. Find the heat required to raise the temperature to 400 K. The molar heat capacity of hydrogen at constant volume is 5 cal/mol K.

Solution

$$(\Delta Q)_V = nC_V \Delta T$$

Here $\quad C_V = \dfrac{5 \text{ cal}}{\text{mol K}} \quad$ and $\quad \Delta T = (400 - 300) = 100\ K$

and as for ideal gas $PV = nRT$

$$n = \frac{\left(10^5\right) \times (0.2)}{8.31 \times 300} = 8 \text{ mol}$$

$$(\Delta Q)_V = 8 \times 5 \times 100 = 4000 \text{ cal} = 4 \text{ kcal}$$

Illustration 3

A gas has molar heat capacity $C = 37.35$ J mol^{-1} K^{-1} in the process PT = constant. Find the number of degrees of freedom of molecules in the gas.

Solution

For one mole of an ideal gas,

$PV = RT \qquad$ and $PT = K$ (constant)

$$\therefore \qquad P = \frac{K}{T} \quad \text{and} \quad V = \frac{RT}{P} = \frac{RT^2}{K}$$

So, $\qquad dV = \dfrac{2RT}{K} dT$

$\therefore$ Work done, when temperature changes from T to $(T + dT)$,

$$dW = \int P dV = \int_T^{T+dT} \frac{K}{T} \cdot \frac{2RT}{K} dT$$

$$= 2R\, dT$$

Now, $\quad dQ = dU + dW$

$$CdT = C_V dT + 2RdT$$

$$\therefore \qquad C_V = C - 2R = \frac{37.35R}{8.3} - 2R$$

$$= \frac{5}{2} R$$

Hence, degree of freedom $n = 5 \quad \left[\because C_V = \dfrac{n}{2} R\right]$

Illustration 4

Two identical containers A and B with frictionless pistons contain the same ideal gas at the same temperature and the same volume V. The mass of the gas in A is m_A, and that in B is m_B. The gas in each cylinder is now allowed to expand isothermally to the same final volume $2V$. The changes in the pressure in A and B are found to be ΔP and 1.5 ΔP respectively. Then,

(a) $4m_A = 9m_B$ (b) $2m_A = 3m_B$
(c) $3m_A = 2m_B$ (d) $9m_A = 4m_B$

Solution

As temperature of the gas is same in the two containers, therefore,

$$P \propto 1/V$$

As volume of the gas in the cylinder increases, the pressure decreases. In container A,

$$\Delta P = (P_A)_i - (P_A)_f$$

$$= \frac{n_A RT}{V} - \frac{n_A RT}{2V} = \frac{n_A RT}{2V} \quad \text{...(i)}$$

In container B,

$$1.5\Delta P = (P_B)_i - (P_B)_f$$

$$= \frac{n_B RT}{V} - \frac{n_B RT}{2V} = \frac{n_B RT}{2V} \quad \ldots\text{(ii)}$$

From (i) and (ii), $\frac{n_A}{n_B} = \frac{1}{1.5} = \frac{2}{3}$.

As $n = \frac{m}{M}$, therefore, $n \propto m$.

$$\therefore \quad \frac{m_A}{m_B} = \frac{n_A}{n_B} = \frac{2}{3} \text{ or } 3m_A = 2m_B.$$

Ans. (c)

Illustration 5

A gas mixture consists of 2 moles of oxygen and 4 moles of argon at temperature T. Neglecting all vibrational modes, the total internal energy of the system is

(a) $4RT$ (b) $15RT$
(c) $9RT$ (d) $11RT$

Solution

Oxygen is diatomic, the total number of degrees of freedom for 2 moles of oxygen is $2N_A \times 5$.

Argon is monatomic, for which the number of degrees of freedom for 4 moles is $4N_A \times 3$.

$\therefore$ Total number of degrees of freedom $= 22\, N_A$

According to the law of equipartition of energy law, the total internal energy

$$U = 22N_A \times \frac{1}{2} k_B$$

$$= 11N_A \times \frac{R}{N_A} T$$

$$= 11\, RT$$

Ans. (d)

Mean Free Path

Mean free path of a molecule in a gas is the average distance travelled by the molecule between two successive collisions. It is generally denoted by λ.

The molecules are in random motion. Therefore the molecules frequently collide against each other. In each collision, there is a change in the magnitude and direction of the velocity of molecules. Between two successive collisions, a molecule moves along a straight path with uniform velocity. This path is known as **free path.**

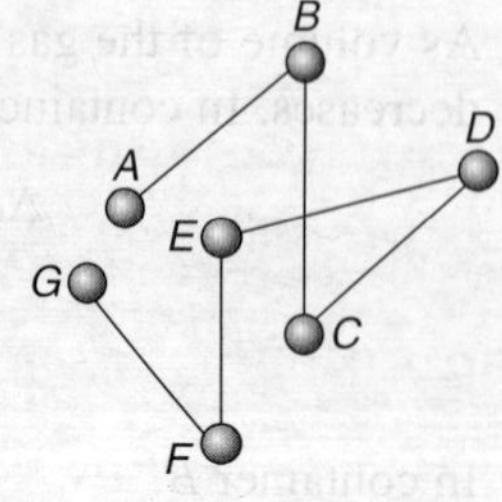

Imagine a molecule starting from A and suffering collisions at B, C, D, E, F, G.

Let $\lambda_1, \lambda_2, \lambda_3 \ldots, \lambda_N$ be the free paths travelled by the molecule in N successive collisions. Then, mean free path λ is given by:

$$\lambda = \frac{\lambda_1 + \lambda_2 + \lambda_3 + \ldots + \lambda_N}{N}$$

Expression for mean free path

Assume that only one molecule is in motion and all other molecules are at rest. Let d be the diameter of each molecule.

The moving molecule shall collide against all those molecules whose centres lie within a distance d from the centre of the moving molecule.

Suppose l is the distance travelled by the 'moving molecule'. The moving molecule will make a collision with all other molecules whose centres lie inside a volume $\pi d^2 l$.

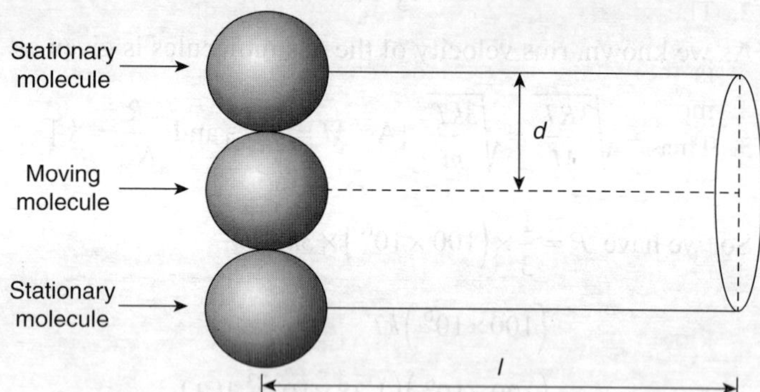

Suppose n is the number of molecules per unit volume in the gas.

Then, it will collide with all the molecules in volume $\pi d^2 l$.
Number of collisions $= n\pi d^2 l$

$$\text{or} \quad \lambda = \frac{1}{n\,\pi\, d^2} \quad \ldots(1)$$

In the above derivation, we have assumed that all, except one, molecules are stationary. All the molecules are in random motion. So, the chances of a collision by a molecule are more. Taking this into account, the mean free path can be shown to be $\sqrt{2}$ times less than that given in Eq. (1).

$$\therefore \quad \lambda = \frac{1}{\sqrt{2}\, n\, \pi\, d^2} \quad \ldots(2)$$

This equation shows that the mean free path is inversely proportional to the number of molecules per unit volume.

From Eq. 2, $\lambda = \frac{m}{\sqrt{2}\, \pi d^2 mn}$

Let m be the mass of a molecule.

But mn = Mass per unit volume of the gas = Density ρ of the gas

$$\therefore \quad \lambda = \frac{m}{\sqrt{2}\, \pi d^2 \rho} \quad \ldots(3)$$

Ideal gas equation $PV = \frac{N}{N_0}RT$

Where, N = Total number of molecules

N_0 = Avogadro's Number

$$P = \frac{N}{V}\frac{R}{N_0}T$$

$$= nkT$$

$$\Rightarrow n = \frac{P}{kT}$$

Putting the value in Eq. (2),

$$\lambda = \frac{kT}{\sqrt{2}\,\pi d^2 P}$$

POINTS TO REMEMBER

1. An ideal gas is one that would have the same simple behaviour under all conditions of temperature and pressure, i.e., which obeys the perfect gas equation, $PV = nRT$
2. The pressure exerted by an ideal gas is $P = \left(\frac{1}{3}\right)\rho v_{\text{rms}}^2$, v_{rms} is the square root of the average of the square of the molecular speeds.
3. The total translational kinetic energy per mole of the molecules of an ideal gas is proportional to temperature and is $\left(\frac{3}{2}\right)RT$. Similarly, the average translational kinetic energy per molecule is $\left(\frac{3}{2}\right)kT$.
4. $C_p - C_v = R$, where C_p and C_v are the molar heat capacities at constant temperature and at constant volume (in J mol^{-1} K^{-1}).
5. The degree of freedom of a system is defined as the number of independent modes of absorption of energy by the particles of the system.
   The energy associated with each degree of freedom per molecule is equal to $\left(\frac{1}{2}\right)kT$.
6. The mean free path of a gas molecule (that is, its average path length between collisions) is given by, $l = \left(\frac{l}{\sqrt{2}\pi n d^2}\right)$, where n is the number of molecules per unit volume and d is the molecular diameter.
7. At a given temperature, the specific heat of a gas at constant pressure is always greater than its specific heat at constant volume.
8. For monatomic gases, the molecules have only translational motion, thus $C_v = \left(\frac{3}{2}\right)R$ and $C_p = \left(\frac{5}{2}\right)R$ and the ratio of specific heat is $\gamma = \left(\frac{C_p}{C_v}\right) = 1.67$.
9. Absolute zero degree temperature is not zero energy temperature.
10. For a molecule with n degree of freedom,

$$U = \left(\frac{1}{2}\right)nRT$$

$$C_v = \left(\frac{dU}{dT}\right) = \left(\frac{n}{2}\right)R$$

$$C_p = \left[\left(\frac{1}{2}\right)n + 1\right]R$$

$$\gamma = \left(\frac{C_p}{C_v}\right) = \left[1 + \left(\frac{2}{n}\right)\right]$$

11. Degree of freedom and principal specific heats of gases.

Molecules type	n_{vib}	n_{raf}	n_{trans}	n_{total}	C_v	C_p	γ
Monoatomic	0	0	3	3	$\left(\frac{3}{2}\right)R$	$\left(\frac{5}{2}\right)R$	1.67
Diatomic	0	2	3	5	$\left(\frac{5}{2}\right)R$	$\left(\frac{7}{2}\right)R$	1.40
Triatomic							
(i) Linear molecule	2	2	3	7	$\left(\frac{7}{2}\right)R$	$\left(\frac{9}{2}\right)R$	1.28
(ii) Non-linear molecule	0	3	3	6	$3R$	$4R$	1.33

EXERCISES

1. An insulated container of gas has two chambers separated by an insulating partition. One of the chambers has volume V_1 and contains ideal gas at pressure P_1 and temperature T_1. The other chamber has volume V_2 and contains ideal gas at pressure P_2 and temperature T_2. If the partition is removed without doing any work on the gas, the final equilibrium temperature of the gas in the container will be
(a) $\dfrac{P_1V_1T_1 + P_2V_2T_2}{P_1V_1 + P_2V_2}$ (b) $\dfrac{P_1V_1T_2 + P_2V_2T_1}{P_1V_1 + P_2V_2}$
(c) $\dfrac{T_1T_2(P_1V_1 + P_2V_2)}{P_1V_1T_1 + P_2V_2T_2}$ (d) $\dfrac{T_1T_2(P_1V_1 + P_2V_2)}{P_1V_1T_2 + P_2V_2T_1}$

2. Two containers A and B are partly filled with water and closed. The volume of A is twice that of B. If both containers are at the same temperature, the water vapour in the containers will have pressure in the ratio of
(a) 1 : 2 (b) 1 : 1
(c) 2 : 1 (d) 4 : 1

3. Let A and B the two gases. $\dfrac{T_A}{M_A} = 4\dfrac{T_B}{M_B}$, where T is temperature and M is molecular mass. If C_A and C_B are the rms speed, then the ratio $\dfrac{C_A}{C_B}$ will be equal to
(a) 2 (b) 4
(c) 0.5 (d) 0.25

4. If $P = 10^6\ kT$, then the number of molecules per unit volume of the gas is:
(a) 1 (b) 10^2
(c) 10^3 (d) 10^6

5. The mean kinetic energy per mole of an ideal gas at 0° C is approximately:
(a) 3.4 J (b) 3.4×10^2 J
(c) 3.4×10^3 J (d) 207 J

6. At what temperature is the KE of gas molecule half that of its value at 27° C?
(a) 13.5° C (b) 150° C
(c) 150 K (d) –123 K

7. A mixture of two gases A and B is in a container at a constant temperature. Gas A is diatomic and B is monatomic. The ratio of molecular masses of A and B is 4, the ratio of the rms speeds of A and B is:
(a) 1 : 1 (b) $\dfrac{1}{2\sqrt{3}}$
(c) $\sqrt{2}:1$ (d) 1 : 2

8. A gas in a vessel is at a pressure P_0. If the masses of all the molecules are made half and their speed made double, then the resultant pressure will be:
(a) $4P_0$ (b) $2P_0$
(c) P_0 (d) $P_0/2$

9. At a certain temperature, hydrogen molecules have rms velocity 2 km/s. The rms velocity of the oxygen molecules at the same temperature is:
(a) 2 km/s (b) 8 km/s
(c) 0.5 km/s (d) 1 km/s

10. The molecular weight of a gas is 44. The volume occupied by 2.2 g of this gas at 0°C and 2 atm, pressure, will be:
(a) 0.56 litre (b) 1.2 litre
(c) 2.4 litre (d) 5.6 litre

11. A mixture of 2 moles of helium gas (atomic mass = 4 amu), and 1 mole of argon gas (atomic mass = 40 amu) is kept at 300 K in a container. The ratio of the rms speeds $\left[\dfrac{V_{rms}(\text{helium})}{V_{rms}(\text{argon})}\right]$ is
(a) 0.32 (b) 0.45
(c) 2.24 (d) 3.16

12. 12 g of a gas occupies a volume of 4×10^{-3} m^3 at a temperature of 7° C. After the gas is heated at constant pressure, its density becomes 6×10^{-4} g/cm^3. What is the temperature to which the gas was heated?
(a) 1000 K (b) 1400 K
(c) 1200 K (d) 800 K

13. The figure shows, the P-V diagram of two different masses m_1 and m_2 drawn at constant temperature T. Then,

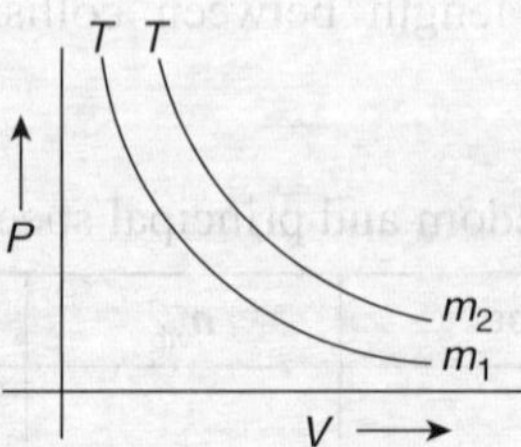

(a) $m_1 > m_2$ (b) $m_2 > m_1$
(c) $m_1 = m_2$ (d) Insufficient data

14. At what temperature is the effective speed of gaseous hydrogen molecules equal to that of oxygen molecules at 47° C?
(a) 200 K (b) 20 K
(c) 40 K (d) 100 K

15. The resultant temperature when two perfectly mono-atomic gases at absolute temperatures T_1 and T_2 and number of moles n_1 and n_2, respectively are mixed, is: (Assume no loss of energy)

(a) $T = \dfrac{n_1T_2 + n_2T_1}{n_1 + n_2}$ (b) $T = \dfrac{n_1T_1 - n_2T_2}{n_1 + n_2}$

(c) $T = \dfrac{n_1T_1 + n_2T_2}{n_1 + n_2}$ (d) $T = \dfrac{n_1T_1 - n_2T_2}{n_1 - n_2}$

16. For a gas, $\dfrac{R}{C_V} = 0.67$. This gas is made up of molecules which are

(a) diatomic
(b) mixture of diatomic and polyatomic molecules
(c) monoatomic
(d) polyatomic.

17. One mole of ideal monoatomic gas ($\gamma = 5/3$) is mixed with one mole of diatomic gas ($\gamma = 7/5$). What is γ for the mixture? γ denotes the ratio of specific heat at constant pressure, to that at constant volume:

(a) 3/2 (b) 23/15
(c) 35/23 (d) 4/3

18. Two kg of a monatomic gas is at a pressure of 4×10^4 N/m^2. The density of the gas is 8 kg/m^3. What is the order of energy of the gas due to its thermal motion

(a) 10^3 J (b) 10^5 J
(c) 10^6 J (d) 10^4 J

19. The density (ρ) versus pressure (P) of a given mass of an ideal gas is shown at two temperatures T_1 and T_2 Then relation between T_1 and T_2 may be

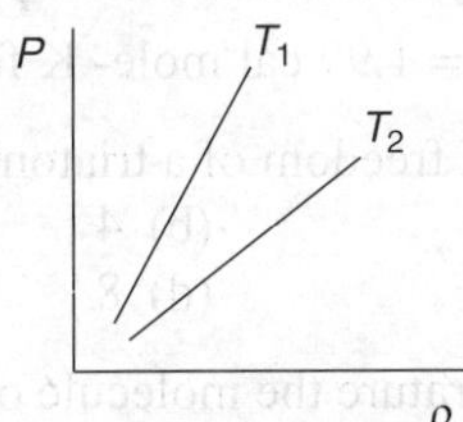

(a) $T_1 > T_2$ (b) $T_2 > T_1$
(c) $T_1 = T_2$ (d) All the three are possible.

20. The pressure of a gas kept in an isothermal container is 200 kPa. If half of the gas is removed from it the pressure will be:

(a) 100 kPa (b) 200 kPa
(c) 400 kPa (d) 800 kPa

21. How many cylinders of hydrogen at atmospheric pressure are required to fill a balloon whose volume is 500 m^3, if hydrogen is stored in cylinders of volume 0.05 m^3 at an absolute pressure of 15×10^5 Pa?

(a) 700 (b) 675
(c) 605 (d) 710

22. The value of critical temperature in terms of van der Waals' constant a and b is given by

(a) $T_C = \dfrac{8a}{27Rb}$ (b) $T_C = \dfrac{27a}{8Rb}$

(c) $T_C = \dfrac{a}{2Rb}$ (d) $T_C = \dfrac{a}{27Rb}$

23. If the ratio of specific heat of a gas at constant pressure to that at constant volume is γ, the change in internal energy of one mole of gas, when the volume changes from V to $2V$ at constant pressure P, is

(a) $\dfrac{PV}{(Y-1)}$ (b) PV

(c) $\dfrac{R}{(Y-1)}$ (d) $\dfrac{YPV}{(Y-1)}$

24. At constant volume, for a fixed number of moles of a gas, the pressure of the gas increases with rise of temperature due to

(a) increase in average molecular sped
(b) increase rate of collisions amongst molecules
(c) increase in molecular attraction
(d) decrease in mean free path.

25. A 15 g mass of nitrogen gas is enclosed in a vessel at a temperature 27°C. Amount of heat transferred to the gas, so that rms velocity of molecules is doubled, is about
[Take $R = 8.3$ J / K mole]

(a) 10 kJ (b) 0.9 kJ
(c) 6 kJ (d) 14 kJ

26. The kinetic energy of one gram molecule of a gas at normal temperature and pressure is

(a) 0.56×10^4 J
(b) 1.3×10^2 J
(c) 2.7×10^2 J
(d) 3.4×10^3 J

27. The law of equipartition of energy was postulated by

(a) Maxwell (b) Planck
(c) Bohr (d) Boltzmann.

28. Two gases are at absolute temperature 300 K and 350 K respectively. The ratio of average kinetic energy of their molecules is

(a) 7 : 6 (b) 6 : 7
(c) 36 : 49 (d) 49 : 36

29. The mean distance between molecules of a gas

(a) increases as density of the gas increases
(b) decreases as density increases
(c) increases with increasing temperature
(d) decreases with increasing temperature.

30. When attractive forces between molecules in a gas are taken into account, the volume occupied by the gas for a given external pressure
(a) decreases (b) increases
(c) does not change (d) cannot be estimated.

31. Real gases obey ideal gas laws at
(a) very low temperature (b) very high temperature
(c) very low densities (d) very high densities.

32. The quantity PV/RT represents
(a) mass of the gas
(b) kinetic energy of the gas
(c) number of moles of the gas
(d) number of molecules of the gas.

33. The molar heat capacity at constant volume of oxygen gas at STP is nearly $\frac{5R}{2}$ and it approaches $\frac{7R}{2}$ as the temperature is increased. This happens because at higher temperature,
(a) oxygen becomes triatomic
(b) oxygen does not behave as an ideal gas
(c) oxygen molecules rotate more vigorously
(d) oxygen molecules start vibrating.

34. According to kinetic theory of gases, the rms velocity of the gas molecules is directly proportional to
(a) $\sqrt{T}$ (b) T
(c) T^2 (d) T^4

35. The speed of sound in a gas is V. The rms velocity of the gas molecules is C. The ratio of V to C is
(a) $3/\gamma$ (b) $\gamma/3$
(c) $\sqrt{(3/\gamma)}$ (d) $\sqrt{(\gamma/3)}$

36. To find out degree of freedom, the expression is
(a) $f=\frac{2}{\gamma-1}$ (b) $f=\frac{\gamma+1}{2}$
(c) $f=\frac{2}{\gamma+1}$ (d) $f=\frac{1}{\gamma+1}$

37. Which of the following statement is incorrect?
Statement: The root mean square value of the speed of the molecules in a fixed mass of an ideal gas is increased by increasing
(a) the temperature while keeping the volume constant
(b) the pressure while keeping the volume constant
(c) the temperature while keeping the pressure constant
(d) the pressure while keeping the temperature constant.

38. At room temperature, the r.m.s speed of the molecules of a certain diatomic gas is found to be 1930 m/s. The gas is
(a) H_2 (b) F_2
(c) O_2 (d) Cl_2

39. The equation of state for 5 g of oxygen at a pressure P and temperature T, when occupying a volume V, will be
(a) $PV = (5/32)RT$ (b) $PV = 5RT$
(c) $PV = (5/2)RT$ (d) $PV = (5/16)RT$
(where R is the gas constant)

40. Oxygen and hydrogen are at the same temperature T. The kinetic energy of the oxygen molecule will be
(a) 16 times
(b) 4 times
(c) equal
(d) one-fourth the KE of hydrogen molecule.

41. If a gas has n degrees of freedom, the value of γ is
(a) n (b) $(n+2)/n$
(c) $n/2$ (d) $2n$

42. The temperature of an ideal gas is increased from 120 K to 480 K. If at 120 K the root-mean square velocity of the gas molecules is v, at 480 K it becomes
(a) $4v$ (b) $2v$
(c) $v/2$ (d) $v/4$

43. Absolute temperature can be calculated by
(a) mean square velocity (b) motion of the molecule
(c) Both (a) and (b) (d) None of these.

44. For the specific heat of 1 mole of an ideal gas at constant pressure (C_P) and at constant volume (C_V), which is correct
(a) C_P of hydrogen gas is $\frac{5}{2}R$
(b) C_V of hydrogen gas is $\frac{7}{2}R$
(c) H_2 has very small values of C_P and C_V
(d) $C_P - C_V = 1.99$ cal/mole- K for H_2

45. The degrees of freedom of a triatomic gas is
(a) 6 (b) 4
(c) 2 (d) 8

46. At what temperature the molecule of nitrogen will have the same r.m.s velocity as the molecules of oxygen at 127°C?
(a) 77°C (b) 350°C
(c) 273°C (d) 557°C

47. A gas mixture consists of 2 moles of oxygen and 4 moles of argon at temperature T. Neglecting all vibrational modes, the total internal energy of the system is
(a) 4 RT (b) 15 RT
(c) 9 RT (d) 11 RT

48. The number of translational degree of freedom for monoatomic and diatomic gases are
(a) 3, 5 (b) 5, 3
(c) 3, 3 (d) 3, 7

49. An ideal gas at 27°C is compressed adiabatically to 8/27 of its original volume. The rise in temperature is (Take $\gamma = 5/3$)
(a) 275 K (b) 375 K
(c) 475 K (d) 175 K

50. The equation of state corresponding to 8 g of O_2 is
(a) $PV = RT$ (b) $PV = 8RT$
(c) $PV = \frac{RT}{2}$ (d) $PV = \frac{RT}{4}$

51. Out of the following assumptions of the kinetic theory of gases, which one is incorrect?
(a) The gas contains large number of molecules.
(b) All the molecules of the gas are moving with the same speed.
(c) The collision between the molecules are elastic.
(d) There is no preferred direction of motion for the molecules.

52. According to kinetic theory of gases, the temperature of a perfect gas is
(a) directly proportional to kinetic energy of the molecules
(b) inversely proportional to kinetic energy of the molecules
(c) independent of the kinetic energy of the molecules
(d) directly proportional to both kinetic energy and potential energy of the molecules.

53. According to the main postulate of kinetic theory of gases, there is
(a) a force of repulsion between molecules
(b) a force of attraction between molecules
(c) no force between molecules
(d) a small force between the molecules.

54. The mean free path is inversely proportional to
(a) molecular diameter
(b) square of the molecular diameter
(c) square root of molecular diameter
(d) fourth power of the molecular diameter.

55. According to kinetic interpretation of temperature, the average kinetic energy of the molecules of a gas at absolute zero of temperature is
(a) positive (b) cannot be estimated
(c) zero (d) None of the above.

56. For a monatomic gas, the ratio of specific heats is
(a) 1.33 (b) 1.67
(c) 1.40 (d) None of the above.

57. When gas is in equilibrium, its molecules
(a) have the same energy
(b) have different energies, which remain constant
(c) have a certain constant average energy
(d) do not collide with each another.

58. At constant volume temperature is increased, then
(a) collision on walls will be less
(b) number of collisions per unit time will increase
(c) collisions will be in straight lines
(d) collisions will not change.

59. How much should the pressure be increased in order to decrease the volume of a gas by 5% at a constant temperature?
(a) 5% (b) 10%
(c) 5.26% (d) 4.26%

60. For an ideal gas $PV = xT$, where x is a constant. Then x must be proportional to
(a) mass of the gas molecule
(b) number of the gas molecules in the vessel
(c) absolute temperature
(d) None of the above.

61. Tiny smoke particles in air are seen through a microscope as flashes of light when illuminated correctly. The irregular movement in it is due to
(a) gravitational attraction between the particles
(b) energy given by light source
(c) continuous bombardment by air molecules
(d) current produced by the lamp used for illumination.

62. For a certain mass of gas, the relations between P and V are shown by the graphs at Kelvin temperature respectively then,

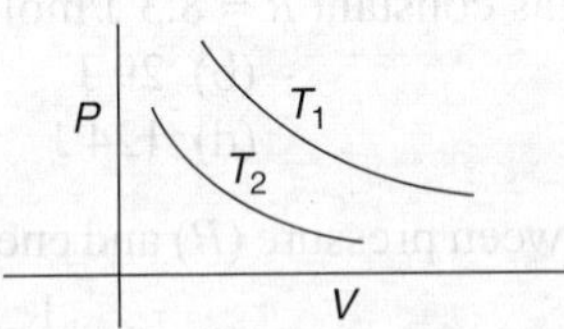

(a) $T_1 = T_2$ (b) $T_1 < T_2$
(c) $T_1 > T_2$ (d) Data incomplete.

63. A polyatomic gas with n degrees of freedom has a mean energy per molecule given by
(a) $\frac{nkT}{N}$ (b) $\frac{nkT}{2}$
(c) $\frac{nkT}{N}$ (d) $\frac{3kT}{N}$

64. For a certain gas, the ratio of specific heats is given to be $\gamma = 1.5$. For this gas
(a) $C_v = 3R/J$ (b) $C_p = 3R/J$
(c) $C_p - 5R/J$ (d) $C_v - 5R/J$

65. If pressure of a gas contained in a closed vessel is increased by 0.4% when heated by 1° C, the initial temperature must be
(a) 250 K (b) 1200 K
(c) 500 K (d) 2500 K

66. According to kinetic theory of gases, at absolute zero of temperature
(a) water freezes
(b) liquid helium freezes
(c) molecular motion stops
(d) liquid hydrogen freezes.

67. A sample of gas is at 0°C. The temperature at which rms speed of its molecules will be doubled is
(a) 103°C (b) 273°C
(c) 723°C (d) 819°C

68. The root mean square value of the speed of the molecules in a fixed mass of an ideal gas is increased by increasing
(a) the temperature while keeping the volume constant
(b) the pressure while keeping the volume constant
(c) the temperature while keeping the pressure constant
(d) the pressure while keeping the temperature constant.

69. In a cubic container of side 5 cm, a gas molecule whose velocity is 50 cm/s make successive collisions on a face. Then the number of collisions per second is:
(a) 2 (b) 5
(c) 10 (d) 250

70. One mole of an ideal gas requires 207 J heat to rise the temperature by 10 K when heated at constant pressure. If the same gas is heated at constant volume to raise the temperature by the same 10 K, the heat required is (Given the gas constant $R = 8.3$ J/mole K)
(a) 198.7 J (b) 29 J
(c) 215.3 J (d) 124 J

71. Relation between pressure (P) and energy (E) of a gas is
(a) $P = \frac{2}{3}E$ (b) $P = \frac{1}{3}E$
(c) $P = E$ (d) $P = 3E$

72. In a gas, according to kinetic theory, which speed among those given below corresponds to a molecule having average kinetic energy of translation?
(a) Most probable speed (b) Root mean square speed
(c) Average speed (d) None of the above.

73. Three containers of the same volume contain three different gases. The masses of the molecules are m_1, m_2 and m_3 and the number of molecules in their respective containers are N_1, N_2 and N_3. The gas pressure in the containers is P_1, P_2 and P_3 respectively. All the gases are now mixed and put in one of these containers. The pressure P of the mixture will be
(a) $P < (P_1 + P_2 + P_3)$ (b) $P = \frac{P_1 + P_2 + P_3}{3}$
(c) $P = P_1 + P_2 + P_3$ (d) $P > (P_1 + P_2 + P_3)$

74. According to kinetic theory of matter, a molecule is the smallest particle of a substance and it possesses
(a) all the properties of the substance
(b) some of the properties of the substance
(c) none of the properties of the substance
(d) Both (a) and (b) are true.

75. The absolute temperature of a gas is determined by
(a) the average momentum of the molecule
(b) the velocity of sound in the gas
(c) the number of molecules in the gas
(d) the mean square velocity of the molecules.

76. For hydrogen gas $C_p - C_v = a$ and for oxygen gas $C_p - C_v = b$, so the relation between a and b is given by
(a) $a = 16b$ (b) $16b = a$
(c) $a = 4b$ (d) $a = b$

77. The molar heat capacity of CO_2 at constant pressure in cal/mole-K is
(a) 8 (b) 6.74
(c) 5.03 (d) 7.5

78. If a gas enclosed in a closed vessel is heated by 10°C the pressure of the gas increases by 2% of its original pressure. The original temperature of the gas
(a) 500°C (b) 327°C
(c) 373°C (d) 227°C

79. At a given temperature, the force between molecules of a gas as a function of intermolecular distance
(a) is always constant
(b) decreases
(c) increases
(d) first decreases and then increases.

80. At room temperature, the rms speed of the molecules of a certain diatomic gas in found to be 1930 m/s. The gas is
(a) H_2 (b) F_2
(c) O_2 (d) Cl_2

81. For an ideal gas, if the pressure is increased at constant volume, the
(a) intermolecular forces will increase
(b) temperature will increase
(c) temperature will decrease
(d) number of collision with the walls of the container will increase.

82. All real gases tend to behave like ideal gases exactly under the conditions of
(a) high temperature and low pressure
(b) high temperature and high pressure
(c) low temperature and high pressure
(d) low temperature and low pressure.

83. The molecules of an ideal gas will
(a) experience a force during collisions with the walls of the container
(b) never experience any force
(c) experience a force during the collisions of the walls as well
(d) always experience due to intermolecular attractions.

84. Velocity of any body depends upon the reference frame from which its motion is viewed. The randomness of motion of the gas molecules is valid in
(a) any reference frame fixed with respect to earth
(b) any reference frame fixed with respect to stars
(c) any uniformly accelerating reference frame
(d) only that reference frame in which the centre of mass is at rest.

85. A sample of an ideal gas is placed in a container of fixed volume. If the average velocity of the molecules is doubled, the new temperature
(a) increases by $\sqrt{2}$ times
(b) increases by 2 times
(c) increases by 4 times
(d) remains constant.

86. A mixture of two gases A and B is enclosed at constant temperature. The molecular weight of A, (which is diatomic) is 8 times that of B (which is monatomic). What is the ratio of the rms speed of B to A?
(a) 2 (b) $2\sqrt{2}$
(c) 4 (d) 8

87. A room contains oxygen gas, at 0°C and 1 atm pressure. If the absolute temperature of the gas is doubled, the oxygen gas dissociated, becoming molecular oxygen. If the rms speed initially was V, what is it after dissociation?
(a) $\left(\frac{1}{2}\right)V$ (b) V
(c) $\sqrt{2}V$ (d) $2V$

88. If N gas molecules per second strike A m^2 of a wall, at an angle θ with the normal, when moving with s speed of v m/s, the pressure that they exert on the wall is:
(a) $\left[\frac{Nmv\cos\theta}{A}\right]$ (b) $\left[\frac{Nmv\sin\theta}{A}\right]$
(c) $\left[\frac{2Nmv\cos\theta}{A}\right]$ (d) $\left[\frac{2mv\cos\theta}{A}\right]$

89. Given 1 cc of hydrogen and 1 cc of oxygen, both at NTP, which sample has larger number of molecules?
(a) Oxygen
(b) Nitrogen
(c) Both of them have equal number.
(d) It depends on which one is more reactive.

90. If for a gas, $\frac{R}{C_V} = 0.67$, this gas is made up of molecules which are
(a) diatomic
(b) mixture of diatomic and polyatomic molecules
(c) monatomic
(d) polyatomic.

91. The total number of degrees of freedom for monatomic and diatomic gases are, respectively,
(a) 3, 5 (b) 5, 3
(c) 3, 3 (d) 3, 7

92. An ideal diatomic gas is at pressure P and volume V. The internal energy of the gas which is at a temperature T is
(a) $\left(\frac{3}{2}\right)PV$ (b) $\left(\frac{5}{2}\right)PV$
(c) $\left(\frac{7}{2}\right)PV$ (d) $3PV$

93. The C_p of gas is greater than C_v because, at constant pressure
(a) the coefficient of expansion is different
(b) work is done by the expanding gas
(c) the molecular attraction is greater
(d) the molecules themselves expand.

94. The internal energy of an ideal monatomic gas at 20° C is
(a) partly potential and partly kinetic
(b) purely kinetic
(c) purely rotational
(d) purely potential.

95. The number of translational degrees of freedom for a diatomic gas is
(a) 2 (b) 3
(c) 5 (d) 6

96. A gas deviates most from ideal gas behaviour at
(a) high temperature and high pressure
(b) low temperature and low pressure
(c) high temperature and low pressure
(d) low temperature and high pressure.

97. 70 calories of heat is required to raise the temperature of 2 moles of an ideal gas at constant pressure from 30° C to 35°C. The amount of heat required (in calories) to raise the temperature of the same gas through the same range (30°C to 35°C) at constant volume is ($R = 2$ cal/g mole^{-1} K^{-1})
(a) 30 (b) 50
(c) 70 (d) 90

Data for Questions 98 to 100

One mole of a monatomic ideal gas is mixed with three moles of a diatomic ideal gas at the same temperature.

98. The molar specific heat at constant volume of the mixture is

(a) $\left(\frac{3}{2}\right)R$ (b) $\left(\frac{5}{2}\right)R$

(c) $\left(\frac{9}{4}\right)R$ (d) $\left(\frac{15}{16}\right)R$

99. The molar specific heat at constant pressure of the mixture is

(a) $\left(\frac{9}{4}\right)R$ (b) $\left(\frac{5}{2}\right)R$

(c) $\left(\frac{15}{17}\right)R$ (d) $\left(\frac{13}{4}\right)R$

100. The γ value of the mixture is

(a) $\left(\frac{5}{3}\right)$ (b) $\left(\frac{13}{9}\right)$

(c) $\left(\frac{7}{5}\right)$ (d) $\left(\frac{4}{3}\right)$

101. According to kinetic theory of gases, the average translational kinetic energy of gas molecules which are at temperature T is proportional to

(a) $(T)^{(1/2)}$ (b) $(T)^1$

(c) $(T)^2$ (d) $(T)^{(-1/2)}$

102. For oxygen gas at 300 K, the rms velocity is

(a) directly proportional to the square root of pressure

(b) inversely proportional to the square root of pressure

(c) inversely proportional to the square root of density

(d) is a constant, and is independent of pressure and density.

103. One mole of a monatomic, ideal gas undergoes the following process A→B as in the figure. The molar specific heat of the gas during the process is

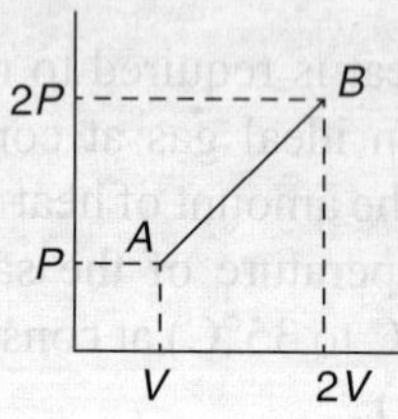

(a) Zero (b) $3R$

(c) $2R$ (d) $5R$

104. The sum of molar specific heats for a monatomic gas is more than the sum for diatomic gas by _____ cal/mole-K.

(a) 2 (b) 4

(c) 8.3 (d) 16.6

105. The ratio $\frac{C_P}{C_v}$ for a gas with n degree of freedom is equal to

(a) $n+1$ (b) $\left(\frac{2-n}{n}\right)$

(c) $n-1$ (d) $\left(\frac{2+n}{n}\right)$

106. V_m, V_{rms} and V_p respectively denote the mean speed, root mean square speed and most probable speed of the molecules in an ideal monoatomic gas at absolute temperature T. Then,

(a) no molecule can have a speed greater than $\sqrt{2}\,V_{rms}$

(b) no molecule can have a speed less than $\frac{V_p}{\sqrt{2}}$

(c) $V_p > V_m > V_{rms}$

(d) the average kinetic energy of a molecule is $\frac{3}{4}mV_p^2$

107. We have a jar A filled with a gas characterized by parameters P, V and T and another jar B filled with gas a parameters $2P$, $\left(\frac{V}{4}\right)$ and $2T$, where the symbols have the usual meanings. The ratio of the number of moles of jar A to those of jar B is

(a) 1 : 1 (b) 2 : 1

(c) 1 : 2 (d) 4 : 1

108. In a cubical vessel, n molecules of a gas are enclosed, each having a mass m and an average speed v. If l is the length of each edge of the cube, the pressure exerted by the gas will be

(a) $\left[\frac{nmv^2}{l^3}\right]$ (b) $\left[\frac{nm^2v}{2l^3}\right]$

(c) $\left[\frac{mnv^2}{3l^3}\right]$ (d) $\left[\frac{mnv}{2l}\right]$

109. The average kinetic energy of a molecule of a perfect gas is

(a) $\left(\frac{2}{3}\right)kT$ (b) $\left(\frac{3}{2}\right)kT$

(c) $\left(\frac{5}{2}\right)kT$ (d) None of these.

110. The speed of sound in a gas is v. The rms velocity of the gas molecules is c. The ratio of v to c is

(a) $\sqrt{\left(\frac{3}{\gamma}\right)}$ (b) $\sqrt{\left(\frac{\gamma}{3}\right)}$

(c) $\left(\frac{3}{\gamma}\right)$ (d) $\left(\frac{\gamma}{3}\right)$

111. In an adiabatic change, the pressure and temperature of a monatomic gas are related as $P \propto T^C$, where C equals

(a) 3/5 (b) 5/3
(c) 2/5 (d) 5/2

112. If one mole of argon is mixed with one mole of oxygen, the value of γ for the mixture will be

(a) 1.40 (b) 1.50
(c) 1.53 (d) 3.07

113. The rms velocity of oxygen molecule at 16°C is 474 m/s. The rms velocity in m/s of hydrogen molecular at 127°C is

(a) 1603 (b) 1896
(c) 2240 (d) 2730

114. The ratio of the velocity of sound in a monatomic gas to the root mean square velocity of its molecules is

(a) $\sqrt{\left(\frac{5}{9}\right)}$ (b) $\sqrt{\left(\frac{5}{3}\right)}$

(c) $\sqrt{\left(\frac{\gamma}{9}\right)}$ (d) $\sqrt{\left(\frac{\gamma}{3}\right)}$

115. 2 mole of an ideal gas at temperature 27°C is mixed with 1 mole of an ideal gas at 57°C. There is no loss of energy. The temperature of the mixture is

(a) $\left(\frac{100\sqrt{2}}{3}\right)$ (b) $100\sqrt{2}$

(c) $\frac{400}{\sqrt{3}}$ (d) None of these.

116. The value of $\gamma = \frac{C_p}{C_v}$ for hydrogen, helium and another ideal diatomic gas X (whose molecules are not rigid but have an additional mode), are respectively equal to

(a) $\frac{7}{5}, \frac{5}{3}, \frac{9}{7}$ (b) $\frac{5}{3}, \frac{7}{5}, \frac{9}{7}$

(c) $\frac{5}{3}, \frac{7}{5}, \frac{9}{7}$ (d) $\frac{7}{5}, \frac{5}{3}, \frac{7}{5}$

117. A sealed container with negligible thermal coefficient of expansion contains helium (a monatomic gas). When it is heated from 300 to 600 K, the average kinetic energy of the helium atom is

(a) halved (b) left unchanged
(c) doubled (d) becomes $\sqrt{2}$ times more.

118. The root mean square speed of hydrogen and oxygen at a given temperature is in the ratio

(a) 16 : 1 (b) 4 : 1
(c) 1 : 2 (d) 1 : 16

119. The equation of state for 4 g of nitrogen at a pressure P and temperature T, when occupying a volume V will be

(a) $PV = \frac{4}{7}RT$ (b) $PV = 7RT$

(c) $PV = \frac{1}{7}RT$ (d) $PV = \frac{1}{8}RT$

120. The molecules of a given mass of gas have rms velocity 200 m/s at 27 °C and 10^5 N/m^2 pressure. When the temperature is doubled and the pressure is halved, the rms speed of the molecules of the same gas is

(a) 200 m/s (b) 400 m/s
(c) $200\sqrt{2}$ m/s (d) $\frac{200}{\sqrt{2}}$ m/s

121. The temperature of a gas is held constant while its volume is increased. The pressure exerted on the walls of the gas container decreases because its molecules

(a) have less kinetic energy
(b) strike the container walls with smaller force
(c) strike the container walls at less frequency
(d) move more randomly.

122. N molecules, each of mass m, of gas A and 2 N molecules, each of mass 2 m, of gas B are contained in the same vessel which is maintained at a temperature T. The mean square of the velocity of molecules of B type is denoted by v^2 and the mean square of x component of the velocity of A type is denoted by w^2, then $\frac{w^2}{v^2}$ is

(a) 2 (b) 1
(c) $\frac{1}{3}$ (d) $\frac{2}{3}$

123. Two vessels having equal volume contain molecular hydrogen at one atmosphere and helium at two atmosphere pressure respectively. If both samples are at the same temperature, the mean velocity of hydrogen molecule is

(a) equal to that of helium
(b) twice that of helium
(c) half that of helium
(d) $\sqrt{2}$ times that of helium.

124. Oxygen and hydrogen gases are at the same temperature T. The kinetic energy of an oxygen molecule will be equal to
(a) 16 times the kinetic energy of a hydrogen molecule
(b) 4 times the kinetic energy of a hydrogen molecule
(c) the kinetic energy of a hydrogen molecule
(d) one-fourth the kinetic energy of a hydrogen molecule.

125. At 0 K, which of the following properties of a gas will be zero?
(a) vibrational energy (b) density
(c) kinetic energy (d) potential energy.

126. The average kinetic energy of a gas molecule at 27 °C is 6.21×10^{-21} joule. The average kinetic energy at 227 °C will be
(a) 9.35×10^{21} joule (b) 10.35×10^{-21} joule
(c) 11.35×10^{-21} joule (d) 12.35×10^{-21} joule.

127. Two identical containers A and B with frictionless pistons contain the same ideal gas at the same temperature and the same volume V. The mass of the gas in A is m_A and that in B is m_B. The gas in each cylinder is allowed to expand isothermally to the same final value $2V$. The change in pressure in A and B are found to be ΔP and $1.5\,\Delta P$ respectively. Then
(a) $41\,m_A = 9\,m_B$ (b) $2\,m_A = 3\,m_B$
(c) $3\,m_A = 2\,m_B$ (d) $9\,m_A = 4\,m_B$

128. An ideal gas occupies a volume of 2m^3 at a pressure of 3×10^6 Pa. The energy of the gas is
(a) 3×10^2 (b) 10^8 J
(c) 6×10^4 J (d) 9×10^6 J

129. A box contains n molecules of a gas. How will be the pressure of the gas affected if the number of molecules is made $2n$?
(a) Pressure will decrease
(b) Pressure will remain unchanged
(c) Pressure will be doubled
(d) Pressure will become three times more.

130. To what temperature should the hydrogen at 327 °C be cooled at constant pressure so that the root mean square velocity of its molecules become half of its previous value
(a) – 123 ° C (b) 123 ° C
(c) – 100 ° C (d) 0 ° C

131. A gas at a pressure P_O is contained in a vessel. If the masses of all the molecules are halved and their velocities doubled, the resulting pressure P would be equal to
(a) $4\,P_O$ (b) $2\,P_O$
(c) P_O (d) $P_O\sqrt{2}$

132. A mixture of two different ideal gases in a container is separated by a porous wall from an evacuated chamber. After instantaneous diffusion, the evacuated chamber contains
(a) large number of atoms of the lighter gas
(b) large number of atoms of the heavier gas
(c) equal number of atoms of the both gases
(d) any number of atoms of either gas without following any specific rule.

133. A what temperature is the root mean square velocity of gaseous hydrogen molecules equal to that of oxygen molecules at 47 °C ?
(a) 20 K (b) –73 K
(c) 80 K (d) 3 K

134. The average velocity of the molecules in a gas in equilibrium is
(a) proportional to $\sqrt{T}$ (b) proportional to T
(c) proportional to T^2 (d) equal to zero.

135. Match the following columns.

(1) Pressure exerted by the gas	(a) $\left(\frac{3}{2}\right)P$
(2) Mean kinetic energy	(b) $\sqrt{\frac{2}{3}}\overline{C}$
(3) Most probable velocity	(c) $\frac{1}{3}\frac{mn\overline{C^2}}{V}$
(4) Ratio of two specific heats	(d) $\left(\frac{3}{2}\right)kT$
(5) Mean kinetic energy per unit volume	(e) $\frac{1}{n\pi d^2}$
(6) Mean free path	(f) γ

The correct alternative is
(a) (1) d (2) a (3) b (4) f (5) c (6) e
(b) (1) d (2) e (3) b (4) f (5) a (6) c
(c) (1) c (2) d (3) b (4) f (5) a (6) e
(d) (1) a (2) b (3) c (4) d (5) e (6) f.

136. A gas can be liquified by application of pressure when
(a) its temperature is greater than the critical temperature
(b) its volume is very low
(c) its temperature is below the critical temperature
(d) its pressure is low.

137. A diatomic gas initially at 18°C is compressed adiabatically to one-eight of its original volume. The temperature after compression will be
(a) 395.4°C (b) 144°C
(c) 18°C (d) 887.4°C

138. The pressure P of a gas and its mean kinetic energy E per unit volume are related as

(a) $P = \left(\frac{1}{2}\right)E$ (b) $P = E$

(c) $P = \left(\frac{3}{2}\right)E$ (d) $P = \left(\frac{2}{3}\right)E$

139. The mass of an oxygen molecule is about 16 times that of hydrogen molecule. At room temperature, the rms speed of oxygen molecule is v. The rms speed of the hydrogen molecule at the same temperature will be

(a) $v/16$ (b) $v/4$

(c) 4 v (d) 16 v

140. The kinetic energy of one mole of argon at 127°C is (Boltzmann's constant $k = 1.381 \times 10^{-23}$J)

(a) 4000 J (b) 5982 J

(c) 4988 J (d) None of the above.

141. At what temperature, pressure remaining constant, will the rms velocity of a gas be half of its value at 0 °C?

(a) 0 °C (b) 32 °C

(c) –273 °C (d) –204 °C

142. The density of a gas is 6×10^{-2} kg/m³ and the root mean square velocity of the gas molecules is 500 m/s. The pressure exerted by the gas on the walls of the vessel is

(a) 5×10^3 N/m² (b) 1.2×10^{-4} N/m²

(c) 0.83×10^{-4} N/m² (d) 30 N/m²

143. Five molecules have speeds 1 km/s, 2 km/s, 3 km/s, 4 km/s, 5 km/s. The root mean square speed of the molecules is approximately

(a) 3.5 m/s (b) 5.5 m/s

(c) 2.8 m/s (d) 4.2 m/s

144. A gas molecule of mass m and velocity v is incident normally on a wall. After undergoing elastic collision, its momentum will be

(a) 2 mv (b) mv

(c) –mv (d) –2mv

145. When water solidifies to ice at 273 K, the kinetic energy of molecules

(a) increases (b) decreases

(c) remains the same (d) becomes zero.

146. In the perfect gas equation, $PV = nRT$, V stands for the volume of

(a) any quantity of gas

(b) one molecule of gas

(c) one gram molecule of gas

(d) one gram of gas.

147. The quantity of heat required to raise one mole through one degree for a monatomic gas at constant volume is

(a) 1.5 R (b) 3.5 R

(c) 2.5 R (d) 0.5 R

148. Which of the following number of molecules is contained in 6 gm of hydrogen at N.T.P.? (N = Avogadro's numbers)

(a) N (b) 2 N

(c) 3N (d) $\frac{N}{3}$

149. If γ be the ratio of specific heats of a perfect gas, the number of degrees of freedom of the molecule of the gas is

(a) $\frac{5}{2}(\gamma - 1)$ (b) $\frac{2}{\gamma - 1}$

(c) $\frac{\gamma - 1}{2}$ (d) $\frac{2}{5}(\gamma - 1)$

150. The average kinetic energy of a molecule of a gas at a temperature of 300 K is

(a) 6.21×10^{-21} J (b) 6.21×10^{-21} J

(c) 7.21×10^{-21} J (d) 7.21×10^{-21} J

151. If the speed of sound in a gas is the same as the root mean square speed of the molecules, then the speed of sound v_1 at a temperature T_1 is related to the speed of sound v_2 in the same gas at a temperature T_2 by

(a) $\frac{v_1}{v_2} = \frac{T_1}{T_2}$ (b) $\frac{v_2}{v_1} = \frac{T_2}{T_1}$

(c) $\frac{v_1}{v_2} = \left(\frac{\sqrt{T_1}}{T_2}\right)^{1/2}$ (d) $\frac{v_2}{v_1} = \left(\frac{\sqrt{T_1}}{T_2}\right)^{1/2}$

152. A gas is enclosed in a vessel at a constant temperature and at a pressure of 2.5 atmosphere and volume 4 litre. Due to leakage in the vessel after some time, the pressure is reduced to 2 atmosphere. As a result,

(a) $\frac{2}{5}$ of the gas has escaped out

(b) $\frac{4}{5}$ of the gas has escaped out

(c) $\frac{4}{5}$ of the gas remained in the vessel

(d) $\frac{1}{5}$ of the gas remained in the vessel.

153. An ideal gas A and a real gas B have their volumes increased from V to $2V$ under isothermal conditions. The increase in internal energy

(a) will be same in both A and B

(b) will be zero in both the gases

(c) of B will be more than that of A

(d) of A will be more than that of B.

154. Which of the following quantities remain same for ideal gases at the same temperature?
(a) The kinetic energy of 1 mole
(b) The kinetic energy of 1 gram
(c) The number molecules in 1 gram
(d) The volume of 1 gram of the gas.

155. Which of the following quantities is zero on an average for the molecules of an ideal gas in equilibrium?
(a) Kinetic energy (b) Momentum
(c) Density (d) Speed

156. The root mean square velocity of hydrogen molecules at 300 K is 1930 m/s. Then the rms velocity oxygen molecules at 1200 K will be
(a) 482 m/s (b) 965 m/s
(c) 1930 m/s (d) 3860 m/s

157. Temperature of an ideal gas is T K and average kinetic energy is 2.07×10^{23} J/molecule. Number of molecules in 1 litre gas at NTP will be
(a) 2.68×10^{22} (b) 2.68×10^{25}
(c) 2.68×10^{28} (d) 2.68×10^{18}

158. A sample of an ideal gas occupies a volume V at a pressure P and absolute temperature T. The mass of each molecule is m. The expression for the density of the gas is
(a) mkT (b) P/kT
(c) $\frac{P}{kT}$ (d) $\frac{Pm}{kT}$

159. Pressure of a gas decreases with increasing volume because
(a) molecules have less momentum
(b) molecules strike with less velocity
(c) molecules strike less frequency
(d) kinetic energy of molecules decrease.

160. The temperature of hydrogen gas at which the average speed of its molecules is equal to that of oxygen gas molecules at a temperature of 31°C is
(a) –216°C (b) –235°C
(c) –254°C (d) –264°C

161. A vessel contains a mixture of one mole of oxygen and two moles of nitrogen at 300 K. The ratio of the average rotational kinetic energy per O_2 molecule to that per N_2 molecule is
(a) 1 : 1
(b) 1 : 2
(c) 2 : 1
(d) It depends upon the moment of inertia of the two molecules.

162. The temperature of a mixture of one mole of helium and one mole of hydrogen is increased from 0°C to 100°C at constant pressure. The amount of heat delivered will be
(a) 600 cal (b) 1200 cal
(c) 1800cal (d) 3600 cal

163. The temperature of a gas consisting of rigid diatomic molecules is t °C. The angular velocity of a rotating molecule, if its moment of inertia is I can be written as
(a) $\left(\frac{3kt}{I}\right)^{1/2}$ (b) $\left[k\frac{(273+t)}{I}\right]^{1/2}$
(c) $\left[\frac{k(273+t)}{I}\right]^{1/2}$ (d) $\left[\frac{2k(273+t)}{I}\right]^{1/2}$

164. Which of the following gasses possess maximum rms velocity, all being at the same temperature?
(a) Oxygen (b) Air
(c) Carbon-dioxide (d) Hydrogen

165. A closed compartment containing gas is moving with some acceleration in horizontal direction.
Neglect effect of gravity. The pressure in the compartment is
(a) same everywhere (b) lower in the front side
(c) lower in the rear side (d) lower in the upper side.

166. A vessel contains 1 mole of O_2 gas at a temperature T. The pressure of the gas is P. An identical vessel containing one mole of He gas at temperature $2T$ has a pressure
(a) $\frac{P}{8}$ (b) P
(c) $8P$ (d) $2P$

167. Four molecules of a gas are having speeds 1, 4, 8 and 16 ms^{-1}. The root mean square velocity of the gas molecules is:
(a) 7.25 ms^{-1} (b) 52.56 ms^{-1}
(c) 84.25 ms^{-1} (d) 9.2 ms^{-1}

168. Temperature of an ideal gas is 300 K. The change in temperature of the gas when its volume changes from V to $2V$ in the process $P = aV$ (Here, a is a positive constant) is
(a) 900 K (b) 1200 K
(c) 600 K (d) 300 K

169. A gaseous mixture consists of 16 g of helium and 16 g of oxygen. The ratio C_p / C_v of the mixture is
(a) 1.59 (b) 1.62
(c) 1.4 (d) 1.54

170. Two thermally insulated vessel 1 and 2 are filled with air at temperature (T_1, T_2), volume (V_1, V_2) and pressure (P_1, P_2) respectively. If the valve joining the two vessels is opened, the temperature inside the vessel at equilibrium will be

(a) $T_1 + T_2$

(b) $\frac{(T_1 + T_2)}{2}$

(c) $\frac{T_1T_2(P_1V_1 + P_2V_2)}{P_1V_1T_1 + P_2V_2T_2}$

(d) $\frac{T_1T_2(P_1V_1 + P_2V_2)}{P_1V_1T_2 + P_2V_2T_1}$

171. The equation of state for 5 g of oxygen at a pressure P and temperature T, when occupying a volume V will be
(a) $PV = (5/32)\ RT$ (b) $PV = 5\ RT$
(c) $PV = (5/2)\ RT$ (d) $PV = (5/16)\ RT$

172. The gas having average speed four times as that of SO_2 (molecular mass 64) is
(a) He (molecular mass 4)
(b) O_2 (molecular mass 32)
(c) H_2 (molecular mass 2)
(d) CH_4 (molecular mass 16)

173. The temperature of a given mass is increased from 27°C to 327°C.The rms velocity of the molecules increases
(a) $\sqrt{2}$ times (b) 2 times
(c) $2\sqrt{2}$ times (d) 4 times

174. An ideal gas undergoes a quasi-static, reversible process in which its molar heat capacity C remains constant. If during this process, the relation of pressure P and volume V is given by PV^n = Constant, then n is given by

(a) $n = \frac{C - C_p}{C - C_v}$ (b) $n = \frac{C_p - C}{C - C_v}$

(c) $n = \frac{C - C_v}{C - C_p}$ (d) $n = \frac{C_p}{C_v}$

175. What is an ideal gas?
(a) One that consists of molecules
(b) A gas satisfying the assumptions of kinetic theory
(c) A gas obeying Maxwell distribution law in the distribution of speed of its molecules
(d) A gas consisting of massless particles.

176. An ideal gas, undergoing adiabatic change, has which of the following pressure temperature relationship?
(a) $P^{\gamma} T^{1-\gamma}$ = Constant (b) $P^{1-\gamma} T^{\gamma}$ = Constant
(c) $P^{\gamma-1} T^{\gamma}$ = Constant (d) $P^{\gamma} T^{\gamma-1}$ = Constant.

177. Pressure of an ideal gas is increased by keeping temperature constant. What is the effect on kinetic energy of molecules?
(a) Increase (b) Decrease
(c) No change (d) Cannot be determined.

178. Two balloons are filled, one with pure He-gas and the other with air. If the pressure and temperature of these balloons are same, then the number of molecules per unit volume is
(a) more in the He gas filled balloon
(b) same in both balloons
(c) more in air filled balloon
(d) in the ratio 1 : 4.

179. The velocity of the molecules of a gas at temperature 120 K is v. At what temperature will the velocity be $2v$?
(a) 120 K (b) 240 K
(c) 480 K (d) 1120 K

180. At what temperature the molecules of nitrogen will have same rms velocities as the molecules of oxygen at 127°C?
(a) 457°C (b) 273°C
(c) 350°C (d) 77°C

181. An ideal gas expands in such a manner that its pressure and volume comply with the condition PV^2 = constant. During this process, the gas is
(a) heated
(b) cooled
(c) neither heated nor cooled
(d) first heated then cooled.

182. Energy of all molecules of a monatomic gas have a volume V and pressure P is (3/2) PV. The total translational kinetic energy of all molecules of a diatomic gas at the same pressure and volume is
(a) (1/2) PV (b) (3/2) PV
(c) (5/2) PV (d) 3 PV

183. Air is pumped into an automobile tube up to a pressure of 200 KPa in the morning when the air temperature is 22°C. During the day the temperature rises to 42°C and the tube expands by 2%. The pressure of the air in the tube at this temperature will be approximately
(a) 212 kPa (b) 209 kPa
(c) 206 kPa (d) 200 kPa.

184. An ideal gas at 27°C is compressed adiabatically to 8/27 of its original volume. If $\gamma = 5/3$, then the rise in temperature is
(a) 575 K (b) 450 K
(c) 225 K (d) 375 K.

185. For an ideal gas of diatomic molecules, where C_P and C_V are the two specific heats of the gas and R is gas constant then

(a) $C_p = \frac{5}{2}R$ (b) $C_v = \frac{3}{2}R$

(c) $C_p - C_v = 2R$ (d) $C_p = \frac{7}{2}R$

186. Oxygen is 16 times heavier than hydrogen. Equal volumes of hydrogen and oxygen are mixed. The ratio of speed of sound in the mixture to that in hydrogen

(a) $\sqrt{8}$ (b) $\sqrt{\frac{2}{17}}$

(c) $\sqrt{\frac{1}{8}}$ (d) $\sqrt{\frac{32}{17}}$

187. Which one of the following graphs represents the behaviour or an ideal gas at constant temperature?

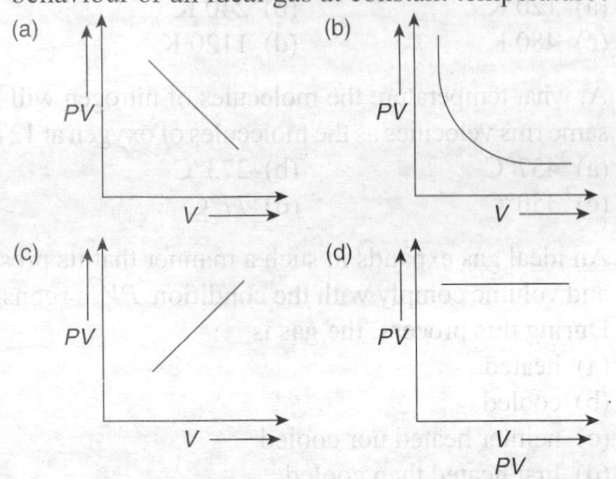

188. The figure below shows the plot of $\frac{PV}{nT}$ versus P for oxygen gas at two different temperatures.

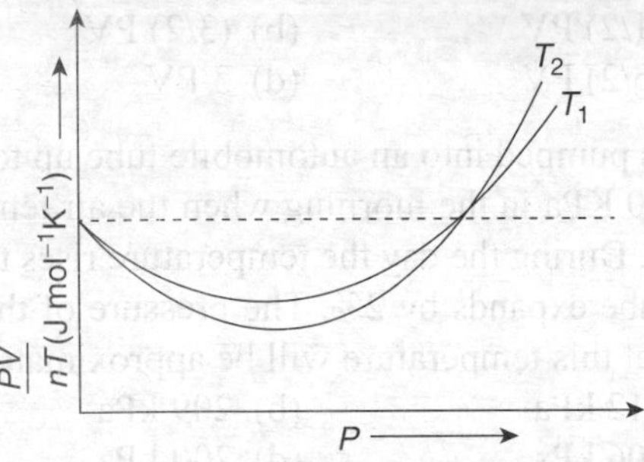

Read the following statements concerning the above curves:

(I) The dotted line corresponds to the 'ideal' gas behavior.

(II) $T_1 > T_2$

(III) The value of $\frac{PV}{nT}$ at the point where the curves meet on the y-axis is the same for all gases.

Which of the above statements is true?

(a) (i) only (b) (i) and (ii) only

(c) All the above. (d) None of the above.

189. At which of the following temperatures would the molecules of a gas have twice the average kinetic energy they have at 20°C

(a) 313°C (b) 373°C

(c) 393°C (d) 586°C

190. If a give mass of a gas occupies a volume 100 m^3 at one atmospheric pressure and temperature of 100°C, what will be its volume at 4 atmospheric pressure, the temperature being the same?

(a) 25 cm^3 (b) 0

(c) 50 cm^3 (d) 5 cm^3

191. Two gas A and B having same pressure P, volume V and absolute temperature T are mixed. If the mixture has the volume and temperature as V and T respectively, then the pressure of the mixture is

(a) 2 P (b) P

(c) P/2 (d) 4 P

192. The tyre of a motor car contains air at 15°C. If the temperature increases to 35°C, the approximate percentage increase in pressure is (ignore to expansion of tyre)

(a) 7 (b) 9

(c) 11 (d) 13

193. If 4 moles of an ideal monatomic gas at temperature 400 K is mixed with 2 moles of another ideal monatomic gas at temperature 700 K, the temperature of the mixture is

(a) 550 °C (b) 500 °C

(c) 550 K (d) 500 K

194. Assertion (A): At 300 K, kinetic energy of 16 g of methane is equal to the kinetic energy of 32 of oxygen.

Reason (R): At constant temperature, kinetic energy of one mole of all gases is equal.

The correct answer is

(a) Both (A) and (R) are true and (R) is the correct explanation of A.

(b) Both (A) and (R) are true and (R) is not the correct explanation of A.

(c) (A) is true but (R) is not true.

(d) (A) is not true but (R) is true.

195. The van der Waal's equation for a gas is $\left(P+\frac{a}{V^2}\right)(V-b)=nRT$, where P, V, R, T and n represent the pressure, volume, universal gas constant, absolute temperature and number of moles of a gas respectively. a and b are constants. The ratio b/a will have the following dimensional formula

(a) $\left[M^{-1}L^{-2}T^{-2}\right]$ (b) $\left[M^{-1}L^{-1}T^{-1}\right]$

(c) $\left[ML^{2}T^{2}\right]$ (d) $\left[MLT^{-2}\right]$

196. The mass of oxygen gas occupying a volume of 11.2 litres at a temperature 27°C and pressure of 760 mm of mercury in kilogram is (Molecular weight of oxygen = 32)

(a) 0.001456 (b) 0.01456

(c) 0.1456 (d) 0.1456

197. During an adiabatic process, the pressure of a gas is found to be proportional to the cube of its absolute temperature. The ratio $\frac{C_p}{C_v}$ for the gas is

(a) $\frac{4}{3}$ (b) 2
(c) $\frac{5}{3}$ (d) $\frac{3}{2}$

198. If average velocity becomes 4 times, then what will be the effect on rms velocity at that temperature?
(a) 1.4 times (b) 4 times
(c) 2 times (d) 1/4 times

199. The phenomenon of Brownian movement may be taken as evidence of
(a) kinetic theory of matter
(b) corpuscular theory of light
(c) electromagnetic theory of radiation
(d) photoelectric phenomenon.

200. If the mass of a He molecule is 4 times that of hydrogen molecule, then the mean speed of He is
(a) 2 times of H mean value
(b) 1/2 times of H mean value
(c) 4 times of H mean value
(d) same as that of H.

201. In gases of diatomic molecules, the ratio of two specific heats of gases C_p / C_v is
(a) 1.21 (b) 1.33
(c) 2.80 (d) 1.40

202. The pressure and density of a diatomic gas $\left(\gamma = \frac{7}{5}\right)$ change adiabatically from (P_1, ρ_1) to (P_2, ρ_2). If $\frac{\rho_1}{\rho_2} = 32$ then $\frac{P_1}{P_2}$ should be
(a) 16 (b) 32
(c) 64 (d) (128)

Questions Asked in NEET/CBSE Medicals

203. Increase in temperature of a gas filled in a container would lead to
(a) decrease in intermolecular distance
(b) increase in its mass
(c) increase in its kinetic energy
(d) decrease in its pressure.

204. At what temperature will the rms speed of oxygen molecules become just sufficient for escaping from the Earth's atmosphere? (Given: Mass of oxygen molecule $(m) = 2.76 \times 10^{-26}$ kg, Boltzmann's constant $k_B = 1.38 \times 10^{-23}$ J)
(a) 2.508×10^4 K (b) 8.360×10^4 K
(c) 5.016×10^4 K (d) 1.254×10^4 K

205. A gas mixture consists of 2 moles of O_2 and 4 moles of Ar at temperature T. Neglecting all vibrational modes, the total internal energy of the system is
(a) 15 RT (b) 9 RT
(c) 11 RT (d) 4 RT

206. One mole of an ideal monatomic gas undergoes a process described by the equation PV^3 = constant. The heat capacity of the gas during this process is
(a) $\frac{3}{2}R$ (b) $\frac{5}{2}R$
(c) $2R$ (d) R

207. A given sample of an ideal gas occupies a volume V at a pressure P and absolute temperature T. The mass of each molecule of the gas is m. Which of the following gives the density of the gas?
(a) $P/(kT)$ (b) $Pm/(kT)$
(c) $P/(kTV)$ (d) mkt

208. The molecules of a given mass of a gas have r.m.s. velocity of 200 m s^{-1} at 27°C and 1.0×10^5 N m^{-2} pressure. When the temperature and pressure of the gas are respectively, 127°C and 0.05×10^5 N m^{-2}, the r.m.s. velocity of its molecules in m s^{-1} is
(a) $\frac{100\sqrt{2}}{3}$ (b) $\frac{100}{3}$
(c) $100\sqrt{2}$ (d) $\frac{400}{\sqrt{3}}$

209. Two vessels separately contain two ideal gases A and B at the same temperature, the pressure of A being twice that of B. Under such conditions, the density of A is found to be 1.5 times the density of B, The ratio of molecular weight of A and B is
(a) 2 (b) 1 / 2
(c) 2 / 3 (d) 3 / 4

210. A monatomic gas at a pressure P, having a volume V expands isothermally to a volume $2V$ and then adiabatically, to a volume $16V$. The final pressure of the gas is (Take γ= 5/3)
(a) 64 P (b) 32 P
(c) P/64 (d) 16 P

211. The mean free path of molecules of a gas, (radius r) is inversely proportional to
(a) r^3 (b) r^2
(c) r (d) $\sqrt{r}$

212. During an adiabatic process, the pressure of a gas is found to be proportional to the cube of its temperature. The ratio C_P/C_V for the gas is
(a) 5/3 (b) 3/2
(c) 4/3 (d) 2

213. The amount of heat energy required to raise the temperature of 1 g of Helium at NTP, from T_1K to T_2K is
(a) $\frac{3}{4}N_a k_B(T_2-T_1)$ (b) $\frac{3}{4}N_a k_B(T_2/T_1)$
(c) $\frac{3}{8}N_a k_B(T_2-T_1)$ (d) $\frac{3}{2}N_a k_B(T_2-T_1)$

214. Which of the following relations does not give the equation of an adiabatic process, where terms have their usual meaning?
(a) $P^{1-\gamma}T^{\gamma}$= constant (b) PV^{γ}= constant
(c) $TV^{\gamma-1}$ = constant (d) $P^{\gamma}T^{1-\gamma}$= constant

215. In a vessel, the gas is at pressure P. If the mass of all the molecules is halved and their speed is doubled, then the resultant pressure will be
(a) $2P$ (b) P
(c) $P/2$ (d) $4P$

216. A mass of diatomic gas (γ = 1.4) at a pressure of 2 atmospheres is compressed adiabatically so that its temperature rises from 27°C to 927°C. The pressure of the gas in the final state is
(a) 8 atm (b) 28 atm
(c) 68.7 atm (d) 256 atm.

217. A monatomic gas at pressure P_1 and volume V_1 is compressed adiabatically $\frac{1}{8}$th of its original volume. What is the final pressure of the gas?
(a) $64P_1$ (b) P_1
(c) $16P_1$ (d) $32P_1$

218. At 10°C the value of the density of a fixed mass of an ideal gas divided by its pressure is x. At 110°C this ratio is
(a) $\frac{10}{110}x$ (b) $\frac{283}{383}x$
(c) x (d) $\frac{383}{283}x$

219. The molar specific heat at constant pressure of an ideal gas is (7/2)R. The ratio of specific heat at constant pressure to that at constant volume is
(a) 9/7 (b) 7/5
(c) 8/7 (d) 5/7

Answer Keys

EXERCISES

1. (d)	2. (b)	3. (a)	4. (d)	5. (c)	6. (c)	7. (d)	8. (b)	9. (b)	10. (a)
11. (d)	12. (b)	13. (b)	14. (b)	15. (c)	16. (c)	17. (a)	18. (d)	19. (b)	20. (c)
21. (b)	22. (a)	23. (a)	24. (a)	25. (a)	26. (d)	27. (a)	28. (b)	29. (b)	30. (a)
31. (b)	32. (c)	33. (d)	34. (a)	35. (d)	36. (a)	37. (d)	38. (a)	39. (a)	40. (c)
41. (b)	42. (b)	43. (a)	44. (d)	45. (a)	46. (a)	47. (c)	48. (c)	49. (b)	50. (d)
51. (b)	52. (a)	53. (c)	54. (c)	55. (c)	56. (b)	57. (a)	58. (b)	59. (c)	60. (b)
61. (c)	62. (c)	63. (b)	64. (b)	65. (a)	66. (c)	67. (d)	68. (b)	69. (b)	70. (d)
71. (a)	72. (b)	73. (c)	74. (a)	75. (d)	76. (d)	77. (b)	78. (d)	79. (d)	80. (a)
81. (b)	82. (a)	83. (a)	84. (d)	85. (c)	86. (b)	87. (d)	88. (c)	89. (c)	90. (b)
91. (c)	92. (b)	93. (b)	94. (b)	95. (b)	96. (d)	97. (b)	98. (c)	99. (d)	100. (b)
101. (b)	102. (b)	103. (c)	104. (a)	105. (c)	106. (d)	107. (d)	108. (c)	109. (b)	110. (b)
111. (d)	112. (b)	113. (c)	114. (a)	115. (b)	116. (a)	117. (c)	118. (b)	119. (c)	120. (c)
121. (c)	122. (a)	123. (d)	124. (c)	125. (c)	126. (b)	127.(c)	128. (d)	129. (c)	130. (a)
131. (b)	132. (a)	133. (a)	134. (a)	135. (c)	136. (c)	137. (a)	138. (d)	139. (c)	140.(c)
141. (d)	142. (a)	143. (a)	144. (c)	145.(c)	146. (a)	147. (a)	148. (c)	149. (b)	150. (b)
151. (c)	152. (c)	153. (b)	154. (a)	155. (a)	156. (b)	157. (a)	158. (d)	159. (c)	160. (c)
161. (a)	162. (b)	163. (c)	164. (d)	165. (a)	166. (d)	167. (d)	168. (a)	169. (b)	170. (a)
171. (a)	172. (a)	173. (a)	174. (a)	175. (b)	176. (b)	177. (c)	178. (b)	179. (c)	180. (d)
181. (b)	182. (b)	183. (b)	184. (d)	185. (d)	186. (b)	187. (d)	188. (c)	189. (a)	190. (a)
191. (a)	192. (a)	193. (d)	194. (a)	195. (a)	196. (b)	197. (d)	198. (b)	199. (a)	200. (b)
201. (d)	202. (d)								

Questions Asked in NEET/CBSE Medicals

203. (c)	204. (b)	205. (c)	206. (d)	207. (b)	208. (d)	209. (d)	210. (c)	211. (b)	212. (b)
213. (c)	214. (d)	215. (a)	216. (d)	217. (d)	218. (b)	219. (b)			

Solutions

1. (d)

The system is thermally isolated; therefore, the total internal energy in the two partitions remains constant. Let n_1 be the number of moles of the gas in one partition and n_2 the number of moles in other partition. Then

$$U = \frac{3}{2}(n_1 + n_2)RT$$

$$\frac{3}{2}(n_1 + n_2)RT = \frac{3}{2}n_1RT_1 + \frac{3}{2}n_2RT_2$$

$$(n_1 + n_2)T = n_1T_1 + n_2T_2$$

$$T = \frac{n_1T_1 + n_2T_2}{n_1 + n_2}$$

Substituting the value of

$$n_1 = \frac{P_1V_1}{RT_1} \text{ and } n_2 = \frac{P_2V_2}{RT_2}$$

$$T = \frac{(P_1V_1 + P_2V_2)/R}{\frac{P_1V_1}{RT_1} + \frac{P_2V_2}{RT_2}}$$

$$T = \frac{\frac{(P_1V_1 + P_2V_2)}{R}}{\frac{1}{R}\left(\frac{P_1V_1}{T_1} + \frac{P_2V_2}{T_2}\right)}$$

$$T_1T_2 = \frac{T(P_1V_1T_2 + P_2V_2T_1)}{P_1V_1 + P_2V_2}$$

2. (b)

Vapour pressure does not depend on the amount of substance. It depends on the temperature alone.

3. (a)

KE for one mole of a gas $= \frac{3}{2}RT$

$$\frac{1}{2}M_AC_A^2 = \frac{3}{2}RT_A$$

$$\frac{1}{2}M_BC_B^2 = \frac{3}{2}RT_B$$

$$\left(\frac{C_A}{C_B}\right)^2 = \frac{T_A}{T_B}\cdot\frac{M_B}{M_A} = 4$$

$$\frac{C_A}{C_B} = 2$$

4. (d)

$PV = nRT$

$$PV = \frac{N}{N_0}RT$$

where N is number of molecules of the gas.

N_0 = Avogadro's number

$$P\frac{V}{N} = \frac{R}{N_0}T = kT$$

$$\frac{V}{N} = \frac{kT}{P}$$

$$\frac{N}{V} = \frac{P}{kT} = \frac{10^6 \times kT}{kT} = 10^6$$

5. (c)

$$\text{KE per mole} = \frac{3}{2}RT$$

$$= \frac{3}{2}\times 8.3 \times 273$$

$$= 3.4 \times 10^3 \text{ J}$$

6. (c)

$$KE_1 = \frac{3}{2}RT_1$$

$$KE_2 = \frac{3}{2}RT_2$$

$$\frac{KE_2}{KE_1} = \frac{T_2}{T_1}$$

$$\frac{T_2}{T_1} = \frac{1}{2}$$

$$T_2 = \frac{T_1}{2} = \frac{(273 + 27)}{2} = 150 \text{ K}$$

7. (d)

Since the gases are at constant temperature, they have the same mean translational kinetic energy.

$$\frac{1}{2}m_Ac_A^2 = \frac{1}{2}m_Bc_B^2$$

$$\frac{c_A}{c_B} = \sqrt{\frac{m_B}{m_A}} = \sqrt{4} = 2$$

8. (b)

$$P_1 = \frac{1}{3}\frac{M_1C_1^2}{V}$$

$$P_2 = \frac{1}{3}\frac{M_2C_2^2}{V}$$

$$\frac{P_2}{P_1} = \frac{M_2}{M_1}\left(\frac{C_2}{C_1}\right)^2$$

$$= \frac{1}{2}\times(2)^2 = 2$$

$$P_2 = 2P_1 = 2P_0$$

9. (b)

$$\frac{C_O}{C_H} = \sqrt{\frac{M_H}{M_O}} = \sqrt{\frac{2}{32}} = \frac{1}{4}$$

$$C_O = 2 \times \frac{1}{4} = 0.5 \text{ km/s}$$

10. (a)

$PV = nRT$

$$n = \frac{2.2}{44} = \frac{1}{20}$$

According to Avogadro's law 1 mole of every gas has a volume of 22.4 litre.

Hence, the volume $= \dfrac{22.4}{20} = 1.12$

Now $\quad P_2V_2 = P_1V_1$

$$2V_2 = 1 \times 1.12$$

$$V_2 = 0.56 \text{ litre}$$

11. (d)

$$\frac{V_{\text{rms}_{\text{He}}}}{V_{\text{rms}_{\text{Ar}}}} = \frac{\sqrt{\dfrac{3RT}{m_{\text{He}}}}}{\sqrt{\dfrac{3RT}{m_{\text{Ar}}}}} = \sqrt{\frac{m_{\text{Ar}}}{m_{\text{He}}}} = \sqrt{\frac{40}{4}} = \sqrt{10} \approx 3.16$$

12. (b)

Volume V_2 after heating

$$V_2 = \frac{m}{\rho} = \frac{12}{6 \times 10^{-4}} = 2 \times 10^4 \text{ cc}$$

Now, $\dfrac{V_2}{V_1} = \dfrac{T_2}{T_1}$

$$\frac{2 \times 10^4}{4 \times 10^{-3} \times 10^6} = \frac{T_2}{290}$$

$$5 = \frac{T_2}{290}$$

$$\Rightarrow \quad T_2 = 5 \times 290$$

$$= 1450 \text{ K}$$

The nearest option is b.

13. (b)

14. (b)

$$\frac{C_H}{C_O} = \sqrt{\frac{T_H}{T_O} \times \frac{M_O}{M_H}}$$

$$1 = \sqrt{\frac{T_H}{320} \times \frac{32}{2}}$$

$$= \sqrt{\frac{T_H}{20}}$$

$$\Rightarrow T_H = 20$$

15. (c)

KE per mole at T_1 temperature $= \dfrac{3}{2}RT_1$

For n_1 mole $= \dfrac{3}{2}RT_1 \cdot n_1$

Similarly KE of n_2 mole of the other gas $= \dfrac{3}{2}RT_2 \cdot n_2$

Total energy for $(n_1 + n_2)$ mole $= \dfrac{3}{2}R\,(n_1\,T_1 + n_2\,T_2)$

Let equilibrium temperature be T.

Then total energy

$$\frac{3}{2}RT(n_1 + n_2)$$

$$\Rightarrow \frac{3}{2}RT(n_1 + n_2) = \frac{3}{2}R(n_1T_1 + n_2T_2)$$

$$\Rightarrow \quad T = \frac{n_1T_1 + n_2T_2}{n_1 + n_2}$$

16. (c)

$$C_V = \frac{R}{0.67} = 1.5R = \frac{3}{2}R.$$

17. (a)

$$\gamma = \frac{n_1c_{p_1} + n_2c_{p_2}}{n_1c_{v_1} + n_2c_{v_2}}$$

$$= \frac{\frac{5}{2}R + \frac{7}{2}R}{\frac{3}{2}R + \frac{5}{2}R}$$

$$= \frac{3}{2}$$

18. (d)

Thermal energy of N molecule

$$= N\left(\frac{3}{2}kT\right)$$

$$= \frac{N}{N_A}\frac{3}{2}RT$$

$$= \frac{3}{2}(nRT) = \frac{3}{2}PV$$

$$= \frac{3}{2}P\left(\frac{m}{\rho}\right) = \frac{3}{2} \times 4 \times 10^4 \times \frac{2}{8} = 1.5 \times 10^4 \text{ J}$$

Order will be 10^4.

19. (b)

According to ideal gas equation $PV = nRT$

$$PV = \frac{m}{M}RT, P = \frac{\rho}{M}RT \text{ or } \frac{\rho}{P} = \frac{M}{RT} \text{ or } \frac{\rho}{P} \propto \frac{1}{T}$$

Here, $\dfrac{P}{\rho}$ represents the slope of the graph.

Hence $T_1 > T_2$.

20. (c)

$$P_1V_1 = P_2V_2$$

$$P_2 = 200 \times \frac{V_1}{V_2}$$

$$= 200 \times \frac{2V_1}{V_1}$$

$$= 400 kP_a$$

21. (b)

$$P_1V_1 = P_2V_2$$

$$15 \times 10^5 \times .05 = 1.01 \times 10^5 \times V_2$$

$$V_2 = \frac{15 \times .05}{1.01} \simeq 0.75 \text{ m}^3$$

Number of cylinders required

$$= \frac{500}{0.75} \simeq 666$$

Closest option is (b).

22. (a)

23. (a)

Change in internal energy,

$$\Delta U = nC_V \Delta T$$

$$= \frac{nR\Delta T}{(\gamma - 1)} = \frac{nP\Delta V}{(\gamma - 1)} = \frac{nP(2V - V)}{(\gamma - 1)}$$

For one mole, $n = 1$.

$$\Delta U = PV/(\gamma - 1)$$

24. (a)

25. (a)

$Q = nC_V\Delta T$ as gas in closed vessel

$$Q = \frac{15}{28} \times \frac{5 \times R}{2} \times (4T - T)$$

$$= \frac{15}{28} \times \frac{5 \times R}{2} \times 3(300)$$

$$\Rightarrow Q = 10021.33 \simeq 10 \text{ kJ}$$

26. (d)

Kinetic energy per g mole $E = \frac{f}{2} RT$

If nothing is said about gas then we should calculate the translational kinetic energy

i.e., $E_{Trans} - \frac{3}{2} RT = \frac{3}{2} \times 8.31 \times (273 + 0) = 3.4 \times 10^3$ J

27. (a)

28. (b)

Average kinetic energy is proportional to temperature. Therefore,

$$\frac{K_1}{K_2} = \frac{300}{350} = \frac{6}{7} = 6:7$$

29. (b)

As the density increases, the intermolecular distance decreases.

30. (a)

Vander Wall's equation is $\left(P + \frac{a}{V^2}\right)(V - b) = \text{constant}.$

$\frac{a}{V^2}$ factor is due to inter-molecular forces; as pressure increases, the volume decreases.

31. (b)

32. (c)

$$\frac{PV}{RT} = n$$

33. (d)

34. (a)

35. (d)

$$V = \sqrt{\frac{\gamma P}{\rho}}$$

$$P = \frac{1}{3} \frac{mnc^2}{V}$$

$$= \frac{1}{3} \frac{M}{V} c^2$$

$$= \frac{1}{3} \rho c^2$$

$$c = \sqrt{\frac{3P}{\rho}}$$

$$\frac{v}{c} = \sqrt{\frac{\gamma P}{\rho} \times \frac{\rho}{3P}} = \sqrt{\gamma/3}$$

36. (a)

Here, $\gamma = 1 + \frac{2}{f}$

where f is the degree of freedom

$$\frac{2}{f} = \gamma - 1 \text{ or } f = \frac{2}{\gamma - 1}$$

37. (d)

38. (a)

39. (a)

As $PV = nRT$

$$n = \frac{m}{\text{molecular mass}} = \frac{5}{32}$$

$$\Rightarrow PV = \left(\frac{5}{32}\right) RT$$

40. (c)

$$KE \propto T$$

41. (b)

$$C_v = \frac{1}{2}nR$$

$$C_p = \frac{1}{2}nR + R$$

$$\gamma = \frac{C_p}{C_v} = \frac{\frac{1}{2}nR + R}{\frac{1}{2}nR} = \frac{n+2}{n}$$

42. (b)

$$c \propto T$$

43. (a)

44. (d)

45. (a) 3 translational, 3 rotational.

46. (a)

$$\frac{C_N}{C_O} = \sqrt{\frac{T_N}{T_O} \times \frac{M_O}{M_N}}$$

$$1 = \sqrt{\frac{T_N}{400} \times \frac{32}{28}}$$

$$T_N = \frac{400 \times 28}{32} = 350 \text{ K}$$

In °C = 350 − 273 = 77°C

47. (c)

For argon (Monatomic)

$$U_1 = \frac{3}{2}RT \times 4 = 6RT$$

For oxygen (Diatomic)

$$U_2 = 2 \times \frac{5}{2}RT = 5RT$$

$$\text{Total } U_1 + U_2 = 6RT + 5RT = 11RT$$

48. (c)

Translational degree of freedom corresponds to motion along three axes X, Y and Z, which is same for all gases.

49. (b)

$TV^{\gamma-1}$ = Constant (adiabatic).

$$\therefore\ (300)(V_0)^{3/2} = (V_f)^{2/3} T$$

$$T = 300\left(\frac{27}{8}\right)^{2/3} = 300 \times \left(\frac{3}{2}\right)^{3\times\frac{2}{3}} = \frac{300 \times 9}{4} = 675 \text{ K}$$

Temperature rise = 675 − 300 = 375 K

50. (d)

51. (b)

The molecules of the gas are moving with different velocities in all directions.

52. (a)

The temperature of the gas is directly proportional to the mean translation energy of gas molecules.

$$\frac{1}{2}mv_{rms}^2 \propto T$$

53. (c)

No intermolecular force between the molecules. It is one of the assumptions of kinetic theory of gases.

54. (c)

Mean free path $l = \frac{1}{\sqrt{2}\pi n v d^2}$

$\Rightarrow\ l \propto \frac{1}{d^2}$ where d is molecular diameter

55. (c)

$$\frac{1}{2}m\,V_{rms}^2 \propto T$$

56. (b)

$$\gamma = \frac{C_p}{C_v} = \frac{\frac{5}{2}R}{\frac{3}{2}R} = \frac{5}{3} = 1.67$$

57. (a)

All the molecules are at the same temperature T.

Therefore, $\frac{1}{2}mv_{rms}^2 \propto T$

58. (b)

As the temperature increases, the average velocity increases. So, the collisions are faster.

59. (c)

At constant temperature,

$$P_1V_1 = P_2V_2$$

$$P_2 = \frac{PV \times 100}{95V} = \frac{20P}{19}$$

$$\Delta P = \frac{20P}{19} - P = \frac{1}{19}P$$

$$\% \text{ increase} = \frac{P \times 100}{19 \times P} = 5.26\%$$

60. (b)

Ideal gas equation is

$$PV = nRT$$

$$x = nR$$

where x is proportional to number of moles ∝ number of molecules.

61. (c)

Brownian motion

62. (c)
Comparing the two isotherms at T_1 and T_2

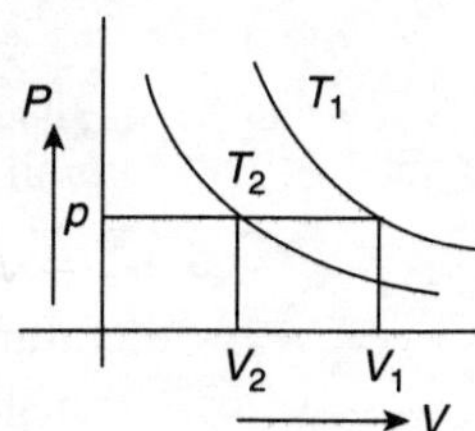

$$\frac{PV_2}{T_2} = \frac{PV_1}{T_1}$$

$$V_1 > V_2 \Rightarrow T_1 > T_2$$

63. (b)
According to law of equipartition of energy, the energy per degree of freedom is $\frac{1}{2}kT$. For a polyatomic gas with n degrees of freedom, the mean energy per molecule $= \frac{1}{2}nkT$

64. (b)

$$\gamma = \frac{C_P}{C_V} = \frac{3}{2} \Rightarrow C_V = \frac{2}{3}C_P$$

$$C_P - C_V = R\ \text{J}^{-1} \quad \text{or} \quad C_P - \frac{2}{3}C_P = R\ \text{J}^{-1}$$

or $$\frac{C_P}{3} = R\ \text{J}^{-1} \quad \text{or} \quad C_P = 3R\ \text{J}^{-1}$$

65. (a)
Volume is constant. Therefore, we have the relation

$$\frac{P}{T} = \text{constant}$$

$$\Rightarrow \quad \frac{P}{T} = \frac{P\left(1 + \frac{0.4}{100}\right)}{T+1}$$

$$\Rightarrow \quad T + 1 = T + \frac{0.4}{100} \times T$$

$$\Rightarrow \quad T = \frac{100}{0.4} = 250\ \text{K}$$

$$T = \frac{\Delta T}{\Delta P} \times P$$

$$= \frac{1}{\frac{0.4}{100}} \times 1$$

$$= \frac{1000}{4} = 250\ \text{K}$$

66. (c)

67. (d)

$$T_1 = 273 \quad T_2 = ?$$

$$\frac{v_2}{v_1} = \sqrt{\frac{T_2}{273}}$$

$$2 = \sqrt{\frac{T_2}{273}}$$

$$\Rightarrow \quad T_2 = 4 \times 273 = 1092\ \text{K}$$

$$1092 - 273 = 819°\text{C}$$

68. (b)

$$P = \frac{1}{3}\frac{mn}{V}v_{rms}^2$$

If volume is constant, then

$$P \propto v_{rms}^2$$

69. (b)
The collisions take place after the molecules travel a distance $2l$

$$2l = 5 \times 2 = 10\ \text{cm}$$

The number of collisions /second

$$= \frac{50}{10} = 5$$

70. (d)
Heat Q required to raise the temperature by 10 k is equal to
$Q_p = C_p.10k$
Similarly $Q_v = C_v.10k$

$$Q_p - Q_v = (C_p - C_v)10$$

$$= R \cdot 10$$

$$Q_v = Q_p - 10R$$

$$= 207 - 10 \times 8.3$$

$$= 124\ \text{J}$$

71. (a)

$$\frac{1}{3}Nmc^2 = \frac{2}{3}\left(\frac{1}{2}Nm\right)c^2 = \frac{2}{3}E$$

72. (b)
The average translational kinetic energy is $\frac{1}{2}mv_{\text{rms}}^2$.

73. (c)
According to Dalton's law of partial pressure, we have
$P = P_1 + P_2 + P_3$

74. (a)

75. (d)

$$\frac{1}{2}mv_{\text{rms}}^2 \propto T$$

76. (d)

$C_P - C_V = R$ for all gases.

77. (b)
C_p for diatomic gas is

$$\frac{7}{2}R = \frac{7}{2} \times 8.31\ \text{J/mol K}$$

$$= \frac{7 \times 8.31}{2 \times 4.2} \approx 6.74\ \text{cal/mol K}$$

78. (d)
Volume is constant

Therefore $\frac{P}{T}$ = Constant

$$T = \frac{\Delta T}{\Delta P} \times P = \frac{10}{2/100} = 500 \text{ K}$$

In °C, $T = 500 - 273 = 227°C$

79. (d)
See the inter-molecular graph (F vs. r)

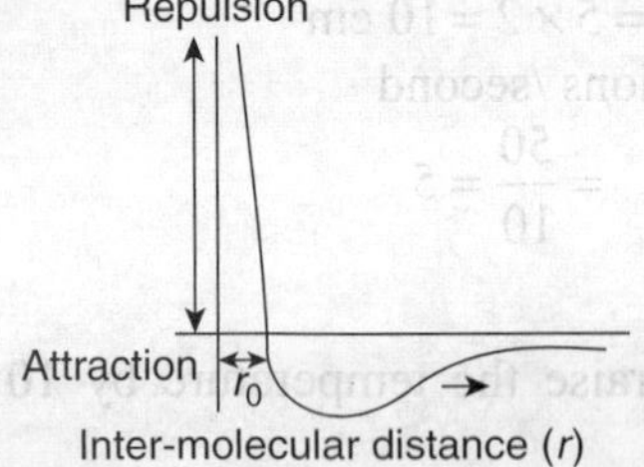

80. (a)

$$C_{max} = \sqrt{\frac{3RT}{M}} = \sqrt{\frac{3 \times 8.31 \times 300}{2 \times 10^{-3}}}$$

$$= \sqrt{3739.5} = 1933.78 \text{ m/s}$$

$$C_{rms} \propto \frac{1}{\sqrt{M}}$$

81. (b)
At constant volume

$$P \propto T$$

82. (a)

83. (a)
Assumption of kinetic theory of gases.

84. (d)

85. (c)

$$\frac{v_2}{v_1} = \sqrt{\frac{T_2}{T_1}}$$

$$(2)^2 = \frac{T_2}{T_1} \Rightarrow T_2 = 4T_1$$

86. (b)

$$\frac{1}{2}M_A v_A^2 = 3/2RT$$

$$\frac{1}{2}M_B v_B^2 = \frac{3}{2}RT$$

$$\Rightarrow \quad \frac{v_B}{v_A} = \sqrt{\frac{M_A}{M_B}} = \sqrt{8} = 2\sqrt{2}$$

87. (d)
Let the molecular mass of the oxygen gas be M_1.
The molecular mass of oxygen be M_2.

Oxygen is diatomic, therefore $M_2 = \frac{M_1}{2}$

$$\frac{V_2}{V_1} = \sqrt{\frac{M_1}{M_2} \times \frac{T_2}{T_1}}$$

$$= \sqrt{2 \times 2} = 2 \quad \{\text{Let } V_1 = V\}$$

$$V_2 = 2V_1 = 2V$$

88. (c)
Momentum imparted by N molecules (initial momentum) = Nmv
The final momentum is also equal to Nmv (in reflected direction)
Resolving along the wall and perpendicular to the wall, respectively
The components are $Nmv \sin \theta$; $Nmv \cos \theta$

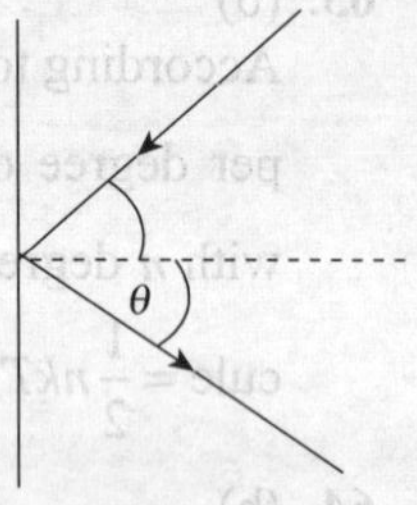

The component of reflected molecules are $Nmv \sin \theta$; $-Nmv \cos \theta$
Considering the component perpendicular to the wall,
Change in momentum per second

$$= Nmv\cos\theta - (-Nmv\cos\theta)$$

$$= 2Nmv\cos\theta$$

$$\text{Pressure} \quad = \frac{2Nmv\cos\theta}{A}$$

89. (c)
Avogadro's Law → Equal volume of all gases have equal number of molecules under the similar conditions of temperature and pressure.

90. (b)
Mean free path $\lambda \,\alpha \frac{1}{P}$; If λ is doubled then P becomes half.

91. (c)
Temperature of the gas is concerned only with its disordered motion. It is no way concerned with its ordered motion.

92. (b)

$$dU = \frac{5}{2}RT$$

$$= \frac{5}{2}PV$$

(Ideal gas equation for one mole is $PV = RT$)

93. (b)
In case of heat supplied to gas at constant pressure, it is spent in two ways:
(i) It increases the temperature of the gas.
(ii) It does work against external pressure.
However, the heat supplied at constant volume is spent in increasing the temperature only
Therefore, $C_p > C_v$

94. (b)
The three degrees of freedom of an ideal monoatomic gas corresponds to kinetic energy.

95. (b)
Numbers of translational degrees of freedom are same for all types of gases, that is, 3.

96. (d)

97. (b)

$$Q_1 = nC_P\Delta T$$
$$70 = 2C_p(35-30)$$
$$\Rightarrow \quad C_p = 7\text{cal mol}^{-1}K^{-1}$$
$$C_p - C_v = R$$
$$C_v = C_p - R = 7-2 = 5$$

Now,
$$Q_2 = 2\times C_v\Delta T$$
$$= 2\times 5\times 5$$
$$= 50 \text{ cal}$$

98. (c)

$$C_v = \frac{n_1C_{v_1} + n_2C_{v_2}}{n_1+n_2}$$
$$= \frac{1\times\frac{3}{2}R + 3\times\frac{5}{2}R}{1+3} = \frac{9}{4}R$$

99. (d)

$$C_p = C_v + R$$
$$= \frac{9}{4}R + R = \frac{13}{4}R$$

100. (b)

$$\gamma = \frac{C_p}{C_v} = \frac{\frac{13}{4}R}{\frac{9}{4}R} = \frac{13}{9}$$

101. (b)
Mean translational kinetic energy is directly proportional to temperature.

102. (b)

$$v_{rms} = \sqrt{\frac{3P}{\rho}}$$

103. (c)

104. (a)

$$\frac{5}{2}R - \frac{3}{2}R = R$$

The value of R is 2 cal/mol K.

105. (c)
Kinetic energy of an ideal gas depends only on its temperature. Hence, it remains constant whether the pressure increases or decreases

106. (d)

$$V_{rms} = \sqrt{\frac{3RT}{M}}$$
$$V_{av} = \sqrt{\frac{8RT}{\pi M}}$$
$$V_p = \sqrt{\frac{2RT}{M}}$$

The average kinetic energy of the gas molecule is

$$= \frac{1}{2}mV_{rms}^2 = \frac{1}{2}m\left(\frac{\sqrt{3}}{2}V_p\right)^2$$
$$= \frac{3}{4}mV_p^2$$

107. (d)

$$PV = nRT$$

For jar A;
$$PV = n_1RT \quad \ldots\text{(i)}$$
For jar B;
$$2P\times\frac{V}{4} = n_2\times R\times 2T \quad \ldots\text{(ii)}$$
Dividing Eq. (i) by Eq. (ii),
$$\Rightarrow \quad \frac{n_1}{n_2} = 4$$

108. (c)
Change in momentum per collision $= mv_x - (-mv_x)$ $= 2mv_x$
A molecule suffers a collision after a distance of $2l$.
For one collision, time taken is $\frac{2l}{v_x}$

$$\frac{\Delta P}{\Delta t} = \frac{2mv_x\times v_x}{2l} = \frac{mv_x^2}{l}$$

Total
$$\frac{\Delta P}{\Delta t} = \frac{mnv_x^2}{l}$$
$$F = \frac{mnv^2}{l}$$
$$P_{yz} = \frac{F}{A} = \frac{mnv_x^2}{l\times l^2} = \frac{mnv_x^2}{l^3}$$
$$P_{xz} = \frac{mnv_y^2}{l^3}$$
$$P_{xy} = \frac{mnv_z^2}{l^3}$$
$$\bar{P} = \frac{1}{3}\frac{mn}{l^3}(v_x^2 + v_y^2 + v_z^2)$$
$$= \frac{1}{3}\frac{mnv^2}{l^3}$$

109. (b)

$$\frac{3}{2}kT$$

110. (b)

$$c=\sqrt{\frac{3P}{\rho}}$$

$$v=\sqrt{\frac{\gamma P}{\rho}}$$

$$\Rightarrow \quad \frac{v}{c}=\sqrt{\frac{\gamma}{3}}$$

111. (d)

For adiabatic change, PV^{γ} = constant

$\Rightarrow P^{1-\gamma}T^{\gamma}$ = constant $\Rightarrow P \propto T^{\frac{-\gamma}{1-\gamma}}$

Therefore the value of constant, $C=\dfrac{\gamma}{(\gamma-1)}$

For monatomic gas, $\gamma = 5/3$

Therefore $C=\dfrac{5/3}{\frac{5}{3}-1}$

$$=\frac{5}{3}\times\frac{3}{2}=\frac{5}{2}$$

112. (b)

$$C_v=\frac{\frac{3}{2}R+\frac{5}{2}R}{2}=2R$$

$$C_p=2R+R=3R$$

$$\gamma=\frac{3R}{2R}=\frac{3}{2}$$

113. (c)

$$\frac{V_2}{V_1}=\sqrt{\frac{M_1}{M_2}\times\frac{T_2}{T_1}}$$

$$=\sqrt{\frac{32}{2}\times\frac{400}{289}}=\frac{4\times 20}{17}$$

$$V_2=\frac{474\times 4\times 20}{17}$$

$$\approx 2240 \text{ m/s}$$

114. (a)

$$\sqrt{\frac{\gamma}{3}}=\sqrt{\frac{5}{3\times 3}}=\sqrt{\frac{5}{9}}$$

115. (b)

116. (a)

$$\gamma=1+\frac{2}{n}$$

For H_2, $\gamma=1+\frac{2}{5}=\frac{7}{5}$

For He, $\gamma=1+\frac{2}{3}=\frac{5}{3}$

For X, $\gamma=1+\frac{2}{7}=\frac{9}{7}$

117. (c)

Mean translational kinetic energy is proportional to absolute temperature.

118. (b)

$$\frac{V_2}{V_1}=\sqrt{\frac{M_1}{M_2}}=\sqrt{\frac{32}{2}}=4$$

119. (c)

$$n=\frac{4}{28}=\frac{1}{7}$$

$$PV=nRT$$

$$PV=\frac{1}{7}RT$$

120. (c)

$V_{rms}\propto\sqrt{T}$ (Independent of pressure)

121. (c)

$$P=\frac{1}{3}\rho c^2$$

122. (a)

Both the gases are at same temperature T.

$$\frac{1}{2}m\omega^2=\frac{1}{2}2mv^2 \quad (M_A=m; M_B=2m)$$

$$\Rightarrow \quad \frac{\omega^2}{v^2}=2$$

123. (d)

$$\frac{V_2}{V_1}=\sqrt{\frac{M_1}{M_2}}=\sqrt{\frac{4}{2}}=\sqrt{2}$$

$$V_2=\sqrt{2}V_1$$

124. (c)

Mean translational kinetic energy of the molecules $=\frac{3}{2}kT$. If the temperature is same, then $\frac{3}{2}kT$ for all the gas molecules is same.

125. (c)

126. (b)

$$\frac{\text{KE}_2}{\text{KE}_1}=\frac{(273+227)}{(273+27)}=\frac{500}{300}=\frac{5}{3}$$

$$KE_2 = 6.21\times10^{-21}\times\frac{5}{3}$$

$$= 10.35\times10^{-21}\text{ J}$$

127. (c)

$$PV = nRT$$

For gas A

$$\Delta P = P_1 - P_2 = \frac{n_1RT}{V} - \frac{n_1RT}{2V}$$

$$= \frac{n_1RT}{2V} \quad \ldots\text{(i)}$$

For gas B

$$1.5\Delta P = P_1 - P_2 = \frac{n_2RT}{V} - \frac{n_2RT}{2V}$$

$$= \frac{n_2RT}{2V} \quad \ldots\text{(ii)}$$

From Eqs. (i) and (ii),

$$\frac{\Delta P}{1.5\Delta P} = \frac{n_1}{n_2}$$

$$\frac{n_1}{n_2} = \frac{1}{1.5} = \frac{2}{3}$$

Let M be the molecular mass of the gas. Then,

$$\frac{m_1/M}{m_2/M} = \frac{2}{3}$$

$$\Rightarrow \frac{m_1}{m_2} = \frac{2}{3}$$

$$3m_1 = 2m_2 \text{ or } 3m_A = 2m_B$$

128. (d)

$$\text{Energy} = \frac{f}{2}nRT = \frac{f}{2}PV$$

$$= \frac{f}{2}(3\times10^6)(2)$$

$$= f\times3\times10^6$$

Considering the gas is monoatomic, i.e., f=3

$$\Rightarrow U = 3\times3\times10^6 = 9\times10^6 \text{ J}$$

129. (c)

$$P = \frac{1}{3}\frac{mn\bar{c}^2}{V}$$

$$P \propto n$$

130. (a)

$$\frac{V_2}{V_1} = \sqrt{\frac{T}{600}}$$

$$\left(\frac{1}{2}\right)^2 = \frac{T}{600}$$

$$\Rightarrow \quad T = 150 \text{ K}$$

$$= 150 - 273$$

$$= -123^\circ \text{ C}$$

131. (b)

$$P_1 = \frac{1}{3}\frac{mnv_1^2}{V}$$

$$P_2 = \frac{1}{3}\frac{m}{2}\frac{n}{V}\times4v_2^2$$

$$\frac{P_2}{P_1} = 2 \text{ or } P = 2P_0$$

132. (a)

The rate of diffusion in inversely proportional to the square root of density of the gas.

133. (a)

$$\frac{V_2}{V_1} = \sqrt{\frac{32}{2}\times\frac{T_2}{320}}$$

$$1 = \sqrt{\frac{T_2}{20}}$$

$$\Rightarrow \quad T_2 = 20 \text{ K}$$

134. (a)

$$\frac{1}{2}mv^2 \propto T$$

$$v \propto \sqrt{T}$$

135. (c)

$$\text{Pressure} = \frac{1}{3}\frac{mnc^{-2}}{V}$$

$$\text{Mean kinetic energy} = \frac{3}{2}kT$$

$$\text{Most probable velocity} = \sqrt{\frac{2}{3}}c$$

$$\text{Ratio of two specific heats} = \gamma$$

$$\text{Mean KE} = \frac{3}{2}PV$$

$$\text{Mean kinetic energy per unit volume} = \frac{3}{2}P$$

$$\text{Mean free path} = \frac{1}{nd^2}$$

136. (c)

137. (a)

Initial temperature (T_1) = 18 °C

$$= (273+18) = 291\text{ K and } V_2 = (1/8)V_1$$

For adiabatic compression, $TV^{\gamma-1}$ = Constant

or $T_1V^{\gamma-1} = T_2V_2^{\gamma-1}$

Therefore $T_2 = T_1\left(\frac{V_1}{V_2}\right)^{\gamma-1}$

$= 291\times(8)^{1.4-1} = 291\times(8)^{0.4}$

$= 291\times 2.297 = 668.4 \text{ K} = 395.4° \text{ C}$

138. (d)

$$P = \frac{1}{3}\frac{mn\bar{c}^2}{V} = \frac{2}{3V}\times\frac{1}{2}mn\bar{c}^2$$

$$P = \frac{2}{3V}\times KE$$

$$P = \frac{2}{3}E$$

139. (c)

$$\frac{V_2}{V_1} = \sqrt{\frac{M_1}{M_2}} = \sqrt{16} = 4v$$

140. (c)

KE per mole $= \frac{3}{2}kT\times N_0$

where N_0 is Avogadro's number

$= \frac{3}{2}\times 1.38\times 10^{-23}\times 6.023\times 10^{23}\times 400$

$= 4988$ J

141. (d)

$$\frac{1}{2} = \sqrt{\frac{T}{273}}$$

$$T = \frac{273}{4} = 68.25 \text{ K}$$

In °C $= 68.25 - 273 \approx -204$

142. (a)

$$P = \frac{1}{2}\rho C^2$$

$$= \frac{1}{3}\times 6\times 10^{-2}\times(500)^2$$

$$= 5\times 10^3 \text{ N/m}^2$$

143. (a)

$$\bar{c} = \sqrt{\frac{1^2+2^2+3^2+4^2+5^2}{5}}$$

$$= \sqrt{\frac{55}{5}} = 3.3 \text{ m/s}$$

144. (c)
Due to change in direction of gas molecule after elastic collision.

145. (c)
Potential energy increases, kinetic energy remains same.

146. (a)
n, the number of moles is for the given quantity of the gas.

147. (a)

$$C_v = \frac{3}{2}R$$

148. (c)

Number of moles $= \frac{6}{2} = 3$

Three moles $= 3N$

149. (b)

$$C_v = N\times\frac{1}{2}R$$

$$C_p = \frac{N}{2}R + R$$

$$\gamma = \frac{C_p}{C_v} = \frac{\frac{N}{2}R + R}{\frac{N}{2}R}$$

$$\gamma = \frac{\frac{N}{2}+1}{N/2} = \frac{N+2}{N}$$

$$\gamma = 1 + \frac{2}{N}$$

$$\Rightarrow \quad N = \frac{2}{\gamma - 1}$$

150. (b)

$$\text{KE} = \frac{3}{2}kT$$

$$= \frac{3}{2}\times 1.38\times 10^{-23}\times 300$$

$$= 6.21\times 10^{-21} \text{ J}$$

151. (c)

$$\frac{V_{rms_1}}{V_{rms_2}} = \sqrt{\frac{T_1}{T_2}} = \frac{v_1}{v_2}$$

152. (c)
Using ideal gas equation,

$$PV = nRT$$

$$2.5\times 4 = n_1 RT$$
$$2\times 4 = n_2 RT$$
$$\Rightarrow \frac{n_2}{n_1} = \frac{2}{2.5} = \frac{4}{5}$$

153. (b)
Under isothermal conditions, there is no change in internal energy.

154. (a)
1 mole of every gas at a temperature has same kinetic energy.

Kinetic energy /mole $= \frac{3}{2}RT$

155. (a)

Kinetic energy $\propto T$

156. (b)

$$\frac{V_2}{V_1} = \sqrt{\frac{M_1}{M_2}\times\frac{T_2}{T_1}}$$
$$= \sqrt{\frac{2}{32}\times\frac{1200}{300}}$$
$$\frac{V_2}{V_1} = \frac{1}{2}$$
$$V_2 = \frac{1930}{2} = 965 \text{ m/s}$$

157. (a)

Number of molecules $= \frac{6.023\times 10^{23}}{22.4}$
$$= 2.68\times 10^{22}$$

158. (d)
$$PV = nRT$$
$$PV = \frac{M}{M_0}RT$$

where M_0 is gram molecular mass. Let N be the number of molecules in 1 mole of the gas, then

$$\frac{M_0}{N} = m$$
$$PV = \frac{M}{Nm}kNT$$
$$P = \frac{M}{V}\frac{kT}{m}$$
$$P = \rho\frac{kT}{m}$$
$$\rho = \frac{Pm}{kT}$$

159. (c)
Because of increased dimension of the vessel.

160. (c)
$$\frac{V_2}{V_1} = \sqrt{\frac{T_2}{T_1}\times\frac{M_1}{M_2}}$$
$$1 = \sqrt{\frac{T_2}{304}\times\frac{32}{2}}$$
$$T_2 = \frac{304}{16} = 19K$$
$$= 19 - 273 = -254^\circ \text{ C}$$

161. (a)
There are two degrees of freedom of a gas for rotational kinetic energy.

Per mole for oxygen $= 2\times\frac{1}{2}RT = RT$

Per mole for nitrogen $= 2\times\frac{1}{2}RT = RT$

The ratio is 1: 1 (one mole of every gas has the same number of molecules).

162. (b)
$$(C_P)_{mix} = \frac{n_1 C_{P_1} + n_2 C_{P_2}}{n_1 + n_2} (C_{P_1}(\text{He}) = \frac{5}{2}R \text{ and } C_{P_2}(\text{H}_2) = \frac{7}{2}R)$$
$$= \frac{1\times\frac{5}{2}R + 1\times\frac{7}{2}R}{1+1} = 3R = 3\times 2 = 6 \text{ cal/mol.°C}$$

∴ Amount of heat needed to raise the temperature from 0 °C to 100 °C
$$(\Delta Q)_P = nC_p\Delta T = 2\times 6\times 100 = 1200 \text{ cal}$$

163. (c)
Rotational kinetic energy of diatomic molecule
$$= 2\times\frac{1}{2}I\omega^2$$
$$= I\omega^2$$
Rotational kinetic energy of diatomic molecule
$$= 2\times\frac{1}{2}KT$$
$$I\omega^2 = k(t+273)$$
$$\omega = \left(\frac{k(273+t)}{I}\right)^{1/2}$$

164. (d)
$$V_{rms} \propto \frac{1}{\sqrt{M_0}}$$
where M_0 = Molecular mass

165. (a)
The large number of gas molecules exerts a steady force on the walls. The force measured per unit area of the wall gives pressure, so it remains same everywhere.

166. (d)

$$P \propto T$$

167. (d)

$$\bar{c} = \frac{\sqrt{1^2 + 4^2 + 8^2 + 16^2}}{5} = 9.2 \text{ ms}^{-1}$$

168. (a)
Using ideal gas equation

$$PV = nRT$$

$$T_1 = \frac{P_1V_1}{nR} = \frac{aV_1V_1}{nR}$$

$$T_2 = \frac{P_2V_2}{nR} = \frac{aV_1V_2}{nR}$$

$$\frac{T_2}{T_1} = \frac{V_2^2}{V_1^2} = (2)^2 = 4$$

$$T_2 = 4T_1 = 4 \times 300 = 1200 \text{ K}$$

$$T_2 - T_1 = 1200 - 300 = 900 \text{ K}$$

169. (b)

$$C_v = \frac{n_1C_{v_1} + n_2C_{v_2}}{n_1 + n_2}$$

$$n_1 = \frac{16}{4} = 4 \text{ and } n_2 = \frac{16}{32} = \frac{1}{2}$$

$$C_v = \frac{4 \times \frac{3}{2}R + \frac{1}{2} \times \frac{5}{2}R}{\left(4 + \frac{1}{2}\right)}$$

$$C_v = \frac{29R}{18}$$

$$C_p = \frac{29R}{18} + R = \frac{47R}{18}$$

$$\gamma = \frac{C_p}{C_v} = \frac{47R}{18} / \frac{29R}{18} = \frac{47}{29} = 1.62$$

170. (a)

$$t \propto \frac{\text{Volume}}{\text{Velocity}}$$

$$\text{Volume} \propto \frac{T}{P}$$

$$t \alpha \frac{\sqrt{T}}{P}$$

$$\frac{t_1}{6 \times 10^{-8}} = \frac{\sqrt{500}}{2P} \times \frac{P}{\sqrt{300}}$$

$$t_1 = 3.8 \times 10^{-8}$$

$$= 4 \times 10^{-8}$$

171. (a)

$$n = \frac{5}{32}$$

Equation of state is

$$PV = nRT$$

$$PV = \frac{5}{32}RT$$

172. (a)

$$\frac{V_2}{V_1} = \sqrt{\frac{M_1}{M_2}}$$

$$4 = \sqrt{\frac{64}{M_2}}$$

$$\Rightarrow \quad M_2 = 4$$

173. (a)

$$\frac{V_2}{V_1} = \sqrt{\frac{T_2}{T_1}} = \sqrt{\frac{273 + 327}{273 + 27}}$$

$$\frac{V_2}{V_1} = \sqrt{2} \text{ times}$$

174. (a)

$$C = C_v + \frac{R}{1-n}$$

$$C - C_v = \frac{C_p - C_v}{1-n}; 1 - n = \frac{C_p - C_v}{C - C_v}$$

$$n = 1 - \frac{C_p - C_v}{C - C_v} = \frac{C - C_p}{C - C_v}.$$

175. (b)

176. (b)
For an adiabatic change,

$$PV^\gamma = K = \text{Constant} \quad \ldots\text{(i)}$$

$$PV = nRT \quad \ldots\text{(ii)}$$

Substitute the value of V from Eq. (ii) in Eq. (i)

$$P\left(\frac{nRT}{P}\right)^\gamma = K$$

$$\frac{P}{P^\gamma} \cdot T^\gamma (nR)^\gamma = K$$

$$P^{1-\gamma}T^\gamma = \frac{K}{(nR)^\gamma} = K'$$

$$P^{1-\gamma}T^\gamma = \text{Constant}$$

177. (c)

$$KE \propto T$$

Temperature is constant, therefore kinetic energy is constant.

178. (b)

$$PV = nRT$$

$$PV_1 = n_1RT \quad \ldots(i)$$

$$PV_2 = n_2RT \quad \ldots(ii)$$

By Eqs. (i) and (ii),

$$\Rightarrow \quad \frac{n_1}{V_1} = \frac{n_2}{V_2}$$

179. (c)

$$\frac{V_2}{V_1} = \sqrt{\frac{T_2}{T_1}}$$

$$2 = \sqrt{\frac{T_2}{120}} \Rightarrow T_2 = 4 \times 120$$

$$= 480 \text{ K}$$

180. (d)

$$\frac{V_2}{V_1} = \sqrt{\frac{T_2}{T_1} \times \frac{M_1}{M_2}}$$

$$1 = \sqrt{\frac{T_2}{400} \times \frac{32}{28}}$$

$$T_2 = 400 \times \frac{28}{32} = 350 \text{ K}$$

In °C = 350 − 273 = 77°C

181. (b)

$$PV = nRT$$

$$PV^2 = nRTV$$

$$\frac{PV^2}{nR} = TV$$

$$\Rightarrow \quad TV = \text{Constant}$$

$$T \propto \frac{1}{V}$$

182. (b)

Mean translational kinetic energy corresponds to translational motion.

KE for monatomic gas $= \frac{3}{2}PV$

KE for diatomic gas $= \frac{3}{2}PV$

183. (b)

$$\frac{P_2V_2}{T_2} = \frac{P_1V_1}{T_1}$$

$$P_2 = \frac{P_1V_1}{T_1} \times \frac{T_2}{V_2}$$

$$P_2 = \frac{200V \times 100 \times 315}{295 \times 102V} = 209 \text{ kPa}$$

184. (d)

$$T_2V_2^{\gamma-1} = T_1V_1^{\gamma-1}$$

$$T_2 = T_1\left(\frac{V_1}{V_2}\right)^{\gamma-1} = 300 \times \left(\frac{27}{8}\right)^{\frac{5}{3}-1}$$

$$= 300 \times \frac{9}{4} = 675$$

$$\Delta T = 675 - 300 = 375 \text{ K}$$

185. (d)

$$C_p = \left(\frac{f}{2}+1\right)R$$

where f = number of degrees of freedom

For diatomic gases,

$f = 5$ (3 translational and 2 rotational)

$$C_p = \left(\frac{5}{2}+1\right)R = \frac{7}{2}R$$

186. (b)

Equal volumes mean equal number of moles.

Therefore C_v for the mixture

$$C_v = \frac{n \times \frac{5}{2}R + n \times \frac{5}{2}R}{n+n} = \frac{5}{2}R$$

$$C_p = \frac{5}{2}R + R = \frac{7}{2}R$$

$$\gamma = \frac{7}{5}$$

Speed of sound in hydrogen

$$V_H = \sqrt{\frac{\gamma P}{\rho m}}$$

Speed of sound in mixture

$$V_m = \sqrt{\frac{\gamma P}{\rho m}}$$

Let there be one mole of each gas having volume V.

$$\rho_m = \frac{\text{Mass of } H_2 + \text{Mass of } O_2}{\text{Total volume}}$$

$$= \frac{2+32}{2V} = \frac{17}{V}$$

$$\rho_H = \frac{2}{V}$$

$$\frac{V_m}{V_H} = \sqrt{\frac{\gamma P}{\rho m} \times \frac{\rho_H}{\gamma P}}$$

$$= \sqrt{\frac{2 \times V}{V \times 17}} = \sqrt{\frac{2}{17}}$$

187. (d)
PV = constant at constant temperature. So the graph is straight line for all volumes.

188. (c)
(i) The dotted line indicates that $\frac{PV}{nT}$ for the temperature T_1 and T_2, and also for changing pressure is constant, so it corresponds to ideal gas behaviour.
(ii) At high temperature the deviation from ideal gas behavior is less. Therefore, $T_1 > T_2$.
(iii) The two curves meet at the same point.

189. (a)
Mean KE $\propto T$
Therefore $T_2 = 2T_1 = 2(273 + 20°) = 586$ K

In °C $= 586 - 273 = 313$°C

190. (a)

$$\frac{P_1V_1}{T_1} = \frac{P_2V_2}{T_2}$$

$$V_2 = \frac{P_1V_1}{P_2} \times \frac{T_2}{T_1} = \frac{1 \times 100}{4}$$

$$= 25 \text{ cm}^3$$

191. (a)
Dalton's Law of partial pressure for two non-reacting gases: The total pressure exerted by gas is equal to sum of the individual pressure which the gas exerts for the same volume and at same temperature.

$$P = P + P = 2P$$

192. (a)

$$\frac{P_1}{P_2} = \frac{T_1}{T_2}$$

$$\frac{P_2}{P_1} = \frac{T_2}{T_1} = \frac{308}{288}$$

$$\% \text{ increase} = \frac{P_2 - P_1}{P_1} \times 100$$

$$= \frac{308 - 288}{288} \times 100 = 7\%$$

193. (d)

Total energy $= 4 \times \frac{3}{2} R \times 400 + 2 \times \frac{3}{2} R \times 700$

Let the temperature of the mixture be T.

$$6 \times \frac{3}{2} RT = 4 \times \frac{3}{2} R \times 400 + 2 \times \frac{3}{2} R \times 700$$

$$6T = 1600 + 1400$$

$$\Rightarrow \quad T = \frac{3000}{6} = 500 \text{ K}$$

194. (a)

$$CH_4 = 12 + 4 \times 1 = 16g$$

$$O_2 = 32g$$

1 mole of methane and 1 mole of oxygen have the same kinetic energy.

195. (a)
The dimensions of b is the dimension of volume and the dimension of $\frac{a}{V^2}$ is the dimension of pressure.

The dimensions of $b = L^3$
The dimensions of $a = L^6 \times ML^{-1}T^{-2}$
Therefore, the dimension of b/a

$$= \frac{L^3}{L^6ML^{-1}T^{-2}} = \left[M^{-1}L^{-2}T^2\right]$$

196. (b)
At STP, the volume of 1 mole of gas is 22.4 litre. Let n_1 be the number of moles at STP and n_2, at the given pressure and temperature, then

$$n_2 = \frac{m_2}{32}$$

$$\frac{P_1V_1}{n_1T_1} = \frac{P_2V_2}{n_2T_2}$$

$$\frac{760 \times 22.4}{273} = \frac{760 \times 11.2 \times 32}{m_2 \times 300}$$

$$\Rightarrow \quad m^2 = \frac{273 \times 32}{300 \times 2} = 14.56 \text{ gm}$$

$$= 0.01456 \text{ kg}$$

197. (d)

$$P = KT^3 \quad \text{...(i)}$$

$$PT^{-3} = K$$

Adiabatic relation between P and T is

$$T^{\gamma}P^{\gamma-1} = K$$

$$PT^{\frac{\gamma}{1-\gamma}} = K \quad \text{...(ii)}$$

Comparing (i) and (ii)

$$\frac{\gamma}{1-\gamma} = -3$$

$$\Rightarrow \quad \frac{1-\gamma}{\gamma} = -\frac{1}{3}$$

$$\frac{1}{\gamma} - 1 = -\frac{1}{3}$$

$$\frac{1}{\gamma}=1-\frac{1}{3}=\frac{2}{3}$$

$$\gamma=\frac{3}{2}$$

198. (b)

$$v_{rms}\propto\sqrt{T}$$

199. (a)

200. (b)

$$\frac{V_2}{V_1}=\sqrt{\frac{M_1}{M_2}}=\sqrt{\frac{1}{4}}$$

$$V_2=\frac{1}{2}V_1$$

201. (d)

$$\gamma=\frac{7}{5}=1.4$$

202. (d)

$$PV^{\gamma}=\text{constant}$$

$$P\left(\frac{m}{\rho}\right)^{\gamma}=\text{constant}$$

$$\Rightarrow\quad \frac{P_1}{\rho_1^{\gamma}}=\frac{P_2}{\rho_2^{\gamma}}$$

$$\frac{P_1}{P_2}=\left(\frac{\rho_1}{\rho_2}\right)^{\gamma}=(32)^{7/5}=128$$

Questions Asked in NEET/CBSE Medicals

203. (c)
As per kinetic theory of gases, kinetic energy of gas molecules is directly proportional to the temperature of the gas.

204. (b)
Escape velocity from the Earth's surface is

$$v_{escape}=11200\text{ ms}^{-1}$$

Say at temperature T, oxygen molecule attains escape velocity.

So, $$v_{escape}=\sqrt{\frac{3k_BT}{m_{O_2}}}$$

$$\Rightarrow\quad 11200=\sqrt{\frac{3\times1.38\times10^{-23}\times T}{2.76\times10^{-26}}}$$

205. (c)
The internal energy of 2 moles of O_2 atom is

$$U_{O_2}=\frac{n_1f_1}{2}RT=2\times\frac{5}{2}\times RT$$

$$U_{O_2}=5RT$$

The internal energy of 4 moles of Ar atom is

$$U_{Ar}=\frac{n_2f_2RT}{2}=4\times\frac{3}{2}\times RT=6RT$$

$\therefore$ The total internal energy of the system is

$$U=U_{O_2}+U_{Ar}=5RT+6RT=11RT$$

206. (d)
PV^3 = constant
For a polytropic process, PV^{α} = constant

$$C=C_V+\frac{R}{1-\alpha}=\frac{3}{2}R+\frac{R}{1-3}=R$$

207. (b)

$$PV=nRT$$

$$\text{or } n=\frac{PV}{RT}=\frac{\text{Mass}}{\text{Molar mass}}$$

Density, $$\rho=\frac{\text{Mass}}{\text{Volume}}=\frac{(\text{Molar mass})P}{RT}$$

$$=\frac{(mN_A)P}{RT}$$

$$\rho=\frac{mP}{kT}\quad(\because R=N_Ak)$$

208. (d)

As, $$v_{rms}=\sqrt{\frac{3k_BT}{m}}$$

$$\therefore\quad \frac{v_{27}}{v_{127}}=\sqrt{\frac{27+273}{127+273}}=\sqrt{\frac{300}{400}}=\frac{\sqrt{3}}{2}$$

or $$v_{127}=\frac{2}{\sqrt{3}}\times v_{27}=\frac{2}{\sqrt{3}}\times200\text{ ms}^{-1}=\frac{400}{\sqrt{3}}\text{ ms}^{-1}$$

209. (d)
According to ideal gas equation, the molecular weight of an ideal gas is

$$M=\frac{\rho RT}{P}$$

$\therefore$ The molecular weight of A is

$$M_A=\frac{\rho_ART_A}{P_A}\text{ and that of B is } M_B=\frac{\rho_BRT_B}{P_B}$$

$$\frac{M_A}{M_B}=\left(\frac{\rho_A}{\rho_B}\right)\left(\frac{T_A}{T_B}\right)\left(\frac{P_A}{P_B}\right)$$

given, $$\frac{\rho_A}{\rho_B}=1.5=\frac{3}{2},\ \frac{T_A}{T_B}=1\text{ and }\frac{P_A}{P_B}=2$$

$$\frac{M_A}{M_B}=\left(\frac{3}{2}\right)(1)\left(\frac{1}{2}\right)=\frac{3}{4}$$

210. (c)
First, isothermal expansion

$$PV = P'(2V);\ P' = \frac{P}{2}$$

Then, adiabatic expansion

$$P'(2V)^{\gamma} = P'(16V)^{\gamma}$$

$$\frac{P}{2}(2V)^{5/3} = P_f(16V)^{5/3}$$

$$P_f = \frac{P}{2}\left(\frac{2V}{16V}\right)^{5/3} = \frac{P}{2}\left(\frac{1}{8}\right)^{5/3} = \frac{P}{2}\left(\frac{1}{2^3}\right)^{5/3}$$

$$= \frac{P}{2}\left(\frac{1}{2^5}\right) = \frac{P}{64}$$

211. (b)

Mean free path, $\lambda = \frac{1}{\sqrt{2}n\pi d^2}$

where n is the number density and d is the diameter of the molecule.

As $d = 2r$, $\quad \therefore \lambda = \frac{1}{4\sqrt{2}n\pi r^2}$ or $\lambda \propto \frac{1}{r^2}$

212. (b)

$P \propto T^3; PT^{-3} = \text{Constant}$ (i)

For an adiabatic process; $PT^{\gamma/1-\gamma}$ = Constant (ii)

Comparing (i) and (ii), we get

$$\frac{\gamma}{1-\gamma} = -3; \gamma = -3 + 3\gamma$$

$$-2\gamma = -3 \text{ or } \gamma = \frac{3}{2}$$

As $\quad \gamma = \frac{C_P}{C_v} \therefore \frac{C_p}{C_v} = \frac{3}{2}$

213. (c)

$\Delta Q = nC_V\Delta T$

Here, number of moles, $n = \frac{1}{4}$

$C_V = \frac{3}{2}R \quad (\because \text{He is a monatomic})$

$\Delta T = T_2 - T_1$

$$\therefore \Delta Q = \frac{1}{4} \times \frac{3}{2}R(T_2 - T_1)$$

$$= \frac{3}{8}N_a k_B(T_2 - T_1)$$

214. (d)

For an adiabatic process,

$$PV^{\gamma} = \text{Constant} \qquad \text{(i)}$$

According to ideal gas equation,

$$PV = nRT \Rightarrow P = \frac{nRT}{V}$$

Putting value of P in (i), we get

$$\frac{nRT}{V}V^{\gamma} = \text{Constant}; \therefore TV^{\gamma-1} = \text{Constant}$$

Again, from the ideal gas equation

$$V = \frac{nRT}{P}$$

Putting value of V in Eq. (i)

$$P\left(\frac{nRT}{P}\right)^{\gamma} = \text{Constant}; \ P^{1-\gamma}T^{\gamma} = \text{Constant}$$

215. (a)

As $P = \frac{1}{3}\frac{mN}{V}v_{rms}^2$ (i)

$$P' = \frac{1}{3}\left(\frac{m}{2}\right) \times \frac{N}{V} \times (2v_{rms})^2 = \frac{2}{3}\frac{mN}{V}v_{rms}^2 = 2P$$

(Using (i))

216. (d)

For an adiabatic process, $\frac{T^{\gamma}}{P^{\gamma-1}}$ = constant

$$\left(\frac{T_i}{T_f}\right)^{\gamma} = \left(\frac{P_i}{P_f}\right)^{\gamma-1}; P_f = P_i\left(\frac{T_f}{T_i}\right)^{\frac{\gamma}{\gamma-1}} \qquad \text{(i)}$$

Here, $T_i = 27°\text{ C} = 300\text{ K}$, $T_f = 927°\text{ C} = 1200\text{ K}$

$P_i = 2\text{ atm}$, $\gamma = 1.4$

Substituting these values in Eq. (i), we get

$$P_f = (2)\left(\frac{1200}{300}\right)^{\frac{1.4}{1.4-1}}$$

$$P_f = (2)(4)^{1.4/0.4} = 2(2^2)^{7/2} = (2)(2)^7 = 2^8$$

$= 256$ atm

217. (d)

Ideal gas equation for an adiabatic process is

PV^{γ} = constant, or $P_1V_1^{\gamma} = P_2V_2^{\gamma}$

For monatomic gas, $\gamma = \frac{5}{3}$

$$P_1V_1^{5/3} = P_2\left(\frac{V_1}{8}\right)^{5/3}$$

$$\Rightarrow \qquad P_2 = P_1 \times (2)^2 = 32P_1$$

218. (b)

Let fixed mass be M, then

$$\rho = \frac{M}{V} \qquad \text{...(i)}$$

$$\frac{\rho}{P} = x \quad \text{...(ii)}$$

Substitute the value of ρ from Eq. (i) into Eq. (ii)

$$\frac{M}{PV} = x$$

$$\frac{M}{nRT} = x$$

$$xT = \frac{M}{nRT} = \text{Constant}$$

$$x_1T_1 = x_2T_2$$

$$x_2 = x_1\frac{T_1}{T_2}$$

$$= x_1\left(\frac{283}{383}\right)$$

219. (b)

$$C_P = \frac{7}{2}R$$

$$C_V = C_P - R = \frac{7}{2}R - R = \frac{5}{2}R.$$

$$\frac{C_P}{C_V} = \frac{\frac{7}{2}}{\frac{5}{2}} = \frac{7}{5}$$

QUICK QUESTIONS

Quiz

1. Can two gases having same number of moles at the same temperature have different internal energy?

2. Name the various speeds which characterize the motion of a molecule in a gas and write them in increasing order.

3. How many degrees of freedom are associated with
(i) an ideal gas (ii) an atom in a solid
(iii) a diatomic molecule free to vibrate?

4. When will a real gas behave like an ideal gas?

5. What are the different ways of increasing the number of collisions per second in molecules of a gas?

6. Two rooms, equal in size, are connected through an open doorway, the average temperature in the two rooms are maintained at different values. Which room contains more air, i.e., the one which has higher temperature or the one which is at lower temperature.

7. Two identical cylinders are at the same temperature. One contains helium at 2 temperature and the other argon at 1 temperature. If both the gases are filled in one of the cylinders then what will be the pressure?

8. Small planets tend to have little or no atmosphere. Why?

9. Why do you feel cold near a pond or river?

10. What is the average velocity of the molecules of a gas in equilibrium?

11. A vessel is filled with a mixture of two different gases. Will the mean kinetic energy of the two gases will be same or different?

12. Can we say that absolute zero temperature is zero energy temperature?

13. The radius of a bubble is doubled as it rises from the bottom of a lake to its surface. What is the depth of the lake?

Think and grow

14. How would the pressure inside a liquid change if the forces of attraction between the molecules suddenly disappear?

15. A group of astronauts with a scientist lands on an inhabited planet. They develop a conversation with the inhabitants regarding temperature scale. During the conversation they came to know they also have a thermometer based on the freezing and boiling point of water which are also separated by 100 degrees based on the inhabitants' scale. Would the measure of alien's degrees be the same as ours on Earth? Suppose they also have a temperature scale similar to our kelvin scale. Would their absolute zero be same as ours?

16. Two vessels are connected by tubes with cocks and filled with water. The air has been pumped out of both the vessels.

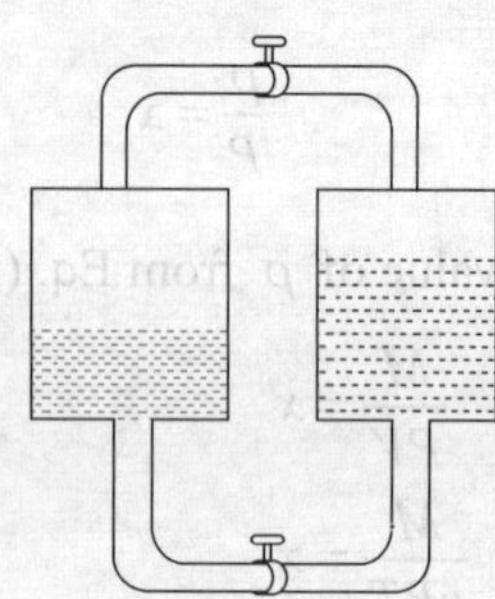

What will happen to the water level if
(i) the cock in the lower tube is opened?
(ii) the cock in the upper tube is opened?

Explanations

Quiz

1. Yes, $U = \frac{1}{2} nfRT$
where f is the degree of freedom. So if gases have different degrees of freedom, their internal energy or KE will be different but translational kinetic energy will be same. Translational kinetic energy $= \frac{3}{2} nRT$.

2. These are most probable (v_m), average (u_v) and root mean square (v_{rms}) velocities. $v_{mp} < v_{av} < v_{rm}$

3. (i) 3 (ii) 6 (iii) 7

4. High temperature and low pressure.

5. According to kinetic theory of gases $P = \frac{1}{3}\frac{mNc^{-2}}{V}$, so the pressure or number of collision per second can be increased by the following three independent sources.
(i) By increasing the mass of the gas, i.e., the number of molecules.
(ii) $P \propto c^{-2} \propto T$ if the temperature of the gas is increased
(iii) $P \propto \frac{1}{V}$ If volume decreases the number of collisions per second will increase due to lesser effective distance between the walls, resulting in greater pressure.

6. Let us apply ideal gas equation for the two rooms $P_1V_1 = n_1RT_1$ and $P_2V_2 = n_2RT_2$ given $P_1 = P_2$ and $V_1 = V_2$. Therefore, $n_1T_1 = n_2T_2$.
Lower the temperature, more will be the mass of the gas. The room at lower temperature has more mass.

7. According to Dalton's law of partial pressure.

$$P = P_1 + P_2 = 2 + 1 = 30 \text{ atm}$$

8. The presence of atmosphere on a planet is due to its gravity. On small planets the gravitational force is small and correspondingly the escape velocity is also small. Let us consider a small planet that has atmosphere at its beginning. The molecules of the gas, as per kinetic theory of gases, have a wide range of speeds. Some of the molecules in this speed distribution may have speed of the planet and will leave the planet. As a result the planet has less atmosphere.

9. According to kinetic energy of gases, at a given temperature in a liquid, molecules are moving randomly in all possible directions with all possible speeds. At all temperature, high-speed molecules may escape. Therefore the temperature of the remaining molecules will decrease. This process is known as evaporation.

10. Zero

11. Equal because the mean translational kinetic energy per molecules is equal to $\frac{3}{2}kT$ which depends only on temperature.

12. No, at absolute zero of temperature, the molecules may possess potential energy. So the absolute zero may more precisely be called zero kinetic energy temperature.

13. The radius is doubled. It means volume becomes eight times more. It means the pressure at the bottom of the lake is 8 times the atmosphere pressure. Let the pressure difference be 8 Pa – Pa = 7 Pa.
The atmospheric pressure in equivalent mercury column is 0.76 m. Therefore in equivalent water column, it is 0.76 × 13.6.
$\therefore D = 7 \times 0.76 \times 13.6 = 72 \text{ m}$

Think and grow

14.

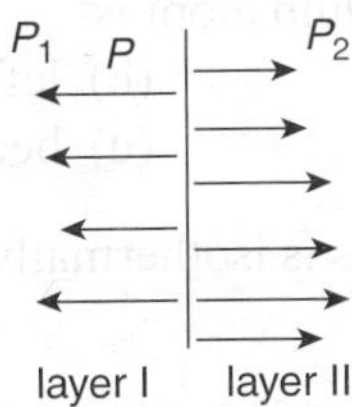

If the attraction between the molecules suddenly disappeared, the pressure would increase. Let's reason by considering two layers I and II inside a fluid. The molecules penetrating from layer I into layer II, because of thermal motion, collide with the molecules in layer II and as a result this layer is acted upon by the pressure force P_2 that depends upon temperature. The force of attraction P_1 acts on layer II from the side of molecule in layer I in opposite direction. The resultant pressure of layer I on layer II $P = P_2 - P_1$. When P_1 disappears P_2 grows.

15. The two temperatures, zero degree (freezing point) and 100 degree (boiling point) of water has been defined at normal atmospheric pressure on the planet. It is less probable that their pressure is same as on earth. The inhabitants may define the two points zero degree and 100 degree for their temperature scale but their degree size would not be the same as ours unless the atmospheric pressure is same.

The Kelvin scale of the temperature may be the same as ours because it is based on a universal definition rather than being defined based on a particular substance at a given pressure.

16. (i) The water levels will become the same as in communicating vessels. The water vapors in the left hand vessels will partly condense and some water will evaporate in the right hand vessels.

(ii) The levels well become the same because the vapors will flow from one vessel into other.

CHAPTER TEST

1. According to kinetic theory of gases, a molecule is the smallest particle of a substance and it possesses
   (a) all the properties of the substance
   (b) some of the properties of the substance
   (c) None of the properties of the substance.
   (d) Both (a) and (b) are true.

2. The internal energy of a gram-molecule of an ideal gas depends on
   (a) pressure alone
   (b) volume alone
   (c) temperature alone
   (d) Both on pressure as well as temperature.

3. The absolute temperature of a gas is determined by
   (a) the average momentum of the molecule
   (b) the velocity of sound in the gas
   (c) the number of molecules in the gas
   (d) the mean square velocity of the molecules.

4. Two vessels having equal volume contain molecular hydrogen at one atmosphere and helium at two atmospheres respectively. If both samples are at the same temperature, the mean velocity of hydrogen molecules is
   (a) equal to that of helium
   (b) twice that of helium
   (c) half that of helium
   (d) $\sqrt{2}$ times that of helium.

5. The root mean square velocity of the molecules in a sample of helium is (5/7)th that of molecules in a sample of hydrogen. If the temperature of hydrogen gas is 0° C then that of the helium sample is about
   (a) 0° C (b) 4 K
   (c) 213° C (d) 100° C

6. A gas at a pressure P_0 is contained in a vessel. If the masses of all the molecules are halved and their velocities doubled, the resulting pressure P would be equal to
   (a) 4 P_0 (b) 2 P_0
   (c) P_0 (d) $P_0/2$

7. At a given temperature, the force between molecules of a gas as a function of intermolecular distance
   (a) is always constant
   (b) always decreases
   (c) always increases
   (d) first decreases and then increases.

8. At a given volume and temperature, the pressure of a gas
   (a) varies inversely as its mass
   (b) varies inversely as the square of its mass
   (c) varies linearly as its mass
   (d) is independent of its mass.

9. At 0° K which of the following properties of a gas will be zero?
   (a) Kinetic energy (b) Potential energy
   (c) Vibrational energy (d) Density

10. If the Avogadro's number were to tend to infinity, the phenomenon of Brownian motion would
(a) remain completely unaffected
(b) become more vigorous, than that observed with present finite values of Avogadro's number, for all sizes of the Brownian particles
(c) become more vigorous than that observed with the present finite value of Avogadro's number, only for relatively large Brownian particles
(d) become practically unobservable as the molecular impact would tend to balance one another, for practically all sizes of Brownian particles.

11. We have a jar A filled with a gas characterized by parameters P, V and T and another jar B filled with a gas with parameters $2P$, $V/4$ and T, where the symbols have their usual meanings. The ratio of the number of molecules of jar A to those of jar B is
(a) 1 : 1 (b) 1 : 2
(c) 2 : 1 (d) 4 : 1

12. A gas in a container A is in thermal equilibrium with another gas of the same mass in container B. If we denote the corresponding pressures and volumes by the suffixes A and B, then which of the following statements is most likely to be true.
(a) $P_A = P_B,\ V_A \neq V_B$ (b) $P_A \neq P_B,\ V_A = V_B$
(c) $P_A / V_A = P_B / V_B$ (d) $P_A V_A = P_B V_B$

13. At what temperature is the root mean square velocity of gaseous hydrogen molecules equal to that of oxygen molecules at 47° C ?
(a) 20° K (b) 80° K
(c) –73° K (d) 3° K

14. In a cubical vessel are enclosed n molecules of a gas each having a mass m and an average speed v. If l is the length of each edge of the cube, the pressure exerted by the gas will be
(a) $\frac{nmv^2}{l^3}$ (b) $\frac{nm^2r}{2l^3}$
(c) $\frac{nm\bar{v}^2}{3l^3}$ (d) $\frac{nmv}{2l^3}$

15. If C_s be the velocity of sound in air and C be the r.m.s velocity, then
(a) $C_s < C$ (b) $C_s = C$
(c) $C_s = C(\gamma / 3)^{1/2}$ (d) None of the above.

16. On any planet, the presence of atmosphere implies: [v_{rms} = root mean square velocity of molecules and v_e = escape velocity]
(a) $v_{rms} << v_e$ (b) $v_{rms} > v_e$
(c) $v_{rms} = v_e$ (d) $v_{rms} = 0$

17. A sealed container with negligible thermal coefficient of expansion contains helium (a monatomic gas). When it is heated from 300 to 600 K, the average kinetic energy of the helium atom is
(a) halved (b) left unchanged
(c) doubled (d) becomes $\sqrt{2}$ times.

18. When an ideal gas is isothermally expanded, its internal energy will
(a) increase
(b) decrease
(c) increase or decrease depending upon the nature of the gas
(d) remain the same.

19. Oxygen and hydrogen gases are at the same temperature T. The kinetic energy of an oxygen molecule will be equal to
(a) 16 times the kinetic energy of a hydrogen molecule
(b) 4 times the kinetic energy of a hydrogen molecule
(c) the kinetic energy of a hydrogen molecule
(d) one-fourth the kinetic energy of a hydrogen molecule.

20. A vessel contains a mixture of different types of gases. Which of the following statements is correct?
(a) On an average, the heavier molecules have higher speed.
(b) The average speed of different molecules is the same.
(c)The average translational energy of different types of molecules is the same.
(d) On an average, the heavier molecules have higher translational energy.

21. According to the law of equipartition of energy, the average KE per degree of freedom per molecule of an ideal gas at absolute temperature T is
(a) $kT/2$ (b) $3\,kT/4$
(c) kT (d) $3\,kT/2$

22. A vessel containing 0.1 m^3 of air at a pressure of 76 cm of Hg is connected to an evacuated vessel of capacity 0.09 m^3. The resultant air pressure is
(a) 20 cm of Hg (b) 30 cm of Hg
(c) 40 cm of Hg (d) 50 cm of Hg.

23. If one mole of a monatomic gas ($\gamma = 5/3$) is mixed with one mole of a diatomic gas ($\gamma = 7/3$), the value of γ for the mixture is
(a) 1.40 (b) 1.50
(c) 1.53 (d) 3.07

24. The law of equipartition is energy was postulated by
(a) Maxwell (b) Planck
(c) Bohr (d) Boltzmann.

25. The pressure–temperature relationship for an ideal gas undergoing adiabatic change is
(a) $P^{1-\gamma}T^{\gamma}$ = constant (b) $P^{\gamma-1}T^{\gamma}$ = constant
(c) $P^{\gamma}T^{1-\gamma}$ = constant (d) $P^{\gamma}T^{\gamma-1}$ = constant

26. When a gas expands adiabatically,
(a) no energy is required for expansion
(b) energy is required and it comes from the wall of the container of the gas
(c) internal energy of the gas is used in doing work
(c) law of conservation of energy does not hold.

27. A sample of gas expands from volume V_1 to V_2. The amount of work done by the gas is greatest, when the expansion is
(a) isothermal (b) isobaric
(c) adiabatic (d) equal in all cases.

28. If a van der Waal gas expands freely, then final temperature is
(a) less than the initial temperature
(b) equal to the initial temperature
(c) more than the initial temperature
(d) less or more than the initial temperature depending on the nature of the gas.

29. A gas is being compressed adiabatically. The specific heat of the gas during compression is
(a) zero
(b) infinite
(c) finite but not zero
(d) undefined.

30. A cycle tyre bursts suddenly. This represents an
(a) isothermal process (b) isobaric process
(c) isochoric process (d) adiabatic process.

31. For adiabatic expansion of a perfect gas, the value of $\Delta P / P$ is equal to
(a) $-\gamma^{1/2}(\Delta V / V)$ (b) $-\Delta V / V$
(c) $-\gamma(\Delta V / V)$ (d) $\gamma^2(\Delta V / V)$

32. Two samples, A and B, of a gas initially of the same temperature and pressure are compressed from a volume V to a volume $V/2$ such that A is compressed isothermally and B adiabatically. The final pressure of
(a) A is greater than that of B
(b) A is equal to that of B
(c) A is less than that of B
(d) A is twice the pressure of B.

33. If a gas heated is at constant pressure, its isothermal compressibility
(a) remains constant
(b) increases linearly with temperature
(c) decreases linearly with temperature
(d) decreases inversely with temperature.

34. Four curves A, B, C and D are drawn in figure for a given amount of gas. The curves which represent adiabatic and isothermal change are

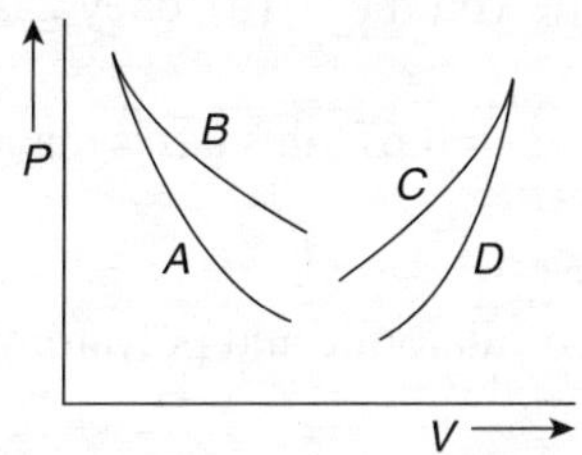

(a) C and D respectively
(b) D and C respectively
(c) A and B respectively
(d) B and A respectively.

35. A box contains n molecules of gas. How will the pressure of the gas be affected if the number of molecules is made $2n$
(a) Pressure will decrease
(b) Pressure will remain unchanged
(c) Pressure will be doubled
(d) Pressure will become three times.

36. To what temperature should the hydrogen at 327° C be cooled at constant pressure so that the root mean square velocity of its molecules becomes half of its previous value
(a) –123° C (b) 123° C
(c) –100° C (d) 0° C

37. Consider a monatomic ideal gas occupying a volume of 22.4 litre at 0^0 C and 1 atm pressure. The work required to compress this gas isothermally to a volume of 16.8 litre is
(a) 285 J (b) 74 J
(c) 1570 J (d) 646 J

38. If a gas has n degree of freedom, the ratio of the specific heats γ of the gas is
(a) $\frac{1+n}{2}$ (b) $1+\frac{n}{2}$
(c) $1+\frac{1}{n}$ (d) $1+\frac{2}{n}$

39. At a given temperature, which of the following gases possesses maximum r.m.s. velocity?
(a) Hydrogen (b) Oxygen
(c) Nitrogen (d) Carbon dioxide

40. The temperature of gas is produced by
(a) the potential energy of its molecules
(b) the kinetic energy of its molecules
(c) the attractive force between its molecules
(d) the repulsive force between its molecules.

41. Gases exert pressure on the walls of the containing vessel because the gas molecules
(a) possess momentum (b) collide with each other
(c) have finite volume (d) obey gas laws.

42. If for a gas, $\frac{R}{C_V} = 0.67$, this gas is made up of molecules which are
(a) diatomic
(b) mixture of diatomic and polyatomic molecules
(c) monatomic
(d) polyatomic.

43. If masses of all molecules of a gas are halved and their speed doubled, then the ratio of initial and final pressure will be
(a) 2 : 1 (b) 1 : 2
(c) 4 : 1 (d) 1 : 4

44. Consider sample of oxygen at 300 K. Find the average time taken by a gas molecule to travel a distance equal to the radius of the earth.
(a) 8 min (b) 8 hr
(c) 800 min (d) 80 hr

45. The graph between (P/T) and T for a gas at constant volume will be

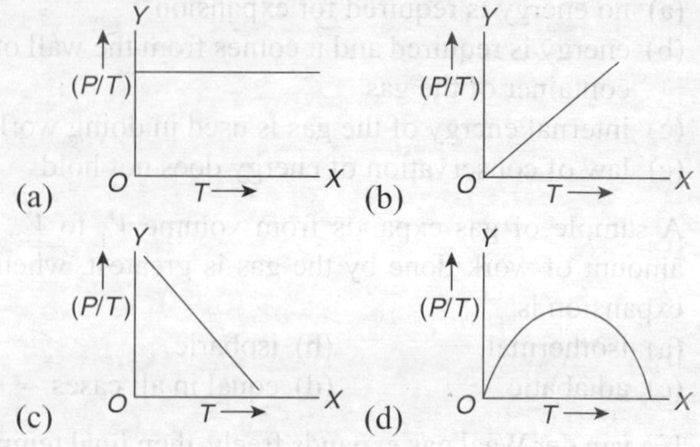

Answer Keys

1. (a)	2. (c)	3. (d)	4. (d)	5. (a)	6. (b)	7. (d)	8. (c)	9. (a)	10. (a)
11. (d)	12. (d)	13. (a)	14. (c)	15. (c)	16. (a)	17. (c)	18. (d)	19. (c)	20. (c)
21. (a)	22. (c)	23. (c)	24. (a)	25. (a)	26. (c)	27. (a)	28. (a)	29. (a)	30. (d)
31. (c)	32. (c)	33. (a)	34. (c)	35. (c)	36. (a)	37. (d)	38. (d)	39. (a)	40. (b)
41. (a)	42. (c)	43. (b)	44. (b)	45. (a)					

Solutions

1. (a)

2. (c)

3. (d)

4. (d)

5. (a)

6. (b)

$$P_0 = \frac{1}{3}\frac{mn}{V}[v_1^2 + v_2^2 + \ldots\ldots\ldots + v_n^2]$$

and $P = \frac{1}{3}(\frac{m}{2})\frac{n}{V}[(2v_1)^2 + (2v_2)^2 + \ldots\ldots\ldots + (2v_n)^2]$

$$= \frac{4}{2}[P_0] = 2P_0$$

7. (d)

8. (c)

$$P = \frac{1}{3}\left(\frac{mn}{V}\right)v^2$$

At constant temperature, v is constant. Further, the volume V is constant. Hence

$$P \propto m$$

9. (a)
The kinetic energy proportional to temperature

10. (a)
Brownian motion is independent of Avogadro's number

11. (d)
We know that $PV = (m/M)RT$
Where m = mass of the gas and M = molecular mass
Let m_0 be the mass of one molecule and N be the number of molecules, then

$$m = m_0 N,$$

$$\therefore PV = \left(\frac{m_0}{M}\right)RT$$

For jar A, $PV = \left(\frac{m_0 N_1}{M}\right)RT$ ...(1)

For jar B

$$(2P)\left(\frac{V}{4}\right) = \left(\frac{m_0 N_2}{M}\right)R(2T) \quad \ldots\ldots (2)$$

From Eqs. (1) and (2), we get

$$\frac{N_1}{N_2}=\frac{4}{1} \text{ or } N_1:N_2=4:1$$

12. (d)

13. (a)

14. (c)

15. (c)

16. (a)
When $v_{rms} << v_e$ the molecules do not escape out.

17. (c)

18. (d)

19. (c)

20. (c)

21. (a)

22. (c)

23. (c)
Let γ_1 and γ_2 denote the ratio of two specific heat of gas 1 and gas 2 respectively when they are mixed in the ratio of n_1 and n_2 mole. The resultant value of γ is weight mean of γ_1 and γ_2, i.e.,

$$\gamma=\frac{n_1\gamma_1+n_2\gamma_2}{n_1+n_2}=\frac{1(5/3)+1(7/5)}{1+1}$$

$$=\frac{25+21}{15\times 2}=\frac{46}{30}=1.53 \text{ J}$$

24. (a)

25. (a)

26. (c)

27. (a)
We know that the *P-V* diagram for an isothermal process is less steep than that for an adiabatic process. Hence, the amount of work done by the gas is greatest when the expansion is isothermal.

28. (a)
As the gas is real, work is done in separating the molecules. So, heat is taken from the gas and as a result, the temperature falls

29. (a)

$$dQ=msdT$$

During adiabatic compression, dQ = 0

$$\therefore \quad 0=msdT \text{ or } s=0$$

30. (d)
As the process is fast, it represents an adiabatic process

31. (c)

32. (c)
For isothermal compression,
$P_2=PV/(V/2)=2P$
For adiabatic compression,
$P_2'=P[V/(V/2)\gamma]=2\gamma P$

33. (a)

34. (c)

35. (c)

36. (a)

$$V_{rms}\,\alpha\sqrt{600} \text{ and } (V_{rms}/2)\,\alpha\sqrt{T}$$

$$\frac{1}{2}=\frac{\sqrt{T}}{\sqrt{600}} \text{ or } \frac{1}{4}=\frac{T}{600} \text{ or } T=150 \text{ K}$$

$$T=150-273=-123°\text{C}$$

37. (d)

38. (d)

39. (a)

40. (b)

41. (a)
Gas molecules possess momentum after collision; hence, there is a change in momentum. We know that rate of change of momentum is force. The force per unit area is pressure.

42. (c)

$$C_V=\frac{R}{0.67}\simeq 1.5R=\frac{3}{2}R$$

This is the case with monatomic gas.

43. (b)

$$P=\frac{1}{3}\times\frac{M}{V}\times\overline{v^2} \text{ and } P'=\frac{1}{3}\frac{(M/2)}{V}4\overline{v^2}$$

$$\frac{P}{P'}=\frac{1}{2} \text{ or } P:P'=1:2$$

44. (b)

$$v_{rms}=\sqrt{\left(\frac{3RT}{M}\right)}$$

$$=\sqrt{\left(\frac{3\times 8.3\times 300}{32\times 10^3}\right)}$$

$$\approx 0.48$$

Now $$t_{av}=\frac{2\times(6400\times 10^3)}{v_{rms}}=8 \text{ hr}$$

45. (a)
According to Gay–Lussac's law, at constant volume, pressure of a gas is directly proportional to *T*. Hence $P \propto T$ or (P/T) = constant.

Oscillations and Waves

Introduction

When particles of a medium move back and forth on the same path, they are said to oscillate and it creates a disturbance (or wave) which propagates energy and momentum from one place to another without the transfer of matter. In this chapter we shall study about the nature of oscillations and waves in detail.

SECTION - I

OSCILLATIONS

Periodic Motion

A motion that repeats itself after a regular time interval is known as periodic motion. It can be rectilinear, closed, open or curvilinear.

Oscillatory or Vibratory Motion

It is a to-and-fro motion. It is a periodic motion between precisely fixed limits. All oscillatory motions are periodic but all periodic motions are not oscillatory. For example, a particle executing circular motion is periodic but not oscillatory. However, projection of this particle on any diameter is both oscillatory and periodic.

Simple Harmonic Motion (SHM)

An oscillatory motion in which the force F (or torque) acting on the particle is always directed towards a mean position (called centre) and is always proportional to displacement x (or angle θ) but opposite in direction is called SHM.

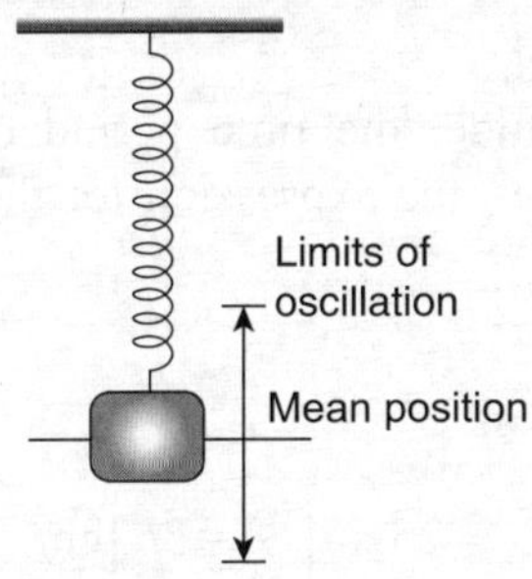

Example of Linear SHM is a mass M attached to a vertical helical spring of spring constant k.

Example of Angular SHM is a disc attached to a rod of stiffness constant C. When the disc is slightly twisted and then released. It executes a SHM.

In a SHM, total force or torque at mean position is always zero. That is, if due to damping (resistive forces) the body comes to rest, it will be at the mean position or at the centre of oscillation.

Equation of Simple Harmonic Motion

A body or particle executing SHM satisfies the condition:

Force $\quad F = -kx \quad$ (for Linear SHM)

or $\quad \tau = -C\theta \quad$ (for Angular SHM)

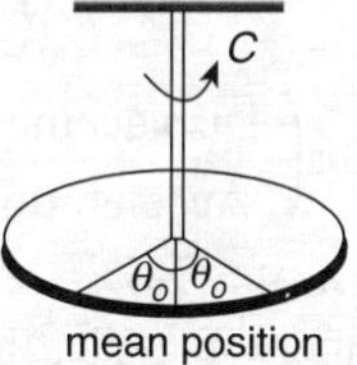

Linear SHM

Force, $M\dfrac{d^2x}{dt^2} = -kx$

or
$$\frac{d^2x}{dt^2} = -(k/M)x \quad (1)$$

Assume a trial solution

$$x = A\sin(\omega t + \varphi) \quad (2)$$

First differentiation gives velocity

$$v = \frac{dx}{dt} = A\omega\cos(\omega t + \varphi)$$

One more differentiation gives acceleration

$$a = \frac{d^2x}{dt^2} = -A\omega^2\sin(\omega t + \varphi)$$

or

$$\frac{d^2x}{dt^2} = -\omega^2 x \qquad (3)$$

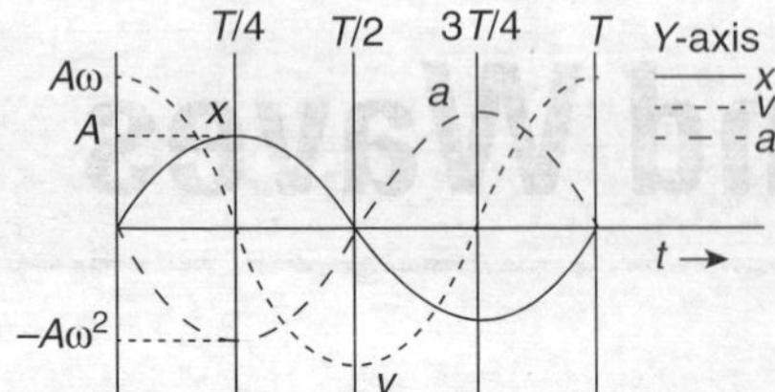

where $\omega = 2\pi\nu$ Comparing (1) and (3), we find that (2) is a solution if

$$\omega = \left(\frac{k}{M}\right)^{1/2}$$

or

$$\nu = \frac{1}{2\pi}\sqrt{\frac{k}{M}}$$

with solution

$$x = A\sin(\omega t + \varphi)$$

Graph of displacement x, velocity v and acceleration a as a function of time: Let us consider the special case when $\varphi = 0$ i.e.

$$x = A\sin\omega t$$

Conclusions from the Graph

(a) **Time period:** The motion repeats after a time interval

$T = \frac{1}{\nu}$ hence

$T = 2\pi\left(\frac{M}{k}\right)$ or $T = 2\pi\left(\frac{x}{kx/M}\right)^{1/2}$ i.e.,

Time period $T = 2\pi\left(\frac{\text{Displacement}}{\text{Acceleration}}\right)^{1/2}$ and is independent of amplitude A of SHM.

(b) **Frequency:** Number of cycles per second – it is the frequency ν and ω is the angular frequency.

(c) **Amplitude:** The maximum value of the displacement is A and it is called amplitude. The system oscillates between the extreme displacements $x = +A$ and $x = -A$.

(d) **Phase:** $(\omega t + \phi)$ is called the phase of motion. The phase of an oscillation system describes the state as regards the position, and the direction of motion at an instant t.

(e) **Initial phase or phase constant (ϕ):** This is the value of the phase when time $t = 0$. Hence phase constant tells us the position and direction of motion at the starting of oscillation. In our examples $\phi = 0$ means, $x = 0$ and $v > 0$, while $\phi = \pi$ means $x = 0$ and $v < 0$.

It can be further inferred from the displacement–velocity, acceleration graph that

(i) **Velocity** is out-of-phase of displacement by 90° or $\pi/2$ radians, i.e., when displacement is zero (at the centre of oscillation), velocity is either a maximum or a minimum. A negative velocity means particle is moving in –ve x direction. Also, it can be shown that

$$v = \omega\sqrt{A^2 - x^2} \text{ and } v_{max} = \omega A$$

(ii) **Acceleration** is always out of phase of displacement by 180° or π radians. Also acceleration, hence force or restoring force is zero at the centre of oscillation and has maximum magnitude at the extreme displacements. Maximum value of acceleration $a_{max} = \omega^2 A$

Angular SHM

Torque $\quad I\frac{d^2\theta}{dt^2} = -C\theta$

or $\quad \frac{d^2\theta}{dt^2} = -\left(\frac{C}{I}\right)\theta$

where I is moment of inertia of the mass about axis of rotation.

This equation has mathematically the same form as that of Linear SHM, hence solution will be once again

$$\theta = \theta_0 \sin(\omega t + \varphi)$$

where θ_0 = max angular displacement and $\omega = \left(\frac{C}{I}\right)^{1/2}$ is angular frequency.

Angular velocity $\frac{d\theta}{dt} = \theta_0\,\omega\cos(\omega t + \varphi)$

and angular acceleration $\frac{d^2\theta}{dt^2} = -\theta_0\,\omega^2\sin(\omega t + \varphi)$

All the conclusions that we have drawn about linear SHM also hold good for angular SHM when the terms *force* is replaced by *torque*, *linear displacement* by θ and *mass* by *moment of inertia I*.

Illustration 1

A particle undergoes SHM and its displacement, x is given by:

$$x = 25\sin\left(\frac{\pi}{3} - 3t\right)\text{ cm},$$

where t is in second.

Find the amplitude, the time period of the SHM, the epoch (initial phase), the expression for the velocity and its amplitude.

Solution

The displacement x is:

$$x = 25\sin\left(\frac{\pi}{3} - 3t\right)\text{cm}$$

$$= 25\sin\left(3t - \frac{\pi}{3} + \pi\right)\text{cm}$$

$$= 25\sin\left(3t + \frac{2\pi}{3}\right)\text{cm}.$$

The amplitude is 25 cm, the time period is given by,

$$T = \frac{2\pi}{\omega}$$

$$= \frac{2\pi}{3} \text{ s.}$$

The epoch (initial phase) is given by: $\frac{2\pi}{3}$. Note that this can be specified up to an integral multiple of 2π.

The velocity is given by, $v = \frac{dx}{dt}$

$$= 75\cos\left(3t + \frac{2\pi}{3}\right) \text{cm s}^{-1}.$$

and the velocity amplitude is 75 cm s^{-1}.

Illustration 2

A point describes SHM along a line 4 cm long. Its velocity, when passing through the centre of the line, is 12 cm/s. Find its frequency.

(a) 0.955 Hz (b) 1.230 Hz
(c) 2.955 Hz (d) 3.955 Hz

Ans. (a)

Solution

Amplitude of SHM will be

$$A = 2 \text{ cm}$$

Hence $\omega A = 12$ cm/s and $\omega = 6$

$$f = \frac{6}{2\pi} = 0.955 \text{ Hz}$$

Illustration 3

In what minimum time after its motion begins, will a particle oscillating according to the equation $y = 7 \sin 0.5\,\pi t$ move from the mean position to maximum displacement?

(a) 1 s (b) 2 s
(c) 3 s (d) 4 s

Ans. (a)

Solution

Comparing with $y = a \sin \omega t$

$$y_{max} = 7 = 7 \sin 0.5\,\pi t$$

$$0.5\pi t = \frac{\pi}{2}$$

or $t = 1$ second

Illustration 4

The equation of a SHM is given as $x = 6 \sin 10\,\pi t + 8 \cos 10\,\pi t$, where x is in cm and t in sec. Find the amplitude

(a) 100 cm (b) 10 cm
(c) 1 cm (d) 0.1 cm

Ans. (b)

Solution

Standard equation is $x = a \sin(\omega t + \phi)$

or $x = a \sin \omega t. \cos\phi + a \cos \omega t \sin\phi$

$$a \cos\phi = 6 \quad \text{(i)}$$

and $a \sin \phi = 8$ (ii)

Squaring and adding (i) and (ii)

$$a = 10 \text{ cm}$$

Illustration 5

A body is describing SHM of amplitude 1 cm. Its velocity while passing through the neutral point is 10 cm/s. What is its frequency?

(a) 2.6 s^{-1} (b) 3.6 Hz
(c) 1.6 s^{-1} (d) 5.2 Hz

Ans. (c)

Solution

$$A = 1 \text{ cm}$$

$$V_{max} = 10 \text{ cm/s}$$

$$V_{max} = \omega A$$

$$\Rightarrow \quad f = \frac{10}{2 \times 3.14 \times 1} = 1.6 \text{ s}^{-1}$$

Illustration 6

A particle executing linear SHM has a maximum velocity of 40 cm/s and a maximum acceleration of 50 cm/s^2. Find its time period.

(a) 5 sec (b) 10 sec
(c) 15 sec (d) 20 sec

Ans. (a)

Solution

$$v_{max} : \omega A = 40 \quad \text{(i)}$$

$$a_{max} : \omega^2 A = 50 \quad \text{(ii)}$$

Dividing (ii) by (i)

$$\omega = \frac{0}{40} = \frac{5}{4} \text{ rad/s}$$

$$\frac{T}{2\pi} = \frac{4}{5} \Rightarrow T = 5 \text{ sec}.$$

Illustration 7

A particle executing a SHM has a maximum displacement of 4 cm and its acceleration at a distance of 1 cm from its mean position is 3 cm/s^2. What will be its velocity when it is at a distance of 2 cm from its mean position?

(a) 2 cm/s (b) 5 cm/s
(c) 6 cm/s (d) 8 cm/s

Ans. (c)

Solution

Acceleration at any time t: $a = -\omega^2 y$

for $y = 1$ cm; $a = 3$ cm/s^2

gives $\omega = \sqrt{3}$ rad/s

Velocity at any time t: $v = \omega\sqrt{A^2 - y^2}$

$$\omega = \sqrt{3}\sqrt{4^2 - 2^2} = 6 \text{ cm/s}$$

Here A is amplitude, ω is angular frequency and y is displacement.

Illustration 8

A body is vibrating with SHM with amplitude 15 cm and frequency 4 Hz. Calculate the maximum acceleration when the amplitude is 10 cm.

(a) 64.28 m/s^2 (b) 73.82 m/s^2
(c) 94.82 m/s^2 (d) 83.28 m/s^2

Ans. (c)

Solution

$\text{Accel}_{max} = \omega^2 A$

$= (2\pi f)^2 a$

$f = 4$ Hz.

$$\text{Accel}_{max} = 4 \times \left(\frac{22}{7}\right)^2 \times (4)^2 \times 0.15$$

$= 94.82$ m/s^2

Illustration 9

A spring compressed by 0.1 m develops a restoring force of 10 N. A body of mass 4 kg is placed on it. Find the period of oscillations, if the body is disturbed slightly.

(a) 1.56 s (b) 2.56 s
(c) 1.26 s (d) 3.54 s

Ans. (c)

Solution

$$k = \frac{10}{0.1} = 100 \text{ N/m}$$

$$T = 2\pi\sqrt{\frac{m}{k}}$$

$$T = 2 \times \frac{22}{7} \times \sqrt{\frac{4}{100}} = 1.26 \text{ s}$$

Energy of an Oscillatory SHM

An oscillating SHM system has kinetic energy of motion (1/2 mv^2) and potential energy of the system (1/2 kx^2). Potential energy is the work done by the external agency against the restoring force ($-kx$) of the system.

The work done in displacement dx; $dW = (kx)\,dx$

(On integration, we get work done from 0 to x displacement, $W = 1/2\, kx^2$ which is also the PE)

Here at $x = 0$ PE is taken to be zero.

Total Energy *E* of the System

$E = \text{KE} + \text{PE}$

or $E = 1/2\, mv^2 + 1/2\, kx^2$

or $E = 1/2\, ma^2\, \omega^2 \cos^2(\omega t + \phi) + 1/2\, k\, a^2 \sin^2(\omega t + \phi)$

$$E = 1/2\, ma^2\omega^2 = \frac{1}{2}ka^2$$

That is total energy E remains constant during oscillations. The figure shows a plot of KE, PE and total Energy E as a function of displacement for one cycle.

Total energy E

Kinetic Energy of mass

$$\text{KE} = 1/2\, ma^2\, \omega^2 \cos^2(\omega t + \phi)$$

$$= 1/2\, k\sqrt{A^2 - x^2}$$

Potential energy of the system

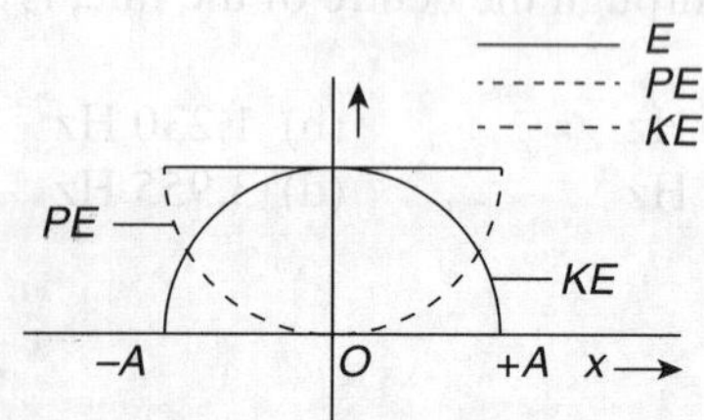

$$\text{PE} = \frac{ma^2}{2}\omega^2 \sin^2(\omega t + \varphi) = 1/2\, kx^2$$

PE is out of phase by $\frac{\pi}{2}$ with KE.

Both PE and KE oscillate with frequency 2ω.

Illustration 10

A particle of mass 5g executes SHM which is given by:

$$x = 10\cos\left(2\pi t - \frac{\pi}{4}\right) \text{cm}.$$

Find the force acting on the particle at the time $t = 2$s, as well as its momentum at the time $t = 4$s. Find the kinetic energy of the particle as a function of time t.

Solution

The force acting on the particle is given by:

$$F = ma$$

$$= m\left(\frac{d^2x}{dt^2}\right)_{t=2}$$

$$= -(5\times10^{-3})(2\pi)^2 \times 10\cos\left(4\pi - \frac{\pi}{4}\right) \times 10^{-2} N$$

$$= -50\times10^{-5} \times 4\pi^2 \times \frac{1}{\sqrt{2}} N$$

$$= 1.414\pi^2 \times 10^{-3}\,\text{N}.$$

The momentum of the particle is given by

$$p = mv$$

$$= (5\times10^{-3})(-2\pi)10\sin\left(8\pi - \frac{\pi}{4}\right)\times10^{-2}$$

$$= \frac{\pi}{\sqrt{2}}\times10^{-3} \text{ kg ms}^{-1}.$$

The kinetic energy of the particle is given by:

$$k = \frac{1}{2}mv^2$$

$$= \frac{1}{2}m\left(\frac{dx}{dt}\right)^2$$

$$= \frac{1}{2}\times5\times10^{-3}\times(20\pi)^2\left[\sin\left(2\pi t - \frac{\pi}{4}\right)\right]^2\times10^{-4}$$

$$= \left[\sin\left(2\pi t - \frac{\pi}{4}\right)\right]^2\times10^{-4}\text{ J}$$

Illustration 11

At a time when the displacement is half the amplitude, what fraction of the total energy is kinetic and what fraction is potential in a SHM?

(a) $\frac{3}{4}E, \frac{1}{4}E$ (b) $\frac{1}{4}E, \frac{3}{4}E$

(c) $\frac{1}{2}E, \frac{1}{2}E$ (d) $\frac{1}{5}E, \frac{4}{5}E$

Ans. (a)

Solution

$$E = \frac{1}{2}m\omega^2a^2$$

$$\text{KE} = \frac{1}{2}m\omega^2(a^2 - y^2)$$

$$= \frac{1}{2}m\omega^2\left(a^2 - \frac{a^2}{4}\right)$$

$$= \frac{3}{4}\cdot\frac{1}{2}m\omega^2a^2 = \frac{3}{4}E$$

$$\text{PE} = \frac{1}{2}m\omega^2y^2$$

$$= \frac{1}{2}m\omega^2\frac{a^2}{4} = \frac{1}{4}E$$

Illustration 12

A body of mass 5 kg executes a SHM of amplitude 0.5 m. If the force constant be 100 N/m calculate its KE when it is half way between the equilibrium and extreme positions.

(a) 1.4 J (b) 2.47 J

(c) 9.375 J (d) 8.125 J

Ans. (c)

Solution

$m = 5$ kg, $A = 0.5$ m, $k = 100$ N/m

$$\omega = \left(\frac{k}{m}\right)^{1/2}$$

$$\omega = \left(\frac{100}{5}\right)^{1/2} = \sqrt{20}\text{ rad/s}$$

Also displacement $y = \frac{0.5}{2} = 0.25$ m

$$\text{KE} = \frac{1}{2}m\omega^2(A^2 - y^2)$$

$$= \frac{1}{2}\times5\times(\sqrt{20})^2(0.5^2 - 0.25^2)$$

$$= 9.375\text{ J}$$

Illustration 13

A particle is executing SHM of amplitude A. At what displacement from the mean position, is the energy half kinetic and half potential?

(a) $\frac{a}{2}$ (b) $\frac{a}{\sqrt{2}}$

(c) $\frac{3a}{\sqrt{2}}$ (d) a

Ans. (b)

Solution

KE = PE

$$\frac{1}{2}m\omega^2(A^2 - y^2) = \frac{1}{2}m\omega^2y^2$$

or $$A^2 - y^2 = y^2$$

gives $$y = \frac{A}{\sqrt{2}}$$

∴ Energy will be half kinetic and half potential at displacements $\frac{A}{\sqrt{2}}$ on either side of the mean position.

Addition of Two SHMs in Perpendicular Directions

Suppose that there are two SHMs which are superposed in perpendicular directions on a particle:

$$x = A\sin\omega t$$

$$y = A\cos\omega t \quad \text{(i)}$$

Eliminating the time t between the two equations above, we get,

$$x^2 + y^2 = A^2 \quad \text{(ii)}$$

This is a circle, and it represents the equation of trajectory of the particle.

If the two SHMs are, however,

$$x = A_1\sin\omega t$$

$$y = A_2\sin\omega t \quad \text{(iii)}$$

The trajectory is a straight line:

$$\frac{x}{A_1} = \frac{y}{A_2} \quad \text{(iv)}$$

It is clear that the trajectory depends on the relative phases, the amplitudes and the individual frequencies (or the ratio of the frequencies).

Consider two SHMs with a phase difference of $\pi/4$, but with identical amplitudes, superposed in perpendicular directions.

$$x = A\sin\omega t,$$

$$y = A\sin(\omega t + \pi/4) \quad \text{(v)}$$

Eliminating ωt between the two equations, we get,

$$y = \frac{1}{\sqrt{2}}\left[x + \sqrt{A^2 - x^2}\right]$$

or, $2x^2 + 2y^2 - 2\sqrt{2}xy = A^2$

Substituting, $X = \frac{x+y}{\sqrt{2}}$ and $Y = \frac{x-y}{\sqrt{2}}$,

or, $x = \frac{X+Y}{\sqrt{2}}$, $y = \frac{X-Y}{\sqrt{2}}$

we get, $$\frac{X^2}{A'^2} + \frac{Y^2}{B'^2} = 1 \quad \text{(vi)}$$

where $A' = \frac{A}{\sqrt{2-\sqrt{2}}}$ and $B' = \frac{A}{\sqrt{2+\sqrt{2}}}$.

Illustration 14

Suppose that a particle is acted upon by two independent SHMs in two mutually perpendicular directions:

$$x = A\sin\omega t$$

$$y = 2A\sin\left(\omega t + \frac{\pi}{4}\right)$$

Find the path of the particle.

Solution

It is given that,

$$x = A\sin\omega t$$

$$y = 2A\sin\left(\omega t + \frac{\pi}{4}\right)$$

$$= \sqrt{2}A\sin\omega t + \sqrt{2}A\cos\omega t \quad \text{(i)}$$

The locus is given by the equation,

$$(y - \sqrt{2}x)^2 = (\sqrt{2}A\cos\omega t)^2$$

$$= 2(A^2 - x^2)$$

or, $$y^2 - 2\sqrt{2}xy + 4x^2 = 2A^2 \quad \text{(ii)}$$

This is the equation of an ellipse.

Important Simple Harmonic Systems

Horizontal Mass-Spring System

The mass-spring system shown in figure has resting position at $x = 0$. That is, the spring is relaxed and here it does not exert any force on the Mass M. At a displacement x of Mass M to the right, spring is stretched it exerts a force along $-x$ direction, i.e.,

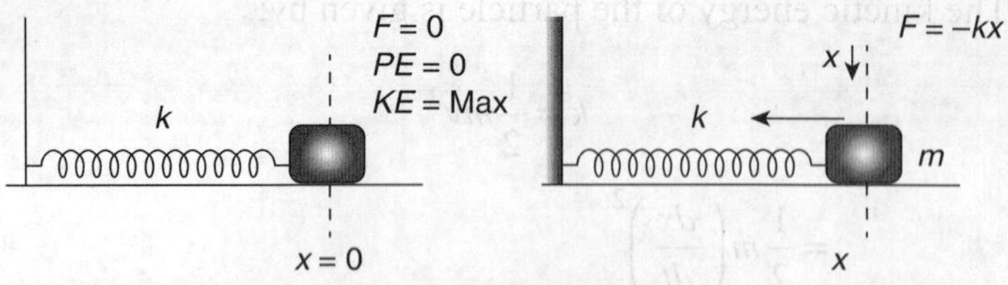

$$F = -kx$$

or $$\frac{d^2x}{dt^2} = -(k/M)x$$

Hence it will execute a SHM.

Here period of oscillation T,

$$T = 2\pi\sqrt{\frac{M}{k}}$$

Remark

Acceleration due to gravity does not appear in the time period. Hence, it has no effect on SHM. If the system is kept in a horizontally accelerated reference frame, the mean position will change but time period or frequency of oscillation will remain unchanged.

Vertical Mass-spring System

At position A gravitational force Mg is being balanced by stretching of length of spring (y_0). Hence total force on mass is zero, i.e., this becomes the mean or central position and the system will oscillate about it.

At position A', the net force due to additional stretching of spring by y will be

$$F = -ky$$

or $$\frac{d^2y}{dt^2} = -\left(\frac{k}{M}\right)y$$

The time period of oscillation will be

$$T = 2\pi\sqrt{\frac{M}{k}}$$

Remarks

(i) If the system oscillates in vertical accelerated reference frame, the mean position of oscillation will change but time period or frequency remains unaltered.

(ii) In the vertical mass-spring system, if the spring is not mass-less but has a spring mass M_s then the effective mass is

$M_{eff} = M + M_s/3$ and $T = 2\pi\left(\frac{M_{\text{eff}}}{k}\right)^{1/2}$ if only the spring oscillates, i.e., $M = 0$

$$T = 2\pi\left(\frac{M_s}{3k}\right)^{1/2}$$

(iii) If both ends of a spring of spring constant k are connected to masses M_1 and M_2 and the spring oscillates on a horizontal surface, then

$$T = 2\pi\left(\frac{M_R}{k}\right)^{1/2}$$

where $\frac{1}{M_R} = \frac{1}{M_1} + \frac{1}{M_2}$, here M_R is called the reduced mass.

Illustration 15

A block of mass m hangs from a spring of spring constant k under gravity. Find the frequency of natural oscillations of this pendulum, and its mean position. Express the displacement of the block as a function of time.

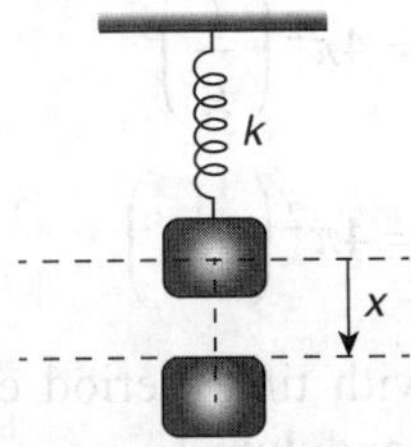

Solution

Suppose that the spring is extended by x from its un-extended position. The total energy of the system is given by:

$$E + \frac{1}{2}kx^2 = \frac{1}{2}m\left(\frac{dx}{dt}\right)^2 - mg\,x, \quad \text{(i)}$$

where the last term is the gravitational potential energy of the spring.

Since energy is conserved, we can write,

$$\frac{dE}{dt} = 0 = m\cdot\left(\frac{dx}{dt}\right)\left(\frac{d^2x}{dt^2}\right) + kx\left(\frac{dx}{dt}\right) - \text{mg}\left(\frac{dx}{dt}\right)$$

or, $$\left[m\frac{d^2x}{dt^2} + kx - \text{mg}\right]\left(\frac{dx}{dt}\right) = 0$$

or, $$m\frac{d^2x}{dt^2} + kx = \text{mg}$$

or, $$\frac{d^2x}{dt^2} + \omega^2 x = g \quad \text{(ii)}$$

where, $\omega^2 = \frac{k}{m}$, gives the angular frequency.

The linear frequency, $f = \frac{\omega}{2\pi}$

$$= \frac{1}{2\pi}\sqrt{\frac{k}{m}}.$$

The mean (or equilibrium) position is given by,

$$x_0 = \frac{g}{\omega^2} = \frac{\text{mg}}{k} \quad \text{(iii)}$$

The displacement is, $x = \frac{\text{mg}}{k} + A\sin(\omega t + \phi)$.

Simple Pendulum

It is a mechanical system which exhibits periodic motion. It consists of a point object of mass m suspended by a light string of length L, the upper end of the string being fixed to a support as shown in the figure.

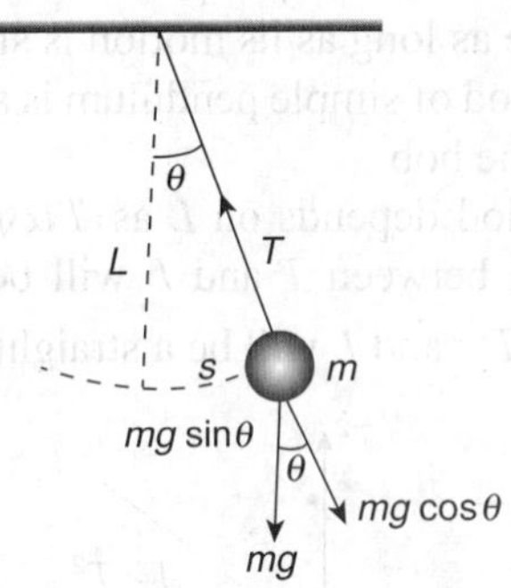

If the size of the object is small as compared to length of the strings, the pendulum can be considered as simple pendulum. When the object is pulled aside and released, it oscillates about its mean position or equilibrium position.

The force acting on the object are the tension force T along the string and gravitational force mg. the tangential component of gravitational force mg Sinθ always acts along its mean position $\theta = 0$, and opposite to the displacement. We can say that the tangential component of force is a restoring force.

Mathematically, the equation of motion in tangential direction is

$$F = -mg\sin\theta$$

$$= m\frac{d^2s}{dt^2}$$

where s is the distance measured along circular path as shown in figure.

As $s = L\theta$, the equation reduces to

$$\frac{d^2\theta}{dt^2} = -\frac{g}{L}\text{Sin}\,\theta$$

Here, L is constant and for a place g is also constant.

If we assume θ is small (less than about 8° or 0.2 rad). We can use simplified model in which Sin $\theta \to \theta$. Therefore, for small angles, the equation of motion becomes

$$\frac{d^2\theta}{dt^2} = -\frac{g}{L}\theta$$

We can compare this with standard equation of SHM with $\omega^2 = g/L$ and motion is approximated as simple harmonic for small amplitude. The solution of $\frac{d^2\theta}{dt^2} = -\frac{g}{L}\theta$ can be written as $\theta = \theta_{max}$ Cos $(wt + \phi)$

where θ_{max} is the maximum angular position with angular frequency $\omega = \sqrt{\frac{g}{L}}$

$$T = 2\pi\sqrt{\frac{L}{g}}$$

The Following Points Regarding Simple Pendulum Worth Remembering

1. Time period of simple pendulum is independent of amplitude as long as its motion is simple harmonic.
2. Time period of simple pendulum is also independent of mass of the bob.
3. Time period depends on L as $T \alpha \sqrt{L}$, i.e., $T^2 \alpha L$. So the graph between T and L will be a parabola while between T^2 and L will be a straight line.

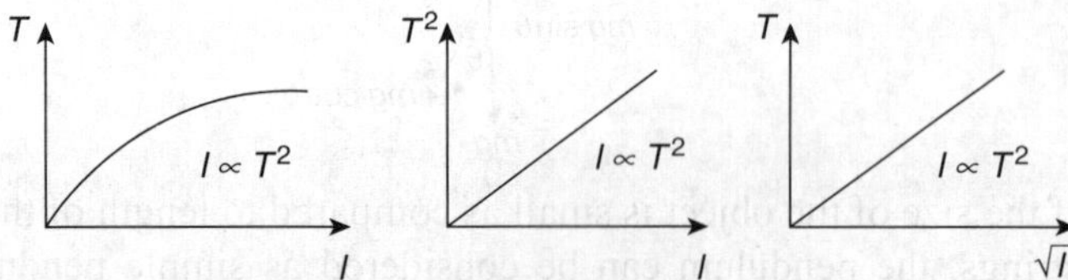

4. If the bob is suspended by the wire, due to change in temperature, length L will change and so will the time period. If $\Delta\theta$ is the increase in temperature and $\frac{\Delta T}{T} \simeq \frac{1}{2}\alpha\Delta\theta$
5. If the clock based on simple pendulum is taken to the moon or a hill, g will decrease and so T will increase. Hence, the clock will take longer time to complete the given number of oscillations and will become slow.
6. If damping is negligible and simple pendulum of density ρ is made to oscillate inside a liquid density $\sigma(<\rho)$, then g will decrease and T will increase.
7. If the time period of a simple pendulum is 2 seconds it is called a seconds pendulum and its length is 1 m.
8. If the length of the pendulum is comparable to the radius of earth, the time period will be approximately 83.8 minutes. This is the maximum time period which an oscillating pendulum can have.
9. If the simple pendulum is made to oscillate in a freely falling lift or a gravity free zone like a satellite i.e., $g = 0$, the pendulum will not oscillate.
10. If a simple pendulum is in an accelerating system like a lift or car, the following results are worth noting.

Motion	g_{eff}	T	Conclusion
Horizontal	$\sqrt{g^2 + a^2}$	$2\pi\sqrt{\frac{L}{(g^2 + a^2)}}$	T Decreases
Upward	$g + a$	$2\pi\sqrt{\frac{L}{g + a}}$	T Decreases
Downward	$g - a$	$2\pi\sqrt{\frac{L}{g - a}}$	T Increases

(Continued)

Illustration 16

What is the length of a simple pendulum whose time period for small oscillations equal 2 seconds?

(a) 0.99 m (b) 1.2 m
(c) 2 m (d) 2.2 m

Solution

Time period of a simple pendulum is given by

$$T = 2\pi\sqrt{\frac{\ell}{g}}$$

$$T^2 = 4\pi^2\left(\frac{\ell}{g}\right)$$

$$T^2 = 4\pi^2\left(\frac{1}{9.8}\right)$$ gives 1 = 1 m

(Simple pendulum with time period equal to 2 seconds is also called seconds pendulum.)

Illustration 17

If pendulum in Example 17 is mounted in a lift which accelerates upwards at 2.8 m/s², by what factor does its period of oscillation change from the original value?

(a) 0.77 (b) 0.88
(c) 0.99 (d) No change

Solution

$$T_1 = 2\pi\sqrt{\frac{\ell}{g}}$$

As the lift accelerates upwards at 2.8 m/s², the effective value of g will become g'

$(= 9.8 + 2.8 = 12.6 \text{ m/s}^2)$ and new time period $T_2 = 2\pi\sqrt{\frac{1}{12.6}}$

$$\frac{T_2}{T_1} = \sqrt{\frac{g}{12.6}} \text{ or } \frac{T_2}{T_1} = 0.88$$

Free, Forced and Damped Oscillations (Qualitative Ideas Only)

Free Oscillations – The oscillation motion which we have considered so far comes under the simplified model of an ideal frictionless system that oscillates indefinitely under a linear restoring force with fundamental frequency. The oscillation of a particle with fundamental frequency, under the effect of restoring force, is known free oscillation and it remains constant.

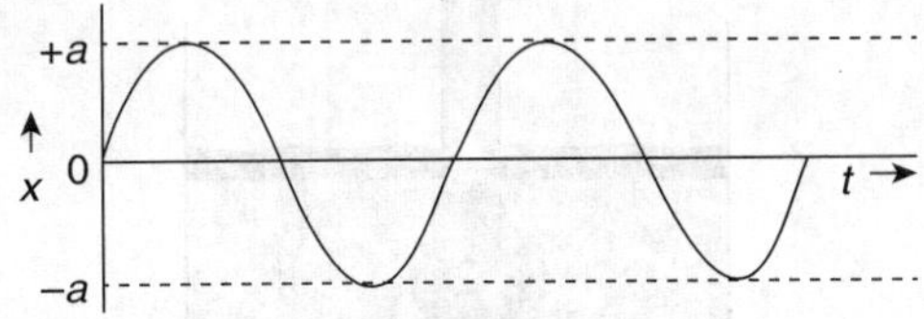

The natural frequency is $v = \frac{1}{2\pi}\sqrt{\frac{K}{m}}$

Damped Simple Harmonic Motion

When a system oscillates in a medium where there is a dissipative force, the oscillations are said to be damped oscillations. The example is motion of a simple pendulum in presence of air resistance. In damped oscillations although the energy of the system is continuously dissipated, the oscillations remain appreciably periodic.

Let us assume a damping force for proportional to velocity

$f_d = -bv$ where b is damping constant

The restoring force on the oscillating system is $f = -kx$

By Newton's second law

$$ma = -kx - bv$$

$$m\frac{d^2x}{dt^2} + b\frac{dx}{dt} + kx = 0$$

The solution of the above differential equations is

$$x = x_m e^{-\frac{bt}{2m}} \cos(\omega' t + \theta)$$

$$\omega' = \sqrt{\frac{k}{m} - \frac{b^2}{4m^2}}$$

where x_m is the amplitude and ω' is angular frequency of the damped oscillator.

Although the cosine function has a period $\frac{2\pi}{\omega'}$ the function x is not strictly periodic because of the term $e^{-\frac{bt}{2m}}$ which decreases continuously with time. If $b = 0$, there is no damping and the function x reduces to $x = x_m \cos(\omega t + \theta)$.

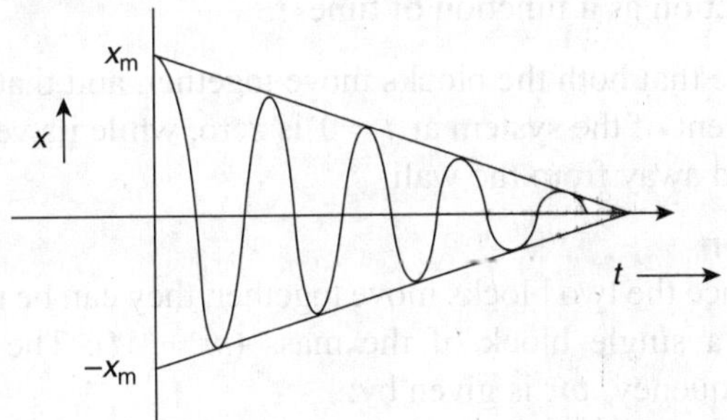

Displacement vs. time graph in damped harmonic oscillations.

Forced Oscillation

When a body oscillates under the influence of an external periodic force with a frequency other than its natural frequency then the oscillations are called forced oscillations. The differential equation of forced oscillations is

$$m\frac{d^2x}{dt^2} + b\frac{dx}{dt} + kx = F_0 \cos \omega_d t \qquad \text{(i)}$$

where ω_d is the frequency of driving force.

The solution of the equation is

$$x = x_m \cos(\omega_d t + \theta) \qquad \text{(ii)}$$

The amplitude x_m is given by

$$x_m = \frac{F_0}{[m^2(\omega_0^2 - \omega_d^2) + \omega_d^2 b^2]^{1/2}} \qquad \text{(iii)}$$

$$\tan\phi = -\frac{v_0}{\omega_d x_0} \qquad \text{(iv)}$$

where v_0 and x_0 are respectively velocity and displacement at $t = 0$

ω_0 is the natural frequency of the driven force.

Resonance

The solution of Eq. (i) gives

$$x_m = \frac{F_0}{[m^2(\omega_0^2 - \omega_d^2) + \omega_d^2 b^2]^{1/2}}$$

and $\omega_0 = \sqrt{k/m}$ is the natural frequency of an undamped oscillator ($b = 0$). The above equation shows that the amplitude of the forced oscillator is constant, for given driving force as it is being driven in steady state by an external force. For small damping, the amplitude becomes large when the frequency of the driven force is very nearly equal to the natural frequency of oscillation, or when $\omega_d = \omega_0$. The dramatic increase in amplitude near the natural frequency is called **resonance.**

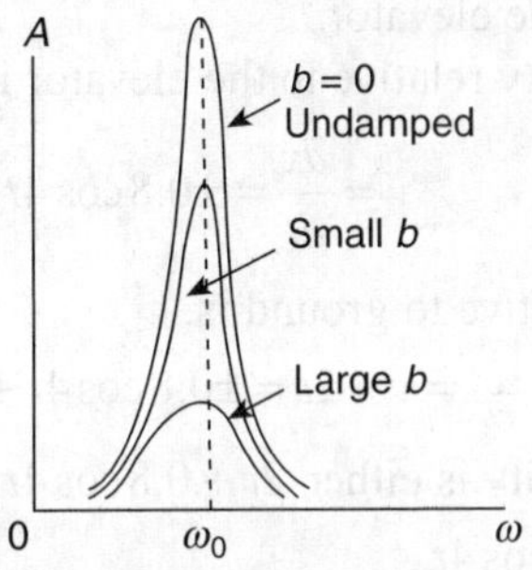

The above is a graph of amplitude vs. frequency for varying resistive forces. The amplitude increases with decreases damping ($b \to 0$) and the resonance curve flattens. In the absence of the damping force ($b = 0$), we see that the steady-state amplitude approaches infinity as $\omega_d = \omega_0$.

SOLVED EXAMPLES

Example 1

A block of mass 60 kg is placed on a stand within an elevator moving upward with an acceleration of $2\,\text{ms}^{-2}$. The stand is loaded on a spring and oscillates with a time period of $\frac{\pi}{2}$ s and amplitude 20 cm.

(a) Find the maximum reaction force acting on the block due to the stand.

(b) The displacement of the stand at the time $t = 0$, measured w.r.t. the elevator, is zero. Find the velocity of the block relative to the ground as a function of time, t. Assume that the elevator is at rest at $t = 0$.

Solution

(a) The reaction force R will be a maximum when the stand is accelerating upward.

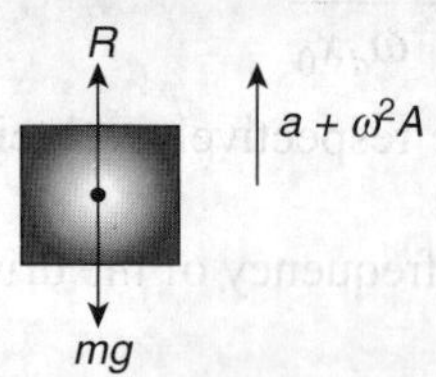

$$R - mg = m(a + \omega^2 A)$$

or, $$R = m(a + g + \omega^2 A) \quad \text{(i)}$$

$\because$ $$m = 60\text{ kg},\ a + g = 2 + 10 = 12\text{ ms}^{-2},$$

$$\omega^2 A = \left(\frac{2\pi}{\pi/2}\right)^2 \times 0.2 = 3.2\text{ ms}^{-2},$$

$$R = 60(12 + 3.2)\text{N}$$

$$= 912\text{ N}.$$

(b) The displacement of the bock may be given by the expression,

$$e\ x = \pm 0.2 \sin 4t$$

relative to the elevator.

$\therefore$ its velocity relative to the elevator is,

$$v = \frac{dx}{dt} = \pm 0.8 \cos 4t$$

and that relative to ground is,

$$v' = v + 2t = \pm 0.8 \cos 4t + 2t \quad \text{(ii)}$$

$\therefore$ the velocity is either $2t + 0.8\cos 4t$

or, $2t - 0.8\cos 4t$,

based on the information given.

Example 2

A hydrometer consists of a bulb attached to a vertical cylindrical tube, the entire apparatus being dipped into a liquid of density ρ_ℓ. If the mass of the hydrometer is m, and a length l of the stem is under the liquid, find the time period of vertical oscillations of the hydrometer. Can this be used to measure the mass? Assume that the stem is cylindrical and its outer radius is r.

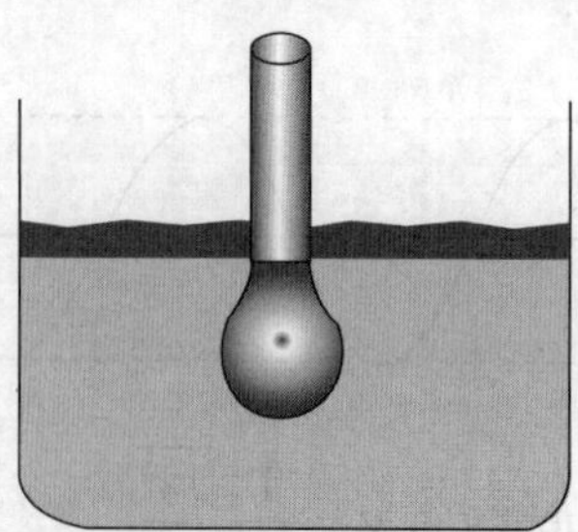

Solution

Suppose that the hydrometer is depressed by x. The restoring force is:

$$F_{\text{rest}} = (B_0 + \delta B) - W,$$

where B_0 is the force of buoyancy acting on it in equilibrium and $W = mg$ is its weight. Thus,

$$B_0 = W$$

$\therefore$ $$F_{rest} = \delta B = \text{excess buoyancy}$$

$$= \rho_\ell \cdot \pi r^2 \times g$$

$\therefore$ the angular frequency, $\omega = \sqrt{\dfrac{\pi r^2 \rho_\ell g}{m}}$

The time period of oscillation, $T = \dfrac{2\pi}{\omega} = 2\pi\sqrt{\dfrac{m}{\pi r^2 \rho_\ell g}}$.

If ρ_ℓ, g are known, m can be measured.

Example 3

The coefficient of friction between the two blocks shown in the figure is μ, and the horizontal plane is smooth. When the system is slightly displaced and released it performs small oscillations. The masses of the blocks are as shown in the diagram.

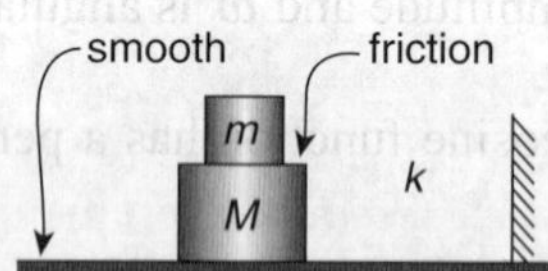

(a) Find the frequency of small oscillations.

(b) The amplitude of oscillations is A. Find the force of friction as a function of time, t.

Assume that both the blocks move together, and that the displacement of the system at $t = 0$ is zero, while its velocity is directed away from the wall.

Solution

(a) Since the two blocks move together, they can be regarded as a single block of the mass ($m + M$). The angular frequency, ω, is given by:

$$\omega = \sqrt{\frac{k}{m + M}}$$

The linear frequency, f, is given by:

$$f = \frac{\omega}{2\pi} = \frac{1}{2\pi}\sqrt{\frac{k}{m+M}} \quad \text{(i)}$$

(b) The displacement, x, is given by:

$$x = A\sin\omega t$$

$\therefore$ the acceleration, a, is given by:

$$a = \ddot{x} = -\omega^2 A\sin\omega t \quad \text{(ii)}$$

Since only the force of static friction acts on the smaller block (of mass m), we can write:

$$F_f = m\ddot{x} = -m\omega^2 A\sin\omega t \quad \text{(iii)}$$

Note: The force of friction is static but does not necessarily reach its limiting value.

Example 4

An ideal monatomic gas is enclosed in a cylinder by a close fitting piston of mass m, and the entire arrangement is placed as shown in the figure. The piston can slide vertically without friction within the cylinder. The sides of the cylinder are insulated to prevent any loss of heat, and the piston is made of thermally insulating material. Find the frequency of small oscillations. Assume that the pressure and temperature of the gas within the cylinder is uniform and that the external atmospheric pressure is p_0, and that the gas undergoes an adiabatic process whose equation is: pV^γ = constant. The volume of the ideal gas, at equilibrium, is V_0 and the cross-sectional area of the piston is A.

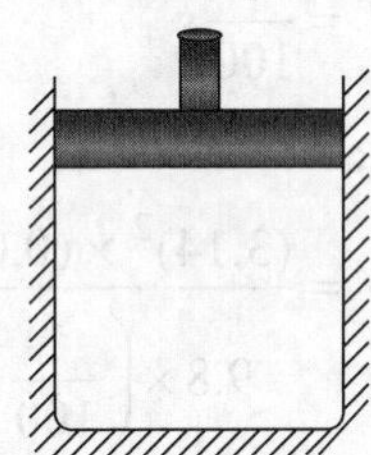

Solution

At equilibrium, the forces acting on the piston are balanced. If the pressure of the gas is p_g,

$$p_g.A = p_0 A + mg$$

$$\therefore \quad p_g = p_0 + \frac{mg}{A} \quad \text{(i)}$$

When the piston is slightly depressed by a small amount x, the pressure of the gas increases and is given by p', where,

$$p'(V_0 - Ax)^\gamma = p_g V_0^\gamma, \qquad \left(\gamma = \frac{C_P}{C_V} = \frac{5}{3}\right)$$

or, $$p' = p_g\left(1 - \frac{Ax}{V_0}\right)^{-5/3}$$

$$\simeq p_g\left(1 + \frac{5Ax}{3V_0}\right) \quad \text{(after binomial expansion and assuming } Ax << V_0)$$

The equation of motion of the piston is given by,

$$m\ddot{x} = (p_0 A + mg) - p'A$$
$$= (p_g - p')A$$
$$\simeq -\frac{5}{3}\frac{A^2 p_g}{V_0}\cdot x$$

or, $$\ddot{x} = -\frac{5}{3}\frac{A^2 p_g}{mV_0}x \quad \text{(ii)}$$

The angular frequency of small oscillations, ω, is given by

$$\omega = \sqrt{\frac{5A^2 p_g}{3mV_0}}$$

$$= \sqrt{\frac{5A}{3mV_0}(p_0 A + mg)}$$

$$\therefore \quad f = \frac{\omega}{2\pi} = \frac{1}{2\pi}\sqrt{\frac{5A}{3mV_0}(p_0 A + mg)} \quad \text{(iii)}$$

Example 5

A thin uniform board of length L and mass M is balanced on a fixed semicircular cylinder of radius R as shown in the figure. If the plank is tilted slightly from its equilibrium position, find the frequency of small oscillations.

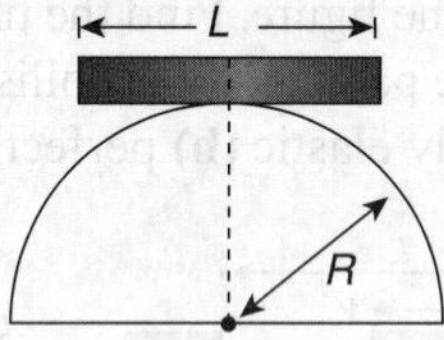

Solution

Suppose that the plank is tilted from its equilibrium position and it is oscillating. Suppose that the angular position of contact of the plank is θ, as shown in the adjacent figure.

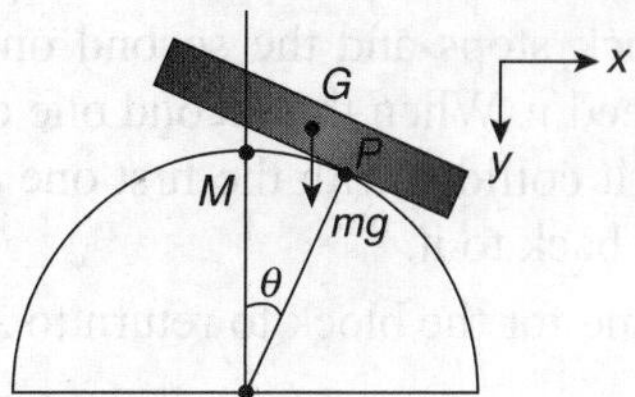

The x, y coordinates of G w.r.t. M as the origin and the directions as shown in the adjacent figure are:

$$x = R\sin\theta - R\theta\cos\theta$$

$$y = R(1 - \cos\theta) - R\theta\sin\theta\,.$$

Then, the velocity components are:

$$\dot{x} = R\theta\,\dot{\theta}\sin\theta$$

$$\dot{y} = -R\theta\,\dot{\theta}\cos\theta\,.$$

The net kinetic energy is,

$$KE = \frac{1}{2}mR^2\theta^2\dot{\theta}^2 + \frac{1}{2}\left(\frac{1}{12}mL^2\right)\dot{\theta}^2$$

$$PE = -mgy$$

$$= -mgR\left[(1-\cos\theta) - \theta\sin\theta\right]$$

$$= \frac{1}{2}mgR\theta^2.$$

∴ The total mechanical energy, for small values of θ is given by,

$$E = \frac{1}{2}\left(\frac{1}{12}mL^2\right)\dot{\theta}^2 + \frac{1}{2}mgR\theta^2$$

Where we have ignored the product $\theta^2\dot{\theta}^2$ since it is smaller in magnitude compared to θ^2.

∴ The frequency of small oscillations is,

$$f = \frac{1}{2\pi}\sqrt{\frac{mgR}{\frac{1}{12}mL^2}} = \frac{\sqrt{3gR}}{\pi L}.$$

Example 6

A block B of mass m is attached to a fixed spring of spring constant k and is free to oscillate on a smooth horizontal plane as shown in the figure. A second identical small block of equal mass starts from A and moves towards B, which is at rest, as shown in the figure. Find the time taken by the free block to return to the point A, if the collision between the two blocks is **(a)** perfectly elastic **(b)** perfectly inelastic .

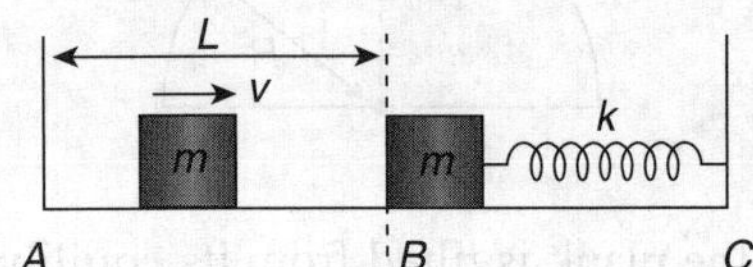

Solution

(a) If the collision between the blocks is perfectly elastic, the first block stops and the second one starts moving with the speed v. When the second one completes a half oscillation, it collides with the first one and transfers its momentum back to it.

The total time for the block to return to A is,

$$t = \frac{2L}{v} + \pi\sqrt{\frac{m}{k}}$$

(b) If the collision is perfectly inelastic, the two blocks move together after the collision. Their common velocity after collision is $\frac{v}{2}$. The time period of a half oscillation is:

$$t_2 = \pi\sqrt{\frac{2m}{k}}$$

When the blocks separate, the velocity of the block is $\frac{v}{2}$.

The total time is, therefore,

$$t = \frac{L}{v} + \pi\sqrt{\frac{2m}{k}} + \frac{L}{v/2}$$

$$= \frac{3L}{v} + \pi\sqrt{\frac{2m}{k}}$$

Example 7

The period of a simple pendulum is increased by $\frac{1}{100}$ s when the length is increased by 1 cm. Find the original length of pendulum.

(a) 1 m (b) 1.5 m
(c) 2 m (d) 2.5 m

Solution

$$T = 2\pi\sqrt{\frac{\ell}{g}}$$

Differentiating,

$$\sqrt{k/m_1} = \frac{\pi(d\ell)}{\sqrt{g\ell}}$$

$$(dT)^2 = \frac{\pi^2(d\ell)^2}{g\ell}$$

or $$1 = \frac{\pi^2(d\ell)^2}{g(dT)^2}$$

$$d\ell = 1 \text{ cm} = 0.01 \text{ m}$$

$$dT = \frac{1}{100}s$$

$$g = 9.8 \text{ m/s}^2$$

$$\ell = \frac{(3.14)^2 \times (0.01)^2}{9.8 \times \left(\frac{1}{100}\right)^2} = 1 \text{ m}$$

Example 8

A small bob of mass 50 g oscillates as a simple pendulum, with an amplitude 5 cm and period 2s. Find the tension in the thread when its velocity is maximum.

(a) 0.29 N (b) 0.39 N
(c) 0.49 N (d) 0.59 N

Solution

$$\omega = \frac{2\pi}{T} = \pi \text{ s}^{-1}$$

Maximum velocity, $v = \omega A = 0.05\,\pi =$ m/s

Tension in the supporting thread $= mg + \frac{mv^2}{\ell}$

Now, $$T = 2\pi = \sqrt{\frac{\ell}{g}}$$

$$\Rightarrow \quad \ell = \frac{T^2 g}{4\pi^2}$$

$$\Rightarrow \quad \text{Tension} = 0.05 \times 9.8 + \frac{0.05 \times (0.05\pi)^2 \times 4\pi^2}{(2)^2 \times 9.8}$$

$$\simeq 0.49 \text{ N}$$

Example 9

Two springs A and B have their force constants k_1 and k_2 ($k_1 > k_2$) respectively. On which spring is more work done when their lengths are increased by the same amount?

(a) On A (b) On B
(c) Can't say (d) Both (a) and (b)

Solution

Work done gets stored in spring as its PE

$$\text{PE} = \frac{1}{2} k x^2$$

$$\text{PE}_A = \frac{1}{2} k_1 x^2, \ \text{PE}_B = \frac{1}{2} k_2 x^2$$

$$\therefore \quad \text{PE}_A > \text{PE}_B$$

Example 10

In Example 9, on which spring is more work done when they are stretched by the same force?

(a) Spring A (b) Spring B
(c) Can't say (d) Both (a) and (b).

Solution

For Spring A, Force F will stretch the spring by an amount x_1 such that

$$x_1 = \frac{F}{k_1}$$

Energy stored in Spring A will be

$$\text{PE}_{(A)} = \frac{1}{2} k_1 x_1^2 = \frac{1}{2} k_1 \frac{F^2}{k_1^2}$$

$$\text{PE}_{(A)} = \frac{F^2}{2k_1}$$

Similarly $\text{PE}(B) = \dfrac{F^2}{2k_2}$ Since $k_1 > k_2$

$$\text{PE}_{(B)} > \text{PE}_{(A)}$$

Example 11

Two identical springs each of force constant k are connected in (i) series (ii) parallel and support a mass. Calculate the ratio of their time periods.

(a) 1 : 2 (b) 1 : 4
(c) 2 : 1 (d) 4 : 1

Solution

In series, resultant force constant will be $\dfrac{k}{2}$

Hence time period $T_1 = 2\pi\sqrt{\dfrac{m}{k/2}} = 2\pi\sqrt{\dfrac{2m}{k}}$

In parallel resultant force constant will be $2k$.

Hence $\quad T_2 = 2\pi\sqrt{\dfrac{m}{2k}}$

$$\frac{T_1}{T_2} = \frac{2}{1}$$

Example 12

The time period of oscillation of a block of mass m attached to a light spring of spring constant k is $2\pi\sqrt{\dfrac{m}{k}}$ in the absence of all other forces. Suppose that a block of mass m is attached to a fixed spring on one side and rests (i.e., is not attached to) against an identical spring on the other side. It is now allowed to oscillate. The frequency of small oscillations is

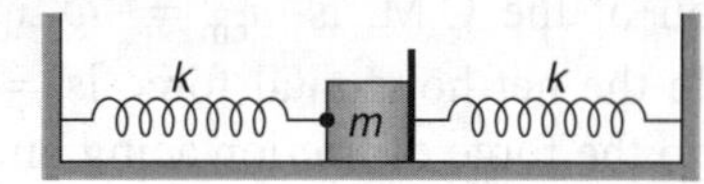

(a) $\dfrac{1}{\pi}\sqrt{\dfrac{k}{m}}$ (b) $\dfrac{(4-2\sqrt{2})}{\pi}\sqrt{\dfrac{k}{m}}$

(c) $\dfrac{1}{\pi}(2-\sqrt{2})\sqrt{\dfrac{k}{m}}$ (d) $\dfrac{(2+\sqrt{2})}{\pi}\sqrt{\dfrac{k}{m}}$

Solution

The effective value of the spring constant during oscillation is $2k$ and during the other half oscillation is k.

$\therefore$ The total time of oscillation is

$$T = \pi\sqrt{\frac{m}{k}} + \pi\sqrt{\frac{m}{2k}}$$

The frequency is $\dfrac{1}{T} = \dfrac{1}{\pi}\sqrt{\dfrac{k}{m}}(2-\sqrt{2})$.

Example 13

A block of mass m, attached to a fixed post O on a smooth inclined wedge of mass M, oscillates with amplitude A and linear frequency f. The wedge is located on a rough horizontal surface. If the angle of the wedge is 60°, then the force of friction acting on the wedge is given by (coefficient of static friction = μ).

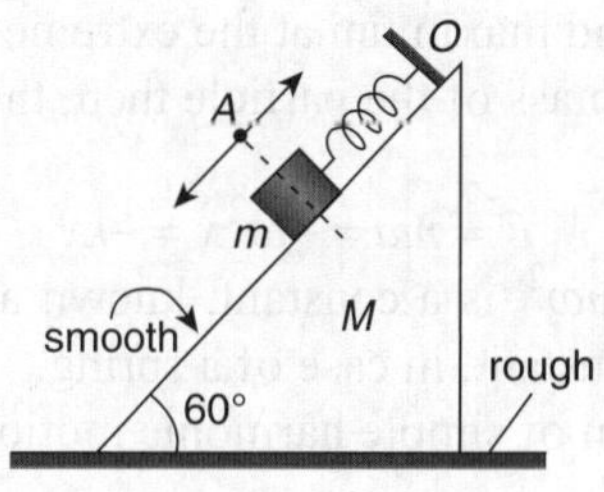

(a) $\mu(M+m)g$

(b) $\frac{1}{2}m\omega^2 A\sin\omega t$

(c) $\mu\left[(M+m)g+\frac{\sqrt{3}}{2}m\omega^2 A\sin\omega t\right]$

(d) $\mu(M+m)\omega^2 A\sin\omega t$.

Solution

The small block oscillates along the inclined plane with an amplitude *A*. As a result the centre of mass of the system undergoes SHM along the horizontal direction:

$$x_{cm}=\frac{mA\sin\omega t}{m+M}\cos 60”$$

$$=\frac{1}{2}\frac{m}{m+M}A\sin\omega t$$

The acceleration of the C.M. is $a_{cm}=-\omega^2 x_{cm}$, along the horizontal while the net horizontal force is $=(M+m)a_{cm}$, which is equal to the force of friction acing on it.

Example 14

A mass *m* is suspended from a spring of force constant *k* and just touches another identical spring fixed to the floor as shown in the figure. The time period of small oscillations is:

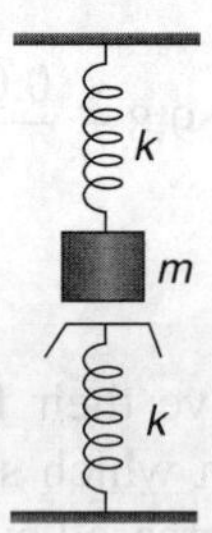

(a) $2\pi\sqrt{\frac{m}{k}}$ (b) $\pi\sqrt{\frac{m}{k}}+\pi\sqrt{\frac{m}{k/2}}$

(c) $\pi\sqrt{\frac{m}{3k/2}}$ (d) $\pi\sqrt{\frac{m}{k}}+\pi\sqrt{\frac{m}{2k}}$

Solution

When the spring undergoes displacement in the downward direction it completes one half oscillation while it completes another half oscillation in the upward direction. The total time period is:

$$T=\pi\sqrt{\frac{m}{k}}+\pi\sqrt{\frac{m}{2k}}$$

POINTS TO REMEMBER

1. To and fro motion of a body on the same path is known as oscillation.
2. When the motion of a body is repeated after a regular interval of time, its motion is said to be harmonic or periodic. The time interval after which it is repeated is known as time period.
3. If the particle moves on the straight path in such a way that its acceleration is always directed towards the mean path the motion is said to be simple harmonic motion (SHM).
4. Taking the mean position as the origin and the line of the motion as x-axis, the defining equation of the SHM can be written as
$$a=-\omega^2 x$$
where *a* is the acceleration of the particle and ω^2 is a constant. Thus, the acceleration of the particle is zero at the origin and maximum at the extreme ends.
5. If *m* be the mass of the particle then, the force acting on it is
$$F=ma=-\omega^2 x=-kx$$
where $k=m\omega^2$ is a constant, known as force constant, or spring constant, in case of a spring.
6. The equation of simple harmonic motion is given by
$$x=A\sin\omega t \quad (1)$$
$$x=A\cos\omega t \quad (2)$$
7. Note the basic difference between the above Eqs. (1) and (2). If the particle starts its motion from either end is the path, we should use Eq. (2) as it gives $x=$ A at $t=0$. On the other hand, if it starts its motion from the mean position, we should use Eq. (1) as it gives $x=0$ at $t=0$.
8. The time period of a particle executing SHM is
$$T=\frac{2\pi}{\omega}=2\pi\sqrt{\frac{m}{k}}=2\pi\sqrt{\frac{\text{Displacement}}{\text{Acceleration}}}$$
9. The velocity of a particle executing SHM at a displacement from the mean position is
$$v=\omega\sqrt{a^2-x^2}$$
This shows that the velocity of the particle executing SHM is maximum at the origin and minimum at the ends (i.e., $x=\pm$A).
10. The kinetic energy of a particle executing SHM at a displacement *x* from the mean position is
$$E_k=\frac{1}{2}m\omega^2(A^2-x^2$$
11. The potential energy of a particle executing SHM at a displacement *x* from the mean position is
$$E_p=\frac{1}{2}m\omega^2 x^2$$

12. The total energy of a particle executing SHM is a constant and is equal to

$$E = E_p + E_k = \frac{1}{2} m\omega^2 A^2$$

13. An SHM is produced when a restoring force proportional to displacement acts on a particle. If a particle is acted simultaneously by two separate forces, each of which can produces an SHM, the resultant motion of the particle is a combination of two SHMs.

14. Let the particle be acted upon by two forces simultaneously along the same line and the equations of the SHMs be

$$x_1 = A_1 \sin \omega t$$

$$x_2 = A_2 \sin(\omega t + \phi)$$

where ϕ is the phase difference between the two SHMs at $t = 0$. The resistant displacement of the particle is given by

$$x = x_1 + x_2 = A \sin(\omega t + \phi)$$

where $A = \sqrt{A_1^2 + A_2^2 + 2A_1 A_2 \text{Cos}\phi}$ is the resultant amplitude of the particle and $\delta = \tan^{-1} . \frac{A_2 \text{Sin}\phi}{A_1 + A_2 \text{C}\,\text{os}\phi}$

A is maximum $\phi = 0$ and $A_{max} = |A_1 + A_2|$

A is minimum when $\phi = \pi$ and $A_{min} = |A_1 - A_2|$

15. Let the particle be acted upon simultaneously by two forces in two mutually perpendicular directions and the equations of SHMs produced by these two forces be given by

$$x = A_1 \sin \omega t$$

and

$$y = A_2 \sin(\omega t + \phi)$$

The resultant motion of the particle will be in a plane and the path of the particle, in general, will be an ellipse. The equation of the path followed by the particle in such a case is

$$\frac{x^2}{A_1^2} + \frac{y^2}{A_2^2} + \frac{2xy}{A_1 A_2} \cos\phi = \sin^2\phi$$

Different values of the angle ϕ result in different path followed by the particle.

SECTION - II

WAVE MOTION

Wave in Deformable or Elastic Media

Unlike electromagnetic waves, mechanical waves require a medium having inertia, elasticity and no friction for their existence and propagation. A mechanical wave is defined as a disturbance in a medium carrying energy and momentum from one part of the medium to another, while the particles of the medium execute simple harmonic motion about their mean positions without moving along with the wave. Mechanical waves originate in the displacement of some portion of an elastic medium from its normal position, causing it to oscillate about an equilibrium position. Because of the elastic forces on adjacent layers, the disturbance is transmitted from one layer to the next through the medium, thus transferring energy to them. The energy in the waves is in the form of both kinetic and potential energy. Thus, mechanical waves are characterized by the transport of energy through matter by the motion of a disturbance in that matter without any corresponding bulk motion of the matter itself.

Mechanical waves can exist in any deformable, elastic medium, be it solid, liquid or gas. The waves are called longitudinal, if the motion of each individual particle of the medium is in the same direction as that of wave propagation through that point. If the motion is perpendicular, the wave is called transverse. The velocity of transverse wave travelling along a string is given by

$$V = \sqrt{(T/m)}$$

where T is the tension applied to the string of mass per unit length m. The velocity of a longitudinal wave (compressional wave) depends on the elasticity and density of the medium

in liquids: $v = \sqrt{\frac{\text{(bulk modulus)}}{\text{density of liquid}}} = \sqrt{\frac{B}{\rho}}$

in solid rods: $v = \sqrt{\frac{\text{(Young's modulus)}}{\text{density of liquid}}} = \sqrt{\frac{Y}{\rho}}$

in gases: $v = \sqrt{\frac{\gamma \times \text{(pressure of gas)}}{\text{density of gas}}} = \sqrt{\frac{\gamma P}{\rho}}$

where γ, the ratio of molar specific heats C_p/C_v is about 1.67 for monoatomic gases such as helium and neon. It is about 1.40 for diatomic gases such as N_2, O_2, and H_2.

Velocity of Sound in an Ideal Gas

The motion of sound wave in air is adiabatic. In the case of an ideal gas, the relation between pressure p and volume V during an adiabatic process is given by

$$pv^\gamma = \text{Constant.}$$

where γ is the ratio of the heat capacity at constant pressure to that at constant volume.

After differentiating, we get

$$\frac{dp}{dV} V^\gamma + \gamma p V^{\gamma-1} = 0$$

Since $\quad B = -V\dfrac{dp}{dV} = \gamma p$

$\therefore \quad v = \sqrt{\dfrac{\gamma p}{\rho}}$

(Laplace correction in contrast to Newton's formula $v = \sqrt{\dfrac{P}{\rho}}$)

Using the gas equation $\dfrac{p}{\rho} = \dfrac{RT}{M}$ where M is the molar mass.

Thus, $v = \sqrt{\dfrac{\gamma RT}{M}}$ $\quad$ (T = Temperature in Kelvin)

Effect of Pressure, Temperature and Humidity on the Speed of Sound in Air

We have stated that for an ideal gas, the pressure, volume and temperature of a given mass satisfy

$$\frac{PV}{T} = \text{Constant.}$$

As the density of a given mass is inversely proportional to its volume, the above equation may also be written as

$$\frac{P}{\rho} = cT,$$

where c is a constant. The speed of sound is

$$v = \sqrt{\frac{\gamma P}{\rho}} = \sqrt{\gamma cT}$$

Thus, if pressure is changed but the temperature is kept constant, the density varies proportionally and P/ρ remains constant. The speed of sound is not affected by the change in pressure provided the temperature is kept constant.

If the temperature of air is changed then the speed of sound is also changed.

From the above result,

$$v \propto \sqrt{T}.$$

At STP, the temperature is 0° C or 273 K. If the speed of sound at 0° C is v_0, its value at the room temperature T (in Kelvin) will satisfy

$$\frac{v}{v_0} = \sqrt{\frac{T}{273}} = \sqrt{\frac{273 + t}{273}},$$

where t is the temperature in °C. This may be approximated as

$$\frac{v}{v_0} = \left(1 + \frac{t}{273}\right)^{1/2} \cong 1 + \frac{t}{546}$$

or, $\quad v = v_0\left(1 + \dfrac{t}{546}\right)$

The density of water vapour is less than that of dry air at the same pressure. Thus, the density of moist air is less than that of dry air. As a result, the speed of sound increases with increasing humidity.

Progressive Wave

The displacement y at time t of a particle vibrating simple harmonically with frequency ν is represented by $y = a \sin 2\pi\nu t$.

Let us consider a wave of frequency ν and wave length λ travelling along +ve x direction.

The particle at origin has $y = 0$ and $t = 0$. The motion of a particle at point p will lag behind the particle at origin O by time $\dfrac{x}{v}$.

Therefore, the displacement at P is given by

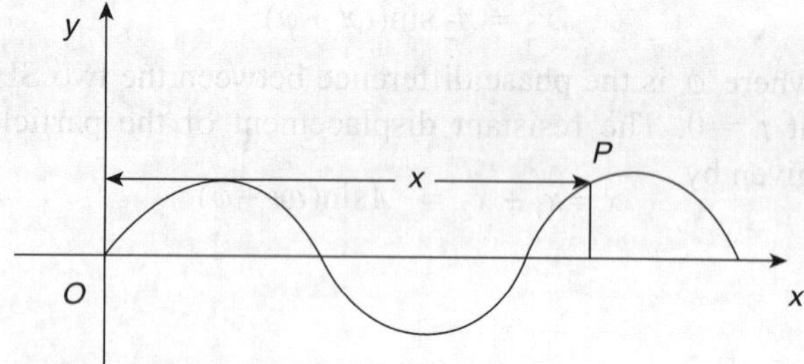

$$y = a\sin\left[2\pi\nu\left(t - \frac{x}{v}\right)\right], \text{ where } v \text{ is wave velocity}$$

Since, $\quad v = \nu\lambda = \dfrac{\lambda}{T}$

The equation may also be written as

$$y = a\sin\left[2\pi\nu\left(t - \frac{xT}{\lambda}\right)\right]$$

$$= a\sin\left(2\pi\nu t - \frac{2\pi\nu T}{\lambda}x\right)$$

$$= a\sin(2\pi\nu t - kx)$$

where $k = \dfrac{2\pi}{\lambda}$ is defined as propagation constant for a progressive wave travelling in −ve x direction The equation takes the form as

$$y = a\sin 2\pi\nu\left(t + \frac{x}{v}\right)$$

$$= a\sin(2\pi\nu t + kx)$$

Differential equation of wave motion is,

$$\frac{d^2y}{dx^2} = \frac{1}{v^2}\frac{d^2y}{dt^2}$$

Principle of Superposition

If two or more waves simultaneously arrive in a medium, the particles of the medium have two or more displacement simultaneously and a new wave is produced. This phenomenon of intermixing of two or more waves is called superposition of waves. Let us have two waves

$$y_1 = a_1 \sin(2\pi v_1 - k_1 x)$$
$$y_2 = a_2 \sin(2\pi v_2 - k_2 x)$$
$$y = y_1 + y_2$$

where y is the resultant displacement.

Reflection of Waves

(i) If a wave goes from rarer to denser medium, the reflected wave suffers a phase change of π.
(ii) There is no phase change when a wave goes from denser to rarer medium.
(iii) Frequency of reflected wave is same as that of incident wave.

Equation of Reflected Wave

(a) If wave goes from rarer to denser ($v_2 < v_1$)

$$y_i = a \sin(kx - \omega t)$$
$$y_r = a \sin(kx_1 + \omega t + \pi)$$
$$= -a \sin(kx + \omega t)$$

(b) If wave goes from denser to rarer medium

$$y_i = a \sin(kx - \omega t)$$
$$y_r = a \sin(kx + \omega t)$$

Stationary Waves

Two waves of same frequency and amplitude travelling along the same straight line in opposite direction with same speed are superposed. They form stationary or standing waves

Let $$y_1 = a \sin(\omega t - kx + \theta_1)$$
$$y_2 = a \sin(\omega t + kx + \theta_2)$$

Then, $$y = y_1 + y_2 = a \sin(\omega t - kx + \theta_1) + a \sin(\omega t + kx + \theta_2)$$
$$= 2a \cos\left(kx + \frac{(\theta_2 - \theta_1)}{2}\right) \sin\left(kt + \frac{(\theta_1 + \theta_2)}{2}\right)$$

The term $2a \cos\left(kx + \frac{(\theta_2 + \theta_1)}{2}\right)$ represents the amplitude at x. Hence, we can conclude the following.

(i) There are positions where the resultant amplitudes are zero called "nodes".
(ii) There are positions where resultant amplitudes are maximum and equal to $2A$. These are called "antinodes".
(iii) Distance between two successive nodes or antinodes are $\lambda/2$.
(iv) Distance between successive nodes and antinodes are $\lambda/4$.

Illustration 1

A transverse harmonic wave on a string is described by

$$y(x,t) = 3.0 \sin\left(36t + 0.018x + \frac{\pi}{4}\right)$$

where x, y are in cm and t in s. The positive direction x is from left to right.

(i) Is this a traveling or a stationary wave? If it is traveling, what are the speed and direction of its propagation?
(ii) What are its amplitude and frequency?
(iii) What is the initial phase at the origin?
(iv) What is the least distance between two successive crests in the wave?

Solution

(i) Travelling, $T = \frac{2\pi}{36}$ and $\lambda = \frac{2\pi}{0.018}$

$$v = \frac{36}{2\pi} \times \frac{2\pi}{0.018}$$

$$\Rightarrow \quad v = 20 \text{ m/s}$$

(ii) $A = 3.0$ cm
(iii) $f = 5.7$ Hz
(iv) $\lambda = 3.47$ m.

Standing Waves in Strings

(a) Fixed at both ends – fixed ends are nodes, so different possibilities are

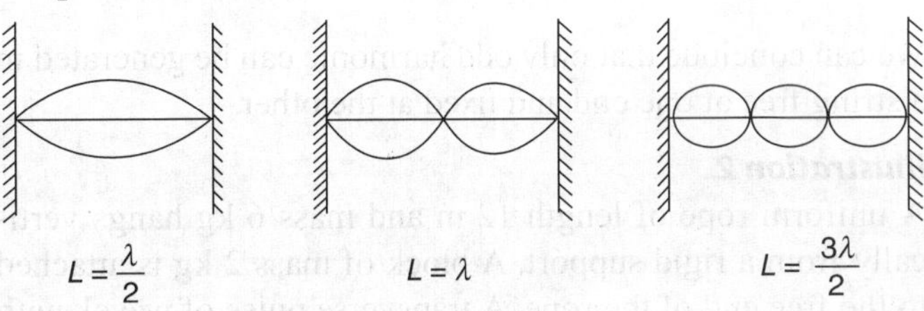

or $L = \frac{n\lambda}{2}$ where $n = 1, 2, 3, \ldots\ldots$

The wave velocity in the string is given by $v = \sqrt{\frac{T}{m}}$, where T is tension and m is mass per unit length of string. The different frequencies are

$$v_1 = \frac{1}{2L}\sqrt{\frac{T}{m}}$$
$$v_2 = \frac{2}{2L}\sqrt{\frac{T}{m}}$$
$$v_3 = \frac{3}{2L}\sqrt{\frac{T}{m}}$$

The frequency, v_1 is called fundamental frequency and corresponding mode of vibration is called fundamental mode of vibration y_2 is called second harmonic or first overtone.

v_3 is third harmonic or second overtone

We can conclude that any harmonic can be generated in strings fixed at both ends.

String Free at One End

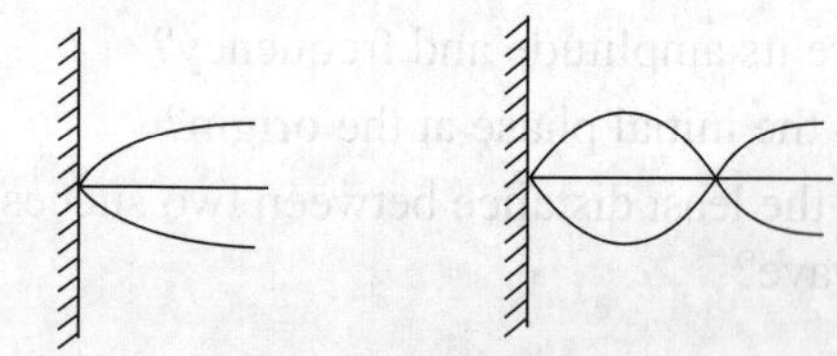

$$L = \frac{\lambda}{4} \quad L = \frac{3\lambda}{4}$$

$$v_1 = \frac{1}{4L}\sqrt{\frac{T}{m}}$$

$$v_2 = \frac{3}{4L}\sqrt{\frac{T}{m}}$$

$$v_3 = \frac{5}{4L}\sqrt{\frac{T}{m}}$$

$$v_n = \frac{2n+1}{4L}\sqrt{\frac{T}{m}}$$

We can conclude that only odd harmonic can be generated in a string free at one end and fixed at the other.

Illustration 2

A uniform rope of length 12 m and mass 6 kg hangs vertically from a rigid support. A block of mass 2 kg is attached to the free end of the rope. A transverse pulse of wavelength 0.06 m is produced at the lower end of the rope. What is the wavelength of the pulse when it reaches the top of the rope?

Solution

As the rope is heavy, its tension will be different at different points. The tension at the free end will be (2 kg)g and that at the upper end it will be (8 kg)g.

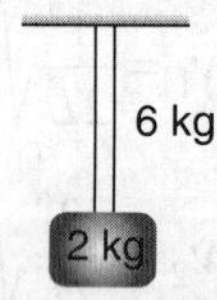

We have, $\upsilon = \nu\lambda$

or, $\sqrt{\frac{F}{\mu}} = \nu\lambda$

or, $\frac{\sqrt{F}}{\lambda} = \nu\sqrt{\mu}$.

The frequency of the wave pulse will be the same everywhere on the rope as it depends only on the frequency of the source. The mass per unit length is also the same throughout the rope as it is uniform.

Thus, by (i) $\frac{\sqrt{F}}{\lambda}$ is constant. Hence, $\frac{\sqrt{(2\text{ kg})g}}{0.06\,m} = \frac{\sqrt{(8\text{ kg})g}}{\lambda_1}$,

where λ_1 is the wavelength at the top of the rope. This gives $\lambda_1 = 0.12$ m.

Illustration 3

The vibrations of a string fixed at both ends are described by the equation $y = (5.00\,\text{mm})\sin(1.57\,\text{cm}^{-1})x]\sin[(314\text{s}^{-1})t]$.

(a) What is the maximum displacement of particle at $x = 5.66$ cm ?

(b) What are the wavelength and the wave speeds of the two transverse waves that combine to give the above vibration?

(c) What is the velocity of the particle at $x = 5.66$ cm at time $t = 2.0$ s?

(d) If the length of the string is 10.0 cm, locate the nodes and the antinodes. How may loops are formed in the vibration?

Solution

(a) The amplitude of the vibration of the particle at position x is

$$A = |(5.00\text{ mm})\sin[(1.57\text{ cm}^{-1})x]|$$

For $x = 5.66$ cm,

$$A = \left|(5.00\text{ mm})\sin\left[\frac{\pi}{2}\times 5.66\right]\right|$$

$$= \left|(5.00\text{ mm})\sin\left(2.5\pi + \frac{\pi}{3}\right)\right|$$

$$= \left|(5.00\text{ mm})\cos\frac{\pi}{3}\right| = 2.50\text{ mm}.$$

(b) From the given equation, the wave number $k = 1.57$ cm^{-1} and the angular frequency $\omega = 314$ s^{-1}. Thus, the wavelength is

$$\lambda = \frac{2\pi}{k} = \frac{2\times 3.14}{1.57\text{cm}^{-1}} = 4.00\text{ cm}$$

and the frequency is $\nu = \frac{\omega}{2\pi} = \frac{314\text{ s}^{-1}}{2\times 3.14} = 50\text{ s}^{-1}$

The wave speed is $\upsilon = \nu\lambda = (50\text{s}^{-1})(4.00\text{cm}) = 2.00$ m/s.

(c) The velocity of the particle at position x at time t is given by

$$\upsilon = \frac{\partial y}{\partial t} = (5.00\text{ mm})\sin[(1.57\text{ cm}^{-1})x]$$

$$[314\text{ s}^{-1}\cos(314\text{ s}^{-1})t]$$

$$= (157\text{ cm/s})\sin(1.57\text{ cm}^{-1})x\cos(314\text{ s}^{-1})t.$$

Putting $x = 5.66$ and $t = 2.00$ s, the velocity of this particle at the given instant is

$$(157\text{ cm/s})\sin\left(\frac{5\pi}{2} + \frac{\pi}{3}\right)\cos(200\pi)$$

$$= (157\text{ cm/s})\times\cos\frac{\pi}{3}\times 1 = 78.5\text{ cm/s}.$$

(d) the nodes occur where the amplitude is zero, i.e.,

$$\sin(1.57\,\text{cm}^{-1})x = 0\,.$$

or, $$\left(\frac{\pi}{2}\text{cm}^{-1}\right)x = 0$$

where n is an integer.

Thus, $x = 2n$ cm

The nodes, therefore, occur at $x = 0$, 2 cm, 4 cm, 6 cm, 8 cm and 10 cm. Antinodes occur in between them i.e., at $x = 1$ cm, 3 cm, 5 cm, 7 cm and 9 cm. The string vibrates in 5 loops.

Vibrations of a Closed Organ Pipe

The waves formed in a vibrating air column of a closed organ pipe are longitudinal stationary waves. The closed end is always a node and open end is always an antinode.

(i) The fundamental frequency or first harmonic $\nu_1 = \dfrac{v}{4L}$

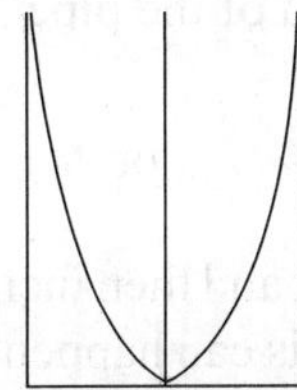

(ii) Third harmonic or first overtone $\nu_3 = \dfrac{3v}{4L}$

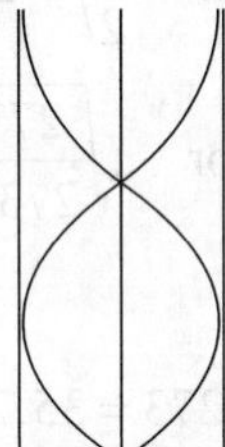

Similarly, $$\nu_5 = \frac{5v}{4L}$$

$$\nu_1 : \nu_3 : \nu_5 = 1 : 3 : 5$$

Vibrations of an Open Organ Pipe

The open ends are always antinodes.

(i) The fundamental frequency or first harmonic $\nu_1 = \dfrac{v}{2L}$

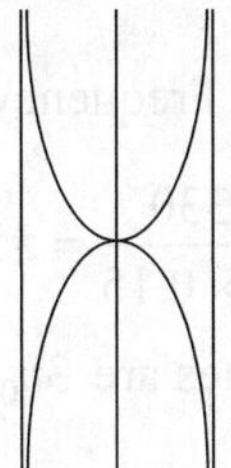

(ii) Second harmonic or first overtone $\nu_2 = \dfrac{2v}{2L}$

Similarly $\nu_3 = \dfrac{3v}{2L}$

$$\nu_1 : \nu_3 : \nu_5 = 1 : 2 : 3$$

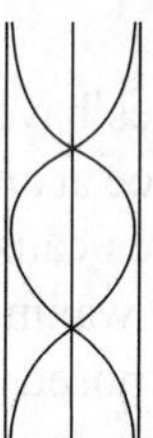

All harmonics are present in an open organ pipe.

Illustration 4

A tube closed at one end has a vibrating diaphragm at the other end, which may be assumed to be displacement node. It is found that when the frequency of the diaphragm is 2000 Hz, a stationary wave pattern is set up in which the distance between adjacent nodes is 8 cm. When the frequency is gradually reduced, the stationary wave pattern disappears but another stationary wave pattern reappears at a frequency of 1600 Hz. Calculate

(i) the speed of sound in air,
(ii) the distance between adjacent nodes at a frequency of 1600 Hz,
(iii) the distance between the diaphragm and the closed end, and
(iv) the next lower frequencies at which stationary wave patterns will be obtained.

Solution

Since the node-to-node distance is $\lambda/2$

$$\lambda/2 = 0.08 \quad \text{or} \quad \lambda = 0.16 \text{ m}$$

(i) $c = n\lambda$

$\therefore c = 2000 \times 0.16 = 320\text{ms}^{-1}$

(ii) $320 = 1600 \times \lambda$ or $\lambda = 0.2$ m

$\therefore$ Distance between nodes = 0.2/2 = 0.1 m = 10 cm

(iii) Since there are nodes at the ends, the distance between the closed end and the membrane must be exact integrals of $\lambda/2$.

$\therefore$ $$0.4 = 2\lambda/2 = n' \times 0.2/2$$

$\Rightarrow$ $$\frac{n}{n'} = \frac{2000}{1600} = \frac{5}{4}$$

When $n = 5, n' = 4$

$$l = n \times 0.16/2 = 0.4 \text{ m} = 40 \text{ cm}.$$

(iv) For the next lower frequency $n = 3, 2, 1$

$\therefore$ $0.4 = 3\lambda/2$ or $\lambda = 0.8/3$

Since $c = n\lambda,\ n = \dfrac{320}{0.8/3} = 1200 \text{ Hz}$

Again $0.4 = 1.\lambda/2$ or $\lambda = 0.4\,\text{m}$

$\therefore$ $n = 320/0.4 = 800\text{ Hz}$

Again $0.4 = 1.\lambda/2$ or $\lambda = 0.8\,\text{m}$

$\therefore$ $n = 320/0.8 = 400\text{ Hz.}$

Beat

When two interacting waves have slightly different frequencies the resultant disturbance at any point due to the superposition periodically fluctuates causing waxing and waning in the resultant intensity. The waxing and waning in the resultant intensity of two superposed waves of slightly different frequencies are known as beats.

Let the displacement produced at a point by one wave be

$$y_1 = A\sin(2\pi f_1 t - \phi_1)$$

And the displacement produced at the point produced by the other wave of equal amplitude as

$$y_2 = A\sin(2\pi f_2 t - \phi_2)$$

By the principle of superposition, the resultant displacement is

$$y = y_1 + y_2 = A\sin(2\pi f_1 t - \phi_1) + A\sin(2\pi f_2 t - \phi_2)$$

$$y = 2A\sin\left\{2\pi\left(\frac{f_1+f_2}{2}\right)t - \left(\frac{\phi_1+\phi_2}{2}\right)\right\}$$

$$\cos 2\pi\left(\frac{f_1-f_2}{2}\right)t$$

$$Y = R\sin\left\{2\pi\left(\frac{f_1+f_2}{2}\right)t - \left(\frac{\phi_1+\phi_2}{2}\right)\right\}$$

where, $R = 2A\cos 2\pi\left(\frac{f_1-f_2}{2}\right)t$

The time for one beat is the time between consecutive maxima or minima.

First maxima would occur when

$$\cos 2\pi\left(\frac{f_1-f_2}{2}\right)t = +1$$

Then $$2\pi\left(\frac{f_1-f_2}{2}\right)t = 0$$

$\therefore$ $$t = 0$$

For second maxima would occur when

$$\cos 2\pi\left(\frac{f_1-f_2}{2}\right)t = -1$$

Then $$2\pi\left(\frac{f_1-f_2}{2}\right)t = \pi$$

or $$t = \frac{1}{f_1-f_2}$$

The time for one beat $= \left(\dfrac{1}{f_1-f_2} - 0\right) = \dfrac{1}{f_1-f_2}$

Similarly it may also be shown that time between two consecutive minima is $\dfrac{1}{f_1-f_2}$.

Hence, frequency of beat i.e., number of beats in one second or

$$\text{Beat frequency} = f_1 \sim f_2.$$

Illustration 5

A tuning fork of frequency 256 Hz and an open organ pipe of slightly lower frequency are at 17° C. When sounded together, they produce 4 beats per second. On altering the temperature of the air in the pipe, it is observed that the number of beats per second first diminishes to zero and then increases again to 4. By how much and in what direction has the temperature of the air in the pipe been altered?

Solution

$n = \dfrac{c_{17}}{2l}$ where l = length of the pipe

$\therefore$ $$256 - \frac{c_{17}}{2l} = 4 \quad \text{or} \quad \frac{c_{17}}{2l} = 252$$

Since beats decrease first and then increase to 4, the frequency of the pipe increases. This can happen only if the temperature increases.

Let t be the final temperature, in Celsius,

Now $$\frac{c_t}{2l} - 256 = 4 \quad \text{or} \quad \frac{c_t}{2l} = 260$$

Dividing $$\frac{c_t}{c_{17}} = \frac{260}{252} \quad \text{or} \quad \sqrt{\frac{273+t}{273+17}} = \frac{260}{252}$$

$$(\because c < \sqrt{T})$$

or $$t = 308.7 - 273 = 35.7°\text{C}$$

$\therefore$ Rise in temperature = 35.7 – 17 = 18.7° C.

Illustration 6

Find the fundamental frequency and the first four overtones of a 15 cm pipe

(a) if the pipe is closed at one end,
(b) if the pipe is open at both ends.
(c) How many overtones may be heard by a person of normal hearing in each of the above cases? Velocity of sound in air = 330 ms^{-1}.

Solution

(a) $n_0 = \dfrac{c}{4l}$ where n_0 = Frequency of the fundamental

$$\Rightarrow \quad n_0 = \frac{330}{4\times 0.15} = 550\,\text{Hz}$$

The first four overtones are $3n_0$, $5n_0$, $7n_0$ and $9n_0$

$\therefore$ So, the required frequencies are 550, 1650, 2750, 3850, 4950 Hz.

(b) $n_0 = \frac{c}{2l} = \frac{330}{2 \times 0.15} = 1100$ Hz.

The first four overtones are $2n_0$, $3n_0$, $4n_0$ and $5n_0$

So, the required frequencies are 1100, 2200, 3300, 4400, 5500 Hz.

(c) The frequency of the nth overtone is $(2n + 1)\ n_0$.

$\therefore \quad (2n+1)\ n_0 = 20000 \quad$ or $\quad (2n+1)\ 550 = 20000$

or $\quad n = 17.68.$

The acceptable value is 17.

The frequency of the nth overtone is $(n+1)\ n_0$.

$\therefore \quad (n+1)\ n_0 = 20000 \quad$ or $\quad (n+1)\ 100 = 20000$

or $\quad n = 17.18$

The acceptable value is 17.

Illustration 7

Two tuning forks A and B, sounded together give 6 beats per second. With an air resonance tube closed at one end, the two forks give resonance when the two air columns are 24 cm and 25 cm respectively. Calculate the frequencies of forks.

Solution

Let the frequency of the first fork be f_1 and that of second be f_2.

We then have,

$$f_1 = \frac{v}{4 \times 24} \quad \text{and} \quad f_2 = \frac{v}{4 \times 25}$$

We also see that $f_1 > f_2$

$$\therefore \quad f_1 - f_2 = 6 \qquad \text{(i)}$$

and
$$\frac{f_1}{f_2} = \frac{24}{25} \qquad \text{(ii)}$$

Solving (i) and (ii), we get

$$f_1 = 150 \text{ Hz}$$

and
$$f_2 = 144 \text{ Hz}$$

Doppler Effect

The apparent shift in frequency of the wave motion when the source of sound or light moves with respect to the observer, is called Doppler Effect.

Calculation of Apparent Frequency

Suppose v is the velocity of sound in air, v_s is the velocity of the source of sound(s), v_0 is the velocity of the observe (O), and f is the frequency of the source.

(i) Source moves towards the stationary observer. It the source were stationary the f waves sent out in one second towards the observer O would occupy a distance v, and the wavelength would be v/f. If S moves with a velocity v_s towards (O), the f waves sent out occupy a distance $(v - v_s)$ because S has moved a distance v_s towards O in 1s. So the apparent wavelength would be

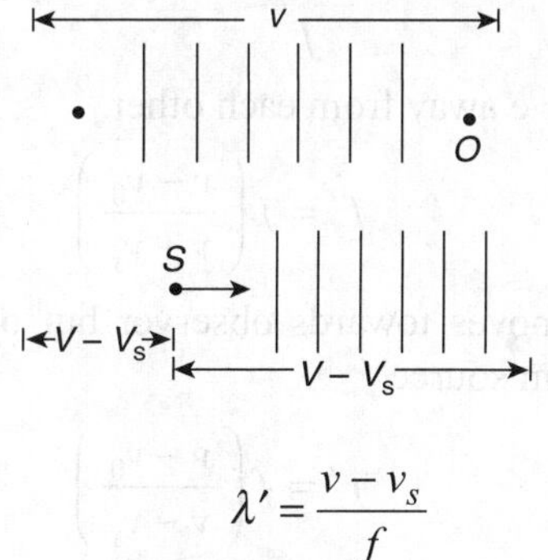

$$\lambda' = \frac{v - v_s}{f}$$

Thus, apparent frequency

$$f' = \frac{\text{Velocity of sound relative to O}}{\text{Wavelenght of wave reaching O}}$$

$$f' = \frac{v}{\lambda'} = f\left(\frac{v}{v - v_s}\right)$$

(ii) Source moves away from stationary observer. Now, apparent wavelength

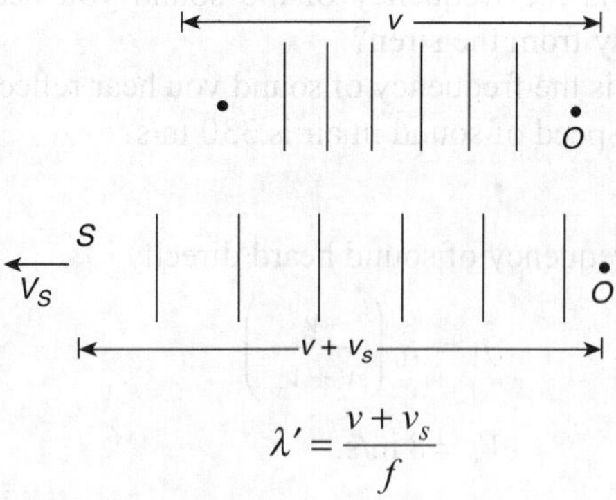

$$\lambda' = \frac{v + v_s}{f}$$

$\therefore$ Apparent frequency

$$f' = v / \lambda'\ n$$

or
$$f'c = f\left(\frac{v}{v + v_s}\right)$$

(iii) Observer moves towards stationary source.

$$f' = \frac{\text{Velocity of sound relative to O}}{\text{Wavelenght of wave reaching O}}$$

Here, velocity of sound relative to O = $v + v_0$

and wavelength of waves reaching O = v/f

$$\therefore \quad f' = \frac{v + v_0}{v / f} = f\left(\frac{v + v_0}{v}\right)$$

(iv) Observer moves away from the stationary source.

$$f' = \frac{v - v_0}{v / f} = f\left(\frac{v - v_0}{v}\right)$$

(v) Source and observer both move toward each other.

$$f' = \frac{v + v_0}{\frac{v - v_s}{f}} = f\left(\frac{v + v_0}{v - v_s}\right)$$

(vi) Both move away from each other.

$$f' = f\left(\frac{v - v_0}{v + v_s}\right)$$

(vii) Source moves towards observer but observer moves away from source

$$f' = f\left(\frac{v - v_0}{v - v_s}\right)$$

(viii) Source moves away from observer but observer moves towards source.

$$f' = f\left(\frac{v + v_0}{v + v_s}\right)$$

Illustration 8

A siren emitting a sound of frequency 2000 Hz moves away from you towards a cliff at a speed of 8 m/s.

(a) What is the frequency of the sound you hear coming directly from the siren?

(b) What is the frequency of sound you hear reflected off the cliff. Speed of sound in air is 330 m/s?

Solution

(a) The frequency of sound heard directly.

$$f_1 = f_0\left(\frac{v}{v + v_s}\right)$$

$$V_s = 8 \text{ m/s.}$$

$$\therefore \quad f_1 = \left(\frac{330}{330 + 8}\right) \times 2000$$

$$f_1 = \frac{330}{338} \times 2000$$

(b) The frequency of the reflected sound is given by

$$f_2 = f_0\left(\frac{v}{v - v_s}\right)$$

$$\therefore \quad f_2 = \left(\frac{330}{330 - 8}\right) \times 2000$$

$$f_2 = \frac{330}{322} \times 2000 = 2050 \text{ Hz}$$

SOLVED EXAMPLES

Example 1

The frequency of sound produced by a bell is 500 Hz. The velocity of the source relative to still air is 60 m/s. An observer moves at 30 m/s along the same line as the source. Calculate the frequency of sound wave measured by the observer. Consider all the possible cases (speed of sound $v = 340$ m/s)

Solution

$\xrightarrow{v_s}$ Source $\xrightarrow{v_0}$ Observer

Case (i) The source and the observer move to the right. Since the observer is moving away from the source, there will be an apparent decrease in frequency. The source is approaching the observer; so there will be apparent increase in frequency. Hence the apparent frequency

$$f' = \left(\frac{v - v_0}{v - v_s}\right) f = \left(\frac{340 - 30}{340 - 60}\right) \times 500\text{Hz} = 553 \text{ Hz}$$

Case (ii) When the source moves to the right and the observer to the left, they approach each other and both contribute to an apparent increase in the frequency. Hence

$$f' = \left(\frac{v + v_0}{v - v_s}\right) f = \frac{340 + 30}{340 - 60} \times 500\text{Hz} = 661 \text{ Hz}$$

Case (iii) When the source moves to the left and the observer to the right so that they recede from each other, consequently both contribute to an apparent decrease in frequency. Hence

$$f' = \left(\frac{v - v_0}{v + v_s}\right) f = \frac{340 - 30}{340 + 60} \times 500\text{Hz} = 387 \text{ Hz}$$

You should repeat the example assuming that the observer is to the left to the source. What conclusions can be derived from this example? Do you get a different set of four apparent frequencies?

Example 2

When 0.98 m long metallic wire is stressed, an extension of 0.02 m is produced. An organ pipe 0.5 m long and open at both ends, when sounded with this stressed metallic wire, produces 8 beats in its fundamental mode. By decreasing the stress in the wire, the frequency of beats is found to decrease. Find the Young's modulus of the wire. The density of metallic wire is 10^4 kg/m³ and speed of sound in air is 292 m/s.

Solution

Frequency of the transverse vibration of the stretched string in

$$f_1 = \frac{1}{2(L + \Delta L)}\sqrt{\frac{T}{\mu}}, \qquad \text{(i)}$$

Here $\quad Y = \frac{T}{A}\frac{L}{\Delta L}$, or $T = \frac{YA\Delta L}{L}$, and $\mu = A\rho$

$$\therefore \quad f_1 = \frac{1}{2(L + \Delta L)}\sqrt{\frac{Y\Delta L}{L\rho}} \qquad \text{(ii)}$$

When the stress in the wire is decreased; f_1 will decrease, consequently the beat frequency will decrease if $f_1 < f_{pipe}$

$$\therefore \quad f_1 - f_{pipe} = 8$$

or $$\frac{1}{2(L+\Delta L)}\sqrt{\frac{Y\Delta L}{L\rho}} - \frac{v}{2\ell} = 8$$

Substituting the value, we get

$$\frac{1}{2(0.98+0.02)}\sqrt{\frac{Y\times 0.02}{0.98\times 10h4}} - \frac{292}{2\times 0.5} = 8$$

Simplifying, we get

$$Y = 1.764\times 10^{11}\ \text{N/m}^2$$

Example 3

A metallic rod of length 1 m is rigidly clamped at its mid-point. Longitudinal stationary waves are set up in the rod in such a way that there are two nodes on either sides of mid-point. The amplitude of antinode is 2×10^{-6} m. Write the equation of motion at a point 2 cm from the mid-point and those of the constituent waves in the rod. (Young's modulus $= 2\times 10^{11}\ \text{Nm}^{-2}$, density = 8000 kg m^{-3}).

Solution

Figure shows the longitudinal displacement y as a function of x from $x = 0$ to $x = L$, where L is length of the rod.

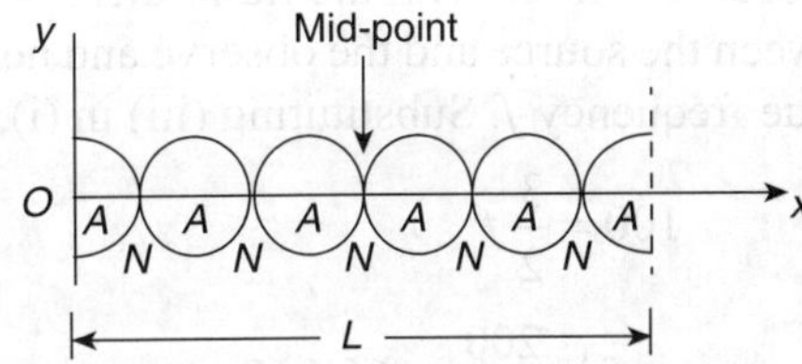

We know that the separation between a node and the next antinode is $\frac{\lambda}{4}$ and that between two consecutive nodes or antinodes is $\frac{\lambda}{4}$. Therefore,

$$L = \frac{\lambda}{4} + \frac{\lambda}{2} + \frac{\lambda}{2} + \frac{\lambda}{2} + \frac{\lambda}{2} + \frac{\lambda}{4} = \frac{5\lambda}{2}$$

Here λ is the wavelength of either of the interfering waves. Thus

$$\lambda = \frac{2L}{5} = \frac{2\times 1}{5} = 0.4\ \text{m}$$

The speed of longitudinal waves in a metal of Young's modulus Y and density ρ is given by

$$\upsilon = \sqrt{\frac{Y}{\rho}} = \left(\frac{2\times 10^{11}}{8000}\right)^{1/2} = 5000\ \text{ms}^{-1}$$

$$\therefore \text{ Frequency } v = \frac{\upsilon}{\lambda} = \frac{5000}{0.4} = 12500\ \text{Hz.}$$

Notice from figure that there is an antinode at the ends $x = 0$ and $x = L$ of the rod. Hence the equation of the stationary wave is (here A is the amplitude of each of the interfering waves)

$$y = 2A\cos\left(\frac{2\pi x}{\lambda}\right)\sin(2\pi vt) \quad \text{(i)}$$

At any instant of time, the amplitude is given by

$$A(x) = 2A\cos\left(\frac{2\pi x}{\lambda}\right) \quad \text{(ii)}$$

At $x = 0$, there is an antinode, i.e. $A(0) = 2\times 10^{-6}$ m (given). Putting $x = 0$ in Eq. (ii) we have

$$A(0) = 2\times 10^{-6}\ \text{m} = 2A$$

or $A = 1\times 10^{-6}$ m. Putting values of A, λ and v in Eq. (i) we get

$$y = 2\times 10^{-6}\cos\left(\frac{2\pi x}{0.4}\right)\sin(2\pi\times 12500\times t)$$

or $$y = 2\times 10^{-6}\cos(5\pi x)\sin(25000\ \pi t) \quad \text{(iii)}$$

where x and y are in metres and t in seconds. For a point at a distance of 2 cm from the mid-point, the value of x is

$$x = 50\ \text{cm} + 2\ \text{cm} = 52\ \text{cm} = 0.52\ \text{m}$$

Using this value of x in Eq. (iii), the required equation of motion at a point 2 cm from the mid-point is

$$y = 2\times 10^{-6}\cos(5\pi\times 0.52)\sin(25000\ \pi t)$$
$$= 2\times 10^{-6}\cos(2.6\pi)\sin(25000\ \pi t)$$

or $$y = 2\times 10^{-6}\cos(0.6\pi)\sin(25000\ \pi t)$$

The equations of the constituent waves (incident wave and reflected wave) in the rod are

$$y_1 = 1\times 10^{-6}\sin 5p(5000t - x)$$

and $$y_2 = 1\times 10^{-6}\sin 5p(5000t + x).$$

Example 4

A uniform rope of mass 0.1 kg and length 2.45 m hangs from the ceiling.

(a) Find the speed of transverse wave in the rope at a point 0.5 m distance from the lower end.

(b) Calculate the time taken by a transverse wave to travel the full length of the rope.

Solution

(a) The speed of the transverse wave is given by

$v = \sqrt{\frac{T}{\mu}}$, where T = tension and μ = linear mass density

At a distance x from the free end of the rope, the tension

$$T = \mu xg$$

$$\therefore \quad v = \sqrt{\frac{\mu xg}{\mu}} = \sqrt{xg} \quad \text{(i)}$$

(i) Here $x = 0.15 \text{ m}, \ g = 9.8 \text{ m/s}^2$

$$\therefore \quad v = \sqrt{(0.5 \text{ m}).(9.8 \text{ m/s}^2)}$$

$$= 2.21 \text{ m/s}$$

(b) The speed of the wave at a distance x from the free end is

$$v = \frac{dx}{dt} = \sqrt{xg} \quad \text{[(from (i)]}$$

or
$$dt = \frac{1}{\sqrt{g}}\frac{dx}{\sqrt{x}}$$

Integrating over the whole length of the rope,

$$t = \int_0^t dt = \frac{1}{\sqrt{g}}\int_0^\ell \frac{dx}{\sqrt{x}} = 2\sqrt{\frac{\ell}{g}}$$

$$= 2\sqrt{\frac{2.45}{9.8}} = 1 \text{ s.}$$

Example 5

A wave pulse starts propagating in the $+x$ direction along a non-uniform wire of length 10 m with mass per unit length given by $\mu = \mu_0 + \alpha x$ and under a tension of 100 N. Find the time taken by a pulse to travel from the lighter end ($x = 0$) to the heavier end. $\left(\mu_0 = 10^{-2} \text{ kg/m and } \alpha = g \times 10^{-3} \text{ kg/m}^2\right)$

Solution

The speed of the wave pulse

$$v = \sqrt{\frac{T}{\mu}} = \sqrt{\frac{T}{\mu_0 + \alpha x}} = \frac{dx}{dt}$$

$$\therefore \quad t = \int_0^t dt = \int_0^\ell \sqrt{\frac{\mu_0 + \alpha x}{T}} dx$$

$$= \frac{2}{3\alpha\sqrt{T}}\left[(\mu_0 + \alpha x)^{3/2}\right]_0^\ell$$

$$= \frac{2}{3}\frac{1}{\alpha\sqrt{T}}\left[(\mu_0 + \alpha\ell)^{3/2} - (\mu_0)^{3/2}\right]$$

Substituting the values, we get

$$t = \frac{2}{3} \times \frac{1}{9\times10^{-3}\times\sqrt{100}}$$

$$\left[\left(10^{-2} + 9\times10^{-3}\times10\right)^{3/2} - \left(10^{-2}\right)^{3/2}\right]$$

$$= \frac{2\times100}{27}\left[(10)^{-3/2} - (10)^{-3}\right] = 0.227\text{s}$$

$$= \frac{200}{27}(0.0316) - 0.001 = 0.227 \text{ s}$$

Example 6

When a train is approaching the observer, the frequency of the whistle is 100 cps. When it has passed the observer, it is 50 cps. Calculate the frequency when the observer moves with the train.

Solution

When the source (the train) moves towards the observer, the apparent frequency is

$$f' = \frac{v}{v - v_s} f \tag{i}$$

while the source is moving away, the apparent frequency

$$f'' = \frac{v}{v + v_s} f \tag{ii}$$

$$\therefore \quad \frac{f'}{f''} = \frac{v + v_s}{v - v_s}$$

or
$$\frac{100}{50} = \frac{2}{1} = \frac{\left(\frac{v}{v_s} + 1\right)}{\left(\frac{v}{v_s} - 1\right)}$$

or
$$\frac{v}{v_s} = 3 \tag{iii}$$

When the observer moves with the train, there is no relative motion between the source and the observe and hence he will listen the true frequency f. Substituting (iii) in (i), we get

$$100 = \frac{3}{2} f$$

or
$$f = \frac{200}{3} = 66.6 \text{ Hz}$$

Example 7

Two metallic strings A and B of different materials are connected in series forming a joint. The strings have equal cross-sectional area. The length of A is $\ell_A = 0.3$ m and that of B is $\ell_B = 0.75$ m. One end of the combined string is tied with a support rigidly and the other end is loaded with a block of mass m passing over a frictionless pulley. Transverse waves are set up in the combined string using an external source of variable frequency. Calculate

(a) the lowest of frequency for which standing waves are observed such that the joint is a node,
(b) the total number of antinodes at this frequency.

The densities of A and B are 6.3×10^3 kg/m³ and 2.8×10^3 kg/m³ respectively.

Solution

(a) Let p and q be the number of loops formed in A and B

$$\therefore \quad n_A = \frac{p}{2\ell_A}\sqrt{\frac{T}{A\rho_A}}, \text{ and}$$

$$n_B = \frac{q}{2\ell_B}\sqrt{\frac{T}{A\rho_B}}$$

$$\therefore \quad \frac{n_A}{n_B} = \frac{p}{q}\cdot\frac{\ell_B}{\ell_A}\cdot\sqrt{\frac{\rho_B}{\rho_A}} = 1[\because n_A = n_B]$$

$$\therefore \quad \frac{p}{q}\cdot\frac{\ell_A}{\ell_B}\cdot\sqrt{\frac{\rho_A}{\rho_B}} = \frac{0.3}{0.75}\sqrt{\frac{6.3}{2.8}} = \frac{3}{5}$$

$\therefore$ The ratio of the number of loops

$$\frac{p}{q} = \frac{3}{5}, \frac{6}{10}, \frac{9}{5}, \ldots$$

For the lowest frequency,

$$\frac{p}{q} = \frac{3}{5}$$

This means that A will have 3 loops and B will have 5 loops. Hence, the required minimum frequency

$$f_{\min} = \frac{3}{2\rho_A}\sqrt{\frac{T}{A\rho_A}} = \frac{5}{2\rho_B}\sqrt{\frac{T}{A\rho_B}}$$

or $$f_{\min} = \frac{3}{2\times 0.3}\sqrt{\frac{mg}{A\times 6.3\times 10^3}}$$

$$= \sqrt{\frac{m}{25.7A}}$$

(b) The total number of antinodes at this minimum frequency is 3 + 5 = 8

Example 8

A source S emitting sound of 300 Hz is fixed on block A which is attached to free end of a spring S_A as shown in the figure.

S D A B S_A S_B

The detector D fixed on block B attached to the free end of spring S_B detects this sound. The blocks A and B are simultaneously displaced towards each other through a distance of 1.0 m and then left to vibrate. Find the maximum and minimum frequencies of sound detected by D of the vibrational frequency of each block is 2 Hz. (velocity of sound = 340 m/s)

Solution

The motion of A and B is synchronized. Hence the maximum and minimum frequencies detected by D will be

$$f_{\max} = \left(\frac{v+v_D}{v-v_S}\right)f,$$

(when S and D an approaching each other while at their equilibrium position.)

$$= \left(\frac{340 + A\omega}{340 - A\omega}\right)f$$

$$= \left(\frac{340+12.56}{340-12.56}\right)\times 300$$

$$= 323 \text{ Hz}$$

Similarly, the apparent frequency will be minimum when the source and the detector recede from each other with maximum speed ($= A\omega$)

$$\therefore \quad f_{\min} = \left(\frac{v - A\omega}{v + A\omega}\right)\cdot f$$

$$= \left(\frac{340-12.56}{340+12.56}\right)\times 300$$

$$= 278.6 \text{ Hz}$$

Example 9

The amplitude of velocity of a particle is given by $V_m = V_0 / (a\omega^2 - b\omega + c)$; the condition for single resonant frequency is

(a) $b^2 < 4ac$ (b) $b^2 > 4ac$

(c) $b^2 = 4ac$ (d) $b^2 = \infty$.

Solution

Determinant of equation $a\omega^2 - b\omega + c$ must be equal to zero for single root, thus $b^2 = 4ac$.

Example 10

The third overtone of a closed pipe is observed to be in unison with the second overtone of an open pipe. The ratio of the lengths of the pipes is

(a) 3/2 (b) 5/3

(c) 7/4 (d) 7/6

Solution

Frequency of the fundamental note in closed pipe is $\frac{v}{4\ell_1} = n_1$.

Only the odd harmonics,

$3n_1$ (First overtone), $5n_1$ (Second overtone) and $7n_1$ (Third overtone) are present.

Frequency of the fundamental note in open pipe is $\frac{v}{2\ell_2} = n_2$.

All the harmonics,

$2n_2$ (Second harmonic or first overtone), $3n_2$ (Third harmonic or second overtone), $4n_2$ (Fourth harmonic or third overtone) are present.

Given that: $7n_1 = 3n_2$

or $$7\frac{v}{4\ell_1} = 3\frac{v}{2\ell_2}$$

$$\therefore \quad \frac{\ell_1}{\ell_2} = \frac{7}{6}$$

Example 11

The speed of sound wave in a mixture of 1 mole of helium and 2 moles of oxygen at 27° C is

(a) 400 m/s (b) 600 m/s

(c) 800 m/s (d) 1200 m/s.

Solution

The speed of sound wave is

$$v=\sqrt{\frac{\gamma RT}{M}},\ \text{where } \gamma=\left(\frac{c_p}{c_v}\right)_{\text{mixture}}$$

The molecular weight of the mixture

$$M=\frac{n_1M_1+n_2M_2}{n_1+n_2}=\frac{1\times 4+2\times 32}{1+2}=\frac{68}{3}\text{g/mol}$$

$$=\frac{68}{3}\times 10^{-3}\,\text{kg/mol}$$

$$(c_v)_{\text{mixture}}=\frac{(n_1c_v)_{He}+(n_1c_v)_0}{n_1+n_2}=\frac{1\times\frac{3}{2}R+2\times\frac{5}{2}R}{1+2}=\frac{13}{6}\text{R}$$

$$(c_p)_{\text{mixture}}=(c_v)_{\text{mixture}}+R=\frac{19}{6}R$$

$$\therefore\ (\gamma)_{\text{mixture}}=\frac{(c_p)_{\text{mixture}}}{(c_v)_{\text{mixture}}}=\frac{19}{13}$$

Substituting these values, we get

$$v=\sqrt{\frac{19}{13}\times 8.31\times(273+27)/\frac{68}{3}\times 10^{-3}}=400.93\,\text{m/s}$$

Example 12

A traveling wave in a stretched string is described by the equation $y=A\sin(kx-\omega t)$. The maximum particle velocity is

(a) $A\omega$ (b) ω/k
(c) $d\omega/dk$ (d) x/t

Solution

Particle velocity $|v|=\left|\frac{dy}{dt}=A\omega\cos(kx-\omega t)\right|$

$\therefore$ Maximum particle velocity $=A\omega$

Example 13

The extension in a string, obeying Hooke's law, is x. The speed of sound in the stretched string is v. If the extension in the string is increased to $1.5x$, the speed of sound will be

(a) 1.22 v (b) 0.61 v
(c) 1.50 v (d) 0.75 v

Solution

Tension $T=kx$

$T'=1.5x$

Since $v\propto\sqrt{T}$

$$\therefore\ \frac{v'}{v}=\sqrt{\frac{T'}{T}}=\sqrt{1.5}=1.22$$

or, $v'=1.22\,v$

Example 14

Two monoatomic ideal gases 1 and 2 of molecular masses m_1 and m_2 respectively are enclosed in separate containers kept at the same temperature. The ratio of the speed of sound in gas 1 to that in gas 2 is given by:

(a) $\sqrt{\frac{m_1}{m_2}}$ (b) $\sqrt{\frac{m_2}{m_1}}$
(c) $\frac{m_1}{m_2}$ (d) $\frac{m_2}{m_1}$.

Solution

Velocity of sound $v-\sqrt{\frac{\gamma RT}{M}}$, where M is the molar mass.

$$\therefore\ \frac{v_1}{v_2}=\sqrt{\frac{\gamma_1T_1}{m_1}\times\frac{m_2}{\gamma_2T_2}}=\sqrt{\frac{m_2}{m_1}}\quad [\because v_1=v_2, T_1=T_2]$$

Example 15

In a wave motion $y=a\sin(kx-\omega t)$, y can represent

(a) electric field (b) magnetic field
(c) displacement (d) pressure.

Solution

All the options represent periodic variation in wave motion.

Example 16

A train moves towards a stationary observer with speed 34 m/s. The train sounds whistle and its frequency registered by the observer is f_1. If the train's speed is reduced to 17 m/s the frequency registered is f_2. If the speed of sound is 340 m/s, then the ratio f_1/f_2 is:

(a) 18/19 (b) 1/2
(c) 2 (d) 19/18

Solution

$$f_1=\frac{v}{v-v_{s_1}}f$$

$$f_2=\frac{v}{v-v_{s_2}}f$$

$$\therefore\ \frac{f_1}{f_2}=\frac{340-17}{340-34}=\frac{323}{306}=\frac{19}{18}$$

Example 17

An open pipe is suddenly closed at one end with the result that the frequency of third harmonic of the closed pipe is found to be higher by 100 Hz than the fundamental frequency of the open pipe. The fundamental frequency of the open pipe is

(a) 200 Hz (b) 300 Hz
(c) 240 Hz (d) 480 Hz.

Solution

Fundamental frequency of open pipe is

$$f_0=\frac{v}{2\ell}$$

Third harmonic of the closed pipe

$$f_c=3\left(\frac{v}{4\ell}\right)$$

Given: $3\frac{v}{4\ell} = \frac{v}{2\ell} + 100$

or, $\frac{v}{2\ell} = 200$ Hz

Example 18

The wires X and Y of a guitar produce 4 beats per second. If the tension of Y is raised , then the number of beats becomes 2 per second. If the frequency of X is 300, then frequency of Y is

(a) 296 (b) 298
(c) 300 (d) 294.

Solution

$$f_X \sim f_Y = 4$$

When the tension of Y is raised, its frequency increases and the difference, $f_X - f_Y$ is reduced.

$\therefore \quad f_Y < f_X$

$\therefore \quad f_Y = f_X - 4 = 300 - 4 = 296$

Example 19

Two pulses in a stretched string whose centres are initially 8 cm apart are moving towards each other as shown in the figure. The speed of each pulse is 2 cm/s. After 2 seconds, the total energy of the pulses will be:

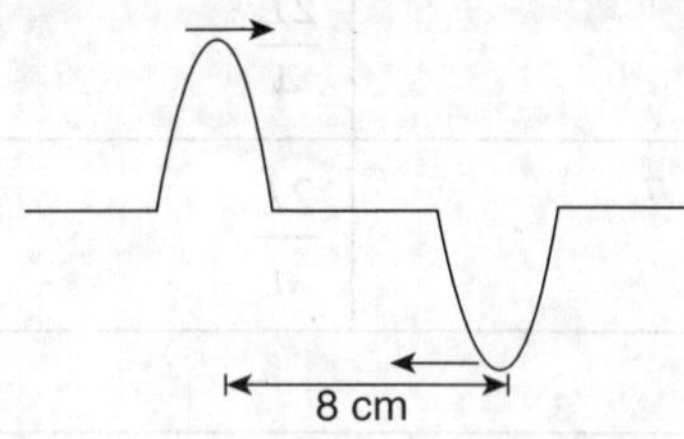

(a) zero
(b) purely kinetic
(c) purely potential
(d) partly kinetic and partly potential.

Solution

Under the given situation, the pulses overlap, the string occupies the equilibrium position. The potential energy will be zero but the total energy will exist in the form of purely kinetic energy.

Example 20

A siren placed at a railway platform is emitting sound of frequency 5 kHz. A passenger sitting in a moving train A records a frequency of 5.5 kHz while the train approaches the siren. During his return journey in a different train B be records a frequency of 6.0 kHz while approaching the same siren. The ratio of the velocity of train B to that of train A is:

(a) 242/252 (b) 2
(c) 5/6 (d) 11/6

Solution

$$f_A = \frac{v + v_A}{v} f, \text{or } v_A = \frac{v}{f}(f_A - f)$$

$$f_B = \frac{v + v_B}{v} f, \text{or } v_B = \frac{v}{f}(f_B - f)$$

$$\therefore \quad \frac{v_B}{v_A} = \frac{f_B - f}{f_A - f} = \frac{6.0 - 5}{5.5 - 5} = \frac{1}{0.5} = 2$$

POINTS TO REMEMBER

1. Wave is a means of transferring energy, in which there is no bulk motion of matter. There are two types of waves.
   (i) Mechanical wave
   (ii) Electromagnetic wave
2. Mechanical wave can be produced and propagated in a material medium which possesses the properties of both elasticity and inertia.
3. Electromagnetic waves do not require any material medium for their production or propagation.
4. Mechanical waves are of two types
   (i) Transverse (ii) Longitudinal
5. Relation between frequency, velocity and wavelength
   $$v = \nu\lambda$$
6. Velocity of transverse waves in a string
   $$v = \sqrt{\frac{T}{m}}$$
   where T is tension in the string and m is mass per unit length.
7. Velocity of longitudinal waves in solids, liquids and gases
   $$v = \sqrt{\frac{E}{\rho}}$$
   where E is elasticity of the medium and τ is the density of undisturbed medium. In the case of solids, E represents Young's modulus of elasticity. In the case of liquids and gases, E represents the bulk modulus of elasticity.
8. Newton's formula for the velocity of sound waves in air
   $v = \sqrt{\frac{P}{\rho}}$ where P is pressure and ρ is density.
9. Laplace correction of Newton's formula
   $v = \sqrt{\frac{\gamma P}{\rho}}$ where γ is the ratio of two specific heat of gases $\gamma = \frac{C_P}{C_V}$

10. Factors affecting the change in velocity of sound in gases

(a) Pressure – Independent

(b) Temperature – $v_t \propto \sqrt{T}$

$$\frac{v_t}{v_0} = (1 + \frac{t}{273})^{1/2}$$

where t is in °C

Also $v_t = v_0 + 0.61\, t$ m/s

(c) Density – $v \propto \frac{1}{\sqrt{\rho}}$

(d) Humidity – Sound travels faster in moist air than dry air.

11. A plane progressive wave travelling from left to right along x axis is given by

$$\varepsilon = a \sin \frac{2\tau}{\lambda}(vt - x)$$

where a is amplitude and v is velocity of the wave.

12. Table for vibrating string

Mode of vibration	Harmonics	Tone	Number of loops	Number of nodes	Number of antinodes	Wavelength	Frequency
First or Fundamental	First	Fundamental	1	2	1	$\frac{2L}{1}$	$\frac{v}{2L} = \nu$
Second	Second	First overtone	2	3	2	$\frac{2L}{2}$	2ν
Third	Third	Second overtone	3	4	3	$\frac{2L}{3}$	3ν
Fourth	Fourth	Third overtone	4	5	4	$\frac{2L}{4}$	4ν
nth	nth	$(n-1)$th overtone	n	$n+1$	n	$\frac{2L}{n}$	$n\nu$

13. Table for vibration of open organ pipe

Mode of vibration	Harmonics	Tone	Number of Nodes	Number of Antinodes	Wavelength	Frequency
First or Fundamental	First	Fundamental	1	2	$\frac{2L}{1}$	$\frac{v}{2L} = \nu$
Second	Second	First overtone	2	3	$\frac{2L}{2}$	2ν
Third	Third	Second overtone	3	4	$\frac{2L}{3}$	3ν
Fourth	Fourth	Third overtone	4	5	$\frac{2L}{4}$	4ν
nth	nth	$(n-1)$th overtone	n	$n+1$	$\frac{2L}{n}$	$n\nu$

14. Table for vibration of closed organ pipe

Mode of Vibration	Harmonics	Tone	Number of Nodes	Number of Antinodes	Wavelength	Frequency
First or Fundamental	First	Fundamental tone	1	1	$\frac{4L}{1}$	$\frac{v}{4L}=\nu$
Second	Third	First overtone	2	2	$\frac{4L}{3}$	3ν
Third	Fifth	Second overtone	3	3	$\frac{4L}{5}$	5ν
Fourth	Seventh	Third overtone	4	4	$\frac{4L}{7}$	7ν
*n*th	(2*n* – 1)th	*n*th overtone	*n*	*n*	$\frac{4L}{2n-1}$	$(2n-1)\nu$

15. The phenomenon of alternate waxing and waning of sound at regular intervals is called beats. Beat frequency $= \nu_1 - \nu_2$

16. Doppler Effect
The apparent change in frequency of sound when the source of sound, the observer and the medium are in relative motion is called Doppler Effect.

17. The formula for apparent frequency ν observed by the observer and the medium moving in the same direction is

$$\nu' = \left(\frac{v + v_m - v_0}{v + v_m - v_s}\right)\nu$$

EXERCISES

SECTION - I

1. A particle executes simple harmonic oscillation with an amplitude *a*. The period of oscillation is *T*. The minimum time taken by the particle to travel half of the amplitude from the equilibrium position is
(a) $\frac{T}{4}$ (b) $\frac{T}{8}$
(c) $\frac{T}{12}$ (d) $\frac{T}{2}$

2. When the maximum kinetic energy of a simple pendulum is *k*, then what is its displacement (in terms of magnitude *a*) when its kinetic energy is $\frac{k}{2}$?
(a) $\frac{a}{\sqrt{2}}$ (b) $\frac{a}{2}$
(c) $\frac{a}{\sqrt{3}}$ (d) $\frac{a}{3}$

3. What is the time period of a pendulum hanged in a satellite? (*T* is time period)
(a) zero (b) *T*
(c) Infinite (d) $\frac{T}{\sqrt{6}}$

4. A particle is executing SHM at about its mid-point of at mean position and extremity. What is the potential energy in terms of total energy *E*?
(a) $\frac{E}{4}$ (b) $\frac{E}{16}$
(c) $\frac{E}{2}$ (d) $\frac{E}{8}$

5. There is a simple pendulum hanging from the ceiling of a lift. When the lift is standstill, the time period of the pendulum is *T*. If the resultant acceleration becomes $\frac{g}{4}$ then the new time period of the pendulum is
(a) 0.8 *T* (b) 0.25 *T*
(c) 2 *T* (d) 4 *T*

6. If the ratio of frequencies of two pendulums are 2 : 3, then their lengths are in the ratio
(a) $\frac{\sqrt{2}}{3}$ (b) $\frac{\sqrt{3}}{2}$
(c) $\frac{4}{9}$ (d) $\frac{9}{4}$

7. What is the maximum acceleration of the particle in SHM? $y = 2\sin\left(\frac{\pi t}{2} + \theta\right)$ where 2 is in cm.
(a) $\frac{\pi}{2}$ cm/s^2 (b) $\frac{\pi^2}{2}$ cm/s^2
(c) $\frac{\pi}{4}$ cm/s^2 (d) $\frac{\pi}{4}$ cm/s^2

8. The mass and the radius of a planet are twice that of earth. Then, period of oscillation of a seconds pendulum on that planet will be
(a) $\frac{1}{\sqrt{2}}$ s (b) $\sqrt{2}$ s
(c) 2s (d) $\frac{1}{2}$ s

9. Which one of the following is a simple harmonic motion?
(a) Ball bouncing between two rigid vertical walls
(b) Particle moving in a circle with uniform speed
(c) Wave moving through a string fixed at both ends
(d) Earth spinning about its own axis.

10. The displacement x (in metres) of a particle in simple harmonic motion is related to time (in seconds) as $x = 0.01\cos\left(\pi t + \frac{\pi}{4}\right)$ The frequency of the motion will be
(a) 0.5 Hz (b) 1.0 Hz
(c) $\frac{\pi}{2}$ (d) π Hz

11. A particle is undergoing a one-dimensional simple harmonic oscillation of amplitude X_M about the origin on X-axis with time period T and is at $-X_M$ at $t = 0$ s. What is the position of the particle after a time interval $t = 3.15\ T$?
(a) Between $-X_M$ and O (b) Between O and $+X_M$
(c) At the origin (d) At $+X_M$

12. Two springs of the same force constant k are joined in series to a black of mass m. What is the frequency of oscillation of the block?
(a) $f = \frac{1}{2\pi}\sqrt{\frac{2k}{m}}$ (b) $f = \frac{1}{2\pi}\sqrt{\frac{k}{2m}}$
(c) $f = \frac{1}{2\pi}\sqrt{\frac{k}{m}}$ (d) $f = \frac{1}{\pi}\sqrt{\frac{k}{m}}$

13. The length of a simple pendulum is about 100 cm known to an accuracy of 1 mm. It's period of oscillation is 2 s determined by measuring the time for 100 oscillations using a clock of 0.01 resolution. What is the accuracy in the determined value of g?
(a) 0.2% (b) 0.5%
(c) 0.1% (d) 2%

14. A particle executes SHM of amplitude 25 cm and time period 3 s. What is the minimum time required for the particle to move between two points 12.5 cm on either side of the mean position?
(a) 0.5 s (b) 1.0 s
(c) 1.5 s (d) 2.0 s

15. The time period of a particle in simple harmonic motion is 8 seconds. At $t = 0$, it is at the mean position. The ratio of the distance travelled by it in the first and second seconds is
(a) $\frac{1}{2}$ (b) $\frac{1}{\sqrt{2}}$
(c) $\frac{1}{\sqrt{2}-1}$ (d) $\frac{1}{\sqrt{3}}$

16. If the displacement (x) and velocity (v) of a particle executing simple harmonic motion are related through the expression $4v^2 = 25 - x^2$, then its time period is
(a) π (b) 2π
(c) 4π (d) 6π

17. A body executes simple harmonic motion under the action of a force F_1 with a time period $\frac{4}{5}$ s. If the force is changed to F_2 it executes SHM with time period $\frac{3}{5}$ s. If both the forces F_1 and F_2 act simultaneously in the same direction on the body, its time period in second is
(a) $\frac{12}{25}$ (b) $\frac{24}{25}$
(c) $\frac{35}{24}$ (d) $\frac{15}{12}$

18. The angular amplitude of a simple pendulum is θ_0. The maximum tension in its string will be
(a) $mg(1-\theta_0)$ (b) $mg(1+\theta_0)$
(c) $mg(1-\theta_0^2)$ (d) $(1+\theta_0^2)$

19. Time period of a simple pendulum length l is T_1 and time period of a uniform rod of the same length l pivoted about one end and oscillating in a vertical plane is T_2. If amplitude of oscillations in both cases is small, then $\frac{T_1}{T_2}$ is
(a) $\frac{1}{\sqrt{3}}$ (b) 1
(c) $\sqrt{\frac{3}{4}}$ (d) $\sqrt{\frac{3}{2}}$

20. What is the velocity of the bobd of a simple pendulum at its mean position, if it is able to rise to vertical height of 10 cm? ($g = 9.8$ m/s^2)?

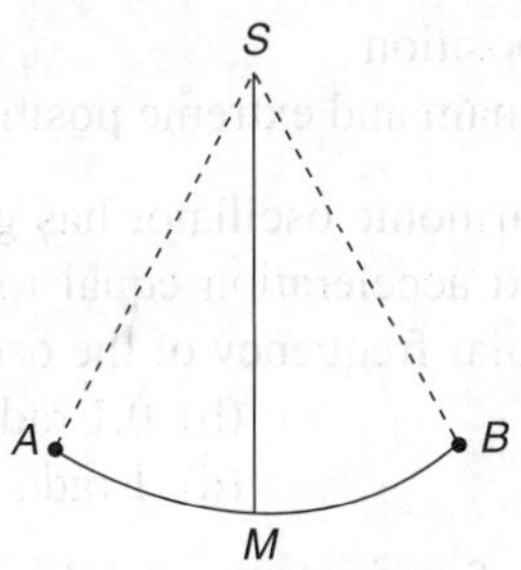

(a) 2.2 m/s (b) 1.8 m/s
(c) 1.4 m/s (d) 0.6 m/s

21. A mass m is suspended from the two coupled springs connected in series. The force constant for the springs are k_1 and k_2. The time period of the suspended mass will be

(a) $T = 2\pi\sqrt{\dfrac{5}{k_1 - k_2}}$ (b) $T = 2\pi\sqrt{\dfrac{mk_1k_2}{k_1 + k_2}}$

(c) $T = 2\pi\sqrt{\dfrac{m}{k_1 + k_2}}$ (d) $T = 2\pi\sqrt{\dfrac{m(k_1 + k_2)}{k_1k_2}}$

22. In SHM the kinetic energy of a particle at some instant is $\dfrac{1}{4}$ of its total energy. Then the displacement of the particle is

(a) $\dfrac{a}{\sqrt{2}}$ (b) $a\sqrt{2}$

(c) $\dfrac{\sqrt{3}a}{2}$ (d) $\dfrac{a}{2}$

23. The time period of a compound pendulum is expressed by

(a) $2\pi\sqrt{\dfrac{l^2 + k^2}{l_g}}$ (b) $2\pi\sqrt{\dfrac{l^2 + k^2}{g}}$

(c) $2\pi\sqrt{\dfrac{l^2 + k}{g}}$ (d) $2\pi\sqrt{\dfrac{l + k^2}{g}}$

24. Which one of the following expressions is that of a simple harmonic progressive wave?

(a) $a\sin\omega t$ (b) $a\sin(\omega t)\cos(kx)$
(c) $a\sin(\omega t - kx)$ (d) $a\cos kx$

25. Two springs of spring constant k_1 and k_2 are arranged in parallel and a body of mass m is attached to it, then calculate the time period of the system.

(a) $2\pi\sqrt{\dfrac{m}{k_1 + k_2}}$ (b) $2\pi\sqrt{\dfrac{2m}{k_1 + k_2}}$

(c) $2\pi\sqrt{\dfrac{mk_1k_2}{k_1 + k_2}}$ (d) $2\pi\sqrt{\dfrac{k_1 + k_2}{m}}$

26. The composition of two simple harmonic motions of equal periods at right angle to each other and with a phase difference of results in the displacement of the particle along

(a) a circle (b) a figure of eight
(c) a straight line (d) an ellipse.

27. Two springs are joined and attached to a mass of 16 kg. The system is then suspended vertically from a rigid support. The spring constant of the two springs are k_1 and k_2 respectively. The period of vertical oscillation of the system will be

(a) $\dfrac{1}{8\pi}\sqrt{k_1 + k_2}$ (b) $8\pi\sqrt{\dfrac{k_1 + k_2}{k_1k_2}}$

(c) $\dfrac{\pi}{2}\sqrt{k_1 - k_2}$ (d) $\dfrac{\pi}{2}\sqrt{\dfrac{k_1}{k_2}}$

28. A particle executes linear simple harmonic motion with an amplitude of 2 cm. When the particle is at 1 cm from the mean position, the magnitude of its velocity is equal to that of its acceleration. Then its time period in second is

(a) $\dfrac{1}{2\pi\sqrt{3}}$ (b) $2\pi\sqrt{3}$

(c) $\dfrac{2\pi}{\sqrt{3}}$ (d) $\dfrac{\sqrt{3}}{2\pi}$

29. A particle starts SHM from the mean position. Its amplitude is a and total energy E. At one instant its kinetic energy $3\dfrac{E}{4}$. Its displacement at that instant is

(a) $\dfrac{a}{\sqrt{2}}$ (b) $\dfrac{a}{2}$

(c) $\dfrac{a}{\sqrt{\left(\dfrac{3}{2}\right)}}$ (d) $\dfrac{a}{\sqrt{3}}$

30. The equation $y = A\sin 2\pi\left(\dfrac{t}{T} - \dfrac{x}{\lambda}\right)$, where the symbols carry their usual meaning and A, T, λ are positive, represents a wave of

(a) amplitude 2A (b) period $\dfrac{T}{\lambda}$

(c) speed × λ (d) speed $= \dfrac{\lambda}{T}$

31. The KE and PE of a particle executing SHM of amplitude a will be equal when displacement is

(a) $\dfrac{a}{2}$ (b) $a\sqrt{2}$

(c) $2a$ (d) $\dfrac{a}{\sqrt{2}}$

32. A particle executes simple harmonic motion when $x = -A$ and $x = +A$. The time taken for it to go from 0 to $\dfrac{A}{2}$ is t_1 and to go from $\dfrac{A}{2}$ to A is t_2, then

(a) $t_1 < t_2$ (b) $t_1 > t_2$
(c) $t_1 = t_2$ (d) $t_1 = 2t_2$

33. A particle executes simple harmonic motion. The potential energy (PE), kinetic energy (KE) and total energy (TE) are measured as a function of displacement y. Which of the following statement is true?
(a) TE is zero when $x = 0$
(b) PE is maximum when $x = 0$
(c) KE is maximum when $x = 0$
(d) KE is maximum when x is maximum.

34. A particle executes SHM of amplitude 25 cm and time period 3s. What is the minimum time required for the particle to move between two points 12.5 cm on either side of the mean position?
(a) 0.5 s (b) 1.0 s
(c) 1.5 s (d) 2.0 s

35. The motion of a particle varies with time according to the equation $y = a(\sin \omega t + \cos \omega t)$. Then
(a) the motion is oscillatory but not SHM
(b) the motion is SHM with amplitude a
(c) the motion is SHM with amplitude $a\sqrt{2}$
(d) the motion is SHM with amplitude $2a$.

36. Two simple harmonic waves of the same amplitude and frequency differ by a phase $\frac{\pi}{2}$ when they are fed simultaneously to the x and y plates of a C.R.O., the screen would display the figure of
(a) a circle (b) an ellipse
(c) straight line (d) a square

37. The maximum velocity of a simple harmonic motion represented by $y = 3\sin\left(100t + \frac{\pi}{6}\right)$ is given by
(a) 300 (b) $\frac{3\pi}{6}$
(c) 100 (d) $\frac{\pi}{6}$

38. Consider the following statement. The total energy of a particle executing simple harmonic motion depends upon (i) amplitude, (ii) period , (iii) displacement.
(a) (i) and (ii) are correct
(b) (ii) and (iii) are correct
(c) (i) and (iii) are correct
(d) (i) (ii) and (iii) are correct.

39. A spring has length 'l' and spring constant 'k'. If the spring is divided into two equal parts, then the spring constant is
(a) k (b) $\frac{k}{2}$
(c) $2k$ (d) $4k$

40. In simple harmonic motion, maximum velocity occurs
(a) at extreme positions
(b) at $\frac{1}{2}$ of extreme positions
(c) at mean position
(d) between main and extreme position.

41. If a simple harmonic oscillator has got a displacement of 0.02 m and acceleration equal to 0.02 m/s^2 at any time, the angular frequency of the oscillator is equal to
(a) 10 rad/s (b) 0.1 rad/s
(c) 100 rod/s (d) 1 rad/s

42. Two springs of spring constant k_1 and k_2 have equal highest velocities, when executing, SHM. Then the ratio of their amplitudes will be
(a) $\frac{k_1}{k_2}$ (b) $\left(\frac{k_1}{k_2}\right)^{1/2}$
(c) $\frac{k_2}{k_1}$ (d) $\left(\frac{k_2}{k_1}\right)^{1/2}$

43. A SHM has amplitude a and time period T. The maximum velocity will be
(a) $\frac{2a}{T}$ (b) $\frac{4a}{T}$
(c) $\frac{2\pi a}{T}$ (d) $\frac{\pi a}{T}$

44. The graph between restoring force and time in case of simple harmonic motion is
(a) a straight line (b) a circle
(c) a parabola (d) a sine curve.

45. The angle between the instantaneous velocity and acceleration of a particle executing simple harmonic motion is
(a) zero (b) $\frac{\pi}{2}$
(c) π (d) zero or π.

46. In simple harmonic motion, the particle is
(a) always accelerated
(b) always retarded
(c) alternately accelerated and retarded
(d) neither accelerated nor retarded.

47. If E is the total energy of a particle executing simple harmonic motion and A be the amplitude of the vibratory motion, then E and A are related as
(a) $E \propto A$ (b) $E \propto \frac{1}{A}$
(c) $E \propto A^2$ (d) $E \propto \frac{1}{A^2}$

48. A mass of 1 kg is suspended from a spring. Its time period of oscillation on earth is T. What will be its time period at the centre of the earth?
(a) Zero (b) T
(c) $2T$ (d) Infinity

49. A mass is attached to a vertically held light spring. The spring extends by 1 mm due to the weight of the mass. The time period (in sec) of oscillation of the mass will be

(a) 1 (b) π
(c) 2π (d) None of these.

50. The dimensions of mass/force constant is

(a) time (b) $(\text{time})^2$
(c) acceleration (d) $\dfrac{1}{\text{acceleration}}$

51. What is the number of degrees of freedom of an oscillating pendulum?

(a) One (b) Two
(c) Three (d) More than three.

52. The angular velocity and the amplitude of a simple pendulum is ω and a respectively. At a displacement x from the mean position if its kinetic energy is T and potential energy is V, then the ratio of T to V is

(a) $\dfrac{(a^2 - x^2\omega^2)}{x^2\omega^2}$ (b) $\dfrac{x^2\omega^2}{(a^2 - x^2\omega^2)}$
(c) $\dfrac{(a^2 - x^2)}{x^2}$ (d) $\dfrac{x^2}{(a^2 - x^2)}$

53. The phase angle between the projections of uniform circular motion on two mutually perpendicular diameters is

(a) zero (b) $\dfrac{\pi}{2}$
(c) $\dfrac{3\pi}{4}$ (d) π

54. In simple harmonic motion the acceleration of the particle is zero, when its

(a) velocity is zero
(b) displacement is zero
(c) both velocity and displacement are zero
(d) both velocity and displacement are maximum.

55. In simple harmonic motion, the variation of which of the following is not a sine curve?

(a) Displacement (b) Velocity
(c) Acceleration (d) Time period

56. The particle executing simple harmonic motion has a kinetic energy $K_0 \cos^2 \omega t$. The maximum values of the potential energy and the total energy are, respectively,

(a) K_0 and $2K_0$ (b) 0 and $2K_0$
(c) K_0 and K_0 (d) $\dfrac{K_0}{2}$ and K_0

57. A particle executing a vibratory motion while passing through the mean position has

(a) maximum potential energy and minimum kinetic energy
(b) maximum potential energy and maximum kinetic energy
(c) maximum kinetic energy and minimum potential energy
(d) minimum kinetic energy and minimum potential energy.

58. A simple pendulum of period T has a metal bob which is negatively charged. If it is allowed to oscillate above a positively charged plate, its period

(a) remains same (b) decreases
(c) increases (d) It will not vibrate.

59. A particle is moving in a circle with uniform speed. Its motion is

(a) periodic and simple harmonic motion
(b) periodic but not simple harmonic motion
(c) a periodic
(d) None of these.

60. A mass m oscillates with simple harmonic motion with a frequency $f = \dfrac{\omega}{2\pi}$ and amplitude A on a spring with constant k. Therefore,

(a) the total energy of the system is $1/2\, kA^2$
(b) its frequency is $\dfrac{1}{2\pi}\left(\dfrac{k}{m}\right)^{1/2}$
(c) the maximum velocity occurs when $y = 0$
(d) All the above are correct.

61. For a particle executing SHM the displacement x is given by $x = a \cos \omega t$. Identify the graph which represents the variation of potential energy (PE) as a function of time t and displacement x.

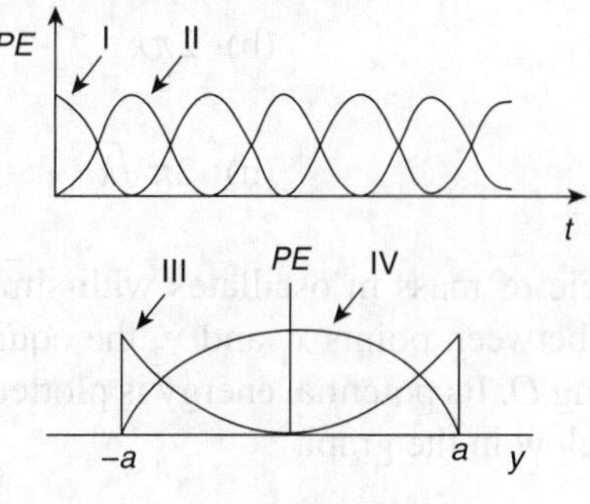

(a) II, IV (b) III, IV
(c) I, III (d) I, II

62. A watch based on an oscillating spring is taken to the moon. It will

(a) become slow
(b) become fast
(c) give the same time as on earth.
(d) None of these.

63. If a hole is bored along the diameter of the earth and a stone is dropped into the hole, the stone
(a) reaches the centre of the earth and stops there
(b) reaches the other side of the earth and stops there
(c) executes simple harmonic motion about the centre of the earth
(d) reaches the other side of the earth and escapes into space.

64. A pendulum suspended from the ceilling of a train has a period T when the train is at rest. When the train is accelerating with uniform acceleration, T will
(a) increase (b) decrease
(c) remain unaffected (d) become infinite.

65. A body of mass 5 gm is executing simple harmonic motion about a point with amplitude of 10 cm. Its maximum velocity is 100 cm/sec. Its velocity will be 50 cm/sec at a distance
(a) 5 (b) $5\sqrt{2}$
(c) $5\sqrt{3}$ (d) $10\sqrt{2}$

66. The time period of a second's pendulum is 2 sec. The spherical bob which is empty from inside has a mass of 50 gm. This is now replaced by another solid bob of same radius but having a different mass of 100 gm. The new time period will be
(a) 4 sec (b) 1 sec
(c) 2 sec (d) 8 sec

67. The equation of motion of a particle is $\frac{d^2y}{dt^2} + ky = 0$ where k is a positive constant. The time period of motion is given by
(a) $\frac{2\pi}{k}$ (b) $2\pi k$
(c) $\frac{2\pi}{\sqrt{k}}$ (d) $2\pi\sqrt{k}$

68. A particle of mass m oscillates with simple harmonic motion between points x_1 and x_2 the equilibrium position being O. Its potential energy is plotted. It will be as given below in the graph
(a)
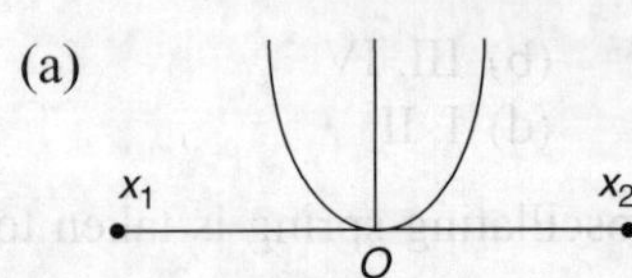

(b)
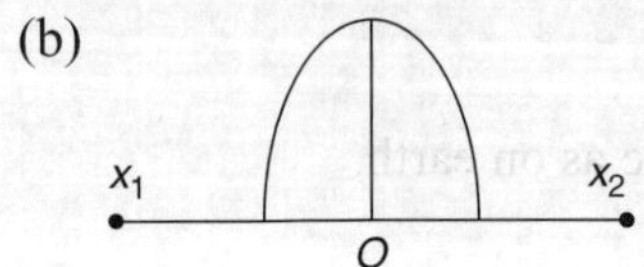

(c)
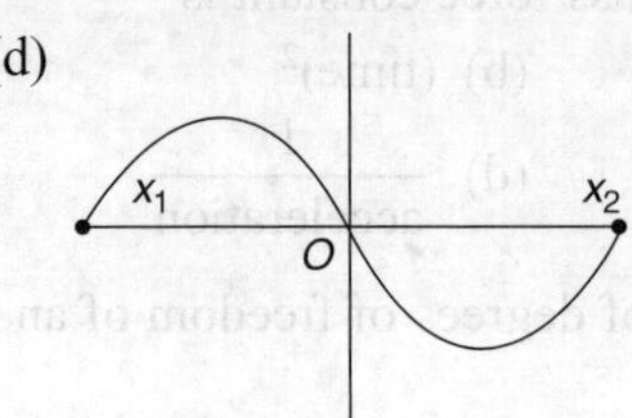

(d)
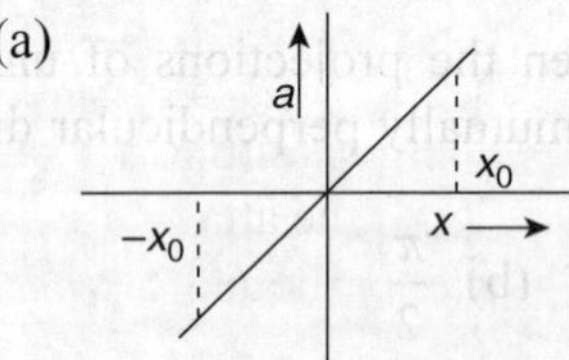

69. A particle of mass m is hanging vertically by an ideal spring of force constant 'k'. If the mass is made to oscillate vertically, its total energy is
(a) maximum at extreme position
(b) maximum near equilibrium
(c) minimum at near equilibrium
(d) same at all positions.

70. The variation of acceleration (a) of the particle executing simple harmonic motion with displacement x is as shown in figure
(a)
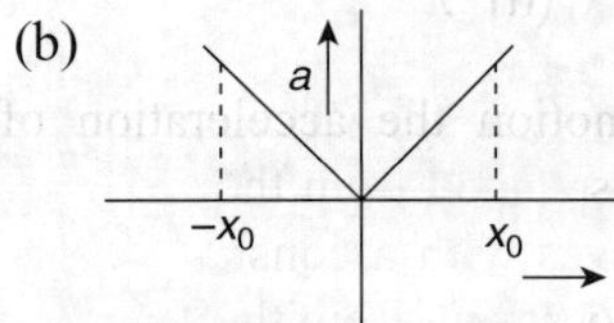

(b)
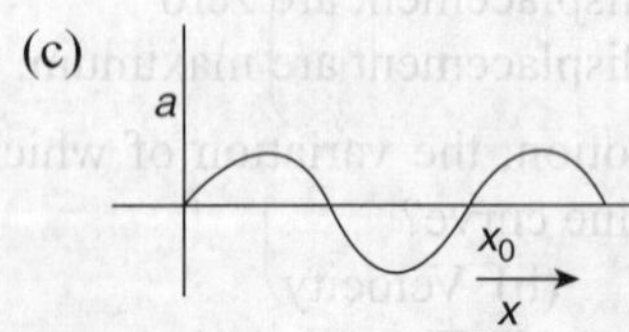

(c)
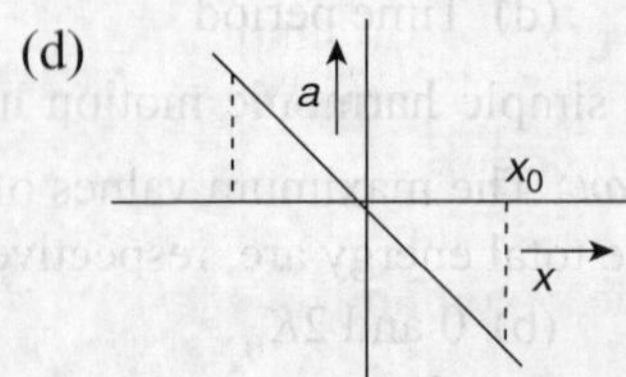

(d)

71. A mass m is suspended from a spring of length l and force constant k. The frequency of vibration of the mass is f_1. The spring is cut into two equal parts and the same mass is suspended from one of the parts. The new frequency of vibration of the mass is f_2. Which of the following relation between frequencies is correct?

(a) $f_2 = \frac{f_1}{\sqrt{2}}$ (b) $f_1 = f_2$
(c) $f_1 = 2f_2$ (d) $f_2 = \sqrt{2}f_1$

72. A man measured the period of a simple pendulum inside a stationary lift and finds it to be T sec. If the lift accelerates upwards with an acceleration of $g/4$, then the period of the pendulum will be
(a) T (b) $T/4$
(c) $\frac{2T}{\sqrt{5}}$ (d) $2T\sqrt{5}$

73. If $y = A\sin^2 \omega t + B\cos^2 \omega t$, then for what value of A, B and C the function y will represent SHM?
(a) For any value of A, B and C
(b) $A = \pm B$, $C = \pm 3B$
(c) $A = C$, $B = 0$
(d) $A = \pm B$, $C = 2B$

74. A simple harmonic oscillator has a period of 0.01 sec and an amplitude of 0.2 m. The magnitude of the velocity in m/sec at the centre of oscillation is
(a) 20π (b) 100
(c) 40π (d) 100π

75. A mass at the end of a spring undergoes simple harmonic motion with frequency of 0.5 Hz. If the attached mass is reduced to one quarter of its value, the new frequency in Hz is
(a) 0.25 (b) 1
(c) 2 (d) 4.5

76. A simple pendulum is suspended from the ceiling of a train. When the train moves with a constant acceleration 'a' the direction of the spring from the vertical is
(a) 0^0 (b) $\tan^{-1}\left(\frac{a}{g}\right)$
(c) $\sin^{-1}\left(\frac{a}{g}\right)$ (d) $\cos^{-1}\left(\frac{a}{g}\right)$

77. The work done by the spring of a simple pendulum during one complete oscillation is equal to
(a) total energy of the pendulum
(b) kinetic energy of the pendulum
(c) potential energy of the pendulum
(d) zero.

78. A simple pendulum is suspended from the roof of the train which is moving with an acceleration of 49 cm/sec^2. By what angle to the vertical will its string will be inclined?
(a) 20° (b) 30°
(c) 0° (d) 3°

79. A simple harmonic oscillator has an amplitude A and time period T. The time required by it to travel from $x = A$ to $x = A/2$ is
(a) $T/6$ (b) $T/4$
(c) $T/3$ (d) $T/2$

80. A particle executes simple harmonic motion with a period of 6 sec and amplitude of 3 cm. Its maximum speed in cm/s is
(a) $\frac{\pi}{2}$ (b) π
(c) 2π (d) 3π

81. The string between mass m and 2 m is inextensible. If the string is cut find the magnitudes of acceleration of mass 2 m and m.

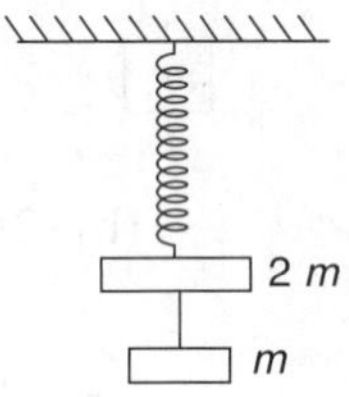

(a) $\frac{g}{2}, g$ (b) g, g
(c) $g, \frac{g}{2}$ (d) $\frac{g}{2}, \frac{g}{2}$

82. Two light springs of force constant k_1 and k_2 are connected to a mass m placed on a horizontal frictionless surface as shown in the figure. The time period of horizontal oscillations will be

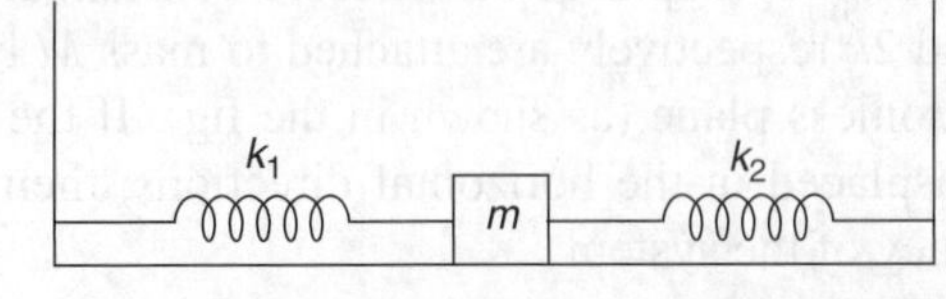

(a) $2\pi\left(\frac{m}{k_1 + k_2}\right)^{1/2}$ (b) $2\pi\left(\frac{mk_2}{k_1^2}\right)^{1/2}$
(c) $2\pi\left(\frac{m(k_1 + k_2)}{k_1k_2}\right)^{1/2}$ (d) $2\pi\left(\frac{m}{k_2 - k_1}\right)^{1/2}$

83. A mass m is suspended from two identical springs, each of force constant k as shown in the figure. The time period of vertical oscillations of the mass will be

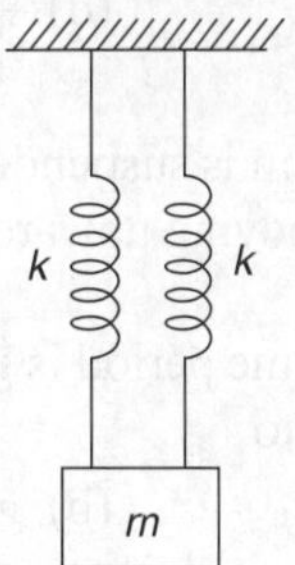

(a) $2\pi\left(\frac{m}{k}\right)^{1/2}$ (b) $2\pi\left(\frac{m}{2k}\right)^{1/2}$

(c) $\frac{1}{2\pi}\left(\frac{2k}{m}\right)^{1/2}$ (d) $\frac{1}{2\pi}\left(\frac{2m}{k}\right)^{1/2}$

84. A mass m is suspended from two identical springs each of force constant k as shown in figure. The frequency of vertical oscillations of the mass will be

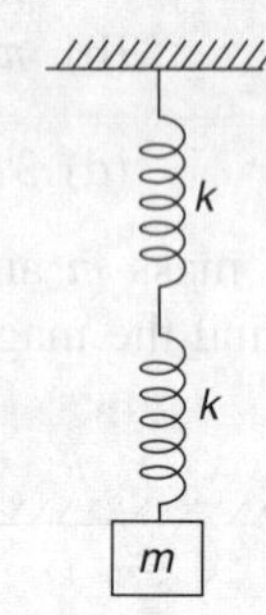

(a) $2\pi\left(\frac{k}{m}\right)^{1/2}$ (b) $2\pi\left(\frac{2m}{k}\right)^{1/2}$

(c) $\frac{1}{2\pi}\left(\frac{2k}{m}\right)^{1/2}$ (d) $\frac{1}{2\pi}\left(\frac{k}{2m}\right)^{1/2}$

85. A body is executing simple harmonic motion. When the displacement from the mean position is 4 cm and 5 cm, the corresponding velocities of the body is 10 cm/sec and 8 cm/sec. Then the time period of the body is

(a) 2π sec (b) $\pi/2$ sec
(c) π sec (d) $3\pi/2$ sec

86. Four mass less springs whose force constant are $2k$, $2k$, k and $2k$ respectively are attached to mass M kept on a frictionless plane (as shown in the fig). If the mass M is displaced in the horizontal directions, then the frequency of the system

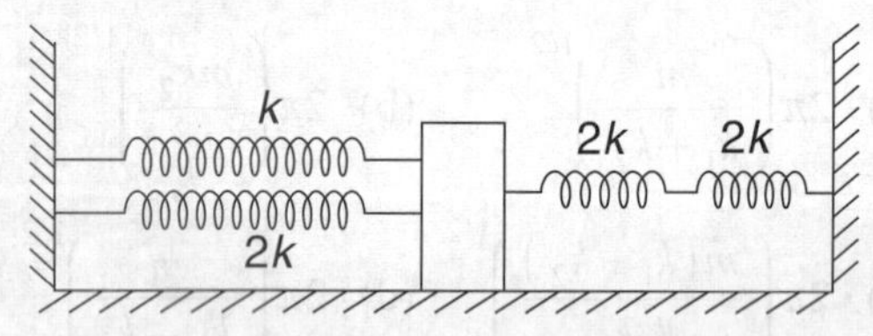

(a) $\frac{1}{2\pi}\sqrt{\frac{k}{4M}}$ (b) $\frac{1}{2\pi}\sqrt{\frac{4k}{M}}$

(c) $\frac{1}{2\pi}\sqrt{\frac{k}{7M}}$ (d) $\frac{1}{2\pi}\sqrt{\frac{7k}{M}}$

87. A simple pendulum is suspended from the roof of a lift. When the lift is moving upwards with an acceleration a ($a < g$), then the time period is given by $T = 2\pi\left(\frac{l}{g'}\right)^{1/2}$ where g' is equal to

(a) g (b) $g - a$
(c) $g + a$ (d) $(g^2 + a^2)^{1/2}$

88. A simple pendulum is suspended from the roof of a trolley. When the trolley moves in a horizontal direction with an acceleration 'a', then the time period is given $T = 2\pi\left(\frac{l}{g'}\right)^{1/2}$ where g' is equal to

(a) g (b) $g - a$
(c) $(g + a)$ (d) $(g^2 + a^2)^{1/2}$

89. A simple pendulum of length L, and mass of bob M is oscillating in a plane about a vertical line between angular limits $(-\phi)$ to $(+\phi)$. For an angular displacement $\theta[|\theta| < \phi]$ the tension in the string and the velocity of the bob are T and v respectively. The relation that holds good under the above conditions is

(a) $T\cos\theta = Mg$
(b) $T - mg\cos\theta = \frac{Mv^2}{L}$
(c) The magnitude of the tangential acceleration of bob $|a_T| = g\cos\theta$
(d) $T = Mg\cos\theta$

90. A particle executes simple harmonic motion with a frequency f. The frequency with which its kinetic energy and the total energy oscillates is

(a) $f/2, f/2$ (b) $2f, 0$
(c) $2f, \infty$ (d) $2f, 2f$

91. A particle executes simple harmonic motion between $x = -A$ and $x = +A$. The time taken for it go from 0 to $\frac{A}{2}$ is T_1 and to go from $\frac{A}{2}$ to A is T_2. Then

(a) $T_1 < T_2$ (b) $T_1 = 2T_2$
(c) $T_1 > T_2$ (d) $T_1 = T_2$

92. The phase difference between two similar pendulums A and B is 90^0. When the bob of pendulum A has maximum velocity v_0, the bob of pendulum B will be

(a) stationary
(b) moving with velocity v_0
(c) moving with a velocity $2v_0$
(d) moving with a velocity of $\frac{v_0}{2}$.

93. A simple pendulum is oscillating without damping. When the displacement of the bob is less than maximum, its acceleration vector $\vec{a}$ is correctly shown in

(a) (b)

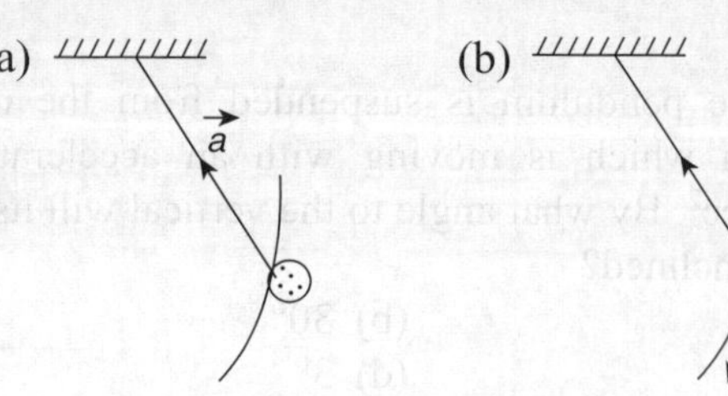

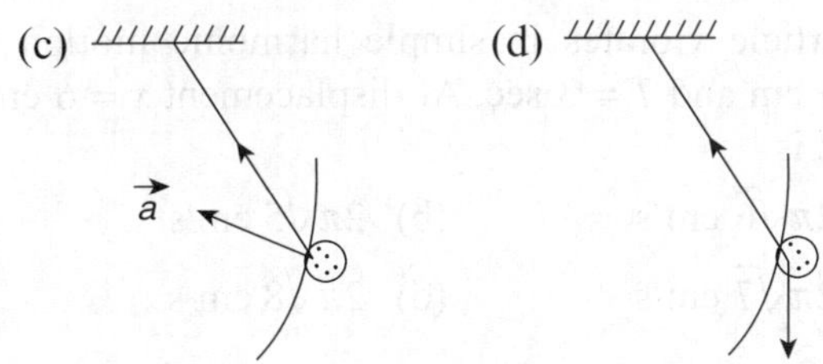

94. A cylindrical piston of mass M slides smoothly inside a long cylinder closed at one end, enclosing a certain mass of gas. The cylinder is kept with its axis horizontal. If the piston is disturbed from its equilibrium position, it oscillates simple harmonically. The period of oscillation will be

(a) $T = 2\pi\left(\frac{MV_0}{PA^2}\right)^{1/2}$ (b) $T = 2\pi\left(\frac{MA^2}{PV_0}\right)^{1/2}$

(c) $T = 2\pi\left(\frac{M}{PA^2V_0}\right)^{1/2}$ (d) $T = 2\pi\sqrt{MPV_0A^2}$

95. A sphere of radius r is kept on a concave mirror of radius of curvature R. The arrangement is kept on a horizontal table. If the sphere is displaced from its equilibrium position and left, then it executes simple harmonic motion. The period of oscillation will be

(a) $2\pi\left(\frac{(R-r)1.4}{g}\right)^{1/2}$ (b) $2\pi\left(\frac{(R-r)}{g}\right)^{1/2}$

(c) $2\pi\left(\frac{rR}{g}\right)^{1/2}$ (d) $2\pi\left(\frac{R}{gr}\right)^{1/2}$

96. A simple pendulum consists of a small sphere of mass m suspended by a thread of length L. The sphere carries a positive charge q. The pendulum is placed in a uniform electric field of strength E directed vertically upwards. With what period will the pendulum oscillate if the electrostatic force on the sphere is less than the gravitational force? Assuming small oscillations, the period is given by

(a) $T = 2\pi\left(\frac{L}{g-E}\right)^{1/2}$ (b) $T = 2\pi\left(\frac{L}{g-\frac{qE}{m}}\right)^{1/2}$

(c) $T = 2\pi\left(\frac{L}{g+\frac{qE}{m}}\right)^{1/2}$ (d) $T = 2\pi\left(\frac{L}{g+qE}\right)^{1/2}$

97. The length of a second's pendulum is increased by 0.1%. The clock
(a) gains 43.2 sec per day
(b) loses 43.2 sec per day
(c) neither loses nor gains
(d) loses 86.4 sec per day.

98. For an oscillating simple pendulum, the tensions in the string is
(a) same through the oscillation
(b) least at the end points and greatest at mean position
(c) least at the mean position and greatest at the end points
(d) greatest at the mid-point between the end point and the mean position.

99. The length of a pendulum is increased by 44%. The percentage increase in the time period is
(a) 10 % (b) 20%
(c) 40% (d) 44%

100. An electrically driven tuning fork with a constant amplitude is an example of
(a) free vibration
(b) damped vibration
(c) maintained vibration
(d) forced vibration.

101. Resonant vibration is a special case of
(a) forced vibrations (b) free vibration
(c) damped vibrations (d) natural vibrations.

102. A test tube of cross-sectional area A has some lead shots in it. The total mass is M. It floats upright in a liquid of density σ. When pushed down a little and released it oscillates up and down with a period given by

(a) $T = 2\pi\left(\frac{MA}{g\sigma}\right)^{1/2}$ (b) $T = 2\pi\left(\frac{Mg}{A\sigma}\right)^{1/2}$

(c) $T = 2\pi\left(\frac{M\sigma}{Ag}\right)^{1/2}$ (d) $T = 2\pi\left(\frac{M}{A\sigma g}\right)^{1/2}$

103. Two equal charges ($-Q$) are fixed at the points (0, a) and (0, $-a$) on the y-axis. A positive charge Q is released from rest at the point ($2a$, 0) on the x-axis. The charge Q will
(a) executes SHM about origin
(b) move to origin and come to rest
(c) move to infinity
(d) executes oscillatory but no SHM period.

104. The time period of a simple pendulum of length equal to the radius of the earth is

(a) $2\pi\left(\frac{R_e}{g}\right)^{1/2}$ (b) $2\pi\left(\frac{2R_e}{g}\right)^{1/2}$

(c) $2\pi\left(\frac{R_e}{2g}\right)^{1/2}$ (d) infinite.

105. The period of oscillation of a physical is (I = moment of inertia about the axis of rotation, M = mass and d is the distance of the centre of gravity from pivot)

(a) $T = 2\pi\left(\frac{I}{Mgd}\right)^{1/2}$ (b) $T = \frac{1}{2\pi}\left(\frac{I}{Mgd}\right)^{1/2}$

(c) $T = 2\pi\left(\frac{Mgd}{I}\right)^{1/2}$ (d) $T = \frac{1}{2\pi}\left(\frac{Mgd}{I}\right)^{1/2}$

106. A point mass M is suspended at the end of a mass less wire of length L and area of cross-section A. If Y is the Young's modulus of elasticity for the wire, the frequency of oscillation for simple harmonic motion along the vertical line is

(a) $n = \frac{1}{2\pi}\left(\frac{ML}{YA}\right)^{1/2}$ (b) $n = \frac{1}{2\pi}\left(\frac{M}{LYA}\right)^{1/2}$

(c) $n = \frac{1}{2\pi}\left(\frac{YA}{ML}\right)^{1/2}$ (d) $n = \frac{1}{2\pi}\left(\frac{Y}{AML}\right)^{1/2}$

107. A particle of mass 10 gm is describing simple harmonic motion along a straight line with a period of 2 sec and an amplitude of 10 cm. The kinetic energy when it is at 2 cm from its equilibrium position is

(a) 2.4×10^{-4} J (b) 1.2×10^{-4} J

(c) 4.8×10^{-4} J (d) 1.6×10^{-24} J

108. A small body of mass 0.1 kg is undergoing simple harmonic motion of amplitude 1 m and period 0.2 s. The maximum value of force acting on it is

(a) π (b) 2π

(c) $2\pi^2$ (d) $10\pi^2$

109. The displacement x (in metres) of a particle in simple harmonic motion is related to time t as $x = 0.01 \cos\left(\pi t + \frac{\pi}{4}\right)$ frequency. Then, frequency is

(a) π Hz (b) $\frac{\pi}{2}$ Hz

(c) 1 Hz (d) 0.5 Hz

110. The displacement of a particle executing simple harmonic motion of $y = 5 \sin 20\pi t$. Its frequency is

(a) 10 Hz (b) 20π Hz

(c) 0.01 Hz (d) 20 Hz

111. A simple harmonic oscillator has amplitude A and time period T. Its maximum speed is

(a) $\frac{4A}{T}$ (b) $\frac{2A}{T}$

(c) $\frac{4\pi A}{T}$ (d) $\frac{2\pi A}{T}$

112. The variation of total energy (E) and kinetic energy (KE) of the particle executing simple harmonic motion with displacement (x) is as shown in figure

(a)
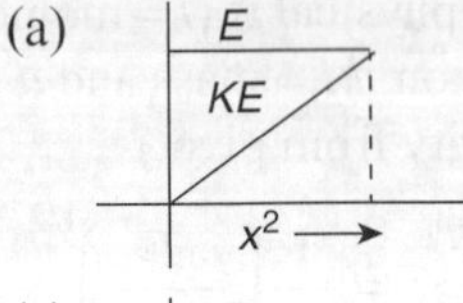

(b)
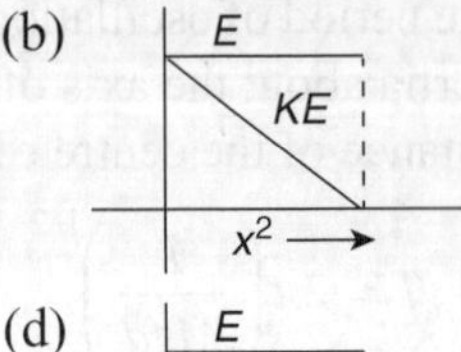

(c)
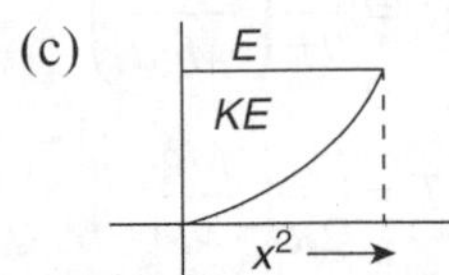

(d)
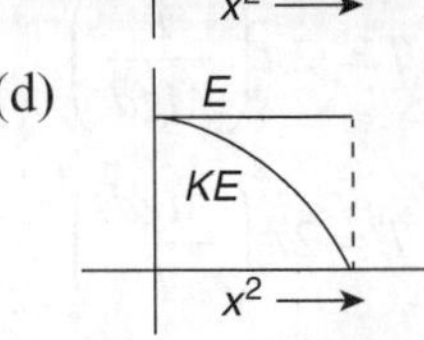

113. A particle vibrates in simple harmonic motion with $A = 9$ cm and $T = 3$ sec. At displacement $x = 6$ cm, its speed is

(a) $2\pi\sqrt{6}$ cm/s (b) $2\pi\sqrt{5}$ cm/s

(c) $2\pi\sqrt{7}$ cm/s (d) $2\pi\sqrt{8}$ cm/s

114. Which one of the following is SHM?

(a) Motion of a particle in a wave moving through a string fixed at both ends
(b) Earth spinning about its own axis
(c) Ball bouncing between two rigid vertical walls
(d) Particle moving in a circle with uniform speed.

115. Which of the following expressions does not represent SHM?

(a) $A \cos \omega t$ (b) $A \sin^2 \omega t$

(c) $A \sin \omega t + B \cos \omega t$ (d) $A (\sin^2 \omega t + \sin^3 \omega t)$

116. Two SHM are given by $y_1 = a \sin\left[(\pi/2)t + \phi\right]$ and $y_2 = b \sin\left[(2\pi t/3) + \phi\right]$. The phase difference between these after 1 sec is

(a) π (b) $\pi/2$

(c) $\pi/4$ (d) $\pi/6$

117. A particle is subjected to two mutually perpendicular SHM such that $x = 2 \sin \omega t$ and $y = 2 \sin(\omega t + \pi/4)$. The path of the particle will be

(a) an ellipse (b) a straight line

(c) a parabola (d) a circle.

118. A simple pendulum of length L has been set up inside a railway wagon sliding down a frictionless inclined plane having an angle of inclination θ with the horizontal. What will be its period of oscillation as recorded by an observer inside the wagon?

(a) $2\pi\sqrt{L/g\cos\theta}$ (b) $2\pi\sqrt{L/g}$

(c) $2\pi\sqrt{L/g\sin\theta}$ (d) $2\pi\sqrt{L\cos\theta/g}$

119. The time period of a simple pendulum is T. The pendulum is charged positively and is made to oscillate while a negatively charged plate is placed below it. The time period of the pendulum is now

(a) $> T$ (b) $< T$

(c) $= T$ (d) ∞

120. The work done by the string of a simple pendulum during one complete oscillation is equal to

(a) total energy of pendulum
(b) KE of pendulum
(c) PE of pendulum
(d) zero.

121. In case of a simple pendulum, time period versus length is depicted by

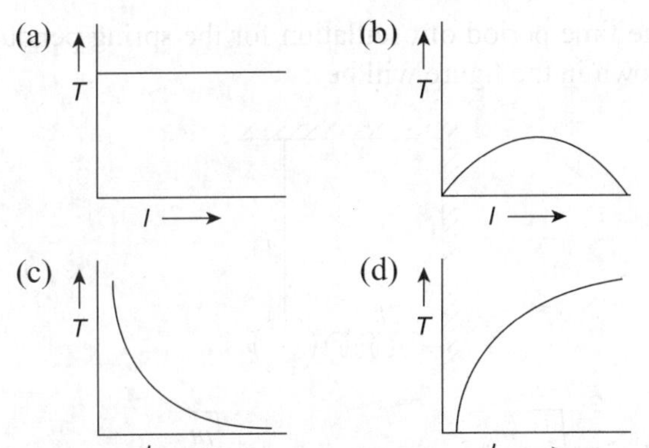

122. Two masses M_A and M_B are hung from two strings of length l_A and l_B respectively. If they are executing SHM with frequency relation $f_A = 2f_B$, then relation

(a) $l_A = 4l_B$, does not depend on mass
(b) $l_A = l_B / 4$, does not depend on mass
(c) $l_A = l_B / 2$ and $M_A = M_B / 2$
(d) $l_A = 2l_B$ and $M_A = 2M_B$

123. Two masses m_1 and m_2 are suspended together by a mass-less spring of spring constant k. When the masses are in equilibrium, m_1 is removed without disturbing the system. The angular frequency of oscillations is

(a) $\sqrt{k / m_1}$ (b) $\sqrt{k / m_2}$
(c) $1 / 2\sqrt{m_1 / k}$ (d) $1 / 2\sqrt{m_2 / k}$

124. M kg weight is suspended from a weightless spring and it has time period T. If now 4 M kg weight is suspended from the same spring, the new time period will be

(a) T (b) $2T$
(c) $T/2$ (d) $4T$

125. A force of 6.4 N stretches a vertical spring by 0.1 m. The mass that must be suspended from the spring so that it oscillates with a period of (π /4) sec is

(a) (π /4) kg (b) 1 kg
(c) (1/π) kg (d) 10 kg

126. A clock S is based on oscillations of a spring and a clock P is based on pendulum motion. Both clocks run at the same rate on earth. On a planet having the same density as earth but twice the radius

(a) S will run faster than P
(b) P will run faster than S
(c) both will run at the same rate as on the earth
(d) both will run at equal rates but not the same as on the earth.

127. Two bodies M and N of equal masses are suspended from two separate mass less spring constant k_1 and k_2 respectively. If the two bodies oscillate vertically such that their maximum velocities are equal, the ratio of the amplitude of vibration of M to that of N is

(a) k_1 / k_2 (b) $\sqrt{k_1 / k_2}$
(c) k_2 / k_1 (d) $\sqrt{k_2 / k_1}$

128. A block P of mass m is placed on a frictionless horizontal surface. Another block Q of same mass is kept on P and connected to the wall with the help of a spring of spring constant k as shown in the figure. μ_s is the coefficient of friction between P and Q. The blocks move together performing SHM of amplitude A. The maximum value of friction force between P and Q is

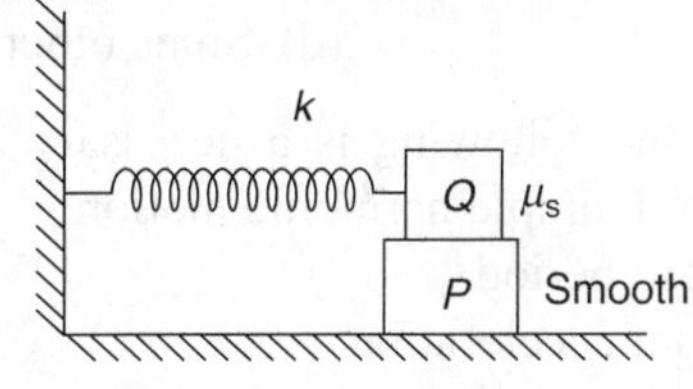

(a) zero (b) kA
(c) $\frac{kA}{2}$ (d) $\mu_s mg$

129. One end of a long metallic wire of length L is tied to the ceiling. The other end is tied to a mass-less with spring constant k. A mass m hangs truly from the free end of the spring. The area of cross-section and the Young's modules of the wire are A and Y respectively. If the mass is slightly pulled down and released, it will oscillate with a time period t equal to

(a) $2\pi\sqrt{mYA / kL}$ (b) $2\pi\sqrt{m(YA / kL) / YAk}$
(c) $2\pi\sqrt{m(YA / kL)}$ (d) $2\pi\sqrt{mL / AY}$

130. A uniform cylinder of length L and mass M having cross-section area A issue suspended with its length vertical, from a fixed point by a mass less spring, such that it is half submerged in a liquid of density ρ at equilibrium position. When the cylinder is given a small downward push and released, it starts oscillating vertically with small amplitude. If the force constant of the spring is k, the frequency of oscillation of the cylinder is

(a) $\frac{1}{2\pi}\left(\frac{k - A\rho g}{M}\right)^{1/2}$ (b) $\frac{1}{2\pi}\left(\frac{k + A\rho g}{M}\right)^{1/2}$
(c) $\frac{1}{2\pi}\left(\frac{k + \rho g L^2}{M}\right)^{1/2}$ (d) $\frac{1}{2\pi}\left(\frac{k + A\rho g}{A\rho g}\right)^{1/2}$

131. A seconds pendulum is mounted in a rocket. Its period of oscillation will decrease when rocket is

(a) moving up with uniform velocity
(b) moving up with uniform acceleration
(c) moving down with uniform acceleration
(d) moving in a geostationary orbit.

132. A simple pendulum performs SHM about $x = 0$ with an amplitude A and time period T. The speed of the pendulum at $x = A/2$ will be

(a) A/T (b) $\left(\sqrt{3}\right)\pi A/T$
(c) $\pi A\left(\sqrt{3}\right)/2T$ (d) $3\pi^2 A/T$

133. A body executes SHM with an amplitude A. At what displacement from the mean position, the potential energy of the body is one fourth of its total energy?
(a) $A/4$ (b) $A/2$
(c) $3A/4$ (d) Some other fraction of A.

134. Which of the following is a necessary and sufficient condition for simple harmonic motion?
(a) Constant period
(b) Constant acceleration
(c) Proportionally between acceleration and restoring force
(d) Proportionality between restoring force and displacement from equilibrium position.

135. A simple harmonic motion having an amplitude A and time period T is represented by the equation $y = 5 \sin \pi (t + 4)$. Then the values of A in m and T in s are
(a) $A = 5; T = 2$ (b) $A = 10; T = 1$
(c) $A = 5; T = 1$ (d) $A = 10; T = 2$

136. A simple pendulum is made of a bob which is a hollow sphere full of mercury suspended by means of a wire. If a little mercury is drained off, the period of the pendulum will
(a) increase (b) decrease
(c) remain the same (d) become erratic.

137. A simple pendulum with a solid metal bob has a period T. The metal bob is now immersed in a liquid of density one-tenth that of the bob. The liquid is non viscous. Now the period of the same pendulum with its bob remaining all the time in the liquid will be
(a) T (b) $9/10\ T$
(c) $\left(\sqrt{10/9}\right)T$ (d) $\sqrt{10/9T}$

138. Two simple pendulums of length 1 m and 16 m respectively are both given small displacement in the same direction at the same instant. They will again be in phase after the shorter pendulum has completed n oscillation, where n is
(a) 1/4 (b) 4/3
(c) 5 (d) 0

139. A light spiral spring supports 200 gm weight at its lower end. It oscillates with a period of 1 sec. How much weight (gm) must be removed from the lower end to reduce the period to 0.5 sec?
(a) 100 gm (b) 50 gm
(c) 150 gm (d) 200 gm

140. The time period of oscillation for the spring pendulum shown in the figure will be

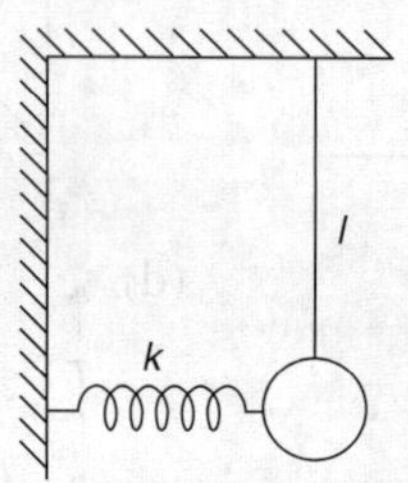

(a) $2\pi\sqrt{\dfrac{l}{g}}$ (b) $2\pi\sqrt{\dfrac{m}{k}}$
(c) $2\pi\sqrt{\dfrac{l}{g}+\dfrac{m}{k}}$ (d) $2\pi\sqrt{\dfrac{lm}{mg+k}}$

141. The simple pendulum is displaced by small angle θ and released. If it makes elastic collision, with the wall, the time period and oscillation will be

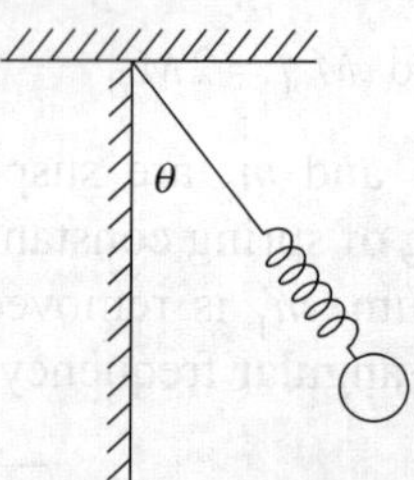

(a) $\pi\sqrt{\dfrac{l}{g}}$ and SHM (b) $2\pi\sqrt{\dfrac{l}{g}}$ and SHM
(c) $\pi\sqrt{\dfrac{l}{g}}$ and oscillatory (d) $\pi\sqrt{\dfrac{l}{g}}$ and periodic

142. A rectangular block of mass m and area of cross-section A floates in a liquid of density ρ. If it is given a small vertical displacement from equilibrium, it undergoes oscillation with a time period T. Then,
(a) $T \propto \dfrac{1}{\rho}$ (b) $T \propto \sqrt{\rho}$
(c) $T \propto \dfrac{1}{\sqrt{m}}$ (d) $T \propto \dfrac{1}{\sqrt{A}}$

143. A simple pendulum has time period T_1. The point of suspension is now moved upward according to the relation $y = Kt^2$, ($K = 1$ m/s^2) where y is the vertical displacement
The time period now becomes T_2. The ratio is $\dfrac{T_1^2}{T_2^2}$ is ($g = 10$ m/s^2)
(a) 1 (b) $\dfrac{4}{5}$
(c) $\dfrac{6}{5}$ (d) $\dfrac{5}{6}$

144. The displacement of an object attached to a spring and executing simple harmonic motion is given by

$x = 2 \times 10^{-2} \cos \pi t$. The time at which the maximum speed first occurs is
(a) 0.25 s (b) 0.5 s
(c) 0.75 s (d) 0.125 s

145. Two springs, of force constant k_1 and k_2, are connected to a mass m as shown. The frequency of oscillation of the mass is f. If both k_1 and k_2 are made four times their original values, the frequency of oscillation becomes

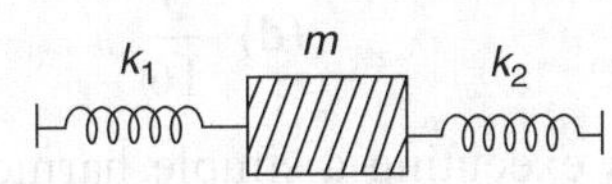

(a) $2f$ (b) $\frac{f}{2}$
(c) $\frac{f}{4}$ (d) $4f$

146. A point mass oscillates along the x-axis according to the law $x = x_0 \cos(\omega t - \pi/4)$. If the acceleration of the particle is written as $a = A\cos(\omega t + \delta)$, then
(a) $A = x_0\omega^2, \delta = \frac{3\pi}{4}$ (b) $A = x_0, \delta = -\frac{\pi}{4}$
(c) $A = x_0\omega^2, \delta = \frac{\pi}{4}$ (d) $A = x_0\omega^2, \delta = -\frac{\pi}{4}$

147. A particle of mass m executes simple harmonic motion with amplitude a and frequency v. The average kinetic energy during its motion from the position of equilibrium to the end is
(a) $2\pi^2 ma^2v^2$ (b) $\pi^2 ma^2v^2$
(c) $1/2ma^2v^2$ (d) $4\pi^2 ma^2v^2$

148. A coin is placed on a horizontal platform which undergoes vertical simple harmonic of angular frequency ω. The amplitude of oscillations is gradually increased. The coin will leave contact with the platform for the first time
(a) at the highest position of the platform
(b) at the mean position of the platform
(c) for an amplitude of $\frac{g}{\omega^2}$
(d) for an amplitude $\frac{g^2}{\omega^2}$

149. Starting from the origin, a body oscillates simple harmonically with a period of 2 s. After what time will its kinetic energy by 75% of the total energy?
(a) $\frac{1}{12}$ s (b) $\frac{1}{6}$ s
(c) $\frac{1}{4}$ s (d) $\frac{1}{3}$ s

150. The maximum velocity of a particle, executing simple harmonic motion with an amplitude 7 mm, is 4.4 m/s. The period of oscillation is
(a) 100 s (b) 0.01 s
(c) 10 s (d) 0.1 s

151. If a simple harmonic motion is represented by $\frac{d^2x}{dt^2} + ax = 0$, its time period is
(a) $2\pi\alpha$ (b) $2\pi\alpha$
(c) $\frac{2\pi}{\alpha}$ (d) $\frac{2\pi}{\sqrt{\alpha}}$

152. The bob of a simple pendulum is a spherical hollow ball filled with water. A plugged hole near the bottom of the oscillating bob gets suddenly unplugged. During observation, till water is coming out, the time period of oscillation, as long as the water continues to come out, would
(a) remain unchanged
(b) increase towards a saturation value
(c) first increase and then decrease to the original value
(d) first decrease and then increase to the original value.

153. Two simple harmonic motions are represented by the equation $y_1 = 0.1 \sin\left(100\pi t \frac{\pi}{3}\right)$ and $y_2 = 0.1 \cos \pi t$. The phase difference of the velocity of particle 1 with respect to the velocity of particle 2 is
(a) $\frac{-\pi}{3}$ (b) $\frac{\pi}{3}$
(c) $\frac{-\pi}{6}$ (d) $\frac{\pi}{6}$

154. The function $\sin^2(\omega t)$ represents
(a) a simple harmonic motion with a period $\frac{2\pi}{\omega}$
(b) a simple harmonic motion with a period $\frac{\pi}{\omega}$
(c) a periodic, but not simple harmonic, motion with a period $\frac{2\pi}{\omega}$
(d) a periodic, but not simple harmonic, motion with a period $\frac{\pi}{\omega}$

155. The total energy of a particle, executing simple harmonic motion is
(a) $\propto x$ (b) $\propto x^{1/2}$
(c) independent of x (d) $\propto x^2$

Questions Asked in NEET/CBSE Medicals

156. The displacement of a particle executing simple harmonic motion is given by $y = A_0 + A \sin \omega t + B \cos \omega t$. Then the amplitude of its oscillation is given by
(a) $A + B$ (b) $A_0 + \sqrt{A^2 + B^2}$
(c) $\sqrt{A^2 + B^2}$ (d) $\sqrt{A_0^2 + (A^2 + B^2)}$

157. The average velocity of a particle executing SHM ion one complete vibration is

(a) zero (b) $\frac{A\omega}{2}$

(c) $A\omega$ (d) $\frac{A\omega^2}{2}$

158. The radius of the circle, the period of revolution, initial position and sense of revolution are indicated in the figure. y-projection of the radius vector of rotating particle P is

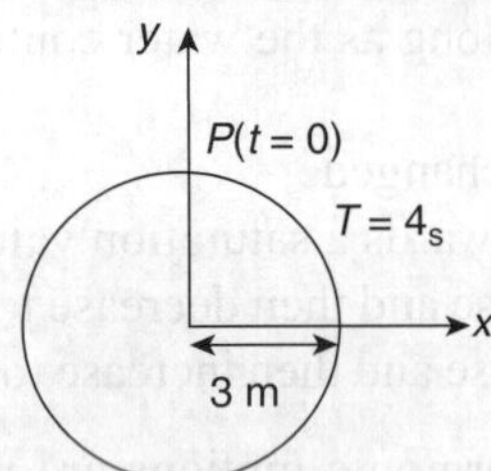

(a) $y(t) = 3\cos\left(\frac{\pi t}{2}\right)$, where y in m

(b) $y(t) = -3\cos 2\pi t$, where y in m

(c) $y(t) = 4\sin\left(\frac{\pi t}{2}\right)$, where y in m

(d) $y(t) = 3\cos\left(\frac{3\pi t}{2}\right)$, where y in m

159. The distance covered by a particle undergoing SHM in one time period is (amplitude = A)

(a) zero (b) A
(c) $2A$ (d) $4A$

160. A pendulum is hung from the roof of a sufficiently high building and is moving freely to and fro like a simple harmonic oscillator. The acceleration of the bob of the pendulum is 20 ms^{-2} at a distance of 5 m from the mean position. The time period of oscillation is

(a) 2π s (b) π s
(c) 2s (d) 1 s

161. A spring of force constant k is cut into length of ratio 1 : 2 : 3. They are connected in series and the new force constant is k'. Then they are connected in parallel and force constant is k''. Then $k' : k''$ is

(a) 1 : 9 (b) 1 : 11
(c) 1 : 14 (d) 1 : 6

162. A particle executes linear simple harmonic motion with an amplitude of 3 cm. When the particle is at 2 cm from the mean position, the magnitude of its velocity is equal to that of its acceleration. Then its time period in seconds is

(a) $\frac{\sqrt{5}}{2\pi}$ (b) $\frac{4\pi}{\sqrt{5}}$

(c) $\frac{2\pi}{\sqrt{3}}$ (d) $\frac{\sqrt{5}}{\pi}$

163. A body of mass m is attached to the lower end of a spring whose upper end is fixed. The spring has negligible mass. When the mass m is slightly pulled down and released, it oscillates with a time period of 3 s. When the mass m is increased by 1 kg, the time period of oscillations becomes 5 s. The value of m in kg is

(a) $\frac{3}{4}$ (b) $\frac{4}{3}$

(c) $\frac{16}{9}$ (d) $\frac{9}{16}$

164. A particle is executing a simple harmonic motion. Its maximum acceleration is α and maximum velocity is β. Then, its time period of vibration will be

(a) $\frac{\beta^2}{\alpha}$ (b) $\frac{2\pi\beta}{\alpha}$

(c) $\frac{\beta^2}{\alpha^2}$ (d) $\frac{\alpha}{\beta}$

165. A particle is executing SHM along a straight line. Its velocities at distances x_1 and x_2 from the mean position are V_1 and V_2, respectively. Its time period is

(a) $2\pi\sqrt{\frac{V_1^2+V_2^2}{x_1^2+x_2^2}}$ (b) $2\pi\sqrt{\frac{V_1^2-V_2^2}{x_1^2-x_2^2}}$

(c) $2\pi\sqrt{\frac{x_1^2+x_2^2}{V_1^2+V_2^2}}$ (d) $2\pi\sqrt{\frac{x_1^2-x_2^2}{V_1^2-V_2^2}}$

166. The oscillation of a body on a smooth horizontal surface is represented by the equation, $X = A\cos(\omega t)$ where X = displacement at time t ω = frequency of oscillation.

Which one of the following graphs shows correctly the variation of a with t?

Here a = acceleration at time t and T = time period

(a) 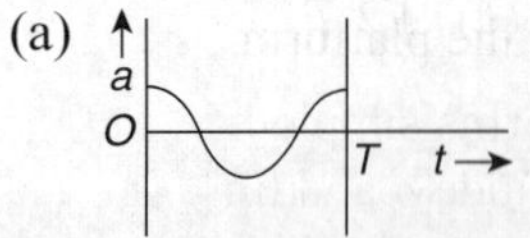

(b) 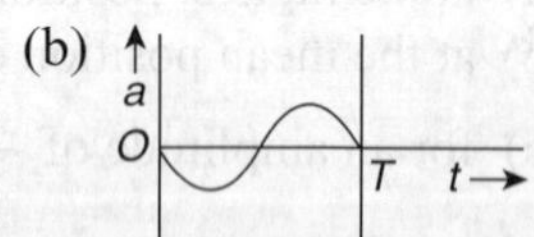

(c) 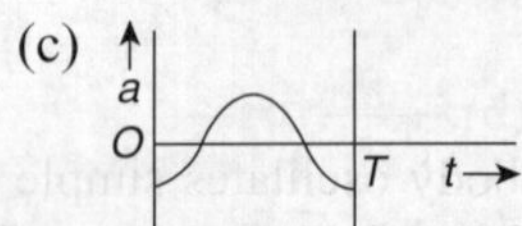

(d) 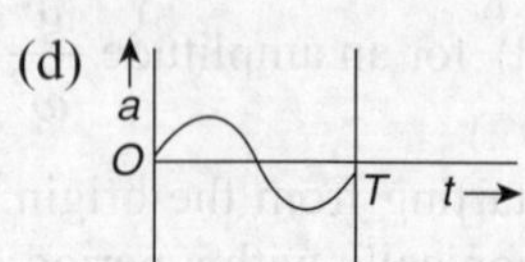

167. A particle of mass m oscillates along x-axis according to equation $x = a\sin\omega t$. The nature of the graph between momentum and displacement of the particle is

(a) Circle
(b) Hyperbola
(c) Ellipse
(d) Straight line passing through origin.

168. Out of the following functions representing motion of a particle, which represents SHM?

(1) $y = \sin \omega t - \cos \omega t$
(2) $y = \sin^3 \omega t$
(3) $y = 5\cos\left(\frac{3\pi}{4} - 3\omega t\right)$
(4) $y = 1 + \omega t + \omega^2 t^2$

(a) Only (1)
(b) Only (4) does not represent SHM
(c) Only (1) to (3)
(d) Only (1) to (2)

169. Two particles are oscillating along two close parallel straight lines side by side, with the same frequency and amplitudes. They pass each other, moving in opposite directions when their displacement is half of the amplitude. The mean position of the two particles lies on a straight line perpendicular to the paths of the two particles. The phase difference is

(a) $\frac{\pi}{6}$ (b) 0
(c) $\frac{2\pi}{3}$ (d) π

170. The displacement of a particle along the x-axis is given by $x = a\sin^2 \omega t$. The motion of the particle corresponds to
(a) simple harmonic motion of frequency ω / π
(b) simple harmonic motion of frequency $3\omega / 2\pi$
(c) non simple harmonic motion
(d) simple harmonic motion of frequency $\omega / 2\pi$.

171. The period of oscillation of a mass M suspended from a spring of negligible mass is T. If along with it another mass M is also suspended, the period of oscillation will now be

(a) T (b) $\frac{T}{\sqrt{2}}$
(c) $2T$ (d) $\sqrt{2}\,T$

172. A simple pendulum performs simple harmonic motion about $x = 0$ with an amplitude a and time period T. The speed of the pendulum at $x = a/2$ will be

(a) $\frac{\pi a}{T}$ (b) $\frac{3\pi^2 a}{T}$
(c) $\frac{\pi a\sqrt{3}}{T}$ (d) $\frac{\pi a\sqrt{3}}{2T}$

173. Which one of the following equations of motion represents simple harmonic motion?
(a) Acceleration $= -k(x + a)$
(b) Acceleration $= k(x + a)$
(c) Acceleration $= -kx$
(d) Acceleration $= -k_0 x + k_1 x^2$
where k, k_0, k_1 and a are all positive.

174. Two simple harmonic motion of angular frequency 100 and 1000 rad s^{-1} have the same displacement amplitude. The ratio of their maximum acceleration is
(a) $1 : 10^3$ (b) $1 : 10^4$
(c) $1 : 10$ (d) $1 : 10^2$

175. A particle executes simple harmonic oscillation with an amplitude a. The period of oscillation is T. The minimum time taken by the particle to travel half of the amplitude from the equilibrium position is
(a) $T/8$ (b) $T/12$
(c) $T/2$ (d) $T/4$

176. A mass of 2.0 kg is put on a flat pan attached to a vertical spring fixed on the ground as shown in the figure. The mass of the spring and the pan is negligible. When pressed slightly and released, the mass executes a simple harmonic motion. The spring constant is 200 N/m. What should be the minimum amplitude of the motion so that the mass gets detached from the pan? (Take $g = 10$ m/s^2.)

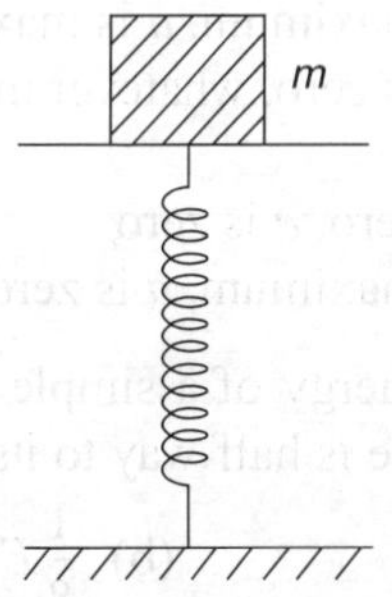

(a) 10.0 cm
(b) Any value less than 12.0 cm
(c) 4.0 cm
(d) 8.0 cm

177. The particle executing simple harmonic motion has a kinetic energy $K_0\cos^2 \omega t$. The maximum values of the potential energy and the total energy are respectively
(a) $K_0/2$ and K_0 (b) K_0 and $2K_0$
(c) K_0 and K_0 (d) 0 and $2K_0$

178. The phase difference between the instantaneous velocity and acceleration of a particle executing simple harmonic motion is
(a) π (b) $0.707\,\pi$
(c) zero (d) $0.5\,\pi$

179. A rectangular block of mass m and area of cross-section A floats in a liquid of density ρ. If it is given a small vertical displacement from equilibrium and it undergoes oscillation with a time period T, then

(a) $T \propto \frac{1}{\sqrt{m}}$ (b) $T \propto \sqrt{\rho}$
(c) $T \propto \frac{1}{\sqrt{A}}$ (d) $T \propto \frac{1}{\rho}$

180. The circular motion of a particle with constant speed is
(a) periodic but not simple harmonic
(b) simple harmonic nut not periodic
(c) period and simple harmonic
(d) neither periodic nor simple harmonic.

181. A particle executing simple harmonic motion of amplitude 5 cm has maximum speed of 31.4 cm/s. The frequency of its oscillation is
(a) 4 Hz (b) 3 Hz
(c) 2 Hz (d) 1 Hz

182. Two springs of spring constant k_1 and k_2 are joined in series. The effective spring constant of the combination is given by
(a) $\sqrt{k_1 k_2}$ (b) $(k_1 + k_2)/2$
(c) $k_1 + k_2$ (d) $k_1 k_2/(k_1 + k_2)$

183. Which one of the following statements is true for the speed v and the acceleration a of a particle executing simple harmonic motion?
(a) When v is maximum, a is maximum
(b) Value of a is zero, whatever may irrespective of the value of v
(c) When v is zero, a is zero
(d) When v is maximum, a is zero.

184. The potential energy of a simple harmonic oscillation when the particle is half-way to its end point is
(a) $\frac{2}{3}E$ (b) $\frac{1}{8}E$
(c) $\frac{1}{4}E$ (d) $\frac{1}{2}E$

185. A particle of mass m oscillates with simple harmonic motion between points x_1 and x_2, the equilibrium position being O. Its potential energy is plotted. It will be as given below in the graph

(a)
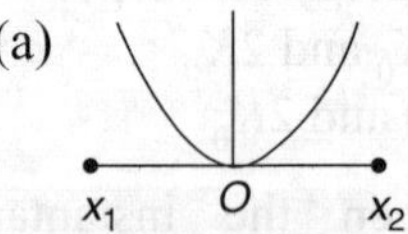

(b)
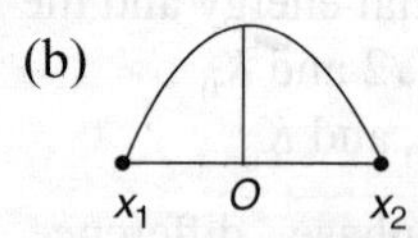

(c)
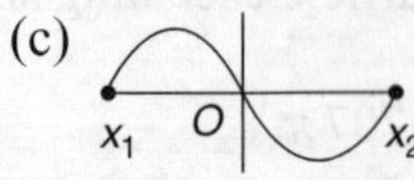

(d)
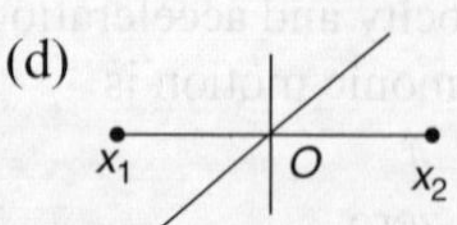

186. The time period of a mass suspended from a spring is T. If the spring is cut into four equal parts and the same mass is suspended from one of the parts, then the new time period will be
(a) $T/4$ (b) T
(c) $T/2$ (d) $2T$

187. In case of a forced vibration, the resonance peak becomes very sharp when the
(a) damping force is small
(b) restoring force is small
(c) applied periodic force is small
(d) quality factor is small.

188. Displacement between maximum potential energy position and maximum kinetic energy position for a particle executing simple harmonic motion is
(a) $\pm a/2$ (b) $+a$
(c) $\pm a$ (d) -1

189. When an oscillator completes 100 oscillations its amplitude reduced to $\frac{1}{3}$ of initial value. What will be its amplitude when it completes 200 oscillations?
(a) $\frac{1}{8}$ (b) $\frac{2}{3}$
(c) $\frac{1}{6}$ (d) $\frac{1}{9}$

190. A mass is suspended separately by two different springs in successive order and the time periods are t_1 and t_2 respectively. If it is connected by both springs as shown in figure, then time period is t_0. Then the correct relation is

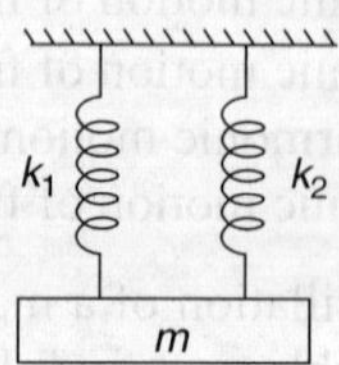

(a) $t_0^2 = t_1^2 + t_2^2$ (b) $t_0^{-2} = t_1^{-2} + t_2^{-2}$
(c) $t_0^{-1} = t_1^{-1} + t_2^{-1}$ (d) $t_0 = t_1 = t_2$

191. The total energy of particle performing SHM depends on
(a) k, a, m (b) k, a
(c) k, a, x (d) k, x

192. Two masses M_A and M_B are hung from two strings of length l_1 and l_2 respectively. They are executing SHM with frequency relation $f_A = 2f_B$, then relation
(a) $l_A = \frac{l_B}{4}$, does not depend on mass
(b) $l_A = 4l_B$, does not depend on mass
(c) $l_A = 2l_B$ and $M_A = 2M_B$
(d) $l_A = \frac{l_B}{2}$ and $M_A = \frac{M_B}{2}$

193. The bob of simple pendulum having length l, is displaced from mean position to an angular position θ with respect to vertical. If it is released, then velocity of bob at equilibrium position is

(a) $\sqrt{2gl(1-\cos\theta)}$ (b) $\sqrt{2gl(1+\cos\theta)}$
(c) $\sqrt{2gl\cos\theta}$ (d) $\sqrt{2gl}$

194. Time period of a simple pendulum is 2 sec. If its length is increased by 4 times, then its time period becomes
(a) 8 sec (b) 12 sec
(c) 16 sec (d) 4 sec

195. A particle, with restoring force proportional to displacement and resisting force proportional to velocity is subjected to a force $F\sin\omega t$. If the amplitude of the particle is maximum for $\omega=\omega_1$ and the energy of the particle is maximum for $\omega=\omega_2$, then
(a) $\omega_1\neq\omega_0$ and $\omega_2=\omega_0$
(b) $\omega_1=\omega_0$ and $\omega_2=\omega_0$
(c) $\omega_1=\omega_0$ and $\omega_2\neq\omega_0$
(d) $\omega_1\neq\omega_0$ and $\omega_2\neq\omega_0$

196. Two simple pendulums of length 5 m and 20 m respectively are given small linear displacement in one direction at the same time. They will again be in the phase when the pendulum of shorter length has complete oscillations.
(a) 2 (b) 1
(c) 5 (d) 3

197. A mass m is vertically suspended from a spring of negligible mass; the system oscillates with a frequency n. What will be the frequency of the system, if a mass 4 m is suspended from the same springs?

(a) $\frac{n}{2}$ (b) $4n$
(c) $\frac{n}{4}$ (d) 2n

198. If the length of a simple pendulum is increased by 2% then the time period
(a) increases by 1% (b) decreases by 1%
(c) increases by 2% (d) decreases by 2%

199. Two SHM's with same amplitude and time period, when acting together in perpendicular directions with a phase difference of $\pi/2$, give rise to
(a) straight motion (b) elliptical motion
(c) circular motion (d) None of these.

200. A particle starts with SHM from the mean position as shown in the figure. Its amplitude is A and its time period is T. At one time, its speed is half that of the maximum speed. What is this displacement?

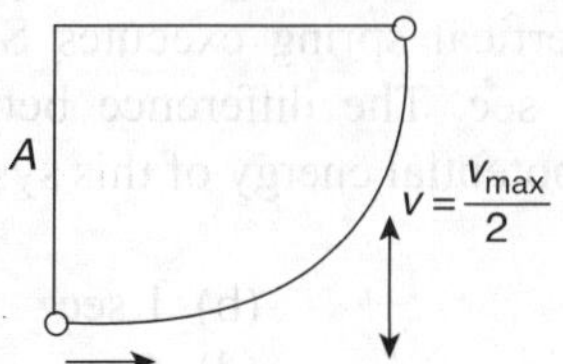

(a) $\frac{2A}{\sqrt{3}}$ (b) $\frac{3A}{\sqrt{2}}$
(c) $\frac{\sqrt{2}A}{3}$ (d) $\frac{\sqrt{3}A}{2}$

201. A linear harmonic oscillator of force constant 2×10^6 N/m and amplitude 0.01 m has a total mechanical energy of 160 J. Its
(a) PE is 160 J (b) PE is zero
(c) PE is 100 J (d) PE is 120 J

202. A simple pendulum with a bob of mass m oscillates from A to C and back to A such that PB is H. If the acceleration due to gravity is g, then the velocity of the bob as it passes through B is

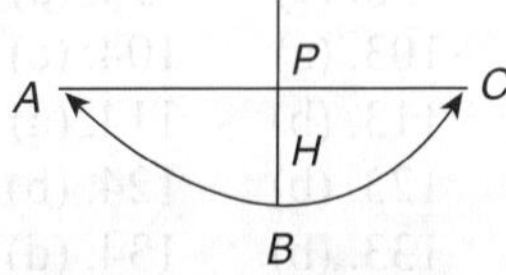

(a) mgH (b) $\sqrt{2gH}$
(c) zero (d) $2gH$

203. In a simple harmonic motion, when the displacement is one-half the amplitude, what fraction of the total energy is kinetic?
(a) 1/2 (b) 3/4
(c) zero (d) 1/4

204. A body of mass 5 kg hangs from a spring and oscillates with a time period of 2π seconds. If the ball is removed, the length of the spring will decrease by
(a) g/k metres (b) k/g metres
(c) 2π metres (d) g metres.

205. A particle executes SHM along x-axis. The force acting on it is given by
(a) $A\cos(kx)$ (b) Ae^{-kx}
(c) Akx (d) $-Akx$

206. A seconds pendulum is mounted in a rocket. Its period of oscillation will decrease when rocket is
(a) moving down with uniform acceleration
(b) moving around the earth in geostationary orbit
(c) moving up with uniform velocity
(d) moving up with uniform acceleration.

207. A loaded vertical spring executes SHM with a time period of 4 sec. The difference between the kinetic energy and potential energy of this system varies with a period of

(a) 2 sec (b) 1 sec
(c) 8 sec (d) 4 sec

208. A body executes simple harmonic motion with amplitude A. At what displacement from the mean position is the potential energy of the body is one-fourth of its total energy?

(a) $A/4$ (b) $A/2$
(c) $3A/4$ (d) Some other fraction of A.

Answer Keys

Exercises

1. (c)	2. (a)	3. (c)	4. (a)	5. (c)	6. (d)	7. (b)	8. (b)	9. (c)	10. (a)
11. (b)	12. (b)	13. (a)	14. (a)	15. (c)	16. (c)	17. (a)	18. (d)	19. (d)	20. (c)
21. (d)	22. (c)	23. (a)	24. (c)	25. (a)	26. (c)	27. (b)	28. (c)	29. (b)	30. (d)
31. (d)	32. (a)	33. (c)	34. (a)	35. (c)	36. (a)	37. (a)	38. (a)	39. (c)	40. (c)
41. (a)	42. (d)	43. (c)	44. (d)	45. (b)	46. (c)	47. (c)	48. (b)	49. (d)	50. (b)
51. (b)	52. (c)	53. (b)	54. (c)	55. (d)	56. (c)	57. (c)	58. (b)	59. (b)	60. (d)
61. (c)	62. (c)	63. (c)	64. (b)	65. (c)	66. (c)	67. (c)	68. (c)	69. (d)	70. (d)
71. (d)	72. (c)	73. (d)	74. (c)	75. (b)	76. (b)	77. (d)	78. (d)	79. (a)	80. (b)
81. (c)	82. (a)	83. (b)	84. (d)	85. (c)	86. (b)	87. (c)	88. (d)	89. (b)	90. (c)
91. (a)	92. (a)	93. (c)	94. (a)	95. (a)	96. (b)	97. (b)	98. (b)	99. (b)	100. (d)
101. (a)	102. (d)	103. (a)	104. (c)	105. (a)	106. (c)	107. (c)	108. (d)	109. (d)	110. (a)
111. (d)	112. (b)	113. (b)	114. (a)	115. (d)	116. (d)	117. (d)	118. (a)	119. (b)	120. (d)
121. (d)	122. (a)	123. (b)	124. (b)	125. (b)	126. (b)	127. (d)	128. (c)	129. (b)	130. (b)
131. (b)	132. (b)	133. (b)	134. (d)	135. (a)	136. (a)	137. (c)	138. (b)	139. (c)	140. (d)
141. (c)	142. (d)	143. (c)	144. (b)	145. (a)	146. (a)	147. (b)	148. (c)	149. (b)	150. (b)
151. (d)	152. (c)	153. (c)	154. (d)	155. (c)					

Question asked in NEET/CBSE medicals

156. (c)	157. (a)	158. (a)	159. (d)	160. (b)	161. (b)	162. (b)	163. (d)	164. (b)	165. (d)
166. (c)	167. (c)	168. (c)	169. (c)	170. (c)	171. (d)	172. (c)	173. (c)	174.(d)	175. (b)
176.(a)	177. (c)	178. (d)	179. (c)	180. (a)	181. (d)	182. (d)	183. (d)	184. (c)	185. (a)
186.(c)	187. (a)	188. (c)	189.(d)	190. (b)	191. (b)	192. (a)	193. (a)	194. (d)	195. (b)
196. (a)	197. (a)	198. (a)	199. (c)	200. (d)	201. (c)	202. (b)	203. (b)	204. (d)	205. (d)
206. (d)	207. (a)	208. (b)							

Solutions

1. (c)

$$y = a \sin \omega t$$

$$\frac{y}{a} = \sin \omega t$$

$$\frac{1}{2} = \sin \omega t \Rightarrow \omega t = \frac{\pi}{6}$$

$$\frac{2\pi t}{T} = \frac{\pi}{6}$$

$$t = \frac{T}{12}$$

2. (a)

$$\text{KE} = \frac{1}{2} m\omega^2 (a^2 - y^2) \quad \text{(i)}$$

$$k = \text{Maximum KE} = \frac{1}{2} m\omega^2 a^2$$

$$\Rightarrow \frac{1}{2} m\omega^2 = \frac{k}{a^2}$$

Substituting the values in Eq. (i),

$$\frac{k}{2} = \frac{k}{a^2}(a^2 - y^2)$$

$$\Rightarrow \quad \frac{1}{2}a^2 = a^2 - y^2$$

$$y^2 = \frac{1}{2}a^2$$

$$y = \frac{a}{\sqrt{2}}$$

3. (c)

In a satellite $g_{eff} = 0$

And $T = 2\pi\sqrt{\frac{l}{g_{eff}}}$

$$\Rightarrow \quad T = \infty$$

4. (a)

$$P.E = \frac{1}{2}ky^2$$

Total energy = Maximum PE = $\frac{1}{2}ka^2$

$$P.E = \frac{1}{2}k\left(\frac{a}{2}\right)^2 = \frac{1}{2}\frac{ka^2}{4} = \frac{E}{4}$$

5. (c)

$$T \propto \sqrt{\frac{l}{g}}$$

If g becomes $\frac{g}{4}$, the time period becomes $2T$.

6. (d)

$$v_1 = \frac{1}{2\pi}\sqrt{\frac{g}{l_1}}$$

$$v_2 = \frac{1}{2\pi}\sqrt{\frac{g}{l_2}}$$

$$\frac{v_1}{v_2} = \frac{2}{3} = \sqrt{\frac{l_2}{l_1}}$$

$$\Rightarrow \quad \frac{l_1}{l_2} = \frac{9}{4}$$

7. (b)

$$a_{max} = \omega^2 A$$

Where A is amplitude and ω = Angular frequency

From the equation

$$y = 2\sin\left(\frac{\pi t}{2} + \theta\right)$$

We find

$\omega = \frac{\pi}{2}$ and $A = 2$ cm

Therefore $a_{max} = \left(\frac{\pi}{2}\right)^2 \times 2 = \frac{\pi^2}{2}$ cm/s^2

8. (b)

The value of gravity g' at other planet is

$$g' = \frac{GM'}{(R')^2} = \frac{G2M}{(2R)^2} = \frac{1}{2}\frac{GM}{R^2} = g/2$$

Time period of second's pendulum is 2 seconds.

$$T' = 2\pi\sqrt{\frac{l}{g'}} = 2\pi\sqrt{\frac{2l}{g}}$$

$$= \sqrt{2}\,2\pi\sqrt{\frac{l}{g}} = \sqrt{2} \times 2 = 2\sqrt{2} \text{ s}$$

9. (c)

In this case only the particles of the string execute SHM.

10. (a)

$x = A\cos(\omega t + \phi)$

Comparing the equation $x = 0.01\cos\left(\pi t + \frac{\pi}{4}\right)$ with general equation,

we find $\omega = \pi \Rightarrow 2\pi v = \pi$ or $v = 0.5$Hz

11. (b)

After a time $2T$, it will be again at $x = -X_M$

$3.15T - 2T = 1.15T$

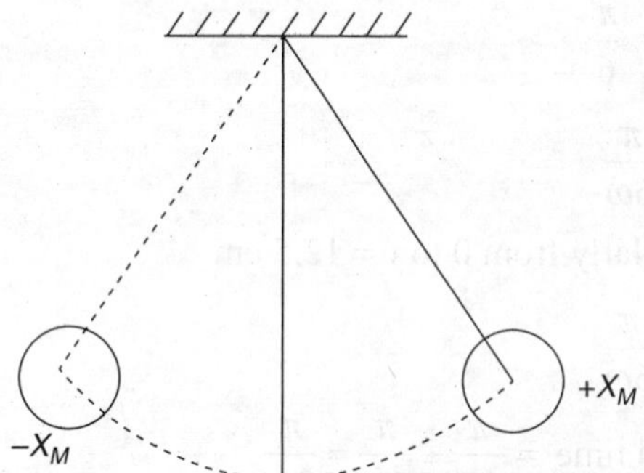

Further after time T it will be at $+x_m$

$1.15T - T = 0.15T$

After 0.15 T from $+x_m$ it will be between O to $+x_m$

12. (b)

In series the equivalent force constant is

$$\frac{1}{k_{eq}} = \frac{1}{k} + \frac{1}{k} = \frac{2}{k}$$

$$K_{eq} = \frac{k}{2}$$

Therefore, $f = \frac{1}{2\pi}\sqrt{\frac{k}{2m}}$

13. (a)

$$T = 2\pi\sqrt{\frac{1}{g}}$$

$$\Rightarrow T^2 = \frac{4\pi^2}{g}\frac{1}{}$$

or $g = \frac{4\pi^2 l}{T^2}$

$$\Delta T = \frac{0.1}{100} = 0.001\text{ s}$$

$$\Delta L = 1\text{ mm} = 0.1\text{ cm}$$

$$\frac{\Delta g}{g} = \frac{\Delta L}{L} + \frac{2\Delta T}{T} = \frac{0.1}{100} + 2\times\frac{0.001}{2}$$

$$= 0.001 + 0.001 = 0.002$$

The maximum % error

$$\frac{\Delta g}{g}\times 100 = 0.002\times 100 = 0.2\%$$

14. (a)

The amplitude $A = 25$ cm

The equation of motion is

$y = A\sin\omega t$

Time taken by the particle to move form $x = -12.5$ to 0.

$12.5 = 25\sin\omega t$

$$\Rightarrow \sin\omega t = \frac{1}{2}$$

$$\omega t = \frac{\pi}{6}$$

$$t = \frac{\pi}{6\omega}$$

Similarly from 0 to $y = 12.5$ cm

$$t = \frac{\pi}{6\omega}$$

$$\text{Total time} = \frac{\pi}{6\omega} + \frac{\pi}{6\omega} = \frac{\pi}{3\omega}$$

$$= \frac{\pi\times T}{3\times 2\pi} = \frac{T}{6} = \frac{3}{6} = 0.5\text{ s}$$

15. (c)

The equation of SHM is

$y = A\sin\omega t$

$y_1 = A\sin\omega t_1 = A\sin\omega$ $\quad t_1 = 1\text{sec}.$

$y_2 = A\sin\omega t_2 = A\sin 2\omega$ $\quad t_2 = 2\text{sec}.$

The distance covered in first two second

$= y_2 - y_1 = A\sin 2\omega - A\sin\omega$

The ratio of distance travelled in first second to that in second

$$= \frac{y_1}{y_2 - y_1} = \frac{A\sin\omega}{A\sin 2\omega - A\sin\omega}$$

$$= \frac{A\sin\frac{2\pi}{T}}{A\sin\left(\frac{2\times 2\pi}{T}\right) - A\sin\left(\frac{2\pi}{T}\right)}$$

$$= \frac{\sin\frac{2\pi}{8}}{\sin\left(\frac{2\times 2\pi}{8}\right) - \sin\frac{2\pi}{8}} = \frac{\sin\frac{\pi}{4}}{\sin\frac{\pi}{2} - \sin\frac{\pi}{4}}$$

$$= \frac{y_1}{y_2 - y_1} = \frac{A\sin\omega}{A\sin 2\omega - A\sin\omega}$$

$$= \frac{A\sin\frac{2\pi}{T}}{A\sin\left(\frac{2\times 2\pi}{T}\right) - A\sin\left(\frac{2\pi}{T}\right)}$$

$$= \frac{\sin\frac{2\pi}{8}}{\sin\left(\frac{2\times 2\pi}{8}\right) - \sin\frac{2\pi}{8}} = \frac{\sin\frac{\pi}{4}}{\sin\frac{\pi}{2} - \sin\frac{\pi}{4}}$$

$$= \frac{\frac{1}{\sqrt{2}}}{1 - \frac{1}{\sqrt{2}}} = \frac{1}{\sqrt{2} - 1}$$

16. (c)

$4v^2 = 25 - x^2$

Differentiating w.r.t to t

$$8v\frac{dv}{dt} = -2x\frac{dx}{dt}$$

$$8va = -2vx$$

$$a = -\frac{2}{8}x$$

$$a = -\frac{1}{4}x$$

$$\omega^2 = \frac{1}{4} \Rightarrow \omega = \frac{1}{2}$$

$$\frac{2\pi}{T} = \frac{1}{2}$$

$$T = 4\pi$$

17. (a)

$F_1 = k_1 x$

$F_2 = k_2 x$

$F = F_1 + F_2 = (k_1 + k_2)x$

$F = kx = k_1 + k_2$

$$k = k_1 + k_2$$

$$T_1 = 2\pi\sqrt{\frac{m}{k_1}}$$

$$T_2 = 2\pi\sqrt{\frac{m}{k_2}}$$

$$k_1 = \frac{4\pi^2 m}{T_1^2}; k_2 = \frac{4\pi^2 m}{T_2^2}$$

$$T = 2\pi\sqrt{\frac{m}{k}} = 2\pi\sqrt{\frac{m}{k_1+k_2}}$$

$$T = \sqrt{\frac{m}{\frac{4\pi^2 m}{T_1^2} + \frac{4\pi^2 m}{T_2^2}}}$$

$$= \sqrt{\frac{T_1^2 T_2^2}{T_1^2 + T_2^2}} = \sqrt{\frac{\left(\frac{4}{5}\right)^2 \times \left(\frac{3}{5}\right)^2}{\left(\frac{4}{5}\right)^2 + \left(\frac{3}{5}\right)^2}}$$

$$\Rightarrow \quad T = \frac{12}{25}$$

18. (d)
The tension T will be maximum at position B (lowest point)

$$T - mg = \frac{mv^2}{r} \quad \text{(i)}$$

Apply conservation of mechanical energy to find velocity at lowest point B

$$\frac{1}{2}mv^2 = mg(l - l\cos\theta) \quad \text{(ii)}$$

Put mv^2 in Eq. (i)

$$T - mg = 2\frac{mg}{l}(l - l\cos\theta_0)$$

$$T = 2mg(1-\cos\theta_0) + mg$$

$$T = 2mg\left(2\sin^2\frac{\theta_0}{2}\right) + mg$$

$$\because \left(1 - \cos\theta_0 = 2\sin^2\frac{\theta}{2}\right) \text{ and } \sin^2\frac{\theta_0}{2} = \left(\frac{\theta_0}{2}\right)^2$$

when θ_0 is small,

$$T = 2mg\left[2\left(\frac{\theta_0}{2}\right)^2\right] + mg$$

$$T = mg(1 + \theta_0^2)$$

19. (d)
For a simple pendulum,

$$T_1 = 2\pi\sqrt{\frac{l}{g}}$$

For a rod, $T_2 = 2\pi\sqrt{\dfrac{I}{mg\times\dfrac{l}{2}}} = 2\pi\sqrt{\dfrac{\frac{1}{3}ml^2}{mg\times\dfrac{l}{2}}}$

$$= 2\pi\sqrt{\frac{2}{3}\frac{l}{g}}$$

$$\frac{T_1}{T_2} = \sqrt{\frac{3}{2}}$$

20. (c)
Let the velocity of the body at the mean position be v.
Loss in KE = Gain in PE

$$\frac{1}{2}mv^2 = mgh$$

$$v = \sqrt{2gh} = \sqrt{2\times10\times0.1}$$

$$= \sqrt{2}\text{m/s} \approx 1.4\text{m/s}$$

21. (d)
The effective spring constant of two springs in series is

$$k = \frac{k_1k_2}{k_1+k_2}$$

Time period, $T = 2\pi\sqrt{\dfrac{m}{k}} = 2\pi\sqrt{\dfrac{m(k_1+k_2)}{k_1k_2}}$

22. (c)

$$\frac{\frac{1}{2}m\omega^2(a^2-y^2)}{\frac{1}{2}m\omega^2a^2} = \frac{1}{4}$$

$$\frac{(a^2-y^2)}{a^2} = \frac{1}{4}$$

$$\Rightarrow \quad y^2 = \frac{3a^2}{4}$$

$$y = \frac{\sqrt{3}a}{2}$$

23. (a)

$$T = 2\pi\sqrt{\frac{l^2+k^2}{I_g}}$$

24. (c)
$y = a\sin(\omega t - kx)$ is the equation of a progressive wave.

25. (a)

$$T = 2\pi\sqrt{\frac{m}{k}}$$

In parallel combination $k = k_1 + k_2$

Therefore, $T = 2\pi\sqrt{\dfrac{m}{k_1 + k_2}}$

26. (c)

$x = a\sin\omega t$ and $y = b\sin(\omega t + \pi) = -b\sin\omega t$

or $\dfrac{x}{a} = -\dfrac{y}{b}$ or $y = -\dfrac{b}{a}x$

It is an equation of straight line.

27. (b)

$$T = 2\pi\sqrt{\frac{m}{k}} = 2\pi\sqrt{\frac{16(k_1 + k_2)}{k_1 k_2}}$$

$$= 8\pi\sqrt{\frac{k_1 + k_2}{k_1 k_2}}$$

(Two springs are in series)

28. (c)

Velocity $= \omega\sqrt{A^2 - y^2}$

Acceleration $= \omega^2 y$

$$\omega^2 y = \omega\sqrt{A^2 - y^2}$$

$$\omega \times 1 = \sqrt{(2)^2 - (1)^2}$$

$$\omega = \sqrt{3}$$

$$\frac{2\pi}{T} = \sqrt{3}$$

$$\Rightarrow \quad T = \frac{2\pi}{\sqrt{3}}$$

29. (b)

$$K.E = \frac{1}{2}m\omega^2(a^2 - y^2)$$

$$\frac{3}{4}\frac{1}{2}m\omega^2 a^2 = \frac{1}{2}m\omega^2(a^2 - y^2)$$

$$\frac{3}{4}a^2 = a^2 - y^2$$

$$\Rightarrow \quad y^2 = \frac{a^2}{2}$$

$$y = \frac{a}{2}$$

30. (d)

Progressive wave equation is

$$y = A\sin\omega\left(t - \frac{x}{v}\right)$$

The given equation is

$$y = A\sin 2\pi\left(\frac{t}{T} - \frac{x}{\lambda}\right)$$

Arranging it in general form

$$y = \sin\frac{2\pi}{T}\left(t - \frac{xT}{\lambda}\right) = \sin\frac{2\pi}{T}\left(t - \frac{x}{\lambda/T}\right) \quad \left(v = \frac{\lambda}{T}\right)$$

Speed $= \dfrac{\lambda}{T}$

31. (d)

$$KE = \frac{1}{2}m\omega^2(a^2 - y^2)$$

$$PE = \frac{1}{2}m\omega^2 y^2$$

$$\frac{1}{2}m\omega^2 y^2 = \frac{1}{2}m\omega^2(a^2 - y^2)$$

$$y^2 = a^2 - y^2$$

$$\Rightarrow y = \frac{a}{\sqrt{2}}$$

32. (a)

From 0 to $\dfrac{A}{2}$

$$y = A\sin\omega t_1$$

$$\frac{A}{2} = A\sin\omega t_1$$

$$\Rightarrow \quad \sin\,\omega t_1 = \frac{1}{2}$$

$$\omega t_1 = \frac{\pi}{6} \Rightarrow t_1 = \frac{\pi}{6\omega}$$

From 0 to A

$$A = A\,\sin\,\omega t_2$$

$$\Rightarrow \quad \omega t_2 = \frac{\pi}{2} \Rightarrow t_2 = \frac{\pi}{2\omega}$$

Therefore, time for $\left(\dfrac{A}{2}\right)$ to $A = t_2 - t$

$$= \frac{\pi}{2\omega} - \frac{\pi}{6\omega} = \frac{2\pi}{6\omega} = \frac{\pi}{3\omega}$$

$$\Rightarrow \quad t_1 < t_2$$

33. (c)

KE is maximum at mean position.

34. (a)

Let time for a displacement of 12.5 cm be t.

$$y = A\sin\,\omega t$$

$$12.5 = 25\sin\,\omega t$$

$$\Rightarrow \quad \sin\,\omega t = \frac{1}{2}$$

$\omega t = \frac{\pi}{6} \Rightarrow \frac{t \times 2\pi}{T} = \frac{\pi}{6}$

$t = \frac{T}{12}$

Similarly time for a displacement 12.5 cm on the other side is also t.

Therefore total time $= \frac{2T}{12} = \frac{3}{6} = 0.5\text{s}$

35. (c)

$y = a\,(\sin\,\omega t + \cos\omega t)$

$= a\sqrt{2}\left(\frac{\sin\omega t}{\sqrt{2}} + \frac{\cos\omega t}{\sqrt{2}}\right)$

$y = a\sqrt{2}\left(\sin\omega t\cos\frac{\pi}{4} + \cos\omega t\sin\frac{\pi}{4}\right)$

$= a\sqrt{2}\sin\left(\omega t + \frac{\pi}{4}\right)$

Therefore, $A = a\sqrt{2}$

And motion is SHM.

36. (a)

$x = A\sin\omega t$

$y = A\sin(\omega t + \pi/2)$

$= A\,\cos\,\omega t$

Now, $x^2 + y^2 = A^2\sin^2\omega t + A^2\sin^2\omega t$

$x^2 + y^2 = A^2(\sin^2\omega t + \cos^2\omega t)$

$x^2 + y^2 = A^2$

This is the equation of a circle.

37. (a)

$y = 3\sin\left(100t + \frac{\pi}{6}\right)$

Here $A = 3;\ \omega = 100$

$v_{\max} = \omega A = 100 \times 3 = 300$

38. (a)

The total energy $= \frac{1}{2}m\omega^2A^2 = \frac{1}{2}m\left(\frac{2\pi}{T}\right)^2A^2$

Hence, it depends upon amplitude and time period.

39. (c)

Restoring force $= AY\frac{\Delta l}{L} = \frac{AY}{L}\Delta l$

$k = \frac{AY}{L}$

In the second case $k' = \frac{AY}{L/2} = \frac{2AY}{L} = 2k$

40. (c)

The maximum velocity is at mean position.

41. (a)

42. (d)

$V_{m_1} = \omega_1 A_1\ ;\ V_{m_2} = \omega_2 A_2$

$1 = \frac{V_{m_1}}{V_{m_2}} = \frac{\omega_1 A_1}{\omega_2 A_2}$

$\frac{A_1}{A_2} = \frac{\omega_2}{\omega_1} = \frac{\sqrt{\frac{k_2}{m}}}{\sqrt{k_1/m}} = \left(\frac{k_2}{k_1}\right)^{1/2}$

43. (c)

$V_{\max} = \omega A = \frac{2\pi}{T}a = \frac{2\pi a}{T}$

44. (d)

The equation is

$a \propto -\text{displacement}$

$ma \propto -mA\sin\omega t$

$F = -\omega^2 mA\sin\omega t$

45. (b)

$y = A\,\sin\omega t$

$v = \frac{dy}{dt} = A\omega\cos\omega t$

$a = \frac{d^2y}{dt^2} = -A\omega^2\,\sin\omega t$

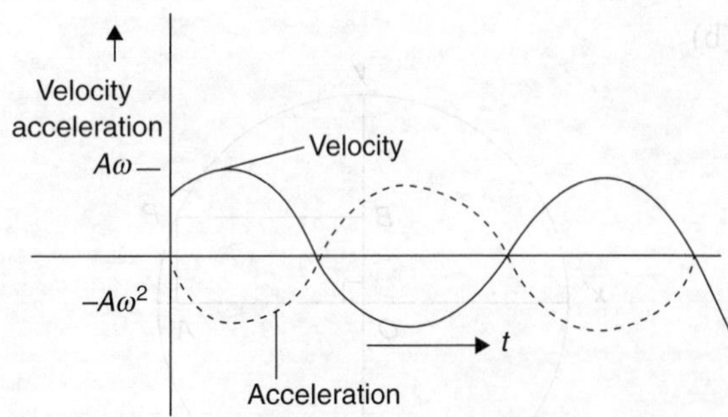

46. (c)

From mean position to one extreme position it is retarded, and from extreme position to equilibrium positions it is accelerated. The similar process is repeated in the other direction also.

47. (c)

The total energy E of a particle in simple harmonic motion is

$E = \frac{1}{2}mA^2\omega^2$

$\Rightarrow \quad E \propto A^2$

48. (b)

In case of oscillating spring, the time period is independent of g.

Note: Had it been simple pendulum its time period would have been infinity as $g = 0$ at the centre of the earth.

49. (d)

$$T = 2\pi\sqrt{\frac{m}{k}}$$

Let the mass be m.

$F = kx$

$$\Rightarrow \quad k = \frac{F}{x} = \frac{mg}{0.001} = \frac{m\times 10}{0.001} = 10000\text{ m}$$

$$T = 2\pi\sqrt{\frac{m}{10000\text{ m}}} = \frac{2\pi}{100}$$

So, the correct option is (d).

50. (b)

$$\frac{m}{\text{Force constant}} = \frac{M}{MLT^{-2}L^{-1}} = T^2 = (\text{time})^2$$

51. (b)

It can oscillate either in y-z plane or in y-x plane independently.

52. (c)

PE $V = \frac{1}{2}m\omega^2x^2$

and KE, $T = \frac{1}{2}m\omega^2\left(a^2 - x^2\right) \quad \therefore \frac{T}{V} = \frac{a^2 - x^2}{x^2}$

53. (b)

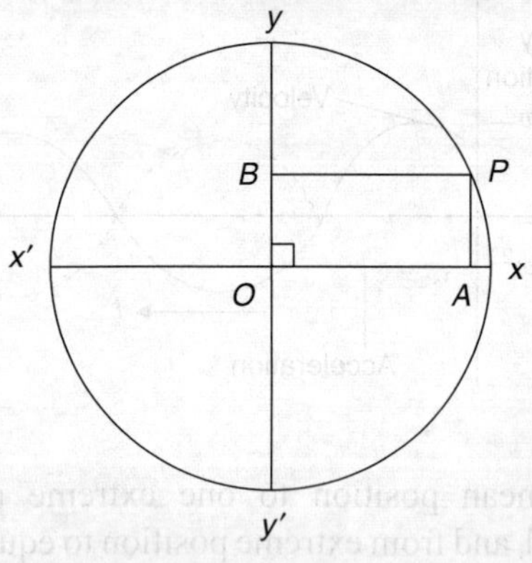

OA and OB are respectively the projections on x and y-axis. $< BOA = \pi/2$

54. (c)

Acceleration $a = -\omega^2 y$

$a = 0$ for $y = 0$

55. (d)

Sine and cosine curves are similar, they only differ in phase.

56. (c)

Kinetic energy at any time is given by

$K = K_0 \cos^2 \omega t$

The maximum value of kinetic energy is K_0. The maximum value of potential energy is also K_0.

57. (c)

$$\text{KE} = \frac{1}{2}m\omega^2\sqrt{A^2 - y^2}$$

$$\text{PE} = \frac{1}{2}m\omega^2 y^2$$

At mean position $y = 0$, kinetic energy is maximum and potential energy is minimum and equal to zero.

58. (b)

In addition to gravity force, there is an electrostatic force on the bob to gravity. Let the equivalent acceleration because of electrostatic force be the effective value of acceleration due to gravity $g' = (g + a)$.

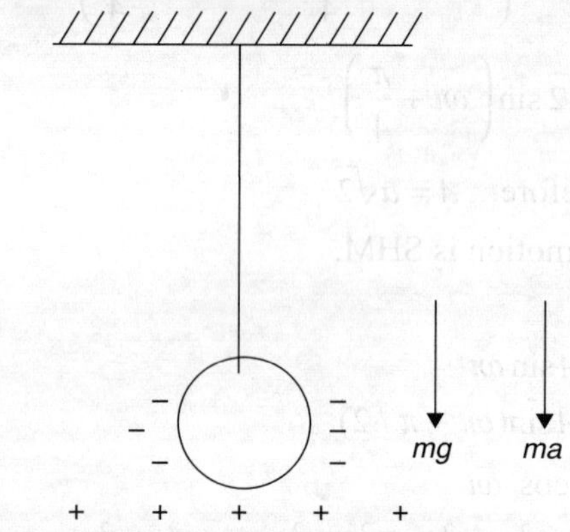

$$T \propto \frac{1}{\sqrt{g}}$$

as g increases T decreases.

59. (b)

The particle will pass a position after at regular intervals but the motion is not directed towards a mean position. It is not to and fro motion.

60. (d)

(i) Total energy in SHM is $\frac{1}{2}m\omega^2A^2$

$$= \frac{1}{2}m\frac{k}{m}A^2 = \frac{1}{2}kA^2$$

(ii) $v = \frac{1}{2\pi}\sqrt{\frac{k}{m}}$

(iii) $v = \frac{1}{2}m\omega^2\sqrt{A^2 - y^2}$

$$v_{\max} = \frac{1}{2}m\omega^2 A \text{ at } y = 0$$

61. (c)

The potential energy PE is given by the equations

$$\text{PE} = \frac{1}{2}m\omega^2A^2\sin^2\omega t \tag{i}$$

$$= \frac{1}{2} m\omega^2 x^2 \quad \text{(ii)}$$

where x is displacement. From the above equations we can conclude that the correct graphs are I and III.

62. (c)
The watch based on oscillating spring is independent of g.

63. (c)
Let the ball be at a distance y from the centre of earth. The restoring force on the ball will act because of gravitation force between ball and earth.
Let M be the mass of earth and M' the effect mass of earth at a distance y from centre of earth.

Then, $M = \frac{4}{3}\pi R^3 \rho$ and $M' = \frac{4}{3}\pi y^3 \rho$

$$F = -\frac{Gm}{y^2} M'$$

$$F = -\frac{Gm}{y^2}\frac{My^3}{R^3} = \frac{GMm}{R^3} y$$

$F \propto -y$ (Characteristic of SHM)

$$T = 2\pi\sqrt{\frac{R^2}{GM}} = 2\pi\sqrt{\frac{R}{g}}$$

64. (b)
In an accelerating train with acceleration a

$$g' = \sqrt{g^2 + a^2}$$

$$T = 2\pi\sqrt{\frac{l}{\left(g^2 + a^2\right)^{1/2}}}$$

65. (c)

$$V_{max} = \omega A = 100$$

$$= \omega \times 10 = 100$$

$$\Rightarrow \quad \omega = 10 s^{-1}$$

$$v = \omega\sqrt{A^2 - y^2}$$

$$\Rightarrow \quad 50 = 10\sqrt{10^2 - y^2}$$

$$\Rightarrow \quad y = 5\sqrt{3} \text{ cm}$$

66. (c)
The time period is independent of mass of the bob.

67. (c)

$$\frac{d^2 y}{dt^2} + ky = 0$$

$$\frac{d^2 y}{dt^2} = -ky$$

$$a = -ky$$

$$\omega^2 = k$$

$$\Rightarrow \quad v = \frac{1}{2\pi}\sqrt{k}$$

$$T = \frac{1}{v} = \frac{2\pi}{\sqrt{k}}$$

68. (c)
Potential energy $= \frac{1}{2} KA^2$

69. (d)
Total energy remains constant.

70. (d)
Acceleration $= -\omega^2 x$
For positive x, acceleration a is –ve, for negative x, a is positive.

71. (d)
When the spring is cut in two halves, the force constant of each half becomes twice that of the original spring.

$$f_1 = \frac{1}{2\pi}\sqrt{\frac{k}{m}}$$

$$f_2 = \frac{1}{2\pi}\sqrt{\frac{2k}{m}}$$

$$f_2 = \sqrt{2} f_1$$

72. (c)

$$T = 2\pi\sqrt{\frac{l}{g}}$$

$$T' = 2\pi\sqrt{\frac{l}{g + \frac{g}{4}}}$$

$$T' = \frac{2T}{\sqrt{5}}$$

73. (d)

$$A = -B$$

$$C = 2B$$

$$y = B\cos 2\omega t + B\sin 2\omega t$$

$$= \sqrt{2}B\sin\left(2\omega t + \frac{\pi}{4}\right)$$

It is an equation of SHM with amplitude $\sqrt{2}B$. If $A = B$ and $C = 2B$ then $y = B + B$ sin. This is also equation of SHM about the point $y = B$.
It oscillates between $x = 0$ and $x = 2B$.

74. (c)
The velocity at the centre of oscillation is maximum.

$$V_{max} = \omega A = \frac{2\pi}{T} A = \frac{2\pi}{0.01} \times 0.2 = 40\pi$$

75. (b)

$$v_1 = 0.5 = \frac{1}{2\pi}\sqrt{\frac{k}{m}}$$

$$v_2 = \frac{1}{2\pi}\sqrt{\frac{k}{m/4}} = \frac{1}{2\pi}\sqrt{\frac{4k}{m}} = 2\times 0.5 = 1$$

76. (b)

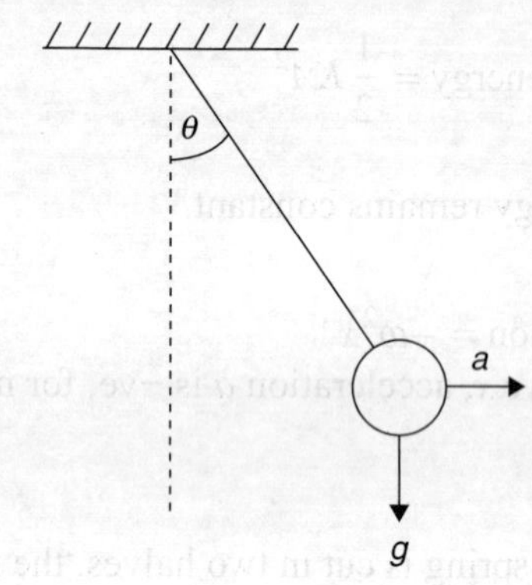

$$\tan\theta = \frac{a}{g}$$

$$\theta = \tan^{-1}\frac{a}{g}$$

77. (d)
The displacement is zero.

78. (d)

$$\tan\theta = \frac{a}{g} = \frac{49}{980} = \frac{1}{20}$$

$$\theta = \tan^{-1}\frac{1}{20} \approx 3°$$

79. (a)

80. (b)

$$V_{max} = \omega A = \frac{2\pi}{T}A = \frac{2\pi}{6}\times 3 = \pi \text{ cm/s}$$

81. (c)
When the string is cut the upward force on 2m is 3 mg and downward force is 2 mg.

$\therefore F_{net} = mg$

$\therefore a = \dfrac{g}{2}$ the acceleration of mass m is g.

82. (a)
The system is a parallel combination of two springs of force constants k_1 and k_2.
Therefore $k = k_1 + k_2$

$$T = 2\pi\sqrt{\frac{m}{k}} = 2\pi\sqrt{\frac{m}{k_1 + k_2}}$$

83. (b)

$$T = 2\pi\sqrt{\frac{m}{k+k}} = 2\pi\sqrt{\frac{m}{2k}}$$

84. (d)
Series combination

$$\frac{1}{k'} = \frac{1}{k} + \frac{1}{k}$$

$$k' = \frac{k}{2}$$

$$v = \frac{1}{2\pi}\sqrt{\frac{k'}{m}}$$

$$= \frac{1}{2\pi}\sqrt{\frac{k}{2m}}$$

85. (c)
For simple harmonic motion velocity

$$v = \omega\sqrt{a^2 - x^2}$$

$$10 = \omega\sqrt{a^2 - 16};$$

$$8 = \omega\sqrt{a^2 - 25}$$

$$\frac{100}{\omega^2} = a^2 - 16;$$

$$\frac{64}{\omega^2} = a^2 - 25$$

Subtract (ii) from (i)

$$36 = \omega^2 \times 9$$

$$\Rightarrow \omega = 2 \text{ rad/s} \quad \text{or} \quad T = \frac{2\pi}{\omega} = \frac{2\pi}{2} = \pi \text{ sec}.$$

86. (b)
The two springs of force constant k and $2k$ are in parallel. The equivalent force constant of these two springs is $k + 2k = 3k$
The other two springs on the right hand side of the mass are equivalent to a spring of force constant k as they are in series. The system reduces to

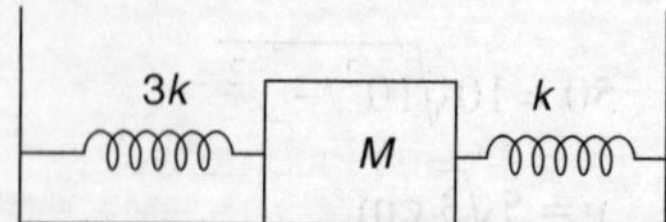

The equivalent force constant
$= 3k + k = 4k$

$$v = \frac{1}{2\pi}\sqrt{\frac{4k}{M}}$$

87. (c)

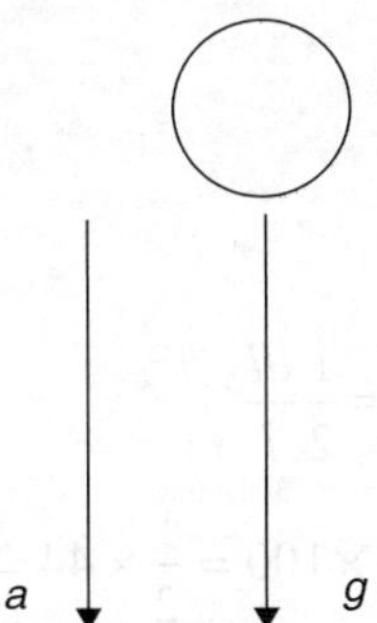

The free-body diagram of the bob is shown in the figure, where a is the pseudo force.

The net acceleration

$$g' = g + a$$

88. (d)

The effective value of acceleration due to gravity is $\sqrt{(a^2 + g^2)}$

89. (b)

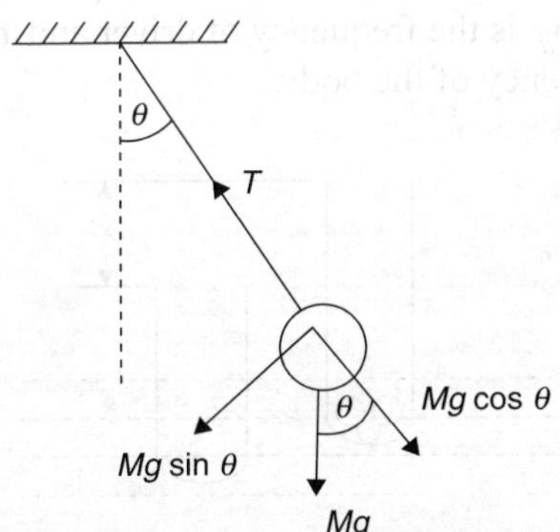

The centripetal force is $= T - Mg\cos\theta$

$$= T - Mg\cos\theta = \frac{Mv^2}{L}$$

where L is the radius of circular motion.

90. (c)

In one oscillation, the particle in SHM has the same kinetic energy twice. But, the total energy remains constant. Therefore if frequency of oscillation is f, the frequency of kinetic energy is $2f$ and that of total energy is infinity.

91. (a)

Acceleration $= -\omega^2$ displacement

$$\omega^2 = \frac{\text{acceleration}}{\text{displacement}} = \frac{2.0}{0.02}$$

$\omega^2 = 100$ or $\omega = 10$ rad/s.

92. (a)

The phase difference is $\pi/2$. Let the equation of pendulum A and B respectively be

$$y_A = A\sin\omega t$$

$$y_B = A\sin(\omega t + \pi/2)$$

$$= A\cos\omega t$$

Then $\dfrac{dy_A}{dt} = A\omega\cos\omega t$

$$V_A = A\omega\cos\omega t$$

$$\frac{dy_B}{dt} = -A\omega\sin\omega t$$

$$V_B = -A\omega\sin\omega t$$

When A has maximum velocity V_0 $\cos\omega t = 1 \Rightarrow \omega t = 0$

At the same instant

$$V_B = A\omega \times 0 = 0$$

Therefore V_B at that instant is stationary.

93. (c)

The acceleration is to be tangential on the path.

94. (a)

Bulk modulus $= B = \dfrac{FV_0}{A.Ax}$

$$M\frac{d^2x}{dt^2} = F = -\frac{BA^2}{V_0}x$$

$$\frac{d^2x}{dt^2} = -\frac{BA^2}{MV_0}x$$

$$\omega^2 = \frac{BA^2}{MV_0}$$

$$T = 2\pi\sqrt{\frac{MV_0}{BA^2}}$$

As motion corresponds to isothermal change, $B = P$

$$T = 2\pi\sqrt{\frac{MV_0}{PA^2}}$$

95. (a)

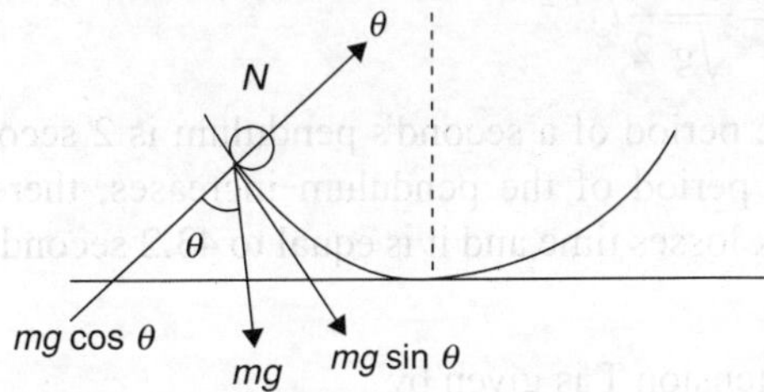

$$E = mgR(1-\cos\theta) + \frac{1}{2}mv^2 + \frac{1}{2}Iw^2$$

∵ The sphere rolls without slipping

$$v = r\omega$$

$$\therefore\ E = mgR(1-\cos\theta) + \frac{1}{2}mv^2 + \frac{1}{2}.\frac{2}{5}mv^2$$

$$= mgR(1-\cos\theta) + \frac{7}{10}mv^2$$

$$\frac{dE}{dt} = mgR\sin\theta.\frac{d\theta}{dt} + 1.4mv\frac{dv}{dt} = 0$$

$$\Rightarrow \quad mg.\frac{xR\omega'}{(R-r)} + 1.4mv\frac{dv}{dt} = 0$$

$$1.4\frac{dv}{dt} = -\frac{g}{(R-r)}x$$

$$\frac{d^2x}{dt^2} = -\frac{g}{1.4(R-r)}x$$

$$\Rightarrow \quad \omega = \sqrt{\frac{g}{1.4(R-r)}}$$

$$\therefore \ T = 2\pi\sqrt{\frac{(R-r)1.4}{g}}$$

96. (b)

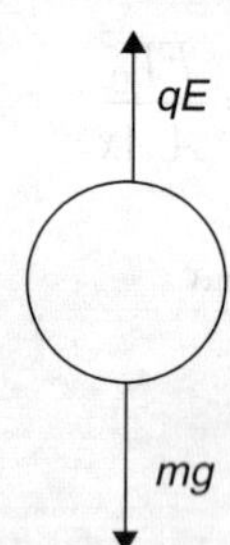

The free-body diagram of sphere is shown in figure

The net force = $mg - qE$

Therefore $g_{eff} = \frac{mg - qE}{m} = g - \frac{qE}{m}$

97. (b)

$$T = 2\pi\sqrt{\frac{l}{g}}$$

Differentiating with respect to l.

$$\frac{dT}{dt} = \frac{2\pi}{\sqrt{g}}\frac{1}{2}l^{-1/2}$$

Time period of a second's pendulum is 2 seconds. The time period of the pendulum increases, therefore the clock losses time and it is equal to 43.2 seconds per day

98. (b)

The tension T is given by

$$T = mg\cos\theta + \frac{mv^2}{r}$$

At end points, velocity is zero and $\theta = \pi/2$, therefore tension is zero.

At mean position, the velocity is maximum and $\theta = 0$

Therefore, $T = mg + \frac{mv_{\max}^2}{r}$ and is maximum.

99. (b)

$$dT = \frac{\pi dl}{\sqrt{gl}}$$

$$T = 2\pi\sqrt{l/g}$$

$$\Rightarrow \quad \frac{dT}{T} = \frac{1}{2}\frac{dl}{l}$$

$$\frac{dT}{T}\times 100 = \frac{1}{2}\frac{dl}{l}\times 100 = \frac{1}{2}\times 44 = 22\%$$

Closest answer is 20%.

100. (d)

When a body oscillates under the influence of an external periodic force with a frequency other than its natural frequency, then oscillations are called forced vibrations. Here the tuning fork oscillates by electrically driven force.

101. (a)

A special case of forced vibration $\omega_d = \omega$ where ω_d is the frequency of driver and ω is the natural frequency of the body.

102. (d)

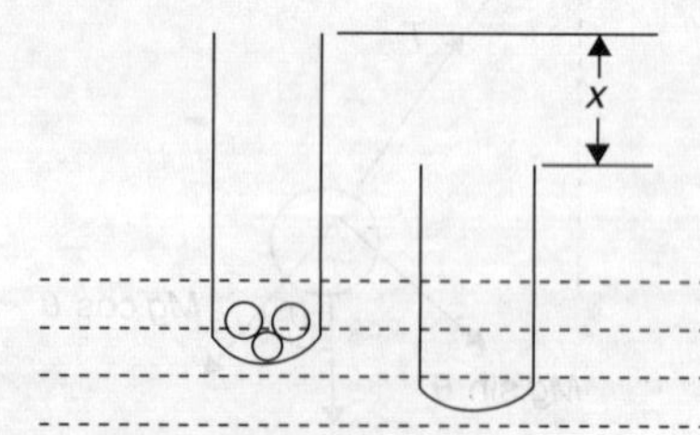

Restoring force $= -Ax\rho g$

Acceleration $= \frac{-A\rho gx}{M}$

$$\omega^2 = \frac{A\rho g}{M}$$

$$\Rightarrow \quad T = 2\pi\sqrt{\frac{M}{A\rho g}}$$

103. (a)

The position of the particle has been shown in the figure.

The forces on charge Q by two charges, each of $-Q$ on Y axis if resolved, cancels out the component of forces along Y axis cancels out. Suppose at any instant the charge Q is at a distance x from 0. Then, net force

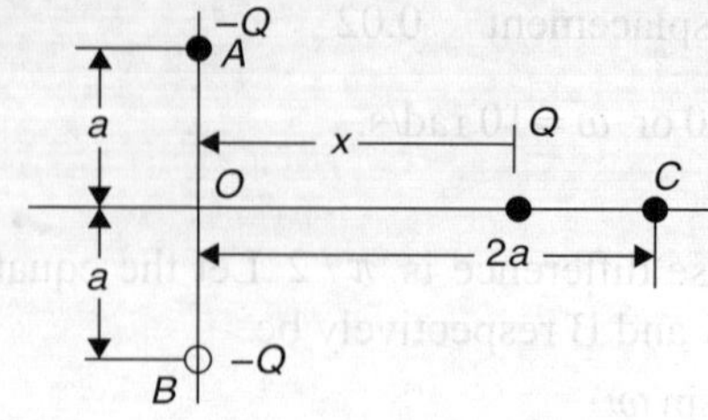

$$F = 2\frac{1}{4\pi\varepsilon_0}\frac{qQ}{\left(a^2+x^2\right)}\cos\theta$$

$$= \frac{2}{4\pi\varepsilon_0}\frac{qQ}{\left(a^2+x^2\right)}\frac{x}{\left(a^2+x^2\right)^{1/2}}$$

$$= \frac{1}{2\pi\varepsilon_0}\frac{qQ}{\left(a^2+x^2\right)^{3/2}}x$$

The restoring force is not proportional to displacement, the motion will not be SHM. It will be oscillatory with amplitude $2a$.

104. (c)

If length of the simple pendulum is comparable or large as compared to radius of the earth 'R' (figure).

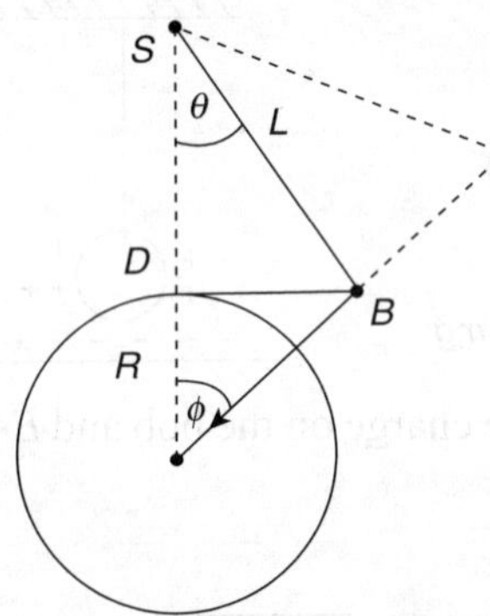

Restoring torque $= mgl\sin(\theta+\phi)$

or $ml^2\dfrac{d^2\theta}{dt^2} = -mgl(\theta+\phi)....\sin(\theta+\phi) = \theta+\phi$

$$\frac{d^2\theta}{dt^2} = \frac{-g}{l}\left(\frac{\theta+\phi}{\theta}\right)\theta$$

$$\tan\phi = \phi = \frac{BD}{R}$$

$$\tan\theta = \theta = \frac{BD}{OD} = \frac{BD}{l}$$

and $\dfrac{\phi}{\theta} = \dfrac{l}{R}$

Here, $\dfrac{d^2\theta}{dt^2} = -g\left(\dfrac{l}{R}+1\right)\theta$

and $T = 2\pi\left(\dfrac{l}{\left(\dfrac{l}{R}+1\right)g}\right)^{\frac{1}{2}} = 2\pi\sqrt{\dfrac{R}{\left(1+\dfrac{R}{l}\right)g}}$

If $l \to \infty$, $T = 2\pi\left(\dfrac{R}{g}\right)^{\frac{1}{2}} = 84.6$ minutes

If $l \to R$, $T = 2\pi\left(\dfrac{R}{2g}\right)^{\frac{1}{2}}$

105. (a)

For SHM, $x = A\sin\left(\dfrac{2\pi}{T}t\right)$

When $x = A$, $A = A\sin\left(\dfrac{2\pi}{T}t\right)$

$$\therefore \sin\left(\frac{2\pi}{T}.t\right) = 1 \Rightarrow \sin\left(\frac{2\pi}{T}.t\right) = \sin\left(\frac{\pi}{2}\right) \Rightarrow t = (T/4)$$

When $x = \dfrac{A}{2}, \dfrac{A}{2} = A\sin\left(\dfrac{2\pi}{T}.t\right)$

or $\sin\dfrac{\pi}{6} = \sin\left(\dfrac{2\pi}{T}t\right)$ or $t = (T/12)$

Now, time taken to travel from $x = A$ to $x = A/2 = T/4 - T/12 = T/6$

106. (c)

Stress = $Y \times$ Strain

Suppose the elongation in the wire is x, then

$$\frac{\textit{Restoring force}}{A} = -Y\times\frac{x}{L}$$

Acceleration $= \dfrac{-YA}{ML}x$

$$\Rightarrow \omega^2 = \frac{YA}{ML} \Rightarrow 2\pi n = \sqrt{\frac{YA}{ML}}$$

$$n = \frac{1}{2\pi}\left(\frac{YA}{ML}\right)^{1/2}$$

107. (c)

$$\text{KE} = \frac{1}{2}m\omega^2\left(A^2-y^2\right)$$

$$= \frac{1}{2}\times 0.01\times\left(\frac{2\pi}{2}\right)^2[(0.1)^2-(0.02)^2]$$

$$= 4.8\times10^{-4}$$

108. (d)

$$|F_{\max}| = m\omega^2 A = 0.1\times\left(\frac{2\pi}{0.2}\right)^2\times 1 = 10\pi^2\ \text{N}$$

109. (d)

$$x = 0.01\cos\left(\pi t+\frac{\pi}{4}\right)$$

Comparing this with general equation $x = A\cos(\omega t+0)$, we get $\omega = \pi$

$2\pi v = \pi$

$v = 0.5$ Hz

110. (a)

$\omega = 20\pi$

$v = 10$ Hz

111. (d)

$$V_{\max} = \omega A = \frac{2\pi A}{T}$$

112. (b)

Total energy E is constant

$$KE = \frac{1}{2} m\omega^2 \left(A^2 - x^2\right)$$

KE vs. x^2 graph is a straight line.

For $x = 0$. KE is maximum and for $x = A$, KE = 0

113. (b)

$$V = \omega\sqrt{\left(A^2 - x^2\right)}$$

$$= \frac{2\pi}{3}\sqrt{\left(9^2 - 6^2\right)}$$

$$= \frac{2\pi}{3}\sqrt{45} = \frac{2\pi}{3} \times 3 \times \sqrt{5}$$

$2\pi\sqrt{5}$ cm s^{-1}

114. (a)

Mechanical wave is a sequence of particles executing SHM.

115. (d)

The options (a), (b) and (c) can be expressed as x = the general equation of SHM except the expression given in option (d).

116. (d)

$$\phi_1 \text{ after 1 sec} = \frac{\pi}{2} + \phi$$

$$\phi_2 \text{ after 1 sec} = \frac{2\pi}{3} + \phi$$

$$\text{Phase difference } \phi_2 - \phi_1 = \frac{2\pi}{3} - \frac{\pi}{2} = \pi/6$$

117. (d)

$x = 2\sin\omega t$ (i)

$$y = 2\sin\left(\omega t + \frac{\pi}{2}\right)$$

$\Rightarrow x^2 + y = 4$ which is the equation of a circle

118. (a)

The free body diagram is shown in the figure $mg\sin\theta$ in upward direction is pseudo force.

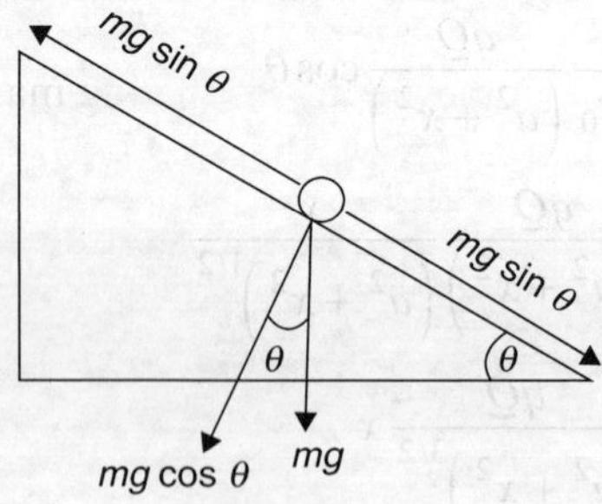

The effective g is $g\cos\theta$

Therefore, $T = 2\pi\sqrt{\dfrac{L}{g\cos\theta}}$

119. (b)

The free body diagram is

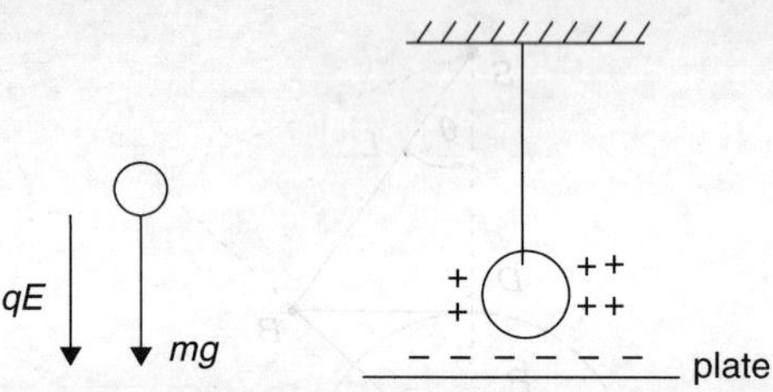

where q is the charge on the bob and E the electric field.

$$g_{eff} = g + \frac{qE}{m}$$

Therefore, $T' = 2\pi\sqrt{\dfrac{l}{g + \dfrac{qE}{m}}}$

and $T = 2\pi\sqrt{\dfrac{l}{g}}$

$T' < T$

120. (d)

This displacement is zero.

121. (d)

122. (a)

$$f_A = \frac{1}{2\pi}\sqrt{\frac{YA}{M_A l_A}}$$

$$f_B = \frac{1}{2\pi}\sqrt{\frac{Y}{M_B l_B}}$$

$$\frac{f_A}{f_B} = \sqrt{\frac{M_B l_B}{M_A l_A}}$$

$$2 = \sqrt{\frac{M_B}{M_A} \times \frac{l_B}{l_A}}$$

This equation is satisfied with $l_A = l_B/2$ and $M_A = M_B/2$

123. (b)
When m_1 is removed the oscillating mass is only m_2

Therefore $\omega = \sqrt{\dfrac{k}{m_2}}$

124. (b)

$$T = 2\pi\sqrt{\frac{M}{K}}$$

$$T' = 2\pi\sqrt{\frac{4M}{K}}$$

$$\Rightarrow \quad T' = 2T$$

125. (b)

$$T = 2\pi\sqrt{\frac{m}{k}}$$

Now, force constant $k = \dfrac{6.4}{0.1} = 64$

$$\frac{\pi}{4} = 2\pi\sqrt{\frac{m}{64}}$$

$$\frac{1}{4} = \frac{2}{8}\sqrt{m}$$

$$\Rightarrow \quad m = 1 \text{ kg}$$

126. (b)
Clock *S* is based on spring. Therefore it will give show the same time as it gives shows on the earth. But, clock *P* is based on pendulum on a planet whose radius is twice the radius of the earth the value of g will increase as

$$g = \frac{GM}{R^2} = \frac{G.\frac{4}{3}\pi R^3\rho}{R^2}$$

$$\Rightarrow \quad g \propto R$$

Accordingly time period *T* of the pendulum will decrease as $T = 2\pi\sqrt{l/g}$. If time period increases the clock runs faster.

127. (d)

$$V_{\max} = \omega A$$

$$\omega_1 A_1 = \omega_2 A_2$$

The ratio of amplitude,

$$\frac{A_1}{A_2} = \frac{\omega_2}{\omega_1} = \frac{\left(\frac{k_2}{m}\right)^{1/2}}{\left(\frac{k_1}{m}\right)^{1/2}} = \left(\frac{k_2}{k_1}\right)^{1/2}$$

128. (c)

$$\omega^2 = \frac{k}{m+m} = \frac{k}{2m}$$

For the blocks to move together the pseudo force because of SHM should be less than or at the most equal to maximum force of friction.

$m\omega^2 A \le$ Maximum force of friction

$$m\frac{k}{2m}A = \frac{kA}{2}$$

129. (b)
A wire of length *L*, area of cross-section *A* and Young's modulus of elasticity *Y* is equivalent to a force constant

$$k_1 = \frac{YA}{L}$$

This could be found out in following way
Stress = $Y \times$ Strain

Restoring force = $-YA\dfrac{x}{L} = \dfrac{-YA}{L}x$

$$k_{eq} = \frac{k_1 k_2}{k_1 + k_2} = \frac{\frac{YAk}{L}}{\frac{YA}{L} + k}$$

$$T = 2\pi\sqrt{\frac{m}{k_{eq}}} = 2\pi\sqrt{\frac{m(YA + kL)}{YAk}}$$

130. (b)
Where the cylinder is given a small downward push '*x*'
Restoring force = Additional spring force + Additional up-thrust force

$$F = -(kx + Ax\rho g)$$

Acceleration $= \dfrac{-(kx + Ax\rho g)}{M}$

$$v = \frac{1}{2\pi}\sqrt{\frac{k + \rho Ag}{M}}$$

131. (b)

$$T = 2\pi\sqrt{\frac{l}{g_{eff}}}$$

Time period *T* decreases when g_{eff} increases. This is possible when the rocket is going up with uniform acceleration.

132. (b)

$$v = \omega\sqrt{A^2 - x^2} = \omega\sqrt{A^2 - \left(\frac{A}{2}\right)^2}$$

$$= \frac{2\pi}{T}\frac{\sqrt{3}}{2}A = \frac{\pi A}{T}\sqrt{3}$$

133. (b)

Total potential energy $= \dfrac{1}{2}kA^2$

Let the displacement be *y*,

then potential energy $= \dfrac{1}{2}ky^2$

$$\frac{1}{2}ky^2 = \left(\frac{1}{2}kA^2\right) \times \frac{1}{4}$$

$$\Rightarrow \quad y = \frac{A}{2}$$

134. (d)

Acceleration $\propto$ −(Displacement from mean position)

135. (a)

$$y = 5\sin\pi(t+4) = 5\sin(\pi t + 4\pi)$$

$$A = 5$$

$$\omega = \frac{2\pi}{T} = \pi$$

$$\Rightarrow T = 2$$

136. (a)

The effective length increases as the centre of gravity shifts downward.

Therefore, time period increases.

137. (c)

The up-thrust force = $V\rho g \; v\rho g$

The downward force = Vdg

$$\frac{\text{Upthrust force}}{\text{Downward force}} = \frac{V\rho g}{Vdg} = \frac{1}{10}$$

Therefore, when metal bob is immersed in a liquid the net downward force

$$= mg - \frac{mg}{10} = \frac{9}{10}mg$$

The effective value of g' is $\frac{9}{10}g$

$$T' = 2\pi\sqrt{\frac{l}{g'}} \text{ and } T = 2\pi\sqrt{\frac{l}{g}}$$

$$T' = T\sqrt{\frac{g}{g'}} = T\sqrt{\frac{10}{9}}$$

138. (b)

Ratio of time period $\Rightarrow T : 4T$

Difference in time period = $4T - T = 3T$

$$n = \frac{4T}{3T} = \frac{4}{3}$$

139. (c)

$$T_1 = 2\pi\sqrt{\frac{0.2}{k}} \quad \text{(i)}$$

$$T_2 = 2\pi\sqrt{\frac{m}{k}} \quad \text{(ii)}$$

From Eqs. (i) and (ii)

$$\frac{0.5}{1} = \sqrt{\frac{m}{0.2}}$$

$$\Rightarrow \quad m = (0.5)^2 \times 0.2 = 0.05 \text{ kg} = 50 \text{ gm}$$

The mass that should be removed

$$= 200 - 50 = 150 \text{ gm}$$

140. (d)

Suppose the pendulum is displaced by small angle θ

Restoring force = $-(kx + mg\sin\theta)$

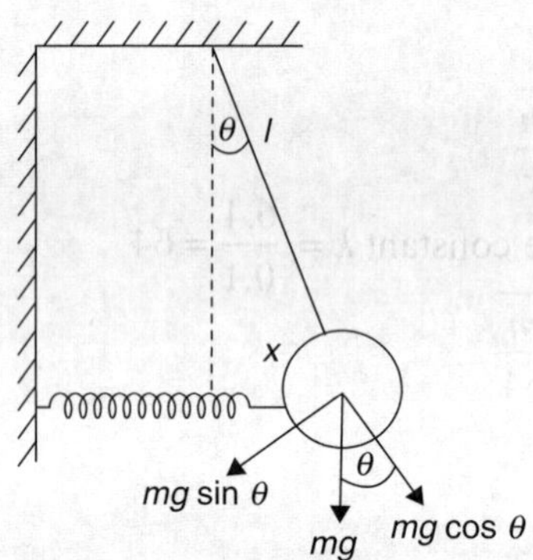

Restoring force = $-(kx + mg\theta)$

$$\text{Restoring force} = -\left(kx + mg\frac{x}{l}\right) = -\left(k + \frac{mg}{l}\right)x$$

$$\text{Acceleration} = -\left(\frac{k}{m} + \frac{g}{l}\right)x$$

$$\omega^2 = \frac{k}{m} + \frac{g}{l}$$

$$T = 2\pi\sqrt{\frac{1}{\frac{k}{m} + \frac{g}{l}}} = 2\pi\sqrt{\frac{ml}{mg + kl}}$$

141. (c)

The motion does not satisfy the criteria of SHM.

The motion is oscillatory.

$$T = \frac{2\pi}{2}\sqrt{\frac{l}{g}}$$

$$= \pi\sqrt{\frac{l}{g}}$$

142. (d)

Suppose the displacement is x

Then up-thrust force $= -Ax\rho g$

Up-thrust force = Restoring force = $-Ax\rho g$

$$\text{Acceleration} = \frac{A\rho g}{m}x$$

$$\omega^2 = \frac{A\rho g}{m}$$

$$T = 2\pi\sqrt{\frac{m}{A\rho g}}$$

$$T \propto \frac{1}{\sqrt{A}}$$

143. (c)

ma mg

$$T_1 = 2\pi\sqrt{\frac{l}{g}}$$

$$y = t^2$$

$$a = \frac{d^2 y}{dt^2} = 2$$

$$g' = (2 + 10)\ \text{m/s}^2$$

$$T_2 = 2\pi\sqrt{\frac{l}{g'}}$$

$$\frac{T_1^2}{T_2^2} = \left(\frac{g_2}{g_1}\right) = \frac{12}{10} = \frac{6}{5}$$

144. (b)

$$x = 2\times10^{-2}\cos\pi t$$

Amplitude $A = 2\times10^{-2}$ m

At $t = 0, x = 2\times10^{-2}$ m the particle is at the extreme position.

The time taken by particle to reach mean position = $\frac{T}{4}$

At mean position, the velocity is maximum.

$$\omega = \pi = \frac{2\pi}{T}$$

$$T = 2\ \text{s}$$

$$\therefore \quad \frac{T}{4} = 0.5\ \text{s}$$

145. (a)

$$f = \frac{1}{2\pi}\sqrt{\frac{k_1 + k_2}{m}}$$

$$f' = \frac{1}{2\pi}\sqrt{\frac{4k_1 + 4k_2}{m}}$$

$$f' = \frac{1}{2\pi}\times 2\sqrt{(k_1 + k_2)/m}$$

$$f' = 2f$$

146. (a)

$$x = x_0\cos\left(\omega t - \frac{\pi}{4}\right)$$

$$\frac{d^2x}{dt^2} = -\omega^2 x_0\cos(\omega t - \pi/4)$$

$$= \omega^2 x_0\cos\left(\omega t - \frac{\pi}{4} + \pi\right)$$

Comparing this with the given equation $a = A\cos(\omega t + \delta)$

$$A = \omega^2 x_0 \text{ and } \delta = \frac{3\pi}{4}$$

147. (b)

Average kinetic energy of the particle

$$= \frac{1}{2}\times\frac{1}{2}ma^2\omega^2 = \frac{1}{4}ma^2(2\pi v)^2$$

$$= \pi^2 ma^2 v^2$$

148. (c)

The acceleration in SHM is $\omega^2 A$, where A is amplitude. Coin will leave contact with the platform when $m\omega^2 A \geq mg$

$$A = \frac{g}{\omega^2}$$

149. (b)

KE in SHM is $\frac{1}{2}m\omega A^2\cos^2\omega t$

$$\frac{\text{KE}}{\text{Total Energy}} = \frac{\frac{1}{2}m\omega A^2\cos^2\omega t}{\frac{1}{2}m\omega^2 A^2}$$

$$\frac{3}{4} = \cos^2\omega t$$

$$\Rightarrow \quad \cos\omega t = \frac{\sqrt{3}}{2}$$

$$\omega t = \frac{\pi}{6}$$

$$\frac{2\pi t}{T} = \frac{\pi}{6}$$

$$\frac{2\pi t}{2} = \frac{\pi}{6}$$

$$\Rightarrow \quad t = 1/6\ \text{s}$$

150. (b)

$$V_{max} = \omega A$$

$$4.4 = \frac{2\pi}{T} \times 7 \times 10^{-3}$$

$$T = \frac{2\pi \times 7 \times 10^{-3}}{4.4} = 0.01 \text{ s}$$

151. (d)

$$\frac{d^2x}{dt^2} + ax = 0$$

$$\frac{d^2x}{dt^2} = -\alpha x$$

$$\Rightarrow \quad \omega^2 = \alpha$$

$$\omega = \sqrt{\alpha}$$

$$\frac{2\pi}{T} = \sqrt{\alpha}$$

$$T = \frac{2\pi}{\sqrt{\alpha}}$$

152. (c)
First the centre of gravity goes down, therefore that effective length increases and since time period $T \propto \sqrt{l}$; the time period also increases. As more and more water has leaked out, the centre of gravity goes up, the effective length decreases and also the time period.

153. (c)

$$y_1 = 0.1 \sin\left(100\pi t + \frac{\pi}{3}\right)$$

$$\frac{dy_1}{dt} = 0.1 \times 100 \cos\left(100\pi t + \frac{\pi}{3}\right)$$

$$v_1 = 0.1 \times 100 \sin\left(100\pi t + \frac{\pi}{3} + \frac{\pi}{2}\right)$$

$$y_2 = 0.1\pi \cos \pi t$$

$$v_2 = \frac{dy_2}{dt} = -0.1\pi \sin \pi t = 0.1\pi \sin(\pi t + \pi)$$

$$\text{Initial phase difference} = \frac{\pi}{3} + \frac{\pi}{2} - \pi = -\frac{\pi}{6}$$

154. (d)

$$y = \sin^2 \omega t$$

$$\frac{dy}{dt} = 2\omega \sin \omega t \cos \omega t = \omega t \sin 2\omega t$$

$$\frac{d^2y}{dt^2} = 2\,\omega^2 t \cos 2\omega t$$

The general equation of SHM is $\frac{d^2y}{dt^2} \propto -y$ on comparison,

We find the given equation is not of this form. Therefore, the function does not represent a simple harmonic motion. The graph of $y = \sin^2 \omega t$ is

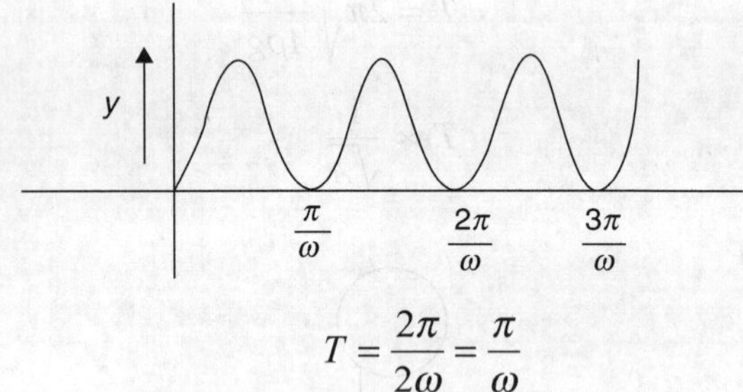

$$T = \frac{2\pi}{2\omega} = \frac{\pi}{\omega}$$

155. (c)
Total energy is constant and is equal to

$$= \frac{1}{2} m\omega^2 A^2$$

Questions Asked in NEET/CBSE Medicals

156. (c)

$$y = A_0 + A \sin \omega t + B \cos \omega t$$

$$\text{or} \quad (y - A_0) = A \sin \omega t + B \cos \omega t$$

$$\text{or} \quad y' = A \sin \omega t + B \cos \omega t$$

$$= A \cos\left(\frac{\pi}{2} - \omega t\right) + B \cos \omega t$$

$$\text{Amplitude} = \sqrt{A^2 + B^2 + 2AB \cos\frac{\pi}{2}} = \sqrt{A^2 + B^2}$$

157. (a)
Since the displacement for a complete vibration is zero, therefore the average velocity will be zero.

158. (a)
Here, $T = 4$ s, $A = 3$ m

Time period $T = \frac{2\pi}{\omega}$

$$4 = \frac{2\pi}{\omega}$$

$$\Rightarrow \omega = \frac{\pi}{2}$$

Since, time period is noted from extreme position it will be cosine function.

So, $y = A \cos(\omega t) \Rightarrow y = 3 \cos\left(\frac{\pi}{2} t\right)$

159. (d)

160. (b)
Magnitude of acceleration of a particle moving in a SHM is, $|a| = \omega^2 y$; where y is amplitude.

$\Rightarrow \quad 20 = \omega^2(5) \Rightarrow \omega = 2 \text{ rad s}^{-1}$

$\therefore$ Time period of oscillation, $T = \dfrac{2\pi}{\omega} = \dfrac{2\pi}{2} = \pi s$

161. (b)

Let us assume, the length of spring be l. When we cut the spring into ratio of length 1 : 2 : 3, we get three springs of lengths $\frac{l}{6}, \frac{2l}{6}$ and $\frac{3l}{6}$ and respective force constants are separate.

$$k_1 = \frac{kl}{l_1} = \frac{kl}{l/6} = 6k,$$

$$k_2 = \frac{kl}{l_2} = \frac{kl}{2l/6} = 3k$$

$$k_3 = \frac{kl}{l_3} = \frac{kl}{3l/6} = 2k$$

In series,

$$\frac{1}{k_s} = \frac{1}{6k} + \frac{1}{3k} + \frac{1}{2k} = \frac{1+2+3}{6k} = \frac{1}{k}$$

$$\therefore k' = k$$

In parallel

$$k_p = 6k + 3k + 2k = 11k$$

$$\therefore \frac{k_s}{k_p} = \frac{k}{11k} = \frac{1}{11}$$

162. (b)

Given $A = 3$ cm, $x = 2$ cm

The velocity of a particle in SHM is

$v = \omega\sqrt{A^2 - x^2}$ and magnitude of its acceleration is $a = \omega^2 x$

Given $|v| = |a|$

$$\therefore \omega\sqrt{A^2 - x^2} = \omega^2 x$$

$$\omega x = \sqrt{A^2 - x^2}$$

$$\omega^2 = \frac{A^2 - x^2}{x^2} = \frac{9-4}{4} = \frac{5}{4}$$

$$\Rightarrow \quad \omega = \frac{\sqrt{5}}{2}$$

Time period, $T = \dfrac{2\pi}{\omega} = 2\pi \cdot \dfrac{2}{\sqrt{5}} = \dfrac{4\pi}{\sqrt{5}}$ s

163. (d)

Time period of spring-mass system,

$$T = 2\pi\sqrt{\frac{m}{k}}$$

$$\frac{T_1}{T_2} = \sqrt{\frac{m_1}{m_2}}$$

Here, $T_1 = 3$ s, $m_1 = m$, $T_2 = 5$ s, $m_2 = m + 1$, $m = ?$

$$\frac{3}{5} = \sqrt{\frac{m}{m+1}}$$

$$\frac{9}{25} = \frac{m}{m+1}$$

$$\therefore m = \frac{9}{16} \text{ kg}$$

164. (b)

If A and ω be the amplitude and angular frequency of vibration, then

$$\alpha = \omega^2 A \quad \text{(i)}$$

and $$\beta = \omega A \quad \text{(ii)}$$

Dividing Eq. (i) by Eq. (ii), we get

$$\frac{\alpha}{\beta} = \frac{\omega^2 A}{\omega A} = \omega$$

$\therefore$ Time period of vibration is

$$T = \frac{2\pi}{\omega} = \frac{2\pi}{(\alpha/\beta)} = \frac{2\pi\beta}{\alpha}$$

165. (d)

In SHM, velocities of a particle at distance x_1 and x_2 from mean position are given by

$$V_1^2 = \omega^2(a^2 - x_1^2) \quad \text{(i)}$$

$$V_2^2 = \omega^2(a^2 - x_2^2) \quad \text{(ii)}$$

From Eqs. (i) and (ii), we get

$$V_1^2 - V_2^2 = \omega^2(x_2^2 - x_1^2)$$

$$\omega = \sqrt{\frac{V_1^2 - V_2^2}{x_2^2 - x_1^2}} \Rightarrow T = 2\pi\sqrt{\frac{x_2^2 - x_1^2}{V_1^2 - V_2^2}}$$

166. (c)

Here, $x = A\cos\omega t$

$\therefore$ Velocity, $v = \dfrac{dx}{dt} = \dfrac{d}{dt}(A\cos\omega t) = -A\omega\sin\omega t$

Acceleration, $a = \dfrac{dv}{dt} = \dfrac{d}{dt}(-A\omega\sin\omega t) = -A\omega^2\cos\omega t$

Hence the variation of a with t is correctly shown by graph (c).

167. (c)

$$x = a\sin\omega t \text{ or } \frac{x}{a} = \sin\omega t \quad \text{(i)}$$

Velocity, $v = \frac{dx}{dt} = a\omega\cos\omega t$

$$\frac{v}{a\omega} = \cos\omega t$$

Squaring and adding (i) and (ii), we get

$$\frac{x^2}{a^2} + \frac{v^2}{a^2\omega^2} = \sin^2\omega t + \cos^2\omega t$$

$$\frac{x^2}{a^2} + \frac{v^2}{a^2\omega^2} = 1$$

It is an equation of ellipse.

This is a graph between velocity and displacement

Momentum of the particle = mv

$\therefore$ The nature of graph of the momentum and displacement is also the same as velocity displacement because mass is constant.

168. (c)

$$y = \sin\omega t - \cos\omega t$$

$$= \sqrt{2}\left[\frac{1}{\sqrt{2}}\sin\omega t - \frac{1}{\sqrt{2}}\cos\omega t\right]$$

$$\sqrt{2}\sin\left(\omega t - \frac{\pi}{4}\right)$$

It represents a SHM with time period, $T = \frac{2\pi}{\omega}$

$$y = \sin^3\omega t = \frac{1}{4}[3\sin\omega t - \sin 3\omega t]$$

It represents a periodic motion with time period $T = \frac{2\pi}{\omega}$ but not SHM.

$$y = 5\cos\left(\frac{3\pi}{4} - 3\omega t\right) = 5\cos\left(3\omega t - \frac{3\pi}{4}\right)$$

It represents a SHM with time period, $T = \frac{2\pi}{3\omega}$

$$y = 1 + \omega t + \omega^2 t^2$$

It represents a non-periodic motion. Also it is not physically acceptable as $y \to \infty$ as $t \to \infty$.

169. (c)

The time taken by the particle to travel from

$x = 0$ to $x = \frac{A}{2}$ is $\frac{T}{6}$

The time taken by the particle to travel from

$x = A$ to $x = \frac{A}{2}$ is $\frac{T}{6}$

Time difference $= \frac{T}{6} + \frac{T}{6} = \frac{T}{3}$

Phase difference, $\Delta d = \frac{2\pi}{T} \times \Delta T = \frac{2\pi}{T} \times \frac{T}{3} = \frac{2\pi}{3}$

170. (c)

$$x = a\sin^2\omega t = a\left(\frac{1 - \cos 2\omega t}{2}\right)$$

$$= \frac{a}{2} - \frac{a\cos 2\omega t}{2}$$

$\therefore$ Velocity, $v = \frac{dx}{dt} = \frac{2\omega a\sin 2\omega t}{2} = \omega a\sin 2\omega t$

Acceleration, $a = \frac{dv}{dt} = 2\omega^2 a\cos 2\omega t$

For the given displacement $x = a\sin^2\omega t$, $a \propto -x$ is not satisfied.

Hence, the motion of the particle is non simple harmonic motion.

171. (d)

$$T = 2\pi\sqrt{\frac{M}{k}} \quad \text{(i)}$$

$$T' = 2\pi\sqrt{\frac{M + M}{k}} = 2\pi\sqrt{\frac{2M}{k}} \quad \text{(ii)}$$

From (i) and (ii)

$$T' = \sqrt{2}T$$

172. (c)

For simple harmonic motion, $v = \omega\sqrt{a^2 - x^2}$

When $x = \frac{a}{2}, v = \omega\sqrt{a^2 - \frac{a^2}{4}} = \omega\sqrt{\frac{3}{4}a^2}$

$$\therefore v = \frac{2\pi}{T} \cdot \frac{\sqrt{3}}{2}a \Rightarrow v = \frac{\pi\sqrt{3}a}{T}$$

173. (c)

174. (d)

$\omega_1 = 100$ rad s^{-1}; $\omega_2 = 1000$ rad s^{-1}

Maximum acceleration of (1) = $\omega_1^2 A$

Maximum acceleration of (2) = $\omega_2^2 A$

$$\therefore \frac{a_1}{a_2} = \frac{\omega_1^2}{\omega_2^2} = \frac{(100)^2}{(1000)^2} = \frac{1}{100}$$

$a_1 : a_2 = 1 : 100.$

175. (b)

$x(t) = a\sin\omega t$ (from the equilibrium position)

At $x(t) = a/2$

$$\frac{a}{2} = a\sin(\omega t)$$

$$\Rightarrow \sin\left(\frac{\pi}{6}\right) = \sin(\omega t) \text{ or, } \frac{\pi}{6} = \frac{2\pi t}{T}\left[\because \omega = \frac{2\pi}{T}\right]$$

or $t = T/12$

176. (a)

The spring has a length l. When m is placed over it, the equilibrium position becomes A. If it is pressed from A (the equilibrium position) to B, AB is the amplitude.

$$\therefore AB = \frac{mg}{k} = \frac{2\times10}{200} = 0.10 \text{ m}$$

$$mg = kx_0$$

If the restoring force $mA\omega^2 > mg$, then the mass will move up with acceleration and be detached from the pan.

$$A \geq \frac{20}{200} = 0.1 \text{ m}$$

Therefore, the amplitude > 10 cm.

177. (c)

Kinetic energy + Potential energy = Total energy

When kinetic energy is maximum, potential energy is zero and vice versa.

Maximum potential energy = Total energy.

$0 + K_0 = K_0$ (KE + PE = Total energy).

178. (d)

Let $y = A\sin\omega t$

$$\frac{dy}{dt} = A\omega\cos\omega t = A\omega\sin\left(\omega t + \frac{\pi}{2}\right)$$

Acceleration = $-A\omega^2\sin\omega t$

The phase difference between acceleration and velocity is $\pi/2$

179. (c)

Let l be the length of block immersed in liquid as shown in the figure. When the block is floating,

$$\therefore mg = Al\rho m$$

If the block is given a vertical displacement y, then the effective restoring force is

$$F = -[A(l+y)\rho g - mg]$$

$$= -[A(l+y)\rho g - Al\rho g] = -Al\rho gy$$

Restoring force = $-[Al\rho g]y$. As this F is directed towards equilibrium position of block, so if the block is left free, it will exeute simple harmonic motion.

Here inertia factor = Mass of block = m

Spring factor = $A\rho g$

$$\therefore \text{Time period, } T = 2\pi\sqrt{\frac{m}{A\rho g}} T \propto \frac{1}{\sqrt{A}}$$

180. (a)

181. (d)

$a = 5$ cm, $v_{max} = 31.4$ cm/s

$$v_{max} = \omega a \Rightarrow 31.4 = 2\pi\upsilon\times5$$

$$\Rightarrow 31.4 = 10\times3.14\times\upsilon \Rightarrow \upsilon = 1 \text{ Hz}$$

182. (d)

When the spring joined in series, the total extension in spring is

$$\Rightarrow y = y_1 + y_2 = \frac{-F}{k_1} - \frac{F}{k_2}$$

$$\Rightarrow y = -F\left[\frac{1}{k_1} + \frac{1}{k_2}\right]$$

Thus spring constant in this case becomes

$$k = \frac{k_1k_2}{k_1 + k_2}$$

183. (d)

In SHM

$$v = A\omega\sin(\omega t + \pi/2)$$

$a = A\omega^2\sin(\omega t + \pi)$. From this we can easily find out that when v is maximum, then a is zero

184. (c)

Potential energy of simple harmonic oscillator = $\frac{1}{2}m\omega^2y^2$

For $y = \frac{a}{2}$, PE = $\frac{1}{2}m\omega^2\frac{a^2}{4}$

$$\Rightarrow P.E. = \frac{1}{4}\left(\frac{1}{2}m\omega^2a^2\right) = \frac{E}{4}$$

185. (a)

Potential energy of particle performing SHM varies parabolically in such a way that at mean position it is zero and maximum at extreme position.

186. (c)

Let k be the force constant of spring. If k' is the force constant of each part, then

$$\frac{1}{k} = \frac{4}{k'} \Rightarrow k' = 4k$$

$$\therefore \text{Time period} = 2\pi\sqrt{\frac{m}{4k}} = \frac{1}{2} \times 2\pi\sqrt{\frac{m}{k}} = \frac{T}{2}$$

187. (a)
Smaller damping gives a taller and narrower resonance peak.

188. (c)

189. (d)
This is a case of damped vibration as the amplitude of vibration is decreasing with time.

Amplitude of vibrations at any instant t is given by a = $a_0 e^{-bt}$, where a_0 is the initial amplitude of vibrations and b is the damping constant.

Now, when $t = 100T$, $a = a_0/3$

Let the amplitude be a' at $t = 200T$ after completing 200 oscillations.

$$\therefore a = a_0 / 3 = a_0 e^{-100Tb} \quad \text{(i)}$$

And $a' = a_0 e^{-200Tb}$ (ii)

From (i) and (ii),

$$a' = a_0 \times \frac{1}{9} = \frac{a_0}{9}$$

190. (b)

$$t_1 = 2\pi\sqrt{\frac{m}{k_1}}$$

$$t_2 = 2\pi\sqrt{\frac{m}{k_2}}$$

$$t_0 = 2\pi\sqrt{\frac{m}{k_1 + k_2}}$$

$$\frac{1}{t_1^2} = \frac{k_1}{4\pi^2 m}$$

$$\frac{1}{t_2^2} = \frac{k_2}{4\pi^2 m}$$

$$\frac{1}{t_0^2} = \frac{k_1 + k_2}{4\pi^2 m}$$

$$\frac{1}{t_1^2} + \frac{1}{t_2^2} = \frac{k_1 + k_2}{4\pi^2 m}$$

$$\frac{1}{t_1^2} + \frac{1}{t_2^2} = \frac{1}{t_0^2}$$

$$\Rightarrow t_1^{-2} + t_2^{-2} = t_0^{-2}$$

191. (b)

$$\text{Energy} = \frac{1}{2} m\omega^2 a^2 = \frac{1}{2} ka^2$$

192. (a)

$$f_A = 2 f_B$$

$$\Rightarrow \frac{1}{2\pi}\sqrt{\frac{g}{l_A}} = 2 \times \frac{1}{2\pi}\sqrt{\frac{g}{l_B}}$$

or, $l_A = \frac{l_B}{4}$ which is independent of mass

193. (a)

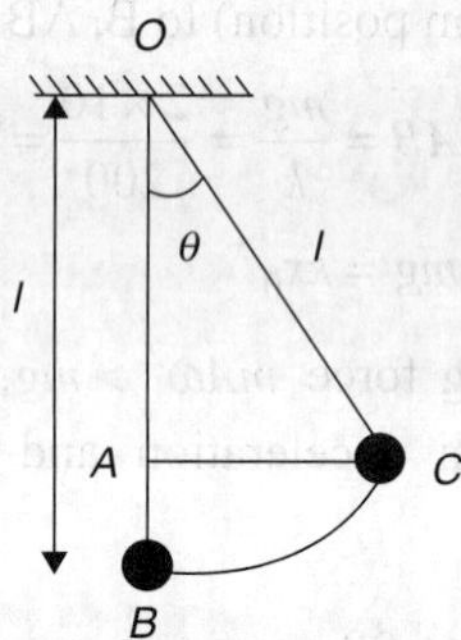

In $\Delta OAC, \cos\theta = A/l$ or, $OA = l\cos\theta$

$$\therefore \quad AB = l(1 - \cos\theta) = h$$

At point C, the velocity of bob = 0. The vertical acceleration = g

$$\therefore v^2 = 2gh$$

or, $v = \sqrt{2gl(1 - \cos\theta)}$

194. (d)
Time period of a simple pendulum is given by $T = 2\pi\sqrt{\frac{l}{g}}$

$$\therefore \frac{T_1}{T_2} = \sqrt{\frac{l_1}{l_2}} = \sqrt{\frac{1}{4}} = \frac{1}{2} \text{ or, } T_2 = 2T_1 = 4 \text{ sec}$$

195. (b)
The amplitude and velocity resonance occurs at the same frequency.

At resonance, $\omega_1 = \omega_0$ and $\omega_2 = \omega_0$, so the amplitude and energy of the particle would be maximum.

196. (a)
Frequency of the pendulum

$$v_1 = \frac{1}{2\pi}\sqrt{\frac{g}{5}}$$

$$v_2 = \frac{1}{2\pi}\sqrt{\frac{g}{20}}$$

$$\therefore \frac{v_1}{v_2} = \sqrt{\frac{20}{5}} = 2$$

197. (a)

$$n = \frac{1}{2\pi}\sqrt{\frac{k}{m}}; n' = \frac{1}{2\pi}\sqrt{\frac{k}{4m}}$$

$$\therefore n' = n/2$$

198. (a)

$$l_2 = 1.02 l_1$$

Time period $(T) = 2\pi \times \sqrt{\frac{l}{g}}$

Therefore $\frac{T_2}{T_1} = \sqrt{\frac{l_2}{l_1}} = \sqrt{\frac{1.02 l_1}{l_1}} = 1.01$

Thus, the time period increases by 1%.

199. (c)

$$x = a\sin\omega t$$

$$y = a\sin(\omega t + \pi/2) = a\cos\omega t$$

or, $\frac{x}{y} = \frac{\sin\omega t}{\cos\omega t} = \tan\omega t$ or, $\frac{x}{y} = \frac{x}{\sqrt{a^2 - x^2}}$ or, $y^2 = a^2 - x^2$

or, $x^2 + y^2 = a^2$

It is an equation of a circle.

200. (d)

Maximum velocity, $v_{\max} = A\omega$

$$\frac{v_{\max}}{2} = \frac{A\omega}{2} = \omega\sqrt{A^2 - y^2}$$

$$\frac{A^2}{4} = A^2 - y^2 \Rightarrow y^2 = A^2 - \frac{A^2}{4} \Rightarrow y = \frac{\sqrt{3}A}{2}$$

201. (c)

$$PE = \frac{1}{2}ky^2$$

$$= \frac{1}{2} \times 2 \times 10^6 \times (0.01)^2$$

$$= 10^6 \times 10^{-4}$$

$$= 100 \text{ J}$$

202. (b)

Potential energy at A (or C) = Kinetic energy at B.

Thus $\frac{1}{2}mv_B^2 = mgH$ or $v_B = \sqrt{2gH}$

203. (b)

Displacement $(x) = \frac{a}{2}$

Total energy $= \frac{1}{2}m\omega^2 a^2$ and kinetic energy when displacement is (x)

$$\frac{1}{2}m\omega^2\left(a^2 - x^2\right)$$

$$= \frac{1}{2}m\omega^2\left(a^2 - \left(\frac{a}{2}\right)^2\right) = \frac{3}{4}\left(\frac{1}{2}m\omega^2 a^2\right)$$

Therefore fraction of the total energy at x,

$$= \frac{\frac{3}{4}\left(\frac{1}{2}m\omega^2 a^2\right)}{\frac{1}{2}m\omega^2 a^2} = \frac{3}{4}$$

204. (d)

Mass $(m) = 5$ kg and time period $(T) = 2\pi$ sec.

$$T = 2\pi = 2\pi\sqrt{\frac{m}{k}}$$

$$\frac{m}{k} = 1$$

$$\Rightarrow \frac{5}{k} = 1$$

or $k = 5$ N/m

$F = -ky$

$5g = 5y$

$\Rightarrow y = g$ metres

205. (d)

For simple harmonic motion

$$\frac{d^2x}{dt^2} \propto -x$$

Therefore force acting on the particle = Akx.

206. (d)

Period of oscillation $T = 2\pi\sqrt{\frac{l}{g}}$.

Therefore T will decreases when acceleration (g) increases. And g will increase when the rocket moves up with a uniform acceleration.

207. (a)

Time-period = 4 sec. In one simple harmonic oscillation, the same kinetic and potential energies are repeated two times. So the difference will be 2 seconds.

208. (b)

$$\text{Potential energy} = \frac{1}{4}\text{ (Total energy)}$$

$$\frac{1}{2}m\omega^2 y^2 = \frac{1}{4}\left(\frac{1}{2}m\omega^2 A^2\right)$$

$$y = \frac{A}{2}$$

EXERCISES

SECTION - II

1. Given that: $y = A\sin\left[\left(\frac{2\pi}{\lambda}\right)(ct - x)\right]$ where y and x are measured in metres. Which of the following statements is true?
(a) The unit of λ is same as that of x and A.
(b) The unit of λ is same as that of x but not A.
(c) The unit of c is same as that of $\frac{2\pi}{\lambda}$.
(d) The unit of $(ct - x)$ is same as that of $\left(\frac{2\pi}{\pi}\right)$.

2. A point source emits sound equally in all directions in a non-absorbing medium. Two points P and Q are at distance of 2 m and 3 m respectively from the source. The ratio of the intensities of the waves at P and Q is
(a) 9 : 4 (b) 2 : 3
(c) 3 : 2 (d) 4 : 9

3. Two tuning forks P and Q when set vibrating, gives 4 beats/s. If a prong of the fork P is filed, the beats are reduced to 2/s. What is frequency of P, if that of Q is 250 Hz?
(a) 246 Hz (b) 250 Hz
(c) 254 Hz (d) 252 Hz

4. The speed (v) of ripples on the surface of water depends on surface tension (σ), density (ρ) and wavelength (λ). The square of speed (v) is proportional to
(a) $\frac{\sigma}{\rho\pi}$ (b) $\frac{\rho}{\sigma\lambda}$
(c) $\frac{\lambda}{\sigma\rho}$ (d) $\rho\lambda\sigma$

5. The source of sound generating a frequency of 3 kHz reaches an observer with a speed of 0.5 times in air. The frequency heard by the observer is
(a) 1 kHz (b) 4 kHz
(c) 4 kHz (d) 6 kHz.

6. A source is moving along a circle of radius 10 metres with angular speed of 20 radius per sec. It is emitting sound of frequency 320 Hz. What is the minimum frequency heard by a stationary observer?
(a) 350 (b) 250
(c) 300 (d) 200

7. A car sounding its horn at 480 Hz moves towards a high wall at a speed of 20 m/s. The frequency of the reflected sound heard by the man sitting in the car will be nearest to
(a) 480 Hz (b) 510 Hz
(c) 540 Hz (d) 570 Hz

8. For production of beats, the two sources must have
(a) different frequencies and same amplitude
(b) different frequencies
(c) different frequencies, same amplitude and same phase
(d) different frequencies and same phase.

9. An organ pipe P_1 closed at one end and vibrating in its first overtone and another pipe P_2 open at both ends vibrating in its second overtone are in resonances. The ratio of length of pipes P_1 and P_2 is
(a) 1 (b) $\frac{1}{2}$
(c) $\frac{1}{3}$ (d) $\frac{1}{4}$

10. A piston fitted in cylindrical pipe is pulled as shown in the figure. A tuning fork is sounded at open end and loudest sound is heard at open length 13 cm, 41 cm and 69 cm, the frequency of tuning fork, if velocity of sound is 350 m/s is

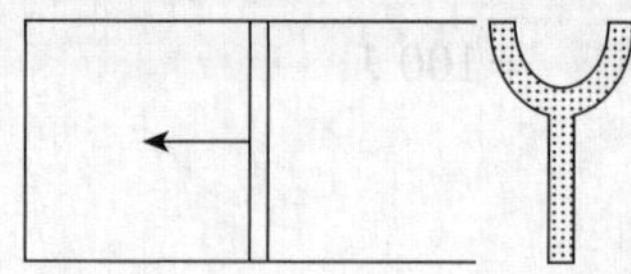

(a) 1250 Hz (b) 625 Hz
(c) 417 Hz (d) 715 Hz

11. A man is standing on the platform and one train is approaching and another train is going away with speed of 4 m/s. Frequency of sound produced by train is 240 Hz. What will be the number of beats heard by him per second?
(a) 12 (b) zero
(c) 6 (d) 3

12. Equation of progressive wave is given by $y = 4 \sin\left[\pi\left(\frac{t}{5} - \frac{x}{9}\right) + \frac{\pi}{6}\right]$ where y, x are in cm and t is in seconds. Then which of the following is correct?
(a) $v = 5$ cm (b) $\lambda = 18$ cm
(c) $a = 0.04$ cm (d) $f = 50$ Hz

13. A boat at anchor is rocket by waves whose crests are 100 m apart and velocity is 25 m/s. The boat bounces up once in every
(a) 2500 s (b) 75 s
(c) 4 s (d) 0.25 s

14. An open pipe is suddenly closed at one end with the result that the frequency of third harmonic of the closed pipe is termed to be higher by 100 Hz than the fundamental frequency of the open pipe. The fundamental frequency of the open pipe is
(a) 200 Hz (b) 150 Hz
(c) 100 Hz (d) 250 Hz

15. The number of beats produced per second by two vibrations:
$x_1 = x_0 \sin 646\pi t$ and $x_2 = x_0 \sin 652\pi t$ is
(a) 2 (b) 3
(c) 4 (d) 6

16. A string vibrates according to the equation
$$Y = 5 \sin\left(\frac{2\pi x}{3}\right)\cos 20\pi t$$
where x and y is in cm and t in second. The distance between two adjacent nodes is
(a) 3 cm (b) 4.5 cm
(c) 6 cm (d) 1.5 cm

17. The ratio of the velocity of sound in oxygen to that in hydrogen at same temperature and pressure approximately
(a) 16 : 1 (b) 1 : 16
(c) 4 : 1 (d) 1 : 4

18. A source of sound of frequency 500 Hz is moving towards a stationary observer with velocity 30 m/s. The speed of sound is 330 m/s. The frequency heard by the observer will be
(a) 545 Hz (b) 580 Hz
(c) 458.3 Hz (d) 550 Hz

19. A wave is represented by the equation $y = a\cos(kx - \omega t)$ is superposed with another wave to form a stationary wave such that the point $x = 0$ is a node. The equation of the other wave is
(a) $a\sin(kx + \omega t)$ (b) $-a\sin(kx - \omega t)$
(c) $-a\cos(kx + \omega t)$ (d) $a\cos(kx + \omega t)$

20. The ratio of fundamental frequency of an organ pipe opened at both ends to that of the organ pipe closed at one end is
(a) 1 : 1 (b) 1.5 : 1
(c) 2 : 1 (d) 3 : 1

21. A police car horn emits a sound at a frequency 240 Hz when the car is at rest. If the speed of sound is 330 m/s the frequency heard by an observer who is approaching the car at a speed of 11 m/s, is
(a) 248 Hz (b) 244 Hz
(c) 240 Hz (d) 230 Hz

22. The ratio of speed of sound in nitrogen and helium gas at 300 K is
(a) $\sqrt{\frac{2}{7}}$ (b) $\sqrt{\frac{1}{7}}$
(c) $\frac{\sqrt{3}}{5}$ (d) $\frac{\sqrt{6}}{5}$

23. If the equation of a wave is given as:
$y = 0.07\sin(12\pi x - 300\pi t)$
where x is in metre and t in sec, then the correct statement is
(a) $\lambda = 1/6$ m, $v = 25$ m/s
(b) $a = 0.07$ m, $v = 300$ m/s
(c) $n = 1500$, $v = 200$ m/s
(d) None of these.

24. A small source of sound moves on a circle as shown in the figure and an observer is sitting on O. Let ν_1, ν_2 and ν_3 be the frequencies heard when the sources are at A, B, and C respectively. Then,

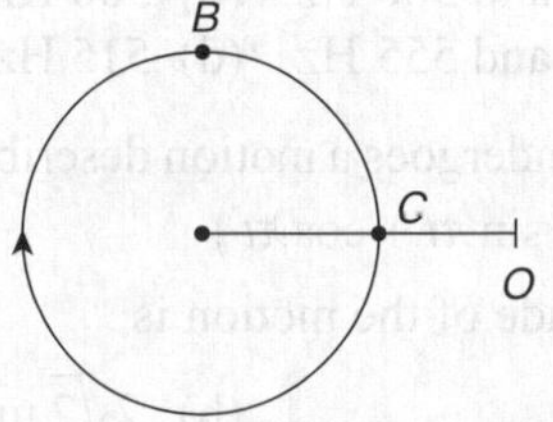

(a) $\nu_1 > \nu_2 > \nu_3$ (b) $\nu_2 > \nu_3 > \nu_1$
(c) $\nu_1 = \nu_2 > \nu_3$ (d) $\nu_2 > \nu_1 > \nu_3$

25. A transverse wave is described by the equation
$$y = y_0 \sin 2\pi\left(ft - \frac{x}{\lambda}\right)$$
The maximum velocity of the particle is equal to four times the wave velocity, if
(a) $\lambda = \pi y_0 / 4$ (b) $\lambda = \pi y_0 / 2$
(c) $\lambda = \pi y_0$ (d) $\lambda = 2\pi y_0$

26. Two identical flutes produce fundamental notes of frequency 300 Hz at 27 °C. If the temperature of air at one flute is increased to 31 °C, the number of the beats heard per second, will be
(a) 1 (b) 2
(c) 3 (d) 4

27. A source of sound of frequency 256 Hz is moving rapidly towards a wall with a velocity of 5 m/s. The speed of sound is 330 m/s. If the observer is in between the wall and the source, then beats per second heard will be
(a) 7.8 Hz (b) 9.7 Hz
(c) 3.9 Hz (d) zero

28. When a sound wave of wave length λ is propagating in a medium, the maximum velocity of the particle is equal to wave velocity. The amplitude of the wave is
(a) λ (b) $\frac{\lambda}{2}$
(c) $\frac{\lambda}{2\pi}$ (d) $\frac{\lambda}{4\pi}$

29. The difference between the apparent frequency of a source of sound as perceived by the observer during its approach and recession is 2% of the frequency of the source. If the speed of sound in air 300 ms^{-1}m, the velocity of the source is
(a) 1.5 ms^{-1} (b) 12 ms^{-1}
(c) 6 ms^{-1} (d) 3 $m\,s^{-1}$

30. A whistle of frequency 540 Hz is moving in a horizontal circle of radius 60 cm at an angular speed of 15 rad/s. The lowest and highest frequencies heard by a listener a long distance away with respect to the centre of the circle is
(a) 520 Hz and 560 Hz (b) 500 Hz and 580 Hz
(c) 526 Hz and 555 Hz (d) 515 Hz and 565 Hz

31. A particle undergoes a motion described by the equation $\xi(x,t) = 2(\sin \pi t + \cos \pi t)$
The amplitude of the motion is
(a) 4 unit (b) $2\sqrt{2}$ unit
(c) 2 unit (d) $\frac{1}{\sqrt{2}}$ unit

32. A tuning fork of frequency 340 Hz is vibrated just above the tube of 120 cm height. Water is poured slowly in the tube. What is the minimum height of water necessary for the resonance?
(Speed of sound in air = 340 m/s)
(a) 45 cm (b) 30 cm
(c) 45 cm (d) 25 cm

33. A stretched wire, 60 cm long, is vibrating with its fundamental frequency 256 Hz. If the length of the wire is decreased to 15 cm and the tension remains the same, then the fundamental frequency of the vibration of the wire will be
(a) 1024 Hz (b) 512 Hz
(c) 64 Hz (d) 256 Hz

34. The equation of a wave travelling in a string is given as $y = 3 \cos \pi(100t - x)$. Its wavelength is
(a) 3 cm (b) 100 cm
(c) 2 cm (d) 5 cm

35. Two sound waves are $y = a \sin(\omega t - kx)$, $y = a \cos(\omega t - kx)$, the phase difference between the two waves is
(a) $\frac{\pi}{2}$ (b) $\frac{\pi}{4}$
(c) π (d) 0

36. The velocity of sound in air is not affected by changes in
(a) moisture contents in air
(b) temperature of air
(c) the atmospheric pressure
(d) the composition of air.

37. Two uniform strings A and B made of steel are made to vibrate under the same tension. If the first overtone of A is equal to the second overtone of B and if the radius of A is twice that of B the ratio of the length of the strings is
(a) 1 : 2 (b) 1 : 3
(c) 1 : 4 (d) 1 : 5

38. When both listener and source are moving towards each other, then which of the wavelength of sound will be observed by the observer?
(a) More frequency, less wavelength
(b) More frequency, more wavelength
(c) Less frequency, less wavelength
(d) More frequency, constant wavelength.

39. A pipe closed at one end produces a fundamental note of 412 Hz. It is cut into two pieces of equal length. The fundamental notes produced by the two pieces of equal length.
(a) 824 Hz, 1648 Hz (b) 412 Hz, 824 Hz
(c) 206 Hz, 412 Hz (d) 216 Hz, 824 Hz

40. Sound waves of $f = 600$ Hz fall normally on a perfectly reflecting wall. The shortest distance from the wall at which all particles will have maximum amplitude of vibration will be (speed of sound = 300 m/s)

(a) $\frac{7}{8}$ m (b) $\frac{3}{8}$ m

(c) $\frac{1}{8}$ (d) $\frac{1}{4}$

41. Compression wave pulses are sent to the bottom of sea from a ship and the echo is heard after 2 s. If bulk modulus of elasticity of water is 2×10^9 N/m^2 and mean temperature is 4 °C, the depth of the sea will be

(a) 1014 m (b) 1414 m

(c) 2828 m (d) None of these.

42. The velocity of sound in any gas depends upon

(a) wavelength of sound only

(b) density and elasticity of gas

(c) intensity of sound waves only

(d) amplitude and frequency of sound.

43. The speed of a wave on a string is 150 m/s when the tension is 120 N. The percentage increase in the tension is 120 N. The percentage increase in the tension in order to raise the wave speed by 20% is

(a) 44 % (b) 40%

(c) 20% (d) 10 %

44. The velocity of sound is generally greater in solids than in gases because

(a) the density of solids is high, but elasticity is low

(b) the density of solids is high, but the elasticity is very high

(c) both the density and elasticity of solids are low.

(d) the density of solids is low but the elasticity is high.

45. Two closed organ pipes A and B, have the same length. A is wider than B. They resonate in the fundamental mode at frequencies n_A and n_B respectively then

(a) $n_A = n_B$

(b) $n_A > n_B$

(c) $n_A < n_B$

(d) either (b) or (c) depending on the ratio of their diameters.

46. A sound wave frequency f propagating through air with a velocity C, is reflected from a surface which is moving away from the source with a constant speed v. The frequency of the reflected wave, measured by the observer at the position of the source is

(a) $\frac{f(C-v)}{C+v}$ (b) $\frac{f(C+v)}{C-v}$

(c) $\frac{f(C+2v)}{C+v}$ (d) $\frac{f(C-2v)}{C-2v}$

47. In an open organ pipe, if fundamental frequency is n, then other frequencies are

(a) $n, 2n, 3n, 4n$ (b) $n, 3n, 5n$

(c) $n, 2n, 4n, 8n$ (d) None of these.

48. The fundamental frequency of a sonometer wire is n. If its radius is doubled and its tension becomes half while the material of the wire remains same, the new fundamental frequency will be

(a) n (b) $\frac{n}{\sqrt{2}}$

(c) $\frac{n}{2}$ (d) $\frac{n}{2\sqrt{2}}$

49. A string of length 2 m is fixed at both ends. If this string vibrates in its fourth normal mode with frequency of 500 Hz, then the waves would travel on it with a velocity of

(a) 125 m/s (b) 250 m/s

(c) 500 m/s (d) 1000 m/s

50. The speed of a wave is 360 m/s and frequency is 500 Hz. Phase difference between two consecutive particles is 60°. The path difference between them will be

(a) 72 cm (b) 120 cm

(c) 12 cm (d) 7.2 cm

51. If the wave motion is $y = 0.08 \sin \frac{2\pi}{\lambda}(200\,t - x)$, then velocity of the wave is

(a) $400\sqrt{2}$ (b) $200\sqrt{2}$

(c) 400 (d) 200

52. The frequency of a tuning fork is 256 Hz. It will not reasonate with a fork of frequency

(a) 512 Hz (b) 786 Hz

(c) 396 Hz (d) None of these.

53. A couple of tuning fork produces 2 beats in time interval of 0.4 sec, so that the beat frequency is

(a) 8 Hz (b) 5 Hz

(c) 2 Hz (d) infinite

54. If the equation of motion of a standing wave be $y = 0.30 \sin (314t - 1.5x)$ where t, x and y are in sec, metre and m respectively, then the velocity of the standing wave will be

(a) 200 m/s (b) 100 m/s

(c) 300 m/sec (d) 400 m/sec.

55. A man is travelling in a train moving away from a station with a speed 30 m/s. Calculate the apparent frequency heard by him of a whistle which is blown at the station with a frequency 1000 c/s.

(a) 909 c/s (b) 1010 c/s
(c) 90.9 c/s (d) 9.09 c/s

56. The ratio of the fundamental frequency of an open and closed organ pipe of same wavelength is
(a) 2 : 1 (b) 1 : 2
(c) 1 : 1 (d) 4 : 1

57. The velocity of sound in air depends upon
(a) density and elasticity of gas
(b) pressure
(c) wavelength of sound
(d) amplitude and frequency of sound.

58. Two wires made up same material are of equal length but their diameters are in the ratio of 1 : 2. On stretching each of these two strings by the same tension, the ratio between the fundamental frequencies of these is
(a) 1 : 2 (b) 2 : 1
(c) 2 : 1 (d) 4 : 9

59. A hollow cylinder with both sides open generates a frequency f in air. When the cylinder is vertically immersed into water by half its length the frequency of the wave is
(a) f (b) $2f$
(c) $\frac{f}{2}$ (d) $\frac{f}{4}$

60. Two stretched strings have length l and $2l$ while tensions are T and $4T$ respectively. If they are made of same material, the ratio of their frequency is
(a) 2 : 1 (b) 1 : 2
(c) 1 : 1 (d) 1 : 4

61. When sound is produced in an aero plane moving with a velocity 200 m/s horizontal, its echo is heard after $10\sqrt{5}$ s. If velocity of the sound in air is 300 ms^{-1}. The elevation of aircraft is
(a) 250 m (b) $250\sqrt{5}$ m
(c) 1250 m (d) 2500 m.

62. Two tuning forks of frequency n_1 and n_2 produces n beats per sec. If n_2 and n are known n_1 may be given by
(a) $\frac{n_2}{n} + n_2$ (b) $n_2 n$
(c) $n_2 \pm n$ (d) $\frac{n_2}{n} - n_2$

63. A pipe open at both ends produces a note of fundamental frequency f_1. When the pipe is kept with $\frac{3}{4}$th of its length in water, it produces a note of fundamental frequency f_2. The ratio of f_1 and f_2 is
(a) $\frac{4}{3}$ (b) $\frac{3}{4}$
(c) 2 (d) $\frac{1}{2}$

64. A closed organ pipe and an open organ pipe are tuned to the same fundamental frequency. The ratio of their lengths is
(a) 1 : 1 (b) 2 : 1
(c) 1 : 4 (d) 1 : 2

65. The equation of a wave $y = 5 \sin\left(\frac{t}{0.04} - \frac{x}{4}\right)$, where x is in cm and t in sec. The maximum velocity of the wave will be
(a) 1 ms^{-1} (b) 2 ms^{-1}
(c) 1.5 ms^{-1} (d) 1.25 ms^{-1}

66. When the two waves $y_2 = a\sin\omega t$ and $y_2 = a\cos\omega t$ are superimposed, then the resultant amplitude is
(a) a (b) $2a$
(c) $\sqrt{2}\,a$ (d) $\frac{a}{\sqrt{2}}$

67. A train moves towards a stationary observer with speed 34 ms^{-1}. The train sounds a whistle and its frequency registered by the observer is v_1. If the trains speed is reduced to 17 ms^{-1}, the frequency registered is v_2. If the speed of the sound is 340 ms^{-1}, then the ratio $\frac{v_1}{v_2}$ is
(a) 2 (b) $\frac{1}{2}$
(c) $\frac{18}{19}$ (d) $\frac{19}{18}$

68. The velocity ratio of sound in air at 4 atm to that at 1 atm pressure would be
(a) : 1 (b) 4 : 1
(c) 1 : 4 (d) 3 : 1

69. A boat at anchor is rocked by waves whose crests are 100 m apart and whose velocity is 25 m per sec. These waves reach the boat once every
(a) 0.25 sec (b) 4 sec
(c) 2500 sec (d) 1500 sec

70. Sound waves are travelling in a medium whose adiabatic elasticity is E and isothermal elasticity is E′. The velocity of sound is proportional to
(a) E' (b) $\sqrt{E}$
(c) $\sqrt{E'}$ (d) $\frac{E}{E'}$

71. The Laplace's correction in the expression for the velocity of sound given by Newton is needed because sound waves
(a) are longitudinal
(b) propagate isothermally
(c) propagate adiabatically
(d) are of long wavelengths.

72. The ratio of velocity of the body to the velocity of sound is
(a) Mach number (b) Wood number
(c) Laplace number (d) Natural number.

73. A wave of frequency 500 Hz has a velocity of 350 m/s. The distance between two nearest points 60^0 out of phase will be
(a) 0.7 cm (b) 12 cm
(c) 70 cm (d) 1.2 cm

74. Sound wave is passing through an air column. During the consequent compression and rarefaction
(a) Boyle's law is obeyed
(b) the total amount of energy remains constant
(c) the density of air remains constant
(d) Bulk Modulus of air remains constant.

75. With sound, one cannot observe the phenomenon of
(a) refraction (b) interference
(c) diffraction (d) polarization.

76. Standing stationary waves can be obtained in air column even if the interfering waves are
(a) of different pitches
(b) of different amplitude
(c) of different qualities
(d) moving with different velocities.

77. In stationary waves, the distance between two consecutive nodes is
(a) $\frac{\lambda}{4}$ (b) $\frac{\lambda}{2}$
(c) $\frac{\lambda}{3}$ (d) λ

78. Two sources of sound are in resonance when
(a) they look alike
(b) they are situated at a particular distance from each other
(c) they produce sound of same frequency
(d) they are excited by the same exciting device.

79. Which one of the following does not represent a travelling wave?
(a) $y = f(x - vt)$ (b) $y = y_m \sin k(x + vt)$
(c) $y = y_m \log(x - vt)$ (d) $y = f(x^2 - vt^2)$

80. The rms velocity of molecules of gas is c and velocity of sound in this gas is v. The relation between v and c is
(a) $\frac{v}{c} = \frac{\gamma}{3}$ (b) $\frac{v}{3} = 3\gamma$
(c) $\frac{v}{c} = \sqrt{3r}$ (d) $\frac{v}{c} = \left(\frac{\gamma}{3}\right)^{1/2}$

81. A stone is dropped into a well. If depth of water below the top be h and velocity of sound is v. If the splash in water is hearted after T sec, then
(a) $T = \left(\frac{2h}{g}\right)^{1/2} + \frac{h}{v}$ (b) $T = \frac{2h}{v}$
(c) $T = 2\left(\frac{2h}{g}\right)^{1/2}$ (d) $T = 2\left(\frac{2h}{g}\right)^{1/2} \times \frac{h}{v}$

82. Ultrasonic, infrasonic and audible waves travel through a medium with speeds v_u, v_i and v_a respectively, then
(a) v_u, v_i and v_a are nearly equal
(b) $v_u \geq v_a \geq v_i$
(c) $v_u \geq v_a \leq v_i$
(d) $v_a \geq v_u$ and $v_u \equiv v_i$

83. The square of the speed of sound in a gas is proportional to
(a) density
(b) pressure
(c) temperature in degree Celsius
(d) temperature in Kelvin.

84. When a tuning fork sounds and pressure of air decreases, then there is a change in
(a) frequency only
(b) velocity only
(c) loudness only
(d) frequency and loudness only.

85. In a stationary wave,
(a) all the particles perform SHM with a frequency which is twice that of the two component waves
(b) all the particles between two adjacent nodes vibrate in phase
(c) the amplitude of particle vibration at an antinodes is equal to that of either component wave
(d) particles on opposite sides of a node vibrate with a phase difference of π.

86. A simple sound wave of frequency 440 Hz is passing through air. An oxygen molecule (mass = 5.3×10^{-26} kg) is set in oscillation with an amplitude of 10^{-6} m. Its speed at the centre of its oscillation is
(a) 1.7×10^{-5} m/s (b) 17×10^{-5} m/s
(c) 2.76×10^{-3} (d) 2.77×10^{-5} m/s

87. A thin plane membrane separates hydrogen at 7°C from hydrogen at 47°C, both being at the same pressure. If a collimated sound beam travelling from colder gas makes an angle of incidence of 30⁰ at the membrane, the angle of refraction is

(a) $\sin^{-1}\left(\frac{7}{32}\right)^{1/2}$ (b) $\sin^{-1}\left(\frac{2}{7}\right)^{1/2}$

(c) $\sin^{-1}\left(\frac{4}{7}\right)^{1/2}$ (d) $\sin^{-1}\left(\frac{1}{8}\right)^{1/2}$

88. A tuning fork whose frequency as given by the manufacture is 514 Hz is being tested with an accurate oscillator. It is found that the fork produces a beat of 2 Hz when the oscillator reads 510 Hz. The actual frequency of the fork is

(a) 508 Hz (b) 512 Hz
(c) 516 Hz (d) 518 Hz

89. When beats are produced by two progressive waves of same amplitude and of nearly the same frequencies, then the maximum loudness heard corresponding to each of its constituent waves is

(a) two times (b) four times
(c) same (d) eight times.

90. The end correction of a resonance column is 1.0 cm. If the shortest length resonating with a turning fork is 15.0 cm, the next resonating length is

(a) 31 cm (b) 45 cm
(c) 46 cm (d) 47 cm

91. A piano wire having a diameter of 0.90 mm is replaced by another wire of the same material but with a diameter 0.93 mm. If the tension of the wire is kept the same, then the percentage change in the frequency of the fundamental tone is

(a) + 3% (b) + 3.2%
(c) – 3.2% (d) –3%

92. A transverse wave is described by the equation $y = y_0 \sin 2\pi\left(ft - \frac{x}{\lambda}\right)$. The maximum particle velocity is equal to four times the wave velocity, if

(a) $\lambda = \frac{\pi y_0}{4}$ (b) $\lambda = \frac{\pi y_0}{2}$

(c) $\lambda = \pi y_0$ (d) $\lambda = 2\pi y_0$

93. A wave equation which gives the displacement along the y-direction is given by $y = 10^{-4} \sin(\cot + 2x)$ (where x and y are distance the metres and t is time in seconds. Which of the following is incorrect?

(a) Wave is travelling with a velocity of 30 m/s in the negative x-direction
(b) Wavelength is π metres
(c) Frequency is $30/\pi$ Hz
(d) The amplitude of the wave is 10^{-4} metres and travelling along positive x-direction.

94. A resonating column of air contains

(a) stationary longitudinal waves
(b) stationary transverse wave
(c) transverse progressive wave
(d) longitudinal progressive waves.

95. With an open end organ pipe of length l, the fundamental tone has frequency

(a) $\frac{v}{2l}$ and only even harmonics are presents
(b) $\frac{v}{2l}$ only odd harmonics are presents
(c) $\frac{v}{2l}$ and even as well as odd harmonics are present
(d) $\frac{v}{4l}$ and only odd harmonics are present.

96. With a closed end organ pipe of length l, the fundamental tone has frequency

(a) $\frac{v}{2l}$ and all harmonics are present
(b) $\frac{v}{4l}$ and all harmonics are present
(c) $\frac{v}{4l}$ and only odd harmonics are present
(d) $\frac{v}{4l}$ and only even harmonics are present.

97. With an increase in temperature, the frequency of sound from an organ pipe

(a) decreases (b) increases
(c) remains unchanged (d) changes erratically.

98. A man moves towards a stationary observer. The apparent pitch of the sound is found to be higher than its actual value. This is because

(a) wavelength of sound decreases
(b) wavelength of sound increases
(c) the number of waves received by the man increases
(d) the number of waves received by the man decreases.

99. A stationary wave is set up in a resonant air column of a glass tube partially filled with water by holding a tuning fork near the open end. The open end of the tube is

(i) always a node
(ii) always an antinodes
(iii) sometimes a node, sometimes an antinodes
(iv) neither a node, nor an antinodes.

100. With an increase in the diameter of an organ pipe, the frequency of sound from the organ
(a) increases slightly (b) decreases slightly
(c) remains unchanged (d) changes erratically.

101. If the velocity of sound in air is 350 m/s, the frequency of the fundamental tone emitted by a tube of length 50 cm, open at both ends, is
(a) 50 Hz (b) 175 Hz
(c) 380 Hz (d) 700 Hz

102. A closed organ pipe and an open organ pipe have their first overtones identical in frequency. Their lengths are in the ratio
(a) 1 : 2 (b) 2 : 3
(c) 3 : 4 (d) 4 :2

103. An empty vessel is filled with water, its frequency
(a) increases (b) decreases
(c) remains the same (d) None of the above.

104. A closed organ pipe is excited so as to support the third overtone. It is found that air in the pipe has
(a) 3 nodes, 3 antinodes (b) 3 nodes, 4 antinodes
(c) 4 nodes, 3 antinodes (d) 4 nodes, 4 antinodes.

105. For a resonance tube, the air columns for the first and second resonance differ by 31.5 cm. The wavelength of the sound waves in the tube is
(a) 31.5 cm (b) 63.0 cm
(c) 126 cm (d) 252 cm

106. A source of sound of frequency 256 Hz is moving rapidly towards a wall with a velocity of 5 m/s. What will be the beat per sec heard, if sound travels at a speed of 330 m/s?
(a) 8 Hz (b) 4 Hz
(c) 16 Hz (d) No beats

107. When both source and listener move in the same direction with a velocity equal to half the velocity of sound, the change in the frequency of sound as detected by the listener is
(a) 0% (b) 25%
(c) 50% (d) None of these.

108. When both source and listener approach each other with a velocity equal to half the velocity of sound, the change in frequency of sound as detected by the listener is
(a) 0% (b) 25%
(c) 50% (d) None of these.

109. The frequency of waves emitted from a radar is 750 MHz. The frequency of the reflected wave from the aero plane, as observed at the radar station, is increased to 2.5 KHz. What is the velocity of the aeroplane?
(a) 4 km/s (b) 3 km/s
(c) 1 km/s (d) 0.5 km/s

110. There are the following four possible relative motions between the source of sound and the listener
I. Source moves towards stationary listener
II. Listener moves towards stationary source moves away from stationary listener
III. Listener moves towards stationary source
IV. Listener moves away from stationary source.

In which of the case, the change of frequency is the same, the magnitude of velocity of source or listener being the same
(a) I and II (b) II and III
(c) III and IV (d) None of these.

111. A source of sound waves moves towards a stationary listener. The apparent pitch of the sound is found to be higher than its actual value. This happens because
(a) wavelength of sound decreases
(b) wavelength of sound increases
(c) the number of wave received by listener increases
(d) the number of waves received by listener decreases.

112. Values of the shortest length L of a closed pipe which resounds to a series of tuning forks giving notes of frequency f are determined. In order to allow the reflection of sound waves a short distance beyond the open end of the tube, the speed of sound in air should be calculated from
(a) the slope of a graph of L against f^1
(b) the slope of a graph of f against L^{-1}
(c) the slope of a graph of f against L
(d) the slope of a graph of f against log 1L

113. A string of length 0.4 m and mass 10^{-2} kg is tightly clamped at its ends. The tension in the string is 1.6 N. Identical wave pulses are produced at one end at equal intervals of time Δt. The minimum value of Δt which allows constructive interference between successive pulses is
(a) 0.05 s (b) 0.10 s
(c) 0.02 s (d) 0.40 s

114. The following two wave trains are approaching each other

$y_1 = a \sin 2000 \pi t$

$y_2 = a \sin 2008 \pi t$

The number of beats heard per second is
(a) 4 (b) zero
(c) 1 (d) 8

115. Two vibrating strings of the same material but lengths L and $2L$ have radii $2r$ and r respectively. They are stretched under the same tension. Both the strings vibrate in their fundamental modes, the one of length L with frequency v_1 and the other with frequency v_2. The ratio $\frac{v_1}{v_2}$ is given by

(a) 4 (b) 8
(c) 2 (d) 1

116. Two trains move towards each other with the same speed. The speed of sound is 340 m/s. If the height of the tone of the whistle of one of them heard on the other changes to 9/8 times, then the speed of each train should be
(a) 20 m/s (b) 2 m/s
(c) 200 m/s (d) 2000 m/s

117. Two stationary sources each emitting waves of wavelength λ. An observer moves from one source to another with velocity U. Then number of beats heard by him is
(a) $\frac{2U}{\lambda}$ (b) $\sqrt{U\lambda}$
(c) $\frac{U}{\lambda}$ (d) $\frac{U}{2\lambda}$

118. With the propagation of a longitudinal wave through a material medium, the quantities transmitted in the propagation direction are
(a) energy, momentum and mass
(b) energy
(c) energy and mass
(d) energy and linear momentum.

119. A siren placed at a railway platform is emitting sound of frequency 5 kHz. A passenger sitting in a moving train A records a frequency of 5.5 kHz while the train approaches the siren. During his return journey in a different train B, he records a frequency of 6.0 kHz while approaching the same siren. The ratio of the velocity of train B to that of train A is
(a) $\frac{242}{252}$ (b) $\frac{11}{6}$
(c) $\frac{5}{6}$ (d) 2

120. The frequency of sinusoidal wave $y = 0.4\cos[2000t + 0.80]$ would be
(a) $1000\,\pi$ Hz (b) 2000 Hz
(c) 20 Hz (d) $\frac{1000}{\pi}$ Hz

121. A sonometer wire resonates with a given tuning fork forming standing waves with five antinodes between the two bridges when a mass of 9 kg is suspended from the wire. When this mass is replaced by a mass M, the wire resonates with the same tuning fork forming three antinodes for the same positions of the bridges. The value of M is
(a) 5 kg (b) $\frac{1}{25}$ kg
(c) 12.5 kg (d) 25 kg

122. A police car moving at 22 m/s chases a motorcyclist. The police man sounds his horn at 176 Hz, while both of them move towards a stationary siren of frequency 165 Hz. Calculate the speed of the motorcycle, if it is given that no beats are heard.

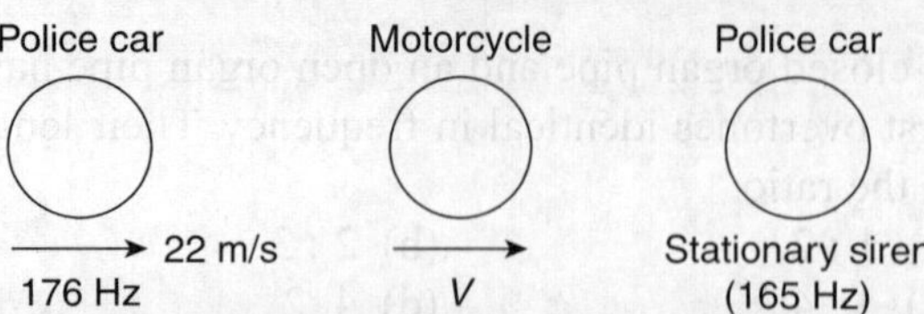

(a) zero (b) 11 m/s
(c) 22 m/s (d) 33 m/s

123. A closed organ pipe of length L and an open organ pipe contain gases of densities ρ_1 and ρ_2 respectively. The compressibility of gases are equal in both the pipes. Both the pipes are vibrating in their first overtone with same frequency. The length of the open organ pipe is
(a) $\frac{4L}{3}\sqrt{\frac{\rho_2}{\rho_1}}$ (b) $\frac{L}{3}$
(c) $\frac{4L}{3}$ (d) $\frac{4L}{3}\sqrt{\frac{\rho_1}{\rho_2}}$

124. If the amplitude of sound is double and the frequency reduced to one fourth, the intensity of sound at the same point will be
(a) increasing by a factor of 2
(b) decreasing by a factor of 2
(c) decreasing by a factor of 4
(d) unchanged.

125. A 5.5 metre length of string has a mass of 0.035 kg. If the tension in the string in 77 N, the speed of a wave on the string is
(a) 110 ms^{-1} (b) 165 ms^{-1}
(c) 77 ms^{-1} (d) 102 ms^{-1}

126. An open pipe is in resonance in 2nd harmonic with frequency f_1. Now one end of the tube is closed and frequency is increased such that the resonance again occurs in nth harmonic. Choose the correct option.
(a) $n = 3,\ f_2 = \frac{5}{4}f_1$ (b) $n = 3,\ f_2 = \frac{3}{4}f_1$
(c) $n = 5,\ f_2 = \frac{5}{4}f_1$ (d) $n = 5,\ f_2 = \frac{3}{4}f_1$

127. The speed of sound in oxygen (O_2) at a certain temperature is 460 ms^{-1}. The speed of sound in helium (He) at the same temperature will be
(Assume both gases to be ideal)
(a) 500 ms^{-1} (b) 1421 ms^{-1}
(c) 330 ms^{-1} (d) 460 ms^{-1}

128. A wave travelling along the *x*-axis is described by the equation $y(x, t) = 0.005 \cos(\alpha x - \beta t)$. If the wavelength and the time period of the wave are 0.08 m and 2.0 s, respectively, then α and β in appropriate units are
(a) $\alpha = \frac{0.08}{\pi}, \beta = \frac{2.0}{\pi}$ (b) $\alpha = \frac{0.04}{\pi}, \beta = \frac{1.0}{\pi}$
(c) $\alpha = 12.50\pi, \beta = \frac{\pi}{2.0}$ (d) $\alpha = 25.00\pi, \beta = \pi$

129. While measuring the speed of sound by performing a resonance column experiment, a student gets the first resonance condition at a column length of 18 cm during winter. Repeating the same experiment during summer, she measure the column length to be *x* cm for the second resonance. Then
(a) $x > 54$ (b) $54 > x > 36$
(c) $36 > x > 18$ (d) $18 > x$

130. A sound absorber attenuates the sound level by 20 dB. The intensity decreases by a factor of
(a) 100 (b) 10000
(c) 10 (d) 1000

131. A whistle producing sound waves of frequencies 9500 Hz and above is approaching a stationary person with speed *v* ms^{-1}. The velocity of sound in air is 300 ms^{-1}. If the person can hear frequencies up to a maximum of 10000 Hz, The maximum value of *v* up to which he can hear the whistle is
(a) $15\sqrt{2}$ ms^{-1} (b) $\frac{15}{\sqrt{2}}$ ms^{-1}
(c) 30 ms^{-1} (d) 15 ms^{-1}

132. A string is stretched between fixed points separated by 75 cm. It is observed to have resonant frequencies of 420 Hz and 315 Hz. There are no other resonant frequencies between these two. Then, the lowest resonant frequency for this string is
(a) 10.5 Hz (b) 1.05 Hz
(c) 105 Hz (d) 1050 Hz

133. When two tuning forks (Fork 1 and Fork 2) are sounded simultaneously, 4 beats per second are heard. Now, some tape is attached on the prong of the Fork 2. When the tuning forks are sounded again, 6 beats per second are heard. If the frequency of Fork 1 is 200 Hz, then what was the original frequency of Fork 2?
(a) 196 Hz (b) 204 Hz
(c) 200 Hz (d) 202 Hz

134. An observer moves towards a stationary source of sound, with a velocity one-fifth of the velocity of sound. What is the percentage increase in the apparent frequency?
(a) 0.5% (b) 20%
(c) 5% (d) zero

135. The displacement *y* of a particle in a medium can be expressed as $y = 10^{-6} \sin(100t + 20x + \pi/4)$ m where *t* is in second and *x* in metres. The speed of the wave is
(a) 200 m/s (b) 5π m/s
(c) 20 m/s (d) 5 m/s

Questions Asked in NEET/CBSE Medicals

136. A tuning fork with frequency 800 Hz produces resonance in a resonance column tube with upper end open and lower end closed by water surface. Successive resonances are observed at length 9.75 cm, 31.25 cm. The speed of sound in air is
(a) 500 m/s (b) 156 m/s
(c) 344 m/s (d) 172 m/s

137. A tuning fork is used to produce resonance in a glass tube. The length of the air column in this tube can be adjusted by a variable piston. At room temperature of 27° C, two successive resonances are produced at 20 cm and 73 cm of column length. If the frequency of the tuning fork is 320 Hz, the velocity of sound in air at 27° C is
(a) 330 ms^{-1} (b) 339 ms^{-1}
(c) 350 ms^{-1} (d) 300 ms^{-1}

138. The fundamental frequency in an open organ pipe is equal to the third harmonic of a closed organ pipe. If the length of the closed organ pipe is 20 cm, the length of the open organ pipe is
(a) 13.2 cm (b) 8 cm
(c) 12.5 cm (d) 16 cm

139. The two nearest harmonics of a tube closed at one end open at other end are 220 Hz and 260 Hz respectively. What is the fundamental frequency of the system?
(a) 20 Hz (b) 30 Hz
(c) 40 Hz (d) 10 Hz

140. Two cars moving in opposite directions approach each other with speed of 22 ms^{-1} and 16.5 ms^{-1} respectively. The driver of the first car blows a horn having a frequency 400 Hz. The frequency heard by the driver of the second car is (velocity of sound is 340 ms^{-1})
(a) 361 Hz (b) 411 Hz
(c) 448 Hz (d) 350 Hz

141. The second overtone of an open organ pipe has the same frequency as the first overtone of a closed pipe *L* metres long. The length of the open pipe will be
(a) *L* (b) 2*L*
(c) $\frac{L}{2}$ (d) 4*L*

142. Three sound waves of equal amplitudes have frequencies $(n-1)$, n, $(n+1)$. They superimpose to give beats. The number of beats produced per second will be

(a) 1 (b) 4
(c) 3 (d) 2

143. A siren emitting a sound of frequency 800 Hz moves away from an observer towards a cliff at a speed of 15 ms^{-1}. Then, the frequency of sound that the observer hears in the echo reflected from the cliff is (Take velocity of sound in air = 330 ms^{-1})

(a) 838 Hz (b) 885 Hz
(c) 765 Hz (d) 800 Hz

144. An air column, closed at one end and open at the other, resonates with a tuning fork when the smallest length of the column is 50 cm. The next larger length of the column resonating with the same tuning fork is

(a) 150 cm (b) 200 cm
(c) 66.7 cm (d) 100 cm

145. A uniform rope of length L and mass m_1 hangs vertically from a rigid support. A block of mass m_2 is attached to the free end of the rope. A transverse pulse of wavelength λ_1 is produced at the lower end of the rope. The wavelength of the pulse when it reaches the top of the rope is λ_2. The ratio λ_2/λ_1 is

(a) $\sqrt{\frac{m_2}{m_1}}$ (b) $\sqrt{\frac{m_1+m_2}{m_1}}$

(c) $\sqrt{\frac{m_1}{m_2}}$ (d) $\sqrt{\frac{m_1+m_2}{m_2}}$

146. A string is stretched between fixed points separated by 75.0 cm. It is observed to have resonant frequencies of 420 Hz and 315 Hz. There are no other resonant frequencies between these two. The lowest resonant frequency for this string is

(a) 10.5 Hz (b) 105 Hz
(c) 155 Hz (d) 205 Hz

147. 4.0 g of a gas occupies 22.4 liters at NTP. The specific heat capacity of the gas at constant volume is 5.0 JK^{-1}mol^{-1}. If the speed of sound in this gas at NTP is 952 ms^{-1}, then the heat capacity at constant pressure is (Take gas constant R = 8.3 J K^{-1}mol^{-1})

(a) 7.0 J K^{-1}mol^{-1} (b) 8.5 J K^{-1}mol^{-1}
(c) 8.0 J K^{-1}mol^{-1} (d) 7.5 J K^{-1}mol^{-1}

148. A source of sound S emitting waves of frequency 100 Hz and an observer O are located at some distance from each other. The source is moving with a speed of 19.4 ms^{-1} at an angle of 60° with the source observer line as shown in the figure. The observer is at rest. The apparent frequency observed by the observer (velocity of sound in air 330 ms^{-1}) is

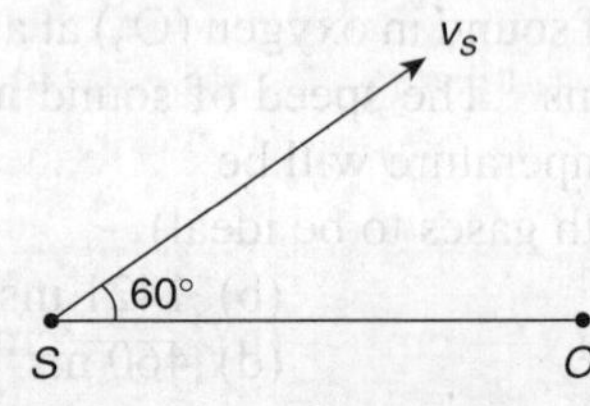

(a) 106 Hz (b) 97 Hz
(c) 100 Hz (d) 103 Hz

149. The fundamental frequency of a closed organ pipe of length 20 cm is equal to the second overtone of an organ pipe open at both the ends. The length of organ pipe open at both the ends is

(a) 120 cm (b) 140 cm
(c) 80 cm (d) 100 cm

150. The number of possible natural oscillations of air column in a pipe closed at one end of length 85 cm whose frequencies lie below 1250 Hz are (Velocity of sound = 340 ms^{-1})

(a) 4 (b) 5
(c) 7 (d) 6

151. A speeding motorcyclist sees traffic jam ahead him. He slows down to 36 km hour^{-1}. He finds that traffic has eased and a car moving ahead of him at 18 km hour^{-1} is honking at a frequency of 1392 Hz. If the speed of sound is 343 ms^{-1}, the frequency of the honk as heard by him will be

(a) 1332 Hz (b) 1372 Hz
(c) 1412 Hz (d) 1454 Hz

152. If we study the vibration of a pipe open at both ends, then the following statements is not true:

(a) All harmonics of the fundamental frequency will be generated
(b) Pressure change will be maximum at both ends
(c) Open end will be actinide
(d) Odd harmonic of the fundamental frequency will be generated.

153. A wave travelling in the positive x-direction having displacement along y-direction as 1 m, wavelength 2 m and frequency of $\frac{1}{\pi}$ Hz is represented by

(a) $y = \sin(10\pi x - 20\pi t)$
(b) $y = \sin(2\pi x + 2\pi t)$
(c) $y = \sin(x - 2t)$
(d) $y = \sin(2\pi x - 2\pi t)$

154. A source of unknown frequency gives 4 beats/s when sounded with a source of known frequency 250 Hz. The second harmonic of the source of unknown frequency gives five beats per second, when sounded with a source of frequency 513 Hz. The unknown frequency is

(a) 240 Hz (b) 260 Hz
(c) 254 Hz (d) 246 Hz

155. The length of the wire between two ends of a sonometer is 100 cm. What should be the positions of two bridges below the wire so that the three segments of the wire have their fundamental frequencies in the ratio 1 : 3 : 5.

(a) $\frac{1500}{23}$ cm, $\frac{500}{23}$ cm (b) $\frac{1500}{23}$ cm, $\frac{300}{23}$ cm

(c) $\frac{300}{23}$ cm, $\frac{1500}{23}$ cm (d) $\frac{1500}{23}$ cm, $\frac{2000}{23}$ cm

156. Two sources P and Q produce notes of frequency 660 Hz each. A listener moves from P to Q with a speed of 1 ms^{-1}. If the speed of sound is 330 m/s, then the number of beats heard by the listener per second will be

(a) 4 (b) 8
(c) 2 (d) zero.

157. When a string is divided into three segments of length l_1, l_2 and l_3 the fundamental frequencies of these three segments are ν_1, ν_2 and ν_3 respectively. The original fundamental frequency (ν) of the string is

(a) $\sqrt{\nu} = \sqrt{\nu_1} + \sqrt{\nu_2} + \sqrt{\nu_3}$

(b) $\nu = \nu_1 + \nu_2 + \nu_3$

(c) $\frac{1}{\nu} = \frac{1}{\nu_1} + \frac{1}{\nu_2} + \frac{1}{\nu_3}$

(d) $\frac{1}{\sqrt{\nu}} = \frac{1}{\sqrt{\nu_1}} + \frac{1}{\sqrt{\nu_2}} + \frac{1}{\sqrt{\nu_3}}$

158. Two sources of sound, placed close to each other, are emitting progressive waves given by

$y_1 = 4\sin 600\pi t$ and $y_2 = 5\sin 60\pi t$

An observer located near these two sources of sound will hear

(a) 4 beats per second with intensity ratio 25 : 16 between waxing and waning
(b) 8 beats per second with intensity ratio 25 : 16 between waxing and waning
(c) 8 beats per second with intensity ratio 81 : 1 between waxing and waning
(d) 4 beats per second with intensity ratio 81 : 1 between waxing and waning.

159. The equation of a simple harmonic wave is given by

$$y = 3\sin\frac{\pi}{2}(50t - x)$$

where x and y are in metres and t is in seconds. The ratio of maximum particle velocity to the wave velocity is

(a) 2π (b) $\frac{3}{2}\pi$

(c) 3π (d) $\frac{2}{3}\pi$

160. A train moving at a speed of 220 ms^{-1} towards a stationary object, emits a sound of frequency 1000 Hz. Some of the sound reaching the object gets reflected back to the train as echo. The frequency of the echo as detected by the driver of the train is
(Speed of sound in air is 330 ms^{-1})

(a) 3500 Hz (b) 4000 Hz
(c) 5000 Hz (d) 3000 Hz

161. Two waves are represented by the equations

$$y_1 = a\sin(\omega t + kx + 0.57)\,\text{m and } y_2 = a\cos(\omega t + kx)$$

where x is in metres and t in sec. The phase difference between them is

(a) 1.0 radian (b) 1.25 radian
(c) 1.57 radian (d) 0.57 radian

162. Sound waves travel at 350 m/s through a warm air and at 3500 m/s through brass. The wavelength of a 700 Hz acoustic wave as it enters brass from warm air

(a) decreases by a factor 10
(b) increases by a factor 20
(c) increases by a factor 10
(d) decreases by a factor 20.

163. Two identical piano wires, kept under the same tension T, have a fundamental frequency of 600 Hz. The fractional increase in the tension of one of the wires which will lead to occurrence of 6 beats/s when both the wires oscillate together would be

(a) 0.01 (b) 0.02
(c) 0.03 (d) 0.04

164. A transverse wave is represented by $y = A\sin(\omega t - kx)$. For what value of the wavelength is the wave velocity equal to the maximum particle velocity?

(a) $\pi A/2$ (b) πA
(c) $2\pi A$ (d) A

165. A tuning fork of frequency 512 Hz makes 4 beats per second with the vibrating string of a piano. The beat frequency decreases to 2 beats per sec when the tension in the piano string is slightly increased. The frequency of the piano string before increasing the tension was

(a) 510 Hz (b) 514 Hz
(c) 516 Hz (d) 508 Hz

166. Each of the two strings of length 51.6 cm and 49.1 cm are tensioned separately by 20 N force. Mass per unit length of both the strings is same and equal to 1 g/m. When both the strings vibrate simultaneously the number of beats is

(a) 7 (b) 8
(c) 3 (d) 5

167. The driver of a car travelling with speed 30 m/s towards a hill sounds a horn of frequency 600 Hz. If the velocity

of sound in air is 330 m/s, the frequency of reflected sound as heard by the driver is
(a) 555.5 Hz (b) 720 Hz
(c) 500 Hz (d) 550 Hz

168. A wave in a string has amplitude of 2 cm. The wave travels in the +ve direction of x axis with a speed of 128 m/s and it is noted that 5 complete waves fit in 4 m length of the string. The equation describing the wave is
(a) $y = (0.02)\text{m} \sin (15.7x - 2010t)$
(b) $y = (0.02)\text{m} \sin (15.7 + 2010t)$
(c) $y = (0.02)\text{m} \sin (7.85x - 1005t)$
(d) $y = (0.02)\text{m} \sin (7.85x + 1005t)$

169. A point performs simple harmonic oscillation of period T and the equation of motion is given by $x = a \sin(\omega t + \pi/6)$. After the elapse of what friction of the time period, the velocity of the point will be equal to half of its maximum velocity?
(a) $T/3$ (b) $T/12$
(c) $T/8$ (d) $T/6$

170. Two periodic waves of intensities I_1 and I_2 pass through a region at the same time in the same direction. The sum of the maximum and minimum intensities is
(a) $\left(\sqrt{I_1} - \sqrt{I_2}\right)^2$ (b) $2(I_1 + I_2)$
(c) $I_1 + I_2$ (d) $\left(\sqrt{I_1} + \sqrt{I_2}\right)^2$

171. The wave described by $y = 0.25 \sin (10\pi x - 2\pi t)$, where x and y are in metres and t in seconds, is a wave travelling along the
(a) +ve x direction with frequency 1 Hz and wavelength $\lambda = 0.2$ m.
(b) -ve x direction with amplitude 0.25 m and wavelength $\lambda = 0.2$ m
(c) –ve x direction with frequency 1 Hz
(d) +ve x direction with frequency π Hz and wavelength $\lambda = 0.2$ m.

172. Two vibrating tuning forks produce waves given by $y_1 = 4 \sin 500\pi t$ and $y_2 = 2 \sin 506\pi t$. Number of beats produced per minute is
(a) 360 (b) 180
(c) 60 (d) 3

173. Two sound waves with wavelengths 5.0 m and 5.5 m respectively, each propagate in a gas with velocity 330 m/s. We expect the following number of beats per second:
(a) 6 (b) 12
(c) 0 (d) 1

174. The time of reverberation of a room A is one second. What will be the time (in seconds) of reverberation of a room, having all the dimensions double of those of room A?
(a) 1 (b) 2
(c) 4 (d) 1/2

175. A transverse wave propagating along x-axis is represented by $y(x, t) = 8.0\sin (0.5\pi x - 4\pi t - \pi/4)$ where x is in metres and t is in seconds. The speed of the wave is
(a) 8 m/s (b) 4 m/s
(c) 0.5 m/s (d) m/s

176. Which one of the following statements is true?
(a) Both light and sound waves can travel in vacuum
(b) Both light and sound waves in air are transverse
(c) The sound waves in air are longitudinal while the light waves are transverse
(d) Both light and sound waves in air are longitudinal.

177. A point source emits sound equally in all directions in a non-absorbing medium. Two points P and Q are at distance of 2 m and 3 m respectively from the source. The ratio of the intensities of the waves at P and Q is
(a) 3 : 2 (b) 2 : 3
(c) 9 : 4 (d) 4 : 9

178. A closed organ pipe (closed at one end) is excited to support the third overtone. It is found that air in the pipe has
(a) three nodes and three antinodes
(b) three nodes and four antinodes
(c) four nodes and three antinodes
(d) four nodes and four antinodes.

179. A car is moving towards a high cliff. The driver sounds a horn of frequency f. The reflected sound heard by the driver has frequency $2f$. If v is the velocity of sound, then the velocity of the car, in the same velocity units, will be
(a) $v/\sqrt{2}$ (b) $v/3$
(c) $v/4$ (d) $v/2$

180. An observer moves towards a stationary source of sound with a speed 1/5th of the speed of sound. The wavelength and frequency of the source emitted are λ and f respectively. The apparent frequency and wavelength recorded by the observer are respectively
(a) $1.2f$, 1.2λ (b) $1.2f$, λ
(c) f, 1.2λ (d) $0.8f$, 0.8λ

181. A whistle revolves in a circle with angular speed ω = 20 rad/s using a string of length 50 cm. If the frequency of sound from the whistle is 385 Hz, then what is the minimum frequency heard by an observer which is far away from the centre (velocity of sound = 340 m/s)
(a) 385 Hz (b) 374 Hz
(c) 394 Hz (d) 333 Hz

182. A wave travelling in positive x-direction with $a = 0.2 \text{ ms}^{-2}$, velocity = 360 ms^{-1} and $\lambda = 60$ m, then correct expression for the wave is

(a) $y = 0.2 \sin\left[2\pi\left(6t + \frac{x}{60}\right)\right]$

(b) $y = 0.2 \sin\left[\pi\left(6t + \frac{x}{60}\right)\right]$

(c) $y = 0.2 \sin\left[2\pi\left(6t - \frac{x}{60}\right)\right]$

(d) $y = 0.2 \sin\left[\pi\left(6t - \frac{x}{60}\right)\right]$

183. If the equation of a wave is represented by $y = 10^{-4} \sin\left(100t - \frac{x}{10}\right)$ m, then the velocity of wave will be

(a) 100 m/s (b) 4 m/s
(c) 1000 m/s (d) 10 m/s

184. Two waves are having equation

$$x_1 = a\sin(\omega t - kx + \phi_1)$$
$$x_2 = a\sin(\omega t - kx + \phi_2)$$

If in the resultant wave, the frequency and amplitude remain equal to amplitude of superimposing waves, the phase difference between them is

(a) $\frac{\pi}{6}$ (b) $\frac{2\pi}{3}$
(c) $\frac{\pi}{4}$ (d) $\frac{\pi}{3}$

185. If the tension and diameter of a sonometer wire of fundamental frequency n is doubled and density is halved then its fundamental frequency will become

(a) $\frac{n}{4}$ (b) $\sqrt{2}n$
(c) n (d) $\frac{n}{\sqrt{2}}$

186. The equations of two waves acting in perpendicular directions are given as

$x = a\cos(\omega t + \delta)$ and $y = a\cos(\omega t + \alpha)$, where $\delta = \alpha + \frac{\pi}{2}$, the resultant wave represents

(a) a parabola (b) a circle
(c) an ellipse (d) a straight line.

187. A string is cut into three parts, having fundamental frequencies n_1, n_2, n_3 respectively. Then the original fundamental frequency n is related by the expression as

(a) $\frac{1}{n} = \frac{1}{n_1} + \frac{1}{n_2} + \frac{1}{n_3}$ (b) $n = n_1 \times n_2 \times n_3$
(c) $n = n_1 + n_2 + n_3$ (d) $n = \frac{n_1 + n_2 + n_3}{3}$

188. Two stationary sources each emitting waves of wavelength λ, an observer moves from one source to another with velocity u. Then number of beats heard by him is

(a) $\frac{2u}{\lambda}$ (b) $\frac{u}{\lambda}$
(c) $\sqrt{u\lambda}$ (d) $\frac{u}{2\lambda}$

189. Two waves of length 50 cm and 51 cm produced 12 beats per sec. The velocity of sound is

(a) 340 m/s (b) 331 m/s
(c) 306 m/s (d) 360 m/s

190. A transverse wave is represented by the equation

$$y = y_0 \sin\frac{2\pi}{\lambda}(vt - x)$$

For what value of λ, is the maximum particle velocity equal to two times the wave velocity?

(a) $\lambda = \frac{\pi y_0}{2}$ (b) $\lambda = \frac{\pi y_0}{3}$
(c) $\lambda = 2\pi y_0$ (d) $\lambda = \pi y_0$

191. A vehicle, with a horn of frequency n is moving with a velocity of 30 m/s in a direction perpendicular to the straight line joining the observer and the vehicle. The observer perceives the sound to have a frequency $n + n_1$. Then (if the sound velocity in air is 300 m/s)

(a) $n_1 = 0.1n$ (b) $n_1 = 0$
(c) $n_1 = 10n$ (d) $n_1 = -0.1n$

192. A standing wave having 3 nodes and 2 antinodes is formed between two atoms having a distance 1.21 Å between them. The wavelength of the standing wave is

(a) 6.05 Å (b) 2.42 Å
(c) 1.21 Å (d) 3.63 Å

193. In a sinusoidal wave, the time required for a particular point to move from maximum displacement to zero displacement is 0.170 s. The frequency of the wave is

(a) 0.73 Hz (b) 0.36 Hz
(c) 1.47 Hz (d) 2.94 Hz

194. Standing waves are produced in a 10 m long stretched string. If the string vibrates in 5 segments and the wave velocity is 20 m/s, the frequency is

(a) 5 Hz (b) 10 Hz
(c) 2 Hz (d) 4 Hz

195. A cylindrical tube, open at both ends has fundamental frequency f in air. The tube is dipped vertically in water, so that half of it is in water. The fundamental frequency of air column is now

(a) $f/2$ (b) $3f/4$
(c) $2f$ (d) f

196. The equation of a sound wave is
$y = 0.0015 \sin (62.4x + 316t)$.
The wavelength of this wave is

(a) 0.3 unit (b) 0.2 unit
(c) 0.1 unit (d) Cannot be calculated.

197. Two sound waves having a phase difference of 60° have path difference of

(a) $\frac{\lambda}{6}$ (b) $\frac{\lambda}{3}$
(c) 2λ (d) $\frac{\lambda}{2}$

198. The length of a sonometer wire AB is 110 cm. Where should the two bridges be placed from A to divide the wire in 3 segments whose fundamental frequencies are in the ratio of 1 : 2 : 3?

(a) 60 cm and 90 cm (b) 30 cm and 60 cm
(c) 30 cm and 90 cm (d) 40 cm and 80 cm

199. A hospital uses an ultrasonic scanner to locate tumors in a tissue. The operating frequency of the scanner is 4.2 MHz. The speed of sound in a tissue is 1.7 km/s. The wavelength of sound in the tissue is close to

(a) 4×10^{-3} m (b) 8×10^{-3} m
(c) 4×10^{-4} m (d) 8×10^{-4} m

200. A source of sound gives 5 beats per second, when sounded with another source of frequency 100 second^{-1}. The second harmonic of the source, together with a source of frequency 205 sec^{-1}, gives 5 beats per second. When is the frequency of the source?

(a) 105 second^{-1} (b) 205 second^{-1}
(c) 95 second^{-1} (d) 100 second^{-1}

201. A star, which is emitting radiation at a wavelength of 5000 Å, is approaching the earth with a velocity of 1.5×10^{4} m/s. The change in wavelength of the radiation, as received on the earth, is

(a) 25 Å (b) 100 Å
(c) zero (d) 2.5 Å

202. Velocity of sound waves in air is 330 m/s. For a particular sound in air, a path difference of 40 cm is equivalent to phase difference of 1.6π. The frequency of this wave is

(a) 165 Hz (b) 150 Hz
(c) 660 Hz (d) 330 Hz

203. A wave of frequency 100 Hz travels along a string towards its fixed end. When this wave travels back, after reflection, a node is formed at a distance of 10 cm from the fixed end. The speed of the wave (incident and reflected) is

(a) 20 m/s (b) 40 m/s
(c) 5 m/s (d) 10 m/s

204. A stationary wave is represented by $y = A \sin (100t) \cos (0.01x)$, where y and A are in millimetres, t is in seconds and x is in metres. The velocity of the wave is

(a) 10^4 m/s (b) not derivable
(c) 1 m/s (d) 10^2 m/s

205. A source of frequency v gives 5 beats/second when sounded with a source of frequency 200 Hz. The second harmonic of frequency $2v$ of source gives 10 beats/second when sounded with a source of frequency 420 Hz. The value of v is

(a) 205 Hz (b) 195 Hz
(c) 200 Hz (d) 210 Hz

206. A wave has simple harmonic motion whose period is 4 seconds while another wave which also possesses simple harmonic motion has its period of 3 seconds. It both are combined, then the resultant wave will have the period equal to

(a) 4 s (b) 5 s
(c) 12 s (d) 3 s

207. A stretched string resonates with tuning fork frequency 512 Hz when length of the string is 0.5 m. The length of the string required to vibrate resonantly with a tuning fork of frequency 256 Hz would be

(a) 0.25 m (b) 0.5 m
(c) 1 m (d) 2 m

208. The temperature at which the speed of sound becomes double as was at 27° C is

(a) 273° C (b) 0° C
(c) 927° C (d) 1027° C

209. For production of beats the two sources must have

(a) different frequencies and same amplitude
(b) different frequencies
(c) different frequencies, same amplitude and same phase
(d) different frequencies and same phase.

Answer Keys

Exercises

1. (a) 2. (a) 3. (a) 4. (a) 5. (d) 6. (d) 7. (c) 8. (b) 9. (b) 10. (b)
11. (c) 12. (b) 13. (c) 14. (a) 15. (b) 16. (d) 17. (d) 18. (d) 19. (c) 20. (c)
21. (a) 22. (c) 23. (a) 24. (b) 25. (b) 26. (b) 27. (a) 28. (c) 29. (d) 30. (c)
31. (b) 32. (a) 33. (a) 34. (c) 35. (a) 36. (c) 37. (b) 38. (a) 39. (a) 40. (c)
41. (b) 42. (b) 43. (a) 44. (b) 45. (c) 46. (a) 47. (a) 48. (d) 49. (c) 50. (c)
51. (d) 52. (c) 53. (b) 54. (a) 55. (a) 56. (a) 57. (a) 58. (b) 59. (a) 60. (c)
61. (d) 62. (c) 63. (d) 64. (d) 65. (d) 66. (c) 67. (d) 68. (a) 69. (b) 70. (b)
71. (c) 72. (a) 73. (b) 74. (b) 75. (d) 76. (b) 77. (d) 78. (c) 79. (d) 80. (d)
81. (a) 82. (a) 83. (d) 84. (c) 85. (b) 86. (c) 87. (b) 88. (c) 89. (b) 90. (d)
91. (c) 92. (b) 93. (a) 94. (a) 95. (c) 96. (c) 97. (b) 98. (c) 99. (b) 100. (b)
101. (c) 102. (c) 103. (a) 104. (d) 105. (b) 106. (a) 107. (a) 108. (d) 109. (d) 110. (d)
111. (c) 112. (a) 113. (b) 114. (a) 115. (d) 116. (a) 117. (a) 118. (b) 119. (b) 120. (d)
121. (d) 122. (c) 123. (d) 124. (c) 125. (a) 126. (c) 127. (b) 128. (d) 129. (a) 130. (a)
131. (d) 132. (c) 133. (a) 134. (b) 135. (d)

Questions asked in NEET/CBSE Medicals

136. (c) 137. (b) 138. (a) 139. (a) 140. (c) 141. (b) 142. (d) 143. (a) 144. (a) 145. (d)
146. (b) 147. (c) 148. (d) 149. (a) 150. (d) 151. (c) 152. (b) 153. (c) 154. (c) 155. (d)
156. (a) 157. (c) 158. (d) 159. (b) 160. (c) 161. (a) 162. (c) 163. (b) 164. (c) 165. (d)
166. (a) 167. (b) 168. (c) 169. (b) 170. (b) 171. (a) 172. (b) 173. (a) 174. (b) 175. (a)
176. (c) 177. (c) 178. (d) 179. (b) 180. (b) 181. (b) 182. (c) 183. (c) 184. (b) 185. (c)
186. (b) 187. (a) 188. (a) 189. (c) 190. (d) 191. (b) 192. (c) 193. (c) 194. (a) 195. (d)
196. (c) 197. (a) 198. (a) 199. (c) 200. (a) 201. (a) 202. (c) 203. (a) 204. (a) 205. (a)
206. (c) 207. (c) 208. (c) 209. (b)

Solutions

1. (a)
λ, x and A all have the dimensions of length.

2. (a)

$$\frac{I_1}{I_2}=\frac{r_2^2}{r_1^2}=\frac{3^2}{2^2}=\frac{9}{4}$$

3. (a)
The frequency of P is 250 ± 4, i.e., 254 or 246
If the prongs of P are filled, the frequency increases, Therefore, there are two possibilities, the frequency of P is either $254+n$ or $246+n$ is ruled out as the number of beats decreases. The frequency of P is 246 Hz.

4. (a)
The option (a) is dimensionally satisfied.

5. (d)

$$v'=v\frac{c}{c-\frac{c}{2}}$$

$$v'=3\times 2=6 \text{ kHz}$$

6. (d)
Source velocity $=\omega r$

$$=20\times 10=200 \text{ rad/s}$$

For minimum frequency the source should move away from the observer.

$$v'=320\times\frac{c}{c+200}=\frac{320\times 330}{530}=220 \text{ Hz}$$

7. (c)
Consider the wall as observer.

$$v''=v'\left(\frac{330+20}{330}\right)$$

$$v''=\frac{480\times 330}{310}\times\frac{350}{330}\approx 540 \text{ Hz}$$

Now consider the wall as a source of frequency v and the man in the car as observer.

8. (b)
Beats are produced when waves having slightly different frequencies travelling in the same direction superpose.

9. (b)

P_1 vibrating in first overtone

$$\nu_2 = \frac{3v}{4l_1}$$

P_2 vibrating in second overtone

$$\nu_3 = \frac{3v}{2l_2}$$

$$\nu_2 = \nu_3$$

$$\Rightarrow \quad \frac{3v}{4l_1} = \frac{3v}{2l_2}$$

$$\frac{l_1}{l_2} = \frac{1}{2}$$

10. (b)

1st case, $\frac{\lambda}{4} = 13$

$\lambda = 52$

2nd case, $\frac{3\lambda}{4} = 41$

$\lambda = \frac{4 \times 41}{3} = 54.6$

3rd case, $\frac{5\lambda}{4} = 69$

$\lambda = \frac{4 \times 69}{5} = 55.2$

$$\text{Average } \lambda = \frac{52 + 54.6 + 55.2}{3} = 53.7 \text{ cm}$$

$$\nu = \frac{v}{\lambda} = \frac{350}{53.7 \times 10^{-2}} \approx 625 \text{ Hz}$$

11. (c)

The apparent frequency observed by the man due to approaching train

$$\nu_1 = 240\left(\frac{330}{330-4}\right)$$

The apparent frequency observed by the man due to receding train

$$\nu_2 = 240\left(\frac{330}{330+4}\right)$$

$$\text{Beats} = \nu_1 - \nu_2 = 240 \times 330\left(\frac{1}{326} - \frac{1}{334}\right) \approx 6$$

12. (b)

The standard equation of a progressive wave is

$$y = a\sin\left[2\pi\left(\frac{t}{T} - \frac{x}{\lambda}\right) + \phi\right]$$

The given equation can be written as

$$y = 4\sin\left[2\pi\left(\frac{t}{10} - \frac{x}{18}\right) + \frac{\pi}{6}\right]$$

$\therefore a = 4$ cm, $T = 10$ s,

$\lambda = 18$ cm

13. (c)

$$v = \frac{\lambda}{T}$$

$$25 = \frac{100}{T}$$

$$\Rightarrow T = \frac{100}{25} = 4 \text{ s}$$

14. (a)

Open pipe,
Fundamental frequency

$$\nu = \frac{v}{2l}$$

Closed pipe,
Third harmonic frequency

$$\nu' = \frac{3v}{4l}$$

$$100 + \nu = \frac{3\nu}{2}$$

$$\Rightarrow \quad \nu = 200 \text{ Hz}$$

15. (b)

$$x_1 = x_0 \sin 2\pi \times 323t$$

$$x_2 = x_0 \sin 2\pi \times 326t$$

$$\nu_1 = 323; \nu_2 = 326$$

$$\text{Beat frequency } = \nu_2 - \nu_1 = 326 - 323 = 3$$

16. (d)

The given equation of standing wave is

$$y = 5\sin\left(\frac{2\pi x}{3}\right)\cos 20\pi t$$

Comparing this with standard standing wave equation

$$y = 2a\sin\frac{2\pi x}{\lambda}\cos\frac{2\pi vt}{\lambda}$$

We find

$$\lambda = 3$$

The distance between two adjacent nodes is

$$\frac{\lambda}{2} = \frac{3}{2} = 1.5 \text{ cm}$$

17. (d)

$$v = \sqrt{\frac{\gamma P}{\rho}}$$

Both the gases are diatomic, therefore λ for the both the gases is same. The pressure is same. Therefore,

$$\frac{v_O}{v_H} = \sqrt{\frac{\rho H}{\rho O}} = \sqrt{\frac{2}{32}} = \frac{1}{4}$$

18. (d)

$$\nu' = \nu\left(\frac{c \pm v_0}{c \pm v_s}\right)$$

$$\nu' = 500\left(\frac{330}{330-30}\right)$$

$$= \frac{500 \times 330}{300} = 550 \text{ Hz}$$

19. (c)

The given wave equation is

$$y = a\ \cos(kx - \omega t) \qquad \text{(i)}$$

Let us assume that the equation of other wave is either

$$y = -a\ \cos(kx + \omega t) \qquad \text{(ii)}$$

or

$$y = a\ \cos(kx + \omega t) \qquad \text{(iii)}$$

Superposing wave given by Eq. (i) with that given by Eq. (ii)

$$y = a\ \cos(kx - \omega t) - a\ \cos(kx + \omega t)$$

$$y = 2a\ \sin kx\ \sin \omega t \qquad \text{(iv)}$$

Similarly superposing the wave given by Eq. (i) with that given by Eq. (iii), on superposing gives

$$y = 2a\ \cos\ kx\ \cos \omega t \qquad \text{(v)}$$

Substituting $x = 0$ in Equation (iv) gives $y = 0$ whereas if we put $x = 0$ in Equation (v), we get

$$y = 2a\ \cos\ \omega t$$

Therefore, the equation (ii) is the equation of unknown wave.

20. (c)

Open organ pipe

$$\nu_1 = \frac{v}{2l}$$

Closed organ pipe

$$\nu_2 = \frac{v}{4l}$$

$$\nu_1 : \nu_2 = \frac{v}{2l} : \frac{v}{4l} = \frac{1}{2} : \frac{1}{4} = 2:1$$

21. (a)

$$\nu' = \nu\frac{(c \pm v_0)}{(c \pm v_s)}$$

$$= 240\left(\frac{330+11}{330}\right) = 240 \times \frac{341}{330} = 248 \text{ Hz}$$

22. (c)

$$v = \sqrt{\frac{\gamma P}{\rho}} = \sqrt{\frac{\gamma RT}{M}}$$

$$v_N = \sqrt{\frac{7RT}{5M_N}}$$

$$v_{He} = \sqrt{\frac{5RT}{3M_{He}}}$$

$$\frac{v_N}{v_{He}} = \sqrt{\frac{7 \times 3}{5 \times 5} \times \frac{4}{28}} = \frac{\sqrt{3}}{5}$$

23. (a)

The standard form of wave equation is

$$y = a \sin \omega\left(t - \frac{x}{v}\right)$$

The given equation is

$$y = 0.07 \sin(12\pi x - 3000\pi t)$$

$$y = 0.07 \sin 3000\pi t\left(\frac{x}{250} - t\right)$$

Comparing this with standard wave equation, we find

$$v = 250 \text{ m/s}$$

$$\omega = 3000\pi$$

$$\nu = 1500 \text{ c/s}$$

$$\lambda = \frac{v}{\nu} = \frac{250}{1500} = \frac{1}{6} \text{ m}$$

24. (b)

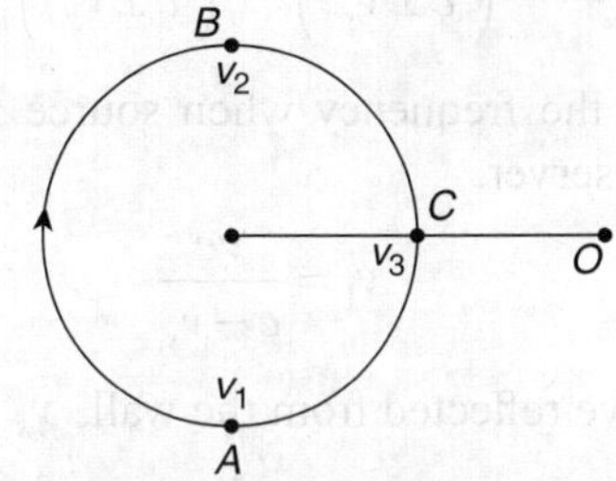

When the source is at B, the source is approaching towards the observer and when it is at A, it is receding from the observer. Therefore $\nu_2\ \nu_1$. In situation C, the source is moving perpendicular to observer

$$\nu_2 > \nu_3 > \nu_1$$

25. (b)

The given equation is $y = y_0 \sin 2\pi\left(ft - \frac{x}{\lambda}\right)$

Converting this in standard form x

$$y = y_0 \sin 2\pi f\left(t - \frac{x}{f\lambda}\right)$$

Wave velocity is $f\lambda$.

Particle velocity is ωA

$$= 2\pi f y_0$$

$$2\pi f y_0 = 4f\lambda$$

$$\Rightarrow \quad \lambda = \frac{\pi y_0}{2}$$

26. (b)

The speed of sound in a gas is given by

$$v = \sqrt{\frac{\gamma P}{\rho}} = \sqrt{\frac{\gamma RT}{M}}$$

$$v_1 = \sqrt{\frac{\gamma R(273+27)}{M}} = v\lambda \qquad \text{(i)}$$

$$v_2 = \sqrt{\frac{\gamma R(273+31)}{M}} = (v+x)\lambda \qquad \text{(ii)}$$

From Eqs. (i) and (ii)

$$\frac{v_1}{v_2} = \sqrt{\frac{300}{304}} = \frac{v}{v+x} = \left(\frac{300}{300+4}\right)^{1/2}$$

$$1 + \frac{x}{v} = \left(1 + \frac{4}{300}\right)^{-1/2}$$

$$x = \frac{4 \times 300}{600} = 2$$

27. (a)

$$v' = v\left(\frac{c \pm v_0}{c \pm v_s}\right) = v\left(\frac{c}{c \pm v_s}\right)$$

Let v_1 be the frequency when source is approaching towards observer.

$$v_1 = \frac{vc}{c - v_s}$$

For the wave reflected from the wall, v_s is negative

$$v_2 = \frac{vc}{c + v_s}$$

Beat frequency $= v_1 - v_2$

$$= vc\left(\frac{1}{c - v_2} - \frac{1}{c + v_2}\right)$$

$$= v_c \frac{(2v_s)}{c^2 - v_s^2}$$

$$= \frac{256 \times 330 \times 2 \times 5}{(330)^2 - 5^2} = 7.8 \text{ Hz}$$

28. (c)

Maximum particle velocity $= \omega A$

Let wave velocity be $v = v\lambda$

$$\omega A = v\lambda$$

$$A = \frac{v\lambda}{\omega} = \frac{v\lambda}{2\pi v} = \frac{\lambda}{2\pi}$$

29. (d)

$$v_1 = v\left(\frac{300}{300 - v_s}\right)$$

$$v_2 = v\left(\frac{300}{300 + v_s}\right)$$

$$\left(\frac{v_1 - v_2}{v}\right) \times 100 = \left(\frac{300}{300 - v_s} - \frac{300}{300 + v_s}\right) \times 100$$

$$2 = 300\left(\frac{1}{300 - v_s} - \frac{1}{300 + v_s}\right) \times 100$$

$$2 = \frac{300 \times 2v_s \times 100}{(300)^2 - (v_s)^2}$$

$$v_s{}^2(300)^2 - v_s{}^2 = 300 \times 100 v_s$$

$$\Rightarrow \quad v_s = 3 \text{ m/s}$$

30. (c)

Frequency $v = 540$ Hz

Source velocity, $v_s = 12 \times 0.6 = 9$ m/s

For highest frequency source should approach towards observer.

$$v' = v\left(\frac{300}{300 - 9}\right)$$

$$= \frac{540 \times 300}{291} = 555 \text{ Hz}$$

For lowest frequency source should move away from observer.

$$v' = 540\left(\frac{300}{300 + 9}\right)$$

$$= \frac{540 \times 300}{309} = 526 \text{ Hz}$$

31. (b)

$$\in(x,t) = 2(\sin\pi + \cos\pi\, t)$$

Converting this in the form $\in(x,t) = a \sin \omega t$

$$\in(x,t)=2\sqrt{2}\left(\frac{1}{\sqrt{2}}\sin\pi t+\cos\pi t\times\frac{1}{\sqrt{2}}\right)$$

$$=2\sqrt{2}\left(\cos\frac{\pi}{4}.\sin\pi t+\cos\pi t\sin\frac{\pi}{4}\right)$$

$$=2\sqrt{2}\sin(\pi t+\pi/4)$$

32. (a)

$$v=\nu\lambda$$

$$\lambda=\frac{340}{340}=1\text{ m}$$

For first resonance, the length of resonance column

$$l_1=\frac{\lambda}{4}=\frac{1}{4}\text{m}=25\text{ cm}$$

For second, resonance the length of resonance column

$$l_2=\frac{3\lambda}{4}=\frac{3}{4}\text{m}=75\text{ cm}$$

For third resonance, the length of resonance column

$$l_3=\frac{5\lambda}{4}=\frac{5\times1\text{ m}}{4}=125\text{ cm}$$

Third resonance is not possible as the length of the tube is 120 cm.
So minimum height of water $=120-75=45$ cm

33. (a)

$$\nu=\frac{1}{2l}\sqrt{\frac{T}{m}}$$

$$256=\frac{1}{2\times0.6}\sqrt{\frac{T}{m}} \quad \text{(i)}$$

$$\nu=\frac{1}{2\times0.15}\sqrt{\frac{T}{m}} \quad \text{(ii)}$$

Substituting the value of $\sqrt{\frac{T}{m}}$, from Eq. (i)

$$\nu=\frac{256\times2\times0.6}{2\times0.15}=1024\text{ Hz.}$$

34. (c)

$$y=\cos\pi(100t-x)$$

$$=100\pi\left(t-\frac{x}{100}\right)$$

$$\omega=100\pi;\ v=100$$

$$2\pi\nu=100\pi$$

$$\nu=50$$

$$v=\nu\lambda$$

$$100=50\lambda$$

$$\Rightarrow \quad \lambda=2\text{ cm}$$

35. (a)
First wave

$$y=a\sin(\omega t-kx)$$

Second wave

$$y'=a\cos(\omega t-kx)$$

$$=a\sin(\omega t-kx+\pi/2)$$

Phase difference $=\pi/2$

36. (c)

$c=\sqrt{\frac{\lambda P}{\rho}}$ with increase in P the value of ρ also increases in the same proportion

$$\sqrt{\frac{P}{\rho}}=\text{Constant}$$

$$c=\text{Constant}$$

37. (b)

$$\nu=\frac{n}{2L}\sqrt{\frac{T}{m}}$$

$$\nu_1=\frac{2}{2L_1}\sqrt{\frac{T}{m_1}}=\frac{1}{L_1}\sqrt{\frac{T}{m_1}}$$

$$\nu_2=\frac{2}{2L_2}\sqrt{\frac{T}{m_2}}$$

$$\sqrt{\frac{T}{m_1}}\frac{2}{2L_1}=\frac{3}{2L_2}\sqrt{\frac{T}{m_2}}$$

$$\Rightarrow \quad \frac{L_1}{L_2}=\frac{2}{3}\sqrt{\frac{m_2}{m_1}}$$

$$\Rightarrow \quad \frac{L_1}{L_2}=\frac{2}{3}\sqrt{\frac{\pi r_2^2}{\pi r_1^2}}$$

$$\frac{L_1}{L_2}=\frac{2}{3}\sqrt{\frac{1}{4}}$$

$$\frac{L_1}{L_2}=\frac{1}{3}$$

38. (a)

$$\nu'=\nu\left(\frac{c+v_0}{c-v_s}\right)$$

$$\Rightarrow \quad \nu'>\nu$$

and $$\lambda\propto\frac{1}{\nu}$$

$$\lambda'<\lambda.$$

39. (a)
Let the length of the pipe be L, then

$$\nu=\frac{v}{4L}=412$$

$$\Rightarrow \quad \frac{v}{L} = 4 \times 412$$

When it is cut into two pieces, we get two pipes one open and other closed each of length $\frac{L}{2}$.

For open pipe fundamental frequency

$$v = \frac{v}{L} = 4 \times 412 = 1648 \text{ Hz}$$

Closed pipe, fundamental frequency

$$v = \frac{v}{4\frac{L}{2}} = \frac{v}{2L} = \frac{4 \times 412}{2}$$

$$= 824 \text{ Hz}$$

40. (c)

$$\lambda = \frac{v}{v} = \frac{300}{600} = \frac{1}{2} \text{ m}$$

Short distance for maximum amplitude

$$= \frac{\lambda}{4} = \frac{1}{2 \times 4} = \frac{1}{8} \text{ m}$$

41. (b)

Let the velocity of sound be v, then

$$\text{Depth } h = \frac{v}{t} = v$$

$$t = \frac{2}{2} = 1 \text{ s}$$

$$\text{Depth } h = \frac{v}{1} = v$$

$$v = \sqrt{\frac{E}{\rho}} = \sqrt{\frac{2 \times 10^9}{1000}}$$

$$= 10^3\sqrt{2} = 1.414 \times 10^3 = 1414 \text{ m/s}$$

(at $4°\text{C}$, the density of water is maximum that is 10^3 kg/m^3)

$$h = v = 1414 \text{ m}$$

42. (b)

$$v = \sqrt{\frac{\gamma P}{\rho}} \text{ or } \sqrt{\frac{B}{\rho}}$$

43. (a)

$$v_1 = \sqrt{\frac{T_1}{m}}$$

$$v_2 = \sqrt{\frac{T_2}{m}}$$

$$\frac{v_2}{v_1} = \sqrt{\frac{T_2}{T_1}}$$

$$\frac{120}{100} = \sqrt{\frac{T_2}{120}}$$

$$\Rightarrow T_2 = \left(\frac{120}{100}\right)^2 \times 120 = \frac{36}{25} \times 120$$

$$= \frac{36 \times 24}{5}$$

$$\text{Change} = \frac{36 \times 24}{5} - 120$$

$$= 24\left(\frac{36}{5} - 5\right) = \frac{24 \times 11}{5}$$

$$\% \text{ increase} = \frac{24 \times 11}{5 \times 120} \times 100 = 44\ \%$$

44. (b)

$$V_g = \sqrt{\frac{B}{\rho}}; V_s = \sqrt{\frac{Y}{\rho}}$$

Y dominates on B as compared to gases.

45. (c)

$$n_A = \frac{c}{4\left(l + \frac{D_1}{3}\right)}$$

$$n_B = \frac{c}{4\left(l + \frac{D_2}{3}\right)}$$

As $\quad D_1 > D_2$

$$n_A < n_B$$

46. (a)

Frequency with which the wave reaches the reflector

$$f' = f\left(\frac{C - v}{C}\right)$$

The frequency of reflected wave received by the observer at position of source is

$$f'' = f'\left(\frac{C}{C + v}\right)$$

$$= f\left(\frac{C - v}{C + v}\right)$$

47. (a)

The ratio of frequencies are 1 : 2 : : 3 : 4

48. (d)

$$n = \frac{1}{2L}\sqrt{\frac{T}{m}}$$

$$n' = \frac{1}{2L}\sqrt{\frac{T/2}{m'}}$$

$$\frac{n'}{n} = \sqrt{\frac{T}{m'} \times \frac{m}{T}}$$

$$= \sqrt{\frac{m}{2m'}} = \sqrt{\frac{\pi r^2 \rho}{2 \times \pi (4r^2)\rho}}$$

$$n' = \frac{n}{2\sqrt{2}}$$

49. (c)

4th normal mode;

$$2\lambda = L$$
$$2\lambda = 2$$
$$\lambda = 1 \text{ m}$$
$$v = \nu\lambda$$
$$= 500 \times 1 = 500 \text{ m/s}$$

50. (c)

$$v = \nu\lambda$$

$$\lambda = \frac{v}{\nu} = \frac{360}{500} = 0.72 \text{ m}$$

Path difference $= \dfrac{\lambda}{2\pi} \times$ phase difference

$$= \frac{0.72}{2\pi} \times \frac{\pi}{3} = 0.12 \text{ m} = 12 \text{ cm}$$

51. (d)

$$y = 0.08 \sin \frac{2\pi}{\lambda}(200t - x)$$

$$= 0.08 \sin \frac{2\pi \times 200}{\lambda}\left(t - \frac{x}{200}\right)$$

$$v = 200 \text{ m/s}$$

52. (c)

Resonance will not occur if frequency is not an integral multiple of fundamental frequency.

53. (b)

Beat frequency $= \dfrac{2}{0.4} = 5$ Hz

54. (a)

$$y = 0.3 \sin(314t - 1.5x)$$

$$= 0.3 \sin 314\left(t - \frac{1.5x}{314}\right)$$

$$v = \frac{314}{1.5} \approx 210 \text{ Hz}$$

55. (a)

$$v' = v\left(\frac{330 - 30}{330}\right)$$

$$= 1000 \times \frac{300}{330}$$

$$= \frac{10000}{11} = 909 \text{ c/s}$$

56. (a)

$$\nu_1 = \frac{v}{2l} \quad \text{(For open pipe)}$$

$$\nu_2 = \frac{v}{4l}$$

$$\nu_1 : \nu_2 = \frac{v}{2l} : \frac{v}{4l} = \frac{1}{2} : \frac{1}{4} = 2:1$$

57. (a)

$$c = \sqrt{\frac{E}{\rho}}$$

58. (b)

$$\nu_1 = \frac{1}{2l}\sqrt{\frac{T}{m_1}}$$

$$\nu_2 = \frac{1}{2l}\sqrt{\frac{T}{m_2}}$$

$$\nu_1 : \nu_2 = \sqrt{\frac{1}{m_1}} : \sqrt{\frac{1}{m_2}}$$

$$= \sqrt{\frac{1}{\frac{\pi d^2}{4}\rho}} : \sqrt{\frac{1}{\frac{\pi 4d^2}{4}\rho}} = \sqrt{4} : 1 = 2:1$$

59. (a)

$$f = \frac{v}{2L}$$

$$f' = \frac{v}{4 \times \frac{L}{2}} = \frac{v}{2L}$$

$$\Rightarrow \quad f' = f$$

60. (c)

$$\nu_1 = \frac{1}{2L}\sqrt{\frac{T}{m}}$$

$$\nu_2 = \frac{1}{2 \times 2L} = \sqrt{\frac{4T}{m}}$$

$$\nu_1 : \nu_2 = \frac{1}{2} : \frac{1}{2} = 1:1$$

61. (d)

Echo time $= 10\sqrt{5}$ s

Plane velocity $= 200$ m/s

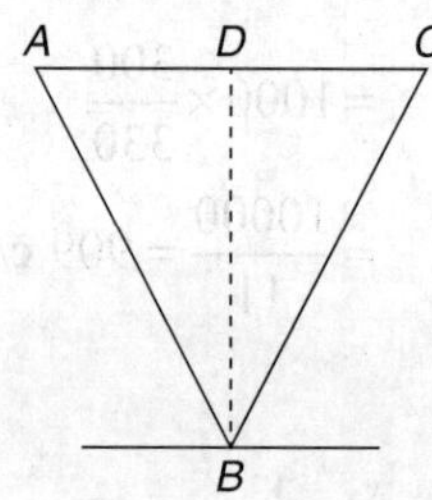

$$DC = 200 \times 5\sqrt{5} = 1000\sqrt{5}$$

$$BC = 300 \times 5\sqrt{5} = 1500\sqrt{5}$$

$$DB = \sqrt{(BC)^2 - (DC)^2}$$

$$= \sqrt{(1500\sqrt{5})^2 - (1000\sqrt{5})^2} \approx 2500 \text{ m}$$

62. (c)

Beat frequency = Difference in the two frequencies

63. (d)

$$f_1 = \frac{v}{2L}$$

$$f_2 = \frac{v}{4 \times \frac{L}{4}} = \frac{v}{L}$$

$$f_1 : f_2 = \frac{v}{2L} : \frac{v}{L} = \frac{1}{2} : 1 = 1 : 2$$

64. (d)

$$v_1 = \frac{v}{4L_1}; \; v_2 = \frac{v}{2L_2}$$

$$\Rightarrow \quad \frac{L_1}{L_2} = 1 : 2$$

65. (d)

$$y = 5\sin\left(\frac{t}{0.04} - \frac{x}{4}\right)$$

$$= 5\sin\frac{1}{0.04}\left(t - \frac{x \times 0.04}{4}\right)$$

$$\omega = \frac{1}{0.04}; A = 5$$

$$\text{Maximum particle velocity} = \omega A = \frac{5}{0.04} = 125 \text{ cm}$$

$$= 1.25 \text{ ms}^{-1}$$

66. (c)

$$y_1 = a\sin\omega t$$

$$y_2 = a\cos\omega t$$

$$y = y_1 + y_2 = a\sin\omega t + a\cos\omega t$$

$$= a\sqrt{2}\left(\frac{1}{\sqrt{2}}\sin\omega t + \frac{1}{\sqrt{2}}\cos\omega t\right)$$

$$= a\sqrt{2}(\cos 45°.\sin\omega t + \sin 45°.\cos\omega t)$$

$$= a\sqrt{2}\sin(\omega t + \pi/4)$$

Resultant amplitude

$$= a\sqrt{2}$$

67. (d)

$$v_1 = v\left(\frac{c}{c-34}\right)$$

$$v_2 = v\left(\frac{c}{c-17}\right)$$

$$\frac{v_1}{v_2} = \frac{c-17}{c-34} = \frac{340-17}{340-34} = \frac{17(19)}{17(18)} = \frac{19}{18}$$

68. (a)

The velocity of sound is independent of pressure.

69. (b)

$$v = v\lambda$$

$$v = \frac{\lambda}{T}$$

$$25 = \frac{100}{T}$$

$$\Rightarrow \quad T = 4 \text{ s}$$

70. (b)

The velocity of sound

$$v = \sqrt{\frac{E_{ad}}{\rho}} = \sqrt{\frac{E}{\rho}}$$

71. (c)

The compression and rarefaction is so fast that the heat exchange with the surrounding is not possible.

72. (a)

$$\text{Match number} = \frac{\text{Velocity of source}}{\text{Velocity of wave}}$$

73. (b)

$$\lambda = \frac{v}{\nu} = \frac{350}{500} = 0.7 \text{ m}$$

$$\text{Phase difference} = 60° = \frac{\pi}{3}$$

$$\text{Path difference} = \text{Phase difference} \times \frac{\lambda}{2\pi}$$

$$= \frac{\pi}{3} \times \frac{\lambda}{2\pi} = \frac{\lambda}{6} = \frac{0.7}{6}$$

$$= 0.12 \text{ m} = 12 \text{ cm}$$

74. (b)

Adiabatic conditions.

75. (d)

∵ The polarization takes place in transverse wave. The sound wave is a longitudinal wave.

76. (b)

Stationary waves are produced by the superposition of waves having equal frequencies in opposite directions.

77. (d)

78. (c)

Resonance occurs when the frequency of driver is equal to the frequency of driven.

79. (d)

Travelling wave is expressed as a periodic function of $(x - vt)$.

80. (d)

$$c = \sqrt{\frac{3P}{\rho}}$$

$$v = \sqrt{\frac{\gamma P}{\rho}}$$

$$\frac{v}{c} = \sqrt{\frac{\gamma}{3}}$$

81. (a)

The time t_1 taken by the stone to reach the bottom,

$$h = \frac{1}{2}gt_1^2$$

$$\Rightarrow \quad t_1 = \sqrt{\frac{2h}{g}}$$

The time t_2 taken by sound to cover a distance h,

$$t_2 = \frac{h}{v}$$

Total time $T = t_1 + t_2$

$$= \left(\frac{2h}{g}\right)^{1/2} + \frac{h}{v}$$

82. (a)

Wave velocity is a property of medium. In a given medium, sound of all frequencies travels with the same velocity.

83. (d)

$$v = \sqrt{\frac{\gamma RT}{M}}$$

$$\Rightarrow \quad v \propto \sqrt{T}$$

84. (c)

∵ Intensity of sound decreases with decrease in pressure.

85. (b)

∵ All the particles attain the same phase at the same instant in stationary waves.

86. (c)

$$v_{max} = \omega A = 2\pi v A$$

$$= 2\times 3.14\times 440\times 10^{-6} \text{ m/s}$$

$$= 2.76\times 10^{-3} \text{ m/s}$$

87. (b)

The ratio of velocity of sound at 7 °C to 47 °C

$$\frac{v_1}{v_2} = \sqrt{\frac{273+7}{273+47}}$$

$$\frac{v_1}{v_2} = \sqrt{\frac{280}{320}}$$

$$\frac{\sin i}{\sin r} = \sqrt{\frac{7}{8}}$$

$$\frac{\sin 30°}{\sin r} = \frac{1}{2}\sqrt{\frac{7}{2}}$$

$$\frac{1}{2\sin r} = \frac{1}{2}\sqrt{\frac{7}{2}}$$

$$\sin r = \sqrt{\frac{2}{7}}$$

$$r = \sin^{-1}\left(\frac{2}{7}\right)^{1/2}$$

88. (c)

Frequency of tuning fork is 514 ± 2, either 516 to 512

Frequency of tuning fork is

510 ± 6 = 516 or 504

Correct frequency is 516 Hz

89. (b)

Intensity (loudness) $\propto A^2$

The maximum amplitude when two waves superpose each other is $2A$.

Intensity $\propto (2A)^2$

$\propto 4A^2$

90. (d)

$$\frac{\lambda}{4} = l + e = 15 + 1$$

$$\lambda = 64 \text{ cm}$$

The next resonating length l' is equal to

$$\frac{3\lambda}{4} = l' + 1$$

$$\frac{3\times 64}{4}=l'+1$$

$$\Rightarrow \quad l'=47 \text{ cm}$$

91. (c)

$$v_1=\frac{1}{\lambda}\sqrt{\frac{T}{m_1}}$$

$$v_2=\frac{1}{\lambda}\sqrt{\frac{T}{m_2}}$$

$$\frac{v_2}{v_1}=\sqrt{\frac{m_1}{m_2}}$$

$$\frac{v_2}{v_1}=\sqrt{\frac{\pi d_1^2\rho}{\pi d_2^2\rho}}=\sqrt{\left(\frac{0.9}{0.93}\right)^2}$$

$$\Rightarrow \quad \frac{v_2}{v_1}=\frac{0.9}{0.93}=\frac{90}{93}$$

$$\frac{v_2-v_1}{v_1}=\frac{90-9}{93}$$

$$\left(\frac{v_2-v_1}{v_1}\right)\times 100=-\left(\frac{3}{93}\times 100\right)$$

$$=-3.2\%$$

92. (b)

The given equation is

$$y=y_0\sin 2\pi\left(ft-\frac{x}{\lambda}\right)$$

$$=y_0\sin\left(2\pi ft-\frac{2\pi}{\lambda}x\right)$$

Maximum particle velocity

$$=\omega A$$

$$=2\pi fy_0$$

Wave velocity $= f\lambda$

$$2\pi fy_0=4f\lambda$$

$$\Rightarrow \quad \lambda=\frac{\pi y_0}{2}$$

93. (a)

$$y=10^{-4}\sin(60t+2x)$$

$$=10^{-4}\sin 60\left(t+\frac{2x}{60}\right)$$

$$y=10^{-4}\sin 60\left(t+\frac{x}{30}\right) \quad \text{(i)}$$

On comparing with wave equation of (–ve) x-direction

$$y=a\sin 2\pi v\left(t+\frac{x}{v}\right)$$

$$a=10^{-4}, 2\pi v=60, v=30 \text{ m/s}$$

So, option 'a' is correct because Eq. (i) is a wave equation of negative x-direction with velocity 30 m/s.

94. (a)

Stationary waves in air columns are longitudinal.

95. (c)

$$v_1=\frac{v}{2l}$$

$$v_2=\frac{2v}{2l}$$

$$v_3=\frac{3v}{2l}$$

$$v_1:v_2:v_3=1:2:3$$

So all odd and even harmonics are present.

96. (c)

$$v_1=\frac{v}{4l}$$

$$v_2=\frac{3v}{4l}$$

$$v_3=\frac{5v}{4l}$$

$$v_1:v_2:v_3=1:3:5$$

97. (b)

$$v=v\lambda$$

With increase in temperature, v increases, therefore frequency 'v' increases

98. (c)
Pitch increases with increase in frequency.

99. (b)
There is always an antinode at open end.

100. (b)

$$v=\frac{c}{\lambda}$$

λ is measured in terms of resonating length of organ pipe. More the diameter of the more the open end the more error in measurement of v for closed pipe

$$v=\frac{C}{\frac{l}{4}+\frac{D}{3}}$$

$\therefore$ As D increases, v decreases.

101. (c)

$$v_1 = \frac{v}{2l} = \frac{350}{2\times 0.5} = 350 \text{ Hz}$$

102. (c)

For closed organ pipe first overtone, $v_2 = \frac{v}{l_1}$

For open organ pipe

$$v_2 = \frac{3v}{4l_2}$$

$$\frac{v}{l_1} = \frac{3v}{4l_2}$$

$$\frac{l_1}{l_2} = \frac{4}{3} = 4:3$$

$$\frac{l_2}{l_1} = \frac{3}{4} = 3:4$$

103. (a)

$$v = \frac{c}{4l}$$

As the vessel fills up the length of air column decreases, v increases.

104. (d)

Third overtone means fourth harmonic.

105. (b)

$$\frac{\lambda}{4} = l_1$$

$$\frac{3\lambda}{4} = l_2$$

$$\frac{3\lambda}{4} - \frac{\lambda}{4} = l_2 - l_1$$

$$\frac{\lambda}{2} = 31.5$$

$$\Rightarrow \quad \lambda = 63 \text{ cm}$$

106. (a)

$$v' = v\frac{c}{c - v_s}$$

$$v'' = v'\left(\frac{c + v_0}{c}\right) = v\frac{c + v_0}{c - v_s}$$

$$= 256\frac{335}{325} \simeq 264$$

∴ No of beats = 264 – 256 = 8 Hz

107. (a)

$$v' = v\left(\frac{c + v_0}{c - v_s}\right)$$

$$= v\frac{c/2}{c/2} = v$$

∴ Percentage change in velocity = 0%

108. (d)

$$v' = v\frac{\left(v + \frac{v}{2}\right)}{v - \frac{v}{2}} = 3v$$

⇒ Percentage change in frequency

$$\frac{\Delta v}{v}\times 100 = \frac{2v}{v}\times 100 = 200\%$$

109. (d)

$$v' = v\left(1 + \frac{2v}{c}\right)$$

$$v = \frac{v' - v}{v}\times\frac{c}{2}$$

$$= \frac{2500\times 3\times 10^8}{750\times 10^6\times 2}$$

$$= 500 \text{ ms}^{-1} = 0.5 \text{ kms}^{-1}$$

110. (d)

$$v' = v\left(\frac{c - v_0}{c - v_s}\right)$$

As v' in all the four cases will have a different value in all the four cases, none of the given options hold good.

111. (c)

$$\because \quad v' = v\left(\frac{c}{c - v}\right)$$

$$v' > v$$

with increase in frequency of pitch increases.

112. (a)

$$f = \frac{c}{4(L + x)}$$

$$\Rightarrow \quad L + x = \frac{c}{4}\times f^{-1}$$

$$L = \frac{c}{4}(f^{-1}) - x$$

The graph between L and f^{-1} will be a straight line and the slope of the graph will give $\frac{c}{4}$.

113. (b)

$$v=\sqrt{\frac{T}{m}} \qquad m=\frac{10^{-2}}{0.4}$$

$$=\sqrt{\frac{1.6\times 0.4}{10^{-2}}}=8 \text{ m/s}$$

$$\lambda=0.4\times 2=0.8 \text{ m}$$

$$\Delta t=\frac{0.8}{8}=0.1 \text{ s}$$

114. (a)

$$y_1=a\sin 2000\pi t$$

$$=a\sin 1000\times 2\pi t$$

$$y_2=a\sin 1004\times 2\pi t$$

Beat frequency $= v_2-v_1$

$= 1004 - 1000 = 4$

115. (d)

$$v_1=\frac{1}{2l_1}\sqrt{\frac{T}{m_1}}$$

$$v_2=\frac{1}{2l_2}\sqrt{\frac{T}{m_2}}$$

$$\frac{v_1}{v_2}=\frac{l_2}{l_1}\sqrt{\frac{m_2}{m_1}}$$

$$=\frac{2L}{L}\sqrt{\frac{\pi r^2\rho 2L/2L}{\pi(2r)^2\rho L/L}}=2\sqrt{\frac{1}{4}}=\frac{2}{2}=1$$

116. (a)

Here, $\upsilon'=\frac{9}{8}\upsilon$

Source and observer are moving in opposite directions, so apparent frequency

$$\upsilon'=\upsilon\times\frac{(v+u)}{(v-u)}$$

$$\frac{9}{8}\upsilon=\upsilon\times\frac{340+u}{340-u}$$

$$\Rightarrow 9\times 340-9u=8\times 340+8u$$

$$\Rightarrow 17u=340\times 1 \Rightarrow u=\frac{340}{17}=20 \text{ m/s}$$

117. (a)

$$\frac{U}{\lambda}-\left(-\frac{U}{\lambda}\right)=\frac{2U}{\lambda}$$

118. (b)

With the propagation of a longitudinal wave, energy alone is propagated.

119. (b)

Let the velocity of trains A and B be respectively v_A and v_B. Then,

$$v'=v\frac{c}{c-v_A}$$

$$5.5=\frac{5c}{c-v_A}$$

$$\Rightarrow \frac{5}{5.5}=\frac{c-v_A}{c}$$

$$\Rightarrow \frac{v_A}{c}=\frac{0.5}{5.5}$$

Similarly,

$$6=5\frac{c}{c-v_B}$$

$$\frac{v_B}{c}=\frac{1}{6}$$

From Eqs. (i) and (ii), we get

$$\frac{v_B}{v_A}=\frac{2\times 5.5}{6}=\frac{11}{6}$$

120. (d)

Compare with the equation,

$$y=a\cos(2\pi\upsilon t+\phi)$$

Thus gives $2\pi\upsilon=2000$

$$\upsilon=\frac{1000}{\pi} \text{ Hz}$$

121. (d)

1 st case

$$\frac{5\lambda_1}{2}=l \text{ or } \lambda_1=\frac{2l}{5}$$

2 nd case $\lambda_2=\frac{2l}{3}$

$$v_1=\frac{1}{\lambda}\sqrt{\frac{T}{m}}$$

$$v_1=\frac{5}{2l}\sqrt{\frac{9\times g}{m}}$$

$$v_2=\frac{3}{2l}\sqrt{\frac{M\times g}{m}}$$

$$v_1=v_2$$

$$\frac{5}{2l}\sqrt{\frac{9\times g}{m}}=\frac{3}{2l}\sqrt{\frac{Mg}{m}}$$

$$\Rightarrow \quad M=\left(\frac{5}{3}\times 3\right)^2=25 \text{ kg}$$

122. (c)

f_1 = frequency of the police car heard by motorcyclist,
f_2 = frequency of the siren heard by motorcyclist

$$f_1 = \frac{330-v}{330-22}\times 176$$

$$f_2 = \frac{330+v}{330}\times 165$$

$$f_1 - f_2 = 0 \Rightarrow v = 22 \text{ m/s}$$

123. (d)

Close organ pipe, second harmonic

$$v_2 = \frac{3v_1}{4L}, \text{ as } \frac{3\lambda}{4} = L$$

Open organ pipe, second harmonic

$$v_2' = \frac{v_2}{l_2}, \text{ as } \lambda = l_2$$

$$\frac{3v_1}{4L} = \frac{v_2}{l_2}$$

$$l_2 = \frac{4L}{3}\frac{v_2}{v_1}$$

$$= \frac{4L}{3}\sqrt{\frac{\gamma P\times\rho_1}{\rho_2\times\gamma P}} = \frac{4L}{3}\sqrt{\frac{\rho_1}{\rho_2}}$$

124. (c)

Intensity $\propto$ (amplitude)^2 and also intensity $\propto$ (frequency)^2

Therefore original $I \propto A^2\omega^2$

$$I' \propto 4A^2\frac{\omega^2}{16} \text{ i.e., } \frac{1}{4}I$$

125. (a)

$$\mu = \frac{0.035}{5.5} \text{ kg/m, T = 77 N}$$

where μ is mass per unit length.

$$v = \sqrt{\frac{T}{\mu}} = \sqrt{\frac{77\times 5.5}{0.035}} = 110 \text{ m/s}$$

126. (c)

$$f_1 = \frac{2v}{2L} = \frac{v}{L}$$

$$f_2 = \frac{nv}{4L} = \frac{n}{4}f_1$$

$$\because \quad f_2 > f_1$$

$$n = 5,\ f_2 = \frac{5}{4}f_1$$

127. (b)

Speed of sound

$$v = \sqrt{\frac{\gamma RT}{M}}$$

Let v_1 and v_2 be respectively the speed of sound in oxygen and helium gas respectively.

$$v_1 = \sqrt{\frac{\frac{7}{5}RT}{32}}$$

$$v_2 = \sqrt{\frac{\frac{5}{3}RT}{4}}$$

$$v_2 = 460\times\sqrt{\frac{5\times 32\times 5}{7\times 3\times 4}} = 1421 \text{ m/s}$$

128. (d)

The given wave equation is

$$y(x,t) = 0.005\cos(\alpha x - \beta t)$$

$$= 0.005\cos\beta\left(\frac{\alpha}{\beta}x - t\right)$$

$$2\pi v = \beta$$

$$\frac{2\pi}{T} = \beta$$

$$\Rightarrow \quad \frac{2\pi}{2} = \beta$$

$$\Rightarrow \quad \beta = \pi$$

$$\text{Now} \quad \frac{\beta}{\alpha} = v$$

$$\frac{\beta}{\alpha} = v\lambda = \frac{\lambda}{T}$$

$$\Rightarrow \quad \frac{\beta}{\alpha} = \frac{0.08}{2} = 0.04$$

$$\Rightarrow \quad \alpha = \frac{\beta}{0.04} = 25\pi$$

129. (a)

$$v = \frac{v_1}{4l_1} \text{ in winter}$$

$$v = \frac{3v_2}{4l_2} \text{ in summer}$$

$$l_1 = 18 \text{ cm}$$

$$\frac{3v_2}{4l_2} = \frac{v_1}{4l_1} \Rightarrow l_2 = 3l_1\frac{v_2}{v_1}$$

$$l_2 = 3 \times 18 \times \frac{v_2}{v_1}$$

$$l_2 = 54 \times \frac{v_2}{v_1}$$

$$v_2 > v_1$$

$$l_2 = 54 \frac{v_2}{v_1}$$

$\Rightarrow \quad l_2 > 54$

According to question $x = l_2$

$\therefore \quad x > 54$

130. (a)

Let intensity of sound be I_1 and I_2.

$$\beta_1 = 10 \log \left(\frac{I_1}{I_0} \right)$$

$$\beta_2 = 10 \log \left(\frac{I_2}{I_0} \right)$$

Given $\quad \beta_1 - \beta_2 = 20$

$$20 = 10 \log \frac{I_1}{I_0} - 10 \log \frac{I_2}{I_0}$$

$$20 = 10 \log \frac{I_1}{I_2}$$

$$\log \frac{I_1}{I_2} = 2$$

$$\Rightarrow \quad I_2 = \frac{I}{(10)^2} = \frac{I}{100}$$

131. (d)

$$v' = \left(\frac{300}{300 - v_s} \right) \times 9500$$

$$\frac{10000}{9500} = \frac{300}{300 - v_s}$$

$$\frac{95}{100} = \frac{300}{300} - \frac{v_s}{300} = 1 - \frac{v_s}{300}$$

$$v_s = \left(1 - \frac{95}{100} \right) \times 300$$

$$= \frac{5}{100} \times 300 = 15 \text{ m/s}$$

132. (c)

The two frequencies are consecutive.

$$315 = \frac{nv}{2L}, \quad 420 = \frac{(n+1)v}{2L}$$

$$\frac{420}{315} = \frac{n+1}{n} \Rightarrow n = 3$$

Lowest resonant frequency

$$= \frac{315}{5} = 105 \text{ Hz}$$

133. (a)

The frequency of Fork 2 is $200 \pm 4 = 204$ or 196 Hz. On loading the Fork 2, its frequency decreases. The number of beats per second is 6. The frequency difference increases. It means the frequency of Fork (2) is less than the frequency of Fork (1). Therefore the frequency of Fork (1) is 196 Hz.

134. (b)

$$v' = v \left(\frac{v + v_0}{v} \right) = v \left(\frac{v + \frac{v}{5}}{v} \right) = v \times \frac{6}{5}$$

$$100 \times \left(\frac{v' - v}{v} \right) = \left(\frac{6}{5} - 1 \right) \times 100 = 20\%$$

135. (d)

$$y = 10^{-4} \sin \left(100t + 20x + \frac{\pi}{4} \right)$$

$$= 10^{-4} \sin 100 \left(t + \frac{x}{5} + \frac{\pi}{400} \right)$$

The general wave equation is

$$y = a \sin \omega \left(t + \frac{x}{v} + \phi_0 \right)$$

Comparing this with the given equation $v = 5$ m/s

Questions Asked in NEET/CBSE Medicals

136. (c)

Frequency (ν) = 800 Hz

As the pipe is closed at one end, so

$$l_3 - l_2 = l_2 - l_1 \frac{\lambda}{2} = 21.5 \text{ cm}$$

$$\therefore \lambda = 43.0 \text{ cm}$$

$$v = \nu\lambda$$

$$= 800 \times 43$$

$$= 344 \text{ m/s}$$

137. (b)

$$\frac{\lambda}{2} = (73 - 20) \text{ cm}$$

$$\lambda = 2(73 - 20) \times 10^{-2} \text{ m}$$

$$v = \nu\lambda$$

$$= 320 \times 2 \times 53 \times 10^{-2}$$

$$= 332 \text{ m/s}$$

Closest option is (b).

138. (a)

For closed organ pipe, third harmonic is $\frac{3v}{4l}$

For open organ pipe, fundamental frequency is $\frac{v}{2l'}$

Given third harmonic for closed organ pipe = Fundamental frequency for open organ pipe.

$$\therefore \frac{3v}{4l} = \frac{v}{2l'} \Rightarrow l' = \frac{4l}{3\times 2} = \frac{2l}{3};$$

where l and l' are the lengths of closed and open organ pipes respectively.

$$\therefore l' = \frac{2\times 20}{3} = 13.33 \text{ cm}$$

139. (a)

The nearest harmonics of an organ pipe closed at one end is differ by twice of its fundamental frequency.

$$\therefore 260 - 220 = 2v, v = 20 \text{ Hz}$$

140. (c)

The required frequency of sound heard by the driver of second car is given as

$$v' = v\left(\frac{v + v_0}{v - v_s}\right)$$

where v = velocity of sound

v_0 = velocity of observer, i.e., second car

v_s = velocity of source, i.e., first car

$$v' = 400\left(\frac{340 + 16.5}{340 - 22}\right) = 400\left(\frac{356.5}{318}\right)$$

$$v' \approx 448 \text{ Hz}$$

141. (b)

Second overtone of an open organ pipe

(Third harmonic) $= 3\times v_0 = 3\times\frac{v}{2L'}$

First overtone of a closed organ pipe

(Third harmonic) $= 3\times v_0 = 3\times\frac{v}{4L}$

According to question,

$$3v_0' = 3v_0 \Rightarrow 3\times\frac{v}{2L'} = 3\times\frac{v}{4L}$$

$$\therefore L' = 2L$$

142. (d)

143. (a)

Here, frequency of sound emitted by siren,

$$v_0 = 800 \text{ Hz}$$

Speed of source, $v_s = 15 \text{ ms}^{-1}$

Speed of sound in air, $v = 330 \text{ ms}^{-1}$

Apparent frequency of sound at the cliff = Frequency heard by observer = v

Using Doppler's effect of sound

$$v = \left(\frac{v}{v - v_s}\right)v_0 = \frac{330}{330 - 15}\times 800$$

$$= \frac{330}{315}\times 800 = 838.09 \text{ Hz} \approx 838 \text{ Hz}$$

144. (a)

First harmonic is obtained at

$$l = \frac{\lambda}{4} = 50 \text{ cm}$$

Third harmonic is obtained for resonance,

$$l' = \frac{3\lambda}{4} = 3\times 50 = 150 \text{ cm}$$

145. (d)

Wavelength of pulse at the lower end,

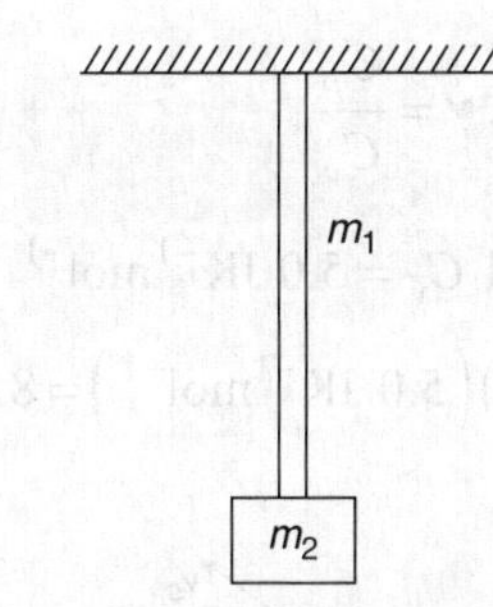

$$\lambda_1 \propto \text{velocity}(v_1) = \sqrt{\frac{T_1}{\mu}}$$

Similarly, $\lambda_2 \propto (v_2) = \sqrt{\frac{T_2}{\mu}}$

$$\therefore \frac{\lambda_2}{\lambda_1} = \sqrt{\frac{T_2}{\mu}} = \sqrt{\frac{(m_1 + m_2)g}{m_2 g}}$$

$$= \sqrt{\frac{m_1 + m_2}{m_2}}$$

146. (b)

For a string fixed at both ends, the resonant frequencies are

$$v_n = \frac{nv}{2L} \text{ where } n = 1, 2, 3, \ldots\ldots$$

The difference between two consecutive resonant frequencies is

$$\Delta v_n = v_{n+1} - v_n = \frac{(n+1)v}{2L} - \frac{nv}{2L} = \frac{v}{2L}$$

This is also the lowest resonant frequency ($n = 1$)
Thus the lowest resonant frequency for the given strings $= 420\text{ Hz} - 315\text{ Hz} = 105\text{ Hz}$

147. (c)
Since 4.0 g of a gas occupies 22.4 liters at NTP, so the molecular mass of the gas is

$$M = 4.0\text{ g mol}^{-1}$$

As the speed of the sound in the gas is

$$v = \sqrt{\frac{\gamma RT}{M}}$$

$$\therefore \gamma = \frac{Mv^2}{RT}$$

Here, $M = 4.0\text{ g mol}^{-1} = 4.0\times10^{-3}\text{ kgmol}^{-1}$

$v = 952\text{ ms}^{-1}$, $R = 8.3\text{ JK}^{-1}\text{mol}^{-1}$ and

$T = 273$ K (at NTP)

$$\therefore \gamma = \frac{(4.0\times10^{-3}\text{ kgmol}^{-1})(952\text{ ms}^{-1})^2}{(8.3\text{ JK}^{-1}\text{mol}^{-1})(273\text{ K})} = 1.6$$

By definition, $\therefore \gamma = \frac{C_p}{C_v}$

But $\gamma = 1.6$ and $C_v = 5.0\text{ JK}^{-1}\text{mol}^{-1}$

$$\therefore C_p = (1.6)(5.0\text{ JK}^{-1}\text{mol}^{-1}) = 8.0\text{ JK}^{-1}\text{mol}^{-1}$$

148. (d)

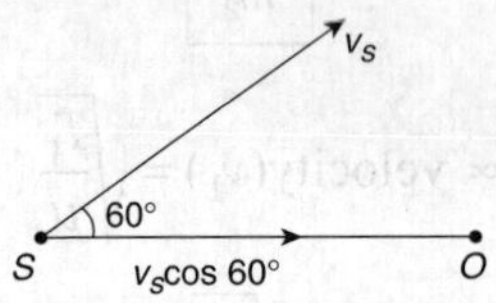

Here,
Frequency of source, $v_0 = 100$ Hz

Velocity of source $v_s = 19.4\text{ ms}^{-1}$

Velocity of sound in air, $v = 330\text{ ms}^{-1}$
As the velocity of source along the source observer line is $v_s = \cos 60°$ and the observer is at rest, so the apparent frequency is

$$v = v_0\left(\frac{v}{v - v_s\cos 60°}\right)$$

$$= (100\text{ Hz})\left(\frac{330\text{ ms}^{-1}}{330\text{ ms}^{-1} - (19.4\text{ ms}^{-1})\left(\frac{1}{2}\right)}\right)$$

$$= (100\text{ Hz})\left(\frac{330\text{ ms}^{-1}}{330\text{ ms}^{-1} - 9.7\text{ ms}^{-1}}\right)$$

$$= (100\text{ Hz})\left(\frac{330\text{ ms}^{-1}}{320.3\text{ ms}^{-1}}\right) = 103\text{ Hz}.$$

149. (a)
For closed organ pipe, fundamental frequency is given by $v_c = \frac{v}{4l_1}$

For open organ pipe, fundamental frequency is given by $v_0 = \frac{v}{2l_2}$

2nd overtone of open organ pipe

$$v_2 = 3v_0; v_2 = \frac{3v}{2l_2}$$

$$\frac{v}{4l_1} = \frac{3v}{2l_2}$$

$$l_2 = 6l_1$$

$$\Rightarrow l_2 = 3\times2\times20 = 120\text{ cm}$$

150. (d)
Fundamental frequency of the closed organ pipe is

$$v = \frac{v}{4L}$$

Here, $v = 340\text{ ms}^{-1}$, $L = 85\text{ cm} = 0.85\text{ m}$

$$\therefore v = \frac{340\text{ ms}^{-1}}{4\times0.85\text{ m}} = 100\text{ Hz}$$

The natural frequency of the closed organ pipe will be

$v_n = (2n-1)v = v, 3v, 5v, 7v, 9v, 11v, 13v, \ldots$
= 100 Hz, 300 Hz, 500 Hz, 700 Hz, 900 Hz, 1100 Hz, 1300 Hz,.... and so on.
Thus the number of natural frequencies that lies below 1250 Hz is 6.

151. (c)
Here, speed of motorcyclist $v_m = 36\text{ km hour}^{-1}$

$$= 36\times\frac{5}{18} = 10\text{ ms}^{-1}$$

Similarly speed of car $v_c = 5$ m/s

Frequency of source, $v_0 = 1392$ Hz

Speed of sound, $v = 343\text{ ms}^{-1}$

The frequency of the honk heard by the motorcyclist is

$$v' = v_0\left(\frac{v+v_m}{v+v_c}\right) = 1392\left(\frac{343+10}{343+5}\right)$$

$$= \frac{1392\times 353}{348} = 1412 \text{ Hz}$$

152. (b)

153. (c)

The standard equation of a wave travelling along +ve x-direction is given by

$$y = A \sin (kx - \omega t)$$

Here, $A = 1$ m, $\lambda = 2\pi, v = \frac{1}{\pi}$ Hz

As $k = \frac{2\pi}{\lambda} = \frac{2\pi}{2\pi} = 1$

$$\omega = 2\pi v = 2\pi \times \frac{1}{\pi} = 2$$

$\therefore$ The equation of the given wave is

$$y = 1 \sin (1x - 2t) = \sin (x - 2t)$$

154. (c)

Let v be frequency of the unknown source.

$\therefore$ $v = 250 \pm 4 = 246$ Hz or 254 Hz

Second harmonic of this unknown source = 492 Hz or 508 Hz which gives 5 beats per second, when sounded with a source of frequency 513 Hz. Therefore unknown frequency, $v = 254$ Hz.

155. (d)

Let L (= 100 cm) be the length of the wire and L_1, L_2 and L_3 are the lengths of the segments as shown in the figure.

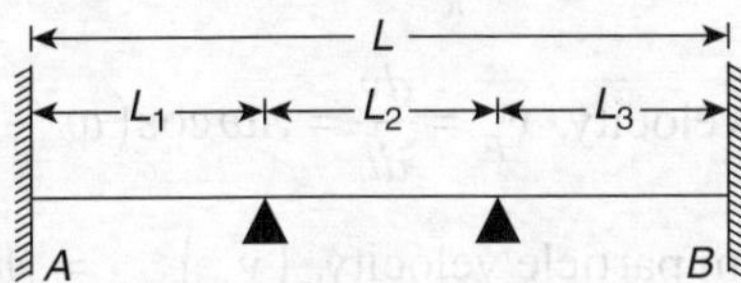

Fundamental frequency, $v \propto \frac{1}{L}$

As the fundamental frequencies are in the ratio of 1 : 3 : 5,

$$\therefore L_1 : L_2 : L_3 = \frac{1}{1} : \frac{1}{3} : \frac{1}{5} = 15 : 5 : 3$$

Position of first bridge

$$L_1 = \frac{100\times 15}{23} = \frac{1500}{23} \text{ cm}$$

Position of second bridge

$$(L_1 + L_2) = \frac{100\times 20}{23} = \frac{2000}{23} \text{ cm}$$

$$15x + 5x + 3x = L = 100$$

$$23x = 100 \text{ or } x = \frac{100}{23}$$

$$\therefore L_1 = 15\times\frac{100}{23} = \frac{1500}{23} \text{ cm}$$

$$L_2 = 5\times\frac{100}{23} = \frac{500}{23}; \; L_3 = 3\times\frac{100}{23} = \frac{300}{23} \text{ cm}$$

$\therefore$ The bridges should be placed from A at $\frac{1500}{23}$ cm and $\frac{2000}{23}$ cm respectively.

156. (a)

Speed of listener, $v_L = 1 \text{ ms}^{-1}$

Speed of sound, $v = 330 \text{ ms}^{-1}$

Frequency of each source, $v = 660$ Hz

Apparent frequency due to P, $v' = \frac{v(v - v_L)}{v}$

Apparent frequency due to Q,

$$v'' = \frac{v(v + v_L)}{v}$$

Number of beats heard by the listener per second is

$$v'' - v = \frac{v(v + v_L)}{v} - v\frac{(v - v_L)}{v}$$

$$= \frac{2vv_L}{v} = \frac{2\times 660\times 1}{330} = 4$$

157. (c)

Let l be the length of the string be l

Fundamental frequency is given by

$$v = \frac{1}{2l}\sqrt{\frac{T}{\mu}}$$

or $\quad v \propto \frac{1}{l} \quad$ ($\because$ T and μ are constants)

Here, $l_1 = \frac{k}{v_1}, l_2 = \frac{k}{v_2}, l_3 = \frac{k}{v_3}$ and $l = \frac{k}{v}$

But $l = l_1 + l_2 + l_3$

$$\therefore \frac{1}{v} = \frac{1}{v_1} + \frac{1}{v_2} + \frac{1}{v_3}$$

158. (d)

Given: $y_1 = 4\sin 600\pi t$, $y_2 = 5\sin 608\pi t$

$\therefore \omega_1 = 600\pi$ or $2\pi v_1 = 600\pi$ or $v_1 = 300$ Hz

$A_1 = 4$

and $\omega_2 = 608$ or $2\pi v_2 = 608\pi$ or $v_2 = 304$ Hz

$A_2 = 5$

Number of beats heard per second

$$= \nu_2 - \nu_1 = 304 - 300 = 4$$

$$\frac{I_{max}}{I_{min}} = \frac{(A_1 + A_2)^2}{(A_1 - A_2)^2} = \frac{(4+5)^2}{(4-5)^2} = \frac{81}{1}$$

159. (b)

The given wave equation is

$$y = 3\sin\frac{\pi}{2}(50t - x)$$

$$y = 3\sin\left(25\pi t - \frac{\pi}{2}x\right) \quad \text{(i)}$$

The standard wave equation is

$$y = A\sin(\omega t - kx) \quad \text{(ii)}$$

Comparing (i) and (ii), we get

$$\omega = 25\pi, k = \frac{\pi}{2}$$

Wave velocity, $v = \frac{\omega}{k} = \frac{25\pi}{(\pi/2)} = 50 \text{ ms}^{-1}$

Particle velocity, $v_p = \frac{dy}{dt} = \frac{d}{dt}\left(3\sin\left(25\pi t - \frac{\pi}{2}\right)\right)$

$$= 75\pi\cos\left(25\pi t - \frac{\pi}{2}\right)$$

Maximum particle velocity, $(v_p)_{max} = 75\pi \text{ ms}^{-1}$

$$\therefore \frac{(v_p)_{max}}{v} = \frac{75\pi}{50} = \frac{3}{2}\pi$$

160. (c)

Here,

Speed of the train, $v_T = 220$ m/s

Speed of sound in air, $v = 330$ m/s

The frequency of the echo detected by the driver of the train is

$$\nu' = \nu\left(\frac{v + v_T}{v - v_T}\right) = 1000\left(\frac{330+220}{330-220}\right)$$

$$= 1000 \times \frac{550}{110} = 5000 \text{ Hz}$$

161. (a)

$$y_1 = a\sin(\omega t + kx + 0.57)$$

$\therefore$ Phase, $\phi_1 = \omega t + kx + 0.57$

$$y_2 = a\cos(\omega t + kx) = a\sin\left(\omega t + kx + \frac{\pi}{2}\right)$$

$\therefore$ Phase, $\phi_2 = \omega t + kx + \frac{\pi}{2}$

Phase difference, $\Delta\phi = \phi_2 - \phi_1$

$$= \left(\omega t + kx + \frac{\pi}{2}\right) - (\omega t + kx + 0.57) = \frac{\pi}{2} - 0.57$$

$$= (1.57 - 0.57) \text{ radian} = 1 \text{ radian}$$

162. (c)

Here, $v_{air} = 350$/ms, $v_{brass} = 3500$/ms

When a sound wave travels from one medium to another medium its frequency remains the same

$\therefore$ Frequency, $\nu = \frac{v}{\lambda}$

Since ν remains the same in both the medium

$$\Rightarrow \frac{v_{air}}{\lambda_{air}} = \frac{v_{brass}}{\lambda_{brass}}$$

$$\lambda_{brass} = \lambda_{air} \times \frac{v_{brass}}{v_{air}} = \lambda_{air} \times \frac{3500}{350} = 10\lambda_{air}$$

163. (b)

As $\nu = \frac{1}{2L}\sqrt{\frac{T}{\mu}} \therefore \frac{\Delta\nu}{\nu} = \frac{1}{2}\frac{\Delta T}{T}$

$$\frac{\Delta T}{T} = 2\frac{\Delta\nu}{\nu} = 2 \times \frac{6}{600} = 0.02$$

164. (c)

The given wave equation is

$$y = A\sin(\omega t - kx)$$

Wave velocity, $v = \frac{\omega}{k}$ (i)

Particle velocity, $v_p = \frac{dy}{dt} = A\omega\cos(\omega t - kx)$

Maximum particle velocity, $(v_p)_{max} = A\omega$ (ii)

$$v = (v_p)_{max}$$

$$\frac{\omega}{k} = A\omega \quad \text{(Using (i) and (ii))}$$

$$\frac{1}{k} = A \text{ or } \frac{\lambda}{2\pi} = A \quad \left(\because k = \frac{2\pi}{\lambda}\right)$$

$$\lambda = 2\pi A$$

165. (d)

Let the frequencies of tuning fork and piano string be ν_1 and ν_2 respectively.

$$\therefore \nu_2 = \nu_1 \pm 4 = 512 \text{ Hz} \pm 4 = 516 \text{ Hz or } 508 \text{ Hz}$$

Using logical contradiction, we can find that the required frequency is 508.

166. (a)

$$l_1 = 0.516 \text{ m}, l_2 = 0.491 \text{ m}, T = 20 \text{ N}$$

Mass per unit length, $\mu = 0.001$ kg/m.

Frequency, $\nu = \frac{1}{2l}\sqrt{\frac{T}{\mu}}$

$$\nu_1 = \frac{1}{2\times 0.516}\sqrt{\frac{20}{0.001}} \text{ Hz}$$

$$\nu_2 = \frac{1}{2\times 0.491}\sqrt{\frac{20}{0.001}} \text{ Hz}$$

$\therefore$ Number of beats $= \nu_1 - \nu_2 = 7$

167. (b)

Frequency heard at the hill, ν_1

$$\therefore \nu_1 = \frac{\nu \times V}{(v - V)} = \frac{600\times 330}{330-30}$$

Now for reflection, the hill is the source and the driver is the observer.

$$\therefore \nu_2 = \nu_1 \times \frac{(330+30)}{330}$$

$$\Rightarrow \nu_2 = \frac{600\times 330}{300}\times\frac{360}{330} \Rightarrow \nu_2 = 720 \text{ Hz}$$

168. (c)

Amplitude = 2 cm = 0.02 m, v = 128 m/s

$$\lambda = \frac{4}{5} = 0.8 \text{ m}; \upsilon = \frac{128}{0.8} = 160 \text{ Hz}$$

$$\omega = 2\pi\upsilon = 2\pi\times 160 = 1005;\ k\frac{2\pi}{\lambda} = \frac{2\pi}{0.8} = 7.85$$

$$\therefore \quad y = (0.02)\text{m} \sin(7.85x - 1005t)$$

169. (b)

$$x = a\sin(\omega t + \pi/6)$$

$$\frac{dx}{dt} = a\omega\cos(\omega t + \pi/6)$$

Max. velocity $= a\omega$

$$\therefore \frac{a\omega}{2} = a\omega\cos(\omega t + \pi/6)$$

$$\therefore \cos(\omega t + \pi/6) = \frac{1}{2}$$

$$\Rightarrow \omega t + \frac{\pi}{6} = \frac{\pi}{3}$$

$$\Rightarrow \frac{2\pi}{T}.t = \frac{2\pi}{6} - \frac{\pi}{6} = \frac{\pi}{6}$$

$$\therefore t = \frac{\pi}{6}\times\frac{T}{2\pi} = \boxed{\frac{T}{12}}$$

170. (b)

Other factors such as ω and v remaining the same, $I = A^2 \times$ constant K, or A $= \sqrt{\frac{I}{K}}$

On superposition,

$$A_{max} = A_1 + A_2 \text{ and } A_{min} = A_1 - A_2$$

$$\therefore A_{max}^2 = A_1^2 + A_2^2 + 2A_1A_2$$

$$\Rightarrow \frac{I_{max}}{K} = \frac{I_1}{K} + \frac{I_2}{K} + \frac{2\sqrt{I_1I_2}}{K}$$

$$A_{min}^2 = A_1^2 + A_2^2 - 2A_1A_2$$

$$\Rightarrow \frac{I_{min}}{K} = \frac{I_1}{K} + \frac{I_2}{K} - \frac{2\sqrt{I_1I_2}}{K}$$

$$\therefore I_{max} + I_{min} = 2I_1 + 2I_2 = 2(I_1 + I_2)$$

171. (a)

$$y = 0.25\sin(10\pi x - 2\pi t)$$

$$y_{max} = 0.25$$

$$k = \frac{2\pi}{\lambda} = 10\pi \Rightarrow \lambda = 0.2 \text{ m}$$

$$\omega = 2\pi f = 2\pi \Rightarrow f = 1 \text{ Hz}$$

The negative sign means that wave travels in +ve x direction

172. (b)

$$Y_1 = 4\sin 500\pi t, Y_2 = 2\sin 506\pi t$$

$$\omega_1 = 500\pi = 2\pi\upsilon_1 \Rightarrow \upsilon_1 = 250 \text{ Hz}$$

$$\omega_2 = 506\pi = 2\pi\upsilon_2 \Rightarrow \upsilon_2 = 253 \text{ Hz}$$

$$\upsilon = \upsilon_2 - \upsilon_1 = 253 - 250 = 3 \text{ beats/s.}$$

Number of beats per minute $= 3\times 60 = 180$

173. (a)

$$\therefore \upsilon_1 = \frac{v}{\lambda_1} = \frac{330}{5} = 66 \text{ Hz}$$

and $$\upsilon_2 = \frac{v}{\lambda_2} = \frac{330}{5.5} = 66 \text{ Hz}$$

Number of beats per second $= \upsilon_1 - \upsilon_2 = 66 - 60 = 6$

174. (b)

Reverberation time, $T = \frac{0.61V}{aS}$

where V is the volume of room in cubic metres, a is the average absorption coefficient of the room, S is the total surface area of room in square metres.

$$\text{or, } T \propto \frac{V}{S} \text{ or, } \frac{T_1}{T_2} = \left(\frac{V_1}{V_2}\right)\left(\frac{S_2}{S_1}\right) = \left(\frac{V}{8V}\right)\left(\frac{4S}{S}\right) = \frac{1}{2}$$

or, $T_2 = 2T_1 = 2\times 1 = 2$ sec. $\quad (\because T_1 = 1 \text{ sec})$

175. (a)

$$y(x, t) = 8.0 \sin\left(0.5\pi x - 4\pi t - \frac{\pi}{4}\right)$$

Compare with a standard wave equation,

$$y = a\sin\left(\frac{2\pi x}{\lambda} - \frac{2\pi t}{T} + \phi\right)$$

we get $\frac{2\pi}{\lambda} = 0.5\pi$ or $\lambda = \frac{2\pi}{0.5\pi} = 4$ m

$\frac{2\pi}{T} = 4\pi$ or, $T = \frac{2\pi}{4\pi} = \frac{1}{2}$ sec

$\upsilon = 1/T = 2$ Hz

Wave velocity, $v = \lambda\upsilon = 4 \times 21 = 8$ m/sec

176. (c)

Light waves are electromagnetic waves. Light waves are transverse in nature and do not require a medium to travel, hence they can travel in vacuum. Sound waves are longitudinal waves and require a medium to travel. They do not travel in vacuum.

177. (c)

$$d_1 = 2\text{m}, d_2 = 3\text{m}$$

$$\text{Intensity} \propto \frac{1}{(\text{distance})^2}$$

$$I_1 \propto \frac{1}{2^2} \text{ and } I_2 \propto \frac{1}{3^2} \qquad \therefore \frac{I_1}{I_2} = \frac{9}{4}$$

178. (d)

Third overtone has a frequency $4n$, 4th harmonic = Three full loops + One half loop, which would make four nodes and four antinodes.

179. (b)

First case: The car is the source and at the cliff, one observes f'

$$\therefore f' = \frac{v}{v - v_s} f$$

Second case: Cliff is now source. It emits frequency f' and the observer is now the driver who observes f''

$$\therefore f'' = \left[\frac{v - v_o}{v}\right] f' \text{ or } 2f = \left[\frac{v + v_o}{v - v_s}\right] f$$

$$\Rightarrow 2v - 2v_o = v + v_o \qquad [\text{as } v_s = v_o]$$

$$\Rightarrow v_o = \frac{v}{3}$$

180. (b)

Apparent frequency,

$$f' = \frac{v + v_o}{v} f = \frac{v + (1/5)v}{v} f = 1.2f$$

Wavelength does not change by motion of observer.

181. (b)

The apparent frequency heard by the observer is then given by $\upsilon' = \upsilon\left(\frac{V}{V + v}\right)$

where V and v are the velocities of sound and source respectively and υ is the actual frequency.

Now, $v = r\omega = 0.5 \times 20 = 10$ m/s

$V = 340$ m/s, $\upsilon = 385$ Hz

$$\therefore \upsilon' = 385 \times \frac{340}{340 + 10} = 374 \text{ Hz}$$

182. (c)

The equation of progressive wave travelling in positive x-direction is given by

$$y = a\sin\frac{2\pi}{\lambda}(vt - x)$$

Here $a = 0.2$ m, $v = 360$ m/sec, $\lambda = 60$ m

$$\therefore y = 0.2\sin\frac{2\pi}{60}(360t - x) = 0.2\sin\left[2\pi\left(6t - \frac{x}{60}\right)\right]$$

183. (c)

Comparing the given equation with the general equation, $y = a\sin 2\pi\left(\frac{t}{T} - \frac{x}{\lambda}\right)$, we get

$$T = \frac{2\pi}{100} \text{ and } \lambda = 20\pi$$

$$\therefore v = \upsilon\lambda = \frac{100}{2\pi} \times 20\pi = 1000 \text{ m/s}$$

184. (b)

Resultant amplitude $= 2a(1 + \cos\phi) = a$

$$\therefore (1 + \cos\phi) = 1/2; \cos\phi = -\frac{1}{2}; \phi = \frac{2\pi}{3}$$

185. (c)

$$n = \frac{1}{2l}\sqrt{\frac{T}{\pi r^2 \rho}}$$

$$\rho_1' = \frac{\rho}{2}; T' = 2T \text{ and } D' = 2D \text{ or } r' = 2r$$

$$n' = \frac{1}{2l}\sqrt{\frac{2T}{\pi(2r)^2\frac{\rho}{2}}}$$

After solving, $n' = \frac{1}{2l}\sqrt{\frac{T}{\pi r^2 \rho}} = n$

186. (b)

Given $x = a\cos(\omega t + \delta)$ and $y = a\cos(\omega t + \alpha)$ (i)

where, $\delta = \alpha + \pi/2$

$$\therefore x = a\cos(\omega t + \alpha + \pi/2) = -a\sin(\omega t + \alpha) + \pi/2) \quad \text{(ii)}$$

Given the two waves are acting in perpendicular directions with the same frequency and phase difference $\pi/2$.

From Eqs. (i) and (ii),

$$x^2 + y^2 = a^2$$

which represents the equation of a circle.

187. (a)

As $n \propto (1/l)$ and $l = l_1 + l_2 + l_3$

$$\therefore \frac{1}{n} = \frac{1}{n_1} + \frac{1}{n_2} + \frac{1}{n_3}$$

188. (a)

$$f' = \frac{v-u}{v}f; f'' = \frac{v+u}{v}f$$

Number of beats = $f'' - f' = \frac{2u}{\lambda}$

189. (c) Number of beats produced per second

$$= \upsilon_1 - \upsilon_2 = \frac{v}{\lambda_1} - \frac{v}{\lambda_2}$$

$$12 = v\left[\frac{1}{50} - \frac{1}{51}\right] \text{ or } 12 = \frac{v \times 1}{50 \times 51}$$

or, $v = 12 \times 50 \times 51$ cm/s = 306 m/s.

190. (d)

The given equation of wave is

$$y = y_0 \sin\frac{2\pi}{\lambda}(vt - x)$$

$$\text{Particle velocity} = \frac{dy}{dt} = y_0 \cos\frac{2\pi}{\lambda}(vt - x)\frac{2\pi v}{\lambda}$$

$$\left(\frac{dy}{dt}\right)_{\max} = y_0\frac{2\pi}{\lambda}v$$

$$\therefore y_0\frac{2\pi}{\lambda}v = 2v \text{ or } \lambda = \pi y_0$$

191. (b)

$$n' = n + n_1 = \frac{nv}{v - v_s\cos\theta} = \frac{nv}{v}$$

$$n' = n \therefore n_1 = 0$$

192. (c)

Distance between a node and adjoining antinodes = $\lambda/4$

From figure, distance between two atoms

$$= 4 \times \frac{\lambda}{4} = 1.21\text{Å} \text{ or, } \lambda = 1.21\text{Å}$$

193. (c)

Displacement, $Y_{\max} = a, Y_{\min} = 0$

$$t = T/4$$

$\therefore T/4 = 0.170$ $\quad \therefore T = 0.68$

The frequency of wave = $1/T = 1.47$ Hz

194. (a)

The frequency of standing wave.

$$\upsilon = \frac{n}{2l}v = \frac{5 \times 20}{2 \times 10} = 5 \text{ Hz}$$

195. (d)

For the cylindrical tube open at both ends, $f = v/2l$.

When half of the tube is in water, it behaves as a closed pipe of length l/2.

$$\therefore f' = \frac{v}{4(l/2)} = \frac{v}{2l} \therefore f' = f$$

196. (c)

Sound wave equation is

$$y = 0.0015 \sin(62.4x + 316t).$$

Comparing it with the general equation of motion

$$y = A\sin 2\pi\left[\frac{x}{\lambda} + \frac{t}{T}\right], \text{ we get } \frac{2\pi}{\lambda} = 62.4$$

$$\text{or } \lambda = \frac{2\pi}{62.4} = 0.1 \text{ unit}$$

197. (a)

$$\text{Phase difference } \theta = 60° = \frac{\pi}{3} \text{ rad}$$

$$\text{Phase difference } (\theta) = \frac{\pi}{3} = \frac{2\pi}{\lambda} \times \text{Path difference}$$

$$\text{Therefore, Path difference} = \frac{\pi}{3} = \frac{\lambda}{2\pi} = \frac{\lambda}{6}$$

198. (a)

Length of sonometer wire (l) = 110 cm and ratio of frequencies = 1 : 2 : 3.

$$\text{Frequency } (\upsilon) \propto \frac{1}{l} \text{ or } l \propto \frac{1}{\upsilon}$$

$$\text{Therefore AC : CD : DB} = \frac{1}{1} : \frac{1}{2} : \frac{1}{3} = 6 : 3 : 2$$

$$\text{AC} = 6 \times \frac{110}{11} \text{ 60 cm and CD} = 3 \times \frac{110}{11} = 30 \text{ cm}$$

AD = 60 + 30 = 90 cm.

199. (c)

Frequency (υ) = 4.2 MHz

$$= 4.2 \times 10^6 \text{ Hz and}$$

Speed of sound (v) = 1.7 km/s = 1.7×10^3 m/s

Wavelength of sound in tissue $(\lambda) = \frac{v}{\upsilon}$

$$= \frac{1.7\times10^3}{4.2\times10^6} = 4\times10^{-4}\text{ m}$$

200. (a)
Frequency of first source with 5 beats/sec = 100 and frequency of second source with 5 beats/sec = 205. The frequency of the first source = 100 ± 5 = 105 or 95 Hz. Therefore frequency of second harmonic source = 210 Hz or 190 Hz.
As the second harmonic gives 5 beats/second with the sound of frequency 205 Hz, the frequency of second harmonic source should be 210 Hz. The frequency of source = 105 Hz.

201. (a)
Wavelength $(\lambda) = 5000$ Å and velocity $(v) = 1.5\times10^4$ m/s.
Wavelength of the approaching star,

$$(\lambda') = \lambda\frac{c-v}{c}$$

$$\text{or } \frac{\lambda'}{\lambda} = 1-\frac{v}{c} \text{ or } \frac{v}{c} = 1-\frac{\lambda'}{\lambda} = \frac{\lambda-\lambda'}{\lambda} = \frac{\Delta\lambda}{\lambda}$$

Therefore $\Delta\lambda = \lambda\times\frac{v}{c} = 5000\text{ Å}\times\frac{1.5\times10^6}{3\times10^8} = 25\text{ Å}$

(where $\Delta\lambda$ is the change in the wavelength)

202. (c)

From $\Delta x = \frac{\lambda}{2\pi}\Delta\phi$

$$\lambda = 2\pi\frac{\Delta x}{\Delta\phi} = \frac{2\pi(0.4)}{1.6\pi} = 0.5\text{ m}$$

$$\upsilon = \frac{v}{\lambda} = \frac{330}{0.5} = 660\text{ Hz}$$

203. (a)
Frequency $(\upsilon) = 100$ Hz and distance from fixed end = 10 cm = 0.1 m. When a stationary wave is produced, the fixed end behaves as a node.

Thus wavelength $(\lambda) = 2\times0.1 = 0.2$ m

Therefore velocity $v = \upsilon\lambda = 100\times0.2 = 20$ m/s

204. (a)

$$y = A\sin(100t)\cos(0.01x)$$

Comparing it with standard equation

$$y = A\sin\left(\frac{2\pi}{T}t\right)\cos\left(\frac{2\pi}{\lambda}x\right)$$

$$T = \frac{\pi}{50} \text{ and } \lambda = 200\pi$$

Therefore velocity, (v)

$$= \frac{\lambda}{T} = \frac{200\pi}{\pi/50} 200\times50 = 1000 = 10^4\text{ m/s.}$$

205. (a)
First case: Frequency = υ;
No. of beats/sec = 5 and
Frequency (sounded with) = 200Hz
Second case: Frequency = 2 υ;
No. of beats/sec = 10 and frequency (sounded with) = 420 Hz.
In the first case, frequency $(\upsilon) = 200\pm5 = 205$ or 195 Hz. And in the second case, frequency $(2v) = 420 \pm 10$ or $\upsilon = 210 \pm 5$=205 or 215. So, common value of υ in both the cases is 205 Hz.

206. (c)
Beats are produced. Frequency of beats will be $\frac{1}{3}-\frac{1}{4} = \frac{1}{12}$
Hence time period = 12 s.

207. (c)

$f = \frac{1}{2l}\left[\frac{T}{\mu}\right]^{\frac{1}{2}}$ when f is halved, the length is doubled.

∴ Length is 1 m.

208. (c)
Velocity of sound, $v \propto \sqrt{T}$

$\therefore \frac{v}{2v} = \frac{\sqrt{273+27}}{\sqrt{T}}$ or $T = 1200\text{ K} = 927°\text{ C}$

209. (b)
For production of beats, different frequencies are essential. The different amplitudes effect the minimum amplitude of the beats. If frequencies are different, phases will be different.

QUICK QUESTIONS

Quiz

1. (i) Can a non-mechanical wave be longitudinal?
   (ii) Can a mechanical wave be transverse?
   (iii) Can the speed of sound wave change with rise in temperature?
   (iv) Can the speed of non-mechanical wave change with temperature?
   (v) Can you find a wave which is neither longitudinal nor transverse?
   (vi) Can transverse mechanical waves be propagated in fluids?
   (vii) Can one listen the song of his companion on the moon?
   (viii) Are the waves in solids always transverse?
   (ix) Do you find a difference between the roar of a lion and humming of mosquito?
   (x) Why is it sometimes difficult to recognize the voice of a person on telephone?
   (xi) A rope of mass M is hanging from a ceiling. A wave is created in the rope. Does the velocity of the wave remain the same throughout the length of the rope?
   (xii) Can a motion be oscillatory beet not simple harmonic?
   (xiii) Can there be a motion which is periodic but not oscillatory?
2. You are very well aware of two kinds of waves – mechanical and electromagnetic waves. Can you name any wave other than mentioned above two?
3. You must have heard that great singers like Tansen and Baijubavra had the power to shatter a glass object by his singing. What is the physics phenomena behind it?
4. Is a bouncing ball an example of simple harmonic motion?
5. A grandfather clock depends on the period of a pendulum to keep correct time. Suppose a grandfather clock is calibrated correctly and then a mischievous child slides the bob of the pendulum downward on the oscillating rod. Does the grandfather clock run (a) slow, (b) fast, and (c) correctly?
6. A long line of people is waiting to buy tickets at a movie theatre. When the first person leaves, a pulse of motion occurs as people step forward to fill in the gap. As each person steps forward, the gap moves through the line. Is the propagation of this gap transverse or longitudinal?
7. In mechanics, a simplification model of massless strings is often adopted. Why is this not a good model when discussing waves on strings?
8. You are driving toward a cliff and you honk your horn. The frequency of the echo you hear will be Doppler shifted (a) up in frequency, equivalent to a moving source; (b) up in frequency, equivalent to a moving observer; (c) up in frequency, equivalent to a moving source and a moving observer; (d) down in frequency, equivalent to moving source; (e) down in frequency, equivalent to a moving observer; (f) down in frequency, equivalent to a moving source and a moving observer. Which of the options mentioned above is correct?

Think and grow

9. We know that the period of oscillation of an object attached to a spring is proportional to the square root of the mass of the object $\left(T = 2\pi\sqrt{\frac{m}{k}}\right)$. Thus, if we perform an experiment in which we place objects with a range of masses on the end of a spring and measure the period of oscillation of each mass-spring system, a graph of the square of the period versus the mass will result in a straight line, as suggested in the figure. But we find that the line does not go through the origin. Why not?

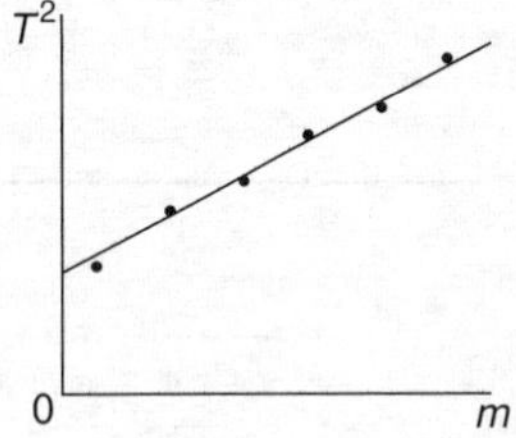

10. Suppose you attach a stapler to a rubber band and allow the stapler to hang in equilibrium from the end of the loop. You now pull the stapler down a short distance and release it, measuring the frequency f of its oscillations. After making this measurement, you cut the rubber band in one place and unfold it so that you have a single rubber strip that is twice as long as the length of the original loop. You attach the stapler to one end of the rubber strip, let it hang while you hold the other end, and again measure the frequency of oscillation. How does this frequency compare with the original frequency?
11. In one of the Moon landings, an astronaut walking on the Moon's surface had a belt hanging from his space suit, and the belt oscillated as a physical pendulum. A scientist on Earth observed this motion on television and from it was able to estimate the free-fall acceleration on the Moon. How do you suppose this calculation was done?

12. Suppose you create a pulse by moving the free end of a taut string up and down once with your hand. The string is attached at its other end to a distant wall. The pulse reaches the wall in a time interval Δt. Which of the following actions, taken by itself, decreases the time interval required for the pulse to reach the wall? More than one choice may be correct.
    (a) Moving your hand more quickly, but still only up and down once by the same amount
    (b) Moving your hand more slowly, but still only up and down once by the same amount
    (c) Moving your hand a greater distance up and down the same amount of time
    (d) Moving your hand a smaller distance up and down the same amount of time
    (e) Using a heavier string of the same length and under the same tension
    (f) Using a lighter string of the same length and under the same tension
    (g) Using a string of the same linear mass density but under decreased tension
    (h) Using a string of the same linear mass density but under increased tension.

    We know that wave velocity in the string $v = \sqrt{\dfrac{T}{m}}$

    where T is tension and m is mass per unit length.

13. A secret agent is trapped in a building on top of an elevator car at a lower floor. He attempts to signal a fellow agent on the roof by tapping a message in Morse code on the elevator cable, so that transverse pulses move upward on the cable. As the pulses move up the cable toward the accomplice, does the speed with which they move stay the same, increase, or decrease? If the pulses are sent 1 s apart, are they received 1 s apart by his partner?

14. You are on a cruise and the person next to you at the deck railing falls overboard. During the entire time of fall, the person screams with a constant frequency. How does the apparent frequency of the sound of the scream change with time?

15. Suppose you run past your stereo speakers, from right to left, or left to right. If you run rapidly enough and have excellent pitch discrimination, you may notice that the music that is playing seems to be out of tune when you are between the speakers. Why?

Explanations

Quiz

1. (i) No
   (ii) Yes
   (iii) Yes
   (iv) No
   (v) Yes, These are called ripples. Waves on the surface of a liquid are example of these types of wave. In these waves, particles of the medium vibrate up and down and simultaneously back and forth describing an ellipse in a vertical plane.
   (vi) No, to propagate a transverse wave, the medium must have elasticity to provide restoring force when acted upon by shearing stress. But, the fluids flow when subjected to shearing stress. They are unable to bear shearing stress and hence cannot provide shearing stress.
   (vii) No, as there is no medium
   (viii) No, it can be either transverse or longitudinal depending upon the mode of exaction
   (ix) Yes. Roar has large loudness but low pitch whereas humming has low loudness but high pitch.
   (x) Due to poor quality of sound
   (xi) It is different; the velocity of sound in a rope is $v = \sqrt{\dfrac{T}{m}}$ where T is tension and m is mass per unit length. The tension is different at different points of the rope. Hence, velocity does not remain same.
   (xii) Yes,
   (xiii) Yes uniform circular motion
2. Yes, matter waves. These waves exhibit wave like properties of particles and governed by the laws of quantum physics.
3. Resonance. When the frequency of his singing coincides with the natural frequency of the glass, the glass vibrates violently because of resonance and may get shattered.
4. The bouncing ball is not an example of simple harmonic motion. The acceleration is not proportional to displacement.

5. (a) With a longer length, the period of the pendulum increases. Thus, it takes longer to execute each swing, so that each second according to the clock takes longer than an actual second. Thus, the clock runs slow.

6. (a) Longitudinal. The movement of the people is parallel to the direction of propagation of the gap.

7. A wave on a massless string would have an infinite speed of propagation.

8. (c). The sound that leaves your horn in the forward direction is Doppler shifted to a higher frequency because it is coming from a moving source. As the sound reflects back and comes toward you, you are a moving observer, so a second Doppler shift occurs to an even higher frequency.

Think and Grow

9. The line does not go through the origin because the spring is not massless. Thus, the resistance to changes in motion of the system is a combination of the mass of the object on the end of the spring and the mass of the oscillating spring coils. The entire mass of the spring is not oscillating in the same way. The coil of the spring attached to the object is oscillating over the same amplitude as the object, but the coil at the opposite end of the spring is not oscillating at all.

10. When the stapler is attached and allowed to hang, the single strip stretches through a distance twice as long as that for the rubber loop, because only one strip is supporting the stapler instead of two in the case of the loop. Now, if the rubber strip is held at the free end and the stapler is hung at the opposite end, each half of the full length of the strip stretches by a distance twice as great as the original stretch of the rubber band. Therefore, the overall stretch will be four times as large as the original rubber band. With four times the stretch for the same hanging weight, the force constant of the rubber strip must be one-fourth as large as that of the original rubber band. Using the equation, $f = \frac{1}{2\pi}\sqrt{\frac{k}{m}}$ we determine that the frequency of the stapler when oscillating must be $\frac{1}{2}f$.

11. The period of oscillation T of the swinging belt can be measured by watching the television coverage. Furthermore, the belt can be geometrically realized as a long thin rod, so that its moment of inertia can be approximated as $\frac{1}{3}mL^2$. Assuming the belt to be of uniform density, the centre of mass is located at $d = L/2$. Thus, the equation becomes

$$T = 2\pi\sqrt{\frac{\frac{1}{3}mL^2}{mg(L/2)}} = 2\pi\sqrt{\frac{2L}{3g}}$$

Upon estimating the length of the swinging belt by comparing it to the height of a typical astronaut this equation can be solved for g.

12. Only choices (f) and (h) are correct. (a) and (b) affect the transverse speed of a particle of the string but not the wave speed along the string. (c) and (d) change the amplitude. (e) and (g) increase the time by decreasing the wave speed.

13. The elevator cable can be realized as a string. The speed of waves on the cable is a function of the tension in the cable. As the waves move higher on the cable, tension increases because each higher point on the cable must support the weight of all the cable below it. Thus, the speed of the pulses increases as they move higher on the cable. The frequency of the pulses will not be affected because each pulse takes the same time to reach the top. They will still arrive at the top of the cable at the intervals of 1 s.

14. The apparent frequency would be determined by Doppler shifted downward in frequency because the screaming person is moving away from you. Because the speed of the source increases due to the gravity, the apparent frequency decreases with time.

15. When you are between the speakers, you are running away from one of them and toward the other. Thus, there is Doppler shift downward from the speaker behind you and a Doppler shift upward from the speaker ahead of you. As a result, the sound from the two speakers will not be in tune.

CHAPTER TEST

1. A tuning fork and a sonometer wire, when sounded together produce 4 beats/s when the length of the wire is 95 cm or 100 cm. The frequency of the tuning fork is:
(a) 156 Hz (b) 148 Hz
(c) 152 Hz (d) 160 Hz

2. Sound travels fastest in:
(a) granite (b) iron
(c) aluminium (d) lead

3. A 120 Hz wave on a string has an amplitude of 0.160 mm. How much energy exists in 80 cm length of string?
(a) 0.1 mJ (b) 0.2 mJ
(c) 0.3 mJ (d) Data is inadequate.

4. An engine running at speed $\frac{v}{10}$ sounds a whistle at frequency 600 Hz. A passenger in a train coming from the opposite side at speed $\frac{v}{15}$ experiences this whistle to be of frequency f. If v is speed of sound in air and there is no wind, f is nearest to
(a) 710 Hz (b) 630 Hz
(c) 580 Hz (d) 510 Hz

5. A source of sound produces waves of wavelength 60 cm when it is stationary. If the speed of sound in air is 320 m s^{-1} and source moves with speed 20 m s^{-1}, the wavelength of sound in the forward direction will be nearest to:
(a) 56 cm (b) 60 cm
(c) 64 cm (d) 68 cm

6. A longitudinal wave is represented by $x = x_0 \sin 2\pi\left(nt - \frac{x}{\lambda}\right)$ The maximum particle velocity will be four times the wave velocity if
(a) $\lambda = \frac{\pi x_0}{4}$ (b) $\lambda = 2\pi x_0$
(c) $\lambda = \frac{\pi x_0}{2}$ (d) $\lambda = 4\pi x_0$

7. A person carrying a whistle emitting continuously a note of 272 Hz is running towards a reflecting surface with a speed of 18 km h^{-1}. The speed of sound in air is 345 m s^{-1}. The number of beats heard by him is:
(a) 4 (b) 6
(c) 8 (d) zero.

8. Shock waves are produced by objects
(a) carrying electric charge and vibrating
(b) vibrating with frequency greater than 20000 Hz
(c) vibrating with very large amplitude
(d) moving with a speed greater than that of sound in the medium.

9. A copper wire 2.4 mm in diameter is 3 m long and is used to suspend a 2 kg mass from a beam. If a transverse disturbance is sent along the wire, how fast will the disturbance travel?
(a) 2 m s^{-1} (b) 22 m s^{-1}
(c) 222 m s^{-1} (d) Data is inadequate.

10. A piano wire having a diameter of 0.90 mm is replaced by another wire of the same material but with diameter of 0.93 mm. If the tension of the wire is kept the same, then the preference change in the frequency of the fundamental tone is:
(a) + 3 % (b) + 3.2 %
(c) −3.2% (d) −3 %

11. An open pipe is suddenly closed with the result that the second overtone of the closed pipe is observed to be higher in frequency by 100 Hz than the first overtone of the original pipe. The fundamental frequency of the open pipe will be
(a) 100 Hz (b) 300 Hz
(c) 150 Hz (d) 200 Hz

12. Which of the following is incorrect?
(a) If a wave is longitudinal, it must be mechanical
(b) If a wave is mechanical, it may or may not be transverse
(c) No mechanical wave can exist in vacuum
(d) Diffraction helps us to distinguish mechanical waves from other waves.

13. The distance between two consecutive crests in a wave train produced in a string is 5 cm. If two complete waves pass through any point per second, then the velocity of the wave is
(a) 10 cm s^{-1} (b) 15 cm s^{-1}
(c) 2.5 cm s^{-1} (d) 5 cm s^{-1}

14. Which of the following is a wrong statement?
(a) Sound waves transport energy
(b) Sound waves travel in straight line
(c) Sound waves travel adiabatically
(d) Sound waves travel faster in vacuum than in air.

15. A transverse wave passes through a string with the equation $y = 10 \sin \pi (0.02\, x - 2.00\, t)$ where x is in metre and t in second. The maximum velocity of the particle in wave motion is:
(a) 100 m/s (b) 63 m/s
(c) 120 m/s (d) 161 m/s

16. A wave is represented by the equation
$y = a \sin (0.01\, x - 2t)$
where a and x are in cm and t in seconds. Velocity of propagation of the wave is
(a) 200 cm/s (b) 10 cm/s
(c) 25 cm/s (d) 50 cm/s

17. Two cars A and B approach a stationary observer from opposite sides as shown in figure. Observer hears no beats. If the frequency of the horn of the car B is 504 Hz, the frequency of horn of car A will be

(a) 529.2 Hz (b) 295.2 Hz
(c) 440.5 Hz (d) None of these.

18. An organ pipe emits fundamental tone of frequency 320 hertz at 47 °C. What would be the fundamental tone emitted by pipe at 27 °C.
(a) 165 Hz (b) 310 Hz
(c) 220 Hz (d) 440 Hz

19. A cylindrical tube, open at both ends, has a fundamental frequency f in air. The tube is dipped vertically in water so that half of it is in water. The fundamental frequency of the air column is now
(a) $f/2$ (b) $3f/4$
(c) f (d) $2f$

20. A thin closed organ pipe and a thin open organ pipe are of equal lengths. The ratio of the frequency of their third overtones will be
(a) 7 : 8 (b) 8 : 7
(c) 7 : 4 (d) 1 : 1

21. The length of a sonometer wire AB is 110 cm. Where should the two bridges be placed from A to divide the wire in 3 segments whose fundamental frequencies are in the ratio of 1: 2: 3?
(a) 30 cm and 90 cm
(b) 40 cm and 80 cm
(c) 60 cm and 90 cm
(d) 30 cm and 60 cm

22. The speed of sound in air is 333 m/s. The fundamental frequency of the open pipe is 333 Hz. The second overtone of the open organ pipe can be produced with a pipe of length
(a) 0.5m (b) 1.0 m
(c) 1.5 m (d) 2.0 m

23. The graph between the frequency n and square root of density (ρ) of a wire, keeping its length, radius and tension constant, is

(a) 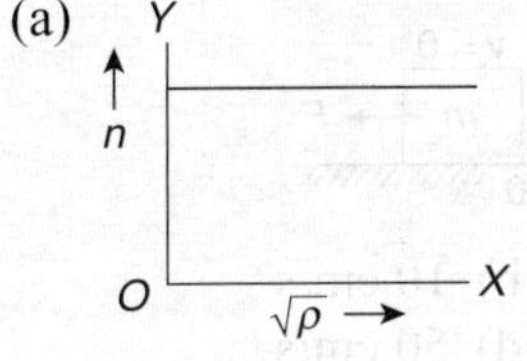

(b) 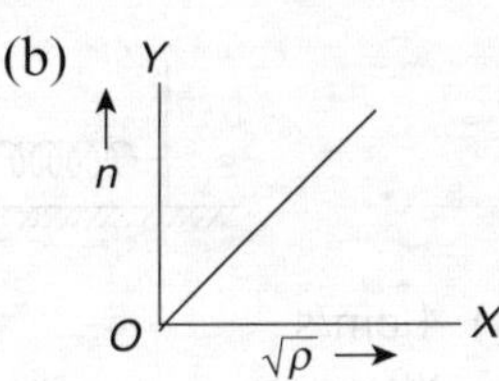

(c) 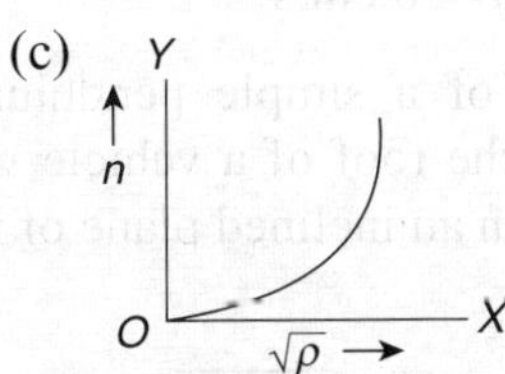

(d) 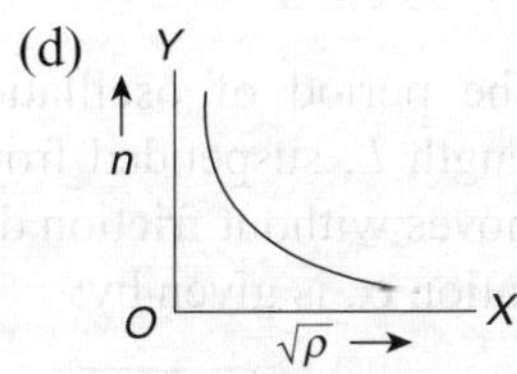

24. A stretched wire of 60 cm vibrates with a fundamental note of frequency 256 Hz. If the length of the wire is reduced to 15 cm while the tension is kept constant, the fundamental frequency of the wire will be
(a) 64 (b) 256
(c) 512 (d) 1024

25. An organ pipe P_1 closed at one end vibrating in its first harmonic and other pipe P_2 open at both ends vibrating in its third harmonic are in resonance with a given tuning fork. The ratio of the length of P_1 to that of P_2 is
(a) 8/3 (b) 3/8
(c) 1/2 (d) 1/3

26. A stretched string of 1 m length fixed at both ends, having a mass of 5×10^{-4} kg is under a tension of 20 N. It is plucked at a point situated at 25 cm from one end. The stretched string would vibrate with a frequency of
(a) 400 Hz (b) 100 Hz
(c) 200 Hz (d) 256 Hz

27. In a closed organ pipe the fundamental frequency is n. What will be the ratio of frequencies of the next three overtones?
(a) 2 : 3 : 4 (b) 3 : 4 : 5
(c) 3 : 7 : 11 (d) 3 : 5 : 7

28. Two particles executing SHM have time periods in ratio of 1:4. Both are given small displacements in the same direction at the same instant. They will again be in phase at the mean position after the second particle has completed n oscillations. The value of n is:
(a) 4 (b) 2
(c) 1/2 (d) zero

29. For a particle executing SHM, the potential energy is given by $U = U_0 \sin^2 \omega t$. The maximum kinetic energy of the particle is:
(a) U_0 (b) $\frac{U_0}{\sqrt{2}}$
(c) $\frac{U_0}{3}$ (d) None of these.

30. The time period of a particle executing SHM is doubled, then
(a) angular frequency becomes half
(b) frequency becomes half
(c) frequency becomes double
(d) None of the above.

31. A particle of mass 10 kg is executing SHM of time period 2 seconds and amplitude 0.25 m. The magnitude of maximum force on the particle is:
(a) 5 N (b) 24.65 N
(c) zero (d) 40.6 N

32. Two SHMs $x = a \sin \omega t$ and $y = b \cos \omega t$ directed along y-axis respectively are acted on a particle. The path of the particle is:
(a) circle (b) straight line
(c) ellipse (d) parabola

33. If the mass of bob of simple pendulum is increased by 50%, the time period of the pendulum:
(a) does not change (b) increases
(c) decreases (d) None of these.

34. The equation of acceleration of a particle is $a = -k(x + b)$, where x is distance along x-axis and k is a positive constant. The motion of particle is
(a) oscillatory (b) periodic
(c) SHM (d) All of these.

35. A particle executes SHM, its time period is 16s. If it passes through the centre of oscillation then its velocity is 2 m/s at time 2 sec. The amplitude will be:
(a) 7.2 m (b) 4 cm
(c) 6 cm (d) 0.72 m

36. A body executing SHM has its velocity 10 cm/s and 7 cm/s, when its displacements from the mean position are 3 cm and 4 cm respectively. The length of path is:
(a) 10 cm (b) 9.5 cm
(c) 4 cm (d) 11.36 cm

37. A particle executes simple harmonic motion. The amplitude of vibration of particle is 2 cm. The displacement of particle in one time period is:
(a) 1 cm (b) 2 cm
(c) 4 cm (d) zero

38. In previous problem, the distance travelled by the particle is:
(a) 8 cm (b) 2 cm
(c) 4 cm (d) zero

39. A particle executes SHM with an amplitude of 10 cm and frequency 2 Hz. At $t = 0$, the particle is at a point where potential energy and kinetic energy are same. The equation of displacement of particle is:
(a) $0.1\sin\left(4\pi t + \dfrac{\pi}{4}\right)$ (b) $0.1\sin 4\pi t$
(c) $0.1\cos\left(4\pi t + \dfrac{\pi}{4}\right)$ (d) None of these.

40. A simple harmonic oscillator has amplitude A, angular velocity ω, and mass m. Then average energy in one time period will be:
(a) $\dfrac{1}{4}m\omega^2A^2$ (b) $\dfrac{1}{2}m\omega^2A^2$
(c) $m\omega^2A^2$ (d) zero

41. A point mass $m = 20$ kg, is suspended by a massless spring of constant 2000 N/m. The point mass is released when elongation in the spring is 15 cm. The equation of displacement of particle as function of time is:

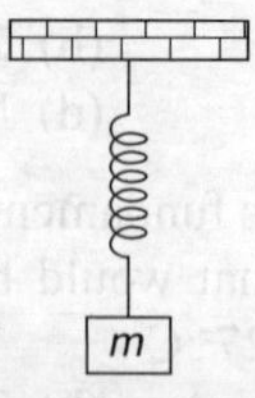

(a) $y = 10\ sin\ 10t$
(b) $y = 10\ \cos 10t$
(c) $y = 10\sin\left(10t + \dfrac{\pi}{6}\right)d$
(d) None of these.

42. A system shown in figure, consists of a massless pulley, a spring of force constant k and a block of mass m. If the block is slightly displaced vertically down from its equilibrium position and released, then the period of vertical oscillations is

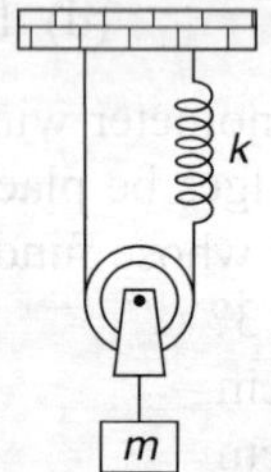

(a) $T = \pi\sqrt{\left(\dfrac{m}{4k}\right)}$ (b) $T = 2\pi\sqrt{\left(\dfrac{m}{4k}\right)}$
(c) $T = 2\pi\sqrt{\left(\dfrac{m}{2k}\right)}$ (d) $T = 2\pi\sqrt{\left(\dfrac{m}{3k}\right)}$

43. A block of mass 1 kg is connected with a massless spring of force constant 100 N/m. At $t = 0$, a constant force $F = 10$N is applied on the block. The spring is in its natural length at $t = 0$. The speed of particle at $x = 6$ cm from mean position is:

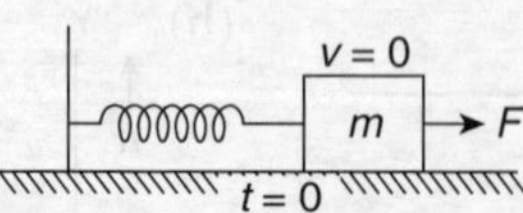

(a) 4 cm/s (b) 10 cm/s
(c) 80 cm/s (d) 50 cm/s

44. The period of oscillation of a simple pendulum of length L, suspended from the roof of a vehicle which moves without friction down an inclined plane of inclination α, is given by:
(a) $T = 2\pi\sqrt{\dfrac{L}{g\cos\alpha}}$ (b) $2\pi\sqrt{\dfrac{L}{g\sin\alpha}}$
(c) $2\pi\sqrt{\dfrac{L}{g}}$ (d) $\sqrt{\dfrac{L}{g\tan\alpha}}$

45. A cylindrical piston of mass M slides smoothly inside a long cylinder closed at one end, enclosing a certain mass of gas. The cylinder is kept with its axis horizontal. If the piston is disturbed from its equilibrium position, it oscillates simple harmonically. Its period is:

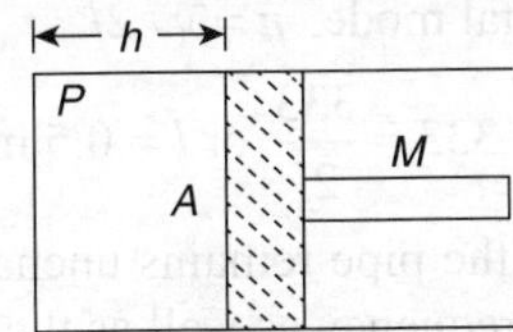

(a) $T = 2\pi\sqrt{\frac{Mh}{PA}}$ (b) $T = 2\pi\sqrt{\frac{MA}{Ph}}$

(c) $T = 2\pi\sqrt{\frac{M}{PAh}}$ (d) $T = 2\pi\sqrt{MPhA}$

Answer Keys

1. (a)	2. (c)	3. (d)	4. (a)	5. (a)	6. (c)	7. (c)	8. (d)	9. (b)	10. (c)
11. (d)	12. (d)	13. (a)	14. (d)	15. (b)	16. (a)	17. (a)	18. (b)	19. (c)	20. (a)
21. (c)	22. (a)	23. (d)	24. (d)	25. (b)	26. (c)	27. (d)	28. (a)	29. (a)	30. (b)
31. (b)	32. (c)	33. (a)	34. (d)	35. (a)	36. (d)	37. (d)	38. (a)	39. (a)	40. (a)
41. (c)	42. (b)	43. (c)	44. (a)	45. (a)					

Solutions

1. (a)

$$\frac{v+4}{v-4} = \frac{100}{95} \Rightarrow V = 156 \text{ Hz}$$

2. (c)

3. (d)

4. (a)

The engine and the train are approaching each another. Hence n' larger than n. This excludes (c) and (d). Source motion alone contributes a factor $\frac{10}{9}$. This excludes (b) $600 \times \frac{16}{15} \times \frac{10}{9} = 710$ Hz

5. (a)

Doppler effect. In forward direction, waves per second spread to 320 – 20 m instead of 320. Hence lower.

6. (c)

7. (c)

Velocity of source = 18 km h^{-1} = 5 m s^{-1}

(i) S moves towards listener (v_S) (ii) Listener moves towards source (v_1)

$$v' = \frac{\upsilon + \upsilon_L}{\upsilon - \upsilon_S} v = 280\text{ , Beats} = v' - v = 8$$

8. (d)

9. (b)

10. (c)

11. (d)

$$\frac{5\upsilon}{4l} - \frac{2\upsilon}{2l} = 100 \Rightarrow \upsilon = 200 \text{ Hz}$$

12. (d)

Diffraction is inherent in wave motion.

13. (a)

$\lambda = 5$ cm, $v = 2$ s^{-1}

14. (d)

15. (b)

Standard equation of a wave travelling with amplitude a and angular frequency ω is given by

$$y = a \sin(\omega t - kx)$$

Given equation of wave is

$$y = 10 \sin \pi (0.02x - 2t)$$

Comparing this equation with standard wave equation, we get

$a = 10$ m, $\omega = 2\pi$ rad/s

Maximum velocity, $v_{max} = a\omega$

$$\therefore \quad v_{max} = 10 \times 2\pi = 20\pi = 20 \times \frac{22}{7} = 63 \text{ m/s}$$

16. (a)

The standard equation to wave is

$$y = a\sin(lx - \omega t) \quad (1)$$

where ω is angular velocity, t is time.

Given equation is

$$y = a\sin(0.01x - 2t) \quad (2)$$

Comparing the given equation with standard wave from, we get

$$k = 0.01 \text{ and } \omega = 2$$

Thus, velocity $= \frac{\omega}{k} = \frac{2}{0.01} = 200$ cm/s.

17. (a)

The apparent frequency of the two cars should be equal for no beats.

From Doppler's effect, the apparent change in frequency of a wave that is perceived by stationary observer is

$$n' = n_0\left(\frac{v}{v - v_s}\right)$$

where v is speed of sound.

Since observer is not hearing any beats, the apparent frequency of the two cars should be equal.

$$n' = n''$$

$$504\left(\frac{v}{v-30}\right) = n\left(\frac{v}{v-15}\right)$$

$$n = \frac{(330-15)\times 504}{(330-30)}$$

$$= \frac{315\times 504}{300} = 529.9 \text{ Hz}$$

18. (b)

$n \propto \sqrt{T}$ (assuming no change in length)

or $\frac{n_1}{n} = \sqrt{\left(\frac{T_1}{T_2}\right)} = \sqrt{\left(\frac{273+27}{273+47}\right)}$

or $\frac{n_1}{320} = \sqrt{\left(\frac{300}{320}\right)}$

$$n_1 = 320\sqrt{\left(\frac{15}{16}\right)} \approx 310 \text{ Hz}$$

19. (c)

For open pipe $f = \frac{v}{2l}$

For closed pipe $f' = \frac{v}{4l'} = \frac{v}{4(l/2)} = \frac{v}{2l}$

$$f' = f$$

20. (a)

21. (c)

$$n \propto \frac{1}{l} \text{ or } l \propto \frac{1}{n}$$

Now AC : CD : DB $= \frac{1}{1} : \frac{1}{2} : \frac{1}{3} = 6:3:2$

$\therefore AC = 60$ cm and $AD = 60 + 30 = 90$ cm

22. (a)

For fundamental mode, $n = v/2l$

$$333 = \frac{333}{2l} \text{ or } l = 0.5 \text{ m}$$

The length of the pipe remains unchanged both for the fundamental frequency as well as the overtones.

23. (d)

$$n = \frac{1}{2l}\sqrt{\left(\frac{T}{m}\right)} = \frac{1}{2l}\sqrt{\left(\frac{T}{\pi r^2 \rho}\right)}$$

or $n \propto 1/\sqrt{\rho}$

So, the graph will be as shown in Fig. (d)

24. (d)

$$n \propto \frac{1}{l}$$

$$256 \propto \frac{1}{60} \text{ and } n_2 \propto \frac{1}{15}$$

$$\therefore \quad \frac{n_2}{256} = \frac{60}{15} \text{ or } n = \frac{256\times 60}{15} = 1024 \text{ Hz}$$

25. (b)

Let l_1 and l_2 be the respective lengths of the closed and open organ pipes. The frequency of vibration of a closed pipe in rth mode of vibration is given by

$$n_r = (2r-1)\frac{v}{4l_1}$$

For first harmonic, $r = 2$

$$n_2 = \frac{3v}{4l_1} \quad (1)$$

For open pipe, $n_r = \frac{rv}{2l_2}$

For third harmonic $n_4 = \frac{4v}{2l_2} \quad (2)$

$$\frac{3v}{4l_1} = \frac{4v}{2l_2}$$

$$\frac{l_1}{l_2} = \frac{3}{8}$$

26. (c)

$$n = \frac{p}{2l}\sqrt{\left(\frac{T}{m}\right)}$$

Fundamental frequency n_1 is given by

$$n_1 = \frac{1}{2\times 1}\left[\frac{20}{5\times 10^{-4}}\right]^{1/2} = 100 \text{ Hz}$$

When the wire is plucked at 25 cm, it will vibrate in two segments.

$\therefore$ Frequency $n_2 = 2n_1 = 200$ Hz

27. (d) **28.** (a)

29. (a)

30. (b)

$$f = \frac{1}{T}$$

$$\frac{f'}{f} = \frac{T}{T'} = \frac{T}{2T}$$

$$\therefore \quad f' = \frac{f}{2}$$

31. (b)

The maximum acceleration is $\alpha_m = -\omega^2 a$

$\therefore$ The magnitude of maximum force is

$$F_{max} = m\omega^2 a = 10\left(\frac{2\pi}{T}\right)^2 \times 0.25$$

$$= 10\left(\frac{2\times 3.14}{2}\right)^2 \times 0.25$$

$$= 24.65 \text{ N}$$

32. (c) **33.** (a)

34. (d)

35. (a)

Given: $t = 2\,\text{sec}, v = 2 \text{ m/s}, T = 16$ s

$$v = a\omega \cos \omega t$$

$$2 = \text{a}\frac{2\pi}{16}\cos\frac{2\pi}{16}.2$$

$$a = \frac{16\sqrt{2}}{\pi} = 7.2 \text{ m}$$

36. (d)

$$v = \omega\sqrt{a^2 - y^2}$$

$$\frac{v^2}{v_1} = \frac{a^2 - 16}{a^2 - 9}$$

$$a = 5.68\,\text{cm}$$

$\therefore$ Length of path $= 2a = 11.36$ cm

37. (d)

$$\because x = A\sin\omega t$$

$$\therefore x = A\sin\left(\frac{2\pi}{T}\times T\right)$$

$$\therefore x = A\sin 2\pi = 0$$

38. (a)

Since, the distance travelled in $T/4$ time is $s_1 = A$

$\therefore$ The distance travelled in one time period is

$$s = 4s_1 = 4A \therefore s = 8 \text{ cm} (\because A = 2 \text{ cm})$$

39. (a)

Let $x = A\sin(\omega t + \phi)$

$$\therefore \quad v = \frac{dx}{dt} = A\omega\cos(\omega t + \phi)$$

$$\therefore \quad \text{KE} = \frac{1}{2}mv^2 = \frac{1}{2}mA^2\omega^2\cos^2(\omega t + \phi)$$

$$\therefore (\text{KE})_{max} = \frac{1}{2}mA^2\omega^2$$

$$\therefore \quad \text{PE} = \frac{1}{2}mA^2\omega^2 - \text{KE}$$

$$= \frac{1}{2}mA^2\omega^2 - \frac{1}{2}mA^2\omega^2\cos^2(\omega t + \phi)$$

$$= \frac{1}{2}mA^2\omega^2\cos^2(\omega t + \phi)$$

According to problem,

KE = PE (at $t = 0$)

$$\therefore \quad \frac{1}{2}mA^2\omega^2\sin^2(\omega t + \phi) = \frac{1}{2}mA^2\omega^2\cos^2(\omega t + \phi)$$

$$\therefore \quad \tan^2(\omega t + \phi) = 1$$

$$\Rightarrow \tan^2(\omega t + \phi) = \tan^2\frac{\pi}{4} \Rightarrow \omega t + \phi = \frac{\pi}{4} \Rightarrow \phi = \frac{\pi}{4} (\because t = 0)$$

$$\because \quad x = A\sin(\omega t + \phi)$$

Here, $A = 10$ cm $= 0.1$ m

$$\omega = 2\pi f = 2\pi\times 2 = 4\pi \text{ rad/s}$$

$$\phi = \pi/4$$

$$x = 0.1\sin\left(4\pi t + \frac{\pi}{4}\right)$$

40. (a)

$$\text{Average energy} = \frac{\int_0^T U dt}{\int_0^T dt} = \frac{1}{T}\int_0^T U dt$$

$$= \frac{1}{2T}\int_0^T m\omega^2 A^2\cos^2(\omega t + \phi)dt$$

$$= \frac{1}{4}m\omega^2 A^2$$

41. (c)

The motion of block is SHM

$$\therefore \quad y = A\sin(\omega t + \phi)$$

Here amplitude is $A = \dfrac{mg}{k}$

$$= \frac{20 \times 10}{2000} \text{ m} = 10 \text{ cm}$$

At $t = 0$, displacement of body with respect to mean position is

$$y = 15 - 10 = 5 \text{ cm}$$

$$5 = 10\sin(\omega \times 0 + \phi)$$

or $$\frac{1}{2} = \sin\phi \Rightarrow \phi = \frac{\pi}{6}$$

$$y = 10\sin\left(10t + \frac{\pi}{6}\right)$$

42. (b)

In this situation, if mass m moves down a distance x from its equilibrium position, the pulley will also move by x and so, the spring will stretch by $2x$.
Therefore, spring force will be $2 \times k$. The restoring force on the block will be $4kx$.

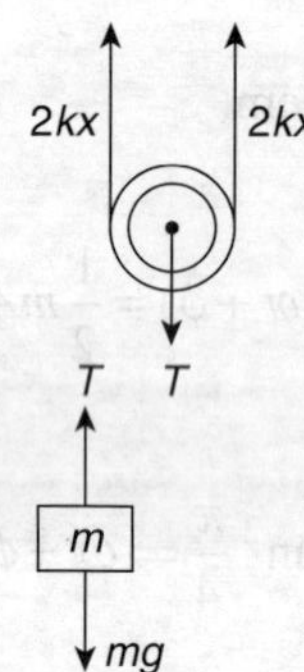

Hence, $F = -4kx$

$$a = -\frac{4kx}{m}$$

$$T = 2\pi\sqrt{\frac{x}{a}}$$

$$T = 2\pi\sqrt{\left(\frac{m}{4k}\right)}$$

43. (c)

At equilibrium, $F = kx_0$

$$x_0 = \frac{F}{k} = \text{Amplitude}$$

$$= \frac{10}{100} = \frac{1}{10} = 0.1 \text{ m}$$

$$\omega = \sqrt{\left(\frac{k}{m}\right)} = \sqrt{\left(\frac{100}{1}\right)} = 10 \text{ rad/s}$$

$$x = A\sin(\omega t + \phi)$$

$$v = \frac{dx}{dt} = A\omega\cos(\omega t + \phi)$$

At $t = 0, v = 0$

$$0 = A\omega\cos(\omega \times 0 + \phi)$$

$$\Rightarrow \quad \cos\phi = 0$$

$$\Rightarrow \quad \phi = \frac{\pi}{2}$$

$$x = A\sin\left(\omega t + \frac{\pi}{2}\right)$$

$$x = A\cos\omega t$$

$$v = \frac{dx}{dt} = A\omega\sin\omega t$$

$$v = A\omega\sqrt{1 - \cos^2\omega t}$$

or $$v = A\omega\sqrt{\left[1 - \left(\frac{x}{A}\right)^2\right]} = \omega\sqrt{A^2 - x^2}$$

$$v = \omega\sqrt{A^2 - x^2} = 10\sqrt{10^2 - x^2}$$

$$v = 10\sqrt{100 - x^2}$$

$$= 10 \times \sqrt{100 - 36}$$

$$= 10 \times \sqrt{64} = 80 \text{ cm/s}$$

44. (a)

The vehicle is moving down an inclined plane with the acceleration

$$a_0 = g\sin\alpha$$

The simple pendulum oscillates in the frame of vehicle. Force $mg\sin\alpha$ on the bob is balanced by pseudo force ma_0.

$$g' = g\cos\alpha$$

$$T = 2\pi\sqrt{\left(\frac{L}{g'}\right)} = 2\pi\sqrt{\left(\frac{L}{g\cos\alpha}\right)}$$

45. (a)

$$f = \frac{1}{2\pi}\sqrt{\left(\frac{BA^2}{MV_0}\right)}$$

B = Bulk modulus of elasticity

$$T = 2\pi\sqrt{\left(\frac{MV_0}{BA^2}\right)} = 2\pi\sqrt{\left(\frac{M(hA)}{PA^2}\right)}$$

$$T = 2\pi\sqrt{\left(\frac{Mh}{PA}\right)}$$